GOODE'S

ATLAS OF Physical Geography

 to accompany titles published by
WILEY John Wiley & Sons, Inc.

Howard Veregin, Ph.D., Editor

Editorial Advisory Board

Byron Augustin, D.A., Texas State University-San Marcos

Joshua Comenetz, Ph.D., University of Florida

Francis Galgano, Ph.D., United States Military Academy

Sallie A. Marston, Ph.D., University of Arizona

Virginia Thompson, Ph.D., Towson University

Abridgement of
21ST Edition

John Wiley & Sons, Inc. and RAND MCNALLY

Working together to bring you the best in geography education

Few publishers can claim as rich a history as John Wiley & Sons, Inc. (publishers since 1807) and Rand McNally & Company (publishers since 1856). Even fewer can claim as long-standing a commitment to geographic education.

Wiley's partnership with the geographic community began at the very beginning of the 20th century with the publication of textbooks on surveying. Rand McNally's partnership began even earlier, with the publication of the first Rand McNally maps in 1872. Since then, both companies have worked in parallel to help students visualize spatial relationships and appreciate the earth's dynamic landscapes and diverse cultures.

Now these two publishers have combined their efforts to bring you this new atlas, which represents the very best in educational resources for geography.

Based on the 21st edition of the *Goode's World Atlas*, the *Goode's Atlas of Physical Geography* features:

- An emphasis on map accuracy and legibility, and the mixture of maps of different types and scales to facilitate interpretation of geographic phenomena.
- World, continental, and regional population density maps, which have been created using LandScan, a digital population database developed using satellite and computer-mapping technology.
- Graphs accompanying many of the maps, to show important statistical information, trends over time, and relationships between variables.
- Maps and graphs that have been updated, based on the most current available data in accordance with the high standards and quality that have always been a defining feature of the *Goode's World Atlas*.

Wiley and Rand McNally are currently offering seven new course-specific atlases, which can be packaged with any of Wiley's best-selling textbooks, or sold separately as stand-alones. These atlases include:

Rand McNally Goode's Atlas of Political Geography	0-471-70694-9
Rand McNally Goode's Atlas of Latin America	0-471-70697-3
Rand McNally Goode's Atlas of North America	0-471-70696-5
Rand McNally Goode's Atlas of Asia	0-471-70699-X
Rand McNally Goode's Atlas of Urban Geography	0-471-70695-7
Rand McNally Goode's Atlas of Physical Geography	0-471-70693-0
Rand McNally Goode's Atlas of Human Geography	0-471-70692-2

This book was set by GGS Book Services and printed and bound by Walsworth Press. The cover was printed by Phoenix Color.

To order books or for customer service please, call 1-800-CALL WILEY (225-5945).

ISBN 0471-70693-0

Printed in the United States

10 9 8 7 6 5 4 3 2 1

Table of Contents

Introduction

Tables and Indexes

Introduction

Basic Earth Properties

The subject matter of **geography** includes people, landforms, climate, and all the other physical and human phenomena that make up the earth's environments and give unique character to different places. Geographers construct maps to visualize the **spatial distributions** of these phenomena: that is, how the phenomena vary over geographic space. Maps help geographers understand and explain phenomena and their interactions.

To better understand how maps portray geographic distributions, it is helpful to have an understanding of the basic properties of the earth.

The earth is essentially **spherical** in shape. Two basic reference points — the **North and South Poles** — mark the locations of the earth's axis of rotation. Equidistant between the two poles and encircling the earth is the **equator**. The equator divides the earth into two halves, called the **northern and southern hemispheres**. (See the figures to the right.)

Latitude and longitude are used to identify the locations of features on the earth's surface. They are measured in degrees, minutes and seconds. There are 60 minutes in a degree and 60 seconds in a minute. Latitude is the angle north or south of the equator. The symbols °, ', and " represent degrees, minutes and seconds, respectively. The N means north of the equator. For latitudes south of the equator, S is used. For example, the Rand McNally head office in Skokie, Illinois, is located at 42°1'51" N. The minimum latitude of 0° occurs at the equator. The maximum latitudes of 90° N and 90° S occur at the North and South Poles.

A **line of latitude** is a line connecting all points on the earth having the same latitude. Lines of latitude are also called **parallels**, as they run parallel to each other. Two parallels of special importance are the **Tropic of Cancer** and the **Tropic of Capricorn**, at approximately 23°30' N and S respectively. This angle coincides with the inclination of the earth's axis relative to its orbital plane around the sun. These tropics are the lines of latitude where the noon sun is directly overhead on the solstices. (See figure on page 66.) Two other important parallels are the **Arctic Circle** and the **Antarctic Circle**, at approximately 66°30' N and S respectively. These lines mark the most northerly and southerly points at which the sun can be seen on the solstices.

While latitude measures locations in a north-south direction, longitude measures them east-west. Longitude is the angle east or west of the **Prime Meridian**. A **meridian** is a line of longitude, a straight line extending from the North Pole to the South Pole. The Prime Meridian is the meridian passing through the Royal Observatory in Greenwich, England. For this reason the Prime Meridian is sometimes referred to as the **Greenwich Meridian**. This location for the Prime Meridian was adopted at the International Meridian Conference in Washington, D.C., in 1884.

Like latitude, longitude is measured in degrees, minutes, and seconds. For example, the Rand McNally head office is located at 87°43'6" W. The qualifiers E and W indicate whether a location is east or west of the Greenwich Meridian. Longitude ranges from 0° at Greenwich to 180° E or W. The meridian at 180° E is the same as the meridian at 180° W. This meridian, together with the Greenwich Meridian, divides the earth into **eastern and western hemispheres**.

Any circle that divides the earth into equal hemispheres is called a **great circle**. The equator is an example. The shortest distance between any two points on the earth is along a great circle. Other circles, including all other lines of latitude, are called **small circles**. Small circles divide the earth into two unequal pieces.

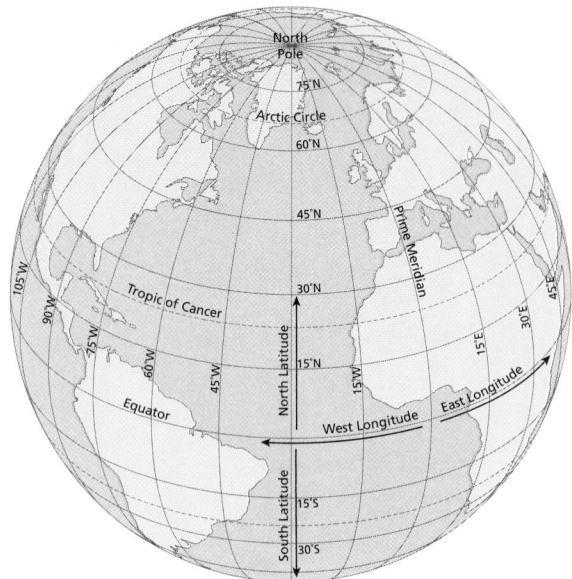

View of earth centered on 30° N, 30° W

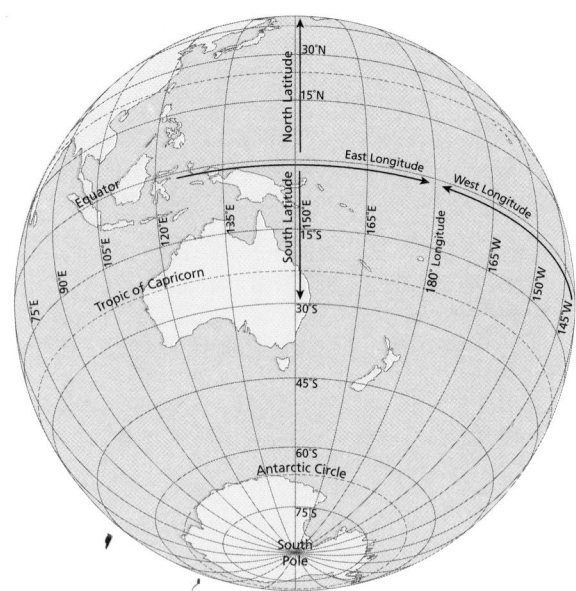

View of earth centered on 30° S, 150° E

The Geographic Grid

The grid of lines of latitude and longitude is known as the **geographic grid**. The following are some important characteristics of the grid.

All lines of longitude are equal in length and meet at the North and South Poles. These lines are called meridians.

All lines of latitude are parallel and equally spaced along meridians. These lines are called parallels.

The length of parallels increases with distance from the poles. For example, the length of the parallel at 60° latitude is one-half the length of the equator.

Meridians get closer together with increasing distance from the equator, and finally converge at the poles.

Parallels and meridians meet at right angles.

Map Scale

To use maps effectively it is important to have a basic understanding of map scale.

Map scale is defined as the ratio of distance on the map to distance on the earth's surface. For example, if a map shows two towns as separated by a distance of 1 inch, and these towns are actually 1 mile apart, then the scale of the map is 1 inch to 1 mile.

The statement "1 inch to 1 mile" is called a **verbal scale**. Verbal scales are simple and intuitive, but a drawback is that they are tied to the specific set of map and real-world units in the numerator and denominator of the ratio. This makes it difficult to compare the scales of different maps.

A more flexible way of expressing scale is as a **representative fraction**. In this case, both the numerator and denominator are converted to the same unit of measurement. For example, since there are 63,360 inches in a mile, the verbal scale "1 inch to 1 mile" can be expressed as the representative fraction 1:63,360. This means that 1 inch on the map represents 63,360 inches on the earth's surface. The advantage of the representative fraction is that it applies to any linear unit of measurement, including inches, feet, miles, meters, and kilometers.

Map scale can also be represented in graphical form. Many maps contain a **graphic scale** (or **bar scale**) showing real-world units such as miles or kilometers. The bar scale is usually subdivided to allow easy calculation of distance on the map.

Map scale has a significant effect on the amount of detail that can be portrayed on a map. This concept is illustrated here using a series of maps of the Washington, D.C., area. (See the figures to the right.) The scales of these maps range from 1:40,000,000 (top map) to 1:4,000,000 (center map) to 1:25,000 (bottom map). The top map has the **smallest scale** of the three maps, and the bottom map has the **largest scale**.

Note that as scale increases, the area of the earth's surface covered by the map decreases. The smallest-scale map covers thousands of square miles, while the largest-scale map covers only a few square miles within the city of Washington. This means that a given feature on the earth's surface will appear larger as map scale increases. On the smallest-scale map, Washington is represented by a small dot. As scale increases the dot becomes an orange shape representing the built-up area of Washington. At the largest scale Washington is so large that only a portion of it fits on the map.

Because small-scale maps cover such a large area, only the largest and most important features can be shown, such as large cities, major rivers and lakes, and international boundaries. In contrast, large-scale maps contain relatively small features, such as city streets, buildings, parks, and monuments.

Small-scale maps depict features in a more simplified manner than large-scale maps. As map scale decreases, the shapes of rivers and other features must be simplified to allow them to be depicted at a highly reduced size. This simplification process is known as **map generalization**.

Maps in *Goode's Atlas of Physical Geography* have a wide range of scales. The smallest scales are used for the world thematic map series, where scales range from approximately 1:200,000,000 to 1:75,000,000. Reference map scales range from a minimum of 1:100,000,000 for world maps to a maximum of 1:1,000,000 for city maps. Most reference maps are regional views with a scale of 1:4,000,000.

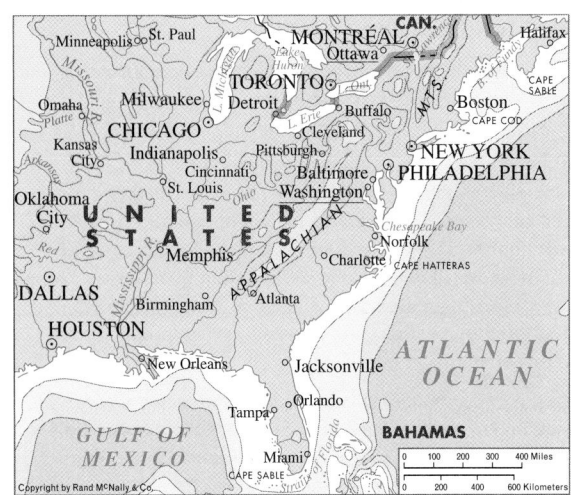

1:40,000,000 scale

1:4,000,000 scale

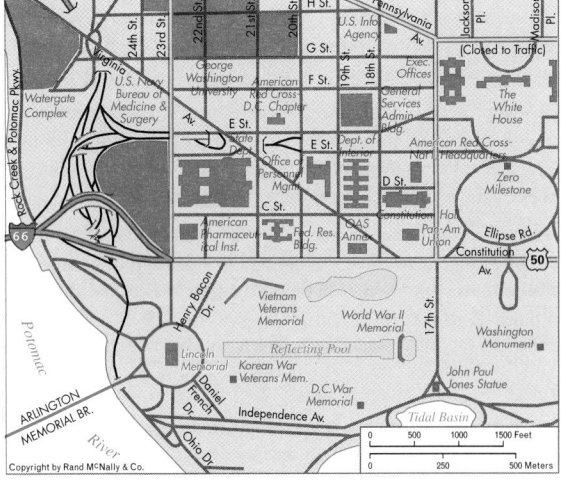

1:25,000 scale

Map Projections

Map projections influence the appearance of features on the map and the ability to interpret geographic phenomena.

A **map projection** is a geometric representation of the earth's surface on a flat or plane surface. Since the earth's surface is curved, a map projection is needed to produce any flat map, whether a page in this atlas or a computer-generated map of driving directions on www.randmcnally.com. Hundreds of projections have been developed since the dawn of mapmaking. A limitation of all projections is that they distort some geometric properties of the earth, such as shape, area, distance, or direction. However, certain properties are preserved on some projections.

If shape is preserved, the projection is called **conformal**. On conformal projections the shapes of features agree with the shapes these features have on the earth. A limitation of conformal projections is that they necessarily distort area, sometimes severely.

Equal-area projections preserve area. On equal area projections the areas of features correspond to their areas on the earth. To achieve this effect, equal-area projections distort shape.

Some projections preserve neither shape nor area, but instead balance shape and area distortion to create an aesthetically-pleasing result. These are often referred to as **compromise** projections.

Distance is preserved on **equidistant** projections, but this can only be achieved selectively, such as along specific meridians or parallels. No projection correctly preserves distance in all directions at all locations. As a result, the stated scale of a map may be accurate for only a limited set of locations. This problem is especially acute for small-scale maps covering large areas.

The projection selected for a particular map depends on the relative importance of different types of distortion, which often depends on the purpose of the map. For example, world maps showing phenomena that vary with area, such as population density or the distribution of agricultural crops, often use an equal-area projection to give an accurate depiction of the importance of each region.

Map projections are created using mathematical procedures. To illustrate the general principles of projections without using mathematics, we can view a projection as the geometric transfer of information from a globe to a flat projection surface, such as a sheet of paper. If we allow the paper to be rolled in different ways, we can derive three basic types of map projections: **cylindrical, conic,** and **azimuthal**. (See the figures to the right.)

For cylindrical projections, the sheet of paper is rolled into a tube and wrapped around the globe so that it is **tangent** (touching) along the equator. Information from the globe is transferred to the tube, and the tube is then unrolled to produce the final flat map.

Conic projections use a cone rather than a cylinder. The figure shows the cone tangent to the earth along a line of latitude with the apex of the cone over the pole. The line of tangency is called the **standard parallel** of the projection.

Azimuthal projections use a flat projection surface that is tangent to the globe at a single point, such as one of the poles.

The figures show the **normal orientation** of each type of surface relative to the globe. The **transverse orientation** is produced when the surface is rotated 90 degrees from normal. For azimuthal projections this orientation is usually called **equatorial** rather than transverse. An **oblique orientation** is created if the projection surface is oriented at an angle between normal and transverse. In general, map distortion increases with distance away from the point or line of tangency. This is why the normal orientations of the cylindrical, conic, and azimuthal projections are often used for mapping equatorial, mid-latitude, and polar regions, respectively.

The projection surface model is a visual tool useful for illustrating how information from the globe can be projected to the map. However, each of the three projection surfaces actually represents scores of individual projections. There are, for example, many projections with the term "cylindrical" in the name, each of which has the same basic rectangular shape, but different spacings of parallels and meridians. The projection surface model does not account for the numerous mathematical details that differentiate one cylindrical, conic, or azimuthal projection from another.

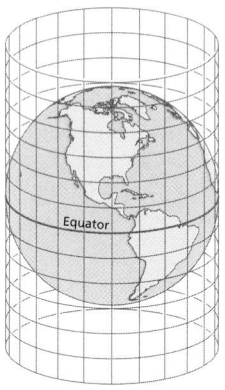
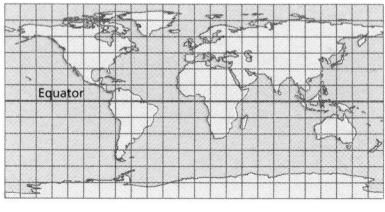

Cylindrical Projection

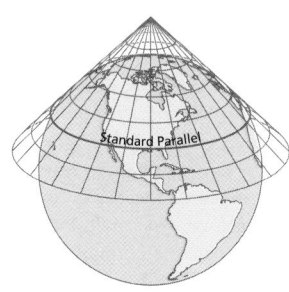
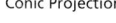
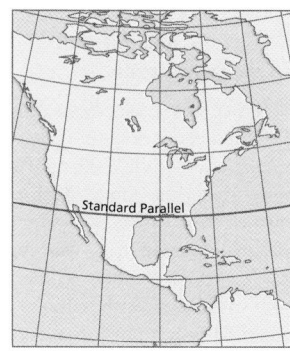

Conic Projection

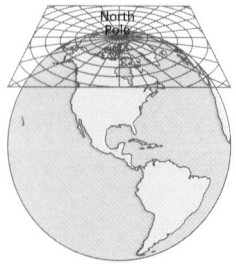
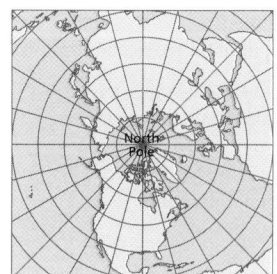

Azimuthal Projection

Map Projections Used in *Goode's Atlas of Physical Geography*

Of the hundreds of projections that have been developed, only a fraction are in everyday use. The main projections used in *Goode's Atlas of Physical Geography* are described below.

Simple Conic

Type: Conic   **Conformal:** No   **Equal-area:** No

Notes: Shape and area distortion on the Simple Conic projection are relatively low, even though the projection is neither conformal nor equal-area. The origins of the Simple Conic can be traced back nearly two thousand years, with the modern form of the projection dating to the 18th century.

Uses in *Goode's Atlas of Physical Geography*: Larger-scale reference maps of North America, Europe, Asia, and other regions.

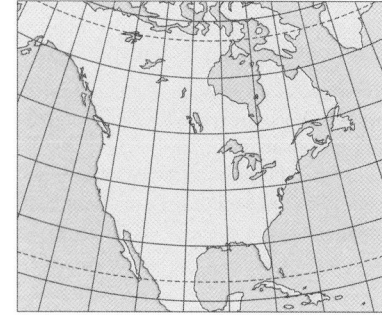
Simple Conic Projection

Lambert Conformal Conic

Type: Conic   **Conformal:** Yes   **Equal-area:** No

Notes: On the Lambert Conformal Conic projection, spacing between parallels increases with distance away from the standard parallel, which allows the property of shape to be preserved. The projection is named after Johann Lambert, an 18th century mathematician who developed some of the most important projections in use today. It became widely used in the United States in the 20th century following its adoption for many statewide mapping programs.

Uses in *Goode's Atlas of Physical Geography*: Thematic maps of the United States and Canada, and reference maps of parts of Asia.

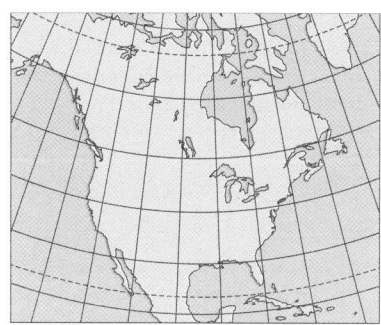
Lambert Conformal Conic Projection

Albers Equal-Area Conic

Type: Conic   **Conformal:** No   **Equal-area:** Yes

Notes: On the Albers Equal-Area Conic projection, spacing between parallels decreases with distance away from the standard parallel, which allows the property of area to be preserved. The projection is named after Heinrich Albers, who developed it in 1805. It became widely used in the 20th century, when the United States Coast and Geodetic Survey made it a standard for equal area maps of the United States.

Uses in *Goode's Atlas of Physical Geography*: Thematic maps of North America and Asia.

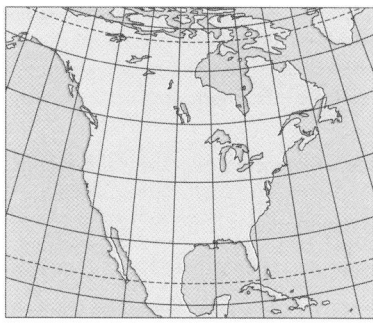
Albers Equal-Area Conic Projection

Polyconic

Type: Conic   **Conformal:** No   **Equal-area:** No

Notes: The term polyconic — literally "many-cones" — refers to the fact that this projection is an assemblage of different cones, each tangent at a different line of latitude. In contrast to many other conic projections, parallels are not concentric, and meridians are curved rather than straight. The Polyconic was first proposed by Ferdinand Hassler, who became Head of the United States Survey of the Coast (later renamed the Coast and Geodetic Survey) in 1807. The United States Geological Survey used this projection exclusively for large-scale topographic maps until the mid-20th century.

Uses in *Goode's Atlas of Physical Geography*: Reference maps of North America and Asia.

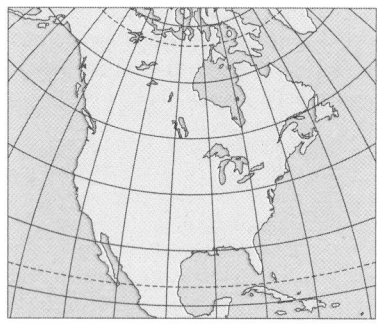
Polyconic Projection

Lambert Azimuthal Equal-Area

Type: Azimuthal   **Conformal:** No   **Equal-area:** Yes

Notes: This projection (another named after Johann Lambert) is useful for mapping large regions, as area is correctly preserved while shape distortion is relatively low. All orientations — polar, equatorial, and oblique — are common.

Uses in *Goode's Atlas of Physical Geography*: Thematic and reference maps of North and South America, Asia, Africa, Australia, and polar regions.

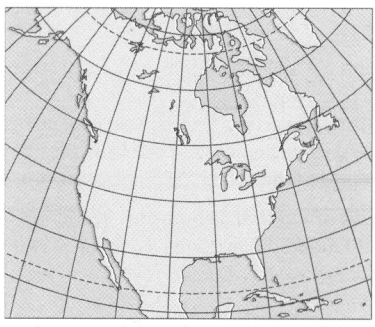
Lambert Azimuthal Equal-Area Projection

Miller Cylindrical

Type: Cylindrical   **Conformal:** No   **Equal-area:** No

Notes: This projection is useful for showing the entire earth in a simple rectangular form. However, polar areas exhibit significant exaggeration of area, a problem common to many cylindrical projections. The projection is named after Osborn Miller, Director of the American Geographical Society, who developed it in 1942 as a compromise projection that is neither conformal nor equal-area.

Uses in *Goode's Atlas of Physical Geography*: World climate and time zone maps.

Sinusoidal

Type: Pseudocylindrical   **Conformal:** No   **Equal-area:** Yes

Notes: The straight, evenly spaced parallels on this projection resemble the parallels on cylindrical projections. Unlike cylindrical projections, however, meridians are curved and converge at the poles. This causes significant shape distortion in polar regions. The Sinusoidal is the oldest-known pseudocylindrical projection, dating to the 16th century.

Uses in *Goode's Atlas of Physical Geography*: Reference maps of equatorial regions.

Mollweide

Type: Pseudocylindrical   **Conformal:** No   **Equal-area:** Yes

Notes: The Mollweide (or Homolographic) projection resembles the Sinusoidal but has less shape distortion in polar areas due to its elliptical (or oval) form. One of several pseudocylindrical projections developed in the 19th century, it is named after Karl Mollweide, an astronomer and mathematician.

Uses in *Goode's Atlas of Physical Geography*: Oceanic reference maps.

Goode's Interrupted Homolosine

Type: Pseudocylindrical   **Conformal:** No   **Equal-area:** Yes

Notes: This projection is a fusion of the Sinusoidal between 40°44'N and S, and the Mollweide between these parallels and the poles. The unique appearance of the projection is due to the introduction of discontinuities in oceanic regions, the goal of which is to reduce distortion for continental landmasses. A condensed version of the projection also exists in which the Atlantic Ocean is compressed in an east-west direction. This modification helps maximize the scale of the map on the page. The Interrupted Homolosine projection is named after J. Paul Goode of the University of Chicago, who developed it in 1923. Goode was an advocate of interrupted projections and, as editor of *Goode's School Atlas*, promoted their use in education.

Uses in *Goode's Atlas of Physical Geography*: Small-scale world thematic and reference maps. Both condensed and non-condensed forms are used. An uninterrupted example is used for the Pacific Ocean map.

Robinson

Type: Pseudocylindrical   **Conformal:** No   **Equal-area:** No

Notes: This projection resembles the Mollweide except that polar regions are flattened and stretched out. While it is neither conformal nor equal-area, both shape and area distortion are relatively low. The projection was developed in 1963 by Arthur Robinson of the University of Wisconsin, at the request of Rand McNally.

Uses in *Goode's Atlas of Physical Geography*: World maps where the interrupted nature of Goode's Homolosine would be inappropriate, such as the World Oceanic Environments map.

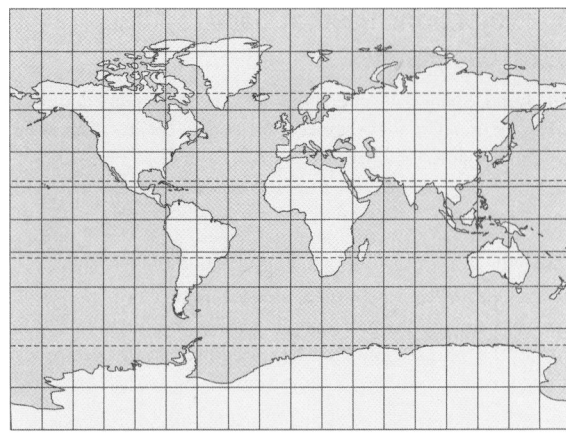

Miller Cylindrical Projection

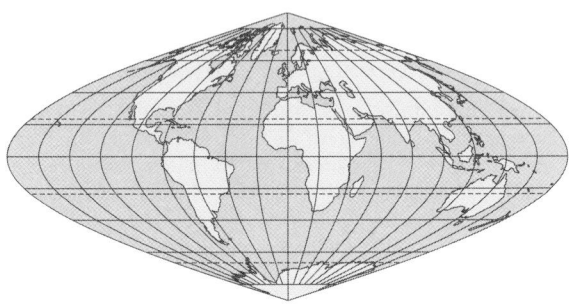

Sinusoidal Projection

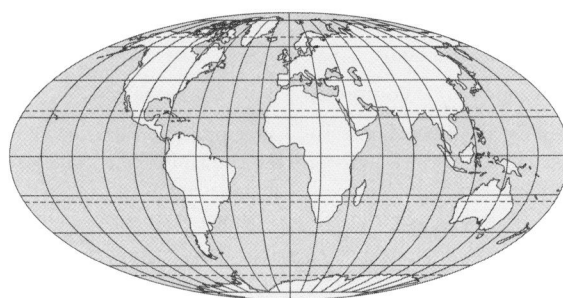

Mollweide Projection

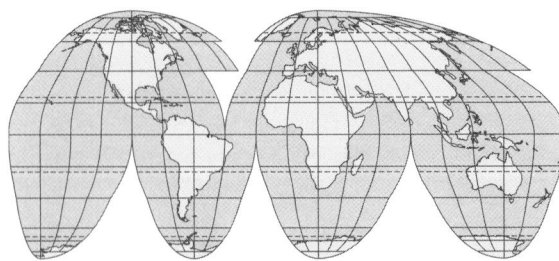

Goode's Interrupted Homolosine Projection

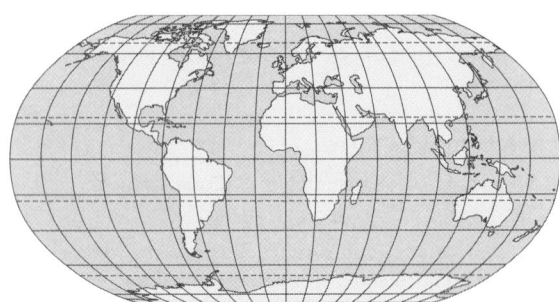

Robinson Projection

Thematic Maps in *Goode's Atlas of Physical Geography*

Thematic maps depict a single "theme" such as population density, agricultural productivity, or annual precipitation. The selected theme is presented on a base of locational information, such as coastlines, country boundaries, and major drainage features. The primary purpose of a thematic map is to convey an impression of the overall geographic distribution of the theme. It is usually not the intent of the map to provide exact numerical values. To obtain such information, the graphs and tables accompanying the map should be used.

Goode's Atlas of Physical Geography contains many different types of thematic maps. The characteristics of each are summarized below.

Point Symbol Maps

Point symbol maps are perhaps the simplest type of thematic map. They show features that occur at discrete locations. Examples include earthquakes, nuclear power plants, and minerals-producing areas. The Precious Metals map is an example of a point symbol map showing the locations of areas producing gold, silver, and platinum. A different color is used for each type of metal, while symbol size indicates relative importance.

Area Symbol Maps

Area symbol maps are useful for delineating regions of interest on the earth's surface. For example, the Tobacco and Fisheries map shows major tobacco-producing regions in one color and important fishing areas in another. On some area symbol maps, different shadings or colors are used to differentiate between major and minor areas.

Dot Maps

Dot maps show a distribution using a pattern of dots, where each dot represents a certain quantity or amount. For example, on the Sugar map, each dot represents 20,000 metric tons of sugar produced. Different dot colors are used to distinguish cane sugar from beet sugar. Dot maps are an effective way of representing the variable density of geographic phenomena over the earth's surface. This type of map is used extensively in *Goode's Atlas of Physical Geography* to show the distribution of agricultural commodities.

Area Class Maps

On area class maps, the earth's surface is divided into areas based on different classes or categories of a particular geographic phenomenon. For example, the Ecoregions map differentiates natural landscape categories, such as Tundra, Savanna, and Prairie. Other examples of area class maps in *Goode's Atlas of Physical Geography* include Landforms, Climatic Regions, Natural Vegetation, Soils, Agricultural Areas, Languages and Religions.

Isoline Maps

Isoline maps are used to portray quantities that vary smoothly over the surface of the earth. These maps are frequently used for climatic variables such as precipitation and temperature, but a variety of other quantities — from crop yield to population density — can also be treated in this way.

An isoline is a line on the map that joins locations with the same value. For example, the Summer (May to October) Precipitation map contains isolines at 5, 10, 20, and 40 inches. On this map, any 10-inch isoline separates areas that have less than 10 inches of precipitation from areas that have more than 10 inches. Note that the areas between isolines are given different colors to assist in map interpretation.

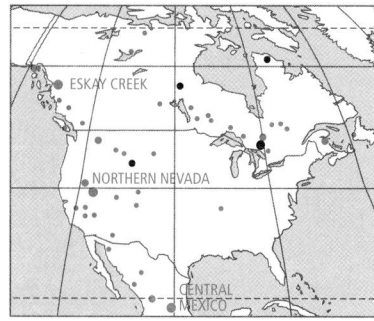

Point symbol map: Detail of Precious Metals

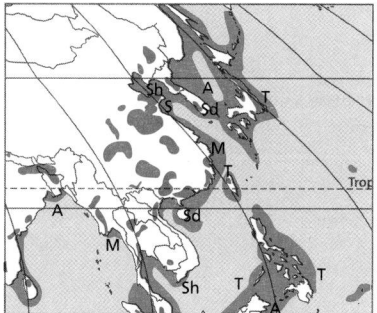

Area symbol map: Detail of Tobacco and Fisheries

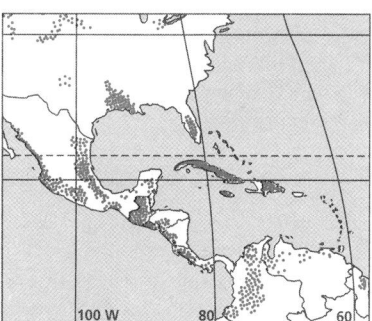

Dot map: Detail of Sugar

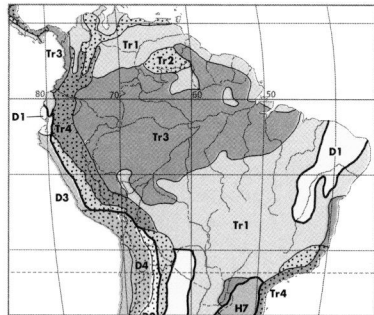

Area class map: Detail of Ecoregions

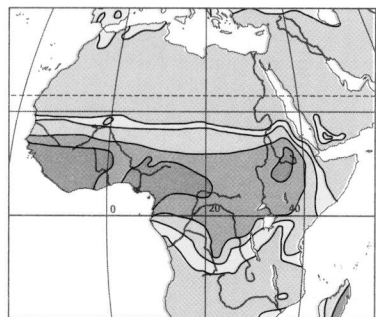

Isoline map: Detail of Precipitation

Proportional Symbol Maps

Proportional symbol maps portray numerical quantities, such as the total population of each state, the total value of agricultural goods produced in different regions, or the amount of hydroelectricity generated in different countries. The symbols on these maps — usually circles — are drawn such that the size of each is proportional to the value at that location. For example the Exports map shows the value of goods exported by each country in the world, in millions of U.S. dollars.

Proportional symbols are frequently subdivided based on the percentage of individual components making up the total. The Exports map uses wedges of different color to show the percentages of various types of exports, such as manufactured articles and raw materials.

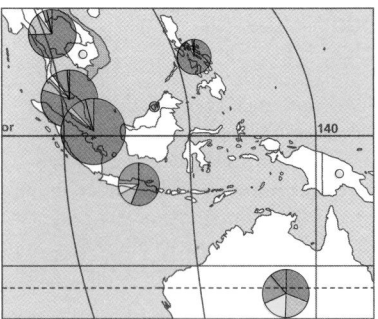
Proportional symbol map: Detail of Exports

Flow Line Maps

Flow line maps show flows between locations. Usually, the thickness of the flow lines is proportional to flow volume. Flows may be physical commodities like petroleum, or less tangible quantities like information. The flow lines on the Mineral Fuels map represent movement of petroleum measured in billions of U.S. dollars. Note that the locations of flow lines may not represent actual physical routes.

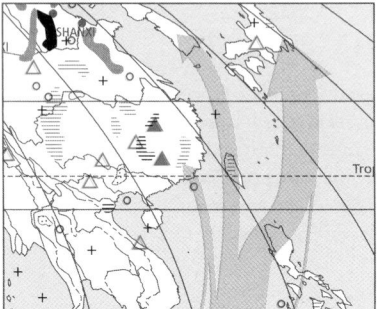
Flow line map: Detail of Mineral Fuels

Choropleth Maps

Choropleth maps apply distinctive colors to predefined areas, such as counties or states, to represent different quantities in each area. The quantities shown are usually rates, percentages, or densities. For example, the Birth Rate map shows the annual number of births per one thousand people for each country.

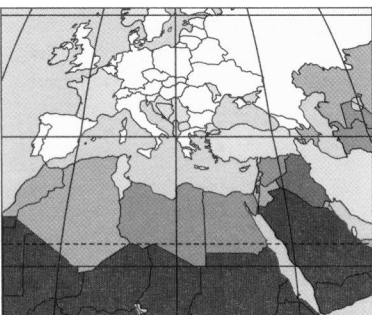
Choropleth map: Detail of Birth Rate

Digital Images

Some maps are actually digital images, analogous to the pictures captured by digital cameras. These maps are created from a very fine grid of cells called **pixels**, each of which is assigned a color that corresponds to a specific value or range of values. The population density maps in this atlas are examples of this type. The effect is much like an isoline map, but the isolines themselves are not shown and the resulting geographic patterns are more subtle and variable. This approach is increasingly being used to map environmental phenomena observable from remote sensing systems.

Cartograms

Cartograms deliberately distort map shapes to achieve specific effects. On **area cartograms**, the size of each area, such as a country, is made proportional to its population. Countries with large populations are therefore drawn larger than countries with smaller populations, regardless of the actual size of these countries on the earth.

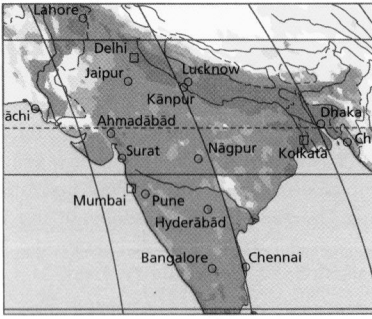
Digital image map: Detail of Population Density

The world cartogram series in this atlas depicts each country as a rectangle. This is a departure from cartograms in earlier editions of the atlas, which attempted to preserve some of the salient shape characteristics for each country. The advantage of the rectangle method is that it is easier to compare the area of countries when their shapes are consistent.

The cartogram series incorporates choropleth shading on top of the rectangular cartogram base. In this way map readers can make inferences about the relationship between population and another thematic variable, such as HIV-infection rates.

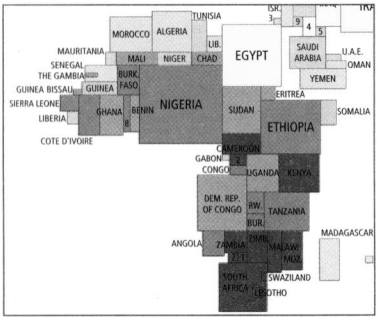
Cartogram: Detail of HIV Infection

Map Legend

Political Boundaries

Political maps	Physical maps	
━━ ━ ━	━━ ━━ ━━	International (Demarcated, Undemarcated, and Administrative)
━━ ━ ━	━━ ━ ━	Disputed de facto
▓▓ ▓ ▓	▓▓ ▓ ▓	Indefinite or Undefined
━━ ━ ━	━━ ━ ━	Secondary, State, Provincial, etc.

⬭	Parks, Indian Reservations
⬭	City Limits
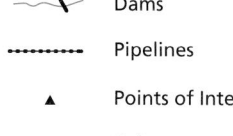	Urbanized Areas

Transportation

Political maps	Physical maps	
───────	───────	Railroads
-------	-------	Railroad Ferries
	───────	Major Roads
	───────	Minor Roads
	....	Caravan Routes
	✈	Airports

Cultural Features

⌁	Dams
••••••••••	Pipelines
▲	Points of Interest
∴	Ruins

Populated Places

◉	1,000,000 and over
◎	250,000 to 1,000,000
⊙	100,000 to 250,000
•	25,000 to 100,000
○	Under 25,000
▫	Neighborhoods, Sections of Cities
TŌKYŌ	National Capitals
Boise	Secondary Capitals

Note: On maps at 1:20,000,000 and smaller, symbols do not follow the population classification shown above. Some other maps use a slightly different classification, which is shown in a separate legend in the map margin. On all maps, type size indicates the relative importance of the city.

Land Features

△	Peaks, Spot Heights
⤬	Passes
▨	Sand
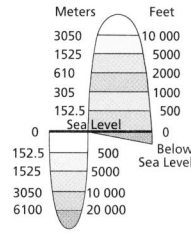	Contours

Elevation

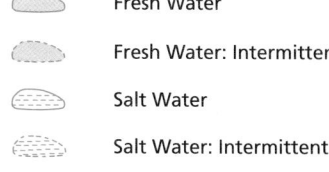

Meters		Feet
3050		10 000
1525		5000
610		2000
305		1000
152.5		500
0	Sea Level	0
152.5	500	Below Sea Level
1525	5000	
3050	10 000	
6100	20 000	

Lakes and Reservoirs

⬭	Fresh Water
⬭	Fresh Water: Intermittent
⬭	Salt Water
⬭	Salt Water: Intermittent

Other Water Features

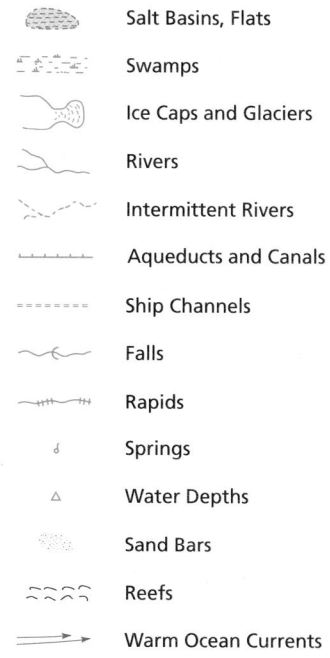

	Salt Basins, Flats
	Swamps
	Ice Caps and Glaciers
	Rivers
	Intermittent Rivers
	Aqueducts and Canals
	Ship Channels
	Falls
	Rapids
	Springs
	Water Depths
	Sand Bars
	Reefs
→	Warm Ocean Currents
→	Cold Ocean Currents

The legend above shows the symbols used for the political and physical reference maps in *Goode's Atlas of Physical Geography*.

To portray relative areas correctly, uniform map scales have been used wherever possible:

Continents – 1:40,000,000
Countries and regions – between 1:4,000,000 and 1:20,000,000
World, polar areas and oceans – between 1:50,000,000 and 1:100,000,000
Urbanized areas – 1:1,000,000

Elevations on the maps are shown using a combination of shaded relief and hypsometric tints. Shaded relief (or hill-shading) gives a three-dimensional impression of the landscape, while hypsometric tints show elevation ranges in different colors.

The choice of names for mapped features is complicated by the fact that a variety of languages and alphabets are used throughout the world. A local-names policy is used in *Goode's Atlas of Physical Geography* for populated places and local physical features. For some major features, an English form of the name is used with the local name given below in parentheses. Examples include Moscow (Moskva), Vienna (Wien) and Naples (Napoli). In countries where more than one official language is used, names are given in the dominant local language. For large physical features spanning international borders, the conventional English form of the name is used. In cases where a non-Roman alphabet is used, names have been transliterated according to accepted practice.

Selected features are also listed in the Index (pp. 217-322), which includes a pronunciation guide. A list of foreign geographic terms is provided in the Glossary (p. 214).

POLITICAL

ARCTIC OCEAN

GREENLAND (Den.)

Baffin Bay

RUS. ALASKA (U.S.)
Nome
Anchorage
Juneau

Reykjavík ○ ICELAND

GREEN

HUDSON BAY

C A N A D A
Edmonton
Vancouver   Winnipeg
Seattle   Ottawa   Québec
Portland   Chicago   Montréal   St. John's
Detroit   Toronto   Halifax
San Francisco   U N I T E D   S T A T E S   Boston
St. Louis   New York
Washington

PORTL
Lisb

Azores (Port.)

Los Angeles   Phoenix
Dallas   New Orleans   BERMUDA (U.K.)
Houston   Miami   BAHAMAS
Atlanta
GULF OF MEXICO   Havana   CUBA

Casablan

Madeira Is. (Port.)

Tropic of Cancer

MIDWAY ISLANDS (U.S.)

MEXICO
Guadalajara
Mexico City

JOHNSTON ATOLL (U.S.)

HAWAII (U.S.)
Honolulu

Canary Is. (Sp.)

W. SAHARA WOR
MAURITANIA
CAPE VERDE   Dakar

A T L A N T I C

HAITI   DOM. REP.   PUERTO RICO (U.S.)
JAMAICA   GUADELOUPE (Fr.)
BELIZE   MARTINIQUE (Fr.)
GUAT.   HOND.   BARBADOS
EL SAL.   NIC.   CARIBBEAN SEA   TRINIDAD AND TOBAGO
COSTA RICA   GUYANA
PANAMA   VENEZUELA   Georgetown   GUYANA
Caracas   SURINAME
Bogotá   FRENCH GUIANA (Fr.)
COLOMBIA

THE GAMBIA   SENEGAL   M
GUINEA-BISSAU   GUINEA   BUR
SIERRA LEONE   CÔTE D'IVOIRE
LIBERIA

P A C I F I C

HOWLAND ISLAND (U.S.)
JARVIS ISLAND (U.S.)
BAKER ISLAND (U.S.)
KIRIBATI

ECUADOR   Quito
Galápagos Is. (Ec.)

Belém   Fortaleza

Equator

Manaus   Amazon
PERU   Lima   BRAZIL   Recife
Brasília

TOKELAU (N.Z.)
SAMOA   AMERICAN SAMOA (U.S.)
COOK ISLANDS (N.Z.)
FRENCH POLYNESIA (Fr.)

O C E A N

Salvador

ST. HEL (U.K.)

O C E A

La Paz   Belo Horizonte
BOLIVIA   Sucre   Rio de Janeiro
PARAGUAY   São Paulo
TONGA

PITCAIRN ISLANDS (U.K.)

Antofagasta   Asunción   Porto Alegre

Trop

Valparaíso   Rosario   URUGUAY
Santiago   ARGENTINA   Montevideo
Buenos Aires

CHILE

FALKLAND ISLANDS (U.K.)
SOUTH GEORGIA AND THE SOUTH SANDWICH ISLANDS (U.K.)

Antarctic Circle

S O U T H E R N   O C E A N
ROSS SEA

WEDDELL SEA

Scale 1 : 100 000 000 (approximate)
One inch to 1,600 miles
0   500   1000   1500   2000 miles
0   500   1000   1500   2000   2500 Kilometers

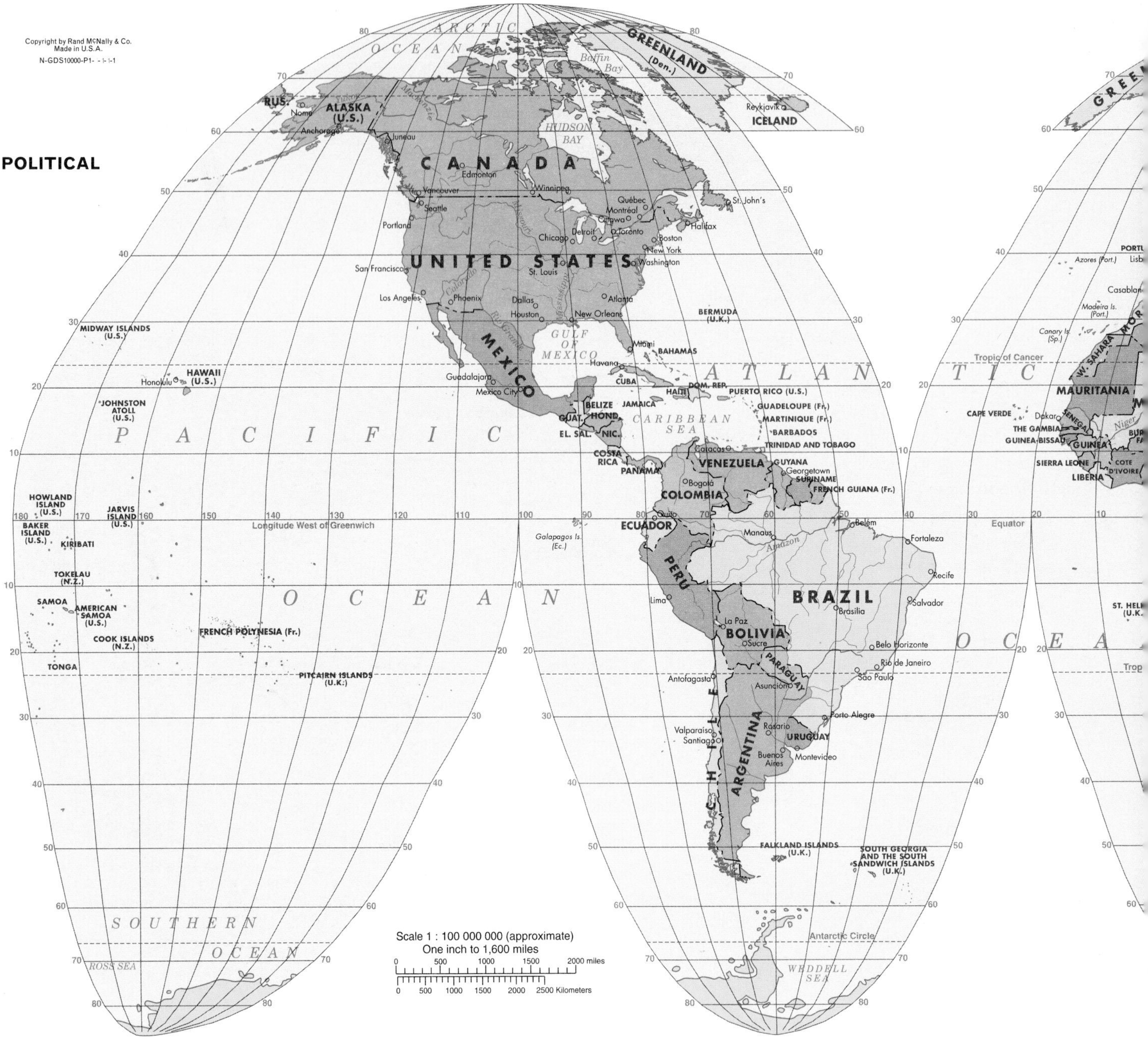

Comparative Land Areas (Land and inland water. Numbers indicate thousands of square miles.)

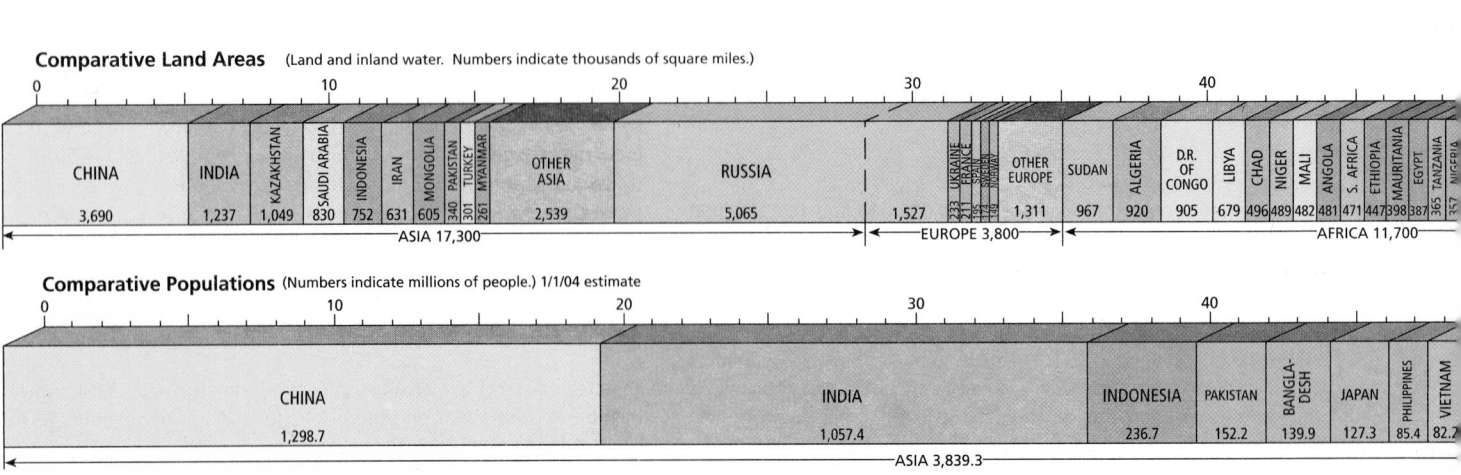

	0		10								20			30				40													

| CHINA | INDIA | KAZAKHSTAN | SAUDI ARABIA | INDONESIA | IRAN | MONGOLIA | PAKISTAN | TURKEY | MYANMAR | OTHER ASIA | RUSSIA | UKRAINE | FRANCE | SPAIN | SWEDEN | GERMANY | OTHER EUROPE | SUDAN | ALGERIA | D.R. OF CONGO | LIBYA | CHAD | NIGER | MALI | ANGOLA | S. AFRICA | ETHIOPIA | MAURITANIA | EGYPT | TANZANIA | NIGERIA |
| 3,690 | 1,237 | 1,049 | 830 | 752 | 631 | 605 | 340 | 301 | 261 | 2,539 | 5,065 | 233 | 211 | 195 | 174 | 138 | 1,311 | 967 | 920 | 905 | 679 | 496 | 489 | 482 | 481 | 471 | 447 | 398 | 387 | 365 | 357 |

←————————————————— ASIA 17,300 ————————————————→ ←—— EUROPE 3,800 ——→ ←——————— AFRICA 11,700

1,527

Comparative Populations (Numbers indicate millions of people.) 1/1/04 estimate

	0		10			20			30				40			

| CHINA | INDIA | INDONESIA | PAKISTAN | BANGLA-DESH | JAPAN | PHILIPPINES | VIETNAM |
| 1,298.7 | 1,057.4 | 236.7 | 152.2 | 139.9 | 127.3 | 85.4 | 82.2 |

←————————————————————————— ASIA 3,839.3 —————————————————————————→

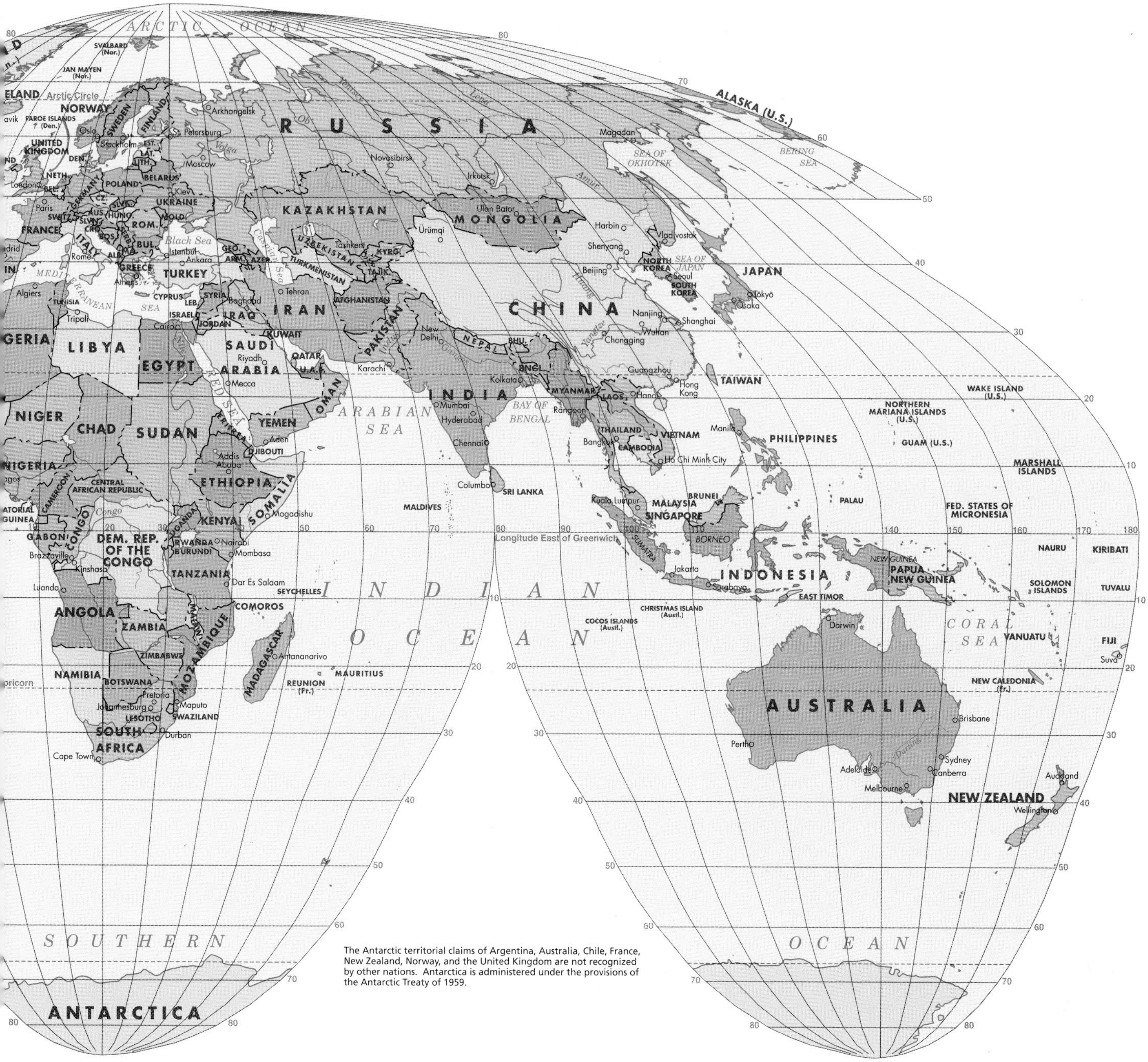

ARCTIC OCEAN

RUSSIA

ALASKA (U.S.)

SVALBARD (Nor.)

JAN MAYEN (Nor.)

ICELAND   Arctic Circle

NORWAY   SWEDEN   FINLAND   Arkhangelsk

FAROE ISLANDS (Den.)   St. Petersburg

UNITED KINGDOM   DEN.   Stockholm   EST.   LAT.   Moscow   Novosibirsk   Irkutsk   Magadan

NETH.   POLAND   BELARUS   SEA OF OKHOTSK   BERING SEA

London   GER.   CZ.   UKRAINE   Kiev   KAZAKHSTAN   MONGOLIA   Ulan Bator   Harbin   Vladivostok

FRANCE   SWITZ.   HUNG.   ROM.   MOLD.   Shenyang   NORTH KOREA   JAPAN

ITALY   BUL.   UZBEKISTAN   KYRG.   Ürümqi   Beijing   SOUTH KOREA   Seoul   Tōkyō

GREECE   TURKEY   ARM.   AZER.   TURKMENISTAN   TAJIK.   Nanjing   Ōsaka

CYPRUS   SYRIA   GEO.   Tehran   AFGHANISTAN   CHINA   Shanghai   Wuhan

ISRAEL   IRAQ   IRAN   PAKISTAN   New Delhi   NEPAL   BHU.   Chongqing   Guangzhou   TAIWAN

LIBYA   EGYPT   SAUDI ARABIA   KUWAIT   QATAR   U.A.E.   Karachi   INDIA   BNGL.   MYANMAR   LAOS   Hong Kong

NIGER   CHAD   SUDAN   YEMEN   OMAN   ARABIAN SEA   Mumbai   Hyderabad   BAY OF BENGAL   THAILAND   VIETNAM   Manila   PHILIPPINES

NIGERIA   CENTRAL AFRICAN REPUBLIC   ETHIOPIA   Chennai   CAMBODIA   Ho Chi Minh City

EQUATORIAL GUINEA   CONGO   KENYA   SOMALIA   Mogadishu   SRI LANKA   MALDIVES   MALAYSIA   BRUNEI   PALAU

GABON   DEM. REP. OF THE CONGO   RWANDA   BURUNDI   Nairobi   TANZANIA   SEYCHELLES   Kuala Lumpur   SINGAPORE   BORNEO

ANGOLA   ZAMBIA   MALAWI   MOZAMBIQUE   MADAGASCAR   COMOROS   Antananarivo   MAURITIUS   REUNION (Fr.)   INDONESIA   Jakarta   EAST TIMOR   PAPUA NEW GUINEA

NAMIBIA   BOTSWANA   ZIMBABWE   SWAZILAND   LESOTHO   SOUTH AFRICA   Durban   Cape Town

INDIAN OCEAN

AUSTRALIA   Darwin   Perth   Adelaide   Melbourne   Sydney   Canberra   Brisbane

NEW ZEALAND   Auckland   Wellington

CORAL SEA   VANUATU   FIJI   Suva   NEW CALEDONIA (Fr.)

SOUTHERN OCEAN

ANTARCTICA

The Antarctic territorial claims of Argentina, Australia, Chile, France, New Zealand, Norway, and the United Kingdom are not recognized by other nations. Antarctica is administered under the provisions of the Antarctic Treaty of 1959.

Goode's Homolosine Equal Area Projection

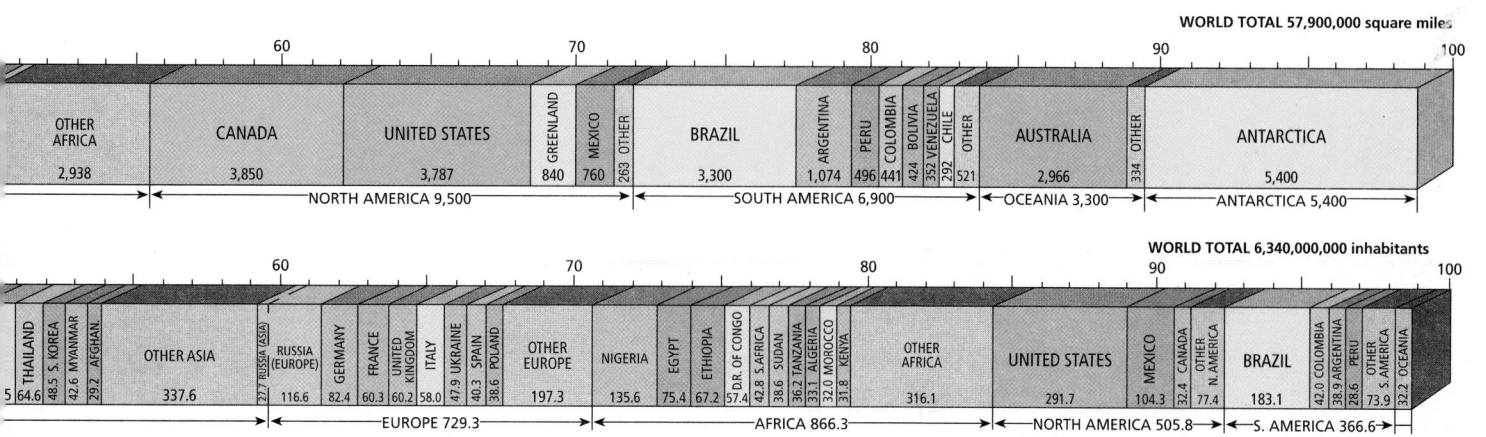

WORLD TOTAL 57,900,000 square miles

| OTHER AFRICA 2,938 | CANADA 3,850 | UNITED STATES 3,787 | GREENLAND 840 | MEXICO 760 | OTHER 263 | BRAZIL 3,300 | ARGENTINA 1,074 | PERU 496 | COLOMBIA 441 | BOLIVIA 424 | VENEZUELA 352 | CHILE 292 | OTHER 521 | AUSTRALIA 2,966 | OTHER 334 | ANTARCTICA 5,400 |

NORTH AMERICA 9,500   SOUTH AMERICA 6,900   OCEANIA 3,300   ANTARCTICA 5,400

WORLD TOTAL 6,340,000,000 inhabitants

| THAILAND 64.6 | S. KOREA 48.5 | MYANMAR 42.6 | AFGHAN. 29.2 | OTHER ASIA 337.6 | RUSSIA (ASIA) 22.7 | RUSSIA (EUROPE) 116.6 | GERMANY 82.4 | FRANCE 60.3 | UNITED KINGDOM 60.2 | ITALY 58.0 | UKRAINE 47.9 | SPAIN 40.3 | POLAND 38.6 | OTHER EUROPE 197.3 | NIGERIA 135.6 | EGYPT 75.4 | ETHIOPIA 67.2 | D.R. OF CONGO 57.4 | S.AFRICA 42.8 | SUDAN 38.6 | TANZANIA 36.2 | ALGERIA 33.1 | MOROCCO 32.0 | KENYA 31.8 | OTHER AFRICA 316.1 | UNITED STATES 291.7 | MEXICO 104.3 | CANADA 32.4 | OTHER N. AMERICA 77.4 | BRAZIL 183.1 | COLOMBIA 42.0 | ARGENTINA 38.9 | PERU 28.6 | OTHER S. AMERICA 73.9 | OCEANIA 32.2 |

EUROPE 729.3   AFRICA 866.3   NORTH AMERICA 505.8   S. AMERICA 366.6

PHYSICAL

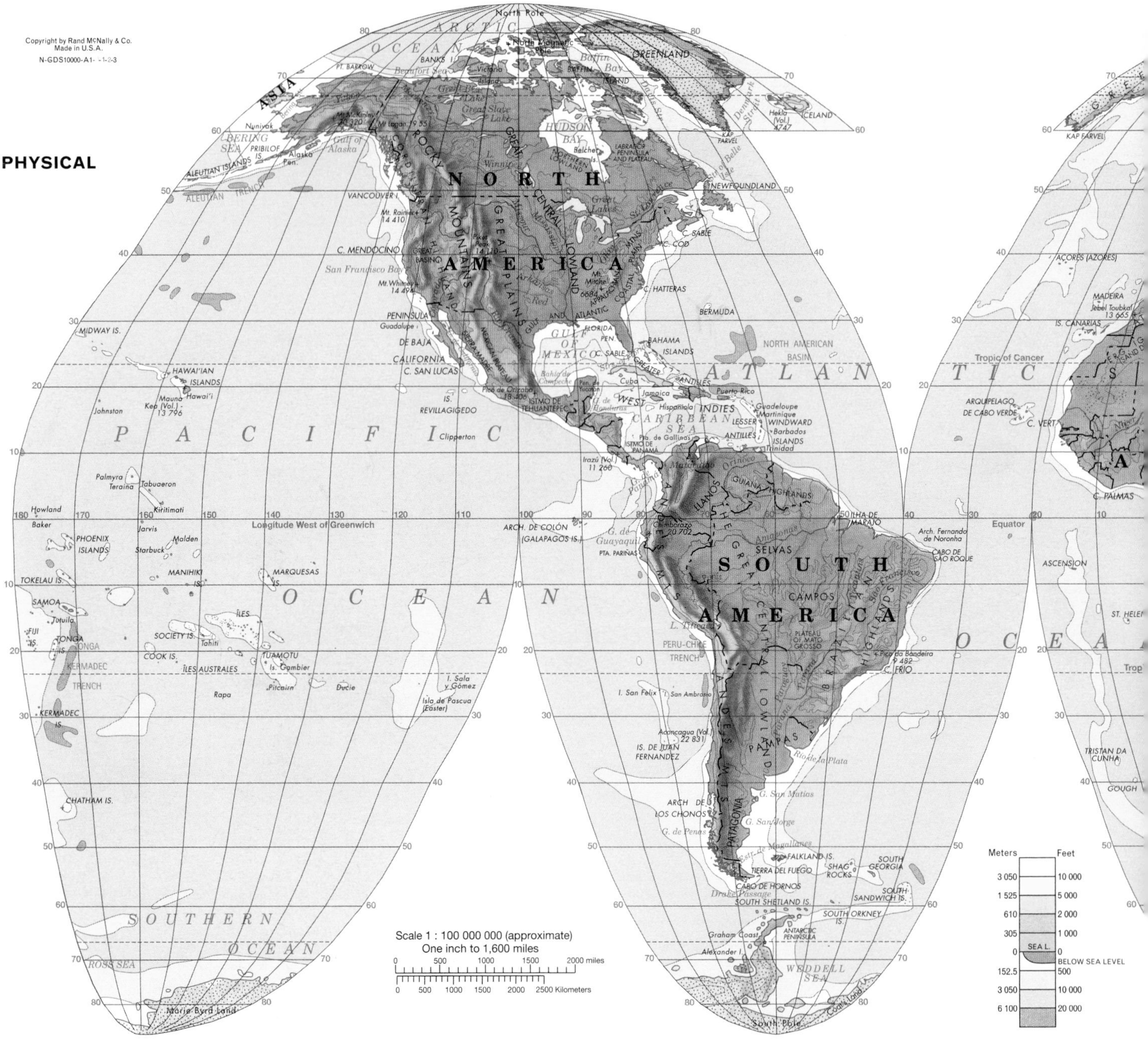

Land Elevations in Profile

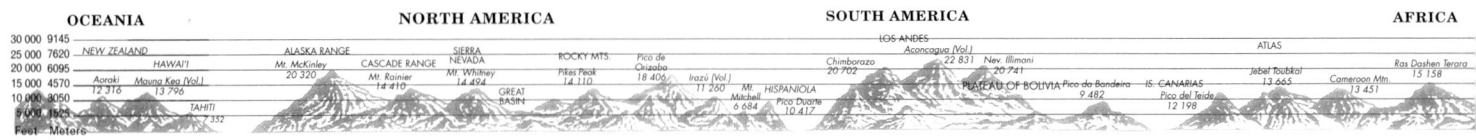

Ocean Depths in Profile

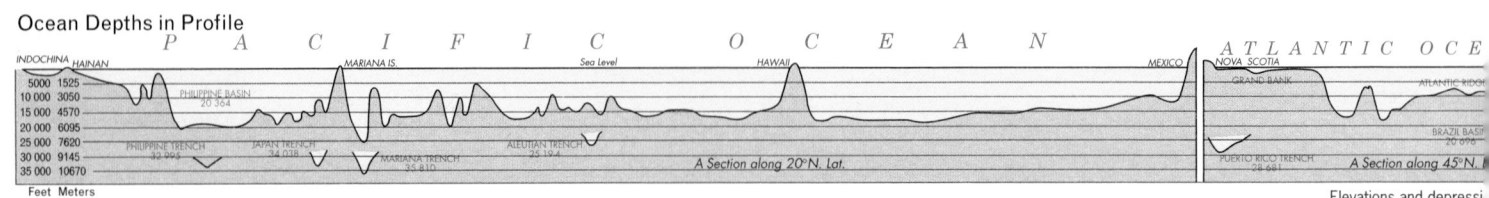

Elevations and depressi

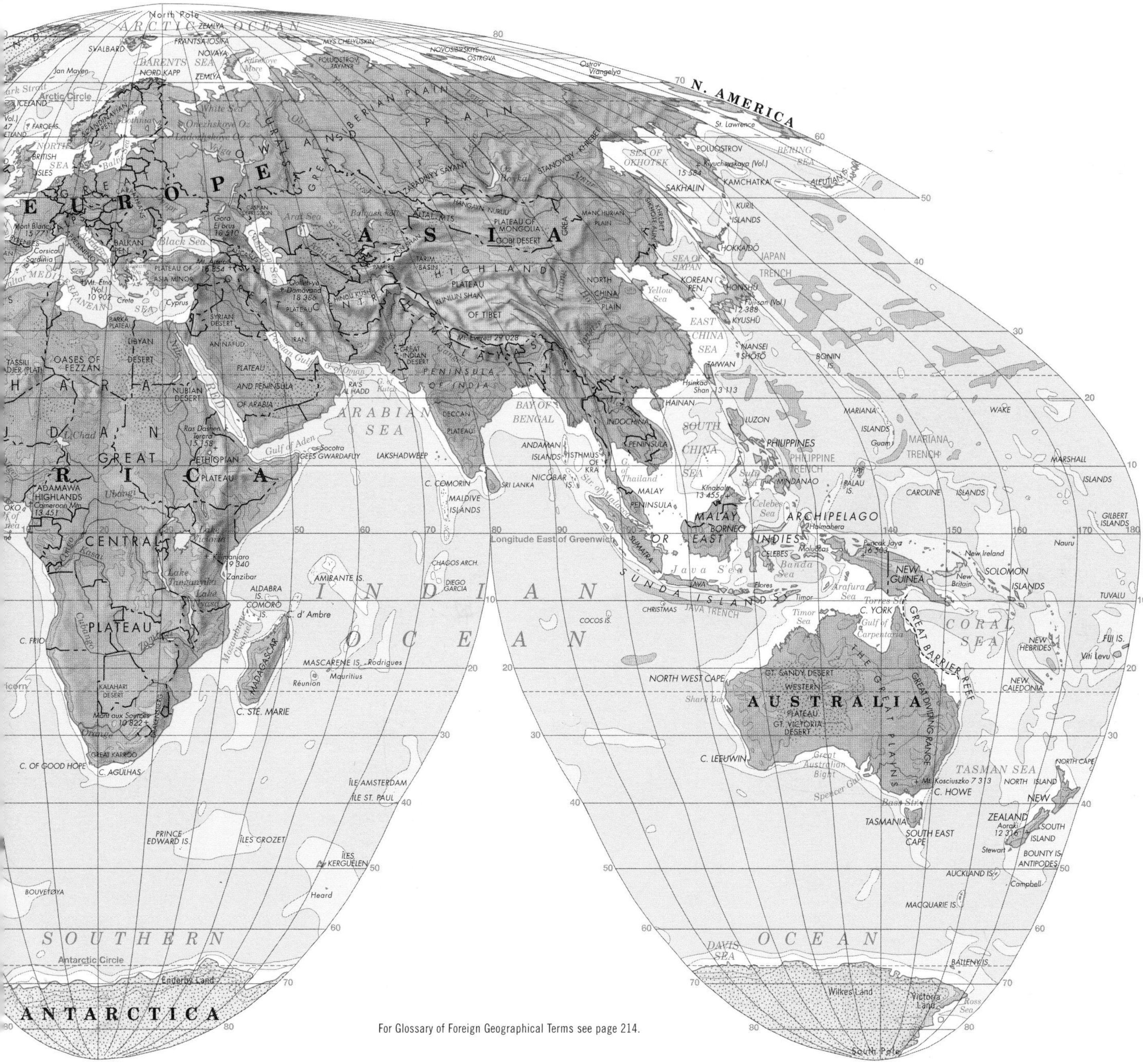

North Pole
ARCTIC OCEAN
ZEMLYA
FRANTSA-IOSIFA
SVALBARD
NOVAYA ZEMLYA
NORD KAPP
Jan Mayen
BARENTS SEA
MYS CHELYUSKIN
NOVOSIBIRSKIYE OSTROVA
N. AMERICA
Ostrov Vrangelya
70
Arctic Circle
ICELAND
FAROES
ST. LAWRENCE
SEA OF OKHOTSK
POLUOSTROV
BERING SEA
GREAT SIBERIAN PLAIN
STANOVOY KHREBET
Klyuchevskaya (Vol.) 15 584
KAMCHATKA
KURIL ISLANDS
ALEUTIAN IS.
50

EUROPE
Mont Blanc 15 771
Gora El'brus 18 510
BALKAN PEN.
Black Sea
Sicily
Corsica
Sardinia
MEDITERRANEAN
Crete
Cyprus
ASIA
Aral Sea
Balkhash Ozero
ALTAI MTS.
TARIM BASIN
PLATEAU OF TIBET
PLATEAU OF MONGOLIA
GOBI DESERT
MANCHURIAN PLAIN
SEA OF JAPAN
HOKKAIDŌ
JAPAN
HONSHŪ
KOREAN PEN.
KYŪSHŪ
Fujisan (Vol.) 12 388
JAPAN TRENCH
40
ASIA MINOR
SYRIAN DESERT
IRAN
HIMALAYAS
Mt. Everest 29 028
KUNLUN SHAN
NORTH CHINA PLAIN
Yellow Sea
EAST CHINA SEA
NANSEI SHŌTŌ
TAIWAN
BONIN IS.
30
SAHARA
LIBYAN DESERT
NUBIAN DESERT
RED SEA
AN NAFUD
PLATEAU AND PENINSULA OF ARABIA
Gulf of Oman
DECCAN PLATEAU
BAY OF BENGAL
INDOCHINA
HAINAN
SOUTH CHINA SEA
LUZON
PHILIPPINES
MARIANA ISLANDS
Guam
MARIANA TRENCH
WAKE
20
JORDAN
Lake Chad
GREAT RIFT VALLEY
Ras Dashen Terara 15 158
ETHIOPIAN PLATEAU
ARABIAN SEA
Socotra
GEES GWARDAFUY
LAKSHADWEEP
C. COMORIN
SRI LANKA
NICOBAR IS.
ANDAMAN ISLANDS
ISTHMUS OF KRA
G. of Thailand
MALAY PENINSULA
Kinabalu 13 455
MINDANAO
PALAU IS.
CAROLINE ISLANDS
MARSHALL ISLANDS
GILBERT ISLANDS
10
ADAMAWA HIGHLANDS
Cameroon Mtn. 13 451
CENTRAL AFRICA
MALDIVE ISLANDS
CHAGOS ARCH.
MALAY ARCHIPELAGO
BORNEO
CELEBES
Celebes Sea
Halmahera
Moluccas
Puncak Jaya 16 503
NEW GUINEA
New Ireland
SOLOMON ISLANDS
New Britain
Nauru
TUVALU
Longitude East of Greenwich
Lake Victoria
Kilimanjaro 19 340
Zanzibar
ALDABRA IS.
COMORO IS.
C. d'Ambre
AMIRANTE IS.
DIEGO GARCIA
INDIAN OCEAN
Java Sea
JAVA
Banda Sea
Flores
Timor
Arafura Sea
Torres Str.
C. YORK
Gulf of Carpentaria
CORAL SEA
NEW HEBRIDES
FIJI IS.
Viti Levu
NEW CALEDONIA
10
MADAGASCAR
MASCARENE IS. Rodrigues
Réunion
Mauritius
COCOS IS.
CHRISTMAS I.
JAVA TRENCH
SUNDA ISLANDS
Timor Sea
20
Mozambique Channel
PLATEAU
C. STE. MARIE
NORTH WEST CAPE
Shark Bay
GT. SANDY DESERT
WESTERN PLATEAU
GT. VICTORIA DESERT
AUSTRALIA
GREAT DIVIDING RANGE
GREAT BARRIER REEF
THE GREAT PLAINS
TASMAN SEA
NORTH CAPE
30
C. FRIO
KALAHARI DESERT
Mont aux Sources 10 822
GREAT KARROO
Orange
C. OF GOOD HOPE
C. AGULHAS
ÎLE AMSTERDAM
ÎLE ST. PAUL
C. LEEUWIN
Great Australian Bight
Spencer Gulf
C. HOWE
Mt. Kosciuszko 7 313
NORTH ISLAND
Aoraki 12 316
NEW ZEALAND
SOUTH ISLAND
40
Tropic of Capricorn
PRINCE EDWARD IS.
ÎLES CROZET
ÎLES KERGUELEN
TASMANIA
SOUTH EAST CAPE
Bass Str.
Stewart
BOUNTY IS.
ANTIPODES
50
BOUVETØYA
Heard
AUCKLAND IS.
Campbell
60
SOUTHERN OCEAN
MACQUARIE IS.
OCEAN
DAVIS SEA
70
Antarctic Circle
Enderby Land
Wilkes Land
Victoria Land
Ross Sea
BALLENY IS.
80
ANTARCTICA
South Pole

For Glossary of Foreign Geographical Terms see page 214.

Goode's Homolosine Equal Area Projection

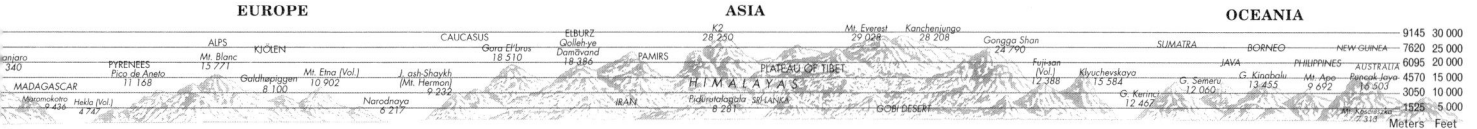

EUROPE	ASIA	OCEANIA

ALPS   KJØLEN   CAUCASUS   ELBURZ   K2 28 250   Mt. Everest 29 028   Kanchenjunga 28 208   Gongga Shan 24 790   SUMATRA   BORNEO   NEW GUINEA   9145   30 000
Gora El'brus 18 510   Qolleh-ye Damāvand 18 386   PAMIRS   7620   25 000
PYRENEES   Mt. Blanc 15 771   Pico de Aneto 11 168   Galdhøpiggen 8 100   Mt. Etna (Vol.) 10 902   J. ash-Shaykh (Mt. Hermon) 9 232   PLATEAU OF TIBET   Fuji-san (Vol.) 12 388   Klyuchevskaya 15 584   JAVA   PHILIPPINES   AUSTRALIA   6095   20 000
HIMALAYAS   G. Kinabalu 13 455   Mt. Apo 9 692   Puncak Jaya 16 503   4570   15 000
MADAGASCAR   Hekla (Vol.) 4 747   Narodnaya 6 217   IRAN   Piduruталagala 8 281   SRI LANKA   G. Semeru 12 060   G. Kerinci 12 467   GOBI DESERT   Mt. Kosciuszko 7 313   3050   10 000   1525   5 000
Meters   Feet

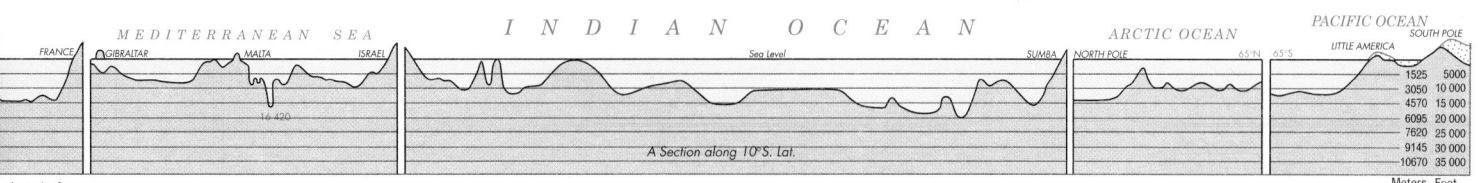

FRANCE
MEDITERRANEAN SEA
GIBRALTAR
MALTA
16 420
ISRAEL
INDIAN OCEAN
Sea Level
SUMBA
ARCTIC OCEAN
NORTH POLE
65°N   65°S
PACIFIC OCEAN
SOUTH POLE
LITTLE AMERICA
1525   5000
3050   10 000
4570   15 000
6095   20 000
7620   25 000
9145   30 000
10670   35 000
A Section along 10°S. Lat.
Meters   Feet

given in feet

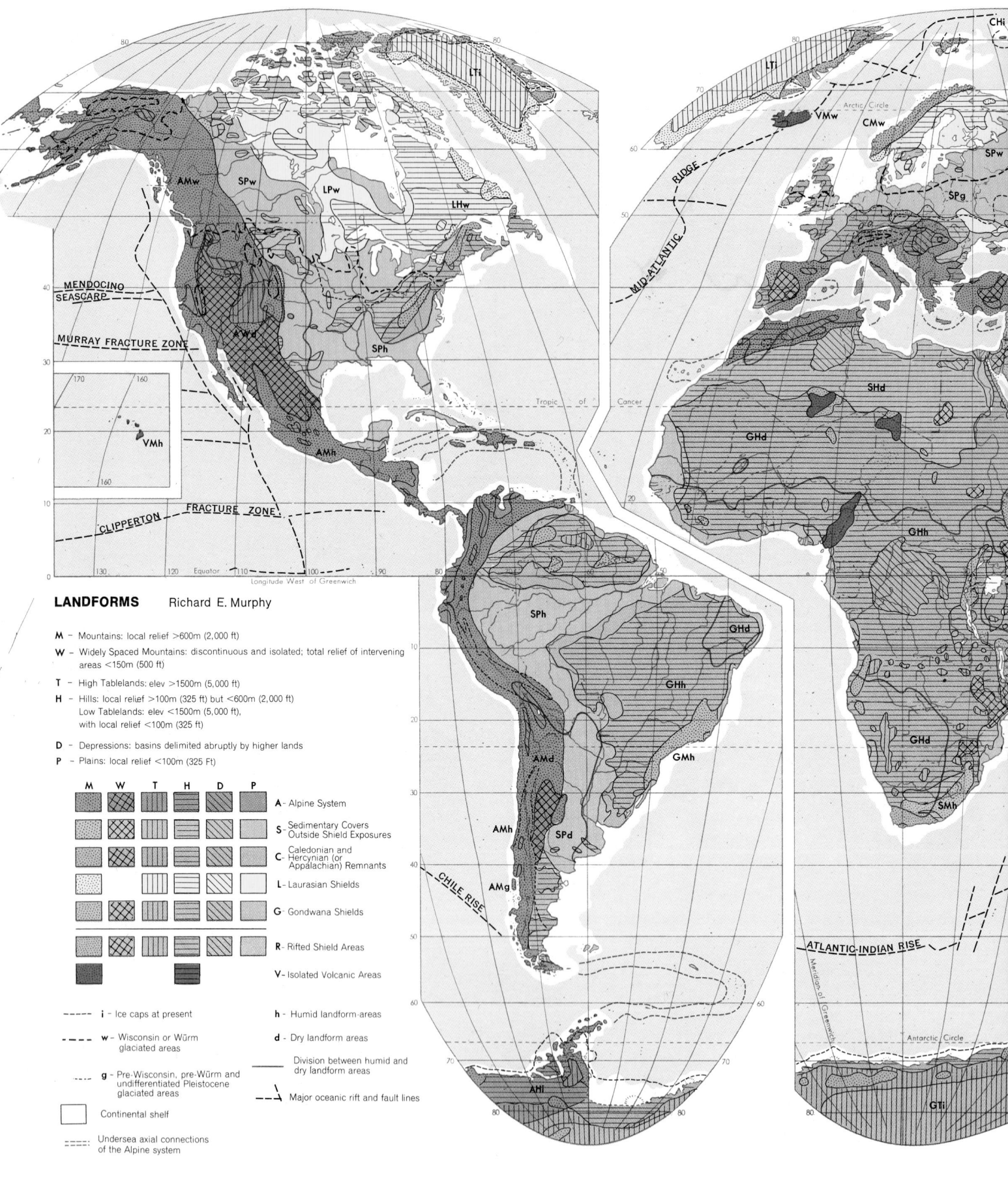

LANDFORMS   Richard E. Murphy

M – Mountains: local relief >600m (2,000 ft)

W – Widely Spaced Mountains: discontinuous and isolated; total relief of intervening areas <150m (500 ft)

T – High Tablelands: elev >1500m (5,000 ft)

H – Hills: local relief >100m (325 ft) but <600m (2,000 ft)
Low Tablelands: elev <1500m (5,000 ft), with local relief <100m (325 ft)

D – Depressions: basins delimited abruptly by higher lands

P – Plains: local relief <100m (325 Ft)

M   W   T   H   D   P

A - Alpine System

S - Sedimentary Covers Outside Shield Exposures

C - Caledonian and Hercynian (or Appalachian) Remnants

L - Laurasian Shields

G - Gondwana Shields

R - Rifted Shield Areas

V - Isolated Volcanic Areas

- - - - **i** - Ice caps at present

- - - - **w** - Wisconsin or Würm glaciated areas

- - - - **g** - Pre-Wisconsin, pre-Würm and undifferentiated Pleistocene glaciated areas

Continental shelf

- - - - Undersea axial connections of the Alpine system

h - Humid landform areas

d - Dry landform areas

———— Division between humid and dry landform areas

- - - - Major oceanic rift and fault lines

SPg
SHh
AMg
SPh
SPd
ADd
AMh
SHd
OWEN FRACTURE ZONE
CARLSBURG RIDGE
GHh
d
GMh
I INDIAN RIDGE
MID-INDIAN RIDGE
Longitude East of Greenwich
Tropic of Cancer
AMh
Equator
Tropic of Capricorn
GHd
SPd
CHh
AMh
AMg
AUSTRALIAN–ANTARCTIC RISE
GTi

Scale 1 : 75 000 000 (approximate)
One inch to 1 200 miles
0   500   1000   1500 Miles
0   500   1000   1500   2000 Kilometers

Goode's Homolosine Equal Area Projection (Condensed)
A-510000- 9A6 -3-5-**7**

CONTINENTAL DRIFT

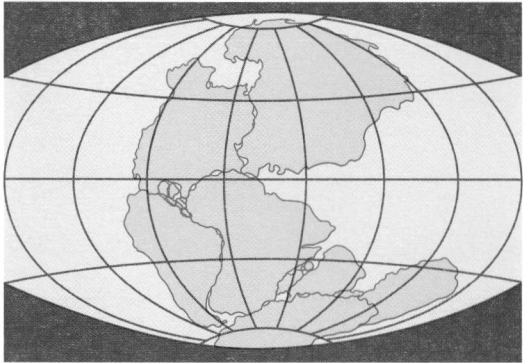

225 million years ago the supercontinent of Pangaea exists and Panthalassa forms the ancestral ocean. Tethys Sea separates Eurasia and Africa.

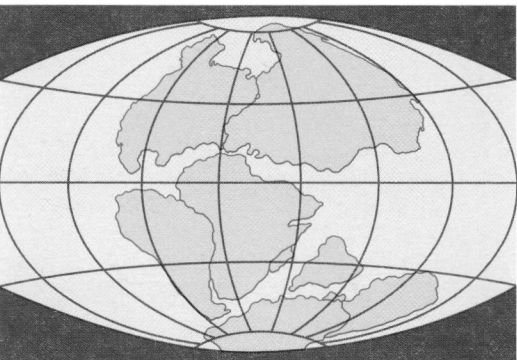

180 million years ago Pangaea splits, Laurasia drifts north. Gondwanaland breaks into South America/Africa, India, and Australia/Antarctica.

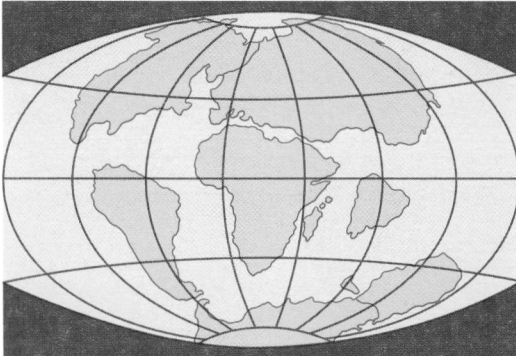

65 million years ago ocean basins take shape as South America and India move from Africa and the Tethys Sea closes to form the Mediterranean Sea.

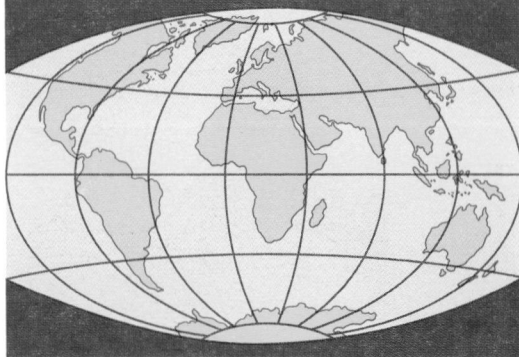

The present day: India has merged with Asia, Australia is free of Antarctica, and North America is free of Eurasia.

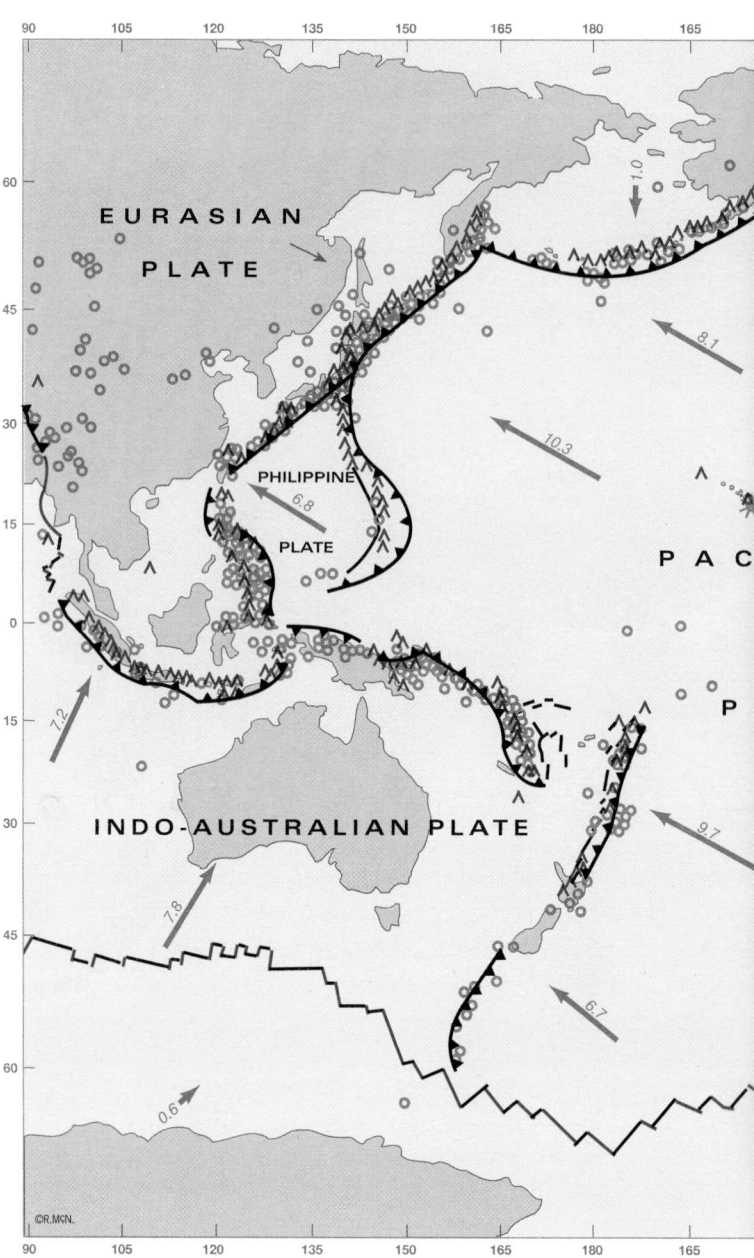

PLATE TECTONICS

Types of plate boundaries

▬▬▬ **Divergent:** magma emerges from the earth's mantle at the mid-ocean ridges forming new crust and forcing the plates to spread apart at the ridges.

▲▲▲ **Convergent:** plates collide at subduction zones where the denser plate is forced back into the earth's mantle forming deep ocean trenches.

▬▬▬ **Transform:** plates slide past one another producing faults and fracture zones.

Other map symbols

→ Direction of plate movement

6.7 ➤ Length of arrow is proportional to the amount of plate movement (number indicates centimeters of movement per year)

○ Earthquake of magnitude 7.5 and above (from 10 A.D. to the present)

∧ Volcano (eruption since 1900)

✳ Selected hot spots

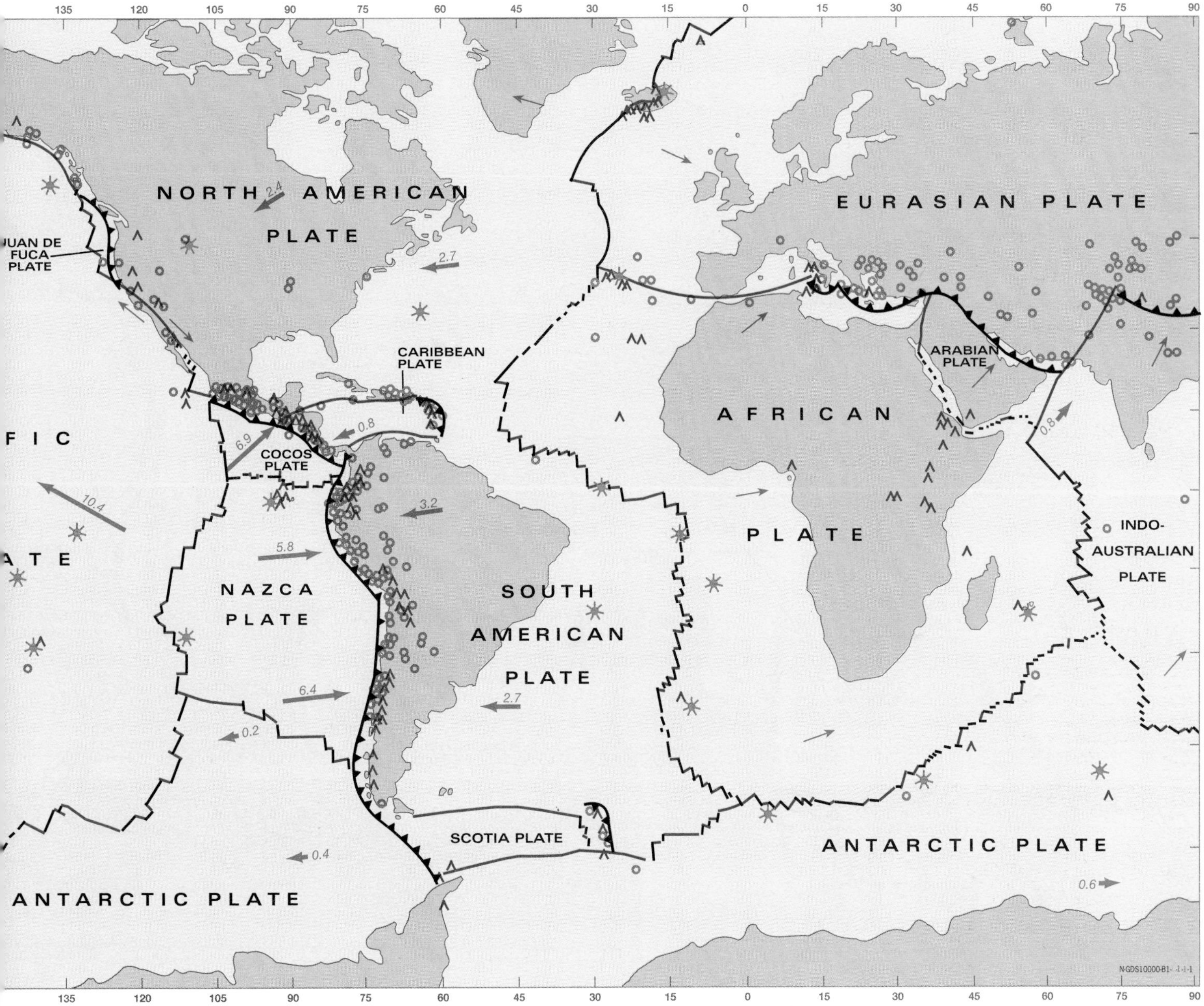

Subduction
Zone

Ocean Ridge
Zone

The plate tectonic theory describes the movement of the earth's surface and subsurface and explains why surface features are where they are.

Stated concisely, the theory presumes the lithosphere - the outside crust and uppermost mantle of the earth - is divided into about a dozen major rigid plates and several smaller platelets that move relative to one another. The position and names of the plates are shown on the map above.

The motor that drives the plates is found deep in the mantle. The theory states that because of temperature differences in the mantle, slow convection currents circulate there. Where two molten currents converge and move upward, they separate, causing the crustal plates to bulge and move apart in mid-ocean regions. Transverse fractures disrupt these broad regions. Lava wells up at these points to cause volcanic activity and to form ridges. The plates grow larger by accretion along these mid-ocean ridges, cause vast regions of the crust to move apart, and force the plates to collide with one another. As the plates do so, they are destroyed at subduction zones, where the plates are consumed downward, back into the earth's mantle, forming deep ocean trenches. The diagrams to the right illustrate the processes.

Most of the earth's volcanic and seismic activities

occur where plates slide past each other at transform boundaries or collide along subduction zones. The friction and heat caused by the grinding motion of the subducted plates causes rock to liquify and rise to the surface as volcanoes and eventually form vast mountain ranges. Strong and deep earthquakes are common here.

Volcanoes and earthquakes also occur at random locations around the earth known as "hot spots". Hot rock from deep in the mantle rises to the surface creating some of the earth's tallest mountains. As the lithospheric plates move slowly over these stationary plumes of magma, island chains (such as the Hawaiian Islands) are formed.

The overall result of tectonic movement is that the crustal plates move slowly and inexorably as relatively rigid entitles, carrying the continents along with them. The history of this continental drifting is illustrated in the four maps to the left. It began with a single landmass called the supercontinent of Pangaea and the ancestral sea, the Panthalassa Ocean. Pangaea first split into a northern landmass called Laurasia and a southern block called Gondwanaland and subsequently into the continents we map today. The map of the future will be significantly different as the continents continue to drift.

Scale 1:72 000 000 at 40° latitude.

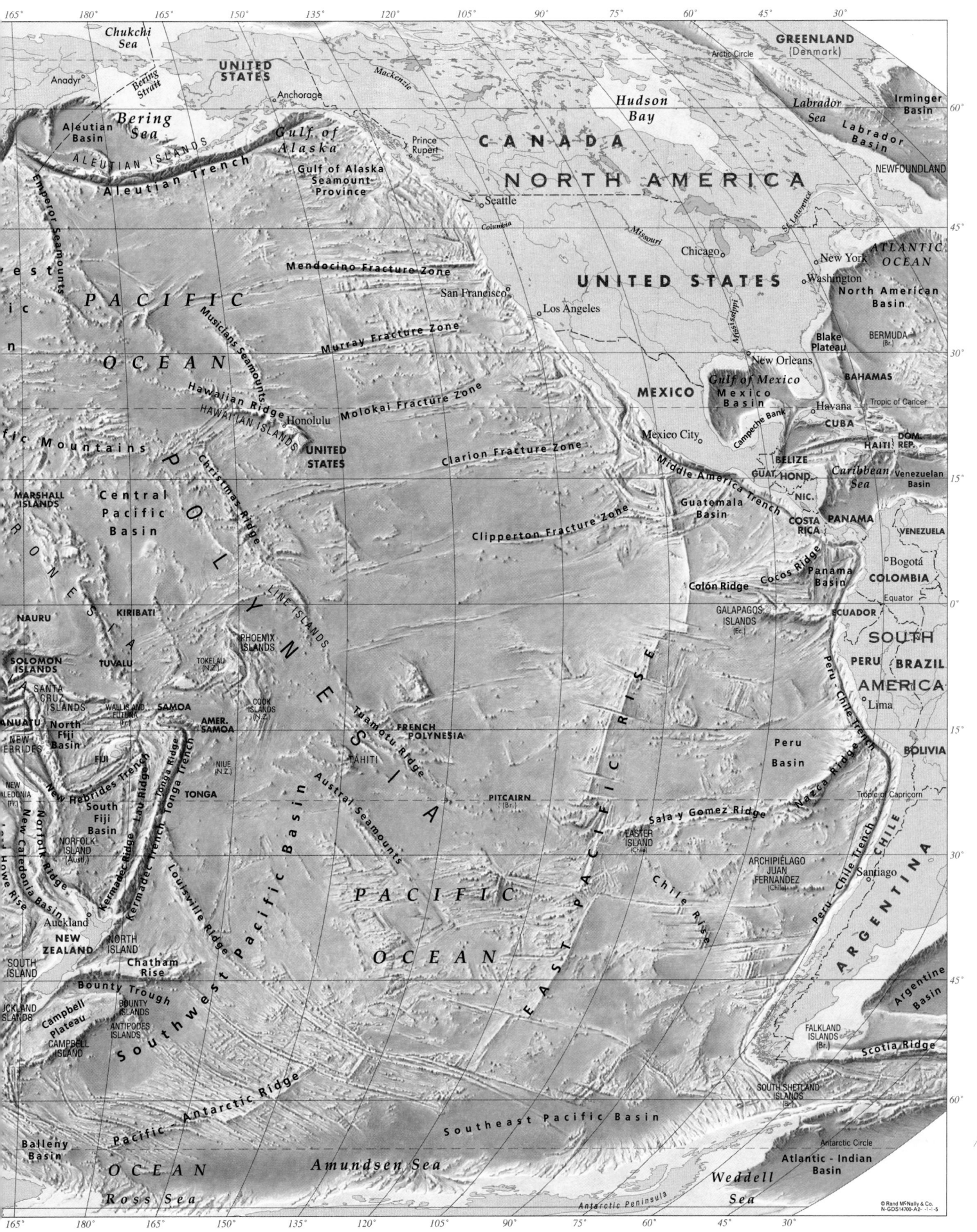

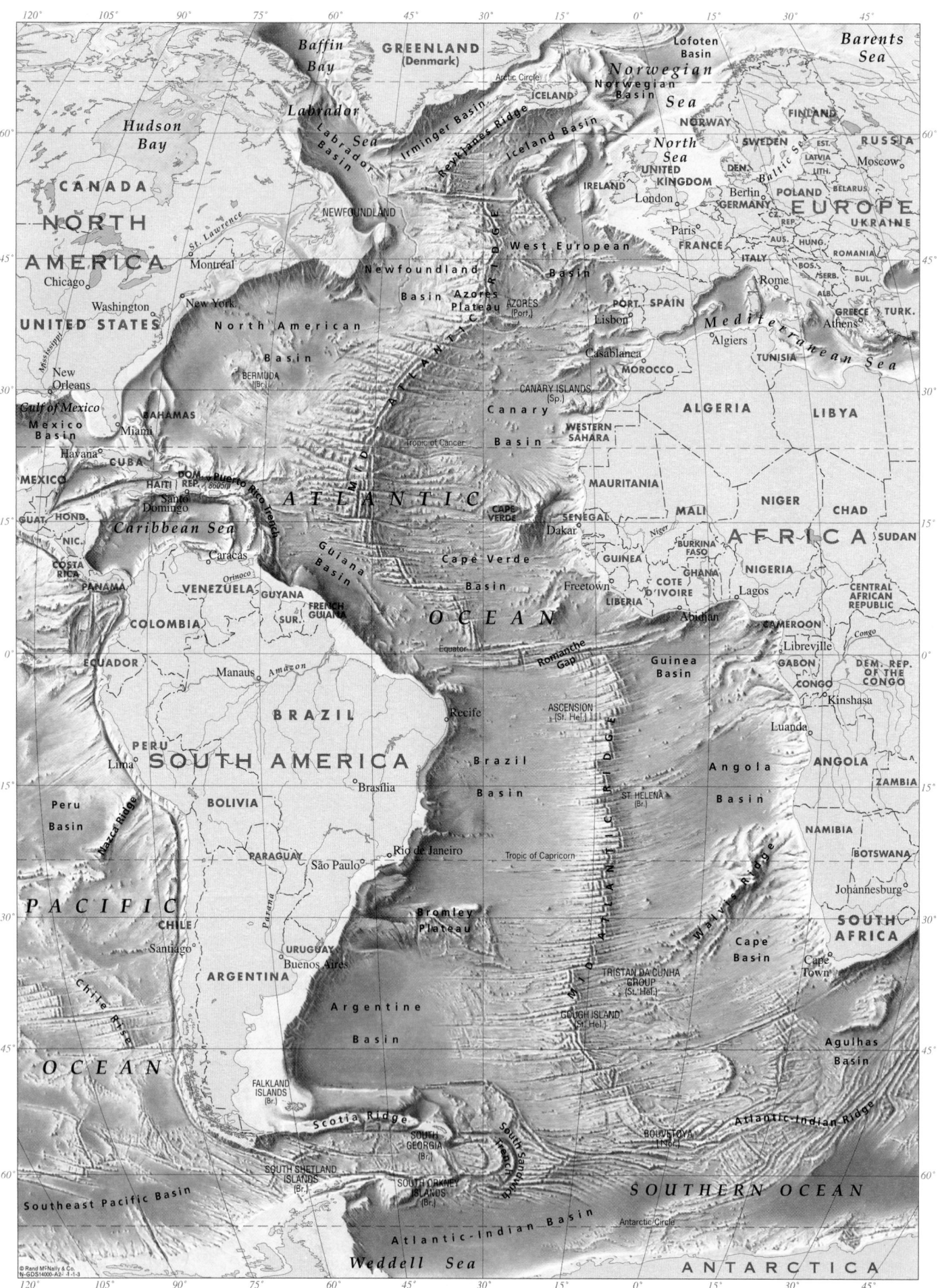

Scale 1:72 000 000 at 40° latitude.   ROBINSON PROJECTION

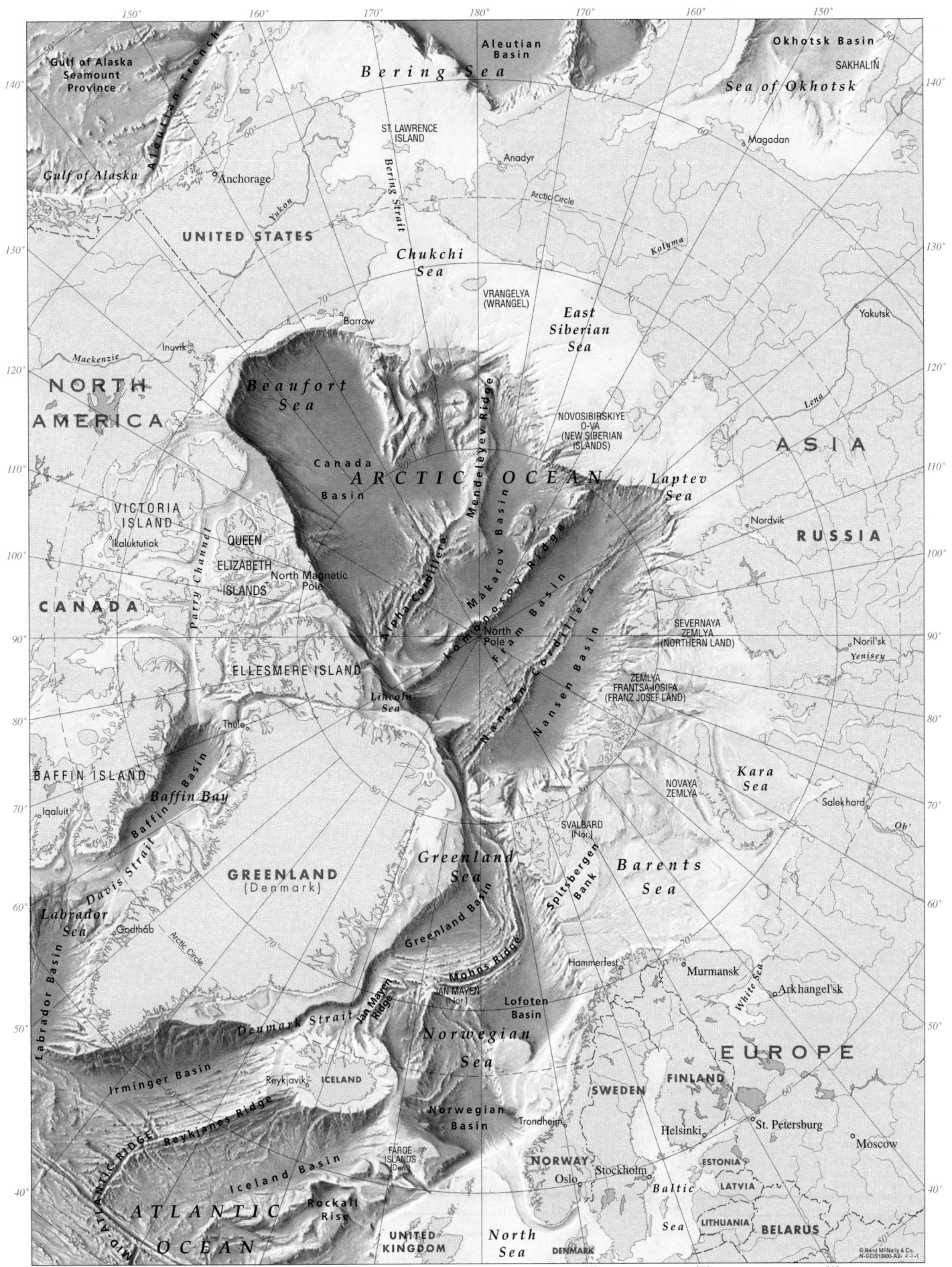

14

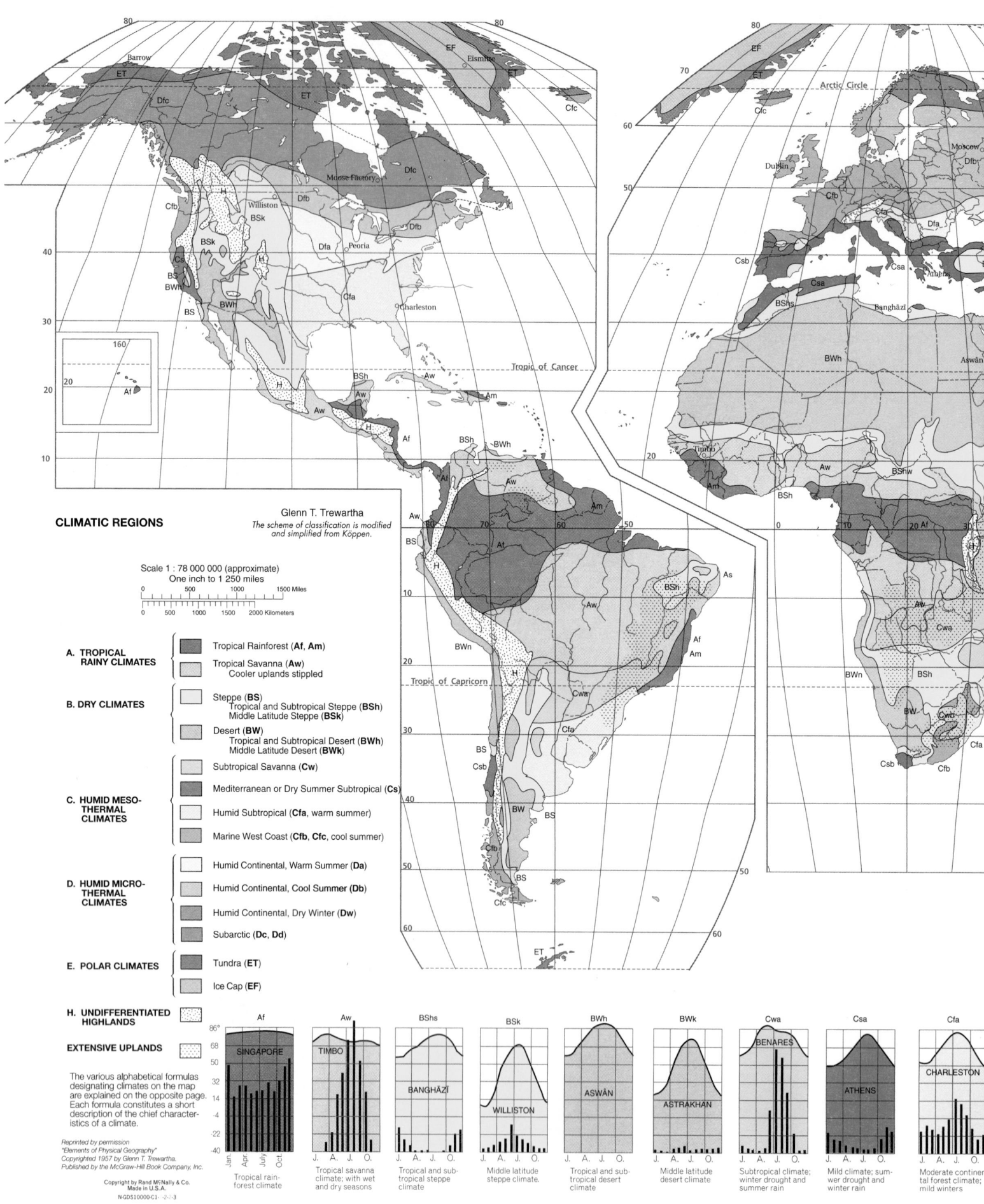

CLIMATIC REGIONS

Glenn T. Trewartha
*The scheme of classification is modified
and simplified from Köppen.*

Scale 1 : 78 000 000 (approximate)
One inch to 1 250 miles

0   500   1000   1500 Miles

0   500   1000   1500   2000 Kilometers

A. TROPICAL RAINY CLIMATES
- Tropical Rainforest (**Af, Am**)
- Tropical Savanna (**Aw**)
  Cooler uplands stippled

B. DRY CLIMATES
- Steppe (**BS**)
  Tropical and Subtropical Steppe (**BSh**)
  Middle Latitude Steppe (**BSk**)
- Desert (**BW**)
  Tropical and Subtropical Desert (**BWh**)
  Middle Latitude Desert (**BWk**)

C. HUMID MESO-THERMAL CLIMATES
- Subtropical Savanna (**Cw**)
- Mediterranean or Dry Summer Subtropical (**Cs**)
- Humid Subtropical (**Cfa**, warm summer)
- Marine West Coast (**Cfb, Cfc**, cool summer)

D. HUMID MICRO-THERMAL CLIMATES
- Humid Continental, Warm Summer (**Da**)
- Humid Continental, Cool Summer (**Db**)
- Humid Continental, Dry Winter (**Dw**)
- Subarctic (**Dc, Dd**)

E. POLAR CLIMATES
- Tundra (**ET**)
- Ice Cap (**EF**)

H. UNDIFFERENTIATED HIGHLANDS

EXTENSIVE UPLANDS

The various alphabetical formulas
designating climates on the map
are explained on the opposite page.
Each formula constitutes a short
description of the chief characteristics of a climate.

Reprinted by permission
"Elements of Physical Geography"
Copyrighted 1957 by Glenn T. Trewartha.
Published by the McGraw-Hill Book Company, Inc.

Copyright by Rand McNally & Co.
Made in U.S.A.
N-GDS10000-C1- -2-3

Af SINGAPORE — Tropical rain-forest climate

Aw TIMBO — Tropical savanna climate; with wet and dry seasons

BShs BANGHĀZĪ — Tropical and subtropical steppe climate

BSk WILLISTON — Middle latitude steppe climate.

BWh ASWĀN — Tropical and subtropical desert climate

BWk ASTRAKHAN — Middle latitude desert climate

Cwa BENARES — Subtropical climate; winter drought and summer rain

Csa ATHENS — Mild climate; summer drought and winter rain

Cfa CHARLESTON — Moderate continental forest climate; mild winters

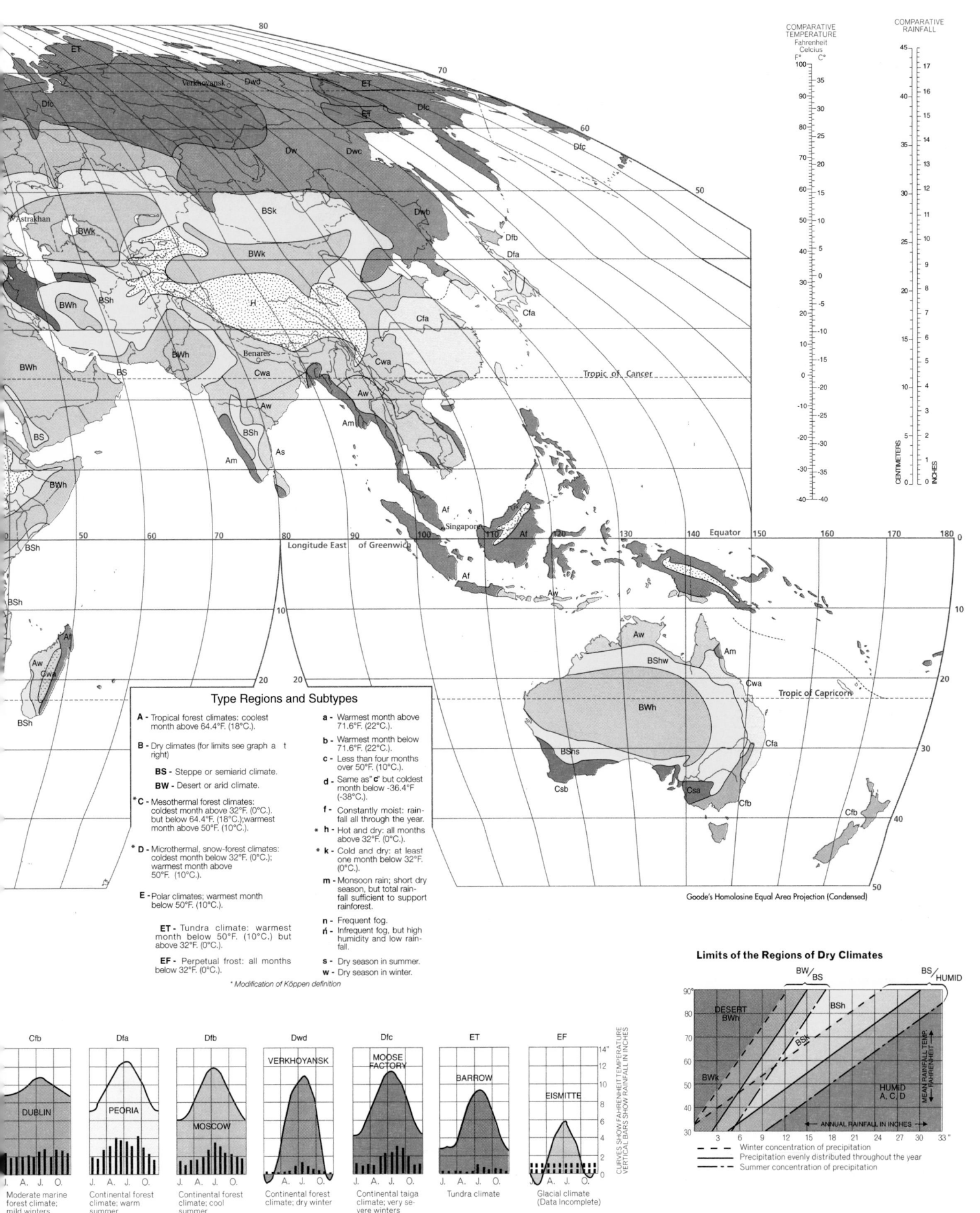

COMPARATIVE
TEMPERATURE
Fahrenheit
Celcius

COMPARATIVE
RAINFALL

Goode's Homolosine Equal Area Projection (Condensed)

Type Regions and Subtypes

A - Tropical forest climates: coolest
month above 64.4°F. (18°C.).

B - Dry climates (for limits see graph at
right)

   BS - Steppe or semiarid climate.

   BW - Desert or arid climate.

*C - Mesothermal forest climates:
coldest month above 32°F. (0°C.).
but below 64.4°F. (18°C.);warmest
month above 50°F. (10°C.).

*D - Microthermal, snow-forest climates:
coldest month below 32°F. (0°C.);
warmest month above
50°F. (10°C.).

E - Polar climates; warmest month
below 50°F. (10°C.).

   ET - Tundra climate: warmest
month below 50°F. (10°C.) but
above 32°F. (0°C.).

   EF - Perpetual frost: all months
below 32°F. (0°C.).

* Modification of Köppen definition

a - Warmest month above
71.6°F. (22°C.).

b - Warmest month below
71.6°F. (22°C.).

c - Less than four months
over 50°F. (10°C.).

d - Same as" **c'** but coldest
month below -36.4°F
(-38°C.).

f - Constantly moist: rainfall all through the year.

* h - Hot and dry: all months
above 32°F. (0°C.).

* k - Cold and dry: at least
one month below 32°F.
(0°C.).

m - Monsoon rain; short dry
season, but total rain-
fall sufficient to support
rainforest.

n - Frequent fog.

ń - Infrequent fog, but high
humidity and low rain-
fall.

s - Dry season in summer.

w - Dry season in winter.

Limits of the Regions of Dry Climates

- - - Winter concentration of precipitation
——— Precipitation evenly distributed throughout the year
—·—·— Summer concentration of precipitation

Cfb	Dfa	Dfb	Dwd	Dfc	ET	EF

DUBLIN | PEORIA | MOSCOW | VERKHOYANSK | MOOSE FACTORY | BARROW | EISMITTE

Moderate marine
forest climate;
mild winters

Continental forest
climate; warm
summer

Continental forest
climate; cool
summer

Continental forest
climate; dry winter

Continental taiga
climate; very se-
vere winters

Tundra climate

Glacial climate
(Data Incomplete)

CURVES SHOW FAHRENHEIT TEMPERATURE
VERTICAL BARS SHOW RAINFALL IN INCHES

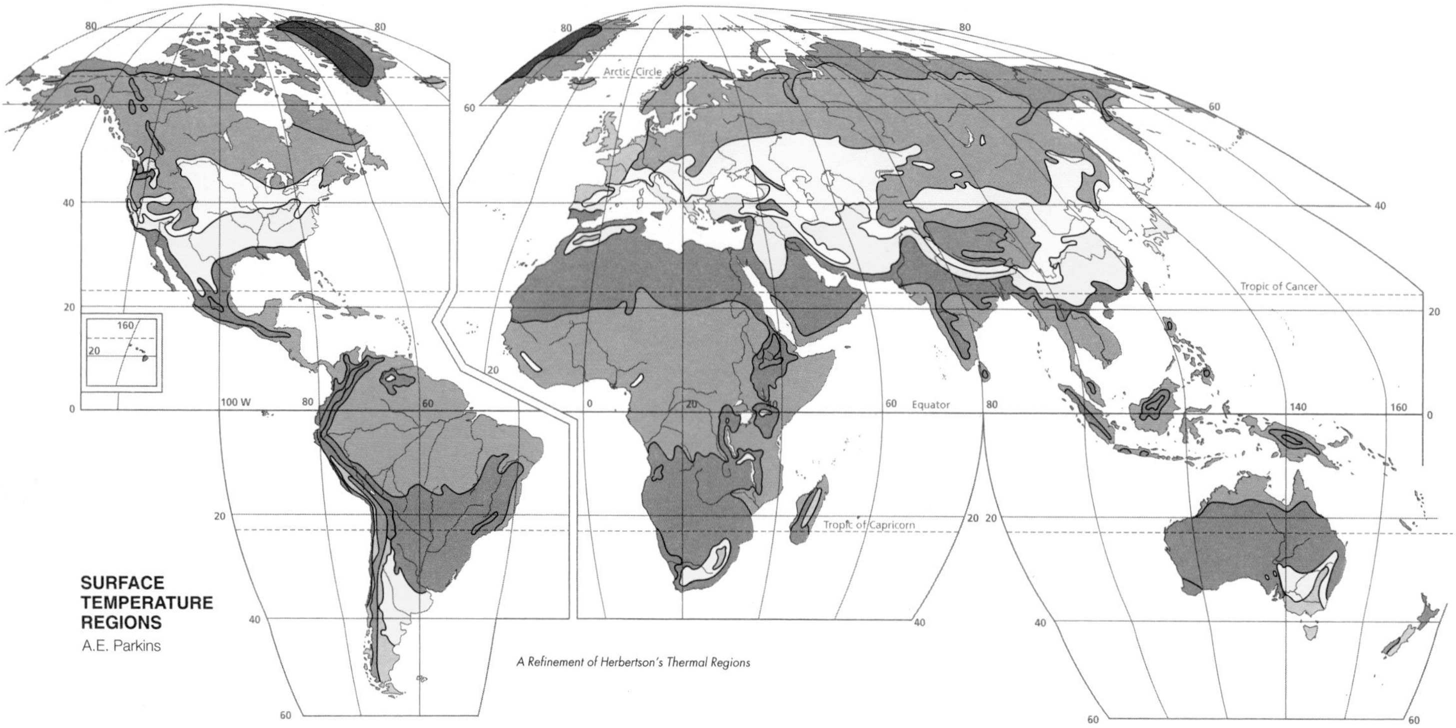

SURFACE TEMPERATURE REGIONS

A.E. Parkins

A Refinement of Herbertson's Thermal Regions

Hot = above 20°C
Mild = 10° to 20°
Cool = 0° to 10°
Cold = below 0°

Always cold; Polar regions and high altitudes	
Cold winter and cool summer; always cool in the Andes	
Cold winter and mild summer	
Cool winter and mild summer	
Hot summer and cold winter	
Hot summer and cool winter	
Hot summer and mild winter	
Always hot	
Always mild	

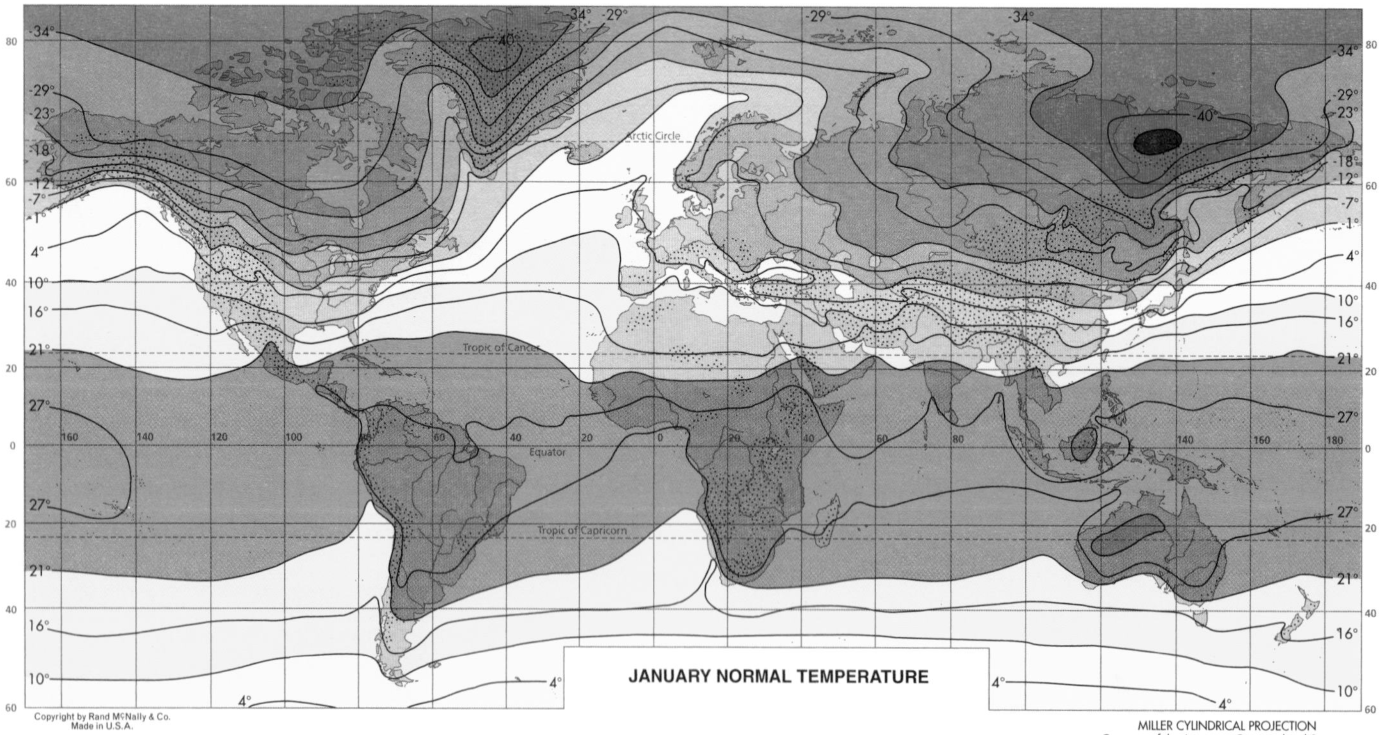

JANUARY NORMAL TEMPERATURE

MILLER CYLINDRICAL PROJECTION
Courtesy of the American Geographical Society.

Copyright by Rand McNally & Co.
Made in U.S.A.
NGDS10000-C4- -2-2-2

Reduced to Sea Level

Below -46°C. (-50°F.)	
-34° to -46° (-30° to -50°)	
-23° to -34° (-10° to -30°)	
-23° to -12° (-10° to +10°)	
-12° to -1° (10° to 30°)	
-1° to 10° (30° to 50°)	
10° to 21° (50° to 70°)	
21° to 32° (70° to 90°)	
Over 32° (90°)	
Highlands above 1000 meters	

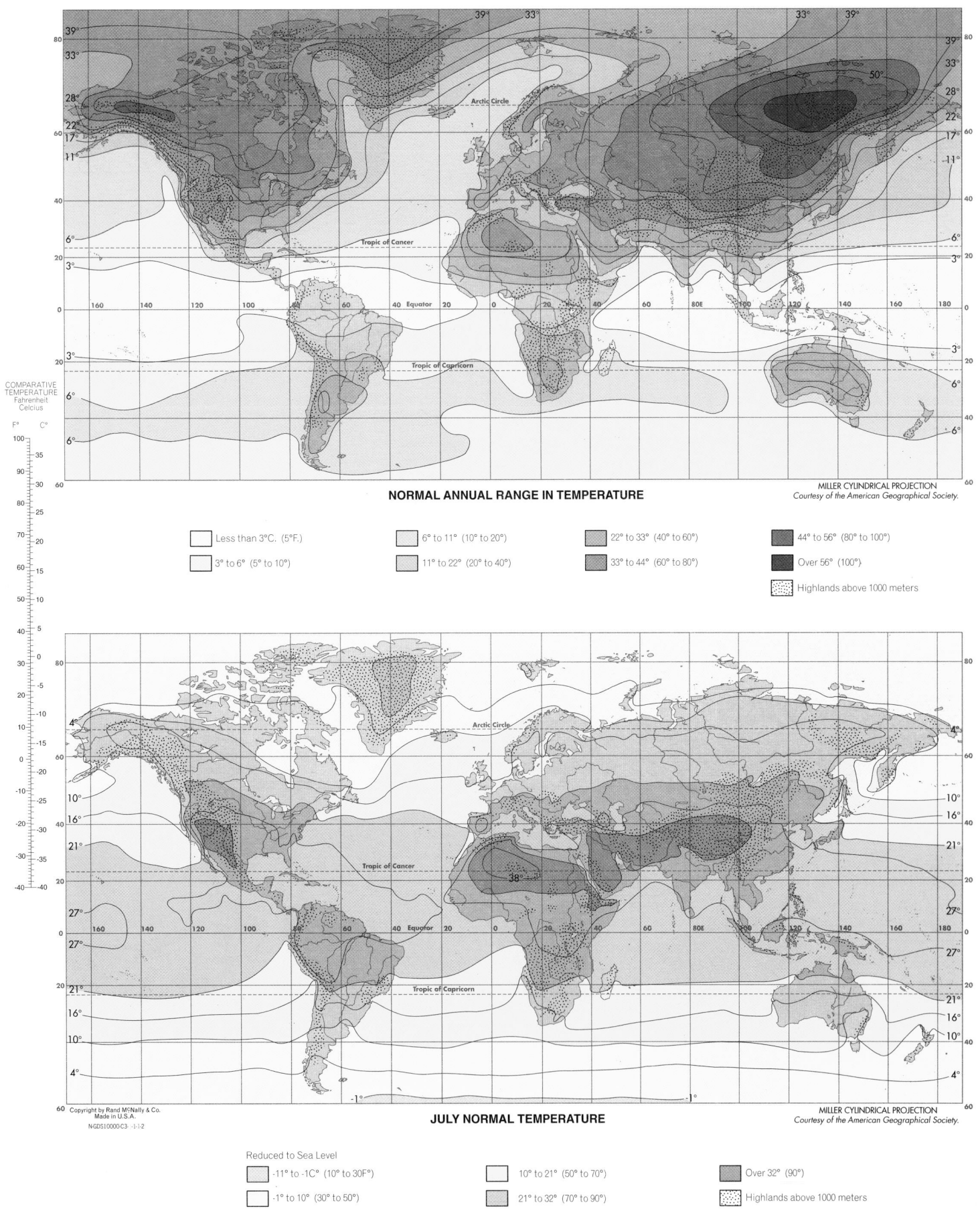

NORMAL ANNUAL RANGE IN TEMPERATURE

MILLER CYLINDRICAL PROJECTION
Courtesy of the American Geographical Society.

COMPARATIVE
TEMPERATURE
Fahrenheit
Celcius

	Less than 3°C. (5°F.)		6° to 11° (10° to 20°)		22° to 33° (40° to 60°)		44° to 56° (80° to 100°)
	3° to 6° (5° to 10°)		11° to 22° (20° to 40°)		33° to 44° (60° to 80°)		Over 56° (100°)

Highlands above 1000 meters

JULY NORMAL TEMPERATURE

MILLER CYLINDRICAL PROJECTION
Courtesy of the American Geographical Society.

Copyright by Rand McNally & Co.
Made in U.S.A.

N-GDS10000-C3- -1-1-2

Reduced to Sea Level

	-11° to -1C° (10° to 30F°)		10° to 21° (50° to 70°)		Over 32° (90°)
	-1° to 10° (30° to 50°)		21° to 32° (70° to 90°)		Highlands above 1000 meters

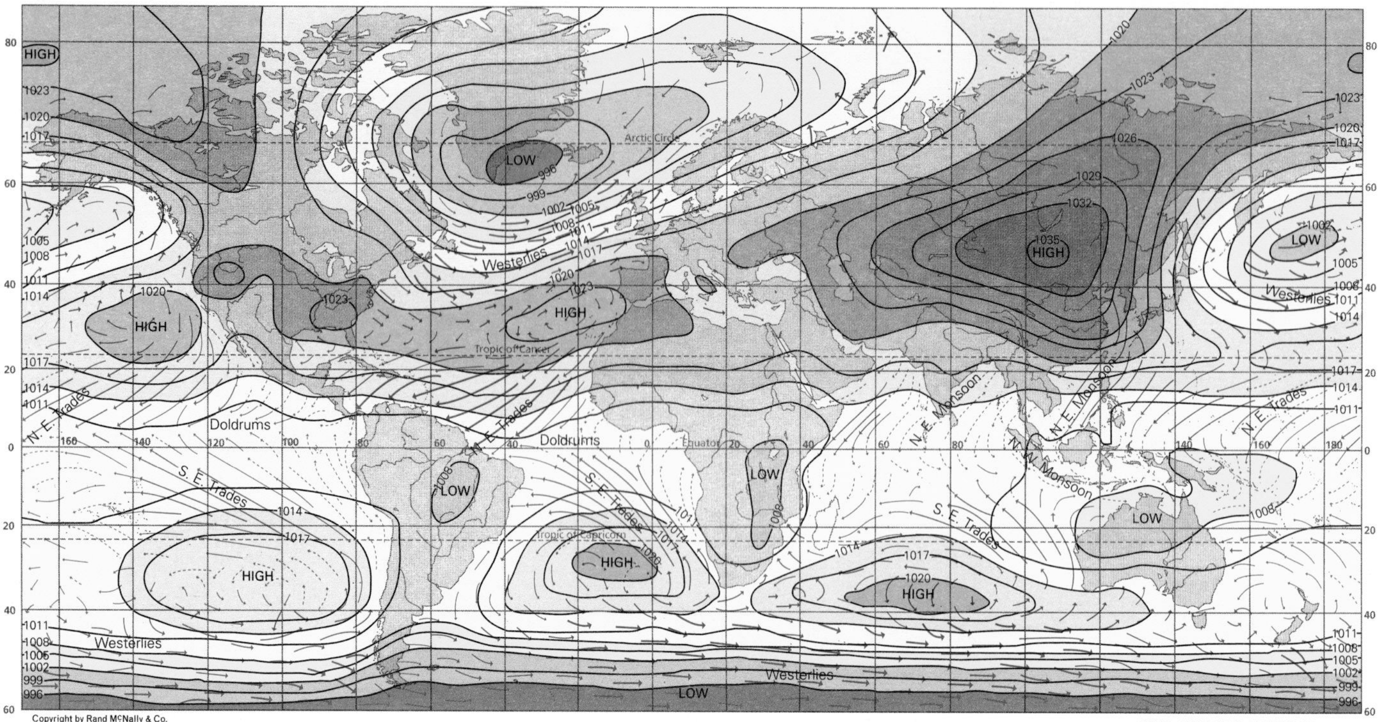

JANUARY PRESSURE AND PREDOMINANT WINDS

MILLER CYLINDRICAL PROJECTION
Courtesy of the American Geographical Society.

Copyright by Rand McNally & Co.
Made in U.S.A.
N- GDS10000-D2- -:-:-2

Low Pressures		High Pressures	
	990 mb.		1014
	996		1020
	1002		1026
	1008		1032
	1014		1038

Isobars on map at intervals of 3 millibars

Arrows fly with the wind. Wind direction determined by the quarter of the compass having highest wind frequency.

Length of arrow indicates the steadiness of the wind. Thickness of shaft indicates wind force.

Dominant Wind Forces

Beaufort Scale	Miles per hour (approx)
0-3	0-10
3-4	10-15
4-5½	15-25
Over 5½	Over 25

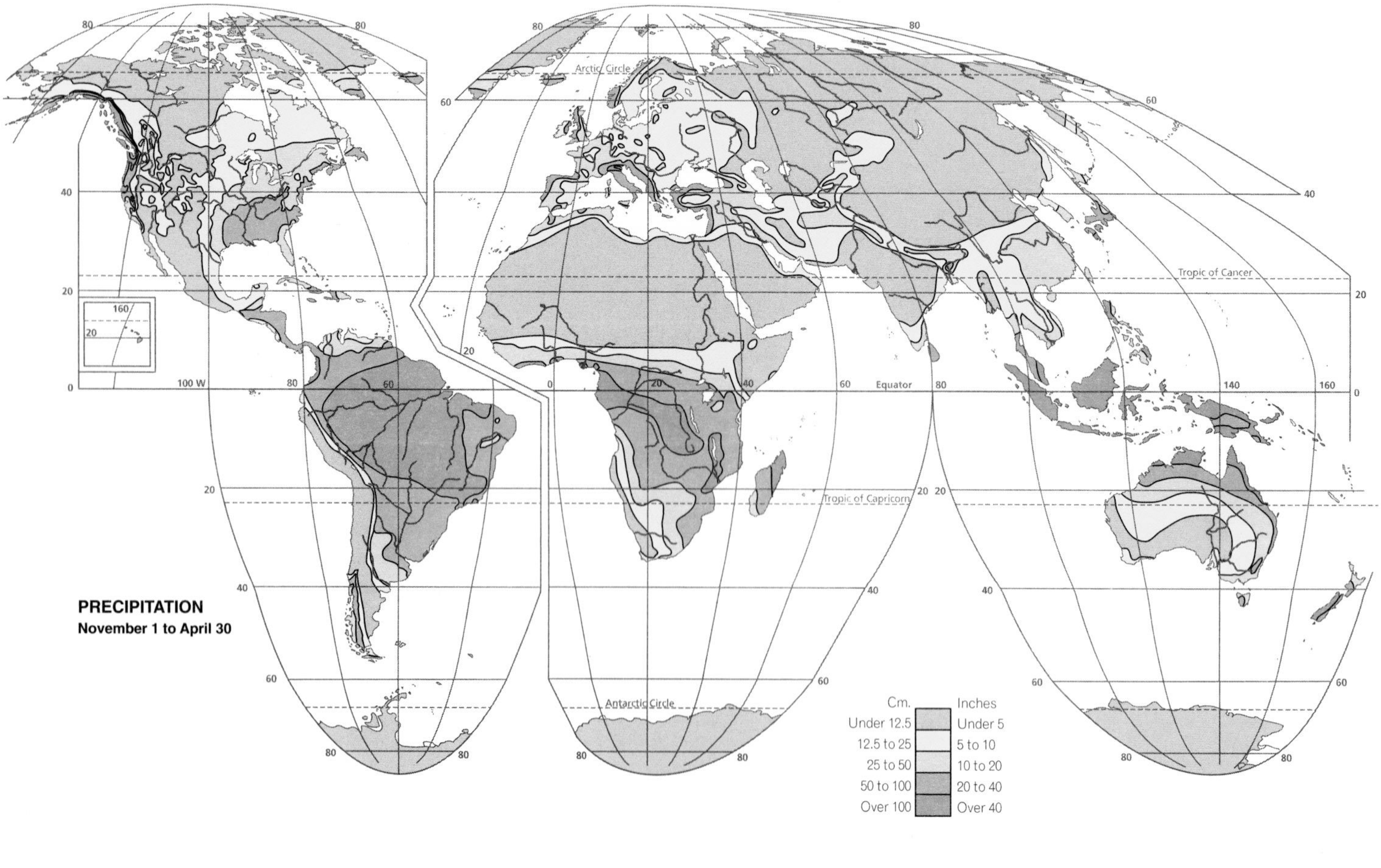

PRECIPITATION
November 1 to April 30

Cm.	Inches
Under 12.5	Under 5
12.5 to 25	5 to 10
25 to 50	10 to 20
50 to 100	20 to 40
Over 100	Over 40

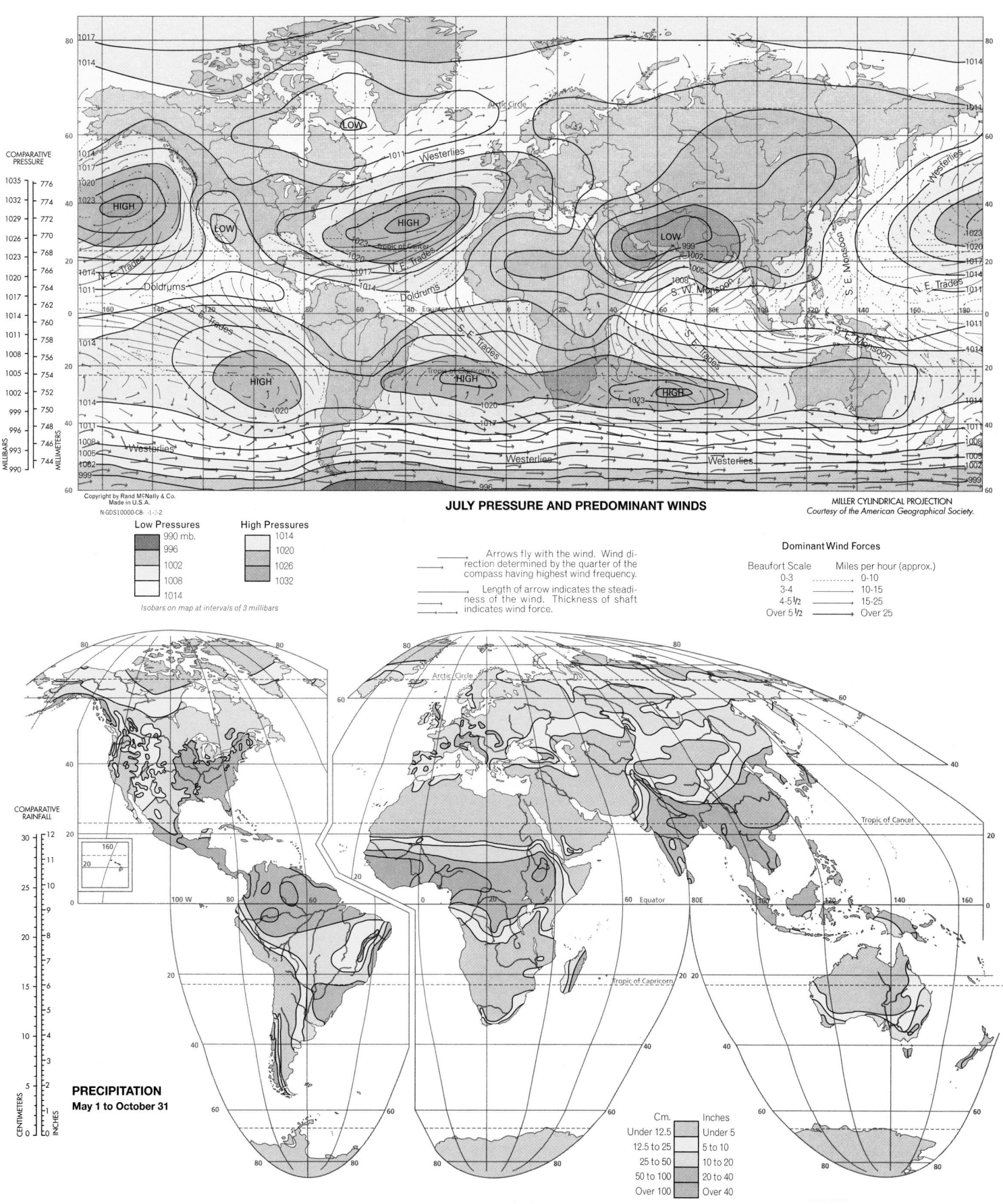

JULY PRESSURE AND PREDOMINANT WINDS

MILLER CYLINDRICAL PROJECTION
Courtesy of the American Geographical Society.

Copyright by Rand McNally & Co.
Made in U.S.A.
N-GDS10000-C8- -1-2-2

COMPARATIVE
PRESSURE

MILLIBARS	MILLIMETERS
1035	776
1032	774
1029	772
1026	770
1023	768
1020	766
1017	764
1014	762
1011	760
0	758
1008	756
1005	754
1002	752
999	750
996	748
993	746
990	744

Low Pressures

	990 mb.
	996
	1002
	1008
	1014

High Pressures

	1014
	1020
	1026
	1032

Isobars on map at intervals of 3 millibars

Arrows fly with the wind. Wind direction determined by the quarter of the compass having highest wind frequency.

Length of arrow indicates the steadiness of the wind. Thickness of shaft indicates wind force.

Dominant Wind Forces

Beaufort Scale	Miles per hour (approx.)
0-3	0-10
3-4	10-15
4-5 ½	15-25
Over 5 ½	Over 25

PRECIPITATION
May 1 to October 31

COMPARATIVE
RAINFALL

CENTIMETERS	INCHES
30	12
	11
25	10
	9
20	8
	7
15	6
	5
10	4
	3
5	2
	1
0	0

Cm.	Inches
Under 12.5	Under 5
12.5 to 25	5 to 10
25 to 50	10 to 20
50 to 100	20 to 40
Over 100	Over 40

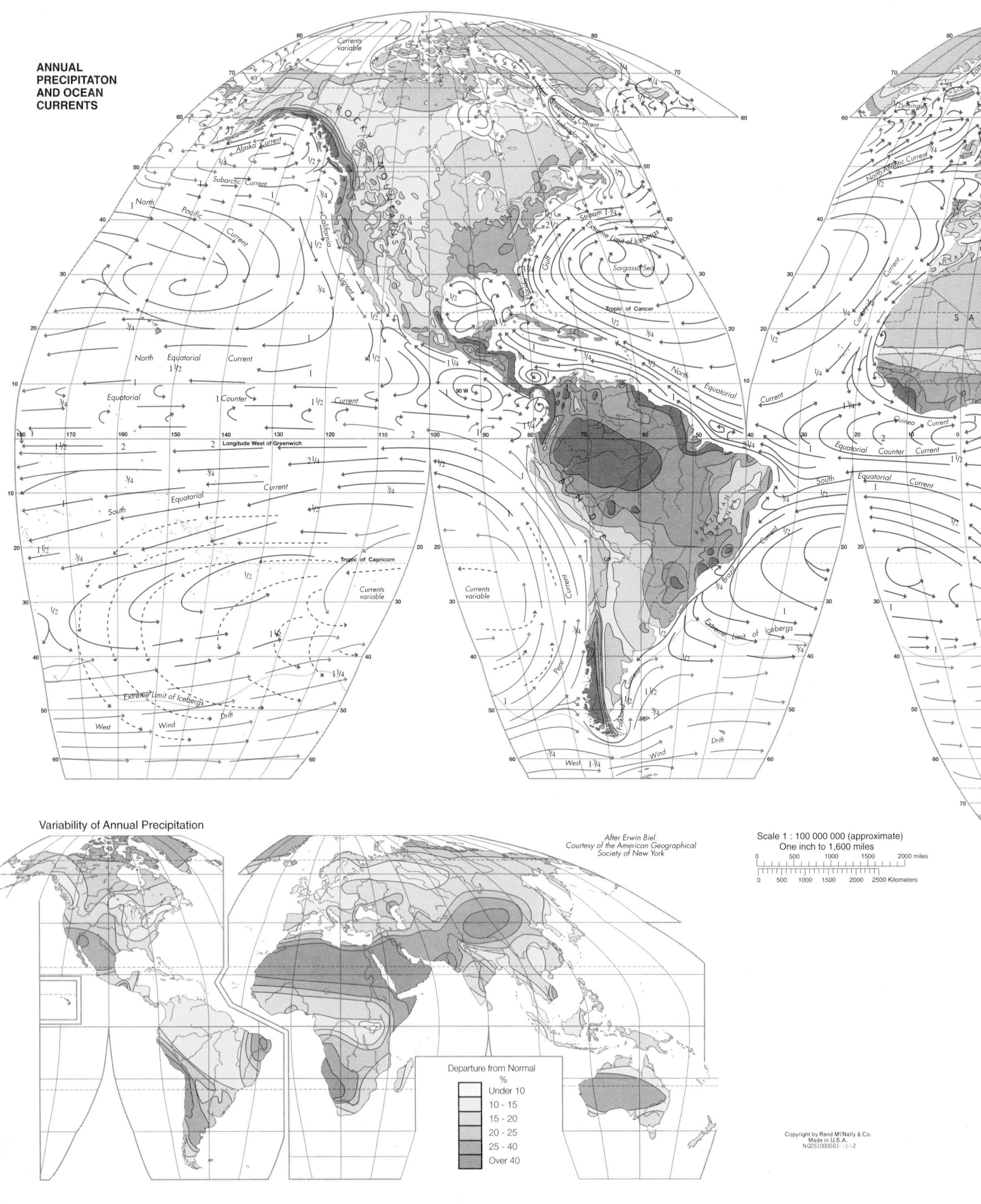

**ANNUAL
PRECIPITATON
AND OCEAN
CURRENTS**

Variability of Annual Precipitation

After Erwin Biel.
Courtesy of the American Geographical
Society of New York

Departure from Normal
%
Under 10
10 - 15
15 - 20
20 - 25
25 - 40
Over 40

Scale 1 : 100 000 000 (approximate)
One inch to 1,600 miles

0   500   1000   1500   2000 miles

0   500   1000   1500   2000   2500 Kilometers

Copyright by Rand McNally & Co.
Made in U.S.A.
NGDS10000-D1- -1-1-2

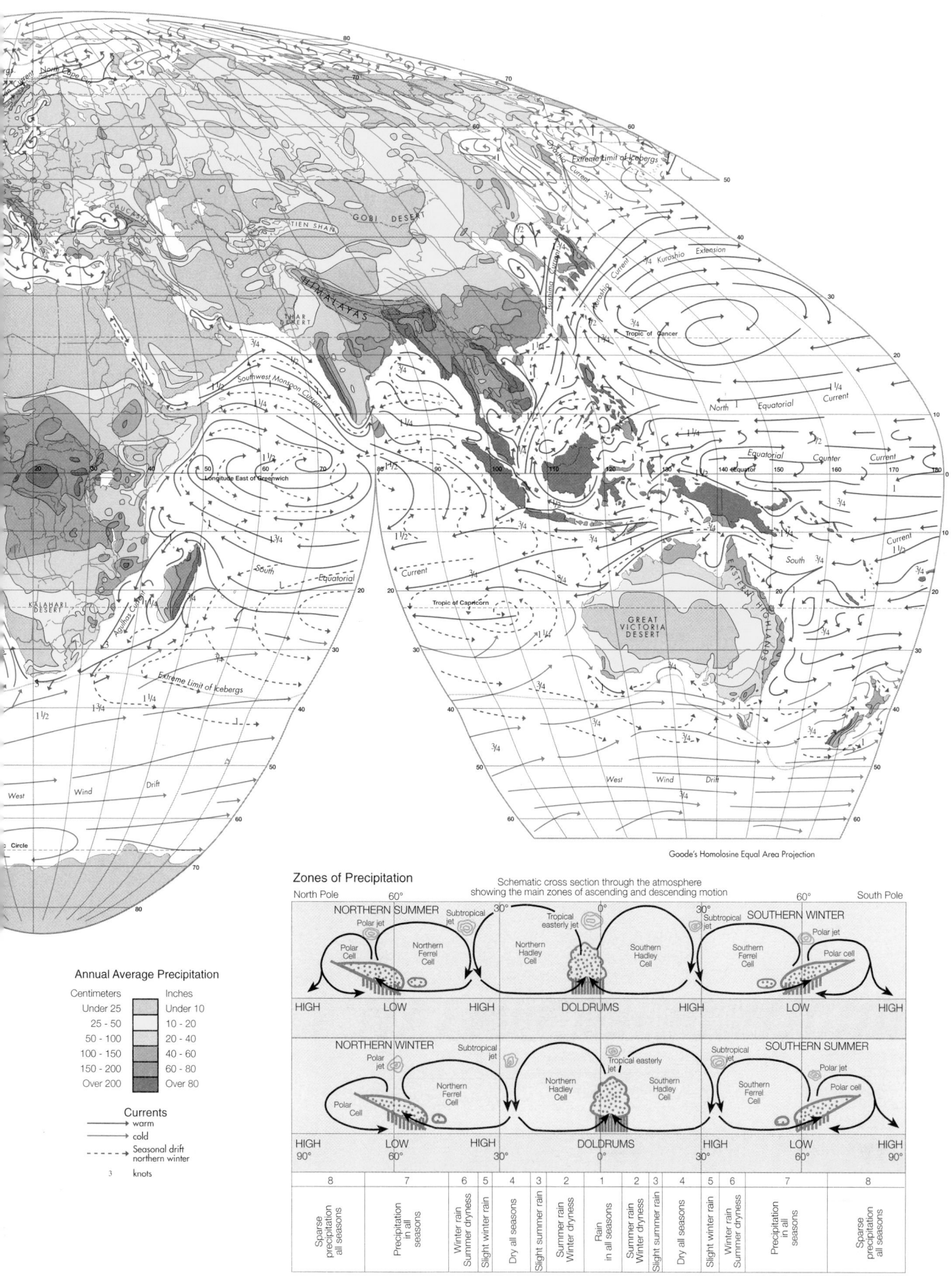

Goode's Homolosine Equal Area Projection

Zones of Precipitation

Schematic cross section through the atmosphere
showing the main zones of ascending and descending motion

Annual Average Precipitation

Centimeters		Inches
Under 25		Under 10
25 - 50		10 - 20
50 - 100		20 - 40
100 - 150		40 - 60
150 - 200		60 - 80
Over 200		Over 80

Currents
→ warm
→ cold
--→ Seasonal drift
northern winter
3   knots

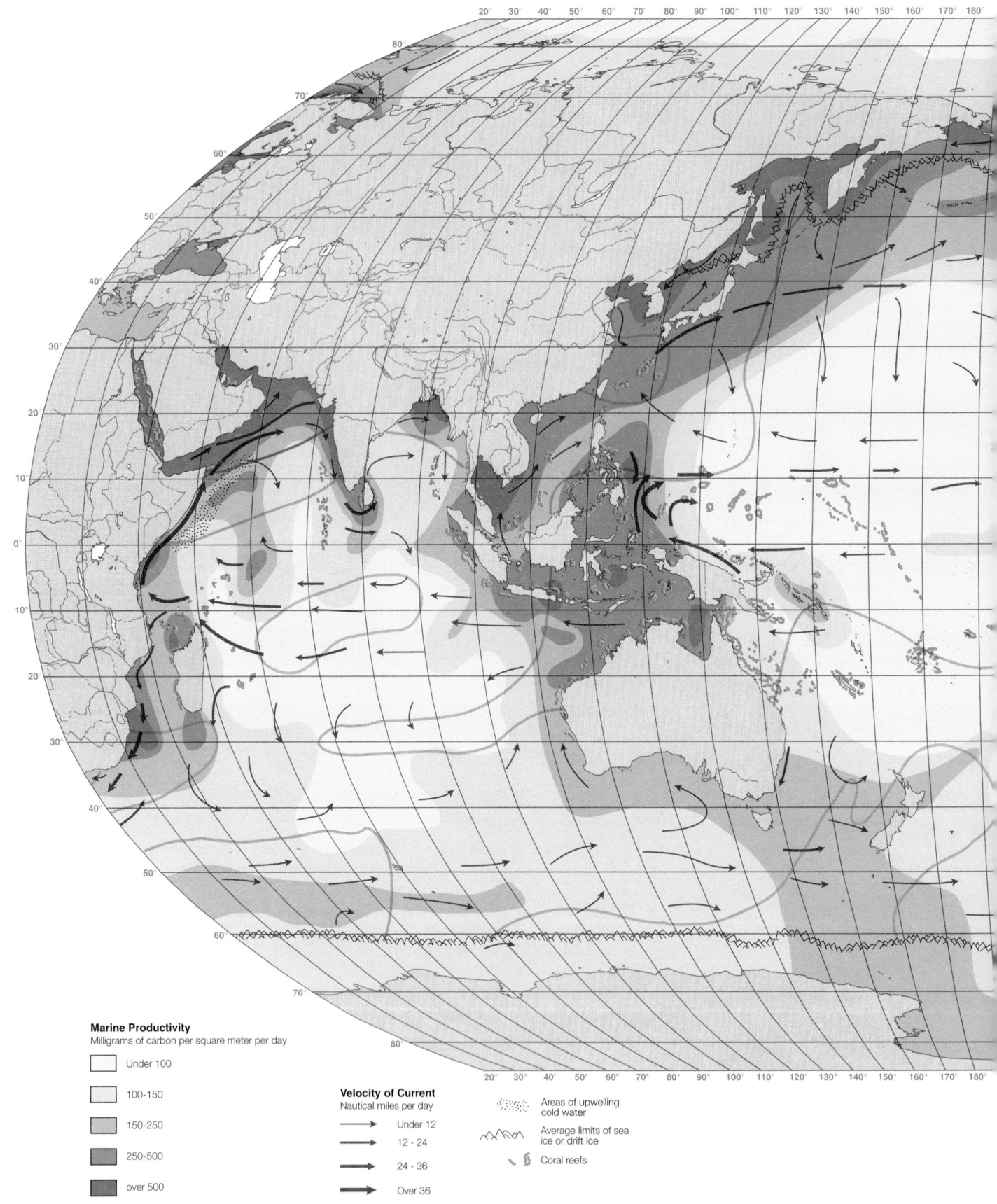

Marine Productivity
Milligrams of carbon per square meter per day

- Under 100
- 100-150
- 150-250
- 250-500
- over 500

Velocity of Current
Nautical miles per day

- Under 12
- 12 - 24
- 24 - 36
- Over 36

Areas of upwelling cold water

Average limits of sea ice or drift ice

Coral reefs

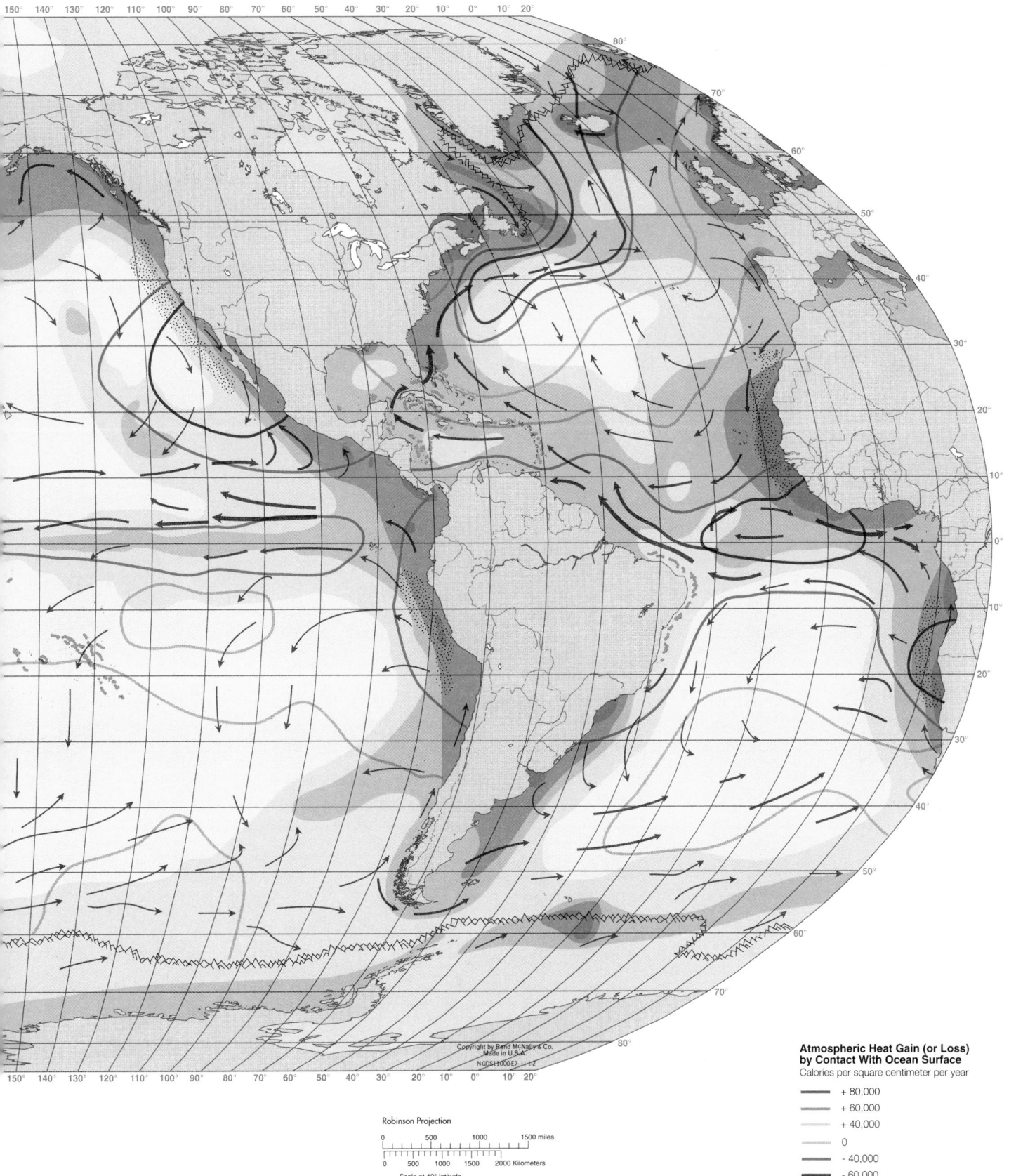

Atmospheric Heat Gain (or Loss) by Contact With Ocean Surface

Calories per square centimeter per year

- + 80,000
- + 60,000
- + 40,000
- 0
- - 40,000
- - 60,000

Robinson Projection

0   500   1000   1500 miles

0   500   1000   1500   2000 Kilometers

Scale at 40° latitude

Copyright by Rand McNally & Co.
Made in U.S.A.
N-00SH000E7-1-12

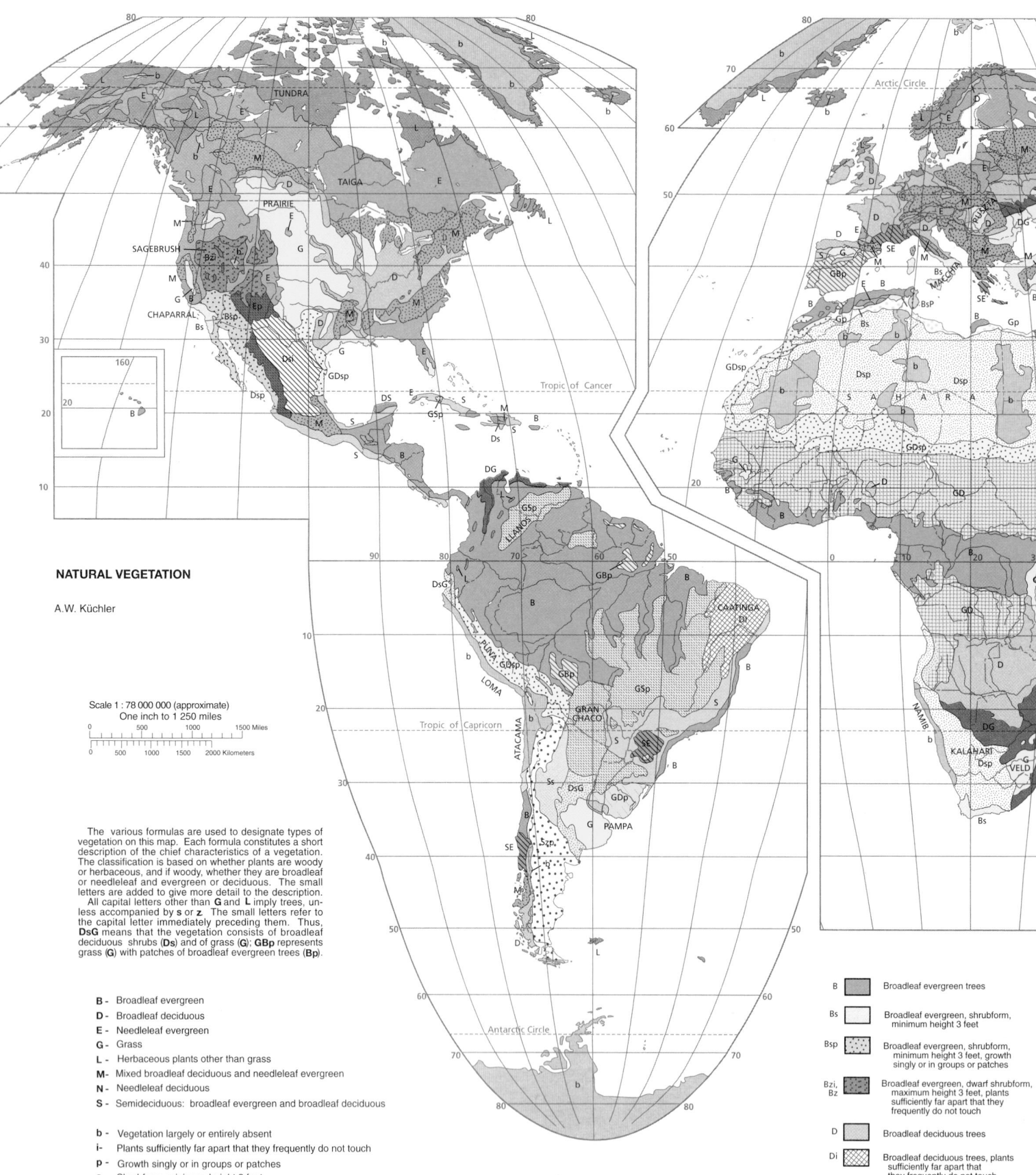

NATURAL VEGETATION

A.W. Küchler

Scale 1 : 78 000 000 (approximate)
One inch to 1 250 miles

```
0     500    1000    1500 Miles
|__|__|__|__|__|__|__|__|__|
0    500   1000   1500   2000 Kilometers
```

The various formulas are used to designate types of vegetation on this map. Each formula constitutes a short description of the chief characteristics of a vegetation. The classification is based on whether plants are woody or herbaceous, and if woody, whether they are broadleaf or needleleaf and evergreen or deciduous. The small letters are added to give more detail to the description.
All capital letters other than **G** and **L** imply trees, unless accompanied by **s** or **z**. The small letters refer to the capital letter immediately preceding them. Thus, **DsG** means that the vegetation consists of broadleaf deciduous shrubs (**Ds**) and of grass (**G**); **GBp** represents grass (**G**) with patches of broadleaf evergreen trees (**Bp**).

B - Broadleaf evergreen
D - Broadleaf deciduous
E - Needleleaf evergreen
G - Grass
L - Herbaceous plants other than grass
M - Mixed broadleaf deciduous and needleleaf evergreen
N - Needleleaf deciduous
S - Semideciduous: broadleaf evergreen and broadleaf deciduous

b - Vegetation largely or entirely absent
i - Plants sufficiently far apart that they frequently do not touch
p - Growth singly or in groups or patches
s - Shrubform, minimum height 3 feet
z - Dwarf shrubform, maximum height 3 feet

B — Broadleaf evergreen trees

Bs — Broadleaf evergreen, shrubform, minimum height 3 feet

Bsp — Broadleaf evergreen, shrubform, minimum height 3 feet, growth singly or in groups or patches

Bzi, Bz — Broadleaf evergreen, dwarf shrubform, maximum height 3 feet, plants sufficiently far apart that they frequently do not touch

D — Broadleaf deciduous trees

Di — Broadleaf deciduous trees, plants sufficiently far apart that they frequently do not touch

Goode's Homolosine Equal Area Projection (Condensed)

Bs	Broadleaf deciduous, shrubform, minimum height 3 feet
Bsi	Broadleaf deciduous, shrubform, minimum height 3 feet, plants sufficiently far apart that they frequently do not touch
Bsp	Broadleaf deciduous, shrubform, minimum height 3 feet, growth singly or in groups or patches
Bzp	Broadleaf deciduous, dwarf shrubform, maximum height 3 feet, growth singly or in groups or patches
BsG	Broadleaf deciduous, shrubform, minimum height 3 feet Grass and other herbaceous plants
BG	Broadleaf deciduous trees Grass and other herbaceous plants
BBs	Broadleaf deciduous trees Broadleaf evergreen, shrubform, minimum height 3 feet
E	Needleleaf evergreen trees
Ep	Needleleaf evergreen trees, growth singly or in groups or patches
G	Grass and other herbaceous plants
Gp	Grass and other herbaceous plants, growth singly or in groups or patches
GBp	Grass and other herbaceous plants Broadleaf evergreen trees, growth singly or in groups or patches
GD	Grass and other herbaceous plants Broadleaf deciduous trees
GDp	Grass and other herbaceous plants Broadleaf deciduous trees, growth singly or in groups or patches
GDsp	Grass and other herbaceous plants Broadleaf deciduous, shrubform, minimum height 3 feet, growth singly or in groups or patches
GSp	Grass and other herbaceous plants Semideciduous: broadleaf evergreen and broadleaf deciduous trees, growth singly or in groups or patches
L	Herbaceous plants other than grass
M	Mixed: broadleaf deciduous and needleleaf evergreen trees
N	Needleleaf deciduous trees
ND	Needleleaf deciduous trees Broadleaf deciduous trees
S	Semideciduous: broadleaf evergreen and broadleaf deciduous trees
Ss	Semideciduous: broadleaf evergreen and broadleaf deciduous, shrubform, minimum height 3 feet
SsG	Semideciduous: broadleaf evergreen and broadleaf deciduous, shrubform, minimum height 3 feet Grass and other herbaceous plants
Szp	Semideciduous: broadleaf evergreen and broadleaf deciduous, dwarf shrubform, maximum height 3 feet, growth singly or in groups or patches
SE	Semideciduous: broadleaf evergreen and broadleaf deciduous trees Needleleaf evergreen trees
b	Vegetation largely or entirely absent

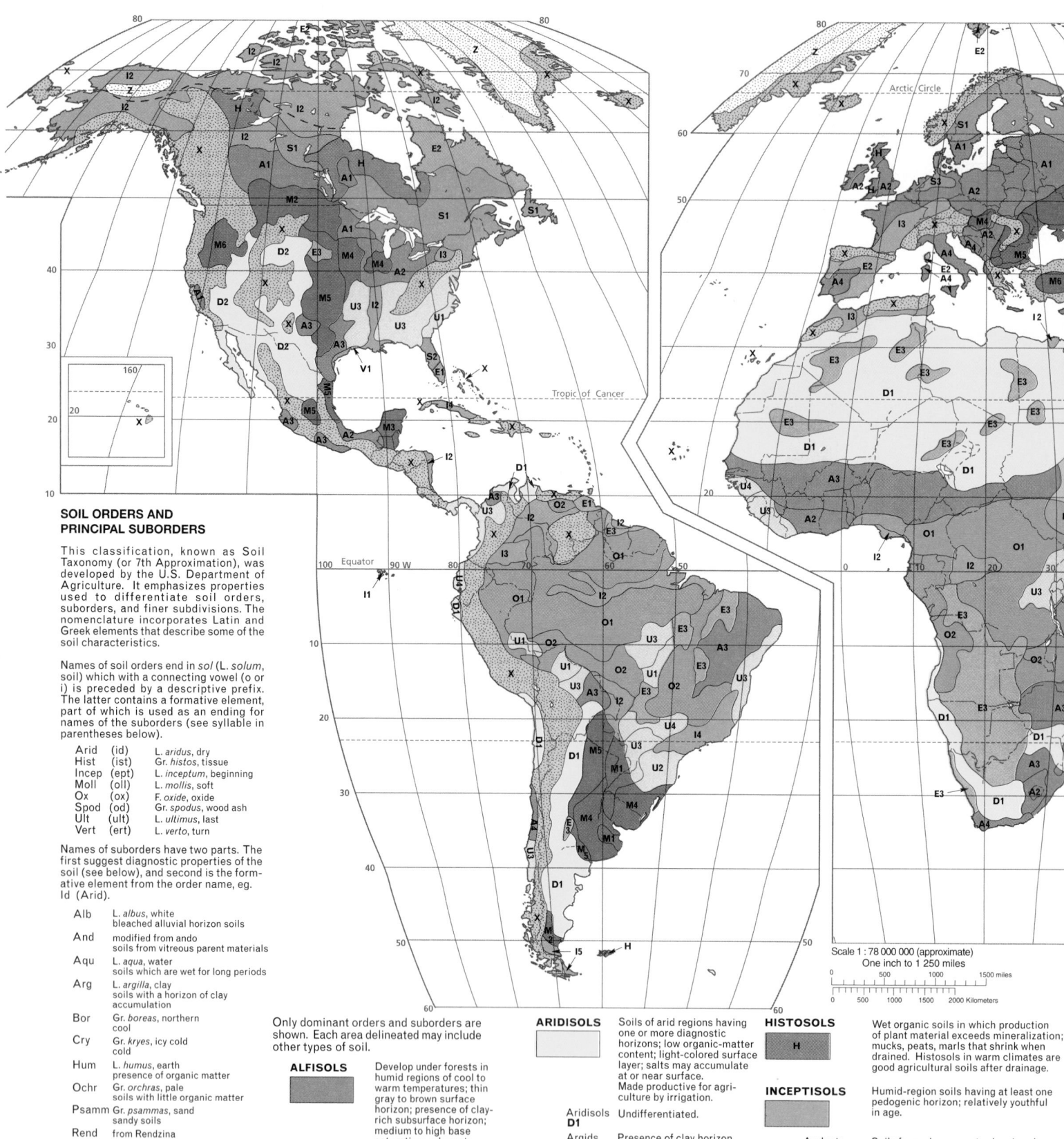

SOIL ORDERS AND PRINCIPAL SUBORDERS

This classification, known as Soil Taxonomy (or 7th Approximation), was developed by the U.S. Department of Agriculture. It emphasizes properties used to differentiate soil orders, suborders, and finer subdivisions. The nomenclature incorporates Latin and Greek elements that describe some of the soil characteristics.

Names of soil orders end in *sol* (L. *solum*, soil) which with a connecting vowel (o or i) is preceded by a descriptive prefix. The latter contains a formative element, part of which is used as an ending for names of the suborders (see syllable in parentheses below).

Arid	(id)	L. *aridus*, dry
Hist	(ist)	Gr. *histos*, tissue
Incep	(ept)	L. *inceptum*, beginning
Moll	(oll)	L. *mollis*, soft
Ox	(ox)	F. *oxide*, oxide
Spod	(od)	Gr. *spodus*, wood ash
Ult	(ult)	L. *ultimus*, last
Vert	(ert)	L. *verto*, turn

Names of suborders have two parts. The first suggest diagnostic properties of the soil (see below), and second is the formative element from the order name, eg. Id (Arid).

Alb	L. *albus*, white bleached alluvial horizon soils
And	modified from ando soils from vitreous parent materials
Aqu	L. *aqua*, water soils which are wet for long periods
Arg	L. *argilla*, clay soils with a horizon of clay accumulation
Bor	Gr. *boreas*, northern cool
Cry	Gr. *kryes*, icy cold cold
Hum	L. *humus*, earth presence of organic matter
Ochr	Gr. *orchras*, pale soils with little organic matter
Psamm	Gr. *psammas*, sand sandy soils
Rend	from Rendzina high carbonate content
Torr	L. *torridus*, hot and dry soils of very dry climate
Ud	L. *udus*, humid soils of humid climate
Umbr	L. *umbra*, shade dark color reflecting relatively high organic matter
Ust	L. *ustus*, burnt soils of dry climates with summer rains
Xer	Gr. *xeros*, dry soils of dry climates with winter rains

Only dominant orders and suborders are shown. Each area delineated may include other types of soil.

ALFISOLS

Develop under forests in humid regions of cool to warm temperatures; thin gray to brown surface horizon; presence of clay-rich subsurface horizon; medium to high base saturation; adequate moisture supply most of year. Generally fertile agricultural soils.

Boralfs A1	Well-drained soils of boreal and subalpine forests.
Udalfs A2	Humid, well-drained, highly fertile soils of warm-summer climates.
Ustalfs A3	Reddish-brown forest and grassland soils of warm, subhumid to semiarid climates.
Xeralfs A4	Reddish soils lacking moisture during summer in Mediterranean climate zones.

ARIDISOLS

Soils of arid regions having one or more diagnostic horizons; low organic-matter content; light-colored surface layer; salts may accumulate at or near surface. Made productive for agriculture by irrigation.

| Aridisols D1 | Undifferentiated. |
| Argids D2 | Presence of clay horizon. |

ENTISOLS

Soils lacking pedogenic horizons; varied in nature.

Aquents E1	Seasonally or perenially wet; bluish or gray and mottled.
Orthents E2	Soils thinning due to erosion or where no sedimentation occurs.
Psamments E3	Sandy texture in all layers below surface; form on dune sands.

HISTOSOLS

Wet organic soils in which production of plant material exceeds mineralization; mucks, peats, marls that shrink when drained. Histosols in warm climates are good agricultural soils after drainage.

INCEPTISOLS

Humid-region soils having at least one pedogenic horizon; relatively youthful in age.

Andepts I1	Soils formed on recent volcanic ash; high organic-matter content.
Aquepts I2	Humid region soils developed on river floodplains. Cryaquepts are tundra soils on permafrost.
Ochrepts I3	Thin, light-colored surface horizons; little organic-matter content.
Tropepts I4	Brownish or reddish soils of tropical environments.
Umbrepts I5	Dark-colored surface layer; high organic-matter content; hilly to mountainous topography.

Scale 1 : 78 000 000 (approximate)
One inch to 1 250 miles

0   500   1000   1500 miles

0   500   1000   1500   2000 Kilometers

Goode's Homolosine Equal Area Projection (Condensed)
Copyright by Rand McNally & Co.
Made in U.S.A.
N-GDS10000-E3- -2-2-5

Longitude East of Greenwich

Tropic of Cancer

Equator

Tropic of Capricorn

— — — — Limit of continuous permafrost

*Terms refer to Great Soils Group terminology.

	OLLISOLS	Deep-profile soils with seasonal moisture deficit associated with grasslands; dark brown to black upper layer; may have subsurface horizon of calcium accumulation; high base saturation. Very productive for grain crops.

Albolls M1	Soils with a grayish subsurface horizon over clay layer and a fluctuating water table.
Borolls M2	Well-drained, fertile grassland soils of cool summers and cold winters.
Rendolls M3	Formed on calcareous limestones.
Udolls M4	Freely drained soils of humid regions with warm summers; excellent agricultural soils.
Ustolls M5	Fertile agricultural soils of subhumid climates.
Xerolls M6	Pronounced soil-moisture deficit during high-sun season; associated with Mediterranean climates.

OXISOLS — Deeply weathered tropical and subtropical soils of low natural fertility; low base saturation; limited ability to hold soil nutrients against leaching; presence of plinthite (laterite) layers. Generally unsuited to large-scale agricultural production.

Orthox O1	Hot and nearly always moist; associated with tropical rainforests.
Ustox O2	Hot to warm forest and savanna soils with a drier season of low soil-moisture availability.

SPODOSOLS — Soils of moist climates ranging from subtropical to cold conditions; include a spodic subsurface horizon incorporating active organic matter beneath a light-colored, leached, sandy horizon. Generally marginal for agriculture.

Spodo-sols S1	Undifferentiated, mostly in high latitudes.
Aquods S2	Seasonally wet developed on sandy parent material.
Humods S3	Considerable organic matter present in subsurface horizon.
Orthods S4	Subsurface accumulations of iron, aluminum, and organic matter.

ULTISOLS — Tropical and subtropical soils with a variety of soil moisture regimes; subsurface clay horizon; low base saturation; very old soils characterized by long weathering of clay minerals; low ability to hold nutrients against leaching. Often marginal for agriculture.

Aquults U1	Seasonally wet with mottled, gray subsurface horizon.
Humults U2	Dark soils with high organic-matter content, warm temperatures.
Udults U3	Low organic-matter content and temperate to hot conditions.
Ustults U4	Seasonally dry, warm to hot conditions.

VERTISOLS — Dark tropical and subtropical soils developed on heavy clays; deep shrinkage cracks appear during dry season which become filled with loose surface materials that absorb moisture and swell during wet season. Generally fertile and well suited to crop production.

Uderts V1	Generally moist with limited period for shrinkage cracks to develop.
Usterts V2	Over three months of shrinkage-crack formation.

MOUNTAIN SOILS

Soils with various moisture and temperature regimes; mainly high altitude soils forming on steep slopes; soils vary greatly within a short distance.

Z — Areas with little or no soils.

APPROXIMATE CORRELATION WITH OTHER SOIL CLASSIFICATION SYSTEMS

Soil Taxonomy	Great Soil Groups (former U.S. system)	Canadian system
Udalfs	Gray-brown Podzolic	Luvisolic Gray-Brown
Ustalfs	Reddish Chestnut; Red and Yellow Podzolic	
Aridisols	Desert and Reddish Desert Solonetz, Solonchak	
Entisols	Lithosols	Regosolic
Histosols	Bog	Organic
Inceptisol		Brunisolic
Orthents	Lithosols	
Aquepts	Humic Gley	Gleysolic
Cryaquept	Tundra	Cryosolic
Boralfs		Luvisolic Gray; Solonetzic
Borolls	Chernozem Chestnut Brown	Chernozemic, Solonetzic
Rendolls	Rendzina	
Udolls	Prairie	
Ustolls	Brown	
Oxisols	Latosols	
Humod		Humic Podzolic
Orthods	Podzols	Podzolic
Udults	Red and Yellow Podzolic Reddish Brown Lateritic	
Vertisols	Rendzina	

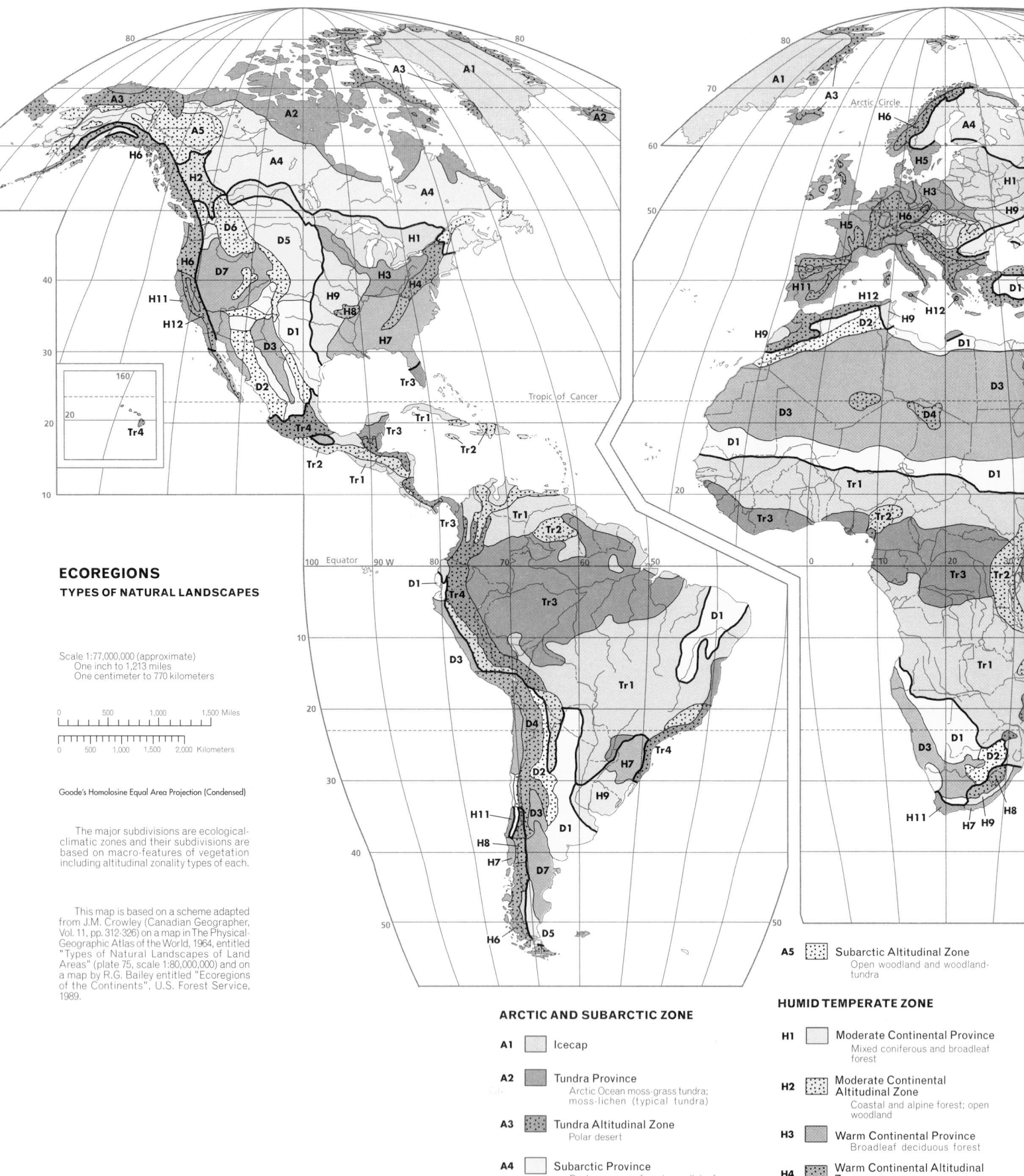

ECOREGIONS

TYPES OF NATURAL LANDSCAPES

Scale 1:77,000,000 (approximate)
One inch to 1,213 miles
One centimeter to 770 kilometers

| 0 | 500 | 1,000 | 1,500 Miles |

| 0 | 500 | 1,000 | 1,500 | 2,000 Kilometers |

Goode's Homolosine Equal Area Projection (Condensed)

The major subdivisions are ecological-climatic zones and their subdivisions are based on macro-features of vegetation including altitudinal zonality types of each.

This map is based on a scheme adapted from J.M. Crowley (Canadian Geographer, Vol. 11, pp. 312-326) on a map in The Physical-Geographic Atlas of the World, 1964, entitled "Types of Natural Landscapes of Land Areas" (plate 75, scale 1:80,000,000) and on a map by R.G. Bailey entitled "Ecoregions of the Continents", U.S. Forest Service, 1989.

A5 [ ] Subarctic Altitudinal Zone
Open woodland and woodland-tundra

ARCTIC AND SUBARCTIC ZONE

A1 [ ] Icecap

A2 [ ] Tundra Province
Arctic Ocean moss-grass tundra; moss-lichen (typical tundra)

A3 [ ] Tundra Altitudinal Zone
Polar desert

A4 [ ] Subarctic Province
Dark evergreen forest; needleleaf taiga; mixed coniferous and small-leafed forest

HUMID TEMPERATE ZONE

H1 [ ] Moderate Continental Province
Mixed coniferous and broadleaf forest

H2 [ ] Moderate Continental Altitudinal Zone
Coastal and alpine forest; open woodland

H3 [ ] Warm Continental Province
Broadleaf deciduous forest

H4 [ ] Warm Continental Altitudinal Zone
Upland broadleaf and alpine needleleaf forest

Copyright by Rand McNally & Co.
Made in U.S.A.
N-GDS10000-E5- -1-2-5

Tropic of Cancer

Equator

Longitude East of Greenwich

Tropic of Capricorn

H5 Marine Province
Lowland, west-coastal humid forest

H6 Marine Altitudinal Zone
Humid coastal and alpine coniferous forest

H7 Humid Subtropical Province
Broadleaf evergreen and broadleaf deciduous forest

H8 Humid Subtropical Altitudinal Zone
Upland, subtropical broadleaf forest

H9 Prairie Province

H10 Prairie Altitudinal Zone
Upland mixed prairie and woodland

H11 Mediterranean Province
Sclerophyll woodland, shrub, and steppe

H12 Mediterranean Altitudinal Zone
Upland shrub and steppe

DRY AND DESERT ZONE

D1 Tropical/Subtropical Steppe Province
Dry steppe, desert shrub, semi-desert savanna

D2 Tropical/Subtropical Steppe Altitudinal Zone
Upland steppe and desert shrub

D3 Tropical/Subtropical Desert Province
Hot, lowland desert at subtropical and coastal locations

D4 Tropical/Subtropical Desert Altitudinal Zone
Desert shrub

D5 Temperate Steppe Province
Medium to short steppe grassland

D6 Temperate Steppe Altitudinal Zone
Alpine meadow and coniferous woodland

D7 Temperate Desert Province
Midlatitude rainshadow desert

D8 Temperate Desert Altitudinal Zone
Extreme continental desert-steppe

HUMID TROPICAL ZONE

Tr1 Savanna Province
Seasonally dry forest, open woodland, tall grass

Tr2 Savanna Altitudinal Zone
Open woodland-steppe

Tr3 Rainforest Province
Constantly humid, broadleaf evergreen forest

Tr4 Rainforest Altitudinal Zone
Broadleaf evergreen and subtropical deciduous forest

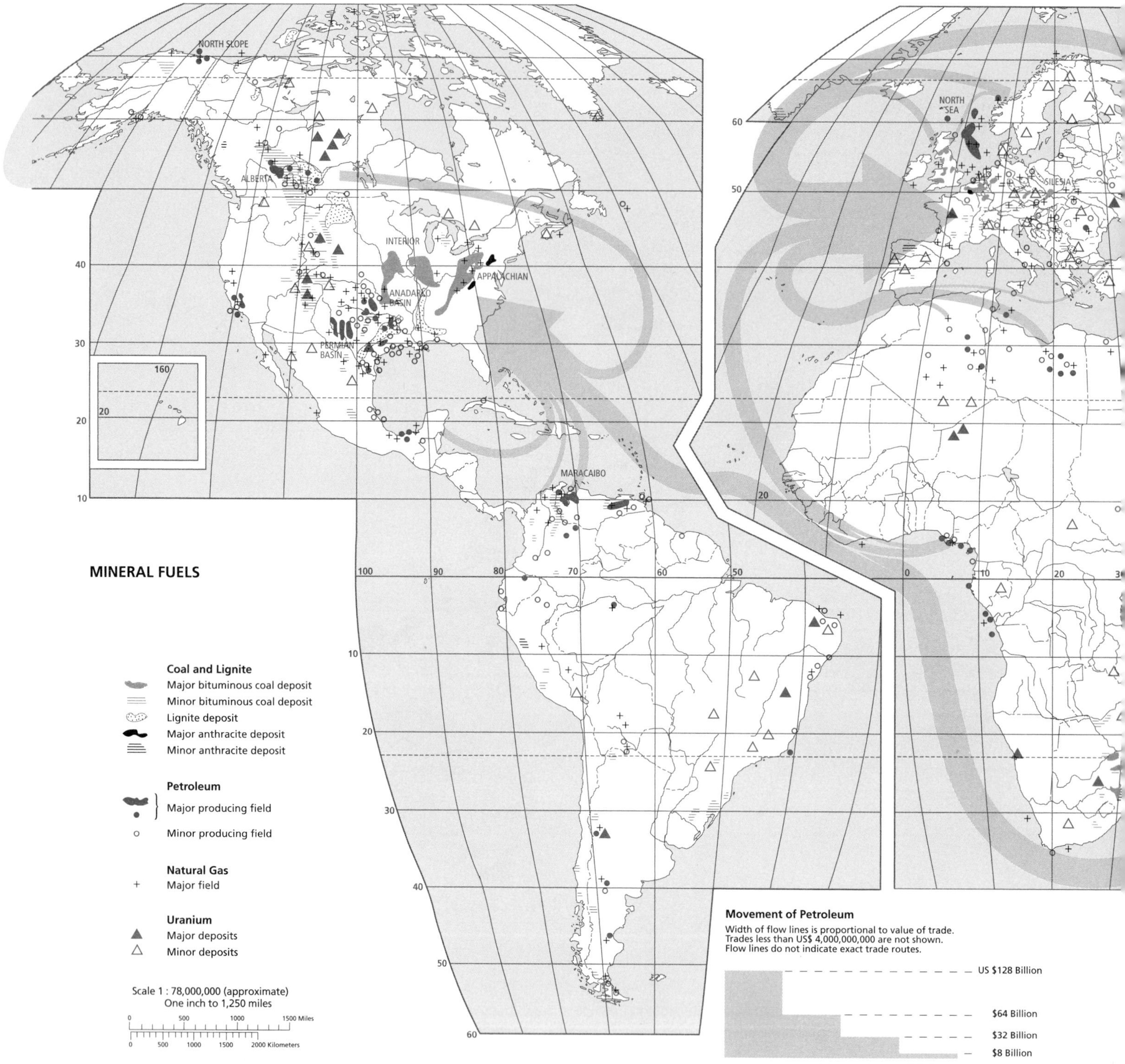

MINERAL FUELS

Coal and Lignite
- Major bituminous coal deposit
- Minor bituminous coal deposit
- Lignite deposit
- Major anthracite deposit
- Minor anthracite deposit

Petroleum
- ⟩ Major producing field
- ○ Minor producing field

Natural Gas
- + Major field

Uranium
- ▲ Major deposits
- △ Minor deposits

Scale 1 : 78,000,000 (approximate)
One inch to 1,250 miles

0   500   1000   1500 Miles

0   500   1000   1500   2000 Kilometers

Movement of Petroleum
Width of flow lines is proportional to value of trade.
Trades less than US$ 4,000,000,000 are not shown.
Flow lines do not indicate exact trade routes.

- — US $128 Billion
- — $64 Billion
- — $32 Billion
- — $8 Billion

Map labels: NORTH SLOPE, ALBERTA, INTERIOR, ANADARKO BASIN, APPALACHIAN, PERMIAN BASIN, MARACAIBO, NORTH SEA, SILESIA

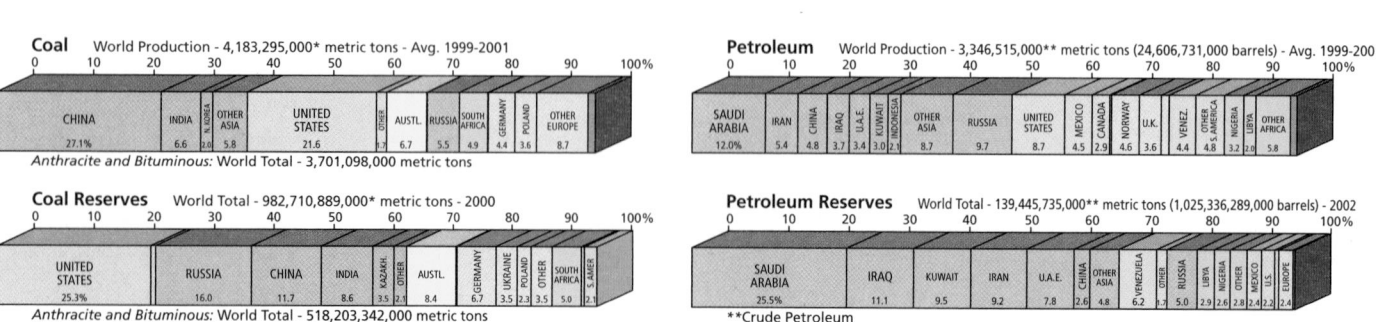

Coal   World Production - 4,183,295,000* metric tons - Avg. 1999-2001

CHINA	INDIA	N. KOREA	OTHER ASIA	UNITED STATES	OTHER	AUSTL.	RUSSIA	SOUTH AFRICA	GERMANY	POLAND	OTHER EUROPE
27.1%	6.6	2.0	5.8	21.6	1.7	6.7	5.5	4.9	4.4	3.6	8.7

Anthracite and Bituminous: World Total - 3,701,098,000 metric tons

Coal Reserves   World Total - 982,710,889,000* metric tons - 2000

UNITED STATES	RUSSIA	CHINA	INDIA	KAZAKH.	OTHER	AUSTL.	GERMANY	UKRAINE	POLAND	OTHER	SOUTH AFRICA	S. AMER.
25.3%	16.0	11.7	8.6	3.5	2.1	8.4	6.7	3.5	2.1	2.3	5.0	

Anthracite and Bituminous: World Total - 518,203,342,000 metric tons
*Includes anthracite, bituminous, and lignite coal

Petroleum   World Production - 3,346,515,000** metric tons (24,606,731,000 barrels) - Avg. 1999-2001

SAUDI ARABIA	IRAN	CHINA	IRAQ	U.A.E.	KUWAIT	INDONESIA	OTHER ASIA	RUSSIA	UNITED STATES	MEXICO	CANADA	NORWAY	U.K.	VENEZ.	OTHER AMERICA	NIGERIA	LIBYA	OTHER AFRICA
12.0%	5.4	4.8	3.7	3.4	3.0	2.1	8.7	9.7	8.7	4.5	2.9	4.6	3.6	4.4	4.8	3.2	2.0	5.8

Petroleum Reserves   World Total - 139,445,735,000** metric tons (1,025,336,289,000 barrels) - 2002

SAUDI ARABIA	IRAQ	KUWAIT	IRAN	U.A.E.	CHINA	OTHER ASIA	VENEZUELA	OTHER	RUSSIA	LIBYA	NIGERIA	MEXICO	U.S.	EUROPE
25.5%	11.1	9.5	9.2	7.8	2.6	4.8	6.2	1.7	5.0	2.9	2.6	2.4	2.2	2.4

**Crude Petroleum

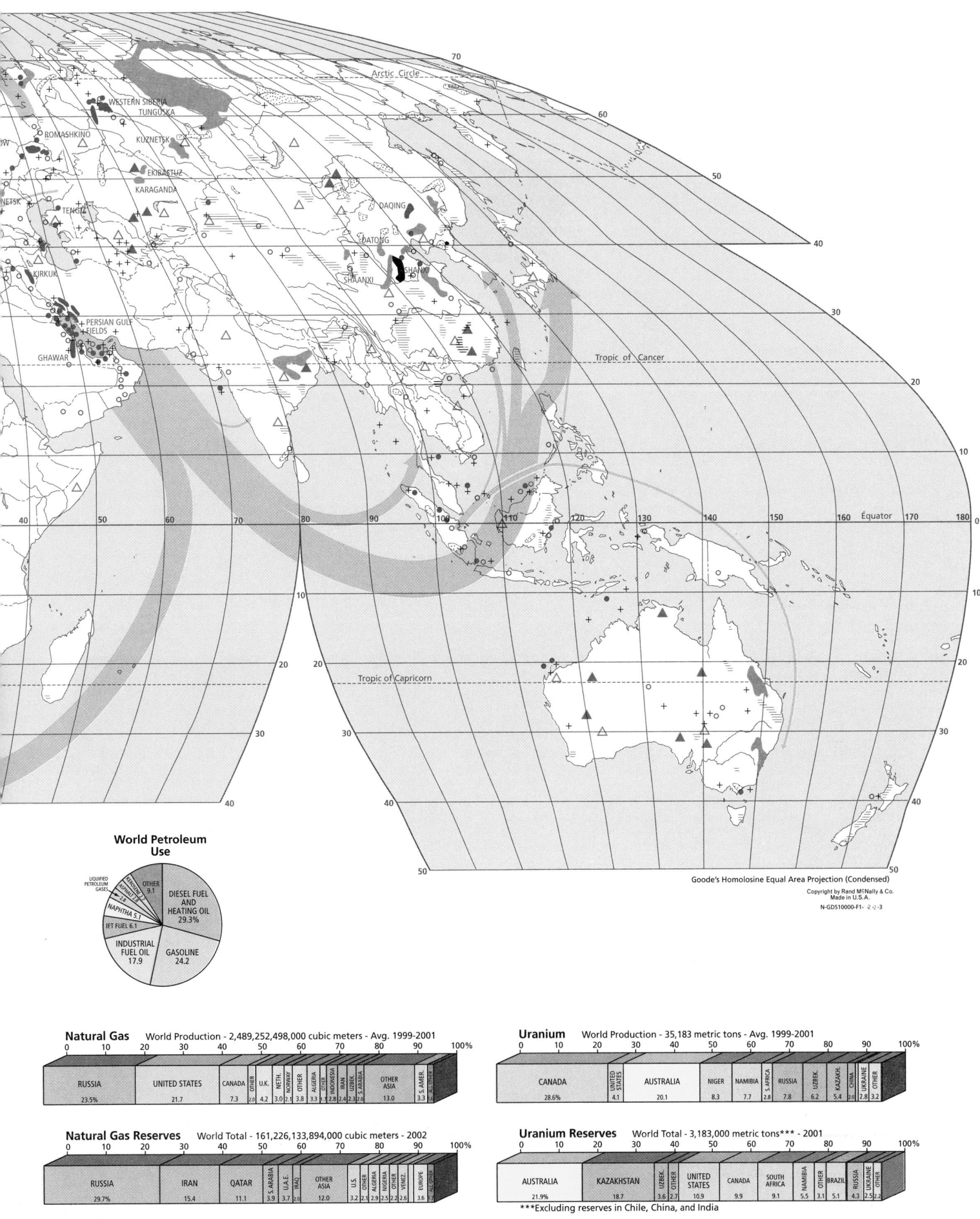

World Petroleum Use

- LIQUIFIED PETROLEUM GASES
- KEROSENE
- ASPHALT 2.3
- OTHER 9.1
- DIESEL FUEL AND HEATING OIL 29.3%
- NAPHTHA 5.1
- JET FUEL 6.1
- INDUSTRIAL FUEL OIL 17.9
- GASOLINE 24.2

Goode's Homolosine Equal Area Projection (Condensed)

Copyright by Rand McNally & Co.
Made in U.S.A.

N-GDS10000-F1- 2-2-3

Natural Gas   World Production - 2,489,252,498,000 cubic meters - Avg. 1999-2001

RUSSIA 23.5%	UNITED STATES 21.7	CANADA 7.3	OTHER 2.0	U.K. 4.2	NETH. 3.0	NORWAY 2.1	OTHER 3.8	ALGERIA 3.3	INDONESIA 2.8	IRAN 2.4	UZBEK. 2.3	S. ARABIA 2.0	OTHER ASIA 13.0	S. AMER. 3.3	ALL OTHER

Natural Gas Reserves   World Total - 161,226,133,894,000 cubic meters - 2002

RUSSIA 29.7%	IRAN 15.4	QATAR 11.1	S. ARABIA 3.9	U.A.E. 3.7	IRAQ 2.0	OTHER ASIA 12.0	U.S. 3.2	OTHER 2.9	ALGERIA 2.5	NIGERIA 2.6	VENEZ. 2.6	EUROPE 3.6	ALL OTHER

Uranium   World Production - 35,183 metric tons - Avg. 1999-2001

CANADA 28.6%	UNITED STATES 4.1	AUSTRALIA 20.1	NIGER 8.3	NAMIBIA 7.7	S. AFRICA 2.8	RUSSIA 7.8	UZBEK. 6.2	KAZAKH. 5.4	CHINA 2.0	UKRAINE 2.8	OTHER 3.2

Uranium Reserves   World Total - 3,183,000 metric tons*** - 2001

AUSTRALIA 21.9%	KAZAKHSTAN 18.7	UZBEK. 3.6	OTHER 2.7	UNITED STATES 10.9	CANADA 9.9	SOUTH AFRICA 9.1	NAMIBIA 5.5	OTHER 3.1	BRAZIL 5.1	RUSSIA 4.3	UKRAINE 2.5	OTHER 2.2

***Excluding reserves in Chile, China, and India

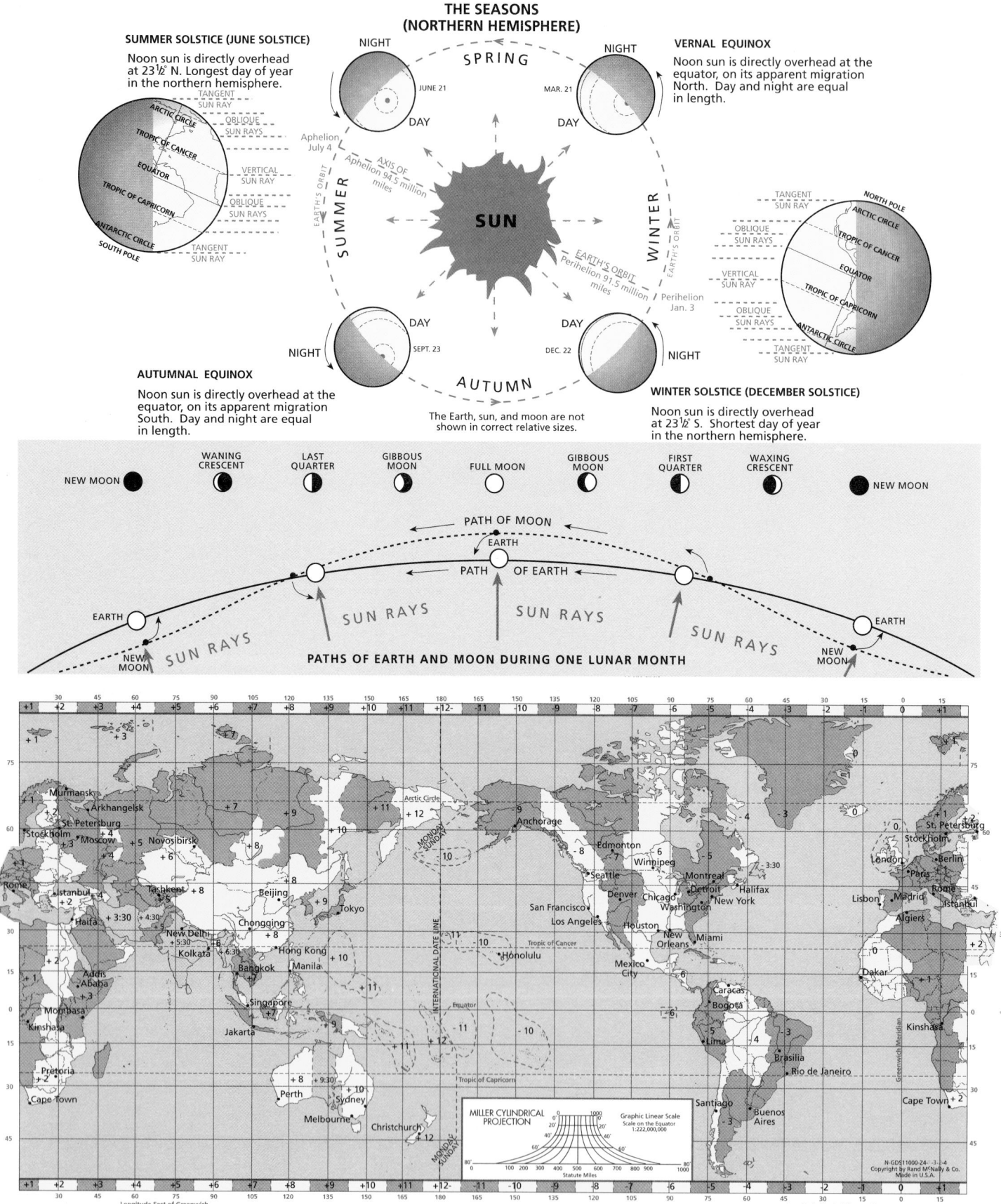

THE SEASONS (NORTHERN HEMISPHERE)

SUMMER SOLSTICE (JUNE SOLSTICE)
Noon sun is directly overhead at 23½° N. Longest day of year in the northern hemisphere.

VERNAL EQUINOX
Noon sun is directly overhead at the equator, on its apparent migration North. Day and night are equal in length.

AUTUMNAL EQUINOX
Noon sun is directly overhead at the equator, on its apparent migration South. Day and night are equal in length.

WINTER SOLSTICE (DECEMBER SOLSTICE)
Noon sun is directly overhead at 23½° S. Shortest day of year in the northern hemisphere.

The Earth, sun, and moon are not shown in correct relative sizes.

PATHS OF EARTH AND MOON DURING ONE LUNAR MONTH

MILLER CYLINDRICAL PROJECTION

Copyright by Rand McNally & Co.
Made in U.S.A.

Time Zones

The surface of the earth is divided into 24 time zones. Each zone represents 15° of longitude or one hour of time. The time of the initial, or zero, zone is based on the Greenwich Meridian and extends eastward and westward for a distance of 7½° of longitude. Each of the zones is designated by a number representing the hours (+ or -) by which its standard time differs from Greenwich mean time. These standard time zones are indicated by bands of orange and yellow. Areas which have a fractional deviation from standard time are shown in an intermediate color. The irregularities in the zones and the fractional deviations are due to political and economic factors.

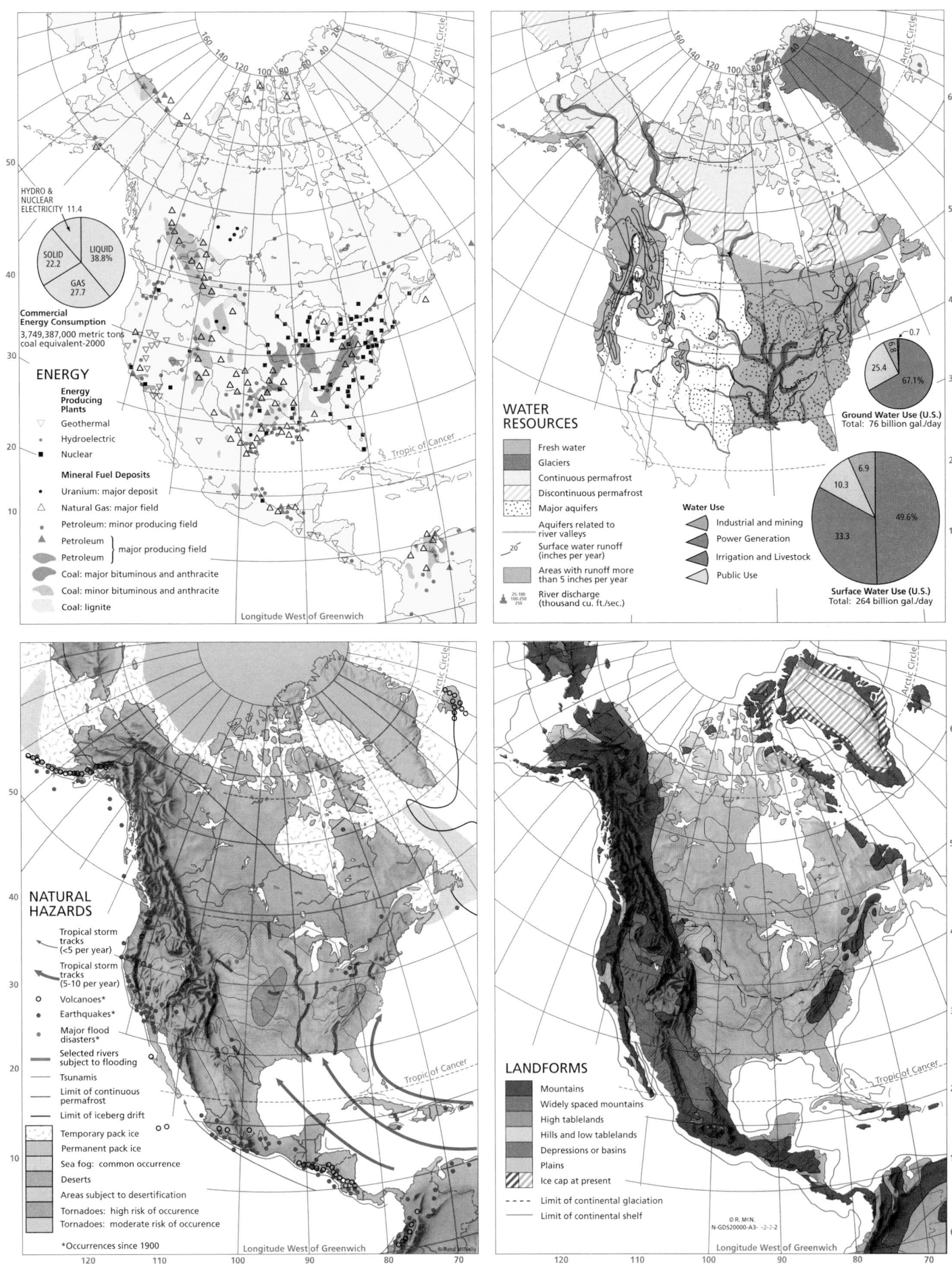

ENERGY

HYDRO & NUCLEAR ELECTRICITY 11.4

SOLID 22.2
LIQUID 38.8%
GAS 27.7

Commercial Energy Consumption
3,749,387,000 metric tons coal equivalent-2000

Energy Producing Plants
▽ Geothermal
• Hydroelectric
■ Nuclear

Mineral Fuel Deposits
• Uranium: major deposit
△ Natural Gas: major field
• Petroleum: minor producing field
▲ Petroleum } major producing field
▬ Petroleum }
▬ Coal: major bituminous and anthracite
▬ Coal: minor bituminous and anthracite
▬ Coal: lignite

Longitude West of Greenwich

WATER RESOURCES

Fresh water
Glaciers
Continuous permafrost
Discontinuous permafrost
Major aquifers
Aquifers related to river valleys
Surface water runoff (inches per year)
Areas with runoff more than 5 inches per year
River discharge (thousand cu. ft./sec.)
25-100
100-250
250

Water Use
Industrial and mining
Power Generation
Irrigation and Livestock
Public Use

Ground Water Use (U.S.)
Total: 76 billion gal./day
0.7
6.8
25.4
67.1%

Surface Water Use (U.S.)
Total: 264 billion gal./day
6.9
10.3
33.3
49.6%

NATURAL HAZARDS

Tropical storm tracks (<5 per year)
Tropical storm tracks (5-10 per year)
○ Volcanoes*
• Earthquakes*
• Major flood disasters*
▬ Selected rivers subject to flooding
Tsunamis
Limit of continuous permafrost
Limit of iceberg drift
Temporary pack ice
Permanent pack ice
Sea fog: common occurrence
Deserts
Areas subject to desertification
Tornadoes: high risk of occurence
Tornadoes: moderate risk of occurence

*Occurrences since 1900

Longitude West of Greenwich

LANDFORMS

Mountains
Widely spaced mountains
High tablelands
Hills and low tablelands
Depressions or basins
Plains
Ice cap at present

- - - Limit of continental glaciation
─── Limit of continental shelf

© R. McN.
N-GD520000-A3- -2-2-2

Longitude West of Greenwich

Tropic of Cancer

Map 1 (top left): Annual Precipitation

WINTER MAXIMUM
FALL MAX
SUMMER MAXIMUM
WINTER MAXIMUM
WINTER MAX.
SUMMER MAXIMUM
SUMMER MAXIMUM
WINTER MAXIMUM
SUMMER MAXIMUM
SUMMER MAXIMUM

Arctic Circle
Tropic of Cancer

ANNUAL PRECIPITATION
Cm. (In.)

- Under 25 (10)
- 25-50 (10-20)
- 50-100 (20-40)
- 100-150 (40-60)
- 150-200 (60-80)
- Over 200 (80)

Longitude West of Greenwich

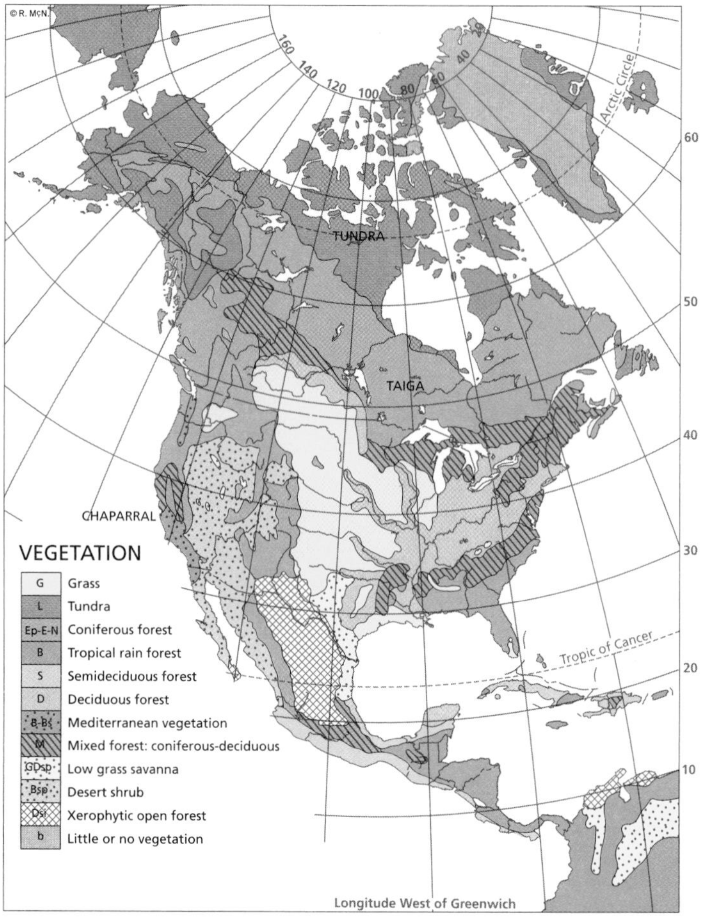

Map 2 (top right): Vegetation

TUNDRA
TAIGA
CHAPARRAL

Arctic Circle
Tropic of Cancer

VEGETATION

G	Grass
L	Tundra
Ep-E-N	Coniferous forest
B	Tropical rain forest
S	Semideciduous forest
D	Deciduous forest
B-B	Mediterranean vegetation
M	Mixed forest: coniferous-deciduous
GDsp	Low grass savanna
BSp	Desert shrub
Dsi	Xerophytic open forest
b	Little or no vegetation

Longitude West of Greenwich

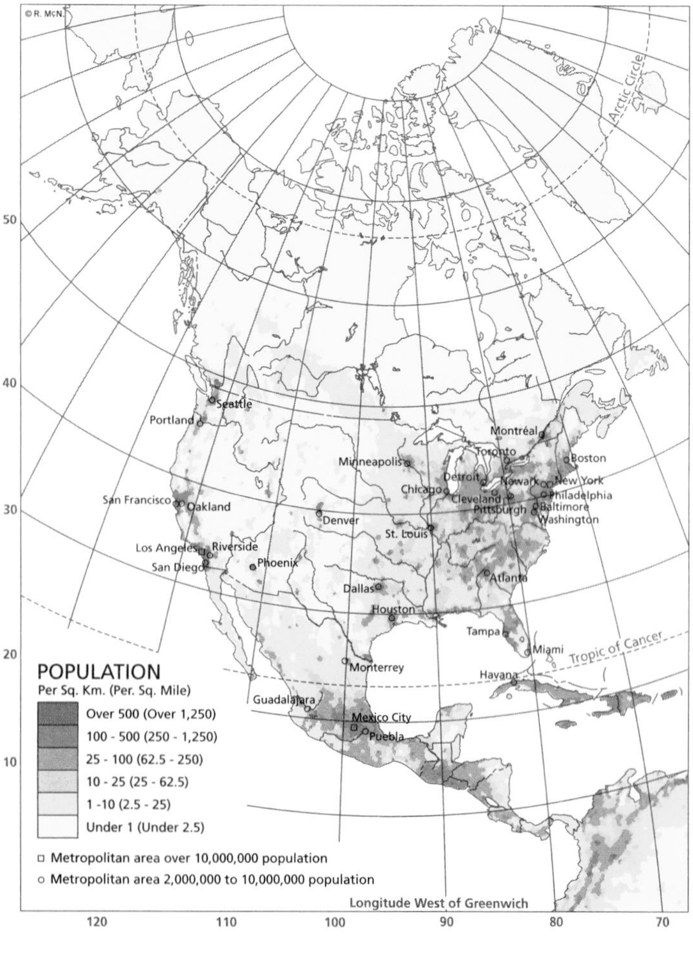

Map 3 (bottom left): Population

Seattle
Portland
Montréal
Toronto
Boston
Minneapolis
Detroit
Chicago
Cleveland New York
San Francisco
Oakland
Pittsburgh Philadelphia
Denver
St. Louis
Baltimore
Washington
Los Angeles Riverside
San Diego
Phoenix
Atlanta
Dallas
Houston
Tampa
Miami
Monterrey
Havana
Guadalajara
Mexico City
Puebla

Arctic Circle
Tropic of Cancer

POPULATION
Per Sq. Km. (Per. Sq. Mile)

- Over 500 (Over 1,250)
- 100 - 500 (250 - 1,250)
- 25 - 100 (62.5 - 250)
- 10 - 25 (25 - 62.5)
- 1 -10 (2.5 - 25)
- Under 1 (Under 2.5)

□ Metropolitan area over 10,000,000 population
○ Metropolitan area 2,000,000 to 10,000,000 population

Longitude West of Greenwich

120   110   100   90   80   70

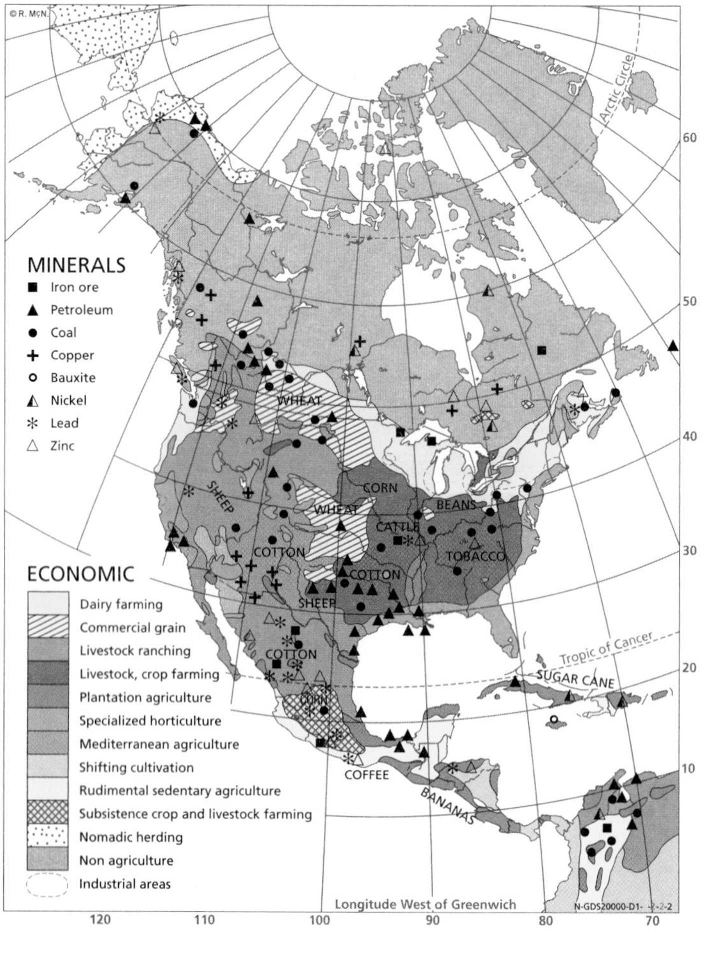

Map 4 (bottom right): Economic, Minerals

MINERALS
- ■ Iron ore
- ▲ Petroleum
- ● Coal
- ✛ Copper
- ○ Bauxite
- ◮ Nickel
- ✳ Lead
- △ Zinc

WHEAT
WHEAT
SHEEP
CORN
BEANS
CATTLE
COTTON
TOBACCO
COTTON
SHEEP
COTTON
CORN
SUGAR CANE
COFFEE
BANANAS

ECONOMIC

- Dairy farming
- Commercial grain
- Livestock ranching
- Livestock, crop farming
- Plantation agriculture
- Specialized horticulture
- Mediterranean agriculture
- Shifting cultivation
- Rudimental sedentary agriculture
- Subsistence crop and livestock farming
- Nomadic herding
- Non agriculture
- Industrial areas

Arctic Circle
Tropic of Cancer

Longitude West of Greenwich

120   110   100   90   80   70

N-GDS20000-D1-

ARCTIC OCEAN

ALEUTIAN ISLANDS

Bering Sea

Bering Strait

Nome

BROOKS RANGE

ALASKA RANGE

Yukon

Fairbanks

Anchorage

Gulf of Alaska

Beaufort Sea

ELLESMERE ISLAND

BANKS ISLAND

MELVILLE ISLAND

VICTORIA ISLAND

DEVON ISLAND

GREENLAND

Baffin Bay

BAFFIN ISLAND

Arctic Circle

Godthab

PACIFIC OCEAN

Juneau

Prince Rupert

Great Slave Lake

Churchill

UNGAVA PENINSULA

Hudson Bay

Labrador Sea

Vancouver

Seattle

Portland

Peace

Edmonton

Calgary

Regina

ROCKY MOUNTAINS

St. Lawrence

St. John's

San Francisco

SIERRA NEVADA

Salt Lake City

GREAT BASIN

Billings

Bismarck

Winnipeg

Minneapolis

Lake Superior

Halifax

Los Angeles

Colorado

Rapid City

Denver

Omaha

Missouri

Mississippi

Lake Michigan

L. Huron

MONTRÉAL

TORONTO

L. Ont.

Albuquerque

Phoenix

Kansas City

CHICAGO

DETROIT

L. Erie

Ohio

Cincinnati

Pittsburgh

BOSTON

NEW YORK

PHILADELPHIA

WASHINGTON

Golfo de California

Dallas

ST. LOUIS

Mississippi

Nashville

APPALACHIAN MOUNTAINS

Chihuahua

Rio Grande

Houston

Atlanta

La Paz

SIERRA MADRE OCCIDENTAL

Mazatlán

Monterrey

New Orleans

Jacksonville

ATLANTIC OCEAN

Guadalajara

SIERRA MADRE ORIENTAL

Gulf of Mexico

MEXICO CITY

SIERRA MADRE DEL SUR

Mérida

Miami

Nassau

BAHAMA ISLANDS

Havana

CUBA

Tropic of Cancer

San Salvador

Port-au-Prince

JAMAICA

Kingston

San Juan

HISPANIOLA

PUERTO RICO

Managua

Caribbean Sea

PACIFIC OCEAN

San Jose

Panamá

Maracaibo

CARACAS

TRINIDAD

Legend:

- Urban
- Cropland
- Cropland & Woodland
- Cropland & Grazing Land
- Grassland, Grazing Land
- Forest, Woodland
- Swamp, Marshland
- Tundra
- Shrub, Sparse Grass, Wasteland
- Barren Land

COPYRIGHT BY RAND McNALLY & COMPANY
MADE IN U.S.A.

A-520000-36 -2-6

Scale 1:36,000,000; one inch to 570 miles. Lambert Azimuthal Equal-Area Projection

0 100 200 400 600 800 Miles

0 150 300 600 900 1200 Kilometers

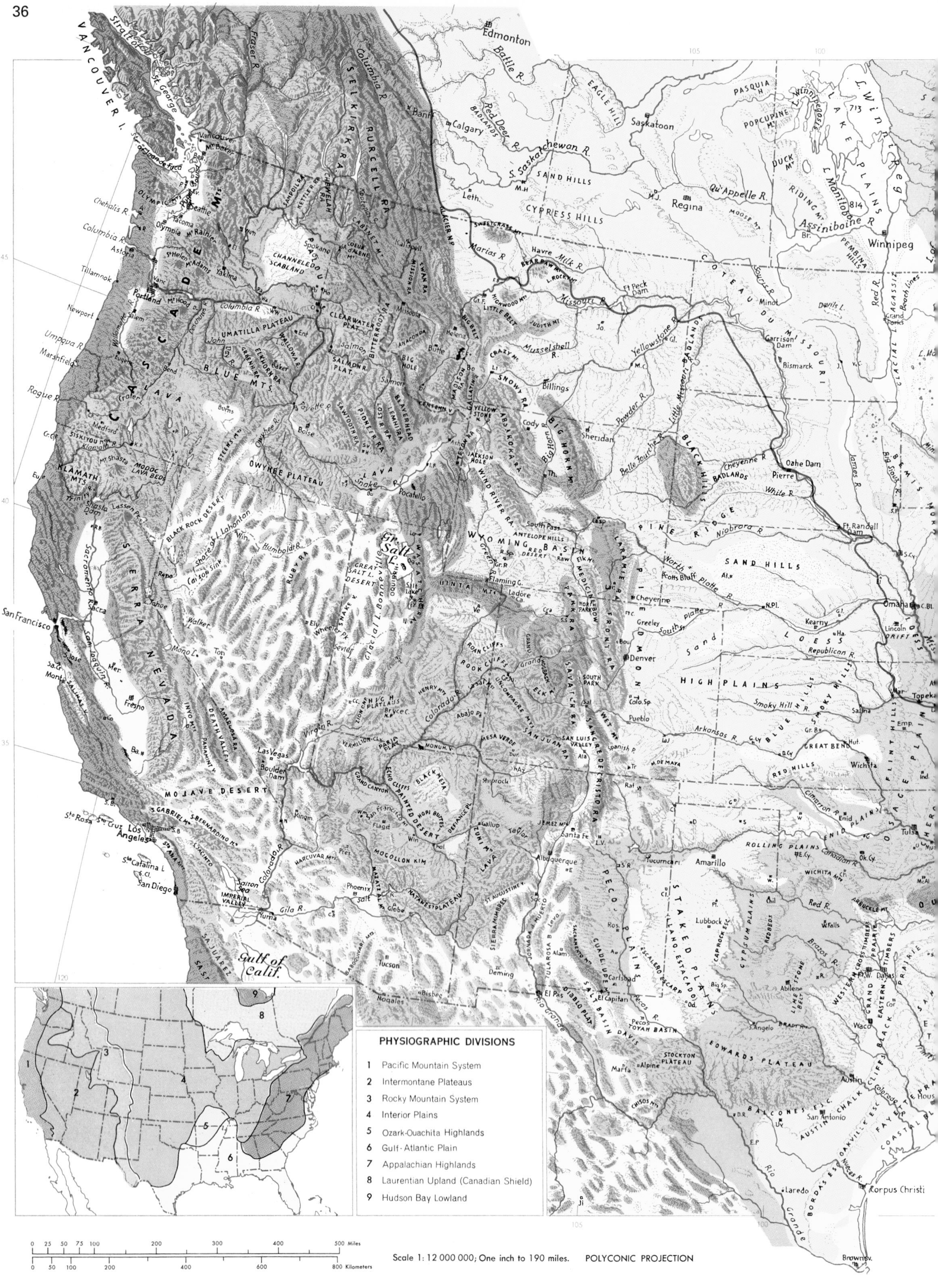

PHYSIOGRAPHIC DIVISIONS
1 Pacific Mountain System
2 Intermontane Plateaus
3 Rocky Mountain System
4 Interior Plains
5 Ozark-Ouachita Highlands
6 Gulf-Atlantic Plain
7 Appalachian Highlands
8 Laurentian Upland (Canadian Shield)
9 Hudson Bay Lowland
0 25 50 75 100 200 300 400 500 Miles
0 50 100 200 400 600 800 Kilometers
Scale 1: 12 000 000; One inch to 190 miles. POLYCONIC PROJECTION
VANCOUVER I.
Edmonton
Calgary
Saskatoon
Regina
Winnipeg
Vancouver
Seattle
Portland
Spokane
CHANNELED SCABLAND
CASCADE MTS.
BLUE MTS.
UMATILLA PLATEAU
OWYHEE PLATEAU
BLACK ROCK DESERT
Gr. Salt L.
GREAT SALT L. DESERT
WYOMING BASIN
CYPRESS HILLS
SAND HILLS
HIGH PLAINS
PINE RIDGE
BLACK HILLS
BADLANDS
Missouri R.
Yellowstone R.
Snake R.
Columbia R.
Colorado R.
Cheyenne
Denver
Omaha
San Francisco
Sacra
Fresno
SIERRA NEVADA
DEATH VALLEY
MOJAVE DESERT
Los Angeles
San Diego
Las Vegas
Boulder Dam
GRAND CANYON
PAINTED DESERT
Phoenix
Tucson
Gulf of Calif.
Santa Fe
Albuquerque
El Paso
STAKED PLAINS (LLANO ESTACADO)
EDWARDS PLATEAU
San Antonio
Austin
Dallas
Corpus Christi
Laredo
Rio Grande

PHYSIOGRAPHY
BY
ERWIN RAISZ

LITHOLOGY AND STRUCTURE

Unconsolidated deposits: alluvium, sands, playa deposits, etc.

Essentially horizontal sedimentary rocks; many partially unconsolidated.

Slightly to moderately tilted, older sedimentary rocks.

Steeply folded or faulted, sedimentary rocks

Volcanics; largely lava flows.

Metamorphic and intrusive igneous rocks; structure complex.

Limits of continental glaciation.

LANDFORMS

PLATEAUS	BASIN RANGES
HILLS	VOLCANO AND LAVA
MOUNTAINS	SAND
MESAS	SINKS
CUESTAS	MORAINES
FOLDED MOUNTAINS	DRUMLINS

A-520500-9A6 -3-3-7
Copyright by Rand McNally & Co.
Made in U.S.A.

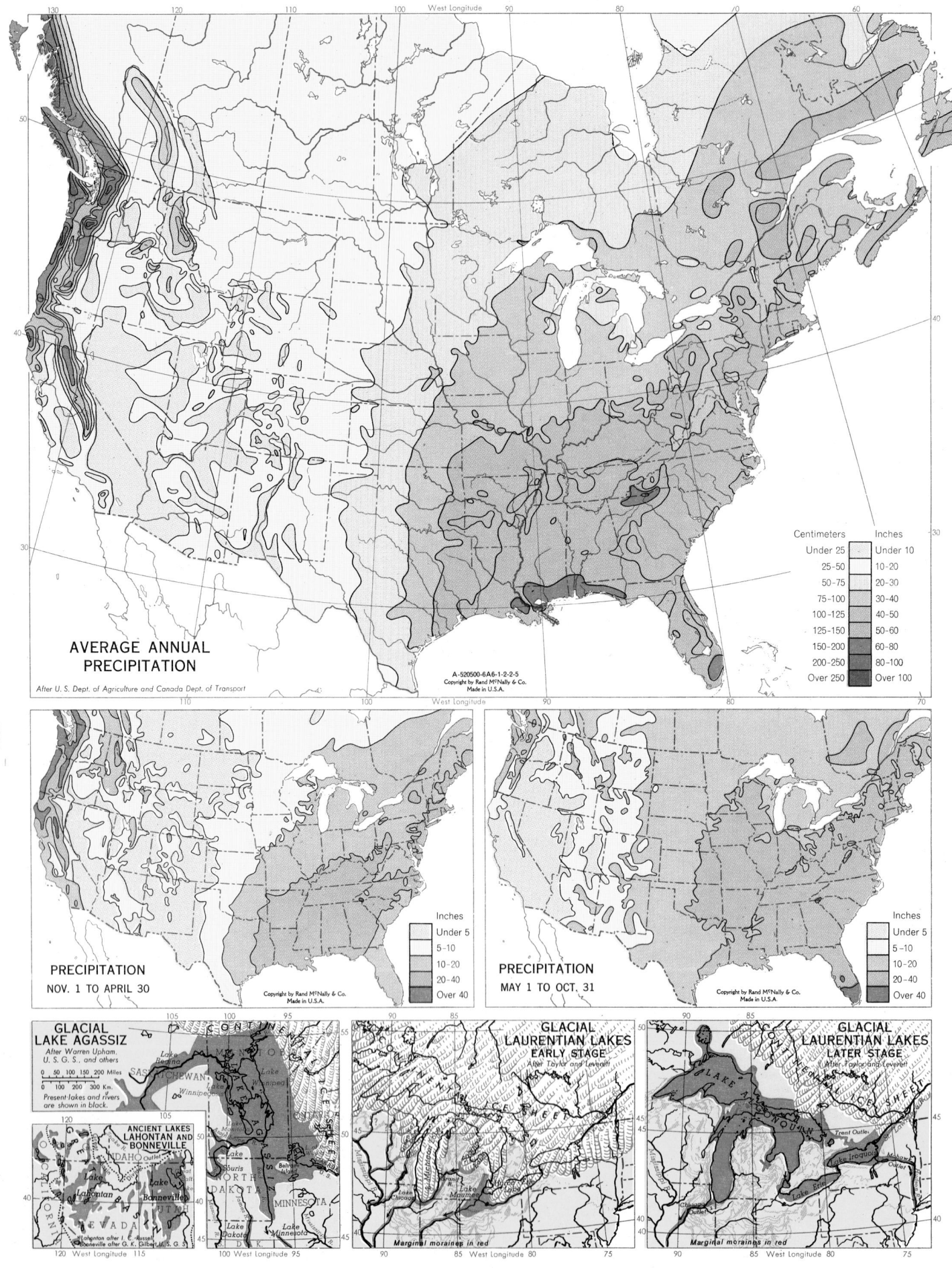

AVERAGE ANNUAL
PRECIPITATION

After U. S. Dept. of Agriculture and Canada Dept. of Transport

A-520500-6A6-1-2-2-5
Copyright by Rand McNally & Co.
Made in U.S.A.

Centimeters	Inches
Under 25	Under 10
25-50	10-20
50-75	20-30
75-100	30-40
100-125	40-50
125-150	50-60
150-200	60-80
200-250	80-100
Over 250	Over 100

PRECIPITATION

NOV. 1 TO APRIL 30

Copyright by Rand McNally & Co.
Made in U.S.A.

Inches
Under 5
5-10
10-20
20-40
Over 40

PRECIPITATION

MAY 1 TO OCT. 31

Copyright by Rand McNally & Co.
Made in U.S.A.

Inches
Under 5
5-10
10-20
20-40
Over 40

GLACIAL
LAKE AGASSIZ

After Warren Upham,
U. S. G. S., and others

0 50 100 150 200 Miles
0 100 200 300 Km.

Present lakes and rivers
are shown in black.

ANCIENT LAKES
LAHONTAN AND
BONNEVILLE

Lahontan after I. C. Russell,
Bonneville after G. K. Gilbert, U. S. G. S.

GLACIAL
LAURENTIAN LAKES
EARLY STAGE

After Taylor and Leverett

Marginal moraines in red

GLACIAL
LAURENTIAN LAKES
LATER STAGE

After Taylor and Leverett

Marginal moraines in red

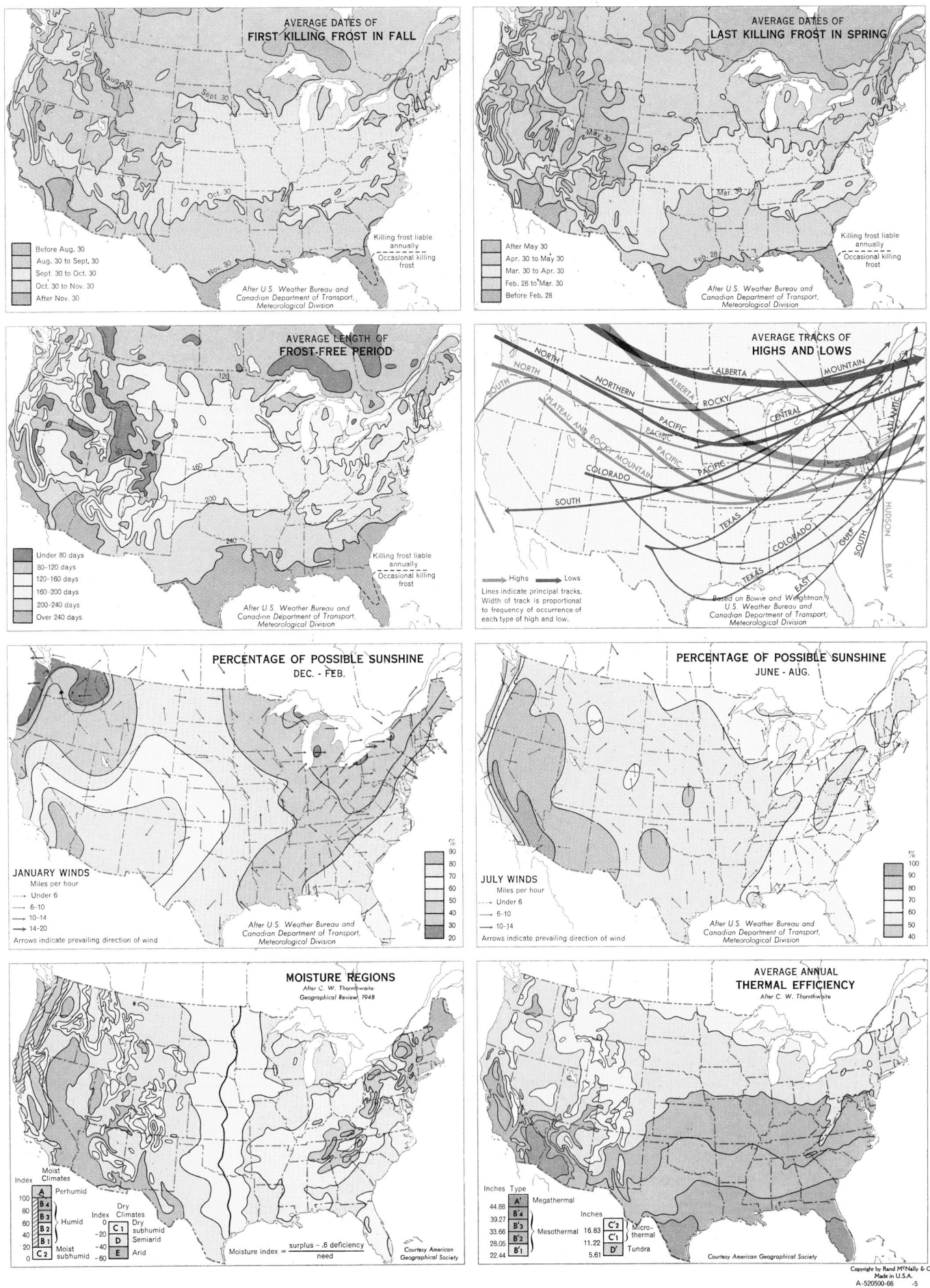

AVERAGE DATES OF
FIRST KILLING FROST IN FALL

Before Aug. 30
Aug. 30 to Sept. 30
Sept. 30 to Oct. 30
Oct. 30 to Nov. 30
After Nov. 30

Killing frost liable
annually
Occasional killing
frost

After U.S. Weather Bureau and
Canadian Department of Transport,
Meteorological Division

AVERAGE DATES OF
LAST KILLING FROST IN SPRING

After May 30
Apr. 30 to May 30
Mar. 30 to Apr. 30
Feb. 28 to Mar. 30
Before Feb. 28

Killing frost liable
annually
Occasional killing
frost

After U.S. Weather Bureau and
Canadian Department of Transport,
Meteorological Division

AVERAGE LENGTH OF
FROST-FREE PERIOD

Under 80 days
80-120 days
120-160 days
160-200 days
200-240 days
Over 240 days

Killing frost liable
annually
Occasional killing
frost

After U.S. Weather Bureau and
Canadian Department of Transport,
Meteorological Division

AVERAGE TRACKS OF
HIGHS AND LOWS

Highs   Lows
Lines indicate principal tracks.
Width of track is proportional
to frequency of occurrence of
each type of high and low.

Based on Bowie and Weightman,
U.S. Weather Bureau and
Canadian Department of Transport,
Meteorological Division

PERCENTAGE OF POSSIBLE SUNSHINE
DEC. - FEB.

JANUARY WINDS
Miles per hour
Under 6
6-10
10-14
14-20
Arrows indicate prevailing direction of wind

%
90
80
70
60
50
40
30
20

After U.S. Weather Bureau and
Canadian Department of Transport,
Meteorological Division

PERCENTAGE OF POSSIBLE SUNSHINE
JUNE - AUG.

JULY WINDS
Miles per hour
Under 6
6-10
10-14
Arrows indicate prevailing direction of wind

%
100
90
80
70
60
50
40

After U.S. Weather Bureau and
Canadian Department of Transport,
Meteorological Division

MOISTURE REGIONS
After C. W. Thornthwaite
Geographical Review, 1948

Index   Moist Climates
100   A   Perhumid
80   B4
60   B3   Humid
40   B2
20   B1   Moist
0   C2   subhumid

Index   Dry Climates
0   C1   Dry subhumid
-20   D   Semiarid
-40   E   Arid
-60

Moisture index = surplus − .6 deficiency / need

Courtesy American
Geographical Society

AVERAGE ANNUAL
THERMAL EFFICIENCY
After C. W. Thornthwaite

Inches   Type
44.88   A'   Megathermal
39.27   B'4
33.66   B'3   Mesothermal   Inches
28.05   B'2   16.83   C'2
22.44   B'1   11.22   C'1   Microthermal
                            5.61   D'   Tundra

Courtesy American Geographical Society

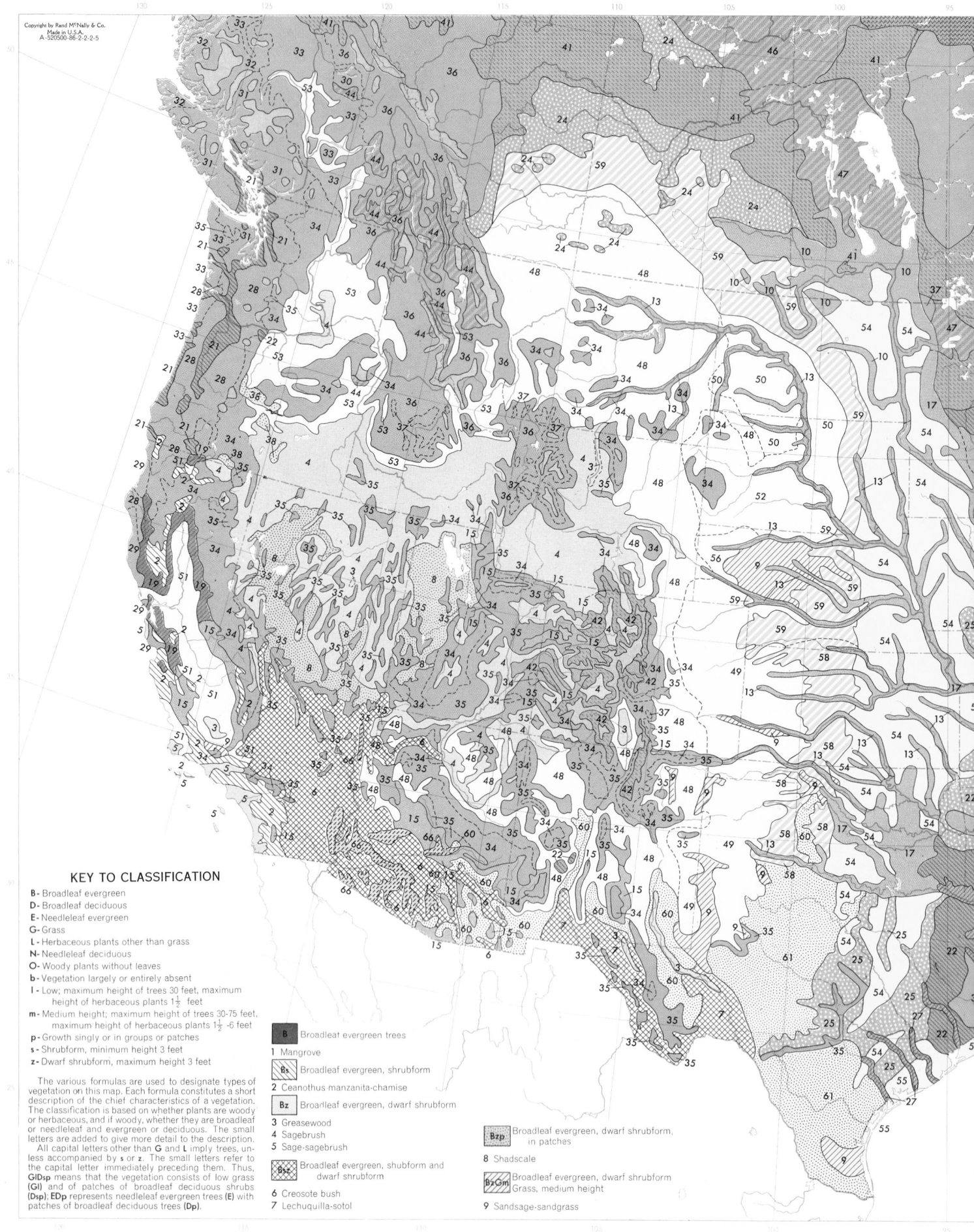

KEY TO CLASSIFICATION

B - Broadleaf evergreen
D - Broadleaf deciduous
E - Needleleaf evergreen
G - Grass
L - Herbaceous plants other than grass
N - Needleleaf deciduous
O - Woody plants without leaves
b - Vegetation largely or entirely absent
l - Low; maximum height of trees 30 feet, maximum
    height of herbaceous plants 1½ feet
m - Medium height; maximum height of trees 30-75 feet,
    maximum height of herbaceous plants 1½ -6 feet
p - Growth singly or in groups or patches
s - Shrubform, minimum height 3 feet
z - Dwarf shrubform, maximum height 3 feet

The various formulas are used to designate types of
vegetation on this map. Each formula constitutes a short
description of the chief characteristics of a vegetation.
The classification is based on whether plants are woody
or herbaceous, and if woody, whether they are broadleaf
or needleleaf and evergreen or deciduous. The small
letters are added to give more detail to the description.
All capital letters other than G and L imply trees, unless accompanied by s or z. The small letters refer to
the capital letter immediately preceding them. Thus,
GlDsp means that the vegetation consists of low grass
(Gl) and of patches of broadleaf deciduous shrubs
(Dsp); EDp represents needleleaf evergreen trees (E) with
patches of broadleaf deciduous trees (Dp).

B Broadleaf evergreen trees

1 Mangrove

Bs Broadleaf evergreen, shrubform

2 Ceanothus-manzanita-chamise

Bz Broadleaf evergreen, dwarf shrubform

3 Greasewood
4 Sagebrush
5 Sage-sagebrush

Bsz Broadleaf evergreen, shubform and
       dwarf shrubform

6 Creosote bush
7 Lechuquilla-sotol

Bzp Broadleaf evergreen, dwarf shrubform,
       in patches

8 Shadscale

BzGm Broadleaf evergreen, dwarf shrubform
        Grass, medium height

9 Sandsage-sandgrass

Copyright by Rand McNally & Co.
Made in U.S.A.
A-520500-86-2-2-5

0 25 50 75 100   200   300   400   500 Miles

0 50 100   200   400   600   800 Kilometers

Scale 1:14 000 000;   One inch to 220 mile

NATURAL VEGETATION

BY A. W. KÜCHLER

Based on "A Physiognomic Classification of Vegetation"
Annals of the Assoc. of American Geographers, Vol. 39, September, 1949

Longitude West of Greenwich

ALBERT CONFORMAL CONIC PROJECTION

D Broadleaf deciduous trees

10 Aspen-oak
11 Beech-maple
12 Beech-tulip tree-maple-basswood
13 Cottonwood-willow
14 Maple-basswood
15 Oak
16 Oak-ash-maple
17 Oak-hickory
18 Oak-tulip tree

DB Broadleaf deciduous trees
Broadleaf evergreen trees

19 Oak-madrone

DE Broadleaf deciduous trees
Needleleaf evergreen trees

20 Maple-yellow birch-hemlock-pine
21 Oak-Douglas fir
22 Oak-pine
23 Maple-beech-hemlock

D / Gmp Broadleaf deciduous trees
Grass, medium height, in patches

24 Aspen-needle grass-wheat grass
25 Oak-hickory-bluestem

DN Broadleaf deciduous trees
Needleleaf deciduous trees

26 Bay trees-bald cypress
27 Tupelo-gum-bald cypress

E Needleleaf evergreen trees

28 Douglas fir
29 Douglas fir-redwood
30 Hemlock-arbor vitae
31 Hemlock-arbor vitae-Douglas fir
32 Hemlock-arbor vitae-fir
33 Hemlock-spruce
34 Pine
35 Pine-juniper
36 Pine-spruce
37 Spruce-fir

Esp Needleleaf evergreen, shrubform,
in patches

38 Juniper

EDp Needleleaf evergreen trees
Broadleaf deciduous trees, in patches

39 Douglas fir-pine-aspen
40 Pine-spruce-birch
41 Spruce-aspen
42 Spruce-fir-aspen
43 Spruce-poplar-birch

EN Needleleaf evergreen trees
Needleleaf deciduous trees

44 Hemlock-arbor vitae-Douglas fir-larch
45 Pine-bald cypress
46 Pine-spruce-larch
47 Spruce-larch

Gl Grass, low

48 Grama grass
49 Grama grass-buffalo grass
50 Grama grass-needle grass
51 Needle grass-blue grass
52 Wheat grass
53 Wheat grass-blue grass

Gm Grass, medium height

54 Bluestem
55 Broom grass-water grass
56 Marsh grass
57 Saw grass

Gml Grass, medium and low height

58 Bluestem-bunch grass
59 Needle grass-wheat grass

Gl / Dsp Grass, low
Broadleaf deciduous, shrubform, in patches

60 Bunch grass-oak

Gm / Dsp Grass, medium height
Broadleaf deciduous, shrubform, in patches

61 Mesquite grass-mesquite

L Herbaceous plants other than grass

62 Lichens, etc.

LEp Herbaceous plants other than grass
Needleleaf evergreen trees, in patches

63 Lichens-spruce

LEp / Np Herbaceous plants other than grass
Needleleaf evergreen trees, in patches
Needleleaf deciduous trees, in patches

64 Lichens-spruce-larch

N Needleleaf deciduous trees

65 Bald cypress

Op Woody plants without leaves, in patches

66 Palo verde-cacti-ocotillo

b Vegetation largely or entirely absent

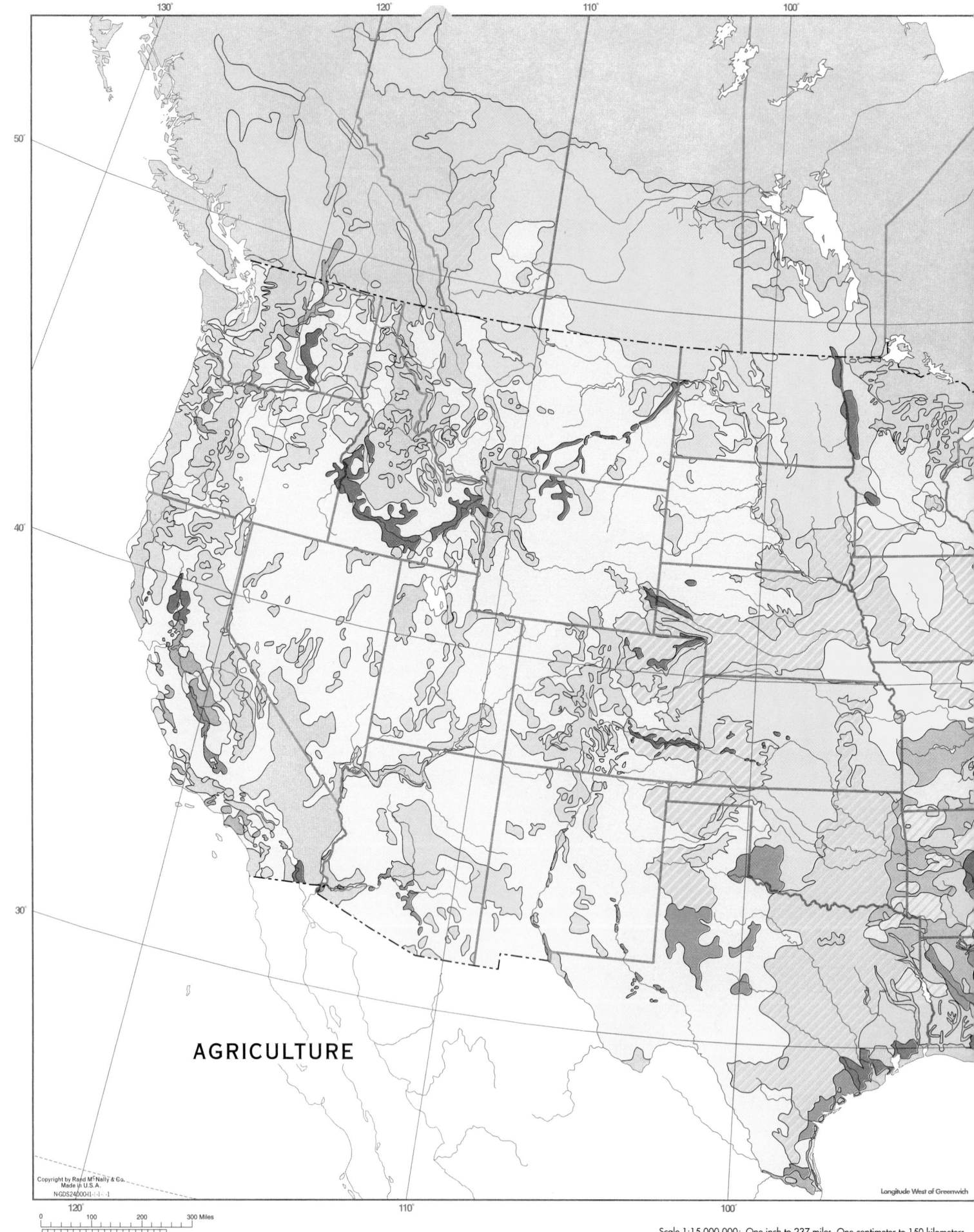

AGRICULTURE

Longitude West of Greenwich

Scale 1:15,000,000; One inch to 237 miles. One centimeter to 150 kilometers.

0   100   200   300 Miles

0   100   200   300   400 Kilometers

Dairying

Fruits and Vegetables

Wheat, Barley, and Oilseeds

Cash Corn and Soybeans

Tobacco

Cotton

Livestock and Feed Grains: Beef

Livestock and Feed Grains: Hogs

Livestock and Feed Grains: Poultry

Livestock and Feed Grains: Mixed

Specialty Crops (Peanuts, Potatoes, Rice, Sugar)

Western Livestock Ranching

Western Feedlots

Agriculture and Forestry

Non-Agricultural Areas

Tropic of Cancer

ALBERS CONIC PROJECTION

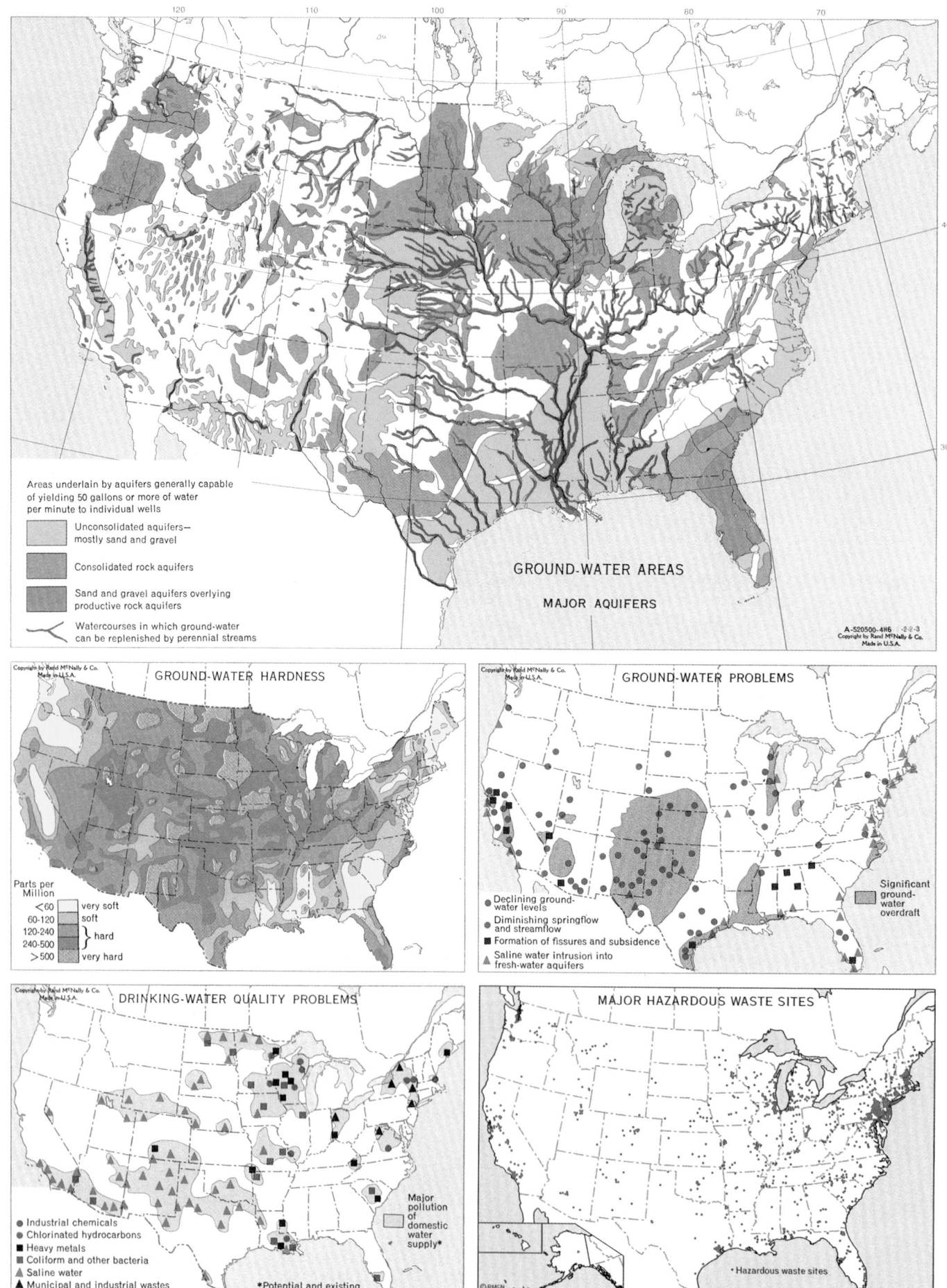

Areas underlain by aquifers generally capable of yielding 50 gallons or more of water per minute to individual wells

Unconsolidated aquifers— mostly sand and gravel

Consolidated rock aquifers

Sand and gravel aquifers overlying productive rock aquifers

Watercourses in which ground-water can be replenished by perennial streams

GROUND-WATER AREAS

MAJOR AQUIFERS

A-520500-4H6   -2-2-3
Copyright by Rand McNally & Co.
Made in U.S.A.

GROUND-WATER HARDNESS

Parts per Million
<60   very soft
60-120   soft
120-240   } hard
240-500
>500   very hard

GROUND-WATER PROBLEMS

● Declining ground-water levels
● Diminishing springflow and streamflow
■ Formation of fissures and subsidence
▲ Saline water intrusion into fresh-water aquifers

Significant ground-water overdraft

DRINKING-WATER QUALITY PROBLEMS

● Industrial chemicals
● Chlorinated hydrocarbons
■ Heavy metals
■ Coliform and other bacteria
▲ Saline water
▲ Municipal and industrial wastes

Major pollution of domestic water supply*

*Potential and existing

MAJOR HAZARDOUS WASTE SITES

• Hazardous waste sites

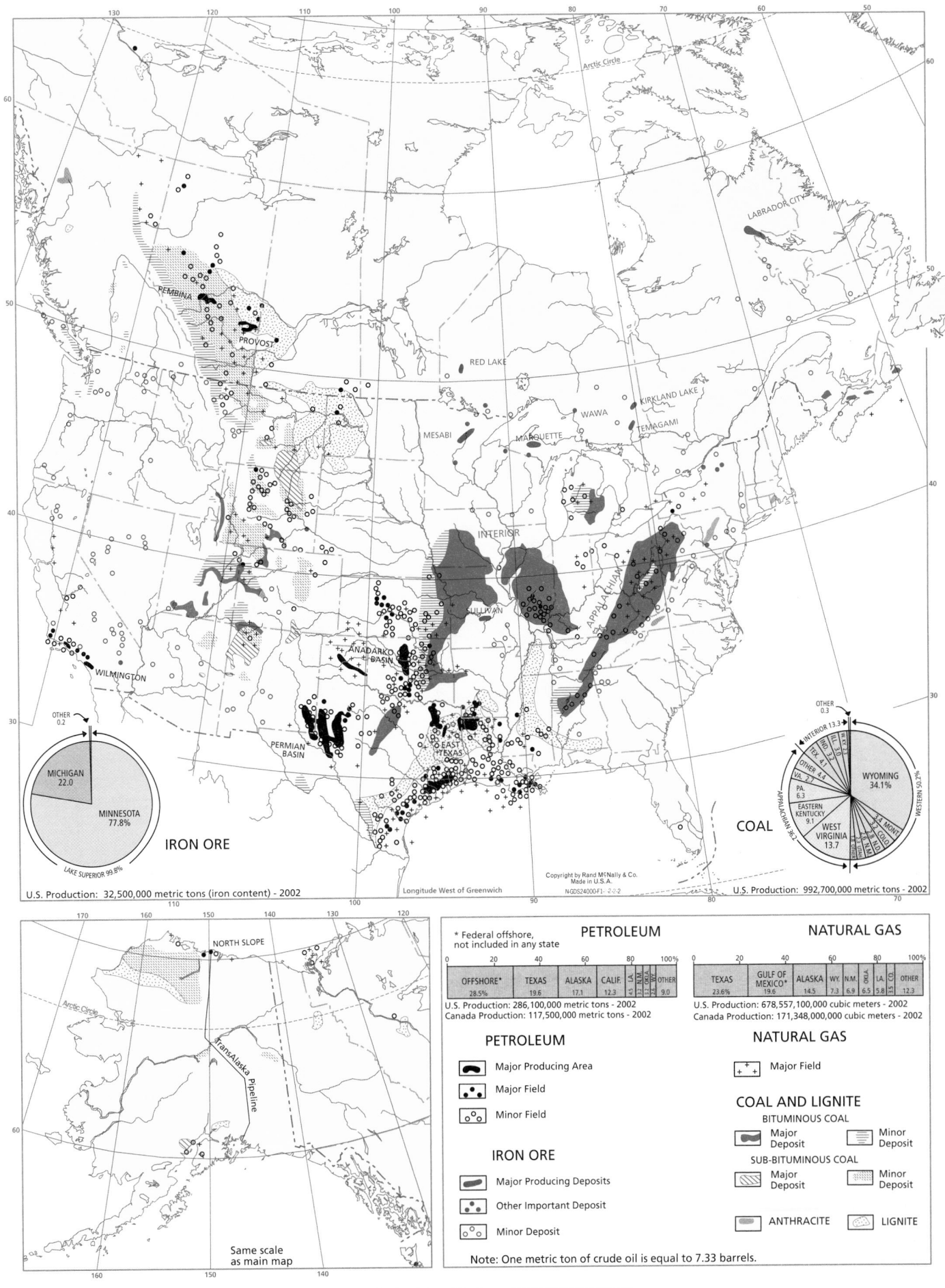

IRON ORE

OTHER 0.2

MICHIGAN 22.0

MINNESOTA 77.8%

LAKE SUPERIOR 99.8%

U.S. Production: 32,500,000 metric tons (iron content) - 2002

COAL

OTHER 0.3

Interior 13.3

ILL. 3.1
IND.
KY. 1.2
TEX. 4.1
OTHER 4.4
VA. 2.3
PA. 6.3
EASTERN KENTUCKY 9.1
WEST VIRGINIA 13.7

WYOMING 34.1%

APPALACHIAN 30.3

WESTERN 50.7%

N. DAK.
MONT.
OKLA
COLO.

U.S. Production: 992,700,000 metric tons - 2002

PEMBINA
PROVOST
RED LAKE
KIRKLAND LAKE
WAWA
MESABI
MARQUETTE
TEMAGAMI
LABRADOR CITY
INTERIOR
APPALACHIAN
SULLIVAN
ANADARKO BASIN
PERMIAN BASIN
EAST TEXAS
WILMINGTON

Copyright by Rand McNally & Co.
Made in U.S.A.
Longitude West of Greenwich
NGDS24000-F1-2-2-2

Arctic Circle

PETROLEUM

* Federal offshore, not included in any state

0	20	40	60	80	100%

| OFFSHORE* 28.5% | TEXAS 19.6 | ALASKA 17.1 | CALIF. 12.3 | LA. | N.M. | OKLA | W.Y. | OTHER 9.0 |

U.S. Production: 286,100,000 metric tons - 2002
Canada Production: 117,500,000 metric tons - 2002

NATURAL GAS

0	20	40	60	80	100%

| TEXAS 23.6% | GULF OF MEXICO* 19.6 | ALASKA 14.5 | W.Y. 7.3 | N.M. 6.9 | OKLA 6.5 | LA. 5.8 | CO. 3.5 | OTHER 12.3 |

U.S. Production: 678,557,100,000 cubic meters - 2002
Canada Production: 171,348,000,000 cubic meters - 2002

PETROLEUM

⬛ Major Producing Area

⬤ Major Field

○ Minor Field

IRON ORE

⬛ Major Producing Deposits

⬤ Other Important Deposit

○ Minor Deposit

NATURAL GAS

+ Major Field

COAL AND LIGNITE

BITUMINOUS COAL

⬛ Major Deposit   Minor Deposit

SUB-BITUMINOUS COAL

Major Deposit   Minor Deposit

ANTHRACITE   LIGNITE

Note: One metric ton of crude oil is equal to 7.33 barrels.

NORTH SLOPE
Arctic Circle
Trans-Alaska Pipeline

Same scale as main map

Scale 1:29,000,000; One inch to 457 miles.   ALBERS CONIC PROJECTION

Scale 1:12,000,000.   One inch to 190 miles.   Albers Conic Projection
                       One centimeter to 120 kilometers.

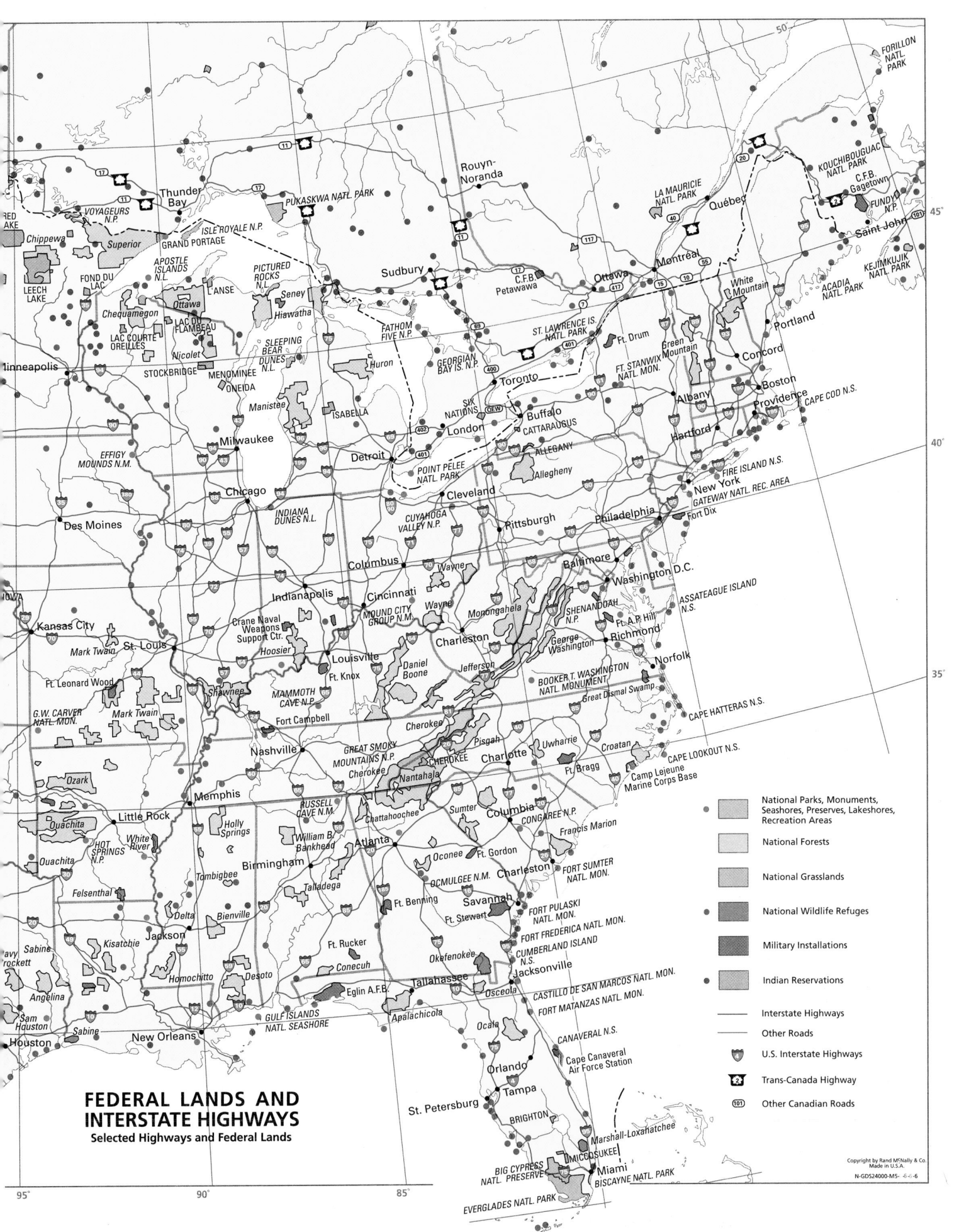

FEDERAL LANDS AND INTERSTATE HIGHWAYS
Selected Highways and Federal Lands

National Parks, Monuments, Seashores, Preserves, Lakeshores, Recreation Areas

National Forests

National Grasslands

National Wildlife Refuges

Military Installations

Indian Reservations

Interstate Highways

Other Roads

U.S. Interstate Highways

Trans-Canada Highway

Other Canadian Roads

Copyright by Rand McNally & Co.
Made in U.S.A.
N-GDS24000-M5- -6-6-6

ASIA
RUSSIA

GREENLAND
SEA

UNITED
KINGDOM

IRELAND

McKinley
Sea

JAN MAYEN
(Nor.)

SHETLAND IS.
(Br.)

FAROE IS.
(Den.)

North Sea

North Pole

Lincoln
Sea

GREENLAND
(Denmark)

10,000

ICELAND

Reykjavík

BERING
SEA

ALEUTIAN ISLANDS

ALEUTIAN TROUGH

POLUOSTROV
KAMCHATKA

INTERNATIONAL DATE LINE

BERING STRAIT

POINT BARROW

ARCTIC OCEAN

BROOKS RANGE

ALASKA

Fairbanks

Anchorage

KLONDIKE REGION

Dawson

Whitehorse

MT. LOGAN

Yukon

Inuvik

Mackenzie

QUEEN ELIZABETH ISLANDS

ELLESMERE I.

North Magnetic Pole

Thule

KAP YORK

Godhavn

DISKO

Egedesminde

Holsteinsborg

Godthåb

Julianehåb

KAP FARVEL

BAFFIN BAY

Davis Strait

ATLANTIC OCEAN

VICTORIA ISLAND

BANKS ISLAND

Viscount Melville Sound

BOOTHIA PEN.

Resolute

FOXE BASIN

BAFFIN ISLAND

Beaufort Sea

Great Bear Lake

Great Slave Lake

Ft. Simpson

Athabasca Lake

Reindeer Lake

CANADA

Churchill

HUDSON BAY

Nelson

SOUTHAMPTON I.

Hudson Strait

CAPE CHIDLEY

UNGAVA BAY

UNGAVA PEN.

LABRADOR

Gulf of Alaska

KODIAK ISLAND

QUEEN CHARLOTTE ISLANDS

Sitka

Prince Rupert

Edmonton

Calgary

Peace

Saskatchewan

Regina

Lake Winnipeg

Winnipeg

Lake of the Woods

JAMES BAY

NEWFOUNDLAND

St. John's

C. RACE

VANCOUVER ISLAND

Vancouver

Seattle

Portland

Columbia River

Spokane

Butte

Nelson

South Saskatchewan

Duluth

Fargo

LAURENTIAN HIGHLANDS

Québec

MONTREAL

Ottawa

Saint John

Halifax

NOVA SCOTIA

CAPE SABLE

CAPE BRETON ISLAND

Gulf of St. Lawrence

PACIFIC OCEAN

CAPE MENDOCINO

San Francisco

Oakland

ROCKY MOUNTAINS

Yellowstone

GREAT PLAINS

St. Paul

Minneapolis

Milwaukee

CHICAGO

L. MICHIGAN

L. SUPERIOR

Toronto

DETROIT

Cleveland

Buffalo

Pittsburgh

Cincinnati

NEW YORK

PHILADELPHIA

Boston

CAPE COD

APPALACHIAN MTS.

Salt Lake City

GREAT BASIN

Denver

Pikes Peak

Kansas City

Omaha

Wichita

St. Louis

Washington

Baltimore

Richmond

Norfolk

CAPE HATTERAS

Chesapeake Bay

UNITED STATES

MT. SHASTA

SIERRA NEVADA

Mt. Whitney 14,494

LOS ANGELES

COAST RANGES

CASCADE RANGE

GUADALUPE (Mex.)

BAJA CALIFORNIA

CABO SAN LUCAS

Memphis

Birmingham

Atlanta

Savannah

Jacksonville

Mobile

New Orleans

Dallas

Fort Worth

El Paso

Rio Grande

San Antonio

Houston

Galveston

Red

Arkansas

Mississippi

Miami

BERMUDA (Br.)

ATLANTIC OCEAN

Tropic of Cancer

GULF OF MEXICO

CAPE SABLE

Florida

SIERRA MADRE ORIENTAL

SIERRA MADRE OCCIDENTAL

MEXICO

Tampico

Guadalajara

MEXICO CITY

Veracruz

Pico de Orizaba 18,406 (Vol.)

Popocatépetl 17,887 (Vol.)

Bahía de Campeche

YUCATÁN PEN.

HAVANA

CUBA

Yucatán Channel

Straits of Florida

BAHAMAS

San Salvador

JAMAICA

Kingston

HAITI

Port-au-Prince

DOM. REP.

Santo Domingo

San Juan

PUERTO RICO (U.S.A.)

GUADELOUPE (Fr.)

MARTINIQUE (Fr.)

PUERTO RICO TRENCH

WINDWARD PASSAGE

WEST INDIES

BARBADOS

TRINIDAD AND TOBAGO

ISLAS REVILLAGIGEDO (Mex.)

BELIZE

GUATEMALA

HONDURAS

Golfo de Honduras

EL SALVADOR

NICARAGUA

CENTRAL AMERICA

COSTA RICA

PANAMA

ISTMO DE PANAMÁ

G. de Panamá

PTA. DE GALLINAS

CARIBBEAN SEA

ISLA DEL COCO (Costa Rica)

ISLA DE MALPELO (Colombia)

Caracas

Bogotá

SOUTH AMERICA

Rio Orinoco

Equator

Quito

Negro

40,000 SQ MI AREA

0   300   600
Miles

A-520000-26 5-5-18
COPYRIGHT BY
RAND McNALLY & COMPANY
MADE IN U.S.A.

0   200   400   600   800   1000 Miles
0   400   800   1200   1600 Kilometers

Longitude   West   of   Greenwich

Scale 1:40 000 000; one inch to 630 miles. Lambert's Azimuthal Equal Area Projection
Elevations and depressions are given in feet

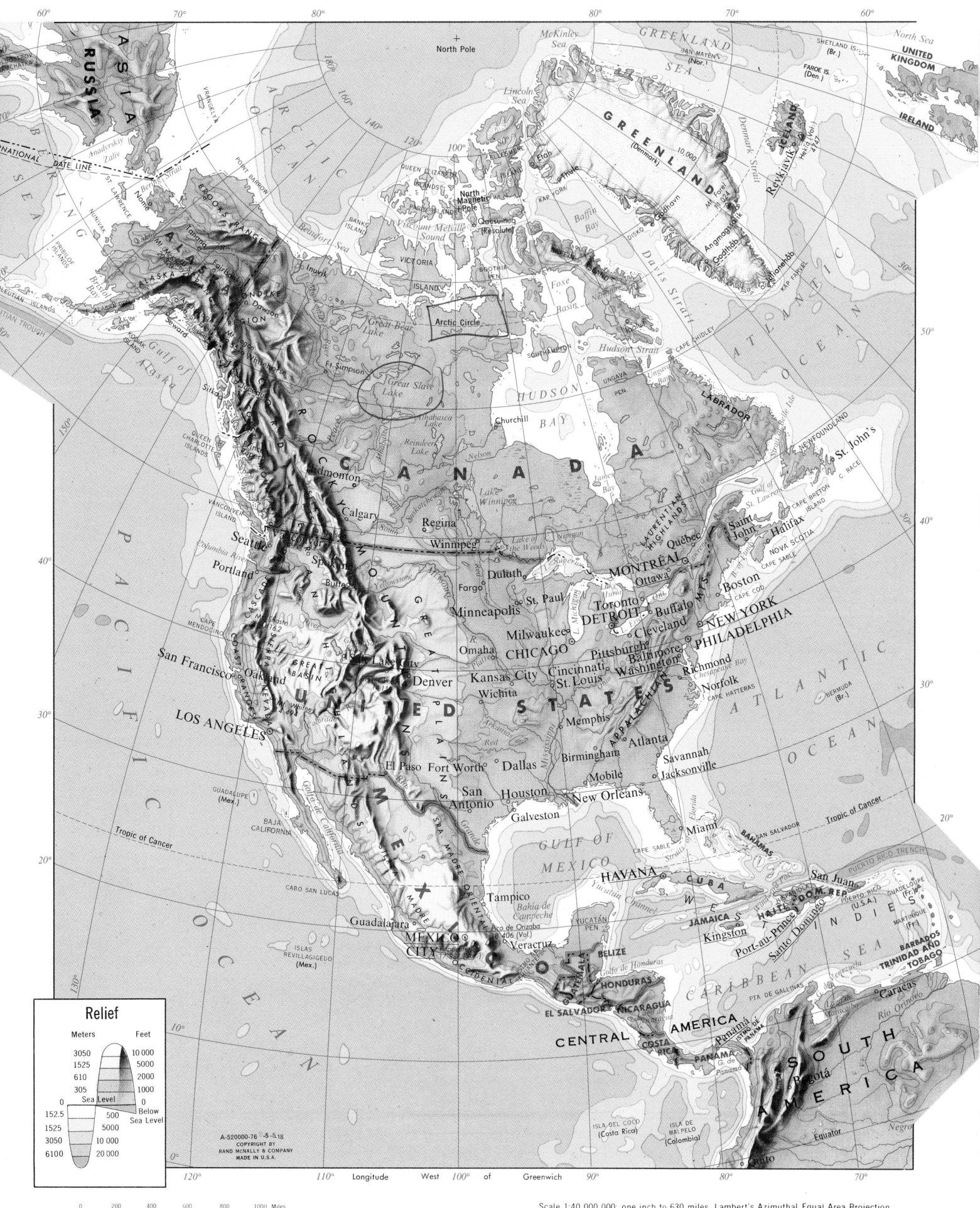

Scale 1:40 000 000; one inch to 630 miles. Lambert's Azimuthal Equal Area Projection
Elevations and depressions are given in feet

Scale 1: 12 000 000; one inch to 190 miles. Conic Projection

Elevations and depressions are given in feet

52

Scale 1: 12 000 000; one inch to 190 miles. Conic Projection

Elevations and depressions are given in feet

Longitude West of Greenwich

Same scale as main map

a

QUEBEC

NEWFOUNDLAND AND LABRADOR

CAPE BAULD

Strait of Belle Isle

LONG RANGE MTS.

Gulf of St. Lawrence

Notre Dame Bay

Twillingate

GROS MORNE NAT'L PARK
Deer Lake
Corner Brook
Stephenville
C. ST. GEORGE
Botwood
Grand Falls
Windsor
Gander
Bonavista

TERRA NOVA NAT'L PARK
Trinity

St. George's
NEWFOUNDLAND
CAPE RAY
Channel-Port-aux-Basques
Grand Bank
Burin
Placentia Bay
St. John's

CAPE NORTH
CAPE BRETON ISLAND
ST. PIERRE AND MIQUELON (Fr.)

ATLANTIC OCEAN

©RMCN

BAFFIN ISLAND NAT'L PARK
BAFFIN ISLAND

MELVILLE PENINSULA

Foxe Basin

Arctic Circle

Pangnirtung
CUMBERLAND PEN.

PRINCE CHARLES ISLAND

Nettilling Lake

C. DE MERCY

Cumberland Sound

N U N A V U T

Igulik

SOUTHAMPTON ISLAND

Iqaluit

Kimmirut
Lake Harbour

EVERETT MTS.

Frobisher Bay

RESOLUTION

FOXE PEN.

Foxe Channel

BELL PEN.

Fisher Strait

C. LOW

COATS

MANSEL

NOTTINGHAM ISLAND

SALISBURY

Hudson

C. DE NOUVELLE-FRANCE

Strait

C. HOPES ADVANCE

AKPATOK

KILLINIQ I.

TORNGAT MTS.

Hebron

HUDSON BAY

All islands within bays and straits lie within Nunavut.

OTTAWA ISLANDS

Povungnituk

Inukjuak

PENINSULE D'UNGAVA

Payne

Ungava Bay

Kuujjuaq

Main

Hopedale

Makkovik

Rigolet

Cartwright

NEWFOUNDLAND

Nakvak

MEALY MTS.

Hamilton Inlet

Bottle Harbour

Michikamau

LABRADOR

Happy Valley-Goose Bay

Churchill Falls

Little Mecatina

LONG RANGE MTS.

Belle Isle

C. HENRIETTA MARIA

BELCHER ISLANDS

Lac Bienville

La Grande

Grande Rivière de la Baleine

Caniapiscau

Schefferville

LABRADOR

Naskaupi

St. Anthony

GROS MORNE NAT'L PARK

Corner Brook
Stephenville
St. George

Ft. Severn

PTE LOUIS-XIV

Chisasibi

Nichicun

MTS. OTISH

Ashuanipi

Natashquan

ÎLE D'ANTICOSTI

James Bay

AKIMISKI I.

Eastmain

R. aux Outardes

Manicouagan

Mingan

Sept-Îles

Gulf of St. Lawrence

Ft. Albany

Winisk

Opinaca

Rivière Rupert

Mistassini

Chibougamau

Betsiamites

Clarke City

MTS. CHIC-CHOCS

Cap Chat

Matane

Gaspé

Moosonee

Missinaibi

Kapuskasing

Chapais

Dolbeau

Alma

Arvida

Rimouski

New Carlisle

Chandler

Rivière-du-Loup

Chaleur Bay

Caraquet

ÎLES DE LA MADELEINE

Coral Rapids
Fraserdale

La Sarre

Rouyn

Senneterre

Parent

La Tuque

Chicoutimi

Jonquière

Edmundston

P.E.I.

PRINCE EDWARD ISLAND NAT'L PARK

Summerside

New Waterford

Sydney Mines

ONTARIO

St. Joseph

Armstrong Sta.

Nakina

Hearst

Cochrane

Iroquois Falls

Timmins

Kirkland Lake

Val-d'Or

Malartic

St. Félicien

Roberval

Chambord

Shawinigan

Grand-Mère

Trois-Rivières

Sorel

Drummondville

Victoriaville

Thetford Mines

Lévis

QUEBEC

NEW BRUNSWICK

Campbellton

Newcastle

Bathurst

Chatham

Richibucto

Moncton

Fredericton

Amherst

Truro

NOVA SCOTIA

New Glasgow

Antigonish

Canso

Lac Seul

Geraldton

Longlac

Obol

Nipigon

Marathon

Chapleau

Cobalt

Ville-Marie

Témiscaming

Sudbury

Sturgeon Falls

Mattawa

North Bay

Pembroke

Renfrew

HULL

MONTRÉAL

St-Hyacinthe

Joliette

Sherbrooke

MAINE

Fredericton

St. Stephen

FUNDY NAT'L PARK

Saint John

St. George

Bay of Fundy

Kentville

Windsor

Springhill

Glasgow

Dartmouth

HALIFAX

Lunenburg

Bridgewater

Liverpool

Shelburne

Yarmouth

CAPE SABLE

Sioux Lookout

Dryden

Thunder Bay

PUKASKWA NAT'L PARK

MICHIPICOTEN I.

Sault Ste. Marie

Thessalon

Blind River

Espanola

Manitoulin I.

Georgian Bay

Parry Sound

Huntsville

Bracebridge

Smiths Falls

Ottawa

Brockville

Kingston

Alexandria Bay

Ogdensburg

VERMONT

NEW HAMPSHIRE

Concord

BOSTON

CAPE COD

Lac Seul

Lake Nipigon

Lake Superior

Marquette

Escanaba

Sault Ste. Marie

Lake Huron

Wiarton

Owen Sound

Midland

Orillia

Barrie

Lindsay

Peterborough

Cobourg

Trenton

Oshawa

Lake Ontario

Rochester

Albany

NEW YORK

Portland

ATLANTIC OCEAN

Duluth

Superior

MINNESOTA

M I C H I G A N

Green Bay

WISCONSIN

Kincardine

Port Elgin

Goderich

London

St. Thomas

Stratford

Kitchener

TORONTO

Hamilton

Niagara Falls

St. Catharines

BUFFALO

PENNSYLVANIA

Scranton

NEW YORK

HARTFORD

MASS.

CONN.

R.I. Providence

N.J.

St. Paul

Madison

MILWAUKEE

Grand Rapids

Lansing

Flint

Saginaw

DETROIT

Windsor

Leamington

Chatham

Sarnia

Port Huron

Lake Erie

OHIO

Toledo

MINNEAPOLIS

ILL.

CHICAGO

Lake Michigan

Relief

Meters		Feet
3050		10 000
1525		5000
610		2000
305		1000
152.5		500
	Sea Level	
152.5		500
1525		5000
3050		10 000

A-520200-76   10-23
COPYRIGHT BY
RAND McNALLY & COMPANY
MADE IN U.S.A.

0   25 50 75 100   200   300   400   500 Miles

0   100   200   400   600   800 Kilometers

Relief

Meters		Feet
3050		10 000
1525		5000
610		2000
305		1000
152.5		500
0	Sea Level	0
152.5		500
1525		5000

A-520220-76 6-69
COPYRIGHT BY
RAND McNALLY & COMPANY
MADE IN U.S.A.

Continued on pages 72-73

Longitude West of Greenwich

Scale 1:4 000 000; one inch to 64 miles. Conic Projection

Elevations and depressions are given in feet.

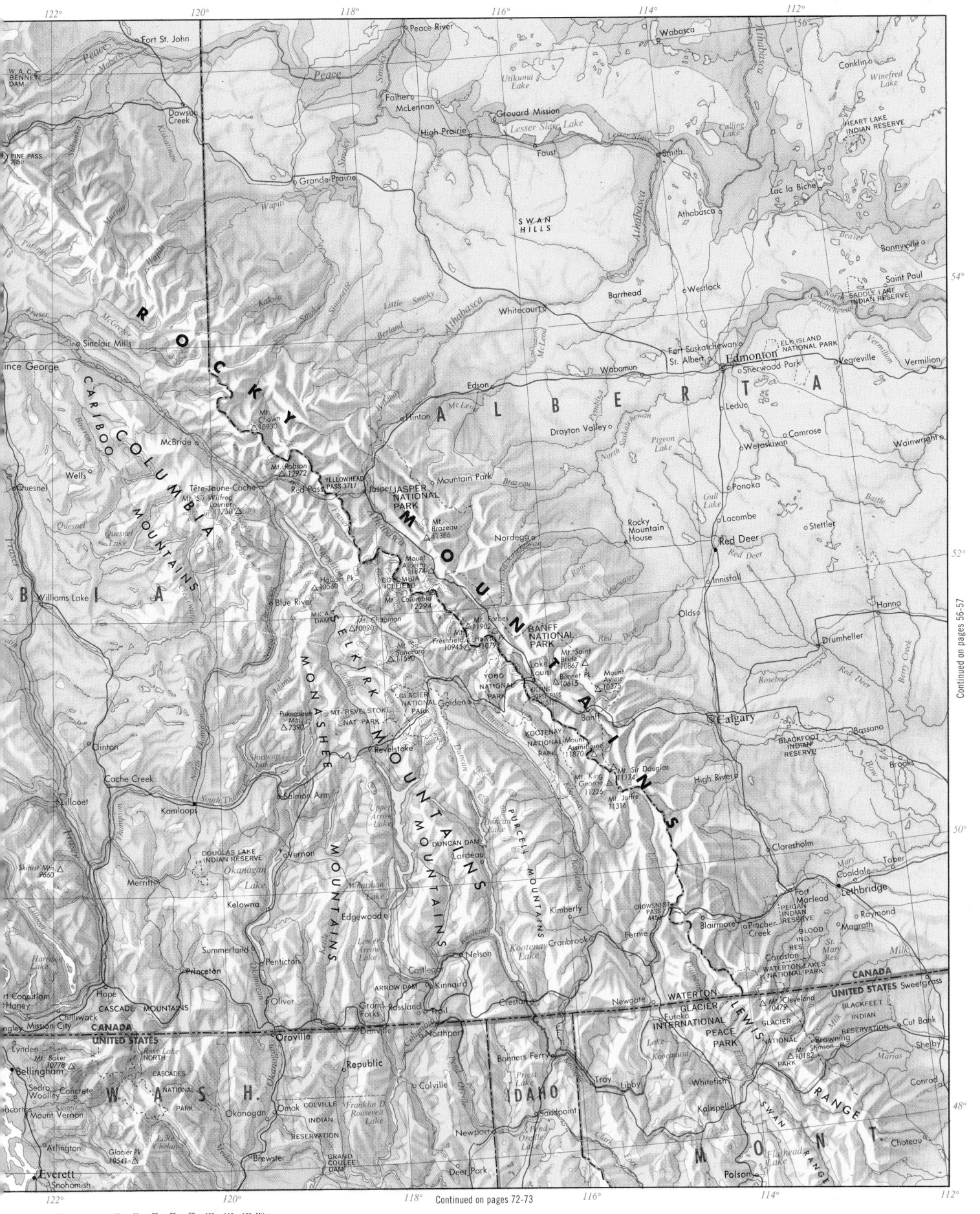

Continued on pages 56-57

Continued on pages 72-73

0 10 20 30 40 50 60 70 80 90 100 110 120 Miles

0   20   40   60   80   100   120   140   160   180   200 Kilometers

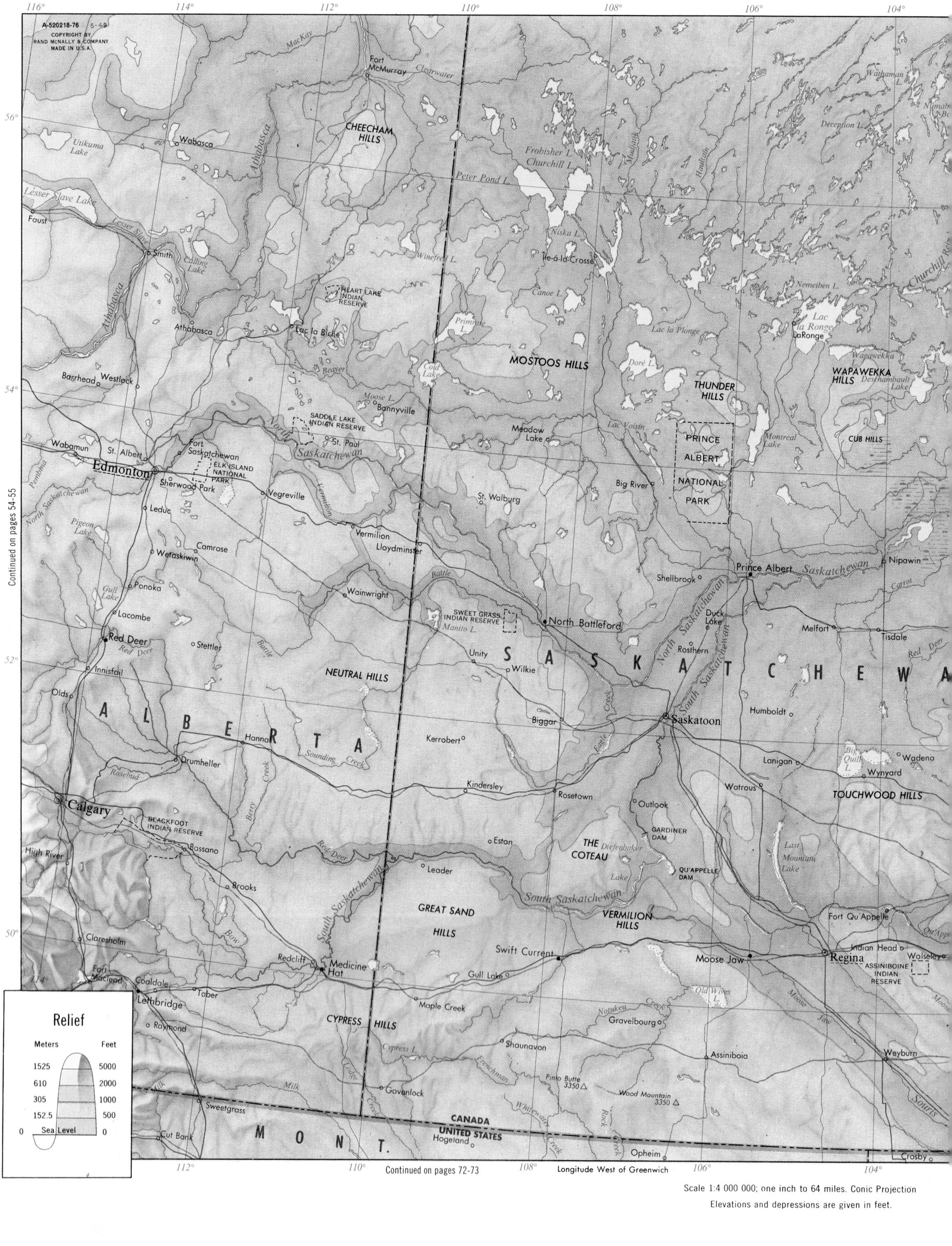

A-520218-76 5-4-9
COPYRIGHT BY
RAND McNALLY & COMPANY
MADE IN U.S.A.

116° 114° 112° 110° 108° 106° 104°

56°

Fort McMurray
Clearwater

CHEECHAM HILLS

Utikuma Lake
Wabasca

Frobisher L.
Churchill L.

Peter Pond L.

MacKay

Athabasca

Lesser Slave Lake

Faust

Smith

Calling Lake

Île-à-la-Crosse

Niska L.

Winefred L.

Canoe L.

Primrose L.

Deception L.

Wathaman L.

Reindeer L.

Nemeiben L.

Lac la Ronge
LaRonge

Wapawekka L.

Lac la Plonge

Doré L.

Montreal Lake

Deschambault Lake

WAPAWEKKA HILLS

54°

Barrhead Westlock

Wabamun St. Albert
Edmonton
Sherwood Park
Leduc

Fort Saskatchewan
ELK ISLAND NATIONAL PARK
Vegreville

Athabasca

Lac la Biche

HEART LAKE INDIAN RESERVE

Beaver

Moose L.
Bonnyville

SADDLE LAKE INDIAN RESERVE

St. Paul

North Saskatchewan

Cold Lake

MOSTOOS HILLS

Meadow Lake

Lac Voisin

THUNDER HILLS

Big River

PRINCE ALBERT NATIONAL PARK

CUB HILLS

St. Walburg

Vermilion
Lloydminster

Wainwright

Battle

Shellbrook

Prince Albert Saskatchewan

Nipawin

Duck Lake
Rosthern

North Saskatchewan

Melfort

Tisdale

Carrot

52°

Wetaskiwin Camrose

Ponoka

Gull Lake
Lacombe

Red Deer

Pigeon Lake

Innisfail

Olds

Stettler

Red Deer

Vermilion

SWEET GRASS INDIAN RESERVE

Manito L.

North Battleford

Unity Wilkie

NEUTRAL HILLS

SASKATCHEWAN

Biggar

Saskatoon

Humboldt

South Saskatchewan

Lanigan

Big Quill L.

Wadena

Wynyard

ALBERTA

Hannah

Kerrobert

Sounding Creek

Kindersley

Rosetown

Berry Creek

Drumheller

Rosebud

Calgary

BLACKFOOT INDIAN RESERVE

Bassano

High River

Brooks

Red Deer

Eston

Leader

Outlook

THE COTEAU

Diefenbaker

GARDINER DAM

QU'APPELLE DAM

Watrous

TOUCHWOOD HILLS

Last Mountain Lake

Fort Qu'Appelle

50°

Claresholm

Fort Macleod

Coaldale Taber

Lethbridge

Raymond

Redcliff Medicine Hat

South Saskatchewan

Bow

Maple Creek

CYPRESS HILLS

Cypress L.

GREAT SAND HILLS

Gull Lake

Swift Current

VERMILION HILLS

South Saskatchewan

Moose Jaw

Regina

Indian Head Wolseley

ASSINIBOINE INDIAN RESERVE

Qu'Appelle

Gravelbourg

Old Wives L.

Notukeu Creek

Weyburn

Shaunavon

Assiniboia

Wood Mountain 3350△

Pinto Butte 3350△

High River

Milk

Sweetgrass

Cut Bank

Govenlock

Frenchman

Whitemud

Moose Jaw

Souris

MONT.

CANADA
UNITED STATES

Hogeland

Opheim

Crosby

Continued on pages 54-55

Continued on pages 72-73

Longitude West of Greenwich

112° 110° 108° 106° 104°

Scale 1:4 000 000; one inch to 64 miles. Conic Projection
Elevations and depressions are given in feet.

Relief

Meters		Feet
1525		5000
610		2000
305		1000
152.5		500
0	Sea Level	0

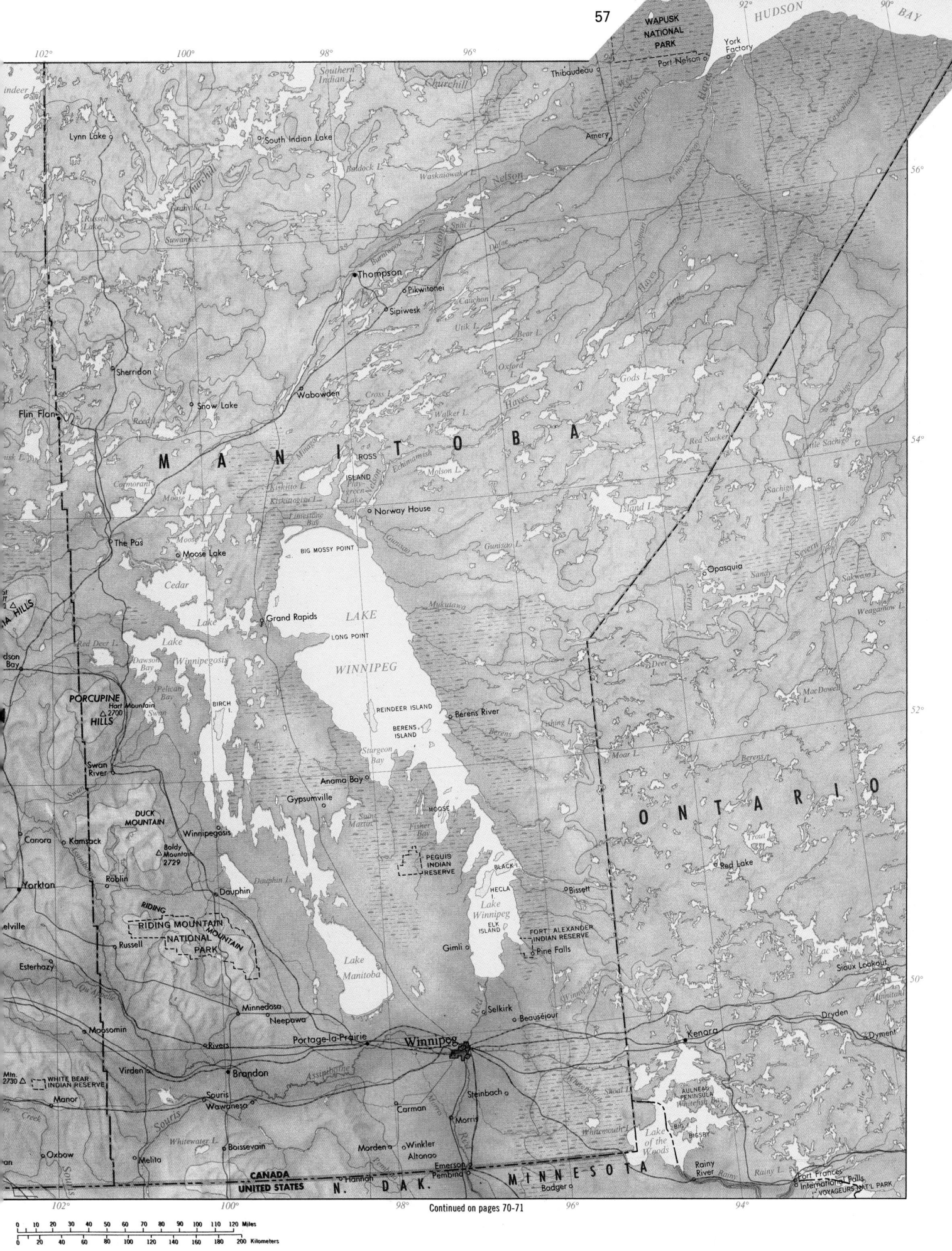

Continued on pages 70-71

0 10 20 30 40 50 60 70 80 90 100 110 120 Miles

0 20 40 60 80 100 120 140 160 180 200 Kilometers

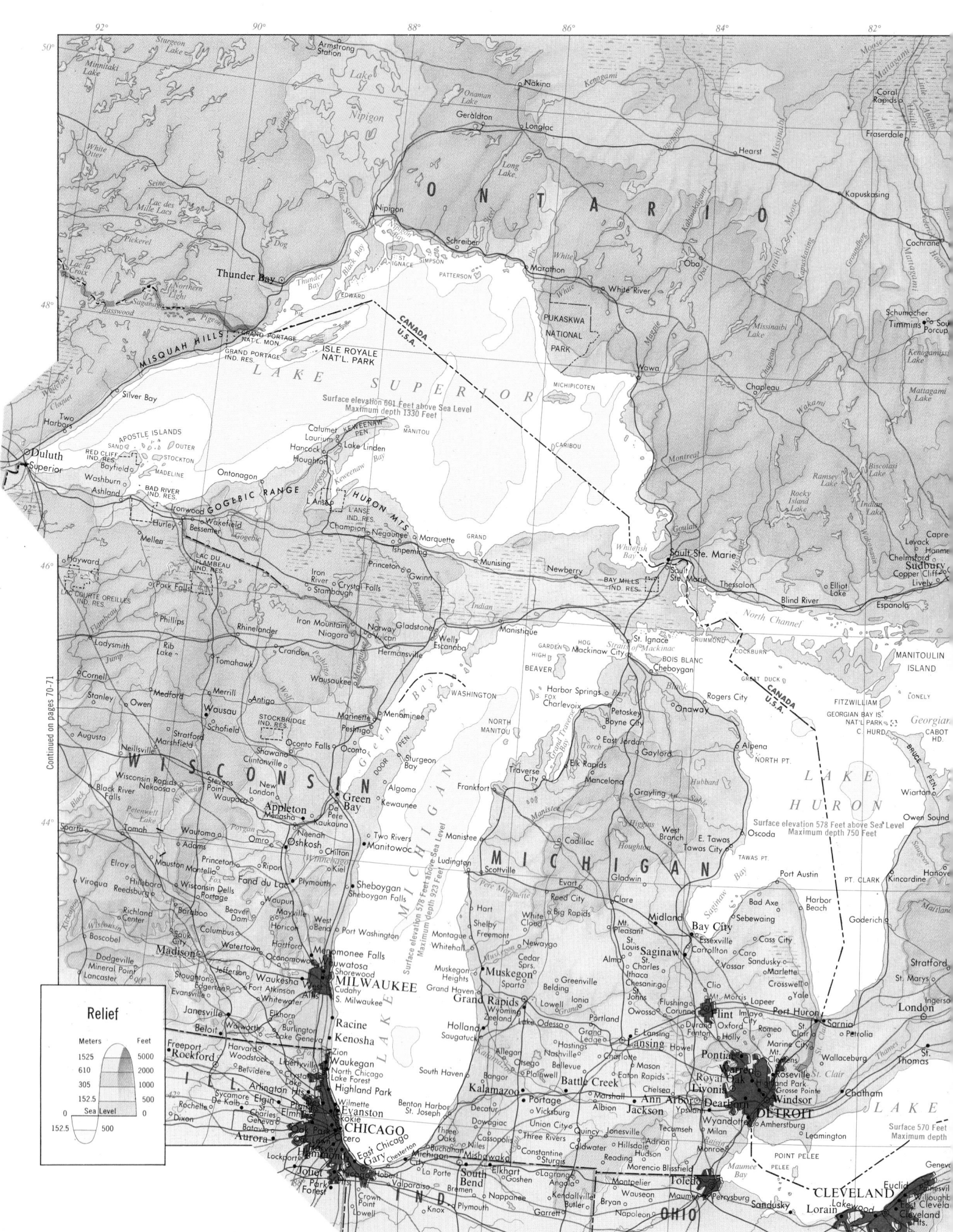

Continued on pages 70-71

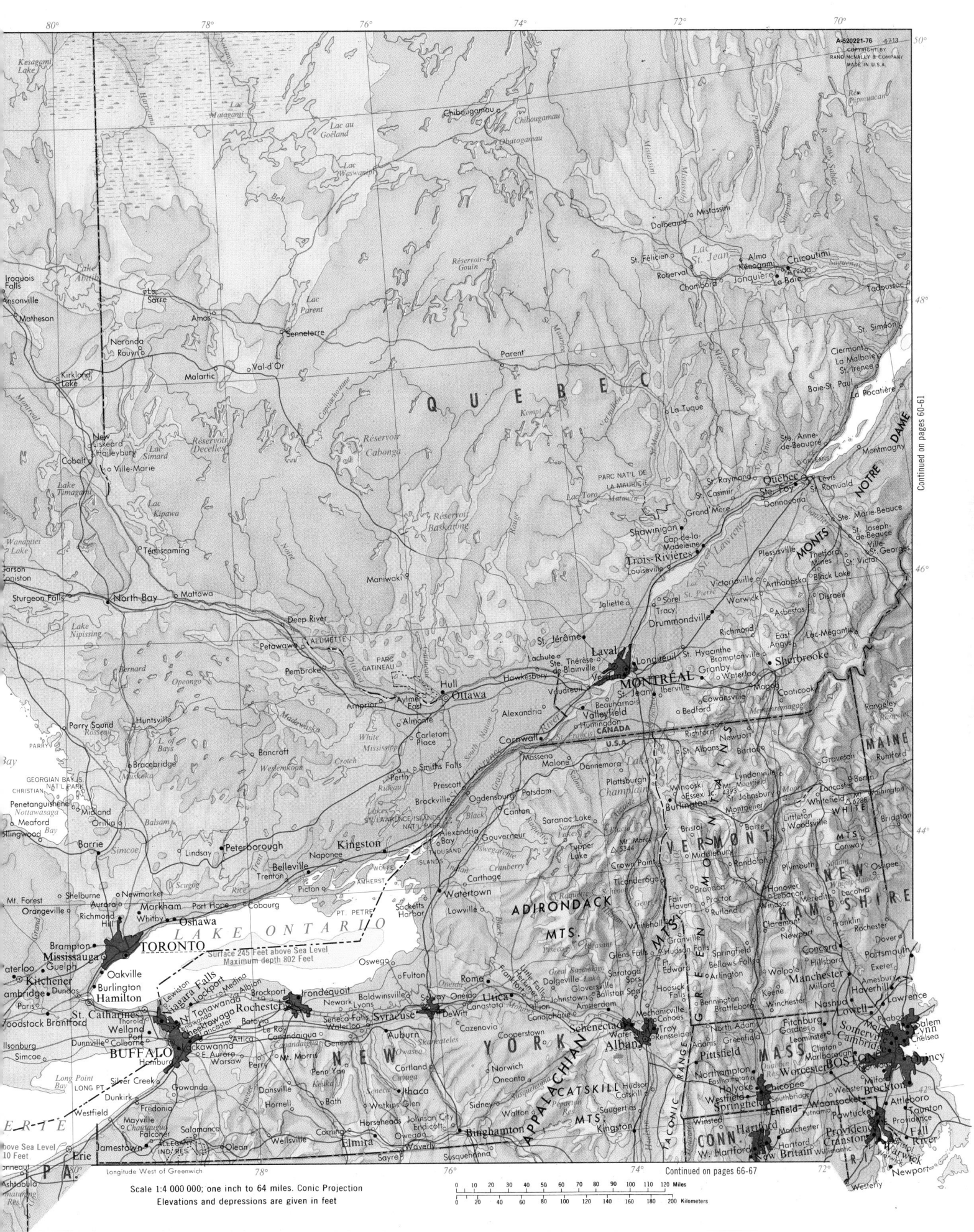

Scale 1:4 000 000; one inch to 64 miles. Conic Projection
Elevations and depressions are given in feet

Continued on pages 60-61

Continued on pages 66-67

Continued on pages 58-59

Continued on pages 66-67

Scale 1:4 000 000; one inch to 64 miles. Conic Projection
Elevations and depressions are given in feet.

Longitude West of Greenwich

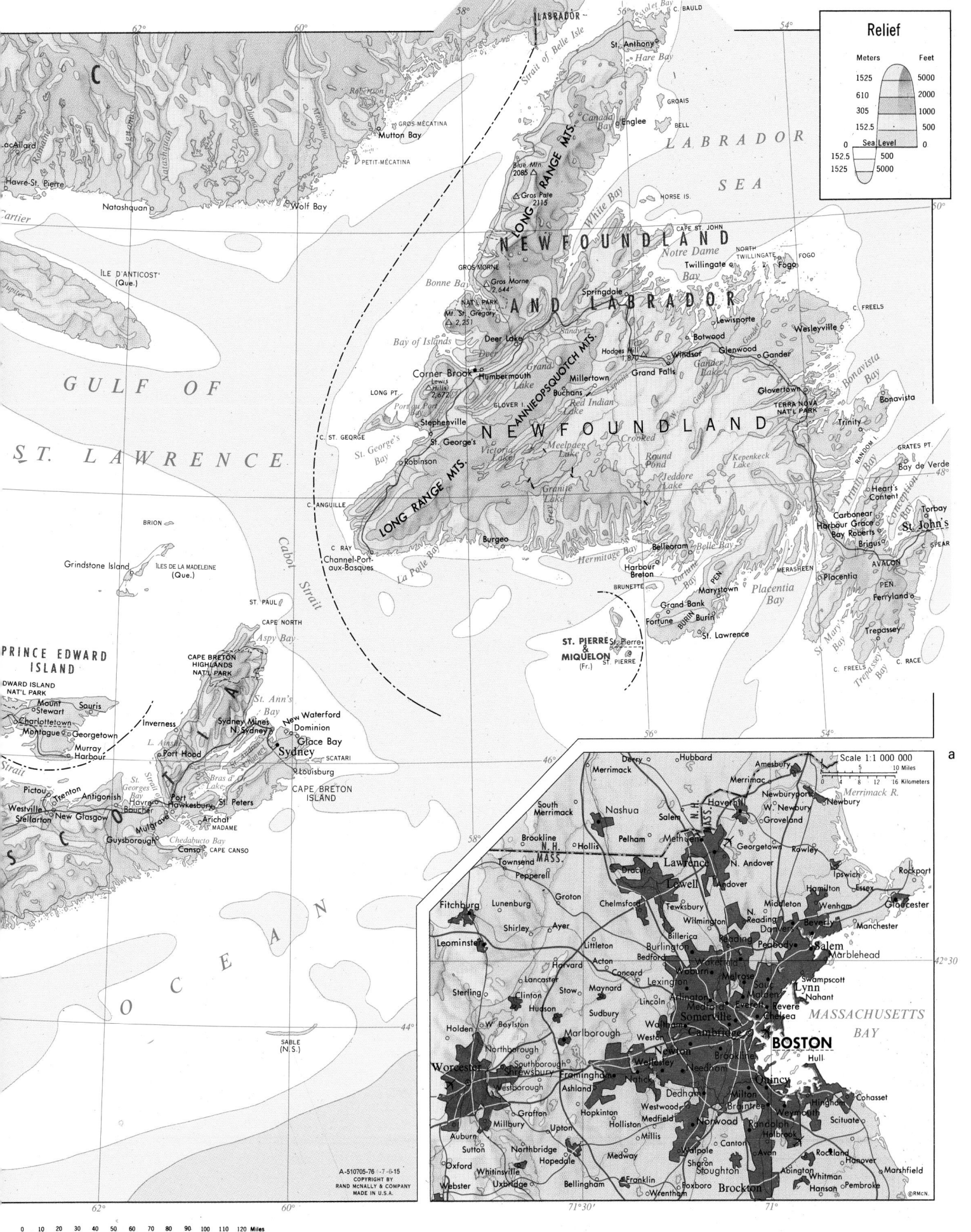

a — Montréal area

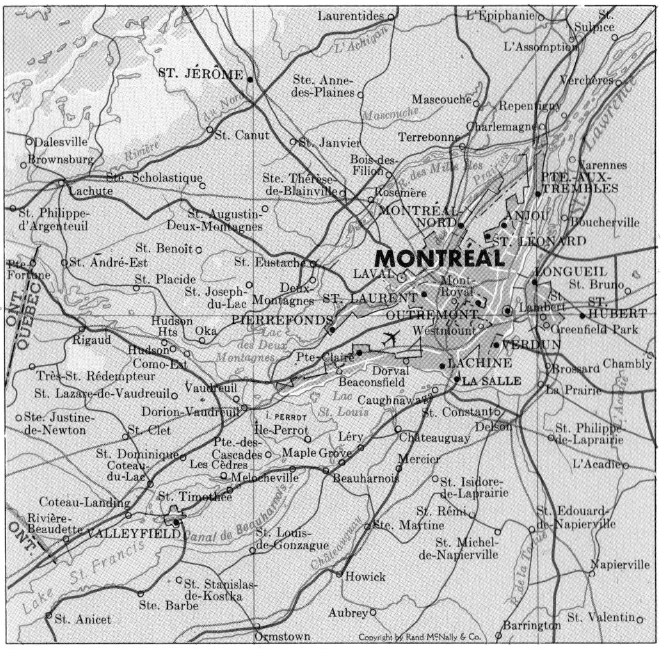

St. Jérôme, Ste. Anne-des-Plaines, L'Épiphanie, St. Sulpice, Laurentides, L'Assomption, Dalesville, Brownsburg, St. Canut, St. Janvier, Mascouche, Repentigny, Charlemagne, Verchères, Lachute, St. Scholastique, Bois-des-Fillion, Terrebonne, Ste. Thérèse-de-Blainville, Rosemère, MONTRÉAL NORD, PTE.-AUX-TREMBLES, Boucherville, St. Philippe-d'Argenteuil, St. Augustin-Deux-Montagnes, St. Benoît, St. André-Est, St. Eustache, Deux-Montagnes, ST. LAURENT, Mont-Royal, ST. LÉONARD, Varennes, LAVAL, St. Placide, St. Joseph-du-Lac, Oka, PIERREFONDS, OUTREMONT, MONTRÉAL, LONGUEUIL, St. Bruno, ST. HUBERT, Hudson Hts., Como-Est, Lac des Deux Montagnes, Westmount, LACHINE, Greenfield Park, Brossard, Chambly, Très-St. Rédempteur, Rigaud, Vaudreuil, Dorval, Beaconsfield, Pte-Claire, VERDUN, LA SALLE, La Prairie, St. Lazare-de-Vaudreuil, Dorion-Vaudreuil, Île-Perrot, Lac St. Louis, St. Constant, St. Philippe-de-Laprairie, Ste. Justine-de-Newton, Pte-des-Cascades, Les Cèdres, Maple Grove, Léry, Châteauguay, Delson, L'Acadie, St. Dominique, Coteau-du-Lac, Melocheville, Mercier, St. Isidore-de-Laprairie, St. Édouard-de-Napierville, Coteau-Landing, St. Timothée, Beauharnois, St. Rémi, St. Michel-de-Napierville, Rivière-Beaudette, VALLEYFIELD, St. Louis-de-Gonzague, Ste. Martine, Napierville, Ste. Justine, St. Stanislas-de-Kostka, Howick, Aubrey, Ste. Barbe, St. Anicet, Ormstown, Barrington, St. Valentin, Lake St. Francis, QUÉBEC, ONT.

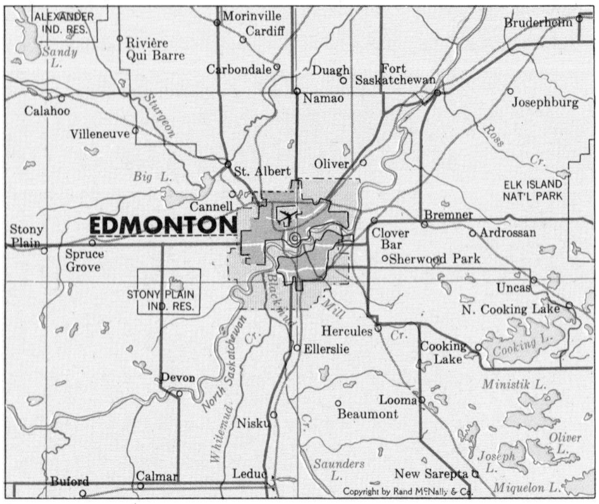

Copyright by Rand McNally & Co.

b — Québec area

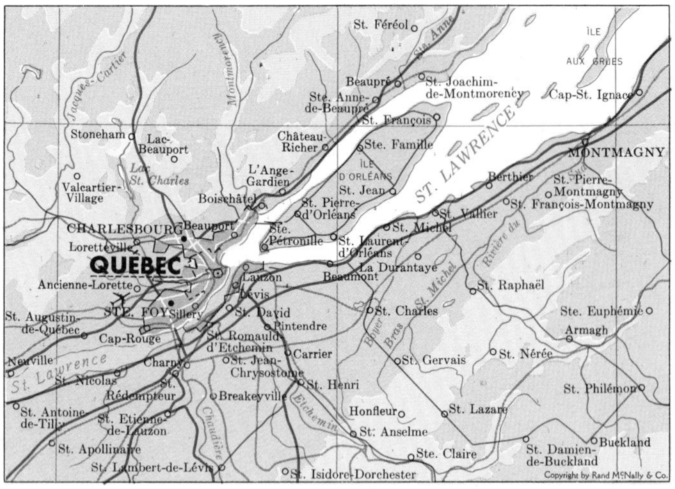

St. Féréol, Ste. Anne-de-Beaupré, St. Joachim-de-Montmorency, Cap-St. Ignace, Beaupré, Stoneham, Lac-Beauport, Château-Richer, Ste. Famille, Berthier, St. Pierre-Montmagny, MONTMAGNY, Valcartier-Village, L'Ange-Gardien, St. Jean, St. François, St. François-Montmagny, CHARLESBOURG, Loretteville, Beauport, Ste. Pétronille, St. Pierre-d'Orléans, St. Michel, St. Vallier, Ancienne-Lorette, St. Augustin-de-Québec, STE. FOY, Sillery, Lévis, St. David, Lauzon, St. Laurent-d'Orléans, La Durantaye, Ste. Euphémie, Cap-Rouge, Charny, St. Romuald-d'Etchemin, Ste. Claire, St. Raphaël, Neuville, St. Jean-Chrysostome, Carrier, St. Gervais, St. Nérée, St. Nicolas, St. Henri, Armagh, St. Antoine-de-Tilly, St. Étienne-de-Lauzon, Breakeyville, Honfleur, St. Lazare, St. Philémon, St. Apollinaire, Ste. Claire, St. Damien-de-Buckland, St. Lambert-de-Lévis, St. Isidore-Dorchester, Buckland, ST. LAWRENCE

Copyright by Rand McNally & Co.

c — Ottawa area

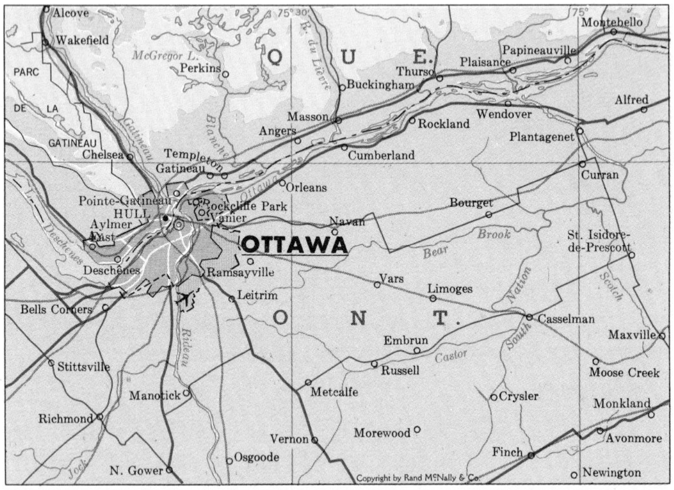

Alcove, Wakefield, Perkins, Papineauville, Montebello, PARC DE LA GATINEAU, McGregor L., Thurso, Buckingham, Plaisance, Chelsea, Masson, Angers, Rockland, Wendover, Alfred, QUE., Templeton, Gatineau, Cumberland, Plantagenet, Pointe-Gatineau, HULL, Rockcliffe Park, Orleans, Curran, Aylmer Dist., Vanier, OTTAWA, Navan, St. Isidore-de-Prescott, Deschênes, Ramsayville, Bourget, Bells Corners, Leitrim, Vars, Limoges, Casselman, Maxville, Stittsville, Russell, Crysler, Moose Creek, Manotick, Embrun, Monkland, Richmond, Metcalfe, Morewood, Avonmore, N. Gower, Osgoode, Vernon, Finch, Newington, ONT.

Copyright by Rand McNally & Co.

d — Toronto area

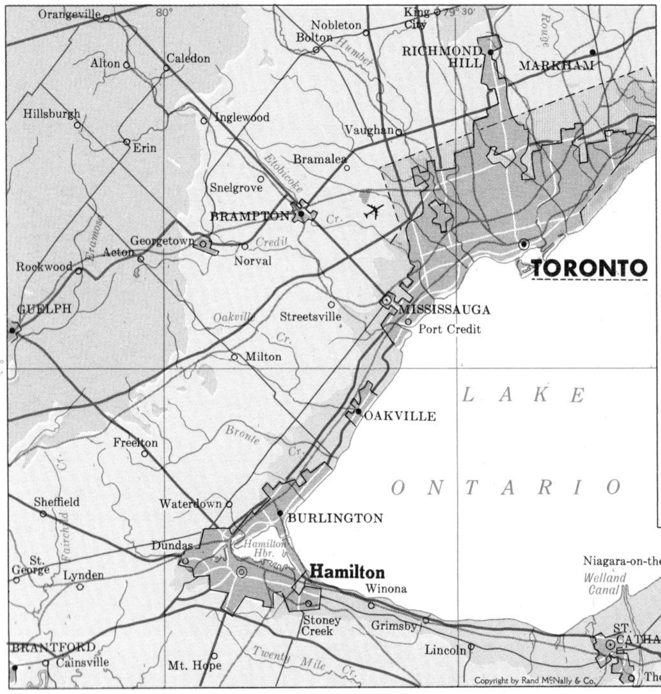

Orangeville, Nobleton, King City, RICHMOND HILL, MARKHAM, Alton, Caledon, Bolton, Inglewood, Vaughan, Hillsburgh, Erin, Bramalea, Snelgrove, BRAMPTON, Georgetown, Rockwood, Acton, Norval, GUELPH, Streetsville, MISSISSAUGA, Port Credit, TORONTO, Milton, LAKE ONTARIO, Sheffield, Waterdown, OAKVILLE, Freelton, St. George, Lynden, Dundas, BURLINGTON, Hamilton Hbr., Hamilton, Winona, Niagara-on-the-Lake, Youngstown, BRANTFORD, Cainsville, Mt. Hope, Stoney Creek, Grimsby, Lincoln, ST. CATHARINES, Lewiston, Thorold, Welland Canal, NEW YORK, U.S. CAN.

Copyright by Rand McNally & Co.

e — Calgary area

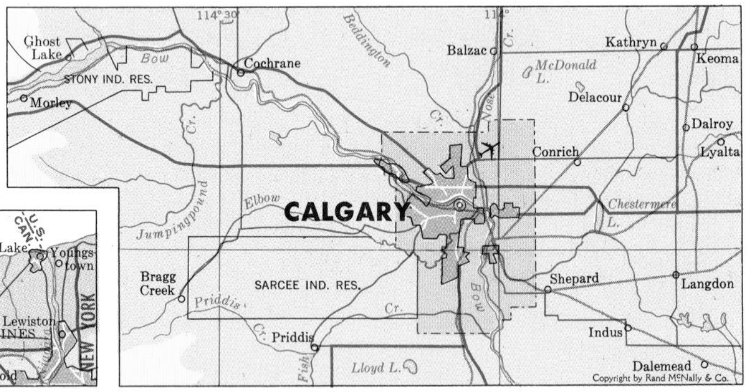

Ghost Lake, STONY IND. RES., Balzac, Kathryn, Keoma, Morley, Cochrane, McDonald L., Delacour, Dalroy, Conrich, Lyalta, CALGARY, Chestermere L., Bragg Creek, Priddis, SARCEE IND. RES., Shepard, Langdon, Jumpingpound, Bragg Creek, Indus, Priddis, Lloyd L., Dalemead

Copyright by Rand McNally & Co.

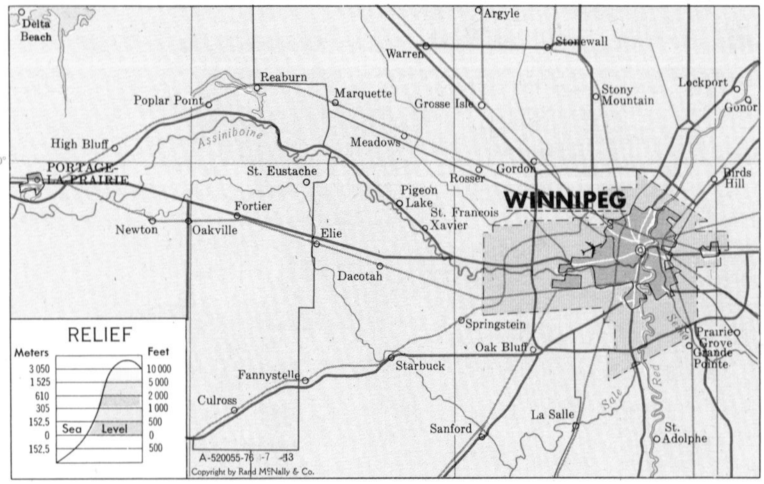

f — Winnipeg area

Delta Beach, Argyle, Stonewall, Warren, Reaburn, Marquette, Grosse Isle, Lockport, Poplar Point, Stony Mountain, Gonor, High Bluff, Meadows, Birds Hill, PORTAGE LA PRAIRIE, St. Eustache, Rosser, Pigeon Lake, Gordon, Fortier, St. François Xavier, WINNIPEG, Newton, Oakville, Elie, Dacotah, Springstein, Prairie Grove, Grande Pointe, Fannystelle, Oak Bluff, Culross, Starbuck, Sanford, La Salle, St. Adolphe

RELIEF

Meters	Feet
3 050	10 000
1 525	5 000
610	2 000
305	1 000
152.5	500
Sea Level	0
152.5	500

A-520055-76 -7 -63

Copyright by Rand McNally & Co.

g — Edmonton area

ALEXANDER IND. RES., Morinville, Cardiff, Bruderheim, Rivière Qui Barre, Carbondale, Duagh, Fort Saskatchewan, Josephburg, Calahoo, Namao, Villeneuve, St. Albert, Oliver, ELK ISLAND NAT'L PARK, Stony Plain, Cannell, EDMONTON, Bremner, Ardrossan, Spruce Grove, Clover Bar, Sherwood Park, Uncas, STONY PLAIN IND. RES., Hercules, N. Cooking Lake, Devon, Ellerslie, Cooking Lake, Nisku, Looma, Beaumont, Ministik L., Buford, Calmar, Leduc, New Sarepta, Miquelon L., Saunders L.

Copyright by Rand McNally & Co.

Scale 1:1 000 000; One inch to 16 miles.
Elevations and depressions are given in feet.

Miles: 0 2 4 6 8 10 12 14 16 18 20 22 24
Kilometers: 0 4 8 12 16 20 24 28 32 36 40

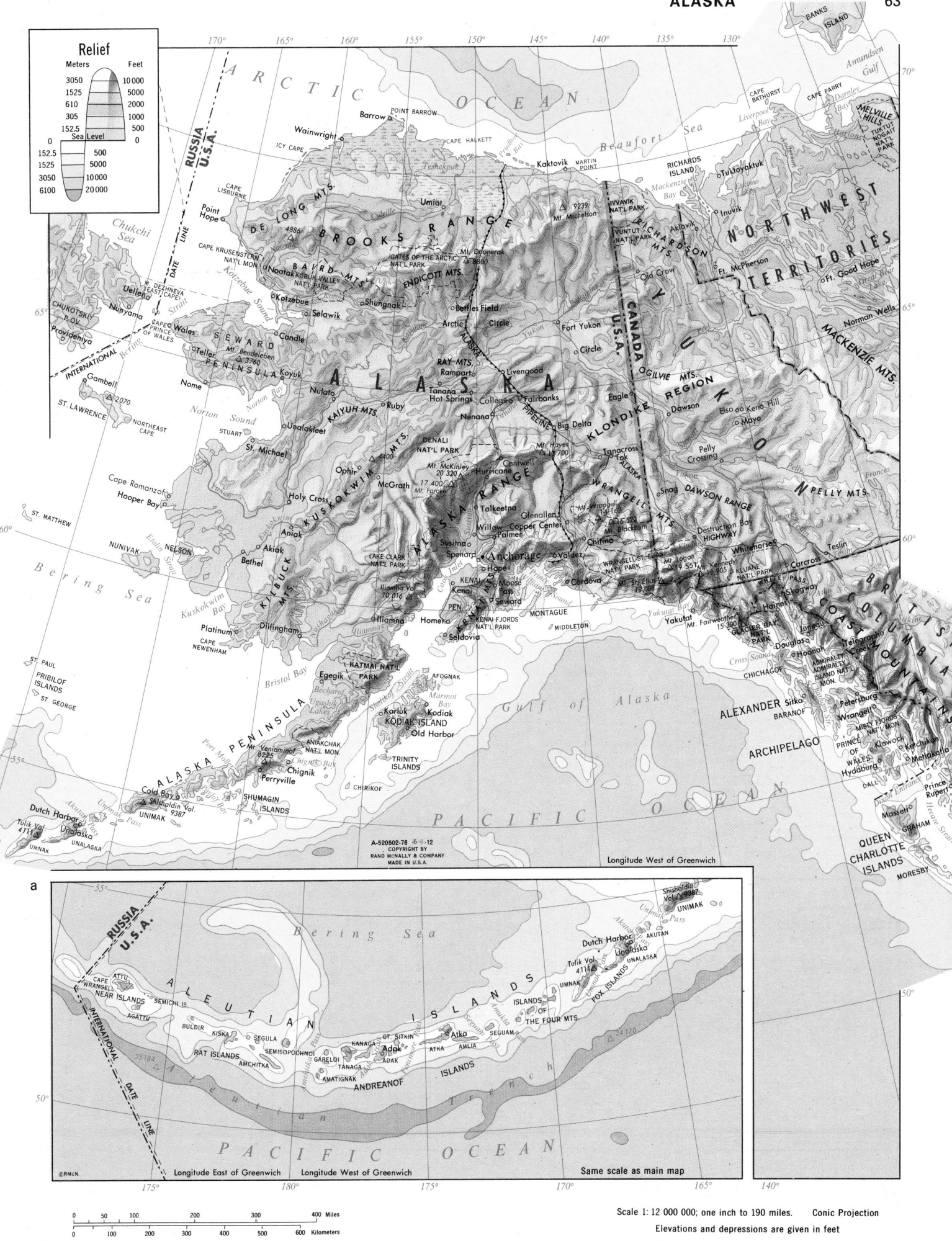

Relief

Meters		Feet
3050		10 000
1525		5000
610		2000
305		1000
152.5		500
Sea Level		0

152.5		500
1525		5000
3050		10 000
6100		20 000

a

Longitude East of Greenwich   Longitude West of Greenwich   Same scale as main map

Scale 1: 12 000 000; one inch to 190 miles.   Conic Projection

Elevations and depressions are given in feet

Longitude West of Greenwich

A-520502-76 -6-6-12
COPYRIGHT BY
RAND McNALLY & COMPANY
MADE IN U.S.A.

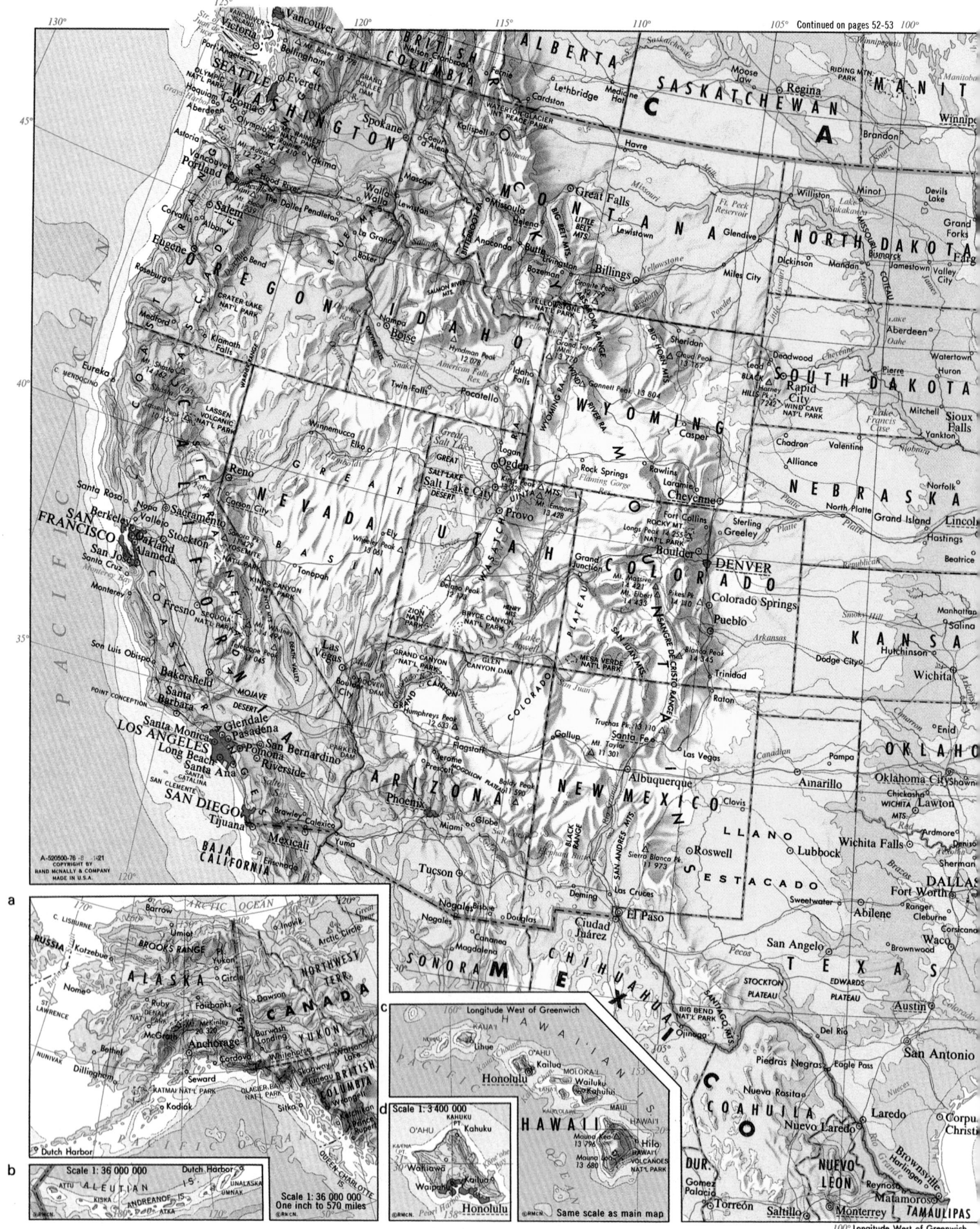

Continued on pages 52-53

Scale 1:12 000 000; one inch to 190 miles. Polyconic Projection
Elevations and depressions are given in feet

100° Longitude West of Greenwich

Continued on pages 70-71

Continued on pages 82-83

Cities and Towns

0 to 50,000 ○

50,000 to 500,000 ⊙

500,000 to 1,000,000 ◉

1,000,000 and over

Longitude West of Greenwich

Scale 1:4 000 000; one inch to 64 miles.   Conic Projection
Elevations and depressions are given in feet

LAKE HURON
Surface 579 Feet above Sea Level
maximum depth 750 Feet

LAKE ERIE
Surface 570 Feet above Sea Level
maximum depth 210 Feet

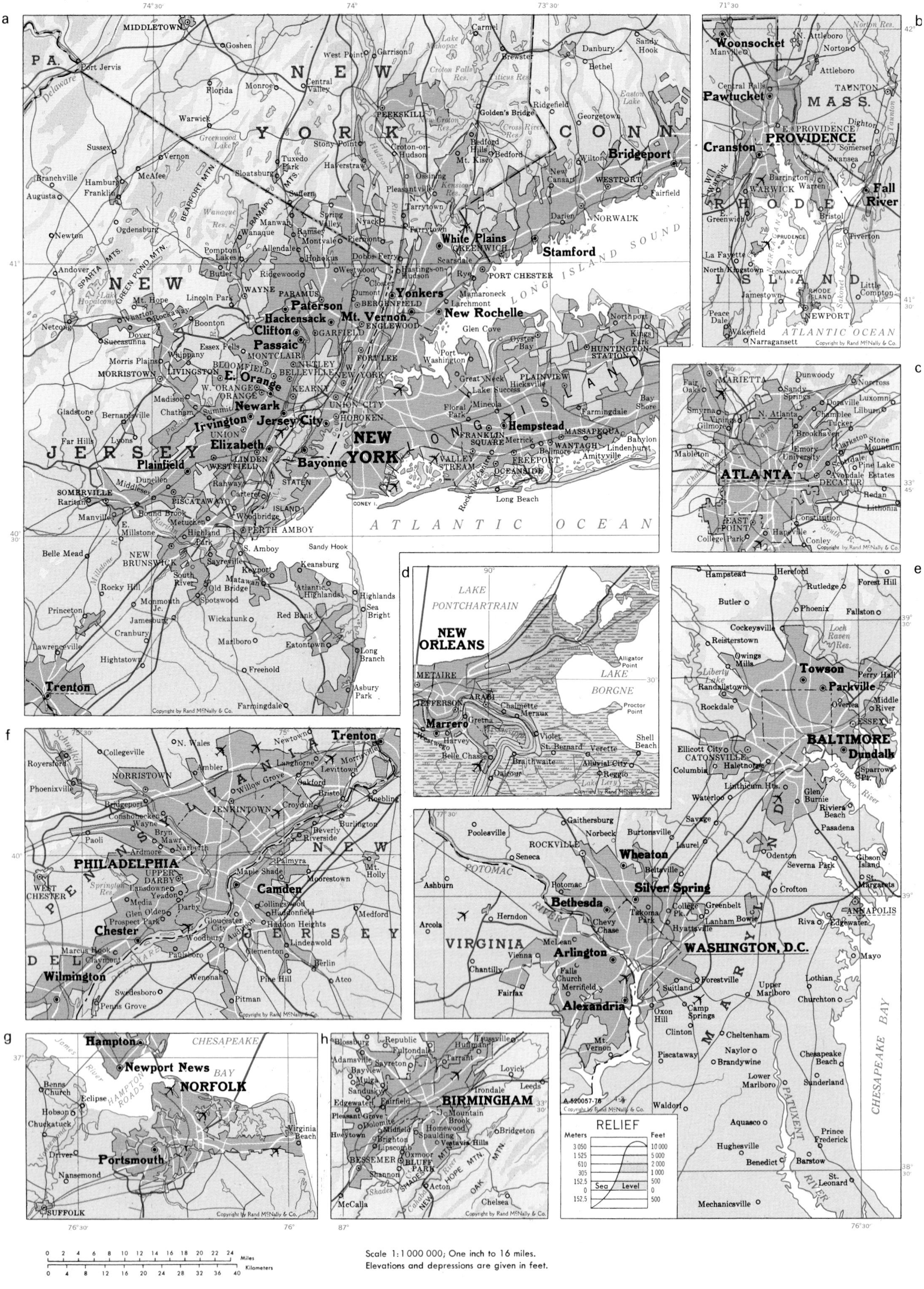

a

MIDDLETOWN

PA.

NEW YORK

Port Jervis
Goshen
Florida
Monroe
Warwick
Sussex
Vernon
McAfee
Branchville
Hamburg
Franklin
Augusta
Newton
Andover
Sparta
Ogdensburg
Netcong
Mt. Hope
Lincoln Park
Wharton
Dover
Rockaway
Boonton
Morris Plains
Whippany
MORRISTOWN
Livingston
Madison
Chatham
Summit
Gladstone
Bernardsville
Far Hills
Lyons
Somerville
Raritan
Manville
Bound Brook
E. Millstone
Metuchen
Highland Park
NEW BRUNSWICK
Rocky Hill
Princeton
Lawrenceville
Hightstown

West Point
Central Valley
Stony Point
Haverstraw
Suffern
Sloatsburg
Tuxedo Park
Ramsey
Spring Valley
Nyack
Dobbs Ferry
Hastings-on-Hudson
Hohokus
Ridgewood
WAYNE
PARAMUS
Paterson
Hackensack
Clifton
Passaic
MONTCLAIR
BLOOMFIELD
NUTLEY
BELLEVILLE
E. Orange
W. ORANGE
ORANGE
KEARNY
Newark
Irvington
UNION
Elizabeth
LINDEN
Jersey City
Bayonne
Plainfield
WESTFIELD
Rahway
Carteret
Woodbridge
PERTH AMBOY
S. Amboy
Sayreville
Keyport
Matawan
Old Bridge
South River
Spotswood
Monmouth Jc.
Jamesburg
Cranbury
Marlboro
Freehold

Carmel
Danbury
Sandy Hook
Bethel
Ridgefield
Georgetown
CONN
Bedford
Mt. Kisco
New Canaan
Wilton
Bridgeport
WESTPORT
Fairfield
NORWALK
Darien
White Plains
GREENWICH
PORT CHESTER
Stamford
Rye
Mamaroneck
Larchmont
Yonkers
Mt. Vernon
New Rochelle
Glen Cove
Oyster Bay
Port Washington
Great Neck
Hicksville
PLAINVIEW
Mineola
HUNTINGTON STATION
Bay Shore
Farmingdale
Hempstead
FRANKLIN SQUARE
MASSAPEQUA
Merrick
Amityville
Lindenhurst
Babylon
Freeport
VALLEY STREAM
OCEANSIDE
NEW YORK
LONG ISLAND
Long Beach
STATEN ISLAND
CONEY ISLAND

Trenton

N E W   J E R S E Y

ATLANTIC OCEAN

Belle Mead
Highlands
Red Bank
Eatontown
Long Branch
Asbury Park
Farmingdale

b
Woonsocket
N. Attleboro
Norton
Attleboro
MASS.
TAUNTON
Pawtucket
Central Falls
E. PROVIDENCE
Dighton
Cranston
PROVIDENCE
Somerset
Swansea
Fall River
Barrington
Warren
Bristol
Tiverton
RHODE ISLAND
La Fayette
North Kingstown
Jamestown
Newport
Little Compton
Peace Dale
Wakefield
Narragansett
ATLANTIC OCEAN

c
MARIETTA
Dunwoody
Norcross
Fair Oaks
Smyrna
Vinings
Sandy Springs
N. Atlanta
Doraville
Chamblee
Lilburn
Luxomni
Mableton
Emory University
Clarkston
Stone Mountain
Avondale Estates
Pine Lake
ATLANTA
DECATUR
Redan
Lithonia
EAST POINT
Constitution
College Park
Hapeville
Conley

d
LAKE PONTCHARTRAIN
NEW ORLEANS
METAIRE
JEFFERSON
ARABI
Chalmette
Meraux
Alligator Point
LAKE BORGNE
Proctor Point
Marrero
Gretna
Harvey
Violet
Verrette
Belle Chasse
Braithwaite
Alluvial City
Shell Beach
Dalcour
Reggio
Copyright by Rand McNally & Co.

e
Hampstead
Hereford
Forest Hill
Butler
Rutledge
Phoenix
Fallston
Cockeysville
Loch Raven Res.
Reisterstown
Owings Mills
Randallstown
Towson
Perry Hall
Rockdale
Parkville
Overlea
Middle River
Essex
Ellicott City
CATONSVILLE
BALTIMORE
Dundalk
Columbia
Sparrows Pt.
Waterloo
Linthicum Hts.
Glen Burnie
Riviera Beach
Savage
Odenton
Pasadena
Severna Park
Gibson Island
Crofton
St. Margarets
ANNAPOLIS
Riva
Edgewater
Mayo

f
Collegeville
N. Wales
Newtown
Trenton
Royersford
NORRISTOWN
Ambler
Langhorne
Morrisville
Phoenixville
Willow Grove
Levittown
Bridgeport
Jenkintown
Croydon
Bristol
Roebling
Conshohocken
Wayne
Bryn Mawr
Narberth
Beverly
Riverside
Burlington
Paoli
Ardmore
Palmyra
Mt. Holly
PHILADELPHIA
UPPER DARBY
Maple Shade
Moorestown
Medford
WEST CHESTER
Lansdowne
Yeadon
Media
Glen Olden
Darby
Collingswood
Camden
Haddonfield
Haddon Heights
Lindenwold
Chester
Prospect Park
Gloucester City
Woodbury
Clementon
Atco
Marcus Hook
Clayton
Paulsboro
Berlin
Wilmington
DEL
Swedesboro
Wenonah
Pine Hill
Penns Grove
Pitman

g
Hampton
CHESAPEAKE BAY
Newport News
Benns Church
Eclipse
HAMPTON ROADS
NORFOLK
Hobson
Chuckatuck
Virginia Beach
Driver
Nansemond
Portsmouth
SUFFOLK

h
Blossburg
Republic
Hayesville
Adamsville
Fultondale
Huffman
Bayview
Sayreton
Tarrant
Lovick
Mulga
Trondale
Leeds
Sandusky
Edgewater
Fairfield
BIRMINGHAM
Bridgeton
Pleasant Grove
Mountain Brook
Dolomite
Midfield
Homewood
Spaulding
Hueytown
Brighton
Lipscomb
Vestavia Hills
BESSEMER
BLUFF PARK
Shannon
Acton
Chelsea
McCalla

Virginia

Poolesville
Gaithersburg
Norbeck
Burtonsville
ROCKVILLE
Laurel
Seneca
Wheaton
Beltsville
POTOMAC
Potomac
Silver Spring
Greenbelt
Odenton
Ashburn
Herndon
Bethesda
Takoma Park
College Pk.
Lanham
Bowie
Arcola
McLean
Chevy Chase
Hyattsville
WASHINGTON, D.C.
Vienna
Arlington
Suitland
Forestville
Chantilly
Falls Church
Oxon Hill
Upper Marlboro
Churchton
Fairfax
Merrifield
Alexandria
Camp Springs
Clinton
Cheltenham
Mt. Vernon
Piscataway
Naylor
Brandywine
Chesapeake Beach
Waldorf
Lower Marlboro
Sunderland
Hughesville
Aquasco
Prince Frederick
Mechanicsville
Benedict
Barstow
St. Leonard

RELIEF

Meters		Feet
3 050		10 000
1 525		5 000
610		2 000
305		1 000
152.5		500
	Sea Level	
152.5		500

0 2 4 6 8 10 12 14 16 18 20 22 24 Miles
0 4 8 12 16 20 24 28 32 36 40 Kilometers

Scale 1:1 000 000; One inch to 16 miles.
Elevations and depressions are given in feet.

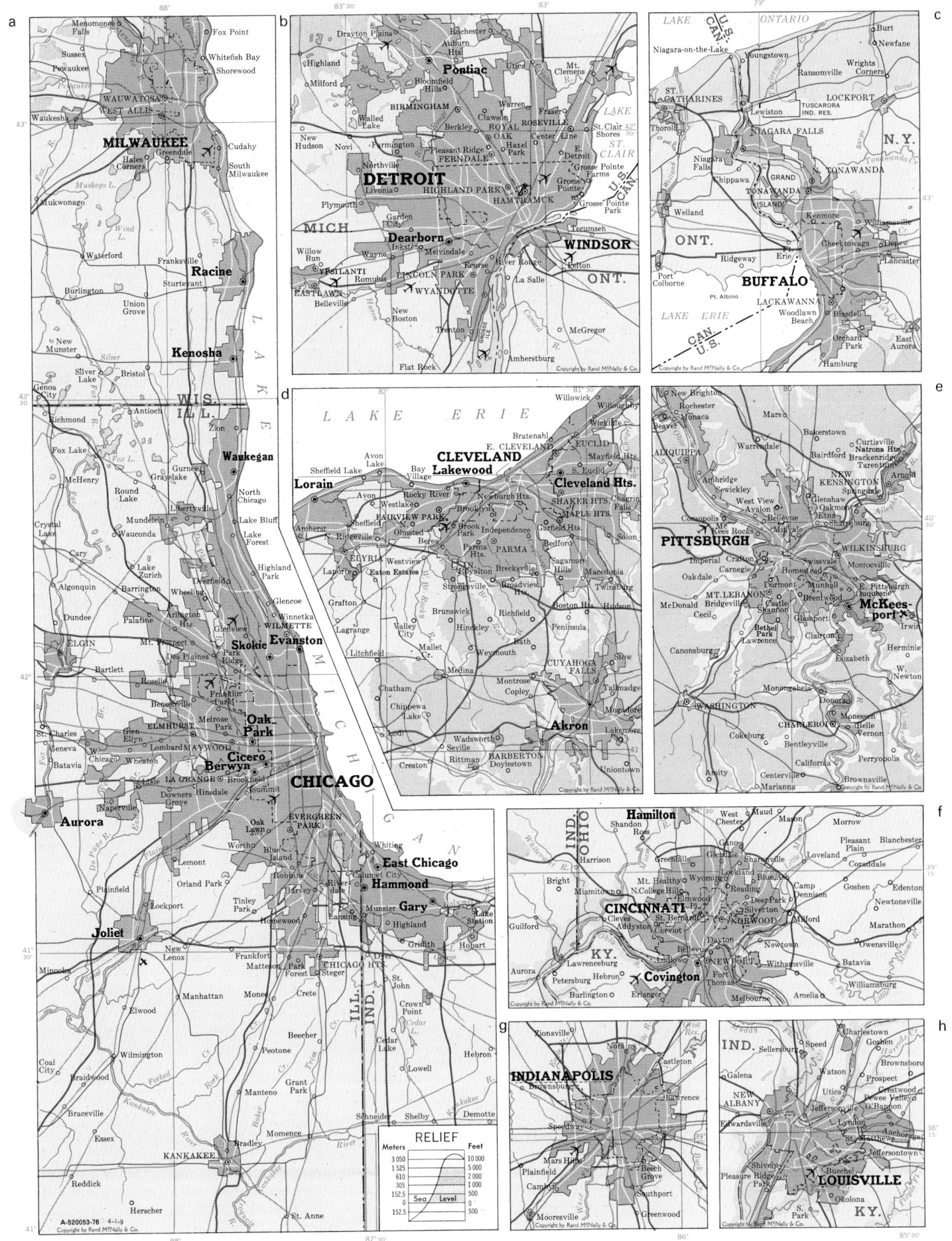

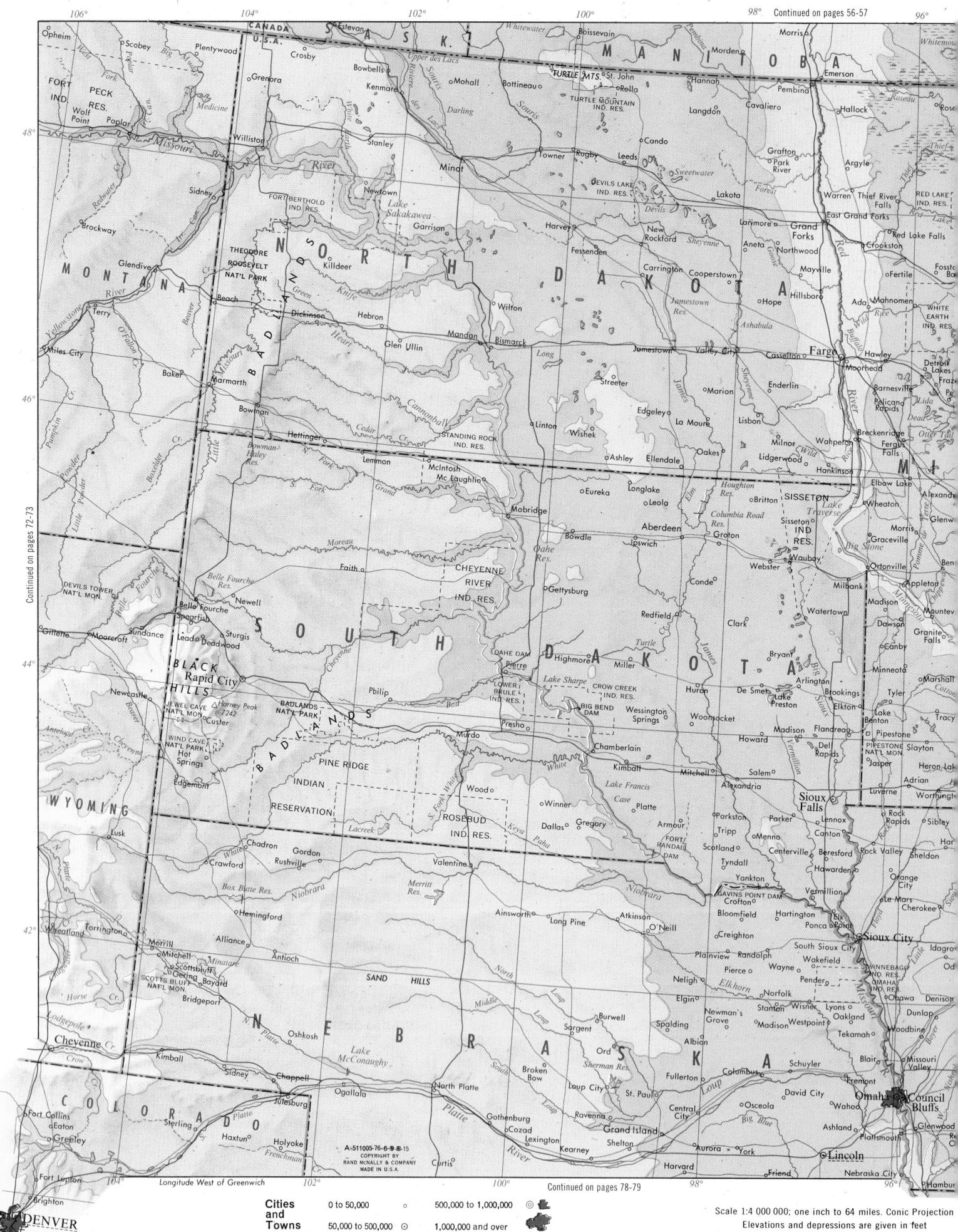

Continued on pages 56-57

Continued on pages 72-73

Continued on pages 78-79

A-511005-76-6-9-8-15
COPYRIGHT BY
RAND McNALLY & COMPANY
MADE IN U.S.A.

Longitude West of Greenwich

**Cities
and
Towns**

| 0 to 50,000 | ○ | 500,000 to 1,000,000 | ◎ |
| 50,000 to 500,000 | ⊙ | 1,000,000 and over | |

Scale 1:4 000 000; one inch to 64 miles. Conic Projection
Elevations and depressions are given in feet

Continued on pages 66-67

Continued on pages 78-79

Relief

Meters	Feet
1525	5000
610	2000
305	1000
152.5	500
0 Sea Level	0
152.5	500

LAKE SUPERIOR
Surface elev. 600 Feet above Sea Level
Maximum depth 1333 Feet

LAKE MICHIGAN
Surface elevation 579 Feet above Sea Level
Maximum depth 870 Feet

0   20   40   60   80   100   120 Miles
0   20   40   60   80   100   120   140   160   180   200 Kilometers

Continued on pages 54-55

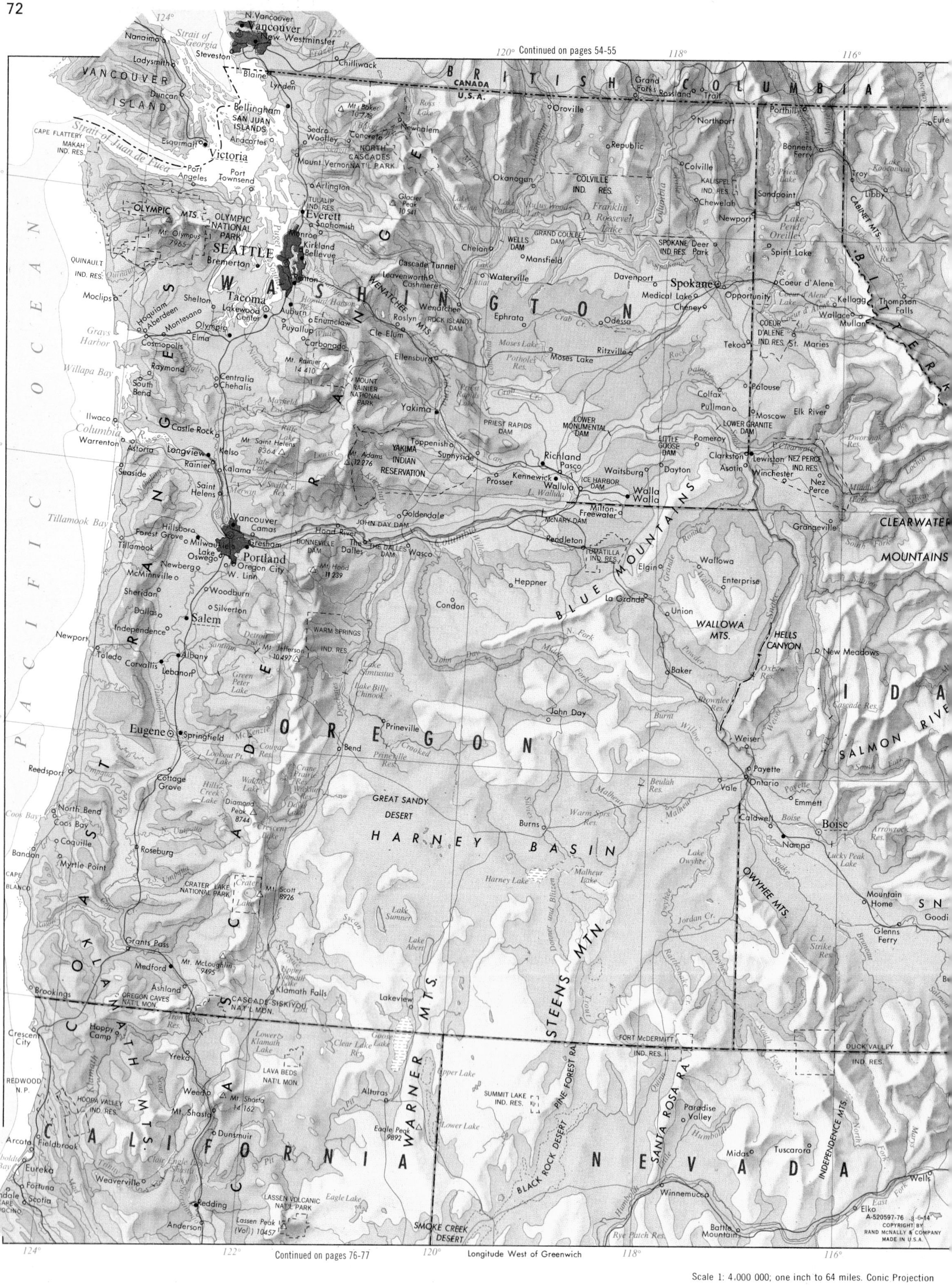

BRITISH COLUMBIA
CANADA
U.S.A.

VANCOUVER ISLAND

WASHINGTON

SEATTLE
Tacoma
Everett
Spokane
Olympia
Yakima
Walla Walla
Wenatchee

OREGON

Portland
Salem
Eugene
Bend
HARNEY BASIN
GREAT SANDY DESERT
BLUE MOUNTAINS
WALLOWA MTS.
HELLS CANYON

IDAHO

Boise
Nampa
SALMON RIVER
CLEARWATER MOUNTAINS

CALIFORNIA

NEVADA

PACIFIC OCEAN

COAST RANGE
CASCADE RANGE
KLAMATH MTS.
STEENS MTN.
WARNER MTS.
SANTA ROSA RA.
OWYHEE MTS.

CRATER LAKE NATIONAL PARK
Mt. Shasta 14,162
Lassen Peak (Vol.) 10,457
Mt. Hood 11,239
Mt. Jefferson 10,497
Mt. Rainier 14,410
Mt. Adams 12,276
Mt. Saint Helens 8,364
Mt. Olympus 7,965
Glacier Peak 10,541
Mt. Baker 10,778

Continued on pages 76-77

Longitude West of Greenwich

Scale 1: 4,000,000; one inch to 64 miles. Conic Projection
Elevations and depressions are given in feet

A-520597-76
COPYRIGHT BY
RAND McNALLY & COMPANY
MADE IN U.S.A.

114° Continued on pages 56-57 112° 110° 108° 106°

ALBERTA CANADA SASKATCHEWAN
U.S.A.

WATERTON-GLACIER INTERNATIONAL PEACE PARK
BLACKFEET IND. RES.
Milk
Morgan
Hogeland
Opheim
Plentywood
Scobey
Grenora

Whitefish
Browning
Sunburst
Cut Bank
Shelby
Chinook
Harlem
Havre
Malta
FORT PECK IND. RES.
Medicine Lake
Williston

Kalispell
Valier
Conrad
Lake Elwell
Marias
FT. BELKNAP IND. RES.
ROCKY BOYS IND. RES.
Glasgow
Wolf Point
Poplar
Sidney
N. DAK.

olson
Choteau
Teton
Fort Benton
Missouri
Fort Peck Lake
Fr. Peck
Brockway
Glendive
Beach

LATHEAD INDIAN RESERVATION
NATIONAL BISON RANGE
Missoula
Lolo
Sun
Great Falls
Belt
Winifred
Lewistown
Winnett
Terry
Miles City
Baker
Marmarth

Stevensville
Hamilton
Helena
East Helena
Townsend
LITTLE BELT MTS.
Neihart
White Sulphur Spgs.
Harlowton
Roundup
Musselshell
Forsyth
Colstrip

Philipsburg
Anaconda
Deer Lodge
Butte
Walkerville
Three Forks
CRAZY MTS.
Big Timber
Bozeman
Livingston
Columbus
Laurel
Billings
Huntley
Hardin
Crow Agency
LITTLE BIGHORN BATTLEFIELD NAT'L MON.
Lame Deer
NORTHERN CHEYENNE IND. RES.

BIG HOLE NAT'L BATTLEFIELD
Homer Youngs Peak 10 621
PIONEER MTS.
Twin Bridges
Dillon
Ennis Lake
Red Lodge
Granite Peak
Bear Creek
CROW IND. RES.
Bighorn Lake
Sheridan
DEVILS TOWER NAT'L MON.

Salmon
Hap Hawkins Lake
Electric Peak 10 992
Gardiner
Mt. Washburn 10 243
ABSAROKA RANGE
Lovell
Powell
BIGHORN MOUNTAINS
Buffalo
Syndance
Moorcroft

Borah Pk. 12 662
LEMHI RANGE
LOST RIVER RA.
BEAVERHEAD MTS.
Lima Res.
Hebgen Lake
YELLOWSTONE NATIONAL PARK
7733 ft above sea level
Cody
Greybull
Basin
Cloud Peak 13 167
Ten Sleep
Gillette

Boulder Peak 10 981
Hyndman Peak 12 009
Mackay
St. Anthony
Ashton
GRAND TETON NAT'L PARK
Grand Teton 13 770
Worland
Gebo
Thermopolis
Kaycee
Antelope

Arco
CRATERS OF THE MOON NAT'L MON.
Rexburg
Rigby
Idaho Falls
WYOMING RANGE
WIND RIVER RANGE
Gannett Peak 13 804
Fremont Peak 13 745
WIND RIVER IND. RES.
Shoshoni
Riverton
Lander
Powder River
Midwest
Glenrock
Douglas
Orin

Shoshone
Shelley
Blackfoot
FORT HALL IND. RES.
Pocatello
Blackfoot Reservoir
Afton
Casper

American Falls Res.
American Falls
Soda Springs
Meade Peak 9957
GREAT DIVIDE BASIN
Pathfinder Res.
Alcova Res.

Rupert
Burley
Lava Hot Sprs.
Montpelier
Green
Fontenelle Res.
Sweetwater
Seminoe Res.
Wheatland

Twin Falls
Oakley
Malad City
Preston
Kemmerer
Granger
Superior
Rawlins
Hanna

GREAT SALT LAKE DESERT
Lucin
Great Salt Lake
Garland
Wallsville
Brigham
Huntsville
Logan
Providence
Smithfield
Richmond
Lewiston
Bear Lake
Green River
Rock Springs
FLAMING GORGE RES.

Wendover
Lake Wendover
Ogden
Morgan
Farmington
Evanston
Flaming Gorge Res.
Craig
Steamboat Spgs.

Tooele
Murray
Midvale
Salt Lake City
Bountiful
Park City
Mt. Emmons 13 440
Kings Peak 13 528
UINTA MTS.
DINOSAUR NAT'L MON.
COLO.
Oak Creek

Heber City
Vernal
UINTAH AND OURAY IND. RES.
PARK RANGE

Relief		
Meters		Feet
3050		10000
1525		5000
610		2000
305		1000
152.5		500
Sea Level		0
1525	500	

114° 112° 110° 108° 106°

Continued on pages 76-77
Continued on pages 70-71

0 20 40 60 80 100 120 Miles
0 20 40 60 80 100 120 140 160 180 200 Kilometers

48°
44°
42°
104°

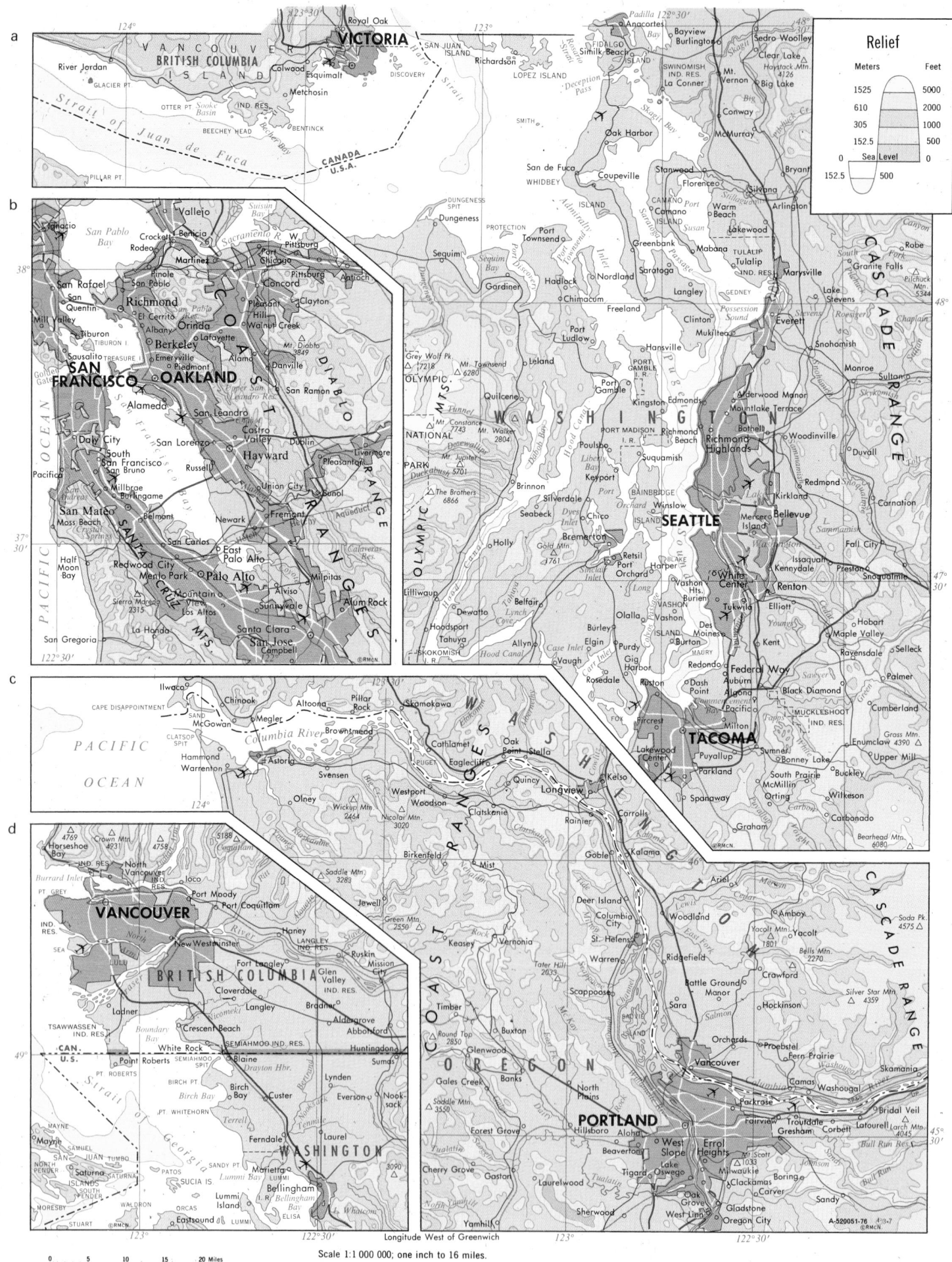

Scale 1:1 000 000; one inch to 16 miles.
Elevations and depressions are given in feet.

Longitude West of Greenwich

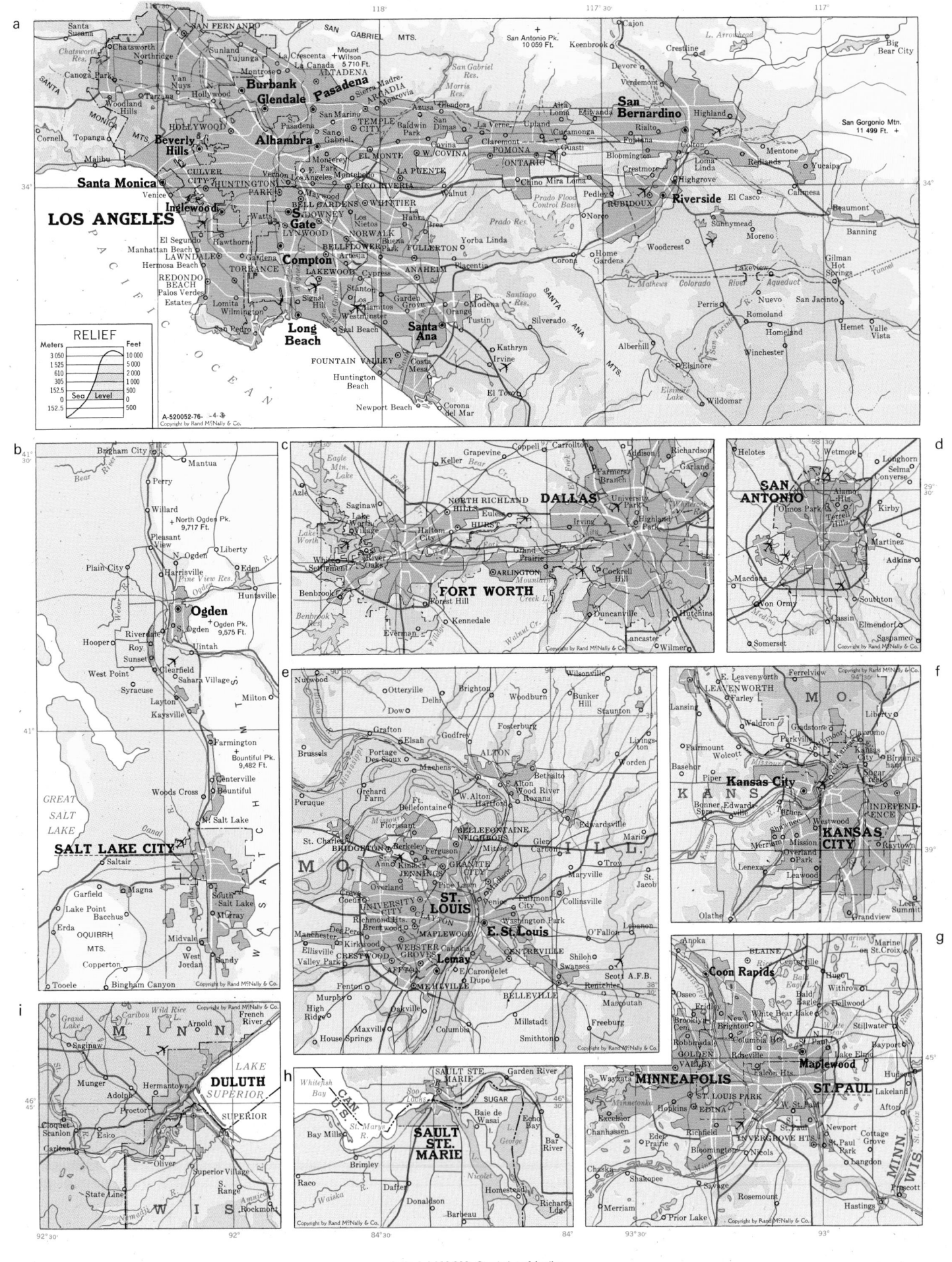

a Los Angeles and Environs

Santa Susana · Chatsworth Res. · Santa Susana · Chatsworth · SAN FERNANDO · SAN GABRIEL MTS. · San Antonio Pk. 10 059 Ft. + · Cajon · L. Arrowhead · Crestline · Big Bear City · Keenbrook · Devore · Verdemont · Canoga Park · Woodland Hills · Northridge · Sunland · Tujunga · La Crescenta · Montrose · Mount Wilson 5 710 Ft. · San Gabriel Res. · Morris Res. · Sierra Madre · Monrovia · Glendora · Alta Loma · Etiwanda · San Bernardino · Highland · San Gorgonio Mtn. 11 499 Ft. + · Cornell · Topanga · Malibu · Van Nuys · N. · Hollywood · HOLLYWOOD · BURBANK · Glendale · PASADENA · ALTADENA · S. Pasadena · San Marino · ARCADIA · TEMPLE CITY · Baldwin Park · San Dimas · La Verne · Upland · Cucamonga · Rialto · Fontana · Colton · Loma Linda · Highland · Highgrove · Redlands · Mentone · Yucaipa · BEVERLY HILLS · ALHAMBRA · El Monte · W. COVINA · POMONA · ONTARIO · Claremont · Guasti · Bloomington · Crestmore · Santa Monica · CULVER CITY · Vernon · Los Angeles · Montebello · PICO RIVERA · WHITTIER · La Habra · Walnut · Chino Mira Loma · Pedley · RUBIDOUX · Riverside · El Casco · Calimesa · Venice · HUNTINGTON PARK · Maywood · BELL GARDENS · Norco · Beaumont · Banning · LOS ANGELES · Inglewood · S. GATE · DOWNEY · Los Nietos · NORWALK · Buena Park · FULLERTON · Yorba Linda · Corona · Home Gardens · Woodcrest · Moreno · Gilman Hot Springs · El Segundo · Hawthorne · LYNWOOD · BELLFLOWER · Anaheim · Placentia · Lakeview · Tunnel · Manhattan Beach · LAWNDALE · Gardena · Compton · Artesia · Cypress · L. Mathews · Colorado River Aqueduct · Hermosa Beach · TORRANCE · LAKEWOOD · Stanton · Garden Grove · Perris · Nuevo · San Jacinto · REDONDO BEACH · Palos Verdes Estates · Lomita · Wilmington · Signal Hill · Los Alamitos · Westminster · Orange · Tustin · Santiago Res. · SANTA ANA MTS. · Alberhill · Homeland · Hemet · Valle Vista · San Pedro · LONG BEACH · Seal Beach · SANTA ANA · El Modena · Silverado · Kathryn · Irvine · Winchester · FOUNTAIN VALLEY · Costa Mesa · El Toro · Elsinore · Elsinore Lake · Wildomar · Huntington Beach · Newport Beach · Corona del Mar · PACIFIC OCEAN

RELIEF
Meters		Feet
3 050		10 000
1 525		5 000
610		2 000
305		1 000
152.5		500
0	Sea Level	0
152.5		500

A-520052-76- -4-3 Copyright by Rand McNally & Co.

b Salt Lake City — Ogden Area
Brigham City · Mantua · Bear River · Perry · Willard · North Ogden Pk. 9,717 Ft. · Pleasant View · N. Ogden · Liberty · Plain City · Harrisville · Pine View Res. · Eden · Huntsville · Ogden · Ogden Pk. 9,575 Ft. · Riverdale · S. Ogden · Uintah · Hooper · Roy · Sunset · Clearfield · Sahara Village · GREAT SALT LAKE · West Point · Syracuse · Layton · Kaysville · Milton · Farmington · Bountiful Pk. 9,482 Ft. · Centerville · Bountiful · Woods Cross · N. Salt Lake · SALT LAKE CITY · Saltair · Garfield · Magna · South Salt Lake · Murray · Lake Point · Bacchus · Erda · OQUIRRH MTS. · Midvale · West Jordan · Sandy · Copperton · Tooele · Bingham Canyon · WASATCH

c Dallas — Fort Worth Area
Eagle Mtn. Lake · Keller · Grapevine · Coppell · Carrollton · Addison · Richardson · Helotes (d) · Azle · Saginaw · NORTH RICHLAND HILLS · Euless · Farmers Branch · Garland · University Park · DALLAS · Lake Worth · Haltom City · HURST · Irving · Highland Park · White Settlement · River Oaks · Grand Prairie · FORT WORTH · ARLINGTON · Mountain Creek L. · Cockrell Hill · Benbrook · Forest Hill · Duncanville · Hutchins · Benbrook Res. · Kennedale · Everman · Crowley · Lancaster · Wilmer · Copyright by Rand McNally & Co.

d San Antonio Area
Helotes · Wetmore · Longhorn · Selma · Converse · SAN ANTONIO · Alamo Hts. · Olmos Park · Terrell Hills · Kirby · Martinez · Adkins · Macdona · Von Ormy · Southton · Cassin · Elmendorf · Somerset · Saspamco · Copyright by Rand McNally & Co.

e St. Louis Area
Nutwood · Otterville · Brighton · Wilsonville · Dow · Delhi · Woodburn · Bunker Hill · Staunton · Livingston · Grafton · Elsah · Godfrey · Fosterburg · Brussels · Portage Des Sioux · Machens · ALTON · Bethalto · Worden · Peruque · Orchard Farm · Ft. Bellefontaine · W. Alton · E. Alton · Wood River · Roxana · St. Charles · Florissant · BELLEFONTAINE NEIGHBORS · Edwardsville · MO. · BRIDGETON · Berkeley · Ferguson · Mitchell · Glen Carbon · Marine · ILL. · Cool Valley · JENNINGS · GRANITE CITY · Troy · St. Jacob · Overland · Pine Lawn · Venice · Maryville · Collinsville · UNIVERSITY CITY · CLAYTON · ST. LOUIS · Fairmont · Washington Park · Richmond Hts. · Brentwood · E. ST. LOUIS · O'Fallon · Lebanon · MAPLEWOOD · Manchester · Des Peres · Kirkwood · WEBSTER GROVES · Cahokia · CENTREVILLE · Ellisville · Valley Park · CRESTWOOD · AFFTON · Lemay · Dupo · Shiloh · Swansea · Fenton · MEHLVILLE · Carondelet · BELLEVILLE · Scott A.F.B. · Rentchler · Murphy · Oakville · Maxville · Columbia · Millstadt · Freeburg · Mascoutah · High Ridge · House Springs · Smithton · Copyright by Rand McNally & Co.

f Kansas City Area
E. Leavenworth · Ferrelview · Copyright by Rand McNally & Co. · LEAVENWORTH · Farley · Lansing · Fairmount · Wolcott · Gladstone · Parkville · Waldron · M O. · Liberty · Birmingham · Sugar Creek · Basehor · Piper · Edwardsville · KANSAS CITY · KANS · Bonner Springs · Shawnee · Merriam · Mission · KANSAS CITY · INDEPENDENCE · Westwood · Overland Park · Lenexa · Leawood · Raytown · Olathe · Grandview · Lee's Summit

g Minneapolis — St. Paul Area
Anoka · BLAINE · Marine on St. Croix · Marine · Coon Rapids · Hugo · Osseo · Fridley · New Brighton · White Bear Lake · Withrow · Brooklyn Center · Columbia Hts. · Stillwater · Robbinsdale · Roseville · Maplewood · Bayport · GOLDEN VALLEY · Falcon Hts. · Lake Elmo · Wayzata · MINNEAPOLIS · ST. PAUL · Lakeland · Minnetonka · ST. LOUIS PARK · W. St. Paul · Afton · Excelsior · Hopkins · EDINA · Richfield · INVER GROVE HTS. · Newport · Cottage Grove · Chanhassen · Eden Prairie · Bloomington · Nicols · St. Paul Park · Chaska · Savage · Langdon · Shakopee · MINN. · WIS. · Merriam · Rosemount · Prior Lake · Hastings · St. Croix R. · Prescott

i Duluth — Superior Area
Grand Lake · Wild Rice L. · Caribou L. · Arnold · French River · Saginaw · MINN. · Copyright by Rand McNally & Co. · Munger · Hermantown · DULUTH · LAKE SUPERIOR · Adolph · Proctor · SUPERIOR · Cloquet · Scanlon · Esko · Oliver · Carlton · State Line · S. Range · Superior Village · WIS. · Rockmont

h Sault Ste. Marie Area
Whitefish Bay · SAULT STE. MARIE · Garden River · CAN. · U.S. · Soo Locks · SUGAR · Bay Mills · St. Marys R. · Baie de Wasai · Echo Bay · Raco · SAULT STE. MARIE · L. George · Bar River · Brimley · Nicolet · Dafter · Homestead · Richards Lda. · Barbeau

Scale 1:1 000 000; One inch to 16 miles.
Elevations and depressions are given in feet.

Miles: 0 2 4 6 8 10 12 14 16 18 20 22 24
Kilometers: 0 4 8 12 16 20 24 28 32 36 40

Continued on pages 72-73

Scale 1:4 000 000; one inch to 64 miles. Conic Projection
Elevations and depressions are given in feet

Continued on pages 78-79

Continued on pages 80-81

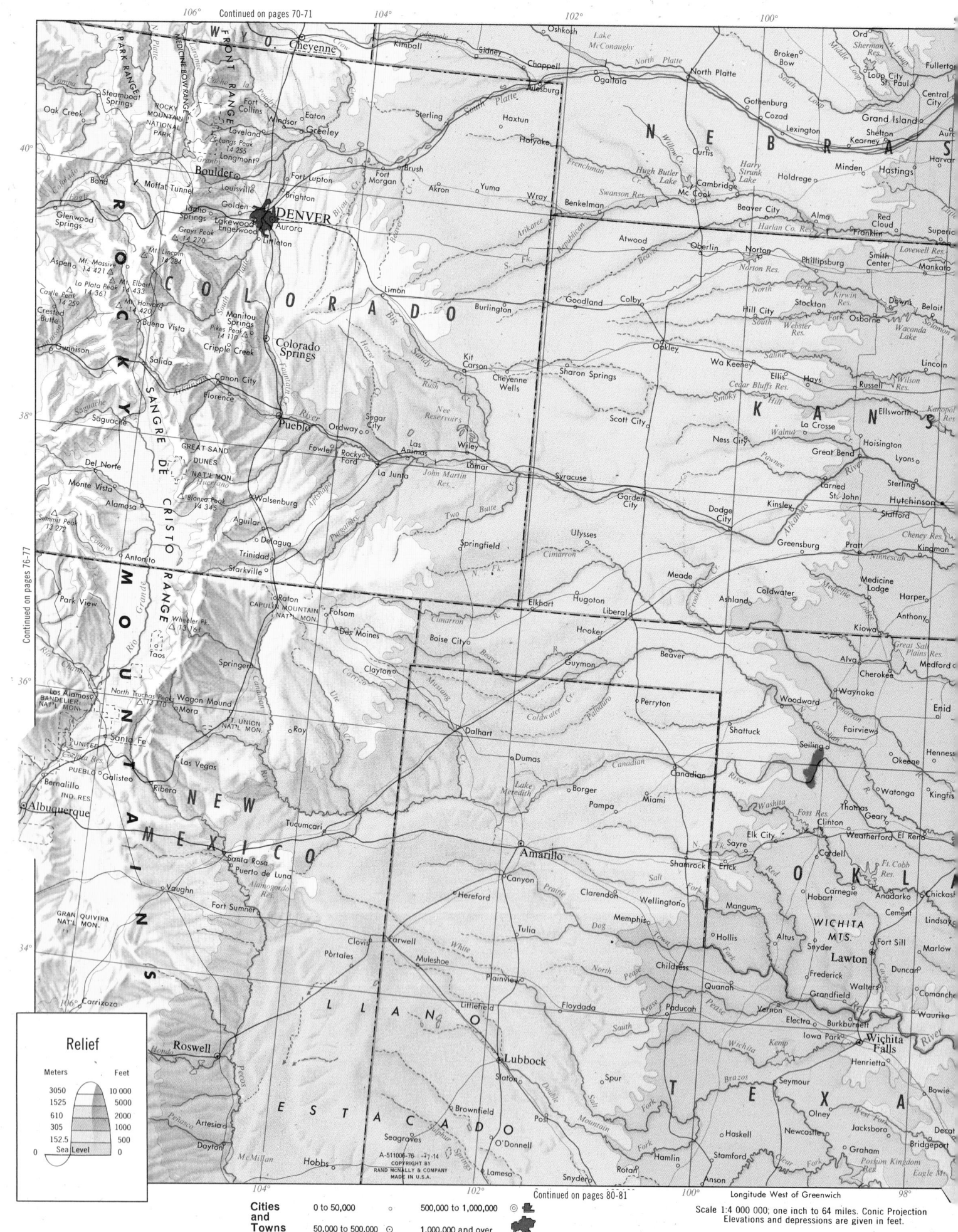

Continued on pages 70-71

Continued on pages 76-77

Continued on pages 80-81

Relief

Meters		Feet
3050		10 000
1525		5000
610		2000
305		1000
152.5		500
0	Sea Level	0

Cities and Towns

| 0 to 50,000 | o | 500,000 to 1,000,000 | ◎ |
| 50,000 to 500,000 | ⊙ | 1,000,000 and over | |

Longitude West of Greenwich

Scale 1:4 000 000; one inch to 64 miles. Conic Projection
Elevations and depressions are given in feet.

A-511006-76 -27-14
COPYRIGHT BY
RAND McNALLY & COMPANY
MADE IN U.S.A.

Continued on pages 70-71

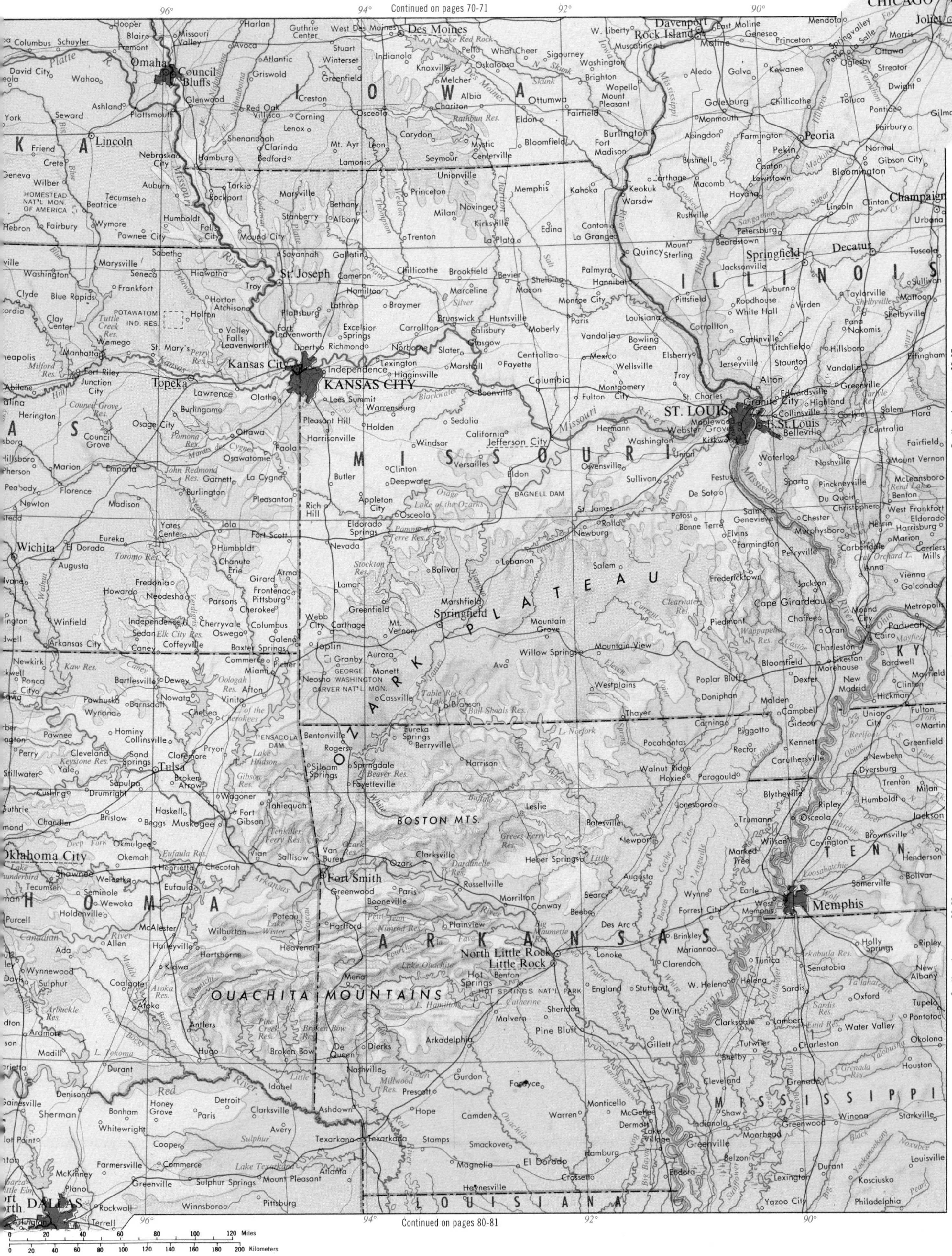

Continued on pages 66-67

Continued on pages 82-83

Continued on pages 80-81

CHICAGO
Aurora
Joliet

IOWA
ILLINOIS
KANSAS
MISSOURI
OKLAHOMA
ARKANSAS
TENN.
KY.
MISSISSIPPI
LOUISIANA

OZARK PLATEAU
BOSTON MTS.
OUACHITA MOUNTAINS

Omaha
Council Bluffs
Lincoln
Des Moines
Davenport
Rock Island
Peoria
Champaign
Decatur
Springfield
St. Joseph
Kansas City
KANSAS CITY
Topeka
Lawrence
Wichita
Tulsa
Oklahoma City
Fort Smith
ST. LOUIS
E. St. Louis
Jefferson City
Springfield
Memphis
North Little Rock
Little Rock
Hot Springs
Pine Bluff
DALLAS

0 20 40 60 80 100 120 Miles
0 20 40 60 80 100 120 140 160 180 200 Kilometers

80

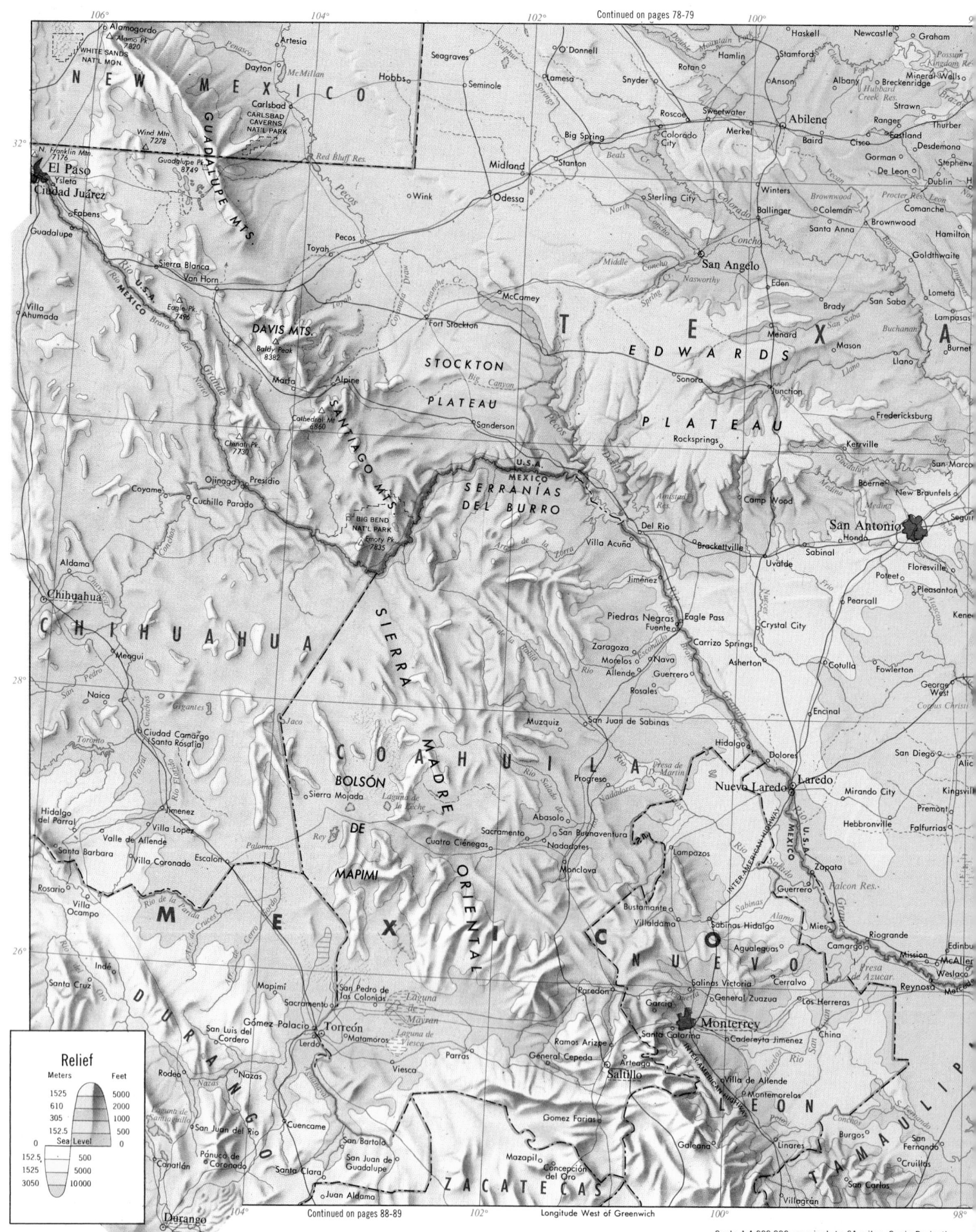

Continued on pages 78-79

106°   104°   102°   100°

NEW MEXICO

White Sands Nat'l Mon.
Alamogordo
Alamo Pk. 820
Penasco
Artesia
Dayton
McMillan
Hobbs
Carlsbad
Carlsbad Caverns Nat'l Park
Wind Mtn. 7278
Guadalupe Pk. 8749
Red Bluff Res.

Haskell
Newcastle
Graham
O'Donnell
Rotan
Hamlin
Stamford
Mineral Wells
Possum Kingdom Res.
Seagraves
Seminole
Lamesa
Snyder
Roscoe
Sweetwater
Anson
Albany
Breckenridge
Strawn
Hubbard Creek Res.
Big Spring
Colorado City
Merkel
Abilene
Baird
Cisco
Eastland
Ranger
Thurber
Desdemona
Gorman
De Leon
Stephenv.
Dublin

N. Franklin Mtn. 7176
El Paso
Ysleta
Ciudad Juárez
Fabens
Guadalupe

Midland
Stanton

Wink
Odessa

Sterling City
North Concho
Winters
Coleman
Brownwood
Comanche
Santa Anna
Brownwood
Hamilton
Goldthwaite

Pecos
Toyah

McCamey

San Angelo
Nasworthy
Eden
Brady
San Saba
Lometa
Lampasas
Burnet
Buchanan

T E X A S

Villa Ahumada
Sierra Blanca
Van Horn
Eagle Pk. 7496

DAVIS MTS.
Baldy Peak 8382
Marfa
Alpine

STOCKTON PLATEAU
Fort Stockton
Big Canyon
Sanderson

EDWARDS PLATEAU
Sonora
Menard
Mason
Llano
Junction
Rocksprings
Kerrville
Fredericksburg

Guadalupe
Rio Grande del Norte
Chinati Pk. 7730
Cathedral Mt. 6860

SANTIAGO MTS.
Ojinaga
Presidio

SERRANÍAS DEL BURRO
U.S.A.
MEXICO

San Marcos
Boerne
New Braunfels
Seguin

Coyame
Cuchillo Parado

BIG BEND NAT'L PARK
Emory Pk. 7835

Del Rio
Villa Acuña
Brackettville
Camp Wood
Amistad Res.

San Antonio
Hondo
Sabinal
Uvalde
Floresville
Pleasanton

Aldama

C H I H U A H U A

Meoqui

Chihuahua

Jiménez
Piedras Negras
Fuente
Eagle Pass
Crystal City
Pearsall
Kene

Naica
San Pedro
Gigantes
Toronto

Zaragoza
Morelos
Nava
Allende
Guerrero
Rosales
Carrizo Springs
Asherton
Cotulla
Fowlerton
Encinal
George West
Corpus Christi

SIERRA

Jaco
Ciudad Camargo (Santa Rosalía)

Muzquiz
San Juan de Sabinas
Hidalgo
Dolores
San Diego
Alic

M A D R E

C O A H U I L A
BOLSÓN
Sierra Mojada
Laguna de la Leche
Progreso
Presa de Martín
Nuevo Laredo
Laredo
Mirando City
Kingsvill
Premont
Falfurrias

Hidalgo del Parral
Jimenez
Villa Lopez

DE

Sacramento
San Buenaventura
Nadadores
Abasolo
Monclova

Lampazos
Sabinas
Presa de Azucar
Guerrero
Zapata
Falcon Res.

O R I E N T A L

Santa Barbara
Valle de Allende
Villa Coronado
Escalón

MAPIMI
Cuatro Ciénegas
Sacramento

Bustamante
Villaldama
Sabinas Hidalgo
Mier
Camargo
Riogrande

Rosario
Villa Ocampo

M E X I C O

Aguileguas
Cerralvo
Mission
McAller
Edinbu
Reynosa

N U E V O

Indé
Santa Cruz
Mapimí
San Pedro de las Colonias
Laguna de Mayran
Paredon
Salinas Victoria
General Zuazua
Los Herreras

D U R A N G O

Gómez Palacio
San Luis del Cordero
Torreón
Matamoros
Lerdo
Laguna de Viesca
Viesca
Parras
Ramos Arizpe
Santa Catarina
Monterrey
Cadereyta Jiménez
China

Rodeo
Nazas
General Cepeda
Arteaga
Villa de Allende
Montemorelos

L E Ó N

San Juan del Rio
Cuencamé
San Bartolo
Gómez Farías
Galeana
Burgos
San Fernando

Pánuco de Coronado
Conatlán
Santa Clara
San Juan de Guadalupe
Mazapil
Concepción del Oro
Linares
Cruillas

Juan Aldama
Z A C A T E C A S
Villagrán
San Carlos

T A M A U L I P

Durango

Relief

Meters	Feet
1525	5000
610	2000
305	1000
152.5	500
Sea Level	0
152.5	500
1525	5000
3050	10000

Continued on pages 88-89

104°   102°   Longitude West of Greenwich   100°   98°

Scale 1:4 000 000; one inch to 64 miles. Conic Projection
Elevations and depressions are given in feet

Continued on pages 78-79

Continued on pages 82-83

Scale 1:1 000 000

A-511007-76
COPYRIGHT BY
RAND McNALLY & COMPANY
MADE IN U.S.A.

Cities	0 to 50,000	○	500,000 to 1,000,000	◉
and				
Towns	50,000 to 500,000	⊙	1,000,000 and over	

Continued on pages 66-67

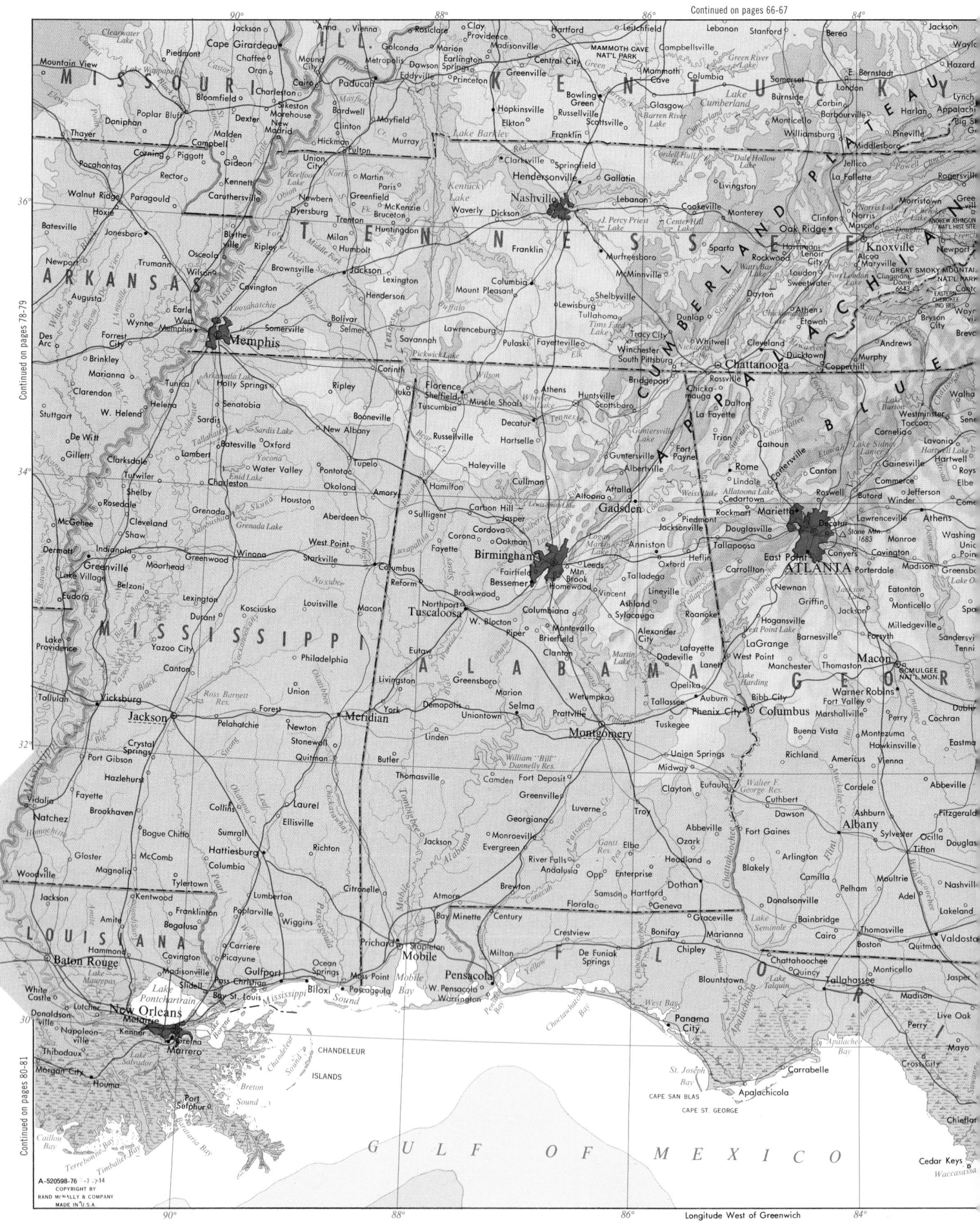

Continued on pages 78-79

Continued on pages 80-81

Scale 1:4 000 000; one inch to 64 miles. Conic Projection
Elevations and depressions are given in feet

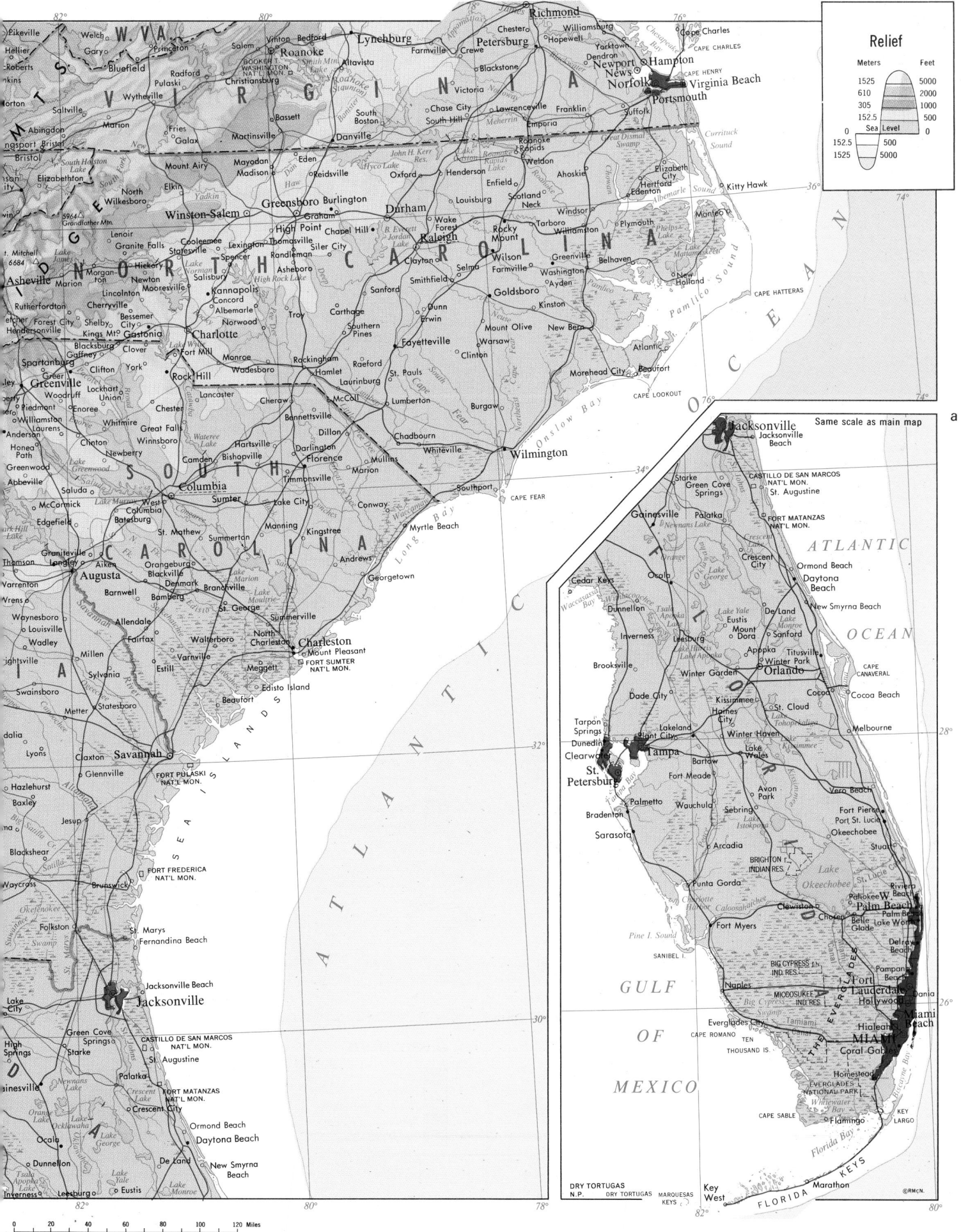

Relief

Meters		Feet
1525		5000
610		2000
305		1000
152.5		500
0	Sea Level	0
152.5		500
1525		5000

Same scale as main map

a

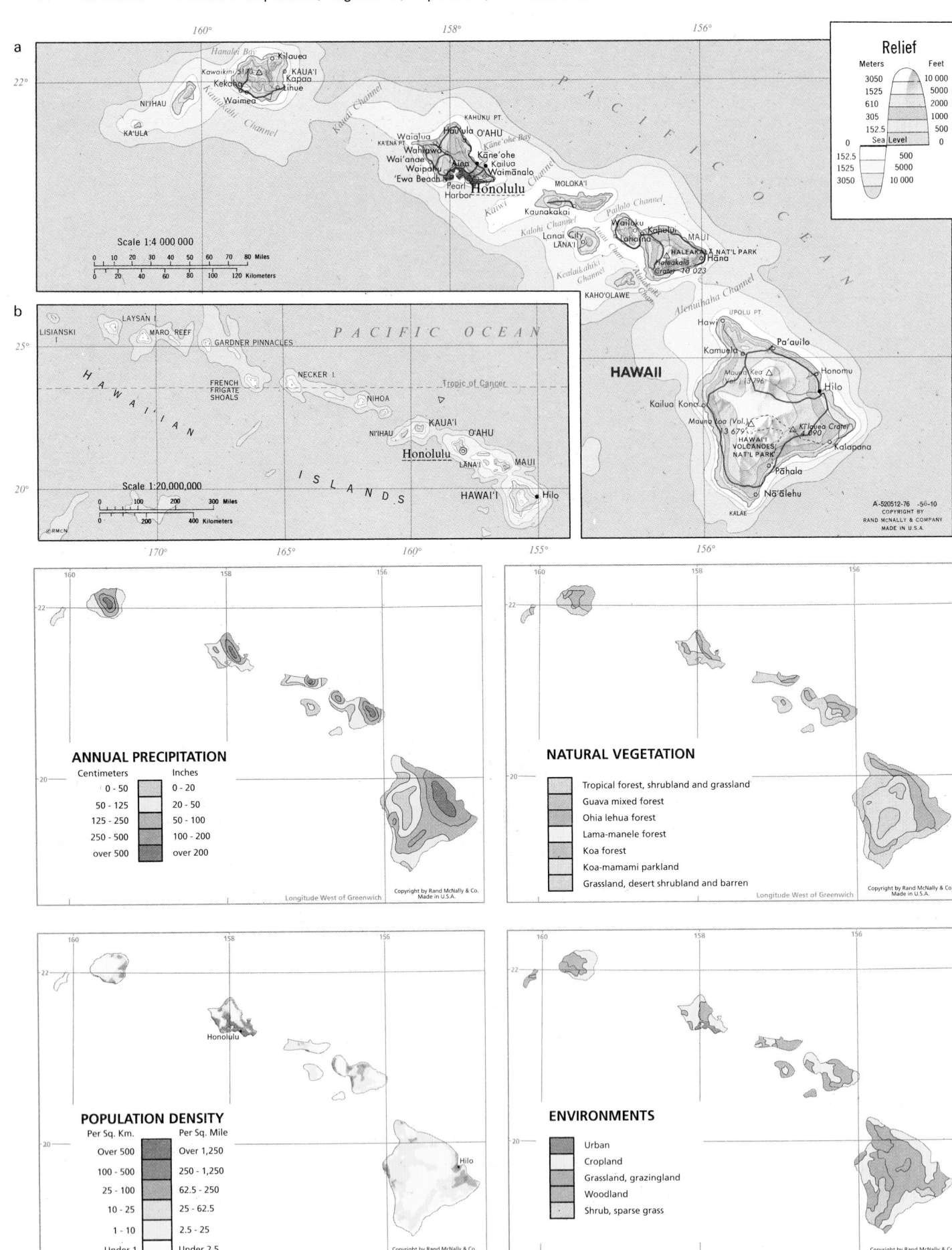

Relief

Meters		Feet
3050		10 000
1525		5000
610		2000
305		1000
152.5		500
0	Sea Level	0
152.5		500
1525		5000
3050		10 000

a

Scale 1:4 000 000

b

PACIFIC OCEAN

HAWAIIAN ISLANDS

HAWAII

Scale 1:20,000,000

ANNUAL PRECIPITATION

Centimeters	Inches
0 - 50	0 - 20
50 - 125	20 - 50
125 - 250	50 - 100
250 - 500	100 - 200
over 500	over 200

Longitude West of Greenwich
Copyright by Rand McNally & Co.
Made in U.S.A.

NATURAL VEGETATION

Tropical forest, shrubland and grassland
Guava mixed forest
Ohia lehua forest
Lama-manele forest
Koa forest
Koa-mamami parkland
Grassland, desert shrubland and barren

Longitude West of Greenwich
Copyright by Rand McNally & Co.
Made in U.S.A.

POPULATION DENSITY

Per Sq. Km.	Per Sq. Mile
Over 500	Over 1,250
100 - 500	250 - 1,250
25 - 100	62.5 - 250
10 - 25	25 - 62.5
1 - 10	2.5 - 25
Under 1	Under 2.5

Longitude West of Greenwich
Copyright by Rand McNally & Co.
Made in U.S.A.
A-GDS25200-T1- -;-1-1

ENVIRONMENTS

Urban
Cropland
Grassland, grazingland
Woodland
Shrub, sparse grass

Longitude West of Greenwich
Copyright by Rand McNally & Co.

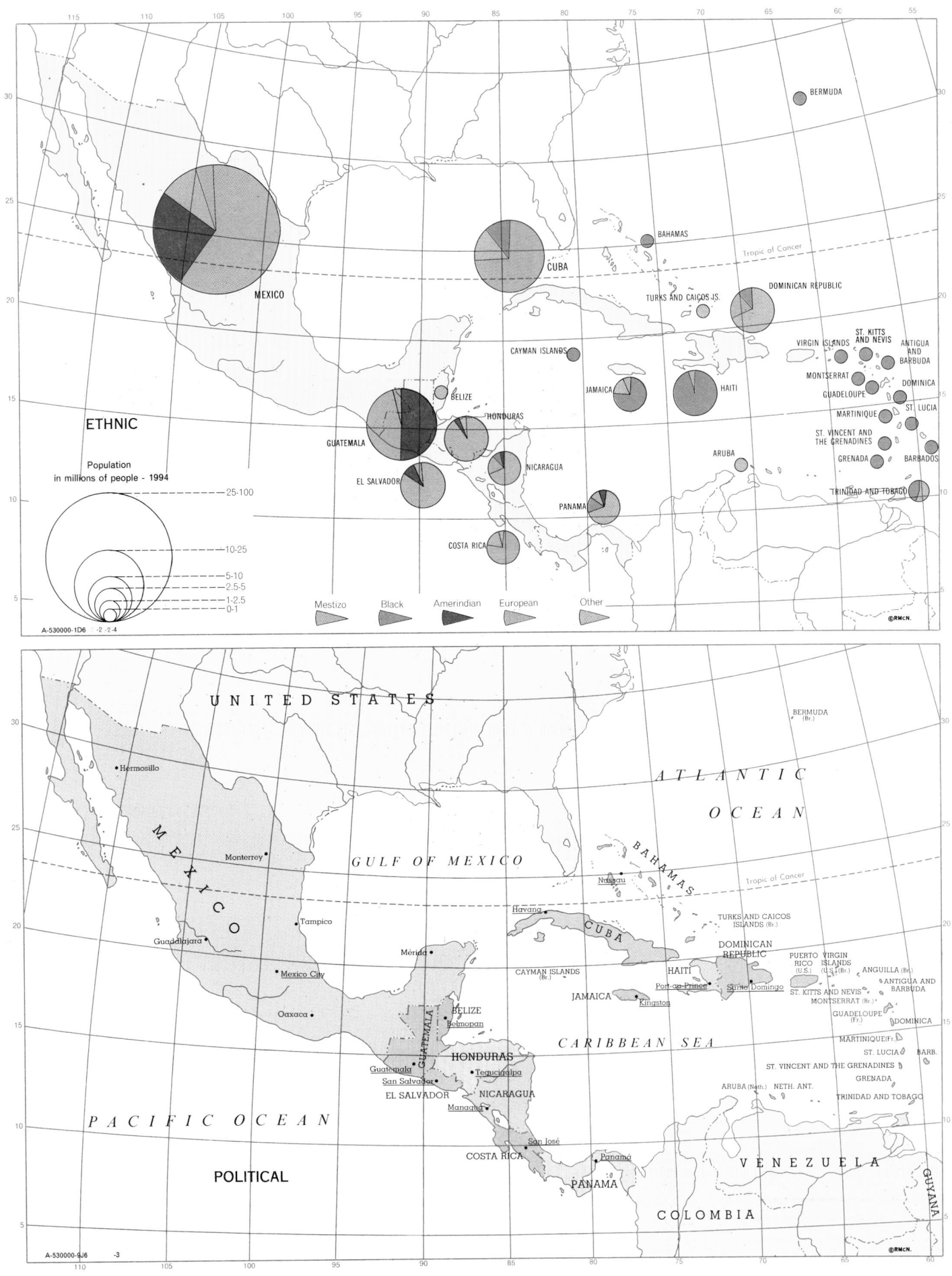

ETHNIC

Population
in millions of people - 1994

25-100

10-25

5-10
2.5-5
1-2.5
0-1

A-530000-1D6   -2 -2 -4

Mestizo   Black   Amerindian   European   Other

©RMCN.

POLITICAL

A-530000-9J6   -3

©RMCN.

Scale 1:16 000 000; one inch to 250 miles. Polyconic Projection
Elevations and depressions are given in feet

a

PANAMA

Scale 1:1 000 000

A-530000-76g   -9-27
COPYRIGHT BY
RAND McNALLY & COMPANY
MADE IN U.S.A.

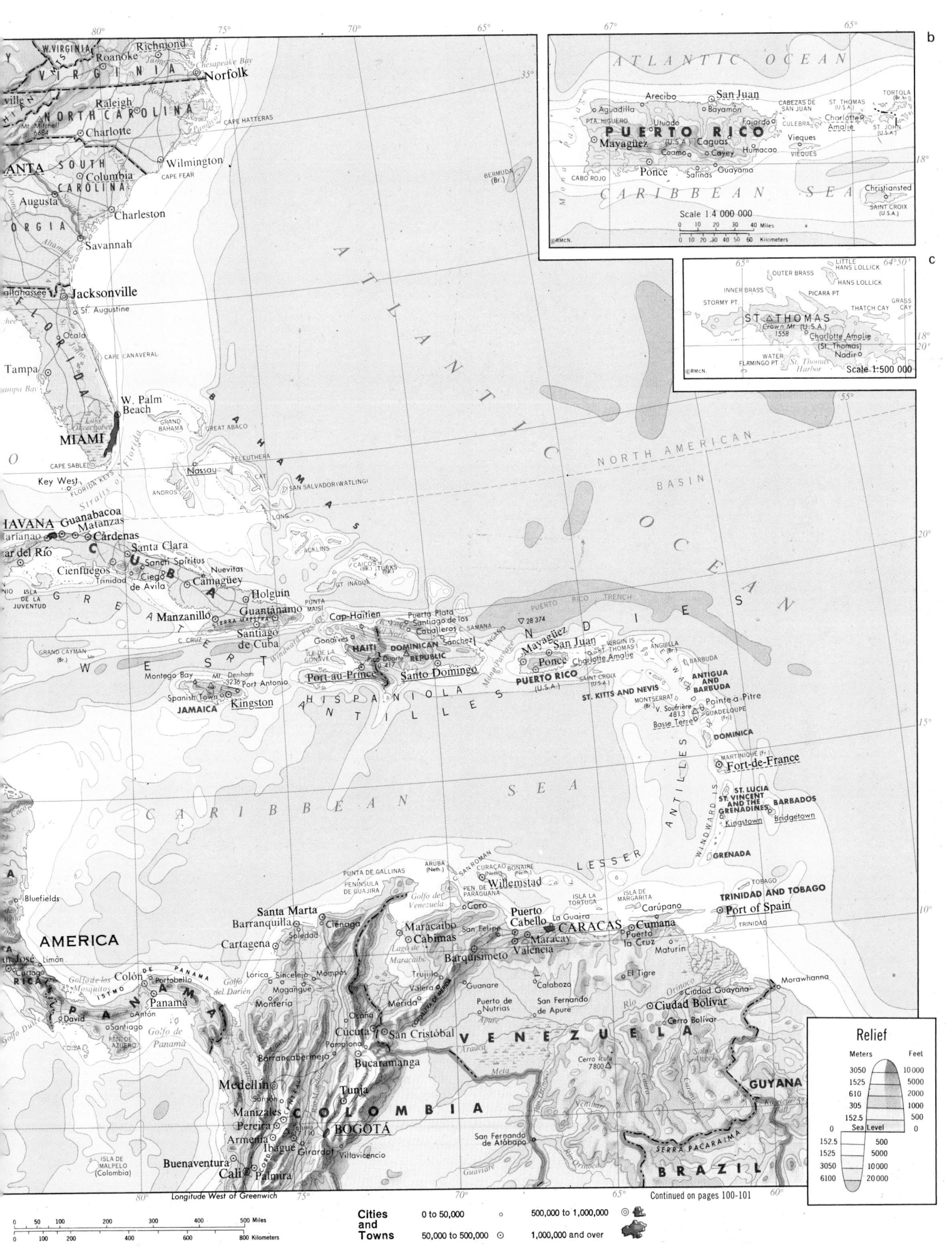

b

ATLANTIC OCEAN

Aguadilla • Arecibo • San Juan
PTA. HIGUERO • Utuado • Bayamón CABEZAS DE SAN JUAN CULEBRA ST. THOMAS TORTOLA (Br.)
PUERTO RICO Caguas • Fajardo • Charlotte (U.S.A.) Amalie ST. JOHN (U.S.A.)
Mayagüez • Coamo • Cayey • Humacao Vieques VIEQUES
CABO ROJO • Ponce • Salinas • Guayama 18°
CARIBBEAN SEA Christiansted
SAINT CROIX (U.S.A.)

Scale 1:4 000 000
0 10 20 30 40 Miles
0 10 20 30 40 50 60 Kilometers
©RMCN

c

LITTLE HANS LOLLICK 64°50'
OUTER BRASS HANS LOLLICK 65°
INNER BRASS PICARA PT GRASS
STORMY PT PICARA PT THATCH CAY CAY
ST △ THOMAS
Crown Mt (U.S.A.) 18°
1558 Charlotte Amalie 20'
WATER (St. Thomas) Nadir
FLAMINGO PT St. Thomas Harbor
©RMCN Scale 1:500 000

Continued on pages 100-101

Relief

Meters	Feet
3050	10 000
1525	5000
610	2000
305	1000
152.5	500
Sea Level	0
152.5	500
1525	5000
3050	10 000
6100	20 000

Longitude West of Greenwich

0 50 100 200 300 400 500 Miles
0 100 200 400 600 800 Kilometers

Cities and Towns
0 to 50,000
50,000 to 500,000
500,000 to 1,000,000
1,000,000 and over

Continued on pages 80-81

PACIFIC OCEAN

Relief

Meters	Feet
3050	10 000
1525	5000
610	2000
305	1000
152.5	500
Sea Level	0
152.5	500
1525	5000
3050	10 000

A-531695-76
COPYRIGHT BY
RAND McNALLY & COMPANY
MADE IN U.S.A.

Cities and Towns

0 to 50,000 • 500,000 to 1,000,000 ◎
50,000 to 500,000 ⊙ 1,000,000 and over

Longitude West of Greenwich

Scale 1:4 000 000; one inch to 64 miles. Conic Projection
Elevations and depressions are given in feet

a

Inset map (Mexico City area)

96°   99°30'   99°   98°30'

24°

Morelos
Nicolás Romero
Cahuacán
Cuautitlán
Tutitlán
Coacalco
Tecamac
Teotihuacán
Acolman
Chiconautla
Tepexpan
Otumba
▲ Pyramids of Teotihuacán
HIDALGO
Apán

San Bartolo
Ixtlahuaca
Jiquipilco
Cerro La Catedral 13 000 △
Atizapán
Mazatla
Tlalnepantla
Tepetlaoxtoc
Tulpetlac
San Jerónimo
Texcoco
Calpulalpan
TLAXCALA
Nanacamilpa

19°30'

Atzcapotzalco
Naucalpan de Juárez
Temoaya
Mimiapan
Chimalpa
Gustavo A. Madero
Texcoco (Dry Lake)
Lago de
Coatlinchán
Chicoloapan

MEXICO CITY
Nezahualcóyotl
Río Frío
HY
PUEBLA

Toluca
Lerma
Río Lerma
Huixquilucan
Cuajimalpa
Iztacalco
Ixtapalapa
Los Reyes
Ayotla
Ixtapaluca
INTER-AMERICAN
Texmelucan

Capultitlán
Metepec
Mexicalcingo
Villa Obregón
Contreras
Tláhuac
Xochimilco
Chalco
Tlalmanalco

Cerro Muneco 12 655 △
San Andrés
Tláhuac
Tecómitl
Iztaccíhuatl 17 343 △
19°

Almoloya
Cerro Ajusco 12 850 △
Ajusco
Topilejo
Milpa Alta
Oxtotepec
Tenango
Amecameca

Nevado de Toluca △ 14 409
Coatepec
DISTRITO FEDERAL

Tenango
Huitzilac
Tres Cumbres
Volcán Popocatépetl 17 887 △
Ozumba

Scale 1:1 000 000

0   10 Miles
0   4   8   12   16 Kilometers

©RMcN.

Tepoztlán
Tlalnepantla
MORELOS
Tlayacapan

Cuernavaca

Main map

90°

20°

GULF OF MEXICO

22°

Laguna Almagre
Tropic of Cancer

PTA. JEREZ

Laguna de San Andrés

amira
iudad Madero
Tampico
Villa Cuauhtémoc
Tampico Alto

Laguna Tamiahua
CABO ROJO
ARRECIFE BLANQUILLA
ISLA DE LOBOS

Tancoco
Alamo
Túxpan
Tamiahua
ARRECIFE TANQUIJO
ARRECIFE TÚXPAN

BAHÍA DE CAMPECHE

YUCATÁN
Sisal
Hunucmá
Maxcanú
Halachó
Calkiní
Dzitbalché
Hecelchakán

Tihuatlán
Poza Rica
Tecolutla
Gutiérrez Zamora
Furbero
Coyutla
Nautla
Coxquihui
Vega de Alatorre
Cuetzalan
del Progreso
Tlapacoyan
Misantla
Atempan
Jalacingo
Altotonga
Teziutlán
Naolinco
Las Vigas
Perote
△14 048
Xalapa
Nauchampatepetl
Coatepec
Libres
Teocelo
Antigua Veracruz
PUNTA ZEMPOALA

20°
Lerma
Campeche

Seybaplaya
Champotón
Pustunich

Sabancuy
Chicbul
CAMPECHE
Mamantel

ISLA DEL CARMEN
Ciudad del Carmen
Laguna de Términos

Veracruz
ARRECIFE CABEZA

San Juan Ixtenco
Huatusco
Ciudad Serdán
Coscomatepec
Pico de Orizaba (Vol.) 18 406
Medellín
Orizaba
Córdoba
Heroica
Nogales
Omealca
Cotaxtla
Maltrata

18°

PUNTA FONTERA
San Pedro
Frontera
Paraíso
Allende
Comalcalco
Jalpa
Cunduacán
Paradise
Palizada
MEXICO
GUATEMALA

Tlalixcoyan
Alvarado
Laguna de Alvarado

Tehuacán
Ajalpan
Zoquitlán
Zinacatepec
San Gabriel Chilac
Chazumba
San Miguel
Huatla de Jiménez
Teotitlán del Camino
Ojitlán (S. Lucas)
Jalapa de Díaz (San Felipe)
Tuxtepec
Tierra Blanca
Cosamaloapan
Chacaltianguis
Santiago Tuxtla
San Andrés Tuxtla
Catemaco
Pajapan
San Martín (Vol.) 6000
PTA. ZAPOTLÁN
Tesechoacan
Coatzacoalcos (Puerto México)
Cárdenas
Villahermosa
San Carlos
Balancán
Emiliano Zapata
Palenque
Tenosique

Petlalcingo
Tepelmeme
Huajuapan de León
Coixtlahuaca
Cuicatlán
Playa Vicente
San Juan Evangelista
Acayucan
Sayula
Jaltipan
Cosoleacaque
Minatitlán
Texistepec
Laguna Rosario
Huimanguillo
Teapa
Tacotalpa
Pichucalco
Chapultenango
Yajalón
Bachajón
Ococingo

nazulapam
l Progreso
Tejupan (Santiago)
Teposcolula (Asunción)
Nochixtlán (Asunción)
Ixtlán de Juárez
Talea de Castro (San Miguel)
Villa Alta (San Ildefonso)
Jesús Carranza
Pueblo Viejo
Presa de Mateos

ISTMO
DE
TEHUANTEPEC

San Pedro y San Pablo
Tlaxiaco
Sta. María Asunción
Hidalgo Yalalag
Zacatepec (Santiago)
Tecpatán
Pantepec
Simojovel
Compañata
Jitotol
MESETA DE AGUA ESCONDIDA

Chalcatongo
Yosondua (Sta. Catarina)
San Mateo (Etlatongo)
Zaachila
Zimatlán de Alvarez
Oaxaca
Tlacolula de Matamoros
Ocotlán de Morelos
Táviche
Mazatlán (San Juan)
Guichicovi (San Juan)
Berriozábal
Ocozocoautla
Tuxtla Gutiérrez
Bohom
Oxchuc
San Cristóbal de las Casas
Cancuc
Chiapa de Corzo
Acala
Teopisca
Amatenango
Las Rosas
Comitán

INTER-AMERICAN HY

Sola de Vega (S. Miguel)
Ejutla de Crespo
Jalapa del Márques
Ixtepec
Ixtaltepec (Asunción)
Zanatepec (Sto. Domingo)
Cintalapa
Chiapa
Suchiapa
Venustiano Carranza
Socoltenango
La Concordia
Trinitaria

huazolotitlán (Sta. María)
omiltepec
Miahuatlán
Las Vacas
Tehuantepec (Sto. Domingo)
Juchitán de Zaragoza
Ixhuatán (San Francisco)
Tapanatepec
Villa Flores
Arriaga
Tonalá
SA. CUCHUMATANES
GUATEMALA

Loxicha (Sta. Catarina)
Pluma Hidalgo
Pochutla (San Pedro)
Puerto Ángel
Salina Cruz
Laguna Superior
Laguna Inferior
Mar Muerto
Golfo de Tehuantepec
CORD. DE CHIAPAS
SIERRA MADRE
Cuauhtemoc
Jacatenango

SIERRA DE OAXACA

DISTRITO FEDERAL

MÉXICO

TABASCO

CHIAPAS

OAXACA

PUEBLA

96°   94°   92°

16°

Continued on pages 90-91

0   20   40   60   80   100   120 Miles
0   20   40   60   80   100   120   140   160   180   200 Kilometers

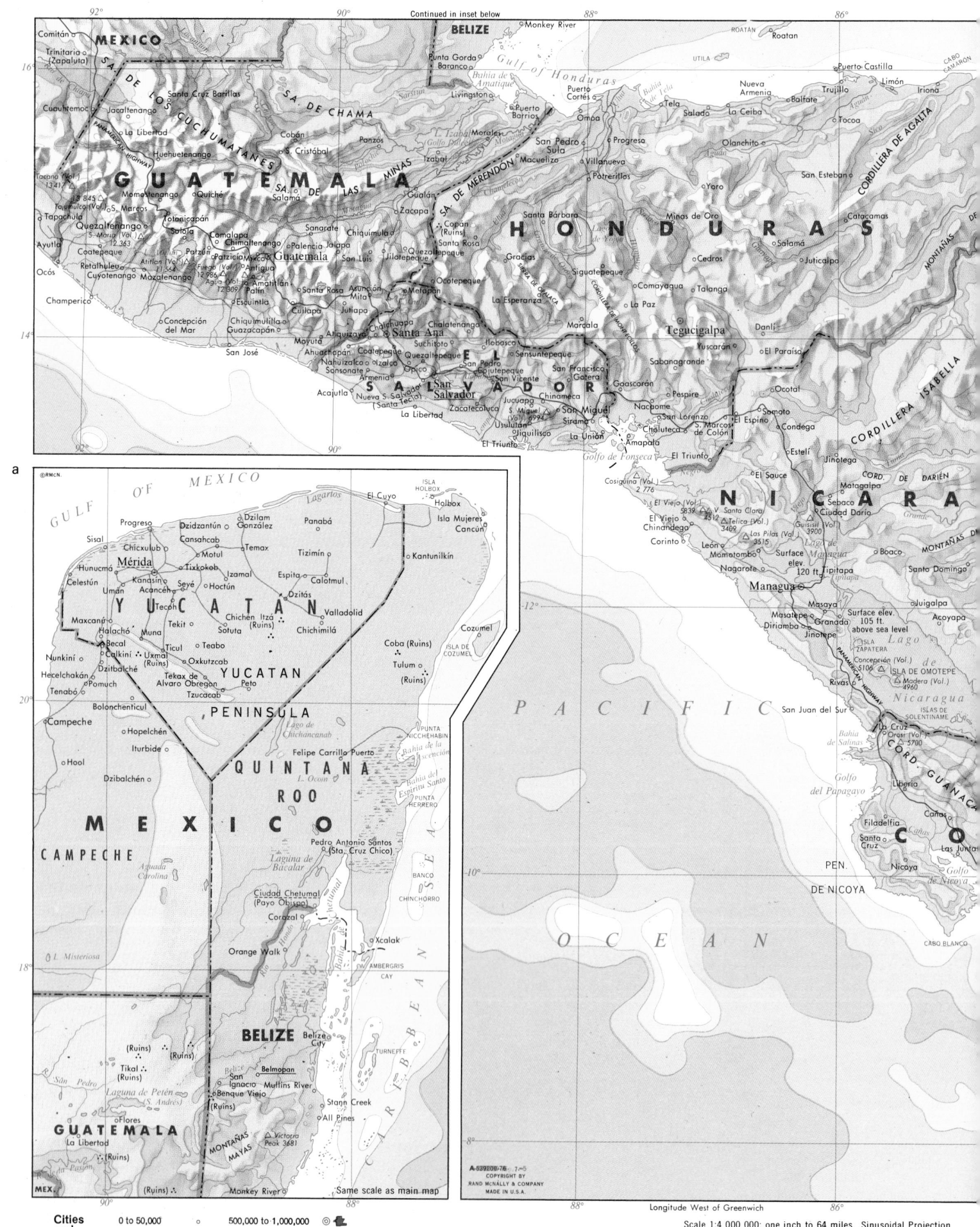

Cities and Towns

	0 to 50,000	○	500,000 to 1,000,000	◎
	50,000 to 500,000	⊙	1,000,000 and over	

Scale 1:4 000 000; one inch to 64 miles. Sinusoidal Projection

Elevations and depressions are given in feet

PUNTA PATUCA

ÓLON

Cabo Gracias a Dios

Coco

Segovia

CAYOS
MISKITO

M O S Q U I T O S

Lone Star

Laguna Carata

Puerto Cabezas

Laguna Huaunta

Huaunta

Prinzapolca

C A R I B B E A N

ISLA DE PROVIDENCIA
(Colombia)

C O S T A   D E

*Laguna
las Perlas*

Rama

Bluefields

ISLA DE LA CIERVO

LITTLE CORN

SAN ANDRÉS
(Colombia)
CAYOS-DE-ESE

GREAT CORN
(Nicaragua)

CAYOS DE ALBUQUERQUE
(Colombia)

PUNTA MICO

S E A

CORD. DE YOLAINA

Río Punta Gorda

*Bahía
de San Juan
del Norte*

an Carlos

San Juan del Norte
(Greytown)

Carlos

ESTA

San Ramón Guapiles oCaño

Esparta Alajuela

Intarenas

San José Cartago Paraíso

R I C A Turrialba

Heredia Irazú Vol. 11760

Matina

Limón

PUNTA CAHUITA

Parrita

Quepos

CORDILLERA DE

Cerro Chirripó
12 530

Guabito

Bocas del Toro

Bahía de Almirante

*Golfo
de los Mosquitos*

PUNTA QUEPOS

San Isidro

Cerro Kámok △
11 696

Almirante

Bocas del Toro

PUNTA CHIRIQUI

Buenos Aires

Cerro Echandí
10 394

TALAMANCA

*Laguna
de Chiriquí*

ESCUDO
DE VERAGUAS

Bahía

Puerto Cortés

Boquete

Chiriquí Grande

de Coronada

Volcán Barú
11 401

ISLA DE CAÑO

PENÍNSULA
Puerto Jiménez

DE OSA

Golfito

*Golfo
Dulce*

La Cuesta

Concepción

David

C. de Santa
Catalina
5249

Horconcitos

Remedios

SERRANIA
DE TABASARÁ

C. Negro 4429

CABO MATAPALO

PUERTO Armuelles

*Bahía Charco
de Azul*

PUNTA BURICA

Golfo

de

Las Palmas

Santiago

Chiriquí

Soná

PENÍNSULA

DE AZUERO

ISLA COIBA

Bahía Mon

ISLA CEBACO

PUNTA MALA

PUNTA MARIATO

ISLA JICARÓN

Nombre
de Dios El
PUNTA MANZANILLO Porvenir PUNTA SAN BLAS

Portobelo Mandinga *Golfo de San Blas*

Colón

Silver City

Gatún

*Lago
Gatún*

North Gamboa

A

Chepo

C. Brewster
3018

CORD. DE SAN BLAS

Balboa Heights

Balboa **Panamá**

Chorrera *Bahía de Panamá*

Chepo

M

PUNTA CHAME

ISTMO DE PANAMÁ

N

Bejuco

ARCHIPIÉLAGO
DE LAS PERLAS

SERRANIA DEL DARIEN

Penonomé

Antón

Río Hato

Natá

ISLA DE SAN JOSÉ

San Miguel

ISLA
DEL REY

*Bahía
San Miguel*

La Palma

CABO
TIBURÓN

Aguadulce

*Golfo
de Parita*

Chitré Los Santos

Río de Jesús

Las Tablas

PENÍNSULA

Golfo de Panamá

PUNTA GARACHINÉ

San Miguel

El Real

Garachiné

COLOMBIA

Relief

Meters	Feet
3050	10 000
1525	5000
610	2000
305	1000
152.5	500
Sea Level	
152.5	500
1525	5000
3050	10 000

b

Longitude West of Greenwich

ANGUILLA
(Br.)
ST. MARTIN
(Neth. and Fr.)

ST. BARTHÉLEMY
(Fr.)

SABA
(Neth.)

Codrington BARBUDA

ST. EUSTATIUS
(Neth.) ST KITTS

Mt. Misery
3792
Basseterre

ST. KITTS AND NEVIS

Charlestown Nevis Peak
3596

NEVIS

St. Johns

Boggy Peak
1319

**ANTIGUA
AND
BARBUDA**

REDONDA

L E E W A R D

MONTSERRAT
(Br.)

Plymouth Chances Pk.
3000

Guadeloupe Passage

POINTE DE
LA GRANDE VIGIE

GRANDE TERRE

Ste. Rose Le Moule

Pointe-à-Pitre DÉSIRADE
(Fr.)

Ste. Anne PETITE TERRE
(Fr.)

BASSE TERRE

Soufrière
4813

GUADELOUPE
Capesterre (Fr.)

Basse Terre

MARIE GALANTE

Grand Bourg
(Fr.)

LES SAINTES IS.

I S .

Portsmouth Morne Diablotins
4747

St. Joseph **DOMINICA**

Roseau

Dominica Channel

Mt. Pelée (Vol.)
4583 Trinité

St. Pierre Pitons du Carbet
3960

Fort-de-France Le François

MARTINIQUE
(Fr.)

Le Marin

POINTE D'ENFER

St. Lucia Channel

Castries

Morne Gimie
3117 **ST. LUCIA**

Soufrière

St. Vincent Passage

I S .

W I N D W A R D

Soufrière
4048

**ST. VINCENT
AND THE
GRENADINES**

Kingstown

BEQUIA

MUSTIQUE

CANOUAN

THE GRENADINES

©RMCN

CARRIACOU

Mt. St. Catherine
2757 Grenville

St.
George's **GRENADA**

C A R I B B E A N   S E A

BARBADOS

NORTH POINT

Mt. Hillaby
1115 Bathsheba

Bridgetown

SOUTH POINT

A T L A N T I C   O C E A N

Same scale as main map

---

0 20 40 60 80 100 120 Miles

0 20 40 60 80 100 120 140 160 180 200 Kilometers

HAVANA
(La Habana)

Scale 1:1 000 000

GULF OF MEXICO

Cojimar
Playa de Guanabo
Guanabacoa
Regla
Campo Florido
Playa de Santa Fé
San Francisco de Paula
Baracoa
Marianao
Cotorro
Arroya Arena
Rancho Bayeros
Cuatro Caminos
Calabazar
Managua
Bauta
Caimito del Guayabal
Santiago de las Vegas
San José de las Lajas
Bejucal
La Sabina
Buenaventura
Ceiba del Agua
San Antonio de los Baños
San Antonio de las Vegas
△ 950
©RMcN

ATLANTIC

OCEAN

Tropic of Cancer

JAMES PT.
Governor's Harbour
PALMETTO PT.
ELEUTHERA
Arthur's Town
NORTHEAST PT.
LITTLE SAN SALVADOR
CAT
Old Bight
HAWKS NEST PT.
COLUMBUS PT.
SAN SALVADOR
(WATLING)
(Columbus, Oct. 12, 1492)
SOUTHWEST PT.
GREAT GUANA CAY
CONCEPTIÓN
LEE STOCKING
Rolleville
CAPE STA MARIA
RUM CAY
GREAT EXUMA
George Town
HOG CAY
LITTLE EXUMA
LONG
SAMANA OR ATWOOD CAY
Clarence Town
JUMENTO CAYS
WATER CAY
CAP VERDE
BIRD ROCK
CROOKED
NORTHEAST PT.
FLAMINGO CAY
PLANA OR FLAT CAYS
Man of War Channel
JAMAICA CAY
FORTUNE
The Bight of Acklins
SEAL CAYS
DIANA BANK
FISH CAY
ACKLINS
MAYAGUANA
NURSE CAY
SALINA PT.
RACCOON CAY
CASTLE
OCHINOS BANKS
GREAT RAGGED
MIRA POR VOS ISLETS
COLUMBUS BANK
CAY VERDE
HOGSTY REEF
CAY STA. DOMINGO
BROWN BANK
LITTLE INAGUA
NORTHEAST PT.
PALMETTO PT.
Ocean Bight
The Lake
WEST SAND SPIT
Man of War Bay
Matthew Town
GREAT INAGUA
South Bay

CAICOS Passage
PROVIDENCIALES
NORTH CAICOS
GRAND CAICOS
CAPE COMETE
EAST CAICOS
WEST CAICOS
CAICOS IS.
(Br.)
SOUTH CAICOS
GRAND TURK
Grand Turk
TURKS IS. (Br.)
SALT CAY
AMBERGRIS CAYS
SEAL CAYS
Turks I. Passage
MOUCHOIR Passage
MOUCHOIR BANK
Silver Bank Passage
SILVER BANK
NAVIDAD BANK

Gibara
CABO LUCRECIA
Banes
Holguín
Bahía de Nipe
Antilla
OLGUÍN
Mayari
Sagua de Tánamo
CUCHILLAS DE TOA
Baracoa
SANTIAGO DE CUBA
GUANTANAMO
DE PURIAL
PUNTA MAISÍ
Soriano
Alto Songo
Bahía de Ovando
San Luis
Coney
Gran Piedra
Guantanamo
ILE DE LA TORTUE
CABO ISABELA
Caimanera
Yateras
Canal de la Tortue
Monte Cristi
Puerto Plata
ESTRA
Santiago de Cuba
Naval Station (U.S.A.)
Port de Paix
CORDILLERA SEPTENTRIONAL
CABO FRANCÉS VIEJO
Bahía de Guantánamo
Windward Passage
CAP ST. NICOLAS
Le Môle
Le Borgne
Cap-Haïtien
Guayubin
Puerto Plata
PTE. PLATEFORME
Imbé
Fort Liberté
Dajabón
Santiago Rodríguez
Gasper Hernández
Bahía Escocesa
Grande Rivière du Nord
Ouanaminthe
Moca
Nagua
Gonaïves
St-Michel de l'Atalaye
Vallière
Santiago de los Caballeros
La Vega
Sánchez
CABO SAMANA
GOLFE DES GONAÏVES
Hinche
DOMINICAN
Samaná
Bahía de Samaná
St. Marc
Bonhomme
CORDILLERA
Riva
CABO SAN RAFAEL
HAITI
Bánica
Pico Duarte
Cotui
Sabana de la Mar
Miches
POINT OUEST
Mirebalais
San Juan
CENTRAL
Miches
ÎLE DE LA GONÂVE
Lascahobas
Mte. Tina
Yamasá
CORDILLERA ORIENTAL
Jérémie
CAP DAME MARIE
Petit Goâve
Port-au-Prince
Pétionville
SIERRA DE NEIBA
Hato Mayor
Bayaguana
Los Llanos
Seibo
CAP DES IROIS
Miragoane
Léogane
Neiba
REPUBLIC
Higuey
FORMIGAS BANK
Anse d'Hainault
Lago Enriquillo
Azua
San Cristóbal
Santo Domingo
La Romana
NAVASSA (U.S.A.)
Pico de Moya
MASSIF DE LA HOTTE
Aquin
MASSIF DE LA SELLE
Barahona
PTA. PALENQUE
CATALINA
Tiburon
Roche à Bateau
Les Cayes
Jacmel
Belle-Anse
SIERRA DE BAHORUCO
Bahía de Neiba
S. Pedro de Macorís
SAONA
HISPANIOLA
Port Antonio
ÎLE À VACHE
Enriquillo
POINTE À GRAVOIS
Oviedo
CABO FALSO
MORANT PT.
CABO BEATA
ALTO VELO
BEATA

0 10 20 30 40 50 60 70 80 90 100 110 120 Miles
0 20 40 60 80 100 120 140 160 180 200 Kilometers

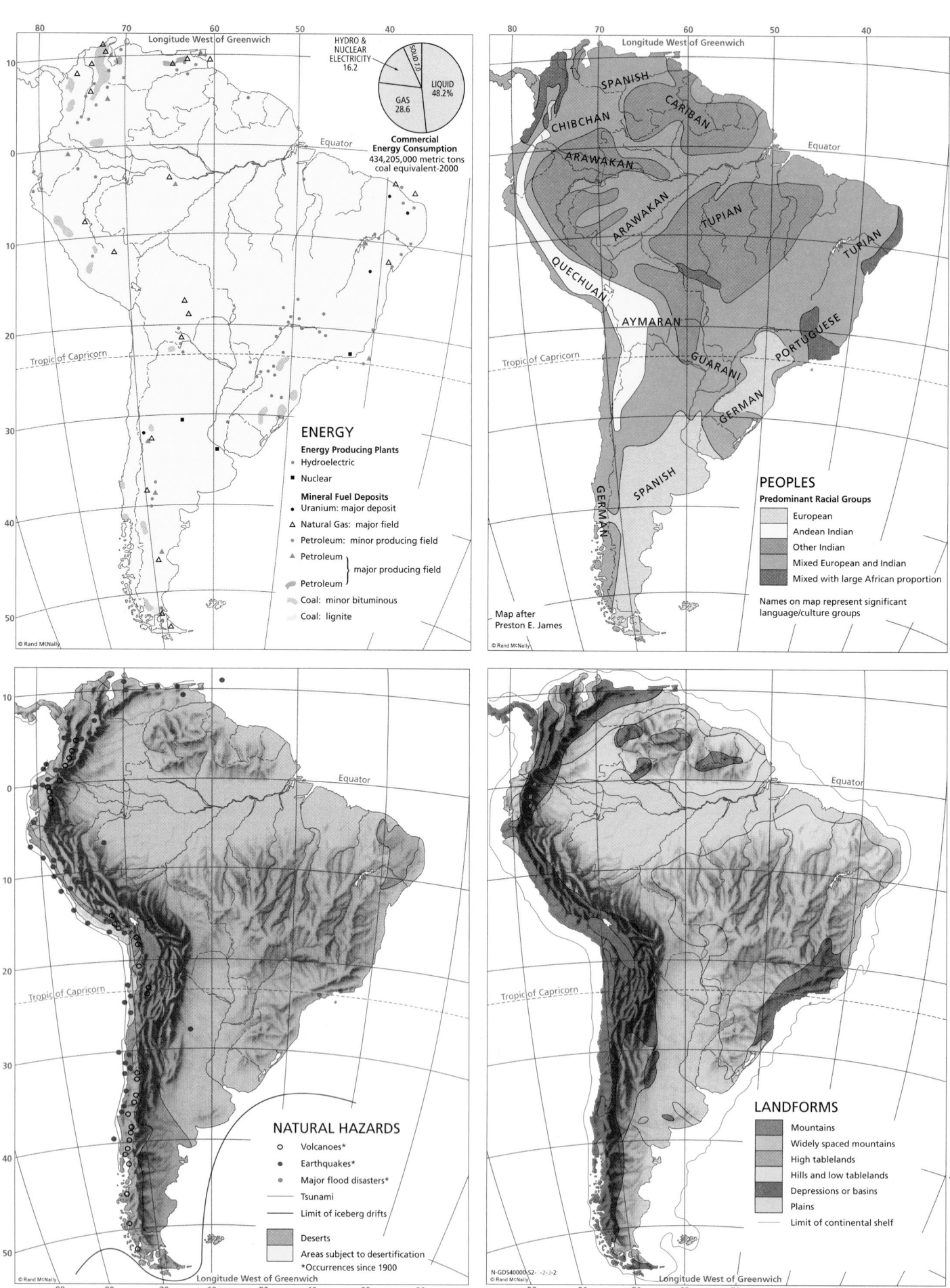

ENERGY

Commercial Energy Consumption
434,205,000 metric tons
coal equivalent-2000

HYDRO & NUCLEAR ELECTRICITY 16.2
SOLID 7.0
LIQUID 48.2%
GAS 28.6

Energy Producing Plants
- Hydroelectric
■ Nuclear

Mineral Fuel Deposits
• Uranium: major deposit
△ Natural Gas: major field
• Petroleum: minor producing field
▲ Petroleum }
Petroleum } major producing field
Coal: minor bituminous
Coal: lignite

PEOPLES

Predominant Racial Groups
- European
- Andean Indian
- Other Indian
- Mixed European and Indian
- Mixed with large African proportion

Names on map represent significant
language/culture groups

Map after
Preston E. James

SPANISH
CHIBCHAN
CARIBAN
ARAWAKAN
ARAWAKAN
QUECHUAN
TUPIAN
TUPIAN
AYMARAN
GUARANI
PORTUGUESE
GERMAN
GERMAN
SPANISH

NATURAL HAZARDS

○ Volcanoes*
● Earthquakes*
● Major flood disasters*
— Tsunami
— Limit of iceberg drifts
Deserts
Areas subject to desertification
*Occurrences since 1900

LANDFORMS

- Mountains
- Widely spaced mountains
- High tablelands
- Hills and low tablelands
- Depressions or basins
- Plains
— Limit of continental shelf

N-GDS40000-S2- -2-2-2
© Rand McNally

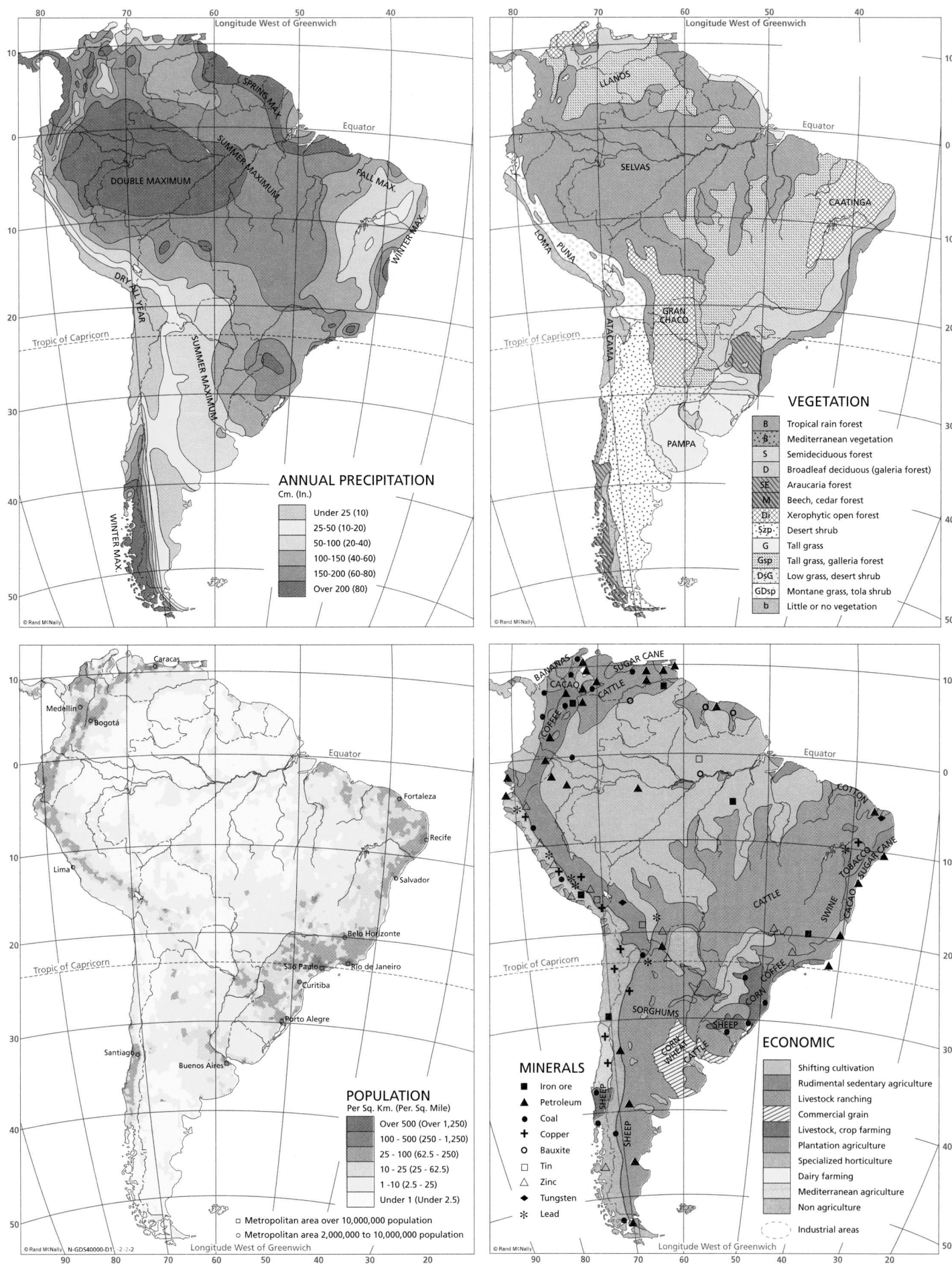

ANNUAL PRECIPITATION
Cm. (In.)

	Under 25 (10)
	25-50 (10-20)
	50-100 (20-40)
	100-150 (40-60)
	150-200 (60-80)
	Over 200 (80)

DOUBLE MAXIMUM
SPRING MAX.
SUMMER MAXIMUM
FALL MAX.
WINTER MAX.
DRY ALL YEAR
SUMMER MAXIMUM
WINTER MAX.
Equator
Tropic of Capricorn

VEGETATION

B	Tropical rain forest
B	Mediterranean vegetation
S	Semideciduous forest
D	Broadleaf deciduous (galeria forest)
SE	Araucaria forest
M	Beech, cedar forest
Di	Xerophytic open forest
Szp	Desert shrub
G	Tall grass
Gsp	Tall grass, galleria forest
DsG	Low grass, desert shrub
GDsp	Montane grass, tola shrub
b	Little or no vegetation

LLANOS
SELVAS
CAATINGA
LOMA
PUNA
ATACAMA
GRAN CHACO
PAMPA

POPULATION
Per Sq. Km. (Per. Sq. Mile)

	Over 500 (Over 1,250)
	100 - 500 (250 - 1,250)
	25 - 100 (62.5 - 250)
	10 - 25 (25 - 62.5)
	1 -10 (2.5 - 25)
	Under 1 (Under 2.5)

☐ Metropolitan area over 10,000,000 population
○ Metropolitan area 2,000,000 to 10,000,000 population

Caracas
Medellín
Bogotá
Lima
Santiago
Fortaleza
Recife
Salvador
Belo Horizonte
São Paulo
Rio de Janeiro
Curitiba
Porto Alegre
Buenos Aires

MINERALS

■	Iron ore
▲	Petroleum
●	Coal
+	Copper
○	Bauxite
☐	Tin
△	Zinc
◆	Tungsten
✳	Lead

ECONOMIC

	Shifting cultivation
	Rudimental sedentary agriculture
	Livestock ranching
	Commercial grain
	Livestock, crop farming
	Plantation agriculture
	Specialized horticulture
	Dairy farming
	Mediterranean agriculture
	Non agriculture
	Industrial areas

BANANAS
CACAO
COFFEE
SUGAR CANE
CATTLE
COTTON
TOBACCO
CACAO SUGAR CANE
SWINE
CATTLE
COFFEE
CORN
SHEEP
SORGHUMS
CORN WHEAT
CATTLE
SHEEP

© Rand McNally

40,000 SQ MI AREA

0   300   600
Miles

Scale 1:40 000 000; one inch to 630 miles. Lambert's Azimuthal, Equal Area Projection
Elevations and depressions are given in feet

Scale 1:40 000 000; one inch to 630 miles. Lambert's Azimuthal, Equal Area Projection
Elevations and depressions are given in feet

Relief

Meters		Feet
3050		10 000
1525		5000
610		2000
305		1000
0	Sea Level	0
152.5		500
1525		5000
3050		10 000
6100		20 000

20°
80° 70° 60° 50° 40°
CUBA
JAMAICA San Juan
Kingston HISPANIOLA PUERTO
RICO
Caribbean Sea
ATLANTIC
OCEAN
Barranquilla
Panamá Maracaibo CARACAS Port of Spain
TRINIDAD
LLANOS *Orinoco*
Georgetown
BOGOTÁ
Quito *Negro* Equator
Belém
Amazon Manaus
Iquitos
S E L V A S Fortaleza
A N D E S
Rio Branco
São Francisco
Recife
LIMA
La Paz Cuiaba
M A T O Brasília
Iquique G R O S S O
Belo Horizonte
C H A C O *Paraná*
Tropic of Capricorn G R A N SÃO
PAULO RIO DE JANEIRO
Asunción
San Miguel
de Tucumán
Porto Alegre
Córdoba
PACIFIC SANTIAGO BUENOS AIRES
ATLANTIC
Montevideo
OCEAN P A M P A *OCEAN*
Bahía Blanca
P A T A G O N I A Puerto Montt
FALKLAND
ISLANDS
Punta Arenas TIERRA
DEL FUEGO
SOUTH
GEORGIA
Drake Passage

| Urban |
| Cropland |
| Cropland & Woodland |
| Cropland & Grazing Land |
| Grassland, Grazing Land |
| Forest, Woodland |
| Swamp, Marshland |
| Shrub, Sparse Grass, Wasteland |
| Barren Land |

A-540000-36
COPYRIGHT BY
RAND McNALLY & COMPANY
MADE IN U.S.A.

Scale 1:36,000,000; one inch to 570 miles Lambert Azimuthal Equal-Area Projection

0 100 200   400   600   800 Miles
0 150 300   600   900   1200 Kilometers

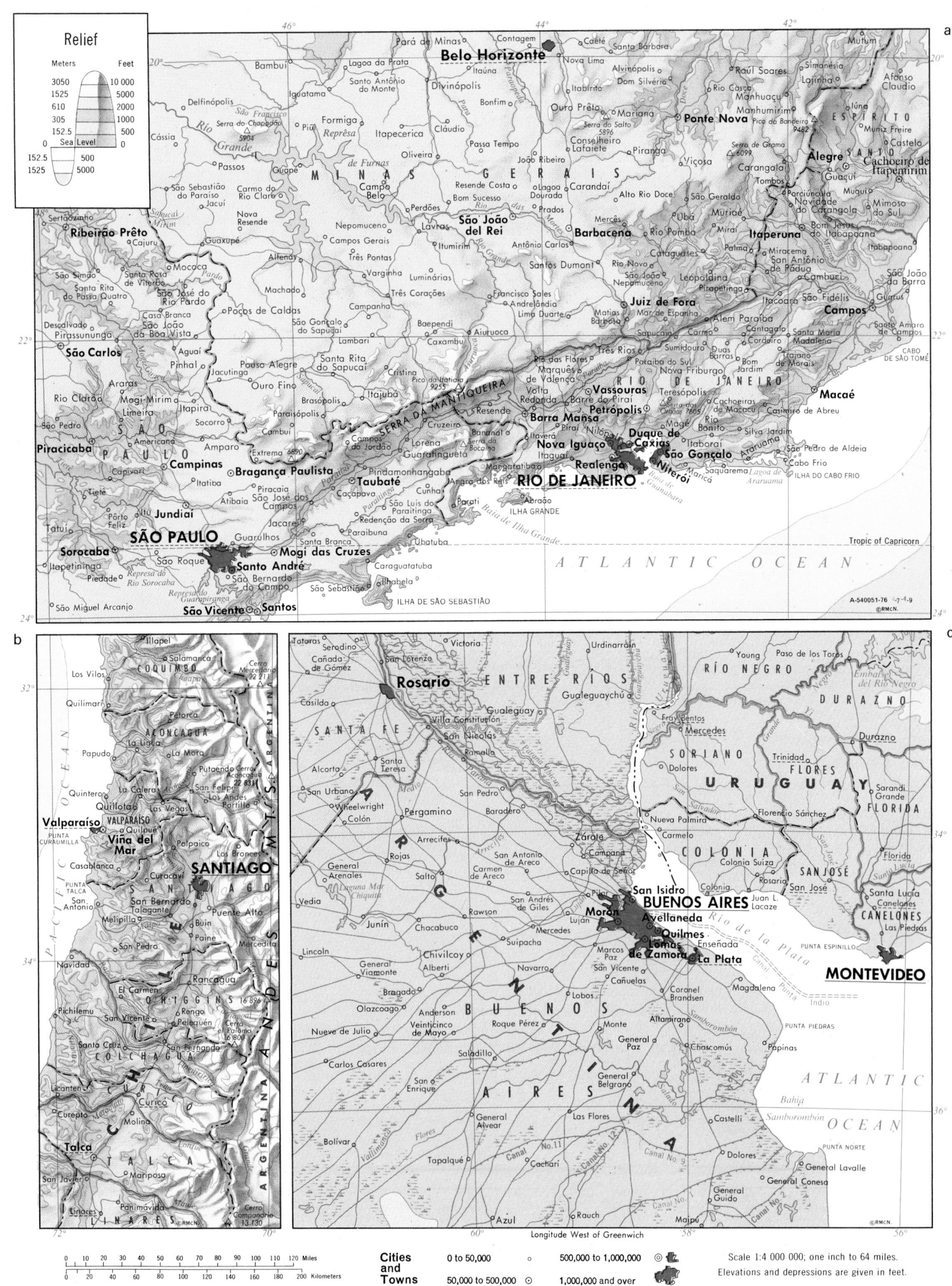

a

Relief

Meters	Feet
3050	10 000
1525	5000
610	2000
305	1000
152.5	500
Sea Level	0
152.5	500
1525	5000

b

c

Cities and Towns

0 to 50,000
50,000 to 500,000
500,000 to 1,000,000
1,000,000 and over

Scale 1:4 000 000; one inch to 64 miles.
Elevations and depressions are given in feet.

Longitude West of Greenwich

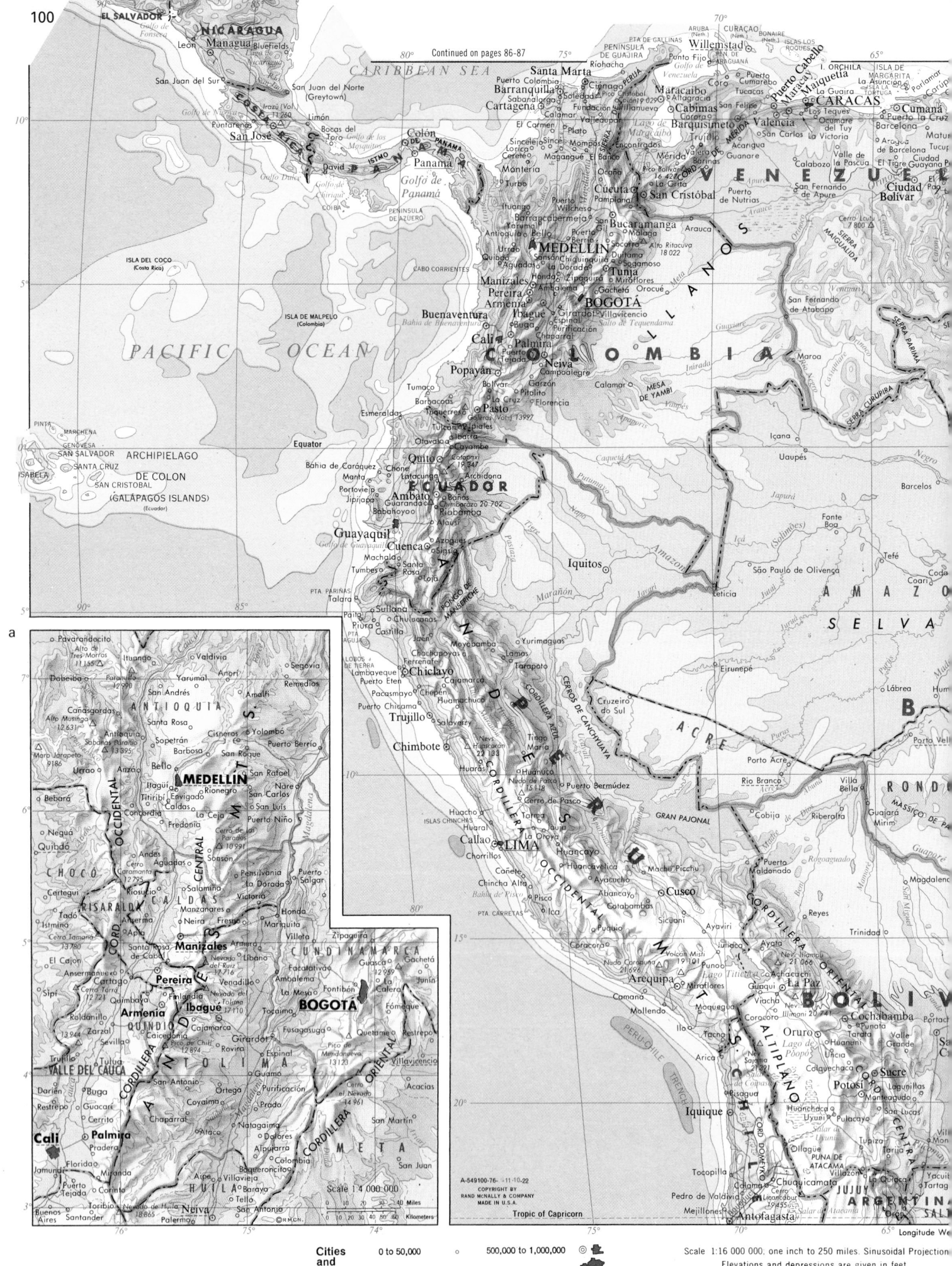

ATLANTIC OCEAN

Inset map (top right):

CARIBBEAN SEA

Tocuyo de la Costa
Chichiriviche
Cayo Sombrero
Tucacas
Golfo Triste
Maiquetía · Guaira · Naiguatá · La Sabana
Carayaca
Puerto Cabello
Morón
El Cambur
San Joaquín
Montalbán · Guacara
Miranda · Guacara
Valencia
Güigüe
Villa de Cura
San Sebastián
Tinaquillo
COJEDES
CARABOBO
Lago de Valencia
CARACAS
Petare · Santa Lucia
Maracay
La Victoria
Cagua
Ocumare del Tuy
San Francisco de Macaira
MIRANDA
Río Chico
Higuerote
Boca de Uchire
Laguna de la Tacarigua
El Hatillo
Puerto Píritu
Clarines · San Miguel
El Pilar
Barcelona
Bergantín
ANZOÁTEGUI
San Mateo
Santa Rosa
GUÁRICO
Dos Caminos
Barbacoas
Libertad de Orituco
Aragua de Barcelona

ISLA DE MARGARITA
Boca del Pozo
Punta de Piedras
NUEVA ESPARTA
ISLA CUBAGUA
LA TORTUGA
ISLA LA BORRACHA
Punta de Araya
Cumaná
SUCRE
Manicuare
Las Vegas
Puerto La Cruz
Guanta

Scale 1:4 000 000
Miles 0 10 20 30 40
Kilometers 0 10 20 30 40 50 60
©RMCN.

Main map:

TRINIDAD AND TOBAGO
TOBAGO
Port of Spain
TRINIDAD
Boca Grande
Morawhanna
Georgetown
Bartica · Rosignol · New Amsterdam
Wismar
Rockstone
Skeldon
Nieuw Nickerie · Paranam
Totness · Paramaribo
Maengo · Albina · St. Laurent
SURINAME
FRENCH GUIANA
ILE DU DIABLE (DEVIL'S I.)
Cayenne
Saint-Georges
CABO ORANGE

MERUME MTS.
WILHELMINA GEBERGTE
ACARAI MTS.
TUMUC-HUMAC MTS.

AMAPÁ
Amapá
Macapá
Mazagão

Manaus (Manáos)
Itacoatiara
ILHA TUPINAMBARANAS
Maués
Borba
Parintins
Óbidos · Alenquer
Faro
Santarém
Altamira
Itaituba
Brasília Legal (Fordlândia)

PARÁ
Belém (Pará)
Abaetetuba
Cametá
Breves
ILHA DE MARAJO
Gurupá
Curuçá
Bragança
Marapanim
ILHA CAVIANA

São Luís (Maranhão)
Alcântara
Rosário
Viana
Tutóia
Parnaíba
Camocim
Acaraú
FORTALEZA (Ceará)
Maranguape
Baturité
Sobral
Ipu
Crateús
Russas
Aracati
Areia Branca
Macau
Ceará-Mirim
Natal
Nova Cruz
RIO GRANDE DO NORTE
Currais Novos

MARANHÃO
Teresina
Barra do Corda
Grajaú
Codó
Caxias
Pedreiras
Campo Maior
Quixadá
CEARÁ
Senador Pompeu
Iguatu
Pedro II

Tucuruí
SERRA DOS CARAJÁS
São João do Araguaia
Araguaína
Tocantinópolis
SERRA DO GURUPÍ
Miradora
Loreto
Riachão
Carolina
Balsas
Floriano
Oeiras
Picos
Amarante

BRAZIL
SERRA DO RONCADOR
SERRA DO ESTRONDO
CHAP. DAS MANGABEIRAS
Santa Filomena
São Raimundo Nonato
Paulistana
Granito
Cabrobó

TOCANTINS
Miracema do Tocantins
Porto Nacional
Palmas
Natividade
Parnaguá
Barra
Correntina
Carinhanha
Barreiras

Juazeiro do Norte
Crato
Flores
Sertânia
PLANALTO DA BORBOREMA
Caruaru
PERNAMBUCO
Campina Grande
João Pessoa (Paraíba)
Nazaré da Mata
Jaboatão
Olinda
RECIFE (Pernambuco)
Palmares
Pôrto de Pedras
Maceió
ALAGOAS
Penedo
SERGIPE
Propriá
Aracaju
Coruripe

São Francisco
Petrolina
Juazeiro
Jeremoabo
Senhor do Bonfim
Jacobina
Serrinha
Feira de Santana
Catú
Santo Amaro
SALVADOR (Bahia)
BAHIA
Morro do Chapéu
Lençóis
Cachoeiro
Nazaré
Valença
Itabuna
Ilhéus
Mucugé
Caetité
Condeúba
Vitória da Conquista
Jequié

SERRA DO NORTE
SERRA DO TOMBADOR
CHAPADA DE MATO GROSSO
SERRA DOS PARECIS
MATO GROSSO
Diamantino
SERRA DA CHAPADA
SA. DA TAQUARA
Cuiabá
Cáceres
Barão de Melgaço
Rosário Oeste
Mato Grosso

GRAN CHACO
PARAGUAY
Bahía Negra
Fuerte Olimpo
Porto Murtinho
Mariscal Estigarribia
Puerto Casado
Concepción
Belén
Horqueta
Ponta Porã

MATO GROSSO DO SUL
Aquidauana
Corumbá
Campo Grande
Três Lagoas
Presidente Epitácio
Bela Vista

GOIÁS
Goiás
Pirenópolis
Anápolis
D.F. Brasília
Formosa
Goiânia
Luziânia
Silvânia
Bela Vista de Goiás
Cavalcante
SERRA GERAL DE GOIÁS
Januária
São Francisco
Pedra Azul

MINAS GERAIS
Paracatu
Patos de Minas
Araguari
Uberlândia
Uberaba
Araxá
Ituiutaba
Pará de Minas
SA. DE CANASTRA
Formiga
Divinópolis
Sete Lagoas
BELO HORIZONTE
Curvelo
Diamantina
Teófilo Otoni
Gov. Valadares
Governador Valadares
Caravelas
Pirapora
Montes Claros
Grão Mogol
Araçuaí
SERRA DO ESPINHAÇO
Pôrto Seguro
Belmonte
Canavieiras

SÃO PAULO
São José do Rio Prêto
Barretos
Franca
Ribeirão Prêto
Araçatuba
Araraquara
São Carlos
Marília
Bauru
Piracicaba
Campinas
Jundiaí
Sorocaba
SÃO PAULO
Santos
São Vicente
Mogi das Cruzes
Jundiaí
Itapetininga

ESPÍRITO SANTO
Aracruz
Cachoeiro do Itapemirim
Campos
RIO DE JANEIRO
Niterói
Nova Friburgo
Petrópolis
Juiz de Fora
Barbacena
Lafaiete
RIO DE JANEIRO
CABO FRIO
ARQUIPÉLAGO DOS ABROLHOS

PARANÁ
Londrina
Porto Mendes
Tibagi
Castro
Guarapuava
Ponta Grossa
Curitiba

Equator
Tropic of Capricorn

Continued on page 102

Relief

Meters	Feet
3050	10 000
1525	5000
610	2000
305	1000
152.5	500
Sea Level	0
152.5	500
1525	5000
3050	10 000
6100	20 000

Miles 0 50 100 200 300 400 500 Miles
Kilometers 0 100 200 400 600 800 Kilometers

Continued on pages 100-101

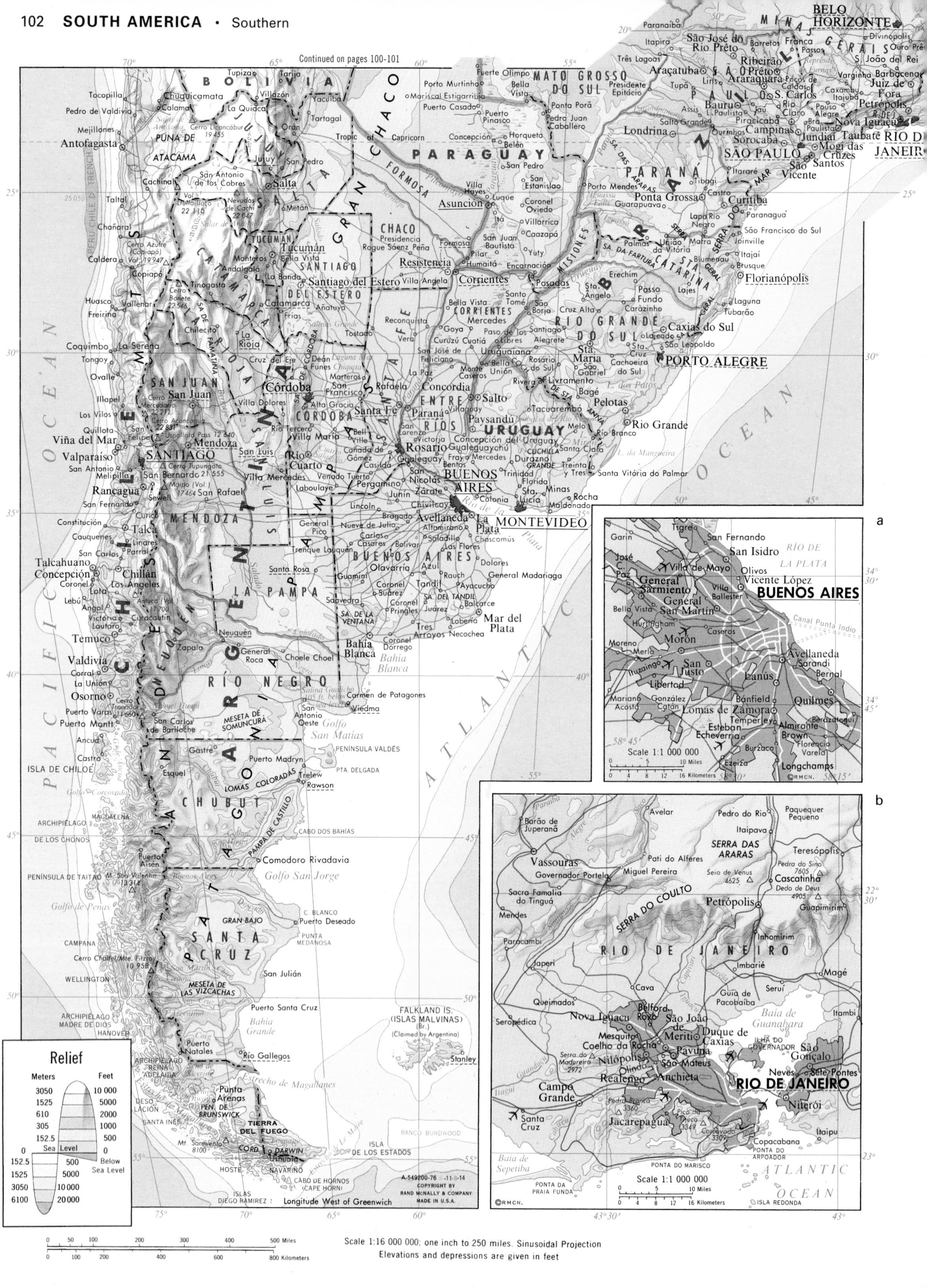

Scale 1:16 000 000; one inch to 250 miles. Sinusoidal Projection
Elevations and depressions are given in feet

Relief

Meters		Feet
3050		10 000
1525		5000
610		2000
305		1000
152.5		500
0	Sea Level	Sea Level
152.5		500
1525		5000
3050		10 000
6100		20 000
		Below Sea Level

a

BUENOS AIRES

Scale 1:1 000 000

b

RIO DE JANEIRO

Scale 1:1 000 000

A-549200-76 -11-9-14
COPYRIGHT BY
RAND McNALLY & COMPANY
MADE IN U.S.A.

Longitude West of Greenwich

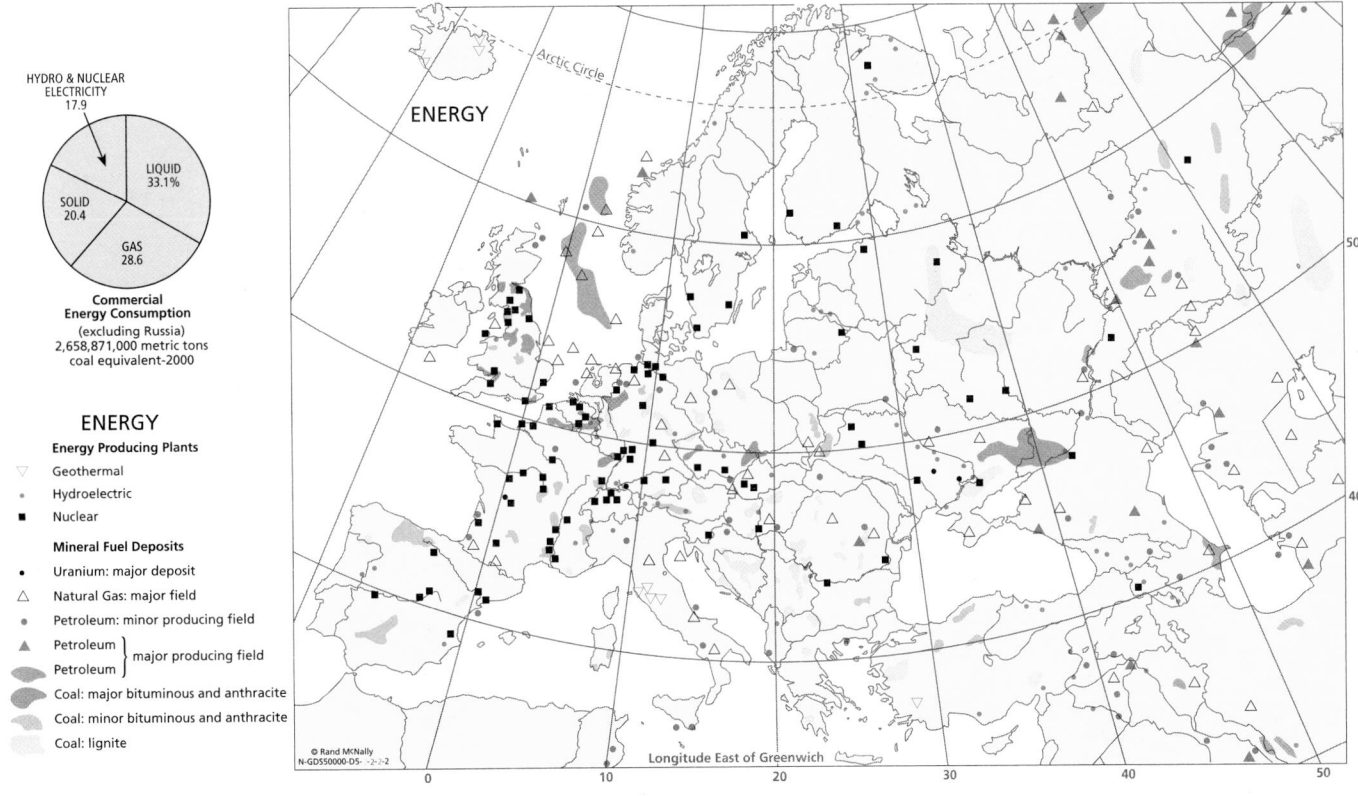

ENERGY

HYDRO & NUCLEAR
ELECTRICITY
17.9

LIQUID
33.1%

SOLID
20.4

GAS
28.6

**Commercial
Energy Consumption**
(excluding Russia)
2,658,871,000 metric tons
coal equivalent-2000

ENERGY

Energy Producing Plants

▽ Geothermal

• Hydroelectric

■ Nuclear

Mineral Fuel Deposits

• Uranium: major deposit

△ Natural Gas: major field

• Petroleum: minor producing field

▲ Petroleum } major producing field

     Petroleum }

     Coal: major bituminous and anthracite

     Coal: minor bituminous and anthracite

     Coal: lignite

© Rand McNally
N-GDS50000-D5- -2-/-/-2

Longitude East of Greenwich

NATURAL HAZARDS

NATURAL HAZARDS

○ Volcanoes*

• Earthquakes*

• Major flood disasters*

—— Tsunamis

—— Limit of iceburg drift

     Temporary pack ice

     Areas subject to desertification

*Occurrences since 1900

© Rand McNally

Longitude East of Greenwich

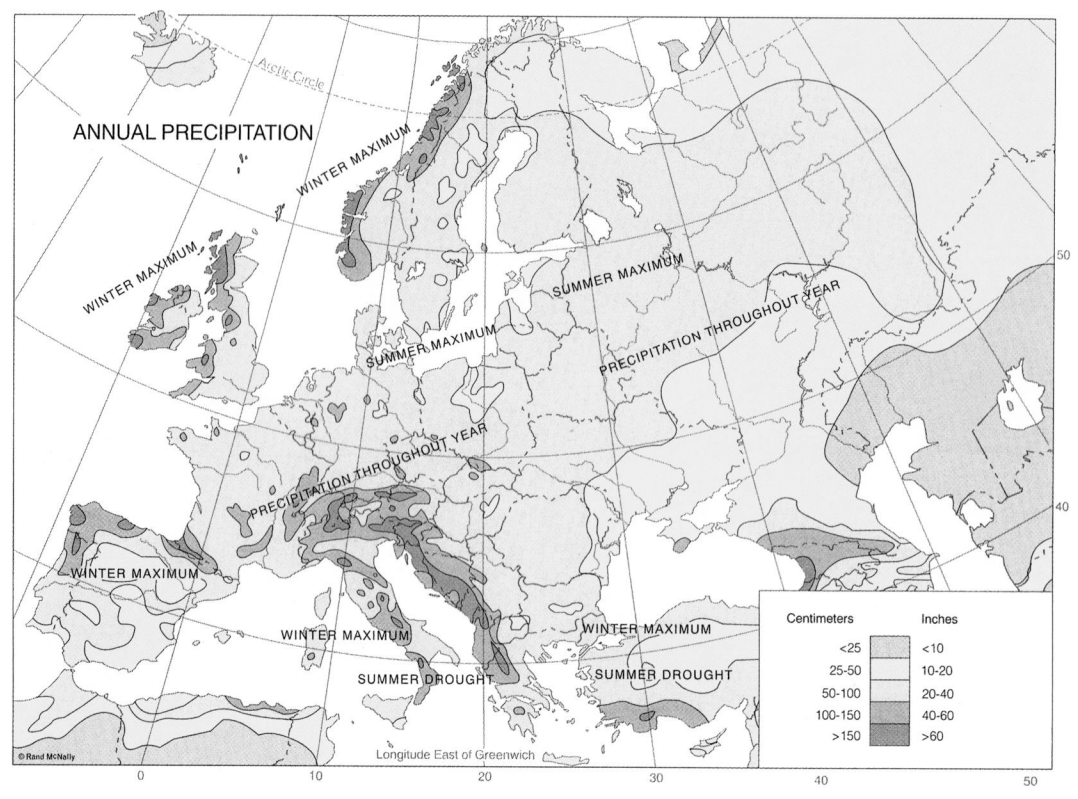

ANNUAL PRECIPITATION

Centimeters	Inches
<25	<10
25-50	10-20
50-100	20-40
100-150	40-60
>150	>60

Longitude East of Greenwich

© Rand McNally

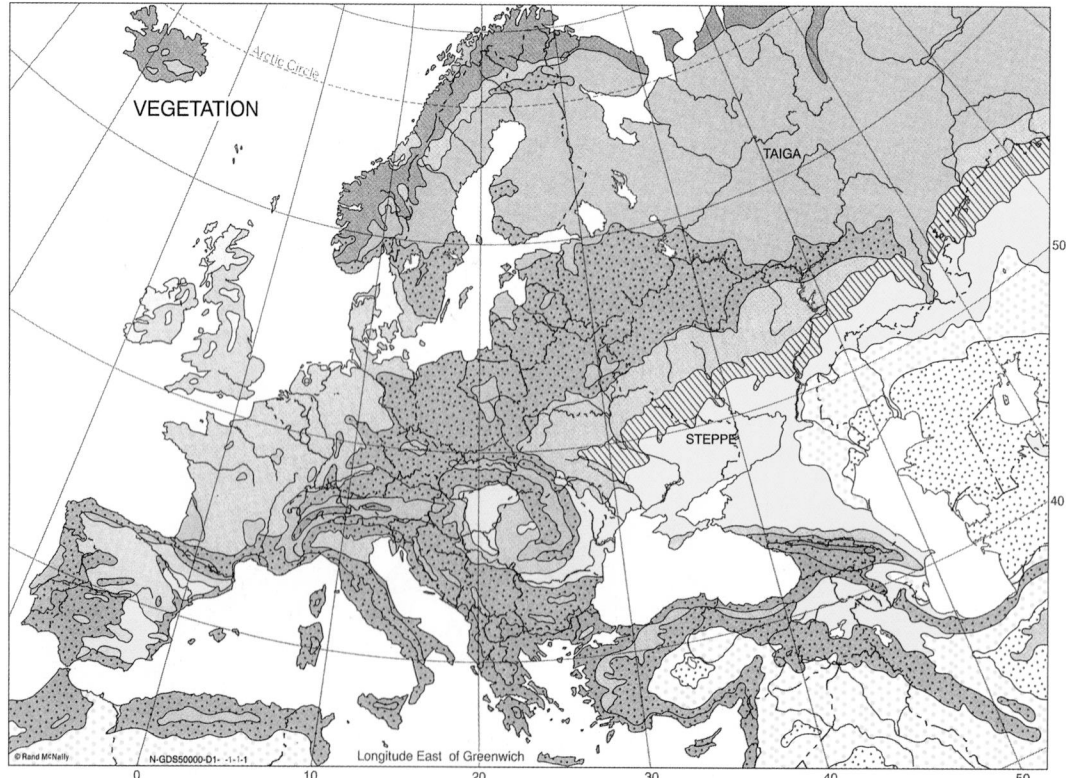

VEGETATION

TAIGA

STEPPE

Longitude East of Greenwich

© Rand McNally N-GDS50000-D1- -1-1-1

VEGETATION

E	Coniferous forest
B, Bs	Mediterranean vegetation
M	Mixed forest: coniferous-deciduous
S	Semi-deciduous forest
D	Deciduous forest
DG	Wooded steppe
G	Grass (steppe)
Gp	Short grass
Den	Desert shrub
L	Heath and moor
L	Alpine vegetation, tundra
b	Little or no vegetation

For explanation of letters in boxes,
see Natural Vegetation Map
by A. W. Kuchler, p. 24

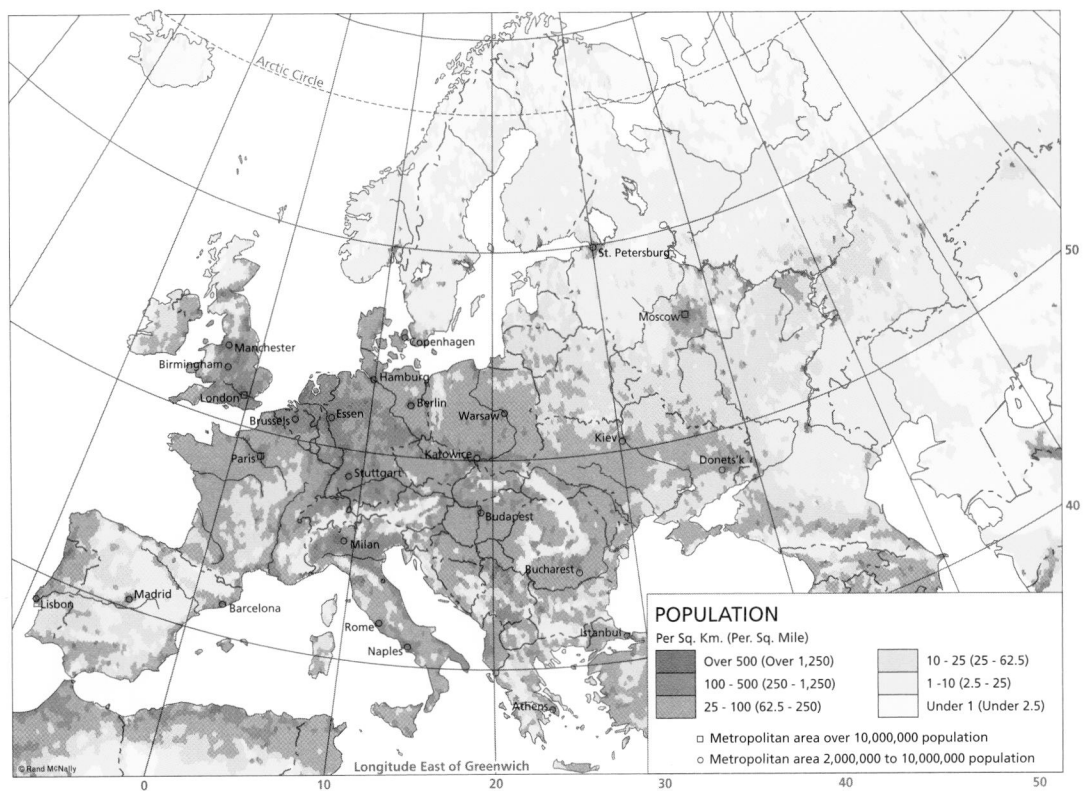

POPULATION

Per Sq. Km. (Per. Sq. Mile)

- Over 500 (Over 1,250)
- 100 - 500 (250 - 1,250)
- 25 - 100 (62.5 - 250)
- 10 - 25 (25 - 62.5)
- 1 -10 (2.5 - 25)
- Under 1 (Under 2.5)

□ Metropolitan area over 10,000,000 population
○ Metropolitan area 2,000,000 to 10,000,000 population

Longitude East of Greenwich

© Rand McNally

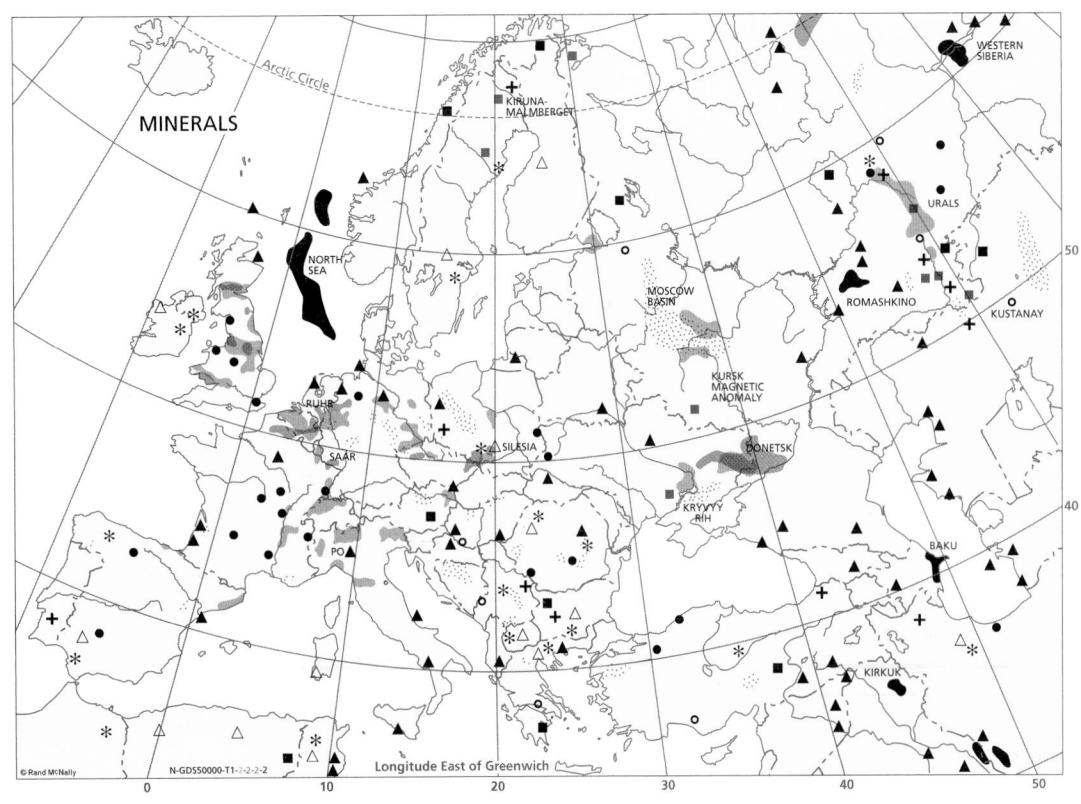

MINERALS

- Industrial areas
- Major coal deposits
- Major petroleum deposits
- Lignite deposits
- Minor petroleum deposits
- Minor coal deposits
- Major iron ore
- Minor iron ore
- ✳ Lead
- ○ Bauxite
- △ Zinc
- + Copper

© Rand McNally   N-GDS50000-T1-2-2-2-2   Longitude East of Greenwich

106

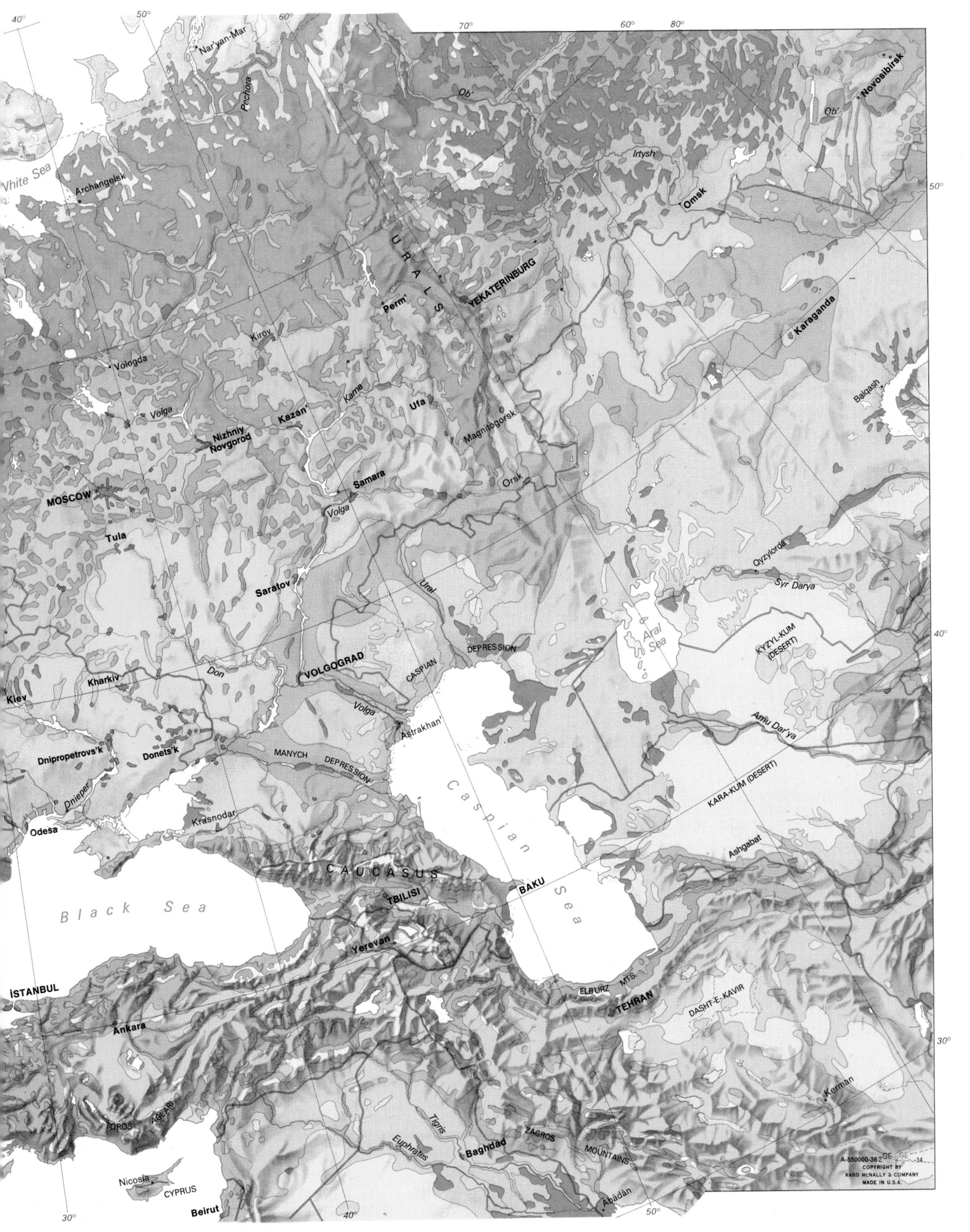

White Sea

Nar'yan-Mar

Pechora

Archangelsk

Ob'

Irtysh

Novosibirsk

Ob'

Omsk

Vologda

U R A L S

YEKATERINBURG

Perm'

Karaganda

Kiev

Kazan'

Kama

Ufa

Magnitogorsk

Balqash

Nizhniy
Novgorod

Volga

MOSCOW

Samara

Orsk

Qyzylorda

Syr Darya

Volga

Tula

Saratov

Ural

Aral
Sea

KYZYL-KUM
(DESERT)

40°

Kharkiv

DEPRESSION

CASPIAN

Kiev

Don

VOLGOGRAD

Volga

Amu Dar'ya

Dnipropetrovs'k

Donets'k

Astrakhan'

MANYCH

DEPRESSION

KARA-KUM (DESERT)

Dnieper

Odesa

Krasnodar

C a s p i a n

Ashgabat

Black   Sea

C A U C A S U S

BAKU

S e a

İSTANBUL

TBILISI

Yerevan

ELBURZ   MTS.

DASHT-E-KAVIR

Ankara

TEHRAN

30°

TOROS

AĞLARI

Tigris

ZAGROS

Kerman

Euphrates

Baghdad

MOUNTAINS

Nicosia

CYPRUS

Beirut

Abādān

A-550000-36.2

COPYRIGHT BY
RAND McNALLY & COMPANY
MADE IN U.S.A.

Scale 1:16 000 000; one inch to 250 miles. Conic Projection
Elevations and depressions are given in feet.

40,000 SQ MI
AREA

0   100   200
Miles

Scale 1: 16 000 000; one inch to 250 miles. Conic Projection

Elevations and depressions are given in feet

Longitude West of Greenwich   Longitude East of Greenwich

0 50 100   200   300   400   500 Miles

0   100   200   400   600   800 Kilometers

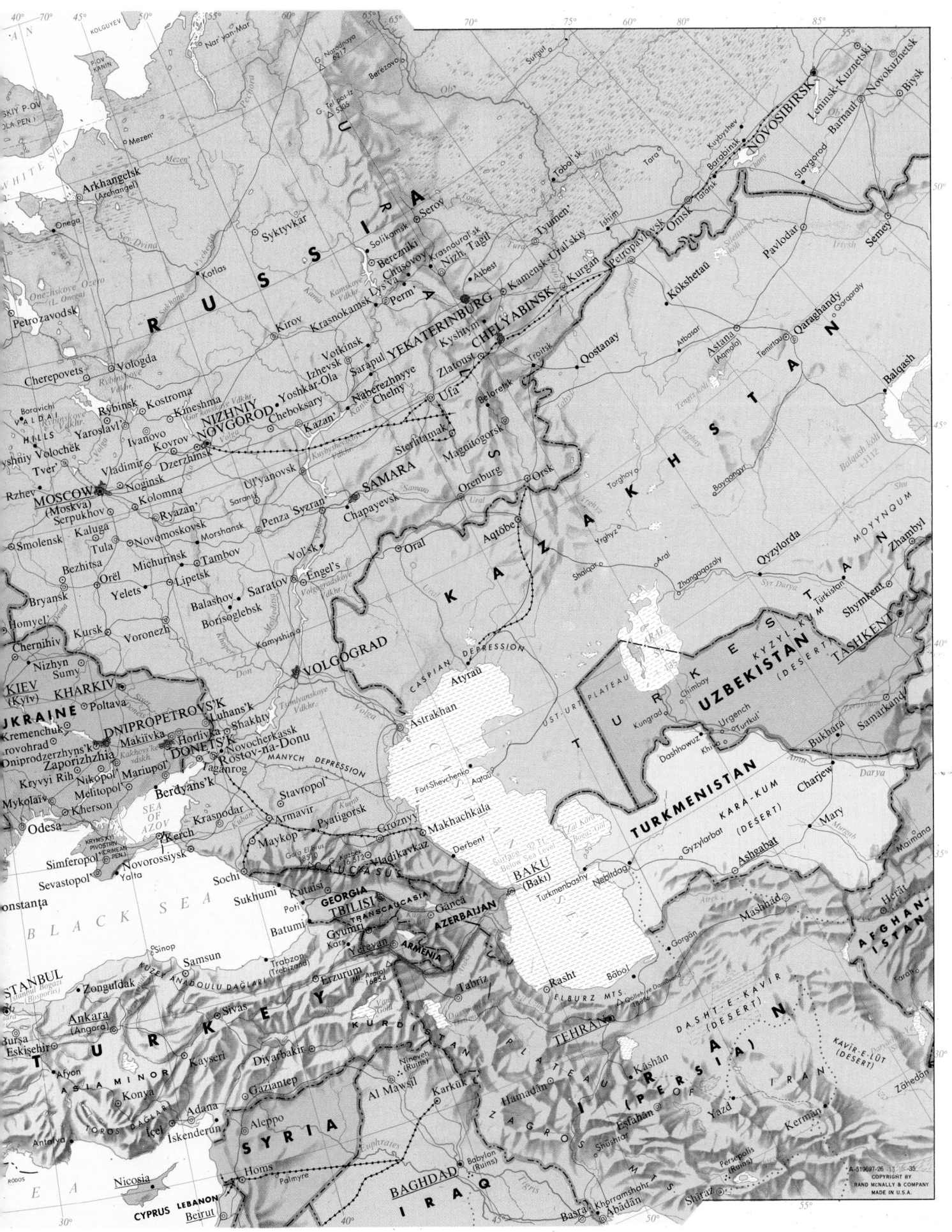

Relief

Meters		Feet
3050		10 000
1525		5000
610		2000
305		1000
152.5		500
0	Sea Level	0
152.5		500
1525		5000
3050		10 000

Below Sea Level

Scale 1: 16 000 000; one inch to 250 miles. Conic Projection

Elevations and depressions are given in feet

Continued on pages 184–185

Longitude West of Greenwich   Longitude East of Greenwich

| 0 | 50 | 100 | 200 | 300 | 400 | 500 Miles |
| 0 | 100 | 200 | 400 | 600 | 800 Kilometers |

Continued on pages 140-141

Continued on pages 154-155

A-519697-76 · 13 10-35
COPYRIGHT BY
RAND McNALLY & COMPANY
MADE IN U.S.A.

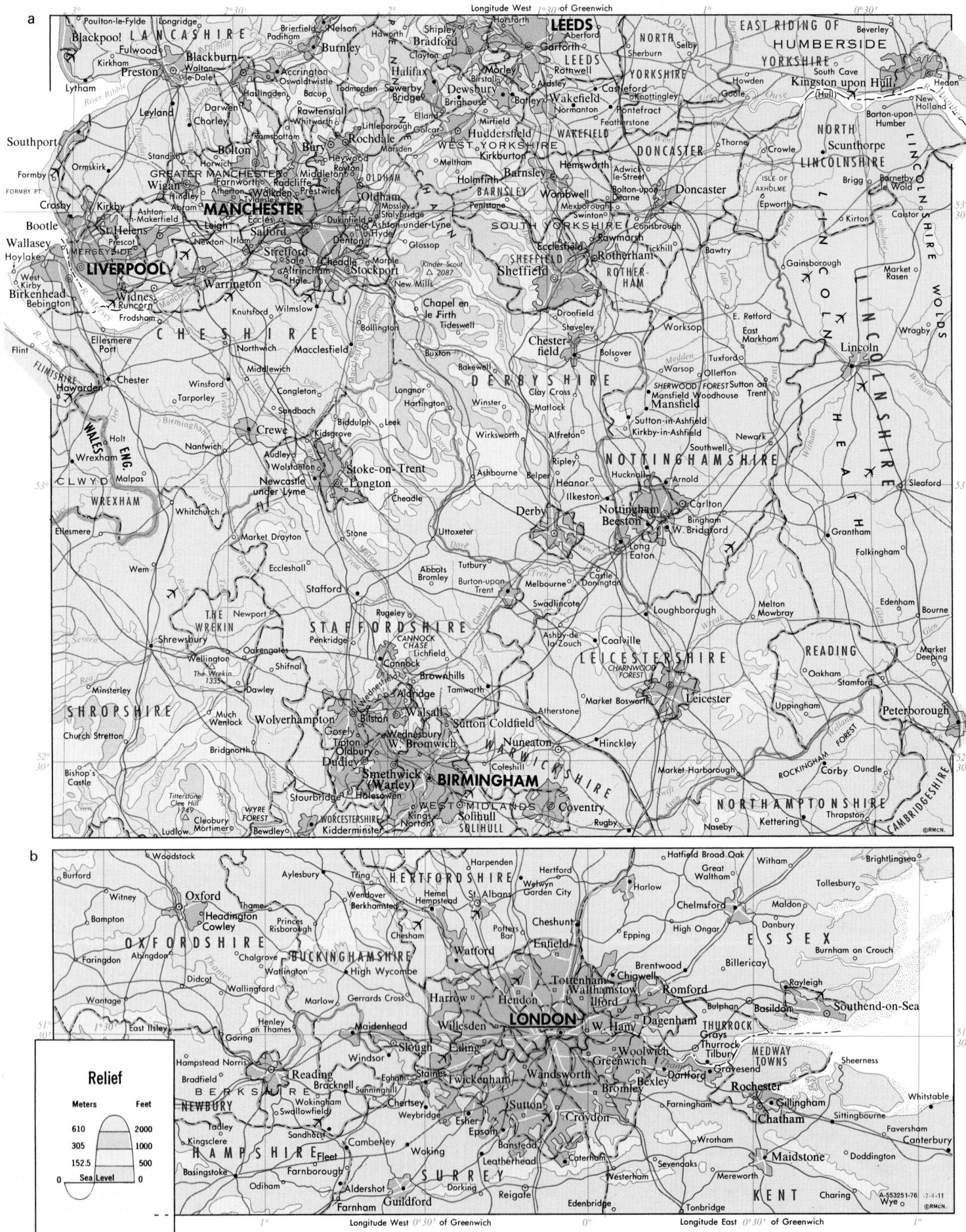

Scale 1:1 000 000; one inch to 16 miles.
Elevations and depressions are given in feet.

A-553251-76   -7-4-11
©RMcN.

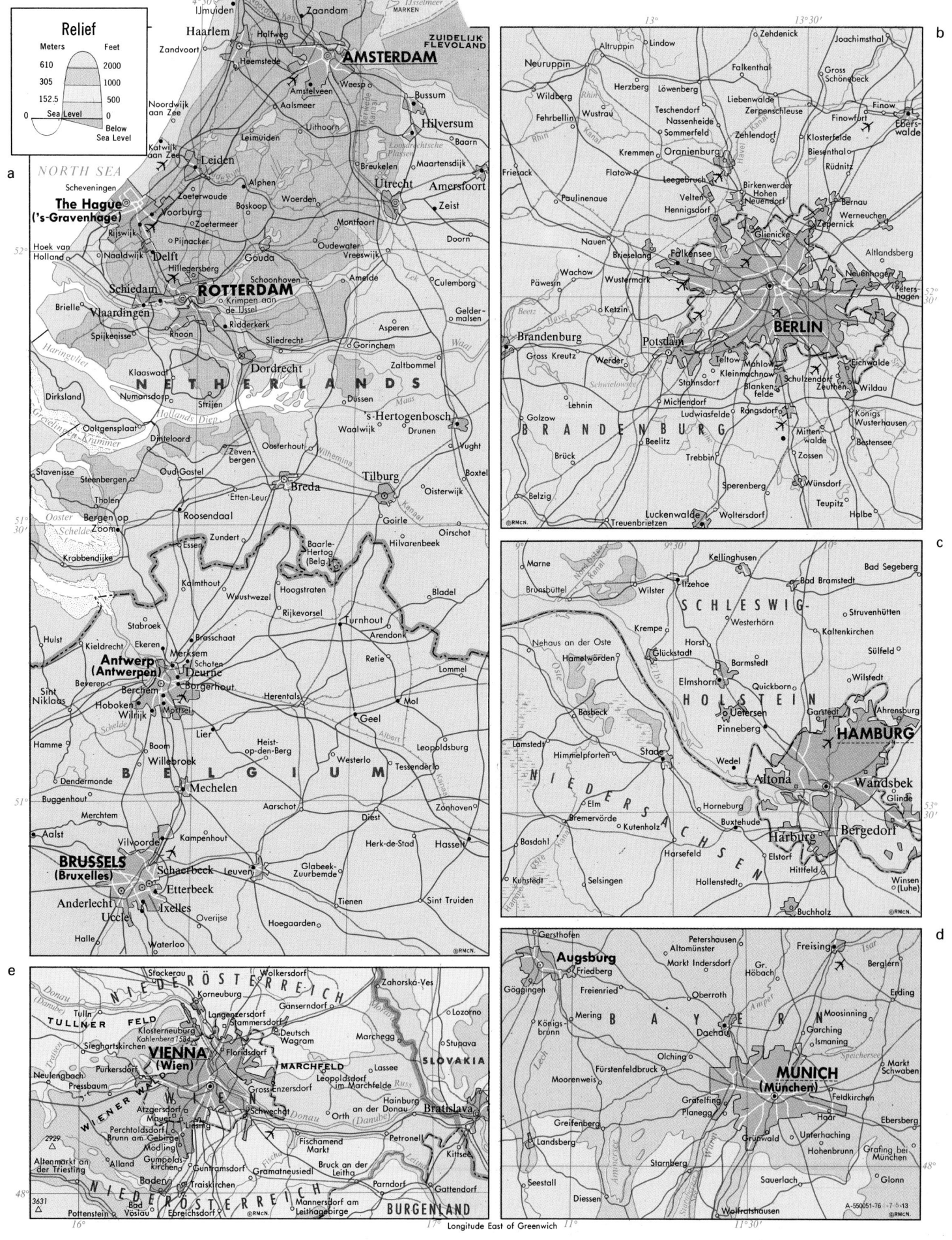

Relief

Meters	Feet
610	2000
305	1000
152.5	500
0 Sea Level	0 Sea Level
	Below Sea Level

a

NORTH SEA

Scheveningen

The Hague
('s-Gravenhage)

Hoek van
Holland

IJmuiden
Haarlem
Zandvoort
Zandaam
Halfweg
Heemstede
Weesp
AMSTERDAM
Noordwijk
aan Zee
Amstelveen
Aalsmeer
Leimuiden
Uithoorn
Bussum
Hilversum
Baarn
Katwijk
aan Zee
Leiden
Alphen
Zoeterwoude
Voorburg
Zoetermeer
Woerden
Boskoop
Gouda
Utrecht
Amersfoort
Zeist
Doorn
Montfoort
Oudewater
Vreeswijk
Rijswijk
Delft
Pijnacker
Hillegersberg
Schoonhoven
Culemborg
Schiedam
ROTTERDAM
Krimpen aan
de IJssel
Amelde
Gelder-
malsen
Brielle
Vlaardingen
Ridderkerk
Asperen
Spijkenisse
Rhoon
Sliedrecht
Gorinchem
Zaltbommel
Klaaswaal
Dordrecht
NETHERLANDS
Dussen
's-Hertogenbosch
Dirksland
Numansdorp
Strijen
Vught
Oud Gastel
Oosterhout
Waalwijk
Drunen
Boxtel
Stavenisse
Steenbergen
Zeven-
bergen
Wilhelmina
Oisterwijk
Tholen
Bergen op
Zoom
Roosendaal
Breda
Tilburg
Goirle
Oirschot
Krabbendijke
Zundert
Essen
Baarle-
Hertog
(Belg)
Hilvarenbeek
Kalmthout
Hoogstraten
Bladel
Wuustwezel
Rijkevorsel
Turnhout
Arendonk
Stabroek
Brasschaat
Retie
Kieldrecht
Ekeren
Schoten
Antwerp
(Antwerpen)
Merksem
Deurne
Borgerhout
Herentals
Lommel
Beveren
Berchem
Mortsel
Mol
Sint
Niklaas
Hoboken
Wilrijk
Lier
Geel
Hamme
Boom
Heist-
op-den-Berg
Leopoldsburg
Willebroek
Westerlo
Tessenderlo
Dendermonde
BELGIUM
Mechelen
Buggenhout
Aarschot
Diest
Zonhoven
Aalst
Merchtem
Kampenhout
Herk-de-Stad
Hasselt
Vilvoorde
BRUSSELS
(Bruxelles)
Schaerbeek
Leuven
Glabeek-
Zuurbemde
Anderlecht
Etterbeek
Sint Truiden
Uccle
Ixelles
Tienen
Halle
Overijse
Hoegaarden
Waterloo

b

Neuruppin
Altruppin
Lindow
Zehdenick
Joachimsthal
Wildberg
Herzberg
Löwenberg
Falkenthal
Gross
Schönebeck
Fehrbellin
Wustrau
Teschendorf
Nassenheide
Liebenwalde
Zerpenschleuse
Finow
Finowfurt
Ebers-
walde
Kremmen
Sommerfeld
Zehlendorf
Klosterfelde
Biesenthal
Rüdnitz
Friesack
Oranienburg
Flatow
Leegebruch
Birkenwerder
Hohen
Neuendorf
Bernau
Werneuchen
Nauen
Brieselang
Velten
Hennigsdorf
Glienicke
Altlandsberg
Päwesin
Wachow
Falkensee
Neuenhagen
Wustermark
Petters-
hagen
Ketzin
Brandenburg
BERLIN
Gross Kreutz
Werder
Potsdam
Teltow
Mahlow
Eichwalde
Golzow
Lehnin
Stahnsdorf
Kleinmachnow
Blanken-
felde
Schulzendorf
Zeuthen
Wildau
BRANDENBURG
Michendorf
Ludwigsfelde
Rangsdorf
Königs
Wusterhausen
Brück
Beelitz
Mitten-
walde
Bestensee
Belzig
Zossen
Wünsdorf
Sperenberg
Teupitz
Halbe
Treuenbrietzen
Luckenwalde
Woltersdorf

c

Marne
Kellinghusen
Bad Segeberg
Brunsbüttel
Wilster
Itzehoe
Bad Bramstedt
Struvenhütten
SCHLESWIG-
Westerhörn
Kaltenkirchen
Nehaus an der Oste
Krempe
Barmstedt
Sülfeld
Hamelwörden
Glückstadt
Horst
Quickborn
Wilstedt
Basbeck
Elmshorn
Pinneberg
Garstedt
Ahrensburg
Lamstedt
HOLSTEIN
Uetersen
HAMBURG
Himmelpforten
Stade
Wedel
Wandsbek
Elm
Altona
Glinde
Bremervörde
Horneburg
Harburg
Bergedorf
Basdahl
Buxtehude
Kuhstedt
Harsefeld
Elstorf
Hittfeld
Winsen
(Luhe)
Selsingen
Hollenstedt
Buchholz
NIEDERSACHSEN

d

Gersthofen
Petershausen
Freising
Altomünster
Augsburg
Friedberg
Markt Indersdorf
Gr.
Höbach
Berglern
Göggingen
Freienried
Oberroth
Erding
Königs
brunn
Mering
BAYERN
Moosinning
Garching
Olching
Dachau
Ismaning
Moorenweis
Fürstenfeldbruck
MUNICH
(München)
Markt
Schwaben
Gräfelfing
Planegg
Haar
Feldkirchen
Greifenberg
Grünwald
Unterhaching
Ebersberg
Landsberg
Hohenbrunn
Grafing bei
München
Starnberg
Sauerlach
Glonn
Seestall
Diessen
Wolfratshausen

e

NIEDERÖSTERREICH
Stockerau
Wolkersdorf
Zahorska-Ves
Tulln
Korneuburg
Gänserndorf
TULLNER FELD
Langenzersdorf
Klosterneuburg
Stammersdorf
Deutsch
Wagram
Marchegg
Stupava
Lozorno
Sieghartskirchen
Kahlenberg 1584
Floridsdorf
MARCHFELD
VIENNA
(Wien)
Lassee
Neulengbach
Pressbaum
Purkersdorf
WIEN
Gross Enzersdorf
Leopoldsdorf
im Marchfelde
SLOVAKIA
Atzgersdorf
Hainburg
an der Donau
Bratislava
Perchtoldsdorf
Liesing
Schwechat
Orth
Mauer
Mödling
Fischamend
Markt
Petronell
Altenmarkt an
der Triesting
Brunn am Gebirge
Gumpolds-
kirchen
Kittsee
2929
Alland
Guntramsdorf
Bruck an der
Leitha
Baden
Traiskirchen
Gramatneusiedl
Parndorf
Gattendorf
Bad
Vöslau
Mannersdorf am
Leithagebirge
BURGENLAND
3631
Pottenstein
Ebreichsdorf
NIEDERÖSTERREICH

Longitude East of Greenwich

Scale 1:1 000 000; one inch to 16 miles.

Elevations and depressions are given in feet.

A-550051-76 -7-5-13

Continued on pages 136-137

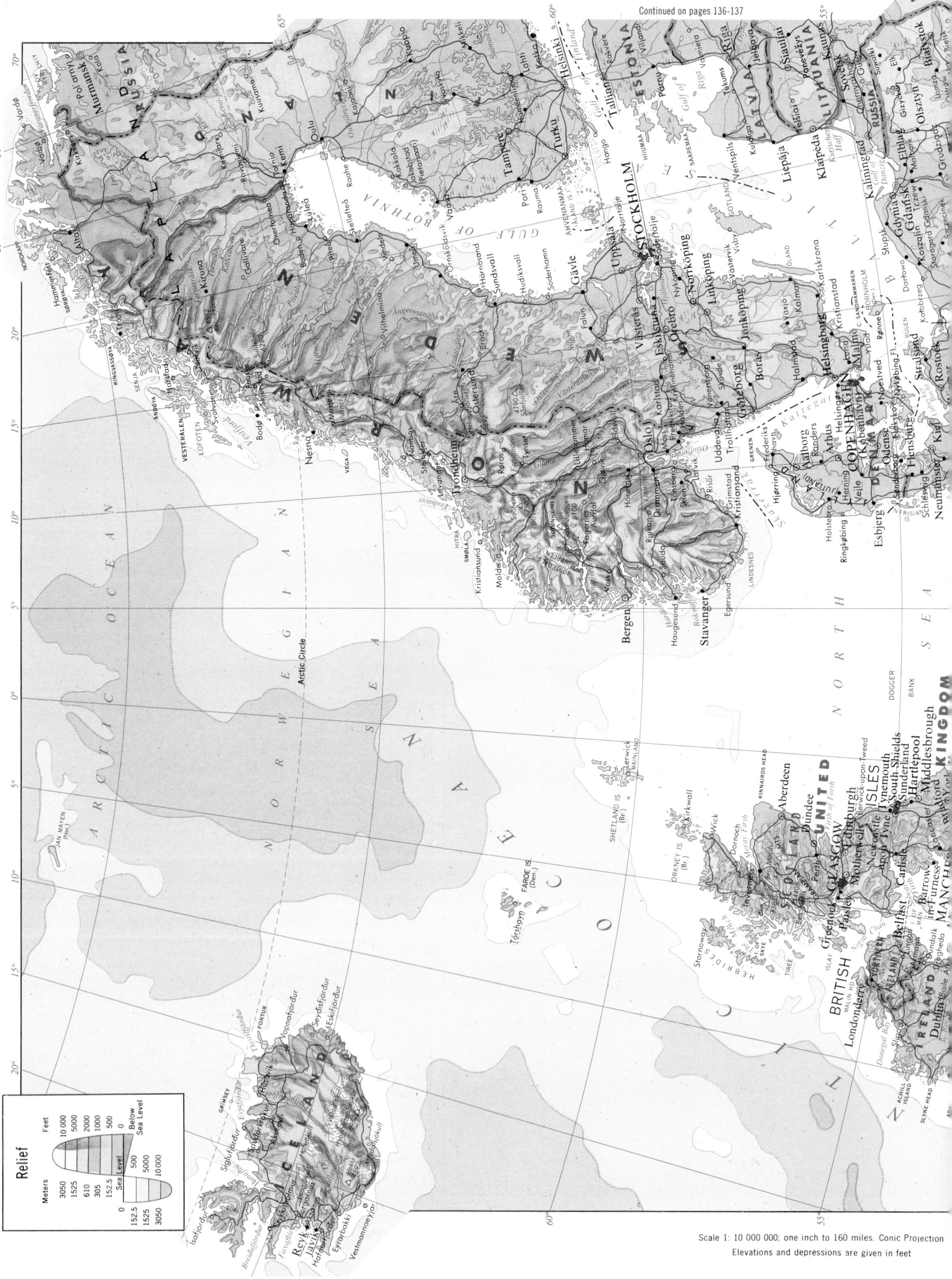

BELARUS

RUSSIA

LAPLAND

Murmansk
Pol'arnyj
Kola
Varde
Vadso
Hammerfest
Alta
Kirkenes

FINLAND

Oulu
Kemi
Tornio
Kuopio
Mikkeli
Lahti
HELSINKI
Tampere
Turku
Pori
Rauma
Hango
Kotka

ESTONIA
Tallinn
Pärnu
Tartu
HIIUMAA
SAAREMAA

Gulf of Riga
Riga
LATVIA
Ventspils
Liepāja

LITHUANIA
Klaipėda
Šiauliai
Kaunas

RUSSIA
Kaliningrad

SWEDEN
NORWAY

Kiruna
Narvik
Boden
Luleå
Skellefteå
Umeå
Örnsköldsvik
Sundsvall
Härnösand
Hudiksvall
Söderhamn
Gävle
Falun
Uppsala
STOCKHOLM
Västerås
Eskilstuna
Örebro
Norrköping
Linköping
Jönköping
Borås
GÖTEBORG
Halmstad
Helsingborg
Malmö
Kalmar
Karlskrona
Kristianstad
Visby
GOTLAND
ÖLAND

GULF OF BOTHNIA

Trondheim
OSLO
Bergen
Stavanger
Haugesund
Egersund
Kristiansand
Arendal
Skien
Drammen
Hamar
Lillehammer
Molde
Ålesund
Kristiansund

DENMARK
COPENHAGEN
Esbjerg
Odense
Ålborg
Århus
Randers
Helsingør
Frederikshavn

RÜGEN
Stralsund
Rostock

NORTH SEA
NORWEGIAN SEA
ARCTIC OCEAN

Arctic Circle

JAN MAYEN
(Nor.)

FAROE IS.
(Den.)
Tórshavn

SHETLAND IS.
(Br.)
Lerwick
MAINLAND

ORKNEY IS.
(Br.)
Kirkwall

UNITED KINGDOM
SCOTLAND
Aberdeen
Dundee
Perth
EDINBURGH
GLASGOW
Greenock
Paisley
Motherwell
Wick
Stornoway
HEBRIDES
SKYE
ISLAY
TIREE

Newcastle-upon-Tyne
Tynemouth
South Shields
Sunderland
Hartlepool
Middlesbrough
Carlisle
Barrow-in-Furness
MANCHESTER
Lancaster

IRELAND
BELFAST
DUBLIN
Londonderry
Sligo
Donegal Bay
ISLE OF MAN
ANGLESEY

BRITISH ISLES

DOGGER BANK

ICELAND
Reykjavík
Akureyri

Relief

Meters	Feet
3050	10 000
1525	5000
610	2000
305	1000
152.5	500
0	0 Sea Level

Below Sea Level
152.5	500
1525	5000
3050	10 000

118

Continued on pages 116–117

10°   5°   0°   5°   10°   15°

ATLANTIC
OCEAN

BAY OF BISCAY

45°

FRANCE

GERMANY
Continued on pages 116–117
PRAGUE
(Praha)
FRANKFURT
MANNHEIM
STUTTGART
MUNICH
AUSTRIA

Cherbourg · Le Havre · Amiens · St. Quentin
Brest · Morlaix · Caen · Rouen · Reims
Quimper · Rennes · Le Mans · Orléans · Paris
St. Nazaire · Nantes · Tours · Blois · Bourges
Poitiers · Châteauroux · Nevers · Dijon
La Rochelle · Limoges · Clermont-Ferrand · Lyon
Bordeaux · Périgueux · St. Étienne
Bayonne · Toulouse · Montpellier · Marseille
SWITZERLAND
Bern · Zürich · Luzern
MILAN · TURIN · Genoa · Venice

C. ORTEGAL
A Coruña · Ferrol
Santiago de Compostela · Oviedo · Gijón · Santander · Bilbao
Vigo · Porto · León · Burgos · Zaragoza
40°
Coimbra · SPAIN · MADRID · BARCELONA
LISBON · Badajoz · Valencia
Sevilla · Córdoba · Murcia · Alacant
Cádiz · Málaga · Almería

LIGURIAN SEA
CORSICA (Fr.) · Ajaccio
SARDINIA (It.) · Cagliari
ROME (Roma)
NAPLES (Napoli)
TYRRHENIAN SEA

MEDITERRANEAN

35°
Tanger · Ceuta (Sp.) · Tetouan · Melilla
Oran (Wahran) · Algiers (El Djazair) · Constantine · Annaba (Bône)
MOROCCO · Rabat · Fès · Meknès · Oujda
MOYEN ATLAS · HAUT ATLAS
ALGERIA
SAHARAN ATLAS · MONTS DES KSOUR
Laghouat
TUNISIA · Sfax · Gabès
Palermo · SICILY · Catania
MALTA · Valletta

30°
GRAND ERG OCCIDENTAL
Ghardaïa · Wargla
GRAND ERG ORIENTAL
Tripoli (Tarabulus)
Misratah
TARABULUS (TRIPOLITANIA)
Ghudamis

Longitude West of Greenwich 0° Longitude East of Greenwich

Relief

Meters		Feet
3050		10000
1525		5000
610		2000
305		1000
152.5		500
0	Sea Level	0
152.5	Below Sea Level	500
1525		5000
3050		10000

A-55630-76 117-138
COPYRIGHT BY
RAND M NALLY & COMPANY
MADE IN U.S.A.

Scale 1:10 000 000; one inch to 160 miles. Bonne's Projection
Elevations and depressions are given in feet

Continued on pages 136-137

Zabrze · KATOWICE · Rzeszów · Jarosław · Brody · Khmel'nyts'kyi · Smila · Pavlohrad · DONETS'K
Ostrava · Kraków · POLAND · Tarnów · L'viv · Ternopil' · Vinnytsia · Zvenyhorodka · Shpola · Kremenchuk · Synel'nykove · Novocherkassk · Rostov-na-Donu
ouc · Žilina · BESKID MTS · CARPATHIANS · Ivano-Frankivs'k · Haisyn · Uman' · Kirovohrad · Dniprodzerzhyns'k · Zaporizhzhia · Taganrog · Yeysk · RUSSIA
SLOVAKIA · Košice · Prešov · Uzhhorod · Mukacheve · UKRAINE · Novoukrainka · Novovorontsovka · Parvomais'k · Nikopol' · DNIPROPETROVS'K · Mariupol' · Tikhoretsk · Kropotkin
NNA · Banská Bystrica · Miskolc · Khust · Chernivtsi · Balta · Soroka · Kryvyi Rih · Berdians'k · Primorsko-Akhtarskaya · Stavropol'
Bratislava · Nové Zámky · Nyíregyháza · Satu Mare · Bălți · Kherson · KOSA BYRIUCHI O. · Timashevskaya · Temryuk · Armavir · Lobinsk
Wiener Neustadt · Győr · Debrecen · Hajdúszoboszló · Baia Mare · MOLDOVA · Iași · Tighina · Odesa · Mykolaïv · Melitopol' · Heniches'k · SEA OF AZOV · Kerch · Krasnodar · Maykop
BUDAPEST · Hajdúszoboszló · Dej · Bistrița · Piatra Neamț · Orhei · Chișinău · KRYMS'KYI PIVOSTRIV (CRIMEAN PEN.) · Feodosiia · Novorossiysk · Tuapse · Sukhumi · GEORGIA
Szombathely · Székesfehérvár · Kecskemét · Cluj-Napoca · Bilhorod-Dnistrovs'kyy · Ievpatoriia · Simferopol' · Yalta
HUNGARY · Szeged · Arad · Târgu Mureș · Sighișoara · Gheorghe · Bolgrad · Izmyil · Sulina · Sevastopol' · M. SARYCH · Sochi

50 100 150 200 250 300 Miles
100 200 300 400 500 Kilometers

a

Same scale as main map

ATLANTIC

OCEAN

SHETLAND
ISLANDS
(Br.)

Lerwick

ORKNEY
ISLANDS
(Br.)

Kirkwall

SCOTLAND

ATLANTIC

OCEAN

NORTHERN
IRELAND
ULSTER

Belfast

IRELAND

CONNACHT

Dublin
Baile Átha Cliath

LEINSTER

MUNSTER

Cork

SCOTLAND

GRAMPIAN MTS

Aberdeen

Dundee

GLASGOW

Edinburgh

UNITED

NEWCASTLE UPON
TYNE

Middlesbrough

LIVERPOOL

MANCHESTER

Sheffield

LEEDS

KINGDOM

BIRMINGHAM

Coventry

Nottingham

Cardiff

Bristol

LONDON

ENGLISH

Relief

Meters		Feet
610		2000
305		1000
152.5		500
0	Sea Level	0
152.5		500
1525		5000

A-559700-76-...g...-7-17
COPYRIGHT BY
RAND McNALLY & COMPANY
MADE IN U.S.A.

Longitude West of Greenwich

Scale 1: 4 000 000; one inch to 64 miles. Conic Projection

Elevations and depressions are given in feet

Continued on pages 122-123

Continued on pages 124-125

Continued on pages 126-127

Longitude East of Greenwich

10   20   30   40   50   60   70   80   90   100   110   120 Miles
20   40   60   80   100   120   140   160   180   200 Kilometers

NORWEGIAN SEA

SMØLA
Kristiansund
Trondheim
Stjørdalshalsen
Selbusjøen
12°
Orkanger
Averøya
GURSKØY
Molde
Halsafjorden
Støren
14°
16°
18°
Ålesund
Hustadvika
Romsdal
TROLLHEIMEN
Støren
Gaula
Sylarna 5781
Storsjön
Östersund
Sollefteå
4°
Åndalsnes
Oppdal
Røros
Helagsfjället 5892
Ragunda
Kramfors
BREMANGERLANDET
Nordfjord
Snøhatta 7500
Aursunden
Storsjö
HEMSÖN
Florø
62°
Tynset
Fermunden
Bräcke
Härnösand
DOVRE FJELL
Savalen
NATIONAL PARK
Ånge
Fränsta
Stöde
Sundsvall
JOTUNHEIMEN
Galdhøpiggen 8100
Glittertinden 8984
Sønfjället 4790
TÖFSINGDALENS NATIONAL PARK
ALNÖN
Njurunda
JOSTEDALSBREEN
Sveg
Stjørdalshalsen
Z
Ramsjö
G
Hjerkinn
Slädjön 3711
Ljusnan
Hudiksvall
Leikanger
Sognefjorden
Vikøyri
Lærdalsøyri
Lillehammer
Rena
Älvdalen
Ljusdal
BO
Gudvangen
Flåm
Fagernes
Aurdal
Orsa
Söderhamn
Dale
Voss
Eidfjord
NORWAY
Gol
Gjøvik
Moelv
Elverum
Lima
Mora
Ockelbo
Enånger
Bollnäs
Bergen
Osøyra
Odda
Gulsvik
Raufoss
Hamar
Leksand
Rättvik
Gävle
60°
STORA SOTRA
Skreia
Filsa
Appelbo
Falun
Siljan
Borlänge
Gävle-bukten
BØMLO
Honefoss
Kongsvinger
Torsby
Ludvika
Storvik
GRÅSÖ
STORD
Rjukan
Vickersund
Oslo
Lillestrøm
Charlottenberg
Smedjebacken
Säter
Hedemora
Tierp
Östhammar
KARMØY
Sauda
Tinnoset
Notodden
Drammen
Øyeren
Arvika
Sunne
Filipstad
Kopparberg
Avesta
Krylbo
Vattholma
Haugesund
Kopervik
Dalen
Kongsberg
Svelvik
Drøbak
Holmstrand
Moss
Kil
Forshaga
Nora
Sala
Heby
Uppsala
Rimbo
Norrtälje
Skudeneshavn
Stavanger
Skien
Porsgrunn
Horten
Tønsberg
Mysen
Karlstad
Karlskoga
Arboga
Tillberga
Enköping
Sigtuna
Västerås
Sandnes
Tau
Fyresvatn
Brevik
Sandefjord
Larvik
Sarpsborg
Fredrikstad
Kristinehamn
Örebro
Eskilstuna
Strängnäs
Mariefred
Sundbyberg
STOCKHOLM
Egersund
Byglandsfjord
Risør
Langesund
Kragerø
Halden
Åmål
Säffle
Hallsberg
Hjälmaren
Malmköping
Södertälje
Saltsjöbad
58°
Flekkefjord
Tvedestrand
Strömstad
Grebbestad
Askersund
Katrineholm
Trosa
Nynäshamn
ORNÖ
Farsund
Mandal
Arendal
Grimstad
Lillesand
Fjällbacka
Mellerud
Mariestad
Töreboda
Motala
Vadstena
Söderköping
Nyköping
LINDESNES
Kristiansand
Uddevalla
Lidköping
Skara
Skövde
Skänninge
Norrköping
Bråviken
Lysekil
Vänersborg
Falköping
Hjo
Gränna
Linköping
Trollhättan
Falköping
Tidaholm
Tranås
Åtvidaberg
Valdemarsvik
Marstrand
Alingsås
Ulricehamn
Huskvarna
Vimmerby
Gamleby
Kungälv
Borås
Jönköping
Nässjö
Eksjö
Västervik
GRENEN
Skagen
Göteborg
Mölndal
Nässjö
Vetlanda
Virserum
Figeholm
GOTLAND
Hjørring
Frederikshavn
Kungsbacka
Vimmerby
Oskarshamn
Klintehamn
Visby
Brønderslev
Sæby
LÆSØ
Varberg
Värnamo
Mönsterås
ÖLAND
Aalborg
Nørresundby
Falkenberg
Oskarström
Alvesta
Växjö
Nybro
Bergholm
Kalmar
Thisted
Løgstør
Nibe
ANHOLT
Halmstad
Ljungby
Almhult
NORTH SEA
Nykøbing
Hobro
Mariager
Laholm
Markaryd
Tingsryd
Mörbylånga
Nissum Fjord
Struer
Skive
Viborg
Randers
Grenaa
Bästad
Ängelholm
Klippan
Hässleholm
Karlshamn
Ronneby
Karlskrona
Ringkøbing
Herning
Silkeborg
Ebeltoft
Helsingør
HELSINGBORG
Kristianstad
Sölvesborg
Åhus
Hanö-bukten
Ringkøbing Fjord
JYLLAND
Århus
Skanderborg
Nykøbing
Landskrona
Eslöv
Härby
Varde
Horsens
SAMSØ
COPENHAGEN
Hillerød
Lund
Simrishamn
Esbjerg
Fredericia
Kolding
Middelfart
Bogense
Holbæk
København
Øresund
Roskilde
Malmö
Svedala
Skurup
Tomelilla
FANØ
Ribe
DENMARK
Odense
FYN
Slagelse
Ringsted
Køge Bugt
Skånör
Falsterbo
Trelleborg
Ystad
SANDHAMMAREN
Haderslev
Åbenrå
ALS
Assens
Faaborg
Nyborg
Korsør
Næstved
BORNHOLM (Den.)
Allinge
Svaneke
RØMØ
Tønder
Sønderborg
ÆRØ
Svendborg
Rudkøbing
LANGELAND
Nakskov
Vordingborg
MØN
Rønne
Neksø
SYLT
FÖHR
Flensburg
Schleswig
LOLLAND
Maribo
Nykøbing
FALSTER
RÜGEN
KAP ARKONA
Łeba
Ustka
Wejherowo
Gdynia
Sopot
FRISIAN ISLANDS
Husum
SCHLESWIG
Kiel
FEHMARN
Gedser
Barth
Sassnitz
B
Dartowo
Lębork
Gdańsk
Tønning
Eckernförde
Kiel Bay
Warnemünde
Stralsund
Bergen
Pomeranian Bay
Kołobrzeg
Słupsk
POLAND
Heide
Rendsburg
Neustadt in Holstein
Lübecker Bucht
Greifswald
Wolgast
Longitude East of Greenwich
18°
HOLSTEIN
Neumünster
Lübeck
Rostock
Wismar
Wolin
Kamień Pomorski
Gdań
54°
Cuxhaven
Elbe
GERMANY
Świnoujście
8°

A-559195-76 -13-9 18
COPYRIGHT BY
RAND MCNALLY & COMPANY
MADE IN U.S.A.

Relief

Meters		Feet
1525		5000
610		2000
305		1000
152.5		500
0	Sea Level	0
152.5		500
		Below Sea Level

Skagerrak
Kattegat
BALTIC SEA
Jammerbugten
Hanöbukten

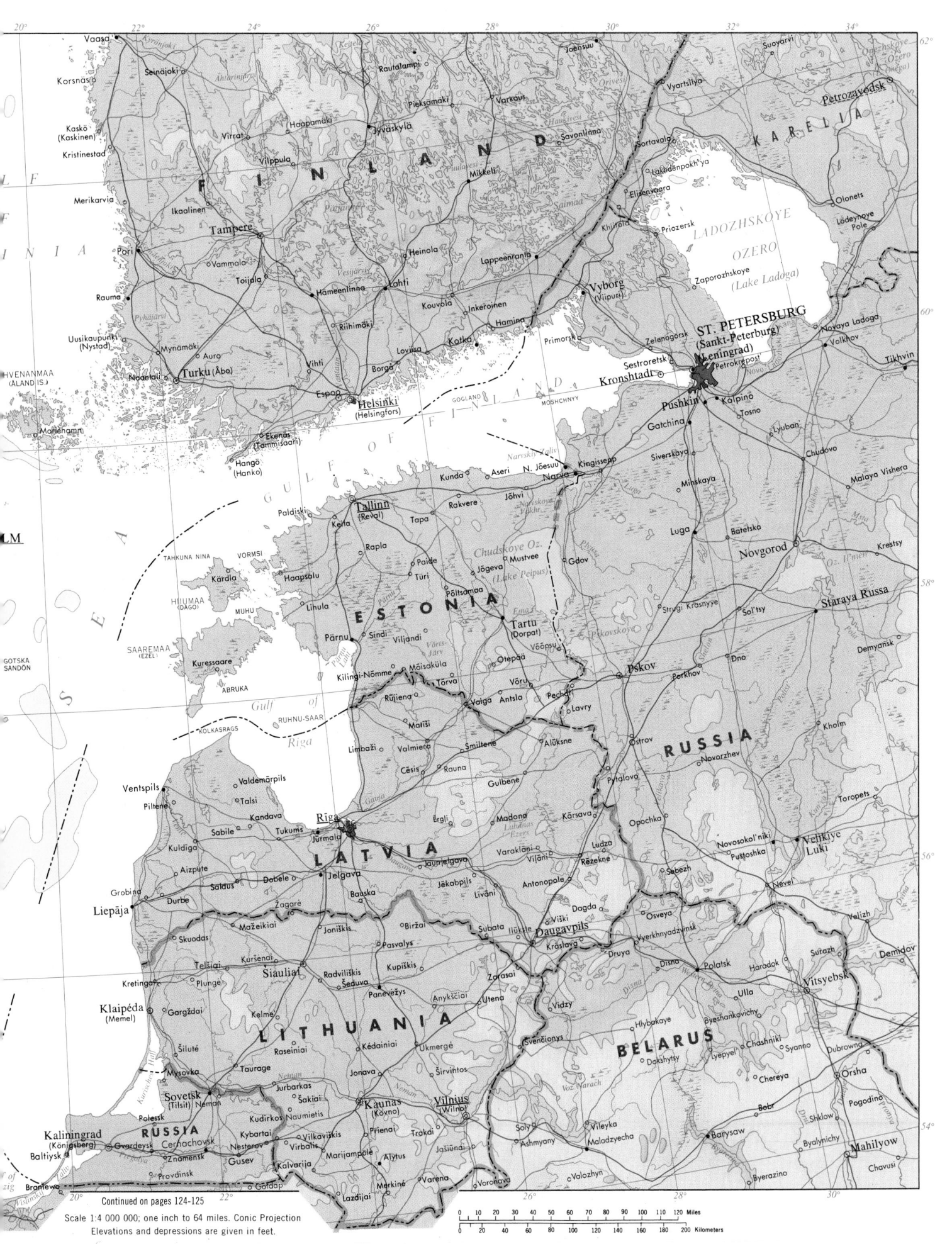

Continued on pages 124-125

Scale 1:4 000 000; one inch to 64 miles. Conic Projection
Elevations and depressions are given in feet.

Continued on pages 122-123

NORTH SEA

DENMARK

BALTIC

FRISIAN ISLANDS

NETHERLANDS

AMSTERDAM

GERMANY

HAMBURG

BREMEN

HANNOVER

BERLIN

MECKLENBURG

BRANDENBURG

POMERANIA

POLAND

DÜSSELDORF

COLOGNE (Köln)

DORTMUND

ESSEN

WESTFALEN

HESSEN

THÜRINGEN

LEIPZIG

DRESDEN

FRANKFURT AM MAIN

Wiesbaden

MANNHEIM

NÜRNBERG

BAYERN (BAVARIA)

CZECH REPUBLIC

PRAGUE (Praha)

ČECHY (BOHEMIA)

STUTTGART

MUNICH (München)

Augsburg

FRANCE

SWITZERLAND

Zürich

LIECHTENSTEIN

ALPS

Innsbruck

AUSTRIA

VIENNA (Wien)

Salzburg

Linz

SLOVENIA

CROATIA

Continued on pages 130-131

Longitude East of Greenwich

Scale 1:4 000 000; one inch to 64 miles. Conic Projection
Elevations and depressions are given in feet.

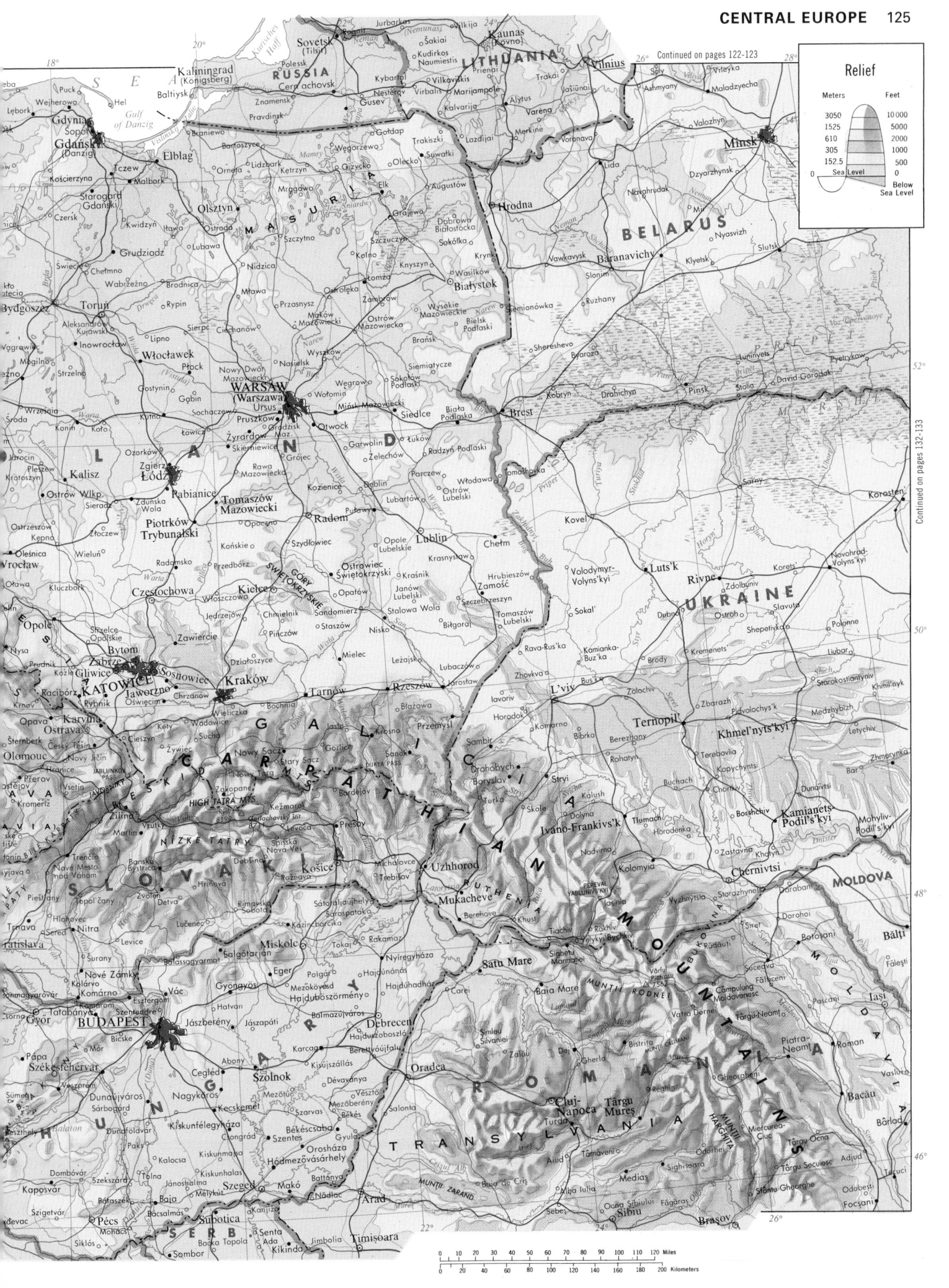

Continued on pages 122-123

Continued on pages 132-133

Relief

Meters	Feet
3050	10 000
1525	5000
610	2000
305	1000
152.5	500
0 Sea Level	0
	Below Sea Level

Continued on pages 120-121

Relief

Meters	Feet
3050	10 000
1525	5000
610	2000
305	1000
152.5	500
0	Sea Level	0
152.5	500
1525	5000

a

A-550900-76 —7-6-14
COPYRIGHT BY
RAND McNALLY & COMPANY
MADE IN U.S.A.

Scale 1:1 000 000

10 Miles

0 4 8 12 16 Kilometers

®RMcN

Continued on pages 128-129

Scale 1:4 000 000; one inch to 64 miles. Conic Projection
Elevations and depressions are given in feet

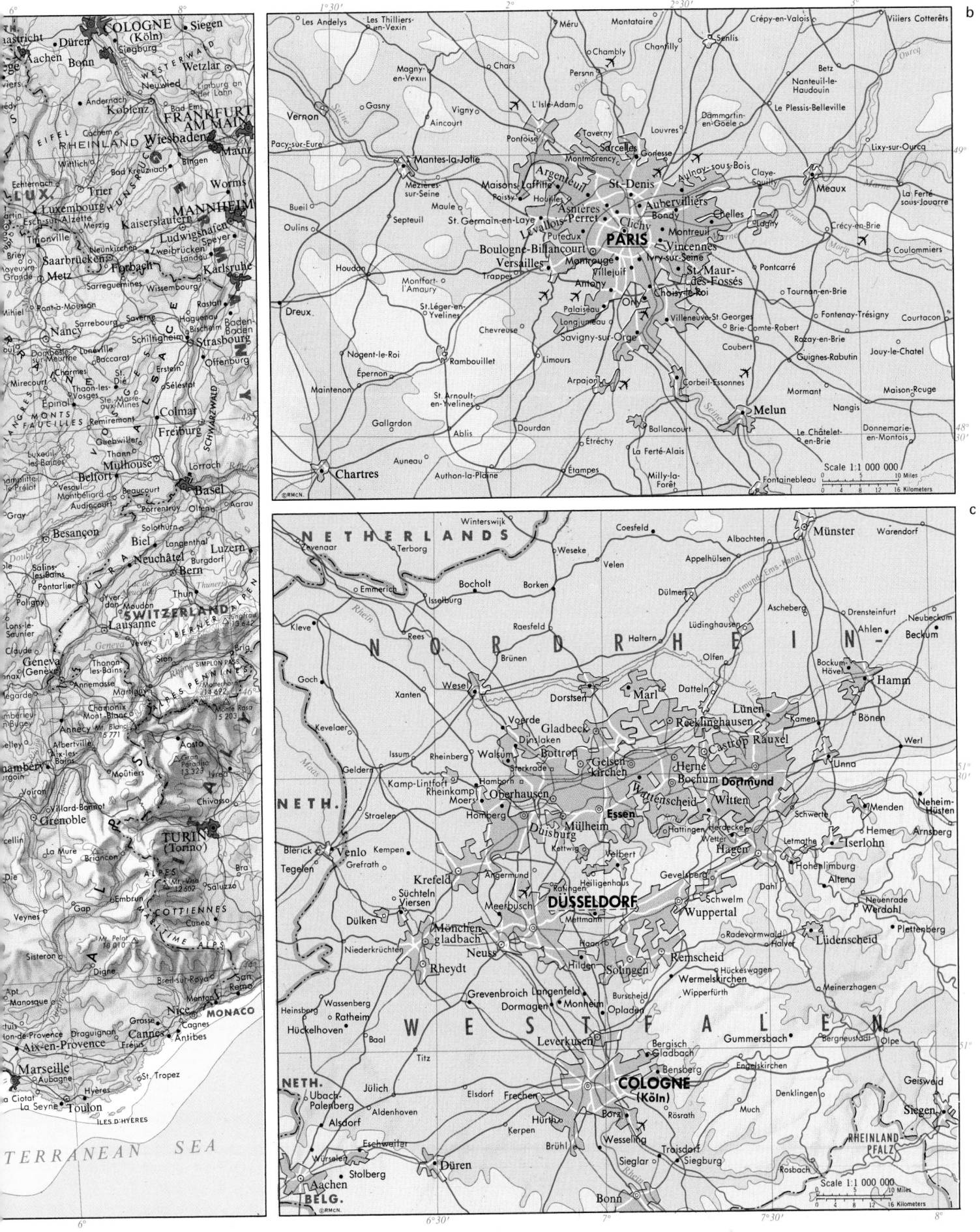

Scale 1:4 000 000, one inch to 64 miles. Conic Projection
Elevations and depressions are given in feet

Longitude West of Greenwich

Continued on pages 126-127

FRANCE

Pau
Tarbes
Toulouse
Verdun
Condom
de-Marsan
St-l'Adour
Olaron
Ste Marie
Lourdes
St. Gaudens
Muret
Pamiers
Foix
Tarascon
Aulus-les Thermes
Prades
Ceret
Perpignan
Narbonne
Béziers
Montpellier
Sète
Agde
Pézenas
Bédarieux
Castres
Albi
Gaillac
Mazamet
Carcassonne
Limoux
Quillan
Rivesaltes
Port Vendres

Golfe du Lion

CAP DE CREUS

PYRENEES
ANDORRA
Andorra
Pico de Aneto 11 168
Pica d'Estats 10,320
Monte Perdido 11 007
Monte Bañona
SA. DE GUARA

ARAGON

Huesca
Barbastro
Monzón
Lleida
Fraga
Sariñena
Balaguer
Tárrega
Cervera
Igualada
Manresa
Vic
Ripoll
Olot
Berga
Montblanc
Valls
Reus
Tarragona
Tortosa
Amposta
Gandesa
Caspe
Alcañiz
Calanda
Morella
Alcanar
Vinaròs
Benicarló
Torreblanca

CATALUNYA

Terrassa
Sabadell
Badalona
BARCELONA
Granollers
Mataró
Catella
Sant Feliu de Guixols
La Bisbal
Girona
Manlleu
Vilafranca del Penedès
Vilanova i la Geltrú

Sant Mateu
DE GUDAR
Penagolosa 5952
Lucena
Ares
Onda
Vila-real
Borriana
La Vall d'Uixò
Segorbe
Iva
Llíria
Sagunt
Torrent
Catarroja
VALÈNCIA
Alzira
Sueca
Cullera
Xàtiva
Gandia
Oliva
Ontinyent
Pego
Dénia
Xàbia
Alcoi
Cocentaina
Xixona
La Vila Joiosa
Monóvar
velda
Alacant
Elx
villent
Oriola
Segura
Torrevieja

Castelló de la Plana

COLUMBRETES

Golf de València

BALEARIC SEA

ILLES BALEARS (BALEARIC ISLANDS)
(Sp)

MENORCA (MINORCA)
Ciutadella
Maó
Es Port de Pollença
Sóller
Inca
Sa Pobla
Ba. d'Alcúdia
Manacor
Felanitx
Santanyí
Llucmajor
Palma
Ba. de Palma
MALLORCA (MAJORCA)
CAP DE SES SALINES
ILLA DE CABRERA

EIVISSA (IBIZA)
Sant Antoni de Portmany
Santa Eulària del Riu
Eivissa
FORMENTERA

CAP DE TORTOSA
CAP DE LA NAU

Mar Menor
La Unión
CABO DE PALOS
artagena

MEDITERRANEAN SEA

Algiers (El Djazaïr)
Dalles
Boudouaou
Cherchell
Boufarik
El Arba
El Boulaïda
Bouïra
Ténès
El Affroun
Meftah
Lakhdaria
Carnot
Ech Cheliff
Oued Rhiou
Qasr el Boukhari
Sidi Aïssa
Sour el Ghozlane

ATLAS MOUNTAINS

Mestghanem
CAP FERRAT
Arzew
Oran (Wahran)
Oued Tlelat
El Mohammadia
Ghilizane
Mouaskar
Stizel
in Temouchent
Sebha Azzel
Ksar Chellala

ALGERIA

Aïn Wessara
Bouira-Sahary
Djelfa
Zahrez Chergui
Tiaret

0° 2° 4°
Longitude East of Greenwich

0 20 40 60 80 100 120 Miles
0 20 40 60 80 100 120 140 160 180 200 Kilometers

S. Lorenzo de El Escorial
SA. DEL HOYO 4606
Colmenar Viejo
Fuente el Saz
El Escorial
Galapagar
El Pardo
Canal Bajo
Algete
Valdemorillo
Las Rozas de Madrid
Fuencarral
Barajas de Madrid
Torrejón de Ardoz
S. Sebastián de los Reyes
Alcobendas
Alcalá de Henares
Pozuelo de Alarcón
Brunete
MADRID
Vicálvaro
S. Fernando de Henares
Alcorcón
Villaviciosa de Odón
Leganés
Getafe
Vallecas
Loeches
Campo Real
Móstoles
Canal del Manzanares
Valdilecha
Arganda
Navalcarnero
Pinto
Carabaña
Parla
S. Martín de la Vega
Marata de Tajuña
Tielmes
Perales de Tajuña

Scale 1:1 000 000
0   5   10 Miles
0   4   8   12   16 Kilometers
©RMCN.

40°30'
4° 3°30'

Mafra
Cheleiros
Alhandra
Alverca
Samora Correia
São João das Lampas
Montelavar
Loures
Almargem do Bispo
Colares
Sintra
Odivelas
Sacavém
Moscavide
Alcochete
CABO DA ROCA
Queluz
Barcarena
Amadora
Alcabideche
Carnaxide
Oeiras
Cascais
Estoril
LISBON (Lisboa)
Montijo
Costa de Caparica
Almada
Barreiro
Moita
Seixal
Albos Vedros
Pinhal Novo
Coina
Palmela
Setúbal
Sesimbra

ATLANTIC OCEAN

CABO ESPICHEL
Ba. de Setúbal
Rio Sado
Comporta

Scale 1:1 000 000
0   5   10 Miles
0   4   8   12   16 Kilometers
©RMCN.

9°30' 9° 38°30'

Frattamaggiore
Afragola
Acerra
Nola
Marano di Napoli
Pomigliano d'Arco
Monteforte Irpino
Avellino
Somma Vesuviana
NAPLES (Napoli)
Vesuvius 3710
Pozzuoli
Vesuvio 4190
S. Giuseppe Vesuviano
Mercato S. Severino
Bacoli
C. MISENO
Portici
Torre del Greco
Sarno
I. DI PROCIDA
Procida
Torre Annunziata
Pompeii Ruins
Nocera Inf.
Cava de' Tirreni
Forio 2585
Ischia
Castellammare di Stabia
Gragnano
Angri
Salerno
I. D'ISCHIA
Golfo di Napoli
Amalfi
Sorrento
TYRRHENIAN SEA
Golfo di Salerno
I. DI CAPRI 1932
Capri
PUNTA CAMPANELLA

Scale 1:1 000 000
0   5   10 Miles
0   4   8   12   16 Kilometers
©RMCN.

14° 14°30' 40°30'

Monterotondo
Pyrgi
Caere
Mentana
Veio
Guidonia
Cerveteri
Tivoli
Ladispoli
ROME (Roma)
Villa Adriana
VATICAN CITY
Zagarolo
Frascati
Fregene
Marino
COLLI ALBANI 3114
Fiumicino
Ostia Antica
Albano Laziale
Genzano di Roma
Lido di Roma
Laurentum
Velletri
Pomezia
Lanuvio
Cisterna di Latina
AGRO PONTINO
Aprilia
Anzio
Nettuno

TYRRHENIAN SEA

Scale 1:1 000 000
0   5   10 Miles
0   4   8   12   16 Kilometers
©RMCN.

12° 12°30' 42° 41°30'

Continued on pages 124-125

Continued on pages 126-127

AUSTRIA

SWITZERLAND

FRANCE

MONACO

Nice
S. Remo
Ventimiglia
Imperia

Brenner Pass
Merano
Bolzano
Bressanone
Pieve di Cadore
Lienz
Carnic Alps
Villach
Klagenfurt
Maribor
Murska Sobota
Čakovec
Varaždin
Koprivnica
Szigetvár

RHAETIAN ALPS
DOLOMITI
Trento
TRENTINO-ALTO ADIGE
FRIULI
VENEZIA GIULIA
Udine
Tolmezzo
Triglav
Kranj
Škofja Loka
Celje
Kobarid
Tolmin
SLOVENIA
Ljubljana
Zagreb
Bjelovar
Daruvar

Sion
Brig
Simplon Pass
St. Moritz
Alpi Lepontine
Locarno
Lugano
Bellinzona
Sondrio
Tirano
Schio
Feltre
Belluno
Conegliano
Vittorio
Pordenone
Idrija
Cerknica
Novo Mesto
Karlovac
Kočevje
Sisak
Petrinja

Como
Lecco
Bergamo
Varese
Monza
Brescia
Riva
Lago di Garda
Verona
Vicenza
Cittadella
Treviso
Mestre
San Donà di Piave
Portogruaro
Monfalcone
Trieste
Rijeka (Fiume)
Pazin
Poreč
Rovinj
Ogulin
Karlobag
Gospić
CROATIA

Gallarate
Legnano
Busto Arsizio
Magenta
Treviglio
MILAN (Milano)
Lodi
Chiari
LOMBARDY
Cremona
Mantova (Mantua)
Padova (Padua)
Este
Rovigo
Adria
Venice (Venezia)
Chioggia
Cavarzere
Gulf of Venice
Pula
Istra
ADRIATIC

TURIN (Torino)
Chivasso
Vercelli
Novara
Abbiategrasso
Vigevano
Pavia
Piacenza
PIEDMONT
Casale Monferrato
Alessandria
Asti
Chieri
Tortona
Voghera
Parma
Reggio nell' Emilia
Modena
Carpi
EMILIA
Bologna
Ferrara
Comacchio
Ravenna

Cuneo
Mondovì
Bra
Alba
Acqui
Novi Ligure
Genoa (Genova)
LIGURIA
Savona
Albenga
La Spezia
Carrara
Massa
Viareggio
Lucca
Pisa
ROMAGNA
Imola
Faenza
Forlì
Cesena
Rimini
Pesaro
Fano
San Marino
SAN MARINO
Ancona
Senigallia

LIGURIAN SEA

CORSICA (Fr.)
Bastia
Calvi
Corte
Ajaccio
Sartène
Bonifacio
Strait of Bonifacio
Porto-Vecchio

ISOLA DI GORGONA
CAPRAIA
C. CORSE
Piombino
Portoferraio
ISOLA D'ELBA
PIANOSA
I. DI MONTECRISTO
I. DEL GIGLIO
I. DI GIANNUTRI

Pistoia
Prato
Florence (Firenze)
Empoli
Pontedera
Poggibonsi
Volterra
Siena
Montepulciano
Massa Marittima
Grosseto
Orbetello
Arezzo
Cortona
Perugia
Assisi
Foligno
Spoleto
TUSCANY
UMBRIA
Città di Castello
Gubbio
Fabriano
Macerata
Fermo
MARCHE
Urbino
Fossombrone
Cagli
Recanati
Ascoli Piceno
San Benedetto del Tronto
Teramo

Lago di Bolsena
Viterbo
Orvieto
Terni
Rieti
Civitavecchia
Tarquinia
Corneto
VATICAN CITY
ROME (Roma)
Frascati
Albano Laziale
Tivoli
LATIUM
L'Aquila
Penne
Mt. Corno 9554
Chieti
Pescara
Ortona
ABRUZZI
Avezzano
Sulmona
MOLISE
Vasto
Termoli

TYRRHENIAN SEA

ISOLA D'ISCHIA
NAPLES (Napoli)
Pozzuoli
Torre del Greco
Vesuvio 4190
Sorrento
Salerno
CAMPANIA
Caserta
Benevento
Avellino
Aversa
Santa Maria
Capua
Gaeta
Golfo di Gaeta
ISOLE PONZIANE
Terracina
Anzio
Nettuno
Aprilia
Velletri
Frosinone
Cassino
Isernia
Campobasso
Lucera
Foggia
San Severo
Manfredonia
Monte Sant'Angelo
Golfo di Manfredonia
Vieste
TESTA DEL GARGANO
ISOLE TREMITI

Barletta
Trani
Andria
Corato
Molfetta
Bitonto
Cerignola
Canosa
Ruvo
Minervino
Gravina
Altamura
Matera
Potenza
APULIA
BASILICATA
CAMPANIA

SARDINIA (It.)
Sassari
Alghero
Porto Torres
Olbia
Golfo dell' Asinara
ASINARA
CAPRERA
CAPRARA PT.
Tempio Pausania
Ozieri
Nuoro
Dorgali
Golfo di Orosei
Bosa
Bonorva
Oristano
Arborea
Golfo di Oristano
Punta la Marmora 6017
Lanusei
Villacidro
Iglesias
CARBONARA
Quartu Sant'Elena
Cagliari
Golfo di Cagliari
Carloforte
I. DI S. PIETRO
I. DI S. ANTIOCO
C. SPARTIVENTO

Golfo di Salerno
Golfo di Policastro
Castrovillari
Rossano
Coriglia
CALABRIA
Cosenza
Golfo di Sant' Eufemia
Nicastro
Vibo Valentia
Polistena
Palmi
Bagnara
Reggio di Calabria
Messina
Milazzo

Palermo
Monreale
Bagheria
Cefalù
Termini
Trapani
ISOLE EGADI
Alcamo
Marsala
Castelvetrano
Mazara del Vallo
Salemi
Sciacca
Agrigento
Caltanissetta
Enna
Catania
Gela
Augusta
Taormina
Acireale
Mt. Etna 10902
SICILY
ISOLE EOLIE
STROMBOLI
Lipari
Vulcano
I. DI USTICA
PALAGRUŽA (Cro.)

BOSNIA
Banja Luka
Bosanski Novi
Bosanska Gradiška
Prijedor
Bosanski Petrovac
Ključ
Donji Vakuf
Bugojno
Livno
Knin
DINARIC ALPS
DALMATIA
Zadar
Šibenik
Split
Trogir
Makarska
DUGI OTOK
KORNAT
HVAR
BRAČ
KORČULA
LASTOVO
VIS
BIŠEVO

AEGEAN SEA
AKRA SPATHA
Kissamos
Khaniá
Kálpos Khaniōn
Kálpos Almirón
Réthimnon
Iráklion (Candia)
DIA
AKRA SIDHEROS
Neápoli
Sitía
CRETE (Greece)
Khóra Sfakíon
Khóra
Ano Viánnos
Ierápetra
AKRA LITHINON
MEDITERRANEAN SEA
GÁVDHOS

Same scale as main map

Scale 1:4 000 000; one inch to 64 miles. Conic Projection
Elevations and depressions are given in feet

Continued on pages 122-123

Cities and Towns

0 to 50,000	○
50,000 to 500,000	⊙
500,000 to 1,000,000	◎
1,000,000 and over	🟦

Scale 1:4 000 000; one inch to 64 miles. Conic Projection
Elevations and depressions are given in feet

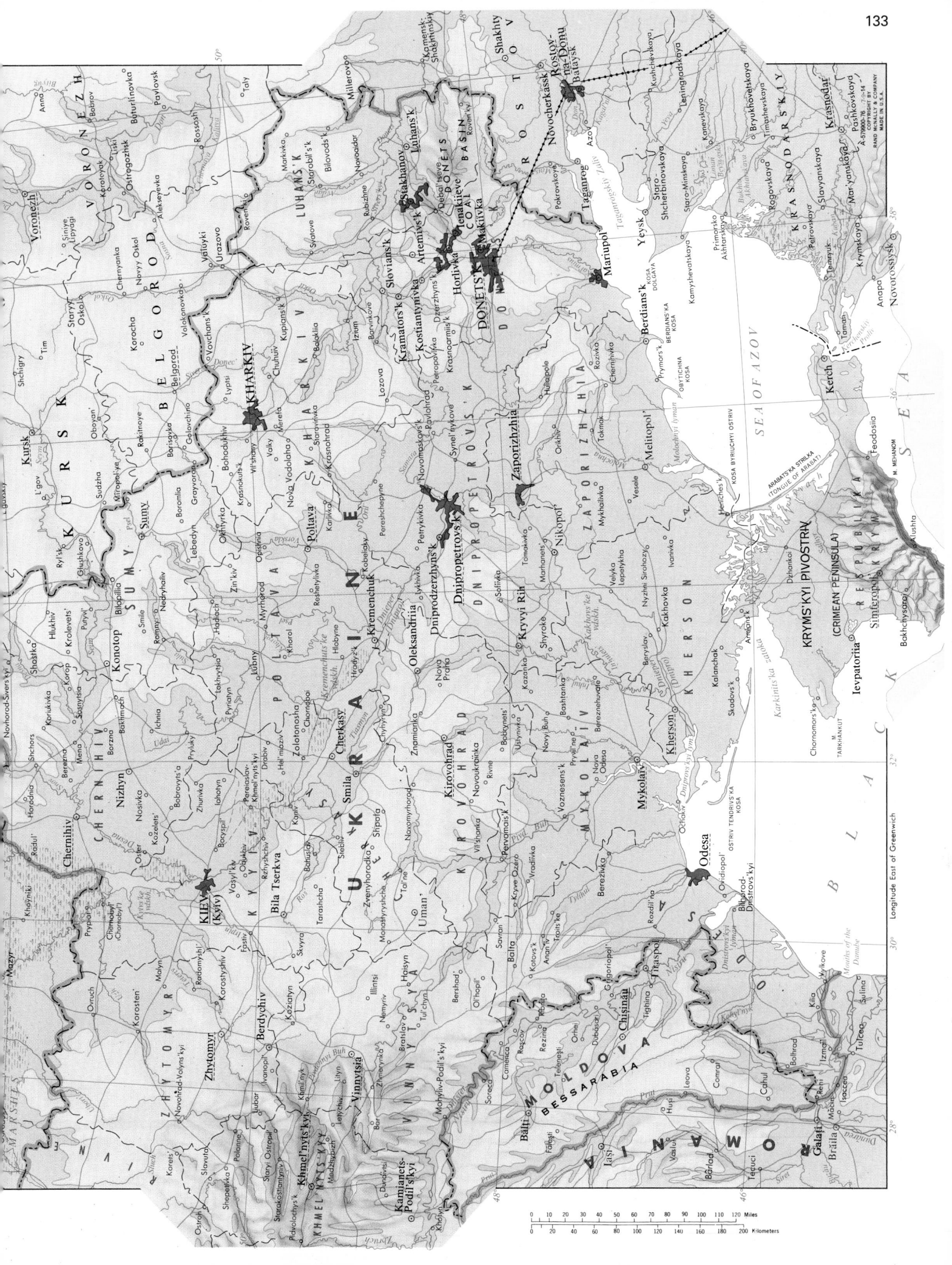

Scale 1:20 000 000; one inch to 315 mile
Lambert's Azimuthal, Equal Area Projecti
Elevations and depressions are given in f

Cities and Towns

0 to 50,000	○
50,000 to 500,000	◉
500,000 to 1,000,000	◎
1,000,000 and over	◆

Relief

Meters	Feet
3050	10000
1525	5000
610	2000
305	1000
152.5	500
Sea Level	Sea Level
152.5	500
1525	5000
3050	10000

Below Sea Level

0	50	100	150	200	250	300 Miles
0	100	200	300	400		500 Kilometers

Continued on pages 116-117

Scale 1:10 000 000; one inch to 160 miles. Conic Projection
Elevations and depressions are given in feet.

Continued on pages 118-119

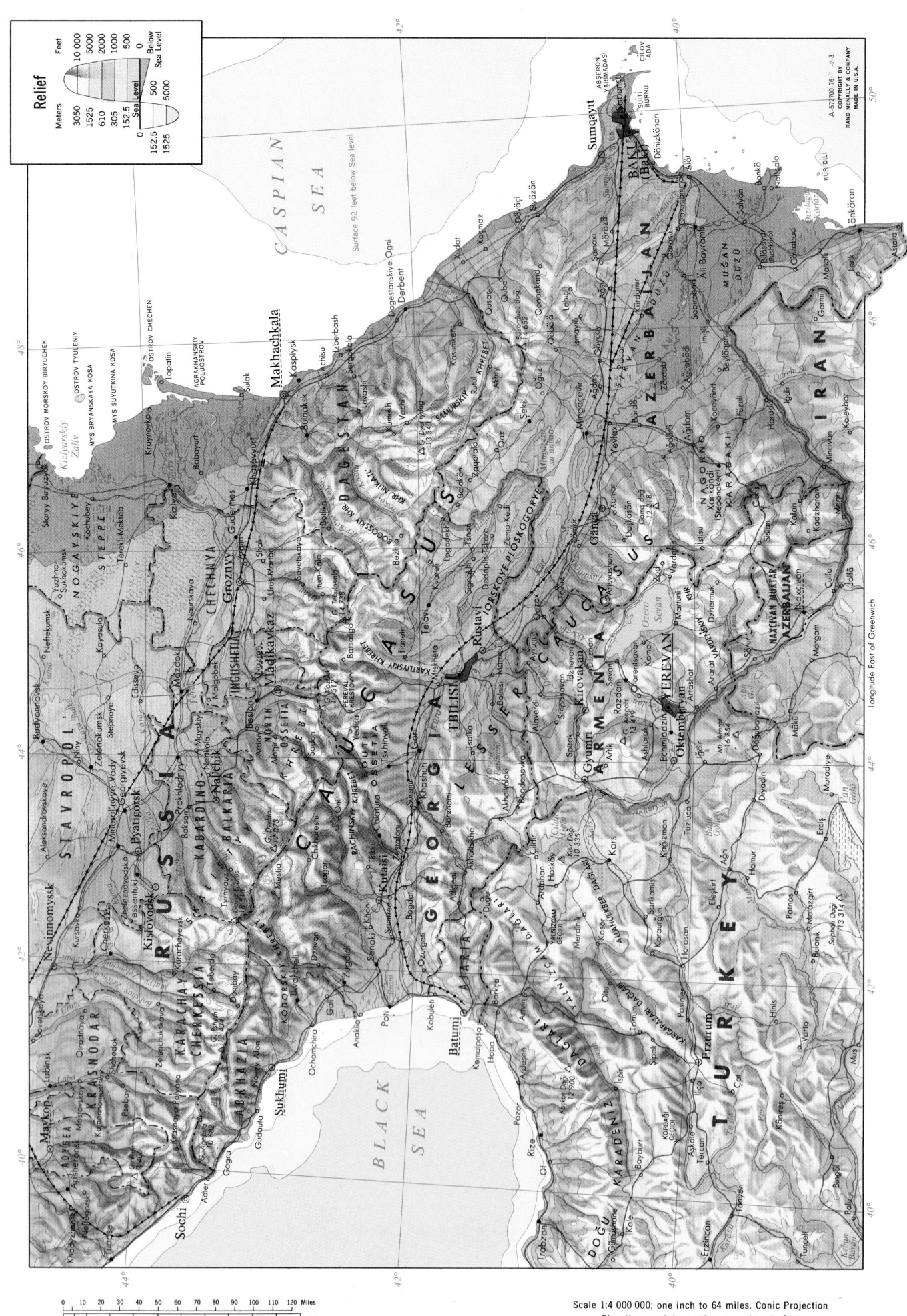

Relief

Meters	Feet
3050	10 000
1525	5000
610	2000
305	1000
152.5	500
0	Sea Level
152.5	500
1525	5000

Below Sea Level

Scale 1:4 000 000; one inch to 64 miles. Conic Projection
Elevations and depressions are given in feet

0 10 20 30 40 50 60 70 80 90 100 110 120 Miles
0 20 40 60 80 100 120 140 160 180 200 Kilometers

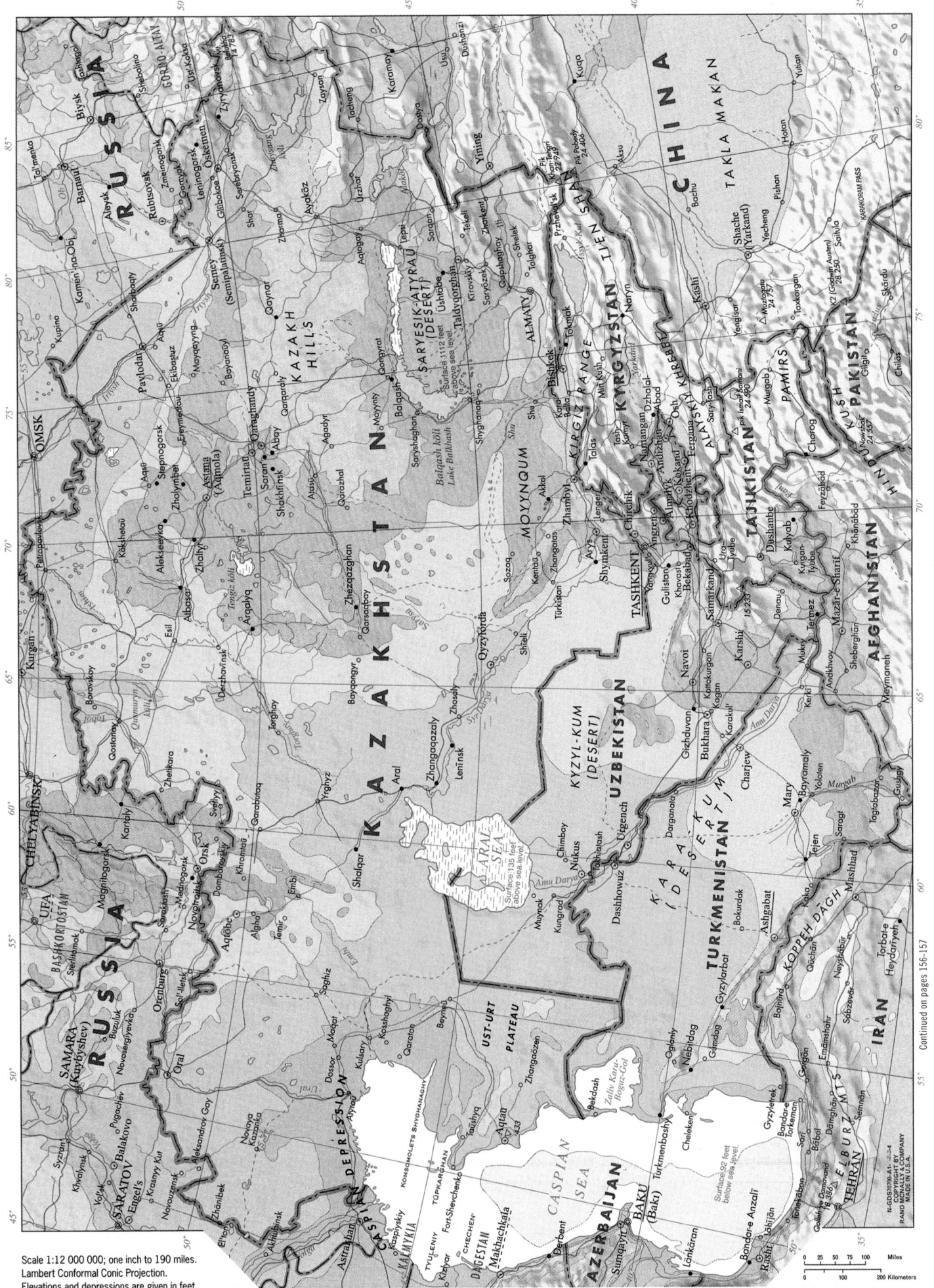

Continued on pages 156-157

Scale 1:12 000 000; one inch to 190 miles.
Lambert Conformal Conic Projection.
Elevations and depressions are given in feet.

| 0 | 25 | 50 | 75 | 100 | Miles |
| 0 | | 100 | | 200 | Kilometers |

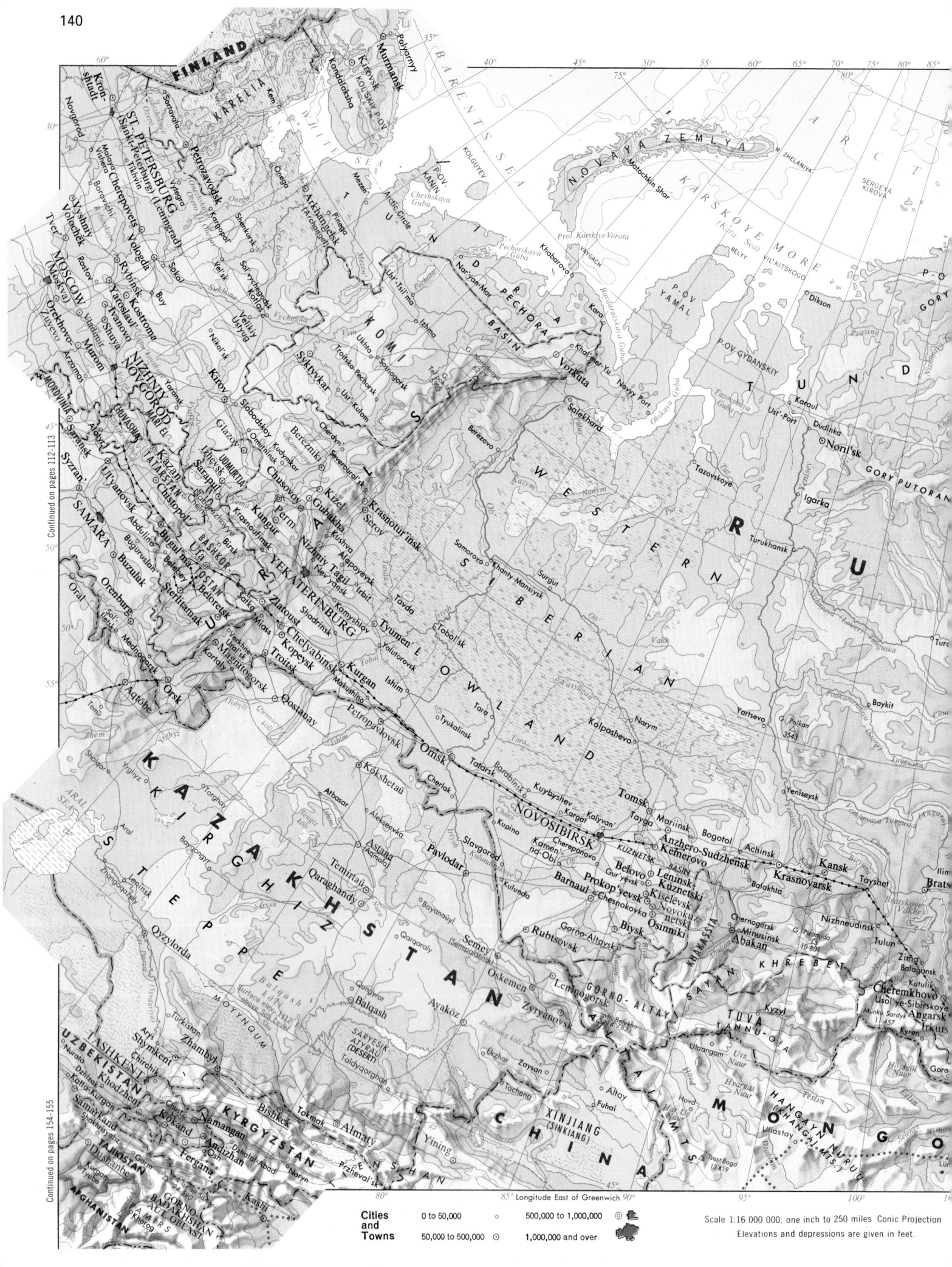

Continued on pages 112-113

Continued on pages 154-155

Cities and Towns

0 to 50,000 ○
500,000 to 1,000,000 ◎
50,000 to 500,000 ⊙
1,000,000 and over

Scale 1:16 000 000; one inch to 250 miles Conic Projection
Elevations and depressions are given in feet.

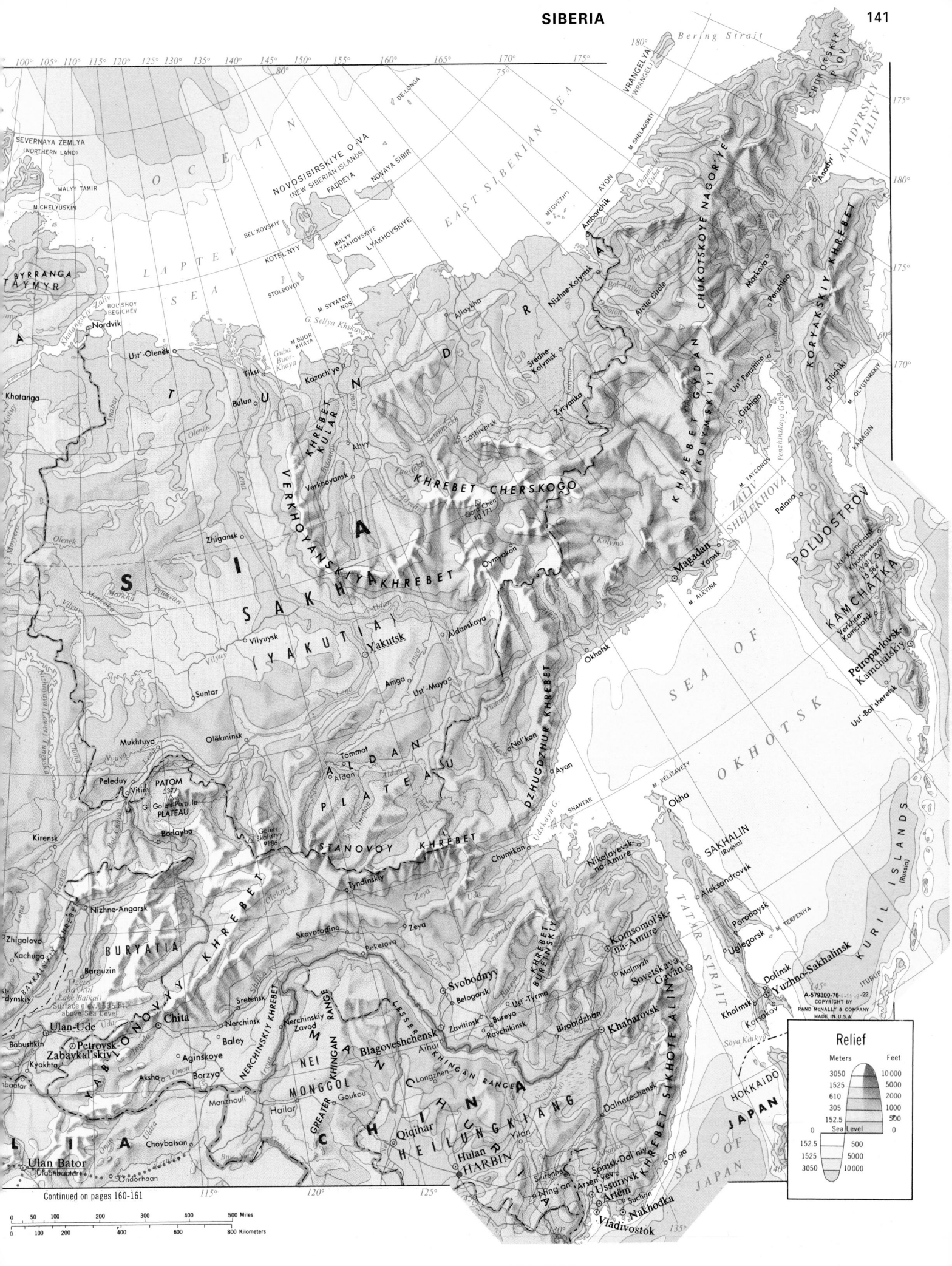

Continued on pages 160-161

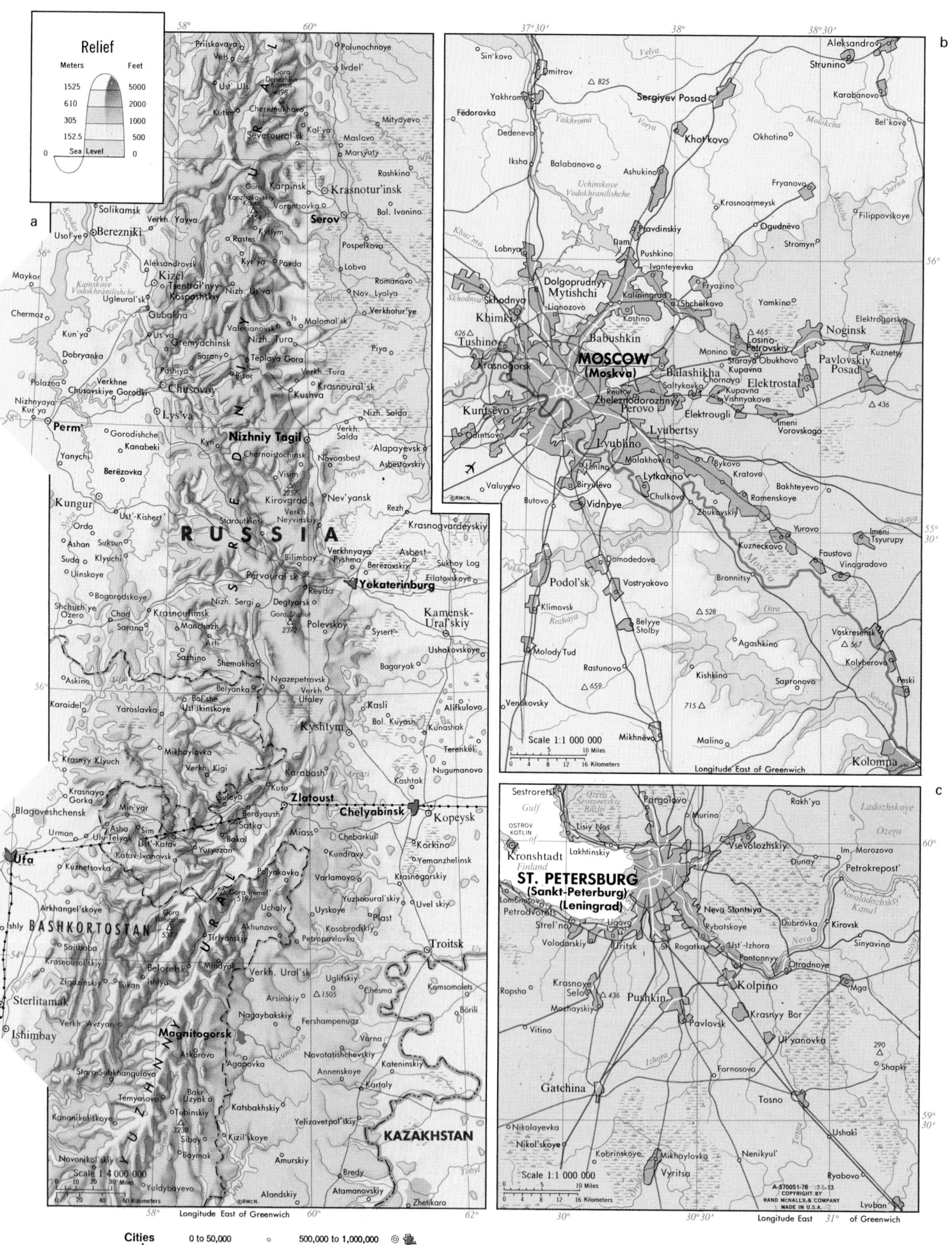

Relief

Meters	Feet
1525	5000
610	2000
305	1000
152.5	500
Sea Level	0

a

b

c

Scale 1:1 000 000

0   4   8   12   16 Kilometers

Longitude East of Greenwich

RUSSIA

BASHKORTOSTAN

KAZAKHSTAN

Scale 1:4 000 000

Longitude East of Greenwich

Scale 1:1 000 000

Longitude East 31° of Greenwich

A-570051-76 -7 5 -13
COPYRIGHT BY
RAND M?NALLY & COMPANY
MADE IN U.S.A.

Cities	0 to 50,000	500,000 to 1,000,000
and		
Towns	50,000 to 500,000	1,000,000 and over

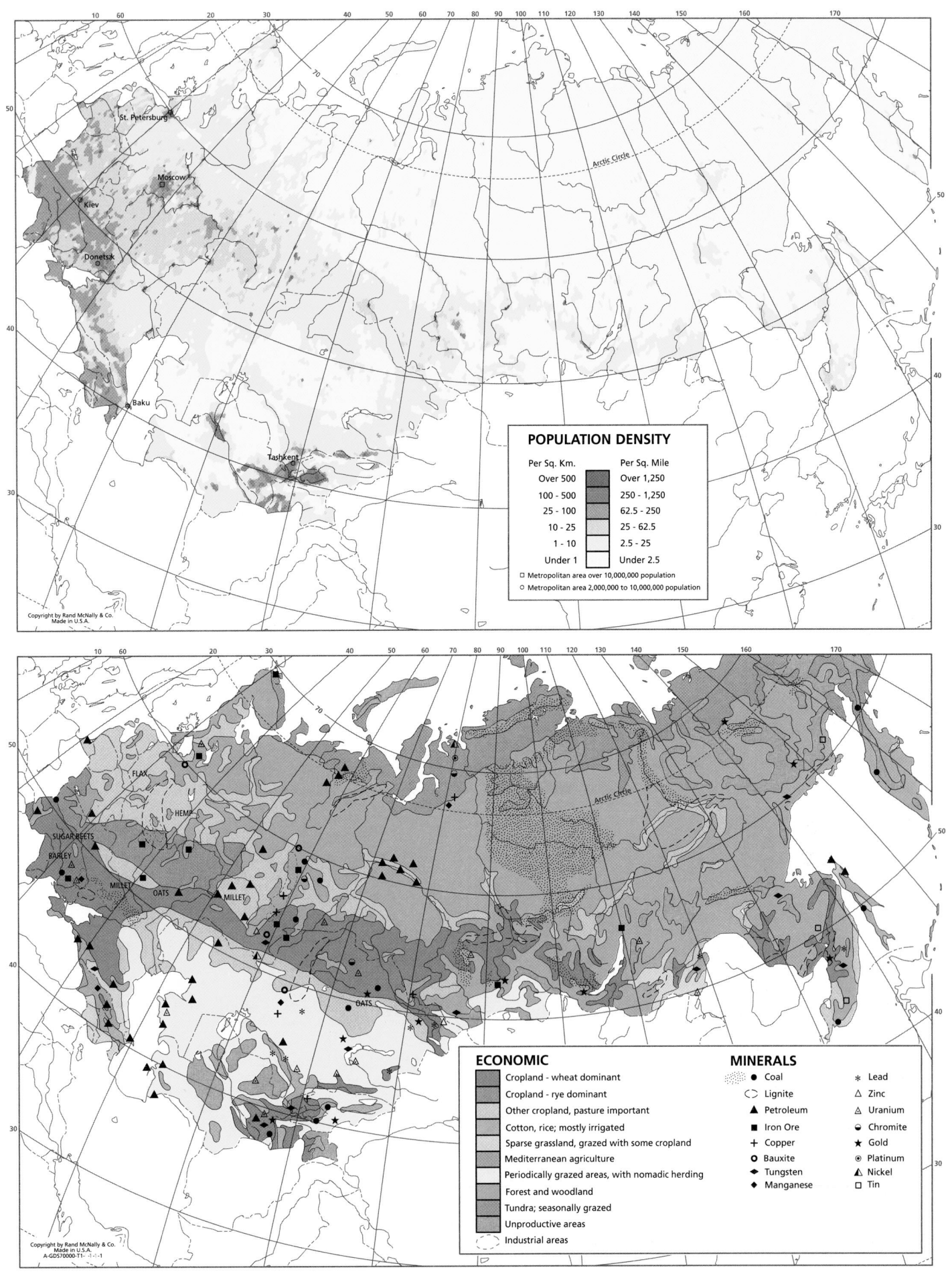

POPULATION DENSITY

Per Sq. Km.	Per Sq. Mile
Over 500	Over 1,250
100 - 500	250 - 1,250
25 - 100	62.5 - 250
10 - 25	25 - 62.5
1 - 10	2.5 - 25
Under 1	Under 2.5

□ Metropolitan area over 10,000,000 population
○ Metropolitan area 2,000,000 to 10,000,000 population

Copyright by Rand McNally & Co.
Made in U.S.A.

ECONOMIC

- Cropland - wheat dominant
- Cropland - rye dominant
- Other cropland, pasture important
- Cotton, rice; mostly irrigated
- Sparse grassland, grazed with some cropland
- Mediterranean agriculture
- Periodically grazed areas, with nomadic herding
- Forest and woodland
- Tundra; seasonally grazed
- Unproductive areas
- Industrial areas

MINERALS

⬟	Coal	✳	Lead
◖	Lignite	△	Zinc
▲	Petroleum	△	Uranium
■	Iron Ore	◠	Chromite
✛	Copper	★	Gold
○	Bauxite	⊙	Platinum
◆	Tungsten	◬	Nickel
◆	Manganese	▢	Tin

Copyright by Rand McNally & Co.
Made in U.S.A.
A-GDS70000-T1- -1-1-1

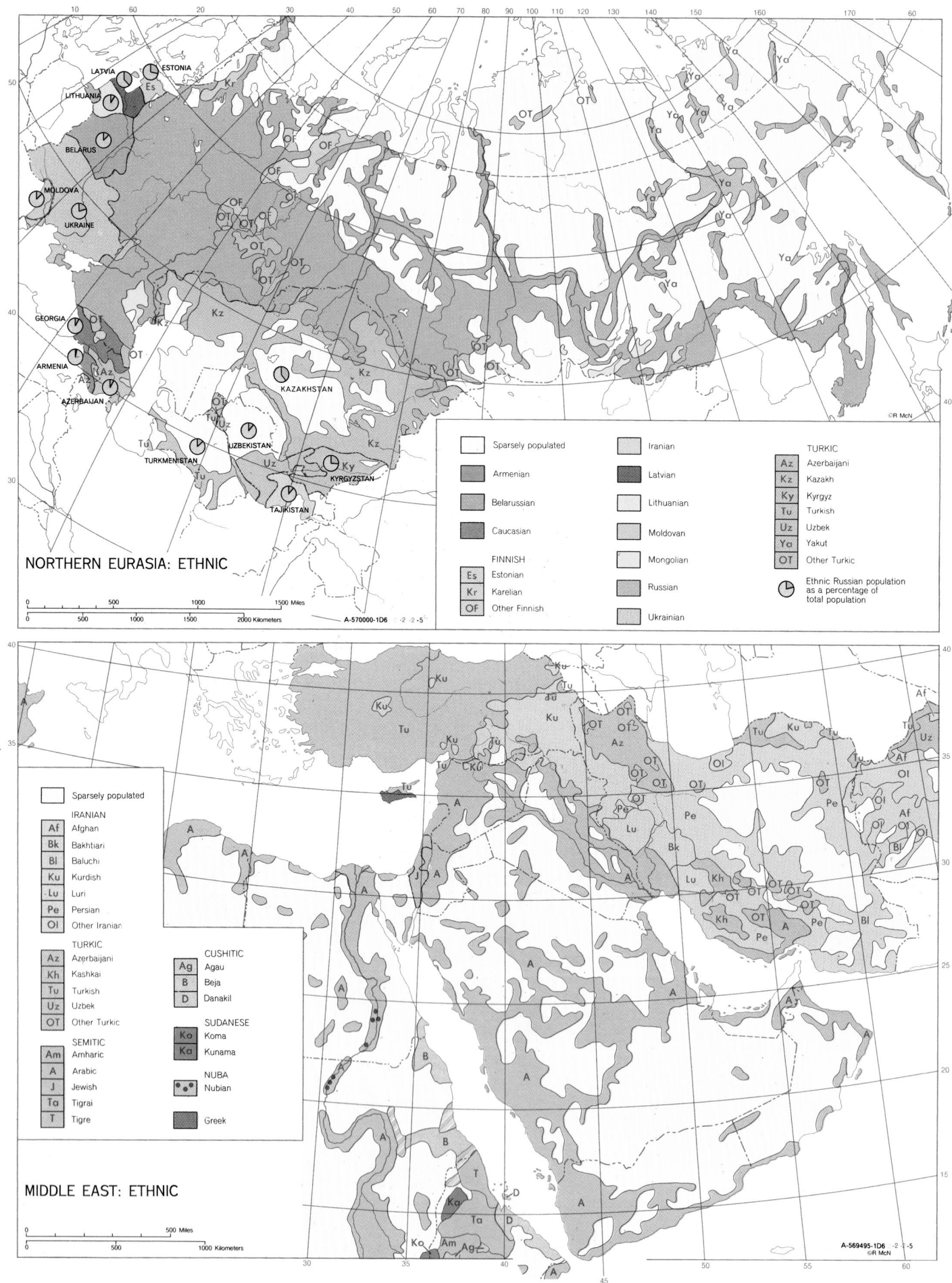

NORTHERN EURASIA: ETHNIC

0 500 1000 1500 Miles
0 500 1000 1500 2000 Kilometers

A-570000-1D6 -2 -2 -5

Sparsely populated	Iranian	TURKIC
Armenian	Latvian	**Az** Azerbaijani
Belarussian	Lithuanian	**Kz** Kazakh
Caucasian	Moldovan	**Ky** Kyrgyz
FINNISH	Mongolian	**Tu** Turkish
Es Estonian	Russian	**Uz** Uzbek
Kr Karelian	Ukrainian	**Ya** Yakut
OF Other Finnish		**OT** Other Turkic

Ethnic Russian population as a percentage of total population

MIDDLE EAST: ETHNIC

0 500 Miles
0 500 1000 Kilometers

IRANIAN		
Af Afghan	CUSHITIC	
Bk Bakhtiari	**Ag** Agau	
Bl Baluchi	**B** Beja	
Ku Kurdish	**D** Danakil	
Lu Luri	SUDANESE	
Pe Persian	**Ko** Koma	
OI Other Iranian	**Ka** Kunama	
TURKIC	NUBA	
Az Azerbaijani	Nubian	
Kh Kashkai		
Tu Turkish	Greek	
Uz Uzbek		
OT Other Turkic		
SEMITIC		
Am Amharic		
A Arabic		
J Jewish		
Ta Tigrai		
T Tigre		

A-569495-1D6 -2 -2 -5
©R McN

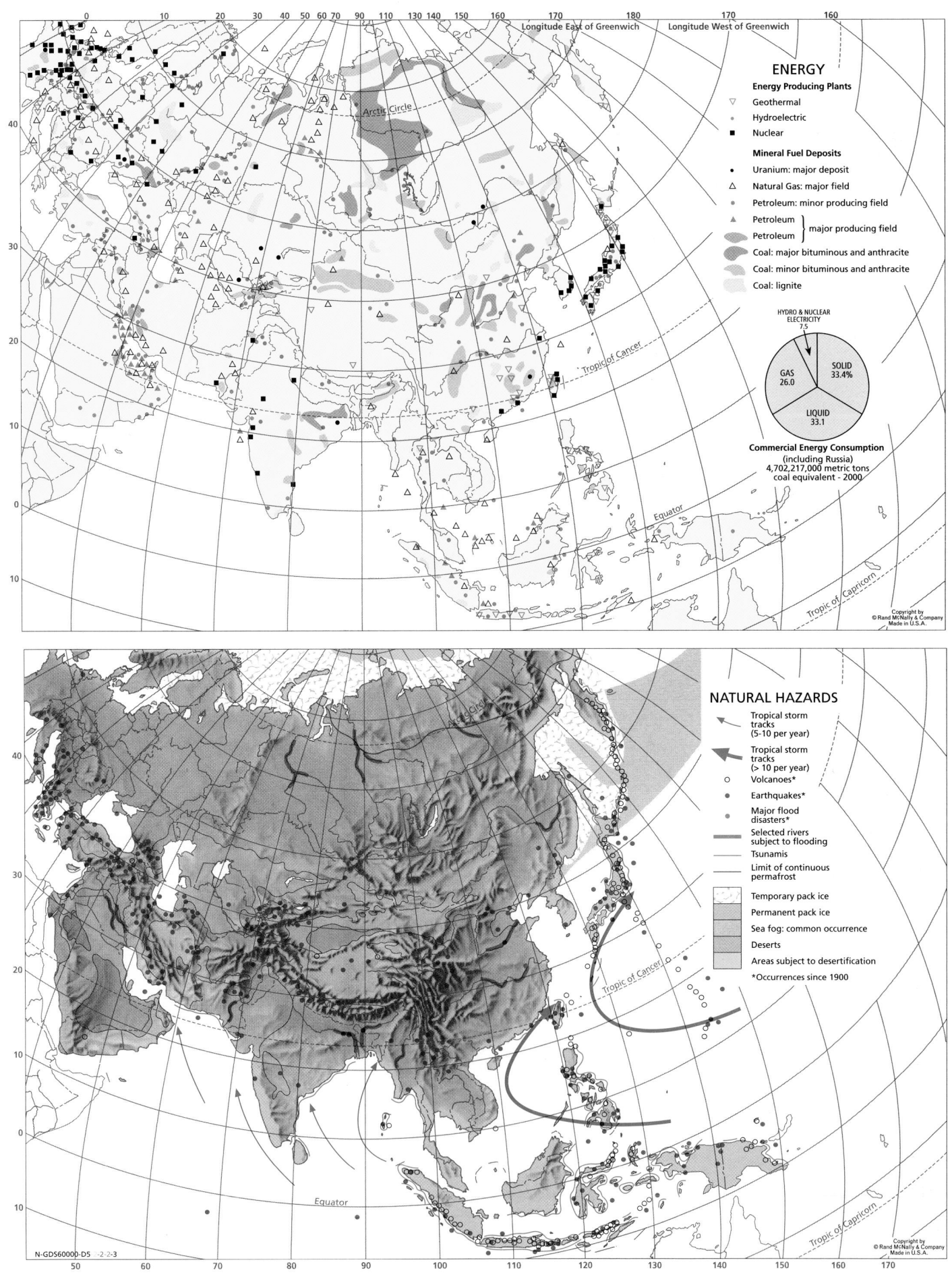

ENERGY

Energy Producing Plants

▽ Geothermal
· Hydroelectric
■ Nuclear

Mineral Fuel Deposits

· Uranium: major deposit
△ Natural Gas: major field
● Petroleum: minor producing field
▲ Petroleum }
Petroleum } major producing field

Coal: major bituminous and anthracite
Coal: minor bituminous and anthracite
Coal: lignite

HYDRO & NUCLEAR
ELECTRICITY
7.5

GAS
26.0

SOLID
33.4%

LIQUID
33.1

Commercial Energy Consumption
(including Russia)
4,702,217,000 metric tons
coal equivalent - 2000

Arctic Circle

Tropic of Cancer

Equator

Tropic of Capricorn

NATURAL HAZARDS

Tropical storm
tracks
(5-10 per year)

Tropical storm
tracks
(> 10 per year)

○ Volcanoes*
● Earthquakes*
· Major flood
disasters*

Selected rivers
subject to flooding

Tsunamis

Limit of continuous
permafrost

Temporary pack ice
Permanent pack ice
Sea fog: common occurrence
Deserts
Areas subject to desertification

*Occurrences since 1900

Tropic of Cancer

Equator

Tropic of Capricorn

N-GDS60000-D5   -2-2-3

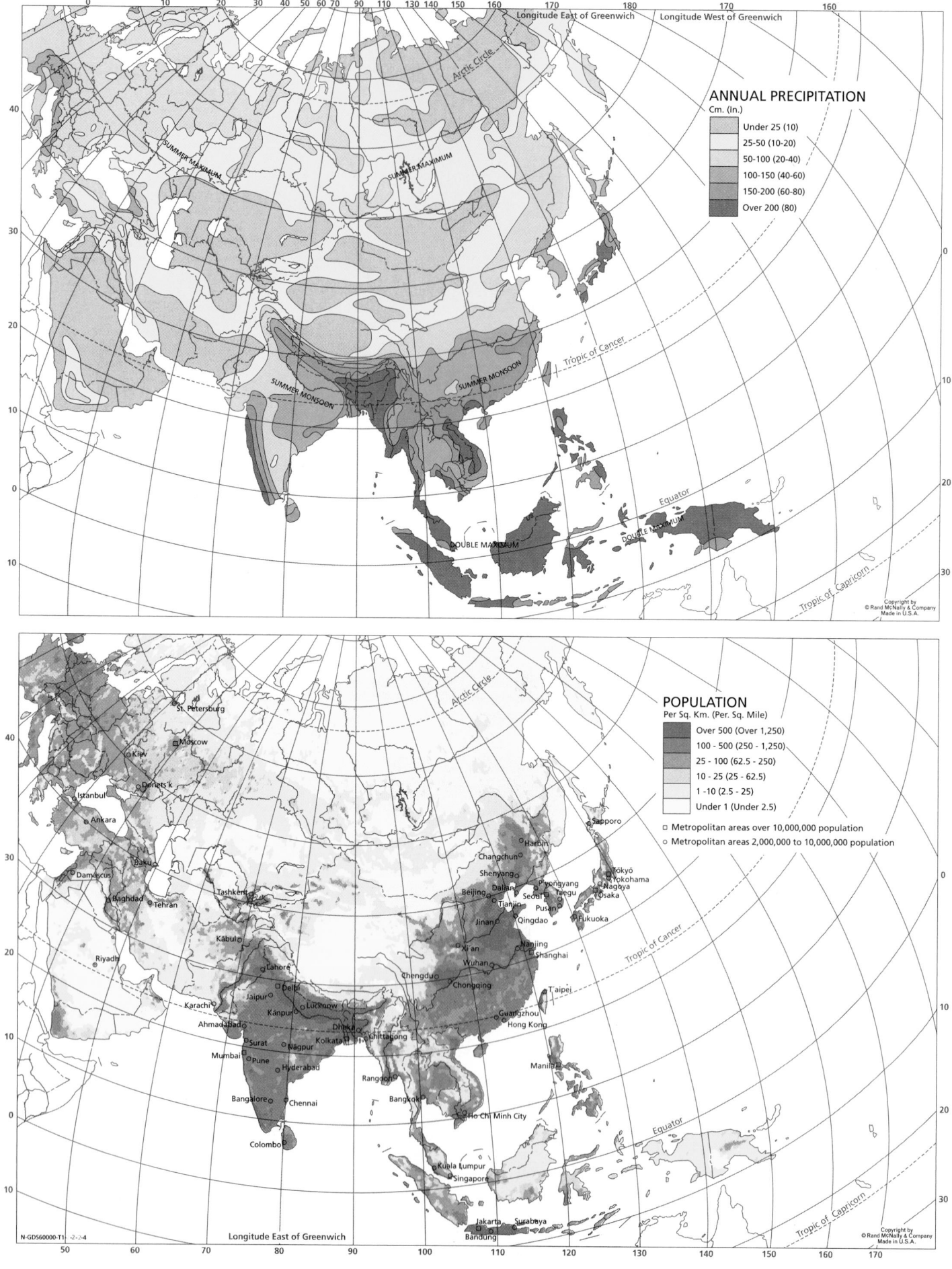

ANNUAL PRECIPITATION
Cm. (In.)

	Under 25 (10)
	25-50 (10-20)
	50-100 (20-40)
	100-150 (40-60)
	150-200 (60-80)
	Over 200 (80)

POPULATION
Per Sq. Km. (Per. Sq. Mile)

	Over 500 (Over 1,250)
	100 - 500 (250 - 1,250)
	25 - 100 (62.5 - 250)
	10 - 25 (25 - 62.5)
	1 -10 (2.5 - 25)
	Under 1 (Under 2.5)

□ Metropolitan areas over 10,000,000 population
○ Metropolitan areas 2,000,000 to 10,000,000 population

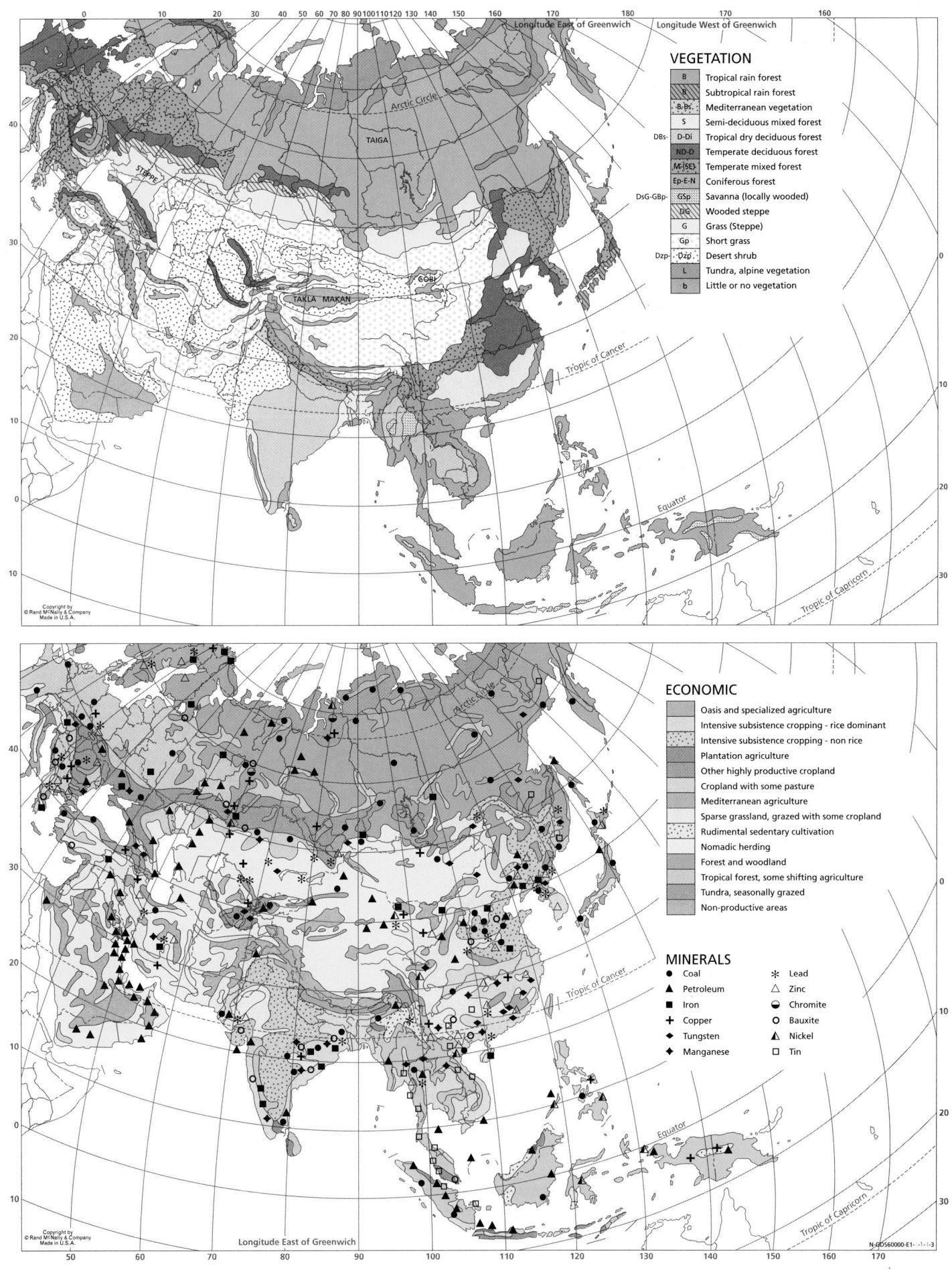

VEGETATION

B	Tropical rain forest
B	Subtropical rain forest
B-Bs.	Mediterranean vegetation
S	Semi-deciduous mixed forest
D-Di	Tropical dry deciduous forest
ND-D	Temperate deciduous forest
M-(SE)	Temperate mixed forest
Ep-E-N	Coniferous forest
GSp	Savanna (locally wooded)
DG	Wooded steppe
G	Grass (Steppe)
Gp	Short grass
Dzp-Dzp	Desert shrub
L	Tundra, alpine vegetation
b	Little or no vegetation

DBs-

DsG-GBp-

Dzp-

ECONOMIC

- Oasis and specialized agriculture
- Intensive subsistence cropping - rice dominant
- Intensive subsistence cropping - non rice
- Plantation agriculture
- Other highly productive cropland
- Cropland with some pasture
- Mediterranean agriculture
- Sparse grassland, grazed with some cropland
- Rudimental sedentary cultivation
- Nomadic herding
- Forest and woodland
- Tropical forest, some shifting agriculture
- Tundra, seasonally grazed
- Non-productive areas

MINERALS

●	Coal	✳	Lead
▲	Petroleum	△	Zinc
■	Iron	⊖	Chromite
✛	Copper	○	Bauxite
◆	Tungsten	▲	Nickel
◆	Manganese	☐	Tin

N-GDS60000-E1- -1-1-3

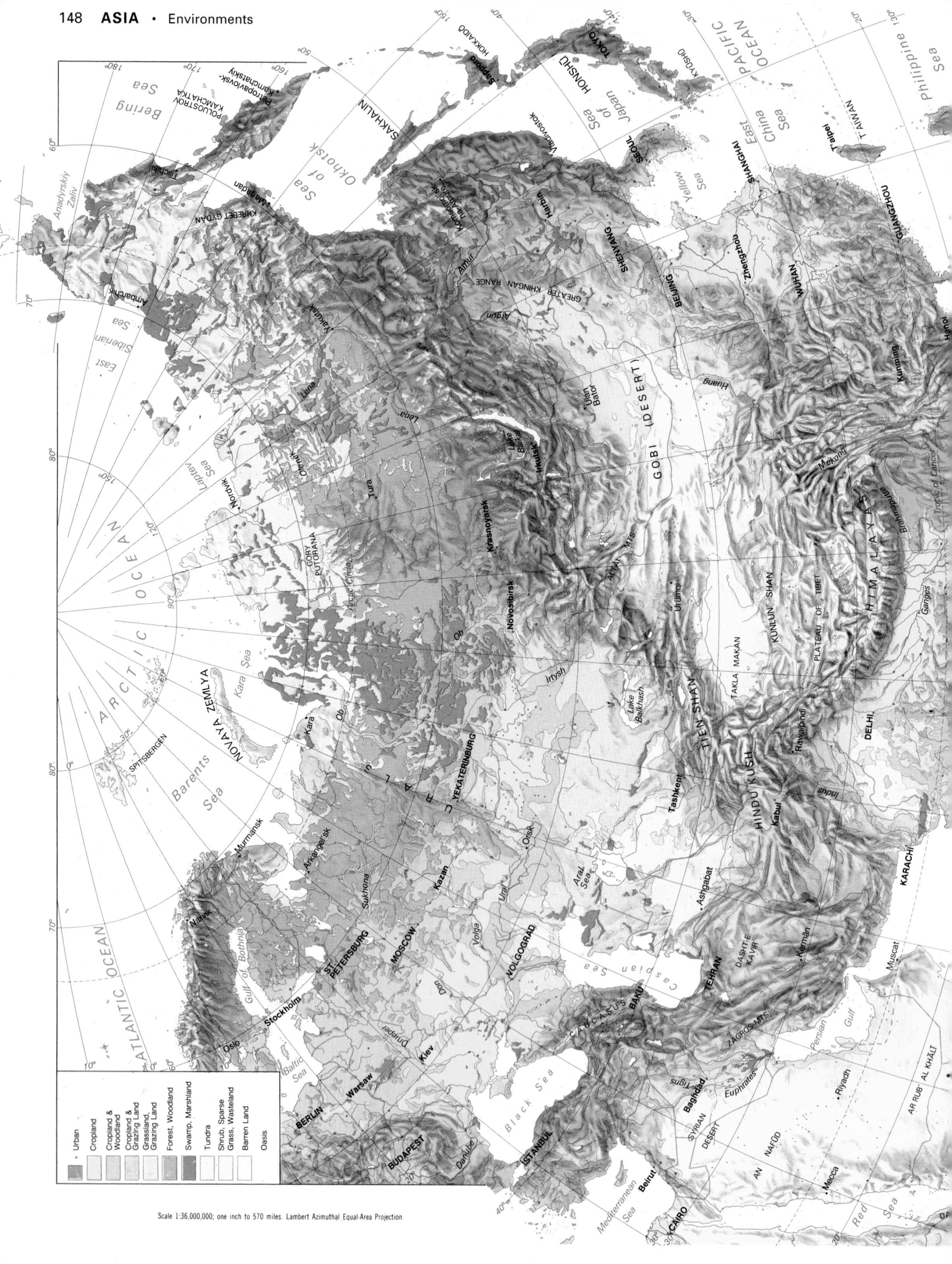

Scale 1:36,000,000; one inch to 570 miles. Lambert Azimuthal Equal-Area Projection

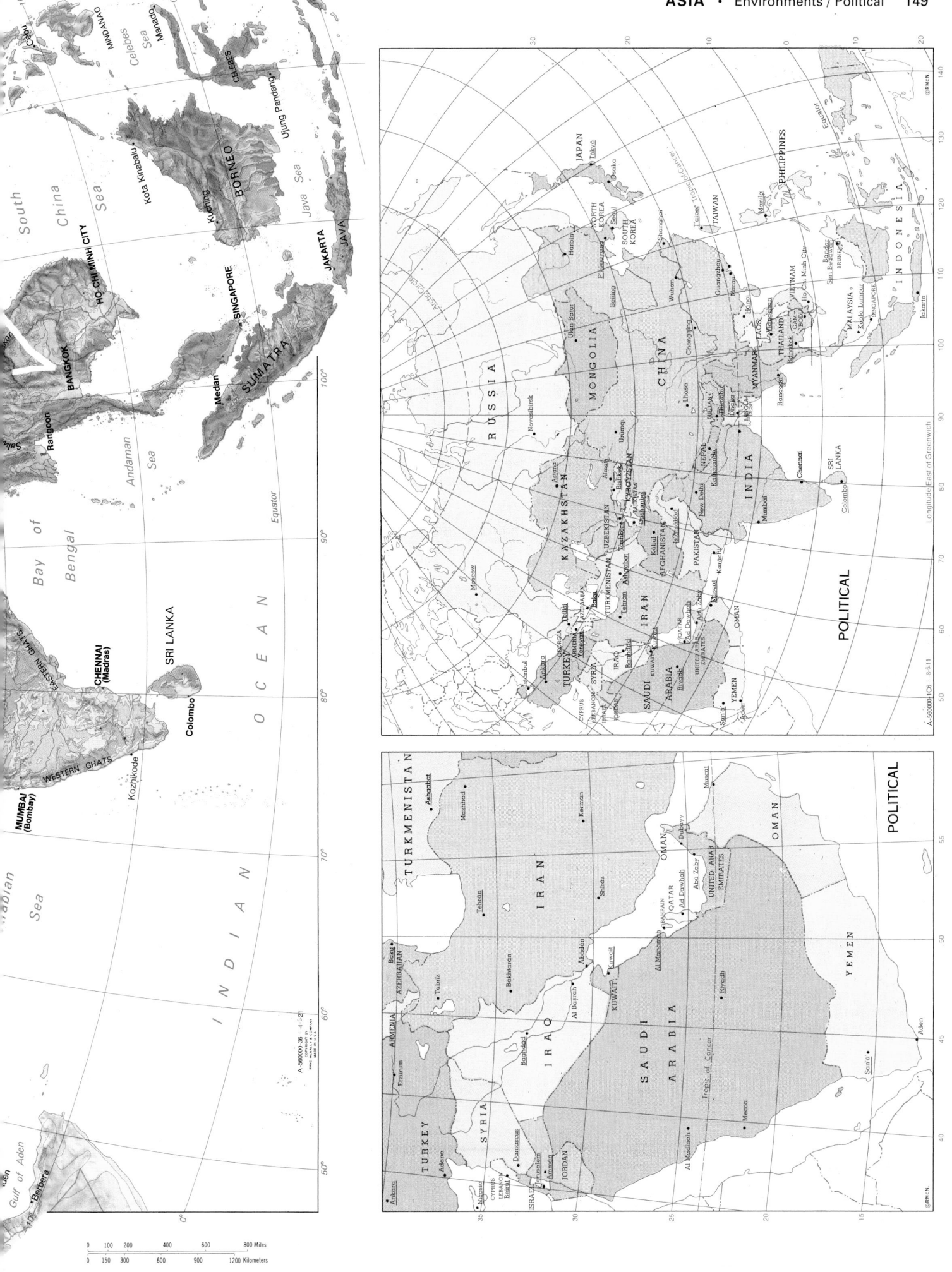

POLITICAL

POLITICAL

Scale 1:40 000 000; one inch to 630 miles. Lambert's Azimuthal, Equal Area Projection
Elevations and depressions are given in feet

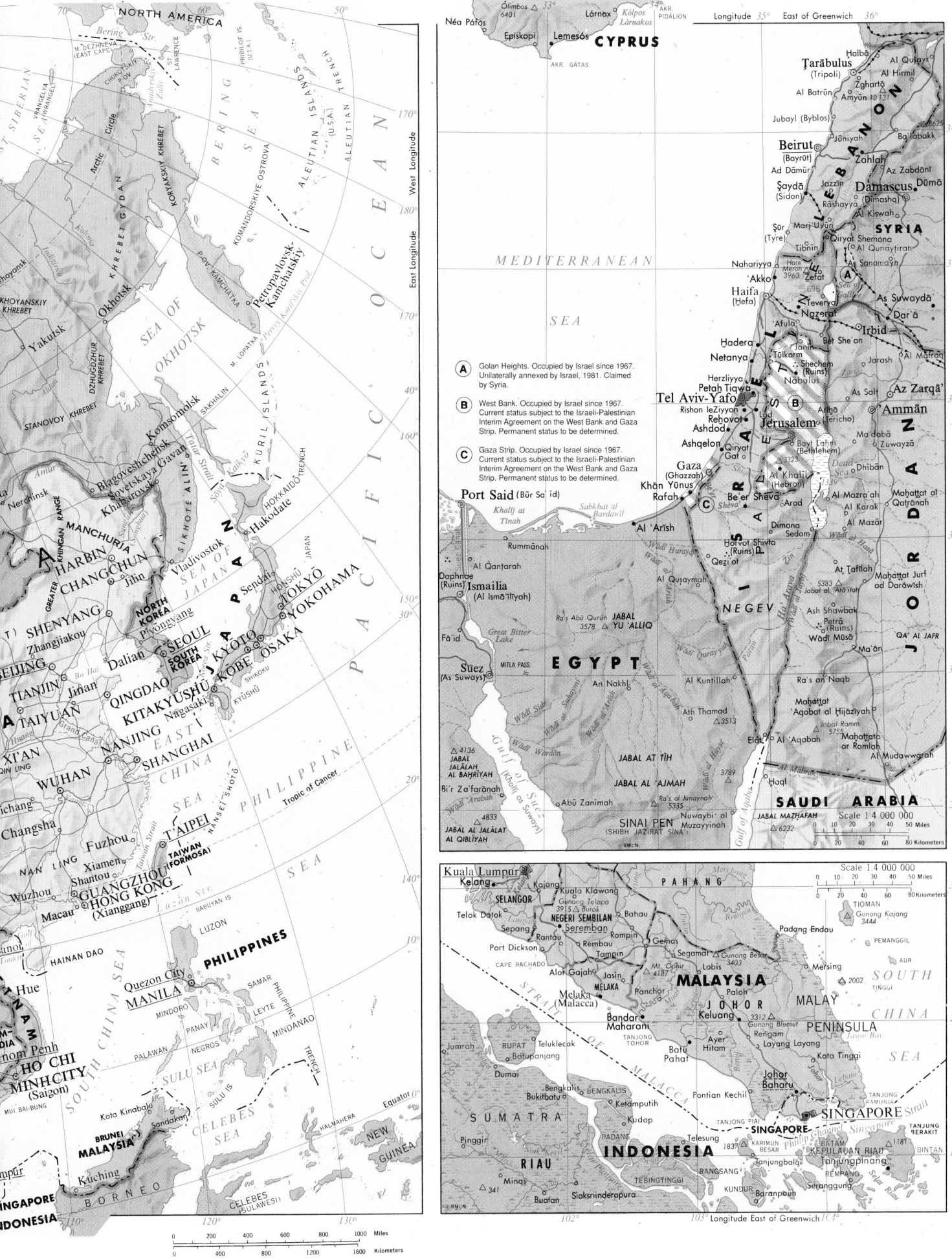

Main map (Middle East / Levant)

Néa Páfos
Ólimbos 6401
Lárnax Kólpos Lárnakos
AKR. PIDÁLION
Episkopi Lemesós
CYPRUS
AKR. GÁTAS

Longitude 35° East of Greenwich 36°

Ţarābulus (Tripoli) Halbā Al Quşayr
Al Batrūn Zgharţā Al Hirmil
Amyūn 10 131
Jubayl (Byblos) Jūniyah Ba'labakk

Beirut (Bayrūt) Zaḥlah Az Zabdānī
Ad Dāmūr Shtawrā
Şaydā (Sidon) Jazzīn **Damascus (Dimashq)** Dūmā
Rāshayyā Al Kiswah
Şūr (Tyre) Marj 'Uyūn
Qiryat Shemona Al Qunayţirah **SYRIA**
Tibnīn
Nahariyya Sanamayn
Hare Meron 3963 Zefat As Suwaydā'
'Akko Teverya Dar'ā
Haifa (Hefa) Nazerat Irbid
'Afula Al Mafraq
Hadera Bet She'an
Netanya Jenin Jarash
Tulkarm
Herzliyya Shechem (Ruins) As Salt
Petah Tiqwa Nābulus **Az Zarqā'**
Tel Aviv-Yafo
Rishon leZiyyon **Amman**
Rehovot Lod Arḍha (Jericho)
Ashdod **Jerusalem** Ma'dabā
Ashqelon Qiryat Gat Bayt Laḥm (Bethlehem) Zuwayzā
3323
Gaza (Ghazzah) Al Khalīl (Hebron) Dhiban
Khān Yūnus Be'er Sheva Al Mazra'ah
Rafah Arad Al Karak
Dimona Al Mazar
Sedom
Port Said (Būr Sa'īd) Ḥorvot Shivta (Ruins) Maḥaţţat al Qaţrānah
Khalīj at Tīnah Sabkhat al Bardawil Qeẕi'ot
Al 'Arīsh Aţ Ţafīlah
Rummānah Ash Shawbak
Al Qanţarah 5383 Petra (Ruins)
Daphnae (Ruins) Al Qusaymah Jabal al 'Atā'itah Wādī Mūsā
Ismailia (Al Ismā'īlīyah) **NEGEV** Ma'ān
Fā'id
Ra's Abū Qurūn JABAL YU 'ALLIQ QA' AL JAFR
Great Bitter Lake 3578
An Nakhl Maḥaţţat 'Aqabat al Ḥijāzīyah
Suez (As Suways) MITLA PASS **EGYPT**
Al Kuntillah Ra's an Naqb
Ath Thamad Elat Al 'Aqabah
3513 Jabal Ramm 575 Al Mudawwarah
JABAL AT TĪH 3789
4136 Ḥaql
JABAL JALĀLAH AL BAḤRĪYAH JABAL AL 'AJMAH
Bi'r Za'farānah **SAUDI ARABIA**
Abū Zanimah Nuwaybi' al Muzayyinah JABAL MAZḤAFAH
4833 Ra's al Junaynah 5335 6232
JABAL AL JALĀLAT AL QIBLĪYAH SINAI PEN (SHIBH JAZĪRAT SĪNĀ)

Scale 1:4 000 000
0 10 20 30 40 50 Miles
0 20 40 60 80 Kilometers

Legend

A Golan Heights. Occupied by Israel since 1967. Unilaterally annexed by Israel, 1981. Claimed by Syria.

B West Bank. Occupied by Israel since 1967. Current status subject to the Israeli-Palestinian Interim Agreement on the West Bank and Gaza Strip. Permanent status to be determined.

C Gaza Strip. Occupied by Israel since 1967. Current status subject to the Israeli-Palestinian Interim Agreement on the West Bank and Gaza Strip. Permanent status to be determined.

Left map (East Asia / Pacific)

NORTH AMERICA
M. DEZHNEVA (EAST CAPE)
Arctic Circle
PRIBILOF IS. (U.S.A.)
ALEUTIAN ISLANDS (U.S.A.)
West Longitude
ALEUTIAN TRENCH
BERING SEA
East Longitude
VRANGELYA (WRANGEL)
Zil
KHREBET GYDAN
KOMANDORSKIYE OSTROVA
P-OV KAMCHATKA
Petropavlovsk-Kamchatskiy
Kuril'skiy Prol.
Khoyonsk
KHOYANSKIY KHREBET
Okhotsk
SEA OF OKHOTSK
M. LOPATKA
Yakutsk
DZHUGDZHUR KHREBET
SAKHALIN
KURIL ISLANDS
STANOVOY KHREBET
HOKKAIDO TRENCH
Nerchinsk
Amur
Blagoveshchensk
Komsomolsk
Sovetskaya Gavan
Khabarovsk
Tatar Strait
Kaikyō
Hakodate
MANCHURIA
GREATER KHINGAN RANGE
SIKHOTE ALIN'
HONSHŪ
JAPAN
A
HARBIN
CHANGCHUN
Vladivostok
SEA OF JAPAN
Sendai
Jilin
TOKYO
YOKOHAMA
SHENYANG
NORTH KOREA
Zhangjiakou
P'yongyang
KYOTO
OSAKA
SEOUL
SHIKOKU
BEIJING
SOUTH KOREA
KOBE
Dalian
KITAKYUSHU
Nagasaki
KYŪSHŪ
TIANJIN
Jinan
QINGDAO
TAIYUAN
Bo Hai
Grand Canal
XI'AN
QIN LING
Huang
NANJING
SHANGHAI
WUHAN
EAST CHINA SEA
NANSEI SHOTŌ
PHILIPPINE SEA
ichang
Changsha
Tropic of Cancer
NAN LING
Fuzhou
T'AIPEI
Wuzhou
Xiamen
Shantou
TAIWAN (FORMOSA)
Macau
GUANGZHOU
HONG KONG (Xianggang)
Luzon Str.
BABUYAN IS.
LUZON
HAINAN DAO
PHILIPPINES
Quezon City
MANILA
MINDORO
SAMAR
PANAY
LEYTE
PHILIPPINE TRENCH
Hue
VIETNAM
PALAWAN
NEGROS
MINDANAO
nom Penh
HO CHI MINH CITY (Saigon)
SULU SEA
SULU IS.
MUI BAI-BUNG
Kota Kinabalu
Sandakan
CELEBES SEA
BRUNEI
MALAYSIA
NEW GUINEA
HALMAHERA
Equator
Kuching
BORNEO
CELEBES (SULAWESI)
umpur
SINGAPORE
INDONESIA
SOUTH CHINA SEA
PACIFIC OCEAN

0 200 400 600 800 1000 Miles
0 400 800 1200 1600 Kilometers

Bottom-right map (Malaysia / Singapore)

Scale 1:4 000 000
0 10 20 30 40 50 Miles
0 20 40 60 80 Kilometers

Kuala Lumpur
Kelang
Kajang
PAHANG
Merchong
SELANGOR
Kuala Klawang
Gunong Telapa 3915 Burok
Bahau
TIOMAN
Telok Datok
Lampat
Sepang
NEGERI SEMBILAN
Rompin
Gunong Kajang 3444
Port Dickson
Rantau
Seremban
Gemas
PEMANGGIL
CAPE RACHADO
Rembau
Tampin
Gunong Besar 3403
Padang Endau
Mt. Ophir 4187
Segamat
Gunong Blumut
AUR
Alor Gajah
Jasin
Labis
Mersing
Sungei
MELAKA
Panchor
MALAYSIA
SOUTH
Melaka (Malacca)
Paloh
MALAY
TINGGI
Bandar Maharani
JOHOR
Keluang 3312
Rengam
Jason Bay
PENINSULA
CHINA
Jumrah
RUPAT
Teluklecak
Batu Pahat
Ayer Hitam
Layang Layang
Dumai
Batupanjang
Pontian Kechil
Kota Tinggi
SEA
Bengkalis
BENGKALIS
Ketamputih
Johor Baharu
TANJONG RAMUNIA
SUMATRA
Kudap
TANJONG PIAI
TANJUNG BERAKIT
Pinggir
Telesung
SINGAPORE
PADANG
1837 KARIMUN
BATAM
KEPULAUAN RIAU
BINTAN
RIAU
Tanjungbalai
Tanjungpinang
REMPANG
INDONESIA
341 Minas
Buatan
1181
Siaksriinderapura
Rangsang
SERANGGUNG
TEBINGTINGGI
KUNDUR
Baranpauh
Sedari Riau

Continued on page 183

Scale 1:40 000 000; one inch to 630 miles. Lambert's Azimuthal, Equal Area Projection
Elevations and depressions are given in feet

Relief

Meters	Feet
3050	10 000
1525	5000
610	2000
305	1000
0 Sea Level	Sea Level
	0 Below Sea Level
152.5	500
1525	5000
3050	10 000
6100	20 000

A-519695-76 · 24 · 2646
COPYRIGHT BY
RAND McNALLY & COMPANY
MADE IN U.S.A.

Longitude East of Greenwich

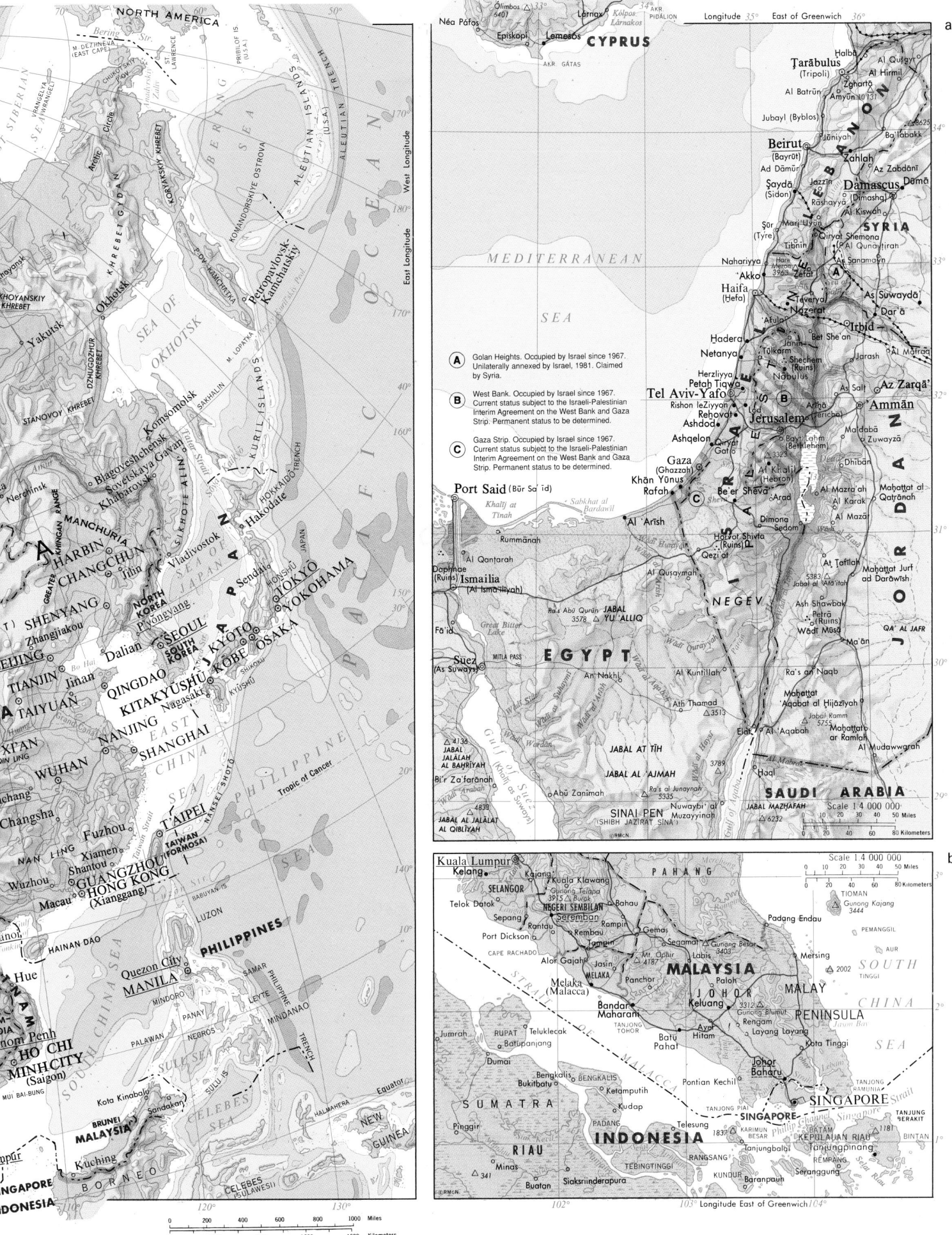

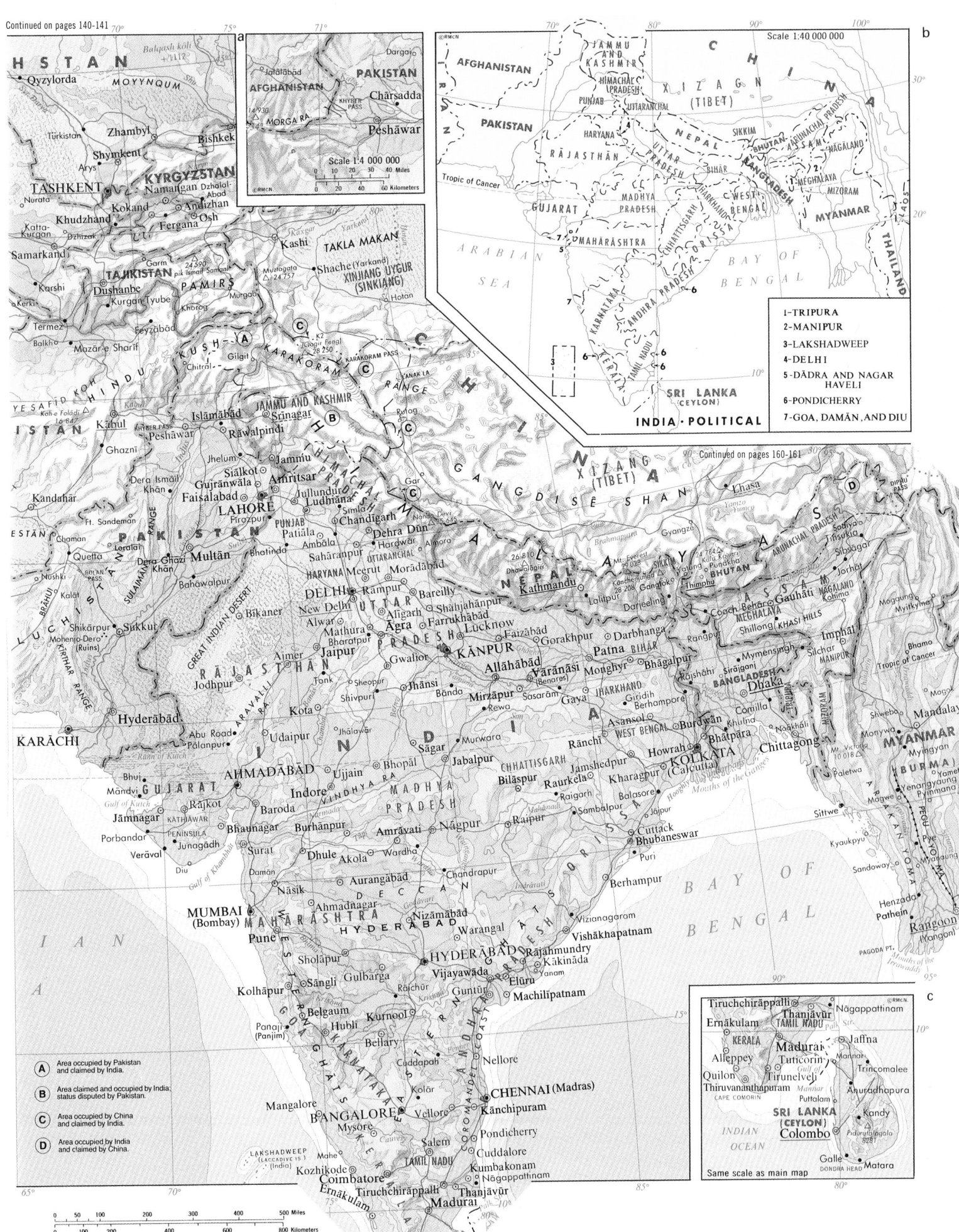

Continued on pages 140-141

Continued on pages 160-161

Scale 1:4 000 000

Scale 1:40 000 000

INDIA · POLITICAL

1-TRIPURA
2-MANIPUR
3-LAKSHADWEEP
4-DELHI
5-DĀDRA AND NAGAR HAVELI
6-PONDICHERRY
7-GOA, DAMĀN, AND DIU

(A) Area occupied by Pakistan and claimed by India.

(B) Area claimed and occupied by India; status disputed by Pakistan.

(C) Area occupied by China and claimed by India.

(D) Area occupied by India and claimed by China.

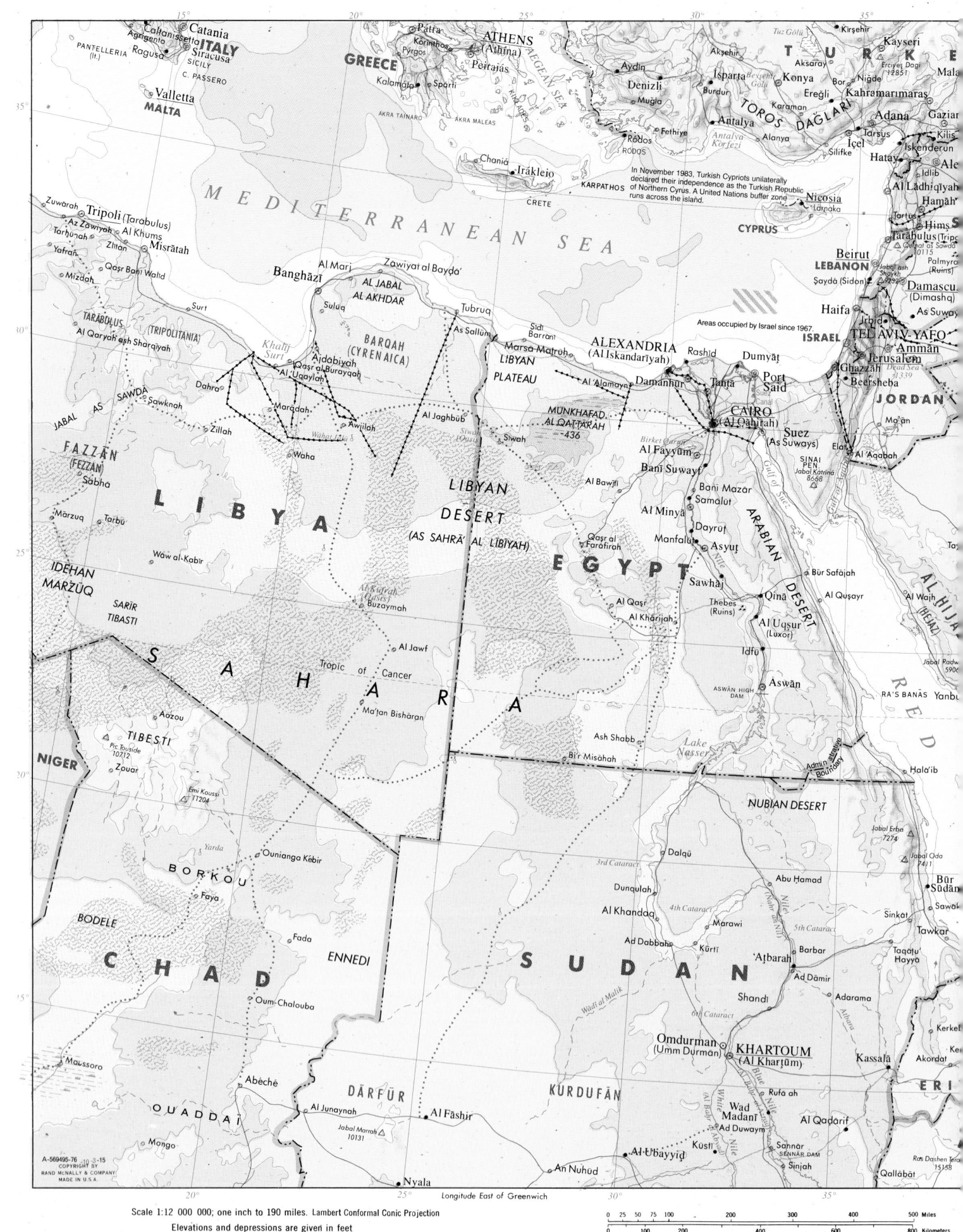

Scale 1:12 000 000; one inch to 190 miles. Lambert Conformal Conic Projection

Elevations and depressions are given in feet

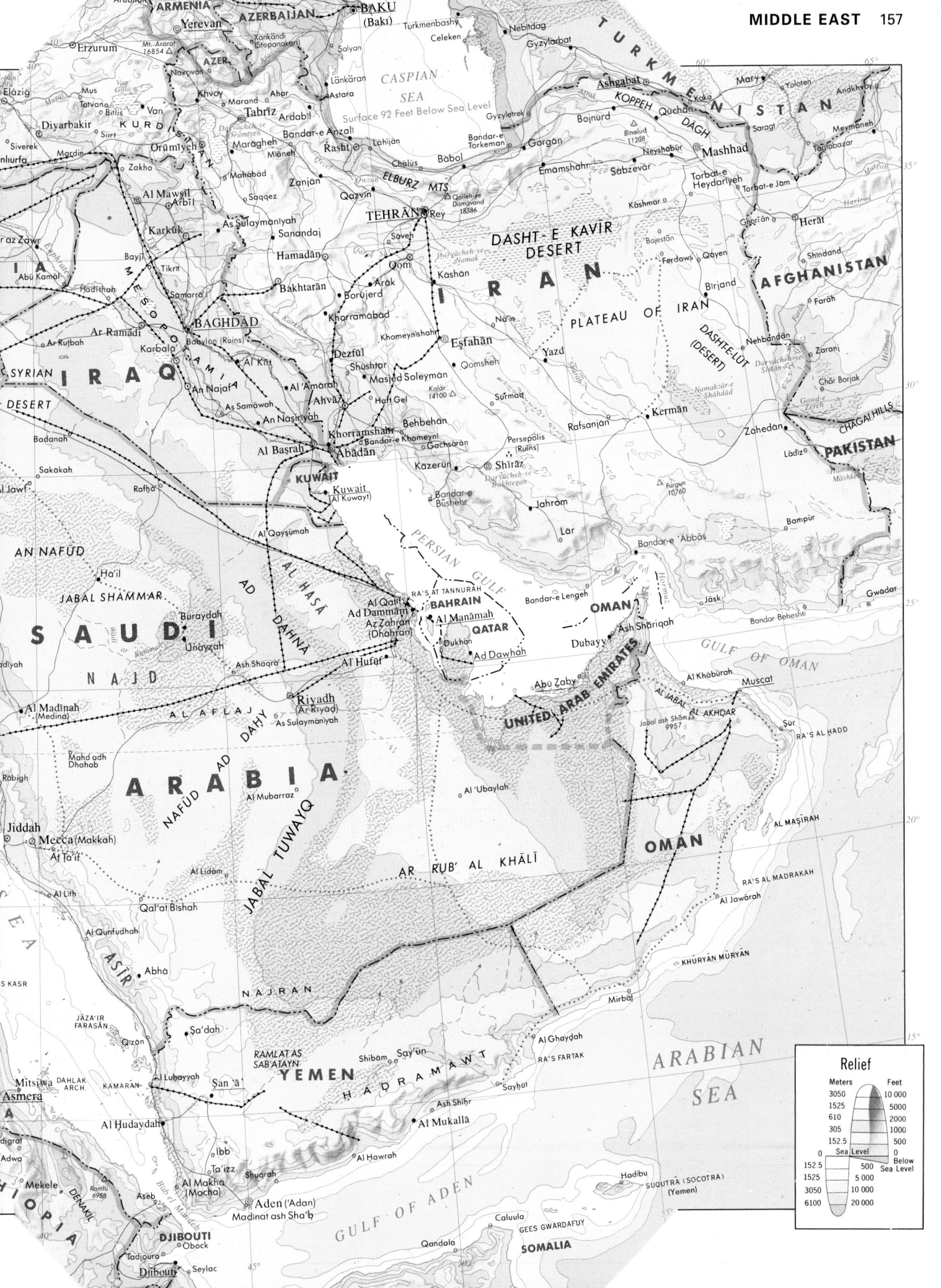

158

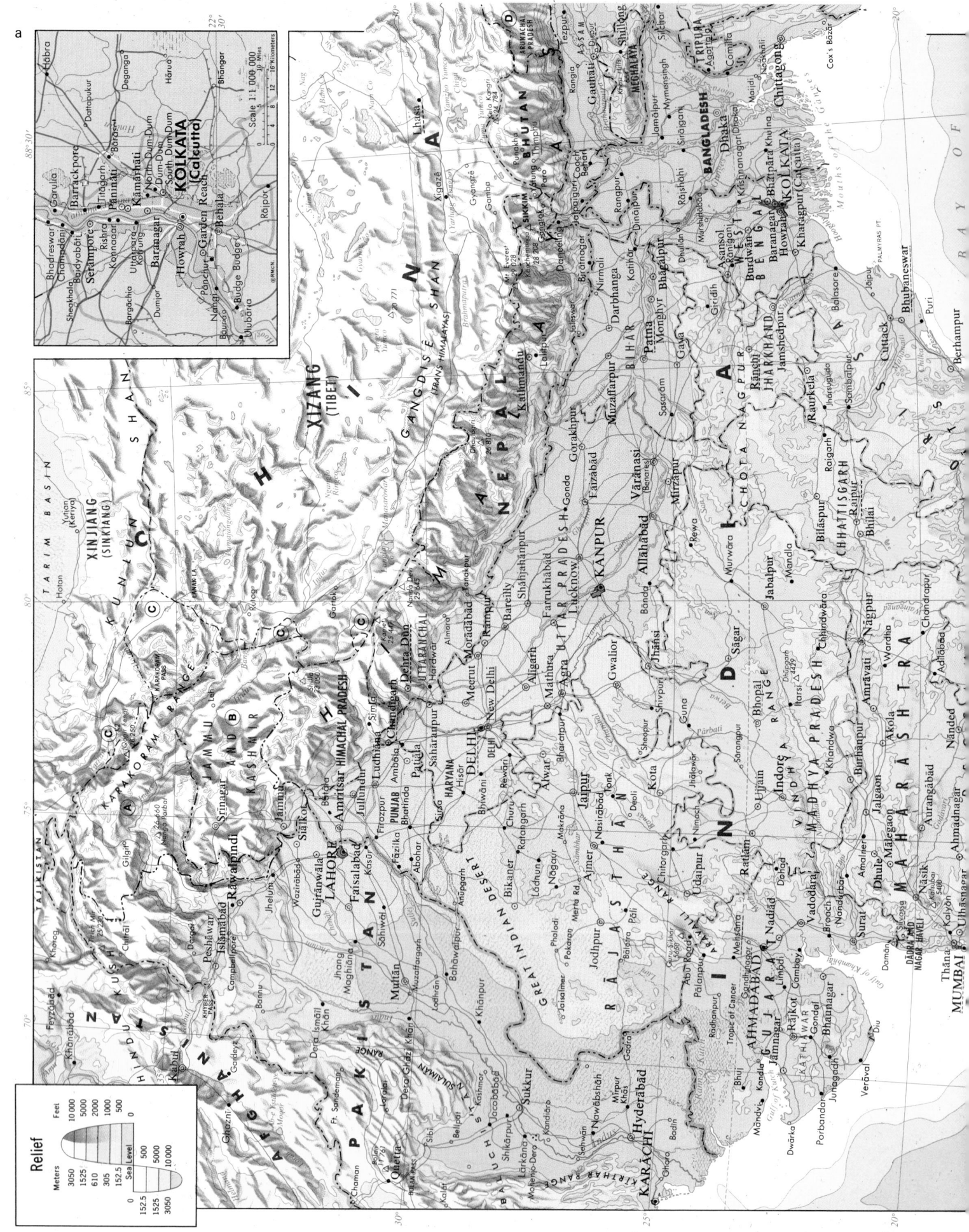

POPULATION DENSITY

Per Sq. Km.	Population	Per Sq. Mile
Over 500		Over 1,250
100 - 500		250 - 1,250
25 - 100		62.5 - 250
10 - 25		25 - 62.5
1 - 10		2.5 - 25
Under 1		Under 2.5

□ Metropolitan area over 10,000,000 population
○ Metropolitan area 2,000,000 to 10,000,000 population

Copyright by Rand McNally & Co.
Made in U.S.A.
A-GO562000-T1-1-11

ECONOMIC AND MINERALS

Copyright by Rand McNally & Co.
Made in U.S.A.

ECONOMIC AND MINERALS

Economic

Rice
Wheat
Sorghum
Woodlands
Scrub and Pasture
Non-productive
Industrial Areas

J Jute
τ Tea
Sc Sugarcane
c Coffee
Co Cotton
R Rubber

Minerals

● Coal
▲ Iron Ore
■ Manganese
✦ Copper
○ Bauxite
◖ Chromite
◀ Petroleum

Longitude East of Greenwich

A-561000-76
COPYRIGHT BY
RAND McNALLY & COMPANY
MADE IN U.S.A.

Ⓐ Area occupied by Pakistan and claimed by India.
Ⓑ Area claimed and occupied by India; status disputed by Pakistan.
Ⓒ Area occupied by China and claimed by India.
Ⓓ Area occupied by India and claimed by China.

MUMBAI (Bombay)

Scale 1:1 000 000

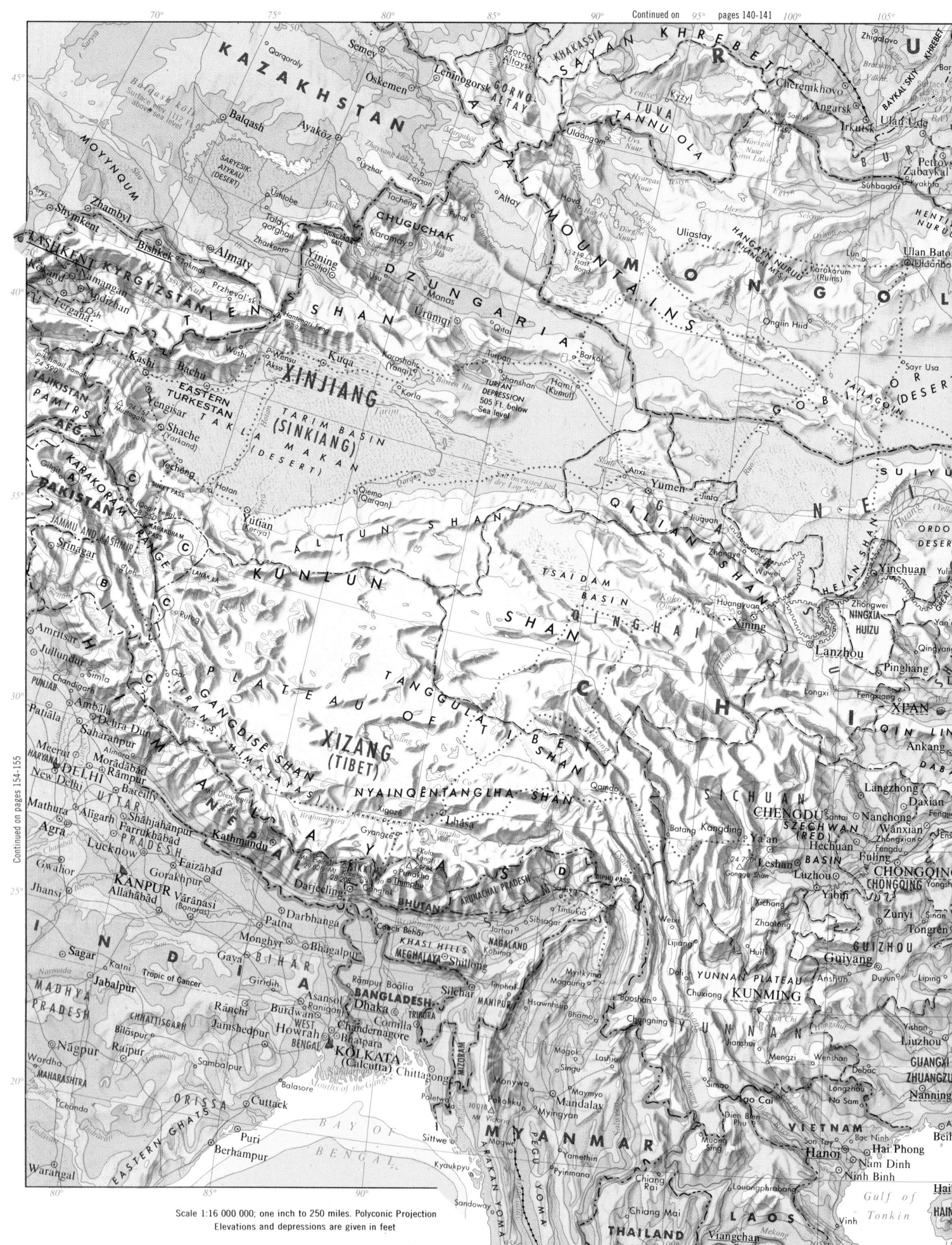

Continued on 95° pages 140-141

Continued on pages 154-155

Scale 1:16 000 000; one inch to 250 miles. Polyconic Projection
Elevations and depressions are given in feet

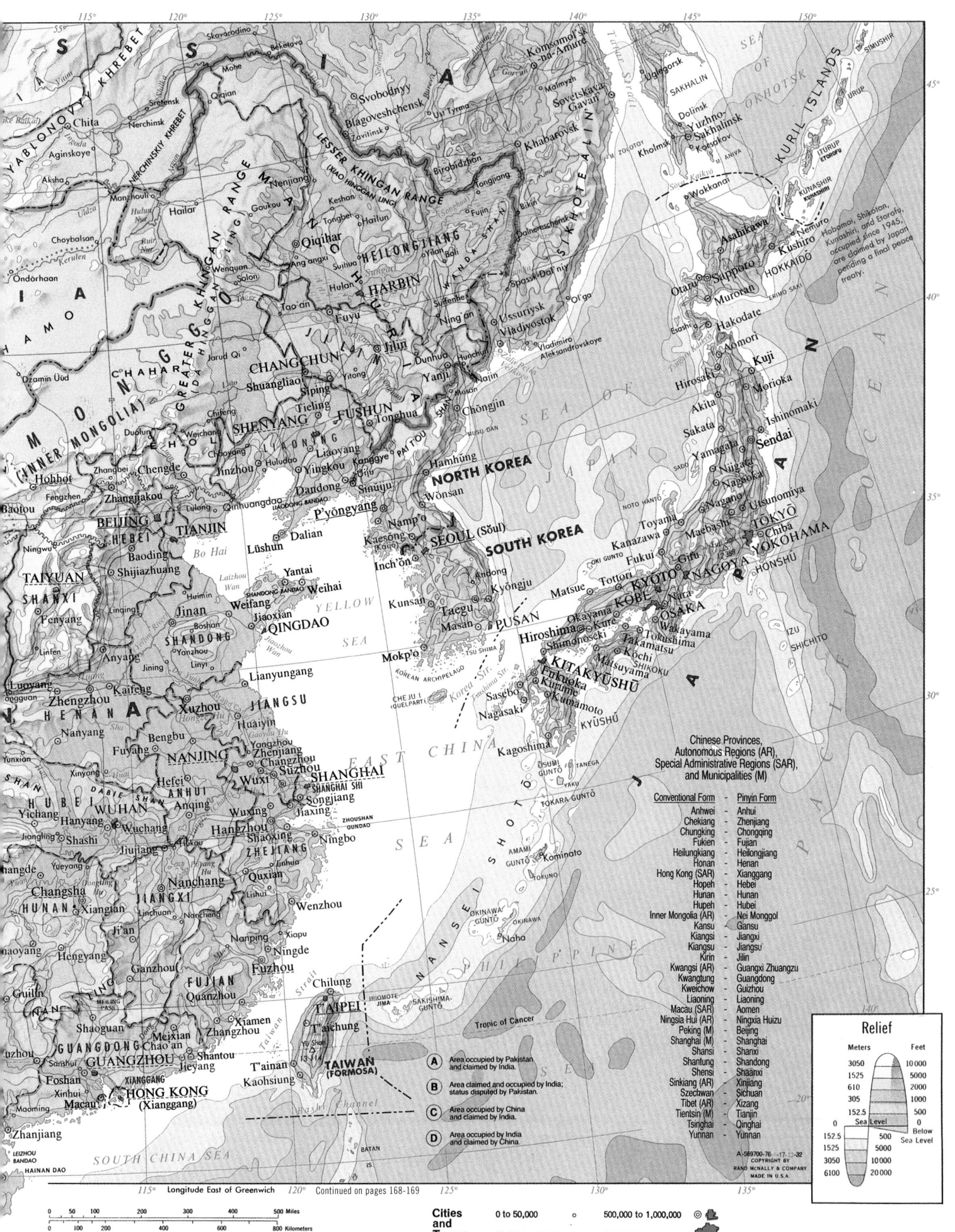

Chinese Provinces,
Autonomous Regions (AR),
Special Administrative Regions (SAR),
and Municipalities (M)

Conventional Form	Pinyin Form
Anhwei	Anhui
Chekiang	Zhenjiang
Chungking	Chongqing
Fukien	Fujian
Heilungkiang	Heilongjiang
Honan	Henan
Hong Kong (SAR)	Xianggang
Hopeh	Hebei
Hunan	Hunan
Hupeh	Hubei
Inner Mongolia (AR)	Nei Monggol
Kansu	Gansu
Kiangsi	Jiangxi
Kiangsu	Jiangsu
Kirin	Jilin
Kwangsi (AR)	Guangxi Zhuangzu
Kwangtung	Guangdong
Kweichow	Guizhou
Liaoning	Liaoning
Macau (SAR)	Aomen
Ningsia Hui (AR)	Ningxia Huizu
Peking (M)	Beijing
Shanghai (M)	Shanghai
Shansi	Shanxi
Shantung	Shandong
Shensi	Shaanxi
Sinkiang (AR)	Xinjiang
Szechwan	Sichuan
Tibet (AR)	Xizang
Tientsin (M)	Tianjin
Tsinghai	Qinghai
Yunnan	Yunnan

(A) Area occupied by Pakistan and claimed by India.

(B) Area claimed and occupied by India; status disputed by Pakistan.

(C) Area occupied by China and claimed by India.

(D) Area occupied by India and claimed by China.

A-569700-76 -17-13-32
COPYRIGHT BY
RAND McNALLY & COMPANY
MADE IN U.S.A.

Relief

Meters	Feet
3050	10 000
1525	5000
610	2000
305	1000
152.5	500
Sea Level	Sea Level
152.5	500
1525	5000
3050	10 000
6100	20 000

Continued on pages 168-169

Longitude East of Greenwich

0   50 100   200   300   400   500 Miles
0 100 200   400   600   800 Kilometers

Cities and Towns

0 to 50,000 ○   500,000 to 1,000,000 ◎

50,000 to 500,000 ⊙   1,000,000 and over

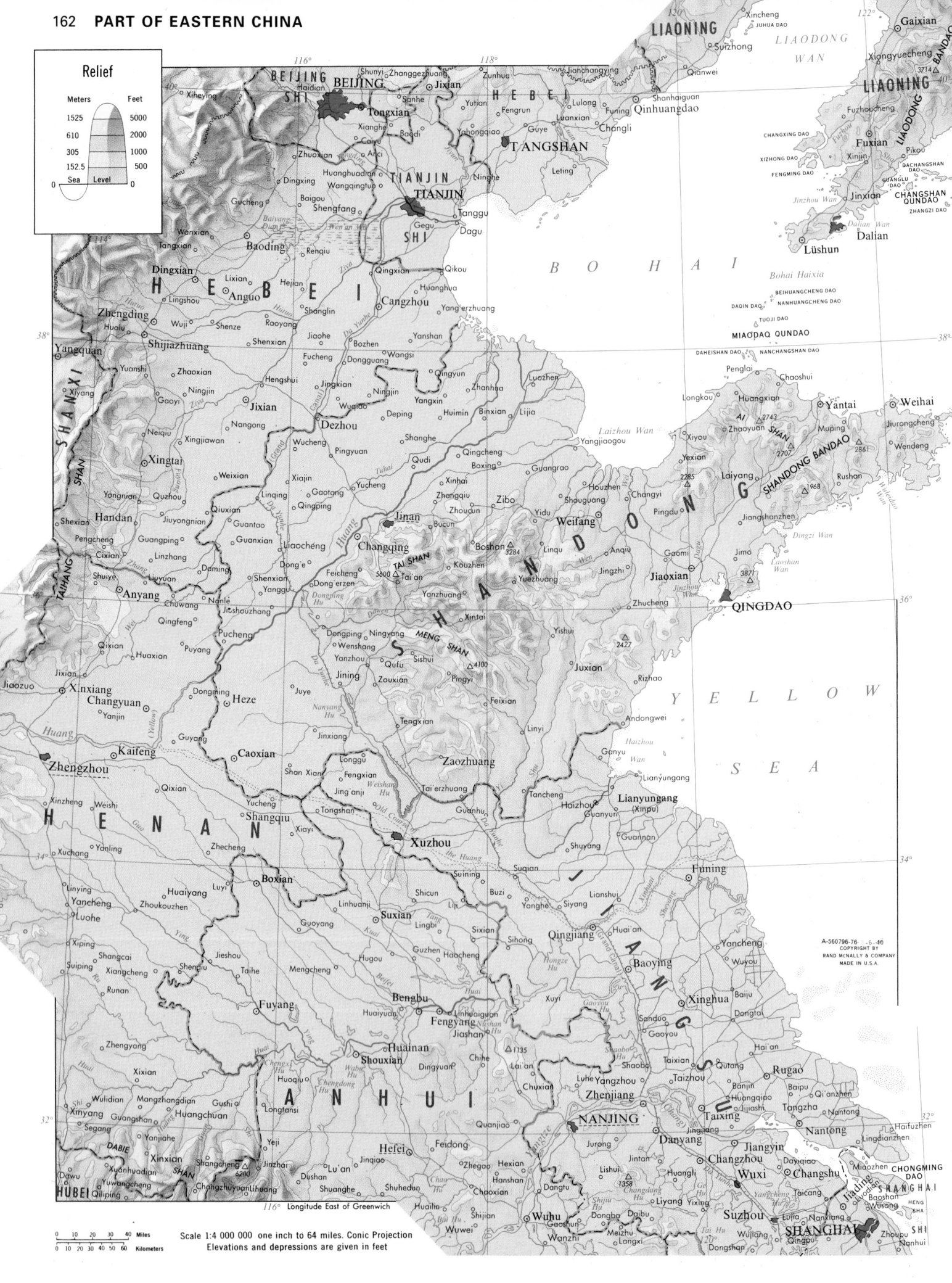

Relief

Meters	Feet
1525	5000
610	2000
305	1000
152.5	500
Sea Level	

Scale 1:4 000 000 one inch to 64 miles. Conic Projection
Elevations and depressions are given in feet

Miles 0 10 20 30 40
Kilometers 0 10 20 30 40 50 60

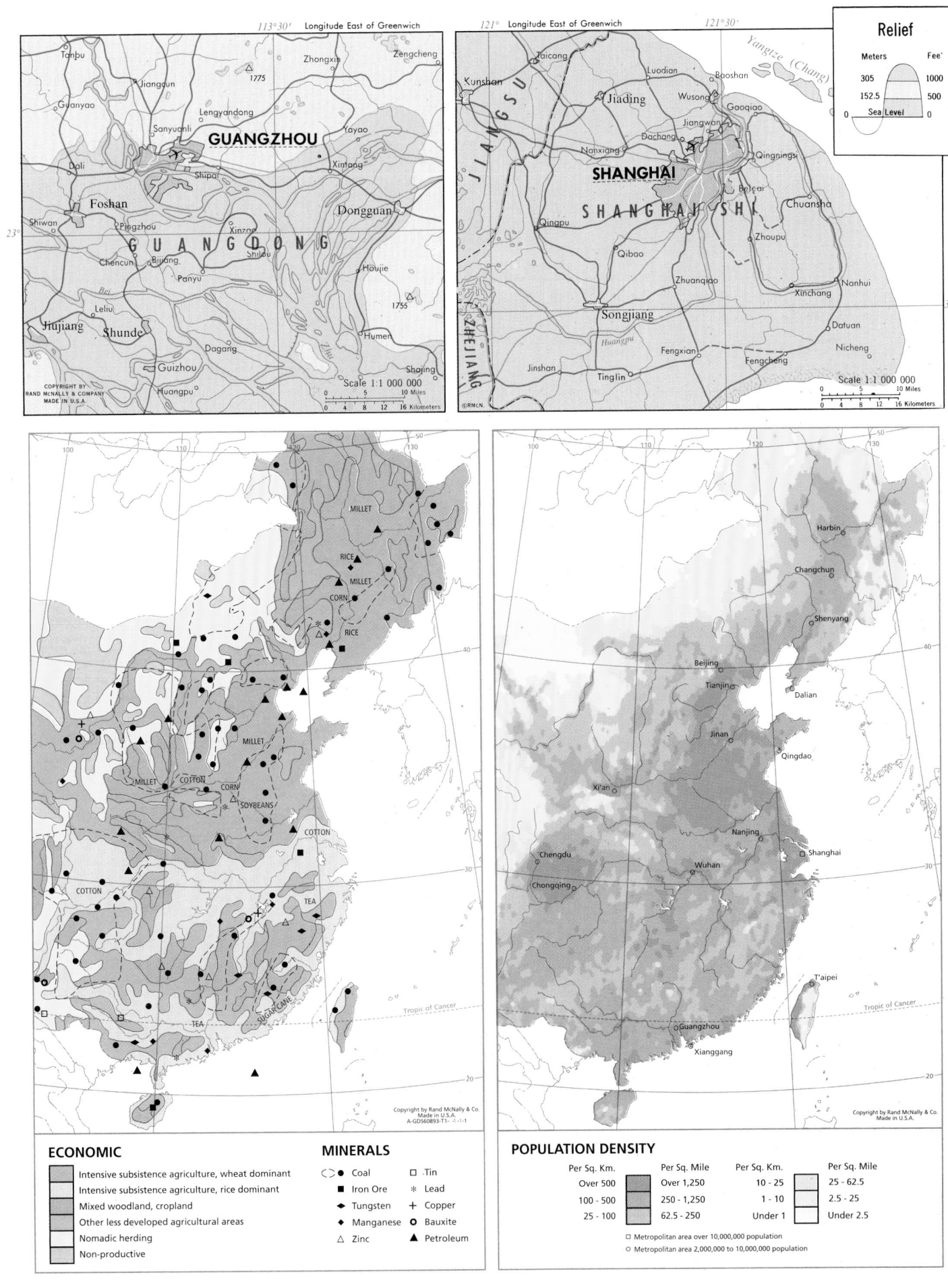

Relief

Meters	Feet
305	1000
152.5	500
0 Sea Level	0

Scale 1:1 000 000

Scale 1:1 000 000

COPYRIGHT BY
RAND MCNALLY & COMPANY
MADE IN U.S.A.

©RMcN.

Copyright by Rand McNally & Co.
Made in U.S.A.
A-GDS60893-T1- -l- -l-1

Copyright by Rand McNally & Co.
Made in U.S.A.

ECONOMIC

- Intensive subsistence agriculture, wheat dominant
- Intensive subsistence agriculture, rice dominant
- Mixed woodland, cropland
- Other less developed agricultural areas
- Nomadic herding
- Non-productive

MINERALS

- ● Coal
- ■ Iron Ore
- ◆ Tungsten
- ◆ Manganese
- △ Zinc
- □ Tin
- ✳ Lead
- + Copper
- ○ Bauxite
- ▲ Petroleum

POPULATION DENSITY

Per Sq. Km.	Per Sq. Mile	Per Sq. Km.	Per Sq. Mile
Over 500	Over 1,250	10 - 25	25 - 62.5
100 - 500	250 - 1,250	1 - 10	2.5 - 25
25 - 100	62.5 - 250	Under 1	Under 2.5

□ Metropolitan area over 10,000,000 population
○ Metropolitan area 2,000,000 to 10,000,000 population

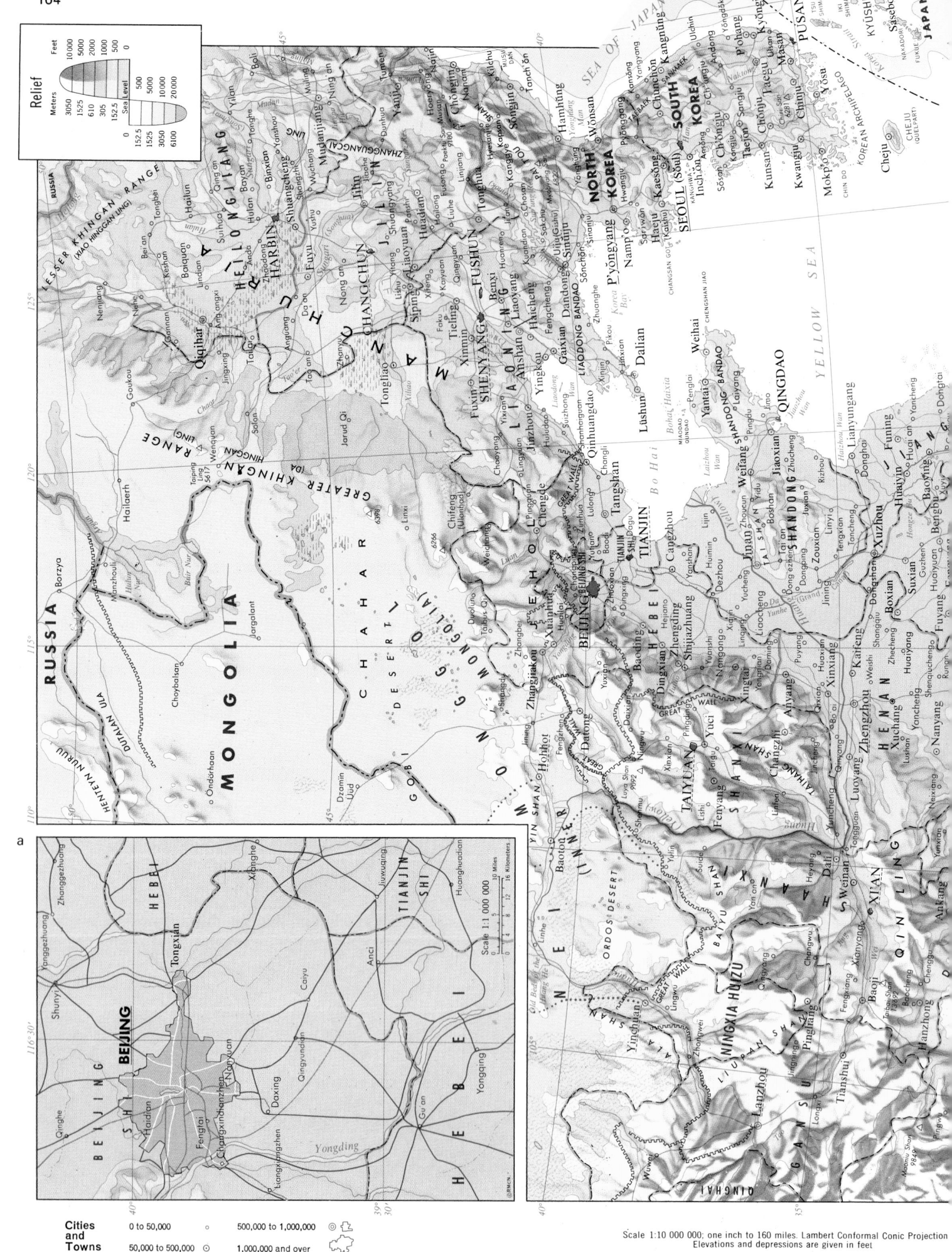

164

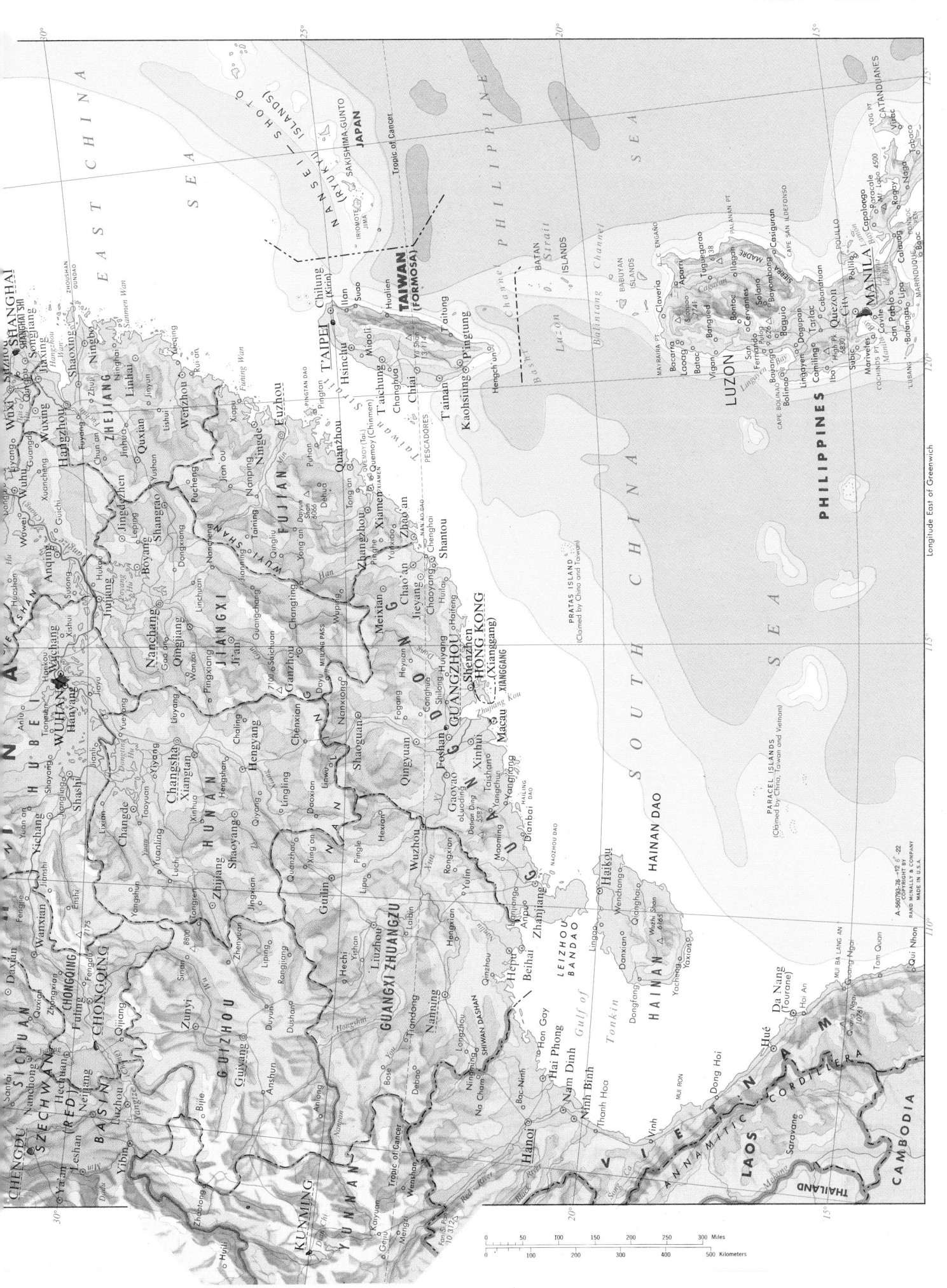

Continued on pages 164–165

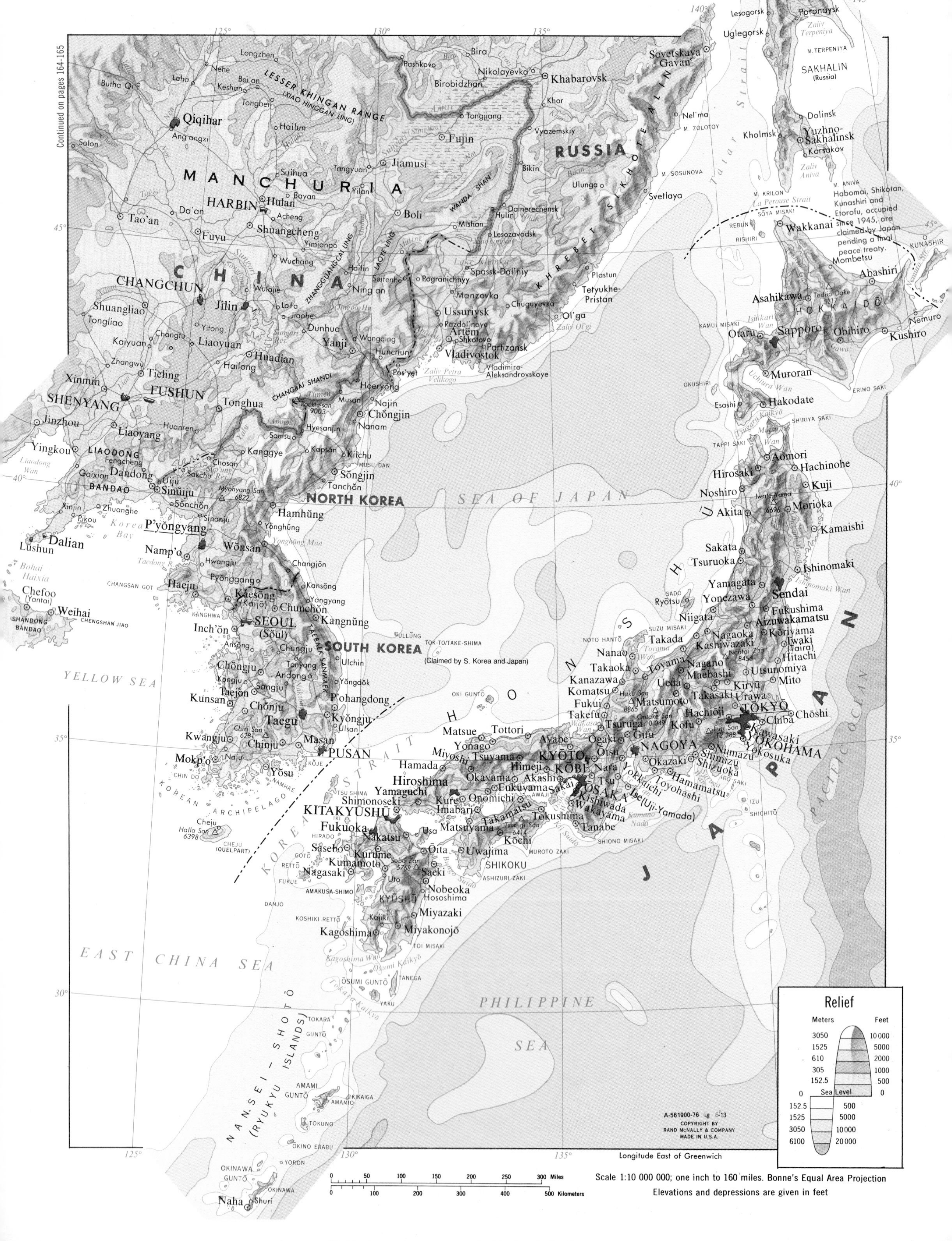

MANCHURIA

CHINA

RUSSIA

LESSER KHINGAN RANGE (XIAO HINGGAN LING)

Qiqihar
Nehe
Longzhen
Laha
Bei'an
Keshan
Tongbei
Butha Qi
Solon
Ang'angxi
Hailun
Suihua
Bayan
Tangyuan
Yilan
Jiamusi
Fujin
Tongjiang
Bira
Pashkovo
Nikolayevka
Birobidzhan
Khabarovsk
Khor
Tonghe
Bikin

HARBIN
Hulan
Acheng
Shuangcheng
Boli
Mishan
Hulin
Lesozavodsk
Dalnerechensk
Vyazemskiy

Tao'an
Da'an
Fuyu

CHANGCHUN
Yitong
Wuchang
Lafa
Jiaohe
Dunhua
Hailin
Ning'an
Suifenhe
Pogranichnyy
Spassk-Dal'niy
Lake Khanka

Shuangliao
Tongliao
Changtu
Liaoyuan
Jilin
KHREBET SIKHOTE ALIN

Kaiyuan
Zhangwu
Tieling
Huadian
Hailong
Yanji
Wangqing
Hunchun
Ussuriysk
Razdol'noye
Artëm
Shkotovo
Chuguyevka
Tetyukhe-Pristan
Plastun

Xinmin
SHENYANG
FUSHUN
Huanren
CHANGBAI SHANDI
Baekdusan 9003
Musan
Pos'yet
Vladivostok
Vladimiro-Aleksandrovskoye
Ol'ga

Jinzhou
Liaoyang
LIAODONG
Dandong
Fengcheng
Uiju
Sakchu
Hoeryŏng
Samsu
Chŏngjin
Nanam

Yingkou
Gaixian
BANDAO
Sinŭiju
Sŏnchŏn
Kanggye
Hyesanjin
Kapsan
Kilchu
Tanchŏn

Xinjin
Zhuanghe
Pikou
Sinanju
Myohyang San 6822
Sŏngjin

Lüshun
Dalian
P'YŎNGYANG
Namp'o
Wŏnsan
Hamhŭng
Yŏnghŭng

Bohai Haixia
Chefoo (Yantai)
Weihai

NORTH KOREA

SEA OF JAPAN

YELLOW SEA

SOUTH KOREA

SEOUL (Sŏul)
Inch'ŏn
Kaesŏng (Kaijō)
Chunchŏn
Kangnŭng
Wŏnju
Ch'ŏngju
Taejŏn
Chŏnju
Kunsan
Kwangju
Mokp'o
Chinju
Masan
PUSAN
Taegu
Kyŏngju
Ulsan
P'ohangdong

TOK-TO/TAKE-SHIMA
(Claimed by S. Korea and Japan)

ULLŬNG

Cheju
Halla San 6398
CHEJU (QUELPART)

KOREAN ARCHIPELAGO

KOREA STRAIT

HOKKAIDŌ

Wakkanai
Asahikawa
Sapporo
Otaru
Obihiro
Kushiro
Muroran
Hakodate
Abashiri
Nemuro

Habomai, Shikotan, Kunashiri and Etorofu, occupied since 1945, are claimed by Japan pending a final peace treaty.

SAKHALIN (Russia)

Poronaysk
Uglegorsk
Kholmsk
Yuzhno-Sakhalinsk
Korsakov
Dolinsk

Aomori
Hirosaki
Hachinohe
Kuji
Noshiro
Akita
Morioka
Kamaishi
Sakata
Tsuruoka
Ishinomaki
Yamagata
Yonezawa
Sendai
Fukushima
Aizuwakamatsu
Niigata
Nagaoka
Kōriyama
Iwaki (Taira)
Hitachi
Takada
Nagano
Mito
Utsunomiya
Maebashi
Kiryū
Takasaki
Urawa
TOKYO
Chiba
Chōshi
YOKOHAMA
Kawasaki
Yokosuka

JAPAN

HONSHU

Matsue
Tottori
Yonago
Ayabe
KYOTO
Ōtsu
Nara
KŌBE
OSAKA
NAGOYA
Gifu
Fukui
Kanazawa
Komatsu
Takaoka
Toyama
Hiroshima
Yamaguchi
Okayama
Akashi
Himeji
Fukuyama
Onomichi
Kure
Imabari

SHIKOKU

Takamatsu
Tokushima
Matsuyama
Kōchi
Uwajima

KYŪSHŪ

KITAKYŪSHŪ
Fukuoka
Nakatsu
Ōita
Saeki
Nobeoka
Miyazaki
Kagoshima
Miyakonojō
Kumamoto
Nagasaki
Sasebo
Kurume

EAST CHINA SEA

PHILIPPINE SEA

PACIFIC OCEAN

NANSEI SHOTŌ (RYUKYU ISLANDS)

AMAMI GUNTŌ
OKINAWA GUNTŌ
Naha
Shuri

Relief

Meters		Feet
3050		10 000
1525		5000
610		2000
305		1000
152.5		500
Sea Level		0
152.5		500
1525		5000
3050		10 000
6100		20 000

A-561900-76
COPYRIGHT BY
RAND McNALLY & COMPANY
MADE IN U.S.A.

Longitude East of Greenwich

0   50   100   150   200   250   300 Miles
0   100   200   300   400   500 Kilometers

Scale 1:10 000 000; one inch to 160 miles. Bonne's Equal Area Projection
Elevations and depressions are given in feet

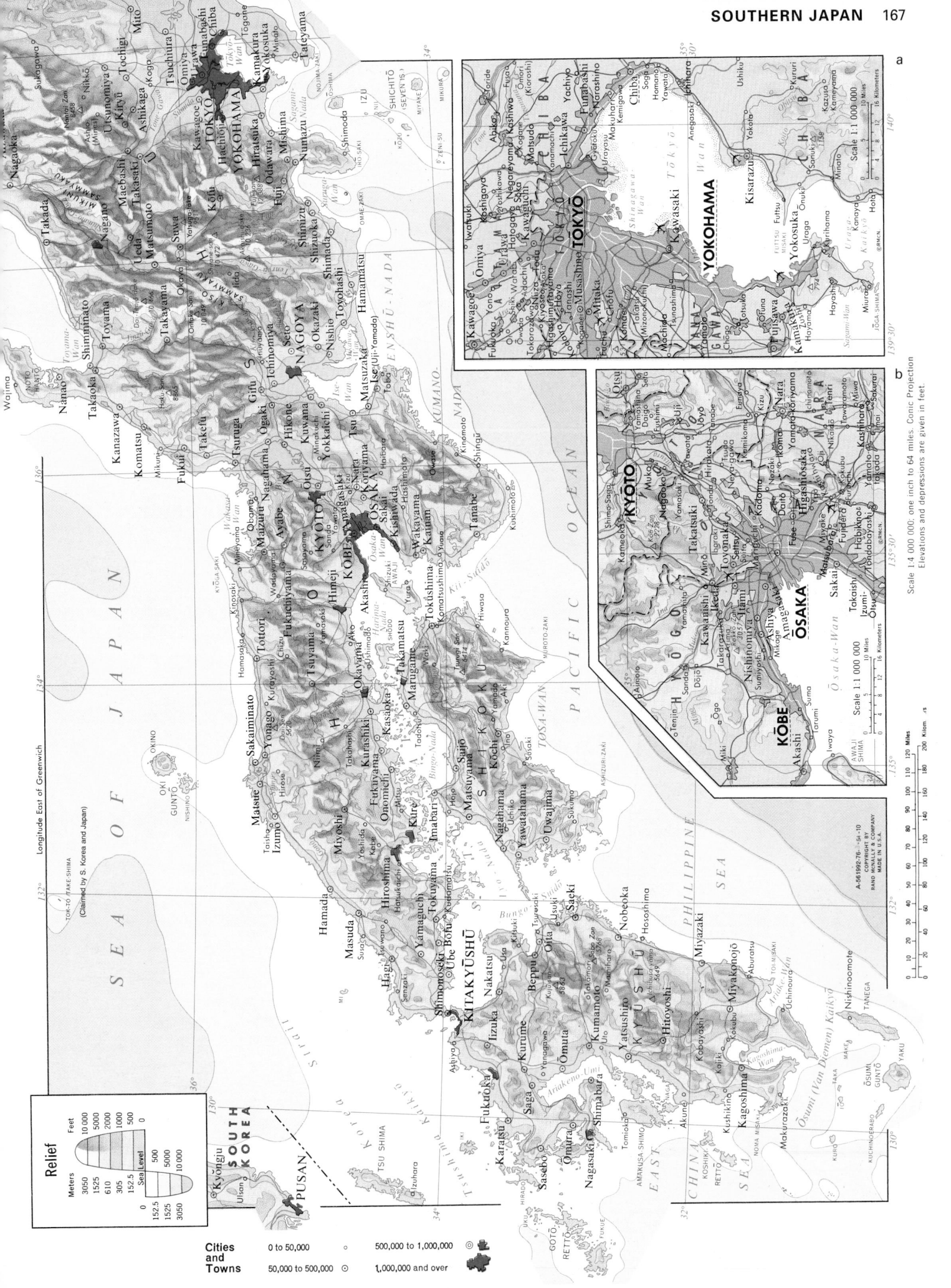

a

b

Scale 1:4 000 000: one inch to 64 miles. Conic Projection
Elevations and depressions are given in feet.

Scale 1:1 000 000

Scale 1:1 000 000

Relief

Meters	Feet
3050	10 000
1525	5000
610	2000
305	1000
152.5	500
0	Sea level
152.5	500
1525	5000
3050	10 000

Cities and Towns

0 to 50,000

50,000 to 500,000

500,000 to 1,000,000

1,000,000 and over

Longitude East of Greenwich

A-561992-76-I-54-10
COPYRIGHT BY
RAND McNALLY & COMPANY
MADE IN U.S.A.

Scale 1:16 000 000; one inch to 250 miles. Polyconic Projection
Elevations and depressions are given in feet

a

Continued on pages 160-161

Scale 1:4 000 000

0   10   20   30   40 Miles

0 10 20 30 40 50 60 Kilometers

©RMCN.

PHILIPPINE

PHILIPPINES

PALAU

SOUTH

CHINA

SEA

PHILIPPINE

SEA

LUZON

MANILA
Quezon
City
Pasig

PHILIPPINES

MINDORO

SIBUYAN
SEA

MASBATE

SONSOROL
ISLANDS

KEPULAUAN
TALAUD

Laut
Maluku
(Molucca Sea)

Laut
Halmahera
(Halmahera Sea)

MALUKU
(MOLUCCAS)

CERAM
(SERAM)

LAUT BANDA
(BANDA SEA)

S   I   A

Equator

JAZIRAH
DOBERAI

PEGUNUNGAN VAN REES

PEGUNUNGAN MAOKE

NEW GUINEA

NINIGO GROUP

HERMIT IS.

ADMIRALTY ISLANDS

MUSSAU
ISLAND

EMIRA
ISLAND

MANUS
ISLAND

NEW HANOVER

BISMARCK

NEW
IRELAND

ARCH.

PAPUA
NEW GUINEA

NEW BRITAIN

NEW BRITAIN
TRENCH

EAST TIMOR

TIMOR

SEA

ARAFURA   SEA

Gulf
of Papua

CORAL SEA

OWEN STANLEY RA.

TROBRIAND IS.

WOODLARK
ISLAND

TIMOR
SEA

AUSTRALIA

Gulf of Carpentaria

Torres
Strait

Continued on pages 174-175

0 50 100   200   300   400   500 Miles

0   100   200   400   600   800 Kilometers

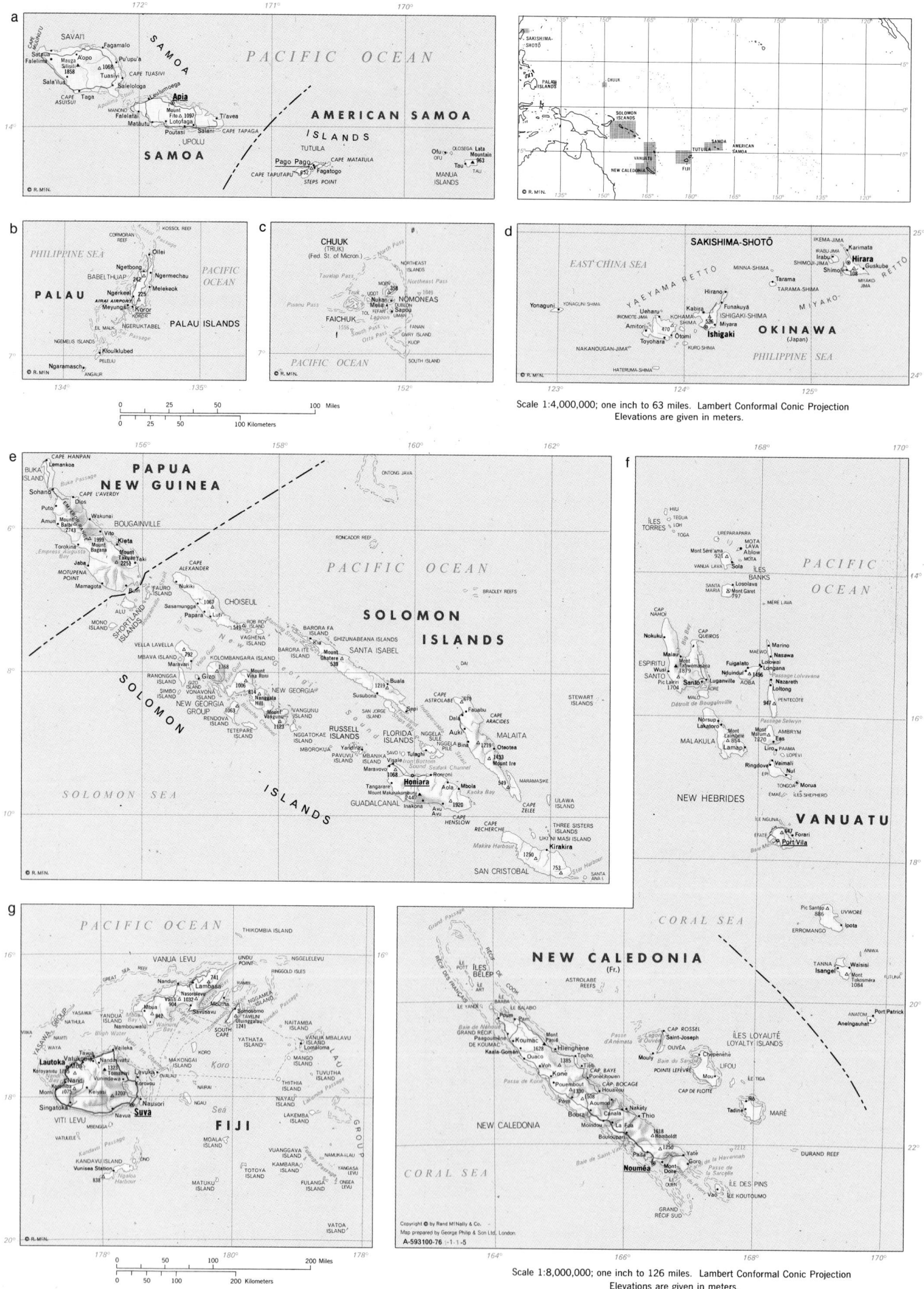

a

PACIFIC OCEAN

CAPE MULINU'U
SAVAI'I
Fagamalo
Mauga
Satauta
Falelima
Sili
Tuasivi CAPE TUASIVI
Silisili
1858
1068
A'opo
Pu'upu'a
Sala'ilua
Leulumoega
CAPE ASUISUI
Taga
Salelologa
Apia
CAPE TAPAGA

SAMOA

MANONO
APOLIMA
Faleolo
Falelatai
Matautu
Poutasi
Fito
1097
Salani
Lefaga
932
Fagatoga
UPOLU
Ti'avea

AMERICAN SAMOA
ISLANDS

Pago Pago
CAPE MATATULA
Olosega Lata Mountain 963
Ofu
Tau
TAU

CAPE TAPUTAPU
STEPS POINT

TUTUILA

MANU'A
ISLANDS

172° 171° 170°
14°

SAKISHIMA-SHOTO
PALAU ISLANDS
CHUUK
SOLOMON ISLANDS
NEW CALEDONIA
VANUATU
FIJI
SAMOA
AMERICAN SAMOA
TUTUILA
PACIFIC OCEAN

b

PHILIPPINE SEA

CORMORAN REEF
KOSSOL REEF
Ollei
BABELTHUAP
Ngermechau
Melekeok
Ngerkeel
AIRAI AIRPORT 225
Meyuns
Koror
EL MALK
NGERUKTABEL
Kloulklubed
PELELIU
NGEMELIS ISLANDS
Ngaramasch
ANGAUR

PALAU

PACIFIC OCEAN

PALAU ISLANDS

134° 135°
7°

c

CHUUK
(TRUK)
(Fed. St. of Micron.)

North Pass
NORTHEAST ISLANDS
NORTHEAST PASS
Tauilap Pass
MOEN
Truk
UDOT
Mesa 358
Nukan
DUBLON
NOMONEAS
TOL
KEFAR
Sapou
South Pass
FAICHUK
1556
Piaanu Pass
FANAN
Otta Pass
GARY ISLAND
KUOP
SOUTH ISLAND

PACIFIC OCEAN

152°
7°

d

SAKISHIMA-SHOTO

EAST CHINA SEA

IKEMA-JIMA
Karimata
IRABU-JIMA
Irabu
SHIMOJI-JIMA Shimoji
Hirara
Guskube
108
MIYAKO-JIMA

MINNA-SHIMA
Tarama
TARAMA-SHIMA
YAEYAMA RETTO

Yonaguni
YONAGUNI-SHIMA
Hirano
Kabira
Funakuya
IRIOMOTE-JIMA
Ueharu
KOHAMA-SHIMA
ISHIGAKI-SHIMA
Amitori
470
526
Miyara
Otomi
Ishigaki
Toyohara
KURO-SHIMA

NAKANOUGAN-JIMA

HATERUMA-SHIMA

OKINAWA
(Japan)

PHILIPPINE SEA

123° 124° 125°
25°
24°

Scale 1:4,000,000; one inch to 63 miles. Lambert Conformal Conic Projection
Elevations are given in meters.

0   25   50   100 Miles
0   25   50   100 Kilometers

e

CAPE HANPAN
Lemankoa
BUKA ISLAND
Buka Passage
Sohano

PAPUA
NEW GUINEA

CAPE L'AVERDY
Dios
Puto
Amun
Mount Balbi 2743
Wakunai
BOUGAINVILLE
Vito
Kieta

Torokina
Jaba
Mount Bagana 1999
Mount Takuan 2251
Buin
Taki
MOTUPENA POINT
Mamagota

ONTONG JAVA

PACIFIC OCEAN

RONCADOR REEF

CAPE ALEXANDER

FAURO ISLAND
Nukiki
ALU
SHORTLAND ISLANDS
MONO ISLAND
Sasamungga
Papara
Luti
CHOISEUL
549
VAGHENA ISLAND
BARORA FA ISLAND
GHIZUNABEANA ISLANDS
BARORA ITE ISLAND
Mount Ghatere 539
SANTA ISABEL
DAI

BRADLEY REEFS

1067

VELLA LAVELLA
792
KOLOMBANGARA ISLAND
1768
RANONGGA ISLAND
Gizo
VONAVONA ISLAND
814
NEW GEORGIA
1006
SIMBO ISLAND
Nangat.
Hill
Buala
Susubona
Zepi
Dala

SOLOMON
ISLANDS

STEWART ISLANDS

NEW GEORGIA GROUP
1063
RENDOVA ISLAND
Mount Vangunu 1123
VANGUNU ISLAND
NGATOTAKAE ISLAND
NGGELA SULE
NGGELA PILE
CAPE ASTROLABE
2679
Fauabu
CAPE ARACIDES
Auki
Oloetea
1219
MALAITA

TETEPARE ISLAND
MBOROKUA ISLAND
RUSSELL ISLANDS
Yandina
PAVUVU ISLAND
SAVO
MBANIKA ISLAND
Visale
Maravovo
FLORIDA ISLANDS
Tulaghi
1433
Mount Ire
Bina
1219

SOLOMON SEA

Tangarare
Mount Makarakomburu 2447
GUADALCANAL
Inakona
Aola
Avu Avu
Mbola
549
KAOKA BAY
MARAMASIKE
ULAWA ISLAND

SOLOMON
ISLANDS

1068
Roroni
Honiara
1920

CAPE HENSLOW
NI MASI ISLAND
THREE SISTERS ISLAND
Kirakira
1250
753
SAN CRISTOBAL
Star Harbour
SANTA ANA ISLAND

156° 158° 160° 162°
6°
8°
10°

f

HIU
TEGUA
ÎLES TORRES
LOH
TOGA
UREPARAPARA
MOTA LAVA
Mont Seré'ama 921
Ablow
MOTA
VANUA LAVA
Sola
ÎLES BANKS
SANTA MARIA
Losolava
Mont Garet 797
MERE LAVA

CAP NAHOI
Nokuku
CAP QUEIROS
Marino
Malau
ESPÍRITU SANTO
Mont Tabwémasana 1879
Wusi
Fuigalato
Nduindui
1496
Nasawa
Longana
MAEWO
Santo
Luganville
AORE
AOBA
Nazareth
Loltong
Pic Lalm 1704
MALO

Détroit de Bougainville

PENTECÔTE
947
Passage Lolvavana

PACIFIC OCEAN

Nórsup
Lakatoro
AMBRYM
Mont Maturn 1270
Eas
Liro
PAAMA
Lamap
LOPEVI
Vaimali
MALAKULA
Ringdove
EPI
Nul
TONGOA Morua
EMAE
ÎLES SHEPHERD

NEW HEBRIDES

VANUATU

ÎLE NGUNA
ÉFATÉ
647
Forari
Port Vila

168° 170°
14°
16°
18°

g

PACIFIC OCEAN

THIKOMBIA ISLAND

VANUA LEVU
UNDU POINT
RAMBI
NGGELELEVU
RINGGOLD ISLS
Nandi
Lambasa
741
Nasorolevu
1032
Vatia
904
YASAWA GROUP
Mbua
842
Savusavu

GREAT SEA REEF
NAITAMBA ISLAND

YASAWA ISLAND
NATHULA
NAVITI
YANDUA ISLAND
Nambouwalu
WAYA
Somosomo
TAVEUNI
Ulunggalau 1241
SOUTH CAPE
YATHATA ISLAND
VANUA MBALAVU
Lomaloma
MANGO ISLAND

Bligh Water

VITI LEVU
Vaileka
KORO
MAKONGAI ISLAND
OVALAU
NAIRAI
THITHIA ISLAND

Lautoka
Vatukoula
Koroyanitu 1195
Nandarivatu
1323
Tomanivi 1322
Levuka
Koro Sea
LAKEMBA ISLAND

Mba
Singatoka 1075
Keiyasi
1207
Nandi
Korovou
Nausori
Suva
MBENGGA
MOALA ISLAND

FIJI

Mbulia
Singatoka
Navua
VITI LEVU
NGAU

YASAWA GROUP
YANUYANU
VATULELE
KANDAVU ISLAND
ONO
Vunisea Station
834
Ngaloa Harbour

VUANGGAVA ISLAND
KAMBARA ISLAND
NAMUKA-ILAU
ONGEA LEVU
TOTOYA ISLAND
MATUKU ISLAND
FULANGA ISLAND

Kandavu Passage

VATOA ISLAND

178° 180° 178°
16°
18°
20°

NEW CALEDONIA
(Fr.)

CORAL SEA

Grand Passage
RÉCIF
RÉCIFS DES FRANÇAIS
ÎLE POTT
ÎLES BÉLEP
ÎLE ART
ÎLE YANDÉ
COOK
ÎLE BALABIO
Poum
Pam
ASTROLABE REEFS

CAP NDOUA
GRAND RÉCIF DE KOUMAC
Paagoumène
Koumac
Mont Panié
Hienghene
Kaala-Gomen
Ouaco
Voh
Tiéta
Tipo
Koné
1385
Touho
Pouembout
Poindimié
CAP BAYE
Pouébo
CAP BOCAGE
Bourail
1330
Houailou
Aoumeu
Nakéty
Canala
Thio
Moindou
Boulouparis
La Foa
1618 Mont Humboldt
Païta
1250
Yaté
CAP ROSSEL
Saint-Joseph

Passe d'Amédéa

Lagon d'Ouvéa
OUVÉA
Mouli
Chépénéhé
LIFOU
ÎLE TIGA
Rô
Tadine
MARÉ

ÎLES LOYAUTÉ
LOYALTY ISLANDS

POINTE LEFÈVRE
CAP DE FLOTTE

Nouméa
Mont Dore
Gorgo
ÎLE OUEN
ÎLE DES PINS
Vao
ÎLE KOUTOUMO
GRAND RÉCIF SUD

DURAND REEF

Pic Santo 886
UVWORÉ
Ipota
ERROMANGO
ANIWA
TANNA
Waisisi
Isangel
Mont Tukosméra 1084
FUTUNA
ANATOM
Port Patrick
Aneingauha

CORAL SEA

NEW CALEDONIA

BAIE DE SAINT-VINCENT
Baie de Kuné
Passe de Kuné
Baie de la Havannah
Passe de Sarcelle
Passe de Prony

Copyright © by Rand McNally & Co.
Map prepared by George Philip & Son Ltd, London.
A-593100-76 -1-1-5

164° 166° 168° 170°
16°
18°
20°
22°

Scale 1:8,000,000; one inch to 126 miles. Lambert Conformal Conic Projection
Elevations are given in meters.

0   50   100   200 Miles
0   50   100   200 Kilometers

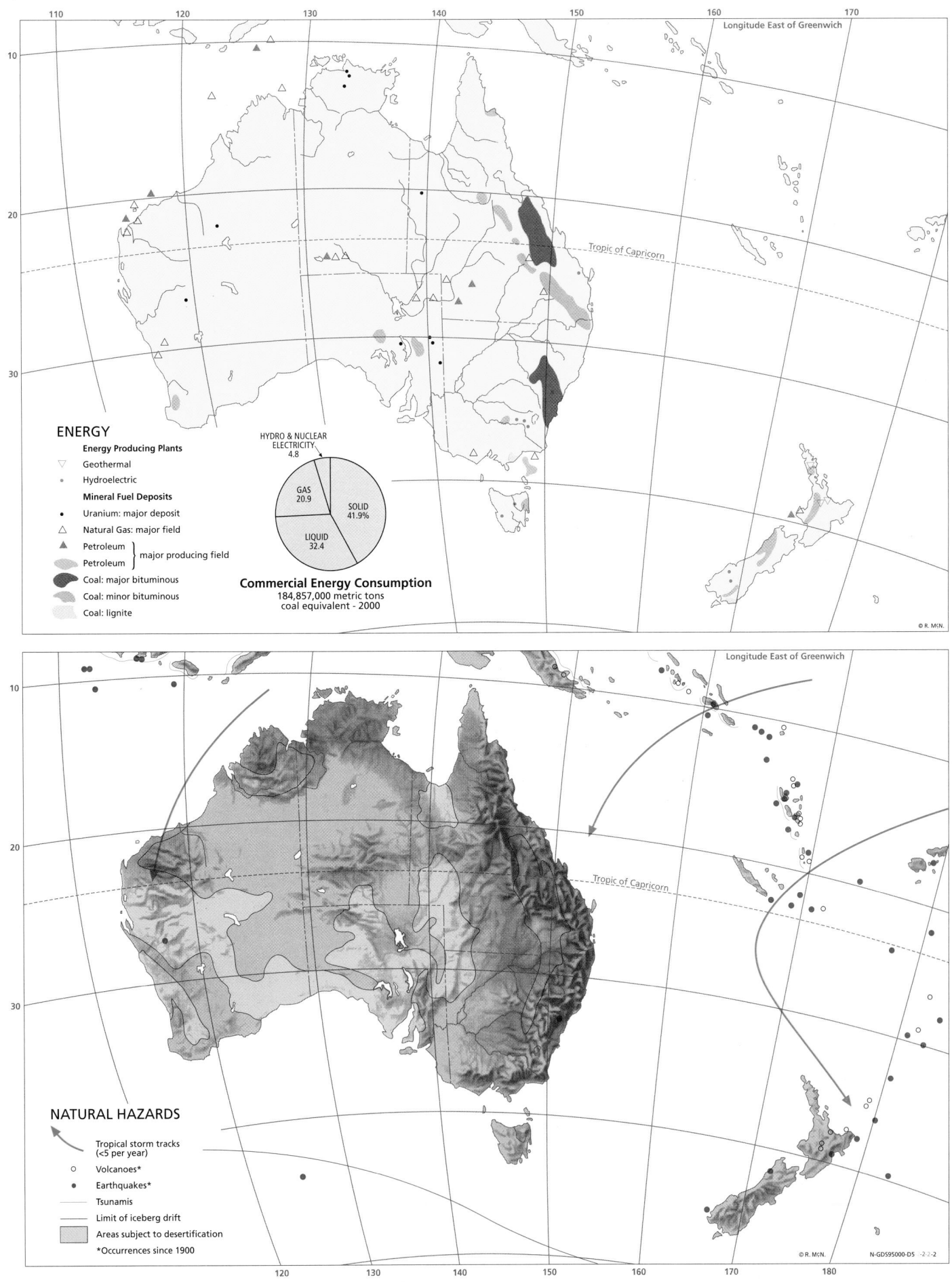

Longitude East of Greenwich

Tropic of Capricorn

ENERGY

Energy Producing Plants

▽ Geothermal

• Hydroelectric

Mineral Fuel Deposits

• Uranium: major deposit

△ Natural Gas: major field

▲ Petroleum } major producing field

▨ Petroleum

■ Coal: major bituminous

▨ Coal: minor bituminous

▨ Coal: lignite

HYDRO & NUCLEAR
ELECTRICITY
4.8

GAS
20.9

SOLID
41.9%

LIQUID
32.4

Commercial Energy Consumption
184,857,000 metric tons
coal equivalent - 2000

© R. McN.

Longitude East of Greenwich

Tropic of Capricorn

NATURAL HAZARDS

↖ Tropical storm tracks
(<5 per year)

○ Volcanoes*

● Earthquakes*

— Tsunamis

— Limit of iceberg drift

▨ Areas subject to desertification

*Occurrences since 1900

© R. McN.   N-GDS95000-D5 -2-2-2

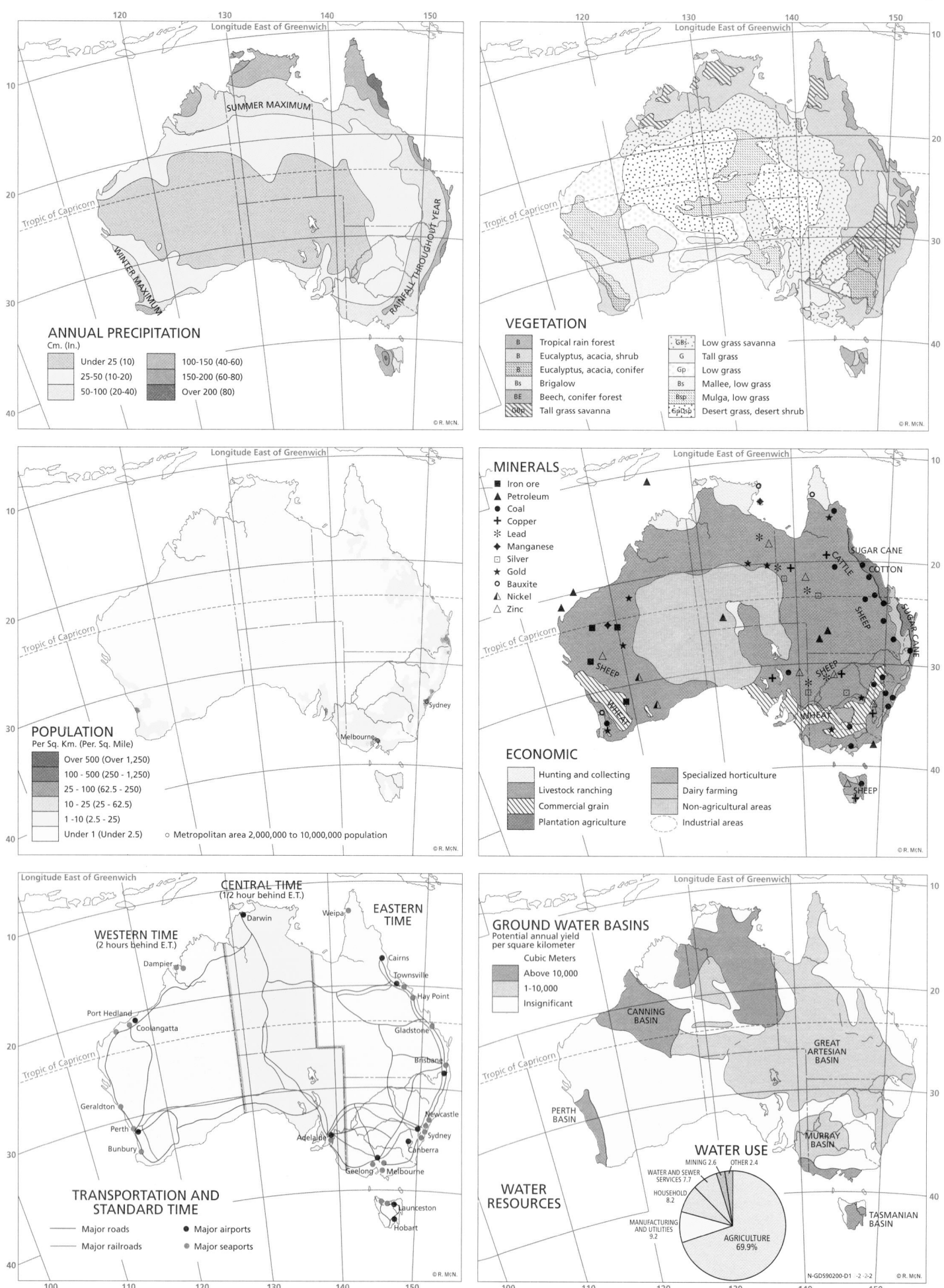

ANNUAL PRECIPITATION
Cm. (In.)

- Under 25 (10)
- 25-50 (10-20)
- 50-100 (20-40)
- 100-150 (40-60)
- 150-200 (60-80)
- Over 200 (80)

SUMMER MAXIMUM
WINTER MAXIMUM
RAINFALL THROUGHOUT YEAR

VEGETATION

B	Tropical rain forest	GBs	Low grass savanna
B	Eucalyptus, acacia, shrub	G	Tall grass
B	Eucalyptus, acacia, conifer	Gp	Low grass
Bs	Brigalow	Bs	Mallee, low grass
BE	Beech, conifer forest	Bsp	Mulga, low grass
GGp	Tall grass savanna	GpBsp	Desert grass, desert shrub

POPULATION
Per Sq. Km. (Per. Sq. Mile)

- Over 500 (Over 1,250)
- 100 - 500 (250 - 1,250)
- 25 - 100 (62.5 - 250)
- 10 - 25 (25 - 62.5)
- 1 - 10 (2.5 - 25)
- Under 1 (Under 2.5)

○ Metropolitan area 2,000,000 to 10,000,000 population

Sydney
Melbourne

MINERALS

- ■ Iron ore
- ▲ Petroleum
- ● Coal
- ✚ Copper
- ✳ Lead
- ◆ Manganese
- ☐ Silver
- ★ Gold
- ○ Bauxite
- △ Nickel
- △ Zinc

SUGAR CANE
CATTLE
COTTON
SHEEP
SUGAR CANE
SHEEP
WHEAT
WHEAT
SHEEP
SHEEP

ECONOMIC

- Hunting and collecting
- Livestock ranching
- Commercial grain
- Plantation agriculture
- Specialized horticulture
- Dairy farming
- Non-agricultural areas
- Industrial areas

TRANSPORTATION AND STANDARD TIME

CENTRAL TIME
(1/2 hour behind E.T.)
EASTERN TIME
WESTERN TIME
(2 hours behind E.T.)

Darwin
Weipa
Cairns
Townsville
Hay Point
Dampier
Gladstone
Port Hedland
Coolangatta
Brisbane
Geraldton
Newcastle
Sydney
Perth
Canberra
Bunbury
Adelaide
Geelong Melbourne
Launceston
Hobart

- —— Major roads
- —— Major railroads
- ● Major airports
- ● Major seaports

GROUND WATER BASINS
Potential annual yield per square kilometer

Cubic Meters
- Above 10,000
- 1-10,000
- Insignificant

CANNING BASIN
GREAT ARTESIAN BASIN
PERTH BASIN
MURRAY BASIN
TASMANIAN BASIN

WATER RESOURCES

WATER USE

MINING 2.6   OTHER 2.4
WATER AND SEWER SERVICES 7.7
HOUSEHOLD 8.2
MANUFACTURING AND UTILITIES 9.2
AGRICULTURE 69.9%

N-GDS90200-D1 -2 -2-2   © R. McN.

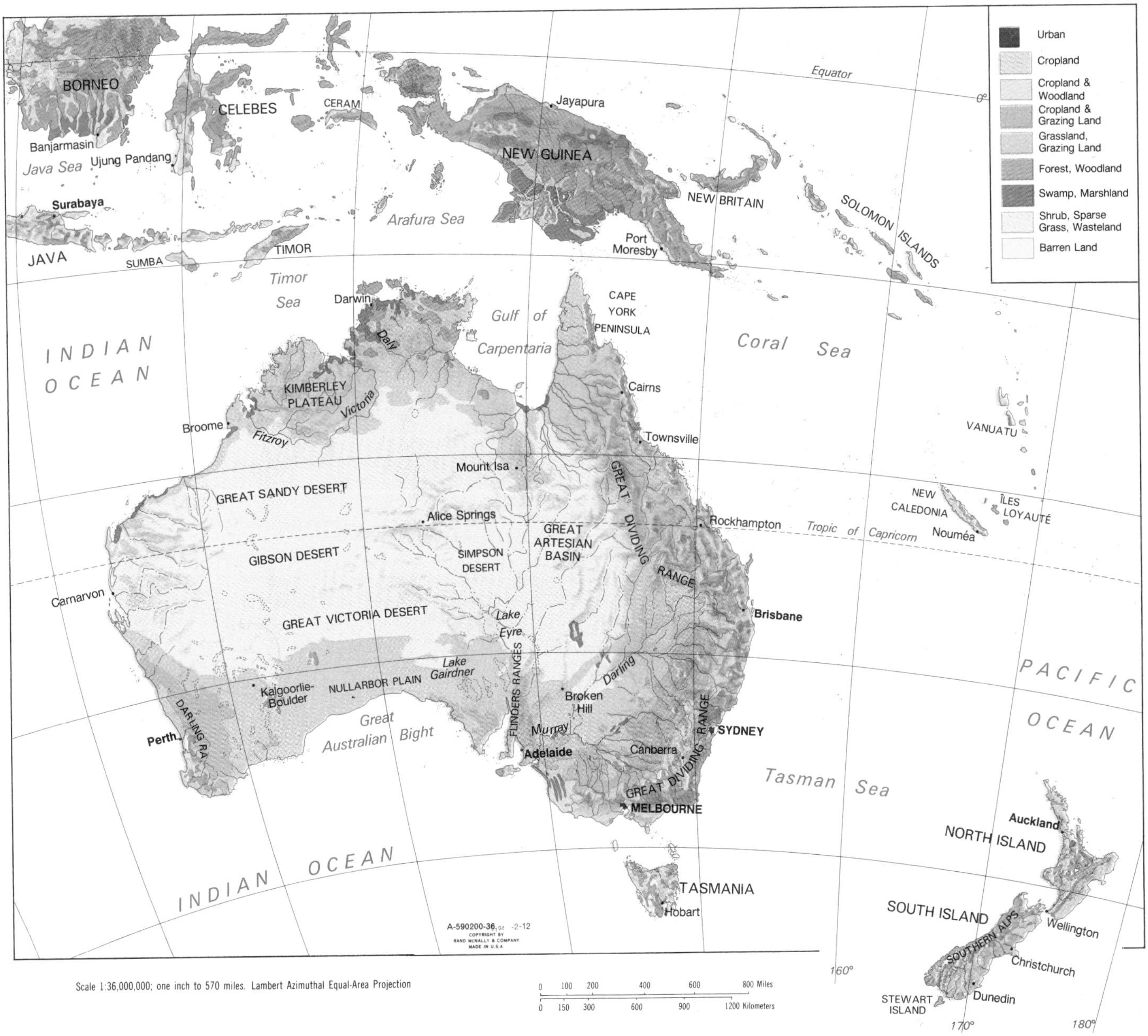

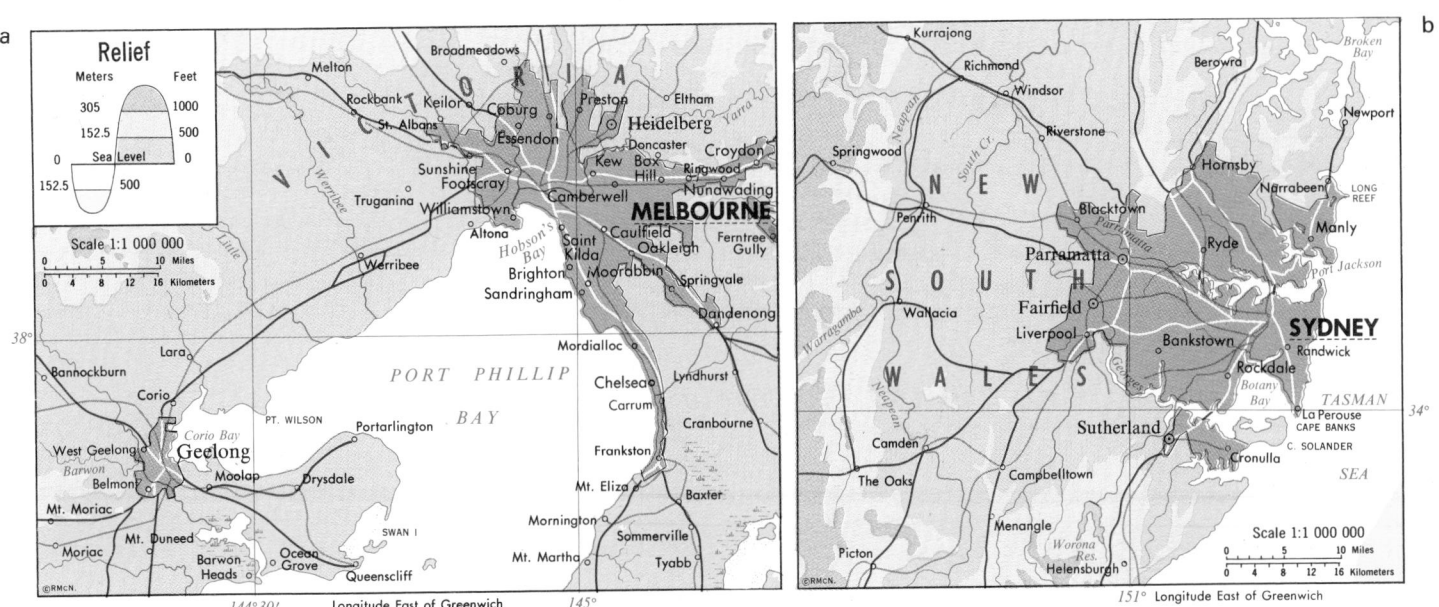

	Legend
	Urban
	Cropland
	Cropland & Woodland
	Cropland & Grazing Land
	Grassland, Grazing Land
	Forest, Woodland
	Swamp, Marshland
	Shrub, Sparse Grass, Wasteland
	Barren Land

BORNEO
CELEBES
CERAM
Banjarmasin
Java Sea Ujung Pandang
Surabaya
JAVA
SUMBA
TIMOR
Timor Sea
Jayapura
NEW GUINEA
NEW BRITAIN
Port Moresby
SOLOMON ISLANDS
Arafura Sea
Coral Sea
Equator
0°

Darwin
Daly
Gulf of Carpentaria
CAPE YORK PENINSULA
Cairns
Townsville
VANUATU
NEW CALEDONIA
Nouméa
ÎLES LOYAUTÉ

INDIAN OCEAN
KIMBERLEY PLATEAU
Victoria
Broome
Fitzroy
GREAT SANDY DESERT
Mount Isa
Alice Springs
SIMPSON DESERT
GREAT DIVIDING RANGE
Rockhampton
Tropic of Capricorn

GIBSON DESERT
GREAT ARTESIAN BASIN
Lake Eyre
Brisbane

Carnarvon
GREAT VICTORIA DESERT
Lake Gairdner
FLINDERS RANGES
Darling
PACIFIC OCEAN

DARLING RA.
Kalgoorlie-Boulder
NULLARBOR PLAIN
Broken Hill
Murray
SYDNEY
Perth
Great Australian Bight
Adelaide
Canberra
GREAT DIVIDING RANGE
Tasman Sea

INDIAN OCEAN
MELBOURNE
TASMANIA
Hobart

Auckland
NORTH ISLAND
SOUTH ISLAND
SOUTHERN ALPS
Wellington
Christchurch
STEWART ISLAND
Dunedin

A-590200-36
COPYRIGHT BY
RAND McNALLY & COMPANY
MADE IN U.S.A.

Scale 1:36,000,000; one inch to 570 miles. Lambert Azimuthal Equal-Area Projection

0 100 200 400 600 800 Miles
0 150 300 600 900 1200 Kilometers

160°
170°
180°

a

Relief

Meters	Feet
305	1000
152.5	500
0 Sea Level	0
152.5	500

Scale 1:1 000 000
0 5 10 Miles
0 4 8 12 16 Kilometers

VICTORIA
Melton
Broadmeadows
Rockbank
Keilor
St. Albans
Coburg
Preston
Eltham
Heidelberg
Yarra
Essendon
Doncaster
Croydon
Kew
Box Hill
Ringwood
Nunawading
Sunshine
Footscray
Camberwell
MELBOURNE
Truganina
Williamstown
Caulfield
Oakleigh
Ferntree Gully
Altona
Hobson's Bay
Saint Kilda
Moorabbin
Springvale
Werribee
Brighton
Sandringham
Dandenong
Mordialloc
Lara
Chelsea
Lyndhurst
Bannockburn
Corio
PORT PHILLIP BAY
Carrum
Cranbourne
Portarlington
Corio Bay
PT. WILSON
Frankston
West Geelong
Geelong
Drysdale
Mt. Eliza
Baxter
Barwon
Belmont
Moolap
Mornington
Sommerville
Mt. Moriac
Ocean Grove
SWAN I
Mt. Martha
Tyabb
Moriac
Mt. Duneed
Barwon Heads
Queenscliff

144°30'
Longitude East of Greenwich
145°

b

Kurrajong
Richmond
Windsor
Berowra
Broken Bay
Newport
Springwood
NEW
Riverstone
Hornsby
Narrabeen
LONG REEF
Penrith
Blacktown
Ryde
Manly
SOUTH
Parramatta
Port Jackson
Wallacia
Fairfield
SYDNEY
Liverpool
Bankstown
Randwick
WALES
Rockdale
Botany Bay
Sutherland
TASMAN
Camden
La Perouse
CAPE BANKS
The Oaks
Campbelltown
Cronulla
C. SOLANDER
Menangle
Worona Res.
SEA
Picton
Helensburgh

Scale 1:1 000 000
0 5 10 Miles
0 4 8 12 16 Kilometers

151° Longitude East of Greenwich
34°

Continued on pages 168-169

INDONESIA

SELARU

TANJUNG VALS

Pasuruan

Singaraja Rinjani Rapa FLORES Lomblen Pantar ALOR Dili **EAST TIMOR**

G. Mahameru BALI Sumbawa-Besar Rapa Dili

12 060 LOMBOK SUMBAWA SAVU SEA Waingapu TIMOR

SUMBA SAWU ROTI Kupang

A R A F U R A S E A

C. VAN DIEMEN CROKER WESSEL IS.

BATHURST MELVILLE CAPE ARNHEM

Van Diemen Gulf CAPE ARNHEM

Clarence Str. Darwin

Anson Bay Blue Mud Bay **GULF OF**

S U N D A   I S L A N D S

T I M O R   S E A

CAPE LONDONDERRY Joseph Bonaparte Gulf

GROOTE EYLANDT

Pine Creek

Katherine Limmen Bight **CARPENTARIA**

ARNHEM LAND

SIR EDWARD PELLEW GROUP

WELLESLEY IS.

I N D I A N

Wyndham Birdum Borroloola

Victoria River Downs Daly Waters Burketown

BUCCANEER ARCH. Mt. Hann Newcastle Waters

CAPE LEVEQUE 2800 KING LEOPOLD RANGES

King Sd. GEIKIE RANGE **N O R T H E R N**

DAMPIER Derby Fitzroy Crossing Halls Creek Alexandria

BROOME LAND Fitzroy Surf Cr. Camooweal

Roebuck Bay LaGrange Tanami Tennant Creek Dobbyn

O C E A N EIGHTY MILE BEACH Mount Isa Malbon

LARREY POINT **T E R R I T O R Y** Dajarra Duchess

RIPON De Grey Barrow Creek

DAMPIER Port Hedland Arltunga **Q U**

MONTE BELLO IS. ARCH. Roebourne **G R E A T   S A N D Y   D E S E R T** Mt. Ziel Alice Springs

BARROW Marble Bar 4955 MACDONNELL RANGES

Millstream Nullagine Mackay JAMES RANGE **SIMPSON**

NORTH WEST CAPE **HAMERSLEY RANGE** Jiggalong Macdonald MACDONNELL **DESERT** **A**

Onslow Mt. Bruce Disappointment Amadeus Charlotte Waters Birdsville

POINT CLOATES 4052 **W E S T E R N** **GIBSON DESERT** Uluru **B**

Tropic of Capricorn Peak Hill (Ayers Rock)

CAPE FARQUHAR Carnarvon Nabberu Carnegie Gillen MUSGRAVE RANGES

BERNIER Gascoyne Wells Mt. Woodroffe Oodnadatta

DORRE Shark Bay **A U S T R A L I A** 4724 EVERARD RANGES The Alberga

DIRK HARTOG Meekatharra Wiluna EVERARD RANGES

STEEP POINT Nannine Yeo STUART RANGE William Creek

Cue Sandstone Laverton Marree

Ajana Mount Magnet **G R E A T   V I C T O R I A   D E S E R T** **S O U T H   A U S T R A L I A** Farina

HOUTMAN ROCKS Northampton Ballard Oldea Station Pimba **FLINDERS RANGES**

Geraldton Menzies Rawlinna Hughes Woomera Paratchilna **FLIND**

Dongara Minginew Coolgardie Kalgoorlie-Boulder Eucla Penong Ceduna Whyalla Port Augusta

Pithara Lefroy Goddards Soak POINT FOWLER EYRE PENINSULA Port Pirie

Miling Lake Brown **NULLARBOR PLAIN** Eyre Moonta Gladstone

Moora Southern Cross Cowan Norseman Wallaroo Peterborough

SWANLAND Dundas Salmon Gums Port Wakefield

Perth Northam Gawler

Fremantle York **G R E A T   A U S T R A L I A N   B I G H T** Port Lincoln Murray Bridge

DARLING RANGE Narrogin Ravensthorpe Esperance Adelaide

Collie Hopetoun KANGAROO

Geographe Bay Bunbury ARCHIPELAGO OF THE RECHERCHE Nardoorte

CAPE NATURALISTE Katanning Kingston

Busselton CAPE JAFFA

CAPE LEEUWIN Normalup Albany Mt. Gambier

PT. D'ENTRECASTEAUX WEST CAPE HOWE King George Sd.

I N D I A N **O C E A N**

Longitude 115° East of Greenwich

Scale 1:16 000 000; one inch to 250 miles. Lambert's Azimuthal, Equal Area Projection
Elevations and depressions are given in feet

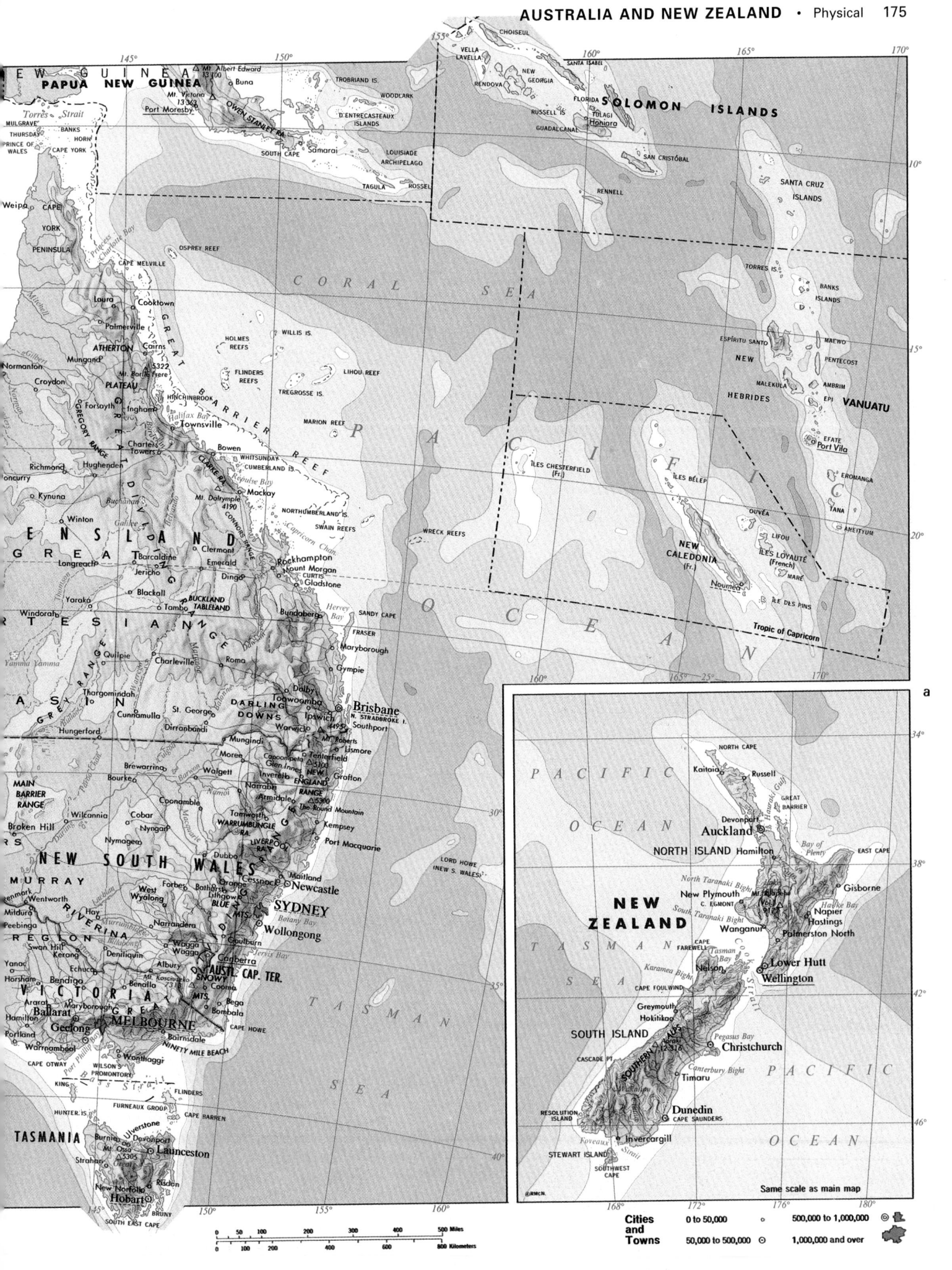

Cities and Towns

0 to 50,000 ∘

50,000 to 500,000 ⊙

500,000 to 1,000,000

1,000,000 and over

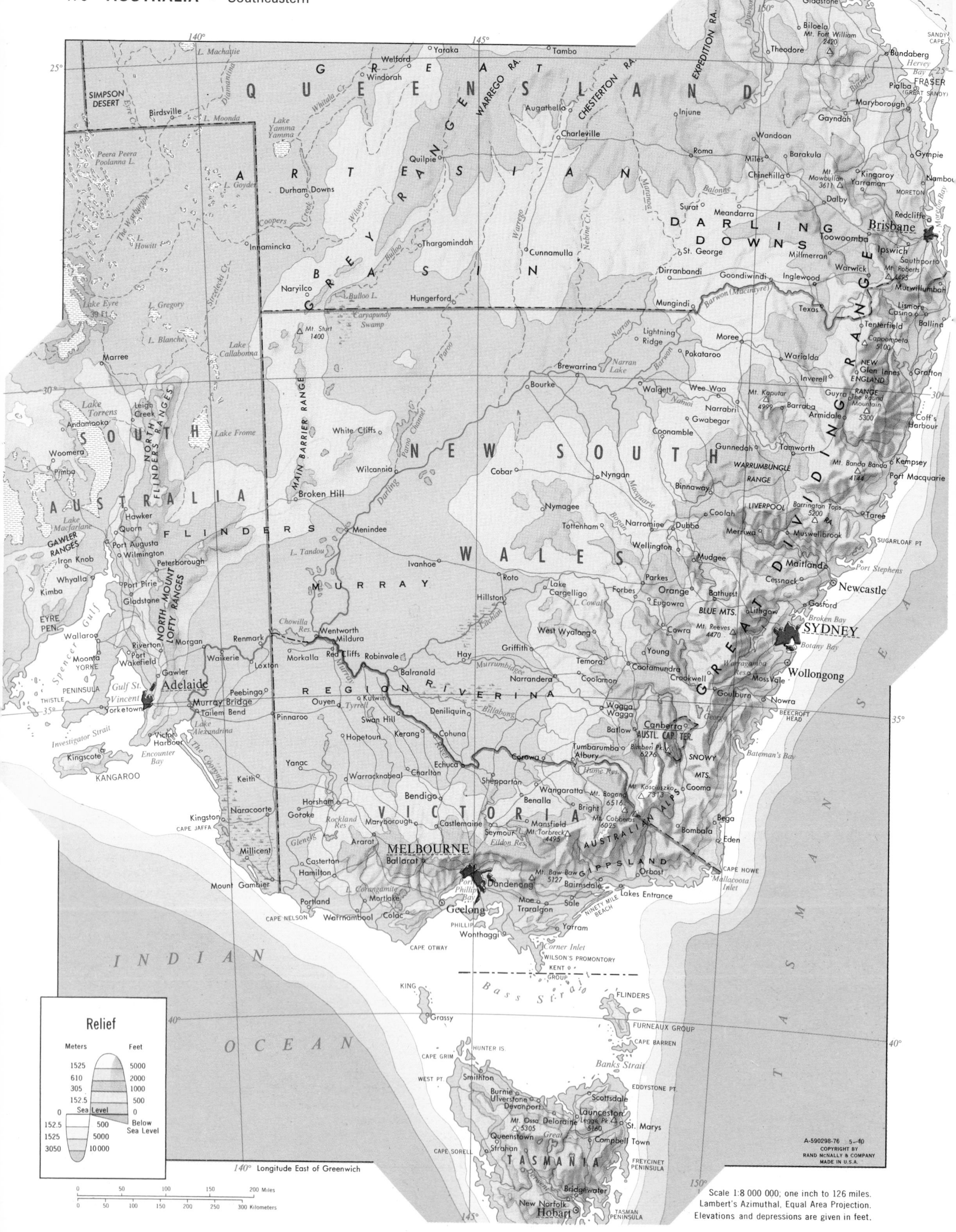

QUEENSLAND

GREAT DIVIDING RANGE

GREY RANGE

ARTESIAN BASIN

GREAT

SIMPSON DESERT

Welford
Yaraka
Tambo
Windorah
Augathella
Charleville
Injune
Roma
Miles
Barakula
Wandoan
Chinchilla
Dalby
Surat
Meandarra
Millmerran
Toowoomba
Warwick
Ipswich
Southport
Brisbane
Redcliffe
Gayndah
Maryborough
Gympie
Nambour
MORETON
Hervey Bay
FRASER (GREAT SANDY)
SANDY CAPE
Bundaberg
Pialba (GREAT SANDY)
Biloela
Theodore
Gladstone
Mt. Fort William 2420
Barcaldine
Mt. Mowbullan 3611
Kingaroy
Yarraman

Birdsville
L. Machattie
Diamantina
L. Yamma Yamma
Durham Downs
Coopers Creek
Innamincka
Thargomindah
Quilpie
Cunnamulla
St. George
Dirranbandi
Goondiwindi
Inglewood
Texas
Tenterfield
Casino
Lismore
Ballina
Grafton
Coff's Harbour
Glen Innes
NEW ENGLAND RANGE
The Round Mountain 5300
Capoompeta 5100
Mutwillumbah
Murwillumbah
Mt. Roberts 4495

Naryilco
Hungerford
Bulloo L.
Caryapundy Swamp
Mt. Sturt 1400
Brewarrina
Narran Lake
Bourke
Walgett
Wee Waa
Moree
Warialda
Inverell
Guyra
Barraba
Armidale
Lightning Ridge
Pokataroo
Narrabri
Gwabegar
Coonabarabran
Gunnedah
Tamworth
Mt. Kaputar 4999
WARRUMBUNGLE RANGE
Mt. Banda Banda 4744
Kempsey
Port Macquarie

Marree
L. Gregory
L. Blanche
Lake Callabonna
Lake Frome
Lake Eyre 39 FT

SOUTH AUSTRALIA
FLINDERS RANGES
NORTH FLINDERS RANGES
Leigh Creek
Hawker
Quorn

Andamooka
Lake Torrens
Woomera
Pimba
GAWLER RANGES
Iron Knob
Whyalla
Kimba
EYRE PEN.
Wallaroo
Moonta
YORKE PENINSULA
Port Augusta
Wilmington
Peterborough
Port Pirie
Gladstone
Crystal Brook
Jamestown
Orroroo

NEW SOUTH WALES

White Cliffs
Wilcannia
MAIN BARRIER RANGE
Broken Hill
Menindee
L. Tandou
Ivanhoe
Cobar
Nymagee
Nyngan
Tottenham
Narromine
Dubbo
Wellington
Coolah
Merriwa
Muswellbrook
Singleton
Maitland
Cessnock
Newcastle
Gosford
Broken Bay
LIVERPOOL RANGE
Barrington Tops 5200
Taree
SUGARLOAF PT.
Port Stephens
BEECROFT HEAD

MURRAY
Chowilla Res.
Wentworth
Mildura
Red Cliffs
Robinvale
Balranald
Hay
Hillston
Roto
Lake Cargelligo
L. Cowal
Condobolin
Forbes
Parkes
Orange
Eugowra
Bathurst
Mudgee
BLUE MTS.
Mt. Reeves 4470
Lithgow
SYDNEY
Botany Bay
Wollongong
Warragamba
Moss Vale
Nowra

Renmark
Morgan
Waikerie
Loxton
Morkalla
Peebinga
Pinnaroo
Ouyen
Swan Hill
Kerang
Kutwin
Balranald
Deniliquin
Hay
Griffith
West Wyalong
Temora
Young
Cootamundra
Coolamon
Junee
Wagga Wagga
Narrandera
RIVERINA REGION
Murrumbidgee R.
Crookwell
Goulburn
Canberra
AUSTL. CAP. TER.

Adelaide
Gawler
Murray Bridge
Tailem Bend
Lake Alexandrina
Victor Harbour
Kingscote
KANGAROO ISLAND
Encounter Bay
Gulf St. Vincent
Investigator Strait
THISTLE
Riverton
Port Wakefield

VICTORIA
Yanac
Warracknabeal
Charlton
Echuca
Cohuna
Shepparton
Benalla
Wangaratta
Bright
Mansfield
Mt. Bogong 6516
Mt. Kosciusko 7313
Mt. Cobberas 6025
AUSTRALIAN ALPS
SNOWY MTS.
Cooma
Bombala
Bega
Eden
CAPE HOWE
Mallacoota Inlet
Albury
Corowa
Tumbarumba
Bimberi Pk. 6276
Batlow
Tumut

Kingston
Naracoorte
Keith
Horsham
Rockland Res.
Maryborough
Castlemaine
Seymour
Eildon Res.
Mt. Torbreck 4495
Mansfield
Bairnsdale
Sale
Orbost
GIPPSLAND
Lakes Entrance
NINETY MILE BEACH
Yarram
CAPE OTWAY
Corner Inlet
WILSON'S PROMONTORY

Millicent
CAPE JAFFA
Mount Gambier
Casterton
Hamilton
Portland
CAPE NELSON
Warrnambool
Colac
Camperdown
L. Corangamite
Ararat
Ballarat
MELBOURNE
Port Phillip Bay
Dandenong
Moe
Traralgon
Geelong
PHILLIP I.
Wonthaggi
Mt. Baw Baw 5127

INDIAN OCEAN

Bass Strait
KING I.
Grassy
CAPE GRIM
WEST PT.
Smithton
Burnie
Ulverstone
Devonport
Deloraine
Mt. Ossa 5305
Launceston
Legges Pk. 5160
St. Marys
Scottsdale
EDDYSTONE PT.
HUNTER IS.
Banks Strait
FLINDERS
FURNEAUX GROUP
CAPE BARREN
KENT GROUP

TASMANIA
Queenstown
Strahan
CAPE SORELL
New Norfolk
Hobart
Bridgewater
TASMAN PENINSULA
FREYCINET PENINSULA
Campbell Town

TASMAN SEA

Relief

Meters		Feet
1525		5000
610		2000
305		1000
152.5		500
0	Sea Level	0
152.5		500
1525		5000 Below Sea Level
3050		10 000

140° Longitude East of Greenwich

0 50 100 150 200 Miles
0 50 100 150 200 250 300 Kilometers

Scale 1:8 000 000; one inch to 126 miles.
Lambert's Azimuthal, Equal Area Projection.
Elevations and depressions are given in feet.

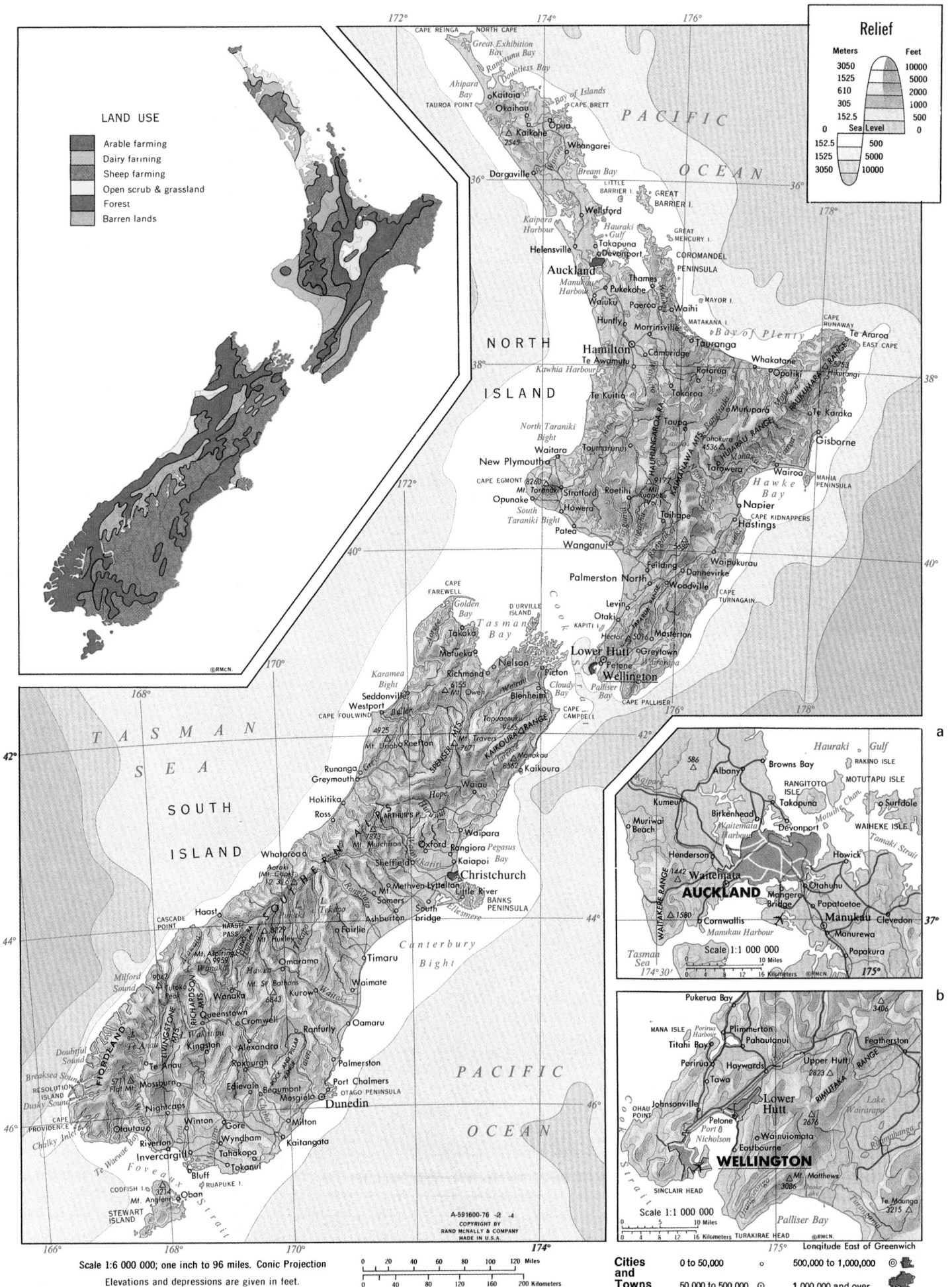

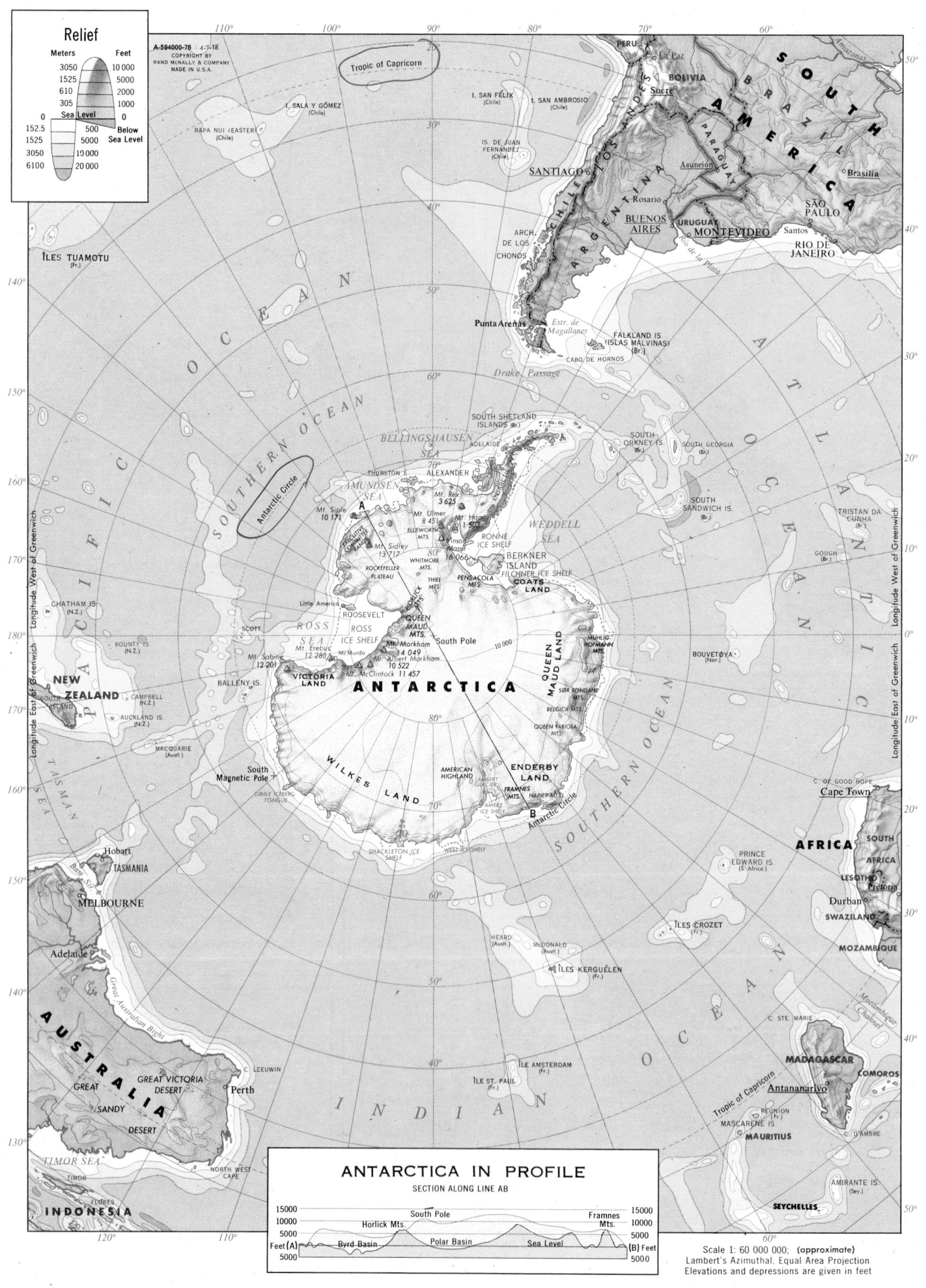

Relief

Meters	Feet
3050	10 000
1525	5000
610	2000
305	1000
Sea Level	Sea Level
0	0
152.5	500
1525	5000
3050	19 000
6100	20 000
	Below Sea Level

A-594000-76 4-7-18
COPYRIGHT BY
RAND McNALLY & COMPANY
MADE IN U.S.A.

ANTARCTICA IN PROFILE
SECTION ALONG LINE AB

Scale 1: 60 000 000; (approximate)
Lambert's Azimuthal, Equal Area Projection
Elevations and depressions are given in feet

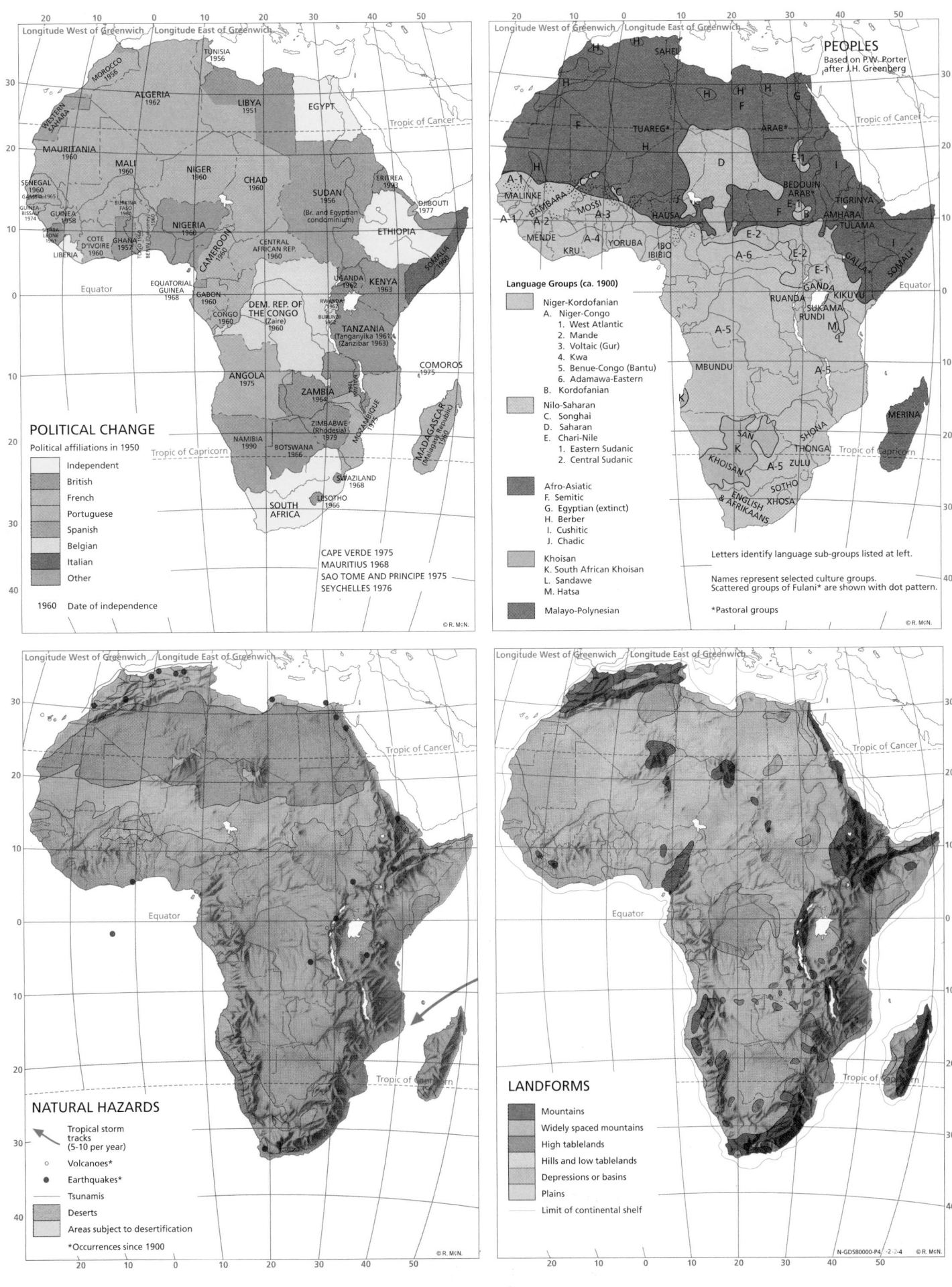

POLITICAL CHANGE

Political affiliations in 1950

- Independent
- British
- French
- Portuguese
- Spanish
- Belgian
- Italian
- Other

1960   Date of independence

CAPE VERDE 1975
MAURITIUS 1968
SAO TOME AND PRINCIPE 1975
SEYCHELLES 1976

MOROCCO 1956
TUNISIA 1956
ALGERIA 1962
LIBYA 1951
EGYPT
WESTERN SAHARA
MAURITANIA 1960
MALI 1960
NIGER 1960
CHAD 1960
SUDAN 1956 (Br. and Egyptian condominium)
ERITREA 1993
DJIBOUTI 1977
SENEGAL 1960
GAMBIA 1965
GUINEA BISSAU 1974
GUINEA 1958
BURKINA FASO 1960
NIGERIA 1960
CAMEROON 1960
CENTRAL AFRICAN REP. 1960
ETHIOPIA
SOMALIA 1960
SIERRA LEONE 1961
COTE D'IVOIRE 1960
GHANA 1957
TOGO 1960
BENIN (Dahomey) 1960
LIBERIA
EQUATORIAL GUINEA 1968
GABON 1960
CONGO 1960
DEM. REP. OF THE CONGO (Zaire) 1960
RWANDA 1962
BURUNDI 1962
UGANDA 1962
KENYA 1963
TANZANIA (Tanganyika 1961/ Zanzibar 1963)
COMOROS 1975
ANGOLA 1975
ZAMBIA 1964
MALAWI 1964
MOZAMBIQUE 1975
MADAGASCAR (Malagasy Republic) 1960
NAMIBIA 1990
ZIMBABWE (Rhodesia) 1979
BOTSWANA 1966
SWAZILAND 1968
SOUTH AFRICA
LESOTHO 1966

© R. McN.

PEOPLES

Based on P.W. Porter
after J.H. Greenberg

Language Groups (ca. 1900)

- Niger-Kordofanian
  - A. Niger-Congo
    1. West Atlantic
    2. Mande
    3. Voltaic (Gur)
    4. Kwa
    5. Benue-Congo (Bantu)
    6. Adamawa-Eastern
  - B. Kordofanian
- Nilo-Saharan
  - C. Songhai
  - D. Saharan
  - E. Chari-Nile
    1. Eastern Sudanic
    2. Central Sudanic
- Afro-Asiatic
  - F. Semitic
  - G. Egyptian (extinct)
  - H. Berber
  - I. Cushitic
  - J. Chadic
- Khoisan
  - K. South African Khoisan
  - L. Sandawe
  - M. Hatsa
- Malayo-Polynesian

Letters identify language sub-groups listed at left.

Names represent selected culture groups.
Scattered groups of Fulani* are shown with dot pattern.

*Pastoral groups

© R. McN.

SAHEL
TUAREG*
ARAB*
MALINKE
BAMBARA
MOSSI
HAUSA
BEDDUIN ARAB*
TIGRINYA
AMHARA
TULAMA
MENDE
KRU
YORUBA
IBO
IBIBIO
GALLA
SOMALI
GANDA
KIKUYU
RUANDA
RUNDI
SUKAMA
MBUNDU
SHONA
SAN
THONGA
ZULU
KHOISAN
SOTHO
XHOSA
ENGLISH & AFRIKAANS
MERINA

NATURAL HAZARDS

- Tropical storm tracks (5-10 per year)
- ○ Volcanoes*
- ● Earthquakes*
- Tsunamis
- Deserts
- Areas subject to desertification

*Occurrences since 1900

LANDFORMS

- Mountains
- Widely spaced mountains
- High tablelands
- Hills and low tablelands
- Depressions or basins
- Plains
- Limit of continental shelf

N-GDS80000-P4 /-2-2-4   © R. McN.

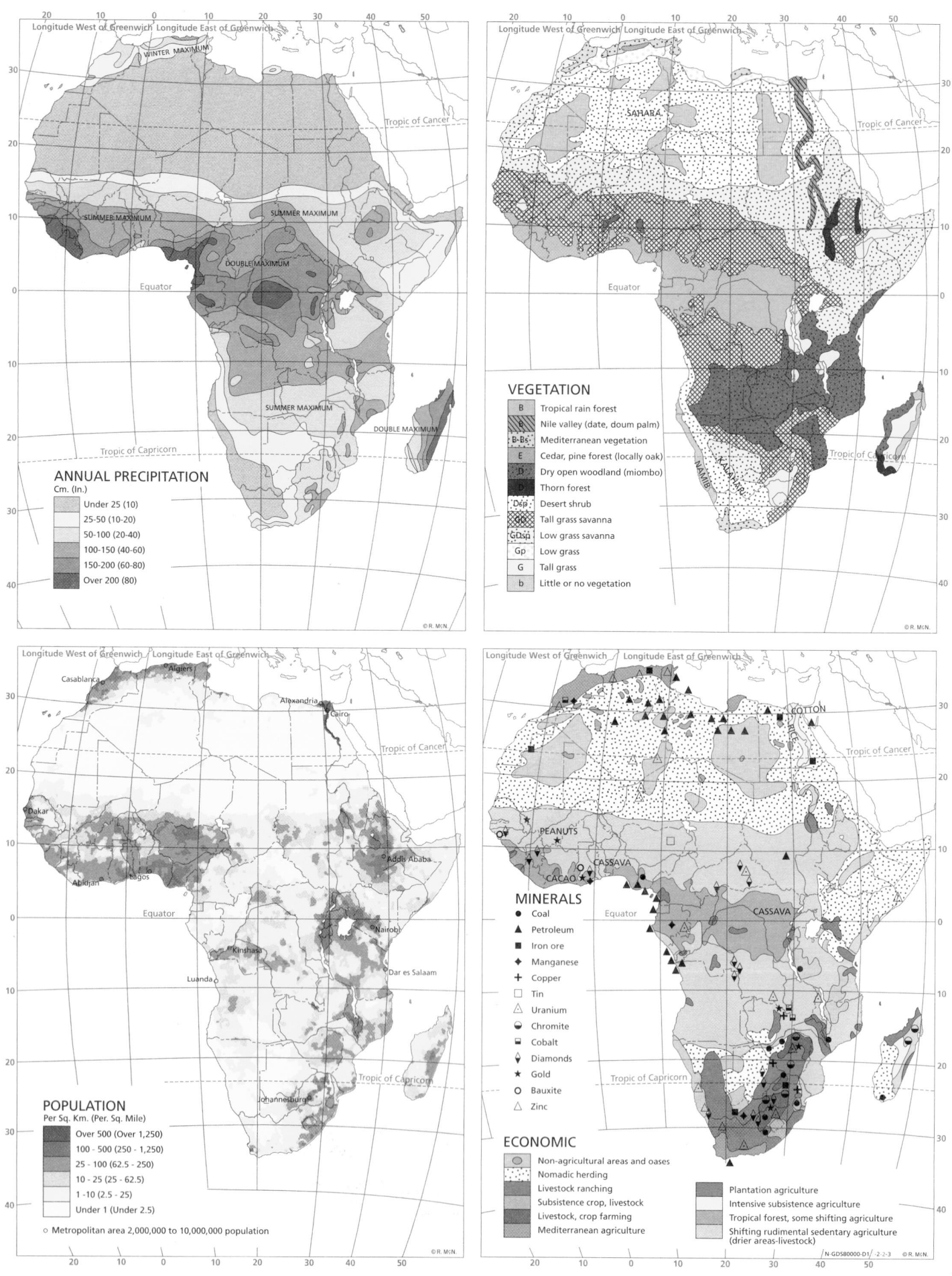

ANNUAL PRECIPITATION
Cm. (In.)

	Under 25 (10)
	25-50 (10-20)
	50-100 (20-40)
	100-150 (40-60)
	150-200 (60-80)
	Over 200 (80)

© R. McN.

VEGETATION

B	Tropical rain forest
B	Nile valley (date, doum palm)
B-Bs	Mediterranean vegetation
E	Cedar, pine forest (locally oak)
D	Dry open woodland (miombo)
D	Thorn forest
Dsp	Desert shrub
GD	Tall grass savanna
GDsp	Low grass savanna
Gp	Low grass
G	Tall grass
b	Little or no vegetation

© R. McN.

POPULATION
Per Sq. Km. (Per. Sq. Mile)

	Over 500 (Over 1,250)
	100 - 500 (250 - 1,250)
	25 - 100 (62.5 - 250)
	10 - 25 (25 - 62.5)
	1 - 10 (2.5 - 25)
	Under 1 (Under 2.5)

○ Metropolitan area 2,000,000 to 10,000,000 population

© R. McN.

MINERALS
- ● Coal
- ▲ Petroleum
- ■ Iron ore
- ◆ Manganese
- ✚ Copper
- □ Tin
- △ Uranium
- ◑ Chromite
- ▣ Cobalt
- ◈ Diamonds
- ★ Gold
- ○ Bauxite
- △ Zinc

ECONOMIC

	Non-agricultural areas and oases
	Nomadic herding
	Livestock ranching
	Subsistence crop, livestock
	Livestock, crop farming
	Mediterranean agriculture
	Plantation agriculture
	Intensive subsistence agriculture
	Tropical forest, some shifting agriculture
	Shifting rudimental sedentary agriculture (drier areas-livestock)

N-GDS80000-D1/ -2-2-3   © R. McN.

ATLANTIC OCEAN

CORSICA
MADRID
SARDINIA
ROME
SICILY
Algiers
Tunis
MALTA
Casablanca
ATLAS MOUNTAINS
Tripoli
Banghāzī
CANARY ISLANDS
El Aaíun
Tropic of Cancer
GRAND ERG OCCIDENTAL
GRAND ERG ORIENTAL
EL DJOUF
AHAGGAR
Tamenghest
S A H A R A
ADRAR DES ÍFÔGHAS
TIBESTI
S U D A N
ENNEDI
Tombouctou
Dakar
Niger
Bamako
Kano
Lake Chad
N'Djamena
Al-Fāshir
Freetown
Niger
Lake Volta
Lagos
Abidjan
Gulf of Guinea
Yaoundé
Bangui
Uele
Equator
Ubangi
Congo
Kisangani
Congo
Kasai
Kinshasa
Luanda
ATLANTIC OCEAN
Congo
Lake Tanganyika
Lubumbashi
Lake Nyasa
Lusaka
Zambezi
Blantyre
Harare
Moçambique
MADAGASCAR
NAMIB DESERT
Windhoek
KALAHARI DESERT
Limpopo
Antananarivo
Orange
Johannesburg
Orange
Durban
Cape Town
INDIAN OCEAN

Mediterranean Sea
Black Sea
İSTANBUL
BAKU
Caspian Sea
Athens
CRETE
CYPRUS
TEHRAN
Tigris
Beirut
Baghdad
Alexandria
SYRIAN DESERT
Euphrates
CAIRO
AN NAFŪD
Nile
ARABIAN DESERT
Riyadh
Lake Nasser
Mecca
NUBIAN DESERT
Red Sea
Nile
Khartoum
Asmera
White Nile
Aden
Gulf of Aden
DANAKIL
Berbera
Blue Nile
Addis Ababa
Mountain Nile
Mogadishu
Lake Victoria
Nairobi
INDIAN OCEAN
Dar es Salaam
COMORO ISLANDS
Mozambique Channel

Urban
Cropland
Cropland & Woodland
Cropland & Grazing Land
Grassland, Grazing Land
Forest, Woodland
Swamp, Marshland
Shrub, Sparse Grass, Wasteland
Barren Land
Oasis

A-580000-36 -2 3-13
COPYRIGHT BY
RAND MCNALLY & COMPANY
MADE IN U.S.A.

Tropic of Capricorn

Scale 1:36,000,000; one inch to 570 miles.   Lambert Azimuthal Equal-Area Projection

0   100   200   400   600   800 Miles
0 150 300   600   900   1200 Kilometers

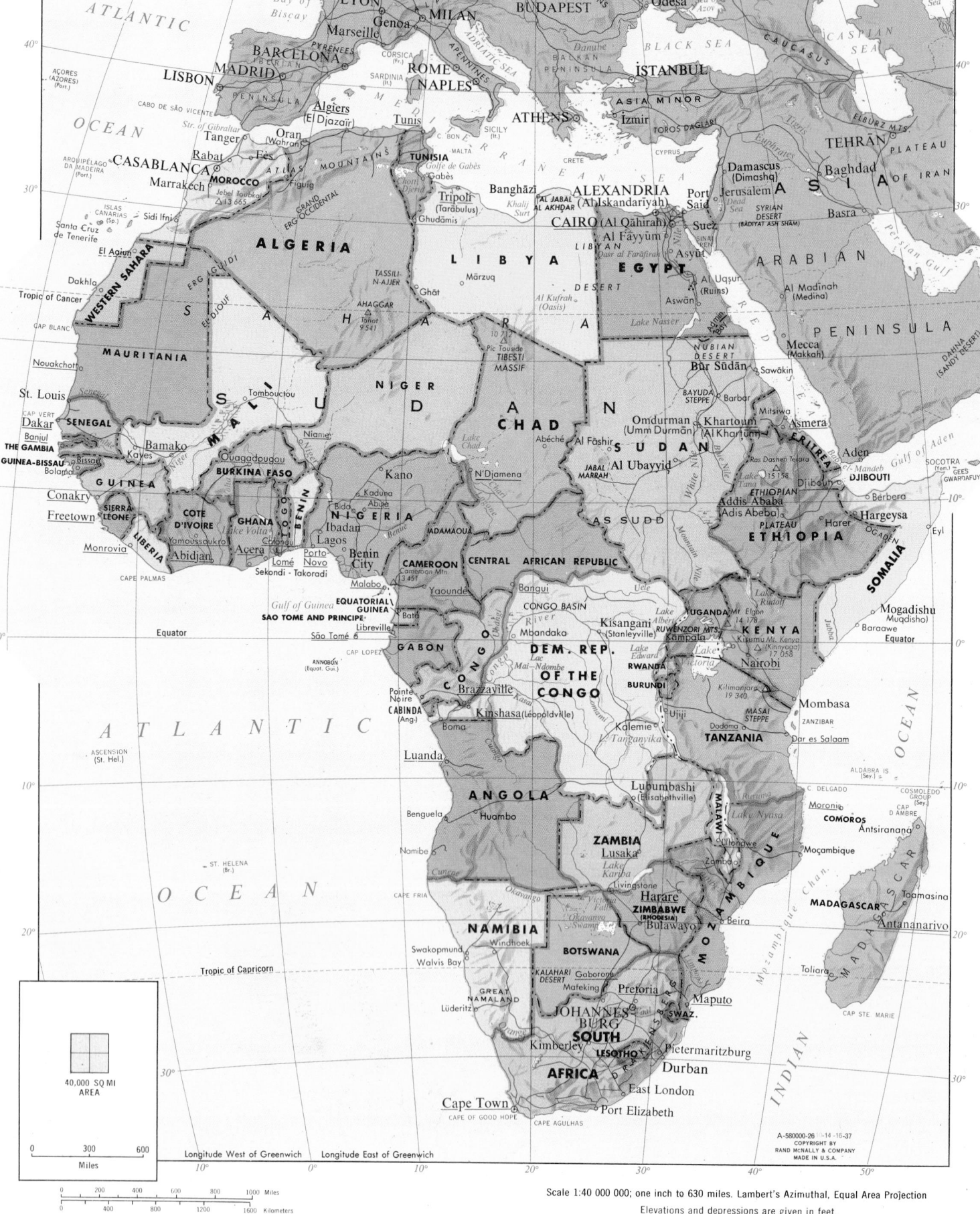

Scale 1:40 000 000; one inch to 630 miles. Lambert's Azimuthal, Equal Area Projection
Elevations and depressions are given in feet.

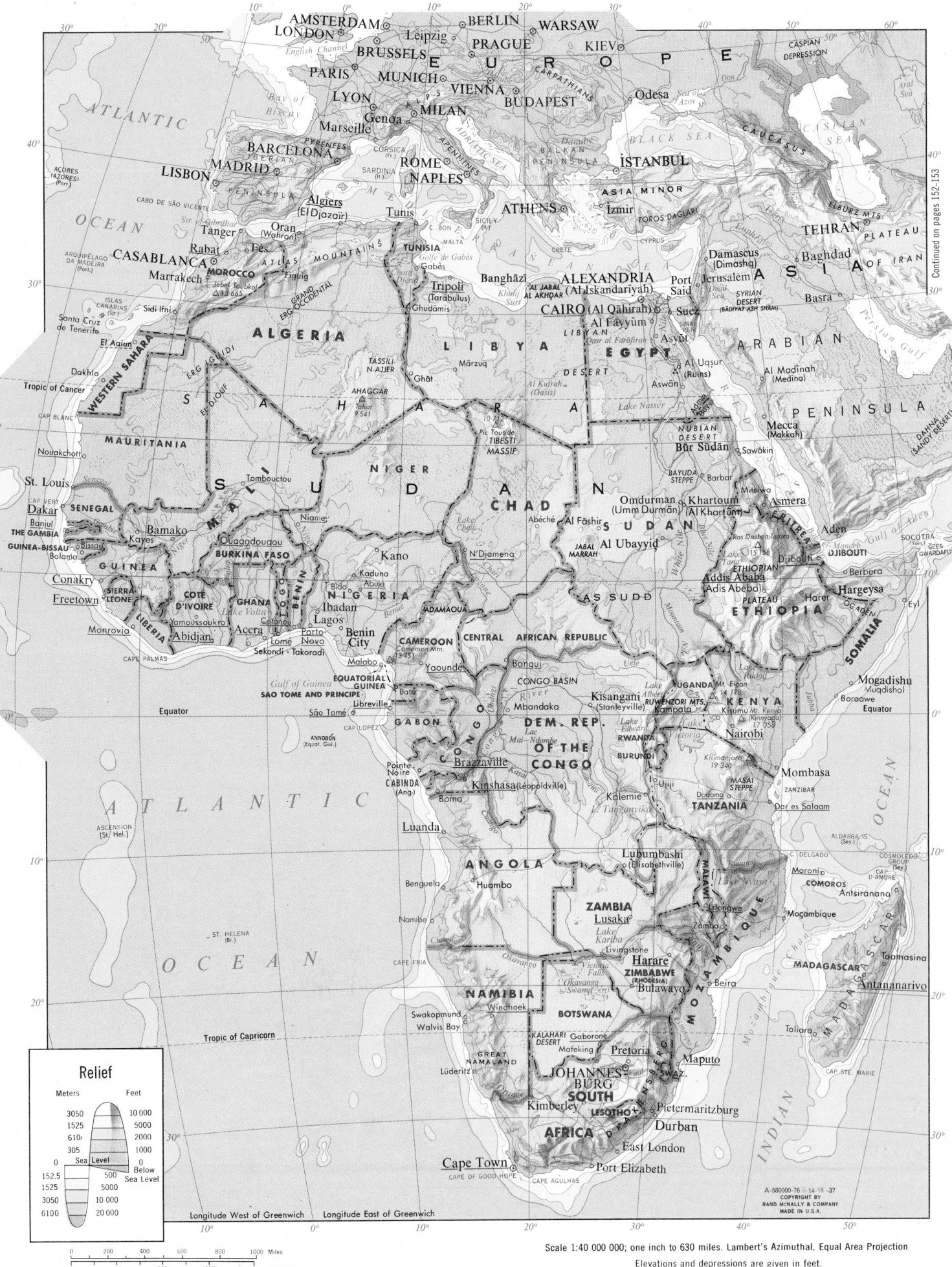

EUROPE

AMSTERDAM
LONDON
BERLIN
WARSAW
Leipzig
PRAGUE
KIEV
BRUSSELS
PARIS
MUNICH
VIENNA
BUDAPEST
LYON
MILAN
Genoa
Marseille
BARCELONA
ROME
NAPLES
LISBON
MADRID
Algiers
(El Djazaïr)
Tunis
TUNISIA
ATHENS
Izmir
ISTANBUL
ASIA MINOR
TEHRAN
Damascus
(Dimashq)
Baghdad
OF IRAN
ALEXANDRIA
(Al Iskandarīyah)
CAIRO (Al Qāhirah)
Jerusalem
Basra
ASIA
Tanger
CASABLANCA
Marrakech
MOROCCO
ATLAS MOUNTAINS
Tripoli
(Tarābulus)
Banghāzi
Al Fayyūm
Suez
ARABIAN
Port Said
Al Madīnah
(Medina)

ALGERIA
LIBYA
EGYPT
PENINSULA
Mecca
(Makkah)

WESTERN SAHARA
S A H A R A

MAURITANIA
Nouakchott

MALI
NIGER
CHAD
SUDAN
ERITREA
Asmera
Aden

SENEGAL
Dakar
THE GAMBIA
GUINEA-BISSAU
GUINEA
SIERRA LEONE
Freetown
LIBERIA
Monrovia
CÔTE D'IVOIRE
BURKINA FASO
GHANA
Accra
TOGO
BENIN
NIGERIA
Lagos
CAMEROON
Yaoundé
EQUATORIAL GUINEA
SAO TOME AND PRINCIPE
GABON
CONGO
Brazzaville
Kinshasa
CABINDA
(Ang.)
Luanda
ANGOLA
Benguela
Namibe
NAMIBIA
Windhoek
BOTSWANA
SOUTH AFRICA
Cape Town

CENTRAL AFRICAN REPUBLIC
Bangui
DEM. REP. OF THE CONGO
ETHIOPIA
Addis Ababa
(Adīs Abeba)
SOMALIA
Mogadishu
(Muqdisho)

UGANDA
Kampala
KENYA
Nairobi
RWANDA
BURUNDI
TANZANIA
Dar es Salaam
Mombasa

MALAWI
ZAMBIA
Lusaka
ZIMBABWE
Harare
Bulawayo
MOZAMBIQUE
MADAGASCAR
Antananarivo
COMOROS

Johannesburg
Pretoria
Maputo
SWAZ.
LESOTHO
Durban

ATLANTIC OCEAN

INDIAN OCEAN

Tropic of Cancer

Equator

Tropic of Capricorn

Relief

Meters		Feet
3050		10 000
1525		5000
610		2000
305		1000
0	Sea Level	0
152.5		500 Below Sea Level
1525		5000
3050		10 000
6100		20 000

Longitude West of Greenwich
Longitude East of Greenwich

0   200   400   600   800   1000 Miles
0   400   800   1200   1600 Kilometers

Scale 1:40 000 000; one inch to 630 miles. Lambert's Azimuthal, Equal Area Projection
Elevations and depressions are given in feet.

a

30° 28° 26°

©RMCN.

FAIAL
GRACIOSA
TERCEIRA
PICO
SÃO JORGE

AÇORES (AZORES)
(Port.)
SÃO MIGUEL
Ponta Delgada
STA. MARIA

Same scale as main map

38°

Continued on pages 112-113

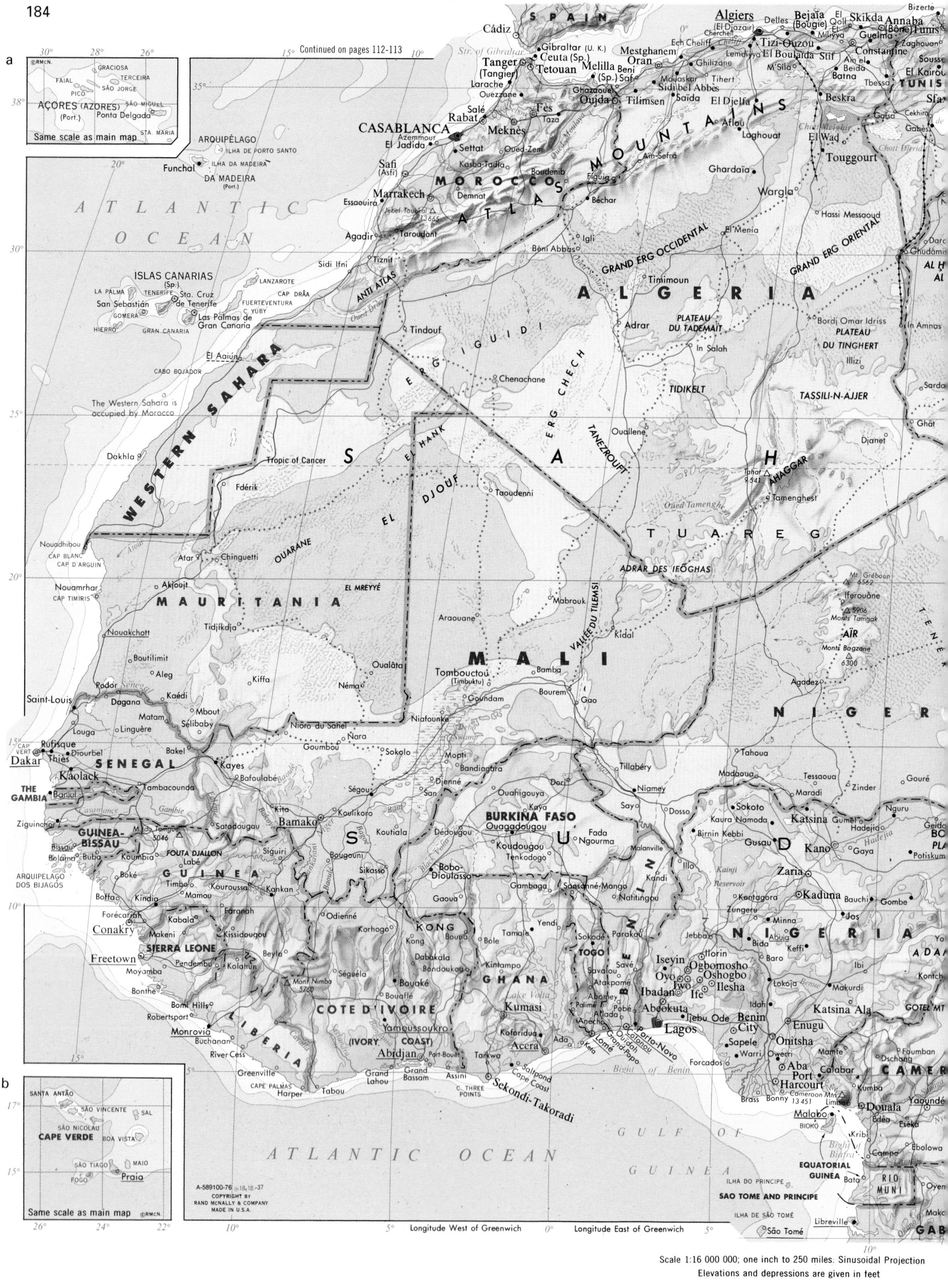

15° 10°

SPAIN

Cádiz
Gibraltar (U.K.)
Str. of Gibraltar
Ceuta (Sp.)
Tanger
(Tangier)
Larache
Melilla
(Sp.)
Beni

35°

Algiers
(El Djazair)
Delles
Béjaïa
(Bougie)
El
Qoll
Skikda
Annaba
(Bône)
Bizerte
Tunis
TUNIS

Ech Cheliff
Mestghanem
Oran
Ghilizane
Tizi-Ouzou
Cherchell
Lemdiyya
El Boulaida
Stif
Guelma
Souk
Ahras
Sousse
Sfax

Tetouan
Ouezzane
Fès
Taza
Sidi bel Abbès
Saïda
El Djelfa
Batna
El Kairo

Sale
Rabat
Meknès

Tilimsen
Aflou
Laghouat
Beskra
El Wad

CASABLANCA
El Jadida
Settat
Kasba-Tadla
Oued-Zem
Boudenib
Figuig
Ghardaïa
Touggourt

MOROCCO
ATLAS
MOUNTAINS
El Menia

Marrakech
Demnat
Béchar
Hassi Messaoud

Essaouira
Agadir
Taroudant
Jebel Toubkal △ 13 665
Igli
Béni Abbès
GRAND ERG OCCIDENTAL
GRAND ERG ORIENTAL
Ghdâmis

30°

Sidi Ifni
Tiznit
ANTI ATLAS
Adrar
In Salah
PLATEAU
DU TADEMAÏT
Bordj Omar Idriss
In Amnas
PLATEAU
DU TINGHERT
Illizi

ATLANTIC

OCEAN

ISLAS CANARIAS
(Sp.)
LA PALMA
TENERIFE
Sta. Cruz
de Tenerife
LANZAROTE
FUERTEVENTURA
CAP DRÂA
C. YUBY

San Sebastián
GOMERA
HIERRO
Las Palmas de
Gran Canaria
GRAN CANARIA

El Aaiún

CABO BOJADOR

WESTERN SAHARA

The Western Sahara is
occupied by Morocco

Tindouf
ERG IGUIDI
Timimoun
ALGERIA

Chenachane
ERG CHECH
TIDIKELT
TASSILI-N-AJJER

Tropic of Cancer
25°

Dakhla

Fdérik

S
A
H
A
R
A

EL HANK

EL DJOUF

Taoudenni

Ouallene
TANEZROUFT

Ghat
Djanet

Tahat △ 9 541
AHAGGAR
Tamenghest

Nouadhibou
CAP BLANC
CAP D'ARGUIN
Atar
Chinguetti
OUARANE
EL MREYYÉ

Mabrouk

TUAREG

ADRAR DES IFÔGHAS

Oued Tamenghest
Mt Gréboun △ 4562
Iferouâne
△ 5906
Monts Tamgak

20°

Nouâmrhâr
CAP TIMIRIS
Akjoujt

MAURITANIA

Tidjikdja

Araouane

Kidal

VALLÉE DU TILEMSI

AÏR
Monts Bagzane △ 6300

Nouakchott
Boutilimit
Aleg

Kiffa
Oualâta
Néma

MALI

Tombouctou
(Timbuktu)
Bamba
Gao

Agadez

NIGER

Saint-Louis
Podor
Dagana
Matam
Sélibaby
Nioro du Sahel
Goundam
Bourem

Tahoua

15°
CAP
VERT
Rufisque
Dakar
Thiès
Diourbel
Louga
Linguère
Kaédi
Mbout

Bakel

Goumbou
Nara

Niafounke
Débo
Swamp
Say
Niamey
Tessaoua
Zinder
Gouré
Nguru

SENEGAL
Kaolack
Kayes
Sokolo

Tillabéry
Maradi
Sokoto

THE
GAMBIA
Banjul
Tambacounda
Bafoulabé
Ségou
Mopti
Bandiagara
Dori
Dosso
Kaura Namoda
Katsina
Gumel
Gusau

Ziguinchor
Kita
Satadougou
Koulikoro
Djenné
San
Ouahigouya
Kaya
Birnin Kebbi
Kano
Gaya

GUINEA-
BISSAU
Bissau
Bolama
FOUTA DJALLON
Labé
Bamako
BURKINA FASO
Ouagadougou
SOUDAN
Malanville
Zaria

ARQUIPÉLAGO
DOS BIJAGÓS
Boké
Mont Tamgué △ 5046
Siguiri
Dédougou
Koudougou
Tenkodogo
Kandi
Kaduna
Gombe

Buba
Timbo
Mamou
Bougouni
Bobo-
Dioulasso
Fada
Ngourma
Gambaga
Sansanné-Mango
Kontagora
Zungeru
Bauchi
Jos

10°
Conakry
Makeni
Forécariah
Kindia
Kissidougou
Kabala
Faranah
Kankan
Odienné
Korhogo
Boua
Yendi
Sokodé
Parakou
Minna
Abuja
NIGERIA
Keffi

GUINEA
SIERRA LEONE
Beyla
Séguéla
KONG
Kong
Tamale
TOGO
Bida
Baro
Ibi

Freetown
Pendembu
Kolahun
Dabakala
Bondoukou
Kintampo
Savalou
Iseyin
Ilorin
Oyo
Ogbomosho
Oshogbo
Lokoja
Idah
Makurdi
Katsina Ala

Moyamba
Bonthe
Bomi Hills
Robertsport
Mont Nimba △ 5740
Bouaké
Bouaflé
GHANA
Kumasi
Abomey
Pobé
Ibadan
Iwo
Ife
Ilesha
Benin
City
Enugu
Onitsha
Sapele
Warri

LIBERIA
COTE D'IVOIRE
(IVORY COAST)
Yamoussoukro
Koforidua
Accra
Anécho
Lomé
Porto-Novo
Lagos
Ijebu Ode
Aba
Oweri
Mamfe

Monrovia
Buchanan
River Cess
Abidjan
Grand
Lahou
Grand
Bassam
Assini
Tarkwa
Saltpond
Cape Coast
Sekondi-Takoradi
Forcados
Port
Harcourt
Calabar

15°
Greenville
CAPE PALMAS
Harper
Tabou
C. THREE
POINTS
Bight of Benin
Brass
Bonny
Cameroon Mtn △ 13 451
Malabo
BIOKO
Douala
Yaoundé
CAMER

GULF OF

GUINEA

ATLANTIC   OCEAN

Bight of
Biafra
EQUATORIAL
GUINEA
Bata
RIO
MUNI

ILHA DO PRINCIPE
SÃO TOME AND PRINCIPE
ILHA DE SÃO TOMÉ
São Tomé
Libreville
GAB

b

SANTA ANTÃO
SÃO VICENTE
SAL
SÃO NICOLAU
BOA VISTA
CAPE VERDE
SÃO TIAGO
MAIO
FOGO
Praia

Same scale as main map

26° 24° 22°

A-589100-76 D-18-18-37
COPYRIGHT BY
RAND McNALLY & COMPANY
MADE IN U.S.A.

10° 5° Longitude West of Greenwich 0° Longitude East of Greenwich 5° 10°

Scale 1:16 000 000; one inch to 250 miles. Sinusoidal Projection
Elevations and depressions are given in feet

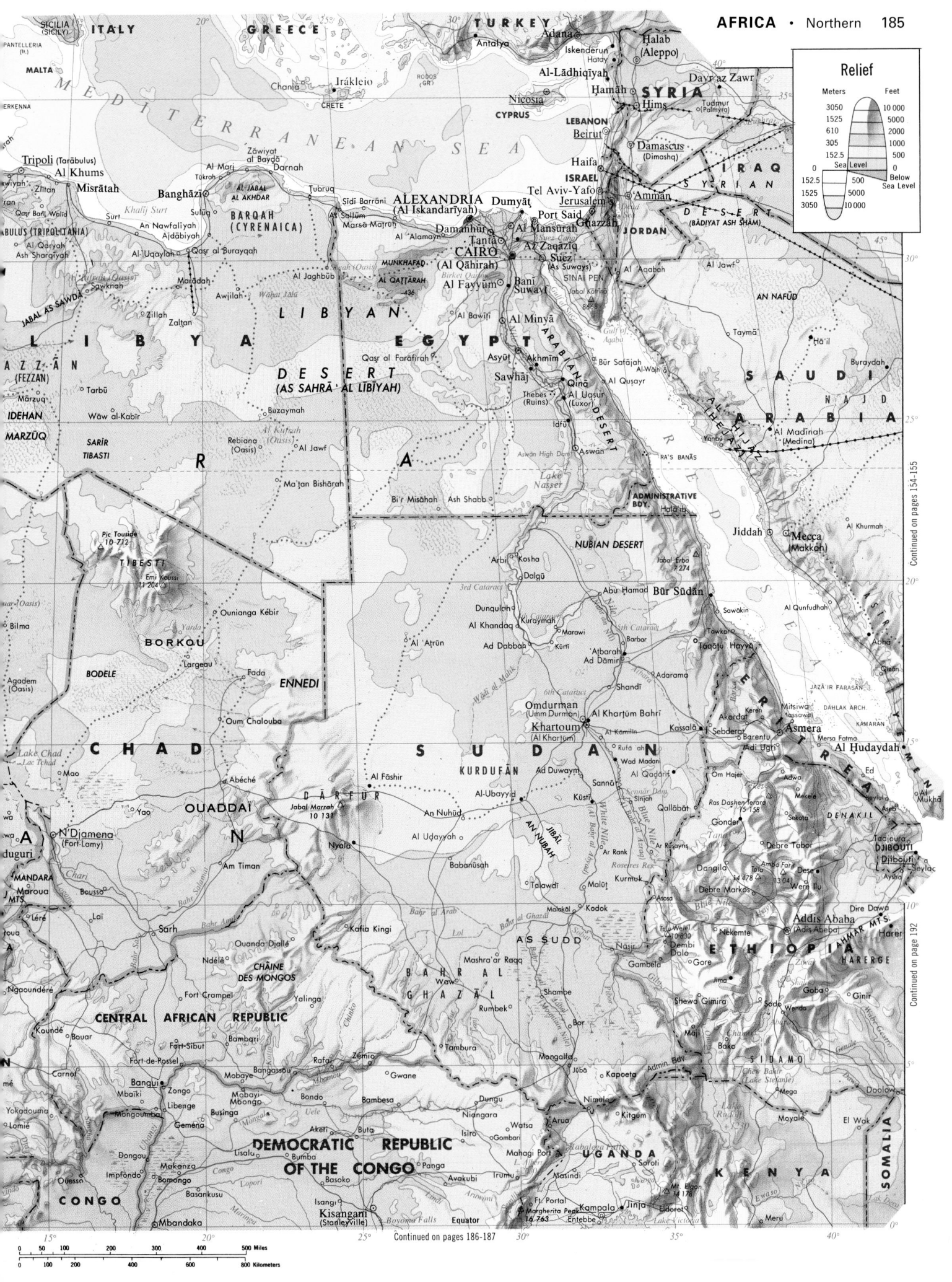

Relief

Meters	Feet
3050	10 000
1525	5000
610	2000
305	1000
152.5	500
0 Sea Level	0 Sea Level
152.5	500 Below Sea Level
1525	5000
3050	10 000

SICILIA (SICILY)
ITALY
PANTELLERIA (It.)
MALTA
KERKENA

GREECE
TURKEY
Adana
Antalya
Iskenderun
Hatay
RODOS (GR)
CRETE
Chania
Irákleio
Nicosia
CYPRUS

Ḥalab (Aleppo)
Al-Lādhiqīyah
Ḥamāh
Ḥimṣ
Dayr az Zawr
Tudmur (Palmyra)
SYRIA
LEBANON
Beirut
Damascus (Dimashq)
IRAQ
SYRIAN DESERT (BĀDIYAT ASH SHĀM)

MEDITERRANEAN SEA

Tripoli (Tarābulus)
Al Khums
Zliten
Miṣrātah
Qaṣr Banī Walīd
BULUS (TRIPOLITANIA)
Al Qaryah Ash Sharqīyah

Zāwiyat al Bayḍā
Darnah
Al Marj
Banghāzī
AL JABAL AL AKHDAR
BARQAH (CYRENAICA)
Surt
An Nawfalīyah
Ajdābiyah
Al 'Uqaylah
Qaṣr Burayqah

Tūkrah
Sūluq
Tubruq
Sīdī Barrānī
As Sallūm
Marsā Maṭrūḥ
Al 'Alamayn
Damanhūr
ALEXANDRIA (Al Iskandarīyah)
Dumyāṭ
Port Said
Al Manṣūrah
Tanta
Al Zaqāzīq
CAIRO (Al Qāhirah)
Suez (As Suways)
Ghazzah
ISRAEL
Tel Aviv-Yafo
Jerusalem
JORDAN
Amman
Al 'Aqabah
Al Jawf
Gulf of Aqaba

Haifa
SINAI PEN.
Jabal Kātrīnā 8668

Khalīj Surt
Rafah (Oasis)
Sawknah
JABAL AS SAWDA

Marādah
Awjilah
Wāḥat Jalū
Al Jaghbūb
MUNKHAFAD AL QATTĀRAH -436
Al Fayyūm
Banī Suwayf
Al Bawītī
Al Minyā
Asyūṭ
Birket Qārūn

ʿAZZĀN (FEZZAN)
Tarbū
Mārzuq
IDEHAN MARZŪQ
Wāw al-Kabīr
Zillah
Zaltan

LIBYA
EGYPT
LIBYAN DESERT (AS SAHRĀ AL LĪBĪYAH)
Qaṣr al Farāfirah
Qaṣr al Farāfirah
Rebiana (Oasis)
Al Jawf
Buzaymah
Maʿtan Bishārah
Bi'r Misāḥah
Ash Shabb

SARĪR TIBESTI
BORKOU
Ounianga Kébir
Largeau
Fada
ENNEDI
Oum Chalouba
Pic Touside 10 712
Emi Koussi 11 204
TIBESTI
Yarda
BODELE

Sawḥāj
Akhmīm
Qinā
Al Uqṣur (Luxor)
Thebes (Ruins)
Idfū
Aswān High Dam
Aswān
Lake Nasser
ADMINISTRATIVE BDY.
Halaib
NUBIAN DESERT
Jabal Erba 7 274
Arbi
Kosha
Dalgū
3rd Cataract
Dunqulah
Al Khandaq
Kuraymah
Marawi
Kürtī
Ad Dabbah
Al 'Aṭrūn
4th Cataract
Abu Ḥamad
Bür Sūdān
Sawākin

ARABIAN DESERT
Būr Safājah
Al Qusayr
Al Wajh
RED SEA
AL HEJAZ
RAS BĀNĀS
Yanbu'
Jiddah
Mecca (Makkah)
Al Khurmah
Al Qunfudhah
Abhā
SAUDI ARABIA
NAJD
Taymā
Ha'il
Buraydah
Al Madīnah (Medina)
AN NAFŪD

JAZĀ'IR FARASAN
DAHLAK ARCH.
KAMARAN
YEMEN
Al Mukhā
Al Ḥudaydah
Ed

CHAD
Lake Chad / Lac Tchad
Mao
N'Djamena (Fort-Lamy)
Yao
OUADDAĪ
Abéché
Am Timan
Ndélé
CHAÎNE DES MONGOS
Fort Crampel
MANDARA MTS.
Maroua
Bousso
Léré
Lai
Sarh
Baïbokoum

SUDAN
DARFUR
Jabal Marrah 10 131
Al Fāshir
Nyala
KURDUFAN
An Nuhūd
Al Uḍayyah
Babanūsah
Talawdī
Malūṭ
Kafia Kingi
BAHR AL GHAZĀL
Wāw
Tambura
Rumbek
Bor
AS SUDD
Mashra'ar Raqq
Shambe
Gambela
Maji
Bako
Juba
Kapoeta
Nimule
Torit

Atbarah
Barbar
Ad Dāmir
Shandī
6th Cataract
Omdurman (Umm Durman)
Al Khartūm Bahrī
KHARTOUM (Al Khartūm)
Rufā'a
Wad Madani
Al Kāmilīn
Kassalā
Sebderat
Adarama
Taqāṭu Hayyā
Tawkar
AN NUBAH
JIBĀL
Ad Duwaym
Al-Ubayyid
Sannār
Al Qaḍārif
Om Hajer
Kūstī
Sinjah
Qallābāt
Sennar Dam
Roseires Res.
Ar Rank
Kurmuk
Asosa
White Nile (Al Baḥr al Abyaḍ)
Blue Nile (Al Baḥr al Azraq)
Malakal
Kodok
Nāṣir

ERITREA
Mitsiwa
Massawa
Akordat
Keren
Asmera
Adi Ugrī
Adwa
Mekele
DENAKIL
Ras Dashen Terara 15 158
Gonder
Debre Tabor
Dangila
Debre Markos
Amba Farīt 13 041
Tana Hāyk 14 478
Dese
Were Ilu
Dire Dawa
Addis Ababa (Adis Abeba)
Nekemte
Dembi Dolo
Gore
Jima
Shewa Gimira
Sodo
Wenda
Chew Bahir Lake Stefanie
SIDAMO
Ginir
Goba
Maydī
Aseb
Tadjoura
DJIBOUTI
Djibouti
Zeyla
Aysha
AHMAR MTS.
HARERGE
Harer
SOMALIA
El Wak
Doolow
Mega
Moyale

CENTRAL AFRICAN REPUBLIC
Koundé
Bauar
Ngaoundéré
Fort-Sibut
Fort-de-Possel
Bambari
Bangui
Mbaïki
Libenge
Zongo
Bangassou
Rafai
Zemio
Gwane
Mobaye
Mobayi-Mbongo
Bondo
Bambesa
Dungu
Niangara
Watsa
Arua
Kitgum
Gulu

DEMOCRATIC REPUBLIC OF THE CONGO
Dongou
Impfondo
Ouésso
Makanza
Bomongo
Mbandaka
Basankusu
Bumba
Basoko
Isangi
Kisangani (Stanleyville)
Boyoma Falls
Avakubi
Isiro
Gombari
Mahagi Port
Panga
Irumu
Fort Portal
Margherita Peak 16 763
Mt. Elgon
Kampala
Entebbe
Lake Victoria
Equator
UGANDA
Masindi
Soroti
Jinja
KENYA
Eldoret
Meru
CONGO
Yokadouma
Lomié
Mongoumba
Gemena
Akéti
Buta

Continued on pages 154-155
Continued on page 192
Continued on pages 186-187

0 50 100 200 300 400 500 Miles
0 100 200 400 600 800 Kilometers

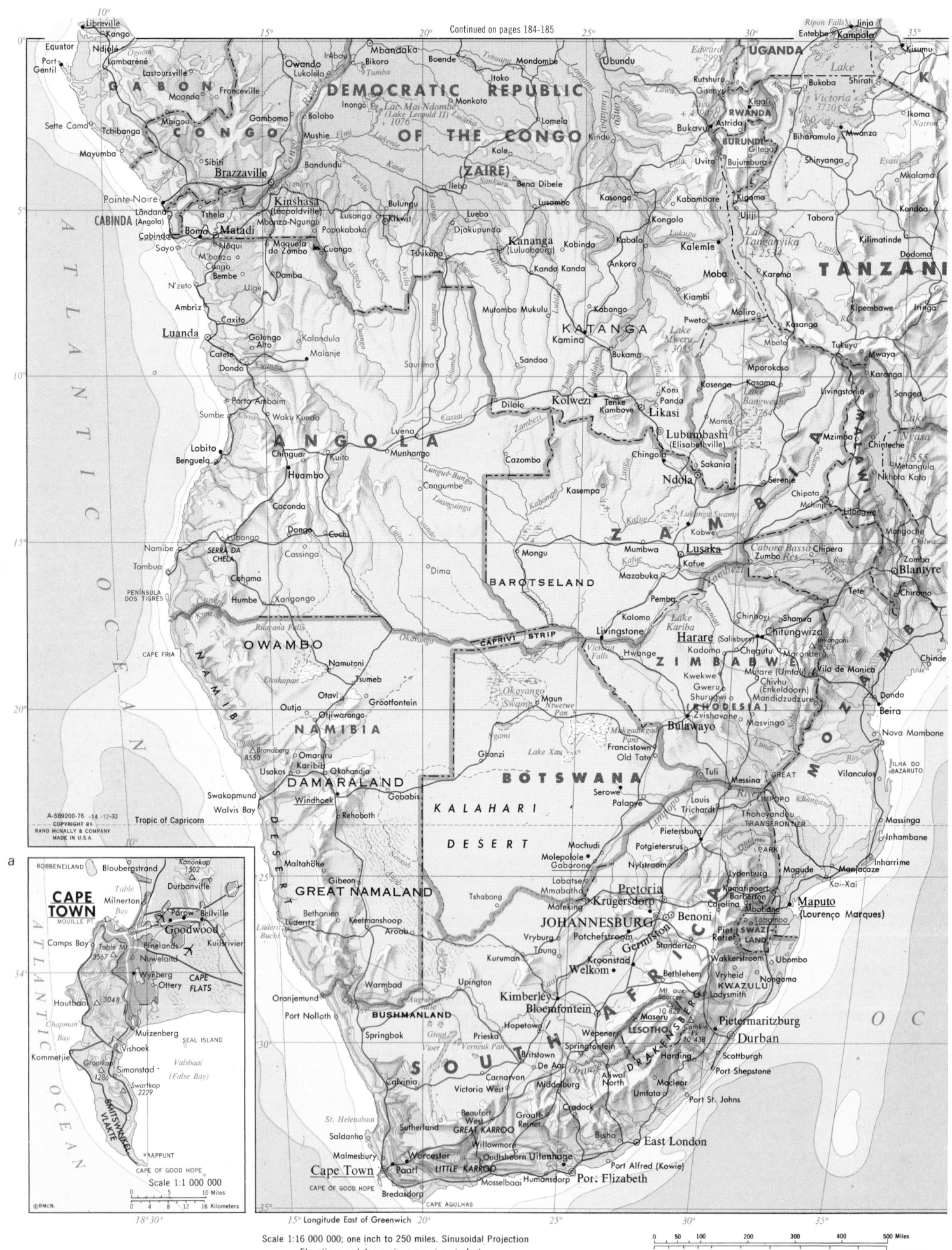

Continued on pages 184-185

Scale 1:16 000 000; one inch to 250 miles. Sinusoidal Projection
Elevations and depressions are given in feet

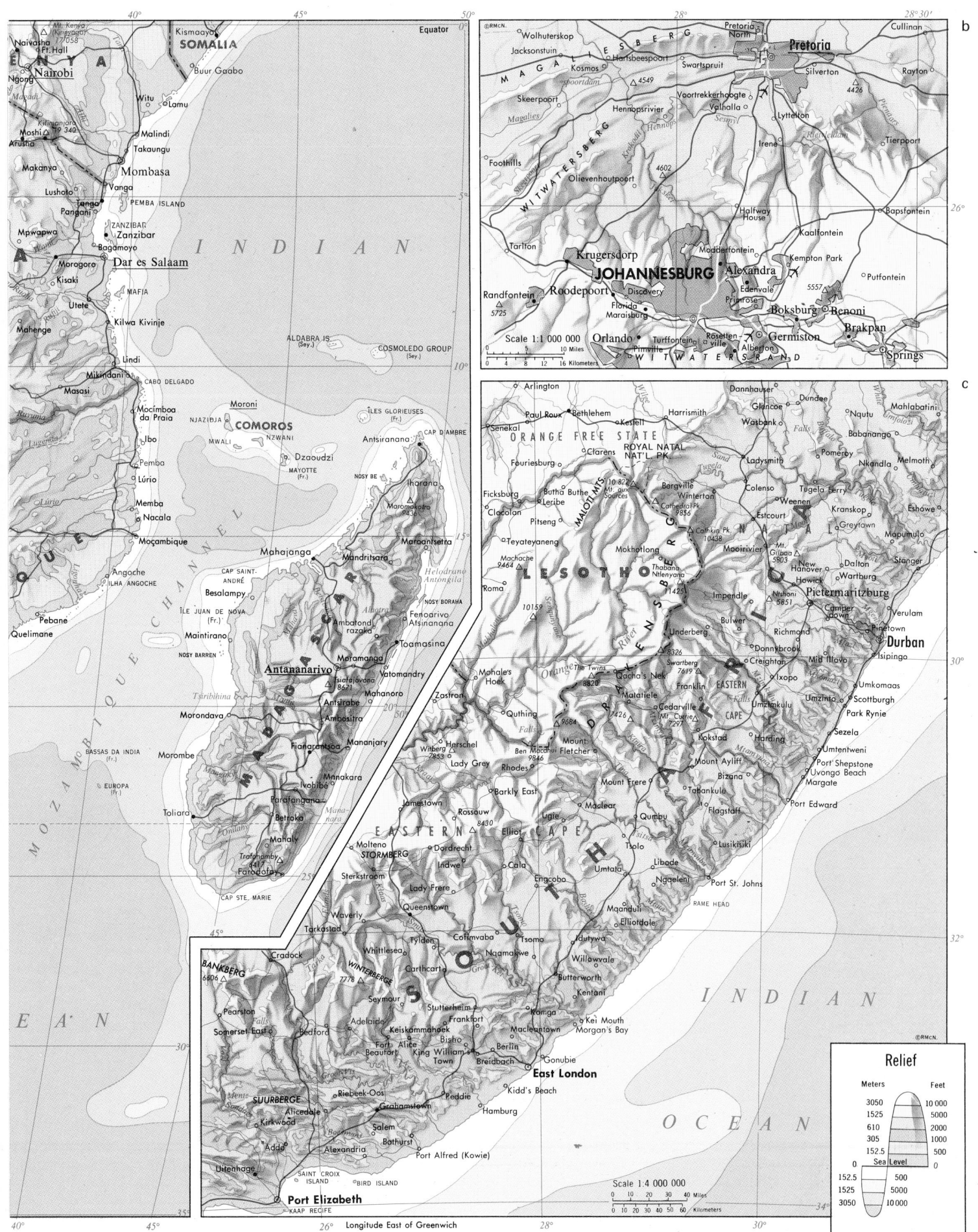

KENYA
Mt. Kenya (Kirinyaga) 17,058
Ngong
Naivasha
Ft. Hall
Nairobi
Machakos
Moshi
Kilimanjaro 19,340
Arusha
Mpwapwa
Makanya
Lushoto
Korogwe
Tanga
Pangani
TANZANIA
Morogoro
Kisaki
Utete
Mahenge

SOMALIA
Kismaayo
Buur Gaabo
Equator
Witu
Lamu
Malindi
Takaungu
Vanga
Mombasa
PEMBA ISLAND
ZANZIBAR
Zanzibar
Bagamoyo
Dar es Salaam
MAFIA

INDIAN

Mogincual
Masasi
Mikindani
Lindi
Kilwa Kivinje
CABO DELGADO
Moçimboa da Praia
Ibo
Pemba
Lúrio
Memba
Nacala
Moçambique
Angoche
ILHA ANGOCHE
Pebane
Quelimane

Moroni
NJAZIDJA
MWALI
COMOROS
NZWANI
MAYOTTE (Fr.)
Dzaoudzi
ALDABRA IS. (Sey.)
COSMOLEDO GROUP (Sey.)
ÎLES GLORIEUSES (Fr.)
CAP D'AMBRE
Antsiranana
NOSY BE
Ihorana
Maromokotro 9436

OCEAN

MOZAMBIQUE CHANNEL
BASSAS DA INDIA (Fr.)
EUROPA (Fr.)
ÎLE JUAN DE NOVA (Fr.)
NOSY BARREN

MADAGASCAR
Mahajanga
Mandritsara
Maroantsetra
CAP SAINT-ANDRÉ
Besalampy
Helodrano Antongila
NOSY BORAHA
Fenoarivo Atsinanana
Maintirano
Ambatondrazaka
Toamasina
Moramanga
Antananarivo
Tsiafajavona 8671
Antsirabe
Vatomandry
Mahanoro
Morondava
Ambositra
Mananjary
Fianarantsoa
Ivohibe
Manakara
Morombe
Ambalavao
Farafangana
Betroka
Mahaly
Trotrohamby 4412
Toliara
Farodofay
CAP STE. MARIE

b
RMcN
Wolhuterskop
Jacksonstuin
MAGALIESBERG
Kosmos
Hartbeespoort
Skeerpoort
spoortdam
Foothills
Magalies
Hennopsrivier
Valhalla
Tarlton
Krugersdorp
Randfontein
5725
JOHANNESBURG
Roodepoort
Discovery
Florida
Maraisburg
WITWATERSBERG
Orlando
Scale 1:1 000 000
0   10 Miles
0   4   8   12   16 Kilometers
Pretoria North
Pretoria
Swartspruit
4549
Voortrekkerhoogte
Silverton
4426
Rayton
Cullinan
Lyttelton
Tierpoort
Irene
Halfway House
Kaalfontein
Bapsfontein
Modderfontein
Kempton Park
Putfontein
Alexandra
Edenvale
5557
Boksburg
Benoni
Primrose
Turffontein
Rösettenville
Germiston
Brakpan
Pimville
Alberton
Springs
WITWATERSRAND

c
Arlington
Dannhauser
Glencoe
Dundee
Nqutu
Mahlabatini
Paul Roux
Bethlehem
Harrismith
Wasbank
ORANGE FREE STATE
Senekal
Kestell
Ladysmith
Pomeroy
Nkandla
Melmoth
Fouriesburg
Clarens
ROYAL NATAL NAT'L. PK.
Colenso
Babanango
Ficksburg
Butha Buthe
10,822 Mt. aux Sources
Bergville
Winterton
Cathedral Pk. 9856
Estcourt
Kranskop
Eshowe
Clocolan
Leribe
Pitseng
Mokhotlong
Catkin Pk. 10,438
Mooirivier
Greytown
Mapumulo
Teyateyaneng
Machache 9464
Thabana Ntlenyana
Mt. Gilboa 5803
Stanger
Roma
LESOTHO
Impendle
New Hanover
Howick
Dalton
Wartburg
10,152
1425
Ntshoni 5851
Pietermaritzburg
Underberg
Bulwer
Richmond
Camperdown
Verulam
8326
The Twins
Qacha's Nek
Creighton
Donnybrook
Pinetown
Durban
Mohale's Hoek
Swartberg 7619
Mid Illovo
Isipingo
Zastron
Matatiele
Franklin
EASTERN CAPE
Ixopo
Umkomaas
Quthing
7426
Cedarville
Mt. Currie 7297
Kokstad
Harding
Sezela
9684
Umzimkulu
Scottburgh
Park Rynie
Witberg 2853
Herschel
Mount Fletcher
Mount Frere
Mount Ayliff
Bizana
Port Shepstone
Uvongo Beach
Margate
Ben Macdhui 9846
Lady Grey
Rhodes
Maclear
Qumbu
Flagstaff
Port Edward
Barkly East
Jamestown
Rossouw
Ugie
Maclear
Tsolo
Lusikisiki
Elliot
Umtata
Libode
RAME HEAD
8430
Dordrecht
Indwe
Cala
Engcobo
Ngqeleni
Port St. Johns
Molteno
STORMBERG
Lady Frere
Ugu
Mqanduli
Elliotdale
Sterkstroom
Queenstown
Tsomo
Idutywa
Waverly
Tarkastad
Tylden
Cofimvaba
Nqamakwe
Willowvale
Cradock
Whittlesea
Carthcart
Butterworth
BANKBERG
6606
WINTERBERG
7778
SOUTH
Tsomo
Kentani
Seymour
Stutterheim
Komga
Adelaide
Frankfort
Maclaentown
Kei Mouth
Morgan's Bay
Pearston
Somerset East
Bedford
Keiskammahoek
Bisho
Berlin
Breidbach
Fort Beaufort
Fort Alice
King William's Town
Gonubie
SUURBERGE
Riebeek-Oos
Grahamstown
East London
Kidd's Beach
Alicedale
Peddie
Hamburg
Kirkwood
Addo
Salem
Bathurst
Uitenhage
Alexandria
Port Alfred (Kowie)
SAINT CROIX ISLAND
BIRD ISLAND
Port Elizabeth
KAAP RECIFE
Scale 1:4 000 000
0 10 20 30 40 Miles
0 10 20 30 40 50 60 Kilometers

Longitude East of Greenwich

Relief

Meters		Feet
3050		10 000
1525		5000
610		2000
305		1000
152.5		500
Sea Level		0
152.5		500
1525		5000
3050		10 000

Relief

Meters	Feet
3050	10 000
1525	5000
610	2000
305	1000
152.5	500
0	Sea Level
152.5	500
1525	5000
3050	10 000

Copyright by Rand McNally & Co.
Made in U.S.A.
A-589400-76   2-2-13

Scale 1:10,000,000; one inch to 160 miles. Lambert Azimuthal Equal Area Projection
Elevations and depressions are given in feet.

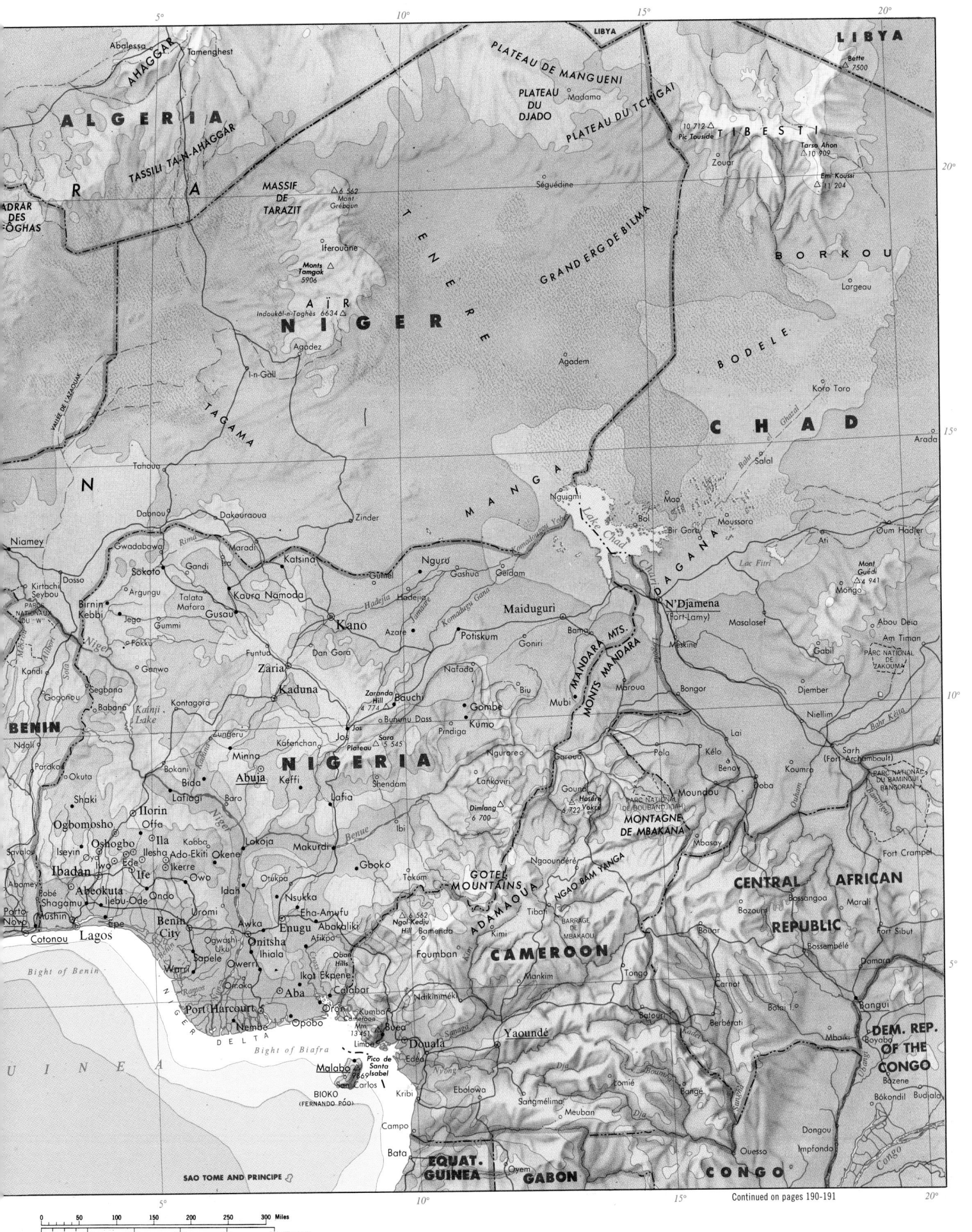

Continued on pages 190-191

0   50   100   150   200   250   300 Miles

0   100   200   300   400   500 Kilometers

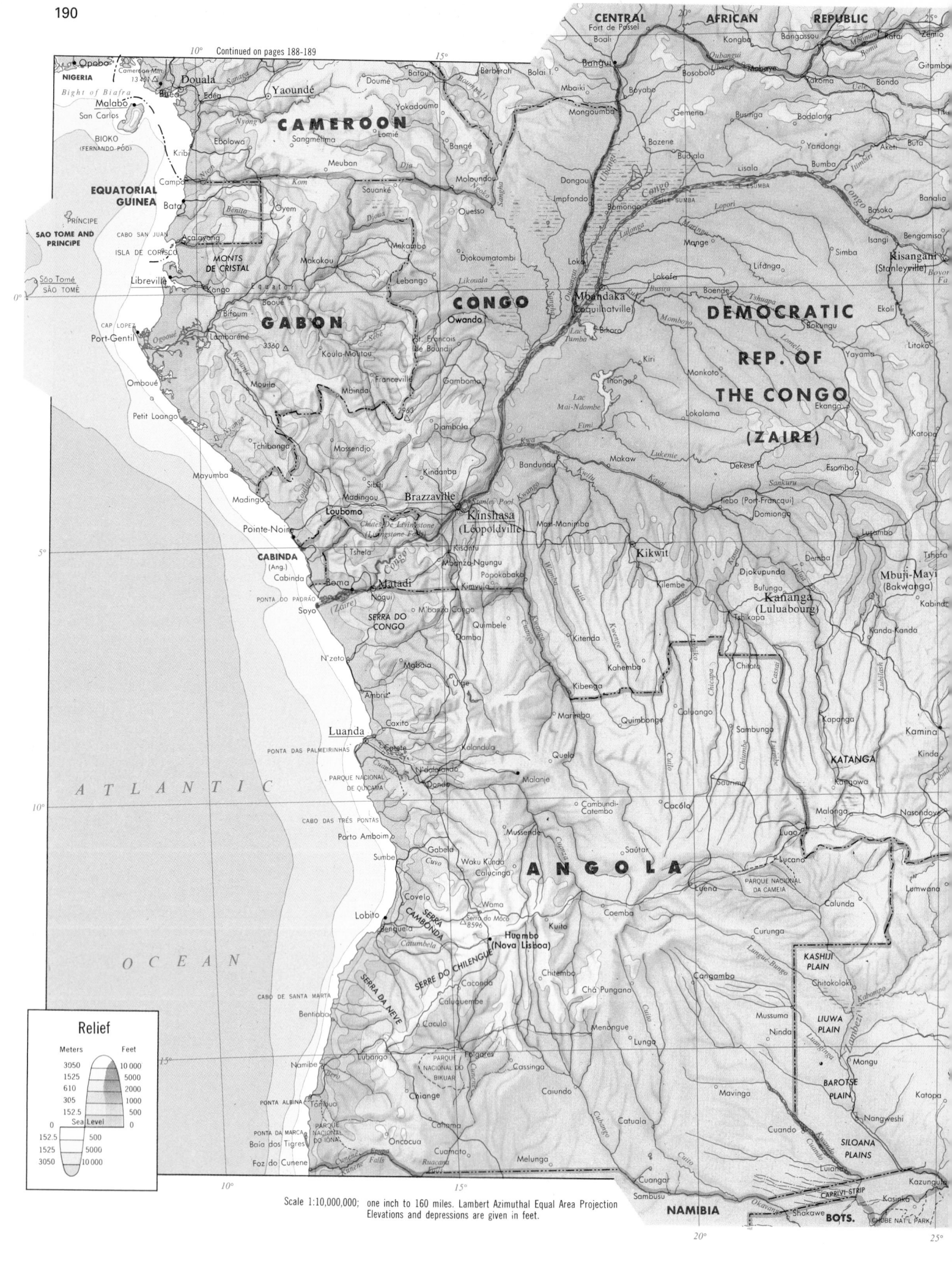

Continued on pages 188-189

NIGERIA

Opobo
Cameroon Mtn.
13 451
Buea
Douala
Edéa
Yaoundé

Bight of Biafra

Malabo
San Carlos

BIOKO
(FERNANDO PÓO)

Doumé
Batouri
Berberati
Bolai T.

CENTRAL   AFRICAN   REPUBLIC
Fort de Possel
Kongba
Bangassou
Zémio
Katai

Yokadouma
Bangé
Berberati
Bangui
Bosobolo
Ubangi
Mobaye
Bondo
Gitamba
Boyaba
Boali T.
Mbaiki
Boyabu
Gemena
Businga
Bodalang
Yandongi
Buta
Aket

CAMEROON

Ebolowa
Sangmelima
Meuban
Lomié
Kom
Souanké
Ouesso

Dongou
Impfondo
Bozene
Budiala
Lisala
Bumba
Banalia

EQUATORIAL
GUINEA

Campo
Bata
Oyem
Benito
Djoum
Mekambo

MONTS
DE CRISTAL

Acalayong

Makokou
Lebango

Djokoumatombi
Likouala

CONGO

Owando

Mbandaka
(Coquilhatville)

DEMOCRATIC

REP. OF

THE CONGO

(ZAIRE)

SÃO TOMÉ AND
PRINCIPE

PRÍNCIPE

CABO SAN JUAN

ISLA DE CORISCO

São Tomé
SÃO TOMÉ

Libreville

Kango

Equator

Booué

GABON

Bifoum
3360

Lac
Tumba
Bikoro

Loka
Lakofa
Boende

Lifanga
Simba
Isangi
Bengamisa
Kisangani
(Stanleyville)

CAP LOPEZ
Port-Gentil

Ogooué
Lambaréné
Koula-Moutou
Francevillé

St. François
de Boundji

Gamboma

Kiri

Monkoto

Lokolama

Ekanga

Yayama
Litoko
Katopa

Omboué
Mouila
Mbinda

Djambala

Lac
Mai-Ndombe

Fimi

Ekoli

Petit Loango

Tchibanga
Mossendjo

Kindamba

Bandundu

Makaw
Lukenie

Dekese

Esambo

Mayumba
Madingo

Sibiti
Madingou

Brazzaville

Kwa
Stanley Pool
Masi-Manimba

Kasai
Sankuru
Tiebo (Port-Francqui)
Domionga

Lumbo

Loubomo

Chutes De Livingstone
(Livingstone Falls)

Kinshasa
(Léopoldville)

Kikwit

Djokupunda
Bulunga

Kananga
(Luluabourg)

Mbuji-Mayi
(Bakwanga)

Tshofa

Pointe-Noire

CABINDA
(Ang.)

Tshela

Congo

Kisantu
Mbanza-Ngungu
Popokabaka

Kilembe

Tshikapa
Damba

Kanda-Kanda
Kabinda

Cabinda
Boma
Matadi

Kimvula

Kitenda

Kahemba

Chitats

Kamina

PONTA DO PADRÃO

Nóqui

Soyo
Zaire

SERRA DO
CONGO

M'banza Congo

Quimbele

Damba

Kibenga

Marimba

Quimbonge

Kaluango
Sambungo

Kapanga
Kamina
Kinda

N'zeto

Mbaia
Uíge

Caluango

Cuilo

KATANGA
Kamgowa

Ambriz

Caxito

Kalandula

Quela

Luena

Malanga
Nasondoye

Luanda

PONTA DAS PALMEIRINHAS

Catete
N'dalatando
Dondo

Malanje

Cacólo

Luao
Lucano

Calunda
Lumwana

PARQUE NACIONAL
DE QUIÇAMA

Cambundi-
Catembo

Cuanza

ATLANTIC

CABO DAS TRÊS PONTAS
Porto Amboim

Mussende

Saútar

PARQUE NACIONAL
DA CAMEIA

KASHIJI
PLAIN

Chitokoloki

Sumbe

Gabela
Waku Kunda
Calucinga

ANGOLA

Coemba

Curunga

Cangamba

Mussuma

LIUWA
PLAIN

Cuvo

Wamo

Kuito

Ninda

OCEAN

Coyelo

SERRA
CAMBONDA

Serra do Moco
8596

Huambo
(Nova Lisboa)

Chitembo

Cuíto

BAROTSE
PLAIN

Katopa

Lobito
Benguela
Catumbela

SERRE DO CHILENGUE

Caconda

Chá Pungana

Mongu

Nangweshi

SERRA DA NEIVE

Caluquembe

Cariango

Menongue

Mavinga

Lungue-Bungo

SILOANA
PLAINS

CABO DE SANTA MARTA
Bentiaba

Cacula

Lunga

Cassinga

Caiundo

Cuando

Katopa

Namibe

Lubango

PARQUE
NACIONAL DO
BIKUAR

Folgares

Cuito

Catuala

Kasinka

PONTA ALBINA
Tômbua

Chiange

Cahama

Cuando

Okavango

CAPRIVI STRIP
Shakawe
CHOBE NAT'L PARK

PONTA DA MARCA
Baía dos Tigres

PARQUE
NACIONAL
DO IONA

Cunene
Epupa Falls
Ruacana Falls

Oncócua
Cuamato

Melunga

NAMIBIA
BOTS.

Foz do Cunene
Kenene

Cuando

Relief

Meters	Feet
3050	10 000
1525	5000
610	2000
305	1000
152.5	500
Sea Level	0
152.5	500
1525	5000
3050	10 000

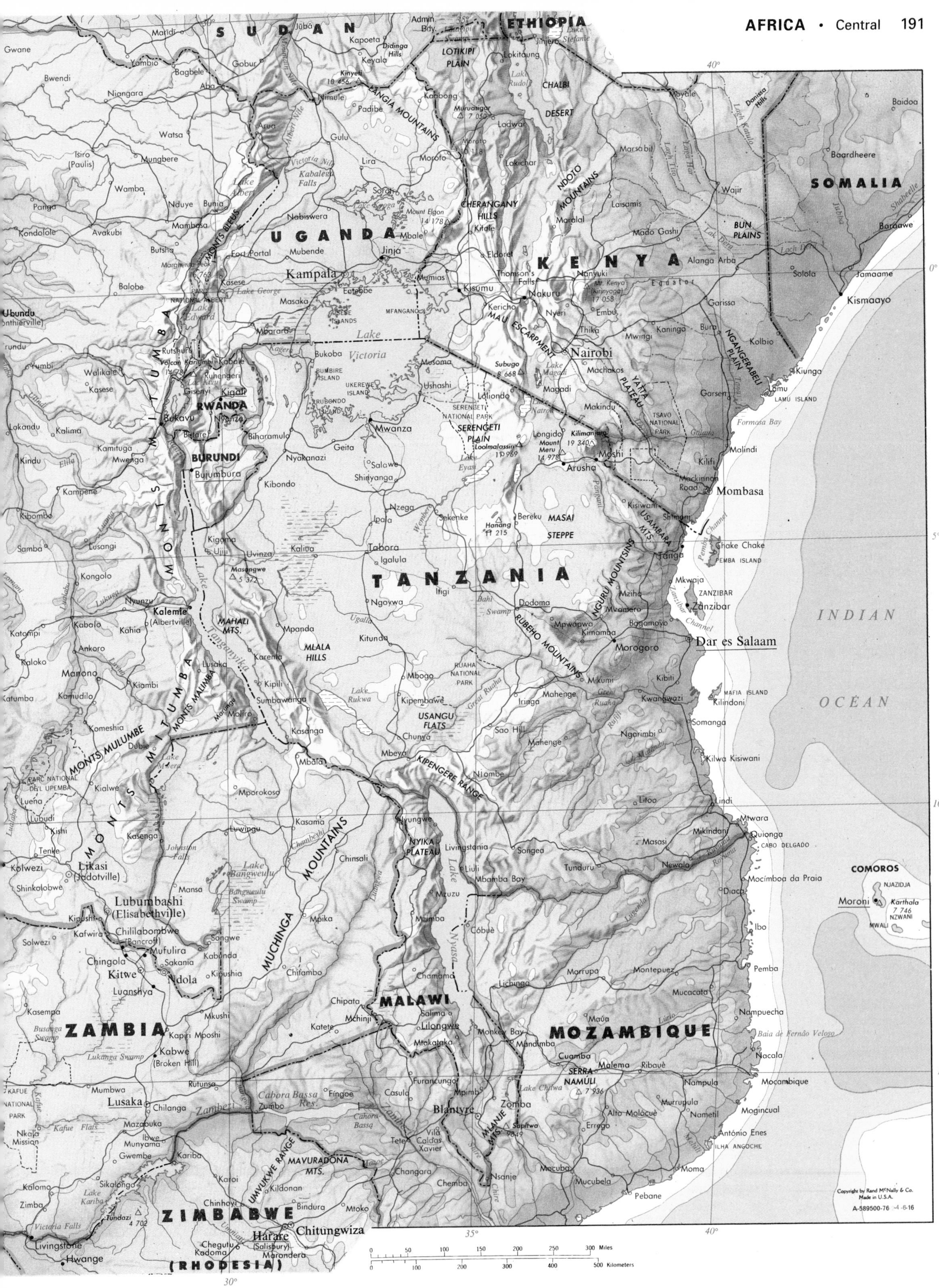

SUDAN
ETHIOPIA
UGANDA
KENYA
SOMALIA
RWANDA
BURUNDI
TANZANIA
ZAMBIA
MALAWI
MOZAMBIQUE
ZIMBABWE
(RHODESIA)
COMOROS

INDIAN OCEAN

Kampala
Nairobi
Kismaayo
Baardheere
Baraawe
Mombasa
Zanzibar
Dar es Salaam
Dodoma
Arusha
Moshi
Kigali
Bujumbura
Lubumbashi (Elisabethville)
Lusaka
Kitwe
Ndola
Blantyre
Lilongwe
Zomba
Harare (Salisbury)
Chitungwiza
Moroni

Lake Victoria
Lake Tanganyika
Lake Nyasa
Lake Edward
Lake Kivu
Lake Rukwa
Lake Bangweulu
Lake Malawi
Lake Kariba
Victoria Falls

CHALBI DESERT
SERENGETI PLAIN
MASAI STEPPE
NYIKA PLATEAU
MALI MTS.
MLALA HILLS
USANGU FLATS
KIPENGERE RANGE
MUCHINGA MOUNTAINS
MONTS MITUMBA
MONTS MULUMBE
RUBEHO MOUNTAINS
NGURU MOUNTAINS
USAMBARA MTS.
MAU ESCARPMENT
CHERANGANY HILLS
NDOTO MOUNTAINS
YATTA PLATEAU
MAVURADONA MTS.
UNYUKWE RANGE
SERRA NAMÚLI

Kilimanjaro 19 340
Mount Kenya (Kirinyaga) 17 058
Mount Elgon 14 178
Mount Meru 14 978

Equator 0°
5°
10°
15°
30°
35°
40°

ZANZIBAR
PEMBA ISLAND
MAFIA ISLAND
LAMU ISLAND
NJAZIDJA
Karthala 7 746
NZWANI
MWALI

Copyright by Rand McNally & Co.
Made in U.S.A.
A-589500-76 -4 -6-16

0 50 100 150 200 250 300 Miles
0 100 200 300 400 500 Kilometers

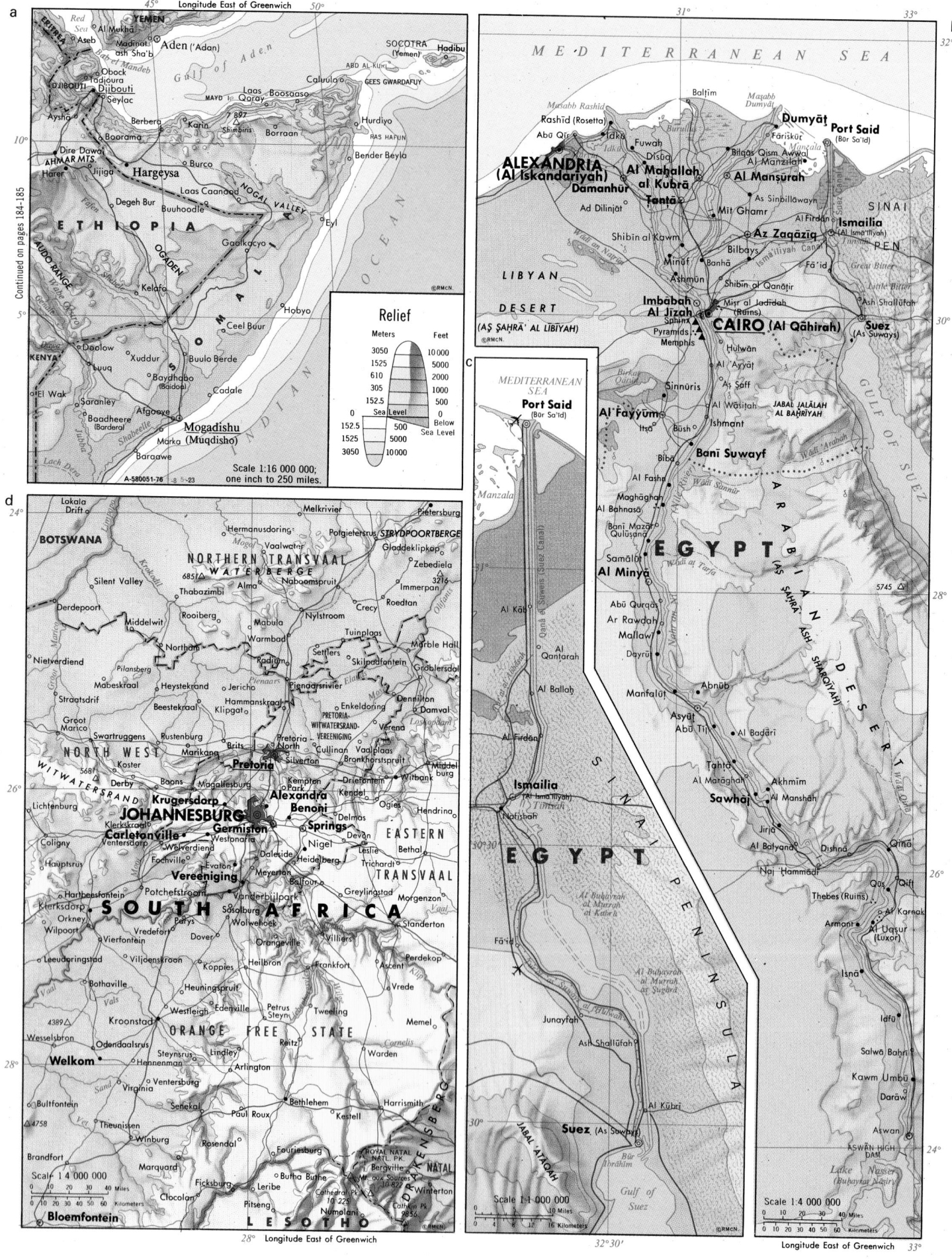

a

Red Sea
Al Mukhā
YEMEN
Aden ('Adan)
SOCOTRA (Yemen)
Hadibu
Madīnat ash Sha'b
Gulf of Aden
ABD AL-KŪRĪ
GEES GWARDAFUY
Obock
Tadjoura
DJIBOUTI
Djibouti
Seylac
MAYD Is.
Laas Qoray
Boosaaso
Caluula
ERITREA
Aseb
Aysha
Berbera
Karin
Hurdiyo
RAS HAFUN
Dire Dawa
AHMAR MTS.
Jijiga
Boorama
Laas Caanood
Borraan
Bender Beyla
Harer
Hargeysa
7 897
Shimbiris
10°
Degeh Bur
Buuhoodle
Eyl
Continued on pages 184–185
ETHIOPIA
Gaalkacyo
AUDO RANGE
OGADEN
NOGAL VALLEY
Kelafo
S
O
Ceel Buur
Hobyo
5°
KENYA
Doolow
Xuddur
Buulo Berde
El Wak
Luuq
Baydhabo (Baidoa)
Cadale
Saranley
Afgooye
Baadheere (Barderel)
INDIAN OCEAN
Mogadishu (Muqdisho)
Marka
Baraawe
Lach Dera
Shabeelle
Jubba

Relief
Meters		Feet
3050		10 000
1525		5000
610		2000
305		1000
152.5		500
0	Sea Level	0
152.5		Below Sea Level
1525		500
3050		5000
		10000

©RMCN.
Scale 1:16 000 000;
one inch to 250 miles.
A-580051-76 -8 5-23

b
MEDITERRANEAN SEA
32°
Baltīm
Maşabb Dumyāţ
Maşabb Rashīd
Rashīd (Rosetta)
Dumyāţ
Port Said
(Būr Sa'īd)
Abū Qīr
Idkū
Fuwah
Burullus
Fāriskūr
Manzala
Bilqās Qism Awwal
Al Manzilah
ALEXANDRIA
(Al Iskandarīyah)
Disūq
Al Maḥallah al Kubrā
Al Manṣūrah
Damanhūr
Ţanţā
As Sinbillāwayn
SINAI PEN.
Ad Dilinjāt
Mīt Ghamr
Al Firdān
Ismailia
(Al Ismā'īlīyah)
Shibīn al Kawm
Az Zaqāzīq
Az Zaqāzīq
Banhā
Fā'id
Great Bitter
Minūf
Ashmūn
Shibīn al Qanāţir
Bilbays
Ismā'īlīyah Canal
Little Bitter
LIBYAN
Imbābah
Al Jīzah
Mişr al Jadīdah (Ruins)
Ash Shallūfah
30°
DESERT
Sphinx
CAIRO (Al Qāhirah)
Suez
(AŞ ŞAḤRĀ' AL LĪBĪYAH)
Pyramids
Memphis
(As Suways)
©RMCN.
Ḥulwān
Al 'Ayyāţ
GULF OF SUEZ
Birkat Qārūn
As Saff
c
Sinnūris
Al Fayyūm
Itsā
Būsh
Ishmant
JABAL JALĀLAH AL BAḤRĪYAH
Biba
Banī Suwayf
Wādī Sannūr
MEDITERRANEAN SEA
Port Said
(Būr Sa'īd)
Al Fashn
Wādī 'Arabah
Maghāghah
Al Bahnāsā
Banī Mazār
Qulūşnā
Manzala
Al Minyā
Samālūṭ
EGYPT
Al Kāb
Abū Qurqāş
Ar Rawdah
Mallawī
Dayrūţ
5745
28°
Al Qantarah
Al Ballaḥ
Manfalūt
Abnūb
Asyūţ
Abū Tīj
Al Badārī
Al Firdān
Ismailia
(Al Ismā'īlīyah)
Nafishah
Al Marāghah
Akhmīm
Ţahţā
Al Manshāh
Sawhāj
SINAI PENINSULA
Jirjā
Al Balyanā
Dishnā
26°
EGYPT
Qinā
Al Buḥayrah al Murrah al Kubrā
Naj Ḥammādī
Fā'id
Qūş
Qift
Thebes (Ruins)
Al Karnak
Armant
Al Uqşur (Luxor)
Al Buḥayrah al Murrah aş Şughrā
Isnā
Junayfah
Ash Shallūfah
Al Kūbrī
Idfū
Būr Ibrāhīm
Suez (As Suways)
30°
JABAL 'ATAQAH
Salwā Baḥrī
Kawm Umbū
Darāw
Gulf of Suez
Aswan
24°
©RMCN.
ASWĀN HIGH DAM
Lake Nasser
(Buḥayrat Nāşir)
Scale 1:4 000 000
0 10 20 30 40 50 60 Miles
0 10 20 30 40 50 60 Kilometers

d
Lokala Drift
Melkrivier
Pietersburg
BOTSWANA
24°
Hermanusdoring
Potgietersrus
STRYDPOORTBERGE
NORTHERN TRANSVAAL
Gladdeklipkop
Zebediela
Silent Valley
6851
WATERBERGE
Vaalwater
Alma
3216
Naboomspruit
Immerpan
Derdepoort
Thabazimbi
Roedtan
Crecy
Middelwit
Mabula
Nylstroom
Rooiberg
Warmbad
Northam
Settlers
Tuinplaas
Marble Hall
Radium
Skilpadfontein
Gröblersdal
Nietverdiend
Pilansberg
Mabeskraal
Heystekrand
Jericho
Klipgat
Settlers
Enkeldoring
Damval
Straatsdrif
Beestekraal
Hammanskraal
Dennilton
Groot Marico
Swartruggens
Rustenburg
PRETORIA-WITWATERSRAND-VEREENIGING
Verena
Middelburg
NORTH WEST
Marikana
Brits
Pretoria North
Cullinan
Witbank
5681
Koster
Magaliesburg
Silverton
Vaalplaas
Bronkhorstspruit
Derby
Boons
Pretoria
Kempton Park
Driefontein
Kendal
Lichtenburg
Krugersdorp
Alexandra
Ogies
Hendrina
Coligny
Klerkskraal
JOHANNESBURG
Benoni
EASTERN
26°
WITWATERSRAND
Carletonville
Germiston
Springs
Delmas
Hauptrus
Ventersdorp
Fochville
Wolverdiend
Nigel
Devon
Bethal
TRANSVAAL
Westonaria
Vereeniging
Evaton
Daleside
Heidelberg
Leslie
Trichardt
Hartbeesfontein
Meyerton
Balfour
Greylingstad
Morgenzon
Potchefstroom
Vanderbijlpark
SOUTH AFRICA
Klerksdorp
Parys
Dover
Orangeville
Villiers
Standerton
Vaal
Orkney
Vredefort
Wilpoort
Vierfontein
Heilbron
Frankfort
Ascent
Perdekop
4389
Leeudoringstad
Viljoenskroon
Koppies
Vrede
Bothaville
Vals
Edenville
Tweeling
Memel
Wesselsbron
Kroonstad
ORANGE FREE STATE
Reitz
Warden
Welkom
Steynsrus
Lindley
Arlington
Hennenman
Cornelis
Ventersburg
28°
Virginia
Bultfontein
Senekal
Paul Roux
Bethlehem
Harrismith
4758
Theunissen
Winburg
ROYAL NATAL NATL. PK.
Vet
Rosendal
Fouriesburg
Bergville
NATAL
Brandfort
Marquard
Ficksburg
Leribe
Cathedral Pk.
10 226
Winterton
Sand
Clocolan
Butha Buthe
DRAKENSBERG
Sources 10 822
Cathkin Pk.
9856
Bloemfontein
LESOTHO
Scale 1:4 000 000
0 10 20 30 40 Miles
0 10 20 30 40 Kilometers
©RMCN.
28° Longitude East of Greenwich

Scale 1:1 000 000
0 10 Miles
0 16 Kilometers
32°30'
Longitude East of Greenwich
33°

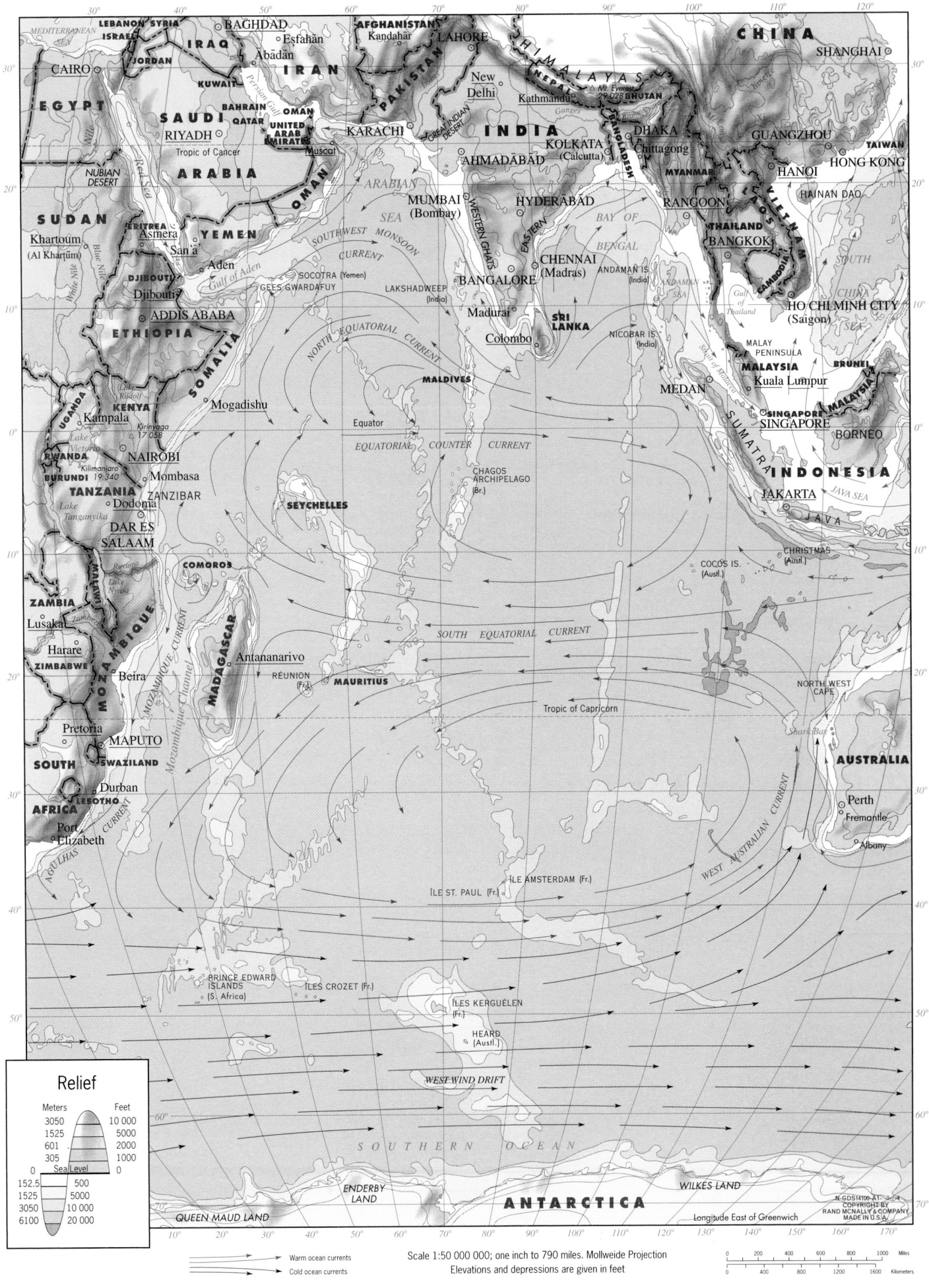

Relief

Meters	Feet
3050	10 000
1525	5000
601	2000
305	1000
0	Sea Level	0
152.5	500
1525	5000
3050	10 000
6100	20 000

Warm ocean currents
Cold ocean currents

Scale 1:50 000 000; one inch to 790 miles. Mollweide Projection
Elevations and depressions are given in feet

N-GDS14100-AT-3-2-4
COPYRIGHT BY
RAND McNALLY & COMPANY
MADE IN U.S.A.

Longitude East of Greenwich

0 200 400 600 800 1000 Miles
0 400 800 1200 1600 Kilometers

194

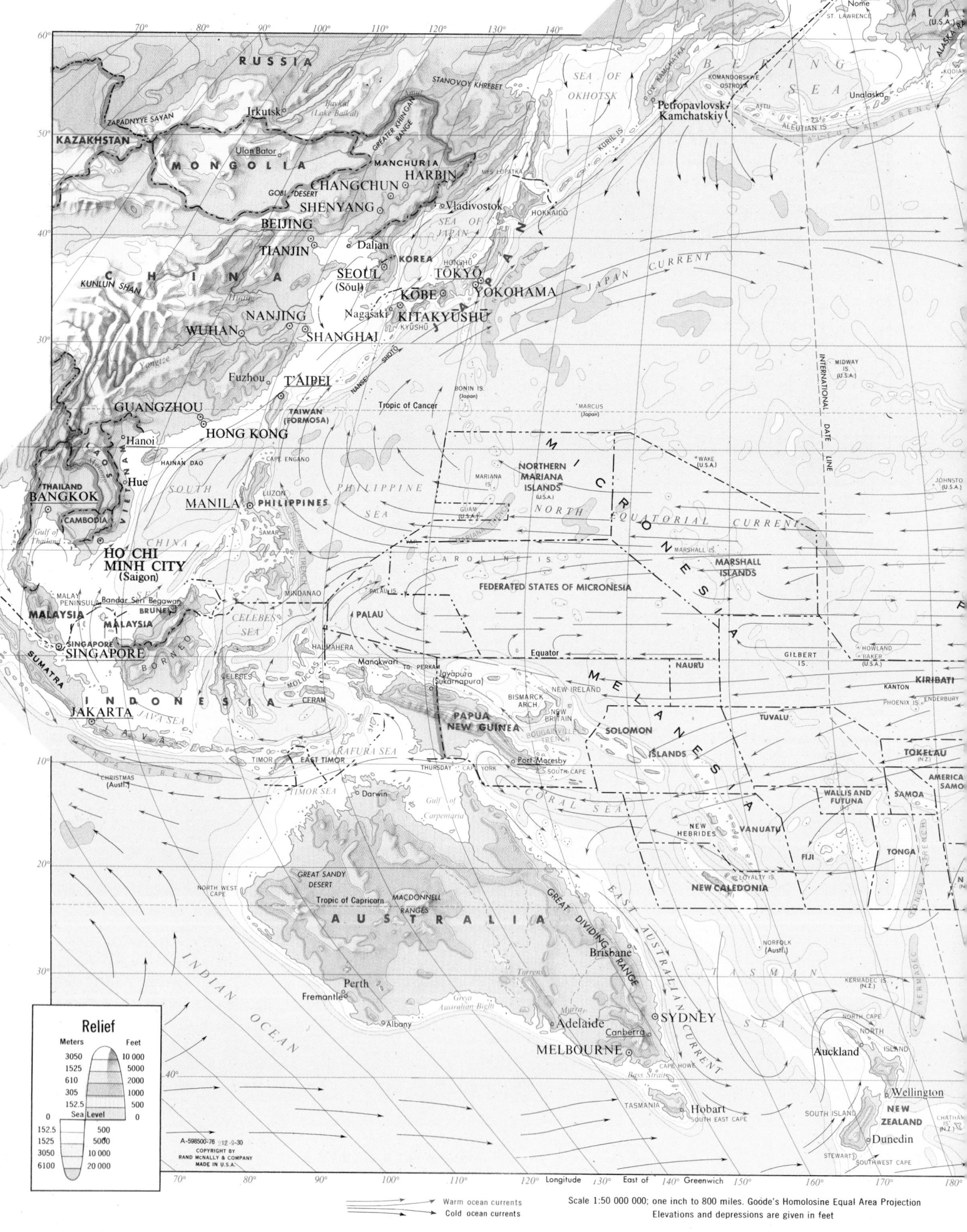

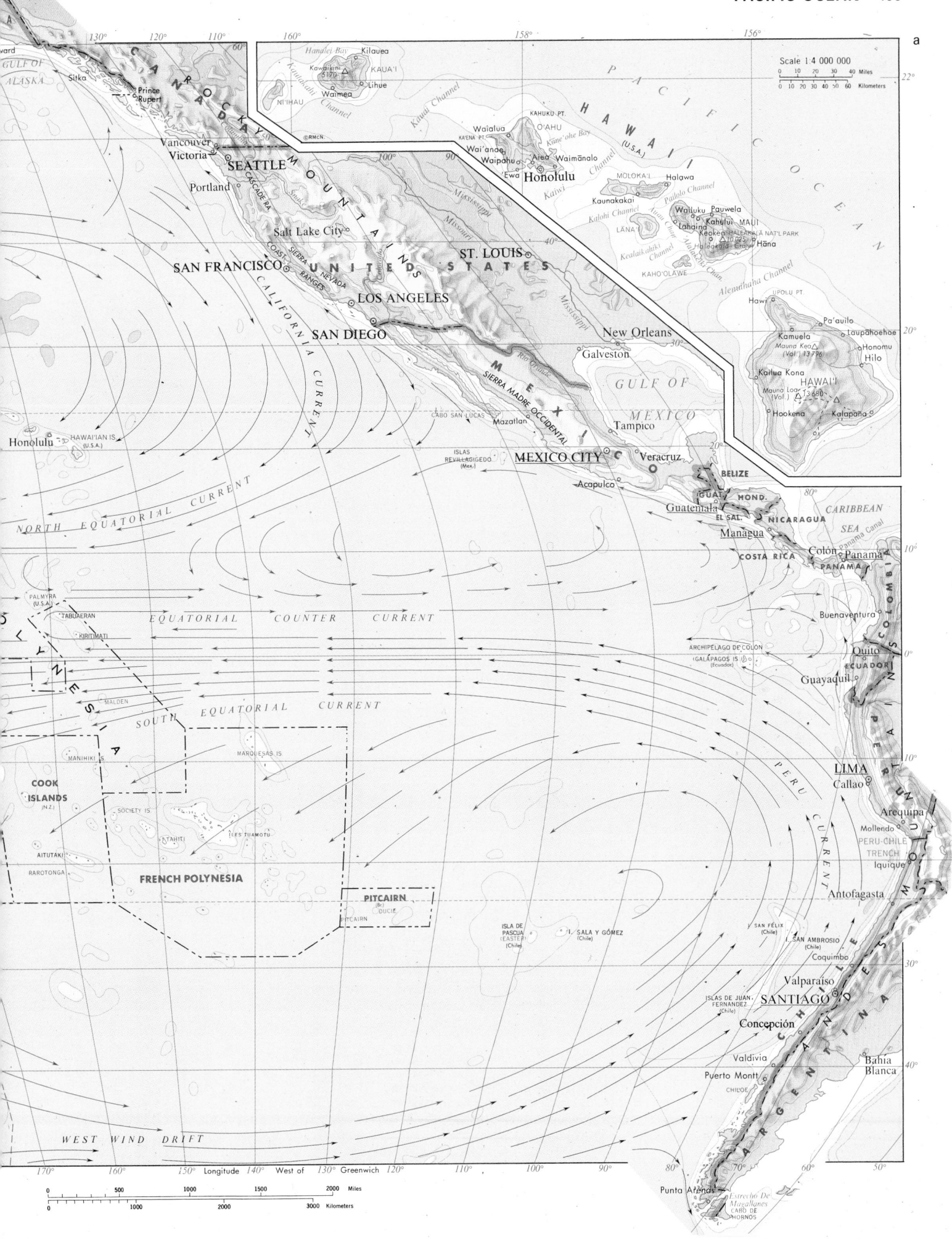

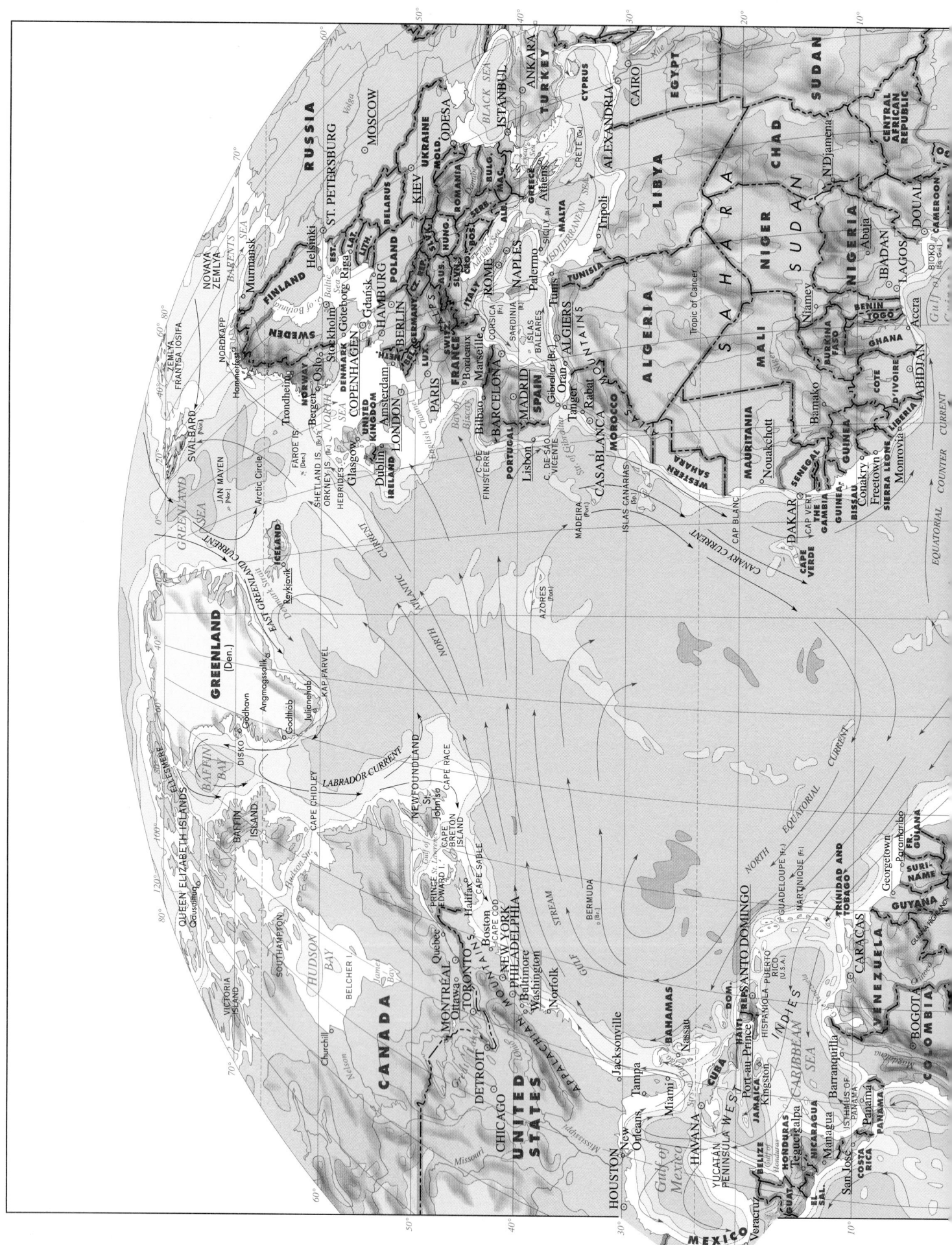

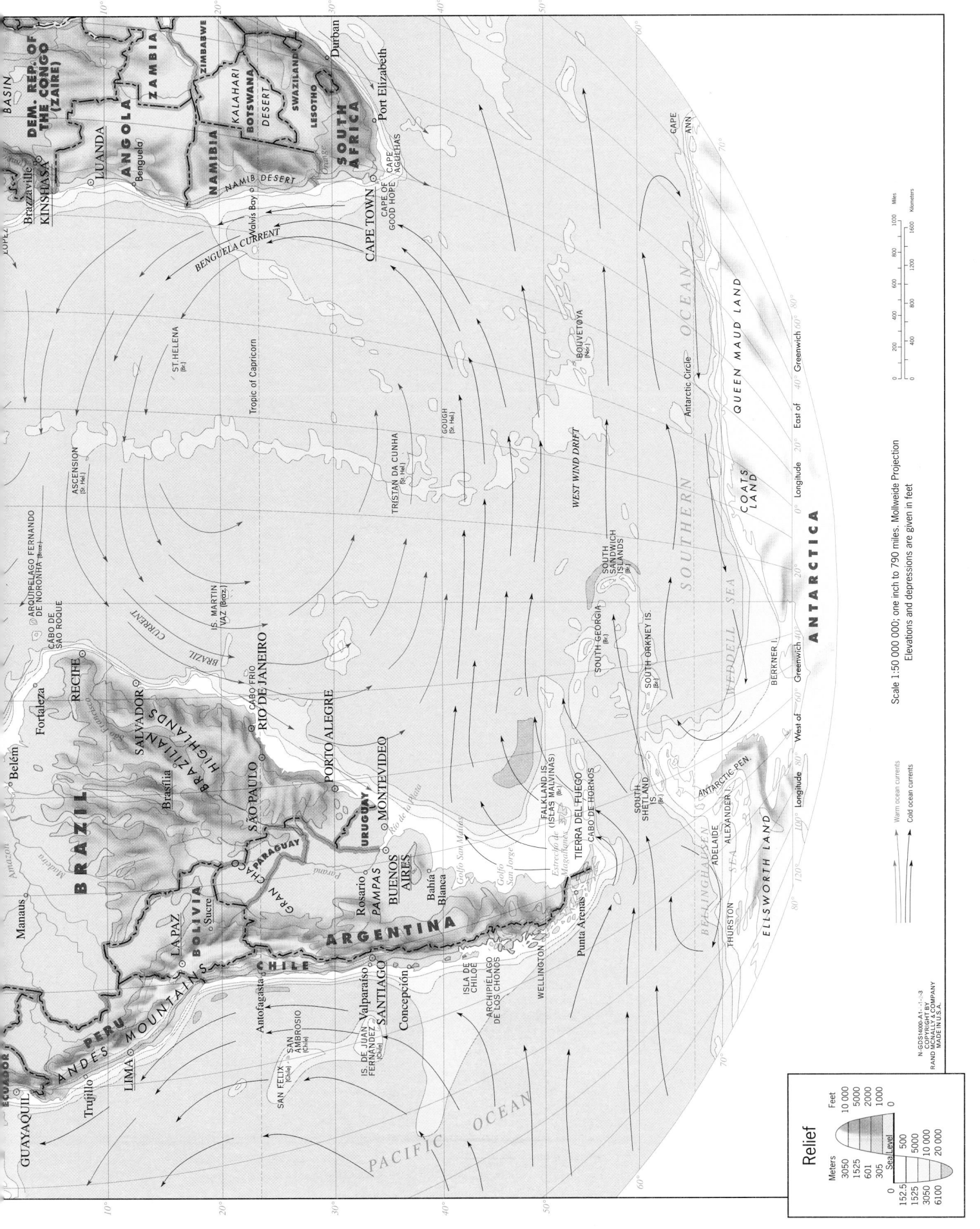

Relief

Meters	Feet
3050	10 000
1525	5000
601	2000
305	1000
0	Sea Level
152.5	500
1525	5000
3050	10 000
6100	20 000

Scale 1:50 000 000; one inch to 790 miles. Mollweide Projection
Elevations and depressions are given in feet

Warm ocean currents
Cold ocean currents

N-GDS14000-A1---2--3
COPYRIGHT BY
RAND McNALLY & COMPANY
MADE IN U.S.A.

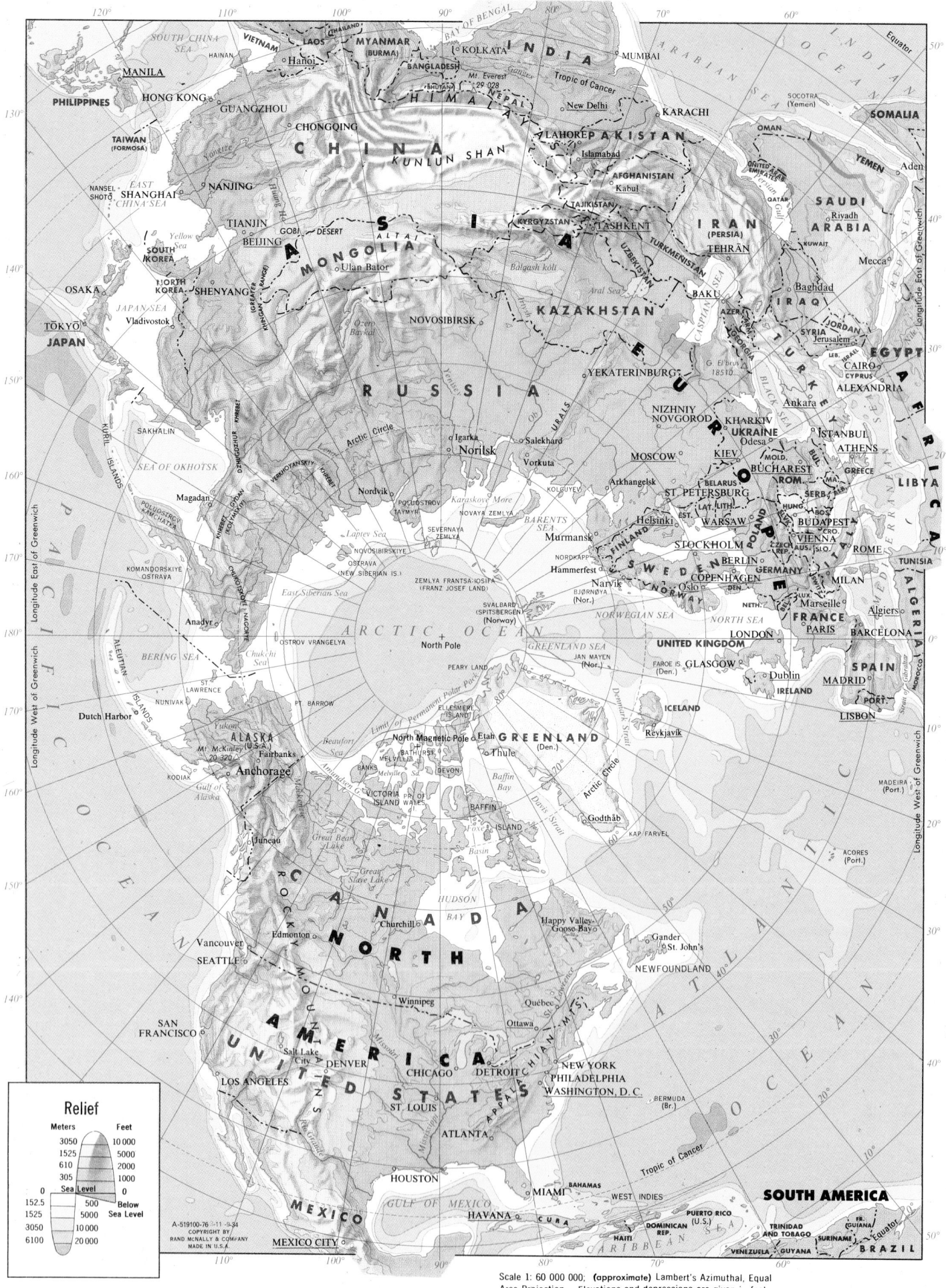

Relief

Meters	Feet
3050	10 000
1525	5000
610	2000
305	1000
0	Sea Level
Sea Level	
152.5	500
1525	5000
3050	10 000
6100	20 000

Below Sea Level

A-519100-76 7-11-9-34
COPYRIGHT BY
RAND McNALLY & COMPANY
MADE IN U.S.A.

Scale 1: 60 000 000; (approximate) Lambert's Azimuthal, Equal
Area Projection   Elevations and depressions are given in feet

WORLD POLITICAL INFORMATION TABLE

This table gives the area, population, population density, political status, capital, and predominant languages for every country in the world. The political units listed are categorized by political status in the form of government column of the table, as follows: A—independent countries; B—internally independent political entities which are under the protection of another country in matters of defense and foreign affairs; C—colonies and other dependent political units; and D—the major administrative subdivisions of Australia, Canada, China, the United Kingdom, and the United States. For comparison, the table also includes the continents and the world. All footnotes appear at the end of the table.

The populations are estimates for January 1, 2004, made by Rand McNally on the basis of official data, United States Census Bureau estimates, and other available information. Area figures include inland water.

REGION OR POLITICAL DIVISION	Area Sq. Mi.	Est. Pop. 1/1/04	Pop. Per Sq. Mi.	Form of Government and Ruling Power	Capital	Predominant Languages	International Organizations
Afars and Issas see Djibouti							
Afghanistan	251,773	29,205,000	116	Transitional ... A	Kābul	Dari, Pashto, Uzbek, Turkmen	UN
Africa	11,700,000	866,305,000	74				
Alabama	52,419	4,515,000	86	State (U.S.) ... D	Montgomery	English	
Alaska	663,267	650,000	1.0	State (U.S.) ... D	Juneau	English, indigenous	
Albania	11,100	3,535,000	318	Republic ... A	Tiranë	Albanian, Greek	NATO/PP, UN
Alberta	255,541	3,215,000	13	Province (Canada) ... D	Edmonton	English	
Algeria	919,595	33,090,000	36	Republic ... A	Algiers (El Djazair)	Arabic, Berber dialects, French	AL, AU, OPEC, UN
American Samoa	77	58,000	753	Unincorporated territory (U.S.) ... C	Pago Pago	Samoan, English	
Andorra	181	70,000	387	Parliamentary co-principality (Spanish and French) ... B	Andorra	Catalan, Spanish (Castilian), French, Portuguese	UN
Angola	481,354	10,875,000	23	Republic ... A	Luanda	Portuguese, indigenous	AU, COMESA, UN
Anguilla	37	13,000	351	Overseas territory (U.K.) ... C	The Valley	English	
Anhui	53,668	61,215,000	1,141	Province (China) ... D	Hefei	Chinese (Mandarin)	
Antarctica	5,400,000	(')					
Antigua and Barbuda	171	68,000	398	Parliamentary state ... A	St. John's	English, local dialects	OAS, UN
Aomen (Macau)	6.9	445,000	64,493	Special administrative region (China) ... D	Macau (Aomen)	Chinese (Cantonese), Portuguese	
Argentina	1,073,519	38,945,000	36	Republic ... A	Buenos Aires	Spanish, English, Italian, German, French	MERCOSUR, OAS, UN
Arizona	113,998	5,600,000	49	State (U.S.) ... D	Phoenix	English	
Arkansas	53,179	2,735,000	51	State (U.S.) ... D	Little Rock	English	
Armenia	11,506	3,325,000	289	Republic ... A	Yerevan	Armenian, Russian	CIS, NATO/PP, UN
Aruba	75	71,000	947	Self-governing territory (Netherlands protection) ... B	Oranjestad	Dutch, Papiamento, English, Spanish	
Ascension	34	1,000	29	Dependency (St. Helena) ... C	Georgetown	English	
Asia	17,300,000	3,839,320,000	222				
Australia	2,969,910	19,825,000	6.7	Federal parliamentary state ... A	Canberra	English, indigenous	ANZUS, UN
Australian Capital Territory	911	325,000	357	Territory (Australia) ... D	Canberra	English	
Austria	32,378	8,170,000	252	Federal republic ... A	Vienna (Wien)	German	EU, NATO/PP, UN
Azerbaijan	33,437	7,850,000	235	Republic ... A	Baku (Bakı)	Azeri, Russian, Armenian	CIS, NATO/PP, UN
Bahamas	5,382	300,000	56	Parliamentary state ... A	Nassau	English, Creole	OAS, UN
Bahrain	267	675,000	2,528	Monarchy ... A	Al Manāmah	Arabic, English, Persian, Urdu	AL, UN
Bangladesh	55,598	139,875,000	2,516	Republic ... A	Dkaha (Dacca)	Bangla, English	UN
Barbados	166	280,000	1,687	Parliamentary state ... A	Bridgetown	English	OAS, UN
Beijing (Peking)	6,487	14,135,000	2,179	Autonomous city (China) ... D	Beijing (Peking)	Chinese (Mandarin)	
Belarus	80,155	10,315,000	129	Republic ... A	Minsk	Belarussian, Russian	CIS, NATO/PP, UN
Belau see Palau							
Belgium	11,787	10,340,000	877	Constitutional monarchy ... A	Brussels (Bruxelles)	Dutch (Flemish), French, German	EU, NATO, UN
Belize	8,867	270,000	30	Parliamentary state ... A	Belmopan	English, Spanish, Mayan, Garifuna, Creole	OAS, UN
Benin	43,484	7,145,000	164	Republic ... A	Porto-Novo and Cotonou	French, Fon, Yoruba, indigenous	AU, UN
Bermuda	21	65,000	3,095	Overseas territory (U.K. protection) ... B	Hamilton	English, Portuguese	
Bhutan	17,954	2,160,000	120	Monarchy (Indian protection) ... B	Thimphu	Dzongkha, Tibetan and Nepalese dialects	UN
Bolivia	424,165	8,655,000	20	Republic ... A	La Paz and Sucre	Aymara, Quechua, Spanish	OAS, UN
Bosnia and Herzegovina	19,767	4,000,000	202	Republic ... A	Sarajevo	Bosnian, Serbian, Croatian	UN
Botswana	224,607	1,570,000	7.0	Republic ... A	Gaborone	English, Tswana	AU, UN
Brazil	3,300,172	183,080,000	55	Federal republic ... A	Brasília	Portuguese, Spanish, English, French	MERCOSUR, OAS, UN
British Columbia	364,764	4,245,000	12	Province (Canada) ... D	Victoria	English	
British Indian Ocean Territory	23	(')		Overseas territory (U.K.) ... C		English	
British Virgin Islands	58	22,000	379	Overseas territory (U.K.) ... C	Road Town	English	
Brunei	2,226	360,000	162	Monarchy ... A	Bandar Seri Begawan	Malay, English, Chinese	ASEAN, UN
Bulgaria	42,855	7,550,000	176	Republic ... A	Sofia (Sofiya)	Bulgarian, Turkish	NATO, UN
Burkina Faso	105,869	13,400,000	127	Republic ... A	Ouagadougou	French, indigenous	AU, UN
Burma see Myanmar							
Burundi	10,745	6,165,000	574	Republic ... A	Bujumbura	French, Kirundi, Swahili	AU, COMESA, UN
California	163,696	35,590,000	217	State (U.S.) ... D	Sacramento	English	
Cambodia	69,898	13,245,000	189	Constitutional monarchy ... A	Phnom Penh (Phnum Pénh)	Khmer, French, English	ASEAN, UN
Cameroon	183,568	15,905,000	87	Republic ... A	Yaoundé	English, French, indigenous	AU, UN
Canada	3,855,103	32,360,000	8.4	Federal parliamentary state ... A	Ottawa	English, French, other	NAFTA, NATO, OAS, UN
Cape Verde	1,557	415,000	267	Republic ... A	Praia	Portuguese, Crioulo	AU, UN
Cayman Islands	102	43,000	422	Overseas territory (U.K.) ... C	George Town	English	
Central African Republic	240,536	3,715,000	15	Republic ... A	Bangui	French, Sango, indigenous	AU, UN
Ceylon see Sri Lanka							
Chad	495,755	9,395,000	19	Republic ... A	N'Djamena	Arabic, French, indigenous	AU, UN
Channel Islands	75	155,000	2,067	Two crown dependencies (U.K. protection)		English, French	
Chile	291,930	15,745,000	54	Republic ... A	Santiago	Spanish	OAS, UN
China (excl. Taiwan)	3,690,045	1,298,720,000	352	Socialist republic ... A	Beijing (Peking)	Chinese dialects	UN
Chongqing	31,815	31,600,000	993	Autonomous city (China) ... D	Chongqing (Chungking)	Chinese (Mandarin)	
Christmas Island	52	400	7.7	External territory (Australia) ... C	Settlement	English, Chinese, Malay	
Cocos (Keeling) Islands	5.4	600	111	External territory (Australia) ... C	West Island	English, Cocos-Malay	
Colombia	439,737	41,985,000	95	Republic ... A	Bogotá	Spanish	OAS, UN
Colorado	104,094	4,565,000	44	State (U.S.) ... D	Denver	English	
Comoros (excl. Mayotte)	863	640,000	742	Republic ... A	Moroni	Arabic, French, Shikomoro	AL, AU, COMESA, UN
Congo	132,047	2,975,000	23	Republic ... A	Brazzaville	French, Lingala, Monokutuba, indigenous	AU, UN
Congo, Democratic Republic of the (Zaire)	905,446	57,445,000	63	Republic ... A	Kinshasa	French, Lingala, indigenous	AU, COMESA, UN
Connecticut	5,543	3,495,000	631	State (U.S.) ... D	Hartford	English	

REGION OR POLITICAL DIVISION	Area Sq. Mi.	Est. Pop. 1/1/04	Pop. Per Sq. Mi.	Form of Government and Ruling Power	Capital	Predominant Languages	International Organizations
Cook Islands	91	21,000	231	Self-governing territory (New Zealand protection) . . . B	Avarua	English, Maori	
Costa Rica	19,730	3,925,000	199	Republic . . . A	San José	Spanish, English	OAS, UN
Cote d'Ivoire (Ivory Coast)	124,504	17,145,000	138	Republic . . . A	Abidjan and Yamoussoukro. .	French, Dioula and other indigenous . .	AU, UN
Croatia	21,829	4,430,000	203	Republic . . . A	Zagreb	Croatian	NATO/PP, UN
Cuba	42,804	11,290,000	264	Socialist republic . . . A	Havana (La Habana)	Spanish	OAS, UN
Cyprus	3,572	775,000	217	Republic . . . A	Nicosia	Greek, Turkish, English	EU, UN
Czech Republic	30,450	10,250,000	337	Republic . . . A	Prague (Praha)	Czech	EU, NATO, UN
Delaware	2,489	820,000	329	State (U.S.) . . . D	Dover	English	
Denmark	16,640	5,405,000	325	Constitutional monarchy . . . A	Copenhagen (København) . . .	Danish	EU, NATO, UN
District of Columbia	68	565,000	8,309	Federal district (U.S.) . . . D	Washington	English	
Djibouti	8,958	460,000	51	Republic . . . A	Djibouti	French, Arabic, Somali, Afar	AL, AU, COMESA, UN
Dominica	290	69,000	238	Republic . . . A	Roseau	English, French	OAS, UN
Dominican Republic	18,730	8,775,000	468	Republic . . . A	Santo Domingo	Spanish	OAS, UN
East Timor	5,743	1,010,000	176	Republic . . . A	Dili	Portuguese, Tetum, Bahasa Indonesia (Malay), English	UN
Ecuador	109,484	13,840,000	126	Republic . . . A	Quito	Spanish, Quechua, indigenous	OAS, UN
Egypt	386,662	75,420,000	195	Republic . . . A	Cairo (Al Qāhirah)	Arabic	AL, AU, CAEU, COMESA, UN
Ellice Islands see Tuvalu							
El Salvador	8,124	6,530,000	804	Republic . . . A	San Salvador	Spanish, Nahua	OAS, UN
England	50,356	50,360,000	1,000	Administrative division (U.K.) . . . D	London	English	
Equatorial Guinea	10,831	515,000	48	Republic . . . A	Malabo	French, Spanish, indigenous, English . .	AU, UN
Eritrea	45,406	4,390,000	97	Republic . . . A	Asmera	Afar, Arabic, Tigre, Kunama, Tigrinya, other	AU, COMESA, UN
Estonia	17,462	1,405,000	80	Republic . . . A	Tallinn	Estonian, Russian, Ukrainian, Finnish, other	EU, NATO, UN
Ethiopia	426,373	67,210,000	158	Federal republic . . . A	Addis Ababa (Adis Abeba) . . .	Amharic, Tigrinya, Orominga, Guaraginga, Somali, Arabic	AU, COMESA, UN
Europe	3,800,000	729,330,000	192				
Falkland Islands (²)	4,700	3,000	0.6	Overseas territory (U.K.) . . . C	Stanley	English	
Faroe Islands	540	47,000	87	Self-governing territory (Danish protection). . . . B	Tórshavn	Danish, Faroese	
Fiji	7,056	875,000	124	Republic . . . A	Suva	English, Fijian, Hindustani	UN
Finland	130,559	5,210,000	40	Republic . . . A	Helsinki (Helsingfors)	Finnish, Swedish, Sami, Russian	EU, NATO/PP, UN
Florida	65,755	17,070,000	260	State (U.S.) . . . D	Tallahassee	English	
France (excl. Overseas Departments) . . .	208,482	60,305,000	289	Republic . . . A	Paris	French	EU, NATO, UN
French Guiana	32,253	190,000	5.9	Overseas department (France) . . . C	Cayenne	French	
French Polynesia	1,544	265,000	172	Overseas territory (France) . . . C	Papeete	French, Tahitian	
Fujian	46,332	35,495,000	766	Province (China) . . . D	Fuzhou	Chinese dialects	
Gabon	103,347	1,340,000	13	Republic . . . A	Libreville	French, Fang, indigenous	AU, UN
Gambia, The	4,127	1,525,000	370	Republic . . . A	Banjul	English, Malinke, Wolof, Fula, indigenous	AU, UN
Gansu	173,746	26,200,000	151	Province (China) . . . D	Lanzhou	Chinese (Mandarin), Mongolian, Tibetan dialects	
Gaza Strip	139	1,300,000	9,353	Israeli territory with limited self-government. . . .		Arabic, Hebrew	(⁴)
Georgia	59,425	8,710,000	147	State (U.S.) . . . D	Atlanta	English	
Georgia	26,911	4,920,000	183	Republic . . . A	Tbilisi	Georgian, Russian, Armenian, Azeri, other	NATO/PP, UN
Germany	137,847	82,415,000	598	Federal republic . . . A	Berlin	German	EU, NATO, UN
Ghana	92,098	20,615,000	224	Republic . . . A	Accra	English, Akan and other indigenous	AU, UN
Gibraltar (¹)	2.3	28,000	12,174	Overseas territory (U.K.) . . . C	Gibraltar	English, Spanish, Italian, Portuguese . .	
Gilbert Islands see Kiribati							
Golan Heights	454	37,000	81	Occupied by Israel . . .		Arabic, Hebrew	
Great Britain see United Kingdom							
Greece	50,949	10,635,000	209	Republic . . . A	Athens (Athina)	Greek, English, French	EU, NATO, UN
Greenland	836,331	56,000	0.07	Self-governing territory (Danish protection). . . B	Godthåb (Nuuk)	Danish, Greenlandic, English	
Grenada	133	89,000	669	Parliamentary state. . . . A	St. George's	English, French	OAS, UN
Guadeloupe (incl. Dependencies)	687	440,000	640	Overseas department (France) . . . C	Basse-Terre	French, Creole	
Guam	212	165,000	778	Unincorporated territory (U.S.). . . . C	Hagåtña (Agana)	English, Chamorro, Japanese	
Guangdong	68,649	88,375,000	1,287	Province (China) . . . D	Guangzhou (Canton)	Chinese dialects, Miao-Yao	
Guangxi Zhuangzu	91,236	45,905,000	503	Autonomous region (China) . . . D	Nanning	Chinese dialects, Thai, Miao-Yao	
Guatemala	42,042	14,095,000	335	Republic . . . A	Guatemala	Spanish, indigenous	OAS, UN
Guernsey (incl. Dependencies)	30	65,000	2,167	Crown dependency (U.K. protection) . . . B	St. Peter Port	English, French	
Guinea	94,926	9,135,000	96	Republic . . . A	Conakry	French, indigenous	AU, UN
Guinea-Bissau	13,948	1,375,000	99	Republic . . . A	Bissau	Portuguese, Crioulo, indigenous	AU, UN
Guizhou	65,637	36,045,000	549	Province (China) . . . D	Guiyang	Chinese (Mandarin), Thai, Miao-Yao . .	
Guyana	83,000	705,000	8.5	Republic . . . A	Georgetown	English, indigenous, Creole, Hindi, Urdu	OAS, UN
Hainan	13,205	8,050,000	610	Province (China) . . . D	Haikou	Chinese, Min, Tai	
Haiti	10,714	7,590,000	708	Republic . . . A	Port-au-Prince	Creole, French	OAS, UN
Hawaii	10,931	1,260,000	115	State (U.S.) . . . D	Honolulu	English, Hawaiian, Japanese	
Hebei	73,359	68,965,000	940	Province (China) . . . D	Shijiazhuang	Chinese (Mandarin)	
Heilongjiang	181,082	37,725,000	208	Province (China) . . . D	Harbin	Chinese dialects, Mongolian, Tungus. . .	
Henan	64,479	94,655,000	1,468	Province (China) . . . D	Zhengzhou	Chinese (Mandarin)	
Holland see Netherlands							
Honduras	43,277	6,745,000	156	Republic . . . A	Tegucigalpa	Spanish, indigenous	OAS, UN
Hubei	72,356	61,645,000	852	Province (China) . . . D	Wuhan	Chinese dialects	
Hunan	81,082	65,855,000	812	Province (China) . . . D	Changsha	Chinese dialects, Miao-Yao	
Hungary	35,919	10,045,000	280	Republic . . . A	Budapest	Hungarian	EU, NATO, UN
Iceland	39,769	280,000	7.0	Republic . . . A	Reykjavik	Icelandic, English, other	EFTA, NATO, UN
Idaho	83,570	1,370,000	16	State (U.S.) . . . D	Boise	English	
Illinois	57,914	12,690,000	219	State (U.S.) . . . D	Springfield	English	
India (incl. part of Jammu and Kashmir)	1,222,510	1,057,415,000	865	Federal republic . . . A	New Delhi	English, Hindi, Telugu, Bengali, indigenous	UN
Indiana	36,418	6,215,000	171	State (U.S.) . . . D	Indianapolis	English	
Indonesia	735,310	236,680,000	322	Republic . . . A	Jakarta	Bahasa Indonesia (Malay), English, Dutch, indigenous	ASEAN, OPEC, UN
Iowa	56,272	2,955,000	53	State (U.S.) . . . D	Des Moines	English	
Iran	636,372	68,650,000	108	Islamic republic . . . A	Tehrān	Persian, Turkish dialects, Kurdish, other	OPEC, UN
Iraq	169,235	25,025,000	148	Republic . . . A	Baghdād	Arabic, Kurdish, Assyrian, Armenian . . .	AL, CAEU, OPEC, UN
Ireland	27,133	3,945,000	145	Republic . . . A	Dublin (Baile Átha Cliath) . . .	English, Irish Gaelic	EU, NATO/PP, UN
Isle of Man	221	74,000	335	Crown dependency (U.K. protection) . . . B	Douglas	English, Manx Gaelic	

REGION OR POLITICAL DIVISION	Area Sq. Mi.	Est. Pop. 1/1/04	Pop. Per Sq. Mi.	Form of Government and Ruling Power	Capital	Predominant Languages	International Organizations
Israel (excl. Occupied Areas)	8,019	6,160,000	768	Republic A	Jerusalem (Yerushalayim)	Hebrew, Arabic	UN
Italy	116,342	58,030,000	499	Republic A	Rome (Roma)	Italian, German, French, Slovene	EU, NATO, UN
Ivory Coast see Cote d'Ivoire	...	...	...	...	...	...	...
Jamaica	4,244	2,705,000	637	Parliamentary state A	Kingston	English, Creole	OAS, UN
Japan	145,850	127,285,000	873	Constitutional monarchy A	Tōkyō	Japanese	UN
Jersey	45	90,000	2,000	Crown dependency (U.K. protection) B	St. Helier	English, French	...
Jiangsu	39,614	76,065,000	1,920	Province (China) D	Nanjing (Nanking)	Chinese dialects	...
Jiangxi	64,325	42,335,000	658	Province (China) D	Nanchang	Chinese dialects	...
Jilin	72,201	27,895,000	386	Province (China) D	Changchun	Chinese (Mandarin), Mongolian, Korean	...
Jordan	34,495	5,535,000	160	Constitutional monarchy A	'Ammān	Arabic	AL, CAEU, UN
Kansas	82,277	2,730,000	33	State (U.S.) D	Topeka	English	...
Kazakhstan	1,049,156	16,780,000	16	Republic A	Astana (Aqmola)	Kazakh, Russian	CIS, NATO/PP, UN
Kentucky	40,409	4,130,000	102	State (U.S.) D	Frankfort	English	...
Kenya	224,961	31,840,000	142	Republic A	Nairobi	English, Swahili, indigenous	AU, COMESA, UN
Kiribati	313	100,000	319	Republic A	Bairiki	English, I-Kiribati	UN
Korea, North	46,540	22,585,000	485	Socialist republic A	P'yŏngyang	Korean	UN
Korea, South	38,328	48,450,000	1,264	Republic A	Seoul (Sŏul)	Korean	UN
Kuwait	6,880	2,220,000	323	Constitutional monarchy A	Kuwait (Al Kuwayt)	Arabic, English	AL, CAEU, OPEC, UN
Kyrgyzstan	77,182	4,930,000	64	Republic A	Bishkek	Kirghiz, Russian	CIS, NATO/PP, UN
Laos	91,429	5,995,000	66	Socialist republic A	Viangchan (Vientiane)	Lao, French, English	ASEAN, UN
Latvia	24,942	2,340,000	94	Republic A	Riga	Latvian, Lithuanian, Russian, other	EU, NATO, UN
Lebanon	4,016	3,755,000	935	Republic A	Beirut (Bayrūt)	Arabic, French, Armenian, English	AL, UN
Lesotho	11,720	1,865,000	159	Constitutional monarchy A	Maseru	English, Sesotho, Zulu, Xhosa	AU, UN
Liaoning	56,255	43,340,000	770	Province (China) D	Shenyang (Mukden)	Chinese (Mandarin), Mongolian	...
Liberia	43,000	3,345,000	78	Republic A	Monrovia	English, indigenous	AU, UN
Libya	679,362	5,565,000	8.2	Socialist republic A	Tripoli (Ţarābulus)	Arabic	AL, AU, CAEU, OPEC, UN
Liechtenstein	62	33,000	532	Constitutional monarchy A	Vaduz	German	EFTA, UN
Lithuania	25,213	3,590,000	142	Republic A	Vilnius	Lithuanian, Polish, Russian	EU, NATO, UN
Louisiana	51,840	4,510,000	87	State (U.S.) D	Baton Rouge	English	...
Luxembourg	999	460,000	460	Constitutional monarchy A	Luxembourg	French, Luxembourgish, German	EU, NATO, UN
Macedonia	9,928	2,065,000	208	Republic A	Skopje	Macedonian, Albanian, other	NATO/PP, UN
Madagascar	226,658	17,235,000	76	Republic A	Antananarivo	French, Malagasy	AU, COMESA, UN
Maine	35,385	1,310,000	37	State (U.S.) D	Augusta	English	...
Malawi	45,747	11,780,000	258	Republic A	Lilongwe	Chichewa, English, indigenous	AU, COMESA, UN
Malaysia	127,320	23,310,000	183	Federal constitutional monarchy A	Kuala Lumpur and Putrajaya (5)	Bahasa Melayu, Chinese dialects, English, other	ASEAN, UN
Maldives	115	335,000	2,913	Republic A	Male'	Dhivehi	UN
Mali	478,841	11,790,000	25	Republic A	Bamako	French, Bambara, indigenous	AU, UN
Malta	122	400,000	3,279	Republic A	Valletta	English, Maltese	EU, UN
Manitoba	250,116	1,190,000	4.8	Province (Canada) D	Winnipeg	English	...
Marshall Islands	70	57,000	814	Republic (U.S. protection) B	Majuro (island)	English, indigenous, Japanese	UN
Martinique	425	430,000	1,012	Overseas department (France) C	Fort-de-France	French, Creole	...
Maryland	12,407	5,525,000	445	State (U.S.) D	Annapolis	English	...
Massachusetts	10,555	6,455,000	612	State (U.S.) D	Boston	English	...
Mauritania	397,956	2,955,000	7.4	Republic A	Nouakchott	Arabic, Wolof, Pular, Soninke, French	AL, AU, CAEU, UN
Mauritius (incl. Dependencies)	788	1,215,000	1,542	Republic A	Port Louis	English, French, Creole, other	AU, COMESA, UN
Mayotte (6)	144	180,000	1,250	Departmental collectivity (France) C	Mamoutzou	French, Swahili (Mahorian)	...
Mexico	758,452	104,340,000	138	Federal republic A	Mexico City (Ciudad de México)	Spanish, indigenous	NAFTA, OAS, UN
Michigan	96,716	10,110,000	105	State (U.S.) D	Lansing	English	...
Micronesia, Federated States of	271	110,000	406	Republic (U.S. protection) B	Palikir	English, indigenous	UN
Midway Islands	2.0	(1)	...	Unincorporated territory (U.S.) C	...	English	...
Minnesota	86,939	5,075,000	58	State (U.S.) D	St. Paul	English	...
Mississippi	48,430	2,890,000	60	State (U.S.) D	Jackson	English	...
Missouri	69,704	5,720,000	82	State (U.S.) D	Jefferson City	English	...
Moldova	13,070	4,440,000	340	Republic A	Chişinău (Kishinev)	Romanian (Moldovan), Russian, Gagauz	CIS, NATO/PP, UN
Monaco	0.8	32,000	40,000	Constitutional monarchy A	Monaco	French, English, Italian, Monegasque	UN
Mongolia	604,829	2,730,000	4.5	Republic A	Ulan Bator (Ulaanbaatar)	Khalkha Mongol, Turkish dialects, Russian	UN
Montana	4,095	920,000	225	State (U.S.) D	Helena	English	...
Montserrat	39	9,000	231	Overseas territory (U.K.) C	Plymouth	English	...
Morocco (excl. Western Sahara)	172,414	31,950,000	185	Constitutional monarchy A	Rabat	Arabic, Berber dialects, French	AL, UN
Mozambique	309,496	18,695,000	60	Republic A	Maputo	Portuguese, indigenous	AU, UN
Myanmar (Burma)	261,228	42,620,000	163	Provisional military government A	Rangoon (Yangon)	Burmese, indigenous	ASEAN, UN
Namibia	317,818	1,940,000	6.1	Republic A	Windhoek	English, Afrikaans, German, indigenous	AU, COMESA, UN
Nauru	8.1	13,000	1,605	Republic A	Yaren District	Nauruan, English	UN
Nebraska	77,354	1,745,000	23	State (U.S.) D	Lincoln	English	...
Nei Mongol (Inner Mongolia)	456,759	24,295,000	53	Autonomous region (China) D	Hohhot	Mongolian	...
Nepal	56,827	26,770,000	471	Constitutional monarchy A	Kathmandu	Nepali, indigenous	UN
Netherlands	16,164	16,270,000	1,007	Constitutional monarchy A	Amsterdam and The Hague ('s-Gravenhage)	Dutch, Frisian	EU, NATO, UN
Netherlands Antilles	309	215,000	696	Self-governing territory (Netherlands protection) B	Willemstad	Dutch, Papiamento, English, Spanish	...
Nevada	110,561	2,250,000	20	State (U.S.) D	Carson City	English	...
New Brunswick	28,150	770,000	27	Province (Canada) D	Fredericton	English, French	...
New Caledonia	7,172	210,000	29	Territorial collectivity (France) C	Nouméa	French, indigenous	...
Newfoundland and Labrador	156,453	535,000	3.4	Province (Canada) D	St. John's	English	...
New Hampshire	9,350	1,290,000	138	State (U.S.) D	Concord	English	...
New Hebrides see Vanuatu	...	...	...	...	...	...	...
New Jersey	8,721	8,665,000	994	State (U.S.) D	Trenton	English	...
New Mexico	121,590	1,880,000	15	State (U.S.) D	Santa Fe	English, Spanish	...
New South Wales	309,129	6,665,000	22	State (Australia) D	Sydney	English	...
New York	54,556	19,245,000	353	State (U.S.) D	Albany	English	...
New Zealand	104,454	3,975,000	38	Parliamentary state A	Wellington	English, Maori	ANZUS, UN
Nicaragua	50,054	5,180,000	103	Republic A	Managua	Spanish, English, indigenous	OAS, UN
Niger	489,192	11,210,000	23	Repubic A	Niamey	French, Hausa, Djerma, indigenous	AU, UN
Nigeria	356,669	135,570,000	380	Transitional military government A	Abuja	English, Hausa, Fulani, Yorbua, Ibo, indigenous	AU, OPEC, UN
Ningxia Huizu	25,637	5,745,000	224	Autonomous region (China) D	Yinchuan	Chinese (Mandarin)	...
Niue	100	2,000	20	Self-governing territory (New Zealand protection) B	Alofi	Niuean, English	...
Norfolk Island	14	2,000	143	External territory (Australia) C	Kingston	English, Norfolk	...

REGION OR POLITICAL DIVISION	Area Sq. Mi.	Est. Pop. 1/1/04	Pop. Per Sq. Mi.	Form of Government and Ruling Power		Capital	Predominant Languages	International Organizations
North America	9,500,000	505,780,000	53					
North Carolina	53,819	8,430,000	157	State (U.S.)	D	Raleigh	English	
North Dakota	70,700	635,000	9.0	State (U.S.)	D	Bismarck	English	
Northern Ireland	5,242	1,725,000	329	Administrative division (U.K.)	D	Belfast	English	
Northern Mariana Islands	179	77,000	430	Commonwealth (U.S. protection)	B	Saipan (island)	English, Chamorro, Carolinian	
Northern Territory	520,902	200,000	0.4	Territory (Australia)	D	Darwin	English, indigenous	
Northwest Territories	519,735	43,000	0.08	Territory (Canada)	D	Yellowknife	English, indigenous	
Norway (incl. Svalbard and Jan Mayen)	125,050	4,565,000	37	Constitutional monarchy	A	Oslo	Norwegian, Sami, Finnish	EFTA, NATO, UN
Nova Scotia	21,345	965,000	45	Province (Canada)	D	Halifax	English	
Nunavut	808,185	30,000	0.04	Territory (Canada)	D	Iqaluit	English, indigenous	
Oceania (incl. Australia)	3,300,000	32,170,000	9.7					
Ohio	44,825	11,470,000	256	State (U.S.)	D	Columbus	English	
Oklahoma	69,898	3,520,000	50	State (U.S.)	D	Oklahoma City	English	
Oman	119,499	2,855,000	24	Monarchy	A	Muscat (Masqat)	Arabic, English, Baluchi, Urdu, Indian dialects	AL, UN
Ontario	415,599	12,495,000	30	Province (Canada)	D	Toronto	English	
Oregon	98,381	3,570,000	36	State (U.S.)	D	Salem	English	
Pakistan (incl. part of Jammu and Kashmir)	339,732	152,210,000	448	Federal Islamic republic	A	Islāmābād	English, Urdu, Punjabi, Sindhi, Pashto, other	UN
Palau (Belau)	188	20,000	106	Republic (U.S. protection)	B	Koror and Melekeok (¹)	Angaur, English, Japanese, Palauan, Sonsorolese, Tobi	UN
Panama	29,157	2,980,000	102	Republic	A	Panamá	Spanish, English	OAS, UN
Papua New Guinea	178,704	5,360,000	30	Parliamentary state	A	Port Moresby	English, Motu, Pidgin, indigenous	UN
Paraguay	157,048	6,115,000	39	Republic	A	Asunción	Guarani, Spanish	MERCOSUR, OAS, UN
Pennsylvania	46,055	12,400,000	269	State (U.S.)	D	Harrisburg	English	
Peru	496,225	28,640,000	58	Republic	A	Lima	Quechua, Spanish, Aymara	OAS, UN
Philippines	115,831	85,430,000	738	Republic	A	Manila	English, Filipino, indigenous	ASEAN, UN
Pitcairn Islands (incl. Dependencies)	19	100	5.3	Overseas territory (U.K.)	C	Adamstown	English, Pitcairnese	
Poland	120,728	38,625,000	320	Republic	A	Warsaw (Warszawa)	Polish	EU, NATO, UN
Portugal	35,516	10,110,000	285	Republic	A	Lisbon (Lisboa)	Portuguese, Mirandese	EU, NATO, UN
Prince Edward Island	2,185	140,000	64	Province (Canada)	D	Charlottetown	English	
Puerto Rico	3,515	3,890,000	1,107	Commonwealth (U.S. protection)	B	San Juan	Spanish, English	
Qatar	4,412	830,000	188	Monarchy	A	Ad Dawḩah (Doha)	Arabic	AL, OPEC, UN
Qinghai	277,994	5,295,000	19	Province (China)	D	Xining	Tibetan dialects, Mongolian, Turkish dialects, Chinese (Mandarin)	
Quebec	595,391	7,675,000	13	Province (Canada)	D	Québec	French, English	
Queensland	668,208	3,785,000	5.7	State (Australia)	D	Brisbane	English	
Reunion	969	760,000	784	Overseas department (France)	C	Saint-Denis	French, Creole	
Rhode Island	1,545	1,080,000	699	State (U.S.)	D	Providence	English	
Rhodesia see Zimbabwe								
Romania	91,699	22,370,000	244	Republic	A	Bucharest (Bucureşti)	Romanian, Hungarian, German	NATO, UN
Russia	6,592,849	144,310,000	22	Federal republic	A	Moscow (Moskva)	Russian, other	CIS, NATO/PP, UN
Rwanda	10,169	7,880,000	775	Republic	A	Kigali	English, French, Kinyarwanda, Kiswahili	AU, COMESA, UN
St. Helena (incl. Dependencies)	121	7,500	62	Overseas territory (U.K.)	C	Jamestown	English	
St. Kitts and Nevis	101	39,000	386	Parliamentary state	A	Basseterre	English	OAS, UN
St. Lucia	238	165,000	693	Parliamentary state	A	Castries	English, French	OAS, UN
St. Pierre and Miquelon	93	7,000	75	Territorial collectivity (France)	C	Saint-Pierre	French	
St. Vincent and the Grenadines	150	115,000	767	Parliamentary state	A	Kingstown	English, French	OAS, UN
Samoa	1,093	180,000	165	Constitutional monarchy	A	Apia	English, Samoan	UN
San Marino	24	28,000	1,167	Republic	A	San Marino	Italian	UN
Sao Tome and Principe	372	180,000	484	Republic	A	São Tomé	Portuguese	AU, UN
Saskatchewan	251,366	1,025,000	4.1	Province (Canada)	D	Regina	English	
Saudi Arabia	830,000	24,690,000	30	Monarchy	A	Riyadh (Ar Riyāḑ)	Arabic	AL, OPEC, UN
Scotland	30,167	5,135,000	170	Administrative division (U.K.)	D	Edinburgh	English, Scots Gaelic	
Senegal	75,951	10,715,000	141	Republic	A	Dakar	French, Wolof and other indigenous	AU, UN
Serbia and Montenegro (Yugoslavia)	39,449	10,660,000	270	Republic	A	Belgrade (Beograd)	Serbian, Albanian	UN
Seychelles	176	81,000	460	Republic	A	Victoria	English, French, Creole	AU, COMESA, UN
Shaanxi	79,151	36,865,000	466	Province (China)	D	Xi'an (Sian)	Chinese (Mandarin)	
Shandong	59,074	92,845,000	1,572	Province (China)	D	Jinan	Chinese (Mandarin)	
Shanghai	2,394	17,120,000	7,151	Autonomous city (China)	D	Shanghai	Chinese (Wu)	
Shanxi	60,232	33,715,000	560	Province (China)	D	Taiyuan	Chinese (Mandarin)	
Sichuan	188,263	85,175,000	452	Province (China)	D	Chengdu	Chinese (Mandarin), Tibetan dialects, Miao-Yao	
Sierra Leone	27,699	5,815,000	210	Republic	A	Freetown	English, Krio, Mende, Temne, indigenous	AU, UN
Singapore	264	4,685,000	17,746	Republic	A	Singapore	Chinese (Mandarin), English, Malay, Tamil	ASEAN, UN
Slovakia	18,924	5,420,000	286	Republic	A	Bratislava	Slovak, Hungarian	EU, NATO, UN
Slovenia	7,821	1,935,000	247	Republic	A	Ljubljana	Slovenian, Croatian, Serbian	EU, NATO, UN
Solomon Islands	10,954	515,000	47	Parliamentary state	A	Honiara	English, indigenous	UN
Somalia	246,201	8,165,000	33	Transitional	A	Mogadishu (Muqdisho)	Arabic, Somali, English, Italian	AL, AU, CAEU, UN
South Africa	470,693	42,770,000	91	Republic	A	Pretoria, Cape Town, and Bloemfontein	Afrikaans, English, Xhosa, Zulu, other indigenous	AU, UN
South America	6,900,000	366,600,000	53					
South Australia	379,724	1,525,000	4.0	State (Australia)	D	Adelaide	English	
South Carolina	32,020	4,160,000	130	State (U.S.)	D	Columbia	English	
South Dakota	77,117	765,000	9.9	State (U.S.)	D	Pierre	English	
South Georgia and the South Sandwich Islands (²)	1,450	(¹)	.....	Overseas territory (U.K.)	C		English	
South West Africa see Namibia								
Spain	194,885	40,250,000	207	Constitutional monarchy	A	Madrid	Spanish (Castilian), Catalan, Galician, Basque	EU, NATO, UN
Spanish North Africa (¹)	12	140,000	11,667	Five possessions (Spain)	C		Spanish, Arabic, Berber dialects	
Spanish Sahara see Western Sahara								
Sri Lanka	25,332	19,825,000	783	Socialist republic	A	Colombo and Sri Jayewardenepura Kotte	English, Sinhala, Tamil	UN
Sudan	967,500	38,630,000	40	Provisional military government	A	Khartoum (Al Kharṭūm)	Arabic, Nubian, and other indigenous, English	AL, AU, CAEU, COMESA, UN
Suriname	63,037	435,000	6.9	Republic	A	Paramaribo	Dutch, Sranan Tongo, English, Hindustani, Javanese	OAS, UN

REGION OR POLITICAL DIVISION	Area Sq. Mi.	Est. Pop. 1/1/04	Pop. Per Sq. Mi.	Form of Government and Ruling Power	Capital	Predominant Languages	International Organizations
Swaziland	6,704	1,165,000	174	Monarchy ... A	Mbabane and Lobamba	English, siSwati	AU, COMESA, UN
Sweden	173,732	8,980,000	52	Constitutional monarchy ... A	Stockholm	Swedish, Sami, Finnish	EU, NATO/PP, UN
Switzerland	15,943	7,430,000	466	Federal republic ... A	Bern (Berne)	German, French, Italian, Romansch	EFTA, NATO/PP, UN
Syria	71,498	17,800,000	249	Republic ... A	Damascus (Dimashq)	Arabic, Kurdish, Armenian, Aramaic, Circassian	AL, CAEU, UN
Taiwan	13,901	22,675,000	1,631	Republic ... A	T'aipei	Chinese (Mandarin), Taiwanese (Min), Hakka	
Tajikistan	55,251	6,935,000	126	Republic ... A	Dushanbe	Tajik, Russian	CIS, NATO/PP, UN
Tanzania	364,900	36,230,000	99	Republic ... A	Dar es Salaam and Dodoma	English, Swahili, indigenous	AU, UN
Tasmania	26,409	475,000	18	State (Australia) ... D	Hobart	English	
Tennessee	42,143	5,860,000	139	State (U.S.) ... D	Nashville	English	
Texas	268,581	22,185,000	83	State (U.S.) ... D	Austin	English, Spanish	
Thailand	198,115	64,570,000	326	Constitutional monarchy ... A	Bangkok (Krung Thep)	Thai, indigenous	ASEAN, UN
Tianjin (Tientsin)	4,363	10,235,000	2,346	Autonomous city (China) ... D	Tianjin (Tientsin)	Chinese (Mandarin)	
Togo	21,925	5,495,000	251	Republic ... A	Lomé	French, Ewe, Mina, Kabye, Dagomba	AU, UN
Tokelau	4.6	1,500	326	Island territory (New Zealand) ... C		English, Tokelauan	
Tonga	251	110,000	438	Constitutional monarchy ... A	Nuku'alofa	Tongan, English	UN
Trinidad and Tobago	1,980	1,100,000	556	Republic ... A	Port of Spain	English, Hindi, French, Spanish, Chinese	OAS, UN
Tristan da Cunha	40	300	7.5	Dependency (St. Helena) ... C	Edinburgh	English	
Tunisia	63,170	9,980,000	158	Republic ... A	Tunis	Arabic, French	AL, AU, UN
Turkey	302,541	68,505,000	226	Republic ... A	Ankara	Turkish, Kurdish, Arabic, Armenian, Greek	NATO, UN
Turkmenistan	188,457	4,820,000	26	Republic ... A	Ashgabat (Ashkhabad)	Turkmen, Russian, Uzbek	CIS, NATO/PP, UN
Turks and Caicos Islands	166	20,000	120	Overseas territory (U.K.) ... C	Grand Turk	English	
Tuvalu	10	11,000	1,100	Parliamentary state ... A	Funafuti	Tuvaluan, English, Samoan, I-Kiribati	UN
Uganda	93,065	26,010,000	279	Republic ... A	Kampala	English, Luganda, Swahili, indigenous, Arabic	AU, COMESA, UN
Ukraine	233,090	47,890,000	205	Republic ... A	Kiev (Kyïv)	Ukrainian, Russian, Romanian, Polish, Hungarian	CIS, NATO/PP, UN
United Arab Emirates	32,278	2,505,000	78	Federation of monarchs ... A	Abū Ẓaby (Abu Dhabi)	Arabic, Persian, English, Hindi, Urdu	AL, CAEU, OPEC, UN
United Kingdom	93,788	60,185,000	642	Constitutional monarchy ... A	London	English, Welsh, Scots Gaelic	EU, NATO, UN
United States	3,794,083	291,680,000	77	Federal republic ... A	Washington	English, Spanish	ANZUS, NAFTA, NATO, OAS, UN
Upper Volta see Burkina Faso							
Uruguay	67,574	3,425,000	51	Republic ... A	Montevideo	Spanish	MERCOSUR, OAS, UN
Utah	84,899	2,360,000	28	State (U.S.) ... D	Salt Lake City	English	
Uzbekistan	172,742	26,195,000	152	Republic ... A	Tashkent (Toshkent)	Uzbek, Russian, Tajik	CIS, NATO/PP, UN
Vanuatu	4,707	200,000	42	Republic ... A	Port Vila	Bislama, English, French	UN
Vatican City	0.2	900	4,500	Ecclesiastical state ... A	Vatican City	Italian, Latin, French, other	
Venezuela	352,145	24,835,000	71	Federal republic ... A	Caracas	Spanish, indigenous	OAS, OPEC, UN
Vermont	9,614	620,000	64	State (U.S.) ... D	Montpelier	English	
Victoria	87,807	4,905,000	56	State (Australia) ... D	Melbourne	English	
Vietnam	128,066	82,150,000	641	Socialist republic ... A	Hanoi	Vietnamese, English, French, Chinese, Khmer, indigenous	ASEAN, UN
Virginia	42,774	7,410,000	173	State (U.S.) ... D	Richmond	English	
Virgin Islands (U.S.)	134	110,000	821	Unincorporated territory (U.S.) ... C	Charlotte Amalie	English, Spanish, Creole	
Wake Island	3.0	(¹)		Unincorporated territory (U.S.) ... C		English	
Wales	8,023	2,965,000	370	Administrative division (U.K.) ... D	Cardiff	English, Welsh Gaelic	
Wallis and Futuna	99	16,000	162	Overseas territory (France) ... C	Mata-Utu	French, Wallisian	
Washington	71,300	6,150,000	86	State (U.S.) ... D	Olympia	English	
West Bank (incl. Jericho and East Jerusalem)	2,263	2,275,000	1,005	Israeli territory with limited self-government		Arabic, Hebrew	(⁴)
Western Australia	976,792	1,945,000	2.0	State (Australia) ... D	Perth	English	
Western Sahara	102,703	265,000	2.6	Occupied by Morocco ... C		Arabic	
West Virginia	24,230	1,815,000	75	State (U.S.) ... D	Charleston	English	
Wisconsin	65,498	5,490,000	84	State (U.S.) ... D	Madison	English	
Wyoming	97,814	505,000	5.2	State (U.S.) ... D	Cheyenne	English	
Xianggang (Hong Kong)	425	7,440,000	17,506	Special administrative region (China) ... D	Hong Kong (Xianggang)	Chinese (Cantonese), English	
Xinjiang Uygur (Sinkiang)	617,764	19,685,000	32	Autonomous region (China) ... D	Ürümqi	Turkish dialects, Mongolian, Tungus, English	
Xizang (Tibet)	471,045	2,680,000	5.7	Autonomous region (China) ... D	Lhasa	Tibetan dialects	
Yemen	203,850	19,680,000	97	Republic ... A	Ṣan'ā' (Sanaa)	Arabic	AL, CAEU, UN
Yugoslavia see Serbia and Montenegro							
Yukon Territory	186,272	32,000	0.2	Territory (Canada) ... D	Whitehorse	English, Inuktitut, indigenous	
Yunnan	152,124	43,850,000	288	Province (China) ... D	Kunming	Chinese (Mandarin), Tibetan dialects, Khmer, Miao-Yao	
Zaire see Congo, Democratic Republic of the							
Zambia	290,586	10,385,000	36	Republic ... A	Lusaka	English, indigenous	AU, COMESA, UN
Zhejiang	39,305	47,830,000	1,217	Province (China) ... D	Hangzhou	Chinese dialects	
Zimbabwe	150,873	12,630,000	84	Republic ... A	Harare (Salisbury)	English, indigenous	AU, COMESA, UN
WORLD	57,900,000	6,339,505,000	109				

... None, or not applicable
(1) No permanent population
(2) Claimed by Argentina
(3) Claimed by Spain
(4) The Palestinian Liberation Organization (PLO) is a member of AL and CAEU
(5) Future capital
(6) Claimed by Comoros
(7) Comprises Ceuta, Melilla, and several small islands

AL	Arab League (League of Arab States)
ANZUS	Australia-New Zealand-U.S. Security Treaty
ASEAN	Association of Southeast Asian Nations
AU	African Union
CAEU	Council of Arab Unity
CIS	Commonwealth of Independent States
COMESA	Common Market for Eastern and Southern Africa
EFTA	European Free Trade Association
EU	European Union
MERCOSUR	Southern Common Market
NAFTA	North American Free Trade Agreement
NATO	North Atlantic Treaty Organization
NATO/PP	NATO-Partnership for Peace Program
OAS	Organization of American States
OPEC	Organization of Petroleum Exporting Countries

WORLD DEMOGRAPHIC TABLE

CONTINENT/Country	Population Estimate 2004	Pop. Per Sq. Mile 2004	Percent Urban[1] 2001	Crude Birth Rate per 1,000[2] 2003	Crude Death Rate per 1,000[2] 2003	Natural Increase Percent[2] 2003	Fertility Rate (Children born/Woman)[3] 2003	Infant Mortality Rate per 1,000[3] 2003	Median Age[2] 2002	Life Expectancy Male[2] 2003	Life Expecta Female 2003
NORTH AMERICA											
Bahamas	300,000	56	64.7	19	9	1.0%	2	26	27	62	69
Belize	270,000	30	48.1	30	6	2.4%	4	27	19	65	70
Canada	32,360,000	8	78.9	11	8	0.3%	2	5	38	76	83
Costa Rica	3,925,000	199	59.5	19	4	1.5%	2	11	25	74	79
Cuba	11,290,000	264	75.5	12	7	0.5%	2	7	35	75	79
Dominica	69,000	238	71.4	17	7	1.0%	2	15	28	71	77
Dominican Republic	8,775,000	468	66.0	24	7	1.7%	3	34	24	66	70
El Salvador	6,530,000	804	61.5	28	6	2.2%	3	27	21	67	74
Guatemala	14,095,000	335	39.9	35	7	2.8%	5	38	18	64	66
Haiti	7,590,000	708	36.3	34	13	2.1%	5	76	18	50	53
Honduras	6,745,000	156	53.7	32	6	2.5%	4	30	19	65	68
Jamaica	2,705,000	637	56.6	17	5	1.2%	2	13	27	74	78
Mexico	104,340,000	138	74.6	22	5	1.7%	3	22	24	72	78
Nicaragua	5,180,000	103	56.5	26	5	2.2%	3	31	20	68	72
Panama	2,980,000	102	56.5	21	6	1.5%	3	21	26	70	75
St. Lucia	165,000	693	38.0	21	5	1.6%	2	14	24	70	77
Trinidad and Tobago	1,100,000	556	74.5	13	9	0.4%	2	25	30	67	72
United States	291,680,000	77	77.4	14	8	0.6%	2	7	36	74	80
SOUTH AMERICA											
Argentina	38,945,000	36	88.3	17	8	1.0%	2	16	29	72	79
Bolivia	8,655,000	20	62.9	26	8	1.8%	3	56	21	62	67
Brazil	183,080,000	55	81.7	18	6	1.2%	2	32	27	67	75
Chile	15,745,000	54	86.1	16	6	1.0%	2	9	30	73	80
Colombia	41,985,000	95	75.5	22	6	1.6%	3	22	26	67	75
Ecuador	13,840,000	126	63.4	25	5	2.0%	3	32	23	69	75
Guyana	705,000	9	36.7	18	9	0.9%	2	38	26	61	66
Paraguay	6,115,000	39	56.7	30	5	2.6%	4	28	21	72	77
Peru	28,640,000	58	73.1	23	6	1.7%	3	37	24	68	73
Suriname	435,000	7	74.8	19	7	1.3%	2	25	26	67	72
Uruguay	3,425,000	51	92.1	17	9	0.8%	2	14	32	73	79
Venezuela	24,835,000	71	87.2	20	5	1.5%	2	24	25	71	77
EUROPE											
Albania	3,535,000	318	42.9	15	5	1.0%	2	23	27	74	80
Austria	8,170,000	252	67.4	9	9	0%	1	5	39	76	82
Belarus	10,315,000	129	69.6	10	14	-0.4%	1	14	37	63	75
Belgium	10,340,000	877	97.4	11	10	0.1%	2	5	40	75	82
Bosnia and Herzegovina	4,000,000	202	43.4	13	8	0.4%	2	23	36	70	75
Bulgaria	7,550,000	176	67.4	10	14	-0.5%	1	22	41	68	75
Croatia	4,430,000	203	58.1	13	11	0.2%	2	7	39	71	78
Czech Republic	10,250,000	337	74.5	9	11	-0.1%	1	4	38	72	79
Denmark	5,405,000	325	85.1	12	11	0.1%	2	5	39	75	80
Estonia	1,405,000	80	69.4	9	13	-0.4%	1	12	38	64	77
Finland	5,210,000	40	58.5	11	10	0.1%	2	4	40	75	82
France	60,305,000	289	75.5	13	9	0.3%	2	4	38	76	83
Germany	82,415,000	598	87.7	9	10	-0.2%	1	4	41	75	82
Greece	10,635,000	209	60.3	10	10	0%	1	6	40	76	81
Hungary	10,045,000	280	64.8	10	13	-0.3%	1	9	38	68	77
Iceland	280,000	7	92.7	14	7	0.7%	2	4	34	78	82
Ireland	3,945,000	145	59.3	14	8	0.6%	2	6	33	75	80
Italy	58,030,000	499	67.1	9	10	-0.1%	1	6	41	76	83
Latvia	2,340,000	94	59.8	9	15	-0.6%	1	15	39	63	75
Lithuania	3,590,000	142	68.6	10	13	-0.2%	1	14	37	64	76
Luxembourg	460,000	460	91.9	12	8	0.4%	2	5	38	75	82
Macedonia	2,065,000	208	59.4	13	8	0.5%	2	12	33	72	77
Moldova	4,440,000	340	41.4	14	13	0.2%	2	42	32	61	69
Netherlands	16,270,000	1,007	89.6	12	9	0.3%	2	5	39	76	81
Norway	4,565,000	37	75.0	12	10	0.3%	2	4	38	77	82
Poland	38,625,000	320	62.5	10	10	0.1%	1	9	36	70	78
Portugal	10,110,000	285	65.8	11	10	0.1%	1	6	38	73	80
Romania	22,370,000	244	55.2	11	12	-0.1%	1	28	35	67	75
Serbia and Montenegro	10,660,000	270	51.7	13	11	0.2%	2	17	36	71	77
Slovakia	5,420,000	286	57.6	10	10	0.1%	1	8	35	70	78
Slovenia	1,935,000	247	49.1	9	10	-0.1%	1	4	39	72	80
Spain	40,250,000	207	77.8	10	9	0.1%	1	5	39	76	83
Sweden	8,980,000	52	83.3	11	10	0%	2	3	40	78	83
Switzerland	7,430,000	466	67.3	10	8	0.1%	1	4	40	77	83
Ukraine	47,890,000	205	68.0	10	16	-0.7%	1	21	38	61	72
United Kingdom	60,185,000	642	89.5	11	10	0.1%	2	5	38	76	81
Russia	144,310,000	22	72.9	10	14	-0.4%	1	20	38	62	73
ASIA											
Afghanistan	29,205,000	116	22.3	41	17	2.3%	6	142	19	48	46
Armenia	3,325,000	289	67.2	13	10	0.2%	2	41	32	62	71
Azerbaijan	7,850,000	235	51.8	19	10	1.0%	2	82	27	59	68
Bahrain	675,000	2,528	92.5	19	4	1.5%	3	19	29	71	76
Bangladesh	139,875,000	2,516	25.6	30	9	2.1%	3	66	21	61	61
Brunei	360,000	162	72.8	20	3	1.6%	2	14	26	72	77
Cambodia	13,245,000	189	17.5	27	9	1.8%	4	76	19	55	60
China	1,298,720,000	352	37.1	13	7	0.6%	2	25	32	70	74
Cyprus	775,000	217	70.2	13	8	0.5%	2	8	34	75	80
East Timor	1,010,000	176	7.5	28	6	2.1%	4	50	20	63	68
Georgia	4,920,000	183	56.5	12	15	-0.3%	2	51	35	61	68
India	1,057,415,000	865	27.9	23	8	1.5%	3	60	24	63	64
Indonesia	236,680,000	322	42.1	21	6	1.5%	3	38	26	67	71
Iran	68,650,000	108	64.7	17	6	1.2%	2	44	23	68	71
Iraq	25,025,000	148	67.4	34	6	2.8%	5	55	19	67	69
Israel	6,160,000	768	91.8	19	6	1.2%	3	7	29	77	81
Japan	127,285,000	873	78.9	10	9	0.1%	1	3	42	78	84
Jordan	5,535,000	160	78.7	24	4	2.1%	3	19	22	75	81
Kazakhstan	16,780,000	16	55.8	18	11	0.8%	2	59	28	58	69
Korea, North	22,585,000	485	60.5	18	7	1.1%	2	26	31	68	74
Korea, South	48,450,000	1,264	82.5	13	6	0.7%	2	7	33	72	79
Kuwait	2,220,000	323	96.1	22	2	1.9%	3	11	26	76	78

CONTINENT/Country	Population Estimate 2004	Pop. Per Sq. Mile 2004	Percent Urban[1] 2001	Crude Birth Rate per 1,000[2] 2003	Crude Death Rate per 1,000[2] 2003	Natural Increase Percent[2] 2003	Fertility Rate (Children born/Woman)[3] 2003	Infant Mortality Rate per 1,000[3] 2003	Median Age[2] 2002	Life Expectancy Male[2] 2003	Life Expectancy Female[2] 2003
Kyrgyzstan	4,930,000	64	34.3	26	9	1.7%	3	75	23	59	68
Laos	5,995,000	66	19.7	37	12	2.5%	5	89	19	52	56
Lebanon	3,755,000	935	90.1	20	6	1.3%	2	26	26	70	75
Malaysia	23,310,000	183	58.1	24	5	1.9%	3	19	24	69	75
Mongolia	2,730,000	5	56.6	21	7	1.4%	2	57	24	62	66
Myanmar	42,620,000	163	28.1	19	12	0.7%	2	70	25	54	58
Nepal	26,770,000	471	12.2	32	10	2.3%	4	71	20	59	59
Oman	2,855,000	24	76.5	37	4	3.4%	6	21	19	70	75
Pakistan	152,210,000	448	33.4	30	9	2.1%	4	77	20	61	63
Philippines	85,430,000	738	59.4	26	6	2.1%	3	25	22	66	72
Qatar	830,000	188	92.9	16	4	1.1%	3	20	31	71	76
Saudi Arabia	24,690,000	30	86.7	37	6	3.1%	6	48	19	67	71
Singapore	4,685,000	17,746	100.0	13	4	0.8%	1	4	35	77	84
Sri Lanka	19,825,000	783	23.1	16	6	1.0%	2	15	29	70	75
Syria	17,800,000	249	51.8	30	5	2.5%	4	32	20	68	71
Taiwan	22,675,000	1,631	[5]	13	6	0.7%	2	7	33	74	80
Tajikistan	6,935,000	126	27.7	33	8	2.4%	4	113	19	61	68
Thailand	64,570,000	326	20.0	16	7	1.0%	2	22	30	69	74
Turkey	68,505,000	226	66.2	18	6	1.2%	2	44	27	69	74
Turkmenistan	4,820,000	26	44.9	28	9	1.9%	4	73	21	58	65
United Arab Emirates	2,505,000	78	87.2	18	4	1.4%	3	16	28	72	77
Uzbekistan	26,195,000	152	36.6	26	8	1.8%	3	72	22	61	68
Vietnam	82,150,000	641	24.5	20	6	1.3%	2	31	25	68	73
Yemen	19,680,000	97	25.0	43	9	3.4%	7	65	16	59	63
AFRICA											
Algeria	33,090,000	36	57.7	22	5	1.7%	3	38	23	69	72
Angola	10,875,000	23	34.9	46	26	2.0%	6	194	18	36	38
Benin	7,145,000	164	43.0	43	14	3.0%	6	87	16	50	52
Botswana	1,570,000	7	49.4	26	31	-0.6%	3	67	19	32	32
Burkina Faso	13,400,000	127	16.9	45	19	2.6%	6	100	17	43	46
Burundi	6,165,000	574	9.3	40	18	2.2%	6	72	16	43	44
Cameroon	15,905,000	87	49.7	35	15	2.0%	5	70	18	47	49
Cape Verde	415,000	267	63.5	27	7	2.0%	4	51	19	67	73
Central African Republic	3,715,000	15	41.7	36	20	1.6%	5	93	18	40	43
Chad	9,395,000	19	24.1	47	16	3.1%	6	96	16	47	50
Comoros	640,000	742	33.8	39	9	3.0%	5	80	19	59	64
Congo	2,975,000	23	66.1	29	14	1.5%	4	95	20	49	51
Congo, Democratic Republic of the	57,445,000	63	30.7	45	15	3.0%	7	97	16	47	51
Cote d'Ivoire	17,145,000	138	44.0	40	18	2.2%	6	98	17	40	45
Djibouti	460,000	51	84.2	41	19	2.1%	6	107	18	42	44
Egypt	75,420,000	195	42.7	24	5	1.9%	3	35	23	68	73
Equatorial Guinea	515,000	48	49.3	37	13	2.4%	5	89	19	53	57
Eritrea	4,390,000	97	19.1	39	13	2.6%	6	76	18	51	55
Ethiopia	67,210,000	158	15.9	40	20	2.0%	6	103	17	40	42
Gabon	1,340,000	13	82.3	37	11	2.5%	5	55	19	55	59
Gambia, The	1,525,000	370	31.3	41	12	2.8%	6	75	17	52	56
Ghana	20,615,000	224	36.4	26	11	1.5%	3	53	20	56	57
Guinea	9,135,000	96	27.9	43	16	2.7%	6	93	18	48	51
Guinea-Bissau	1,375,000	99	32.3	38	17	2.2%	5	110	19	45	49
Kenya	31,840,000	142	34.4	29	16	1.3%	3	63	18	45	45
Lesotho	1,865,000	159	28.8	27	25	0.3%	4	86	20	37	37
Liberia	3,345,000	78	45.5	45	18	2.7%	6	132	18	47	49
Libya	5,565,000	8	88.0	27	3	2.4%	3	27	22	74	78
Madagascar	17,235,000	76	30.1	42	12	3.0%	6	80	17	54	59
Malawi	11,780,000	258	15.1	45	23	2.2%	6	105	16	38	38
Mali	11,790,000	25	30.9	48	19	2.9%	7	119	16	45	46
Mauritania	2,955,000	7	59.1	42	13	2.9%	6	74	17	50	54
Mauritius	1,215,000	1,542	41.6	16	7	0.9%	3	16	30	68	76
Morocco	31,950,000	185	56.1	23	6	1.7%	3	45	23	68	72
Mozambique	18,695,000	60	33.3	37	23	1.4%	5	138	19	39	37
Namibia	1,940,000	6	31.4	34	19	1.5%	5	68	18	44	41
Niger	11,210,000	23	21.1	50	22	2.8%	7	124	16	42	42
Nigeria	135,570,000	380	44.9	39	14	2.5%	5	71	18	51	51
Rwanda	7,880,000	775	6.3	40	22	1.8%	6	103	18	39	40
Sao Tome and Principe	180,000	484	47.7	42	7	3.5%	6	46	16	65	68
Senegal	10,715,000	141	48.2	36	11	2.5%	5	58	18	55	58
Sierra Leone	5,815,000	210	37.3	44	21	2.3%	6	147	18	40	45
Somalia	8,165,000	33	27.9	46	18	2.9%	7	120	18	46	49
South Africa	42,770,000	91	57.7	19	18	0%	2	61	25	47	47
Sudan	38,630,000	40	37.1	36	10	2.7%	5	66	18	57	59
Swaziland	1,165,000	174	26.7	29	21	0.8%	4	67	19	41	38
Tanzania	36,230,000	99	33.3	40	17	2.2%	5	104	18	43	46
Togo	5,495,000	251	33.9	35	12	2.4%	5	69	17	51	55
Tunisia	9,980,000	158	66.2	17	5	1.2%	2	27	26	73	76
Uganda	26,010,000	279	14.5	47	17	3.0%	7	88	15	43	46
Zambia	10,385,000	36	39.8	40	24	1.5%	5	99	17	35	35
Zimbabwe	12,630,000	84	36.0	30	22	0.8%	4	66	19	40	38
OCEANIA											
Australia	19,825,000	7	91.2	13	7	0.5%	2	5	36	77	83
Fiji	875,000	124	50.2	23	6	1.7%	3	13	24	66	71
Kiribati	100,000	319	38.6	31	9	2.3%	4	51	20	58	64
Micronesia, Federated States of	110,000	406	28.6	26	5	2.1%	4	32	19[4]	67	71
New Zealand	3,975,000	38	85.9	14	8	0.7%	2	6	33	75	81
Papua New Guinea	5,360,000	30	17.6	31	8	2.3%	4	55	21	62	66
Samoa	180,000	165	22.3	15	6	0.9%	3	30	24	67	73
Solomon Islands	515,000	47	20.2	32	4	2.8%	4	23	18	70	75
Tonga	110,000	438	33.0	25	6	1.9%	3	13	20	66	71
Vanuatu	200,000	42	22.1	24	8	1.6%	3	58	22	60	63

This table presents data for most independent nations having an area greater than 200 square miles
(1) Source: United Nations World Urbanization Prospects
(2) Source: United States Census Bureau International Database
(3) Source: United States Central Intelligence Agency World Factbook
(4) 2000 Census preliminary count from www.fsmgov.org/info/people.html
(5) Data for Taiwan is included with China

WORLD AGRICULTURE TABLE

CONTINENT/Country	Agricultural Area 2001 Total Area Sq. Miles	Cropland Area[1] Sq. Miles	Cropland Area[1] %	Pasture Area[1] Sq. Miles	Pasture Area[1] %	Average Production 1999-2001 Wheat[1] 1,000 metric tons	Rice[1] 1,000 metric tons	Corn[1] 1,000 metric tons	Average 1999-2001 Cattle[1] 1,000	Pigs[1] 1,000	Sheep[1] 1,000
NORTH AMERICA											
Bahamas	5,382	46	0.9%	8	0.1%	-	-	-	1	5	6
Belize	8,867	402	4.5%	193	2.2%	-	12	36	52	25	4
Canada	3,855,103	177,144	4.6%	111,970	2.9%	24,676	-	8,168	13,340	12,970	819
Costa Rica	19,730	2,027	10.3%	9,035	45.8%	-	267	20	1,358	438	3
Cuba	42,804	17,239	40.3%	8,494	19.8%	-	342	207	4,305	2,600	310
Dominica	290	77	26.6%	8	2.7%	-	-	-	13	5	8
Dominican Republic	18,730	6,162	32.9%	8,108	43.3%	-	615	30	2,026	548	106
El Salvador	8,124	3,514	43.2%	3,066	37.7%	-	47	605	1,190	195	5
Guatemala	42,042	7,355	17.5%	10,046	23.9%	9	46	1,057	2,500	1,417	270
Haiti	10,714	4,247	39.6%	1,892	17.7%	-	111	211	1,390	934	147
Honduras	43,277	5,514	12.7%	5,822	13.5%	1	9	509	1,737	474	14
Jamaica	4,244	1,097	25.8%	884	20.8%	-	-	2	400	180	1
Mexico	758,452	105,406	13.9%	308,882	40.7%	3,263	324	18,466	30,428	16,112	6,048
Nicaragua	50,054	8,382	16.7%	18,591	37.1%	-	234	374	2,008	402	4
Panama	29,157	2,683	9.2%	5,927	20.3%	-	237	71	1,348	279	-
St. Lucia	238	69	29.2%	8	3.2%	-	-	-	12	15	13
Trinidad and Tobago	1,980	471	23.8%	42	2.1%	-	13	5	36	41	12
United States	3,794,083	684,401	18.0%	903,479	23.8%	58,862	9,222	244,296	98,197	60,229	7,071
SOUTH AMERICA											
Argentina	1,073,519	135,136	12.6%	548,265	51.1%	15,642	1,140	15,217	49,299	4,200	13,588
Bolivia	424,165	11,973	2.8%	130,618	30.8%	121	281	607	6,715	2,786	8,743
Brazil	3,300,172	256,623	7.8%	760,621	23.0%	2,461	10,998	35,119	170,295	30,608	14,728
Chile	291,930	8,880	3.0%	49,942	17.1%	1,490	113	685	4,117	2,395	4,153
Colombia	439,737	16,405	3.7%	161,391	36.7%	37	2,262	1,128	25,274	2,726	2,247
Ecuador	109,484	11,525	10.5%	19,653	18.0%	19	1,340	483	5,261	2,654	2,214
Guyana	83,000	1,969	2.4%	4,749	5.7%	-	560	3	220	20	130
Paraguay	157,048	12,008	7.6%	83,784	53.3%	256	112	804	9,758	2,633	402
Peru	496,225	16,255	3.3%	104,634	21.1%	180	1,963	1,205	4,936	2,795	14,414
Suriname	63,037	259	0.4%	81	0.1%	-	178	-	128	22	8
Uruguay	67,574	5,174	7.7%	52,290	77.4%	284	1,189	190	10,446	375	13,257
Venezuela	352,145	13,158	3.7%	70,425	20.0%	1	696	1,547	14,620	5,555	780
EUROPE											
Albania	11,100	2,699	24.3%	1,699	15.3%	298	-	203	719	96	1,929
Austria	32,378	5,676	17.5%	7,413	22.9%	1,412	-	1,774	2,166	3,556	357
Belarus	80,155	24,151	30.1%	11,564	14.4%	903	-	13	4,411	3,565	96
Belgium	11,787	3,344[2]	26.2%[2]	2,618[2]	20.5%[2]	1,535	-	420	3,165	7,462	150
Bosnia and Herzegovina	19,767	3,243	16.4%	4,633	23.4%	289	-	656	448	345	645
Bulgaria	42,855	17,900	41.8%	6,236	14.6%	3,071	8	1,137	664	1,459	2,536
Croatia	21,829	6,124	28.1%	6,035	27.6%	852	-	1,958	435	1,276	519
Czech Republic	30,450	12,788	42.0%	3,730	12.2%	4,196	-	324	1,604	3,761	87
Denmark	16,640	8,880	53.4%	1,452	8.7%	4,683	-	-	1,887	12,052	147
Estonia	17,462	2,691	15.4%	745	4.3%	123	-	-	276	304	29
Finland	130,559	8,490	6.5%	77	0.1%	427	-	-	1,060	1,303	101
France	208,482	75,618	36.3%	38,788	18.6%	35,327	110	15,928	20,377	14,693	9,754
Germany	137,847	46,409	33.7%	19,355	14.0%	21,358	-	3,362	14,723	26,021	2,746
Greece	50,949	14,873	29.2%	17,954	35.2%	2,111	153	2,007	584	925	8,977
Hungary	35,919	18,548	51.6%	4,097	11.4%	3,843	9	6,664	845	5,216	991
Iceland	39,769	27	0.1%	8,780	22.1%	-	-	-	72	44	477
Ireland	27,133	4,050	14.9%	12,934	47.7%	688	-	-	6,613	1,765	5,311
Italy	116,342	42,379	36.4%	16,907	14.5%	7,239	1,310	10,222	7,167	8,356	11,000
Latvia	24,942	7,220	28.9%	2,355	9.4%	410	-	-	393	407	28
Lithuania	25,213	11,541	45.8%	1,923	7.6%	1,062	-	-	856	984	14
Luxembourg	999	[3]	[3]	[3]	[3]	-	-	2	134	-	-
Macedonia	9,928	2,363	23.8%	2,432	24.5%	308	20	135	267	209	1,285
Moldova	13,070	8,398	64.3%	1,483	11.3%	902	-	1,096	423	646	929
Netherlands	16,164	3,622	22.4%	3,834	23.7%	995	-	148	4,108	13,253	1,335
Norway	125,050	3,398	2.7%	625	0.5%	265	-	-	1,017	414	2,342
Poland	120,728	55,267	45.8%	15,745	13.0%	8,946	-	962	6,124	17,588	366
Portugal	35,516	10,444	29.4%	5,548	15.6%	295	146	907	1,415	2,346	4,337
Romania	91,699	38,305	41.8%	19,039	20.8%	5,610	3	8,317	3,021	5,946	8,062
Serbia and Montenegro	39,449	14,394	36.5%	7,197	18.2%	2,207	-	5,013	1,550	4,012	1,853
Slovakia	18,924	6,085	32.2%	3,375	17.8%	1,445	-	612	671	1,548	344
Slovenia	7,821	784	10.0%	1,185	15.2%	153	-	283	473	585	80
Spain	194,885	69,298	35.6%	44,209	22.7%	5,785	844	4,208	6,140	22,079	24,185
Sweden	173,732	10,413	6.0%	1,726	1.0%	2,135	-	-	1,683	1,975	440
Switzerland	15,943	1,683	10.6%	4,417	27.7%	535	-	214	1,603	1,499	421
Ukraine	233,090	129,321	55.5%	30,541	13.1%	15,043	74	3,075	10,591	9,270	1,074
United Kingdom	93,788	22,019	23.5%	43,440	46.3%	14,380	-	-	11,052	6,537	41,205
Russia	6,592,849	485,400	7.4%	351,905	5.3%	37,455	509	1,133	27,936	17,076	12,954
ASIA											
Afghanistan	251,773	31,097	12.4%	115,831	46.0%	1,821	205	172	2,600	-	12,762
Armenia	11,506	2,162	18.8%	3,089	26.8%	211	-	9	478	75	515
Azerbaijan	33,437	7,471	22.3%	10,039	30.0%	1,172	19	107	1,965	21	5,321
Bahrain	267	23	8.7%	15	5.8%	-	-	-	12	-	17
Bangladesh	55,598	32,761	58.9%	2,317	4.2%	1,807	36,909	8	23,817	-	1,128
Brunei	2,226	27	1.2%	23	1.0%	-	-	-	2	6	2
Cambodia	69,898	14,699	21.0%	5,792	8.3%	-	4,035	146	2,896	2,079	-
China	3,690,045	599,520[4]	16.2%[4]	1,544,412[4]	41.9%[4]	102,463[4]	189,840[4]	116,240[4]	104,179[4]	440,384[4]	130,536[4]
Cyprus	3,572	436	12.2%	15	0.4%	12	-	-	55	419	240
East Timor	5,743	309	5.4%	579	10.1%	-	33	93	173	300	36
Georgia	26,911	4,104	15.3%	7,490	27.8%	207	-	358	1,117	433	541
India	1,222,510	655,987	53.7%	42,124	3.4%	72,140	132,818	12,285	217,773	17,000	57,900
Indonesia	735,310	129,730	17.6%	43,155	5.9%	-	50,953	9,409	11,370	6,098	7,316
Iran	636,372	63,892	10.0%	169,885	26.7%	8,740	2,103	1,113	8,273	-	53,900
Iraq	169,235	23,514	13.9%	15,444	9.1%	667	110	73	1,342	-	6,770
Israel	8,019	1,637	20.4%	548	6.8%	94	-	73	393	138	373
Japan	145,850	18,510	12.7%	1,564	1.1%	657	11,551	-	4,592	9,823	11
Jordan	34,495	1,544	4.5%	2,865	8.3%	18	-	13	66	-	1,900
Kazakhstan	1,049,156	83,672	8.0%	714,667	68.1%	10,938	225	256	4,021	984	8,785
Korea, North	46,540	10,811	23.2%	193	0.4%	88	2,031	1,253	575	3,076	186
Korea, South	38,328	7,293	19.0%	208	0.5%	4	7,204	67	2,191	8,266	1
Kuwait	6,880	58	0.8%	525	7.6%	-	-	-	19	-	543

CONTINENT/Country	Agricultural Area 2001 Total Area Sq. Miles	Cropland Area[1] Sq. Miles	Cropland Area[1] %	Pasture Area[1] Sq. Miles	Pasture Area[1] %	Average Production 1999-2001 Wheat[1] 1,000 metric tons	Rice[1] 1,000 metric tons	Corn[1] 1,000 metric tons	Average 1999-2001 Cattle[1] 1,000	Pigs[1] 1,000	Sheep[1] 1,000
Kyrgyzstan	77,182	5,664	7.3%	35,873	46.5%	1,113	17	363	942	98	3,101
Laos	91,429	3,699	4.0%	3,390	3.7%	-	2,213	108	1,106	1,390	-
Lebanon	4,016	1,208	30.1%	62	1.5%	60	-	4	76	63	354
Malaysia	127,320	29,286	23.0%	1,100	0.9%	-	2,170	63	744	1,943	167
Mongolia	604,829	4,633	0.8%	499,230	82.5%	148	-	-	2,997	17	14,587
Myanmar	261,228	41,023	15.7%	1,212	0.5%	105	20,683	413	10,974	3,923	390
Nepal	56,827	12,324	21.7%	6,784	11.9%	1,143	4,137	1,528	7,012	872	852
Oman	119,499	313	0.3%	3,861	3.2%	1	-	-	299	-	342
Pakistan	339,732	85,560	25.2%	19,305	5.7%	19,319	6,920	1,653	22,007	-	24,067
Philippines	115,831	41,120	35.5%	4,942	4.3%	-	12,377	4,540	2,467	10,724	30
Qatar	4,412	81	1.8%	193	4.4%	-	-	1	15	-	214
Saudi Arabia	830,000	14,649	1.8%	656,373	79.1%	1,871	-	5	304	-	7,848
Singapore	264	4	1.5%	-	0.0%	-	-	-	-	190	-
Sri Lanka	25,332	7,378	29.1%	1,699	6.7%	-	2,804	30	1,580	71	12
Syria	71,498	21,043	29.4%	31,942	44.7%	3,514	-	196	933	-	13,288
Taiwan	13,901	(5)	(5)	(5)	(5)	(5)	(5)	(5)	(5)	(5)	(5)
Tajikistan	55,251	4,093	7.4%	13,514	24.5%	375	67	38	1,045	1	1,481
Thailand	198,115	70,657	35.7%	3,089	1.6%	1	25,578	4,405	4,973	6,539	40
Turkey	302,541	101,757	33.6%	47,792	15.8%	19,341	350	2,266	10,949	4	29,394
Turkmenistan	188,457	7,008	3.7%	118,533	62.9%	1,472	33	9	863	46	5,750
United Arab Emirates	32,278	919	2.8%	1,178	3.6%	-	-	-	94	-	504
Uzbekistan	172,742	18,649	10.8%	88,031	51.0%	3,637	219	133	5,279	83	7,980
Vietnam	128,066	32,579	25.4%	2,479	1.9%	-	31,964	1,961	4,029	20,273	-
Yemen	203,850	6,158	3.0%	62,027	30.4%	145	-	48	1,320	-	4,758
AFRICA											
Algeria	919,595	31,861	3.5%	122,780	13.4%	1,414	-	1	1,667	6	19,000
Angola	481,354	12,741	2.6%	208,495	43.3%	4	16	417	3,995	800	345
Benin	43,484	8,745	20.1%	2,124	4.9%	-	46	740	1,486	463	650
Botswana	224,607	1,440	0.6%	98,842	44.0%	1	-	8	2,035	6	347
Burkina Faso	105,869	15,444	14.6%	23,166	21.9%	-	102	500	4,767	621	6,722
Burundi	10,745	4,865	45.3%	3,610	33.6%	7	57	124	321	67	215
Cameroon	183,568	27,645	15.1%	7,722	4.2%	-	69	759	5,761	1,232	3,734
Cape Verde	1,557	158	10.2%	97	6.2%	-	-	27	22	195	9
Central African Republic	240,536	7,799	3.2%	12,066	5.0%	-	23	101	3,096	669	218
Chad	495,755	14,016	2.8%	173,746	35.0%	3	114	88	5,852	22	2,374
Comoros	863	510	59.1%	58	6.7%	-	17	4	51	-	21
Congo	132,047	849	0.6%	38,610	29.2%	-	1	6	87	46	102
Congo, Democratic Republic of the	905,446	30,425	3.4%	57,915	6.4%	9	338	1,184	823	1,050	925
Cote d'Ivoire	124,504	28,958	23.3%	50,193	40.3%	-	1,217	693	1,398	333	1,439
Djibouti	8,958	4	0.0%	5,019	56.0%	-	-	-	269	-	465
Egypt	386,662	12,888	3.3%	-	0.0%	6,388	5,681	6,487	3,583	29	4,510
Equatorial Guinea	10,831	888	8.2%	402	3.7%	-	-	-	5	6	37
Eritrea	45,406	1,942	4.3%	26,900	59.2%	32	-	13	2,150	-	1,570
Ethiopia	426,373	44,255	10.4%	77,220	18.1%	1,340	-	2,938	35,025	25	22,333
Gabon	103,347	1,911	1.8%	18,012	17.4%	-	1	26	36	213	197
Gambia, The	4,127	985	23.9%	1,772	42.9%	-	28	24	350	12	115
Ghana	92,098	22,780	24.7%	32,240	35.0%	-	244	988	1,297	327	2,715
Guinea	94,926	5,888	6.2%	41,313	43.5%	-	830	96	2,576	93	824
Guinea-Bissau	13,948	2,116	15.2%	4,170	29.9%	-	95	26	509	347	283
Kenya	224,961	19,923	8.9%	82,240	36.6%	184	58	2,419	13,229	311	7,000
Lesotho	11,720	1,290	11.0%	7,722	65.9%	39	-	128	547	63	839
Liberia	43,000	2,317	5.4%	7,722	18.0%	-	188	-	36	127	210
Libya	679,362	8,301	1.2%	51,352	7.6%	128	-	-	207	-	5,100
Madagascar	226,658	13,707	6.0%	92,664	40.9%	9	2,412	175	10,339	1,267	793
Malawi	45,747	9,035	19.7%	7,143	15.6%	2	86	2,190	741	450	110
Mali	478,841	18,147	3.8%	115,831	24.2%	8	801	378	6,594	72	6,282
Mauritania	397,956	1,931	0.5%	151,545	38.1%	-	65	7	1,470	-	7,437
Mauritius	788	409	51.9%	27	3.4%	-	-	-	27	12	10
Morocco	172,414	37,529	21.8%	81,081	47.0%	2,284	33	95	2,629	8	17,059
Mozambique	309,496	16,351	5.3%	169,885	54.9%	1	168	1,136	1,317	179	125
Namibia	317,818	3,166	1.0%	146,719	46.2%	4	-	26	2,436	21	2,330
Niger	489,192	17,375	3.6%	46,332	9.5%	10	66	5	2,217	39	4,386
Nigeria	356,669	120,464	33.8%	151,352	42.4%	75	3,109	4,734	19,677	5,000	20,833
Rwanda	10,169	5,019	49.4%	2,124	20.9%	6	13	66	766	172	264
Sao Tome and Principe	372	205	55.0%	4	1.0%	-	-	2	4	2	3
Senegal	75,951	9,653	12.7%	21,815	28.7%	-	229	84	3,076	263	4,619
Sierra Leone	27,699	2,178	7.9%	8,494	30.7%	-	215	9	413	52	365
Somalia	246,201	4,135	1.7%	166,024	67.4%	1	2	188	5,133	4	13,100
South Africa	470,693	60,664	12.9%	324,048	68.8%	2,200	3	9,147	13,594	1,542	28,677
Sudan	967,500	64,298	6.6%	452,434	46.8%	230	8	48	37,081	-	45,980
Swaziland	6,704	734	10.9%	4,633	69.1%	-	-	94	613	32	27
Tanzania	364,900	19,112	5.2%	135,136	37.0%	87	509	2,567	17,350	449	3,513
Togo	21,925	10,154	46.3%	3,861	17.6%	-	69	480	277	287	1,528
Tunisia	63,170	18,954	30.0%	15,792	25.0%	1,111	-	-	760	6	6,862
Uganda	93,065	27,799	29.9%	19,738	21.2%	12	106	1,108	5,977	1,540	1,065
Zambia	290,586	20,386	7.0%	115,831	39.9%	80	11	768	2,709	324	137
Zimbabwe	150,873	12,934	8.6%	66,410	44.0%	282	-	1,698	5,840	494	602
OCEANIA											
Australia	2,969,910	195,368	6.6%	1,563,327	52.6%	23,654	1,417	363	27,645	2,607	116,736
Fiji	7,056	1,100	15.6%	676	9.6%	-	16	1	335	139	7
Kiribati	313	151	48.1%	-	0.0%	-	-	-	-	10	-
Micronesia, Federated States of ...	271	139	51.3%	42	15.7%	-	-	-	14	32	-
New Zealand	104,454	13,019	12.5%	53,525	51.2%	337	-	185	9,025	364	45,114
Papua New Guinea	178,704	3,320	1.9%	676	0.4%	-	1	7	87	1,583	6
Samoa	1,093	498	45.6%	8	0.7%	-	-	-	28	179	-
Solomon Islands	10,954	286	2.6%	154	1.4%	-	5	-	11	63	-
Tonga	251	185	73.8%	15	6.2%	-	-	-	11	81	-
Vanuatu	4,707	463	9.8%	162	3.4%	-	-	1	151	62	-

This table presents data for most independent nations having an area greater than 200 square miles
- Zero, insignificant, or not available
(1) Source: United Nations Food and Agriculture Organization
(2) Includes data for Luxembourg
(3) Data for Luxembourg is included with Belgium
(4) Includes data for Taiwan
(5) Data for Taiwan is included with China

WORLD ECONOMIC TABLE

CONTINENT/Country	Total GDP[1]	GDP Per Capita[1]	Value of Exports[1]	Value of Imports[1]	Total (1,000 Metric Tons of Coal Equiv.)	Solid %	Liquid %	Gas %	Hydro & Nuclear %	Coal[3]	Petroleum[3]	Iron Ore[4]	Bauxite
	GDP 2002		Trade		Commercial Energy Production Avg. 2000[2]					Average Production 1999-2001 in Metric Tons			
NORTH AMERICA													
Bahamas	$4,590,000,000	$17,000	$560,700,000	$1,860,000,000	-	-	-	-	-	-	-	-	-
Belize	$1,280,000,000	$4,900	$290,000,000	$430,000,000	12	-	-	-	100%	-	-	-	-
Canada	$934,100,000,000	$29,400	$260,500,000,000	$229,000,000,000	507,218	10%	33%	43%	14%	70,711,084	97,834,913	20,527,000	-
Costa Rica	$32,000,000,000	$8,500	$5,100,000,000	$6,400,000,000	1,937	-	-	-	100%	-	-	-	-
Cuba	$30,690,000,000	$2,300	$1,800,000,000	$4,800,000,000	4,626	-	83%	17%	-	-	2,134,520	-	-
Dominica	$380,000,000	$5,400	$50,000,000	$135,000,000	4	-	-	-	100%	-	-	-	-
Dominican Republic	$53,780,000,000	$6,100	$5,300,000,000	$8,700,000,000	115	-	-	-	100%	-	-	-	-
El Salvador	$29,410,000,000	$4,700	$3,000,000,000	$4,900,000,000	1,110	-	-	-	100%	-	-	-	-
Guatemala	$53,200,000,000	$3,700	$2,700,000,000	$5,600,000,000	1,822	-	81%	1%	18%	-	1,076,526	9,000	-
Haiti	$10,600,000,000	$1,700	$298,000,000	$1,140,000,000	33	-	-	-	100%	-	-	-	-
Honduras	$16,290,000,000	$2,600	$1,300,000,000	$2,700,000,000	347	-	-	-	100%	-	-	-	-
Jamaica	$10,080,000,000	$3,900	$1,400,000,000	$3,100,000,000	18	-	-	-	100%	-	-	-	11,728,000
Mexico	$924,400,000,000	$9,000	$158,400,000,000	$168,400,000,000	340,594	1%	79%	16%	4%	11,097,943	150,165,451	6,860,000	-
Nicaragua	$11,160,000,000	$2,500	$637,000,000	$1,700,000,000	706	-	-	-	100%	-	-	-	-
Panama	$18,060,000,000	$6,000	$5,800,000,000	$6,700,000,000	418	-	-	-	100%	-	-	-	-
St. Lucia	$866,000,000	$5,400	$68,300,000	$319,400,000	-	-	-	-	-	-	-	-	-
Trinidad and Tobago	$11,070,000,000	$9,500	$4,200,000,000	$3,800,000,000	22,768	-	39%	61%	-	-	5,964,991	-	-
United States	$10,450,000,000,000	$37,600	$733,900,000,000	$1,194,100,000,000	2,342,228	33%	22%	30%	14%	996,498,186	289,640,487	35,178,000	-
SOUTH AMERICA													
Argentina	$403,800,000,000	$10,200	$25,300,000,000	$9,000,000,000	118,739	-	50%	45%	5%	260,299	38,783,798	-	-
Bolivia	$21,150,000,000	$2,500	$1,300,000,000	$1,600,000,000	7,732	-	33%	64%	3%	-	1,599,401	-	-
Brazil	$1,376,000,000,000	$7,600	$59,400,000,000	$46,200,000,000	143,640	3%	63%	6%	28%	4,446,477	61,155,586	124,667,000	13,654,000
Chile	$156,100,000,000	$10,000	$17,800,000,000	$15,600,000,000	6,180	6%	11%	45%	38%	475,484	349,201	5,523,000	-
Colombia	$251,600,000,000	$6,500	$12,900,000,000	$12,500,000,000	99,513	36%	52%	9%	4%	38,112,136	34,896,672	348,000	-
Ecuador	$42,650,000,000	$3,100	$4,900,000,000	$6,000,000,000	32,171	-	94%	3%	3%	-	19,520,185	-	-
Guyana	$2,628,000,000	$4,000	$500,000,000	$575,000,000	1	-	-	-	100%	-	-	-	2,272,000
Paraguay	$25,190,000,000	$4,200	$2,000,000,000	$2,400,000,000	6,577	-	-	-	100%	-	-	-	-
Peru	$138,800,000,000	$4,800	$7,600,000,000	$7,300,000,000	10,933	-	73%	9%	18%	52,297	4,932,561	2,701,000	-
Suriname	$1,469,000,000	$3,500	$445,000,000	$300,000,000	1,022	-	84%	-	16%	-	496,400	-	3,946,000
Uruguay	$26,820,000,000	$7,800	$2,100,000,000	$1,870,000,000	867	-	-	-	100%	-	-	-	-
Venezuela	$131,700,000,000	$5,500	$28,600,000,000	$18,800,000,000	311,899	3%	81%	14%	2%	7,482,998	146,621,238	10,497,000	4,309,000
EUROPE													
Albania	$15,690,000,000	$4,500	$340,000,000	$1,500,000,000	1,089	1%	42%	2%	55%	32,666	284,321	-	-
Austria	$227,700,000,000	$27,700	$70,000,000,000	$74,000,000,000	9,611	5%	15%	24%	56%	1,197,660	921,120	525,000	-
Belarus	$90,190,000,000	$8,200	$7,700,000,000	$8,800,000,000	3,644	18%	73%	9%	-	-	1,830,872	-	-
Belgium	$299,700,000,000	$29,000	$162,000,000,000	$152,000,000,000	18,451	2%	-	-	98%	318,998	-	-	-
Bosnia and Herzegovina	$7,300,000,000	$1,900	$1,150,000,000	$2,800,000,000	6,553	90%	-	-	10%	8,414,623	-	50,000	75,000
Bulgaria	$49,230,000,000	$6,600	$5,300,000,000	$6,900,000,000	13,500	46%	-	-	53%	28,841,963	37,048	310,000	-
Croatia	$43,120,000,000	$8,800	$4,900,000,000	$10,700,000,000	4,962	-	42%	43%	15%	5,104	1,191,360	-	-
Czech Republic	$157,100,000,000	$15,300	$40,800,000,000	$43,200,000,000	39,843	85%	1%	1%	14%	63,466,671	283,097	-	-
Denmark	$155,300,000,000	$29,000	$56,300,000,000	$47,900,000,000	36,502	-	70%	29%	2%	-	16,701,163	-	-
Estonia	$15,520,000,000	$10,900	$3,400,000,000	$4,400,000,000	3,892	100%	-	-	-	-	-	-	-
Finland	$133,800,000,000	$26,200	$40,100,000,000	$31,800,000,000	11,933	15%	-	-	85%	-	-	-	-
France	$1,558,000,000,000	$25,700	$307,800,000,000	$303,700,000,000	175,306	2%	4%	1%	93%	3,616,981	1,446,228	12,000	-
Germany	$2,160,000,000,000	$26,600	$608,000,000,000	$487,300,000,000	181,697	47%	2%	13%	38%	204,685,080	3,044,206	5,000	-
Greece	$203,300,000,000	$19,000	$12,600,000,000	$31,400,000,000	12,988	92%	3%	1%	4%	64,503,999	166,807	583,000	1,975,000
Hungary	$134,000,000,000	$13,300	$31,400,000,000	$33,900,000,000	16,319	25%	19%	24%	32%	14,796,257	1,301,710	-	994,000
Iceland	$8,444,000,000	$25,000	$2,300,000,000	$2,100,000,000	1,638	-	-	-	100%	-	-	-	-
Ireland	$113,700,000,000	$30,500	$86,600,000,000	$48,600,000,000	3,232	47%	-	47%	6%	-	-	-	-
Italy	$1,455,000,000,000	$25,000	$259,200,000,000	$238,200,000,000	40,332	-	16%	54%	30%	47,666	4,144,278	-	-
Latvia	$20,990,000,000	$8,300	$2,300,000,000	$3,900,000,000	369	6%	-	-	94%	-	-	-	-
Lithuania	$30,080,000,000	$8,400	$5,400,000,000	$6,800,000,000	3,677	-	12%	-	87%	-	251,824	-	-
Luxembourg	$21,940,000,000	$44,000	$10,100,000,000	$13,250,000,000	113	-	-	-	100%	-	-	-	-
Macedonia	$10,570,000,000	$5,000	$1,100,000,000	$1,900,000,000	3,038	95%	-	-	5%	7,463,628	-	9,000	-
Moldova	$11,510,000,000	$2,500	$590,000,000	$980,000,000	7	-	-	-	100%	-	-	-	-
Netherlands	$437,800,000,000	$26,900	$243,300,000,000	$201,100,000,000	87,974	-	4%	94%	2%	-	1,437,293	-	-
Norway	$149,100,000,000	$31,800	$68,200,000,000	$37,300,000,000	324,396	-	72%	22%	5%	847,996	154,419,533	355,000	-
Poland	$373,200,000,000	$9,500	$32,400,000,000	$43,400,000,000	108,277	94%	1%	5%	-	164,737,813	645,072	-	-
Portugal	$195,200,000,000	$18,000	$25,900,000,000	$39,000,000,000	1,560	-	-	-	100%	-	-	6,000	-
Romania	$169,300,000,000	$7,400	$13,700,000,000	$16,700,000,000	37,598	19%	24%	46%	10%	27,392,191	6,038,110	24,000	-
Serbia and Montenegro	$23,150,000,000	$2,370	$2,400,000,000	$6,300,000,000	14,188	74%	8%	8%	10%	34,480,488	810,787	10,000	580,000
Slovakia	$67,340,000,000	$12,200	$12,900,000,000	$15,400,000,000	8,813	17%	1%	2%	79%	3,606,648	48,134	200,000	-
Slovenia	$37,060,000,000	$18,000	$10,300,000,000	$11,100,000,000	3,644	38%	-	-	62%	4,391,644	991	-	-
Spain	$850,700,000,000	$20,700	$122,200,000,000	$156,600,000,000	40,444	28%	2%	1%	68%	23,479,212	296,665	-	-
Sweden	$230,700,000,000	$25,400	$80,600,000,000	$68,600,000,000	31,413	1%	-	-	99%	-	-	12,114,000	-
Switzerland	$233,400,000,000	$31,700	$100,300,000,000	$94,400,000,000	14,710	-	-	-	100%	-	-	-	-
Ukraine	$218,000,000,000	$4,500	$18,100,000,000	$18,000,000,000	118,973	50%	5%	20%	25%	81,998,575	3,747,936	28,933,000	-
United Kingdom	$1,528,000,000,000	$25,300	$286,300,000,000	$330,100,000,000	397,906	7%	47%	38%	8%	32,758,497	119,820,635	1,000	-
Russia	$1,409,000,000,000	$9,300	$104,600,000,000	$60,700,000,000	1,412,286	10%	33%	52%	5%	253,376,954	324,436,632	48,300,000	3,983,000
ASIA													
Afghanistan	$19,000,000,000	$700	$1,200,000,000	$1,300,000,000	195	1%	-	79%	20%	1,000	-	-	-
Armenia	$12,130,000,000	$3,800	$525,000,000	$991,000,000	901	-	-	-	100%	-	-	-	-
Azerbaijan	$28,610,000,000	$3,500	$2,000,000,000	$1,800,000,000	27,748	-	72%	27%	1%	-	14,183,985	-	-
Bahrain	$9,910,000,000	$14,000	$5,800,000,000	$4,200,000,000	14,442	-	22%	78%	-	-	1,827,397	-	-
Bangladesh	$238,200,000,000	$1,700	$6,200,000,000	$8,500,000,000	11,713	-	-	99%	1%	-	120,476	-	-
Brunei	$6,500,000,000	$18,600	$3,000,000,000	$1,400,000,000	27,922	-	49%	51%	-	-	9,435,323	-	-
Cambodia	$20,420,000,000	$1,500	$1,380,000,000	$1,730,000,000	10	-	-	-	100%	-	-	-	-
China	$5,989,000,000,000	$4,400	$658,260,000,000	$618,930,000,000	1,023,314[5]	70%[5]	23%[5]	4%[5]	3%[5]	1,251,423,183	161,226,848	72,967,000	9,000,000
Cyprus	$9,400,000,000	$15,000	$1,030,000,000	$3,900,000,000	-	-	-	-	-	-	-	-	-
East Timor	$440,000,000	$500	$8,000,000	$237,000,000	-	-	-	-	-	-	-	-	-
Georgia	$16,050,000,000	$3,100	$515,000,000	$750,000,000	963	1%	16%	8%	75%	10,000	102,258	-	-
India	$2,664,000,000,000	$2,540	$44,500,000,000	$53,800,000,000	367,807	73%	14%	8%	4%	304,842,421	32,123,682	48,080,000	7,554,000
Indonesia	$714,200,000,000	$3,100	$52,300,000,000	$32,100,000,000	279,695	27%	45%	26%	2%	79,664,587	70,565,213	282,000	1,168,000
Iran	$458,300,000,000	$7,000	$24,800,000,000	$21,800,000,000	350,729	-	77%	23%	-	1,376,993	181,632,777	5,367,000	136,000
Iraq	$58,000,000,000	$2,400	$13,000,000,000	$7,800,000,000	186,519	-	97%	3%	-	-	124,281,583	-	-
Israel	$117,400,000,000	$19,000	$28,100,000,000	$30,800,000,000	334	94%	2%	4%	1%	-	5,957	-	-
Japan	$3,651,000,000,000	$28,000	$383,800,000,000	$292,100,000,000	142,731	2%	1%	2%	95%	3,286,983	351,650	1,000	-
Jordan	$22,630,000,000	$4,300	$2,500,000,000	$4,400,000,000	316	-	1%	97%	2%	-	1,986	-	-
Kazakhstan	$120,000,000,000	$6,300	$10,300,000,000	$9,600,000,000	113,390	40%	45%	14%	1%	70,311,969	30,508,827	7,467,000	3,668,000
Korea, North	$22,260,000,000	$1,000	$842,000,000	$1,314,000,000	65,932	96%	-	-	4%	94,174,845	-	3,000,000	-
Korea, South	$941,500,000,000	$19,400	$162,600,000,000	$148,400,000,000	43,892	6%	-	-	94%	4,054,646	-	175,000	-
Kuwait	$36,850,000,000	$15,000	$16,000,000,000	$7,300,000,000	161,322	-	92%	8%	-	-	98,844,823	-	-

CONTINENT/Country	GDP 2002 Total GDP[1]	GDP Per Capita[1]	Trade Value of Exports[1]	Value of Imports[1]	Commercial Energy Production Avg. 2000[2] Total (1,000 Metric Tons of Coal Equiv.)	Solid %	Liquid %	Gas %	Hydro & Nuclear %	Average Production 1999-2001 in Metric Tons Coal[3]	Petroleum[3]	Iron Ore[4]	Bauxite[4]
Kyrgyzstan	$13,880,000,000	$2,800	$488,000,000	$587,000,000	2,026	9%	5%	2%	83%	423,664	91,503	-	-
Laos	$10,400,000,000	$1,700	$345,000,000	$555,000,000	146	1%	-	-	99%	1,000	-	-	-
Lebanon	$17,610,000,000	$5,400	$1,000,000,000	$6,000,000,000	55	-	-	-	100%	-	-	-	-
Malaysia	$198,400,000,000	$9,300	$95,200,000,000	$76,800,000,000	110,069	-	41%	58%	1%	314,332	33,792,132	208,000	137,000
Mongolia	$5,060,000,000	$1,840	$501,000,000	$659,000,000	2,212	100%	-	-	-	5,099,640	-	-	-
Myanmar	$73,690,000,000	$1,660	$2,700,000,000	$2,500,000,000	9,297	3%	6%	88%	2%	358,331	587,374	-	-
Nepal	$37,320,000,000	$1,400	$720,000,000	$1,600,000,000	172	10%	-	-	90%	9,667	-	-	-
Oman	$22,400,000,000	$8,300	$10,600,000,000	$5,500,000,000	74,376	-	92%	8%	-	-	46,989,489	-	-
Pakistan	$295,300,000,000	$2,100	$9,800,000,000	$11,100,000,000	33,773	6%	12%	74%	7%	3,247,391	2,768,108	-	10,000
Philippines	$379,700,000,000	$4,200	$35,100,000,000	$33,500,000,000	16,244	6%	-	-	94%	1,306,993	173,128	-	-
Qatar	$15,910,000,000	$21,500	$10,900,000,000	$3,900,000,000	92,237	-	57%	43%	-	-	35,018,538	-	-
Saudi Arabia	$268,900,000,000	$10,500	$71,000,000,000	$39,500,000,000	736,996	-	91%	9%	-	-	401,559,222	-	-
Singapore	$112,400,000,000	$24,000	$127,000,000,000	$113,000,000,000	-	-	-	-	-	-	-	-	-
Sri Lanka	$73,700,000,000	$3,700	$4,600,000,000	$5,400,000,000	394	-	-	-	100%	-	-	-	-
Syria	$63,480,000,000	$3,500	$6,200,000,000	$4,900,000,000	47,898	-	83%	15%	2%	-	26,119,029	-	-
Taiwan	$406,000,000,000	$18,000	$130,000,000,000	$113,000,000,000	(6)	(6)	(6)	(6)	(6)	58,284	38,686	-	-
Tajikistan	$8,476,000,000	$1,250	$710,000,000	$830,000,000	1,790	-	1%	3%	95%	20,667	16,613	-	-
Thailand	$445,800,000,000	$6,900	$67,700,000,000	$58,100,000,000	44,127	25%	24%	50%	2%	18,551,756	5,080,720	20,000	-
Turkey	$489,700,000,000	$7,000	$35,100,000,000	$50,800,000,000	28,167	69%	14%	3%	14%	65,334,995	2,642,106	2,300,000	303,000
Turkmenistan	$31,340,000,000	$5,500	$2,970,000,000	$2,250,000,000	71,764	-	15%	85%	-	-	7,139,688	-	-
United Arab Emirates	$53,970,000,000	$22,000	$44,900,000,000	$30,800,000,000	199,656	-	83%	17%	-	-	112,737,023	-	-
Uzbekistan	$66,060,000,000	$2,500	$2,800,000,000	$2,500,000,000	85,806	1%	13%	85%	1%	2,736,319	4,419,300	-	-
Vietnam	$183,800,000,000	$2,250	$16,500,000,000	$16,800,000,000	39,300	30%	59%	5%	7%	9,688,950	15,926,911	-	-
Yemen	$15,070,000,000	$840	$3,400,000,000	$2,900,000,000	30,622	-	100%	-	-	-	21,304,264	-	-
AFRICA													
Algeria	$173,800,000,000	$5,300	$19,500,000,000	$10,600,000,000	222,648	-	47%	53%	-	24,000	61,651,110	757,000	-
Angola	$18,360,000,000	$1,600	$8,600,000,000	$4,100,000,000	53,315	-	98%	1%	-	-	36,961,745	-	-
Benin	$7,380,000,000	$1,070	$207,000,000	$479,000,000	69	-	100%	-	-	-	39,547	-	-
Botswana	$13,480,000,000	$9,500	$2,400,000,000	$1,900,000,000	(7)	(7)	(7)	(7)	(7)	956,767	-	-	-
Burkina Faso	$14,510,000,000	$1,080	$250,000,000	$525,000,000	15	-	-	-	100%	-	-	-	-
Burundi	$3,146,000,000	$600	$26,000,000	$135,000,000	21	29%	-	-	71%	-	-	-	-
Cameroon	$26,840,000,000	$1,700	$1,900,000,000	$1,700,000,000	10,722	-	96%	-	4%	1,000	4,326,440	-	-
Cape Verde	$600,000,000	$1,400	$30,000,000	$220,000,000	-	-	-	-	-	-	-	-	-
Central African Republic	$4,296,000,000	$1,300	$134,000,000	$102,000,000	10	-	-	-	100%	-	-	-	-
Chad	$9,297,000,000	$1,100	$197,000,000	$570,000,000	-	-	-	-	-	-	-	-	-
Comoros	$441,000,000	$720	$16,300,000	$39,800,000	-	-	-	-	-	-	-	-	-
Congo	$2,500,000,000	$900	$2,400,000,000	$73,000,000	19,097	-	99%	1%	-	-	13,651,000	-	-
Congo, Democratic Republic of the	$34,000,000,000	$610	$1,200,000,000	$890,000,000	2,630	4%	71%	-	25%	96,000	1,194,669	-	-
Cote d'Ivoire	$24,030,000,000	$1,500	$4,400,000,000	$2,500,000,000	4,439	-	50%	45%	5%	-	620,450	-	-
Djibouti	$619,000,000	$1,300	$70,000,000	$255,000,000	-	-	-	-	-	-	-	-	-
Egypt	$289,800,000,000	$3,900	$7,000,000,000	$15,200,000,000	86,315	-	65%	32%	2%	-	38,024,058	1,283,000	-
Equatorial Guinea	$1,270,000,000	$2,700	$2,500,000,000	$562,000,000	7,531	-	100%	-	-	-	7,461,521	-	-
Eritrea	$3,300,000,000	$740	$20,000,000	$500,000,000	-	-	-	-	-	-	-	-	-
Ethiopia	$48,530,000,000	$750	$433,000,000	$1,630,000,000	211	-	-	-	100%	-	-	-	-
Gabon	$8,354,000,000	$5,700	$2,600,000,000	$1,100,000,000	23,273	-	95%	5%	-	-	15,674,359	-	-
Gambia, The	$2,582,000,000	$1,800	$138,000,000	$225,000,000	-	-	-	-	-	-	-	-	-
Ghana	$41,250,000,000	$2,100	$2,200,000,000	$2,800,000,000	830	-	2%	-	98%	-	-	330,933	525,000
Guinea	$18,690,000,000	$2,000	$835,000,000	$670,000,000	25	-	-	-	100%	-	-	-	15,663,000
Guinea-Bissau	$901,400,000	$800	$71,000,000	$59,000,000	-	-	-	-	-	-	-	-	-
Kenya	$32,890,000,000	$1,020	$2,100,000,000	$3,000,000,000	642	-	-	-	100%	-	-	-	-
Lesotho	$5,106,000,000	$2,700	$422,000,000	$738,000,000	(7)	(7)	(7)	(7)	(7)	-	-	-	-
Liberia	$3,116,000,000	$1,100	$110,000,000	$165,000,000	24	-	-	-	100%	-	-	-	-
Libya	$33,360,000,000	$7,600	$11,800,000,000	$6,300,000,000	103,205	-	92%	8%	-	-	67,767,436	-	-
Madagascar	$12,590,000,000	$760	$700,000,000	$985,000,000	64	-	-	-	100%	-	-	-	-
Malawi	$6,811,000,000	$670	$435,000,000	$505,000,000	107	-	-	-	100%	-	-	-	-
Mali	$9,775,000,000	$860	$680,000,000	$630,000,000	29	-	-	-	100%	-	-	-	-
Mauritania	$4,891,000,000	$1,900	$355,000,000	$360,000,000	4	-	-	-	100%	-	-	7,492,000	-
Mauritius	$12,150,000,000	$11,000	$1,600,000,000	$1,800,000,000	12	-	-	-	100%	-	-	-	-
Morocco	$121,800,000,000	$3,900	$7,500,000,000	$10,400,000,000	201	14%	9%	33%	43%	61,000	15,223	4,000	-
Mozambique	$19,520,000,000	$1,000	$680,000,000	$1,180,000,000	874	2%	-	-	98%	18,667	-	-	8,000
Namibia	$13,150,000,000	$6,900	$1,210,000,000	$1,380,000,000	(7)	(7)	(7)	(7)	(7)	-	-	-	-
Niger	$8,713,000,000	$830	$293,000,000	$368,000,000	175	100%	-	-	-	151,666	-	-	-
Nigeria	$112,500,000,000	$875	$17,300,000,000	$13,600,000,000	172,641	-	90%	10%	-	61,000	108,397,478	-	-
Rwanda	$8,920,000,000	$1,200	$68,000,000	$253,000,000	20	-	-	-	100%	-	-	-	-
Sao Tome and Principe	$200,000,000	$1,200	$5,500,000	$24,800,000	1	-	-	-	100%	-	-	-	-
Senegal	$15,640,000,000	$1,500	$1,150,000,000	$1,460,000,000	1	-	-	100%	-	-	-	-	-
Sierra Leone	$2,826,000,000	$580	$35,000,000	$190,000,000	(7)	(7)	(7)	(7)	(7)	-	-	-	-
Somalia	$4,270,000,000	$550	$126,000,000	$343,000,000	-	-	-	-	-	-	-	-	-
South Africa	$427,700,000,000	$10,000	$31,800,000,000	$26,600,000,000	245,195(8)	92%(8)	5%(8)	1%(8)	2%(8)	224,286,505	1,277,485	20,751,000	-
Sudan	$52,900,000,000	$1,420	$1,800,000,000	$1,500,000,000	13,436	-	99%	-	1%	-	7,679,837	-	-
Swaziland	$5,542,000,000	$4,400	$820,000,000	$938,000,000	(7)	(7)	(7)	(7)	(7)	288,665	-	-	-
Tanzania	$20,420,000,000	$630	$863,000,000	$1,670,000,000	343	23%	-	-	77%	5,000	-	-	-
Togo	$7,594,000,000	$1,500	$449,000,000	$561,000,000	-	-	-	-	-	-	-	-	-
Tunisia	$67,130,000,000	$6,500	$6,800,000,000	$8,700,000,000	8,065	-	66%	34%	-	-	3,826,400	105,000	-
Uganda	$30,490,000,000	$1,260	$476,000,000	$1,140,000,000	193	-	-	-	100%	-	-	3,000	-
Zambia	$8,240,000,000	$890	$709,000,000	$1,123,000,000	1,117	15%	-	-	85%	192,358	-	-	-
Zimbabwe	$26,070,000,000	$2,400	$1,570,000,000	$1,739,000,000	4,801	92%	-	-	8%	4,508,643	-	237,000	-
OCEANIA													
Australia	$525,500,000,000	$27,000	$66,300,000,000	$68,000,000,000	331,923	71%	14%	14%	1%	307,176,075	31,728,994	104,014,000	51,834,000
Fiji	$4,822,000,000	$5,500	$442,000,000	$642,000,000	53	-	-	-	100%	-	-	-	-
Kiribati	$79,000,000	$840	$6,000,000	$44,000,000	-	-	-	-	-	-	-	-	-
Micronesia, Federated States of	$277,000,000	$2,000	$22,000,000	$149,000,000	-	-	-	-	-	-	-	-	-
New Zealand	$78,400,000,000	$20,200	$15,000,000,000	$12,500,000,000	19,812	14%	13%	40%	33%	3,452,315	1,839,394	660,000	-
Papua New Guinea	$10,860,000,000	$2,300	$1,800,000,000	$1,100,000,000	5,864	-	96%	2%	2%	-	3,874,601	-	-
Samoa	$1,000,000,000	$5,600	$15,500,000	$130,100,000	3	-	-	-	100%	-	-	-	-
Solomon Islands	$800,000,000	$1,700	$47,000,000	$82,000,000	-	-	-	-	-	-	-	-	-
Tonga	$236,000,000	$2,200	$8,900,000	$70,000,000	-	-	-	-	-	-	-	-	-
Vanuatu	$563,000,000	$2,900	$22,000,000	$93,000,000	-	-	-	-	-	-	-	-	-

This table presents data for most independent nations having an area greater than 200 square miles
- Zero, insignificant, or not available
(1) Source: United States Central Intelligence Agency World Factbook
(2) Source: United Nations Energy Statistics Yearbook
(3) Source: United States Energy Information Administration International Energy Annual
(4) Source: United States Geological Survey Minerals Yearbook
(5) Includes data for Taiwan
(6) Data for Taiwan is included with China
(7) Data for countries in the South Africa Customs Union are included with South Africa
(8) Includes data for countries in the South Africa Customs Union

WORLD ENVIRONMENT TABLE

CONTINENT/Country	Total Area Sq. Miles	Protected Area 2002[1,2] Sq. Miles	%	Mammal	Bird	Endangered Species 2003[3] Reptile	Amphib.	Fish	Invrt.	Forest Cover[4] Sq. Miles 2000	Percent Chan 1990-2000
NORTH AMERICA											
Bahamas	5,382	-	-	5	4	6	0	15	1	3,251	-
Belize	8,867	3,999	45.1%	5	2	4	0	17	1	5,205	-20.9%
Canada	3,855,103	427,916	11.1%	16	8	2	1	25	11	944,294	-
Costa Rica	19,730	4,538	23.0%	13	13	7	1	13	9	7,598	-7.4%
Cuba	42,804	29,578	69.1%	11	18	7	0	23	3	9,066	13.4%
Dominica	290	-	-	1	3	4	0	11	0	178	-8.0%
Dominican Republic	18,730	9,721	51.9%	5	15	10	1	10	2	5,313	-
El Salvador	8,124	33	0.4%	2	0	4	0	5	1	467	-37.3%
Guatemala	42,042	8,408	20.0%	7	6	8	0	14	8	11,004	-15.9%
Haiti	10,714	43	0.4%	4	14	8	1	12	2	340	-44.3%
Honduras	43,277	2,770	6.4%	10	5	6	0	14	2	20,784	-9.9%
Jamaica	4,244	3,590	84.6%	5	12	8	4	12	5	1,255	-14.2%
Mexico	758,452	77,362	10.2%	72	40	18	4	106	41	213,148	-10.3%
Nicaragua	50,054	8,910	17.8%	6	5	7	0	17	2	12,656	-26.3%
Panama	29,157	6,327	21.7%	17	16	7	0	17	2	11,104	-15.3%
St. Lucia	238	-	-	2	5	6	0	10	0	35	-35.7%
Trinidad and Tobago	1,980	119	6.0%	1	1	5	0	15	0	1,000	-7.8%
United States	3,794,083	982,668	25.9%	39	56	27	25	155	557	872,563	1.7%
SOUTH AMERICA											
Argentina	1,073,519	70,852	6.6%	32	39	5	5	9	10	133,777	-7.6%
Bolivia	424,165	56,838	13.4%	25	28	2	1	0	1	204,897	-2.9%
Brazil	3,300,172	221,112	6.7%	74	113	22	6	33	34	2,100,028	-4.1%
Chile	291,930	55,175	18.9%	21	22	0	3	9	0	59,985	-1.3%
Colombia	439,737	44,853	10.2%	39	78	14	0	23	0	191,510	-3.7%
Ecuador	109,484	20,036	18.3%	34	62	10	0	11	48	40,761	-11.5%
Guyana	83,000	249	0.3%	13	2	6	0	13	1	65,170	-2.8%
Paraguay	157,048	5,497	3.5%	10	26	2	0	0	0	90,240	-5.0%
Peru	496,225	30,270	6.1%	46	76	6	1	8	2	251,796	-4.0%
Suriname	63,037	3,089	4.9%	12	1	6	0	12	0	54,491	-
Uruguay	67,574	203	0.3%	6	11	3	0	8	1	4,988	63.3%
Venezuela	352,145	224,669	63.8%	26	24	13	0	19	1	191,144	-4.2%
EUROPE											
Albania	11,100	422	3.8%	3	3	4	0	16	4	3,826	-7.3%
Austria	32,378	10,685	33.0%	7	3	0	0	7	44	15,004	2.0%
Belarus	80,155	5,050	6.3%	7	3	0	0	0	5	36,301	37.5%
Belgium	11,787	-	-	11	2	0	0	7	11	2,811	-1.8%
Bosnia and Herzegovina	19,767	99	0.5%	10	3	1	1	10	10	8,776	-
Bulgaria	42,855	1,928	4.5%	14	10	2	0	10	9	14,247	5.9%
Croatia	21,829	1,637	7.5%	9	4	1	1	26	11	6,884	1.1%
Czech Republic	30,450	4,902	16.1%	8	2	0	0	7	19	10,162	0.2%
Denmark	16,640	5,658	34.0%	5	1	0	0	7	11	1,757	2.2%
Estonia	17,462	2,061	11.8%	5	3	0	0	1	4	7,954	6.5%
Finland	130,559	12,142	9.3%	4	3	0	0	1	10	84,691	0.4%
France	208,482	27,728	13.3%	18	5	3	2	15	65	59,232	4.2%
Germany	137,847	43,973	31.9%	11	5	0	0	12	31	41,467	-
Greece	50,949	1,834	3.6%	13	7	6	1	26	11	13,896	9.1%
Hungary	35,919	2,514	7.0%	9	8	1	0	8	25	7,104	4.1%
Iceland	39,769	3,897	9.8%	7	0	0	0	8	0	120	24.0%
Ireland	27,133	461	1.7%	6	1	0	0	6	3	2,544	34.8%
Italy	116,342	9,191	7.9%	14	5	4	4	16	58	38,622	3.0%
Latvia	24,942	3,342	13.4%	5	3	0	0	3	8	11,286	4.5%
Lithuania	25,213	2,597	10.3%	6	4	0	0	3	5	7,699	2.5%
Luxembourg	999	-	-	3	1	0	0	0	4	-	-
Macedonia	9,928	705	7.1%	11	3	2	0	4	5	3,498	-
Moldova	13,070	183	1.4%	6	5	1	0	9	5	1,255	2.2%
Netherlands	16,164	2,295	14.2%	10	4	0	0	7	7	1,448	2.7%
Norway	125,050	8,503	6.8%	10	2	0	0	7	9	34,240	3.6%
Poland	120,728	14,970	12.4%	14	4	0	0	3	15	34,931	2.0%
Portugal	35,516	2,344	6.6%	17	7	0	1	19	82	14,154	18.4%
Romania	91,699	4,310	4.7%	17	8	2	0	10	22	24,896	2.3%
Serbia and Montenegro	39,449	1,302	3.3%	12	5	1	0	19	19	11,147	-0.5%
Slovakia	18,924	4,315	22.8%	9	4	1	0	8	19	8,405	9.0%
Slovenia	7,821	469	6.0%	9	1	0	1	15	42	4,274	2.0%
Spain	194,885	16,565	8.5%	24	7	7	3	23	63	55,483	6.4%
Sweden	173,732	15,810	9.1%	6	2	0	0	6	13	104,765	-
Switzerland	15,943	4,783	30.0%	5	2	0	0	4	30	4,629	3.7%
Ukraine	233,090	9,091	3.9%	16	8	2	0	11	14	37,004	3.3%
United Kingdom	93,788	19,602	20.9%	12	2	0	0	11	10	10,788	6.5%
Russia	6,592,849	514,242	7.8%	45	38	6	0	18	30	3,287,242	0.2%
ASIA											
Afghanistan	251,773	755	0.3%	13	11	1	1	0	1	5,216	-
Armenia	11,506	874	7.6%	11	4	5	0	1	7	1,355	13.6%
Azerbaijan	33,437	2,040	6.1%	13	8	5	0	5	6	4,224	13.5%
Bahrain	267	-	-	1	6	0	0	6	0	-	-
Bangladesh	55,598	445	0.8%	22	23	20	0	8	0	5,151	14.1%
Brunei	2,226	-	-	11	14	4	0	6	0	1,707	-2.2%
Cambodia	69,898	12,931	18.5%	24	19	10	0	11	0	36,043	-5.7%
China	3,690,045	287,824	7.8%	81	75	31	1	46	4	631,200	12.4%
Cyprus	3,572	-	-	3	3	3	0	6	0	664	44.5%
East Timor	5,743	-	-	0	6	0	0	2	0	1,958	-6.3%
Georgia	26,911	619	2.3%	13	3	7	1	6	10	11,537	-
India	1,222,510	63,571	5.2%	86	72	25	3	27	23	247,542	0.6%
Indonesia	735,310	151,474	20.6%	147	114	28	0	91	31	405,353	-11.1%
Iran	636,372	30,546	4.8%	22	13	8	2	14	3	28,182	-
Iraq	169,235			11	11	2	0	3	2	3,085	-
Israel	8,019	1,267	15.8%	15	12	4	0	10	10	510	61.0%
Japan	145,850	9,918	6.8%	37	35	11	10	27	45	92,977	0.1%
Jordan	34,495	1,173	3.4%	9	8	1	0	5	3	332	-
Kazakhstan	1,049,156	28,327	2.7%	17	15	2	1	7	4	46,904	24.5%
Korea, North	46,540	1,210	2.6%	13	19	0	0	5	1	31,699	-
Korea, South	38,328	2,645	6.9%	13	25	0	0	7	1	24,124	-0.8%

CONTINENT/Country	Total Area Sq. Miles	Protected Area 2002[1,2] Sq. Miles	%	Endangered Species 2003[3] Mammal	Bird	Reptile	Amphib.	Fish	Invrt.	Forest Cover[4] Sq. Miles 2000	Percent Change 1990-2000
Kuwait	6,880	103	1.5%	1	7	1	0	6	0	19	66.7%
Kyrgyzstan	77,182	2,779	3.6%	7	4	2	0	0	3	3,873	29.4%
Laos	91,429	11,429	12.5%	31	20	11	0	6	0	48,498	-4.0%
Lebanon	4,016	20	0.5%	6	7	1	0	8	1	139	-2.7%
Malaysia	127,320	7,257	5.7%	50	37	21	0	34	3	74,487	52.4%
Mongolia	604,829	69,555	11.5%	14	16	0	0	1	3	41,101	-5.3%
Myanmar	261,228	784	0.3%	39	35	20	0	7	2	132,892	-13.1%
Nepal	56,827	5,058	8.9%	29	25	6	0	0	1	15,058	-16.7%
Oman	119,499	16,730	14.0%	11	10	4	0	17	1	4	
Pakistan	339,732	16,647	4.9%	17	17	9	0	14	0	9,116	-14.3%
Philippines	115,831	6,602	5.7%	50	67	8	23	48	19	22,351	-13.3%
Qatar	4,412	-	-	0	6	1	0	4	0	-	
Saudi Arabia	830,000	317,890	38.3%	9	15	2	0	8	1	5,807	-
Singapore	264	13	4.9%	3	7	3	0	12	1	8	
Sri Lanka	25,332	3,420	13.5%	22	14	8	0	22	2	7,490	-15.2%
Syria	71,498	-	-	4	8	3	0	8	3	1,780	
Taiwan	13,901	-	-	12	21	8	0	23	0	-	
Tajikistan	55,251	2,321	4.2%	9	7	1	0	3	2	1,544	5.3%
Thailand	198,115	27,538	13.9%	37	37	19	0	35	1	56,996	-7.1%
Turkey	302,541	4,841	1.6%	17	11	12	3	29	13	39,479	2.2%
Turkmenistan	188,457	7,915	4.2%	13	6	2	0	8	5	14,498	-
United Arab Emirates	32,278	-	-	4	8	1	0	6	0	1,239	32.1%
Uzbekistan	172,742	3,455	2.0%	9	9	2	0	4	1	7,602	2.4%
Vietnam	128,066	4,738	3.7%	42	37	24	1	22	0	37,911	5.5%
Yemen	203,850	-	-	6	12	2	0	10	2	1,734	-17.0%
AFRICA											
Algeria	919,595	45,980	5.0%	13	6	2	0	9	12	8,282	14.2%
Angola	481,354	31,769	6.6%	19	15	4	0	8	6	269,329	-1.7%
Benin	43,484	4,957	11.4%	9	2	1	0	7	0	10,232	-20.9%
Botswana	224,607	41,552	18.5%	7	7	0	0	0	0	47,981	-8.7%
Burkina Faso	105,869	12,175	11.5%	7	2	1	0	0	0	27,371	-2.1%
Burundi	10,745	612	5.7%	6	7	0	0	0	3	363	-61.0%
Cameroon	183,568	8,261	4.5%	38	15	1	1	34	4	92,116	-8.5%
Cape Verde	1,557	-	-	3	2	0	0	13	0	328	142.9%
Central African Republic	240,536	20,927	8.7%	14	3	1	0	0	0	88,444	-1.3%
Chad	495,755	45,114	9.1%	15	5	1	0	0	1	49,004	-6.0%
Comoros	863	-	-	2	9	2	0	3	4	31	-33.3%
Congo	132,047	6,602	5.0%	15	3	1	0	9	1	85,174	-0.8%
Congo, Democratic Republic of the	905,446	58,854	6.5%	40	28	2	0	9	45	522,037	-3.8%
Cote d'Ivoire	124,504	7,470	6.0%	19	12	2	1	10	1	27,479	-27.1%
Djibouti	8,958	-	-	5	5	0	0	9	0	23	-
Egypt	386,662	37,506	9.7%	13	7	6	0	13	1	278	38.5%
Equatorial Guinea	10,831	-	-	16	5	2	1	7	2	6,765	-5.7%
Eritrea	45,406	1,952	4.3%	12	7	6	0	8	0	6,120	-3.3%
Ethiopia	426,373	72,057	16.9%	35	16	1	0	0	4	17,734	-8.1%
Gabon	103,347	723	0.7%	14	5	1	0	11	1	84,271	-0.5%
Gambia, The	4,127	95	2.3%	3	2	1	0	10	0	1,857	10.3%
Ghana	92,098	5,157	5.6%	14	8	2	0	7	0	24,460	-15.9%
Guinea	94,926	664	0.7%	12	10	1	1	7	3	26,753	-4.8%
Guinea-Bissau	13,948	-	-	3	0	1	0	9	1	8,444	-9.0%
Kenya	224,961	17,997	8.0%	50	24	5	0	27	15	66,008	-5.2%
Lesotho	11,720	23	0.2%	6	7	0	0	1	1	54	-
Liberia	43,000	731	1.7%	16	11	2	0	7	2	13,440	-17.9%
Libya	679,362	679	0.1%	8	1	3	0	8	0	1,382	15.1%
Madagascar	226,658	9,746	4.3%	50	27	18	2	25	32	45,278	-9.1%
Malawi	45,747	5,124	11.2%	8	11	0	0	0	8	9,892	-21.6%
Mali	478,841	17,717	3.7%	13	4	1	0	1	0	50,911	-7.0%
Mauritania	397,956	6,765	1.7%	10	2	2	0	10	1	1,224	-23.6%
Mauritius	788	-	-	3	9	4	0	7	32	62	-5.9%
Morocco	172,414	1,207	0.7%	16	9	2	0	10	8	11,680	-0.4%
Mozambique	309,496	25,998	8.4%	15	16	5	0	19	7	118,151	-2.0%
Namibia	317,818	43,223	13.6%	14	11	3	1	11	1	31,043	-8.4%
Niger	489,192	37,668	7.7%	11	3	0	0	0	1	5,127	-31.7%
Nigeria	356,669	11,770	3.3%	27	9	2	0	11	1	52,189	-22.8%
Rwanda	10,169	630	6.2%	8	9	0	0	0	2	1,185	-32.8%
Sao Tome and Principe	372	-	-	3	9	1	0	6	2	104	-
Senegal	75,951	8,810	11.6%	12	4	6	0	17	0	23,958	-6.8%
Sierra Leone	27,699	582	2.1%	12	10	3	0	7	4	4,073	-25.5%
Somalia	246,201	1,970	0.8%	19	10	2	0	16	1	29,016	-9.3%
South Africa	470,693	25,888	5.5%	36	28	19	9	47	113	34,429	-0.9%
Sudan	967,500	50,310	5.2%	22	6	2	0	7	1	237,943	-13.5%
Swaziland	6,704	-	-	5	5	0	0	0	0	2,015	12.5%
Tanzania	364,900	108,740	29.8%	41	33	5	0	26	47	149,850	-2.3%
Togo	21,925	1,732	7.9%	9	0	2	0	7	0	1,969	-29.1%
Tunisia	63,170	190	0.3%	11	5	3	0	8	5	1,969	2.2%
Uganda	93,065	22,894	24.6%	20	13	0	0	27	10	16,178	-17.9%
Zambia	290,586	92,697	31.9%	11	11	0	0	0	6	120,641	-21.4%
Zimbabwe	150,873	18,256	12.1%	11	10	0	0	0	2	73,514	-14.4%
OCEANIA											
Australia	2,969,910	397,968	13.4%	63	35	38	35	74	282	596,678	-1.8%
Fiji	7,056	78	1.1%	5	13	6	1	8	2	3,147	-2.0%
Kiribati	313	-	-	0	4	1	0	4	1	108	-
Micronesia, Federated States of	271	-	-	6	5	2	0	6	4	58	-37.5%
New Zealand	104,454	30,918	29.6%	8	63	11	1	16	13	30,680	5.2%
Papua New Guinea	178,704	4,110	2.3%	58	32	9	0	31	12	118,151	-3.6%
Samoa	1,093	-	-	3	8	1	0	4	1	405	-19.2%
Solomon Islands	10,954	33	0.3%	20	23	4	0	4	6	9,792	-1.7%
Tonga	251	-	-	2	3	2	0	3	2	15	-
Vanuatu	4,707	-	-	5	8	2	0	4	0	1,726	1.4%

This table presents data for most independent nations having an area greater than 200 square miles
- Zero, insignificant, or not available

(1) Source: World Resources Institute, 2003. Earth Trends: The Environmental Information Portal. Available at http://earthtrends.wri.org. Washington D. C. World Resources Institute
(2) Source: United Nations Environment Programme - World Conservation Monitoring Centre (UNEP-WCMC); World Database on Protected Areas
(3) Source: International Union of Conservation of Nature and Natural Resources; IUCN 2003 Red List of Threatened Species <www.redlist.org>
(4) Source: United Nations Food and Agriculture Organization; Global Forest Resources Assessment 2000

WORLD COMPARISONS

General Information

Equatorial diameter of the earth, 7,926.38 miles.
Polar diameter of the earth, 7,899.80 miles.
Mean diameter of the earth, 7,917.52 miles.
Equatorial circumference of the earth, 24,901.46 miles.
Polar circumference of the earth, 24,855.34 miles.
Mean distance from the earth to the sun, 93,020,000 miles.
Mean distance from the earth to the moon, 238,857 miles.
Total area of the earth, 197,000,000 sq. miles.

Highest elevation on the earth's surface, Mt. Everest, Asia, 29,028 ft.
Lowest elevation on the earth's land surface, shores of the Dead Sea, Asia, 1,339 ft. below sea level.
Greatest known depth of the ocean, southwest of Guam, Pacific Ocean, 35,810 ft.
Total land area of the earth (incl. inland water and Antarctica), 57,900,000 sq. miles.

Area of Africa, 11,700,000 sq. miles.
Area of Antarctica, 5,400,000 sq. miles.
Area of Asia, 17,300,000 sq. miles.
Area of Europe, 3,800,000 sq. miles.
Area of North America, 9,500,000 sq. miles.
Area of Oceania (incl. Australia) 3,300,000 sq. miles.
Area of South America, 6,900,000 sq. miles.
Population of the earth (est. 1/1/04), 6,339,505,000.

Principal Islands and Their Areas

ISLAND	Area (Sq. Mi.)
Baffin I., Canada	195,928
Banks I., Canada	27,038
Borneo (Kalimantan), Asia	287,300
Bougainville, Papua New Guinea	3,591
Cape Breton I., Canada	3,981
Celebes (Sulawesi), Indonesia	73,057
Ceram (Seram), Indonesia	7,191
Corsica, France	3,367
Crete, Greece	3,189
Cuba, N. America	42,780
Cyprus, Asia	3,572
Devon I., Canada	21,331
Ellesmere I., Canada	75,767
Flores, Indonesia	5,502
Great Britain, U.K.	88,795
Greenland, N. America	840,000
Guadalcanal, Solomon Is.	2,060
Hainan Dao, China	13,127
Hawaii, U.S.	4,028
Hispaniola, N. America	29,300
Hokkaidō, Japan	32,245
Honshū, Japan	89,176
Iceland, Europe	39,769
Ireland, Europe	32,587
Jamaica, N. America	4,247
Java (Jawa), Indonesia	51,038
Kodiak I., U.S.	3,670
Kyūshū, Japan	17,129
Lyete, Philippines	2,785
Long Island, U.S.	1,377
Luzon, Philippines	40,420
Madagascar, Africa	226,642
Melville I., Canada	16,274
Mindanao, Philippines	36,537
Mindoro, Philippines	3,759
Negros, Philippines	4,907
New Britain, Papua New Guinea	14,093
New Caledonia, Oceania	6,252
Newfoundland, Canada	42,031
New Guinea, Asia-Oceania	308,882
New Ireland, Papua New Guinea	3,475
North East Land, Norway	6,350
North I., New Zealand	44,333
Novaya Zemlya, Russia	31,892
Palawan, Philippines	4,550
Panay, Philippines	4,446
Prince of Wales I., Canada	12,872
Puerto Rico, N. America	3,514
Sakhalin, Russia	29,498
Samar, Philippines	5,050
Sardinia, Italy	9,301
Shikoku, Japan	7,258
Sicily, Italy	9,926
Somerset I., Canada	9,570
Southampton I., Canada	15,913
South I., New Zealand	57,708
Spitsbergen, Norway	15,260
Sri Lanka, Asia	24,942
Sumatra (Sumatera), Indonesia	182,860
Taiwan, Asia	13,900
Tasmania, Australia	26,178
Tierra del Fuego, S. America	18,600
Timor, Asia	5,743
Vancouver I., Canada	12,079
Victoria I., Canada	83,897
Vrangelya (Wrangel), Russia	2,819

Principal Lakes, Oceans, Seas, and Their Areas

LAKE Country	Area (Sq. Mi.)
Arabian Sea	1,492,000
Aral Sea, Kazakhstan-Uzbekistan	13,000
Arctic Ocean	5,400,000
Athabasca, L., Canada	3,064
Atlantic Ocean	29,600,000
Balqash köli (L. Balkhash), Kazakhstan	7,027
Baltic Sea, Europe	163,000
Baykal, Ozero (L. Baikal), Russia	12,162
Bering Sea, Asia-N.A.	876,000
Black Sea, Europe-Asia	178,000
Caribbean Sea, N.A.-S.A.	1,063,000
Caspian Sea, Asia-Europe	144,402
Chad, L., Cameroon-Chad-Nigeria	595
Erie, L., Canada-U.S.	9,910
Eyre, L., Australia	3,668
Gairdner, L., Australia	1,076
Great Bear Lake, Canada	12,096
Great Salt Lake, U.S.	1,700
Great Slave Lake, Canada	11,030
Hudson Bay, Canada	475,000
Huron, L., Canada-U.S.	23,000
Indian Ocean	26,500,000
Japan, Sea of, Asia	389,000
Koko Nor (Qinghai Hu), China	1,722
Ladozhskoye Ozero (L. Ladoga), Russia	7,002
Manitoba, L., Canada	1,785
Mediterranean Sea, Europe-Africa-Asia	967,000
Mexico, Gulf of, N. America	596,000
Michigan, L., U.S.	22,300
Nicaragua, Lago de, Nicaragua	3,147
North Sea, Europe	222,000
Nyasa, L., Malawi-Mozambique-Tanzania	11,120
Onezhskoye Ozero (L. Onega), Russia	3,819
Ontario, L., Canada-U.S.	7,340
Pacific Ocean	60,100,000
Red Sea, Africa-Asia	169,000
Rudolf, L., Ethiopia-Kenya	2,471
Southern Ocean	7,800,000
Superior, L., Canada-U.S.	31,700
Tanganyika, L., Africa	12,355
Titicaca, Lago, Bolivia-Peru	3,232
Torrens, L., Australia	1,076
Vänern (L.), Sweden	2,181
Van Gölü (L.), Turkey	1,434
Victoria, L., Kenya-Tanzania-Uganda	26,564
Winnipeg, L., Canada	9,416
Winnipegosis, L., Canada	2,075
Yellow Sea, China-Korea	480,000

Principal Mountains and Their Heights

MOUNTAIN Country	Elev. (Ft.)
Aconcagua, Cerro, Argentina	22,831
Annapurna, Nepal	26,504
Aoraki, New Zealand	12,316
Api, Nepal	23,399
Apo, Philippines	9,692
Ararat, Mt., Turkey	16,854
Barú, Volcán, Panama	11,401
Bangueta, Mt., Papua New Guinea	13,520
Belukha, Mt., Kazakhstan-Russia	14,783
Bia, Phou, Laos	9,249
Blanc, Mont (Monte Bianco), France-Italy	15,771
Blanca Pk., Colorado, U.S.	14,345
Bolívar, Pico, Venezuela	16,427
Bonete, Cerro, Argentina	22,546
Borah Pk., Idaho, U.S.	12,662
Boundary Pk., Nevada, U.S.	13,140
Cameroon Mtn., Cameroon	13,451
Carrauntoohil, Ireland	3,406
Chaltel, Cerro (Monte Fitzroy), Argentina-Chile	10,958
Chimborazo, Ecuador	20,702
Chirripó, Cerro, Costa Rica	12,530
Colima, Nevado de, Mexico	13,911
Cotopaxi, Ecuador	19,347
Cristóbal Colón, Pico, Colombia	19,029
Damāvand, Qolleh-ye, Iran	18,386
Dhawalāgiri, Nepal	26,810
Duarte, Pico, Dominican Rep.	10,417
Dufourspitze (Monte Rosa), Italy-Switzerland	15,203
Elbert, Mt., Colorado, U.S.	14,433
El'brus, Gora, Russia	18,510
Elgon, Mt., Kenya-Uganda	14,178
Erciyeş, Dağı, Turkey	12,848
Etna, Mt., Italy	10,902
Everest, Mt., China-Nepal	29,028
Fairweather, Mt., Alaska-Canada	15,300
Folādi, Koh-e, Afghanistan	16,847
Foraker, Mt., Alaska, U.S.	17,400
Fuji San, Japan	12,388
Galdhøpiggen, Norway	8,100
Gannett Pk., Wyoming, U.S.	13,804
Gasherbrum, China-Pakistan	26,470
Gerlachovský štít, Slovakia	8,711
Giluwe, Mt., Papua New Guinea	14,331
Gongga Shan, China	24,790
Grand Teton, Wyoming, U.S.	13,770
Grossglockner, Austria	12,457
Hadūr Shu'ayb, Yemen	12,008
Haleakalā Crater, Hawaii, U.S.	10,023
Hekla, Iceland	4,892
Hood, Mt., Oregon, U.S.	11,239
Huascarán, Nevado, Peru	22,133
Huila, Nevado de, Colombia	18,865
Hvannadalshnúkur, Iceland	6,952
Illampu, Nevado, Bolivia	21,066
Illimani, Nevado, Bolivia	20,741
Ismail Samani, pik, Tajikistan	24,590
Iztaccíhuatl, Mexico	17,159
Jaya, Puncak, Indonesia	16,503
Jungfrau, Switzerland	13,642
K2 (Qogir Feng), China-Pakistan	28,250
Kāmet, China-India	25,447
Kānchenjunga, India-Nepal	28,208
Kātrīnā, Jabal, Egypt	8,668
Kebnekaise, Sweden	6,926
Kenya, Mt. (Kirinyaga), Kenya	17,058
Kerinci, Gunung, Indonesia	12,467
Kilimanjaro, Tanzania	19,340
Kinabalu, Gunong, Malaysia	13,455
Klyuchevskaya, Russia	15,584
Kosciuszko, Mt., Australia	7,313
Koussi, Emi, Chad	11,204
Kula Kangri, Bhutan	24,784
La Selle, Massif de, Haiti	8,793
Lassen Pk., California, U.S.	10,457
Llullaillaco, Volcán, Argentina-Chile	22,110
Logan, Mt., Canada	19,551
Longs Pk., Colorado, U.S.	14,255
Makalu, China-Nepal	27,825
Margherita Peak, Dem. Rep. of the Congo-Uganda	16,763
Markham, Mt., Antarctica	14,049
Maromokotro, Madagascar	9,436
Massive, Mt., Colorado, U.S.	14,421
Matterhorn, Italy-Switzerland	14,692
Mauna Kea, Hawaii, U.S.	13,796
Mauna Loa, Hawaii, U.S.	13,679
Mayon Volcano, Philippines	8,077
McKinley, Mt., Alaska, U.S.	20,320
Meron, Hare, Israel	3,963
Meru, Mt., Tanzania	14,978
Misti, Volcán, Peru	19,101
Mitchell, Mt., North Carolina, U.S.	6,684
Môco, Serra do, Angola	8,596
Moldoveanu, Romania	8,346
Mulhacén, Spain	11,424
Musala, Bulgaria	9,596
Muztag, China	25,338
Muztagata, China	24,757
Namjagbarwa Feng, China	25,446
Nanda Devi, India	25,645
Nanga Parbat, Pakistan	26,660
Narodnaya, Gora, Russia	6,217
Nevis, Ben, United Kingdom	4,406
Ojos del Salado, Nevado, Argentina-Chile	22,615
Ólimbos, Cyprus	6,401
Ólympos, Greece	9,570
Olympus, Mt., Washington, U.S.	7,965
Orizaba, Pico de, Mexico	18,406
Paektu San, North Korea-China	9,003
Paricutín, Mexico	9,186
Parnassós, Greece	8,061
Pelée, Montagne, Martinique	4,583
Pidurutalagala, Sri Lanka	8,281
Pikes Pk., Colorado, U.S.	14,110
Pobedy, pik, China-Kyrgyzstan	24,406
Popocatépetl, Volcán, Mexico	17,930
Pulog, Mt., Philippines	9,626
Rainier, Mt., Washington, U.S.	14,410
Ramm, Jabal, Jordan	5,755
Ras Dashen Terara, Ethiopia	15,158
Rinjani, Gunung, Indonesia	12,224
Robson, Mt., Canada	12,972
Roraima, Mt., Brazil-Guyana-Venezuela	9,432
Ruapehu, Mt., New Zealand	9,177
St. Elias, Mt., Alaska, U.S.-Canada	18,008
Sajama, Nevado, Bolivia	21,391
Semeru, Gunung, Indonesia	12,060
Shām, Jabal ash, Oman	9,957
Shasta, Mt., California, U.S.	14,162
Snowdon, United Kingdom	3,560
Tahat, Algeria	9,541
Tajumulco, Guatemala	13,845
Taranaki, Mt., New Zealand	8,260
Tirich Mīr, Pakistan	25,230
Tomanivi (Victoria), Fiji	4,341
Toubkal, Jebel, Morocco	13,665
Triglav, Slovenia	9,396
Trikora, Puncak, Indonesia	15,584
Tupungato, Cerro, Argentina-Chile	21,555
Turquino, Pico, Cuba	6,470
Uluru (Ayers Rock), Australia	2,844
Uncompahgre Pk, Colorado, U.S.	14,309
Vesuvio (Vesuvius), Italy	4,190
Victoria, Mt., Papua New Guinea	13,238
Vinson Massif, Antarctica	16,066
Waddington, Mt., Canada	13,163
Washington, Mt., New Hampshire, U.S.	6,288
Whitney, Mt., California, U.S.	14,494
Wilhelm, Mt., Papua New Guinea	14,793
Wrangell, Mt., Alaska, U.S.	14,163
Xixabangma Feng (Gosainthan), China	26,286
Yü Shan, Taiwan	13,114
Zugspitze, Austria-Germany	9,718

Principal Rivers and Their Lengths

RIVER Continent	Length (Mi.)
Albany, N. America	610
Aldan, Asia	1,412
Amazonas-Ucayali, S. America	4,000
Amu Darya, Asia	1,578
Amur, Asia	1,752
Araguaia, S. America	1,367
Arkansas, N. America	1,460
Atchafalaya, N. America	1,420
Athabasca, N. America	765
Brahmaputra, Asia	1,770
Brazos, N. America	1,280
Canadian, N. America	906
Churchill, N. America	1,000
Colorado, N. America (U.S.-Mexico)	1,450
Colorado, N. America (Texas)	862
Columbia, N. America	1,240
Congo (Zaïre), Africa	2,715
Danube, Europe	1,777
Darling, Australia	864
Dnieper (Dnipro), Europe	1,367
Don, Europe	1,162
Elbe, Europe	690
Essequibo, S. America	603
Euphrates, Asia	1,510
Fraser, N. America	851
Ganges, Asia	1,560
Gila, N. America	649
Godāvari, Asia	932
Huang (Yellow), Asia	2,902
Indigirka, Asia	1,072
Indus, Asia	1,118
Irrawaddy, Asia	1,300
Juruá, S. America	1,250
Kama, Europe	1,122
Kasai, Africa	1,338
Kolyma, Asia	1,323
Lena, Asia	2,734
Limpopo, Africa	1,100
Loire, Europe	690
Mackenzie, N. America	2,635
Madeira, S. America	2,013
Magdalena, S. America	951
Marañón, S. America	1,000
Mekong, Asia	2,796
Meuse, Europe	575
Mississippi, N. America	2,340
Mississippi-Missouri, N. America	3,710
Missouri, N. America	2,540
Murray-Darling, Australia	2,169
Negro, S. America	1,305
Nelson, N. America	1,600
Niger, Africa	2,585
Nile, Africa	4,132
Ob', Asia	2,268
Oder, Europe	565
Ohio, N. America	1,310
Oka, Europe	932
Orange, Africa	1,300
Orinoco, S. America	1,703
Ottawa, N. America	790
Paraguay, S. America	1,610
Parnaíba, S. America	901
Peace, N. America	1,195
Pechora, Europe	1,125
Pecos, N. America	926
Pilcomayo, S. America	1,550
Plata-Paraná, S. America	2,920
Platte, N. America	990
Purús, S. America	1,860
Red, N. America	1,290
Rhine, Europe	820
Rhône, Europe	503
Rio Grande, N. America	1,900
Roosevelt, S. America	950
St. Lawrence, N. America	1,900
Salado, S. America	870
Salween (Nu), Asia	1,750
São Francisco, S. America	1,740
Saskatchewan-Bow, N. America	1,205
Severnaya Dvina (Northern Dvina), Europe	462
Snake, N. America	1,040
Sungari (Songhua), Asia	1,140
Syr Darya, Asia	1,370
Tagus, Europe	625
Tarim, Asia	1,328
Tennessee, N. America	886
Tigris, Asia	1,180
Tisa, Europe	607
Tocantins, S. America	1,640
Ucayali, S. America	1,220
Ural, Asia	1,509
Uruguay, S. America	1,025
Verkhnyaya Tunguska (Angara), Asia	1,105
Vilyuy, Asia	1,647
Volga, Europe	2,082
Volta, Africa	994
Wisła (Vistula), Europe	630
Xiang, Asia	930
Xingu, S. America	1,230
Yangtze (Chang), Asia	3,915
Yellowstone, N. America	692
Yenisey, Asia	2,169
Yukon, N. America	1,980
Zambezi, Africa	1,653

PRINCIPAL CITIES OF THE WORLD

Abidjan, Cote d'Ivoire ... 1,929,079
Abū Ẓaby (Abu Dhabi), United Arab Emirates ... 242,975
Accra, Ghana (1,390,000) ... 949,113
Addis Ababa, Ethiopia ... 2,424,000
Ahmadābād, India (4,519,278) ... 3,515,361
Aleppo (Ḥalab), Syria (1,640,000) ... 1,591,400
Alexandria (Al Iskandarīyah), Egypt (3,350,000) ... 3,339,076
Algiers (El Djazaïr), Algeria (2,547,983) ... 1,507,241
Al Jīzah (Giza), Egypt (*Al Qāhirah) ... 2,221,817
Almaty, Kazakhstan (1,190,000) ... 1,129,356
'Ammān, Jordan (1,500,000) ... 1,147,447
Amsterdam, Netherlands (1,121,303) ... 727,053
Ankara, Turkey (3,294,220) ... 2,984,099
Antananarivo, Madagascar ... 1,250,000
Antwerp (Antwerpen), Belgium (1,135,000) ... 453,030
Ashgabat (Ashkhabad), Turkmenistan ... 557,600
Asmera, Eritrea ... 358,100
Astana (Aqmola), Kazakhstan (319,324) ... 312,965
Asunción, Paraguay (700,000) ... 546,637
Athens (Athína), Greece (3,150,000) ... 772,072
Atlanta, Georgia, U.S. (4,112,198) ... 416,474
Auckland, New Zealand (1,074,510) ... 367,737
Baghdād, Iraq ... 3,841,268
Baku (Bakı), Azerbaijan (2,020,000) ... 1,792,300
Bamako, Mali ... 658,275
Bandung, Indonesia ... 5,919,400
Bangalore, India (5,686,844) ... 4,292,223
Banghāzī, Libya ... 800,000
Bangkok (Krung Thep), Thailand (7,060,000) ... 5,620,591
Bangui, Central African Republic ... 451,690
Barcelona, Spain (4,000,000) ... 1,496,266
Beijing, China (7,320,000) ... 6,690,000
Beirut (Bayrūt), Lebanon (1,675,000) ... 509,000
Belfast, N. Ireland, U.K. (730,000) ... 297,300
Belgrade (Beograd), Serbia and Montenegro ... 1,594,483
Belo Horizonte, Brazil (4,055,000) ... 1,366,301
Berlin, Germany (4,220,000) ... 3,386,667
Birmingham, England, U.K. (2,705,000) ... 1,020,589
Bishkek, Kyrgyzstan ... 753,400
Bogotá, Colombia ... 6,422,198
Bonn, Germany (600,000) ... 301,048
Boston, Massachusetts, U.S. (5,819,100) ... 589,141
Brasília, Brazil ... 1,947,133
Bratislava, Slovakia ... 451,395
Brazzaville, Congo ... 693,712
Brisbane, Australia (1,627,535) ... 888,449
Brussels (Bruxelles), Belgium (2,390,000) ... 133,845
Bucharest (București), Romania (2,300,000) ... 2,016,131
Budapest, Hungary (2,450,000) ... 1,825,153
Buenos Aires, Argentina (11,000,000) ... 2,960,976
Cairo (Al Qāhirah), Egypt (9,300,000) ... 6,800,992
Calgary, Alberta, Canada (951,395) ... 878,866
Cali, Colombia ... 2,128,920
Canberra, Australia (342,798) ... 311,518
Cape Town, South Africa (1,900,000) ... 854,616
Caracas, Venezuela (4,000,000) ... 1,822,465
Cardiff, Wales, U.K. (645,000) ... 315,040
Casablanca, Morocco (3,400,000) ... 3,022,000
Changchun, China ... 2,470,000
Chelyabinsk, Russia (1,320,000) ... 1,086,300
Chengdu, China ... 2,760,000
Chennai (Madras), India (6,424,624) ... 4,216,268
Chicago, Illinois, U.S. (9,157,540) ... 2,896,016
Chișinău (Kishinev), Moldova (746,500) ... 658,300
Chittagong, Bangladesh (2,342,662) ... 1,566,070
Chongqing, China ... 3,870,000
Cincinnati, Ohio, U.S. (1,979,202) ... 331,285
Cleveland, Ohio, U.S. (2,945,831) ... 478,403
Cologne (Köln), Germany (1,830,000) ... 962,507
Colombo, Sri Lanka (2,050,000) ... 615,000
Conakry, Guinea ... 950,000
Copenhagen (København), Denmark (2,030,000) ... 499,148
Córdoba, Argentina (1,260,000) ... 1,179,067
Cotonou, Benin ... 650,660
Curitiba, Brazil (2,595,000) ... 1,586,848
Dakar, Senegal (1,976,533) ... 879,703
Dalian, China ... 2,400,000
Dallas, Texas, U.S. (5,221,801) ... 1,188,580
Damascus (Dimashq), Syria (2,230,000) ... 1,549,932
Dar es Salaam, Tanzania ... 1,360,850
Delhi, India (12,791,458) ... 9,817,439
Denver, Colorado, U.S. (2,581,506) ... 554,636
Detroit, Michigan, U.S. (5,456,428) ... 951,270
Dhaka (Dacca), Bangladesh (6,537,308) ... 3,637,892
Djibouti, Djibouti ... 329,337
Dnipropetrovs'k, Ukraine (1,590,000) ... 1,108,682
Donets'k, Ukraine (2,090,000) ... 1,050,369
Douala, Cameroon ... 712,251
Dublin (Baile Átha Cliath), Ireland (1,175,000) ... 481,854
Durban, South Africa (1,740,000) ... 669,242
Dushanbe, Tajikistan (700,000) ... 528,600
Düsseldorf, Germany (1,200,000) ... 568,855
Edinburgh, Scotland, U.K. (640,000) ... 448,850
Edmonton, Alberta, Canada (937,845) ... 666,104
Eşfahān, Iran (1,525,000) ... 1,266,072
Essen, Germany (5,040,000) ... 599,515
Fortaleza, Brazil (2,780,000) ... 788,956
Frankfurt am Main, Germany (1,960,000) ... 643,821
Fukuoka, Japan (2,000,000) ... 1,341,489
Geneva (Génève), Switzerland (450,592) ... 172,598
Glasgow, Scotland, U.K. (1,870,000) ... 616,430
Goiânia, Brazil ... 1,075,761
Guadalajara, Mexico (3,669,021) ... 1,646,183
Guangzhou (Canton), China ... 3,750,000
Guatemala, Guatemala (1,500,000) ... 1,006,954
Guayaquil, Ecuador ... 2,117,553
Halifax, Nova Scotia, Canada (359,183) ... 119,300
Hamburg, Germany (2,460,000) ... 1,704,735
Hannover, Germany (1,015,000) ... 514,718
Hanoi, Vietnam (1,275,000) ... 1,073,760
Harare, Zimbabwe (1,470,000) ... 1,189,103
Harbin, China ... 3,120,000
Havana (La Habana), Cuba (2,285,000) ... 2,189,716
Helsinki, Finland (939,697) ... 548,720
Hiroshima, Japan (1,600,000) ... 1,126,282
Ho Chi Minh City (Saigon), Vietnam (3,300,000) ... 3,015,743
Hong Kong (Xianggang), China (4,770,000) ... 1,250,993
Honolulu, Hawaii, U.S. (876,156) ... 371,657
Houston, Texas, U.S. (4,669,571) ... 1,953,631
Hyderābād, India (5,533,640) ... 3,449,878
Ibadan, Nigeria ... 1,144,000
Islāmābād, Pakistan (*Rāwalpindi) ... 529,180
İstanbul, Turkey (8,506,026) ... 8,260,438
İzmir, Turkey (2,554,363) ... 2,081,556
Jaipur, India ... 2,324,319
Jakarta, Indonesia (10,200,000) ... 9,373,900
Jerusalem (Yerushalayim), Israel (685,000) ... 633,700
Jiddah, Saudi Arabia ... 1,450,000
Jinan, China ... 2,150,000
Johannesburg, South Africa (4,000,000) ... 752,349
Kābul, Afghanistan ... 1,424,400
Kampala, Uganda ... 773,463
Kānpur, India (2,690,486) ... 2,540,069
Kaohsiung, Taiwan (1,845,000) ... 1,468,586
Karāchi, Pakistan ... 9,339,023
Katowice, Poland (2,755,000) ... 343,158
Kharkiv, Ukraine (1,950,000) ... 1,494,235
Khartoum (Al Kharṭūm), Sudan (1,450,000) ... 947,483
Kiev (Kyïv), Ukraine (3,250,000) ... 2,589,541
Kingston, Jamaica (830,000) ... 516,500
Kinshasa, Dem. Rep. of the Congo ... 3,000,000
Kitakyūshū, Japan (1,550,000) ... 1,011,491
Kolkata (Calcutta), India (13,216,546) ... 4,580,544
Kuala Lumpur, Malaysia (2,500,000) ... 1,297,526
Kuwait (Al Kuwayt), Kuwait (1,126,000) ... 28,859
Lagos, Nigeria (3,800,000) ... 1,213,000
Lahore, Pakistan ... 5,143,495
La Paz, Bolivia (1,487,854) ... 792,611
Libreville, Gabon (418,616) ... 362,386
Lilongwe, Malawi ... 435,964
Lima, Peru (6,321,173) ... 340,422
Lisbon (Lisboa), Portugal (2,350,000) ... 563,210
Liverpool, England, U.K. (1,515,000) ... 467,995
Ljubljana, Slovenia ... 263,832
Lomé, Togo ... 450,000
London, England, U.K. (12,000,000) ... 7,074,265
Los Angeles, California, U.S. (16,373,645) ... 3,694,820
Luanda, Angola ... 1,459,900
Lucknow, India (2,266,933) ... 2,207,340
Lusaka, Zambia ... 1,269,848
Lyon, France (1,648,216) ... 445,452
Madrid, Spain (4,690,000) ... 2,882,860
Managua, Nicaragua ... 864,201
Manaus, Brazil ... 1,394,724
Manchester, England, U.K. (2,760,000) ... 430,818
Manila, Philippines (11,200,000) ... 1,654,761
Mannheim, Germany (1,525,000) ... 307,730
Maputo, Mozambique ... 966,837
Maracaibo, Venezuela ... 1,249,670
Marseille, France (1,516,340) ... 798,430
Mashhad, Iran ... 1,887,405
Mecca (Makkah), Saudi Arabia ... 630,000
Medan, Indonesia ... 1,988,200
Medellín, Colombia (2,290,000) ... 1,885,001
Melbourne, Australia (3,366,542) ... 67,784
Mexico City (Ciudad de México), Mexico (17,786,983) ... 8,605,239
Miami, Florida, U.S. (3,876,380) ... 362,470
Milan (Milano), Italy (3,790,000) ... 1,305,591
Milwaukee, Wisconsin, U.S. (1,689,572) ... 596,974
Minneapolis, Minnesota, U.S. (2,968,806) ... 382,618
Minsk, Belarus (1,680,567) ... 1,677,137
Mogadishu (Muqdisho), Somalia ... 600,000
Monrovia, Liberia ... 465,000
Monterrey, Mexico (3,236,604) ... 1,110,909
Montevideo, Uruguay (1,650,000) ... 1,303,182
Montréal, Quebec, Canada (3,426,350) ... 1,039,534
Moscow (Moskva), Russia (12,850,000) ... 8,389,700
Mumbai (Bombay), India (16,368,084) ... 11,914,398
Munich (München), Germany (1,930,000) ... 1,194,560
Nagoya, Japan (5,250,000) ... 2,171,378
Nāgpur, India (2,122,965) ... 2,051,320
Nairobi, Kenya ... 2,143,254
Nanjing, China ... 2,490,000
Naples (Napoli), Italy (3,150,000) ... 1,046,987
N'Djamena, Chad ... 546,572
Newcastle upon Tyne, England, U.K. (1,350,000) ... 282,338
New Delhi, India (*Delhi) ... 294,783
New York, New York, U.S. (21,199,865) ... 8,008,278
Niamey, Niger ... 392,165
Nizhniy Novgorod, Russia (1,950,000) ... 1,364,900
Nouakchott, Mauritania ... 393,325
Novosibirsk, Russia (1,505,000) ... 1,402,400
Nürnberg, Germany (1,065,000) ... 486,628
Odesa, Ukraine (1,150,000) ... 1,002,246
Omsk, Russia (1,190,000) ... 1,157,600
Ōsaka, Japan (17,050,000) ... 2,598,589
Oslo, Norway (773,498) ... 504,040
Ottawa, Ontario, Canada (1,063,664) ... 774,072
Ouagadougou, Burkina Faso ... 634,479
Palembang, Indonesia ... 1,415,500
Panamá, Panama (995,000) ... 415,964
Paris, France (11,174,743) ... 2,125,246
Patna, India (1,707,429) ... 1,376,950
Perm', Russia (1,110,000) ... 1,017,100
Perth, Australia (1,244,320) ... 10,195
Philadelphia, Pennsylvania, U.S. (6,188,463) ... 1,517,550
Phnom Penh (Phnum Pénh), Cambodia ... 570,155
Phoenix, Arizona, U.S. (3,251,876) ... 1,321,045
Port Moresby, Papua New Guinea ... 246,664
Port-au-Prince, Haiti (1,425,594) ... 990,558
Portland, Oregon, U.S. (2,265,223) ... 529,121
Porto, Portugal (1,230,000) ... 273,060
Porto Alegre, Brazil (3,375,000) ... 1,304,998
Prague (Praha), Czech Republic (1,328,000) ... 1,193,270
Pretoria, South Africa (1,100,000) ... 692,348
Pune, India (3,755,525) ... 2,540,069
Pusan, South Korea ... 3,814,325
P'yŏngyang, North Korea ... 2,741,260
Qingdao, China ... 2,300,000
Québec, Quebec, Canada (682,757) ... 169,076
Quezon City, Philippines (*Manila) ... 1,989,419
Quito, Ecuador ... 1,615,809
Rabat, Morocco (1,200,000) ... 717,000
Rangoon (Yangon), Myanmar (2,800,000) ... 2,705,039
Recife, Brazil (3,160,000) ... 1,421,993
Regina, Saskatchewan, Canada (192,800) ... 178,225
Reykjavík, Iceland (166,015) ... 107,684
Rīga, Latvia (1,000,000) ... 792,508
Rio de Janeiro, Brazil (10,465,000) ... 5,851,914
Riyadh (Ar Riyāḍ), Saudi Arabia ... 1,800,000
Rome (Roma), Italy (3,235,000) ... 2,649,765
Rosario, Argentina (1,190,000) ... 894,645
Rostov-na-Donu, Russia (1,160,000) ... 1,017,300
Rotterdam, Netherlands (1,089,979) ... 539,000
Sacramento, California, U.S. (1,796,857) ... 407,018
St. Louis, Missouri, U.S. (2,603,607) ... 348,189
St. Petersburg (Leningrad), Russia (6,000,000) ... 4,728,200
Salvador, Brazil (2,855,000) ... 2,439,823
Samara, Russia (1,450,000) ... 1,168,000
San Diego, California, U.S. (2,813,833) ... 1,223,400
San Francisco, California, U.S. (7,039,362) ... 776,733
San José, Costa Rica (996,194) ... 309,672
San Juan, Puerto Rico (1,967,627) ... 421,958
San Salvador, El Salvador (1,908,921) ... 473,372
Santiago, Chile ... 4,788,543
Santo Domingo, Dominican Republic ... 2,677,056
São Paulo, Brazil (17,380,000) ... 9,713,692
Sapporo, Japan (2,000,000) ... 1,822,300
Sarajevo, Bosnia and Herzegovina ... 367,703
Saratov, Russia (1,135,000) ... 881,000
Seattle, Washington, U.S. (3,554,760) ... 563,374
Seoul (Sŏul), South Korea (15,850,000) ... 10,231,217
Shanghai, China (11,010,000) ... 8,930,000
Shenyang (Mukden), China ... 4,050,000
Singapore, Singapore (4,400,000) ... 4,017,700
Skopje, Macedonia ... 440,577
Sofia (Sofiya), Bulgaria (1,189,794) ... 1,138,629
Stockholm, Sweden (1,643,366) ... 743,703
Stuttgart, Germany (2,020,000) ... 582,443
Surabaya, Indonesia ... 2,801,300
Sūrat, India (2,811,466) ... 2,433,787
Sydney, Australia (3,741,290) ... 11,115
T'aipei, Taiwan (6,200,000) ... 2,640,322
Tallinn, Estonia ... 403,981
Tashkent (Toshkent), Uzbekistan (2,325,000) ... 2,142,700
Tbilisi, Georgia (1,460,000) ... 1,279,000
Tegucigalpa, Honduras ... 576,661
Tehrān, Iran (8,800,000) ... 6,758,845
Tel Aviv-Yafo, Israel (1,890,000) ... 348,100
Tianjin (Tientsin), China ... 5,000,000
Tiranë, Albania ... 244,153
Tōkyō, Japan (30,300,000) ... 8,130,408
Toronto, Ontario, Canada (4,682,897) ... 2,481,494
Tripoli (Ṭarābulus), Libya ... 1,500,000
Tunis, Tunisia (1,300,000) ... 702,330
Turin (Torino), Italy (1,550,000) ... 921,485
Ufa, Russia (1,110,000) ... 1,088,900
Ulan Bator (Ulaanbaatar), Mongolia ... 672,882
Ürümqi, China ... 1,130,000
València, Spain (1,340,000) ... 739,014
Vancouver, British Columbia, Canada (1,986,965) ... 545,671
Viangchan (Vientiane), Laos ... 464,000
Vienna (Wien), Austria (1,950,000) ... 1,609,631
Vilnius, Lithuania ... 578,334
Volgograd (Stalingrad), Russia (1,358,000) ... 1,000,000
Warsaw (Warszawa), Poland (2,300,000) ... 1,615,369
Washington, D.C., U.S. (7,608,070) ... 572,059
Wellington, New Zealand (346,500) ... 167,400
Winnipeg, Manitoba, Canada (671,274) ... 619,544
Wuhan, China ... 3,870,000
Xi'an, China ... 2,410,000
Yekaterinburg, Russia (1,530,000) ... 1,272,900
Yerevan, Armenia (1,315,000) ... 1,249,202
Yokohama, Japan (*Tōkyō) ... 3,426,506
Zagreb, Croatia ... 867,865
Zürich, Switzerland (932,681) ... 337,553

Metropolitan area populations are shown in parentheses.
* City is located within the metropolitan area of another city; for example, Yokohama, Japan is located in the Tōkyō metropolitan area.

GLOSSARY OF FOREIGN GEOGRAPHICAL TERMS

Annam	Annamese
Arab	Arabic
Bantu	Bantu
Bur	Burmese
Camb	Cambodian
Celt	Celtic
Chn	Chinese
Czech	Czech
Dan	Danish
Du	Dutch
Fin	Finnish
Fr	French
Ger	German
Gr	Greek
Hung	Hungarian
Ice	Icelandic
India	India
Indian	American Indian
Indon	Indonesian
It	Italian
Jap	Japanese
Kor	Korean
Mal	Malayan
Mong	Mongolian
Nor	Norwegian
Per	Persian
Pol	Polish
Port	Portuguese
Rom	Romanian
Rus	Russian
Siam	Siamese
So. Slav	Southern Slavonic
Sp	Spanish
Swe	Swedish
Tib	Tibetan
Tur	Turkish
Yugo	Yugoslav

å, Nor., Swe	brook, river
aa, Dan., Nor	brook
aas, Dan., Nor	ridge
âb, Per	water, river
abad, India, Per	town, city
ada, Tur	island
adrar, Berber	mountain
air, Indon	stream
akrotírion, Gr	cape
älf, Swe	river
alp, Ger	mountain
altipiano, It	plateau
alto, Sp	height
archipel, Fr	archipelago
archipiélago, Sp	archipelago
arquipélago, Port	archipelago
arroyo, Sp	brook, stream
ås, Nor., Swe	ridge
austral, Sp	southern
baai, Du	bay
bab, Arab	gate, port
bach, Ger	brook, stream
backe, Swe	hill
bad, Ger	bath, spa
bahía, Sp	bay, gulf
bahr, Arab	river, sea, lake
baia, It	bay, gulf
baía, Port	bay
baie, Fr	bay, gulf
bajo, Sp	depression
bak, Indon	stream
bakke, Dan., Nor	hill
balkan, Tur	mountain range
bana, Jap	point, cape
banco, Sp	bank
bandar, Mal., Per.	town, port, harbor
bang, Siam	village
bassin, Fr	basin
batang, Indon., Mal	river
ben, Celt	mountain, summit
bender, Arab	harbor, port
bereg, Rus	coast, shore
berg, Du., Ger., Nor., Swe.	mountain, hill
bir, Arab	well
birkat, Arab	lake, pond, pool
bit, Arab	house
bjaerg, Dan., Nor	mountain
bocche, It	mouth
boğazı, Tur	strait
bois, Fr	forest, wood
boloto, Rus	marsh
bolsón, Sp.	flat-floored desert valley
boreal, Sp	northern
borg, Dan., Nor., Swe	castle, town
borgo, It	town, suburb
bosch, Du	forest, wood
bouche, Fr	river mouth
bourg, Fr	town, borough
bro, Dan., Nor., Swe	bridge
brücke, Ger	bridge
bucht, Ger	bay, bight
bugt, Dan., Nor., Swe	bay, gulf
bulu, Indon	mountain
burg, Du., Ger	castle, town
buri, Siam	town
burun, burnu, Tur	cape
by, Dan., Nor., Swe	village
caatinga, Port. (Brazil)	open brushland
cabezo, Sp	summit
cabo, Port., Sp	cape
campo, It., Port., Sp	plain, field
campos, Port. (Brazil)	plains
cañón, Sp	canyon
cap, Fr	cape
capo, It	cape
casa, It., Port., Sp	house
castello, It., Port	castle, fort
castillo, Sp	castle
càte, Fr	hill
çay, Tur	stream, river
cayo, Sp	rock, shoal, islet
cerro, Sp	mountain, hill
champ, Fr	field
chang, Chn	village, middle
château, Fr	castle
chen, Chn	market town
chiang, Chn	river
chott, Arab	salt lake
chou, Chn.	capital of district; island
chu, Tib	water, stream
cidade, Port	town, city
cima, Sp	summit, peak
città, It	town, city
ciudad, Sp	town, city
cochilha, Port	ridge
col, Fr	pass
colina, Sp	hill
cordillera, Sp	mountain chain
costa, It., Port., Sp	coast
côte, Fr	coast
cuchilla, Sp	mountain ridge
dağ, Tur	mountain(s)
dake, Jap	peak, summit
dal, Dan., Du., Nor., Swe	valley
dan, Kor	point, cape
danau, Indon	lake
dar, Arab	house, abode, country
darya, Per	river, sea
dasht, Per	plain, desert
deniz, Tur	sea
désert, Fr	desert
deserto, It	desert
desierto, Sp	desert
détroit, Fr	strait
dijk, Du	dam, dike
djebel, Arab	mountain
do, Kor	island
dorf, Ger	village
dorp, Du	village
duin, Du	dune
dzong, Tib.	fort, administrative capital
eau, Fr	water
ecuador, Sp	equator
eiland, Du	island
elv, Dan., Nor	river, stream
embalse, Sp	reservoir
erg, Arab	dune, sandy desert
est, Fr., It	east
estado, Sp	state
este, Port., Sp	east
estrecho, Sp	strait
étang, Fr	pond, lake
état, Fr	state
eyjar, Ice	islands
feld, Ger	field, plain
festung, Ger	fortress
fiume, It	river
fjäll, Swe	mountain
fjärd, Swe	bay, inlet
fjeld, Nor	mountain, hill
fjord, Dan., Nor	fiord, inlet
fjördur, Ice	fiord, inlet
fleuve, Fr	river
flod, Dan., Swe	river
flói, Ice	bay, marshland
fluss, Ger	river
foce, It	river mouth
fontein, Du	a spring
forêt, Fr	forest
fors, Swe	waterfall
forst, Ger	forest
fos, Dan., Nor	waterfall
fu, Chn	town, residence
fuente, Sp	spring, fountain
fuerte, Sp	fort
furt, Ger	ford
gang, Kor	stream, river
gangri, Tib	mountain
gat, Dan., Nor	channel
gàve, Fr	stream
gawa, Jap	river
gebergte, Du	mountain range
gebiet, Ger	district, territory
gebirge, Ger	mountains
ghat, India	pass, mountain range
gobi, Mong	desert
gol, Mong	river
göl, gölü, Tur	lake
golf, Du., Ger	gulf, bay
golfe, Fr	gulf, bay
golfo, It., Port., Sp	gulf, bay
gomba, gompa, Tib	monastery
gora, Rus., So. Slav	mountain
góra, Pol	mountain
gorod, Rus	town
grad, Rus., So. Slav	town
guba, Rus	bay, gulf
gundung, Indon	mountain
guntō, Jap	archipelago
gunung, Mal	mountain
haf, Swe	sea, ocean
hafen, Ger	port, harbor
haff, Ger	gulf, inland sea
hai, Chn	sea, lake
hama, Jap	beach, shore
hamada, Arab	rocky plateau
hamn, Swe	harbor
hāmūn, Per	swampy lake, plain
hantō, Jap	peninsula
hassi, Arab	well, spring
haus, Ger	house
haut, Fr	summit, top
hav, Dan., Nor	sea, ocean
havn, Dan., Nor	harbor, port
havre, Fr	harbor, port
háza, Hung	house, dwelling of
heim, Ger	hamlet, home
hem, Swe	hamlet, home
higashi, Jap	east
hisar, Tur	fortress
hissar, Arab	fort
ho, Chn	river
hoek, Du	cape
hof, Ger	court, farmhouse
höfn, Ice	harbor
hoku, Jap	north
holm, Dan., Nor., Swe	island
hora, Czech	mountain
horn, Ger	peak
hoved, Dan., Nor	cape
hsien, Chn	district, district capital
hu, Chn	lake
hügel, Ger	hill
huk, Dan., Swe	point
hus, Dan., Nor., Swe	house
île, Fr	island
ilha, Port	island
indsö, Dan., Nor	lake
insel, Ger	island
insjö, Swe	lake
irmak, irmagi, Tur	river
isla, Sp	island
isola, It	island
istmo, It., Sp	isthmus
järvi, jaur, Fin	lake
jebel, Arab	mountain
jima, Jap	island
jökel, Nor	glacier
joki, Fin	river
jökull, Ice	glacier
kaap, Du	cape
kai, Jap	bay, gulf, sea
kaikyō, Jap	channel, strait
kalat, Per	castle, fortress
kale, Tur	fort
kali, Mal	creek, river
kand, Per	village
kang, Chn	mountain ridge; village
kap, Dan., Ger	cape
kapp, Nor., Swe	cape
kasr, Arab	fort, castle
kawa, Jap	river
kefr, Arab	village
kei, Jap	creek, river
ken, Jap	prefecture
khor, Arab	bay, inlet
khrebet, Rus	mountain range
kiang, Chn	large river
king, Chn	capital city, town
kita, Jap	north
ko, Jap	lake
köbstad, Dan	market-town
kol, Mong	lake
kólpos, Gr	gulf
kong, Chn	river
kopf, Ger	head, summit, peak
köpstad, Swe	market-town
körfezi, Tur	gulf
kosa, Rus	spit
kou, Chn	river mouth
köy, Tur	village
kraal, Du. (Africa)	native village
ksar, Arab	fortified village
kuala, Mal	bay, river mouth
kuh, Per	mountain
kum, Tur	sand
kuppe, Ger	summit
küste, Ger	coast
kyo, Jap	town, capital
la, Tib	mountain pass
labuan, Mal	anchorage, port
lac, Fr	lake
lago, It., Port., Sp	lake
lagoa, Port	lake, marsh
laguna, It., Port., Sp	lagoon, lake
lahti, Fin	bay, gulf
län, Swe	county
landsby, Dan., Nor	village
liehtao, Chn	archipelago
liman, Tur	bay, port
ling, Chn	pass, ridge, mountain
llanos, Sp	plains
loch, Celt. (Scotland)	lake, bay
loma, Sp	long, low hill
lough, Celt. (Ireland)	lake, bay
machi, Jap	town
man, Kor	bay
mar, Port., Sp	sea
mare, It., Rom	sea
marisma, Sp	marsh, swamp
mark, Ger	boundary, limit
massif, Fr	block of mountains
mato, Port	forest, thicket
me, Siam	river
meer, Du., Ger	lake, sea
mer, Fr	sea
mesa, Sp	flat-topped mountain
meseta, Sp	plateau
mina, Port., Sp	mine
minami, Jap	south
minato, Jap	harbor, haven
misaki, Jap	cape, headland
mont, Fr	mount, mountain
montagna, It	mountain
montagne, Fr	mountain
montaña, Sp	mountain
monte, It., Port., Sp.	mount, mountain
more, Rus., So. Slav	sea
morro, Port., Sp	hill, bluff
mühle, Ger	mill
mund, Ger	mouth, opening
mündung, Ger	river mouth
mura, Jap	township
myit, Bur	river
mys, Rus	cape
nada, Jap	sea
nadi, India	river, creek
naes, Dan., Nor	cape
nafud, Arab	desert of sand dunes
nagar, India	town, city
nahr, Arab	river
nam, Siam	river, water
nan, Chn., Jap	south
näs, Nor., Swe	cape
nez, Fr	point, cape
nishi, nisi, Jap	west
njarga, Fin	peninsula
nong, Siam	marsh
noord, Du	north
nor, Mong	lake
nord, Dan., Fr., Ger., It., Nor., Swe	north
norte, Port., Sp	north
nos, Rus	cape
nyasa, Bantu	lake
ö, Dan., Nor., Swe	island
occidental, Sp	western
ocna, Rom	salt mine
odde, Dan., Nor	point, cape
oeste, Port., Sp	west
oka, Jap	hill
oost, Du	east
oriental, Sp	eastern
óros, Gr	mountain
ost, Ger., Swe	east
öster, Dan., Nor., Swe	eastern
ostrov, Rus	island
oued, Arab	river, stream
ouest, Fr	west
ozero, Rus	lake
pää, Fin	mountain
padang, Mal	plain, field
pampas, Sp. (Argentina)	grassy plains
pará, Indian (Brazil)	river
pas, Fr	channel, passage
paso, Sp	mountain pass, passage
passo, It., Port.	mountain pass, passage, strait
patam, India	city, town
pei, Chn	north
pélagos, Gr	open sea
pegunungan, Indon	mountains
peña, Sp	rock
peresheyek, Rus	isthmus
pertuis, Fr	strait
peski, Rus	desert
pic, Fr	mountain peak
pico, Port., Sp	mountain peak
piedra, Sp	stone, rock
ping, Chn	plain, flat
planalto, Port	plateau
planina, Yugo	mountains
playa, Sp	shore, beach
pnom, Camb	mountain
pointe, Fr	point
polder, Du., Ger	reclaimed marsh
polje, So. Slav	plain, field
poluostrov, Rus	peninsula
pont, Fr	bridge
ponta, Port	point, headland
ponte, It., Port	bridge
pore, India	city, town
porthmós, Gr	strait
porto, It., Port	port, harbor
potamós, Gr	river
p'ov, Rus	peninsula
prado, Sp	field, meadow
presqu'île, Fr	peninsula
proliv, Rus	strait
pu, Chn	commercial village
pueblo, Sp	town, village
puerto, Sp	port, harbor
pulau, Indon	island
punkt, Ger	point
punt, Du	point
punta, It., Sp	point
pur, India	city, town
puy, Fr	peak
qal'a, qal'at, Arab	fort, village
qasr, Arab	fort, castle
rann, India	wasteland
ra's, Arab	cape, head
reka, Rus., So. Slav	river
represa, Port	reservoir
rettō, Jap	island chain
ría, Sp	estuary
ribeira, Port	stream
riberão, Port	river
rio, It., Port	stream, river
río, Sp	river
rivière, Fr	river
roca, Sp	rock
rt, Yugo	cape
rūd, Per	river
saari, Fin	island
sable, Fr	sand
sahara, Arab	desert, plain
saki, Jap	cape
sal, Sp	salt
salar, Sp	salt flat, salt lake
salto, Sp	waterfall
san, Jap., Kor	mountain, hill
sat, satul, Rom	village
schloss, Ger	castle
sebkha, Arab	salt marsh
see, Ger	lake, sea
şehir, Tur	town, city
selat, Indon	stream
selvas, Port. (Brazil)	tropical rain forests
seno, Sp	bay
serra, Port	mountain chain
serranía, Sp	mountain ridge
seto, Jap	strait
severnaya, Rus	northern
shahr, Per	town, city
shan, Chn	mountain, hill, island
shatt, Arab	river
shi, Jap	city
shima, Jap	island
shōtō, Jap	archipelago
si, Chn	west, western
sierra, Sp	mountain range
sjö, Nor., Swe	lake, sea
sö, Dan., Nor	lake, sea
söder, södra, Swe	south
song, Annam	river
sopka, Rus	peak, volcano
source, Fr	a spring
spitze, Ger	summit, point
staat, Ger	state
stad, Dan., Du., Nor., Swe.	city, town
stadt, Ger	city, town
stato, It	state
step', Rus	treeless plain, steppe
straat, Du	strait
strand, Dan., Du., Ger., Nor., Swe	shore, beach
stretto, It	strait
strom, Ger	river, stream
ström, Dan., Nor., Swe.	stream, river
stroom, Du	stream, river
su, suyu, Tur	water, river
sud, Fr., Sp	south
süd, Ger	south
suidō, Jap	channel
sul, Port	south
sund, Dan., Nor., Swe	sound
sungai, sungei, Indon., Mal	river
sur, Sp	south
syd, Dan., Nor., Swe	south
tafelland, Ger	plateau
take, Jap	peak, summit
tal, Ger	valley
tanjung, tanjong, Mal	cape
tao, Chn	island
târg, târgul, Rom	market, town
tell, Arab	hill
teluk, Indon	bay, gulf
terra, It	land
terre, Fr	earth, land
thal, Ger	valley
tierra, Sp	earth, land
tō, Jap	east; island
tonle, Camb	river, lake
top, Du	peak
torp, Swe	hamlet, cottage
tsangpo, Tib	river
tsi, Chn	village, borough
tso, Tib	lake
tsu, Jap	harbor, port
tundra, Rus	treeless arctic plains
tung, Chn	east
tuz, Tur	salt
udde, Swe	cape
ufer, Ger	shore, riverbank
ujung, Indon	point, cape
umi, Jap	sea, gulf
ura, Jap	bay, coast, creek
ust'ye, Rus	river mouth
valle, It., Port., Sp	valley
vallée, Fr	valley
valli, It	lake
vár, Hung	fortress
város, Hung	town
varoš, So. Slav	town
veld, Du	open plain, field
verkh, Rus	top, summit
ves, Czech	village
vest, Dan., Nor., Swe	west
vik, Swe	cove, bay
vila, Port	town
villa, Sp	town
villar, Sp	village, hamlet
ville, Fr	town, city
vostok, Rus	east
wad, wādī, Arab.	intermittent stream
wald, Ger	forest, woodland
wan, Chn., Jap	bay, gulf
weiler, Ger	hamlet, village
westersch, Du	western
wüste, Ger	desert
yama, Jap	mountain
yarimada, Tur	peninsula
yug, Rus	south
zaki, Jap	cape
zaliv, Rus	bay, gulf
zapad, Rus	west
zee, Du	sea
zemlya, Rus	land
zuid, Du	south

ABBREVIATIONS OF GEOGRAPHICAL NAMES AND TERMS

Afg.	Afghanistan
Afr.	Africa
Ak., U.S.	Alaska, U.S.
Al., U.S.	Alabama, U.S.
Alb.	Albania
Alg.	Algeria
Am. Sam.	American Samoa
And.	Andorra
Ang.	Angola
Ant.	Antarctica
Antig.	Antigua and Barbuda
aq.	Aqueduct
Ar., U.S.	Arkansas, U.S.
Arg.	Argentina
Arm.	Armenia
arpt.	Airport
Aus.	Austria
Austl.	Australia
Az., U.S.	Arizona, U.S.
Azer.	Azerbaijan
b.	Bay, Gulf, Inlet, Lagoon
Bah.	Bahamas
Bahr.	Bahrain
Barb.	Barbados
Bdi.	Burundi
Bel.	Belgium
Bela.	Belarus
Ber.	Bermuda
Bhu.	Bhutan
bk.	Undersea Bank
bldg.	Building
Blg.	Bulgaria
Bngl.	Bangladesh
Bol.	Bolivia
Bos.	Bosnia and Herzegovina
Bots.	Botswana
Braz.	Brazil
Bru.	Brunei
Br. Vir. Is.	British Virgin Islands
bt.	Bight
Burkina	Burkina Faso
c.	Cape, Point
Ca., U.S.	California, U.S.
Cam.	Cameroon
Camb.	Cambodia
can.	Canal
Can.	Canada
C.A.R.	Central African Republic
Cay. Is.	Cayman Islands
C. Iv.	Cote d'Ivoire
clf.	Cliff, Escarpment
co.	County, Parish
Co., U.S.	Colorado, U.S.
Col.	Colombia
Com.	Comoros
cont.	Continent
Cook Is.	Cook Islands
C.R.	Costa Rica
Cro.	Croatia
cst.	Coast, Beach
Ct., U.S.	Connecticut, U.S.
C.V.	Cape Verde
Cyp.	Cyprus
Czech Rep.	Czech Republic
d.	Delta
D.C., U.S.	District of Columbia, U.S.
De., U.S.	Delaware, U.S.
Den.	Denmark
dep.	Dependency, Colony
depr.	Depression
dept.	Department, District
des.	Desert
Dji.	Djibouti
Dom.	Dominica
Dom. Rep.	Dominican Republic
D.R.C.	Democratic Republic of the Congo
Ec.	Ecuador
educ.	Educational Facility
El Sal.	El Salvador
Eng., U.K.	England, U.K.
Eq. Gui.	Equatorial Guinea
Erit.	Eritrea
Est.	Estonia
est.	Estuary
Eth.	Ethiopia
E. Timor	East Timor
Eur.	Europe
Falk. Is.	Falkland Islands
Far. Is.	Faroe Islands
Fin.	Finland
fj.	Fjord
Fl., U.S.	Florida, U.S.
for.	Forest, Moor
Fr.	France
Fr. Gu.	French Guiana
Fr. Poly.	French Polynesia
Ga., U.S.	Georgia, U.S.
Gam.	The Gambia
Gaza	Gaza Strip
Geor.	Georgia
Ger.	Germany
Grc.	Greece
Gren.	Grenada
Grnld.	Greenland
Guad.	Guadeloupe
Guat.	Guatemala
Guern.	Guernsey
Gui.	Guinea
Gui.-B.	Guinea-Bissau
Guy.	Guyana
Hi., U.S.	Hawaii, U.S.
hist.	Historic Site, Ruins
hist. reg.	Historic Region
Hond.	Honduras
Hung.	Hungary
i.	Island
Ia., U.S.	Iowa, U.S.
ice	Ice Feature, Glacier
Ice.	Iceland
Id., U.S.	Idaho, U.S.
Il., U.S.	Illinois, U.S.
In., U.S.	Indiana, U.S.
Indon.	Indonesia
I. of Man	Isle of Man
I.R.	Indian Reservation
Ire.	Ireland
is.	Islands
Isr.	Israel
isth.	Isthmus
Jam.	Jamaica
Jord.	Jordan
Kaz.	Kazakhstan
Kir.	Kiribati
Kor., N.	Korea, North
Kor., S.	Korea, South
Ks., U.S.	Kansas, U.S.
Kuw.	Kuwait
Ky., U.S.	Kentucky, U.S.
Kyrg.	Kyrgyzstan
l.	Lake, Pond
La., U.S.	Louisiana, U.S.
Lat.	Latvia
Leb.	Lebanon
Leso.	Lesotho
Lib.	Liberia
Liech.	Liechtenstein
Lith.	Lithuania
Lux.	Luxembourg
Ma., U.S.	Massachusetts, U.S.
Mac.	Macedonia
Madag.	Madagascar
Malay.	Malaysia
Mald.	Maldives
Marsh. Is.	Marshall Islands
Mart.	Martinique
Maur.	Mauritania
May.	Mayotte
Md., U.S.	Maryland, U.S.
Me., U.S.	Maine, U.S.
Mex.	Mexico
Mi., U.S.	Michigan, U.S.
Micron.	Micronesia, Federated States of
Mn., U.S.	Minnesota, U.S.
Mo., U.S.	Missouri, U.S.
Mol.	Moldova
Mong.	Mongolia
Monts.	Montserrat
Mor.	Morocco
Moz.	Mozambique
Ms., U.S.	Mississippi, U.S.
Mt., U.S.	Montana, U.S.
mth.	River Mouth or Channel
mtn.	Mountain
mts.	Mountains
Mwi.	Malawi
Mya.	Myanmar
N.A.	North America
N.C., U.S.	North Carolina, U.S.
N. Cal.	New Caledonia
N.D., U.S.	North Dakota, U.S.
Ne., U.S.	Nebraska, U.S.
neigh.	Neighborhood
Neth.	Netherlands
Neth. Ant.	Netherlands Antilles
N.H., U.S.	New Hampshire, U.S.
Nic.	Nicaragua
Nig.	Nigeria
N. Ire., U.K.	Northern Ireland, U.K.
N.J., U.S.	New Jersey, U.S.
N.M., U.S.	New Mexico, U.S.
N. Mar. Is.	Northern Mariana Islands
Nmb.	Namibia
Nor.	Norway
Nv., U.S.	Nevada, U.S.
N.Y., U.S.	New York, U.S.
N.Z.	New Zealand
o.	Ocean
Oc.	Oceania
Oh., U.S.	Ohio, U.S.
Ok., U.S.	Oklahoma, U.S.
Or., U.S.	Oregon, U.S.
p.	Pass
Pa., U.S.	Pennsylvania, U.S.
Pak.	Pakistan
Pan.	Panama
Pap. N. Gui.	Papua New Guinea
Para.	Paraguay
pen.	Peninsula
Phil.	Philippines
Pit.	Pitcairn
pl.	Plain, Flat
plat.	Plateau, Highland
Pol.	Poland
Port.	Portugal
P.R.	Puerto Rico
prov.	Province, Region
pt. of i.	Point of Interest
r.	River, Creek
Reu.	Reunion
rec.	Recreational Site, Park
reg.	Physical Region
rel.	Religious Institution
res.	Reservoir
rf.	Reef, Shoal
R.I., U.S.	Rhode Island, U.S.
Rom.	Romania
Rw.	Rwanda
S.A.	South America
S. Afr.	South Africa
Sau. Ar.	Saudi Arabia
S.C., U.S.	South Carolina, U.S.
sci.	Scientific Station
Scot., U.K.	Scotland, U.K.
S.D., U.S.	South Dakota, U.S.
sea feat.	Undersea Feature
Sen.	Senegal
Serb.	Serbia and Montenegro
Sey.	Seychelles
S. Geor.	South Georgia
Sing.	Singapore
S.L.	Sierra Leone
Slvk.	Slovakia
Slvn.	Slovenia
S. Mar.	San Marino
Sol. Is.	Solomon Islands
Som.	Somalia
Sp. N. Afr.	Spanish North Africa
Sri L.	Sri Lanka
St. Hel.	St. Helena
St. K./N.	St. Kitts and Nevis
St. Luc.	St. Lucia
St. P./M.	St. Pierre and Miquelon
strt.	Strait, Channel, Sound
S. Tom./P.	Sao Tome and Principe
St. Vin.	St. Vincent and the Grenadines
Sur.	Suriname
Sval.	Svalbard
sw.	Swamp, Marsh
Swaz.	Swaziland
Swe.	Sweden
Switz.	Switzerland
Tai.	Taiwan
Taj.	Tajikistan
Tan.	Tanzania
T./C. Is.	Turks and Caicos Islands
ter.	Territory
Thai.	Thailand
Tn., U.S.	Tennessee, U.S.
trans.	Transportation Facility
Trin.	Trinidad and Tobago
Tun.	Tunisia
Tur.	Turkey
Turkmen.	Turkmenistan
Tx., U.S.	Texas, U.S.
U.A.E.	United Arab Emirates
Ug.	Uganda
U.K.	United Kingdom
Ukr.	Ukraine
Ur.	Uruguay
U.S.	United States
Ut., U.S.	Utah, U.S.
Uzb.	Uzbekistan
Va., U.S.	Virginia, U.S.
val.	Valley, Watercourse
Ven.	Venezuela
Viet.	Vietnam
V.I.U.S.	Virgin Islands (U.S.)
vol.	Volcano
Vt., U.S.	Vermont, U.S.
Wa., U.S.	Washington, U.S.
W.B.	West Bank
Wi., U.S.	Wisconsin, U.S.
W. Sah.	Western Sahara
wtfl.	Waterfall
W.V., U.S.	West Virginia, U.S.
Wy., U.S.	Wyoming, U.S.
Zam.	Zambia
Zimb.	Zimbabwe

PRONUNCIATION OF GEOGRAPHICAL NAMES

Key to the Sound Values of Letters and Symbols Used in the Index to Indicate Pronunciation

ă-ăt; băttle
ă-fină̆l; appeă̆l
ā-rāte; elāte
å-senåte; inanimåte
ä-ärm; cälm
à-àsk; bàth
ȧ-sofȧ; mȧrine (short neutral or indeterminate sound)
â-fâre; prepâre
ch-choose; church
dh-as th in other; either
ē-bē; ēve
ḗ-ḗvent; crḗate
ĕ-bĕt; ĕnd
ĕ̆-recĕ̆nt (short neutral or indeterminate sound)
ẽ-cratẽr; cindẽr
g-gō; gāme
gh-guttural g
ĭ-bĭt; wĭll
ĭ-(short neutral or indeterminate sound)
ī-rīde; bīte
ĸ-gutteral k as *ch* in German *ich*
ng-sing
ŋ-baŋk; liŋger
ɴ-indicates nasalized
ŏ-nŏd; ŏdd
ŏ̆-cŏ̆mmit; cŏ̆nnect
ō-ōld; bōld
ô-ôbey; hôtel
ô-ôrder; nôrth
oi-boil
o͞o-fo͞od; ro͞ot
ò-as oo in foot; wood
ou-out; thou
s-soft; so; sane
sh-dish; finish
th-thin; thick
ū-pūre; cūre
ů-ůnite; ůsůrp
û-ûrn; fûr
ŭ-stŭd; ŭp
ŭ̆-circŭ̆s; sŭ̆bmit
ü-as in French *tu*
zh-as *z* in azure
'-indeterminate vowel sound

In many cases the spelling of foreign geographical names does not even remotely indicate the pronunciation to an American, i.e., Słupsk in Poland is pronounced swôpsk; Jujuy in Argentina is pronounced ho͞ohwē', La Spezia in Italy is lä-spě'zyä.

This condition is hardly surprising, however, when we consider that in our own language Worcester, Massachusetts, is pronounced wŏs'tẽr; Sioux City, Iowa, so͞o sĭ'tĭ; Schuylkill Haven, Pennsylvania, sko͞ol'kĭl hā-vĕn; Poughkeepsie, New York, pŏ̆-kĭp'sḗ.

The indication of pronunciation of geographic names presents several peculiar problems:

1. Many foreign tongues use sounds that are not present in the English language and which an American cannot normally articulate. Thus, though the nearest English equivalent sound has been indicated, only approximate results are possible.

2. There are several dialects in each foreign tongue which cause variation in the local pronunciation of names. This also occurs in identical names in the various divisions of a great language group, as the Slavic or the Latin.

3. Within the United States there are marked differences in pronunciation, not only of local geographic names, but also of common words, indicating that the sound and tone values for letters as well as the placing of the emphasis vary considerably from one part of the country to another.

4. A number of different letters and diacritical combinations could be used to indicate essentially the same or approximate pronunciations.

Some variation in pronunciation other than that indicated in this index may be encountered, but such a difference does not necessarily indicate that either is in error, and in many cases it is a matter of individual choice as to which is preferred. In fact, an exact indication of pronunciation of many foreign names using English letters and diacritical marks is extremely difficult and sometimes impossible.

PRONOUNCING INDEX

This universal index includes in a single alphabetical list approximately 30,000 names of features that appear on the reference maps. Each name is followed by a page reference and geographical coordinates.

Abbreviation and Capitalization Abbreviations of names on the maps have been standardized as much as possible. Names that are abbreviated on the maps are generally spelled out in full in the index. Periods are used after all abbreviations regardless of local practice. The abbreviation "St." is used only for "Saint". "Sankt" and other forms of this term are spelled out.

Most initial letters of names are capitalized, except for a few Dutch names, such as "s-Gravenhage". Capitalization of noninitial words in a name generally follows local practice.

Alphabetization Names are alphabetized in the order of the letters of the English alphabet. Spanish *ll* and *ch*, for example, are not treated as direct letters. Furthermore, diacritical marks are disregarded in alphabetization — German or Scandinavian *ä* or *ö* are treated as *a* or *o*.

The names of physical features may appear inverted, since they are always alphabetized under the proper, not the generic, part of the name, thus: "Gibraltar, Strait of". Otherwise every entry, whether consisting of one word or more, is alphabetized as a single continuous entity. "Lakeland", for example, appears after "La Crosse" and before "La Salle". Names beginning with articles (Le Harve, Den Helder, Al Manāmah, Ad Dawhah) are not inverted.

In the case of identical names, towns are listed first, then political divisions, then physical features.

Generic Terms Except for cities, the names of all features are followed by terms that represent broad classes of features, for example, Mississippi, r. or Alabama, state. A list of all abbreviations used in the index is on page 215.

Country names and the names of features that extend beyond the boundaries of one county are followed by the name of the continent in which each is located. Country designations follow the names of all other places in the index. The locations of places in the United States and the United Kingdom are further defined by abbreviations that include the state or political division in which each is located.

Pronunciations Pronunciations are included for most names listed. An explanation of the pronunciation system used appears on page 215.

Page References and Geographical Coordinates The geographical coordinates and page references are found in the last columns of each entry.

If a page contains several maps or insets, a lowercase letter identifies the specific map or inset.

Latitude and longitude coordinates for point features, such as cities and mountain peaks, indicate the location of the symbols. For extensive areal features, such as countries or mountain ranges, or linear features, such as canals and rivers, locations are given for the position of the type as it appears on the map.

ăt; finăl; rāte; senăte; ärm; åsk; sofá; fâre; ch-choose; dh-as th in other; bē; ĕvent; bĕt; recĕnt; cratĕr; g-gō; gh-guttural g; bĭt; ĭ-short neutral; rīde; к-guttural k as ch in German ich;

PLACE (Pronunciation)	PAGE	LAT.	LONG.
Adams, r., Can. (ăd′ămz)	55	51°30'N	119°20'W
Adams, Mount, mtn., Wa., U.S. (ăd′ămz)	64	46°15'N	121°19'W
Adamsville, Al., U.S. (ăd′ămz-vĭl)	68h	33°36'N	86°57'W
Adana, Tur. (ä′dä-nä)	154	37°05'N	35°20'E
Adapazarı, Tur. (ä-dä-pä-zä′rē)	119	40°45'N	30°20'E
Adarama, Sudan (ä-dä-rä′mä)	185	17°11'N	34°56'E
Adda, r., Italy (äd′dä)	130	45°43'N	9°31'E
Ad Dabbah, Sudan	185	18°04'N	30°58'E
Ad Dahnā, des., Sau. Ar.	154	26°05'N	47°15'E
Ad-Dāmir, Sudan (ad-dä′mĕr)	185	17°38'N	33°57'E
Ad Dammām, Sau. Ar.	154	26°27'N	49°59'E
Ad Dāmūr, Leb.	153a	33°44'N	35°27'E
Ad Dawhah, Qatar	154	25°02'N	51°28'E
Ad Dilam, Sau. Ar.	154	23°47'N	47°03'E
Ad Dilinjāt, Egypt	192b	30°48'N	30°32'E
Addis Ababa, Eth.	185	9°00'N	38°44'E
Addison, Tx., U.S. (ă′dĭ-sŭn)	75c	32°58'N	96°50'W
Addo, S. Afr. (ădō)	187c	33°33'S	25°43'E
Ad Duwaym, Sudan (ad-dō-ām′)	185	13°56'N	32°22'E
Addyston, Oh., U.S. (ăd′ē-stŭn)	69f	39°09'N	84°42'W
Adel, Ga., U.S. (ā-dĕl′)	82	31°08'N	83°55'W
Adelaide, Austl. (ăd′ē-lād)	174	34°46'S	139°08'E
Adelaide, S. Afr. (ăd-ĕl′ād)	187c	32°41'S	26°07'E
Adelaide Island, i., Ant. (ăd′ē-lād)	178	67°15'S	68°40'W
Aden ('Adan), Yemen (ä′dĕn)	154	12°48'N	45°00'E
Aden, Gulf of, b.	154	11°45'N	45°45'E
Adi, Pulau, i., Indon. (ä′dē)	169	4°25'S	133°52'E
Adige, r., Italy (ä′dē-jā)	118	46°38'N	10°43'E
Adigrat, Eth.	157	14°17'N	39°28'E
Adilābād, India (ŭ-dĭl-ä-bäd′)	158	19°47'N	78°30'E
Adirondack Mountains, mts., N.Y., U.S. (ăd-ĭ-rŏn′dăk)	65	43°45'N	74°40'W
Adis Abeba see Addis Ababa, Eth.	185	9°00'N	38°44'E
Adi Ugri, Erit. (ä-dē ōō′grē)	185	14°54'N	38°52'E
Adjud, Rom. (äd′zhōd)	125	46°05'N	27°12'E
Adkins, Tx., U.S.	75d	29°22'N	98°18'W
Admiralty, i., Ak., U.S. (ăd′mĭ-răl-tē)	63	57°50'N	133°50'W
Admiralty Inlet, Wa., U.S. (ăd′mĭ-răl-tē)	74a	48°10'N	122°45'W
Admiralty Island National Monument, rec., Ak., U.S. (ăd′mĭ-răl-tē)	63	57°50'N	137°30'W
Admiralty Islands, is., Pap. N. Gui. (ăd′mĭ-răl-tē)	169	1°40'S	146°45'E
Ado-Ekiti, Nig.	189	7°38'N	5°12'E
Adolph, Mn., U.S. (ā′dolf)	75h	46°47'N	92°17'W
Ādoni, India	159	15°42'N	77°18'E
Adour, r., Fr. (á-dōōr′)	117	43°43'N	0°38'W
Adra, Spain (ä′drä)	128	36°45'N	3°02'W
Adrano, Italy (ä-drä′nō)	130	37°42'N	14°52'E
Adrar, Alg.	184	27°53'N	0°15'W
Adria, Italy (ä′drē-ä)	130	45°03'N	12°01'E
Adrian, Mi., U.S. (ā′drĭ-ăn)	66	41°55'N	84°00'W
Adrian, Mn., U.S. (ā′drĭ-ăn)	70	43°39'N	95°56'W
Adrianople see Edirne, Tur.	110	41°41'N	26°35'E
Adriatic Sea, sea, Eur.	112	43°30'N	14°27'E
Adwa, Eth.	185	14°02'N	38°58'E
Adwick-le-Street, Eng., U.K. (ăd′wĭk-lē-strēt′)	114a	53°35'N	1°11'W
Adycha, r., Russia (ä′dĭ-chà)	141	66°11'N	136°45'E
Adygea, prov., Russia	136	45°00'N	40°00'E
Adz′va, r., Russia (ädz′vá)	136	67°00'N	59°20'E
Aegean Sea, sea, (ē-jē′ăn)	112	39°04'N	24°56'E
A Estrada, Spain	128	42°42'N	8°29'W
Affton, Mo., U.S.	75e	38°33'N	90°20'W
Afghanistan, nation, Asia (ăf-găn-ĭ-stăn′)	154	33°00'N	63°00'E
Afgooye, Som. (äf-gō′ĭ)	192a	2°08'N	45°08'E
Afikpo, Nig.	189	5°53'N	7°56'E
Aflou, Alg. (ä-flōō′)	184	33°59'N	2°04'E
Afognak, i., Ak., U.S. (ä-fŏg-nák′)	63	58°28'N	151°35'W
A Fonsagrada, Spain	128	43°08'N	7°07'W
Afonso Claudio, Braz. (äl-fōn′sō-klou′dēō)	99a	20°05'S	41°05'W
Afragola, Italy (ä-frä′gō-lä)	129c	40°40'N	14°19'E
Africa, cont.	183	10°00'N	22°00'E
Afton, Mn., U.S. (ăf′tŭn)	75g	44°54'N	92°47'W
Afton, Ok., U.S. (ăf′tŭn)	79	36°42'N	94°56'W
Afton, Wy., U.S. (ăf′tŭn)	73	42°42'N	110°52'W
'Afula, Isr. (ä-fōō′lá)	153a	32°36'N	35°17'E
Afyon, Tur. (ä-fē-ōn)	154	38°45'N	30°20'E
Agadem, Niger (ä′gá-dĕm)	185	16°50'N	13°17'E
Agadez, Niger (ä′gá-dĕs)	184	16°58'N	7°59'E
Agadir, Mor. (ä-gá-dēr)	184	30°30'N	9°37'W
Agalta, Cordillera de, mts., Hond. (kōr-dēl-yĕ′rä-dĕ-ä-gä′l-tä)	90	15°15'N	85°42'W
Agapovka, Russia (ä-gä-pôv′kä)	142a	53°18'N	59°10'E
Agartala, India	158	23°53'N	91°22'E
Agāshi, India	159b	19°28'N	72°46'E
Agashkino, Russia (á-gäsh′kĭ-nô)	142b	55°18'N	38°13'E
Agattu, i., Ak., U.S. (ä′gä-tōō)	63a	52°14'N	173°40'E
Agboville, C. Iv.	188	5°56'N	4°13'W
Ağdam, Azer. (äg′däm)	137	40°00'N	47°00'E
Agde, Fr. (ägd)	126	43°19'N	3°30'E
Agen, Fr. (á-zhäN′)	117	44°13'N	0°31'E
Agiásos, Grc.	131	39°06'N	26°25'E
Aginskoye, Russia (ä-hĭn′skō-yĕ)	135	51°15'N	113°15'E
Ágios Efstrátios, i., Grc.	119	39°30'N	24°58'E
Agíou Órous, Kólpos, b., Grc.	131	40°15'N	24°00'E
Agno, Phil. (äg′nō)	169a	16°07'N	119°49'E
Agno, r., Phil.	169a	15°42'N	120°28'E
Agnone, Italy (än-yō′nā)	130	41°49'N	14°23'E
Agogo, Ghana	188	6°47'N	1°04'W
Agra, India (ä′grä)	155	27°18'N	78°00'E
Ağrı, Tur.	137	39°50'N	43°10'E
Agri, r., Italy (ä′grē)	130	40°15'N	16°21'E
Agrínio, Grc.	119	38°38'N	21°06'E
Agua, vol., Guat. (ä′gwä)	90	14°28'N	90°43'W
Agua Blanca, Río, r., Mex. (rē′ō-ä-gwä-blä′n-kä)	88	21°46'N	102°54'W
Agua Brava, Laguna de, l., Mex.	88	22°04'N	105°40'W
Agua Caliente Indian Reservation, I.R., Ca., U.S. (ä′gwä kal-yĕn′tä)	76	33°50'N	116°24'W
Aguada, Cuba (ä-gwä′dä)	92	22°25'N	80°50'W
Aguada, l., Mex. (ä-gwä′dä)	90a	18°46'N	89°40'W
Aguadas, Col. (ä-gwä′dás)	100	5°37'N	75°27'W
Aguadilla, P.R. (ä-gwä-dēl′yä)	87b	18°27'N	67°10'W
Aguadulce, Pan. (ä-gwä-dōōl′sā)	91	8°15'N	80°33'W
Agua Escondida, Meseta de, plat., Mex.	89	16°54'N	91°35'W
Agua Fria, r., Az., U.S. (ä′gwä frē-ä)	77	33°43'N	112°22'W
Agua Fria National Monument, rec., Az., U.S.	77	34°13'N	112°03'W
Aguai, Braz. (ägwä-ē′)	99a	22°04'S	46°57'W
Agualeguas, Mex. (ä-gwä-lä′gwäs)	80	26°19'N	99°33'W
Aguán, r., Hond. (ä-gwä′n)	90	15°22'N	87°00'W
Aguanaval, r., Mex. (á-guä-nä-väl′)	80	25°12'N	103°28'W
Aguanus, r., Can. (ä-gwä′nŭs)	61	50°45'N	62°03'W
Aguascalientes, Mex. (ä′gwäs-käl-yĕn′tās)	86	21°52'N	102°17'W
Aguascalientes, state, Mex. (ä′gwäs-käl-yĕn′tās)	88	22°00'N	102°18'W
Águeda, Port. (ä-gwä′dä)	128	40°36'N	8°26'W
Agueda, r., Eur. (ä-gĕ-dä)	128	40°50'N	6°44'W
Aguelhok, Mali	188	19°28'N	0°52'E
Aguilar, Spain	128	37°32'N	4°39'W
Aguilar, Co., U.S. (ä-gē-lär′)	78	37°24'N	104°38'W
Aguilas, Spain (ä-gē-läs)	118	37°26'N	1°35'W
Aguililla, Mex. (ä-gē-lēl-yä)	88	18°44'N	102°44'W
Aguililla, r., Mex. (ä-gē-lēl-yä)	88	18°30'N	102°48'W
Aguja, Punta, c., Peru (pŭn′tá à-gōō′ hä)	100	6°00'S	81°15'W
Agulhas, Cape, c., S. Afr. (ä-gōōl′yäs)	186	34°47'S	20°00'E
Agusan, r., Phil. (ä-gōō′sän)	169	8°12'N	126°07'E
Ahaggar, mts., Alg. (á-há-gär′)	184	23°14'N	6°00'E
Ahar, Iran	157	38°28'N	47°04'E
Ahlen, Ger. (ä′lĕn)	124	51°45'N	7°52'E
Ahmadābād, India (ŭ-mĕd-ä-bäd′)	155	23°04'N	72°38'E
Ahmadnagar, India (ä′mûd-nû-gûr)	155	19°09'N	74°45'E
Ahmar Mountains, mts., Eth.	185	9°22'N	42°00'E
Ahoskie, N.C., U.S. (ā-hŏs′kē)	83	36°15'N	77°00'W
Ahrensburg, Ger. (ä′rĕns-bōrg)	115c	53°40'N	10°14'E
Ahrweiler, Ger. (är′vī-lĕr)	124	50°34'N	7°05'E
Ähtärinjärvi, l., Fin.	123	62°46'N	24°25'E
Ahuacatlán, Mex. (ä-wä-kät-län′)	88	21°05'N	104°28'W
Ahuachapán, El Sal. (ä-wä-chä-pän′)	90	13°57'N	89°53'W
Ahualulco, Mex. (ä-wä-lōōl′kō)	88	20°43'N	103°57'W
Ahuatempan, Mex. (ä-wä-tĕm-pän)	88	18°11'N	98°02'W
Åhus, Swe. (ō′hōs)	122	55°56'N	14°19'E
Ahvāz, Iran	154	31°15'N	48°54'E
Ahvenanmaa (Åland), is., Fin. (ä′vĕ-nán-mô) (ō′länd)	116	60°36'N	19°55'E
'Aiea, Hi., U.S.	84a	21°18'N	157°52'W
Aígina, Grc. (ägwä-ē′)	131	37°43'N	23°35'E
Aígina, i., Grc.	131	37°43'N	23°35'E
Aígio, Grc.	131	38°13'N	22°04'E
Aiken, S.C., U.S. (ā′kĕn)	83	33°32'N	81°43'W
Aimorès, Serra dos, mts., Braz. (sĕ′r-rä-dôs-ī-mō-rĕ′s)	101	17°40'S	42°38'W
Aimoto, Japan (ī-mō-tō)	167b	34°59'N	135°09'E
Aincourt, Fr. (ân-kōō′r)	127b	49°04'N	1°47'E
Aïn el Beïda, Alg.	184	35°57'N	7°25'E
Ainsworth, Ne., U.S. (ānz′wûrth)	70	42°32'N	99°51'W
Aïn Témouchent, Alg. (ä′ēntĕ-mōō-shaN′)	118	35°20'N	1°23'W
Aïn Wessara, Alg. (ĕn ōō-sä-rá)	129	35°25'N	2°50'E
Aipe, Col. (ī′pĕ)	100a	3°13'N	75°15'W
Aïr, mts., Niger	184	18°00'N	8°30'E
Aire, r., Eng., U.K.	114a	53°42'N	1°00'W
Aire-sur-l'Adour, Fr. (âr)	126	43°42'N	0°17'W
Airhitam, Selat, strt., Indon.	153b	0°58'N	102°38'E
Ai Shan, mts., China (äī′shän)	162	37°27'N	120°35'E
Aisne, r., Fr. (ĕn)	117	49°28'N	3°32'E
Aitape, Pap. N. Gui. (ä-ē-tä′pā)	169	3°00'S	142°10'E
Aitkin, Mn., U.S. (āt′kĭn)	71	46°32'N	93°43'W
Aitolikó, Grc.	131	38°27'N	21°21'E
Aitos, Blg. (ä-ē′tōs)	131	42°42'N	27°17'E
Aitutaki, i., Cook Is. (ī-tōō-tä′kē)	195	19°00'S	162°00'W
Aiud, Rom. (ä′ē-ŏd)	119	46°19'N	23°40'E
Aiuruoca, Braz. (äē′ōō-rōōō′-ká)	99a	21°57'S	44°36'W
Aiuruoca, r., Braz.	99a	22°11'S	44°35'W
Aix-en-Provence, Fr. (ĕks-prō-văns)	117	43°32'N	5°27'E
Aix-les-Bains, Fr. (ĕks′-lā-baN′)	127	45°42'N	5°56'E
Aizpute, Lat. (ä′ēz-pōō-tĕ)	123	56°44'N	21°37'E
Aizuwakamatsu, Japan	166	37°27'N	139°51'E
Ajaccio, Fr. (ä-yät′chō)	110	41°55'N	8°42'E
Ajalpan, Mex. (ä-häl′pän)	89	18°21'N	97°14'W
Ajana, Austl. (ăj-ăn′ĕr)	174	28°00'S	114°45'E
Ajaria, state, Geor.	138	41°40'N	42°00'E
Ajdābiyah, Libya	185	30°56'N	20°16'E
Ajjer, Tassili-n-, plat., Alg.	184	25°40'N	6°57'E
Ajmah, Jabal al, mts., Egypt	153a	29°12'N	34°03'E
Ajman, U.A.E.	154	25°15'N	54°30'E
Ajmer, India (ŭj-mĕr′)	155	26°26'N	74°42'E
Ajo, Az., U.S. (ä′hō)	77	32°20'N	112°55'W
Ajuchitlán del Progreso, Mex. (ä-hōō-chet-län)	88	18°11'N	100°32'W
Ajusco, Mex. (ä-hōō′s-kō)	89a	19°13'N	99°12'W
Ajusco, Cerro, mtn., Mex. (sĕ′r-rō-ä-hōō′s-kō)	89a	19°12'N	99°16'W
Akaishi-dake, mtn., Japan (ä-kī-shē dä′kä)	167	35°30'N	138°00'E
Akashi, Japan (ä′kä-shē)	166	34°38'N	134°59'E
Aketi, D.R.C. (ä-kå-tē)	185	2°44'N	23°46'E
Akhaltsikhe, Geor. (äkà′l-tsĭ-kĕ)	137	41°40'N	42°50'E
Akhdar, Al Jabal al, mts., Libya	185	32°00'N	22°00'E
Akhḑar, Al Jabal al, mts., Oman	154	23°30'N	56°43'W
Akhisar, Tur. (äk-hĭs-sär′)	119	38°58'N	27°58'E
Akhtarskaya, Bukhta, b., Russia (bōōk′tä ák-tär′ská-yá)	133	45°53'N	38°22'E
Akhtopol, Blg. (äĸ′tō-pōl)	131	42°08'N	27°54'E
Akhunovo, Russia (ä-kû′nô-vô)	142a	54°13'N	59°36'E
Aki, Japan (ä′kē)	167	33°31'N	133°51'E
Akiak, Ak., U.S. (äk′yák)	63	61°00'N	161°02'W
Akimiski, i., Can. (ä-kĭ-mĭ′skĭ)	53	52°54'N	80°22'W
Akita, Japan (ä′kē-tä)	161	39°40'N	140°12'E
Akjoujt, Maur.	184	19°45'N	14°23'W
'Akko, Isr.	153a	32°56'N	35°05'E
Aklavik, Can. (äk′lä-vĭk)	50	68°28'N	135°26'W
Aklé'Aouâna, dunes, Afr.	188	18°07'N	6°00'W
Ako, Japan (ä′kō)	167	34°44'N	134°22'E
Akola, India (á-kō′lä)	155	20°47'N	77°00'E
Akordat, Erit.	185	15°34'N	37°54'E
Akpatok, i., Can. (äk′pá-tŏk)	53	60°30'N	67°10'W
Akranes, Ice.	116	64°18'N	21°40'W
Akron, Co., U.S. (ăk′rŭn)	78	40°09'N	103°14'W
Akron, Oh., U.S. (ăk′rŭn)	65	41°05'N	81°30'W
Aksaray, Tur. (äk-sá-rī′)	119	38°30'N	34°05'E
Akşehir, Tur. (äk′shå-hĕr)	119	38°20'N	31°20'E
Akşehir Gölü, l., Tur. (äk′shå-hĕr)	154	38°40'N	31°30'E
Aksha, Russia (äk′shá)	135	50°28'N	113°00'E
Aksu, China (ä-kŭ-sōō)	160	41°29'N	80°15'E
Akune, Japan (ä′kō-nå)	167	32°03'N	130°16'E
Akureyri, Ice. (ä-kō-rå′rē)	116	65°39'N	18°01'W
Akutan, i., Ak., U.S. (ä-kōō-tän′)	63a	53°58'N	169°54'W
Akwatia, Ghana	188	6°04'N	0°49'W
Alabama, state, U.S. (ăl-á-băm′á)	65	32°50'N	87°30'W
Alabama, r., Al., U.S. (ăl-á-băm′á)	65	31°20'N	87°39'W
Alabat, i., Phil. (ä-lä-bät′)	169a	14°14'N	122°05'E
Alacam, Tur. (ä-lä-chäm′)	137	41°30'N	35°40'E
Alacant, Spain	118	38°20'N	0°30'W
Alacranes, Cuba (ä-lä-krä′nås)	92	22°45'N	81°35'W
Al Aflaj, des., Sau. Ar.	154	24°00'N	44°47'E
Alagôas, state, Braz. (ä-lä-gō′äzh)	101	9°50'S	36°33'W
Alagoinhas, Braz. (ä-lä-gō-ēn′yäzh)	101	12°13'S	38°12'W
Alagón, Spain (ä-lä-gōn′)	128	41°46'N	1°07'W
Alagón, r., Spain (ä-lä-gōn′)	128	39°53'N	6°42'W
Alahuatán, r., Mex. (ä-lä-wä-tá′n)	88	18°30'N	100°00'W
Alajuela, C.R. (ä-lä-hwä′lä)	91	10°01'N	84°14'W
Alajuela, Lago, l., Pan. (ä-lä-hwä′lä)	86a	9°15'N	79°34'W
Alaköl, l., Kaz.	139	45°45'N	81°13'E
'Alalakeiki Channel, strt., Hi., U.S. (ä-lä-lä-kā′kē)	84a	20°40'N	156°30'W
Al 'Alamayn, Egypt	185	30°53'N	28°52'E
Al 'Amārah, Iraq	157	31°50'N	47°09'E
Alameda, Ca., U.S. (ăl-á-mā′dá)	64	37°46'N	122°15'W
Alameda, r., Ca., U.S. (ăl-á-mā′dá)	74b	37°36'N	122°02'W
Alaminos, Phil. (ä-lä-mē′nōs)	169a	16°09'N	119°58'E
Al 'Amiriyah, Egypt	119	31°01'N	29°52'E
Alamo, Mex. (ä′lä-mō)	89	20°55'N	97°41'W
Alamo, Ca., U.S. (ă′lá-mō)	74b	37°51'N	122°02'W
Alamo, Nv., U.S. (ă′lá-mō)	76	37°22'N	115°10'W
Alamo, r., Mex. (ä′lä-mō)	80	26°33'N	99°35'W
Alamogordo, N.M., U.S. (ăl-á-mō-gôr′dō)	77	32°55'N	106°00'W
Alamo Heights, Tx., U.S. (ă′lá-mō)	75d	29°28'N	98°27'W
Alamo Indian Reservation, I.R., N.M., U.S.	77	34°30'N	107°30'W
Alamo Peak, mtn., N.M., U.S. (ă′lá-mō pēk)	80	32°50'N	105°55'W
Alamosa, Co., U.S. (ăl-á-mō′sá)	77	37°25'N	105°50'W
Åland see Ahvenanmaa, is., Fin.	116	60°36'N	19°55'E
Alandskiy, Russia (ä-länt′skĭ)	142a	52°14'N	59°48'E
Alanga Arba, Kenya	191	0°07'N	40°25'E
Alanya, Tur.	119	36°40'N	32°10'E
Alaotra, l., Madag. (ä-lä-ō′trá)	187	17°15'S	48°17'E
Alapayevsk, Russia (ä-lä-pä′yĕfsk)	134	57°50'N	61°35'E
Al 'Aqabah, Jord.	154	29°32'N	35°00'E
Alaquines, Mex. (ä-lá-kē′nås)	88	22°07'N	99°35'W
Al 'Arish, Egypt (ä-rēsh′)	153a	31°08'N	33°48'E
Alaska, state, U.S. (á-lăs′ká)	64a	64°00'N	150°00'W
Alaska, Gulf of, b., Ak., U.S. (á-lăs′ká)	63	57°42'N	147°40'W
Alaska Highway, Ak., U.S. (á-lăs′ká)	63	63°00'N	142°00'W
Alaska Peninsula, pen., Ak., U.S. (á-lăs′ká)	63	55°50'N	162°10'W
Alaska Range, mts., Ak., U.S. (á-lăs′ká)	63	62°00'N	152°18'W
Al 'Atrūn, Sudan	185	18°13'N	26°44'E
Alatyr′, Russia (ä′lä-tür)	134	54°55'N	46°30'E
Alazani, r., Asia	138	41°05'N	46°40'E
Alba, Italy (äl′bä)	130	44°41'N	8°02'E
Albacete, Spain (äl-bä-thā′tā)	118	39°00'N	1°49'W
Albachten, Ger. (äl-bä′ĸ-tĕn)	127c	51°55'N	7°31'E
Alba de Tormes, Spain (äl-bá dā tôr′mās)	128	40°48'N	5°28'W
Alba Iulia, Rom. (äl-bä yōō′lyá)	119	46°05'N	23°32'E
Albani, Colli, hills, Italy	129d	41°46'N	12°45'E
Albania, nation, Eur. (ăl-bā′nĭ-á)	110	41°45'N	20°00'E
Albano, Lago, l., Italy (lä′-gō äl-bä′nō)	129d	41°45'N	12°44'E
Albano Laziale, Italy (äl-bä′nō lät-zē-ä′lā)	130	41°44'N	12°43'E
Albany, Austl. (ôl′bá nĭ)	174	35°00'S	118°00'E
Albany, Ca., U.S. (ôl′bá-nĭ)	74b	37°54'N	122°18'W
Albany, Ga., U.S. (ôl′bá-nĭ)	65	31°35'N	84°10'W
Albany, Mo., U.S. (ôl′bá-nĭ)	79	40°14'N	94°18'W
Albany, N.Y., U.S. (ôl′bá-nĭ)	65	42°40'N	73°50'W
Albany, Or., U.S. (ôl′bá-nĭ)	64	44°38'N	123°06'W
Albany, r., Can. (ôl′bá-nĭ)	53	51°45'N	83°30'W
Al Başrah, Iraq	154	30°35'N	47°59'E

ăt; finăl; rāte; senāte; ärm; ásk; sofá; fâre; ch-choose; dh-as th in other; bē; ēvent; bĕt; recĕnt; cratẽr; g-gō; gh-guttural g; bĭt; ĭ-short neutral; rīde; ĸ-guttural k as ch in German ich;

ng-sing; ŋ-baŋk; N-nasalized n; nŏd; cŏmmit; ōld; ôbey; ôrder; oi-boil; fōōd; ò-as oo in foot; ou-out; s-soft; sh-dish; th-thin; pūre; ûnite; ûrn; stŭd; circŭs; ü-as in French tu; ′-indeterminate vowel.

220 Almonte – Angeles PRONOUNCING INDEX

PLACE (Pronunciation)	PAGE	LAT.	LONG.
Almonte, r., Spain (äl-mōn′tā)	128	39°35′N	5°50′W
Almora, India	155	29°20′N	79°40′E
Al Mubarraz, Sau. Ar.	154	22°31′N	46°27′E
Al Mudawwarah, Jord.	153a	29°20′N	36°01′E
Al Mukhā (Mocha), Yemen	154	13°11′N	43°20′E
Almuñécar, Spain (äl-mo͞on-yā′kär)	128	36°44′N	3°43′W
Almyrós, Grc.	131	39°13′N	22°47′E
Alnön, i., Swe.	122	62°20′N	17°39′E
Aloha, Or., U.S. (ä′lồ-hā)	74c	45°29′N	122°52′W
Alor, Pulau, i., Indon. (ä′lồr)	169	8°07′S	125°00′E
Álora, Spain (ä′lồ-rä)	128	36°49′N	4°42′W
Alor Gajah, Malay.	153b	2°23′N	102°13′E
Alor Setar, Malay. (ä′lồr stär)	168	6°10′N	100°16′E
Alouette, r., Can. (ä-lo͞o-ĕt′)	74d	49°16′N	122°32′W
Alpena, Mi., U.S. (ăl-pē′nȧ)	65	45°05′N	83°30′W
Alpes Cotiennes, mts., Eur.	127	44°46′N	7°02′E
Alphen, Neth.	115a	52°07′N	4°38′E
Alpiarça, Port. (äl-pyär′sȧ)	128	39°38′N	8°37′W
Alpine, Tx., U.S. (ăl′pīn)	80	30°21′N	103°41′W
Alps, mts., Eur. (ălps)	112	46°18′N	8°42′E
Alpujarra, Col. (ȧl-po͞o-κȧ′rȧ)	100a	3°23′N	74°56′W
Al Qaḑārif, Sudan	185	14°03′N	35°11′E
Al Qāhirah *see* Cairo, Egypt	185	30°00′N	31°17′E
Al Qanţarah, Egypt	192d	30°51′N	32°20′E
Al Qaryah Ash Sharqīyah, Libya	185	30°36′N	13°13′E
Al Qaşr, Egypt	156	25°42′N	28°53′E
Al Qaţīf, Sau. Ar.	154	26°30′N	50°00′E
Al Qayşūmah, Sau. Ar.	154	28°15′N	46°20′E
Al Qunaytirah, Syria	153a	33°09′N	35°49′E
Al Qunfudhah, Sau. Ar.	154	19°08′N	41°05′E
Al Quşaymah, Egypt	153a	30°40′N	34°23′E
Al Quşayr, Egypt	185	26°14′N	34°11′E
Al Qusayr, Syria	153a	34°32′N	36°33′E
Als, i., Den. (äls)	122	55°06′N	9°40′E
Alsace, hist. reg., Fr. (äl-sȧ′s)	127	48°25′N	7°24′E
Altadena, Ca., U.S. (ăl-tä-dē′nä)	75a	34°12′N	118°08′W
Alta Gracia, Arg. (äl′tä grä′sề-a)	102	31°41′S	64°19′W
Altagracia, Ven.	100	10°42′N	71°34′W
Altagracia de Orituco, Ven.	101b	9°53′N	66°22′W
Altai Mountains, mts., Asia (äl′tī′)	160	49°11′N	87°15′E
Alta Loma, Ca., U.S. (ăl′tă lō′mä)	75a	34°07′N	117°35′W
Alta Loma, Tx., U.S. (ăl′tȧ lō-mȧ)	81a	29°22′N	95°05′W
Altamaha, r., Ga., U.S. (ôl-tȧ-mȧ-hô′)	83	31°50′N	82°00′W
Altamira, Braz. (äl-tä-mē′rä)	101	3°13′S	52°14′W
Altamira, Mex.	89	22°25′N	97°55′W
Altamirano, Arg. (äl-tä-mề-rā′nō)	102	35°26′S	58°12′W
Altamura, Italy (äl-tä-mo͞o′rä)	119	40°40′N	16°35′E
Altavista, Va., U.S. (äl-tä-vĭs′tȧ)	83	37°08′N	79°14′W
Altay, China (äl-tā)	160	47°52′N	86°50′E
Altenburg, Ger. (äl-tĕn-bo͝orgh)	124	50°59′N	12°27′E
Altenmarkt an der Triesting, Aus.	115e	48°02′N	16°00′E
Alter do Chão, Port. (äl-tĕr′dȯ shäɴ′ōɴ)	128	39°13′N	7°38′W
Altiplano, pl., Bol. (äl-tē-plȧ′nō)	100	18°38′S	68°20′W
Altlandsberg, Ger. (ält länts′bĕrgh)	115b	52°34′N	13°44′E
Alto, La., U.S. (ăl′tō)	81	32°21′N	91°52′W
Alto Marañón, r., Peru (ȧl′tô-mä-rän-yŏ′n)	100	8°18′S	77°13′W
Altomünster, Ger. (äl′tō-mün′stĕr)	115d	48°24′N	11°16′E
Alton, Can. (ôl′tŭn)	62d	43°52′N	80°05′W
Alton, Il., U.S. (ôl′tŭn)	65	38°53′N	90°11′W
Altona, Austl.	173a	37°52′S	144°50′E
Altona, Can.	57	49°06′N	97°33′W
Altona, Ger. (äl′tō-nȧ)	115c	53°33′N	9°54′E
Altoona, Al., U.S. (ăl-to͞o′nȧ)	82	34°01′N	86°15′W
Altoona, Pa., U.S. (ăl-to͞o′nȧ)	65	40°25′N	78°25′W
Altoona, Wa., U.S. (ăl-to͞o′nȧ)	74c	46°16′N	123°39′W
Alto Rio Doce, Braz. (äl′tồ-rē′ồ-dō′sĕ)	99a	21°02′S	43°23′W
Alto Songo, Cuba (äl-fō-sôŋ′gō)	93	20°10′N	75°45′W
Altotonga, Mex. (ȧl-tō-tôŋ′gä)	89	19°44′N	97°13′W
Alto Velo, i., Dom. Rep. (äl-tồ-vĕ′lō)	93	17°30′N	71°35′W
Altrincham, Eng., U.K. (ôl′trĭng-ăm)	114a	53°18′N	2°21′W
Altruppin, Ger. (ält ro͞o′ppēn)	115b	52°56′N	12°50′E
Altun Shan, mts., China (äl-tȯn shän)	160	36°58′N	85°09′E
Alturas, Ca., U.S. (ăl-to͞o′rȧs)	72	41°29′N	120°33′W
Altus, Ok., U.S. (ăl′tŭs)	78	34°38′N	99°20′W
Al ′Ubaylah, Sau. Ar.	157	21°59′N	50°57′E
Al-Uḑayyah, Sudan	185	12°06′N	28°16′E
Alūksne, Lat. (ä′lȯks-nĕ)	136	57°24′N	27°04′E
Alumette Island, i., Can. (ȧ-lü-mĕt′)	59	45°50′N	77°00′W
Alum Rock, Ca., U.S.	74b	37°23′N	121°50′W
Al ′Uqaylah, Libya	185	30°15′N	19°07′E
Al Uqşur, Egypt	185	25°38′N	32°59′E
Alushta, Ukr. (ä′lshȯ-tȧ)	133	44°39′N	34°23′E
Alva, Ok., U.S. (ăl′vȧ)	78	36°46′N	98°41′W
Alvarado, Mex. (äl-vä-rä′dhō)	89	18°48′N	95°45′W
Alvarado, Laguna de, l., Mex. (lä-gȯ′nä-dĕ-äl-vä-rȧ′dồ)	89	18°44′N	95°45′W
Älvdalen, Swe. (ĕlv′dä-lĕn)	122	61°14′N	14°04′E
Alverca, Port. (al-vĕr′kȧ)	129b	38°53′N	9°02′W
Alvesta, Swe. (äl-vĕs′tä)	122	56°55′N	14°29′E
Alvin, Tx., U.S. (ăl′vĭn)	81a	29°25′N	95°14′W
Alvinópolis, Braz. (äl-vēnō′pō-lềs)	99a	20°07′S	43°03′W
Alviso, Ca., U.S. (ăl-vī′sō)	74b	37°26′N	121°59′W
Al Wajh, Sau. Ar.	154	26°15′N	36°32′E
Alwar, India (ŭl′wŭr)	155	27°39′N	76°39′E
Al Wāsiţah, Egypt	192b	29°21′N	31°15′E
Alytus, Lith. (ä′lề-tȯs)	123	54°25′N	24°05′E
Amacuzac, r., Mex. (ä-mä-ko͞o-zȧk)	88	18°00′N	99°03′W
Amadeus, l., Austl. (ăm-ȧ-dē′ŭs)	174	24°30′S	131°25′E
Amadjuak, l., Can. (ä-mädj′wäk)	53	64°50′N	69°20′W
Amadora, Port.	129b	38°45′N	9°14′W
Amagasaki, Japan (ä′mä-gä-sä′kề)	167	34°43′N	135°25′E
Amakusa-Shimo, i., Japan (ämä-ko͞o′sä shē-mō)	166	32°24′N	129°35′E
Åmål, Swe. (ồ′mồl)	122	59°05′N	12°40′E
Amalfi, Col. (ä′mȧ′l-fē)	100a	6°55′N	75°04′W
Amalfi, Italy (ä-mä′l-fề)	129c	40°23′N	14°36′E
Amaliáda, Grc.	131	37°48′N	21°23′E
Amalner, India	158	21°07′N	75°06′E
Amambai, Serra de, mts., S.A.	101	20°06′S	57°08′W
Amami, i., Japan	161	28°10′N	129°55′E
Amapala, Hond. (ä-mä-pä′lä)	90	13°16′N	87°39′W
Amarante, Braz. (ä-mä-rän′tā)	101	6°17′S	42°43′W
Amargosa, r., Ca., U.S. (ȧ′mär-gō′sȧ)	76	35°55′N	116°45′W
Amarillo, Tx., U.S. (ăm-ȧ-rĭl′ō)	64	35°14′N	101°49′W
Amaro, Mount, mtn., Italy (ä-mä′rō)	118	42°07′N	14°07′E
Amasya, Tur. (ä-mä′sề-ȧ)	119	40°40′N	35°50′E
Amatenango, Mex. (ä-mä-tā-naŋ′gō)	89	16°30′N	92°29′W
Amatignak, i., Ak., U.S. (ä-mȧ′tề-näk)	63a	51°12′N	178°30′W
Amatique, Bahía de, b., N.A. (bä-ē′ä-dĕ-ä-mä-tē′kā)	90	15°58′N	88°50′W
Amatitlán, Guat. (ä-mä-tề-tlän′)	90	14°27′N	90°39′W
Amatlán de Cañas, Mex. (ä-mät-län′dā kän-yäs)	88	20°50′N	104°22′W
Amazon (Amazonas) (Solimões), r., S.A.	101	2°03′S	53°18′W
Amazonas, state, Braz. (ä-mä-thō′näs)	100	4°15′S	64°30′W
Ambāla, India (ŭm-bä′lŭ)	155	30°31′N	76°48′E
Ambalema, Col. (äm-bä-lā′mä)	100	4°47′N	74°45′W
Ambarchik, Russia (ŭm-bär′chĭk)	135	69°39′N	162°18′E
Ambarnāth, India	159b	19°12′N	73°10′E
Ambato, Ec. (äm-bä′tō)	100	1°15′S	78°30′W
Ambatondrazaka, Madag.	187	17°58′S	48°43′E
Amberg, Ger. (äm′bĕrgh)	124	49°26′N	11°51′E
Ambergris Cay, i., Belize (ăm′bẽr-grēs kāz)	90a	18°04′N	87°43′W
Ambergris Cays, is., T./C. Is.	93	21°20′N	71°40′W
Ambérieu-en-Bugey, Fr. (äɴ-bā-rề-u′)	127	45°57′N	5°21′E
Ambert, Fr. (äɴ-bĕr′)	126	45°32′N	3°41′E
Ambil Island, i., Phil. (äm′bēl)	169a	13°51′N	120°25′E
Ambler, Pa., U.S. (ăm′blẽr)	68f	40°09′N	75°13′W
Amboise, Fr. (äɴ-bwăz′)	126	47°25′N	0°56′E
Ambon, Indon.	169	3°45′S	128°17′E
Ambon, Pulau, i., Indon.	169	4°50′S	128°45′E
Ambositra, Madag. (äɴ-bồ-sē′trä)	187	20°31′S	47°28′E
Amboy, Il., U.S. (ăm′boi)	66	41°41′N	89°15′W
Amboy, Wa., U.S. (ăm′boi)	74c	45°55′N	122°27′W
Ambre, Cap d′, c., Madag.	187	12°06′S	49°15′E
Ambridge, Pa., U.S. (ăm′brĭdj)	69e	40°36′N	80°13′W
Ambrim, i., Vanuatu	175	16°25′S	168°15′E
Ambriz, Ang.	186	7°50′S	13°06′E
Amchitka, i., Ak., U.S. (ăm-chĭt′kȧ)	63a	51°25′N	178°10′E
Amchitka Passage, strt., Ak., U.S. (ăm-chĭt′kȧ)	63a	51°30′N	179°36′W
Amealco, Mex. (ä-mā-äl′kō)	88	20°12′N	100°08′W
Ameca, Mex. (ä-mĕ′kä)	86	20°34′N	104°02′W
Amecameca, Mex. (ä-mȧ-kä-mā′kä)	88	19°06′N	98°46′W
Ameide, Neth.	115a	51°57′N	4°57′E
Ameland, i., Neth.	121	53°29′N	5°54′E
Amelia, Oh., U.S. (ȧ-mēl′yā)	69f	39°01′N	84°12′W
American, South Fork, r., Ca., U.S. (ȧ-mĕr′ĭ-kăn)	76	38°43′N	120°45′W
Americana, Braz. (ä-mĕ-rề-kȧ′nȧ)	99a	22°46′S	47°19′W
American Falls, Id., U.S. (ȧ-mĕr′ĭ-kăn-fāls′)	73	42°45′N	112°53′W
American Falls Reservoir, res., Id., U.S. (ȧ-mĕr′ĭ-kăn-fāls′)	64	42°56′N	113°18′W
American Fork, Ut., U.S.	77	40°20′N	111°50′W
American Highland, plat., Ant.	178	72°00′S	79°00′E
American Samoa, dep., Oc.	2	14°20′S	170°00′W
Americus, Ga., U.S. (ȧ-mĕr′ĭ-kŭs)	65	32°04′N	84°15′W
Amersfoort, Neth. (ä′mĕrz-fōrt)	115a	52°08′N	5°23′E
Amery, Can. (ā′mẽr-ề)	51	56°34′N	94°03′W
Amery, Wi., U.S.	71	45°19′N	92°24′W
Ames, Ia., U.S. (āmz)	71	42°00′N	93°36′W
Amesbury, Ma., U.S. (āmz′bẽr-ề)	61a	42°51′N	70°56′W
Amfissa, Grc. (ȧm-fĭ′sȧ)	131	38°32′N	22°26′E
Amga, Russia (ŭm-gä′)	135	61°08′N	132°09′E
Amga, r., Russia	141	61°41′N	133°11′E
Amgun′, r., Russia	141	52°30′N	138°00′E
Amherst, Can. (ăm′hẽrst)	51	45°49′N	64°14′W
Amherst, Oh., U.S.	69d	41°24′N	82°13′W
Amherst, i., Can. (ăm′hẽrst)	59	44°08′N	76°45′W
Amiens, Fr. (ä-myäɴ′)	117	49°54′N	2°18′E
Amirante Islands, is., Sey.	5	6°02′S	52°30′E
Amisk Lake, l., Can.	57	54°35′N	102°13′W
Amistad Reservoir, res., N.A.	80	29°20′N	101°00′W
Amite, La., U.S. (ă-mēt′)	81	30°43′N	90°32′W
Amite, r., La., U.S.	81	30°45′N	90°48′W
Amity, Pa., U.S. (ăm′ĭ-tĭ)	69e	40°02′N	80°11′W
Amityville, N.Y., U.S. (ăm′ĭ-tĭ-vĭl)	68a	40°41′N	73°24′W
Amlia, i., Ak., U.S. (ȧ′mlề-ä)	63a	52°00′N	173°28′W
′Ammān, Jord. (äm′mȧn)	154	31°57′N	35°57′E
Ammersee, l., Ger. (äm′mĕr)	115d	48°00′N	11°08′E
Amnicon, r., Wi., U.S. (ăm′nề-kŏn)	75h	46°35′N	91°56′W
Amorgós, i., Grc. (ä-môr′gōs)	119	36°47′N	25°47′E
Amory, Ms., U.S. (āmồ-rē)	82	33°58′N	88°27′W
Amos, Can. (ā′mŭs)	51	48°31′N	78°04′W
Amoy *see* Xiamen, China	161	24°30′N	118°10′E
Amparo, Braz. (äm-pȧ′-rô)	99a	22°43′S	46°44′W
Amper, r., Ger. (äm′pĕr)	115d	48°18′N	11°32′E
Amposta, Spain (äm-pōs′tä)	129	40°42′N	0°34′E
Amqui, Can.	60	48°28′N	67°28′W
Amrāvati, India	155	20°58′N	77°47′E
Amritsar, India (ŭm-rĭt′sŭr)	155	31°43′N	74°52′E
Amstelveen, Neth.	115a	52°18′N	4°51′E
Amsterdam, Neth. (äm-stẽr-däm′)	110	52°21′N	4°52′E
Amsterdam, N.Y., U.S. (ăm′stĕr-dăm)	67	42°55′N	74°10′W
Amsterdam, Île, i., Afr.	178	37°52′S	77°32′E
Amstetten, Aus. (äm′stĕt-ĕn)	124	48°09′N	14°53′E
Am Timan, Chad (äm′tề-män′)	185	11°18′N	20°30′E
Amu Darya, r., Asia (ä-mȯ-dä′rēä)	134	38°30′N	64°00′E
Amukta Passage, strt., Ak., U.S. (ä-mo͞ok′tȧ)	63a	52°30′N	172°00′W
Amundsen Gulf, b., Can. (ä′mŭn-sĕn-gŭlf′)	52	70°17′N	123°28′W
Amundsen Sea, sea, Ant. (ä′mŭn-sĕn-sē′)	178	72°00′S	110°00′W
Amungen, l., Swe.	122	61°07′N	16°00′E
Amur, r., Asia	135	49°00′N	136°00′E
Amurskiy, Russia (ä-mûr′skĭ)	142a	52°35′N	59°36′E
Amurskiy, Zaliv, b., Russia (zä′lĭf ä-mȯr′skĭ)	166	43°20′N	131°40′E
Amusgos, Mex.	88	16°39′N	98°09′W
Amuyao, Mount, mtn., Phil. (ä-mo͞o-yä′ō)	169a	17°04′N	121°09′E
Amvrakikos Kólpos, b., Grc.	131	39°00′N	21°00′E
Amyun, Leb.	153a	34°18′N	35°48′E
Anabar, r., Russia (ăn-ȧ-bär′)	141	71°15′N	113°00′E
Anaco, Ven. (ä-nȧ′kô)	101b	9°29′N	64°27′W
Anaconda, Mt., U.S. (ăn-ȧ-kŏn′dȧ)	64	46°07′N	112°55′W
Anacortes, Wa., U.S. (ăn-ȧ-kôr′tĕz)	74a	48°30′N	122°37′W
Anadarko, Ok., U.S. (ăn-ȧ-där′kō)	78	35°05′N	98°14′W
Anadyr′, Russia (ŭ-nȧ-dĭr′)	135	64°47′N	177°01′E
Anadyr, r., Russia	141	65°30′N	172°45′E
Anadyrskiy Zaliv, b., Russia	134	64°10′N	178°00′W
Anaheim, Ca., U.S. (ăn′ȧ-hīm)	75a	33°50′N	117°55′W
Anahuac, Tx., U.S. (ä-nä′wäk)	81a	29°46′N	94°41′W
Ānai Mudi, mtn., India	159	10°10′N	77°00′E
Anama Bay, Can.	57	51°56′N	98°05′W
Ana María, Cayos, is., Cuba	92	21°25′N	78°50′W
Anambas, Kepulauan, is., Indon. (ä-näm-bäs)	168	2°41′N	106°38′E
Anamosa, Ia., U.S. (ăn-ȧ-mō′sȧ)	71	42°06′N	91°18′W
Anan′iv, Ukr.	137	47°43′N	29°59′E
Anapa, Russia (ȧ-nä′pä)	137	44°54′N	37°19′E
Anápolis, Braz. (ä-nȧ′pō-lềs)	101	16°17′S	48°47′W
Añatuya, Arg. (ȧ-nyä-to͞o′yȧ)	102	28°22′S	62°45′W
Anchieta, Braz. (ȧn-chyĕ′tä)	102b	22°49′S	43°24′W
Ancholme, r., Eng., U.K. (ăn′chŭm)	114a	53°28′N	0°27′W
Anchorage, Ak., U.S. (ăŋ′kẽr-åj)	64a	61°12′N	149°48′W
Anchorage, Ky., U.S.	69h	38°16′N	85°32′W
Anci, China (än-tsŭ)	162	39°31′N	116°41′E
Ancienne-Lorette, Can. (äɴ-syĕn′ lō-rĕt′)	62b	46°48′N	71°21′W
Ancón, Pan. (ăŋ-kōn′)	86a	8°55′N	79°32′W
Ancona, Italy (än-kō′nä)	110	43°37′N	13°32′E
Ancud, Chile (äŋ-ko͞odh′)	102	41°52′S	73°45′W
Ancud, Golfo de, b., Chile (gồl-fô-dĕ-äŋ-ko͞odh′)	102	41°15′S	73°00′W
Anda, China	164	46°20′N	125°20′E
Åndalsnes, Nor.	122	62°33′N	7°46′E
Andalucia, hist. reg., Spain (än-dä-lo͞o-sề′ă)	128	37°35′N	5°40′W
Andalusia, Al., U.S. (ăn-dȧ-lo͞o′zhĭȧ)	82	31°19′N	86°19′W
Andaman Islands, is., India (ăn-dȧ-măn′)	168	11°38′N	92°17′E
Andaman Sea, sea, Asia	168	12°44′N	95°45′E
Andarax, r., Spain	128	37°00′N	2°40′W
Anderlecht, Bel. (än′dĕr-lĕkt)	115a	50°49′N	4°16′E
Andernach, Ger. (än′dẽr-näk)	124	50°25′N	7°23′E
Anderson, Arg. (ȧ′n-dĕr-sōn)	99c	35°15′S	60°15′W
Anderson, Ca., U.S. (ăn′dẽr-sŭn)	72	40°28′N	122°19′W
Anderson, In., U.S. (ăn′dẽr-sŭn)	66	40°05′N	85°50′W
Anderson, S.C., U.S. (ăn′dẽr-sŭn)	65	34°30′N	82°40′W
Anderson, r., Can. (ăn′dẽr-sŭn)	52	68°32′N	125°12′W
Andes Mountains, mts., S.A. (ăn′dēz) (än′dās)	97	13°00′S	75°00′W
Andheri, neigh., India	159b	19°08′N	72°50′E
Andhra Pradesh, state, India	155	16°00′N	79°00′E
Andikýthira, i., Grc.	119	35°50′N	23°20′E
Andizhan, Uzb. (än-dề-zhän′)	139	40°45′N	72°22′E
Andong, Kor., S. (än′dŭng′)	161	36°31′N	128°42′E
Andongwei, China (än-dồŋ-wā)	162	35°08′N	119°19′E
Andorra, And. (än-dồr′rä)	129	42°38′N	1°30′E
Andorra, nation, Eur. (än-dồr′rä)	110	42°30′N	2°00′E
Andover, Ma., U.S. (ăn′dồ-vẽr)	61a	42°39′N	71°08′W
Andover, N.J., U.S. (ăn′dồ-vẽr)	68a	40°59′N	74°45′W
Andøya, i., Nor. (änd-ûề)	116	69°12′N	14°58′E
Andreanof Islands, is., Ak., U.S. (än-drå-ä′nồf-ī′ăndz)	64b	51°10′N	177°00′W
Andrelândia, Braz. (än-drĕ-lȧ′n-dyȧ)	99a	21°45′S	44°18′W
Andrew Johnson National Historic Site, rec., Tn., U.S. (ăn′dro͞o jŏn′sŭn)	83	36°15′N	82°55′W
Andrews, N.C., U.S. (ăn′dro͞oz)	82	35°12′N	83°48′W
Andrews, S.C., U.S. (ăn′dro͞oz)	83	33°25′N	79°32′W
Andria, Italy (än′drề-ä)	119	41°17′N	15°55′E
Andros, Grc. (ȧn′dhrồs)	131	37°50′N	24°54′E
Ándros, i., Grc. (ȧn′drồs)	119	37°59′N	24°55′E
Androscoggin, r., Me., U.S. (ăn-drŭs-kŏg′ĭn)	60	44°25′N	70°45′W
Andros Island, i., Bah. (ăn′drōs)	87	24°30′N	78°00′W
Anefis i-n-Darane, Mali	188	18°03′N	0°36′E
Anegasaki, Japan (ä′nå-gä-sä′kề)	167a	35°29′N	140°02′E
Aneityum, i., Vanuatu (ä-nå-ē′tề-ŭm)	175	20°15′S	169°49′E
Aneta, N.D., U.S. (ă-nē′tȧ)	70	47°41′N	97°57′W
Aneto, Pico de, mtn., Spain (pề′kō-dĕ-ä-nĕ′tồ)	112	42°35′N	0°38′E
Angamacutiro, Mex. (än′gä-mă-ko͞o-tē′rồ)	88	20°08′N	101°44′W
Angangueo, Mex. (än-gäŋ′gwå-ō)	88	19°36′N	100°18′W
Ang′angxi, China (äŋ-äŋ-shyē)	161	47°05′N	123°58′E
Angarsk, Russia	135	52°48′N	104°15′E
Ånge, Swe. (ồng′ā)	122	62°31′N	15°39′E
Ángel, Salto, wtfl., Ven. (säl′tō-ȧ′n-hĕl)	100	5°44′N	62°27′W
Ángel de la Guarda, i., Mex. (ȧ′n-hĕl-dĕ-lä-gwä′r-dä)	86	29°30′N	113°00′W
Angeles, Phil. (än′hå-lās)	169a	15°09′N	120°35′E

ăt; finăl; rāte; senåte; ärm; ȧsk; sofȧ; fâre; ch-choose; dh-as th in other; bē; ềvent; bĕt; recĕnt; cratẽr; g-gō; gh-guttural g; bĭt; ĭ-short neutral; rīde; κ-guttural k as ch in German ich;

PRONOUNCING INDEX

Ängelholm – Argentino 221

PLACE (Pronunciation)	PAGE	LAT.	LONG.
Ängelholm, Swe. (ĕng′ĕl-hôlm)	122	56°14′N	12°50′E
Angelina, r., Tx., U.S. (ăn-jė lē′nȧ)	81	31°30′N	94°53′W
Angels Camp, Ca., U.S. (ān′jĕls kămp′)	76	38°03′N	120°33′W
Ångermanälven, r., Swe.	116	64°10′N	17°30′E
Angermund, Ger. (ȧn′ngĕr-mŭnd)	127c	51°20′N	6°47′E
Angermünde, Ger. (äng′ĕr-mûn-dĕ)	124	53°02′N	14°00′E
Angers, Can. (äɴ-zhā′)	62c	45°31′N	75°29′W
Angers, Fr.	126	47°29′N	0°36′W
Angkor, hist., Camb. (äng′kôr)	168	13°52′N	103°50′E
Anglesey, i., Wales, U.K. (ăŋ′g′l-sė)	120	53°35′N	4°28′W
Angleton, Tx., U.S. (aŋ′g′l-tŭn)	81a	29°10′N	95°25′W
Angmagssalik, Grnld. (ȧŋ-mȧ′sȧ-lĭk)	49	65°40′N	37°40′W
Angoche, Ilha, i., Moz. (ė′lä-ăn-gō′chå)	187	16°20′S	40°00′E
Angol, Chile (aŋ-gōl′)	102	37°47′S	72°43′W
Angola, In., U.S. (ăŋ-gō′lȧ)	66	41°35′S	85°00′W
Angola, nation, Afr. (ăŋ-gō′lȧ)	186	14°15′S	16°00′E
Angora *see* Ankara, Tur.	154	39°55′N	32°50′E
Angoulême, Fr. (äŋ′gōō-lâm′)	126	45°40′N	0°09′E
Angra dos Reis, Braz. (aŋ′grä dōs rā′ės)	99a	23°01′S	44°17′W
Angri, Italy (ä′n-grė)	129c	40°30′N	14°35′E
Anguang, China (än-gůäŋ)	164	45°28′N	123°42′E
Anguilla, dep., N.A.	87	18°15′N	62°54′W
Anguilla Cays, is., Bah. (ăŋ-gwĭl′ȧ)	92	23°30′N	79°35′W
Anguille, Cape, c., Can. (kāp′-äŋ-gē′yĕ)	61	47°55′N	59°25′W
Anguo, China (än-gwô)	162	38°27′N	115°19′E
Anholt, i., Den. (än′hŏlt)	122	56°43′N	11°34′E
Anhui, prov., China (än-hwā)	161	31°30′N	117°15′E
Aniak, Ak., U.S. (ä-nyȧ′k)	63	61°32′N	159°35′W
Aniakchak National Monument, rec., Ak., U.S.	63	56°50′N	157°50′W
Animas, r., Co., U.S. (ä′nė-mȧs)	77	37°03′N	107°50′W
Anina, Rom. (ä-nē′nä)	131	45°03′N	21°50′E
Anita, Pa., U.S. (ȧ-nē′ȧ)	67	41°05′N	79°00′W
Aniva, Mys, c., Russia (mĭs ȧ-nē′vȧ)	166	46°08′N	143°13′E
Aniva, Zaliv, b., Russia (zä′lĭf ȧ-nē′vȧ)	166	46°30′N	143°00′E
Anjou, Can.	62a	45°37′N	73°33′W
Ankang, China (än-ĸäŋ)	160	32°38′N	109°10′E
Ankara, Tur. (än′kȧ-rȧ)	154	39°55′N	32°50′E
Anklam, Ger. (än′kläm)	124	53°52′N	13°43′E
Ankoro, D.R.C. (ăŋ-kō′rō)	186	6°45′S	26°57′E
Anloga, Ghana	188	5°47′N	0°50′E
Anlong, China (än-loŋ)	165	25°01′N	105°32′E
Anlu, China (än′lōō′)	165	31°18′N	113°40′E
Ann, Cape, c., Ma., U.S. (kāp′ăn′)	67	42°40′N	70°40′W
Anna, Russia (än′ä)	133	51°31′N	40°27′E
Anna, Il., U.S. (ăn′ȧ)	79	37°28′N	89°15′W
Annaba, Alg.	184	36°57′N	7°39′E
Annaberg-Bucholz, Ger. (än′ä-bĕrgh)	124	50°35′N	13°02′E
An Nafūd, des., Sau. Ar.	154	28°30′N	40°30′E
An Najaf, Iraq (än nä-jäf′)	154	32°00′N	44°25′E
An Nakhl, Egypt	153a	29°55′N	33°45′E
Annamese Cordillera, mts., Asia	168	17°34′N	105°38′E
Annapolis, Md., U.S. (ă-năp′ô-lĭs)	65	39°00′N	76°25′W
Annapolis Royal, Can.	60	44°45′N	65°31′W
Ann Arbor, Mi., U.S. (ăn är′bĕr)	65	42°15′N	83°45′W
An Nāşirīyah, Iraq	154	31°08′N	46°15′E
An Nawfalīyah, Libya	185	30°57′N	17°38′E
Annecy, Fr. (ȧn sē′)	127	45°54′N	6°07′E
Annemasse, Fr. (än′mäs′)	127	46°09′N	6°13′E
Annette Island, i., Ak., U.S.	54	55°13′N	131°30′W
An Nhon, Viet.	168	13°55′N	109°00′E
Annieopsquotch Mountains, mts., Can.	61	48°37′N	57°17′W
Anniston, Al., U.S. (ăn′ĭs-tŭn)	65	33°39′N	85°47′W
Annobón, i., Eq. Gui.	183	2°00′S	3°30′E
Annonay, Fr. (ȧn′ĭs-tsiŭn)	126	45°16′N	4°36′E
Annotto Bay, Jam. (än-nō′tō)	92	18°15′N	76°45′W
An Nuhūd, Sudan	185	12°39′N	28°18′E
Anoka, Mn., U.S. (å-nō′ká)	75g	45°12′N	93°24′W
Anori, Col. (ȧ-nō′rė)	100a	7°01′N	75°09′W
Áno Viánnos, Grc.	130a	35°02′N	25°26′E
Anpu, China (än-pōō)	160	21°28′N	110°00′E
Anqiu, China (än-chyô)	162	36°26′N	119°12′E
Ansbach, Ger. (äns′bäk)	124	49°18′N	10°35′E
Anse à Veau, Haiti (äɴs′ ä-vō′)	93	18°30′N	73°25′W
Anse d'Hainault, Haiti (äɴs′dĕnō)	93	18°30′N	74°25′W
Anserma, Col. (ȧ′n-sĕ′r-mȧ)	100a	5°13′N	75°47′W
Ansermanuevo, Col. (ȧ′n-sĕ′r-mȧ-nwĕ′vō)	100a	4°47′N	75°59′W
Anshan, China	164	41°00′N	123°00′E
Anshun, China (än-shōōn′)	160	26°12′N	105°50′E
Anson, Tx., U.S. (ăn′sŭn)	80	32°45′N	99°52′W
Anson Bay, b., Austl.	174	13°10′S	130°00′E
Ansŏng, Kor., S. (än′sŭng′)	166	37°00′N	127°12′E
Ansongo, Mali	188	15°40′N	0°30′E
Ansonia, Ct., U.S. (ăn-sōnĭ-ȧ)	67	41°20′N	73°05′W
Antalya, Tur. (än-tä′lė-ä) (ä-dä′lė-ä)	119	37°00′N	30°50′E
Antalya Körfezi, b., Tur.	119	36°40′N	31°20′E
Antananarivo, Madag.	187	18°51′S	47°40′E
Antarctica, cont.	178	80°15′S	127°00′E
Antarctic Peninsula, pen., Ant.	178	70°00′S	65°00′W
Antelope Creek, r., Wy., U.S. (ăn′tė-lōp)	73	43°29′N	105°42′W
Antequera, Spain (än-tĕ-kĕ′rä)	118	37°01′N	4°34′W
Anthony, Ks., U.S. (ăn′thô-nė)	78	37°08′N	98°01′W
Anthony Peak, mtn., Ca., U.S.	76	39°51′N	122°58′W
Anti Atlas, mts., Mor.	184	28°45′N	9°30′W
Antibes, Fr. (äɴ-tēb′)	127	43°36′N	7°12′E
Anticosti, Île d', i., Can. (ăn-tĭ-kŏs′tė)	53	49°30′N	62°00′W
Antigo, Wi., U.S. (ăn′tĭ-gō)	71	45°09′N	89°11′W
Antigonish, Can. (ăn-tĭ-gô-nėsh′)	61	45°35′N	61°55′W
Antigua, Guat. (än-tē′gwä)	86	14°32′N	90°43′W
Antigua, r., Mex.	89	19°16′N	96°36′W
Antigua and Barbuda, nation, N.A.	87	17°15′N	61°15′W
Antigua Veracruz, Mex. (än-tē′gwä vā-rä-krōōz′)	89	19°18′N	96°17′W
Antilla, Cuba (än-tē′lyä)	93	20°50′N	75°50′W
Antioch, Ca., U.S. (ăn′tĭ-ŏk)	74b	38°00′N	121°48′W
Antioch, Il., U.S.	69a	42°29′N	88°06′W
Antioch, Ne., U.S.	70	42°05′N	102°36′W
Antioquia, Col. (än-tė-ō′kėä)	100	6°34′N	75°49′W
Antioquia, dept., Col.	100a	6°48′N	75°42′W
Antlers, Ok., U.S. (ănt′lẽrz)	79	34°14′N	95°38′W
Antofagasta, Chile (än-tô-fä-gäs′tä)	102	23°32′S	70°21′W
Antofalla, Salar de, pl., Arg. (sȧ-lȧr′de ȧn′tō-fä′lä)	102	26°00′S	67°52′W
Antón, Pan. (än-tōn′)	87	8°24′N	80°15′W
Antongila, Helodrano, b., Madag.	187	16°15′S	50°15′E
Antônio Carlos, Braz. (än-tō′nėö-kȧ′r-lôs)	99a	21°19′S	43°45′W
António Enes, Moz. (ăn-to′nyŏ ĕn′ēs)	187	16°14′S	39°58′E
Antonito, Co., U.S. (ăn-tô-nē′tō)	78	37°04′N	106°01′W
Antonopole, Lat. (än′tô-nô-pō lyĕ)	123	56°19′N	27°11′E
Antony, Fr.	127b	48°45′N	2°18′E
Antsirabe, Madag. (änt-sė-rä′bā)	187	19°49′S	47°16′E
Antsiranana, Madag.	187	12°18′S	49°16′E
Antsla, Est. (änt′slȧ)	123	57°49′N	26°29′E
Antuco, vol., S.A. (än-tōō′kō)	102	37°30′S	72°30′W
Antwerp, Bel.	110	51°13′N	4°24′E
Antwerpen *see* Antwerp, Bel.	110	51°13′N	4°24′E
Anūpgarh, India (ŭ-nŏp′gŭr)	158	29°22′N	73°20′E
Anuradhapura, Sri L. (ŭ-nōō′rä-dŭ-pōō′rŭ)	159	8°24′N	80°25′E
Anxi, China (än-shyē)	160	40°36′N	95°49′E
Anyang, China (än′yäng)	161	36°05′N	114°22′E
Anykščiai, Lith. (anĭksh-chȧ′ė)	123	55°34′N	25°04′E
Anzhero-Sudzhensk, Russia (än′zhâ-rô-sŏd′zhĕnsk)	134	56°08′N	86°08′E
Anzio, Italy (änt′zė-ō)	130	41°28′N	12°39′E
Anzoátegui, dept., Ven. (ȧn-zôȧ′tĕ-gė)	101b	9°38′N	64°45′W
Aoba, i., Vanuatu	170f	15°25′S	167°50′E
Aomori, Japan (äô-mō′rė)	161	40°45′N	140°52′E
Aoraki (Cook, Mount), mtn., N.Z.	175a	43°27′S	170°13′E
Aosta, Italy (ä-ôs′tä)	130	45°45′N	7°20′E
Aouk, Bahr, r., Afr. (ä-ŏk′)	185	9°30′N	20°45′E
Aoukâr, reg., Maur.	188	18°00′N	9°40′W
Apalachicola, Fl., U.S. (ăp-ȧ-lăch-ĭ-kō′lȧ)	82	29°43′N	84°59′W
Apalachicola, r., Fl., U.S. (ăpȧ-lăch′ĭ-cōlā)	65	30°11′N	85°00′W
Apan, Mex. (ä-pȧ′n)	88	19°43′N	98°27′W
Apango, Mex. (ä-päŋ′gō)	88	17°41′N	99°22′W
Apaporis, r., S.A. (ä-pä-pô′rĭs)	100	0°48′N	72°32′W
Aparri, Phil. (ä-pär′rē)	168	18°15′N	121°40′E
Apasco, Mex. (ä-pȧ′s-kō)	88	20°33′N	100°43′W
Apatin, Serb. (ŏ′pŏ-tĭn)	131	45°40′N	19°00′E
Apatzingán de la Constitución, Mex.	88	19°07′N	102°21′W
Apeldoorn, Neth. (ä′pĕl-dōōrn)	117	52°14′N	5°55′E
Apennines *see* Appennino, mts., Italy	112	43°48′N	11°06′E
Apía, Col. (ȧ-pē′ä)	100a	5°07′N	75°58′W
Apia, Samoa	170a	13°50′S	171°44′W
Apipilulco, Mex. (ä-pĭ-pĭ-lōōl′kō)	88	18°09′N	99°40′W
Apishapa, r., Co., U.S. (ăp-ĭ-shä′pȧ)	78	37°40′N	104°08′W
Apizaco, Mex. (ä-pė-zä′kō)	88	19°18′N	98°11′W
Apo, Mount, mtn., Phil. (ä′pō)	169	6°56′N	125°05′E
Apopka, Fl., U.S. (ä-pŏp′kȧ)	83a	28°37′N	81°30′W
Apopka, Lake, l., Fl., U.S.	83a	28°38′N	81°50′W
Apostle Islands, is., Wi., U.S. (ă-pŏs′l)	71	47°05′N	90°55′W
Appalachia, Va., U.S. (ăpȧ-lăch′ĭ-ȧ)	83	36°54′N	82°49′W
Appalachian Mountains, mts., N.A. (ăp-ȧ-lăch′ĭ-ȧn)	65	37°20′N	82°00′W
Äppelbo, Swe. (ĕp-ĕl-bōō)	122	60°30′N	14°02′E
Appelhülsen, Ger. (ä′pĕl-hül′sĕn)	127c	51°55′N	7°26′E
Appennino, mts., Italy (äp-pĕn-nė′nô)	112	43°48′N	11°06′E
Appleton, Mn., U.S. (ăp′l-tŭn)	70	45°10′N	96°01′W
Appleton, Wi., U.S.	65	44°14′N	88°27′W
Appleton City, Mo., U.S.	79	38°10′N	94°02′W
Appomattox, r., Va., U.S. (ăp-ô-măt′ŭks)	83	37°22′N	78°09′W
Aprília, Italy (ȧ-prē′lyȧ)	130	41°36′N	12°40′E
Apsheronsk, Russia	138	44°28′N	39°44′E
Apt, Fr. (äpt)	127	43°54′N	5°19′E
Apure, r., Ven. (ä-pōō′rā)	100	8°08′N	68°46′W
Apurimac, r., Peru (ä-pōō-rė-mäk′)	100	11°39′S	73°48′W
Aqaba, Gulf of, b., (ä′kȧ-bȧ)	154	28°30′N	34°40′E
Aqabah, Wādī al, r., Egypt	153a	29°48′N	34°05′E
Aqmola *see* Astana, Kaz.	139	51°10′N	71°43′E
Aqtaū, Kaz.	139	43°35′N	51°05′E
Aqtöbe, Kaz.	139	50°20′N	57°00′E
Aquasco, Md., U.S. (ȧ′gwȧ′scô)	68e	38°35′N	76°44′W
Aquidauana, Braz. (ä-kē-däwä′nä)	101	20°24′S	55°46′W
Aquin, Haiti (ä-kăn′)	93	18°20′N	73°25′W
Ara, r., Japan (ä-rä)	167a	35°40′N	139°52′E
Arab, Baḩr al, r., Sudan	185	9°46′N	26°52′E
'Arabah, Wādī, val., Egypt	192b	29°02′N	32°10′E
Arabats'ka Strilka (Tongue of Arabat), spit, Ukr.	133	45°50′N	35°05′E
Arabi, La., U.S.	68d	29°58′N	90°01′W
Arabian Desert, des., Egypt (ȧ-rā′bĭ-ăn)	185	27°06′N	32°49′E
Arabian Sea, sea, (ȧ-rā′bĭ-ăn)	152	16°00′N	65°15′E
Aracaju, Braz. (ä-rä′kä-zhōō′)	101	11°00′S	37°01′W
Aracati, Braz. (ä-rä′kä-tē′)	101	4°31′S	37°41′W
Araçatuba, Braz. (ä-rä-sȧ-tōō′bä)	101	21°14′S	50°19′W
Aracena, Spain	128	37°53′N	6°34′W
Arachthos, r., Grc. (är′äĸ-thôs)	131	39°10′N	21°05′E
Aracruz, Braz. (ä-rä-krōō′s)	101	19°58′S	40°11′W
'Arad, Isr.	153a	31°20′N	35°15′E
Arad, Rom. (ŏ′rŏd)	119	46°10′N	21°18′E
Arafura Sea, sea, (ä-rä-fōō′rä)	169	8°40′S	130°00′E
Aragats, Gora, mtn., Arm.	138	40°32′N	44°14′E
Aragon, hist. reg., Spain (ä-rä-gōn′)	129	40°55′N	0°45′W
Aragón, r., Spain	128	42°35′N	1°10′W
Aragua, dept., Ven. (ä-rä′gwä)	101b	10°00′N	67°05′W
Aragua de Barcelona, Ven.	100	9°29′N	64°48′W
Araguaía, r., Braz. (ä-rä-gwä′yä)	101	8°37′S	49°43′W
Araguari, Braz. (ä-rä-gwä′rė)	101	18°43′S	48°03′W
Araguatins, Braz. (ä-rä-gwä-tēns)	101	5°41′S	48°04′W
Aragüita, Ven. (ärȧ-gwė′tȧ)	101b	10°13′N	66°28′W
Araj, oasis, Egypt (ä-räj′)	119	29°05′N	26°51′E
Arāk, Iran	154	34°08′N	49°57′E
Arakan Yoma, mts., Mya. (ŭ-rŭ-kŭn′yō′mȧ)	155	19°51′N	94°13′E
Aral, Kaz.	139	46°47′N	62°00′E
Aral Sea, sea, Asia	134	45°17′N	60°02′E
Aralsor köli, l., Kaz. (ȧ-räl′sôr′)	137	49°00′N	48°20′E
Aramberri, Mex. (ä-rȧm-bĕr-rē′)	88	24°05′N	99°47′W
Arana, Sierra, mts., Spain	128	37°17′N	3°28′W
Aranda de Duero, Spain (ä-rän′dä dā dwā′rō)	128	41°43′N	3°45′W
Arandas, Mex. (ä-rän′däs)	88	20°43′N	102°18′W
Aran Island, i., Ire. (ăr′ăn)	120	54°58′N	8°33′W
Aran Islands, is., Ire.	116	53°04′N	9°59′W
Aranjuez, Spain (ä-rän-hwāth′)	118	40°02′N	3°24′W
Aransas Pass, Tx., U.S. (ȧ-răn′sȧs pȧs)	81	27°55′N	97°09′W
Araouane, Mali	184	18°54′N	3°33′W
Arapkir, Tur. (ä-räp-kēr′)	119	39°00′N	38°10′E
Araraquara, Braz. (ä-rä-rä-kwȧ′rä)	101	21°47′S	48°08′W
Araras, Braz. (ä-rȧ′răs)	99a	22°21′S	47°22′W
Araras, Serra das, mts., Braz. (sĕ′r-rä-däs-ä-rä′räs)	101	18°03′S	53°23′W
Araras, Serra das, mts., Braz. (sĕ′r-rä-däs-ä-rä′räs)	102	23°30′S	53°00′W
Araras, Serra das, mts., Braz.	102b	22°24′S	43°15′W
Ararat, Austl. (ăr′ȧrăt)	175	37°17′S	142°56′E
Ararat, Mount, mtn., Tur.	154	39°50′N	44°20′E
Arari, l., Braz. (ä-rȧ′rē)	101	0°30′S	48°50′W
Araripe, Chapada do, hills, Braz. (shä-pä′dä-dô-ä-rä-rē′pĕ)	101	5°55′S	40°42′W
Araruama, Braz. (ä-rä-rōō-ä′mä)	99a	22°53′S	42°19′W
Araruama, Lagoa de, l., Braz.	99a	23°00′S	42°15′W
Aras, r., Asia (ȧ-räs)	154	39°15′N	47°10′E
Aratuípe, Braz. (ä-rä-tōō-ē′pĕ)	101	13°12′S	38°58′W
Arauca, Col. (ä-rou′kä)	100	6°56′N	70°45′W
Arauca, r., S.A.	100	7°13′N	68°43′W
Aravalli Range, mts., India (ä-rä′vŭ-lė)	155	24°15′N	72°40′E
Araya, Punta de, c., Ven. (pŭn′tȧ-dĕ-ä-rä′yȧ)	101b	10°40′N	64°15′W
Arayat, Phil. (ä-rä′yät)	169a	15°10′N	120°44′E
'Arbi, Sudan	185	20°36′N	29°57′E
Arbīl, Iraq	154	36°10′N	44°00′E
Arboga, Swe. (är-bō′gä)	122	59°26′N	15°50′E
Arborea, Italy (är-bō-rĕ′ä)	130	39°50′N	8°36′E
Arbroath, Scot., U.K. (är-brōth′)	120	56°36′N	2°25′W
Arcachon, Fr. (är-ĸä-shôɴ′)	117	44°39′N	1°12′W
Arcachon, Bassin d', Fr. (bä-sĕn′ där-kä-shôn′)	126	44°42′N	1°50′W
Arcadia, Ca., U.S. (är-kā′dĭ-ȧ)	75a	34°08′N	118°02′W
Arcadia, Fl., U.S.	83a	27°12′N	81°51′W
Arcadia, La., U.S.	81	32°33′N	92°56′W
Arcadia, Wi., U.S.	71	44°15′N	91°30′W
Arcata, Ca., U.S. (är-kȧ′tȧ)	72	40°54′N	124°05′W
Arc Dome Mountain, mtn., Nv., U.S. (ärk dōm)	76	38°51′N	117°21′W
Arcelia, Mex. (är-sā′lė-ä)	88	18°19′N	100°14′W
Archbald, Pa., U.S. (ärch′bôld)	67	41°30′N	75°35′W
Arches National Park, rec., Ut., U.S. (är′ches)	77	38°45′N	109°35′W
Archidona, Ec. (är-chė-do′nä)	100	1°01′S	77°49′W
Archidona, Spain (är-chē-dô′nä)	128	37°08′N	4°24′W
Arcis-sur-Aube, Fr. (är-sēs′sûr-ōb′)	126	48°31′N	4°04′E
Arco, Id., U.S. (är′kô)	73	43°39′N	113°15′W
Arcola, Tx., U.S.	81a	29°30′N	95°28′W
Arcola, Va., U.S. (ăr′cōlă)	68e	38°57′N	77°32′W
Arcos de la Frontera, Spain (är′kōs-dĕ-lä-frôn-tĕ′rä)	128	36°44′N	5°48′W
Arctic Ocean, o.	198	85°00′N	170°00′E
Arda, r., Blg. (är′dä)	131	41°36′N	25°18′E
Ardabīl, Iran	154	38°15′N	48°00′E
Ardahan, Tur. (är-dȧ-hän′)	137	41°10′N	42°40′E
Ardatov, Russia (är-dȧ-tôf′)	136	54°58′N	46°10′E
Ardennes, mts., Eur. (ȧr-dĕn′)	117	50°01′N	5°12′E
Ardila, r., Eur. (är-dē′lȧ)	128	38°10′N	7°15′W
Ardmore, Ok., U.S. (ärd′mōr)	64	34°10′N	97°08′W
Ardmore, Pa., U.S.	68f	40°01′N	75°18′W
Ardrossan, Can. (är-dros′ȧn)	62g	53°33′N	113°08′W
Ardsley, Eng., U.K. (ärdz′lė)	114a	53°43′N	1°33′W
Åre, Swe.	116	63°12′N	13°12′E
Arecibo, P.R. (ä-rå-sē′bō)	87b	18°28′N	66°45′W
Areia Branca, Braz. (ä-rĕ′yä-brȧ′n-kă)	101	4°58′S	37°02′W
Arena, Point, c., Ca., U.S. (ä-rā′nȧ)	76	38°57′N	123°40′W
Arenas, Punta, c., Ven. (pȯn′tä-rĕ′näs)	101b	10°57′N	64°24′W
Arenas de San Pedro, Spain	128	40°12′N	5°04′W
Arendal, Nor. (ä′rĕn-däl)	122	58°29′N	8°44′E
Arendonk, Bel.	115a	51°19′N	5°07′E
Arequipa, Peru (ä-rå-kē′pä)	100	16°27′S	71°30′W
Arezzo, Italy (ä-rĕt′sō)	118	43°28′N	11°54′E
Arga, r., Spain (är′gä)	128	42°35′N	1°55′W
Arganda, Spain (är-gän′dä)	129a	40°18′N	3°27′W
Argazi, l., Russia (är′gä-zĭ)	142a	55°24′N	60°37′E
Argazi, r., Russia	142a	55°33′N	57°30′E
Argentan, Fr. (ȧr-zhäɴ-täɴ′)	126	48°45′N	0°01′W
Argentat, Fr. (ȧr-zhäɴ-tä′)	126	45°07′N	1°57′E
Argenteuil, Fr. (är-zhäɴ-tû′y′)	126	48°56′N	2°15′E
Argentina, nation, S.A. (är-jĕn-tē′nȧ)	102	35°30′S	67°00′W
Argentino, l., Arg. (är-ĸĕn-tē′nō)	102	50°15′S	72°45′W

ng-sing; ŋ-baŋk; ɴ-nasalized n; nŏd; cŏmmit; ōld; ôbey; ôrder; oi-boil; fōōd; ȯ-as oo in foot; ou-out; s-soft; sh-dish; th-thin; pūre; ůnite; ûrn; stŭd; circŭs; ü-as in French tu; ′-indeterminate vowel.

222 Argenton-sur-Creuse – Athy PRONOUNCING INDEX

PLACE (Pronunciation)	PAGE	LAT.	LONG.
Argenton-sur-Creuse, Fr. (är-zhän'tôN-sür-krôs)	126	46°34'N	1°28'E
Argolikos Kólpos, b., Grc.	131	37°20'N	23°00'E
Argonne, mts., Fr. (ä'r-gôn)	127	49°21'N	5°54'E
Argos, Grc. (är'gŏs)	131	37°38'N	22°45'E
Argostóli, Grc.	131	38°10'N	20°30'E
Arguello, Point, c., Ca., U.S. (är-gwāl'yō)	76	34°35'N	120°40'W
Arguin, Cap d', c., Maur.	184	20°28'N	17°46'W
Argun', r., Asia (ȧr-gōōn')	135	50°00'N	119°00'E
Argungu, Nig.	189	12°45'N	4°31'E
Argyle, Can. (är'gīl)	62f	50°11'N	97°27'W
Argyle, Mn., U.S.	70	48°21'N	96°48'W
Århus, Den. (ôr'hōōs)	116	56°09'N	10°10'E
Ariakeno-Umi, b., Japan (ä-rē'ä-kä'nō ōō'nē)	167	33°03'N	130°18'E
Ariake-Wan, b., Japan (ä'rē-ä'kå wän)	167	31°19'N	131°15'E
Ariano, Italy (ä-rē-ä'nō)	130	41°09'N	15°11'E
Ariari, r., Col. (ä-ryà'rē)	100a	3°34'N	73°42'W
Aribinda, Burkina	188	14°14'N	0°52'W
Arica, Chile (ä-rē'kä)	100	18°34'S	70°14'W
Arichat, Can. (ä-rĭ-shät')	61	45°31'N	61°01'W
Ariège, r., Fr. (à-rê-ězh')	126	43°26'N	1°29'E
Ariel, Wa., U.S. (ā'rĭ-ĕl)	74c	45°57'N	122°34'W
Arieş, r., Rom.	125	46°25'N	23°15'E
Ariguanabo, Lago de, l., Cuba (lä'gô-dĕ-ä-rē-gwä-nä'bô)	93a	22°52'N	82°33'W
Arikaree, r., Co., U.S. (ä-rĭ-kȧ-rē')	78	39°51'N	102°18'W
Arima, Japan (ä'rē-mä')	167b	34°48'N	135°16'E
Aringay, Phil. (ä-rĭŋ-gä'ê)	169a	16°25'N	120°20'E
Arinos, r., Braz. (ä-rē'nôzh)	101	12°09'S	56°49'W
Aripuanã, r., Braz. (à-rê-pwän'yȧ)	101	7°06'S	60°29'W
'Arīsh, Wādī al, r., Egypt (à-rēsh')	153a	30°36'N	34°07'E
Aristazabal Island, i., Can.	54	52°30'N	129°20'W
Arizona, state, U.S. (ăr-ĭ-zō'nȧ)	64	34°00'N	113°00'W
Arjona, Spain (är-hō'nä)	128	37°58'N	4°03'W
Arka, r., Russia	141	60°45'N	142°30'E
Arkabutla Lake, res., Ms., U.S. (är-kȧ-bŭt'lȧ)	82	34°48'N	90°00'W
Arkadelphia, Ar., U.S. (är-kȧ-dĕl'fĭ-ȧ)	79	34°06'N	93°05'W
Arkansas, state, U.S. (är'kăn-sô) (är-kăn'sȧs)	65	34°50'N	93°40'W
Arkansas, r., U.S.	64	37°30'N	97°00'W
Arkansas City, Ks., U.S.	79	37°04'N	97°02'W
Arkhangelsk (Archangel), Russia (ár-kän'gĕlsk)	134	64°30'N	40°25'E
Arkhangel'skoye, Russia (är-kän-gĕl'skô-yĕ)	142a	54°25'N	56°48'E
Arklow, Ire. (ärk'lō)	120	52°47'N	6°10'W
Arkonam, India (är-kō-näm')	159	13°05'N	79°43'E
Arlanza, r., Spain (är-län-thä')	128	42°08'N	3°45'W
Arlanzón, r., Spain (är-län-thōn')	128	42°12'N	3°58'W
Arlberg Tunnel, trans., Aus. (ärl'bĕrgh)	124	47°05'N	10°15'E
Arles, Fr. (ärl)	126	43°42'N	4°38'E
Arlington, S. Afr.	192c	28°02'S	27°52'E
Arlington, Ga., U.S. (är'lĭng-tun')	82	31°25'N	84°42'W
Arlington, Ma., U.S.	61a	42°26'N	71°13'W
Arlington, S.D., U.S. (är'lĕng-tŭn)	70	44°23'N	97°09'W
Arlington, Tx., U.S. (är'lĭng-tŭn)	75c	32°44'N	97°07'W
Arlington, Va., U.S.	68e	38°55'N	77°10'W
Arlington, Vt., U.S.	67	43°05'N	73°05'W
Arlington, Wa., U.S.	74a	48°11'N	122°08'W
Arlington Heights, Il., U.S. (är'lĕng-tŭn-hī'ts)	69a	42°05'N	87°59'W
Arltunga, Austl. (ärl-tŏŋ'gȧ)	174	23°19'S	134°45'E
Arma, Ks., U.S. (är'mȧ)	79	37°34'N	94°43'W
Armagh, Can. (är-mä') (är-mäk')	62b	46°45'N	70°36'W
Armagh, N. Ire., U.K.	116	54°21'N	6°25'W
Armant, Egypt (är-mänt')	192b	25°37'N	32°32'E
Armaro, Col. (är-má'rō)	100a	4°58'N	74°54'W
Armavir, Russia (ár-má-vĭr')	134	45°00'N	41°00'E
Armenia, Col. (ȧr-mĕ'nē-ä)	100	4°33'N	75°40'W
Armenia, El Sal. (är-mā'nē-ä)	90	13°44'N	89°31'W
Armenia, nation, Asia	134	41°00'N	44°39'E
Armentières, Fr. (ár-mäN-tyâr')	126	50°43'N	2°53'E
Armeria, Río de, r., Mex. (rē'ō-dĕ-är-mā-rē'ä)	88	19°36'N	104°10'W
Armherstburg, Can. (ärm'hĕrst-bōōrgh)	58	42°06'N	83°06'W
Armians'k, Ukr.	133	46°06'N	33°42'E
Armidale, Austl. (är'mĭ-dāl)	175	30°27'S	151°50'E
Armour, S.D., U.S. (är'mẽr)	70	43°18'N	98°21'W
Armstrong Station, Can. (ärm'strŏng)	51	50°21'N	89°00'W
Arnedo, Spain (är-nā'dō)	128	42°12'N	2°03'W
Arnhem, Neth. (ärn'hĕm)	117	51°58'N	5°56'E
Arnhem, Cape, c., Austl.	174	12°15'S	137°00'E
Arnhem Land, reg., Austl. (ärn'hĕm-lănd)	174	13°15'S	133°00'E
Arno, r., Italy (ä'r-nō)	118	43°30'N	11°00'E
Arnold, Eng., U.K. (är'nŭld)	114a	53°00'N	1°08'W
Arnold, Mn., U.S. (är'nŭld)	75h	46°53'N	92°06'W
Arnold, Pa., U.S.	69e	40°35'N	79°45'W
Arnprior, Can. (ärn-prī'ẽr)	59	45°25'N	76°20'W
Arnsberg, Ger. (ärns'bĕrgh)	127c	51°25'N	8°02'E
Arnstadt, Ger. (ärn'shtät)	124	50°51'N	10°57'E
Aroab, Nmb. (är'ō-ȧb)	186	25°40'S	19°45'E
Aroostook, r., Me., U.S. (ȧ-rōōs'tŏŏk)	60	46°44'N	68°15'W
Aroroy, Phil. (ä-rō-rō'ē)	169a	12°30'N	123°24'E
Arpajon, Fr. (ăr-pä-jô'n)	127b	48°35'N	2°15'E
Arpoador, Ponta do, c., Braz. (pō'n-tä-dō-är'pōä-dō'r)	102b	22°59'S	43°11'W
Arraiolos, Port. (är-rī-ō'lōzh)	128	38°47'N	7°59'W
Ar Ramādī, Iraq	154	33°26'N	43°19'E
Arran, Island of, Scot., U.K. 7	120	55°25'N	5°25'W
Ar Rank, Sudan	185	11°45'N	32°53'E
Arras, Fr. (à-räs')	117	50°21'N	2°40'E
Ar Rawḍah, Egypt	192b	27°47'N	30°52'E
Arrecifes, Arg. (är-rå-sē'fås)	99c	34°03'S	60°05'W
Arrecifes, r., Arg.	99c	34°07'S	59°50'W
Arrée, Monts d', mts., Fr. (är-rā')	126	48°27'N	4°00'W
Arriaga, Mex. (är-rēä'gä)	89	16°15'N	93°54'W
Arrone, r., Italy	129d	41°57'N	12°17'E
Arrow Creek, r., Mt., U.S. (ăr'ō)	73	47°29'N	109°53'W
Arrowhead, Lake, l., Ca., U.S. (lăk ăr'ōhĕd)	75a	34°17'N	117°13'W
Arrowrock Reservoir, res., Id., U.S. (ăr'ō-rŏk)	72	43°40'N	115°30'W
Arroya Arena, Cuba (är-rō'yä-rĕ'nä)	93a	23°01'N	82°30'W
Arroyo de la Luz, Spain (är-rō'yō-dĕ-lä-lōō'z)	128	39°39'N	6°46'W
Arroyo Seco, Mex. (är-rō'yō sä'kō)	88	21°31'N	99°44'W
Ar Rub' al Khālī, des., Asia	154	20°00'N	51°00'E
Ar Ruṭbah, Iraq	157	33°02'N	40°17'E
Arsen'yev, Russia	135	44°13'N	133°32'E
Arsinskiy, Russia (är-sĭn'skĭ)	142a	53°46'N	59°54'E
Árta, Grc. (är'tā)	119	39°08'N	21°02'E
Arteaga, Mex. (är-tā-ä'gä)	80	25°28'N	100°50'W
Artëm, Russia (ȧr-tyôm')	135	43°28'N	132°29'E
Artemisa, Cuba (är-tå-mē'sä)	92	22°50'N	82°45'W
Artemivs'k, Ukr.	137	48°37'N	38°00'E
Artesia, N.M., U.S. (är-tē'sĭ-ȧ)	78	32°44'N	104°23'W
Arthabaska, Can.	59	46°03'N	71°54'W
Arthur's Town, Bah.	93	24°40'N	75°40'W
Arti, Russia (är'tĭ)	142a	56°20'N	58°38'E
Artibonite, r., N.A. (är-tē-bō-nē'tä)	93	19°00'N	72°25'W
Aru, Kepulauan, is., Indon.	169	6°20'S	133°00'E
Arua, Ug. (ä'rōō-ä)	185	3°01'N	30°55'E
Aruba, dep., N.A. (ä-rōō'bȧ)	87	12°29'N	70°00'W
Arunachal Pradesh, state, India	155	27°35'N	92°56'E
Arusha, Tan. (ȧ-rōō'shä)	186	3°22'S	36°41'E
Arvida, Can.	51	48°26'N	71°11'W
Arvika, Swe. (är-vē'kȧ)	122	59°41'N	12°35'E
Arzamas, Russia (är-zä-mäs')	136	55°20'N	43°52'E
Arziw, Alg.	118	35°50'N	0°20'W
Arzúa, Spain	128	42°54'N	8°19'W
Aš, Czech Rep. (äsh')	124	50°12'N	12°13'E
Asahi-Gawa, r., Japan (ä-sä'hē-gä'wä)	167	35°01'N	133°40'E
Asahikawa, Japan	161	43°50'N	142°09'E
Asaka, Japan (ä-sä'kä)	167a	35°47'N	139°36'E
Asansol, India	155	23°45'N	86°58'E
Asbest, Russia (äs-bĕst')	136	57°02'N	61°28'E
Asbestos, Can. (ăs-bĕs'tōs)	59	45°49'N	71°52'W
Asbestovskiy, Russia	142a	57°46'N	61°23'E
Asbury Park, N.J., U.S. (ăz'bẽr-ĭ)	68a	40°13'N	74°01'W
Ascención, Bahía de la, b., Mex.	90a	19°39'N	87°30'W
Ascensión, Mex. (äs-sĕn-sē-ōn')	88	24°21'N	99°54'W
Ascension, i., St. Hel. (ȧ-sĕn'shŭn)	183	8°00'S	13°00'W
Ascent, S. Afr. (ăs-ĕnt')	192c	27°14'S	29°06'E
Aschaffenburg, Ger. (ä-shäf'ĕn-bōrgh)	124	49°58'N	9°12'E
Ascheberg, Ger. (ä'shĕ-bĕrg)	127c	51°47'N	7°38'E
Aschersleben, Ger. (äsh'ĕrs-lā-bĕn)	124	51°46'N	11°28'E
Ascoli Piceno, Italy (äs'kô-lēpē-chā'nō)	130	42°50'N	13°55'E
Aseb, Erit.	185	12°52'N	43°39'E
Asenovgrad, Blg.	131	42°00'N	24°49'E
Aseri, Est. (ȧ'sĕ-rĭ)	123	59°26'N	26°58'E
Asha, Russia (ä'shä)	142a	55°01'N	57°17'E
Ashabula, l., N.D., U.S. (ăsh'ȧ-bū-lä)	70	47°07'N	97°51'W
Ashan, Russia (ä'shän)	142a	57°08'N	56°25'E
Ashbourne, Eng., U.K. (ăsh'bŭrn)	114a	53°01'N	1°44'W
Ashburn, Ga., U.S. (ăsh'bŭrn)	82	31°42'N	83°42'W
Ashburn, Va., U.S.	68e	39°02'N	77°30'W
Ashburton, r., Austl. (ăsh'bûr-tŭn)	174	22°30'S	115°30'E
Ashby-de-la-Zouch, Eng., U.K. (ăsh'bĭ-dĕ-lȧ zōōsh')	114a	52°44'N	1°23'W
Ashdod, Isr.	153a	31°46'N	34°39'E
Ashdown, Ar., U.S. (ăsh'doun)	79	33°41'N	94°07'W
Asheboro, N.C., U.S. (ăsh'bŭr-ō)	83	35°41'N	79°50'W
Asherton, Tx., U.S. (ăsh'ẽr-tŭn)	80	28°26'N	99°45'W
Asheville, N.C., U.S. (ăsh'vĭl)	65	35°35'N	82°35'W
Ash Fork, Az., U.S.	77	35°13'N	112°29'W
Ashgabat, Turkmen.	139	37°57'N	58°23'E
Ashikaga, Japan (ä'shē-kä'gȧ)	167	36°22'N	139°26'E
Ashiya, Japan (ä'shē-yä')	167	33°54'N	130°40'E
Ashiya, Japan	167b	34°44'N	135°18'E
Ashizuri-Zaki, c., Japan (ä-shē-zò-rē zä-kē)	166	32°43'N	133°04'E
Ashland, Al., U.S. (ăsh'lȧnd)	82	33°15'N	85°50'W
Ashland, Ks., U.S.	78	37°11'N	99°46'W
Ashland, Ky., U.S.	66	38°25'N	82°40'W
Ashland, Ma., U.S.	61a	42°16'N	71°28'W
Ashland, Me., U.S.	60	46°37'N	68°26'W
Ashland, Ne., U.S.	70	41°02'N	96°23'W
Ashland, Oh., U.S.	66	40°50'N	82°15'W
Ashland, Or., U.S.	72	42°12'N	122°42'W
Ashland, Pa., U.S.	67	40°45'N	76°20'W
Ashland, Wi., U.S.	65	46°34'N	90°55'W
Ashley, N.D., U.S. (ăsh'lē)	70	46°03'N	99°23'W
Ashley, Pa., U.S.	67	41°15'N	75°55'W
Ashmūn, Egypt (ȧsh-mōōn')	192b	30°19'N	30°57'E
Ashmyany, Bela.	123	54°27'N	25°55'E
Ashqelon, Isr. (äsh'kĕ-lōn)	153a	31°40'N	34°36'E
Ash Shabb, Egypt (shĕb)	185	22°34'N	29°52'E
Ash Shallūfah, Egypt (shäl'lō-fȧ)	192b	30°09'N	32°33'E
Ash Shaqrā', Sau. Ar.	154	25°10'N	45°08'E
Ash Shāriqah, U.A.E.	157	25°22'N	55°23'E
Ash Shawbak, Jord.	153a	30°31'N	35°35'E
Ash Shiḥr, Yemen	154	14°45'N	49°32'E
Ashtabula, Oh., U.S. (ăsh-tȧ-bū'lȧ)	65	41°55'N	80°50'W
Ashton, Id., U.S. (ăsh'tŭn)	73	44°04'N	111°28'W
Ashton-in-Makerfield, Eng., U.K. (ăsh'tŭn-ĭn-māk'ẽr-fēld)	114a	53°29'N	2°39'W
Ashton-under-Lyne, Eng., U.K. (ăsh'tŭn-ŭn-dẽr-līn')	114a	53°29'N	2°04'W
Ashuanipi, l., Can. (ăsh-wȧ-nĭp'ĭ)	53	52°40'N	67°42'W
Ashukino, Russia (ȧ-shōō'kinô)	142b	56°10'N	37°57'E
Asia, cont.	152	50°00'N	100°00'E
Asia Minor, reg., Tur. (ā'zhȧ)	113	38°18'N	31°18'E
Asientos, Mex. (ä-sē-ĕn'tōs)	88	22°13'N	102°05'W
Asilah, Mor.	128	35°30'N	6°05'W
Asinara, i., Italy	130	41°02'N	8°22'E
Asinara, Golfo dell', b., Italy (gôl'fō-dĕl-ä-sē-nä'rä)	130	40°58'N	8°28'E
Asīr, reg., Sau. Ar. (ä-sēr')	154	19°30'N	42°00'E
Askarovo, Russia (äs-kä-rō'vô)	142a	53°21'N	58°32'E
Askersund, Swe. (äs'kĕr-sònd)	122	58°43'N	14°53'E
Askino, Russia (äs'kĭ-nô)	142a	56°06'N	56°29'E
Asmara *see* Asmera, Erit.	184	15°17'N	38°56'E
Asmera, Erit. (äs-mä'rä)	185	15°17'N	38°56'E
Asnieres, Fr. (ä-nyâr')	127b	48°55'N	2°18'E
Asosa, Eth.	185	10°13'N	34°28'E
Asotin, Wa., U.S. (ȧ-sō'tĭn)	72	46°19'N	117°01'W
Aspen, Co., U.S. (ăs'pĕn)	77	39°15'N	106°55'W
Asperen, Neth.	115a	51°52'N	5°07'E
Aspy Bay, b., Can. (ăs'pē)	61	46°55'N	60°25'W
Aş Şaff, Egypt	192b	29°33'N	31°23'E
As Sallūm, Egypt	185	31°35'N	25°05'E
As Salt, Jord.	153a	32°02'N	35°44'E
Assam, state, India (ăs-săm')	155	26°00'N	91°00'E
As Samāwah, Iraq	157	31°18'N	45°17'E
Assens, Den. (äs'sĕns)	122	55°16'N	9°54'E
As Sinbillāwayn, Egypt	192b	30°53'N	31°37'E
Assini, C. Iv. (ȧ-sē-nē')	184	4°52'N	3°16'W
Assiniboia, Can.	50	49°38'N	105°59'W
Assiniboine, r., Can. (ȧ-sĭn'ĭ-boin)	57	50°03'N	97°57'W
Assiniboine, Mount, mtn., Can.	55	50°52'N	115°39'W
Assis, Braz. (ä-sē's)	101	22°39'S	50°21'W
Assisi, Italy	118	43°04'N	12°37'E
As-Sudd, reg., Sudan	185	8°45'N	30°45'E
As Sulaymānīyah, Iraq	154	35°47'N	45°23'E
As Sulaymānīyah, Sau. Ar.	157	24°09'N	46°19'E
As Suwaydā', Syria	154	32°41'N	36°41'E
Astakós, Grc. (äs'tä-kôs)	131	38°42'N	21°00'E
Astana (Aqmola), Kaz.	139	51°10'N	71°43'E
Astara, Azer.	137	38°30'N	48°50'E
Asti, Italy (äs'tē)	118	44°54'N	8°12'E
Astorga, Spain (äs-tôr'gä)	128	42°28'N	6°03'W
Astoria, Or., U.S. (ăs-tō'rĭ-ȧ)	64	46°11'N	123°51'W
Astrakhan', Russia (äs-trȧ-kän')	134	46°15'N	48°00'E
Astrida, Rw. (äs-trē'dȧ)	186	2°37'S	29°48'E
Asturias, hist. reg., Spain (äs-tōō'ryäs)	128	43°21'N	6°00'W
Astypalaia, i., Grc.	119	36°31'N	26°19'E
Asunción *see* Ixtaltepec, Mex.	89	16°33'N	95°04'W
Asunción *see* Nochistlán, Mex.	88	21°23'N	102°52'W
Asunción, Para. (ä-sōōn-syōn')	102	25°25'S	57°30'W
Asunción Mita, Guat. (ä-sōōn-syō'n-mē'tä)	90	14°19'N	89°43'W
Aswān, Egypt (ä-swän')	185	24°05'N	32°57'E
Aswān High Dam, dam, Egypt	185	23°58'N	32°53'E
Atacama, Desierto de, des., Chile (dĕ-syĕ'r-tô-dĕ-ä-tä-kä'mä)	97	23°50'S	69°00'W
Atacama, Puna de, plat., Bol. (pōō'nä-dĕ-ä-tä-kà'mä)	100	21°35'S	66°58'W
Atacama, Puna de, reg., Chile (pōō'nä-dĕ-ätä-kà'mä)	102	23°15'S	68°45'W
Atacama, Salar de, l., Chile (sá-lär'dĕ-ätä-kà'mä)	102	23°38'S	68°15'W
Ataco, Col. (ä-tà'kō)	100a	3°36'N	75°22'W
Atacora, Chaîne de l', mts., Benin	188	10°15'N	1°15'E
Atā 'itah, Jabal al, mtn., Jord.	153a	30°48'N	35°19'E
Atamanovskiy, Russia (ä-tä-mä'nôv-skĭ)	142a	52°15'N	60°47'E
'Atāqah, Jabal, mts., Egypt	192d	29°59'N	32°20'E
Atar, Maur. (ä-tär')	184	20°45'N	13°16'W
Atascadero, Ca., U.S. (ăt-ăs-kȧ-dâ'rō)	76	35°29'N	120°40'W
Atascosa, r., Tx., U.S. (ăt-ăs-kō'sȧ)	80	28°50'N	98°17'W
Atauro, Ilha de, i., E. Timor (dĕ-ä-tä'ōō-rō)	169	8°20'S	126°15'E
Atbara, r., Afr.	185	17°14'N	34°27'E
'Aṭbarah, Sudan (ät'bȧ-rä)	185	17°45'N	33°15'E
Atbasar, Kaz. (ät'bä-sär')	139	51°42'N	68°28'E
Atchafalaya, r., La., U.S.	81	30°53'N	91°51'W
Atchafalaya Bay, b., La., U.S. (ăch-ȧ-fȧ-lī'ȧ)	81	29°25'N	91°30'W
Atchison, Ks., U.S. (ăch'ĭ-sŭn)	65	39°33'N	95°08'W
Atco, N.J., U.S. (ăt'kō)	68f	39°46'N	74°53'W
Atempan, Mex. (ä-tĕm-pȧ'n)	89	19°49'N	97°25'W
Atenguillo, r., Mex. (ä-tĕn-gē'l-yō)	88	20°18'N	104°35'W
Athabasca, Can. (ăth-ȧ-băs'kȧ)	50	54°43'N	113°17'W
Athabasca, l., Can.	52	59°04'N	109°10'W
Athabasca, r., Can.	52	57°30'N	112°00'W
Athens (Athína), Grc.	131	38°00'N	23°38'E
Athens, Al., U.S. (ăth'ĕnz)	82	34°47'N	86°58'W
Athens, Ga., U.S.	65	33°55'N	83°24'W
Athens, Oh., U.S.	66	39°20'N	82°10'W
Athens, Pa., U.S.	67	42°00'N	76°30'W
Athens, Tn., U.S.	82	35°26'N	84°36'W
Athens, Tx., U.S.	81	32°13'N	95°51'W
Atherstone, Eng., U.K. (ăth'ẽr-stŭn)	114a	52°34'N	1°33'W
Atherton, Eng., U.K. (ăth'ẽr-tŭn)	114a	53°32'N	2°29'W
Atherton Plateau, plat., Austl. (ădh-ẽr-tŏn)	175	17°00'S	144°30'E
Athi, r., Kenya (ä'tē)	187	2°43'S	38°30'E
Athína *see* Athens, Grc.	110	38°00'N	23°38'E
Athlone, Ire. (ăth-lōn')	116	53°24'N	7°30'W
Áthos, mtn., Grc. (ăth'ŏs)	131	40°10'N	24°15'E
Ath Thamad, Egypt	153a	29°41'N	34°17'E
Athy, Ire. (ȧ-thī)	120	52°59'N	7°08'W

ăt; finăl; rāte; senāte; ärm; ȧsk; sofȧ; fâre; ch-choose; dh-as th in other; bē; êvent; bĕt; recĕnt; cratẽr; g-gō; gh-guttural g; bĭt; ĭ-short neutral; rīde; K-guttural k as ch in German ich;

PLACE (Pronunciation)	PAGE	LAT.	LONG.
Ati, Chad	189	13°13'N	18°20'E
Atibaia, Braz. (ä-tē-bá'yă)	99a	23°08'S	46°32'W
Atikonak, l., Can.	53	52°34'N	63°49'W
Atimonan, Phil. (ä-tē-mō'nän)	169a	13°59'N	121°56'E
Atiquizaya, El Sal. (ä'tē-kē-zä'yä)	90	14°00'N	89°42'W
Atitlan, vol., Guat. (ä-tē-tlän')	90	14°35'N	91°11'W
Atitlan, Lago, l., Guat. (ä-tē-tlän')	90	14°38'N	91°23'W
Atizapán, Mex. (ä'tē-zà-pän')	89a	19°33'N	99°16'W
Atka, Ak., U.S. (ăt'kă)	63a	52°18'N	174°18'W
Atka, i., Ak., U.S.	64b	51°58'N	174°30'W
Atkarsk, Russia (àt-kärsk')	137	51°50'N	45°00'E
Atkinson, Ne., U.S. (ăt'kĭn-sŭn)	70	42°32'N	98°58'W
Atlanta, Ga., U.S. (ăt-lăn'tă)	65	33°45'N	84°23'W
Atlanta, Tx., U.S.	79	33°09'N	94°09'W
Atlantic, Ia., U.S. (ăt-lăn'tĭk)	71	41°23'N	94°58'W
Atlantic, N.C., U.S.	83	34°54'N	76°20'W
Atlantic City, N.J., U.S.	65	39°20'N	74°30'W
Atlantic Highlands, N.J., U.S.	68a	40°25'N	74°04'W
Atlantic Ocean, o.	4	5°00'S	25°00'W
Atlas Mountains, mts., Afr. (ăt'lăs)	184	31°22'N	4°57'W
Atliaca, Mex. (ät-lē-ä'kä)	88	17°38'N	99°24'W
Atlin, l., Can. (ăt'lĭn)	52	59°34'N	133°20'W
Atlixco, Mex. (ät-lēz'kō)	88	18°52'N	98°27'W
Atmore, Al., U.S. (ăt'mōr)	82	31°01'N	87°31'W
Atoka, Ok., U.S. (à-tō'kă)	79	34°23'N	96°07'W
Atoka Reservoir, res., Ok., U.S.	79	34°30'N	96°05'W
Atotonilco el Alto, Mex.	88	20°35'N	102°32'W
Atotonilco el Grande, Mex.	88	20°17'N	98°41'W
Atoui, r., Afr. (à-tōō-ē')	184	21°00'N	15°32'W
Atoyac, Mex. (ä-tō-yäk')	88	20°01'N	103°28'W
Atoyac, r., Mex.	88	18°35'N	98°16'W
Atoyac, r., Mex.	89	16°27'N	97°28'W
Atoyac de Alvarez, Mex. (ä-tō-yäk'dā äl'vä-räz)	88	17°13'N	100°29'W
Atoyatempan, Mex. (ä-tō'yá-těm-pän')	89	18°47'N	97°54'W
Atrak, r., Asia	154	37°45'N	56°30'E
Ätran, r., Swe.	122	57°02'N	12°43'E
Atrato, Río, r., Col. (rē'ō-ä-trä'tō)	100	7°15'N	77°18'W
Aṭ Ṭafīlah, Jord. (tä-fē'la)	153a	30°50'N	35°36'E
Aṭ Ṭā'if, Sau. Ar.	154	21°03'N	41°00'E
Attalla, Al., U.S. (ă-tăl'yă)	82	34°01'N	86°05'W
Attawapiskat, r., Can. (ăt'à-wà-pĭs'kăt)	53	52°31'N	86°22'W
Attersee, l., Aus.	124	47°57'N	13°25'E
Attica, N.Y., U.S. (ăt'ĭ-kă)	67	42°55'N	78°15'W
Attleboro, Ma., U.S. (ăt''l-bŭr-ō)	68b	41°56'N	71°15'W
Attow, Ben, mtn., Scot., U.K. (běn ăt'tō)	120	57°15'N	5°25'W
Attoyac Bay, Tx., U.S. (ă-toi'yăk)	81	31°45'N	94°23'W
Attu, i., Ak., U.S. (ăt-tōō')	64b	53°08'N	173°18'E
Aṭ Ṭūr, Egypt	119	28°09'N	33°47'E
Aṭ Ṭurayf, Sau. Ar.	154	31°32'N	38°30'E
Åtvidaberg, Swe. (ôt-vē'dá-běrgh)	122	58°12'N	15°55'E
Atwood, Ks., U.S. (ăt'wŏŏd)	78	39°48'N	101°06'W
Atyraū, Kaz.	139	47°10'N	51°50'E
Atzcapotzalco, Mex. (ät'zkä-pō-tzäl'kō)	88	19°29'N	99°11'W
Atzgersdorf, Aus.	115e	48°10'N	16°17'E
Auau Channel, strt., Hi., U.S. (ä'ō-ä'ō)	84a	20°55'N	156°50'W
Aubagne, Fr. (ō-bän'y')	127	43°18'N	5°34'E
Aube, r., Fr. (ōb)	126	48°42'N	3°49'E
Aubenas, Fr. (ōb-nä')	126	44°37'N	4°22'E
Aubervilliers, Fr. (ō-bĕr-vē-yā')	127b	48°54'N	2°23'E
Aubin, Fr. (ō-băn')	126	44°29'N	2°12'E
Aubrey, Can. (ô-brē')	62a	45°08'N	73°47'W
Auburn, Al., U.S. (ô'bŭrn)	82	32°35'N	85°26'W
Auburn, Ca., U.S.	76	38°52'N	121°05'W
Auburn, Il., U.S.	79	39°36'N	89°46'W
Auburn, In., U.S.	66	41°20'N	85°05'W
Auburn, Ma., U.S.	61a	42°11'N	71°51'W
Auburn, Me., U.S.	65	44°04'N	70°24'W
Auburn, Ne., U.S.	79	40°23'N	95°50'W
Auburn, N.Y., U.S.	67	42°55'N	76°35'W
Auburn, Wa., U.S.	74a	47°18'N	122°14'W
Auburn Heights, Mi., U.S.	69b	42°37'N	83°13'W
Aubusson, Fr. (ō-bü-sôN')	126	45°57'N	2°10'E
Auch, Fr. (ōsh)	117	43°38'N	0°35'E
Aucilla, r., Fl., U.S. (ô-sĭl'ă)	82	30°15'N	83°55'W
Auckland, N.Z. (ôk'lănd)	175a	36°53'S	174°45'E
Auckland Islands, is., N.Z.	3	50°30'S	166°30'E
Aude, r., Fr. (ōd)	126	42°55'N	2°08'E
Audierne, Fr. (ō-dyĕrn')	126	48°02'N	4°31'W
Audincourt, Fr. (ō-dăn-kōōr')	127	47°30'N	6°49'E
Audley, Eng., U.K. (ôd'lĭ)	114a	53°03'N	2°18'W
Audo Range, mts., Eth.	192a	6°58'N	41°18'E
Audubon, Ia., U.S. (ô'dŏŏ-bŏn)	71	41°43'N	94°57'W
Audubon, N.J., U.S.	68f	39°54'N	75°04'W
Aue, Ger. (ou'ē)	124	50°35'N	12°44'E
Augathella, Austl. (ôr'gá'thē-lă)	176	25°49'S	146°40'E
Augrabiesvalle, wtfl., S. Afr.	186	28°30'S	20°00'E
Augsburg, Ger. (ouks'bŏŏrgh)	117	48°23'N	10°55'E
Augusta, Ar., U.S. (ô-gŭs'tă)	79	35°16'N	91°21'W
Augusta, Ga., U.S.	65	33°26'N	82°00'W
Augusta, Ks., U.S.	79	37°41'N	96°58'W
Augusta, Ky., U.S.	66	38°45'N	84°00'W
Augusta, Me., U.S.	65	44°19'N	69°42'W
Augusta, N.J., U.S.	68a	41°07'N	74°44'W
Augusta, Wi., U.S.	71	44°40'N	91°09'W
Augustow, Pol. (ou-gŏŏs'tōf)	125	53°52'N	23°00'E
Auki, Sol. Is.	170e	8°46'S	160°42'E
Aulnay-sous-Bois, Fr. (ō-nĕ'sōō-bwä')	127b	48°56'N	2°30'E
Aulne, r., Fr. (ōn)	126	48°08'N	3°53'W
Auneau, Fr. (ō-nĕü)	127b	48°28'N	1°45'E
Auob, r., Afr. (ä'wŏb)	186	25°00'S	19°00'E
Aur, i., Malay.	153b	2°27'N	104°51'E
Aura, Fin.	123	60°38'N	22°32'E
Aurangābād, India (ou-rŭŋ-gä-bäd')	155	19°56'N	75°19'E

PLACE (Pronunciation)	PAGE	LAT.	LONG.
Aurdal, Nor. (äŭr-däl)	116	60°54'N	9°24'E
Aurès, Massif de l', mts., Alg.	118	35°16'N	5°53'E
Aurillac, Fr. (ō-rē-yák')	117	44°57'N	2°27'E
Aurora, Can.	59	43°59'N	79°25'W
Aurora, Co., U.S.	78	39°44'N	104°50'W
Aurora, Il., U.S. (ō-rō'ră)	65	41°45'N	88°18'W
Aurora, In., U.S.	69f	39°04'N	84°55'W
Aurora, Mn., U.S.	71	47°31'N	92°17'W
Aurora, Mo., U.S.	79	36°58'N	93°42'W
Aurora, Ne., U.S.	78	40°54'N	98°01'W
Aursunden, l., Nor. (äûr-sûndĕn)	122	62°42'N	11°10'E
Au Sable, r., Mi., U.S. (ô-sä'b'l)	66	44°40'N	84°25'W
Ausable, r., N.Y., U.S.	67	44°25'N	73°50'W
Austin, Mn., U.S. (ôs'tĭn)	71	43°40'N	92°58'W
Austin, Nv., U.S.	76	39°30'N	117°05'W
Austin, Tx., U.S.	64	30°15'N	97°42'W
Austin, l., Austl.	174	27°45'S	117°30'E
Austin Bayou, Tx., U.S. (ôs'tĭn bī-ōō')	81a	29°17'N	95°21'W
Australia, nation, Oc.	174	25°00'S	135°00'E
Australian Alps, mts., Austl.	176	37°10'S	147°55'E
Australian Capital Territory, ter., Austl. (ôs-trā'lĭ-ăn)	175	35°30'S	148°40'E
Austria, nation, Eur. (ôs'trĭ-ă)	110	47°15'N	11°53'E
Authon-la-Plaine, Fr. (ō-tôN-lă-plĕ'n)	127b	48°27'N	1°58'E
Autlán, Mex. (ä-ōōt-län')	86	19°47'N	104°24'W
Autun, Fr. (ō-tŭN')	126	46°58'N	4°14'E
Auvergne, mts., Fr. (ō-vĕrn'y')	126	45°12'N	2°31'E
Auxerre, Fr. (ō-sâr')	117	47°48'N	3°32'E
Ava, Mo., U.S. (ā'vă)	79	36°56'N	92°40'W
Avakubi, D.R.C. (ä-vá-kōō'bē)	185	1°20'N	27°34'E
Avallon, Fr. (à-vá-lôN')	126	47°30'N	3°58'E
Avalon, Ca., U.S.	76	33°21'N	118°22'W
Avalon, Pa., U.S. (ăv'à-lŏn)	69e	40°31'N	80°05'W
Aveiro, Port. (ä-vā'rō)	118	40°38'N	8°38'W
Avelar, Braz. (ä'vē-lá'r)	102b	22°20'S	43°25'W
Avellaneda, Arg. (ä-vĕl-yä-nä'dhä)	102	34°40'S	58°23'W
Avellino, Italy (ä-vĕl-lē'nō)	130	40°40'N	14°46'E
Averøya, i., Nor. (ävĕr-ûē)	122	63°40'N	7°16'E
Aversa, Italy (ä-vĕr'sä)	130	40°58'N	14°13'E
Avery, Tx., U.S. (ā'vĕr-ĭ)	79	33°34'N	94°46'W
Avesta, Swe. (ä-vĕs'tä)	122	60°16'N	16°09'E
Aveyron, r., Fr. (ä-vâ-rôN)	126	44°07'N	1°45'E
Avezzano, Italy (ä-vät-sä'nō)	130	42°03'N	13°27'E
Avigliano, Italy (ä-vēl-yä'nō)	130	40°45'N	15°44'E
Avignon, Fr. (à-vē-nyôN')	117	43°55'N	4°50'E
Ávila, Spain (ä-vē-lä)	128	40°39'N	4°42'W
Avilés, Spain (ä-vē-lās')	118	43°33'N	5°55'W
Aviño, Spain	128	43°36'N	8°05'W
Avoca, Ia., U.S. (à-vō'kă)	79	41°29'N	95°16'W
Avon, Ct., U.S. (ā'vŏn)	67	41°40'N	72°50'W
Avon, Ma., U.S. (ā'vŏn)	61a	42°08'N	71°03'W
Avon, Oh., U.S.	69d	41°27'N	82°02'W
Avon, r., Eng., U.K. (ā'vŭn)	120	52°05'N	1°55'W
Avondale, Ga., U.S.	68c	33°47'N	84°16'W
Avon Lake, Oh., U.S.	69d	41°31'N	82°01'W
Avonmore, Can. (ä'vôN-mōr)	62c	45°11'N	74°58'W
Avon Park, Fl., U.S. (ā'vŏn pärk')	83a	27°35'N	81°29'W
Avranches, Fr. (à-vräNsh')	126	48°43'N	1°34'W
Awaji-Shima, i., Japan	166	34°32'N	135°02'E
Awe, Loch, l., Scot., U.K. (lŏk ôr)	120	56°22'N	5°04'W
Awjilah, Libya	185	29°07'N	21°21'E
Ax-les-Thermes, Fr. (äks'lā tĕrm')	126	42°43'N	1°50'E
Axochiapan, Mex. (äks-ō-chyä'pän)	88	18°29'N	98°49'W
Ay, r., Russia	136	55°55'N	57°55'E
Ayabe, Japan (ä'yä-bĕ)	166	35°16'N	135°17'E
Ayachi, Arin', mtn., Mor.	118	32°29'N	4°57'W
Ayacucho, Arg. (ä-yä-kōō'chō)	102	37°05'S	58°30'W
Ayacucho, Peru	100	13°12'S	74°03'W
Ayaköz, Kaz.	139	48°00'N	80°12'E
Ayamonte, Spain (ä-yä-mô'n-tĕ)	118	37°14'N	7°28'W
Ayan, Russia (ä-yän')	135	56°26'N	138°18'E
Ayata, Bol. (ä-yä'tä)	100	15°17'S	68°43'W
Ayaviri, Peru (ä-yä-vē'rē)	100	14°46'S	70°38'W
Aydar, r., Eur. (ī-där')	133	49°15'N	38°48'E
Ayden, N.C., U.S. (ā'dĕn)	83	35°27'N	77°25'W
Aydın, Tur. (āīy-dĕn)	154	37°40'N	27°40'E
Ayer, Ma., U.S. (âr)	61a	42°33'N	71°36'W
Ayer Hitam, Malay.	153b	1°55'N	103°11'E
Ayers Rock *see* Uluru, mtn., Austl.	174	25°23'S	131°05'E
Aylesbury, Eng., U.K. (ālz'bĕr-ĭ)	120	51°47'N	0°49'W
Aylmer, l., Can. (āl'mēr)	52	64°27'N	108°22'W
Aylmer, Mount, mtn., Can.	55	51°19'N	115°26'W
Aylmer East, Can. (āl'mēr)	59	45°24'N	75°50'W
Ayo el Chico, Mex. (ä'yō el chē'kō)	88	20°31'N	102°21'W
Ayon, i., Russia (ī-ôn')	135	69°50'N	168°40'E
Ayorou, Niger	188	14°44'N	0°55'E
Ayotla, Mex. (ä-yōt'lä)	89a	19°18'N	98°55'W
Ayoun el Atrous, Maur.	188	16°40'N	9°37'W
Ayr, Scot., U.K. (âr)	120	55°27'N	4°40'W
Aysha, Eth.	185	10°48'N	42°32'E
Ayutla, Guat. (à-yōōt'lä)	90	14°44'N	92°11'W
Ayutla, Mex.	88	16°50'N	99°16'W
Ayutla, Mex.	88	20°09'N	104°20'W
Ayvalık, Tur. (āīy-wä-lĭk)	119	39°19'N	26°40'E
Azaouad, reg., Mali	188	18°00'N	3°20'W
Azaouak, Vallée de l', val., Afr.	189	15°50'N	3°10'E
Azare, Nig.	189	11°40'N	10°11'E
Azemmour, Mor. (á-zē-mōōr')	184	33°20'N	8°21'W
Azerbaijan, nation, Asia	134	40°30'N	47°30'E
Azle, Tx., U.S. (áz'lē)	75c	35°54'N	97°33'W
Azogues, Ec. (ä-sō'gās)	100	2°47'S	78°45'W
Azores *see* Açores, is., Port.	183	37°44'N	29°25'W
Azov, Russia (á-zôf') (ä-zôf)	137	47°07'N	39°19'E
Azov, Sea of, sea, Eur.	134	46°00'N	36°20'E
Aztec, N.M., U.S. (ăz'tĕk)	77	36°40'N	108°00'W
Aztec Ruins National Monument, rec., N.M., U.S.	77	36°50'N	108°00'W

PLACE (Pronunciation)	PAGE	LAT.	LONG.
Azua, Dom. Rep. (ä'swä)	93	18°30'N	70°45'W
Azuaga, Spain (ä-thwä'gä)	128	38°15'N	5°42'W
Azucar, Presa de, res., Mex.	80	26°06'N	98°44'W
Azuero, Península de, pen., Pan.	87	7°30'N	80°34'W
Azufre, Cerro (Copiapó), mtn., Chile	102	27°10'S	69°00'W
Azul, Arg. (ä-sōōl')	102	36°46'S	59°51'W
Azul, Cordillera, mts., Peru	100	7°15'S	75°30'W
Azul, Sierra, mts., Mex.	88	23°20'N	98°28'W
Azusa, Ca., U.S. (á-zōō'sá)	75a	34°08'N	117°55'W
Aẓ Ẓahrān (Dhahran), Sau. Ar.	154	26°13'N	50°00'E
Az Zaqāzīq, Egypt	185	30°36'N	31°36'E
Az Zarqā', Jord.	153a	32°03'N	36°07'E
Az Zāwiyah, Libya	184	32°28'N	11°55'E

B

PLACE (Pronunciation)	PAGE	LAT.	LONG.
Baadheere (Bardera), Som.	192a	2°13'N	42°24'E
Baal, Ger. (bäl)	127c	51°02'N	6°17'E
Baao, Phil. (bä'ō)	169a	13°27'N	123°22'E
Baarle-Hertog, Bel.	115a	51°26'N	4°57'E
Baarn, Neth.	115a	52°12'N	5°18'E
Babaeski, Tur. (bä'bä-ĕs'kĭ)	131	41°25'N	27°05'E
Babahoyo, Ec. (bä-bä-ō'yō)	100	1°56'S	79°24'W
Babana, Nig.	189	10°36'N	3°50'E
Babanango, S. Afr.	187c	28°24'S	31°11'E
Babanūsah, Sudan	185	11°30'N	27°55'E
Babar, Pulau, i., Indon. (bä'bär)	169	7°50'S	129°15'E
Bab-el-Mandeb *see* Mandeb, Bab-el-, strt.,	154	13°17'N	42°49'E
Babelthuap, i., Palau	170b	7°30'N	134°36'E
Babia, Arroyo de la, r., Mex.	80	28°26'N	101°50'W
Babine, r., Can.	54	55°10'N	127°00'W
Babine Lake, l., Can. (băb'ēn)	52	54°45'N	126°00'W
Bābol, Iran	154	36°30'N	52°48'E
Babruysk, Bela.	136	53°07'N	29°13'E
Babushkin, Russia (bä'bŏŏsh-kĭn)	140	51°47'N	106°08'W
Babushkin, Russia	132	55°52'N	37°42'E
Babuyan Islands, is., Phil. (bä-bōō-yän')	168	19°30'N	122°38'E
Babyak, Blg. (băb'zhák)	131	41°59'N	23°42'E
Babylon, N.Y., U.S. (băb'ĭ-lŏn)	68a	40°42'N	73°19'W
Babylon, hist., Iraq	154	32°15'N	45°23'E
Bacalar, Laguna de, l., Mex. (lä-gōō-nä-dĕ-bä-kä-lär')	90a	18°50'N	88°31'W
Bacan, Pulau, i., Indon.	169	0°30'S	127°00'E
Bacarra, Phil. (bä-kär'rä)	165	18°22'N	120°40'E
Bacău, Rom.	119	46°34'N	27°00'E
Baccarat, Fr. (bá-ká-rá')	127	48°29'N	6°42'E
Bacchus, Ut., U.S. (băk'ŭs)	75b	40°40'N	112°06'W
Bachajón, Mex. (bä-chä-hōn')	89	17°08'N	92°18'W
Bachu, China (bä-chōō)	160	39°50'N	78°23'E
Back, r., Can.	52	65°30'N	104°15'W
Bačka Palanka, Serb. (bäch'kä pälän-kä)	131	45°14'N	19°24'E
Bačka Topola, Serb. (bäch'kä tô'pô-lä')	131	45°48'N	19°38'E
Back Bay, India (băk)	159b	18°55'N	72°45'E
Backstairs Passage, strt., Austl. (băk-stârs')	174	35°50'S	138°15'E
Bac Lieu, Viet.	168	9°45'N	105°50'E
Bac Ninh, Viet. (bäk'nēn'')	165	21°10'N	106°02'E
Baco, Mount, mtn., Phil. (bä'kō)	169a	12°50'N	121°11'E
Bacoli, Italy (bä-kō-lē')	129c	40°33'N	14°05'E
Bacolod, Phil. (bä-kō'lōd)	169	10°42'N	123°03'E
Bácsalmás, Hung. (bäch'ôl-mäs)	125	46°07'N	19°18'E
Bacup, Eng., U.K. (băk'ŭp)	114a	53°42'N	2°12'W
Bad, r., S.D., U.S. (băd)	70	44°04'N	100°58'W
Badajoz, Spain (bä-dhä-hōth')	118	38°52'N	6°56'W
Badalona, Spain (bä-dhä-lō'nä)	129	41°27'N	2°15'E
Badanah, Sau. Ar.	154	30°49'N	40°45'E
Bad Axe, Mi., U.S. (băd' ăks)	66	43°50'N	82°55'W
Bad Bramstedt, Ger. (bät bräm'shtĕt)	115c	53°55'N	9°53'E
Baden, Aus. (bä'dĕn)	124	48°00'N	16°14'E
Baden, Switz.	124	47°28'N	8°17'E
Baden-Baden, Ger. (bä'dĕn-bä'dĕn)	117	48°46'N	8°11'E
Bad Freienwalde, Ger. (bät frī'ĕn-väl'dĕ)	124	52°47'N	14°00'E
Bad Hersfeld, Ger. (bät hĕrsh'fĕlt)	124	50°53'N	9°43'E
Badīn, Pak.	158	24°47'N	69°51'E
Bad Ischl, Aus. (bät ĭsh''l)	124	47°46'N	13°37'E
Bad Kissingen, Ger. (bät kĭs'ĭng-ĕn)	124	50°12'N	10°05'E
Bad Kreuznach, Ger. (bät kroits'näk)	124	49°52'N	7°53'E
Badlands, reg., N.D., U.S. (băd' lănds)	70	46°43'N	103°22'W
Badlands, reg., S.D., U.S.	70	43°43'N	102°36'W
Badlands National Park, rec., S.D., U.S.	70	43°56'N	102°37'W
Badlāpur, India	159b	19°12'N	73°12'E
Badogo, Mali	188	11°02'N	8°13'W
Bad Oldesloe, Ger. (bät ōl'dĕs-lōē)	124	53°48'N	10°21'E
Bad Reichenhall, Ger. (bät rī'kĕn-häl)	124	47°43'N	12°53'E
Bad River Indian Reservation, I.R., Wi., U.S. (băd)	71	46°41'N	90°36'W
Bad Segeberg, Ger. (bät sĕ'gĕ-bōōrgh)	115c	53°56'N	10°18'E
Bad Tölz, Ger. (bät tûltz)	124	47°46'N	11°35'E
Badulla, Sri L.	159	6°55'N	81°07'E
Bad Vöslau, Aus.	115e	47°58'N	16°13'E
Badwater Creek, r., Wy., U.S. (băd'wô-tēr)	73	43°13'N	107°55'W
Baena, Spain (bä-ā'nä)	118	37°38'N	4°20'W
Baependi, Braz. (bä-ĕ-pĕn'dĭ)	99a	21°57'S	44°51'W
Baffin Bay, b., N.A. (băf'ĭn)	49	72°00'N	65°00'W

ng-sing; ŋ-baŋk; N-nasalized n; nŏd; cŏmmit; ōld; ŏbey; ôrder; oi-boil; fōōd; ŏŏ-as oo in foot; ou-out; s-soft; sh-dish; th-thin; pūre; ûnite; ûrn; stŭd; circŭs; ü-as in French tu; '-indeterminate vowel.

åt; finǎl; rāte; senǎte; ärm; àsk; sofà; fâre; ch-choose; dh-as th in other; bē; ĕvent; bĕt; recĕnt; crātēr; g-gō; gh-guttural g; bĭt; ĭ-short neutral; rīde; ĸ-guttural k as ch in German ich;

PRONOUNCING INDEX
Barão de Melgaço – Beatty 225

PLACE (Pronunciation)	PAGE	LAT.	LONG.
Barão de Melgaço, Braz. (bä-roun-dĕ-mĕl-gä′sô)	101	16°12′S	55°48′W
Bărăsat, India	158a	22°42′N	88°29′E
Barataria Bay, b., La., U.S.	81	29°13′N	89°50′W
Baraya, Col. (bä-rá′yä)	100a	3°10′N	75°04′W
Barbacena, Braz. (bär-bä-sā′nȧ)	101	21°15′S	43°46′W
Barbacoas, Col. (bär-bä-kō′äs)	100	1°39′N	78°12′W
Barbacoas, Ven. (bä-bä-kô′äs)	101b	9°30′N	66°58′W
Barbados, nation, N.A. (bär-bā′dōz)	87	13°30′N	59°00′W
Barbar, Sudan	185	18°11′N	34°00′E
Barbastro, Spain (bär-bäs′trō)	129	42°05′N	0°05′E
Barbeau, Mi., U.S. (bȧr-bō′)	75k	46°17′N	84°16′W
Barberton, S. Afr.	186	25°48′S	31°04′E
Barberton, Oh., U.S. (bär′bẽr-tŭn)	69d	41°01′N	81°37′W
Barbezieux, Fr. (bärb′zyû′)	126	45°30′N	0°11′W
Barbosa, Col. (bär-bô′-sä)	100a	6°26′N	75°19′W
Barboursville, W.V., U.S. (bär′bẽrs-vĭl)	66	38°20′N	82°20′W
Barbourville, Ky., U.S.	82	36°52′N	83°58′W
Barbuda, i., Antig. (bär-bōō′dȧ)	87	17°45′N	61°15′W
Barcaldine, Austl. (bär′kôl-dĭn)	175	23°33′S	145°17′E
Barcarrota, Spain (bär-kär-rō′tä)	128	38°31′N	6°50′W
Barcellona, Italy (bä-chĕl-lō′nä)	130	38°07′N	15°15′E
Barcelona, Spain (bär-thå-lō′nä)	110	41°25′N	2°08′E
Barcelona, Ven. (bär-så-lō′nä)	100	10°09′N	64°41′W
Barcelos, Braz. (bär-sĕ′lôs)	100	1°04′S	63°00′W
Barcelos, Port. (bär-thå′lōs)	128	41°34′N	8°39′W
Bardawīl, Sabkhat al, b., Egypt	153a	31°20′N	33°24′E
Bardejov, Czech Rep. (bär′dyĕ-yôf)	125	49°18′N	21°18′E
Bardsey Island, i., Wales, U.K. (bärd′sē)	120	52°45′N	4°50′W
Bardstown, Ky., U.S. (bärds′toun)	66	37°50′N	85°30′W
Bardwell, Ky., U.S. (bärd′wĕl)	82	36°51′N	88°57′W
Bareilly, India	155	28°21′N	79°25′E
Barents Sea, sea, Eur. (bä′rĕnts)	134	72°14′N	37°28′E
Barentu, Erit. (bä-rĕn′tōō)	185	15°06′N	37°39′E
Barfleur, Pointe de, c., Fr. (bȧr-flûr′)	126	49°43′N	1°17′W
Barguzin, Russia (bär′gōō-zĭn)	135	53°44′N	109°28′E
Bar Harbor, Me., U.S. (bär här′bẽr)	60	44°22′N	68°13′W
Bari, Italy (bä′rē)	110	41°08′N	16°53′E
Barinas, Ven. (bä-rē′näs)	100	8°36′N	70°14′W
Baring, Cape, c., Can. (bâr′ĭng)	52	70°07′N	119°48′W
Barisan, Pegunungan, mts., Indon. (bä-rē-sän′)	168	2°38′S	101°45′E
Barito, r., Indon. (bä-rē′tō)	168	2°10′S	114°38′E
Barka, r., Afr.	185	16°44′N	37°34′E
Barkley Sound, strt., Can.	54	48°53′N	125°20′W
Barkly East, S. Afr. (bärk′lē ēst)	187c	30°58′S	27°37′E
Barkly Tableland, plat., Austl. (bär′klē)	174	18°15′S	137°05′E
Barkol, China (bär-kŭl)	160	43°43′N	92°50′E
Bârlad, Rom.	119	46°15′N	27°43′E
Bar-le-Duc, Fr. (bär-lē-dük′)	127	48°47′N	5°05′E
Barlee, l., Austl. (bär-lē′)	174	29°45′S	119°00′E
Barletta, Italy (bär-lĕt′tä)	119	41°19′N	16°20′E
Barmstedt, Ger. (bärm′shtĕt)	115c	53°47′N	9°46′E
Barnaul, Russia (bär-nä-ōl′)	134	53°18′N	83°23′E
Barnesboro, Pa., U.S. (bärnz′bẽr-ō)	67	40°45′N	78°50′W
Barnesville, Ga., U.S. (bärnz′vĭl)	82	33°03′N	84°10′W
Barnesville, Mn., U.S.	70	46°38′N	96°25′W
Barnesville, Oh., U.S.	66	39°55′N	81°10′W
Barnet, Vt., U.S. (bär′nĕt)	67	44°20′N	72°00′W
Barnetby le Wold, Eng., U.K. (bär′nĕt-bī)	114a	53°34′N	0°26′W
Barnett Harbor, b., Bah.	92	25°40′N	79°20′W
Barnsdall, Ok., U.S. (bärnz′dôl)	79	36°38′N	96°14′W
Barnsley, Eng., U.K. (bärnz′lĭ)	114a	53°33′N	1°29′W
Barnsley, co., Eng., U.K.	114a	53°33′N	1°30′W
Barnstaple, Eng., U.K. (bärn′stā-p′l)	120	51°06′N	4°05′W
Barnwell, S.C., U.S. (bärn′wĕl)	83	33°14′N	81°23′W
Baro, Nig. (bä′rō)	184	8°37′N	6°25′E
Baroda, India (bär-rō′dä)	155	22°21′N	73°12′E
Barotse Plain, pl., Zam.	190	15°50′S	22°55′E
Barqah (Cyrenaica), hist. reg., Libya	185	31°09′N	21°45′E
Barquisimeto, Ven. (bär-kē-sē-mä′tō)	100	10°04′N	69°16′W
Barra, Braz. (bär′rä)	101	11°04′S	43°11′W
Barraba, Austl.	176	30°22′S	150°36′E
Barrackpore, India	158a	22°46′N	88°21′E
Barra do Corda, Braz. (bär′rä dō cōr-dä)	101	5°33′S	45°13′W
Barra Mansa, Braz. (bär′rä män′sä)	99a	22°35′S	44°09′W
Barrancabermeja, Col. (bär-rän′kä-bēr-mā′hä)	100	7°06′N	73°49′W
Barranquilla, Col. (bär-rän-kēl′yä)	100	10°57′N	75°00′W
Barras, Braz. (bȧ′r-räs)	101	4°13′S	42°14′W
Barre, Vt., U.S. (băr′ē)	67	44°15′N	72°30′W
Barreiras, Braz. (bär-rā′räs)	101	12°13′S	44°59′W
Barreiro, Port. (bär-rĕ′ē-rō)	118	38°39′N	9°05′W
Barren, r., Ky., U.S.	82	37°00′N	86°20′W
Barren, Cape, c., Austl. (băr′ĕn)	175	40°20′S	149°00′E
Barren, Nosy, is., Madag.	187	18°18′S	43°57′E
Barren River Lake, res., Ky., U.S.	82	36°45′N	86°02′W
Barretos, Braz. (bär-rā′tōs)	101	20°40′S	48°36′W
Barrhead, Can. (băr′ĭd)	50	54°08′N	114°24′W
Barrie, Can. (băr′ĭ)	51	44°25′N	79°45′W
Barrington, Can. (bă-rĕng-tŏn)	62a	45°07′N	73°35′W
Barrington, Il., U.S.	69a	42°09′N	88°08′W
Barrington, R.I., U.S.	68b	41°44′N	71°16′W
Barrington Tops, mtn., Austl.	176	32°00′S	151°25′E
Bar River, Can. (bär)	75k	46°27′N	84°02′W
Barron, Wi., U.S. (băr′ŭn)	71	45°24′N	91°51′W
Barrow, Ak., U.S. (băr′ō)	64a	71°20′N	156°00′W
Barrow, i., Austl.	174	20°50′S	115°00′E
Barrow, r., Ire. (bȧ-rȯ)	120	52°35′N	7°05′W
Barrow, Point, c., Ak., U.S.	63	71°20′N	156°00′W
Barrow Creek, Austl.	174	21°23′S	133°55′E
Barrow-in-Furness, Eng., U.K.	116	54°10′N	3°15′W
Barstow, Ca., U.S. (bär′stō)	76	34°53′N	117°03′W
Barstow, Md., U.S.	68e	38°32′N	76°37′W
Barth, Ger. (bärt)	124	54°20′N	12°43′E
Bartholomew Bayou, r., U.S. (bär-thŏl′ō-mū bī-ōō′)	79	33°53′N	91°45′W
Barthurst, Can. (bär-thŭrst′)	51	47°38′N	65°40′W
Bartica, Guy. (bär′tĭ-kȧ)	101	6°23′N	58°32′W
Bartın, Tur. (bär′tĭn)	119	41°35′N	32°12′E
Bartle Frere, Mount, mtn., Austl. (bärt′′l frēr′)	175	17°30′S	145°46′E
Bartlesville, Ok., U.S. (bär′tlz-vĭl)	79	36°44′N	95°58′W
Bartlett, Il., U.S. (bärt′lĕt)	69a	41°59′N	88°11′W
Bartlett, Tx., U.S.	81	30°48′N	97°25′W
Barton, Vt., U.S. (bär′tŭn)	67	44°45′N	72°05′W
Barton-upon-Humber, Eng., U.K. (bär′tŭn-ŭp′ŏn-hŭm′bẽr)	114a	53°41′N	0°26′W
Bartoszyce, Pol. (bär-tō-shĭ′tsȧ)	125	54°15′N	20°50′E
Bartow, Fl., U.S. (bär′tō)	83a	27°51′N	81°50′W
Barvinkove, Ukr.	133	48°55′N	36°59′E
Barwon, r., Austl. (bär′wŭn)	175	30°00′S	147°30′E
Barwon Heads, Austl.	173a	38°17′S	144°29′E
Barycz, r., Pol. (bä′rĭch)	124	51°30′N	16°38′E
Barysaw, Bela.	136	54°16′N	28°33′E
Basankusu, D.R.C. (bä-sän-kōō′sōō)	185	1°14′N	19°45′E
Basbeck, Ger. (bäs′bĕk)	115c	53°40′N	9°11′E
Basdahl, Ger. (bäs′däl)	115c	53°27′N	9°00′E
Basehor, Ks., U.S. (bās′hôr)	75f	39°08′N	94°55′W
Basel, Switz. (bä′z′l)	117	47°32′N	7°35′E
Bashee, r., S. Afr. (bă-shē′)	187c	31°47′S	28°25′E
Bashi Channel, strt., Asia (bäsh′ē)	161	21°20′N	120°22′E
Bashkortostan, prov., Russia	136	54°12′N	57°15′E
Bashtanka, Ukr. (bȧsh-tän′kȧ)	133	47°32′N	32°31′E
Basilan Island, i., Phil.	168	6°37′N	122°07′E
Basildon, Eng., U.K.	121	51°35′N	0°25′E
Basilicata, hist. reg., Italy (bä-zē-lē-kä′tä)	130	40°30′N	15°55′E
Basin, Wy., U.S. (bā′sĭn)	73	44°22′N	108°02′W
Basingstoke, Eng., U.K. (bā′zĭng-stōk)	114b	51°14′N	1°06′W
Baška, Cro. (bäsh′kȧ)	130	44°58′N	14°44′E
Baskale, Tur. (bäsh-kä′lē)	137	38°10′N	44°00′E
Baskatong, Réservoir, res., Can.	59	46°50′N	75°50′W
Baskunchak, l., Russia	137	48°20′N	46°40′E
Basoko, D.R.C. (bȧ-sō′kō)	185	0°52′N	23°50′E
Basque Provinces, hist. reg., Spain	128	43°00′N	2°46′W
Basra *see* Al Başrah, Iraq	154	30°35′N	47°59′E
Bassano, Can. (bäs-sän′ō)	50	50°47′N	112°28′W
Bassano del Grappa, Italy	130	45°46′N	11°44′E
Bassari, Togo	188	9°15′N	0°47′E
Bassas da India, i., Reu. (bäs′säs dä ēn′dē-ȧ)	187	21°23′S	39°42′E
Basse Terre, Guad. (bȧs′ târ′)	87	16°00′N	61°43′W
Basseterre, St. K./N.	91b	17°20′N	62°42′W
Basse Terre, i., Guad.	91b	16°10′N	62°14′W
Bassett, Va., U.S. (băs′sĕt)	83	36°45′N	81°58′W
Bass Islands, is., Oh., U.S. (băs)	66	41°40′N	82°50′W
Bass Strait, strt., Austl.	175	39°40′S	145°40′E
Basswood, l., N.A. (băs′wŏŏd)	71	48°10′N	91°36′W
Båstad, Swe. (bô′stät)	122	56°26′N	12°46′E
Bastia, Fr. (bäs′tē-ä)	117	42°43′N	9°27′E
Bastogne, Bel. (bäs-tôn′y′)	121	50°02′N	5°45′E
Bastrop, La., U.S. (băs′trŭp)	81	32°47′N	91°55′W
Bastrop, Tx., U.S.	81	30°08′N	97°18′W
Bastrop Bayou, Tx., U.S.	81a	29°07′N	95°22′W
Bata, Eq. Gui. (bä′tä)	184	1°51′N	9°45′E
Batabano, Golfo de, b., Cuba (gôl-fō-dĕ-bä-tä-bá′nō)	92	22°10′N	83°05′W
Batāla, India	158	31°54′N	75°18′E
Batam, i., Indon. (bä-täm′)	153b	1°03′N	104°00′E
Batang, China (bä-täŋ)	160	30°08′N	99°00′E
Batangas, Phil. (bä-tän′gäs)	168	13°45′N	121°04′E
Batan Islands, is., Phil. (bä-tän′)	168	20°58′N	122°20′E
Bátaszék, Hung. (bä′tá-sĕk)	125	46°07′N	18°40′E
Batavia, Il., U.S. (bȧ-tā′vĭ-ȧ)	69a	41°51′N	88°18′W
Batavia, N.Y., U.S.	67	43°00′N	78°15′W
Batavia, Oh., U.S.	69f	39°05′N	84°10′W
Bataysk, Russia (bȧ-tīsk′)	137	47°08′N	39°44′E
Bătdâmbâng, Camb. (bȧt-tȧm-bāng′)	168	13°14′N	103°15′E
Batesburg, S.C., U.S. (bāts′bûrg)	83	33°53′N	81°34′W
Batesville, Ar., U.S. (bāts′vĭl)	79	35°46′N	91°39′W
Batesville, In., U.S.	66	39°15′N	85°15′W
Batesville, Ms., U.S.	82	34°17′N	89°55′W
Batetska, Russia (bȧ-tĕ′tskȧ)	132	58°36′N	30°21′E
Bath, Can. (bȧth)	60	46°31′N	67°36′W
Bath, Eng., U.K.	117	51°24′N	2°20′W
Bath, Me., U.S.	60	43°54′N	69°50′W
Bath, N.Y., U.S.	67	42°25′N	77°20′W
Bath, Oh., U.S.	69d	41°11′N	81°38′W
Bathsheba, Barb.	91b	13°13′N	60°30′W
Bathurst, Austl. (băth′ŭrst)	175	33°28′S	149°30′E
Bathurst *see* Banjul, Gam.	184	13°28′N	16°39′W
Bathurst, S. Afr. (băt-hûrst)	187c	33°26′S	26°53′E
Bathurst, i., Austl.	174	11°19′S	130°13′E
Bathurst, Cape, c., Can. (bath′-ŭrst)	52	70°33′N	127°55′W
Bathurst Inlet, b., Can.	52	68°10′N	108°00′W
Batia, Benin	188	10°54′N	1°29′E
Batley, Eng., U.K. (băt′lĭ)	114a	53°43′N	1°37′W
Batna, Alg. (bät′nä)	184	35°41′N	6°12′E
Baton Rouge, La., U.S. (băt′ŭn rōōzh′)	65	30°28′N	91°10′W
Batticaloa, Sri L.	159	7°40′N	81°10′E
Battle, r., Can.	56	52°20′N	111°59′W
Battle Creek, Mi., U.S. (băt′′l krēk′)	65	42°20′N	85°15′W
Battle Ground, Wa., U.S. (băt′′l ground)	74c	45°47′N	122°32′W
Battle Harbour, Can. (băt′′l här′bẽr)	51	52°17′N	55°33′W
Battle Mountain, Nv., U.S.	72	40°40′N	116°56′W
Battonya, Hung. (bät-tô′nyä)	125	46°17′N	21°00′E
Batu, Kepulauan, is., Indon. (bä′tōō)	168	0°10′S	98°00′E
Batumi, Geor. (bŭ-tōō′mē)	134	41°40′N	41°30′E
Batu Pahat, Malay.	168	1°51′N	102°56′E
Batupanjang, Indon.	153b	1°42′N	101°35′E
Bauang, Phil. (bä′wäng)	169a	16°31′N	120°19′E
Bauchi, Nig. (bȧ-ōō′chē)	184	10°19′N	9°50′E
Bauld, Cape, c., Can.	53a	51°38′N	55°25′W
Bāuria, India	158a	22°29′N	88°08′E
Bauru, Braz. (bou-rōō′)	101	22°21′S	48°57′W
Bauska, Lat. (bou′skȧ)	123	56°24′N	24°12′E
Bauta, Cuba (bá′ōō-tä)	93a	22°59′N	82°33′W
Bautzen, Ger. (bout′sĕn)	117	51°11′N	14°27′E
Bavaria *see* Bayern, hist. reg., Ger.	124	49°00′N	11°16′E
Baw Baw, Mount, mtn., Austl.	176	37°50′S	146°17′E
Bawean, Pulau, i., Indon. (bá′vē-än)	168	5°50′S	112°40′E
Bawtry, Eng., U.K. (bôtrĭ)	114a	53°26′N	1°01′W
Baxley, Ga., U.S. (băks′lĭ)	83	31°47′N	82°22′W
Baxter, Austl.	173a	38°12′S	145°10′E
Baxter Springs, Ks., U.S. (băks′tẽr sprĭngs′)	79	37°01′N	94°44′W
Bay, Laguna de, l., Phil. (lä-gōō′nä dȧ bä′ē)	169a	14°24′N	121°13′E
Bayaguana, Dom. Rep. (bä-yä-gwä′nä)	93	18°45′N	69°40′W
Bay al Kabīr, Wādī, val., Libya	118	29°52′N	14°28′E
Bayambang, Phil. (bä-yäm-bäng′)	169a	15°50′N	120°26′E
Bayamo, Cuba (bä-yä′mō)	92	20°25′N	76°35′W
Bayamón, P.R.	87b	18°27′N	66°13′W
Bayan, China (bä-yän)	164	46°00′N	127°20′E
Bayanaūyl, Kaz.	139	50°43′N	75°37′E
Bayard, Ne., U.S. (bā′ẽrd)	70	41°45′N	103°20′W
Bayard, N.M., U.S.	77	32°45′N	108°07′W
Bayard, W.V., U.S.	67	39°15′N	79°20′W
Bayburt, Tur. (bä′ĭ-bȯrt)	137	40°15′N	40°10′E
Bay City, Mi., U.S. (bā)	65	43°35′N	83°55′W
Bay City, Tx., U.S.	81	28°59′N	95°58′W
Baydaratskaya Guba, b., Russia	136	69°20′N	66°10′E
Bay de Verde, Can.	61	48°05′N	52°54′W
Baydhabo (Baidoa), Som.	192a	3°19′N	44°20′E
Baydrag, r., Mong.	160	46°09′N	98°52′E
Bayern, state, Ger.	115d	48°05′N	11°30′E
Bayern (Bavaria), hist. reg., Ger. (bī′ẽrn) (bȧ-vâ-rĭ-ȧ)	124	49°00′N	11°16′E
Bayeux, Fr. (bȧ-yû′)	117	49°19′N	0°41′W
Bayfield, Wi., U.S. (bā′fēld)	71	46°48′N	90°51′W
Baykal, Ozero (Lake Baikal), l., Russia	135	53°00′N	109°28′E
Baykal′skiy Khrebet, mts., Russia	135	53°30′N	107°30′E
Baykit, Russia (bī-kēt′)	135	61°43′N	96°39′E
Baymak, Russia (báy′mäk)	142a	52°35′N	58°21′E
Bay Mills, Mi., U.S. (bā mĭlls)	75k	46°27′N	84°36′W
Bay Mills Indian Reservation, I.R., Mi., U.S.	71	46°19′N	85°03′W
Bay Minette, Al., U.S. (bā′mĭn-ĕt′)	82	30°52′N	87°44′W
Bayombong, Phil. (bä-yôm-bŏng′)	169a	16°28′N	121°09′E
Bayonne, Fr. (bȧ-yŏn′)	110	43°28′N	1°30′W
Bayonne, N.J., U.S. (bā-yōn′)	68a	40°40′N	74°07′W
Bayou Bodcau Reservoir, res., La., U.S. (bī′yōō bŏd′kō)	65	32°49′N	93°22′W
Bayport, Mn., U.S. (bā′pōrt)	75g	45°02′N	92°46′W
Bayqongyr, Kaz.	139	47°46′N	66°11′E
Bayramiç, Tur.	131	39°48′N	26°35′E
Bayreuth, Ger. (bī-roit′)	124	49°56′N	11°35′E
Bay Roberts, Can. (bā rŏb′ẽrts)	61	47°36′N	53°16′W
Bays, Lake of, l., Can. (bās)	59	45°15′N	79°00′W
Bay Saint Louis, Ms., U.S. (bā′ sȧnt lōō′ĭs)	82	30°19′N	89°20′W
Bay Shore, N.Y., U.S. (bā′ shôr)	68a	40°44′N	73°15′W
Bayt Lahm, W.B. (bĕth′lē-hĕm)	153a	31°42′N	35°13′E
Baytown, Tx., U.S. (bā′town)	81a	29°44′N	95°01′W
Bayview, Al., U.S. (bā′vū)	68h	33°34′N	86°59′W
Bayview, Wa., U.S.	74a	48°29′N	122°28′W
Bay Village, Oh., U.S. (bā)	69d	41°29′N	81°56′W
Baza, Spain (bä′thä)	118	37°29′N	2°46′W
Baza, Sierra de, mts., Spain	128	37°19′N	2°48′W
Bazar-Dyuzi, mtn., Azer. (bä′zär-dyōōz′ĕ)	137	41°20′N	47°40′E
Bazaruto, Ilha do, i., Moz. (bȧ-zȧ-rōō′tō)	186	21°42′S	36°10′E
Bazièges, Fr.	126	43°25′N	1°41′E
Be, Nosy, i., Madag.	187	13°14′S	47°28′E
Beach, N.D., U.S. (bēch)	70	46°55′N	104°00′W
Beachy Head, c., Eng., U.K. (bēchē hĕd)	121	50°40′N	0°25′E
Beacon, N.Y., U.S. (bē′kŭn)	67	41°30′N	73°55′W
Beaconsfield, Can. (bē′kŭnz-fēld)	62a	45°26′N	73°51′W
Beals Creek, r., Tx., U.S. (bēls)	80	32°10′N	101°14′W
Bear, r., Ut., U.S.	75b	41°28′N	112°10′W
Bear, r., U.S.	73	42°17′N	111°42′W
Bear Brook, r., Can.	62c	45°24′N	75°15′W
Bear Creek, Mt., U.S. (bâr krēk)	73	45°11′N	109°07′W
Bear Creek, r., Al., U.S. (bâr)	82	34°27′N	88°00′W
Bear Creek, r., Tx., U.S.	75c	32°56′N	97°09′W
Beardstown, Il., U.S. (bērds′toun)	79	40°01′N	90°26′W
Bearfort Mountain, mtn., N.J., U.S. (bē′fōrt)	68a	41°08′N	74°23′W
Bearhead Mountain, mtn., Wa., U.S. (bâr′hĕd)	74a	47°01′N	121°49′W
Bear Lake, l., Can.	57	55°08′N	96°00′W
Bear Lake, l., Id., U.S.	73	41°56′N	111°10′W
Bear River Range, mts., U.S.	73	41°50′N	111°30′W
Beas de Segura, Spain (bā′äs dā sä-gōō′rä)	128	38°16′N	2°53′W
Beata, i., Dom. Rep. (bĕ-ä′tä)	93	17°40′N	71°40′W
Beata, Cabo, c., Dom. Rep. (ká′bō-bĕ-ä′tä)	93	17°40′N	71°20′W
Beatrice, Ne., U.S. (bē′ȧ-trĭs)	64	40°16′N	96°45′W
Beatty, Nv., U.S. (bēt′ē)	76	36°58′N	116°48′W

ng-sing; ŋ-baŋk; N-nasalized n; nŏd; cŏmmit; ōld; ôbey; ôrder; oi-boil; fōōd; ȯ-as oo in foot; ou-out; s-soft; sh-dish; th-thin; pūre; ůnite; ûrn; stŭd; circŭs; ü-as in French tu; ′-indeterminate vowel.

226 Beattyville – Bergedorf PRONOUNCING INDEX

PLACE (Pronunciation)	PAGE	LAT.	LONG.
Beattyville, Ky., U.S. (bēt'ē-vĭl)	66	37°35'N	83°40'W
Beaucaire, Fr. (bō-kâr')	126	43°49'N	4°37'E
Beaucourt, Fr. (bō-ko͞or')	127	47°30'N	6°54'E
Beaufort, N.C., U.S. (bō'fĕrt)	83	34°43'N	76°40'W
Beaufort, S.C., U.S.	83	32°25'N	80°40'W
Beaufort Sea, sea, N.A.	63	70°30'N	138°40'W
Beaufort West, S. Afr.	186	32°20'S	22°45'E
Beauharnois, Can. (bō-är-nwä')	59	45°23'N	73°52'W
Beaumont, Can.	62b	46°50'N	71°01'W
Beaumont, Can.	62g	53°22'N	113°18'W
Beaumont, Ca., U.S. (bō'mŏnt)	75a	33°57'N	116°57'W
Beaumont, Tx., U.S.	65	30°05'N	94°06'W
Beaune, Fr. (bōn)	126	47°02'N	4°49'E
Beauport, Can. (bō-pôr')	62b	46°52'N	71°11'W
Beauséjour, Can.	50	50°04'N	96°33'W
Beauvais, Fr. (bō-vĕ')	126	49°25'N	2°05'E
Beaver, Ok., U.S. (bē'vẽr)	78	36°46'N	100°31'W
Beaver, Pa., U.S.	69e	40°42'N	80°18'W
Beaver, Ut., U.S.	77	38°15'N	112°40'W
Beaver, i., Mi., U.S.	66	45°40'N	85°30'W
Beaver, r., Can.	52	54°20'N	111°10'W
Beaver City, Ne., U.S.	78	40°08'N	99°52'W
Beaver Creek, r., Co., U.S.	78	39°42'N	103°37'W
Beaver Creek, r., Ks., U.S.	78	39°44'N	101°05'W
Beaver Creek, r., Mt., U.S.	70	46°45'N	104°18'W
Beaver Creek, r., Wy., U.S.	70	43°46'N	104°25'W
Beaver Dam, Wi., U.S.	71	43°29'N	88°50'W
Beaverhead, r., Mt., U.S.	73	45°25'N	112°35'W
Beaverhead Mountains, mts., Mt., U.S. (bē'vẽr-hĕd)	73	44°33'N	112°59'W
Beaver Indian Reservation, I.R., Mi., U.S.	66	45°40'N	85°30'W
Beaverton, Or., U.S. (bē'vẽr-tŭn)	74c	45°29'N	122°49'W
Bebington, Eng., U.K. (bē'bĭng-tŭn)	114a	53°20'N	2°59'W
Bečej, Serb. (bĕ'chā)	131	45°36'N	20°03'E
Béchar, Alg.	184	31°39'N	2°14'W
Becharof, l., Ak., U.S. (bĕk-à-rôf)	63	57°58'N	156°58'W
Becher Bay, b., Can. (bēch'ẽr)	74a	48°18'N	123°37'W
Beckley, W.V., U.S. (bĕk'lĭ)	66	37°40'N	81°15'W
Bédarieux, Fr. (bā-dà-ryû')	126	43°36'N	3°11'E
Beddington Creek, r., Can. (bĕd'ĕng tŭn)	62e	51°14'N	114°13'W
Bedford, Can. (bĕd'fẽrd)	59	45°10'N	73°00'W
Bedford, S. Afr.	187c	32°43'S	26°19'E
Bedford, Eng., U.K.	117	52°10'N	0°25'W
Bedford, Ia., U.S.	71	40°40'N	94°41'W
Bedford, In., U.S.	66	38°50'N	86°30'W
Bedford, Ma., U.S.	61a	42°30'N	71°17'W
Bedford, N.Y., U.S.	68a	41°12'N	73°38'W
Bedford, Oh., U.S.	69d	41°23'N	81°32'W
Bedford, Pa., U.S.	67	40°05'N	78°20'W
Bedford, Va., U.S.	83	37°19'N	79°27'W
Bedford Hills, N.Y., U.S.	68a	41°14'N	73°41'W
Beebe, Ar., U.S. (bē'bē)	79	35°04'N	91°54'W
Beecher, Il., U.S. (bē'chŭr)	69a	41°20'N	87°38'W
Beechey Head, c., Can. (bē'chĭ hĕd)	74a	48°19'N	123°40'W
Beech Grove, In., U.S. (bēch grōv)	69g	39°43'N	86°05'W
Beecroft Head, c., Austl. (bē'krŭft)	176	35°03'S	151°15'E
Beelitz, Ger. (bĕ'lētz)	115b	52°14'N	12°59'E
Be'er Sheva', Isr. (bēr-shē'bà)	153a	31°15'N	34°48'E
Be'er Sheva', r., Isr.	153a	31°23'N	34°30'E
Beestekraal, S. Afr.	192c	25°22'S	27°34'E
Beeston, Eng., U.K. (bēs't'n)	114a	52°55'N	1°11'W
Beetz, r., Ger. (bĕtz)	115b	52°28'N	12°37'E
Beeville, Tx., U.S. (bē'vĭl)	81	28°24'N	97°44'W
Bega, Austl. (bā'gà)	175	36°50'S	149°49'E
Beggs, Ok., U.S. (bĕgz)	79	35°46'N	96°06'W
Bégles, Fr. (bĕ'gl')	126	44°47'N	0°34'W
Begoro, Ghana	188	6°23'N	0°23'W
Behala, India	158a	22°31'N	88°19'E
Behbehān, Iran	157	30°35'N	50°14'E
Behm Canal, can., Ak., U.S.	54	55°41'N	131°35'W
Bei, r., China (bā)	163a	22°54'N	113°08'E
Bei'an, China (bā-än)	164	48°05'N	126°26'E
Beicai, China (bā-tsī)	163b	31°12'N	121°33'E
Beifei, r., China (bā-fā)	162	33°14'N	117°03'E
Beihai, China (bā-hī)	160	21°30'N	109°10'E
Beihuangcheng Dao, i., China (bā-hûäŋ-chŭŋ dou)	162	38°23'N	120°55'E
Beijing, China	161	39°55'N	116°23'E
Beijing Shi, prov., China (bā-jyĭŋ shr)	164	40°07'N	116°00'E
Beira, Moz. (bā'rà)	186	19°45'N	34°58'E
Beira, hist. reg., Port. (bĕ'y-rä)	128	40°38'N	8°00'W
Beirut, Leb. (bā-ro͞ot')	154	33°53'N	35°30'E
Beja, Port. (bā'zhä)	118	38°03'N	7°53'W
Béja, Tun.	118	36°52'N	9°20'E
Bejaïa (Bougie), Alg.	184	36°46'N	5°00'E
Bejar, Spain	128	40°25'N	5°43'W
Bejestān, Iran	154	34°30'N	58°22'E
Bejucal, Cuba (bā-ho͞o-käl')	92	22°56'N	82°23'W
Bejuco, Pan. (bĕ-ko͞o'kō)	91	8°37'N	79°54'W
Békés, Hung. (bā'kāsh)	125	46°45'N	21°08'E
Békéscsaba, Hung. (bā'kāsh-chô'bô)	119	46°39'N	21°06'E
Beketova, Russia (bĕkĕ-to'và)	141	53°23'N	125°21'E
Bela Crkva, Serb. (bĕ'lä tsĕrk'vä)	131	44°53'N	21°25'E
Belalcázar, Spain (bāl-à-kä'thär)	128	38°35'N	5°12'W
Belarus, nation, Eur.	134	53°30'N	25°33'E
Belau *see* Palau, nation, Oc.	3	7°15'N	134°30'E
Bela Vista de Goiás, Braz.	101	16°57'S	48°47'W
Belawan, Indon. (bå-lä'wän)	168	3°43'N	98°43'E
Belaya, r., Russia (byĕ'lī-yà)	137	52°30'N	56°15'E
Belcher Islands, is., Can. (bĕl'chĕr)	53	56°20'N	80°40'W
Belding, Mi., U.S. (bĕl'dĭng)	66	43°05'N	85°25'W
Belebey, Russia (byĕ'lĕ-bă'ĭ)	136	54°00'N	54°10'E
Belém, Braz. (bå-lĕN')	101	1°18'S	48°27'W
Belén, Para. (bā-lān')	102	23°30'S	57°09'W
Belen, N.M., U.S. (bĕ-lān')	77	34°40'N	106°45'W
Bélep, Îles, is., N. Cal.	175	19°30'S	164°00'E
Belëv, Russia (byĕl'yĕf)	136	53°49'N	36°06'E
Belfair, Wa., U.S. (bĕl'far)	74a	47°27'N	122°50'W
Belfast, N. Ire., U.K.	110	54°36'N	5°45'W
Belfast, Me., U.S. (bĕl'fàst)	60	44°25'N	69°01'W
Belfast, Lough, b., N. Ire., U.K. (lŏĸ bĕl'fàst)	120	54°45'N	6°00'W
Belford Roxo, Braz.	102b	22°46'S	43°24'W
Belfort, Fr. (bā-fôr')	117	47°40'N	7°50'E
Belgaum, India	155	15°57'N	74°32'E
Belgium, nation, Eur. (bĕl'jĭ-ŭm)	110	51°00'N	2°52'E
Belgorod, Russia (byĕl'gŭ-rut)	137	50°36'N	36°32'E
Belgorod, prov., Russia	133	50°40'N	36°42'E
Belgrade (Beograd), Serb.	110	44°48'N	20°32'E
Belhaven, N.C., U.S. (bĕl'hā-vĕn)	83	35°33'N	76°37'W
Belington, W.V., U.S. (bĕl'ĭng-tŭn)	67	39°00'N	79°55'W
Belitung, i., Indon.	168	3°30'S	107°30'E
Belize, nation, N.A.	86	17°00'N	88°40'W
Belize, r., Belize	90a	17°16'N	88°56'W
Belize City, Belize (bĕ-lēz')	86	17°31'N	88°10'W
Bel'kovo, Russia (byĕl'kô-vô)	142b	56°15'N	38°49'E
Bel'kovskiy, i., Russia (byĕl-kôf'skĭ)	141	75°45'N	137°00'E
Bell, i., Can. (bĕl)	61	50°45'N	55°35'W
Bell, r., Can.	59	49°25'N	77°15'W
Bella Bella, Can.	54	52°10'N	128°07'W
Bella Coola, Can.	54	52°22'N	126°46'W
Bellaire, Oh., U.S. (bĕl-âr')	66	40°00'N	80°45'W
Bellaire, Tx., U.S.	81a	29°43'N	95°28'W
Bellary, India (bĕl-lä'rē)	155	15°15'N	76°56'E
Bella Union, Ur. (bĕ'l-yà-o͞o-nyō'n)	102	30°18'S	57°26'W
Bella Vista, Arg. (bā'lyà vēs'tà)	102	27°07'S	65°14'W
Bella Vista, Arg.	102	28°35'S	58°53'W
Bella Vista, Arg.	102a	34°35'S	58°41'W
Bella Vista, Para.	101	22°16'S	56°14'W
Belle-Anse, Haiti	93	18°15'N	72°00'W
Belle Bay, b., Can. (bĕl)	61	47°35'N	55°15'W
Belle Chasse, La., U.S. (bĕl shäs')	68d	29°52'N	90°00'W
Bellefontaine, Oh., U.S. (bel-fōn'tân)	66	40°25'N	83°50'W
Bellefontaine Neighbors, Mo., U.S.	75e	38°46'N	90°13'W
Belle Fourche, S.D., U.S. (bĕl' fo͞orsh')	70	44°28'N	103°50'W
Belle Fourche, r., Wy., U.S.	70	44°29'N	104°40'W
Belle Fourche Reservoir, res., S.D., U.S.	70	44°51'N	103°44'W
Bellegarde, Fr. (bĕl-gärd')	127	46°06'N	5°50'E
Belle Glade, Fl., U.S. (bĕl glād)	83a	26°39'N	80°37'W
Belle-Île, i., Fr. (bĕlēl')	117	47°15'N	3°30'W
Belle Isle, Strait of, strt., Can.	53	51°35'N	56°30'W
Belle Mead, N.J., U.S. (bĕl mēd)	68a	40°28'N	74°40'W
Belleoram, Can.	61	47°31'N	55°25'W
Belle Plaine, Ia., U.S. (bĕl plān')	71	41°52'N	92°19'W
Belle Vernon, Pa., U.S. (bĕl vŭr'nŭn)	69e	40°08'N	79°52'W
Belleville, Can. (bĕl'vĭl)	59	44°15'N	77°25'W
Belleville, Il., U.S.	75e	38°31'N	89°59'W
Belleville, Ks., U.S.	79	39°49'N	97°37'W
Belleville, Mi., U.S.	69b	42°12'N	83°29'W
Belleville, N.J., U.S.	68a	40°47'N	74°09'W
Bellevue, Ia., U.S. (bĕl'vū)	71	42°14'N	90°26'W
Bellevue, Ky., U.S.	69f	39°06'N	84°29'W
Bellevue, Mi., U.S.	66	42°30'N	85°00'W
Bellevue, Oh., U.S.	66	41°15'N	82°45'W
Bellevue, Pa., U.S.	69e	40°30'N	80°04'W
Bellevue, Wa., U.S.	74a	47°37'N	122°12'W
Belley, Fr. (bĕ-lĕ')	127	45°46'N	5°41'E
Bellflower, Ca., U.S. (bĕl-flou'ẽr)	75a	33°53'N	118°08'W
Bell Gardens, Ca., U.S.	75a	33°59'N	118°11'W
Bellingham, Ma., U.S. (bĕl'ĭng-hăm)	61a	42°05'N	71°28'W
Bellingham, Wa., U.S.	64	48°46'N	122°29'W
Bellingham Bay, b., Wa., U.S.	74d	48°44'N	122°34'W
Bellingshausen Sea, sea, Ant. (bĕl'ĭngz houz'n)	178	72°00'S	80°30'W
Bellinzona, Switz. (bĕl-ĭn-tsō'nä)	124	46°10'N	9°09'E
Bellmore, N.Y., U.S. (bĕl-mōr)	68a	40°40'N	73°31'W
Bello, Col. (bĕ'l-yô)	100	6°20'N	75°33'W
Bellow Falls, Vt., U.S. (bĕl'ōz fôls)	67	43°10'N	72°30'W
Bellpat, Pak.	158	29°08'N	68°00'E
Bell Peninsula, pen., Can.	53	63°50'N	81°16'W
Bells Corners, Can.	62c	45°20'N	75°49'W
Bells Mountain, mtn., Wa., U.S. (bĕls)	74c	45°50'N	122°21'W
Belluno, Italy (bĕl-lo͞o'nō)	130	46°08'N	12°14'E
Bell Ville, Arg. (bĕl vēl')	102	32°33'S	62°36'W
Bellville, S. Afr.	186a	33°54'S	18°38'E
Bellville, Tx., U.S. (bĕl'vĭl)	81	29°57'N	96°15'W
Bélmez, Spain (bĕl'mĕth)	128	38°17'N	5°17'W
Belmond, Ia., U.S. (bĕl'mŏnd)	71	42°49'N	93°37'W
Belmont, Ca., U.S.	74b	37°34'N	122°18'W
Belmonte, Braz. (bĕl-mōn'tå)	101	15°58'S	38°47'W
Belmopan, Belize	86	17°15'N	88°47'W
Belogorsk, Russia	135	51°09'N	128°32'E
Belo Horizonte, Braz. (bĕ'lôre-sô'n-tĕ)	101	19°54'S	43°56'W
Beloit, Ks., U.S. (bĕ-loit')	78	39°26'N	98°06'W
Beloit, Wi., U.S.	65	42°31'N	89°04'W
Belomorsk, Russia (byĕl-ô-môrsk')	136	64°30'N	34°42'E
Beloretsk, Russia (byĕ'lō-rĕtsk)	136	53°58'N	58°25'E
Belosarayskaya, Kosa, c., Ukr.	133	46°43'N	37°18'E
Belovo, Russia (byĕ'lŭ-vû)	140	54°25'N	86°18'E
Beloye, l., Russia	136	60°10'N	38°05'E
Belozersk, Russia (byĕ-lŭ-zyôrsk')	136	60°00'N	38°00'E
Belper, Eng., U.K. (bĕl'pẽr)	114a	53°01'N	1°28'W
Belt, Mt., U.S. (bĕlt)	73	47°11'N	110°58'W
Belt Creek, r., Mt., U.S.	73	47°19'N	110°58'W
Belton, Tx., U.S. (bĕl'tŭn)	81	31°04'N	97°27'W
Belton Lake, l., Tx., U.S.	81	31°15'N	97°35'W
Beltsville, Md., U.S. (belts-vĭl)	68e	39°03'N	76°56'W
Belukha, Mount, mtn., Asia	134	49°47'N	86°23'E
Belvidere, Il., U.S. (bĕl-vĕ-dēr')	71	42°14'N	88°52'W
Belvidere, N.J., U.S.	67	40°50'N	75°05'W
Belyando, r., Austl. (bĕl-yän'dō)	175	22°09'S	146°48'E
Belyanka, Russia (byĕl'yàn-kà)	142a	56°04'N	59°16'E
Belyy, Russia (byĕ'lē)	136	55°52'N	32°58'E
Belyy, i., Russia	134	73°19'N	72°00'E
Belyye Stolby, Russia (byĕ'lī-ye stôl'bĭ)	142b	55°20'N	37°52'E
Belzig, Ger. (bĕl'tsēg)	115b	52°08'N	12°35'E
Belzoni, Ms., U.S. (bĕl-zō'nē)	82	33°09'N	90°30'W
Bembe, Ang. (bĕn'bĕ)	186	7°00'S	14°20'E
Bembézar, r., Spain (bĕm-bā-thär')	128	38°00'N	5°18'W
Bemidji, Mn., U.S. (bē-mĭj'ĭ)	71	47°28'N	94°54'W
Bena Dibele, D.R.C. (bĕn'à dē-bĕ'lĕ)	186	4°00'S	22°49'E
Benalla, Austl. (bĕn-ăl'à)	175	36°30'S	146°00'E
Benares *see* Vārānasi, India	155	25°25'N	83°00'E
Benavente, Spain (bā-nä-vĕn'tä)	118	42°01'N	5°43'W
Benbrook, Tx., U.S. (bĕn'bro͝ok)	75c	32°41'N	97°27'W
Benbrook Reservoir, res., Tx., U.S.	75c	32°35'N	97°30'W
Bend, Or., U.S. (bĕnd)	64	44°04'N	121°17'W
Bendeleben, Mount, mtn., Ak., U.S. (bĕn-dĕl-bĕn)	63	65°18'N	163°45'W
Bender Beyla, Som.	192a	9°40'N	50°45'E
Bendigo, Austl. (bĕn'dĭ-gō)	175	36°39'S	144°20'E
Benedict, Md., U.S. (bĕnē'dĭct)	68e	38°31'N	76°41'W
Benešov, Czech Rep. (bĕn'ĕ-shôf)	124	49°48'N	14°40'E
Benevento, Italy (bā-nā-vĕn'tō)	118	41°08'N	14°46'E
Bengal, Bay of, b., Asia (bĕn-gôl')	152	17°30'N	87°00'E
Bengamisa, D.R.C.	191	0°57'N	25°10'E
Bengbu, China (bŭŋ-bo͞o)	161	32°52'N	117°22'E
Benghazi *see* Banghāzī, Libya	184	32°07'N	20°04'E
Bengkalis, Indon. (bĕng-kä'lĭs)	168	1°29'N	102°06'E
Bengkulu, Indon.	168	3°46'S	102°18'E
Benguela, Ang. (bĕn-gĕl'à)	186	12°35'S	13°25'E
Beni, r., Bol. (bā'nē)	100	13°41'S	67°30'W
Béni-Abbas, Alg. (bā'nē à-bĕs')	184	30°11'N	2°13'W
Benicia, Ca., U.S. (bĕ-nĭsh'ĭ-à)	74b	38°03'N	122°09'W
Benin, nation, Afr.	184	8°00'N	2°00'E
Benin, r., Nig. (bĕn-ēn')	189	5°55'N	5°15'E
Benin, Bight of, b., Afr.	184	5°30'N	3°00'E
Benin City, Nig.	184	6°19'N	5°41'E
Beni Saf, Alg. (bā'nē säf')	184	35°23'N	1°20'W
Benito, r., Eq. Gui.	190	1°35'N	10°45'E
Benkelman, Ne., U.S. (bĕn-kĕl-màn)	78	40°05'N	101°35'W
Benkovac, Cro. (bĕn'kō-vàts)	130	44°02'N	15°41'E
Bennettsville, S.C., U.S. (bĕn'ĕts vĭl)	83	34°35'N	79°41'W
Bennington, Vt., U.S. (bĕn'ĭng-tŭn)	67	42°55'N	73°15'W
Benns Church, Va., U.S. (bĕnz' chûrch')	68g	36°47'N	76°35'W
Benoni, S. Afr. (bĕ-nō'nī)	186	26°11'S	28°19'E
Benoy, Chad	189	8°59'N	16°19'E
Benque Viejo, Belize (bĕn-kĕ biĕ'hō)	90a	17°07'N	89°07'W
Bensberg, Ger.	127c	50°58'N	7°09'E
Bensenville, Il., U.S. (bĕn'sĕn-vĭl)	69a	41°57'N	87°56'W
Bensheim, Ger. (bĕns-hīm)	124	49°42'N	8°38'E
Benson, Az., U.S. (bĕn-sŭn)	77	32°00'N	110°20'W
Benson, Mn., U.S.	70	45°18'N	95°36'W
Bentiaba, Ang.	190	14°15'S	12°21'E
Bentleyville, Pa., U.S. (bent'lē vĭl)	69e	40°07'N	80°01'W
Benton, Can.	60	45°59'N	67°36'W
Benton, Ar., U.S. (bĕn'tŭn)	79	34°34'N	92°34'W
Benton, Ca., U.S.	76	37°44'N	118°22'W
Benton, Il., U.S.	66	38°00'N	88°55'W
Benton Harbor, Mi., U.S. (bĕn'tŭn här'bẽr)	66	42°05'N	86°30'W
Bentonville, Ar., U.S. (bĕn'tŭn-vĭl)	79	36°22'N	94°11'W
Benue, r., Afr. (bā'no͞o-à)	184	8°00'N	8°00'E
Benut, r., Malay.	153b	1°43'N	103°20'E
Benwood, W.V., U.S. (bĕn-wo͝od)	66	39°55'N	80°45'W
Benxi, China (bŭn-shyē)	164	41°25'N	123°50'E
Beograd *see* Belgrade, Serb.	110	44°48'N	20°32'E
Beppu, Japan (bĕ'po͞o)	167	33°16'N	131°30'E
Bequia Island, i., St. Vin. (bĕk-ē'ä)	91b	13°00'N	61°08'W
Berakit, Tanjung, c., Indon.	153b	1°16'N	104°44'E
Berat, Alb. (bĕ-rät')	131	40°43'N	19°59'E
Berau, Teluk, b., Indon.	169	2°22'S	131°40'E
Berazategui, Arg. (bĕ-rä-zá'tē-gē)	102a	34°46'S	58°14'W
Berbera, Som. (bûr'bûr-à)	192a	10°25'N	45°05'E
Berbérati, C.A.R.	189	4°16'N	15°47'E
Berck, Fr. (bĕrk)	126	50°26'N	1°36'E
Berdians'k, Ukr.	137	46°45'N	36°47'E
Berdians'ka kosa, c., Ukr.	133	46°38'N	36°42'E
Berdyaush, Russia (bĕr'dyàûsh)	142a	55°10'N	59°12'E
Berdychiv, Ukr.	134	49°53'N	28°32'E
Berea, Ky., U.S. (bĕ-rē'à)	82	37°30'N	84°19'W
Berea, Oh., U.S.	69d	41°22'N	81°51'W
Berehove, Ukr.	125	48°13'N	22°40'E
Bereku, Tan.	191	4°27'S	35°44'E
Berens, r., Can. (bĕr'ĕnz)	57	52°15'N	96°30'W
Berens Island, i., Can.	57	52°18'N	97°40'W
Berens River, Can.	50	52°22'N	97°02'W
Beresford, S.D., U.S. (bĕr'ĕs-fẽrd)	70	43°05'N	96°46'W
Berettyóújfalu, Hung. (bĕ'rĕt-tyō-o͞o'y'fō-lo͞o)	125	47°14'N	21°33'E
Berezhany, Ukr. (bĕr-yĕ'zhà-nē)	125	49°25'N	24°58'E
Berezivka, Ukr.	133	47°12'N	30°56'E
Berezna, Ukr. (bĕr-yôz'nà)	133	51°32'N	31°47'E
Bereznehuvate, Ukr.	133	47°19'N	32°58'E
Berezniki, Russia (bĕr-yôz'nyē-kē)	136	59°25'N	56°46'E
Berëzovka, Russia	142a	57°35'N	57°19'E
Berëzovo, Russia (bĭr-yô'zĕ-vû)	134	64°10'N	65°10'E
Berëzovskiy, Russia (bĕr-yô'zôf-skĭ)	142a	56°54'N	60°47'E
Berga, Spain (bĕr'gä)	129	42°05'N	1°52'E
Bergama, Tur. (bĕr'gä-mä)	154	39°08'N	27°09'E
Bergamo, Italy (bĕr'gä-mō)	118	45°43'N	9°41'E
Bergantin, Ven. (bĕr-gän-tē'n)	101b	10°04'N	64°23'W
Bergara, Spain	128	43°08'N	2°23'W
Bergedorf, Ger. (bĕr'gĕ-dôrf)	115c	53°29'N	10°12'E

ăt; finăl; rāte; senāte; ärm; åsk; sofà; fâre; ch-choose; dh-as th in other; bē; ēvent; bĕt; recĕnt; cratẽr; g-gō; gh-guttural g; bĭt; ĭ-short neutral; rīde; ĸ-guttural k as ch in German ich;

PLACE (Pronunciation)	PAGE	LAT.	LONG.
Bergen, Ger. (bĕr′gĕn)	124	54°26′N	13°26′E
Bergen, Nor.	110	60°24′N	5°20′E
Bergenfield, N.J., U.S.	68a	40°55′N	73°59′W
Bergen op Zoom, Neth.	121	51°29′N	4°16′E
Bergerac, Fr. (bĕr-zhĕ-răk′)	117	44°49′N	0°28′E
Bergisch Gladbach, Ger. (bĕrg′ĭsh-glät′bäk)	127c	50°59′N	7°08′E
Berglern, Ger. (bĕrgh′lĕrn)	115d	48°24′N	11°55′E
Bergneustadt, Ger.	127c	51°01′N	7°39′E
Bergville, S. Afr. (bĕrg′vĭl)	187c	28°46′S	29°22′E
Berhampur, India	155	19°19′N	84°48′E
Bering Sea, sea, (bē′rĭng)	194	58°00′N	175°00′W
Bering Strait, strt.,	64a	64°50′N	169°50′W
Berja, Spain (bĕr′hä)	128	36°50′N	2°56′W
Berkeley, Ca., U.S. (bûrk′lĭ)	64	37°52′N	122°17′W
Berkeley, Mo., U.S.	75e	38°45′N	90°20′W
Berkeley Springs, W.V., U.S. (bûrk′lĭ springz)	67	39°40′N	78°10′W
Berkhamsted, Eng., U.K. (bĕk′hăm′stĕd)	114b	51°44′N	0°34′W
Berkley, Mi., U.S. (bûrk′lĭ)	69b	42°30′N	83°10′W
Berkovitsa, Blg. (bĕ-kō′vē-tsȧ)	131	43°14′N	23°08′E
Berkshire, hist. reg., Eng., U.K.	114b	51°23′N	1°07′W
Berland, r., Can.	55	54°00′N	117°10′W
Berlenga, is., Port. (bĕr-lĕn′gäzh)	128	39°25′N	9°33′W
Berlin, Ger. (bĕr-lēn′)	110	52°31′N	13°28′E
Berlin, S. Afr. (bĕr-lĭn)	187c	32°53′S	27°36′E
Berlin, N.H., U.S. (bûr-lĭn)	67	44°25′N	71°10′W
Berlin, N.J., U.S.	68f	39°47′N	74°56′W
Berlin, Wi., U.S. (bûr-lĭn′)	71	43°58′N	88°58′W
Bermejo, r., S.A. (bĕr-mā′hō)	102	25°05′S	61°00′W
Bermeo, Spain (bĕr-mā′yō)	128	43°23′N	2°43′W
Bermuda, dep., N.A.	87	32°20′N	65°45′W
Bern, Switz. (bĕrn)	110	46°55′N	7°25′E
Bernal, Arg. (bĕr-näl′)	102a	34°43′S	58°17′W
Bernalillo, N.M., U.S. (bĕr-nä-lē′yō)	77	35°20′N	106°30′W
Bernard, l., Can. (bĕr-närd′)	67	45°45′N	79°25′W
Bernardsville, N.J., U.S. (bŭr nârds′vĭl)	68a	40°43′N	74°34′W
Bernau, Ger. (bĕr′nou)	124	52°40′N	13°35′E
Bernburg, Ger. (bĕrn′bŏorgh)	124	51°48′N	11°43′E
Berndorf, Aus. (bĕrn′dôrf)	124	47°57′N	16°05′E
Berne, In., U.S. (bûrn)	66	40°40′N	84°55′W
Berner Alpen, mts., Switz.	124	46°29′N	7°30′E
Bernier, i., Austl. (bĕr-nēr′)	174	24°58′S	113°15′E
Bernina, Pizzo, mtn., Eur.	124	46°23′N	9°58′E
Bero, r., Ang.	190	15°10′S	12°20′E
Beroun, Czech Rep. (bā′rōn)	124	49°57′N	14°03′E
Berounka, r., Czech Rep. (bĕ-rōn′kȧ)	124	49°53′N	13°40′E
Berowra, Austl.	173b	33°36′S	151°10′E
Berre, Étang de, l., Fr. (ā-tôN′ dĕ bâr′)	126a	43°27′N	5°07′E
Berre-l'Étang, Fr. (bâr′lā-tôN′)	126a	43°28′N	5°11′E
Berriozabal, Mex. (bă′rēō-zä-bäl′)	89	16°47′N	93°16′W
Berriyyane, Alg.	118	32°50′N	3°49′E
Berry Creek, r., Can.	56	51°15′N	111°40′W
Berryessa, r., Ca., U.S. (bĕ′rĭ ĕs′ȧ)	76	38°35′N	122°33′W
Berry Islands, is., Bah.	92	25°40′N	77°50′W
Berryville, Ar., U.S. (bĕr′ē-vĭl)	79	36°21′N	93°34′W
Bershad', Ukr. (byĕr′shät)	133	48°22′N	29°31′E
Berthier, Can.	62b	46°56′N	70°44′W
Bertrand, r., Wa., U.S. (bûr′trȧnd)	74d	48°58′N	122°31′W
Berwick, Pa., U.S. (bûr′wĭk)	67	41°05′N	76°10′W
Berwick-upon-Tweed, Eng., U.K. (bûr′ĭk)	116	55°45′N	2°01′W
Berwyn, Il., U.S. (bûr′wĭn)	69a	41°49′N	87°47′W
Beryslav, Ukr.	133	46°49′N	33°24′E
Besalampy, Madag. (bĕz-ȧ-lȧm-pē′)	187	16°48′S	44°40′E
Besançon, Fr. (bĕ-säN-sôN)	117	47°14′N	6°02′E
Besar, Gunong, mtn., Malay.	153b	2°31′N	103°09′E
Besed', r., Eur. (byĕ′syĕt)	132	52°58′N	31°36′E
Beskid Mountains, mts., Eur.	125	49°23′N	19°00′E
Beskra, Alg.	184	34°52′N	5°39′E
Beslan, Russia	138	43°12′N	44°33′E
Bessarabia, hist. reg., Mol.	133	47°00′N	28°30′E
Bességes, Fr. (bĕ-sĕzh′)	126	44°20′N	4°07′E
Bessemer, Al., U.S. (bĕs′ē-mĕr)	68h	33°24′N	86°58′W
Bessemer, Mi., U.S.	71	46°29′N	90°04′W
Bessemer City, N.C., U.S.	83	35°16′N	81°17′W
Bestensee, Ger. (bĕs′tĕn-zā)	115b	52°15′N	13°39′E
Betanzos, Spain (bĕ-tän′thōs)	128	43°18′N	8°14′W
Betatakin Ruin, Az., U.S. (bĕt-ȧ-tăk′ĭn)	77	36°40′N	110°29′W
Bethal, S. Afr. (bĕth′ăl)	192c	26°27′S	29°28′E
Bethalto, Il., U.S. (bȧ-thăl′tō)	75e	38°54′N	90°03′W
Bethanien, Nmb.	186	26°20′S	16°10′E
Bethany, Mo., U.S.	79	40°15′N	94°04′W
Bethel, Ak., U.S. (bĕth′ĕl)	64	60°50′N	161°50′W
Bethel, Ct., U.S.	68a	41°22′N	73°24′W
Bethel, Vt., U.S.	67	43°50′N	72°40′W
Bethel Park, Pa., U.S.	69e	40°19′N	80°02′W
Bethesda, Md., U.S. (bĕ-thĕs′dȧ)	68e	39°00′N	77°10′W
Bethlehem, S. Afr.	186	28°14′S	28°18′E
Bethlehem, Pa., U.S. (bĕth′lē-hĕm)	67	40°40′N	75°25′W
Bethlehem *see* Bayt Lahm, W.B.	153a	31°42′N	35°13′E
Béthune, Fr. (bā-tün′)	126	50°32′N	2°37′E
Betroka, Madag. (bĕ-trōk′ȧ)	187	23°13′S	46°17′E
Bet She'an, Isr.	153a	32°30′N	35°30′E
Betsiamites, Can.	51	48°57′N	68°36′W
Betsiamites, r., Can.	60	49°11′N	69°20′W
Betsiboka, r., Madag. (bĕt-sĭ-bō′kȧ)	187	16°47′S	46°45′E
Bettles Field, Ak., U.S. (bĕt′ŭls)	63	66°58′N	151°48′W
Betwa, r., India (bĕt′wȧ)	155	25°00′N	78°00′E
Betz, Fr. (bĕ)	127b	49°09′N	2°58′E
Beveren, Bel.	115a	51°13′N	4°14′E
B. Everett Jordan Lake, res., N.C., U.S.	83	35°45′N	79°00′W
Beverly, Ma., U.S.	61a	42°34′N	70°53′W
Beverly, N.J., U.S.	68f	40°03′N	74°56′W
Beverly Hills, Ca., U.S.	75a	34°05′N	118°24′W
Bevier, Mo., U.S. (bē-vēr′)	79	39°44′N	92°36′W
Bewdley, Eng., U.K. (būd′lĭ)	114a	52°22′N	2°19′W
Bexhill, Eng., U.K. (bĕks′hĭl)	121	50°49′N	0°25′E
Bexley, Eng., U.K. (bĕks′ly)	114b	51°26′N	0°09′E
Beyla, Gui. (bā′lȧ)	184	8°41′N	8°37′W
Beylul, Erit.	185	13°15′N	42°21′E
Beypazari, Tur. (bā-pȧ-zä′rĭ)	119	40°10′N	31°40′E
Beyşehir, Tur.	137	38°00′N	31°45′E
Beysugskiy, Liman, b., Russia (lĭ-män′ bĕy-sōōg′skĭ)	133	46°07′N	38°35′E
Bezhetsk, Russia (byĕ-zhĕtsk′)	136	57°46′N	36°40′E
Bezhitsa, Russia (byĕ-zhĭ′tsȧ)	136	53°19′N	34°18′E
Béziers, Fr. (bā-zyā′)	117	43°21′N	3°12′E
Bhadreswar, India	158a	22°49′N	88°22′E
Bhāgalpur, India (bä′gŭl-pŏor)	155	25°15′N	86°59′E
Bhamo, Mya. (bŭ-mō′)	155	24°00′N	96°15′E
Bhāngar, India	158a	22°30′N	88°36′E
Bharatpur, India (bĕrt′pŏor)	155	27°21′N	77°33′E
Bhatinda, India (bŭ-tĭn-dȧ)	155	30°19′N	74°56′E
Bhātpāra, India	155	22°52′N	88°24′E
Bhaunagar, India (bäv-nŭg′ŭr)	155	21°45′N	72°58′E
Bhayandar, India	159b	19°20′N	72°50′E
Bhilai, India	158	21°14′N	81°23′E
Bhīma, r., India (bē′mȧ)	155	18°00′N	74°45′E
Bhiwandi, India	159b	19°18′N	73°03′E
Bhiwāni, India	158	28°53′N	76°08′E
Bhopāl, India (bō-päl)	155	23°20′N	77°25′E
Bhubaneswar, India (bō-bû-nāsh′vûr)	155	20°21′N	85°53′E
Bhuj, India (bōōj)	155	23°22′N	69°39′E
Bhutan, nation, Asia (bōō-tän′)	155	27°15′N	90°30′E
Biafra, Bight of, b., Afr.	184	4°05′N	7°10′E
Biak, i., Indon. (bē′äk)	169	1°00′S	136°00′E
Biała Podlaska, Pol. (byä′wä pōd-läs′kä)	125	52°01′N	23°08′E
Białogard, Pol.	124	54°00′N	16°01′E
Bialystok, Pol. (byä-wĭs′tôk)	110	53°08′N	23°12′E
Biankouma, C. Iv.	188	7°44′N	7°37′W
Biarritz, Fr. (byä-rēts′)	117	43°27′N	1°39′W
Bibb City, Ga., U.S. (bĭb′ sĭ′tē)	82	32°31′N	84°56′W
Biberach, Ger. (bē′bēräk)	124	48°06′N	9°49′E
Bibiani, Ghana	188	6°28′N	2°20′W
Bic, Can. (bĭk)	60	48°22′N	68°42′W
Bicknell, In., U.S. (bĭk′nĕl)	66	38°45′N	87°20′W
Bicske, Hung. (bĭsh′kē)	125	47°29′N	18°38′E
Bida, Nig. (bē′dä)	184	9°05′N	6°01′E
Biddeford, Me., U.S. (bĭd′ē-fērd)	60	43°29′N	70°29′W
Biddulph, Eng., U.K. (bĭd′ŭlf)	114a	53°07′N	2°10′W
Biebrza, r., Pol. (byĕb′zhȧ)	125	53°18′N	22°25′E
Biel, Switz. (bēl)	124	47°09′N	7°12′E
Bielefeld, Ger. (bē′lē-fĕlt)	117	52°01′N	8°35′E
Biella, Italy (byĕl′lä)	130	45°34′N	8°05′E
Bielsk Podlaski, Pol. (byĕlsk pŭd-lä′skĭ)	117	52°47′N	23°14′E
Bien Hoa, Viet.	168	10°59′N	106°49′E
Bienville, Lac, l., Can.	53	55°32′N	72°45′W
Biesenthal, Ger. (bē′sĕn-täl)	115b	52°46′N	13°38′E
Biferno, r., Italy (bē-fĕr′nō)	130	41°49′N	14°46′E
Bifoum, Gabon	190	0°22′S	10°23′E
Biga, Tur. (bē′ghȧ)	131	40°13′N	27°14′E
Big Bay de Noc, Mi., U.S. (bĭg bā dē nok′)	71	45°48′N	86°41′W
Big Bayou, Ar., U.S. (bĭg′bī′yōō)	79	33°04′N	91°28′W
Big Bear City, Ca., U.S. (bĭg bâr)	75a	34°16′N	116°51′W
Big Belt Mountains, mts., Mt., U.S. (bĭg bĕlt)	64	46°53′N	111°43′W
Big Bend Dam, S.D., U.S. (bĭg bĕnd)	70	44°11′N	99°33′W
Big Bend National Park, rec., Tx., U.S.	64	29°15′N	103°15′W
Big Black, r., Ms., U.S. (bĭg blăk)	82	32°05′N	90°49′W
Big Blue, r., Ne., U.S. (bĭg blōō)	79	40°53′N	97°00′W
Big Canyon, Tx., U.S. (bĭg kăn′yŭn)	80	30°27′N	102°19′W
Big Cypress Indian Reservation, I.R., Fl., U.S.	83a	26°19′N	81°11′W
Big Cypress Swamp, sw., Fl., U.S. (bĭg sī′prĕs)	83a	26°02′N	81°20′W
Big Delta, Ak., U.S. (bĭg dĕl′tȧ)	63	64°08′N	145°48′W
Big Fork, r., Mn., U.S. (bĭg fôrk)	71	48°08′N	93°47′W
Biggar, Can.	50	52°04′N	108°00′W
Big Hole, r., Mt., U.S. (bĭg hōl)	73	45°53′N	113°15′W
Big Hole National Battlefield, Mt., U.S. (bĭg hōl băt′'l-fēld)	73	45°44′N	113°35′W
Bighorn, r., U.S.	64	45°30′N	108°00′W
Bighorn Lake, res., Mt., U.S.	73	45°00′N	108°10′W
Bighorn Mountains, mts., U.S. (bĭg hôrn)	64	44°47′N	107°40′W
Big Island, i., Can.	57	49°10′N	94°40′W
Big Lake, Wa., U.S. (bĭg lāk)	74a	48°24′N	122°14′W
Big Lake, l., Can.	62g	53°35′N	113°47′W
Big Lake, l., Wa., U.S.	74a	48°24′N	122°14′W
Big Lost, r., Id., U.S. (lôst)	73	43°56′N	113°38′W
Big Mossy Point, c., Can.	57	53°45′N	97°50′W
Big Muddy, r., Il., U.S.	66	37°50′N	89°00′W
Big Muddy Creek, r., Mt., U.S. (bĭg mud′ĭ)	73	48°53′N	105°02′W
Bignona, Sen.	188	12°49′N	16°14′W
Big Porcupine Creek, r., Mt., U.S. (pôr′kū-pīn)	73	46°38′N	107°04′W
Big Quill Lake, l., Can.	52	51°55′N	104°22′W
Big Rapids, Mi., U.S. (bĭg răp′ĭdz)	66	43°40′N	85°30′W
Big River, Can.	50	53°50′N	107°01′W
Big Sandy, r., Az., U.S. (bĭg sănd′ē)	77	34°59′N	113°36′W
Big Sandy, r., Ky., U.S.	66	38°15′N	82°35′W
Big Sandy, r., Wy., U.S.	73	42°08′N	109°35′W
Big Sandy Creek, r., Co., U.S.	78	39°08′N	103°36′W
Big Sandy Creek, r., Mt., U.S.	73	48°20′N	110°08′W
Bigsby Island, i., Can.	57	49°04′N	94°35′W
Big Sioux, r., U.S. (bĭg sōō)	70	44°34′N	97°00′W
Big Spring, Tx., U.S. (bĭg sprĭng)	80	32°15′N	101°28′W
Big Stone, l., Mn., U.S. (bĭg stōn)	70	45°29′N	96°40′W
Big Stone Gap, Va., U.S.	83	36°50′N	82°50′W
Big Sunflower, r., Ms., U.S. (sŭn-flou′ēr)	82	32°57′N	90°40′W
Big Timber, Mt., U.S. (bĭg′tĭm-bēr)	73	45°50′N	109°57′W
Big Wood, r., Id., U.S. (bĭg wŏod)	73	43°02′N	114°30′W
Bihār, state, India (bē-här′)	155	25°30′N	87°00′E
Biharamulo, Tan. (bē-hä-rä-mōō′lō)	186	2°38′S	31°20′E
Bihorului, Munţii, mts., Rom.	125	46°37′N	22°37′E
Bijagós, Arquipélago dos, is., Gui.-B.	184	11°20′N	17°10′W
Bijāpur, India	159	16°53′N	75°42′E
Bijeljina, Bos.	131	44°44′N	19°15′E
Bijelo Polje, Serb. (bē′yĕ-lō pō′lyĕ)	131	43°02′N	19°48′E
Bijiang, China (bē-jyän)	163a	22°57′N	113°15′E
Bijie, China (bē-jyĕ)	165	27°20′N	105°18′E
Bijou Creek, r., Co., U.S. (bē′zhōō)	78	39°41′N	104°13′W
Bīkaner, India (bĭ-kă′nûr)	155	28°07′N	73°19′E
Bikin, Russia (bē-kēn′)	166	46°41′N	134°29′E
Bikin, r., Russia	166	46°37′N	135°55′E
Bikoro, D.R.C. (bē-kō′rō)	186	0°45′S	18°07′E
Bikuar, Parque Nacional do, rec., Ang.	190	15°07′S	14°40′E
Bilāspur, India (bē-läs′pōōr)	155	22°08′N	82°12′E
Bila Tserkva, Ukr.	137	49°48′N	30°09′E
Bilauktaung, mts., Asia	168	14°40′N	98°50′E
Bilbao, Spain (bĭl-bä′ō)	110	43°12′N	2°48′W
Bilbays, Egypt	192b	30°26′N	31°37′E
Bileća, Bos. (bē′lĕ-chä)	131	42°52′N	18°26′E
Bilecik, Tur. (bē-lĕd-zhēk′)	119	40°10′N	29°58′E
Bílé Karpaty, mts., Eur.	125	48°53′N	17°35′E
Biłgoraj, Pol. (bēw-gō′rĭ)	125	50°31′N	22°43′E
Bilhorod-Dnistrovs'kyi, Ukr.	137	46°09′N	30°19′E
Bilimbay, Russia (bē′lĭm-bȧy)	142a	56°59′N	59°53′E
Billabong, r., Austl. (bĭl′ȧ-bŏng)	175	35°15′S	145°20′E
Billerica, Ma., U.S. (bĭl′rĭk-ȧ)	61a	42°33′N	71°16′W
Billericay, Eng., U.K.	114b	51°38′N	0°25′E
Billings, Mt., U.S. (bĭl′ĭngz)	64	45°47′N	108°29′W
Bill Williams, r., Az., U.S. (bĭl-wĭl′yumz)	77	34°10′N	113°50′W
Bilma, Niger (bēl′mä)	185	18°41′N	13°20′E
Bilopillia, Ukr.	137	51°10′N	34°19′E
Bilovods'k, Ukr.	133	49°12′N	39°36′E
Biloxi, Ms., U.S. (bĭ-lŏk′sĭ)	65	30°24′N	88°50′W
Bilqās Qism Awwal, Egypt	192b	31°14′N	31°25′E
Bimberi Peak, mtn., Austl. (bĭm′bĕrĭ)	176	35°45′S	148°50′E
Binalonan, Phil. (bē-nä-lō′nän)	169a	16°03′N	120°35′E
Bingen, Ger. (bĭn′gĕn)	124	49°57′N	7°54′E
Bingham, Eng., U.K. (bĭng′ȧm)	114a	52°57′N	0°57′W
Bingham, Me., U.S.	60	45°03′N	69°51′W
Bingham Canyon, Ut., U.S.	75b	40°33′N	112°09′W
Binghamton, N.Y., U.S. (bĭng′ȧm-tŭn)	65	42°05′N	75°55′W
Bingo-Nada, b., Japan (bĭn′gō nä-dä)	167	34°06′N	133°14′E
Binjai, Indon.	168	3°59′N	108°00′E
Binnaway, Austl. (bĭn′ȧ-wā)	176	31°42′S	149°22′E
Bintan, i., Indon. (bĭn′tän)	153b	1°09′N	104°43′E
Bintimani, mtn., S.L.	188	9°13′N	11°07′W
Bintulu, Malay. (bēn′tōō-lōō)	168	3°07′N	113°06′E
Binxian, China	164	45°40′N	127°20′E
Binxian, China (bĭn-shyän)	162	37°27′N	117°58′E
Bio Gorge, val., Ghana	188	8°30′N	2°05′W
Bioko (Fernando Póo), i., Eq. Gui.	184	3°35′N	7°45′E
Bira, Russia (bē′rȧ)	166	49°00′N	133°18′E
Bira, r., Russia	166	48°55′N	132°25′E
Birātnagar, Nepal (bĭ-rät′nŭ-gŭr)	158	26°35′N	87°18′E
Birbka, Ukr.	125	49°36′N	24°18′E
Birch Bay, Wa., U.S. (bûrch)	74d	48°55′N	122°45′W
Birch Bay, b., Wa., U.S.	74d	48°55′N	122°52′W
Birch Island, i., Can.	57	52°25′N	99°55′W
Birch Mountains, mts., Can.	52	57°36′N	113°10′W
Birch Point, c., Wa., U.S.	74d	48°57′N	122°50′W
Bird Island, i., S. Afr. (bĕrd)	187c	33°51′S	26°21′E
Bird Rock, i., Bah. (bûrd)	93	22°50′N	74°20′W
Birds Hill, Can. (bûrds)	62f	49°58′N	97°00′W
Birdsville, Austl. (bûrd′vĭl)	174	25°50′S	139°31′E
Birdum, Austl. (bûrd′ŭm)	174	15°45′S	133°25′E
Birecik, Tur. (bē-rĕd-zhēk′)	119	37°10′N	37°50′E
Bir Gara, Chad	189	13°11′N	15°58′E
Bīrjand, Iran (bēr′jänd)	154	33°07′N	59°16′E
Birkenfeld, Or., U.S.	74c	45°59′N	123°20′W
Birkenhead, Eng., U.K. (bûr′kĕn-hĕd)	120	53°23′N	3°02′W
Birkenwerder, Ger. (bēr′kĕn-vĕr-dĕr)	115b	52°41′N	13°22′E
Birmingham, Eng., U.K.	110	52°29′N	1°53′W
Birmingham, Al., U.S. (bûr′mĭng-häm)	65	33°31′N	86°49′W
Birmingham, Mi., U.S.	69b	42°32′N	83°13′W
Birmingham, Mo., U.S.	75f	39°10′N	94°22′W
Birmingham Canal, can., Eng., U.K.	114a	53°07′N	2°40′W
Bi'r Misāhah, Egypt	185	22°16′N	28°04′E
Birnin Kebbi, Nig.	184	12°32′N	4°12′E
Birobidzhan, Russia (bē′rō-bē-jän′)	135	48°42′N	133°28′E
Birsk, Russia (bĭrsk)	134	55°25′N	55°30′E
Birstall, Eng., U.K. (bûr′stôl)	114a	53°44′N	1°39′W
Biryulëvo, Russia (bēr-yōōl′yô-vô)	142b	55°35′N	37°39′E
Biryusa, r., Russia (bēr-yōō′sä)	140	56°43′N	97°30′E
Bi'r Za'farānah, Egypt	153a	29°07′N	32°38′E
Biržai, Lith. (bēr-zhä′ē)	123	56°11′N	24°45′E
Bisbee, Az., U.S. (bĭz′bē)	64	31°30′N	109°55′W
Biscay, Bay of, b., Eur. (bĭs′kā′)	112	45°19′N	3°51′W
Biscayne Bay, b., Fl., U.S. (bĭs-kān′)	83a	25°22′N	80°15′W
Bischeim, Fr. (bĭsh′hīm)	127	48°40′N	7°48′E
Biscotasi Lake, l., Can.	58	47°20′N	81°55′W
Biser, Russia (bē′sĕr)	142a	58°24′N	58°54′E
Biševo, is., Serb. (bē′shĕ-vō)	130	42°58′N	15°50′E

ng-sing; ŋ-baŋk; N-nasalized n; nŏd; cŏmmit; ōld; ôbey; ôrder; oi-boil; fōōd; ŏ-as oo in foot; ou-out; s-soft; sh-dish; th-thin; pūre; ûnite; ûrn; stŭd; circŭs; ü-as in French tu; ′-indeterminate vowel.

ăt; finăl; rāte; senăte; ärm; åsk; sofá; fâre; ch-choose; dh-as th in other; bē; ěvent; bět; recěnt; cratẽr; g-gō; gh-guttural g; bǐt; ĭ-short neutral; rīde; ĸ-guttural k as ch in German ich;

PRONOUNCING INDEX

Bolton-upon-Dearne – Bratsk 229

PLACE (Pronunciation)	PAGE	LAT.	LONG.
Bolton-upon-Dearne, Eng., U.K. (bōl′tŭn-ŭp′ŏn-dûrn)	114a	53°31′N	1°19′W
Bolu, Tur. (bō′lò)	119	40°45′N	31°45′E
Bolva, r., Russia (bôl′vä)	132	53°30′N	34°30′E
Bolvadin, Tur. (bŏl-vä-dēn′)	119	38°50′N	30°50′E
Bolzano, Italy (bôl-tsä′nō)	118	46°31′N	11°22′E
Boma, D.R.C. (bō′mä)	186	5°51′S	13°03′E
Bombala, Austl. (bŭm-bä′lă)	175	36°55′S	149°07′E
Bombay *see* Mumbai, India	155	18°58′N	72°50′E
Bombay Harbour, b., India	159b	18°55′N	72°52′E
Bomi Hills, Lib.	184	7°00′N	11°00′W
Bom Jardim, Braz. (bôn zhär-dēn′)	99a	22°10′S	42°25′W
Bom Jesus do Itabapoana, Braz.	99a	21°08′S	41°51′W
Bømlo, i., Nor. (bûmlò)	122	59°47′N	4°57′E
Bomongo, D.R.C.	185	1°22′N	18°21′E
Bom Sucesso, Braz. (bôn-so͞o-sĕ′sō)	99a	21°02′S	44°44′W
Bomu *see* Mbomou, r., Afr.	185	4°50′N	24°00′E
Bon, Cap, c., Tun. (bôn)	118	37°04′N	11°13′E
Bonaire, i., Neth. Ant. (bò-nâr′)	100	12°10′N	68°15′W
Bonavista, Can. (bŏ-nȧ-vĭs′tȧ)	53a	48°39′N	53°07′W
Bonavista Bay, b., Can.	53a	48°45′N	53°20′W
Bond, Co., U.S. (bŏnd)	78	39°53′N	106°40′W
Bondo, D.R.C. (bôn′dō)	140	3°49′N	23°40′E
Bondoc Peninsula, pen., Phil. (bôn-dŏk′)	169a	13°24′N	122°30′E
Bondoukou, C. Iv. (bŏn-do͞o′ko͞o)	184	8°02′N	2°48′W
Bonds Cay, i., Bah. (bŏnds kē)	92	25°30′N	77°45′W
Bondy, Fr.	127b	48°54′N	2°28′E
Bône *see* Annaba, Alg.	184	36°57′N	7°39′E
Bone, Teluk, b., Indon.	168	4°09′S	121°00′E
Bonete, Cerro, mtn., Arg. (bô′nětěh çĕrrô)	102	27°50′S	68°35′W
Bonfim, Braz. (bôn-fē′ɴ)	99a	20°20′S	44°15′W
Bongor, Chad	189	10°17′N	15°22′E
Bonham, Tx., U.S. (bŏn′ăm)	79	33°35′N	96°09′W
Bonhomme, Pic, mtn., Haiti	93	19°10′N	72°20′W
Bonifacio, Fr. (bō-nē-fä′chō)	130	41°23′N	9°10′E
Bonifacio, Strait of, strt., Eur.	118	41°14′N	9°02′E
Bonifay, Fl., U.S. (bŏn-ĭ-fā′)	82	30°46′N	85°40′W
Bonin Islands, is., Japan (bō′nĭn)	195	26°30′N	141°00′E
Bonn, Ger. (bōn)	110	50°44′N	7°06′E
Bonne Bay, b., Can. (bŏn)	61	49°33′N	57°55′W
Bonners Ferry, Id., U.S. (bŏnẽrz fĕr′ĭ)	72	48°41′N	116°19′W
Bonner Springs, Ks., U.S. (bŏn′ẽr sprĭngz)	75f	39°04′N	94°52′W
Bonne Terre, Mo., U.S. (bŏn târ′)	79	37°55′N	90°32′W
Bonnet Peak, mtn., Can. (bŏn′ĭt)	55	51°26′N	115°53′W
Bonneville Dam, dam, U.S. (bŏn′ē-vĭl)	72	45°37′N	121°57′W
Bonny, Nig. (bŏn′ē)	184	4°29′N	7°13′E
Bonny Lake, Wa., U.S. (bŏn′ē lăk)	74a	47°11′N	122°11′W
Bonnyville, Can. (bŏnē-vĭl)	55	54°16′N	110°44′W
Bonorva, Italy (bō-nôr′vä)	130	40°26′N	8°46′E
Bonthain, Indon. (bŏn-tīn′)	168	5°30′S	119°52′E
Bonthe, S.L.	184	7°32′N	12°30′W
Bontoc, Phil. (bŏn-tŏk′)	169a	17°10′N	121°01′E
Booby Rocks, is., Bah. (bo͞o′bĭ rŏks)	92	23°55′N	77°00′W
Booker T. Washington National Monument, rec., Va., U.S. (bo͝ok′ẽr tē wŏsh′ĭng-tŭn)	83	37°07′N	79°45′W
Boom, Bel.	115a	51°05′N	4°22′E
Boone, Ia., U.S. (bo͞on)	71	42°04′N	93°51′W
Booneville, Ar., U.S. (bo͞on′vĭl)	79	35°09′N	93°54′W
Booneville, Ky., U.S.	66	37°25′N	83°40′W
Booneville, Ms., U.S.	82	34°37′N	88°35′W
Boons, S. Afr.	192c	25°59′S	27°15′E
Boonton, N.J., U.S. (bo͞on′tŭn)	68a	40°54′N	74°24′W
Boonville, In., U.S.	66	38°00′N	87°15′W
Boonville, Mo., U.S.	79	38°57′N	92°44′W
Boorama, Som.	192a	10°05′N	43°08′E
Boosaaso, Som.	192a	11°19′N	49°10′E
Boothbay Harbor, Me., U.S. (bo͞oth′bā här′bẽr)	60	43°51′N	69°39′W
Boothia, Gulf of, b., Can. (bo͞o′thĭ-ȧ)	53	69°04′N	86°04′W
Boothia Peninsula, pen., Can.	49	73°30′N	95°00′W
Bootle, Eng., U.K. (bo͞ot′l)	114a	53°29′N	3°02′W
Bor, Sudan (bôr)	185	6°13′N	31°35′E
Bor, Tur. (bōr)	137	37°50′N	34°40′E
Boraha, Nosy, i., Madag.	187	16°58′S	50°15′E
Borah Peak, mtn., Id., U.S. (bō′rä)	73	44°12′N	113°47′W
Borås, Swe. (bò′rōs)	116	57°43′N	12°55′E
Borāzjān, Iran (bō-räz-jän′)	154	29°13′N	51°13′E
Borba, Braz. (bôr′bä)	101	4°23′S	59°31′W
Borborema, Planalto da, plat., Braz. (plä-näl′tō-dä-bōr-bō-rĕ′mä)	101	7°35′S	36°40′W
Bordeaux, Fr. (bôr-dō′)	110	44°50′N	0°37′W
Bordentown, N.J., U.S. (bôr′dĕn-toun)	67	40°05′N	74°40′W
Bordj-bou-Arréridj, Alg. (bôrj-bo͞o-ȧ-rā-rēj′)	118	36°03′N	4°48′E
Bordj Omar Idriss, Alg.	184	28°06′N	6°34′E
Borgarnes, Ice.	116	64°31′N	21°40′W
Borger, Tx., U.S. (bôr′gẽr)	78	35°40′N	101°23′W
Borgholm, Swe. (bôrg-hŏlm′)	122	56°52′N	16°40′E
Borgne, l., La., U.S. (bôrn′y′)	81	30°03′N	89°36′W
Borgomanero, Italy (bôr′gō-mä-nâ′rō)	130	45°40′N	8°28′E
Borgo Val di Taro, Italy (bô′r-zhō-väl-dē-tá′rō)	130	44°29′N	9°44′E
Börili, Kaz.	142a	53°36′N	61°55′E
Boring, Or., U.S. (bōring)	74c	45°26′N	122°22′W
Borisoglebsk, Russia (bŏ-rē sŏ-glyĕpsk′)	134	51°20′N	42°00′E
Borisovka, Russia (bŏ-rē-sôf′kȧ)	137	50°38′N	36°00′E
Borivli, India	159b	19°15′N	72°48′E
Borja, Spain (bôr′hä)	128	41°50′N	1°33′W
Borken, Ger. (bōr′kĕn)	127c	51°50′N	6°51′E
Borkou, reg., Chad (bôr-ko͞o′)	185	18°11′N	18°28′E
Borkum, i., Ger. (bōr′ko͞om)	124	53°31′N	6°50′E
Borlänge, Swe. (bôr-lĕŋ′gĕ)	122	60°30′N	15°24′E
Borneo, i., Asia	168	0°25′N	112°39′E
Bornholm, i., Den. (bôrn-hôlm)	112	55°16′N	15°15′E
Boromlia, Ukr.	133	50°36′N	34°58′E
Boromo, Burkina	188	11°45′N	2°56′W
Borovan, Blg. (bō-rō-vän′)	131	43°24′N	23°47′E
Borovichi, Russia (bō-rô-vē′chē)	134	58°22′N	33°56′E
Borovsk, Russia (bô′rŏvsk)	132	55°13′N	36°26′E
Borraan, Som.	192a	10°38′N	48°30′E
Borracha, Isla la, i., Ven. (ē′s-lä-lä-bôr-rä′chä)	101b	10°18′N	64°44′W
Borriana, Spain	118	39°53′N	0°05′W
Borroloola, Austl. (bŏr-rō-lo͞o′lȧ)	174	16°15′S	136°19′E
Borshchiv, Ukr.	125	48°47′N	26°04′E
Bort-les-Orgues, Fr. (bôr-lā-zôrg)	126	45°26′N	2°26′E
Borūjerd, Iran	154	33°45′N	48°53′E
Boryslav, Ukr.	125	49°17′N	23°24′E
Boryspil′, Ukr.	133	50°17′N	30°54′E
Borzna, Ukr. (bôrz′nȧ)	137	51°15′N	32°26′E
Borzya, Russia (bôrz′yȧ)	135	50°37′N	116°53′E
Bosa, Italy (bō′sä)	130	40°18′N	8°34′E
Bosanska Dubica, Bos. (bō′sän-skä do͞o′bĭt-sä)	130	45°10′N	16°49′E
Bosanska Gradiška, Bos. (bō′sän-skä grä-dĭsh′kä)	131	45°08′N	17°15′E
Bosanski Novi, Bos. (bō′s sän-skĭ nō′vē)	130	45°00′N	16°22′E
Bosanski Petrovac, Bos. (bō′sän-skĭ pĕt′rō-väts)	130	44°33′N	16°23′E
Bosanski Šamac, Bos. (bō′sän-skĭ shä′mȧts)	131	45°03′N	18°30′E
Boscobel, Wi., U.S. (bŏs′kō-bĕl)	71	43°08′N	90°44′W
Bose, China (bwo-sŭ)	165	24°00′N	106°38′E
Boshan, China (bwo-shan)	161	36°32′N	117°51′E
Boskoop, Neth.	115a	52°04′N	4°39′E
Boskovice, Czech Rep. (bŏs′kō-vē-tsĕ)	124	49°26′N	16°37′E
Bosna, r., Serb.	131	44°19′N	17°54′E
Bosnia and Herzegovina, nation, Eur.	131	44°15′N	17°30′E
Bosobolo, D.R.C.	190	4°11′N	19°54′E
Bosporus *see* İstanbul Boğazı, strt., Tur.	154	41°10′N	29°10′E
Bossangoa, C.A.R.	189	6°29′N	17°27′E
Bossier City, La., U.S. (bŏsh′ẽr)	81	32°31′N	93°42′W
Bosten Hu, l., China (bwo-stŭn ho͞o)	160	42°06′N	88°01′E
Boston, Ga., U.S. (bôs′tŭn)	82	30°47′N	83°47′W
Boston, Ma., U.S.	65	42°15′N	71°07′W
Boston Heights, Oh., U.S.	69d	41°15′N	81°30′W
Boston Mountains, mts., Ar., U.S.	65	35°46′N	93°32′W
Botany Bay, b., Austl. (bŏt′ȧ-nĭ)	175	33°58′S	151°11′E
Botevgrad, Blg.	131	42°54′N	23°41′E
Bothaville, S. Afr. (bō′tȧ-vĭl)	192c	27°24′S	26°38′E
Bothell, Wa., U.S. (bŏth′ĕl)	74a	47°46′N	122°12′W
Bothnia, Gulf of, b., Eur. (bŏth′nĭ-ȧ)	112	63°40′N	21°30′E
Botoşani, Rom. (bō-tō-shan′ĭ)	125	47°46′N	26°40′E
Botswana, nation, Afr. (bŏtswănă)	186	22°10′S	23°13′E
Bottineau, N.D., U.S. (bŏt-ĭ-nō′)	70	48°48′N	100°28′W
Bottrop, Ger. (bōt′trōp)	124	51°31′N	6°56′E
Botwood, Can. (bŏt′wo͝od)	53a	49°08′N	55°21′W
Bouafle, C. Iv. (bò-ȧ-flā′)	184	6°59′N	5°45′W
Bouar, C.A.R. (bo͞o-är)	185	5°57′N	15°36′E
Bou Areg, Sebkha, Mor.	128	35°09′N	3°02′W
Boubandjidah, Parc National de, rec., Cam.	189	8°20′N	14°40′E
Boucherville, Can. (bo͞o-shā-vēl′)	62a	45°37′N	73°27′W
Boudenib, Mor. (bo͞o-dē-nĕb′)	184	32°14′N	3°04′W
Boudette, Mn., U.S. (bo͞o-dĕt)	71	48°42′N	94°34′W
Boudouaou, Alg.	129	36°44′N	3°25′E
Boufarik, Alg. (bo͞o-fä-rēk′)	129	36°35′N	2°55′E
Bougainville, i., Pap. N. Gui.	170e	6°00′S	155°00′E
Bougainville Trench, deep, (bo͞o-găɴ-vēl′)	195	7°00′S	152°00′E
Bougie *see* Bejaïa, Alg.	184	36°46′N	5°00′E
Bougouni, Mali (bo͞o-go͞o-nē′)	184	11°27′N	7°30′W
Bouïra, Alg. (boo-ē′rȧ)	118	36°25′N	3°55′E
Bouïra-Sahary, Alg. (bwē-rȧ sȧ′ȧ-rē)	129	35°16′N	3°23′E
Bouka, r., Gui.	188	11°05′N	10°40′W
Boulder, Co., U.S.	64	40°02′N	105°19′W
Boulder, r., Mt., U.S.	73	46°10′N	112°07′W
Boulder City, Nv., U.S.	64	35°57′N	114°50′W
Boulder Peak, mtn., Id., U.S.	73	43°53′N	114°33′W
Boulogne-Billancourt, Fr. (bo͞o-lôn′y′-bē-yäɴ-ko͞or′)	126	48°50′N	2°14′E
Boulogne-sur-Mer, Fr. (bo͞o-lôn′y-sür-mâr′)	117	50°44′N	1°37′E
Boumba, r., Cam.	189	3°20′N	14°40′E
Bouna, C. Iv. (bo͞o-nä′)	184	9°16′N	3°00′W
Bouna, Parc National de, rec., C. Iv.	188	9°20′N	3°35′W
Boundary Bay, b., N.A. (boun′dȧ-rĭ)	74d	49°03′N	122°59′W
Boundary Peak, mtn., Nv., U.S.	76	37°52′N	118°20′W
Bound Brook, N.J., U.S. (bound brŏk)	68a	40°34′N	74°32′W
Bountiful, Ut., U.S. (boun′tĭ-fōl)	75b	40°55′N	111°53′W
Bountiful Peak, mtn., Ut., U.S. (boun′tĭ-fōl)	75b	40°58′N	111°49′W
Bounty Islands, is., N.Z.	5	47°42′S	179°05′E
Bourail, N. Cal.	170f	21°34′S	165°30′E
Bourem, Mali (bo͞o-rĕm′)	184	16°43′N	0°15′W
Bourg-en-Bresse, Fr. (bo͞or-gĕɴ-brĕs′)	117	46°12′N	5°13′E
Bourges, Fr. (bo͞orzh)	117	47°06′N	2°22′E
Bourget, Can. (bo͞or-zhĕ′)	62c	45°26′N	75°09′W
Bourgoin, Fr. (bo͞or-gwăɴ′)	127	45°46′N	5°17′E
Bourke, Austl. (bŭrk)	175	30°10′S	146°00′E
Bourne, Eng., U.K. (bōrn)	114a	52°46′N	0°22′W
Bournemouth, Eng., U.K. (bôrn′mŭth)	120	50°44′N	1°55′W
Bou Saâda, Alg. (bo͞o-sä′dä)	118	35°13′N	4°17′E
Bousso, Chad (bo͞o-sô′)	185	10°33′N	16°45′E
Boutilimit, Maur.	184	17°30′N	14°54′W
Bouvetøya, i., Ant.	3	55°00′S	3°00′E
Bow, r., Can. (bō)	52	50°35′N	112°15′W
Bowbells, N.D., U.S. (bō′bĕls)	70	48°50′N	102°16′W
Bowdle, S.D., U.S. (bōd′'l)	70	45°28′N	99°42′W
Bowen, Austl. (bō′ĕn)	175	20°02′S	148°14′E
Bowie, Md., U.S. (bo͞o′ĭ) (bō′ē)	68e	38°59′N	76°47′W
Bowie, Tx., U.S.	79	33°34′N	97°50′W
Bowling Green, Ky., U.S. (bōlĭng grēn)	65	37°00′N	86°26′W
Bowling Green, Mo., U.S.	79	39°19′N	91°09′W
Bowling Green, Oh., U.S.	66	41°25′N	83°40′W
Bowman, N.D., U.S. (bō′măn)	70	46°11′N	103°23′W
Bowron, r., Can. (bō′rŭn)	55	53°20′N	121°10′W
Boxelder Creek, r., Mt., U.S. (bŏks′ĕl-dẽr)	70	45°35′N	104°28′W
Box Elder Creek, r., Mt., U.S.	73	47°17′N	108°37′W
Box Hill, Austl.	173a	37°49′S	145°08′E
Boxian, China (bwo shyĕn)	164	33°52′N	115°47′E
Boxing, China (bwo-shyĭŋ)	162	37°09′N	118°08′E
Boxtel, Neth.	115a	51°40′N	5°21′E
Boyabo, D.R.C.	190	3°43′N	18°46′E
Boyang, China (bwo-yäŋ)	165	29°00′N	116°42′E
Boyer, r., Can. (boi′ẽr)	62b	46°45′N	70°56′W
Boyer, r., Ia., U.S.	70	41°45′N	95°36′W
Boyle, Ire. (boil)	120	53°59′N	8°15′W
Boyne, r., Ire. (boin)	120	53°40′N	6°40′W
Boyne City, Mi., U.S.	66	45°15′N	85°05′W
Boyoma Falls, wtfl., D.R.C.	185	0°30′N	25°12′E
Boysen Reservoir, res., Wy., U.S.	73	43°19′N	108°11′W
Bozcaada, Tur. (bōz-cä′dä)	131	39°50′N	26°05′E
Bozca Ada, i., Tur.	131	39°50′N	26°00′E
Bozeman, Mt., U.S. (bōz′măn)	64	45°41′N	111°00′W
Bozene, D.R.C.	190	2°56′N	19°12′E
Bozhen, China (bwo-jŭn)	162	38°05′N	116°35′E
Bozoum, C.A.R.	189	6°19′N	16°23′E
Bra, Italy (brä)	130	44°41′N	7°52′E
Bracciano, Lago di, l., Italy (lä′gō-dē-brä-chä′nō)	130	42°05′N	12°00′E
Bracebridge, Can. (brās′brĭj)	59	45°05′N	79°20′W
Braceville, Il., U.S. (brās′vĭl)	69a	41°13′N	88°16′W
Bräcke, Swe. (brĕk′kĕ)	116	62°44′N	15°28′E
Brackenridge, Pa., U.S. (brăk′ĕn-rĭj)	69e	40°37′N	79°44′W
Brackettville, Tx., U.S. (brăk′ĕt-vĭl)	80	29°19′N	100°24′W
Braço Maior, mth., Braz.	101	11°00′S	51°00′W
Braço Menor, mth., Braz. (brä′zō-mē-nō′r)	101	11°38′S	50°00′W
Bradano, r., Italy (brä-dä′nō)	130	40°43′N	16°22′E
Bradenton, Fl., U.S. (brā′dĕn-tŭn)	83a	27°28′N	82°35′W
Bradfield, Eng., U.K. (brăd′fēld)	114b	51°25′N	1°08′W
Bradford, Eng., U.K. (brăd′fẽrd)	116	53°47′N	1°44′W
Bradford, Oh., U.S.	66	40°10′N	84°30′W
Bradford, Pa., U.S.	67	42°00′N	78°40′W
Bradley, Il., U.S. (brăd′lĭ)	69a	41°09′N	87°52′W
Bradner, Can. (brăd′nẽr)	74d	49°05′N	122°26′W
Brady, Tx., U.S. (brā′dĭ)	80	31°09′N	99°21′W
Braga, Port. (brä′gä)	118	41°20′N	8°25′W
Bragado, Arg. (brä-gä′dō)	102	35°07′S	60°28′W
Bragança, Braz. (brä-gän′sä)	101	1°02′S	46°50′W
Bragança, Port.	128	41°48′N	6°46′W
Bragança Paulista, Braz. (brä-gän′sä-pä′o͞o-lē′s-tá)	102	22°58′S	46°31′W
Bragg Creek, Can. (brăg)	62e	50°57′N	114°35′W
Brahmaputra, r., Asia (brä′mȧ-po͞o′trȧ)	155	26°45′N	92°45′E
Brāhui, mts., Pak.	155	28°32′N	66°15′E
Braidwood, Il., U.S. (brād′wŏd)	69a	41°16′N	88°13′W
Brăila, Rom. (brē′ēlä)	110	45°15′N	27°58′E
Brainerd, Mn., U.S. (brān′ẽrd)	71	46°20′N	94°09′W
Braintree, Ma., U.S. (brān′trē)	61a	42°14′N	71°00′W
Braithwaite, La., U.S. (brīth′wīt)	68d	29°52′N	89°57′W
Brakpan, S. Afr. (brăk′păn)	187b	26°15′S	28°22′E
Bralorne, Can. (brā′lôrn)	55	50°47′N	122°49′W
Bramalea, Can.	62d	43°48′N	79°41′W
Brampton, Can. (brămp′tŭn)	59	43°41′N	79°46′W
Branca, Pedra, mtn., Braz. (pĕ′drä-brá′ɴ-kä)	102b	22°55′S	43°28′W
Branchville, N.J., U.S. (brănch′vĭl)	68a	41°09′N	74°44′W
Branchville, S.C., U.S.	83	33°17′N	80°48′W
Branco, r., Braz. (brăŋ′kō)	101	2°21′N	60°38′W
Brandberg, mtn., Nmb.	186	21°15′S	14°15′E
Brandenburg, Ger. (brän′dĕn-bōrgh)	117	52°25′N	12°33′E
Brandenburg, state, Ger.	115b	52°15′N	13°00′E
Brandenburg, hist. reg., Ger.	124	52°12′N	13°31′E
Brandfort, S. Afr. (brăn′d-fŏrt)	192c	28°42′S	26°29′E
Brandon, Can. (brăn′dŭn)	50	49°50′N	99°57′W
Brandon, Vt., U.S.	67	43°45′N	73°05′W
Brandon Mountain, mtn., Ire. (brăn-dŏn)	120	52°15′N	10°12′W
Brandywine, Md., U.S. (brăndĭ′wīn)	68e	38°42′N	76°51′W
Branford, Ct., U.S. (brăn′fẽrd)	67	41°15′N	72°50′W
Braniewo, Pol. (brä-nyĕ′vô)	125	54°23′N	19°50′E
Brańsk, Pol. (brän′ sk)	125	52°44′N	22°51′E
Branson, Mo., U.S.	79	36°39′N	93°13′W
Brantford, Can. (brănt′fẽrd)	59	43°09′N	80°17′W
Bras d'Or Lake, l., Can. (brä-dōr′)	61	45°52′N	60°50′W
Brasília, Braz. (brä-sē′lvä)	101	15°49′S	47°39′W
Brasilia Legal, Braz.	101	3°45′S	55°46′W
Brasópolis, Braz. (brä-sô′pô-lēs)	99a	22°30′S	45°36′W
Braşov, Rom.	119	45°39′N	25°35′E
Brass, Nig. (brăs)	184	4°28′N	6°28′E
Brasschaat, Bel. (bräs′kät)	115a	51°19′N	4°30′E
Bratenahl, Oh., U.S. (brā′tĕn-ôl)	69d	41°34′N	81°36′W
Bratislava, Slvk. (brä′tĭs-lä-vä)	110	48°09′N	17°07′E
Bratsk, Russia (brätsk)	135	56°10′N	102°04′E

ng-sing; ŋ-baŋk; ɴ-nasalized n; nŏd; cŏmmit; ōld; ôbey; ôrder; oi-boil; fo͞od; o͝o-as oo in foot; ou-out; s-soft; sh-dish; th-thin; pūre; ûnite; ûrn; stŭd; circŭs; ü-as in French tu; ′-indeterminate vowel.

230 Bratskoye Vodokhranilishche – Buenos Aires PRONOUNCING INDEX

PLACE (Pronunciation)	PAGE	LAT.	LONG.
Bratskoye Vodokhranilishche, res., Russia	135	56°10'N	102°05'E
Bratslav, Ukr. (brät'släf)	133	48°48'N	28°59'E
Brattleboro, Vt., U.S. (brăt''l-bŭr-ō)	67	42°50'N	72°35'W
Braunau, Aus. (brou'nou)	124	48°15'N	13°05'E
Braunschweig, Ger. (broun'shvīgh)	117	52°16'N	10°32'E
Bråviken, r., Swe.	122	58°40'N	16°40'E
Brawley, Ca., U.S. (brô'lĭ)	64	32°59'N	115°32'W
Bray, Ire. (brā)	120	53°10'N	6°05'W
Braymer, Mo., U.S. (brā'mẽr)	79	39°34'N	93°47'W
Brays Bay, Tx., U.S. (brās'bī'yo͞o)	81a	29°41'N	95°33'W
Brazeau, r., Can.	55	52°55'N	116°10'W
Brazeau, Mount, mtn., Can. (brä-zō')	55	52°33'N	117°21'W
Brazil, In., U.S. (brá-zĭl')	66	39°30'N	87°00'W
Brazil, nation, S.A.	101	9°00'S	53°00'W
Brazilian Highlands, mts., Braz. (brá zĭl yăn hī-lándz)	97	14°00'S	48°00'W
Brazos, r., Tx., U.S. (brä'zōs)	64	33°10'N	98°50'W
Brazos, Clear Fork, r., Tx., U.S.	80	32°56'N	99°14'W
Brazos, Double Mountain Fork, r., Tx., U.S.	78	33°23'N	101°21'W
Brazos, Salt Fork, r., Tx., U.S. (sôlt fôrk)	78	33°20'N	101°57'W
Brazzaville, Congo (brá-zá-vēl')	186	4°16'S	15°17'E
Brčko, Bos. (bĕrch'kō)	131	44°54'N	18°46'E
Brda, r., Pol. (bĕr-dä)	125	53°18'N	17°55'E
Brea, Ca., U.S. (brē'á)	75a	33°55'N	117°54'W
Breakeyville, Can.	62b	46°40'N	71°13'W
Breckenridge, Mn., U.S. (brĕk'ĕn-rĭj)	70	46°17'N	96°35'W
Breckenridge, Tx., U.S.	80	32°46'N	98°53'W
Brecksville, Oh., U.S. (brĕks'vĭl)	69d	41°19'N	81°38'W
Břeclav, Czech Rep. (brzhĕl'läf)	124	48°46'N	16°54'E
Breda, Neth. (brā-dä')	121	51°35'N	4°47'E
Bredasdorp, S. Afr. (brā'das-dôrp)	186	34°15'S	20°00'E
Bredy, Russia (brĕ'dĭ)	142a	52°25'N	60°23'E
Bregenz, Aus. (brā'gĕnts)	124	47°30'N	9°46'E
Bregovo, Blg. (brĕ'gô-vô)	131	44°07'N	22°45'E
Breidafjördur, b., Ice.	116	65°15'N	22°50'W
Breidbach, S. Afr. (brĕd'băk)	187c	32°54'S	27°26'E
Breil-sur-Roya, Fr. (brĕ'y')	127	43°57'N	7°36'E
Brejo, Braz. (brá'zhō)	101	3°33'S	42°46'W
Bremangerlandet, i., Nor.	122	61°51'N	4°25'E
Bremen, Ger. (brā-mĕn)	110	53°05'N	8°50'E
Bremen, In., U.S. (brē'mĕn)	66	41°25'N	86°05'W
Bremerhaven, Ger. (brām-ẽr-hä'fĕn)	116	53°33'N	8°38'E
Bremerton, Wa., U.S. (brĕm'ẽr-tŭn)	72	47°34'N	122°38'W
Bremervörde, Ger. (brĕ'mĕr-fûr-dĕ)	115c	53°29'N	9°09'E
Bremner, Can. (brĕm'nẽr)	62g	53°34'N	113°14'W
Bremond, Tx., U.S. (brĕm'ŭnd)	81	31°11'N	96°40'W
Brenham, Tx., U.S. (brĕn'ăm)	81	30°10'N	96°24'W
Brenner Pass, p., Eur. (brĕn'ẽr)	117	47°00'N	11°30'E
Brentwood, Eng., U.K. (brĕnt'wo͝od)	121	51°37'N	0°18'E
Brentwood, Md., U.S.	67	39°00'N	76°55'W
Brentwood, Mo., U.S.	75e	38°37'N	90°21'W
Brentwood, Pa., U.S.	69e	40°22'N	79°59'W
Brescia, Italy (brā'shä)	118	45°33'N	10°15'E
Bressanone, Italy (brĕs-sä-nō'nā)	130	46°42'N	11°40'E
Bressuire, Fr. (grĕ-swēr')	126	46°49'N	0°14'W
Brest, Bela.	134	52°06'N	23°43'E
Brest, Fr. (brĕst)	110	48°24'N	4°30'W
Brest, prov., Bela.	132	52°30'N	26°50'E
Bretagne, hist. reg., Fr. (brĕ-tăn'yĕ)	126	48°00'N	3°00'W
Breton, Pertuis, strt., Fr. (pâr-twĕ'brĕ-tôN')	126	46°18'N	1°43'W
Breton Sound, strt., La., U.S. (brĕt'ŭn)	82	29°38'N	89°15'W
Breukelen, Neth.	115a	52°09'N	5°00'E
Brevard, N.C., U.S. (brē-värd')	83	35°14'N	82°45'W
Breves, Braz. (brā'vĕzh)	101	1°32'S	50°13'W
Brevik, Nor. (brĕ'vĕk)	122	59°04'N	9°39'E
Brewarrina, Austl. (bro͞o-ẽr-rē'ná)	175	29°54'S	146°50'E
Brewer, Me., U.S. (bro͞o'ẽr)	60	44°46'N	68°46'W
Brewerville, Lib.	188	6°26'N	10°47'W
Brewster, N.Y., U.S. (bro͞o'stẽr)	68a	41°23'N	73°38'W
Brewster, Cerro, mtn., Pan. (sĕ'r-rô-bro͞o'stẽr)	91	9°19'N	79°15'W
Brewton, Al., U.S. (bro͞o'tŭn)	82	31°06'N	87°04'W
Brežice, Slvn. (brĕ'zhē-tsĕ)	130	45°55'N	15°37'E
Breznik, Blg. (brĕs'nĕk)	131	42°44'N	22°55'E
Briancon, Fr. (brē-äN-sôN')	127	44°54'N	6°39'E
Briare, Fr. (brē-är')	126	47°40'N	2°46'E
Bridal Veil, Or., U.S. (brīd'ál vāl)	74c	45°33'N	122°10'W
Bridge Point, c., Bah. (brĭj)	92	25°35'N	76°40'W
Bridgeport, Al., U.S. (brĭj'pôrt)	82	34°55'N	85°42'W
Bridgeport, Ct., U.S.	65	41°12'N	73°12'W
Bridgeport, Il., U.S.	66	38°40'N	87°45'W
Bridgeport, Ne., U.S.	70	41°40'N	103°06'W
Bridgeport, Oh., U.S.	66	40°00'N	80°45'W
Bridgeport, Pa., U.S.	68f	40°06'N	75°21'W
Bridgeport, Tx., U.S.	79	33°13'N	97°46'W
Bridgeton, Al., U.S. (brĭj'tŭn)	68h	33°27'N	86°39'W
Bridgeton, Mo., U.S.	75e	38°45'N	90°23'W
Bridgeton, N.J., U.S.	67	39°30'N	75°15'W
Bridgetown, Barb. (brĭj' toun)	87	13°08'N	59°37'W
Bridgetown, Can.	60	44°51'N	65°18'W
Bridgeville, Pa., U.S. (brĭj'vĭl)	69e	40°22'N	80°07'W
Bridgewater, Austl. (brĭj'wô-tẽr)	176	42°50'S	147°28'E
Bridgewater, Can.	51	44°23'N	64°31'W
Bridgnorth, Eng., U.K. (brĭj'nôrth)	114a	52°32'N	2°25'W
Bridgton, Me., U.S. (brĭj'tŭn)	60	44°04'N	70°45'W
Bridlington, Eng., U.K. (brĭd'lĭng-tŭn)	120	54°06'N	0°10'W
Brie-Comte-Robert, Fr. (brē-kôNt-ĕ-rō-bâr')	127b	48°42'N	2°37'E
Brielle, Neth.	115a	51°54'N	4°08'E
Brierfield, Eng., U.K. (brī'ẽr fēld)	114a	53°49'N	2°14'W
Brierfield, Al., U.S. (brī'ẽr-fēld)	82	33°01'N	86°55'W
Brier Island, i., Can. (brī'ẽr)	60	44°16'N	66°24'W
Brieselang, Ger. (brē'zĕ-läng)	115b	52°36'N	12°59'E
Briey, Fr. (brē-ĕ')	127	49°15'N	5°57'E
Brig, Switz. (brēg)	117	46°17'N	7°59'E
Brigg, Eng., U.K. (brĭg)	114a	53°33'N	0°29'W
Brigham City, Ut., U.S. (brĭg'ăm)	75b	41°31'N	112°01'W
Brighouse, Eng., U.K. (brĭg'hous)	114a	53°42'N	1°47'W
Bright, Austl. (brīt)	176	36°43'S	147°00'E
Bright, In., U.S. (brīt)	69f	39°13'N	84°51'W
Brightlingsea, Eng., U.K. (brī't-lĭng-sē)	114b	51°50'N	1°00'E
Brighton, Austl.	173a	37°55'S	145°00'E
Brighton, Eng., U.K.	117	50°47'N	0°07'W
Brighton, Al., U.S. (brī't'ŭn)	68h	33°27'N	86°56'W
Brighton, Co., U.S.	78	39°58'N	104°49'W
Brighton, Ia., U.S.	71	41°11'N	91°47'W
Brighton, Il., U.S.	75e	39°03'N	90°08'W
Brighton Indian Reservation, I.R., Fl., U.S.	83a	27°05'N	81°25'W
Brihuega, Spain (brē-wā'gä)	128	40°32'N	2°52'W
Brimley, Mi., U.S. (brĭm'lē)	75k	46°24'N	84°34'W
Brindisi, Italy (brēn'dē-zē)	110	40°38'N	17°57'E
Brinje, Cro. (brēn'yĕ)	130	45°00'N	15°08'E
Brinkley, Ar., U.S. (brĭŋk'lĭ)	79	34°52'N	91°12'W
Brinnon, Wa., U.S. (brĭn'ŭn)	74a	47°41'N	122°54'W
Brion, i., Can. (brē-ôN')	61	47°47'N	61°29'W
Brioude, Fr. (brē-o͞od')	126	45°18'N	3°22'E
Brisbane, Austl. (brĭz'bån)	176	27°30'S	153°10'E
Bristol, Eng., U.K.	117	51°29'N	2°39'W
Bristol, Ct., U.S. (brĭs'tŭl)	67	41°40'N	72°55'W
Bristol, Pa., U.S.	68f	40°06'N	74°51'W
Bristol, R.I., U.S.	68b	41°41'N	71°14'W
Bristol, Tn., U.S.	65	36°35'N	82°10'W
Bristol, Va., U.S.	65	36°36'N	82°00'W
Bristol, Vt., U.S.	67	44°10'N	73°00'W
Bristol, Wi., U.S.	69a	42°32'N	88°04'W
Bristol Bay, b., Ak., U.S.	63	58°05'N	158°54'W
Bristol Channel, strt., Eng., U.K.	117	51°20'N	3°47'W
Bristow, Ok., U.S. (brĭs'tō)	79	35°50'N	96°25'W
British Columbia, prov., Can. (brĭt'ĭsh kŏl'ŭm-bĭ-á)	50	56°00'N	124°53'W
British Indian Ocean Territory, dep., Afr.	2	7°00'S	72°00'E
British Isles, is., Eur.	112	54°00'N	4°00'W
Brits, S. Afr.	192c	25°39'S	27°47'E
Britstown, S. Afr. (brĭts'toun)	186	30°30'S	23°40'E
Britt, Ia., U.S. (brĭt)	71	43°05'N	93°47'W
Brittany see Bretagne, hist. reg., Fr.	126	48°00'N	3°00'W
Britton, S.D., U.S. (brĭt'ŭn)	70	45°47'N	97°44'W
Brive-la-Gaillarde, Fr. (brēv-lä-gī-yärd'ĕ)	117	45°10'N	1°31'E
Briviesca, Spain (brē-vyās'ká)	128	42°34'N	3°21'W
Brno, Czech Rep. (b'r'nô)	110	49°18'N	16°37'E
Broa, Ensenada de la, b., Cuba	92	22°30'N	82°00'W
Broach, India	158	21°47'N	72°58'E
Broad, r., Ga., U.S. (brôd)	82	34°15'N	83°14'W
Broad, r., N.C., U.S.	83	35°38'N	82°40'W
Broadmeadows, Austl. (brôd'mĕd-ōz)	173a	37°40'S	144°53'E
Broadview Heights, Oh., U.S. (brôd'vū)	69d	41°18'N	81°41'W
Brockport, N.Y., U.S. (brŏk'pōrt)	67	43°15'N	77°55'W
Brockton, Ma., U.S. (brŏk'tŭn)	61a	42°04'N	71°01'W
Brockville, Can. (brŏk'vĭl)	51	44°35'N	75°40'W
Brockway, Mt., U.S. (brŏk'wā)	73	47°24'N	105°41'W
Brodnica, Pol. (brôd'nĭt-sá)	125	53°16'N	19°26'E
Brody, Ukr. (brô'dĭ)	137	50°05'N	25°10'E
Broken Arrow, Ok., U.S. (brō'kĕn ăr'ō)	79	36°03'N	95°48'W
Broken Bay, b., Austl.	176	33°34'S	151°20'E
Broken Bow, Ne., U.S. (brō'kĕn bō)	70	41°24'N	99°37'W
Broken Bow, Ok., U.S.	79	34°02'N	94°43'W
Broken Hill, Austl. (brōk'ĕn)	175	31°55'S	141°35'E
Broken Hill see Kabwe, Zam.	186	14°27'S	28°27'E
Bromley, Eng., U.K. (brŭm'lĭ)	114b	51°23'N	0°01'E
Bromptonville, Can. (brŭmp'tŭn-vĭl)	59	45°30'N	72°00'W
Brønderslev, Den. (brûn'dẽr-slĕv)	122	57°15'N	9°56'E
Bronkhorstspruit, S. Afr.	192c	25°50'S	28°48'E
Bronnitsy, Russia (brô-nyĭ'tsĭ)	132	55°26'N	38°16'E
Bronson, Mi., U.S. (brŏn'sŭn)	66	41°55'N	85°15'W
Bronte Creek, r., Can.	62d	43°25'N	79°53'W
Brood, r., S.C., U.S. (bro͞od)	83	34°46'N	81°25'W
Brookfield, Il., U.S. (bro͝ok'fēld)	69a	41°49'N	87°51'W
Brookfield, Mo., U.S.	79	39°45'N	93°04'W
Brookhaven, Ga., U.S. (bro͝ok'hāv'n)	68c	33°52'N	84°21'W
Brookhaven, Ms., U.S.	82	31°35'N	90°26'W
Brookings, Or., U.S. (bro͝ok'ĭngs)	72	42°04'N	124°16'W
Brookings, S.D., U.S.	70	44°18'N	96°47'W
Brookline, Ma., U.S. (bro͝ok'līn)	61a	42°20'N	71°08'W
Brookline, N.H., U.S.	61a	42°44'N	71°37'W
Brooklyn, Oh., U.S. (bro͝ok'lĭn)	69d	41°26'N	81°44'W
Brooklyn Center, Mn., U.S.	75g	45°05'N	93°21'W
Brook Park, Oh., U.S. (bro͝ok)	69d	41°24'N	81°50'W
Brooks, Can.	55	50°35'N	111°53'W
Brooks Range, mts., Ak., U.S. (bro͝oks)	64a	68°20'N	159°00'W
Brooksville, Fl., U.S. (bro͝oks'vĭl)	83a	28°32'N	82°28'W
Brookville, In., U.S. (bro͝ok'vĭl)	66	39°20'N	85°00'W
Brookville, Pa., U.S.	67	41°10'N	79°00'W
Brookwood, Al., U.S. (bro͝ok'wo͝od)	82	33°15'N	87°17'W
Broome, Austl. (bro͞om)	174	18°00'S	122°15'E
Brossard, Can.	62a	45°26'N	73°28'W
Brothers, is., Bah. (brŭd'hẽrs)	92	26°05'N	79°00'W
Broumov, Czech Rep. (bro͞o'môf)	124	50°33'N	15°55'E
Brown Bank, bk.,	93	21°30'N	74°35'W
Brownfield, Tx., U.S. (broun'fēld)	78	33°11'N	102°16'W
Browning, Mt., U.S. (broun'ĭng)	73	48°37'N	113°05'W
Brownsboro, Ky., U.S. (brounz'bô-rô)	69h	38°22'N	85°30'W
Brownsburg, Can. (brouns'bûrg)	62a	45°40'N	74°24'W
Brownsburg, In., U.S.	69g	39°51'N	86°23'W
Brownsmead, Or., U.S. (brounz'-mēd)	74c	46°13'N	123°33'W
Brownstown, In., U.S. (brounz'toun)	66	38°50'N	86°00'W
Brownsville, Pa., U.S. (brounz'vĭl)	69e	40°01'N	79°53'W
Brownsville, Tn., U.S.	82	35°35'N	89°15'W
Brownsville, Tx., U.S.	64	25°55'N	97°30'W
Brownville Junction, Me., U.S. (broun'vĭl)	60	45°20'N	69°04'W
Brownwood, Tx., U.S. (broun'wo͝od)	64	31°44'N	98°58'W
Brownwood, l., Tx., U.S.	80	31°55'N	99°15'W
Brozas, Spain (brō'thäs)	128	39°37'N	6°44'W
Bruce, Mount, mtn., Austl. (bro͞os)	174	22°35'S	118°15'E
Bruce Peninsula, pen., Can.	58	44°50'N	81°20'W
Bruceton, Tn., U.S. (bro͞os'tŭn)	82	36°02'N	88°14'W
Bruchsal, Ger. (bro͝ok'zäl)	124	49°08'N	8°34'E
Bruck, Aus. (bro͝ok)	124	47°25'N	15°14'E
Bruck, Aus.	124	48°01'N	16°47'E
Brück, Ger. (brük)	115b	52°12'N	12°45'E
Bruderheim, Can. (bro͞o'dẽr-hīm)	62g	53°47'N	112°56'W
Brugge, Bel.	117	51°13'N	3°05'E
Brühl, Ger. (brül)	127c	50°49'N	6°54'E
Bruneau, r., Id., U.S. (bro͞o-nō')	72	42°47'N	115°43'W
Brunei, nation, Asia (brô-nī')	168	4°52'N	113°38'E
Brünen, Ger. (brü'nĕn)	127c	51°43'N	6°41'E
Brunete, Spain (bro͞o-nā'tā)	129a	40°24'N	4°00'W
Brunette, i., Can. (brô-nĕt')	61	47°16'N	55°54'W
Brunn am Gebirge, Aus. (bro͝on'äm gĕ-bĭr'gĕ)	115e	48°07'N	16°18'E
Brunsbüttel, Ger. (brôns'büt-tĕl)	115c	53°58'N	9°10'E
Brunswick, Ga., U.S. (brŭnz'wĭk)	65	31°08'N	81°30'W
Brunswick, Md., U.S.	67	39°20'N	77°35'W
Brunswick, Me., U.S.	60	43°54'N	69°57'W
Brunswick, Mo., U.S.	79	39°25'N	93°07'W
Brunswick, Oh., U.S.	69d	41°14'N	81°50'W
Brunswick, Península de, pen., Chile	102	53°25'S	71°15'W
Bruny, i., Austl. (bro͞o'nē)	175	43°30'S	147°50'E
Brush, Co., U.S. (brŭsh)	78	40°14'N	103°40'W
Brusque, Braz. (bro͞o's-ko͞oĕ)	102	27°15'S	48°45'W
Brussels, Bel.	110	50°51'N	4°21'E
Brussels, Il., U.S. (brŭs'ĕls)	75e	38°57'N	90°36'W
Bruxelles see Brussels, Bel.	110	50°51'N	4°21'E
Bryan, Oh., U.S. (brī'ăn)	66	41°25'N	84°30'W
Bryan, Tx., U.S.	81	30°40'N	96°22'W
Bryansk, Russia	134	53°15'N	34°22'E
Bryansk, prov., Russia	132	52°43'N	32°25'E
Bryant, S.D., U.S. (brī'ănt)	70	44°35'N	97°29'W
Bryant, Wa., U.S.	74a	48°14'N	122°10'W
Bryce Canyon National Park, rec., Ut., U.S. (brīs)	64	37°35'N	112°15'W
Bryn Mawr, Pa., U.S. (brĭn mår')	68f	40°02'N	75°20'W
Bryson City, N.C., U.S. (brīs'ŭn)	82	35°25'N	83°25'W
Bryukhovetskaya, Russia (b'ryŭk'ô-vyĕt-skä'yä)	133	45°56'N	38°58'E
Buala, Sol. Is.	170e	8°08'S	159°35'E
Buatan, Indon.	153b	0°45'N	101°49'E
Buba, Gui.-B. (bo͞o'bá)	184	11°39'N	14°58'W
Bucaramanga, Col. (bo͞o-kä'rä-mäŋ'gä)	100	7°12'N	73°14'W
Buccaneer Archipelago, is., Austl. (bŭk-á-nēr')	174	16°05'S	122°00'E
Buchach, Ukr. (bô'chách)	125	49°04'N	25°25'E
Buchanan, Lib. (bû-kăn'ăn)	184	5°57'N	10°02'W
Buchanan, Mi., U.S.	66	41°50'N	86°25'W
Buchanan, l., Austl. (bû-kăn'nŏn)	175	21°40'S	145°00'E
Buchanan, l., Tx., U.S. (bû-kăn'ăn)	80	30°55'N	98°40'W
Buchans, Can.	61	48°49'N	56°52'W
Bucharest, Rom.	110	44°23'N	26°10'E
Buchholz, Ger. (bo͞ok'hōltz)	115c	53°19'N	9°53'E
Buck Creek, r., In., U.S. (bŭk)	69g	39°43'N	85°58'W
Buckhannon, W.V., U.S. (bŭk-hăn'ŭn)	66	39°00'N	80°10'W
Buckhaven, Scot., U.K. (bŭk-hā'v'n)	120	56°10'N	3°10'W
Buckie, Scot., U.K. (bŭk'ĭ)	120	57°40'N	2°50'W
Buckingham, Can. (bŭk'ĭng-ăm)	62c	45°35'N	75°25'W
Buckingham, can., India (bŭk'ĭng-ăm)	159	15°18'N	79°50'E
Buckinghamshire, co., Eng., U.K.	114b	51°45'N	0°48'W
Buckland, Can. (bŭk'lănd)	62b	46°37'N	70°33'W
Buckland Tableland, reg., Austl.	175	24°31'S	148°00'E
Buckley, Wa., U.S. (bŭk'lē)	74a	47°10'N	122°02'W
Bucksport, Me., U.S. (bŭks'pôrt)	60	44°35'N	68°47'W
Bucouche, Can. (bü-ko͞osh')	60	46°28'N	64°43'W
Bucun, China (bo͞o-tsôn)	162	36°38'N	117°26'E
București see Bucharest, Rom.	110	44°23'N	26°10'E
Bucyrus, Oh., U.S. (bû-sī'rŭs)	66	40°50'N	82°55'W
Budapest, Hung. (bo͞o'dá-pĕsht')	110	47°30'N	19°05'E
Budge Budge, India	158a	22°28'N	88°08'E
Budjala, D.R.C.	190	2°39'N	19°42'E
Budyonnovsk, Russia	138	44°46'N	44°09'E
Buea, Cam.	189	4°09'N	9°14'E
Buechel, Ky., U.S. (bē-chŭl')	69h	38°12'N	85°38'W
Bueil, Fr. (bwā')	127b	48°55'N	1°27'E
Buena Park, Ca., U.S. (bwā'ná pärk)	75a	33°52'N	118°00'W
Buenaventura, Col. (bwā'nä-vĕn-to͞o'rä)	100	3°46'N	77°09'W
Buenaventura, Cuba	93a	22°53'N	82°22'W
Buenaventura, Bahía de, b., Col.	100	3°45'N	79°23'W
Buena Vista, Co., U.S. (bū'ná vĭs'tá)	78	38°51'N	106°07'W
Buena Vista, Ga., U.S.	82	32°15'N	84°30'W
Buena Vista, Va., U.S.	67	37°45'N	79°20'W
Buena Vista, Bahía, b., Cuba (bä-ē'ä-bwĕ-nä-vē's-tä)	92	22°30'N	79°10'W
Buena Vista Lake Bed, l., Ca., U.S. (bū'ná vĭs'tá)	76	35°14'N	119°17'W
Buendia, Embalse de, res., Spain	128	40°30'N	2°45'W
Buenos Aires, Arg. (bwā'nōs ī'rās)	102	34°20'S	58°30'W
Buenos Aires, Col.	100a	3°01'N	76°34'W
Buenos Aires, C.R.	91	9°10'N	83°21'W
Buenos Aires, prov., Arg.	102	36°15'S	61°45'W

ăt; finăl; rāte; senåte; ärm; åsk; sofá; fâre; ch-choose; dh-as th in other; bē; ēvent; bĕt; recĕnt; cratẽr; g-gō; gh-guttural g; bĭt; ĭ-short neutral; rīde; K-guttural k as ch in German ich;

PRONOUNCING INDEX

Buenos Aires – Cahama 231

PLACE (Pronunciation)	PAGE	LAT.	LONG.
Buenos Aires, l., S.A.	102	46°30'S	72°15'W
Buffalo, Mn., U.S. (bŭf′*a* lō)	71	45°10'N	93°50'W
Buffalo, N.Y., U.S.	65	42°54'N	78°51'W
Buffalo, Tx., U.S.	81	31°28'N	96°04'W
Buffalo, Wy., U.S.	73	44°19'N	106°42'W
Buffalo, r., S. Afr.	187c	28°35'S	30°27'E
Buffalo, r., Ar., U.S.	79	35°56'N	92°58'W
Buffalo, r., Tn., U.S.	82	35°24'N	87°10'W
Buffalo Bayou, Tx., U.S.	81a	29°46'N	95°32'W
Buffalo Creek, r., Mn., U.S.	71	44°46'N	94°28'W
Buffalo Head Hills, hills, Can.	52	57°16'N	116°18'W
Buford, Can. (bū′fŭrd)	62g	53°15'N	113°55'W
Buford, Ga., U.S. (bū′fērd)	82	34°05'N	84°00'W
Bug (Zakhidnyy Buh), r., Eur.	125	52°29'N	21°20'E
Buga, Col. (bo͞o′gä)	100	3°54'N	76°17'W
Buggenhout, Bel.	115a	51°01'N	4°10'E
Buglandsfjorden, l., Nor.	122	58°53'N	7°55'E
Bugojno, Bos. (bo̤-gō ĭ nō)	131	44°03'N	17°28'E
Bugul'ma, Russia (bo̤-gōl′má)	134	54°40'N	52°40'E
Buguruslan, Russia (bo̤-go̤-ro̤s-lán′)	134	53°30'N	52°32'E
Buhi, Phil. (bo͞o′ē)	169a	13°26'N	123°31'E
Buhl, Id., U.S. (būl)	73	42°36'N	114°45'W
Buhl, Mn., U.S.	71	47°28'N	92°49'W
Buin, Chile (bo̤-ēn′)	99b	33°44'S	70°44'W
Buinaksk, Russia (bo̤′ē-näksk)	137	42°40'N	47°20'E
Buir Nur, l., Asia (bo͞o-ēr no͞or)	161	47°50'N	117°00'E
Bujalance, Spain (bo͞o-hä-län′thā)	128	37°54'N	4°22'W
Bujumbura, Bdi.	191	3°23'S	29°22'E
Buka Island, i., Pap. N. Gui.	170e	5°15'S	154°35'E
Bukama, D.R.C. (bo͞o-kä′mä)	186	9°08'S	26°00'E
Bukavu, D.R.C.	186	2°30'S	28°52'E
Bukhara, Uzb. (bo̤-kä′rä)	139	39°31'N	64°22'E
Bukitbatu, Indon.	153b	1°25'N	101°58'E
Bukittinggi, Indon.	168	0°25'S	100°28'E
Bukoba, Tan.	186	1°20'S	31°49'E
Bukovina, hist. reg., Eur. (bo̤-kô-vĭ-ná)	125	48°06'N	25°20'E
Bula, Indon. (bo͞o′lä)	169	3°00'S	130°30'E
Bulalacao, Phil. (bo͞o-lä-lä′kä-ô)	169a	12°30'N	121°20'E
Bulawayo, Zimb. (bo͞o-lȧ-wā′yō)	186	20°12'S	28°43'E
Buldir, i., Ak., U.S. (bŭl dĭr)	63a	52°22'N	175°50'E
Bulgaria, nation, Eur. (bo̤l-gā′rĭ-ă)	110	42°12'N	24°13'E
Bulkley Ranges, mts., Can. (bŭlk′lē)	54	54°30'N	127°30'W
Bullaque, r., Spain (bo̤-lä′kå)	128	39°15'N	4°13'W
Bullas, Spain (bo͞ol′yäs)	128	38°07'N	1°48'W
Bullfrog Creek, r., Ut., U.S.	77	37°45'N	110°55'W
Bull Harbour, Can. (hăr′bēr)	54	50°45'N	127°55'W
Bull Head, mtn., Jam.	92	18°10'N	77°15'W
Bull Run, r., Or., U.S. (bo̤l)	74c	45°26'N	122°11'W
Bull Run Reservoir, res., Or., U.S.	74c	45°29'N	122°11'W
Bull Shoals Reservoir, res., U.S. (bo̤l shōlz)	65	36°35'N	92°57'W
Bulpham, Eng., U.K. (bo͞ol′făn)	114b	51°33'N	0°21'E
Bultfontein, S. Afr. (bo̤lt′fŏn-tān′)	192c	28°18'S	26°10'E
Bulun, Russia (bo͞o-lo̤n′)	135	70°48'N	127°27'E
Bulungu, D.R.C. (bo͞o-lo̤ŋ′go͞o)	190	6°04'S	21°54'E
Bulwer, S. Afr. (bo̤l-wēr)	187c	29°49'S	29°48'E
Bumba, D.R.C. (bo̤m′bä)	185	2°11'N	22°28'E
Bumbire Island, i., Tan.	191	1°40'S	32°05'E
Buna, Pap. N. Gui. (bo͞o′nä)	169	8°58'S	148°38'E
Bunbury, Austl. (bŭn′b*ŭ*rĭ)	174	33°25'S	115°45'E
Bundaberg, Austl. (bŭn′d*a*-bûrg)	175	24°45'S	152°18'E
Bunguran Utara, Kepulauan, is., Indon.	168	3°22'N	108°00'E
Bunia, D.R.C.	191	1°34'N	30°15'E
Bunker Hill, Il., U.S. (bŭnk′ēr hĭl)	75e	39°03'N	89°57'W
Bunkie, La., U.S. (bŭŋ′kĭ)	81	30°55'N	92°10'W
Bun Plains, pl., Kenya	191	0°55'N	40°35'E
Bununu Dass, Nig.	189	10°00'N	9°31'E
Buor-Khaya, Guba, b., Russia	141	71°45'N	131°00'E
Buor Khaya, Mys, c., Russia	135	71°47'N	133°22'E
Bura, Kenya	191	1°06'S	39°57'E
Buraydah, Sau. Ar.	154	26°23'N	44°14'E
Burbank, Ca., U.S. (bûr′bănk)	75a	34°11'N	118°19'W
Burco, Som.	192a	9°20'N	45°45'E
Burdekin, r., Austl. (bûr′dē-kĭn)	175	19°22'S	145°07'E
Burdur, Tur. (bo͞or-do̤r′)	119	37°50'N	30°15'E
Burdwān, India (bo̤d-wän′)	155	23°29'N	87°53'E
Bureinskiy, Khrebet, mts., Russia	135	51°15'N	133°30'E
Bureya, Russia (bo̤rā′*a*)	135	49°55'N	130°00'E
Bureya, r., Russia (bo̤-rā′yä)	141	51°00'N	131°15'E
Burford, Eng., U.K. (bûr-fērd)	114b	51°46'N	1°38'W
Burgas, Blg. (bo̤r-gäs′)	119	42°29'N	27°30'E
Burgas, Gulf of, b., Blg.	119	42°30'N	27°40'E
Burgaw, N.C., U.S. (bûr′gô)	83	34°31'N	77°56'W
Burgdorf, Switz. (bo̤rg′dôrf)	124	47°04'N	7°37'E
Burgenland, state, Aus.	115e	47°58'N	16°57'E
Burgeo, Can.	61	47°36'N	57°34'W
Burgess, Va., U.S.	67	37°53'N	76°21'W
Burgo de Osma, Spain	128	41°35'N	3°02'W
Burgos, Mex. (bo̤r′gōs)	80	24°57'N	98°47'W
Burgos, Phil.	169a	16°03'N	119°52'E
Burgos, Spain (bo͞o′r-gōs)	118	42°20'N	3°44'W
Burgsvik, Swe. (bo̤rgs′vĭk)	122	57°04'N	18°18'E
Burhānpur, India (bo̤r′hán-po͞or)	155	21°26'N	76°08'E
Burias Island, i., Phil. (bo͞o′rē-äs)	169a	12°56'N	122°56'E
Burias Pass, strt., Phil. (bo͞o′rē-äs)	169a	13°04'N	123°11'E
Burica, Punta, c., N.A. (po͞o′n-tä-bo͞o′rē-kä)	91	8°02'N	83°12'W
Burien, Wa., U.S. (bū′rĭ-ĕn)	74a	47°28'N	122°20'W
Burin, Can. (bûr′ĭn)	53a	47°02'N	55°10'W
Burin Peninsula, pen., Can.	61	47°00'N	55°40'W
Burkburnett, Tx., U.S. (bûrk-bûr′nĕt)	78	34°04'N	98°35'W
Burke, Vt., U.S. (bûrk)	67	44°40'N	72°00'W
Burke Channel, strt., Can.	54	52°07'N	127°38'W
Burketown, Austl. (bûrk′toun)	174	17°50'S	139°30'E
Burkina Faso, nation, Afr.	184	13°00'N	2°00'W
Burley, Id., U.S. (bûr′lĭ)	73	42°31'N	113°48'W
Burley, Wa., U.S.	74a	47°25'N	122°38'W
Burlingame, Ca., U.S. (bûr′lĭn-găm)	74b	37°35'N	122°22'W
Burlingame, Ks., U.S.	79	38°45'N	95°49'W
Burlington, Can. (bûr′lĭng-t*ŭ*n)	59	43°19'N	79°48'W
Burlington, Co., U.S.	78	39°17'N	102°26'W
Burlington, Ia., U.S.	65	40°48'N	91°05'W
Burlington, Ks., U.S.	79	38°10'N	95°46'W
Burlington, Ky., U.S.	69f	39°01'N	84°44'W
Burlington, Ma., U.S.	61a	42°31'N	71°13'W
Burlington, N.C., U.S.	83	36°05'N	79°26'W
Burlington, N.J., U.S.	68f	40°04'N	74°52'W
Burlington, Vt., U.S.	65	44°30'N	73°15'W
Burlington, Wa., U.S.	74a	48°28'N	122°20'W
Burlington, Wi., U.S.	69a	42°41'N	88°16'W
Burma *see* Myanmar, nation, Asia	150	21°00'N	95°15'E
Burnaby, Can.	50	49°14'N	122°58'W
Burnet, Tx., U.S. (bûrn′ĕt)	80	30°46'N	98°14'W
Burnham on Crouch, Eng., U.K. (bûrn′*ă*m-ŏn-krouch)	114b	51°38'N	0°48'E
Burnie, Austl. (bûr′nē)	175	41°15'S	146°05'E
Burnley, Eng., U.K. (bûrn′lē)	120	53°47'N	2°19'W
Burns, Or., U.S. (bûrnz)	72	43°35'N	119°05'W
Burnside, Ky., U.S. (bûrn′sīd)	82	36°57'N	84°33'W
Burns Lake, Can. (bûrnz lăk)	50	54°14'N	125°46'W
Burnsville, Can. (bûrnz′vĭl)	60	47°44'N	65°07'W
Burnt, r., Or., U.S. (bûrnt)	72	44°26'N	117°53'W
Burntwood, r., Can.	57	55°53'N	97°30'W
Burrard Inlet, b., Can. (bûr′*a*rd)	74d	49°19'N	123°15'W
Burr Gaabo, Som.	187	1°14'N	51°47'E
Burro, Serranías del, mts., Mex. (sĕr-rä-nē′äs dĕl bo͞o′r-rō)	80	29°39'N	102°07'W
Bursa, Tur. (bo͞or′sȧ)	154	40°10'N	28°10'E
Būr Safājah, Egypt	185	26°57'N	33°56'E
Burscheid, Ger. (bo͞or′shīd)	127c	51°05'N	7°07'E
Būr Sūdān, Sudan (so͞o-dán′)	185	19°30'N	37°10'E
Burt, N.Y., U.S. (bŭrt)	69c	43°19'N	78°45'W
Burt, l., Mi., U.S. (bŭrt)	66	45°25'N	84°45'W
Burton, Wa., U.S. (bûr′t*ŭ*n)	74a	47°24'N	122°28'W
Burton, Lake, res., Ga., U.S.	82	34°46'N	83°40'W
Burtonsville, Md., U.S. (bûrtŏns-vĭl)	68e	39°07'N	76°57'W
Burton-upon-Trent, Eng., U.K. (bŭr′t*ŭ*n-ŭp′-ŏn-trĕnt)	120	52°48'N	1°37'W
Buru, i., Indon.	169	3°30'S	126°30'E
Burullus, l., Egypt	192b	31°20'N	30°58'E
Burundi, nation, Afr.	186	3°00'S	29°30'E
Burwell, Ne., U.S. (bûr′w*ĕ*l)	70	41°46'N	99°08'W
Bury, Eng., U.K. (bĕr′ĭ)	114a	53°36'N	2°17'W
Buryatia, prov., Russia	141	55°15'N	112°00'E
Bury Saint Edmunds, Eng., U.K. (bĕr′ĭ-sănt ĕd′m*ŭ*ndz)	121	52°14'N	0°44'E
Burzaco, Arg. (bo͞or-zá′kô)	102a	34°50'S	58°23'W
Busanga Swamp, sw., Zam.	191	14°10'S	25°50'E
Būsh, Egypt (bo͞osh)	192b	29°13'N	31°08'E
Bushmanland, hist. reg., S. Afr. (bo̤sh-m*ă*n lănd)	186	29°15'S	18°45'E
Bushnell, Il., U.S. (bo̤sh′nĕl)	79	40°33'N	90°28'W
Businga, D.R.C. (bo̤-sĭŋ′gȧ)	185	3°20'N	20°53'E
Busira, r., D.R.C.	190	0°05'S	19°20'E
Bus'k, Ukr.	125	49°58'N	24°39'E
Busselton, Austl. (bûs′l-t*ŭ*n)	174	33°40'S	115°30'E
Bussum, Neth.	115a	52°16'N	5°10'E
Bustamante, Mex. (bo͞os-tä-män′tå)	80	26°34'N	100°30'W
Busto Arsizio, Italy (bo͞os′tō är-sēd′zē-ō)	130	45°47'N	8°51'E
Busuanga, i., Phil. (bo͞o-swäŋ′gä)	169a	12°20'N	119°43'E
Buta, D.R.C. (bo͞o′tȧ)	185	2°48'N	24°44'E
Butha Buthe, Leso. (bo͞o-thȧ-bo͞o′thȧ)	187c	28°49'S	28°16'E
Butler, Al., U.S. (bŭt′lēr)	82	32°05'N	88°10'W
Butler, In., U.S.	66	41°25'N	84°50'W
Butler, Md., U.S.	68e	39°32'N	76°46'W
Butler, N.J., U.S.	68a	41°00'N	74°20'W
Butler, Pa., U.S.	67	40°50'N	79°55'W
Butovo, Russia (bo̤-tô′vô)	142b	55°33'N	37°36'E
Butsha, D.R.C.	191	0°57'N	29°13'E
Buttahatchee, r., Al., U.S. (bŭt-ȧ-hăch′ē)	82	34°02'N	88°05'W
Butte, Mt., U.S. (būt)	64	46°00'N	112°31'W
Butterworth, S. Afr. (bŭ tēr′wûrth)	187c	32°20'S	28°09'E
Butt of Lewis, c., Scot., U.K. (bŭt ŏv lū′ĭs)	120	58°34'N	6°15'W
Butuan, Phil. (bo͞o-to͞o′än)	169	8°40'N	125°33'E
Buturlinovka, Russia (bo͞o-to͞o′lē-nôf′ka)	137	50°47'N	40°35'E
Buuhoodle, Som.	192a	8°15'N	46°20'E
Buulo Berde, Som.	192a	3°53'N	45°30'E
Buxtehude, Ger.	115c	53°29'N	9°42'E
Buxton, Eng., U.K. (bŭks′t′n)	114a	53°15'N	1°55'W
Buxton, Or., U.S.	74c	45°41'N	123°11'W
Buy, Russia (bwē)	134	58°30'N	41°48'E
Büyükmenderes, r., Tur.	154	37°50'N	28°20'E
Buzău, Rom. (bo͞o-z*ĕ*′o̤)	131	45°09'N	26°51'E
Buzău, r., Rom.	133	45°17'N	27°22'E
Buzaymah, Libya	185	25°14'N	22°13'E
Buzi, China (bo͞o-dz)	162	33°48'N	118°13'E
Buzuluk, Russia (bo̤-zó-lo̤k′)	134	52°50'N	52°10'E
Bwendi, D.R.C.	191	4°01'N	26°41'E
Byala, Blg.	131	43°26'N	25°44'E
Byala Slatina, Blg. (byä′la slä′tēnä)	131	43°26'N	23°56'E
Byalynichy, Bela. (byĕl-ĭ-nĭ′chĭ)	132	54°02'N	29°42'E
Byarezina, r., Bela. (bĕr-yĕ′zē-nȧ)	132	53°20'N	29°05'E
Byaroza, Bela.	125	52°29'N	24°59'E
Byblos *see* Jubayl, Leb.	153a	34°07'N	35°38'E
Bydgoszcz, Pol. (bĭd′gôshch)	116	53°07'N	18°00'E
Byelorussia *see* Belarus, nation, Eur.	134	53°30'N	25°33'E
Byerazino, Bela. (bĕr-yä′zē-nô)	132	53°51'N	28°54'E
Byeshankovichy, Bela.	132	55°04'N	29°29'E
Byesville, Oh., U.S. (bīz-vĭl)	66	39°55'N	81°35'W
Bygdin, l., Nor. (bügh-dēn′)	122	61°24'N	8°31'E
Byglandsfjord, Nor. (bügh′länds-fyôr)	122	58°40'N	7°49'E
Bykhaw, Bela.	132	53°32'N	30°15'E
Bykovo, Russia (bĭ-kô′vô)	142b	55°38'N	38°05'E
Byrranga, Gory, mts., Russia	140	74°15'N	94°28'E
Bytantay, r., Russia (byän′täy)	141	68°15'N	132°15'E
Bytom, Pol. (bĭ′tŭm)	117	50°21'N	18°55'E
Bytosh', Russia (bĭ-tôsh′)	132	53°48'N	34°06'E
Bytow, Pol. (bĭ′tŭf)	125	54°10'N	17°30'E

C

PLACE (Pronunciation)	PAGE	LAT.	LONG.
Cabagan, Phil. (kä-bä-gän′)	169a	17°27'N	121°50'E
Cabalete, i., Phil. (kä-bä-lā′tå)	169a	14°19'N	122°00'E
Caballones, Canal de, strt., Cuba (kä-nä′l-dĕ-kä-bäl-yô′nĕs)	92	20°45'N	79°20'W
Caballo Reservoir, res., N.M., U.S. (kä-bä-lyō′)	77	33°00'N	107°20'W
Cabanatuan, Phil. (kä-bä-nä-twän′)	169a	15°30'N	120°56'E
Cabano, Can. (kä-bä-nō′)	60	47°41'N	68°54'W
Cabarruyan, i., Phil. (kä-bä-ro͞o′yän)	169a	16°21'N	120°10'E
Cabedelo, Braz. (kä-bē-dā′lo̤)	101	6°58'S	34°49'W
Cabeza, Arrecife, i., Mex.	89	19°07'N	95°52'W
Cabeza del Buey, Spain (kä-bā′thä dĕl bwā′)	128	38°43'N	5°18'W
Cabimas, Ven. (kä-bē′mäs)	100	10°21'N	71°27'W
Cabinda, Ang.	186	5°33'S	12°12'E
Cabinda, hist. reg., Ang. (kä-bĭn′dȧ)	186	5°10'S	10°00'E
Cabinet Mountains, mts., Mt., U.S. (kăb′ĭ-nĕt)	72	48°13'N	115°52'W
Cabo Frio, Braz. (kä′bô-frē′ô)	99a	22°53'S	42°02'W
Cabo Frio, Ilha do, Braz. (ē′lä-dô-kä′bô frē′ô)	99a	23°01'S	42°00'W
Cabo Gracias a Dios, Hond. (kȧ′bô-grä-syäs-ä-dyô′s)	91	15°00'N	83°13'W
Cabonga, Réservoir, res., Can.	59	47°25'N	76°35'W
Cabora Bassa Reservoir, res., Moz.	186	15°45'S	32°00'E
Cabot Head, c., Can. (kăb′*ŭ*t)	58	45°15'N	81°20'W
Cabot Strait, strt., Can. (kăb′*ŭ*t)	53a	47°35'N	60°00'W
Cabra, Spain (käb′rä)	128	37°28'N	4°29'W
Cabra, i., Phil.	169a	13°55'N	119°55'E
Cabrera, Illa de, i., Spain	129	39°08'N	2°57'E
Cabrera, Sierra de la, mts., Spain	128	42°15'N	6°45'W
Cabriel, r., Spain (kä-brē-ĕl′)	128	39°25'N	1°20'W
Cabrillo National Monument, rec., Ca., U.S. (kä-brēl′yō)	76a	32°41'N	117°03'W
Cabuçu, r., Braz. (kä-bo͞o′-so͞o)	102b	22°57'S	43°36'W
Cabugao, Phil. (kä-bo͞o′gä-ô)	169a	17°48'N	120°28'E
Čačak, Serb. (chä′chák)	131	43°51'N	20°22'E
Caçapava, Braz. (kä′sä-pá′vä)	99a	23°05'S	45°52'W
Cáceres, Braz. (kȧ′sĕ-rĕs)	101	16°11'S	57°32'W
Cáceres, Spain (kȧ′thå-rās)	118	39°28'N	6°20'W
Cachapoal, r., Chile (kä-chä-pô-ä′l)	99b	34°23'S	70°19'W
Cache, r., Ar., U.S. (kȧsh)	79	35°24'N	91°12'W
Cache Creek, Can.	55	50°48'N	121°19'W
Cache Creek, r., Ca., U.S. (kăsh)	76	38°53'N	122°24'W
Cache la Poudre, r., Co., U.S. (kåsh lȧ po͞od′r′)	78	40°43'N	105°39'W
Cachi, Nevados de, mtn., Arg. (nĕ-vá′dôs-dĕ-kȧ′chē)	102	25°05'S	66°40'W
Cachinal, Chile (kä-chē-näl′)	102	24°57'S	69°33'W
Cachoeira, Braz. (kä-shô-ā′rä)	101	12°32'S	38°47'W
Cachoeirá do Sul, Braz. (kä-shô-ā′rä-dô-so͞o′l)	102	30°02'S	52°49'W
Cachoeiras de Macacu, Braz. (kä-shô-ā′räs-dĕ-mä-kȧ′ko͞o)	99a	22°28'S	42°39'W
Cachoeiro de Itapemirim, Braz.	101	20°51'S	41°06'W
Cacólo, Ang.	190	10°07'S	19°17'E
Caconda, Ang. (kä-kōn′dä)	186	13°43'S	15°06'E
Cacouna, Can.	60	47°54'N	69°31'W
Cacula, Ang.	190	14°29'S	14°10'E
Cadale, Som.	192a	2°45'N	46°15'E
Caddo, l., La., U.S. (kăd′ō)	81	32°37'N	94°15'W
Cadereyta, Mex. (kä-dȧ-rā′tä)	88	20°42'N	99°47'W
Cadereyta Jimenez, Mex. (kä-dä-rā′tä hē-mā′nāz)	80	25°36'N	99°59'W
Cadí, Sierra de, mts., Spain (sē-ĕ′r-rä-dĕ-kä′dē)	129	42°17'N	1°34'E
Cadillac, Mi., U.S. (kăd′ĭ-lăk)	66	44°15'N	85°25'W
Cádiz, Spain (kȧ′dēz)	110	36°34'N	6°20'W
Cadiz, Ca., U.S. (kā′dĭz)	76	34°33'N	115°30'W
Cadiz, Oh., U.S.	66	40°15'N	81°00'W
Cádiz, Golfo de, b., Spain (gôl-fô-dĕ-kȧ′dēz)	118	36°50'N	7°00'W
Caen, Fr. (kän)	117	49°13'N	0°22'W
Caernarfon, Wales, U.K.	116	53°08'N	4°17'W
Caernarfon Bay, b., Wales, U.K.	120	53°09'N	4°56'W
Cagayan, Phil. (kä-gä-yän′)	169	8°13'N	124°30'E
Cagayan, r., Phil.	168	16°45'N	121°55'E
Cagayan Islands, is., Phil.	168	9°40'N	120°30'E
Cagayan Sulu, i., Phil. (kä-gä-yän so͞o′lo͞o)	168	7°00'N	118°30'E
Cagli, Italy (käl′yē)	130	43°35'N	12°40'E
Cagliari, Italy (käl′yä-rē)	110	39°16'N	9°08'E
Cagliari, Golfo di, b., Italy (gôl-fô-dē-käl′yä-rē)	118	39°08'N	9°12'E
Cagnes, Fr. (kän′y′)	127	43°40'N	7°14'E
Cagua, Ven. (kä′gwä)	101b	10°12'N	67°27'W
Caguas, P.R. (kä′gwäs)	87b	18°12'N	66°01'W
Cahaba, r., Al., U.S. (kȧ hä-bä)	82	32°50'N	87°15'W
Cahama, Ang. (kä-á′mä)	186	16°17'S	14°19'E

PLACE (Pronunciation)	PAGE	LAT.	LONG.
Cahokia, Il., U.S. (kȧ-hō′kĭ-ȧ)	75e	38°34′N	90°11′W
Cahora-Bassa, wtfl., Moz.	191	15°40′S	32°50′E
Cahors, Fr. (kȧ-ŏr′)	117	44°27′N	1°27′E
Cahuacán, Mex. (kä-wä-kä′n)	89a	19°38′N	99°25′W
Cahuita, Punta, c., C.R. (po͞o′n-tä-kä-wē′tȧ)	91	9°47′N	82°41′W
Cahul, Mol.	133	45°49′N	28°17′E
Caibarién, Cuba (kī-bä-rē-ĕn′)	92	22°35′N	79°30′W
Caicedonia, Col. (kī-sĕ-dŏ-nēä)	100a	4°21′N	75°48′W
Caicos Bank, bk., (kī′kōs)	93	21°35′N	72°00′W
Caicos Islands, is., T./C. Is.	87	21°45′N	71°50′W
Caicos Passage, strt., N.A.	93	21°55′N	72°45′W
Caillou Bay, b., La., U.S. (kȧ-yo͞o′)	81	29°07′N	91°00′W
Caimanera, Cuba (kī-mä-nā′rä)	93	20°00′N	75°10′W
Caiman Point, c., Phil. (kī′mȧn)	169a	15°56′N	119°33′E
Caimito, r., Pan. (kä-ē-mē′tô)	86a	8°50′N	79°45′W
Caimito del Guayabal, Cuba (kä-ē-mē′tō-dĕl-gwä-yä-bä′l)	93a	22°57′N	82°36′W
Cairns, Austl. (kârnz)	175	17°02′S	145°49′E
Cairo, C.R. (kī′rō)	91	10°06′N	83°47′W
Cairo, Egypt	185	30°00′N	31°17′E
Cairo, Ga., U.S. (kā′rō)	82	30°48′N	84°12′W
Cairo, Il., U.S.	65	36°59′N	89°11′W
Caistor, Eng., U.K. (kâs′tēr)	114a	53°30′N	0°20′W
Caiundo, Ang.	190	15°46′S	17°28′E
Caiyu, China (tsī-yo͞o)	162	39°39′N	116°36′E
Cajamarca, Col. (kä-κä-mȧ′r-kä)	100a	4°25′N	75°25′W
Çajamarca, Peru (kä-hä-mär′kä)	100	7°16′S	78°30′W
Čajniče, Bos. (chī′nĭ-chĕ)	131	43°32′N	19°04′E
Cajon, Ca., U.S. (kȧ-hōn′)	75a	34°18′N	117°28′W
Çajuru, Braz. (kȧ-zho͞o′ro͞o)	99a	21°17′S	47°17′W
Čakovec, Cro. (chȧ′kō-vĕts)	130	46°23′N	16°27′E
Cala, S. Afr. (cä-lȧ)	187c	31°33′S	27°41′E
Calabar, Nig. (kăl-ȧ-bär′)	184	4°57′N	8°19′E
Calabazar, Cuba (kä-lä-bä-zä′r)	93a	23°02′N	82°25′W
Calabozo, Ven. (kä-lä-bō′zō)	100	8°48′N	67°27′W
Calabria, hist. reg., Italy (kä-lä′brē-ä)	130	39°26′N	16°23′E
Calafat, Rom. (kȧ-lȧ-făt′)	131	43°59′N	22°56′E
Calaguas Islands, is., Phil. (kä-läg′wäs)	169a	14°30′N	123°06′E
Calahoo, Can. (kă-lȧ-ho͞o′)	62g	53°42′N	113°58′W
Calahorra, Spain (kä-lä-ŏr′rä)	118	42°18′N	1°58′W
Calais, Fr. (kȧ-lĕ′)	110	50°56′N	1°51′E
Calais, Me., U.S.	65	45°11′N	67°15′W
Calama, Chile (kä-lä′mä)	102	22°17′S	68°58′W
Calamar, Col. (kä-lä-mär′)	100	10°24′N	75°00′W
Calamar, Col.	100	1°55′N	72°33′W
Calamba, Phil. (kä-läm′bä)	169a	14°12′N	121°10′E
Calamian Group, is., Phil. (kä-lä-myän′)	168	12°14′N	118°38′E
Calañas, Spain (kä-län′yäs)	128	37°41′N	6°52′W
Calanda, Spain	129	40°53′N	0°20′W
Calapan, Phil. (kä-lä-pän′)	169a	13°25′N	121°11′E
Călăraşi, Rom. (kŭ-lŭ-rȧsh′ĭ)	119	44°09′N	27°20′E
Calatayud, Spain (kä-lä-tä-yo͞odh′)	118	41°23′N	1°37′W
Calauag Bay, b., Phil.	169a	14°07′N	122°10′E
Calaveras Reservoir, res., Ca., U.S. (kăl-ȧ-vẽr′ăs)	74b	37°29′N	121°47′W
Calavite, Cape, c., Phil. (kä-lä-vē′tā)	169a	13°29′N	120°00′E
Calcasieu, r., La., U.S. (kăl′kȧ-shū)	81	30°22′N	93°08′W
Calcasieu Lake, l., La., U.S.	81	29°58′N	93°08′W
Calcutta *see* Kolkata, India (kăl-kŭt′ȧ)	155	22°32′N	88°22′E
Caldas, Col. (kȧ′l-däs)	100a	6°06′N	75°38′W
Caldas, dept., Col.	100a	5°20′N	75°38′W
Caldas da Rainha, Port. (käl′däs dä rīn′yă)	128	39°25′N	9°08′W
Calder, r., Eng., U.K. (kôl′dẽr)	114a	53°39′N	1°30′W
Caldera, Chile (käl-dā′rä)	102	27°02′S	70°53′W
Calder Canal, can., Eng., U.K.	114a	53°48′N	2°25′W
Caldwell, Id., U.S. (kôld′wĕl)	72	43°40′N	116°43′W
Caldwell, Ks., U.S.	79	37°04′N	97°36′W
Caldwell, Oh., U.S.	66	39°40′N	81°30′W
Caldwell, Tx., U.S.	81	30°30′N	96°40′W
Caledon, Can. (kăl′ē-dŏn)	62d	43°52′N	79°59′W
Caledonia, Mn., U.S. (kăl-ē-dō′nĭ-ȧ)	71	43°38′N	91°31′W
Calella, Spain (kä-lĕl′yä)	129	41°37′N	2°39′E
Calera Victor Rosales, Mex. (kä-lā′rä-vē′k-tŏr-rŏ-sȧ′lĕs)	88	22°57′N	102°42′W
Calexico, Ca., U.S. (kȧ-lĕk′sĭ-kō)	64	32°41′N	115°30′W
Calgary, Can. (kăl′gȧ-rĭ)	50	51°03′N	114°05′W
Calhoun, Ga., U.S. (kăl-ho͞on′)	82	34°30′N	84°56′W
Cali, Col. (kä′lē)	100	3°26′N	76°30′W
Caliente, Nv., U.S. (käl-yĕn′tā)	77	37°38′N	114°30′W
California, Mo., U.S. (kăl-ĭ-fôr′nĭ-ȧ)	79	38°38′N	92°38′W
California, Pa., U.S.	69e	40°03′N	79°53′W
California, state, U.S.	64	38°10′N	121°20′W
California, Golfo de, b., Mex. (gŏl-fô-dĕ-kä-lē-fôr-nyä)	86	30°30′N	113°45′W
California Aqueduct, aq., Ca., U.S.	76	37°10′N	121°10′W
Călimani, Munţii, mts., Rom.	125	47°05′N	24°47′E
Calimere, Point, c., India	159	10°20′N	80°20′E
Calimesa, Ca., U.S. (kă-lĭ-mā′sȧ)	75a	34°00′N	117°04′W
Calipatria, Ca., U.S. (kăl-ĭ-păt′rĭ-ȧ)	76	33°03′N	115°30′W
Calkini, Mex. (käl-kē-nē′)	89	20°21′N	90°06′W
Callabonna, Lake, l., Austl. (călă′bŏnȧ)	176	29°35′S	140°28′E
Callao, Peru (käl-yä′ō)	100	12°02′S	77°07′W
Calling, l., Can. (kôl′ĭng)	55	55°15′N	113°12′W
Calmar, Can. (kăl′mär)	62g	53°16′N	113°49′W
Calmar, Ia., U.S.	71	43°12′N	91°54′W
Caloosahatchee, r., Fl., U.S. (kȧ-loo-sȧ-hăch′ē)	83a	26°45′N	81°41′W
Calotmul, Mex. (kä-lôt-mo͞ol)	90a	20°58′N	88°11′W
Calpulalpan, Mex. (käl-po͞o-lȧl′pän)	88	19°35′N	98°33′W
Caltagirone, Italy (käl-tä-jē-rō′nȧ)	118	37°14′N	14°32′E
Caltanissetta, Italy (käl-tä-nē-sĕt′tä)	118	37°30′N	14°02′E
Caluango, Ang.	190	8°21′S	19°40′E
Calucinga, Ang.	190	11°18′S	16°12′E
Calumet, Mi., U.S. (kă-lū-mĕt′)	71	47°15′N	88°29′W
Calumet, Lake, l., Il., U.S.	69a	41°43′N	87°36′W
Calumet City, Il., U.S.	69a	41°37′N	87°33′W
Calunda, Ang.	190	12°06′S	23°23′E
Caluquembe, Ang.	190	13°47′S	14°44′E
Caluula, Som.	192a	11°53′N	50°40′E
Calvert, Tx., U.S. (kăl′vẽrt)	81	30°59′N	96°41′W
Calvert Island, i., Can.	52	51°35′N	128°00′W
Calvi, Fr. (käl′vē)	130	42°33′N	8°35′E
Calvillo, Mex. (käl-vēl′yō)	89	21°51′N	102°44′E
Calvinia, S. Afr. (kăl-vĭn′ĭ-ȧ)	186	31°20′S	19°50′E
Cam, r., Eng., U.K. (kăm)	121	52°15′N	0°05′E
Camagüey, Cuba (kä-mä-gwā′)	87	21°25′N	78°00′W
Camagüey, prov., Cuba	92	21°30′N	78°10′W
Camajuani, Cuba (kä-mä-hwä′nē)	92	22°25′N	79°50′W
Camano, Wa., U.S. (kä-mä′no)	74a	48°10′N	122°32′W
Camano Island, i., Wa., U.S.	74a	48°11′N	122°29′W
Camargo, Mex. (kä-mär gō)	80	26°19′N	98°49′W
Camarón, Cabo, c., Hond. (kȧ′bō-kä-mä-rōn′)	90	16°06′N	85°05′W
Camas, Wa., U.S. (kăm′ȧs)	74c	45°36′N	122°24′W
Camas Creek, r., Id., U.S.	73	44°10′N	112°09′W
Camatagua, Ven. (kä-mä-tȧ′gwä)	101b	9°49′N	66°55′W
Ca Mau, Mui, c., Viet.	168	8°36′N	104°43′E
Cambay, India (kăm-bā′)	158	22°22′N	72°39′E
Cambodia, nation, Asia	168	12°15′N	104°00′E
Cambonda, Serra, mts., Ang.	190	12°10′S	14°15′E
Camborne, Eng., U.K. (kăm′bôrn)	120	50°15′N	5°28′W
Cambrai, Fr. (käɴ-brĕ′)	117	50°10′N	3°15′E
Cambrian Mountains, mts., Wales, U.K. (kăm′brĭ-ăn)	120	52°05′N	4°05′W
Cambridge, Can.	59	43°22′N	80°19′W
Cambridge, Eng., U.K. (kām′brĭj)	117	52°12′N	0°11′E
Cambridge, Ma., U.S.	61a	42°23′N	71°07′W
Cambridge, Md., U.S.	67	38°35′N	76°10′W
Cambridge, Mn., U.S.	71	45°35′N	93°14′W
Cambridge, Ne., U.S.	78	40°17′N	100°10′W
Cambridge, Oh., U.S.	66	40°00′N	81°35′W
Cambridge Bay *see* Kaluktutiak, Can.	52	69°15′N	105°00′W
Cambridge City, In., U.S.	66	39°45′N	85°15′W
Cambridgeshire, co., Eng., U.K.	114a	52°26′N	0°19′W
Cambuci, Braz. (käm-bo͞o′sē)	99a	21°35′S	41°54′W
Cambundi-Catembo, Ang.	190	10°09′S	17°31′E
Camby, In., U.S. (kăm′bē)	69g	39°40′N	86°19′W
Camden, Austl.	173b	34°03′S	150°42′E
Camden, Al., U.S. (kăm′dĕn)	82	31°58′N	87°15′W
Camden, Ar., U.S.	79	33°36′N	92°49′W
Camden, Me., U.S.	60	44°11′N	69°05′W
Camden, N.J., U.S.	65	39°56′N	75°06′W
Camden, S.C., U.S.	83	34°14′N	80°37′W
Cameia, Parque Nacional da, rec., Ang.	190	11°40′S	21°20′E
Camenca, Mol.	133	48°02′N	28°43′E
Cameron, Mo., U.S. (kăm′ẽr-ŭn)	79	39°44′N	94°14′W
Cameron, Tx., U.S.	81	30°52′N	96°57′W
Cameron, W.V., U.S.	66	39°40′N	80°35′W
Cameron Hills, hills, Can.	52	60°13′N	120°20′W
Cameroon, nation, Afr.	184	5°48′N	11°00′E
Cameroon Mountain, mtn., Cam.	184	4°12′N	9°11′E
Camiling, Phil. (kä-mē-lĭng′)	169a	15°42′N	120°24′E
Camilla, Ga., U.S. (kȧ-mĭl′ȧ)	82	31°13′N	84°12′W
Caminha, Port. (kä-mĭn′yȧ)	128	41°52′N	8°44′W
Camoçim, Braz. (kä-mô-sēɴ′)	101	2°56′S	40°55′W
Camooweal, Austl.	174	20°00′S	138°13′E
Campana, Arg. (käm-pä′nä)	99c	34°10′S	58°58′W
Campana, i., Chile (käm-pän′yä)	102	48°20′S	75°15′W
Campanario, Spain (kä-pä-nä′rē-ō)	128	38°51′N	5°36′W
Campanella, Punta, c., Italy (pŏ′n-tä-käm-pä-nĕ′lä)	129c	40°20′N	14°21′E
Campanha, Braz. (käm-pän-yäɴ′)	99a	21°51′S	45°24′W
Campania, hist. reg., Italy (käm-pän′yä)	130	41°00′N	14°40′E
Campbell, Ca., U.S. (kăm′bĕl)	74b	37°17′N	121°57′W
Campbell, Mo., U.S.	79	36°29′N	90°04′W
Campbell, is., N.Z.	3	52°30′S	169°00′E
Campbellpore, Pak.	158	33°49′N	72°24′E
Campbell River, Can.	50	50°01′N	125°15′W
Campbellsville, Ky., U.S. (kăm′bĕlz-vĭl)	82	37°19′N	85°20′W
Campbellton, Can. (kăm′bĕl-tŭn)	51	48°00′N	66°40′W
Campbelltown, Austl. (kăm′bĕl-toun)	173b	34°04′S	150°49′E
Campbelltown, Scot., U.K. (kăm′b′l-toun)	120	55°25′N	5°50′W
Camp Dennison, Oh., U.S. (dĕ′nĭ-sŏn)	69f	39°12′N	84°17′W
Campeche, Mex. (käm-pā′chȧ)	86	19°51′N	90°32′W
Campeche, state, Mex.	86	18°55′N	90°20′W
Campeche, Bahía de, b., Mex. (bä-ē′ä-dĕ-käm-pā′chā)	86	19°30′N	93°40′W
Campechuela, Cuba (käm-pȧ-chwā′lä)	92	20°15′N	77°15′W
Camperdown, S. Afr. (kăm′pẽr-doun)	187c	29°44′S	30°33′E
Câmpina, Rom.	131	45°08′N	25°47′E
Campina Grande, Braz. (käm-pē′nä grän′dĕ)	101	7°15′S	35°49′W
Campinas, Braz. (käm-pē′näzh)	101	22°53′S	47°03′W
Camp Indian Reservation, I.R., Ca., U.S. (kămp)	76	32°39′N	116°26′W
Campo, Cam. (käm′pō)	184	2°22′N	9°49′E
Campoalegre, Col. (kä′m-pō-ȧlĕ′grĕ)	100	2°34′N	75°20′W
Campobasso, Italy (käm′pō-bäs′sō)	130	41°35′N	14°39′E
Campo Belo, Braz.	99a	20°52′S	45°15′W
Campo de Criptana, Spain (käm′pō dā krēp-tä′nä)	128	39°24′N	3°09′W
Campo Florido, Cuba (kä′m-pō flō-rē′dō)	93a	23°07′N	82°07′W
Campo Grande, Braz. (käm-pȯ grän′dĕ)	101	20°28′S	54°32′W
Campo Grande, Braz.	102b	22°54′S	43°33′W
Campo Maior, Braz. (käm-pȯ mä-yôr′)	101	4°48′S	42°12′W
Campo Maior, Port.	128	39°03′N	7°06′W
Campo Real, Spain (käm′pō rå-äl′)	129a	40°21′N	3°23′W
Campos, Braz. (kä′m-pōs)	101	21°46′S	41°19′W
Campos do Jordão, Braz. (kä′m-pōs-dō-zhôr-dou′ɴ)	99a	22°45′S	45°35′W
Campos Gerais, Braz. (kä′m-pōs-zhĕ-rȧēs)	99a	21°17′S	45°43′W
Camps Bay, S. Afr. (kămps)	186a	33°57′S	18°22′E
Camp Springs, Md., U.S. (cămp sprĭngz)	68e	38°48′N	76°55′W
Câmpulung, Rom.	119	45°15′N	25°03′E
Câmpulung Moldovenesc, Rom.	125	47°31′N	25°36′E
Camp Wood, Tx., U.S. (kămp wŏd)	80	29°39′N	100°02′W
Camrose, Can. (kăm-rōz)	50	53°01′N	112°50′W
Camu, r., Dom. Rep. (kä′mo͞o)	93	19°05′N	70°15′W
Canada, nation, N.A. (kăn′ȧ-dȧ)	50	50°00′N	100°00′W
Canada Bay, b., Can.	61	50°43′N	56°10′W
Cañada de Gómez, Arg. (kä-nyä′dä-dĕ-gô′mĕz)	102	32°49′S	61°24′W
Canadian, Tx., U.S. (kȧ-nā′dĭ-ăn)	78	35°54′N	100°24′W
Canadian, r., U.S.	64	35°30′N	102°30′W
Canajoharie, N.Y., U.S. (kăn-ȧ-jō-hăr′ē)	67	42°55′N	74°35′W
Çanakkale, Tur. (chä-näk-kä′lĕ)	119	40°10′N	26°26′E
Çanakkale Boğazi (Dardanelles), strt., Tur.	119	40°05′N	25°50′E
Canandaigua, N.Y., U.S. (kăn-ăn-dā′gwȧ)	67	42°55′N	77°20′W
Canandaigua, l., N.Y., U.S.	67	42°45′N	77°20′W
Cananea, Mex. (kä-nä-nĕ′ä)	86	31°00′N	110°20′W
Canarias, Islas (Canary Is.), is., Spain (ē′s-läs-kä-nä′ryäs)	183	29°15′N	16°30′W
Canarreos, Archipiélago de los, is., Cuba	92	21°35′N	82°20′W
Canary Islands *see* Canarias, Islas, is., Spain	183	29°15′N	16°30′W
Cañas, C.R. (kä′-nyäs)	90	10°26′N	85°06′W
Cañas, r., C.R.	90	10°20′N	85°21′W
Cañasgordas, Col. (kä′nyäs-gō′r-däs)	100a	6°44′N	76°01′W
Canastota, N.Y., U.S. (kăn-ȧs-tō′tȧ)	67	43°05′N	75°45′W
Canastra, Serra de, mts., Braz. (sĕ′r-rä-dĕ-kä-nä′s-trä)	101	19°53′S	46°57′W
Canatlán, Mex. (kä-nät-län′)	80	24°30′N	104°45′W
Canaveral, Cape, c., Fl., U.S.	65	28°30′N	80°23′W
Canavieiras, Braz. (kä-nȧ-vē-ā′räs)	101	15°40′S	38°49′W
Canberra, Austl. (kăn′bĕr-ȧ)	175	35°21′S	149°10′E
Canby, Mn., U.S. (kăn′bĭ)	70	44°43′N	96°15′W
Canchyuaya, Cerros de, mts., Peru (sĕ′r-rōs-dĕ-kän-cho͞o-ȧ′ĭä)	100	7°30′S	74°30′W
Cancuc, Mex. (kän-ko͞ok)	89	16°58′N	92°17′W
Cancún, Mex.	90a	21°25′N	86°50′W
Candelaria, Cuba (kän-dĕ-lä′ryä)	92	22°45′N	82°55′W
Candelaria, Phil. (kän-dȧ-lä′rē-ä)	169a	15°39′N	119°55′E
Candelaria, r., Mex. (kän-dĕ-lä-ryä)	89	18°25′N	91°21′W
Candeleda, Spain (kän-dhȧ-lā′dhä)	128	40°09′N	5°18′W
Candia *see* Irákleio, Grc.	110	35°20′N	25°10′E
Candle, Ak., U.S. (kăn′d′l)	63	65°00′N	162°04′W
Cando, N.D., U.S. (kăn′dō)	70	48°27′N	99°13′W
Candon, Phil. (kän-dōn′)	169a	17°13′N	120°26′E
Canelones, Ur. (kä-nĕ-lô-nĕs)	99c	34°32′S	56°19′W
Canelones, dept., Ur.	99c	34°34′S	56°15′W
Cañete, Peru (kän-yā′tȧ)	100	13°06′S	76°17′W
Caney, Cuba (kä-nā′) (kā′nĭ)	93	20°05′N	75°45′W
Caney, Ks., U.S. (kā′nĭ)	79	37°00′N	95°57′W
Caney Fork, r., Tn., U.S.	82	36°10′N	85°50′W
Cangamba, Ang.	186	13°40′S	19°54′E
Cangas, Spain (kän′gäs)	128	42°15′N	8°43′W
Cangas de Narcea, Spain (kä′n-gäs-dĕ-när-sĕ-ä)	128	43°08′N	6°36′W
Cangzhou, China (tsäŋ-jō)	164	38°21′N	116°53′E
Caniapiscau, l., Can.	53	54°10′N	71°13′E
Caniapiscau, r., Can.	53	57°00′N	68°45′W
Canicatti, Italy (kä-nē-kät′tē)	130	37°18′N	13°58′E
Cañitas, Mex. (kän-yē′täs)	88	23°38′N	102°44′W
Cannell, Can.	62g	53°35′N	113°38′W
Cannelton, In., U.S. (kăn′ĕl-tŭn)	66	37°55′N	86°45′W
Cannes, Fr. (kȧn)	117	43°34′N	7°05′E
Canning, Can. (kăn′ĭng)	60	45°09′N	64°25′W
Cannock, Eng., U.K. (kăn′ŭk)	114a	52°41′N	2°02′W
Cannock Chase, reg., Eng., U.K. (kăn′ŭk chās)	114a	52°43′N	1°54′W
Cannon, r., Mn., U.S. (kăn′ŭn)	71	44°18′N	93°24′W
Cannonball, r., N.D., U.S. (kăn′ŭn-bäl)	70	46°17′N	101°35′W
Caño, Isla de, i., C.R. (ē′s-lä-dĕ-kä′nō)	91	8°38′N	84°00′W
Canoga Park, Ca., U.S. (kä-nō′gȧ)	75a	34°07′N	118°36′W
Canoncito Indian Reservation, I.R., N.M., U.S.	77	35°00′N	107°05′W
Canon City, Co., U.S. (kăn′yŭn)	78	38°27′N	105°16′W
Canonsburg, Pa., U.S. (kăn′ŭnz-bûrg)	69e	40°16′N	80°11′W
Canoochee, r., Ga., U.S. (kȧ-no͞o′chē)	83	32°25′N	82°11′W
Canora, Can. (kȧ-nōrȧ)	50	51°37′N	102°26′W
Canosa, Italy (kä-nō′sä)	130	41°14′N	16°03′E
Canouan, i., St. Vin.	91b	12°44′N	61°10′W
Cansahcab, Mex.	90a	21°11′N	89°05′W
Canso, Can. (kăn′sō)	61	45°20′N	61°00′W
Canso, Cape, c., Can.	61	45°21′N	60°46′W
Canso, Strait of, strt., Can.	61	45°37′N	61°25′W
Cantabrica, Cordillera, mts., Spain	112	43°05′N	6°05′W
Cantagalo, Braz. (kän-tä-gȧ′lo)	99a	21°59′S	42°22′W
Cantanhede, Port. (kän-täɴ-yā′dȧ)	128	40°22′N	8°35′W

ăt; finăl; rāte; senåte; ärm; ȧsk; sofȧ; fâre; ch-choose; dh-as th in other; bē; ēvent; bĕt; recĕnt; cratẽr; g-gō; gh-guttural g; bĭt; ĭ-short neutral; rīde; ĸ-guttural k as ch in German ich;

PRONOUNCING INDEX

Canterbury – Castellammare di Stabia 233

PLACE (Pronunciation)	PAGE	LAT.	LONG.
Canterbury, Eng., U.K. (kăn'tĕr-bĕr-ē)	121	51°17'N	1°06'E
Canterbury Bight, b., N.Z.	175a	44°15'S	172°08'E
Cantiles, Cayo, i., Cuba (ky-ō-kän-tē'lås)	92	21°40'N	82°00'W
Canton see Guangzhou, China	161	23°07'N	113°15'E
Canton, Ga., U.S.	82	34°13'N	84°29'W
Canton, Il., U.S.	79	40°34'N	90°02'W
Canton, Ma., U.S.	61a	42°09'N	71°09'W
Canton, Mo., U.S.	79	40°08'N	91°33'W
Canton, Ms., U.S.	82	32°36'N	90°01'W
Canton, N.C., U.S.	83	35°32'N	82°50'W
Canton, Oh., U.S.	65	40°50'N	81°25'W
Canton, Pa., U.S.	67	41°50'N	76°45'W
Canton, S.D., U.S.	70	43°17'N	96°37'W
Cantu, Italy (kän-tō')	130	45°43'N	9°09'E
Cañuelas, Arg. (kä-nyōĕ'-läs)	99c	35°03'S	58°45'W
Canyon, Tx., U.S. (kăn'yŭn)	78	34°59'N	101°57'W
Canyon, r., Wa., U.S.	74a	48°09'N	121°48'W
Canyon de Chelly National Monument, rec., Az., U.S.	77	36°14'N	110°00'W
Canyon Ferry Lake, res., Mt., U.S.	73	46°33'N	111°37'W
Canyonlands National Park, rec., Ut., U.S.	77	38°10'N	110°00'W
Canyons of the Ancients National Monument, rec., Co., U.S.	77	37°30'N	108°50'W
Caoxian, China (tsou shyĕn)	162	34°48'N	115°33'E
Capalonga, Phil. (kä-pä-lôŋ'gä)	169a	14°20'N	122°30'E
Capannori, Italy (kä-pän'nō-rē)	130	43°50'N	10°30'E
Capaya, r., Ven. (kä-pä-ïä)	101b	10°28'N	66°15'W
Cap-Chat, Can. (kȧp-shä')	51	48°02'N	65°20'W
Cap-de-la-Madeleine, Can. (kȧp dē lȧ mȧ-d'lĕn')	59	46°23'N	72°30'W
Cape Breton, i., Can. (kāp brĕt'ŭn)	61	45°48'N	59°50'W
Cape Breton Highlands National Park, rec., Can.	51	46°45'N	60°45'W
Cape Charles, Va., U.S. (kāp chärlz)	83	37°13'N	76°02'W
Cape Coast, Ghana	184	5°05'N	1°15'W
Cape Fear, r., N.C., U.S. (kāp fēr)	65	35°00'N	79°00'W
Cape Flats, pl., S. Afr. (kāp flăts)	186a	34°01'S	18°37'E
Cape Girardeau, Mo., U.S. (jē-rär-dō')	65	37°17'N	89°32'W
Cape Krusenstern National Monument, rec., Ak., U.S.	63	67°30'N	163°40'W
Cape May, N.J., U.S. (kāp mā)	67	38°55'N	74°50'W
Cape May Court House, N.J., U.S.	67	39°05'N	75°00'W
Cape Romanzof, Ak., U.S. (rō' män zôf)	63	61°50'N	165°45'W
Capesterre, Guad.	91b	16°02'N	61°37'W
Cape Tormentine, Can.	60	46°08'N	63°47'W
Cape Town, S. Afr. (kāp toun)	186	33°48'S	18°28'E
Cape Verde, nation, Afr.	184b	15°48'N	26°02'W
Cape York Peninsula, pen., Austl. (kāp yôrk)	175	12°30'S	142°35'E
Cap-Haïtien, Haiti (kȧp ȧ-ē-syăn')	87	19°45'N	72°15'W
Capilla de Señor, Arg. (kä-pēl'yä dā sān-yôr')	99c	34°18'S	59°07'W
Capitachouane, r., Can.	59	47°50'N	76°45'W
Capitol Reef National Park, rec., Ut., U.S. (kăp'ĭ-tŏl)	77	38°15'N	111°10'W
Capivari, Braz. (kä-pē-vä'rē)	99a	22°59'S	47°29'W
Capivari, r., Braz.	102b	22°39'S	43°19'W
Capoompeta, mtn., Austl. (kȧ-pōōm-pē'tȧ)	175	29°15'S	152°12'E
Capraia, i., Italy (kä-prä'yä)	118	43°02'N	9°51'E
Caprara Point, c., Italy (kä-prä'rä)	130	41°08'N	8°20'E
Capreol, Can.	59	46°43'N	80°56'W
Caprera, i., Italy (kä-prā'rä)	130	41°12'N	9°28'E
Capri, Italy	129c	40°18'N	14°16'E
Capri, Isola di, i., Italy (ē'-sō-lä-dē-kä'prē)	129c	40°19'N	14°10'E
Capricorn Channel, strt., Austl. (kăp'rĭ-kôrn)	175	22°27'S	151°24'E
Caprivi Strip, hist. reg., Nmb.	186	18°00'S	22°00'E
Cap-Rouge, Can. (kȧp rōōzh')	62b	46°45'N	71°21'W
Cap-Saint Ignace, Can. (kĭp săN-tē-nyås')	62b	47°02'N	70°27'W
Capua, Italy (kä'pwä)	118	41°07'N	14°14'E
Capulhuac, Mex. (kä-pōōl-hwäk')	88	19°33'N	99°43'W
Capulin Mountain National Monument, rec., N.M., U.S. (kȧ-pū'lĭn)	78	36°15'N	103°58'W
Capultitlán, Mex. (kä-pōō'l-tē-tlä'n)	89a	19°15'N	99°40'W
Caquetá (Japurá), r., S.A.	100	0°20'S	73°00'W
Carabaña, Spain (kä-rä-bän'yä)	129a	40°16'N	3°15'W
Carabelle, Fl., U.S. (kăr'ȧ-bĕl)	82	29°50'N	84°40'W
Carabobo, dept., Ven. (kä-rä-bô'-bô)	101b	10°07'N	68°06'W
Caracal, Rom. (kȧ-rȧ-kȧl')	131	44°06'N	24°22'E
Caracas, Ven. (kä-rä'käs)	100	10°30'N	66°58'W
Carácuaro de Morelos, Mex. (kä-rä'kwä-rō-dē-mô-rē-lôs)	88	18°44'N	101°04'W
Caraguatatuba, Braz. (kä-rä-gwä-tȧ-tōō'bä)	99a	23°37'S	45°26'W
Carajás, Serra dos, mts., Braz. (sĕ'r-rä-dôs-kä-rä-zhå's)	101	5°58'S	51°45'W
Caramanta, Cerro, mtn., Col. (sĕ'r-rō-kä-rä-mä'n-tä)	100a	5°29'N	76°01'W
Carangola, Braz. (kä-rán'gô'lä)	99a	20°46'S	42°02'W
Caraquet, Can. (kä-rä-kĕt')	51	47°48'N	64°57'W
Carata, Laguna, l., Nic. (lä-gô'nä-kä-rä'tä)	91	13°59'N	83°41'W
Caratasca, Laguna, l., Hond. (lä-gô'nä-kä-rä-täs'kä)	91	15°20'N	83°45'W
Caravaca, Spain (kä-rä-vä'kä)	128	38°05'N	1°51'W
Caravelas, Braz. (kȧ-rä-vĕl'äzh)	101	17°46'S	39°06'W
Carayaca, Ven. (kȧ-rä-ïă'kä)	101b	10°32'N	67°07'W
Caràzinho, Braz. (kä-rä'zē-nyō)	102	28°22'S	52°33'W
Carballiño, Spain	118	42°26'N	8°04'W
Carballo, Spain (kär-bäl'yō)	128	43°13'N	8°40'W

PLACE (Pronunciation)	PAGE	LAT.	LONG.
Carbet, Pitons du, mtn., Mart.	91b	14°40'N	61°05'W
Carbon, r., Wa., U.S. (kär'bŏn)	74a	47°06'N	122°08'W
Carbonado, Wa., U.S. (kȧr-bō-nā'dō)	74a	47°05'N	122°03'W
Carbonara, Cape, c., Italy (kär-bō-nä'rä)	118	39°08'N	9°33'E
Carbondale, Can. (kär'bŏn-dāl)	62g	53°45'N	113°32'W
Carbondale, Il., U.S.	66	37°42'N	89°12'W
Carbondale, Pa., U.S.	67	41°35'N	75°30'W
Carbonear, Can. (kär-bō-nēr')	61	47°45'N	53°14'W
Carbon Hill, Al., U.S. (kär'bŏn hĭl)	82	33°53'N	87°34'W
Carcaixent, Spain	129	39°09'N	0°29'W
Carcans, Étang de, l., Fr. (ā-taN-dĕ-kär-kän)	126	45°12'N	1°00'W
Carcassonne, Fr. (kȧr-kȧ-sôn')	117	43°12'N	2°23'E
Carcross, Can. (kär'krôs)	50	60°18'N	134°54'W
Cárdenas, Cuba (kär'dā-näs)	87	23°00'N	81°10'W
Cárdenas, Mex. (kȧ'r-dĕ-näs)	89	17°59'N	93°23'W
Cárdenas, Mex.	88	22°01'N	99°38'W
Cárdenas, Bahía de, b., Cuba (bä-ē'ä-dĕ-kär'dā-näs)	92	23°10'N	81°10'W
Cardiff, Can. (kär'dĭf)	62g	53°46'N	113°36'W
Cardiff, Wales, U.K.	117	51°30'N	3°18'W
Cardigan, Wales, U.K. (kär'dĭ-găn)	117	52°05'N	4°40'W
Cardigan Bay, b., Wales, U.K.	117	52°35'N	4°40'W
Cardston, Can. (kärds'tŭn)	50	49°12'N	113°18'W
Carei, Rom. (kä-rē')	125	47°42'N	22°28'E
Carentan, Fr. (kä-rôN-täN')	126	49°19'N	1°14'W
Carey, Oh., U.S. (kā'rē)	66	40°55'N	83°25'W
Carey, l., Austl. (kâr'ē)	174	29°20'S	123°35'E
Carhaix-Plouguer, Fr. (kär-ĕ')	126	48°17'N	3°37'W
Caribbean Sea, sea (kăr-ĭ-bē'ăn)	87	14°30'N	75°30'W
Caribe, Arroyo, r., Mex. (är-ro'ï-kä-rē'bē)	89	18°18'N	90°38'W
Cariboo Mountains, mts., Can. (kă'rĭ-bōō)	52	53°00'N	121°00'W
Caribou, Me., U.S.	60	46°51'N	68°01'W
Caribou, i., Can.	58	47°22'N	85°42'W
Caribou Lake, l., Mn., U.S.	75h	46°54'N	92°16'W
Caribou Mountains, mts., Can.	52	59°20'N	115°30'W
Carinhanha, Braz. (kä-rĭ-nyän'yä)	101	14°14'S	43°44'W
Carini, Italy (kä-rē'nē)	130	38°09'N	13°10'E
Carleton Place, Can. (kärl'tŭn)	59	45°15'N	76°10'W
Carletonville, S. Afr.	192c	26°20'S	27°23'E
Carlinville, Il., U.S. (kär'lĭn-vĭl)	79	39°16'N	89°52'W
Carlisle, Eng., U.K. (kär-līl')	110	54°54'N	3°03'W
Carlisle, Ky., U.S.	66	38°20'N	84°00'W
Carlisle, Pa., U.S.	67	40°10'N	77°15'W
Carloforte, Italy (kär'lō-fôr-tȧ)	130	39°11'N	8°28'E
Carlos Casares, Arg. (kär-lôs-kä-sȧ'rĕs)	102	35°38'S	61°17'W
Carlow, Ire. (kär'lō)	120	52°50'N	7°00'W
Carlsbad, N.M., U.S. (kärlz'băd)	80	32°24'N	104°12'W
Carlsbad Caverns National Park, rec., N.M., U.S.	80	32°08'N	104°30'W
Carlton, Eng., U.K. (kärl'tŭn)	114a	52°58'N	1°05'W
Carlton, Mn., U.S.	75h	46°40'N	92°26'W
Carlton Center, Mi., U.S. (kärl'tŭn sĕn'tēr)	66	42°45'N	85°20'W
Carlyle, Il., U.S. (kärlīl')	79	38°37'N	89°23'W
Carmagnola, Italy (kär-mä-nyô'lä)	130	44°52'N	7°48'E
Carman, Can. (kär'mȧn)	50	49°32'N	98°00'W
Carmarthen, Wales, U.K. (kär-mär'thĕn)	120	51°50'N	4°20'W
Carmaux, Fr. (kȧr-mō')	126	44°05'N	2°09'E
Carmel, N.Y., U.S. (kär'mĕl)	68a	41°25'N	73°42'W
Carmelo, Ur. (kär-mĕ'lo)	99c	33°59'S	58°15'W
Carmen, Isla del, i., Mex. (ē's-lä-dĕl-kȧ'r-mĕn)	89	18°43'N	91°40'W
Carmen, Laguna del, l., Mex. (lä-gô'nä-dĕl-kȧ'r-mĕn)	89	18°15'N	93°26'W
Carmen de Areco, Arg. (kär'mĕn' dā ä-rā'kō)	99c	34°21'S	59°50'W
Carmen de Patagones, Arg. (kȧ'r-mĕn-dĕ-pä-tä-gô'nĕs)	102	41°00'S	63°00'W
Carmi, Il., U.S. (kär'mī)	66	38°05'N	88°10'W
Carmo, Braz. (kȧ'r-mô)	99a	21°57'S	42°45'W
Carmo do Rio Clara, Braz. (kȧ'r-mô-dô-rē'ō-klä'rä)	99a	20°57'S	46°04'W
Carmona, Spain	128	37°28'N	5°38'W
Carnarvon, Austl. (kär-när'vŭn)	174	24°45'S	113°45'E
Carnarvon, S. Afr.	186	31°00'S	22°15'E
Carnation, Wa., U.S. (kär-nā'shŭn)	74a	47°39'N	121°55'W
Carnaxide, Port. (kär-nä-shē'dĕ)	129b	38°44'N	9°15'W
Carndonagh, Ire. (kärn-dō-nä')	120	55°15'N	7°15'W
Carnegie, Ok., U.S. (kär-nĕg'ĭ)	78	35°06'N	98°38'W
Carnegie, Pa., U.S.	69e	40°24'N	80°06'W
Carneys Point, N.J., U.S. (kär'nēs)	67	39°45'N	75°25'W
Carnic Alps, mts., Eur.	117	46°43'N	12°38'E
Carnot, Alg. (kär nō')	129	36°15'N	1°40'E
Carnot, C.A.R.	185	5°00'N	15°52'E
Carnsore Point, c., Ire. (kärn'sôr)	120	52°10'N	6°16'W
Caro, Mi., U.S. (kȧ'rō)	66	43°30'N	83°25'W
Carolina, Braz. (kä-rô-lē'nä)	101	7°26'S	47°16'W
Carolina, S. Afr. (kăr-ō-lī'nȧ)	186	26°07'S	30°09'E
Carolina, l., Mex. (kä-rō-lē'nä)	90a	18°41'N	89°40'W
Caroline Islands, is., Oc.	5	8°00'N	140°00'E
Caroni, r., Ven. (kä-rō'nē)	100	5°49'N	62°57'W
Carora, Ven. (kä-rō'rä)	100	10°09'N	70°12'W
Carpathians, mts., Eur. (kär-pā'thĭ-ăn)	112	49°23'N	20°14'E
Carpaţii Meridionali (Transylvanian Alps), mts., Rom.	112	45°30'N	23°30'E
Carpentaria, Gulf of, b., Austl. (kär-pĕn-târ'ĭȧ)	174	14°45'S	138°50'E
Carpentras, Fr. (kär-päN-träs')	127	44°04'N	5°01'E
Carpi, Italy	130	44°48'N	10°54'E
Carrara, Italy (kä-rä'rä)	118	44°05'N	10°05'E
Carrauntoohil, Ire. (kă-răn-tōō'ĭl)	120	52°01'N	9°48'W

PLACE (Pronunciation)	PAGE	LAT.	LONG.
Carretas, Punta, c., Peru (pōō'n-tä-kär-rĕ'tĕ'räs)	100	14°15'S	76°25'W
Carriacou, i., Gren.	91b	12°28'N	61°20'W
Carrick-on-Sur, Ire. (kăr'-ĭk)	120	52°20'N	7°35'W
Carrier, Can. (kăr'ĭ-ēr)	62b	46°43'N	71°05'W
Carriere, Ms., U.S. (kȧ-rēr')	82	30°37'N	89°37'W
Carriers Mills, Il., U.S. (kăr'ĭ-ērs)	66	37°40'N	88°40'W
Carrington, N.D., U.S. (kăr'ĭng-tŭn)	70	47°26'N	99°06'W
Carr Inlet, Wa., U.S. (kăr ĭn'lĕt)	74a	47°20'N	122°42'W
Carrion Crow Harbor, b., Bah. (kăr'ĭŭn krō)	92	26°35'N	77°55'W
Carrión de los Condes, Spain (kär-rē-ōn' dā lōs kōn'dās)	128	42°20'N	4°35'W
Carrizo Creek, r., N.M., U.S. (kär-rē'zō)	78	36°22'N	103°39'W
Carrizo Springs, Tx., U.S.	80	28°32'N	99°51'W
Carrizozo, N.M., U.S. (kär-rē-zō'zō)	77	33°40'N	105°55'W
Carroll, Ia., U.S. (kăr'ŭl)	71	42°03'N	94°51'W
Carrollton, Ga., U.S. (kăr-ŭl-tŭn)	82	33°35'N	85°05'W
Carrollton, Il., U.S.	79	39°18'N	90°22'W
Carrollton, Ky., U.S.	66	38°45'N	85°15'W
Carrollton, Mi., U.S.	66	43°30'N	83°55'W
Carrollton, Mo., U.S.	79	39°21'N	93°29'W
Carrollton, Oh., U.S.	66	40°35'N	81°10'W
Carrollton, Tx., U.S.	75c	32°58'N	96°53'W
Carrols, Wa., U.S. (kăr'ŭlz)	74c	46°05'N	122°51'W
Carrot, r., Can.	56	53°12'N	103°50'W
Carry-le-Rouet, Fr. (kä-rē'lē-rōō-ā')	126a	43°20'N	5°10'E
Carsamba, Tur. (chär-shäm'bä)	119	41°05'N	36°40'E
Carson, r., Nv., U.S. (kär'sŭn)	76	39°15'N	119°25'W
Carson City, Nv., U.S.	64	39°10'N	119°45'W
Carson Sink, Nv., U.S.	76	39°51'N	118°25'W
Cartagena, Col. (kär-tä-hā'nä)	100	10°30'N	75°40'W
Cartagena, Spain (kär-tä-ĸĕ'nä)	110	37°46'N	1°00'W
Cartago, Col. (kär-tä'gō)	100a	4°44'N	75°54'W
Cartago, C.R.	87	9°52'N	83°56'W
Cartaxo, Port. (kär-tä'shō)	128	39°10'N	8°48'W
Carteret, N.J., U.S. (kär'tē-ret)	68a	40°35'N	74°13'W
Cartersville, Ga., U.S. (kär'tērs-vĭl)	82	34°09'N	84°47'W
Carthage, Tun.	184	37°04'N	10°18'E
Carthage, Il., U.S. (kär'thāj)	79	40°27'N	91°09'W
Carthage, Mo., U.S.	79	37°10'N	94°18'W
Carthage, N.C., U.S.	83	35°22'N	79°25'W
Carthage, N.Y., U.S.	67	44°00'N	75°45'W
Carthage, Tx., U.S.	81	32°09'N	94°20'W
Carthcart, S. Afr. (cärth-cȧ't)	187c	32°18'S	27°11'E
Cartwright, Can. (kärt'rīt)	51	53°36'N	57°00'W
Caruaru, Braz. (kä-rô-ä-rōō')	101	8°19'S	35°52'W
Carúpano, Ven. (kä-rōō'pä-nō)	100	10°45'N	63°21'W
Caruthersville, Mo., U.S. (kȧ-rŭdh'ērz-vĭl)	79	36°09'N	89°41'W
Carver, Or., U.S. (kärv'ēr)	74c	45°24'N	122°30'W
Carvoeiro, Cabo, c., Port. (kȧ'bō-kär-vô-ē'y-rō)	128	39°22'N	9°24'W
Cary, Il., U.S. (kā'rē)	69a	42°13'N	88°14'W
Casablanca, Chile (kä-sä-bläŋ'kä)	99b	33°19'S	71°24'W
Casablanca, Mor.	184	33°32'N	7°41'W
Casa Branca, Braz. (kȧ'sä-brä'N-kä)	99a	21°47'S	47°04'W
Casa Grande, Az., U.S. (kä'sä grän'dā)	77	32°50'N	111°45'W
Casa Grande Ruins National Monument, rec., Az., U.S.	77	33°00'N	111°33'W
Casale Monferrato, Italy (kä-sä'lā)	130	45°08'N	8°26'E
Casalmaggiore, Italy (kä-säl-mäd-jō'rā)	130	45°00'N	10°24'E
Casamance, r., Sen. (kä-sä-mäNs')	184	12°30'N	15°00'W
Cascade Mountains, mts., N.A.	55	49°10'N	121°00'W
Cascade Point, c., N.Z. (kăs-kād')	175a	43°59'S	168°23'E
Cascade Range, mts., N.A.	64	42°50'N	122°20'W
Cascade-Siskiyou National Monument, rec., Or., U.S.	72	42°05'N	122°30'W
Cascade Tunnel, trans., Wa., U.S.	72	47°41'N	120°53'W
Cascais, Port. (käs-kȧ-ēzh)	128	38°42'N	9°25'W
Case Inlet, Wa., U.S. (kās)	74a	47°22'N	122°47'W
Caseros, Arg. (kä-sā'rôs)	102a	34°35'S	58°34'W
Caserta, Italy (kä-zĕr'tä)	130	41°04'N	14°21'E
Casey, Il., U.S. (kā'sĭ)	66	39°20'N	88°00'W
Cashmere, Wa., U.S. (kăsh'mĭr)	72	47°30'N	120°28'W
Casiguran, Phil. (käs-sē-gōō'rän)	169a	16°15'N	122°10'E
Casiguran Sound, strt., Phil.	169a	16°02'N	121°51'E
Casilda, Arg. (kä-sē'l-dä)	102	33°02'S	61°11'W
Casilda, Cuba	92	21°50'N	80°00'W
Casimiro de Abreu, Braz. (kä'sē-mē'ro-dē-ä-brĕ'ōō)	99a	22°30'S	42°11'W
Casino, Austl. (kȧ-sē'nō)	176	28°35'S	153°10'E
Casiquiare, r., Ven. (kä-sē-kyä'rä)	100	2°11'N	66°15'W
Caspe, Spain (käs'på)	129	41°18'N	0°02'W
Casper, Wy., U.S. (kăs'pēr)	64	42°51'N	106°18'W
Caspian Depression, depr., (kăs'pĭ-ăn)	134	47°40'N	52°35'E
Caspian Sea, sea	134	40°00'N	52°00'E
Cass, W.V., U.S. (kăs)	67	38°25'N	79°55'W
Cass, l., Mn., U.S.	71	47°23'N	94°28'W
Cassai (Kasai), r., Afr. (kä-sä'ē)	186	11°30'S	21°00'E
Cass City, Mi., U.S. (kăs)	66	43°35'N	83°10'W
Casselman, Can. (kăs''l-mȧn)	62c	45°18'N	75°05'W
Casselton, N.D., U.S. (kăs''l-tŭn)	70	46°53'N	97°14'W
Cássia, Braz. (kȧ'syä)	99a	20°36'S	46°53'W
Cassin, Tx., U.S. (kăs'ĭn)	75d	29°16'N	98°29'W
Cassinga, Ang.	186	15°05'S	16°15'E
Cassino, Italy (käs-sē'nō)	118	41°30'N	13°50'E
Cass Lake, Mn., U.S. (kăs)	71	47°23'N	94°37'W
Cassopolis, Mi., U.S. (kăs-ŏ'pō-lĭs)	66	41°55'N	86°00'W
Cassville, Mo., U.S. (kăs'vĭl)	79	36°41'N	93°52'W
Castanheira de Pêra, Port. (käs-tän-yā'rä-dĕ-pē'rä)	128	40°00'N	8°07'W
Castellammare di Stabia, Italy	129c	40°26'N	14°29'E

ng-sing; ŋ-baŋk; N-nasalized n; nŏd; cŏmmit; ōld; ōbey; ôrder; oi-boil; fōōd; ŏŏ-as oo in foot; ou-out; s-soft; sh-dish; th-thin; pūre; ûnite; ûrn; stŭd; circŭs; ü-as in French tu; '-indeterminate vowel.

234 Castelli - Chame

PRONOUNCING INDEX

PLACE (Pronunciation)	PAGE	LAT.	LONG.
Castelli, Arg. (kàs-tĕ'zhē)	99c	36°07'S	57°48'W
Castelló de la Plana, Spain	118	39°59'N	0°05'W
Castelnaudary, Fr. (kȧs'tĕl-nō-dȧ-rē')	126	43°20'N	1°57'E
Castelo, Braz. (käs-tĕ'lô)	99a	20°37'S	41°13'W
Castelo Branco, Port. (käs-tā'lò brän'kò)	118	39°48'N	7°37'W
Castelo de Vide, Port. (käs-tā'lò dī vē'dī)	128	39°25'N	7°25'W
Castelsarrasin, Fr. (kȧs'tĕl-sȧ-rȧ-zăN')	126	44°03'N	1°05'E
Castelvetrano, Italy (käs'tĕl-vē-trä'nō)	130	37°43'N	12°50'E
Castilla, Peru (kȧs-tē'l-yä)	100	5°18'S	80°40'W
Castilla La Nueva, hist. reg., Spain (käs-tē'lyä lä nwä'vä)	128	39°15'N	3°55'W
Castilla La Vieja, hist. reg., Spain (käs-tēl'yä lä vyä'hä)	128	40°48'N	4°24'W
Castillo de San Marcos National Monument, rec., Fl., U.S. (käs-tē'lyä de-sän mär-kōs)	83	29°55'N	81°25'W
Castle, i., Bah. (kås''l)	93	22°05'N	74°20'W
Castlebar, Ire. (kås''l-bär)	120	53°55'N	9°15'W
Castle Dale, Ut., U.S. (kås'l dāl)	77	39°15'N	111°00'W
Castle Donington, Eng., U.K. (dŏn'ĭng-tŭn)	114a	52°50'N	1°21'W
Castleford, Eng., U.K. (kås'l-fērd)	114a	53°43'N	1°21'W
Castlegar, Can. (kås''l-gär)	55	49°19'N	117°40'W
Castlemaine, Austl. (kăs''l-mān)	176	37°05'S	144°10'E
Castle Peak, mtn., Co., U.S.	77	39°00'N	106°50'W
Castle Rock, Wa., U.S. (kȧs''l-rŏk)	72	46°17'N	122°53'W
Castle Rock Flowage, res., Wi., U.S.	71	44°03'N	89°48'W
Castle Shannon, Pa., U.S. (shăn'ŭn)	69e	40°22'N	80°02'W
Castleton, In., U.S. (kȧs''l-tŭn)	69g	39°54'N	86°03'W
Castor, r., Can. (kăs'tôr)	62c	45°16'N	75°14'W
Castor, r., Mo., U.S.	79	36°59'N	89°53'W
Castres, Fr. (kȧs'tr')	126	43°36'N	2°13'E
Castries, St. Luc. (kȧs-trē')	91b	14°01'N	61°00'W
Castro, Braz. (käs'trò)	101	24°56'S	50°00'W
Castro, Chile (käs'tro)	102	42°27'S	73°48'W
Castro Daire, Port. (käs'trò dīr'ĭ)	128	40°56'N	7°57'W
Castro del Río, Spain (käs-trō-dĕl rē'ō)	128	37°42'N	4°28'W
Castrop Rauxel, Ger. (käs'trōp rou'ksĕl)	127c	51°33'N	7°19'E
Castro-Urdiales, Spain	118	43°23'N	3°11'W
Castro Valley, Ca., U.S.	74b	37°42'N	122°05'W
Castro Verde, Port. (käs-trō vĕr'dĕ)	128	37°43'N	8°05'W
Castrovillari, Italy (käs'trō-vēl-lyä'rē)	130	39°48'N	16°11'E
Castuera, Spain (käs-tò-ā'rä)	128	38°43'N	5°33'W
Casula, Moz.	191	15°25'S	33°40'E
Cat, i., Bah.	93	24°30'N	75°30'W
Catacamas, Hond. (kä-tä-kȧ'mäs)	90	14°52'N	85°55'W
Cataguases, Braz. (kä-tä-gwȧ'sĕs)	99a	21°23'S	42°42'W
Catahoula, l., La., U.S. (kăt-ȧ-hò'lȧ)	81	31°35'N	92°20'W
Catalão, Braz. (kä-tä-louN')	101	18°09'S	47°42'W
Catalina, i., Dom. Rep. (kä-tä-lē'nä)	93	18°20'N	69°00'W
Catalunya, hist. reg., Spain	129	41°23'N	0°50'E
Catamarca, Arg. (kä-tä-mä'r-kä)	102	28°29'S	65°45'W
Catamarca, prov., Arg. (kä-tä-mär'kä)	102	27°15'S	67°15'W
Catanaun, Phil. (kä-tä-nä'wän)	169a	13°36'N	122°20'E
Catanduanes Island, i., Phil. (kä-tän-dwä'nĕs)	169	13°55'N	125°00'E
Catanduva, Braz. (kä-tän-doo'vä)	101	21°12'S	48°47'W
Catania, Italy (kä-tä'nyä)	110	37°30'N	15°09'E
Catania, Golfo di, b., Italy (gôl-fô-dē-kȧ-tä'nyä)	130	37°24'N	15°28'E
Catanzaro, Italy (kä-tän-dzä'rō)	119	38°53'N	16°34'E
Catarroja, Spain (kä-tär-rō'hä)	129	39°24'N	0°25'W
Catawba, r., N.C., U.S. (kȧ-tô'bȧ)	83	35°25'N	80°55'W
Catbalogan, Phil. (kät-bä-lō'gän)	169	11°45'N	124°52'E
Catemaco, Mex. (kä-tä-mä'kō)	89	18°26'N	95°06'W
Catemaco, Lago, l., Mex. (lä'gô-kä-tä-mä'kō)	89	18°23'N	95°04'W
Caterham, Eng., U.K. (kā'tēr-ŭm)	114b	51°16'N	0°04'W
Catete, Ang. (kä-tĕ'tĕ)	186	9°06'S	13°43'E
Cathedral Mountain, mtn., Tx., U.S. (kȧ-thē'drȧl)	80	30°09'N	103°46'W
Cathedral Peak, mtn., Afr. (kȧ-thē'drȧl)	187c	28°53'S	29°04'E
Catherine, Lake, l., Ar., U.S. (kă-thēr-ĭn)	79	34°26'N	92°47'W
Cathkin Peak, mtn., Afr. (kăth'kĭn)	186	29°08'S	29°22'E
Cathlamet, Wa., U.S. (kăth-lăm'ĕt)	74c	46°12'N	123°22'W
Catlettsburg, Ky., U.S. (kăt'lĕts-bûrg)	66	38°20'N	82°35'W
Catoche, Cabo, c., Mex. (kä-tō'chĕ)	86	21°30'N	87°15'W
Catonsville, Md., U.S. (kā'tŭnz-vĭl)	68e	39°16'N	76°45'W
Catorce, Mex. (kä-tôr'sȧ)	88	23°41'N	100°51'W
Catskill, N.Y., U.S. (kăts'kĭl)	67	42°15'N	73°50'W
Catskill Mountains, mts., N.Y., U.S.	65	42°20'N	74°35'W
Cattaraugus Indian Reservation, I.R., N.Y., U.S. (kăt'tȧ-rȧ-gŭs)	67	42°30'N	79°05'W
Catu, Braz. (kȧ-too)	101	12°26'S	38°12'W
Catuala, Ang.	190	16°29'S	19°03'E
Catumbela, r., Ang. (kä'tŏm-bĕl'ȧ)	190	12°40'S	14°10'E
Cauayan, Phil. (kou-ä'yän)	169a	16°56'N	121°46'E
Cauca, r., Col. (kou'kä)	100	7°30'N	75°26'W
Caucagua, Ven. (käò-kä'gwä)	101b	10°17'N	66°22'W
Caucasus, mts.	134	43°20'N	42°00'E
Cauchon Lake, l., Can. (kô-shŏn')	57	55°25'N	96°30'W
Caughnawaga, Can.	62a	45°24'N	73°41'W
Caulfield, Austl.	173a	37°53'S	145°03'E
Caulonia, Italy (kou-lō'nyä)	130	38°24'N	16°22'E
Cauquenes, Chile (kou-kā'nās)	102	35°54'S	72°14'W
Caura, r., Ven. (kou'rä)	100	6°48'N	64°40'W
Causapscal, Can.	60	48°22'N	67°14'W
Caution, Cape, c., Can. (kô'shŭn)	54	51°10'N	127°47'W
Cauto, r., Cuba (kou'tō)	92	20°33'N	76°20'W
Cauvery, r., India	155	12°00'N	77°00'E
Cava, Braz. (kȧ'vä)	102b	22°41'S	43°26'W
Cava de' Tirreni, Italy (kä'vä-dē-tēr-rē'nē)	129c	40°27'N	14°43'E
Cávado, r., Port. (kä-vä'dō)	128	41°43'N	8°08'W
Cavalcante, Braz. (kä-väl-kän'tā)	101	13°45'S	47°33'W
Cavalier, N.D., U.S. (kăv-ȧ-lēr')	70	48°45'N	97°39'W
Cavally, r., Afr.	188	4°40'N	7°30'W
Cavan, Ire. (kăv'ăn)	120	54°01'N	7°00'W
Cavarzere, Italy (kä-vär'dzā-rā)	130	45°08'N	12°06'E
Cavendish, Vt., U.S. (kăv'ĕn-dĭsh)	67	43°25'N	72°35'W
Caviana, Ilha, i., Braz. (kä-vyä'nä)	101	0°45'N	49°33'W
Cavite, Phil. (kä-vē'tā)	169a	14°30'N	120°54'E
Caxambu, Braz. (kä-shä'm-boo)	101	22°00'S	44°45'W
Caxias, Braz. (kä'shē-äzh)	101	4°48'S	43°16'W
Caxias do Sul, Braz. (kä'shē-äzh-dò-sool)	102	29°13'S	51°03'W
Caxito, Ang. (kä-shē'tō)	186	8°33'S	13°36'E
Cayambe, Ec. (kȧ-īä'm-bĕ)	100	0°03'N	79°09'W
Cayenne, Fr. Gu. (kā-ĕn')	101	4°56'N	52°18'W
Cayetano Rubio, Mex. (kä-yĕ-tä-nò-roo'byô)	88	20°37'N	100°21'W
Cayey, P.R.	87b	18°05'N	66°12'W
Cayman Brac, i., Cay. Is. (kī-män' bräk)	92	19°45'N	79°50'W
Cayman Islands, dep., N.A.	92	19°30'N	80°30'W
Cay Sal Bank, bk., (kē-sȧl)	92	23°55'N	80°20'W
Cayuga, l., N.Y., U.S. (kā-yoo'gȧ)	67	42°35'N	76°35'W
Cazalla de la Sierra, Spain	128	37°55'N	5°48'W
Cazaux, Étang de, l., Fr. (ā-tăN' dĕ kä-zō')	126	44°32'N	0°59'W
Cazenovia, N.Y., U.S. (kăz-ē-nō'vĭ-ȧ)	67	42°55'N	75°50'W
Çazenovia Creek, r., N.Y., U.S.	69c	42°49'N	78°45'W
Čazma, Cro. (chäz'mä)	130	45°44'N	16°39'E
Cazombo, Ang. (kä-zô'm-bō)	186	11°54'S	22°52'E
Cazones, r., Mex. (kä-zō'nĕs)	89	20°37'N	97°28'W
Cazones, Ensenada de, b., Cuba (ĕn-sĕ-nä-dä-dĕ-kä-zō'nås)	92	22°05'N	81°30'W
Cazones, Golfo de, b., Cuba (gôl-fô-dĕ-kä-zō'nås)	92	21°55'N	81°15'W
Cazorla, Spain (kä-thôr'lä)	128	37°55'N	2°58'W
Cea, r., Spain (thā'ä)	128	42°18'N	5°10'W
Ceará-Mirim, Braz. (sā-ä-rä'mē-rē'N)	101	6°00'S	35°13'W
Cebaco, Isla, i., Pan. (ē's-lä-sȧ-bä'kō)	91	7°27'N	81°08'W
Cebolla Creek, r., Co., U.S. (sē-bōl'yä)	77	38°15'N	107°10'W
Cebreros, Spain (sĕ-brĕ'rôs)	128	40°28'N	4°28'W
Cebu, Phil. (sā-boo')	169	10°22'N	123°49'E
Čechy (Bohemia), hist. reg., Czech Rep.	124	49°51'N	13°55'E
Cecil, Pa., U.S. (sē'sĭl)	69e	40°20'N	80°10'W
Cedar, r., Ia., U.S.	71	42°23'N	92°07'W
Cedar, r., Wa., U.S.	74c	45°56'N	122°32'W
Cedar, West Fork, r., Ia., U.S.	71	42°49'N	93°10'W
Cedar Bayou, Tx., U.S.	81a	29°54'N	94°58'W
Cedar Breaks National Monument, rec., Ut., U.S.	77	37°35'N	112°55'W
Cedarburg, Wi., U.S. (sē'dēr bûrg)	71	43°23'N	88°00'W
Cedar City, Ut., U.S.	77	37°40'N	113°10'W
Cedar Creek, r., N.D., U.S.	70	46°05'N	102°10'W
Cedar Falls, Ia., U.S.	71	42°31'N	92°29'W
Cedar Keys, Fl., U.S.	82	29°06'N	83°03'W
Cedar Lake, In., U.S.	69a	41°22'N	87°27'W
Cedar Lake, l., In., U.S.	69a	41°23'N	87°25'W
Cedar Lake, res., Can.	52	53°10'N	100°00'W
Cedar Rapids, Ia., U.S.	65	42°00'N	91°43'W
Cedar Springs, Mi., U.S.	66	43°15'N	85°40'W
Cedartown, Ga., U.S. (sē'dēr-toun)	82	34°00'N	85°15'W
Cedarville, S. Afr. (cĕdȧr'vĭl)	187c	30°23'S	29°04'E
Cedral, Mex. (sā-dräl')	88	23°47'N	100°42'W
Cedros, Hond. (sā'drōs)	90	14°36'N	87°07'W
Cedros, i., Mex.	86	28°10'N	115°10'W
Ceduna, Austl. (sē-dò'nȧ)	174	32°15'S	133°55'E
Ceel Buur, Som.	192a	4°35'N	46°40'E
Cega, r., Spain (thā'gä)	128	41°25'N	4°27'W
Cegléd, Hung. (tsā'glād)	125	47°10'N	19°49'E
Ceglie, Italy (chĕ'lyĕ)	131	40°39'N	17°32'E
Cehegín, Spain (thā-å-hēn')	128	38°05'N	1°48'W
Ceiba del Agua, Cuba (sā'bä-dĕl-ā'gwä)	93a	22°53'N	82°38'W
Cekhira, Tun.	184	34°17'N	10°00'E
Celaya, Mex. (sā-lä'yä)	86	20°33'N	100°49'W
Celebes (Sulawesi), i., Indon.	168	2°15'S	120°30'E
Celebes Sea, sea, Asia	168	3°45'N	121°52'E
Celestún, Mex. (sĕ-lĕs-too'n)	90a	20°57'N	90°18'W
Celina, Oh., U.S. (sĕlī'nȧ)	66	40°30'N	84°35'W
Celje, Slvn. (tsĕl'yĕ)	130	46°13'N	15°17'E
Celle, Ger. (tsĕl'ē)	117	52°37'N	10°05'E
Cement, Ok., U.S. (sē-mĕnt')	78	34°56'N	98°07'W
Cenderawasih, Teluk, b., Indon.	169	2°20'S	135°30'E
Ceniza, Pico, mtn., Ven. (pē'kô-sē-nē'zä)	101b	10°24'N	67°26'W
Center, Tx., U.S. (sĕn'tēr)	81	31°50'N	94°10'W
Center Hill Lake, res., Tn., U.S. (sĕn'tēr-hĭl)	82	36°02'N	86°00'W
Center Line, Mi., U.S. (sĕn'tēr līn)	69b	42°29'N	83°01'W
Centerville, Ia., U.S. (sĕn'tēr-vĭl)	71	40°44'N	92°48'W
Centerville, Mn., U.S.	75g	45°10'N	93°03'W
Centerville, Pa., U.S.	69e	40°02'N	79°58'W
Centerville, S.D., U.S.	70	43°07'N	96°56'W
Centerville, Ut., U.S.	75b	40°55'N	111°53'W
Central, Cordillera, mts., Bol. (kôr-dēl-yĕ'rä-sĕn-trä'l)	100	19°18'S	65°29'W
Central, Cordillera, mts., Col.	100a	3°58'N	75°55'W
Central, Cordillera, mts., Dom. Rep.	93	19°05'N	71°30'W
Central, Cordillera, mts., Phil. (kôr-dēl-yĕ'rä-sĕn'träl)	169a	17°05'N	120°55'E
Central African Republic, nation, Afr.	185	7°50'N	21°00'E
Central America, reg., N.A. (ä-mĕr'ĭ-kȧ)	86	10°45'N	87°15'W
Central City, Ky., U.S. (sĕn'trȧl)	82	37°15'N	87°09'W
Central City, Ne., U.S. (sĕn'trȧl sĭ'tĭ)	70	41°07'N	98°00'W
Central Falls, R.I., U.S. (sĕn'trȧl fôlz)	68b	41°54'N	71°23'W
Centralia, Il., U.S. (sĕn-trā'lĭ-ȧ)	66	38°35'N	89°05'W
Centralia, Mo., U.S.	79	39°11'N	92°07'W
Centralia, Wa., U.S.	72	46°42'N	122°58'W
Central Plateau, plat., Russia	136	55°00'N	33°30'E
Central Valley, N.Y., U.S.	68a	41°19'N	74°07'W
Centreville, Il., U.S. (sĕn'tēr-vĭl)	75e	38°33'N	90°06'W
Centreville, Md., U.S.	67	39°05'N	76°05'W
Century, Fl., U.S. (sĕn'tū-rĭ)	82	30°57'N	87°15'W
Ceram (Seram), i., Indon.	169	2°45'S	129°30'E
Céret, Fr.	126	42°29'N	2°47'E
Cerignola, Italy (chā-rē-nyô'lä)	130	41°16'N	15°55'E
Cerknica, Slvn. (tsĕr'knē-tsȧ)	130	45°48'N	14°21'E
Cern'achovsk, Russia (chĕr-nyä'kôfsk)	136	54°38'N	21°49'E
Cerralvo, Mex. (sĕr-räl'vō)	80	26°05'N	99°37'W
Cerralvo, i., Mex.	86	24°00'N	109°59'W
Cerrito, Col. (sĕr-rē'tō)	100a	3°41'N	76°17'W
Cerritos, Mex. (sĕr-rē'tôs)	88	22°26'N	100°16'W
Cerro de Pasco, Peru (sĕr'rō dā päs'kō)	100	10°45'S	76°14'W
Cerro Gordo, Arroyo de, r., Mex. (är-rô-yō-dĕ-sĕ'r-rô-gôr-dō)	80	26°12'N	104°06'W
Certegui, Col. (sĕr-tĕ'gē)	100a	5°21'N	76°35'W
Cervantes, Phil. (sĕr-vän'tās)	169a	16°59'N	120°42'E
Cervera del Río Alhama, Spain	128	42°02'N	1°55'W
Cerveteri, Italy (chĕr-vĕ'tĕ-rē)	129d	42°00'N	12°06'E
Cesena, Italy (chĕ'sĕ-nä)	130	44°08'N	12°16'E
Çēsis, Lat. (sā'sĭs)	123	57°19'N	25°17'E
Česká Lípa, Czech Rep. (chĕs'kä lē'pa)	124	50°41'N	14°31'E
České Budějovice, Czech Rep. (chĕs'kā boo'dyĕ-yô-vēt-sĕ)	117	49°00'N	14°30'E
Českomoravská Vysočina, hills, Czech Rep.	124	49°21'N	15°40'E
Český Těšín, Czech Rep.	125	49°43'N	18°22'E
Çeşme, Tur. (chĕsh'mĕ)	131	38°20'N	26°20'E
Cessnock, Austl.	175	32°58'S	151°15'E
Cestos, r., Lib.	188	5°40'N	9°25'W
Cetinje, Serb. (tsĕt'in-yĕ)	110	42°23'N	18°55'E
Ceuta, Sp. N. Afr. (thā-oo'tä)	184	36°04'N	5°36'W
Cévennes, reg., Fr. (sā-vĕn')	117	44°20'N	3°48'E
Ceylon see Sri Lanka, nation, Asia	159	8°45'N	82°30'E
Chabot, Lake, l., Ca., U.S. (sha'bŏt)	74b	37°44'N	122°06'W
Chacabuco, Arg. (chä-kä-boo'kō)	99c	34°37'S	60°27'W
Chacaltianguis, Mex. (chä-käl-tē-äŋ'gwēs)	89	18°18'N	95°50'W
Chachapoyas, Peru (chä-chä-poi'yäs)	100	6°16'S	77°48'W
Chaco, prov., Arg. (chä'kō)	102	26°00'S	60°45'W
Chaco Culture National Historic Park, rec., N.M., U.S. (chä'kō)	77	36°05'N	108°00'W
Chad, Russia (chäd)	142a	56°33'N	57°11'E
Chad, nation, Afr.	185	17°48'N	19°00'E
Chad, Lake, l., Afr.	185	13°55'N	13°40'E
Chadbourn, N.C., U.S. (chăd'bŭrn)	83	34°19'N	78°55'W
Chadron, Ne., U.S. (chăd'rŭn)	64	42°50'N	103°10'W
Chafarinas, Islas, is., Sp. N. Afr.	128	35°08'N	2°20'W
Chaffee, Mo., U.S. (chăf'ē)	79	37°10'N	89°39'W
Chāgal Hills, hills, Afg.	154	29°15'N	63°28'E
Chagodoshcha, r., Russia (chä-gō-dôsh-chä)	132	59°08'N	35°13'E
Chagres, r., Pan. (chä'grĕs)	91	9°18'N	79°22'W
Chagrin, r., Oh., U.S. (shȧ'grĭn)	69d	41°34'N	81°24'W
Chagrin Falls, Oh., U.S. (shȧ'grĭn fôls)	69d	41°26'N	81°23'W
Chahar, hist. reg., China (chä-här)	161	44°25'N	115°00'E
Chake Chake, Tan.	191	5°15'S	39°46'E
Chalatenango, El Sal. (chäl-ä-tĕ-näŋ'gō)	90	14°04'N	88°54'W
Chalbi Desert, des., Kenya	191	3°40'N	36°50'E
Chalcatongo, Mex. (chäl-kä-tôŋ'gō)	89	17°04'N	97°41'W
Chalchihuites, Mex. (chäl-chē-wē'tås)	88	23°28'N	103°57'W
Chalchuapa, El Sal. (chäl-chwä'pä)	90	14°01'N	89°39'W
Chalco, Mex. (chäl-kō)	89a	19°15'N	98°54'W
Chaleur Bay, b., Can. (shȧ-lûr')	53	47°58'N	65°33'W
Chalgrove, Eng., U.K. (chăl'grôv)	114b	51°38'N	1°05'W
Chaling, China (chä'lĭng)	165	27°00'N	113°31'E
Chalkída, Grc.	119	38°28'N	23°38'E
Chalmette, La., U.S. (shăl-mĕt')	68d	29°57'N	89°57'W
Châlons-sur-Marne, Fr. (shȧ-lôN'sür-märn)	117	48°57'N	4°23'E
Chalon-sur-Saône, Fr.	117	46°47'N	4°54'E
Chaltel, Cerro (Monte Fitzroy), mtn., S.A. (sĕ'r-rô-chäl'tĕl)	102	48°10'S	73°18'W
Chālūs, Iran	157	36°38'N	51°26'E
Chama, Rio, r., N.M., U.S. (chä'mä)	77	36°19'N	106°31'W
Chama, Sierra de, mts., Guat. (sē-ĕ'r-rä-dĕ-chä-mä)	90	15°48'N	90°20'W
Chamama, Mwi.	191	12°55'S	33°43'E
Chaman, Pak. (chŭm-än')	155	30°58'N	66°21'E
Chambal, r., India (chŭm-bäl')	155	24°30'N	75°30'E
Chamberlain, S.D., U.S. (chām'bēr-lĭn)	70	43°48'N	99°21'W
Chamberlain, l., Me., U.S.	60	46°15'N	69°10'W
Chambersburg, Pa., U.S. (chām'bērz-bûrg)	67	40°00'N	77°40'W
Chambéry, Fr. (shäm-bā-rē')	117	45°35'N	5°54'E
Chambeshi, r., Zam.	191	10°35'S	31°20'E
Chamblee, Ga., U.S. (chăm-blē')	68c	33°55'N	84°18'W
Chambly, Can. (shäN-blē')	62a	45°27'N	73°17'W
Chambly, Fr.	127b	49°11'N	2°14'E
Chambord, Can.	51	48°22'N	72°01'W
Chame, Punta, c., Pan. (pô'n-tä-chä'mȧ)	91	8°41'N	79°27'W

ăt; finăl; rāte; senåte; ärm; ȧsk; sofȧ; fâre; ch-choose; dh-as th in other; bē; ēvent; bĕt; recĕnt; cratēr; g-gō; gh-guttural g; bĭt; ĭ-short neutral; rīde; K-guttural k as ch in German ich;

PRONOUNCING INDEX

Chamelecón – Chesnokovka 235

PLACE (Pronunciation)	PAGE	LAT.	LONG.
Chamelecón, r., Hond. (chä-mĕ-lĕ-kô'n)	90	15°09'N	88°42'W
Chamo, l., Eth.	185	5°58'N	37°00'E
Chamonix-Mont-Blanc, Fr. (shå-mō-nē')	127	45°55'N	6°50'E
Champagne, reg., Fr. (shäm-pän'yē)	126	48°53'N	4°48'E
Champaign, Il., U.S. (shăm-pān')	65	40°10'N	88°15'W
Champdāni, India	158a	22°48'N	88°21'E
Champerico, Guat. (chăm-pä-rē'kō)	90	14°18'N	91°55'W
Champion, Mi., U.S. (chăm'pĭ-ŭn)	71	46°30'N	87°59'W
Champlain, Lake, l., N.A. (shăm-plān')	65	44°45'N	73°20'W
Champlitte-et-le-Prálot, Fr. (shäɴ-plēt')	127	47°38'N	5°28'E
Champotón, Mex. (chäm-pō-tōn')	89	19°21'N	90°43'W
Champotón, r., Mex.	89	19°19'N	90°15'W
Chañaral, Chile (chän-yä-räl')	102	26°20'S	70°46'W
Chances Peak, vol., Monts.	91b	16°43'N	62°10'W
Chandeleur Islands, is., La., U.S. (shăn-dē-lo͞or')	82	29°53'N	88°35'W
Chandeleur Sound, strt., La., U.S.	82	29°47'N	89°08'W
Chandīgarh, India	155	30°51'N	77°13'E
Chandler, Can. (chăn'dlēr)	51	48°21'N	64°41'W
Chandler, Ok., U.S.	79	35°42'N	96°52'W
Chandrapur, India	155	19°58'N	79°21'E
Chang *see* Yangtze, r., China	161	30°30'N	117°25'E
Changane, r., Moz.	186	22°42'S	32°46'E
Changara, Moz.	191	16°54'S	33°14'E
Changchun, China (chäŋ-chòn)	161	43°55'N	125°25'E
Changdang Hu, l., China (chäŋ-däŋ ho͞o)	162	31°37'N	119°29'E
Changde, China (chäŋ-dŭ)	161	29°00'N	111°38'E
Changhua, Tai. (chäng'hwä')	165	24°02'N	120°32'E
Changjŏn, Kor., N. (chäng'jŭn')	166	38°40'N	128°05'E
Changli, China (chäŋ-lē)	164	39°46'N	119°10'E
Changning, China (chäŋ-nĭŋ)	160	24°34'N	99°49'E
Changping, China (chäŋ-pĭŋ)	164	40°12'N	116°10'E
Changqing, China (chäŋ-chyĭŋ)	162	36°33'N	116°42'E
Changsan Got, c., Kor., N.	166	38°06'N	124°50'E
Changsha, China (chäŋ-shä)	161	28°20'N	113°00'E
Changshan Qundao, is., China (chäŋ-shän chyòn-dou)	162	39°08'N	122°26'E
Changshu, China (chäŋ-sho͞o)	162	31°40'N	120°45'E
Changting, China	165	25°50'N	116°18'E
Changwu, China (chäng'wo͞o')	164	35°12'N	107°45'E
Changxindianzhen, China (chäŋ-shyĭn-dĭĕn-jŭn)	164a	39°49'N	116°12'E
Changxing Dao, i., China (chäŋ-shyĭŋ dou)	162	39°38'N	121°10'E
Changyi, China (chäŋ-yē)	162	36°51'N	119°23'E
Changyuan, China (chyäŋ-yuän)	162	35°10'N	114°41'E
Changzhi, China (chäŋ-jr)	164	35°58'N	112°58'E
Changzhou, China (chäŋ-jō)	161	31°47'N	119°56'E
Changzhuyuan, China (chäŋ-jo͞o-yuän)	162	31°33'N	115°17'E
Chanhassen, Mn., U.S. (shăn'hăs-sĕn)	75g	44°52'N	93°32'W
Chaniá, Grc.	118	35°31'N	24°01'E
Channel Islands, is., Eur. (chăn'ĕl)	112	49°15'N	3°30'W
Channel Islands, is., Ca., U.S.	76	33°30'N	119°15'W
Channel-Port-aux-Basques, Can.	51	47°35'N	59°11'W
Channelview, Tx., U.S. (chănĕlvū)	81a	29°46'N	95°07'W
Chantada, Spain (chän-tä'dä)	128	42°38'N	7°36'W
Chanthaburi, Thai.	168	12°37'N	102°04'E
Chantilly, Fr. (shäɴ-tē-yē')	127b	49°12'N	2°30'E
Chantilly, Va., U.S. (shăn'tĭlē)	68e	38°53'N	77°26'W
Chantrey Inlet, b., Can. (chăn-trē)	52	67°49'N	95°00'W
Chanute, Ks., U.S. (shȧ-no͞ot')	65	37°41'N	95°27'W
Chany, l., Russia (chä'nē)	134	54°15'N	77°31'E
Chao'an, China (chou-än)	161	23°48'N	116°35'E
Chao Hu, l., China	165	31°45'N	116°59'E
Chao Phraya, r., Thai.	168	16°13'N	99°33'E
Chaor, r., China (chou-r)	164	47°20'N	121°40'E
Chaoshui, China (chou-shwä)	162	37°43'N	120°56'E
Chaoxian, China (chou shyĕn)	162	31°37'N	117°50'E
Chaoyang, China	161	41°32'N	120°20'E
Chaoyang, China (chou-yäŋ)	165	23°18'N	116°32'E
Chapada, Serra da, mts., Braz. (sĕ'r-rä-dä-shä-pä'dä)	101	14°57'S	54°34'W
Chapadão, Serra do, mtn., Braz. (sĕ'r-rä-dô-shä-pȧ-dou'ɴ)	99a	20°31'S	46°20'W
Chapala, Mex. (chä-pä'lä)	88	20°18'N	103°10'W
Chapala, Lago de, l., Mex. (lä'gô-dē-chä-pä'lä)	86	20°14'N	103°02'W
Chapalagana, r., Mex. (chä-pä-lä-gȧ'nä)	88	22°11'N	104°09'W
Chaparral, Col. (chä-pär-rá'l)	100	3°44'N	75°28'W
Chapayevsk, Russia (châ-pī'ĕfsk)	136	53°00'N	49°30'E
Chapel Hill, N.C., U.S. (chăp''l hĭl)	83	35°55'N	79°05'W
Chaplain, l., Wa., U.S. (chăp'lĭn)	74a	47°58'N	121°50'W
Chapleau, Can. (chăp-lō')	51	47°43'N	83°28'W
Chapman, Mount, mtn., Can. (chăp'măn)	55	51°50'N	118°20'W
Chapman's Bay, b., S. Afr. (chăp'măns bā)	186a	34°06'S	18°17'E
Chappell, Ne., U.S. (chă-pĕl')	70	41°06'N	102°29'W
Chapultenango, Mex. (chä-pòl-tē-näŋ'gō)	89	17°19'N	93°08'W
Chá Pungana, Ang.	190	13°44'S	18°39'E
Chār Borjak, Afg.	157	30°17'N	62°03'E
Charcas, Mex. (chär'käs)	88	23°09'N	101°09'W
Charco de Azul, Bahía, b., Pan.	91	8°14'N	82°45'W
Charente, r., Fr. (shȧ-räɴt')	126	45°48'N	0°28'W
Chari, r., Afr. (shä-rē')	189	12°45'N	14°55'E
Charing, Eng., U.K. (chă'rĭng)	114b	51°13'N	0°49'E
Chariton, Ia., U.S. (chăr'ĭ-tŭn)	71	41°02'N	93°16'W
Chariton, r., Mo., U.S.	79	40°24'N	92°38'W

PLACE (Pronunciation)	PAGE	LAT.	LONG.
Charjew, Turkmen.	139	38°52'N	63°37'E
Charlemagne, Can. (shärl-mäny')	62a	45°43'N	73°29'W
Charleroi, Bel. (shȧr-lē-rwä')	117	50°25'N	4°35'E
Charleroi, Pa., U.S. (shär'lē-roi)	69e	40°08'N	79°54'W
Charles, Cape, c., Va., U.S. (chärlz)	67	37°05'N	75°48'W
Charlesbourg, Can. (shȧrl-bo͞or')	62b	46°51'N	71°16'W
Charles City, Ia., U.S. (chärlz)	71	43°03'N	92°40'W
Charleston, Il., U.S. (chärlz'tŭn)	66	39°30'N	88°10'W
Charleston, Mo., U.S.	79	36°53'N	89°20'W
Charleston, Ms., U.S.	82	34°00'N	90°02'W
Charleston, S.C., U.S.	65	32°47'N	79°56'W
Charleston, W.V., U.S.	65	38°20'N	81°35'W
Charlestown, St. K./N.	91b	17°10'N	62°32'W
Charlestown, In., U.S. (chärlz'toun)	69h	38°46'N	85°39'W
Charleville, Austl. (chär'lē-vĭl)	175	26°16'S	146°28'E
Charleville Mézières, Fr. (shärl-vēl')	126	49°48'N	4°41'E
Charlevoix, Mi., U.S. (shär'lē-voi)	66	45°20'N	85°15'W
Charlevoix, Lake, l., Mi., U.S.	71	45°17'N	85°43'W
Charlotte, Mi., U.S. (shär'lŏt)	66	42°35'N	84°50'W
Charlotte, N.C., U.S.	65	35°15'N	80°50'W
Charlotte Amalie, V.I.U.S. (shär-lŏt'ĕ ä-mä'lĭ-ȧ)	87	18°21'N	64°54'W
Charlotte Harbor, b., Fl., U.S.	83a	26°49'N	82°00'W
Charlotte Lake, l., Can.	54	52°07'N	125°30'W
Charlottenberg, Swe. (shär-lŭt'ĕn-bĕrg)	122	59°53'N	12°17'E
Charlottesville, Va., U.S. (shär'lŏtz-vĭl)	65	38°00'N	78°25'W
Charlottetown, Can. (shär'lŏt-toun)	51	46°14'N	63°08'W
Charlotte Waters, Austl. (shär'lŏt)	174	26°00'S	134°50'E
Charmes, Fr. (shärm)	127	48°23'N	6°19'E
Charnwood Forest, for., Eng., U.K. (chärn'wòd)	114a	52°42'N	1°15'W
Charny, Can. (shär-nē')	62b	46°43'N	71°16'W
Chars, Fr. (shär)	127b	49°09'N	1°57'E
Chārsadda, Pak. (chŭr-sä'dä)	155a	34°17'N	71°43'E
Charters Towers, Austl. (chär'tērz)	175	20°03'S	146°20'E
Chartres, Fr. (shärt'r')	117	48°26'N	1°29'E
Chascomús, Arg. (chäs-kō-mo͞os')	102	35°32'S	58°01'W
Chase City, Va., U.S. (chās)	83	36°45'N	78°27'W
Chashniki, Bela. (chäsh'nyē-kē)	132	54°51'N	29°08'E
Chaska, Mn., U.S. (chăs'kȧ)	75g	44°48'N	93°36'W
Châteaudun, Fr. (shä-tō-dăɴ')	126	48°04'N	1°23'E
Château-Gontier, Fr. (chȧ-tō'gôɴ'tyā')	126	47°48'N	0°43'W
Châteauguay, Can. (chȧ-tō-gā)	62a	45°22'N	73°45'W
Châteauguay, r., N.A.	62a	45°13'N	73°51'W
Châteauneuf, Fr.	126a	43°23'N	5°11'E
Château-Renault, Fr. (shȧ-tō-rĕ-nō')	126	47°36'N	0°57'E
Château-Richer, Can. (shȧ-tō-rē-shā')	62b	47°00'N	71°01'W
Châteauroux, Fr. (shȧ-tō-ro͞o')	117	46°47'N	1°39'E
Château-Thierry, Fr. (shȧ-tō'ty-ĕr-rē')	126	49°03'N	3°22'E
Châtellerault, Fr. (shȧ-tĕl-rō')	117	46°48'N	0°31'E
Chatfield, Mn., U.S. (chăt'fēld)	71	43°50'N	92°10'W
Chatham, Can. (chăt'ăm)	51	42°25'N	82°10'W
Chatham, Can.	51	47°02'N	65°28'W
Chatham, Eng., U.K. (chăt'ŭm)	121	51°23'N	0°32'E
Chatham, N.J., U.S. (chăt'ăm)	68a	40°44'N	74°23'W
Chatham, Oh., U.S.	69d	41°06'N	82°01'W
Chatham Islands, is., N.Z.	2	44°00'S	178°00'W
Chatham Sound, strt., Can.	54	54°32'N	130°35'W
Chatham Strait, strt., Ak., U.S.	63	57°00'N	134°40'W
Chatsworth, Ca., U.S. (chătz'wûrth)	75a	34°16'N	118°36'W
Chatsworth Reservoir, res., Ca., U.S.	75a	34°15'N	118°41'W
Chattahoochee, Fl., U.S. (chăt-tȧ-ho͞o' cheē)	82	30°42'N	84°47'W
Chattahoochee, r., U.S.	65	32°00'N	85°10'W
Chattanooga, Tn., U.S. (chăt-ȧ-no͞o'gȧ)	65	35°01'N	85°15'W
Chattooga, r., Ga., U.S. (chă-to͞o'gȧ)	82	34°47'N	83°13'W
Chaudière, r., Can. (shō-dyĕr')	59	46°26'N	71°10'W
Chaumont, Fr. (shō-môɴ')	117	48°08'N	5°07'E
Chaunskaya Guba, b., Russia	141	69°15'N	170°00'E
Chauny, Fr. (shō-nē')	126	49°40'N	3°09'E
Chau-phu, Viet.	168	10°49'N	104°57'E
Chautauqua, l., N.Y., U.S. (shȧ-tô'kwȧ)	67	42°10'N	79°25'W
Chavaniga, Russia	136	66°02'N	37°50'E
Chaves, Port. (chä'vĕzh)	128	41°44'N	7°30'W
Chavinda, Mex. (chä-vē'n-dä)	88	20°01'N	102°27'W
Chavusi, Bela.	132	53°57'N	30°58'E
Chazumba, Mex. (chä-zòm'bä)	89	18°11'N	97°41'W
Cheadle, Eng., U.K. (chē'd'l)	114a	52°59'N	1°59'W
Cheat, W.V., U.S. (chēt)	67	39°35'N	79°40'W
Cheb, Czech Rep. (kĕb)	124	50°05'N	12°23'E
Chebarkul', Russia (chĕ-bár-kûl')	142a	54°59'N	60°22'E
Cheboksary, Russia (chyĕ-bôk-sä'rĕ)	136	56°00'N	47°20'E
Cheboygan, Mi., U.S. (shē-boi'găn)	66	45°40'N	84°30'W
Chech, Erg, des., Alg.	184	24°45'N	2°07'W
Chechen', i., Russia (chyĕch'ĕn)	137	44°00'N	48°10'E
Chechnya, prov., Russia	138	43°30'N	45°50'E
Checotah, Ok., U.S. (chĕ-kō'tȧ)	79	35°27'N	95°32'W
Chedabucto Bay, b., Can. (chĕd-ȧ-bŭk-tō)	61	45°23'N	61°10'W
Cheduba Island, i., Mya.	168	18°45'N	93°01'E
Cheecham Hills, hills, Can. (chēē'hăm)	56	56°20'N	111°10'W
Cheektowaga, N.Y., U.S. (chēk-tō-wä'gȧ)	69c	42°54'N	78°46'W
Chefoo *see* Yantai, China	161	37°32'N	121°22'E
Chegutu, Zimb.	186	18°18'S	30°10'E
Chehalis, Wa., U.S. (chē-hā'lĭs)	72	46°39'N	122°58'W
Chehalis, r., Wa., U.S.	72	46°47'N	123°17'W
Cheju, Kor., S. (chĕ'jo͞o')	166	33°29'N	126°40'E
Cheju (Quelpart), i., Kor., S.	166	33°20'N	126°25'E
Chekalin, Russia (chĕ-kä'lĭn)	132	54°05'N	36°13'E
Chela, Serra da, mts., Ang. (sĕr'rȧ dä shä'lȧ)	186	15°30'S	13°30'E

PLACE (Pronunciation)	PAGE	LAT.	LONG.
Chelan, Wa., U.S. (chē-lăn')	72	47°51'N	119°59'W
Chelan, Lake, l., Wa., U.S.	72	48°09'N	120°20'W
Cheleiros, Port. (shĕ-la'rōzh)	129b	38°54'N	9°19'W
Chéliff, r., Alg. (shā-lēf)	184	36°00'N	2°00'E
Chelles, Fr.	127b	48°53'N	2°36'E
Chełm, Pol. (ĸĕlm)	117	51°08'N	23°30'E
Chełmno, Pol. (ĸĕlm'nō)	125	53°20'N	18°25'E
Chelmsford, Can.	58	46°35'N	81°12'W
Chelmsford, Eng., U.K. (chĕlm's-fērd)	121	51°44'N	0°28'E
Chelmsford, Ma., U.S.	61a	42°36'N	71°21'W
Chelsea, Austl.	173a	38°05'S	145°08'E
Chelsea, Can.	62c	45°30'N	75°46'W
Chelsea, Al., U.S. (chĕl'sē)	68h	33°20'N	86°38'W
Chelsea, Ma., U.S.	61a	42°23'N	71°02'W
Chelsea, Mi., U.S.	66	42°20'N	84°00'W
Chelsea, Ok., U.S.	79	36°32'N	95°23'W
Cheltenham, Eng., U.K. (chĕlt'nŭm)	120	51°57'N	2°06'W
Cheltenham, Md., U.S. (chĕltĕn-hăm)	68e	38°45'N	76°50'W
Chelyabinsk, Russia (chĕl-yä-bênsk')	134	55°10'N	61°25'E
Chelyuskin, Mys, c., Russia (chĕl-yòs'-kĭn)	135	77°45'N	104°45'E
Chemba, Moz.	191	17°08'S	34°52'E
Chemnitz, Ger.	117	50°48'N	12°53'E
Chemung, r., N.Y., U.S. (shē-mŭng)	67	42°20'N	77°25'W
Chĕn, Gora, mtn., Russia	135	65°13'N	142°12'E
Chenāb, r., Asia (chē-näb)	155	30°30'N	71°30'E
Chenachane, Alg. (shē-ná-shän')	184	26°14'N	4°14'W
Chencun, China (chŭn-tsòn)	163a	22°58'N	113°14'E
Cheney, Wa., U.S. (chē'nā)	72	47°29'N	117°34'W
Chengde, China (chŭŋ-dŭ)	161	40°50'N	117°50'E
Chengdong Hu, l., China (chŭŋ-dôŋ ho͞o)	162	32°22'N	116°32'E
Chengdu, China (chŭŋ-do͞o)	160	30°30'N	104°10'E
Chenggu, China (chŭŋ-go͞o)	164	33°05'N	107°25'E
Chenghai, China (chŭŋ-hī)	165	23°22'N	116°40'E
Chengshan Jiao, c., China (jyou chŭŋ-shän)	164	37°28'N	122°40'E
Chengxi Hu, l., China (chŭŋ-shyē ho͞o)	162	32°31'N	116°04'E
Chennai (Madras), India	155	13°08'N	80°15'E
Chenxian, China (chŭn-shyĕn)	165	25°40'N	113°00'E
Chepén, Peru (chĕ-pĕ'n)	100	7°17'S	79°24'W
Chepo, Pan. (chā'pō)	91	9°12'N	79°06'W
Chepo, r., Pan.	91	9°10'N	78°36'W
Cher, r., Fr. (shâr)	117	47°14'N	1°34'E
Cherán, Mex. (chā-rän')	88	19°41'N	101°54'W
Cherangany Hills, hills, Kenya	191	1°25'N	35°20'E
Cheraw, S.C., U.S. (chē'rō)	83	34°40'N	79°52'W
Cherbourg, Fr. (shär-bòr')	110	49°39'N	1°43'W
Cherdyn', Russia (chĕr-dyēn')	134	60°25'N	56°32'E
Cheremkhovo, Russia (chĕr'yĕm-kô-vō)	135	52°58'N	103°18'E
Cherëmukhovo, Russia (chĕr-yĕ-mû-kô-vô)	142a	60°20'N	60°00'E
Cherepanovo, Russia (chĕr'yĕ pä-nô'vō)	134	54°13'N	83°22'E
Cherepovets, Russia (chĕr-yĕ-pô'vyĕtz)	134	59°08'N	37°59'E
Chereya, Bela. (chĕr-ā'yä)	132	54°38'N	29°16'E
Chergui, i., Tun.	118	34°50'N	11°40'E
Chergui, Chott ech, l., Alg. (chĕr gē)	118	34°12'N	0°10'W
Cherkasy, Ukr.	133	49°26'N	32°03'E
Cherkasy, prov., Ukr.	133	48°58'N	30°55'E
Cherkessk, Russia	138	44°14'N	42°04'E
Cherlak, Russia (chĭr-läk')	134	54°04'N	74°28'E
Chermoz, Russia (chĕr-môz')	136	58°47'N	56°08'E
Chern', Russia (chĕrn)	132	53°28'N	36°49'E
Chërnaya Kalitva, r., Russia (chôr'ná yá ká-lēt'vá)	133	50°15'N	39°16'E
Chernihiv, Ukr.	137	51°23'N	31°15'E
Chernihiv, prov., Ukr.	133	51°28'N	31°18'E
Chernihivka, Ukr.	133	47°08'N	36°20'E
Chernivtsi, Ukr.	134	48°18'N	25°56'E
Chernobyl' *see* Chornobai, Ukr.	132	51°17'N	30°14'E
Chernogorsk, Russia (chĕr-nŏ-gôrsk')	140	54°01'N	91°07'E
Chernoistochinsk, Russia (chĕr-nôy-stô'chĭnsk)	142a	57°44'N	59°55'E
Chernyanka, Russia (chĕrn-yäŋ'kä)	133	50°56'N	37°48'E
Cherokee, Ia., U.S. (chĕr-ô-kē')	70	42°43'N	95°33'W
Cherokee, Ks., U.S.	79	37°21'N	94°50'W
Cherokee, Ok., U.S.	78	36°44'N	98°22'W
Cherokee Lake, res., Tn., U.S.	82	36°22'N	83°22'W
Cherokees, Lake of the, res., Ok., U.S. (chĕr-ô-kēz')	65	36°32'N	95°14'W
Cherokee Sound, Bah.	92	26°15'N	76°55'W
Cherryfield, Me., U.S. (chĕr'ĭ-fēld)	60	44°37'N	67°56'W
Cherry Grove, Or., U.S.	74c	45°27'N	123°15'W
Cherryvale, Ks., U.S.	79	37°16'N	95°33'W
Cherryville, N.C., U.S. (chĕr'ĭ-vĭl)	83	35°32'N	81°22'W
Cherskogo, Khrebet, mts., Russia	135	67°15'N	140°00'E
Chertsey, Eng., U.K.	114b	51°24'N	0°30'W
Chervonoye, Vozyera, l., Bela. (chĕr-vô'nô-yĕ)	132	52°24'N	28°00'E
Chervyen', Bela. (chĕr'vyĕn)	132	53°43'N	28°26'E
Cherykaw, Bela.	132	53°34'N	31°22'E
Chesaning, Mi., U.S. (chĕs'ȧ-nĭng)	66	43°10'N	84°10'W
Chesapeake, Va., U.S. (chĕs'ȧ-pēk)	68g	36°48'N	76°16'W
Chesapeake Bay, b., U.S.	65	38°20'N	76°15'W
Chesapeake Beach, Md., U.S.	68e	38°42'N	76°33'W
Chesham, Eng., U.K. (chĕsh'ŭm)	114b	51°41'N	0°37'W
Cheshire, Mi., U.S. (chĕsh'ĭr)	66	42°25'N	86°00'W
Cheshire, co., Eng., U.K.	114a	53°16'N	2°30'W
Chëshskaya Guba, b., Russia	134	67°25'N	46°00'E
Cheshunt, Eng., U.K.	114b	51°43'N	0°02'W
Chesma, Russia (chĕs'má)	142a	53°50'N	60°42'E
Chesnokovka, Russia (chĕs-nŏ-kôf'kȧ)	134	53°28'N	83°41'E

236 Chester – Chuuk (Truk) PRONOUNCING INDEX

PLACE (Pronunciation)	PAGE	LAT.	LONG.
Chester, Eng., U.K. (chĕs′tẽr)	120	53°12′N	2°53′W
Chester, Il., U.S.	79	37°54′N	89°48′W
Chester, Pa., U.S.	68f	39°51′N	75°22′W
Chester, S.C., U.S.	83	34°42′N	81°11′W
Chester, Va., U.S.	83	37°20′N	77°24′W
Chester, W.V., U.S.	66	40°35′N	80°30′W
Chesterfield, Eng., U.K. (chĕs′tẽr-fēld)	120	53°14′N	1°26′W
Chesterfield, Îles, is., N. Cal.	175	19°38′S	160°08′E
Chesterfield Inlet *see* Igluligaarjuk, Can.	52	63°19′N	91°11′W
Chesterfield Inlet, b., Can.	53	63°59′N	92°09′W
Chestermere Lake, l., Can. (chĕs′tē-mēr)	62e	51°03′N	113°45′W
Chesterton, In., U.S. (chĕs′tẽr-tŭn)	66	41°35′N	87°05′W
Chestertown, Md., U.S. (chĕs′tẽr-toun)	67	39°15′N	76°05′W
Chesuncook, l., Me., U.S. (chĕs′ŭn-ko̅o̅k)	60	46°03′N	69°40′W
Chetek, Wi., U.S. (chĕ′tĕk)	71	45°18′N	91°41′W
Chetumal, Bahía de, b., N.A. (bä-ē′ä dĕ chĕt-o̅o̅-mäl′)	86	18°07′N	88°05′W
Chevelon Creek, r., Az., U.S. (shĕv′á-lŏn)	77	34°35′N	111°00′W
Cheviot, Oh., U.S. (shĕv′ĭ-ŭt)	69f	39°10′N	84°37′W
Chevreuse, Fr. (shĕ-vrûz′)	127b	48°42′N	2°02′E
Chevy Chase, Md., U.S. (shĕvĭ chās)	68e	38°58′N	77°06′W
Chew Bahir, Afr. (stĕf-a-nē)	185	4°46′N	37°31′E
Chewelah, Wa., U.S. (chē-wē′lă)	72	48°17′N	117°42′W
Cheyenne, Wy., U.S. (shī-ĕn′)	64	41°10′N	104°49′W
Cheyenne, r., U.S.	64	44°20′N	102°15′W
Cheyenne River Indian Reservation, I.R., S.D., U.S.	70	45°07′N	100°46′W
Cheyenne Wells, Co., U.S.	78	38°46′N	102°21′W
Chhattisgarh, state, India	155	23°00′N	83°00′E
Chhindwāra, India	158	22°08′N	78°57′E
Chiai, Tai. (chī′ī′)	165	23°28′N	120°28′E
Chiange, Ang.	190	15°45′S	13°48′E
Chiang Mai, Thai.	168	18°38′N	98°44′E
Chiang Rai, Thai.	168	19°53′N	99°48′E
Chiapa, Río de, r., Mex.	90	16°00′N	92°20′W
Chiapa de Corzo, Mex. (chê-ä′pä dā kôr′zō)	89	16°44′N	93°01′W
Chiapas, state, Mex. (chê-ä′päs)	86	17°10′N	93°00′W
Chiapas, Cordillera de, mts., Mex. (kôr-dēl-yĕ′rä-dĕ-chyá′räs)	89	15°55′N	93°15′W
Chiari, Italy (kyä′rē)	130	45°31′N	9°57′E
Chiasso, Switz.	124	45°50′N	8°57′E
Chiatura, Geor.	138	42°17′N	43°17′E
Chiautla, Mex. (chyä-o̅o̅t′lä)	88	18°16′N	98°37′W
Chiavari, Italy (kyä-vä′rē)	130	44°18′N	9°21′E
Chiba, Japan (chē′bá)	161	35°37′N	140°08′E
Chiba, dept., Japan	167a	35°47′N	140°02′E
Chibougamau, Can. (chē-bo̅o̅′gä-mou)	51	49°57′N	74°23′W
Chibougamau, l., Can.	59	49°53′N	74°21′W
Chicago, Il., U.S. (shĭ-kô-gō) (chĭ-kä′gō)	65	41°49′N	87°37′W
Chicago Heights, Il., U.S.	69a	41°30′N	87°38′W
Chicapa, r., Afr. (chē-kä′pä)	186	7°45′S	20°25′E
Chicbul, Mex. (chēk-bo̅o̅l)	89	18°45′N	90°56′W
Chic-Chocs, Monts, mts., Can.	53	48°38′N	66°37′W
Chichagof, i., Ak., U.S. (chê-chä′gôf)	63	57°50′N	137°00′W
Chichancanab, Lago de, l., Mex. (lä′gô-dĕ-chē-chän-kä-nä′b)	90a	19°50′N	88°28′W
Chichén Itzá, hist., Mex.	90a	20°40′N	88°35′W
Chichester, Eng., U.K. (chĭch′ĕs-tẽr)	120	50°50′N	0°55′W
Chichimilá, Mex. (chē-chē-mê′lä)	90a	20°36′N	88°14′W
Chichiriviche, Ven. (chē-chē-rē-vē-chĕ)	101b	10°56′N	68°17′W
Chickamauga, Ga., U.S. (chĭk-*á*-mô′g*á*)	82	34°50′N	85°15′W
Chickamauga Lake, res., Tn., U.S.	82	35°18′N	85°22′W
Chickasawhay, r., Ms., U.S. (chĭk-*á*-sô′wā)	82	31°45′N	88°45′W
Chickasha, Ok., U.S. (chĭk′*á*-shā)	64	35°04′N	97°56′W
Chiclana de la Frontera, Spain (chê-klä′nä)	128	36°25′N	6°09′W
Chiclayo, Peru (chê-klä′yō)	100	6°46′S	79°50′W
Chico, Ca., U.S. (chē′kō)	76	39°43′N	121°51′W
Chico, Wa., U.S.	74a	47°37′N	122°43′W
Chico, r., Arg.	102	44°30′S	66°00′W
Chico, r., Arg.	102	49°15′S	69°30′W
Chico, r., Phil.	169a	17°33′N	121°24′E
Chicoloapan, Mex. (chê-kō-lwä′pän)	89a	19°24′N	98°54′W
Chiconautla, Mex.	89a	19°39′N	99°01′W
Chicontepec, Mex. (chê-kōn′tē-pĕk′)	88	20°58′N	98°08′W
Chicopee, Ma., U.S. (chĭk′ō-pē)	67	42°10′N	72°35′W
Chicoutimi, Can. (shē-ko̅o̅′tē-mē′)	51	48°26′N	71°04′W
Chicxulub, Mex. (chēk-so̅o̅-lo̅o̅′b)	90a	21°10′N	89°30′W
Chiefland, Fl., U.S. (chēf′l*ă*nd)	83	29°30′N	82°50′W
Chiemsee, l., Ger. (kēm zā)	124	47°58′N	12°20′E
Chieri, Italy (kyâ′rē)	130	45°03′N	7°48′E
Chieti, Italy (kyĕ′tē)	118	42°22′N	14°22′E
Chifeng, China (chr-fŭŋ)	161	42°18′N	118°52′E
Chignanuapan, Mex. (chē′g-nä-nwä-pá′n)	88	19°49′N	98°02′W
Chignecto Bay, b., Can. (shĭg-nĕk′tō)	60	45°33′N	64°50′W
Chignik, Ak., U.S. (chĭg′nĭk)	63	56°14′N	158°12′W
Chignik Bay, b., Ak., U.S.	63	56°18′N	157°22′W
Chigu Co, l., China (chr-go̅o̅ tswo)	158	28°55′N	91°47′E
Chigwell, Eng., U.K.	114b	51°38′N	0°05′E
Chihe, China (chr-hŭ)	162	32°32′N	117°57′E
Chihuahua, Mex. (chê-wä′wä)	86	28°37′N	106°06′W
Chihuahua, state, Mex.	86	29°00′N	107°30′W
Chikishlyar, Turkmen. (chê-kĕsh-lyär′)	139	37°40′N	53°50′E
Chilanga, Zam.	191	15°34′S	28°17′E
Chilapa, Mex. (chê-lä′pä)	88	17°34′N	99°14′W
Chilchota, Mex. (chêl-chō′tä)	88	19°40′N	102°04′W
Chilcotin, r., Can. (chĭl-kō′tĭn)	54	52°20′N	124°15′W
Childress, Tx., U.S. (chĭld′rĕs)	78	34°26′N	100°11′W
Chile, nation, S.A. (chē′lā)	102	35°00′S	72°00′W
Chilecito, Arg. (chê-lå-sē′tō)	102	29°06′S	67°25′W
Chilengue, Serra do, mts., Ang.	190	13°20′S	15°00′E
Chilibre, Pan. (chē-lē′brĕ)	86a	9°09′N	79°37′W
Chililabombwe, Zam.	191	12°18′S	27°43′E
Chilka, l., India	158	19°26′N	85°42′E
Chilko, r., Can. (chĭl′kō)	54	51°53′N	123°53′W
Chilko Lake, l., Can.	54	51°20′N	124°05′W
Chillán, Chile (chêl-yän′)	102	36°44′S	72°06′W
Chillicothe, Il., U.S. (chĭl-ĭ-kŏth′ē)	66	41°55′N	89°30′W
Chillicothe, Mo., U.S.	79	39°46′N	93°32′W
Chillicothe, Oh., U.S.	66	39°20′N	83°00′W
Chilliwack, Can. (chĭl′ĭ-wăk)	50	49°10′N	121°57′W
Chiloé, Isla de, i., Chile	102	42°30′S	73°55′W
Chilpancingo de los Bravo, Mex.	86	17°32′N	99°30′W
Chilton, Wi., U.S. (chĭl′tŭn)	71	44°00′N	88°12′W
Chilung, Tai. (chī′lung)	161	25°02′N	121°48′E
Chilwa, Lake, l., Afr.	186	15°12′S	36°30′E
Chimacum, Wa., U.S. (chĭm′ä-kŭm)	74a	48°01′N	122°47′W
Chimalpa, Mex. (chē-mäl′pä)	89a	19°26′N	99°22′W
Chimaltenango, Guat. (chê-mäl-tå-nän′gō)	90	14°39′N	90°48′W
Chimaltitan, Mex. (chēmäl-tē-tän′)	88	21°36′N	103°50′W
Chimbay, Uzb. (chĭm-bī′)	139	43°00′N	59°44′E
Chimborazo, mtn., Ec. (chêm-bô-rä′zō)	100	1°35′S	78°45′W
Chimbote, Peru (chêm-bō′tå)	100	9°02′S	78°33′W
China, Mex. (chē′nä)	80	25°43′N	99°13′W
China, nation, Asia (chī′n*á*)	160	36°45′N	93°00′E
Chinameca, El Sal. (Chē-nä-mä′kä)	90	13°31′N	88°18′W
Chinandega, Nic. (chē-nän-dā′gä)	90	12°38′N	87°08′W
Chinati Peak, mtn., Tx., U.S. (chĭ-nä′tē)	80	29°56′N	104°29′W
Chincha Alta, Peru (chĭn′chä äl′tä)	100	13°24′S	76°04′W
Chinchas, Islas, is., Peru (ē′s-läs-chē′n-chäs)	100	11°27′S	79°05′W
Chinchilla, Austl. (chĭn-chĭl′*á*)	176	26°44′S	150°36′E
Chinchorro, Banco, bk., Mex. (bä′n-kô-chĕn-chô′r-rō)	90a	18°43′N	87°25′W
Chincilla de Monte Aragon, Spain	128	38°54′N	1°43′W
Chinde, Moz. (shên′dĕ)	186	17°39′S	36°34′E
Chin Do, i., Kor., S.	166	34°30′N	125°43′E
Chindwin, r., Mya. (chĭn-dwĭn)	155	23°30′N	94°34′E
Chingola, Zam. (chĭng-gōlä)	186	12°32′S	27°52′E
Chinguar, Ang. (chĭng-gär)	186	12°35′S	16°15′E
Chinguetti, Maur. (chēŋ-gĕt′ē)	184	20°34′N	12°34′W
Chinhoyi, Zimb.	186	17°22′S	30°12′E
Chinju, Kor., S. (chĭn′jo̅o̅)	166	35°13′N	128°10′E
Chinko, r., C.A.R. (shĭn′kô)	185	6°37′N	24°31′E
Chinmen *see* Quemoy, Tai.	165	24°30′N	118°20′E
Chino, Ca., U.S. (chē′nō)	75a	34°01′N	117°42′W
Chinon, Fr. (shē-nôn′)	126	47°09′N	0°13′E
Chinook, Mt., U.S. (shĭn-ŏk′)	73	48°35′N	109°15′W
Chinsali, Zam.	191	10°34′S	32°03′E
Chinteche, Mwi. (chĭn-tē′chĕ)	186	11°48′S	34°14′E
Chioggia, Italy (kyôd′jä)	130	45°12′N	12°17′E
Chíos, Grc. (kē′ŏs)	119	38°23′N	26°09′E
Chíos, i., Grc.	119	38°20′N	25°45′E
Chipata, Zam.	186	13°39′S	32°40′E
Chipera, Moz. (zhê-pē′rä)	186	15°16′S	32°30′E
Chipley, Fl., U.S. (chĭp′lĭ)	82	30°45′N	85°33′W
Chipman, Can. (chĭp′m*ă*n)	60	46°11′N	65°53′W
Chipola, r., Fl., U.S. (chĭ-pō′l*á*)	82	30°40′N	85°14′W
Chippawa, Can. (chĭp′ê-wä)	69c	43°03′N	79°03′W
Chippewa, r., Mn., U.S. (chĭp′ê-wä)	70	45°07′N	95°41′W
Chippewa, r., Wi., U.S.	71	45°07′N	91°19′W
Chippewa Falls, Wi., U.S.	71	44°55′N	91°26′W
Chippewa Lake, Oh., U.S.	69d	41°04′N	81°54′W
Chiputneticook Lakes, l., N.A. (chĭ-pôt-nĕt′ĭ-ko̅o̅k)	60	45°47′N	67°45′W
Chiquimula, Guat. (chē-kê-mo̅o̅′lä)	90	14°47′N	89°31′W
Chiquimulilla, Guat. (chē-kē-mo̅o̅-lē′l-yä)	90	14°08′N	90°23′W
Chiquinquira, Col. (chê-kēŋ′kê-rä′)	100	5°33′N	73°49′W
Chirala, India	159	15°52′N	80°22′E
Chirchik, Uzb. (chĭr-chĕk′)	139	41°28′N	69°18′E
Chire (Shire), r., Afr.	191	17°15′S	35°25′E
Chiricahua National Monument, rec., Az., U.S. (chĭ-rä-cä′hwä)	77	32°02′N	109°18′W
Chirikof, i., Ak., U.S. (chĭ′rĭ-kôf)	63	55°50′N	155°35′W
Chiriquí, Punta, c., Pan. (pŏ′n-tä-chē-rē-kē′)	91	9°13′N	81°39′W
Chiriquí Grande, Pan. (chē-rê-kē′ grän′dā)	91	8°57′N	82°08′W
Chiri San, mtn., Kor., S. (chĭ′rĭ-sän′)	166	35°20′N	127°39′E
Chiromo, Mwi.	186	16°34′S	35°13′E
Chirpan, Blg.	119	42°12′N	25°19′E
Chirripó, Río, r., C.R.	91	9°50′N	83°20′W
Chisasibi, Can.	51	53°40′N	78°58′W
Chisholm, Mn., U.S. (chĭz′*ŭ*m)	71	47°28′N	92°53′W
Chişinău, Mol.	134	47°02′N	28°52′E
Chistopol′, Russia (chĭs-tô′pôl-y′)	134	55°21′N	50°37′E
Chita, Russia (chê-tá′)	135	52°09′N	113°39′E
Chitambo, Zam.	191	12°55′S	30°39′E
Chitato, Ang.	190	7°20′S	20°47′E
Chitembo, Ang.	190	13°34′S	16°40′E
Chitina, Ak., U.S. (chĭ-tē′ná)	63	61°28′N	144°35′W
Chitokoloki, Zam.	190	13°50′S	23°13′E
Chitorgarh, India	158	24°59′N	74°42′E
Chitrāl, Pak. (chē-träl′)	155	35°58′N	71°48′E
Chittagong, Bngl. (chĭt-á-gông′)	155	22°26′N	90°51′E
Chitungwiza, Zimb.	186	17°51′S	31°05′E
Chiumbe, r., Afr. (chê-ŏm′bå)	186	9°45′S	21°00′E
Chivasso, Italy (kê-väs′sō)	130	45°13′N	7°52′E
Chivhu, Zimb.	186	18°59′S	30°58′E
Chivilcoy, Arg. (chē-vêl-koi′)	102	34°51′S	60°03′W
Chixoy, r., Guat. (chē-кoi′)	90	15°40′N	90°35′W
Chizu, Japan (chê-zo̅o̅′)	167	35°16′N	134°15′E
Chloride, Az., U.S. (klō′rīd)	77	35°25′N	114°15′W
Chmielnik, Pol. (кmyĕl′nĕк)	125	50°36′N	20°46′E
Choapa, r., Chile (chô-á′pä)	99b	31°56′S	70°48′W
Choctawhatchee, r., Fl., U.S.	82	30°37′N	85°56′W
Choctawhatchee Bay, b., Fl., U.S. (chôk-tô-hăch′ē)	82	30°15′N	86°32′W
Chodziez, Pol. (кōj′yĕsh)	124	52°59′N	16°55′E
Choele Choel, Arg. (chô-ĕ′lĕ-chôĕ′l)	102	39°14′S	65°46′W
Chōfu, Japan (chō′fo̅o̅′)	167a	35°39′N	139°33′E
Chōgo, Japan (chō-gō)	167a	35°25′N	139°28′E
Choiseul, i., Sol. Is. (shwä-zûl′)	175	7°30′S	157°30′E
Choisy-le-Roi, Fr.	127b	48°46′N	2°25′E
Chojnice, Pol. (кōĭ-nē-tsĕ)	125	53°41′N	17°34′E
Cholet, Fr. (shô-lĕ′)	117	47°06′N	0°54′W
Cholula, Mex. (chō-lo̅o̅′lä)	88	19°04′N	98°19′W
Choluteca, Hond. (chō-lo̅o̅-tā′kä)	90	13°18′N	87°12′W
Choluteco, r., Hond.	90	13°34′N	86°59′W
Chomutov, Czech Rep. (kō′mò-tôf)	124	50°27′N	13°23′E
Chona, r., Russia (chô′nä)	141	60°45′N	109°15′E
Chone, Ec. (chô′nĕ)	100	0°48′S	80°06′W
Chŏngjin, Kor., N. (chŭng-jĭn′)	161	41°48′N	129°46′E
Chŏngju, Kor., S. (chŭng-jo̅o̅′)	166	36°35′N	127°30′E
Chongming Dao, i., China (chôŋ-mĭŋ dou)	165	31°40′N	122°30′E
Chongqing, China (chôŋ-chyĭŋ)	160	29°38′N	107°30′E
Chongqing, prov., China	160	30°00′N	108°00′E
Chŏnju, Kor., S. (chŭn-jo̅o̅′)	166	35°48′N	127°08′E
Chonos, Archipiélago de los, is., Chile	102	44°35′N	76°15′W
Chorley, Eng., U.K. (chôr′lĭ)	114a	53°40′N	2°38′W
Chornaya, neigh., Russia	142b	55°45′N	38°04′E
Chornobai, Ukr.	133	51°17′N	30°14′E
Chornobay, Ukr. (chĕr-nō-bī′)	133	49°41′N	32°24′E
Chornomors′ke, Ukr.	137	45°29′N	32°43′E
Chorrillos, Peru (chôr-rē′l-yōs)	100	12°17′S	76°55′W
Chortkiv, Ukr.	125	49°01′N	25°48′E
Chosan, Kor., N. (chō-sän′)	166	40°44′N	125°48′E
Chosen, Fl., U.S. (chō′z′n)	83a	26°41′N	80°41′W
Chōshi, Japan (chō′shē)	166	35°40′N	140°55′E
Choszczno, Pol. (chôsh′chnô)	124	53°10′N	15°25′E
Chota Nagpur, plat., India	158	23°40′N	82°50′E
Choteau, Mt., U.S. (shō′tō)	73	47°51′N	112°10′W
Chowan, r., N.C., U.S. (chô-wän′)	83	36°13′N	76°46′W
Chowilla Reservoir, res., Austl.	176	34°05′S	141°20′E
Chown, Mount, mtn., Can. (choun)	55	53°24′N	119°22′W
Choybalsan, Mong.	161	47°50′N	114°15′E
Christchurch, N.Z. (krīst′chûrch)	175a	43°30′S	172°38′E
Christian, i., Can. (krĭs′ch*ă*n)	59	44°50′N	80°00′W
Christiansburg, Va., U.S. (krĭs′ch*ă*nz-bûrg)	83	37°08′N	80°25′W
Christiansted, V.I.U.S.	87b	17°45′N	64°44′W
Christmas Island, dep., Oc.	168	10°35′S	105°40′E
Christopher, Il., U.S. (krĭs′tô-fẽr)	79	37°58′N	89°04′W
Chrudim, Czech Rep. (кro̅o̅′dyĕm)	124	49°57′N	15°46′E
Chrzanów, Pol. (кzhä′nôf)	125	50°08′N	19°24′E
Chuansha, China (chůän-shä)	163b	31°12′N	121°41′E
Chubut, prov., Arg. (chò-bo̅o̅t′)	102	44°00′S	69°15′W
Chubut, r., Arg. (chò-bo̅o̅t′)	102	43°05′S	69°00′W
Chuckatuck, Va., U.S. (chŭck *á*-tŭck)	68g	36°51′N	76°35′W
Chucunaque, r., Pan. (cho̅o̅-ko̅o̅-nä′kå)	91	8°36′N	77°48′W
Chudovo, Russia (chò′dô-vô)	132	59°03′N	31°56′E
Chudskoye Ozero, l., Eur. (chòt′skô-yĕ)	136	58°43′N	26°45′E
Chuguchak, hist. reg., China (cho̅o̅′go̅o̅-chäk′)	160	46°09′N	83°58′E
Chuguyevka, Russia (chò-go̅o̅′yĕf-kå)	166	43°58′N	133°49′E
Chugwater Creek, r., Wy., U.S. (chŭg′wô-tẽr)	70	41°43′N	104°54′W
Chuhuïv, Ukr.	137	49°52′N	36°40′E
Chukotskiy Poluostrov, pen., Russia	135	66°12′N	175°00′W
Chukotskoye Nagor′ye, mts., Russia	135	66°00′N	166°00′E
Chula Vista, Ca., U.S. (cho̅o̅′lá vĭs′tá)	76a	32°38′N	117°05′W
Chulkovo, Russia (cho̅o̅l-kô vô)	142b	55°33′N	38°04′E
Chulucanas, Peru	100	5°13′S	80°13′W
Chulum, r., Russia	140	57°52′N	84°45′E
Chumikan, Russia (cho̅o̅-mē-kän′)	135	54°47′N	135°09′E
Chun′an, China (chòn-än)	165	29°38′N	119°00′E
Chunchŏn, Kor., S. (chòn-chŭn′)	166	37°51′N	127°46′E
Chungju, Kor., S. (chŭng′jo̅o̅′)	166	37°00′N	128°19′E
Chungking *see* Chongqing, China	160	29°38′N	107°30′E
Chunya, Tan.	191	8°32′S	33°25′E
Chunya, r., Russia (chòn′yä′)	140	61°45′N	101°28′E
Chuquicamata, Chile (cho̅o̅-kê-kä-mä′tä)	102	22°08′S	68°57′W
Chur, Switz. (ko̅o̅r)	117	46°51′N	9°32′E
Churchill, Can. (chûrch′ĭl)	51	58°50′N	94°10′W
Churchill, r., Can.	52	58°00′N	95°00′W
Churchill, Cape, c., Can.	53	59°07′N	93°50′W
Churchill Falls, wtfl., Can.	53	53°35′N	64°27′W
Churchill Lake, l., Can.	56	56°12′N	108°40′W
Churchill Peak, mtn., Can.	52	58°10′N	125°14′W
Church Stretton, Eng., U.K. (chûrch strĕt′*ŭ*n)	114a	52°32′N	2°49′W
Churchton, Md., U.S.	68e	38°49′N	76°33′W
Churu, India	158	28°22′N	75°00′E
Churumuco, Mex. (cho̅o̅-ro̅o̅-mo̅o̅′kō)	88	18°39′N	101°40′W
Chuska Mountains, mts., Az., U.S. (chŭs-k*á*)	77	36°21′N	109°11′W
Chusovaya, r., Russia (cho̅o̅-sô-vä′yá)	136	58°08′N	58°35′E
Chusovoy, Russia (cho̅o̅-sô-vôy′)	134	58°18′N	57°50′E
Chust, Uzb. (chòst)	139	41°05′N	71°28′E
Chuuk (Truk), is., Micron.	170c	7°25′N	151°47′E

PRONOUNCING INDEX Chuvashia – Codogno 237

PLACE (Pronunciation)	PAGE	LAT.	LONG.
Chuvashia, prov., Russia	136	55°45′N	46°00′E
Chuviscar, r., Mex. (chōō-vēs-kär´)	80	28°34′N	105°36′W
Chuwang, China (chōō-wäŋ)	162	36°08′N	114°53′E
Chuxian, China (chōō shyĕn)	164	32°19′N	118°19′E
Chuxiong, China (chōō-shyôŋ)	160	25°19′N	101°34′E
Chyhyryn, Ukr.	133	49°02′N	32°39′E
Cicero, Il., U.S. (sĭs´ẽr-ō)	69a	41°50′N	87°46′W
Cide, Tur. (jē´dĕ)	119	41°50′N	33°00′E
Ciechanów, Pol. (tsyĕ-kä´nôf)	125	52°52′N	20°39′E
Ciego de Avila, Cuba (syä´gō dā ä´vē-lä)	87	21°50′N	78°45′W
Ciego de Avila, prov., Cuba	92	22°00′N	78°40′W
Ciempozuelos, Spain (thyĕm-pō-thwā´lōs)	128	40°09′N	3°36′W
Ciénaga, Col. (syā´nä-gä)	100	11°01′N	74°15′W
Cienfuegos, Cuba (syĕn-fwā´gōs)	87	22°10′N	80°30′W
Cienfuegos, prov., Cuba	92	22°15′N	80°40′W
Cienfuegos, Bahía, b., Cuba (bä-ē´ä-syĕn-fwā´gōs)	92	22°00′N	80°35′W
Ciervo, Isla de la, i., Nic. (ē´s-lä-dĕ-lä-syĕ´r-vô)	91	11°56′N	83°20′W
Cieszyn, Pol. (tsyĕ´shĕn)	125	49°47′N	18°45′E
Cieza, Spain (thyā´thä)	128	38°13′N	1°25′W
Cigüela, r., Spain	128	39°53′N	2°54′W
Cihuatlán, Mex. (sē-wä-tlä´n)	88	19°13′N	104°36′W
Cihuatlán, r., Mex.	88	19°11′N	104°30′W
Cijara, Embalse de, res., Spain	128	39°25′N	5°00′W
Cilician Gates, p., Tur.	137	37°30′N	35°30′E
Cimarron, r., Co., U.S.	78	37°13′N	102°30′W
Cimarron, r., U.S. (sĭm-ȧ-rōn´)	64	36°26′N	98°27′W
Cinca, r., Spain (thēŋ´kä)	129	42°09′N	0°08′E
Cincinnati, Oh., U.S. (sĭn-sĭ-nát´ĭ)	65	39°08′N	84°30′W
Cinco Balas, Cayos, is., Cuba (kä´yōs-thēŋ´kō bä´läs)	92	21°05′N	79°25′W
Cintalapa, Mex. (sēn-tä-lä´pä)	89	16°41′N	93°44′W
Cinto, Monte, mtn., Fr. (chēn´tō)	117	42°24′N	8°54′E
Circle, Ak., U.S. (sûr´k'l)	64a	65°49′N	144°22′W
Circleville, Oh., U.S. (sûr´k'lvĭl)	66	39°35′N	83°00′W
Cirebon, Indon.	168	6°50′S	108°33′E
Ciri Grande, r., Pan. (sē´rē-grä´n´dĕ)	86a	8°55′N	80°04′W
Cisco, Tx., U.S. (sĭs´kō)	80	32°23′N	98°57′W
Cisneros, Col. (sēs-nĕ´rōs)	100a	6°33′N	75°05′W
Cisterna di Latina, Italy (chēs-tĕ´r-nä-dē-lä-tē´nä)	129d	41°36′N	12°53′E
Cistierna, Spain (thēs-tyĕr´nä)	128	42°48′N	5°08′W
Citronelle, Al., U.S. (cĭt-rō´nĕl)	82	31°05′N	88°15′W
Cittadella, Italy (chēt-tä-dĕl´lä)	130	45°39′N	11°51′E
Città di Castello, Italy (chēt-tä´dē käs-tĕl´lō)	130	43°27′N	12°17′E
Ciudad Altamirano, Mex. (syōō-dä´d-äl-tä-mē-rä´nō)	88	18°24′N	100°38′W
Ciudad Bolívar, Ven. (syōō-dhädh´ bō-lē´vär)	100	8°07′N	63°41′W
Ciudad Camargo, Mex.	86	27°42′N	105°10′W
Ciudad Chetumal, Mex.	86	18°30′N	88°17′W
Ciudad Darío, Nic. (syōō-dhädh´dä´rē-ō)	90	12°44′N	86°08′W
Ciudad de la Habana, prov., Cuba	92	23°20′N	82°10′W
Ciudad del Carmen, Mex. (syōō-dä´d-dĕl-kä´r-mĕn)	86	18°39′N	91°49′W
Ciudad del Maíz, Mex. (syōō-dhädh´del mä-ēz´)	88	22°24′N	99°37′W
Ciudad Fernández, Mex. (syōō-dhädh´fĕr-nän´dĕz)	88	21°56′N	100°03′W
Ciudad García, Mex. (syōō-dhädh´gär-sē´ä)	86	22°39′N	103°02′W
Ciudad Guayana, Ven.	100	8°30′N	62°45′W
Ciudad Guzmán, Mex. (syōō-dhädh´gōz-män)	86	19°40′N	103°29′W
Ciudad Hidalgo, Mex. (syōō-dä´d-ē-dä´l-gō)	88	19°41′N	100°35′W
Ciudad Juárez, Mex. (syōō-dhädh hwä´rāz)	86	31°44′N	106°28′W
Ciudad Madero, Mex. (syōō-dä´d-mä-dĕ´rô)	89	22°16′N	97°52′W
Ciudad Mante, Mex. (syōō-dä´d-män´tĕ)	86	22°34′N	98°58′W
Ciudad Manuel Doblado, Mex. (syōō-dä´d-män-wāl´dō-blä´dō)	88	20°43′N	101°57′W
Ciudad Obregón, Mex. (syōō-dhädh-ô-brĕ-gô´n)	86	27°40′N	109°58′W
Ciudad Real, Spain (thyōō-dhädh´rā-äl´)	128	38°59′N	3°55′W
Ciudad Rodrigo, Spain (thyōō-dhädh´rō-drē´gō)	118	40°38′N	6°34′W
Ciudad Serdán, Mex. (syōō-dä´d-sĕr-dá´n)	89	18°58′N	97°26′W
Ciudad Victoria, Mex. (syōō-dhädh´vĕk-tō´rē-ä)	86	23°43′N	99°09′W
Ciutadella, Spain	129	40°00′N	3°52′E
Civitavecchia, Italy (chē´vē-tä-vĕk´kyä)	130	42°06′N	11°49′E
Cixian, China (tsē shyĕn)	162	36°22′N	114°23′E
Clackamas, Or., U.S. (klăc-kȧ´măs)	74c	45°25′N	122°34′W
Claire, l., Can. (klâr)	52	58°33′N	113°16′W
Clair Engle Lake, l., Ca., U.S.	72	40°51′N	122°41′W
Clairton, Pa., U.S. (klârtŭn)	69e	40°17′N	79°53′W
Clanton, Al., U.S. (klăn´tŭn)	82	32°50′N	86°38′W
Clare, Mi., U.S. (klâr)	66	43°50′N	84°45′W
Clare Island, i., Ire.	120	53°46′N	10°00′W
Claremont, Ca., U.S. (klâr´mŏnt)	75a	34°06′N	117°43′W
Claremont, N.H., U.S. (klâr´mŏnt)	67	43°20′N	72°20′W
Claremont, W.V., U.S.	66	37°55′N	81°00′W
Claremore, Ok., U.S. (klâr´mōr)	79	36°16′N	95°37′W
Claremorris, Ire. (klâr-mŏr´ĭs)	120	53°46′N	9°05′W
Clarence Strait, strt., Austl. (klăr´ĕns)	174	12°15′S	130°05′E
Clarence Strait, strt., Ak., U.S.	54	55°25′N	132°00′W
Clarence Town, Bah.	93	23°05′N	75°00′W
Clarendon, Ar., U.S. (klâr´ĕn-dŭn)	79	34°42′N	91°17′W
Clarendon, Tx., U.S.	78	34°55′N	100°52′W
Clarens, S. Afr. (clâ-rĕns)	187c	28°34′S	28°26′E
Claresholm, Can. (klâr´ĕs-hōlm)	50	50°02′N	113°35′W
Clarinda, Ia., U.S. (klȧ-rĭn´dȧ)	70	40°42′N	95°00′W
Clarines, Ven. (klä-rē´nĕs)	101b	9°57′N	65°10′W
Clarion, Ia., U.S. (klăr´ĭ-ŭn)	71	42°43′N	93°45′W
Clarion, Pa., U.S.	67	41°10′N	79°25′W
Clark, S.D., U.S. (klärk)	70	44°52′N	97°45′W
Clark, Point, c., Can.	58	44°05′N	81°50′W
Clarkdale, Az., U.S. (klärk-dāl)	77	34°45′N	112°05′W
Clarke City, Can.	51	50°12′N	66°38′W
Clarke Range, mts., Austl.	175	20°30′S	148°00′E
Clark Fork, r., Mt., U.S.	72	47°50′N	115°35′W
Clarksburg, W.V., U.S. (klärkz´bûrg)	65	39°15′N	80°20′W
Clarksdale, Ms., U.S. (klärks-dāl)	82	34°10′N	90°31′W
Clark's Harbour, Can. (klärks)	60	43°26′N	65°38′W
Clarks Hill Lake, res., U.S. (klärk-hĭl)	65	33°50′N	82°35′W
Clarkston, Ga., U.S. (klärks´tŭn)	68c	33°49′N	84°15′W
Clarkston, Wa., U.S.	72	46°24′N	117°01′W
Clarksville, Ar., U.S. (klärks-vĭl)	79	35°28′N	93°26′W
Clarksville, Tn., U.S.	82	36°30′N	87°23′W
Clarksville, Tx., U.S.	79	33°37′N	95°02′W
Clatskanie, Or., U.S.	74c	46°04′N	123°11′W
Clatskanie, r., Or., U.S. (klăt-skā´nē)	74c	46°06′N	123°11′W
Clatsop Spit, Or., U.S. (klăt-sŏp)	74c	46°13′N	124°04′W
Cláudio, Braz. (klou´-dēō)	99a	20°26′S	44°44′W
Claveria, Phil. (klä-vȧ-rē´ä)	165	18°38′N	121°08′E
Clawson, Mi., U.S. (klô´s'n)	69b	42°32′N	83°09′W
Claxton, Ga., U.S. (klăks´tŭn)	83	32°07′N	81°54′W
Clay, Ky., U.S. (klā)	82	37°28′N	87°50′W
Clay Center, Ks., U.S. (klā sĕn´tẽr)	79	39°23′N	97°08′W
Clay City, Ky., U.S. (klā sĭ´tĭ)	66	37°50′N	83°55′W
Claycomo, Mo., U.S. (kla-kō´mo)	75f	39°12′N	94°30′W
Clay Cross, Eng., U.K. (klā krŏs)	114a	53°10′N	1°25′W
Claye-Souilly, Fr. (klĕ-sōō-yē´)	127b	48°56′N	2°43′E
Claymont, De., U.S. (klā-mŏnt)	68f	39°48′N	75°28′W
Clayton, Eng., U.K.	114a	53°47′N	1°49′W
Clayton, Al., U.S. (klā´tŭn)	82	31°52′N	85°25′W
Clayton, Ca., U.S.	74b	37°56′N	121°56′W
Clayton, Mo., U.S.	75e	38°39′N	90°20′W
Clayton, N.C., U.S.	83	35°40′N	78°27′W
Clayton, N.M., U.S.	78	36°26′N	103°12′W
Clear, l., Ca., U.S.	76	39°05′N	122°50′W
Clear Boggy Creek, r., Ok., U.S. (klẽr bŏg´ĭ krēk)	79	34°21′N	96°22′W
Clear Creek, r., Az., U.S.	77	34°40′N	111°05′W
Clear Creek, r., Tx., U.S.	81a	29°34′N	95°13′W
Clear Creek, r., Wy., U.S.	73	44°35′N	106°20′W
Clearfield, Pa., U.S. (klẽr-fēld)	67	41°00′N	78°25′W
Clearfield, Ut., U.S.	75b	41°07′N	112°01′W
Clear Hills, Can.	50	57°11′N	119°20′W
Clear Lake, Ia., U.S.	71	43°09′N	93°23′W
Clear Lake, Wa., U.S.	74a	48°27′N	122°14′W
Clear Lake Reservoir, res., Ca., U.S.	72	41°53′N	121°00′W
Clearwater, Fl., U.S. (klẽr-wô´tẽr)	83a	27°43′N	82°45′W
Clearwater, r., Can.	55	52°22′N	114°57′W
Clearwater, r., Can.	55	52°00′N	120°10′W
Clearwater, r., Can.	56	56°10′N	110°40′W
Clearwater, r., Id., U.S.	72	46°27′N	116°33′W
Clearwater, Middle Fork, r., Id., U.S.	72	46°10′N	115°48′W
Clearwater, North Fork, r., Id., U.S.	72	46°34′N	116°08′W
Clearwater, South Fork, r., Id., U.S.	72	45°46′N	115°53′W
Clearwater Mountains, mts., Id., U.S.	72	45°56′N	115°15′W
Cleburne, Tx., U.S. (klē´bŭrn)	64	32°21′N	97°23′W
Cle Elum, Wa., U.S. (klē ĕl´ŭm)	72	47°12′N	120°55′W
Clementon, N.J., U.S. (klē´mĕn-tŭn)	68f	39°49′N	75°00′W
Cleobury Mortimer, Eng., U.K. (klēō-bĕr´ĭ môr´tĭ-mẽr)	114a	52°22′N	2°29′W
Clermont, Austl. (klẽr´mŏnt)	175	23°02′S	147°46′E
Clermont, Can.	59	47°45′N	70°20′W
Clermont-Ferrand, Fr. (klĕr-môn´fĕr-räN´)	110	45°47′N	3°03′E
Cleveland, Ms., U.S. (klēv´lȧnd)	82	33°45′N	90°42′W
Cleveland, Oh., U.S.	65	41°30′N	81°42′W
Cleveland, Ok., U.S.	79	36°18′N	96°28′W
Cleveland, Tn., U.S.	82	35°09′N	84°52′W
Cleveland, Tx., U.S.	81	30°18′N	95°05′W
Cleveland Heights, Oh., U.S.	69d	41°30′N	81°35′W
Cleveland Peninsula, pen., Ak., U.S.	54	55°45′N	132°00′W
Cleves, Oh., U.S. (klē´vĕs)	69f	39°10′N	84°45′W
Clew Bay, b., Ire. (klōō)	120	53°47′N	9°45′W
Clewiston, Fl., U.S. (klē´wis-tŭn)	83a	26°44′N	80°55′W
Clichy, Fr. (klē-shē´)	126	48°54′N	2°18′E
Clifden, Ire. (klĭf´dĕn)	120	53°31′N	10°04′W
Clifton, Az., U.S. (klĭf´tŭn)	77	33°05′N	109°20′W
Clifton, N.J., U.S.	68a	40°52′N	74°09′W
Clifton, S.C., U.S.	83	35°00′N	81°47′W
Clifton, Tx., U.S.	81	31°45′N	97°31′W
Clifton Forge, Va., U.S.	67	37°50′N	79°50′W
Clinch, r., Tn., U.S. (klĭnch)	82	36°30′N	83°19′W
Clingmans Dome, mtn., U.S. (klĭng´măns dōm)	82	35°37′N	83°26′W
Clinton, Can. (klĭn-´tŭn)	50	51°05′N	121°35′W
Clinton, Ia., U.S.	71	41°50′N	90°13′W
Clinton, Il., U.S.	66	40°10′N	88°55′W
Clinton, In., U.S.	66	39°40′N	87°25′W
Clinton, Ky., U.S.	82	36°39′N	88°56′W
Clinton, Ma., U.S.	61a	42°25′N	71°41′W
Clinton, Md., U.S.	68e	38°46′N	76°54′W
Clinton, Mo., U.S.	79	38°23′N	93°46′W
Clinton, N.C., U.S.	83	34°58′N	78°20′W
Clinton, Ok., U.S.	78	35°31′N	98°56′W
Clinton, S.C., U.S.	83	34°27′N	81°53′W
Clinton, Tn., U.S.	82	36°05′N	84°08′W
Clinton, Wa., U.S.	74a	47°59′N	122°22′W
Clinton, r., Mi., U.S.	69b	42°36′N	83°00′W
Clinton-Colden, l., Can.	52	63°58′N	106°34′W
Clintonville, Wi., U.S. (klĭn´tŭn-vĭl)	71	44°37′N	88°46′W
Clio, Mi., U.S. (klē´ō)	66	43°10′N	83°45′W
Cloates, Point, c., Austl. (klōts)	174	22°47′S	113°45′E
Clocolan, S. Afr.	192c	28°56′S	27°35′E
Clonakilty Bay, b., Ire. (klŏn-ȧ-kĭltē)	120	51°30′N	8°50′W
Cloncurry, Austl. (klŏn-kŭr´ē)	174	20°58′S	140°42′E
Clonmel, Ire. (klŏn-mĕl)	120	52°21′N	7°45′W
Cloquet, Mn., U.S. (klō-kā´)	75h	46°42′N	92°28′W
Closter, N.J., U.S. (klōs´tẽr)	68a	40°58′N	73°57′W
Cloud Peak, mtn., Wy., U.S. (kloud)	64	44°23′N	107°11′W
Clover, S.C., U.S. (klō´vẽr)	83	35°08′N	81°08′W
Clover Bar, Can. (klō´vẽr bär)	62g	53°34′N	113°20′W
Cloverdale, Can.	74d	49°06′N	122°44′W
Cloverdale, Ca., U.S. (klō´vẽr-dāl)	76	38°47′N	123°03′W
Cloverport, Ky., U.S. (klō´vẽr pōrt)	66	37°50′N	86°35′W
Clovis, N.M., U.S. (klō´vĭs)	64	34°24′N	103°11′W
Cluj-Napoca, Rom.	110	46°46′N	23°34′E
Clun, r., Eng., U.K. (klŭn)	114a	52°25′N	2°56′W
Cluny, Fr. (klü-nē´)	126	46°27′N	4°40′E
Clutha, r., N.Z. (klōō´thȧ)	175a	45°52′S	169°30′E
Clwyd, hist. reg., Wales, U.K.	114a	53°01′N	2°59′W
Clyde, Ks., U.S.	79	39°34′N	97°23′W
Clyde, Oh., U.S.	66	41°15′N	83°00′W
Clyde, r., Scot., U.K.	120	55°35′N	3°50′W
Clyde, Firth of, b., Scot., U.K. (fûrth ŏv klīd)	120	55°28′N	5°01′W
Côa, r., Port. (kō´ä)	128	40°28′N	6°55′W
Coacalco, Mex. (kō-ä-käl´kō)	89a	19°37′N	99°06′W
Coachella, Canal, can., Ca., U.S. (kō´chĕl-lȧ)	76	33°15′N	115°25′W
Coahuayana, Río de, r., Mex. (rē´ō-dĕ-kô-ä-wä-yá´nä)	88	19°00′N	103°33′W
Coahuayutla, Mex. (kō´ä-wī-yōōt´lä)	88	18°19′N	101°44′W
Coahuila, state, Mex. (kō-ä-wē´lä)	86	27°30′N	103°00′W
Coal City, Il., U.S. (kōl sĭ´tĭ)	69a	41°17′N	88°17′W
Coalcomán, Río de, r., Mex. (rē´ō-dĕ-kō-äl-kō-män´)	88	18°45′N	103°15′W
Coalcomán, Sierra de, mts., Mex.	88	18°30′N	102°45′W
Coalcomán de Matamoros, Mex.	88	18°46′N	103°10′W
Coaldale, Can. (kōl´dāl)	55	49°43′N	112°37′W
Coalgate, Ok., U.S. (kōl´gāt)	79	34°44′N	96°13′W
Coal Grove, Oh., U.S. (kōl grōv)	66	38°20′N	82°40′W
Coalinga, Ca., U.S. (kō-ȧ-lĭŋ´gȧ)	76	36°09′N	120°23′W
Coalville, Eng., U.K. (kōl´vĭl)	114a	52°43′N	1°21′W
Coamo, P.R. (kō-ä´mō)	87b	18°05′N	66°21′W
Coari, Braz. (kō-är´ē)	100	4°06′S	63°10′W
Coast Mountains, mts., N.A. (kōst)	52	54°10′N	128°00′W
Coast Ranges, mts., U.S.	64	41°28′N	123°30′W
Coatepec, Mex. (kō-ä-tā-pĕk)	88	19°23′N	98°44′W
Coatepec, Mex.	89	19°26′N	96°56′W
Coatepec, Mex.	89a	19°08′N	99°25′W
Coatepeque, El Sal.	90	13°56′N	89°30′W
Coatepeque, Guat. (kō-ä-tā-pä´kā)	90	14°40′N	91°52′W
Coatesville, Pa., U.S. (kōts´vĭl)	67	40°00′N	75°50′W
Coatetelco, Mex. (kō-ä-tâ-tĕl´kō)	88	18°43′N	99°17′W
Coaticook, Can. (kō´tĭ-kŏŏk)	59	45°10′N	71°55′W
Coatlinchán, Mex. (kô-ä-tlĕ´n-chä´n)	89a	19°26′N	98°52′W
Coats, i., Can. (kōts)	53	62°23′N	82°11′W
Coats Land, reg., Ant.	178	74°00′S	30°00′W
Coatzacoalcos, Mex.	86	18°09′N	94°26′W
Coatzacoalcos, r., Mex.	89	17°40′N	94°41′W
Coba, hist., Mex. (kô´bä)	90a	20°23′N	87°23′W
Cobalt, Can. (kō´bôlt)	51	47°21′N	79°40′W
Cobán, Guat. (kō-bän´)	86	15°28′N	90°19′W
Cobar, Austl.	175	31°28′S	145°50′E
Cobberas, Mount, mtn., Austl. (cŏ-bĕr-ăs)	176	36°45′S	148°15′E
Cobequid Mountains, mts., Can.	60	45°35′N	64°10′W
Cobh, Ire. (kôv)	110	51°52′N	8°09′W
Cobija, Bol. (kô-bē´hä)	100	11°12′S	68°49′W
Cobourg, Can. (kō´bôrgh)	51	43°55′N	78°05′W
Cobre, r., Jam. (kō´brä)	92	18°05′N	77°00′W
Coburg, Austl.	173a	37°45′S	144°58′E
Coburg, Ger. (kō´bōōrg)	124	50°16′N	10°57′E
Cocentaina, Spain (kō-thân-tä-ē´nȧ)	129	38°44′N	0°27′W
Cochabamba, Bol.	100	17°24′S	66°09′W
Cochinos, Bahía, b., Cuba (bä-ē´ä-kô-chē´nōs)	92	22°05′N	81°10′W
Cochinos Banks, bk.	92	22°20′N	76°15′W
Cochiti Indian Reservation, I.R., N.M., U.S.	77	35°37′N	106°20′W
Cochran, Ga., U.S. (kŏk´răn)	82	32°23′N	83°23′W
Cochrane, Can. (kŏk´răn)	51	49°01′N	81°06′W
Cochrane, Can.	62e	51°11′N	114°28′W
Cockburn, i., Can. (kŏk-bûrn)	58	45°55′N	83°25′W
Cockeysville, Md., U.S. (kŏk´ĭz-vĭl)	68e	39°30′N	76°40′W
Cockrell Hill, Tx., U.S. (kŏk´rĕl)	75c	32°44′N	96°53′W
Coco, r., N.A.	87	14°55′N	83°45′W
Coco, Cayo, i., Cuba (kä´-yō-kō´kô)	92	22°30′S	78°30′W
Coco, Isla del, i., C.R. (ē´s-lä-dĕl-kô-kô)	86	5°33′N	87°02′W
Cocoa, Fl., U.S. (kō´kō)	83a	28°21′N	80°44′W
Cocoa Beach, Fl., U.S.	83a	28°20′N	80°35′W
Cocoli, Pan. (kŏ-kŏ´lē)	86a	8°58′N	79°36′W
Coconino, Plateau, plat., Az., U.S. (kō kō nē´nō)	77	35°45′N	112°28′W
Cocos (Keeling) Islands, is., Oc. (kō´kōs) (kē´lĭng)	3	11°50′S	90°50′E
Coco Solito, Pan. (kô-kô-sŏ-lē´tō)	86a	9°21′N	79°53′W
Cocula, Mex. (kō-kōō´lä)	88	20°23′N	103°47′W
Cocula, r., Mex.	88	18°17′N	99°45′W
Cod, Cape, pen., Ma., U.S.	65	41°42′N	70°15′W
Codajás, Braz. (kō-dä-häzh´)	100	3°44′S	62°09′W
Codera, Cabo, c., Ven. (kȧ´bô-kō-dĕ´rä)	101b	10°35′N	66°06′W
Codogno, Italy (kô-dō´nyō)	130	45°08′N	9°43′E

ng-sing; ŋ-baŋk; N-nasalized n; nŏd; cŏmmit; ōld; ôbey; ôrder; oi-boil; fōōd; ŏ-as oo in foot; ou-out; s-soft; sh-dish; th-thin; pūre; ûnite; ûrn; stŭd; circŭs; ü-as in French tu; ´-indeterminate vowel.

238 Codrington – Copano Bay

PRONOUNCING INDEX

PLACE (Pronunciation)	PAGE	LAT.	LONG.
Codrington, Antig. (kŏd′rĭng-tŭn)	91b	17°39′N	61°49′W
Cody, Wy., U.S. (kō′dĭ)	73	44°31′N	109°02′W
Coelho da Rocha, Braz.	102b	22°47′S	43°23′W
Coemba, Ang.	190	12°08′S	18°05′E
Coesfeld, Ger. (kûs′fĕld)	127c	51°56′N	7°10′E
Coeur d'Alene, Id., U.S. (kûr dȧ-lān′)	64	47°43′N	116°35′W
Coeur d'Alene, r., Id., U.S.	72	47°26′N	116°35′W
Coeur d'Alene Indian Reservation, I.R., Id., U.S.	72	47°18′N	116°45′W
Coeur d'Alene Lake, l., Id., U.S.	72	47°32′N	116°39′W
Coffeyville, Ks., U.S. (kŏf′ĭ-vĭl)	65	37°01′N	95°38′W
Coff's Harbour, Austl.	176	30°20′S	153°10′E
Cofimvaba, S. Afr. (căfĭm′vä-bȧ)	187c	32°01′S	27°37′E
Coghinas, r., Italy (kō′gē-nȧs)	130	40°31′N	9°00′E
Cognac, Fr. (kŏn-yak′)	117	45°41′N	0°22′W
Cohasset, Ma., U.S. (kŏ-hăs′ĕt)	61a	42°14′N	70°48′W
Cohoes, N.Y., U.S. (kô-hōz′)	67	42°50′N	73°40′W
Coig, r., Arg. (kŏ′ĕk)	102	51°15′N	71°00′W
Coimbatore, India (kŏ-ēm-bȧ-tŏr′)	155	11°03′N	76°56′E
Coimbra, Port. (kŏ-ēm′brä)	110	40°14′N	8°23′W
Coín, Spain (kŏ-ēn′)	128	36°40′N	4°45′W
Coina, Port. (kŏ-ē′nȧ)	129b	38°35′N	9°03′W
Coina, r., Port. (kŏ′y-nä)	129b	38°35′N	9°02′W
Coipasa, Salar de, pl., Bol. (sä-lä′r-dĕ-koi-pä′-sä)	100	19°12′S	69°13′W
Coixtlahuaca, Mex. (kō-ēks′tlä-wä′kä)	89	17°42′N	97°17′W
Cojedes, dept., Ven. (kŏ-kĕ′dĕs)	101b	9°50′N	68°21′W
Cojimar, Cuba (kō-hĕ-mär′)	93a	23°10′N	82°19′W
Cojutepeque, El Sal. (kŏ-hò-tĕ-pā′kå)	90	13°45′N	88°50′W
Cokato, Mn., U.S. (kŏ-kā′tō)	71	45°03′N	94°11′W
Cokeburg, Pa., U.S. (kōk bŭgh)	69e	40°06′N	80°03′W
Colac, Austl. (kŏ′lac)	176	38°25′S	143°40′E
Colares, Port. (kŏ-lä′rĕs)	129b	38°47′N	9°27′W
Colatina, Braz. (kô-lä-tĕ′nä)	101	19°33′S	40°42′W
Colby, Ks., U.S. (kōl′bĭ)	78	39°23′N	101°04′W
Colchagua, prov., Chile (kŏl-chá′gwä)	99b	34°42′S	71°24′W
Colchester, Eng., U.K. (kōl′chĕs-tēr)	121	51°52′N	0°50′E
Cold Lake, l., Can. (kōld)	56	54°33′N	110°05′W
Coldwater, Ks., U.S. (kōld′wô-tēr)	78	37°14′N	99°21′W
Coldwater, Mi., U.S.	66	41°55′N	85°00′W
Coldwater, r., Ms., U.S.	82	34°25′N	90°12′W
Coldwater Creek, r., Tx., U.S.	78	36°10′N	101°45′W
Coleman, Tx., U.S. (kōl′mȧn)	80	31°50′N	99°25′W
Colenso, S. Afr. (kŏ-lĕnz′ŏ)	187c	28°48′S	29°49′E
Coleraine, N. Ire., U.K.	120	55°08′N	6°40′W
Coleraine, Mn., U.S. (kōl-rān′)	71	47°16′N	93°29′W
Coleshill, Eng., U.K. (kōlz′hĭl)	114a	52°30′N	1°42′W
Colfax, Ia., U.S. (kōl′făks)	71	41°40′N	93°13′W
Colfax, La., U.S.	81	31°31′N	92°42′W
Colfax, Wa., U.S.	72	46°53′N	117°21′W
Colhué Huapi, l., Arg. (kŏl-wā′oȧ′pĕ)	102	45°30′S	68°45′W
Coligny, S. Afr.	192c	26°20′S	26°18′E
Colima, Mex. (kōlē′mä)	86	19°13′N	103°45′W
Colima, state, Mex.	88	19°10′N	104°00′W
Colima, Nevado de, mtn., Mex. (nĕ-vä′dŏ-dĕ-kŏ-lē′mä)	86	19°30′N	103°38′W
Coll, i., Scot., U.K. (kōl)	120	56°42′N	6°23′W
College, Ak., U.S.	63	64°43′N	147°50′W
College Park, Ga., U.S. (kŏl′ĕj)	68c	33°39′N	84°27′W
College Park, Md., U.S.	68e	38°59′N	76°58′W
Collegeville, Pa., U.S. (kŏl′ĕj-vĭl)	68f	40°11′N	75°27′W
Collie, Austl. (kŏl′ĕ)	174	33°20′S	116°20′E
Collier Bay, b., Austl. (kŏl-yēr)	174	15°30′S	123°30′E
Collingswood, N.J., U.S. (kŏl′ĭngz-wŏd)	68f	39°54′N	75°04′W
Collingwood, Can.	59	44°30′N	80°20′W
Collins, Ms., U.S. (kŏl′ĭns)	82	31°40′N	89°34′W
Collinsville, Il., U.S. (kŏl′ĭnz-vĭl)	75e	38°41′N	89°59′W
Collinsville, Ok., U.S.	79	36°21′N	95°50′W
Colmar, Fr. (kŏl′mär)	117	48°03′N	7°25′E
Colmenar de Oreja, Spain (kŏl-mā-när′dāōrā′hä)	128	40°06′N	3°25′W
Colmenar Viejo, Spain (kŏl-mā-när′vyā′hō)	128	40°40′N	3°46′W
Cologne, Ger.	110	50°56′N	6°57′E
Colombia, Col. (kŏ-lŏm′bĕ-ä)	100a	3°23′N	74°48′W
Colombia, nation, S.A.	100	3°30′N	72°30′W
Colombo, Sri L. (kŏ-lŏm′bō)	159	6°58′N	79°52′E
Colón, Arg. (kō-lōn′)	99c	33°55′S	61°08′W
Colón, Cuba (kŏ-lŏ′n)	92	22°45′N	80°55′W
Colón, Mex. (kō-lōn′)	88	20°46′N	100°02′W
Colón, Pan. (kŏ-lô′n)	87	9°22′N	79°54′W
Colón, Archipiélago de, is., Ec.	100	0°10′S	87°45′W
Colón, Montañas de, mts., Hond. (môn-tä′n-yäs-dĕ-kŏ-lô′n)	91	14°58′N	84°39′W
Colonia, Ur. (kō-lō′nĕ-ä)	102	34°27′S	57°50′W
Colonia, dept., Ur.	99c	34°08′S	57°50′W
Colonia Suiza, Ur. (kŏ-lô′nĕä-sòē′zä)	99c	34°17′S	57°15′W
Colonna, Capo, c., Italy	131	39°02′N	17°15′E
Colonsay, i., Scot., U.K. (kŏl-ŏn-sā′)	121	56°08′N	6°08′E
Coloradas, Lomas, Arg. (lŏ′mäs-kō-lō-rä′däs)	102	43°30′S	68°00′W
Colorado, state, U.S.	64	39°30′N	106°55′W
Colorado, r., Arg.	102	38°30′S	66°00′W
Colorado, r., N.A.	64	36°00′N	113°30′W
Colorado, r., Tx., U.S.	64	30°08′N	97°33′W
Colorado City, Tx., U.S. (kŏl-ŏ-rä′dō sĭ′tĭ)	80	32°24′N	100°50′W
Colorado National Monument, rec., Co., U.S.	77	39°00′N	108°40′W
Colorado Plateau, plat., U.S.	64	36°20′N	109°25′W
Colorado River Aqueduct, aq., Ca., U.S.	76	33°38′N	115°43′W
Colorado River Indian Reservation, I.R., Az., U.S.	77	34°03′N	114°02′W
Colorados, Archipiélago de los, is., Cuba	92	22°25′N	84°25′W

PLACE (Pronunciation)	PAGE	LAT.	LONG.
Colorado Springs, Co., U.S. (kŏl-ŏ-rä′dō)	64	38°49′N	104°48′W
Colotepec, r., Mex. (kô-lŏ′tĕ-pĕk)	89	15°56′N	96°57′W
Colotlán, Mex. (kŏ-lŏ-tlän′)	88	22°06′N	103°14′W
Colotlán, r., Mex.	88	22°09′N	103°17′W
Colquechaca, Bol. (kŏl-kā-chä′kä)	100	18°47′S	66°02′W
Colstrip, Mt., U.S. (kōl′strip)	73	45°54′N	106°38′W
Colton, Ca., U.S. (kōl′tŭn)	75a	34°04′N	117°20′W
Columbia, Il., U.S. (kŏ-lŭm′bĭ-ȧ)	75e	38°26′N	90°12′W
Columbia, Ky., U.S.	82	37°06′N	85°15′W
Columbia, Md., U.S.	68e	39°15′N	76°51′W
Columbia, Mo., U.S.	65	38°55′N	92°19′W
Columbia, Ms., U.S.	82	31°15′N	89°49′W
Columbia, Pa., U.S.	67	40°00′N	76°25′W
Columbia, S.C., U.S.	65	34°00′N	81°00′W
Columbia, Tn., U.S.	82	35°36′N	87°02′W
Columbia, r., N.A.	52	46°00′N	120°00′W
Columbia, Mount, mtn., Can.	55	52°09′N	117°25′W
Columbia City, In., U.S.	66	41°10′N	85°30′W
Columbia City, Or., U.S.	74c	45°53′N	112°49′W
Columbia Heights, Mn., U.S.	75g	45°03′N	93°15′W
Columbia Icefield, ice, Can.	55	52°08′N	117°26′W
Columbia Mountains, mts., N.A.	55	51°30′N	118°30′W
Columbiana, Al., U.S. (kô-ŭm-bĭ-ă′nȧ)	82	33°10′N	86°35′W
Columbretes, is., Spain (kŏ-lōōm-brĕ′tĕs)	129	39°54′N	0°54′E
Columbus, Ga., U.S. (kŏ-lŭm′bŭs)	65	32°29′N	84°56′W
Columbus, In., U.S.	66	39°15′N	85°55′W
Columbus, Ks., U.S.	79	37°10′N	94°50′W
Columbus, Ms., U.S.	82	33°30′N	88°25′W
Columbus, Mt., U.S.	73	45°39′N	109°15′W
Columbus, Ne., U.S.	70	41°25′N	97°25′W
Columbus, N.M., U.S.	77	31°50′N	107°40′W
Columbus, Oh., U.S.	65	40°00′N	83°00′W
Columbus, Tx., U.S.	81	29°44′N	96°34′W
Columbus, Wi., U.S.	71	43°20′N	89°01′W
Columbus Bank, bk., (kŏ-lŭm′bŭs)	93	22°05′N	75°30′W
Columbus Grove, Oh., U.S.	66	40°55′N	84°05′W
Columbus Point, c., Bah.	93	24°10′N	75°15′W
Colusa, Ca., U.S. (kŏ-lū′sȧ)	76	39°12′N	122°01′W
Colville, Wa., U.S. (kŏl′vĭl)	72	48°33′N	117°53′W
Colville, r., Ak., U.S.	63	69°00′N	156°25′W
Colville Indian Reservation, I.R., Wa., U.S.	72	48°15′N	119°00′W
Colville R, Wa., U.S.	72	48°25′N	117°58′W
Colvos Passage, strt., Wa., U.S. (kōl′vōs)	74a	47°24′N	122°32′W
Colwood, Can. (kŏl′wòd)	74a	48°26′N	123°30′W
Comacchio, Italy (kŏ-mäk′kyō)	130	44°42′N	12°12′E
Comala, Mex. (kō-mä-lä′)	88	19°22′N	103°47′W
Comalapa, Guat. (kō-mä-lä′-pä)	90	14°43′N	90°56′W
Comalcalco, Mex. (kō-mäl-käl′kō)	89	18°16′N	93°13′W
Comanche, Ok., U.S. (kŏ-măn′chĕ)	79	34°20′N	97°58′W
Comanche, Tx., U.S.	80	31°54′N	98°37′W
Comanche Creek, r., Tx., U.S.	80	31°02′N	102°47′W
Comayagua, Hond. (kō-mä-yä′gwä)	86	14°24′N	87°36′W
Combahee, r., S.C., U.S. (kŏm-bȧ-hē′)	83	32°42′N	80°40′W
Comer, Ga., U.S. (kŭm′ēr)	82	34°02′N	83°07′W
Comete, Cape, c., T./C. Is. (kŏ-mā′tå)	93	21°45′N	71°25′W
Comilla, Bngl. (kŏ-mĭl′ä)	155	23°33′N	91°17′E
Comino, Cape, c., Italy (kŏ-mē′nō)	130	40°30′N	9°48′E
Comitán, Mex. (kŏ-mē-tän′)	86	16°16′N	92°09′W
Commencement Bay, b., Wa., U.S. (kŏ-mĕns′mĕnt bā)	74a	47°17′N	122°21′W
Commentry, Fr. (kŏ-mäN-trē′)	126	46°16′N	2°44′E
Commerce, Ga., U.S. (kŏm′ērs)	82	34°10′N	83°27′W
Commerce, Ok., U.S.	79	36°57′N	94°54′W
Commerce, Tx., U.S.	79	33°15′N	95°52′W
Como, Italy (kō′mō)	118	45°48′N	9°03′E
Como, Lago di, l., Italy (lä′gō-dē-kō′mō)	118	46°00′N	9°30′E
Comodoro Rivadavia, Arg.	102	45°47′S	67°31′W
Como-Est, Can.	62a	45°27′N	74°08′W
Comonfort, Mex. (kŏ-mōn-fô′rt)	88	20°43′N	100°47′W
Comorin, Cape, c., India (kŏ′mŏ-rĭn)	159	8°05′N	78°05′E
Comoros, nation, Afr.	187	12°30′S	42°45′E
Comox, Can. (kō′mŏks)	54	49°40′N	124°55′W
Companario, Cerro, mtn., S.A. (sĕ′r-rŏ-kôm-pä-nä′ryŏ)	99b	35°54′S	70°23′W
Compiègne, Fr. (kôN-pyĕn′y′)	117	49°25′N	2°49′E
Comporta, Port. (kŏm-pŏr′tȧ)	129b	38°24′N	8°48′W
Compostela, Mex. (kōm-pŏ-stā′lä)	88	21°14′N	104°54′W
Compton, Ca., U.S. (kŏmpt′tŭn)	75a	33°54′N	118°14′W
Comrat, Mol. (kŏm-rät′)	137	46°17′N	28°38′E
Conakry, Gui. (kŏ-nȧ-krē′)	184	9°31′N	13°43′W
Conanicut, i., R.I., U.S. (kŏn′ȧ-nĭ-kŭt)	68b	41°34′N	71°20′W
Conasauga, r., Ga., U.S. (kŏ-nȧ)	82	34°40′N	84°51′W
Concarneau, Fr. (kŏN-kär-nō′)	126	47°54′N	3°52′W
Concepción, Bol. (kŏn-sĕp′syōn′)	101	15°47′S	61°08′W
Concepción, Chile	102	36°51′S	72°59′W
Concepción, Pan.	91	8°31′N	82°38′W
Concepción, Para.	102	23°29′S	57°18′W
Concepcion, Phil.	169a	15°19′N	120°40′E
Concepción, vol., Nic.	90	11°36′N	85°43′W
Concepción, r., Mex.	86	30°25′N	112°20′W
Concepción del Mar, Guat. (kŏn-sĕp-syōn′dĕl mär′)	90	14°07′N	91°23′W
Concepción del Oro, Mex. (kŏn-sĕp-syōn′ dĕl ō′rō)	86	24°39′N	101°24′W
Concepción del Uruguay, Arg. (kŏn-sĕp-syô′n-dĕl-ōō-rōō-gwī′)	102	32°31′S	58°10′W
Conception, i., Bah.	93	23°50′N	75°05′W
Conception, Point, c., Ca., U.S.	64	34°27′N	120°28′W
Conception Bay, b., Can. (kŏn-sĕp′shŭn)	61	47°50′N	52°50′W
Concho, r., Tx., U.S. (kŏn′chō)	80	31°34′N	100°00′W

PLACE (Pronunciation)	PAGE	LAT.	LONG.
Conchos, r., Mex.	86	29°30′N	105°00′W
Conchos, r., Mex. (kōn′chōs)	80	25°03′N	99°00′W
Concord, Ca., U.S. (kŏŋ′kŏrd)	74b	37°58′N	122°02′W
Concord, Ma., U.S.	61a	42°28′N	71°21′W
Concord, N.C., U.S.	83	35°23′N	80°11′W
Concord, N.H., U.S.	65	43°10′N	71°30′W
Concordia, Arg. (kŏn-kôr′dĭ-ȧ)	102	31°18′S	57°59′W
Concordia, Col.	100a	6°04′N	75°54′W
Concordia, Mex. (kŏn-kô′r-dyä)	88	23°17′N	106°06′W
Concordia, Ks., U.S.	79	39°32′N	97°39′W
Concrete, Wa., U.S. (kŏn-′krēt)	72	48°33′N	121°44′W
Conde, Fr.	126	48°50′N	0°36′W
Conde, S.D., U.S. (kŏn-dē′)	70	45°10′N	98°06′W
Condega, Nic. (kŏn-dĕ′gä)	90	13°20′N	86°27′W
Condeúba, Braz. (kōn-dā-ōō′bä)	101	14°47′S	41°44′W
Condom, Fr.	126	43°58′N	0°22′E
Condon, Or., U.S. (kŏn′dŭn)	72	45°14′N	120°10′W
Conecun, r., Al., U.S. (kŏ-nē′kŭ)	82	31°05′N	86°52′W
Conegliano, Italy (kō-nål-yä′nō)	130	45°59′N	12°17′E
Conejos, r., Co., U.S. (kŏ-nā′hōs)	77	37°07′N	106°19′W
Conemaugh, Pa., U.S. (kŏn′ĕ-mô)	67	40°25′N	78°50′W
Coney Island, i., N.Y., U.S. (kō′nĭ)	68a	40°34′N	73°27′W
Confolens, Fr. (kôN-fä-läN′)	126	46°01′N	0°41′E
Congaree, r., S.C., U.S. (kŏŋ-gȧ-rē′)	83	33°53′N	80°55′W
Conghua, China (tsŏŋ-hwä)	165	23°30′N	113°40′E
Congleton, Eng., U.K. (kŏŋ′g′l-tŭn)	114a	53°10′N	2°13′W
Congo, nation, Afr. (kŏn′gō)	186	3°00′S	13°48′E
Congo (Zaire), r., Afr. (kŏn′gō)	183	2°00′S	17°00′E
Congo, Democratic Republic of the (Zaire), nation, Afr.	186	1°00′S	22°15′E
Congo, Serra do, mts., Ang.	190	6°25′S	13°30′E
Congo Basin, basin, D.R.C.	183	2°47′N	20°58′E
Conisbrough, Eng., U.K. (kŏn′ĭs-bŭr-ŏ)	114a	53°29′N	1°13′W
Coniston, Can.	59	46°29′N	80°51′W
Conklin, Can. (kŏŋk′lĭn)	55	55°38′N	111°05′W
Conley, Ga., U.S. (kŏn′lĭ)	68c	33°38′N	84°19′W
Conn, Lough, l., Ire. (lŏk kŏn)	120	53°56′N	9°25′W
Connacht, hist. reg., Ire. (cŏn′åt)	120	53°50′N	8°45′W
Conneaut, Oh., U.S. (kŏn-ĕ-ôt′)	66	41°55′N	80°35′W
Connecticut, state, U.S. (kŏ-nĕt′ĭ-kŭt)	65	41°40′N	73°10′W
Connecticut, r., U.S.	65	43°55′N	72°15′W
Connellsville, Pa., U.S. (kŏn′nĕlz-vĭl)	67	40°00′N	79°40′W
Connemara, mts., Ire. (kŏn-nĕ-mȧ′rȧ)	120	53°30′N	9°54′W
Connersville, In., U.S. (kŏn′ērz-vĭl)	66	39°35′N	85°10′W
Connors Range, mts., Austl. (kŏn′nŏrs)	175	22°15′S	149°00′E
Conrad, Mt., U.S. (kŏn′răd)	73	48°11′N	111°56′W
Conrich, Can. (kŏn′rĭch)	62e	51°06′N	113°51′W
Conroe, Tx., U.S. (kŏn′rō)	81	30°18′N	95°23′W
Conselheiro Lafaiete, Braz.	101	20°40′S	43°46′W
Conshohocken, Pa., U.S. (kŏn-shŏ-hŏk′ĕn)	68f	40°04′N	75°18′W
Consolación del Sur, Cuba (kōn-sŏ-lä-syōn′)	92	22°30′N	83°55′W
Con Son, is., Viet.	168	8°30′N	106°28′E
Constance, Mount, mtn., Wa., U.S. (kŏn′stăns)	74a	47°46′N	123°08′W
Constanţa, Rom. (kōn-stȧn′tsȧ)	110	44°12′N	28°36′E
Constantina, Spain (kŏn-stän-tē′nä)	128	37°52′N	5°39′W
Constantine, Alg. (kôN-stäN′tēn′)	184	36°28′N	6°38′E
Constantine, Mi., U.S. (kŏn′stăn-tēn)	66	41°50′N	85°40′W
Constitución, Chile (kōn′stĭ-tōō-syōn′)	102	35°24′S	72°25′W
Constitution, Ga., U.S. (kŏn-stĭ-tū′shŭn)	68c	33°41′N	84°20′W
Contagem, Braz. (kōn-tȧ′zhĕm)	99a	19°54′S	44°05′W
Contepec, Mex. (kōn-tĕ-pĕk′)	88	20°04′N	100°07′W
Contreras, Mex. (kôn-trĕ′räs)	89a	19°18′N	99°14′W
Contwoyto, l., Can.	52	65°42′N	110°50′W
Converse, Tx., U.S. (kōn′vērs)	75d	29°31′N	98°17′W
Conway, Ar., U.S. (kŏn′wā)	79	35°06′N	92°27′W
Conway, N.H., U.S.	67	44°00′N	71°10′W
Conway, S.C., U.S.	83	33°49′N	79°01′W
Conway, Wa., U.S.	74a	48°20′N	122°20′W
Conyers, Ga., U.S. (kŏn′yōrz)	82	33°41′N	84°01′W
Cooch Behār, India (kòch bĕ-här′)	155	26°25′N	89°34′E
Cook, Cape, c., Can. (kòk)	54	50°08′N	127°55′W
Cook, Mount *see* Aoraki, mtn., N.Z.	175a	43°27′S	170°13′E
Cookeville, Tn., U.S. (kòk′vĭl)	82	36°07′N	85°30′W
Cooking Lake, Can. (kōōk′ĭng)	62g	53°25′N	113°08′W
Cooking Lake, l., Can.	62g	53°25′N	113°02′W
Cook Inlet, b., Ak., U.S.	63	60°50′N	151°38′W
Cook Islands, dep., Oc.	2	20°00′S	158°00′W
Cook Strait, strt., N.Z.	175a	40°37′S	174°15′E
Cooktown, Austl. (kòk′toun)	175	15°40′S	145°20′E
Cooleemee, N.C., U.S. (kōō-lē′mē)	83	35°50′N	80°32′W
Coolgardie, Austl. (kōōl-gär′dĕ)	174	31°00′S	121°25′E
Cooma, Austl. (kōō′mȧ)	175	36°22′S	149°10′E
Coonamble, Austl. (kōō-năm′b′l)	175	31°00′S	148°30′E
Coonoor, India	159	10°22′N	76°15′E
Coon Rapids, Mn., U.S. (kòn)	75g	45°09′N	93°17′W
Cooper, Tx., U.S. (kōōp′ēr)	79	33°23′N	95°40′W
Cooper Center, Ak., U.S.	63	61°54′N	15°30′W
Coopers Creek, r., Austl. (kōō′pērz)	175	27°32′N	141°19′E
Cooperstown, N.D., U.S.	70	47°26′N	98°07′W
Cooperstown, N.Y., U.S. (kōōp′ērs-toun)	67	42°45′N	74°55′W
Coosa, Al., U.S. (kōō′sȧ)	82	32°43′N	86°25′W
Coosa, r., U.S.	65	34°00′N	86°00′W
Coosawattee, r., Ga., U.S.	82	34°37′N	84°45′W
Coos Bay, Or., U.S. (kōōs)	72	43°21′N	124°12′W
Coos Bay, b., Or., U.S.	72	43°19′N	124°40′W
Cootamundra, Austl. (kòtă-mŭnd′ră)	176	34°25′S	148°00′E
Copacabana, Braz. (kô′pä-kȧ-bȧ′nä)	102b	22°57′S	43°11′W
Copalita, r., Mex. (kŏ-pä-lē′tä)	89	15°55′N	96°06′W
Copán, hist., Hond. (kō-pän′)	90	14°50′N	89°10′W
Copano Bay, b., Tx., U.S. (kō-pän′ō)	81	28°08′N	97°25′W

ăt; finăl; rāte; senåte; ärm; ȧsk; sofȧ; fâre; ch-choose; dh-as th in other; bē; ĕvent; bĕt; recĕnt; cratēr; g-gō; gh-guttural g; bĭt; ĭ-short neutral; rīde; K-guttural k as ch in German ich;

PRONOUNCING INDEX

Copenhagen – Crofton 239

PLACE (Pronunciation)	PAGE	LAT.	LONG.
Copenhagen (København), Den.	110	55°43'N	12°27'E
Copiapó, Chile (kō-pyä-pō')	102	27°16'S	70°28'W
Copley, Oh., U.S. (kŏp'lē)	69d	41°06'N	81°38'W
Copparo, Italy (kôp-pä'rō)	130	44°53'N	11°50'E
Coppell, Tx., U.S. (kŏp'pĕl)	75c	32°57'N	97°00'W
Copper, r., Ak., U.S. (kŏp'ẽr)	63	62°38'N	145°00'W
Copper Cliff, Can.	58	46°28'N	81°04'W
Copper Harbor, Mi., U.S.	71	47°27'N	87°53'W
Copperhill, Tn., U.S. (kŏp'ẽr hĭl)	82	35°00'N	84°22'W
Coppermine *see* Kugluktuk, Can.	52	67°46'N	115°19'W
Coppermine, r., Can.	52	66°48'N	114°59'W
Copper Mountain, mtn., Ak., U.S.	54	55°14'N	132°36'W
Copperton, Ut., U.S. (kŏp'ẽr-tŭn)	75b	40°34'N	112°06'W
Coquilee, Or., U.S. (kô-kēl')	72	43°11'N	124°11'W
Coquilhatville *see* Mbandaka, D.R.C.	186	0°04'N	18°16'E
Coquimbo, Chile (kô-kēm'bō)	102	29°58'S	71°31'W
Coquimbo, prov., Chile	99b	31°50'S	71°05'W
Coquitlam Lake, l., Can. (kō-kwĭt-lám)	74d	49°23'N	122°44'W
Corabia, Rom. (kô-rä'bĭ-á)	119	43°45'N	24°29'E
Coracora, Peru (kō'rä-kō'rä)	100	15°12'S	73°42'W
Coral Gables, Fl., U.S.	83a	25°43'N	80°14'W
Coral Rapids, Can. (kŏr'ál)	51	50°18'N	81°49'W
Coral Sea, sea, Oc. (kŏr'ál)	175	13°30'S	150°00'E
Coralville Reservoir, res., Ia., U.S.	71	41°45'N	91°50'W
Corangamite, Lake, l., Austl. (cŏr-ăng'á-mīt)	176	38°05'S	142°55'E
Coraopolis, Pa., U.S. (kō-rä-ŏp'ō-lĭs)	69e	40°30'N	80°09'W
Corato, Italy (kô'rä-tô)	130	41°08'N	16°28'E
Corbeil-Essonnes, Fr. (kôr-bâ'yĕ-sŏn')	126	48°31'N	2°29'E
Corbett, Or., U.S. (kôr'bĕt)	74c	45°31'N	122°17'W
Corbie, Fr. (kôr-bē')	126	49°55'N	2°27'E
Corbin, Ky., U.S. (kôr'bĭn)	82	36°55'N	84°06'W
Corby, Eng., U.K. (kôr'bĭ)	114a	52°29'N	0°38'W
Corcovado, mtn., Braz. (kôr-kô-vä'dò)	102b	22°57'S	43°13'W
Corcovado, Golfo, b., Chile (kôr-kô-vä'dhō)	102	43°40'S	75°00'W
Cordeiro, Braz. (kôr-dā'rō)	99a	22°03'S	42°22'W
Cordele, Ga., U.S. (kôr-dēl')	82	31°55'N	83°50'W
Cordell, Ok., U.S. (kôr-dĕl')	78	35°19'N	98°58'W
Córdoba, Arg. (kôr'dō-vä)	102	30°20'S	64°03'W
Córdoba, Mex. (kô'r-dō-bä)	86	18°53'N	96°54'W
Córdoba, Spain (kô'r-dô-bä)	128	37°55'N	4°45'W
Córdoba, prov., Arg. (kôr'dô-vä)	102	32°00'S	64°00'W
Córdoba, Sierra de, mts., Arg.	102	31°15'S	64°30'W
Cordova, Ak., U.S. (kôr'dô-vä)	64a	60°34'N	145°38'W
Cordova, Al., U.S. (kôr'dô-á)	82	33°45'N	86°22'W
Cordova Bay, b., Ak., U.S.	54	54°55'N	132°35'W
Corfu *see* Kérkyra, i., Grc.	112	39°33'N	19°36'E
Corigliano, Italy (kō-rē-lyä'nō)	130	39°35'N	16°30'E
Corinth *see* Kórinthos, Grc.	110	37°56'N	22°54'E
Corinth, Ms., U.S. (kŏr'ĭnth)	82	34°55'N	88°30'W
Corinto, Braz. (kô-rē'n-tō)	101	18°20'S	44°16'W
Corinto, Col.	100a	3°09'N	76°12'W
Corinto, Nic. (kŏr-ĭn'to)	90	12°30'N	87°12'W
Corio, Austl.	173a	38°05'S	144°22'E
Corio Bay, b., Austl.	173a	38°07'S	144°25'E
Corisco, Isla de, i., Eq. Gui.	190	0°50'N	8°40'E
Cork, Ire. (kôrk)	110	51°54'N	8°25'W
Cork Harbour, b., Ire.	120	51°44'N	8°15'W
Corleone, Italy (kôr-lå-ō'nä)	130	37°48'N	13°18'E
Cormorant Lake, l., Can.	57	54°13'N	100°47'W
Cornelia, Ga., U.S. (kôr-nē'lyá)	82	34°31'N	83°30'W
Cornelis, r., S. Afr. (kôr-nē'lĭs)	192c	27°48'S	29°15'E
Cornell, Ca., U.S. (kôr-nĕl')	75a	34°06'N	118°46'W
Cornell, Wi., U.S.	71	45°10'N	91°10'W
Corner Brook, Can. (kôr'nẽr)	51	48°57'N	57°57'W
Corner Inlet, b., Austl.	176	38°55'S	146°45'E
Corning, Ar., U.S. (kôr'nĭng)	79	36°26'N	90°35'W
Corning, Ia., U.S.	71	40°58'N	94°40'W
Corning, N.Y., U.S.	67	42°10'N	77°05'W
Corno, Monte, mtn., Italy (kôr'nō)	118	42°28'N	13°37'E
Cornwall, Bah.	92	25°55'N	77°15'W
Cornwall, Can. (kôrn'wôl)	59	45°05'N	74°35'W
Coro, Ven. (kō'rō)	100	11°22'N	69°43'W
Corocoro, Bol. (kō-rō-kō'rō)	100	17°15'S	68°21'W
Coromandel Coast, cst., India (kŏr-ô-man'dĕl)	155	13°30'N	80°30'E
Coromandel Peninsula, pen., N.Z.	177	36°50'S	176°00'E
Corona, Al., U.S. (kô-rō'ná)	82	33°42'N	87°28'W
Corona, Ca., U.S.	75a	33°52'N	117°34'W
Coronada, Bahía de, b., C.R. (bä-ē'ä-dĕ-kô-rô-nä'dō)	91	8°47'N	84°04'W
Corona del Mar, Ca., U.S. (kô-rō'ná dĕl mär)	75a	33°36'N	117°53'W
Coronado, Ca., U.S. (kŏr-ō-nä'dō)	76a	32°42'N	117°12'W
Coronation Gulf, b., Can. (kŏr-ô-nā'shŭn)	52	68°07'N	112°50'W
Coronel, Chile (kō-rō-nĕl')	102	37°00'S	73°10'W
Coronel Brandsen, Arg. (kŏ-rō-nĕl-brá'nd-sĕn)	99c	35°09'S	58°15'W
Coronel Dorrego, Arg. (kŏ-rō-nĕl-dôr-rĕ'gō)	102	38°43'S	61°16'W
Coronel Oviedo, Para. (kŏ-rō-nĕl-ō-vê-ĕ'dō)	102	25°28'S	56°22'W
Coronel Pringles, Arg. (kŏ-rō-nĕl-prēn'glĕs)	102	37°54'S	61°22'W
Coronel Suárez, Arg. (kŏ-rō-nĕl-swä'rās)	102	37°27'S	61°49'W
Corowa, Austl. (cŏr-ôwá)	176	36°02'S	146°23'E
Corozal, Belize (cŏr-ŏth-äl')	90a	18°25'N	88°23'W
Corpus Christi, Tx., U.S. (kôr'pŭs krĭs'tē)	64	27°48'N	97°24'W
Corpus Christi Bay, b., Tx., U.S.	81	27°47'N	97°14'W
Corpus Christi Lake, l., Tx., U.S.	80	28°08'N	98°20'W
Corral, Chile (kô-räl')	102	39°57'S	73°15'W
Corral de Almaguer, Spain (kô-räl'dä äl-mä-gâr')	128	39°45'N	3°10'W

PLACE (Pronunciation)	PAGE	LAT.	LONG.
Corralillo, Cuba (kô-rä-lē-yō)	92	23°00'N	80°40'W
Corregidor Island, i., Phil. (kō-rā-hē-dōr')	169a	14°21'N	120°25'E
Correntina, Braz. (kô-rĕn-tē-ná)	101	13°18'S	44°33'W
Corrib, Lough, l., Ire. (lŏk kŏr'ĭb)	120	53°25'N	9°19'W
Corrientes, Arg. (kō-ryĕn'tās)	102	27°25'S	58°39'W
Corrientes, prov., Arg.	102	28°45'S	58°00'W
Corrientes, Cabo, c., Col. (ká'bô-kō-ryĕn'tās)	100	5°34'N	77°35'W
Corrientes, Cabo, c., Cuba (ká'bô-kôr-rē-ĕn'tĕs)	92	21°50'N	84°25'W
Corrientes, Cabo, c., Mex.	86	20°25'N	105°41'W
Corry, Pa., U.S. (kŏr'ĭ)	67	41°55'N	79°40'W
Corse, Cap, c., Fr. (kôrs)	117	42°59'N	9°19'E
Corsica, i., Fr. (kôr'sĭ-kä)	112	42°10'N	8°55'E
Corsicana, Tx., U.S. (kôr-sĭ-kăn'á)	64	32°06'N	96°28'W
Cortazar, Mex. (kôr-tä-zär)	88	20°30'N	100°57'W
Corte, Fr. (kôr'tā)	130	42°18'N	9°10'E
Cortegana, Spain (kôr-tâ-gä'nä)	128	37°54'N	6°48'W
Cortés, Ensenada de, b., Cuba (ĕn-sĕ-nä-dä-dĕ-kôr-tās')	92	22°05'N	83°45'W
Cortez, Co., U.S.	77	37°21'N	108°35'W
Cortland, N.Y., U.S. (kôrt'lǎnd)	67	42°35'N	76°10'W
Cortona, Italy (kôr-tō'nä)	130	43°16'N	12°00'E
Corubal, r., Gui.-B.	188	11°43'N	14°40'W
Coruche, Port. (kô-rōō'she)	128	38°58'N	8°34'W
Çoruh, r., Asia (chō-rōōk')	137	40°30'N	41°10'E
Çorum, Tur. (chô-rōōm')	154	40°34'N	34°45'E
Corunna, Mi., U.S. (kô-rŭn'á)	66	43°00'N	84°05'W
Coruripe, Braz. (kō-rò-rē'pĭ)	101	10°09'S	36°13'W
Corvallis, Or., U.S. (kôr-văl'ĭs)	64	44°34'N	123°17'W
Corve, r., Eng., U.K. (kôr'vē)	114a	52°28'N	2°43'W
Corydon, Ia., U.S.	71	40°45'N	93°20'W
Corydon, In., U.S. (kŏr'ĭ-dŭn)	66	38°10'N	86°05'W
Corydon, Ky., U.S.	66	37°45'N	87°40'W
Cosamaloápan, Mex. (kô-sä-mä-lwä'pän)	89	18°21'N	95°48'W
Coscomatepec, Mex. (kôs'kōmä-tē-pĕk')	89	19°04'N	97°03'W
Cosenza, Italy (kô-zĕnt'sä)	119	39°18'N	16°15'E
Coshocton, Oh., U.S. (kō-shŏk'tŭn)	66	40°15'N	81°55'W
Cosigüina, vol., Nic.	90	12°59'N	87°35'W
Cosmoledo Group, is., Sey. (kŏs-mō-lā'dō)	187	9°42'S	47°45'E
Cosmopolis, Wa., U.S. (kŏz-mŏp'ô-lĭs)	72	46°58'N	123°47'W
Cosne-sur-Loire, Fr. (kōn-sür-lwär')	126	47°25'N	2°57'E
Cosoleacaque, Mex. (kō sō lā-ä-kä'kĕ)	89	18°01'N	94°38'W
Costa de Caparica, Port.	129b	38°40'N	9°12'W
Costa Mesa, Ca., U.S. (kŏs'tá mā'sá)	75a	33°39'N	118°54'W
Costa Rica, nation, N.A. (kŏs'tá rē'ká)	87	10°30'N	84°30'W
Cosumnes, r., Ca., U.S. (kō-sŭm'nĕz)	76	38°21'N	121°17'W
Cotabambas, Peru (kô-tä-bàm'bäs)	100	13°49'S	72°17'W
Cotabato, Phil. (kō-tä-bä'tō)	169	7°06'N	124°13'E
Cotaxtla, Mex. (kō-täs'tlä)	89	18°49'N	96°22'W
Cotaxtla, r., Mex.	89	18°54'N	96°21'W
Coteau-du-Lac, Can. (cō-tō'dü-läk)	62a	45°17'N	74°11'W
Coteau-Landing, Can.	62a	45°15'N	74°13'W
Coteaux, Haiti	93	18°15'N	74°05'W
Cote d'Ivoire (Ivory Coast), nation, Afr.	184	7°43'N	6°30'W
Côte d'Or, reg., Fr.	126	47°02'N	4°35'E
Cotija de la Paz, Mex. (kô-tē'-kä-dĕ-lä-pá'z)	88	19°46'N	102°43'W
Cotonou, Benin (kô-tô-nōō')	184	6°21'N	2°26'E
Cotopaxi, mtn., Ec. (kō-tô-păk'sē)	100	0°40'S	78°26'W
Cotorro, Cuba (kô-tôr-rō)	93a	23°03'N	82°17'W
Cotswold Hills, hills, Eng., U.K. (kŭtz'wōld)	120	51°35'N	2°16'W
Cottage Grove, Mn., U.S. (kŏt'āj grōv)	75g	44°50'N	92°52'W
Cottage Grove, Or., U.S.	72	43°48'N	123°04'W
Cottbus, Ger. (kŏtt'bōōs)	117	51°47'N	14°20'E
Cottonwood, r., Mn., U.S. (kŏt'ŭn-wŏod)	70	44°25'N	95°35'W
Cotulla, Tx., U.S. (kô-tŭl'lá)	80	28°26'N	99°14'W
Coubert, Fr. (kōō-bâr')	127b	48°40'N	2°43'E
Coudersport, Pa., U.S. (kou'dẽrz-port)	67	41°45'N	78°00'W
Coudres, Île aux, i., Can.	60	47°17'N	70°12'W
Coulommiers, Fr. (kōō-lô-myā')	127b	48°49'N	3°05'E
Coulto, Serra do, mts., Braz. (sĕ'r-rä-dō-kō-ò'tō)	102b	22°33'S	43°27'W
Council Bluffs, Ia., U.S. (koun'sĭl blŭf)	65	41°16'N	95°53'W
Council Grove, Ks., U.S. (koun'sĭl grōv)	79	38°39'N	96°30'W
Coupeville, Wa., U.S. (kōōp'vĭl)	74a	48°13'N	122°41'W
Courantyne, r., S.A. (kôr'ăntĭn)	101	4°28'N	57°42'W
Courtenay, Can. (cōōrt-nā')	50	49°41'N	125°00'W
Coushatta, La., U.S. (kou-shăt'á)	81	32°02'N	93°21'W
Coutras, Fr. (kōō-trá')	126	45°02'N	0°07'W
Covelo, Ang.	190	12°06'S	13°55'E
Coventry, Eng., U.K. (kŭv'ĕn-trĭ)	120	52°25'N	1°29'W
Covina, Ca., U.S. (kô-vē'ná)	75a	34°06'N	117°54'W
Covington, Ga., U.S. (kŭv'ĭng-tŭn)	82	33°36'N	83°50'W
Covington, In., U.S.	66	40°10'N	87°15'W
Covington, Ky., U.S.	65	39°05'N	84°31'W
Covington, La., U.S.	81	30°30'N	90°06'W
Covington, Oh., U.S.	66	40°10'N	84°20'W
Covington, Ok., U.S.	79	36°18'N	97°32'W
Covington, Tn., U.S.	82	35°33'N	89°40'W
Covington, Va., U.S.	66	37°50'N	80°00'W
Cowal, Lake, l., Austl. (kou'ăl)	176	33°40'S	147°10'E
Cowan, l., Austl. (kou'án)	174	32°00'S	122°30'E
Cowansville, Can.	59	45°13'N	72°47'W
Cow Creek, r., Or., U.S. (kou)	72	42°45'N	123°35'W

PLACE (Pronunciation)	PAGE	LAT.	LONG.
Cowes, Eng., U.K. (kouz)	120	50°43'N	1°25'W
Cowichan Lake, l., Can.	54	48°54'N	124°20'W
Cowlitz, r., Wa., U.S. (kou'lĭts)	72	46°30'N	122°45'W
Cowra, Austl. (kou'rá)	176	33°50'S	148°33'E
Coxim, Braz. (kō-shēn')	101	18°32'S	54°43'W
Coxquihui, Mex. (kōz-kē-wē')	89	20°10'N	97°34'W
Cox's Bāzār, Bngl.	158	21°32'N	92°00'E
Coyaima, Col. (kō-yáē'mä)	100a	3°48'N	75°11'W
Coyame, Mex. (kō-yä'mā)	80	29°26'N	105°05'W
Coyanosa Draw, Tx., U.S. (kŏ yá-nō'sä)	80	30°55'N	103°07'W
Coyoacán, Mex. (kô-yō-á-kän')	88	19°21'N	99°10'W
Coyote, r., Ca., U.S. (kī'ōt)	74b	37°37'N	121°57'W
Coyuca de Benítez, Mex. (kō-yōō'kä dā bā-nē'tāz)	88	17°04'N	100°06'W
Coyuca de Catalán, Mex. (kō-yōō'kä dā kä-tä-län')	88	18°19'N	100°41'W
Coyutla, Mex. (kō-yōō'tlä)	89	20°13'N	97°40'W
Cozad, Ne., U.S. (kō'zăd)	78	40°53'N	99°59'W
Cozaddale, Oh., U.S. (kō-zăd-dāl)	69f	39°16'N	84°09'W
Cozoyoapan, Mex. (kô-zō-yô-ä-pá'n)	88	16°45'N	98°17'W
Cozumel, Mex. (kō-zōō-mĕ'l)	90a	20°31'N	86°55'W
Cozumel, Isla de, i., Mex. (ē's-lä-dĕ-kō-zōō-mĕ'l)	86	20°26'N	87°10'W
Crab Creek, r., Wa., U.S.	72	47°21'N	119°09'W
Crab Creek, r., Wa., U.S. (krăb)	72	46°47'N	119°43'W
Cradock, S. Afr. (kră'dŭk)	186	32°12'S	25°38'E
Crafton, Pa., U.S. (krăf'tŭn)	69e	40°26'N	80°04'W
Craig, Co., U.S. (krāg)	73	40°32'N	107°31'W
Craiova, Rom. (krá-yō'vá)	119	44°18'N	23°50'E
Cranberry, l., N.Y., U.S. (krăn'bĕr-ĭ)	67	44°10'N	74°50'W
Cranbourne, Austl.	173a	38°07'S	145°16'E
Cranbrook, Can. (krăn'brŏok)	50	49°31'N	115°46'W
Cranbury, N.J., U.S. (krăn'bē-rĭ)	68a	40°19'N	74°31'W
Crandon, Wi., U.S. (krăn'dŭn)	71	45°35'N	88°55'W
Crane Prairie Reservoir, res., Or., U.S.	72	43°50'N	121°55'W
Cranston, R.I., U.S. (krăns'tŭn)	68b	41°46'N	71°25'W
Crater Lake, l., Or., U.S. (krā'tẽr)	72	43°00'N	122°08'W
Crater Lake National Park, rec., Or., U.S.	72	42°58'N	122°40'W
Craters of the Moon National Monument, rec., Id., U.S. (krā'tẽr)	73	43°28'N	113°15'W
Crateús, Braz. (krä-tâ-ōōzh')	101	5°09'S	40°35'W
Crato, Braz. (krä'tò)	101	7°19'S	39°13'W
Crawford, Ne., U.S. (krô'fẽrd)	70	42°41'N	103°25'W
Crawford, Wa., U.S.	74c	45°49'N	122°24'W
Crawfordsville, In., U.S. (krô'fẽrdz-vĭl)	66	40°00'N	86°55'W
Crazy Mountains, mts., Mt., U.S. (krā'zĭ)	73	46°11'N	110°25'W
Crazy Woman Creek, r., Wy., U.S.	73	44°08'N	106°40'W
Crecy, S. Afr. (krĕ-sē)	192c	24°38'S	28°52'E
Crécy-en-Brie, Fr. (krā-sē'-ĕN-brē')	127b	48°52'N	2°55'E
Crécy-en-Ponthieu, Fr.	126	50°13'N	1°48'E
Credit, r., Can.	62d	43°41'N	79°55'W
Cree, l., Can. (krē)	52	57°35'N	107°52'W
Creighton, S. Afr. (cre-tŏn)	187c	30°02'S	29°52'E
Creighton, Ne., U.S. (krā'tŭn)	70	42°27'N	97°54'W
Creil, Fr. (krĕ'y')	126	49°18'N	2°28'E
Crema, Italy (krā'mä)	130	45°21'N	9°53'E
Cremona, Italy (krā-mō'nä)	118	45°09'N	10°02'E
Crépy-en-Valois, Fr. (krā-pē'ĕN-vä-lwä')	127b	49°14'N	2°53'E
Cres, Cro. (tsrĕs)	130	44°58'N	14°21'E
Crescent Beach, Can.	74d	49°03'N	122°58'W
Crescent City, Ca., U.S. (krĕs'ĕnt)	72	41°46'N	124°13'W
Crescent City, Fl., U.S.	83	29°26'N	81°35'W
Crescent Lake, l., Fl., U.S. (krĕs'ĕnt)	83	29°33'N	81°30'W
Crescent Lake, l., Or., U.S.	72	43°25'N	121°58'W
Cresco, Ia., U.S. (krĕs'kō)	71	43°23'N	92°07'W
Crested Butte, Co., U.S. (krĕst'ĕd būt)	77	38°50'N	107°00'W
Crestline, Ca., U.S. (krĕst-līn)	75a	34°15'N	117°17'W
Crestline, Oh., U.S.	66	40°50'N	82°40'W
Crestmore, Ca., U.S. (krĕst'môr)	75a	34°02'N	117°23'W
Creston, Can. (krĕs'tŭn)	50	49°06'N	116°31'W
Creston, Ia., U.S.	71	41°04'N	94°22'W
Creston, Oh., U.S.	69d	40°59'N	81°54'W
Crestview, Fl., U.S. (krĕst'vū)	82	30°44'N	86°35'W
Crestwood, Ky., U.S. (krĕst'wŏod)	69h	38°20'N	85°28'W
Crestwood, Mo., U.S.	75e	38°33'N	90°23'W
Crete, Il., U.S. (krēt)	69a	41°26'N	87°38'W
Crete, Ne., U.S.	79	40°38'N	96°56'W
Crete, i., Grc.	112	35°15'N	24°30'E
Creus, Cap de, c., Spain (krē-ōōs)	129	42°16'N	3°18'E
Creuse, r., Fr. (krûz)	126	46°51'N	0°49'E
Creve Coeur, Mo., U.S. (krēv kōr)	75e	38°40'N	90°27'W
Crevillent, Spain	129	38°12'N	0°48'W
Crewe, Eng., U.K. (krōō)	120	53°06'N	2°27'W
Crewe, Va., U.S.	83	37°09'N	78°08'W
Crimean Peninsula *see* Kryms'kyi Pivostriv, pen., Ukr.	137	45°18'N	33°30'E
Crimmitschau, Ger. (krĭm'ĭt-shou)	124	50°49'N	12°22'E
Cripple Creek, Co., U.S. (krĭp''l)	78	38°44'N	105°12'W
Crisfield, Md., U.S. (krĭs-fēld)	67	38°00'N	75°50'W
Cristal, Monts de, mts., Gabon	190	0°50'N	10°30'E
Cristina, Braz. (krês-tē'-nä)	99a	22°13'S	45°15'W
Cristóbal Colón, Pico, mtn., Col. (pē'kô-krēs-tô'bäl-kō-lôn')	100	11°00'N	74°00'W
Crişul Alb, r., Rom. (krē'shōōl älb)	125	46°20'N	22°15'E
Crna, r., Serb. (ts'r'nà)	131	41°03'N	21°46'E
Crna Gora (Montenegro), state, Serb.	131	42°55'N	18°52'E
Črnomelj, Slvn. (ch'r'nō-māl')	130	45°35'N	15°11'E
Croatia, nation, Eur.	130	45°24'N	15°18'E
Crockett, Ca., U.S. (krŏk'ĕt)	74b	38°03'N	122°14'W
Crockett, Tx., U.S.	81	31°19'N	95°28'W
Crofton, Md., U.S.	68e	39°01'N	76°43'W

ng-sing; ŋ-baŋk; N-nasalized n; nŏd; cŏmmit; ōld; ōbey; ôrder; oi-boil; fōōd; ò-as oo in foot; ou-out; s-soft; sh-dish; th-thin; pūre; ûnite; ûrn; stŭd; circŭs; ü-as in French tu; '-indeterminate vowel.

240 Crofton – Dall Island

PRONOUNCING INDEX

PLACE (Pronunciation)	PAGE	LAT.	LONG.
Crofton, Ne., U.S.	70	42°44'N	97°32'W
Croix, Lac la, l., N.A. (läk lä krōō-ä')	71	48°19'N	91°53'W
Croker, i., Austl. (krō'kà)	174	10°45'S	132°25'E
Cronulla, Austl. (krō-nŭl'à)	173b	34°03'S	151°09'E
Crooked, i., Bah.	93	22°45'N	74°10'W
Crooked, l., Can.	61	48°25'N	56°05'W
Crooked, r., Can.	55	54°30'N	122°55'W
Crooked, r., Or., U.S.	72	44°07'N	120°30'W
Crooked Creek, r., Il., U.S. (krŏŏk'ĕd)	79	40°21'N	90°49'W
Crooked Island Passage, strt., Bah.	93	22°40'N	74°50'W
Crookston, Mn., U.S. (krŏŏks'tŭn)	70	47°44'N	96°35'W
Crooksville, Oh., U.S. (krŏŏks'vĭl)	66	39°45'N	82°05'W
Crosby, Eng., U.K.	114a	53°30'N	3°02'W
Crosby, Mn., U.S. (krôz'bĭ)	71	46°29'N	93°58'W
Crosby, N.D., U.S.	70	48°55'N	103°18'W
Crosby, Tx., U.S.	81a	29°55'N	95°04'W
Cross, l., La., U.S.	81	32°33'N	93°58'W
Cross, r., Nig.	189	5°35'N	8°05'E
Cross City, Fl., U.S.	82	29°55'N	83°25'W
Crossett, Ar., U.S. (krôs'ĕt)	79	33°08'N	92°00'W
Cross Lake, l., Can.	52	54°45'N	97°30'W
Cross River Reservoir, res., N.Y., U.S. (krôs)	68a	41°14'N	73°34'W
Cross Sound, strt., Ak., U.S. (krŏs)	63	58°12'N	137°20'W
Crosswell, Mi., U.S. (krŏz'wĕl)	66	43°15'N	82°35'W
Croswell, i., Serb.	130	44°50'N	14°31'E
Crotch, l., Can.	59	44°55'N	76°55'W
Crotone, Italy (krō-tô'nĕ)	131	39°05'N	17°08'E
Croton Falls Reservoir, res., N.Y., U.S. (krōtŭn)	68a	41°22'N	73°44'W
Croton-on-Hudson, N.Y., U.S. (krō'tŭn-ŏn hŭd'sŭn)	68a	41°12'N	73°53'W
Crow, l., Can.	71	49°13'N	93°29'W
Crow Agency, Mt., U.S.	73	45°36'N	107°27'W
Crow Creek, r., Co., U.S.	78	41°08'N	104°25'W
Crow Creek Indian Reservation, I.R., S.D., U.S.	70	44°17'N	99°17'W
Crow Indian Reservation, I.R., Mt., U.S. (krō)	73	45°26'N	108°12'W
Crowle, Eng., U.K. (kroul)	114a	53°36'N	0°49'W
Crowley, La., U.S. (krou'lē)	81	30°13'N	92°22'W
Crown Mountain, mtn., Can. (kroun)	74d	49°24'N	123°05'W
Crown Mountain, mtn., V.I.U.S.	87c	18°22'N	64°58'W
Crown Point, In., U.S. (kroun point')	69a	41°25'N	87°22'W
Crown Point, N.Y., U.S.	67	44°00'N	73°25'W
Crowsnest Pass, p., Can.	55	49°39'N	114°45'W
Crow Wing, r., Mn., U.S. (krō)	71	44°50'N	94°01'W
Crow Wing, r., Mn., U.S.	71	46°42'N	94°48'W
Crow Wing, North Fork, r., Mn., U.S.	71	45°16'N	94°28'W
Crow Wing, South Fork, r., Mn., U.S.	71	44°59'N	94°42'W
Croydon, Austl. (kroi'dŭn)	175	18°15'S	142°15'E
Croydon, Austl.	173a	37°48'S	145°17'E
Croydon, Eng., U.K.	117	51°22'N	0°06'W
Croydon, Pa., U.S.	68f	40°05'N	74°55'W
Crozet, Îles, is., Afr. (krô-zĕ')	3	46°20'S	51°30'E
Cruces, Cuba (krōō'sås)	92	22°20'N	80°20'W
Cruces, Arroyo de, r., Mex. (är-rō'yô-dĕ-krōō'sĕs)	80	26°17'N	104°32'W
Cruillas, Mex. (krōō-ēl'yäs)	80	24°45'N	98°31'W
Cruz, Cabo, c., Cuba (kà'-bô-krōōz)	87	19°50'N	77°45'W
Cruz, Cayo, i., Cuba (kä'yō-krōōz)	92	22°15'N	77°50'W
Cruz Alta, Braz. (krōōz äl'tä)	102	28°41'S	54°02'W
Cruz del Eje, Arg. (krōō's-dĕl-ĕ-kĕ)	102	30°46'S	64°45'W
Cruzeiro, Braz. (krōō-zā'rò)	99a	22°36'S	44°57'W
Cruzeiro do Sul, Braz. (krōō-zā'rò dò sōōl)	100	7°34'S	72°40'W
Crysler, Can.	62c	45°13'N	75°09'W
Crystal City, Tx., U.S. (krĭs'tăl sĭ'tĭ)	80	28°40'N	99°50'W
Crystal Falls, Mi., U.S. (krĭs'tăl fôls)	71	46°06'N	88°21'W
Crystal Lake, Il., U.S. (krĭs'tăl lăk)	69a	42°15'N	88°18'W
Crystal Springs, Ms., U.S. (krĭs'tăl sprĭngz)	82	31°58'N	90°20'W
Crystal Springs, oasis, Ca., U.S.	74b	37°31'N	122°26'W
Csongrád, Hung. (chôn'gräd)	125	46°42'N	20°09'E
Csorna, Hung. (chôr'nä)	125	47°39'N	17°11'E
Cúa, Ven. (kōō'ä)	101b	10°10'N	66°54'W
Cuajimalpa, Mex. (kwä-hē-mäl'pä)	89a	19°21'N	99°18'W
Cuale, Sierra del, mts., Mex. (sē-ĕ'r-rä-dĕl-kwä'lĕ)	88	20°20'N	104°58'W
Cuamato, Ang. (kwä-mä'tō)	190	17°05'S	15°09'E
Cuamba, Moz.	191	14°49'S	36°33'E
Cuando, Ang. (kwän'dō)	190	16°32'S	22°07'E
Cuando, r., Afr.	186	14°30'S	20°00'E
Cuangar, Ang.	190	17°36'S	18°39'E
Cuango, r., Afr.	186	9°00'S	18°00'E
Cuanza, r., Ang. (kwän'zä)	186	9°45'S	15°00'E
Cuarto, r., Arg.	102	33°00'S	63°25'W
Cuatro Caminos, Cuba (kwä'trô-kä-mē'nōs)	93a	23°01'N	82°13'W
Cuatro Ciénegas, Mex. (kwä'trô syā'nâ-gäs)	80	26°59'N	102°03'W
Cuauhtemoc, Mex. (kwä-ōō-tĕ-mōk')	89	15°43'N	91°57'W
Cuautepec, Mex. (kwä-ōō-tĕ-pĕk)	88	16°41'N	99°04'W
Cuautepec, Mex.	88	20°01'N	98°19'W
Cuautitlán, Mex. (kwä-ōō-tĕt-län')	89a	19°40'N	99°12'W
Cuautla, Mex. (kwä-ōō'tlà)	88	18°47'N	98°57'W
Cuba, Port. (kōō'bà)	128	38°10'N	7°55'W
Cuba, nation, N.A. (kū'bà)	87	22°00'N	79°00'W
Cubagua, Isla, i., Ven. (ē's-lä-kōō-bä'gwä)	101b	10°48'N	64°10'W
Cubango (Okavango), r., Afr. (kōō-bäŋ'gō)	186	17°10'S	18°20'E
Cub Hills, hills, Can. (kŭb)	56	54°20'N	104°30'W
Cucamonga, Ca., U.S. (kōō-kà-mōŋ'gà)	75a	34°05'N	117°35'W
Cuchi, Ang.	186	14°40'S	16°50'E
Cuchillo Parado, Mex. (kōō-chē'lyô pä-rä'dō)	80	29°26'N	104°52'W
Cuchumatanes, Sierra de los, mts., Guat.	90	15°35'N	91°10'W
Cúcuta, Col. (kōō'kōō-tä)	100	7°56'N	72°30'W
Cudahy, Wi., U.S. (kŭd'à-hī)	69a	42°57'N	87°52'W
Cuddalore, India (kŭd à-lōr')	155	11°49'N	79°46'E
Cuddapah, India (kŭd'à-pä)	155	14°31'N	78°52'E
Cue, Austl. (kū)	174	27°30'S	118°10'E
Cuéllar, Spain (kwā'lyär')	128	41°24'N	4°15'W
Cuenca, Ec. (kwĕn'kä)	100	2°52'S	78°54'W
Cuenca, Spain	118	40°05'N	2°07'W
Cuenca, Sierra de, mts., Spain (sē-ĕ'r-rä-dĕ-kwĕ'n-kä)	128	40°02'N	1°50'W
Cuencame, Mex. (kwĕn-kä-mā')	80	24°52'N	103°42'W
Cuerámaro, Mex. (kwā-rä'mä-rô)	88	20°39'N	101°44'W
Cuernavaca, Mex. (kwĕr-nä-vä'kä)	86	18°55'N	99°15'W
Cuero, Tx., U.S. (kwā'rō)	81	29°05'N	97°16'W
Cuetzalá del Progreso, Mex. (kwĕt-zä-lä dĕl prō-grā'sō)	88	18°07'N	99°51'W
Cuetzalan del Progreso, Mex.	89	20°02'N	97°33'W
Cuevas del Almanzora, Spain (kwĕ'väs-dĕl-äl-män-zō-rä)	118	37°19'N	1°54'W
Cuglieri, Italy (kōō-lyâ'rē)	130	40°11'N	8°37'E
Cuicatlán, Mex. (kwē-kä-tlän')	89	17°46'N	96°57'W
Cuilapa, Guat. (kò-ē-lä'pä)	90	14°16'N	90°20'W
Cuilo (Kwilu), r., Afr.	190	9°15'S	19°30'E
Cuito, r., Ang. (kōō-ē'tō)	186	14°45'S	19°00'E
Cuitzeo, Mex. (kwēt'zā-ō)	88	19°57'N	101°11'W
Cuitzeo, Laguna de, l., Mex. (lä-ò'nä-dĕ-kwēt'zā-ō)	88	19°58'N	101°05'W
Cul de Sac, pl., Haiti (kōō'l-dĕ-sä'k)	93	18°35'N	72°05'W
Culebra, i., P.R. (kōō-lā'brä)	87b	18°19'N	65°32'W
Culebra, Sierra de la, mts., Spain (sē-ĕ'r-rä-dĕ-lä-kōō-lĕ-brä)	128	41°52'N	6°21'W
Culemborg, Neth.	115a	51°57'N	5°14'E
Culfa, Azer.	138	38°58'N	45°38'E
Culgoa, r., Austl. (kŭl-gō'à)	175	29°21'S	147°00'E
Culiacán, Mex. (kōō-lyä-kà'n)	86	24°45'N	107°30'W
Culion, Phil. (kōō-lē-ōn')	168	11°43'N	119°58'E
Cúllar de Baza, Spain (kōō'l-yär-dĕ-bä'zä)	128	37°36'N	2°35'W
Cullera, Spain (kōō-lyā'rä)	118	39°12'N	0°15'W
Cullinan, S. Afr. (kŏ'lĭ-nán)	192c	25°41'S	28°32'E
Cullman, Al., U.S. (kŭl'măn)	82	34°10'N	86°50'W
Culpeper, Va., U.S. (kŭl'pĕp-ĕr)	67	38°30'N	77°55'W
Culross, Can. (kŭl'rôs)	62f	49°43'N	97°54'W
Culver, In., U.S. (kŭl'vĕr)	66	41°15'N	86°25'W
Culver City, Ca., U.S.	75a	34°00'N	118°23'W
Cumaná, Ven.	100	10°28'N	64°10'W
Cumberland, Can. (kŭm'bĕr-lănd)	62c	45°31'N	75°25'W
Cumberland, Md., U.S.	65	39°40'N	78°40'W
Cumberland, Wa., U.S.	74a	47°17'N	121°55'W
Cumberland, Wi., U.S.	71	45°31'N	92°01'W
Cumberland, r., U.S.	82	36°45'N	85°33'W
Cumberland, Lake, res., Ky., U.S.	65	36°55'N	85°20'W
Cumberland Islands, is., Austl.	175	20°20'S	149°46'E
Cumberland Peninsula, pen., Can.	53	65°59'N	64°05'W
Cumberland Plateau, plat., U.S.	82	35°25'N	85°30'W
Cumberland Sound, strt., Can.	53	65°27'N	65°44'W
Cundinamarca, dept., Col. (kōōn-dē-nä-mä'r-kä)	100a	4°57'N	74°27'W
Cunduacán, Mex. (kòn-dōō-à-kän')	89	18°04'N	93°23'W
Cunene (Kunene), r., Afr.	186	17°05'S	12°35'E
Cuneo, Italy (kōō'nā-ō)	130	44°24'N	7°31'E
Cunha, Braz. (kōō'nyà)	99a	23°05'S	44°56'W
Cunnamulla, Austl. (kŭn-à-mŭl-à)	175	28°00'S	145°55'E
Cupula, Pico, mtn., Mex. (pē'kō-kōō'pōō-lä)	86	24°45'N	111°10'W
Cuquío, Mex. (kōō-kē'ō)	88	20°55'N	103°03'W
Curaçao, i., Neth. Ant. (kōō-rä-sä'ō)	100	12°12'N	68°58'W
Curacautín, Chile (kä-rä-käōō-tē'n)	102	38°25'S	71°53'W
Curaumilla, Punta, c., Chile (kōō-rou-mē'lyä)	99b	33°05'S	71°44'W
Curepto, Chile (kōō-rĕp-tô)	99b	35°06'S	72°02'W
Curitiba, Braz. (kōō-rē-tē'bà)	101	25°20'S	49°15'W
Curly Cut Cays, is., Bah.	92	23°40'N	77°40'W
Currais Novos, Braz. (kōōr-rä'ēs nō-vōs)	101	6°02'S	36°39'W
Curran, Can. (kü-rän')	62c	45°30'N	74°59'W
Current, i., Bah. (kŭ-rĕnt)	92	25°20'N	76°50'W
Current, r., Mo., U.S. (kûr'ĕnt)	79	37°18'N	91°21'W
Currie, Mount, mtn., S. Afr. (kŭ-rē)	187c	30°28'S	29°23'E
Currituck Sound, strt., N.C., U.S. (kûr'ĭ-tŭk)	83	36°27'N	75°42'W
Curtis, Ne., U.S. (kûr'tĭs)	78	40°36'N	100°29'W
Curtis, i., Austl.	175	23°38'S	151°43'E
Curtisville, Pa., U.S. (kûr'tĭs-vĭl)	69e	40°38'N	79°50'W
Ćurug, Serb. (chōō'ròg)	131	45°27'N	20°03'E
Curunga, Ang.	190	12°51'S	21°12'E
Curupira, Serra, mts., S.A. (sĕr'rà kōō-rōō-pē'rà)	100	1°00'N	65°30'W
Cururupu, Braz. (kōō-rò-rò-pōō')	101	1°40'S	44°56'W
Curvelo, Braz. (kòr-vĕl'ō)	101	18°47'S	44°14'W
Cusco, Peru	100	13°36'S	71°52'W
Cushing, Ok., U.S. (kŭsh'ĭng)	79	35°58'N	96°46'W
Custer, S.D., U.S. (kŭs'tĕr)	70	43°46'N	103°36'W
Custer, Wa., U.S.	74d	48°55'N	122°39'W
Cut Bank, Mt., U.S. (kŭt bănk)	73	48°38'N	112°19'W
Cuthbert, Ga., U.S. (kŭth'bĕrt)	82	31°47'N	84°48'W
Cuttack, India (kŭ-tăk')	155	20°38'N	85°53'E
Cutzamalá, r., Mex. (kōō-tzä-mä-lä')	88	18°57'N	100°41'W
Cutzamalá de Pinzón, Mex. (kōō-tzä-mä-lä'dĕ-pēn-zō'n)	88	18°28'N	100°36'W
Cuvo, r., Ang. (kōō'vō)	186	11°00'S	14°30'E
Cuxhaven, Ger. (kòks'hä-fĕn)	116	53°51'N	8°43'E
Cuyahoga, r., Oh., U.S. (kī-à-hō'gà)	69d	41°22'N	81°38'W
Cuyahoga Falls, Oh., U.S.	69d	41°08'N	81°29'W
Cuyapaire Indian Reservation, I.R., Ca., U.S. (kū-yà-pâr)	76	32°46'N	116°20'W
Cuyo Islands, is., Phil. (kōō'yō)	168	10°54'N	120°08'E
Cuyotenango, Guat. (kōō-yô-tĕ-näŋ'gô)	90	14°30'N	91°35'W
Cuyuni, r., S.A. (kōō-yōō'nē)	101	6°40'N	60°44'W
Cuyutlán, Mex. (kōō-yōō-tlän')	88	18°54'N	104°04'W
Cyclades *see* Kikládes, is., Grc.	112	37°30'N	24°45'E
Cynthiana, Ky., U.S. (sĭn-thĭ-ăn'à)	66	38°20'N	84°20'W
Cypress, Ca., U.S. (sī'prĕs)	75a	33°50'N	118°03'W
Cypress Hills, hills, Can.	56	49°40'N	110°20'W
Cypress Lake, l., Can.	56	49°28'N	109°43'W
Cyprus, nation, Asia (sī'prŭs)	154	35°00'N	31°00'E
Cyrenaica *see* Barqah, hist. reg., Libya	185	31°09'N	21°45'E
Czech Republic, nation, Eur.	110	50°00'N	15°00'E
Czersk, Pol. (chĕrsk)	125	53°47'N	17°58'E
Częstochowa, Pol. (chăn-stô κô'và)	117	50°49'N	19°10'E

D

PLACE (Pronunciation)	PAGE	LAT.	LONG.
Da'an, China (dä-än)	164	45°25'N	124°22'E
Dabakala, C. Iv. (dä-bä-kä'lä)	184	8°16'N	4°36'W
Daba Shan, mts., China (dä-bä shän)	160	32°25'N	108°20'E
Dabeiba, Col. (dà-bā'bä)	100a	7°01'N	76°16'W
Dabie Shan, mts., China (dä-bĭē shän)	161	31°40'N	114°50'E
Dabnou, Niger	189	14°09'N	5°22'E
Dabob Bay, b., Wa., U.S. (dā'bŏb)	74a	47°50'N	122°50'W
Dabola, Gui.	188	10°45'N	11°07'W
Dąbrowa Białostocka, Pol.	125	53°37'N	23°18'E
Dacca *see* Dhaka, Bngl.	155	23°45'N	90°29'E
Dachang, China (dä-chäŋ)	163b	31°18'N	121°25'E
Dachangshan Dao, i., China (dä-chäŋ-shän dou)	162	39°21'N	122°31'E
Dachau, Ger. (dä'κou)	124	48°16'N	11°26'E
Dacotah, Can. (dà-kō'tà)	62f	49°52'N	97°38'W
Dade City, Fl., U.S. (dād)	83a	28°22'N	82°09'W
Dadeville, Al., U.S. (dād'vĭl)	82	32°48'N	85°44'W
Dādra & Nagar Haveli, India	155	20°00'N	73°00'E
Dadu, r., China (dä-dōō)	165	29°20'N	103°03'E
Daet, mtn., Phil. (dä'ăt)	169a	14°07'N	122°59'E
Dafoe, r., Can.	57	55°50'N	95°50'W
Dafter, Mi., U.S. (dăf'tĕr)	75k	46°21'N	84°26'W
Dagana, Sen. (dä-gä'nä)	184	16°31'N	15°30'W
Dagana, reg., Chad	189	12°20'N	15°15'E
Dagang, China (dä-gäŋ)	163a	22°48'N	113°24'E
Dagda, Lat. (dåg'dà)	123	56°04'N	27°30'E
Dagenham, Eng., U.K. (dăg'ĕn-ăm)	114b	51°32'N	0°09'E
Dagestan, prov., Russia (dä-gĕs-tän')	137	43°40'N	46°10'E
Daggett, Ca., U.S. (dăg'ĕt)	76	34°50'N	116°52'W
Dagu, China (dä-gōō)	164	39°00'N	117°42'E
Dagu, r., China	162	36°29'N	120°06'E
Dagupan, Phil. (dä-gōō'pän)	169a	16°02'N	120°20'E
Daheishan Dao, i., China (dä-hā-shän dou)	162	37°57'N	120°37'E
Dahl, Ger. (däl)	127c	51°18'N	7°33'E
Dahlak Archipelago, is., Erit.	185	15°45'N	40°30'E
Dahomey *see* Benin, nation, Afr.	184	8°00'N	2°00'E
Dahra, Libya	156	29°34'N	17°50'E
Daibu, China (dī-bōō)	162	31°22'N	119°29'E
Daigo, Japan (dī-gō)	167b	34°57'N	135°49'E
Daimiel Manzanares, Spain (dī-myĕl'män-zä-nä'rĕs)	128	39°05'N	3°36'W
Dairen *see* Dalian, China	161	38°54'N	121°35'E
Dairy, r., Or., U.S. (dâr'ĭ)	74c	45°33'N	123°04'W
Dai-Sen, mtn., Japan (dī'sĕn')	167	35°22'N	133°35'E
Dai-Tenjo-dake, mtn., Japan (dī-tĕn'jō dä-κä)	167	36°21'N	137°38'E
Daiyun Shan, mtn., China (dī-yòn shän)	165	25°40'N	118°08'E
Dajabón, Dom. Rep. (dä-κä-bô'n)	93	19°35'N	71°40'W
Dajarra, Austl. (dà-jăr'à)	174	21°45'S	139°30'E
Dakar, Sen. (dà-kär')	184	14°40'N	17°26'W
Dakhla, W. Sah.	184	23°45'N	16°04'W
Dakouraoua, Niger	189	13°58'N	6°15'E
Dakovica, Serb.	131	42°33'N	20°28'E
Dalälven, r., Swe.	112	60°26'N	15°50'E
Dalby, Austl. (dôl'bē)	175	27°10'S	151°15'E
Dalcour, La., U.S. (dăl-kour)	68d	29°49'N	89°59'W
Dale, Nor. (dä'lĕ)	122	60°35'N	5°55'E
Dale Hollow Lake, res., Tn., U.S. (dăl hŏl'ō)	65	36°33'N	85°03'W
Dalemead, Can. (dä'lĕ-mēd)	62e	50°53'N	113°38'W
Dalen, Nor. (dä'lĕn)	122	59°28'N	8°01'E
Daleside, S. Afr. (dāl'sīd)	192c	26°30'S	28°03'E
Dalesville, Can. (dālz'vĭl)	62a	45°42'N	74°23'W
Daley Waters, Austl. (dā lē)	174	16°15'S	133°30'E
Dalhart, Tx., U.S. (dăl härt)	78	36°04'N	102°32'W
Dalhousie, Can. (dăl-hōō'zē)	60	48°04'N	66°23'W
Dali, China (dä-lē)	163a	23°07'N	113°06'E
Dali, China	160	26°00'N	100°08'E
Dali, China	160	35°00'N	109°38'E
Dalian, China (lů-dä)	161	38°54'N	121°35'E
Dalian Wan, b., China (dä-lĭĕn wän)	162	38°55'N	121°50'E
Dalías, Spain (dä-lē'ás)	128	36°49'N	2°50'W
Dall, i., Ak., U.S. (dâl)	63	54°50'N	133°10'W
Dallas, Or., U.S. (dăl'lás)	72	44°55'N	123°20'W
Dallas, S.D., U.S.	70	43°13'N	99°34'W
Dallas, Tx., U.S.	64	32°45'N	96°48'W
Dalles Dam, Or., U.S.	72	45°36'N	121°08'W
Dall Island, i., Ak., U.S.	54	54°50'N	132°55'W

PRONOUNCING INDEX

Dalmacija – Denmark Strait 241

PLACE (Pronunciation)	PAGE	LAT.	LONG.
Dalmacija, hist. reg., Serb. (däl-mä'tsē-yä)	130	43°25'N	16°37'E
Dalnerechensk, Russia	135	46°07'N	133°21'E
Daloa, C. Iv.	188	6°53'N	6°27'W
Dalroy, Can. (dăl'roi)	62e	51°07'N	113°39'W
Dalrymple, Mount, mtn., Austl. (dăl'rĭm-p'l)	175	21°14'S	148°46'E
Dalton, S. Afr. (dôl'tŏn)	187c	29°21'S	30°41'E
Dalton, Ga., U.S. (dôl'tŭn)	82	34°46'N	84°58'W
Daly, r., Austl. (dā'lĭ)	174	14°15'S	131°15'E
Daly City, Ca., U.S. (dā'lē)	74b	37°42'N	122°27'W
Damān, India	155	20°32'N	72°53'E
Damanhūr, Egypt (dä-män-hōōr')	185	30°59'N	30°31'E
Damar, Pulau, i., Indon.	169	7°15'S	129°15'E
Damara, C.A.R.	189	4°58'N	18°42'E
Damaraland, hist. reg., Nmb. (dă'nȧ-rȧ-lănd)	186	22°15'S	16°15'E
Damas Cays, is., Bah. (dä'mäs)	92	23°50'N	79°50'W
Damascus, Syria	154	33°30'N	36°18'E
Damāvand, Qolleh-ye, mtn., Iran	154	36°05'N	52°05'E
Damba, Ang. (däm'bä)	186	6°41'S	15°08'E
Dâmbovița, r., Rom.	131	44°43'N	25°41'E
Dame Marie, Cap, c., Haiti (dām mȧrē')	93	18°35'N	74°50'W
Dāmghān, Iran (dām-gän')	154	35°50'N	54°15'E
Daming, China (dä-mĭŋ)	164	36°15'N	115°09'E
Dammartin-en-Goële, Fr. (däɴ-mär-tăɴ-äɴ-gô-ĕl')	127b	49°03'N	2°40'E
Dampier, Selat, strt., Indon. (dăm'pēr)	169	0°40'S	131°15'E
Dampier Archipielago, is., Austl.	174	20°15'S	116°25'E
Dampier Land, reg., Austl.	174	17°30'S	122°25'E
Dan, r., N.C., U.S. (dăn)	83	36°26'N	79°40'W
Dana, Mount, mtn., Ca., U.S.	76	37°54'N	119°13'W
Da Nang, Viet.	168	16°08'N	108°22'E
Danbury, Eng., U.K.	114b	51°42'N	0°34'E
Danbury, Ct., U.S. (dăn'bēr-ĭ)	68a	41°23'N	73°27'W
Danbury, Tx., U.S.	81a	29°14'N	95°22'W
Dandenong, Austl. (dăn'dē-nông)	176	37°59'S	145°13'E
Dandong, China (dän-dôŋ)	161	40°10'N	124°30'E
Dane, r., Eng., U.K. (dān)	114a	53°11'N	2°14'W
Danea, Gui.	188	11°27'N	13°12'W
Danforth, Me., U.S.	60	45°38'N	67°53'W
Dan Gora, Nig.	189	11°30'N	8°09'E
Dangtu, China (däŋ-tōō)	165	31°35'N	118°28'E
Dani, Burkina	184	13°43'N	0°10'W
Dania, Fl., U.S. (dā'nĭ-ȧ)	83a	26°01'N	80°10'W
Danilov, Russia (dä'nē-lôf)	136	58°12'N	40°08'E
Danissa Hills, hills, Kenya	191	3°20'N	40°55'E
Dänizkänarı, Azer.	138	40°13'N	49°33'E
Dankov, Russia (däŋ'kôf)	136	53°17'N	39°09'E
Dannemora, N.Y., U.S. (dăn-ē-mō'rȧ)	67	44°45'N	73°45'W
Dannhauser, S. Afr. (dăn'hou-zẽr)	187c	28°07'S	30°04'E
Dansville, N.Y., U.S. (dănz'vĭl)	67	42°30'N	77°40'W
Danube, r., Eur.	112	43°00'N	24°00'E
Danube, Mouths of the, mth., Rom. (dăn'ūb)	133	45°13'N	29°37'E
Danvers, Ma., U.S. (dăn'vẽrz)	61a	42°34'N	70°57'W
Danville, Ca., U.S. (dăn'vĭl)	74b	37°49'N	122°00'W
Danville, Il., U.S.	66	40°10'N	87°35'W
Danville, In., U.S.	66	39°45'N	86°30'W
Danville, Ky., U.S.	66	37°35'N	84°50'W
Danville, Pa., U.S.	67	41°00'N	76°35'W
Danville, Va., U.S.	65	36°35'N	79°24'W
Danxian, China (dän shyĕn)	165	19°30'N	109°38'E
Danyang, China (dän-yäŋ)	162	32°01'N	119°32'E
Danzig *see* Gdańsk, Pol.	110	54°20'N	18°40'E
Danzig, Gulf of, b., Eur. (dăn'tsĭk)	116	54°41'N	19°01'E
Daoxian, China (dou shyĕn)	165	25°35'N	111°27'E
Dapango, Togo	188	10°52'N	0°12'E
Daphnae, hist., Egypt	153a	30°43'N	32°12'E
Daqin Dao, i., China (dä-chyĭn dou)	162	38°18'N	120°50'E
Darabani, Rom. (dă-rä-băn'ĭ)	125	48°13'N	26°38'E
Daraj, Libya	184	30°12'N	10°14'E
Darāw, Egypt (dȧ-rä'ōō)	192b	24°24'N	32°56'E
Darbhanga, India (dŭr-bŭŋ'gä)	155	26°03'N	85°09'E
Darby, Pa., U.S. (där'bĭ)	68f	39°55'N	75°16'W
Darby, i., Bah.	92	23°50'N	76°20'W
Dardanelles *see* Çanakkale Boğazi, strt., Tur.	119	40°05'N	25°50'E
Dar es Salaam, Tan. (där ĕs sȧ-läm')	187	6°48'S	39°17'E
Dārfūr, hist. reg., Sudan (där-fōōr')	185	13°21'N	23°46'E
Dargai, Pak. (dŭr-gä'ē)	158	34°35'N	72°00'E
Darien, Col. (dā-rĭ-ĕn')	100a	3°56'N	76°30'W
Darien, Ct., U.S. (dâ-rē-ĕn')	68a	41°04'N	73°28'W
Darién, Cordillera de, mts., Nic.	90	13°00'N	85°42'W
Darien, Serranía del, mts.	91	8°13'N	77°28'W
Darjeeling, India (dŭr-jē'lĭng)	155	27°05'N	88°16'E
Darling, r., Austl.	175	31°50'S	143°20'E
Darling Downs, reg., Austl.	175	27°22'S	150°00'E
Darling Range, mts., Austl.	174	30°30'S	115°45'E
Darlington, Eng., U.K. (där'lĭng-tŭn)	120	54°32'N	1°35'W
Darlington, S.C., U.S.	83	34°15'N	79°52'W
Darlington, Wi., U.S.	71	42°41'N	90°06'W
Darłowo, Pol. (där-lô'vô)	124	54°26'N	16°23'E
Darmstadt, Ger. (därm'shtät)	117	49°53'N	8°40'E
Darnah, Libya	185	32°44'N	22°41'E
Darnley Bay, b., Ak., U.S. (därn'lē)	63	70°00'N	124°00'W
Daroca, Spain (dä-rō-kä)	128	41°08'N	1°24'W
Dartford, Eng., U.K.	114b	51°27'N	0°14'E
Dartmoor, for., Eng., U.K. (därt'mōōr)	120	50°35'N	4°05'W
Dartmouth, Can. (därt'mŭth)	51	44°40'N	63°34'W
Dartmouth, Eng., U.K.	120	50°33'N	3°28'W
Daru, Pap. N. Gui. (dä'rōō)	169	9°04'S	143°21'E
Daruvar, Cro. (där'rōō-vär)	131	45°37'N	17°16'E
Darwen, Eng., U.K. (där'wĕn)	114a	53°42'N	2°28'W
Darwin, Austl. (där'wĭn)	174	12°25'S	131°00'E
Darwin, Cordillera, mts., Chile (kôr-dēl-yĕ'rä-där'wēn)	102	54°40'S	69°30'W
Dashhowuz, Turkmen.	139	41°50'N	59°45'E
Dash Point, Wa., U.S. (dăsh)	74a	47°19'N	122°25'W
Dasht, r., Pak. (dŭsht)	154	25°30'N	62°30'E
Dasol Bay, b., Phil. (dä-sōl')	169a	15°53'N	119°40'E
Datian Ding, mtn., China (dä-tĭĕn dĭŋ)	165	22°25'N	111°20'E
Datong, China (dä-tôŋ)	164	40°00'N	113°30'E
Dattapukur, India	158a	22°45'N	88°32'E
Datteln, Ger. (dät'tĕln)	127c	51°39'N	7°20'E
Datu, Tandjung, c., Asia	168	2°08'N	110°15'E
Datuan, China (dä-tůän)	163b	30°57'N	121°43'E
Daugava (Zapadnaya Dvina), r., Eur.	123	56°40'N	24°40'E
Daugavpils, Lat. (dä'ŏ-gäv-pĭls)	136	55°52'N	26°32'E
Dauphin, Can. (dô'fĭn)	50	51°09'N	100°00'W
Dauphin Lake, l., Can.	57	51°17'N	99°48'W
Dāvangere, India	159	14°30'N	75°55'E
Davao, Phil. (dä'vä-ō)	169	7°05'N	125°30'E
Davao Gulf, b., Phil.	169	6°30'N	125°45'E
Davenport, Ia., U.S. (dăv'ĕn-pōrt)	65	41°34'N	90°38'W
Davenport, Wa., U.S.	72	47°39'N	118°07'W
David, Pan. (dȧ-vēdh')	87	8°27'N	82°27'W
David City, Ne., U.S. (dā'vĭd)	70	41°15'N	97°10'W
David-Gorodok, Bela. (dȧ-vēt' gô-rō'dôk)	137	52°02'N	27°14'E
Davis, Ok., U.S. (dā'vĭs)	79	34°34'N	97°08'W
Davis, W.V., U.S.	67	39°15'N	79°25'W
Davis Lake, l., Or., U.S.	72	43°38'N	121°43'W
Davis Mountains, mts., Tx., U.S.	80	30°45'N	104°17'W
Davis Strait, strt., N.A.	49	66°00'N	60°00'W
Davlekanovo, Russia	136	54°15'N	55°05'E
Davos, Switz. (dä'vōs)	124	46°47'N	9°50'E
Dawa, r., Afr.	185	4°30'N	40°30'E
Dawāsir, Wādī ad, val., Sau. Ar.	154	20°48'N	44°07'E
Dawei, Mya.	168	14°04'N	98°19'E
Dawen, r., China (dä-wŭn)	162	35°58'N	116°53'E
Dawley, Eng., U.K. (dô'lĭ)	114a	52°38'N	2°28'W
Dawna Range, mts., Mya. (dô'nȧ)	168	17°02'N	98°01'E
Dawson, Can. (dô'sŭn)	50	64°04'N	139°22'W
Dawson, Ga., U.S.	82	31°45'N	84°29'W
Dawson, Mn., U.S.	70	44°54'N	96°03'W
Dawson, r., Austl.	175	24°20'S	149°45'E
Dawson Bay, b., Can.	57	52°55'N	100°50'W
Dawson Creek, Can.	50	55°46'N	120°14'W
Dawson Range, mts., Can.	63	62°15'N	138°10'W
Dawson Springs, Ky., U.S.	82	37°10'N	87°40'W
Dawu, China (dä-wōō)	162	31°33'N	114°07'E
Dax, Fr. (dȧks)	117	43°42'N	1°06'W
Daxian, China (dä-shyĕn)	160	31°12'N	107°30'E
Daxing, China (dä-shyĭŋ)	164a	39°44'N	116°19'E
Dayiqiao, China (dä-yē-chyou)	162	31°43'N	120°40'E
Dayr az Zawr, Syria (dä-ērĕz-zôr')	154	35°15'N	40°01'E
Dayton, Ky., U.S. (dā'tŭn)	69f	39°07'N	84°28'W
Dayton, N.M., U.S.	78	32°44'N	104°23'W
Dayton, Oh., U.S.	65	39°54'N	84°15'W
Dayton, Tn., U.S.	82	35°30'N	85°00'W
Dayton, Tx., U.S.	81	30°03'N	94°53'W
Dayton, Wa., U.S.	72	46°18'N	117°59'W
Daytona Beach, Fl., U.S. (dā-tō'nȧ)	65	29°11'N	81°02'W
Dayu, China (dä-yōō)	165	25°20'N	114°20'E
Da Yunhe (Grand Canal), can., China (dä yòn-hŭ)	161	35°00'N	117°00'E
Dayville, Ct., U.S. (dā'vĭl)	67	41°50'N	71°55'W
De Aar, S. Afr. (dē-är')	186	30°45'S	24°05'E
Dead, l., Mn., U.S. (dĕd)	70	46°28'N	96°00'W
Dead Sea, l., Asia	154	31°30'N	35°30'E
Deadwood, S.D., U.S. (dĕd'wŏŏd)	64	44°23'N	103°43'W
Deal Island, Md., U.S. (dēl-ī'lănd)	67	38°10'N	75°55'W
Dean, r., Can. (dēn)	54	52°45'N	125°30'W
Dean Channel, strt., Can.	54	52°33'N	127°13'W
Deán Funes, Arg. (dĕ-ä'n-fōō-nĕs)	102	30°26'S	64°12'W
Dearborn, Mi., U.S. (dẽr'bŭrn)	69b	42°18'N	83°15'W
Dearg, Ben, mtn., Scot., U.K. (bĕn dŭrg)	120	57°48'N	4°59'W
Dease Strait, strt., Can. (dēz)	52	68°50'N	108°20'W
Death Valley, Ca., U.S.	76	36°18'N	116°26'W
Death Valley, val., Ca., U.S.	64	36°30'N	117°00'W
Death Valley National Park, rec., U.S.	76	36°34'N	117°00'W
Debal'tseve, Ukr.	133	48°23'N	38°29'E
Debao, China (dŭ-bou)	160	23°18'N	106°40'E
Debar, Mac. (dĕ'bär) (dä'brä)	131	41°31'N	20°32'E
Dęblin, Pol. (dăN'blĭn)	125	51°34'N	21°49'E
Dębno, Pol. (dĕb-nô')	124	52°47'N	13°43'E
Debo, Lac, l., Mali	188	15°15'N	4°40'W
Debrecen, Hung. (dĕ'brĕ-tsĕn)	110	47°32'N	21°40'E
Debre Markos, Eth.	185	10°15'N	37°45'E
Debre Tabor, Eth.	185	11°57'N	38°09'E
Decatur, Al., U.S. (dē-kā'tŭr)	82	34°35'N	87°00'W
Decatur, Ga., U.S.	68c	33°47'N	84°18'W
Decatur, Il., U.S.	65	39°50'N	88°59'W
Decatur, In., U.S.	66	40°50'N	84°55'W
Decatur, Mi., U.S.	66	42°10'N	86°00'W
Decatur, Tx., U.S.	79	33°14'N	97°33'W
Decazeville, Fr. (dē-käz'vēl')	117	44°33'N	2°16'E
Deccan, plat., India (dĕk'ăn)	155	19°05'N	76°40'E
Deception Lake, l., Can.	56	56°33'N	104°15'W
Deception Pass, p., Wa., U.S. (dē-sĕp'shŭn)	74a	48°24'N	122°44'W
Děčín, Czech Rep. (dyĕ'chēn)	124	50°47'N	14°14'E
Decorah, Ia., U.S. (dē-kō'rȧ)	71	43°18'N	91°48'W
Dedenevo, Russia (dyĕ-dyĕ'nyĕ-vô)	142b	56°14'N	37°31'E
Dedham, Ma., U.S. (dĕd'ăm)	61a	42°15'N	71°11'W
Dedo do Deus, mtn., Braz. (dĕ-dô-dô-dĕ'ōōs)	102b	22°30'S	43°02'W
Dédougou, Burkina (dā-dŏ-gōō')	184	12°38'N	3°28'W
Dee, r., Scot., U.K.	120	57°05'N	2°25'W
Dee, r., U.K.	114a	53°15'N	3°05'E
Deep, r., N.C., U.S. (dēp)	83	35°36'N	79°32'W
Deep Fork, r., Ok., U.S.	79	35°35'N	96°42'W
Deep River, Can.	59	46°06'N	77°20'W
Deepwater, Mo., U.S. (dep-wô-tẽr)	79	38°15'N	93°46'W
Deer, i., Me., U.S.	60	44°07'N	68°38'W
Deerfield, Il., U.S. (dẽr'fēld)	69a	42°10'N	87°51'W
Deer Island, Or., U.S.	74c	45°56'N	122°51'W
Deer Lake, Can.	53a	49°10'N	57°25'W
Deer Lake, l., Can.	57	52°40'N	94°30'W
Deer Lodge, Mt., U.S. (dẽr lŏj)	73	46°23'N	112°42'W
Deer Park, Oh., U.S.	69f	39°12'N	84°24'W
Deer Park, Wa., U.S.	72	47°58'N	117°28'W
Deer River, Mn., U.S.	71	47°20'N	93°49'W
Defiance, Oh., U.S. (dē-fī'ăns)	66	41°15'N	84°20'W
DeFuniak Springs, Fl., U.S. (dē fū'nĭ-ăk)	82	30°42'N	86°06'W
Deganga, India	158a	22°41'N	88°41'E
Degeh Bur, Eth.	192a	8°10'N	43°25'E
Deggendorf, Ger. (dĕ'ghĕn-dôrf)	124	48°50'N	12°59'E
Degollado, Mex. (dā-gô-lyä'dō)	88	20°27'N	102°11'W
DeGrey, r., Austl. (dē grā')	174	20°20'S	119°25'E
Degtyarsk, Russia (dĕg-ty'arsk)	142a	56°42'N	60°05'E
Dehiwala-Mount Lavinia, Sri L.	159	6°47'N	79°55'E
Dehra Dūn, India (dā'rŭ)	155	30°09'N	78°07'E
Dehua, China (dŭ-hwä)	165	25°30'N	118°15'E
Dej, Rom. (dāzh)	119	47°09'N	23°53'E
De Kalb, Il., U.S. (dē kălb')	66	41°54'N	88°46'W
Dekese, D.R.C.	190	3°27'S	21°24'E
Delacour, Can. (dĕ-lä-kōōr')	62e	51°09'N	113°45'W
Delagua, Co., U.S. (dĕl-ä'gwä)	78	37°19'N	104°42'W
De Land, Fl., U.S. (dē lănd')	83	29°00'N	81°19'W
Delano, Ca., U.S. (dĕl'ȧ-nō)	76	35°47'N	119°15'W
Delano Peak, mtn., Ut., U.S.	64	38°25'N	112°25'W
Delavan, Wi., U.S. (dĕl'ȧ-văn)	71	42°39'N	88°38'W
Delaware, Oh., U.S. (dĕl'ȧ-wâr)	66	40°15'N	83°05'W
Delaware, state, U.S.	65	38°40'N	75°30'W
Delaware, r., Ks., U.S.	79	39°45'N	95°47'W
Delaware, r., U.S.	67	41°50'N	75°20'W
Delaware Bay, b., U.S.	65	39°05'N	75°10'W
Delaware Reservoir, res., Oh., U.S.	67	40°30'N	83°05'E
Delémont, Switz. (dē-lā-môn')	124	47°21'N	7°18'E
De Leon, Tx., U.S. (dē lē-ŏn')	80	32°06'N	98°33'W
Delft, Neth. (dĕlft)	121	52°01'N	4°20'E
Delfzijl, Neth.	121	53°20'N	6°50'E
Delgada, Punta, c., Arg. (pōō'n-tä-dĕl-gä'dä)	102	43°46'S	63°46'W
Delgado, Cabo, c., Moz. (kä'bō-dĕl-gä'dō)	187	10°40'S	40°35'E
Delhi, India	155	28°54'N	77°13'E
Delhi, Il., U.S. (dĕl'hī)	75e	39°03'N	90°16'W
Delhi, La., U.S.	81	32°26'N	91°29'W
Delhi, state, India	155	28°30'N	76°50'E
Delitzsch, Ger. (dā'lĭch)	124	51°32'N	12°18'E
Dellansjöarna, l., Swe.	122	61°57'N	16°25'E
Delles, Alg. (dĕ'lĕs')	184	36°59'N	3°40'E
Dell Rapids, S.D., U.S. (dĕl)	70	43°50'N	96°43'W
Dellwood, Mn., U.S. (dĕl'wŏŏd)	75g	45°05'N	92°58'W
Del Mar, Ca., U.S. (dĕl mär')	76a	32°57'N	117°16'W
Delmas, S. Afr. (dĕl'mȧs)	192c	26°08'S	28°43'E
Delmenhorst, Ger. (dĕl'mĕn-hôrst)	124	53°03'N	8°38'E
Del Norte, Co., U.S. (dĕl nôrt')	77	37°40'N	106°25'W
De-Longa, i., Russia	135	76°21'N	148°56'E
De Long Mountains, mts., Ak., U.S. (dē'lŏng)	63	68°38'N	162°30'W
Deloraine, Austl. (dē-lŭ-rān)	176	41°30'S	146°40'E
Delphi, In., U.S. (dĕl'fī)	66	40°35'N	86°40'W
Delphos, Oh., U.S. (dĕl'fŏs)	66	40°50'N	84°20'W
Delray Beach, Fl., U.S. (dĕl-rā')	83a	26°27'N	80°05'W
Del Rio, Tx., U.S. (dĕl rē'ō)	64	29°21'N	100°52'W
Delson, Can. (dĕl'sŭn)	62a	45°24'N	73°32'W
Delta, Co., U.S.	77	38°45'N	108°05'W
Delta, Ut., U.S.	77	39°20'N	112°35'W
Delta Beach, Can.	62f	50°10'N	98°20'W
Delvine, Alb. (dĕl'vē-nȧ)	131	39°58'N	20°10'E
Dëma, r., Russia (dyĕm'ä)	136	53°40'N	54°30'E
Demba, D.R.C.	190	5°30'S	22°16'E
Dembi Dolo, Eth.	185	8°46'N	34°46'E
Demidov, Russia (dzyĕ'mē-dô'f)	132	55°16'N	31°32'E
Deming, N.M., U.S. (dĕm'ĭng)	64	32°15'N	107°45'W
Demmin, Ger. (dĕm'mēn)	124	53°54'N	13°04'E
Demnat, Mor. (dĕm-nät)	184	31°58'N	7°03'W
Demopolis, Al., U.S. (dē-mŏp'ō-lĭs)	82	32°30'N	87°50'W
Demotte, In., U.S. (dē'mŏt)	69a	41°12'N	87°13'W
Dempo, Gunung, mtn., Indon. (dĕm'pô)	168	4°04'S	103°11'E
Dem'yanka, r., Russia (dyĕm-yän'kä)	140	59°07'N	72°58'E
Demyansk, Russia (dyĕm-yänsk')	132	57°39'N	32°26'E
Denain, Fr. (dē-năN')	126	50°23'N	3°21'E
Denakil Plain, pl., Eth.	185	12°45'N	41°01'E
Denali National Park, rec., Ak., U.S.	64a	63°48'N	153°02'W
Denbigh, Wales, U.K. (dĕn'bĭ)	120	53°15'N	3°25'W
Dendermonde, Bel.	115a	51°02'N	4°04'E
Dendron, Va., U.S. (dĕn'drŭn)	83	37°02'N	76°53'W
Denezhkin Kamen, Gora, mtn., Russia (dzyĕ-ŋĕ'zhkĕn kămĭēn)	142a	60°26'N	59°35'E
Denham, Mount, mtn., Jam.	87	18°20'N	77°30'W
Den Helder, Neth. (dĕn hĕl'dĕr)	121	52°55'N	5°45'E
Dénia, Spain	129	38°48'N	0°06'E
Deniliquin, Austl. (dĕ-nĭl'ĭ-kwĭn)	175	35°20'S	144°52'E
Denison, Ia., U.S. (dĕn'ĭ-sŭn)	70	42°01'N	95°22'W
Denison, Tx., U.S.	64	33°45'N	97°02'W
Denizli, Tur. (dĕn-ĭz-lē')	119	37°40'N	29°10'E
Denklingen, Ger. (dĕn'klĕn-gĕn)	127c	50°54'N	7°40'E
Denmark, S.C., U.S. (dĕn'märk)	83	33°18'N	81°09'W
Denmark, nation, Eur.	110	56°14'N	8°30'E
Denmark Strait, strt., Eur.	49	66°30'N	27°00'W

242 Dennilton – Dominion PRONOUNCING INDEX

PLACE (Pronunciation)	PAGE	LAT.	LONG.
Dennilton, S. Afr. (děn-ĭl-tǔn)	192c	25°18'S	29°13'E
Dennison, Oh., U.S. (děn'ĭ-sǔn)	66	40°25'N	81°20'W
Denpasar, Indon.	168	8°35'S	115°10'E
Denton, Eng., U.K. (děn'tǔn)	114a	53°27'N	2°07'W
Denton, Md., U.S.	67	38°55'N	75°50'W
Denton, Tx., U.S.	79	33°12'N	97°06'W
D'Entrecasteaux, Point, c., Austl. (däɴ-tr'-kȧs-tō')	174	34°50'S	114°45'E
D'Entrecasteaux Islands, is., Pap. N. Gui. (däɴ-tr'-kȧs-tō')	169	9°45'S	152°00'E
Denver, Co., U.S. (děn'vẽr)	64	39°44'N	104°59'W
Deoli, India	158	25°52'N	75°23'E
De Pere, Wi., U.S. (dē pēr')	71	44°25'N	88°04'W
Depew, N.Y., U.S. (dē-pū')	69c	42°55'N	78°43'W
Deping, China (dŭ-pĭŋ)	162	37°28'N	116°57'E
Depue, Il., U.S. (dē pū)	66	41°15'N	89°55'W
De Queen, Ar., U.S. (dē kwēn')	79	34°02'N	94°21'W
De Quincy, La., U.S. (dē kwĭn'sĭ)	81	30°27'N	93°27'W
Dera, Lach, r., Afr. (läk dä'rä)	192a	0°45'N	41°26'E
Dera, Lak, r., Afr.	185	0°45'N	41°30'E
Dera Ghāzi Khān, Pak. (dā'rŭ gä-zē' ĸän')	155	30°09'N	70°39'E
Dera Ismāīl Khān, Pak. (dā'rŭ ĭs-mä-ēl' ĸän')	158	31°55'N	70°51'E
Derbent, Russia (dĕr-bĕnt')	137	42°00'N	48°10'E
Derby, Austl. (där'bē) (dûr'bē)	174	17°20'S	123°40'E
Derby, S. Afr. (där'bĭ)	192c	25°55'S	27°02'E
Derby, Eng., U.K. (där'bē)	117	52°55'N	1°29'W
Derby, Ct., U.S. (dûr'bē)	67	41°20'N	73°05'W
Derbyshire, co., Eng., U.K.	114a	53°11'N	1°30'W
Derdepoort, S. Afr.	192c	24°39'S	26°21'E
Derg, Lough, l., Ire. (lŏk dĕrg)	120	53°00'N	8°09'W
De Ridder, La., U.S. (dē rĭd'ẽr)	81	30°50'N	93°18'W
Dermott, Ar., U.S. (dûr'mŏt)	79	33°32'N	91°24'W
Derry, N.H., U.S. (dâr'ĭ)	61a	42°53'N	71°22'W
Derventa, Bos. (dĕr'ven-tȧ)	131	44°58'N	17°58'E
Derwent, r., Austl. (dẽr'wĕnt)	176	42°21'S	146°30'E
Derwent, r., Eng., U.K.	114a	52°54'N	1°24'W
Des Arc, Ar., U.S. (dāz ärk')	79	34°59'N	91°31'W
Descalvado, Braz. (dĕs-käl-vȧ-dō)	99a	21°55'S	47°37'W
Descartes, Fr.	126	46°58'N	0°42'E
Deschambault Lake, l., Can.	56	54°40'N	103°35'W
Deschênes, Can.	62c	45°23'N	75°47'W
Deschenes, Lake, l., Can.	62c	45°25'N	75°53'W
Deschutes, r., Or., U.S. (dā-shōōt')	72	44°25'N	121°21'W
Desdemona, Tx., U.S. (dĕz-dē-mō'nȧ)	80	32°16'N	98°33'W
Dese, Eth.	185	11°00'N	39°51'E
Deseado, r., Arg. (dā-sā-ä'dhō)	102	46°50'S	67°45'W
Desirade Island, i., Guad. (dā-zē-räs')	91b	16°21'N	60°51'W
De Smet, S.D., U.S. (dē smĕt')	70	44°23'N	97°33'W
Des Moines, Ia., U.S. (dē moin')	65	41°35'N	93°37'W
Des Moines, N.M., U.S.	78	36°42'N	103°48'W
Des Moines, Wa., U.S.	74a	47°24'N	122°20'W
Des Moines, r., U.S.	65	42°30'N	94°20'W
Desna, r., Eur. (dyĕs-nȧ')	137	51°55'N	31°45'E
Desolación, i., Chile (dĕ-sō-lä-syō'n)	102	53°05'S	74°00'W
De Soto, Mo., U.S. (dē sō'tō)	79	38°07'N	90°32'W
Des Peres, Mo., U.S. (dĕs pĕr'ĕs)	75e	38°36'N	90°26'W
Des Plaines, Il., U.S. (dĕs plānz')	69a	42°02'N	87°54'W
Des Plaines, r., U.S.	69a	41°39'N	87°56'W
Dessau, Ger. (dĕsou)	117	51°50'N	12°15'E
Detmold, Ger. (dĕt'mōld)	124	51°57'N	8°55'E
Detroit, Mi., U.S. (dē-troit')	65	42°22'N	83°10'W
Detroit, Tx., U.S.	79	33°41'N	95°16'W
Detroit Lake, res., Or., U.S.	72	44°42'N	122°10'W
Detroit Lakes, Mn., U.S. (dē-troit'lăkz)	70	46°48'N	95°51'W
Detva, Slvk. (dyĕt'vȧ)	125	48°32'N	19°21'E
Deurne, Bel.	115a	51°13'N	4°27'E
Deutsch Wagram, Aus.	115e	48°19'N	16°34'E
Deux-Montagnes, Can.	62a	45°33'N	73°53'W
Deux Montagnes, Lac des, l., Can.	62a	45°28'N	74°00'W
Deva, Rom. (dā'vä)	119	45°52'N	22°52'E
Dévaványa, Hung. (dā'vô-vän-yô)	125	47°01'N	20°58'E
Develi, Tur. (dĕ'vȧ-lē)	137	38°20'N	35°10'E
Deventer, Neth. (dĕv'ĕn-tẽr)	121	52°14'N	6°07'E
Devils, r., Tx., U.S.	80	29°55'N	101°10'W
Devils Island see Diable, Île du, i., Fr. Gu.	101	5°15'N	52°40'W
Devils Lake, N.D., U.S.	64	48°10'N	98°55'W
Devils Lake, l., N.D., U.S. (dĕv''lz)	70	47°57'N	99°04'W
Devils Lake Indian Reservation, I.R., N.D., U.S.	70	48°08'N	99°40'W
Devils Postpile National Monument, rec., Ca., U.S.	76	37°42'N	119°12'W
Devils Tower National Monument, rec., Wy., U.S.	73	44°38'N	105°07'W
Devoll, r., Alb.	131	40°55'N	20°10'E
Devon, Can.	62g	53°23'N	113°43'W
Devon, S. Afr. (dĕv'ŭn)	192c	26°23'S	28°47'E
Devonport, Austl. (dĕv'ŭn-pôrt)	175	41°20'S	146°30'E
Devonport, N.Z.	175a	36°50'S	174°45'E
Devore, Ca., U.S. (dē-vôr')	75a	34°13'N	117°24'W
Dewatto, Wa., U.S. (dē-wăt'ō)	74a	47°27'N	123°04'W
Dewey, Ok., U.S. (dū'ĭ)	79	36°48'N	95°55'W
De Witt, Ar., U.S. (dē wĭt')	79	34°17'N	91°22'W
De Witt, Ia., U.S.	71	41°46'N	90°34'W
Dewsbury, Eng., U.K. (dūz'bẽr-ĭ)	114a	53°42'N	1°39'W
Dexter, Me., U.S. (dĕks'tẽr)	60	45°01'N	69°19'W
Dexter, Mo., U.S.	79	36°46'N	89°56'W
Dezfūl, Iran	154	32°14'N	48°37'E
Dezhnëva, Mys, c., Russia (dyĕzh'nyĭf)	152	68°00'N	172°00'W
Dezhou, China (dŭ-jō)	164	37°28'N	116°17'E
Dhahran see Az̧ Z̧ahrān, Sau. Ar.	154	26°13'N	50°00'E
Dhaka, Bngl. (dä'kä) (dăk'ȧ)	155	23°45'N	90°29'E
Dharamtar Creek, r., India	159b	18°49'N	72°54'E

PLACE (Pronunciation)	PAGE	LAT.	LONG.
Dharmavaram, India	159	14°32'N	77°43'E
Dhawalāgiri, mtn., Nepal	155	28°42'N	83°31'E
Dhibān, Jord.	153a	31°30'N	35°46'E
Dhidhimótikhon, Grc.	131	41°20'N	26°27'E
Dhule, India	155	20°58'N	74°43'E
Día, i., Grc. (dē'ä)	130a	35°27'N	25°17'E
Diable, Île du, i., Fr. Gu.	101	5°15'N	52°40'W
Diablo, Mount, mtn., Ca., U.S. (dyä'blō)	74b	37°52'N	121°55'W
Diablo Heights, Pan. (dyȧ'blō)	86a	8°58'N	79°34'W
Diablo Range, mts., Ca., U.S.	74b	37°47'N	121°50'W
Diablotins, Morne, mtn., Dom.	91b	15°31'N	61°24'W
Diaca, Moz.	191	11°30'S	39°59'E
Diaka, r., Mali	189	14°40'N	5°00'E
Diamantina, Braz.	101	18°14'S	43°32'W
Diamantina, r., Austl. (dī'man-tē'nȧ)	174	25°38'S	139°53'E
Diamantino, Braz. (dē-ȧ-män-tē'nō)	101	14°22'S	56°23'W
Diamond Peak, mtn., Or., U.S.	72	43°32'N	122°08'W
Diana Bank, bk., (dī'ăn'ȧ)	93	22°30'N	74°45'W
Dianbai, China (dĭĕn-bī)	165	21°30'N	111°20'E
Dian Chi, l., China (dĭĕn chē)	160	24°58'N	103°18'E
Dickinson, N.D., U.S. (dĭk'ĭn-sŭn)	64	46°52'N	102°49'W
Dickinson, Tx., U.S. (dĭk'ĭn-sŭn)	81a	29°28'N	95°02'W
Dickinson Bayou, Tx., U.S.	81a	29°26'N	95°08'W
Dickson, Tn., U.S. (dĭk'sŭn)	82	36°03'N	87°24'W
Dickson City, Pa., U.S.	67	41°25'N	75°40'W
Didcot, Eng., U.K. (dĭd'cŏt)	114b	51°35'N	1°15'W
Didiéni, Mali	188	13°53'N	8°06'W
Die, Fr. (dē)	127	44°45'N	5°22'E
Diefenbaker, res., Can.	52	51°20'N	108°10'W
Diego de Ocampo, Pico, mtn., Dom. Rep. (pē'-kō-dyĕ'gō-dē-ō-kä'm-pō)	93	19°40'N	70°45'W
Diego Ramirez, Islas, is., Chile (dē ā'gō rä-mē'rāz)	102	56°15'S	70°15'W
Diéma, Mali	188	14°32'N	9°12'W
Dien Bien Phu, Viet.	160	21°38'N	102°49'E
Dieppe, Can. (dē-ĕp')	60	46°06'N	64°45'W
Dieppe, Fr.	117	49°54'N	1°05'E
Dierks, Ar., U.S. (dērks)	79	34°06'N	94°02'W
Diessen, Ger. (dēs'sĕn)	115d	47°57'N	11°06'E
Diest, Bel.	115a	50°59'N	5°05'E
Digby, Can. (dĭg'bĭ)	51	44°37'N	65°46'W
Dighton, Ma., U.S. (dī-tŭn)	68b	41°49'N	71°05'W
Digne, Fr. (dēn'y')	127	44°07'N	6°16'E
Digoin, Fr. (dē-gwăɴ')	126	46°28'N	4°06'E
Digul, r., Indon.	169	7°00'S	140°27'E
Dijohan Point, c., Phil. (dē-kō-än)	169a	16°24'N	122°25'E
Dijon, Fr. (dē-zhŏɴ')	110	47°21'N	5°02'E
Dikson, Russia (dĭk'sŏn)	134	73°30'N	80°35'E
Dikwa, Nig. (dē'kwȧ)	185	12°06'N	13°53'E
Dili, E. Timor (dĭl'ē)	169	8°35'S	125°35'E
Di Linosa Island, i., Italy (dē-lē-nō'sä)	118	36°01'N	12°43'E
Dilizhan, Arm.	137	40°45'N	45°00'E
Dillingham, Ak., U.S. (dĭl'ĕng-hăm)	64a	59°10'N	158°38'W
Dillon, Mt., U.S. (dĭl'ŭn)	73	45°12'N	112°40'W
Dillon, S.C., U.S.	83	34°24'N	79°28'W
Dillon Reservoir, res., Oh., U.S.	66	40°05'N	82°05'W
Dilolo, D.R.C. (dē-lō'lō)	186	10°19'S	22°23'E
Dimashq see Damascus, Syria	154	33°31'N	36°18'E
Dimbokro, C. Iv.	188	6°39'N	4°42'W
Dimitrovo see Pernik, Blg.	119	42°36'N	23°04'E
Dimlang, mtn., Nig.	189	8°24'N	11°47'E
Dimona, Isr.	153a	31°03'N	35°01'E
Dinagat Island, i., Phil.	169	10°15'N	126°15'E
Dinājpur, Bngl.	158	25°38'N	87°39'E
Dinan, Fr. (dē-näɴ')	126	48°27'N	2°03'W
Dinant, Bel. (dē-näɴ')	121	50°17'N	4°50'E
Dinara, mts., Serb. (dē'nä-rä)	119	43°50'N	16°15'E
Dinard, Fr.	126	48°38'N	2°04'W
Dindigul, India	159	10°25'N	78°03'E
Dingalan Bay, b., Phil. (dĭŋ-gä'län)	169a	15°19'N	121°33'E
Dingle, Ire. (dĭng''l)	120	52°10'N	10°13'W
Dingle Bay, b., Ire.	117	52°02'N	10°15'W
Dingo, Austl. (dĭŋ'gō)	175	23°45'S	149°26'E
Dinguiraye, Gui.	188	11°18'N	10°43'W
Dingwall, Scot., U.K. (dĭng'wôl)	120	57°37'N	4°23'W
Dingxian, China (dĭŋ shyĕn)	164	38°30'N	115°00'E
Dingxing, China (dĭŋ-shyĭŋ)	164	39°18'N	115°50'E
Dingyuan, China (dĭŋ-yüän)	162	32°32'N	117°40'E
Dingzi Wan, b., China	162	36°33'N	121°06'E
Dinosaur National Monument, rec., Co., U.S. (dī'nō-sôr)	73	40°45'N	109°17'W
Dinslaken, Ger. (dēns'lä-kĕn)	127c	51°33'N	6°44'E
Dinteloord, Neth.	115a	51°38'N	4°21'E
Dinuba, Ca., U.S. (dĭ-nū'bȧ)	76	36°33'N	119°29'W
Dios, Cayo de, i., Cuba (kä'yō-dĕ-dē-ōs')	92	22°05'N	83°05'W
Diourbel, Sen. (dē-ōōr-bĕl')	184	14°40'N	16°15'W
Diphu Pass, p., Asia (dĭ-pōō)	160	28°15'N	96°45'E
Diquis, r., C.R. (dē-kēs')	91	8°59'N	83°24'W
Dire Dawa, Eth.	185	9°40'N	41°47'E
Diriamba, Nic. (dēr-yäm'bä)	90	11°52'N	86°15'W
Dirk Hartog, i., Austl.	174	26°25'S	113°15'E
Dirksland, Neth.	115a	51°45'N	4°04'E
Dirranbandi, Austl. (dĭ-rȧ-băn'dē)	175	28°24'S	148°29'E
Dirty Devil, r., Ut., U.S. (dûr'tĭ dĕv''l)	77	38°20'N	110°30'W
Disappointment, l., Austl.	174	23°20'S	123°00'E
Disappointment, Cape, c., Wa., U.S. (dĭs'ȧ-point'ment)	74c	46°16'N	124°11'W
Discovery, S. Afr. (dĭs-kŭv'ẽr-ĭ)	187b	26°10'S	27°53'E
Discovery, is., Can. (dĭs-kŭv'ẽr-ē)	74a	48°25'N	123°13'W
Disko, i., Grnld. (dĭs'kō)	49	70°00'N	54°00'W
Disna, Bela. (dēs'nȧ)	136	55°34'N	28°15'E
Dispur, India	158	26°00'N	91°50'E
Disraëli, Can. (dĭs-rā'lĭ)	59	45°53'N	71°23'W
District of Columbia, state, U.S.	65	38°50'N	77°00'W

PLACE (Pronunciation)	PAGE	LAT.	LONG.
Distrito Federal, state, Braz. (dēs-trē'tô-fĕ-dĕ-rä'l)	101	15°49'S	47°39'W
Distrito Federal, state, Mex.	88	19°14'N	99°08'W
Disūq, Egypt (dē-sōōk')	192b	31°07'N	30°41'E
Diu, India (dē'ōō)	155	20°48'N	70°58'E
Divilacan Bay, b., Phil. (dē-vē-lä'kän)	169a	17°26'N	122°25'E
Divinópolis, Braz. (dē-vē-nô'pō-lēs)	101	20°10'S	44°53'W
Divo, C. Iv.	188	5°50'N	5°22'W
Dixon, Il., U.S. (dĭks'ŭn)	71	41°50'N	89°30'W
Dixon Entrance, strt., N.A.	52	54°25'N	132°00'W
Diyarbakir, Tur. (dē-yär-bĕk'ĭr)	154	38°00'N	40°10'E
Dja, r., Afr.	185	2°30'N	14°00'E
Djambala, Congo	190	2°33'S	14°45'E
Djanet, Alg.	184	24°29'N	9°26'E
Djebobo, mtn., Ghana	188	8°20'N	0°37'E
Djedi, Oued, r., Alg.	118	34°18'N	4°39'E
Djember, Chad	189	10°25'N	17°50'E
Djerba, Île de, i., Tun.	118	33°53'N	11°26'E
Djerid, Chott, l., Tun. (jĕr'ĭd)	184	33°15'N	8°29'E
Djibasso, Burkina	188	13°07'N	4°10'W
Djibo, Burkina	188	14°06'N	1°38'W
Djibouti, Dji. (jē-bōō-tē')	192a	11°34'N	43°00'E
Djibouti, nation, Afr.	192a	11°35'N	48°08'E
Djokoumatombi, Congo	190	0°47'N	15°22'E
Djokupunda, D.R.C.	186	5°27'S	20°58'E
Djoua, r., Afr.	190	1°25'N	13°40'E
Djursholm, Swe. (djōōrs'hōlm)	122	59°26'N	18°01'E
Dmitriyev-L'govskiy, Russia (d'mē'trĭ-yĕf l'gôf'skĭ)	132	52°07'N	35°05'E
Dmitrov, Russia (d'mē'trôf)	132	56°21'N	37°32'E
Dmitrovsk, Russia (d'mē'trôfsk)	132	52°30'N	35°10'E
Dmytrivka, Ukr.	133	47°57'N	38°56'E
Dnepropetrovsk see Dnipropetrovs'k, Ukr.	134	48°15'N	34°08'E
Dnieper (Dnipro), r., Eur.	137	46°45'N	33°40'E
Dniester, r., Eur.	137	48°21'N	28°10'E
Dniprodzerzhyns'k, Ukr.	137	48°32'N	34°38'E
Dniprodzerzhyns'ke vodoskhovyshche, res., Ukr.	134	49°00'N	34°10'E
Dnipropetrovs'k, Ukr.	134	48°15'N	34°08'E
Dnipropetrovs'k, prov., Ukr.	133	48°15'N	34°10'E
Dniprovs'kyi lyman, b., Ukr.	133	46°33'N	31°45'E
Dnistrovs'kyi lyman, l., Ukr.	133	46°13'N	29°50'E
Dno, Russia (d'nô')	132	57°49'N	29°59'E
Do, Lac, l., Mali	188	15°50'N	2°20'W
Doba, Chad	189	8°39'N	16°51'E
Dobbs Ferry, N.Y., U.S. (dŏbz'fĕ'rē)	68a	41°01'N	73°53'W
Dobbyn, Austl. (dôb'ĭn)	174	19°45'S	140°02'E
Dobele, Lat. (dô'bĕ-lĕ)	123	56°37'N	23°18'E
Doberai, Jazirah, pen., Indon.	169	1°25'S	133°15'E
Dobo, Indon.	169	6°00'S	134°18'E
Doboj, Bos. (dō'boi)	131	44°42'N	18°04'E
Dobrich, Blg.	119	43°33'N	27°52'E
Dobryanka, Russia (dôb-ryän'kȧ)	142a	58°27'N	56°26'E
Dobšina, Slvk. (dôp'shē-nä)	125	48°48'N	20°25'E
Doce, r., Braz. (dō'sȧ)	101	19°01'S	42°14'W
Doce, Canal Numero, can., Arg.	99c	36°47'S	59°00'W
Doce Leguas, Cayos de las, is., Cuba	92	20°55'N	79°05'W
Doctor Arroyo, Mex. (dōk-tōr' är-rō'yō)	88	23°41'N	100°10'W
Doddington, Eng., U.K. (dŏd'dĭng-tŏn)	114b	51°17'N	0°47'E
Dodecanese see Dodekanisoy, is., Grc.	131	38°00'N	26°10'E
Dodekanisoy (Dodecanese), is., Grc.	131	38°00'N	26°10'E
Dodge City, Ks., U.S. (dŏj)	64	37°44'N	100°01'W
Dodgeville, Wi., U.S. (dŏj'vĭl)	71	42°58'N	90°07'W
Dodoma, Tan. (dō'dō-mȧ)	186	6°11'S	35°45'E
Dog, l., Can. (dŏg)	58	48°42'N	89°24'W
Dogger Bank, bk., (dŏg'gẽr)	121	55°07'N	2°25'E
Dogubayazit, Tur.	137	39°35'N	44°00'E
Doha see Ad Dawhah, Qatar	154	25°02'N	51°28'E
Dohad, India	158	22°52'N	74°18'E
Dokshytsy, Bela. (dôk-shĕtsĕ)	132	54°53'N	27°49'E
Dolbeau, Can.	51	48°52'N	72°16'W
Dole, Fr. (dōl)	117	47°07'N	5°28'E
Dolgaya, Kosa, c., Russia (kô'sȧ dôl-gä'yȧ)	133	46°42'N	37°42'E
Dolgeville, N.Y., U.S.	67	43°10'N	74°45'W
Dolgiy, i., Russia	136	69°20'N	59°20'E
Dolgoprudnyy, Russia	142b	55°57'N	37°33'E
Dolinsk, Russia (dȧ-lēnsk')	141	47°29'N	142°31'E
Dollar Harbor, b., Bah.	92	25°30'N	79°15'W
Dolomite, Al., U.S. (dŏl'ō-mīt)	68h	33°28'N	86°57'W
Dolomiti, mts., Italy	130	46°16'N	11°43'E
Dolores, Arg. (dō-lō'rĕs)	102	36°20'S	57°42'W
Dolores, Col.	100a	3°33'N	74°54'W
Dolores, Ur.	99c	33°32'S	58°15'W
Dolores, Tx., U.S. (dō-lō'rĕs)	80	27°42'N	99°47'W
Dolores, r., Co., U.S.	77	38°35'N	108°50'W
Dolores Hidalgo, Mex. (dō-lō'rĕs-ē-däl'gō)	88	21°09'N	100°56'W
Dolphin and Union Strait, strt., Can. (dŏl'fĭn ūn'yŭn)	52	69°22'N	117°10'W
Dolyna, Ukr.	125	48°57'N	24°01'E
Domažlice, Czech Rep. (dō'mäzh-lē-tsĕ)	124	49°27'N	12°55'E
Dombasle-sur-Meurthe, Fr. (dôɴ-bäl')	127	48°38'N	6°18'E
Dombóvár, Hung. (dŏm'bō-vär)	125	46°22'N	18°08'E
Domeyko, Cordillera, mts., Chile (kôr-dēl-yĕ'rä-dō-mā'kō)	100	20°50'S	69°02'W
Dominica, nation, N.A. (dŏ-mĭ-nē'kȧ)	87	15°30'N	60°45'W
Dominica Channel, strt., N.A.	91b	15°00'N	61°30'W
Dominican Republic, nation, N.A. (dō-mĭn'ĭ-kăn)	87	19°00'N	70°45'W
Dominion, Can. (dō-mĭn'yŭn)	61	46°13'N	60°01'W

244 Durango – El Cajon PRONOUNCING INDEX

PLACE (Pronunciation)	PAGE	LAT.	LONG.
Durango, Mex. (do͞o-rä′n-gō)	86	24°02′N	104°42′W
Durango, Co., U.S. (dò-răn′gō)	77	37°15′N	107°55′W
Durango, state, Mex.	86	25°00′N	106°00′W
Durant, Ms., U.S. (dů-rănt′)	82	33°05′N	89°50′W
Durant, Ok., U.S.	79	33°59′N	96°23′W
Duratón, r., Spain (do͞o-rä-tōn′)	128	41°30′N	3°55′W
Durazno, Ur. (do͞o-räz′nō)	102	33°21′S	56°31′W
Durazno, dept., Ur.	99c	33°00′S	56°35′W
Durban, S. Afr. (dûr′băn)	186	29°48′S	31°00′E
Durbanville, S. Afr. (dûr-băn′vĭl)	186a	33°50′S	18°39′E
Durbe, Lat. (do͞or′bĕ)	123	56°36′N	21°24′E
Đurđevac, Cro.	119	46°03′N	17°03′E
Düren, Ger. (dü′rĕn)	127c	50°48′N	6°30′E
Durham, Eng., U.K. (dûr′ăm)	120	54°47′N	1°46′W
Durham, N.C., U.S.	65	36°00′N	78°55′W
Durham Downs, Austl.	176	27°30′S	141°55′E
Durrës, Alb. (dòr′ĕs)	110	41°19′N	19°27′E
Duryea, Pa., U.S. (do͞or-yā′)	67	41°20′N	75°50′W
Dushan, China	162	31°38′N	116°16′E
Dushan, China (do͞o-shän)	165	25°50′N	107°42′E
Dushanbe, Taj.	139	38°30′N	68°45′E
Düsseldorf, Ger. (düs′ĕl-dôrf)	117	51°14′N	6°47′E
Dussen, Neth.	115a	51°43′N	4°58′E
Dutalan Ula, mts., Mong.	164	49°25′N	112°40′E
Dutch Harbor, Ak., U.S. (dŭch här′bẽr)	64a	53°58′N	166°30′W
Duvall, Wa., U.S. (do͞o′vâl)	74a	47°44′N	121°59′W
Duwamish, r., Wa., U.S. (do͞o-wăm′ĭsh)	74a	47°24′N	122°18′W
Duyun, China (do͞o-yòn)	160	26°18′N	107°40′E
Dvinskaya Guba, b., Russia	136	65°10′N	38°40′E
Dwārka, India	158	22°18′N	68°59′E
Dwight, Il., U.S. (dwīt)	66	41°00′N	88°20′W
Dworshak Res, Id., U.S.	72	46°45′N	115°50′W
Dyat′kovo, Russia (dyät′kō-vō)	132	53°36′N	34°19′E
Dyer, In., U.S. (dī′ẽr)	69a	41°30′N	87°31′W
Dyersburg, Tn., U.S. (dī′ẽrz-bûrg)	82	36°02′N	89°23′W
Dyersville, Ia., U.S. (dī′ẽrz-vĭl)	71	42°28′N	91°09′W
Dyes Inlet, Wa., U.S. (dīz)	74a	47°37′N	122°45′W
Dykhtau, Gora, mtn., Russia	138	43°03′N	43°08′E
Dyment, Can. (dī′mĕnt)	57	49°37′N	92°19′W
Dzamïn Üüd, Mong.	161	44°38′N	111°32′E
Dzaoudzi, May. (dzou′dzĭ)	187	12°44′S	45°15′E
Dzavhan, r., Mong.	160	48°19′N	94°08′E
Dzerzhinsk, Russia	136	56°20′N	43°50′E
Dzerzhyns′k, Ukr.	133	48°26′N	37°50′E
Dzhalal-Abad, Kyrg. (jà-läl′à-bät′)	139	40°56′N	73°00′E
Dzhambul *see* Zhambyl, Kaz.	139	42°51′N	71°29′E
Dzhankoi, Ukr.	137	45°43′N	34°22′E
Dzhizak, Uzb. (dzhĭ′zäk)	139	40°13′N	67°58′E
Dzhugdzhur Khrebet, mts., Russia (jòg-jo͞or′)	135	56°15′N	137°00′E
Działoszyce, Pol. (jyä-wō-shĭ′tsĕ)	125	50°21′N	20°22′E
Dzibalchén, Mex. (zē-bäl-chĕ′n)	90a	19°25′N	89°39′W
Dzidzantún, Mex. (zēd-zän-to͞o′n)	90a	21°18′N	89°00′W
Dzierżoniów, Pol. (dzyĕr-zhōn′yŭf)	124	50°44′N	16°38′E
Dzilam González, Mex. (zē-lä′m-gōn-zä′lĕz)	90a	21°21′N	88°53′W
Dzitás, Mex. (zē-tá′s)	90a	20°47′N	88°32′W
Dzungaria, reg., China (dzòn-gä′rĭ-à)	160	44°39′N	86°13′E
Dzungarian Gate, p., Asia	160	45°00′N	88°00′E
Dzyarzhynsk, Bela.	132	53°41′N	27°14′E

E

PLACE (Pronunciation)	PAGE	LAT.	LONG.
Eagle, W.V., U.S.	66	38°10′N	81°20′W
Eagle, r., Co., U.S.	77	39°32′N	106°28′W
Eaglecliff, Wa., U.S. (ē′gl-klĭf)	74c	46°10′N	123°13′W
Eagle Creek, r., In., U.S.	69g	39°54′N	86°17′W
Eagle Grove, Ia., U.S.	71	42°39′N	93°55′W
Eagle Lake, Me., U.S.	60	47°03′N	68°38′W
Eagle Lake, Tx., U.S.	81	29°37′N	96°20′W
Eagle Lake, l., Ca., U.S.	72	40°45′N	120°52′W
Eagle Mountain, Ca., U.S.	76	33°49′N	115°27′W
Eagle Mountain L, Tx., U.S.	75c	32°56′N	97°27′W
Eagle Pass, Tx., U.S.	64	28°49′N	100°30′W
Eagle Pk, Ca., U.S.	72	41°18′N	120°11′W
Ealing, Eng., U.K. (ē′lĭng)	114b	51°29′N	0°19′W
Earle, Ar., U.S. (ûrl)	79	35°14′N	90°28′W
Earlington, Ky., U.S. (ûr′lĭng-tŭn)	82	37°15′N	87°31′W
Easley, S.C., U.S. (ēz′lĭ)	83	34°48′N	82°37′W
East, Mount, mtn., Pan.	86a	9°09′N	79°46′W
East Alton, Il., U.S. (ôl′tŭn)	75e	38°53′N	90°08′W
East Angus, Can. (ăŋ′gŭs)	59	45°35′N	71°40′W
East Aurora, N.Y., U.S. (ô-rō′rà)	69c	42°46′N	78°38′W
East Bay, b., Tx., U.S.	81a	29°30′N	94°41′W
East Bernstadt, Ky., U.S. (bûrn′stát)	82	37°09′N	84°08′W
Eastbourne, Eng., U.K. (ēst′bôrn)	121	50°48′N	0°16′E
East Caicos, i., T./C. Is. (kī′kōs)	93	21°40′N	71°35′W
East Cape, c., N.Z.	175a	37°37′S	178°33′E
East Cape *see* Dezhnëva, Mys, c., Russia	152	68°00′N	172°00′W
East Carondelet, Il., U.S. (kà-rŏn′dĕ-lĕt)	75e	38°33′N	90°14′W
East Cherokee Indian Reservation, I.R., N.C., U.S.	82	35°33′N	83°12′W
East Chicago, In., U.S. (shĭ-kô′gō)	69a	41°39′N	87°29′W
East China Sea, sea, Asia	161	30°28′N	125°52′E
East Cleveland, Oh., U.S. (klēv′lănd)	69d	41°33′N	81°35′W
East Cote Blanche Bay, b., La., U.S. (kōt blănsh′)	81	29°30′N	92°07′W
East Des Moines, r., Ia., U.S. (dē moin′)	71	42°57′N	94°17′W
East Detroit, Mi., U.S. (dē-troit′)	69b	42°28′N	82°57′W
Easter Island *see* Pascua, Isla de, i., Chile	195	26°50′S	109°00′W
Eastern Ghāts, mts., India	155	13°50′N	78°45′E
Eastern Turkestan, hist. reg., China (tòr-kĕ-stän′)(tûr-kĕ-stăn′)	160	39°40′N	78°20′E
East Grand Forks, Mn., U.S. (grănd fôrks)	70	47°56′N	97°02′W
East Greenwich, R.I., U.S. (grĭn′ĭj)	68b	41°40′N	71°27′W
Easthampton, Ma., U.S. (ēst-hămp′tŭn)	67	42°15′N	72°45′W
East Hartford, Ct., U.S. (härt′fẽrd)	67	41°45′N	72°35′W
East Helena, Mt., U.S. (hĕ-lē′nà)	73	46°31′N	111°50′W
East Ilsley, Eng., U.K. (ĭl′slē)	114b	51°30′N	1°18′W
East Jordan, Mi., U.S. (jôr′dăn)	66	45°05′N	85°05′W
East Kansas City, Mo., U.S. (kăn′zàs)	75f	39°09′N	94°30′W
Eastland, Tx., U.S. (ēst′lănd)	80	32°24′N	98°47′W
East Lansing, Mi., U.S. (lăn′sĭng)	66	42°45′N	84°30′W
Eastlawn, Mi., U.S.	69b	42°15′N	83°35′W
East Leavenworth, Mo., U.S. (lĕv′ĕn-wûrth)	75f	39°18′N	94°50′W
East Liverpool, Oh., U.S. (lĭv′ẽr-po͞ol)	66	40°40′N	80°35′W
East London, S. Afr. (lŭn′dŭn)	186	33°02′S	27°54′E
East Los Angeles, Ca., U.S. (lōs ăŋ′há-lās)	75a	34°01′N	118°09′W
Eastmain, r., Can. (ēst′mān)	53	52°12′N	73°19′W
Eastman, Ga., U.S. (ēst′măn)	82	32°10′N	83°11′W
East Millstone, N.J., U.S. (mĭl′stōn)	68a	40°30′N	74°35′W
East Moline, Il., U.S. (mō-lēn′)	71	41°31′N	90°28′W
East Nishnabotna, r., Ia., U.S. (nĭsh-nà-bŏt′nà)	70	40°53′N	95°23′W
Easton, Md., U.S. (ēs′tŭn)	67	38°45′N	76°05′W
Easton, Pa., U.S.	67	40°45′N	75°15′W
Easton L, Ct., U.S.	68a	41°18′N	73°17′W
East Orange, N.J., U.S. (ŏr′ĕnj)	68a	40°46′N	74°12′W
East Pakistan *see* Bangladesh, nation, Asia	155	24°15′N	90°00′E
East Palo Alto, Ca., U.S.	74b	37°27′N	122°07′W
East Peoria, Il., U.S. (pē-ō′rĭ-à)	66	40°40′N	89°30′W
East Pittsburgh, Pa., U.S. (pĭts′bûrg)	69e	40°24′N	79°50′W
East Point, Ga., U.S.	68c	33°41′N	84°27′W
Eastport, Me., U.S. (ēst′pōrt)	60	44°53′N	67°01′W
East Providence, R.I., U.S. (prŏv′ĭ-dĕns)	68b	41°49′N	71°22′W
East Retford, Eng., U.K. (rĕt′fẽrd)	114a	53°19′N	0°56′W
East Riding of Yorkshire, co., Eng., U.K.	114a	53°45′N	0°40′W
East Rochester, N.Y., U.S. (rŏch′ĕs-tẽr)	67	43°10′N	77°30′W
East Saint Louis, Il., U.S.	65	38°38′N	90°10′W
East Siberian Sea, sea, Russia (sī-bĭr′y′n)	135	73°00′N	153°28′E
Eastsound, Wa., U.S. (ēst-sound)	74d	48°42′N	122°42′W
East Stroudsburg, Pa., U.S. (stroudz′bûrg)	67	41°00′N	75°10′W
East Syracuse, N.Y., U.S. (sĭr′à-kūs)	67	43°05′N	76°00′W
East Tavaputs Plateau, plat., Ut., U.S. (tă-vă′-pŭts)	77	39°25′N	109°45′W
East Tawas, Mi., U.S. (tô′wăs)	66	44°15′N	83°30′W
East Timor, nation, Asia	169	9°00′S	125°30′E
East Walker, r., U.S. (wôk′ẽr)	76	38°36′N	119°02′W
Eaton, Co., U.S. (ē′tŭn)	78	40°31′N	104°42′W
Eaton, Oh., U.S.	66	39°45′N	84°40′W
Eaton Estates, Oh., U.S.	69d	41°19′N	82°01′W
Eaton Rapids, Mi., U.S. (răp′ĭdz)	66	42°30′N	84°40′W
Eatonton, Ga., U.S. (ē′tŭn-tŭn)	82	33°20′N	83°24′W
Eatontown, N.J., U.S. (ē′tŭn-toun)	68a	40°18′N	74°04′W
Eau Claire, Wi., U.S. (ō klâr′)	65	44°47′N	91°32′W
Ebeltoft, Den. (ĕ′bĕl-tŭft)	122	56°11′N	10°39′E
Ebensburg, Pa., U.S.	67	40°29′N	78°44′W
Ebersberg, Ger. (ĕ′bĕrs-bĕrgh)	115d	48°05′N	11°58′E
Ebingen, Ger. (ā′bĭng-ĕn)	124	48°13′N	9°04′E
Eboli, Italy (ĕb′ō-lē)	130	40°38′N	15°04′E
Ebolowa, Cam.	184	2°54′N	11°09′E
Ebreichsdorf, Aus.	115e	47°58′N	16°24′E
Ebrié, Lagune, b., C. Iv.	188	5°20′N	4°50′W
Ebro, r., Spain (ā′brō)	112	42°00′N	2°00′W
Eccles, Eng., U.K. (ĕk′′lz)	114a	53°29′N	2°20′W
Eccles, W.V., U.S.	66	37°45′N	81°10′W
Eccleshall, Eng., U.K.	114a	52°51′N	2°15′W
Eceabat, Tur.	131	40°10′N	26°21′E
Echague, Phil. (ā-chä′gwä)	169a	16°43′N	121°40′E
Echandi, Cerro, mtn., N.A. (sĕ′r-rô-ĕ-chä′nd)	91	9°05′N	82°51′W
Ech Cheliff, Alg.	184	36°14′N	1°32′E
Echimamish, r., Can.	57	54°15′N	97°30′W
Echmiadzin, Arm.	138	40°10′N	44°18′E
Echo Bay, Can. (ĕk′ō)	75k	46°29′N	84°04′W
Echoing, r., Can. (ĕk′ō-ĭng)	57	55°15′N	91°30′W
Echternach, Lux. (ĕk′tĕr-näk)	127	49°48′N	6°25′E
Echuca, Austl. (ē-chò′kà)	175	36°10′S	144°47′E
Écija, Spain (ā′thē-hä)	118	37°20′N	5°07′W
Eckernförde, Ger.	124	54°27′N	9°51′E
Eclipse, Va., U.S. (ē-klĭps′)	68g	36°55′N	76°29′W
Ecorse, Mi., U.S. (ē-kôrs′)	69b	42°15′N	83°09′W
Ecuador, nation, S.A. (ĕk′wà-dôr)	100	0°00′N	78°30′W
Ed, Erit.	185	13°57′N	41°37′E
Eddyville, Ky., U.S. (ĕd′ĭ-vĭl)	82	37°03′N	88°03′W
Ede, Nig.	189	7°44′N	4°27′E
Edéa, Cam. (ĕ-dā′ä)	184	3°48′N	10°08′E
Eden, Tx., U.S. (ē′dĕn)	80	31°13′N	99°51′W
Eden, Ut., U.S.	75b	41°18′N	111°49′W
Eden, r., Eng., U.K. (ē′dĕn)	120	54°40′N	2°35′W
Edenbridge, Eng., U.K. (ē′dĕn-brĭj)	114b	51°11′N	0°05′E
Edenham, Eng., U.K. (ē′d′n-ăm)	114a	52°46′N	0°25′W
Eden Prairie, Mn., U.S. (prâr′ĭ)	75g	44°51′N	93°29′W
Edenton, N.C., U.S. (ē′dĕn-tŭn)	83	36°02′N	76°37′W
Edenton, Oh., U.S.	69f	39°14′N	84°02′W
Edenvale, S. Afr. (ĕd′ĕn-văl)	187b	26°09′S	28°10′E
Edenville, S. Afr. (ĕ′d′n-vĭl)	192c	27°33′S	27°42′E
Eder, r., Ger. (ā′dẽr)	124	51°05′N	8°52′E
Édessa, Grc.	119	40°48′N	22°04′E
Edgefield, S.C., U.S. (ĕj′fēld)	83	33°52′N	81°55′W
Edgeley, N.D., U.S. (ĕj′lĭ)	70	46°24′N	98°43′W
Edgemont, S.D., U.S. (ĕj′mŏnt)	70	43°19′N	103°50′W
Edgerton, Wi., U.S. (ĕj′ẽr-tŭn)	71	42°49′N	89°06′W
Edgewater, Al., U.S. (ĕj-wô-tẽr)	68h	33°31′N	86°52′W
Edgewater, Md., U.S.	68e	38°58′N	76°35′W
Edgewood, Can. (ĕj′wòd)	55	49°47′N	118°08′W
Edina, Mn., U.S. (ē-dī′nà)	75g	44°55′N	93°20′W
Edina, Mo., U.S.	79	40°10′N	92°11′W
Edinburg, In., U.S. (ĕd′′n-bûrg)	66	39°20′N	85°55′W
Edinburg, Tx., U.S.	80	26°18′N	98°08′W
Edinburgh, Scot., U.K. (ĕd′′n-bŭr-ō)	110	55°57′N	3°10′W
Edirne, Tur.	131	41°41′N	26°35′E
Edisto, r., S.C., U.S. (ĕd′ĭs-tō)	83	33°10′N	80°50′W
Edisto, North Fork, r., S.C., U.S.	83	33°42′N	81°24′W
Edisto, South Fork, r., S.C., U.S.	83	33°43′N	81°35′W
Edisto Island, S.C., U.S.	83	32°32′N	80°20′W
Edmond, Ok., U.S. (ĕd′mŭnd)	79	35°39′N	97°29′W
Edmonds, Wa., U.S. (ĕd′mŭndz)	74a	47°49′N	122°23′W
Edmonton, Can.	50	53°33′N	113°28′W
Edmundston, Can. (ĕd′mŭn-stŭn)	51	47°22′N	68°20′W
Edna, Tx., U.S. (ĕd′nà)	81	28°59′N	96°39′W
Edremit, Tur. (ĕd-rĕ-mēt′)	119	39°35′N	27°00′E
Edremit Körfezi, b., Tur.	131	39°28′N	26°35′E
Edson, Can. (ĕd′sŭn)	50	53°35′N	116°26′W
Edward, i., Can. (ĕd′wẽrd)	58	48°21′N	88°29′W
Edward, l., Afr.	186	0°25′S	29°40′E
Edwardsville, Il., U.S. (ĕd′wẽrdz-vĭl)	75e	38°49′N	89°58′W
Edwardsville, In., U.S.	69h	38°17′N	85°53′W
Edwardsville, Ks., U.S.	75f	39°04′N	94°49′W
Eel, r., Ca., U.S. (ēl)	72	40°39′N	124°15′W
Eel, r., In., U.S.	66	40°50′N	85°55′W
Efate, i., Vanuatu (å-fä′tä)	175	18°02′S	168°29′E
Effigy Mounds National Monument, rec., Ia., U.S. (ĕf′ĭ-jŭ mounds)	71	43°04′N	91°15′W
Effingham, Il., U.S. (ĕf′ĭng-hăm)	66	39°05′N	88°30′W
Ega, r., Spain (ā′gä)	128	42°40′N	2°20′W
Egadi, Isole, is., Italy (ĕ′gä-dē)	118	38°01′N	12°00′E
Egegik, Ak., U.S. (ĕg′ĕ-jĭt)	63	58°10′N	157°22′W
Eger, Hung. (ĕ′gĕr)	125	47°53′N	20°24′E
Egersund, Nor. (ĕ′ghĕr-sòn′)	116	58°29′N	6°01′E
Egg Harbor, N.J., U.S. (ĕg här′bẽr)	67	39°30′N	74°35′W
Egham, Eng., U.K. (ĕg′ŭm)	114b	51°24′N	0°33′W
Egiyn, r., Mong.	160	49°41′N	100°40′E
Egmont, Cape, c., N.Z. (ĕg′mŏnt)	175a	39°18′S	173°49′E
Egypt, nation, Afr. (ē′jĭpt)	185	26°58′N	27°01′E
Eha-Amufu, Nig.	189	6°40′N	7°46′E
Eibar, Spain (ā′ē-bär)	128	43°12′N	2°20′W
Eichstätt, Ger. (īк′shtät)	124	48°54′N	11°14′E
Eichwalde, Ger. (īк′väl-dĕ)	115b	52°22′N	13°37′E
Eidfjord, Nor. (ĕĭd′fyôr)	122	60°28′N	7°04′E
Eidsvoll, Nor. (ĭdhs′vôl)	116	60°19′N	11°15′E
Eifel, mts., Ger. (ī′fĕl)	124	50°08′N	6°30′E
Eighty Mile Beach, cst., Austl.	174	19°00′S	121°00′E
Eilenburg, Ger. (ī′lĕn-bòrgh)	124	51°27′N	12°38′E
Einbeck, Ger. (īn′bĕk)	124	51°49′N	9°52′E
Eindhoven, Neth. (īnd′hō-vĕn)	121	51°29′N	5°20′E
Eisenach, Ger. (ī′zĕn-äк)	117	50°58′N	10°18′E
Eisenhüttenstadt, Ger.	124	52°08′N	14°40′E
Eivissa, Spain	129	38°55′N	1°24′E
Eivissa, i., Spain	112	38°55′N	1°24′E
Ejea de los Caballeros, Spain	128	42°07′N	1°05′W
Ejura, Ghana	188	7°23′N	1°22′W
Ejutla de Crespo, Mex. (å-hòt′lä dā kräs′pō)	89	16°34′N	96°44′W
Ekanga, D.R.C.	190	2°23′S	23°14′E
Ekenäs, Fin. (ĕ′kĕ-nâs)	123	59°59′N	23°25′E
Ekeren, Bel.	115a	51°17′N	4°27′E
Ekoli, D.R.C.	190	0°23′S	24°16′E
El Aaiún, W. Sah.	184	26°45′N	13°15′W
El Affroun, Alg. (ĕl áf-froun′)	129	36°28′N	2°38′E
Elands, r., S. Afr. (ĕlănds)	187c	31°48′S	26°09′E
Elands, r., S. Afr.	192c	25°11′S	28°52′E
El Arahal, Spain (ĕl ä-rä-äl′)	128	37°17′N	5°32′W
El Arba, Alg.	129	36°35′N	3°10′E
Elat, Isr.	154	29°34′N	34°57′E
Elazığ, Tur. (ĕl-ä′zĕz)	154	38°40′N	39°00′E
Elba, Al., U.S. (ĕl′bà)	82	31°25′N	86°01′W
Elba, Isola d′, i., Italy (ē-sō lä-d-ĕl′bà)	118	42°42′N	10°25′E
El Banco, Col. (ĕl băn′cō)	100	8°58′N	74°01′W
Elbansan, Alb. (ĕl-bä-sän′)	119	41°08′N	20°05′E
Elbe (Labe), r., Eur. (ĕl′bĕ)(lä′bĕ)	112	52°30′N	11°30′E
Elbert, Mount, mtn., Co., U.S. (ĕl′bẽrt)	64	39°05′N	106°25′W
Elberton, Ga., U.S. (ĕl′bẽr-tŭn)	83	34°05′N	82°53′W
Elbeuf, Fr. (ĕl-bûf′)	117	49°16′N	0°59′E
El Beyadh, Alg.	118	33°42′N	1°06′E
Elbistan, Tur. (ĕl-bē-stän′)	119	38°20′N	37°10′E
Elbląg, Pol. (ĕl′blăŋg)	116	54°11′N	19°25′E
El Bonillo, Spain (ĕl bō-nēl′yō)	128	38°56′N	2°31′W
El Boulaïda, Alg.	184	36°33′N	2°45′E
Elbow, r., Can. (ĕl′bō)	62e	51°03′N	114°24′W
Elbow Cay, i., Bah.	92	26°25′N	76°55′W
Elbow Lake, Mn., U.S.	70	46°00′N	95°59′W
El′brus, Gora, mtn., Russia (ĕl′bròs′)	134	43°20′N	42°25′E
Elbrus, Mount *see* El′brus, Gora, mtn., Russia	134	43°20′N	42°25′E
Elburz Mountains, mts., Iran (ĕl′bòrz′)	154	36°30′N	51°00′E
El Cajon, Col. (ĕl-kä-kŏ′n)	100a	4°50′N	76°35′W

ăt; finăl; rāte; senăte; ärm; ásk; sofá; fâre; ch-choose; dh-as th in other; bē; ēvent; bĕt; recĕnt; cratẽr; g-gō; gh-guttural g; bĭt; ĭ-short neutral; rīde; к-guttural k as ch in German ich;

PRONOUNCING INDEX

El Cajon – Erlanger 245

PLACE (Pronunciation)	PAGE	LAT.	LONG.
El Cajon, Ca., U.S.	76a	32°48'N	116°58'W
El Cambur, Ven. (käm-bōōr')	101b	10°24'N	68°06'W
El Campo, Tx., U.S. (kăm'pō)	81	29°13'N	96°17'W
El Carmen, Chile (kà'r-měn)	99b	34°14'S	71°23'W
El Carmen, Col. (kà'r-měn)	100	9°54'N	75°12'W
El Casco, Ca., U.S. (kăs'kô)	75a	33°59'N	117°08'W
El Centro, Ca., U.S. (sěn'trô)	76	32°47'N	115°33'W
El Cerrito, Ca., U.S. (sěr-rē'tō)	74b	37°55'N	122°19'W
El Cuyo, Mex.	90a	21°30'N	87°42'W
Elda, Spain (ĕl'dä)	129	38°28'N	0°44'W
El Djelfa, Alg.	184	34°40'N	3°17'E
El Djouf, des., Afr. (ĕl djōōf)	184	21°45'N	7°05'W
Eldon, Ia., U.S. (ĕl-dŭn)	71	40°55'N	92°15'W
Eldon, Mo., U.S.	79	38°21'N	92°36'W
Eldora, Ia., U.S. (ĕl-dō'rà)	71	42°21'N	93°08'W
El Dorado, Ar., U.S. (ĕl dô-rä'dō)	65	33°13'N	92°39'W
Eldorado, Il., U.S.	66	37°50'N	88°30'W
El Dorado, Ks., U.S.	79	37°49'N	96°51'W
Eldorado Springs, Mo., U.S. (sprĭngz)	79	37°51'N	94°02'W
Eldoret, Kenya (ĕl-dô-rĕt')	191	0°31'N	35°17'E
El Ebano, Mex. (ā-bä'nō)	88	22°13'N	98°26'W
Electra, Tx., U.S. (ê-lĕk'trà)	78	34°02'N	98°54'W
Electric Peak, mtn., Mt., U.S. (ê-lĕk'trĭk)	73	45°03'N	110°52'W
Elek, r.,	137	51°20'N	53°10'E
Elektrogorsk, Russia (ĕl-yĕk'trô-gôrsk)	142b	55°53'N	38°48'E
Elektrostal', Russia (ĕl-yĕk'trô-stál)	142b	55°47'N	38°27'E
Elektrougli, Russia	142b	55°43'N	38°13'E
Elephant Butte Reservoir, res., N.M., U.S. (ĕl'ê-fănt būt)	64	33°25'N	107°10'W
El Escorial, Spain (ĕl-ĕs-kō-ryä'l)	129a	40°38'N	4°08'W
El Espino, Nic. (ĕl-ĕs-pē'nō)	90	13°26'N	86°48'W
Eleuthera, i., Bah. (ê-lū'thēr-à)	87	25°05'N	76°10'W
Eleuthera Point, c., Bah.	92	24°35'N	76°05'W
Eleven Point, r., Mo., U.S. (ê-lĕv'ĕn)	79	36°53'N	91°39'W
Elgin, Scot., U.K.	120	57°40'N	3°30'W
Elgin, Il., U.S. (ĕl'jĭn)	69a	42°03'N	88°16'W
Elgin, Ne., U.S.	70	41°58'N	98°04'W
Elgin, Or., U.S.	72	45°34'N	117°58'W
Elgin, Tx., U.S.	81	30°21'N	97°22'W
Elgin, Wa., U.S.	74a	47°23'N	122°42'W
Elgon, Mount, mtn., Afr. (ĕl'gŏn)	185	1°00'N	34°25'E
El Grara, Alg.	118	32°50'N	4°26'E
El Grullo, Mex. (grōōl-yô)	88	19°46'N	104°10'W
El Guapo, Ven. (gwà'pô)	101b	10°07'N	66°00'W
El Hank, reg., Afr.	184	23°44'N	6°45'W
El Hatillo, Ven. (ä-tē'l-yô)	101b	10°08'N	65°13'W
Elie, Can. (ē'lē)	62f	49°55'N	97°45'W
Elila, r., D.R.C. (ê-lē'là)	186	3°30'S	28°00'E
Elisa, i., Wa., U.S. (ê-lī'sà)	74d	48°43'N	122°37'W
Élisabethville see Lubumbashi, D.R.C.	186	11°40'S	27°28'E
Elisenvaara, Russia (ā-lē'sĕn-vä'rà)	123	61°25'N	29°46'E
Elizabeth, La., U.S. (ê-lĭz'à-bĕth)	81	30°50'N	92°47'W
Elizabeth, N.J., U.S.	68a	40°40'N	74°13'W
Elizabeth, Pa., U.S.	69e	40°16'N	79°53'W
Elizabeth City, N.C., U.S.	83	36°15'N	76°15'W
Elizabethton, Tn., U.S. (ê-lĭz-à-bĕth'tŭn)	83	36°19'N	82°12'W
Elizabethtown, Ky., U.S. (ê-lĭz'à-bĕth-toun)	66	37°40'N	85°55'W
El Jadida, Mor.	184	33°14'N	8°34'W
Elk, Pol.	116	53°53'N	22°23'E
Elk, r., Can.	55	50°00'N	115°00'W
Elk, r., Tn., U.S.	82	35°05'N	86°36'W
Elk, r., W.V., U.S.	66	38°30'N	81°05'W
El Kairouan, Tun. (kĕr-ô-än)	184	35°46'N	10°04'E
Elk City, Ok., U.S. (ĕlk)	78	35°23'N	99°23'W
El Kef, Tun. (xĕf')	118	36°14'N	8°42'E
Elkhart, In., U.S. (ĕlk'härt)	66	41°40'N	86°00'W
Elkhart, Ks., U.S.	78	37°00'N	101°54'W
Elkhart, Tx., U.S.	81	31°38'N	95°35'W
Elkhorn, Wi., U.S. (ĕlk'hôrn)	71	42°39'N	88°32'W
Elkhorn, r., Ne., U.S.	70	42°06'N	97°46'W
Elkin, N.C., U.S. (ĕl'kĭn)	83	36°15'N	80°50'W
Elk Island, i., Can.	57	50°45'N	96°32'W
Elk Island National Park, rec., Can. (ĕlk ī'lănd)	52	53°37'N	112°45'W
Elko, Nv., U.S. (ĕl'kō)	64	40°51'N	115°46'W
Elk Point, S.D., U.S.	70	42°41'N	96°41'W
Elk Rapids, Mi., U.S. (răp'ĭdz)	66	44°55'N	85°25'W
Elk River, Id., U.S. (rĭv'ēr)	72	46°47'N	116°11'W
Elk River, Mn., U.S.	71	45°17'N	93°33'W
Elkton, Ky., U.S. (ĕlk'tŭn)	82	36°47'N	87°08'W
Elkton, Md., U.S.	67	39°35'N	75°50'W
Elkton, S.D., U.S.	70	44°15'N	96°28'W
Elland, Eng., U.K. (ĕl'ănd)	114a	53°41'N	1°50'W
Ellen, Mount, mtn., Ut., U.S. (ĕl'ĕn)	77	38°05'N	110°50'W
Ellendale, N.D., U.S. (ĕl'ĕn-dāl)	70	46°01'N	98°33'W
Ellensburg, Wa., U.S. (ĕl'ĕnz-bûrg)	72	47°00'N	120°31'W
Ellenville, N.Y., U.S. (ĕl'ĕn-vĭl)	67	41°40'N	74°25'W
Ellerslie, Can. (ĕl'ērz-lê)	62g	53°25'N	113°30'W
Ellesmere, Eng., U.K. (ĕlz'mēr)	114a	52°55'N	2°54'W
Ellesmere Island, i., Can.	49	81°00'N	80°00'W
Ellesmere Port, Eng., U.K.	114a	53°17'N	2°54'W
Ellice Islands see Tuvalu, nation, Oc.	3	5°20'S	174°00'E
Ellicott City, is., Md., U.S. (ĕl'ĭ-kŏt sĭ'tê)	68e	39°16'N	76°48'W
Ellicott Creek, r., N.Y., U.S.	69c	43°00'N	78°46'W
Elliot, S. Afr.	187c	31°19'S	27°52'E
Elliot, Wa., U.S. (ĕl'ĭ-ŭt)	74a	47°28'N	122°08'W
Elliotdale, S. Afr. (ĕl-ĭ-ŏt'dāl)	187c	31°58'S	28°42'E
Elliot Lake, Can.	58	46°23'N	82°39'W
Ellis, Ks., U.S. (ĕl'ĭs)	78	38°56'N	99°34'W
Ellisville, Mo., U.S.	75e	38°35'N	90°35'W
Ellisville, Ms., U.S. (ĕl'ĭs-vĭl)	82	31°37'N	89°10'W
Ellsworth, Ks., U.S. (ĕlz'wûrth)	78	38°43'N	98°14'W
Ellsworth, Me., U.S.	60	44°33'N	68°26'W
Ellsworth Mountains, mts., Ant.	178	77°00'S	90°00'W
Ellwangen, Ger. (ĕl'väŋ-gĕn)	124	48°47'N	10°08'E
Elm, Ger. (ĕlm)	115c	53°31'N	9°13'E
Elm, r., S.D., U.S.	70	45°47'N	98°28'W
Elm, r., W.V., U.S.	66	38°30'N	81°05'W
Elma, Wa., U.S. (ĕl'mà)	72	47°02'N	123°20'W
El Mahdia, Tun. (mä-dēä)(mä'dê-à)	118	35°30'N	11°09'E
Elmendorf, Tx., U.S. (ĕl'mĕn-dôrf)	75d	29°16'N	98°20'W
El Menia, Alg.	184	30°39'N	2°52'E
Elm Fork, Tx., U.S. (ĕlm fôrk)	75c	32°55'N	96°56'W
Elmhurst, Il., U.S. (ĕlm'hûrst)	69a	41°54'N	87°56'W
El Miliyya, Alg. (mē'ä)	184	36°30'N	6°16'E
Elmira, N.Y., U.S. (ĕl-mī'rà)	67	42°05'N	76°50'W
Elmira Heights, N.Y., U.S.	67	42°10'N	76°50'W
El Modena, Ca., U.S. (mô-dē'nô)	75a	33°47'N	117°48'W
El Mohammadia, Alg.	129	35°35'N	0°05'E
El Monte, Ca., U.S. (mōn'tā)	75a	34°04'N	118°02'W
El Morro National Monument, rec., N.M., U.S.	77	35°05'N	108°20'W
Elmshorn, Ger. (ĕlms'hôrn)	124	53°45'N	9°39'E
Elmwood Place, Oh., U.S. (ĕlm'wŏŏd plās)	69f	39°11'N	84°30'W
Elokomin, r., Wa., U.S. (ê-lō'kô-mĭn)	74c	46°16'N	123°16'W
El Oro, Mex. (ô-rô)	88	19°49'N	100°04'W
El Pao, Ven. (ĕl pà'ô)	100	8°08'N	62°37'W
El Paraíso, Hond. (pä-rä-ē'sō)	90	13°55'N	86°35'W
El Pardo, Spain (pä'r-dô)	129a	40°31'N	3°47'W
El Paso, Tx., U.S. (pas'ō)	64	31°47'N	106°27'W
El Pilar, Ven. (pē-lä'r)	101b	9°56'N	64°48'W
El Porvenir, Pan. (pôr-vä-nēr')	91	9°34'N	78°55'W
El Puerto de Santa María, Spain	128	36°36'N	6°18'W
El Qala, Alg.	118	36°52'N	8°23'E
El Qoll, Alg.	184	37°02'N	6°29'E
El Real, Pan. (rā-äl)	91	8°07'N	77°43'W
El Reno, Ok., U.S. (rē'nō)	79	35°31'N	97°57'W
Elroy, Wi., U.S. (ĕl'roi)	71	43°44'N	90°17'W
Elsa, Can.	63	63°55'N	135°25'W
Elsah, Il., U.S. (ĕl'zà)	75e	38°57'N	90°22'W
El Salto, Mex. (säl'tō)	88	23°48'N	105°22'W
El Salvador, nation, N.A.	86	14°00'N	89°30'W
El Sauce, Nic. (ĕl-sà'ō-sĕ)	90	13°00'N	86°40'W
Elsberry, Mo., U.S. (ĕlz'bĕr-ĭ)	79	39°09'N	90°44'W
Elsdorf, Ger. (ĕls'dôrf)	127c	50°56'N	6°35'E
El Segundo, Ca., U.S. (sĕgŭn'dō)	75a	33°55'N	118°24'W
Elsinore, Ca., U.S. (ĕl'sĭ-nôr)	75a	33°40'N	117°19'W
Elsinore Lake, l., Ca., U.S.	75a	33°38'N	117°21'W
Elstorf, Ger. (ĕls'tōrf)	115c	53°25'N	9°48'E
Eltham, Austl. (ĕl'thăm)	173a	37°43'S	145°08'E
El Tigre, Ven. (tē'grĕ)	100	8°49'N	64°15'W
El'ton, l., Russia	137	49°10'N	47°00'E
El Toro, Ca., U.S. (tō'rō)	75a	33°37'N	117°42'W
El Triunfo, El Sal.	90	13°17'N	88°32'W
El Triunfo, Hond. (ĕl-trē-ōō'n-fō)	90	13°06'N	87°00'W
Elūru, India	155	16°44'N	80°09'E
El Vado Res, N.M., U.S.	77	36°37'N	106°30'W
Elvas, Port. (ĕl'väzh)	118	38°53'N	7°11'W
Elverum, Nor. (ĕl'vĕ-rŏŏm)	122	60°53'N	11°33'E
El Viejo, Nic. (ĕl-vyĕ'κō)	90	12°10'N	87°10'W
El Viejo, vol., Nic.	90	12°44'N	87°03'W
Elvins, Mo., U.S. (ĕl'vĭnz)	79	37°49'N	90°31'W
El Wad, Alg.	184	33°23'N	6°49'E
El Wak, Kenya (wäk')	185	3°00'N	41°00'E
Elwell, Lake, res., Mt., U.S.	73	48°22'N	111°17'W
Elwood, Il., U.S. (ĕl'wŏŏd)	69a	41°24'N	88°07'W
Elwood, In., U.S.	66	40°15'N	85°50'W
Elx, Spain	129	38°15'N	0°42'W
Ely, Eng., U.K. (ē'lĭ)	121	52°25'N	0°17'E
Ely, Mn., U.S.	71	47°54'N	91°53'W
Ely, Nv., U.S.	64	39°16'N	114°53'W
Elyria, Oh., U.S. (ê-lĭr'ĭ-à)	69d	41°22'N	82°07'W
Ema, r., Est. (â'mà)	123	58°25'N	27°00'E
Emāmshahr, Iran	154	36°25'N	55°01'E
Emån, r., Swe.	122	57°15'N	15°46'E
Embarrass, r., Il., U.S. (ĕm-băr'ăs)	66	39°15'N	88°05'W
Embrun, Can. (ĕm'brŭn)	62c	45°16'N	75°17'W
Embrun, Fr. (äɴ-brûɴ')	127	44°35'N	6°32'E
Embu, Kenya	191	0°32'S	37°27'E
Emden, Ger. (ĕm'dĕn)	124	53°21'N	7°15'E
Emerson, Can. (ĕm'ēr-sŭn)	50	49°00'N	97°12'W
Emeryville, Ca., U.S. (ĕm'ēr-ĭ-vĭl)	74b	37°50'N	122°17'W
Emi Koussi, mtn., Chad (ā'mê κōō-sē')	185	19°50'N	18°30'E
Emiliano Zapata, Mex. (ĕ-mē-lyà'nô-zä-pà'tà)	89	17°45'N	91°46'W
Emilia-Romagna, hist. reg., Italy (ē-mēl'yä rô-mà'n-yä)	130	44°35'N	10°48'E
Eminence, Ky., U.S. (ĕm'ĭ-nĕns)	66	38°25'N	85°15'W
Emira Island, i., Pap. N. Gui. (ā-mê-rä')	169	1°40'S	150°28'E
Emmen, Neth. (ĕm'ĕn)	121	52°48'N	6°55'E
Emmerich, Ger. (ĕm'ēr-ĭk)	127c	51°51'N	6°16'E
Emmetsburg, Ia., U.S. (ĕm'ĕts-bûrg)	71	43°07'N	94°41'W
Emmett, Id., U.S. (ĕm'ĕt)	72	43°53'N	116°30'W
Emmons, Mount, mtn., Ut., U.S. (ĕm'ŭnz)	64	40°43'N	110°20'W
Emory Peak, mtn., Tx., U.S. (ĕ'mŏ-rē pēk)	80	29°13'N	103°20'W
Empoli, Italy (ām'pô-lē)	130	43°43'N	10°55'E
Emporia, Ks., U.S. (ĕm-pō'rĭ-à)	64	38°24'N	96°11'W
Emporia, Va., U.S.	83	37°40'N	77°34'W
Emporium, Pa., U.S. (ĕm-pō'rĭ-ŭm)	67	41°30'N	78°15'W
Empty Quarter see Ar Rub' al Khālī, des., Asia	154	20°00'N	51°00'E
Ems, r., Ger. (ĕms)	124	52°52'N	7°16'E
Ems-Weser Kanal, can., Ger.	124	52°23'N	8°11'E
Enånger, Swe. (ĕn-ôŋ'gĕr)	122	61°36'N	16°55'E
Encantada, Cerro de la, mtn., Mex. (sĕ'r-rô-dĕ-lä-ĕn-kän-tä'dä)	86	31°58'N	115°15'W
Encanto, Cape, c., Phil. (ĕn-kän'tō)	169a	15°44'N	121°46'E
Encarnación, Para. (ĕn-kär-nä-syōn')	102	27°26'S	55°52'W
Encarnación de Díaz, Mex. (ĕn-kär-nä-syōn dà dē'äz)	88	21°34'N	102°15'W
Encinal, Tx., U.S. (ĕn'sĭ-nôl)	80	28°02'N	99°22'W
Encontrados, Ven. (ĕn-kōn-trä'dōs)	100	9°01'N	72°10'W
Encounter Bay, b., Austl. (ĕn-koun'tēr)	174	35°50'S	138°45'E
Endako, r., Can.	54	54°05'N	125°30'W
Endau, r., Malay.	153b	2°29'N	103°40'E
Enderbury, i., Kir. (ĕn'dĕr-bûrĭ)	194	2°00'S	171°00'W
Enderby Land, reg., Ant. (ĕn'dĕr bĭ)	178	72°00'S	52°00'E
Enderlin, N.D., U.S. (ĕn'dĕr-lĭn)	70	46°38'N	97°37'W
Endicott, N.Y., U.S. (ĕn'dĭ-kŏt)	67	42°05'N	76°00'W
Endicott Mountains, mts., Ak., U.S.	63	67°30'N	153°45'W
Enez, Tur.	131	40°42'N	26°05'E
Enfer, Pointe d', c., Mart.	91b	14°21'N	60°48'W
Enfield, Eng., U.K.	114b	51°38'N	0°06'W
Enfield, Ct., U.S. (ĕn'fēld)	67	41°55'N	72°35'W
Enfield, N.C., U.S.	83	36°10'N	77°41'W
Engaño, Cabo, c., Dom. Rep. (kà'-bô- ĕn-gä-nô)	87	18°40'N	68°30'W
Engcobo, S. Afr. (ĕng-cô-bô)	187c	31°41'S	27°59'E
Engel's, Russia (ĕn'gĕls)	137	51°20'N	45°40'E
Engelskirchen, Ger. (ĕn'gĕls-kēr'κĕn)	127c	50°59'N	7°25'E
Enggano, Pulau, i., Indon. (ĕng-gä'nō)	168	5°22'S	102°18'E
England, Ar., U.S. (ĭŋ'glănd)	79	34°33'N	91°58'W
England, state, U.K. (ĭŋ'glănd)	110	51°35'N	1°40'W
Englewood, Co., U.S. (ĕn'g'l-wŏŏd)	78	39°39'N	105°00'W
Englewood, N.J., U.S.	68a	40°54'N	73°59'W
English, In., U.S. (ĭn'glĭsh)	66	38°15'N	86°25'W
English, r., Can.	53	50°31'N	94°12'W
English Channel, strt., Eur.	112	49°45'N	3°06'W
Énguera, Spain (ān'gārä)	129	38°58'N	0°42'W
Enid, Ok., U.S. (ē'nĭd)	64	36°25'N	97°52'W
Enid Lake, res., Ms., U.S.	82	34°13'N	89°47'W
Enkeldoring, S. Afr. (ĕŋ'k'l-dôr-ĭng)	192c	25°24'S	28°43'E
Enköping, Swe. (ĕn'kû-pĭng)	122	59°39'N	17°05'E
Ennedi, mts., Chad (ĕn-nĕd'ē)	185	16°45'N	22°45'E
Ennis, Ire. (ĕn'ĭs)	120	52°54'N	9°05'W
Ennis, Tx., U.S.	81	32°20'N	96°38'W
Enniscorthy, Ire. (ĕn-ĭs-kôr'thĭ)	120	52°33'N	6°27'W
Enniskillen, N. Ire., U.K. (ĕn-ĭs-kĭl'ĕn)	120	54°20'N	7°25'W
Ennis Lake, res., Mt., U.S.	73	45°15'N	111°30'W
Enns, r., Aus. (ĕns)	117	47°37'N	14°35'E
Enoree, S.C., U.S. (ê-nō'rē)	83	34°43'N	81°58'W
Enoree, r., S.C., U.S.	83	34°35'N	81°55'W
Enriquillo, Dom. Rep. (ĕn-rê-kē'l-yō)	93	17°55'N	71°15'W
Enriquillo, Lago, l., Dom. Rep. (lä'gô-ĕn-rê-kē'l-yô)	93	18°35'N	71°35'W
Enschede, Neth. (ĕns'κä-dĕ)	117	52°10'N	6°50'E
Ensenada, Arg.	99c	34°50'S	57°55'W
Ensenada, Mex. (ĕn-sĕ-nä'dä)	86	32°00'N	116°30'W
Enshi, China (ŭn-shr)	160	30°18'N	109°25'E
Enshū-Nada, b., Japan (ĕn'shōō nä-dä)	167	34°25'N	137°14'E
Entebbe, Ug.	185	0°04'N	32°28'E
Enterprise, Al., U.S. (ĕn'tēr-prīz)	82	31°20'N	85°50'W
Enterprise, Or., U.S.	72	45°25'N	117°16'W
Entiat, L, Wa., U.S.	72	45°43'N	120°11'W
Entraygues, Fr. (ĕɴ-trĕg')	126	44°39'N	2°33'E
Entre Rios, prov., Arg.	102	31°30'S	59°00'W
Enugu, Nig. (ĕ-nōō'gōō)	184	6°27'N	7°27'E
Enumclaw, Wa., U.S. (ĕn'ŭm-klô)	74a	47°12'N	121°59'W
Envigado, Col. (ĕn-vē-gä'dô)	100a	6°10'N	75°34'W
Eolie, Isole, is., Italy (ĕ'sô-lĕ-ĕ-ô'lyĕ)	118	38°43'N	14°43'E
Epe, Nig.	189	6°37'N	3°59'E
Épernay, Fr. (ā-pĕr-nĕ')	117	49°02'N	3°54'E
Épernon, Fr. (ā-pĕr-nôɴ')	127b	48°36'N	1°41'E
Ephraim, Ut., U.S. (ē'frâ-ĭm)	77	39°20'N	111°40'W
Ephrata, Wa., U.S. (ĕfrā'tà)	72	47°18'N	119°35'W
Epi, Vanuatu (ā'pê)	175	16°59'S	168°29'E
Épila, Spain (ā'pê-lä)	128	41°38'N	1°15'W
Épinal, Fr. (ā-pē-nál')	117	48°11'N	6°27'E
Episkopi, Cyp.	153a	34°38'N	32°55'E
Epping, Eng., U.K. (ĕp'ĭng)	114b	51°41'N	0°06'E
Epsom, Eng., U.K.	114b	51°20'N	0°16'W
Epupa Falls, wtfl., Afr.	190	17°00'S	13°05'E
Epworth, Eng., U.K. (ĕp'wûrth)	114a	53°31'N	0°50'W
Equatorial Guinea, nation, Afr.	184	2°00'N	7°15'E
Équilles, Fr.	126a	43°34'N	5°21'E
Eramosa, r., Can. (ĕr-à-mō'sà)	62d	43°39'N	80°08'W
Erba, Jabal, mtn., Sudan (ĕr-bà)	185	20°53'N	36°45'E
Erciyeş Daği, mtn., Tur.	119	38°30'N	35°36'E
Erding, Ger. (ĕr'dĕng)	115d	48°19'N	11°54'E
Erechim, Braz. (ĕ-rĕ-shē'ɴ)	102	27°43'S	52°11'W
Ereğli, Tur. (ĕ-rā'ĭ-le)	119	37°40'N	34°00'E
Ereğli, Tur.	119	41°15'N	31°25'E
Erfurt, Ger. (ĕr'fŏŏrt)	117	50°59'N	11°04'E
Ergene, r., Tur. (ĕr'gĕ-nĕ)	131	41°17'N	26°50'E
Erges, r., Eur. (ĕr'-zhĕs)	128	39°45'N	7°01'W
Ergļi, Lat.	123	56°54'N	25°38'E
Eria, r., Spain (ā-rē'ä)	128	42°10'N	6°08'W
Erick, Ok., U.S. (âr'ĭk)	78	35°14'N	99°51'W
Erie, Ks., U.S. (ē'rĭ)	79	37°35'N	95°17'W
Erie, Pa., U.S.	65	42°05'N	80°05'W
Erie, Lake, l., N.A.	65	42°15'N	81°25'W
Erimo Saki, c., Japan (ā'rē-mō sä-kē)	161	41°53'N	143°20'E
Erin, Can. (ĕ'rĭn)	62d	43°46'N	80°04'W
Eritrea, nation, Afr. (ā-rê-trā'à)	185	16°15'N	38°30'E
Erlangen, Ger. (ĕr'läng-ĕn)	124	49°36'N	11°03'E
Erlanger, Ky., U.S. (ĕr'lăng-ēr)	69f	39°01'N	84°36'W

246 Ermoúpoli – Farmers Branch PRONOUNCING INDEX

PLACE (Pronunciation)	PAGE	LAT.	LONG.
Ermoúpoli, Grc.	131	37°30'N	24°56'E
Ernākulam, India	155	9°58'N	76°23'E
Erne, Lower Lough, l., N. Ire., U.K.	120	54°30'N	7°40'W
Erne, Upper Lough, l., N. Ire., U.K. (lŏk ûrn)	120	54°20'N	7°24'W
Erode, India	159	11°20'N	77°45'E
Eromanga, i., Vanuatu	175	18°58'S	169°18'E
Eros, La., U.S. (ē'rōs)	81	32°23'N	92°22'W
Errego, Moz.	191	16°02'S	37°14'E
Errigal, mtn., Ire. (ĕr-ĭ-gôl')	120	55°02'N	8°07'W
Errol Heights, Or., U.S.	74c	45°29'N	122°38'W
Erstein, Fr. (ĕr'shtīn)	127	48°27'N	7°40'E
Erwin, N.C., U.S. (ûr'wĭn)	83	35°16'N	78°40'W
Erwin, Tn., U.S.	83	36°07'N	82°25'W
Erzgebirge, mts., Eur. (ĕrts'gĕ-bē'gĕ)	112	50°29'N	12°40'E
Erzincan, Tur. (ĕr-zĭn-jän')	154	39°50'N	39°30'E
Erzurum, Tur. (ĕrz'rōōm')	154	39°55'N	41°10'E
Esambo, D.R.C.	190	3°40'S	23°24'E
Esashi, Japan (ĕs'ä-shē)	161	41°50'N	140°10'E
Esbjerg, Den. (ĕs'byĕrgh)	116	55°29'N	8°25'E
Escalante, Ut., U.S. (ĕs-kȧ-lăn'tē)	77	37°50'N	111°40'W
Escalante, r., Ut., U.S.	77	37°40'N	111°20'W
Escalón, Mex.	80	26°45'N	104°20'W
Escambia, r., Fl., U.S. (ĕs-kăm'bĭ-ȧ)	82	30°38'N	87°20'W
Escanaba, Mi., U.S. (ĕs-kȧ-nô'bȧ)	65	45°44'N	87°05'W
Escanaba, r., Mi., U.S.	71	46°10'N	87°22'W
Escarpada Point, Phil.	168	18°40'N	122°45'E
Esch-sur-Alzette, Lux.	127	49°32'N	6°21'E
Eschwege, Ger. (ĕsh'vā-gē)	124	51°11'N	10°02'E
Eschweiler, Ger. (ĕsh'vī-lĕr)	127c	50°49'N	6°15'E
Escondido, Ca., U.S. (ĕs-kŏn-dē'dō)	76	33°07'N	117°00'W
Escondido, r., Nic.	91	12°04'N	84°09'W
Escondido, Río, r., Mex. (rē'ō-ĕs-kōn-dē'dô)	80	28°30'N	100°45'W
Escudo de Veraguas, i., Pan. (ĕs-kōō'dä dä vä-rä'gwäs)	91	9°07'N	81°25'W
Escuinapa, Mex. (ĕs-kwē-nä'pä)	86	22°49'N	105°44'W
Escuintla, Guat. (ĕs-kwēn'tlä)	90	14°16'N	90°47'W
Ese, Cayos de, i., Col.	91	12°24'N	81°07'W
Eşfahān, Iran	154	32°38'N	51°30'E
Esgueva, r., Spain (ĕs-gĕ'vä)	128	41°48'N	4°10'W
Esher, Eng., U.K.	114b	51°23'N	0°22'W
Eshowe, S. Afr. (ĕsh'ô-wê)	187c	28°54'S	31°28'E
Esiama, Ghana	188	4°56'N	2°21'W
Eskdale, W.V., U.S. (ĕsk'dāl)	66	38°05'N	81°25'W
Eskifjördur, Ice. (ĕs'kē-fyŭr'dōōr)	110	65°04'N	14°01'W
Eskilstuna, Swe. (å'shĕl-stü-na)	116	59°23'N	16°28'E
Eskimo Lakes, l., Can. (ĕs'kĭ-mō)	52	69°40'N	130°10'W
Eskişehir, Tur. (ĕs-kē-shĕ'h'r)	154	39°40'N	30°20'E
Esko, Mn., U.S. (ĕs'kô)	75h	46°27'N	92°22'W
Esla, r., Spain (ĕs-lä)	128	41°50'N	5°48'W
Eslöv, Swe. (ĕs'lûv)	122	55°50'N	13°17'E
Esmeraldas, Ec. (ĕs-må-räl'däs)	100	0°58'N	79°45'W
Espanola, Can. (ĕs-pȧ-nō'lȧ)	51	46°11'N	81°59'W
Esparta, C.R. (ĕs-pär'tä)	91	9°59'N	84°40'W
Esperance, Austl. (ĕs'pĕ-răns)	174	33°45'S	122°07'E
Esperanza, Cuba (ĕs-pĕ-rä'n-zä)	92	22°30'N	80°10'W
Espichel, Cabo, c., Port. (kä'bō-ĕs-pē-shĕl')	128	38°25'N	9°13'W
Espinal, Col. (ĕs-pē-näl')	100	4°10'N	74°53'W
Espinhaço, Serra do, mts., Braz. (sĕ'r-rä-dô-ĕs-pē-ná-sō)	101	16°00'S	44°00'W
Espinillo, Punta, c., Ur. (pōō'n-tä-ĕs-pē-nē'l-yô)	99c	34°49'S	56°27'W
Espírito Santo, Braz. (ĕs-pē'rē-tô-sän'tô)	101	20°27'S	40°18'W
Espírito Santo, state, Braz.	101	19°57'S	40°58'W
Espiritu Santo, i., Vanuatu (ĕs-pē'rē-tōō sän'tō)	175	15°45'S	166°50'E
Espíritu Santo, Bahía del, b., Mex.	90a	19°25'N	87°28'W
Espita, Mex. (ĕs-pē'tä)	90a	20°57'N	88°22'W
Espoo, Fin.	123	60°13'N	24°41'E
Es Port de Pollença, Spain	129	39°50'N	3°00'E
Esposende, Port. (ĕs-pō-zĕn'dā)	128	41°33'N	8°45'W
Esquel, Arg. (ĕs-kĕ'l)	102	42°47'S	71°22'W
Esquimalt, Can. (ĕs-kwī'mŏlt)	54	48°26'N	123°24'W
Essaouira, Mor.	184	31°34'N	9°44'W
Essen, Bel.	115a	51°28'N	4°27'E
Essen, Ger. (ĕs'sĕn)	110	51°26'N	6°59'E
Essendon, Austl.	173a	37°46'S	144°55'E
Essequibo, r., Guy. (ĕs-ä-kē'bō)	101	4°26'N	58°17'W
Essex, Il., U.S.	69a	41°11'N	88°11'W
Essex, Ma., U.S.	61a	42°38'N	70°47'W
Essex, Md., U.S.	68e	39°19'N	76°29'W
Essex, Vt., U.S.	67	44°30'N	73°05'W
Essex Fells, N.J., U.S. (ĕs'ĕks fĕlz)	68a	40°50'N	74°16'W
Essexville, Mi., U.S. (ĕs'ĕks-vĭl)	66	43°35'N	83°50'W
Esslingen, Ger. (ĕs'slēn-gĕn)	124	48°45'N	9°19'E
Estacado, Llano, pl., U.S. (yá-nō ĕs-tȧcȧ-dō')	64	33°50'N	103°20'W
Estância, Braz. (ĕs-tän'sĭ-ä)	101	11°17'S	37°18'W
Estarreja, Port. (ĕ-tär-rä'zhä)	128	40°44'N	8°39'W
Estats, Pique d', mtn., Eur.	129	42°43'N	1°30'E
Estcourt, S. Afr. (ĕst-coort)	187c	29°04'S	29°53'E
Este, Italy (ĕs'tā)	130	45°13'N	11°40'E
Estella, Spain (ĕs-tāl'yä)	128	42°40'N	2°01'W
Estepa, Spain (ĕs-tā'pä)	128	37°18'N	4°54'W
Estepona, Spain (ĕs-tā-pō'nä)	128	36°26'N	5°08'W
Esterhazy, Can. (ĕs'tĕr-hä-zē)	57	50°40'N	102°08'W
Estero Bay, b., Ca., U.S. (ĕs-tā'rōs)	76	35°22'N	121°04'W
Estevan, Can. (ĕ-stē'vȧn)	50	49°07'N	103°05'W
Estevan Group, is., Can.	54	53°05'N	129°40'W
Estherville, Ia., U.S. (ĕs'tĕr-vĭl)	71	43°24'N	94°49'W
Estill, S.C., U.S. (ĕs'tĭl)	83	32°46'N	81°15'W
Eston, Can.	56	51°10'N	108°45'W
Estonia, nation, Eur.	134	59°10'N	25°00'E
Estoril, Port. (ĕs-tô-rĕl')	129b	38°45'N	9°24'W
Estrêla, mtn., Port. (mäl-you'n-dä-ĕs-trĕ'lä)	128	40°20'N	7°38'W
Estrêla, r., Braz. (ĕs-trĕ'lä)	102b	22°39'S	43°16'W
Estrêla, Serra da, mts., Port. (sĕr'rȧ dä ĕs-trä'lȧ)	128	40°25'N	7°45'W
Estremadura, hist. reg., Port. (ĕs-trā-mä-dōō'rȧ)	128	39°00'N	8°36'W
Estremoz, Port. (ĕs-trā-mōzh')	128	38°50'N	7°35'W
Estrondo, Serra do, mts., Braz. (sĕr'-rȧ dô ĕs-trôn'-dô)	101	9°52'S	48°56'W
Esumba, Île, i., D.R.C.	190	2°00'N	21°12'E
Esztergom, Hung. (ĕs'tĕr-gōm)	125	47°46'N	18°45'E
Etah, Grnld. (ē'tȧ)	49	78°20'N	72°42'W
Étampes, Fr. (ā-tänp')	126	48°26'N	2°09'E
Étaples, Fr. (ā-täp'l')	126	50°32'N	1°38'E
Etchemin, r., Can. (ĕch'ĕ-mĭn)	62b	46°39'N	71°03'W
Ethiopia, nation, Afr. (ê-thē-ō'pē-ȧ)	185	7°53'N	37°55'E
Eticoga, Gui.-B.	188	11°09'N	16°08'W
Etiwanda, Ca., U.S. (ē-tĭ-wän'dȧ)	75a	34°07'N	117°31'W
Etna, Pa., U.S. (ĕt'nȧ)	69e	40°30'N	79°55'W
Etna, Mount, vol., Italy	112	37°48'N	15°00'E
Etobicoke Creek, r., Can.	62d	43°44'N	79°48'W
Etolin Strait, strt., Ak., U.S. (ĕt ō lĭn)	63	60°35'S	165°40'W
Etoshapan, pl., Nmb. (ĕtō'shä)	186	19°07'S	15°30'E
Etowah, Tn., U.S. (ĕt'ô-wä)	82	35°18'N	84°31'W
Etowah, r., Ga., U.S.	82	34°23'N	84°19'W
Étréchy, Fr. (ā-trā-shē')	127b	48°29'N	2°12'E
Etten-Leur, Neth.	115a	51°34'N	4°38'E
Etterbeek, Bel. (ĕt'ĕr-bāk)	115a	50°51'N	4°24'E
Etzatlán, Mex. (ĕt-zä-tlän')	88	20°44'N	104°04'W
Eucla, Austl. (ū'klȧ)	174	31°45'S	128°50'E
Euclid, Oh., U.S. (ū'klĭd)	69d	41°34'N	81°32'W
Eudora, Ar., U.S. (u-dō'rȧ)	79	33°07'N	91°16'W
Eufaula, Al., U.S. (ū-fô'lȧ)	82	31°53'N	85°09'W
Eufaula, Ok., U.S.	79	35°16'N	95°35'W
Eufaula Reservoir, res., Ok., U.S.	79	35°00'N	94°45'W
Eugene, Or., U.S. (û-jēn')	64	44°02'N	123°06'W
Euless, Tx., U.S. (ū'lĕs)	75c	32°50'N	97°05'W
Eunice, La., U.S. (ū'nĭs)	81	30°30'N	92°25'W
Eupen, Bel. (oi'pĕn)	121	50°39'N	6°05'E
Euphrates, r., Asia (û-frā'tēz)	154	36°00'N	40°00'E
Eure, r., Fr. (ûr)	126	49°03'N	1°22'E
Eureka, Ca., U.S. (û-rē'kȧ)	64	40°45'N	124°10'W
Eureka, Ks., U.S.	79	37°48'N	96°17'W
Eureka, Mt., U.S.	72	48°53'N	115°07'W
Eureka, Nv., U.S.	76	39°33'N	115°58'W
Eureka, S.D., U.S.	70	45°46'N	99°38'W
Eureka, Ut., U.S.	77	39°55'N	112°10'W
Eureka Springs, Ar., U.S.	79	36°24'N	93°43'W
Europe, cont. (ū'rŭp)	112	50°00'N	15°00'E
Eustis, Fl., U.S. (ūs'tĭs)	83	28°50'N	81°41'W
Eutaw, Al., U.S. (ū-tȧ)	82	32°48'N	87°50'W
Eutsuk Lake, l., Can. (ōōt'sŭk)	54	53°20'N	126°44'W
Evanston, Il., U.S. (ĕv'ăn-stŭn)	65	42°03'N	87°41'W
Evanston, Wy., U.S.	73	41°17'N	111°02'W
Evansville, In., U.S. (ĕv'ănz-vĭl)	65	38°00'N	87°30'W
Evansville, Wi., U.S.	71	42°46'N	89°19'W
Evart, Mi., U.S. (ĕv'ērt)	66	43°55'N	85°10'W
Evaton, S. Afr. (ĕv'ȧ-tŏn)	192c	26°32'S	27°53'E
Eveleth, Mn., U.S. (ĕv'ē-lĕth)	71	47°27'N	92°35'W
Everard, l., Austl. (ĕv'ĕr-ärd)	174	31°20'S	134°10'E
Everard Ranges, mts., Austl.	174	27°15'S	132°00'E
Everest, Mount, mtn., Asia (ĕv'ĕr-ĕst)	155	28°00'N	86°57'E
Everett, Ma., U.S. (ĕv'ĕr-ĕt)	61a	42°24'N	71°03'W
Everett, Wa., U.S. (ĕv'ĕr-ĕt)	64	47°59'N	122°11'W
Everett Mountains, mts., Can.	53	62°34'N	68°00'W
Everglades, The, sw., Fl., U.S.	83a	25°35'N	80°55'W
Everglades City, Fl., U.S. (ĕv'ĕr-glādz)	83a	25°50'N	81°25'W
Everglades National Park, rec., Fl., U.S.	65	25°39'N	80°57'W
Evergreen, Al., U.S. (ĕv'ĕr-grēn)	82	31°25'N	87°56'W
Evergreen Park, Il., U.S.	69a	41°44'N	87°42'W
Everman, Tx., U.S. (ĕv'ĕr-măn)	75c	32°38'N	97°17'W
Everson, Wa., U.S. (ĕv'ĕr-sŭn)	74d	48°55'N	122°21'W
Évora, Port. (ĕv'ô-rä)	118	38°35'N	7°54'W
Évreux, Fr. (ā-vrû')	117	49°02'N	1°11'E
Evrótas, r., Grc. (ĕv-rō'täs)	131	37°15'N	22°17'E
Évvoia, i., Grc.	119	38°38'N	23°45'E
'Ewa Beach, Hi., U.S. (ē'wä)	84a	21°17'N	158°03'W
Ewaso Ng'iro, r., Kenya	185	0°59'N	37°47'E
Excelsior, Mn., U.S. (ĕk-sel'sĭ-ŏr)	75g	44°54'N	93°35'W
Excelsior Springs, Mo., U.S.	79	39°20'N	94°13'W
Exe, r., Eng., U.K. (ĕks)	120	50°57'N	3°37'W
Exeter, Eng., U.K.	117	50°45'N	3°33'W
Exeter, Ca., U.S. (ĕk'sē-tēr)	76	36°18'N	119°09'W
Exeter, N.H., U.S.	67	43°00'N	71°00'W
Exmoor, for., Eng., U.K. (ĕks'mōr)	120	51°10'N	3°55'W
Exmouth, Eng., U.K. (ĕks'mŭth)	120	50°40'N	3°20'W
Exmouth Gulf, b., Austl.	174	21°45'S	114°30'E
Exploits, r., Can. (ĕks-ploits')	61	48°50'N	56°15'W
Extórrax, r., Mex. (ĕx-tó'ràx)	88	21°04'N	99°39'W
Extrema, Braz. (ĕsh-trē'mä)	99a	22°52'S	46°19'W
Extremadura, hist. reg., Spain (ĕks-trā-mä-doo'rä)	128	38°43'N	6°30'W
Exuma Sound, strt., Bah. (ĕk-sōō'mä)	92	24°20'N	76°20'W
Eyasi, Lake, l., Tan. (å-yä'sē)	186	3°25'S	34°55'E
Eyjafjördur, b., Ice.	116	66°21'N	18°20'W
Eyl, Som.	192a	7°53'N	49°45'E
Eyrarbakki, Ice.	116	63°51'N	20°52'W
Eyre, Austl. (âr)	174	32°15'S	126°20'E
Eyre, l., Austl.	174	28°43'S	137°50'E
Eyre Peninsula, pen., Austl.	174	33°30'S	136°00'E
Ezeiza, Arg. (ĕ-zā'zä)	102a	34°52'S	58°31'W
Ezine, Tur. (â'zĭ-nå)	131	39°47'N	26°18'E
F			
Faaborg, Den. (fō'bôrg)	122	55°06'N	10°19'E
Fabens, Tx., U.S. (fā'bĕnz)	80	31°30'N	106°07'W
Fabriano, Italy (fä-brē-ä'nō)	130	43°20'N	12°55'E
Fada, Chad (fä'dä)	185	17°06'N	21°18'E
Fada Ngourma, Burkina (fä'dä''n gōōr'mä)	184	12°04'N	0°21'E
Faddeya, i., Russia (fȧd-yä')	135	76°12'N	145°00'E
Faenza, Italy (fä-ĕnd'zä)	130	44°16'N	11°53'E
Fafe, Port. (fä'fä)	128	41°30'N	8°10'W
Fafen, r., Eth.	192a	8°15'N	42°40'E
Făgăras, Rom. (fȧ-gä'räsh)	131	45°50'N	24°55'E
Fagerness, Nor. (fä'ghĕr-nĕs)	116	61°00'N	9°10'E
Fagnano, l., S.A. (fäk-nä'nō)	102	54°35'S	68°20'W
Faguibine, Lac, l., Mali	188	16°50'N	4°20'W
Faial, i., Port. (fä-yä'l)	184a	38°40'N	29°19'W
Fā'id, Egypt (fä-yēd')	192d	30°19'N	32°18'E
Fairbanks, Ak., U.S. (fâr'bănks)	64a	64°50'N	147°48'W
Fairbury, Il., U.S. (fâr'bĕr-ĭ)	66	40°45'N	88°25'W
Fairbury, Ne., U.S.	79	40°09'N	97°11'W
Fairchild Creek, r., Can. (fâr'chīld)	62d	43°18'N	80°10'W
Fairfax, Mn., U.S. (fâr'făks)	71	44°29'N	94°44'W
Fairfax, S.C., U.S.	83	32°29'N	81°13'W
Fairfax, Va., U.S.	68e	38°51'N	77°20'W
Fairfield, Austl.	173b	33°52'S	150°57'E
Fairfield, Al., U.S. (fâr'fēld)	68h	33°30'N	86°50'W
Fairfield, Ct., U.S.	68a	41°08'N	73°22'W
Fairfield, Ia., U.S.	71	41°00'N	91°59'W
Fairfield, Il., U.S.	66	38°25'N	88°20'W
Fairfield, Me., U.S.	60	44°35'N	69°38'W
Fairhaven, Ma., U.S. (fâr-hā'vĕn)	67	41°35'N	70°55'W
Fair Haven, Vt., U.S.	67	43°35'N	73°15'W
Fair Island, i., Scot., U.K. (fâr)	120a	59°34'N	1°41'W
Fairmont, Mn., U.S. (fâr'mŏnt)	71	43°39'N	94°26'W
Fairmont, W.V., U.S.	66	39°30'N	80°10'W
Fairmont City, Il., U.S.	75e	38°39'N	90°05'W
Fairmount, In., U.S.	66	40°25'N	85°45'W
Fairmount, Ks., U.S.	75f	39°12'N	95°55'W
Fair Oaks, Ga., U.S. (fâr ōks)	68c	33°56'N	84°33'W
Fairport, N.Y., U.S. (fâr'pōrt)	67	43°05'N	77°30'W
Fairport Harbor, Oh., U.S.	66	41°45'N	81°15'W
Fairview, Ok., U.S. (fâr'vū)	78	36°16'N	98°28'W
Fairview, Or., U.S.	74c	45°32'N	112°26'W
Fairview, Ut., U.S.	77	39°35'N	111°30'W
Fairview Park, Oh., U.S.	69d	41°27'N	81°52'W
Fairweather, Mount, mtn., N.A. (fâr-wĕdh'ĕr)	63	59°12'N	137°22'W
Faisalabad, Pak.	155	31°29'N	73°06'E
Faith, S.D., U.S. (fāth)	70	45°02'N	120°02'W
Faizābād, India	155	26°50'N	82°17'E
Fajardo, P.R.	87b	18°20'N	65°40'W
Fakfak, Indon.	169	2°56'S	132°25'E
Faku, China (fä-kōō)	164	42°28'N	123°20'E
Falcón, dept., Ven. (fäl-kô'n)	101b	11°00'N	68°28'W
Falconer, N.Y., U.S. (fô'k'n-ēr)	67	42°10'N	79°10'W
Falcon Heights, Mn., U.S. (fô'k'n)	75g	44°59'N	93°10'W
Falcon Reservoir, res., N.A. (fôk'n)	80	26°47'N	99°03'W
Făleşti, Mol.	133	47°33'N	27°46'E
Falfurrias, Tx., U.S. (făl'fōō-rē'ás)	80	27°15'N	98°08'W
Falher, Can. (făl'ĕr)	55	55°44'N	117°12'W
Falkenberg, Swe. (fäl'kĕn-bĕrgh)	122	56°54'N	12°25'E
Falkensee, Ger. (fäl'kĕn-zā)	115b	52°34'N	13°05'E
Falkenthal, Ger. (fäl'kĕn-täl)	115b	52°54'N	13°18'E
Falkirk, Scot., U.K. (fôl'kûrk)	120	55°59'N	3°55'W
Falkland Islands, dep., S.A. (fôk'lȧnd)	102	50°45'S	61°00'W
Falköping, Swe. (fäl'chûp-ĭng)	122	58°09'N	13°30'E
Fall City, Wa., U.S.	74a	47°34'N	121°53'W
Fall Creek, r., In., U.S. (fôl)	69g	39°52'N	86°04'W
Fallon, Nv., U.S. (făl'ŭn)	76	39°30'N	118°48'W
Fall River, Ma., U.S.	65	41°42'N	71°07'W
Falls Church, Va., U.S. (fâlz chûrch)	68e	38°53'N	77°10'W
Falls City, Ne., U.S.	79	40°04'N	95°37'W
Fallston, Md., U.S. (fâls'ton)	68e	39°32'N	76°26'W
Falmouth, Jam.	92	18°30'N	77°40'W
Falmouth, Eng., U.K. (făl'mŭth)	120	50°08'N	5°04'W
Falmouth, Ky., U.S.	66	38°40'N	84°20'W
False Divi Point, c., India	159	15°45'N	80°50'E
Falster, i., Den. (fäls'tĕr)	122	54°48'N	11°58'E
Fălticeni, Rom. (fŭl-tē-chān'y')	125	47°27'N	26°17'E
Falun, Swe. (fä-lōōn')	116	60°38'N	15°35'E
Famagusta, Cyp. (fä-mä-gōōs'tä)	119	35°08'N	33°59'E
Famatina, Sierra de, mts., Arg.	102	29°00'S	67°50'W
Fangxian, China (fäŋ-shyĕn)	164	32°05'N	110°45'E
Fanning, i., Can.	62f	49°45'N	97°46'W
Fano, Italy (fä'nō)	130	43°49'N	13°01'E
Fanø, i., Den. (fän'û)	122	55°24'N	8°10'E
Fan Si Pan, mtn., Viet.	165	22°25'N	103°50'E
Farafangana, Madag. (fä-rä-fäŋ-gä'nä)	187	23°18'S	47°59'E
Farāh, Afg. (fä-rä')	154	32°15'N	62°13'E
Farallón, Punta, c., Mex. (pô'n-tä-fä-rä-lōn)	88	19°21'N	105°03'W
Faranah, Gui. (fä-rä'nä)	184	10°02'N	10°44'W
Farasān, Jaza'ir, is., Sau. Ar.	154	16°45'N	41°08'E
Fārigh, Wādī al, r., Libya (wädē ĕl fä-rĕg')	119	30°10'N	19°34'E
Farewell, Cape, c., N.Z. (fâr-wĕl')	175a	40°37'S	172°40'E
Fargo, N.D., U.S. (fär'gō)	64	46°53'N	96°48'W
Far Hills, N.J., U.S. (fär hĭlz)	68a	40°41'N	74°38'W
Faribault, Mn., U.S. (fâ'rĭ-bō)	71	44°19'N	93°16'W
Farilhões, is., Port. (fä-rē-lyônzh')	128	39°28'N	9°32'W
Faringdon, Eng., U.K. (fă'rĭng-dŏn)	114b	51°38'N	1°35'W
Fāriskūr, Egypt (fä-rēs-kōōr')	192b	31°19'N	31°46'E
Farit, Amba, mtn., Eth.	185	10°51'N	37°52'E
Farley, Mo., U.S. (fär'lē)	75f	39°16'N	94°49'W
Farmers Branch, Tx., U.S.	75c	32°56'N	96°53'W

ăt; finȧl; rāte; senåte; ärm; ȧsk; sofȧ; fâre; ch-choose; dh-as th in other; bē; ēvent; bĕt; recĕnt; cratĕr; g-gō; gh-guttural g; bĭt; ĭ-short neutral; rīde; K-guttural k as ch in German ich;

PRONOUNCING INDEX Farmersburg – Fonseca 247

PLACE (Pronunciation)	PAGE	LAT.	LONG.
Farmersburg, In., U.S. (fär'mẽrz-bûrg)	66	39°15'N	87°25'W
Farmersville, Tx., U.S. (fär'mẽrz-vĭl)	79	33°11'N	96°22'W
Farmingdale, N.J., U.S. (färm'ĕng-dāl)	68a	40°11'N	74°10'W
Farmingdale, N.Y., U.S.	68a	40°44'N	73°26'W
Farmingham, Ma., U.S. (färm-ĭng-hăm)	61a	42°17'N	71°25'W
Farmington, Il., U.S. (färm-ĭng-tŭn)	79	40°42'N	90°01'W
Farmington, Me., U.S.	60	44°40'N	70°10'W
Farmington, Mi., U.S.	69b	42°28'N	83°23'W
Farmington, Mo., U.S.	79	37°46'N	90°26'W
Farmington, N.M., U.S.	77	36°40'N	108°10'W
Farmington, Ut., U.S.	75b	40°59'N	111°53'W
Farmville, N.C., U.S. (färm-vĭl)	83	35°35'N	77°35'W
Farmville, Va., U.S.	83	37°15'N	78°23'W
Farnborough, Eng., U.K. (färn'bŭr-ô)	114b	51°15'N	0°45'W
Farne Islands, is., Eng., U.K. (färn)	120	55°40'N	1°32'W
Farnham, Can. (făr'năm)	67	45°15'N	72°55'W
Farningham, Eng., U.K. (fär'nĭng-ŭm)	114b	51°22'N	0°14'E
Farnworth, Eng., U.K. (färn'wŭrth)	114a	53°34'N	2°24'W
Faro, Braz. (fä'rò)	101	2°05'S	56°32'W
Faro, Port.	118	37°01'N	7°57'W
Farodofay, Madag.	187	24°59'S	46°58'E
Faroe Islands, is., Eur.	112	62°00'N	5°45'W
Fårön, i., Swe.	123	57°57'N	19°10'E
Farquhar, Cape, c., Austl. (fär'kwȧr)	174	23°50'S	112°55'E
Farrell, Pa., U.S. (făr'ĕl)	66	41°10'N	80°30'W
Farrukhābād, India (fŭ-ròk-hä-bäd')	155	27°29'N	79°35'E
Fársala, Grc.	131	39°18'N	22°25'E
Farsund, Nor. (fär'sòn)	122	58°05'N	6°47'E
Fartak, Ra's, c., Yemen	154	15°43'N	52°17'E
Fartura, Serra da, mts., Braz. (sĕ'r-rä-dá-fär-tōō'rä)	102	26°40'S	53°15'W
Farvel, Kap, c., Grnld.	49	60°00'N	44°00'W
Farwell, Tx., U.S. (fär'wĕl)	78	34°24'N	103°03'W
Fasano, Italy (fä-zä'nō)	131	40°50'N	17°22'E
Fastiv, Ukr.	133	50°04'N	29°57'E
Fatëzh, Russia	132	52°06'N	35°51'E
Fatima, Port.	129	39°36'N	9°36'E
Fatsa, Tur. (fät'sä)	119	40°50'N	37°30'E
Faucilles, Monts, mts., Fr. (môn' fō-sēl')	127	48°07'N	6°13'E
Fauske, Nor.	116	67°15'N	15°24'E
Faust, Can. (foust)	55	55°19'N	115°38'W
Faustovo, Russia	142b	55°27'N	38°29'E
Faversham, Eng., U.K. (fă'vẽr-sh'm)	114b	51°19'N	0°54'E
Faxaflói, b., Ice.	116	64°33'N	22°40'W
Fayette, Al., U.S. (fā-yĕt')	82	33°40'N	87°54'W
Fayette, Ia., U.S.	71	42°49'N	91°49'W
Fayette, Mo., U.S.	79	39°09'N	92°41'W
Fayette, Ms., U.S.	82	31°43'N	91°00'W
Fayetteville, Ar., U.S. (fā-yĕt'vĭl)	79	36°03'N	94°08'W
Fayetteville, N.C., U.S.	83	35°02'N	78°54'W
Fayetteville, Tn., U.S.	82	35°10'N	86°33'W
Fazao, Forêt Classée du, for., Togo	188	8°50'N	0°40'E
Fazilka, India	158	30°30'N	74°02'E
Fazzān (Fezzan), hist. reg., Libya	185	26°45'N	13°01'E
Fdérik, Maur.	184	22°45'N	12°38'W
Fear, Cape, c., N.C., U.S. (fēr)	83	33°52'N	77°48'W
Feather, r., Ca., U.S. (fĕth'ẽr)	76	38°56'N	121°41'W
Feather, Middle Fork of, r., Ca., U.S.	76	39°49'N	121°10'W
Feather, North Fork of, r., Ca., U.S.	76	40°00'N	121°20'W
Featherstone, Eng., U.K. (fĕdh'ẽr stŭn)	114a	53°39'N	1°21'W
Fécamp, Fr. (fā-käN')	117	49°45'N	0°20'E
Federal, Distrito, dept., Ven. (dēs-trē'tô-fĕ-dĕ-rä'l)	101b	10°34'N	66°55'W
Federal Way, Wa., U.S.	74a	47°20'N	122°20'W
Fëdorovka, Russia (fyô'dō-rôf-kȧ)	142b	56°15'N	37°14'E
Fehmarn, i., Ger. (fā'märn)	124	54°28'N	11°15'E
Fehrbellin, Ger. (fĕr'bĕl-lēn)	115b	52°49'N	12°46'E
Feia, Logoa, l., Braz. (lô-gôä-fĕ'yä)	99a	21°54'S	41°15'W
Feicheng, China (fā-chŭŋ)	162	36°18'N	116°45'E
Feidong, China (fā-dôŋ)	162	31°53'N	117°28'E
Feira de Santana, Braz. (fĕ'ē-rä dä sänt-än'ä)	101	12°16'S	38°46'W
Feixian, China (fā-shyĕn)	162	35°17'N	117°59'E
Felanitx, Spain (fā-lä-nēch')	118	39°29'N	3°09'E
Feldkirch, Aus. (fĕlt'kĭrk)	124	47°15'N	9°36'E
Feldkirchen, Ger. (fĕld'kĕr-kĕn)	115d	48°09'N	11°44'E
Felipe Carrillo Puerto, Mex.	90a	19°36'N	88°04'W
Feltre, Italy (fĕl'trā)	130	46°02'N	11°56'E
Femunden, l., Nor.	116	62°17'N	11°40'E
Fengcheng, China (fŭŋ-chŭŋ)	164	40°28'N	124°03'E
Fengcheng, China	163b	30°55'N	121°38'E
Fengdu, China (fŭŋ-dōō)	160	29°58'N	107°50'E
Fengjie, China (fŭŋ-jyĕ)	160	31°02'N	109°30'E
Fengming Dao, i., China (fŭŋ-mĭŋ dou)	162	39°19'N	121°15'E
Fengrun, China (fŭŋ-ròn)	162	39°51'N	118°06'E
Fengtai, China (fŭŋ-tī)	164a	39°51'N	116°19'E
Fengxian, China (fŭŋ-shyĕn)	163b	30°55'N	121°26'E
Fengxian, China	162	34°41'N	116°36'E
Fengxiang, China (fŭŋ-shyäŋ)	160	34°25'N	107°20'E
Fengyang, China (fŭng'yăng')	164	32°55'N	117°32'E
Fengzhen, China (fŭŋ-jŭn)	161	40°28'N	113°20'E
Fennimore Pass, strt., Ak., U.S. (fĕn-ĭ-mōr)	63a	51°40'N	175°38'W
Fenoarivo Atsinanana, Madag.	187	17°30'S	49°31'E
Fenton, Mi., U.S. (fĕn-tŭn)	66	42°50'N	83°40'W
Fenton, Mo., U.S.	75e	38°31'N	90°27'W
Fenyang, China	161	37°20'N	111°48'E
Feodosiia, Ukr.	137	45°02'N	35°21'E
Ferdows, Iran	154	34°00'N	58°13'E
Ferentino, Italy (fā-rĕn-tē'nō)	130	41°42'N	13°18'E
Fergana, Uzb.	139	40°23'N	71°46'E
Fergus Falls, Mn., U.S. (fûr'gŭs)	65	46°17'N	96°03'W
Ferguson, Mo., U.S. (fûr-gŭ-sŭn)	75e	38°45'N	90°18'W
Ferkéssédougou, C. Iv.	188	9°36'N	5°12'W
Fermo, Italy (fĕr'mō)	130	43°10'N	13°43'E
Fermoselle, Spain (fĕr-mō-säl'yå)	128	41°20'N	6°23'W
Fermoy, Ire. (fûr-moi')	120	52°05'N	8°06'W
Fernandina Beach, Fl., U.S. (fûr-năn-dē'nȧ)	83	30°38'N	81°29'W
Fernando de Noronha, Arquipélago, is., Braz.	101	3°51'S	32°25'W
Fernando Póo *see* Bioko, i., Eq. Gui.	184	3°35'N	7°45'E
Fernán-Núñez, Spain (fĕr-nän'nōōn'yâth)	128	37°42'N	4°43'W
Fernâo Veloso, Baia de, b., Moz.	191	14°20'S	40°55'E
Ferndale, Ca., U.S. (fûrn'dāl)	72	40°34'N	124°18'W
Ferndale, Mi., U.S.	69b	42°27'N	83°08'W
Ferndale, Wa., U.S.	74d	48°51'N	122°36'W
Fernie, Can. (fûr'nĭ)	50	49°30'N	115°03'W
Fern Prairie, Wa., U.S. (fûrn prâr'ĭ)	74c	45°38'N	122°25'W
Ferrara, Italy (fĕr-rä'rä)	118	44°50'N	11°37'E
Ferrat, Cap, c., Alg. (kăp fĕr-rät)	129	35°49'N	0°29'W
Ferreira do Alentejo, Port.	128	38°03'N	8°06'W
Ferreira do Zezere, Port. (fĕr-rĕ'ē-rä dò zā-zā'rĕ)	128	39°49'N	8°17'W
Ferrelview, Mo., U.S. (fĕr'rĕl-vū)	75f	39°18'N	94°40'W
Ferreñafe, Peru (fĕr-rĕn-yä'fĕ)	100	6°38'S	79°48'W
Ferriday, La., U.S. (fĕr'ĭ-dā)	81	31°38'N	91°33'W
Ferrol, Spain	110	43°30'N	8°12'W
Fershampenuaz, Russia (fĕr-shám'pĕn-wäz)	142a	53°32'N	59°50'E
Fertile, Mn., U.S. (fur'tĭl)	70	47°33'N	96°18'W
Fès, Mor. (fĕs)	184	34°08'N	5°00'W
Fessenden, N.D., U.S. (fĕs'ĕn-dĕn)	70	47°39'N	99°40'W
Festus, Mo., U.S. (fĕst'ŭs)	79	38°12'N	90°22'W
Fethiye, Tur. (fĕt-hē'yĕ)	119	36°40'N	29°05'E
Feuilles, Rivière aux, r., Can.	53	58°30'N	70°50'W
Ffestiniog, Wales, U.K.	120	52°59'N	3°58'W
Fianarantsoa, Madag. (fyȧ-nä'rȧn-tsō'ȧ)	187	21°21'S	47°15'E
Ficksburg, S. Afr. (fĭks'bûrg)	192c	28°53'S	27°53'E
Fidalgo Island, i., Wa., U.S. (fĭ-dăl'gō)	74a	48°28'N	122°39'W
Fieldbrook, Ca., U.S. (fēld'bròk)	72	40°59'N	124°02'W
Fier, Alb. (fyĕr)	131	40°43'N	19°34'E
Fife Ness, c., Scot., U.K. (fīf'nes')	120	56°15'N	2°19'W
Fifth Cataract, wtfl., Sudan	185	18°27'N	33°38'E
Figeac, Fr. (fē-zhák')	126	44°37'N	2°02'E
Figeholm, Swe. (fē-ghĕ-hōlm)	122	57°24'N	16°33'E
Figueira da Foz, Port. (fē-gwĕy-rä-dá-fō'z)	128	40°10'N	8°50'W
Figuig, Mor.	184	32°20'N	1°30'W
Fiji, nation, Oc. (fē'jē)	3	18°40'S	175°00'E
Filadelfia, C.R. (fĭl-ȧ-dĕl'fĭ-ȧ)	90	10°26'N	85°37'W
Filatovskoye, Russia (fĭ-lä'tôf-skô-yĕ)	142a	56°49'N	62°20'E
Filchner Ice Shelf, ice, Ant. (fĭlk'nẽr)	178	80°00'S	35°00'W
Filicudi, i., Italy (fē'le-kōō'dē)	130	38°34'N	14°39'E
Filippovskoye, Russia (fĭ-lĭ-pôf'skô-yĕ)	142b	56°06'N	38°38'E
Filipstad, Swe. (fĭl'ĭps-städh)	122	59°44'N	14°09'E
Fillmore, Ut., U.S. (fĭl'môr)	77	39°00'N	112°20'W
Filsa, Nor.	122	60°35'N	12°03'E
Fimi, r., D.R.C.	186	2°43'S	17°50'E
Finch, Can. (fĭnch)	62c	45°09'N	75°06'W
Findlay, Oh., U.S. (fĭnd'lå)	66	41°05'N	83°40'W
Fingoe, Moz.	191	15°12'S	31°50'E
Finke, r., Austl. (fĭŋ'kē)	174	25°25'S	134°30'E
Finland, nation, Eur. (fĭn'lănd)	110	62°45'N	26°13'E
Finland, Gulf of, b., Eur. (fĭn'lănd)	112	59°35'N	23°35'E
Finlandia, Col. (fēn-lä'n-dēä)	100a	4°38'N	75°39'W
Finlay, r., Can. (fĭn'lå)	52	57°45'N	125°30'W
Finow, Ger. (fē'nōv)	115b	52°50'N	13°44'E
Finowfurt, Ger. (fē'nō-fōōrt)	115b	52°50'N	13°41'E
Fircrest, Wa., U.S. (fûr'krĕst)	74a	47°14'N	122°31'W
Firenze *see* Florence, Italy	110	43°47'N	11°15'E
Firenzuola, Italy (fē-rĕnt-swô'lä)	130	44°08'N	11°21'E
Firozpur, India	155	30°58'N	74°39'E
Fischa, r., Aus.	115e	48°04'N	16°33'E
Fischamend Markt, Aus.	115e	48°07'N	16°37'E
Fish, r., Nmb. (fĭsh)	186	28°00'S	17°30'E
Fish Cay, i., Bah.	93	22°30'N	74°20'W
Fish Creek, r., Can. (fĭsh)	62e	50°52'N	114°21'W
Fisher, La., U.S. (fĭsh'ẽr)	81	31°28'N	93°30'W
Fisher Bay, b., Can.	57	51°30'N	97°16'W
Fisher Channel, strt., Can.	54	52°10'N	127°42'W
Fisher Strait, strt., Can.	53	62°43'N	84°28'W
Fisterra, Cabo de, c., Spain	112	42°52'N	9°48'W
Fitchburg, Ma., U.S. (fĭch'bûrg)	67	42°35'N	71°48'W
Fitri, Lac, l., Chad	189	12°50'N	17°28'E
Fitzgerald, Ga., U.S. (fĭts-jĕr'ăld)	82	31°42'N	83°17'W
Fitz Hugh Sound, strt., Can. (fĭts hū)	54	51°40'N	127°57'W
Fitzroy, r., Austl. (fĭts-roi')	174	18°00'S	124°05'E
Fitzroy, r., Austl.	175	23°45'S	150°02'E
Fitzroy, Monte (Cerro Chaltel), mtn., S.A.	102	48°10'S	73°18'W
Fitzroy Crossing, Austl.	174	18°08'S	126°00'E
Fitzwilliam, i., Can. (fĭts-wĭl'yŭm)	58	45°30'N	81°45'W
Fiume *see* Rijeka, Cro.	118	45°22'N	14°24'E
Fiumicino, Italy (fyōō-mē-chē'nô)	129d	41°47'N	12°19'E
Fjällbacka, Swe. (fyĕl'bäk-å)	122	58°37'N	11°17'E
Flagstaff, S. Afr. (flăg'stäf)	187c	31°06'S	29°31'E
Flagstaff, Az., U.S. (flăg-stáf)	64	35°15'N	111°40'W
Flagstaff, l., Me., U.S. (flăg-stáf)	67	45°05'N	70°30'W
Flåm, Nor. (flôm)	122	60°50'N	7°00'E
Flambeau, r., Wi., U.S. (flăm-bō')	71	45°32'N	91°05'W
Flaming Gorge Reservoir, res., U.S.	64	41°13'N	109°30'W
Flamingo, Fl., U.S. (flȧ-mĭŋ'gō)	83	25°10'N	80°55'W
Flamingo Cay, i., Bah. (flȧ-mĭn'gō)	93	22°50'N	75°50'W
Flamingo Point, c., V.I.U.S.	87c	18°19'N	65°00'W
Flanders, hist. reg., Fr. (flăn'dẽrz)	121	50°53'N	2°29'E
Flandreau, S.D., U.S. (flăn'drō)	70	44°02'N	96°35'W
Flathead, r., N.A.	55	49°30'N	114°30'W
Flathead, Middle Fork, r., Mt., U.S.	73	48°30'N	113°47'W
Flathead, North Fork, r., N.A.	73	48°45'N	114°20'W
Flathead, South Fork, r., Mt., U.S.	73	48°05'N	113°45'W
Flathead Indian Reservation, I.R., Mt., U.S.	73	47°30'N	114°25'W
Flathead Lake, l., Mt., U.S. (flăt'hĕd)	64	47°57'N	114°20'W
Flatow, Ger.	115b	52°44'N	12°58'E
Flat Rock, Mi., U.S. (flăt rŏk)	69b	42°06'N	83°17'W
Flattery, Cape, c., Wa., U.S. (flăt'ẽr-ĭ)	72	48°22'N	124°45'W
Flatwillow Creek, r., Mt., U.S. (flat wĭl'ō)	73	46°45'N	108°47'W
Flekkefjord, Nor. (fläk'kē-fyôr)	122	58°19'N	6°38'E
Flemingsburg, Ky., U.S. (flĕm'ĭngz-bûrg)	66	38°25'N	83°45'W
Flensburg, Ger. (flĕns'borgh)	116	54°48'N	9°27'E
Flers, Fr. (flĕr)	117	48°43'N	0°37'W
Fletcher, N.C., U.S.	83	35°26'N	82°30'W
Flinders, i., Austl.	175	39°35'S	148°10'E
Flinders, r., Austl.	175	18°48'S	141°07'E
Flinders, reg., Austl. (flĭn'dẽrz)	174	32°15'S	138°45'E
Flinders Reefs, rf., Austl.	175	17°30'S	149°02'E
Flin Flon, Can. (flĭn flŏn)	50	54°46'N	101°53'W
Flint, Wales, U.K.	114a	53°15'N	3°07'W
Flint, Mi., U.S.	65	43°00'N	83°45'W
Flint, r., Ga., U.S. (flĭnt)	65	31°25'N	84°15'W
Flintshire, co., Wales, U.K.	114a	53°13'N	3°00'W
Flora, Il., U.S. (flō'rȧ)	66	38°40'N	88°25'W
Flora, In., U.S.	66	40°25'N	86°30'W
Florala, Al., U.S. (flŏr-ăl'ȧ)	82	31°01'N	86°19'W
Floral Park, N.Y., U.S. (flôr'ăl pärk)	68a	40°42'N	73°42'W
Florence, Italy	110	43°47'N	11°15'E
Florence, Al., U.S. (flŏr'ĕns)	65	34°46'N	87°40'W
Florence, Az., U.S.	77	33°00'N	111°25'W
Florence, Co., U.S.	78	38°23'N	105°08'W
Florence, Ks., U.S.	79	38°14'N	96°56'W
Florence, S.C., U.S.	83	34°10'N	79°45'W
Florence, Wa., U.S.	74a	48°13'N	122°21'W
Florencia, Col. (flō-rĕn'sē-ȧ)	100	1°31'N	75°13'W
Florencio Sánchez, Ur. (flō-rĕn'sēō-sá'n-chĕz)	99c	33°52'S	57°24'W
Florencio Varela, Arg. (flō-rĕn'sē-o vä-rā'lä)	102a	34°50'S	58°16'W
Flores, Braz. (flō'rĕzh)	101	7°57'S	37°48'W
Flores, Guat.	90a	16°53'N	89°54'W
Flores, dept., Ur.	99c	33°33'S	57°00'W
Flores, i., Indon.	168	8°14'S	121°08'E
Flores, r., Arg.	99c	36°13'S	60°28'W
Flores, Laut (Flores Sea), sea, Indon.	168	7°09'S	120°30'E
Floresville, Tx., U.S. (flō'rĕs-vĭl)	80	29°10'N	98°08'W
Floriano, Braz. (flō-rá-ä'nò)	101	6°17'S	42°58'W
Florianópolis, Braz. (flō-rē-ä-nō'pò-lēs)	102	27°30'S	48°30'W
Florida, Col. (flō-rē'dä)	100a	3°20'N	76°12'W
Florida, Cuba	92	22°10'N	79°50'W
Florida, S. Afr.	187b	26°11'S	27°56'E
Florida, Ur. (flò-rē-dhä)	102	34°06'S	56°14'W
Florida, N.Y., U.S. (flŏr'ĭ-dȧ)	68a	41°20'N	74°21'W
Florida, state, U.S. (flŏr'ĭ-dȧ)	65	30°30'N	84°40'W
Florida, dept., Ur. (flô-rē'dhä)	99c	33°48'S	56°15'W
Florida, i., Sol. Is.	175	8°56'S	159°45'E
Florida, Straits of, strt., N.A.	87	24°10'N	81°00'W
Florida Bay, b., Fl., U.S. (flŏr'ĭ-dȧ)	83a	24°55'N	80°55'W
Florida Keys, is., Fl., U.S.	65	24°33'N	81°20'W
Florida Mountains, mts., N.M., U.S.	77	32°10'N	107°35'W
Florido, Río, r., Mex. (flō-rē'dō)	80	27°21'N	104°48'W
Floridsdorf, Aus. (flō'rĭds-dôrf)	115e	48°16'N	16°25'E
Florina, Grc. (flô-rē'nä)	119	40°48'N	21°24'E
Florissant, Mo., U.S. (flŏr'ĭ-sănt)	75e	38°47'N	90°20'W
Floyd, r., Ia., U.S. (floid)	70	42°38'N	96°15'W
Floydada, Tx., U.S. (floi-dā'dȧ)	78	33°59'N	101°19'W
Floyds Fork, r., Ky., U.S. (floi-dz)	69h	38°08'N	85°30'W
Flumendosa, r., Italy	130	39°45'N	9°18'E
Flushing, Mi., U.S. (flûsh'ĭng)	66	43°05'N	83°50'W
Fly, r., Pap. N. Gui. (flī)	169	8°00'S	141°45'E
Foča, Bos. (fō'chä)	131	43°29'N	18°48'E
Fochville, S. Afr. (fŏk'vĭl)	192c	26°29'S	27°29'E
Focşani, Rom. (fòk-shä'nē)	125	45°41'N	27°17'E
Fogang, China (fwo-gäŋ)	165	23°50'N	113°35'E
Foggia, Italy (fôd'jä)	119	41°30'N	15°34'E
Fogo, Can. (fō'gō)	61	49°43'N	54°17'W
Fogo, i., Can.	59	49°40'N	54°13'W
Fogo, i., C.V.	184b	14°46'N	24°51'W
Fohnsdorf, Aus. (fōns'dôrf)	124	47°13'N	14°40'E
Föhr, i., Ger. (fûr)	124	54°47'N	8°30'E
Foix, Fr. (fwä)	126	42°58'N	1°34'E
Fokku, Nig.	189	11°40'N	4°31'E
Foladi, Koh-e, mtn., Afg.	155	34°38'N	67°32'E
Folgares, Ang.	190	14°54'S	15°08'E
Foligno, Italy (fō-lēn'yō)	130	42°58'N	12°41'E
Folkeston, Eng., U.K.	121	51°05'N	1°18'E
Folkingham, Eng., U.K. (fō'kĭng-ăm)	114a	52°53'N	0°24'W
Folkston, Ga., U.S.	83	30°50'N	82°01'W
Folsom, Ca., U.S.	76	38°40'N	121°10'W
Folsom, N.M., U.S. (fōl'sŭm)	78	36°47'N	103°56'W
Fomento, Cuba (fō-mĕ'n-tō)	92	21°35'N	78°20'W
Fómeque, Col. (fō'mĕ-kĕ)	100a	4°29'N	73°52'W
Fonda, Ia., U.S. (fŏn'dȧ)	71	42°33'N	94°51'W
Fond du Lac, Wi., U.S. (fŏn dū läk')	65	43°47'N	88°29'W
Fond du Lac Indian Reservation, I.R., Mn., U.S.	71	46°44'N	93°04'W
Fondi, Italy (fōn'dē)	130	41°23'N	13°25'E
Fonseca, Golfo de, b., N.A. (gôl-fô-dĕ-fōn-sä'kä)	86	13°09'N	87°55'W

ng-sing; ŋ-baŋk; N-nasalized n; nŏd; cŏmmit; ōld; ôbey; ôrder; oi-boil; fōōd; ò-as oo in foot; ou-out; s-soft; sh-dish; th-thin; pūre; ůnite; ûrn; stŭd; circŭs; ü-as in French tu; '-indeterminate vowel.

248 Fontainebleau – Freienried PRONOUNCING INDEX

PLACE (Pronunciation)	PAGE	LAT.	LONG.
Fontainebleau, Fr. (fôN-tĕn-blō′)	117	48°24′N	2°42′E
Fontana, Ca., U.S. (fŏn-tă′nȧ)	75a	34°06′N	117°27′W
Fonte Boa, Braz. (fōn′tå bō′ă)	100	2°32′S	66°05′W
Fontenay-le-Comte, Fr. (fôNt-nĕ′lē-kôNt′)	126	46°28′N	0°53′W
Fontenay-Trésigny, Fr. (fôN-te-nā′ tra-sēn-yē′)	127b	48°43′N	2°53′E
Fontenelle Reservoir, res., Wy., U.S.	73	42°05′N	110°05′W
Fontera, Punta, c., Mex. (po͞o′n-tä-fôn-tĕ′rä)	89	18°36′N	92°43′W
Fontibón, Col. (fōn-tē-bôn′)	100a	4°42′N	74°09′W
Fontur, c., Ice.	112	66°21′N	14°02′W
Foothills, S. Afr. (fŏt-hĭls)	187b	25°55′S	27°36′E
Footscray, Austl.	173a	37°48′S	144°54′E
Foraker, Mount, mtn., Ak., U.S. (fŏr′ā-kēr)	63	62°40′N	152°40′W
Forbach, Fr. (fôr′bäk)	127	49°12′N	6°54′E
Forbes, Austl. (fôrbz)	175	33°24′S	148°05′E
Forbes, Mount, mtn., Can.	55	51°52′N	116°56′W
Forchheim, Ger. (fôrk′hīm)	124	49°43′N	11°05′E
Fordyce, Ar., U.S. (fôr′dīs)	79	33°48′N	92°24′W
Forécariah, Gui. (fōr-ká-rē′ä′)	184	9°26′N	13°06′W
Forel, Mont, mtn., Grnld.	49	65°50′N	37°41′W
Forest, Ms., U.S. (fŏr′ĕst)	82	32°22′N	89°29′W
Forest, r., N.D., U.S.	70	48°08′N	97°45′W
Forest City, Ia., U.S.	71	43°14′N	93°40′W
Forest City, N.C., U.S.	83	35°20′N	81°52′W
Forest City, Pa., U.S.	67	41°35′N	75°30′W
Forest Grove, Or., U.S. (grōv)	74c	45°31′N	123°07′W
Forest Hill, Md., U.S.	68e	39°35′N	76°26′W
Forest Hill, Tx., U.S.	75c	32°40′N	97°16′W
Forestville, Can. (fŏr′ĕst-vĭl)	60	48°45′N	69°06′W
Forestville, Md., U.S.	68e	38°51′N	76°55′W
Forez, Monts du, mts., Fr. (môN dü fō-rā′)	126	44°55′N	3°43′E
Forfar, Scot., U.K. (fôr′fȧr)	120	57°10′N	2°55′W
Forillon, Parc National, rec., Can.	60	48°50′N	64°05′W
Forio, mtn., Italy (fô′ryō)	129c	40°29′N	13°55′E
Forked Creek, r., Il., U.S. (fôrk′d)	69a	41°16′N	88°01′W
Forked Deer, r., Tn., U.S.	82	35°53′N	89°29′W
Forli, Italy (fôr-lē′)	118	44°13′N	12°03′E
Formby, Eng., U.K. (fôrm′bē)	114a	53°34′N	3°04′W
Formby Point, c., Eng., U.K.	114a	53°33′N	3°06′W
Formentera, Isla de, i., Spain (ē′s-lä-dĕ-fōr-mĕn-tā′rä)	118	38°43′N	1°25′E
Formiga, Braz. (fōr-mē′gȧ)	101	20°27′S	45°25′W
Formigas Bank, bk., (fōr-mē′gäs)	93	18°30′N	75°40′W
Formosa, Arg. (fōr-mō′sä)	102	27°25′S	58°12′W
Formosa, Braz.	101	15°32′S	47°10′W
Formosa, prov., Arg.	102	24°30′S	60°45′W
Formosa, Serra, mts., Braz. (sĕ′r-rä)	101	12°59′S	55°11′W
Formosa Bay, b., Kenya	191	2°45′S	40°30′E
Formosa Strait *see* Taiwan Strait, strt., Asia	161	24°30′N	120°00′E
Fornosovo, Russia (fôr-nô′sô vô)	142c	59°35′N	30°34′E
Forrest City, Ar., U.S. (fŏr′ĕst sĭ′tĭ)	79	35°00′N	90°46′W
Forsayth, Austl. (fôr-sīth′)	175	18°33′S	143°42′E
Forshaga, Swe. (fōrs′hä′gä)	122	59°34′N	13°25′E
Forst, Ger. (fôrst)	117	51°45′N	14°38′E
Forsyth, Ga., U.S. (fôr-sīth′)	82	33°02′N	83°56′W
Forsyth, Mt., U.S.	73	46°15′N	106°41′W
Fort Albany, Can. (fôrt ôl′bȧ nĭ)	51	52°20′N	81°30′W
Fort Alexander Indian Reserve, I.R., Can.	57	50°27′N	96°15′W
Fortaleza, Braz. (fôr′tä-lā′zȧ)	101	3°35′S	38°31′W
Fort Apache Indian Reservation, I.R., Az., U.S. (ă-păch′ē)	77	34°02′N	110°27′W
Fort Atkinson, Wi., U.S. (ăt′kĭn-sŭn)	71	42°55′N	88°46′W
Fort Beaufort, S. Afr. (bō′fôrt)	187c	32°47′S	26°39′E
Fort Belknap Indian Reservation, I.R., Mt., U.S.	73	48°16′N	108°38′W
Fort Bellefontaine, Mo., U.S. (bĕl-fŏn-tān′)	75f	38°50′N	90°15′W
Fort Benton, Mt., U.S. (bĕn′tŭn)	73	47°51′N	110°40′W
Fort Berthold Indian Reservation, I.R., N.D., U.S. (bērth′ōld)	70	47°47′N	103°28′W
Fort Bragg, Ca., U.S.	76	39°26′N	123°48′W
Fort Branch, In., U.S. (brănch)	66	38°15′N	87°35′W
Fort Chipewyan, Can.	50	58°46′N	111°15′W
Fort Cobb Reservoir, res., Ok., U.S.	78	35°12′N	98°28′W
Fort Collins, Co., U.S. (kŏl′ĭns)	64	40°36′N	105°04′W
Fort Crampel, C.A.R. (krȧm-pĕl′)	185	6°59′N	19°11′E
Fort-de-France, Mart. (dē frăns)	87	14°37′N	61°06′W
Fort Deposit, Al., U.S. (dē-pŏz′ĭt)	82	31°58′N	86°35′W
Fort-de-Possel, C.A.R. (dē pō-sĕl′)	185	5°03′N	19°11′E
Fort Dodge, Ia., U.S. (dŏj)	65	42°31′N	94°10′W
Fort Edward, N.Y., U.S. (wẽrd)	67	43°15′N	73°30′W
Fort Erie, Can. (ē′rĭ)	69c	42°55′N	78°56′W
Fortescue, r., Austl. (fôr′tĕs-kū)	174	21°25′S	116°50′E
Fort Fairfield, Me., U.S. (fâr′fēld)	60	46°46′N	67°53′W
Fort Fitzgerald, Can. (fĭts-jĕr′ăld)	50	59°48′N	111°50′W
Fort Frances, Can. (frăn′sĕs)	51	48°36′N	93°24′W
Fort Frederica National Monument, rec., Ga., U.S. (frĕd′ē-rĭ-kȧ)	82	31°13′N	85°25′W
Fort Gaines, Ga., U.S. (gānz)	82	31°35′N	85°03′W
Fort Gibson, Ok., U.S. (gĭb′sŭn)	79	35°50′N	95°13′W
Fort Good Hope, Can. (go͝od hōp)	50	66°19′N	128°52′W
Forth, Firth of, b., Scot., U.K. (fûrth ŏv fōrth)	112	56°04′N	3°03′W
Fort Hall, Kenya (hôl)	187	0°47′S	37°13′E
Fort Hall Indian Reservation, I.R., Id., U.S.	73	43°02′N	112°21′W
Fort Huachuca, Az., U.S. (wä-cho͞o′kä)	77	31°30′N	110°25′W
Fortier, Can. (fôr′tyā′)	62f	49°56′N	97°55′W
Fort Kent, Me., U.S. (kĕnt)	60	47°14′N	68°37′W
Fort Langley, Can. (lăng′lĭ)	74d	49°10′N	122°35′W

PLACE (Pronunciation)	PAGE	LAT.	LONG.
Fort Lauderdale, Fl., U.S. (lô′dēr-dāl)	83a	26°07′N	80°09′W
Fort Lee, N.J., U.S.	68a	40°50′N	73°58′W
Fort Liard, Can.	50	60°16′N	123°34′W
Fort Loudoun Lake, res., Tn., U.S. (fôrt lou′dĕn)	82	35°52′N	84°10′W
Fort Lupton, Co., U.S. (lŭp′tŭn)	78	40°04′N	104°54′W
Fort Macleod, Can. (mȧ-kloud′)	50	49°43′N	113°25′W
Fort Madison, Ia., U.S. (măd′ĭ-sŭn)	71	40°40′N	91°17′W
Fort Matanzas, Fl., U.S. (mä-tän′zäs)	83	29°39′N	81°17′W
Fort McDermitt Indian Reservation, I.R., Or., U.S. (măk dẽr′mĭt)	72	42°04′N	118°07′W
Fort McMurray, Can. (mȧk-mŭr′ĭ)	50	56°44′N	111°23′W
Fort McPherson, Can. (mȧk-fŭr′s′n)	50	67°37′N	134°59′W
Fort Meade, Fl., U.S. (mēd)	83a	27°45′N	81°48′W
Fort Mill, S.C., U.S. (mĭl)	83	35°03′N	80°57′W
Fort Mojave Indian Reservation, I.R., Ca., U.S. (mō-hä′vâ)	76	34°59′N	115°02′W
Fort Morgan, Co., U.S. (môr′gȧn)	78	40°14′N	103°49′W
Fort Myers, Fl., U.S. (mī′ẽrz)	83a	26°36′N	81°45′W
Fort Nelson, Can. (nĕl′sŭn)	50	58°57′N	122°30′W
Fort Nelson, r., Can. (nĕl′sŭn)	52	58°44′N	122°20′W
Fort Payne, Al., U.S. (pān)	82	34°26′N	85°41′W
Fort Peck, Mt., U.S. (pĕk)	73	47°58′N	106°30′W
Fort Peck Indian Reservation, I.R., Mt., U.S.	70	48°22′N	105°40′W
Fort Peck Lake, res., Mt., U.S.	64	47°52′N	106°59′W
Fort Pierce, Fl., U.S. (pērs)	83a	27°25′N	80°20′W
Fort Portal, Ug. (pôr′tăl)	185	0°40′N	30°16′E
Fort Providence, Can. (prŏv′ĭ-dĕns)	50	61°27′N	117°59′W
Fort Pulaski National Monument, rec., Ga., U.S. (pu-lăs′kĭ)	83	31°59′N	80°56′W
Fort Qu'Appelle, Can.	56	50°46′N	103°55′W
Fort Randall Dam, dam, S.D., U.S.	70	42°48′N	98°35′W
Fort Resolution, Can. (rĕz′ō-lū′shŭn)	50	61°08′N	113°42′W
Fort Riley, Ks., U.S. (rī′lĭ)	79	39°05′N	96°46′W
Fort Saint James, Can. (fôrt sānt jāmz)	50	54°26′N	124°15′W
Fort Saint John, Can. (sānt jŏn)	50	56°15′N	120°51′W
Fort Sandeman, Pak. (săn′da-măn)	155	31°28′N	69°29′E
Fort Saskatchewan, Can. (săs-kăt′cho͞o-ån)	62g	53°43′N	113°13′W
Fort Scott, Ks., U.S. (skŏt)	65	37°50′N	94°43′W
Fort Severn, Can. (sĕv′ẽrn)	51	55°58′N	87°50′W
Fort-Shevchenko, Kaz. (shĕv-chĕn′kô)	139	44°30′N	50°18′E
Fort Sibut, C.A.R. (fôr sē-bü′)	185	5°44′N	19°05′E
Fort Sill, Ok., U.S. (fôrt sĭl)	78	34°41′N	98°25′W
Fort Simpson, Can. (sĭmp′sŭn)	50	61°52′N	121°48′W
Fort Smith, Can.	50	60°09′N	112°08′W
Fort Smith, Ar., U.S. (smĭth)	65	35°23′N	94°24′W
Fort Stockton, Tx., U.S. (stŏk′tŭn)	80	30°54′N	102°51′W
Fort Sumner, N.M., U.S. (sŭm′nẽr)	78	34°30′N	104°17′W
Fort Sumter National Monument, rec., S.C., U.S. (sŭm′tẽr)	83	32°43′N	79°54′W
Fort Thomas, Ky., U.S. (tŏm′ăs)	69f	39°05′N	84°27′W
Fortuna, Ca., U.S. (fôr-tū′nȧ)	72	40°36′N	124°10′W
Fortune, Can. (fôr′tûn)	61	47°04′N	55°51′W
Fortune, i., Bah.	93	22°35′N	74°20′W
Fortune Bay, b., Can.	53a	47°25′N	55°25′W
Fort Union National Monument, rec., N.M., U.S. (ūn′yŭn)	78	35°51′N	104°57′W
Fort Valley, Ga., U.S. (văl′ĭ)	82	32°33′N	83°53′W
Fort Vermilion, Can. (vẽr-mĭl′yŭn)	50	58°23′N	115°50′W
Fort Victoria *see* Masvingo, Zimb.	186	20°07′S	30°47′E
Fort Wayne, In., U.S. (wān)	65	41°00′N	85°10′W
Fort William, Scot., U.K. (wĭl′yŭm)	120	56°50′N	3°00′W
Fort William, Mount, mtn., Austl. (wĭ′ĭ-ăm)	176	24°45′S	151°15′E
Fort Worth, Tx., U.S. (wûrth)	64	32°45′N	97°20′W
Fort Yukon, Ak., U.S. (yo͞o′kŏn)	64a	66°30′N	145°00′W
Fort Yuma Indian Reservation, I.R., Ca., U.S. (yo͞o′mä)	77	32°54′N	114°47′W
Foshan, China	161	23°02′N	113°07′E
Fossano, Italy (fōs-sä′nō)	130	44°34′N	7°42′E
Fossil Creek, r., Tx., U.S. (fŏs-ĭl)	75c	32°53′N	97°19′W
Fossombrone, Italy (fōs-sôm-brō′nā)	130	43°41′N	12°48′E
Foss Res, Ok., U.S.	78	35°38′N	99°11′W
Fosston, Mn., U.S. (fŏs′tŭn)	70	47°34′N	95°44′W
Fosterburg, Il., U.S. (fŏs′tẽr-bûrg)	75e	38°58′N	90°04′W
Fostoria, Oh., U.S. (fŏs-tō′rĭ-ȧ)	66	41°10′N	83°20′W
Fougéres, Fr. (fo͞o-zhâr′)	117	48°23′N	1°14′W
Foula, i., Scot., U.K. (fou′lȧ)	120a	60°08′N	2°04′W
Foulwind, Cape, c., N.Z. (foul′wĭnd)	175a	41°45′S	171°00′E
Foumban, Cam. (fo͞om-bän′)	184	5°43′N	10°55′E
Fountain Creek, r., Co., U.S. (foun′tĭn)	78	38°36′N	104°37′W
Fountain Valley, Ca., U.S.	75a	33°42′N	117°57′W
Fourche la Fave, r., Ar., U.S. (fo͞orsh lä fàv′)	79	34°46′N	93°45′W
Fouriesburg, S. Afr. (fō′rēz-bûrg)	192c	28°38′S	28°13′E
Fourmies, Fr. (fo͞or-mē′)	126	50°01′N	4°01′E
Four Mountains, Islands of the, is., Ak., U.S.	63a	52°58′N	170°40′W
Fourth Cataract, wtfl., Sudan	185	18°52′N	32°07′E
Fouta Djallon, mts., Gui. (fo͞o′tä jä-lôN)	184	11°37′N	12°29′W
Foveaux Strait, strt., N.Z. (fô-vō′)	175a	46°30′S	167°43′E
Fowler, Co., U.S. (foul′ẽr)	78	38°04′N	104°02′W
Fowler, In., U.S.	66	40°35′N	87°20′W
Fowlerton, Point, c., Austl.	174	32°05′S	132°30′E
Fowlerton, Tx., U.S. (foul′ẽr-tŭn)	80	28°26′N	98°48′W
Fox, i., Wa., U.S. (fŏks)	74a	47°15′N	122°08′W
Fox, r., Il., U.S.	71	41°35′N	88°43′W
Fox, r., Wi., U.S.	71	44°18′N	88°23′W
Foxboro, Ma., U.S. (fŏks′bŭr-ō)	61a	42°04′N	71°15′W
Foxe Basin, b., Can. (fŏks)	53	67°35′N	79°21′W
Foxe Channel, strt., Can.	53	64°30′N	79°23′W
Foxe Peninsula, pen., Can.	53	64°57′N	77°26′W
Fox Islands, is., Ak., U.S. (fŏks)	63a	53°04′N	167°30′W

PLACE (Pronunciation)	PAGE	LAT.	LONG.
Fox Lake, Il., U.S. (lăk)	69a	42°24′N	88°11′W
Fox Lake, l., Il., U.S.	69a	42°24′N	88°07′W
Fox Point, Wi., U.S.	69a	43°10′N	87°54′W
Foyle, Lough, b., Eur. (lŏk foil′)	120	55°07′N	7°08′W
Foz do Cunene, Ang.	190	17°16′S	11°50′E
Fraga, Spain (frä′gä)	129	41°31′N	0°20′E
Fragoso, Cayo, i., Cuba (kä′yō-frä-gō′sō)	92	22°45′N	79°30′W
Framnes Mountains, mts., Ant.	178	67°50′S	62°35′E
Franca, Braz. (frä′n-kä)	101	20°28′S	47°20′W
Francavilla, Italy (fräŋ-kä-vēl′lä)	131	40°32′N	17°37′E
France, nation, Eur. (fråns)	110	46°39′N	0°47′E
Frances, l., Can. (frăn′sĭs)	52	61°27′N	128°28′W
Frances, Cabo, c., Cuba (kä′bô-frän-sĕ′s)	92	21°55′N	84°05′W
Frances, Punta, c., Cuba (po͞o′n-tä-frän-sĕ′s)	92	21°45′N	83°10′W
Francés Viejo, Cabo, c., Dom. Rep. (ká′bô-frän′sås vyā′hô)	93	19°40′N	69°35′W
Franceville, Gabon (fräns-vēl′)	186	1°38′S	13°35′E
Francis Case, Lake, res., S.D., U.S. (frăn′sĭs)	64	43°15′N	99°00′W
Francisco Sales, Braz. (frän-sē′s-kô-sá′lĕs)	99a	21°42′S	44°26′W
Francistown, Bots. (frăn′sĭs-toun)	186	21°17′S	27°28′E
Frankfort, S. Afr. (frănk′fŏrt)	187c	32°43′S	27°28′E
Frankfort, S. Afr.	192c	27°17′S	28°30′E
Frankfort, Il., U.S. (frăŋk′fŭrt)	69a	41°30′N	87°51′W
Frankfort, In., U.S.	66	40°15′N	86°30′W
Frankfort, Ks., U.S.	79	39°42′N	96°27′W
Frankfort, Ky., U.S.	65	38°10′N	84°55′W
Frankfort, Mi., U.S.	66	44°40′N	86°15′W
Frankfort, N.Y., U.S.	67	43°05′N	75°05′W
Frankfurt am Main, Ger.	110	50°07′N	8°40′E
Frankfurt an der Oder, Ger.	117	52°20′N	14°31′E
Franklin, S. Afr.	187c	30°19′S	29°28′E
Franklin, In., U.S. (frănk′lĭn)	66	39°25′N	86°00′W
Franklin, Ky., U.S.	82	36°42′N	86°34′W
Franklin, La., U.S.	81	29°47′N	91°31′W
Franklin, Ma., U.S.	61a	42°05′N	71°24′W
Franklin, Ne., U.S.	78	40°06′N	99°01′W
Franklin, N.H., U.S.	67	43°25′N	71°40′W
Franklin, N.J., U.S.	68a	41°08′N	74°35′W
Franklin, Oh., U.S.	66	39°30′N	84°20′W
Franklin, Pa., U.S.	67	41°25′N	79°50′W
Franklin, Tn., U.S.	82	35°54′N	86°54′W
Franklin, Va., U.S.	83	36°41′N	76°57′W
Franklin, l., Nv., U.S.	76	40°23′N	115°10′W
Franklin D. Roosevelt Lake, res., Wa., U.S.	72	48°12′N	118°43′W
Franklin Mountains, mts., Can.	52	65°36′N	125°55′W
Franklin Park, Il., U.S.	69a	41°56′N	87°53′W
Franklin Square, N.Y., U.S.	68a	40°43′N	73°40′W
Franklinton, La., U.S. (frăŋk′lĭn-tŭn)	81	30°49′N	90°09′W
Frankston, Austl.	173a	38°09′S	145°08′E
Franksville, Wi., U.S. (frănkz′vĭl)	69a	42°46′N	87°55′W
Fransta, Swe.	122	62°30′N	16°04′E
Franz Josef Land *see* Zemlya Frantsa-Iosifa, is., Russia	134	81°32′N	40°00′E
Frascati, Italy (fräs-kä′tē)	130	41°49′N	12°45′E
Fraser, Mi., U.S. (frā′zẽr)	69b	42°32′N	82°57′W
Fraser, i., Austl.	175	25°12′S	153°00′E
Fraser, r., Can.	52	51°30′N	122°00′W
Fraserburgh, Scot., U.K. (frā′zẽr-bûrg)	120	57°40′N	2°01′W
Fraser Plateau, plat., Can.	55	51°30′N	122°00′W
Frattamaggiore, Italy (frät-tä-mäg-zhyô′rĕ)	129c	40°41′N	14°16′E
Fray Bentos, Ur. (frī bĕn′tōs)	102	33°10′S	58°19′W
Frazee, Mn., U.S. (frȧ-zē′)	70	46°42′N	95°43′W
Fraziers Hog Cay, i., Bah.	92	25°25′N	77°55′W
Frechen, Ger. (frĕ′κĕn)	127c	50°54′N	6°49′E
Fredericia, Den. (frĕdh-ĕ-rē′tsē-ȧ)	122	55°35′N	9°45′E
Frederick, Md., U.S. (frĕd′ẽr-ĭk)	65	39°25′N	77°25′W
Frederick, Ok., U.S.	78	34°23′N	99°01′W
Frederick House, r., Can.	58	49°05′N	81°20′W
Fredericksburg, Tx., U.S. (frĕd′ẽr-ĭkz-bûrg)	80	30°16′N	98°52′W
Fredericksburg, Va., U.S.	67	38°20′N	77°30′W
Fredericktown, Mo., U.S. (frĕd′ẽr-ĭk-toun)	79	37°32′N	90°16′W
Fredericton, Can. (frĕd′-ẽr-ĭk-tŭn)	51	45°48′N	66°39′W
Frederikshavn, Den. (frĕdh′ĕ-rĕks-houn)	116	57°27′N	10°31′E
Frederikssund, Den. (frĕdh′ĕ-rĕks-sōn)	122	55°51′N	12°04′E
Fredonia, Col. (frĕ-dō′nyȧ)	100a	5°55′N	75°40′W
Fredonia, Ks., U.S. (frē-dō′nĭ-ȧ)	79	36°31′N	95°50′W
Fredonia, N.Y., U.S.	67	42°25′N	79°20′W
Fredrikstad, Nor. (frådh′rĕks-ståd)	116	59°14′N	10°58′E
Freeburg, Il., U.S. (frē′bûrg)	75e	38°26′N	89°59′W
Freehold, N.J., U.S. (frē′hōld)	68a	40°15′N	74°16′W
Freeland, Pa., U.S. (frē′lănd)	67	41°00′N	75°50′W
Freeland, Wa., U.S.	74a	48°01′N	122°32′W
Freels, Cape, c., Can. (frēlz)	61	46°37′N	53°45′W
Freelton, Can. (frēl′tŭn)	62d	43°24′N	80°02′W
Freeport, Bah.	92	26°30′N	78°45′W
Freeport, Il., U.S. (frē′pōrt)	65	42°19′N	89°30′W
Freeport, N.Y., U.S.	68a	40°39′N	73°35′W
Freeport, Tx., U.S.	81	28°56′N	95°21′W
Freetown, S.L. (frē′toun)	184	8°30′N	13°15′W
Fregenal de la Sierra, Spain (frā-hå-näl′ dā lä syĕr′rä)	128	38°09′N	6°40′W
Fregene, Italy (frĕ-zhĕ′-nĕ)	129d	41°52′N	12°12′E
Freiberg, Ger. (frī′bĕrgh)	117	50°54′N	13°18′E
Freiburg, Ger.	117	48°00′N	7°50′E
Freienried, Ger. (frī′ĕn-rēd)	115d	48°20′N	11°08′E

ăt; finăl; rāte; senāte; ärm; àsk; sofȧ; fâre; ch-choose; dh-as th in other; bē; ēvent; bĕt; recĕnt; cratẽr; g-gō; gh-guttural g; bĭt; ĭ-short neutral; rīde; κ-guttural k as ch in German ich;

ng-sing; ŋ-baŋk; N-nasalized n; nŏd; cŏmmit; ōld; ŏbey; ôrder; oi-boil; fōōd; ò-as oo in foot; ou-out; s-soft; sh-dish; th-thin; pūre; ûnite; ûrn; stŭd; circŭs; ü-as in French tu; ′-indeterminate vowel.

250 Gardner Pinnacles – Gladstone PRONOUNCING INDEX

PLACE (Pronunciation)	PAGE	LAT.	LONG.
Gardner Pinnacles, Hi., U.S.	84b	25°10'N	167°00'W
Gareloi, i., Ak., U.S. (gär-lo͞o-ä')	63a	51°40'N	178°48'W
Garfield, N.J., U.S. (gär'fēld)	68a	40°53'N	74°06'W
Garfield, Ut., U.S.	75b	40°45'N	112°10'W
Garfield Heights, Oh., U.S.	69d	41°25'N	81°36'W
Gargaliánoi, Grc. (gär-gä-lyä'nē)	131	37°07'N	21°50'E
Gargždai, Lith. (gärgzh'dī)	123	55°43'N	20°09'E
Garibaldi, Mount, mtn., Can. (gâr-ĭ-băl'dē)	54	49°51'N	123°01'W
Garin, Arg. (gä-rē'n)	102a	34°25'S	58°44'W
Garissa, Kenya	191	0°28'S	39°38'E
Garland, Tx., U.S. (gär'lănd)	75c	32°55'N	96°39'W
Garland, Ut., U.S.	73	41°45'N	112°10'W
Garm, Taj.	139	39°12'N	70°28'E
Garmisch-Partenkirchen, Ger. (gär'mēsh pär'tĕn-kēr'kĕn)	124	47°38'N	11°10'E
Garnett, Ks., U.S. (gär'nĕt)	79	38°16'N	95°15'W
Garonne, r., Fr. (gȧ-rôn)	112	44°00'N	1°00'E
Garoua, Cam. (gär'wä)	185	9°18'N	13°24'E
Garrett, In., U.S. (găr'ĕt)	66	41°20'N	85°10'W
Garrison, N.D., U.S.	70	47°38'N	101°24'W
Garrison, N.Y., U.S. (găr'ĭ-sŭn)	68a	41°23'N	73°57'W
Garrovillas, Spain (gä-rô-vēl'yäs)	128	39°42'N	6°30'W
Garry, l., Can. (găr'ĭ)	52	66°16'N	99°23'W
Garsen, Kenya	191	2°16'S	40°07'E
Garson, Can.	59	46°34'N	80°52'W
Garstedt, Ger. (gär'shtĕt)	115c	53°40'N	9°58'E
Garulia, India	158a	22°48'N	88°23'E
Garwolin, Pol. (gär-vō'lēn)	125	51°54'N	21°40'E
Gary, In., U.S. (gā'rĭ)	65	41°35'N	87°21'W
Gary, W.V., U.S. (fĭl'bĕrt)	83	37°21'N	81°33'W
Garzón, Col. (gär-thōn')	100	2°13'N	75°44'W
Gasan, Phil. (gä-sän')	169a	13°19'N	121°52'E
Gasan-Kuli, Turkmen.	139	37°25'N	53°55'E
Gas City, In., U.S. (găs)	66	40°30'N	85°40'W
Gascogne, reg., Fr. (gäs-kôn'yĕ)	126	43°45'N	1°49'W
Gasconade, r., Mo., U.S. (găs-kô-nād')	79	37°46'N	92°15'W
Gascoyne, r., Austl. (găs-koin')	174	25°15'S	117°00'E
Gashland, Mo., U.S. (găsh'-lănd)	75f	39°15'N	94°35'W
Gashua, Nig.	189	12°54'N	11°00'E
Gasny, Fr. (gäs-nē')	127b	49°05'N	1°36'E
Gaspé, Can.	51	48°50'N	64°29'W
Gaspé, Péninsule de, pen., Can.	53	48°30'N	65°00'W
Gasper Hernández, Dom. Rep. (gäs-pär' ĕr-nän'dāth)	93	19°40'N	70°15'W
Gassaway, W.V., U.S. (găs'ȧ-wā)	66	38°40'N	80°45'W
Gaston, Or., U.S. (găs'tŭn)	74c	45°26'N	123°08'W
Gastonia, N.C., U.S. (găs-tô'nĭ-ȧ)	83	35°15'N	81°14'W
Gastre, Arg. (gäs-trĕ)	102	42°12'S	68°50'W
Gata, Cabo de, c., Spain (kȧ'bô-dĕ-gä'tä)	118	36°42'N	2°00'W
Gata, Sierra de, mts., Spain (syĕr'rä dä gä'tä)	118	40°12'N	6°39'W
Gatchina, Russia (gä-chē'nȧ)	136	59°33'N	30°08'E
Gátes, Akrotírion, c., Cyp.	153a	34°30'N	33°15'E
Gateshead, Eng., U.K. (gāts'hĕd)	120	54°56'N	1°38'W
Gates of the Arctic National Park, rec., Ak., U.S.	63	67°45'N	153°30'W
Gatesville, Tx., U.S. (gāts'vĭl)	81	31°26'N	97°34'W
Gâtine, Hauteurs de, hills, Fr.	126	46°40'N	0°50'W
Gatineau, Can. (gȧ'tē-nō)	62c	45°29'N	75°38'W
Gatineau, r., Can.	59	45°45'N	75°50'W
Gatineau, Parc de la, rec., Can.	59	45°32'N	75°53'W
Gattendorf, Aus.	115e	48°01'N	17°00'E
Gatun, Pan. (gä-to͞on')	91	9°16'N	79°25'W
Gatun, r., Pan.	86a	9°21'N	79°40'W
Gatún, Lago, l., Pan.	91	9°13'N	79°24'W
Gatun Locks, trans., Pan.	86a	9°16'N	79°57'W
Gauhāti, India	155	26°09'N	91°51'E
Gauja, r., Lat. (gȧ'ō-yȧ)	123	57°10'N	24°30'E
Gaula, r., Nor.	122	62°55'N	10°45'E
Gávdos, i., Grc. (gäv'dôs)	119	34°48'N	24°08'E
Gavins Point Dam, Ne., U.S. (gă'-vĭns)	70	42°47'N	97°47'W
Gāvkhūnī, Bātlāq-e, l., Iran	154	31°40'N	52°48'E
Gävle, Swe. (yĕv'lĕ)	110	60°40'N	17°07'E
Gävlebukten, b., Swe.	122	60°45'N	17°30'E
Gavrilov Posad, Russia (gȧ'vrē-lôf'ka po-sȧt)	132	56°34'N	40°09'E
Gavrilov-Yam, Russia (gȧ'vrē-lôf yäm')	132	57°17'N	39°49'E
Gawler, Austl. (gô'lẽr)	174	34°35'S	138°47'E
Gawler Ranges, mts., Austl.	176	32°35'S	136°30'E
Gaya, India (gŭ'yä)(gī'ȧ)	155	24°53'N	85°00'E
Gaya, Nig. (gä'yä)	184	11°58'N	9°05'E
Gaylord, Mi., U.S. (gā'lôrd)	66	45°00'N	84°35'W
Gayndah, Austl. (gān'dăh)	176	25°43'S	151°33'E
Gaza, Gaza	154	31°30'N	34°29'E
Gaziantep, Tur. (gä-zē-än'tĕp)	154	37°10'N	37°30'E
Gbarnga, Lib.	188	7°00'N	9°29'W
Gdańsk, Pol. (g'dänsk)	110	54°20'N	18°40'E
Gdov, Russia (g'dôf')	136	58°44'N	27°51'E
Gdynia, Pol. (g'dēn'yȧ)	116	54°29'N	18°30'E
Geary, Ok., U.S. (gē'rĭ)	78	35°36'N	98°19'W
Géba, r., Gui.-B.	188	12°25'N	14°35'W
Gebo, Wy., U.S. (gĕb'ō)	73	43°49'N	108°13'W
Ged, La., U.S. (gĕd)	81	30°07'N	93°36'W
Gediz, r., Tur.	119	38°44'N	28°45'E
Gedney, i., Wa., U.S. (gĕd-nē)	74a	48°01'N	122°18'W
Gedser, Den.	122	54°35'N	12°08'E
Geel, Bel.	115a	51°09'N	5°01'E
Geelong, Austl. (jē-lông')	175	38°06'S	144°13'E
Gegu, China (gŭ-go͞o)	162	39°00'N	117°30'E
Ge Hu, l., China (gŭ ho͞o)	162	31°37'N	119°57'E
Geidam, Nig.	184	12°57'N	11°57'E
Geikie Range, mts., Austl. (gē'kē)	174	17°35'S	125°32'E
Geislingen, Ger. (gis'lĭng-ĕn)	124	48°37'N	9°52'E
Geist Reservoir, res., In., U.S. (gēst)	69g	39°57'N	85°59'W
Geita, Tan.	191	2°52'S	32°10'E
Gejiu, China (gŭ-jĭō)	165	23°32'N	102°50'E
Geldermalsen, Neth.	115a	51°53'N	5°18'E
Geldern, Ger. (gĕl'dĕrn)	127c	51°31'N	6°20'E
Gelibolu, Tur. (gĕ-lĭb'ō-lò)	119	40°25'N	26°40'E
Gelsenkirchen, Ger. (gĕl-zĕn-kĭrk-ĕn)	124	51°31'N	7°05'E
Gemas, Malay. (jĕm'ȧs)	153b	2°35'N	102°37'E
Gemena, D.R.C.	185	3°15'N	19°46'E
Gemlik, Tur. (gĕm'lĭk)	119	40°30'N	29°10'E
Genale (Jubba), r., Afr.	192a	5°15'N	41°00'E
General Alvear, Arg. (gĕ-nĕ-răl'ăl-vĕ-ä'r)	99c	36°04'S	60°02'W
General Arenales, Arg. (ä-rĕ-nä'lĕs)	99c	34°19'S	61°16'W
General Belgrano, Arg. (bĕl-grȧ'nô)	99c	35°45'S	58°32'W
General Cepeda, Mex. (sĕ-pĕ'dä)	80	25°24'N	101°29'W
General Conesa, Arg. (kô-nĕ'sä)	99c	36°30'S	57°19'W
General Guido, Arg. (gĕ'dō)	99c	36°41'S	57°48'W
General Lavalle, Arg. (lȧ-vȧ'l-yĕ)	99c	36°25'S	56°55'W
General Madariaga, Arg. (män-dȧ-rēȧ'gä)	102	36°59'S	57°14'W
General Paz, Arg. (pȧ'z)	99c	35°30'S	58°20'W
General Pedro Antonio Santos, Mex.	88	21°37'N	98°58'W
General Pico, Arg. (pē'kô)	102	36°46'S	63°44'W
General Roca, Arg. (rô-kä)	102	39°01'S	67°31'W
General San Martín, Arg. (sän-mȧr-tē'n)	102a	34°35'S	58°32'W
General Sarmiento (San Miguel), Arg.	102a	34°33'S	58°43'W
General Viamonte, Arg. (vēȧ'môn-tĕ)	99c	35°01'S	60°59'W
General Zuazua, Mex. (zwä'zwä)	80	25°54'N	100°07'W
Genesee, r., N.Y., U.S. (jĕn-ē-sē')	67	42°25'N	78°10'W
Geneseo, Il., U.S. (jē-nĕsēō)	66	41°28'N	90°11'W
Geneva (Genève), Switz.	110	46°14'N	6°04'E
Geneva, Al., U.S. (jē-nē'vȧ)	82	31°03'N	85°50'W
Geneva, Il., U.S.	69a	41°53'N	88°18'W
Geneva, Ne., U.S.	79	40°32'N	97°37'W
Geneva, N.Y., U.S.	67	42°50'N	77°00'W
Geneva, Oh., U.S.	66	41°45'N	80°55'W
Geneva, Lake, l., Switz.	117	46°28'N	6°30'E
Genève *see* Geneva, Switz.	110	46°14'N	6°04'E
Genil, r., Spain (hå-nēl')	128	37°15'N	4°05'W
Genoa, Italy	110	44°23'N	9°52'E
Genoa, Ne., U.S. (jen'ō-ȧ)	79	41°26'N	97°43'W
Genoa City, Wi., U.S.	69a	42°31'N	88°19'W
Genova, Golfo di, b., Italy (gôl-fô-dē-jĕn'ō-vä)	112	44°10'N	8°45'E
Genovesa, i., Ec. (ĕ's-lä-gē-nō-vē-sä)	100	0°08'N	90°15'W
Gent, Bel.	117	51°05'N	3°40'E
Genthin, Ger. (gĕn-tēn')	124	52°24'N	12°10'E
Genzano di Roma, Italy (gzhĕnt-zä'-nō-dē-rō'mä)	129d	41°43'N	12°49'E
Geographe Bay, b., Austl. (jē-ō-graf')	174	33°00'S	114°00'E
Geographe Channel, strt., Austl. (jēō'grȧ-fĭk)	174	24°15'S	112°50'E
George, l., N.Y., U.S. (jôrj)	67	43°40'N	73°30'W
George, Lake, l., N.A. (jôrg)	75k	46°26'N	84°09'W
George, Lake, l., Ug.	191	0°02'N	30°25'E
George, Lake, l., Fl., U.S. (jôr-ĭj)	83	29°10'N	81°50'W
George, Lake, l., In., U.S.	69a	41°31'N	87°17'W
Georges, r., Austl.	173b	33°57'S	151°00'E
George Town, Bah.	93	23°30'N	75°50'W
Georgetown, Can. (jôrg-toun)	62d	43°39'N	79°56'W
Georgetown, Can. (jôr-ĭj-toun)	61	46°11'N	62°32'W
George Town, Cay. Is.	92	19°20'N	81°20'W
Georgetown, Guy. (jôrj'toun)	101	7°45'N	58°04'W
George Town, Malay.	168	5°21'N	100°09'E
Georgetown, Ct., U.S.	68a	41°15'N	73°25'W
Georgetown, De., U.S.	67	38°40'N	75°20'W
Georgetown, Il., U.S.	66	40°00'N	87°40'W
Georgetown, Ky., U.S.	66	38°10'N	84°35'W
Georgetown, Ma., U.S. (jôrg-toun)	61a	42°43'N	71°00'W
Georgetown, Md., U.S.	67	39°25'N	75°55'W
Georgetown, S.C., U.S. (jôr-ĭj-toun)	83	33°22'N	79°17'W
Georgetown, Tx., U.S. (jôrg-toun)	81	30°37'N	97°40'W
George Washington Birthplace National Monument, rec., Va., U.S. (jôrj wŏsh'ĭng-tŭn)	67	38°10'N	77°00'W
George Washington Carver National Monument, rec., Mo., U.S. (jôrg wăsh-ĭng-tŭn kär'vẽr)	79	36°58'N	94°21'W
George West, Tx., U.S.	80	28°20'N	98°07'W
Georgia, nation, Asia	134	42°17'N	43°00'E
Georgia, state, U.S. (jôr'jĭ-ȧ)	65	32°40'N	83°50'W
Georgia, Strait of, strt., N.A.	54	49°20'N	124°00'W
Georgiana, Al., U.S. (jôr-jē-ăn'ȧ)	82	31°39'N	86°44'W
Georgian Bay, b., Can.	53	45°15'N	80°50'W
Georgian Bay Islands National Park, rec., Can.	58	45°20'N	81°40'W
Georgina, r., Austl. (jôr-jē'nȧ)	174	22°00'S	138°15'E
Georgiyevsk, Russia (gyôr-gyĕfsk')	137	44°05'N	43°30'E
Gera, Ger. (gā'rä)	117	50°52'N	12°06'E
Geral, Serra, mts., Braz. (sĕr'rȧ zhä-räl')	102	28°30'S	51°00'W
Geral de Goiás, Serra, mts., Braz. (zhä-räl'-dĕ-gô-yȧ's)	101	14°22'S	45°40'W
Geraldton, Austl. (jĕr'ăld-tŭn)	174	28°40'S	114°35'E
Geraldton, Can.	51	49°43'N	87°00'W
Gérgal, Spain (gĕr'gäl)	128	37°08'N	2°29'W
Gering, Ne., U.S. (gē'rĭng)	70	41°49'N	103°41'W
Gerlachovský štít, mtn., Slvk.	125	49°12'N	20°08'E
Germantown, Oh., U.S. (jûr'măn-toun)	66	39°35'N	84°25'W
Germany, nation, Eur. (jûr'mȧ-nĭ)	110	51°00'N	10°00'E
Germiston, S. Afr. (jûr'mĭs-tŭn)	186	26°19'S	28°11'E
Gerona, Phil. (hā-rō'nä)	169a	15°36'N	120°36'E
Gerrards Cross, Eng., U.K. (jĕrårds krŏs)	114b	51°34'N	0°33'W
Gers, r., Fr. (zhĕr)	129	43°25'N	0°30'E
Gersthofen, Ger. (gĕrst-hō'fĕn)	115d	48°26'N	10°54'E
Getafe, Spain (hā-tä'fā)	128	40°19'N	3°44'W
Gettysburg, Pa., U.S. (gĕt'ĭs-bûrg)	67	39°50'N	77°15'W
Gettysburg, S.D., U.S.	70	45°01'N	99°59'W
Gevelsberg, Ger. (gĕ-fĕls'bĕrgh)	127c	51°18'N	7°20'E
Ghāghra, r., India	155	26°00'N	83°00'E
Ghana, nation, Afr. (gän'ä)	184	8°00'N	2°00'W
Ghanzi, Bots. (gän'zē)	186	21°30'S	22°00'E
Ghardaïa, Alg. (gär-dä'ē-ä)	184	32°29'N	3°38'E
Gharo, Pak.	158	24°50'N	68°35'E
Ghāt, Libya	184	24°52'N	10°16'E
Ghazāl, Bahr al-, r., Sudan	185	9°30'N	30°00'E
Ghazal, Bahr el, r., Chad (bär ĕl ghä-zäl')	189	14°30'N	17°00'E
Ghazzah *see* Gaza, Gaza	154	31°30'N	34°29'E
Gheorgheni, Rom.	119	46°48'N	25°30'E
Gherla, Rom. (gĕr'lä)	125	47°01'N	23°55'E
Ghilizane, Alg.	184	35°43'N	0°43'E
Ghorīān, Afg.	157	34°21'N	61°30'E
Ghost Lake, Can.	62e	51°15'N	114°46'W
Ghudāmis, Libya	184	30°07'N	9°26'E
Giannitsá, Grc.	131	40°47'N	22°26'E
Giannutri, Isola di, i., Italy (jän-no͞o'trē)	130	42°15'N	11°06'E
Giant Sequoia National Monument, rec., Ca., U.S.	76	36°10'N	118°35'W
Gibara, Cuba (hē-bä'rä)	92	21°05'N	76°10'W
Gibeon, Nmb. (gĭb'ē-ŭn)	186	25°15'S	17°30'E
Gibraleón, Spain (hē-brä-lå-ōn')	128	37°24'N	7°00'W
Gibraltar, dep., Eur. (gĭ-bräl-tä'r)	110	36°08'N	5°22'W
Gibraltar, Strait of, strt.	112	35°55'N	5°45'W
Gibson City, Il., U.S. (gĭb'sŭn)	66	40°25'N	88°20'W
Gibson Desert, des., Austl.	174	24°45'S	123°15'E
Gibson Island, Md., U.S.	68e	39°05'N	76°26'W
Gibson Reservoir, res., Ok., U.S.	79	36°07'N	95°08'W
Giddings, Tx., U.S. (gĭd'ĭngz)	81	30°11'N	96°55'W
Gideon, Mo., U.S. (gĭd'ē-ŭn)	79	36°27'N	89°56'W
Gien, Fr. (zhē-ăn')	117	47°43'N	2°37'E
Giessen, Ger. (gēs'sĕn)	124	50°35'N	8°40'E
Gifu, Japan (gē'fo͞o)	161	35°25'N	136°45'E
Gig Harbor, Wa., U.S. (gĭg)	74a	47°20'N	122°36'W
Giglio, Isola del, i., Italy (jēl'yō)	130	42°23'N	10°55'E
Gijón, Spain (hē-hōn')	110	43°33'N	5°37'W
Gila, r., U.S. (hē'lȧ)	64	33°00'N	110°00'W
Gila Bend, Az., U.S.	77	32°59'N	112°41'W
Gila Cliff Dwellings National Monument, rec., N.M., U.S.	77	33°15'N	108°20'W
Gila River Indian Reservation, I.R., Az., U.S.	77	33°11'N	112°38'W
Gilbert, Mn., U.S. (gĭl'bẽrt)	71	47°27'N	92°29'W
Gilbert, r., Austl. (gĭl-bẽrt)	175	17°15'S	142°09'E
Gilbert, Mount, mtn., Can.	54	50°51'N	124°20'W
Gilbert Islands, is., Kir.	195	0°30'S	174°00'E
Gilboa, Mount, mtn., S. Afr. (gĭl-bōȧ)	187c	29°13'S	30°17'E
Gilford Island, i., Can. (gĭl'fẽrd)	54	50°45'N	126°25'W
Gilgit, Pak. (gĭl'gĭt)	155	35°58'N	73°48'E
Gil Island, i., Can. (gĭl)	54	53°13'N	129°15'W
Gillen, l., Austl. (jĭl'ĕn)	174	26°15'S	125°15'E
Gillett, Ar., U.S. (jĭ-lĕt')	79	34°07'N	91°22'W
Gillette, Wy., U.S.	73	44°17'N	105°30'W
Gillingham, Eng., U.K. (gĭl'ĭng ăm)	121	51°23'N	0°33'E
Gilman, Il., U.S. (gĭl'măn)	66	40°45'N	87°55'W
Gilman Hot Springs, Ca., U.S.	75a	33°49'N	116°57'W
Gilmer, Tx., U.S. (gĭl'mẽr)	81	32°43'N	94°57'W
Gilmore, Ga., U.S. (gĭl'môr)	68c	33°51'N	84°29'W
Gilo, r., Eth.	185	7°40'N	34°17'E
Gilroy, Ca., U.S. (gĭl-roi')	76	37°00'N	121°34'W
Giluwe, Mount, mtn., Pap. N. Gui.	169	6°04'S	144°00'E
Gimli, Can. (gĭm'lē)	57	50°39'N	97°00'W
Gimone, r., Fr. (zhē-mōn')	126	43°26'N	0°36'E
Ginir, Eth.	185	7°13'N	40°44'E
Ginosa, Italy (jē-nō'zä)	130	40°35'N	16°48'E
Gioia del Colle, Italy (jô'yä dĕl kôl'lā)	130	40°48'N	16°55'E
Girard, Ks., U.S. (jĭ-rärd')	79	37°30'N	94°50'W
Girardot, Col. (hē-rär-dōt')	100	4°19'N	74°47'W
Giresun, Tur. (ghĕr'ĕ-sŏŏn')	154	40°55'N	38°20'E
Giridih, India (jē-rē-dē)	155	24°12'N	86°18'E
Girona, Spain	118	41°55'N	2°48'E
Gironde, r., Fr. (zhē-rônd')	112	45°31'N	1°00'W
Girvan, Scot., U.K. (gûr'văn)	120	55°15'N	5°01'W
Gisborne, N.Z. (gĭz'bŭrn)	175a	38°40'S	178°08'E
Gisenyi, Rw.	186	1°43'S	29°15'E
Gisors, Fr. (zhē-zôr')	126	49°19'N	1°47'E
Gitambo, D.R.C.	190	4°21'N	24°45'E
Gitega, Bdi.	186	3°39'S	30°05'E
Giurgiu, Rom. (jôr'jō)	131	43°53'N	25°58'E
Givet, Fr. (zhē-vĕ')	126	50°08'N	4°47'E
Givors, Fr. (zhē-vôr')	126	45°35'N	4°46'E
Giza *see* Al Jīzah, Egypt	192b	30°01'N	31°12'E
Gizhiga, Russia (gē'zhi-gȧ)	135	61°59'N	160°46'E
Gizo, Sol. Is.	170e	8°06'S	156°51'E
Gizycko, Pol. (gĭ'zhĭ-ko)	116	54°03'N	21°48'E
Gjirokastër, Alb.	119	40°04'N	20°10'E
Gjøvik, Nor. (gyû'vĕk)	116	60°47'N	10°36'E
Glabeek-Zuurbemde, Bel.	115a	50°52'N	4°59'E
Glace Bay, Can. (gläs bā)	61	46°12'N	59°57'W
Glacier Bay National Park, rec., Ak., U.S. (glā'shẽr)	64a	58°40'N	136°50'W
Glacier National Park, rec., Can.	52	51°45'N	117°35'W
Glacier Peak, mtn., Wa., U.S.	72	48°07'N	121°10'W
Glacier Point, c., Can.	74a	48°24'N	123°59'W
Gladbeck, Ger. (gläd'bĕk)	124	51°35'N	6°59'E
Gladdeklipkop, S. Afr.	192c	24°17'S	29°36'E
Gladstone, Austl. (glăd'stōn)	175	23°45'S	152°00'E

PRONOUNCING INDEX Gladstone – Grande 251

PLACE (Pronunciation)	PAGE	LAT.	LONG.
Gladstone, Austl.	174	33°15'S	138°20'E
Gladstone, Mi., U.S.	71	45°50'N	87°04'W
Gladstone, N.J., U.S.	68a	40°43'N	74°39'W
Gladstone, Or., U.S.	74c	45°23'N	122°36'W
Gladwin, Mi., U.S. (glăd'wĭn)	66	44°00'N	84°25'W
Glåma, r., Nor.	112	61°30'N	10°30'E
Glarus, Switz. (glä'rōs)	124	47°02'N	9°03'E
Glasgow, Scot., U.K. (glȧs'gō)	110	55°54'N	4°25'W
Glasgow, Ky., U.S.	82	37°00'N	85°55'W
Glasgow, Mo., U.S.	79	39°14'N	92°48'W
Glasgow, Mt., U.S.	73	48°14'N	106°39'W
Glassport, Pa., U.S. (glȧs'pōrt)	69e	40°19'N	79°53'W
Glauchau, Ger. (glou'kou)	124	50°51'N	12°28'E
Glazov, Russia (glä'zôf)	134	58°05'N	52°52'E
Glen, r., Eng., U.K. (glĕn)	114a	52°44'N	0°18'W
Glénan, Îles de, is., Fr. (ĕl-dĕ-glā-näN')	126	47°43'N	4°42'W
Glen Burnie, Md., U.S. (bûr'nē)	68e	39°10'N	76°38'W
Glen Canyon, p., Ut., U.S.	77	37°10'N	110°50'W
Glen Canyon Dam, dam, Az., U.S. (glĕn kăn'yŭn)	64	36°57'N	111°25'W
Glen Canyon National Recreation Area, rec., U.S.	77	37°00'N	111°20'W
Glen Carbon, Il., U.S. (kär'bŏn)	75e	38°45'N	89°59'W
Glencoe, S. Afr. (glĕn-cô)	187c	28°14'S	30°09'E
Glencoe, Il., U.S.	69a	42°08'N	87°45'W
Glencoe, Mn., U.S. (glĕn'kō)	71	44°44'N	94°07'W
Glen Cove, N.Y., U.S. (kōv)	68a	40°51'N	73°38'W
Glendale, Az., U.S. (glĕn'dāl)	77	33°30'N	112°15'W
Glendale, Ca., U.S.	64	34°09'N	118°15'W
Glendale, Oh., U.S.	69f	31°16'N	84°22'W
Glendive, Mt., U.S. (glĕn'dīv)	64	47°08'N	104°41'W
Glendo, Wy., U.S.	73	42°32'N	104°54'W
Glendora, Ca., U.S. (glĕn-dō'rȧ)	75a	34°08'N	117°52'W
Glenelg, r., Austl.	176	37°20'S	141°30'E
Glen Ellyn, Il., U.S. (glĕn ĕl'-lĕn)	69a	41°53'N	88°04'W
Glen Innes, Austl. (ĭn'ĕs)	175	29°45'S	152°02'E
Glenns Ferry, Id., U.S. (fĕr'ĭ)	72	42°58'N	115°21'W
Glen Olden, Pa., U.S. (ōl'd'n)	68f	39°54'N	75°17'W
Glenomra, La., U.S. (glĕn-mō'rȧ)	81	30°58'N	92°36'W
Glenrock, Wy., U.S. (glĕn'rŏk)	73	42°50'N	105°53'W
Glens Falls, N.Y., U.S. (glĕnz fôlz)	67	43°20'N	73°40'W
Glenshaw, Pa., U.S. (glĕn'shô)	69e	40°33'N	79°57'W
Glen Valley, Can.	74d	49°09'N	122°30'W
Glenview, Il., U.S. (glĕn'vū)	69a	42°04'N	87°48'W
Glenville, Ga., U.S. (glĕn'vĭl)	83	31°55'N	81°56'W
Glenwood, Ia., U.S.	70	41°03'N	95°44'W
Glenwood, Mn., U.S.	70	45°39'N	95°23'W
Glenwood, N.M., U.S.	77	33°19'N	108°52'W
Glenwood Springs, Co., U.S.	77	39°35'N	107°20'W
Glienicke, Ger. (glē'nē-kĕ)	115b	52°38'N	13°19'E
Glinde, Ger. (glĕn'dĕ)	115c	53°32'N	10°13'E
Glittertinden, mtn., Nor.	122	61°39'N	8°33'E
Gliwice, Pol. (gwĭ-wĭt'sĕ)	117	50°18'N	18°40'E
Globe, Az., U.S. (glōb)	64	33°20'N	110°50'W
Głogów, Pol. (gwō'gōōv)	117	51°40'N	16°04'E
Glommen, r., Nor. (glôm'ĕn)	122	60°03'N	11°15'E
Glonn, Ger. (glônn)	115d	47°59'N	11°52'E
Glorieuses, Îles, is., Reu.	187	11°28'S	47°50'E
Glossop, Eng., U.K. (glŏs'ŭp)	114a	53°26'N	1°57'W
Gloster, Ms., U.S. (glŏs'tēr)	82	31°10'N	91°00'W
Gloucester, Eng., U.K. (glŏs'tēr)	117	51°54'N	2°11'W
Gloucester, Ma., U.S.	61a	42°37'N	70°40'W
Gloucester City, N.J., U.S.	68f	39°53'N	75°08'W
Glouster, Oh., U.S. (glŏs'tēr)	66	39°35'N	82°05'W
Glover Island, i., Can. (glŭv'ēr)	61	48°44'N	57°45'W
Gloversville, N.Y., U.S. (glŭv'ērz-vĭl)	67	43°05'N	74°20'W
Glovertown, Can. (glŭv'ēr-toun)	61	48°41'N	54°02'W
Glückstadt, Ger. (glük-shtät)	115c	53°47'N	9°25'E
Glushkovo, Russia (glōsh'kô-vō)	133	51°21'N	34°43'E
Gmünden, Aus. (g'mòn'dĕn)	124	47°57'N	13°47'E
Gniezno, Pol. (g'nyåz'nô)	117	52°32'N	17°34'E
Gnjilane, Serb. (gnyē'lá-nĕ)	131	42°28'N	21°27'E
Goa, state, India (gō'ȧ)	155	15°45'N	74°00'E
Goascorán, Hond. (gō-äs'kō-rän')	90	13°37'N	87°43'W
Goba, Eth. (gō'bȧ)	185	7°17'N	39°58'E
Gobabis, Nmb. (gō-bä'bĭs)	186	22°25'S	18°50'E
Gobi, des., Asia (gō'be)	160	43°29'N	103°15'E
Goble, Or., U.S. (gō'b'l)	74c	46°01'N	122°53'W
Goch, Ger. (gōk)	127c	51°35'N	6°10'E
Godāvari, r., India (gō-dä'vŭ-rē)	155	19°00'N	78°30'E
Goddards Soak, sw., Austl. (gŏd'ȧrdz)	174	31°20'S	123°30'E
Goderich, Can. (gōd'rĭch)	58	43°45'N	81°45'W
Godfrey, Il., U.S. (gŏd'frē)	75e	38°57'N	90°12'W
Godhavn, Grnld. (gōdh'håvn)	49	69°15'N	53°30'W
Gods, r., Can. (gŏdz)	57	55°17'N	93°35'W
Gods Lake, Can.	51	54°40'N	94°09'W
Godthåb, Grnld. (gŏt'hòb)	49	64°10'N	51°32'W
Goéland, Lac au, l., Can.	59	49°47'N	76°41'W
Goffs, Ca., U.S. (gôfs)	76	34°57'N	115°06'W
Gogebic, l., Mi., U.S. (gô-gē'bĭk)	71	46°24'N	89°25'W
Gogebic Range, mts., Mi., U.S.	71	46°37'N	89°48'W
Göggingen, Ger. (gûg'gĕn-gĕn)	115d	48°21'N	10°53'E
Gogland, i., Russia	123	60°04'N	26°55'E
Gogonou, Benin	189	10°50'N	2°50'E
Gogorrón, Mex. (gō-gō-rōn')	88	21°51'N	100°54'W
Goiânia, Braz. (gô-yä'nyä)	101	16°41'S	48°57'W
Goiás, Braz. (gô-yà's)	101	15°57'S	50°10'W
Goiás, state, Braz.	101	16°00'S	48°00'W
Goirle, Neth.	115a	51°31'N	5°06'E
Gökçeada, i., Tur.	131	40°10'N	25°27'E
Göksu, r., Tur. (gûk'sōō')	137	36°40'N	33°30'E
Gol, Nor. (gûl)	122	60°58'N	8°54'E
Golax, Va., U.S. (gō'lăks)	83	36°41'N	80°56'W
Golcar, Eng., U.K. (gŏl'kȧr)	114a	53°38'N	1°52'W
Golconda, Il., U.S. (gŏl-kŏn'dȧ)	79	37°21'N	88°32'W
Gołdap, Pol. (gôl'dăp)	125	54°17'N	22°17'E

PLACE (Pronunciation)	PAGE	LAT.	LONG.
Golden, Can.	55	51°18'N	116°58'W
Golden, Co., U.S.	78	39°44'N	105°15'W
Goldendale, Wa., U.S. (gōl'dĕn-dāl)	72	45°49'N	120°48'W
Golden Gate, strt., Ca., U.S. (gōl'dĕn gāt)	74b	37°48'N	122°32'W
Golden Hinde, mtn., Can. (hīnd)	54	49°40'N	125°45'W
Golden's Bridge, N.Y., U.S.	68a	41°17'N	73°41'W
Golden Valley, Mn., U.S.	75g	44°58'N	93°23'W
Goldfield, Nv., U.S. (gōld'fēld)	76	37°42'N	117°15'W
Gold Hill, mtn., Pan.	86a	9°03'N	79°08'W
Gold Mountain, mtn., Wa., U.S. (gōld)	74a	47°33'N	122°48'W
Goldsboro, N.C., U.S. (gōldz-bûr'ô)	83	35°23'N	77°59'W
Goldthwaite, Tx., U.S. (gōld'thwāt)	80	31°27'N	98°34'W
Goleniów, Pol. (gô-lĕ-nyŭf')	124	53°33'N	14°51'E
Golets-Purpula, Gora, mtn., Russia	135	59°08'N	115°22'E
Golfito, C.R. (gôl-fē'tō)	91	8°40'N	83°12'W
Goliad, Tx., U.S. (gō-lĭ-ăd')	81	28°40'N	97°12'W
Golo, r., Fr.	130	42°28'N	9°18'E
Golo Island, i., Phil. (gō'lō)	169a	13°38'N	120°17'E
Golovchino, Russia (gō-lôf'chĕ-nō)	133	50°34'N	35°52'E
Golyamo Konare, Blg. (gō'lā-mō-kō'nä-rĕ)	131	42°16'N	24°33'E
Golzow, Ger. (gōl'tsōv)	115b	52°17'N	12°36'E
Gombe, Nig.	184	10°19'N	11°02'E
Gomera Island, i., Spain (gō-mä'rä)	184	28°00'N	18°01'W
Gomez Farias, Mex. (gō'māz fä-rē'äs)	80	24°59'N	101°02'W
Gómez Palacio, Mex. (pä-lä'syō)	86	25°35'N	103°30'W
Gonaïves, Haiti (gō-nà-ēv')	87	19°25'N	72°45'W
Gonaïves, Golfe des, b., Haiti (gō-nà-ēv')	93	19°20'N	73°20'W
Gonâve, Île de la, i., Haiti (gō-nàv')	87	18°50'N	73°30'W
Gonda, India	158	27°13'N	82°00'E
Gondal, India	158	22°02'N	70°47'E
Gonder, Eth.	185	12°39'N	37°30'E
Gonesse, Fr. (gô-nĕs')	127b	48°59'N	2°28'E
Gongga Shan, mtn., China (gōŋ-gä shän)	160	29°16'N	101°46'E
Goniri, Nig.	189	11°30'N	12°20'E
Gonō, r., Japan (gō'nō)	167	35°00'N	132°25'E
Gonor, Can. (gō'nŏr)	62f	50°04'N	96°57'W
Gonubie, S. Afr. (gŏn'ōō-bē)	187c	32°56'S	28°02'E
Gonzales, Mex. (gôn-zá'lĕs)	88	22°47'N	98°26'W
Gonzales, Tx., U.S. (gŏn-zä'lĕz)	81	29°31'N	97°25'W
González Catán, Arg. (gōn-zá'lĕz-kä-tà'n)	102a	34°47'S	58°39'W
Good Hope, Cape of, c., S. Afr. (kāp ov gŏŏd hōp)	186	34°21'S	18°29'E
Good Hope Mountain, mtn., Can.	54	51°09'N	124°10'W
Gooding, Id., U.S. (gòd'ĭng)	73	42°55'N	114°43'W
Goodland, In., U.S. (gòd'lănd)	66	40°50'N	87°15'W
Goodland, Ks., U.S.	78	39°19'N	101°43'W
Goodwood, S. Afr. (gòd'wòd)	186a	33°54'S	18°33'E
Goole, Eng., U.K. (gōōl)	114a	53°42'N	0°52'W
Goose, r., N.D., U.S.	70	47°40'N	97°41'W
Gooseberry Creek, r., Wy., U.S. (gōōs-bĕr'ĭ)	73	44°04'N	108°35'W
Goose Creek, r., Id., U.S. (gōōs)	73	42°07'N	113°53'W
Goose Lake, l., Ca., U.S.	72	41°56'N	120°35'W
Gorakhpur, India (gō'rŭk-pōōr)	155	26°45'N	82°39'E
Gorda, Punta, c., Cuba (pōō'n-tä-gôr-dä)	92	22°25'N	82°10'W
Gorda Cay, i., Bah. (gôr'dä)	92	26°05'N	77°30'W
Gordon, Can. (gôr'dŭn)	62f	50°00'N	97°20'W
Gordon, Ne., U.S.	70	42°47'N	102°14'W
Gore, Eth. (gō'rĕ)	185	8°12'N	35°34'E
Gorgān, Iran	154	36°44'N	54°30'E
Gorgona, Isola di, Italy (gôr-gō'nä)	118	43°27'N	9°55'E
Gori, Geor. (gō'rĕ)	137	42°00'N	44°08'E
Gorinchem, Neth. (gō'rĭn-kĕm)	115a	51°50'N	4°59'E
Goring, Eng., U.K. (gôr'ĭng)	114b	51°30'N	1°08'W
Gorizia, Italy (gō-rē'tsē-yä)	130	45°56'N	13°40'E
Gor'kiy *see* Nizhniy Novgorod, Russia	134	56°15'N	44°05'E
Gor'kovskoye, res., Russia	134	56°38'N	43°40'E
Gorlice, Pol. (gôr-lē'tsĕ)	125	49°38'N	21°11'E
Görlitz, Ger. (gûr'lĭts)	117	51°10'N	15°01'E
Gorman, Tx., U.S. (gôr'măn)	80	32°13'N	98°40'W
Gorna Oryakhovitsa, Blg. (gôr'nä-ôr-yĕk'ô-vē-tsá)	131	43°08'N	25°40'E
Gornji Milanovac, Serb. (gôrn'yē-mē'la-nô-väts)	131	44°02'N	20°29'E
Gorno-Altay, prov., Russia	140	51°00'N	86°00'E
Gorno-Altaysk, Russia (gôr'nŭ'ŭl-tīsk')	134	51°58'N	85°58'E
Gorodishche, Russia (gô-rô'dĭsh-chĕ)	142a	57°57'N	57°03'E
Gorodok, Russia	135	50°30'N	103°58'E
Gorontalo, Indon. (gō-rōn-tä'lo)	169	0°40'N	123°04'E
Gorzów Wielkopolski, Pol. (gō-zhōōv' vyĕl-ko-pōl'skē)	116	53°44'N	15°15'E
Gosely, Eng., U.K.	114a	52°33'N	2°10'W
Goshen, In., U.S. (gō'shĕn)	66	41°35'N	85°50'W
Goshen, Ky., U.S.	69h	38°24'N	85°34'W
Goshen, N.Y., U.S.	68a	41°24'N	74°19'W
Goshen, Oh., U.S.	69f	39°14'N	84°09'W
Goshute Indian Reservation, I.R., Ut., U.S. (gō-shōōt')	77	39°50'N	114°00'W
Goslar, Ger. (gôs'lär)	124	51°55'N	10°25'E
Gospa, r., Ven. (gôs-pä)	101b	9°43'N	64°23'W
Gostivar, Mac. (gos'tē-vär)	131	41°46'N	20°58'E
Gostynin, Pol. (gôs-tē'nĭn)	125	52°24'N	19°30'E
Göta, r., Swe. (gōē'tä)	122	58°11'N	12°03'E
Göta Kanal, can., Swe. (yû'tȧ)	122	58°35'N	15°24'E
Göteborg, Swe. (yû'tĕ-bôrgh)	110	57°39'N	11°56'E
Gotel Mountains, mts., Afr.	189	7°05'N	11°20'E
Gotera, El Sal. (gō-tä'rä)	90	13°41'N	88°06'W
Gotha, Ger. (gō'tȧ)	117	50°47'N	10°43'E

PLACE (Pronunciation)	PAGE	LAT.	LONG.
Gothenburg *see* Göteborg, Swe.	110	57°39'N	11°56'E
Gothenburg, Ne., U.S. (gŏth'ĕn-bûrg)	78	40°57'N	100°08'W
Gotland, i., Swe.	112	57°35'N	17°35'E
Gotska Sandön, i., Swe.	123	58°24'N	19°15'E
Göttingen, Ger. (gŭt'ĭng-ĕn)	124	51°32'N	9°57'E
Gouda, Neth. (gou'dä)	115a	52°00'N	4°42'E
Gough, i., St. Hel. (gŏf)	2	40°00'S	10°00'W
Gouin, Réservoir, res., Can.	53	48°15'N	74°15'W
Goukou, China (gō-kō)	161	48°45'N	121°42'E
Goulais, r., Can.	58	46°45'N	84°10'W
Goulburn, Austl. (gōl'bŭrn)	175	34°47'S	149°40'E
Goumbati, mtn., Sen.	188	13°08'N	12°06'W
Goumbou, Mali (gōō-bōō')	184	14°59'N	7°27'W
Gouna, Cam.	189	8°32'N	13°34'E
Goundam, Mali (gōōn-däN')	184	16°29'N	3°37'W
Gouverneur, N.Y., U.S. (gŭv-ēr-nōōr')	67	44°20'N	75°25'W
Govenlock, Can. (gŭvĕn-lŏk)	50	49°15'N	109°48'W
Governador, Ilha do, i., Braz. (gô-vĕr-nä-dō-'r-ē-lá'dō)	102b	22°48'S	43°13'W
Governador Portela, Braz. (pōr-tĕ'là)	102b	22°28'S	43°30'W
Governador Valadares, Braz. (vä-lä-dä'rĕs)	101	18°47'S	41°45'W
Governor's Harbour, Bah.	92	25°15'N	76°15'W
Gowanda, N.Y., U.S. (gô-wŏn'dȧ)	67	42°30'N	78°55'W
Goya, Arg. (gō'yä)	102	29°06'S	59°12'W
Göyçay, Azer. (gĕ-ôk'chī)	137	40°40'N	47°40'E
Goyt, r., Eng., U.K. (goit)	114a	53°19'N	2°03'W
Graaff-Reinet, S. Afr. (gräf'rī'nĕt)	186	32°10'S	24°40'E
Gračac, Cro. (grä'chäts)	130	44°16'N	15°50'E
Gračanica, Bos.	131	44°42'N	18°18'E
Graceville, Fl., U.S. (grās'vĭl)	82	30°57'N	85°30'W
Graceville, Mn., U.S.	70	45°33'N	96°25'W
Gracias, Hond. (grä'sē-äs)	90	14°35'N	88°37'W
Graciosa Island, i., Port. (grä-syō'sä)	184a	39°07'N	27°30'W
Gradačac, Bos. (gra-dä'chats)	119	44°50'N	18°28'E
Grado, Spain (grä'dō)	128	43°24'N	6°04'W
Gräfelfing, Ger. (grä'fĕl-fēng)	115d	48°07'N	11°27'E
Grafing bei München, Ger. (grä'fēng)	115d	48°03'N	11°58'E
Grafton, Austl. (grȧf'tŭn)	175	29°38'S	153°05'E
Grafton, Il., U.S.	75e	38°58'N	90°26'W
Grafton, Ma., U.S.	61a	42°13'N	71°41'W
Grafton, N.D., U.S.	70	48°24'N	97°25'W
Grafton, Oh., U.S.	69d	41°16'N	82°04'W
Grafton, W.V., U.S.	66	39°20'N	80°00'W
Gragnano, Italy (grän-yä'nô)	129c	40°27'N	14°32'E
Graham, N.C., U.S. (grā'ăm)	83	36°03'N	79°23'W
Graham, Tx., U.S.	78	33°07'N	98°34'W
Graham, Wa., U.S.	74a	47°03'N	122°18'W
Graham, i., Can.	52	53°50'N	132°40'W
Grahamstown, S. Afr. (grā'ăms'toun)	187c	33°19'S	26°33'E
Grajewo, Pol. (grä-yä'vo)	125	53°38'N	22°28'E
Grama, Serra de, mtn., Braz. (sĕ'r-rä-dĕ-grá'mà)	99a	20°42'S	42°28'W
Gramada, Blg. (grä'mä-dä)	131	43°46'N	22°41'E
Gramatneusiedl, Aus.	115e	48°02'N	16°29'E
Grampian Mountains, mts., Scot., U.K. (grăm'pĭ-ăn)	112	56°30'N	4°55'W
Granada, Nic. (grä-nä'dhä)	86	11°55'N	85°58'W
Granada, Spain (grä-nä'dä)	118	37°13'N	3°37'W
Gran Bajo, reg., Arg. (grán'bä'kō)	102	47°35'S	68°45'W
Granbury, Tx., U.S. (grăn'bĕr-ĭ)	81	32°26'N	97°45'W
Granby, Can. (grăn'bĭ)	51	45°30'N	72°40'W
Granby, Mo., U.S.	79	36°54'N	94°15'W
Granby, l., Co., U.S.	78	40°07'N	105°40'W
Gran Canaria Island, i., Spain (grän'kä-nä'rē-ä)	184	27°39'N	15°39'W
Gran Chaco, reg., S.A. (grän'chá'kō)	102	25°30'S	62°15'W
Grand, i., Mi., U.S.	71	46°37'N	86°38'W
Grand, l., Can.	60	45°59'N	66°15'W
Grand, l., Me., U.S.	60	45°17'N	67°42'W
Grand, r., Can.	59	43°45'N	80°20'W
Grand, r., Mi., U.S.	66	42°58'N	85°13'W
Grand, r., Mo., U.S.	79	39°50'N	93°52'W
Grand, r., S.D., U.S.	70	45°40'N	101°55'W
Grand, North Fork, r., U.S.	70	45°52'N	102°49'W
Grand, South Fork, r., S.D., U.S.	70	45°38'N	102°56'W
Grand Bahama, i., Bah.	87	26°35'N	78°30'W
Grand Bank, Can. (grănd băngk)	53a	47°06'N	55°47'W
Grand Bassam, C. Iv. (grän bá-säN')	184	5°12'N	3°44'W
Grand Bourg, Guad. (grän bōōr')	91b	15°54'N	61°20'W
Grand Caicos, i., T./C. Is. (grănd kä-ē'kōs)	93	21°45'N	71°50'W
Grand Canal *see* Da Yunhe, can., China	161	35°00'N	117°00'E
Grand Canal, can., Ire.	120	53°21'N	7°15'W
Grand Canyon, Az., U.S.	77	36°05'N	112°10'W
Grand Canyon, p., Az., U.S.	64	35°50'N	113°16'W
Grand Canyon National Park, rec., Az., U.S.	64	36°15'N	112°20'W
Grand Canyon-Parashant National Monument, rec., Az., U.S.	77	36°25'N	113°45'W
Grand Cayman, i., Cay. Is. (kā'măn)	87	19°15'N	81°15'W
Grand Coulee Dam, dam, Wa., U.S. (kōō'lē)	64	47°58'N	119°28'W
Grande, r., Arg.	99b	35°25'S	70°14'W
Grande, r., Bol.	100	16°49'S	63°19'W
Grande, r., Braz.	101	19°48'S	49°54'W
Grande, r., Mex.	89	17°37'N	96°41'W
Grande, r., Nic. (grän'dĕ)	91	13°01'N	84°21'W
Grande, r., Ur.	99c	33°19'S	57°15'W
Grande, Arroyo, r., Mex. (är-rō'yō-grä'n-dĕ)	88	23°30'N	98°45'W
Grande, Bahía, b., Arg. (bä-ē'ä-grän'dĕ)	102	50°45'S	68°00'W
Grande, Boca, mth., Ven. (bō'kä-grä'n-dĕ)	101	8°46'N	60°17'W

252 Grande – Gros Morne PRONOUNCING INDEX

PLACE (Pronunciation)	PAGE	LAT.	LONG.
Grande, Cuchilla, mts., Ur. (kōō-chē´l-yä)	102	33°00′S	55°15′W
Grande, Ilha, i., Braz. (grän´dē)	99a	23°11′S	44°14′W
Grande, Rio, r., N.A. (grän´dä)	64	26°50′N	99°10′W
Grande, Salinas, l., Arg. (sä-lē´näs)	102	29°45′S	65°00′W
Grande, Salto, wtfl., Braz. (säl-tô)	101	16°18′S	39°38′W
Grande Cayemite, Île, i., Haiti	93	18°45′N	73°45′W
Grande de Otoro, r., Hond. (grä´dą dā ô-tō´rô)	90	14°42′N	88°21′W
Grande de Santiago, Río, r., Mex. (rē̂ō-grä´n-dĕ-dĕ-sän-tyá´gô)	86	20°30′N	104°00′W
Grande Pointe, Can. (grănd point´)	62f	49°47′N	97°03′W
Grande Prairie, Can. (prâr´ĭ)	50	55°10′N	118°48′W
Grand Erg Occidental, des., Alg.	184	30°00′N	1°00′E
Grand Erg Oriental, des., Alg.	184	30°00′N	7°00′E
Grande Rivière du Nord, Haiti (rē-vyâr´ dü nôr´)	93	19°35′N	72°10′W
Grande Ronde, r., Or., U.S. (rŏnd´)	72	45°32′N	117°52′W
Gran Desierto, des., Mex. (grän-dē-syĕ´r-tô)	77	32°14′N	114°28′W
Grande Terre, i., Guad.	91b	16°28′N	61°13′W
Grande Vigie, Pointe de la, c., Guad. (gräɴd vē-gē´)	91b	16°32′N	61°25′W
Grand Falls, Can. (fôlz)	53a	48°56′N	55°40′W
Grandfather Mountain, mtn., N.C., U.S. (grănd-fä-thẽr)	83	36°07′N	81°48′W
Grandfield, Ok., U.S. (grănd´fēld)	78	34°13′N	98°39′W
Grand Forks, Can. (fôrks)	50	49°02′N	118°27′W
Grand Forks, N.D., U.S.	64	47°55′N	97°05′W
Grand Haven, Mi., U.S. (hā´v´n)	66	43°05′N	86°15′W
Grand Island, Ne., U.S. (ī´lănd)	64	40°56′N	98°20′W
Grand Island, i., N.Y., U.S.	69c	43°03′N	78°58′W
Grand Junction, Co., U.S. (jŭngk´shŭn)	64	39°05′N	108°35′W
Grand Lake, l., Can. (lāk)	53a	49°00′N	57°10′W
Grand Lake, l., La., U.S.	81	29°57′N	91°25′W
Grand Lake, l., Mn., U.S.	75h	46°54′N	92°26′W
Grand Ledge, Mi., U.S. (lĕj)	66	42°45′N	84°50′W
Grand Lieu, Lac de, l., Fr. (gräɴ´-lyû)	126	47°00′N	1°45′W
Grand Manan, i., Can. (mă-năn)	60	44°40′N	66°50′W
Grand Mère, Can. (gräɴ mâr´)	51	46°36′N	72°43′W
Grândola, Port. (grăɴ´dô-lă)	128	38°10′N	8°36′W
Grand Portage Indian Reservation, I.R., Mn., U.S. (pōr´tĭj)	71	47°54′N	89°34′W
Grand Portage National Monument, rec., Mi., U.S.	71	47°59′N	89°47′W
Grand Prairie, Tx., U.S. (prē´rē)	75c	32°45′N	97°00′W
Grand Rapids, Can.	57	53°08′N	99°20′W
Grand Rapids, Mi., U.S. (răp´ĭdz)	65	43°00′N	85°45′W
Grand Rapids, Mn., U.S.	71	47°16′N	93°33′W
Grand-Riviere, Can.	60	48°26′N	64°30′W
Grand Staircase-Escalante National Monument, rec., Ut., U.S.	77	37°25′N	111°30′W
Grand Teton, mtn., Wy., U.S.	64	43°46′N	110°50′W
Grand Teton National Park, rec., Wy., U.S. (tē´tŏn)	73	43°54′N	110°15′W
Grand Traverse Bay, b., Mi., U.S. (trăv´ẽrs)	66	45°00′N	85°30′W
Grand Turk, T./C. Is. (tûrk)	93	21°30′N	71°10′W
Grand Turk, i., T./C. Is.	93	21°30′N	71°10′W
Grandview, Mo., U.S. (grănd´vyōō)	75f	38°53′N	94°32′W
Granger, Wy., U.S. (grān´jẽr)	73	41°37′N	109°58′W
Grangeville, Id., U.S. (grānj´vĭl)	72	45°56′N	116°08′W
Granite City, Il., U.S. (grăn´ĭt sĭt´ĭ)	75e	38°42′N	90°09′W
Granite Falls, Mn., U.S. (fôlz)	70	44°46′N	95°34′W
Granite Falls, N.C., U.S.	83	35°49′N	81°25′W
Granite Falls, Wa., U.S.	74a	48°05′N	121°59′W
Granite Lake, l., Can.	61	48°01′N	57°00′W
Granite Peak, mtn., Mt., U.S.	64	45°13′N	109°48′W
Graniteville, S.C., U.S. (grăn´ĭt-vĭl)	83	33°35′N	81°50′W
Granito, Braz. (grä-nē´tô)	101	7°39′S	39°34′W
Granma, prov., Cuba	92	20°10′N	76°50′W
Gränna, Swe. (grĕn´à)	122	58°02′N	14°38′E
Granollers, Spain (grä-nôl-yĕrs´)	129	41°36′N	2°19′E
Gran Pajonal, reg., Peru (grä´n-pä-κô-näl´)	100	11°14′S	71°45′W
Gran Paradiso, mtn., Italy	130	45°32′N	7°16′E
Gran Piedra, mtn., Cuba (grän-pyĕ´drä)	93	20°00′N	75°40′W
Grantham, Eng., U.K. (grăn´tăm)	120	52°54′N	0°38′W
Grant Park, Il., U.S. (grănt pärk)	69a	41°14′N	87°39′W
Grants Pass, Or., U.S. (grănts pȧs)	72	42°26′N	123°20′W
Granville, Fr. (gräɴ-vēl´)	117	48°52′N	1°35′W
Granville, N.Y., U.S. (grăn´vĭl)	67	43°25′N	73°15′W
Granville, l., Can.	52	56°18′N	100°30′W
Grão Mogol, Braz. (groun´ mô-gôl´)	101	16°34′S	42°35′W
Grapevine, Tx., U.S. (grăp´vīn)	75c	32°56′N	97°05′W
Gräso, i., Swe.	122	60°30′N	18°35′E
Grass, r., N.Y., U.S.	67	44°45′N	75°10′W
Grass Cay, i., V.I.U.S.	87c	18°22′N	64°50′W
Grasse, Fr. (gräs)	127	43°39′N	6°57′E
Grass Mountain, mtn., Wa., U.S. (grȧs)	74a	47°13′N	121°48′W
Grates Point, c., Can. (grāts)	61	48°09′N	52°57′W
Gravelbourg, Can. (grăv´ĕl-bôrg)	50	49°53′N	106°34′W
Gravesend, Eng., U.K. (grāvz´ĕnd´)	114b	51°26′N	0°22′E
Gravina, Italy (grä-vē´nä)	130	40°48′N	16°27′E
Gravois, Pointe à, c., Haiti (grȧ-vwä´)	93	18°00′N	74°20′W
Gray, Fr. (grå)	127	47°26′N	5°35′E
Grayling, Mi., U.S. (grā´lĭng)	66	44°40′N	84°40′W
Grays Harbor, b., Wa., U.S. (grās)	64	46°55′N	124°23′W
Grayslake, Il., U.S. (grāz´lāk)	69a	42°20′N	88°20′W
Grays Peak, mtn., Co., U.S. (grāz)	78	39°29′N	105°52′W
Grays Thurrock, Eng., U.K. (thŭ´rŏk)	114b	51°28′N	0°19′E
Grayvoron, Russia (grá-ē´vô-rôn)	133	50°28′N	35°41′E
Graz, Aus. (gräts)	110	47°05′N	15°26′E
Great Abaco, i., Bah. (ä´bä-kō)	87	26°30′N	77°05′W
Great Artesian Basin, basin, Austl. (är-tēzh-ăn bā-sĭn)	175	23°16′S	143°37′E
Great Australian Bight, b., Austl. (ôs-trā´lĭ-ăn bĭt)	174	33°30′S	127°00′E
Great Bahama Bank, bk., (bȧ-hā´mȧ)	92	25°00′N	78°50′W
Great Barrier, i., N.Z. (băr´ĭ-ẽr)	175a	36°10′S	175°30′E
Great Barrier Reef, rf., Austl. (bȧ-rĭ-ẽr rēf)	175	16°43′S	146°34′E
Great Basin, basin, U.S. (grāt bā´s´n)	64	40°08′N	117°10′W
Great Bear Lake, l., Can. (bâr)	52	66°10′N	119°53′W
Great Bend, Ks., U.S. (bĕnd)	78	38°41′N	98°46′W
Great Bitter Lake, l., Egypt	192b	30°24′N	32°27′E
Great Blasket Island, i., Ire. (bläs´kĕt)	120	52°05′N	10°55′W
Great Corn Island, i., Nic.	91	12°10′N	82°54′W
Great Dismal Swamp, sw., U.S. (dĭz´măl)	83	36°35′N	76°34′W
Great Divide Basin, basin, Wy., U.S. (dĭ-vīd´ bā´s´n)	73	42°10′N	108°10′W
Great Dividing Range, mts., Austl. (dĭ-vī-dĭng rānj)	175	35°16′S	146°38′E
Great Duck, i., Can. (dŭk)	58	45°40′N	83°22′W
Greater Antilles, is., N.A.	87	20°30′N	79°15′W
Greater Khingan Range, mts., China (dä hĭŋ-gän lĭŋ)	161	46°30′N	120°00′E
Greater Leech Indian Reservation, I.R., Mn., U.S. (grāt´ẽr lēch)	71	47°39′N	94°27′W
Greater Manchester, hist. reg., Eng., U.K.	114a	53°34′N	2°41′W
Greater Sunda Islands, is., Asia	168	4°00′S	108°00′E
Great Exuma, i., Bah. (ĕk-sōō´mä)	92	23°35′N	76°00′W
Great Falls, Mt., U.S. (fôlz)	64	47°30′N	111°15′W
Great Falls, S.C., U.S.	83	34°32′N	80°53′W
Great Guana Cay, i., Bah. (gwä´nä)	92	24°00′N	76°20′W
Great Harbor Cay, i., Bah. (kē)	92	25°45′N	77°50′W
Great Inagua, i., Bah. (ē-nä´gwä)	87	21°00′N	73°15′W
Great Indian Desert, des., Asia	155	27°35′N	71°37′E
Great Isaac, i., Bah. (ī´zăk)	92	26°05′N	79°05′W
Great Karroo, plat., S. Afr. (grāt kă´rōō)	186	32°45′S	22°00′E
Great Limpopo Transfrontier Park, rec., Afr.	186	22°00′S	31°30′E
Great Namaland, hist. reg., Nmb.	186	25°45′S	16°15′E
Great Neck, N.Y., U.S. (nĕk)	68a	40°48′N	73°44′W
Great Nicobar Island, i., India (nĭk-ô-bär´)	168	7°00′N	94°18′E
Great Pedro Bluff, c., Jam.	92	17°50′N	78°05′W
Great Pee Dee, r., S.C., U.S. (pē-dē´)	65	34°01′N	79°26′W
Great Plains, pl., N.A. (plānz)	49	45°00′N	104°00′W
Great Ragged, i., Bah.	93	22°10′N	75°45′W
Great Ruaha, r., Tan.	186	7°30′S	37°00′E
Great Salt Lake, l., Ut., U.S. (sôlt lāk)	64	41°19′N	112°48′W
Great Salt Lake Desert, des., Ut., U.S.	64	41°00′N	113°30′W
Great Salt Plains Reservoir, res., Ok., U.S.	78	36°56′N	98°14′W
Great Sand Dunes National Monument, rec., Co., U.S.	78	37°56′N	105°25′W
Great Sand Hills, hills, Can. (sănd)	56	50°35′N	109°05′W
Great Sandy Desert, des., Austl. (săn´dē)	174	21°50′S	123°10′E
Great Sandy Desert, des., Or., U.S. (săn´dĭ)	72	43°43′N	120°44′W
Great Sitkin, i., Ak., U.S. (sĭt-kĭn)	63a	52°18′N	176°22′W
Great Slave Lake, l., Can. (slāv)	52	61°37′N	114°58′W
Great Smoky Mountains National Park, rec., U.S. (smōk-ē)	65	35°43′N	83°20′W
Great Stirrup Cay, i., Bah. (stĭr-ŭp)	92	25°50′N	77°55′W
Great Victoria Desert, des., Austl. (vĭk-tō´rĭ-ȧ)	174	29°45′S	124°30′E
Great Wall, hist., China	160	38°00′N	109°00′E
Great Waltham, Eng., U.K. (wôl´thŭm)	114b	51°47′N	0°27′E
Great Yarmouth, Eng., U.K. (yär-mŭth)	117	52°35′N	1°45′E
Grebbestad, Swe. (grĕb-bĕ-städh)	122	58°42′N	11°15′E
Gréboun, Mont, mtn., Niger	184	20°00′N	8°35′E
Gredos, Sierra de, mts., Spain (syĕr´rä dā grā´dōs)	128	40°13′N	5°30′W
Greece, nation, Eur. (grēs)	110	39°00′N	21°30′E
Greeley, Co., U.S. (grē´lĭ)	64	40°25′N	104°41′W
Green, r., Ky., U.S. (grēn)	82	37°13′N	86°30′W
Green, r., N.D., U.S.	70	47°05′N	103°05′W
Green, r., Ut., U.S.	77	38°30′N	110°05′W
Green, r., Wa., U.S.	74a	47°17′N	121°57′W
Green, r., Wy., U.S.	73	41°08′N	110°27′W
Green, r., U.S.	64	38°30′N	110°10′W
Greenbank, Wa., U.S. (grēn´bănk)	74a	48°06′N	122°35′W
Green Bay, Wi., U.S.	65	44°30′N	88°04′W
Green Bay, b., U.S.	65	44°55′N	87°40′W
Green Bayou, Tx., U.S.	81a	29°53′N	95°13′W
Greenbelt, Md., U.S. (grēn´bĕlt)	68e	38°59′N	76°53′W
Greencastle, In., U.S. (grēn-kȧs´´l)	66	39°40′N	86°50′W
Green Cay, i., Bah.	92	24°05′N	77°10′W
Green Cove Springs, Fl., U.S. (kōv)	83	29°56′N	81°42′W
Greendale, Wi., U.S. (grēn´dāl)	69a	42°56′N	87°59′W
Greenfield, Ia., U.S.	71	41°16′N	94°30′W
Greenfield, In., U.S. (grēn´fēld)	66	39°45′N	85°40′W
Greenfield, Ma., U.S.	67	42°35′N	72°35′W
Greenfield, Mo., U.S.	79	37°23′N	93°48′W
Greenfield, Oh., U.S.	66	39°15′N	83°25′W
Greenfield, Tn., U.S.	82	36°08′N	88°45′W
Greenfield Park, Can.	62a	45°29′N	73°29′W
Greenhills, Oh., U.S. (grēn-hĭls)	69f	39°16′N	84°31′W
Greenland, dep., N.A. (grēn´lănd)	49	74°00′N	40°00′W
Greenland Sea, sea,	198	77°00′N	1°00′W
Green Mountain, mtn., Or., U.S.	74c	45°52′N	123°24′W
Green Mountain Reservoir, res., Co., U.S.	77	39°50′N	106°20′W
Green Mountains, mts., N.A.	65	43°10′N	73°05′W
Greenock, Scot., U.K. (grēn´ŭk)	116	55°55′N	4°45′W
Green Peter Lake, res., Or., U.S.	72	44°28′N	122°30′W
Green Pond Mountain, mtn., N.J., U.S. (pŏnd)	68a	41°00′N	74°32′W
Greenport, N.Y., U.S.	67	41°06′N	72°22′W
Green River, Ut., U.S. (grēn rĭv´ẽr)	77	39°00′N	110°05′W
Green River, Wy., U.S.	73	41°32′N	109°26′W
Green River Lake, res., Ky., U.S.	82	37°15′N	85°15′W
Greensboro, Al., U.S. (grēnz´bŭro)	82	32°42′N	87°36′W
Greensboro, Ga., U.S. (grēns-bûr´ô)	82	33°34′N	83°11′W
Greensboro, N.C., U.S.	65	36°04′N	79°45′W
Greensburg, In., U.S. (grēnz´bûrg)	66	39°20′N	85°30′W
Greensburg, Ks., U.S. (grēns-bûrg)	78	37°36′N	99°17′W
Greensburg, Pa., U.S.	67	40°20′N	79°30′W
Greenville, Lib.	184	5°01′N	9°03′W
Greenville, Al., U.S. (grēn´vĭl)	82	31°49′N	86°39′W
Greenville, Il., U.S.	79	38°52′N	89°22′W
Greenville, Ky., U.S.	82	37°11′N	87°11′W
Greenville, Me., U.S.	60	45°26′N	69°35′W
Greenville, Mi., U.S.	66	43°10′N	85°25′W
Greenville, Ms., U.S.	65	33°25′N	91°00′W
Greenville, N.C., U.S.	83	35°35′N	77°22′W
Greenville, Oh., U.S.	66	40°05′N	84°35′W
Greenville, Pa., U.S.	66	41°20′N	80°25′W
Greenville, S.C., U.S.	65	34°50′N	82°25′W
Greenville, Tn., U.S.	83	36°08′N	82°50′W
Greenville, Tx., U.S.	81	33°09′N	96°07′W
Greenwich, Eng., U.K.	114b	51°28′N	0°00′
Greenwich, Ct., U.S.	68a	41°01′N	73°37′W
Greenwood, Ar., U.S. (grēn-wŏŏd)	79	35°13′N	94°15′W
Greenwood, In., U.S.	69g	39°37′N	86°07′W
Greenwood, Ms., U.S.	82	33°30′N	90°09′W
Greenwood, S.C., U.S.	83	34°10′N	82°10′W
Greenwood, Lake, res., S.C., U.S.	83	34°17′N	81°55′W
Greenwood Lake, l., N.Y., U.S.	68a	41°13′N	74°20′W
Greer, S.C., U.S. (grēr)	83	34°55′N	81°56′W
Grefrath, Ger. (grĕf´rät)	127c	51°20′N	6°21′E
Gregory, S.D., U.S. (grĕg´ô-rĭ)	70	43°12′N	99°27′W
Gregory, Lake, l., Austl. (grĕg´ô-rē)	174	28°47′S	139°15′E
Gregory Range, mts., Austl.	175	19°23′S	143°45′E
Greifenberg, Ger. (grī´fĕn-bĕrgh)	115d	48°04′N	11°06′E
Greifswald, Ger. (grīfs´vält)	124	54°05′N	13°24′E
Greiz, Ger. (grīts)	124	50°39′N	12°14′E
Gremyachinsk, Russia (grâ´myá-chĭnsk)	142a	58°35′N	57°53′E
Grenada, Ms., U.S. (grē-nä´da)	82	33°45′N	89°47′W
Grenada, nation, N.A.	87	12°02′N	61°15′W
Grenada Lake, res., Ms., U.S.	82	33°52′N	89°30′W
Grenadines, The, is., N.A. (grĕn´ȧ-dēnz)	91b	12°37′N	61°35′W
Grenen, c., Den.	116	57°43′N	10°31′E
Grenoble, Fr. (grẽ-nô´bl´)	117	45°14′N	5°45′E
Grenora, N.D., U.S. (grē-nō´rȧ)	70	48°38′N	103°55′W
Grenville, Can. (grĕn´vĭl)	67	45°40′N	74°35′W
Grenville, Gren.	91b	12°07′N	61°38′W
Gresham, Or., U.S. (grĕsh´ăm)	74c	45°30′N	122°25′W
Gretna, La., U.S. (grĕt´nȧ)	68d	29°56′N	90°03′W
Grevelingen Krammer, r., Neth.	115a	51°42′N	4°03′E
Grevenbroich, Ger. (grĕ´fĕn-broik)	127c	51°05′N	6°36′E
Grey, r., Can. (grā)	61	47°53′N	57°00′W
Grey, Point, c., Can.	74d	49°22′N	123°16′W
Greybull, Wy., U.S. (grā´bŏŏl)	73	44°28′N	108°05′W
Greybull, r., Wy., U.S.	73	44°13′N	108°43′W
Greylingstad, S. Afr. (grā-lĭng´shtät)	192c	26°40′S	29°13′E
Greymouth, N.Z. (grā´mouth)	175a	42°27′S	171°17′E
Grey Range, mts., Austl.	175	28°40′S	142°05′E
Greytown, S. Afr. (grā´toun)	187c	29°07′S	30°38′E
Grey Wolf Peak, mtn., Wa., U.S. (grā wŏŏlf)	74a	48°53′N	123°12′W
Gridley, Ca., U.S. (grĭd´lĭ)	76	39°22′N	121°43′W
Griffin, Ga., U.S. (grĭf´ĭn)	82	33°15′N	84°16′W
Griffith, Austl. (grĭf-ĭth)	176	34°16′S	146°10′E
Griffith, In., U.S.	69a	41°31′N	87°26′W
Grigoriopol', Mol. (grĭ´gor-i-ŏ´pôl)	133	47°09′N	29°18′E
Grijalva, r., Mex. (grē-häl´vä)	89	17°25′N	93°23′W
Grim, Cape, c., Austl. (grĭm)	176	40°43′S	144°30′E
Grimma, Ger. (grĭm´ä)	124	51°14′N	12°43′E
Grimsby, Can. (grĭmz´bĭ)	62d	43°11′N	79°33′W
Grimsby, Eng., U.K.	116	53°35′N	0°05′W
Grímsey, i., Ice. (grĭms´ȧ)	116	66°30′N	17°50′W
Grimstad, Nor. (grĭm-städh)	116	58°21′N	8°30′E
Grindstone Island, Can.	61	47°25′N	61°51′W
Grinnel, Ia., U.S. (grĭ-nĕl´)	71	41°44′N	92°44′W
Griswold, Ia., U.S. (grĭz´wŭld)	70	41°11′N	95°05′W
Groais Island, i., Can.	61	50°57′N	55°35′W
Grobina, Lat. (grō´bĭnĭa)	123	56°35′N	21°10′E
Groblersdal, S. Afr.	192c	25°11′S	29°25′E
Grodzisk, Pol. (grō´jēsk)	124	52°14′N	16°22′E
Grodzisk Masowiecki, Pol. (grō´jēsk mä-zō-vyĕts´ke)	125	52°06′N	20°40′E
Groesbeck, Tx., U.S. (grōs´bĕk)	81	31°32′N	96°31′W
Groix, Île de, i., Fr. (ēl dē grwä´)	126	47°39′N	3°28′W
Grójec, Pol. (grō´yĕts)	125	51°53′N	20°52′E
Gronau, Ger. (grō´nou)	124	52°12′N	7°05′E
Groningen, Neth. (grō´nĭng-ĕn)	116	53°13′N	6°30′E
Groote Eylandt, i., Austl. (grō´tē ī´länt)	174	13°50′S	137°30′E
Grootfontein, Nmb. (grōt´fōn-tān´)	186	19°30′S	18°15′E
Groot-Kei, r., Afr. (kē)	187c	32°17′S	27°30′E
Grootkop, mtn., S. Afr.	186a	34°11′S	18°23′E
Groot Marico, S. Afr.	192c	25°36′S	26°23′E
Groot Marico, r., Afr.	192c	25°13′S	26°20′E
Groot-Vis, r., S. Afr.	187c	33°04′S	26°08′E
Groot Vloer, pl., S. Afr. (grōt´ vlōr´)	186	30°00′S	21°00′E
Gros-Mécatina, i., Can.	61	50°50′N	58°33′W
Gros Morne, mtn., Can. (grō môrn´)	61	49°36′N	57°48′W

ăt; finăl; rāte; senâte; ärm; åsk; sofȧ; fâre; ch-choose; dh-as th in other; bē; ēvent; bĕt; recĕnt; cratẽr; g-gō; gh-guttural g; bĭt; ĭ-short neutral; rīde; κ-guttural k as ch in German ich;

PRONOUNCING INDEX

Gros Morne National Park – Hadlock 253

PLACE (Pronunciation)	PAGE	LAT.	LONG.
Gros Morne National Park, rec., Can.	53a	49°45'N	59°15'W
Gros Pate, mtn., Can.	61	50°16'N	57°25'W
Grosse Island, i., Mi., U.S. (grōs)	69b	42°08'N	83°09'W
Grosse Isle, Can. (īl')	62f	50°04'N	97°27'W
Grossenhain, Ger. (grōs'ĕn-hīn)	124	51°17'N	13°33'E
Gross-Enzersdorf, Aus.	115e	48°13'N	16°33'E
Grosse Pointe, Mi., U.S. (point')	69b	42°23'N	82°54'W
Grosse Pointe Farms, Mi., U.S. (färm)	69b	42°25'N	82°53'W
Grosse Pointe Park, Mi., U.S. (pärk)	69b	42°23'N	82°55'W
Grosseto, Italy (grŏs-sā'tō)	130	42°46'N	11°09'E
Grossglockner, mtn., Aus.	117	47°06'N	12°45'E
Gross Höbach, Ger. (hû'bäk)	115d	48°21'N	11°36'E
Gross Kreutz, Ger. (kroitz)	115b	52°24'N	12°47'E
Gross Schönebeck, Ger. (shö'nĕ-bĕk)	115b	52°54'N	13°32'E
Gros Ventre, r., Wy., U.S. (grōvĕn't'r)	73	43°38'N	110°34'W
Groton, Ct., U.S. (grŏt'ŭn)	67	41°20'N	72°00'W
Groton, Ma., U.S.	61a	42°37'N	71°34'W
Groton, S.D., U.S.	70	45°25'N	98°04'W
Grottaglie, Italy (grŏt-täl'yā)	131	40°32'N	17°26'E
Grouard Mission, Can.	50	55°31'N	116°09'W
Groveland, Ma., U.S. (grōv'land)	61a	42°25'N	71°02'W
Groveton, N.H., U.S. (grōv'tŭn)	67	44°35'N	71°30'W
Groveton, Tx., U.S.	81	31°04'N	95°07'W
Groznyy, Russia (grôz'nĭ)	134	43°20'N	45°40'E
Grudziądz, Pol. (grŏ'jyŏNts)	116	53°30'N	18°48'E
Grues, Île aux, i., Can. (ō grü)	62b	47°05'N	70°32'W
Grundy Center, Ia., U.S. (grŭn'dĭ sĕn'tẽr)	71	42°22'N	92°45'W
Gruñidora, Mex. (grōō-nyē-dō'rō)	88	24°10'N	101°49'W
Grünwald, Ger. (grōōn'väld)	115d	48°04'N	11°34'E
Gryazi, Russia (gryä'zĭ)	132	52°31'N	39°59'E
Gryazovets, Russia (gryä'zō-vĕts)	136	58°52'N	40°14'E
Gryfice, Pol. (grĭ'fĭ-tsĕ)	124	53°55'N	15°11'E
Gryfino, Pol. (grĭ'fē-nō)	124	53°16'N	14°30'E
Guabito, Pan. (gwä-bē'tō)	91	9°30'N	82°33'W
Guacanayabo, Golfo de, b., Cuba (gôl-fô-dĕ-gwä-kä-nä-yä'bō)	92	20°30'N	77°40'W
Guacara, Ven. (gwà'kä-rä)	101b	10°16'N	67°48'W
Guadalajara, Mex. (gwä-dhä-lä-hä'rä)	86	20°41'N	103°21'W
Guadalajara, Spain (gwä-dä-lä-kä'rä)	118	40°37'N	3°10'W
Guadalcanal, Spain (gwä-dhäl-kä-näl')	128	38°05'N	5°48'W
Guadalcanal, i., Sol. Is.	175	9°48'S	158°43'E
Guadalcázar, Mex. (gwä-dhäl-kä'zär)	88	22°38'N	100°24'W
Guadalete, r., Spain (gwä-dhä-lā'tå)	128	36°53'N	5°38'W
Guadalhorce, r., Spain (gwä-dhäl-ôr'thā)	128	37°05'N	4°50'W
Guadalimar, r., Spain (gwä-dhä-lē-mär')	128	38°29'N	2°53'W
Guadalope, r., Spain (gwä-dä-lō-pĕ)	129	40°48'N	0°10'W
Guadalquivir, Río, r., Spain (rē'ō-gwä-dhäl-kē-vēr')	112	37°30'N	5°00'W
Guadalupe, Mex.	80	31°23'N	106°06'W
Guadalupe, i., Mex.	86	29°00'N	118°45'W
Guadalupe, r., Tx., U.S. (gwä-dhä-lōō'pā)	80	29°54'N	99°03'W
Guadalupe, Sierra de, mts., Spain (syĕr'rä dā gwä-dhä-lōō'pā)	118	39°30'N	5°25'W
Guadalupe Mountains, mts., N.M., U.S.	80	32°00'N	104°55'W
Guadalupe Peak, mtn., Tx., U.S.	80	31°55'N	104°55'W
Guadarrama, r., Spain (gwä-dhär-rä'mä)	129a	40°34'N	3°58'W
Guadarrama, Sierra de, mts., Spain (gwä-dhär-rä'mä)	112	41°00'N	3°40'W
Guadatentin, r., Spain	128	37°43'N	1°58'W
Guadeloupe, dep., N.A. (gwä-dē-lōōp)	87	16°40'N	61°10'W
Guadeloupe Passage, strt., N.A.	91b	16°26'N	62°00'W
Guadiana, r., Eur. (gwä-dvä'nä)	112	39°00'N	6°00'W
Guadiana, Bahía de, b., Cuba (bä-ē'ä-dĕ-gwä-dhē-ä'nä)	92	22°10'N	84°35'W
Guadiana Alto, r., Spain (äl'tō)	128	39°02'N	2°52'W
Guadiana Menor, r., Spain (mā'nôr)	128	37°43'N	2°45'W
Guadiaro, r., Spain (gwä-dhē-ä'rō)	128	36°38'N	5°25'W
Guadiela, r., Spain (gwä-dhē-ā'lä)	128	40°27'N	2°05'W
Guadix, Spain (gwä-dēsh')	128	37°18'N	3°09'W
Guaira, Braz. (gwä-ē-rä)	101	24°03'S	54°02'W
Guaire, r., Ven. (gwī'rĕ)	101b	10°25'N	66°43'W
Guajaba, Cayo, i., Cuba (kä'yō-gwä-hä'bä)	92	21°50'N	77°35'W
Guajará Mirim, Braz. (gwä-zhä-rä'mē-rēN')	100	10°58'S	65°12'W
Guajira, Península de, pen., S.A.	100	12°35'N	73°00'W
Gualán, Guat. (gwä-län')	90	15°08'N	89°21'W
Gualeguay, Arg. (gwä-lĕ-gwä'y)	102	33°10'S	59°20'W
Gualeguay, r., Arg.	102	32°49'S	59°05'W
Gualicho, Salina, l., Arg. (sä-lē'nä-gwä-lē'chō)	102	40°20'S	65°15'W
Guam, i., Oc. (gwäm)	3	14°00'N	143°20'E
Guamo, Col. (gwà'mō)	100a	4°02'N	74°58'W
Gu'an, China (gōō-än)	164a	39°25'N	116°18'E
Guan, r., China (gůän)	162	31°56'N	115°19'E
Guanabacoa, Cuba (gwä-nä-bä-kō'ä)	87	23°08'N	82°19'W
Guanabara, Baía de, b., Braz.	99a	22°44'S	43°09'W
Guanacaste, Cordillera, mts., C.R.	90	10°54'N	85°27'W
Guanacevi, Mex. (gwä-nä-sĕ-vē')	86	25°30'N	105°45'W
Guanahacabibes, Península de, pen., Cuba	92	21°55'N	84°35'W
Guanajay, Cuba (gwänä-hī')	92	22°55'N	82°40'W
Guanajuato, Mex. (gwä-nä-hwä'tō)	86	21°01'N	101°16'W
Guanajuato, state, Mex.	86	21°00'N	101°00'W
Guanape, Ven. (gwä-nä'pĕ)	101b	9°55'N	65°32'W
Guanape, r., Ven.	101b	9°52'N	65°20'W
Guanare, Ven. (gwä-nä'rå)	100	8°57'N	69°47'W
Guanduçu, r., Braz. (gwä'n-dōō'sōō)	102b	22°50'S	43°40'W
Guane, Cuba (gwä'nå)	92	22°10'N	84°05'W
Guangchang, China (gůäŋ-chäŋ)	165	26°50'N	116°18'E
Guangde, China (gůäŋ-dŭ)	165	30°40'N	119°20'E
Guangdong, prov., China (gůäŋ-dôŋ)	161	23°45'N	113°15'E
Guanglu Dao, i., China (gůäŋ-lōō dou)	162	39°13'N	122°21'E
Guangping, China (gůäŋ-pĭŋ)	162	36°30'N	114°57'E
Guangrao, China (gůaŋ-rou)	162	37°04'N	118°24'E
Guangshan, China (gůäŋ-shän)	162	32°02'N	114°53'E
Guangxi Zhuangzu, prov., China (gůäŋ-shyē)	160	24°00'N	108°30'E
Guangzhou (Canton), China	160	23°07'N	113°15'W
Guanhu, China (gůän-hōō)	162	34°26'N	117°59'E
Guannan, China (gůän-nän)	162	34°17'N	119°17'E
Guanta, Ven. (gwän'tä)	101b	10°15'N	64°35'W
Guantánamo, Cuba (gwän-tä'nä-mō)	93	20°10'N	75°10'W
Guantánamo, prov., Cuba	93	20°10'N	75°05'W
Guantánamo, Bahía de, b., Cuba	93	19°35'N	75°35'W
Guantao, China (gůän-tou)	162	36°39'N	115°25'E
Guanxian, China (gůän-shyĕn)	162	36°30'N	115°28'E
Guanyao, China (gůän-you)	163a	23°13'N	113°04'E
Guanyun, China (gůän-yòn)	162	34°28'N	119°16'E
Guapiles, C.R. (gwä-pē-lĕs)	91	10°05'N	83°54'W
Guapimirim, Braz. (gwä-pē-mē-rē'N)	102b	22°31'S	42°59'W
Guaporé, r., S.A. (gwä-pō-rā')	100	12°11'S	63°47'W
Guaqui, Bol. (guä'kē)	100	16°42'S	68°47'W
Guara, Sierra de, mts., Spain (sē-ĕ'r-rä-dĕ-gwä'rä)	129	42°24'N	0°15'W
Guarabira, Braz. (gwä-rä-bē'rà)	101	6°49'S	35°27'W
Guaranda, Ec. (gwä-rán'dä)	100	1°39'S	78°57'W
Guarapari, Braz. (gwä-rä-pä'rē)	101	20°34'S	40°31'W
Guarapiranga, Represa do, res., Braz.	99a	23°45'S	46°44'W
Guarapuava, Braz. (gwä-rä-pwä'vá)	102	25°29'S	51°26'W
Guarda, Port. (gwär'dä)	128	40°32'N	7°17'W
Guardiato, r., Spain	128	38°10'N	5°05'W
Guarena, Spain (gwä-rā'nyä)	128	38°52'N	6°08'W
Guaribe, r., Ven. (gwä-rē'bĕ)	101b	9°48'N	65°17'W
Guárico, dept., Ven.	101b	9°42'N	67°25'W
Guarulhos, Braz. (gwá-rō'l-yôs)	99a	23°28'S	46°30'W
Guarus, Braz. (gwá'rōōs)	99a	21°44'S	41°19'W
Guasca, Col. (gwäs'kä)	100a	4°52'N	73°52'W
Guasipati, Ven. (gwä-sē-pä'tē)	101	7°26'N	61°57'W
Guastalla, Italy (gwäs-täl'lä)	130	44°53'N	10°39'E
Guasti, Ca., U.S. (gwäs'tĭ)	75a	34°04'N	117°35'W
Guatemala, Guat. (guä-tå-mä'lä)	86	14°37'N	90°32'W
Guatemala, nation, N.A.	86	15°45'N	91°45'W
Guatire, Ven. (gwä-tē'rē)	101b	10°28'N	66°34'W
Guaviare, r., Col.	100	3°35'N	69°28'W
Guayabal, Cuba (gwä-yä-bä'l)	92	20°40'N	77°40'W
Guayalejo, r., Mex. (gwä-yä-lĕ'hō)	88	23°24'N	99°09'W
Guayama, P.R. (gwä-yä'mä)	87b	18°00'N	66°08'W
Guayamouc, r., Haiti	93	19°05'N	72°00'W
Guayaquil, Ec. (gwī-ä-kēl')	100	2°16'S	79°53'W
Guayaquil, Golfo de, b., Ec. (gôl-fô-dĕ)	100	3°03'S	82°12'W
Guaymas, Mex. (gwä'y-mäs)	86	27°49'N	110°58'W
Guayubin, Dom. Rep. (gwä-yōō-bĕ'n)	93	19°40'N	71°25'W
Guazacapán, Guat. (gwä-zä-kä-pän')	90	14°04'N	90°26'W
Gubakha, Russia (gōō-bä'kå)	134	58°53'N	57°35'E
Gubbio, Italy (gōōb'byō)	130	43°23'N	12°36'E
Guben, Ger.	124	51°57'N	14°43'E
Gucheng, China (gōō-chŭŋ)	162	39°09'N	115°43'E
Gúdar, Sierra de, mts., Spain	129	40°28'N	0°47'W
Gudena, r., Den.	122	56°20'N	9°47'E
Gudermes, Russia	138	43°20'N	46°08'E
Gudvangen, Nor. (gōōdh'väŋ-gĕn)	122	60°52'N	6°45'E
Guebwiller, Fr. (gĕb-vē-lâr')	127	47°53'N	7°10'E
Guédi, Mont, mtn., Chad	189	12°14'N	18°58'E
Guelma, Alg. (gwĕl'mä)	184	36°32'N	7°17'E
Guelph, Can. (gwĕlf)	59	43°33'N	80°15'W
Güere, r., Ven. (gwĕ'rĕ)	101b	9°39'N	65°00'W
Guéret, Fr. (gā-rĕ')	126	46°09'N	1°52'E
Guernsey, dep., Eur.	126	49°28'N	2°35'W
Guernsey, i., Guern. (gûrn'zĭ)	117	49°27'N	2°36'W
Guerrero, Mex. (gĕr-rä'rō)	80	26°47'N	99°20'W
Guerrero, Mex.	80	28°20'N	100°24'W
Guerrero, state, Mex.	86	17°45'N	100°15'W
Gueydan, La., U.S. (gā'dăn)	81	30°01'N	92°31'W
Guia de Pacobaíba, Braz. (gwē'ä-dē-pä'kō-bī'bä)	102b	22°42'S	43°10'W
Guiana Highlands, mts., S.A.	97	3°20'N	60°00'W
Guichi, China (gwā-chr)	165	30°35'N	117°28'E
Guichicovi, Mex. (gwē-chē-kō'vē)	89	16°58'N	95°10'W
Guidonia, Italy (gwē-dō'nyá)	130	42°00'N	12°45'E
Guiglo, C. Iv.	188	6°33'N	7°29'W
Guignes-Rabutin, Fr. (gēn'yĕ)	127b	48°38'N	2°48'E
Güigüe, Ven. (gwē'gwĕ)	101b	10°05'N	67°48'W
Guija, Lago, l., N.A. (gē'hä)	90	14°16'N	89°21'W
Guildford, Eng., U.K. (gĭl'fẽrd)	120	51°13'N	0°34'W
Guilford, In., U.S. (gĭl'fẽrd)	69f	39°10'N	84°55'W
Guilin, China (gwā-lĭn)	161	25°18'N	110°22'E
Guimarães, Port. (gē-mä-rănsh')	128	41°27'N	8°22'W
Guinea, nation, Afr. (gĭn'ē)	184	10°48'N	12°28'W
Guinea, Gulf of, b., Afr.	184	2°00'N	1°00'E
Guinea-Bissau, nation, Afr. (gĭn'ē)	184	12°00'N	20°00'W
Guingamp, Fr. (găn-gän')	126	48°35'N	3°10'W
Guir, r., Mor.	118	31°55'N	2°48'W
Güira de Melena, Cuba (gwē'rä dā mā-lā'nä)	92	22°45'N	82°30'W
Güiria, Ven. (gwē-rē'ä)	100	10°43'N	62°16'W
Guise, Fr. (guēz)	126	49°54'N	3°37'E
Guisisil, vol., Nic. (gē-sē-sēl')	90	12°40'N	86°11'W
Guiyang, China (gwā-yäŋ)	160	26°45'N	107°00'E
Guizhou, China (gwā-jō)	163a	22°46'N	113°15'E
Guizhou, prov., China	160	27°00'N	106°10'E
Gujānwāla, Pak. (gōj-rän'va-lá)	155	32°08'N	74°14'E
Gujarat, India	155	22°54'N	72°00'E
Gulbarga, India (gōl-bûr'gá)	155	17°25'N	76°52'E
Gulbene, Lat. (gōl-bā'nē)	123	57°09'N	26°49'E
Gulfport, Ms., U.S. (gŭlf'pōrt)	82	30°24'N	89°05'W
Gulja *see* Yining, China	160	43°58'N	80°40'E
Gull Lake, Can.	56	50°10'N	108°25'W
Gull Lake, l., Can.	55	52°35'N	114°00'W
Gulu, Ug.	191	2°47'N	32°18'E
Gumaca, Phil. (gōō-mä-kä')	169a	13°55'N	122°06'E
Gumbeyka, r., Russia (gòm-bĕy'ká)	142a	53°20'N	59°42'E
Gumel, Nig.	184	12°39'N	9°22'E
Gummersbach, Ger. (gòm'ĕrs-bäk)	124	51°02'N	7°34'E
Gummi, Nig.	189	12°09'N	5°09'E
Gumpoldskirchen, Aus.	115e	48°04'N	16°15'E
Guna, India	158	24°44'N	77°17'E
Gunisao, r., Can. (gŭn-ĭ-sā'ō)	57	53°40'N	97°35'W
Gunisao Lake, l., Can.	57	53°35'N	96°10'W
Gunnedah, Austl. (gŭ'nē-dä)	176	31°00'S	150°10'E
Gunnison, Co., U.S. (gŭn'ĭ-sŭn)	77	38°33'N	106°56'W
Gunnison, Ut., U.S.	77	39°10'N	111°50'W
Gunnison, r., Co., U.S.	77	38°45'N	108°20'W
Guntersville, Al., U.S. (gŭn'tẽrz-vĭl)	82	34°20'N	86°19'W
Guntersville Lake, res., Al., U.S.	82	34°30'N	86°20'W
Guntramsdorf, Aus.	115e	48°04'N	16°19'E
Guntūr, India (gòn'tōōr)	155	16°22'N	80°29'E
Guoyang, China (gwô-yäŋ)	162	33°32'N	116°10'E
Gurdon, Ar., U.S. (gûr'dŭn)	79	33°56'N	93°10'W
Gurgueia, r., Braz.	101	8°12'S	43°49'W
Guri, Embalse, res., Ven.	100	7°30'N	63°00'W
Gurnee, Il., U.S. (gûr'nē)	69a	42°22'N	87°55'W
Gurskøy, i., Nor. (gōōrskûê)	122	62°18'N	5°20'E
Gurupi, Serra do, mts., Braz. (sĕ'r-rä-dô-gōō-rōō-pē')	101	5°32'S	47°02'W
Guru Sikhar, mtn., India	158	29°42'N	72°50'E
Gur'yevsk, Russia (gōōr-yĭfsk')	134	54°17'N	85°56'E
Gusau, Nig. (gōō-zä'ōō)	184	12°12'N	6°40'E
Gusev, Russia (gōō'sĕf)	123	54°35'N	22°15'E
Gushi, China (gōō-shr)	162	32°11'N	115°39'E
Gushiago, Ghana	188	9°55'N	0°12'W
Gusinje, Serb. (gōō-sēn'yĕ)	131	42°34'N	19°54'E
Gus'-Khrustal'nyy, Russia (gōōs-ĸrōō-stäl'ny')	136	55°39'N	40°41'E
Gustavo A. Madero, Mex. (gōōs-tä'vô-à-mä-dĕ'rō)	88	19°29'N	99°07'W
Güstrow, Ger. (güs'trō)	124	53°48'N	12°12'E
Gütersloh, Ger. (gü'tĕrs-lo)	124	51°54'N	8°22'E
Guthrie, Ok., U.S. (gŭth'rĭ)	79	35°52'N	97°26'W
Guthrie Center, Ia., U.S.	71	41°41'N	94°33'W
Gutiérrez Zamora, Mex. (gōō-tĭ-âr'räz zä-mō'rä)	89	20°27'N	97°17'W
Guttenberg, Ia., U.S. (gŭt'ĕn-bûrg)	71	42°48'N	91°09'W
Guyana, nation, S.A. (güy'änä)	101	7°45'N	59°00'W
Guyang, China (gōō-yäŋ)	162	34°56'N	114°57'E
Guye, China (gōō-yŭ)	162	39°46'N	118°23'E
Guymon, Ok., U.S. (gī'mŏn)	78	36°41'N	101°29'W
Guysborough, Can. (gīz'bŭr-ō)	61	45°23'N	61°30'W
Guzhen, China (gōō-jŭn)	164	33°20'N	117°18'E
Gvardeysk, Russia (gvár-dĕysk')	123	54°39'N	21°11'E
Gwadabawa, Nig.	189	13°20'N	5°15'E
Gwādar, Pak. (gwä'dŭr)	154	25°15'N	62°29'E
Gwalior, India	155	26°13'N	78°10'E
Gwane, D.R.C. (gwän)	185	4°43'N	25°50'E
Gwardafuy, Gees, c., Som.	192a	11°55'N	51°30'E
Gwda, r., Pol.	124	53°27'N	16°52'E
Gwembe, Zam.	191	16°30'S	27°35'E
Gweru, Zimb.	186	19°15'S	29°48'E
Gwinn, Mi., U.S. (gwĭn)	71	46°15'N	87°30'W
Gyaring Co, l., China	158	30°37'N	88°33'E
Gydan, Khrebet (Kolymskiy), mts., Russia	135	61°45'N	155°00'E
Gydanskiy Poluostrov, pen., Russia	134	70°42'N	76°03'E
Gympie, Austl. (gĭm'pē)	175	26°20'S	152°50'E
Gyöngyös, Hung. (dyûn'dyûsh)	119	47°47'N	19°55'E
Györ, Hung. (dyûr)	119	47°40'N	17°37'E
Gyōtoku, Japan (gyō'tō-kōō')	167a	35°42'N	139°56'E
Gypsumville, Can. (jĭp'sŭm'vĭl)	50	51°45'N	98°35'W
Gýtheio, Grc.	131	36°50'N	22°37'E
Gyula, Hung. (dyò'lä)	125	46°38'N	21°18'E
Gyumri, Arm.	137	40°40'N	43°50'E
Gyzylarbat, Turkmen.	139	38°55'N	56°33'E

H

PLACE (Pronunciation)	PAGE	LAT.	LONG.
Haan, Ger. (hän)	127c	51°12'N	7°00'E
Haapamäki, Fin. (häp'ä-mĕ-kē)	123	62°16'N	24°20'E
Haapsalu, Est. (häp'sä-lò)	123	58°56'N	23°33'E
Haar, Ger. (här)	115d	48°06'N	11°44'E
Ha'Arava (Wādī al Jayb), val., Asia	153a	30°33'N	35°10'E
Haarlem, Neth. (här'lĕm)	121	52°22'N	4°37'E
Habana, prov., Cuba (hä-vä'nä)	92	22°45'N	82°25'W
Häbra, India	158a	22°49'N	88°38'E
Hachinohe, Japan (hä'chē-nō'hå)	166	40°29'N	141°40'E
Hachiōji, Japan (hä'chē-ō'jē)	166	35°39'N	139°18'E
Hackensack, N.J., U.S. (hăk'ĕn-săk)	68a	40°54'N	74°03'W
Ḥadd, Ra's al, c., Oman	154	22°29'N	59°46'E
Haddonfield, N.J., U.S. (hăd'ŭn-fēld)	68f	39°53'N	75°02'W
Haddon Heights, N.J., U.S. (hăd'ŭn hīts)	68f	39°53'N	75°03'W
Hadejia, Nig. (hä-dā'jä)	184	12°30'N	9°59'E
Hadejia, r., Nig.	184	12°15'N	10°00'E
Ḥadera, Isr. (ĸå-dĕ'rá)	153a	32°26'N	34°55'E
Haderslev, Den. (hä'dhĕrs-lĕv)	122	55°17'N	9°28'E
Hadiach, Ukr.	137	50°22'N	33°59'E
Ḥadīdū, Yemen	154	12°40'N	53°50'E
Hadlock, Wa., U.S. (hăd'lŏk)	74a	48°02'N	122°46'W

PLACE (Pronunciation)	PAGE	LAT.	LONG.
Haḑramawt, reg., Yemen	154	15°22'N	48°40'E
Hadūr Shu'ayb, mtn., Yemen	154	15°45'N	43°45'E
Haeju, Kor., N. (hä'ĕ-jŭ)	166	38°03'N	125°42'E
Hafnarfjördur, Ice.	116	64°02'N	21°32'W
Haft Gel, Iran	157	31°27'N	49°27'E
Hafun, Ras, c., Som. (hä-fōōn')	192a	10°15'N	51°35'E
Hageland, Mt., U.S. (hăge'lănd)	73	48°53'N	108°43'W
Hagen, Ger. (hä'gĕn)	124	51°21'N	7°29'E
Hagerstown, In., U.S. (hā'gẽrz-toun)	66	39°55'N	85°10'W
Hagerstown, Md., U.S.	65	39°40'N	77°45'W
Hagi, Japan (hä'gĭ)	167	34°25'N	131°25'E
Hague, Cap de la, c., Fr. (dē là åg')	126	49°44'N	1°55'W
Haguenau, Fr. (åg'nō')	127	48°47'N	7°48'E
Hai'an, China (hī-än)	162	32°35'N	120°25'E
Haibara, Japan (hä'ê-bä'rä)	167	34°29'N	135°57'E
Haicheng, China (hī-chŭŋ)	164	40°58'N	122°45'E
Haidian, China (hī-dĭĕn)	162	39°59'N	116°17'E
Haifa, Isr. (hä'ē-fà)	154	32°48'N	35°00'E
Haifeng, China (hä'ê-fĕng')	165	23°00'N	115°20'E
Haifuzhen, China (hī-fōō-jŭn)	162	31°57'N	121°48'E
Haikou, China (hī-kō)	165	20°00'N	110°20'E
Ḩa'il, Sau. Ar.	154	27°30'N	41°47'E
Hailar, China	161	49°10'N	118°40'E
Hailey, Id., U.S. (hā'lĭ)	73	43°31'N	114°19'W
Haileybury, Can.	59	47°27'N	79°38'W
Haileyville, Ok., U.S. (hā'lĭ-vĭl)	79	34°51'N	95°34'W
Hailing Dao, i., China (hī-lĭŋ dou)	165	21°30'N	112°15'E
Hailong, China (hī-loŋ)	164	42°32'N	125°52'E
Hailun, China (hä'ê-lōōn')	161	47°18'N	126°50'E
Hainan, prov., China	160	19°00'N	109°30'E
Hainan Dao, i., China (hī-nän dou)	161	19°00'N	111°10'E
Hainburg, Aus.	124	48°09'N	16°57'E
Haines, Ak., U.S. (hānz)	63	59°10'N	135°38'W
Haines City, Fl., U.S.	83a	28°05'N	81°38'W
Hai Phong, Viet. (hī'fông')(hä'êp-hŏng)	168	20°52'N	106°40'E
Haisyn, Ukr.	137	48°46'N	29°22'E
Haiti, nation, N.A. (hā'tĭ)	87	19°00'N	72°15'W
Haizhou, China	162	34°34'N	119°11'E
Haizhou Wan, b., China	164	34°49'N	120°35'E
Hajduböszormény, Hung. (hô'dò-bû'sûr-mān')	125	47°41'N	21°30'E
Hajdúhadház, Hung. (hô'ĭ-dò-hôd'häz)	125	47°32'N	21°32'E
Hajdúnánás, Hung. (hô'ĭ-dò-nä'näsh)	125	47°52'N	21°27'E
Hakodate, Japan (hä-kō-dä't å)	161	41°46'N	140°42'E
Haku-San, mtn., Japan (hä'kōō-sän')	166	36°11'N	136°45'E
Halā'ib, Egypt (hä-lä'ēb)	185	22°10'N	36°40'E
Halbe, Ger. (häl'bē)	115b	52°07'N	13°43'E
Halberstadt, Ger. (häl'bĕr-shtät)	124	51°54'N	11°07'E
Halcon, Mount, mtn., Phil. (häl-kōn')	169a	13°19'N	120°55'E
Halden, Nor. (häl'dĕn)	116	59°10'N	11°21'E
Haldensleben, Ger.	124	52°18'N	11°23'E
Hale, Eng., U.K. (hāl)	114a	53°22'N	2°20'W
Haleakalā Crater, depr., Hi., U.S. (hä'lā-ä'kä-lä)	84a	20°44'N	156°15'W
Haleakalā National Park, rec., Hi., U.S.	84a	20°46'N	156°00'W
Hales Corners, Wi., U.S. (hālz kŏr'nẽrz)	69a	42°56'N	88°03'W
Halesowen, Eng., U.K. (hālz'ô-wĕn)	114a	52°26'N	2°03'W
Halethorpe, Md., U.S. (hāl-thôrp)	68e	39°15'N	76°40'W
Haleyville, Al., U.S. (hā'lĭ-vĭl)	82	34°11'N	87°36'W
Half Moon Bay, Ca., U.S. (hăf'mōōn)	74b	37°28'N	122°26'W
Halfway House, S. Afr. (häf-wā hous)	187b	26°00'S	28°08'E
Halfweg, Neth.	115a	52°23'N	4°45'E
Halifax, Can. (hăl'ĭ-făks)	51	44°39'N	63°36'W
Halifax, Eng., U.K.	120	53°44'N	1°52'W
Halifax Bay, b., Austl. (hăl'ĭ-făx)	175	18°56'S	147°07'E
Halifax Harbour, b., Can.	60	44°35'N	63°31'W
Halkett, Cape, c., Ak., U.S.	63	70°50'N	151°15'W
Hallam Peak, mtn., Can.	55	52°11'N	118°46'E
Halla San, mtn., Kor., S. (häl'lä-sän)	166	33°20'N	126°37'E
Halle, Bel. (häl'lē)	115a	50°45'N	4°13'E
Halle, Ger.	117	51°30'N	11°59'E
Hallettsville, Tx., U.S. (hăl'ĕts-vĭl)	81	29°26'N	96°55'W
Hallock, Mn., U.S. (hăl'ŭk)	70	48°46'N	96°57'W
Hall Peninsula, pen., Can. (hôl)	53	63°14'N	65°40'W
Halls Bayou, Tx., U.S.	81a	29°55'N	95°23'W
Hallsberg, Swe. (häls'bĕrgh)	122	59°04'N	15°04'E
Halls Creek, Austl. (hôlz)	174	18°15'S	127°45'E
Halmahera, i., Indon. (häl-mä-hä'rä)	169	0°45'N	128°45'E
Halmahera, Laut, Indon.	169	1°00'S	129°00'E
Halmstad, Swe. (hälm'städ)	116	56°40'N	12°46'E
Halsafjorden, b., Nor. (häl'sĕ fyôrd)	122	63°03'N	8°23'E
Halstead, Ks., U.S. (hôl'stĕd)	79	38°02'N	97°36'W
Haltern, Ger. (häl'tĕrn)	127c	51°45'N	7°10'E
Haltom City, Tx., U.S. (hôl'tŏm)	75c	32°48'N	97°13'W
Halver, Ger.	127c	51°11'N	7°30'E
Hamada, Japan	166	34°53'N	132°05'E
Hamadān, Iran (hŭ-mŭ-dän')	154	34°45'N	48°07'E
Ḩamāh, Syria (hä'mä)	154	35°08'N	36°53'E
Hamamatsu, Japan (hä'mä-mät'sò)	166	34°41'N	137°43'E
Hamar, Nor. (hä'mär)	116	60°49'N	11°05'E
Hamasaka, Japan (hä'mà-sä'kà)	167	35°57'N	134°27'E
Hamborn, Ger. (häm'bōrn)	127c	51°30'N	6°43'E
Hamburg, Ger. (häm'bōōrgh)	110	53°34'N	10°02'E
Hamburg, S. Afr. (häm'bûrg)	187c	33°18'S	27°28'E
Hamburg, Ar., U.S. (hăm'bûrg)	79	33°15'N	91°49'W
Hamburg, N.J., U.S.	68a	41°09'N	74°35'W
Hamburg, N.Y., U.S.	69c	42°44'N	78°51'W
Hamden, Ct., U.S. (hăm'dĕn)	67	41°20'N	72°55'W
Hämeenlinna, Fin. (hĕ'măn-lĭn-nà)	116	61°00'N	24°29'E
Hameln, Ger. (hä'mĕln)	124	52°06'N	9°23'E
Hamelwörden, Ger. (hä'mĕl-vûr-dĕn)	115c	53°47'N	9°19'E
Hamersley Range, mts., Austl. (hăm'ẽrz-lê)	174	22°15'S	117°50'E
Hamhŭng, Kor., N. (häm'hòng')	161	39°57'N	127°35'E
Hami, China (hä-mē)	160	42°58'N	93°14'E
Hamilton, Austl. (hăm'ĭl-tŭn)	175	37°50'S	142°10'E
Hamilton, Can.	51	43°15'N	79°52'W
Hamilton, N.Z.	175a	37°45'S	175°28'E
Hamilton, Al., U.S.	82	34°09'N	88°01'W
Hamilton, Ma., U.S.	61a	42°37'N	70°52'W
Hamilton, Mo., U.S.	79	39°43'N	93°59'W
Hamilton, Mt., U.S.	73	46°15'N	114°09'W
Hamilton, Oh., U.S.	65	39°22'N	84°33'W
Hamilton, Tx., U.S.	80	31°42'N	98°07'W
Hamilton, Lake, l., Ar., U.S.	79	34°25'N	93°32'W
Hamilton Harbour, b., Can.	62d	43°17'N	79°50'W
Hamilton Inlet, b., Can.	53	54°20'N	56°57'W
Hamina, Fin. (hà'mē-nà)	123	60°34'N	27°15'E
Hamlet, N.C., U.S. (hăm'lĕt)	83	34°53'N	79°42'W
Hamlin, Tx., U.S. (hăm'lĭn)	78	32°54'N	100°08'W
Hamm, Ger. (häm)	124	51°40'N	7°48'E
Hammanskraal, S. Afr. (hä-mäns-kräl')	192c	25°24'S	28°17'E
Hamme, Bel.	115a	51°06'N	4°07'E
Hamme-Oste Kanal, can., Ger. (hä'mĕ-ōs'tĕ kä-näl)	115c	53°20'N	8°59'E
Hammerfest, Nor. (hä'mĕr-fĕst)	110	70°38'N	23°59'E
Hammond, In., U.S. (hăm'ŭnd)	65	41°37'N	87°31'W
Hammond, La., U.S.	81	30°30'N	90°28'W
Hammond, Or., U.S.	74c	46°12'N	123°57'W
Hammonton, N.J., U.S. (hăm'ŭn-tŭn)	67	39°40'N	74°45'W
Hampden, Me., U.S. (hăm'dĕn)	60	44°44'N	68°51'W
Hampstead, Md., U.S.	68e	39°36'N	76°54'W
Hampstead Norris, Eng., U.K. (hămp-stĕd nŏ'rĭs)	114b	51°27'N	1°14'W
Hampton, Can. (hămp'tŭn)	60	45°32'N	65°51'W
Hampton, Ia., U.S.	71	42°43'N	93°15'W
Hampton, Va., U.S.	67	37°02'N	76°21'W
Hampton Roads, b., Va., U.S.	68g	36°56'N	76°23'W
Hams Fork, r., Wy., U.S.	73	41°55'N	110°40'W
Hamtramck, Mi., U.S. (hăm-trăm'ĭk)	69b	42°24'N	83°03'W
Han, r., China (hän)	165	25°00'N	116°35'E
Han, r., China	161	31°40'N	112°04'E
Han, r., Kor., S.	166	37°10'N	127°40'E
Hāna, Hi., U.S. (hä'nä)	84a	20°43'N	155°59'W
Hanábana, r., Cuba (hä-nä-bä'nä)	92	22°30'N	80°55'W
Hanalei Bay, b., Hi., U.S. (hä-nä-lā'ē)	84a	22°15'N	159°40'W
Hanang, mtn., Tan.	191	4°26'S	35°24'E
Hanau, Ger. (hä'nou)	124	50°08'N	8°56'E
Hancock, Mi., U.S. (hăn'kŏk)	71	47°08'N	88°37'W
Handan, China (hän-dän)	162	36°37'N	114°30'E
Haney, Can. (hā-nē)	55	49°13'N	122°36'W
Hanford, Ca., U.S. (hăn'fẽrd)	76	36°20'N	119°38'W
Hangayn Nuruu, mts., Mong.	160	48°03'N	99°45'E
Hango, Fin. (häŋ'gû)	110	59°49'N	22°56'E
Hangzhou, China (hăng'chō')	161	30°17'N	120°12'E
Hangzhou Wan, b., China (häŋ-jō wän)	165	30°20'N	121°25'E
Hankamer, Tx., U.S. (hăn'kà-mẽr)	81a	29°52'N	94°42'W
Hankinson, N.D., U.S. (häŋ'kĭn-sŭn)	70	46°04'N	96°54'W
Hankou, China (hän-kō)	165	30°42'N	114°22'E
Hann, Mount, mtn., Austl. (hän)	174	16°05'S	126°07'E
Hanna, Can. (hăn'à)	50	51°38'N	111°54'W
Hanna, Wy., U.S.	73	41°51'N	106°34'W
Hannah, N.D., U.S.	70	48°58'N	98°42'W
Hannibal, Mo., U.S. (hăn'ĭ băl)	65	39°42'N	91°22'W
Hannover, Ger. (hän-ō'vĕr)	110	52°22'N	9°45'E
Hannover, hist. reg., Ger.	124	52°52'N	8°27'E
Hanöbukten, b., Swe.	122	55°54'N	14°55'E
Hanoi, Viet. (hä-noi')	168	21°04'N	105°50'E
Hanover, Can. (hăn'ô-vẽr)	58	44°10'N	81°05'W
Hanover, Ma., U.S.	61a	42°07'N	70°49'W
Hanover, N.H., U.S.	67	43°45'N	72°15'W
Hanover, Pa., U.S.	67	39°50'N	77°00'W
Hanover, i., Chile	102	51°00'S	74°45'W
Hanshan, China (hän'shän')	162	31°43'N	118°06'E
Hans Lollick, i., V.I.U.S. (häns'lôl'ĭk)	87c	18°24'N	64°55'W
Hanson, Ma., U.S. (hăn'sŭn)	61a	42°04'N	70°53'W
Hansville, Wa., U.S. (häns'-vĭl)	74a	47°55'N	122°33'W
Hantengri Feng, mtn., Asia (hän-tŭŋ-rē fŭŋ)	160	42°10'N	80°20'E
Hantsport, Can. (hănts'pōrt)	60	45°04'N	64°11'W
Hanyang, China (han'yäng')	161	30°30'N	114°10'E
Hanzhong, China (hän-jôŋ)	164	33°02'N	107°00'E
Haocheng, China (hou-chŭŋ)	162	33°19'N	117°33'E
Haparanda, Swe. (hà-pa-rän'dä)	116	65°54'N	23°57'E
Hapeville, Ga., U.S. (hāp'vĭl)	68c	33°39'N	84°25'W
Happy Camp, Ca., U.S.	72	41°47'N	123°22'W
Happy Valley-Goose Bay, Can.	51	53°19'N	60°33'W
Ḩaql, Sau. Ar.	153a	29°15'N	34°57'E
Har, Laga, r., Kenya	191	2°15'N	39°30'E
Haradok, Bela.	132	55°27'N	29°58'E
Harare, Zimb.	186	17°50'S	31°03'E
Harbin, China	161	45°40'N	126°30'E
Harbor Beach, Mi., U.S. (här'bẽr bēch)	66	43°50'N	82°40'W
Harbor Springs, Mi., U.S.	66	45°25'N	85°05'W
Harbour Breton, Can. (brĕt'ŭn) (brē-tôn')	61	47°29'N	55°48'W
Harbour Grace, Can. (grās)	61	47°32'N	53°13'W
Harburg, Ger. (här-bòrgh)	115c	53°28'N	9°58'E
Hardangerfjorden, Nor. (här-däng'ẽr fyôrd)	116	59°58'N	6°30'E
Hardin, Mt., U.S. (här'dĭn)	73	45°44'N	107°36'W
Harding, S. Afr. (här'dĭng)	186	30°34'S	29°54'E
Harding, Lake, res., U.S.	82	32°43'N	85°00'W
Hardwār, India (hŭr'dvär)	155	29°56'N	78°06'E
Hardy, r., Mex. (här'dĭ)	76	32°04'N	115°10'W
Hare Bay, b., Can. (hâr)	61	51°18'N	55°50'W
Harer, Eth.	185	9°43'N	42°10'E
Harerge, hist. reg., Eth.	185	8°15'N	41°00'E
Hargeysa, Som. (här-gä'ē-sà)	192a	9°20'N	43°57'E
Harghita, Munţii, mts., Rom.	125	46°25'N	25°40'E
Harima-Nada, b., Japan (hä'rē-mä nä-dä)	167	34°34'N	134°37'E
Haringvliet, r., Neth.	115a	51°49'N	4°03'E
Harīrūd, r., Asia	154	34°29'N	61°16'E
Harlan, Ia., U.S. (här'lăn)	79	41°40'N	95°10'W
Harlan, Ky., U.S.	82	36°50'N	83°19'W
Harlan County Reservoir, res., Ne., U.S.	78	40°03'N	99°51'W
Harlem, Mt., U.S. (hàr'lĕm)	73	48°33'N	108°50'W
Harlingen, Neth. (här'lĭng-ĕn)	121	53°10'N	5°24'E
Harlingen, Tx., U.S.	64	26°12'N	97°42'W
Harlow, Eng., U.K. (här'lō)	114b	51°46'N	0°08'E
Harlowton, Mt., U.S. (hàr'lô-tun)	73	46°26'N	109°50'W
Harmony, In., U.S. (här'mô-nĭ)	66	39°35'N	87°00'W
Harney Basin, Or., U.S. (här'nĭ)	72	43°26'N	120°19'W
Harney Lake, l., Or., U.S.	72	43°11'N	119°23'W
Harney Peak, mtn., S.D., U.S.	64	43°52'N	103°32'W
Härnosand, Swe. (hĕr-nû-sänd)	116	62°37'N	17°54'E
Haro, Spain (ä'rō)	128	42°35'N	2°49'W
Haro Strait, strt., N.A. (hă'rō)	74a	48°27'N	123°11'W
Harpenden, Eng., U.K. (här'pĕn-d'n)	114b	51°48'N	0°22'W
Harper, Lib.	184	4°25'N	7°43'W
Harper, Ks., U.S. (här'pẽr)	78	37°17'N	98°02'W
Harper, Wa., U.S.	74a	47°31'N	122°32'W
Harpers Ferry, W.V., U.S. (här'pẽrz)	67	39°20'N	77°45'W
Harricana, r., Can.	59	50°10'N	78°50'W
Harriman, Tn., U.S. (hă'ĭ-măn)	82	35°55'N	84°34'W
Harrington, De., U.S. (hăr'ĭng-tŭn)	67	38°55'N	75°35'W
Harris, i., Scot., U.K. (hăr'ĭs)	120	57°55'N	6°40'W
Harris, Lake, l., Fl., U.S.	83a	28°43'N	81°40'W
Harrisburg, Il., U.S. (hăr'ĭs-bûrg)	66	37°45'N	88°35'W
Harrisburg, Pa., U.S.	65	40°15'N	76°50'W
Harrismith, S. Afr. (hă-rĭs'mĭth)	192c	28°17'S	29°08'E
Harrison, Ar., U.S. (hăr'ĭ-sŭn)	79	36°13'N	93°06'W
Harrison, Oh., U.S.	69f	39°16'N	84°45'W
Harrisonburg, Va., U.S. (hăr'ĭ-sŭn-bûrg)	67	38°30'N	78°50'W
Harrison Lake, l., Can.	55	49°31'N	121°59'W
Harrisonville, Mo., U.S. (hăr-ĭ-sŭn-vĭl)	79	38°39'N	94°21'W
Harrisville, Ut., U.S. (hăr'ĭs-vĭl)	75b	41°17'N	112°00'W
Harrisville, W.V., U.S.	66	39°10'N	81°05'W
Harrodsburg, Ky., U.S. (hăr'ŭdz-bûrg)	66	37°45'N	84°50'W
Harrods Creek, r., Ky., U.S. (hăr'ŭdz)	69h	38°24'N	35°33'W
Harrow, Eng., U.K. (hăr'ō)	114b	51°34'N	0°21'W
Harsefeld, Ger. (här'zĕ-fĕld')	115c	53°27'N	9°30'E
Harstad, Nor. (här'städh)	116	68°49'N	16°10'E
Hart, Mi., U.S. (härt)	66	43°40'N	86°25'W
Hartbeesfontein, S. Afr.	192c	26°46'S	26°25'E
Hartbeespoortdam, res., S. Afr.	187b	25°47'S	27°43'E
Hartford, Al., U.S. (härt'fẽrd)	82	31°05'N	85°42'W
Hartford, Ar., U.S.	79	35°01'N	94°21'W
Hartford, Ct., U.S.	65	41°45'N	72°40'W
Hartford, Il., U.S.	75e	38°50'N	90°06'W
Hartford, Ky., U.S.	82	37°25'N	86°50'W
Hartford, Mi., U.S.	66	42°15'N	86°15'W
Hartford, Wi., U.S.	71	43°19'N	88°25'W
Hartford City, In., U.S.	66	40°35'N	85°25'W
Hartington, Eng., U.K. (härt'ĭng-tŭn)	114a	53°08'N	1°48'W
Hartington, Ne., U.S.	70	42°37'N	97°18'W
Hartland Point, c., Eng., U.K.	120	51°03'N	4°40'W
Hartlepool, Eng., U.K. (här't'l-pōōl)	116	54°40'N	1°12'W
Hartley, Ia., U.S. (härt'lĭ)	70	43°12'N	95°29'W
Hartley Bay, Can.	54	53°25'N	129°15'W
Hart Mountain, mtn., Can. (härt)	57	52°25'N	101°30'W
Hartsbeespoort, S. Afr.	187b	25°44'S	27°51'E
Hartselle, Al., U.S. (härt'sĕl)	82	34°24'N	86°55'W
Hartshorne, Ok., U.S. (härts'hôrn)	79	34°49'N	95°34'W
Hartsville, S.C., U.S. (härts'vĭl)	83	34°20'N	80°04'W
Hartwell, Ga., U.S. (härt'wĕl)	83	34°21'N	82°56'W
Hartwell Lake, res., U.S.	65	34°30'N	83°00'W
Hārua, India	158a	22°36'N	88°40'E
Harvard, Il., U.S. (här'vȧrd)	71	42°25'N	88°39'W
Harvard, Ma., U.S.	61a	42°30'N	71°35'W
Harvard, Ne., U.S.	78	40°36'N	98°08'W
Harvard, Mount, mtn., Co., U.S.	77	38°55'N	106°20'W
Harvey, Can.	60	45°44'N	64°46'W
Harvey, Il., U.S.	69a	41°37'N	87°39'W
Harvey, La., U.S.	68d	29°54'N	90°05'W
Harvey, N.D., U.S.	70	47°46'N	99°55'W
Harwich, Eng., U.K. (här'wĭch)	121	51°53'N	1°13'E
Haryana, state, India	155	29°00'N	75°45'E
Harz Mountains, mts., Ger. (härts)	124	51°42'N	10°50'E
Hashimoto, Japan (hä'shē-mō'tō)	167	34°19'N	135°37'E
Haskell, Ok., U.S. (hăs'kĕl)	79	35°49'N	95°41'W
Haskell, Tx., U.S.	78	33°09'N	99°43'W
Haslingden, Eng., U.K. (hăz'lĭng dĕn)	114a	53°43'N	2°19'W
Hassi Messaoud, Alg.	184	31°17'N	6°13'E
Hässleholm, Swe. (hĕs'lĕ-hōlm)	122	56°10'N	13°44'E
Hastings, N.Z.	175a	39°33'S	176°53'E
Hastings, Eng., U.K. (hās'tĭngz)	117	50°52'N	0°28'E
Hastings, Mi., U.S.	66	42°40'N	85°20'W
Hastings, Mn., U.S.	75g	44°44'N	92°51'W
Hastings, Ne., U.S.	64	40°34'N	98°42'W
Hastings-on-Hudson, N.Y., U.S. (ŏn-hŭd'sŭn)	68a	40°59'N	75°53'W
Hatay, Tur.	154	36°20'N	36°10'E
Hatchie, r., Tn., U.S. (hăch'ē)	82	35°28'N	89°14'W
Haţeg, Rom. (kät-säg')	131	45°35'N	22°57'E
Hatfield Broad Oak, Eng., U.K. (hăt-fēld brôd ōk)	114b	51°50'N	0°14'E
Hatogaya, Japan (hä'tō-gä-yä)	167a	35°50'N	139°45'E

ăt; finăl; rāte; senāte; ärm; åsk; sofȧ; fâre; ch-choose; dh-as th in other; bē; ēvent; bĕt; recĕnt; cratẽr; g-gō; gh-guttural g; bĭt; ĭ-short neutral; rīde; κ-guttural k as ch in German ich;

PRONOUNCING INDEX
Hatsukaichi – Hillsboro 255

PLACE (Pronunciation)	PAGE	LAT.	LONG.
Hatsukaichi, Japan (hät′sōō-kȧ′ē-chē)	167	34°22′N	132°19′E
Hatteras, Cape, c., N.C., U.S. (hăt′ẽr-ȧs)	65	35°15′N	75°24′W
Hattiesburg, Ms., U.S. (hăt′ĭz-bûrg)	65	31°20′N	89°18′W
Hattingen, Ger. (hä′tĕn-gĕn)	127c	51°24′N	7°11′E
Hatvan, Hung. (hôt′vôn)	125	47°39′N	19°44′E
Hat Yai, Thai.	168	7°01′N	100°29′E
Haugesund, Nor. (hou′gĕ-soon′)	116	59°26′N	5°20′E
Haukivesi, l., Fin. (hou′kē-vĕ′sē)	123	62°02′N	29°02′E
Haultain, r., Can.	56	56°15′N	106°35′W
Hauptsrus, S. Afr.	192c	26°35′S	26°16′E
Hauraki Gulf, b., N.Z. (hä-ōō-rä′kē)	175a	36°30′S	175°00′E
Haut, Isle au, Me., U.S. (hō)	60	44°03′N	68°13′W
Haut Atlas, mts., Mor.	118	32°10′N	5°49′W
Hauterive, Can.	60	49°11′N	68°16′W
Hau′ula, Hi., U.S.	84a	21°37′N	157°45′W
Havana, Cuba	87	23°08′N	82°23′W
Havana, Il., U.S. (hȧ-vă′nȧ)	79	40°17′N	90°02′W
Havasu, Lake, res., U.S. (hăv′ȧ-sōō)	77	34°26′N	114°09′W
Havel, r., Ger. (hä′fĕl)	124	53°09′N	13°10′E
Havel-Kanal, can., Ger.	115b	52°36′N	13°12′E
Haverhill, Ma., U.S. (hā′vẽr-hĭl)	61a	42°46′N	71°05′W
Haverhill, N.H., U.S.	67	44°00′N	72°05′W
Haverstraw, N.Y., U.S. (hā′vẽr-strô)	68a	41°11′N	73°58′W
Havlíčkův Brod, Czech Rep.	117	49°38′N	15°34′E
Havre, Mt., U.S. (hăv′ẽr)	64	48°34′N	109°42′W
Havre-Boucher, Can. (hăv′rà-bōō-shā′)	61	45°42′N	61°30′W
Havre de Grace, Md., U.S. (hăv′ẽr dē gràs′)	67	39°35′N	76°05′W
Havre-Saint Pierre, Can.	60	50°15′N	63°36′W
Haw, r., N.C., U.S. (hô)	83	36°17′N	79°46′W
Hawaii, state, U.S.	64c	20°00′N	157°40′W
Hawai′i, i., Hi., U.S. (häw wī′ē)	64c	19°30′N	155°30′W
Hawai′ian Islands, is., Hi., U.S. (hä-wī′ȧn)	64c	22°00′N	158°00′W
Hawai′i Volcanoes National Park, rec., Hi., U.S.	64c	19°30′N	155°25′W
Hawarden, Ia., U.S. (hā′wàr-dĕn)	70	43°00′N	96°28′W
Hawi, Hi., U.S. (hä′wē)	84a	20°16′N	155°48′W
Hawick, Scot., U.K. (hô′ĭk)	120	55°25′N	2°55′W
Hawke Bay, b., N.Z. (hôk)	175a	39°17′S	177°20′E
Hawker, Austl. (hô′kẽr)	176	31°58′S	138°12′E
Hawkesbury, Can. (hôks′bẽr-ĭ)	59	45°35′N	74°35′W
Hawkinsville, Ga., U.S. (hô′kĭnz-vĭl)	82	32°15′N	83°30′W
Hawks Nest Point, c., Bah.	93	24°05′N	75°30′W
Hawley, Mn., U.S. (hô′lĭ)	70	46°52′N	96°18′W
Haworth, Eng., U.K. (hā′wûrth)	114a	53°50′N	1°57′W
Hawthorne, Ca., U.S. (hô′thôrn)	75a	33°55′N	118°22′W
Hawthorne, Nv., U.S.	76	38°33′N	118°39′W
Haxtun, Co., U.S. (hăks′tŭn)	78	40°39′N	102°38′W
Hay, r., Austl. (hā)	174	23°00′S	136°45′E
Hay, r., Can.	52	60°21′N	117°14′W
Hayama, Japan (hä-yä′mä)	167a	35°16′N	139°35′E
Hayashi, Japan (hä-yä′shē)	167a	35°13′N	139°38′E
Hayden, Az., U.S. (hā′dĕn)	77	33°00′N	110°50′W
Hayes, r., Can.	53	55°25′N	93°55′W
Hayes, Mount, mtn., Ak., U.S. (hāz)	63	63°32′N	146°40′W
Haynesville, La., U.S. (hānz′vĭl)	81	32°55′N	93°08′W
Hayrabolu, Tur.	131	41°14′N	27°05′E
Hay River, Can.	50	60°50′N	115°53′W
Hays, Ks., U.S. (hāz)	78	38°51′N	99°20′W
Haystack Mountain, mtn., Wa., U.S. (hā-stăk′)	74a	48°26′N	122°07′W
Hayward, Ca., U.S. (hā′wẽrd)	74b	37°40′N	122°06′W
Hayward, Wi., U.S.	71	46°01′N	91°31′W
Hazard, Ky., U.S. (hăz′ȧrd)	82	37°13′N	83°10′W
Hazelhurst, Ga., U.S. (hā′z′l-hûrst)	83	31°50′N	82°36′W
Hazelhurst, Ms., U.S.	82	31°52′N	90°23′W
Hazel Park, Mi., U.S.	69b	42°28′N	83°06′W
Hazelton, Can. (hā′z′l-tŭn)	50	55°15′N	127°40′W
Hazelton Mountains, mts., Can.	54	55°00′N	128°00′W
Hazleton, Pa., U.S.	67	41°00′N	76°00′W
Headland, Al., U.S. (hĕd′lănd)	82	31°22′N	85°20′W
Healdsburg, Ca., U.S. (hēldz′bûrg)	76	38°37′N	122°52′W
Healdton, Ok., U.S. (hēld′tŭn)	79	34°13′N	97°28′W
Heanor, Eng., U.K. (hēn′ŏr)	114a	53°01′N	1°22′W
Heard Island, i., Austl. (hûrd)	3	53°10′S	74°35′E
Hearne, Tx., U.S. (hûrn)	81	30°53′N	96°35′W
Hearst, Can. (hûrst)	51	49°36′N	83°40′W
Heart, r., N.D., U.S. (härt)	70	46°46′N	102°34′W
Heart Lake Indian Reserve, I.R., Can.	55	55°02′N	111°30′W
Heart′s Content, Can. (härts kŏn′tĕnt)	61	47°52′N	53°22′W
Heavener, Ok., U.S. (hēv′nẽr)	79	34°52′N	94°36′W
Hebbronville, Tx., U.S. (hē′brŭn-vĭl)	80	27°18′N	98°40′W
Hebei, prov., China (hŭ-bā)	161	39°15′N	115°40′E
Heber City, Ut., U.S. (hē′bẽr)	77	40°30′N	111°25′W
Heber Springs, Ar., U.S.	79	35°28′N	91°59′W
Hebgen Lake, res., Mt., U.S. (hĕb′gĕn)	73	44°47′N	111°38′W
Hebrides, is., Scot., U.K.	112	57°00′N	6°30′W
Hebrides, Sea of the, sea, Scot., U.K.	120	57°00′N	7°00′W
Hebron, Can. (hĕb′rŭn)	51	58°11′N	62°56′W
Hebron, In., U.S.	69a	41°19′N	87°13′W
Hebron, Ky., U.S.	69f	39°04′N	84°43′W
Hebron, N.D., U.S.	70	46°54′N	102°04′W
Hebron, Ne., U.S.	79	40°11′N	97°36′W
Hebron *see* Al Khalīl, W.B.	153a	31°31′N	35°07′E
Heby, Swe. (hĭ′bü)	122	59°56′N	16°48′E
Hecate Strait, strt., Can. (hĕk′ȧ-tē)	52	53°00′N	131°00′W
Hecelchakán, Mex. (ā-sĕl-chä-kän′)	89	20°10′N	90°09′W
Hechi, China (hŭ-chr)	165	24°50′N	108°18′E
Hechuan, China (hŭ-chyuän)	160	30°00′N	106°20′E
Hecla Island, i., Can.	57	51°08′N	96°45′W
Hedemora, Swe. (hĭ-dĕ-mō′rä)	122	60°16′N	15°55′E
Hedon, Eng., U.K. (hĕ-dŭn)	114a	53°44′N	0°12′W
Heemstede, Neth.	115a	52°20′N	4°36′E
Heerlen, Neth.	121	50°55′N	5°58′E

PLACE (Pronunciation)	PAGE	LAT.	LONG.
Hefei, China (hŭ-fā)	161	31°51′N	117°15′E
Heflin, Al., U.S. (hĕf′lĭn)	82	33°40′N	85°33′W
Heide, Ger. (hī′dē)	124	54°13′N	9°06′E
Heidelberg, Austl. (hī′dĕl-bûrg)	173a	37°45′S	145°04′E
Heidelberg, Ger. (hīdĕl-bĕrgh)	117	49°24′N	8°43′E
Heidelberg, S. Afr.	192c	26°32′S	28°22′E
Heidenheim, Ger. (hī′dĕn-hīm)	124	48°41′N	10°09′E
Heilbron, S. Afr. (hīl′brōn)	192c	27°17′S	27°58′E
Heilbronn, Ger. (hīl′brōn)	117	49°09′N	9°16′E
Heiligenhaus, Ger. (hī′lē-gĕn-houz)	127c	51°19′N	6°58′E
Heiligenstadt, Ger. (hī′lē-gĕn-shtät)	124	51°21′N	10°10′E
Heilongjiang, prov., China (hā-lŏŋ-jyäŋ)	161	46°36′N	128°07′E
Heinola, Fin. (hå-nō′lȧ)	123	61°13′N	26°03′E
Heinsberg, Ger. (hīnz′bĕrgh)	127c	51°04′N	6°07′E
Heist-op-den-Berg, Bel.	115a	51°05′N	4°14′E
Hejaz *see* Al Ḥijāz, reg., Sau. Ar.	154	23°45′N	39°08′E
Hejian, China (hŭ-jyĕn)	164	38°28′N	116°05′E
Hekla, vol., Ice.	112	63°53′N	19°37′W
Hel, Pol. (hāl)	125	54°37′N	18°53′E
Helagsfjället, mtn., Swe.	116	62°54′N	12°24′E
Helan Shan, mts., China (hŭ-län shän)	160	38°02′N	105°20′E
Helena, Ar., U.S. (hē-lē′nȧ)	65	34°33′N	90°35′W
Helena, Mt., U.S. (hĕ-lē′nȧ)	64	46°35′N	112°01′W
Helensburgh, Austl. (hĕl′ĕnz-bŭr-ō)	173b	34°11′S	150°59′E
Helensburgh, Scot., U.K.	120	56°01′N	4°53′W
Helgoland, i., Ger. (hĕl′gō-länd)	124	54°13′N	7°30′E
Hellier, Ky., U.S. (hĕl′yẽr)	83	37°16′N	82°27′W
Hellín, Spain (ĕl-yēn′)	118	38°30′N	1°40′W
Hells Canyon, p., U.S.	72	45°20′N	116°45′W
Helmand, r., Afg. (hĕl′mŭnd)	154	31°00′N	63°48′E
Hel′miaziv, Ukr.	133	49°49′N	31°54′E
Helmond, Neth. (hĕl′mōnt) (ĕl′môN′)	121	51°35′N	5°04′E
Helmstedt, Ger. (hĕlm′shtĕt)	124	52°14′N	11°03′E
Helotes, Tx., U.S. (hē′lōts)	75d	29°35′N	98°41′W
Helper, Ut., U.S. (hĕlp′ẽr)	77	39°40′N	110°55′W
Helsingborg, Swe. (hĕl′sĭng-bôrgh)	116	56°04′N	12°40′E
Helsingfors *see* Helsinki, Fin.	110	60°10′N	24°53′E
Helsingør, Den. (hĕl-sĭng-ûr′)	116	56°03′N	12°33′E
Helsinki, Fin. (hĕl′sēn-kē)	110	60°10′N	24°53′E
Hemel Hempstead, Eng., U.K. (hĕm′ĕl hĕmp′stĕd)	114b	51°43′N	0°29′W
Hemer, Ger.	127c	51°22′N	7°46′E
Hemet, Ca., U.S. (hĕm′ĕt)	75a	33°45′N	116°57′W
Hemingford, Ne., U.S. (hĕm′ĭng-fẽrd)	70	42°21′N	103°30′W
Hemphill, Tx., U.S. (hĕmp′hĭl)	81	31°20′N	93°48′W
Hempstead, N.Y., U.S. (hĕmp′stĕd)	68a	40°42′N	73°37′W
Hempstead, Tx., U.S.	81	30°07′N	96°05′W
Hemse, Swe. (hĕm′sĕ)	122	57°15′N	18°25′E
Hemsön, i., Swe.	122	62°43′N	18°22′E
Henan, prov., China (hŭ-nän)	161	33°58′N	112°33′E
Henares, r., Spain (å-nä′rås)	128	40°50′N	2°55′W
Henderson, Ky., U.S. (hĕn′dẽr-sŭn)	66	37°50′N	87°30′W
Henderson, N.C., U.S.	83	36°18′N	78°24′W
Henderson, Nv., U.S.	76	36°09′N	115°04′W
Henderson, Tn., U.S.	82	35°25′N	88°40′W
Henderson, Tx., U.S.	81	32°09′N	94°48′W
Hendersonville, N.C., U.S. (hĕn′dẽr-sŭn-vĭl)	83	35°17′N	82°28′W
Hendersonville, Tn., U.S.	82	36°18′N	86°37′W
Hendon, Eng., U.K. (hĕn′dŭn)	114b	51°34′N	0°13′W
Hendrina, S. Afr. (hĕn-drē′nȧ)	192c	26°10′S	29°44′E
Hengch′un, Tai. (hĕng′chŭn′)	165	22°00′N	120°42′E
Hengelo, Neth. (hĕng′ĕ-lō)	121	52°20′N	6°45′E
Hengshan, China (hĕng′shän′)	165	27°20′N	112°40′E
Hengshui, China (hĕng′shōō-ē′)	162	37°43′N	115°42′E
Hengxian, China (hŭŋ shyĕn)	165	22°40′N	109°20′E
Hengyang, China	161	26°58′N	112°30′E
Heniches′k, Ukr.	137	46°11′N	34°47′E
Henley on Thames, Eng., U.K. (hĕn′lē ŏn tĕmz)	114b	51°31′N	0°54′W
Henlopen, Cape, c., De., U.S. (hĕn-lō′pĕn)	67	38°45′N	75°05′W
Hennebont, Fr. (ĕn-bôN′)	126	47°47′N	3°16′W
Hennenman, S. Afr.	192c	27°59′S	27°03′E
Hennessey, Ok., U.S. (hĕn′ĕ-sĭ)	79	36°04′N	97°53′W
Hennigsdorf, Ger. (hĕ′nĕngz-dōrf))	115b	52°39′N	13°12′E
Hennops, r., S. Afr. (hĕn′ŏps)	187b	25°51′S	27°57′E
Hennopsrivier, S. Afr.	187b	25°50′S	27°59′E
Henrietta, Ok., U.S. (hĕn-rĭ-ĕt′ȧ)	79	35°25′N	95°58′W
Henrietta, Tx., U.S. (hen-rĭ-ĕ′tȧ)	78	33°47′N	98°11′W
Henrietta Maria, Cape, c., Can. (hĕn-rĭ-ĕt′ȧ)	53	55°10′N	82°20′W
Henry Mountains, mts., Ut., U.S. (hĕn′rĭ)	64	37°55′N	110°45′W
Henrys Fork, r., Id., U.S.	73	43°52′N	111°55′W
Henteyn Nuruu, mtn., Russia	164	49°40′N	111°00′E
Hentiyn Nuruu, mts., Mong.	160	49°25′N	107°51′E
Henzada, Mya.	155	17°38′N	95°28′E
Heppner, Or., U.S. (hĕp′nẽr)	72	45°21′N	119°33′W
Hepu, China (hŭ-pōō)	165	21°28′N	109°10′E
Herāt, Afg. (hĕ-rät′)	154	34°28′N	62°13′E
Hercules, Can.	62g	53°27′N	113°20′W
Herdecke, Ger. (hĕr′dĕ-kĕ)	127c	51°24′N	7°26′E
Heredia, C.R. (ā-rā′dhē-ä)	91	10°04′N	84°06′W
Hereford, Eng., U.K. (hĕrĕ′fẽrd)	120	52°05′N	2°44′W
Hereford, Md., U.S.	68e	39°35′N	76°42′W
Hereford, Tx., U.S. (hĕr′ĕ-fẽrd)	78	34°47′N	102°25′W
Hereford and Worcester, co., Eng., U.K.	114a	52°24′N	2°15′W
Herencia, Spain (å-rān′thē-ä)	128	39°23′N	3°22′W
Herentals, Bel.	115a	51°10′N	4°51′E
Herford, Ger. (hĕr′fōrt)	124	52°06′N	8°42′E
Herington, Ks., U.S. (hĕr′ĭng-tŭn)	79	38°41′N	96°57′W
Herisau, Switz. (hā′rē-zou)	124	47°23′N	9°18′E
Herk-de-Stad, Bel.	115a	50°56′N	5°13′E
Herkimer, N.Y., U.S. (hûr′kĭ-mẽr)	67	43°05′N	75°00′W

PLACE (Pronunciation)	PAGE	LAT.	LONG.
Hermansville, Mi., U.S. (hûr′măns-vĭl)	66	45°40′N	87°35′W
Hermantown, Mn., U.S. (hĕr′măn-toun)	75h	46°46′N	92°12′W
Hermanusdorings, S. Afr.	192c	24°08′S	27°46′E
Herminie, Pa., U.S. (hûr-mĭ′nē)	69e	40°16′N	79°45′W
Hermitage Bay, b., Can. (hûr′mĭ-tĕj)	61	47°35′N	56°05′W
Hermit Islands, is., Pap. N. Gui. (hûr′mĭt)	169	1°48′S	144°55′E
Hermosa Beach, Ca., U.S. (hĕr-mō′sȧ)	75a	33°51′N	118°24′W
Hermosillo, Mex. (ĕr-mô-sē′l-yō)	86	29°00′N	110°57′W
Herndon, Va., U.S. (hĕrn′don)	68e	38°58′N	77°22′W
Herne, Ger. (hĕr′nĕ)	127c	51°32′N	7°13′E
Herning, Den. (hĕr′nĭng)	116	56°08′N	8°55′E
Heron, l., Mn., U.S. (hĕr′ŭn)	70	43°42′N	95°23′W
Heron Lake, Mn., U.S.	70	43°48′N	95°20′W
Herrero, Punta, Mex. (pŏ′n-tä-ĕr-rĕ′rô)	90a	19°18′N	87°24′W
Herrin, Il., U.S. (hĕr′ĭn)	66	37°50′N	89°00′W
Herschel, S. Afr. (hĕr′-shĕl)	187c	30°37′S	27°12′E
Herscher, Il., U.S. (hĕr′shẽr)	69a	41°03′N	88°06′W
Herstal, Bel. (hĕr′stäl)	121	50°42′N	5°32′E
Hertford, Eng., U.K.	120	51°48′N	0°05′W
Hertford, N.C., U.S. (hûrt′fẽrd)	83	36°10′N	76°30′W
Hertfordshire, co., Eng., U.K.	114b	51°46′N	0°05′W
Hertzberg, Ger. (hĕrtz′bĕrgh)	115b	52°54′N	12°58′E
Hervás, Spain	128	40°16′N	5°51′W
Herzliyya, Isr.	153a	32°10′N	34°49′E
Hessen, hist. reg., Ger. (hĕs′ĕn)	124	50°42′N	9°00′E
Hetch Hetchy Aqueduct, Ca., U.S. (hĕtch hĕt′chĭ ăk′wē-dŭkt)	76	37°27′N	120°54′W
Hettinger, N.D., U.S. (hĕt′ĭn-jẽr)	70	45°58′N	102°36′W
Heuningspruit, S. Afr.	192c	27°28′S	27°26′E
Hexian, China (hŭ shyĕn)	165	24°20′N	111°28′E
Hexian, China	162	31°44′N	118°20′E
Heyang, China (hŭ-yäŋ)	164	35°18′N	110°18′E
Heystekrand, S. Afr.	192c	25°16′S	27°14′E
Heyuan, China (hŭ-yüän)	165	23°48′N	114°45′E
Heywood, Eng., U.K. (hā′wŏd)	114a	53°36′N	2°12′W
Heze, China (hŭ-dzŭ)	162	35°13′N	115°28′E
Hialeah, Fl., U.S. (hī-ȧ-lē′äh)	83a	25°49′N	80°18′W
Hiawatha, Ks., U.S. (hī-ȧ-wô′thȧ)	79	39°50′N	95°33′W
Hiawatha, Ut., U.S.	77	39°25′N	111°05′W
Hibbing, Mn., U.S. (hĭb′ĭng)	65	47°26′N	92°58′W
Hickman, Ky., U.S. (hĭk′măn)	82	34°33′N	89°10′W
Hickory, N.C., U.S. (hĭk′ô-rĭ)	83	35°43′N	81°21′W
Hicksville, N.Y., U.S.	66	41°15′N	84°45′W
Hicksville, N.Y., U.S. (hĭks′vĭl)	68a	40°47′N	73°25′W
Hico, Tx., U.S. (hī′kō)	80	32°00′N	98°02′W
Hidalgo, Mex. (ē-dhäl′gō)	88	24°14′N	99°25′W
Hidalgo, Mex.	80	27°49′N	99°53′W
Hidalgo, state, Mex.	86	20°45′N	99°30′W
Hidalgo del Parral, Mex. (ē-dä′l-gō-dĕl-pär-rá′l)	86	26°55′N	105°40′W
Hidalgo Yalalag, Mex. (ē-dhäl′gō-yä-lä-läg)	89	17°12′N	96°11′W
Hierro Island, i., Spain (yĕ′r-rô)	184	27°37′N	18°29′W
Higashimurayama, Japan	167a	35°46′N	139°28′E
Higashiōsaka, Japan	167b	34°40′N	135°44′E
Higgins, l., Mi., U.S. (hĭg′ĭnz)	66	44°20′N	84°45′W
Higginsville, Mo., U.S. (hĭg′ĭnz-vĭl)	79	39°05′N	93°44′W
High, i., Mi., U.S.	66	45°45′N	85°45′W
High Bluff, Can.	62f	50°01′N	98°08′W
Highborne Cay, i., Bah. (hī′bôrn kē)	92	24°45′N	76°50′W
Highgrove, Ca., U.S. (hī′grōv)	75a	34°01′N	117°20′W
High Island, Tx., U.S.	81a	29°34′N	94°24′W
Highland, Ca., U.S. (hī′lănd)	75a	34°08′N	117°13′W
Highland, Il., U.S.	79	38°44′N	89°41′W
Highland, In., U.S.	69a	41°33′N	87°28′W
Highland, Mi., U.S.	69b	42°38′N	83°37′W
Highland Park, Il., U.S.	69a	42°11′N	87°47′W
Highland Park, Mi., U.S.	69b	42°24′N	83°06′W
Highland Park, N.J., U.S.	68a	40°30′N	74°25′W
Highland Park, Tx., U.S.	75c	32°49′N	96°48′W
Highlands, N.J., U.S. (hī-lăndz)	68a	40°24′N	73°59′W
Highlands, Tx., U.S.	81a	29°49′N	95°01′W
Highmore, S.D., U.S. (hī′mōr)	70	44°30′N	99°26′W
High Ongar, Eng., U.K. (ŏn′gẽr)	114b	51°43′N	0°15′E
High Peak, mtn., Phil.	169a	15°38′N	120°05′E
High Point, N.C., U.S.	83	35°55′N	80°00′W
High Prairie, Can.	50	55°26′N	116°29′W
High Ridge, Mo., U.S.	75e	38°27′N	90°32′W
High River, Can.	50	50°35′N	113°52′W
High Rock Lake, res., N.C., U.S. (hī′-rŏk)	83	35°40′N	80°15′W
High Springs, Fl., U.S.	83	29°48′N	82°38′W
High Tatra Mountains, mts., Eur.	125	49°15′N	19°40′E
Hightstown, N.J., U.S. (hīts-toun)	68a	40°16′N	74°32′W
High Wycombe, Eng., U.K. (wĭ-kŭm)	120	51°36′N	0°45′W
Higuero, Punta, c., P.R.	87b	18°21′N	67°11′W
Higuerote, Ven. (ē-gĕ-rô′tĕ)	101b	10°29′N	66°06′W
Higüey, Dom. Rep. (ē-gwĕ′y)	93	18°40′N	68°45′W
Hiiumaa, i., Est. (hē′ŏm-ô)	136	58°47′N	22°05′E
Hikone, Japan (hē′kō-nĕ)	167	35°15′N	136°15′E
Hildburghausen, Ger. (hĭld′bŏrg hou-zĕn)	124	50°26′N	10°45′E
Hilden, Ger. (hĕl′dĕn)	127c	51°10′N	6°56′E
Hildesheim, Ger. (hĭl′dĕs-hīm)	117	52°08′N	9°56′E
Hillaby, Mount, mtn., Barb. (hĭl′ȧ-bĭ)	91b	13°15′N	59°35′W
Hill City, Ks., U.S. (hĭl)	78	39°22′N	99°54′W
Hill City, Mn., U.S.	71	46°58′N	93°38′W
Hillegersberg, Neth.	115a	51°57′N	4°29′E
Hillerød, Den. (hē′lē-rûdh)	122	55°56′N	12°17′E
Hillsboro, Il., U.S. (hĭlz′bŭr-ō)	79	39°09′N	89°28′W
Hillsboro, Ks., U.S.	78	38°22′N	97°11′W
Hillsboro, N.D., U.S.	70	47°23′N	97°05′W

ng-sing; ŋ-baŋk; N-nasalized n; nŏd; cŏmmit; ōld; ôbey; ôrder; oi-boil; fōōd; ŏ-as oo in foot; ou-out; s-soft; sh-dish; th-thin; pūre; ûnite; ûrn; stŭd; circŭs; ü-as in French tu; ′-indeterminate vowel.

256 Hillsboro – Hrodna PRONOUNCING INDEX

PLACE (Pronunciation)	PAGE	LAT.	LONG.
Hillsboro, N.H., U.S.	67	43°05'N	71°55'W
Hillsboro, Oh., U.S.	66	39°10'N	83°40'W
Hillsboro, Or., U.S.	74c	45°31'N	122°59'W
Hillsboro, Tx., U.S.	81	32°01'N	97°06'W
Hillsboro, Wi., U.S.	71	43°39'N	90°20'W
Hillsburgh, Can. (hĭlz'bûrg)	62d	43°48'N	80°09'W
Hills Creek Lake, res., Or., U.S.	72	43°41'N	122°26'W
Hillsdale, Mi., U.S. (hĭls-dāl)	77	41°55'N	84°35'W
Hilo, Hi., U.S. (hē'lō)	64c	19°44'N	155°01'W
Hilvarenbeek, Neth.	115a	51°29'N	5°10'E
Hilversum, Neth. (hĭl'vĕr-sŭm)	115a	52°13'N	5°10'E
Himachal Pradesh, India	155	32°00'N	77°30'E
Himalayas, mts., Asia	155	29°30'N	85°02'E
Himeji, Japan (hē'må-jė)	166	34°50'N	134°42'E
Himmelpforten, Ger. (hē'mĕl-pfōr-tĕn)	115c	53°37'N	9°19'E
Ḥimṣ, Syria	154	34°44'N	36°43'E
Hinche, Haiti (hēn'chå) (ăɴsh)	93	19°10'N	72°05'W
Hinchinbrook, i., Austl. (hĭn-chĭn-brōōk)	174	18°23'S	146°57'W
Hinckley, Eng., U.K. (hĭnk'lĭ)	114a	52°32'N	1°21'W
Hindley, Eng., U.K. (hĭnd'lĭ)	114a	53°32'N	2°35'W
Hindu Kush, mts., Asia (hĭn'dōō kōōsh')	155	35°15'N	68°44'E
Hindupur, India (hĭn'dōō-pōōr)	159	13°52'N	77°34'E
Hingham, Ma., U.S. (hĭng'ăm)	61a	42°14'N	70°53'W
Hinkley, Oh., U.S. (hĭnk'-lĭ)	69d	41°14'N	81°45'W
Hinojosa del Duque, Spain (ē-nō-kō'sä)	128	38°30'N	5°09'W
Hinsdale, Il., U.S. (hĭnz'dāl)	69a	41°48'N	87°56'W
Hinton, Can. (hĭn'tŭn)	55	53°25'N	117°34'W
Hinton, W.V., U.S. (hĭn'tŭn)	66	37°40'N	80°55'W
Hirado, i., Japan (hē'rä-dō)	166	33°19'N	129°18'E
Hirakata, Japan (hē'rä-kä'tä)	167b	34°49'N	135°40'E
Hirara, Japan	170d	24°48'N	125°17'E
Hiratsuka, Japan (hē-rät-sōō'kå)	167	35°20'N	139°19'E
Hirosaki, Japan (hē'rō-sä'kē)	161	40°31'N	140°38'E
Hirose, Japan (hē'rō-sā)	167	35°20'N	133°11'E
Hiroshima, Japan (hē-rō-shē'mȧ)	161	34°22'N	132°25'E
Hirson, Fr. (ēr-sôɴ')	126	49°54'N	4°00'E
Hisar, India	158	29°15'N	75°47'E
Hispaniola, i., N.A. (hĭ'spăn-ĭ-ō-lȧ)	87	17°30'N	73°15'W
Hitachi, Japan (hē-tä'chē)	166	36°42'N	140°47'E
Hitchcock, Tx., U.S. (hĭch'kŏk)	81a	29°21'N	95°01'W
Hitoyoshi, Japan (hē'tō-yō'shē)	167	32°13'N	130°45'E
Hitra, i., Nor. (hĭträ)	116	63°34'N	7°37'E
Hittefeld, Ger. (hē'tĕ-fĕld)	115c	53°23'N	9°59'E
Hiwasa, Japan (hē'wȧ-sä)	167	33°44'N	134°31'E
Hiwassee, r., Tn., U.S. (hī-wôs'sē)	82	35°10'N	84°35'W
Hjälmaren, l., Swe.	116	59°07'N	16°05'E
Hjo, Swe. (yō)	122	58°19'N	14°11'E
Hjørring, Den. (jûr'ĭng)	116	57°27'N	9°59'E
Hlobyne, Ukr.	133	49°22'N	33°17'E
Hlohovec, Slvk. (hlō'ho-vĕts)	125	48°24'N	17°49'E
Hlukhiv, Ukr.	137	51°42'N	33°52'E
Hlybokaye, Bela.	136	55°08'N	27°44'E
Hobart, Austl. (hō'bȧrt)	175	43°00'S	147°30'E
Hobart, In., U.S.	69a	41°31'N	87°15'W
Hobart, Ok., U.S.	78	35°02'N	99°06'W
Hobart, Wa., U.S.	74a	47°25'N	121°58'W
Hobbs, N.M., U.S. (hŏbs)	78	32°41'N	103°15'W
Hoboken, Bel. (hō'bō-kĕn)	115a	51°11'N	4°20'E
Hoboken, N.J., U.S.	68a	40°43'N	74°03'W
Hobro, Den. (hô-brô')	122	56°38'N	9°47'E
Hobson, Va., U.S. (hŏb'sŭn)	68g	36°54'N	76°31'W
Hobson's Bay, b., Austl. (hŏb'sŭnz)	173a	37°54'S	144°45'E
Hobyo, Som.	192a	5°24'N	48°28'E
Ho Chi Minh City, Viet.	168	10°46'N	106°34'E
Hockinson, Wa., U.S. (hŏk'ĭn-sŭn)	74c	45°44'N	122°29'W
Hoctún, Mex. (ôk-tōō'n)	90a	20°52'N	89°10'W
Hodgenville, Ky., U.S. (hŏj'ĕn-vĭl)	66	37°35'N	85°45'W
Hodges Hill, mtn., Can. (hŏj'ĕz)	61	49°04'N	55°53'W
Hódmezővásárhely, Hung. (hōd'mĕ-zû-vô'shôr-hĕl-y')	125	46°24'N	20°21'E
Hodna, Chott el, l., Alg.	118	35°20'N	3°27'E
Hodonin, Czech Rep. (hē'dô-nēn)	125	48°50'N	17°06'E
Hoegaarden, Bel.	115a	50°46'N	4°55'E
Hoek van Holland, Neth.	115a	51°59'N	4°05'E
Hoeryŏng, Kor., N. (hwĕr'yŭng)	166	42°28'N	129°39'E
Hof, Ger. (hōf)	124	50°19'N	11°55'E
Hofsjökull, ice, Ice. (hôfs'yü'kōōl)	116	64°55'N	18°40'W
Hog, i., Mi., U.S.	66	45°50'N	85°20'W
Hogansville, Ga., U.S. (hō'gănz-vĭl)	82	33°10'N	84°54'W
Hog Cay, i., Bah.	93	23°35'N	75°30'W
Hogsty Reef, rf., Bah.	93	21°45'N	73°50'W
Hohenbrunn, Ger. (hō'hĕn-brōōn)	115d	48°03'N	11°42'E
Hohenlimburg, Ger. (hō'hĕn lēm'bōōrg)	127c	51°20'N	7°35'E
Hohen Neuendorf, Ger. (hō'hĕn noi'ĕn-dōrf)	115b	52°40'N	13°22'E
Hohe Tauern, mts., Aus. (hō'ĕ tou'ērn)	124	47°11'N	12°12'E
Hohhot, China (hŭ-hōō-tŭ)	161	41°05'N	111°50'E
Hohoe, Ghana	188	7°09'N	0°28'E
Hohokus, N.J., U.S. (hō-hō-kŭs)	68a	41°01'N	74°08'W
Hoi An, Viet.	165	15°48'N	108°30'E
Hoisington, Ks., U.S. (hoi'zĭng-tŭn)	78	38°30'N	98°46'W
Hojo, Japan (hō'jō)	167	33°58'N	132°50'E
Hokitika, N.Z. (hō-kĭ-tē'kä)	175a	42°43'S	170°59'E
Hokkaidō, i., Japan (hŏk'kī-dō)	166	43°30'N	142°45'E
Holbaek, Den. (hôl'bĕk)	122	55°42'N	11°40'E
Holbox, Mex. (ôl-bō'x)	90a	21°33'N	87°19'W
Holbox, Isla, i., Mex. (ē's-lä-ôl-bō'x)	90a	21°40'N	87°21'W
Holbrook, Az., U.S. (hōl'brŏŏk)	77	34°55'N	110°15'W
Holbrook, Ma., U.S.	61a	42°10'N	71°01'W
Holden, Ma., U.S. (hōl'dĕn)	61a	42°21'N	71°51'W
Holden, Mo., U.S.	79	38°42'N	94°00'W
Holden, W.V., U.S.	66	37°45'N	82°05'W
Holdenville, Ok., U.S. (hōl'dĕn-vĭl)	79	35°05'N	96°25'W
Holdrege, Ne., U.S. (hōl'drĕj)	78	40°25'N	99°28'W
Holguín, Cuba (ôl-gēn')	87	20°55'N	76°15'W
Holguín, prov., Cuba	92	20°40'N	76°15'W
Holidaysburg, Pa., U.S. (hŏl'ĭ-dāz-bûrg)	67	40°30'N	78°30'W
Hollabrunn, Aus.	124	48°33'N	16°04'E
Holland, Mi., U.S. (hŏl'ănd)	66	42°45'N	86°10'W
Hollands Diep, strt., Neth.	115a	51°43'N	4°25'E
Hollenstedt, Ger. (hō'lĕn-shtĕt)	115c	53°22'N	9°43'E
Hollis, N.H., U.S. (hŏl'ĭs)	61a	42°30'N	71°29'W
Hollis, Ok., U.S.	78	34°39'N	99°56'W
Hollister, Ca., U.S. (hŏl'ĭs-tēr)	76	36°50'N	121°25'W
Holliston, Ma., U.S. (hŏl'ĭs-tŭn)	61a	42°12'N	71°25'W
Holly, Mi., U.S. (hŏl'ĭ)	66	42°45'N	83°30'W
Holly, Wa., U.S.	74a	47°34'N	122°58'W
Holly Springs, Ms., U.S. (hŏl'ĭ sprĭngz)	82	34°45'N	89°28'W
Hollywood, Ca., U.S. (hŏl'ē-wŏŏd)	75a	34°06'N	118°20'W
Hollywood, Fl., U.S.	83a	26°00'N	80°11'W
Holmes Reefs, rf., Austl. (hōmz)	175	16°33'S	148°43'E
Holmestrand, Nor. (hôl'mĕ-strän)	122	59°29'N	10°17'E
Holmsbu, Nor. (hôlms'bōō)	122	59°36'N	10°26'E
Holmsjön, l., Swe.	122	62°23'N	15°43'E
Holstebro, Den. (hôl'stĕ-brô)	116	56°22'N	8°39'E
Holstein, hist. reg., Ger.	124	54°10'N	9°40'E
Holston, r., Tn., U.S. (hōl'stŭn)	82	36°02'N	83°42'W
Holt, Eng., U.K. (hōlt)	114a	53°05'N	2°53'W
Holton, Ks., U.S. (hōl'tŭn)	79	39°27'N	95°43'W
Holy Cross, Ak., U.S. (hō'lĭ krôs)	63	62°10'N	159°40'W
Holyhead, Wales, U.K. (hŏl'ē-hĕd)	120	53°18'N	4°45'W
Holy Island, i., Eng., U.K.	120	55°43'N	1°48'W
Holy Island, i., Wales, U.K. (hō'lĭ)	120	53°15'N	4°45'W
Holyoke, Co., U.S. (hōl'yōk)	78	40°36'N	102°18'W
Holyoke, Ma., U.S.	67	42°10'N	72°40'W
Homano, Japan (hō-mä'nō)	167a	35°33'N	140°08'E
Homberg, Ger. (hōm'bĕrgh)	127c	51°27'N	6°42'E
Hombori, Mali	188	15°17'N	1°42'W
Home Gardens, Ca., U.S. (hōm gär'd'nz)	75a	33°53'N	117°32'W
Homeland, Ca., U.S. (hōm'lănd)	75a	33°44'N	117°07'W
Homer, Ak., U.S. (hō'mēr)	63	59°42'N	151°30'W
Homer, La., U.S.	81	32°46'N	93°05'W
Homer Youngs Peak, mtn., Mt., U.S.	73	45°19'N	113°41'W
Homestead, Fl., U.S. (hōm'stĕd)	83a	25°27'N	80°28'W
Homestead, Mi., U.S.	75k	46°20'N	84°07'W
Homestead, Pa., U.S.	69e	40°29'N	79°55'W
Homestead National Monument of America, rec., Ne., U.S.	79	40°16'N	96°51'W
Homewood, Al., U.S. (hōm'wŏŏd)	68h	33°28'N	86°48'W
Homewood, Il., U.S.	69a	41°34'N	87°40'W
Hominy, Ok., U.S. (hŏm'ĭ-nĭ)	79	36°25'N	96°24'W
Homochitto, r., Ms., U.S. (hō-mō-chĭt'ō)	82	31°23'N	91°15'W
Homyel', Bela.	136	52°20'N	31°03'E
Homyel', prov., Bela.	132	52°18'N	29°00'E
Honda, Col. (hōn'dȧ)	100	5°13'N	74°45'W
Honda, Bahía, b., Cuba (bä-ē'ä-ô'n-dä)	92	23°10'N	83°20'W
Hondo, Tx., U.S.	80	29°20'N	99°08'W
Hondo, r., N.M., U.S.	78	33°22'N	105°06'W
Hondo, Río, r., N.A. (hon-dō')	90a	18°16'N	88°32'W
Honduras, nation, N.A. (hŏn-dōō'rȧs)	86	14°30'N	88°00'W
Honduras, Gulf of, b., N.A.	86	16°30'N	87°30'W
Honea Path, S.C., U.S. (hŭn'ĭ păth)	83	34°25'N	82°16'W
Hönefoss, Nor. (hē'nē-fôs)	116	60°10'N	10°15'E
Honesdale, Pa., U.S. (hōnz'dāl)	67	41°30'N	75°15'W
Honey Grove, Tx., U.S. (hŭn'ĭ grōv)	79	33°35'N	95°54'W
Honey Lake, l., Ca., U.S. (hŭn'ĭ)	76	40°11'N	120°34'W
Honfleur, Can. (ôɴ-flûr')	62b	46°39'N	70°53'W
Honfleur, Fr. (ôɴ-flûr')	126	49°26'N	0°13'E
Hon Gay, Viet.	165	20°58'N	107°10'E
Hong Kong (Xianggang), China	161	21°45'N	115°00'E
Hongshui, r., China (hôŋ-shwā)	160	24°30'N	105°00'E
Honguedo, Détroit d', strt., Can.	60	49°08'N	63°45'W
Hongze Hu, l., China	161	33°17'N	118°37'E
Honiara, Sol. Is.	175	9°26'S	159°57'E
Honiton, Eng., U.K. (hŏn'ĭ-tŏn)	120	50°49'N	3°10'W
Honolulu, Hi., U.S. (hŏn-ō-lōō'lōō)	64c	21°18'N	157°50'W
Honomu, Hi., U.S. (hŏn'ō-mōō)	84a	19°50'N	155°04'W
Honshū, i., Japan	161	36°00'N	138°00'E
Hood, Mount, mtn., Or., U.S.	64	45°20'N	121°43'W
Hood Canal, b., Wa., U.S. (hŏŏd)	74a	47°45'N	122°45'W
Hood River, Or., U.S.	64	45°42'N	121°30'W
Hoodsport, Wa., U.S. (hŏŏdz'pōrt)	74a	47°25'N	123°09'W
Hoogly, r., India (hōōg'lĭ)	155	21°35'N	87°50'E
Hoogstraten, Bel.	115a	51°24'N	4°46'E
Hooker, Ok., U.S. (hŏŏk'ēr)	78	36°49'N	101°13'W
Hool, Mex. (ōō'l)	90a	19°32'N	90°22'W
Hoonah, Ak., U.S. (hōō'nȧ)	63	58°05'N	135°25'W
Hoopa Valley Indian Reservation, I.R., Ca., U.S.	72	41°18'N	123°35'W
Hooper, Ne., U.S. (hŏŏp'ēr)	79	41°37'N	96°31'W
Hooper, Ut., U.S.	75b	41°10'N	112°08'W
Hooper Bay, Ak., U.S.	63	61°32'N	166°02'W
Hoopeston, Il., U.S. (hōōps'tŭn)	66	40°35'N	87°40'W
Hoosick Falls, N.Y., U.S. (hōō'sĭk)	67	42°55'N	73°15'W
Hoover Dam, Nv., U.S. (hōō'vēr)	76	36°00'N	115°06'W
Hoover Dam, dam, U.S.	64	36°00'N	114°27'W
Hopatcong, Lake, l., N.J., U.S. (hō-păt'kong)	68a	40°57'N	74°38'W
Hope, Ak., U.S. (hōp)	63	60°54'N	149°48'W
Hope, Ar., U.S.	79	33°41'N	93°35'W
Hope, N.D., U.S.	70	47°17'N	97°45'W
Hope, Ben, mtn., Scot., U.K. (bĕn hōp)	120	58°25'N	4°25'W
Hopedale, Can. (hōp'dāl)	51	55°26'N	60°11'W
Hopedale, Ma., U.S. (hōp'dāl)	61a	42°08'N	71°33'W
Hopelchén, Mex. (o-pĕl-chĕ'n)	90a	19°47'N	89°51'W
Hopes Advance, Cap, c., Can. (hōps ăd-vans')	53	61°05'N	69°35'W
Hopetoun, Austl. (hōp'toun)	174	33°50'S	120°15'E
Hopetown, S. Afr. (hōp'toun)	186	29°35'S	24°10'E
Hopewell, Va., U.S. (hōp'wĕl)	83	37°14'N	77°15'W
Hopewell Culture National Historical Park, rec., Oh., U.S.	66	39°25'N	83°00'W
Hopi Indian Reservation, I.R., Az., U.S. (hō'pē)	77	36°20'N	110°30'W
Hopkins, Mn., U.S. (hŏp'kĭns)	75g	44°55'N	93°24'W
Hopkinsville, Ky., U.S. (hŏp'kĭns-vĭl)	65	36°50'N	87°28'W
Hopkinton, Ma., U.S. (hŏp'kĭn-tŭn)	61a	42°14'N	71°31'W
Hoquiam, Wa., U.S. (hō'kwĭ-ăm)	64	47°00'N	123°53'W
Horconcitos, Pan. (ŏr-kôn-sē'-tōs)	91	8°18'N	82°11'W
Horgen, Switz. (hôr'gĕn)	124	47°16'N	8°35'E
Horicon, Wi., U.S. (hŏr'ĭ-kŏn)	71	43°26'N	88°40'W
Horlivka, Ukr.	137	48°17'N	38°03'E
Hormuz, Strait of, strt., Asia (hôr'mŭz')	154	26°30'N	56°30'E
Horn, i., Austl. (hôrn)	175	10°30'S	143°30'E
Horn, Cape *see* Hornos, Cabo de, c., Chile	102	56°00'S	67°00'W
Hornavan, l., Swe.	116	65°54'N	16°17'E
Horneburg, Ger. (hōr'nĕ-bōrgh)	115c	53°30'N	9°35'E
Hornell, N.Y., U.S. (hôr-nĕl')	67	42°20'N	77°40'W
Hornos, Cabo de, c., Chile	102	56°00'S	67°00'W
Horn Plateau, plat., Can.	52	62°12'N	120°29'W
Hornsby, Austl. (hôrnz'bĭ)	173b	33°43'S	151°06'E
Horodenka, Ukr.	125	48°40'N	25°30'E
Horodnia, Ukr.	133	51°54'N	31°31'E
Horodok, Ukr.	125	49°47'N	23°39'E
Horqueta, Para. (ŏr-kĕ'tä)	102	23°20'S	57°00'W
Horse Creek, r., Co., U.S. (hôrs)	78	38°49'N	103°48'W
Horse Creek, r., Wy., U.S.	70	41°33'N	104°39'W
Horse Islands, is., Can.	61	50°11'N	55°45'W
Horsens, Den. (hôrs'ĕns)	122	55°50'N	9°49'E
Horseshoe Bay, Can. (hôrs-shōō)	74d	49°23'N	123°16'W
Horsforth, Eng., U.K. (hôrs'fŭrth)	114a	53°50'N	1°38'W
Horsham, Austl. (hôr'shăm) (hôrs'ăm)	175	36°42'S	142°17'E
Horst, Ger. (hôrst)	115c	53°49'N	9°37'E
Horten, Nor. (hôr'tĕn)	122	59°26'N	10°27'E
Horton, Ks., U.S. (hôr'tŭn)	79	39°38'N	95°32'W
Horton, r., Ak., U.S. (hôr'tŭn)	63	68°38'N	122°00'W
Horwich, Eng., U.K. (hŏr'ĭch)	114a	53°36'N	2°33'W
Horyn', r., Eur. (gō'rĕn')	125	50°55'N	26°07'E
Hososhima, Japan (hō'sô-shē'mä)	166	32°25'N	131°40'E
Hoste, i., Chile (ôs'tā)	102	55°20'S	70°45'W
Hostotipaquillo, Mex. (ôs-tō'tĭ-pä-kēl'yō)	88	21°09'N	104°05'W
Hota, Japan (hō'tä)	167a	35°08'N	139°50'E
Hotan, China (hwō-tän)	160	37°11'N	79°50'E
Hotan, r., China	160	39°09'N	81°08'E
Hoto Mayor, Dom. Rep. (ô-tô-mä-yō'r)	93	18°45'N	69°10'W
Hot Springs, Ak., U.S. (hŏt sprĭngs)	63	65°00'N	150°20'W
Hot Springs, Ar., U.S.	65	34°29'N	93°02'W
Hot Springs, S.D., U.S.	70	43°28'N	103°32'W
Hot Springs, Va., U.S.	67	38°00'N	79°55'W
Hot Springs National Park, rec., Ar., U.S.	65	34°30'N	93°00'W
Hotte, Massif de la, mts., Haiti	93	18°25'N	74°00'W
Hotville, Ca., U.S. (hŏt'vĭl)	76	32°50'N	115°24'W
Houdan, Fr. (ōō-däɴ')	127b	48°47'N	1°36'E
Houghton, Mi., U.S. (hō'tŭn)	71	47°06'N	88°36'W
Houghton, l., Mi., U.S.	66	44°20'N	84°45'W
Houilles, Fr. (ōō-yēs')	127b	48°55'N	2°11'E
Houjie, China (hwô-jyĕ)	163a	22°58'N	113°39'E
Houlton, Me., U.S. (hōl'tŭn)	60	46°07'N	67°50'W
Houma, La., U.S. (hōō'mȧ)	81	29°36'N	90°43'W
Housatonic, r., U.S. (hōō-sȧ-tŏn'ĭk)	67	41°50'N	73°25'W
House Springs, Mo., U.S. (hous sprĭngs)	75e	38°24'N	90°34'W
Houston, Ms., U.S. (hūs'tŭn)	82	33°53'N	89°00'W
Houston, Tx., U.S.	65	29°46'N	95°21'W
Houston Ship Channel, strt., Tx., U.S.	81a	29°38'N	94°57'W
Houtbaai, S. Afr.	186a	34°03'S	18°22'E
Houtman Rocks, is., Austl. (hout'män)	174	28°15'S	112°45'E
Houzhen, China (hwō-jŭn)	162	36°59'N	118°59'E
Hovd, Mong.	160	48°08'N	91°40'E
Hovd Gol, r., Mong.	160	49°06'N	91°16'E
Hove, Eng., U.K. (hōv)	120	50°50'N	0°09'W
Hövsgöl Nuur, l., Mong.	160	51°11'N	99°11'E
Howard, Ks., U.S. (hou'ărd)	79	37°27'N	96°10'W
Howard, S.D., U.S.	70	44°01'N	97°31'W
Howden, Eng., U.K. (hou'dĕn)	114a	53°44'N	0°52'W
Howe, Cape, c., Austl. (hou)	175	37°30'S	150°40'E
Howell, Mi., U.S. (hou'ĕl)	66	42°40'N	84°00'W
Howe Sound, strt., Can.	54	49°22'N	123°18'W
Howick, Can. (hou'ĭk)	62a	45°11'N	73°51'W
Howick, S. Afr.	187c	29°29'S	30°16'E
Howland, i., Oc. (hou'lănd)	2	1°00'N	176°00'W
Howrah, India (hou'rä)	155	22°33'N	88°20'E
Howse Peak, mtn., Can.	55	51°30'N	116°40'W
Howson Peak, mtn., Can.	54	54°25'N	127°45'W
Hoxie, Ar., U.S. (kŏh'sĭ)	79	36°03'N	91°00'W
Hoy, i., Scot., U.K. (hoi)	120a	58°53'N	3°10'W
Höya, Japan	167a	35°45'N	139°35'E
Hoylake, Eng., U.K. (hoi-lāk')	114a	53°23'N	3°11'W
Hoyo, Sierra del, mts., Spain (sē-ĕ'r-rä-dĕl-ō'yō)	129a	40°39'N	3°56'W
Hradec Králové, Czech Rep.	117	50°12'N	15°50'E
Hradyz'k, Ukr.	133	49°12'N	33°06'E
Hranice, Czech Rep. (hrän'yĕ-tsĕ)	125	49°33'N	17°45'E
Hröby, Swe. (hûr'bü)	122	55°50'N	13°41'E
Hrodna, Bela.	136	53°40'N	23°49'E

ăt; finăl; rāte; senāte; ärm; åsk; sofȧ; fâre; ch-choose; dh-as th in other; bē; ēvent; bĕt; recĕnt; cratēr; g-gō; gh-guttural g; bĭt; ĭ-short neutral; rīde; ĸ-guttural k as ch in German ich;

PLACE (Pronunciation)	PAGE	LAT.	LONG.
Hron, r., Slvk.	125	48°22'N	18°42'E
Hrubieszów, Pol. (hrōō-byā'shōōf)	125	50°48'N	23°54'E
Hsawnhsup, Mya.	160	24°29'N	94°45'E
Hsinchu, Tai. (hsĭn'chōō')	165	24°48'N	121°00'E
Huadian, China (hwä-dĭĕn)	164	42°38'N	126°45'E
Huai, r., China (hwī)	161	32°07'N	114°38'E
Huai'an, China (hwī-än)	164	33°31'N	119°11'E
Huailai, China	164	40°20'N	115°45'E
Huailin, China (hwī-lĭn)	162	31°27'N	117°36'E
Huainan, China	162	32°38'N	117°02'E
Huaiyang, China (hōōäi'yang)	164	33°45'N	114°54'E
Huaiyuan, China (hwī-yůän)	164	32°53'N	117°13'E
Huajicori, Mex. (wä-jē-kō'rē)	88	22°41'N	105°24'W
Huajuapan de León, Mex. (wäj-wä'päm dā lā-ón')	89	17°46'N	97°45'W
Hualapai Indian Reservation, I.R., Az., U.S. (wälåpī)	77	35°41'N	113°38'W
Hualapai Mountains, mts., Az., U.S.	77	34°53'N	113°54'W
Hualien, Tai. (hwä'lyĕn')	165	23°58'N	121°58'E
Huallaga, r., Peru (wäl-yä'gä)	100	8°12'S	76°34'W
Huamachuco, Peru (wä-mä-chōō'kō)	100	7°52'S	78°11'W
Huamantla, Mex. (wä-män'tlà)	89	19°18'N	97°54'W
Huambo, Ang.	186	12°44'S	15°47'E
Huamuxtitlán, Mex. (wä-mōōs-tē-tlän')	88	17°49'N	98°38'W
Huancavelica, Peru (wän'kä-vā-lē'kä)	100	12°47'S	75°02'W
Huancayo, Peru (wän-kä'yō)	100	12°09'S	75°04'W
Huanchaca, Bol. (wän-chä'kä)	100	20°09'S	66°40'W
Huang (Yellow), r., China (hůäŋ)	161	35°06'N	113°39'E
Huang, Old Beds of the, mth., China	160	40°28'N	106°34'E
Huang, Old Course of the, r., China	162	34°28'N	116°59'E
Huangchuan, China (hůäŋ-chůän)	164	32°07'N	115°01'E
Huanghua, China (hůäŋ-hwä)	162	38°28'N	117°18'E
Huanghuadian, China (hůäŋ-hwä-dĭĕn)	162	39°22'N	116°53'E
Huangli, China (hōōäng'lē)	162	31°39'N	119°42'E
Huangpu, China (hůäŋ-pōō)	163a	22°44'N	113°20'E
Huangpu, r., China	163b	30°56'N	121°16'E
Huangqiao, China (hůän-chyou)	162	32°15'N	120°13'E
Huangxian, China (hůäŋ shyĕn)	162	37°39'N	120°32'E
Huangyuan, China (hůäŋ-yůän)	160	37°00'N	101°01'E
Huanren, China (hůän-rŭn)	164	41°10'N	125°30'E
Huánuco, Peru (wä-nōō'kō)	100	9°50'S	76°17'W
Huánuni, Bol. (wä-nōō'nē)	100	18°11'S	66°43'W
Huaquechula, Mex. (wä-kē-chōō'lä)	88	18°44'N	98°37'W
Huaral, Peru (wä-rä'l)	100	11°28'S	77°11'W
Huarás, Peru (ōä'rà's)	100	9°32'S	77°29'W
Huascarán, Nevados, mts., Peru (wäs-kä-rän')	100	9°05'S	77°50'W
Huasco, Chile (wäs'kō)	102	28°32'S	71°16'W
Huatla de Jiménez, Mex. (wà'tlä-dĕ-ĸe-mĕ'nĕz)	89	18°08'N	96°49'W
Huatlatlauch, Mex. (wä'tlä-tlä-ōō'ch)	88	18°40'N	98°04'W
Huatusco, Mex. (wä-tōōs'kō)	89	19°09'N	96°57'W
Huauchinango, Mex. (wä-ōō-chē-näŋ'gō)	88	20°09'N	98°03'W
Huaunta, Nic. (wä-ō'n-tä)	91	13°30'N	83°32'W
Huaunta, Laguna, l., Nic. (lä-gò'nä-wä-ō'n-tä)	91	13°35'N	83°46'W
Huautla, Mex. (wä-ōō'tlä)	88	21°04'N	98°13'W
Huaxian, China (hwä shyĕn)	164	35°34'N	114°32'E
Huaynamota, Río de, r., Mex. (rē'ō-dĕ-wäy-nä-mō'tä)	88	22°10'N	104°36'W
Huazolotitlán, Mex. (wäzō-lō-tlē-tlän')	89	16°18'N	97°55'W
Hubbard, N.H., U.S. (hŭb'ērd)	61a	42°53'N	71°12'W
Hubbard, Tx., U.S.	81	31°53'N	96°46'W
Hubbard, l., Mi., U.S.	66	44°45'N	83°30'W
Hubbard Creek Reservoir, res., Tx., U.S.	80	32°50'N	98°55'W
Hubei, prov., China (hōō-bā)	161	31°20'N	111°58'E
Hubli, India (hò'blē)	155	15°25'N	75°09'E
Hückeswagen, Ger. (hü'kĕs-vä'gĕn)	127c	51°09'N	7°20'E
Hucknall, Eng., U.K. (hŭk'năl)	114a	53°02'N	1°12'W
Huddersfield, Eng., U.K. (hŭd'ērz-fēld)	120	53°39'N	1°47'W
Hudiksvall, Swe. (hōō'dĭks-väl)	116	61°44'N	17°05'E
Hudson, Can. (hŭd'sŭn)	62a	45°26'N	74°08'W
Hudson, Ma., U.S.	61a	42°24'N	71°34'W
Hudson, Mi., U.S.	66	41°50'N	84°15'W
Hudson, N.Y., U.S.	67	42°15'N	73°45'W
Hudson, Oh., U.S.	69d	41°15'N	81°27'W
Hudson, Wi., U.S.	75g	44°59'N	92°45'W
Hudson, r., U.S.	65	42°30'N	73°55'W
Hudson Bay, Can.	57	52°52'N	102°25'W
Hudson Bay, b., Can.	53	60°15'N	85°30'W
Hudson Falls, N.Y., U.S.	67	43°20'N	73°30'W
Hudson Heights, Can.	62a	45°28'N	74°09'W
Hudson Strait, strt., Can.	53	63°25'N	74°05'W
Hue, Viet. (ü-ā')	168	16°28'N	107°42'E
Huebra, r., Spain (wĕ'brä)	128	40°44'N	6°17'W
Huehuetenango, Guat. (wā-wå-tå-näŋ'gō)	90	15°19'N	91°26'W
Huejotzingo, Mex. (wā-hò-tzĭŋ'gō)	88	19°09'N	98°24'W
Huejúcar, Mex. (wā-hōō'kär)	88	22°26'N	103°12'W
Huejuquilla el Alto, Mex. (wā-hōō-kēl'yä ĕl äl'tō)	88	22°42'N	103°54'W
Huejutla, Mex. (wā-hōō'tlä)	88	21°08'N	98°26'W
Huelma, Spain (wĕl'mä)	128	37°39'N	3°36'W
Huelva, Spain (wĕl'vä)	118	37°16'N	6°58'W
Huércal-Overa, Spain (wĕr-käl' ō-vā'rä)	128	37°12'N	1°58'W
Huerfano, r., Co., U.S. (wâr'fá-nō)	78	37°41'N	105°13'W
Huésca, Spain (wĕs-kä)	118	42°07'N	0°25'W
Huéscar, Spain (wās'kär)	128	37°50'N	2°34'W
Huetamo de Núñez, Mex.	88	18°34'N	100°53'W
Huete, Spain (wā'tå)	128	40°09'N	2°42'W
Hueycatenango, Mex. (wĕy-kà-tĕ-nä'n-gô)	88	17°31'N	99°10'W
Hueytlalpan, Mex. (wā'ĭ-tläl'pän)	89	20°03'N	97°41'W
Hueytown, Al., U.S.	68h	33°28'N	86°59'W
Huffman, Al., U.S. (hŭf'măn)	68h	33°36'N	86°42'W
Hugh Butler, l., Ne., U.S.	78	40°21'N	100°40'W
Hughenden, Austl. (hū'ĕn-dĕn)	175	20°58'S	144°13'E
Hughes, Austl. (hūz)	174	30°45'S	129°30'E
Hughesville, Md., U.S.	68e	38°32'N	76°48'W
Hugo, Mn., U.S. (hū'gō)	75g	45°10'N	93°00'W
Hugo, Ok., U.S.	79	34°01'N	95°32'W
Hugoton, Ks., U.S. (hū'gō-tŭn)	78	37°10'N	101°28'W
Hugou, China (hōō-gō)	162	33°22'N	117°07'E
Huichapan, Mex. (wē-chä-pän')	88	20°22'N	99°39'W
Huila, dept., Col. (wē'lä)	100a	3°10'N	75°20'W
Huila, Nevado de, mtn., Col. (nĕ-vä-dô-de-wē'lä)	100a	2°59'N	76°01'W
Huilai, China	165	23°02'N	116°18'E
Huili, China	160	26°48'N	102°20'E
Huimanguillo, Mex. (wē-män-gēl'yō)	89	17°50'N	93°16'W
Huimin, China (hōōĭ mĭn)	161	37°29'N	117°32'E
Huitzilac, Mex. (ōē't-zē-lä'k)	89a	19°01'N	99°16'W
Huitzitzilingo, Mex. (wē-tzē-tzē-lē'n-go)	88	21°11'N	98°42'W
Huitzuco, Mex. (wē-tzōō'kō)	88	18°16'N	99°20'W
Huixquilucan, Mex. (ōē'x-kē-lōō-kä'n)	89a	19°21'N	99°22'W
Huiyang, China	165	23°05'N	114°25'E
Hukou, China (hōō-kō)	161	29°58'N	116°20'E
Hulan, China (hōō'län')	161	45°58'N	126°32'E
Hulan, r., China	164	47°20'N	126°30'E
Huliaipole, Ukr.	133	47°39'N	36°12'E
Hulin, China (hōō'lĭn')	166	45°45'N	133°25'E
Hull, Can. (hŭl)	51	45°26'N	75°43'W
Hull, Ma., U.S.	61a	42°18'N	70°54'W
Hull, r., Eng., U.K.	114a	53°47'N	0°20'W
Hulst, Neth. (hòlst)	115a	51°17'N	4°01'E
Huludao, China (hōō-lōō-dou)	161	40°40'N	120°55'E
Hulun Nur, l., China (hōō-lòn nòr)	161	48°50'N	116°45'E
Humacao, P.R. (ōō-mä-kä'ō)	87b	18°09'N	65°49'W
Humansdorp, S. Afr. (hōō'mäns-dòrp)	186	33°57'S	24°45'E
Humbe, Ang. (hòm'bâ)	186	16°50'S	14°55'E
Humber, r., Can.	62d	43°53'N	79°40'W
Humber, r., Eng., U.K. (hŭm'bēr)	116	53°30'N	0°30'E
Humbermouth, Can. (hŭm'bēr-mŭth)	61	48°58'N	57°55'W
Humberside, hist. reg., Eng., U.K.	114a	53°47'N	0°36'W
Humble, Tx., U.S. (hŭm'b'l)	81	29°58'N	95°15'W
Humboldt, Can. (hŭm'bōlt)	50	52°12'N	105°07'W
Humboldt, Ia., U.S.	71	42°43'N	94°11'W
Humboldt, Ks., U.S.	79	37°48'N	95°26'W
Humboldt, Ne., U.S.	79	40°10'N	95°57'W
Humboldt, r., Nv., U.S.	64	40°30'N	116°50'W
Humboldt, East Fork, r., Nv., U.S.	72	40°59'N	115°21'W
Humboldt, North Fork, r., Nv., U.S.	72	41°25'N	115°45'W
Humboldt Bay, b., Ca., U.S.	72	40°48'N	124°25'W
Humboldt Range, mts., Nv., U.S.	76	40°12'N	118°16'W
Humbolt, Tn., U.S.	82	35°47'N	88°55'W
Humbolt Salt Marsh, Nv., U.S.	76	39°49'N	117°41'W
Humbolt Sink, Nv., U.S.	76	39°58'N	118°54'W
Humen, China (hōō-mŭn)	163a	22°49'N	113°39'E
Humphreys Peak, mtn., Az., U.S. (hŭm'frĭs)	64	35°20'N	111°40'W
Humpolec, Czech Rep. (hòm'pō-lĕts)	124	49°33'N	15°21'E
Humuya, r., Hond. (ōō-mōō'yä)	90	14°38'N	87°36'W
Hunaflói, b., Ice. (hōō'nä-flō'ĭ)	116	65°41'N	20°44'W
Hunan, prov., China (hōō'nän')	161	28°08'N	111°25'E
Hunchun, China (hòn-chůn)	161	42°53'N	130°34'E
Hunedoara, Rom. (kōō'nĕd-wä'rà)	131	45°45'N	22°54'E
Hungary, nation, Eur. (hŭŋ'gá-rĭ)	110	46°44'N	17°55'E
Hungerford, Austl. (hŭn'gēr-fērd)	175	28°50'S	144°32'E
Hungry Horse Reservoir, res., Mt., U.S. (hŭŋ'gá-rĭ hôrs)	73	48°11'N	113°30'W
Hunsrück, mts., Ger. (hōōns'rûk)	124	49°43'N	7°12'E
Hunte, r., Ger. (hòn'tĕ)	124	52°45'N	8°26'E
Hunter Islands, is., Austl. (hŭn-tēr)	175	40°33'S	143°36'E
Huntingburg, In., U.S. (hŭnt'ĭng-bûrg)	66	38°15'N	86°55'W
Huntingdon, Can. (hŭnt'ĭng-dŭn)	59	45°10'N	74°05'W
Huntingdon, Can.	74d	49°00'N	122°16'W
Huntingdon, Tn., U.S.	82	36°00'N	88°23'W
Huntington, In., U.S.	66	40°55'N	85°30'W
Huntington, Pa., U.S.	67	40°30'N	78°00'W
Huntington, W.V., U.S.	65	38°25'N	82°25'W
Huntington Beach, Ca., U.S.	75a	33°39'N	118°00'W
Huntington Park, Ca., U.S.	75a	33°59'N	118°14'W
Huntington Station, N.Y., U.S.	68a	40°51'N	73°25'W
Huntley, Mt., U.S.	73	45°54'N	108°01'W
Huntsville, Can.	51	45°20'N	79°15'W
Huntsville, Al., U.S. (hŭnts'vĭl)	82	34°44'N	86°36'W
Huntsville, Mo., U.S.	79	39°24'N	92°32'W
Huntsville, Tx., U.S.	81	30°44'N	95°34'W
Huntsville, Ut., U.S.	75b	41°16'N	111°46'W
Huolu, China (hòŭ lōō)	162	38°05'N	114°20'E
Huon Gulf, b., Pap. N. Gui.	169	7°15'S	147°45'E
Huoqiu, China (hwô-chyō)	162	32°19'N	116°17'E
Huoshan, China (hwô-shän)	165	31°30'N	116°25'E
Ḩuraydin, Wādī, r., Egypt	153a	30°55'N	34°12'E
Hurd, Cape, c., Can. (hûrd)	58	45°15'N	81°45'W
Hurdiyo, Som.	192a	10°43'N	51°05'E
Hurley, Wi., U.S. (hûr'lĭ)	71	46°26'N	90°11'W
Hurlingham, Arg. (ōō'r-lēn-gäm)	102a	34°36'S	58°38'W
Huron, Oh., U.S. (hū'rŏn)	66	41°20'N	82°35'W
Huron, S.D., U.S.	64	44°22'N	98°15'W
Huron, r., Mi., U.S.	69b	42°12'N	83°26'W
Huron, Lake, l., N.A. (hū'rŏn)	65	45°15'N	82°40'W
Huron Mountains, mts., Mi., U.S. (hū'rŏn)	71	46°47'N	87°52'W
Hurricane, Ak., U.S. (hŭr'ĭ-kān)	63	63°00'N	149°30'W
Hurricane, Ut., U.S.	77	37°10'N	113°20'W
Hurricane Flats, bk., (hŭ-rĭ-kăn flăts)	92	23°35'N	78°30'W
Hurst, Tx., U.S.	75c	32°48'N	97°12'W
Húsavik, Ice.	116	66°00'N	17°10'W
Huşi, Rom. (kòsh')	133	46°52'N	28°04'E
Huskvarna, Swe. (hòsk-vär'nà)	122	57°48'N	14°16'E
Husum, Ger. (hōō'zòm)	124	54°29'N	9°04'E
Hutchins, Tx., U.S. (hŭch'ĭnz)	75c	32°38'N	96°43'W
Hutchinson, Ks., U.S. (hŭch'ĭn-sŭn)	64	38°02'N	97°56'W
Hutchinson, Mn., U.S.	71	44°53'N	94°23'W
Hutuo, r., China	164	38°10'N	114°00'E
Huy, Bel. (û-ē') (hû'ē)	121	50°33'N	5°14'E
Hvannadalshnúkur, mtn., Ice.	116	64°09'N	16°46'W
Hvar, i., Serb. (khvär)	130	43°08'N	16°28'E
Hwange, Zimb.	186	18°22'S	26°29'E
Hwangju, Kor., N. (hwäng'jōō')	166	38°39'N	125°49'E
Hyargas Nuur, l., Mong.	160	48°00'N	92°32'E
Hyattsville, Md., U.S. (hī'ăt's-vil)	68e	38°57'N	76°58'W
Hyco Lake, res., N.C., U.S. (rōks' bŭr-ō)	83	36°22'N	78°58'W
Hydaburg, Ak., U.S. (hī-dă'bûrg)	63	55°12'N	132°49'W
Hyde, Eng., U.K. (hīd)	114a	53°27'N	2°05'W
Hyderābād, India (hī-dēr-å-băd')	155	17°29'N	78°28'E
Hyderābād, India	155	18°30'N	76°50'E
Hyderābād, Pak.	155	25°29'N	68°28'E
Hyéres, Fr. (ē-âr')	117	43°09'N	6°08'E
Hyéres, Îles d', is., Fr. (ēl'dyâr')	117	42°57'N	6°17'E
Hyesanjin, Kor., N. (hyē'sän-jĭn')	166	41°11'N	128°12'E
Hymera, In., U.S. (hī-mē'rá)	66	39°10'N	87°20'W
Hyndman Peak, mtn., Id., U.S. (hīnd'măn)	64	43°38'N	114°04'W
Hyōgo, dept., Japan (hĭ'yō'gō)	167b	34°54'N	135°15'E

I

PLACE (Pronunciation)	PAGE	LAT.	LONG.
Ia, r., Japan (ē'ä)	167b	34°54'N	135°34'E
Iahotyn, Ukr.	133	50°18'N	31°46'E
Ialomiţa, r., Rom.	131	44°37'N	26°42'E
Iaşi, Rom. (yä'shē)	110	47°10'N	27°40'E
Iasinia, Ukr.	125	48°17'N	24°21'E
Iavoriv, Ukr.	125	49°56'N	23°24'E
Iba, Phil. (ē'bä)	169a	15°20'N	119°59'E
Ibadan, Nig. (ē-bä'dän)	184	7°17'N	3°30'E
Ibagué, Col.	100	4°27'N	75°14'W
Ibar, r., Serb. (ē'bär)	131	43°22'N	20°35'E
Ibaraki, Japan (ē-bä'rä-gē)	167b	34°49'N	135°35'E
Ibarra, Ec. (ē-bär'rä)	100	0°19'N	78°08'W
Ibb, Yemen	157	14°01'N	44°10'E
Iberville, Can. (ē-bår-vēl') (ī'bēr-vĭl)	59	45°14'N	73°01'W
Ibi, Nig. (ē'bē)	184	8°12'N	9°45'E
Ibiapaba, Serra da, mts., Braz. (sĕ'r-rä-dä-ē-byä-pá'bä)	101	3°30'S	40°55'W
Ibiza *see* Eivissa, i., Spain	112	38°55'N	1°24'E
Ibo, Moz. (ē'bō)	187	12°20'S	40°35'E
Ibrāhīm, Būr, b., Egypt	192d	29°57'N	32°33'E
Ibrahim, Jabal, mtn., Sau. Ar.	154	20°31'N	41°17'E
Ibwe Munyama, Zam.	191	16°09'S	28°34'E
Ica, Peru (ē'kå)	100	14°09'S	75°42'W
Icá (Putumayo), r., S.A.	100	3°00'S	69°00'W
Içana, Braz. (ē-sä'nä)	100	0°15'N	67°19'W
Ice Harbor Dam, Wa., U.S.	72	46°15'N	118°54'W
İçel, Tur.	154	37°00'N	34°40'E
Iceland, nation, Eur. (īs'lănd)	110	65°12'N	19°45'W
Ichibusayama, mtn., Japan (ē'chē-bōō'sá-yä'mä)	167	32°19'N	131°08'E
Ichihara, Japan	167a	35°31'N	140°05'E
Ichikawa, Japan (ē'chē-kä'wä)	167a	35°44'N	139°54'E
Ichinomiya, Japan (ē'chē-nō-mē'yá)	167	35°19'N	136°49'E
Ichinomoto, Japan (ē-chē'nō-mō-tō)	167b	34°37'N	135°50'E
Ichnia, Ukr.	137	50°47'N	32°23'E
Icy Cape, c., Ak., U.S. (ī'sĭ)	63	70°20'N	161°40'W
Idabel, Ok., U.S. (ī'dá-bĕl)	79	33°52'N	94°47'W
Idagrove, Ia., U.S. (ī'dá-grōv)	70	42°22'N	95°29'W
Idah, Nig. (ē'dä)	184	7°07'N	6°43'E
Idaho, state, U.S. (ī'dá-hō)	64	44°00'N	115°10'W
Idaho Falls, Id., U.S.	64	43°30'N	112°01'W
Idaho Springs, Co., U.S.	78	39°43'N	105°32'W
Idanha-a-Nova, Port. (ē-dän'yá-ä-nō'vá)	128	39°58'N	7°13'W
Ider, r., Mong.	160	48°58'N	98°38'E
Idi, Indon. (ē'dē)	168	4°58'N	97°47'E
Idkū Lake, l., Egypt	192b	31°13'N	30°22'E
Idle, r., Eng., U.K. (īd''l)	114a	53°22'N	0°56'W
Idlib, Syria	156	35°55'N	36°38'E
Idriaj, Slvn. (ē'drē-å)	130	46°01'N	14°01'E
Idutywa, S. Afr. (ē-dò-tī'wá)	187c	32°06'S	28°18'E
Ienakiieve, Ukr.	133	48°14'N	38°12'E
Ieper, Bel.	121	50°50'N	2°53'E
Ierápetra, Grc.	130a	35°01'N	25°48'E
Iesi, Italy (yā'sē)	130	43°37'N	13°20'E
Ievpatoriia, Ukr.	137	45°13'N	33°22'E
Ife, Nig.	184	7°30'N	4°30'E
Iferouâne, Niger (ēf'rōō-än')	184	19°04'N	8°24'E
Ifôghas, Adrar des, plat., Afr.	184	19°55'N	2°00'E
Igalula, Tan.	191	5°14'S	33°00'E
Igarka, Russia (ē-gär'kà)	134	67°22'N	86°16'E
Iglesias, Italy (ē-lĕ'syôs)	118	39°20'N	8°34'E
Igli, Alg. (ē-glē')	184	30°32'N	2°15'W
Igluligaarjuk (Chesterfield Inlet), Can.	51	63°19'N	91°11'W
Iglulik, Can.	51	69°33'N	81°18'W
Ignacio, Ca., U.S. (ĭg-nā'cĭ-ō)	74b	38°05'N	122°32'W
Iguaçu, r., Braz. (ē-gwä-sōō')	102b	22°42'S	43°19'W

ng-sing; ŋ-baŋk; ɴ-nasalized n; nŏd; cŏmmit; ōld; ôbey; ôrder; oi-boil; fōōd; ò-as oo in foot; ou-out; s-soft; sh-dish; th-thin; pūre; ûnite; ûrn; stŭd; circŭs; ü-as in French tu; '-indeterminate vowel.

PLACE (Pronunciation)	PAGE	LAT.	LONG.
Iguala, Mex. (ė-gwä′lä)	88	18°18′N	99°34′W
Igualada, Spain (ė-gwä-lä′dä)	129	41°35′N	1°38′E
Iguassu, r., S.A. (ė-gwä-sōō′)	102	25°45′S	52°30′W
Iguassu Falls, wtfl., S.A.	101	25°40′S	54°16′W
Iguatama, Braz. (ē-gwä-tȧ′mä)	99a	20°13′S	45°40′W
Iguatu, Braz. (ė-gwä-tōō′)	101	6°22′S	39°17′W
Iguidi, Erg, Afr.	184	26°22′N	6°53′W
Iguig, Phil. (ē-gēg′)	169a	17°46′N	121°44′E
Iharana, Madag.	187	13°35′S	50°05′E
Ihiala, Nig.	189	5°51′N	6°51′E
Iida, Japan (ē′ė-dȧ)	167	35°39′N	137°53′E
Iijoki, r., Fin. (ē′yō′kĭ)	136	65°28′N	27°00′E
Iizuka, Japan (ē′ė-zȯ-kȧ)	167	33°39′N	130°39′E
Ijebu-Ode, Nig. (ė-jĕ′bōō ōdå)	184	6°50′N	3°56′E
IJmuiden, Neth.	115a	52°27′N	4°36′E
IJsselmeer, l., Neth. (ī′sĕl-mār)	121	52°46′N	5°14′E
Ikaalinen, Fin. (ē′kä-lĭ-nĕn)	123	61°47′N	22°55′E
Ikaría, i., Grc. (ē-kä′ryȧ)	131	37°43′N	26°07′E
Ikeda, Japan (ē′kå-dä)	167b	34°49′N	135°26′E
Ikerre, Nig.	189	7°31′N	5°14′E
Ikhtiman, Blg. (ėk′tė-män)	131	42°26′N	23°49′E
Iki, i., Japan (ē′kė)	166	33°46′N	129°44′E
Ikoma, Japan	167b	34°41′N	135°43′E
Ikoma, Tan. (ė-kō′mä)	186	2°08′S	34°47′E
Iksha, Russia (ĭk′shȧ)	142b	56°10′N	37°30′E
Ila, Nig.	189	8°01′N	4°55′E
Ilagan, Phil.	169a	17°09′N	121°52′E
Ilan, Tai. (ē′län′)	165	24°50′N	121°42′E
Iława, Pol. (ė-lä′vȧ)	125	53°35′N	19°36′E
Île-à-la-Crosse, Can.	56	55°34′N	108°00′W
Ilebo, D.R.C.	186	4°19′S	20°35′E
Ilek, Russia (ē′lyĕk)	137	51°30′N	53°10′E
Île-Perrot, Can. (yl-pĕ-rōt′)	62a	45°21′N	73°54′W
Ilesha, Nig.	184	7°38′N	4°45′E
Ilford, Eng., U.K. (ĭl′fẽrd)	114b	51°33′N	0°06′E
Ilfracombe, Eng., U.K. (ĭl-frȧ-kōōm′)	120	51°13′N	4°08′W
Ilhabela, Braz. (ē′lä-bĕ′lä)	99a	23°47′S	45°21′W
Ilha Grande, Baía de, b., Braz. (ēl′yȧ grän′dĕ)	99a	23°17′S	44°25′W
Ílhavo, Port. (ēl′yȧ-vȯ)	118	40°36′N	8°41′W
Ilhéus, Braz. (ē-lĕ′ōōs)	101	14°52′S	39°00′W
Ili, r., Asia	140	44°30′N	76°45′E
Iliamna, Ak., U.S. (ē-lė-ăm′nȧ)	63	59°45′N	155°05′W
Iliamna, Ak., U.S.	63	60°18′N	153°25′W
Iliamna, l., Ak., U.S.	63	59°25′N	155°30′W
Ilim, r., Russia (ė-lyėm′)	140	57°28′N	103°00′E
Ilimsk, Russia (ė-lyėmsk′)	135	56°47′N	103°43′E
Ilin Island, i., Phil. (ė-lyēn′)	169a	12°16′N	120°57′E
Ilion, N.Y., U.S. (ĭl′ĭ-ŭn)	67	43°00′N	75°05′W
Ilkeston, Eng., U.K. (ĭl′kĕs-tŭn)	114a	52°58′N	1°19′W
Illampu, Nevado, mtn., Bol. (nĕ-vȧ′dȯ-ėl-yäm-pōō′)	100	15°50′S	68°15′W
Illapel, Chile (ē-zhä-pĕ′l)	102	31°37′S	71°10′W
Iller, r., Ger. (ĭlẽr)	124	47°52′N	10°06′E
Illimani, Nevado, mtn., Bol. (nĕ-vȧ′dȯ-ēl-yė-mä′nė)	100	16°50′S	67°38′W
Illinois, state, U.S. (ĭl-ĭ-noi′) (ĭl-ĭ-noiz′)	65	40°25′N	90°40′W
Illinois, r., Il., U.S.	65	39°00′N	90°30′W
Illintsi, Ukr.	133	49°07′N	29°13′E
Illizi, Alg.	184	26°35′N	8°24′E
Il′men, l., Russia (ô′zĕ-rȯ el′′men′′) (ĭl′mĕn)	136	58°18′N	32°00′E
Ilo, Peru	100	17°46′S	71°13′W
Ilobasco, El Sal. (ė-lȯ-bäs′kȯ)	90	13°57′N	88°46′W
Iloilo, Phil. (ē-lȯ-ē′lō)	168	10°49′N	122°33′E
Ilopango, Lago, l., El Sal. (ē-lȯ-päŋ′gō)	90	13°48′N	88°50′W
Ilorin, Nig. (ē-lȯ-rēn′)	184	8°30′N	4°32′E
Ilūkste, Lat.	123	55°59′N	26°20′E
Ilwaco, Wa., U.S. (ĭl-wā′kȯ)	74c	46°19′N	124°02′W
Ilych, r., Russia (ė′l′ĭch)	136	62°30′N	57°30′E
Imabari, Japan (ē′mä-bä′rė)	166	34°05′N	132°58′E
Imai, Japan (ė-mī′)	167b	34°30′N	135°47′E
Iman, r., Russia (ė-män′)	166	45°40′N	134°31′E
Imandra, l., Russia (ē-män′drȧ)	136	67°40′N	32°30′E
Imbābah, Egypt (ēm-bä′bȧ)	192b	30°06′N	31°09′E
Imeni Morozova, Russia (ĭm-yĕ′nyĭ mô rô′zô vȧ)	142c	59°58′N	31°02′E
Imeni Moskvy, Kanal (Moscow Canal), can., Russia (kȧ-näl′ĭm-yă′nĭ mŏs-kvĭ)	132	56°33′N	37°15′E
Imeni Tsyurupy, Russia	142b	55°30′N	38°39′E
Imeni Vorovskogo, Russia	142b	55°43′N	38°21′E
Imlay City, Mi., U.S. (ĭm′lā)	66	43°00′N	83°15′W
Immenstadt, Ger. (ĭm′ĕn-shtät)	124	47°34′N	10°12′E
Immerpan, S. Afr. (ĭmẽr-pän)	192c	24°29′S	29°14′E
Imola, Italy (ē′mȯ-lä)	130	44°19′N	11°43′E
Imotski, Cro. (ė-môts′kė)	131	43°25′N	17°15′E
Impameri, Braz.	101	17°44′S	48°03′W
Impendle, S. Afr. (ĭm-pĕnd′lå)	187c	29°38′S	29°54′E
Imperia, Italy (ėm-pā′rė-ä)	118	43°52′N	8°00′E
Imperial, Pa., U.S. (ĭm-pē′rĭ-ăl)	69e	40°27′N	80°15′W
Imperial Beach, Ca., U.S.	76a	32°34′N	117°08′W
Imperial Valley, Ca., U.S.	76	33°00′N	115°22′W
Impfondo, Congo (ĭmp-fōn′dȯ)	185	1°37′N	18°04′E
Imphāl, India (ĭmp′hŭl)	155	24°42′N	94°00′E
Ina, r., Japan (ė-nä′)	167b	34°56′N	135°21′E
Inaja Indian Reservation, I.R., Ca., U.S. (ė-nä′hä)	76	32°56′N	116°37′W
Inari, l., Fin.	116	69°02′N	26°22′E
Inca, Spain (ėŋ′kä)	129	39°43′N	2°53′E
Ince Burun, c., Tur. (ĭn′jå)	119	42°00′N	35°00′E
Inch′ŏn, Kor., S. (ĭn′chŭn)	161	37°26′N	126°46′E
Incudine, Monte, mtn., Fr. (ēn-kōō-dē′nå) (ăɴ-kü-dēn′)	130	41°53′N	9°17′E
Indalsälven, r., Swe.	116	62°50′N	16°50′E
Independence, Ks., U.S. (ĭn-dė-pĕn′dĕns)	79	37°14′N	95°42′W
Independence, Mo., U.S.	75f	39°06′N	94°26′W
Independence, Oh., U.S.	69d	41°23′N	81°39′W
Independence, Or., U.S.	72	44°49′N	123°13′W
Independence Mountains, mts., Nv., U.S.	72	41°15′N	116°02′W
Īnder köli, l., Kaz.	137	48°20′N	52°10′E
India, nation, Asia (ĭn′dĭ-ȧ)	155	23°00′N	77°30′E
Indian, l., Mi., U.S. (ĭn′dĭ-ăn)	71	46°04′N	86°34′W
Indian, r., N.Y., U.S.	67	44°05′N	75°45′W
Indiana, Pa., U.S. (ĭn-dĭ-än′ȧ)	67	40°40′N	79°10′W
Indiana, state, U.S.	65	39°50′N	86°45′W
Indianapolis, In., U.S. (ĭn-dĭ-ăn-ăp′ȯ-lĭs)	65	39°45′N	86°08′W
Indian Arm, b., Can. (ĭn′dĭ-ăn ärm)	74d	49°21′N	122°55′W
Indian Head, Can.	50	50°29′N	103°44′W
Indian Lake, l., Can.	58	47°00′N	82°00′W
Indian Ocean, o.	5	10°00′S	70°00′E
Indianola, Ia., U.S. (ĭn-dĭ-ăn-ō′lȧ)	71	41°22′N	93°33′W
Indianola, Ms., U.S.	82	33°29′N	90°35′W
Indigirka, r., Russia (ėn-dė-gēr′kȧ)	141	67°45′N	145°45′E
Indio, r., Pan. (ē′n-dyô)	86a	9°13′N	79°28′W
Indochina, reg., Asia (ĭn-dȯ-chī′nȧ)	168	17°22′N	105°18′E
Indonesia, nation, Asia (ĭn′dȯ-nė-zhȧ)	168	4°38′S	118°45′E
Indore, India (ĭn-dōr′)	155	22°48′N	76°51′E
Indragiri, r., Indon. (ĭn-drȧ-jē′rė)	168	0°27′S	102°05′E
Indrāvati, r., India (ĭn-drŭ-vä′tė)	155	19°00′N	82°00′E
Indre, r., Fr. (ăɴ′dr′)	126	47°13′N	0°29′E
Indus, Can. (ĭn′dŭs)	62e	50°55′N	113°45′W
Indus, r., Asia	155	26°43′N	67°41′E
Indwe, S. Afr. (ĭnd′wå)	187c	31°30′S	27°21′E
Inebolu, Tur. (ė-nå-bō′lōō)	119	41°50′N	33°40′E
Inego, Tur. (ė′nå-gŭ)	137	40°05′N	29°20′E
Infanta, Phil. (ėn-fän′tä)	169a	14°44′N	121°39′E
Infanta, Phil.	169a	15°50′N	119°53′E
Inferror, Laguna, l., Mex. (lä-gȯ′nä-ėn-fẽr-rôr)	89	16°18′N	94°40′W
Infiernillo, Presa de, res., Mex.	88	18°50′N	101°50′W
Infiesto, Spain (ēn-fyĕ′s-tȯ)	128	43°21′N	5°24′W
I-n-Gall, Niger	189	16°47′N	6°56′E
Ingersoll, Can. (ĭn′gẽr-sŏl)	58	43°05′N	81°00′W
Ingham, Austl. (ĭng′ăm)	175	18°45′S	146°14′E
Ingles, Cayos, is., Cuba (kä-yōs-ė′n-glĕ′s)	92	21°55′N	82°35′W
Inglewood, Can.	62d	43°48′N	79°56′W
Inglewood, Ca., U.S. (ĭn′g′l-wȯd)	75a	33°57′N	118°22′W
Ingoda, r., Russia (ėn-gō′dȧ)	141	51°29′N	112°32′E
Ingolstadt, Ger. (ĭŋ′gȯl-shtät)	124	48°46′N	11°27′E
Ingur, r., Geor. (ėn-gȯr′)	137	42°30′N	42°00′E
Ingushetia, prov., Russia	138	43°15′N	45°00′E
Inhambane, Moz. (ėn-äm-bä′-nĕ)	186	23°47′S	35°28′E
Inhambupe, Braz. (ėn-yäm-bōō′pä)	101	11°47′S	38°13′W
Inharrime, Moz. (ėn-yär-rē′mä)	186	24°17′S	35°07′E
Inhomirim, Braz. (ē-nô-mė-rē′ɴ)	102b	22°34′S	43°11′W
Inhul, r., Ukr.	133	47°22′N	32°52′E
Inhulets′, r., Ukr.	133	47°12′N	33°12′E
Inírida, r., Col. (ē-nė-rē′dä)	100	2°25′N	70°38′W
Injune, Austl. (ĭn′jȯn)	176	25°52′S	148°30′E
Inkeroinem, Fin. (ĭn′kĕr-oi-nĕn)	123	60°42′N	26°50′E
Inkster, Mi., U.S. (ĭngk′stẽr)	69b	42°18′N	83°19′W
Inn, r., Eur. (ĭn)	117	48°00′N	12°00′E
Innamincka, Austl. (ĭnn-á′mĭn-ká)	176	27°50′S	140°48′E
Inner Brass, i., V.I.U.S. (bräs)	87c	18°23′N	64°58′W
Inner Hebrides, is., Scot., U.K.	120	57°20′N	6°20′W
Inner Mongolia *see* Nei Monggol, prov., China	160	40°15′N	105°00′E
Innisfail, Can.	50	52°02′N	113°57′W
Innsbruck, Aus. (ĭns′brȯk)	117	47°15′N	11°25′E
Ino, Japan (ē′nȯ)	167	33°34′N	133°23′E
Inongo, D.R.C. (ė-nôŋ′gō)	186	1°57′S	18°16′E
Inowrocław, Pol. (ē-nô-vrŏts′läf)	125	52°48′N	18°16′E
In Salah, Alg.	184	27°13′N	2°22′E
Inscription House Ruin, Az., U.S. (ĭn′skrĭp-shŭn hous rōō′ĭn)	77	36°45′N	110°47′W
International Falls, Mn., U.S. (ĭn′tẽr-năsh′ŭn-ăl fôlz)	65	48°34′N	93°26′W
Inuvik, Can.	50	68°40′N	134°10′W
Inuyama, Japan (ē′nōō-yä′mä)	167	35°24′N	137°01′E
Invercargil, N.Z. (ĭn-vẽr-kär′gĭl)	177	46°25′S	168°27′E
Inverell, Austl. (ĭn-vẽr-el′)	175	29°50′S	151°32′E
Invergrove Heights, Mn., U.S. (ĭn′vẽr-grōv)	75g	44°51′N	93°01′W
Inverness, Can. (ĭn-vẽr-nĕs′)	61	46°14′N	61°18′W
Inverness, Scot., U.K.	116	57°30′N	4°07′W
Inverness, Fl., U.S.	83	28°48′N	82°22′W
Investigator Strait, strt., Austl. (ĭn-vĕst′ĭ′gā-tôr)	176	35°33′S	137°00′E
Inyangani, mtn., Zimb. (ėn-yän-gä′nė)	186	18°06′S	32°37′E
Inyokern, Ca., U.S.	76	35°39′N	117°51′W
Inyo Mountains, mts., Ca., U.S. (ĭn′yō)	64	36°55′N	118°04′W
Inzer, r., Russia (ĭn′zĕr)	142a	54°24′N	57°17′E
Inzia, r., D.R.C.	190	5°55′S	17°50′E
Ioánnina, Grc. (yȯ-ä′nė-nȧ)	119	39°39′N	20°52′E
Ioco, Can.	74d	49°18′N	122°53′W
Iola, Ks., U.S. (ī-ō′lȧ)	79	37°55′N	95°23′W
Iôna, Parque Nacional do, rec., Ang.	190	16°35′S	12°00′E
Ionia, Mi., U.S. (ī-ō′nĭ-ȧ)	66	43°00′N	85°10′W
Ionian Islands, is., Grc. (ī-ō′nĭ-ăn)	119	39°10′N	20°05′E
Ionian Sea, sea, Eur.	112	38°59′N	18°48′E
Iori, r., Asia	138	41°03′N	46°17′E
Íos, i., Grc. (ĭ′ŏs)	131	36°48′N	25°25′E
Iowa, state, U.S. (ī′ȯ-wȧ)	65	42°05′N	94°20′W
Iowa, r., Ia., U.S.	71	41°55′N	92°20′W
Iowa City, Ia., U.S.	65	41°39′N	91°31′W
Iowa Falls, Ia., U.S.	71	42°32′N	93°16′W
Iowa Park, Tx., U.S.	78	33°57′N	98°39′W
Ipala, Tan.	191	4°30′S	32°53′E
Ípeirus, hist. reg., Grc.	131	39°35′N	20°45′E
Ipel′, r., Eur. (ė′pĕl)	125	48°08′N	19°00′E
Ipiales, Col. (ē-pė-ä′lås)	100	0°48′N	77°45′W
Ipoh, Malay.	168	4°45′N	101°05′E
Ipswich, Austl. (ĭps′wĭch)	175	27°40′S	152°50′E
Ipswich, Eng., U.K.	117	52°05′N	1°05′E
Ipswich, Ma., U.S.	61a	42°41′N	70°50′W
Ipswich, S.D., U.S.	70	45°26′N	99°01′W
Ipu, Braz. (ē-pōō)	101	4°11′S	40°45′W
Iput′, r., Eur. (ė-pȯt′)	137	52°53′N	31°57′E
Iqaluit, Can.	51	63°48′N	68°31′W
Iquique, Chile (ē-kė′kĕ)	100	20°16′S	70°07′W
Iquitos, Peru (ė-kē′tōs)	100	3°39′S	73°18′W
Irákleio, Grc.	110	35°20′N	25°10′E
Iran, nation, Asia (ē-rän′)	154	31°15′N	53°30′E
Iran, Plateau of, plat., Iran	154	32°28′N	58°00′E
Iran Mountains, mts., Asia	168	2°30′N	114°30′E
Irapuato, Mex. (ē-rä-pwä′tō)	88	20°41′N	101°24′W
Iraq, nation, Asia (ē-räk′)	154	32°00′N	42°30′E
Irazú, vol., C.R. (ē-rä-zōō′)	91	9°58′N	83°54′W
Irbid, Jord. (ėr-bēd′)	156	32°33′N	35°51′E
Irbit, Russia (ėr-bėt′)	134	57°40′N	63°10′E
Irébou, D.R.C. (ē-rā′bōō)	186	0°40′S	17°48′E
Ireland, nation, Eur. (īr-lănd)	110	53°33′N	8°00′W
Iremel′, Gora, mtn., Russia (gȧ-rä′ĭ-rĕ′mĕl)	142a	54°32′N	58°52′E
Irene, S. Afr. (ĭ-rē-nē)	187b	25°53′S	28°13′E
Irîgui, reg., Mali	188	16°45′N	5°35′W
Iriklinskoye Vodokhranilishche, res., Russia	137	52°20′N	58°50′E
Iringa, Tan. (ė-rĭŋ′gä)	186	7°46′S	35°42′E
Iriomote Jima, i., Japan (ērē′-ō-mō-tā)	161	24°20′N	123°30′E
Iriona, Hond. (ė-rė-ō′nä)	90	15°53′N	85°12′W
Irish Sea, sea, Eur. (ī′rĭsh)	112	53°55′N	5°25′W
Irkutsk, Russia (ĭr-kȯtsk′)	135	52°16′N	104°00′E
Irlam, Eng., U.K. (ûr′lăm)	114a	53°26′N	2°26′W
Irois, Cap des, c., Haiti	93	18°25′N	74°50′W
Iron Bottom Sound, strt., Sol. Is.	170e	9°15′S	160°00′E
Irondale, Al., U.S. (ī′ẽrn-dăl)	68h	33°32′N	86°43′W
Iron Gate, val., Eur.	131	44°43′N	22°32′E
Iron Knob, Austl. (ĭ-ăn nŏb)	176	32°47′S	137°10′E
Iron Mountain, Mi., U.S. (ī′ẽrn)	71	45°49′N	88°04′W
Iron River, Mi., U.S.	71	46°09′N	88°39′W
Ironton, Oh., U.S. (ī′ẽrn-tŭn)	66	38°30′N	82°45′W
Ironwood, Mi., U.S. (ī′ẽrn-wȯd)	71	46°28′N	90°10′W
Ironwood Forest National Monument, rec., Az., U.S.	77	32°30′N	111°25′W
Iroquois, r., Il., U.S. (ĭr′ȯ-kwoi)	66	40°55′N	87°20′W
Iroquois Falls, Can.	51	48°41′N	80°39′W
Irō-Saki, c., Japan (ē′rō sä′kē)	166	34°35′N	138°54′E
Irpin, r., Ukr.	133	50°13′N	29°55′E
Irrawaddy, r., Mya. (ĭr-ȧ-wäd′ė)	155	23°27′N	96°25′E
Irtysh, r., Asia (ĭr-tĭsh′)	134	59°00′N	69°00′E
Irumu, D.R.C. (ė-rȯ′mōō)	185	1°30′N	29°52′E
Irun, Spain (ė-rōōn′)	128	43°20′N	1°47′W
Irvine, Scot., U.K.	120	55°39′N	4°40′W
Irvine, Ca., U.S. (ûr′vĭn)	75a	33°40′N	117°45′W
Irvine, Ky., U.S.	66	37°40′N	84°00′W
Irving, Tx., U.S. (ûr′vēng)	75c	32°49′N	96°57′W
Irvington, N.J., U.S. (ûr′vēng-tŭn)	68a	40°43′N	74°15′W
Irwin, Pa., U.S. (ûr′wĭn)	69e	40°19′N	79°42′W
Is, Russia (ēs)	142a	58°48′N	59°44′E
Isa, Nig.	189	13°14′N	6°24′E
Isaacs, Mount, mtn., Pan. (ē-sä-ȧ′ks)	86a	9°22′N	79°31′W
Isabela, i., Ec. (ē-sä-bā′lä)	100	0°47′S	91°35′W
Isabela, Cabo, c., Dom. Rep. (kȧ′bȯ-ė-sä-bĕ′lä)	93	20°00′N	71°00′W
Isabella, Cordillera, mts., Nic. (kȯr-dēl-yĕ′rä-ē-sä-bĕlä)	90	13°20′N	85°37′W
Isabella Indian Reservation, I.R., Mi., U.S. (ĭs-ȧ-bĕl′-lå)	66	43°35′N	84°55′W
Isaccea, Rom. (ė-säk′chä)	133	45°16′N	28°26′E
Ísafjördur, Ice. (ēs′ȧ-fŷr-dȯr)	116	66°09′N	22°39′W
Isangi, D.R.C. (ė-säŋ′gē)	160	0°46′N	24°15′E
Isar, r., Ger. (ē′zär)	117	48°30′N	12°30′E
Isarco, r., Italy (ė-sär′kō)	130	46°37′N	11°25′E
Isarog, Mount, mtn., Phil. (ė-sä-rȯ-g)	169a	13°40′N	123°23′E
Ischia, Italy (ēs′kyä)	129c	40°29′N	13°58′E
Ischia, Isola d′, i., Italy (dė′sh-kyä)	118	40°26′N	13°55′E
Ise, Japan (ĭs′hĕ) (û′gē-yă′mă′dȧ)	166	34°30′N	136°43′E
Iseo, Lago d′, l., Italy (lä-′gō-dē-ė-zĕ′ō)	130	45°50′N	9°55′E
Isére, r., Fr. (ė-zâr′)	117	45°15′N	5°15′E
Iserlohn, Ger. (ē′zĕr-lōn)	127c	51°22′N	7°42′E
Isernia, Italy (ė-zĕr′nyä)	130	41°35′N	14°14′E
Ise-Wan, b., Japan (ē′sĕ wän)	166	34°49′N	136°44′E
Iseyin, Nig.	184	7°58′N	3°36′E
Ishigaki, Japan	170d	24°20′N	124°09′E
Ishikari Wan, b., Japan (ē′shē-kä-rē wän)	166	43°30′N	141°05′E
Ishim, Russia (ĭsh-ėm′)	134	56°07′N	69°13′E
Ishim, r., Asia	134	53°17′N	67°45′E
Ishimbay, Russia (ē-shēm-bī′)	142a	53°28′N	56°02′E
Ishinomaki, Japan (ĭsh-nō-mä′kē)	161	38°22′N	141°22′E
Ishinomaki Wan, b., Japan (ē-shē-nō-mä′kē wän)	166	38°10′N	141°40′E
Ishly, Russia (ĭsh′lĭ)	142a	54°13′N	55°55′E
Ishlya, Russia (ĭsh′lyȧ)	142a	53°54′N	57°48′E
Ishmant, Egypt	192b	29°17′N	31°15′E
Ishpeming, Mi., U.S. (ĭsh′pė-mĭng)	71	46°28′N	87°42′W
Isipingo, S. Afr. (ĭs-ĭ-pĭŋ-gȯ)	187c	29°59′S	30°58′E
Isiro, D.R.C.	185	2°47′N	27°37′E
İskenderun, Tur. (ĭs-kĕn′dẽr-ōōn)	154	36°45′N	36°15′E
İskenderun Körfezi, b., Tur.	119	36°22′N	35°25′E
İskilip, Tur. (ės′kĭ-lėp′)	119	40°40′N	34°30′E

ăt; finăl; rāte; senåte; ärm; ȧsk; sofȧ; fâre; ch-choose; dh-as th in other; bē; ėvent; bĕt; recĕnt; cratẽr; g-gō; gh-guttural g; bĭt; ĭ-short neutral; rīde; ĸ-guttural k as ch in German ich;

PLACE (Pronunciation)	PAGE	LAT.	LONG.
Iskūr, r., Blg. (ĭs′k′r)	131	43°05′N	23°37′E
Isla-Cristina, Spain (ī′lä-krē-stē′nä)	128	37°13′N	7°20′W
Islāmābād, Pak.	155	33°55′N	73°05′E
Isla Mujeres, Mex. (ē′s-lä-mōō-kĕ′rĕs)	90a	21°25′N	86°53′W
Island Lake, l., Can.	53	53°47′N	94°25′W
Islands, Bay of, b., Can. (ī′lăndz)	61	49°10′N	58°15′W
Islay, i., Scot., U.K. (ī′lā)	116	55°55′N	6°35′W
Isle, r., Fr. (ēl)	126	45°02′N	0°29′E
Isle of Axholme, reg., Eng., U.K. (āks′-hōm)	114a	53°33′N	0°48′W
Isle of Man, dep., Eur. (măn)	120	54°26′N	4°21′W
Isle Royale National Park, rec., Mi., U.S. (īl′roi-ăl′)	65	47°57′N	88°37′W
Isleta, N.M., U.S. (ēs-lā′tȧ) (ī-lē′tȧ)	77	34°55′N	106°45′W
Isleta Indian Reservation, I.R., N.M., U.S.	77	34°55′N	106°45′W
Ismailia, Egypt (ēs-mā-ēl′ēä)	192b	30°35′N	32°17′E
Ismā′īlīyah Canal, can., Egypt	192b	30°25′N	31°45′E
Ismail Samani, pik, mtn., Taj.	139	38°57′N	72°01′E
Ismaning, Ger. (ēz′mä-nēng)	115d	48°14′N	11°41′E
Isparta, Tur. (ē-spär′tȧ)	154	37°50′N	30°40′E
Israel, nation, Asia	154	32°40′N	34°00′E
Issaquah, Wa., U.S. (īz′să-kwăh)	74a	47°32′N	122°02′W
Isselburg, Ger. (ē′sĕl-bōōrg)	127c	51°50′N	6°28′E
Issoire, Fr. (ē-swȧr′)	126	45°32′N	3°13′E
Issoudun, Fr. (ē-sōō-dăN′)	126	46°56′N	2°00′E
Issum, Ger. (ē′sōōm)	127c	51°32′N	6°24′E
Issyk-Kul, Ozero, l., Kyrg.	139	42°13′N	76°12′E
İstanbul, Tur. (ē-stän-bōōl′)	154	41°02′N	29°00′E
İstanbul Boğazı (Bosporus), strt., Tur.	154	41°10′N	29°10′E
Istiaía, Grc. (ĭs-tyī′yä)	131	38°58′N	23°11′E
Istmina, Col. (ēst-mē′nä)	100a	5°10′N	76°40′W
Istokpoga, Lake, l., Fl., U.S. (īs-tōk-pō′gȧ)	83a	27°20′N	81°33′W
Istra, pen., Serb. (ē-strä)	130	45°18′N	13°48′E
Istranca Dağlari, mts., Eur. (ī-strän′jȧ)	131	41°50′N	27°25′E
Istres, Fr. (ēs′tr′)	126a	43°30′N	5°00′E
Itabaiana, Braz. (ē-tä-bä-yȧ-nä)	101	10°42′S	37°17′W
Itabapoana, Braz. (ē-tä′-bä-pôä′nä)	99a	21°19′S	40°58′W
Itabapoana, r., Braz.	99a	21°11′S	41°18′W
Itabirito, Braz. (ē-tä-bē-rē′tô)	99a	20°15′S	43°46′W
Itabuna, Braz. (ē-tä-bōō′nȧ)	101	14°47′S	39°17′W
Itacoara, Braz. (ē-tä-kô′ä-rä)	99a	21°41′S	42°04′W
Itacoatiara, Braz. (ē-tȧ-kwȧ-tyä′rȧ)	101	3°03′S	58°18′W
Itagüí, Col. (ē-tä′gwē)	100a	6°11′N	75°36′W
Itagui, r., Braz.	102b	22°53′S	43°43′W
Itaipava, Braz. (ē-tī-pȧ′-vä)	102b	22°23′S	43°09′W
Itaipu, Braz. (ē-tī′pōō)	102b	22°58′S	43°02′W
Itaituba, Braz. (ē-tä′ī-tōō′bȧ)	101	4°12′S	56°00′W
Itajái, Braz. (ē-tä-zhī′)	102	26°52′S	48°39′W
Italy, Tx., U.S.	81	32°11′N	96°51′W
Italy, nation, Eur. (ĭt′ȧ-lē)	110	43°58′N	11°14′E
Itambi, Braz. (ē-tä′m-bē)	102b	22°44′S	42°57′W
Itami, Japan (ē′tä′mē′)	167b	34°47′N	135°25′E
Itapecerica, Braz. (ē-tä-pĕ-sĕ-rē′kȧ)	99a	20°29′S	45°08′W
Itapecuru-Mirim, Braz. (ē-tä-pĕ′kōō-rōō-mē-rēN′)	101	3°17′S	44°15′W
Itaperuna, Braz. (ē-tȧ′pȧ-rōō′nä)	101	21°12′S	41°53′W
Itapetininga, Braz. (ē-tä-pĕ-tē-nē′N-gä)	101	23°37′S	48°03′W
Itapira, Braz. (ē-tȧ-pē′rȧ)	101	20°42′S	51°19′W
Itapira, Braz.	99a	22°27′S	46°47′W
Itarsi, India	158	22°43′N	77°45′E
Itasca, Tx., U.S. (ī-tăs′kȧ)	81	32°09′N	97°08′W
Itasca, l., Mn., U.S.	70	47°13′N	95°14′W
Itatiaia, Pico da, mtn., Braz. (pē′-kô-dȧ-ē-tä-tyȧ′ēä)	101	22°18′S	44°41′W
Itatiba, Braz. (ē-tä-tē′bä)	99a	23°01′S	46°48′W
Itaúna, Braz. (ē-tä-ōō′nä)	99a	20°05′S	44°35′W
Ithaca, Mi., U.S. (ĭth′ȧ-kȧ)	66	43°20′N	84°35′W
Ithaca, N.Y., U.S.	65	42°25′N	76°30′W
Itháka, i., Grc. (ē′thä-kē)	131	38°27′N	20°48′E
Itigi, Tan.	191	5°42′S	34°29′E
Itimbiri, r., D.R.C.	190	2°40′N	23°30′E
Itoko, D.R.C. (ē-tō′kō)	186	1°13′S	22°07′E
Itu, Braz. (ē-tōō′)	99a	23°16′S	47°16′W
Ituango, Col. (ē-twäŋ′gō)	100	7°07′N	75°44′W
Ituiutaba, Braz. (ē-tōō-ēōō-tä′bä)	101	18°56′S	49°17′W
Itumirim, Braz. (ē-tōō-mē-rē′N)	99a	21°20′S	44°51′W
Itundujia Santa Cruz, Mex. (ē-tōōn-dōō-hē′ȧ sä′n-tä krōō′z)	89	16°50′N	97°43′W
Iturbide, Mex. (ē′tōōr-bē′dhā)	90a	19°38′N	89°31′W
Iturup, i., Russia (ē-tōō-rōōp′)	141	45°35′N	147°15′E
Ituzaingo, Arg. (ē-tōō-zä-ē′n-gō)	102a	34°40′S	58°40′W
Itzehoe, Ger. (ē′tzĕ-hō)	124	53°55′N	9°31′E
Iuka, Ms., U.S. (ī-ū′kȧ)	82	34°47′N	88°10′W
Iúna, Braz. (ē-ōō′-nä)	99a	20°22′S	41°32′W
Ivanhoe, Austl. (ĭv′ăn-hō)	176	32°53′S	144°10′E
Ivanivka, Ukr.	132	46°43′N	34°33′E
Ivano-Frankivs′k, Ukr.	137	48°53′N	24°46′E
Ivanopil′, Ukr.	133	49°51′N	28°11′E
Ivanovo, Russia (ē-vä′nô-vō)	134	57°02′N	41°54′E
Ivanovo, prov., Russia	132	56°55′N	40°30′E
Ivanteyevka, Russia (ē-vȧn-tyĕ′yĕf-kȧ)	142b	55°58′N	37°56′E
Ivdel′, Russia (ĭv′dyĕl)	142a	60°42′N	60°27′E
Iviza see Eivissa, i., Spain	112	38°55′N	1°24′E
Ivohibé, Madag. (ē-vō-hē-bā′)	187	22°28′S	46°59′E
Ivory Coast see Cote d′Ivoire, nation, Afr.	184	7°43′N	6°30′W
Ivrea, Italy (ē-vrĕ′ä)	118	45°25′N	7°54′E
Ivry-sur-Seine, Fr.	127b	48°49′N	2°23′E
Ivujivik, Can.	51	62°17′N	77°52′W
Ivvavik National Park, rec., Can.	63	69°10′N	139°30′W
Iwaki, Japan	166	37°03′N	140°57′E
Iwate Yama, mtn., Japan (ē-wä-tē-yä′mä)	166	39°50′N	140°56′E
Iwatsuki, Japan	167a	35°48′N	139°43′E
Iwaya, Japan (ē′wȧ-yȧ)	167b	34°35′N	135°01′E
Iwo, Nig.	184	7°38′N	4°11′E
Ixcateopán, Mex. (ēs-kä-tä-ō-pän′)	88	18°29′N	99°49′W
Ixelles, Bel.	115a	50°49′N	4°23′E
Ixhautlán, Mex. (ēs-wat-län′)	88	20°41′N	98°01′W
Ixhuatán, Mex. (ēs-hwä-tän′)	89	16°19′N	94°30′W
Ixmiquilpan, Mex. (ēs-mē-kēl′pän)	88	20°30′N	99°12′W
Ixopo, S. Afr.	187c	30°10′S	30°04′E
Ixtacalco, Mex. (ēs-tȧ-käl′kō)	89a	19°23′N	99°07′W
Ixtaltepec, Mex. (ēs-täl-tē-pĕk′)	89	16°33′N	95°04′W
Ixtapalapa, Mex. (ēs′tä-pä-lä′pä)	89a	19°21′N	99°06′W
Ixtapaluca, Mex. (ēs′tä-pä-lōō′kä)	89a	19°18′N	98°53′W
Ixtepec, Mex. (ēks-tĕ′pĕk)	89	16°37′N	95°09′W
Ixtlahuaca, Mex. (ēs-tlä-wä′kä)	88	19°34′N	99°46′W
Ixtlán de Juárez, Mex. (ēs-tlän′ dā hwä′rȧz)	89	17°20′N	96°29′W
Ixtlán del Río, Mex. (ēs-tlän′dĕl rē′ō)	88	21°05′N	104°22′W
Iya, r., Russia	140	53°45′N	99°30′E
Iyo-Nada, b., Japan (ē′yō nä-dä)	167	33°33′N	132°07′E
Izabal, Guat. (ē′zä-bäl′)	90	15°23′N	89°10′W
Izabal, Lago, l., Guat.	90	15°30′N	89°04′W
Izalco, El Sal. (ē-zäl′kō)	90	13°50′N	89°40′W
Izamal, Mex. (ē-zä-mä′l)	90a	20°55′N	89°00′W
Izberbash, Russia	138	42°33′N	47°52′E
Izhevsk, Russia (ē-zhyĕfsk′)	134	56°50′N	53°15′E
Izhma, Russia (ĭzh′mȧ)	136	65°00′N	54°05′E
Izhma, r., Russia	136	64°00′N	53°00′E
Izhora, r., Russia (ēz′hô-rȧ)	142c	59°36′N	30°20′E
Izmaïl, Ukr.	137	45°00′N	28°49′E
İzmir, Tur. (ĭz-mēr′)	154	38°25′N	27°05′E
İzmit, Tur. (ĭz-mēt′)	119	40°45′N	29°45′E
Iznajar, Embalse de, res., Spain	128	37°15′N	4°30′W
Iztaccíhuatl, mtn., Mex.	88	19°10′N	98°38′W
Izuhara, Japan (ē′zōō-hä′rä)	167	34°11′N	129°18′E
Izumi-Ōtsu, Japan (ē′zōō-mōō ō′tsōō)	167b	34°30′N	135°24′E
Izumo, Japan (ē′zōō-mō)	167	35°22′N	132°45′E
Izu Shichitō, is., Japan	161	34°32′N	139°25′E

J

PLACE (Pronunciation)	PAGE	LAT.	LONG.
Jabal, Bahr al, r., Sudan	185	7°30′N	31°00′E
Jabalpur, India	155	23°18′N	79°59′E
Jablonec nad Nisou, Czech Rep. (yäb′lô-nyĕts)	124	50°43′N	15°12′E
Jablunkov Pass, p., Eur. (yäb′lôn-kôf)	125	49°31′N	18°35′E
Jaboatão, Braz. (zhä-bô-ȧ-touN)	101	8°14′S	35°08′W
Jaca, Spain (hä′kä)	129	42°35′N	0°30′W
Jacala, Mex. (hä-kȧ′lä)	88	21°01′N	99°11′W
Jacaltenango, Guat. (hä-käl-tĕ-näŋ′gō)	90	15°39′N	91°41′W
Jacarézinho, Braz. (zhä-kä-rĕ′zē-nyō)	101	23°13′S	49°58′W
Jachymov, Czech Rep. (yä′chĭ-môf)	124	50°22′N	12°51′E
Jacinto City, Tx., U.S. (hä-sēn′tō) (jȧ-sĭn′tō)	81a	29°45′N	95°14′W
Jacksboro, Tx., U.S. (jăks′bŭr-ō)	78	33°13′N	98°11′W
Jackson, Al., U.S. (jăk′sŭn)	82	31°31′N	87°52′W
Jackson, Ca., U.S.	76	38°22′N	120°47′W
Jackson, Ga., U.S.	82	33°19′N	83°55′W
Jackson, Ky., U.S.	82	37°32′N	83°17′W
Jackson, La., U.S.	81	30°50′N	91°13′W
Jackson, Mi., U.S.	65	42°15′N	84°25′W
Jackson, Mn., U.S.	70	43°37′N	95°00′W
Jackson, Mo., U.S.	79	37°23′N	89°40′W
Jackson, Ms., U.S.	65	32°17′N	90°10′W
Jackson, Oh., U.S.	66	39°00′N	82°40′W
Jackson, Tn., U.S.	65	35°37′N	88°49′W
Jackson, Port, b., Austl.	173b	33°50′S	151°18′E
Jackson Lake, l., Wy., U.S.	73	43°57′N	110°28′W
Jacksonville, Al., U.S. (jăk′sŭn-vĭl)	82	33°52′N	85°45′W
Jacksonville, Fl., U.S.	65	30°20′N	81°40′W
Jacksonville, Il., U.S.	65	39°43′N	90°12′W
Jacksonville, Tx., U.S.	81	31°58′N	95°18′W
Jacksonville Beach, Fl., U.S.	83	31°18′N	81°25′W
Jacmel, Haiti (zhȧk-mĕl′)	93	18°15′N	72°30′W
Jaco, l., Mex. (hä′kō)	80	27°51′N	103°50′W
Jacobābād, Pak.	158	28°22′N	68°30′E
Jacobina, Braz. (zhä-kô-bē′nȧ)	101	11°13′S	40°30′W
Jacques-Cartier, r., Can.	62b	47°04′N	71°28′W
Jacques Cartier, Détroit de, strt., Can.	60	50°07′S	63°58′W
Jacques-Cartier, Mont, mtn., Can.	60	48°59′N	66°00′W
Jacquet River, Can. (zhȧ-kĕ′) (jăk′ĕt)	60	47°55′N	66°00′W
Jacutinga, Braz. (zhȧ-kōō-tēn′gä)	99a	22°17′S	46°36′W
Jadebusen, b., Ger.	124	53°28′N	8°17′E
Jadotville see Likasi, D.R.C.	186	10°59′S	26°44′E
Jaén, Peru (kä-ĕ′n)	100	5°38′S	78°49′W
Jaen, Spain	118	37°45′N	3°48′W
Jaffa, Cape, c., Austl. (jăf′ȧ)	174	36°58′S	139°29′E
Jaffna, Sri L. (jăf′nȧ)	159	9°44′N	80°09′E
Jagüey Grande, Cuba (hä′gwā grän′dā)	92	22°35′N	81°05′W
Jahore Strait, strt., Asia	153b	1°22′N	103°37′E
Jahrom, Iran	154	28°30′N	53°28′E
Jaibo, r., Cuba (hä-ē′bō)	93	20°10′N	75°20′W
Jaipur, India	155	27°00′N	75°50′E
Jaisalmer, India	158	27°00′N	70°54′E
Jajce, Bos. (yī′tsĕ)	131	44°20′N	17°19′E
Jajpur, India	155	20°49′N	86°37′E
Jakarta, Indon. (yä-kär′tä)	168	6°17′S	106°45′E
Jakobstad, Fin. (yȧ′kôb-stádh)	116	63°33′N	22°31′E
Jalacingo, Mex. (hä-lä-sĭŋ′gō)	89	19°47′N	97°16′W
Jalālābād, Afg. (jŭ-lä-lȧ-bäd)	155a	34°25′N	70°27′E
Jalālah al Baḩrīyah, Jabal, mts., Egypt	192b	29°20′N	32°00′E
Jalapa, Guat. (hä-lä′pȧ)	90	14°38′N	89°58′W
Jalapa de Díaz, Mex.	89	18°06′N	96°33′W
Jalapa del Marqués, Mex. (dĕl mär-kās′)	89	16°30′N	95°29′W
Jaleswar, Nepal	158	26°50′N	85°55′E
Jalgaon, India	158	21°08′N	75°33′E
Jalisco, Mex. (hä-lēs′kō)	88	21°27′N	104°54′W
Jalisco, state, Mex.	86	20°07′N	104°45′W
Jalón, r., Spain (hä-lōn′)	128	41°22′N	1°46′W
Jalostotitlán, Mex. (hä-lōs-tē-tlän′)	88	21°09′N	102°30′W
Jalpa, Mex. (häl′pä)	89	18°12′N	93°06′W
Jalpa, Mex. (häl′pä)	88	21°40′N	103°04′W
Jalpan, Mex. (häl′pän)	88	21°13′N	99°31′W
Jaltepec, Mex. (häl-tȧ-pĕk′)	89	17°20′N	95°15′W
Jaltipan, Mex. (häl-tȧ-pän′)	89	17°59′N	94°42′W
Jaltocan, Mex. (häl-tô-kän′)	88	21°08′N	98°32′W
Jamaare, r., Nig.	189	11°50′N	10°10′E
Jamaica, nation, N.A.	87	17°45′N	78°00′W
Jamaica Cay, i., Bah.	93	22°45′N	75°55′W
Jamālpur, Bngl.	158	24°56′N	89°58′E
Jamay, Mex. (hä-mī′)	88	20°16′N	102°43′W
Jambi, Indon. (mäm′bē)	168	1°45′S	103°28′E
James, r., Mo., U.S.	79	36°51′N	93°22′W
James, r., Va., U.S.	65	37°35′N	77°50′W
James, r., U.S.	64	46°25′N	98°55′W
James, Lake, res., N.C., U.S.	83	36°07′N	81°48′W
James Bay, b., Can. (jāmz)	53	53°53′N	80°40′W
Jamesburg, N.J., U.S. (jāmz′bûrg)	68a	40°21′N	74°26′W
James Point, c., Bah.	92	25°20′N	76°30′W
James Range, mts., Austl.	174	24°15′S	133°30′E
James Ross, i., Ant.	97	64°20′S	58°20′W
Jamestown, S. Afr.	187c	31°07′S	26°49′E
Jamestown, N.D., U.S.	64	46°54′N	98°42′W
Jamestown, N.Y., U.S. (jāmz′toun)	65	42°05′N	79°15′W
Jamestown, R.I., U.S.	68b	41°30′N	71°21′W
Jamestown Reservoir, res., N.D., U.S.	70	47°16′N	98°40′W
Jamiltepec, Mex. (hä-mēl-tȧ-pĕk)	89	16°16′N	97°54′W
Jammerbugten, b., Den.	122	57°20′N	9°28′E
Jammu, India	155	32°50′N	74°52′E
Jammu and Kashmīr, state, India (kȧsh-mēr′)	155	34°30′N	76°00′E
Jammu and Kashmīr, hist. reg., Asia (kȧsh-mēr′)	155	39°10′N	75°05′E
Jāmnagar, India (jäm-nû′gŭr)	155	22°33′N	70°03′E
Jamshedpur, India (jäm′shād-pōōr)	155	22°52′N	86°11′E
Jándula, r., Spain (hän′dōō-lȧ)	128	38°28′N	3°52′W
Janesville, Wi., U.S. (jānz′vĭl)	71	42°41′N	89°03′W
Janin, W.B.	153a	32°27′N	35°19′E
Jan Mayen, i., Nor. (yän mī′ĕn)	116	70°59′N	8°05′W
Jánoshalma, Hung. (yä′nôsh-hôl-mô)	125	46°17′N	19°18′E
Janów Lubelski, Pol. (yä′nōōf lū-bĕl′skĭ)	125	50°40′N	22°25′E
Januária, Braz. (zhä-nwä′rē-ä)	101	15°31′S	44°17′W
Japan, nation, Asia (jȧ-păn′)	161	36°30′N	133°30′E
Japan, Sea of, sea, Asia (jȧ-păn′)	161	40°08′N	132°55′E
Japeri, Braz. (zhȧ-pĕ′rē)	102b	22°38′S	43°40′W
Japurá (Caquetá), r., S.A.	100	2°00′S	68°00′W
Jarabacoa, Dom. Rep. (kä-rä-bä-kô′ä)	93	19°05′N	70°40′W
Jaral del Progreso, Mex. (hä-räl dĕl prô-grā′sō)	88	20°21′N	101°05′W
Jarama, r., Spain (hä-rä′mä)	128	40°33′N	3°30′W
Jarash, Jord.	153a	32°17′N	35°53′E
Jardines, Banco, bk., Cuba (bä′n-kō-här-dē′nås)	92	21°45′N	81°40′W
Jargalant, Mong.	164	46°28′N	115°10′E
Jari, r., Braz. (zhä-rē)	101	0°28′N	53°00′W
Jarocin, Pol. (yä-rō′tsyēn)	125	51°58′N	17°31′E
Jarosław, Pol. (yä-rôs-wȧf)	117	50°01′N	22°41′E
Jarud Qi, China (jya-lōō-tŭ shyē)	161	44°35′N	120°40′E
Jasin, Malay.	153b	2°19′N	102°26′E
Jašiūnai, Lith. (dzȧ-shōō-nȧ′yē)	123	54°27′N	25°25′E
Jāsk, Iran (jäsk)	154	25°46′N	57°48′E
Jasło, Pol. (yäs′wō)	125	49°44′N	21°28′E
Jason Bay, b., Malay.	153b	1°53′N	104°14′E
Jasonville, In., U.S. (jā′sŭn-vĭl)	66	39°10′N	87°15′W
Jasper, Can.	50	52°53′N	118°05′W
Jasper, Al., U.S. (jăs′pēr)	82	33°50′N	87°17′W
Jasper, Fl., U.S.	83	30°30′N	82°56′W
Jasper, In., U.S.	66	38°20′N	86°55′W
Jasper, Mn., U.S.	70	43°51′N	96°22′W
Jasper, Tx., U.S.	81	30°55′N	93°59′W
Jasper National Park, rec., Can.	52	53°09′N	117°45′W
Jászapáti, Hung. (yäs′ô-pä-tē)	125	47°29′N	20°10′E
Jászberény, Hung.	125	47°30′N	19°56′E
Jatibonico, Cuba (hä-tē-bô-nē′kō)	92	22°00′N	79°15′W
Jauja, Peru (kä-ō′κ)	100	11°43′S	75°32′W
Jaumave, Mex. (hou-mä′vå)	88	23°23′N	99°24′W
Jaunjelgava, Lat. (youn′yĕl′gȧ-vȧ)	136	56°37′N	25°06′E
Java (Jawa), i., Indon.	168	8°35′S	111°11′E
Javari, r., S.A. (kä-vä-rē)	100	4°25′S	72°07′W
Java Trench, deep,	168	9°45′S	107°30′E
Jawa, Laut (Java Sea), sea, Indon.	168	5°10′S	110°30′E
Jawor, Pol. (yä′vôr)	124	51°04′N	16°12′E
Jaworzno, Pol. (yä-vôzh′nô)	125	50°11′N	19°18′E
Jaya, Puncak, mtn., Indon.	169	4°00′S	137°00′E
Jayapura, Indon.	168	2°30′S	140°45′W
Jayb, Wādī al (Ha′Arava), val., Asia	153a	30°33′N	35°10′E
Jazzīn, Leb.	153a	33°34′N	35°37′E
Jeanerette, La., U.S. (jĕn-ĕr-et′) (zhän-rĕt′)	81	29°54′N	91°41′W
Jebba, Nig. (jĕb′ȧ)	184	9°07′N	4°46′E
Jeddore Lake, l., Can.	61	48°07′N	55°35′W
Jędrzejów, Pol. (yăN-dzhā′yôf)	125	50°38′N	20°18′E
Jefferson, Ga., U.S. (jĕf′ēr-sŭn)	82	34°05′N	83°35′W
Jefferson, Ia., U.S.	71	42°10′N	94°22′W

ng-sing; ŋ-baŋk; N-nasalized n; nŏd; cŏmmit; ōld; ôbey; ôrder; oi-boil; fōōd; ŏŏ-as oo in foot; ou-out; s-soft; sh-dish; th-thin; pūre; ůnite; ûrn; stŭd; circŭs; ü-as in French tu; ′-indeterminate vowel.

260 Jefferson – Jylland PRONOUNCING INDEX

PLACE (Pronunciation)	PAGE	LAT.	LONG.
Jefferson, La., U.S.	68d	29°57'N	90°04'W
Jefferson, Tx., U.S.	81	32°47'N	94°21'W
Jefferson, Wi., U.S.	71	42°59'N	88°45'W
Jefferson, r., Mt., U.S.	73	45°37'N	112°22'W
Jefferson, Mount, mtn., Or., U.S.	72	44°41'N	121°50'W
Jefferson City, Mo., U.S.	65	38°34'N	92°10'W
Jeffersontown, Ky., U.S. (jĕf'ẽr-sŭn-toun)	69h	38°11'N	85°34'W
Jeffersonville, In., U.S. (jĕf'ẽr-sŭn-vĭl)	69h	38°17'N	85°44'W
Jega, Nig.	189	12°15'N	4°23'E
Jehol, hist. reg., China (jĕ-hŏl)	161	42°31'N	118°12'E
Jēkabpils, Lat. (yĕk'áb-pĭls)	136	56°29'N	25°50'E
Jelenia Góra, Pol. (yĕ-lĕn'yà gó'rà)	124	50°53'N	15°43'E
Jelgava, Lat.	123	56°39'N	23°42'E
Jellico, Tn., U.S. (jĕl'ĭ-kō)	82	36°34'N	84°06'W
Jemez Indian Reservation, I.R., N.M., U.S.	77	35°35'N	106°45'W
Jena, Ger. (yā'nä)	117	50°55'N	11°37'E
Jenkins, Ky., U.S. (jĕŋ'kĭnz)	83	37°09'N	82°38'W
Jenkintown, Pa., U.S. (jĕŋ'kĭn-toun)	68f	40°06'N	75°08'W
Jennings, La., U.S. (jĕn'ĭngz)	81	30°14'N	92°40'W
Jennings, Mi., U.S.	66	44°20'N	85°20'W
Jennings, Mo., U.S.	75e	38°43'N	90°16'W
Jequitinhonha, r., Braz. (zhĕ-kē-tēŋ-ō'n-yä)	101	16°47'S	41°19'W
Jérémie, Haiti (zhā-rà-mē')	93	18°40'N	74°10'W
Jeremoabo, Braz. (zhĕ-rä-mō-á'bō)	101	10°03'S	38°13'W
Jerez, Punta, c., Mex. (pōō'n-tä-кĕ-räz')	89	23°04'N	97°44'W
Jerez de la Frontera, Spain	118	36°42'N	6°09'W
Jerez de los Caballeros, Spain	128	38°20'N	6°45'W
Jericho, Austl. (jĕr'ĭ-kō)	175	23°38'S	146°24'E
Jericho, S. Afr. (jĕr-ĭkô)	192c	25°16'N	27°47'E
Jericho see Arīḥā, W.B.	153a	31°51'N	35°28'E
Jerome, Az., U.S. (jē-rōm')	64	34°45'N	112°10'W
Jerome, Id., U.S.	73	42°44'N	114°31'W
Jersey, dep., Eur.	126	49°15'N	2°10'W
Jersey, i., Jersey (jûr'zĭ)	117	49°13'N	2°07'W
Jersey City, N.J., U.S.	65	40°43'N	74°05'W
Jersey Shore, Pa., U.S.	67	41°10'N	77°15'W
Jerseyville, Il., U.S. (jẽr'zē-vĭl)	79	39°07'N	90°18'W
Jerusalem, Isr. (jē-rōō'sá-lĕm)	154	31°46'N	35°14'E
Jesup, Ga., U.S. (jĕs'ŭp)	83	31°36'N	81°53'W
Jesús Carranza, Mex. (hĕ-sōō's-kär-rá'n-zä)	89	17°26'N	95°01'W
Jewel, Or., U.S. (jū'ĕl)	74c	45°56'N	123°30'W
Jewel Cave National Monument, rec., S.D., U.S.	70	43°44'N	103°52'W
Jhālawār, India	155	24°30'N	76°00'E
Jhang Maghiāna, Pak.	158	31°21'N	72°19'E
Jhānsi, India (jän'sē)	155	25°29'N	78°32'E
Jharkhand, state, India	155	23°30'N	85°00'E
Jhārsuguda, India	158	22°51'N	84°13'E
Jhelum, Pak.	155	32°59'N	73°43'E
Jhelum, r., Asia (jā'lŭm)	155	31°40'N	71°51'E
Jiading, China (jyä-dĭŋ)	162	31°23'N	121°15'E
Jialing, r., China (jyä-lĭŋ)	160	32°30'N	105°30'E
Jiamusi, China	166	46°50'N	130°21'E
Ji'an, China (jyē-än)	161	27°15'N	115°10'E
Ji'an, China	164	41°00'N	126°04'E
Jianchangying, China (jyĕn-chäŋ-yĭŋ)	162	40°09'N	118°47'E
Jiangcun, China (jyän-tsòn)	163a	23°16'N	113°14'E
Jiangling, China (jyäŋ-lĭŋ)	161	30°30'N	112°10'E
Jiangshanzhen, China (jyäŋ-shän-jŭn)	162	36°39'N	120°31'E
Jiangsu, prov., China (jyäŋ-sōō)	161	33°45'N	120°30'E
Jiangwan, China (jyäŋ-wän)	163b	31°18'N	121°29'E
Jiangxi, prov., China (jyäŋ-shyē)	161	28°15'N	116°00'E
Jiangyin, China (jyäŋ-yĭn)	165	31°54'N	120°15'E
Jianli, China (jyĕn-lē)	165	29°50'N	112°52'E
Jianning, China (jyĕn-nĭŋ)	165	26°50'N	116°55'E
Jian'ou, China (jyĕn-ō)	165	27°10'N	118°18'E
Jianshi, China (jyĕn-shr)	165	30°40'N	109°45'E
Jiaohe, China	162	38°03'N	116°18'E
Jiaohe, China (jyou-hŭ)	164	43°40'N	127°20'E
Jiaoxian, China (jyou shyĕn)	161	36°18'N	120°01'E
Jiaozuo, China (jyou-dzwô)	162	35°15'N	113°18'E
Jiashan, China (jyä-shän)	162	32°41'N	118°00'E
Jiaxing, China (jyä-shyĭŋ)	161	30°45'N	120°50'E
Jiayu, China (jyä-yōō)	165	30°00'N	114°00'E
Jiazhou Wan, b., China (jyä-jō wän)	161	36°10'N	119°55'E
Jicarilla Apache Indian Reservation, I.R., N.M., U.S. (кē-kà-rēl'yà)	77	36°45'N	107°00'W
Jicarón, Isla, i., Pan. (кē-kä-rōn')	91	7°14'N	81°41'W
Jiddah, Sau. Ar.	154	21°30'N	39°15'E
Jieshou, China	162	33°17'N	115°20'E
Jieyang, China (jyĕ-yäŋ)	161	23°38'N	116°20'E
Jiggalong, Austl. (jĭg'á-lông)	174	23°20'S	120°45'E
Jiguani, Cuba (кē-gwä-nē')	92	20°20'N	76°30'W
Jigüey, Bahía, b., Cuba (bä-ē'ä-кē'gwā)	92	22°15'N	78°10'W
Jihlava, Czech Rep. (yē'hlà-và)	117	49°23'N	15°33'E
Jijel, Alg.	117	36°49'N	5°47'E
Jijia, r., Rom.	125	47°35'N	27°02'E
Jijiashi, China (jyē-jyä-shr)	162	32°10'N	120°17'E
Jijiga, Eth.	192a	9°15'N	42°48'E
Jilin, China (jyē-lĭn)	161	43°58'N	126°40'E
Jilin, prov., China	161	44°20'N	124°50'E
Jiloca, r., Spain (кē-lō'kà)	128	41°13'N	1°30'W
Jilotepeque, Guat. (кē-lô-tĕ-pĕ'kĕ)	90	14°39'N	89°36'W
Jima, Eth.	185	7°41'N	36°52'E
Jimbolia, Rom. (zhĭm-bô'lyä)	131	45°45'N	20°44'E
Jiménez, Mex. (кē-mā'nāz)	88	24°12'N	98°29'W
Jiménez, Mex.	80	29°03'N	100°42'W
Jiménez, Mex.	80	27°09'N	104°55'W
Jiménez del Téul, Mex. (tĕ-ōō'l)	88	21°28'N	103°51'W
Jimo, China (jyē-mwo)	164	36°22'N	120°28'E

PLACE (Pronunciation)	PAGE	LAT.	LONG.
Jim Thorpe, Pa., U.S. (jĭm' thôrp')	67	40°50'N	75°45'W
Jinan, China (jyē-nän)	161	36°40'N	117°01'E
Jincheng, China (jyĭn-chŭŋ)	164	35°30'N	112°50'E
Jindřichův Hradec, Czech Rep. (yĕn'd'r-zhĭ-kōōf hrä'dĕts)	124	49°09'N	15°02'E
Jing, r., China (jyĭŋ)	164	34°40'N	108°20'E
Jing'anji, China (jyĭŋ-än-jē)	162	34°30'N	116°55'E
Jingdezhen, China (jyĭn-dŭ-jŭn)	165	29°18'N	117°18'E
Jingjiang, China (jyĭŋ-jyäŋ)	162	32°02'N	120°15'E
Jingning, China (jyĭŋ-nĭŋ)	164	35°28'N	105°50'E
Jingpo Hu, l., China (jyĭŋ-pwo hōō)	164	44°10'N	129°00'E
Jingxian, China (jyĭŋ shyĕn)	165	26°32'N	109°45'E
Jingxian, China	162	37°43'N	116°17'E
Jingxing, China (jyĭŋ-shyĭŋ)	164	47°00'N	123°00'E
Jingzhi, China (jyĭŋ-jr)	162	36°19'N	119°23'E
Jinhua, China (jyĭn-hwä)	161	29°10'N	119°42'E
Jining, China (jyē-nĭŋ)	161	35°26'N	116°34'E
Jining, China	164	41°00'N	113°10'E
Jinja, Ug. (jĭn'jä)	185	0°26'N	33°12'E
Jinotega, Nic. (кē-nō-tā'gä)	90	13°07'N	86°00'W
Jinotepe, Nic. (кē-nō-tā'pā)	90	11°52'N	86°12'W
Jinqiao, China (jyĭn-chyou)	162	31°46'N	116°46'E
Jinshan, China (jyĭn-shän)	163b	30°53'N	121°09'E
Jinta, China (jyĭn-tä)	160	40°11'N	98°45'E
Jintan, China (jyĭn-tän)	162	31°47'N	119°34'E
Jin Xian, China (jyĭn shyĕn)	164	39°04'N	121°40'E
Jinxiang, China (jyĭn-shyäŋ)	162	35°03'N	116°20'E
Jinyun, China (jyĭn-yòn)	165	28°40'N	120°08'E
Jinzhai, China (jyĭn-jī)	162	31°41'N	115°51'E
Jinzhou, China (jyĭn-jō)	161	41°00'N	121°00'E
Jinzhou Wan, b., China (jyĭn-jō wän)	162	39°07'N	121°17'E
Jinzū-Gawa, r., Japan (jēn'zōō gä'wä)	167	36°26'N	137°18'E
Jipijapa, Ec. (кē-pē-hä'pä)	100	1°36'S	80°52'W
Jiquilisco, El Sal. (кē-kē-lē's-kô)	90	13°18'N	88°32'W
Jiquilpan de Juárez, Mex. (кē-kēl'pän dā hwä'räz)	88	20°00'N	102°43'W
Jiquipilco, Mex. (hē-kē-pē'l-kô)	89a	19°32'N	99°37'W
Jitotol, Mex. (кē-tô-tōl')	89	17°03'N	92°54'W
Jiu, r., Rom.	131	44°45'N	23°17'E
Jiujiang, China (jyô-jyän)	163a	22°50'N	113°02'E
Jiujiang, China	161	29°43'N	116°00'E
Jiuquan, China (jyô-chyän)	160	39°46'N	98°26'E
Jiurongcheng, China (jyô-rôŋ-chŭŋ)	162	37°23'N	122°31'E
Jiushouzhang, China (jyô-shō-jäŋ)	162	35°59'N	115°52'E
Jiuwuqing, China (jyô-wōō-chyĭŋ)	164a	32°31'N	116°51'E
Jiuyongnian, China (jyô-yôŋ-nĭĕn)	162	36°41'N	114°46'E
Jixian, China (jyē shyĕn)	162	35°25'N	114°03'E
Jixian, China	162	37°37'N	115°33'E
Jixian, China	162	40°03'N	117°25'E
Jiyun, r., China (jyē-yōōm)	162	39°35'N	117°34'E
Joachimsthal, Ger.	115b	52°58'N	13°45'E
João Pessoa, Braz.	101	7°09'S	34°45'W
João Ribeiro, Braz. (zhô-un-rē-bā'rō)	99a	20°42'S	44°03'W
Jobabo, r., Cuba (hō-bä'bä)	92	20°50'N	77°15'W
Jock, r., Can. (jŏk)	62c	45°08'N	75°51'W
Jocotepec, Mex. (jô-kō-tà-pĕk')	88	20°17'N	103°26'W
Jodar, Spain (hō'där)	128	37°54'N	3°20'W
Jodhpur, India (hōd'pōōr)	155	26°23'N	73°00'E
Joensuu, Fin. (yô-ĕn'sōō)	123	62°35'N	29°46'E
Joffre, Mount, mtn., Can. (jŏ'f'r)	55	50°32'N	115°13'W
Jõgeva, Est. (yû'gĕ-vá)	123	58°45'N	26°23'E
Joggins, Can. (jŏ'gĭnz)	60	45°42'N	64°27'W
Johannesburg, S. Afr. (yô-hän'ĕs-bŏorgh)	186	26°08'S	27°54'E
John Day, r., Or., U.S. (jŏn'dā)	72	44°46'N	120°15'W
John Day, Middle Fork, r., Or., U.S.	72	44°53'N	119°04'W
John Day, North Fork, r., Or., U.S.	72	45°03'N	118°50'W
John Day Dam, Or., U.S.	72	45°40'N	120°15'W
John H. Kerr Reservoir, res., U.S.	65	36°30'N	78°38'W
John Martin Reservoir, res., Co., U.S. (jŏn mär'tĭn)	78	37°57'N	103°04'W
Johnson, r., Or., U.S. (jŏn'sŭn)	74c	45°27'N	122°20'W
Johnsonburg, Pa., U.S. (jŏn'sŭn-bûrg)	67	41°30'N	78°40'W
Johnson City, Il., U.S. (jŏn'sŭn)	66	37°50'N	88°55'W
Johnson City, N.Y., U.S.	67	42°10'N	76°00'W
Johnson City, Tn., U.S.	65	36°17'N	82°23'W
Johnston, i., Oc. (jŏn'stŭn)	2	17°00'N	168°00'W
Johnstone Strait, strt., Can.	54	50°25'N	126°00'W
Johnston Falls, wtfl., Afr.	191	10°35'S	28°50'E
Johnstown, N.Y., U.S. (jonz'toun)	67	43°00'N	74°20'W
Johnstown, Pa., U.S.	65	40°20'N	78°50'W
Johor, r., Malay. (jŭ-hōr')	153b	1°39'N	103°52'E
Johor Baharu, Malay.	168	1°28'N	103°46'E
Jõhvi, Est. (yû'vĭ)	123	59°21'N	27°21'E
Joigny, Fr. (zhwán-yē')	126	47°58'N	3°26'E
Joinville, Braz. (zhwăN-vēl')	102	26°18'S	48°47'W
Joinville, Fr.	127	48°28'N	5°05'E
Joinville, i., Ant.	97	63°00'S	53°30'W
Jojutla, Mex. (hō-hōō'tlä)	88	18°39'N	99°11'W
Jola, Mex. (kō'lä)	88	21°08'N	104°26'W
Joliet, Il., U.S. (jō-lĭ-ĕt')	69a	41°32'N	88°05'W
Joliette, Can. (zhō-lyĕt')	51	46°01'N	73°30'W
Jolo, Phil. (hō-lô)	168	5°59'N	121°05'E
Jolo Island, i., Phil.	168	5°55'N	121°15'E
Jomalig, i., Phil. (hô-mä'lēg)	169a	14°44'N	122°34'E
Jomulco, Mex. (hô-mōōl'kō)	88	21°08'N	104°24'W
Jonacatepec, Mex.	88	18°39'N	98°46'W
Jonava, Lith. (yō-nä'vá)	123	55°05'N	24°15'E
Jones, Phil. (jōnz)	169a	12°56'N	122°05'E
Jones, Phil.	169a	16°35'N	121°39'E
Jonesboro, Ar., U.S. (jōnz'bŭro)	65	35°49'N	90°42'W
Jonesboro, La., U.S.	81	32°14'N	92°43'W
Jonesville, La., U.S. (jōnz'vĭl)	81	31°35'N	91°50'W
Jonesville, Mi., U.S.	66	42°00'N	84°45'W
Jong, r., S.L.	188	8°10'N	12°10'W

PLACE (Pronunciation)	PAGE	LAT.	LONG.
Joniškis, Lith. (yō'nĭsh-kĭs)	123	56°14'N	23°36'E
Jönköping, Swe. (yûn'chû-pĭng)	116	57°47'N	14°10'E
Jonquière, Can. (zhôN-kyâr')	51	48°25'N	71°15'W
Jonuta, Mex. (hō-nōō'tä)	89	18°07'N	92°09'W
Jonzac, Fr. (zhôN-zák')	126	45°27'N	0°27'W
Joplin, Mo., U.S. (jŏp'lĭn)	65	37°05'N	94°31'W
Jordan, nation, Asia (jôr'dăn)	154	30°15'N	38°00'E
Jordan, r., Asia	153a	32°05'N	35°35'E
Jordan, r., Ut., U.S.	75b	40°42'N	111°56'W
Jorhāt, India (jôr-hät')	155	26°43'N	94°16'E
Jorullo, Volcán de, vol., Mex. (vôl-ká'n-dĕ-hō-rōōl'yō)	88	18°54'N	101°38'W
José C. Paz, Arg.	102a	34°32'S	58°44'W
Joseph Bonaparte Gulf, b., Austl. (jō'sĕf bō'nȧ-pärt)	174	13°30'S	128°40'E
Josephburg, Can.	62g	53°45'N	113°06'W
Joseph Lake, l., Can. (jō'sĕf läk)	62g	53°18'N	113°06'W
Joshua Tree National Park, rec., Ca., U.S. (jō'shū-á trē)	76	34°02'N	115°53'W
Jos Plateau, plat., Nig. (jŏs)	189	9°53'N	9°05'E
Jostedalsbreen, ice, Nor. (yôstĕ-däls-brĕĕn)	116	61°40'N	6°55'E
Jotunheimen, mts., Nor.	116	61°44'N	8°11'E
Joulter's Cays, is., Bah. (jōl'tẽrz)	92	25°20'N	78°10'W
Jouy-le-Chatel, Fr. (zhwē-lĕ-shä-tĕl')	127b	48°40'N	3°07'E
Jovellanos, Cuba (hō-vĕl-yä'nōs)	92	22°50'N	81°10'W
J. Percy Priest Lake, res., Tn., U.S.	82	36°00'N	86°45'W
Juan Aldama, Mex. (kòà'n-äl-dá'mä)	88	24°16'N	103°21'W
Juan de Fuca, Strait of, strt., N.A. (hwän' dā fōō'kä)	52	48°25'N	124°37'W
Juan de Nova, Île, i., Reu.	187	17°18'S	43°07'E
Juan Diaz, r., Pan. (кōōà'n-dē'äz)	86a	9°05'N	79°30'W
Juan Fernández, Islas de, is., Chile	97	33°30'S	79°00'W
Juan L. Lacaze, Ur. (hōōà'n-ĕ'lĕ-lä-kä'zĕ)	99c	34°25'S	57°28'W
Juan Luis, Cayos de, is., Cuba (ka-yōs-dĕ-hwän lōō-ēs')	92	22°15'N	82°00'W
Juárez, Arg. (hōōá'rĕz)	102	37°42'S	59°46'W
Juázeiro, Braz. (zhōōá'zā'rō)	101	9°27'S	40°28'W
Juazeiro do Norte, Braz. (zhōōá'zā'rō-dō-nôr-tĕ)	101	7°16'S	38°57'W
Jubayl, Leb. (jōō-bīl')	153a	34°07'N	35°38'E
Jubba (Genale), r., Afr.	192a	1°30'N	42°25'E
Juby, Cap, c., Mor. (yōō'bē)	184	28°01'N	13°21'W
Júcar, r., Spain (hōō'kär)	118	39°10'N	1°22'W
Júcaro, Cuba (hōō'ká-rō)	92	21°40'N	78°50'W
Juchipila, Mex. (hōō-chē-pē'lá)	88	21°26'N	103°09'W
Juchitán, Mex. (hōō-chē-tän')	86	16°15'N	95°00'W
Juchitlán, Mex. (hōō-chē-tlän)	88	20°05'N	104°07'W
Jucuapa, El Sal. (кōō-kwä'pä)	90	13°30'N	88°24'W
Judenburg, Aus. (jōō'dĕn-bûrg)	124	47°10'N	14°40'E
Judith, r., Mt., U.S. (jōō'dĭth)	73	47°20'N	109°36'W
Juhua Dao, i., China (jyōō-hwä dou)	162	40°30'N	120°47'E
Juigalpa, Nic. (hwê-gäl'pä)	90	12°02'N	85°24'W
Juiz de Fora, Braz. (zhò-ēzh' dā fō'rä)	101	21°47'S	43°20'W
Jujuy, Arg. (hōō-hwē')	102	24°14'S	65°15'W
Jujuy, prov., Arg. (hōō-hwē')	102	23°00'S	65°45'W
Jukskei, r., S. Afr.	187b	25°58'S	27°58'E
Julesburg, Co., U.S. (jōōlz'bûrg)	78	40°59'N	102°16'W
Juliaca, Peru (hōō-lē-ä'kä)	100	15°26'S	70°12'W
Julian Alps, mts., Serb.	118	46°05'N	14°05'E
Julianehåb, Grnld.	49	60°07'N	46°20'W
Jülich, Ger. (yü'lĕk)	127c	50°55'N	6°22'E
Jullundur, India	155	31°29'N	75°39'E
Julpaiguri, India	158	26°35'N	88°48'E
Jumento Cays, is., Bah. (hōō-mĕn'tō)	93	23°05'N	75°40'W
Jumilla, Spain (hōō-mēl'yä)	128	38°28'N	1°20'W
Jump, r., Wi., U.S. (jŭmp)	71	45°18'N	90°53'W
Jumpingpound Creek, r., Can. (jŭmp-ĭng-pound)	62e	51°01'N	114°34'W
Jumrah, Indon.	153b	1°48'N	101°04'E
Junagādh, India (jò-nä'gŭd)	155	21°33'N	70°25'E
Junayfah, Egypt	192d	30°11'N	32°26'E
Junaynah, Ra's al, mtn., Egypt	153a	29°02'N	33°58'E
Junction, Tx., U.S. (jŭŋk'shŭn)	80	30°29'N	99°48'W
Junction City, Ks., U.S.	79	39°01'N	96°49'W
Jundiaí, Braz.	101	23°11'S	46°52'W
Juneau, Ak., U.S. (jōō'nō)	64a	58°25'N	134°30'W
Jungfrau, mtn., Switz. (yòng'frou)	124	46°30'N	7°59'E
Junín, Arg. (hōō-nē'n)	102	34°35'S	60°56'W
Junín, Col.	100a	4°47'N	73°39'W
Juniyah, Leb. (jōō-nē'ĕ)	153a	33°59'N	35°38'E
Jupiter, r., Can.	60	49°40'N	63°20'W
Jupiter, Mount, mtn., Wa., U.S.	74a	47°42'N	123°04'W
Jur, r., Sudan (jòr)	185	6°38'N	27°52'E
Jura, mts., Eur. (zhü-rá')	117	46°55'N	6°49'E
Jura, i., Scot., U.K. (jōō'rá)	120	56°09'N	6°45'W
Jura, Sound of, strt., Scot., U.K. (jōō'rȧ)	120	55°45'N	5°55'W
Jurbarkas, Lith. (yōōr-bär'käs)	123	55°06'N	22°50'E
Jūrmala, Lat.	123	56°57'N	23°37'E
Jurong, China (jyōō-roŋ)	162	31°58'N	119°12'E
Juruá, r., S.A.	100	5°30'S	67°30'W
Juruena, r., Braz. (zhōō-rōōĕ'nä)	101	12°22'S	58°34'W
Jutiapa, Guat. (hōō-tē-ä'pä)	90	14°16'N	89°55'W
Juticalpa, Hond. (hōō-tē-käl'pä)	86	14°35'N	86°17'W
Jutland see Jylland, reg., Den.	116	56°04'N	9°00'E
Juventino Rosas, Mex.	88	20°38'N	101°02'W
Juventud, Isla de la, i., Cuba	87	21°40'N	82°45'W
Juxian, China (jyōō shyĕn)	164	35°35'N	118°50'E
Juxtlahuaca, Mex. (hōōs-tlà-hwä'kä)	88	17°20'N	98°02'W
Juye, China (jyōō-yŭ)	162	35°25'N	116°05'E
Južna Morava, r., Serb. (ŭ'zhnä mō'rä-vä)	131	42°30'N	22°00'E
Jylland, reg., Den.	116	56°04'N	9°00'E

ăt; finăl; rāte; senāte; ärm; ásk; sofá; fâre; ch-choose; dh-as th in other; bē; ēvent; bĕt; recĕnt; cratẽr; g-gō; gh-guttural g; bĭt; ĭ-short neutral; rīde; к-guttural k as ch in German ich;

K

PLACE (Pronunciation)	PAGE	LAT.	LONG.
K2 (Qogir Feng), mtn., Asia	155	36°06′N	76°38′E
Kaabong, Ug.	191	3°31′N	34°08′E
Kaalfontein, S. Afr. (kärl-fōn-tān)	187b	26°02′S	28°16′E
Kaappunt, c., S. Afr.	186a	34°21′S	18°30′E
Kabaena, Pulau, i., Indon. (kä-bȧ-ā′nȧ)	168	5°35′S	121°07′E
Kabala, S.L. (kȧ-bȧ′lȧ)	184	9°43′N	11°39′W
Kabale, Ug.	191	1°15′S	29°59′E
Kabalega Falls, wtfl., Ug.	185	2°15′N	31°41′E
Kabalo, D.R.C. (kä-bä′lō)	186	6°03′S	26°55′E
Kabambare, D.R.C. (kä-bäm-bä′rȧ)	186	4°47′S	27°45′E
Kabardino-Balkaria, prov., Russia	136	43°30′N	43°30′E
Kabba, Nig.	189	7°50′N	6°03′E
Kabe, Japan (kä′bā)	167	34°32′N	132°30′E
Kabinakagami, r., Can.	58	49°00′N	84°15′W
Kabinda, D.R.C. (kä-bēn′dä)	186	6°08′S	24°29′E
Kabompo, r., Zam. (kȧ-bôm′pō)	186	14°00′S	23°40′E
Kabongo, D.R.C. (kȧ-bông′ō)	186	7°58′S	25°10′E
Kabot, Gui.	188	10°48′N	14°57′W
Kaboudia, Ra′s, c., Tun.	118	35°17′N	11°28′E
Kābul, Afg. (kä′bōōl)	155	34°39′N	69°14′E
Kabul, r., Asia (kä′bōōl)	155	34°44′N	69°43′E
Kabunda, D.R.C.	191	12°25′S	29°22′E
Kabwe, Zam.	186	14°27′S	28°27′E
Kachuga, Russia (kȧ-chōō-gȧ)	135	54°09′N	105°43′E
Kadei, r., Afr.	189	4°00′N	15°10′E
Kadnikov, Russia (käd′nė-kôf)	136	59°30′N	40°10′E
Kadoma, Japan	167b	34°43′N	135°36′E
Kadoma, Zimb.	186	18°21′S	29°55′E
Kaduna, Nig. (kä-dōō′nä)	184	10°33′N	7°27′E
Kaduna, r., Nig.	189	9°30′N	6°00′E
Kaédi, Maur. (kä-ā-dē′)	184	16°09′N	13°30′W
Ka′ena Point, c., Hi., U.S. (kä′ā-nä)	64d	21°33′N	158°19′W
Kaesŏng, Kor., N. (kä′ē-sŭng) (kī′jō)	161	38°00′N	126°35′E
Kafanchan, Nig.	189	9°36′N	8°17′E
Kafia Kingi, Sudan (kä′fē-ȧ kĭŋ′gē)	185	9°17′N	24°28′E
Kafue, Zam. (kä′fōō)	186	15°45′S	28°17′E
Kafue, r., Zam.	186	15°45′S	26°30′E
Kafue Flats, sw., Zam.	191	16°15′S	26°30′E
Kafue National Park, rec., Zam.	191	15°00′S	25°35′E
Kafwira, D.R.C.	191	12°10′S	27°33′E
Kagal′nik, r., Russia (kä-gäl′′nėk)	133	46°58′N	39°25′E
Kagera, r., Afr. (kä-gā′rȧ)	186	1°10′S	31°10′E
Kagoshima, Japan (kä′gô-shē′mȧ)	161	31°35′N	130°31′E
Kagoshima-Wan, b., Japan (kä′gô-shē′mä wän)	166	31°24′N	130°39′E
Kahayan, r., Indon.	168	1°45′S	113°40′E
Kahemba, D.R.C.	190	7°17′S	19°00′E
Kahia, D.R.C.	191	6°21′S	28°24′E
Kahoka, Mo., U.S. (kȧ-hō′kȧ)	79	40°26′N	91°42′W
Kaho′olawe, i., Hi., U.S. (kä-hōō-lä′wē)	64c	20°28′N	156°48′W
Kahramanmaraş, Tur.	154	37°40′N	36°50′W
Kahshahpiwi, r., Can.	71	48°24′N	90°56′W
Kahuku Point, c., Hi., U.S. (kä-hōō′kōō)	64d	21°50′N	157°50′W
Kahului, Hi., U.S.	64c	20°53′N	156°28′W
Kai, Kepulauan, is., Indon.	169	5°35′S	132°45′E
Kaiang, Malay.	153b	3°00′N	101°47′E
Kaiashk, r., Can.	58	49°40′N	89°30′W
Kaibab Indian Reservation, I.R., Az., U.S. (kä′ē-bȧb)	77	36°55′N	112°45′W
Kaibab Plat, Az., U.S.	77	36°30′N	112°10′W
Kaidu, r., China (kī-dōō)	160	42°35′N	84°04′E
Kaieteur Fall, wtfl., Guy. (kī-ē-tōōr′)	101	4°48′N	59°24′W
Kaifeng, China (kī-fŭŋ)	161	34°48′N	114°22′E
Kai Kecil, i., Indon.	169	5°45′S	132°40′E
Kailua, Hi., U.S. (kä′ē-lōō′ä)	64c	21°18′N	157°43′W
Kailua Kona, Hi., U.S.	84a	19°49′N	155°59′W
Kaimana, Indon.	169	3°32′S	133°47′E
Kaimanawa Mountains, mts., N.Z.	177	39°10′S	176°00′E
Kainan, Japan (kä′ē-nän′)	167	34°09′N	135°14′E
Kainji Lake, res., Nig.	184	10°25′N	4°50′E
Kaiserslautern, Ger. (kī-zĕrs-lou′tĕrn)	117	49°26′N	7°46′E
Kaitaia, N.Z. (kä-ē-tä′ē-ä)	175a	35°30′S	173°28′E
Kaiwi Channel, strt., Hi., U.S. (käē-wē)	64c	21°10′N	157°38′W
Kaiyuan, China (kī-yuän)	165	23°42′N	103°20′E
Kaiyuan, China	164	42°30′N	124°00′E
Kaiyuh Mountains, mts., Ak., U.S. (kī-yōō′)	63	64°25′N	157°38′W
Kajaani, Fin. (kä′yä-nė)	116	64°15′N	27°16′E
Kajang, Gunong, mtn., Malay.	153b	2°47′N	104°05′E
Kajiki, Japan (kä′jē-kē)	166	31°44′N	130°41′E
Kakhovka, Ukr. (kä-kôf′kȧ)	133	46°46′N	33°32′E
Kakhovs′ke vodoskhovyshche, res., Ukr.	134	47°21′N	33°33′E
Kākināda, India	155	16°58′N	82°18′E
Kaktovik, Ak., U.S. (käk-tō′vĭk)	63	70°08′N	143°51′W
Kakwa, r., Can. (käk′wȧ)	55	54°00′N	118°55′W
Kalach, Russia (kȧ-lách′)	137	50°15′N	40°55′E
Kaladan, r., Asia	160	21°07′N	93°04′E
Kalae, c., Hi., U.S.	84a	18°55′N	155°41′W
Kalahari Desert, des., Afr. (kä-lä-hä′rē)	186	23°00′S	22°03′E
Kalama, Wa., U.S. (kȧ-lăm′ȧ)	74c	46°01′N	122°50′W
Kalama, r., Wa., U.S.	74c	46°03′N	122°47′W
Kalamáta, Grc.	110	37°04′N	22°08′E
Kalamazoo, Mi., U.S. (kăl-ȧ-mȧ-zōō′)	65	42°20′N	85°40′W
Kalamazoo, r., Mi., U.S.	66	42°35′N	86°00′W
Kalanchak, Ukr. (kä-län-chäk′)	133	46°17′N	33°14′E
Kalandula, Ang. (dōō′kȧ dȧ brä-gäN′sä)	186	9°06′S	15°57′E
Kalaotoa, Pulau, i., Indon.	168	7°22′S	122°30′E
Kalapana, Hi., U.S. (kä-lä-pä′nä)	84a	19°25′N	155°00′W
Kalar, mtn., Iran	154	31°43′N	51°41′E
Kalāt, Pak. (kŭ-lät′)	155	29°05′N	66°36′E
Kalemie, D.R.C.	186	5°56′S	29°12′E
Kalgan see Zhangjiakou, China	161	40°45′N	114°58′E
Kalgoorlie-Boulder, Austl. (kăl-gōōr′lė)	174	30°45′S	121°35′E
Kaliakra, Nos, c., Blg.	119	43°25′N	28°42′E
Kalima, D.R.C.	191	2°34′S	26°37′E
Kaliningrad, Russia	134	54°42′N	20°32′E
Kaliningrad, Russia (kä-lė-nēn′grät)	142b	55°55′N	37°49′E
Kalinkavichy, Bela.	132	52°07′N	29°19′E
Kalispel Indian Reservation, I.R., Wa., U.S. (kăl-ĭ-spĕl′)	72	48°25′N	117°30′W
Kalispell, Mt., U.S. (kăl′ĭ-spĕl)	64	48°12′N	114°18′W
Kalisz, Pol. (kä′lēsh)	117	51°45′N	18°05′E
Kaliua, Tan.	191	5°04′S	31°48′E
Kalixälven, r., Swe.	116	67°12′N	22°00′E
Kalmar, Swe. (käl′mär)	116	56°40′N	16°19′E
Kalmarsund, strt., Swe. (käl′mär)	122	56°30′N	16°17′E
Kal′mius, r., Ukr. (käl′′myōōs)	133	47°15′N	37°38′E
Kalmykia, prov., Russia	137	46°56′N	46°00′E
Kalocsa, Hung. (kä′lô-chä)	125	46°32′N	19°00′E
Kalohi Channel, strt., Hi., U.S. (kä-lō′hĭ)	84a	20°55′N	157°15′W
Kaloko, D.R.C.	191	6°47′S	25°48′E
Kalomo, Zam. (kä-lō′mō)	186	17°02′S	26°30′E
Kalsubai Mount, mtn., India	158	19°43′N	73°47′E
Kaltenkirchen, Ger. (käl′tĕn-kēr-кĕn)	115c	53°50′N	9°57′E
Kālu, r., India	159b	19°18′N	73°14′E
Kaluga, Russia (kȧ-lō′gä)	134	54°29′N	36°12′E
Kaluga, prov., Russia	132	54°10′N	35°00′E
Kaluktutiak (Cambridge Bay), Can.	50	69°15′N	105°00′W
Kalundborg, Den. (kȧ-lōn′′bôr′)	122	55°42′N	11°07′E
Kalush, Ukr. (kä′lōsh)	125	49°02′N	24°24′E
Kalvarija, Lith. (käl-vä-rē′yȧ)	123	54°24′N	23°17′E
Kalwa, India	159b	19°12′N	72°59′E
Kal′ya, Russia (käl′yȧ)	142a	60°17′N	59°58′E
Kalyān, India	158	19°16′N	73°07′E
Kalyazin, Russia (käl-yȧ′zēn)	132	57°13′N	37°55′E
Kama, r., Russia (kä′mȧ)	134	56°10′N	53°50′E
Kamaishi, Japan (kä′mä-ē′shė)	166	39°16′N	142°03′E
Kamakura, Japan (kä′mä-kōō′rä)	167	35°19′N	139°33′E
Kamarān, i., Yemen	154	15°19′N	41°47′E
Kāmārhāti, India	158a	22°41′N	88°23′E
Kambove, D.R.C. (käm-bō′vē)	186	10°58′S	26°43′E
Kamchatka, r., Russia	141	54°15′N	158°38′E
Kamchatka, Poluostrov, pen., Russia	141	55°19′N	157°45′E
Kamen, Ger. (kä′mĕn)	127c	51°35′N	7°40′E
Kamenjak, Rt, c., Cro. (kä′mĕ-nyäk)	130	44°45′N	13°57′E
Kamen′-na-Obi, Russia (kä-mĭny′nŭ ô′bē)	134	53°43′N	81°28′E
Kamensk-Shakhtinskiy, Russia (kä′mĕnsk shäk′tĭn-skĭ)	133	48°17′N	40°16′E
Kamensk-Ural′skiy, Russia (kä′mĕnsk ōō-räl′skĭ)	136	56°27′N	61°55′E
Kamenz, Ger. (kä′mĕnts)	124	51°16′N	14°05′E
Kameoka, Japan (kä′mā-ōkä)	167b	35°01′N	135°35′E
Kāmet, mtn., Asia	158	30°50′N	79°42′E
Kamianets′-Podil′s′kyi, Ukr.	137	48°41′N	26°34′E
Kamianka-Buz′ka, Ukr.	125	50°06′N	24°20′E
Kamień Pomorski, Pol.	124	53°57′N	14°48′E
Kamikoma, Japan (kä′mė-kō′mä)	167b	34°45′N	135°50′E
Kamina, D.R.C.	186	8°44′S	25°00′E
Kaministikwia, r., Can. (kȧ-mĭ-nĭ-stĭk′wĭ-ȧ)	71	48°40′N	89°41′W
Kamituga, D.R.C.	191	3°04′S	28°11′E
Kamloops, Can. (kăm′lōōps)	50	50°40′N	120°20′W
Kamp, r., Aus. (kämp)	124	48°30′N	15°45′E
Kampala, Ug. (käm-pä′lä)	185	0°19′N	32°25′E
Kampar, r., Indon. (käm′pär)	168	0°30′N	101°30′E
Kampene, D.R.C.	191	3°36′S	26°40′E
Kampenhout, Bel.	115a	50°56′N	4°33′E
Kamp-Lintfort, Ger. (kämp-lĕnt′fŏrt)	127c	51°30′N	6°34′E
Kâmpóng Saôm, Camb.	168	10°40′N	103°50′E
Kâmpóng Thum, Camb. (kŏm′pŏng-tŏm)	168	12°41′N	104°29′E
Kâmpôt, Camb. (käm′pōt)	168	10°41′N	104°07′E
Kampuchea see Cambodia, nation, Asia	168	12°15′N	104°00′E
Kamsack, Can. (kăm′săk)	50	51°34′N	101°54′W
Kamskoye, res., Russia	134	59°08′N	56°30′E
Kamudilo, D.R.C.	191	7°42′S	27°18′E
Kamuela, Hi., U.S.	84a	20°01′N	155°40′W
Kamui Misaki, c., Japan	166	43°25′N	139°35′E
Kámuk, Cerro, mtn., C.R. (sĕ′r-rō-kä-mōō′k)	91	9°18′N	83°02′W
Kamyshevatskaya, Russia	133	46°24′N	37°58′E
Kamyshin, Russia (kä-mwēsh′ĭn)	134	50°08′N	45°20′E
Kamyshlov, Russia (kä-mēsh′lôf)	134	56°50′N	62°32′E
Kan, r., Russia (kän)	140	56°30′N	94°17′E
Kanab, Ut., U.S. (kăn′ăb)	77	37°00′N	112°30′W
Kanabeki, Russia (kȧ-nä′byĕ-kĭ)	142a	57°48′N	57°16′E
Kanab Plateau, plat., Az., U.S.	77	36°31′N	112°55′W
Kanaga, i., Ak., U.S. (kă-nä′gä)	63a	52°02′N	177°38′W
Kanagawa, dept., Japan (kä′nä-gä′wä)	167a	35°29′N	139°32′E
Kanā′is, Ra′s al, c., Egypt	119	31°14′N	28°08′E
Kanamachi, Japan (kä-nä-mä′chē)	167a	35°46′N	139°52′E
Kananga, D.R.C.	186	6°14′S	22°17′E
Kananikol′skoye, Russia	142a	52°48′N	57°29′E
Kanasín, Mex. (kä-nä-sē′n)	90a	20°54′N	89°31′W
Kanatak, Ak., U.S. (kă-nă′tŏk)	63	57°35′N	155°48′W
Kanawha, r., W.V., U.S. (kȧ-nô′wȧ)	65	37°55′N	81°50′W
Kanaya, Japan (kä-nä′yä)	167a	35°10′N	139°49′E
Kanazawa, Japan (kä′nȧ-zä′wȧ)	161	36°34′N	136°38′E
Kānchenjunga, mtn., Asia (kĭn-chĭn-jŏn′gȧ)	155	27°30′N	88°18′E
Kānchipuram, India	155	12°55′N	79°43′E
Kandahār, Afg.	155	31°43′N	65°58′E
Kanda Kanda, D.R.C. (kän′dȧ kän′dȧ)	186	6°56′S	23°36′E
Kandalaksha, Russia (kȧn-dȧ-lák′shȧ)	134	67°10′N	33°05′E
Kandalakshskiy Zaliv, b., Russia	136	66°20′N	35°00′E
Kandava, Lat. (kän′dȧ-vȧ)	123	57°03′N	22°45′E
Kandi, Benin (käN-dē′)	184	11°08′N	2°56′E
Kandiāro, Pak.	158	27°09′N	68°12′E
Kandla, India (kŭnd′lŭ)	158	23°00′N	70°20′E
Kandy, Sri L. (kän′dė)	159	7°18′N	80°42′E
Kane, Pa., U.S. (kān)	67	41°40′N	78°50′W
Kāne′ohe, Hi., U.S. (kä-nä-ō′hā)	84a	21°25′N	157°47′W
Kāne′ohe Bay, b., Hi., U.S.	64d	21°32′N	157°40′W
Kanevskaya, Russia (kȧ-nyĕf′skȧ)	133	46°07′N	38°58′E
Kangaroo, i., Austl. (kăŋ-gȧ-rō′)	174	36°05′S	137°05′E
Kangāvar, Iran (kŭŋ′gä-vär)	154	34°37′N	46°45′E
Kangean, Kepulauan, is., Indon. (käŋ′gē-än)	168	6°50′S	116°22′E
Kanggye, Kor., N. (käng′gyĕ)	161	40°55′N	126°40′E
Kanghwa, i., Kor., S. (käng′hwä)	166	37°38′N	126°00′E
Kangnŭng, Kor., S. (käng′nŏ ng)	166	37°42′N	128°50′E
Kango, Gabon (käN-gō)	186	0°09′N	10°08′E
Kangowa, D.R.C.	190	9°55′S	22°48′E
Kanin, Poluostrov, pen., Russia	134	68°00′N	45°00′E
Kaningo, Kenya	191	0°49′S	38°32′E
Kanin Nos, Mys, c., Russia	136	68°40′N	44°00′E
Kaniv, Ukr.	133	49°46′N	31°27′E
Kanivs′ke vodoskhovyshche, res., Ukr.	134	50°10′N	30°40′E
Kanjiža, Serb. (kä′nyē-zhä)	131	46°05′N	20°02′E
Kankakee, Il., U.S. (kăŋ-kȧ-kē′)	66	41°07′N	87°53′W
Kankakee, r., Il., U.S.	66	41°15′N	88°15′W
Kankan, Gui. (käN-käN) (kän-kän′)	184	10°23′N	9°18′W
Kannapolis, N.C., U.S. (kăn-ăp′ō-lĭs)	83	35°30′N	80°38′W
Kannoura, Japan (kä′nō-ōō′rä)	167	33°34′N	134°18′E
Kano, Nig. (kä′nō)	184	12°00′N	8°30′E
Kanonkop, mtn., S. Afr.	186a	33°49′S	18°37′E
Kanopolis Reservoir, res., Ks., U.S. (kăn-ŏp′ō-lĭs)	78	38°44′N	98°01′W
Kānpur, India (kän′pûr)	158	26°30′N	80°10′E
Kansas, state, U.S. (kăn′zȧs)	64	38°30′N	99°40′W
Kansas, r., Ks., U.S.	79	39°08′N	95°52′W
Kansas City, Ks., U.S.	65	39°06′N	94°39′W
Kansas City, Mo., U.S.	65	39°05′N	94°35′W
Kansk, Russia	135	56°14′N	95°43′E
Kansŏng, Kor., S.	166	38°09′N	128°29′E
Kantang, Thai. (kän′täng′)	168	7°26′N	99°28′E
Kantchari, Burkina	188	12°29′N	1°31′E
Kanton, i., Kir.	194	3°50′S	174°00′W
Kantunilkin, Mex. (kän-tōō-nēl-kē′n)	90a	21°07′N	87°30′W
Kanzhakovskiy Kamen, Gora, mtn., Russia (kȧn-zhä′kôvs-kēē kä′mĭen)	142a	59°38′N	59°12′E
Kaohsiung, Tai. (kä-ô-syŏng′)	161	22°35′N	120°25′E
Kaolack, Sen.	184	14°09′N	16°04′W
Kaouar, oasis, Niger	185	19°16′N	13°09′E
Kapaa, Hi., U.S.	84a	22°06′N	159°20′W
Kapanga, D.R.C.	190	8°21′S	22°35′E
Kapfenberg, Aus. (käp′fĕn-bĕrgh)	124	47°27′N	15°16′E
Kapiri Mposhi, Zam.	191	13°58′S	28°41′E
Kapoeta, Sudan	185	4°45′N	33°35′E
Kaposvár, Hung. (kô′pôsh-vär)	125	46°21′N	17°45′E
Kapsan, Kor., N. (käp′sän′)	166	40°59′N	128°22′E
Kapuskasing, Can.	51	49°28′N	82°22′W
Kapuskasing, r., Can.	58	48°55′N	82°55′W
Kapustin Yar, Russia (kä′pòs-tėn yär′)	137	48°30′N	45°40′E
Kaputar, Mount, mtn., Austl. (kă-pū-tăr)	176	30°11′S	150°11′E
Kapuvár, Hung. (kô′pōō-vär)	125	47°35′N	17°02′E
Kara, Russia (kärȧ)	134	68°42′N	65°30′E
Kara, r., Russia	136	68°30′N	65°20′E
Karabalā′, Iraq (kŭr′bȧ-lä)	154	32°31′N	43°58′E
Karabanovo, Russia (kä′rȧ-bȧ-nō-vō)	132	56°19′N	38°43′E
Karabash, Russia (kó-rȧ-bäsh′)	142a	55°27′N	60°14′E
Kara-Bogaz-Gol, Zaliv, b., Turkmen. (kȧ-rä′ bŭ-gäs′)	139	41°30′N	53°40′E
Karachay-Cherkessia, prov., Russia	138	44°00′N	42°00′E
Karachev, Russia (kȧ-rȧ-chôf′)	136	53°08′N	34°54′E
Karāchi, Pak.	155	24°59′N	68°56′E
Karaganda see Qaraghandy, Kaz.	139	49°42′N	73°18′E
Karaidel′, Russia (kä′rī-dĕl)	142a	55°52′N	56°54′E
Karakoram Pass, p., Asia	155	35°35′N	77°45′E
Karakoram Range, mts., India (kä′rä kō′rōm)	155	35°24′N	76°38′E
Karakorum, hist., Mong.	160	47°25′N	102°22′E
Kara-Kum, des., Turkmen.	139	40°00′N	57°00′E
Kara Kum Canal, can., Turkmen.	139	37°35′N	61°50′E
Karaman, Tur. (kä-rä-män′)	119	37°10′N	33°00′E
Karamay, China (kär-äm-ā)	160	45°37′N	84°53′E
Karamea Bight, b., N.Z. (kȧ-rä-mē′ȧ bĭt)	175a	41°20′S	171°30′E
Kara Sea see Karskoye More, sea, Russia	134	74°00′N	68°00′E
Karashahr (Yanqui), China (kä-rä-shä-är) (yän-chyē)	160	42°14′N	86°28′E
Karatsu, Japan (kä′rȧ-tsōō)	167	33°28′N	129°59′E
Karaul, Russia (kä-rä-ōl′)	140	70°13′N	83°46′E
Karawanken, mts., Eur.	124	46°32′N	14°07′E
Karcag, Hung. (kär′tsäg)	125	47°18′N	20°58′E
Kárditsa, Grc.	131	39°23′N	21°57′E
Kärdla, Est. (kĕrd′lȧ)	123	58°59′N	22°44′E
Karelia, prov., Russia	140	62°30′N	32°35′E
Karema, Tan.	186	6°49′S	30°26′E
Kargat, Russia (kär-gät′)	134	55°17′N	80°07′E
Karghalik see Yecheng, China	160	37°54′N	77°25′E
Kargopol′, Russia (kär-gō-pōl′′)	134	61°30′N	38°50′E
Kariba, Lake, res., Afr.	186	17°15′S	27°55′E
Karibib, Nmb. (kár′ȧ-bĭb)	186	21°55′S	15°50′E

ng-sing; ŋ-baŋk; N-nasalized n; nŏd; cŏmmit; ōld; ôbey; ôrder; oi-boil; fōōd; ŏ-as oo in foot; ou-out; s-soft; sh-dish; th-thin; pūre; ûnite; ûrn; stŭd; circŭs; ü-as in French tu; ′-indeterminate vowel.

262 Kārikāl – Key West

PRONOUNCING INDEX

PLACE (Pronunciation)	PAGE	LAT.	LONG.
Kārikāl, India (kä-rē-käl′)	159	10°58′N	79°49′E
Karimata, Kepulauan, is., Indon. (kä-rē-mä′tä)	168	1°08′S	108°10′E
Karimata, Selat, strt., Indon.	168	1°00′S	107°10′E
Karimun Besar, i., Indon.	153b	1°10′N	103°28′E
Karimunjawa, Kepulauan, is., Indon. (kä′rē-mōōn-yä′vä)	168	5°36′S	110°15′E
Karin, Som. (kȧr′ĭn)	192a	10°43′N	45°50′E
Karkar Island, i., Pap. N. Gui. (kär′kär)	169	4°50′S	146°45′E
Karkheh, r., Iran	154	32°45′N	47°50′E
Karkinits′ka zatoka, b., Ukr.	133	45°50′N	32°45′E
Karkūk, Iraq	154	35°28′N	44°22′E
Karlivka, Ukr.	133	49°26′N	35°08′E
Karlobag, Cro. (kär-lō-bäg′)	130	44°30′N	15°03′E
Karlovac, Cro. (kär′lō-väts)	119	45°29′N	15°16′E
Karlovo, Blg. (kär′lō-vō)	131	42°39′N	24°48′E
Karlovy Vary, Czech Rep. (kär′lō-vē vä′rē)	117	50°13′N	12°53′E
Karlshamn, Swe. (kärls′häm)	122	56°11′N	14°50′E
Karlskrona, Swe. (kärls′krō-nä)	116	56°10′N	15°33′E
Karlsruhe, Ger. (kärls′rōō-ĕ)	117	49°00′N	8°23′E
Karlstad, Swe. (kärl′städ)	110	59°25′N	13°28′E
Karluk, Ak., U.S. (kär′lŭk)	63	57°30′N	154°22′W
Karmøy, i., Nor. (kärm-ûe)	122	59°14′N	5°00′E
Karnataka, state, India	155	14°55′N	75°00′E
Karnobat, Blg. (kär-nô′bät)	131	42°39′N	26°59′E
Karonga, Mwi. (kȧ-rōŋ′gȧ)	186	9°52′S	33°57′E
Kárpathos, i., Grc.	119	35°34′N	27°26′E
Karpinsk, Russia (kär′pinsk)	142a	59°46′N	60°00′E
Kars, Tur. (kärs)	154	40°35′N	43°00′E
Kārsava, Lat. (kär′sä-vä)	123	56°46′N	27°39′E
Karshi, Uzb. (kär′shē)	139	38°30′N	66°08′E
Karskiye Vorota, Proliv, strt., Russia	134	70°30′N	58°07′E
Karskoye More (Kara Sea), sea, Russia	134	74°00′N	68°00′E
Kartaly, Russia (kȧr′tȧ lē)	134	53°05′N	60°40′E
Karunagapalli, India	159	9°09′N	76°34′E
Karvina, Czech Rep.	125	49°50′N	18°30′E
Kasai (Cassai), r., Afr.	186	3°45′S	19°10′E
Kasama, Zam. (kȧ-sä′mȧ)	186	10°13′S	31°12′E
Kasanga, Tan. (kȧ-säŋ′gȧ)	186	8°28′S	31°09′E
Kasaoka, Japan (kä′sȧ-ō′kȧ)	167	34°33′N	133°29′E
Kasba-Tadla, Mor. (käs′bȧ-täd′lȧ)	184	32°37′N	5°57′W
Kasempa, Zam. (kȧ-sĕm′pȧ)	186	13°27′S	25°50′E
Kasenga, D.R.C. (kȧ-seŋ′gȧ)	186	10°22′S	28°38′E
Kasese, D.R.C.	191	1°38′S	27°07′E
Kasese, Ug.	191	0°10′N	30°05′E
Kāshān, Iran (kä-shän′)	154	33°52′N	51°15′E
Kashgar see Kashi, China	160	39°29′N	76°00′E
Kashi (Kashgar), China (kä-shr) (käsh-gär)	160	39°29′N	76°00′E
Kashihara, Japan (kä′shē-hä′rä)	167b	34°31′N	135°48′E
Kashiji Plain, pl., Zam.	190	13°25′S	22°30′E
Kashin, Russia (kä-shēn′)	132	57°20′N	37°38′E
Kashira, Russia (kä-shē′rȧ)	132	54°49′N	38°11′E
Kashiwa, Japan (kä′shē-wä)	167a	35°51′N	139°58′E
Kashiwara, Japan	167b	34°35′N	135°38′E
Kashiwazaki, Japan (kä′shē-wä-zä′kē)	166	37°06′N	138°17′E
Kāshmar, Iran	157	35°12′N	58°27′E
Kashmīr see Jammu and Kashmīr, state, India	155	34°30′N	76°00′E
Kashmor, Pak.	158	28°33′N	69°34′E
Kashtak, Russia (käsh′tăk)	142a	55°18′N	61°25′E
Kasimov, Russia (kȧ-sē′môf)	136	54°56′N	41°23′E
Kaskanak, Ak., U.S. (kăs-kă′năk)	63	60°00′N	158°00′W
Kaskaskia, r., Il., U.S. (kăs-kăs′kĭ-ȧ)	66	39°10′N	88°50′W
Kaskattama, r., Can. (kăs-kȧ-tă′mȧ)	57	56°28′N	90°55′W
Kaskö (Kaskinen), Fin. (käs′kû) (käs′kē-nĕn)	123	62°24′N	21°18′E
Kasli, Russia (käs′lĭ)	136	55°53′N	60°46′E
Kasongo, D.R.C. (kä-sôŋ′gō)	186	4°31′S	26°42′E
Kásos, i., Grc.	119	35°20′N	26°55′E
Kaspiysk, Russia	138	42°52′N	47°38′E
Kassándras, Kólpos, b., Grc.	131	40°10′N	23°35′E
Kassel, Ger. (käs′ĕl)	117	51°19′N	9°30′E
Kasson, Mn., U.S. (kăs′ŭn)	71	44°01′N	92°45′W
Kastamonu, Tur. (kä-stȧ-mō′nōō)	154	41°20′N	33°50′E
Kastoría, Grc. (kȧs-tō′rĭ-ȧ)	119	40°28′N	21°17′E
Kasūr, Pak.	158	31°10′N	74°29′E
Kataba, Zam.	191	16°05′S	25°10′E
Katahdin, Mount, mtn., Me., U.S. (kȧ-tä′dĭn)	60	45°56′N	68°57′W
Katanga, hist. reg., D.R.C. (kȧ-täŋ′gȧ)	186	8°30′S	25°00′E
Katanning, Austl. (kȧ-tăn′ĭng)	174	33°45′S	117°45′E
Katav-Ivanovsk, Russia (kä′tăf ĭ-vä′nôfsk)	142a	54°46′N	58°13′E
Kateninskiy, Russia (kătyĕ′nĭs-kĭ)	142a	53°12′N	61°05′E
Kateríni, Grc.	131	40°18′N	22°36′E
Katete, Zam.	191	14°05′S	32°07′E
Katherine, Austl. (kăth′ẽr-ĭn)	174	14°15′S	132°20′E
Kāthiāwār, pen., India (kä′tyȧ-wär′)	155	22°10′N	70°20′E
Kathmandu, Nepal (kät-män-dōō′)	155	27°49′N	85°21′E
Kathryn, Can. (kăth′rĭn)	62e	51°13′N	113°42′W
Kathryn, Ca., U.S.	75a	33°42′N	117°45′W
Katihār, India	158	25°39′N	87°39′E
Katiola, C. Iv.	188	8°08′N	5°06′W
Katmai National Park, rec., Ak., U.S. (kăt′mī)	64a	58°38′N	155°00′W
Katompi, D.R.C.	191	6°11′S	26°20′E
Katopa, D.R.C.	191	2°45′S	25°06′E
Katowice, Pol.	110	50°15′N	19°00′E
Katrineholm, Swe. (kä-trē′nĕ-hōlm)	122	59°01′N	16°10′E
Katsbakhskiy, Russia (kȧts-bäk′skĭ)	142a	52°57′N	59°37′E
Katsina, Nig. (kät′sē-nȧ)	184	13°00′N	7°32′E
Katsina Ala, Nig.	184	7°10′N	9°17′E
Katsura, r., Japan (kä′tsō-rä)	167b	34°55′N	135°43′E
Katta-Kurgan, Uzb. (kȧ-tȧ-kȯr-gän′)	139	39°45′N	66°42′E
Kattegat, strt., Eur. (kăt′ĕ-gät)	112	56°57′N	11°25′E
Katumba, D.R.C.	191	7°45′S	25°18′E
Katun′, r., Russia (kȧ-tōn′)	140	51°30′N	86°18′E
Katwijk aan Zee, Neth.	115a	52°12′N	4°23′E
Kaua′i, i., Hi., U.S.	64c	22°09′N	159°15′W
Kauai Channel, strt., Hi., U.S. (kä-ōō-ä′ē)	64c	21°35′N	158°52′W
Kaufbeuren, Ger. (kouf′boi-rĕn)	124	47°52′N	10°38′E
Kaufman, Tx., U.S. (kôf′măn)	81	32°36′N	96°18′W
Kaukauna, Wi., U.S. (kô-kô′nȧ)	71	44°17′N	88°15′W
Kaulakahi Channel, strt., Hi., U.S. (kä′ōō-lȧ-kä′hē)	84a	22°00′N	159°55′W
Kaunakakai, Hi., U.S. (kä′ōō-nȧ-kä′kī)	84a	21°06′N	156°59′W
Kaunas, Lith. (kou′nȧs) (kôv′nô)	134	54°42′N	23°54′E
Kaura Namoda, Nig.	184	12°35′N	6°35′E
Kavála, Grc. (kä-vä′lä)	119	40°55′N	24°24′E
Kavieng, Pap. N. Gui. (kä-vē-ĕng′)	169	2°44′S	151°02′E
Kavīr, Dasht-e, des., Iran (dŭsht-ē-ka-vēr′)	154	34°41′N	53°30′E
Kawagoe, Japan (kä-wä-gō′ȧ)	167	35°55′N	139°29′E
Kawaguchi, Japan (kä-wä-gōō-chē)	167a	35°48′N	139°44′E
Kawaikini, mtn., Hi., U.S. (kä-wä′ē-kĭ-nĭ)	84a	22°05′N	159°33′W
Kawanishi, Japan (kä′wä′nē-shē)	167b	34°49′N	135°26′E
Kawasaki, Japan (kä-wä-sä′kē)	166	35°32′N	139°43′E
Kaxgar, r., China	160	39°30′N	75°00′E
Kaya, Burkina (kä′yä)	184	13°05′N	1°05′W
Kayan, r., Indon.	168	1°45′N	115°38′E
Kaycee, Wy., U.S. (kā-sē′)	73	43°43′N	106°38′W
Kayes, Mali (kāz)	184	14°27′N	11°26′W
Kayseri, Tur. (kī′sĕ-rē)	154	38°45′N	35°20′E
Kazach′ye, Russia	135	70°46′N	135°47′E
Kazakhstan, nation, Asia	134	48°45′N	59°00′E
Kazan′, Russia (kȧ-zän′)	134	55°50′N	49°18′E
Kazanka, Ukr. (kȧ-zän′kȧ)	133	47°49′N	32°50′E
Kazanlŭk, Blg. (kȧ′zȧn-lĕk)	131	42°47′N	25°23′E
Kazbek, Gora, mtn. (kȧz-bĕk′)	137	42°42′N	44°31′E
Kāzerūn, Iran	154	29°37′N	51°44′E
Kazincbarcika, Hung. (kô′zĭnts-bôr-tsĭ-ko)	125	48°15′N	20°39′E
Kazungula, Zam.	191	17°45′S	25°20′E
Kazusa Kameyama, Japan (kä-zōō-sä kä-mä′yä-mä)	167a	35°14′N	140°06′E
Kazym, r., Russia (kä-zĕm′)	140	63°30′N	67°41′E
Kéa, i., Grc.	131	37°36′N	24°13′E
Kealaikahiki Channel, strt., Hi., U.S. (kā-ä′lä-ē-kä-hē′kē)	84a	20°38′N	157°00′W
Keansburg, N.J., U.S. (kēnz′bûrg)	68a	40°26′N	74°08′W
Kearney, Ne., U.S. (kär′nĭ)	70	40°42′N	99°05′W
Kearny, N.J., U.S.	68a	40°46′N	74°09′W
Keasey, Or., U.S. (kēz′ĭ)	74c	45°51′N	123°20′W
Kebnekaise, mtn., Swe. (kĕp′nĕ-kȧ-ēs′ĕ)	112	67°53′N	18°10′E
Kecskemét, Hung. (kĕch′kĕ-māt)	119	46°52′N	19°42′E
Kedah, hist. reg., Malay. (kä′dä)	168	6°00′N	100°31′E
Kédainiai, Lith. (kē-dī′nĭ-ī)	123	55°16′N	23°58′E
Kedgwick, Can. (kĕdj′wĭk)	60	47°39′N	67°21′W
Keenbrook, Ca., U.S. (kēn′brŏŏk)	75a	34°16′N	117°29′W
Keene, N.H., U.S. (kēn)	67	42°55′N	72°15′W
Keetmanshoop, Nmb. (kāt′măns-hōp)	186	26°30′S	18°05′E
Keet Seel Ruin, Az., U.S. (kēt sēl)	77	36°46′N	110°32′W
Keewatin, Mn., U.S. (kē-wä′tĭn)	71	47°24′N	93°03′W
Kefallonía, i., Grc.	119	38°08′N	20°58′E
Keffi, Nig. (kĕf′ē)	184	8°51′N	7°52′E
Ke Ga, Mui, c., Viet.	168	12°58′N	109°50′E
Kei, r., Afr. (kā)	187c	32°57′S	26°50′E
Keila, Est. (kā′lȧ)	123	59°19′N	24°25′E
Keilor, Austl.	173a	37°43′S	144°50′E
Kei Mouth, S. Afr.	187c	32°40′S	28°23′E
Keiskammahoek, S. Afr. (kās′kämä-hōōk′)	187c	32°42′S	27°11′E
Kéita, Bahr, r., Chad	189	9°30′N	19°17′E
Keitele, l., Fin. (kā′tĕ-lĕ)	123	62°50′N	25°40′E
Kekaha, Hi., U.S.	84a	21°57′N	159°42′W
Kelafo, Eth.	192a	5°40′N	44°00′E
Kelang, Malay.	168	3°20′N	101°27′E
Kelang, r., Malay.	153b	3°00′N	101°40′E
Kelkit, r., Tur.	119	40°38′N	37°03′E
Keller, Tx., U.S. (kĕl′ẽr)	75c	32°56′N	97°15′W
Kellinghusen, Ger. (kĕ′lĕng-hōō-zĕn)	115c	53°57′N	9°43′E
Kellogg, Id., U.S. (kĕl′ŏg)	72	47°32′N	116°07′W
Kelme′, Lith. (kĕl-mȧ)	123	55°36′N	22°53′E
Kélo, Chad	189	9°19′N	15°48′E
Kelowna, Can.	50	49°53′N	119°29′W
Kelsey Bay, Can. (kĕl′sē)	54	50°24′N	125°57′W
Kelso, Wa., U.S.	74c	46°09′N	122°54′W
Keluang, Malay.	153b	2°01′N	103°19′E
Kem′, Russia (kĕm)	134	65°00′N	34°48′E
Kemah, Tx., U.S. (kē′mȧ)	81a	29°32′N	95°01′W
Kemerovo, Russia	134	55°31′N	86°05′E
Kemi, Fin. (kā′mē)	116	65°48′N	24°38′E
Kemi, r., Fin.	116	67°02′N	27°50′E
Kemigawa, Japan (kĕ′mē-gä′wä)	167a	35°38′N	140°07′E
Kemijarvi, Fin. (kā′mē-yĕr-vē)	116	66°48′N	27°21′E
Kemi-joki, l., Fin.	116	66°37′N	28°13′E
Kemmerer, Wy., U.S. (kĕm′ẽr-ẽr)	73	41°48′N	110°36′W
Kemp, l., Tx., U.S. (kĕmp)	78	33°55′N	99°22′W
Kempen, Ger. (kĕm′pĕn)	127c	51°22′N	6°25′E
Kempsey, Austl. (kĕmp′sē)	175	30°59′S	152°50′E
Kempt, l., Can. (kĕmpt)	59	47°28′N	74°00′W
Kempten, Ger. (kĕmp′tĕn)	117	47°44′N	10°17′E
Kempton Park, S. Afr. (kĕmp′tŏn pärk)	192c	26°07′S	28°29′E
Ken, r., India	158	25°00′N	79°55′E
Kenai, Ak., U.S. (kē-nī′)	63	60°38′N	151°18′W
Kenai Fjords National Park, rec., Ak., U.S.	63	59°45′N	150°00′W
Kenai Mountains, mts., Ak., U.S.	63	60°00′N	150°00′W
Kenai Pen, Ak., U.S.	63	64°40′N	150°18′W
Kendal, S. Afr.	192c	26°03′S	28°58′E
Kendal, Eng., U.K. (kĕn′dȧl)	120	54°20′N	1°48′W
Kendallville, In., U.S. (kĕn′dȧl-vĭl)	66	41°25′N	85°20′W
Kenedy, Tx., U.S. (kĕn′ĕ-dĭ)	81	28°49′N	97°50′W
Kenema, S.L.	188	7°52′N	11°12′W
Kenitra, Mor. (kĕ-nē′trȧ)	118	34°21′N	6°34′W
Kenmare, N.D., U.S. (kĕn-mâr′)	70	48°41′N	102°05′W
Kenmore, N.Y., U.S. (kĕn′mōr)	69c	42°58′N	78°53′W
Kennebec, r., Me., U.S. (kĕn-ē-bĕk′)	60	44°23′N	69°48′W
Kennebunk, Me., U.S. (kĕn-ē-buŋk′)	60	43°24′N	70°33′W
Kennedale, Tx., U.S. (kĕn′-ē-dāl)	75c	32°38′N	97°13′W
Kennedy, Cape see Canaveral, Cape, c., Fl., U.S.	65	28°30′N	80°23′W
Kennedy, Mount, mtn., Can.	63	60°25′N	138°50′W
Kenner, La., U.S. (kĕn′ẽr)	81	29°58′N	90°15′W
Kennett, Mo., U.S. (kĕn′ĕt)	79	36°14′N	90°01′W
Kennewick, Wa., U.S. (kĕn′ē-wĭk)	72	46°12′N	119°06′W
Kenney Dam, dam, Can.	54	53°37′N	124°58′W
Kennydale, Wa., U.S. (kĕn-nē′dāl)	74a	47°31′N	122°12′W
Kénogami, Can. (kĕn-ŏ′gä-mē)	51	48°26′N	71°14′W
Kenogamissi Lake, l., Can.	58	48°15′N	81°31′W
Keno Hill, Can.	63	63°58′N	135°18′W
Kenora, Can. (kē-nō′rȧ)	51	49°47′N	94°29′W
Kenosha, Wi., U.S. (kē-nō′shȧ)	65	42°34′N	87°50′W
Kenova, W.V., U.S. (kē-nō′vȧ)	66	38°20′N	82°35′W
Kensico Reservoir, res., N.Y., U.S. (kĕn′sĭ-kō)	68a	41°08′N	73°45′W
Kent, Oh., U.S. (kĕnt)	66	41°05′N	81°20′W
Kent, Wa., U.S.	74a	47°23′N	122°14′W
Kentani, S. Afr. (kĕnt-änĭ′)	187c	32°31′S	28°19′E
Kentland, In., U.S. (kĕnt′lănd)	66	40°50′N	87°25′W
Kenton, Oh., U.S. (kĕn′tŭn)	66	40°40′N	83°35′W
Kent Peninsula, pen., Can.	52	68°28′N	108°10′W
Kentucky, state, U.S. (kĕn-tŭk′ĭ)	65	37°30′N	87°35′W
Kentucky, res., U.S.	65	36°20′N	88°50′W
Kentucky, r., Ky., U.S.	65	38°15′N	85°01′W
Kentwood, La., U.S. (kĕnt′wŏŏd)	81	30°56′N	90°31′W
Kenya, nation, Afr. (kĕn′yȧ)	186	1°00′N	36°53′E
Kenya, Mount (Kirinyaga), mtn., Kenya	187	0°10′S	37°20′E
Kenyon, Mn., U.S. (kĕn′yŭn)	71	44°15′N	92°58′W
Keokuk, Ia., U.S. (kē′ō-kŭk)	65	40°24′N	91°34′W
Keoma, Can. (kē-ō′mȧ)	62e	51°13′N	113°39′W
Kepenkeck Lake, l., Can.	61	48°13′N	54°45′W
Kępno, Pol. (kăn′pnō)	125	51°17′N	17°59′E
Kerala, state, India	155	16°38′N	76°00′E
Kerang, Austl. (kē-răng′)	175	35°32′S	143°58′E
Kerch, Ukr.	134	45°20′N	36°26′E
Kerchenskiy Proliv, strt., Eur. (kĕr-chĕn′skĭ prô′lĭf)	133	45°08′N	36°35′E
Kerempe Burun, c., Tur.	119	42°00′N	33°20′E
Keren, Erit.	185	15°46′N	38°28′E
Kerguélen, Îles, is., Afr. (kĕr′gȧ-lĕn)	3	49°50′S	69°30′E
Kericho, Kenya	191	0°22′S	35°17′E
Kerinci, Gunung, mtn., Indon.	168	1°45′S	101°18′E
Keriya see Yutian, China	160	36°55′N	81°39′E
Keriya, r., China (kĕ′rē-yä)	160	37°13′N	81°59′E
Kerkebet, Erit.	156	16°18′N	37°24′E
Kerkenna, Îles, i., Tun. (kĕr′kĕn-nä)	184	34°49′N	11°37′E
Kerki, Turkmen. (kĕr′kē)	139	37°52′N	65°15′E
Kérkyra, Grc.	119	39°36′N	19°56′E
Kérkyra, i., Grc.	118	39°33′N	19°36′E
Kermadec Islands, is., N.Z. (kẽr-măd′ĕk)	3	30°30′S	177°00′E
Kermān, Iran (kĕr-män′)	154	30°23′N	57°08′E
Kermānshāh see Bakhtarān, Iran	154	34°01′N	47°00′E
Kern, r., Ca., U.S.	76	35°31′N	118°37′W
Kern, South Fork, r., Ca., U.S.	76	35°40′N	118°15′W
Kerpen, Ger. (kĕr′pĕn)	127c	50°52′N	6°42′E
Kerrobert, Can.	56	51°53′N	109°13′W
Kerrville, Tx., U.S. (kûr′vĭl)	80	30°02′N	99°07′W
Kerulen, r., Asia (kĕr′ōō-lĕn)	161	47°52′N	113°22′E
Kesagami Lake, l., Can.	59	50°23′N	80°15′W
Keşan, Tur. (kĕ′shȧn)	131	40°50′N	26°37′E
Keshan, China (kŭ-shän)	161	48°00′N	126°30′E
Kesour, Monts des, mts., Alg.	118	32°51′N	0°30′W
Kestell, S. Afr. (kĕs′tĕl)	192c	28°19′N	28°43′E
Keszthely, Hung. (kĕst′hĕl-lĭ)	125	46°46′N	17°12′E
Ket′, r., Russia (kyĕt)	140	58°30′N	84°15′E
Keta, Ghana	184	6°00′N	1°00′E
Ketamputih, Indon.	153b	1°25′N	102°19′E
Ketapang, Indon. (kĕ-tȧ-päng′)	168	2°00′S	109°57′E
Ketchikan, Ak., U.S. (kĕch-ĭ-kăn′)	64a	55°21′N	131°35′W
Kętrzyn, Pol. (kăn′t′r-zĭn)	125	54°04′N	21°24′E
Kettering, Eng., U.K. (kĕt′ẽr-ĭng)	114a	52°23′N	0°43′W
Kettering, Oh., U.S.	66	39°40′N	84°15′W
Kettle, r., Can.	55	49°40′N	119°00′W
Kettle, r., Mn., U.S. (kĕt′'l)	71	46°20′N	92°57′W
Kettwig, Ger. (kĕt′vēg)	127c	51°22′N	6°56′E
Kęty, Pol. (kăn tĭ)	125	49°54′N	19°16′E
Ketzin, Ger. (kĕ′tzēn)	115b	52°29′N	12°51′E
Keuka, l., N.Y., U.S. (kē-ū′kȧ)	67	42°30′N	77°10′W
Kevelaer, Ger. (kĕ′fĕ-lăr)	127c	51°35′N	6°15′E
Kew, Austl.	173a	37°49′S	145°02′E
Kewanee, Il., U.S. (kē-wä′nē)	71	41°15′N	89°55′W
Kewaunee, Wi., U.S. (kē-wô′nē)	71	44°27′N	87°33′W
Keweenaw Bay, b., Mi., U.S. (kē′wē-nô)	71	46°59′N	88°15′W
Keweenaw Peninsula, pen., Mi., U.S.	71	47°28′N	88°12′W
Keya Paha, r., S.D., U.S. (kē-yȧ pä′hä)	70	43°11′N	100°10′W
Key Largo, i., Fl., U.S.	83a	25°11′N	80°15′W
Keyport, N.J., U.S. (kē′pōrt)	68a	40°26′N	74°12′W
Keyport, Wa., U.S.	74a	47°42′N	122°38′W
Keyser, W.V., U.S. (kī′sẽr)	67	39°25′N	79°00′W
Key West, Fl., U.S. (kē wĕst′)	65	24°31′N	81°47′W

ăt; finăl; rāte; senāte; ärm; ȧsk; sofȧ; fâre; ch-choose; dh-as th in other; bē; ēvent; bĕt; recĕnt; cratẽr; g-gō; gh-guttural g; bĭt; ĭ-short neutral; rīde; κ-guttural k as ch in German ich;

ng-sing; ŋ-baŋk; N-nasalized n; nŏd; cŏmmit; ōld; ȯbey; ȯrder; oi-boil; fōōd; ȯ-as oo in foot; ou-out; s-soft; sh-dish; th-thin; pūre; ūnite; ûrn; stŭd; circŭs; ü-as in French tu; ´-indeterminate vowel.

264 Klickitat – Krasnousol'skiy PRONOUNCING INDEX

PLACE (Pronunciation)	PAGE	LAT.	LONG.
Klickitat, r., Wa., U.S.	72	46°01'N	121°07'W
Klimovichi, Bela. (klē-mô-vē'chē)	132	53°37'N	31°21'E
Klimovsk, Russia (klĭ'môfsk)	142b	55°21'N	37°32'E
Klin, Russia (klēn)	132	56°18'N	36°43'E
Klintehamn, Swe. (klēn'tĕ-häm)	122	57°24'N	18°14'E
Klintsy, Russia (klĭn'tsĭ)	137	52°46'N	32°14'E
Klip, r., S. Afr. (klĭp)	192c	27°18'S	29°25'E
Klipgat, S. Afr.	192c	25°26'S	27°57'E
Klippan, Swe. (klyp'pān)	122	56°08'N	13°09'E
Kłodzko, Pol. (klôd'skô)	124	50°26'N	16°38'E
Klondike Region, hist. reg., N.A. (klŏn'dīk)	50	64°12'N	142°38'W
Klosterfelde, Ger. (klōs'tĕr-fĕl-dĕ)	115b	52°47'N	13°29'E
Klosterneuburg, Aus. (klōs-tĕr-noi'bo͞orgh)	115e	48°19'N	16°20'E
Kluane, l., Can.	52	61°15'N	138°40'W
Kluane National Park, rec., Can.	52	60°25'N	137°53'W
Kluczbork, Pol. (klo͞och'bŏrk)	125	50°59'N	18°15'E
Klyaz'ma, r., Russia (klyäz'mà)	132	55°49'N	39°19'E
Klyetsk, Bela. (klĕtsk)	132	53°04'N	26°43'E
Klyuchevskaya, vol., Russia (klyo͞o-chĕfskä'yä)	135	56°13'N	160°00'E
Klyuchi, Russia (klyo͞o'chĭ)	142a	57°03'N	57°20'E
Knezha, Blg. (knyā'zhà)	119	43°27'N	24°03'E
Knife, r., N.D., U.S. (nīf)	70	47°06'N	102°33'W
Knight Inlet, b., Can. (nīt)	54	50°41'N	125°40'W
Knightstown, In., U.S. (nīts'toun)	66	39°45'N	85°30'W
Knin, Cro. (knēn)	130	44°02'N	16°14'E
Knittelfeld, Aus.	117	47°13'N	14°50'E
Knob Peak, mtn., Phil. (nŏb)	169a	12°30'N	121°20'E
Knottingley, Eng., U.K. (nŏt'ĭng-lĭ)	114a	53°42'N	1°14'W
Knox, In., U.S. (nŏks)	66	41°15'N	86°40'W
Knox, Cape, c., Can.	54	54°12'N	133°20'W
Knoxville, Ia., U.S. (nŏks'vĭl)	71	41°19'N	93°05'W
Knoxville, Tn., U.S.	65	35°58'N	83°55'W
Knutsford, Eng., U.K. (nŭts'fērd)	114a	53°18'N	2°22'W
Knyszyn, Pol. (knĭ'shĭn)	125	53°16'N	22°59'E
Kobayashi, Japan (kō'bà-yä'shē)	167	31°58'N	130°59'E
Kōbe, Japan (kō'bĕ)	161	34°30'N	135°10'E
Kobeliaky, Ukr.	137	49°11'N	34°12'E
København *see* Copenhagen, Den.	110	55°43'N	12°27'E
Koblenz, Ger. (kō'blĕntz)	117	50°18'N	7°36'E
Kobozha, r., Russia (kô-bō'zhà)	132	58°55'N	35°18'E
Kobrinskoye, Russia (kô-brĭn'skô-yĕ)	142c	59°25'N	30°07'E
Kobryn, Bela. (kō'brēn')	137	52°13'N	24°23'E
Kobuk, r., Ak., U.S. (kō'bŭk)	63	66°58'N	158°48'W
Kobuk Valley National Park, rec., Ak., U.S.	63	67°20'N	159°00'W
Kobuleti, Geor. (kô-bò-lyā'tē)	137	41°50'N	41°40'E
Kočani, Mac. (kô'chä-nĕ)	131	41°54'N	22°25'E
Kočevje, Slvn. (kô'chāv-ye)	130	45°38'N	14°51'E
Kocher, r., Ger. (kôĸ'ēr)	124	49°00'N	9°52'E
Kochi, India	159	9°58'N	76°19'E
Kōchi, Japan (kō'chē)	161	33°35'N	133°32'E
Kodaira, Japan	167a	35°43'N	139°29'E
Kodiak, Ak., U.S. (kō'dyăk)	64a	57°50'N	152°30'W
Kodiak Island, i., Ak., U.S.	63	57°24'N	153°32'W
Kodok, Sudan (ko'dŏk)	185	9°57'N	32°08'E
Koforidua, Ghana (kō fō-rĭ-do͞o'à)	184	6°03'N	0°17'W
Kōfu, Japan (kō'fo͞o')	166	35°41'N	138°34'E
Koga, Japan (kō'gà)	167	36°13'N	139°40'E
Kogan, r., Gui.	188	11°30'N	14°05'W
Kogane, Japan (kō'gä-nā)	167a	35°50'N	139°56'E
Koganei, Japan (kō'gä-nā)	167a	35°42'N	139°31'E
Køge, Den. (kû'gĕ)	122	55°27'N	12°09'E
Køge Bugt, b., Den.	122	55°30'N	12°25'E
Kogoni, Mali	188	14°44'N	6°02'W
Kohīma, India (kô-ē'mà)	155	25°45'N	94°41'E
Kohyl'nyk, r., Eur.	133	46°08'N	29°10'E
Koito, r., Japan (kô'ē-tō)	167a	35°19'N	139°58'E
Kŏje, i., Kor., S. (kû'jĕ)	166	34°53'N	129°00'E
Kokand, Uzb. (kô-känt')	139	40°27'N	71°07'E
Kokemäenjoki, r., Fin.	123	61°23'N	22°03'E
Kokhma, Russia (kôĸ'mà)	132	56°57'N	41°08'E
Kokkola, Fin. (kô'kô-lå)	116	63°47'N	22°58'E
Kokomo, In., U.S. (kō'kō-mō)	66	40°30'N	86°20'W
Koko Nor (Qinghai Hu), l., China (kō'kō nor) (chyĭŋ-hī ho͞o)	160	37°26'N	98°30'E
Kokopo, Pap. N. Gui. (kô-kô'pō)	169	4°25'S	152°27'E
Kökshetaū, Kaz.	139	53°15'N	69°13'E
Koksoak, r., Can. (kôk'sô-ăk)	53	57°42'N	69°50'W
Kokstad, S. Afr. (kôk'shtät)	187c	30°33'S	29°27'E
Kokubu, Japan (kō'ko͞o-bo͞o)	167	31°42'N	130°46'E
Kokuou, Japan (kō'ko͞o-ō'o͞o)	167b	34°34'N	135°39'E
Kola Peninsula *see* Kol'skiy Poluostrov, pen., Russia	134	67°15'N	37°40'E
Kolār (Kolār Gold Fields), India (kōl-är')	155	13°39'N	78°33'E
Kolárvo, Slvk. (kōl-árōvō)	125	47°54'N	17°59'E
Kolbio, Kenya	191	1°10'S	41°15'E
Kol'chugino, Russia (kôl-chô'gē-nô)	132	56°19'N	39°29'E
Kolda, Sen.	188	12°53'N	14°57'W
Kolding, Den. (kŭl'dĭng)	122	55°29'N	9°24'E
Kole, D.R.C. (kō'lå)	186	3°19'S	22°46'E
Kolguyev, i., Russia (kôl-gô'yĕf)	134	69°00'N	49°00'E
Kolhāpur, India	159	16°48'N	74°15'E
Kolín, Czech Rep. (kō'lēn)	124	50°01'N	15°11'E
Kolkasrags, c., Lat. (kôl-käs'rägz)	123	57°46'N	22°39'E
Kolkata (Calcutta), India	155	22°32'N	88°22'E
Köln *see* Cologne, Ger.	127c	50°56'N	6°57'E
Kolno, Pol. (kôw'nô)	125	53°23'N	21°56'E
Koło, Pol. (kô'wô)	125	52°11'N	18°37'E
Kołobrzeg, Pol. (kô-lôb'zhĕk)	116	54°10'N	15°35'E
Kolomna, Russia (käl-ôm'nà)	136	55°06'N	38°47'E
Kolomyia, Ukr.	125	48°32'N	25°04'E
Kolp', r., Russia (kôlp)	132	59°18'N	35°32'E
Kolpashevo, Russia (kŭl pá shô'và)	134	58°16'N	82°43'E
Kolpino, Russia (kôl'pē-nô)	136	59°45'N	30°37'E
Kolpny, Russia (kôlp'nyê)	132	52°14'N	36°54'E
Kol'skiy Poluostrov, pen., Russia	134	67°15'N	37°40'E
Kolva, r., Russia	136	61°00'N	57°00'E
Kolwezi, D.R.C. (kōl-wĕ'zē)	186	10°43'S	25°28'E
Kolyberovo, Russia (kô-lĭ-byâ'rô-vô)	142b	55°16'N	38°45'E
Kolyma, r., Russia	135	66°30'N	151°45'E
Kolymskiy Mountains *see* Gydan, Khrebet, mts., Russia	135	61°45'N	155°00'E
Kom, r., Afr.	190	2°15'N	12°05'E
Komadugu Gana, r., Nig.	189	12°15'N	11°10'E
Komae, Japan	167a	35°37'N	139°35'E
Komandorskiye Ostrova, is., Russia	153	55°40'N	167°13'E
Komárno, Slvk. (kô'mär-nô)	125	47°46'N	18°08'E
Komarno, Ukr.	125	49°38'N	23°42'E
Komárom, Hung. (kô'mä-rôm)	125	47°45'N	18°06'E
Komatipoort, S. Afr. (kō-mä'tē-pōrt)	186	25°21'S	32°00'E
Komatsu, Japan (kō-mät'so͞o)	166	36°23'N	136°26'E
Komatsushima, Japan (kō-mät'so͞o-shē'mä)	167	34°04'N	134°32'E
Komeshia, D.R.C.	191	8°01'S	27°07'E
Komga, S. Afr. (kŏm'gà)	187c	32°36'S	27°54'E
Komi, prov., Russia (kômē)	140	63°00'N	55°00'E
Kommetijie, S. Afr.	186a	34°09'S	18°19'E
Komoé, r., C. Iv.	188	5°40'N	3°40'W
Komsomolets, Kaz.	142a	53°45'N	62°04'E
Komsomol'sk-na-Amure, Russia	135	50°46'N	137°14'E
Kona, Mali	188	14°57'N	3°53'W
Konda, r., Russia (kŏn'dà)	136	60°50'N	64°00'E
Kondas, r., Russia (kŏn'dàs)	142a	59°30'N	56°28'E
Kondoa, Tan. (kŏn-dō'à)	186	4°52'S	36°00'E
Kondolole, D.R.C.	191	1°20'N	25°58'E
Koné, N. Cal.	170f	21°04'S	164°52'E
Kong, C. Iv. (kông)	184	9°05'N	4°41'W
Kongbo, C.A.R.	190	4°44'N	21°23'E
Kongolo, D.R.C. (kôŋ'gō'lō)	186	5°23'S	27°00'E
Kongsberg, Nor. (kŭngs'bĕrg)	122	59°40'N	9°36'E
Kongsvinger, Nor. (kŭngs'vĭŋ-gĕr)	122	60°12'N	12°00'E
Koni, D.R.C. (kō'nē)	186	10°32'S	27°27'E
Königsberg *see* Kaliningrad, Russia	134	54°42'N	20°32'E
Königsbrunn, Ger. (kû'nēgs-bro͞on)	115d	48°16'N	10°53'E
Königs Wusterhausen, Ger. (kû'nēgs vo͞os'tĕr-hou-zĕn)	115b	52°18'N	13°38'E
Konin, Pol. (kô'nyēn)	117	52°11'N	18°17'E
Kónitsa, Grc. (kô'nyē'tsà)	131	40°03'N	20°46'E
Konjic, Bos. (kôn'yēts)	131	43°38'N	17°59'E
Konju, Kor., S.	166	36°21'N	127°05'E
Konnagar, India	158a	22°41'N	88°22'E
Konotop, Ukr. (kô-nô-tôp')	137	51°13'N	33°14'E
Konpienga, r., Burkina	188	11°15'N	0°35'E
Konqi, r., China (kôn-chyē)	160	41°09'N	87°46'E
Końskie, Pol. (koin''skyĕ)	125	51°12'N	20°26'E
Konstanz, Ger. (kōn'shtänts)	124	47°39'N	9°10'E
Kontagora, Nig. (kōn-tà-gō'rä)	184	10°24'N	5°28'E
Konya, Tur. (kōn'yá)	154	36°55'N	32°25'E
Koocanusa, Lake, res., N.A.	72	49°00'N	115°10'W
Kootenay (Kootenai), r., N.A.	55	49°45'N	117°05'W
Kootenay Lake, l., Can.	55	49°35'N	116°50'W
Kootenay National Park, rec., Can. (ko͞o'tē-nā)	50	51°06'N	117°02'W
Kōō-zan, mtn., Japan (ko͞o'zän)	167b	34°53'N	135°32'E
Kopervik, Nor. (kô'pĕr-vēk)	122	59°18'N	5°20'E
Kopeysk, Russia (kô-pāsk')	140	55°07'N	61°37'E
Köping, Swe. (chû'pĭng)	122	59°32'N	15°58'E
Kopparberg, Swe. (kŏp'pàr-bĕrgh)	122	59°53'N	15°00'E
Koppeh Dāgh, mts., Asia	154	37°28'N	58°29'E
Koppies, S. Afr.	192c	27°15'S	27°35'E
Koprivnica, Cro. (kô'prēv-nē'tsà)	130	46°10'N	16°48'E
Kopychyntsi, Ukr.	125	49°06'N	25°55'E
Korčula, i., Serb. (kôr'cho͞o-là)	131	42°50'N	17°05'E
Korea, North, nation, Asia	161	40°00'N	127°00'E
Korea, South, nation, Asia	161	36°30'N	128°00'E
Korea Bay, b., Asia	164	39°18'N	123°50'E
Korean Archipelago, is., Kor., S.	161	34°05'N	125°35'E
Korea Strait, strt., Asia	161	33°30'N	128°30'E
Korets', Ukr.	125	50°35'N	27°13'E
Korhogo, C. Iv. (kôr-hō'gō)	184	9°27'N	5°38'W
Korinthiakós Kólpos, b., Grc.	119	38°15'N	22°33'E
Kórinthos, Grc. (kô-rēn'thôs) (kŏr'ĭnth)	110	37°56'N	22°54'E
Koriukivka, Ukr.	133	51°44'N	32°24'E
Kōriyama, Japan (kō'rē-yä'mä)	166	37°18'N	140°25'E
Korkino, Russia (kôr'kē-nŭ)	142a	54°53'N	61°25'E
Korla, China (kôr-lä)	160	41°37'N	86°03'E
Körmend, Hung. (kûr'mĕnt)	124	47°02'N	16°36'E
Kornat, i., Serb. (kôr-nät')	130	43°46'N	15°10'E
Korneuburg, Aus. (kôr'noi-bôrgh)	115e	48°22'N	16°21'E
Koro, Mali	188	14°04'N	3°05'W
Korocha, Russia (kô-rō'chà)	133	50°50'N	37°13'E
Korop, Ukr. (kō'rôp)	133	51°33'N	32°54'E
Koro Sea, sea, Fiji	170g	18°00'S	179°50'E
Korosten', Ukr. (kô'rôs-tĕn)	137	50°51'N	28°39'E
Korostyshiv, Ukr.	133	50°19'N	29°05'E
Koro Toro, Chad	189	16°05'N	18°30'E
Korotoyak, Russia (kô'rô-tô-yák')	133	51°00'N	39°06'E
Korsakov, Russia (kôr'sá-kôf')	135	46°42'N	143°16'E
Korsnäs, Fin. (kôrs'nĕs)	123	62°51'N	21°17'E
Korsør, Den. (kôrs'ûr')	122	55°19'N	11°08'E
Kortrijk, Bel.	121	50°49'N	3°10'E
Koryakskiy Khrebet, mts., Russia	135	62°00'N	168°45'E
Kosa Byriuchyi ostriv, i., Ukr.	133	46°07'N	35°12'E
Kościan, Pol. (kŭsh'tsyán)	124	52°05'N	16°38'E
Kościerzyna, Pol. (kŭsh-tsyĕ-zhē'nà)	125	54°08'N	17°59'E
Kosciusko, Ms., U.S. (kŏs-ĭ-ŭs'kō)	82	33°04'N	89°35'W
Kosciuszko, Mount, mtn., Austl.	175	36°26'S	148°20'E
Kosha, Sudan	185	20°49'N	30°27'E
Koshigaya, Japan (kō'shē-gä'yä)	167a	35°53'N	139°48'E
Kōshim, r., Kaz.	137	50°30'N	50°40'E
Kosi, r., India (kō'sē)	158	26°00'N	86°20'E
Košice, Slvk. (kō'shē-tsĕ')	117	48°43'N	21°17'E
Kosmos, S. Afr. (kŏz'mŏs)	187b	25°45'S	27°51'E
Kosobrodskiy, Russia (kä-sô'brôd-skĭ)	142a	54°14'N	60°53'E
Kosovo, hist. reg., Serb.	131	42°35'N	21°00'E
Kosovska Mitrovica, Serb. (kō'sôv-skä' mē'trô-vē-tsä')	131	42°51'N	20°50'E
Kostajnica, Cro. (kôs'tä-ē-nē'tsà)	130	45°14'N	16°32'E
Koster, S. Afr.	192c	25°52'S	26°52'E
Kostiantynivka, Ukr.	133	48°33'N	37°42'E
Kostino, Russia (kôs'tĭ-nô)	142b	55°54'N	37°51'E
Kostroma, Russia (kôs-trô-má')	134	57°46'N	40°55'E
Kostroma, prov., Russia	132	57°50'N	41°10'E
Kostrzyn, Pol. (kôst'chēn)	117	52°35'N	14°38'E
Kos'va, r., Russia (kôs'vá)	142a	58°44'N	57°08'E
Koszalin, Pol. (kô-shä'lĭn)	116	54°12'N	16°10'E
Kőszeg, Hung. (kû'sĕg)	124	47°21'N	16°32'E
Kota, India	155	25°17'N	75°49'E
Kota Baharu, Malay. (kō'tä bä'ro͞o)	168	6°15'N	102°23'E
Kotabaru, Indon.	168	3°22'S	116°15'E
Kota Kinabalu, Malay.	168	5°55'N	116°05'E
Kota Tinggi, Malay.	153b	1°43'N	103°54'E
Kotel, Blg. (kō-tĕl')	131	42°54'N	26°28'E
Kotel'nich, Russia (kô-tyĕl'nēch)	136	58°15'N	48°20'E
Kotel'nyy, i., Russia (kô-tyĕl'nē)	135	74°51'N	134°09'E
Kotka, Fin. (kŏt'kà)	116	60°28'N	26°56'E
Kotlas, Russia (kôt'lås)	136	61°10'N	46°50'E
Kotlin, Ostrov, i., Russia (ôs-trôf' kôt'lĭn)	142c	60°02'N	29°49'E
Kotor, Serb.	131	42°25'N	18°46'E
Kotorosl', r., Russia (kô-tô'rôsl)	132	57°18'N	39°08'E
Kotovs'k, Ukr.	133	47°49'N	29°31'E
Kotto, r., C.A.R.	185	5°17'N	22°04'E
Kotuy, r., Russia (kô-to͞o')	140	71°00'N	103°15'E
Kotzebue, Ak., U.S. (kŏt'sĕ-bo͞o)	64a	66°48'N	162°42'W
Kotzebue Sound, strt., Ak., U.S.	63	67°00'N	164°28'W
Kouchibouguac National Park, rec., Can.	60	46°53'N	65°35'W
Koudougou, Burkina (ko͞o-do͞o'go͞o)	184	12°15'N	2°22'W
Kouilou, r., Congo	186	4°30'S	12°00'E
Koula-Moutou, Gabon	190	1°08'S	12°29'E
Koulikoro, Mali (ko͞o-lē-kō'rō)	184	12°53'N	7°33'W
Koulouguidi, Mali	189	13°27'N	17°33'E
Koumac, N. Cal.	170f	20°33'S	164°17'E
Koumra, Chad	189	8°55'N	17°33'E
Koundara, Gui.	188	12°29'N	13°18'W
Kouroussa, Gui. (ko͞o-ro͞o'sä)	184	10°39'N	9°53'W
Koutiala, Mali (ko͞o-tē-ä'lä)	184	12°29'N	5°29'W
Kouvola, Fin. (kō'ò-vō-lä)	123	60°51'N	26°40'E
Kouzhen, China (kō-jŭn)	162	36°19'N	117°37'E
Kovda, l., Russia (kôv'dá)	136	66°45'N	32°00'E
Kovel', Ukr. (kō'vĕl)	137	51°13'N	24°45'E
Kovno *see* Kaunas, Lith.	134	54°42'N	23°54'E
Kovrov, Russia (kôv-rôf')	136	56°23'N	41°21'E
Koyuk, Ak., U.S. (kô-yo͞ok')	63	65°00'N	161°18'W
Koyukuk, r., Ak., U.S. (kô-yo͞o'kŏk)	63	66°25'N	153°50'W
Kozáni, Grc.	119	40°16'N	21°51'E
Kozelets', Ukr. (kōzĕ-lyĕts)	133	50°53'N	31°07'E
Kozel'sk, Russia (kô-zĕlsk')	132	54°01'N	35°49'E
Kozhikode, India	155	11°19'N	75°49'E
Koziatyn, Ukr.	137	49°43'N	28°50'E
Kozienice, Pol. (kō-zyĕ-nē'tsĕ)	125	51°34'N	21°35'E
Koźle, Pol. (kôzh'lĕ)	125	50°19'N	18°10'E
Kozloduy, Blg. (kŭz'lô-dwē)	131	43°45'N	23°42'E
Kōzu, i., Japan (kō'zo͞o)	167	34°16'N	139°03'E
Kra, Isthmus of, isth., Asia	168	9°30'S	99°45'E
Kraai, r., S. Afr. (krā'ē)	187c	30°50'S	27°03'E
Krabbendijke, Neth.	115a	51°26'N	4°05'E
Krâchéh, Camb.	168	12°28'N	106°06'E
Kragujevac, Serb. (krä'go͞o'yĕ-váts)	119	44°01'N	20°55'E
Kraków, Pol. (krä'kŏf)	110	50°05'N	20°00'E
Kraljevo, Serb. (kräl'ye-vô)	119	43°39'N	20°48'E
Kramators'k, Ukr.	133	48°43'N	37°32'E
Kramfors, Swe. (kräm'fōrs)	122	62°54'N	17°49'E
Kranj, Slvn. (krän')	118	46°16'N	14°23'E
Kranskop, S. Afr. (kränz'kŏp)	187c	28°57'S	30°54'E
Krāslava, Lat. (kräs'lä-vä)	123	55°53'N	27°12'E
Kraslice, Czech Rep. (kräs'lē-tsĕ)	124	50°19'N	12°30'E
Krasnaya Gorka, Russia	142a	55°12'N	56°40'E
Krasnaya Sloboda, Russia	137	48°25'N	44°35'E
Kraśnik, Pol. (kräsh'nĭk)	125	50°53'N	22°15'E
Krasnoarmeysk, Russia (kräs'nô-ár-mask')	142b	56°06'N	38°09'E
Krasnoarmiis'k, Ukr.	133	48°19'N	37°04'E
Krasnodar, Russia (kräs'nô-dár)	134	45°03'N	38°55'E
Krasnodarskiy, prov., Russia (kräs-nô-där'skĭ ôb'lást)	133	45°25'N	38°10'E
Krasnogorsk, Russia	142b	55°49'N	37°20'E
Krasnogorskiy, Russia (kräs-nô-gôr'skĭ)	142a	54°36'N	61°15'E
Krasnogvardeyskiy, Russia (krá'sno-gvăr-dzyē ēs-kēē)	142a	57°17'N	62°05'E
Krasnohrad, Ukr.	133	49°23'N	35°26'E
Krasnokamsk, Russia (kräs-nô-kämsk')	136	58°00'N	55°45'E
Krasnokuts'k, Ukr.	133	50°03'N	35°05'E
Krasnoslobodsk, Russia (kräs'nô-slôbôtsk')	136	54°20'N	43°50'E
Krasnotur'insk, Russia (krŭs-nŭ-to͞o-rensk')	134	59°47'N	60°15'E
Krasnoufimsk, Russia (krŭs-nŭ-o͞o-fēmsk')	134	56°38'N	57°46'E
Krasnoural'sk, Russia (kräs'nô-o͞o-rälsk')	136	58°21'N	60°05'E
Krasnousol'skiy, Russia (kräs-nô-ó-sôl'skĭ)	142a	53°54'N	56°27'E

PRONOUNCING INDEX

Krasnovishersk – Ladoga 265

PLACE (Pronunciation)	PAGE	LAT.	LONG.
Krasnovishersk, Russia (kräs-nô-vêshersk′)	136	60°22′N	57°20′E
Krasnoyarsk, Russia (kräs-nô-yársk′)	135	56°13′N	93°12′E
Krasnoye Selo, Russia (kräs′nŭ-yŭ sâ′lô)	142c	59°44′N	30°06′E
Krasny Kholm, Russia (kräs′nê ĸôlm)	132	58°03′N	37°11′E
Krasnystaw, Pol. (kräs-nê-stáf′)	125	50°59′N	23°11′E
Krasnyy Bor, Russia (kräs′nê bôr)	142c	59°41′N	30°40′E
Krasnyy Klyuch, Russia (kräs′nê′klyûch′)	142a	55°24′N	56°43′E
Krasnyy Kut, Russia (kràs-nê ko͞ot′)	137	50°50′N	47°00′E
Kratovo, Mac. (krä′tô-vô)	131	42°04′N	22°12′E
Kratovo, Russia (krä′tô-vô)	142b	55°35′N	38°10′E
Krefeld, Ger. (krā′fĕlt)	127c	51°20′N	6°34′E
Kremenchuk, Ukr.	137	49°04′N	33°26′E
Kremenchuts′ke vodoskhovyshche, res., Ukr.	137	49°20′N	32°45′E
Kremenets′, Ukr.	125	50°06′N	25°43′E
Kremmen, Ger. (krĕ′mĕn)	115b	52°45′N	13°02′E
Krempe, Ger. (krĕm′pĕ)	115c	53°50′N	9°29′E
Krems, Aus. (krĕms)	124	48°25′N	15°36′E
Krestovyy, Pereval, p., Geor.	138	42°32′N	44°28′E
Kresttsy, Russia (kråst′sê)	132	58°16′N	32°25′E
Kretinga, Lith. (krĕ-tĭŋ′gá)	123	55°55′N	21°17′E
Kribi, Cam. (krē′bê)	184	2°57′N	9°55′E
Krilon, Mys, c., Russia (mĭs krĭl′ôn)	166	45°58′N	142°00′E
Krimpen aan de IJssel, Neth.	115a	51°55′N	4°34′E
Krishna, r., India	155	16°00′N	79°00′E
Krishnanagar, India	158	23°29′N	88°33′E
Kristiansand, Nor. (krĭs-tyán-sän′′)	110	58°09′N	7°59′E
Kristianstad, Swe. (krĭs-tyán-städ′)	116	56°02′N	14°09′E
Kristiansund, Nor. (krĭs-tyán-sòn′′)	116	63°07′N	7°49′E
Kristinehamn, Swe. (krês-tē′nĕ-häm′)	116	59°20′N	14°05′E
Kristinestad, Fin. (krĭs-tē′nĕ-städh)	123	62°16′N	21°28′E
Kriva-Palanka, Mac. (krē-vá-pá-län̄′ká)	131	42°12′N	22°21′E
Krivoy Rog *see* Kryvyi Rih, Ukr.	134	47°54′N	33°22′E
Križevci, Cro. (krē′zhĕv-tsĭ)	130	46°02′N	16°30′E
Krk, i., Serb. (k′rk)	130	45°06′N	14°33′E
Krnov, Czech Rep. (k′r′nôf)	125	50°05′N	17°41′E
Krokodil, r., S. Afr. (krŏ′kŏ-dĭl)	192c	24°25′S	27°08′E
Krolevets′, Ukr.	137	51°33′N	33°21′E
Kromy, Russia (krô′mê)	132	52°44′N	35°41′E
Kronshtadt, Russia (krōn′shtät)	136	59°59′N	29°47′E
Kroonstad, S. Afr. (krōn′shtät)	186	27°40′S	27°15′E
Kropotkin, Russia (krá-pôt′kĭn)	137	45°25′N	40°30′E
Krosno, Pol. (krôs′nô)	125	49°41′N	21°46′E
Krotoszyn, Pol. (krô-tō′shĭn)	125	51°41′N	17°25′E
Krško, Slvn. (k′rsh′kô)	130	45°58′N	15°30′E
Krugersdorp, S. Afr. (kro͞o′gẽrz-dôrp)	186	26°06′S	27°46′E
Krung Thep *see* Bangkok, Thai.	168	13°50′N	100°29′E
Kruševac, Serb. (krô′shĕ-váts)	131	43°34′N	21°21′E
Kruševo, Mac.	131	41°20′N	21°15′E
Krychaw, Bela.	132	53°44′N	31°39′E
Krylbo, Swe. (krül′bô)	122	60°07′N	16°14′E
Krym, Respublika, prov., Ukr.	133	45°08′N	34°05′E
Krymskaya, Russia (krĭm′ská-yá)	133	44°58′N	38°01′E
Kryms′kyi Pivostriv (Crimean Peninsula), pen., Ukr.	137	45°18′N	33°30′E
Krynki, Pol. (krĭn′kê)	125	53°15′N	23°47′E
Kryve Ozero, Ukr.	133	47°57′N	30°21′E
Kryvyi Rih, Ukr.	134	47°54′N	33°22′E
Ksar Chellala, Alg.	129	35°12′N	2°20′E
Ksar-el-Kebir, Mor.	118	35°01′N	5°48′W
Ksar-es-Souk, Mor.	118	31°58′N	4°25′W
Kuai, r., China (ko͞o-ī)	162	33°30′N	116°56′E
Kuala Klawang, Malay.	153b	2°57′N	102°04′E
Kuala Lumpur, Malay. (kwä′lä lòm-po͞or′)	168	3°08′N	101°42′E
Kuandian, China (kůän-dĭĕn)	164	40°40′N	124°50′E
Kuban, r., Russia	137	45°20′N	40°05′E
Kubenskoye, l., Russia	136	59°40′N	39°40′E
Kuching, Malay. (ko͞o′chĭng)	168	1°30′N	110°26′E
Kuchinoerabo, i., Japan (ko͞o′chē nō ẽr′á-bō)	167	30°31′N	129°53′E
Kudamatsu, Japan (ko͞o′dá-mä′tso͞o)	167	34°00′N	131°51′E
Kudap, Indon.	153b	1°14′N	102°30′E
Kudat, Malay. (ko͞o-dät′)	168	6°56′N	116°48′E
Kudirkos Naumietis, Lith. (ko͞odĭr-kôs nä′ô-mē′tĭs)	123	54°51′N	23°00′E
Kudymkar, Russia (ko͞o-dĭm-kär′)	134	58°43′N	54°52′E
Kufstein, Aus. (ko͞of′shtīn)	124	47°34′N	12°11′E
Kugluktuk (Coppermine), Can.	50	67°46′N	115°19′W
Kuhstedt, Ger. (ko͞o′shtĕ)	115c	53°23′N	8°58′E
Kuibyshev *see* Kuybyshev, Russia	134	53°10′N	50°05′E
Kuilsrivier, S. Afr.	186a	33°56′S	18°41′E
Kuito, Ang.	186	12°22′S	16°56′E
Kuji, Japan	161	40°11′N	141°46′E
Kujū-san, mtn., Japan (ko͞o′jò-sän′)	167	33°07′N	131°14′E
Kukës, Alb. (ko͞o′kĕs)	131	42°03′N	20°25′E
Kula, Blg. (ko͞o′lá)	131	43°52′N	23°13′E
Kula, Tur.	119	38°32′N	28°30′E
Kula Kangri, mtn., Bhu.	155	33°11′N	90°36′E
Kular, Khrebet, mts., Russia (kò-lär′)	141	69°00′N	131°45′E
Kuldīga, Lat. (kòl′dê-gá)	123	56°59′N	21°59′E
Kulebaki, Russia (ko͞o-lĕ-bäk′ĭ)	136	55°22′N	42°30′E
Kulmbach, Ger. (klòlm′bäĸ)	124	50°07′N	11°28′E
Kulunda, Russia (kò-lòn′dá)	134	52°38′N	79°00′E
Kulundinskoye, l., Russia	140	52°45′N	77°18′E
Kum, r., Kor., S. (kòm)	166	36°50′N	127°30′E
Kuma, r., Russia (ko͞o′mä)	137	44°50′N	45°10′E
Kumamoto, Japan (ko͞o′mä-mō′tô)	161	32°49′N	130°40′E
Kumano-Nada, b., Japan (ko͞o-mä′nō nä-dä)	167	34°03′N	136°36′E
Kumanovo, Mac. (kò-mä′nô-vô)	131	42°10′N	21°41′E
Kumasi, Ghana (ko͞o-mä′sê)	184	6°41′N	1°35′W
Kumba, Cam. (kòm′bá)	184	4°38′N	9°25′E

PLACE (Pronunciation)	PAGE	LAT.	LONG.
Kumbakonam, India (kòm′bŭ-kō′nŭm)	155	10°59′N	79°25′E
Kumkale, Tur.	131	39°59′N	26°10′E
Kumo, Nig.	189	10°03′N	11°13′E
Kumta, India	159	14°19′N	75°28′E
Kumul *see* Hami, China	160	42°58′N	93°14′E
Kunashak, Russia (kû-nä′shák)	142a	55°43′N	61°35′E
Kunashir (Kunashiri), i., Russia (ko͞o-nŭ-shēr′)	161	44°00′N	145°45′E
Kunda, Est.	123	59°30′N	26°28′E
Kundravy, Russia (ko͞on′drá-vĭ)	142a	54°50′N	60°14′E
Kundur, i., Indon.	153b	0°49′N	103°20′E
Kunene (Cunene), r., Afr.	186	17°05′S	12°35′E
Kungälv, Swe. (kŭng′ĕlf)	122	57°53′N	12°01′E
Kungsbacka, Swe. (kŭngs′bä-ká)	122	57°31′N	12°04′E
Kungur, Russia (kòn-go͞or′)	134	57°27′N	56°53′E
Kunlun Shan, mts., China (ko͞on-lo͞on shän)	160	35°26′N	83°09′E
Kunming, China (ko͞on-mĭŋ)	160	25°10′N	102°50′E
Kunsan, Kor., S. (kòn′sän′)	161	35°54′N	126°46′E
Kunshan, China (ko͞onshän)	163b	31°23′N	120°57′E
Kuntsëvo, Russia (kòn-tsyô′vô)	132	55°43′N	37°27′E
Kun′ya, Russia	142a	58°42′N	56°47′E
Kun′ya, r., Russia (kòn′yá)	132	56°45′N	30°53′E
Kuopio, Fin. (kò-ô′pê-ŏ)	116	62°48′N	28°30′E
Kupa, r., Serb.	130	45°32′N	14°50′E
Kupang, Indon.	169	10°14′S	123°37′E
Kupavna, Russia	142b	55°49′N	38°11′E
Kupians′k, Ukr.	137	49°44′N	37°38′E
Kupino, Russia (ko͞o-pĭ′nô)	134	54°00′N	77°47′E
Kupiškis, Lith. (kò-pĭsh′kĭs)	123	55°50′N	24°55′E
Kuqa, China (ko͞o-chyä)	160	41°34′N	82°44′E
Kür, r., Asia	137	41°10′N	45°40′E
Kurashiki, Japan (ko͞o′rä-shē′kê)	167	34°37′N	133°44′E
Kuraymah, Sudan	185	18°34′N	31°49′E
Kurayoshi, Japan (ko͞o′rá-yō′shê)	167	35°25′N	133°49′E
Kurdistan, hist. reg., Asia (kûrd′ĭ-stăn)	154	37°40′N	43°30′E
Kurdufān, hist. reg., Sudan (kôr-dô-fän′)	185	14°08′N	28°39′E
Kŭrdzhali, Blg.	131	41°39′N	25°21′E
Kure, Japan (ko͞o′rĕ)	161	34°17′N	132°35′E
Kuressaare, Est. (kò′rĕ-sä′rĕ)	123	58°15′N	22°26′E
Kurgan, Russia (kòr-găn′)	134	55°28′N	65°14′E
Kurgan-Tyube, Taj. (kòr-gän′ tyò′bĕ)	139	38°00′N	68°49′E
Kurihama, Japan (ko͞o-rē-hä′mä)	167a	35°14′N	139°42′E
Kuril Islands, is., Russia (ko͞o′rĭl)	141	46°20′N	149°30′E
Kurisches Haff, b., Eur.	123	55°10′N	21°08′E
Kurla, neigh., India	159b	19°03′N	72°53′E
Kurmuk, Sudan (kòr′mo͞ok)	185	10°40′N	34°13′E
Kurnool, India (kòr-no͞ol′)	155	16°00′N	78°04′E
Kurrajong, Austl.	173b	33°33′S	150°40′E
Kuršenai, Lith. (kòr′shá-nī)	123	56°01′N	22°56′E
Kursk, Russia (kòrsk)	134	51°44′N	36°08′E
Kuršumlija, Serb. (kòr′shòm′lĭ-yá)	131	43°08′N	21°18′E
Kuruman, S. Afr. (ko͞o-ro͞o-män′)	186	27°25′S	23°30′E
Kurume, Japan (ko͞o′rò-mĕ)	161	33°10′N	130°30′E
Kururi, Japan (ko͞o′rò-rê)	167a	35°17′N	140°05′E
Kusa, Russia (ko͞o′sá)	142a	55°19′N	59°27′E
Kushchëvskaya, Russia	133	46°34′N	39°40′E
Kushikino, Japan (ko͞o′shĭ-kē′nô)	167	31°44′N	130°19′E
Kushimoto, Japan (ko͞o′shĭ-mō′tô)	167	33°29′N	135°47′E
Kushiro, Japan (ko͞o′shē-rō)	161	43°00′N	144°22′E
Kushva, Russia (ko͞osh′vá)	134	58°18′N	59°51′E
Kuskokwim, r., Ak., U.S.	63	61°32′N	160°36′W
Kuskokwim Bay, b., Ak., U.S. (kŭs′kô-kwĭm)	63	59°25′N	163°14′W
Kuskokwim Mountains, mts., Ak., U.S.	63	62°08′N	158°00′W
Kuskovak, Ak., U.S. (kŭs-kō′văk)	63	60°10′N	162°50′W
Kütahya, Tur. (kû-tä′hyá)	154	39°20′N	29°50′E
Kutaisi, Geor. (ko͞o-tŭ-ē′sē)	137	42°15′N	42°40′E
Kutch, Gulf of, b., India	155	22°45′N	68°33′E
Kutch, Rann of, sw., Asia	155	23°59′N	69°13′E
Kutenholz, Ger. (ko͞o′tĕn-hōlts)	115c	53°29′N	9°20′E
Kutim, Russia (ko͞o′tĭm)	142a	60°22′N	58°51′E
Kutina, Cro. (ko͞o′tĕ-ná)	130	45°29′N	16°48′E
Kutno, Pol. (kòt′nô)	117	52°14′N	19°22′E
Kutno, l., Russia	136	65°15′N	31°30′E
Kutulik, Russia (kò to͞o′lyĭk)	135	53°12′N	102°51′E
Kuujjuaq, Can.	51	58°06′N	68°25′W
Kuusamo, Fin. (ko͞o′sá-mô)	116	65°59′N	29°10′E
Kuvshinovo, Russia (kòv-shē′nô-vô)	132	57°01′N	34°09′E
Kuwait *see* Al Kuwayt, Kuw.	154	29°04′N	47°59′E
Kuwait, nation, Asia	154	29°00′N	48°45′E
Kuwana, Japan (ko͞o′wá-ná)	167	35°02′N	136°40′E
Kuybyshev *see* Samara, Russia	136	53°10′N	50°05′E
Kuybyshevskoye, res., Russia	134	53°40′N	49°00′E
Kuzneckovo, Russia	142b	55°29′N	38°22′E
Kuznetsk, Russia (ko͞oz-nyĕtsk′)	136	53°00′N	46°30′E
Kuznetsk Basin, basin, Russia	134	56°30′N	86°15′E
Kuznetsovka, Russia (kòz-nyĕt′sôf-ká)	142a	54°41′N	56°40′E
Kuznetsovo, Russia (kòz-nyĕt-sô′vô)	132	56°39′N	36°55′E
Kuznetsy, Russia	142b	55°50′N	38°39′E
Kvarner Zaliv, b., Serb. (kvär′nĕr)	130	44°41′N	14°05′E
Kwa, r., D.R.C.	190	3°00′S	16°45′E
Kwahu Plateau, plat., Ghana	188	7°00′N	1°35′W
Kwando (Cuando), r., Afr.	190	16°50′S	22°40′E
Kwangju, Kor., S.	166	35°09′N	126°54′E
Kwango (Cuango), r., Afr. (kwäng′ō′)	190	6°35′S	16°50′E
Kwangwazi, Tan.	191	7°47′S	38°15′E
Kwekwe, Zimb.	186	18°49′S	29°45′E
Kwenge, r., Afr. (kwĕŋ′gĕ)	190	6°45′S	18°23′E
Kwilu, r., Afr. (kwē′lo͞o)	186	4°00′S	18°00′E
Kyakhta, Russia (kyäĸ′ta)	135	51°00′N	107°30′E
Kyaukpyu, Mya. (chouk′pyoo′)	155	19°19′N	93°33′E
Kybartai, Lith. (kē′bär-tī′)	123	54°40′N	22°46′E
Kyiv *see* Kiev, Ukr.	134	50°27′N	30°30′E

PLACE (Pronunciation)	PAGE	LAT.	LONG.
Kyïvs′ke vodoskhovyshche, res., Ukr.	134	51°00′N	30°20′E
Kými, Grc.	131	38°38′N	24°05′E
Kyn, Russia (kĭn′)	142a	57°52′N	58°42′E
Kynuna, Austl. (kĭ-no͞o′ná)	175	21°30′S	142°12′E
Kyoga, Lake, l., Ug.	185	1°30′N	32°45′E
Kyōga-Saki, c., Japan (kyō′gä sa′kê)	167	35°46′N	135°14′E
Kyŏngju, Kor., S. (kyŭng′yo͞o)	161	35°48′N	129°12′E
Kyōto, Japan (kyō′tō′)	161	35°00′N	135°46′E
Kyōto, dept., Japan	167b	34°56′N	135°42′E
Kyparissía, Grc.	119	37°17′N	21°43′E
Kyparissiakós Kólpos, b., Grc.	131	37°28′N	21°15′E
Kyren, Russia (kĭ-rĕn′)	135	51°46′N	102°13′E
Kyrgyzstan, nation, Asia	134	41°45′N	74°38′E
Kyrönjoki, r., Fin.	123	63°03′N	22°20′E
Kyrya, Russia (kēr′yá)	142a	59°18′N	59°03′E
Kyshtym, Russia (kĭsh-tĭm′)	136	55°42′N	60°34′E
Kýthira, i., Grc.	119	36°15′N	22°56′E
Kýthnos, i., Grc.	131	37°24′N	24°10′E
Kytlym, Russia (kĭt′lĭm)	142a	59°30′N	59°15′E
Kyūshū, i., Japan	161	33°00′N	131°00′E
Kyustendil, Blg. (kyòs-tĕn-dĭl′)	119	42°16′N	22°39′E
Kyyiv, prov., Ukr.	133	50°05′N	30°40′E
Kyzyl, Russia (kĭ zĭl)	135	51°37′N	93°38′E
Kyzyl-Kum, des., Asia	134	42°47′N	64°45′E

L

PLACE (Pronunciation)	PAGE	LAT.	LONG.
Laa, Aus.	124	48°42′N	16°23′E
La Almunia de Doña Godina, Spain	128	41°29′N	1°22′W
Laas Caanood, Som.	192a	8°24′N	47°20′E
La Asunción, Ven. (lä ä-so͞on-syōn′)	100	11°02′N	63°57′W
La Baie, Can.	59	48°21′N	70°53′W
La Banda, Arg. (lä bän′dä)	102	27°48′S	64°12′W
La Barca, Mex. (lä bär′kä)	88	20°17′N	102°33′W
Laberge, Lake, l., Can. (lá-bẽrzh′)	52	61°08′N	136°42′W
Laberinto de las Doce Leguas, is., Cuba	92	20°40′N	78°35′W
Labinsk, Russia	137	44°30′N	40°40′E
Labis, Malay. (läb′ĭs)	153b	2°23′N	103°01′E
La Bisbal, Spain (lä bēs-bäl′)	129	41°55′N	3°00′E
Labo, Phil. (lä′bô)	169a	14°11′N	122°49′E
Labo, Mount, mtn., Phil.	169a	14°00′N	122°47′E
Labouheyre, Fr. (lá-bo͞o-âr′)	126	44°14′N	0°58′W
Laboulaye, Arg. (lä-bô′o͞o-lä-yĕ)	102	34°01′S	63°10′W
Labrador, reg., Can. (lăb′rá-dôr)	53	53°05′N	63°30′W
Labrador Sea, sea, Can.	61	50°38′N	55°00′W
Lábrea, Braz. (lä-brā′ä)	100	7°28′S	64°39′W
Labuan, Pulau, i., Malay. (lä-bò-än′)	168	5°28′N	115°11′E
Labuha, Indon.	169	0°43′S	127°35′E
L′Acadie, Can. (lá-ká-dē′)	62a	45°18′N	73°22′W
L′Acadie, r., Can.	62a	45°24′N	73°21′W
La Calera, Chile (lä-kä-lĕ-rä)	99b	32°47′S	71°11′W
La Calera, Col.	100a	4°43′N	73°58′W
Lac Allard, Can.	60	50°38′N	63°28′W
La Canada, Ca., U.S. (lä kän-yä′dä)	75a	34°13′N	118°12′W
Lacantum, r., Mex. (lä-kän-to͞o′m)	89	16°13′N	90°52′W
La Carolina, Spain (lä kä-rô-lē′nä)	128	38°16′N	3°48′W
La Catedral, Cerro, mtn., Mex. (sĕ′r-rô-lä-kä-tĕ-drá′l)	89a	19°32′N	99°31′W
Lac-Beauport, Can. (läk-bō-pōr′)	62b	46°58′N	71°17′W
Laccadive Islands *see* Lakshadweep, is., India	155	11°00′N	73°02′E
Laccadive Sea, sea, Asia	159	9°10′N	75°17′E
Lac Court Oreille Indian Reservation, I.R., Wi., U.S.	71	46°04′N	91°18′W
Lac du Flambeau Indian Reservation, I.R., Wi., U.S.	71	46°12′N	89°50′W
La Ceiba, Hond. (lä sēbä)	86	15°45′N	86°52′W
La Ceja, Col. (lä-sĕ-kä)	100a	6°02′N	75°25′W
Lac-Frontière, Can.	51	46°42′N	70°00′W
Lacha, l., Russia (lá′chä)	136	61°15′N	39°05′E
La Chaux de Fonds, Switz. (lá shō dē-fôɴ′)	124	47°07′N	6°47′E
L′Achigan, r., Can. (lä-shē-gäɴ)	62a	45°49′N	73°48′W
Lachine, Can. (lá-shēn′)	62a	45°26′N	73°40′W
Lachlan, r., Austl. (läk′lăn)	175	34°00′S	145°00′E
La Chorrera, Pan. (láchôr-rā′rä)	91	8°54′N	79°47′W
Lachute, Can. (lá-sho͞ot′)	59	45°39′N	74°20′W
La Ciotat, Fr. (lá syō-tá′)	127	43°13′N	5°35′E
Lackawanna, N.Y., U.S. (lak-á-wŏn′á)	69c	42°49′N	78°50′W
Lac La Biche, Can.	50	54°46′N	112°58′W
Lacombe, Can.	50	52°28′N	113°44′W
Laconia, N.H., U.S. (lá-kō′nĭ-á)	67	43°30′N	71°30′W
La Conner, Wa., U.S. (lá kŏn′ẽr)	74a	48°23′N	122°30′W
Lacreek, l., S.D., U.S. (lá′krēk)	70	43°04′N	101°46′W
La Cresenta, Ca., U.S. (lá krĕs′ĕnt-á)	75a	34°14′N	118°13′W
La Cross, Ks., U.S. (lá-krôs′)	78	38°30′N	99°20′W
La Crosse, Wi., U.S.	65	43°48′N	91°14′W
La Cruz, Col. (lá kro͞oz′)	100	1°37′N	77°00′W
La Cruz, C.R. (lä-kro͞o′z)	90	11°05′N	85°37′W
Lacs, Riviere des, r., N.D., U.S. (rē-vyĕr′ dē läk)	70	48°30′N	101°45′W
La Cuesta, C.R. (lä-kwĕ′s-tä)	91	8°32′N	82°51′W
La Cygne, Ks., U.S. (lá-sēn′y′) (lá-sēn′)	79	38°20′N	94°45′W
Ladd, Il., U.S. (lăd)	66	41°25′N	89°25′W
Ladíspoli, Italy (lä-dē′s-pô-lē)	129d	41°57′N	12°05′E
Lādīz, Iran	157	28°56′N	61°19′E
Ladner, Can. (lăd′nẽr)	54	49°05′N	123°05′W
Lādnun, India (läd′nón)	158	27°45′N	74°20′E
Ladoga, Lake *see* Ladozhskoye Ozero, l., Russia	134	60°59′N	31°30′E

266 La Dorado – Larch Mountain PRONOUNCING INDEX

PLACE (Pronunciation)	PAGE	LAT.	LONG.
La Dorado, Col. (lä dô-rä'dä)	100	5°28'N	74°42'W
Ladozhskoye Ozero, l., Russia (là-dôsh'skô-yĕ ô'zĕ-rô)	134	60°59'N	31°30'E
La Durantaye, Can. (lä dü-räN-tā')	62b	46°51'N	70°51'W
Lady Frere, S. Afr. (lā-dē frâ'r')	187c	31°48'S	27°16'E
Lady Grey, S. Afr.	187c	30°44'S	27°17'E
Ladysmith, Can. (lā'dĭ-smĭth)	54	48°58'N	123°49'W
Ladysmith, S. Afr.	186	28°38'S	29°48'E
Ladysmith, Wi., U.S.	71	45°27'N	91°07'W
Lae, Pap. N. Gui. (lä'å)	169	6°15'S	146°57'E
Laerdalsøyri, Nor.	122	61°08'N	7°26'E
La Esperanza, Hond. (lä ĕs-pă-rän'zä)	90	14°20'N	88°21'W
Lafayette, Al., U.S.	82	32°52'N	85°25'W
Lafayette, Ca., U.S.	74b	37°53'N	122°07'W
Lafayette, Ga., U.S. (lä-fā-yĕt')	82	34°41'N	85°19'W
Lafayette, In., U.S.	65	40°25'N	86°55'W
Lafayette, La., U.S.	65	30°15'N	92°02'W
La Fayette, R.I., U.S.	68b	41°34'N	71°29'W
La Ferté-Alais, Fr. (lä-fĕr-tā'ä-lā')	127b	48°29'N	2°19'E
La Ferté-sous-Jouarre, Fr. (là fĕr-tā'sōō-zhōō-är')	127b	48°56'N	3°07'E
Lafia, Nig.	189	8°30'N	8°30'E
Lafiagi, Nig.	189	8°52'N	5°25'E
La Flèche, Fr. (là flâsh')	126	47°43'N	0°03'W
La Follete, Tn., U.S. (lă-fŏl'ĕt)	82	36°23'N	84°07'W
Lafourche, Bayou, r., La., U.S. (bä-yōō'lá-fōōrsh')	81	29°25'N	90°15'W
La Gaiba, Braz. (lä-gī'bä)	101	17°54'S	57°32'W
La Galite, i., Tun. (gä-lēt)	118	37°36'N	8°03'E
Lågan, r., Nor. (lô'ghĕn)	112	61°00'N	10°00'E
Lagan, r., Swe.	122	56°34'N	13°25'E
Lagan, r., N. Ire., U.K. (lā'găn)	120	54°30'N	6°00'W
Lagarto, r., Pan. (lä-gä'r-tô)	86a	9°08'N	80°05'W
Lagartos, l., Mex. (lä-gä'r-tôs)	90a	21°32'N	88°15'W
Laghouat, Alg. (lä-gwät')	184	33°45'N	2°49'E
Lagkadás, Grc.	131	40°44'N	23°10'E
Lagny, Fr. (län-yē')	127b	48°53'N	2°41'E
Lagoa da Prata, Braz. (lä-gô'ä-dä-prä'tä)	99a	20°04'S	45°33'W
Lagoa Dourada, Braz. (lä-gô'ä-dō-rä'dä)	99a	20°55'S	44°03'W
Lagogne, Fr. (lan-gōn'y')	126	44°43'N	3°50'E
Lagonay, Phil.	169a	13°44'N	123°31'E
Lagos, Nig. (lä'gōs)	184	6°27'N	3°24'E
Lagos, Port. (lä'gōzh)	128	37°08'N	8°43'W
Lagos de Moreno, Mex. (lä'gōs dā mō-rā'nō)	86	21°21'N	101°55'W
La Grand' Combe, Fr. (lå grän kanb')	126	44°12'N	4°03'E
La Grande, Or., U.S. (là grănd')	64	45°20'N	118°06'W
La Grande, r., Can.	53	53°55'N	77°30'W
La Grange, Austl. (lä grānj)	174	18°40'S	122°00'E
La Grange, Ga., U.S. (là-grānj')	65	33°01'N	85°00'W
La Grange, Il., U.S.	69a	41°49'N	87°53'W
Lagrange, In., U.S.	66	41°40'N	85°25'W
La Grange, Ky., U.S.	66	38°20'N	85°25'W
La Grange, Mo., U.S.	79	40°04'N	91°30'W
Lagrange, Oh., U.S.	69d	41°14'N	82°07'W
Lagrange, Tx., U.S.	81	29°55'N	96°50'W
La Grita, Ven. (lä grē'tä)	100	8°02'N	71°59'W
La Guaira, Ven. (lä gwä'ē-rä)	100	10°36'N	66°54'W
La Guardia, Spain (lä gwär'dē-ä)	128	41°55'N	8°48'W
Laguna, Braz. (lä-gōō'nä)	102	28°19'S	48°42'W
Laguna, Cayos, is., Cuba (kä'yōs-lä-gō'nä)	92	22°15'N	82°45'W
Laguna Indian Reservation, I.R., N.M., U.S.	77	35°00'N	107°30'W
Lagunillas, Bol. (lä-gōō-nēl'yäs)	100	19°42'S	63°38'W
Lagunillas, Mex. (lä-gōō-nē'l-yäs)	88	21°34'N	99°41'W
La Habana *see* Havana, Cuba	87	23°08'N	82°23'W
La Habra, Ca., U.S. (là hăb'rà)	75a	33°56'N	117°57'W
Lahaina, Hi., U.S. (lä-hä'ē-nä)	84a	20°52'N	156°39'W
Lāhījān, Iran	157	37°12'N	50°01'E
Laholm, Swe. (lä'hôlm)	122	56°30'N	13°00'E
La Honda, Ca., U.S. (lā hŏn'dă)	74b	37°20'N	122°16'W
Lahore, Pak. (lä-hōr')	155	32°00'N	74°18'E
Lahr, Ger. (lär)	124	48°19'N	7°52'E
Lahti, Fin. (lä'tē)	116	60°59'N	27°39'E
Lai, Chad	185	9°29'N	16°18'E
Lai'an, China (lī-än)	162	32°27'N	118°25'E
Laibin, China (lī-bĭn)	165	23°42'N	109°20'E
L'Aigle, Fr. (lĕ'gl')	126	48°45'N	0°37'E
Laisamis, Kenya	191	1°36'N	37°48'E
Laiyang, China (lāī'yäng)	164	36°59'N	120°42'E
Laizhou Wan, b., China (lī-jō wän)	161	37°22'N	119°19'E
Laja, Río de la, r., Mex. (rē'ō-dĕ-lä-lä'Kä)	88	21°17'N	100°57'W
Lajas, Cuba (lä'häs)	92	22°25'N	80°20'W
Lajeado, Braz. (lä-zhĕä'dō)	102	29°24'S	51°46'W
Lajes, Braz. (lá'zhĕs)	102	27°47'S	50°17'W
Lajinha, Braz. (lä-zhē'nyä)	99a	20°08'S	41°36'W
La Jolla, Ca., U.S. (là hoi'yà)	76a	32°51'N	117°16'W
La Jolla Indian Reservation, I.R., Ca., U.S.	76	33°19'N	116°21'W
La Junta, Co., U.S. (là hōōn'tà)	78	37°59'N	103°35'W
Lake Arthur, La., U.S. (är'thŭr)	81	30°06'N	92°40'W
Lake Barkley, res., U.S.	82	36°45'N	88°00'W
Lake Benton, Mn., U.S. (bĕn'tŭn)	70	44°15'N	96°17'W
Lake Bluff, Il., U.S. (blŭf)	69a	42°17'N	87°50'W
Lake Brown, Austl. (broun)	174	31°03'S	118°30'E
Lake Charles, La., U.S. (chärlz')	65	30°15'N	93°14'W
Lake City, Fl., U.S.	83	30°09'N	82°40'W
Lake City, Ia., U.S.	71	42°14'N	94°43'W
Lake City, Mn., U.S.	71	44°28'N	92°19'W
Lake City, S.C., U.S.	83	33°57'N	79°45'W
Lake Clark National Park, rec., Ak., U.S.	63	60°30'N	153°15'W
Lake Cowichan, Can. (kou'ĭ-chăn)	54	48°50'N	124°03'W
Lake Crystal, Mn., U.S. (krĭs'tăl)	71	44°05'N	94°12'W
Lake District, reg., Eng., U.K. (lāk)	120	54°25'N	3°20'W
Lake Elmo, Mn., U.S. (ĕlmō)	75g	45°00'N	92°53'W
Lake Forest, Il., U.S. (fŏr'ĕst)	69a	42°16'N	87°50'W
Lake Fork, r., Ut., U.S.	77	40°30'N	110°25'W
Lake Geneva, Wi., U.S. (jĕ-nē'vå)	71	42°36'N	88°28'W
Lake Havasu City, Az., U.S.	77	34°27'N	114°22'W
Lake June, Tx., U.S. (jōōn)	75c	32°43'N	96°45'W
Lakeland, Fl., U.S. (lāk'lănd)	65	28°02'N	81°58'W
Lakeland, Ga., U.S.	82	31°02'N	83°02'W
Lakeland, Mn., U.S.	75g	44°57'N	92°47'W
Lake Linden, Mi., U.S. (lĭn'dĕn)	71	47°11'N	88°26'W
Lake Louise, Can. (lōō-ēz')	55	51°26'N	116°11'W
Lake Mead National Recreation Area, rec., U.S.	77	36°00'N	114°30'W
Lake Mills, Ia., U.S. (mĭlz')	71	43°25'N	93°32'W
Lakemore, Oh., U.S. (lāk-mōr)	69d	41°01'N	81°24'W
Lake Odessa, Mi., U.S.	66	42°50'N	85°15'W
Lake Oswego, Or., U.S. (ŏs-wē'go)	74c	45°25'N	122°40'W
Lake Placid, N.Y., U.S.	67	44°17'N	73°59'W
Lake Point, Ut., U.S.	75b	40°41'N	112°16'W
Lakeport, Ca., U.S. (lāk'pōrt)	76	39°03'N	122°54'W
Lake Preston, S.D., U.S. (prēs'tŭn)	70	44°21'N	97°23'W
Lake Providence, La., U.S. (prŏv'ĭ-dĕns)	81	32°48'N	91°12'W
Lake Red Rock, res., Ia., U.S.	71	41°30'N	93°15'W
Lake Sharpe, res., S.D., U.S.	70	44°30'N	100°00'W
Lakeside, Ca., U.S. (lāk'sīd)	76a	32°52'N	116°55'W
Lake Station, In., U.S.	69a	41°34'N	87°15'W
Lake Stevens, Wa., U.S.	74a	48°01'N	122°04'W
Lake Success, N.Y., U.S. (sŭk-sĕs')	68a	40°46'N	73°43'W
Lakeview, Or., U.S.	72	42°11'N	120°21'W
Lake Village, Ar., U.S.	79	33°20'N	91°17'W
Lake Wales, Fl., U.S. (wālz')	83a	27°54'N	81°35'W
Lakewood, Ca., U.S. (lāk'wŏod)	75a	33°50'N	118°09'W
Lakewood, Co., U.S.	78	39°44'N	105°06'W
Lakewood, Oh., U.S.	65	41°29'N	81°48'W
Lakewood, Pa., U.S.	67	40°05'N	74°10'W
Lakewood, Wa., U.S.	74a	48°09'N	122°13'W
Lakewood Center, Wa., U.S.	74a	47°10'N	122°31'W
Lake Worth, Fl., U.S. (wûrth')	83a	26°37'N	80°04'W
Lake Worth Village, Tx., U.S.	75c	32°49'N	97°26'W
Lake Zurich, Il., U.S. (tsü'rĭk)	69a	42°11'N	88°05'W
Lakhdenpokh'ya, Russia (l'äk-dĭe'npōkyá)	123	61°33'N	30°10'E
Lakhtinskiy, Russia (làk-tĭn'skĭ)	142c	59°59'N	30°10'E
Lakota, N.D., U.S. (lå-kō'tå)	70	48°04'N	98°21'W
Lakshadweep, state, India	155	10°10'N	72°50'E
Lakshadweep, is., India	155	11°00'N	73°02'E
La Libertad, El Sal.	90	13°29'N	89°20'W
La Libertad, Guat. (lä lē-bĕr-tädh')	90	15°31'N	91°44'W
La Libertad, Guat.	90a	16°46'N	90°12'W
La Ligua, Chile (lä lē'gwä)	99b	32°21'S	71°13'W
Lalín, Spain (lä-lē'n)	128	42°40'N	8°05'W
La Línea, Spain (lä lē'nå-ä)	118	36°11'N	5°22'W
Lalitpur, Nepal	155	27°23'N	85°24'E
La Louviere, Bel. (là lōō-vyår')	121	50°30'N	4°10'E
La Luz, Mex. (lä lōōz')	88	21°04'N	101°19'W
Lama-Kara, Togo	188	9°33'N	1°12'E
La Malbaie, Can. (lä mäl-bå')	51	47°39'N	70°10'W
La Mancha, reg., Spain (lä män'chä)	128	38°55'N	4°20'W
Lamar, Co., U.S. (lå-mär')	78	38°04'N	102°44'W
Lamar, Mo., U.S.	79	37°28'N	94°15'W
La Marmora, Punta, mtn., Italy (lä-mä'r-mô-rä)	118	40°00'N	9°28'E
La Marque, Tx., U.S. (lå-märk)	81a	29°23'N	94°58'W
Lamas, Peru (lä'màs)	100	6°24'S	76°41'W
Lamballe, Fr. (läN-bäl')	126	48°29'N	2°36'W
Lambari, Braz. (läm-bá'rē)	99a	21°58'S	45°22'W
Lambasa, Fiji	170g	16°26'S	179°24'E
Lambayeque, Peru (läm-bä-yā'kå)	100	6°41'S	79°58'W
Lambert, Ms., U.S. (lăm'bērt)	82	34°10'N	90°16'W
Lambertville, N.J., U.S. (lăm'bērt-vĭl)	67	40°20'N	75°00'W
Lame Deer, Mt., U.S. (lām dēr')	73	45°36'N	106°40'W
Lamego, Port. (lä-mā'gō)	128	41°07'N	7°47'W
La Mesa, Col.	100a	4°38'N	74°27'W
La Mesa, Ca., U.S. (lä mā'sä)	76a	32°46'N	117°01'W
Lamesa, Tx., U.S.	78	32°44'N	101°54'W
Lamía, Grc. (lå-mē'å)	119	38°54'N	22°25'E
Lamon Bay, b., Phil. (lä-mōn')	168	14°35'N	121°52'E
La Mora, Chile (lä-mō'rä)	99b	32°28'S	70°56'W
La Moure, N.D., U.S. (lå mōōr')	70	46°23'N	98°17'W
Lampa, r., Chile (lá'm-pä)	99b	33°15'S	70°55'W
Lampasas, Tx., U.S. (lăm-păs'ås)	80	31°06'N	98°10'W
Lampasas, r., Tx., U.S.	80	31°18'N	98°08'W
Lampazos, Mex. (läm-pä'zōs)	86	27°03'N	100°30'W
Lampedusa, i., Italy (läm-pā-dōō'sä)	118	35°29'N	12°58'E
Lamstedt, Ger. (läm'shtĕt)	115c	53°38'N	9°06'E
Lamu, Kenya (lä'mōō)	187	2°16'S	40°54'E
Lamu Island, i., Kenya	191	2°25'S	40°50'E
La Mure, Fr. (lå mür')	127	44°55'N	5°50'E
Lan', r., Bela. (lån')	132	52°38'N	27°05'E
Lāna'i, i., Hi., U.S. (lä-nä'ē)	64c	20°48'N	157°06'W
Lanai City, Hi., U.S.	84a	20°50'N	156°56'W
Lanak La, p., China	160	34°40'N	79°50'E
Lanark, Scot., U.K. (lăn'àrk)	120	55°40'N	3°50'W
Lancashire, co., Eng., U.K. (lăŋ'kå-shĭr)	114a	53°49'N	2°42'W
Lancaster, Eng., U.K.	116	54°04'N	2°55'W
Lancaster, Ky., U.S.	66	37°35'N	84°30'W
Lancaster, Ma., U.S.	61a	42°28'N	71°40'W
Lancaster, N.H., U.S.	67	44°25'N	71°30'W
Lancaster, N.Y., U.S.	69c	42°54'N	78°42'W
Lancaster, Oh., U.S.	66	39°40'N	82°35'W
Lancaster, Pa., U.S.	65	40°05'N	76°20'W
Lancaster, Tx., U.S.	75c	32°36'N	96°45'W
Lancaster, Wi., U.S.	71	42°51'N	90°44'W
Lândana, Ang. (län-dä'nä)	186	5°15'S	12°07'E
Landau, Ger. (län'dou)	124	49°13'N	8°07'E
Lander, Wy., U.S. (lăn'dēr)	73	42°49'N	108°24'W
Landerneau, Fr. (läN-dĕr-nō')	126	48°28'N	4°14'W
Landes, reg., Fr. (läNd)	126	44°22'N	0°52'W
Landsberg, Ger. (länds'bĕrgh)	124	48°03'N	10°53'E
Lands End, c., Eng., U.K.	112	50°03'N	5°45'W
Landshut, Ger. (länts'hōōt)	117	48°32'N	12°09'E
Landskrona, Swe. (läns-krō'nå)	122	55°51'N	12°47'E
Lanett, Al., U.S. (lå-nĕt')	82	32°52'N	85°13'W
Langat, r., Malay.	153b	2°46'N	101°33'E
Langdon, Can. (lăng'dŭn)	62e	50°58'N	113°40'W
Langdon, Mn., U.S.	75g	44°49'N	92°56'W
L'Ange-Gardien, Can. (länzh gàr-dyăN')	62b	46°55'N	71°06'W
Langeland, i., Den.	122	54°52'N	10°46'E
Langenzersdorf, Aus.	115e	48°30'N	16°22'E
Langesund, Nor. (läng'ĕ-sŏn')	122	58°59'N	9°38'E
Langfjorden, b., Nor.	122	62°40'N	7°45'E
Langhorne, Pa., U.S. (lăng'hôrn)	68f	40°10'N	74°55'W
Langia Mountains, mts., Ug.	191	3°35'N	33°35'E
Langjökoll, ice, Ice. (läng-yû'kōōl)	116	64°40'N	20°31'W
Langla Co, l., China (läŋ-lä tswo)	158	30°42'N	80°40'E
Langley, Can. (lăng'lĭ)	55	49°06'N	122°39'W
Langley, S.C., U.S.	83	33°32'N	81°52'W
Langley, Wa., U.S.	74a	48°02'N	122°25'W
Langley Indian Reserve, I.R., Can.	74d	49°12'N	122°31'W
Langnau, Switz. (läng'nou)	124	46°56'N	7°46'E
Langon, Fr. (läN-gôN')	126	44°34'N	0°16'W
Langres, Fr. (läN'gr')	127	47°53'N	5°20'E
Langres, Plateau de, plat., Fr. (plä-tō'dĕ-läN'grĕ)	126	47°39'N	5°00'E
Langsa, Indon. (läng'sá)	168	4°33'N	97°52'E
Lang Son, Viet. (lăng'sŏn')	168	21°52'N	106°42'E
L'Anguille, r., Ar., U.S. (läN-gē'y')	79	35°23'N	90°52'W
Langxi, China (läŋ-shyē)	162	31°10'N	119°09'E
Langzhong, China (läŋ-jôŋ)	160	31°40'N	106°05'E
Lanham, Md., U.S. (lăn'ăm)	68e	38°58'N	76°54'W
Lanigan, Can. (lăn'ĭ-găn)	50	51°52'N	105°02'W
Länkäran, Azer. (lĕn-kô-rän')	134	38°52'N	48°58'E
Lankoviri, Nig.	189	9°00'N	11°25'E
Lansdale, Pa., U.S. (lănz'dāl)	67	40°20'N	75°15'W
Lansdowne, Pa., U.S.	68f	39°57'N	75°17'W
L'Anse, Mi., U.S. (làns)	71	46°43'N	88°28'W
L'Anse and Vieux Desert Indian Reservation, I.R., Mi., U.S.	71	46°41'N	88°12'W
Lansford, Pa., U.S. (lănz'fērd)	67	40°50'N	75°50'W
Lansing, Ia., U.S.	71	43°22'N	91°16'W
Lansing, Il., U.S.	69a	41°34'N	87°33'W
Lansing, Ks., U.S.	75f	39°15'N	94°53'W
Lansing, Mi., U.S.	65	42°45'N	84°35'W
Lanús, Arg. (lä-nōōs')	102a	34°42'S	58°24'W
Lanusei, Italy (lä-nōō-sĕ'y)	130	39°51'N	9°34'E
Lanúvio, Italy (lä-nōō'vyô)	129d	41°41'N	12°42'E
Lanzarote Island, i., Spain (län-zá-rō'tā)	184	29°04'N	13°03'W
Lanzhou, China (län-jō)	160	35°55'N	103°55'E
Laoag, Phil. (lä-wäg')	168	18°13'N	120°38'E
Laon, Fr. (läN)	126	49°36'N	3°35'E
La Oroya, Peru (lä-ô-rō'yä)	100	11°30'S	76°00'W
Laos, nation, Asia (lā-ōs) (là-ōs')	168	20°15'N	102°00'E
Laoshan Wan, b., China (lou-shän wän)	162	36°21'N	120°48'E
La Palma, Pan. (lä-päl'mä)	91	8°25'N	78°07'W
La Palma, Spain	128	37°24'N	6°36'W
La Palma Island, i., Spain	184	28°42'N	19°03'W
La Pampa, prov., Arg.	102	37°25'S	67°00'W
Lapa Rio Negro, Braz. (lä-pä-rē'ō-nĕ'grō)	102	26°12'S	49°56'W
La Paz, Arg. (lä päz')	102	30°48'S	59°47'W
La Paz, Bol.	100	16°31'S	68°03'W
La Paz, Hond.	90	14°15'N	87°40'W
La Paz, Mex. (lä-pá'z)	88	23°39'N	100°44'W
La Paz, Mex.	86	24°00'N	110°15'W
Lapeer, Mi., U.S. (lå-pēr')	66	43°05'N	83°15'W
La-Penne-sur-Huveaune, Fr. (la-pĕn'sür-ü-vōn')	126a	43°18'N	5°33'E
La Perouse, Austl.	173b	33°59'S	151°14'E
La Piedad Cabadas, Mex. (lä pyä-dhädh' kä-bä'dhäs)	88	20°20'N	102°04'W
Lapland, hist. reg., Eur. (lăp'lănd)	110	68°20'N	22°00'E
La Plata, Arg. (lä plä'tä)	102	34°54'S	57°57'W
La Plata, Mo., U.S. (lå plä'tå)	79	40°03'N	92°28'W
La Plata Peak, mtn., Co., U.S.	77	39°00'N	106°25'W
La Pocatière, Can. (lä pô-ká-tyâr')	59	47°24'N	70°01'W
La Poile Bay, b., Can. (lå pwäl')	61	47°38'N	58°20'W
La Porte, In., U.S. (lå pōrt')	66	41°35'N	86°45'W
Laporte, Oh., U.S.	69d	41°19'N	82°05'W
La Porte, Tx., U.S.	81a	29°40'N	95°01'W
La Porte City, Ia., U.S.	71	42°20'N	92°10'W
Lappeenranta, Fin. (lä'pĕn-rän'tä)	123	61°04'N	28°08'E
La Prairie, Can. (lä-prä-rē')	62a	45°24'N	73°30'W
Lâpseki, Tur. (läp'sá-kē)	131	40°20'N	26°41'E
Laptev Sea, sea, Russia (läp'tyĭf)	135	75°39'N	120°00'E
La Puebla de Montalbán, Spain	128	39°54'N	4°21'W
La Puente, Ca., U.S. (pwĕn'tē)	75a	34°01'N	117°57'W
Lapuşul, r., Rom. (lä'pōō-shōōl)	125	47°29'N	23°46'E
La Quiaca, Arg. (lä kē-ä'kä)	102	22°15'S	65°44'W
L'Aquila, Italy (lá'kē-lä)	118	42°22'N	13°24'E
Lār, Iran (lär)	154	27°31'N	54°12'E
Lara, Austl.	173a	38°02'S	144°24'E
Larache, Mor. (lä-räsh')	184	35°15'N	6°09'W
Laramie, Wy., U.S. (lăr'å-mĭ)	64	41°20'N	105°40'W
Laramie, r., Co., U.S.	78	40°56'N	105°55'W
Larchmont, N.Y., U.S. (lärch'mŏnt)	68a	40°56'N	73°46'W
Larch Mountain, mtn., Or., U.S. (lärch)	74c	45°32'N	122°06'W

PLACE (Pronunciation)	PAGE	LAT.	LONG.
Laredo, Spain (lȧ-rā'dhō)	128	43°24'N	3°24'W
Laredo, Tx., U.S.	64	27°31'N	99°29'W
La Réole, Fr. (lä rȧ-ōl')	126	44°37'N	0°03'W
Largeau, Chad (lär-zhō')	185	17°55'N	19°07'E
Largo, Cayo, Cuba (kä'yō-lär'gō)	92	21°40'N	81°30'W
Larimore, N.D., U.S. (lăr'ĭ-môr)	70	47°53'N	97°38'W
Larino, Italy (lä-rē'nô)	130	41°48'N	14°54'E
La Rioja, Arg. (lä rê-ōhä)	102	29°18'S	67°42'W
La Rioja, prov., Arg. (lä-rê-ō'kä)	102	28°45'S	68°00'W
Lárisa, Grc. (lä'rê-sȧ)	119	39°38'N	22°25'E
Lārkāna, Pak.	158	27°40'N	68°12'E
Larnaka, Cyp.	119	34°55'N	33°37'E
Lárnakos, Kólpos, b., Cyp.	153a	36°50'N	33°45'E
Larned, Ks., U.S. (lär'nĕd)	78	38°09'N	99°07'W
La Robla, Spain (lä rōb'lä)	128	42°48'N	5°36'W
La Rochelle, Fr. (lȧ rō-shĕl')	110	46°10'N	1°09'W
La Roche-sur-Yon, Fr. (lȧ rôsh'sûr-yôn')	117	46°39'N	1°27'W
La Roda, Spain (lä rō'dä)	128	39°13'N	2°08'W
La Romana, Dom. Rep. (lä-rä-mô'nä)	93	18°25'N	69°00'W
Larrey Point, c., Austl. (lăr'ê)	174	19°15'S	118°15'E
Laruns, Fr. (lȧ-răNs')	126	42°58'N	0°28'W
Larvik, Nor. (lär'vēk)	116	59°06'N	10°03'E
La Sabana, Ven. (lä-sä-bȧ'nä)	101b	10°38'N	66°24'W
La Sabina, Cuba (lä-sä-bē'nä)	93a	22°51'N	82°05'W
La Sagra, mtn., Spain (lä sä'grä)	118	37°56'N	2°35'W
La Sal, Ut., U.S. (lȧ sȧl')	77	38°10'N	109°20'W
La Salle, Can. (lȧ sȧl')	69b	42°14'N	83°06'W
La Salle, Can.	62a	45°26'N	73°39'W
La Salle, Can.	62f	49°41'N	97°16'W
La Salle, Il., U.S.	66	41°20'N	89°05'W
Las Animas, Co., U.S. (lȧs ă'nĭ-mȧs)	78	38°03'N	103°16'W
La Sarre, Can.	51	48°43'N	79°12'W
Lascahobas, Haiti (läs-kä-ō'bȧs)	93	19°00'N	71°55'W
Las Cruces, Mex. (läs-krōō'sĕs)	89	16°37'N	93°54'W
Las Cruces, N.M., U.S.	64	32°20'N	106°50'W
La Selle, Massif de, mtn., Haiti (lä'sĕl')	93	18°25'N	72°05'W
La Serena, Chile (lä-sĕ-rē'nä)	102	29°55'S	71°24'W
La Seyne, Fr. (lä-sān')	117	43°07'N	5°52'E
Las Flores, Arg. (läs flō'rĕs)	102	36°01'S	59°07'W
Lashio, Mya. (lăsh'ê-ō)	160	22°58'N	98°03'E
Las Juntas, C.R. (läs-kōō'n-täs)	90	10°15'N	85°00'W
Las Maismas, sw., Spain (läs-mī's-mäs)	128	37°05'N	6°25'W
La Solana, Spain (lä-sō-lä-nä)	128	38°56'N	3°13'W
Las Palmas, Pan.	91	8°08'N	81°30'W
Las Palmas de Gran Canaria, Spain (läs päl'mäs)	184	28°07'N	15°28'W
La Spezia, Italy (lä-spĕ'zyä)	110	44°07'N	9°48'E
Las Piedras, Ur. (läs-pyĕ'drȧs)	99c	34°42'S	56°08'W
Las Pilas, vol., Nic. (läs-pē'läs)	90	12°32'N	86°43'W
Las Rosas, Mex. (läs rō thäs)	89	16°24'N	92°23'W
Las Rozas de Madrid, Spain (läs rō'thas dā mä-dhrēd')	129a	40°29'N	3°53'W
Lassee, Aus.	115e	48°14'N	16°50'E
Lassen Peak, mtn., Ca., U.S. (lăs'ĕn)	64	40°30'N	121°32'W
Lassen Volcanic National Park, rec., Ca., U.S.	64	40°43'N	121°35'W
L'Assomption, Can. (lȧs-sôm-syôn)	62a	45°50'N	73°25'W
Lass Qoray, Som.	192a	11°13'N	48°19'E
Las Tablas, Pan. (läs tä'bläs)	91	7°48'N	80°16'W
Last Mountain, l., Can. (lást moun'tĭn)	52	51°05'N	105°10'W
Lastoursville, Gabon (lȧs-tōōr-vēl')	186	1°00'S	12°49'E
Las Tres Vírgenes, Volcán, vol., Mex. (vē'r-hĕ-nĕs)	86	26°00'N	111°45'W
Las Tunas, prov., Cuba	92	21°05'N	77°00'W
Las Vacas, Mex. (läs-vá'kás)	89	16°24'N	95°48'W
Las Vegas, Chile (läs-vĕ'gäs)	99b	32°50'S	70°59'W
Las Vegas, N.M., U.S.	64	35°36'N	105°13'W
Las Vegas, Nv., U.S. (läs vā'gas)	64	36°12'N	115°10'W
Las Vegas, Ven. (läs-vĕ'gäs)	101b	10°26'N	64°08'W
Las Vigas, Mex.	89	19°38'N	97°03'W
Las Vizcachas, Meseta de, plat., Arg.	102	49°35'S	71°00'W
Latacunga, Ec. (lä-tä-kōŋ'gä)	100	1°02'S	78°33'W
Latakia see Al Lādhiqīyah, Syria	154	35°32'N	35°51'E
La Teste-de-Buch, Fr. (lä-tĕst-dē-büsh)	126	44°38'N	1°11'W
Lathrop, Mo., U.S. (lā'thrŭp)	79	39°32'N	94°21'W
La Tortuga, Isla, i., Ven. (ê's-lä-lä-tôr-tōō'gä)	100	10°55'N	65°18'W
Latorytsia, r., Eur.	125	48°27'N	22°30'E
Latourell, Or., U.S. (lȧ-tou'rĕl)	74c	45°32'N	122°13'W
La Tremblade, Fr. (lä-trĕn-bläd')	126	45°45'N	1°12'W
Latrobe, Pa., U.S. (lȧ-trōb')	67	40°25'N	79°15'W
La Tuque, Can. (lȧ'tük')	51	47°27'N	72°49'W
Lātūr, India (lä-tōōr')	158	18°20'N	76°35'E
Latvia, nation, Eur.	134	57°28'N	24°29'E
Lau Group, is., Fiji	170g	18°20'S	178°30'W
Launceston, Austl. (lôn'sĕs-tŭn)	175	41°35'S	147°22'E
Launceston, Eng., U.K. (lôrn'stŏn)	120	50°38'N	4°26'W
La Unión, Chile (lä-ōō-nyō'n)	102	40°15'S	73°04'W
La Unión, El Sal.	90	13°18'N	87°51'W
La Unión, Mex. (lä ōōn-nyōn')	88	17°59'N	101°48'W
La Unión, Spain	118	37°38'N	0°50'W
Laura, Austl. (lôrȧ)	175	15°40'S	144°45'E
Laurel, De., U.S. (lô'rĕl)	67	38°30'N	75°40'W
Laurel, Md., U.S.	68e	39°06'N	76°51'W
Laurel, Ms., U.S.	65	31°42'N	89°07'W
Laurel, Mt., U.S.	73	45°41'N	108°45'W
Laurel, Wa., U.S.	74d	48°52'N	122°29'W
Laurelwood, Or., U.S. (lô'rĕl-wŏŏd)	74c	45°25'N	123°05'W
Laurens, S.C., U.S. (lô'rĕnz)	83	34°29'N	82°03'W
Laurentian Highlands, hills, Can. (lô'rĕn-tĭ-ăn)	49	49°00'N	74°50'W
Laurentides, Can. (lô'rĕn-tīdz)	62a	45°51'N	73°46'W
Lauria, Italy (lou'rē-ä)	130	40°03'N	15°02'E
Laurinburg, N.C., U.S. (lô'rĭn-bûrg)	83	34°45'N	79°27'W

PLACE (Pronunciation)	PAGE	LAT.	LONG.
Laurium, Mi., U.S. (lô'rĭ-ŭm)	71	47°13'N	88°28'W
Lausanne, Switz. (lō-zȧn')	110	46°32'N	6°35'E
Laut, Pulau, i., Indon.	168	3°39'S	116°07'E
Lautaro, Chile (lou-tä'rô)	102	38°40'S	72°24'W
Laut Kecil, Kepulauan, is., Indon.	168	4°44'S	115°43'E
Lautoka, Fiji	170g	17°37'S	177°27'E
Lauzon, Can. (lō-zōN')	62b	46°50'N	71°10'W
Lava Beds National Monument, rec., Ca., U.S. (lä'vȧ bĕds)	72	41°38'N	121°44'W
Lavaca, r., Tx., U.S. (lȧ-văk'ȧ)	81	29°05'N	96°50'W
Lava Hot Springs, Id., U.S.	73	42°37'N	111°58'W
Laval, Can.	51	45°31'N	73°44'W
Laval, Fr. (lä-väl')	117	48°05'N	0°47'W
La Vecilla de Curueño, Spain	128	42°53'N	5°18'W
La Vega, Dom. Rep. (lä-vĕ'gä)	93	19°15'N	70°35'W
Lavello, Italy (lä-vĕl'lō)	130	41°05'N	15°50'E
La Verne, Ca., U.S. (lȧ vûrn')	75a	34°06'N	117°46'W
Laverton, Austl. (lā'vēr-tŭn)	174	28°45'S	122°30'E
La Victoria, Ven. (lä vēk-tō'rê-ä)	100	10°14'N	67°20'W
La Vila Joiosa, Spain	129	38°30'N	0°14'W
Lavonia, Ga., U.S. (lȧ-vō'nĭ-ȧ)	82	34°26'N	83°05'W
Lavon Reservoir, res., Tx., U.S.	81	33°06'N	96°20'W
Lavras, Braz. (lä'vräzh)	99a	21°15'S	44°59'W
Lávrio, Grc.	131	37°44'N	24°05'E
Lavry, Russia (lou'ra)	132	57°35'N	27°28'E
Lawndale, Ca., U.S. (lôn'dāl)	75a	33°54'N	118°22'W
Lawra, Ghana	188	10°39'N	2°52'W
Lawrence, In., U.S. (lô'rĕns)	69g	39°59'N	86°01'W
Lawrence, Ks., U.S.	65	38°57'N	95°13'W
Lawrence, Ma., U.S.	61a	42°42'N	71°09'W
Lawrence, Pa., U.S.	69e	40°18'N	80°07'W
Lawrenceburg, In., U.S. (lô'rĕns-bûrg)	69f	39°06'N	84°47'W
Lawrenceburg, Ky., U.S.	66	38°00'N	85°00'W
Lawrenceburg, Tn., U.S.	82	35°13'N	87°20'W
Lawrenceville, Ga., U.S. (lô-rĕns-vĭl)	82	33°56'N	83°57'W
Lawrenceville, Il., U.S.	66	38°45'N	87°45'W
Lawrenceville, N.J., U.S.	68a	40°17'N	74°44'W
Lawrenceville, Va., U.S.	83	36°43'N	77°52'W
Lawsonia, Md., U.S. (lô-sō'nĭ-ȧ)	67	38°00'N	75°50'W
Lawton, Ok., U.S. (lô'tŭn)	64	34°36'N	98°25'W
Lawz, Jabal al, mtn., Sau. Ar.	154	28°46'N	35°37'E
Layang Layang, Malay. (lä-yäng' lä-yäng')	153b	1°49'N	103°28'E
Laysan, i., Hi., U.S.	84b	26°00'N	171°00'W
Layton, Ut., U.S. (lā'tŭn)	75b	41°04'N	111°58'W
Lazdijai, Lith. (läzh'dē-yī')	123	54°12'N	23°35'E
Lazio (Latium), hist. reg., Italy	130	42°05'N	12°25'E
Lead, S.D., U.S. (lēd)	64	44°22'N	103°47'W
Leader, Can.	56	50°55'N	109°32'W
Leadville, Co., U.S. (lĕd'vĭl)	78	39°14'N	106°18'W
Leaf, r., Ms., U.S. (lēf)	82	31°43'N	89°20'W
League City, Tx., U.S. (lēg)	81a	29°31'N	95°05'W
Leamington, Can. (lĕm'ĭng-tŭn)	58	42°05'N	82°35'W
Leamington, Eng., U.K. (lĕ'mĭng-tŭn)	120	52°17'N	1°25'W
Leatherhead, Eng., U.K. (lĕdh'ēr-hĕd')	114b	51°17'N	0°20'W
Leavenworth, Ks., U.S. (lĕv'ĕn-wûrth)	65	39°19'N	94°54'W
Leavenworth, Wa., U.S.	72	47°35'N	120°39'W
Leawood, Ks., U.S. (lē'wŏŏd)	75f	38°58'N	94°37'W
Łeba, Pol. (lā'bä)	125	54°45'N	17°34'E
Lebam, r., Malay.	153b	1°35'N	104°09'E
Lebango, Congo	190	0°22'N	14°49'E
Lebanon, Il., U.S. (lĕb'ȧ-nŭn)	75e	38°36'N	89°49'W
Lebanon, In., U.S.	66	40°00'N	86°30'W
Lebanon, Ky., U.S.	82	37°32'N	85°15'W
Lebanon, Mo., U.S.	79	37°40'N	92°43'W
Lebanon, N.H., U.S.	67	43°40'N	72°15'W
Lebanon, Oh., U.S.	66	39°25'N	84°10'W
Lebanon, Or., U.S.	72	44°31'N	122°53'W
Lebanon, Pa., U.S.	67	40°20'N	76°20'W
Lebanon, Tn., U.S.	82	36°10'N	86°16'W
Lebanon, nation, Asia	154	34°00'N	34°00'E
Lebedyan', Russia (lyĕ'bĕ-dyän')	136	53°03'N	39°08'E
Lebedyn, Ukr.	137	50°34'N	34°27'E
Le Blanc, Fr. (lē-bläN')	126	46°38'N	0°59'E
Le Borgne, Haiti (lē bōrn'y')	93	19°50'N	72°30'W
Lębork, Pol. (lăN-bork')	125	54°33'N	17°46'E
Lebrija, Spain (lå-brē'hä)	128	36°55'N	6°06'W
Lecce, Italy (lĕt'chā)	119	40°22'N	18°11'E
Lecco, Italy (lĕk'kō)	130	45°52'N	9°28'E
Lech, r., Ger. (lĕk)	124	47°41'N	10°52'E
Le Châtelet-en-Brie, Fr. (lĕ-shä-tĕ-lā'ĕN-brē')	127b	48°29'N	2°50'E
Leche, Laguna de, l., Cuba (lä-gō'nä-dē-lĕ'chĕ)	92	22°10'N	78°30'W
Leche, Laguna de la, l., Mex.	80	27°16'N	102°45'W
Lecompte, La., U.S.	81	31°06'N	92°25'W
Le Creusot, Fr. (lĕkrû-zō)	117	46°48'N	4°23'E
Ledesma, Spain (lå-dĕs'mä)	128	41°05'N	5°59'W
Leduc, Can. (lĕ-dōōk')	55	53°16'N	113°33'W
Leech, l., Mn., U.S. (lēch)	71	47°06'N	94°16'W
Leeds, Eng., U.K.	110	53°48'N	1°33'W
Leeds, Al., U.S. (lēdz)	68h	33°33'N	86°33'W
Leeds, N.D., U.S.	70	48°18'N	99°24'W
Leeds, co., Eng., U.K.	114a	53°50'N	1°30'W
Leeds and Liverpool Canal, can., Eng., U.K. (lĭv'ēr-pōōl)	114a	53°36'N	2°38'W
Leegebruch, Ger. (lĕh'gĕn-brōōk)	115b	52°43'N	13°12'E
Leek, Eng., U.K. (lēk)	114a	53°06'N	2°01'W
Leer, Ger. (lār)	124	53°14'N	7°27'E
Leesburg, Fl., U.S. (lēz'bûrg)	83	28°49'N	81°53'W
Leesburg, Va., U.S.	67	39°10'N	77°30'W
Lees Summit, Mo., U.S.	75f	38°55'N	94°23'W
Lee Stocking, i., Bah.	92	23°45'N	76°05'W
Leesville, La., U.S. (lēz'vĭl)	81	31°09'N	93°17'W
Leetonia, Oh., U.S. (lē-tō'nĭ-ȧ)	66	40°50'N	80°45'W
Leeuwarden, Neth. (lā'wär-dĕn)	117	52°12'N	5°50'E
Leeuwin, Cape, c., Austl. (lōō'wĭn)	174	34°15'S	114°30'E

PLACE (Pronunciation)	PAGE	LAT.	LONG.
Leeward Islands, is., N.A. (lē'wērd)	87	17°00'N	62°15'W
Lefkáda, Grc.	131	38°49'N	20°43'E
Lefkáda, i., Grc.	119	38°42'N	20°22'E
Le François, Mart.	91b	14°37'N	60°55'W
Lefroy, l., Austl. (lē-froi')	174	31°30'S	122°00'E
Leganés, Spain (lå-gä'nås)	129a	40°20'N	3°46'W
Legazpi, Phil. (lå-gäs'pē)	169	13°09'N	123°44'E
Legge Peak, mtn., Austl. (lĕg)	176	41°33'S	148°10'E
Leggett, Ca., U.S.	76	39°51'N	123°42'W
Leghorn see Livorno, Italy	110	43°32'N	11°18'E
Legnano, Italy (lå-nyä'nō)	130	45°35'N	8°53'E
Legnica, Pol. (lĕk-nĭt'sȧ)	117	51°13'N	16°10'E
Leh, India (lā)	158	34°10'N	77°40'E
Le Havre, Fr. (lē áv'r')	110	49°31'N	0°07'E
Lehi, Ut., U.S. (lē'hī)	77	40°25'N	111°55'W
Lehman Caves National Monument, rec., Nv., U.S. (lē'mȧn)	77	38°54'N	114°08'W
Lehnin, Ger. (lĕh'nēn)	115b	52°19'N	12°45'E
Leicester, Eng., U.K. (lĕs'tēr)	110	52°37'N	1°08'W
Leicestershire, co., Eng., U.K.	114a	52°40'N	1°12'W
Leichhardt, r., Austl. (līk'härt)	174	18°30'S	139°45'E
Leiden, Neth. (lī'dĕn)	121	52°09'N	4°29'E
Leigh Creek, Austl. (lē krēk)	176	30°33'S	138°30'E
Leikanger, Nor. (lī'käŋ'gēr)	122	61°11'N	6°51'E
Leimuiden, Neth.	115a	52°13'N	4°40'E
Leine, r., Ger. (lī'nĕ)	124	51°58'N	9°56'E
Leinster, hist. reg., Ire. (lĕn-stēr)	120	52°45'N	7°19'W
Leipsic, Oh., U.S. (līp'sĭk)	66	41°05'N	84°00'W
Leipzig, Ger. (līp'tsĭk)	110	51°20'N	12°24'E
Leiria, Port. (lā-rē'ä)	128	39°45'N	8°50'W
Leitchfield, Ky., U.S. (lēch'fēld)	82	37°28'N	86°20'W
Leitha, r., Aus.	115e	48°04'N	16°57'E
Leitrim, Can.	62c	45°20'N	75°36'W
Leivádia, Grc.	131	38°25'N	22°51'E
Leizhou Bandao, pen., China (lā-jō bän-dou)	160	20°42'N	109°10'E
Leksand, Swe. (lĕk'sänd)	122	60°45'N	14°56'E
Leland, Wa., U.S. (lē'lănd)	74a	47°54'N	122°53'W
Leliu, China (lŭ-lĭō)	163a	22°52'N	113°09'E
Le Locle, Switz. (lē lô'kl')	124	47°03'N	6°43'E
Le Maire, Estrecho de, strt., Arg. (ĕs-trĕ'chô-dĕ-lĕ-mī'rĕ)	102	55°15'S	65°30'W
Le Mans, Fr. (lē mäN')	117	48°01'N	0°12'E
Le Marin, Mart.	91b	14°28'N	60°55'W
Le Mars, Ia., U.S. (lē märz')	70	42°46'N	96°09'W
Lemay, Mo., U.S.	75e	38°32'N	90°17'W
Lemdiyya, Alg.	184	36°18'N	2°40'E
Lemery, Phil. (lā-mā-rē')	169a	13°51'S	120°55'E
Lemhi, r., Id., U.S.	73	44°40'N	113°27'W
Lemhi Range, mts., Id., U.S. (lĕm'hī)	73	44°35'N	113°33'W
Lemmon, S.D., U.S. (lĕm'ŭn)	70	45°55'N	102°10'W
Le Môle, Haiti (lē mōl')	93	19°50'N	73°20'W
Lemon Grove, Ca., U.S. (lĕm'ŭn-grōv)	76a	32°44'N	117°02'W
Le Moule, Guad. (lē mōōl')	91b	16°19'N	61°22'W
Lempa, r., N.A. (lĕm'pä)	90	13°20'N	88°46'W
Lemvig, Den. (lĕm'vēgh)	122	56°33'N	8°16'E
Lena, r., Russia	135	68°00'N	123°00'E
Lençóes Paulista, Braz. (lĕN-sôNs' pou-lēs'tä)	102	22°30'S	48°45'W
Lençóis, Braz. (lĕn-sóis)	101	12°38'S	41°28'W
Lenexa, Ks., U.S. (lē'nĕx-ă)	75f	38°58'N	99°44'W
Lengyandong, China (lŭŋ-yän-dôŋ)	163a	23°12'N	113°21'E
Lenik, r., Malay.	153b	1°59'N	102°51'E
Leningrad see Saint Petersburg, Russia	134	59°57'N	30°20'E
Leningrad, prov., Russia	132	59°15'N	30°30'E
Leningradskaya, Russia (lyĕ-nĭn-gräd'skä-yȧ)	133	46°19'N	39°23'E
Lenino, Russia (lyĕ'nĭ-nô)	142b	55°37'N	37°41'E
Leninogorsk, Kaz.	139	50°29'N	83°25'E
Lenīnsk, Kaz.	139	45°39'N	63°19'E
Leninsk, Russia (lyĕ-nēnsk')	137	48°40'N	45°10'E
Leninsk-Kuznetski, Russia (lyĕ-nēnsk'kōōz-nyĕt'skĭ)	134	54°28'N	86°48'E
Lennox, S.D., U.S. (lĕn'ŭks)	70	43°22'N	96°53'W
Lenoir, N.C., U.S. (lē-nōr')	83	35°54'N	81°35'W
Lenoir City, Tn., U.S.	82	35°47'N	84°16'W
Lenox, Ia., U.S.	71	40°51'N	94°29'W
Léo, Burkina	188	11°06'N	2°06'W
Leoben, Aus. (lå-ō'bĕn)	124	47°22'N	15°09'E
Léogane, Haiti (lā-ō-gän')	93	18°30'N	72°35'W
Leola, S.D., U.S. (lē-ō'lȧ)	70	45°43'N	99°55'W
Leominster, Ma., U.S. (lĕm'ĭn-stēr)	67	42°32'N	71°45'W
León, Mex. (lå-ōn')	86	21°08'N	101°41'W
León, Nic. (lĕ-ô'n)	86	12°28'N	86°53'W
León, Spain (lĕ-ô'n)	118	42°38'N	5°33'W
Leon, Ia., U.S. (lē'ŏn)	71	40°43'N	93°44'W
León, hist. reg., Spain	128	41°18'N	5°50'W
Leon, r., Tx., U.S. (lē'ŏn)	80	31°54'N	98°20'W
Leonforte, Italy (lā-ôn-fôr'tä)	130	37°40'N	14°27'E
Leopold II, Lac see Mai-Ndombe, Lac, l., D.R.C.	186	2°16'S	19°00'E
Leopoldina, Braz. (lā-ô-pōl-dē'nä)	99a	21°32'S	42°38'W
Leopoldsburg, Bel.	115a	51°07'N	5°18'E
Leopoldsdorf im Marchfelde, Aus. (lā'ô-pōlts-dôrf)	115e	48°14'N	16°42'E
Léopoldville see Kinshasa, D.R.C.	186	4°18'S	15°18'E
Leova, Mol.	133	46°30'N	28°16'E
Lepe, Spain (lā'pā)	128	37°15'N	7°12'W
Leping, China (lŭ-pĭŋ)	165	29°02'N	117°12'E
L'Épiphanie, Can. (lā-pē-fä-nē')	62a	45°51'N	73°29'W
Le Plessis-Belleville, Fr. (lĕ-plĕ-sē'bĕl-vēl')	127b	49°05'N	2°46'E
Lepreau, Can. (lē-prō')	60	45°10'N	66°28'W
Le Puy, Fr. (lē pwē')	117	45°02'N	3°54'E
Lercara Friddi, Italy (lĕr-kä'rä)	130	37°47'N	13°36'E
Lerdo, Mex. (lĕr'dō)	86	25°31'N	103°30'W

ng-sing; ŋ-baŋk; N-nasalized n; nŏd; cŏmmit; ōld; ôbey; ôrder; oi-boil; fōōd; ŏŏ-as oo in foot; ou-out; s-soft; sh-dish; th-thin; pūre; ûnite; ûrn; stŭd; circŭs; ü-as in French tu; '-indeterminate vowel.

268 Leribe – Lisiy Nos

PRONOUNCING INDEX

PLACE (Pronunciation)	PAGE	LAT.	LONG.
Leribe, Leso.	187c	28°53'S	28°02'E
Lerma, Mex. (lĕr′mä)	89	19°49'N	90°34'W
Lerma, Mex.	89a	19°17'N	99°30'W
Lerma, Spain (lĕ′r-mä)	128	42°03'N	3°45'W
Lerma, r., Mex.	88	20°14'N	101°50'W
Le Roy, N.Y., U.S. (lē roi′)	67	43°00'N	78°00'W
Lerwick, Scot., U.K. (lĕr′ĭk) (lûr′wĭk)	110	60°08'N	1°27'W
Léry, Can. (lā-rī′)	62a	45°21'N	73°49'W
Lery, Lake, l., La., U.S. (lĕ′rē)	68d	29°48'N	89°45'W
Les Andelys, Fr. (lā-zäN-dē-lē′)	127b	49°15'N	1°25'E
Les Borges Blanques, Spain	129	41°29'N	0°53'E
Lesbos *see* Lésvos, i., Grc.	112	39°15'N	25°40'E
Les Cayes, Haiti	93	18°15'N	73°45'W
Les Cèdres, Can. (lā-sĕdr′)	62a	45°18'N	74°03'W
Lesh, Alb. (lĕshē) (ä-lā′sē-ō)	131	41°47'N	19°40'E
Leshan, China (lŭ-shän)	160	29°40'N	103°40'E
Lésina, Lago di, l., Italy (lă′gō dē lā′zē-nä)	130	41°48'N	15°12'E
Leskovac, Serb. (lĕs′kô-väts)	131	43°00'N	21°58'E
Leslie, S. Afr.	192c	26°23'S	28°57'E
Leslie, Ar., U.S. (lĕz′lĭ)	79	35°49'N	92°32'W
Lesnoy, Russia (lĕs′noi)	136	66°45'N	34°45'E
Lesogorsk, Russia (lyĕs′ō-gôrsk)	166	49°28'N	141°59'E
Lesotho, nation, Afr. (lĕsō′thō)	186	29°45'S	28°07'E
Lesozavodsk, Russia (lyĕ-sô-zá-vôdsk′)	166	45°21'N	133°19'E
Les Sables-d'Olonne, Fr. (lā sá′bl′dō-lŭn′)	117	46°30'N	1°47'W
Les Saintes Islands, is., Guad. (lā-sănt′)	91b	15°50'N	61°40'W
Lesser Antilles, is.	87	12°15'N	65°00'W
Lesser Caucasus, mts., Asia	138	41°00'N	44°35'E
Lesser Khingan Range, mts., China	161	49°50'N	129°26'E
Lesser Slave, r., Can.	55	55°15'N	114°30'W
Lesser Slave Lake, l., Can. (lĕs′ĕr slāv)	52	55°25'N	115°30'W
Lesser Sunda Islands, is., Indon.	168	9°00'S	120°00'E
L'Estaque, Fr. (lĕs-täl)	126a	43°22'N	5°20'E
Les Thilliers-en-Vexin, Fr. (lā-tē-yā′ĕN-vĕ-sáN′)	127b	49°19'N	1°36'E
Le Sueur, Mn., U.S. (lē sōōr′)	71	44°27'N	93°53'W
Lésvos, i., Grc.	112	39°15'N	25°40'E
Leszno, Pol. (lĕsh′nō)	117	51°51'N	16°35'E
Le Teil, Fr. (lē tā′y′)	126	44°34'N	4°39'E
Lethbridge, Can. (lĕth′brĭj)	50	49°42'N	112°50'W
Leticia, Col. (lē-tē′syá)	100	4°04'S	69°57'W
Leting, China (lŭ-tĭŋ)	162	39°26'N	118°53'E
Le Tréport, Fr. (lē-trā′pôr′)	126	50°03'N	1°21'E
Letychiv, Ukr.	133	49°22'N	27°29'E
Leuven, Bel.	121	50°53'N	4°42'E
Levack, Can.	58	46°38'N	81°23'W
Levallois-Perret, Fr. (lē-vál-wä′pĕ-rĕ′)	127b	48°53'N	2°17'E
Levanger, Nor. (lĕ-väng′ĕr)	116	63°42'N	11°01'E
Levanna, mtn., Eur. (lā-vä′nä)	130	45°25'N	7°14'E
Leveque, Cape, c., Austl. (lē-vĕk′)	174	16°26'S	123°08'E
Leverkusen, Ger. (lĕ′fĕr-kōō-zĕn)	127c	51°01'N	6°59'E
Levice, Slvk. (lā′vĕt-sĕ)	125	48°13'N	18°37'E
Levico, Italy (lā′vē-kō)	130	46°02'N	11°20'E
Le Vigan, Fr. (lē vē-gäN′)	126	43°59'N	3°36'E
Lévis, Can. (lā-vē′) (lē′vĭs)	51	46°49'N	71°11'W
Levittown, Pa., U.S. (lĕ′vĭt-toun)	68f	40°08'N	74°50'W
Levoča, Slvk. (lā′vô-chá)	125	49°03'N	20°38'E
Levuka, Fiji	170g	17°41'S	178°50'E
Lewes, Eng., U.K.	121	50°51'N	0°01'E
Lewes, De., U.S. (lōō′ĭs)	67	38°45'N	75°10'W
Lewis, r., Wa., U.S.	72	46°05'N	122°09'W
Lewis, East Fork, r., Wa., U.S.	74c	45°52'N	122°40'W
Lewis, Island of, i., Scot., U.K. (lōō′ĭs)	120	58°05'N	6°07'W
Lewisburg, Tn., U.S. (lū′ĭs-bûrg)	82	35°27'N	86°47'W
Lewisburg, W.V., U.S.	66	37°50'N	80°20'W
Lewis Hills, hills, Can.	61	48°48'N	58°30'W
Lewisporte, Can. (lū′ĭs-pôrt)	61	49°15'N	55°04'W
Lewis Range, mts., Mt., U.S. (lū′ĭs)	73	48°15'N	113°20'W
Lewis Smith Lake, res., Al., U.S.	82	34°05'N	87°07'W
Lewiston, Id., U.S. (lū′ĭs-tŭn)	64	46°24'N	116°59'W
Lewiston, Me., U.S.	65	44°05'N	70°14'W
Lewiston, N.Y., U.S.	69c	43°11'N	79°02'W
Lewiston, Ut., U.S.	73	41°58'N	111°51'W
Lewistown, Il., U.S. (lū′ĭs-toun)	79	40°23'N	90°06'W
Lewistown, Mt., U.S.	64	47°05'N	109°25'W
Lewistown, Pa., U.S.	67	40°35'N	77°30'W
Lexington, Ky., U.S. (lĕk′sĭng-tŭn)	65	38°05'N	84°30'W
Lexington, Ma., U.S.	61a	42°27'N	71°14'W
Lexington, Mo., U.S.	79	39°11'N	93°52'W
Lexington, Ms., U.S.	82	33°08'N	90°02'W
Lexington, N.C., U.S.	83	35°47'N	80°15'W
Lexington, Ne., U.S.	78	40°46'N	99°44'W
Lexington, Tn., U.S.	82	35°37'N	88°24'W
Lexington, Va., U.S.	67	37°45'N	79°20'W
Leyte, i., Phil. (lā′tā)	169	10°35'N	125°35'E
Leżajsk, Pol. (lĕ′zhä-ĭsk)	125	50°14'N	22°25'E
Lezha, r., Russia (lĕ-zhä′)	132	58°59'N	40°27'E
L'gov, Russia (lgôf)	133	51°42'N	35°15'E
Lhasa, China (läs′ä)	160	29°41'N	91°12'E
Liangxiangzhen, China (lĭäŋ-shyäŋ-jŭn)	164a	39°43'N	116°08'E
Lianjiang, China (lĭĕn-jyäŋ)	165	21°38'N	110°15'E
Lianozovo, Russia (lĭ-a-nô′zô-vô)	142b	55°54'N	37°36'E
Lianshui, China (lĭĕn-shwā)	162	33°46'N	119°15'E
Lianyungang, China (lĭĕn-yòn-gäŋ)	161	34°35'N	119°09'E
Liao, r., China	164	41°40'N	122°40'E
Liao, r., China	161	43°37'N	120°05'E
Liaocheng, China (lĭou-chŭŋ)	164	36°27'N	115°56'E
Liaodong Bandao, pen., China (lĭou-dôŋ bän-dou)	161	39°45'N	122°22'E
Liaodong Wan, b., China (lĭou-dôŋ wäŋ)	164	40°25'N	121°15'E
Liaoning, prov., China	161	41°31'N	122°11'E
Liaoyang, China (lyä′ō-yäng′)	161	41°18'N	123°10'E
Liaoyuan, China (lĭou-yůän)	164	43°00'N	124°59'E
Liard, r., Can. (lē-är′)	52	59°43'N	126°42'W
Libano, Col. (lē′bá-nō)	100a	4°55'N	75°05'W
Libby, Mt., U.S. (lĭb′ē)	72	48°27'N	115°35'W
Libenge, D.R.C. (lē-bĕŋ′gā)	185	3°39'N	18°40'E
Liberal, Ks., U.S. (lĭb′ĕr-ăl)	78	37°01'N	100°56'W
Liberec, Czech Rep. (lē′bĕr-ĕts)	117	50°45'N	15°06'E
Liberia, C.R.	90	10°38'N	85°28'W
Liberia, nation, Afr. (lĭ-bē′rĭ-ȧ)	184	6°30'N	9°55'W
Libertad, Arg.	102a	34°42'S	58°42'W
Libertad de Orituco, Ven. (lē-bĕr-tä′d-dĕ-ō-rē-tōō′kô)	101b	9°32'N	66°24'W
Liberty, In., U.S. (lĭb′ĕr-tĭ)	66	39°35'N	84°55'W
Liberty, Mo., U.S.	75f	39°15'N	94°25'W
Liberty, S.C., U.S.	83	34°47'N	82°41'W
Liberty, Tx., U.S.	81	30°03'N	94°46'W
Liberty, Ut., U.S.	75b	41°20'N	111°52'W
Liberty Bay, b., Wa., U.S.	74a	47°43'N	122°41'W
Liberty Lake, l., Md., U.S.	68e	39°25'N	76°56'W
Libertyville, Il., U.S. (lĭb′ĕr-tĭ-vĭl)	69a	42°17'N	87°57'W
Libode, S. Afr. (lĭ-bō′dĕ)	187c	31°33'S	29°03'E
Libón, r., N.A.	93	19°30'N	71°45'W
Libourne, Fr. (lē-bōōrn′)	117	44°55'N	0°12'W
Libres, Mex. (lē′brās)	89	19°26'N	97°41'W
Libreville, Gabon (lē-br′vēl′)	186	0°23'N	9°27'E
Liburn, Ga., U.S. (lĭb′ûrn)	68c	33°53'N	84°09'W
Libya, nation, Afr. (lĭb′ē-ä)	185	27°38'N	15°00'E
Libyan Desert, des., Afr. (lĭb′ē-ăn)	185	28°23'N	23°34'E
Libyan Plateau, plat., Afr.	156	30°58'N	26°20'E
Licancábur, Cerro, mtn., S.A. (sĕ′r-rô-lē-kän-ká′bōōr)	102	22°45'S	67°45'W
Licanten, Chile (lē-kän-tē′n)	99b	34°58'S	72°00'W
Lichfield, Eng., U.K. (lĭch′fēld)	114a	52°41'N	1°49'W
Lichinga, Moz.	191	13°18'S	35°14'E
Lichtenburg, S. Afr. (lĭk′tĕn-bĕrgh)	192c	26°09'S	26°10'E
Lick Creek, r., In., U.S. (lĭk)	69g	39°43'N	86°06'W
Licking, r., Ky., U.S. (lĭk′ĭng)	66	38°30'N	84°10'W
Lida, Bela. (lē′dá)	125	53°53'N	25°19'E
Lidgerwood, N.D., U.S. (lĭj′ĕr-wood)	70	46°04'N	97°10'W
Lidköping, Swe. (lēt′chû-pĭng)	122	58°31'N	13°06'E
Lido di Roma, Italy (lē′dô-dē-rô′mä)	129d	41°19'N	12°17'E
Lidzbark, Pol. (lĭts′bärk)	125	54°07'N	20°36'E
Liebenbergsvlei, r., S. Afr.	192c	27°35'S	28°25'E
Liebenwalde, Ger. (lē′bĕn-väl-dē)	115b	52°52'N	13°24'E
Liechtenstein, nation, Eur. (lĕk′tĕn-shtīn)	117	47°10'N	10°00'E
Liège, Bel.	117	50°38'N	5°34'E
Lienz, Aus. (lē-ĕnts′)	124	46°49'N	12°45'E
Liepāja, Lat. (le′pä-yä′)	136	56°31'N	20°59'E
Lier, Bel.	115a	51°08'N	4°34'E
Liesing, Aus. (lē′sĭng)	115e	48°09'N	16°17'E
Liestal, Switz. (lēs′täl)	124	47°28'N	7°44'E
Lifanga, D.R.C.	190	0°19'N	21°57'E
Lifou, i., N. Cal.	175	21°15'S	167°32'E
Ligao, Phil. (lē-gä′ō)	169a	13°14'N	123°33'E
Lightning Ridge, Austl.	176	29°23'S	147°50'E
Ligonha, r., Moz. (lē-gō′nyȧ)	187	16°14'S	39°00'E
Ligonier, In., U.S. (lĭg-ô-nēr)	66	41°30'N	85°35'W
Ligovo, Russia (lē′gô-vô)	142c	59°51'N	30°13'E
Liguria, hist. reg., Italy (lē-gōō-rē-ä)	130	44°24'N	8°27'E
Ligurian Sea, sea, Eur. (lĭ-gū′rĭ-ăn)	118	43°42'N	8°32'E
Lihou Reef, rf., Austl. (lē-hōō′)	175	17°23'S	152°43'E
Lihuang, China (lē′hōōäng)	162	31°32'N	115°46'E
Lihue, Hi., U.S. (lē-hōō′ä)	64c	21°59'N	159°23'W
Lihula, Est. (lē′hò-lá)	123	58°41'N	23°50'E
Liji, China (lē-jyē)	162	33°47'N	117°47'E
Lijiang, China (lē-jyäŋ)	160	27°00'N	100°08'E
Lijin, China (lē-jyĭn)	164	37°30'N	118°15'E
Likasi, D.R.C.	186	10°59'S	26°44'E
Likhoslavl', Russia (lyē-kôsläv′'l)	132	57°07'N	35°27'E
Likouala, r., Congo	190	0°10'S	16°30'E
Lille, Fr. (lēl)	110	50°38'N	3°01'E
Lille Baelt, strt., Den.	122	55°09'N	9°53'E
Lillehammer, Nor. (lēl′ē-häm′mĕr)	116	61°07'N	10°25'E
Lillesand, Nor. (lēl′ē-sän′)	122	58°16'N	8°19'E
Lilleström, Nor. (lēl′ē-strûm)	122	59°56'N	11°04'E
Lilliwaup, Wa., U.S. (lĭl′ĭ-wŏp)	74a	47°28'N	123°07'W
Lillooet, Can. (lĭ′lōō-ĕt)	50	50°30'N	121°55'W
Lillooet, r., Can.	55	49°50'N	122°10'W
Lilongwe, Mwi. (lē-lô-än)	186	13°59'S	33°44'E
Lima, Peru (lē′mä)	100	12°06'S	76°55'W
Lima, Swe.	122	60°54'N	13°24'E
Lima, Oh., U.S. (lī′mȧ)	65	40°40'N	84°05'W
Lima, r., Eur.	128	41°45'N	8°22'W
Lima Duarte, Braz. (dwä′r-tē)	99a	21°52'S	43°47'W
Lima Reservoir, res., Mt., U.S.	73	44°45'N	112°15'W
Limassol, Cyp.	119	34°39'N	33°02'E
Limay, r., Arg. (lē-mä′ē)	102	39°50'S	69°15'W
Limbaži, Lat. (lēm′bä-zĭ)	123	57°32'N	24°44'E
Limbdi, India	158	22°37'N	71°52'E
Limbe, Cam.	184	4°01'N	9°12'E
Limburg an der Lahn, Ger. (lem-bōrg′)	124	50°22'N	8°03'E
Limeira, Braz. (lē-mā′rä)	99a	22°34'S	47°24'W
Limerick, Ire. (lĭm′nák)	117	52°39'N	8°35'W
Limestone Bay, b., Can. (lĭm′stōn)	57	53°50'N	98°50'W
Limfjorden, Den.	116	56°55'N	8°56'E
Limmen Bight, b., Austl. (lĭm′ĕn)	174	14°45'S	136°00'E
Limnos, i., Grc.	119	39°58'N	24°48'E
Limoges, Can. (lē-mōzh′)	62c	45°20'N	75°15'W
Limoges, Fr.	117	45°50'N	1°15'E
Limón, C.R. (lē-mōn′)	87	10°01'N	83°02'W
Limón, Hond. (lē-mô′n)	90	15°53'N	85°34'W
Limon, Co., U.S. (lĭ′mŏn)	78	39°15'N	103°41'W
Limon, r., Dom. Rep.	93	18°20'N	71°40'W
Limón, Bahía, b., Pan.	86a	9°21'N	79°58'W
Limours, Fr. (lē-mōōr′)	127b	48°39'N	2°05'E
Limousin, Plateaux du, plat., Fr. (plä-tō′ dü lē-mōō-zăN′)	126	45°44'N	1°09'E
Limoux, Fr. (lē-mōō′)	126	43°03'N	2°14'E
Limpopo, r., Afr. (lĭm-pō′pō)	186	23°15'S	27°46'E
Linares, Chile (lē-nä′rās)	102	35°51'S	71°35'W
Linares, Mex.	86	24°53'N	99°34'W
Linares, Spain (lē-nä′rĕs)	118	38°07'N	3°38'W
Linares, prov., Chile	99b	35°53'S	71°30'W
Linaro, Cape, c., Italy (lē-nä′rä)	130	42°02'N	11°53'E
Linchuan, China (lĭn-chůän)	161	27°58'N	116°18'E
Lincoln, Arg. (lĭŋ′kŭn)	102	34°51'S	61°29'W
Lincoln, Can.	62d	43°10'N	79°29'W
Lincoln, Eng., U.K.	116	53°14'N	0°33'W
Lincoln, Ca., U.S.	76	38°51'N	121°19'W
Lincoln, Il., U.S.	79	40°09'N	89°21'W
Lincoln, Ks., U.S.	78	39°02'N	98°08'W
Lincoln, Ma., U.S.	61a	42°25'N	71°19'W
Lincoln, Me., U.S.	60	45°23'N	68°31'W
Lincoln, Ne., U.S.	64	40°49'N	96°43'W
Lincoln, Mount, mtn., Co., U.S.	78	39°20'N	106°19'W
Lincoln Heath, reg., Eng., U.K.	114a	53°23'N	0°39'W
Lincoln Park, Mi., U.S.	69b	42°14'N	83°11'W
Lincoln Park, N.J., U.S.	68a	40°56'N	74°18'W
Lincolnshire, co., Eng., U.K.	114a	53°12'N	0°29'W
Lincolnshire Wolds, Eng., U.K. (woldz′)	120	53°25'N	0°23'W
Lincolnton, N.C., U.S. (lĭŋ′kŭn-tŭn)	83	35°27'N	81°15'W
Lindale, Ga., U.S. (lĭn′dāl)	82	34°10'N	85°10'W
Lindau, Ger. (lĭn′dou)	124	47°33'N	9°40'E
Linden, Al., U.S. (lĭn′dĕn)	82	32°16'N	87°47'W
Linden, Mo., U.S.	75f	39°13'N	94°35'W
Linden, N.J., U.S.	68a	40°39'N	74°14'W
Lindenhurst, N.Y., U.S. (lĭn′dĕn-hûrst)	68a	40°41'N	73°23'W
Lindenwold, N.J., U.S. (lĭn′dĕn-wôld)	68f	39°50'N	75°00'W
Lindesberg, Swe. (lĭn′dĕs-bĕrgh)	122	59°37'N	15°14'E
Lindesnes, c., Nor. (lĭn′ĕs-nĕs)	112	58°00'N	7°05'E
Lindi, Tan. (lĭn′dē)	187	10°00'S	39°43'E
Lindi, r., D.R.C.	185	1°00'N	27°13'E
Lindian, China (lĭn-dĭĕn)	164	47°08'N	124°59'E
Lindley, S. Afr. (lĭnd′lē)	192c	27°52'S	27°55'E
Lindow, Ger. (lēn′dōv)	115b	52°58'N	12°59'E
Lindsay, Can. (lĭn′zē)	59	44°20'N	78°45'W
Lindsay, Ok., U.S.	79	34°50'N	97°38'W
Lindsborg, Ks., U.S. (lĭnz′bôrg)	79	38°34'N	97°42'W
Lineville, Al., U.S. (līn′vĭl)	82	33°18'N	85°45'W
Linfen, China	161	36°00'N	111°38'E
Linga, Kepulauan, is., Indon.	168	0°35'S	105°05'E
Lingao, China (lĭn-gou)	165	19°58'N	109°40'E
Lingayen, Phil. (lĭŋ′gä-yän′)	168	16°01'N	120°13'E
Lingayen Gulf, b., Phil.	169a	16°18'N	120°11'E
Lingdianzhen, China	162	31°52'N	121°28'E
Lingen, Ger. (lĭŋ′gĕn)	124	52°32'N	7°20'E
Lingling, China (lĭŋ-lĭŋ)	165	26°10'N	111°40'E
Lingshou, China (lĭŋ-shō)	162	38°21'N	114°41'E
Linguère, Sen. (lĭŋ-gĕr′)	184	15°24'N	15°07'W
Lingwu, China	164	38°05'N	106°18'E
Lingyuan, China (lĭŋ-yůän)	164	41°12'N	119°20'E
Linhai, China	165	28°52'N	121°08'E
Linhe, China (lĭn-hŭ)	164	40°49'N	107°45'E
Linhuaiguan, China (lĭn-hwī-gůän)	162	32°55'N	117°38'E
Linhuanji, China	162	33°42'N	116°33'E
Linjiang, China (lĭn-jyäŋ)	164	41°45'N	127°00'E
Linköping, Swe. (lĭn′chû-pĭng)	116	58°25'N	15°35'E
Linnhe, Loch, b., Scot., U.K. (lĭn′ē)	120	56°35'N	4°30'W
Linqing, China (lĭn-chyĭŋ)	161	36°49'N	115°42'E
Linqu, China (lĭn-chyōō)	162	36°31'N	118°33'E
Lins, Braz. (lē′Ns)	101	21°42'S	49°41'W
Linthicum Heights, Md., U.S. (lĭn′thĭ-kŭm)	68e	39°12'N	76°39'W
Linton, In., U.S. (lĭn′tŭn)	66	39°05'N	87°15'W
Linton, N.D., U.S.	70	46°16'N	100°15'W
Linwu, China (lĭn′wōō′)	165	25°20'N	112°30'E
Linxi, China (lĭn-shyē)	164	43°30'N	118°02'E
Linyi, China (lĭn-yē)	161	35°04'N	118°21'E
Linying, China (lĭn′yĭng′)	162	33°48'N	113°56'E
Linz, Aus. (lĭnts)	117	48°18'N	14°18'E
Linzhang, China (lĭn-jäŋ)	162	36°19'N	114°40'E
Lion, Golfe du, b., Fin.	112	43°00'N	4°00'E
Lipa, Phil. (lē-pä′)	168	13°55'N	121°10'E
Lipari, Italy (lē′pä-rē)	130	38°29'N	15°00'E
Lipari, i., Italy	130	38°32'N	15°04'E
Lipetsk, Russia (lyē′pĕtsk)	134	52°26'N	39°34'E
Lipetsk, prov., Russia	132	52°18'N	38°30'E
Liping, China (lē-pĭŋ)	160	26°18'N	109°00'E
Lipno, Pol. (lēp′nō)	125	52°50'N	19°12'E
Lippe, r., Ger. (lĭp′ē)	127b	51°36'N	6°45'E
Lippstadt, Ger. (lĭp′shtät)	124	51°39'N	8°20'E
Lipscomb, Al., U.S. (lĭp′skŭm)	68h	33°26'N	86°56'W
Lipu, China (lē-pōō)	165	24°38'N	110°35'E
Lira, Ug.	191	2°15'N	32°54'E
Liri, r., Italy (lē′rē)	130	41°49'N	13°30'E
Lisala, D.R.C. (lē-sä′lä)	185	2°09'N	21°31'E
Lisboa *see* Lisbon, Port.	110	38°42'N	9°05'W
Lisbon (Lisboa), Port.	110	38°42'N	9°05'W
Lisbon, N.D., U.S.	70	46°21'N	97°43'W
Lisbon, Oh., U.S.	66	40°45'N	80°50'W
Lisbon Falls, Me., U.S.	60	43°59'N	70°03'W
Lisburn, N. Ire., U.K. (lĭs′bŭrn)	120	54°35'N	6°05'W
Lisburne, Cape, c., Ak., U.S.	64a	68°20'N	165°40'W
Lishi, China (lē-shr)	164	37°32'N	111°12'E
Lishu, China	164	43°12'N	124°18'E
Lishui, China (lĭ′shwĭ′)	162	31°41'N	119°01'E
Lishui, China	161	28°28'N	120°00'E
Lisianski Island, i., Hi., U.S.	84b	25°30'N	174°00'W
Lisieux, Fr. (lē-zyû′)	126	49°10'N	0°13'E
Lisiy Nos, Russia (lĭ′sĭy-nôs)	142c	60°01'N	30°00'E

ăt; finăl; rāte; senāte; ärm; ásk; sofȧ; fâre; ch-choose; dh-as th in other; bē; ēvent; bĕt; recĕnt; cratĕr; g-gō; gh-guttural g; bĭt; ĭ-short neutral; rīde; K-guttural k as ch in German ich;

PRONOUNCING INDEX

Liski – Loosdrechtsche Plassen 269

PLACE (Pronunciation)	PAGE	LAT.	LONG.
Liski, Russia (lyēs′kė)	133	50°56′N	39°28′E
Lisle, Il., U.S. (līl)	69a	41°48′N	88°04′W
L'Isle-Adam, Fr. (lēl-ädän′)	127b	49°05′N	2°13′E
Lismore, Austl. (lĭz′môr)	175	28°48′S	153°18′E
Litani, r., Leb.	153a	33°28′N	35°42′E
Litchfield, Il., U.S. (lĭch′fēld)	79	39°10′N	89°38′W
Litchfield, Mn., U.S.	71	45°08′N	94°34′W
Litchfield, Oh., U.S.	69d	41°10′N	82°01′W
Lithgow, Austl. (lĭth′gō)	175	33°23′S	149°31′E
Lithinon, Akra, c., Grc.	130a	34°59′N	24°35′E
Lithonia, Ga., U.S. (lĭ-thō′nĭ-ȧ)	68c	33°43′N	84°07′W
Lithuania, nation, Eur. (lĭth-ů-ā′-nĭ-ȧ)	134	55°42′N	23°30′E
Litóchoro, Grc.	131	40°05′N	22°29′E
Litoko, D.R.C.	190	1°13′S	24°47′E
Litoměřice, Czech Rep. (lē′tô-myĕr′zhĭ-tsĕ)	124	50°33′N	14°10′E
Litomyšl, Czech Rep. (lē′tô-mēsh′l)	124	49°52′N	16°14′E
Litoo, Tan.	191	9°45′S	38°24′E
Little, r., Austl.	173a	37°54′S	144°27′E
Little, r., Tn., U.S.	82	36°28′N	89°39′W
Little, r., Tx., U.S.	81	30°48′N	96°50′W
Little Abaco, i., Bah. (ä′bä-kō)	92	26°55′N	77°45′W
Little Abitibi, r., Can.	58	50°15′N	81°30′W
Little America, sci., Ant.	178	78°30′S	161°30′W
Little Andaman, i., India (ăn-dȧ-măn′)	168	10°39′N	93°08′E
Little Bahama Bank, bk., (bȧ-hā′mȧ)	92	26°55′N	78°40′W
Little Belt Mountains, mts., Mt., U.S. (bĕlt)	64	47°00′N	110°50′W
Little Bighorn, r., Mt., U.S. (bĭg-hôrn)	73	45°08′N	107°30′W
Little Bighorn Battlefield National Monument, rec., Mt., U.S. (bĭg-hôrn băt′′l-fēld)	73	45°44′N	107°15′W
Little Bitter Lake, l., Egypt	192b	30°10′N	32°36′E
Little Bitterroot, r., Mt., U.S. (bĭt′ẽr-ōōt)	73	47°45′N	114°45′W
Little Blue, r., Ia., U.S. (blōō)	75f	38°52′N	94°25′W
Little Blue, r., Ne., U.S.	78	40°15′N	98°01′W
Littleborough, Eng., U.K. (lĭt′′l-bŭr-ô)	114a	53°39′N	2°06′W
Little Calumet, r., Il., U.S. (kăl-ů-mĕt′)	69a	41°38′N	87°38′W
Little Cayman, i., Cay. Is. (kā′mȧn)	92	19°40′N	80°05′W
Little Colorado, r., Az., U.S. (kŏl-ô-rä′dō)	64	36°05′N	111°35′W
Little Compton, R.I., U.S. (kŏmp′tŭn)	68b	41°31′N	71°07′W
Little Corn Island, i., Nic.	91	12°19′N	82°50′W
Little Exuma, i., Bah. (ĕk-sōō′mä)	93	23°25′N	75°40′W
Little Falls, Mn., U.S. (fôlz)	71	45°58′N	94°23′W
Little Falls, N.Y., U.S.	67	43°05′N	74°55′W
Littlefield, Tx., U.S. (lĭt′′l-fēld)	78	33°55′N	102°17′W
Little Fork, r., Mn., U.S. (fôrk)	71	48°24′N	93°30′W
Little Goose Dam, dam, Wa., U.S.	72	46°35′N	118°02′W
Little Hans Lollick, i., V.I.U.S. (häns lôl′lĭk)	87c	18°25′N	64°54′W
Little Humboldt, r., Nv., U.S. (hŭm′bōlt)	72	41°10′N	117°40′W
Little Inagua, i., Bah. (ē-nä′gwä)	93	21°30′N	73°00′W
Little Isaac, i., Bah. (ī′zȧk)	92	25°55′N	79°00′W
Little Kanawha, r., W.V., U.S. (kȧ-nô′wȧ)	66	39°05′N	81°30′W
Little Karroo, plat., S. Afr. (kä-rōō)	186	33°50′S	21°02′E
Little Mecatina, r., Can. (mĕ cȧ tĭ nȧ)	53	52°40′N	62°21′W
Little Miami, r., Oh., U.S. (mī-ăm′ĭ)	69f	39°19′N	84°15′W
Little Minch, strt., Scot., U.K.	120	57°35′N	6°45′W
Little Missouri, r., Ar., U.S. (mĭ-sōō′rĭ)	79	34°15′N	93°54′W
Little Missouri, r., U.S.	64	46°00′N	104°00′W
Little Pee Dee, r., S.C., U.S. (pē-dē′)	83	34°35′N	79°21′W
Little Powder, r., Wy., U.S. (pou′dẽr)	73	44°51′N	105°20′W
Little Red, r., Ar., U.S. (rĕd)	79	35°25′N	91°55′W
Little Red, r., Ok., U.S.	79	33°53′N	94°38′W
Little Rock, Ar., U.S. (rôk)	65	34°42′N	92°16′W
Little Sachigo Lake, l., Can. (să′chĭ-gō)	57	54°09′N	92°11′W
Little Salt Lake, l., Ut., U.S.	77	37°55′N	112°53′W
Little San Salvador, i., Bah. (săn săl′vȧ-dôr)	93	24°35′N	75°55′W
Little Satilla, r., Ga., U.S. (sȧ-tĭl′ȧ)	83	31°43′N	82°47′W
Little Sioux, r., Ia., U.S. (sōō)	70	42°22′N	95°47′W
Little Smoky, r., Can. (smōk′ĭ)	55	55°10′N	116°55′W
Little Snake, r., Co., U.S. (snāk)	73	40°40′N	108°21′W
Little Tallapoosa, r., Al., U.S. (tăl-ȧ-pōō′sȧ)	82	32°25′N	85°28′W
Little Tennessee, r., Tn., U.S. (tĕn-ĕ-sē′)	82	35°36′N	84°05′W
Littleton, Co., U.S. (lĭt′′l-tŭn)	78	39°34′N	105°01′W
Littleton, Ma., U.S.	61a	42°32′N	71°29′W
Littleton, N.H., U.S.	67	44°15′N	71°45′W
Little Wabash, r., Il., U.S. (wô′băsh)	66	38°50′N	88°30′W
Little Wood, r., Id., U.S. (wŏŏd)	73	43°00′N	114°08′W
Lityn, Ukr.	133	49°16′N	28°11′E
Liubar, Ukr.	133	49°56′N	27°44′E
Liuhe, China	164	42°10′N	125°38′E
Liuli, Tan.	191	11°05′S	34°38′E
Liupan Shan, mts., China	164	36°20′N	105°30′E
Liuwa Plain, pl., Zam.	190	14°30′S	22°40′E
Liuyang, China (lyōō′yäng′)	165	28°10′N	113°35′E
Liuyuan, China (lĭō-yüän)	162	36°09′N	114°37′E
Liuzhou, China (lĭō-jō)	160	24°25′N	109°30′E
Līvāni, Lat. (lē′vá-nē)	123	56°24′N	26°12′E
Lively, Can.	58	46°26′N	81°09′W
Livengood, Ak., U.S. (lĭv′ĕn-gŏŏd)	63	65°30′N	148°35′W
Live Oak, Fl., U.S. (lĭv′ōk)	82	30°15′N	83°00′W
Livermore, Ca., U.S. (lĭv′ẽr-mōr)	74b	37°41′N	121°46′W
Livermore, Ky., U.S.	66	37°30′N	87°05′W
Liverpool, Austl. (lĭv′ẽr-pōōl)	173b	33°55′S	150°56′E
Liverpool, Can.	51	44°02′N	64°41′W
Liverpool, Eng., U.K.	110	53°25′N	2°52′W
Liverpool, Tx., U.S.	81a	29°18′N	95°17′W
Liverpool Bay, b., Can.	63	69°45′N	130°00′W
Liverpool Range, mts., Austl.	175	31°47′S	151°00′E
Livindo, r., Afr.	185	1°09′N	13°30′E
Livingston, Guat.	90	15°50′N	88°45′W
Livingston, Al., U.S. (lĭv′ĭng-stŭn)	82	32°35′N	88°09′W
Livingston, Il., U.S.	75e	38°58′N	89°51′W
Livingston, Mt., U.S.	64	45°40′N	110°35′W
Livingston, N.J., U.S.	68a	40°47′N	74°20′W
Livingston, Tn., U.S.	82	36°23′N	85°20′W
Livingstone, Zam. (lĭv-ĭng-stōn)	186	17°50′S	25°53′E
Livingstone, Chutes de, wtfl., Afr.	190	4°50′S	14°30′E
Livingstonia, Mwi. (lĭv-ĭng-stō′nĭ-ȧ)	186	10°36′S	34°07′E
Livno, Bos. (lēv′nō)	119	43°50′N	17°03′E
Livny, Russia (lĕv′nė)	137	52°28′N	37°36′E
Livonia, Mi., U.S. (lĭ-vō-nĭ-ȧ)	69b	42°25′N	83°23′W
Livorno, Italy (lė-vôr′nō) (lĕg′hôrn)	110	43°32′N	11°18′E
Livramento, Braz. (lē-vrá-mĕ′n-tô)	102	30°46′S	55°21′W
Lixian, China (lē shyĕn)	165	29°42′N	111°40′E
Lixian, China	162	38°30′N	115°38′E
Liyang, China (lē′yäng′)	165	31°30′N	119°29′E
Lizard Point, c., Eng., U.K. (lĭz′ȧrd)	117	49°55′N	5°09′W
Lizy-sur-Ourcq, Fr. (lēk-sē′sür-ōōrk′)	127b	49°01′N	3°02′E
Ljubljana, Slvn. (lyōō′blyä′na)	110	46°04′N	14°29′E
Ljubuški, Bos. (lyōō′bòsh-kė)	131	43°11′N	17°29′E
Ljungan, r., Swe.	122	62°50′N	13°45′E
Ljungby, Swe. (lyòng′bü)	122	56°49′N	13°56′E
Ljusdal, Swe. (lyōōs′däl)	122	61°50′N	16°11′E
Ljusnan, r., Swe.	116	61°55′N	15°33′E
Llandudno, Wales, U.K. (lăn-düd′nō)	120	53°20′N	3°46′W
Llanelli, Wales, U.K. (lá-nĕl′ĭ)	117	51°44′N	4°09′W
Llanes, Spain (lyä′nås)	118	43°25′N	4°41′W
Llano, Tx., U.S. (lä′nō) (lyä′nō)	80	30°45′N	98°41′W
Llano, r., Tx., U.S.	80	30°38′N	99°04′W
Llanos, reg., S.A. (lyá′nōs)	100	4°00′N	71°15′W
Lleida, Spain	118	41°38′N	0°37′E
Llera, Mex. (lyā′rä)	88	23°16′N	99°03′W
Llerena, Spain (lyā-rā′nä)	128	38°14′N	6°02′W
Lliria, Spain	129	39°35′N	0°34′W
Llobregat, r., Spain (lyô-brē-gät′)	129	41°55′N	1°55′E
Lloyd Lake, l., Can. (loid)	62e	50°52′N	114°13′W
Lloydminster, Can.	50	53°17′N	110°00′W
Llucena, Spain	129	40°08′N	0°18′W
Llucmajor, Spain	129	39°28′N	2°53′E
Llullaillaco, Volcán, vol., S.A. (lyōō-lyī-lyá′kō)	102	24°50′S	68°30′W
Loange, r., Afr. (lō-äŋ′gā)	186	5°00′S	20°15′E
Lobamba, Swaz.	186	26°27′S	31°12′E
Lobatse, Bots. (lō-bä′tsĕ)	186	25°13′S	25°35′E
Lobería, Arg. (lô-bĕ′rē′ä)	102	38°13′S	58°48′W
Lobito, Ang. (lô-bē′tō)	186	12°30′S	13°34′E
Lobnya, Russia (lôb′nyá)	142b	56°01′N	37°29′E
Lobo, Phil.	169a	13°39′N	121°14′E
Lobos, Arg. (lō′bôs)	99c	35°10′S	59°08′W
Lobos, Cayo, i., Bah. (lō′bôs)	92	22°25′N	77°40′W
Lobos, Isla de, i., Mex. (ē′s-lä-dĕ-lô′bōs)	89	21°24′N	97°11′W
Lobos de Tierra, i., Peru (lô′bō-dĕ-tyĕ′r-rä)	100	6°29′S	80°55′W
Lobva, Russia (lôb′vá)	142a	59°12′N	60°28′E
Lobva, r., Russia	142a	59°14′N	60°17′E
Locarno, Switz. (lô-kär′nō)	124	46°10′N	8°43′E
Loches, Fr. (lôsh)	126	47°08′N	0°56′E
Loch Raven Reservoir, res., Md., U.S.	68e	39°28′N	76°38′W
Lockeport, Can.	60	43°42′N	65°07′W
Lockhart, S.C., U.S. (lŏk′härt)	83	34°47′N	81°30′W
Lockhart, Tx., U.S.	81	29°54′N	97°40′W
Lock Haven, Pa., U.S. (lŏk′hā-vĕn)	67	41°05′N	77°30′W
Lockland, Oh., U.S. (lŏk′lănd)	69f	39°14′N	84°27′W
Lockport, Il., U.S.	69a	41°35′N	88°04′W
Lockport, N.Y., U.S.	67	43°11′N	78°43′W
Loc Ninh, Viet. (lōk′nĭng′)	168	12°00′N	106°30′E
Lod, Isr. (lôd)	153a	31°57′N	34°55′E
Lodève, Fr. (lô-dĕv′)	126	43°43′N	3°18′E
Lodeynoye Pole, Russia (lô-dĕy-nô′yĕ)	136	60°43′N	33°24′E
Lodge Creek, r., N.A. (lŏj)	73	49°20′N	110°20′W
Lodge Creek, r., Mt., U.S.	73	48°51′N	109°30′W
Lodgepole Creek, r., Wy., U.S. (lŏj′pōl)	70	41°22′N	104°48′W
Lodhran, Pak.	158	29°40′N	71°39′E
Lodi, Italy (lô′dē)	130	45°18′N	9°30′E
Lodi, Ca., U.S. (lō′dī)	76	38°07′N	121°17′W
Lodi, Oh., U.S. (lō′dī)	69d	41°02′N	82°01′W
Lodosa, Spain (lô-dô′sä)	128	42°27′N	2°04′W
Lodwar, Kenya	191	3°07′N	35°36′E
Łódź, Pol.	110	51°46′N	19°30′E
Loeches, Spain (lô-āch′ĕs)	129a	40°22′N	3°25′W
Loffa, r., Afr.	188	7°10′N	10°35′W
Lofoten, is., Nor. (lô′fō-tĕn)	112	68°26′N	13°42′E
Logan, Oh., U.S. (lō′gȧn)	66	39°35′N	82°25′W
Logan, Ut., U.S.	64	41°46′N	111°51′W
Logan, W.V., U.S.	66	37°50′N	82°00′W
Logan, Mount, mtn., Can.	52	60°54′N	140°33′W
Logansport, In., U.S. (lō′gănz-pōrt)	65	40°45′N	86°25′W
Logone, r., Afr. (lō-gō′nä) (lô-gôn′)	185	10°20′N	15°30′E
Logroño, Spain (lô-grō′nyō)	118	42°28′N	2°25′W
Logrosán, Spain (lô-grô-sän′)	128	39°22′N	5°29′W
Løgstør, Den. (lügh-stûr′)	122	56°56′N	9°15′E
Loir, r., Fr. (lwár)	126	47°40′N	0°07′E
Loire, r., Fr.	112	47°30′N	2°00′E
Loja, Ec. (lō′hä)	100	3°49′S	79°13′W
Loja, Spain (lô′-kä)	128	37°10′N	4°11′W
Loka, D.R.C.	190	0°20′N	17°57′E
Lokala Drift, Bots. (lō′kä-lȧ drĭft)	192c	24°00′S	26°38′E
Lokandu, D.R.C.	191	2°31′S	25°47′E
Lokhvytsia, Ukr.	137	50°21′N	33°16′E
Lokichar, Kenya	191	2°23′N	35°39′E
Lokitaung, Kenya	191	4°16′N	35°45′E
Lokofa-Bokolongo, D.R.C.	190	0°12′N	19°22′E
Lokoja, Nig. (lô-kō′yä)	184	7°47′N	6°45′E
Lokolama, D.R.C.	190	2°34′S	19°53′E
Lokosso, Burkina	188	10°19′N	3°40′W
Lol, r., Sudan (lōl)	185	9°06′N	28°09′E
Loliondo, Tan.	191	2°03′S	35°37′E
Lolland, i., Den. (lôl′än′)	122	54°41′N	11°00′E
Lolo, Mt., U.S.	73	46°45′N	114°05′W
Lom, Blg. (lōm)	119	43°48′N	23°15′E
Loma Linda, Ca., U.S. (lō′mȧ lĭn′dȧ)	75a	34°04′N	117°16′W
Lomami, r., D.R.C.	186	0°50′S	24°40′E
Lomas de Zamora, Arg. (lō′mäs dā zä-mō′rä)	99c	34°46′S	58°24′W
Lombard, Il., U.S. (lŏm-bärd)	69a	41°53′N	88°01′W
Lombardia, hist. reg., Italy (lôm-bär-dē′ä)	130	45°20′N	9°30′E
Lomblen, Pulau, i., Indon. (lŏm-blĕn′)	169	8°08′S	123°45′E
Lombok, i., Indon. (lŏm-bŏk′)	168	9°15′S	116°15′E
Lomé, Togo	184	6°08′N	1°13′E
Lomela, D.R.C. (lô-mā′lä)	186	2°19′S	23°33′E
Lomela, r., D.R.C.	186	0°35′S	21°20′E
Lometa, Tx., U.S. (lô-mē′tȧ)	80	31°10′N	98°25′W
Lomié, Cam. (lô-mē-ā′)	189	3°10′N	13°37′E
Lomita, Ca., U.S. (lō-mē′tȧ)	75a	33°48′N	118°20′W
Lommel, Bel.	115a	51°14′N	5°21′E
Lommond, Loch, l., Scot., U.K. (lŏk lō′mŭnd)	120	56°15′N	4°40′W
Lomonosov, Russia (lô-mô′nô-sof)	142c	59°54′N	29°47′E
Lompoc, Ca., U.S. (lŏm-pōk′)	76	34°39′N	120°30′W
Łomża, Pol. (lôm′zhá)	125	53°11′N	22°04′E
Lonaconing, Md., U.S. (lō-nȧ-kō′nĭng)	67	39°35′N	78°55′W
London, Can. (lŭn′dŭn)	51	43°00′N	81°20′W
London, Eng., U.K.	110	51°30′N	0°07′W
London, Ky., U.S.	82	37°07′N	84°06′W
London, Oh., U.S.	66	39°50′N	83°30′W
Londonderry, Can. (lŭn′dŭn-dĕr-ĭ)	60	45°29′N	63°36′W
Londonderry, N. Ire., U.K.	116	55°00′N	7°19′W
Londonderry, Cape, c., Austl.	174	13°30′S	127°00′E
Londrina, Braz. (lôn-drē′nä)	101	21°53′S	51°17′W
Lonely, i., Can. (lōn′lĭ)	53	45°35′N	81°30′W
Lone Pine, Ca., U.S.	76	36°36′N	118°03′W
Lone Star, Nic.	91	13°58′N	84°25′W
Long, i., Bah.	87	23°25′N	75°10′W
Long, i., Can.	60	44°21′S	66°25′W
Long, l., N.D., U.S.	70	46°47′N	100°14′W
Long, l., Wa., U.S.	74a	47°29′N	122°36′W
Longa, r., Ang. (lôŋ′gȧ)	186	10°20′S	15°15′E
Long Bay, b., S.C., U.S.	83	33°30′N	78°54′W
Long Beach, Ca., U.S. (lông bēch)	64	33°46′N	118°12′W
Long Beach, N.Y., U.S.	68a	40°35′N	73°38′W
Long Branch, N.J., U.S. (lông brȧnch)	68a	40°18′N	73°59′W
Longdon, N.D., U.S. (lông′-dŭn)	70	48°45′N	98°23′W
Long Eaton, Eng., U.K. (ē′tŭn)	114a	52°54′N	1°16′W
Longford, Ire. (lông′fẽrd)	120	53°43′N	7°40′W
Longgu, China (lôŋ-gōō)	162	34°52′N	116°48′E
Longhorn, Tx., U.S. (lông-hôrn)	75d	29°33′N	98°23′W
Longido, Tan.	191	2°44′S	36°41′E
Long Island, i., Pap. N. Gui.	169	5°10′S	147°30′E
Long Island, i., Ak., U.S.	54	54°54′N	132°45′W
Long Island, i., N.Y., U.S. (lông)	65	40°50′N	72°50′W
Long Island Sound, strt., U.S. (lông ī′lănd)	65	41°05′N	72°45′W
Longjumeau, Fr. (lôn-zhü-mō′)	127b	48°42′N	2°17′E
Longkou, China (lôŋ-kō)	162	37°39′N	120°21′E
Longlac, Can. (lông′lák)	51	49°41′N	86°28′W
Longlake, S.D., U.S. (lông-lāk)	70	45°52′N	99°06′W
Long Lake, l., Can.	58	49°10′N	86°45′W
Longmont, Co., U.S. (lông′mŏnt)	78	40°11′N	105°07′W
Longnor, Eng., U.K. (lông-nôr)	114a	53°11′N	1°52′W
Long Pine, Ne., U.S. (lông pīn)	70	42°31′N	99°42′W
Long Point, c., Can.	57	53°02′N	98°40′W
Long Point, c., Can.	61	48°48′N	58°46′W
Long Point, c., Can.	59	42°35′N	80°05′W
Long Point Bay, b., Can.	59	42°40′N	80°10′W
Long Range Mountains, mts., Can.	53a	48°00′N	58°30′W
Longreach, Austl. (lông′rēch)	175	23°32′S	144°17′E
Long Reach, r., Can.	60	45°26′N	66°05′W
Long Reef, c., Austl.	173b	33°45′S	151°22′E
Longridge, Eng., U.K. (lông′rĭj)	114a	53°51′N	2°37′W
Longs Peak, mtn., Co., U.S. (lôngz)	64	40°17′N	105°37′W
Longtansi, China (lôŋ-tä-sz)	162	32°12′N	115°53′E
Longton, Eng., U.K. (lông′tŭn)	114a	52°59′N	2°08′W
Longueuil, Can. (lôn-gû′y′)	59	45°32′N	73°30′W
Longview, Tx., U.S.	81	32°29′N	94°44′W
Longview, Wa., U.S. (lông-vū)	72	46°06′N	123°02′W
Longville, La., U.S. (lông′vĭl)	81	30°36′N	93°14′W
Longwy, Fr. (lôn-wē′)	127	49°32′N	6°14′E
Longxi, China (lôŋ-shyē)	160	35°00′N	104°40′E
Long Xuyen, Viet. (loung′ sōō′yĕn)	168	10°31′N	105°28′E
Longzhou, China (lôŋ-jō)	160	22°20′N	107°02′E
Lonoke, Ar., U.S. (lō′nōk)	79	34°48′N	91°52′W
Lons-le-Saunier, Fr. (lôn-lĕ-sō-nyā′)	127	46°40′N	5°33′E
Lontue, r., Chile (lŏn-tòĕ′)	99b	35°20′S	70°45′W
Looc, Phil. (lō-ōk′)	169a	12°16′N	121°59′E
Loogootee, In., U.S.	66	38°40′N	86°55′W
Lookout, Cape, c., N.C., U.S. (lòk′out)	83	34°34′N	76°38′W
Lookout Point Lake, res., Or., U.S.	72	43°51′N	122°38′W
Loolmalasin, mtn., Tan.	191	3°03′S	35°46′E
Looma, Can. (ō′mä)	62g	53°22′N	113°15′W
Loop Head, c., Ire. (lōōp)	120	52°32′N	9°59′W
Loosahatchie, r., Tn., U.S. (lōz-ȧ-hă′chē)	82	35°20′N	89°45′W
Loosdrechtsche Plassen, l., Neth.	115a	52°11′N	5°09′E

PLACE (Pronunciation)	PAGE	LAT.	LONG.
Lopatka, Mys, c., Russia (lô-pät'kȧ)	153	51°00'N	156°52'E
Lopez, Cap, c., Gabon	190	0°37'N	8°43'E
Lopez Bay, b., Phil. (lō'päz)	169a	14°04'N	122°00'E
Lopez I, Wa., U.S.	74a	48°25'N	122°53'W
Lopori, r., D.R.C. (lō-pō'rê)	185	1°35'N	20°43'E
Lora, Spain (lō'rä)	128	37°40'N	5°31'W
Lorain, Oh., U.S. (lô-rān')	69d	41°28'N	82°10'W
Loralai, Pak. (lō-rŭ-lī')	155	30°31'N	68°35'E
Lorca, Spain (lôr'kä)	118	37°39'N	1°40'W
Lord Howe, i., Austl. (lôrd hou)	174	31°44'S	157°56'W
Lordsburg, N.M., U.S. (lôrdz'bûrg)	77	32°20'N	108°45'W
Lorena, Braz. (lô-rā'nȧ)	99a	22°45'S	45°07'W
Loreto, Braz. (lô-rā'tō)	101	7°09'S	45°10'W
Loretteville, Can. (lô-rĕt-vēl')	62b	46°51'N	71°21'W
Lorica, Col. (lô-rē'kä)	100	9°14'N	75°54'W
Lorient, Fr. (lō-rē'äN')	117	47°45'N	3°22'W
Lorn, Firth of, b., Scot., U.K. (fûrth ŏv lôrn')	120	56°10'N	6°09'W
Lörrach, Ger. (lûr'äK)	124	47°36'N	7°38'E
Lorraine, hist. reg., Fr.	127	49°00'N	6°00'E
Los Alamitos, Ca., U.S. (lōs ăl-ȧ-mē'tōs)	75a	33°48'N	118°04'W
Los Alamos, N.M., U.S. (ăl-ȧ-mōs')	77	35°53'N	106°20'W
Los Altos, Ca., U.S. (ăl-tôs')	74b	37°23'N	122°06'W
Los Andes, Chile (än'dĕs)	99b	32°44'S	70°36'W
Los Angeles, Chile (äŋ'hâ-lās)	102	37°27'S	72°15'W
Los Angeles, Ca., U.S.	64	34°03'N	118°14'W
Los Angeles, Ca., U.S.	76	34°03'N	118°14'W
Los Angeles, r., Ca., U.S.	75a	33°50'N	118°13'W
Los Angeles Aqueduct, Ca., U.S.	76	35°12'N	118°02'W
Los Bronces, Chile (lōs brō'n-sĕs)	99b	33°09'S	70°18'W
Loscha, r., Id., U.S. (lŏs'chă)	72	46°20'N	115°11'W
Los Estados, Isla de, i., Arg. (ē's-lä dĕ lōs ĕs-dôs)	102	54°45'S	64°25'W
Los Gatos, Ca., U.S. (gă'tôs)	76	37°13'N	121°59'W
Los Herreras, Mex. (ĕr-rā-räs)	80	25°55'N	99°23'W
Los Indios, Cayos de, is., Cuba (kä'yōs dĕ lōs ē'n-dyō's)	92	21°50'N	83°10'W
Los Llanos, Dom. Rep. (lōs ê-lä'nōs)	93	18°35'N	69°30'W
Lošinj, i., Serb.	130	44°35'N	14°34'E
Losino Petrovskiy, Russia	142b	55°52'N	38°12'E
Los Nietos, Ca., U.S. (nyā'tôs)	75a	33°57'N	118°05'W
Los Palacios, Cuba	92	22°35'N	83°15'W
Los Pinos, r., Co., U.S. (pē'nôs)	77	36°58'N	107°35'W
Los Reyes, Mex.	86	19°35'N	102°29'W
Los Reyes, Mex.	89a	19°21'N	98°58'W
Los Santos, Pan. (săn'tōs)	91	7°57'N	80°24'W
Los Santos de Maimona, Spain (săn'tōs)	128	38°38'N	6°30'W
Lost, r., Or., U.S.	72	42°07'N	121°30'W
Los Teques, Ven. (tĕ'kĕs)	100	10°22'N	67°04'W
Lost River Range, mts., Id., U.S. (rĭ'vẽr)	73	44°23'N	113°48'W
Los Vilos, Chile (vē'lōs)	102	31°56'S	71°29'W
Lot, r., Fr. (lôt)	117	44°30'N	1°30'E
Lota, Chile (lō'tä)	102	37°11'S	73°14'W
Lothian, Md., U.S. (lōth'ĭăn)	68e	38°50'N	76°38'W
Lotikipi Plain, pl., Afr.	191	4°25'N	34°55'E
Lötschberg Tunnel, trans., Switz.	124	46°26'N	7°54'E
Louangphrabang, Laos (lōō-ang'prä-bäng')	168	19°47'N	102°15'E
Loudon, Tn., U.S. (lou'dŭn)	82	35°43'N	84°20'W
Loudonville, Oh., U.S. (lou'dŭn-vĭl)	66	40°40'N	82°15'W
Loudun, Fr.	126	47°03'N	0°00'
Loughborough, Eng., U.K. (lŭf'bŭr-ō)	114a	52°46'N	1°12'W
Louisa, Ky., U.S. (lōō'ēz-ȧ)	66	38°05'N	82°40'W
Louisade Archipelago, is., Pap. N. Gui.	175	10°44'S	153°58'E
Louisberg, N.C., U.S. (lōō'ĭs-bûrg)	83	36°05'N	79°19'W
Louisburg, Can. (lōō'ĭs-bourg)	61	45°55'N	59°58'W
Louiseville, Can.	59	46°17'N	72°58'W
Louisiana, Mo., U.S. (lōō-ē-zē-ăn'ȧ)	79	39°24'N	91°03'W
Louisiana, state, U.S.	65	30°50'N	92°50'W
Louis Trichardt, S. Afr. (lōō'ĭs trĭchårt)	186	22°52'S	29°53'E
Louisville, Co., U.S. (lōō'ĭs-vĭl) (lōō'ē-vĭl)	78	39°58'N	105°08'W
Louisville, Ga., U.S.	83	33°00'N	82°25'W
Louisville, Ky., U.S.	65	38°15'N	85°45'W
Louisville, Ms., U.S.	82	33°07'N	89°02'W
Louis XIV, Pointe, c., Can.	53	54°35'N	79°51'W
Louny, Czech Rep. (lō'nê)	124	50°20'N	13°47'E
Loup, r., Ne., U.S. (lōōp)	70	41°17'N	97°58'W
Loup City, Ne., U.S.	70	41°15'N	98°59'W
Lourdes, Fr. (lōōrd)	117	43°06'N	0°03'W
Lourenço Marques see Maputo, Moz.	186	26°50'S	32°30'E
Loures, Port. (lō'rĕzh)	129b	38°49'N	9°10'W
Lousa, Port. (lō'zȧ)	128	40°05'N	8°12'W
Louth, Eng., U.K. (louth)	120	53°27'N	0°02'W
Louvain see Leuven, Bel.	121	50°53'N	4°42'E
Louviers, Fr. (lōō-vyā')	126	49°13'N	1°11'E
Lovech, Blg. (lō'vĕts)	131	43°10'N	24°40'E
Loveland, Co., U.S. (lŭv'lănd)	78	40°24'N	105°04'W
Loveland, Oh., U.S.	69f	39°16'N	84°15'W
Lovell, Wy., U.S. (lŭv'ĕl)	73	44°50'N	108°23'W
Lovelock, Nv., U.S. (lŭv'lŏk)	76	40°10'N	118°37'W
Lovick, Al., U.S. (lŭ'vĭk)	68h	33°34'N	86°38'W
Loviisa, Fin. (lō'vē-sȧ)	123	60°28'N	26°10'E
Low, Cape, c., Can. (lō)	53	62°58'N	86°50'W
Lowa, r., D.R.C. (lō'wȧ)	186	1°30'S	27°18'E
Lowell, In., U.S.	69a	41°17'N	87°26'W
Lowell, Ma., U.S.	65	42°38'N	71°18'W
Lowell, Mi., U.S.	66	42°55'N	85°20'W
Löwenberg, Ger. (lû'vĕn-bĕrgh)	115b	52°53'N	13°09'E
Lower Brule Indian Reservation, I.R., S.D., U.S. (brü'lā)	70	44°15'N	100°21'W

PLACE (Pronunciation)	PAGE	LAT.	LONG.
Lower California see Baja California, pen., Mex.	49	28°00'N	113°30'W
Lower Granite Dam, dam, Wa., U.S.	72	46°40'N	117°26'W
Lower Hutt, N.Z. (hŭt)	175a	41°10'S	174°55'E
Lower Klamath Lake, l., Ca., U.S. (klăm'ăth)	72	41°55'N	121°50'W
Lower Lake, l., Ca., U.S.	72	41°21'N	119°53'W
Lower Marlboro, Md., U.S. (lō'ẽr märl'bōrō)	68e	38°40'N	76°42'W
Lower Monumental Dam, dam, Wa., U.S.	72	46°34'N	118°32'W
Lower Otay Lake, res., Ca., U.S. (ō'tā)	76a	32°37'N	116°46'W
Lower Red Lake, l., Mn., U.S.	71	47°58'N	94°31'W
Lower Saxony see Niedersachsen, state, Ger.	115c	53°30'N	9°30'E
Lowestoft, Eng., U.K. (lō'stŏf)	121	52°31'N	1°45'E
Łowicz, Pol. (lô'vĭch)	125	52°06'N	19°57'E
Lowville, N.Y., U.S. (lou'vĭl)	67	43°45'N	75°30'W
Loxicha, Mex.	89	16°03'N	96°46'W
Loxton, Austl. (lôks'tŭn)	176	34°25'S	140°38'E
Loyauté, Îles, is., N. Cal.	175	21°00'S	167°00'E
Loznica, Serb. (lôz'nê-tsȧ)	119	44°31'N	19°16'E
Lozova, Ukr.	137	48°53'N	36°23'E
Luama, r., D.R.C. (lōō'ä-mȧ)	186	4°17'S	27°45'E
Lu'an, China (lōō-än)	165	31°45'N	116°29'E
Luan, r., China	161	41°25'N	117°15'E
Luanda, Ang. (lōō-än'dä)	186	8°48'S	13°14'E
Luanguinga, r., Afr. (lōō-ä-gĭŋ'gȧ)	186	14°00'S	20°45'E
Luanshya, Zam.	191	13°08'S	28°24'E
Luanxian, China (luän shyĕn)	162	39°47'N	118°40'E
Luao, Ang.	190	10°42'S	22°12'E
Luarca, Spain (lwä'kä)	118	43°33'N	6°30'W
Lubaczów, Pol. (lōō-bä'chôf)	125	50°08'N	23°10'E
Lubán, Pol. (lōō'bän')	124	51°08'N	15°17'E
Lubānas Ezers, l., Lat. (lōō-bä'nȧs ă'zĕrs)	123	56°48'N	26°30'E
Lubang, Phil. (lōō-bäng')	169a	13°49'N	120°07'E
Lubang Islands, is., Phil.	168	13°47'N	119°56'E
Lubango, Ang.	186	14°55'S	13°30'E
Lubartów, Pol. (lōō-bär'tŏf)	125	51°27'N	22°37'E
Lubawa, Pol. (lōō-bä'vä)	125	53°31'N	19°47'E
Lübben, Ger. (lüb'ĕn)	124	51°56'N	13°53'E
Lubbock, Tx., U.S.	64	33°35'N	101°50'W
Lubec, Me., U.S. (lū'bĕk)	60	44°49'N	67°01'W
Lübeck, Ger. (lü'bĕk)	110	53°53'N	10°42'E
Lübecker Bucht, b., Ger. (lü'bĕ-kĕr bōōkt)	116	54°10'N	11°20'E
Lubilash, r., D.R.C. (lōō-bê-lȧsh')	186	7°35'S	23°55'E
Lubin, Pol. (lyō'bĭn)	124	51°24'N	16°14'E
Lublin, Pol. (lyō'blēn')	110	51°14'N	22°33'E
Lubny, Ukr. (lōōb'nê)	137	50°01'N	33°02'E
Lubuagan, Phil. (lō-bwä-gä'n)	169a	17°24'N	121°11'E
Lubudi, D.R.C.	191	9°57'S	25°58'E
Lubudi, r., D.R.C. (lō-bō'dê)	186	10°00'S	24°30'E
Lubumbashi, D.R.C.	186	11°40'S	27°28'E
Lucano, Ang.	190	11°16'S	21°38'E
Lucca, Italy (lōōk'kä)	118	43°51'N	10°29'E
Lucea, Jam.	92	18°25'N	78°10'W
Luce Bay, b., Scot., U.K. (lūs)	120	54°45'N	4°45'W
Lucena, Phil. (lōō-sā'nä)	169a	13°55'N	121°36'E
Lucena, Spain (lōō-thā'nä)	118	37°25'N	4°28'W
Lučenec, Slvk. (lōō'châ-nyĕts)	117	48°19'N	19°41'E
Lucera, Italy (lōō-châ'rä)	130	41°31'N	15°22'E
Luchi, China	165	28°18'N	110°10'E
Lucin, Ut., U.S. (lû-sēn')	73	41°23'N	113°59'W
Lucipara, Kepulauan, is., Indon.	169	5°45'S	128°15'E
Luckenwalde, Ger.	124	52°05'N	13°10'E
Lucknow, India (lŭk'nou)	155	26°54'N	80°58'E
Lucky Peak Lake, res., Id., U.S.	72	43°33'N	116°00'W
Luçon, Fr. (lü-sôN')	126	46°27'N	1°12'W
Lucrecia, Cabo, c., Cuba	93	21°05'N	75°30'W
Luda Kamchiya, r., Blg.	131	42°46'N	27°13'E
Lüdenscheid, Ger. (lü'dĕn-shīt)	127c	51°13'N	7°38'E
Lüderitz, Nmb. (lü'dĕr-ĭts) (lü'dĕ-rĭts)	186	26°35'S	15°15'E
Lüderitz Bucht, b., Nmb.	186	26°35'S	14°30'E
Ludhiāna, India	155	31°00'N	75°52'E
Lüdinghausen, Ger.	127c	51°46'N	7°27'E
Ludington, Mi., U.S. (lŭd'ĭng-tŭn)	66	44°00'N	86°25'W
Ludlow, Eng., U.K. (lŭd'lō)	114a	52°22'N	2°43'W
Ludlow, Ky., U.S.	69f	39°05'N	84°33'W
Ludvika, Swe. (loodh-vē'kȧ)	122	60°10'N	15°09'E
Ludwigsburg, Ger.	124	48°53'N	9°14'E
Ludwigsfelde, Ger.	115b	52°18'N	13°16'E
Ludwigshafen, Ger.	124	49°29'N	8°26'E
Ludwigslust, Ger.	124	53°18'N	11°31'E
Ludza, Lat. (lōōd'zȧ)	123	56°33'N	27°45'E
Luebo, D.R.C. (lōō-ā'bō)	186	5°15'S	21°22'E
Luena, Ang.	186	11°45'S	19°55'E
Luena, D.R.C.	191	9°27'S	25°47'E
Lufira, r., D.R.C. (lōō-fē'rä)	186	9°32'S	27°15'E
Lufkin, Tx., U.S. (lŭf'kĭn)	81	31°21'N	94°43'W
Luga, Russia (lōō'gȧ)	136	58°43'N	29°52'E
Luga, r., Russia	132	59°00'N	29°25'E
Lugano, Switz. (lōō-gä'nō)	124	46°01'N	8°52'E
Lugenda, r., Moz.	187	12°05'S	38°15'E
Lugo, Italy (lōō'gô)	130	44°28'N	11°57'E
Lugo, Spain (lōō'gō)	118	43°01'N	7°32'W
Lugoj, Rom.	119	45°51'N	21°56'E
Luhans'k, Ukr.	134	48°34'N	39°18'E
Luhans'k, prov., Ukr.	133	49°30'N	38°35'E
Luhe, China (lōō-hŭ)	162	32°22'N	118°50'E
Luiana, Ang.	190	17°23'S	23°03'E
Luilaka, r., D.R.C. (lōō-ê-lä'kȧ)	186	2°18'S	21°15'E
Luis Moya, Mex. (lōōē's-mô-yä)	88	22°26'N	102°14'W
Luján, Arg. (lōō'hän')	99c	34°36'S	59°07'W
Luján, r., Arg.	99c	34°33'S	58°59'W

PLACE (Pronunciation)	PAGE	LAT.	LONG.
Lujia, China (lōō-jyä)	162	31°17'N	120°54'W
Lukanga Swamp, sw., Zam. (lōō-käŋ'gȧ)	186	14°30'S	27°25'E
Lukenie, r., D.R.C. (lōō-kā'nyȧ)	186	3°10'S	19°05'E
Lukolela, D.R.C.	186	1°03'S	17°01'E
Lukovit, Blg. (lōō'kō-vêt')	131	43°13'N	24°07'E
Łuków, Pol. (wō'kôf)	125	51°57'N	22°25'E
Lukuga, r., D.R.C. (lōō-kōō'gȧ)	186	5°50'S	27°35'E
Lüleburgaz, Tur. (lü'lĕ-bōr-gäs')	131	41°25'N	27°23'E
Luling, Tx., U.S. (lū'lĭng)	81	29°41'N	97°38'W
Lulong, China (lōō-lôŋ)	161	39°54'N	118°53'E
Lulonga, r., D.R.C.	190	1°00'N	18°37'E
Luluabourg see Kananga, D.R.C.	186	6°14'S	22°17'E
Lulu Island, i., Can.	74d	49°09'N	123°05'W
Lulu Island, i., Ak., U.S.	54	55°28'N	133°30'W
Lumajangdong Co, l., China	158	34°00'N	81°47'E
Lumber, r., N.C., U.S. (lŭm'bẽr)	83	34°45'N	79°10'W
Lumberton, Ms., U.S. (lŭm'bẽr-tŭn)	82	31°00'N	89°25'W
Lumberton, N.C., U.S.	83	34°47'N	79°00'W
Luminárias, Braz. (lōō-mē-nä'ryäs)	99a	21°32'S	44°53'W
Lummi, i., Wa., U.S.	74d	48°42'N	122°43'W
Lummi Bay, b., Wa., U.S. (lŭm'ĭ)	74d	48°47'N	122°44'W
Lummi Island, Wa., U.S.	74d	48°44'N	122°42'W
Lumwana, Zam.	191	11°50'S	25°10'E
Lün, Mong.	160	47°58'N	104°52'E
Luna, Phil. (lōō'nä)	169a	16°51'N	120°22'E
Lund, Swe. (lŭnd)	116	55°42'N	13°10'E
Lundy, i., Eng., U.K. (lŭn'dê)	120	51°12'N	4°50'W
Lüneburg, Ger. (lü'nĕ-bŏrgh)	124	53°16'N	10°25'E
Lunel, Fr. (lü-nĕl')	126	43°41'N	4°07'E
Lünen, Ger. (lü'nĕn)	127c	51°36'N	7°30'E
Lunenburg, Can. (lōō'nĕn-bûrg)	51	44°23'N	64°19'W
Lunenburg, Ma., U.S.	61a	42°36'N	71°44'W
Lunéville, Fr. (lü-nȧ-vel')	127	48°37'N	6°29'E
Lunga, Ang.	190	14°42'S	18°32'E
Lungué-Bungo, r., Afr.	186	13°00'S	20°30'E
Lunsar, S.L.	188	8°41'N	12°32'W
Luodian, China (lwô-dĭĕn)	162	31°25'N	121°20'E
Luoding, China (lwô-dĭŋ)	165	23°42'N	111°35'E
Luohe, China (lwô-hŭ)	162	33°35'N	114°02'E
Luoyang, China (lwô-yäŋ)	161	34°45'N	112°32'E
Luozhen, China (lwô-jŭn)	162	37°45'N	118°29'E
Luque, Para. (loo'kä)	102	25°18'S	57°17'W
Luray, Va., U.S. (lū-rā')	67	38°40'N	78°25'W
Lurgan, N. Ire., U.K. (lûr'găn)	116	54°27'N	6°28'W
Lúrio, Moz. (lōō'rē-ō)	187	13°17'S	40°29'E
Lúrio, Moz.	187	14°00'S	38°45'E
Lusaka, D.R.C.	191	7°10'S	29°27'E
Lusaka, Zam. (lō-sä'kȧ)	186	15°25'S	28°17'E
Lusambo, D.R.C. (lōō-säm'bō)	186	4°58'S	23°27'E
Lusanga, D.R.C.	186	5°13'S	18°43'E
Lusangi, D.R.C.	186	4°37'S	27°08'E
Lushan, China	164	33°45'N	113°00'E
Lushiko, r., Afr.	190	6°35'S	19°45'E
Lushoto, Tan. (lōō-shō'tō)	187	4°47'S	38°17'E
Lüshun, China (lü-shŭn)	161	38°49'N	121°15'E
Lusikisiki, S. Afr. (lōō-sê-kê-sē'kê)	187c	31°22'S	29°37'E
Lusk, Wy., U.S. (lŭsk)	70	42°46'N	104°27'W
Lūt, Dasht-e, des., Iran (dă'sht-ē-lōōt)	154	31°47'N	58°38'E
Lutcher, La., U.S. (lŭch'ẽr)	81	30°03'N	90°43'W
Luton, Eng., U.K. (lū'tŭn)	120	51°55'N	0°28'W
Luts'k, Ukr.	137	50°45'N	25°20'E
Luuq, Som.	192a	3°38'N	42°35'E
Luverne, Al., U.S. (lŭ-vûn')	82	31°42'N	86°15'W
Luverne, Mn., U.S.	70	43°40'N	96°13'W
Luwingu, Zam.	191	10°15'S	29°55'E
Luxapallila Creek, r., U.S. (lŭk-sȧ-pôl'ĭ-lȧ)	82	33°36'N	88°08'W
Luxembourg, Lux.	110	49°38'N	6°30'E
Luxembourg, nation, Eur.	110	49°30'N	6°22'E
Luxeuil-les-Baines, Fr.	127	47°49'N	6°19'E
Luxomni, Ga., U.S. (lŭx'ôm-nī)	68c	33°54'N	84°07'W
Luxor see Al Uqşur, Egypt	185	25°38'N	32°59'E
Luya Shan, mtn., China	164	38°50'N	111°40'E
Luyi, China (lōō-yē)	162	33°52'N	115°32'E
Luzern, Switz. (lō-tsĕrn)	117	47°03'N	8°18'E
Luzhou, China (lōō-jō)	160	28°58'N	105°25'E
Luziânia, Braz. (lōō-zyä'nēä)	101	16°17'S	47°44'W
Luzon, i., Phil. (lōō-zōn')	168	17°10'N	119°45'E
Luzon Strait, strt., Asia	165	20°40'N	121°00'E
L'viv, Ukr.	134	49°50'N	24°00'E
L'vov see L'viv, Ukr.	134	49°50'N	24°00'E
Lyalta, Can.	62e	51°07'N	113°36'W
Lyalya, r., Russia (lyá'lyá)	142a	58°58'N	60°17'E
Lyaskovets, Blg.	131	43°07'N	25°41'E
Lydenburg, S. Afr. (lī'dĕn-bûrg)	186	25°06'S	30°21'E
Lyell, Mount, mtn., Ca., U.S. (lī'ĕl)	76	37°44'N	119°22'W
Lyepye', Bela. (lyĕ-pĕl')	132	54°52'N	28°41'E
Lykens, Pa., U.S. (lī'kĕnz)	67	40°35'N	76°45'W
Lykhivka, Ukr.	133	48°52'N	33°57'E
Lyna, r., Eur. (lĭn'ȧ)	125	53°56'N	20°30'E
Lynch, Ky., U.S. (lĭnch)	83	36°56'N	82°55'W
Lynchburg, Va., U.S. (lĭnch'bûrg)	65	37°23'N	79°08'W
Lynch Cove, Wa., U.S. (lĭnch)	74a	47°26'N	122°54'W
Lynden, Can. (lĭn'dĕn)	62d	43°14'N	80°08'W
Lynden, Wa., U.S.	74d	48°56'N	122°27'W
Lyndhurst, Austl.	173a	38°03'S	145°14'E
Lyndon, Ky., U.S. (lĭn'dŭn)	69h	38°15'N	85°36'W
Lyndonville, Vt., U.S. (lĭn'dŭn-vĭl)	67	44°35'N	72°00'W
Lynn, Ma., U.S. (lĭn)	65	42°28'N	70°57'W
Lynn Lake, Can. (lăk)	50	56°51'N	101°05'W
Lynwood, Ca., U.S. (lĭn'wŏŏd)	75a	33°56'N	118°13'W
Lyon, Fr. (lē-ôN')	110	45°44'N	4°52'E
Lyons, Ga., U.S. (lī'ŭnz)	83	32°08'N	82°19'W
Lyons, Ks., U.S.	78	38°20'N	98°11'W
Lyons, Ne., U.S.	70	41°57'N	96°28'W
Lyons, N.J., U.S.	68a	40°41'N	74°33'W

ăt; finȧl; rāte; senȧte; ärm; ȧsk; sofȧ; fâre; ch-choose; dh-as th in other; bē; ēvent; bĕt; recĕnt; cratẽr; g-gō; gh-guttural g; bĭt; ĭ-short neutral; rīde; K-guttural k as ch in German ich;

PRONOUNCING INDEX

Lyons – Maldon 271

PLACE (Pronunciation)	PAGE	LAT.	LONG.
Lyons, N.Y., U.S.	67	43°05'N	77°00'W
Lyptsi, Ukr.	133	50°11'N	36°25'E
Lysefjorden, b., Nor.	122	58°59'N	6°35'E
Lysekil, Swe. (lü'sĕ-kêl)	122	58°17'N	11°22'E
Lys'va, Russia (lĭs'vȧ)	136	58°07'N	57°47'E
Lytham, Eng., U.K. (lĭth'ăm)	114a	53°44'N	2°58'W
Lytkarino, Russia	142b	55°35'N	37°55'E
Lyttelton, S. Afr. (lĭt'l'ton)	187b	25°51'S	28°13'E
Lyuban', Russia (lyōō'bȧn)	132	59°21'N	31°15'E
Lyubertsy, Russia (lyōō'bĕr-tsê)	132	55°40'N	37°55'E
Lyubim, Russia (lyōō-bêm')	132	58°24'N	40°39'E
Lyublino, Russia (lyōōb'lĭ-nô)	142b	55°41'N	37°45'E
Lyudinovo, Russia (lū-dē'novō)	132	53°52'N	34°28'E

M

PLACE (Pronunciation)	PAGE	LAT.	LONG.
Ma'ān, Jord. (mä-än')	154	30°12'N	35°45'E
Maartensdijk, Neth.	115a	52°09'N	5°10'E
Maas (Meuse), r., Eur.	121	51°50'N	5°40'E
Maastricht, Neth. (mäs'trĭkt)	121	50°51'N	5°35'E
Mabaia, Ang.	190	7°13'S	14°03'E
Mabana, Wa., U.S. (mä-bä-nä)	74a	48°06'N	122°25'W
Mabank, Tx., U.S. (mā'bănk)	81	32°21'N	96°05'W
Mabeskraal, S. Afr.	192c	25°12'S	26°47'E
Mableton, Ga., U.S. (mā'b'l-tŭn)	68c	33°49'N	84°34'W
Mabrouk, Mali	184	19°27'N	1°16'W
Mabula, S. Afr. (mă'bōō-la)	192c	24°49'S	27°59'E
Macalelon, Phil. (mä-kä-lā-lōn')	169a	13°46'N	122°09'E
Macau, Braz. (mä-kà'ō)	101	5°12'S	36°34'W
Macau, China	161	22°00'N	113°00'E
Macaya, Pico de, mtn., Haiti	93	18°25'N	74°00'W
Macclesfield, Eng., U.K. (măk''lz-fēld)	114a	53°15'N	2°07'W
Macclesfield Canal, can., Eng., U.K. (măk''lz-fēld)	114a	53°14'N	2°07'W
Macdona, Tx., U.S. (măk-dō'nȧ)	75d	29°20'N	98°42'W
Macdonald, l., Austl. (măk-dŏn'ăld)	174	23°40'S	127°40'E
Macdonnell Ranges, mts., Austl. (măk-dŏn'ĕl)	174	23°40'S	131°30'E
MacDowell Lake, l., Can. (măk-dou ĕl)	57	52°15'N	92°45'W
Macdui, Ben, mtn., Scot., U.K. (bĕn măk-dōō'ê)	116	57°06'N	3°45'W
Macedonia, Oh., U.S. (măs-ê-dō'nĭ-ȧ)	69d	41°19'N	81°30'W
Macedonia, nation, Eur.	131	41°50'N	22°00'E
Macedonia, hist. reg., Eur. (măs-ê-dō'nĭ-ȧ)	119	41°05'N	22°15'E
Maceió, Braz.	101	9°40'S	35°43'W
Macerata, Italy (mä-chå-rä'tä)	130	43°18'N	13°28'E
Macfarlane, Lake, l., Austl. (măc'fär-lān)	176	32°10'S	137°00'E
Machache, mtn., Leso.	187c	29°22'S	27°53'E
Machado, Braz. (mä-shà-dô)	99a	21°42'S	45°55'W
Machakos, Kenya	191	1°31'S	37°16'E
Machala, Ec. (mä-chá'lä)	100	3°18'S	78°54'W
Machens, Mo., U.S. (măk'ĕns)	75e	38°54'N	90°20'W
Machias, Me., U.S. (mȧ-chī'ăs)	60	44°22'N	67°29'W
Machida, Japan (mä-chē'dä)	167a	35°32'N	139°28'E
Machilīpatnam, India	155	16°22'N	81°10'E
Machu Picchu, Peru (mä'chò-pē'k-chò)	100	13°07'S	72°34'W
Măcin, Rom. (mȧ-chēn')	133	45°15'N	28°09'E
Macina, reg., Mali	188	14°50'N	4°40'W
Mackay, Austl. (mȧ-kī')	175	21°15'S	149°08'E
Mackay, Id., U.S. (măk-kā')	73	43°55'N	113°38'W
Mackay, l., Austl. (mȧ-kī')	174	22°30'S	127°45'E
MacKay, l., Can. (măk-kā')	52	64°10'N	112°35'W
Mackenzie, r., Can.	52	63°38'N	124°23'W
Mackenzie Bay, b., Can.	63	69°20'N	137°10'W
Mackenzie Mountains, mts., Can. (mȧ-kĕn'zĭ)	52	63°41'N	129°27'W
Mackinaw, r., Il., U.S.	66	40°35'N	89°25'W
Mackinaw City, Mi., U.S. (măk'ĭ-nô)	66	45°45'N	84°45'W
Mackinnon Road, Kenya	191	3°44'S	39°03'E
Macleantown, S. Afr. (măk-lăn'toun)	187c	32°48'S	27°48'E
Maclear, S. Afr. (mȧ-klēr')	186	31°06'S	28°23'E
Macomb, Il., U.S. (mȧ-kōōm')	79	40°27'N	90°40'W
Mâcon, Fr. (mä-kôN)	117	46°19'N	4°51'E
Macon, Ga., U.S. (mā'kŏn)	65	32°49'N	83°39'W
Macon, Mo., U.S.	79	39°42'N	92°29'W
Macon, Ms., U.S.	82	32°07'N	88°31'W
Macquarie, r., Austl.	175	31°43'S	148°04'E
Macquarie Islands, is., Austl. (mȧ-kwŏr'ê)	3	54°36'S	158°45'E
Macuelizo, Hond. (mä-kwĕ-lē'zô)	90	15°22'N	88°32'W
Mad, r., Ca., U.S. (măd)	72	40°38'N	123°37'W
Madagascar, nation, Afr. (măd-ȧ-găs'kăr)	187	18°05'S	43°12'E
Madame, i., Can. (mȧ-dȧm')	61	45°33'N	61°02'W
Madanapalle, India	159	13°06'N	78°09'E
Madang, Pap. N. Gui. (mä-däng')	169	5°15'S	145°45'E
Madaoua, Niger (mȧ-dou'ȧ)	184	14°04'N	6°03'E
Madawaska, r., Can. (măd-ȧ-wôs'kȧ)	59	45°20'N	77°25'W
Madeira, r., S.A.	100	6°48'S	62°43'W
Madeira, Arquipélago da, is., Port.	183	33°26'N	16°44'W
Madeira, Ilha da, i., Port. (mä-dā'rä)	184	32°41'N	16°15'W
Madeleine, Îles de la, is., Can.	53	47°30'N	61°45'W
Madelia, Mn., U.S. (mȧ-dē'lĭ-ȧ)	71	44°03'N	94°23'W
Madeline, i., Wi., U.S. (măd'ĕ-lĭn)	71	46°47'N	91°30'W
Madera, Ca., U.S. (mȧ-dā'rȧ)	76	36°57'N	120°04'W
Madera, vol., Nic.	90	11°27'N	85°30'W
Madgaon, India	159	15°09'N	73°58'E
Madhya Pradesh, state, India (mŭd'vŭ prŭ-dāsh')	155	22°04'N	77°48'E
Madill, Ok., U.S. (mȧ-dĭl')	79	34°04'N	96°45'W
Madīnat ash Sha'b, Yemen	154	12°45'N	44°00'E
Madingo, Congo	190	4°07'S	11°22'E
Madingou, Congo	190	4°09'S	13°34'E
Madison, Fl., U.S. (măd'ĭ-sŭn)	82	30°28'N	83°25'W
Madison, Ga., U.S.	82	33°34'N	83°29'W
Madison, Il., U.S.	75e	38°40'N	90°09'W
Madison, In., U.S.	66	38°45'N	85°25'W
Madison, Ks., U.S.	79	38°08'N	96°07'W
Madison, Me., U.S.	60	44°47'N	69°52'W
Madison, Mn., U.S.	70	44°59'N	96°13'W
Madison, N.C., U.S.	83	36°22'N	79°59'W
Madison, Ne., U.S.	70	41°49'N	97°27'W
Madison, N.J., U.S.	68a	40°46'N	74°25'W
Madison, S.D., U.S.	70	44°01'N	97°08'W
Madison, Wi., U.S.	65	43°05'N	89°23'W
Madison Res, Mt., U.S.	73	45°25'N	111°28'W
Madisonville, Ky., U.S. (măd'ĭ-sŭn-vĭl)	66	37°20'N	87°30'W
Madisonville, La., U.S.	81	30°22'N	90°10'W
Madisonville, Tx., U.S.	81	30°57'N	95°55'W
Madjori, Burkina	188	11°26'N	1°15'E
Mado Gashi, Kenya	191	0°44'N	39°10'E
Madona, Lat. (mȧ'dō'nȧ)	123	56°50'N	26°14'E
Madrakah, Ra's al, c., Oman	154	18°53'N	57°48'E
Madras *see* Chennai, India	155	13°08'N	80°15'E
Madre, Laguna, l., Mex. (lä-gōō'nä mä'drå)	81	25°08'N	97°41'W
Madre, Sierra, mts., N.A. (sē-ĕ'r-rä-mȧ'drĕ)	89	15°55'N	92°40'W
Madre, Sierra, mts., Phil.	169a	16°40'N	122°10'E
Madre de Dios, r., S.A. (mä'drå dā dē-ōs')	100	12°07'S	68°02'W
Madre de Dios, Archipiélago, is., Chile (mȧ'drå dā dē-ōs')	102	50°40'S	76°30'W
Madre del Sur, Sierra, mts., Mex. (sē-ĕ'r-rä-mä'drå dělsōōr')	86	17°35'N	100°35'W
Madre Occidental, Sierra, mts., Mex.	86	29°30'N	107°30'W
Madre Oriental, Sierra, mts., Mex.	86	25°30'N	100°45'W
Madrid, Spain (mä-drē'd)	110	40°26'N	3°42'W
Madrid, Ia., U.S. (măd'rĭd)	71	41°51'N	93°48'W
Madridejos, Spain (mä-dhrē-dhā'hōs)	128	39°29'N	3°32'W
Madura, i., Indon. (mȧ-dōō'rä)	168	6°45'S	113°30'E
Madurai, India (mä-dōō-rä')	155	9°57'N	78°04'E
Madureira, Serra do, mtn., Braz. (sĕ'r-rä-då-mä-dōō-rā'rȧ)	102b	22°49'S	43°30'W
Maebashi, Japan (mä-ĕ-bä'shė)	161	36°26'N	139°04'E
Maestra, Sierra, mts., Cuba (sē-ĕ'r-rȧ-mä-ās'trä)	87	20°05'N	77°05'W
Maewo, i., Vanuatu	175	15°17'S	168°16'E
Mafeking, S. Afr. (mȧf'ê'kĭng)	186	25°46'S	24°45'E
Mafra, Braz. (mä'frä)	102	26°21'N	49°59'W
Mafra, Port. (mäf'rȧ)	129b	38°56'N	9°20'W
Magadan, Russia (mȧ-gȧ-dän')	135	59°39'N	150°43'E
Magadan Oblast, Russia	141	65°00'N	160°00'E
Magadi, Kenya	191	1°54'S	36°17'E
Magalies, r., S. Afr. (mä-gä'lyĕs)	187b	25°51'S	27°42'E
Magaliesberg, mts., S. Afr.	187b	25°45'S	27°43'E
Magaliesburg, S. Afr.	192c	26°01'S	27°32'E
Magallanes, Estrecho de, strt., S.A.	102	52°30'S	68°45'W
Magat, r., Phil. (mä-gät')	169a	16°45'N	121°16'E
Magdalena, Arg. (mäg-dä-lā'nä)	99c	35°05'S	57°32'W
Magdalena, Bol.	100	13°17'S	63°57'W
Magdalena, Mex.	64	30°34'N	110°50'W
Magdalena, N.M., U.S.	77	34°10'N	107°45'W
Magdalena, i., Chile	102	44°45'S	73°15'W
Magdalena, r., Col.	100	7°45'N	74°04'W
Magdalena, Bahía, b., Mex. (bä-ē'ä-mäg-dä-lā'nä)	86	24°30'N	114°00'W
Magdeburg, Ger. (mäg'dĕ-bòrgh)	110	52°07'N	11°39'E
Magellan, Strait of *see* Magallanes, Estrecho de, strt., S.A.	102	52°30'S	68°45'W
Magenta, Italy (mȧ-jĕn'tȧ)	130	45°26'N	8°53'E
Magerøya, i., Nor.	116	71°10'N	24°11'E
Maggiore, Lago, l., Italy	118	46°03'N	8°25'E
Maghāghah, Egypt	192b	28°38'N	30°50'W
Maghniyya, Alg.	118	34°52'N	1°40'W
Magiscatzin, Mex. (mä-kēs-kät-zēn')	88	22°48'N	98°42'W
Maglaj, Bos. (mȧ'glä-ê)	131	44°34'N	18°12'E
Maglie, Italy (mäl'yā)	131	40°06'N	18°20'E
Magna, Ut., U.S. (măg'nȧ)	75b	40°43'N	112°06'W
Magnitogorsk, Russia (mȧg-nyē'tô-gôrsk)	134	53°26'N	59°05'E
Magnolia, Ar., U.S. (măg-nō'lĭ-ȧ)	79	33°16'N	93°13'W
Magnolia, Ms., U.S.	82	31°08'N	90°27'W
Magny-en-Vexin, Fr. (mä-nyē'ĕN-vĕ-săN')	127b	49°09'N	1°45'E
Magog, Can. (mȧ-gŏg')	59	45°15'N	72°10'W
Magpie, r., Can.	60	50°40'N	64°30'W
Magpie, r., Can.	58	48°13'N	84°50'W
Magpie, Lac, l., Can.	60	50°55'N	64°39'W
Magrath, Can.	50	49°25'N	112°52'W
Magude, Moz. (mä-gōō'dȧ)	186	24°58'S	32°39'E
Magwe, Mya. (mŭg-wā')	155	20°19'N	94°57'E
Mahābād, Iran	157	36°55'N	45°50'E
Mahahi Port, D.R.C. (mä-hä'gê)	185	2°14'N	31°12'E
Mahajanga, Madag.	187	15°12'S	46°26'E
Mahakam, r., Indon.	168	0°30'S	116°15'E
Mahali Mountains, mts., Tan.	191	6°20'S	30°00'E
Mahaly, Madag. (mȧ-hȧl-ē')	187	24°09'S	46°20'E
Mahanoro, Madag. (mȧ-hȧ-nō'rō)	187	19°57'S	48°47'E
Mahanoy City, Pa., U.S. (mä-hȧ-noi')	67	40°50'N	76°10'W
Maḩaţţat al Qaţrānah, Jord.	153a	31°15'N	36°04'E
Maḩaţţat 'Aqabat al Ḩijāzīyah, Jord.	153a	29°45'N	35°55'E
Maḩaţţat ar Ramlah, Jord.	153a	29°31'N	35°57'E
Maḩaţţat Jurf ad Darāwīsh, Jord.	153a	30°41'N	35°51'E
Mahd adh-Dhahab, Sau. Ar.	157	23°30'N	40°52'E
Mahe, India (mä-ā')	155	11°42'N	75°39'E
Mahenge, Tan. (mä-hĕŋ'gȧ)	186	7°38'S	36°16'E
Mahi, r., India	158	23°16'N	73°20'E
Mahilyow, Bela.	136	53°53'N	30°22'E
Mahilyow, prov., Bela.	132	53°28'N	30°15'E
Māhīm Bay, b., India	159b	19°03'N	72°45'E
Mahlabatini, S. Afr. (mä'lȧ-bȧ-tē'nê)	187c	28°15'S	31°29'E
Mahlow, Ger. (mä'lōv)	115b	52°23'N	13°24'E
Mahnomen, Mn., U.S. (mô-nō'mĕn)	70	47°18'N	95°58'W
Mahone Bay, Can. (mȧ-hōn')	60	44°27'N	64°23'W
Mahone Bay, b., Can.	60	44°30'N	64°15'W
Mahopac, Lake, l., N.Y., U.S. (mā-hō'păk)	68a	41°24'N	73°45'W
Mahwah, N.J., U.S. (mȧ-wä')	68a	41°05'N	74°09'W
Maidenhead, Eng., U.K. (mād'ĕn-hĕd)	114b	51°30'N	0°44'W
Maidstone, Eng., U.K.	121	51°17'N	0°32'E
Maiduguri, Nig. (mä'ē-dȧ-gōō'rê)	185	11°51'N	13°10'E
Maigualida, Sierra, mts., Ven. (sē-ĕ'r-rȧ-mī-gwä'lē-dĕ)	100	6°30'N	65°50'W
Maijdi, Bngl.	158	22°59'N	91°08'E
Maikop *see* Maykop, Russia	134	44°35'N	40°07'E
Main, r., Ger. (mīn)	124	49°49'N	9°20'E
Main Barrier Range, mts., Austl. (băr''ēr)	175	31°25'S	141°40'E
Mai-Ndombe, Lac, l., D.R.C.	186	2°16'S	19°00'E
Maine, state, U.S. (mān)	65	45°25'N	69°50'W
Mainland, i., Scot., U.K. (mān-lănd)	116	60°19'N	2°40'W
Maintenon, Fr. (măN-tē-nôN')	127b	48°35'N	1°35'E
Maintirano, Madag. (mä'ēn-tē-rä'nō)	187	18°05'S	44°08'E
Mainz, Ger. (mīnts)	110	49°59'N	8°16'E
Maio, i., C.V. (mä'yo)	184b	15°15'N	22°50'W
Maipo, S.A.	102	34°08'S	69°51'W
Maipo, r., Chile (mī'pô)	99b	33°45'S	71°08'W
Maiquetía, Ven. (mī-kĕ-tē'ä)	100	10°37'N	66°56'W
Maison-Rouge, Fr. (mĕ-zôN-rōōzh')	127b	48°34'N	3°09'E
Maisons-Laffitte, Fr.	127b	48°57'N	2°09'E
Maitland, Austl. (māt'lănd)	175	32°45'S	151°40'E
Maizuru, Japan (mä-ĭ'zōō-rōō)	167	35°26'N	135°15'E
Majene, Indon.	168	3°34'S	119°00'E
Maji, Eth.	185	6°14'N	35°34'E
Majorca *see* Mallorca, i., Spain	112	39°18'N	2°22'E
Makah Indian Reservation, I.R., Wa., U.S.	72	48°17'N	124°52'W
Makanya, Tan. (mä-kän'yä)	187	4°15'S	37°49'E
Makanza, D.R.C.	185	1°42'N	19°08'E
Makarakomburu, Mount, mtn., Sol. Is.	170e	9°43'S	160°02'E
Makarska, Cro. (mȧ'kär-skȧ)	131	43°17'N	17°05'E
Makar'yev, Russia	136	57°50'N	43°48'E
Makasar *see* Ujungpandang, Indon.	168	5°08'S	119°28'E
Makasar, Selat (Makassar Strait), strt., Indon.	168	2°00'S	118°07'E
Makaw, D.R.C.	190	3°29'S	18°19'E
Make, i., Japan (mä'kȧ)	167	30°43'N	130°49'E
Makeni, S.L.	184	8°53'N	12°03'W
Makgadikgadi Pans, pl., Bots.	186	20°38'S	21°31'E
Makhachkala, Russia (mäĸ'äch-kä'lä)	137	43°00'N	47°40'E
Makhaleng, r., Leso.	187c	29°53'S	27°33'E
Makiïvka, Ukr.	137	48°03'N	38°00'E
Makindu, Kenya	191	2°17'S	37°49'E
Makkah *see* Mecca, Sau. Ar.	154	21°27'N	39°45'E
Makkovik, Can.	51	55°01'N	59°10'W
Makokou, Gabon (mȧ-kō-kōō')	184	0°34'N	12°52'E
Maków Mazowiecki, Pol. (mä'kōov mä-zō-vyĕts'kē)	125	52°51'N	21°07'E
Makuhari, Japan (mä-kōō-hä'rē)	167a	35°39'N	140°04'E
Makurazaki, Japan (mä'kò-rä-zä'kê)	167	31°16'N	130°18'E
Makurdi, Nig.	184	7°45'N	8°32'E
Makushin, Ak., U.S. (mȧ-kò'shĭn)	63	53°57'N	166°28'W
Makushino, Russia (mȧ-kò-shē'nô)	134	55°03'N	67°43'E
Mala, Punta, c., Pan. (pò'n-tä-mä'lä)	91	7°32'N	79°44'W
Malabar Coast, cst., India (măl'ȧ-bär)	159	11°19'N	75°33'E
Malabar Point, c., India	159b	18°57'N	72°47'E
Malabo, Eq. Gui.	184	3°45'N	8°47'E
Malabon, Phil.	169a	14°39'N	120°57'E
Malacca, Strait of, strt., Asia (mȧ-lăk'ȧ)	168	4°15'N	99°44'E
Malad City, Id., U.S. (mȧ-lăd')	73	42°11'N	112°15'W
Maladzyecha, Bela.	136	54°18'N	26°57'E
Málaga, Col. (mȧ'lä-gȧ)	100	6°41'N	72°46'W
Málaga, Spain	110	36°45'N	4°25'W
Malagón, Spain (mä-lä-gōn')	128	39°12'N	3°52'W
Malaita, i., Sol. Is. (mȧ-lä'ē-tȧ)	175	8°38'S	161°15'E
Malakāl, Sudan (mȧ-lȧ-käl')	185	9°46'N	31°54'E
Malakhovka, Russia (mȧ-läk'ôf-kȧ)	142b	55°38'N	38°01'E
Malang, Indon.	168	8°06'S	112°50'E
Malanje, Ang. (mä-läŋ-gȧ)	186	9°32'S	16°20'E
Malanville, Benin	184	12°04'N	3°09'E
Mälaren, l., Swe.	116	59°38'N	16°55'E
Malartic, Can.	51	48°07'N	78°11'W
Malatya, Tur. (mȧ-lä'tyȧ)	154	38°30'N	38°15'E
Malawi, nation, Afr.	186	11°15'S	33°45'E
Malawi, Lake *see* Nyasa, Lake, l., Afr.	186	10°45'S	34°30'E
Malaya Vishera, Russia (vē-shȧ'rä)	134	58°51'N	32°13'E
Malay Peninsula, pen., Asia (mȧ-lā') (mā'lā)	168	6°00'N	101°00'E
Malaysia, nation, Asia (mȧ-lā'zhȧ)	168	4°10'N	101°22'E
Malbon, Austl. (măl'bŭn)	174	21°15'S	140°30'E
Malbork, Pol. (mäl'bôrk)	116	54°02'N	19°04'E
Malcabran, r., Port. (mäl-kä-brän')	129b	38°47'N	8°46'W
Malden, Ma., U.S. (môl'dĕn)	61a	42°26'N	71°04'W
Malden, Mo., U.S.	79	36°32'N	89°56'W
Malden, l., Kir.	2	4°20'S	154°30'W
Maldives, nation, Asia	150	4°30'N	71°30'E
Maldon, Eng., U.K. (môrl'dŏn)	114b	51°44'N	0°39'E

ng-sing; ŋ-baŋk; N-nasalized n; nŏd; cŏmmit; ōld; ôbey; ôrder; oi-boil; fōōd; ò-as oo in foot; ou-out; s-soft; sh-dish; th-thin; pūre; ůnite; ûrn; stŭd; circŭs; ü-as in French tu; '-indeterminate vowel.

272 Maldonado – Margate PRONOUNCING INDEX

PLACE (Pronunciation)	PAGE	LAT.	LONG.
Maldonado, Ur. (mäl-dō-nä′dô)	102	34°54′S	54°57′W
Maldonado, Punta, c., Mex. (pōō′n-tä)	88	16°18′N	98°34′W
Maléas, Ákra, c., Grc.	119	36°31′N	23°13′E
Mālegaon, India	158	20°35′N	74°30′E
Malé Karpaty, mts., Slvk.	125	48°31′N	17°15′E
Malekula, i., Vanuatu (mä-lā-kōō′lä)	175	16°44′S	167°45′E
Malema, Moz.	191	14°57′S	37°20′E
Malheur, r., Or., U.S. (má-lōōr′)	72	43°45′N	117°41′W
Malheur Lake, l., Or., U.S. (má-lōōr′)	72	43°16′N	118°37′W
Mali, nation, Afr.	184	15°45′N	0°15′W
Malibu, Ca., U.S. (măl′ĭ-bōō)	75a	34°03′N	118°38′W
Malik, Wādī al, r., Sudan	185	16°48′N	29°30′E
Malimba, Monts, mts., D.R.C.	191	7°45′S	29°15′E
Malinalco, Mex. (mä-lē-näl′kō)	88	18°54′N	99°31′W
Malinaltepec, Mex. (mä-lē-näl-tå-pĕk′)	88	17°01′N	98°41′W
Malindi, Kenya (mä-lēn′dē)	187	3°14′S	40°04′E
Malin Head, c., Ire.	116	55°23′N	7°24′W
Malino, Russia (mä′lĭ-nô)	142b	55°07′N	38°12′E
Malkara, Tur. (mäl′kȧ-rȧ)	131	40°51′N	26°52′E
Malko Tŭrnovo, Blg. (mäl′kō-t′r′nō-vȧ)	131	41°59′N	27°28′E
Mallaig, Scot., U.K.	120	56°59′N	5°55′W
Mallet Creek, Oh., U.S. (măl′ĕt)	69d	41°10′N	81°55′W
Mallorca, i., Spain	112	39°30′N	3°00′E
Mallow, Ire. (măl′ō)	120	52°07′N	9°04′W
Malmédy, Bel. (mål-mā-dē′)	121	50°25′N	6°01′E
Malmesbury, S. Afr. (mämz′bēr-ĭ)	186	33°30′S	18°35′E
Malmköping, Swe. (mälm′chû′pĭng)	122	59°09′N	16°39′E
Malmö, Swe.	110	55°36′N	13°00′E
Malmyzh, Russia (mål-mêzh′)	135	49°58′N	137°07′E
Malmyzh, Russia	136	56°30′N	50°48′E
Maloarkhangelsk, Russia (mä′lô-ár-κäŋ′gĕlsk)	132	52°26′N	36°29′E
Malolos, Phil. (mä-lō′lōs)	169a	14°51′N	120°49′E
Malomal′sk, Russia (mȧ-lô-mälsk′′)	142a	58°47′N	59°55′E
Malone, N.Y., U.S. (mȧ-lōn′)	67	44°50′N	74°20′W
Malonga, D.R.C.	190	10°24′S	23°10′E
Maloti Mountains, mts., Leso.	187c	29°00′S	28°29′E
Maloyaroslavets, Russia (mä′lō-yä-rô-slä-vyĕts)	132	55°01′N	36°25′E
Malozemel′skaya Tundra, reg., Russia	136	67°30′N	50°00′E
Malpas, Eng., U.K. (măl′păz)	114a	53°01′N	2°46′W
Malpelo, Isla de, i., Col. (mäl-pā′lō)	100	3°55′N	81°30′W
Malpeque Bay, b., Can. (môl-pĕk′)	60	46°30′N	63°47′W
Malta, Mt., U.S. (môl′tȧ)	73	48°20′N	107°50′W
Malta, nation, Eur.	110	35°52′N	13°30′E
Maltahöhe, Nmb. (mäl′tä-hō′ē)	186	24°45′S	16°45′E
Maltrata, Mex. (mäl-trä′tä)	89	18°48′N	97°16′W
Maluku (Moluccas), is., Indon.	169	2°22′S	128°25′E
Maluku, Laut (Molucca Sea), sea, Indon.	169	0°15′N	125°41′E
Malūţ, Sudan	185	10°30′N	32°17′E
Mālvan, India	159	16°08′N	73°32′E
Malvern, Ar., U.S. (măl′vērn)	79	34°21′N	92°47′W
Malyn, Ukr.	133	50°44′N	29°15′E
Malynivka, Ukr.	133	49°50′N	36°43′E
Malyy Anyuy, r., Russia	141	67°52′N	164°30′E
Malyy Tamir, i., Russia	141	78°10′N	107°30′E
Mamantel, Mex. (mä-mán-tĕl′)	89	18°36′N	91°06′W
Mamaroneck, N.Y., U.S. (măm′ȧ-rō-nĕk)	68a	40°57′N	73°44′W
Mambasa, D.R.C.	191	1°21′N	29°03′E
Mamburao, Phil. (mäm-bōō′rä-ō)	169a	13°14′N	120°35′E
Mamfe, Cam. (mäm′fĕ)	184	5°46′N	9°17′E
Mamihara, Japan (mä′mê-hä-rä)	167	32°41′N	131°12′E
Mammoth Cave, Ky., U.S. (măm′ŏth)	82	37°10′N	86°04′W
Mammoth Cave National Park, rec., Ky., U.S.	65	37°20′N	86°21′W
Mammoth Hot Springs, Wy., U.S. (măm′ŭth hôt sprĭngz)	73	44°55′N	110°50′W
Mamnoli, India	159b	19°17′N	73°15′E
Mamoré, r., S.A.	100	13°00′S	65°20′W
Mamou, Gui.	184	10°26′N	12°07′W
Mampong, Ghana	188	7°04′N	1°24′W
Mamry, Jezioro, l., Pol. (mäm′rĭ)	125	54°10′N	21°28′E
Man, C. Iv.	188	7°24′N	7°33′W
Manacor, Spain (mä-nä-kôr′)	129	39°35′N	3°15′E
Manado, Indon.	169	1°29′N	124°50′E
Managua, Cuba (mä-nä′gwä)	93a	22°58′N	82°17′W
Managua, Nic.	86	12°10′N	86°16′W
Managua, Lago de, l., Nic. (lá′gô-dĕ)	90	12°28′N	86°10′W
Manakara, Madag. (mä-nä-kä′rŭ)	187	22°17′S	48°06′E
Manama *see* Al Manāmah, Bahr.	154	26°01′N	50°33′E
Mananara, r., Madag. (mä-nä-nä′rŭ)	187	23°15′S	48°15′E
Mananjary, Madag. (mä-nän-zhä′rē)	187	20°16′S	48°13′E
Manas, China	160	44°30′N	86°00′E
Manassas, Va., U.S. (mȧ-năs′ȧs)	67	38°45′N	77°30′W
Manaus, Braz. (mä-nä′ōōzh)	101	3°01′S	60°00′W
Mancelona, Mi., U.S. (măn-sê-lō′nȧ)	66	44°50′N	85°05′W
Mancha Real, Spain (män′chä rā-äl′)	128	37°48′N	3°37′W
Manchazh, Russia (män′chásh)	142a	56°30′N	58°10′E
Manchester, Eng., U.K.	110	53°28′N	2°14′W
Manchester, Ct., U.S. (măn′chĕs-tẽr)	67	41°45′N	72°30′W
Manchester, Ga., U.S.	82	32°50′N	84°37′W
Manchester, Ia., U.S.	71	42°30′N	91°30′W
Manchester, Ma., U.S.	61a	42°35′N	70°47′W
Manchester, Mo., U.S.	75e	38°36′N	90°31′W
Manchester, N.H., U.S.	65	43°00′N	71°30′W
Manchester, Oh., U.S.	66	38°40′N	83°35′W
Manchester Ship Canal, Eng., U.K.	114a	53°20′N	2°40′W
Manchuria, hist. reg., China (măn-chōō′rē-ȧ)	161	48°00′N	124°58′E
Mandal, Nor. (män′däl)	122	58°03′N	7°28′E
Mandalay, Mya. (măn′dȧ-lā)	155	22°00′N	96°08′E
Mandalselva, r., Nor.	122	58°25′N	7°30′E
Mandan, N.D., U.S. (măn′dăn)	64	46°49′N	100°54′W
Mandara Mountains, mts., Afr. (män-dä′rä)	185	10°15′N	13°23′E
Mandau Siak, r., Indon.	153b	1°03′N	101°25′E
Mandeb, Bab-el-, strt., (bäb′ĕl män-dĕb′)	154	13°17′N	42°49′E
Mandimba, Moz.	191	14°21′S	35°39′E
Mandinga, Pan. (män-dĭŋ′gä)	91	9°32′N	79°04′W
Mandla, India	158	22°43′N	80°23′E
Mándra, Grc. (män′drä)	131	38°06′N	23°32′E
Mandritsara, Madag. (mån-drēt-sä′rȧ)	187	15°49′S	48°47′E
Manduria, Italy (män-dōō′rē-ä)	131	40°23′N	17°41′E
Mandve, India	159b	18°47′N	72°52′E
Māndvi, India (mŭnd′vē)	159b	19°29′N	72°53′E
Māndvi, India (mŭnd′vē)	155	22°54′N	69°23′E
Mandya, India	159	12°40′N	77°00′E
Manfredonia, Italy (män-frå-dô′nyä)	130	41°39′N	15°55′E
Manfredónia, Golfo di, b., Italy (gôl-fô-dē)	130	41°34′N	16°05′E
Mangabeiras, Chapada das, pl., Braz.	101	8°05′S	47°32′W
Mangalore, India (mŭŋ-gŭ-lōr′)	155	12°53′N	74°52′E
Mangaratiba, Braz. (män-gä-rä-tē′bȧ)	99a	22°56′S	44°03′W
Mangatarem, Phil. (män′gȧ-tä′rĕm)	169a	15°48′N	120°18′E
Mange, D.R.C.	190	0°54′N	20°30′E
Mangkalihat, Tanjung, c., Indon.	168	1°25′N	119°55′E
Mangles, Islas de, Cuba (ē′s-läs-dĕ-män′glās) (măn′g′lz)	92	22°05′N	82°50′W
Mangoche, Mwi.	186	14°16′S	35°14′E
Mangoky, r., Madag. (män-gō′kē)	187	22°02′S	44°11′E
Mangole, Pulau, i., Indon.	169	1°35′S	126°22′E
Mangualde, Port. (män-gwäl′dĕ)	128	40°38′N	7°44′W
Mangueira, Lagoa da, l., Braz.	102	33°15′S	52°45′W
Mangum, Ok., U.S. (măŋ′gŭm)	78	34°52′N	99°31′W
Mangzhangdian, China (mäŋ-jäŋ-dĭĕn)	162	32°07′N	114°44′E
Manhattan, Il., U.S.	69a	41°25′N	87°29′W
Manhattan, Ks., U.S. (măn-hăt′ăn)	64	39°11′N	96°34′W
Manhattan Beach, Ca., U.S.	75a	33°53′N	118°24′W
Manhuaçu, Braz. (män-ōȧ′sōō)	99a	20°17′S	42°01′W
Manhumirim, Braz. (män-ōō-mē-rē′ɴ)	99a	22°30′S	41°57′W
Manicouagane, r., Can.	53	50°00′N	68°35′W
Manicouagane, Lac, res., Can.	53	51°30′N	68°19′W
Manicuare, Ven. (mä-nē-kwä′rē)	101b	10°35′N	64°10′W
Manihiki Islands, is., Cook Is. (mä′nē-hē′kē)	195	9°40′S	158°00′W
Manila, Phil.	168	14°37′N	121°00′E
Manila Bay, b., Phil. (mȧ-nĭl′ȧ)	169a	14°38′N	120°46′E
Manisa, Tur. (mä′nē-sȧ)	119	38°40′N	27°30′E
Manistee, Mi., U.S. (măn-ĭs-tē′)	66	44°15′N	86°20′W
Manistee, r., Mi., U.S.	66	44°25′N	85°45′W
Manistique, Mi., U.S. (măn-ĭs-tēk′)	71	45°58′N	86°16′W
Manistique, l., Mi., U.S.	71	46°14′N	85°30′W
Manistique, r., Mi., U.S.	71	46°05′N	86°09′W
Manitoba, prov., Can. (măn-ĭ-tō′bȧ)	50	55°12′N	97°29′W
Manitoba, Lake, l., Can.	52	51°00′N	98°45′W
Manito Lake, l., Can. (măn′ĭ-tō)	56	52°45′N	109°45′W
Manitou, i., Mi., U.S. (măn′ĭ-tōō)	71	47°21′N	87°33′W
Manitou, l., Can.	71	49°21′N	93°01′W
Manitou Islands, is., Mi., U.S.	66	45°05′N	86°00′W
Manitoulin Island, i., Can. (măn-ĭ-tōō′lĭn)	53	45°45′N	81°30′W
Manitou Springs, Co., U.S.	78	38°51′N	104°58′W
Manitowoc, Wi., U.S. (măn-ĭ-tō-wŏk′)	71	44°05′N	87°42′W
Manitqueira, Serra da, mts., Braz.	99a	22°40′S	45°12′W
Maniwaki, Can.	59	46°23′N	76°00′W
Manizales, Col. (mä-nē-zä′läs)	100	5°05′N	75°31′W
Manjacaze, Moz. (man′yä-kä′zĕ)	186	24°37′S	33°49′E
Mankato, Ks., U.S. (măn-kā′tō)	78	39°45′N	98°12′W
Mankato, Mn., U.S.	65	44°10′N	93°59′W
Mankim, Cam.	189	5°01′N	12°00′E
Manlléu, Spain (män-lyā′ōō)	129	42°00′N	2°16′E
Mannar, Sri L. (mȧ-när′)	159	9°48′N	80°03′E
Mannar, Gulf of, b., Asia	155	8°47′N	78°33′E
Mannheim, Ger. (män′hīm)	117	49°30′N	8°31′E
Manning, Ia., U.S. (măn′ĭng)	70	41°53′N	95°04′W
Manning, S.C., U.S.	83	33°41′N	80°12′W
Mannington, W.V., U.S. (măn′ĭng-tŭn)	66	39°30′N	80°55′W
Mano, r., Afr.	188	7°00′N	11°25′W
Man of War Bay, b., Bah.	93	21°05′N	74°05′W
Man of War Channel, strt., Bah.	92	22°45′N	76°10′W
Manokwari, Indon. (mȧ-nŏk-wä′rē)	169	0°56′S	134°10′E
Manono, D.R.C.	191	7°18′S	27°25′E
Manor, Can. (măn′ẽr)	57	49°36′N	102°05′W
Manor, Wa., U.S.	74c	45°45′N	122°36′W
Manori, neigh., India	159b	19°13′N	72°43′E
Manosque, Fr. (mȧ-nôsh′)	127	43°51′N	5°48′E
Manotick, Can.	62c	45°13′N	75°41′W
Manouane, r., Can.	59	50°15′N	70°30′W
Manouane, Lac, l., Can. (mä-nōō′än)	60	50°36′N	70°50′W
Manresa, Spain (män-rā′sä)	118	41°44′N	1°52′E
Mansa, Zam.	186	11°12′S	28°53′E
Mansel, i., Can. (măn′sĕl)	53	61°56′N	81°10′W
Manseriche, Pongo de, reg., Peru (pô′n-gô-dĕ-män-sĕ-rē′chĕ)	100	4°15′S	77°45′W
Mansfield, Eng., U.K. (mănz′fēld)	114a	53°08′N	1°12′W
Mansfield, La., U.S.	81	32°02′N	93°43′W
Mansfield, Oh., U.S.	66	40°45′N	82°30′W
Mansfield, Wa., U.S.	72	47°48′N	119°39′W
Mansfield, Mount, mtn., Vt., U.S.	67	44°30′N	72°45′W
Mansfield Woodhouse, Eng., U.K. (wŏŏd-hous)	114a	53°08′N	1°12′W
Manta, Ec. (män′tä)	100	1°03′S	80°16′W
Manteno, Il., U.S. (măn-tē-nō)	69a	41°15′N	87°50′W
Manteo, N.C., U.S.	83	35°55′N	75°40′W
Mantes-la-Jolie, Fr. (mäɴt-ĕ-lä-zhō-lē′)	126	48°59′N	1°42′E
Manti, Ut., U.S. (măn′tī)	77	39°15′N	11°40′W
Mantova, Italy (män′tô-vä) (măn′tū-ȧ)	118	45°09′N	10°47′E
Mantua, Cuba (män-tōō′ȧ)	92	22°20′N	84°15′W
Mantua *see* Mantova, Italy	118	45°09′N	10°47′E
Mantua, Ut., U.S. (măn′tū-ȧ)	75b	41°30′N	111°57′W
Manua Islands, is., Am. Sam.	170a	14°13′S	169°35′W
Manui, Pulau, i., Indon. (mä-nōō′ē)	169	3°35′S	123°38′E
Manus Island, i., Pap. N. Gui. (mä′nōōs)	169	2°22′S	146°22′E
Manvel, Tx., U.S. (măn′vel)	81a	29°28′N	95°22′W
Manville, N.J., U.S. (măn′vĭl)	68a	40°33′N	74°36′W
Manville, R.I., U.S.	68b	41°57′N	71°27′W
Manzala Lake, l., Egypt	192b	31°14′N	32°04′E
Manzanares, Col. (män-sä-nä′rĕs)	100a	5°15′N	75°09′W
Manzanares, r., Spain (mänz-nä′rĕs)	129a	40°36′N	3°48′W
Manzanares, Canal del, Spain (kä-nä′l-dĕl-män-thä-nä′rĕs)	129a	40°20′N	3°38′W
Manzanillo, Cuba (män′zä-nēl′yō)	87	20°20′N	77°05′W
Manzanillo, Mex.	86	19°02′N	104°21′W
Manzanillo, Bahía de, b., Mex. (bä-ē′ä-dĕ-män-zä-nē′l-yō)	88	19°00′N	104°38′W
Manzanillo, Bahía de, b., N.A.	93	19°55′N	71°50′W
Manzanillo, Punta, c., Pan.	91	9°40′N	79°33′W
Manzhouli, China (män-jō-lē)	161	49°25′N	117°15′E
Manzovka, Russia (mȧn-zhō′f-kȧ)	166	44°16′N	132°13′E
Mao, Chad (mä′ō)	185	14°07′N	15°19′E
Mao, Dom. Rep.	93	19°35′N	71°10′W
Maó, Spain	118	39°52′N	4°15′E
Maoke, Pegunungan, mts., Indon.	169	4°00′S	138°00′E
Maoming, China	161	21°55′N	110°40′E
Maoniu Shan, mtn., China (mou-nĭō shän)	164	32°45′N	104°09′E
Mapastepec, Mex. (ma-päs-tå-pĕk′)	89	15°24′N	92°52′W
Mapia, Kepulauan, i., Indon.	169	0°57′N	134°22′E
Mapimí, Mex. (mä-pê-mē′)	80	25°50′N	103°50′W
Mapimí, Bolsón de, des., Mex. (bôl-sŏ′n-dĕ-mä-pē′mē)	80	27°27′N	103°20′W
Maple Creek, Can. (mā′p′l) (crēk)	50	49°55′N	109°27′W
Maple Grove, Can. (grōv)	62a	45°19′N	73°51′W
Maple Heights, Oh., U.S.	69d	41°25′N	81°34′W
Maple Shade, N.J., U.S. (shād)	68f	39°57′N	75°01′W
Maple Valley, Wa., U.S. (văl′ē)	74a	47°24′N	122°02′W
Maplewood, Mn., U.S. (wŏd)	75g	45°00′N	93°03′W
Maplewood, Mo., U.S.	75e	38°37′N	90°20′W
Mapumulo, S. Afr. (mä-pä-mōō′lō)	187c	29°12′S	31°05′E
Maputo, Moz.	186	26°50′S	32°30′E
Maquela do Zombo, Ang. (mȧ-kā′lȧ dō zôm′bō)	186	6°08′S	15°15′E
Maquoketa, Ia., U.S. (mȧ-kō-kē-tȧ)	71	42°04′N	90°42′W
Maquoketa, r., Ia., U.S.	71	42°08′N	90°40′W
Mar, Serra do, mts., Braz. (sĕr′rȧ dō mär′)	102	26°30′S	49°15′W
Maracaibo, Ven. (mä-rä-kī′bō)	100	10°38′N	71°45′W
Maracaibo, Lago de, l., Ven. (lä′gô-dĕ-mä-rä-kī′bō)	100	9°55′N	72°13′W
Maracay, Ven. (mä-rä-käy′)	100	10°15′N	67°35′W
Marādah, Libya	185	29°10′N	19°07′E
Maradi, Niger (mȧ-rȧ-dē′)	184	13°29′N	7°06′E
Marāgheh, Iran	157	37°20′N	46°10′E
Maraisburg, S. Afr.	187b	26°12′S	27°57′E
Marais des Cygnes, r., Ks., U.S.	79	38°30′N	95°30′W
Marajó, Ilha de, i., Braz.	101	1°00′S	49°30′W
Maralal, Kenya	191	1°06′N	36°42′E
Marali, C.A.R.	189	6°01′N	18°24′E
Marand, Iran	157	38°26′N	45°46′E
Maranguape, Braz. (mä-rȧŋ-gwä′pĕ)	101	3°48′S	38°38′W
Maranhão, state, Braz. (mä-rän-youɴ)	101	5°15′S	45°52′W
Maranoa, r., Austl. (mä-rä-nō′ä)	175	27°01′S	148°03′E
Marano di Napoli, Italy (mä-rä′nô-dē-nä′pô-lē)	129c	40°39′N	14°12′E
Marañón, r., Peru (mä-rä-nyōn′)	100	4°26′S	75°08′W
Marapanim, Braz. (mä-rä-pä-nē′ɴ)	101	0°45′S	47°42′W
Marathon, Can.	51	48°50′N	86°10′W
Marathon, Fl., U.S. (măr′ȧ-thŏn)	83a	24°41′N	81°06′W
Marathon, Oh., U.S.	69f	39°09′N	83°59′W
Maravatío, Mex. (mä-rȧ-vä′tē-ō)	88	19°54′N	100°25′W
Marawi, Sudan	185	18°07′N	31°57′E
Marble Bar, Austl. (märb′l bär)	174	21°15′S	119°15′E
Marble Canal, can., Az., U.S. (mär′b′l)	77	36°21′N	111°48′W
Marblehead, Ma., U.S. (mär′b′l-hĕd)	61a	42°30′N	70°51′W
Marburg an der Lahn, Ger.	124	50°49′N	8°46′E
Marca, Ponta da, c., Ang.	190	16°31′S	11°42′E
Marcala, Hond. (mär-kä-lä)	90	14°08′N	88°01′W
Marceline, Mo., U.S. (mär-sê-lēn′)	79	39°42′N	92°56′W
Marche, hist. reg., Italy (mär′kā)	130	43°35′N	12°33′E
Marchegg, Aus.	115e	48°18′N	16°55′E
Marchena, Spain (mär-chā′nä)	118	37°20′N	5°25′W
Marchena, i., Ec. (ē′s-lä-mär-chĕ′nä)	100	0°29′N	90°31′W
Marchfeld, reg., Aus.	115e	48°14′N	16°37′E
Mar Chiquita, Laguna, l., Arg. (lä-gōō′nä-mär-chē-kē′tä)	99c	34°25′S	61°10′W
Marcos Paz, Arg. (mär-kōs′ päz)	99c	34°49′S	58°51′W
Marcus, i., Japan (mär′kŭs)	195	24°00′N	155°00′E
Marcus Hook, Pa., U.S. (mär′kŭs hŏŏk)	68f	39°49′N	75°25′W
Marcy, Mount, mtn., N.Y., U.S. (mär′sē)	67	44°10′N	73°55′W
Mar de Espanha, Braz. (mär-dĕ-ĕs-pá′nyá)	99a	21°53′S	43°00′W
Mar del Plata, Arg. (mär dĕl- plä′ta)	102	37°59′S	57°35′W
Mardin, Tur. (mär-dēn′)	154	37°25′N	40°40′E
Maré, i., N. Cal. (mȧ-rā′)	175	21°53′S	168°30′E
Maree, Loch, b., Scot., U.K. (mȧ-rē′)	120	57°40′N	5°44′W
Marengo, Ia., U.S. (mȧ-rĕŋ′gō)	71	41°47′N	92°04′W
Marennes, Fr. (mȧ-rĕn′)	126	45°49′N	1°08′W
Marfa, Tx., U.S. (mär′fȧ)	80	30°19′N	104°01′W
Margarita, Pan. (mär-gōō-rē′tä)	86a	9°20′N	79°55′W
Margarita, Isla de, i., Ven. (mä-gȧ-rē′tȧ)	100	11°00′N	64°15′W
Margate, S. Afr. (mä-gāt′)	187c	30°52′S	30°21′E

PLACE (Pronunciation)	PAGE	LAT.	LONG.
Margate, Eng., U.K. (mär′gāt)	121	51°21'N	1°17'E
Margherita Peak, mtn., Afr.	185	0°22'N	29°51'E
Marguerite, r., Can.	60	50°39'N	66°42'W
Marhanets', Ukr.	133	47°41'N	34°33'E
Maria, Can. (mȧ-rē′ȧ)	60	48°10'N	66°04'W
Mariager, Den. (mä-rê-ägh′ĕr)	122	56°38'N	10°00'E
Mariana, Braz. (mä-ryá′nä)	99a	20°23'S	43°24'W
Mariana Islands, is., Oc.	5	16°00'N	145°30'E
Marianao, Cuba (mä-rê-ä-nä′ō)	87	23°05'N	82°26'W
Mariana Trench, deep,	195	12°00'N	144°00'E
Marianna, Ar., U.S. (mā-rĭ-ăn′ȧ)	79	34°45'N	90°45'W
Marianna, Fl., U.S.	82	30°46'N	85°14'W
Marianna, Pa., U.S.	69e	40°01'N	80°05'W
Mariano Acosta, Arg. (mä-rēä′nō-ȧ-kōs′tä)	102a	34°28'S	58°48'W
Mariánské Lázně, Czech Rep. (mär′yàn-skĕ′läz′nyĕ)	124	49°58'N	12°42'E
Marias, r., Mt., U.S. (mȧ-rī′ȧz)	73	48°15'N	110°50'W
Marias, Islas, is., Mex. (mä-rē′äs)	86	21°30'N	106°40'W
Mariato, Punta, c., Pan.	91	7°17'N	81°09'W
Maribo, Den. (mä′rê-bō)	122	54°46'N	11°29'E
Maribor, Slvn. (mä′re-bôr)	110	46°33'N	15°37'E
Maricaban, i., Phil. (mä-rê-kä-bän′)	169a	13°40'N	120°44'E
Mariefred, Swe. (mä-rē′ĕ-frĭd)	122	59°17'N	17°09'E
Marie Galante, i., Guad. (mȧ-rē′ gȧ-länt′)	91b	15°58'N	61°05'W
Mariehamn, Fin. (mȧ-rē′ê-häm′'n)	123	60°07'N	19°57'E
Mari El, prov., Russia	136	56°30'N	48°00'E
Mariestad, Swe. (mä-rē′ĕ-städ′)	122	58°43'N	13°45'E
Marietta, Ga., U.S. (mā-rĭ-ĕt′ȧ)	68c	33°57'N	84°33'W
Marietta, Oh., U.S.	66	39°25'N	81°30'W
Marietta, Ok., U.S.	79	33°53'N	97°07'W
Marietta, Wa., U.S.	74d	48°48'N	122°35'W
Mariinsk, Russia (mȧ-rē′ĭnsk)	140	56°15'N	87°28'E
Marijampole, Lith. (mä-rê-yäm-pô′lĕ)	123	54°33'N	23°26'E
Marikana, S. Afr. (mă′-rĭ-kä-nȧ)	192c	25°40'S	27°28'E
Marília, Braz. (mä-rē′lyȧ)	101	22°02'S	49°48'W
Marimba, Ang.	190	8°28'S	17°08'E
Marín, Spain	128	42°24'N	8°40'W
Marinduque Island, i., Phil. (mä-rên-dōō′kā)	169a	13°14'N	121°45'E
Marine, Il., U.S. (mȧ-rēn′)	75e	38°48'N	89°47'W
Marine City, Mi., U.S.	66	42°45'N	82°30'W
Marine Lake, l., Mn., U.S.	75g	45°13'N	92°55'W
Marine on Saint Croix, Mn., U.S.	75g	45°11'N	92°47'W
Marinette, Wi., U.S. (măr-ĭ-nĕt′)	65	45°04'N	87°40'W
Maringa, r., D.R.C. (mȧ-rĭŋ′gä)	185	0°30'N	21°00'E
Marinha Grande, Port. (mä-rēn′yȧ grän′dē)	128	39°49'N	8°53'W
Marion, Al., U.S. (măr′ĭ-ŭn)	82	32°36'N	87°19'W
Marion, Ia., U.S.	71	42°01'N	91°39'W
Marion, Il., U.S.	66	37°40'N	88°55'W
Marion, In., U.S.	65	40°35'N	85°45'W
Marion, Ks., U.S.	79	38°21'N	97°02'W
Marion, Ky., U.S.	82	37°19'N	88°05'W
Marion, N.C., U.S.	83	35°40'N	82°00'W
Marion, N.D., U.S.	70	46°37'N	98°20'W
Marion, Oh., U.S.	66	40°35'N	83°10'W
Marion, S.C., U.S.	83	34°08'N	79°23'W
Marion, Va., U.S.	83	36°48'N	81°33'W
Marion, Lake, res., S.C., U.S.	83	33°25'N	80°35'W
Marion Reef, rf., Austl.	175	18°57'S	151°31'E
Mariposa, Chile (mä-rê-pō′sä)	99b	35°33'S	71°21'W
Mariposa Creek, r., Ca., U.S.	76	37°14'N	120°30'W
Mariquita, Col. (mä-rē-kē′tä)	100a	5°13'N	74°52'W
Mariscal Estigarribia, Para.	102	22°03'S	60°28'W
Marisco, Ponta do, c., Braz. (pô′n-tä-dô-mä-rē′s-kö)	102b	23°01'S	43°17'W
Maritime Alps, mts., Eur. (mă′rĭ-tīm ălps)	117	44°20'N	7°02'E
Mariupol', Ukr.	134	47°07'N	37°32'E
Mariveles, Phil.	169a	14°27'N	120°29'E
Marj Uyan, Leb.	153a	33°21'N	35°36'E
Marka, Som.	192a	1°45'N	44°47'E
Markaryd, Swe. (mär′kä-rüd)	122	56°30'N	13°34'E
Marked Tree, Ar., U.S. (märkt trē)	79	35°31'N	90°26'W
Marken, i., Neth.	115a	52°26'N	5°08'E
Market Bosworth, Eng., U.K. (bŏz′wûrth)	114a	52°37'N	1°23'W
Market Deeping, Eng., U.K. (dēp′ĭng)	114a	52°40'N	0°19'W
Market Drayton, Eng., U.K. (drā′tŭn)	114a	52°54'N	2°29'W
Market Harborough, Eng., U.K. (här′bŭr-ō)	114a	52°28'N	0°55'W
Market Rasen, Eng., U.K. (rā′zĕn)	114a	53°23'N	0°21'W
Markham, Can. (märk′ăm)	59	43°53'N	79°15'W
Markham, Mount, mtn., Ant.	178	82°59'S	159°30'E
Markivka, Ukr.	133	49°32'N	39°34'E
Markovo, Russia (mär′kô-vô)	135	64°46'N	170°48'E
Markrāna, India	158	27°08'N	74°43'E
Marks, Russia	137	51°42'N	46°46'E
Marksville, La., U.S. (märks′vĭl)	81	31°09'N	92°05'W
Markt Indersdorf, Ger. (märkt ēn′dĕrs-dōrf)	115d	48°22'N	11°23'E
Marktredwitz, Ger. (märk-rĕd′vēts)	124	50°02'N	12°05'E
Markt Schwaben, Ger. (märkt shvä′bĕn)	115d	48°12'N	11°52'E
Marl, Ger. (märl)	127c	51°40'N	7°05'E
Marlboro, N.J., U.S.	68a	40°18'N	74°15'W
Marlborough, Ma., U.S.	61a	42°21'N	71°33'W
Marlette, Mi., U.S. (mär-lĕt′)	66	43°25'N	83°05'W
Marlin, Tx., U.S. (mär′lĭn)	81	31°18'N	96°52'W
Marlinton, W.V., U.S. (mär′lĭn-tŭn)	66	38°15'N	80°10'W
Marlow, Eng., U.K. (mär′lō)	114b	51°33'N	0°46'W
Marlow, Ok., U.S.	79	34°38'N	97°56'W
Marls, The, b., Bah. (märls)	92	26°30'N	77°15'W
Marmande, Fr. (mȧr-mäNd′)	126	44°30'N	0°10'E
Marmara Denizi, sea, Tur.	154	40°40'N	28°00'E
Marmarth, N.D., U.S. (mär-märth)	70	46°19'N	103°57'W
Mar Muerto, l., Mex. (mär-mòĕ′r-tô)	89	16°13'N	94°22'W
Marne, Ger. (mär′nĕ)	115c	53°57'N	9°01'E
Marne, r., Fr. (märn)	117	49°00'N	4°30'E
Maroa, Ven. (mä-rō′ä)	100	2°43'N	67°37'W
Maroantsetra, Madag. (mȧ-rō-äŋ-tsä′trȧ)	187	15°18'S	49°48'E
Maro Jarapeto, mtn., Col. (mä-rō-hä-rä-pĕ′tō)	100a	6°29'N	76°39'W
Maromokotro, mtn., Madag.	187	14°00'S	49°11'E
Marondera, Zimb.	186	18°10'S	31°36'E
Maroni, r., S.A. (mȧ-rō′nē)	101	3°02'N	53°54'W
Maro Reef, rf., Hi., U.S.	84b	25°15'N	170°00'W
Maroua, Cam. (mär′wä)	185	10°36'N	14°20'E
Marple, Eng., U.K. (mär′p'l)	114a	53°24'N	2°04'W
Marquard, S. Afr.	192c	28°41'S	27°26'E
Marquesas Islands, is., Fr. Poly. (mär-kĕ′säs)	2	8°50'S	141°00'W
Marquesas Keys, is., Fl., U.S. (mȧr-kē′zȧs)	83a	24°37'N	82°15'W
Marquês de Valença, Braz. (mär-kĕ′s-dĕ-vä-lĕ′n-sä)	99a	22°16'S	43°42'W
Marquette, Can. (mȧr-kĕt′)	62f	50°04'N	97°43'W
Marquette, Mi., U.S.	65	46°32'N	87°25'W
Marquez, Tx., U.S. (mär-kāz′)	81	31°14'N	96°15'W
Marra, Jabal, mtn., Sudan (jĕb′ĕl mär′ä)	185	13°00'N	23°47'E
Marrakech, Mor. (mȧr-rä′kĕsh)	184	31°38'N	8°00'W
Marree, Austl. (măr′rē)	174	29°38'S	137°55'E
Marrero, La., U.S.	68d	29°55'N	90°06'W
Marrupa, Moz.	191	13°08'S	37°30'E
Mars, Pa., U.S. (märz)	69e	40°42'N	80°01'W
Marsabit, Kenya	191	2°20'N	37°59'E
Marsala, Italy (mär-sä′lä)	118	37°48'N	12°28'E
Marsden, Eng., U.K. (märz′dĕn)	114a	53°36'N	1°55'W
Marseille, Fr. (mȧr-sâ′y')	110	43°18'N	5°25'E
Marseilles, Il., U.S. (mär-sĕlz′)	66	41°20'N	88°40'W
Marshall, Il., U.S. (mär′shăl)	66	39°20'N	87°40'W
Marshall, Mi., U.S.	66	42°20'N	84°55'W
Marshall, Mn., U.S.	70	44°28'N	95°49'W
Marshall, Mo., U.S.	79	39°07'N	93°12'W
Marshall, Tx., U.S.	65	32°33'N	94°22'W
Marshall Islands, nation, Oc.	3	10°00'N	165°00'E
Marshalltown, Ia., U.S. (mär′shȧl-toun)	71	42°02'N	92°55'W
Marshallville, Ga., U.S. (mär′shȧl-vĭl)	82	32°29'N	83°55'W
Marshfield, Ma., U.S. (märsh′fēld)	61a	42°06'N	70°43'W
Marshfield, Mo., U.S.	79	37°20'N	92°53'W
Marshfield, Wi., U.S.	71	44°40'N	90°10'W
Marsh Harbour, Bah.	92	26°30'N	77°00'W
Mars Hill, In., U.S. (märz′hĭl′)	69g	39°43'N	86°15'W
Mars Hill, Me., U.S.	60	46°34'N	67°54'W
Marstrand, Swe. (mär′stränd)	122	57°54'N	11°33'E
Marsyaty, Russia (märs′yȧ-tĭ)	142a	60°03'N	60°28'E
Mart, Tx., U.S. (märt)	81	31°32'N	96°49'W
Martaban, Gulf of, b., Mya. (mär-tŭ-bän′)	168	16°34'N	96°58'E
Martapura, Indon.	168	3°19'S	114°45'E
Martha's Vineyard, i., Ma., U.S. (mär′thȧz vĭn′yȧrd)	67	41°25'N	70°35'W
Martigny, Switz. (mȧr-tē-nyē′)	124	46°06'N	7°00'E
Martigues, Fr.	127	43°24'N	5°05'E
Martin, Tn., U.S. (mär′tĭn)	82	36°20'N	88°45'W
Martina Franca, Italy (mär-tē′nä fräŋ′kä)	131	40°43'N	17°21'E
Martinez, Ca., U.S. (mär-tē′nĕz)	74b	38°01'N	122°08'W
Martinez, Tx., U.S.	75d	29°25'N	98°20'W
Martinique, dep., N.A. (mȧr-tē-nēk′)	87	14°50'N	60°40'W
Martin Lake, res., Al., U.S.	82	32°40'N	86°05'W
Martin Point, c., Ak., U.S.	63	70°10'N	142°00'W
Martinsburg, W.V., U.S. (mär′tĭnz-bûrg)	67	39°30'N	78°00'W
Martins Ferry, Oh., U.S. (mär′tĭnz)	66	40°05'N	80°45'W
Martinsville, In., U.S. (mär′tĭnz-vĭl)	66	39°25'N	86°25'W
Martinsville, Va., U.S.	83	36°40'N	79°53'W
Martos, Spain (mär′tōs)	128	37°43'N	3°58'W
Martre, Lac la, l., Can. (läk la märtr)	52	63°24'N	119°58'W
Marugame, Japan (mä′rōō-gä′mä)	167	34°19'N	133°48'E
Marungu, mts., D.R.C.	191	7°50'S	29°50'E
Marve, neigh., India	159b	19°12'N	72°43'E
Mary, Turkmen. (mä′rê)	139	37°45'N	61°47'E
Mar'yanskaya, Russia (mȧr-yän′skȧ-yȧ)	133	45°04'N	38°39'E
Maryborough, Austl. (mâ′rĭ-bŭr-ô)	175	25°35'S	152°40'E
Maryborough, Austl.	175	37°00'S	143°50'E
Maryland, state, U.S. (mĕr′ĭ-lănd)	65	39°10'N	76°25'W
Marys, r., Nv., U.S. (mā′rĭz)	72	41°25'N	115°10'W
Marystown, Can. (mâr′ĭz-toun)	61	47°11'N	55°10'W
Marysville, Can.	60	45°59'N	66°35'W
Marysville, Ca., U.S.	76	39°09'N	121°37'W
Marysville, Oh., U.S.	66	40°15'N	83°25'W
Marysville, Wa., U.S.	74a	48°03'N	122°11'W
Maryville, Il., U.S. (mā′rĭ-vĭl)	75e	38°44'N	89°57'W
Maryville, Mo., U.S.	79	40°21'N	94°51'W
Maryville, Tn., U.S.	82	35°44'N	83°59'W
Mārzuq, Libya	185	26°00'N	14°09'E
Marzūq, Idehan, des., Libya	184	24°30'N	13°00'E
Masai Steppe, plat., Tan.	191	4°30'S	36°40'E
Masaka, Ug.	191	0°20'S	31°44'E
Masalasef, Chad	189	11°43'N	17°08'E
Masalembo-Besar, i., Indon.	168	5°40'S	114°28'E
Masan, Kor., S. (mä-sän′)	161	35°10'N	128°31'E
Masangwe, Tan.	191	5°28'S	30°05'E
Masasi, Tan. (mä-sä′sē)	187	10°43'S	38°48'E
Masatepe, Nic. (mä-sä-tĕ′pĕ)	90	11°57'N	86°10'W
Masaya, Nic. (mä-sä′yä)	90	11°58'N	86°05'W
Masbate, Phil. (mäs-bä′tā)	169a	12°21'N	123°38'E
Masbate, i., Phil.	169	12°19'N	123°03'E
Mascarene Islands, is., Afr.	5	20°20'S	56°40'E
Mascot, Tn., U.S. (măs′kŏt)	82	36°04'N	83°45'W
Mascota, Mex. (mäs-kō′tä)	88	20°33'N	104°45'W
Mascota, r., Mex.	88	20°33'N	104°52'W
Mascouche, Can. (màs-kōōsh′)	62a	45°45'N	73°36'W
Mascouche, r., Can.	62a	45°44'N	73°45'W
Mascoutah, Il., U.S. (măs-kū′tā)	75e	38°29'N	89°48'W
Maseru, Leso. (măz′ēr-ōō)	186	29°09'S	27°11'E
Mashhad, Iran	154	36°17'N	59°30'E
Māshkel, Hāmūn-i-, l., Asia (hä-mōōn′ē māsh-kĕl′)	154	28°28'N	64°13'E
Mashra'ar Raqq, Sudan	185	8°28'N	29°15'E
Masi-Manimba, D.R.C.	190	4°46'S	17°55'E
Masindi, Ug. (mä-sēn′dē)	185	1°44'N	31°43'E
Masjed Soleymān, Iran	154	31°45'N	49°17'E
Mask, Lough, b., Ire. (lŏk mȧsk)	120	53°35'N	9°23'W
Maslovo, Russia (mäs′lô-vô)	142a	60°08'N	60°28'E
Mason, Mi., U.S. (mā′sŭn)	66	42°35'N	84°25'W
Mason, Oh., U.S.	69f	39°22'N	84°18'W
Mason, Tx., U.S.	80	30°46'N	99°14'W
Mason City, Ia., U.S.	65	43°08'N	93°14'W
Massa, Italy (mäs′sä)	130	44°02'N	10°08'E
Massachusetts, state, U.S. (măs-ȧ-chōō′sĕts)	65	42°20'N	72°30'W
Massachusetts Bay, b., Ma., U.S.	60	42°26'N	70°20'W
Massafra, Italy (mäs-sä′frä)	131	40°35'N	17°05'E
Massa Marittima, Italy	130	43°03'N	10°55'E
Massapequa, N.Y., U.S.	68a	40°41'N	73°28'W
Massaua *see* Mitsiwa, Erit.	185	15°40'N	39°19'E
Massena, N.Y., U.S. (mă-sē′nȧ)	67	44°55'N	74°55'W
Masset, Can. (măs′ĕt)	50	54°02'N	132°09'W
Masset Inlet, b., Can.	55	53°42'N	132°20'E
Massif Central, Fr. (mȧ-sēf′ sän-trȧl′)	110	45°12'N	3°02'E
Massillon, Oh., U.S. (măs′ĭ-lŏn)	66	40°50'N	81°35'W
Massinga, Moz. (mä-sĭn′gä)	186	23°18'S	35°18'E
Massive, Mount, mtn., Co., U.S. (mȧs′ĭv)	64	39°05'N	106°30'W
Masson, Can. (măs-sŭn)	62c	45°33'N	75°25'W
Masuda, Japan (mä-sōō′dȧ)	167	34°42'N	131°53'E
Masuria, reg., Pol.	125	53°40'N	21°10'E
Masvingo, Zimb.	186	20°07'S	30°47'E
Matadi, D.R.C. (mȧ-tä′dē)	186	5°49'S	13°27'E
Matagalpa, Nic. (mä-tä-gäl′pä)	86	12°52'N	85°57'W
Matagami, l., Can. (mȧ-tä-gä′mē)	53	50°10'N	78°28'W
Matagorda Bay, b., Tx., U.S. (măt-ȧ-gôr′dȧ)	81	28°32'N	96°13'W
Matagorda Island, i., Tx., U.S.	81	28°13'N	96°27'W
Matam, Sen. (mä-täm′)	184	15°40'N	13°15'W
Matamoros, Mex. (mä-tä-mō′rōs)	80	25°32'N	103°13'W
Matamoros, Mex.	86	25°52'N	97°30'W
Matane, Can. (mȧ-tȧn′)	51	48°51'N	67°32'W
Matanzas, Cuba (mä-tän′zäs)	87	23°05'N	81°35'W
Matanzas, prov., Cuba	92	22°45'N	81°20'W
Matanzas, Bahía, b., Cuba (bä-ē′ä)	92	23°10'N	81°30'W
Matapalo, Cabo, c., C.R. (kä′bō-mä-tä-pä′lō)	91	8°22'N	83°25'W
Matapédia, Can. (mä-tȧ-pā′dē-ȧ)	60	47°58'N	66°56'W
Matapédia, l., Can.	60	48°33'N	67°32'W
Matapédia, r., Can.	60	48°10'N	67°10'W
Mataquito, r., Chile (mä-tä-kē′tô)	99b	35°08'S	71°35'W
Matara, Sri L. (mä-tä′rä)	159	5°59'N	80°35'E
Mataram, Indon.	168	8°45'S	116°15'E
Matatiele, S. Afr. (mä-tä-tyā′lā)	187c	30°21'S	28°49'E
Matawan, N.J., U.S.	68a	40°24'N	74°13'W
Matehuala, Mex. (mä-tā-wä′lä)	86	23°38'N	100°39'W
Matera, Italy (mä-tā′rä)	130	40°42'N	16°37'E
Mateur, Tun. (mȧ-tûr′)	118	37°09'N	9°43'E
Mātherān, India	159b	18°58'N	73°16'E
Matheson, Can.	59	48°35'N	80°33'W
Mathews, Lake, l., Ca., U.S. (măth ūz)	75a	33°50'N	117°24'W
Mathura, India (mu-tŏ′rŭ)	155	27°39'N	77°39'E
Matias Barbosa, Braz. (mä-tē′äs-bär-bô-sä)	99a	21°53'S	43°19'W
Matillas, Laguna, l., Mex. (lä-gò′nä-mä-tē′l-yäs)	89	18°02'N	92°36'W
Matina, C.R. (mä-tē′nä)	91	10°06'N	83°20'W
Matiši, Lat. (mä′tē-sē)	123	57°43'N	25°09'E
Matlalcueyetl, Cerro, mtn., Mex. (sĕ′r-rä-mä-tläl-kwĕ′yĕtl)	88	19°13'N	98°02'W
Matlock, Eng., U.K. (măt′lŏk)	114a	53°08'N	1°33'W
Matochkin Shar, Russia (mä′tŏch-kĭn)	134	73°57'N	56°16'E
Mato Grosso, Braz. (mät′ō grōs′ō)	101	15°04'S	59°58'W
Mato Grosso, state, Braz.	101	14°38'S	55°36'W
Mato Grosso, Chapada de, hills, Braz. (shä-pä′dä-dĕ)	101	13°39'S	55°42'W
Mato Grosso do Sul, state, Braz.	101	20°00'S	56°00'W
Matosinhos, Port.	128	41°10'N	8°48'W
Maţraḩ, Oman (mȧ-trä′)	154	23°36'N	58°27'E
Matsubara, Japan	167b	34°34'N	135°34'E
Matsudo, Japan (mät′sò-dō)	167a	35°48'N	139°55'E
Matsue, Japan (mät′sò-ĕ)	161	35°29'N	133°04'E
Matsumoto, Japan (mät′sò-mō′tō)	166	36°15'N	137°59'E
Matsuyama, Japan (mät′sò-yä′mä)	161	33°48'N	132°45'E
Matsuzaka, Japan (mät′sò-zä′kä)	167	34°35'N	136°34'E
Mattamuskeet, Lake, l., N.C., U.S. (măt-tȧ-mŭs′kēt)	83	35°34'N	76°03'W
Mattaponi, r., Va., U.S. (măt′ȧ-pō-nī′)	67	37°45'N	77°00'W
Mattawa, Can. (măt′ȧ-wȧ)	51	46°15'N	78°49'W
Matterhorn, mtn., Eur. (mät′ēr-hôrn)	124	45°57'N	7°36'E
Matteson, Il., U.S. (mătt′ē-sŭn)	69a	41°30'N	87°42'W
Matthew Town, Bah. (măth′ū toun)	93	21°00'N	73°40'W
Mattoon, Il., U.S. (mă-tōōn′)	65	39°30'N	88°20'W
Maturín, Ven. (mä-tōō-rēn′)	100	9°48'N	63°16'W
Maúa, Moz.	191	13°51'S	37°10'E
Mauban, Phil. (mä′ōō-bän′)	169a	14°11'N	121°44'E

274 Maubeuge – Mendota PRONOUNCING INDEX

PLACE (Pronunciation)	PAGE	LAT.	LONG.
Maubeuge, Fr. (mô-bûzh′)	126	50°18′N	3°57′E
Maud, Oh., U.S. (môd)	69f	39°21′N	84°23′W
Mauer, Aus. (mou′ẽr)	115e	48°09′N	16°16′E
Maués, Braz. (mà-wĕ′s)	101	3°34′S	57°30′W
Mau Escarpment, clf., Kenya	191	0°45′S	35°50′E
Maui, i., Hi., U.S. (mä′ōō-ē)	64c	20°52′N	156°02′W
Maule, r., Chile (má′ô-lĕ)	99b	35°45′S	70°50′W
Maumee, Oh., U.S. (mô-mē′)	66	41°30′N	83°40′W
Maumee, r., In., U.S.	66	41°10′N	84°50′W
Maumee Bay, b., Oh., U.S.	66	41°50′N	83°20′W
Maun, Bots. (mä-ön′)	186	19°52′S	23°40′E
Mauna Kea, mtn., Hi., U.S. (mä′ó-näkä′ä)	64c	19°52′N	155°30′W
Mauna Loa, mtn., Hi., U.S. (mä′ó-nälō′ä)	64c	19°28′N	155°38′W
Maurepas Lake, l., La., U.S. (mō-rē-pä′)	81	30°18′N	90°40′W
Mauricie, Parc National de la, rec., Can.	59	46°46′N	73°00′W
Mauritania, nation, Afr. (mô-rē-tā′nĭ-*à*)	184	19°38′N	13°30′W
Mauritius, nation, Afr. (mô-rĭsh′ĭ-*ŭs*)	3	20°18′S	57°36′E
Maury, Wa., U.S. (mô′rĭ)	74a	47°22′N	122°23′W
Mauston, Wi., U.S. (môs′*tŭn*)	71	43°46′N	90°05′W
Maverick, r., Az., U.S. (mā-vûr′ĭk)	77	33°40′N	109°30′W
Mavinga, Ang.	190	15°50′S	20°21′E
Mawlamyine, Mya.	168	16°30′N	97°39′E
Maxville, Can. (măks′vĭl)	62c	45°17′N	74°52′W
Maxville, Mo., U.S.	75e	38°26′N	90°24′W
Maya, r., Russia (mä′yä)	141	58°00′N	135°45′E
Mayaguana, i., Bah.	93	22°25′N	73°00′W
Mayaguana Passage, strt., Bah.	93	22°20′N	73°25′W
Mayagüez, P.R. (mä-yä-gwāz′)	87	18°12′N	67°10′W
Mayari, r., Cuba	93	20°25′N	75°35′W
Mayas, Montañas, mts., N.A. (mŏntăn′ăs mä′äs)	90a	16°43′N	89°00′W
Mayd, i., Som.	192a	11°24′N	46°38′E
Mayen, Ger. (mī′ĕn)	124	50°19′N	7°14′E
Mayenne, r., Fr. (mà-yĕn)	126	48°14′N	0°45′W
Mayfield, Ky., U.S. (mā′fēld)	82	36°44′N	88°19′W
Mayfield Creek, r., Ky., U.S.	82	36°54′N	88°47′W
Mayfield Heights, Oh., U.S.	69d	41°31′N	81°26′W
Mayfield Lake, res., Wa., U.S.	72	46°31′N	122°34′W
Maykop, Russia	134	44°35′N	40°07′E
Maykor, Russia (mī-kôr′)	142a	59°01′N	55°52′E
Maymyo, Mya. (mī′myō)	160	22°14′N	96°32′E
Maynard, Ma., U.S. (mā′*nàrd*)	61a	42°25′N	71°27′W
Mayne, Can. (mān)	74d	48°51′N	123°18′W
Mayne, i., Can.	74d	48°52′N	123°14′W
Mayo, Can. (mā-yō′)	50	63°40′N	135°51′W
Mayo, Fl., U.S.	82	30°02′N	83°08′W
Mayo, Md., U.S.	68e	38°54′N	76°31′W
Mayodan, N.C., U.S. (mā-yō′*dăn*)	83	36°25′N	79°59′W
Mayon Volcano, vol., Phil. (mä-yōn′)	169a	13°21′N	123°43′E
Mayotte, dep., Afr. (mà-yŏt′)	187	13°07′S	45°32′E
May Pen, Jam.	92	18°00′N	77°25′W
Mayraira Point, c., Phil.	165	18°40′N	120°45′E
Mayran, Laguna de, l., Mex. (lä-ō′nä-dĕ-mī-rän′)	86	25°40′N	102°35′W
Mayskiy, Russia	138	43°38′N	44°04′E
Maysville, Ky., U.S. (māz′vĭl)	66	38°35′N	83°45′W
Mayumba, Gabon	186	3°25′S	10°39′E
Mayville, N.D., U.S.	70	47°30′N	97°20′W
Mayville, N.Y., U.S. (mā′vĭl)	67	42°15′N	79°30′W
Mayville, Wi., U.S.	71	43°30′N	88°45′W
Maywood, Ca., U.S. (mā′wŏŏd)	75a	33°59′N	118°11′W
Maywood, Il., U.S.	69a	41°53′N	87°51′W
Mazabuka, Zam. (mä-zä-bōō′kä)	186	15°51′S	27°46′E
Mazagão, Braz. (mä-zá-gouN)	101	0°05′S	51°27′W
Mazapil, Mex. (mä-zä-pēl′)	80	24°40′N	101°30′W
Mazara del Vallo, Italy (mät-sä′rä dĕl väl′lō)	130	37°40′N	12°37′E
Mazār-i-Sharīf, Afg. (mà-zär′-ē-shà-rēf′)	155	36°48′N	67°12′E
Mazarrón, Spain (mä-zär-rô′n)	128	37°37′N	1°29′W
Mazatenango, Guat. (mä-zä-tå-näŋ′gō)	86	14°30′N	91°30′W
Mazatla, Mex.	89a	19°30′N	99°24′W
Mazatlán, Mex.	86	23°14′N	106°27′W
Mazatlán (San Juan), Mex. (mä-zä-tlän′) (sän hwän′)	89	17°05′N	95°26′W
Mažeikiai, Lith. (mà-zhā′kē-ĭ)	123	56°19′N	22°24′E
Mazhafah, Jabal, mtn., Sau. Ar.	153a	28°56′N	35°05′E
Mazyr, Bela.	137	52°03′N	29°14′E
Mbabane, Swaz. (m′bä-bä′nĕ)	186	26°18′S	31°14′E
Mbaiki, C.A.R. (m′bà-ē′kē)	185	3°53′N	18°00′E
Mbakana, Montagne de, mts., Cam.	189	7°55′N	14°40′E
Mbakaou, Barrage de, dam, Cam.	189	6°10′N	12°55′E
Mbala, Zam.	186	8°50′S	31°22′E
Mbale, Ug.	191	1°05′N	34°10′E
Mbamba Bay, Tan.	191	11°17′S	34°46′E
Mbandaka, D.R.C.	186	0°04′N	18°16′E
M'banza Congo, Ang.	186	6°30′S	14°10′E
Mbanza-Ngungu, D.R.C.	186	5°20′S	10°55′E
Mbarara, Ug.	191	0°37′S	30°39′E
Mbasay, Chad	189	7°39′N	15°40′E
Mbigou, Gabon (m-bē-gōō′)	186	2°07′S	11°30′E
Mbinda, Congo	190	2°00′S	12°55′E
Mbogo, Tan.	191	7°26′S	33°26′E
Mbomou (Bomu), r., Afr. (m′bō′mōō)	185	4°50′N	24°00′E
Mbout, Maur. (m′bōō′)	184	16°03′N	12°31′W
Mbuji-Mayi, D.R.C.	190	6°09′S	23°38′E
McAdam, Can. (măk-ăd′ăm)	60	45°36′N	67°20′W
McAfee, N.J., U.S. (măk-ă′fē)	68a	41°10′N	74°32′W
McAlester, Ok., U.S. (m*ă*k ăl′ĕs-tẽr)	65	34°55′N	95°45′W
McAllen, Tx., U.S. (m*ă*k-ăl′ĕn)	80	26°12′N	98°14′W
McBride, Can. (m*ă*k-brīd′)	50	53°18′N	120°10′W
McCalla, Al., U.S. (m*ă*k-kăl′lä)	68h	33°20′N	87°00′W
McCamey, Tx., U.S. (m*ă*-kā′mĭ)	80	31°08′N	102°13′W
McColl, S.C., U.S. (m*à*-kól′)	83	34°40′N	79°34′W
McComb, Ms., U.S. (m*à*-kōm′)	82	31°14′N	90°27′W
McConaughy, Lake, l., Ne., U.S. (m*ă*k kō′nō ĭ′)	70	41°24′N	101°40′W
McCook, Ne., U.S. (mà-kók′)	78	40°13′N	100°37′W
McCormick, S.C., U.S. (m*à*-kôr′mĭk)	83	33°56′N	82°20′W
McDonald, Pa., U.S. (măk-dŏn′*ă*ld)	69e	40°22′N	80°13′W
McDonald Island, i., Austl.	178	53°00′S	72°45′E
McDonald Lake, l., Can. (m*ă*k-dŏn-ăld)	62e	51°12′N	113°53′W
McGehee, Ar., U.S. (m*à*-gē′)	79	33°39′N	91°22′W
McGill, Nv., U.S. (m*à*-gĭl′)	77	39°25′N	114°47′W
McGowan, Wa., U.S. (m*ă*k-gou′ăn)	74c	46°15′N	123°55′W
McGrath, Ak., U.S. (măk′gràth)	64a	62°58′N	155°20′W
McGregor, Can. (m*ă*k-grĕg′ẽr)	69b	42°08′N	82°58′W
McGregor, Ia., U.S.	71	42°58′N	91°12′W
McGregor, Tx., U.S.	81	31°26′N	97°23′W
McGregor, r., Can.	55	54°10′N	121°00′W
McGregor Lake, l., Can. (măk-grĕg′ẽr)	62c	45°38′N	75°44′W
McHenry, Il., U.S. (măk-hĕn′rĭ)	69a	42°21′N	88°16′W
Mchinji, Mwi.	186	13°42′S	32°50′E
McIntosh, S.D., U.S. (măk′ĭn-tŏsh)	70	45°54′N	101°22′W
McKay, r., Or., U.S.	74c	45°43′N	123°00′W
McKeesport, Pa., U.S. (m*à*-kez′pōrt)	69e	40°21′N	79°51′W
McKees Rocks, Pa., U.S. (m*à*-kēz′ rŏks)	69e	40°29′N	80°05′W
McKenzie, Tn., U.S. (mà-kĕn′zĭ)	82	36°07′N	88°30′W
McKenzie, r., Or., U.S.	72	44°07′N	122°20′W
McKinley, Mount, mtn., Ak., U.S. (m*à*-kĭn′lĭ)	64a	63°00′N	151°02′W
McKinney, Tx., U.S. (m*à*-kĭn′ĭ)	79	33°12′N	96°35′W
McLaughlin, S.D., U.S. (m*ă*k-lŏf′lĭn)	70	45°48′N	100°45′W
McLean, Va., U.S. (m*ă*c′lān)	68e	38°56′N	77°11′W
McLeansboro, Il., U.S. (m*à*-klānz′bŭr-ō)	66	38°10′N	88°35′W
McLennan, Can. (măk-lĭn′nán)	50	55°42′N	116°54′W
McLeod, r., Can.	55	53°45′N	115°55′W
McLeod Lake, Can.	54	54°59′N	123°02′W
McLoughlin, Mount, mtn., Or., U.S. (màk-lŏk′lĭn)	72	42°27′N	122°20′W
McMillan Lake, l., Tx., U.S. (m*ă*k-mĭl′*ă*n)	80	32°40′N	104°09′W
McMillin, Wa., U.S. (m*à*k-mĭl′*ĭ*n)	74a	47°08′N	122°14′W
McMinnville, Or., U.S. (m*ă*k-mĭn′vĭl)	72	45°13′N	123°13′W
McMinnville, Tn., U.S.	82	35°41′N	85°47′W
McMurray, Wa., U.S. (m*à*k-mŭr′ĭ)	74a	48°19′N	122°15′W
McNary, Az., U.S. (m*à*k-nâr′ē)	77	34°10′N	109°55′W
McNary, La., U.S.	81	30°58′N	92°32′W
McNary Dam, Or., U.S.	72	45°57′N	119°15′W
McPherson, Ks., U.S. (m*à*k-fŭr′s′n)	79	38°21′N	97°41′W
McRae, Ga., U.S. (m*à*k-rā′)	83	32°02′N	82°55′W
McRoberts, Ky., U.S. (m*à*k-rŏb′ẽrts)	83	37°12′N	82°40′W
Mead, Ks., U.S. (mēd)	78	37°17′N	100°21′W
Mead, Lake, l., U.S.	64	36°20′N	114°14′W
Meade Peak, mtn., Id., U.S.	72	42°19′N	111°16′W
Meadow Lake, Can. (mĕd′ō lăk)	50	54°08′N	108°26′W
Meadows, Can. (mĕd′ōz)	62f	50°02′N	97°35′W
Meadville, Pa., U.S. (mēd′vĭl)	66	41°40′N	80°10′W
Meaford, Can. (mē′fẽrd)	59	44°35′N	80°40′W
Mealy Mountains, mts., Can. (mē′lē)	53	53°32′N	57°58′W
Meandarra, Austl. (mē-ăn-dă′rá)	176	27°47′S	149°40′E
Meaux, Fr. (mō)	126	48°58′N	2°53′E
Mecapalapa, Mex. (mā-kä-pä-lä′pä)	89	20°32′N	97°52′W
Mecatina, r., Can. (mā-k*à*-tē′n*à*)	61	50°50′N	59°45′W
Mecca (Makkah), Sau. Ar. (mĕk′à)	154	21°27′N	39°45′E
Mechanic Falls, Me., U.S. (mē-kăn′ĭk)	60	44°05′N	70°23′W
Mechanicsburg, Pa., U.S. (mē-kăn′ĭks-bŭrg)	67	40°15′N	77°00′W
Mechanicsville, Md., U.S. (mē-kăn′ĭks-vĭl)	68e	38°27′N	76°45′W
Mechanicville, N.Y., U.S. (mēkăn′ĭk-vĭl)	67	42°55′N	73°45′W
Mechelen, Bel.	121	51°01′N	4°28′E
Mechriyya, Alg.	118	33°30′N	0°13′W
Mecklenburg, hist. reg., Ger.	124	53°30′N	13°00′E
Medan, Indon. (mà-dän′)	168	3°35′N	98°35′E
Medanosa, Punta, c., Arg. (pōō′n-tä-mē-dä-nô′sä)	102	47°50′S	65°53′W
Medden, r., Eng., U.K. (mĕd′*ĕ*n)	114a	53°14′N	1°05′W
Medellín, Col. (mā-dhĕl-yēn′)	100	6°15′N	75°34′W
Medellín, Mex. (mĕ-dĕl-yē′n)	89	19°03′N	96°08′W
Medenine, Tun. (mā-dē-nēn′)	118	33°22′N	10°33′E
Medfeld, Ma., U.S. (mĕd′fēld)	61a	42°11′N	71°19′W
Medford, Ma., U.S. (mĕd′fẽrd)	61a	42°25′N	71°07′W
Medford, N.J., U.S.	68f	39°54′N	74°50′W
Medford, Ok., U.S.	79	36°47′N	97°44′W
Medford, Or., U.S.	64	42°19′N	122°52′W
Medford, Wi., U.S.	71	45°09′N	90°22′W
Media, Pa., U.S. (mē′dĭ-*à*)	68f	39°55′N	75°24′W
Mediaş, Rom. (mĕd-yäsh′)	125	46°09′N	24°21′E
Medical Lake, Wa., U.S. (mĕd′ĭ-*kăl*)	72	47°34′N	117°40′W
Medicine Bow, r., Wy., U.S.	73	41°58′N	106°30′W
Medicine Bow Range, mts., Co., U.S. (mĕd′ĭ-sĭn bō)	78	40°55′N	106°02′W
Medicine Hat, Can. (mĕd′ĭ-sĭn hăt)	50	50°03′N	110°40′W
Medicine Lake, l., Mt., U.S. (mĕd′ĭ-sĭn)	73	48°24′N	104°15′W
Medicine Lodge, Ks., U.S.	78	37°17′N	98°37′W
Medicine Lodge, r., Ks., U.S.	78	37°20′N	98°57′W
Medina *see* Al Madīnah, Sau. Ar.	154	24°26′N	39°42′E
Medina, N.Y., U.S. (mē-dī′n*à*)	67	43°15′N	78°20′W
Medina, Oh., U.S.	69d	41°08′N	81°52′W
Medina, r., Tx., U.S.	80	29°45′N	99°13′W
Medina del Campo, Spain (mā-dē′nä dĕl käm′pō)	118	41°18′N	4°54′W
Medina de Ríoseco, Spain (mā-dē′nä dā rē-ô-sā′kō)	128	41°53′N	5°05′W
Medina Lake, l., Tx., U.S.	80	29°36′N	98°47′W
Medina Sidonia, Spain	128	36°28′N	5°58′W
Mediterranean Sea, sea (mĕd-ĭ-tẽr-ā′nē-*ă*n)	118	36°22′N	13°25′E
Medjerda, Oued, r., Afr.	118	36°43′N	9°54′E
Mednogorsk, Russia	134	51°27′N	57°22′E
Medveditsa, r., Russia (mĕd-vyĕ′dē tsà)	137	50°10′N	43°40′E
Medvezhegorsk, Russia (mĕd-vyĕzh′yĕ-gôrsk′)	136	63°00′N	34°20′E
Medway, Ma., U.S. (mĕd′wā)	61a	42°08′N	71°23′W
Medway Towns, co., Eng., U.K.	114b	51°27′N	0°30′E
Medyn′, Russia (mĕ-dēn′)	132	54°58′N	35°53′E
Medzhybizh, Ukr.	133	49°23′N	27°29′E
Meekatharra, Austl. (mē-k*à*-thăr′*à*)	174	26°30′S	118°38′E
Meeker, Co., U.S. (mēk′ẽr)	77	40°00′N	107°55′W
Meelpaeg Lake, l., Can. (mēl′p*à*-ĕg)	61	48°22′N	56°52′W
Meerane, Ger. (mā-rä′nĕ)	124	50°51′N	12°27′E
Meerbusch, Ger.	127c	51°15′N	6°41′E
Meerut, India (mē′rŏt)	155	28°59′N	77°43′E
Megalópoli, Grc.	131	37°22′N	22°08′E
Mégara, Grc. (mĕg′*á*-r*à*)	131	37°59′N	23°21′E
Megget, S.C., U.S. (mĕg′ĕt)	83	32°44′N	80°15′W
Megler, Wa., U.S. (mĕg′lẽr)	74c	46°15′N	123°52′W
Mehanom, Mys, c., Ukr.	133	44°48′N	35°17′E
Meherrin, r., Va., U.S. (mē-hĕr′ĭn)	83	36°40′N	77°49′W
Mehlville, Mo., U.S.	75e	38°30′N	90°19′W
Mehsāna, India	158	23°42′N	72°23′E
Mehun-sur-Yèvre, Fr. (mē-ŭN-sür-yĕvr′)	126	47°11′N	2°14′E
Meiling Pass, p., China (mā′lĭng′)	161	25°22′N	115°00′E
Meinerzhagen, Ger. (mī′nĕrts-hä-gĕn)	127c	51°06′N	7°39′E
Meiningen, Ger. (mī′nĭng-*ĕ*n)	124	50°35′N	10°25′E
Meiringen, Switz.	124	46°45′N	8°11′E
Meissen, Ger.	124	51°11′N	13°28′E
Meizhu, China (mā-jōō)	162	31°17′N	119°12′E
Mejillones, Chile (må-кē-lyō′nås)	102	23°07′S	70°31′W
Mekambo, Gabon	190	1°01′N	13°56′E
Mekele, Eth.	185	13°31′N	39°19′E
Meknés, Mor. (mĕk′nĕs) (mĕk-nĕs′)	184	33°56′N	5°44′W
Mekong, r., Asia	168	18°00′N	104°30′E
Melaka, Malay.	168	2°11′N	102°15′E
Melaka, state, Malay.	153b	2°19′N	102°09′E
Melanesia, is., Oc.	194	13°00′S	164°00′E
Melbourne, Austl. (mĕl′b*ŭ*rn)	175	37°52′S	145°08′E
Melbourne, Eng., U.K.	114a	52°49′N	1°26′W
Melbourne, Fl., U.S.	83a	28°05′N	80°37′W
Melbourne, Ky., U.S.	69f	39°02′N	84°22′W
Melcher, Ia., U.S. (mĕl′chẽr)	71	41°13′N	93°11′W
Melekess, Russia (mĕl-yĕk-ĕs)	136	54°14′N	49°39′E
Melenki, Russia (mĕ-lyĕŋ′kē)	136	55°25′N	41°34′E
Melfort, Can. (mĕl′fôrt)	50	52°52′N	104°36′W
Melghir, Chott, l., Alg.	184	33°52′N	5°22′E
Melilla, Sp. N. Afr. (mā-lēl′yä)	184	35°24′N	3°30′W
Melipilla, Chile (må-lē-pē′lyä)	102	33°40′S	71°12′W
Melita, Can.	57	49°11′N	101°09′W
Melitopol′, Ukr. (mā-lē-tô′pôl-y′)	137	46°49′N	35°19′E
Melívoia, Grc.	131	39°42′N	22°47′E
Melkrivier, S. Afr.	192c	24°01′S	28°23′E
Mellen, Wi., U.S. (mĕl′*ĕ*n)	71	46°20′N	90°40′W
Mellerud, Swe. (mål′ĕ-rōōdh)	122	58°43′N	12°25′E
Melmoth, S. Afr.	187c	28°38′S	31°26′E
Melo, Ur. (mā′lō)	102	32°18′S	54°07′W
Melocheville, Can. (m*ĕ*-lôsh-vĕl′)	62a	45°24′N	73°56′W
Melozha, r., Russia (myĕ′lô-zhà)	142b	56°06′N	38°34′E
Melrose, Ma., U.S. (mĕl′rōz)	61a	42°29′N	71°06′W
Melrose, Mn., U.S.	71	45°39′N	94°49′W
Melrose Park, Il., U.S.	69a	41°54′N	87°52′W
Meltham, Eng., U.K. (mĕl′th*ă*m)	114a	53°35′N	1°51′W
Melton, Austl. (mĕl′t*ŭ*n)	173a	37°41′S	144°35′E
Melton Mowbray, Eng., U.K. (mō′brå)	114a	52°45′N	0°52′W
Melúli, r., Moz.	191	16°10′S	39°30′E
Melun, Fr. (mē-lŭN′)	117	48°32′N	2°40′E
Melunga, Ang.	190	17°16′S	16°24′E
Melville, Can. (mĕl′vĭl)	50	50°55′N	102°48′W
Melville, La., U.S.	81	30°39′N	91°45′W
Melville, i., Austl.	174	11°30′S	131°12′E
Melville, l., Can.	53	53°46′N	59°31′W
Melville, Cape, c., Austl.	175	14°15′S	145°50′E
Melville Hills, hills, Can.	52	69°18′N	124°57′W
Melville Peninsula, pen., Can.	53	67°44′N	84°09′W
Melvindale, Mi., U.S. (mĕl′vĭn-dāl)	69b	42°17′N	83°11′W
Melyana, Alg.	117	36°19′N	1°56′E
Mélykút, Hung. (mā′l′kōōt)	125	46°14′N	19°21′E
Memba, Moz. (mĕm′b*à*)	187	14°12′N	40°35′E
Memel *see* Klaipėda, Lith.	136	55°43′N	21°10′E
Memel, S. Afr. (mĕ′m*ĕ*l)	192c	27°42′S	29°35′E
Memmingen, Ger. (mĕm′ĭng-*ĕ*n)	124	47°59′N	10°10′E
Memo, r., Ven. (mĕ′mō)	101b	9°32′N	66°30′W
Memphis, Mo., U.S. (mĕm′fĭs)	79	40°27′N	92°11′W
Memphis, Tn., U.S. (mĕm′fĭs)	65	35°07′N	90°03′W
Memphis, Tx., U.S.	78	34°42′N	100°33′W
Memphis, hist., Egypt	192b	29°50′N	31°12′E
Mena, Ukr. (mĕ-n*á*′)	133	51°31′N	32°14′E
Mena, Ar., U.S. (mē′n*à*)	79	34°35′N	94°09′W
Menangle, Austl.	173b	34°08′S	150°48′E
Menard, Tx., U.S. (mē-närd′)	80	30°56′N	99°48′W
Menasha, Wi., U.S. (mē-năsh′*à*)	71	44°12′N	88°29′W
Mende, Fr. (mänd)	126	44°31′N	3°30′E
Menden, Ger. (mĕn′dĕn)	127c	51°26′N	7°47′E
Mendes, Braz. (mĕ′n-dĕs)	102b	22°32′S	43°44′W
Mendocino, Ca., U.S.	76	39°18′N	123°47′W
Mendocino, Cape, c., Ca., U.S. (mĕn′dô-sē′nō)	65	40°25′N	12°42′W
Mendota, Il., U.S. (mĕn-dō′t*à*)	71	41°34′N	89°06′W
Mendota, l., Wi., U.S.	71	43°09′N	89°41′W

ăt; finăl; rāte; senåte; ärm; åsk; sofá; fâre; ch-choose; dh-as th in other; bē; ēvent; bĕt; recĕnt; cratẽr; g-gō; gh-guttural g; bĭt; ĭ-short neutral; rīde; к-guttural k as ch in German ich;

PRONOUNCING INDEX Mendoza – Millicent 275

PLACE (Pronunciation)	PAGE	LAT.	LONG.
Mendoza, Arg. (mĕn-dō'sä)	102	32°48'S	68°45'W
Mendoza, prov., Arg.	102	35°10'S	69°00'W
Mengcheng, China (mŭŋ-chŭŋ)	162	33°15'N	116°34'E
Meng Shan, mts., China (mŭŋ shän)	162	35°47'N	117°23'E
Mengzi, China	160	23°22'N	103°20'E
Menindee, Austl. (mĕ-nĭn-dē)	176	32°23'S	142°30'E
Menlo Park, Ca., U.S. (mĕn'lō pärk)	74b	37°27'N	122°11'W
Menno, S.D., U.S. (mĕn'ō)	70	43°14'N	97°34'W
Menominee, Mi., U.S. (mê-nŏm'ĭ-nē)	71	45°08'N	87°40'W
Menominee, r., Mi., U.S.	71	45°37'N	87°54'W
Menominee Falls, Wi., U.S. (fôls)	69a	43°11'N	88°06'W
Menominee Range, Mi., U.S.	71	46°07'N	88°53'W
Menomonee, r., Wi., U.S.	69a	43°09'N	88°06'W
Menomonie, Wi., U.S.	71	44°53'N	91°55'W
Menongue, Ang.	190	14°36'S	17°48'E
Menorca (Minorca), i., Spain (mĕ-nô'r-kä)	112	40°05'N	3°58'E
Mentana, Italy (mĕn-tá'nä)	129d	42°02'N	12°40'E
Mentawai, Kepulauan, is., Indon. (mĕn-tä-vī')	168	1°08'S	98°10'E
Menton, Fr. (mäN-tôN')	127	43°46'N	7°37'E
Mentone, Ca., U.S. (mĕn'tōne)	75a	34°05'N	117°08'W
Mentz, l., S. Afr. (mĕnts)	187c	33°13'S	25°15'E
Menzel Bourguiba, Tun.	118	37°12'N	9°51'E
Menzelinsk, Russia (mĕn'zyĕ-lênsk')	136	55°40'N	53°15'E
Menzies, Austl. (mĕn'zēz)	174	29°45'S	122°15'E
Meogui, Mex. (må-ō'gē)	80	28°17'N	105°28'W
Meppel, Neth. (mĕp'ĕl)	121	52°41'N	6°08'E
Meppen, Ger. (mĕp'ĕn)	124	52°40'N	7°18'E
Merabéllou, Kólpos, b., Grc.	130a	35°16'N	25°55'E
Meramec, r., Mo., U.S. (mĕr'á-mĕk)	79	38°06'N	91°06'W
Merano, Italy (må-rä'nō)	118	46°39'N	11°10'E
Merasheen, i., Can. (mē'rá-shēn)	61	47°30'N	54°15'W
Merauke, Indon. (må-rou'kä)	169	8°32'S	140°17'E
Meraux, La., U.S. (mē-ro')	68d	29°56'N	89°56'W
Mercato San Severino, Italy	129c	40°34'N	14°38'E
Merced, Ca., U.S. (mĕr-sĕd')	76	37°17'N	120°30'W
Merced, r., Ca., U.S.	76	37°25'N	120°31'W
Mercedario, Cerro, mtn., Arg. (mĕr-så-dhä'rē-ō)	102	31°58'S	70°07'W
Mercedes, Arg.	99c	34°41'S	59°26'W
Mercedes, Arg. (mĕr-sā'dhās)	102	29°04'S	58°01'W
Mercedes, Ur.	102	33°17'S	58°04'W
Mercedes, Tx., U.S.	81	26°09'N	97°55'W
Mercedita, Chile (mĕr-sĕ-dē'tä)	99b	33°51'S	71°10'W
Mercer Island, Wa., U.S. (mûr'sēr)	74a	47°35'N	122°15'W
Mercês, Braz. (mĕ-sĕ's)	99a	21°13'S	43°20'W
Merchtem, Bel.	115a	50°57'N	4°13'E
Mercier, Can.	62a	45°19'N	73°45'W
Mercy, Cape, c., Can.	53	64°48'N	63°22'W
Meredith, N.H., U.S. (mĕr'ē-dĭth)	67	43°35'N	71°35'W
Merefa, Ukr. (må-rĕf'á)	133	49°49'N	36°04'E
Merendón, Serranía de, mts., Hond.	90	15°01'N	89°05'W
Mereworth, Eng., U.K. (mĕ-rĕ wûrth)	114b	51°15'N	0°23'E
Mergui, Mya. (mĕr-gē')	168	12°29'N	98°39'E
Mergui Archipelago, is., Mya.	168	12°04'N	97°02'E
Meric (Maritsa), r., Eur.	123	40°43'N	26°19'E
Mérida, Mex.	86	20°58'N	89°37'W
Mérida, Ven.	100	8°30'N	71°15'W
Mérida, Cordillera de, mts., Ven. (mĕ'rē-dhä)	100	8°30'N	70°45'W
Meriden, Ct., U.S. (mĕr'ĭ-dĕn)	67	41°30'N	72°50'W
Meridian, Ms., U.S. (mē-rĭd-ĭ-ăn)	65	32°21'N	88°41'W
Meridian, Tx., U.S.	81	31°56'N	97°37'W
Mérignac, Fr.	126	44°50'N	0°40'W
Merikarvia, Fin. (mā'rē-kàr'vē-à)	123	61°51'N	21°30'E
Mering, Ger. (mĕ'rēng)	115d	48°16'N	11°00'E
Merkel, Tx., U.S. (mûr'kĕl)	80	32°26'N	100°02'W
Merkinė, Lith.	123	54°10'N	24°10'E
Merksem, Bel.	115a	51°15'N	4°27'E
Merkys, r., Lith. (mär'kĭs)	125	54°23'N	25°00'E
Merlo, Arg. (mĕr-lô)	102a	34°40'S	58°44'W
Meron, Hare, mtn., Isr.	153a	32°58'N	35°25'E
Merriam, Ks., U.S. (mĕr-rĭ-yàm)	75f	39°01'N	94°42'W
Merriam, Mn., U.S.	75g	44°44'N	93°36'W
Merrick, N.Y., U.S. (mĕr'ĭk)	68a	40°40'N	73°33'W
Merrifield, Va., U.S. (mĕr'ĭ-fēld)	68e	38°50'N	77°12'W
Merrill, Wi., U.S. (mĕr'ĭl)	71	45°11'N	89°42'W
Merrimac, Ma., U.S. (mĕr'ĭ-măk)	61a	45°20'N	71°00'W
Merrimack, N.H., U.S.	61a	42°51'N	71°25'W
Merrimack, r., Ma., U.S. (mĕr'ĭ-măk)	67	43°10'N	71°30'W
Merritt, Can. (mĕr'ĭt)	50	50°07'N	120°47'W
Merryville, La., U.S. (mĕr'ĭ-vĭl)	81	30°46'N	93°34'W
Mersa Fatma, Erit.	185	14°54'N	40°14'E
Merseburg, Ger. (mĕr'zĕ-bŏŏrgh)	124	51°21'N	11°59'E
Mersey, r., Eng., U.K. (mûr'zē)	114a	53°20'N	2°55'W
Merseyside, hist. reg., Eng., U.K.	114a	53°29'N	2°59'W
Mersing, Malay.	153b	2°25'N	103°51'E
Merta Road, India (mār'tŭ rōd)	158	26°50'N	73°54'E
Merthyr Tydfil, Wales, U.K. (mûr'thēr tĭd'vĭl)	120	51°46'N	3°30'W
Mértola Almodóvar, Port. (mĕr-tô-lá-äl-mô-dô'vär)	128	37°39'N	8°04'W
Méru, Fr. (mā-rü')	126	49°14'N	2°08'E
Meru, Kenya (mā'rōō)	185	0°01'N	37°45'E
Meru, Mount, mtn., Tan.	191	3°15'S	36°43'E
Merume Mountains, mts., Guy. (mĕr-ü'mē)	101	5°45'N	60°15'W
Merwede Kanaal, can., Neth.	115a	52°15'N	5°01'E
Merwin, l., Wa., U.S. (mĕr'wĭn)	74c	45°58'N	122°27'W
Merzifon, Tur. (mĕr'ze-fōn)	154	40°50'N	35°30'E
Mesa, Az., U.S. (mā'sá)	77	33°25'N	111°50'W
Mesabi Range, mts., Mn., U.S. (mā-sŏb'bē)	71	47°17'N	93°04'W
Mesagne, Italy (mā-sän'yå)	131	40°34'N	17°51'E
Mesa Verde National Park, rec., Co., U.S. (vĕr'dē)	64	37°22'N	108°27'W
Mescalero Apache Indian Reservation, I.R., N.M., U.S. (mĕs-kä-lā'rō)	77	33°10'N	105°45'W
Meshchovsk, Russia (myĕsh'chĕfsk)	132	54°17'N	35°19'E
Mesilla, N.M., U.S. (må-sē'yä)	77	32°15'N	106°45'W
Meskine, Chad	189	11°25'N	15°21'E
Mesolóngi, Grc.	131	38°23'N	21°28'E
Mesopotamia, hist. reg., Asia	157	34°00'N	44°00'E
Mesquita, Braz.	102b	22°48'S	43°26'W
Messina, Italy (mĕ-sē'ná)	110	38°11'N	15°34'E
Messina, S. Afr.	186	22°17'S	30°13'E
Messina, Stretto di, strt., Italy (stĕ't-tô dē)	119	38°10'N	15°34'E
Messíni, Grc.	131	37°05'N	22°00'E
Mestaganem, Alg.	184	36°04'N	0°11'E
Mestre, Italy (mĕs'trā)	130	45°29'N	12°15'E
Meta, dept., Col. (mĕ'tä)	100a	3°28'N	74°07'W
Meta, r., S.A.	100	4°33'N	72°09'W
Métabetchouane, r., Can. (mĕ-tá-bĕt-chōō-än')	59	47°45'N	72°00'W
Metairie, La., U.S.	81	30°00'N	90°11'W
Metán, Arg. (mĕ-tá'n)	102	25°32'S	64°51'W
Metangula, Moz.	186	12°42'S	34°48'E
Metapán, El Sal. (må-täpän')	90	14°21'N	89°26'W
Metcalfe, Can. (mĕt-käf)	62c	45°14'N	75°27'W
Metchosin, Can.	74a	48°22'N	123°33'W
Metepec, Mex. (må-tĕ-pĕk')	88	18°56'N	98°31'W
Metepec, Mex.	88	19°15'N	99°36'W
Methow, r., Wa., U.S. (mĕt'hou) (mĕt hou')	72	48°26'N	120°15'W
Methuen, Ma., U.S. (mê-thū'ĕn)	61a	42°44'N	71°11'W
Metković, Cro. (mĕt'kô-vĭch)	131	43°02'N	17°40'E
Metlakatla, Ak., U.S. (mĕt-lá-kăt'lá)	63	55°08'N	131°35'W
Metropolis, Il., U.S. (mê-trŏp'ô-lĭs)	79	37°09'N	88°46'W
Metter, Ga., U.S. (mĕt'ēr)	83	32°21'N	82°05'W
Mettmann, Ger. (mĕt'män)	127c	51°15'N	6°58'E
Metuchen, N.J., U.S. (mĕ-tŭ'chĕn)	68a	40°32'N	74°21'W
Metz, Fr. (mĕtz)	117	49°08'N	6°10'E
Metztitlán, Mex. (mĕtz-tĕt-län)	88	20°36'N	98°45'W
Meuban, Cam.	189	2°27'N	12°41'E
Meuse (Maas), r., Eur. (mûz) (müz)	121	50°32'N	5°22'E
Mexborough, Eng., U.K. (mĕks'bŭr-ô)	114a	53°30'N	1°17'W
Mexia, Tx., U.S. (må-hē'ä)	81	31°32'N	96°29'W
Mexian, China	161	24°20'N	116°10'E
Mexicalcingo, Mex. (mĕ-kē-käl-sĕn'go)	89a	19°13'N	99°34'W
Mexicali, Mex. (māk-sē-kä'lē)	86	32°28'N	115°29'W
Mexicana, Altiplanicie, plat., Mex.	88	22°38'N	102°33'W
Mexican Hat, Ut., U.S. (mĕk'sĭ-kăn hăt)	77	37°10'N	109°55'W
Mexico, Me., U.S. (mĕk'sĭ-kō)	60	44°34'N	70°33'W
Mexico, Mo., U.S.	79	39°09'N	91°51'W
Mexico, nation, N.A.	86	23°45'N	104°00'W
Mexico, Gulf of, b., N.A.	86	25°15'N	93°45'W
Mexico City, Mex. (mĕk'sĭ-kō)	86	19°28'N	99°09'W
Mexticacán, Mex. (mĕs'tē-kä-kän')	88	21°12'N	102°43'W
Meyersdale, Pa., U.S. (mī'ērz-dāl)	67	39°55'N	79°00'W
Meyerton, S. Afr. (mī'ēr-tŭn)	192c	26°35'S	28°01'E
Meymaneh, Afg.	154	35°53'N	64°38'E
Mezen', Russia	134	65°50'N	44°05'E
Mezen', r., Russia	136	65°20'N	44°45'E
Mézenc, Mont, mtn., Fr. (mōN-mä-zĕN')	126	44°55'N	4°12'E
Mezha, r., Eur. (myā'zhá)	132	55°53'N	31°44'E
Mézieres-sur-Seine, Fr. (mā-zyâr'sür-sån')	127b	48°58'N	1°49'E
Mezőkövesd, Hung. (mĕ'zû-kû'vĕsht)	125	47°49'N	20°36'E
Mezőtúr, Hung. (mĕ'zû-tōōr)	125	47°00'N	20°36'E
Mezquital, Mex. (måz-kē-täl')	88	23°30'N	104°20'W
Mezquitic, Mex. (måz-kê-tēk')	88	22°25'N	103°43'W
Mezquitic, r., Mex.	88	22°25'N	103°45'W
Mfangano Island, i., Kenya	191	0°28'S	33°35'E
Mga, Russia (m'gá)	142c	59°45'N	31°04'E
Mglin, Russia (m'glēn')	132	53°03'N	32°52'E
Mia, Oued, r., Alg.	118	29°26'N	3°15'E
Miacatlán, Mex. (mê'ä-kä-tlän')	88	18°42'N	99°17'W
Miahuatlán, Mex. (mê'ä-wä-tlän')	89	16°20'N	96°38'W
Miajadas, Spain (mê-ä-hä'däs)	128	39°10'N	5°53'W
Miami, Az., U.S.	64	33°20'N	110°55'W
Miami, Fl., U.S.	65	25°45'N	80°11'W
Miami, Ok., U.S.	79	36°51'N	94°51'W
Miami, Tx., U.S.	78	35°41'N	100°39'W
Miami Beach, Fl., U.S.	83a	25°47'N	80°07'W
Miamisburg, Oh., U.S. (mī-ăm'ĭz-bûrg)	66	39°40'N	84°20'W
Miamitown, Oh., U.S. (mī-ăm'ĭ-toun)	69f	39°13'N	84°43'W
Mīāneh, Iran	154	37°15'N	47°13'E
Miangas, Pulau, i., Indon.	169	5°30'N	127°00'E
Miaoli, Tai. (mê-ou'lĭ)	165	24°30'N	120°48'E
Miaozhen, China (miou-jŭn)	162	31°44'N	121°28'E
Miass, Russia (mī-äs')	140	54°59'N	60°06'E
Miastko, Pol. (myäst'kô)	124	54°01'N	17°00'E
Miccosukee Indian Reservation, I.R., Fl., U.S.	83a	26°10'N	80°50'W
Michalovce, Slvk. (mē'ĸä-lôf'tsĕ)	125	48°44'N	21°56'E
Michel Peak, mtn., Can.	54	53°35'N	126°25'W
Michelson, Mount, mtn., Ak., U.S. (mĭch'ĕl-sŭn)	63	69°11'N	144°12'W
Michendorf, Ger. (mē'kĕn-dōrf)	115b	52°19'N	13°02'E
Miches, Dom. Rep. (mē'chĕs)	93	19°00'N	69°05'W
Michigan, state, U.S. (mĭsh-'ĭ-găn)	65	45°55'N	87°00'W
Michigan, Lake, l., U.S.	65	43°20'N	87°10'W
Michigan City, In., U.S.	66	41°40'N	86°55'W
Michipicoten, r., Can.	71	47°56'N	84°42'W
Michipicoten Harbour, Can.	71	47°58'N	84°58'W
Michurinsk, Russia (mĭ-chōō-rĭnsk')	137	52°53'N	40°32'E
Mico, Punta, c., Nic. (pōō'n-tä-mē'kô)	91	11°38'N	83°24'W
Micronesia, is., Oc.	194	11°00'N	159°00'E
Micronesia, Federated States of, nation, Oc.	3	5°00'N	152°00'E
Midas, Nv., U.S. (mī'dàs)	72	41°15'N	116°50'W
Middelfart, Den. (mĕd'ĭl-färt)	122	55°30'N	9°45'E
Middle, r., Can.	54	55°00'N	125°50'W
Middle Andaman, i., India (ăn-dá-măn')	168	12°44'N	93°21'E
Middle Bayou, Tx., U.S.	81a	29°38'N	95°06'W
Middleburg, S. Afr. (mĭd'ĕl-bûrg)	186	31°30'S	25°00'E
Middleburg, S. Afr.	192c	25°47'S	29°30'E
Middlebury, Vt., U.S. (mĭd''l-bĕr-ĭ)	67	44°00'N	73°10'W
Middle Concho, Tx., U.S. (kŏn'chō)	80	31°21'N	100°50'W
Middle River, Md., U.S.	68e	39°20'N	76°27'W
Middlesboro, Ky., U.S. (mĭd''lz-bŭr-ô)	82	36°36'N	83°42'W
Middlesbrough, Eng., U.K. (mĭd''lz-brŭ)	116	54°35'N	1°18'W
Middlesex, N.J., U.S. (mĭd''l-sĕks)	68a	40°34'N	74°30'W
Middleton, Can. (mĭd''l-tŭn)	60	44°57'N	65°04'W
Middleton, Eng., U.K.	114a	53°34'N	2°12'W
Middletown, Ct., U.S.	67	41°35'N	72°40'W
Middletown, De., U.S.	67	39°30'N	75°40'W
Middletown, Ma., U.S.	61a	42°35'N	71°01'W
Middletown, N.Y., U.S.	67	41°26'N	74°25'W
Middletown, Oh., U.S.	66	39°30'N	84°25'W
Middlewich, Eng., U.K. (mĭd''l-wĭch)	114a	53°11'N	2°27'W
Middlewit, S. Afr. (mĭd'l'wĭt)	192c	24°50'S	27°00'E
Midfield, Al., U.S.	68h	33°28'N	86°54'W
Midi, Canal du, Fr. (kä-näl-dü-mē-dē')	117	43°22'N	1°35'E
Mid Illovo, S. Afr. (mĭd ĭl'ô-vō)	187c	29°59'S	30°32'E
Midland, Can. (mĭd'lănd)	51	44°45'N	79°50'W
Midland, Mi., U.S.	66	43°40'N	84°20'W
Midland, Tx., U.S.	80	32°05'N	102°05'W
Midvale, Ut., U.S. (mĭd'vāl)	75b	40°37'N	111°54'W
Midway, Al., U.S. (mĭd'wā)	82	32°03'N	85°30'W
Midway Islands, is., Oc.	2	28°00'N	179°00'W
Midwest, Wy., U.S. (mĭd-wĕst')	73	43°25'N	106°15'W
Midye, Tur. (mēd'yĕ)	137	41°35'N	28°10'E
Międzyrzecz, Pol. (myăn-dzû'zhĕch)	124	52°26'N	15°35'E
Mielec, Pol. (myĕ'lĕts)	125	50°17'N	21°27'E
Mier, Mex. (myâr)	80	26°26'N	99°08'W
Mieres, Spain (myā'rās)	128	43°14'N	5°45'W
Mier y Noriega, Mex. (myâr'ē nô-rē-ā'gä)	88	23°28'N	100°08'W
Miguel Auza, Mex.	88	24°17'N	103°27'W
Miguel Pereira, Braz.	102b	22°27'S	43°28'W
Mijares, r., Spain	129	39°55'N	0°01'W
Mikage, Japan (mē'kà-gå)	167b	34°42'N	135°15'E
Mikawa-Wan, b., Japan (mē'kä-wä wän)	167	34°43'N	137°09'E
Mikhaylov, Russia (mē-käy'lôf)	136	54°14'N	39°03'E
Mikhaylovka, Russia	142a	55°35'N	57°57'E
Mikhaylovka, Russia	142c	59°20'N	30°21'E
Mikhaylovka, Russia	137	50°05'N	43°10'E
Mikhnëvo, Russia (mĭk-nyô'vô)	142b	55°08'N	37°57'E
Miki, Japan (mē'kê)	167b	34°47'N	134°59'E
Mikindani, Tan. (mê-kên-dä'nê)	187	10°17'S	40°07'E
Mikkeli, Fin. (mĕk'ĕ-lĭ)	116	61°42'N	27°14'E
Mikulov, Czech Rep. (mĭ'kōō-lôf)	124	48°47'N	16°39'E
Mikumi, Tan.	191	7°24'S	36°59'E
Mikuni, Japan (mē'kōō-nê)	167	36°09'N	136°14'E
Mikuni-Sammyaku, mts., Japan (säm'myä-kōō)	167	36°51'N	138°38'E
Mikura, i., Japan (mē'kōō-rà)	167	33°53'N	139°26'E
Milaca, Milaca, Mn., U.S. (mĭ-lăk'á)	71	45°45'N	93°41'W
Milan (Milano), Italy (mê-lä'nō)	130	45°29'N	9°12'E
Milan, Mi., U.S. (mī'lăn)	66	42°05'N	83°40'W
Milan, Mo., U.S.	79	40°13'N	93°07'W
Milan, Tn., U.S.	82	35°54'N	88°47'W
Milâs, Tur. (mê'läs)	119	37°10'N	27°25'E
Milazzo, Italy	130	38°13'N	15°17'E
Milbank, S.D., U.S. (mĭl'băŋk)	70	45°13'N	96°38'W
Mildura, Austl. (mĭl-dū'rá)	175	34°10'S	142°18'E
Miles City, Mt., U.S. (mīlz)	64	46°24'N	105°50'W
Milford, Ct., U.S. (mĭl'fērd)	67	41°15'N	73°05'W
Milford, De., U.S.	67	38°55'N	75°25'W
Milford, Ma., U.S.	61a	42°09'N	71°31'W
Milford, Mi., U.S.	69b	42°35'N	83°36'W
Milford, N.H., U.S.	67	42°50'N	71°40'W
Milford, Oh., U.S.	69f	39°11'N	84°18'W
Milford, Ut., U.S.	77	38°20'N	113°05'W
Milford Sound, strt., N.Z.	177	44°35'S	167°47'E
Miling, Austl. (mĭl''ng)	174	30°30'S	116°25'E
Milipitas, Ca., U.S. (mĭl-ĭ-pĭ'tàs)	74b	37°26'N	121°54'W
Milk, r., N.A.	64	48°30'N	107°00'W
Millau, Fr. (mē-yō')	117	44°06'N	3°04'E
Millbrae, Ca., U.S. (mĭl'brā)	74b	37°36'N	122°23'W
Millbury, Ma., U.S. (mĭl'bĕr-ĭ)	61a	42°12'N	71°46'W
Mill Creek, r., Can. (mĭl)	62g	53°28'N	113°25'W
Mill Creek, r., Ca., U.S.	76	40°07'N	121°55'W
Milledgeville, Ga., U.S. (mĭl'ĕj-vĭl)	82	33°05'N	83°15'W
Mille Îles, Rivière des, r., Can. (rê-vyâr' dā mĭl'ĭl')	62a	45°41'N	73°40'W
Mille Lac Indian Reservation, I.R., Mn., U.S. (mĭl lăk')	71	46°14'N	94°13'W
Mille Lacs, l., Mn., U.S.	71	46°25'N	93°22'W
Mille Lacs, Lac des, l., Can. (läk dē mēl läks)	58	48°52'N	90°53'W
Millen, Ga., U.S. (mĭl'ĕn)	83	32°47'N	81°55'W
Miller, S.D., U.S. (mĭl'ēr)	70	44°31'N	99°00'W
Millerovo, Russia (mĭl'ĕ-rô-vô)	137	48°58'N	40°27'E
Millersburg, Ky., U.S. (mĭl'ērz-bûrg)	66	38°15'N	84°10'W
Millersburg, Oh., U.S.	66	40°35'N	81°55'W
Millersburg, Pa., U.S.	67	40°35'N	76°55'W
Millerton, Can. (mĭl'ēr-tŭn)	60	46°56'N	65°40'W
Millertown, Can. (mĭl'ēr-toun)	61	48°49'N	56°32'W
Millicent, Austl. (mĭl-ĭ-sĕnt)	176	37°30'S	140°20'E

276 Millinocket – Molody Tud PRONOUNCING INDEX

PLACE (Pronunciation)	PAGE	LAT.	LONG.
Millinocket, Me., U.S. (mĭl-ĭ-nŏk'ĕt)	60	45°40'N	68°44'W
Millis, Ma., U.S. (mĭl-ĭs)	61a	42°10'N	71°22'W
Millstadt, Il., U.S. (mĭl'stăt)	75e	38°27'N	90°06'W
Millstone, r., N.J., U.S. (mĭl'stōn)	68a	40°27'N	74°38'W
Millstream, Austl. (mĭl'strēm)	174	21°45'S	117°10'E
Milltown, Can. (mĭl'toun)	60	45°13'N	67°19'W
Mill Valley, Ca., U.S. (mĭl)	74b	37°54'N	122°32'W
Millwood Reservoir, res., Ar., U.S.	79	33°00'N	94°00'W
Milly-la-Forêt, Fr. (mē-yē'-la-fō-rĕ')	127b	48°24'N	2°28'E
Milnerton, S. Afr. (mĭl'nĕr-tŭn)	186a	33°52'S	18°30'E
Milnor, N.D., U.S. (mĭl'nēr)	70	46°17'N	97°29'W
Milo, Me., U.S.	60	44°16'N	69°01'W
Milos, i., Grc. (mē'lōs)	119	36°45'N	24°35'E
Milpa Alta, Mex. (mē'l-pä-ä'l-tä)	89a	19°11'N	99°01'W
Milton, Can.	62d	43°31'N	79°53'W
Milton, Fl., U.S. (mĭl'tŭn)	82	30°37'N	87°02'W
Milton, Pa., U.S.	67	41°00'N	76°50'W
Milton, Ut., U.S.	75b	41°04'N	111°44'W
Milton, Wa., U.S.	74a	47°15'N	122°20'W
Milton, Wi., U.S.	71	42°45'N	89°00'W
Milton-Freewater, Or., U.S.	72	45°57'N	118°25'W
Milvale, Pa., U.S. (mĭl'văl)	69e	40°29'N	79°58'W
Milville, N.J., U.S. (mĭl'vĭl)	67	39°25'N	75°00'W
Milwaukee, Wi., U.S.	65	43°03'N	87°55'W
Milwaukee, r., Wi., U.S.	69a	43°10'N	87°56'W
Milwaukie, Or., U.S. (mĭl-wô'kē)	72	45°27'N	122°38'W
Mimiapan, Mex. (mē-myä-pån')	89a	19°26'N	99°28'W
Mimoso do Sul, Braz. (mē-mô'sō-dô-sōō'l)	99a	21°03'S	41°21'W
Min, r., China (mēn)	161	26°03'N	118°30'E
Min, r., China	165	29°30'N	104°00'E
Mina, r., Alg. (mē'nå)	129	35°24'N	0°51'E
Minago, r., Can. (mĭ-nä'gō)	57	54°25'N	98°45'W
Minakuchi, Japan (mē'nä-kōō'chê)	167	34°59'N	136°06'E
Minas, Cuba (mē'näs)	92	21°30'N	77°35'W
Minas, Indon.	153b	0°52'N	101°29'E
Minas, Ur. (mē'näs)	102	34°18'S	55°12'W
Minas, Sierra de las, mts., Guat. (syĕr'rä dā läs mē'näs)	90	15°08'N	90°25'W
Minas Basin, b., Can. (mĭ'nås)	60	45°20'N	64°00'W
Minas Channel, strt., Can.	60	45°15'N	64°45'W
Minas de Oro, Hond. (mē'näs-dĕ-dĕ-ō-rō)	90	14°52'N	87°19'W
Minas de Riotinto, Spain (mē'näs dā rē-ō-tēn'tō)	128	37°43'N	6°35'W
Minas Novas, Braz. (mē'näzh nō'väzh)	101	17°20'S	42°19'W
Minatare, l., Ne., U.S. (mĭn'å-târ)	70	41°56'N	103°07'W
Minatitlán, Mex. (mê-nä-tê-tlän')	86	17°59'N	94°33'W
Minatitlán, Mex.	88	19°21'N	104°02'W
Minato, Japan (mē'nå-tô)	167	35°13'N	139°52'E
Minch, The, strt., Scot., U.K.	112	58°04'N	6°04'W
Mindanao, i., Phil.	169	8°00'N	125°00'E
Mindanao Sea, sea, Phil.	169	8°55'N	124°00'E
Minden, Ger. (mĭn'dĕn)	124	52°17'N	8°58'E
Minden, La., U.S.	81	32°36'N	93°19'W
Minden, Ne., U.S.	78	40°30'N	98°54'W
Mindoro, i., Phil.	168	12°50'N	121°05'E
Mindoro Strait, strt., Phil.	169a	12°28'N	120°33'E
Mindyak, Russia (mēn'dyák)	142a	54°01'N	58°48'E
Mineola, N.Y., U.S. (mĭn-ē-ō'lå)	68a	40°43'N	73°38'W
Mineola, Tx., U.S.	81	32°39'N	95°31'W
Mineral del Chico, Mex. (mê-nå-räl'dĕl chē'kō)	88	20°13'N	98°46'W
Mineral del Monte, Mex. (mê-nå-räl dĕl mōn'tå)	88	20°18'N	98°39'W
Mineral'nyye Vody, Russia	137	44°10'N	43°15'E
Mineral Point, Wi., U.S. (mĭn'ēr-ăl)	71	42°50'N	90°10'W
Mineral Wells, Tx., U.S. (mĭn'ēr-ăl wĕlz)	80	32°48'N	98°06'W
Minerva, Oh., U.S. (mĭ-nur'vå)	66	40°45'N	81°10'W
Minervino, Italy (mē-nĕr-vē'nô)	130	41°07'N	16°05'E
Mineyama, Japan (mē-nĕ-yä'mä)	167	35°38'N	135°05'E
Mingäçevir, Azer.	138	40°45'N	47°03'E
Mingäçevir su anbarı, res., Azer.	138	40°50'N	46°50'E
Mingan, Can.	51	50°18'N	64°02'W
Mingenew, Austl. (mĭn'gĕ-nû)	174	29°15'S	115°45'E
Mingo Junction, Oh., U.S. (mĭŋ'gō)	66	40°15'N	80°40'W
Minho, hist. reg., Port. (mēn yô)	128	41°32'N	8°13'W
Minho (Miño), r., Eur. (mē'n-yô)	128	41°28'N	9°05'W
Ministik Lake, l., Can. (mĭ-nĭs'tĭk)	62g	53°23'N	113°05'W
Minna, Nig. (mĭn'å)	184	9°37'N	6°33'E
Minneapolis, Ks., U.S. (mĭn-ē-ăp'ō-lĭs)	79	39°07'N	97°41'W
Minneapolis, Mn., U.S.	65	44°58'N	93°15'W
Minnedosa, Can. (mĭn-ē-dō'så)	50	50°14'N	99°51'W
Minneota, Mn., U.S. (mĭn-ē-ō'tå)	70	44°34'N	95°59'W
Minnesota, state, U.S. (mĭn-ē-sō'tå)	65	46°10'N	90°20'W
Minnesota, r., Mn., U.S.	65	44°30'N	95°00'W
Minnetonka, l., Mn., U.S. (mĭn-ē-tŏŋ'kå)	71	44°52'N	93°34'W
Minnitaki Lake, l., Can. (mĭ'nĭ-tä'kê)	57	49°58'N	92°00'W
Mino, r., Japan	167b	34°56'N	135°06'E
Minonk, Il., U.S. (mĭ'nŏnk)	66	40°55'N	89°00'W
Minooka, Il., U.S. (mĭ-nōō'kå)	69a	41°27'N	88°15'W
Minot, N.D., U.S.	64	48°13'N	101°17'W
Minsk, Bela. (mênsk)	134	53°54'N	27°35'E
Minsk, prov., Bela.	132	53°50'N	27°43'E
Mińsk Mazowiecki, Pol. (mēn'sk mä-zō-vyĕt'skĭ)	125	52°10'N	21°35'E
Minsterley, Eng., U.K. (mĭnstēr-lē)	114a	52°38'N	2°55'W
Minto, Can.	60	46°05'N	66°05'W
Minto, l., Can.	53	57°18'N	75°50'W
Minturno, Italy (mēn-tōōr'nō)	130	41°17'N	13°44'E
Minūf, Egypt (mê-nōōf')	192b	30°26'N	30°55'E
Minusinsk, Russia (mē-nō-sēnsk')	135	53°47'N	91°45'E
Min'yar, Russia	142a	55°06'N	57°33'E

PLACE (Pronunciation)	PAGE	LAT.	LONG.
Miquelon Lake, l., Can. (mĭ'kê-lôn)	62g	53°16'N	112°55'W
Miquihuana, Mex. (mê-kê-wä'nä)	88	23°36'N	99°45'W
Mir, Bela. (mēr)	125	53°27'N	26°25'E
Miracema, Braz. (mē-rä-sĕ'mä)	99a	21°24'S	42°10'W
Miracema do Tocantins, Braz.	101	9°34'S	48°24'W
Mirador, Braz. (mê-rå-dōr')	101	6°19'S	44°12'W
Miraflores, Col. (mē-rä-flō'räs)	100	5°10'N	73°13'W
Miraflores, Peru	100	16°19'S	71°20'W
Miraflores Locks, trans., Pan.	86a	9°00'N	79°35'W
Miragoâne, Haiti (mê-rå-gwän')	93	18°25'N	73°05'W
Mira Loma, Ca., U.S. (mĭ'rå lō'må)	75a	34°01'N	117°32'W
Miramar, Ca., U.S. (mĭr'ă-mär)	76a	32°53'N	117°08'W
Miramas, Fr.	126	43°35'N	5°00'E
Miramichi Bay, b., Can. (mĭr'å-mê-shē')	60	47°08'N	65°08'W
Miranda, Col. (mē-rä'n-dä)	100a	3°14'N	76°11'W
Miranda, Ca., U.S.	76	40°14'N	123°49'W
Miranda, Ven.	101b	10°09'N	68°24'W
Miranda, dept., Ven.	101b	10°17'N	66°41'W
Miranda de Ebro, Spain (mē-rä'n-dä-dĕ-ĕ'brō)	128	42°42'N	2°59'W
Miranda do Douro, Port. (mê-rän'dä dô-dwĕ'rō)	128	41°30'N	6°17'W
Mirandela, Port. (mê-rän-dä'lå)	128	41°28'N	7°10'W
Mirando City, Tx., U.S. (mĭr-ăn'dō)	80	27°25'N	99°03'W
Mira Por Vos Islets, is., Bah. (mē'rä pōr vōs)	93	22°05'N	74°30'W
Mira Por Vos Pass, strt., Bah.	93	22°10'N	74°35'W
Mirbāţ, Oman	154	16°58'N	54°42'E
Mirebalais, Haiti (mēr-bå-lĕ')	93	18°50'N	72°05'W
Mirecourt, Fr. (mēr-kōōr')	127	48°20'N	6°08'E
Mirfield, Eng., U.K. (mûr'fēld)	114a	53°41'N	1°42'W
Miri, Malay. (mē'rē)	168	4°13'N	113°56'E
Mirim, Lagoa, l., S.A. (mê-rēɴ')	102	33°00'S	53°15'W
Miropol'ye, Ukr. (mē-rō-pôl'yĕ)	133	51°02'N	35°13'E
Mīrpur Khās, Pak. (mēr'pōōr käs)	158	25°36'N	69°10'E
Mirzāpur, India (mēr'zä-pōōr)	155	25°12'N	82°38'E
Misantla, Mex. (mê-sän'tlä)	89	19°55'N	96°49'W
Miscou, i., Can. (mĭs'kō)	60	47°58'N	64°35'W
Miscou Point, c., Can.	60	48°04'N	64°32'W
Miseno, Cape, c., Italy (mê-zĕ'nō)	129c	40°33'N	14°12'E
Misery, Mount, mtn., St. K./N. (mĭz'rē-ĭ)	91b	17°28'N	62°47'W
Mishan, China (mĭ'shän)	166	45°32'N	132°19'E
Mishawaka, In., U.S. (mĭsh-å-wôk'å)	66	41°45'N	86°15'W
Mishina, Japan (mē'shē-mä)	167	35°09'N	138°56'E
Misiones, prov., Arg. (mê-syō'näs)	102	27°00'S	54°30'W
Miskito, Cayos, is., Nic.	91	14°34'N	82°30'W
Miskolc, Hung. (mĭsh'kōlts)	110	48°07'N	20°50'E
Misool, Pulau, i., Indon. (mê-sōl')	169	2°00'S	130°05'E
Misquah Hills, Mn., U.S. (mĭs-kwä' hĭlz)	71	47°50'N	90°30'W
Mişr al Jadīdah, Egypt	192b	30°06'N	31°35'E
Misrātah, Libya	185	32°23'N	14°58'E
Missinaibi, r., Can. (mĭs'ĭn-ä'ē-bê)	53	50°27'N	83°01'W
Missinaibi Lake, l., Can.	58	48°23'N	83°40'W
Mission, Ks., U.S. (mĭsh'ŭn)	75f	39°02'N	94°39'W
Mission, Tx., U.S.	80	26°14'N	98°19'W
Mission City, Can. (sĭ'tĭ)	55	49°08'N	112°18'W
Mississagi, r., Can.	58	46°35'N	83°30'W
Mississauga, Can.	59	43°34'N	79°37'W
Mississippi, state, U.S. (mĭs-ĭ-sĭp'ē)	65	32°30'N	89°45'W
Mississippi, l., Can.	59	45°05'N	76°15'W
Mississippi, r., U.S.	65	32°00'N	91°30'W
Mississippi Sound, strt., Ms., U.S.	82	34°16'N	89°10'W
Missoula, Mt., U.S. (mĭ-zōō'lå)	64	46°55'N	114°00'W
Missouri, state, U.S. (mĭ-sōō'rē)	65	38°00'N	93°40'W
Missouri, r., U.S.	64	40°40'N	96°00'W
Missouri City, Tx., U.S.	81a	29°37'N	95°32'W
Missouri Coteau, hills, U.S.	64	47°30'N	101°00'W
Missouri Valley, Ia., U.S.	70	41°35'N	95°53'W
Mist, Or., U.S. (mĭst)	74c	46°00'N	123°15'W
Mistassini, Can. (mĭs-tå-sĭ'nē)	59	48°56'N	71°55'W
Mistassini, l., Can. (mĭs-tå-sĭ'nê)	53	50°48'N	73°30'W
Mistelbach, Aus. (mĭs'tĕl-bäk)	124	48°34'N	16°33'E
Misteriosa, Lago, l., Mex. (mēs-tĕ-ryô'sä)	90a	18°05'N	90°15'W
Misti, Volcán, vol., Peru	100	16°04'S	71°20'W
Mistretta, Italy (mê-strĕt'tä)	130	37°54'N	14°22'E
Misty Fjords National Monument, rec., Ak., U.S.	63	51°00'N	131°00'W
Mita, Punta de, c., Mex. (pōō'n-tä-dĕ-mē'tä)	88	20°44'N	105°34'W
Mitaka, Japan (mē'tä-kä)	167a	35°42'N	139°34'E
Mitchell, Il., U.S. (mĭch'ĕl)	75e	38°46'N	90°05'W
Mitchell, In., U.S.	66	38°45'N	86°25'W
Mitchell, Ne., U.S.	70	41°56'N	103°49'W
Mitchell, S.D., U.S.	64	43°42'N	98°01'W
Mitchell, Mount, mtn., N.C., U.S.	65	35°47'N	82°15'W
Mit Ghamr, Egypt	192b	30°43'N	31°20'E
Mitla Pass, p., Egypt	153a	30°03'N	32°40'E
Mito, Japan (mē'tō)	166	36°20'N	140°23'E
Mitsiwa, Erit.	185	15°40'N	39°19'E
Mitsu, Japan (mēt'sò)	167	34°21'N	132°49'E
Mittelland Kanal, can., Ger. (mĭt'ĕl-lănd)	124	52°18'N	10°42'E
Mittenwalde, Ger. (mē'tĕn-väl-dĕ)	115b	52°16'N	13°33'E
Mittweida, Ger. (mĭt-vī'dä)	124	50°59'N	12°58'E
Mitumba, Monts, mts., D.R.C.	191	10°50'S	27°00'E
Mityayevo, Russia (mĭt-yä'yĕ-vô)	142a	60°17'N	61°02'E
Miura, Japan	167a	35°08'N	139°37'E
Miwa, Japan (mē'wä)	167b	34°32'N	135°51'E
Mixico, Guat. (mēs'kô)	90	14°37'N	90°37'W
Mixquiahuala, Mex. (mēs-kê-wä'lä)	88	20°12'N	99°13'W
Mixteco, r., Mex. (mês-tā'kō)	88	17°45'N	98°10'W
Miyake, Japan (mē'yä-kå)	167b	34°35'N	135°34'E
Miyake, i., Japan (mē'yä-kå)	167	34°06'N	139°21'E
Miyakonojō, Japan	166	31°44'N	131°04'E

PLACE (Pronunciation)	PAGE	LAT.	LONG.
Miyazaki, Japan (mē'yä-zä'kê)	166	31°55'N	131°27'E
Miyoshi, Japan (mê-yō'shê')	166	34°48'N	132°49'E
Mizdah, Libya (mēz'dä)	156	31°29'N	13°09'E
Mizil, Rom. (mē'zêl)	131	45°01'N	26°30'E
Mizoram, state, India	155	23°25'N	92°45'E
Mjölby, Swe. (myûl'bü)	122	58°20'N	15°09'E
Mjörn, l., Swe.	122	57°55'N	12°22'E
Mjösa, l., Nor. (myûsä)	116	60°41'N	11°25'E
Mkalama, Tan.	186	4°07'S	34°38'E
Mkushi, Zam.	191	13°40'S	29°20'E
Mkwaja, Tan.	191	5°47'S	38°51'E
Mladá Boleslav, Czech Rep. (mlä'dä bô'lĕ-slåf)	124	50°26'N	14°52'E
Mlala Hills, hills, Tan.	191	6°47'S	31°45'E
Mlanje Mountains, mts., Mwi.	191	15°55'S	35°30'E
Mława, Pol. (mwä'vå)	116	53°07'N	20°25'E
Mmabatho, S. Afr.	186	25°42'S	25°43'E
Moa, r., Afr.	188	7°40'N	11°15'W
Moa, Pulau, i., Indon.	169	8°30'S	128°30'E
Moab, Ut., U.S. (mō'ăb)	77	38°35'N	109°35'W
Moanda, Gabon	186	1°37'S	13°09'E
Moar Lake, l., Can. (môr)	57	52°00'N	95°09'W
Moba, D.R.C.	186	7°12'S	29°39'E
Mobaye, C.A.R. (mô-bä'y')	185	4°19'N	21°11'E
Mobayi-Mbongo, D.R.C.	185	4°14'N	21°11'E
Moberly, Mo., U.S. (mō'bēr-lĭ)	65	39°24'N	92°25'W
Mobile, Al., U.S. (mô-bēl')	65	30°42'N	88°03'W
Mobile, r., Al., U.S.	82	31°15'N	88°00'W
Mobile Bay, b., Al., U.S.	65	30°26'N	87°56'W
Mobridge, S.D., U.S. (mō'brĭj)	70	45°32'N	100°26'W
Moca, Dom. Rep. (mô'kä)	93	19°25'N	70°35'W
Moçambique, Moz. (mō-säɴ-bē'kĕ)	191	15°03'S	40°42'E
Moçâmedes, Ang. (mô-zå-mĕ-dĕs)	186	15°10'S	12°09'E
Moçâmedes, hist. reg., Ang.	186	16°00'S	12°15'E
Mochitlán, Mex. (mô-chê-tlän')	88	17°10'N	99°19'W
Mochudi, Bots. (mō-chōō'dê)	186	24°13'S	26°07'E
Mocímboa da Praia, Moz. (mô-sē'ēm-bô-å prä'ēå)	187	11°20'S	40°21'E
Moclips, Wa., U.S.	72	47°14'N	124°13'W
Môco, Serra do, mtn., Ang.	190	12°25'S	15°10'E
Mococa, Braz. (mô-kô'kå)	99a	21°29'S	46°58'W
Moctezuma, Mex. (mōk'tå-zōō'mä)	88	22°44'N	101°06'W
Mocuba, Moz.	191	16°50'S	36°59'E
Modderfontein, S. Afr.	187b	26°06'S	28°10'E
Modena, Italy (mô'dĕ-nä)	118	44°38'N	10°54'E
Modesto, Ca., U.S. (mô-dĕs'tō)	76	37°39'N	121°00'W
Mödling, Aus. (mûd'lĭng)	115e	48°06'N	16°17'E
Moelv, Nor.	122	60°55'N	10°40'E
Moengo, Sur.	101	5°43'N	54°19'W
Moenkopi, Az., U.S.	77	36°07'N	111°13'W
Moers, Ger. (mûrs)	127c	51°27'N	6°38'E
Moffat Tunnel, trans., Co., U.S. (môf'ăt)	78	39°52'N	106°20'W
Mogadishu (Muqdisho), Som.	192a	2°08'N	45°22'E
Mogadore, Oh., U.S. (mŏg-å-dōr')	69d	41°04'N	81°23'E
Mogaung, Mya. (mô-gä'ông)	155	25°30'N	96°52'E
Mogi das Cruzes, Braz. (mô-gê-däs-krōō'sĕs)	101	23°33'S	46°10'W
Mogi-Guaçu, r., Braz. (mô-gê-gwä'sōō)	99a	22°06'S	47°12'W
Mogilno, Pol. (mô-gēl'nô)	124	52°38'N	17°58'W
Mogi-Mirim, Braz. (mô-gê-mē-rē'ɴ)	99a	22°26'S	46°57'W
Mogok, Mya. (mô-gōk')	155	23°14'N	96°38'E
Mogol, r., S. Afr. (mô-gōl)	192c	24°12'S	27°55'E
Mogollon Plateau, plat., Az., U.S.	64	34°15'N	110°45'W
Mogollon Rim, clf., Az., U.S. (mō-gô-yōn')	77	34°26'N	111°17'W
Moguer, Spain (mô-gĕr')	128	37°15'N	6°50'W
Mohács, Hung. (mô'häch)	125	45°59'N	18°38'E
Mohale's Hoek, Leso.	187c	30°09'S	27°28'E
Mohall, N.D., U.S. (mō'hôl)	70	48°46'N	101°29'W
Mohave, l., Nv., U.S. (mô-hä'vä)	77	35°23'N	114°40'W
Mohe, China (mwo-hŭ)	161	53°33'N	122°30'E
Mohenjo-Dero, hist., Pak.	155	27°20'N	68°10'E
Mohyliv-Podil's'kyi, Ukr.	137	48°27'N	27°51'E
Mõisaküla, Est. (mē̃'sä-kü'lä)	123	58°07'N	25°12'E
Moissac, Fr. (mwä-såk')	126	44°07'N	1°05'E
Moita, Port. (mô-ē'tå)	129b	38°39'N	9°00'W
Mojave, Ca., U.S.	76	35°06'N	118°09'W
Mojave, r., Ca., U.S. (mô-hä'vä)	76	34°46'N	117°24'W
Mojave Desert, Ca., U.S.	76	35°05'N	117°30'W
Mojave Desert, des., Ca., U.S.	64	35°00'N	117°00'W
Mokhotlong, Leso.	187c	29°18'S	29°06'E
Mokp'o, Kor., S. (mōk'pō')	161	34°50'N	126°30'E
Mol, Bel.	115a	51°21'N	5°09'E
Moldavia *see* Moldova, nation, Eur.	134	48°00'N	28°00'E
Moldavia, hist. reg., Rom.	125	47°20'N	27°12'E
Molde, Nor. (môl'dĕ)	116	62°44'N	7°15'E
Moldova, nation, Eur.	134	48°00'N	28°00'E
Moldova, r., Rom.	125	47°17'N	26°27'E
Moldoveanu, Vârful, mtn., Rom.	131	45°33'N	24°38'E
Molepolole, Bots. (mō-lå-pô-lō'lå)	186	24°15'S	25°33'W
Molfetta, Italy (môl-fĕt'tä)	119	41°11'N	16°38'E
Molina, Chile (mō-lē'nä)	99b	35°07'S	71°17'W
Molina de Aragón, Spain (mô-lē'nä dĕ ä-rä-gô'n)	128	40°40'N	1°54'W
Molína de Segura, Spain (mô-lē'nä dĕ sĕ-gōō'rä)	128	38°03'N	1°07'W
Moline, Il., U.S. (mô-lēn')	79	41°31'N	90°34'W
Moliro, D.R.C.	186	8°13'S	30°34'E
Moliterno, Italy (mōl-ē-tĕr'nō)	130	40°13'N	15°54'W
Mollendo, Peru (mô-lyĕn'dō)	100	17°02'S	71°59'W
Moller, Port, Ak., U.S. (pōrt mōl'ēr)	63	56°18'N	161°30'W
Mölndal, Swe. (mûln'däl)	122	57°39'N	12°01'E
Molochna, r., Ukr.	133	47°05'N	35°22'E
Molochnyĭ lyman, l., Ukr.	133	46°35'N	35°32'E
Molody Tud, Russia (mō-lō-dô'ē tōō'd)	142b	55°17'N	37°31'E

ăt; finăl; rāte; senåte; ärm; åsk; sofå; fâre; ch-choose; dh-as th in other; bē; ēvent; bĕt; recĕnt; cratēr; g-gō; gh-guttural g; bĭt; ĭ-short neutral; rīde; ĸ-guttural k as ch in German ich;

PRONOUNCING INDEX Moloka'i – Moreton Bay 277

PLACE (Pronunciation)	PAGE	LAT.	LONG.
Moloka'i, i., Hi., U.S. (mō-lō kä'ē)	64c	21°15'N	157°05'W
Molokcha, r., Russia (mô'lôk-chä)	142b	56°15'N	38°29'E
Molopo, r., Afr. (mō-lō-pō)	186	27°45'S	20°45'E
Molson Lake, l., Can. (mōl'sŭn)	57	54°12'N	96°45'W
Molteno, S. Afr. (môl-tā'nō)	187c	31°24'S	26°23'E
Moluccas *see* Maluku, is., Indon.	169	2°22'S	128°25'E
Moma, Moz.	191	16°44'S	39°14'E
Mombasa, Kenya (mōm-bä'sä)	187	4°03'S	39°40'E
Mombetsu, Japan (mōm'bĕt-sōō')	166	44°21'N	142°48'E
Momence, Il., U.S. (mō-mĕns')	69a	41°09'N	87°40'W
Momostenango, Guat. (mō-mōs-tā-näŋ'gō)	90	15°02'N	91°25'W
Momotombo, Nic.	90	12°25'N	86°43'W
Mompog Pass, strt., Phil. (mōm-pōg')	169a	13°35'N	122°09'E
Mompos, Col. (mōm-pōs')	100	9°05'N	74°30'W
Momtblanc, Spain	129	41°20'N	1°08'E
Møn, i., Den. (mûn)	122	54°54'N	12°30'E
Monaca, Pa., U.S. (mō-nă'kō)	69e	40°41'N	80°17'W
Monaco, nation, Eur. (mŏn'ă-kō)	110	43°43'N	7°47'E
Monaghan, Ire. (mŏ'ă-găn)	120	54°16'N	7°20'W
Mona Passage, strt., N.A. (mō'nä)	87	18°00'N	68°10'W
Monarch Mountain, mtn., Can. (mŏn'ĕrk)	54	51°41'N	125°53'W
Monashee Mountains, mts., Can. (mŏ-nă'shē)	55	50°30'N	118°30'W
Monastir *see* Bitola, Mac.	130	41°02'N	21°22'E
Monastir, Tun. (mŏn-ăs-tēr')	118	35°49'N	10°56'E
Monastyrshchina, Russia (mô-nás-tērsh'chĭ-nȧ)	132	54°19'N	31°49'E
Monastyrshche, Ukr.	133	48°57'N	29°53'E
Monção, Braz. (mon-souN')	101	3°39'S	45°23'W
Moncayo, mtn., Spain (mȯn-kä'yō)	128	41°44'N	1°48'W
Monchegorsk, Russia (mōn'chĕ-gôrsk)	136	69°00'N	33°35'E
Mönchengladbach, Ger. (mûn'kĕn gläd'bäk)	124	51°12'N	6°28'E
Moncique, Serra de, mts., Port. (sĕr'rä dä mōn-chē'kĕ)	128	37°22'N	8°37'W
Monclova, Mex. (mōn-klō'vä)	86	26°53'N	101°25'W
Moncton, Can. (mŭngk'tŭn)	51	46°06'N	64°47'W
Mondêgo, r., Port. (mōn-dĕ'gō)	128	40°10'N	8°36'W
Mondego, Cabo, c., Port. (kä'bō mōn-dā'gō)	128	40°12'N	8°55'W
Mondombe, D.R.C. (mōn-dôm'bå)	186	0°45'S	23°06'E
Mondoñedo, Spain (mōn-dō-nyā'dō)	128	43°35'N	7°18'W
Mondovi, Wi., U.S. (mŏn-dō'vĭ)	71	44°35'N	91°42'W
Monee, Il., U.S. (mō-nī)	69a	41°25'N	87°45'W
Monessen, Pa., U.S. (mō-nĕs'sen)	69e	40°09'N	79°53'W
Monett, Mo., U.S. (mō-nĕt')	79	36°55'N	93°55'W
Monfalcone, Italy	130	45°49'N	13°30'E
Monforte de Lemos, Spain (mōn-fōr'tä dĕ lĕ'mōs)	128	42°30'N	7°30'W
Mongala, r., D.R.C. (mōn-gäl'ȧ)	185	3°20'N	21°30'E
Mongalla, Sudan	185	5°11'N	31°46'E
Monghyr, India (mōn-gēr')	155	25°23'N	86°34'E
Mongo, r., Afr.	188	9°50'N	11°50'W
Mongolia, nation, Asia (mŏŋ-gō'lĭ-ȧ)	160	46°00'N	100°00'E
Mongos, Chaîne des, mts., C.A.R.	185	8°04'N	21°59'E
Mongoumba, C.A.R. (mōŋ-gōōm'bå)	185	3°38'N	18°36'E
Mongu, Zam. (mŏŋ-gōō')	186	15°15'S	23°09'E
Monkey Bay, Mwi.	191	14°05'S	34°55'E
Monkey River, Belize (mŭŋ'kĭ)	90a	16°22'N	88°33'W
Monkland, Can. (mŭngk-lănd)	62c	45°12'N	74°52'W
Monkoto, D.R.C. (mōn-kō'tō)	186	1°38'S	20°39'E
Monmouth, Il., U.S. (mŏn'mŭth) (mŏn'mouth)	79	40°54'N	90°38'W
Monmouth Junction, N.J., U.S. (mŏn'mouth jŭngk'shŭn)	68a	40°23'N	74°33'W
Monmouth Mountain, mtn., Can. (mŏn'mŭth)	54	51°00'N	123°47'W
Mono, r., Afr.	188	7°20'N	1°25'E
Mono Lake, l., Ca., U.S. (mō'nō)	76	38°04'N	119°00'W
Monon, In., U.S. (mō'nŏn)	66	40°55'N	86°55'W
Monongah, W.V., U.S. (mō-nŏŋ'gȧ)	66	39°25'N	80°10'W
Monongahela, Pa., U.S. (mō-nŏn-gȧ-hē'lȧ)	69a	40°11'N	79°55'W
Monongahela, r., W.V., U.S.	66	39°30'N	80°10'W
Monopoli, Italy (mō-nō'pō-lē)	131	40°55'N	17°17'E
Monóvar, Spain (mō-nō'vär)	129	38°26'N	0°50'W
Monreale, Italy (mōn-rā-ä'lā)	130	38°04'N	13°15'E
Monroe, Ga., U.S. (mŭn-rō')	82	33°47'N	83°43'W
Monroe, La., U.S.	65	32°30'N	92°06'W
Monroe, Mi., U.S.	66	41°55'N	83°25'W
Monroe, N.C., U.S.	83	34°58'N	80°34'W
Monroe, N.Y., U.S.	68a	41°19'N	74°11'W
Monroe, Ut., U.S.	77	38°35'N	112°10'W
Monroe, Wa., U.S.	74a	47°52'N	121°58'W
Monroe, Wi., U.S.	71	42°35'N	89°40'W
Monroe, Lake, l., Fl., U.S.	83	28°50'N	81°15'W
Monroe City, Mo., U.S.	79	39°38'N	91°41'W
Monroeville, Al., U.S. (mŭn-rō'vĭl)	82	31°33'N	87°19'W
Monroeville, Pa., U.S.	69e	40°26'N	79°46'W
Monrovia, Lib.	184	6°18'N	10°47'W
Monrovia, Ca., U.S. (mōn-rō'vĭ-ȧ)	75a	34°09'N	118°00'W
Mons, Bel. (mōN')	117	50°29'N	3°55'E
Monson, Me., U.S. (mŏn'sŭn)	60	45°17'N	69°28'W
Mönsterås, Swe. (mŭn'stĕr-ōs)	122	57°04'N	16°24'E
Montagne Tremblant Provincial Park, rec., Can.	65	46°30'N	75°51'W
Montague, Can. (mŏn'tȧ-gū)	61	46°10'N	62°39'W
Montague, Mi., U.S.	66	43°30'N	86°25'W
Montague, i., Ak., U.S.	63	60°10'N	147°00'W
Montalbán, Ven. (mōnt-äl-bän)	101b	10°14'N	68°19'W
Montalegre, Port. (mōn-tä-lā'grĕ)	128	41°49'N	7°48'W
Montana, state, U.S. (mŏn-tăn'ȧ)	64	47°10'N	111°50'W
Montánchez, Spain (mōn-tän'chāth)	128	39°18'N	6°09'W
Montargis, Fr. (môN-tár-zhē')	117	47°59'N	2°42'E
Montataire, Fr. (môN-tȧ-târ)	127b	49°15'N	2°26'E
Montauban, Fr. (môN-tō-bäN')	117	44°01'N	1°22'E
Montauk, N.Y., U.S.	67	41°03'N	71°57'W
Montauk Point, c., N.Y., U.S. (mŏn-tôk')	67	41°05'N	71°55'W
Montbard, Fr. (mōN-bár')	126	47°40'N	4°19'E
Montbéliard, Fr. (môN-bā-lyár')	127	47°32'N	6°45'E
Mont Belvieu, Tx., U.S. (mŏnt bĕl'vū)	81a	29°51'N	94°53'W
Montbrison, Fr. (môN-brē-zoN')	126	45°38'N	4°06'E
Montceau, Fr. (môN-sō')	126	46°39'N	4°22'E
Montclair, N.J., U.S. (mŏnt-klâr')	68a	40°49'N	74°13'W
Mont-de-Marsan, Fr. (môN-dē-már-säN')	117	43°54'N	0°32'W
Montdidier, Fr. (môN-dē-dyā')	126	49°42'N	2°33'E
Monte, Arg. (mō'n-tĕ)	99c	35°25'S	58°49'W
Monteagudo, Bol. (mōn'tā-ä-gōō'dhō)	100	19°49'S	63°48'W
Montebello, Can.	62c	45°40'N	74°56'W
Montebello, Ca., U.S. (mŏn-tē-bĕl'ō)	75a	34°01'N	118°06'W
Monte Bello Islands, is., Austl.	174	20°30'S	114°10'E
Monte Caseros, Arg. (mō'n-tĕ-kä-sĕ'rôs)	102	30°16'S	57°39'W
Montecillos, Cordillera de, mts., Hond.	90	14°19'N	87°52'W
Monte Cristi, Dom. Rep. (mŏ'n-tĕ-krē's-tē)	93	19°50'N	71°40'W
Montecristo, Isola di, i., Italy (mōn'tā-krēs'tō)	130	42°20'N	10°19'E
Monte Escobedo, Mex. (mōn'tā ĕs-kō-bā'dhō)	88	22°18'N	103°34'W
Monteforte Irpino, Italy (mōn-tĕ-fô'r-tĕ ē'r-pē'nō)	129c	40°39'N	14°42'E
Montefrío, Spain (mōn-tå-frē'ō)	128	37°20'N	4°02'W
Montego Bay, Jam. (mŏn-tē'gō)	87	18°30'N	77°55'W
Montelavar, Port. (mōn-tĕ-lá-vär')	129b	38°51'N	9°20'W
Montélimar, Fr. (môN-tā-lē-mär')	117	44°33'N	4°47'E
Montellano, Spain (mōn-tå-lyä'nō)	128	37°00'N	5°34'W
Montello, Wi., U.S. (mŏn-tĕl'ō)	71	43°47'N	89°20'W
Montemorelos, Mex. (mōn'tā-mō-rā'lōs)	86	25°14'N	99°50'W
Montemor-o-Novo, Port. (mōN-tĕ-môr'ō-nō'vō)	128	38°39'N	8°11'W
Montenegro *see* Crna Gora, state, Serb.	131	42°55'N	18°52'E
Montenegro, reg., Moz.	191	13°07'S	39°00'E
Montepulciano, Italy (mōn'tā-pōōl-chä'nō)	130	43°05'N	11°48'E
Montereau-faut-Yonne, Fr. (môN-t'rō'fō-yōn')	126	48°24'N	2°57'E
Monterey, Ca., U.S. (mŏn-tĕ-rā')	64	36°36'N	121°53'W
Monterey, Tn., U.S.	82	36°06'N	85°15'W
Monterey Bay, b., Ca., U.S.	64	36°48'N	122°01'W
Monterey Park, Ca., U.S.	75a	34°04'N	118°08'W
Montería, Col. (mōn-tā-rā'ä)	100	8°47'N	75°57'W
Monteros, Arg. (mōn-tĕ'rôs)	102	27°14'S	65°29'W
Monterotondo, Italy (mōn-tĕ-rô-tô'n-dō)	129d	42°03'N	12°39'E
Monterrey, Mex. (mōn-tĕr-rā')	86	25°43'N	100°19'W
Montesano, Wa., U.S. (mŏn-tĕ-sä'nō)	72	46°59'N	123°35'W
Monte Sant'Angelo, Italy (mô'n-tĕ sän ä'n-gzhĕ-lō)	119	41°43'N	15°59'E
Montes Claros, Braz. (môn-tĕs-klä'rôs)	101	16°44'S	43°41'W
Montevallo, Al., U.S. (mŏn-tē-văl'ō)	82	33°05'N	86°49'W
Montevarchi, Italy (mōn-tå-vär'kē)	130	43°30'N	11°45'E
Montevideo, Ur. (mōn'tå-vē-dhā'ō)	102	34°50'S	56°10'W
Montevideo, Mn., U.S. (mŏn'tå-vē-dhā'ō)	70	44°56'N	95°42'W
Monte Vista, Co., U.S. (mŏn'tē vĭs'tȧ)	77	37°35'N	106°10'W
Montezuma, Ga., U.S. (mŏn-tē-zōō'mȧ)	82	32°17'N	84°00'W
Montezuma Castle National Monument, rec., Az., U.S.	77	34°38'N	111°50'W
Montfoort, Neth.	115a	52°02'N	4°56'E
Montfor-l'Amaury, Fr. (môN-fōr'lä-mō-rē')	127b	48°47'N	1°49'E
Montfort, Fr. (mōn-fōr)	126	48°09'N	1°58'W
Montgomery, Al., U.S. (mŏnt-gŭm'ĕr-ĭ)	65	32°23'N	86°17'W
Montgomery, W.V., U.S.	66	38°10'N	81°25'W
Montgomery City, Mo., U.S.	79	38°58'N	91°29'W
Monticello, Ar., U.S. (mŏn-tĭ-sĕl'ō)	79	33°38'N	91°47'W
Monticello, Fl., U.S.	82	30°32'N	83°53'W
Monticello, Ga., U.S.	82	33°00'N	83°11'W
Monticello, Ia., U.S.	71	42°14'N	91°13'W
Monticello, Il., U.S.	66	40°05'N	88°35'W
Monticello, In., U.S.	66	40°40'N	86°50'W
Monticello, Ky., U.S.	82	36°47'N	84°50'W
Monticello, Me., U.S.	60	46°19'N	67°53'W
Monticello, Mn., U.S.	71	45°18'N	93°48'W
Monticello, N.Y., U.S.	67	41°35'N	74°40'W
Monticello, Ut., U.S.	77	37°55'N	109°25'W
Montijo, Port. (mōn-tē'zhō)	129b	38°42'N	8°58'W
Montijo, Spain (mōn-tē'hō)	128	38°55'N	6°35'W
Montijo, Bahía, b., Pan. (bä-ē'ä mōn-tē'hō)	87	7°36'N	81°11'W
Mont-Joli, Can. (môN zhō-lē')	51	48°35'N	68°11'W
Montluçon, Fr. (môN-lü-sôN')	117	46°20'N	2°35'E
Montmagny, Can. (môN-mán-yē')	59	46°59'N	70°33'W
Montmorency, Fr. (môN'mō-rän-sē')	127b	48°59'N	2°19'E
Montmorency, r., Can. (mŏnt-mō-rĕn'sĭ)	62b	47°03'N	71°10'W
Montmorillon, Fr. (môN'mō-rē-yôN')	126	46°26'N	0°50'E
Montone, r., Italy (mōn-tō'nĕ)	130	44°03'N	11°45'E
Montoro, Spain (mōn-tō'rō)	128	38°01'N	4°22'W
Montpelier, Id., U.S. (mŏnt-pēl'yẽr)	73	42°19'N	111°19'W
Montpelier, In., U.S. (mŏnt-pēl'yẽr)	66	40°35'N	85°20'W
Montpelier, Oh., U.S.	66	41°35'N	84°35'W
Montpelier, Vt., U.S.	65	44°20'N	72°35'W
Montpellier, Fr. (môN-pĕ-lyā')	117	43°38'N	3°53'E
Montréal, Can. (mŏn-trē-ôl')	51	45°30'N	73°35'W
Montreal, r., Can.	59	47°50'N	80°30'W
Montreal, r., Can.	58	47°15'N	84°20'W
Montreal Lake, l., Can.	56	54°20'N	105°40'W
Montréal-Nord, Can.	62a	45°36'N	73°38'W
Montreuil, Fr.	127b	48°52'N	2°27'E
Montreux, Switz. (môN-trû')	124	46°26'N	6°52'E
Montrose, Scot., U.K.	120	56°45'N	2°25'W
Montrose, Ca., U.S. (mŏnt-rōz)	75a	34°13'N	118°13'W
Montrose, Co., U.S. (mŏn-trōz')	77	38°30'N	107°55'W
Montrose, Oh., U.S.	69d	41°08'N	81°38'W
Montrose, Pa., U.S. (mŏnt-rōz')	67	41°50'N	75°50'W
Montrouge, Fr.	127b	48°49'N	2°19'E
Mont-Royal, Can.	62a	47°31'N	73°39'W
Monts, Pointe des, c., Can. (pwăNt' dā môN')	60	49°19'N	67°22'W
Mont Saint Martin, Fr. (môN săN mär-tăN')	127	49°34'N	6°13'E
Montserrat, dep., N.A. (mŏnt-sĕ-răt')	87	16°48'N	63°15'W
Montvale, N.J., U.S. (mŏnt-vāl')	68a	41°02'N	74°01'W
Monywa, Mya. (mŏn'yōō-wȧ)	155	22°02'N	95°16'E
Monza, Italy (mōn'tsä)	130	45°34'N	9°17'E
Monzón, Spain (mōn-thōn')	129	41°54'N	0°09'E
Moody, Tx., U.S. (mōō'dĭ)	81	31°18'N	97°20'W
Mooi, r., S. Afr. (mōō'ĭ)	192c	26°34'S	27°03'E
Mooi, r., S. Afr.	187c	29°00'S	30°15'E
Mooirivier, S. Afr.	187c	29°14'S	29°59'E
Moolap, Austl.	173a	38°11'S	144°26'E
Moonta, Austl. (mōōn'tȧ)	174	34°05'S	137°42'E
Moora, Austl. (mōr'ȧ)	174	30°35'S	116°12'E
Moorabbin, Austl.	173a	37°56'S	145°02'E
Moore, l., Austl. (mōr)	174	29°50'S	118°12'E
Moorenweis, Ger. (mō'rĕn-vīz)	115d	48°10'N	11°05'E
Moore Reservoir, res., Vt., U.S.	67	44°20'N	72°10'W
Moorestown, N.J., U.S. (morz'toun)	68f	39°58'N	74°56'W
Mooresville, In., U.S. (mōrz'vĭl)	69g	39°37'N	86°22'W
Mooresville, N.C., U.S.	83	35°34'N	80°48'W
Moorhead, Mn., U.S. (mōr'hĕd)	70	46°52'N	96°44'W
Moorhead, Ms., U.S.	82	33°25'N	90°30'W
Moose, r., Can.	53	51°01'N	80°42'W
Moose Creek, Can.	62c	45°16'N	74°58'W
Moosehead, Me., U.S. (mōōs'hĕd)	60	45°37'N	69°15'W
Moose Island, i., Can.	57	51°50'N	97°09'W
Moose Jaw, Can. (mōōs jô)	50	50°23'N	105°32'W
Moose Jaw, r., Can.	56	50°34'N	105°17'W
Moose Lake, Can.	57	53°40'N	100°28'W
Moose Mountain, mtn., Can.	57	49°45'N	102°37'W
Moose Mountain Creek, r., Can.	57	49°12'N	102°10'W
Moosilauke, mtn., N.H., U.S. (mōō-sĭ-lȧ'kē)	67	44°00'N	71°50'W
Moosinning, Ger. (mō'zē-nēng)	115d	48°17'N	11°51'E
Moosomin, Can. (mōō'sō-mĭn)	57	50°07'N	101°40'W
Moosonee, Can. (mōō'sō-nē)	51	51°20'N	80°44'W
Mopti, Mali (mŏp'tē)	184	14°30'N	4°12'W
Moquegua, Peru (mō-kā'gwä)	100	17°15'S	70°54'W
Mór, Hung. (mōr)	125	47°25'N	18°14'E
Mora, India	159b	18°54'N	72°56'E
Mora, Spain (mō-rä)	128	39°42'N	3°45'W
Mora, Swe. (mō'rä)	122	61°00'N	14°29'E
Mora, Mn., U.S. (mō'rȧ)	71	45°52'N	93°18'W
Mora, N.M., U.S.	78	35°58'N	105°17'W
Morādābād, India (mō-rä-dä-bäd')	155	28°57'N	78°48'E
Morales, Guat. (mō-rä'lĕs)	90	15°29'N	88°46'W
Moramanga, Madag. (mō-rä-mäŋ'gä)	187	18°48'S	48°09'E
Morant Point, c., Jam. (mō-rănt')	92	17°55'N	76°10'W
Morata de Tajuña, Spain (mō-rä'tä dā tä-hōō'nyä)	129a	40°14'N	3°27'W
Moratuwa, Sri L.	159	6°35'N	79°59'E
Morava (Moravia), hist. reg., Czech Rep.	124	49°21'N	16°57'E
Morava, r., Eur.	117	49°00'N	17°30'E
Moravia *see* Morava, hist. reg., Czech Rep.	124	49°21'N	16°57'E
Morawhanna, Guy. (mō-rá-hwä'nȧ)	101	8°12'N	59°33'W
Moray Firth, b., Scot., U.K. (mŭr'å)	112	57°41'N	3°55'W
Mörbylånga, Swe. (mûr'bü-lôŋ'gȧ)	122	56°32'N	16°23'E
Morden, Can. (môr'dĕn)	50	49°11'N	98°05'W
Mordialloc, Austl. (môr-dĭ-ăl'ŏk)	173a	38°00'S	145°05'E
Mordvinia, prov., Russia	136	54°18'N	43°50'E
More, Ben, mtn., Scot., U.K. (bĕn môr)	120	58°09'N	5°01'W
Moreau, r., S.D., U.S. (mō-rō')	70	45°13'N	102°22'W
Moree, Austl. (mō'rē)	175	29°20'S	149°50'E
Morehead, Ky., U.S.	66	38°10'N	83°25'W
Morehead City, N.C., U.S. (mōr'hĕd)	83	34°43'N	76°43'W
Morehouse, Mo., U.S. (mōr'hous)	79	36°49'N	89°41'W
Morelia, Mex. (mō-rā'lyä)	86	19°43'N	101°12'W
Morella, Spain (mō-rāl'yä)	129	40°38'N	0°07'W
Morelos, Mex. (mō-rā'lōs)	88	22°46'N	102°36'W
Morelos, Mex.	89a	19°41'N	99°29'W
Morelos, Mex.	80	28°24'N	100°51'W
Morelos, r., Mex.	80	25°27'N	99°35'W
Morena, Sierra, mtn., Ca., U.S. (syĕr'rä mō-rā'nä)	74b	37°24'N	122°19'W
Morena, Sierra, mts., Spain (syĕr'rä mō-rā'nä)	112	38°15'N	5°45'W
Morenci, Az., U.S. (mō-rĕn'sĭ)	77	33°05'N	109°25'W
Morenci, Mi., U.S.	66	41°50'N	84°50'W
Moreno, Arg. (mō-rĕ'nō)	102a	34°39'S	58°47'W
Moreno, Ca., U.S.	75a	33°55'N	117°09'W
Mores, i., Bah. (mōrz)	92	26°20'N	77°35'W
Moresby, i., Can. (mōrz'bĭ)	74d	48°43'N	123°15'W
Moresby Island, i., Can.	52	52°50'N	131°55'W
Moreton, i., Austl. (môr'tŭn)	176	26°53'S	152°42'E
Moreton Bay, b., Austl. (môr'tŭn)	176	27°12'S	153°10'E

278 Morewood – Muqui PRONOUNCING INDEX

PLACE (Pronunciation)	PAGE	LAT.	LONG.
Morewood, Can. (mōr′wo͝od)	62c	45°11′N	75°17′W
Morgan, Mt., U.S. (môr′găn)	73	48°55′N	107°56′W
Morgan, Ut., U.S.	73	41°04′N	111°42′W
Morgan City, La., U.S.	81	29°41′N	91°11′W
Morganfield, Ky., U.S. (môr′găn-fēld)	66	37°40′N	87°55′W
Morgan's Bay, S. Afr.	187c	32°42′S	28°19′E
Morganton, N.C., U.S. (môr′găn-tŭn)	83	35°44′N	81°42′W
Morgantown, W.V., U.S. (môr′găn-toun)	67	39°40′N	79°55′W
Morga Range, mts., Afg.	155a	34°02′N	70°38′E
Morgenzon, S. Afr. (môr′gănt-sōn)	192c	26°44′S	29°39′E
Moriac, Austl.	173a	38°15′S	144°20′E
Morice Lake, l., Can.	54	54°00′N	127°37′W
Moriguchi, Japan (mō′rē-go͞o′chē)	167b	34°44′N	135°34′E
Morinville, Can. (mō′rĭn-vĭl)	62g	53°48′N	113°39′W
Morioka, Japan (mō′rē-ō′kȧ)	161	39°40′N	141°21′E
Morkoka, r., Russia (môr-kô′kȧ)	141	65°35′N	111°00′E
Morlaix, Fr. (mŏr-lĕ′)	117	48°36′N	3°48′W
Morley, Can. (mŏr′lē)	62e	51°10′N	114°51′W
Mormant, Fr.	127b	48°35′N	2°54′E
Morne Gimie, St. Luc. (môrn′ zhē-mē′)	91b	13°53′N	61°03′W
Mornington, Austl.	173a	38°13′S	145°02′E
Morobe, Pap. N. Gui.	169	8°03′S	147°45′E
Morocco, nation, Afr. (mô-rŏk′ō)	184	32°00′N	7°00′W
Morogoro, Tan. (mō-rō-gō′rō)	187	6°49′S	37°40′E
Moroleón, Mex. (mō-rō-lā-ōn′)	88	20°07′N	101°15′W
Morombe, Madag. (mo͞o-ro͞om′bā)	187	21°39′S	43°34′E
Morón, Arg. (mo-rŏ′n)	99c	34°39′S	58°37′W
Morón, Cuba (mô-rōn′)	92	22°05′N	78°35′W
Morón, Ven. (mô-rŏ′n)	101b	10°29′N	68°11′W
Morondava, Madag. (mô-rōn-dä′vȧ)	187	20°17′S	44°18′E
Morón de la Frontera, Spain (mô-rōn′dā läf rôn-tā′rä)	128	37°08′N	5°20′W
Morongo Indian Reservation, I.R., Ca., U.S. (mō-rôŋ′gō)	76	33°54′N	116°47′W
Moroni, Com.	187	11°41′S	43°16′E
Moroni, Ut., U.S. (mô-rō′nĭ)	77	39°30′N	111°40′W
Morotai, i., Indon. (mō-rō-tä′ē)	169	2°12′N	128°30′E
Moroto, Ug.	191	2°32′N	34°39′E
Morozovsk, Russia	137	48°20′N	41°50′E
Morrill, Ne., U.S. (mŏr′ĭl)	70	41°59′N	103°54′W
Morrilton, Ar., U.S. (mŏr′ĭl-tŭn)	79	35°09′N	92°42′W
Morrinhos, Braz. (mô-rēn′yōzh)	101	17°45′S	48°56′W
Morris, Can. (mŏr′ĭs)	50	49°21′N	97°22′W
Morris, Il., U.S.	66	41°20′N	88°25′W
Morris, Mn., U.S.	70	45°35′N	95°53′W
Morris, r., Can.	57	49°30′N	97°30′W
Morrison, Il., U.S. (mŏr′ĭ-sŭn)	71	41°48′N	89°58′W
Morris Reservoir, res., Ca., U.S.	75a	34°11′N	117°49′W
Morristown, N.J., U.S. (mŏr′rĭs-toun)	68a	40°48′N	74°29′W
Morristown, Tn., U.S.	82	36°10′N	83°18′W
Morrisville, Pa., U.S. (mŏr′ĭs-vĭl)	68f	40°12′N	74°46′W
Morro do Chapéu, Braz. (môr-ō dô-shä-pĕ′o͞o)	101	11°34′S	41°03′W
Morrow, Oh., U.S. (mŏr′ō)	69f	39°21′N	84°07′W
Mors, i., Den.	122	56°46′N	8°38′E
Morshansk, Russia (môr-shänsk′)	136	53°25′N	41°35′E
Mortara, Italy (môr-tä′rä)	130	45°13′N	8°47′E
Morteros, Arg. (môr-tĕ′tôs)	102	30°47′S	62°00′W
Mortes, Rio das, r., Braz. (rē′ō-däs-mô′r-tĕs)	99a	21°04′S	44°29′W
Morton Indian Reservation, I.R., Mn., U.S. (mŏr′tŭn)	71	44°35′N	94°48′W
Mortsel, Bel. (môr-sĕl′)	115a	51°10′N	4°28′E
Morvan, mts., Fr. (môr-vän′)	126	47°11′N	4°10′E
Morzhovets, i., Russia (môr′zhô-vyĕts′)	136	66°40′N	42°30′E
Mosal'sk, Russia (mō-zälsk′)	132	54°27′N	34°57′E
Moscavide, Port.	129b	38°47′N	9°06′W
Moscow (Moskva), Russia	134	55°45′N	37°37′E
Moscow, Id., U.S. (mŏs′kō)	64	46°44′N	116°57′W
Mosel (Moselle), r., Eur. (mō′sĕl) (mō-zĕl′)	124	49°49′N	7°00′E
Moses, r., S. Afr.	192c	25°17′S	29°04′E
Moses Lake, Wa., U.S.	72	47°08′N	119°15′W
Moses Lake, l., Wa., U.S. (mō′zĕz)	72	47°09′N	119°30′W
Moshchnyy, is., Russia (môsh′chnĭ)	123	59°56′N	28°07′E
Moshi, Tan. (mō′shē)	187	3°21′S	37°20′E
Mosjøen, Nor.	116	65°50′N	13°10′E
Moskva *see* Moscow, Russia	134	55°45′N	37°37′E
Moskva, prov., Russia	132	55°38′N	36°48′E
Moskva, r., Russia	136	55°30′N	37°05′E
Mosonmagyaróvár, Hung.	125	47°51′N	17°16′E
Mosquitos, Costa de, cst., Nic. (kôs-tä-dĕ-mōs-kē′tō)	91	12°05′N	83°49′W
Mosquitos, Gulfo de los, b., Pan. (go͞o′l-fô-dĕ-lôs-mōs-kē′tōs)	87	9°17′N	80°59′W
Moss, Nor. (môs)	116	59°29′N	10°39′E
Moss Beach, Ca., U.S. (môs bēch)	74b	37°32′N	122°31′W
Mosselbaai, S. Afr. (mô′sŭl bā)	186	34°06′S	22°23′E
Mossendjo, Congo	190	2°57′S	12°44′E
Mossley, Eng., U.K. (môs′lĭ)	114a	53°31′N	2°02′W
Moss Point, Ms., U.S. (môs)	82	30°25′N	88°32′W
Most, Czech Rep. (mōst)	124	50°32′N	13°37′E
Mostar, Bos. (môs′tär)	119	43°20′N	17°51′E
Móstoles, Spain (môs-tō′läs)	129a	40°19′N	3°52′W
Mostoos Hills, hills, Can. (mŏs′to͞os)	56	54°50′N	108°45′W
Mosvatnet, l., Nor.	122	59°55′N	7°50′E
Motagua, r., N.A. (mô-tä′gwä)	90	15°29′N	88°39′W
Motala, Swe. (mô-tō′lä)	122	58°34′N	15°00′E
Motherwell, Scot., U.K. (mŭdh′ēr-wĕl)	116	55°45′N	4°05′W
Motril, Spain (mô-trēl′)	118	36°44′N	3°32′W
Motul, Mex. (mō-to͞o′l)	90a	21°07′N	89°14′W
Mouaskar, Alg.	184	35°25′N	0°08′E
Mouchoir Bank, bk., (mo͞o-shwȧr′)	93	21°35′N	70°40′W
Mouchoir Passage, strt., T./C. Is.	93	21°05′N	71°05′W
Moudjéria, Maur.	188	17°53′N	12°20′W
Mouila, Gabon	190	1°52′S	11°01′E
Mouille Point, c., S. Afr.	186a	33°54′S	18°19′E
Moulins, Fr. (mo͞o-lăn′)	117	46°34′N	3°19′E
Moulouya, Oued, r., Mor. (mo͞o-lo͞o′yȧ)	184	34°00′N	4°00′W
Moultrie, Ga., U.S. (mōl′trĭ)	82	31°10′N	83°48′W
Moultrie, Lake, l., S.C., U.S.	83	33°12′N	80°00′W
Mound City, Il., U.S.	79	37°06′N	89°13′W
Mound City, Mo., U.S.	79	40°08′N	95°13′W
Moundou, Chad	189	8°34′N	16°05′E
Moundsville, W.V., U.S. (moundz′vĭl)	66	39°50′N	80°50′W
Mount, Cape, c., Lib.	188	6°47′N	11°20′W
Mountain Brook, Al., U.S. (moun′tĭn bro͝ok)	68h	33°30′N	86°45′W
Mountain Creek Lake, l., Tx., U.S.	75c	32°43′N	97°03′W
Mountain Grove, Mo., U.S. (grōv)	79	37°07′N	92°16′W
Mountain Home, Id., U.S. (hōm)	72	43°08′N	115°43′W
Mountain Park, Can. (pärk)	50	52°55′N	117°14′W
Mountain View, Ca., U.S. (moun′tĭn vū)	74b	37°25′N	122°07′W
Mountain View, Mo., U.S.	79	36°59′N	91°46′W
Mount Airy, N.C., U.S. (âr′ĭ)	83	36°28′N	80°37′W
Mount Ayliff, S. Afr. (ā′lĭf)	187c	30°48′S	29°24′E
Mount Ayr, Ia., U.S. (âr)	71	40°43′N	94°06′W
Mount Carmel, Il., U.S. (kär′mĕl)	66	38°25′N	87°45′W
Mount Carmel, Pa., U.S.	67	40°50′N	76°25′W
Mount Carooll, Il., U.S.	71	42°05′N	89°55′W
Mount Clemens, Mi., U.S. (klĕm′ĕnz)	69b	42°36′N	82°52′W
Mount Desert, i., Me., U.S. (dē-zûrt′)	60	44°15′N	68°08′W
Mount Dora, Fl., U.S. (dō′rȧ)	83a	28°45′N	81°38′W
Mount Duneed, Austl.	173a	38°15′S	144°20′E
Mount Eliza, Austl.	173a	38°11′S	145°05′E
Mount Fletcher, S. Afr. (flĕ′chēr)	187c	30°42′S	28°32′E
Mount Forest, Can. (fŏr′ĕst)	59	44°00′N	80°45′W
Mount Frere, S. Afr. (frâr′)	187c	30°54′S	29°02′E
Mount Gambier, Austl. (găm′bēr)	174	37°30′S	140°53′E
Mount Gilead, Oh., U.S. (gĭl′ēăd)	66	40°30′N	82°50′W
Mount Healthy, Oh., U.S. (hĕlth′ē)	69f	39°14′N	84°32′W
Mount Holly, N.J., U.S. (hŏl′ĭ)	68f	39°59′N	74°47′W
Mount Hope, Can.	62d	43°09′N	79°55′W
Mount Hope, N.J., U.S. (hōp)	68a	40°55′N	74°32′W
Mount Hope, W.V., U.S.	66	37°55′N	81°10′W
Mount Isa, Austl. (ī′zȧ)	174	21°00′S	139°45′E
Mount Kisco, N.Y., U.S. (kĭs′ko)	68a	41°12′N	73°44′W
Mountlake Terrace, Wa., U.S. (mount lāk tĕr′ĭs)	74a	47°48′N	122°19′W
Mount Lebanon, Pa., U.S. (lĕb′ȧ-nŭn)	69e	40°22′N	80°03′W
Mount Magnet, Austl. (măg-nĕt)	174	28°00′S	118°00′E
Mount Martha, Austl.	173a	38°17′S	145°01′E
Mount Morgan, Austl. (môr-găn)	175	23°42′S	150°45′E
Mount Moriac, Austl.	173a	38°13′S	144°12′E
Mount Morris, Mi., U.S. (mŏr′ĭs)	66	43°10′N	83°45′W
Mount Morris, N.Y., U.S.	67	42°45′N	77°50′W
Mount Nimba National Park, rec., C. Iv.	188	7°35′N	8°10′W
Mount Olive, N.C., U.S. (ŏl′ĭv)	83	35°11′N	78°05′W
Mount Peale, Ut., U.S.	77	38°26′N	109°16′W
Mount Pleasant, Ia., U.S. (plĕz′ănt)	71	40°59′N	91°34′W
Mount Pleasant, Mi., U.S.	66	43°35′N	84°45′W
Mount Pleasant, S.C., U.S.	83	32°46′N	79°51′W
Mount Pleasant, Tn., U.S.	82	35°31′N	87°12′W
Mount Pleasant, Tx., U.S.	81	33°10′N	94°56′W
Mount Pleasant, Ut., U.S.	77	39°35′N	111°20′W
Mount Prospect, Il., U.S. (prŏs′pĕkt)	69a	42°03′N	87°56′W
Mount Rainier National Park, rec., Wa., U.S. (rā-nēr′)	64	46°47′N	121°17′W
Mount Revelstoke National Park, rec., Can. (rĕv′ĕl-stōk)	50	51°22′N	120°15′W
Mount Savage, Md., U.S. (săv′āj)	67	39°45′N	78°55′W
Mount Shasta, Ca., U.S. (shăs′tȧ)	72	41°18′N	122°17′W
Mount Sterling, Il., U.S. (stûr′lĭng)	79	39°59′N	90°44′W
Mount Sterling, Ky., U.S.	66	38°05′N	84°00′W
Mount Stewart, Can. (stū′ărt)	61	46°22′N	62°52′W
Mount Union, Pa., U.S. (ūn′yŭn)	67	40°25′N	77°50′W
Mount Vernon, Il., U.S. (vûr′nŭn)	66	38°20′N	88°50′W
Mount Vernon, In., U.S.	66	37°55′N	87°50′W
Mount Vernon, Mo., U.S.	79	37°09′N	93°48′W
Mount Vernon, N.Y., U.S.	68a	40°55′N	73°51′W
Mount Vernon, Oh., U.S.	66	40°25′N	82°30′W
Mount Vernon, Va., U.S.	68e	38°43′N	77°06′W
Mount Vernon, Wa., U.S.	72	48°25′N	122°20′W
Moura, Braz. (mō′rȧ)	101	1°33′S	61°38′W
Moura, Port.	128	38°08′N	7°28′W
Mourne Mountains, mts., N. Ire., U.K. (môrn)	120	54°10′N	6°09′W
Moussoro, Chad	189	13°39′N	16°29′E
Moûtiers, Fr. (mo͞o-tyâr′)	127	45°31′N	6°34′E
Mowbullan, Mount, mtn., Austl.	176	26°50′S	151°34′E
Moyahua, Mex. (mô-yä′wä)	88	21°16′N	103°10′W
Moyale, Kenya (mô-yä′lä)	185	3°28′N	39°04′E
Moyamba, S.L. (mô-yäm′bä)	184	8°10′N	12°26′W
Moyen Atlas, mts., Mor.	118	32°49′N	5°28′W
Moyeuvre-Grande, Fr.	127	49°15′N	6°26′E
Moyie, r., Id., U.S. (moi′yē)	72	38°50′N	116°10′W
Moyobamba, Peru (mō-yō-bäm′bä)	100	6°12′S	76°56′W
Moyuta, Guat. (mô-ē-o͞o′tä)	90	14°01′N	90°05′W
Moyyero, r., Russia	140	67°15′N	104°10′E
Moyynqum, des., Kaz.	139	44°30′N	70°00′E
Mozambique, nation, Afr. (mō-zăm-bēk′)	186	20°15′S	33°53′E
Mozambique Channel, strt., Afr. (mō-zăm-bek′)	187	24°00′S	38°00′E
Mozdok, Russia (môz-dôk′)	137	43°45′N	44°35′E
Mozhaysk, Russia (mô-zhäysk′)	132	55°31′N	36°02′E
Mozhayskiy, Russia (mô-zhäy′skĭ)	142c	59°42′N	30°08′E
Mpanda, Tan.	191	6°22′S	31°02′E
Mpika, Zam.	191	11°54′S	31°26′E
Mpimbe, Mwi.	191	15°18′S	35°04′E
Mporokoso, Zam. (′m-pō-rō-kō′sō)	186	9°23′S	30°05′E
Mpwapwa, Tan. (′m-pwä′pwä)	186	6°21′S	36°29′E
Mqanduli, S. Afr. (′m-kän′do͞o-lē)	187c	31°50′S	28°42′E
Mrągowo, Pol. (mräŋ′gô-vô)	125	53°52′N	21°18′E
M'Sila, Alg. (m′sē′lȧ)	184	35°47′N	4°34′E
Msta, r., Russia (m′stȧ′)	136	58°30′N	33°00′E
Mstsislaw, Bela.	132	54°01′N	31°42′E
Mtakataka, Mwi.	191	14°12′S	34°32′E
Mtamvuna, r., Afr.	187c	30°43′S	29°53′E
Mtata, r., S. Afr.	187c	31°48′S	29°03′E
Mtsensk, Russia (m′tsĕnsk)	136	53°17′N	36°33′E
Mtwara, Tan.	191	10°16′S	40°11′E
Muar, r., Malay.	153b	2°18′N	102°43′E
Mubende, Ug.	191	0°35′N	31°23′E
Mubi, Nig.	189	10°18′N	13°20′E
Mucacata, Moz.	191	13°20′S	39°59′E
Much, Ger. (mo͞ok)	127c	50°54′N	7°24′E
Muchinga Mountains, mts., Zam.	191	12°40′S	30°50′E
Much Wenlock, Eng., U.K. (mŭch wĕn′lŏk)	114a	52°35′N	2°33′W
Muckalee Creek, r., Ga., U.S. (mŭk′ă lē)	82	31°55′N	84°10′W
Muckleshoot Indian Reservation, I.R., Wa., U.S. (mŭck′′l-sho͞ot)	74a	47°21′N	122°04′W
Mucubela, Moz.	191	16°55′S	37°52′E
Mud, l., Mi., U.S. (mŭd)	71	46°12′N	84°32′W
Mudan, r., China (mo͞o-dän)	164	45°30′N	129°40′E
Mudanjiang, China (mo͞o-dän-jyäŋ)	164	44°28′N	129°38′E
Muddy, r., Nv., U.S. (mŭd′ĭ)	77	36°56′N	114°42′W
Muddy Boggy Creek, r., Ok., U.S. (mud′ĭ bôg′ĭ)	79	34°42′N	96°11′W
Muddy Creek, r., Ut., U.S. (mŭd′ĭ)	77	38°45′N	111°10′W
Mudgee, Austl. (mŭ-jē)	176	32°47′S	149°10′E
Mudjatik, r., Can.	56	56°23′N	107°40′W
Mufulira, Zam.	191	12°33′S	28°14′E
Muğla, Tur. (mo͞og′lä)	154	37°10′N	28°20′E
Mühldorf, Ger. (mül-dôrf)	124	48°15′N	12°33′E
Mühlhausen, Ger. (mül′hou-zĕn)	124	51°13′N	10°25′E
Muhu, i., Est. (mo͞o′ho͞o)	123	58°41′N	22°55′E
Muir Woods National Monument, rec., Ca., U.S. (mūr)	76	37°54′N	123°22′W
Muizenberg, S. Afr. (mwĭz-ĕn-bûrg′)	186a	34°07′S	18°28′E
Mukacheve, Ukr.	125	48°25′N	22°43′E
Mukden *see* Shenyang, China	161	41°45′N	123°22′E
Mukhtuya, Russia (mōk-to͞o′yȧ)	135	61°00′N	113°00′E
Mukilteo, Wa., U.S. (mū-kĭl-tā′ō)	74a	47°57′N	122°18′W
Muko, Japan (mo͞o′kō)	167b	34°57′N	135°43′E
Muko, r., Japan (mo͞o′kō)	167b	34°52′N	135°17′E
Mukutawa, r., Can.	57	53°10′N	97°28′W
Mukwonago, Wi., U.S. (mū-kwō-nȧ′gō)	69a	42°52′N	88°19′W
Mula, Spain (mo͞o′lä)	128	38°05′N	1°12′W
Mula, Al., U.S. (mŭl′gȧ)	68h	33°33′N	86°59′W
Mulde, r., Ger. (mol′dĕ)	124	50°30′N	12°30′E
Muleros, Mex. (mo͞o-lā′rōs)	88	23°44′N	104°00′W
Muleshoe, Tx., U.S.	78	34°13′N	102°43′W
Mulgrave, Can. (mŭl′grāv)	61	45°37′N	61°23′W
Mulhacén, mtn., Spain	118	37°04′N	3°18′W
Mülheim, Ger. (mül′hīm)	127c	51°25′N	6°53′E
Mulhouse, Fr. (mü-lo͞oz′)	117	47°46′N	7°20′E
Muling, China (mo͞o-lĭŋ)	164	44°32′N	130°18′E
Muling, r., China	164	44°40′N	130°30′E
Mull, Island of, i., Scot., U.K. (mŭl)	120	56°40′N	6°19′W
Mullan, Id., U.S. (mŭl′ăn)	72	47°26′N	115°50′W
Müller, Pegunungan, mts., Indon. (mül′ēr)	168	0°22′N	113°05′E
Mullingar, Ire. (mŭl-ĭn-gär)	120	53°31′N	7°26′W
Mullins, S.C., U.S. (mŭl′ĭnz)	83	34°11′N	79°13′W
Mullins River, Belize	90a	17°08′N	88°18′W
Multān, Pak. (mo͞ol-tän′)	155	30°17′N	71°13′E
Multnomah Channel, strt., Or., U.S. (mŭl nō mȧ)	74c	45°41′N	122°53′W
Mulumbe, Monts, mts., D.R.C.	191	8°47′S	27°20′E
Mulvane, Ks., U.S. (mŭl-vān′)	79	37°30′N	97°13′W
Mumbai (Bombay), India	155	18°58′N	72°50′E
Mumbwa, Zam. (mōm′bwä)	186	14°59′S	27°04′E
Mumias, Kenya	191	0°20′N	34°29′E
Muna, Mex. (mo͞o′nȧ)	90a	20°28′N	89°42′W
München *see* Munich, Ger.	110	48°08′N	11°35′E
Muncie, In., U.S. (mŭn′sĭ)	65	40°10′N	85°30′W
Mundelein, Il., U.S. (mŭn-dĕ-līn′)	69a	42°16′N	88°00′W
Mundonueva, Pico de, mtn., Col. (pē′kō-dĕ-mo͞o′n-dō-nwĕ′vä)	100a	4°18′N	74°12′W
Muneco, Cerro, mtn., Mex. (sĕ′r-rō-mo͞o-nĕ′kō)	89a	19°13′N	99°20′W
Mungana, Austl. (mŭn-găn′ȧ)	175	17°15′S	144°18′E
Mungbere, D.R.C.	191	2°38′N	28°30′E
Munger, Mn., U.S. (mŭn′gēr)	75h	46°48′N	92°20′W
Mungindi, Austl. (mŭn-gĭn′dē)	175	29°00′S	148°45′E
Munhall, Pa., U.S. (mŭn′hôl)	69e	40°24′N	79°53′W
Munhango, Ang. (mōn-häŋ′gȧ)	186	12°15′S	18°55′E
Munich, Ger.	110	48°08′N	11°35′E
Munising, Mi., U.S. (mū′nĭ-sĭng)	71	46°24′N	86°41′W
Muniz Freire, Braz.	99a	20°29′S	41°25′W
Munku Sardyk, mtn., Asia (mōn′kō sär-dĭk′)	135	51°45′N	100°30′E
Muñoz, Phil. (mo͞on-nyōth′)	169a	15°44′N	120°53′E
Münster, Ger. (mün′stēr)	117	51°57′N	7°38′E
Munster, In., U.S. (mŭn′stēr)	69a	41°34′N	87°31′W
Munster, hist. reg., Ire. (mŭn-stēr)	120	52°30′N	9°24′W
Muntok, Indon. (mōn-tŏk′)	168	2°05′S	105°11′E
Muong Sing, Laos (mo͞o′ông-sĭng′)	168	21°06′N	101°17′E
Muping, China (mo͞o-pĭŋ)	162	37°23′N	121°36′E
Muqui, Braz. (mo͞o-kôē)	99a	20°56′S	41°20′W

ăt; finăl; rāte; senåte; ärm; åsk; sofȧ; fâre; ch-choose; dh-as th in other; bē; ēvent; bĕt; recĕnt; cratēr; g-gō; gh-guttural g; bĭt; ĭ-short neutral; rīde; ĸ-guttural k as ch in German ich;

PLACE (Pronunciation)	PAGE	LAT.	LONG.
Mur, r., Eur. (mo͞or)	117	47°00'N	15°00'E
Muradiye, Tur. (mo͞o-rä'dē-yĕ)	137	39°00'N	43°40'E
Murat, Fr. (mü-rä')	126	45°05'N	2°56'E
Murat, r., Tur. (mo͞o-rät')	154	39°00'N	42°00'E
Murchison, r., Austl. (mûr'chĭ-sŭn)	174	26°45'S	116°15'E
Murcia, Spain (mo͞or'thyä)	110	38°00'N	1°10'W
Murcia, hist. reg., Spain	128	38°35'N	1°51'W
Murdo, S.D., U.S. (mûr'dô)	70	43°53'N	100°42'W
Mureş, r., Rom. (mo͞o'rĕsh)	119	46°02'N	21°50'E
Muret, Fr. (mü-rĕ')	126	43°28'N	1°17'E
Murfreesboro, Tn., U.S. (mûr'frēz-bŭr-ô)	82	35°50'N	86°19'W
Murgab, Taj.	139	38°10'N	73°59'E
Murgab, r., Asia (mo͞or-gäb')	154	37°07'N	62°32'E
Muriaé, r., Braz.	99a	21°20'S	41°40'W
Murino, Russia (mo͞o'rĭ-nô)	142c	60°03'N	30°28'E
Müritz, l., Ger. (mür'its)	124	53°20'N	12°33'E
Murmansk, Russia (mo͞or-mänsk')	134	69°00'N	33°20'E
Murom, Russia (mo͞o'rôm)	134	55°30'N	42°00'W
Muroran, Japan (mo͞o'rô-rän)	161	42°21'N	141°05'E
Muros, Spain (mo͞o'rōs)	128	42°48'N	9°00'W
Muroto-Zaki, c., Japan (mo͞o'rô-tō zä'kē)	166	33°14'N	134°12'E
Murphy, Mo., U.S. (mŭr'fĭ)	75e	38°29'N	90°29'W
Murphy, N.C., U.S.	82	35°05'N	84°00'W
Murphysboro, Il., U.S. (mŭr'fĭz-bŭr-ô)	79	37°46'N	89°21'W
Murray, Ky., U.S. (mŭr'ĭ)	82	36°39'N	88°17'W
Murray, Ut., U.S.	75b	40°40'N	111°53'W
Murray, r., Austl.	174	34°20'S	140°00'E
Murray, r., Can.	55	55°00'N	121°00'W
Murray, Lake, res., S.C., U.S. (mŭr'ĭ)	83	34°07'N	81°18'W
Murray Bridge, Austl.	174	35°10'S	139°35'E
Murray Harbour, Can.	61	46°00'N	62°31'W
Murray Region, reg., Austl. (mŭ'rē)	175	33°20'S	142°30'E
Murrumbidgee, r., Austl. (mŭr-ŭm-bĭd'jē)	175	34°30'S	145°20'E
Murrupula, Moz.	191	15°27'S	38°47'E
Murshidābād, India (mo͝or'shē-dä-bäd')	158	24°08'N	88°11'E
Murska Sobota, Slvn. (mo͞or'skä sô'bô-tä)	130	46°40'N	16°14'E
Muruasigar, mtn., Kenya	191	3°08'N	35°02'E
Murwāra, India	155	23°54'N	80°23'E
Murwillumbah, Austl. (mŭr-wĭl'lŭm-bŭ)	176	28°15'S	153°30'E
Mürz, r., Aus. (mürts)	124	47°30'N	15°21'E
Mürzzuschlag, Aus. (mürts'tso͞o-shläg)	124	47°37'N	15°41'E
Mus, Tur. (mo͞osh)	137	38°55'N	41°30'E
Musala, mtn., Blg.	131	42°05'N	23°24'E
Musan, Kor., N. (mo͞o'sän)	161	41°11'N	129°10'E
Musashino, Japan (mo͞o-sä'shē-nō)	167a	35°43'N	139°35'E
Muscat, Oman (mŭs-kăt')	154	23°23'N	58°30'E
Muscat and Oman *see* Oman, nation, Asia	154	20°00'N	57°45'E
Muscatine, Ia., U.S. (mŭs-kȧ-tēn)	71	41°26'N	91°00'W
Muscle Shoals, Al., U.S. (mŭs''l shōlz)	82	34°44'N	87°38'W
Musgrave Ranges, mts., Austl. (mŭs'grāv)	174	26°15'S	131°15'E
Mushie, D.R.C. (mŭsh'ē)	186	3°04'S	16°50'E
Mushin, Nig.	189	6°32'N	3°22'E
Musi, r., Indon. (mo͞o'sē)	168	2°40'S	103°42'E
Musinga, Alto, mtn., Col. (ä'l-tô-mo͞o-sē'n-gä)	100a	6°40'N	76°13'W
Muskego Lake, l., Wi., U.S. (mŭs-kē'gō)	69a	42°53'N	88°10'W
Muskegon, Mi., U.S. (mŭs-kē'gŭn)	65	43°15'N	86°20'W
Muskegon, r., Mi., U.S.	66	43°20'N	85°55'W
Muskegon Heights, Mi., U.S.	66	43°10'N	86°20'W
Muskingum, r., Oh., U.S. (mŭs-kĭŋ'gŭm)	66	39°45'N	81°55'W
Muskogee, Ok., U.S. (mŭs-kō'gē)	65	35°44'N	95°21'W
Muskoka, l., Can. (mŭs-kō'kȧ)	59	45°00'N	79°30'W
Musoma, Tan.	191	1°30'S	33°48'E
Mussau Island, i., Pap. N. Gui. (mo͞o-sä'o͞o)	169	1°30'S	149°32'E
Musselshell, r., Mt., U.S. (mŭs''l-shĕl)	73	46°25'N	108°20'W
Mussende, Ang.	190	10°32'S	16°05'E
Mussuma, Ang.	190	14°14'S	21°59'E
Mustafakemalpaşa, Tur.	119	40°05'N	28°30'E
Mustang Bayou, Tx., U.S.	81a	29°22'N	95°12'W
Mustang Creek, r., Tx., U.S. (mŭs'tăng)	78	36°22'N	102°46'W
Mustang Island, i., Tx., U.S.	81	27°43'N	97°00'W
Mustique, i., St. Vin. (müs-tēk')	91b	12°53'N	61°03'W
Mustvee, Est. (mo͞ost'vē-ē)	123	58°50'N	26°54'E
Musu Dan, c., Kor., N. (mo͞o'so͞o dän)	161	40°51'N	130°00'E
Muswellbrook, Austl. (mŭs'wŭnl-brŏk)	176	32°15'S	150°50'E
Mutare, Zimb.	186	18°49'S	32°39'E
Mutombo Mukulu, D.R.C. (mo͞o-tôm'bō mo͞o-ko͞o'lo͞o)	186	8°12'S	23°56'E
Mutsu Wan, b., Japan (mo͞ot'so͞o wän)	166	41°20'N	140°55'E
Mutton Bay, Can. (mŭt''n)	61	50°48'N	59°02'W
Mutum, Braz. (mo͞o-to͞o'm)	99a	19°48'S	41°24'W
Muzaffargarh, Pak.	158	30°09'N	71°15'E
Muzaffarpur, India	158	26°13'N	85°20'E
Muzon, Cape, c., Ak., U.S.	54	54°41'N	132°44'W
Múzquiz, Mex. (mo͞oz'kēz)	80	27°53'N	101°31'W
Muztagata, mtn., China	160	38°20'N	75°28'E
Mvomero, Tan.	191	6°20'S	37°25'E
Mvoti, r., S. Afr.	187c	29°18'S	30°52'E
Mwali, i., Com.	187	12°15'S	43°45'E
Mwanza, Tan. (mwän'zä)	186	2°31'S	32°54'E
Mwaya, Tan. (mwä'yä)	186	9°19'S	33°51'E
Mwenga, D.R.C.	191	3°02'S	28°26'E
Mweru, l., Afr.	186	8°50'S	28°50'E
Mwingi, Kenya	191	0°56'S	38°04'E
Myanmar (Burma), nation, Asia	150	21°00'N	95°15'E
Myingyan, Mya. (myĭng-yŭn')	155	21°37'N	95°26'E
Myitkyina, Mya. (myĭ'chē-nȧ)	155	25°33'N	97°25'E
Myjava, Slvk. (mûē'yȧ-vä)	125	48°45'N	17°33'E
Mykhailivka, Ukr.	133	47°16'N	35°12'E
Mykolaïv, Ukr.	134	46°58'N	32°02'E
Mykolaïv, prov., Ukr.	133	47°27'N	31°25'E
Mýkonos, i., Grc.	131	37°26'N	25°30'E
Mymensingh, Bngl.	155	24°48'N	90°28'E
Mynämäki, Fin.	123	60°41'N	21°58'E
Myohyang San, mtn., Kor., N. (myō'hyang)	166	40°00'N	126°12'E
Mýrdalsjökull, ice, Ice. (mür'däls-yû'kôl)	116	63°34'N	18°04'W
Myrhorod, Ukr.	137	49°56'N	33°36'E
Mýrina, Grc.	131	39°52'N	25°01'E
Myrtle Beach, S.C., U.S. (mûr't'l)	83	33°42'N	78°53'W
Myrtle Point, Or., U.S.	72	43°04'N	124°08'W
Mysen, Nor.	122	59°32'N	11°16'E
Myshikino, Russia (mēsh'kē-nô)	132	57°48'N	38°21'E
Mysore, India (mī-sōr')	155	12°31'N	76°42'E
Mysovka, Russia (mē' sôf-kȧ)	123	55°11'N	21°17'E
Mystic, Ia., U.S. (mĭs'tĭk)	71	40°47'N	92°54'W
Mytilíni, Grc.	119	39°09'N	26°35'E
Mytishchi, Russia (mē-tēsh'chi)	142b	55°55'N	37°46'E
Mziha, Tan.	191	5°54'S	37°47'E
Mzimba, Mwi. ('m-zĭm'bä)	186	11°52'S	33°34'E
Mzimkulu, r., Afr.	187c	30°12'S	29°57'E
Mzimvubu, r., S. Afr.	187c	31°22'S	29°20'E
Mzuzu, Mwi.	191	11°30'S	34°10'E

N

PLACE (Pronunciation)	PAGE	LAT.	LONG.
Naab, r., Ger. (näp)	124	49°38'N	12°15'E
Naaldwijk, Neth.	115a	52°00'N	4°11'E
Nā'ālehu, Hi., U.S.	84a	19°00'N	155°35'W
Naantali, Fin. (nän'tȧ-lē)	123	60°29'N	22°03'E
Nabberu, l., Austl. (năb'ēr-o͞o)	174	26°05'S	120°35'E
Naberezhnyye Chelny, Russia	134	55°42'N	52°19'E
Nabeul, Tun. (nā-bûl')	184	36°34'N	10°45'E
Nabiswera, Ug.	191	1°28'N	32°16'E
Naboomspruit, S. Afr.	192c	24°32'S	28°43'E
Nābulus, W.B.	153a	32°13'N	35°16'E
Nacala, Moz. (nä-kȧ'lȧ)	187	14°34'S	40°41'E
Nacaome, Hond. (nä-kä-ō'mȧ)	90	13°32'N	87°28'W
Na Cham, Viet. (nä chäm')	165	22°02'N	106°30'E
Naches, r., Wa., U.S. (năch'ĕz)	72	46°51'N	121°03'W
Náchod, Czech Rep. (näk'ôt)	124	50°25'N	16°08'E
Nacimiento, Lake, res., Ca., U.S. (nȧ-sĭ-myĕn'tô)	76	35°50'N	121°00'W
Nacogdoches, Tx., U.S. (năk'ô-dō'chĕz)	81	31°36'N	94°40'W
Nadadores, Mex. (nä-dä-dō'räs)	80	27°04'N	101°36'W
Nadiād, India	158	22°45'N	72°51'E
Nadir, V.I.U.S.	87c	18°19'N	64°53'W
Nădlac, Rom.	131	46°09'N	20°52'E
Nadvirna, Ukr.	125	48°37'N	24°35'E
Nadym, r., Russia (nȧ'dĭm)	140	64°30'N	72°48'E
Naestved, Den. (nĕst'vĭdh)	116	55°14'N	11°46'E
Nafada, Nig.	189	11°08'N	11°20'E
Nafishah, Egypt	192d	30°34'N	32°15'E
Náfplio, Grc.	131	37°33'N	22°46'E
Nafūd ad Dahy, des., Sau. Ar.	154	22°15'N	44°15'E
Nag, Co, l., China	158	31°38'N	91°18'E
Naga, Phil. (nä'gä)	169	13°37'N	123°12'E
Naga, i., Japan	167	32°09'N	130°16'E
Nagahama, Japan (nä'gä-hä'mä)	167	33°32'N	132°29'E
Nagahama, Japan	167	35°23'N	136°16'E
Nagaland, India	155	25°47'N	94°15'E
Nagano, Japan (nä'gä-nô)	161	36°42'N	138°12'E
Nagaoka, Japan (nä'gȧ-ō'kȧ)	161	37°22'N	138°49'E
Nagaoka, Japan	167b	34°54'N	135°42'E
Nāgappattinam, India	155	10°48'N	79°51'E
Nagarote, Nic. (nä-gä-rō'tĕ)	90	12°17'N	86°35'W
Nagasaki, Japan (nä'gȧ-sä'kē)	161	32°48'N	129°53'E
Nāgaur, India	158	27°19'N	73°41'E
Nagaybakskiy, Russia (nȧ-gáy-bäk'skĭ)	142a	53°33'N	59°33'E
Nagcarlan, Phil. (näg-kär-län')	169a	14°07'N	121°24'E
Nāgercoil, India	159	8°15'N	77°29'E
Nagorno Karabakh, hist. reg., Azer. (nu-gôr'nŭ-kŭ-rŭ-bäk')	137	40°10'N	46°50'E
Nagoya, Japan	161	35°09'N	136°53'E
Nāgpur, India (näg'po͞or)	155	21°12'N	79°09'E
Nagua, Dom. Rep. (nä'gwä)	93	19°20'N	69°40'W
Nagykanizsa, Hung. (nôd'y'kô'nē-shô)	119	46°27'N	17°00'E
Nagykőrös, Hung. (nôd'y'kŭ-rŭsh)	125	47°02'N	19°46'E
Naha, Japan (nä'hä)	161	26°02'N	127°43'E
Nahanni National Park, rec., Can.	52	62°10'N	125°15'W
Nahant, Ma., U.S. (nȧ-hănt)	61a	42°26'N	70°55'W
Nahariyya, Isr.	153a	33°01'N	35°06'E
Nahuel Huapi, l., Arg. (nä'wĕl wä'pē)	102	41°00'S	71°30'W
Nahuizalco, El Sal. (nä-wē-zäl'kō)	90	13°50'N	89°43'W
Naic, Phil. (nä-ēk)	169a	14°20'N	120°46'E
Naica, Mex. (nä-ē'kä)	80	27°53'N	105°30'W
Naiguata, Pico, mtn., Ven. (pē'kô)	101b	10°32'N	66°44'W
Nain, Can. (nīn)	51	56°29'N	61°52'W
Nā'īn, Iran	157	32°52'N	53°05'E
Nairn, Scot., U.K. (nârn)	120	57°35'N	3°54'W
Nairobi, Kenya (nī-rō'bē)	186	1°17'S	36°49'E
Naivasha, Kenya (nī-vä'shȧ)	186	0°47'S	36°29'E
Najd, hist. reg., Sau. Ar.	154	25°18'N	42°38'E
Najin, Kor., N. (nä'jĭn)	161	42°04'N	130°35'E
Najran, des., Sau. Ar. (nŭj-rän')	154	17°29'N	45°30'E
Naju, Kor., S. (nä'jo͞o')	166	35°02'N	126°42'E
Najusa, r., Cuba (nä-ho͞o'sä)	92	20°55'N	77°55'W
Nakatsu, Japan (nä'käts-o͞o)	166	33°34'N	131°10'E
Nakhodka, Russia (nŭ-ĸôt'kŭ)	135	43°03'N	133°08'E
Nakhon Ratchasima, Thai.	168	14°56'N	102°14'E
Nakhon Sawan, Thai.	168	15°42'N	100°06'E
Nakhon Si Thammarat, Thai.	168	8°27'N	99°58'E
Nakło nad Notecia, Pol.	125	53°10'N	17°35'E
Nakskov, Den. (näk'skou)	116	54°51'N	11°06'E
Naktong, r., Kor., S. (näk'tŭng)	166	36°10'N	128°30'E
Nal'chik, Russia (nál-chĕk')	137	43°30'N	43°35'E
Nalón, r., Spain (nä-lōn')	128	43°15'N	5°38'W
Nālūt, Libya (nä-lo͞ot')	184	31°51'N	10°49'E
Namak, Daryacheh-ye, l., Iran	154	34°58'N	51°33'E
Namakan, l., Mn., U.S. (nă'mȧ-kăn)	71	48°20'N	92°43'W
Namangan, Uzb. (nȧ-màn-gän')	139	41°08'N	71°59'E
Namao, Can.	62g	53°43'N	113°30'W
Namatanai, Pap. N. Gui. (nä'mä-tȧ-nä'ē)	169	3°43'S	152°26'E
Nambour, Austl. (năm'bŏr)	176	26°48'S	153°00'E
Nam Co, l., China (näm tswo)	160	30°30'N	91°10'E
Nam Dinh, Viet. (näm dĕnĸ')	168	20°30'N	106°10'E
Nametil, Moz.	191	15°43'S	39°21'E
Namhae, i., Kor., S. (näm'hī')	166	34°23'N	128°05'E
Namib Desert, des., Nmb. (nä-mēb')	186	18°45'S	12°45'E
Namibia, nation, Afr.	186	19°30'S	16°13'E
Namoi, r., Austl. (nămoi)	175	30°10'S	148°43'E
Namous, Oued en, r., Alg. (nȧ-mo͞os')	118	31°48'N	0°19'W
Nampa, Id., U.S. (năm'pȧ)	64	43°35'N	116°35'W
Namp'o, Kor., N.	161	38°47'N	125°28'E
Nampuecha, Moz.	191	13°59'S	40°18'E
Nampula, Moz.	191	15°07'S	39°15'E
Namsos, Nor. (näm'sŏs)	116	64°28'N	11°14'E
Namu, Can.	54	51°53'N	127°50'W
Namuli, Serra, mts., Moz.	191	15°05'S	37°05'E
Namur, Bel. (nȧ-mür')	117	50°29'N	4°55'E
Namutoni, Nmb. (nȧ-mo͞o-tō'nē)	186	18°45'S	17°00'E
Nan, r., Thai.	168	18°11'N	100°29'E
Nanacamilpa, Mex. (nä-nä-kä-mē'l-pä)	89a	19°30'N	98°33'W
Nanaimo, Can. (nȧ-nī'mō)	50	49°10'N	123°56'W
Nanam, Kor., N. (nä'nän')	166	41°38'N	129°37'E
Nanao, Japan (nä'nä-ō)	166	37°03'N	136°59'E
Nan'ao Dao, i., China (nän-ou dou)	165	23°30'N	117°30'E
Nanchang, China (nän'chäng')	161	28°38'N	115°48'E
Nanchangshan Dao, i., China (nän-chäŋ-shän dou)	162	37°56'N	120°42'E
Nancheng, China (nän-chŭŋ)	161	26°50'N	116°40'E
Nanchong, China (nän-chôŋ)	160	30°45'N	106°05'E
Nancy, Fr. (näɴ-sē')	117	48°42'N	6°11'E
Nancy Creek, r., Ga., U.S. (năn'cē)	68c	33°51'N	84°25'W
Nanda Devi, mtn., India (nän'dä dā'vē)	155	30°30'N	80°25'E
Nānded, India	158	19°13'N	77°21'E
Nandurbār, India	158	21°29'N	74°13'E
Nandyāl, India	159	15°54'N	78°09'E
Nanga Parbat, mtn., Pak.	158	35°20'N	74°35'E
Nangi, India	158a	22°30'N	88°14'E
Nangis, Fr. (näɴ-zhē')	127b	48°33'N	3°01'E
Nangong, China (nän-gôŋ)	164	37°22'N	115°22'E
Nangweshi, Zam.	190	16°26'S	23°17'E
Nanhuangcheng Dao, i., China (nän-hůäŋ-chŭŋ dou)	162	38°22'N	120°54'E
Nanhui, China	162	31°03'N	121°45'E
Nanjing, China (nän-jyĭŋ)	161	32°04'N	118°46'E
Nanjuma, r., China (nän-jyo͞o-mä)	162	39°37'N	115°45'E
Nanking *see* Nanjing, China	161	32°04'N	118°46'E
Nanle, China (nän-lŭ)	162	36°03'N	115°13'E
Nan Ling, mts., China	161	25°15'N	111°40'E
Nanliu, r., China (nän-lĭō)	165	22°00'N	109°18'E
Nannine, Austl. (nă-nēn')	174	25°50'S	118°30'E
Nanning, China (nän'nĭng')	160	22°56'N	108°10'E
Nanpan, r., China (nän-pän)	165	24°50'N	105°30'E
Nanping, China (nän-pĭŋ)	161	26°40'N	118°05'E
Nansei-shotō, is., Japan	161	27°30'N	127°00'E
Nansemond, Va., U.S. (năn'sē-mŭnd)	68g	36°46'N	76°32'W
Nantai Zan, mtn., Japan (nän-täē zän)	166	36°47'N	139°28'E
Nantes, Fr. (näɴt')	110	47°13'N	1°37'W
Nanteuil-le-Haudouin, Fr. (näɴ-tû-lē-ō-dwăɴ')	127b	49°08'N	2°49'E
Nanticoke, Pa., U.S. (năn'tĭ-kōk)	67	41°10'N	76°00'W
Nantong, China (nän-tôŋ)	162	32°02'N	120°51'E
Nantong, China	162	32°08'N	121°06'E
Nantucket, i., Ma., U.S. (năn-tŭk'ĕt)	65	41°15'N	70°05'W
Nantwich, Eng., U.K. (nănt'wĭch)	114a	53°04'N	2°31'W
Nanxiang, China (nän-shyäŋ)	162	31°17'N	121°17'E
Nanxiong, China (nän-shôŋ)	165	25°10'N	114°20'E
Nanyang, China	161	33°00'N	112°42'E
Nanyang Hu, l., China (nän-yäŋ ho͞o)	162	35°14'N	116°24'E
Nanyuan, China (nän-yůän)	164a	39°48'N	116°24'E
Naolinco, Mex. (nä-o-lēŋ'kō)	89	19°39'N	96°50'W
Náousa, Grc. (nä'o͞o-sȧ)	131	40°38'N	22°05'E
Naozhou Dao, i., China (nou-jō dou)	165	20°58'N	110°58'E
Napa, Ca., U.S. (năp'ȧ)	64	38°20'N	122°17'W
Napanee, Can. (năp'ȧ-nē)	59	44°15'N	77°00'W
Naperville, Il., U.S. (nā'pēr-vĭl)	69a	41°46'N	88°09'W
Napier, N.Z. (nā'pĭ-ēr)	175a	39°30'S	177°00'E
Napierville, Can. (nā'pĭ-ē-vĭl)	62a	45°11'N	73°24'W
Naples (Napoli), Italy	110	40°37'N	14°12'E
Naples, Fl., U.S. (nā'p'lz)	83a	26°07'N	81°46'W
Napo, r., S.A. (nä'pō)	100	1°49'S	74°20'W

ng-sing; ŋ-baŋk; ɴ-nasalized n; nŏd; cŏmmit; ōld; ôbey; ôrder; oi-boil; fo͞od; ŏ-as oo in foot; ou-out; s-soft; sh-dish; th-thin; pūre; ûnite; ûrn; stŭd; circŭs; ü-as in French tu; '-indeterminate vowel.

PLACE (Pronunciation)	PAGE	LAT.	LONG.
Napoleon, Oh., U.S. (nȧ-pō'lē-ŭn)	66	41°20'N	84°10'W
Napoleonville, La., U.S. (nȧ-pō'lē-ŭn-vĭl)	81	29°56'N	91°03'W
Napoli *see* Naples, Italy	110	40°37'N	14°12'E
Napoli, Golfo di, b., Italy	118	40°29'N	14°08'E
Nappanee, In., U.S. (năp'ȧ-nē)	66	41°30'N	86°00'W
Nara, Japan (nä'rä)	161	34°41'N	135°50'E
Nara, Mali	184	15°09'N	7°27'W
Nara, dept., Japan	167b	34°36'N	135°49'E
Nara, r., Russia	132	55°05'N	37°16'E
Narach, Vozyera, l., Bela.	132	54°51'N	27°00'E
Naracoorte, Austl. (nȧ-rȧ-kōōn'tē)	174	36°50'S	140°50'E
Narashino, Japan	167a	35°41'N	140°01'E
Naraspur, India	159	16°32'N	81°43'E
Narberth, Pa., U.S. (när'bûrth)	68f	40°01'N	75°17'W
Narbonne, Fr. (nȧr-bôn')	117	43°12'N	3°00'E
Nare, Col. (nä'rē)	100a	6°12'N	74°37'W
Narew, r., Pol. (när'ĕf)	125	52°43'N	21°19'E
Narmada, r., India	155	22°30'N	75°30'E
Narodnaya, Gora, mtn., Russia (nȧ-rôd'nȧ-yȧ)	134	65°10'N	60°10'E
Naro-Fominsk, Russia (nä'rô-mēnsk')	136	55°23'N	36°43'E
Narrabeen, Austl. (năr-ȧ-bĭn)	173b	33°44'S	151°18'E
Narragansett, R.I., U.S. (năr-ȧ-găn'sĕt)	68b	41°26'N	71°27'W
Narragansett Bay, b., R.I., U.S.	67	41°20'N	71°15'W
Narrandera, Austl. (nȧ-rȧn-dē'rȧ)	175	34°40'S	146°40'E
Narrogin, Austl. (năr'ô-gĭn)	174	33°00'S	117°15'E
Narva, Est. (nȧr'vȧ)	136	59°24'N	28°12'E
Narvacan, Phil. (när-vä-kän')	169a	17°27'N	120°29'E
Narva Jõesuu, Est. (nȧr'vȧ ò-ô-ä'sōō-ò)	123	59°26'N	28°02'E
Narvik, Nor. (när'vêk)	110	68°21'N	17°18'E
Narvskiy Zaliv, b., Eur. (nȧr'vskĭ zä'lĭf)	123	59°35'N	27°25'E
Narvskoye, res., Eur.	123	59°18'N	28°14'E
Nar'yan-Mar, Russia (när'yȧn mär')	134	67°42'N	53°30'E
Naryilco, Austl. (năr-ĭl'kô)	176	28°40'S	141°50'E
Narym, Russia (nä-rēm')	134	58°47'N	82°05'E
Naryn, r., Asia (nŭ-rĭn')	140	41°20'N	76°00'E
Naseby, Eng., U.K. (nāz'bĭ)	114a	52°23'N	0°59'W
Nashua, Mo., U.S. (năsh'ū-ȧ)	75f	39°18'N	94°34'W
Nashua, N.H., U.S.	65	42°47'N	71°23'W
Nashville, Ar., U.S. (năsh'vĭl)	79	33°56'N	93°50'W
Nashville, Ga., U.S.	82	31°12'N	83°15'W
Nashville, Il., U.S.	79	38°21'N	89°42'W
Nashville, Mi., U.S.	66	42°35'N	85°50'W
Nashville, Tn., U.S.	65	36°10'N	86°48'W
Nashwauk, Mn., U.S. (năsh'wôk)	71	47°21'N	93°12'W
Näsi, l., Fin.	116	61°42'N	24°05'E
Našice, Cro. (nä'shē-tsĕ)	119	45°29'N	18°06'E
Nasielsk, Pol. (nä'syĕlsk)	125	52°35'N	20°50'E
Nāsik, India (nä'sĭk)	155	20°02'N	73°49'E
Nāṣir, Sudan (nä-zēr')	185	8°30'N	33°06'E
Nasirabād, India	158	26°13'N	74°48'E
Naskaupi, r., Can. (năs'kô-pĭ)	53	53°59'N	61°10'W
Nasondoye, D.R.C.	191	10°22'S	25°06'E
Nass, r., Can. (năs)	54	55°00'N	129°30'W
Nassau, Bah. (năs'ô)	87	25°05'N	77°20'W
Nassenheide, Ger. (nä'sĕn-hī-dĕ)	115b	52°49'N	13°13'E
Nasser, Lake, res., Egypt	185	23°50'N	32°50'E
Nasugbu, Phil. (nȧ-sòg-bōō')	169a	14°05'N	120°37'E
Nasworthy Lake, l., Tx., U.S. (năz'wûr-thē)	80	31°17'N	100°30'W
Natagaima, Col. (nä-tä-gī'mä)	100a	3°38'N	75°07'W
Natal, Braz. (nä-täl')	101	6°00'S	35°13'W
Natashquan, Can. (nä-tăsh'kwăn)	51	50°11'N	61°49'W
Natashquan, r., Can.	61	50°35'N	61°35'W
Natchez, Ms., U.S. (năch'ĕz)	65	31°35'N	91°20'W
Natchitoches, La., U.S. (năk'ĭ-tŏsh)(nȧch-ĭ-tōsh')	81	31°46'N	93°06'W
Natick, Ma., U.S. (nā'tĭk)	61a	42°17'N	71°21'W
National Bison Range, I.R., Mt., U.S. (năsh'ŭn-ăl bī's'n)	73	47°18'N	113°58'W
National City, Ca., U.S.	76a	32°38'N	117°01'W
Natitingou, Benin	184	10°19'N	1°22'E
Natividade, Braz. (nä-tē-vē-dä'dĕ)	101	11°43'S	47°34'W
Natron, Lake, l., Tan. (nā'trŏn)	186	2°17'S	36°10'E
Natrona Heights, Pa., U.S. (nā'trō nȧ)	69e	40°38'N	79°43'W
Naṭrūn, Wādī an, val., Egypt	192b	30°33'N	30°12'E
Natuna Besar, i., Indon.	168	4°00'N	106°50'E
Natural Bridges National Monument, rec., Ut., U.S. (năt'ū-răl brĭj'ĕs)	77	37°20'N	110°20'W
Naturaliste, Cape, c., Austl. (năt-ū-rȧ-lĭst')	174	33°30'S	115°10'E
Nau, Cap de la, c., Spain	112	38°43'N	0°14'E
Naucalpan de Juárez, Mex.	89a	19°28'N	99°14'W
Nauchampatepetl, mtn., Mex. (näōō-chäm-pä-tĕ'pĕtl)	89	19°32'N	97°09'W
Nauen, Ger. (nou'ĕn)	115b	52°36'N	12°53'E
Naugatuck, Ct., U.S. (nô'gȧ-tŭk)	67	41°25'N	73°05'W
Naujan, Phil. (nä-ò-hän')	169a	13°19'N	121°17'E
Naumburg, Ger. (noum'bòrgh)	124	51°10'N	11°50'E
Nauru, nation, Oc.	3	0°30'S	167°00'E
Nautla, Mex. (nä-ōōt'lä)	86	20°14'N	96°44'W
Nava, Mex. (nä'vä)	80	28°25'N	100°44'W
Nava del Rey, Spain (nä-vä dĕl rā'ē)	128	41°22'N	5°04'W
Navahermosa, Spain (nä-vä-ĕr-mō'sä)	128	39°39'N	4°28'W
Navajas, Cuba (nä-vä-häs')	92	22°40'N	81°20'W
Navajo Hopi Joint Use Area, I.R., Az., U.S.	77	36°15'N	110°30'W
Navajo Indian Reservation, I.R., U.S. (năv'ȧ-hō)	77	36°31'N	109°24'W
Navajo National Monument, rec., Az., U.S.	77	36°43'N	110°39'W
Navajo Reservoir, res., N.M., U.S.	77	36°57'N	107°26'W
Navalcarnero, Spain (nä-väl'kär-nā'rō)	129a	40°17'N	4°05'W
Navalmoral de la Mata, Spain	128	39°53'N	5°32'W
Navan, Can. (nă'vȧn)	62c	45°25'N	75°26'W
Navarino, i., Chile (nȧ-vä-rē'nō)	102	55°30'S	68°15'W
Navarra, hist. reg., Spain (nä-vär'rä)	128	42°40'N	1°35'W
Navarro, Arg. (nä-vä'r-rō)	99c	35°00'S	59°16'W
Navasota, Tx., U.S. (năv-ȧ-sō'tȧ)	81	30°24'N	96°05'W
Navasota, r., Tx., U.S.	81	31°03'N	96°11'W
Navassa, i., N.A. (nȧ-văs'ȧ)	93	18°25'N	75°15'W
Navia, r., Spain (nä-vē'ä)	128	43°10'N	6°45'W
Navidad, Chile (nä-vē-dä'd)	99b	33°57'S	71°51'W
Navidad Bank, bk., (nä-vē-dädh')	93	20°05'N	69°00'W
Navidade do Carangola, Braz. (nä-vē-dä'dô-kȧ-rän-gô'lȧ)	99a	21°04'S	41°58'W
Navojoa, Mex. (nä-vô-kô'ä)	86	27°00'N	109°40'W
Nawābshāh, Pak. (nȧ-wäb'shä)	158	26°20'N	68°30'E
Naxçıvan, Azer.	137	39°10'N	45°30'E
Naxçıvan Muxtar, state, Azer.	138	39°20'N	45°30'E
Náxos, i., Grc. (năk'sŏs)	119	37°15'N	25°20'E
Nayarit, state, Mex. (nä-yä-rēt')	86	22°00'N	105°15'W
Nayarit, Sierra de, mts., Mex. (sē-ĕ'r-rä-dĕ)	88	23°20'N	105°07'W
Naye, Sen.	188	14°25'N	12°12'W
Naylor, Md., U.S. (nā'lŏr)	68e	38°43'N	76°46'W
Nazaré da Mata, Braz. (dä-mä-tä)	101	7°46'S	35°13'W
Nazas, Mex. (nä'zäs)	80	25°14'N	104°08'W
Nazas, r., Mex.	86	25°30'N	104°40'W
Naẕerat, Isr.	153a	32°43'N	35°19'E
Nazilli, Tur. (nȧ-zĭ-lē')	137	37°40'N	28°10'E
Naziya, r., Russia (nȧ-zē'yȧ)	142c	59°48'N	31°18'E
Nazko, r., Can.	54	52°35'N	123°10'W
N'dalatando, Ang.	190	9°18'S	14°54'E
Ndali, Benin	189	9°51'N	2°43'E
Ndikiniméki, Cam.	189	4°46'N	10°50'E
N'Djamena, Chad	185	12°07'N	15°03'E
Ndola, Zam. (n'dō'lä)	186	12°58'S	28°38'E
Ndoto Mountains, mts., Kenya	191	1°55'N	37°05'E
Ndrhamcha, Sebkha de, l., Maur.	188	18°50'N	15°15'W
Nduye, D.R.C.	191	1°50'N	29°01'E
Neagh, Lough, l., N. Ire., U.K. (lŏk nā)	116	54°40'N	6°47'W
Néa Páfos, Cyp.	153a	34°46'N	32°27'E
Neapean, r., Austl.	173b	33°40'S	150°39'E
Neápoli, Grc.	131	36°35'N	23°08'E
Neápolis, Grc.	130a	35°17'N	25°37'E
Near Islands, is., Ak., U.S. (nēr)	63a	52°20'N	172°40'E
Neath, Wales, U.K. (nēth)	120	51°41'N	3°50'W
Nebine Creek, r., Austl. (nĕ-bēne')	176	27°50'S	147°00'E
Nebitdag, Turkmen.	139	39°30'N	54°20'E
Nebraska, state, U.S. (nē-brăs'kȧ)	64	41°45'N	101°30'W
Nebraska City, Ne., U.S.	79	40°40'N	95°50'W
Nechako, r., Can.	54	53°45'N	124°55'W
Nechako Plateau, plat., Can. (nĭ-chă'kō)	54	54°00'N	124°30'W
Nechako Range, mts., Can.	54	53°20'N	124°30'W
Nechako Reservoir, res., Can.	54	53°25'N	125°10'W
Neches, r., Tx., U.S. (nĕch'ĕz)	81	31°03'N	94°40'W
Neckar, r., Ger. (nĕk'är)	124	49°16'N	9°06'E
Necker Island, i., Hi., U.S.	84b	24°00'N	164°00'W
Necochea, Arg. (nā-kô-chā'ä)	102	38°30'S	58°45'W
Nedryhailiv, Ukr.	133	50°49'N	33°52'E
Needham, Ma., U.S. (nēd'ăm)	61a	42°17'N	71°14'W
Needles, Ca., U.S. (nē'd'lz)	77	34°51'N	114°39'W
Neenah, Wi., U.S. (nē'nȧ)	71	44°10'N	88°30'W
Neepawa, Can.	50	50°13'N	99°29'W
Nee Reservoir, res., Co., U.S. (nee)	78	38°26'N	102°56'W
Negareyama, Japan (nā'gä-rā-yä'mä)	167a	35°52'N	139°54'E
Negaunee, Mi., U.S. (nē-gô'nē)	71	46°30'N	87°37'W
Negeri Sembilan, state, Malay. (nā'grē-sĕm-bē-län')	153b	2°46'N	101°54'E
Negev, des., Isr. (nĕ'gĕv)	153a	30°34'N	34°43'E
Negombo, Sri L.	159	7°39'N	79°49'E
Negotin, Serb. (nĕ'gô-tēn)	131	44°13'N	22°33'E
Negro, r., Arg.	102	39°50'S	65°00'W
Negro, r., N.A.	90	13°01'N	87°10'W
Negro, r., S.A.	99c	33°17'S	58°18'W
Negro, r., S.A. (nā'grō)	100	0°18'S	63°21'W
Negro, Cerro, mtn., Pan. (sĕ'-rrô-nā'grô)	91	8°44'N	80°37'W
Negros, i., Phil. (nā'grōs)	168	9°50'N	121°45'E
Nehalem, r., Or., U.S. (nē-hăl'ĕm)	72	45°52'N	123°37'W
Nehaus an der Oste, Ger. (noi'houz)(ōz'tē)	115c	53°48'N	9°02'E
Nehbandān, Iran	157	31°32'N	60°02'E
Nehe, China (nŭ-hŭ)	164	48°23'N	124°58'E
Neheim-Hüsten, Ger. (nĕ'hīm)	127c	51°28'N	7°58'E
Neiba, Dom. Rep. (nā-ē'bä)	93	18°30'N	71°20'W
Neiba, Bahía de, b., Dom. Rep.	93	18°10'N	71°00'W
Neiba, Sierra de, mts., Dom. Rep. (sē-ĕ'r-rä-dĕ)	93	18°40'N	71°40'W
Neihart, Mt., U.S. (nī'härt)	73	46°54'N	110°39'W
Neijiang, China (nā-jyäŋ)	165	29°38'N	105°01'E
Neillsville, Wi., U.S. (nēlz'vĭl)	71	44°35'N	90°37'W
Nei Monggol (Inner Mongolia), prov., China	160	40°15'N	105°00'E
Neiqiu, China (nā-chyō)	162	37°17'N	114°32'E
Neira, Col. (nā'rä)	100a	5°10'N	75°32'W
Neisse, r., Eur. (nēs)	124	51°30'N	15°00'E
Neiva, Col. (nā-ē'vä)(nā'vä)	100	2°55'N	75°16'W
Neixiang, China (nā-shyäŋ)	164	33°00'N	111°38'E
Nekemte, Eth.	185	9°09'N	36°29'E
Nekoosa, Wi., U.S. (nē-kōō'sȧ)	71	44°19'N	89°54'W
Neligh, Ne., U.S. (nē'-lē)	70	42°06'N	98°02'W
Nel'kan, Russia (nĕl-kän')	135	57°45'N	136°36'E
Nellore, India (nĕl-lōr')	155	14°28'N	79°59'E
Nel'ma, Russia (nĕl-mä')	166	47°34'N	139°05'E
Nelson, Can. (nĕl'sŭn)	50	49°29'N	117°17'W
Nelson, N.Z.	175a	41°15'S	173°22'E
Nelson, Eng., U.K.	114a	53°50'N	2°13'W
Nelson, i., Ak., U.S.	63	60°38'N	164°42'W
Nelson, r., Can.	57	56°50'N	93°40'W
Nelson, Cape, c., Austl.	176	38°29'S	141°20'E
Nelsonville, Oh., U.S. (nĕl'sŭn-vĭl)	66	39°30'N	82°15'W
Néma, Maur. (nā'mä)	184	16°37'N	7°15'W
Nemadji, r., Wi., U.S. (nē-măd'jē)	75h	46°33'N	92°16'W
Neman, Russia (nĕ'-män)	123	55°02'N	22°01'E
Neman, r., Eur.	136	53°28'N	24°45'E
Nembe, Nig.	189	4°35'N	6°26'E
Nemeiben Lake, l., Can. (nĕ-mē'bȧn)	56	55°20'N	105°20'W
Nemours, Fr.	126	48°16'N	2°41'E
Nemuro, Japan (nā'mò-rō)	161	43°13'N	145°10'E
Nemuro Strait, strt., Asia	166	43°07'N	145°10'E
Nemyriv, Ukr.	133	48°56'N	28°51'E
Nen, r., China (nŭn)	161	47°07'N	123°28'E
Nen, r., Eng., U.K. (nĕn)	114a	52°32'N	0°19'W
Nenagh, Ire. (nē'nȧ)	120	52°50'N	8°05'W
Nenana, Ak., U.S. (nā-nä'nȧ)	63	64°28'N	149°18'W
Nenikyul', Russia (nĕ-nyē'kyûl)	142c	59°26'N	30°40'E
Nenjiang, China (nŭn-jyäŋ)	161	49°02'N	125°15'E
Neodesha, Ks., U.S. (nē-ô-dē-shô')	79	37°24'N	95°41'W
Neosho, Mo., U.S.	79	36°51'N	94°22'W
Neosho, r., Ks., U.S. (nē-ō'shō)	79	38°07'N	95°40'W
Nepal, nation, Asia (nĕ-pôl')	155	28°45'N	83°00'E
Nephi, Ut., U.S. (nē'fī)	77	39°40'N	111°50'W
Nepomuceno, Braz. (nĕ-pô-mōō-sĕ'no)	99a	21°15'S	45°13'W
Nera, r., Italy (nā'rä)	130	42°45'N	12°54'E
Nérac, Fr. (nā-răk')	126	44°08'N	0°19'E
Nerchinsk, Russia (nyĕr'chēnsk)	135	51°47'N	116°17'E
Nerchinskiy Khrebet, mts., Russia	135	50°30'N	118°30'E
Nerchinskiy Zavod, Russia (nyĕr'chēn-skĭzȧ-vôt')	135	51°35'N	119°46'E
Nerekhta, Russia (nyĕ-rĕκ'tȧ)	132	57°29'N	40°34'E
Neretva, r., Serb. (nĕ'rĕt-vȧ)	131	43°08'N	17°50'E
Nerja, Spain (nĕr'hä)	128	36°45'N	3°53'W
Nerl', r., Russia (nyĕrl)	132	56°59'N	37°57'E
Nerskaya, r., Russia (nyĕr'skȧ-yȧ)	142b	55°31'N	38°46'E
Nerussa, r., Russia (nyȧ-rōō'sȧ)	132	52°24'N	34°20'E
Ness, Loch, l., Scot., U.K. (lŏκ nĕs)	120	57°23'N	4°20'W
Ness City, Ks., U.S. (nĕs)	78	38°27'N	99°55'W
Nesterov, Russia (nyĕs-tā'rôf)	123	54°39'N	22°38'E
Néstos (Mesta), r., Eur. (nās'tōs)	131	41°25'N	24°12'E
Netanya, Isr.	153a	32°19'N	34°52'E
Netcong, N.J., U.S. (nĕt'cŏnj)	68a	40°54'N	74°42'W
Netherlands, nation, Eur. (nĕdh'ẽr-lăndz)	110	53°01'N	3°57'E
Netherlands Guiana *see* Suriname, nation, S.A.	101	4°00'N	56°00'W
Nettilling, l., Can.	53	66°30'N	70°40'W
Nett Lake Indian Reservation, I.R., Mn., U.S. (nĕt lāk)	71	48°23'N	93°19'W
Nettuno, Italy (nĕt-tōō'nô)	129d	41°28'N	12°40'E
Neubeckum, Ger. (noi'bĕ-kōōm)	127c	51°48'N	8°01'E
Neubrandenburg, Ger. (noi-brän'dĕn-bòrgh)	124	53°33'N	13°16'E
Neuburg, Ger. (noi'bòrgh)	124	48°43'N	11°12'E
Neuchâtel, Switz. (nû-shȧ-tĕl')	117	47°00'N	6°52'E
Neuchâtel, Lac de, l., Switz.	124	46°48'N	6°53'E
Neuenhagen, Ger. (noi'ĕn-hä-gĕn)	115b	52°31'N	13°41'E
Neuenrade, Ger. (noi'ĕn-rä-dĕ)	127c	51°17'N	7°47'E
Neufchâtel-en-Bray, Fr. (nû-shä-tĕl'ĕn-brā')	126	49°43'N	1°25'E
Neulengbach, Aus.	115e	48°13'N	15°55'E
Neumarkt, Ger. (noi'märkt)	124	49°17'N	11°30'E
Neumünster, Ger. (noi'münstẽr)	116	54°04'N	10°00'E
Neunkirchen, Aus. (noin'kĭrκ-ĕn)	124	47°43'N	16°05'E
Neuquén, Arg. (nĕ-ò-kān')	102	38°52'S	68°12'W
Neuquén, prov., Arg.	102	39°40'S	70°45'W
Neuquén, r., Arg.	102	38°45'S	69°00'W
Neuruppin, Ger. (noi'rōō-pēn)	124	52°55'N	12°48'E
Neuse, r., N.C., U.S. (nūz)	83	36°12'N	78°50'W
Neusiedler See, l., Eur. (noi-zēd'lẽr)	124	47°54'N	16°31'E
Neuss, Ger. (nois)	127c	51°12'N	6°41'E
Neustadt, Ger. (noi'shtät)	124	49°21'N	8°08'E
Neustadt bei Coburg, Ger. (bī kō'bōōrgh)	124	50°20'N	11°09'E
Neustadt in Holstein, Ger.	124	54°06'N	10°50'E
Neustrelitz, Ger. (noi-strā'lĭts)	124	53°21'N	13°05'E
Neutral Hills, hills, Can. (nū'trăl)	56	52°10'N	110°50'W
Neu Ulm, Ger. (noi ò lm')	124	48°23'N	10°01'E
Neuville, Can. (nū'vĭl)	62b	46°39'N	71°35'W
Neuwied, Ger. (noi'vēdt)	124	50°26'N	7°28'E
Neva, r., Russia (nyĕ-vä')	132	59°49'N	30°54'E
Nevada, Ia., U.S. (nē-vä'dȧ)	71	42°01'N	93°27'W
Nevada, Mo., U.S.	79	37°49'N	94°21'W
Nevada, state, U.S. (nē vä'dȧ)	64	39°30'N	117°00'W
Nevada, Sierra, mts., Spain (syĕr'rä nä-vä'dhä)	112	37°01'N	3°28'W
Nevada, Sierra, mts., U.S. (sē-ĕ'r-rä nē-vă'dȧ)	64	39°20'N	120°05'W
Nevado, Cerro el, mtn., Col. (sĕ'r-rô-ĕl-nĕ-vä'dô)	100a	4°02'N	74°08'W
Neva Stantsiya, Russia (nyĕ-vä' stän'tsĭ-yȧ)	142c	59°53'N	30°30'E
Neve, Serra da, mts., Ang.	190	13°40'S	13°20'E
Nevel', Russia (nyĕ'vĕl)	136	56°03'N	29°57'E
Neveri, r., Ven. (nĕ-vĕ-rē)	101b	10°13'N	64°18'W
Nevers, Fr. (nē-vâr')	117	46°59'N	3°10'E
Neves, Braz.	102b	22°51'S	43°06'W
Nevesinje, Bos. (nĕ-vĕ'sēn-yĕ)	131	43°15'N	18°08'E
Nevinnomyssk, Russia	138	44°38'N	41°56'E
Nevis, i., St. K./N. (nē'vĭs)	87	17°05'N	62°38'W
Nevis, Ben, mtn., Scot., U.K. (bĕn)	116	56°47'N	5°00'W

ăt; finăl; rāte; senåte; ärm; ȧsk; sofȧ; fâre; ch-choose; dh-as th in other; bē; ėvent; bĕt; recĕnt; cratẽr; g-gō; gh-guttural g; bĭt; ĭ-short neutral; rīde; κ-guttural k as ch in German ich;

PRONOUNCING INDEX

Nevis Peak – Nipissing 281

PLACE (Pronunciation)	PAGE	LAT.	LONG.
Nevis Peak, mtn., St. K./N.	91b	17°11'N	62°33'W
Nevşehir, Tur. (nĕv-shĕ'hĕr)	119	38°40'N	34°35'E
Nev'yansk, Russia (nĕv-yänsk')	134	57°29'N	60°14'E
New, r., Va., U.S. (nū)	83	37°20'N	80°35'W
Newala, Tan.	191	10°56'S	39°18'E
New Albany, In., U.S. (nû ôl'bá-nĭ)	69h	38°17'N	85°49'W
New Albany, Ms., U.S.	83	34°28'N	39°00'W
New Amsterdam, Guy. (ăm'stẽr-dăm)	101	6°14'N	57°30'W
Newark, Eng., U.K. (nū'ẽrk)	114a	53°04'N	0°49'W
Newark, Ca., U.S. (nū'ẽrk)	74b	37°32'N	122°02'W
Newark, De., U.S. (nōō'ärk)	67	39°40'N	75°45'W
Newark, N.J., U.S. (nōō'ûrk)	65	40°44'N	74°10'W
Newark, N.Y., U.S. (nū'ẽrk)	67	43°05'N	77°10'W
Newark, Oh., U.S.	66	40°05'N	82°25'W
Newaygo, Mi., U.S. (nū'wā-go)	66	43°25'N	85°50'W
New Bedford, Ma., U.S. (bĕd'fẽrd)	65	41°35'N	70°55'W
Newberg, Or., U.S. (nū'bûrg)	66	45°17'N	122°58'W
New Bern, N.C., U.S. (bûrn)	65	35°05'N	77°05'W
Newbern, Tn., U.S.	82	36°05'N	89°12'W
Newberry, Mi., U.S. (nū'bĕr-ĭ)	71	46°22'N	85°31'W
Newberry, S.C., U.S.	83	34°15'N	81°40'W
New Boston, Mi., U.S. (bôs'tŭn)	69b	42°10'N	83°24'W
New Boston, Oh., U.S.	66	38°45'N	82°55'W
New Braunfels, Tx., U.S. (nū broun'fĕls)	80	29°43'N	98°07'W
New Brighton, Mn., U.S. (brī'tŭn)	75g	45°04'N	93°12'W
New Brighton, Pa., U.S.	69e	40°34'N	80°18'W
New Britain, Ct., U.S. (brĭt''n)	67	41°40'N	72°45'W
New Britain, i., Pap. N. Gui.	169	6°45'S	149°38'E
New Brunswick, N.J., U.S. (brŭnz'wĭk)	68a	40°29'N	74°27'W
New Brunswick, prov., Can.	51	47°14'N	66°30'W
Newburg, In., U.S.	66	38°00'N	87°25'W
Newburg, Mo., U.S.	79	37°54'N	91°53'W
Newburgh, N.Y., U.S.	67	41°30'N	74°00'W
Newburgh Heights, Oh., U.S.	69d	41°27'N	81°40'W
Newbury, Eng., U.K. (nū'bẽr-ĭ)	120	51°24'N	1°26'W
Newbury, Ma., U.S.	61a	42°48'N	70°52'W
Newbury, co., Eng., U.K.	114b	51°25'N	1°15'W
Newburyport, Ma., U.S. (nū'bẽr-ĭ-pōrt)	61a	42°48'N	70°53'W
New Caledonia, dep., Oc.	175	21°28'S	164°40'E
New Canaan, Ct., U.S. (kā-nán)	68a	41°06'N	73°30'W
New Carlisle, Can. (kär-līl')	51	48°01'N	65°20'W
Newcastle, Austl. (nū-kás''l)	176	33°00'S	151°55'E
Newcastle, Can.	51	47°00'N	65°34'W
New Castle, De., U.S.	67	39°40'N	75°35'W
New Castle, In., U.S.	66	39°55'N	85°25'W
New Castle, Oh., U.S.	66	40°20'N	82°10'W
New Castle, Pa., U.S.	66	41°00'N	80°25'W
Newcastle, Tx., U.S.	78	33°13'N	98°44'W
Newcastle, Wy., U.S.	70	43°51'N	104°11'W
Newcastle under Lyme, Eng., U.K. (nů-kás''l) (nŭ-kás''l)	114a	53°01'N	2°14'W
Newcastle, Eng., U.K.	110	55°00'N	1°45'W
Newcastle Waters, Austl. (wô'tẽrz)	174	17°10'S	133°25'E
Newcomerstown, Oh., U.S. (nū'kŭm-ẽrz-toun)	66	40°15'N	81°40'W
New Croton Reservoir, res., N.Y., U.S. (krō'tôn)	68a	41°15'N	73°47'W
New Delhi, India (dĕl'hī)	155	28°43'N	77°18'E
Newell, S.D., U.S. (nū'ĕl)	70	44°43'N	103°26'W
New England Range, mts., Austl. (nů ĭŋ'glănd)	175	29°32'S	152°30'E
Newenham, Cape, c., Ak., U.S. (nū-ĕn-hăm)	63	58°40'N	162°32'W
Newfane, N.Y., U.S. (nū-fān)	69c	43°17'N	78°44'W
Newfoundland, i., Can.	53a	48°30'N	56°00'W
Newfoundland and Labrador, prov., Can.	51	48°15'N	56°53'W
Newgate, Can. (nū'gāt)	55	49°01'N	115°10'W
New Georgia, i., Sol. Is. (jôr'jĭ-á)	175	8°08'S	158°00'E
New Georgia Group, is., Sol. Is.	170e	8°30'S	157°20'E
New Georgia Sound, strt., Sol. Is.	170e	8°00'S	158°10'E
New Glasgow, Can. (glás'gō)	51	45°35'N	62°36'W
New Guinea, i., (gĭne)	169	5°45'S	140°00'E
Newhalem, Wa., U.S. (nū hā'lŭm)	72	48°44'N	121°11'W
New Hampshire, state, U.S. (hămp'shĭr)	65	43°55'N	71°40'W
New Hampton, Ia., U.S. (hămp'tŭn)	71	43°03'N	92°20'W
New Hanover, S. Afr. (hăn'ôvẽr)	187c	29°23'S	30°32'E
New Hanover, i., Pap. N. Gui.	169	2°37'S	150°15'E
New Harmony, In., U.S. (nū här'mô-nĭ)	66	38°10'N	87°55'W
New Haven, Ct., U.S. (hā'vĕn)	65	41°20'N	72°55'W
New Haven, In., U.S. (nū hāv''n)	66	41°05'N	85°00'W
New Hebrides, is., Vanuatu	175	16°00'S	167°00'E
New Holland, Eng., U.K. (hŏl'ănd)	114a	53°42'N	0°21'W
New Holland, N.C., U.S.	83	35°27'N	76°14'W
New Hope Mountain, mtn., Al., U.S. (hōp)	68h	33°23'N	86°45'W
New Hudson, Mi., U.S. (hŭd'sŭn)	69b	42°30'N	83°36'W
New Iberia, La., U.S. (ī-bē'rĭ-á)	81	30°00'N	91°50'W
Newington, Can. (nū'ĕng-tŏn)	62c	45°07'N	75°00'W
New Ireland, i., Pap. N. Gui. (īr'lănd)	169	3°15'S	152°30'E
New Jersey, state, U.S. (jûr'zĭ)	65	40°30'N	74°50'W
New Kensington, Pa., U.S. (kĕn'zĭng-tŭn)	69e	40°34'N	79°35'W
Newkirk, Ok., U.S. (nū'kûrk)	79	36°52'N	97°03'W
New Lenox, Il., U.S. (lĕn'ŭk)	69a	41°31'N	87°58'W
New Lexington, Oh., U.S. (lĕk'sĭng-tŭn)	66	39°40'N	82°10'W
New Lisbon, Wi., U.S. (lĭz'bŭn)	71	43°52'N	90°11'W
New Liskeard, Can.	59	47°30'N	79°40'W
New London, Ct., U.S. (lŭn'dŭn)	67	41°20'N	72°05'W
New London, Wi., U.S.	71	44°24'N	88°45'W
New Madrid, Mo., U.S. (măd'rĭd)	79	36°34'N	89°31'W
Newman's Grove, Ne., U.S. (nū'măn grōv)	70	41°46'N	97°44'W
Newmarket, Can. (nū'mär-kĕt)	59	44°00'N	79°30'W
New Martinsville, W.V., U.S. (mär'tĭnz-vĭl)	66	39°35'N	80°50'W
New Meadows, Id., U.S.	72	44°58'N	116°20'W
New Mexico, state, U.S. (mĕk'sĭ-kō)	64	34°30'N	107°10'W
New Mills, Eng., U.K. (mĭlz)	114a	53°22'N	2°00'W
New Munster, Wi., U.S. (mŭn'stẽr)	69a	42°35'N	88°13'W
Newnan, Ga., U.S. (nū'năn)	82	33°22'N	84°47'W
New Norfolk, Austl. (nôr'fŏk)	175	42°50'S	147°17'E
New Orleans, La., U.S. (ôr'lē-ănz)	65	30°00'N	90°05'W
New Philadelphia, Oh., U.S. (fĭl-á-dĕl'fĭ-á)	66	40°30'N	81°30'W
New Plymouth, N.Z. (plĭm'ŭth)	175a	39°04'S	174°13'E
Newport, Austl.	173b	33°39'S	151°19'E
Newport, Eng., U.K. (nū-pôrt)	120	50°41'N	1°25'W
Newport, Eng., U.K.	114a	52°46'N	2°22'W
Newport, Wales, U.K.	117	51°36'N	3°05'W
Newport, Ar., U.S. (nū'pōrt)	79	35°35'N	91°16'W
Newport, Ky., U.S.	65	39°05'N	84°30'W
Newport, Me., U.S.	60	44°49'N	69°20'W
Newport, Mn., U.S.	75g	44°52'N	92°59'W
Newport, N.H., U.S.	67	43°20'N	72°10'W
Newport, Or., U.S.	72	44°39'N	124°02'W
Newport, R.I., U.S.	67	41°29'N	71°16'W
Newport, Tn., U.S.	82	35°55'N	83°12'W
Newport, Vt., U.S.	67	44°55'N	72°15'W
Newport, Wa., U.S.	72	48°12'N	117°01'W
Newport Beach, Ca., U.S. (bēch)	75a	33°36'N	117°55'W
Newport News, Va., U.S.	65	36°59'N	76°24'W
New Prague, Mn., U.S. (nū prāg)	71	44°33'N	93°35'W
New Providence, i., Bah. (prŏv'ĭ-dĕns)	92	25°00'N	77°25'W
New Richmond, Oh., U.S. (rĭch'mŭnd)	66	38°55'N	84°15'W
New Richmond, Wi., U.S.	71	45°07'N	92°34'W
New Roads, La., U.S. (rōds)	81	30°42'N	91°26'W
New Rochelle, N.Y., U.S. (rū-shĕl')	68a	40°55'N	73°47'W
New Rockford, N.D., U.S. (rŏk'fōrd)	70	47°40'N	99°08'W
New Ross, Ire. (rôs)	120	52°25'N	6°55'W
New Sarepta, Can.	62g	53°17'N	113°09'W
New Siberian Islands *see* Novosibirskiye Ostrova, is., Russia	135	74°00'N	140°30'E
New Smyrna Beach, Fl., U.S. (smûr'ná)	83	29°00'N	80°57'W
New South Wales, state, Austl. (wālz)	175	32°45'S	146°14'E
Newton, Can. (nū'tŭn)	62f	49°56'N	98°04'W
Newton, Eng., U.K.	114a	53°27'N	2°37'W
Newton, Ia., U.S.	71	41°42'N	93°04'W
Newton, Il., U.S.	66	39°00'N	88°10'W
Newton, Ks., U.S.	79	38°03'N	97°22'W
Newton, Ma., U.S.	61a	42°21'N	71°13'W
Newton, Ms., U.S.	82	32°18'N	89°10'W
Newton, N.C., U.S.	83	35°40'N	81°19'W
Newton, N.J., U.S.	68a	41°03'N	74°45'W
Newton, Tx., U.S.	81	30°47'N	93°45'W
Newtonsville, Oh., U.S. (nū'tŭnz-vĭl)	69f	39°11'N	84°04'W
Newtown, N.D., U.S. (nū'toun)	70	47°57'N	102°25'W
Newtown, Oh., U.S.	69f	39°08'N	84°22'W
Newtown, Pa., U.S.	68f	40°13'N	74°56'W
Newtownards, N. Ire., U.K. (nu-t'n-ardz')	120	54°35'N	5°39'W
New Ulm, Mn., U.S. (ŭlm)	71	44°18'N	94°27'W
New Waterford, Can. (wô'tẽr-fẽrd)	51	46°15'N	60°05'W
New Westminster, Can. (wĕst'mĭn-stẽr)	55	49°12'N	122°55'W
New York, N.Y., U.S. (yôrk)	65	40°40'N	73°58'W
New York, state, U.S.	65	42°45'N	78°05'W
New Zealand, nation, Oc. (zē'lănd)	175a	42°00'S	175°00'E
Nexapa, r., Mex. (nĕks-ä'pä)	88	18°32'N	98°29'W
Neya-gawa, Japan (nā'yä gä'wä)	167b	34°47'N	135°38'E
Neyshābūr, Iran	154	36°06'N	58°45'E
Neyva, r., Russia (nĕy'vá)	142a	57°39'N	60°37'E
Nezahualcóyotl, Mex.	89a	19°27'N	99°03'W
Nez Perce, Id., U.S. (nĕz' pûrs')	72	46°16'N	116°15'W
Nez Perce Indian Reservation, I.R., Id., U.S.	72	46°20'N	116°30'W
Ngami, l., Bots. (n'gä'mē)	186	20°56'S	22°31'E
Ngangerabeli Plain, pl., Kenya	191	1°20'S	40°10'E
Ngangla Ringco, l., China (ŋäŋ-lä rĭŋ-tswo)	158	31°42'N	82°53'E
Ngarimbi, Tan.	191	8°28'S	38°36'E
Ngoko, r., Afr.	190	1°55'N	15°53'E
Ngol-Kedju Hill, mtn., Cam.	189	6°20'N	9°45'E
Ngong, Kenya ('n-gông)	186	1°27'S	36°39'E
Ngounié, r., Gabon	190	1°15'S	10°43'E
Ngoywa, Tan.	191	5°56'S	32°48'E
Ngqeleni, S. Afr. ('ng-kĕ-lā'nē)	187c	31°41'S	29°04'E
Nguigmi, Niger ('n-gĕg'mē)	185	14°15'N	13°07'E
Ngurore, Nig.	189	9°18'N	12°14'E
Nguru, Nig. ('n-gōō'rōō)	184	12°53'N	10°26'E
Nguru Mountains, mts., Tan.	191	6°10'S	37°35'E
Nha Trang, Viet. (nyä-träng')	168	12°08'N	108°56'E
Niafounke, Mali	184	16°03'N	4°17'W
Niagara, Wi., U.S. (nī-ăg'á-rá)	71	45°45'N	88°05'W
Niagara, r., N.A.	69c	43°12'N	79°03'W
Niagara Falls, Can.	69c	43°05'N	79°05'W
Niagara Falls, N.Y., U.S.	65	43°06'N	79°02'W
Niagara-on-the-Lake, Can.	62d	43°16'N	79°05'W
Niakaramandougou, C. Iv.	188	8°40'N	5°17'W
Niamey, Niger (nē-ä-mâ')	184	13°31'N	2°07'E
Niamtougou, Togo	188	9°46'N	1°06'E
Niangara, D.R.C. (nē-äŋ-gá'rá)	185	3°42'N	27°52'E
Niangua, r., Mo., U.S. (nī-ăŋ'gwá)	79	37°30'N	93°05'W
Nias, Pulau, i., Indon. (nē'äs')	168	0°58'N	97°43'E
Nibe, Den. (nē'bĕ)	122	56°57'N	9°36'E
Nicaragua, nation, N.A. (nĭk-á-rä'gwá)	86	12°45'N	86°15'W
Nicaragua, Lago de, l., Nic. (lä'gō dĕ)	86	11°45'N	85°28'W
Nicastro, Italy (nē-käs'trō)	119	38°39'N	16°15'E
Nicchehabin, Punta, c., Mex. (pōō'n-tä-nĕk-chĕ-ä-bē'n)	90a	19°50'N	87°20'W
Nice, Fr. (nēs)	110	43°42'N	7°21'E
Nicheng, China (nē-chŭŋ)	163b	30°54'N	121°48'E
Nichicun, l., Can. (nĭch'ĭ-kŭn)	53	53°07'N	72°10'W
Nicholas Channel, strt., N.A. (nĭk'ô-lás)	92	23°30'N	80°20'W
Nicholasville, Ky., U.S. (nĭk'ô-lás-vĭl)	66	37°55'N	84°35'W
Nicobar Islands, is., India (nĭk-ô-bär')	168	8°28'N	94°04'E
Nicolai Mountain, mtn., Or., U.S. (nē-cō lī')	74c	46°05'N	123°27'W
Nicolás Romero, Mex. (nē-kô-lá's rô-mĕ'rô)	89a	19°38'N	99°20'W
Nicolet, Lake, l., Mi., U.S. (nĭ'kŏ-lĕt)	75k	46°22'N	84°14'W
Nicolls Town, Bah.	92	25°10'N	78°00'W
Nicols, Mn., U.S. (nĭk'ĕls)	75g	44°50'N	93°12'W
Nicomeki, r., Can.	74d	49°04'N	122°47'W
Nicosia, Cyp. (nē-kô-sē'á)	154	35°10'N	33°22'E
Nicoya, C.R. (nē-kō'yä)	90	10°08'N	85°27'W
Nicoya, Golfo de, b., C.R. (gôl-fô-dĕ)	90	10°03'N	85°04'W
Nicoya, Península de, pen., C.R.	90	10°05'N	86°00'W
Nidzica, Pol. (nē-jēt'sá)	125	53°21'N	20°30'E
Niedere Tauern, mts., Aus.	124	47°15'N	13°41'E
Niederkrüchten, Ger. (nē'dĕr-krük-tĕn)	127c	51°12'N	6°14'E
Niederösterreich, state, Aus.	115e	48°24'N	16°20'E
Niedersachsen (Lower Saxony), state, Ger. (nē'dĕr-zäk-sĕn)	115c	53°30'N	9°30'E
Niellim, Chad	189	9°42'N	17°49'E
Nienburg, Ger. (nē'ĕn-bŏrgh)	124	52°40'N	9°15'E
Nietverdiend, S. Afr.	192c	25°02'S	26°10'E
Nieuw Nickerie, Sur. (nē-nē'kĕ-rē')	101	5°51'N	57°00'W
Nieves, Mex. (nyā'vås)	88	24°00'N	102°57'W
Niğde, Tur. (nĭg'dĕ)	119	37°55'N	34°40'E
Nigel, S. Afr. (nī'jĕl)	192c	26°26'S	28°27'E
Niger, nation, Afr. (nī'jẽr)	184	18°02'N	8°30'E
Niger, r., Afr.	184	8°00'N	6°00'E
Niger Delta, d., Nig.	189	4°45'N	5°20'E
Nigeria, nation, Afr. (nī-jē'rĭ-á)	184	8°57'N	6°30'E
Nihoa, i., Hi., U.S.	64b	23°15'N	161°30'W
Nii, i., Japan (nē)	167	34°26'N	139°23'E
Niigata, Japan (nē'ê-gä'tá)	161	37°47'N	139°04'E
Ni'ihau, i., Hi., U.S. (nē'ê-ha'ōō)	64c	21°50'N	160°05'W
Niimi, Japan (nē'mē)	167	34°59'N	133°28'E
Niiza, Japan	167a	35°48'N	139°34'E
Nijmegen, Neth. (nī'mä-gĕn)	121	51°50'N	5°52'E
Nikitinka, Russia (nê-kĭ'tĭn-ká)	132	55°33'N	33°19'E
Nikolayevka, Russia (nē-kô-lä'yĕf-ká)	142c	59°29'N	29°48'E
Nikolayevka, Russia	166	48°37'N	134°09'E
Nikolayevskiy, Russia	137	50°00'N	45°30'E
Nikolayevsk-na-Amure, Russia	135	53°18'N	140°49'E
Nikol'sk, Russia (nê-kôlsk')	134	59°30'N	45°40'E
Nikol'skoye, Russia (nê-kôl'skô-yĕ)	142c	59°27'N	30°00'E
Nikopol, Blg. (nē'kô-pōl')	119	43°41'N	24°52'E
Nikopol', Ukr.	137	47°36'N	34°24'E
Nilahue, r., Chile (nē-lá'wĕ)	99b	34°36'S	71°50'W
Nile, r., Afr. (nīl)	185	27°30'N	31°00'E
Niles, Mi., U.S. (nīlz)	66	41°50'N	86°15'W
Niles, Oh., U.S.	66	41°15'N	80°45'W
Nileshwar, India	159	12°08'N	74°14'E
Nilgiri Hills, hills, India	159	12°05'N	76°22'E
Nilópolis, Braz. (nē-lô'pô-lēs)	99a	22°48'S	43°25'W
Nīmach, India	158	24°32'N	74°51'E
Nimba, Mont, mtn., Afr. (nĭm'bá)	184	7°40'N	8°33'W
Nimba Mountains, mts., Afr.	188	7°30'N	8°35'W
Nîmes, Fr. (nēm)	110	43°49'N	4°22'E
Nimrod Reservoir, res., Ar., U.S. (nĭm'rŏd)	79	34°58'N	93°46'W
Nimule, Sudan (nê-mōō'lå)	185	3°38'N	32°12'E
Ninda, Ang.	190	14°47'S	21°24'E
Nine Mile Creek, r., Ut., U.S. (mīn'ĭmôd')	77	39°50'N	110°30'W
Ninety Mile Beach, cst., Austl.	175	38°20'S	147°30'E
Nineveh, Iraq (nĭn'ĕ-vá)	154	36°30'N	43°10'E
Ning'an, China (nĭŋ-än)	161	44°20'N	129°20'E
Ningbo, China (nĭŋ-bwo)	161	29°56'N	121°30'E
Ningde, China (nĭŋ-dŭ)	161	26°38'N	119°33'E
Ninghai, China (nĭng'hī')	165	29°20'N	121°20'E
Ninghe, China (nĭŋ-hŭ)	162	39°20'N	117°50'E
Ningjin, China (nĭŋ-jyĭn)	162	37°39'N	116°47'E
Ningjin, China	162	37°37'N	114°55'E
Ningming, China	165	22°22'N	107°06'E
Ningwu, China (nĭng'wōō')	161	39°00'N	112°12'E
Ningxia Huizu, prov., China (nĭŋ-shyä)	160	37°10'N	106°00'E
Ningyang, China (nĭng'yäng')	162	35°46'N	116°48'E
Ninh Binh, Viet. (nēn bĕnk')	168	20°22'N	106°00'E
Ninigo Group, is., Pap. N. Gui.	169	1°15'S	143°30'E
Ninnescah, r., Ks., U.S. (nĭn'ĕs-kä)	78	37°37'N	98°31'W
Nioaque, Braz. (nēō-á'-kĕ)	101	21°14'S	55°41'W
Niobrara, r., U.S. (nī-ô-brâr'á)	64	42°46'N	98°46'W
Niokolo Koba, Parc National du, rec., Sen.	188	13°05'N	13°00'W
Nioro du Sahel, Mali (nê-ō'rō)	184	15°15'N	9°35'W
Nipawin, Can.	50	53°22'N	104°00'W
Nipe, Bahía de, b., Cuba (bä-ē'ä-dĕ-nē'pā)	93	20°50'N	75°30'W
Nipe, Sierra de, mts., Cuba (sē-ĕ'r-rá-dĕ)	93	20°20'N	75°50'W
Nipigon, Can. (nĭp'ĭ-gŏn)	51	48°58'N	88°17'W
Nipigon, l., Can.	53	49°37'N	89°55'W
Nipigon Bay, b., Can.	58	48°56'N	88°00'W
Nipisiguit, r., Can. (nĭ-pĭ'sĭ-kwĭt)	60	47°26'N	66°15'W
Nipissing, l., Can. (nĭp'ĭ-sĭng)	53	45°59'N	80°19'W

ng-sing; ŋ-baŋk; ɴ-nasalized n; nŏd; cŏmmit; ōld; ôbey; ôrder; oi-boil; fōōd; ŏ-as oo in foot; ou-out; s-soft; sh-dish; th-thin; pūre; ůnite; ûrn; stŭd; circŭs; ü-as in French tu; '-indeterminate vowel.

282 Niquero – North York Moors PRONOUNCING INDEX

PLACE (Pronunciation)	PAGE	LAT.	LONG.
Niquero, Cuba (nē-kä′rō)	92	20°00′N	77°35′W
Nirmali, India	158	26°30′N	86°43′E
Niš, Serb.	110	43°19′N	21°54′E
Nisa, Port. (nē′sȧ)	128	39°32′N	7°41′W
Nišava, r., Eur. (nē′shȧ-vȧ)	131	43°17′N	22°17′E
Nishino, i., Japan (nēsh′ē-nô)	167	36°06′N	132°49′E
Nishinomiya, Japan (nēsh′ē-nô-mē′yȧ)	167b	34°44′N	135°21′E
Nishio, Japan (nēsh′ē-ô)	167	34°50′N	137°01′E
Niska Lake, l., Can. (nĭs′kȧ)	56	55°35′N	108°38′W
Nisko, Pol. (nēs′kô)	125	50°30′N	22°07′E
Nisku, Can. (nĭs-kū′)	62g	53°21′N	113°33′W
Nisqually, r., Wa., U.S. (nĭs-kwôl′ĭ)	72	46°51′N	122°33′W
Nissan, r., Swe.	122	57°06′N	13°22′E
Nisser, l., Nor. (nĭs′ĕr)	122	59°14′N	8°35′E
Nissum Fjord, b., Den.	122	56°24′N	7°35′E
Niterói, Braz. (nē-tĕ-rô′ĭ)	101	22°53′S	43°07′W
Nith, r., Scot., U.K. (nĭth)	120	55°13′N	3°55′W
Nitra, Slvk. (nē′trȧ)	125	48°18′N	18°04′E
Nitra, r., Slvk.	125	48°13′N	18°14′E
Nitro, W.V., U.S. (nī′trô)	66	38°25′N	81°50′W
Niue, dep., Oc. (nĭ′ô)	195	19°50′S	167°00′W
Nivelles, Bel. (nē′vĕl′)	121	50°33′N	4°17′E
Nixon, Tx., U.S. (nĭk′sŭn)	81	29°16′N	97°48′W
Nizāmābād, India	155	18°48′N	78°07′E
Nizhne-Angarsk, Russia (nyēzh′nyĭ-ŭngärsk′)	135	55°49′N	108°46′E
Nizhne-Chirskaya, Russia	137	48°20′N	42°50′E
Nizhne-Kolymsk, Russia (kô-lēmsk′)	135	68°32′N	160°56′E
Nizhneudinsk, Russia (nēzh′nyĭ-ōōdēnsk′)	135	54°58′N	99°15′E
Nizhniye Sergi, Russia (nyēzh′ nyē sĕr′gē)	136	56°41′N	59°19′E
Nizhniy Novgorod (Gor'kiy), Russia	134	56°15′N	44°05′E
Nizhniy Tagil, Russia (tŭgēl′)	134	57°54′N	59°59′E
Nizhnyaya Kur'ya, Russia (nyē′zhnyȧ-yȧ koor′yȧ)	142a	58°01′N	56°00′E
Nizhnyaya Salda, Russia (nyē′zhnya′ya säl′da′)	142a	58°05′N	60°43′E
Nizhnyaya Taymyra, r., Russia	140	72°30′N	95°18′E
Nizhnyaya Tunguska, r., Russia	135	64°13′N	91°30′E
Nizhnyaya Tura, Russia (tōō′rȧ)	142a	58°38′N	59°50′E
Nizhnyaya Us'va, Russia (ô′vȧ)	142a	59°05′N	58°53′E
Nizhyn, Ukr.	137	51°03′N	31°52′E
Nízke Tatry, mts., Slvk.	125	48°57′N	19°18′E
Njazidja, i., Com.	187	11°44′S	42°38′E
Njombe, Tan.	191	9°20′S	34°46′E
Njurunda, Swe. (nyōō-rôn′dȧ)	122	62°15′N	17°24′E
Nkala Mission, Zam.	191	15°55′S	26°00′E
Nkandla, S. Afr. ('n-känd′lä)	187c	28°40′S	31°06′E
Nkawkaw, Ghana	188	6°33′N	0°47′W
Nkhota, Mwi. (kō-tȧ kō-tȧ)	186	12°52′S	34°16′E
Noākhāli, Bngl.	155	22°52′N	91°08′E
Noatak, Ak., U.S. (nô-ȧ′tăk)	63	67°22′N	163°28′W
Noatak, r., Ak., U.S.	63	67°58′N	162°15′W
Nobeoka, Japan (nō-bȧ-ō′kȧ)	166	32°36′N	131°41′E
Noblesville, In., U.S. (nō′bl′z-vĭl)	66	40°00′N	86°00′W
Nobleton, Can. (nō′bl′tŭn)	62d	43°54′N	79°39′W
Nocera Inferiore, Italy (ĕn-fē-ryô′rē)	129c	40°30′N	14°38′E
Nochistlán, Mex. (nô-chēs-tlän′)	88	21°23′N	102°52′W
Nochixtlón, Mex. (ä-sôn-syōn′)	89	17°28′N	97°12′W
Nogales, Mex. (nō-gä′lĕs)	89	18°49′N	97°09′W
Nogales, Mex.	86	31°15′N	111°00′W
Nogales, Az., U.S. (nô-gä′lĕs)	64	31°20′N	110°55′W
Nogal Valley, val., Som. (nō′gäl)	192a	8°30′N	47°50′E
Nogent-le-Roi, Fr. (nŏ-zhōn-lĕ-rwä′)	127b	48°39′N	1°32′E
Nogent-le-Rotrou, Fr. (rŏ-trōō′)	126	48°22′N	0°47′E
Noginsk, Russia (nô-gēnsk′)	136	55°52′N	38°28′E
Noguera Pallaresa, r., Spain	129	42°18′N	1°03′E
Noia, Spain	128	42°46′N	8°50′W
Noirmoutier, Île de, i., Fr. (nwȧr-mōō-tyā′)	117	47°03′N	3°08′W
Nojima-Zaki, c., Japan (nō′jē-mä zä-kē)	167	34°54′N	139°48′E
Nokomis, Il., U.S. (nô-kō′mĭs)	66	39°15′N	89°10′W
Nola, Italy (nô′lä)	130	40°41′N	14°32′E
Nolinsk, Russia (nô-lēnsk′)	136	57°32′N	49°50′E
Noma Misaki, c., Japan (nô′mä mē′sä-kē)	167	31°25′N	130°09′E
Nombre de Dios, Mex. (nôm-brĕ-dĕ-dyô′s)	88	23°50′N	104°14′W
Nombre de Dios, Pan. (nô′m-brĕ)	91	9°34′N	79°28′W
Nome, Ak., U.S. (nōm)	64a	64°30′N	165°20′W
Nonacho, l., Can.	52	61°48′N	111°20′W
Nong'an, China (nôŋ-än)	164	44°25′N	125°10′E
Nongoma, S. Afr. (nôn-gō′mȧ)	186	27°48′S	31°45′E
Nooksack, Wa., U.S. (nŏŏk′săk)	74d	48°55′N	122°19′W
Nooksack, r., Wa., U.S.	74d	48°54′N	122°31′W
Noordwijk aan Zee, Neth.	115a	52°14′N	4°25′E
Noordzee Kanaal, can., Neth.	115a	52°27′N	4°42′E
Nootka, i., Can. (nōōt′kȧ)	52	49°32′N	126°42′W
Nootka Sound, strt., Can.	54	49°33′N	126°38′W
Nóqui, Ang. (nô-kē′)	186	5°51′S	13°25′E
Nor, r., China (nou′)	166	46°55′N	132°45′E
Nora, Swe.	122	59°32′N	14°56′E
Nora, In., U.S. (nō′rä)	69g	39°54′N	86°08′W
Noranda, Can.	59	48°15′N	79°01′W
Norbeck, Md., U.S. (nôr′bĕk)	68e	39°06′N	77°05′W
Norborne, Mo., U.S. (nôr′bôrn)	79	39°17′N	93°39′W
Norco, Ca., U.S. (nôr′kô)	75a	33°57′N	117°33′W
Norcross, Ga., U.S. (nôr′krôs)	68c	33°56′N	84°13′W
Nord, Riviere du, Can. (rēv-yĕr′ dü nôr)	62a	45°45′N	74°02′W
Nordegg, Can. (nûr′dĕg)	55	52°28′N	116°04′W
Norden, Ger. (nôr′dĕn)	124	53°35′N	7°14′E
Norderney, i., Ger. (nōr′dĕr-nĕy)	124	53°45′N	6°58′E
Nordfjord, b., Nor. (nô′fyôr)	122	61°50′N	5°35′E
Nordhausen, Ger. (nôrt′hau-zĕn)	117	51°30′N	10°48′E
Nordhorn, Ger. (nôrt′hôrn)	124	52°26′N	7°05′E
Nord Kapp, c., Nor.	136	71°11′N	25°48′E
Nordland, Wa., U.S. (nôrd′lȧnd)	74a	48°03′N	122°41′W
Nördlingen, Ger. (nûrt′lĭng-ĕn)	124	48°51′N	10°30′E
Nord-Ostsee Kanal (Kiel Canal), can., Ger. (nōrd-ōzt-zā) (kēl)	124	54°03′N	9°23′E
Nordrhein-Westfalen (North Rhine-Westphalia), state, Ger. (nōrd′hīn-vĕst-fä-lĕn)	127c	51°40′N	7°00′E
Nordvik, Russia (nôrd′vēk)	135	73°57′N	111°15′E
Nore, r., Ire. (nōr)	120	52°34′N	7°15′W
Norfolk, Ma., U.S. (nôr′fŏk)	61a	42°07′N	71°19′W
Norfolk, Ne., U.S.	64	42°10′N	97°25′W
Norfolk, Va., U.S.	65	36°55′N	76°15′W
Norfolk, i., Oc.	195	27°10′S	166°50′E
Norfork, Lake, l., Ar., U.S.	79	36°25′N	92°09′W
Noril'sk, Russia (nô rēlsk′)	134	69°00′N	87°11′E
Normal, Il., U.S. (nôr′mȧl)	66	40°35′N	89°00′W
Norman, r., Austl.	175	18°27′S	141°29′E
Norman, Lake, res., N.C., U.S.	65	35°30′N	80°53′W
Normandie, hist. reg., Fr. (nōr-mäN-dē′)	126	49°02′N	0°17′E
Normandie, Collines de, hills, Fr. (kô-lēn′dĕ-nōr-män-dē′)	126	48°46′N	0°50′W
Normandy *see* Normandie, hist. reg., Fr.	126	49°02′N	0°17′E
Normanton, Austl. (nôr′mȧn-tŭn)	175	17°45′S	141°10′E
Normanton, Eng., U.K.	114a	53°40′N	1°21′W
Norman Wells, Can.	50	65°26′N	127°00′W
Nornalup, Austl. (nôr-năl′ŭp)	174	35°00′S	117°00′E
Nørresundby, Den. (nû-rĕ-sôn′bü)	122	57°04′N	9°55′E
Norris, Tn., U.S. (nŏr′ĭs)	82	36°09′N	84°05′W
Norris Lake, res., Tn., U.S.	65	36°17′N	84°10′W
Norristown, Pa., U.S. (nŏr′ĭs-town)	68f	40°07′N	75°21′W
Norrköping, Swe. (nôr′chûp′ĭng)	110	58°37′N	16°10′E
Norrtälje, Swe. (nôr-tĕl′yĕ)	116	59°47′N	18°39′E
Norseman, Austl. (nôrs′mȧn)	174	32°15′S	122°00′E
Norte, Punta, c., Arg. (pōō′n-tä-nōr′tĕ)	99c	36°17′S	56°46′W
Norte, Serra do, mts., Braz. (sĕ′r-rä-dō-nôr′tĕ)	101	12°04′S	59°08′W
North, Cape, c., Can.	61	47°02′N	60°25′W
North Adams, Ma., U.S. (ăd′ămz)	67	42°40′N	73°05′W
Northam, Austl. (nôr-dhăm)	174	31°50′S	116°45′E
Northam, S. Afr. (nôr′thăm)	192c	24°52′S	27°16′E
North America, cont.	49	45°00′N	100°00′W
North American Basin, deep, (ȧ-mĕr′ĭ-kȧn)	4	23°45′N	62°45′W
Northampton, Austl. (nôr-thămp′tŭn)	174	28°22′S	114°45′E
Northampton, Eng., U.K. (nôrth-ămp′tŭn)	117	52°14′N	0°56′W
Northampton, Ma., U.S.	67	42°20′N	72°45′W
Northampton, Pa., U.S.	67	40°45′N	75°30′W
Northamptonshire, co., Eng., U.K.	114a	52°25′N	0°47′W
North Andaman Island, i., India (ăn-dȧ-măn′)	168	13°15′N	93°30′E
North Andover, Ma., U.S. (ăn′dō-vĕr)	61a	42°42′N	71°07′W
North Arm, mth., Can. (ärm)	74d	49°13′N	123°01′W
North Atlanta, Ga., U.S. (ăt-lăn′tȧ)	68c	33°52′N	84°20′W
North Attleboro, Ma., U.S. (ăt′'l-bŭr-ô)	68b	41°59′N	71°18′W
North Baltimore, Oh., U.S. (bôl′tĭ-môr)	66	41°10′N	83°40′W
North Basque, Tx., U.S. (băsk)	80	31°56′N	98°01′W
North Battleford, Can. (băt′'l-fērd)	50	52°47′N	108°17′W
North Bay, Can.	51	46°13′N	79°26′W
North Bend, Or., U.S. (bĕnd)	72	43°23′N	124°13′W
North Berwick, Me., U.S. (bûr′wĭk)	60	43°18′N	70°46′W
North Bight, b., Bah. (bīt)	92	24°30′N	77°40′W
North Bimini, i., Bah. (bĭ′mĭ-nē)	92	25°45′N	79°20′W
North Borneo *see* Sabah, hist. reg., Malay.	168	5°10′N	116°25′E
Northborough, Ma., U.S.	61a	42°19′N	71°39′W
Northbridge, Ma., U.S. (nôrth′brĭj)	61a	42°09′N	71°39′W
North Caicos, i., T./C. Is. (kī′kôs)	93	21°55′N	72°00′W
North Cape, c., N.Z.	175a	34°31′S	173°02′E
North Carolina, state, U.S. (kăr-ô-lī′nȧ)	65	35°40′N	81°30′W
North Cascades National Park, rec., Wa., U.S.	72	48°50′N	120°50′W
North Cat Cay, i., Bah.	92	25°35′N	79°20′W
North Channel, strt., Can.	58	46°10′N	83°20′W
North Channel, strt., U.K.	112	55°15′N	7°56′W
North Charleston, S.C., U.S. (chärlz′tŭn)	83	32°49′N	79°57′W
North Chicago, Il., U.S. (shĭ-kô′gō)	69a	42°19′N	87°51′W
North College Hill, Oh., U.S. (kŏl′ĕj hĭl)	69f	39°13′N	84°33′W
North Concho, Tx., U.S. (kŏn′chō)	80	31°40′N	100°48′W
North Cooking Lake, Can. (kŏŏk′ĭng lăk)	62g	53°28′N	112°57′W
North Dakota, state, U.S. (dȧ-kō′tȧ)	64	47°20′N	101°55′W
North Downs, Eng., U.K. (dounz)	120	51°11′N	0°01′W
North Dum-Dum, India	158a	22°38′N	88°23′E
Northeast Cape, c., Ak., U.S. (nôrth-ēst)	63	63°15′N	169°04′W
Northeast Point, c., Bah.	93	21°25′N	73°00′W
Northeast Point, c., Bah.	93	22°45′N	73°50′W
Northeast Providence Channel, strt., Bah. (prŏv′ĭ-dĕns)	92	25°45′N	77°00′W
Northeim, Ger. (nôrt′hīm)	124	51°42′N	9°59′E
North Elbow Cays, is., Bah.	92	23°55′N	80°30′W
Northern Cheyenne Indian Reservation, I.R., Mt., U.S.	73	45°32′N	106°43′W
Northern Dvina *see* Severnaya Dvina, r., Russia	134	63°00′N	42°40′E
Northern Ireland, state, U.K. (īr′lănd)	110	54°48′N	7°00′W
Northern Land *see* Severnaya Zemlya, is., Russia	135	79°33′N	101°15′E
Northern Mariana Islands, dep., Oc. (mä-rē-ä′nȧ)	3	17°20′N	145°00′E
Northern Territory, ter., Austl.	174	18°15′S	133°00′E
Northern Yukon National Park, rec., Can.	63	69°00′N	140°00′W
Northfield, Mn., U.S. (nôrth′fēld)	71	44°28′N	93°11′W
North Flinders Ranges, mts., Austl. (flĭn′dĕrz)	176	31°55′S	138°45′E
North Foreland, Eng., U.K. (nôrth-fōr′lănd)	121	51°20′N	1°30′E
North Franklin Mountain, mtn., Tx., U.S. (frăŋ′klĭn)	80	31°55′N	106°30′W
North Frisian Islands, is., Eur.	116	55°16′N	8°15′E
North Gamboa, Pan. (găm-bô′ä)	91	9°07′N	79°40′W
North Gower, Can. (gōw′ĕr)	62c	45°08′N	75°43′W
North Hollywood, Ca., U.S. (hŏl′ē-wŏd)	75a	34°10′N	118°23′W
North Island, i., N.Z.	175a	37°20′S	173°30′E
North Island, i., Ca., U.S.	76a	32°39′N	117°14′W
North Judson, In., U.S. (jŭd′sŭn)	66	41°15′N	86°50′W
North Kansas City, Mo., U.S. (kăn′zȧs)	75f	39°08′N	94°34′W
North Kingstown, R.I., U.S.	68b	41°34′N	71°26′W
North Lincolnshire, co., Eng., U.K.	114a	53°40′N	0°35′W
North Little Rock, Ar., U.S. (lĭt′'l rŏk)	79	34°46′N	92°13′W
North Loup, r., Ne., U.S. (lōōp)	70	42°05′N	100°10′W
North Magnetic Pole, pt. of. i.,	198	77°19′N	101°49′W
North Manchester, In., U.S. (măn′chĕs-tĕr)	66	41°00′N	85°45′W
Northmoor, Mo., U.S. (nôth′mōōr)	75f	39°10′N	94°37′W
North Moose Lake, l., Can.	57	54°09′N	100°20′W
North Mount Lofty Ranges, mts., Austl.	176	33°50′S	138°30′E
North Ogden, Ut., U.S. (ŏg′dĕn)	75b	41°18′N	111°58′W
North Ogden Peak, mtn., Ut., U.S.	75b	41°23′N	111°59′W
North Olmsted, Oh., U.S. (ōlm-stĕd)	69d	41°25′N	81°55′W
North Ossetia, prov., Russia	136	43°00′N	44°15′E
North Pease, r., Tx., U.S. (pēz)	78	34°19′N	100°58′W
North Pender, i., Can. (pĕn′dĕr)	74d	48°48′N	123°16′W
North Plains, Or., U.S. (plānz)	74c	45°36′N	123°00′W
North Platte, Ne., U.S. (plăt)	64	41°08′N	100°45′W
North Platte, r., U.S.	64	41°20′N	102°40′W
North Point, c., Barb.	91b	13°22′N	59°36′W
North Point, c., Mi., U.S.	66	45°00′N	83°20′W
North Pole, pt. of. i.,	198	90°00′N	0°00′
Northport, Al., U.S. (nôrth′pōrt)	82	33°12′N	87°35′W
Northport, N.Y., U.S.	68a	40°53′N	73°20′W
Northport, Wa., U.S.	72	48°53′N	117°47′W
North Reading, Ma., U.S. (rĕd′ĭng)	61a	42°34′N	71°04′W
North Richland Hills, Tx., U.S.	75c	32°50′N	97°13′W
Northridge, Ca., U.S. (nôrth′rĭdj)	75a	34°14′N	118°32′W
North Ridgeville, Oh., U.S. (rĭj-vĭl)	69d	41°23′N	82°01′W
North Ronaldsay, i., Scot., U.K.	120a	59°21′N	2°23′W
North Royalton, Oh., U.S. (roi′ăl-tŭn)	69d	41°19′N	81°44′W
North Saint Paul, Mn., U.S. (sȧnt pôl′)	71	45°01′N	92°59′W
North Santiam, r., Or., U.S. (săn′tyăm)	72	44°42′N	122°50′W
North Saskatchewan, r., Can. (săn-kăch′ē-wän)	52	54°00′N	111°30′W
North Sea, Eur.	110	56°09′N	3°16′E
North Skunk, r., Ia., U.S. (skŭnk)	71	41°39′N	92°46′W
North Stradbroke Island, i., Austl. (străd′brōk)	175	27°45′S	154°18′E
North Sydney, Can. (sĭd′nē)	61	46°13′N	60°15′W
North Taranaki Bight, N.Z. (tȧ-rȧ-nä′kĭ bīt)	175a	38°40′S	174°00′E
North Tarrytown, N.Y., U.S. (tăr′ĭ-toun)	68a	41°05′N	73°52′W
North Thompson, r., Can.	55	50°50′N	120°10′W
North Tonawanda, N.Y., U.S. (tŏn-ȧ-wŏn′dȧ)	69c	43°02′N	78°53′W
North Truchas Peaks, mtn., N.M., U.S. (trōō′chäs)	64	35°58′N	105°40′W
North Twillingate, i., Can. (twĭl′ĭn-gāt)	60	35°58′N	105°37′W
North Uist, i., Scot., U.K. (û′ĭst)	120	57°37′N	7°22′W
Northumberland, N.H., U.S.	67	44°30′N	71°30′W
Northumberland Islands, is., Austl.	175	21°42′S	151°30′E
Northumberland Strait, strt., Can. (nôr thŭm′bēr-lȧnd)	60	46°25′N	64°20′W
North Umpqua, r., Or., U.S. (ŭmp′kwȧ)	72	43°20′N	122°50′W
North Vancouver, Can. (văn-kōō′vĕr)	50	49°19′N	123°04′W
North Vernon, In., U.S. (vûr′nŭn)	66	39°05′N	85°45′W
Northville, Mi., U.S. (nôrth-vĭl)	69b	42°26′N	83°28′W
North Wales, Pa., U.S. (wālz)	68f	40°12′N	75°16′W
North West Cape, c., Austl. (nôrth′wĕst)	174	21°50′S	112°25′E
Northwest Cape Fear, r., N.C., U.S. (cāp fēr)	83	34°34′N	79°46′W
North West Gander, r., Can. (găn′dĕr)	61	48°40′N	55°15′W
Northwest Providence Channel, strt., Bah. (prŏv′ĭ-dĕns)	92	26°15′N	78°45′W
Northwest Territories, ter., Can. (tĕr′ĭ-tō′rĭs)	50	65°00′N	120°00′W
Northwich, Eng., U.K. (nôrth′wĭch)	114a	53°15′N	2°31′W
North Wilkesboro, N.C., U.S. (wĭlks′bûrô)	83	36°08′N	81°10′W
Northwood, Ia., U.S. (nôrth′wŏd)	71	43°26′N	93°13′W
Northwood, N.D., U.S.	70	47°44′N	97°36′W
North Yamhill, r., Or., U.S. (yăm′ hĭl)	74c	45°22′N	123°21′W
North York, Can.	59	43°47′N	79°25′W
North York Moors, for., Eng., U.K. (yôrk mōrz′)	120	54°20′N	0°40′W

ăt; finăl; rāte; senāte; ärm; ȧsk; sofȧ; fâre; ch-choose; dh-as th in other; bē; ēvent; bĕt; recĕnt; cratēr; g-gō; gh-guttural g; bĭt; ĭ-short neutral; rīde; ĸ-guttural k as ch in German ich;

284 Oceanside – Opava PRONOUNCING INDEX

PLACE (Pronunciation)	PAGE	LAT.	LONG.
Oceanside, N.Y., U.S.	68a	40°38'N	73°39'W
Ocean Springs, Ms., U.S. (sprĭngs)	82	30°25'N	88°49'W
Ochakiv, Ukr.	133	46°38'N	31°33'E
Ochamchira, Geor.	138	42°44'N	41°28'E
Ochlockonee, r., Fl., U.S. (ŏk-lô-kō'nē)	82	30°10'N	84°38'W
Ocilla, Ga., U.S. (ô-sĭl'ȧ)	82	31°36'N	83°15'W
Ockelbo, Swe. (ôk'ĕl-bô)	122	60°54'N	16°35'E
Ocklawaha, Lake, res., Fl., U.S.	83	29°30'N	81°50'W
Ocmulgee, r., Ga., U.S.	82	32°25'N	83°30'W
Ocmulgee National Monument, rec., Ga., U.S. (ôk-mŭl'gē)	82	32°45'N	83°28'W
Ocoa, Bahía de, b., Dom. Rep.	93	18°20'N	70°40'W
Ococingo, Mex. (ô-kô-sē'n-gô)	89	17°03'N	92°18'W
Ocom, Lago, l., Mex. (ô-kô'm)	90a	19°26'N	88°18'W
Oconee, r., Ga., U.S. (ô-kō'nē)	65	32°45'N	83°00'W
Oconee, Lake, res., Ga., U.S.	82	33°30'N	83°15'W
Oconomowoc, Wi., U.S. (ô-kŏn'ô-mô-wŏk')	71	43°06'N	88°24'W
Oconto, Wi., U.S. (ô-kŏn'tō)	71	44°54'N	87°55'W
Oconto, r., Wi., U.S.	71	45°08'N	88°24'W
Oconto Falls, Wi., U.S.	71	44°53'N	88°11'W
Ocós, Guat. (ô-kōs')	90	14°31'N	92°12'W
Ocotal, Nic. (ō-kô-täl')	90	13°36'N	86°31'W
Ocotepeque, Hond. (ô-kō-tā-pā'kå)	90	14°25'N	89°13'W
Ocotlán, Mex. (ō-kô-tlän')	88	20°19'N	102°44'W
Ocotlán de Morelos, Mex. (dā mô-rā'lōs)	89	16°46'N	96°41'W
Ocozocoautla, Mex. (ô-kō'zô-kwä-ōō'tlä)	89	16°44'N	93°22'W
Ocumare del Tuy, Ven. (ō-kōō-mä'rā del twē')	100	10°07'N	66°47'W
Oda, Ghana	188	5°55'N	0°59'W
Odawara, Japan (ō'dá-wä'rä)	167	35°15'N	139°10'E
Odda, Nor. (ôdh-á)	122	60°04'N	6°30'E
Odebolt, Ia., U.S. (ō'dē-bōlt)	70	42°20'N	95°14'W
Odemira, Port. (ō-dȧ-mē'rȧ)	128	37°35'N	8°40'W
Ödemiş, Tur. (û'dē-mēsh)	119	38°12'N	28°00'E
Odendaalsrus, S. Afr. (ō'dĕn-däls-rûs')	192c	27°52'S	26°41'E
Odense, Den. (ō'dhĕn-sĕ)	116	55°24'N	10°20'E
Odenton, Md., U.S. (ō'dĕn-tŭn)	68e	39°05'N	76°43'W
Odenwald, for., Ger. (ō'dĕn-väld)	124	49°39'N	8°55'E
Oder, r., Eur. (ō'dẽr)	112	52°40'N	14°19'E
Oderhaff, l., Eur.	124	53°47'N	14°02'E
Odesa, Ukr.	134	46°28'N	30°44'E
Odesa, prov., Ukr.	133	46°05'N	29°48'E
Odessa, Tx., U.S. (ô-dĕs'á)	80	31°52'N	102°21'W
Odessa, Wa., U.S.	72	47°20'N	118°42'W
Odiel, r., Spain (ō-dē-ĕl')	128	37°47'N	6°42'W
Odiham, Eng., U.K. (ŏd'ē-ȧm)	114b	51°14'N	0°56'W
Odintsovo, Russia (ô-dĕn'tsô-vô)	142b	55°40'N	37°16'E
Odiongan, Phil. (ō-dē-ôŋ'gän)	169a	12°24'N	121°59'E
Odivelas, Port. (ō-dē-vä'lyäs)	129b	38°47'N	9°11'W
Odobeşti, Rom. (ō-dô-bĕsh't')	125	45°46'N	27°08'E
O'Donnell, Tx., U.S. (ō-dŏn'ĕl)	78	32°59'N	101°51'W
Odorhei, Rom. (ō-dôr-hā')	125	46°18'N	25°17'E
Odra see Oder, r., Eur. (ô'drȧ)	112	52°40'N	14°19'E
Oeiras, Braz. (wå-ē-räzh')	101	7°05'S	42°01'W
Oeirás, Port. (ô-ĕ'y-rȧ's)	129b	38°42'N	9°18'W
Oelwein, Ia., U.S. (ōl'wīn)	71	42°40'N	91°56'W
O'Fallon, Il., U.S. (ō-făl'ŭn)	75e	38°36'N	89°55'W
O'Fallon Creek, r., Mt., U.S.	73	46°25'N	104°47'W
Ofanto, r., Italy (ô-fän'tō)	130	41°08'N	15°33'E
Offa, Nig.	189	8°09'N	4°44'E
Offenbach, Ger. (ôf'ĕn-bäk)	124	50°06'N	8°50'E
Offenburg, Ger. (ôf'ĕn-bòrgh)	124	48°28'N	7°57'E
Ofuna, Japan (ō'fōō-nä)	167a	35°21'N	139°32'E
Ogaden Plateau, plat., Eth.	192a	6°45'N	44°53'E
Ogaki, Japan	166	35°21'N	136°36'E
Ogallala, Ne., U.S. (ō-gä-lä'lä)	70	41°08'N	101°44'W
Ogbomosho, Nig. (ŏg-bô-mō'shō)	184	8°08'N	4°15'E
Ogden, Ia., U.S. (ŏg'dĕn)	71	42°10'N	94°20'W
Ogden, Ut., U.S.	64	41°14'N	111°58'W
Ogden, r., Ut., U.S.	75b	41°16'N	111°54'W
Ogden Peak, mtn., Ut., U.S.	75b	41°11'N	111°51'W
Ogdensburg, N.J., U.S. (ŏg'dĕnz-bûrg)	68a	41°05'N	74°36'W
Ogdensburg, N.Y., U.S.	65	44°40'N	75°30'W
Ogeechee, r., Ga., U.S. (ō-gē'chē)	83	32°35'N	81°50'W
Ogies, S. Afr.	192c	26°03'S	29°04'E
Ogilvie Mountains, mts., Can. (ō'g'l-vĭ)	52	64°45'N	138°10'W
Oglesby, Il., U.S. (ō'g'lz-bĭ)	66	41°20'N	89°00'W
Oglio, r., Italy (ôl'yō)	130	45°15'N	10°19'E
Ōgo, Japan (ō'gô)	167b	34°49'N	135°06'E
Ogou, r., Togo	188	8°05'N	1°30'E
Ogudnëvo, Russia (ŏg-ŏd-nyô'vô)	142b	56°04'N	38°17'E
Ogulin, Cro. (ō-gōō-lēn')	130	45°17'N	15°11'E
Ogwashi-Uku, Nig.	189	6°10'N	6°31'E
O'Higgins, prov., Chile (ô-kē'gēns)	99b	34°17'S	70°52'W
Ohio, state, U.S. (ô'hī'ō)	65	40°30'N	83°15'W
Ohio, r., U.S.	65	37°25'N	88°05'W
Ohoopee, r., Ga., U.S. (ô-hōō'pe-mc)	83	32°32'N	82°38'W
Ohře, r., Eur. (ōr'zhě)	124	50°08'N	12°45'E
Ohrid, Mac. (ō'krēd)	131	41°08'N	20°46'E
Ohrid, Lake, l., Eur.	131	40°58'N	20°35'E
Ōi, Japan (oi')	167a	35°51'N	139°31'E
Oi-Gawa, r., Japan (ō'ē-gä'wȧ)	167	35°09'N	138°05'E
Oil City, Pa., U.S. (oil sĭ'tĭ)	67	41°25'N	79°40'W
Oirschot, Neth.	115a	51°30'N	5°20'E
Oise, r., Fr. (wäz)	117	49°30'N	2°56'E
Oisterwijk, Neth.	115a	51°34'N	5°13'E
Oita, Japan (ō'ē-tä)	166	33°14'N	131°38'E
Oji, Japan (ō'jē)	167b	34°36'N	135°43'E
Ojinaga, Mex. (ō-kē-nä'gä)	86	29°34'N	104°26'W
Ojitlán, Mex. (ōkē-tlän') (sän-lōō'käs)	89	18°04'N	96°23'W

PLACE (Pronunciation)	PAGE	LAT.	LONG.
Ojo Caliente, Mex. (ōkō käl-yĕn'tå)	88	21°50'N	100°43'W
Ojocaliente, Mex. (ô-kō-kä-lyĕ'n-tĕ)	88	22°39'N	102°15'W
Ojo del Toro, Pico, mtn., Cuba (pē'kô-ô-kō-dĕl-tô'rô)	92	19°55'N	77°25'W
Oka, Can. (ō-kä)	62a	45°28'N	74°05'W
Oka, r., Russia (ô-kȧ')	136	55°10'N	42°10'E
Oka, r., Russia (ō-kä')	140	53°28'N	101°09'E
Oka, r., Russia (ō-kä')	137	52°10'N	35°20'E
Okahandja, Nmb.	186	21°50'S	16°45'E
Okanagan (Okanogan), r., N.A. (ō'kȧ-näg'ȧn)	55	49°06'N	119°43'W
Okanagan Lake, l., Can.	52	50°00'N	119°28'W
Okano, r., Gabon (ō'kä'nō)	184	0°15'N	11°08'E
Okanogan, Wa., U.S.	72	48°20'N	119°34'W
Okanogan, r., Wa., U.S.	72	48°36'N	119°33'W
Okatibbee, r., Ms., U.S. (ō'kä-tĭb'ē)	82	32°37'N	88°54'W
Okatoma Creek, r., Ms., U.S. (ô-kȧ-tō'mă)	82	31°43'N	89°34'W
Okavango (Cubango), r., Afr.	186	18°00'S	20°00'E
Okavango Swamp, sw., Bots.	186	19°30'S	23°02'E
Okaya, Japan (ō'kȧ-yȧ)	167	36°04'N	138°01'E
Okayama, Japan (ō'kȧ-yä'mȧ)	161	34°39'N	133°54'E
Okazaki, Japan (ō'kȧ-zä'kē)	166	34°58'N	137°09'E
Okeechobee, Fl., U.S. (ō-kē-chō'bē)	83	27°15'N	80°50'W
Okeechobee, Lake, l., Fl., U.S.	65	27°00'N	80°49'W
Okeene, Ok., U.S. (ô-kēn')	78	36°06'N	98°19'W
Okefenokee Swamp, sw., U.S. (ō'kē-fē-nō'kē)	83	30°54'N	82°20'W
Okemah, Ok., U.S. (ô-kē'mä)	79	35°26'N	96°18'W
Okene, Nig.	189	7°33'N	6°15'E
Okha, Russia (ŭ-kä')	135	53°44'N	143°12'E
Okhotino, Russia (ô-kô'tĭ-nô)	142b	56°14'N	38°24'E
Okhotsk, Russia (ô-kôtsk')	135	59°28'N	143°32'E
Okhotsk, Sea of, sea, Asia (ô-kôtsk')	135	56°45'N	146°00'E
Okhtyrka, Ukr.	137	50°18'N	34°53'E
Okinawa, i., Japan	161	26°30'N	128°00'E
Okino, i., Japan (ō'kē-nô)	167	36°22'N	133°27'E
Okino Erabu, i., Japan (ō-kē'nô-å-rä'bōō)	166	27°18'N	129°00'E
Oklahoma, state, U.S. (ô-klȧ-hō'mȧ)	64	36°00'N	98°20'W
Oklahoma City, Ok., U.S.	64	35°27'N	97°32'W
Oklawaha, r., Fl., U.S. (ŏk-lȧ-wô'hô)	83	29°13'N	82°00'W
Okmulgee, Ok., U.S. (ôk-mŭl'gē)	79	35°37'N	95°58'W
Okolona, Ky., U.S. (ō-kô-lō'nȧ)	69h	38°08'N	85°41'W
Okolona, Ms., U.S.	82	33°59'N	88°43'W
Oktemberyan, Arm.	138	40°09'N	44°02'E
Okushiri, i., Japan (ō'koo-shē'rē)	166	42°12'N	139°30'E
Okuta, Nig.	189	9°14'N	3°15'E
Olalla, Wa., U.S. (ō-lä'lȧ)	74a	47°26'N	122°33'W
Olanchito, Hond. (ō'län-chē'tô)	90	15°28'N	86°35'W
Öland, i., Swe. (û-lând')	112	57°03'N	17°15'E
Olathe, Ks., U.S. (ō-lā'thē)	75f	38°53'N	94°49'W
Olavarría, Arg. (ō-lä-vär-rē'ä)	102	36°49'N	60°15'W
Oława, Pol. (ō-lä'vȧ)	125	50°57'N	17°18'E
Olazoago, Arg. (ô-läz-kōä'gô)	99c	35°14'S	60°37'W
Olbia, Italy (ô'l-byä)	130	40°55'N	9°28'E
Olching, Ger. (ōl'kēng)	115d	48°13'N	11°21'E
Old Bahama Channel, strt., N.A. (bȧ-hā'mȧ)	92	22°45'N	78°30'W
Old Bight, Bah.	93	24°15'N	75°20'W
Old Bridge, N.J., U.S. (brĭj)	68a	40°24'N	74°22'W
Old Crow, Can. (crō)	50	67°51'N	139°58'W
Oldenburg, Ger. (ôl'dĕn-bòrgh)	116	53°09'N	8°13'E
Old Forge, Pa., U.S. (fôrj)	67	41°20'N	75°50'W
Oldham, Eng., U.K. (ōld'ȧm)	120	53°32'N	2°07'W
Oldham, co., Eng., U.K.	114a	53°35'N	2°05'W
Old Harbor, Ak., U.S. (här'bẽr)	63	57°18'N	153°20'W
Old Head of Kinsale, c., Ire. (ōld hĕd ŏv kĭn-sāl)	120	51°35'N	8°35'W
Old R, Tx., U.S.	81a	29°54'N	94°52'W
Olds, Can. (ōldz)	50	51°47'N	114°06'W
Old Tate, Bots.	186	21°18'S	27°43'E
Old Town, Me., U.S. (toun)	60	44°55'N	68°42'W
Old Wives Lake, l., Can. (wīvz)	56	50°05'N	106°00'W
Olean, N.Y., U.S. (ō-lē-ăn')	65	42°05'N	78°25'W
Olecko, Pol. (ō-lĕt'skô)	125	54°02'N	22°29'E
Olekma, r., Russia (ô-lyĕk-mȧ')	141	55°41'N	120°33'E
Olëkminsk, Russia (ô-lyĕk-mĕnsk')	135	60°39'N	120°40'E
Oleksandriia, Ukr.	132	48°40'N	33°07'E
Olenëk, r., Russia (ô-lyĕ-nyôk')	135	68°00'N	113°00'E
Oléron Île, d', i., Fr. (ēl' dō lā-rôn')	117	45°52'N	1°58'W
Oleśnica, Pol. (ô-lĕsh-nĭ'tsȧ)	117	51°13'N	17°24'E
Olfen, Ger. (ōl'fĕn)	127c	51°43'N	7°22'E
Ol'ga, Russia (ôl'gȧ)	135	43°48'N	135°44'E
Ol'gi, Zaliv, b., Russia (zä'lĭf ôl'gĭ)	166	43°43'N	135°25'E
Olhão, Port. (ôl-youn')	118	37°02'N	7°54'W
Ol'hopil', Ukr.	133	48°11'N	29°28'E
Olievenhoutpoort, S. Afr.	187b	25°58'S	27°55'E
Ōlimbos, mtn., Cyp.	153a	34°56'N	32°52'E
Olinda, Braz. (ô-lē'n-dä)	101	8°00'S	34°58'W
Olinda, Braz.	102b	22°49'S	43°25'W
Oliva, Spain (ô-lē'vä)	129	38°54'N	0°07'W
Oliva de la Frontera, Spain (ô-lē'vä dā)	128	38°33'N	6°55'W
Olive Hill, Ky., U.S. (ŏl'ĭv)	66	38°15'N	83°10'W
Oliveira, Braz. (ō-lē-vā'rä)	99a	20°42'S	44°49'W
Olivenza, Spain (ō-lē-vĕn'thä)	128	38°42'N	7°06'W
Oliver, Can. (ŏ'lĭ-vẽr)	50	49°11'N	119°33'W
Oliver, Can.	62g	53°38'N	113°21'W
Oliver, Wi., U.S. (ŏ'lĭvẽr)	75h	46°39'N	92°12'W
Oliver Lake, l., Can.	62g	53°19'N	113°00'W
Olivia, Mn., U.S. (ō-lĭv'ē-ȧ)	70	44°46'N	95°00'W
Olivos, Arg. (ōlē'vōs)	102a	34°30'S	58°29'W
Ollagüe, Chile (ō-lyä'gā)	100	21°17'S	68°17'W
Ollerton, Eng., U.K. (ŏl'ẽr-tŭn)	114a	53°12'N	1°02'W
Olmos Park, Tx., U.S. (ŏl'mŭs pärk')	75d	29°27'N	98°32'W
Olney, Il., U.S. (ŏl'nĭ)	66	38°45'N	88°05'W

PLACE (Pronunciation)	PAGE	LAT.	LONG.
Olney, Or., U.S. (ŏl'nē)	74c	46°06'N	123°45'W
Olney, Tx., U.S.	78	33°24'N	98°43'W
Olomane, r., Can. (ō'lô mȧ'nĕ)	61	51°05'N	60°50'W
Olomouc, Czech Rep. (ô'lô-mōts)	117	49°37'N	17°15'E
Olonets, Russia (ô-lô'nĕts)	123	60°58'N	32°54'E
Olongapo, Phil.	168	14°49'S	120°17'E
Oloron, Gave d', r., Fr. (gäv-dō-lō-rôn')	126	43°21'N	0°44'W
Oloron-Sainte Marie, Fr. (ō-lō-rônt'sănt mȧ-rē')	126	43°11'N	1°37'W
Olot, Spain (ô-lōt')	118	42°09'N	2°30'E
Olpe, Ger. (ōl'pĕ)	127c	51°02'N	7°51'E
Olsnitz, Ger. (ōlz'nētz)	124	50°25'N	12°11'E
Olsztyn, Pol. (ōl'shtĭn)	116	53°47'N	20°28'E
Olt, r., Rom.	119	44°09'N	24°40'E
Olten, Switz. (ōl'tĕn)	124	47°20'N	7°53'E
Olteniţa, Rom. (ōl-tā'nĭ-tsȧ)	131	44°05'N	26°39'E
Olvera, Spain (ōl-vĕ'rä)	128	36°55'N	5°16'W
Olympia, Wa., U.S. (ô-lĭm'pĭ-ȧ)	64	47°02'N	122°52'W
Olympic Mountains, mts., Wa., U.S.	72	47°54'N	123°58'W
Olympic National Park, rec., Wa., U.S. (ô-lĭm'pĭk)	64	47°54'N	123°00'W
Ólympos, mtn., Grc.	118	40°05'N	22°21'E
Olympus, Mount, mtn., Wa., U.S. (ô-lĭm'pŭs)	72	47°43'N	123°30'W
Olyphant, Pa., U.S. (ŏl'ĭ-fănt)	67	41°30'N	75°40'W
Olyutorskiy, Mys, c., Russia (ŭl-yōō'tôr-skē)	135	59°49'N	167°16'E
Omae-Zaki, c., Japan (ō'mä-å zä'kē)	167	34°37'N	138°15'E
Omagh, N. Ire., U.K. (ō'mä)	120	54°35'N	7°25'W
Omaha, Ne., U.S. (ō'mȧ-hä)	65	41°18'N	95°57'W
Omaha Indian Reservation, I.R., Ne., U.S.	70	42°09'N	96°08'W
Oman, nation, Asia	154	20°00'N	57°45'E
Oman, Gulf of, b., Asia	154	24°24'N	58°58'E
Omaruru, Nmb. (ō-mä-rōō'rōō)	186	21°25'S	16°50'E
Ombrone, r., Italy (ôm-brō'nä)	130	42°48'N	11°18'E
Omdurman, Sudan	185	15°45'N	32°30'E
Omealca, Mex. (ōmå-äl'kō)	89	18°44'N	96°45'W
Ometepec, Mex. (ô-mā-tå-pĕk')	88	16°41'N	98°27'W
Om Hajer, Eth.	185	14°06'N	36°46'E
Omineca, r., Can. (ô-mĭ-nĕk'ȧ)	54	55°50'N	125°45'W
Omineca Mountains, mts., Can.	54	56°00'N	125°00'W
Ōmiya, Japan (ō'mē-yȧ)	167	35°54'S	139°38'E
Omo, r., Eth. (ō'mō)	185	5°54'N	36°09'E
Omoa, Hond. (ô-mō'rä)	90	15°43'N	88°03'W
Omoko, Nig.	189	5°20'N	6°39'E
Omolon, r., Russia (ō'mō)	141	67°43'N	159°15'E
Ōmori, Japan (ō'mô-rē)	167a	35°50'N	140°09'E
Omotepe, Isla de, i., Nic. (ē's-lä-dĕ-ō-mô-tā'pä)	90	11°32'N	85°30'W
Omro, Wi., U.S. (ŏm'rō)	71	44°01'N	89°46'W
Omsk, Russia (ômsk)	134	55°12'N	73°19'E
Ōmura, Japan (ō'mōō-rȧ)	167	32°56'N	129°57'E
Ōmuta, Japan (ō-mò-tȧ)	167	33°02'N	130°28'E
Omutninsk, Russia (ō'mōō-tnēnsk)	136	58°38'N	52°10'E
Onawa, Ia., U.S. (ŏn-ȧ-wȧ)	70	42°02'N	96°05'W
Onaway, Mi., U.S.	66	45°25'N	84°10'W
Oncócua, Ang.	190	16°34'S	13°28'E
Onda, Spain (ōn'dä)	129	39°58'N	0°13'W
Ondava, r., Slvk. (ōn'dȧ-vȧ)	125	48°51'N	21°40'E
Ondo, Nig.	189	7°04'N	4°47'E
Öndörhaan, Mong.	161	47°20'N	110°40'E
Onega, Russia (ô-nyĕ'gȧ)	134	63°50'N	38°08'E
Onega, r., Russia	136	63°20'N	39°20'E
Onega, Lake see Onezhskoye Ozero, l., Russia	136	62°02'N	34°35'E
Oneida, N.Y., U.S. (ô-nī'dȧ)	67	43°05'N	75°40'W
Oneida, l., N.Y., U.S.	67	43°10'N	76°00'W
O'Neill, Ne., U.S. (ō-nēl')	70	42°28'N	98°38'W
Oneonta, N.Y., U.S. (ō-nē-ŏn'tȧ)	67	42°25'N	75°05'W
Onezhskaja Guba, b., Russia	136	64°30'N	36°00'E
Onezhskiy, Poluostrov, pen., Russia	136	64°30'N	37°40'E
Onezhskoye Ozero, Russia (ô-năsh'skô-yĕ ô'zĕ-rô)	136	62°02'N	34°35'E
Ongiin Hiid, Mong.	160	46°00'N	102°46'E
Ongole, India	159	15°36'N	80°03'E
Onilahy, r., Madag.	187	23°41'S	45°00'E
Onitsha, Nig. (ô-nĭt'shȧ)	184	6°09'N	6°47'W
Onomichi, Japan (ō'nô-mē'chē)	166	34°27'N	133°12'E
Onon, r., Asia (ô'nôn)	135	49°00'N	112°00'E
Onoto, Ven. (ô-nō'tô)	101b	9°38'N	65°03'W
Onslow, Austl. (ŏnz'lō)	174	21°53'S	115°00'E
Onslow, b., N.C., U.S. (ŏnz'lō)	83	34°22'N	77°35'W
Ontake San, mtn., Japan (ôn'tä-kå sän)	166	35°55'N	137°29'E
Ontario, Ca., U.S. (ŏn-tā'rĭ-ō)	75a	34°04'N	117°39'E
Ontario, Or., U.S.	72	44°02'N	116°57'W
Ontario, prov., Can.	51	50°47'N	88°50'W
Ontario, Lake, l., N.A.	65	43°35'N	79°05'W
Ontinyent, Spain	129	38°48'N	0°35'W
Ontonagon, Mi., U.S. (ŏn-tô-năg'ŏn)	71	46°50'N	89°20'W
Ōnuki, Japan (ō'nōō-kē)	167a	35°17'N	139°51'E
Oodnadatta, Austl. (ōōd'nȧ-dȧ'tȧ)	174	27°38'S	135°40'E
Ooldea Station, Austl. (ōōl-dā'ä)	174	30°35'S	132°08'E
Oologah Reservoir, res., Ok., U.S.	65	36°43'N	95°32'W
Ooltgensplaat, Neth.	115a	51°41'N	4°19'E
Oostanaula, r., Ga., U.S. (ōō-stȧ-nô'lȧ)	82	34°25'N	85°10'W
Oostende, Bel. (ōst-ĕn'dĕ)	117	51°14'N	2°55'E
Oosterhout, Neth.	115a	51°38'N	4°52'E
Ooster Schelde, r., Neth.	115a	51°40'N	3°40'E
Ootsa Lake, l., Can.	54	53°49'N	126°18'W
Opalaca, Sierra de, mts., Hond. (sē-sĕ'r-rä-dĕ-ô-pä-lä'kä)	90	14°30'N	88°29'W
Opasquia, Can. (ō-păs'kwē-ȧ)	57	53°16'N	93°53'W
Opatów, Pol. (ô-pä'tôf)	125	50°47'N	21°25'E
Opava, Czech Rep. (ō'pä-vä)	125	49°56'N	17°52'E

ăt; finăl; rāte; senāte; ärm; ȧsk; sofȧ; fâre; ch-choose; dh-as th in other; bē; ēvent; bĕt; recĕnt; cratẽr; g-gō; gh-guttural g; bĭt; ĭ-short neutral; rīde; ĸ-guttural k as ch in German ich;

PRONOUNCING INDEX Opelika – Ouro Fino 285

PLACE (Pronunciation)	PAGE	LAT.	LONG.
Opelika, Al., U.S. (ŏp-ē-lī'kȧ)	82	32°39'N	85°23'W
Opelousas, La., U.S. (ŏp-ē-lōō'sȧs)	81	30°33'N	92°04'W
Opeongo, l., Can. (ŏp-ē-ôŋ'gō)	59	45°40'N	78°20'W
Opheim, Mt., U.S. (ō-fīm')	73	48°51'N	106°19'W
Ophir, Ak., U.S. (ō'fēr)	63	63°10'N	156°28'W
Ophir, Mount, mtn., Malay.	153b	2°22'N	102°37'E
Opico, El Sal. (ô-pē'kō)	90	13°50'N	89°23'W
Opinaca, r., Can. (ŏp-ĭ-nä'kȧ)	53	52°28'N	77°40'W
Opishnia, Ukr.	133	49°57'N	34°34'E
Opladen, Ger. (ŏp'lä-dĕn)	127c	51°04'N	7°00'E
Opobo, Nig.	189	4°34'N	7°27'E
Opochka, Russia (ō-pôch'kȧ)	136	56°43'N	28°39'E
Opoczno, Pol. (ô-pôch'nô)	125	51°22'N	20°18'E
Opole, Pol. (ô-pôl'à)	117	50°42'N	17°55'E
Opole Lubelskie, Pol. (ō-pō'lā lōō-bĕl'skyĕ)	125	51°09'N	21°58'E
Opp, Al., U.S. (ŏp)	82	31°18'N	86°15'W
Oppdal, Nor. (ôp'däl)	122	62°37'N	9°41'E
Opportunity, Wa., U.S. (ŏp-ŏr tū'nĭ tĭ)	72	47°37'N	117°20'W
Oquirrh Mountains, mts., Ut., U.S. (ō'kwēr)	75b	40°38'N	112°11'W
Oradea, Rom. (ô-räd'yä)	110	47°02'N	21°55'E
Oral, Kaz.	139	51°14'N	51°22'E
Oran, Alg. (ō-rän)(ô-rän')	184	35°46'N	0°45'W
Orán, Arg. (ō-rá'n)	102	23°13'S	64°17'W
Oran, Mo., U.S. (ôr'ȧn)	79	37°05'N	89°39'W
Oran, Sebkha d', l., Alg.	129	35°28'N	0°28'W
Orange, Austl. (ŏr'ĕnj)	175	33°15'S	149°08'E
Orange, Fr. (ô-ranzh')	117	44°08'N	4°48'E
Orange, Ca., U.S.	75a	33°48'N	117°51'W
Orange, Ct., U.S.	67	41°15'N	73°00'W
Orange, N.J., U.S.	68a	40°46'N	74°14'W
Orange, Tx., U.S.	79	30°07'N	93°44'W
Orange, r., Afr.	186	29°15'S	17°30'E
Orange, Cabo, c., Braz. (kȧ-bô-rȧ'n-zhĕ)	101	4°25'N	51°30'W
Orangeburg, S.C., U.S. (ŏr'ĕnj-bûrg)	83	33°30'N	80°50'W
Orange Cay, i., Bah. (ŏr-ĕnj kē)	92	24°55'N	79°05'W
Orange City, Ia., U.S.	70	43°01'N	96°06'W
Orange Lake, l., Fl., U.S.	83	29°30'N	82°12'W
Orangeville, Can. (ŏr'ĕnj-vĭl)	59	43°55'N	80°06'W
Orangeville, S. Afr.	192c	27°05'S	28°13'E
Orange Walk, Belize (wôl'k)	90a	18°09'N	88°32'W
Orani, Phil. (ō-rä'nē)	169a	14°47'N	120°32'E
Oranienburg, Ger. (ō-rä'nē-ĕn-bŏorgh)	124	52°45'N	13°14'E
Oranjemund, Nmb.	186	28°33'S	16°20'E
Orăştie, Rom. (ô-rûsh'tyå)	131	45°50'N	23°14'E
Orbetello, Italy (ôr-bā-tĕl'lō)	130	42°27'N	11°15'E
Orbigo, r., Spain (ôr-bē'gō)	128	42°30'N	5°55'W
Orbost, Austl. (ôr'bŭst)	176	37°43'S	148°20'E
Orcas, i., Wa., U.S. (ôr'kȧs)	74d	48°43'N	122°52'W
Orchard Farm, Mo., U.S. (ôr'chērd färm)	75e	38°53'N	90°27'W
Orchard Park, N.Y., U.S.	69c	42°46'N	78°46'W
Orchards, Wa., U.S. (ôr'chērdz)	74c	45°40'N	122°33'W
Orchila, Isla, i., Ven.	100	11°47'N	66°34'W
Ord, Ne., U.S. (ôrd)	70	41°35'N	98°57'W
Ord, r., Austl.	174	17°30'S	128°40'E
Ord, Mount, mtn., Az., U.S.	77	33°55'N	109°40'W
Orda, Kaz. (ôr'dȧ)	137	48°50'N	47°30'E
Orda, Russia (ôr'dȧ)	142a	57°10'N	57°12'E
Ordes, Spain	128	43°00'N	8°24'W
Ordos Desert, des., China	160	39°12'N	108°10'E
Ordu, Tur. (ôr'dô)	119	41°00'N	37°50'E
Ordway, Co., U.S. (ôrd'wā)	78	38°11'N	103°46'W
Örebro, Swe. (û'rē-brō)	116	59°16'N	15°11'E
Oredezh, r., Russia (ô'rĕ-dĕzh)	142c	59°23'N	30°21'E
Oregon, Il., U.S.	71	42°01'N	89°21'W
Oregon, state, U.S.	64	43°40'N	121°50'W
Oregon Caves National Monument, rec., Or., U.S. (cāvz)	72	42°05'N	123°13'W
Oregon City, Or., U.S.	74c	45°21'N	122°36'W
Öregrund, Swe. (û-rē-grŏnd)	122	60°20'N	18°26'E
Orekhovo, Blg.	131	43°43'N	23°59'E
Orekhovo-Zuyevo, Russia (ôr-yě'kô-vô zŏŏ'yĕ-vô)	134	55°46'N	39°00'E
Orël, Russia (ôr-yôl')	134	52°59'N	36°05'E
Orël, prov., Russia	132	52°35'N	36°08'E
Orem, Ut., U.S. (ō'rĕm)	77	40°15'N	111°50'W
Ore Mountains *see* Erzgebirge, mts., Eur.	112	50°29'N	12°40'E
Orenburg, Russia (ō'rĕn-bŏŏrg)	134	51°50'N	55°05'E
Øresund, strt., Eur.	122	55°50'N	12°40'E
Órganos, Sierra de los, mts., Cuba (sē-ĕ'r-rȧ-dĕ-lôs-ŏ'r-gä-nôs)	92	22°20'N	84°10'W
Organ Pipe Cactus National Monument, rec., Az., U.S. (ôr'găn pīp kăk'tŭs)	77	32°14'N	113°05'W
Orgãos, Serra das, mtn., Braz. (sĕ'r-rä-däs-ôr-goun's)	99a	22°30'S	43°01'W
Orhei, Mol.	137	47°27'N	28°49'E
Orhon, r., Mong.	160	48°33'N	103°07'E
Oriental, Cordillera, mts., Col. (kôr-dēl-yĕ'rä)	100a	3°30'N	74°27'W
Oriental, Cordillera, mts., Dom. Rep. (kôr-dēl-yĕ'rä-ô-ryĕ'n-täl)	93	18°55'N	69°40'W
Oriental, Cordillera, mts., S.A. (kôr-dēl-yĕ'rä ō-rē-ĕn-täl')	100	14°00'S	68°33'W
Orikhiv, Ukr.	133	47°34'N	35°51'E
Oril', r., Ukr.	133	49°08'N	34°55'E
Orillia, Can. (ô-rĭl'ĭ-ȧ)	51	44°35'N	79°25'W
Orin, Wy., U.S.	73	42°40'N	105°10'W
Orinda, Ca., U.S.	74b	37°53'N	122°11'W
Orinoco, r., Ven. (ō-rĭ-nō'kō)	100	8°32'N	63°13'W
Oriola, Spain	129	38°04'N	0°55'W
Orion, Phil. (ō-rē-ōn')	169a	14°37'N	120°34'E
Orissa, state, India (ō-rĭs'ȧ)	155	25°09'N	83°50'E
Oristano, Italy (ô-rês-tä'nō)	118	39°53'N	8°38'E
Oristano, Golfo di, b., Italy (gôl-fô-dē-ô-rês-tä'nō)	130	39°53'N	8°12'E
Orituco, r., Ven. (ô-rē-tōō'kō)	101b	9°37'N	66°25'W
Oriuco, r., Ven. (ô-rēōō'kō)	101b	9°36'N	66°25'W
Orivesi, l., Fin.	123	62°15'N	29°55'E
Orizaba, Mex. (ô-rē-zä'bä)	87	18°52'N	97°05'E
Orizaba, Pico de, vol., Mex.	86	19°04'N	97°14'W
Orkanger, Nor.	122	63°19'N	9°54'W
Orkla, r., Nor. (ôr'klȧ)	122	62°55'N	9°50'E
Orkney, S. Afr. (ôrk'nĭ)	192c	26°58'S	26°39'E
Orkney Islands, is., Scot., U.K.	112	59°01'N	2°08'W
Orlando, S. Afr. (ôr-lăn-dô)	187b	26°15'S	27°56'E
Orlando, Fl., U.S. (ôr-lăn'dō)	65	28°32'N	81°22'W
Orland Park, Il., U.S. (ôr-lăn')	69a	41°38'N	87°52'W
Orleans, Can. (ôr-lā-än')	62c	45°28'N	75°31'W
Orléans, Fr. (ôr-lā-än')	110	47°55'N	1°56'E
Orleans, In., U.S. (ôr-lēnz')	66	38°40'N	86°25'W
Orléans, Île d', i., Can.	59	46°56'N	70°57'W
Orly, Fr.	127b	48°45'N	2°24'E
Ormond Beach, Fl., U.S. (ôr'mŏnd)	83	29°15'N	81°05'W
Ormskirk, Eng., U.K. (ôrms'kērk)	114a	53°34'N	2°53'W
Ormstown, Can. (ôrms'toun)	62a	45°07'N	74°00'W
Orneta, Pol. (ôr-nyĕ'tä)	125	54°07'N	20°10'E
Örnsköldsvik, Swe. (ûrn'skōlts-vēk)	116	63°10'N	18°32'E
Oro, Río del, r., Mex. (rē'ō dĕl ō'rō)	88	18°04'N	100°59'W
Oro, Río del, r., Mex.	80	26°04'N	105°40'W
Orobie, Alpi, mts., Italy (äl'pē-ô-rō'byĕ)	130	46°05'N	9°47'E
Oron, Nig.	189	4°48'N	8°14'E
Orosei, Golfo di, b., Italy (gôl-fô-dē-ō-rô-sā'ē)	130	40°12'N	9°45'E
Orosháza, Hung. (ô-rôsh-hä'sô)	125	46°33'N	20°31'E
Orosi, vol., C.R. (ō-rō'sē)	90	11°00'N	85°30'W
Oroville, Ca., U.S. (ôr'ō-vĭl)	76	39°29'N	121°34'W
Oroville, Wa., U.S.	72	48°55'N	119°25'W
Oroville, Lake, res., Ca., U.S.	76	39°32'N	121°25'W
Orreagal, Spain	128	43°00'N	1°17'W
Orrville, Oh., U.S. (ôr'vĭl)	66	40°45'N	81°50'W
Orsa, Swe. (ōr'sä)	122	61°08'N	14°35'E
Orsha, Bela. (ôr'shȧ)	136	54°29'N	30°28'E
Orsk, Russia (ôrsk)	134	51°15'N	58°50'E
Orşova, Rom. (ôr'shô-vä)	131	44°43'N	22°26'E
Ortega, Col. (ôr-tĕ'gä)	100a	3°56'N	75°12'W
Ortegal, Cabo, c., Spain (kȧ'bô-ôr-tȧ-gäl')	118	43°46'N	8°15'W
Orth, Aus.	115e	48°09'N	16°42'E
Orthez, Fr. (ôr-tĕz')	127	43°29'N	0°43'W
Órthrys, Óros, mtn., Grc.	131	39°00'N	22°15'E
Ortigueira, Spain (ôr-tē-gā'ē-rä)	118	43°40'N	7°50'W
Orting, Wa., U.S. (ôrt'ĭng)	74a	47°06'N	122°12'W
Ortona, Italy (ôr-tō'nä)	130	42°22'N	14°22'E
Ortonville, Mn., U.S. (ôr-tŭn-vĭl)	70	45°18'N	96°26'W
Orūmīyeh, Iran	154	37°30'N	45°15'E
Orūmīyeh, Daryacheh-ye, l., Iran	154	38°01'N	45°17'E
Oruro, Bol. (ô-rōō'rō)	100	17°57'S	66°59'W
Orvieto, Italy (ôr-vyå'tō)	130	42°43'N	12°08'E
Osa, Russia (ō'sȧ)	136	57°18'N	55°25'E
Osa, Península de, pen., C.R. (ō'sä)	91	8°30'N	83°25'W
Osage, Ia., U.S. (ō'sāj)	71	43°16'N	92°49'W
Osage, r., Mo., U.S.	79	38°10'N	93°12'W
Osage City, Ks., U.S. (ō'sāj sĭ'tĭ)	79	38°28'N	95°53'W
Ōsaka, Japan (ō'sä-kä)	161	34°40'N	135°27'E
Ōsaka, dept., Japan	167b	34°45'N	135°36'E
Ōsaka-Wan, b., Japan (wän)	166	34°34'N	135°16'E
Osakis, Mn., U.S. (ō-sā'kĭs)	70	45°51'N	95°09'W
Osakis, l., Mn., U.S.	71	45°55'N	94°55'W
Osawatomie, Ks., U.S. (ŏs-ȧ-wăt'ō-mē)	79	38°29'N	94°57'W
Osborne, Ks., U.S. (ŏz'bŭrn)	78	39°25'N	98°42'W
Osceola, Ar., U.S. (ŏs-ē-ō'lȧ)	79	35°42'N	89°58'W
Osceola, Ia., U.S.	71	41°04'N	93°45'W
Osceola, Mo., U.S.	79	38°02'N	93°41'W
Osceola, Ne., U.S.	70	41°11'N	97°34'W
Oscoda, Mi., U.S. (ŏs-kō'dȧ)	66	44°25'N	83°20'W
Osëtr, r., Russia (ō'sĕt'r)	132	54°27'N	38°15'E
Osgood, In., U.S. (ŏz'gŏd)	66	39°10'N	85°20'W
Osgoode, Can.	62c	45°09'N	75°37'W
Osh, Kyrg. (ôsh)	139	40°33'N	72°48'E
Oshawa, Can. (ŏsh'ȧ-wȧ)	51	43°50'N	78°50'W
Ōshima, i., Japan (ō'shē'mä)	167	34°47'N	139°35'E
Oshkosh, Ne., U.S. (ŏsh'kŏsh)	70	41°24'N	102°22'W
Oshkosh, Wi., U.S.	65	44°01'N	88°35'W
Oshogbo, Nig.	184	7°47'N	4°34'E
Osijek, Cro. (ôs'ĭ-yĕk)	119	45°33'N	18°48'E
Osinniki, Russia (ŭ-sē'nyĭ-kē)	140	53°37'N	87°21'E
Oskaloosa, Ia., U.S. (ŏs-kȧ-lōō'sȧ)	71	41°16'N	92°40'W
Oskarshamn, Swe. (ôs'kärs-häm'n)	122	57°16'N	16°24'E
Oskarström, Swe. (ôs'kärs-strûm)	122	56°48'N	12°55'E
Öskemen, Kaz.	139	49°58'N	82°38'E
Oskil, r., Eur.	137	51°00'N	37°41'E
Oslo, Nor. (ôs'lō)	110	59°56'N	10°41'E
Oslofjorden, b., Nor.	122	59°03'N	10°35'E
Osmaniye, Tur.	119	37°10'N	36°30'E
Osnabrück, Ger. (ôs-nä-brük')	124	52°16'N	8°05'E
Osorno, Chile (ō-sō'r-nō)	102	40°42'S	73°13'W
Osøyra, Nor.	122	60°24'N	5°22'E
Osprey Reef, rf., Austl. (ŏs'prå)	175	14°00'S	146°45'E
Ossa, Mount, mtn., Austl. (ŏsȧ)	175	41°45'S	146°05'E
Osseo, Mn., U.S. (ŏs'sē-ō)	75g	45°07'N	93°24'W
Ossining, N.Y., U.S. (ŏs'ĭ-nĭng)	68a	41°09'N	73°51'W
Ossipee, N.H., U.S. (ŏs'ĭ-pē)	60	43°42'N	71°08'W
Ossjøen, l., Nor. (ôs-syûěn)	122	61°20'N	12°00'E
Ostashkov, Russia (ôs-täsh'kôf)	136	57°07'N	33°04'E
Oster, Ukr. (ōs'tĕr)	133	50°55'N	30°52'E
Osterdalälven, r., Swe.	116	61°40'N	13°00'E
Osterfjord, b., Nor. (ûs'tĕr fyôr')	122	60°40'N	5°25'E
Östersund, Swe. (ûs'tĕr-sōōnd)	116	63°09'N	14°49'E
Östhammar, Swe. (ûst'häm'är)	122	60°16'N	18°21'E
Ostrava, Czech Rep.	110	49°51'N	18°18'E
Ostróda, Pol. (ôs'trŏt-ȧ)	125	53°41'N	19°58'E
Ostrogozhsk, Russia (ôs-tr-gôzhk')	137	50°53'N	39°03'E
Ostroh, Ukr.	137	50°21'N	26°40'E
Ostrołęka, Pol. (ôs-trô-wŏn'kȧ)	125	53°04'N	21°35'E
Ostrov, Russia (ôs-trôf')	136	57°21'N	28°22'E
Ostrowiec Świętokrzyski, Pol. (ôs-trō'vyĕts shvyĕN-tō-kzhĭ'ske)	117	50°55'N	21°24'E
Ostrów Lubelski, Pol. (ôs'trôf lōō'bĕl-skĭ)	125	51°32'N	22°49'E
Ostrów Mazowiecka, Pol. (mä-zô-vyĕt'skä)	117	52°47'N	21°54'E
Ostrów Wielkopolski, Pol. (ôs'trōōf vyĕl-kō-pōl'skē)	117	51°38'N	17°49'E
Ostrzeszów, Pol. (ôs-tzhä'shôf)	125	51°26'N	17°56'E
Ostuni, Italy (ôs-tōō'nē)	131	40°44'N	17°35'E
Osum, r., Alb. (ō'sŏm)	131	40°37'N	20°00'E
Osuna, Spain (ô-sōō'nä)	128	37°18'N	5°05'W
Osveya, Bela. (ôs'vĕ-yȧ)	132	56°00'N	28°08'E
Oswaldtwistle, Eng., U.K. (ŏz-wȧld-twĭs''l)	114a	53°44'N	2°23'W
Oswegatchie, r., N.Y., U.S. (ŏs-wē-găch'ĭ)	67	44°15'N	75°20'W
Oswego, Ks., U.S. (ŏs-wē'gō)	79	37°10'N	95°08'W
Oswego, N.Y., U.S.	65	43°25'N	76°30'W
Oświęcim, Pol. (ôsh-vyăN'tsĭm)	125	50°02'N	19°17'E
Otaru, Japan (ō'tȧ-rŏ)	161	43°07'N	141°00'E
Otavalo, Ec. (ōtä-vä'lō)	100	0°14'N	78°16'W
Otavi, Nmb. (ō-tä'vē)	186	19°35'S	17°20'E
Otay, Ca., U.S. (ō'tā)	76a	32°36'N	117°04'W
Otepää, Est.	123	58°03'N	26°30'E
Oti, r., Afr.	188	9°00'N	0°10'E
Otish, Monts, mts., Can. (ô-tĭsh')	53	52°15'N	70°20'W
Otjiwarongo, Nmb. (ŏt-jē-wȧ-rôn'gō)	186	20°20'S	16°25'E
Otočac, Cro. (ō'tō-chȧts)	130	44°53'N	15°15'E
Otra, r., Nor.	122	59°13'N	7°20'E
Otra, r., Russia (ôt'rȧ)	142b	55°22'N	38°20'E
Otradnoye, Russia (ô-trȧ'd-nôyĕ)	142c	59°46'N	30°50'E
Otranto, Italy (ō'trän-tō) (ô-trän'tō)	131	40°07'N	18°30'E
Otranto, Strait of, strt., Eur.	112	40°30'N	18°45'E
Otsego, Mi., U.S. (ŏt-sē'gō)	66	42°25'N	85°45'W
Otsu, Japan (ō'tsōō)	166	35°00'N	135°54'E
Otta, l., Nor. (ôt'tȧ)	122	61°53'N	8°40'E
Ottawa, Can. (ŏt'ȧ-wȧ)	51	45°25'N	75°43'W
Ottawa, Il., U.S.	66	41°20'N	88°50'W
Ottawa, Ks., U.S.	79	38°37'N	95°16'W
Ottawa, Oh., U.S.	66	41°00'N	84°00'W
Ottawa, r., Can.	53	46°05'N	77°20'W
Otter Creek, r., Ut., U.S. (ŏt'ēr)	77	38°20'N	111°55'W
Otter Creek, r., Vt., U.S.	67	44°05'N	73°15'W
Otter Point, c., Can.	74a	48°21'N	123°50'W
Otter Tail, l., Mn., U.S.	70	46°21'N	95°52'W
Otterville, Il., U.S. (ŏt'ēr-vĭl)	75e	39°03'N	90°24'W
Ottery, S. Afr. (ŏt'ēr-ĭ)	186a	34°02'S	18°31'E
Ottumwa, Ia., U.S. (ô-tŭm'wȧ)	65	41°00'N	92°26'W
Otukpa, Nig.	189	7°09'N	7°41'E
Otumba, Mex. (ô-tŭm'bä)	88	19°41'N	98°46'W
Otway, Cape, c., Austl. (ŏt'wā)	175	38°55'S	153°40'E
Otway, Seno, b., Chile (sĕ'nō-ŏ't-wä'y)	102	53°00'S	73°00'W
Otwock, Pol. (ôt'vôtsk)	125	52°05'N	21°18'E
Ouachita, r., U.S.	65	33°25'N	92°30'W
Ouachita Mountains, mts., U.S. (wŏsh'ĭ-tô)	65	34°29'N	95°01'W
Ouagadougou, Burkina (wä'gȧ-dōō'gōō)	184	12°22'N	1°31'W
Ouahigouya, Burkina (wä-ē-gōō'yä)	184	13°35'N	2°25'W
Oualâta, Maur. (wä-lä'tä)	184	17°11'N	6°50'W
Ouallene, Alg. (wäl-lân')	184	24°43'N	1°15'E
Ouanaminthe, Haiti	93	19°35'N	71°45'W
Ouarane, reg., Maur.	184	20°44'N	10°27'W
Ouarkoye, Burkina	188	12°05'N	3°40'W
Ouassel, r., Alg.	129	35°30'N	1°55'E
Oubangui (Ubangi), r., Afr. (ōō-bäN'gē)	190	4°30'N	20°35'E
Oude Rijn, r., Neth.	115a	52°09'N	4°33'E
Oudewater, Neth.	115a	52°01'N	4°52'E
Oud-Gastel, Neth.	115a	51°35'N	4°27'E
Oudtshoorn, S. Afr. (outs'hōrn)	186	33°33'S	23°36'E
Oued Rhiou, Alg.	129	35°55'N	0°57'E
Oued Tlelat, Alg.	129	35°33'N	0°28'W
Oued-Zem, Mor. (wĕd-zĕm')	184	33°05'N	5°49'W
Ouessant, Island d', i., Fr. (ēl-dwĕ-säN')	117	48°28'N	5°00'W
Ouesso, Congo	185	1°37'N	16°04'E
Ouest, Point, c., Haiti	93	19°00'N	73°25'W
Ouezzane, Mor. (wĕ-zan')	184	34°48'N	5°40'W
Ouham, r., Afr.	189	8°30'N	17°50'E
Ouidah, Benin (wē-dä')	184	6°25'N	2°05'E
Oujda, Mor.	184	34°41'N	1°45'W
Oulins, Fr. (ōō-lăN')	127b	48°52'N	1°27'E
Oullins, Fr. (ōō-lăN')	126	45°44'N	4°46'E
Oulu, Fin. (ō'lŏ)	110	64°58'N	25°43'E
Oulujärvi, l., Fin.	116	64°20'N	25°48'E
Oum Chalouba, Chad (ōōm shä-lōō'bä)	185	15°48'N	20°30'E
Oum Hadjer, Chad	189	13°18'N	19°41'E
Ounas, r., Fin. (ō'nȧs)	116	67°46'N	24°40'E
Oundle, Eng., U.K. (ōn'd'l)	114a	52°28'N	0°28'W
Ounianga Kébir, Chad (ōō-nē-äŋ'gȧ kē-bēr')	185	19°04'N	20°22'E
Ouray, Co., U.S. (ōō-rā')	78	38°00'N	107°40'W
Ourense, Spain	128	42°20'N	7°52'W
Ourinhos, Braz. (ôô-rē'nyôs)	101	23°04'S	49°45'W
Ourique, Port. (ō-rē'kĕ)	128	37°39'N	8°10'W
Ouro Fino, Braz. (ōŭ-rô-fē'nō)	99a	22°18'S	46°21'W

286 Ouro Prêto – Paracale PRONOUNCING INDEX

PLACE (Pronunciation)	PAGE	LAT.	LONG.
Ouro Prêto, Braz. (ō′rò prä′tò)	102	20°24′S	43°30′W
Outardes, Rivière aux, r., Can.	53	50°53′N	68°50′W
Outer, i., Wi., U.S. (out′ẽr)	71	47°03′N	90°20′W
Outer Brass, i., V.I.U.S. (bräs)	87c	18°24′N	64°58′W
Outer Hebrides, is., Scot., U.K.	120	57°20′N	7°50′W
Outjo, Nmb. (ōt′yō)	186	20°05′S	17°10′E
Outlook, Can.	56	51°31′N	107°05′W
Outremont, Can. (o͞o-trē-môn′)	62a	45°31′N	73°36′W
Ouvéa, i., N. Cal.	175	20°43′S	166°48′E
Ouyen, Austl. (o͞o-ĕn)	176	35°05′S	142°10′E
Ovalle, Chile (ō-väl′yā)	102	30°43′S	71°16′W
Ovando, Bahía de, b., Cuba (bä-ē′ä-dĕ-ô-vä′n-dō)	93	20°10′N	74°05′W
Ovar, Port. (ô-vär′)	128	40°52′N	8°38′W
Overijse, Bel.	115a	50°46′N	4°32′E
Overland, Mo., U.S. (ō-vẽr-lănd)	75e	38°42′N	90°22′W
Overland Park, Ks., U.S.	75f	38°59′N	94°40′W
Overlea, Md., U.S. (ō′vẽr-lā)(ō′vẽr-lē)	68e	39°21′N	76°31′W
Övertornea, Swe.	116	66°19′N	23°31′E
Ovidiopol′, Ukr.	133	46°15′N	30°28′E
Oviedo, Dom. Rep. (ô-vyĕ′dō)	93	17°50′N	71°25′W
Oviedo, Spain (ō-vê-ā′dhō)	110	43°22′N	5°50′W
Ovruch, Ukr.	133	51°19′N	28°51′E
Owada, Japan (ō′wä-dà)	167a	35°49′N	139°33′E
Owambo, hist. reg., Nmb.	186	18°10′S	15°00′E
Owando, Congo	186	0°29′S	15°55′E
Owasco, l., N.Y., U.S. (ō-wăsk′kō)	67	42°50′N	76°30′W
Owase, Japan (ō′wä-shě)	167	34°03′N	136°12′E
Owego, N.Y., U.S. (ō-wē′gō)	67	42°05′N	76°15′W
Owen, Wi., U.S. (ō′ĕn)	71	44°56′N	90°35′W
Owensboro, Ky., U.S. (ō′ĕnz-bŭr-ô)	65	37°45′N	87°05′W
Owens Lake, l., Ca., U.S.	76	37°13′N	118°20′W
Owen Sound, Can. (ō′ĕn)	51	44°30′N	80°55′W
Owen Stanley Range, mts., Pap. N. Gui. (stăn′lē)	169	9°00′S	147°30′E
Owensville, In., U.S. (ō′ĕnz-vĭl)	66	38°15′N	87°40′W
Owensville, Mo., U.S.	79	38°20′N	91°29′W
Owensville, Oh., U.S.	69f	39°08′N	84°07′W
Owenton, Ky., U.S. (ō′ĕn-tŭn)	66	38°35′N	84°55′W
Owerri, Nig. (ô-wĕr′ē)	184	5°26′N	7°04′E
Owings Mill, Md., U.S. (ōwĭngz mĭl)	68e	39°25′N	76°50′W
Owl Creek, r., Wy., U.S. (oul)	73	43°45′N	108°46′W
Owo, Nig.	189	7°15′N	5°37′E
Owosso, Mi., U.S. (ô-wŏs′ō)	66	43°00′N	84°15′W
Owyhee, r., U.S.	64	43°04′N	117°45′W
Owyhee, Lake, res., Or., U.S.	64	43°27′N	117°30′W
Owyhee, South Fork, r., Id., U.S.	72	42°07′N	116°43′W
Owyhee Mountains, mts., Id., U.S. (ô-wī′hê)	64	43°15′N	116°48′W
Oxbow, Can.	57	49°12′N	102°11′W
Oxchuc, Mex. (ôs-cho͝ok′)	89	16°47′N	92°24′W
Oxford, Can. (ŏks′fẽrd)	60	45°44′N	63°52′W
Oxford, Eng., U.K.	117	51°43′N	1°16′W
Oxford, Al., U.S. (ŏks′fẽrd)	83	33°38′N	80°46′W
Oxford, Ma., U.S.	61a	42°07′N	71°52′W
Oxford, Mi., U.S.	66	42°50′N	83°15′W
Oxford, Ms., U.S.	82	34°22′N	89°30′W
Oxford, N.C., U.S.	83	36°17′N	78°35′W
Oxford, Oh., U.S.	66	39°30′N	84°45′W
Oxford Lake, l., Can.	57	54°51′N	95°37′W
Oxfordshire, co., Eng., U.K.	114b	51°36′N	1°30′W
Oxkutzcab, Mex. (ôx-ko͞o′tz-käb)	90a	20°18′N	89°22′W
Oxmoor, Al., U.S. (ŏks′mŏr)	68h	33°25′N	86°52′W
Oxnard, Ca., U.S. (ŏks′närd)	76	34°08′N	119°12′W
Oxon Hill, Md., U.S. (ŏks′ŏn hĭl)	68e	38°48′N	77°00′W
Oyapock, r., S.A. (ō-yà-pŏk′)	101	2°45′N	52°15′W
Oyem, Gabon	184	1°37′N	11°35′E
Øyeren, l., Nor. (ûĕrĕn)	122	59°50′N	11°25′E
Oymyakon, Russia (oi-myŭ-kôn′)	135	63°14′N	142°58′E
Oyo, Nig. (ō′yō)	184	7°51′N	3°56′E
Oyonnax, Fr. (ô-yô-näks′)	127	46°16′N	5°40′E
Oyster Bay, N.Y., U.S.	68a	40°52′N	73°32′W
Oyster Bayou, Tx., U.S.	81a	29°41′N	94°33′W
Oyster Creek, r., Tx., U.S. (ois′tẽr)	81a	29°13′N	95°29′W
Oyyl, r., Kaz.	137	49°30′N	55°10′E
Ozama, r., Dom. Rep. (ô-zä′mä)	93	18°45′N	69°55′W
Ozamiz, Phil. (ô-zä′mêz)	169	8°06′N	123°43′E
Ozark, Al., U.S. (ō′zärk)	82	31°28′N	85°28′W
Ozark, Ar., U.S.	79	35°29′N	93°49′W
Ozark Plateau, plat., U.S.	65	36°37′N	93°56′W
Ozarks, Lake of the, l., Mo., U.S. (ō′zärksz)	65	38°06′N	93°26′W
Ozëry, Russia (ô-zyô′rê)	132	54°53′N	38°31′E
Ozieri, Italy	118	40°38′N	8°53′E
Ozorków, Pol. (ô-zôr′kôf)	125	51°58′N	19°20′E
Ozuluama, Mex.	89	21°34′N	97°52′W
Ozumba, Mex.	89a	19°02′N	98°48′W
Ozurgeti, Geor.	138	41°56′N	42°00′E
P			
Paarl, S. Afr. (pärl)	186	33°45′S	18°55′E
Pa′auilo, Hi., U.S. (pä-ä-o͞o′ê-lō)	84a	20°03′N	155°25′W
Pabianice, Pol. (pä-byà-nē′tsĕ)	125	51°40′N	19°29′E
Pacaás Novos, Massiço de, mts., Braz.	100	11°03′S	64°02′W
Pacaraima, Serra, mts., S.A. (sĕr′rá pä-kä-rä-ē′mà)	100	3°45′N	62°30′W
Pacasmayo, Peru (pä-käs-mä′yō)	100	7°24′S	79°30′W
Pachuca, Mex. (pä-cho͞o′kä)	86	20°07′N	98°43′W
Pacific, Wa., U.S. (pá-sĭf′ĭk)	74a	47°16′N	122°15′W
Pacifica, Ca., U.S. (pá-sĭf′ĭ-kä)	74b	37°38′N	122°29′W
Pacific Beach, Ca., U.S.	76a	32°47′N	117°22′W
Pacific Grove, Ca., U.S.	76	36°37′N	121°54′W
Pacific Islands, Trust Territory of the *see* Palau, nation, Oc.	3	7°15′N	134°30′E
Pacific Ocean, o.	2	0°00′	170°00′W
Pacific Ranges, mts., Can.	54	51°00′N	125°30′W
Pacific Rim National Park, rec., Can.	54	49°00′N	126°00′W
Pacolet, r., S.C., U.S. (pă′cō-lĕt)	83	34°55′N	81°49′W
Pacy-sur-Eure, Fr. (pä-sē-sür-ûr′)	127b	49°01′N	1°24′E
Padang, Indon. (pä-däng′)	168	1°01′S	100°28′E
Padang, i., Indon.	153b	1°12′N	102°21′E
Padang Endau, Malay.	153b	2°39′N	103°38′E
Paden City, W.V., U.S. (pā′dĕn)	66	39°30′N	80°55′W
Paderborn, Ger. (pä-dĕr-bôrn′)	124	51°43′N	8°46′E
Padibe, Ug.	191	3°28′N	32°50′E
Padiham, Eng., U.K. (păd′ĭ-hăm)	114a	53°48′N	2°19′W
Padilla, Mex. (pä-dēl′yä)	88	24°00′N	98°45′W
Padilla Bay, b., Wa., U.S. (pä-dĕl′lá)	74a	48°31′N	122°34′W
Padova, Italy (pä′dô-vä)(păd′û-á)	118	45°24′N	11°53′E
Padre Island, i., Tx., U.S. (pä′drā)	81	27°09′N	97°15′W
Padua *see* Padova, Italy	118	45°24′N	11°53′E
Paducah, Ky., U.S.	65	37°05′N	88°36′W
Paducah, Tx., U.S.	78	34°01′N	100°18′W
Paektu-san, mtn., Asia (pâk′to͞o-sän′)	166	42°00′N	128°03′E
Pag, i., Serb. (päg)	130	44°30′N	14°48′E
Pagai Selatan, Pulau, i., Indon.	168	2°48′S	100°22′E
Pagai Utara, Pulau, i., Indon.	168	2°45′S	100°02′E
Pagasitikós Kólpos, b., Grc.	131	39°15′N	23°00′E
Page, Az., U.S.	77	36°57′N	111°27′W
Pago Pago, Am. Sam.	170a	14°16′S	170°42′W
Pagosa Springs, Co., U.S. (pá-gō′sá)	78	37°15′N	107°05′W
Pāhala, Hi., U.S. (pä-hä′lä)	84a	19°11′N	155°28′W
Pahang, state, Malay.	153b	3°02′N	102°57′E
Pahang, r., Malay.	168	3°39′N	102°41′E
Pahokee, Fl., U.S. (pá-hō′kē)	83a	26°45′N	80°40′W
Paide, Est. (pī′dĕ)	123	58°54′N	25°30′E
Päijänne, l., Fin. (pĕ′ē-yĕn-nĕ)	116	61°38′N	25°05′E
Pailolo Channel, strt., Hi., U.S. (pä-ê-lō′lō)	84a	21°05′N	156°41′W
Paine, Chile (pī′nĕ)	99b	33°49′S	70°44′W
Painesville, Oh., U.S. (pānz′vĭl)	66	41°40′N	81°15′W
Painted Desert, des., Az., U.S. (pānt′ĕd)	78	36°15′N	111°35′W
Painted Rock Reservoir, res., Az., U.S.	77	33°00′N	113°05′W
Paintsville, Ky., U.S. (pānts′vĭl)	66	37°50′N	82°50′W
Paisley, Scot., U.K. (pāz′lĭ)	116	55°50′N	4°30′W
Paita, Peru (pä-ē′tä)	100	5°11′S	81°12′W
Pai T′ou Shan, mts., Kor., N.	161	40°30′N	127°20′E
Paiute Indian Reservation, I.R., Ut., U.S.	77	38°17′N	113°50′W
Pajápan, Mex. (pä-hä′pän)	89	18°16′N	94°41′W
Pakanbaru, Indon.	168	0°43′N	101°15′E
Pakhra, r., Russia (päk′rá)	142b	55°29′N	37°51′E
Pakistan, nation, Asia	155	28°00′N	67°30′E
Pakokku, Mya. (pá-kŏk′kò)	160	21°29′N	95°00′E
Paks, Hung. (pôksh)	125	46°38′N	18°53′E
Pala, Chad	189	9°22′N	14°54′E
Palacios, Tx., U.S. (pá-lä′syōs)	81	28°42′N	96°12′W
Palagruža, Otoci, is., Cro.	130	42°20′N	16°23′E
Palaiseau, Fr. (pá-lĕ-zō′)	127b	48°44′N	2°16′E
Palana, Russia	135	59°07′N	159°58′E
Palanan Bay, b., Phil. (pä-lä′nän)	169a	17°14′N	122°35′E
Palanan Point, c., Phil.	169a	17°12′N	122°40′E
Pālanpur, India (pä′lŭn-po͞or)	155	24°08′N	73°29′E
Palapye, Bots. (pá-läp′yĕ)	186	22°34′S	27°28′E
Palatine, Il., U.S. (păl′á-tīn)	69a	42°07′N	88°03′W
Palatka, Fl., U.S. (pá-lăt′ká)	83	29°39′N	81°40′W
Palau (Belau), nation, Oc. (pä-lä′ò)	3	7°15′N	134°30′E
Palauig, Phil. (pä-lou′ĕg)	169a	15°27′N	119°54′E
Palawan, i., Phil. (pä-lä′wán)	168	9°50′N	117°38′E
Pālayankottai, India	159	8°50′N	77°50′E
Paldiski, Est. (päl′dĭ-skĭ)	123	59°22′N	24°04′E
Palembang, Indon. (pä-lĕm-bäng′)	168	2°57′S	104°40′E
Palencia, Guat. (pä-lĕn′sē-à)	90	14°40′N	90°22′W
Palencia, Spain (pä-lĕ′n-syä)	118	42°02′N	4°32′W
Palenque, Mex. (pä-lĕŋ′kā)	89	17°34′N	91°58′W
Palenque, Punta, c., Dom. Rep. (po͞o′n-tä)	93	18°10′N	70°10′W
Palermo, Col. (pä-lĕr′mō)	100a	2°53′N	75°26′W
Palermo, Italy	110	38°08′N	13°24′E
Palestine, Tx., U.S.	65	31°46′N	95°38′W
Palestine, hist. reg., Asia (păl′ĕs-tīn)	153a	31°33′N	35°00′E
Paletwa, Mya. (pŭ-lĕt′wä)	155	21°19′N	92°52′E
Palghāt, India	159	10°49′N	76°40′E
Pāli, India	158	25°53′N	73°18′E
Palín, Guat. (pä-lēn′)	90	14°42′N	90°42′W
Palizada, Mex. (pä-lē-zä′dä)	89	18°17′N	92°04′W
Palk Strait, strt., Asia (pôk)	155	10°00′N	79°23′E
Palma, Braz. (päl′mä)	99a	21°23′S	42°18′W
Palma, Spain	110	39°35′N	2°38′E
Palma, Bahía de, b., Spain	129	39°24′N	2°37′E
Palma del Río, Spain	128	37°43′N	5°19′W
Palmares, Braz. (päl-má′rĕs)	101	8°46′S	35°28′W
Palmas, Braz. (päl′mäs)	102	26°20′S	51°56′W
Palmas, Braz.	101	10°08′S	48°18′W
Palmas, Cape, c., Lib.	184	4°22′N	7°44′W
Palma Soriano, Cuba (sô-rē-ä′nō)	92	20°15′N	76°00′W
Palm Beach, Fl., U.S. (päm bēch′)	83a	26°43′N	80°03′W
Palmeira dos Índios, Braz. (pä-mā′rä-dôs-ē′n-dyôs)	101	9°26′S	36°33′W
Palmeirinhas, Ponta das, c., Ang.	190	9°05′S	13°00′E
Palmela, Port. (päl-mā′lä)	128	38°34′N	8°54′W
Palmer, Ak., U.S. (päm′ẽr)	63	61°38′N	149°15′W
Palmer, Wa., U.S.	74a	47°19′N	121°53′W
Palmerston North, N.Z. (päm′ẽr-stŭn)	175a	40°20′S	175°35′E
Palmerville, Austl. (päm′ẽr-vĭl)	175	16°08′S	144°15′E
Palmetto, Fl., U.S. (pál-mĕt′ō)	83a	27°32′N	82°34′W
Palmetto Point, c., Bah.	93	21°15′N	73°25′W
Palmi, Italy (päl′mē)	130	38°21′N	15°54′E
Palmira, Col. (päl-mē′rä)	100	3°33′N	76°17′W
Palmira, Cuba	92	22°15′N	80°25′W
Palmyra, Mo., U.S. (păl-mī′rá)	79	39°45′N	91°32′W
Palmyra, N.J., U.S.	68f	40°01′N	75°00′W
Palmyra, i., Oc.	2	6°00′N	162°20′W
Palmyra, hist., Syria	154	34°25′N	38°28′E
Palmyras Point, c., India	158	20°42′N	87°45′E
Palo Alto, Ca., U.S. (pä′lō äl′tō)	74b	37°27′N	122°09′W
Paloduro Creek, r., Tx., U.S. (pā-lô-do͞o′rô)	78	36°16′N	101°12′W
Paloh, Malay.	153b	2°11′N	103°12′E
Paloma, l., Mex. (pä-lō′mä)	80	26°53′N	104°02′W
Palomo, Cerro el, mtn., Chile (sĕ′r-rô-ĕl-pä-lô′mô)	99b	34°36′S	70°20′W
Palos, Cabo de, c., Spain (ká′bô-dĕ-pä′lôs)	118	39°38′N	0°43′W
Palos Verdes Estates, Ca., U.S. (pă′lŭs vûr′dĭs)	75a	33°48′N	118°24′W
Palouse, Wa., U.S. (pá-lo͞oz′)	72	46°54′N	117°04′W
Palouse, r., Wa., U.S.	72	47°02′N	117°35′W
Palu, Tur. (pä-loo′)	137	38°55′N	40°10′E
Paluan, Phil. (pä-lo͞o′än)	169a	13°25′N	120°29′E
Pamiers, Fr. (pá-myā′)	117	43°07′N	1°34′E
Pamirs, mts., Asia	155	38°14′N	72°27′E
Pamlico, r., N.C., U.S. (păm′lĭ-kō)	83	35°25′N	76°59′W
Pamlico Sound, strt., N.C., U.S.	65	35°10′N	76°10′W
Pampa, Tx., U.S. (păm′pá)	64	35°32′N	100°56′W
Pampa de Castillo, pl., Arg. (pä′m-pä-dĕ-käs-tē′l-yô)	102	45°30′S	67°30′W
Pampana, r., S.L.	188	8°35′N	11°55′W
Pampanga, r., Phil. (päm-päŋ′gä)	169a	15°20′N	120°48′E
Pampas, reg., Arg. (päm′päs)	102	37°00′S	64°30′W
Pampilhosa do Botão, Port. (päm-pê-lyō′sá-dô-bō-toun)	128	40°21′N	8°32′W
Pamplona, Col. (päm-plō′nä)	100	7°19′N	72°41′W
Pamplona, Spain (päm-plô′nä)	118	42°49′N	1°39′W
Pamunkey, r., Va., U.S. (pá-mŭŋ′kĭ)	67	37°40′N	77°20′W
Pana, Il., U.S. (pā′ná)	66	39°25′N	89°05′W
Panagyurishte, Blg. (pá-ná-gyo͞o′rēsh-tĕ)	131	42°30′N	24°11′E
Panaji (Panjim), India	155	15°33′N	73°52′E
Panamá, Pan.	87	8°58′N	79°32′W
Panama, nation, N.A.	87	9°00′N	80°00′W
Panamá, Istmo de, isth., Pan.	87	9°00′N	80°00′W
Panama Canal, can., Pan.	86a	9°20′N	79°55′W
Panama City, Fl., U.S. (păn-á mä′ sĭ′tĭ)	82	30°08′N	85°39′W
Panamint Range, mts., Ca., U.S. (păn-à-mĭnt′)	76	36°40′N	117°30′W
Panarea, i., Italy (pä-nä′rē-a)	130	38°37′N	15°05′E
Panaro, r., Italy (pä-nä′rô)	130	44°47′N	11°06′E
Panay, i., Phil. (pä-nī′)	168	11°15′N	121°38′E
Pančevo, Serb. (pán′chĕ-vô)	119	44°52′N	20°42′E
Panchor, Malay.	153b	2°11′N	102°43′E
Pānchur, India	158a	22°31′N	88°17′E
Panda, D.R.C. (pän′dä′)	186	10°59′S	27°24′E
Pan de Guajaibon, mtn., Cuba (pän dā gwä-jä-bōn′)	92	22°50′N	83°20′W
Panevėžys, Lith. (pä′nyĕ-vāzh′ēs)	136	55°44′N	24°21′E
Panga, D.R.C. (päŋ′gä)	185	1°51′N	26°25′E
Pangani, Tan. (pän-gä′nē)	187	5°28′S	38°58′E
Pangani, r., Tan.	191	4°40′S	37°45′E
Pangkalpinang, Indon. (päng-käl′pē-näng′)	168	2°11′S	106°04′E
Pangnirtung, Can.	51	66°08′N	65°26′W
Panguitch, Ut., U.S. (păn′gwĭch)	77	37°50′N	112°30′W
Panié, Mont, mtn., N. Cal.	170f	20°36′S	164°46′E
Pānihāti, India	158a	22°42′N	88°23′E
Panimávida, Chile (pä-nē-má′vē-dä)	99b	35°44′S	71°26′W
Panshi, China (pän-shē)	164	42°50′N	126°48′E
Pantar, Pulau, i., Indon. (pän′tär)	169	8°40′S	123°45′E
Pantelleria, i., Italy (pän-tĕl-lå-rē′ä)	118	36°43′N	11°59′E
Pantepec, Mex. (pän-tå-pĕk′)	89	17°11′N	93°04′W
Panuco, Mex. (pä′no͞o-kô)	88	22°04′N	98°11′W
Pánuco, Mex. (pä′no͞o-kô)	88	23°25′N	105°55′W
Panuco, r., Mex.	86	21°59′N	98°20′W
Pánuco de Coronado, Mex. (pä′no͞o-kô dā kō-rô-nä′dhō)	88	24°33′N	104°20′W
Panvel, India	159b	18°59′N	73°06′E
Panyu, China (pä-yo͞o)	163a	22°56′N	113°22′E
Panzós, Guat. (pä-zós′)	90	15°26′N	89°40′W
Pao, r., Ven. (pá′ō)	101b	9°52′N	67°57′W
Paola, Ks., U.S. (pā-ō′lá)	79	38°34′N	94°51′W
Paoli, In., U.S. (pā-ō′lī)	66	38°35′N	86°30′W
Paoli, Pa., U.S.	68f	40°03′N	75°29′W
Paonia, Co., U.S. (pā-ō′nyá)	77	38°50′N	107°40′W
Pápa, Hung. (pä′pô)	119	47°18′N	17°27′E
Papagayo, r., Mex. (pä-pä-gä′yō)	88	16°52′N	99°41′W
Papagayo, Golfo del, b., C.R. (gôl-fô-dĕl-pä-pä-gá′yō)	90	10°44′N	85°56′W
Papagayo, Laguna, l., Mex. (lä-ò-nä)	88	16°44′N	99°44′W
Papantla de Olarte, Mex. (pä-pän′tlä dā-ô-lä′r-tĕ)	86	20°30′N	97°15′W
Papatoapan, r., Mex. (pä-pä-tô-ä-pá′n)	89	18°00′N	96°22′W
Papenburg, Ger. (päp′ĕn-bôrgh)	124	53°05′N	7°23′E
Papinas, Arg. (pä-pē′näs)	99c	35°30′S	57°19′W
Papineauville, Can. (pä-pē-nō′vēl)	62c	45°38′N	75°01′W
Papua, Gulf of, b., Pap. N. Gui. (păp-o͞o-á)	169	8°20′S	144°45′E
Papua New Guinea, nation, Oc. (păp-o͞o-á)(gĭne)	169	7°00′S	142°15′E
Papudo, Chile (pä-po͞o′dô)	99b	32°30′S	71°25′W
Paquequer Pequeno, Braz. (pä-kĕ-kĕ′r-pĕ-kĕ′nô)	102b	22°19′S	43°02′W
Para, r., Russia	132	53°45′N	40°58′E
Paracale, Phil. (pä-rä-kä′lå)	169a	14°17′N	122°47′E

ăt; finăl; rāte; senåte; ärm; ȧsk; sofá; fâre; ch-choose; dh-as th in other; bē; ēvent; bĕt; recĕnt; cratẽr; g-gō; gh-guttural g; bĭt; ĭ-short neutral; rīde; ĸ-guttural k as ch in German ich;

PRONOUNCING INDEX

Paracambi – Pelee Island 287

PLACE (Pronunciation)	PAGE	LAT.	LONG.
Paracambi, Braz.	102b	22°36'S	43°43'W
Paracatu, Braz. (pä-rä-kä-to͞o')	101	17°17'S	46°43'W
Paracel Islands, is., Asia	168	16°40'N	113°00'E
Paraćin, Serb. (pá'rä-chėn)	119	43°51'N	21°26'E
Para de Minas, Braz. (pä-rä-dĕ-mē'näs)	101	19°52'S	44°37'W
Paradise, i., Bah.	92	25°05'N	77°20'W
Paradise Valley, Nv., U.S. (păr'á-dīs)	72	41°28'N	117°32'W
Parados, Cerro de los, mtn., Col. (sĕ'r-rô-dĕ-lôs-pä-rä'dôs)	100a	5°44'N	75°13'W
Paragould, Ar., U.S. (păr'á-go͞old)	79	36°03'N	90°29'W
Paraguaçu, r., Braz. (pä-rä-gwä-zo͞o')	101	12°25'S	39°46'W
Paraguay, nation, S.A. (păr'á-gwā)	102	24°00'S	57°00'W
Paraguay, r., S.A. (pä-rä-gwä'y)	102	21°12'S	57°31'W
Paraíba, state, Braz. (pä-rä-ē'bä)	101	7°11'S	37°05'W
Paraíba, r., Braz.	99a	23°02'S	45°43'W
Paraíba do Sul, Braz. (dô-so͞o'l)	99a	22°10'S	43°18'W
Paraibuna, Braz. (pä-räē-bo͞o'nä)	99a	23°23'S	45°38'W
Paraíso, C.R.	91	9°50'N	83°53'W
Paraíso, Mex.	89	18°24'N	93°11'W
Paraíso, Pan. (pä-rä-ē'sō)	86a	9°02'N	79°38'W
Paraisópolis, Braz. (pä-räē-sô'pō-lês)	99a	22°35'S	45°45'W
Paraitinga, r., Braz. (pä-rä-ē-tē'n-gä)	99a	23°15'S	45°24'W
Parakou, Benin (pá-rá-ko͞o')	184	9°21'N	2°37'E
Paramaribo, Sur. (pá-rá-má'rê-bō)	101	5°50'N	55°15'W
Paramatta, Austl. (păr-á-măt'á)	173b	33°49'S	150°59'E
Paramillo, mtn., Col. (pä-rä-mē'l-yō)	100a	7°06'N	75°55'W
Paramus, N.J., U.S.	68a	40°56'N	74°04'W
Paran, r., Asia	153a	30°05'N	34°50'E
Paraná, Arg.	102	31°44'S	60°32'W
Paraná, r., S.A.	102	24°00'S	54°00'W
Paranaíba, Braz. (pä-rä-ná-ē'bá)	101	19°43'S	51°13'W
Paranaíba, r., Braz.	101	18°58'S	50°44'W
Paraná Ibicuy, r., Arg.	99c	33°27'S	59°26'W
Paranam, Sur.	101	5°39'N	55°13'W
Paranápanema, r., Braz. (pä-rä'ná'pä-nĕ-mä)	101	22°28'S	52°15'W
Paraopeda, r., Braz. (pä-rä-o-pĕ'dä)	99a	20°09'S	44°14'W
Parapara, Ven. (pä-rä-pä-rä)	101b	9°44'N	67°17'W
Parati, Braz. (pä-rätē)	99a	23°14'S	44°43'W
Paray-le-Monial, Fr. (pá-rĕ'lē-mô-nyäl')	126	46°27'N	4°14'E
Pārbati, r., India	158	24°50'N	76°44'E
Parchim, Ger. (par'kĭm)	124	53°25'N	11°52'E
Parczew, Pol. (pär'chĕf)	125	51°38'N	22°53'E
Pardo, r., Braz. (pär'dō)	101	15°25'S	39°40'W
Pardo, r., Braz.	99a	21°32'S	46°40'W
Pardubice, Czech Rep. (pär'dô-bĭt-sĕ)	124	50°02'N	15°47'E
Parecis, Serra dos, mts., Braz. (sĕr'rá dōs pä-rå-sēzh')	101	13°45'S	59°28'W
Paredes de Nava, Spain (pä-rā'dās dā nä'vä)	128	42°10'N	4°41'W
Paredón, Mex.	80	25°56'N	100°58'W
Parent, Can.	51	47°59'N	74°30'W
Parent, Lac, l., Can.	59	48°40'N	77°00'W
Parepare, Indon.	168	4°01'S	119°38'E
Pargolovo, Russia (pár-gô'lô vô)	142c	60°04'N	30°18'E
Paria, r., Az., U.S.	77	37°07'N	111°51'W
Paria, Golfo de, b., (gôl-fô-dĕ-br-pä-rē-ä)	100	10°33'N	62°14'W
Paricutín, Volcán, vol., Mex.	88	19°27'N	102°14'W
Parida, Río de la, r., Mex. (rē'ō-dĕ-lä-pä-rē'dä)	80	26°23'N	104°40'W
Parima, Serra, mts., S.A. (sĕr'rá pä-rē'má)	100	3°45'N	64°00'W
Pariñas, Punta, c., Peru (po͞o'n-tä-pä-rē'n-yäs)	100	4°30'S	81°23'W
Parintins, Braz. (pä-rĭn-tĭnzh')	101	2°34'S	56°30'W
Paris, Can.	59	43°15'N	80°23'W
Paris, Fr. (pá-rē')	110	48°51'N	2°20'E
Paris, Ar., U.S. (păr'ĭs)	79	35°17'N	93°43'W
Paris, Il., U.S.	66	39°35'N	87°40'W
Paris, Ky., U.S.	66	38°15'N	84°15'W
Paris, Mo., U.S.	79	39°27'N	91°59'W
Paris, Tn., U.S.	82	36°16'N	88°20'W
Paris, Tx., U.S.	65	33°39'N	95°33'W
Parita, Golfo de, b., Pan. (gôl-fô-dĕ-pä-rē'tä)	91	8°06'N	80°10'W
Park City, Ut., U.S.	73	40°39'N	111°33'W
Parker, S.D., U.S. (pär'kẽr)	70	43°24'N	97°10'W
Parker Dam, dam, U.S.	64	34°20'N	114°00'W
Parkersburg, W.V., U.S. (pär'kẽrz-bûrg)	65	39°15'N	81°35'W
Parkes, Austl. (pärks)	176	33°10'S	148°10'E
Park Falls, Wi., U.S. (pärk)	71	45°55'N	90°29'W
Park Forest, Il., U.S.	69a	41°29'N	87°41'W
Parkland, Wa., U.S. (pärk'lănd)	74a	47°09'N	122°26'W
Park Range, mts., Co., U.S.	73	40°54'N	106°40'W
Park Rapids, Mn., U.S.	70	46°53'N	95°05'W
Park Ridge, Il., U.S.	69a	42°00'N	87°50'W
Park River, N.D., U.S.	70	48°22'N	97°43'W
Parkrose, Or., U.S. (pärk'rōz)	74c	45°33'N	122°33'W
Park Rynie, S. Afr.	187c	30°22'S	30°43'E
Parkston, S.D., U.S. (pärks'tŭn)	70	43°22'N	97°59'W
Parkville, Md., U.S.	68e	39°22'N	76°32'W
Parkville, Mo., U.S.	75f	39°12'N	94°41'W
Parla, Spain (pär'lä)	129a	40°14'N	3°46'W
Parma, Italy (pär'mä)	118	44°48'N	10°20'E
Parma, Oh., U.S.	69d	41°23'N	81°44'W
Parma Heights, Oh., U.S.	69d	41°23'N	81°36'W
Parnaíba, Braz. (pär-nä-ē'bä)	101	3°00'S	41°42'W
Parnaiba, r., Braz.	101	3°57'S	42°30'W
Parnassós, mtn., Grc.	131	38°36'N	22°35'E
Parndorf, Aus.	115e	48°00'N	16°52'E
Pärnu, Est. (pĕr'no͞o)	136	58°24'N	24°29'E
Pärnu, r., Est.	123	58°40'N	25°05'E
Pärnu Laht, b., Est. (läkt)	123	58°15'N	24°17'E

PLACE (Pronunciation)	PAGE	LAT.	LONG.
Paro, Bhu. (pä'rô)	158	27°30'N	89°30'E
Paroo, r., Austl. (pä'ro͞o)	175	30°00'S	144°00'E
Páros, Grc. (pä'rôs) (pā'rŏs)	131	37°05'N	25°14'E
Páros, i., Grc.	119	37°11'N	25°00'E
Parow, S. Afr. (pă'rô)	186a	33°54'S	18°36'E
Parowan, Ut., U.S. (păr'ô-wăn)	77	37°50'N	112°50'W
Parral, Chile (pär-rä'l)	102	36°07'S	71°47'W
Parral, r., Mex.	80	27°25'N	105°08'W
Parramatta, r., Austl. (păr-á-măt'á)	173b	33°42'S	150°58'E
Parras, Mex. (pär-räs')	80	25°28'N	102°08'W
Parrita, C.R. (pär-rē'tä)	91	9°32'N	84°17'W
Parrsboro, Can. (pärz'bŭr-ô)	60	45°24'N	64°20'W
Parry, i., Can. (păr'ĭ)	59	45°15'N	80°00'W
Parry, Mount, mtn., Can.	54	52°53'N	128°45'W
Parry Islands, is., Can.	49	75°30'N	110°00'W
Parry Sound, Can.	51	45°20'N	80°00'W
Parsnip, r., Can. (pärs'nĭp)	55	54°45'N	122°20'W
Parsons, Ks., U.S. (pär's'nz)	65	37°20'N	95°16'W
Parsons, W.V., U.S.	67	39°05'N	79°40'W
Parthenay, Fr. (pȧr-t'nĕ')	126	46°39'N	0°16'W
Partinico, Italy (pär-tē'nē-kô)	130	38°02'N	13°11'E
Partizansk, Russia	135	43°15'N	133°19'E
Parys, S. Afr. (pá-rīs')	192c	26°53'S	27°28'E
Pasadena, Ca., U.S. (păs-á-dē'ná)	64	34°09'N	118°09'W
Pasadena, Md., U.S.	68e	39°06'N	76°35'W
Pasadena, Tx., U.S.	81a	29°43'N	95°13'W
Pascagoula, Ms., U.S. (păs-ká-go͞o'lá)	82	30°22'N	88°33'W
Pascagoula, r., Ms., U.S.	82	30°52'N	88°48'W
Paşcani, Rom. (päsh-kän')	125	47°46'N	26°42'E
Pasco, Wa., U.S. (păs'kō)	72	46°13'N	119°04'W
Pascua, Isla de (Easter Island), i., Chile	195	26°50'S	109°00'W
Pasewalk, Ger. (pä'zē-välk)	124	53°31'N	14°01'E
Pashiya, Russia (pä'shĭ-yȧ)	142a	58°27'N	58°17'E
Pashkovo, Russia (pȧsh-kô'vô)	166	48°52'N	131°09'E
Pashkovskaya, Russia (pȧsh-kôf'ská-yȧ)	133	45°00'N	39°04'E
Pasig, Phil.	169a	14°34'N	121°05'E
Pasión, Río de la, r., Guat. (rē'ō-dĕ-lä-pä-syōn')	90a	16°31'N	90°11'W
Paso de los Libres, Arg. (pä-sô-dĕ-lôs-lē'brĕs)	102	29°33'S	57°05'W
Paso de los Toros, Ur. (tŏ'rôs)	99c	32°43'S	56°33'W
Paso Robles, Ca., U.S. (pă'sō rō'blĕs)	76	35°38'N	120°44'W
Pasquia Hills, hills, Can. (păs'kwē-á)	57	53°13'N	102°37'W
Passaic, N.J., U.S. (pă-sā'ĭk)	68a	40°52'N	74°08'W
Passaic, r., N.J., U.S.	68a	40°42'N	74°26'W
Passamaquoddy Bay, b., N.A. (păs'á-má-kwŏd'ĭ)	60	45°06'N	66°59'W
Passa Tempo, Braz. (pá's-sä-tē'm-pô)	99a	20°40'S	44°29'W
Passau, Ger. (päsou)	117	48°34'N	13°27'E
Pass Christian, Ms., U.S. (pȧs krĭs'tyĕn)	82	30°20'N	89°15'W
Passero, Cape, c., Italy (päs-sĕ'rô)	112	36°34'N	15°13'E
Passo Fundo, Braz. (pä'sō fo͞on'dō)	102	28°16'S	52°13'W
Passos, Braz. (pä's-sōs)	101	20°45'S	46°37'W
Pastaza, r., S.A. (päs-tä'zä)	100	3°05'S	76°18'W
Pasto, Col. (päs'tô)	100	1°15'N	77°19'W
Pastora, Mex. (päs-tô-rä)	88	22°08'N	100°04'W
Pasuruan, Indon.	168	7°45'S	112°50'E
Pasvalys, Lith. (päs-vä-lēs')	123	56°04'N	24°23'E
Patagonia, reg., Arg. (păt-á-gō'nĭ-á)	102	46°45'S	69°30'W
Pātālganga, r., India	159b	18°52'N	73°08'E
Patapsco, r., Md., U.S. (pá-tăps'kō)	68e	39°12'N	76°30'W
Pateros, Lake, res., Wa., U.S.	72	48°05'N	119°45'W
Paterson, N.J., U.S. (păt'ẽr-sŭn)	68a	40°55'N	74°10'W
Pathein, Mya.	155	16°46'N	94°47'E
Pathfinder Reservoir, res., Wy., U.S. (păth'fīn-dẽr)	73	42°22'N	107°10'W
Patiāla, India (pŭt-ê-ä'lŭ)	155	30°25'N	76°28'E
Pati do Alferes, Braz. (pä-tē-dô-äl-fĕ'rĕs)	102b	22°25'S	43°25'W
Patna, India (pŭt'nŭ)	155	25°33'N	85°18'E
Patnanongan, i., Phil. (pät-nä-nôŋ'gän)	169a	14°50'N	122°25'E
Patoka, r., In., U.S. (pá-tō'ká)	66	38°25'N	87°25'W
Patom Plateau, plat., Russia	135	59°30'N	115°00'E
Patos, Braz. (pä'tōzh)	101	7°03'S	37°14'W
Patos, Wa., U.S. (pä'tōs)	74d	48°47'N	122°57'W
Patos, Lagoa dos, l., Braz. (lä'gō-ä dozh pä'tōzh)	102	31°15'S	51°30'W
Patos de Minas, Braz. (dĕ-mē'näzh)	101	18°39'S	46°31'W
Pátra, Grc.	119	38°15'N	21°48'E
Patraïkós Kólpos, b., Grc.	131	38°16'N	21°19'E
Patras see Pátra, Grc.	119	38°15'N	21°48'E
Patrocínio, Braz. (pä-trō-sē'nē-ô)	101	18°48'S	46°47'W
Pattani, Thai. (pät'á-nē)	168	6°56'N	101°13'E
Patten, Me., U.S. (păt''n)	60	45°59'N	68°27'W
Patterson, La., U.S. (păt'ẽr-sŭn)	81	29°41'N	91°20'W
Patterson, i., Can.	58	48°38'N	87°14'W
Patton, Pa., U.S.	67	40°40'N	78°45'W
Patuca, r., Hond.	91	15°22'N	84°31'W
Patuca, Punta, c., Hond. (po͞o'n-tä-pä-to͞o'kä)	91	15°55'N	84°05'W
Patuxent, r., Md., U.S. (pá-tŭk'sĕnt)	67	39°10'N	77°10'W
Pátzcuaro, Mex. (päts'kwä-rô)	88	19°30'N	101°36'W
Pátzcuaro, Lago de, l., Mex. (lä'gô-dĕ)	88	19°36'N	101°38'W
Patzicia, Guat. (pät-zē'syä)	90	14°36'N	90°57'W
Patzún, Guat. (pät-zo͞on')	90	14°40'N	91°00'W
Pau, Fr. (pō)	117	43°18'N	0°23'W
Pau, Gave de, r., Fr. (gäv-dĕ)	126	43°33'N	0°51'W
Paulding, Oh., U.S. (pôl'dĭng)	66	41°05'N	84°35'W
Paulinenaue, Ger. (pou'lē-nĕ-nou-ĕ)	115b	52°40'N	12°43'E
Paulistano, Braz. (pá'ó-lēs-tá-nä)	101	8°13'S	41°06'W
Paulo Afonso, Salto, wtfl., Braz. (säl-tô-pou'lô äf-fôn'sô)	101	9°33'S	38°32'W
Paul Roux, S. Afr. (pôrl ro͞o)	192c	28°18'S	27°57'E

PLACE (Pronunciation)	PAGE	LAT.	LONG.
Paulsboro, N.J., U.S. (pôlz'bē-rô)	68f	39°50'N	75°16'W
Pauls Valley, Ok., U.S. (pôlz văl'ē)	79	34°43'N	97°13'W
Pavarandocito, Col. (pä-vä-rän-dô-sē'tô)	100a	7°18'N	76°32'W
Pavda, Russia (päv'da)	142a	59°16'N	59°32'E
Pavia, Italy (pä-vē'ä)	118	45°12'N	9°11'E
Pavlodar, Kaz. (páv-lô-dár')	139	52°17'N	77°23'E
Pavlof Bay, b., Ak., U.S. (păv-lōf)	63	55°20'N	161°20'W
Pavlohrad, Ukr.	137	48°32'N	35°52'E
Pavlovsk, Russia (páv-lôfsk')	133	50°28'N	40°05'E
Pavlovsk, Russia	142c	59°41'N	30°27'E
Pavlovskiy Posad, Russia (páv-lôf'skĭ pô-sát')	136	55°47'N	38°39'E
Pavuna, Braz. (pä-vo͞o'ná)	102b	22°48'S	43°21'W
Päwesin, Ger. (pă'vĕ-zēn)	115b	52°31'N	12°44'E
Pawhuska, Ok., U.S. (pô-hŭs'ká)	79	36°41'N	96°20'W
Pawnee, Ok., U.S. (pô-nē')	79	36°20'N	96°47'W
Pawnee, r., Ks., U.S.	78	38°18'N	99°42'W
Pawnee City, Ne., U.S.	79	40°08'N	96°09'W
Paw Paw, Mi., U.S. (pô'pô)	66	42°15'N	85°55'W
Paw Paw, r., Mi., U.S.	71	42°14'N	86°21'W
Pawtucket, R.I., U.S. (pô-tŭk'ĕt)	67	41°53'N	71°23'W
Paxoi, i., Grc.	131	39°14'N	20°15'E
Paxton, Il., U.S. (păks'tŭn)	66	40°35'N	88°00'W
Payette, Id., U.S. (pâ-ĕt')	72	44°05'N	116°55'W
Payette, r., Id., U.S.	72	43°57'N	116°26'W
Payette, North Fork, r., Id., U.S.	72	44°10'N	116°10'W
Payette, South Fork, r., Id., U.S.	72	44°07'N	115°43'W
Pay-Khoy, Khrebet, mts., Russia	136	68°08'N	63°04'E
Payne, l., Can. (pān)	53	59°22'N	73°16'W
Paynesville, Mn., U.S. (pānz'vĭl)	71	45°23'N	94°43'W
Paysandú, Ur. (pī-sän-do͞o')	102	32°16'S	57°55'W
Payson, Ut., U.S. (pā's'n)	77	40°05'N	111°45'W
Pazardzhik, Blg. (pä-zár-dzhek')	119	42°10'N	24°22'E
Pazin, Cro. (pä'zēn)	130	45°14'N	13°57'E
Peabody, Ks., U.S. (pē'bŏd-ĭ)	79	38°09'N	97°09'W
Peabody, Ma., U.S.	61a	42°32'N	70°56'W
Peace, r., Can.	52	57°30'N	117°30'W
Peace Creek, r., Fl., U.S. (pēs)	83a	27°16'N	81°53'W
Peace Dale, R.I., U.S. (dāl)	68b	41°27'N	71°30'W
Peace River, Can. (rĭv'ẽr)	50	56°14'N	117°17'W
Peacock Hills, hills, Can. (pē-kŏk' hĭlz)	52	66°08'N	109°55'W
Peak Hill, Austl.	174	25°38'S	118°50'E
Pearl, r., U.S. (pûrl)	65	30°30'N	89°45'W
Pearland, Tx., U.S. (pûrl'ănd)	81a	29°34'N	95°17'W
Pearl Harbor, Hi., U.S.	84a	21°20'N	157°53'W
Pearl Harbor, b., Hi., U.S.	64d	21°22'N	157°58'W
Pearsall, Tx., U.S. (pēr'sôl)	80	28°53'N	99°06'W
Pearse Island, i., Can. (pērs)	54	54°51'N	130°21'W
Pearston, S. Afr. (pē'ẽrstŏn)	187c	32°36'S	25°09'E
Peary Land, reg., Grnld. (pēr'ĭ)	198	82°00'N	40°00'W
Pease, r., Tx., U.S. (pēz)	78	34°07'N	99°53'W
Peason, La., U.S. (pēz''n)	81	31°25'N	93°19'W
Pebane, Moz. (pĕ-bá'nē)	187	17°10'S	38°08'E
Pecan Bay, Tx., U.S. (pê-kăn')	80	32°04'N	99°15'W
Peçanha, Braz. (pā-kän'yá)	101	18°37'S	42°26'W
Pecatonica, r., Il., U.S. (pĕk-á-tŏn-ĭ-ká)	71	42°21'N	89°28'W
Pechenga, Russia (pyĕ'chĕŋ-gȧ)	136	69°30'N	31°10'E
Pechora, r., Russia	134	66°00'N	54°00'E
Pechora Basin, Russia (pyĕ-chô'rá)	134	67°55'N	58°37'E
Pechori, Russia (pĕt'sĕ-rĕ)	132	57°48'N	27°33'E
Pecos, N.M., U.S. (pā'kôs)	77	35°29'N	105°41'W
Pecos, Tx., U.S.	80	31°26'N	103°30'W
Pecos, r., U.S.	64	31°10'N	103°10'W
Pécs, Hung. (pāch)	119	46°04'N	18°15'E
Peddie, S. Afr.	187c	33°13'S	27°09'E
Pedley, Ca., U.S. (pĕd'lē)	75a	33°59'N	117°29'W
Pedra Azul, Braz. (pā'drä-zo͞o'l)	101	16°03'S	41°13'W
Pedreiras, Braz. (pĕ-drā'räs)	101	4°30'S	44°31'W
Pedro, Point, c., Sri L. (pē'drô)	159	9°50'N	80°14'E
Pedro Antonio Santos, Mex.	90a	18°55'N	88°13'W
Pedro Betancourt, Cuba (bā-tāŋ-kōrt')	92	22°40'N	81°15'W
Pedro de Valdivia, Chile (pĕ'drô-dĕ-väl-dē'vē-ä)	102	22°32'S	69°55'W
Pedro do Rio, Braz. (dô-rē'rô)	102b	22°20'S	43°09'W
Pedro II, Braz. (pā'drô sā-gón'dó)	101	4°20'S	41°27'W
Pedro Juan Caballero, Para. (hòá'n-kä-bäl-yĕ'rō)	102	22°40'S	55°42'W
Pedro Miguel, Pan. (mĕ-gāl')	86a	9°01'N	79°36'W
Pedro Miguel Locks, trans., Pan. (mĕ-gāl')	86a	9°01'N	79°36'W
Peebinga, Austl. (pē-bĭng'á)	175	34°43'S	140°55'E
Peebles, Scot., U.K. (pē'b'lz)	120	55°40'N	3°15'W
Peekskill, N.Y., U.S. (pēks'kĭl)	68a	41°17'N	73°55'W
Pegasus Bay, b., N.Z. (pĕg'á-sŭs)	175a	43°18'S	173°25'E
Pegnitz, r., Ger. (pĕgh-nēts)	124	49°38'N	11°40'E
Pego, Spain (pā'gō)	129	38°50'N	0°09'W
Peguis Indian Reserve, I.R., Can.	57	51°20'N	97°35'W
Pegu Yoma, mts., Mya. (pĕ-go͞o'yō'mä)	155	19°16'N	95°59'E
Pehčevo, Mac. (pĕk'chĕ-vô)	131	41°42'N	22°57'E
Peigan Indian Reserve, I.R., Can.	55	49°35'N	113°40'W
Peipus, Lake see Chudskoye Ozero, l., Eur.	136	58°43'N	26°45'E
Peiraiás, Grc.	119	37°57'N	23°38'E
Pekin, Il., U.S. (pē'kĭn)	66	40°35'N	89°30'W
Peking see Beijing, China	161	39°55'N	116°23'E
Pelagie, Isole, is., Italy	118	35°46'N	12°32'E
Pélagos, i., Grc.	131	39°17'N	24°05'E
Pelahatchie, Ms., U.S. (pĕl-á-hăch'ē)	82	32°17'N	89°48'W
Pelat, Mont, mtn., Fr. (pē-lá')	117	44°16'N	6°43'E
Peleduy, Russia (pyĕl-yĭ-do͞o'ē)	135	59°50'N	112°47'E
Pelée, Mont, mtn., Mart. (pē-lā')	91b	14°49'N	61°10'W
Pelee, Point, c., Can.	58	41°55'N	82°30'W
Pelee Island, i., Can. (pē'lē)	58	41°45'N	82°30'W

288 Pelequén – Piedmont PRONOUNCING INDEX

PLACE (Pronunciation)	PAGE	LAT.	LONG.
Pelequén, Chile (pě-lě-kě'n)	99b	34°26'S	71°52'W
Pelham, Ga., U.S. (pěl'hăm)	82	31°07'N	84°10'W
Pelham, N.H., U.S.	61a	42°43'N	71°22'W
Pelican, l., Mn., U.S.	71	46°36'N	94°00'W
Pelican Bay, b., Can.	57	52°45'N	100°20'W
Pelican Harbor, b., Bah. (pěl'ĭ-kăn)	92	26°20'N	76°45'W
Pelican Rapids, Mn., U.S. (pěl'ĭ-kăn)	70	46°34'N	96°05'W
Pella, Ia., U.S. (pěl'à)	71	41°25'N	92°50'W
Pellworm, i., Ger. (pěl'vôrm)	124	54°33'N	8°25'E
Pelly, l., Can.	52	66°08'N	102°57'W
Pelly, r., Can.	52	62°20'N	133°00'W
Pelly Bay, b., Can. (pěl'ĭ)	53	68°57'N	91°05'W
Pelly Crossing, Can.	63	62°50'N	136°50'W
Pelly Mountains, mts., Can.	52	61°50'N	133°05'W
Peloncillo Mountains, mts., Az., U.S. (pěl-ŏn-sĭl'lō)	77	32°40'N	109°20'W
Peloponnisos, pen., Grc.	131	37°28'N	22°14'E
Pelotas, Braz. (på-lō'tăzh)	102	31°45'S	52°18'W
Pelton, Can. (pěl'tŭn)	69b	42°15'N	82°57'W
Pelym, r., Russia	136	60°20'N	63°05'E
Pelzer, S.C., U.S. (pěl'zẽr)	83	34°38'N	82°30'W
Pemanggil, i., Malay.	153b	2°37'N	104°41'E
Pematangsiantar, Indon.	168	2°58'N	99°03'E
Pemba, Moz. (pěm'bà)	187	12°58'S	40°30'E
Pemba, Zam.	186	15°29'S	27°22'E
Pemba Channel, strt., Afr.	191	5°10'S	39°30'E
Pemba Island, i., Tan.	191	5°20'S	39°57'E
Pembina, N.D., U.S. (pěm'bĭ-nà)	70	48°58'N	97°15'W
Pembina, r., Can.	55	53°05'N	114°30'W
Pembina, r., N.A.	57	49°08'N	98°20'W
Pembroke, Can. (pěm' brōk)	51	45°50'N	77°00'W
Pembroke, Wales, U.K.	120	51°40'N	5°00'W
Pembroke, Ma., U.S. (pěm'brōk)	61a	42°05'N	70°49'W
Pen, India	159b	18°44'N	73°06'E
Penafiel, Port. (pā-nà-fyěl')	128	41°12'N	8°19'W
Peñafiel, Spain (pā-nyà-fyěl')	128	41°38'N	4°08'W
Peñalara, mtn., Spain (pā-nyä-lä'rä)	118	40°52'N	3°57'W
Pena Nevada, Cerro, Mex.	88	23°47'N	99°52'W
Peñaranda de Bracamonte, Spain	128	40°54'N	5°11'W
Peñarroya-Pueblonuevo, Spain (pěn-yär-rō'yä-pwě'blō-nwě'vō)	128	38°18'N	5°18'W
Peñas, Cabo de, c., Spain (ká'bō-dě-pā'nyäs)	128	43°42'N	6°12'W
Penas, Golfo de, b., Chile (gōl-fō-dě-pě'n-äs)	102	47°15'S	77°30'W
Penasco, r., Tx., U.S. (pā-nàs'kō)	80	32°50'N	104°45'W
Pendembu, S.L. (pěn-děm'bōō)	184	8°06'N	10°42'W
Pender, Ne., U.S. (pěn'dẽr)	70	42°08'N	96°43'W
Penderisco, r., Col. (pěn-dě-rē's-kô)	100a	6°30'N	76°21'W
Pendjari, Parc National de la, rec., Benin	188	11°25'N	1°30'E
Pendleton, Or., U.S. (pěn'd'l-tŭn)	64	45°41'N	118°47'W
Pend Oreille, r., Wa., U.S.	72	48°44'N	117°20'W
Pend Oreille, Lake, l., Id., U.S. (pŏn-dō-rā') (pěn-dô-rěl')	64	48°09'N	116°38'W
Penedo, Braz. (på-nā'dô)	101	10°17'S	36°28'W
Penetanguishene, Can. (pěn'ě-tăŋ-gĭ-shēn')	59	44°45'N	79°55'W
Pengcheng, China (pŭŋ-chŭŋ)	162	36°24'N	114°11'E
Penglai, China (pŭŋ-lī)	164	37°49'N	120°45'E
Peniche, Port. (pě-nē'chå)	128	39°22'N	9°24'W
Peninsula, Oh., U.S. (pěn-ĭn'sū-là)	69d	41°14'N	81°32'W
Penistone, Eng., U.K. (pěn'ĭ-stŭn)	114a	53°31'N	1°38'W
Penjamillo, Mex. (pěn-hä-mēl'yō)	88	20°06'N	101°56'W
Pénjamo, Mex. (pān'hä-mō)	88	20°27'N	101°43'W
Penk, r., Eng., U.K. (pěnk)	114a	52°41'N	2°10'W
Penkridge, Eng., U.K. (pěnk'rĭj)	114a	52°43'N	2°07'W
Penne, Italy (pěn'nā)	130	42°28'N	13°57'E
Penner, r., India (pěn'ẽr)	155	14°43'N	79°09'E
Pennines, hills, Eng., U.K. (pěn-īn')	120	54°30'N	2°10'W
Pennines, Alpes, mts., Eur.	124	46°02'N	7°07'E
Pennsboro, W.V., U.S. (pěnz'bŭr-ô)	66	39°10'N	81°00'W
Penns Grove, N.J., U.S. (pěnz grōv)	68f	39°44'N	75°28'W
Pennsylvania, state, U.S. (pěn-sĭl-vā'nĭ-à)	65	41°00'N	78°10'W
Penn Yan, N.Y., U.S. (pěn yăn')	67	42°40'N	77°00'W
Pennycutaway, r., Can.	57	56°10'N	93°25'W
Peno, l., Russia (pā'nô)	132	56°55'N	32°28'E
Penobscot, r., Me., U.S.	65	45°00'N	68°36'W
Penobscot Bay, b., Me., U.S. (pě-nŏb'skŏt)	60	44°20'N	69°00'W
Penong, Austl. (pē-nông')	174	32°00'S	133°00'E
Penrith, Austl.	173b	33°45'S	150°42'E
Pensacola, Fl., U.S. (pěn-sà-kō'là)	65	30°25'N	87°13'W
Pensacola Dam, Ok., U.S.	79	36°27'N	95°02'W
Pensilvania, Col. (pěn-sěl-vä'nyä)	100a	5°31'N	75°05'W
Pentecost, i., Vanuatu (pěn'tê-kôst)	175	16°05'S	168°28'E
Penticton, Can.	50	49°30'N	119°35'W
Pentland Firth, strt., Scot., U.K. (pěnt'lănd)	120	58°44'N	3°25'W
Penza, Russia (pěn'zà)	134	53°10'N	45°00'E
Penzance, Eng., U.K. (pěn-zăns')	120	50°07'N	5°40'W
Penzberg, Ger. (pěnts'běrgh)	124	47°43'N	11°21'E
Penzhina, r., Russia (pyĭn-zē-nŭ)	141	62°15'N	166°30'E
Penzhino, Russia	135	63°42'N	168°00'E
Penzhinskaya Guba, b., Russia	141	60°30'N	161°30'E
Peoria, Il., U.S. (pē-ō'rĭ-à)	65	40°45'N	89°35'W
Peotillos, Mex. (pā-ō-tel'yōs)	88	22°30'N	100°39'W
Peotone, Il., U.S. (pē'ŏ-tôn)	69a	41°20'N	87°47'W
Pepacton Reservoir, res., N.Y., U.S. (pěp-ăc'tŭn)	67	42°05'N	74°40'W
Pepe, Cabo, c., Cuba (kä'bō-pě'pě)	92	21°30'N	83°10'W
Pepperell, Ma., U.S. (pěp'ẽr-ěl)	61a	42°40'N	71°36'W
Peqin, Alb. (pě-kēn')	131	41°03'N	19°48'E
Perales, r., Spain (pā-rä'läs)	129a	40°24'N	4°07'W
Perales de Tajuña, Spain (dā tä-hōō'nyä)	129a	40°14'N	3°22'W
Perche, Collines du, hills, Fr.	126	48°25'N	0°40'E
Perchtoldsdorf, Aus. (pěrk'tôlts-dôrf)	115e	48°07'N	16°17'E
Perdekop, S. Afr.	192c	27°11'S	29°38'E
Perdido, r., Al., U.S. (pěr-dī'dō)	82	30°45'N	87°38'W
Perdido, Monte, mtn., Spain (pěr-dē'dō)	129	42°40'N	0°00'
Perdões, Braz. (pěr-dô'ěs)	99a	21°05'S	45°05'W
Pereiaslav-Khmel'nyts'kyi, Ukr.	137	50°05'N	31°25'E
Pereira, Col. (på-rā'rä)	100	4°49'N	75°42'W
Pere Marquette, Mi., U.S.	66	43°55'N	86°10'W
Pereshchepyne, Ukr.	133	49°02'N	35°19'E
Pereslavl'-Zalesskiy, Russia (pě-rå-släv''l zà-lyěs'kĭ)	136	56°43'N	38°52'E
Pergamino, Arg. (pěr-gä-mē'nō)	102	33°53'S	60°36'W
Perham, Mn., U.S. (pěr'hăm)	70	46°37'N	95°35'W
Peribonca, r., Can. (pěr-ĭ-bŏŋ'kà)	53	50°30'N	71°00'W
Périgueux, Fr. (pā-rē-gû')	117	45°12'N	0°43'E
Perija, Sierra de, mts., Col. (sē-ě'r-rà-dě-pě-rē'kä)	100	9°25'N	73°30'W
Perkam, Tanjung, c., Indon.	169	1°20'S	138°45'E
Perkins, Can. (pěr'kĕns)	62c	45°37'N	75°37'W
Perlas, Archipiélago de las, is., Pan.	91	8°29'N	79°15'W
Perlas, Laguna las, l., Nic. (lä-gō'nä-dě-läs)	91	12°34'N	83°19'W
Perleberg, Ger. (pěr'lē-běrg)	124	53°06'N	11°51'E
Perm', Russia (pěrm)	134	58°00'N	56°15'E
Pernambuco see Recife, Braz.	101	8°09'S	34°59'W
Pernambuco, state, Braz. (pěr-näm-bōō'kō)	101	8°08'S	38°54'W
Pernik, Blg. (pěr-něk')	119	42°36'N	23°04'E
Péronne, Fr. (pā-rŏn')	126	49°57'N	2°49'E
Perote, Mex. (pě-rō'tě)	89	19°33'N	97°13'W
Perovo, Russia (pâ'rô-vô)	142b	55°43'N	37°47'E
Perpignan, Fr. (pěr-pē-nyän')	117	42°42'N	2°48'E
Perris, Ca., U.S. (pěr'ĭs)	75a	33°46'N	117°14'W
Perros, Bahía, b., Cuba (bä-ē'ä-pā'rōs)	92	22°25'N	78°35'W
Perrot, Île, i., Can.	62a	45°23'N	73°57'W
Perry, Fl., U.S. (pěr'ĭ)	82	30°06'N	83°35'W
Perry, Ga., U.S.	82	32°27'N	83°44'W
Perry, Ia., U.S.	71	41°49'N	94°40'W
Perry, N.Y., U.S.	67	42°45'N	78°00'W
Perry, Ok., U.S.	79	36°17'N	97°18'W
Perry, Ut., U.S.	75b	41°27'N	112°02'W
Perry Hall, Md., U.S.	68e	39°24'N	76°29'W
Perryopolis, Pa., U.S. (pě-rē-ŏ'pō-lĭs)	69e	40°05'N	79°45'W
Perrysburg, Oh., U.S. (pěr'ĭz-bûrg)	66	41°35'N	83°35'W
Perryton, Tx., U.S. (pěr'ĭ-tŭn)	78	36°23'N	100°48'W
Perryville, Ak., U.S. (pěr-ĭ-vĭl)	63	55°58'N	159°28'W
Perryville, Mo., U.S.	79	37°41'N	89°52'W
Persan, Fr. (pěr-sän')	127b	49°09'N	2°15'E
Persepolis, hist., Iran (pẽr-sěpō-lĭs)	154	30°15'N	53°08'E
Persian Gulf, b., Asia (pûr'zhăn)	154	27°38'N	50°30'E
Perth, Austl. (pûrth)	174	31°50'S	116°10'E
Perth, Can.	59	44°40'N	76°15'W
Perth, Scot., U.K.	116	56°24'N	3°25'W
Perth Amboy, N.J., U.S. (ăm'boi)	68a	40°31'N	74°16'W
Pertuis, Fr. (pěr-tüē')	127	43°43'N	5°29'E
Peru, Il., U.S. (pě-rōō')	66	41°20'N	89°10'W
Peru, In., U.S.	66	40°45'N	86°00'W
Peru, nation, S.A.	100	10°00'S	75°00'W
Peru-Chile Trench, deep,	97	25°00'S	71°30'W
Perugia, Italy (pā-rōō'jä)	118	43°08'N	12°24'E
Peruque, Mo., U.S. (pě rō'kě)	75e	38°52'N	90°36'W
Pervomais'k, Ukr.	137	48°04'N	30°52'E
Pervoural'sk, Russia (pěr-vô-ô-rálsk')	142a	56°54'N	59°58'E
Pesaro, Italy (pā'zä-rō)	118	43°54'N	12°55'E
Pescado, r., Ven. (pěs-kä'dō)	101b	9°33'N	65°32'W
Pescara, Italy (pās-kä'rä)	130	42°26'N	14°15'E
Pescara, r., Italy	130	42°18'N	13°22'E
Peschanyy müyisi, c., Kaz.	137	43°10'N	51°20'E
Pescia, Italy (pā'shä)	130	43°53'N	11°42'E
Peshāwar, Pak. (pě-shä'wŭr)	155	34°01'N	71°34'E
Peshtera, Blg.	131	42°03'N	24°19'E
Peshtigo, Wi., U.S. (pěsh'tē-gō)	71	45°03'N	87°46'W
Peshtigo, r., Wi., U.S.	71	45°15'N	88°14'W
Peski, Russia (pyâs'kĭ)	142b	55°13'N	38°48'E
Pêso da Régua, Port. (pā-sò-dä-rā'gwä)	128	41°09'N	7°47'W
Pespire, Hond. (pås-pē'rå)	90	13°35'N	87°20'W
Pesqueria, r., Mex. (pås-kå-rē'à)	80	25°55'N	100°25'W
Pessac, Fr.	126	44°48'N	0°38'W
Petacalco, Bahía de, b., Mex. (bä-ē'ä-dě-pě-tä-käl'kô)	88	17°55'N	102°00'W
Petah Tiqwa, Isr.	153a	32°05'N	34°53'E
Petaluma, Ca., U.S. (pět-à-lō'mà)	76	38°15'N	122°38'W
Petare, Ven. (pě-tä'rě)	101b	10°28'N	66°48'W
Petatlán, Mex. (pā-tä-tlän')	88	17°31'N	101°17'W
Petawawa, Can.	59	45°54'N	77°17'W
Petén, Laguna de, l., Guat. (lä-gō'nä-dě-på-tān')	90a	17°05'N	89°54'W
Petenwell Reservoir, res., Wi., U.S.	71	44°10'N	89°55'W
Peterborough, Austl.	174	32°53'S	138°58'E
Peterborough, Can. (pē'tẽr-bŭr-ô)	51	44°20'N	78°20'W
Peterborough, Eng., U.K.	120	52°35'N	0°14'W
Peterhead, Scot., U.K. (pē-tẽr-hěd')	120	57°36'N	3°47'W
Peter Pond Lake, l., Can. (pŏnd)	52	55°55'N	108°44'W
Petersburg, Ak., U.S. (pē'tẽrz-bûrg)	63	56°52'N	133°10'W
Petersburg, Il., U.S.	79	40°01'N	89°51'W
Petersburg, In., U.S.	66	38°30'N	87°15'W
Petersburg, Ky., U.S.	69f	39°04'N	84°52'W
Petersburg, Va., U.S.	65	37°12'N	77°30'W
Petershagen, Ger. (pě'těrs-hä-gěn)	115b	52°32'N	13°46'E
Petershausen, Ger. (pě'těrs-hou-zěn)	115d	48°25'N	11°29'E
Pétionville, Haiti	93	18°30'N	72°20'W
Petitcodiac, Can. (pē-tē-kō-dyăk')	60	45°56'N	65°10'W
Petite Terre, i., Guad. (pē-tēt'târ')	91b	16°12'N	61°00'W
Petit Goâve, Haiti (pē-tē' gō-áv')	93	18°25'N	72°50'W
Petit Jean Creek, r., Ar., U.S. (pē-tē'zhän')	79	35°05'N	93°55'W
Petit Loango, Gabon	190	2°16'S	9°35'E
Petlalcingo, Mex. (pě-tläl-sěŋ'gō)	89	18°05'N	97°53'W
Peto, Mex. (pě'tô)	90a	20°07'N	88°49'W
Petorca, Chile (pā-tōr'kà)	99b	32°14'S	70°55'W
Petoskey, Mi., U.S. (pê-tŏs-kĭ)	66	45°25'N	84°55'W
Petra, hist., Jord.	153a	30°21'N	35°25'E
Petra Velikogo, Zaliv, b., Russia	166	42°40'N	131°50'E
Petre, Point, c., Can.	59	43°50'N	77°00'W
Petrich, Blg. (pā'trĭch)	119	41°24'N	23°13'E
Petrified Forest National Park, rec., Az., U.S. (pět'rĭ-fīd fŏr'ěst)	77	34°58'N	109°35'W
Petrinja, Cro. (pā'trēn-yà)	130	45°25'N	16°17'E
Petrodvorets, Russia (pyě-trô-dvô-ryěts')	142c	59°53'N	29°55'E
Petrokrepost', Russia (pyě'trô-krě-pôst)	136	59°56'N	31°03'E
Petrolia, Can. (pē-trō'lĭ-à)	58	42°50'N	82°10'W
Petrolina, Braz. (pě-trō-lē'nà)	101	9°18'S	40°28'W
Petronell, Aus.	115e	48°07'N	16°52'E
Petropavlivka, Ukr.	133	48°24'N	36°23'E
Petropavlovka, Russia	142a	54°10'N	59°50'E
Petropavlovsk, Kaz.	139	54°44'N	69°07'E
Petropavlovsk-Kamchatskiy, Russia (käm-chät'skĭ)	135	53°13'N	158°56'E
Petrópolis, Braz. (på-trô-pō-lēzh')	101	22°31'S	43°10'W
Petroşani, Rom.	131	45°24'N	23°24'E
Petrovsk, Russia (pyě-trôfsk')	137	52°20'N	45°15'E
Petrovskaya, Russia (pyě-trôf'skà-yà)	133	45°25'N	37°50'E
Petrovskoye, Russia	137	45°20'N	43°00'E
Petrovsk-Zabaykal'skiy, Russia (pyě-trôfskzä-bī-käl'skĭ)	135	51°13'N	109°08'E
Petrozavodsk, Russia (pyå'trō-zà-vôtsk')	134	61°46'N	34°25'E
Petrus Steyn, S. Afr.	192c	27°40'S	28°09'E
Petrykivka, Ukr.	133	48°43'N	34°29'E
Pewaukee, Wi., U.S. (pĭ-wô'kē)	69a	43°05'N	88°15'W
Pewaukee Lake, l., Wi., U.S.	69a	43°03'N	88°18'W
Pewee Valley, Ky., U.S. (pe wē)	69h	38°19'N	85°29'W
Peza, r., Russia (pyâ'zà)	136	65°35'N	46°50'E
Pézenas, Fr. (pā-zě-nä')	126	43°26'N	3°24'E
Pforzheim, Ger. (pfôrts'hīm)	117	48°52'N	8°43'E
Phalodi, India	158	27°13'N	72°22'E
Phan Thiet, Viet. (p'hän')	168	11°30'N	108°43'E
Phelps Lake, l., N.C., U.S.	83	35°46'N	76°27'W
Phenix City, Al., U.S. (fē'nĭks)	82	32°29'N	85°00'W
Philadelphia, Ms., U.S. (fĭl-à-děl'phĭ-à)	82	32°45'N	89°07'W
Philadelphia, Pa., U.S.	65	40°00'N	75°13'W
Philip, S.D., U.S. (fĭl'ĭp)	70	44°03'N	101°35'W
Philippeville see Skikda, Alg.	184	36°58'N	6°51'E
Philippines, nation, Asia (fĭl'ĭ-pēnz)	169	14°25'N	125°00'E
Philippine Sea, sea, (fĭl'ĭ-pēn)	195	16°00'N	133°00'E
Philippine Trench, deep,	169	10°30'N	127°15'E
Philipsburg, Pa., U.S. (fĭl'lĭps-běrg)	67	40°55'N	78°10'W
Philipsburg, Wy., U.S.	73	46°19'N	113°19'W
Phillip, i., Austl. (fĭl'ĭp)	176	38°32'S	145°10'E
Phillip Channel, strt., Indon.	153b	1°04'N	103°40'E
Phillipi, W.V., U.S. (fĭ-lĭp'ĭ)	66	39°10'N	80°00'W
Phillips, Wi., U.S. (fĭl'ĭps)	71	45°41'N	90°24'W
Phillipsburg, Ks., U.S. (fĭl'lĭps-běrg)	78	39°44'N	99°19'W
Phillipsburg, N.J., U.S.	67	40°45'N	75°10'W
Phitsanulok, Thai.	168	16°51'N	100°15'E
Phnom Penh (Phnum Pénh), Camb. (nŏm'pěn')	168	11°39'N	104°53'E
Phnum Pénh see Phnom Penh, Camb.	168	11°39'N	104°53'E
Phoenix, Az., U.S. (fē'nĭks)	64	33°30'N	112°00'W
Phoenix, Md., U.S.	68e	39°31'N	76°40'W
Phoenix Islands, is., Kir.	2	4°00'S	174°00'W
Phoenixville, Pa., U.S. (fē'nĭks-vĭl)	68f	40°08'N	75°31'W
Phou Bia, mtn., Laos	168	19°36'N	103°00'E
Phra Nakhon Si Ayutthaya, Thai.	168	14°16'N	100°37'E
Phuket, Thai.	168	7°57'N	98°19'E
Phu Quoc, Dao, i., Viet.	168	10°13'N	104°00'E
Pi, r., China (bē)	162	32°06'N	116°31'E
Piacenza, Italy (pyä-chěnt'sä)	118	45°02'N	9°42'E
Pianosa, i., Italy (pyä-nō'sä)	130	42°13'N	15°45'E
Piave, r., Italy (pyä'vā)	130	45°45'N	12°15'E
Piazza Armerina, Italy (pyät'sä är-må-rē'nä)	130	37°23'N	14°26'E
Pibor, r., Sudan (pē'bôr)	185	7°21'N	32°54'E
Pic, r., Can. (pěk)	58	48°48'N	86°28'W
Picara Point, c., V.I.U.S. (pě-kä'rä)	87c	18°23'N	64°57'W
Picayune, Ms., U.S. (pĭk'à yōōn)	82	30°32'N	89°41'W
Picher, Ok., U.S. (pĭch'ẽr)	79	36°58'N	94°49'W
Pichilemu, Chile (pē-chē-lě'mōō)	99b	34°22'S	72°01'W
Pichucalco, Mex. (pē-chōō-käl'kō)	89	17°34'N	93°06'W
Pickerel, l., Can. (pĭk'ẽr-ěl)	58	48°35'N	91°10'W
Pickwick Lake, res., U.S. (pĭk'wĭck)	82	35°04'N	88°05'W
Pico, Ca., U.S. (pē'kô)	75a	34°01'N	118°05'W
Pico Island, i., Port. (pē'kô)	184a	38°16'N	28°49'W
Pico Rivera, Ca., U.S.	75a	34°01'N	118°05'W
Picos, Braz. (pē'kōzh)	101	7°13'S	41°23'W
Picton, Austl. (pĭk'tŭn)	173b	34°11'S	150°37'E
Picton, Can.	59	44°00'N	77°15'W
Pictou, Can. (pĭk-tōō')	61	45°41'N	62°43'W
Pidálion, Akrotírion, c., Cyp.	153a	34°50'N	34°05'E
Pidurutalagala, mtn., Sri L. (pē'dō-rō-tä'lå-gä'là)	159	7°00'N	80°46'E
Pidvolochys'k, Ukr.	133	49°32'N	26°16'E
Pie, i., Can. (pī)	58	48°10'N	89°07'W
Piedade, Braz. (pyä-dä'dě)	99a	23°42'S	47°25'W
Piedmont, Al., U.S. (pēd'mŏnt)	82	33°54'N	85°36'W
Piedmont, Ca., U.S.	74b	37°50'N	122°14'W
Piedmont, Mo., U.S.	79	37°09'N	90°42'W

ăt; finăl; rāte; senåte; ärm; åsk; sofà; fâre; ch-choose; dh-as th in other; bē; êvent; bět; recĕnt; cratẽr; g-gō; gh-guttural g; bĭt; ĭ-short neutral; rīde; к-guttural k as ch in German ich;

PRONOUNCING INDEX Piedmont – Pointe-des-Cascades 289

PLACE (Pronunciation)	PAGE	LAT.	LONG.
Piedmont, S.C., U.S.	83	34°40'N	82°27'W
Piedmont, W.V., U.S.	67	39°30'N	79°05'W
Piedrabuena, Spain (pyā-drä-bwā'nä)	128	39°01'N	4°10'W
Piedras, Punta, c., Arg. (pōō'n-tä-pyĕ'dräs)	99c	35°25'S	57°10'W
Piedras Negras, Mex. (pyā'dräs nā'gräs)	86	28°41'N	100°33'W
Pieksämäki, Fin. (pyĕk'sĕ-mĕ-kĕ)	123	62°18'N	27°14'E
Piemonte, hist. reg., Italy (pyĕ-mô'n-tĕ)	130	44°30'N	7°42'E
Pienaars, r., S. Afr.	192c	25°13'S	28°05'E
Pienaarsrivier, S. Afr.	192c	25°12'S	28°18'E
Pierce, Ne., U.S. (pērs)	70	42°11'N	97°33'W
Pierce, W.V., U.S.	67	39°15'N	79°30'W
Piermont, N.Y., U.S. (pēr'mŏnt)	68a	41°03'N	73°55'W
Pierre, S.D., U.S. (pēr)	64	44°22'N	100°20'W
Pierrefonds, Can.	62a	45°29'N	73°52'W
Piešt'any, Slvk.	125	48°36'N	17°48'E
Pietermaritzburg, S. Afr. (pē-tēr-mä-rĭts-bûrg)	186	29°36'S	30°23'E
Pietersburg, S. Afr. (pē'tērz-bûrg)	186	23°56'S	29°30'E
Piet Retief, S. Afr. (pēt rĕ-tēf')	186	27°00'S	30°58'E
Pietrosu, Vârful, mtn., Rom.	125	47°35'N	24°49'E
Pieve di Cadore, Italy (pyā'vā dē kä-dō'rå)	118	46°26'N	12°22'E
Pigeon, r., N.A. (pĭj'ŭn)	71	48°05'N	90°13'W
Pigeon Lake, Can.	62f	49°57'N	97°36'W
Pigeon Lake, l., Can.	55	53°00'N	114°00'W
Piggott, Ar., U.S. (pĭg-ŭt)	79	36°22'N	90°10'W
Pijijiapan, Mex. (pēkē-kē-ä'pän)	89	15°40'N	93°12'W
Pijnacker, Neth.	115a	52°01'N	4°25'E
Pikes Peak, mtn., Co., U.S. (pīks)	64	38°49'N	105°03'W
Pikeville, Ky., U.S. (pīk'vĭl)	66	37°28'N	82°31'W
Pikou, China (pē-kō)	164	39°25'N	122°19'E
Pikwitonei, Can. (pĭk'wĭ-tōn)	57	55°35'N	97°09'W
Piła, Pol. (pē'lå)	124	53°09'N	16°44'E
Pilansberg, mtn., S. Afr. (pē'ăns'bûrg)	192c	25°08'S	26°55'E
Pilar, Arg. (pē'lär)	99c	34°27'S	58°55'W
Pilar, Para.	102	27°00'S	58°15'W
Pilar de Goiás, Braz. (dē-gô'yá's)	101	14°47'S	49°33'W
Pilchuck, r., Wa., U.S.	74a	48°03'N	121°58'W
Pilchuck Creek, r., Wa., U.S. (pĭl'chŭck)	74a	48°19'N	122°11'W
Pilchuck Mountain, mtn., Wa., U.S.	74a	48°03'N	121°48'W
Pilcomayo, r., S.A. (pēl-cō-mī'ō)	102	24°45'S	59°15'W
Pili, Phil. (pē'lē)	169a	13°34'N	123°17'E
Pilica, r., Pol. (pē-lēt'sä)	125	51°00'N	19°48'E
Pillar Point, c., Wa., U.S. (pĭl'ăr)	74a	48°14'N	124°06'W
Pillar Rock, Wa., U.S.	74c	46°16'N	123°35'W
Pilón, r., Mex. (pē-lōn')	88	24°13'N	99°03'W
Pilot Point, Tx., U.S. (pī'lŭt)	79	33°24'N	97°00'W
Pilsen see Plzeň, Czech Rep.	110	49°45'N	13°23'E
Piltene, Lat. (pĭl'tĕ-nĕ)	123	57°17'N	21°40'E
Pimal, Cerra, mtn., Mex. (sĕ'r-rä-pē-mäl')	88	22°58'N	104°19'W
Pimba, Austl. (pĭm'bå)	174	31°15'S	137°50'E
Pimville, neigh., S. Afr. (pĭm'vĭl)	187b	26°17'S	27°54'E
Pinacate, Cerro, mtn., Mex. (sĕ'r-rô-pē-nä-kä'tĕ)	86	31°45'N	113°30'W
Pinamalayan, Phil. (pē-nä-mä-lä'yän)	169a	13°04'N	121°31'E
Pinang see George Town, Malay.	168	5°21'N	100°09'E
Pınarbaşı, Tur. (pē'när-bä'shĭ)	119	38°50'N	36°10'E
Pinar del Río, Cuba (pē-när' dĕl rē'ō)	87	22°25'N	83°35'W
Pinar del Río, prov., Cuba	92	22°45'N	83°25'W
Pinatubo, mtn., Phil. (pē-nä-tōō'bō)	169a	15°09'N	120°19'E
Pincher Creek, Can. (pĭn'chēr krēk)	55	49°29'N	113°57'W
Pinckneyville, Il., U.S. (pĭnk'nĭ-vĭl)	79	38°06'N	89°22'W
Pińczów, Pol. (pēn''chôf)	125	50°32'N	20°33'E
Pindamonhangaba, Braz. (pē'n-dä-mōnyä'n-gä-bä)	99a	22°56'S	45°26'W
Pinder Point, c., Bah.	92	26°35'N	78°35'W
Pindiga, Nig.	189	9°59'N	10°54'E
Píndos Óros, mts., Grc.	112	39°48'N	21°19'E
Pine, r., Can. (pīn)	55	55°30'N	122°20'W
Pine, r., Wi., U.S.	71	45°50'N	88°37'W
Pine Bluff, Ar., U.S. (pīn blŭf)	65	34°13'N	92°01'W
Pine City, Mn., U.S. (pīn)	71	45°50'N	93°01'W
Pine Creek, Austl.	174	13°45'S	132°00'E
Pine Creek, r., Nv., U.S.	76	40°15'N	116°17'W
Pine Falls, Can.	57	50°35'N	96°15'W
Pine Flat Lake, res., Ca., U.S.	76	36°52'N	119°18'W
Pine Forest Range, mts., Nv., U.S.	72	41°35'N	118°45'W
Pinega, Russia (pē-nyĕ'gá)	134	64°40'N	43°30'E
Pinega, r., Russia	136	64°10'N	42°30'E
Pine Hill, N.J., U.S. (pīn hĭl)	68f	39°47'N	74°59'W
Pineiós, r., Grc.	131	39°30'N	21°40'E
Pine Island Sound, strt., Fl., U.S.	83a	26°32'N	82°30'W
Pine Lake Estates, Ga., U.S. (lăk ĕs-tāts')	68c	33°47'N	84°13'W
Pinelands, S. Afr. (pīn'lănds)	186a	33°57'S	18°30'E
Pine Lawn, Mo., U.S. (lôn)	75e	38°42'N	90°17'W
Pine Pass, p., Can.	55	55°22'N	122°40'W
Pinerolo, Italy (pē-nå-rô'lō)	130	44°47'N	7°18'E
Pines, Lake o' the, Tx., U.S.	81	32°50'N	94°40'W
Pinetown, S. Afr. (pīn'toun)	187c	29°47'S	30°52'E
Pine View Reservoir, res., Ut., U.S. (vū)	75b	41°17'N	111°54'W
Pineville, Ky., U.S. (pīn'vĭl)	82	36°48'N	83°43'W
Pineville, La., U.S.	81	31°20'N	92°25'W
Ping, r., Thai.	168	17°54'N	98°29'E
Pingding, China (pĭŋ-dĭŋ)	164	37°50'N	113°30'E
Pingdu, China (pĭŋ-dōō)	164	36°46'N	119°57'E
Pinggir, Indon.	153b	1°05'N	101°12'E
Pinghe, China (pĭŋ-hŭ)	165	24°30'N	117°02'E
Pingle, China (pĭŋ-lŭ)	165	24°30'N	110°22'E
Pingliang, China (pĭng'lyäng')	160	35°12'N	106°50'E
Pingquan, China (pĭŋ-chyůän)	164	40°58'N	118°40'E
Pingtan, China (pĭŋ-tän)	165	25°30'N	119°45'E
Pingtan Dao, i., China (pĭŋ-tän dou)	165	25°40'N	119°45'E
P'ingtung, Tai.	165	22°40'N	120°35'E
Pingwu, China (pĭŋ-wōō)	164	32°20'N	104°40'E
Pingxiang, China (pĭŋ-shyäŋ)	165	27°40'N	113°50'E
Pingyi, China (pĭŋ-yē)	162	35°30'N	117°38'E
Pingyuan, China (pĭŋ-yůän)	162	37°11'N	116°26'E
Pingzhou, China (pĭŋ-jō)	163a	23°01'N	113°11'E
Pinhal, Braz. (pē-nyá'l)	99a	22°11'S	46°43'W
Pinhal Novo, Port. (nô vô)	129b	38°38'N	8°54'W
Pinhel, Port. (pēn-yĕl')	128	40°45'N	7°03'W
Pini, Pulau, i., Indon.	168	0°07'S	98°38'E
Pinnacles National Monument, rec., Ca., U.S. (pĭn'á-k'lz)	76	36°30'N	121°00'W
Pinneberg, Ger. (pĭn'ĕ-bĕrg)	115c	53°40'N	9°48'E
Pinole, Ca., U.S. (pĭ-nō'lē)	74b	38°01'N	122°17'W
Pinos-Puente, Spain (pwān'tā)	128	37°15'N	3°43'W
Pinotepa Nacional, Mex. (pē-nô-tā'pä nä-syô-näl')	88	16°21'N	98°04'W
Pins, Île des, i., N. Cal.	175	22°44'S	167°44'E
Pinsk, Bela. (pēn'sk)	134	52°07'N	26°05'E
Pinta, i., Ec.	100	0°41'N	90°47'W
Pintendre, Can. (pēn-täNdr')	62b	46°45'N	71°07'W
Pinto, Spain (pēn'tō)	129a	40°14'N	3°42'W
Pinto Butte, Can. (pĭn'tō)	56	49°22'N	107°25'W
Pioche, Nv., U.S. (pī-ō'chē)	77	37°56'N	114°28'W
Piombino, Italy (pyôm-bē'nō)	118	42°56'N	10°33'E
Pioneer Mountains, mts., Mt., U.S. (pī'ō-nēr')	73	45°23'N	112°51'W
Piotrków Trybunalski, Pol. (pyōtr'kōōv trĭ-bōō-nal'skē)	117	51°23'N	19°44'E
Piper, Al., U.S. (pī'pēr)	82	33°04'N	87°00'W
Piper, Ks., U.S.	75f	39°09'N	94°51'W
Pipe Spring National Monument, rec., Az., U.S. (pīp sprĭng)	77	36°50'N	112°45'W
Pipestone, Mn., U.S. (pīp'stōn)	70	44°00'N	96°19'W
Pipestone National Monument, rec., Mn., U.S.	70	44°03'N	96°24'W
Pipmuacan, Réservoir, res., Can. (pĭp-mä-kän')	59	49°45'N	70°00'W
Piqua, Oh., U.S. (pĭk'wá)	66	40°10'N	84°15'W
Piracaia, Braz. (pē-rä-ká'yä)	99a	23°04'S	46°20'W
Piracicaba, Braz. (pē-rä-sē-kä'bä)	101	22°43'S	47°39'W
Piraíba, r., Braz. (pä-rä-ē'bå)	99a	21°38'S	41°29'W
Piramida, mtn., Russia	135	54°00'N	96°00'E
Piran, Slvn. (pē-rä'n)	130	45°31'N	13°34'E
Piranga, Braz. (pē-rä'n-gä)	99a	20°41'S	43°17'W
Pirapetinga, Braz. (pē-rä-pē-tē'n-gä)	99a	21°40'S	42°20'W
Pirapora, Braz. (pē-rá-pō'rá)	101	17°39'S	44°54'W
Pirassununga, Braz. (pē-rä-sōō-nōō'n-gä)	99a	22°00'S	47°24'W
Pirenópolis, Braz. (pē-rē-nô'pō-lēs)	101	15°56'S	48°49'W
Piritu, Laguna de, l., Ven. (lä-gô'nä-dĕ-pē-rē'tōō)	101b	10°00'N	64°57'W
Pirmasens, Ger. (pĭr-mä-zĕns')	124	49°12'N	7°34'E
Pirna, Ger. (pĭr'nä)	124	50°57'N	13°56'E
Pirot, Serb. (pē'rōt)	119	43°09'N	22°35'E
Pirtleville, Az., U.S. (pûr't'l-vĭl)	77	31°25'N	109°35'W
Piru, Indon. (pē-rōō')	169	3°15'S	128°25'E
Pisa, Italy (pē'sä)	118	43°52'N	10°24'E
Pisagua, Chile (pē-sä'gwä)	100	19°43'S	70°12'W
Piscataway, Md., U.S. (pĭs-kă-tă-wā)	68e	38°42'N	76°59'W
Piscataway, N.J., U.S.	68a	40°35'N	74°27'W
Pisco, Peru (pēs'kō)	100	13°43'S	76°07'W
Pisco, Bahía de, b., Peru	100	13°43'S	77°48'W
Piseco, l., N.Y., U.S. (pĭ-sā'kō)	67	43°25'N	74°35'W
Pisek, Czech Rep. (pē'sĕk)	117	49°18'N	14°08'E
Pisticci, Italy (pēs-tē'chē)	130	40°24'N	16°34'E
Pistoia, Italy (pēs-tô'yä)	118	43°57'N	11°54'E
Pisuerga, r., Spain (pē-swĕr'gä)	128	41°48'N	4°28'W
Pit, r., Ca., U.S. (pĭt)	72	40°58'N	121°42'W
Pitalito, Col. (pē-tä-lē'tō)	100	1°45'N	75°09'W
Pitcairn, dep., Oc.	2	25°04'S	130°05'W
Pitealven, r., Swe.	116	66°08'N	18°51'E
Pitești, Rom. (pē-tĕsht'')	131	44°51'N	24°51'E
Pithara, Austl. (pĭt'ără)	174	30°27'S	116°45'E
Pithiviers, Fr. (pē-tē-vyā')	126	48°12'N	2°14'E
Pitman, N.J., U.S. (pĭt'măn)	68f	39°44'N	75°08'W
Pitseng, Leso.	187c	29°03'S	28°13'E
Pitt, r., Can.	74d	49°19'N	122°39'W
Pitt Island, i., Can.	54	53°35'N	129°45'W
Pittsburg, Ca., U.S. (pĭts'bûrg)	74b	38°01'N	121°52'W
Pittsburg, Ks., U.S.	65	37°25'N	94°43'W
Pittsburg, Tx., U.S.	79	32°00'N	94°57'W
Pittsburgh, Pa., U.S.	65	40°26'N	80°01'W
Pittsfield, Il., U.S. (pĭts'fēld)	79	39°37'N	90°47'W
Pittsfield, Ma., U.S.	67	42°25'N	73°15'W
Pittsfield, Me., U.S.	60	44°45'N	69°44'W
Pittston, Pa., U.S. (pĭts'tŭn)	67	41°20'N	75°50'W
Piũi, Braz. (pē-ōō'ē)	99a	20°27'S	45°57'W
Piura, Peru (pē-ōō'rä)	100	5°13'S	80°46'W
Pivdennyi Buh, r., Ukr.	137	48°12'N	30°13'E
Piya, Russia (pē'yá)	142a	58°34'N	61°12'E
Placentia, Can.	61	47°15'N	53°58'W
Placentia, Ca., U.S. (plă-sĕn'shĭ-á)	75a	33°52'N	117°50'W
Placentia Bay, b., Can.	53a	47°14'N	54°30'W
Placerville, Ca., U.S. (plăs'ēr-vĭl)	76	38°43'N	120°47'W
Placetas, Cuba (plä-thā'täs)	92	22°10'N	79°40'W
Placid, l., N.Y., U.S. (plăs'ĭd)	67	44°20'N	74°00'W
Plain City, Ut., U.S. (plān)	75b	41°18'N	112°06'W
Plainfield, Il., U.S. (plān'fēld)	69a	41°37'N	88°12'W
Plainfield, In., U.S.	69g	39°42'N	86°23'W
Plainfield, N.J., U.S.	68a	40°38'N	74°25'W
Plainview, Ar., U.S. (plān'vū)	79	34°59'N	93°15'W
Plainview, Mn., U.S.	71	44°09'N	93°12'W
Plainview, Ne., U.S.	70	42°20'N	97°47'W
Plainview, Tx., U.S.	78	34°11'N	101°42'W
Plainwell, Mi., U.S. (plan'wĕl)	66	42°25'N	85°40'W
Plaisance, Can. (plĕ-zäns')	62c	45°37'N	75°07'W
Plana or Flat Cays, is., Bah. (plä'nä)	93	22°35'N	73°35'W
Planegg, Ger. (plä'nĕg)	115d	48°06'N	11°27'E
Plano, Tx., U.S. (plā'nō)	79	33°01'N	96°42'W
Plantagenet, Can. (pläN-täzh-nē')	62c	45°33'N	75°00'W
Plant City, Fl., U.S. (plånt sĭ'tĭ)	83a	28°00'N	82°07'W
Plaquemine, La., U.S. (pläk'mēn')	81	30°17'N	91°14'W
Plasencia, Spain (plä-sĕn'thē-ä)	128	40°02'N	6°07'W
Plast, Russia (plást)	136	54°22'N	60°48'E
Plaster Rock, Can. (plås'tēr rŏk)	60	46°54'N	67°24'W
Plastun, Russia (plås-tōōn')	166	44°41'N	136°08'E
Plata, Río de la, est., S.A. (dālä plä'tä)	102	34°35'S	58°15'W
Platani, r., Italy (plä-tä'nē)	130	37°26'N	13°28'E
Plateforme, Pointe, c., Haiti	93	19°35'N	73°50'W
Platinum, Ak., U.S. (plăt'ĭ-nŭm)	63	59°00'N	161°27'W
Plato, Col. (plä'tō)	100	9°49'N	74°48'W
Platón Sánchez, Mex. (plä-tōn' sän'chĕz)	88	21°14'N	98°20'W
Platte, S.D., U.S. (plăt)	70	43°22'N	98°51'W
Platte, r., Mo., U.S.	79	40°09'N	94°40'W
Platte, r., Ne., U.S.	64	40°50'N	100°40'W
Platteville, Wi., U.S. (plăt'vĭl)	71	42°44'N	90°31'W
Plattsburg, Mo., U.S. (plăts'bûrg)	79	39°33'N	94°26'W
Plattsburg, N.Y., U.S.	67	44°40'N	73°30'W
Plattsmouth, Ne., U.S. (plăts'mŭth)	70	41°00'N	95°53'W
Plauen, Ger. (plou'ĕn)	117	50°30'N	12°08'E
Playa de Guanabo, Cuba (plä-yä-dĕ-gwä-nä'bō)	93a	23°10'N	82°07'W
Playa de Santa Fé, Cuba	93a	23°05'N	82°31'W
Playas Lake, l., N.M., U.S. (plä'yås)	77	31°50'N	108°30'W
Playa Vicente, Mex. (vē-sĕn'tå)	89	17°49'N	95°49'W
Playa Vicente, r., Mex.	89	17°36'N	96°13'W
Playgreen Lake, l., Can. (plā'grēn)	57	54°00'N	98°10'W
Pleasant, l., N.Y., U.S. (plĕz'ănt)	67	43°25'N	74°25'W
Pleasant Grove, Al., U.S.	68h	33°29'N	86°57'W
Pleasant Hill, Ca., U.S.	74b	37°57'N	122°04'W
Pleasant Hill, Mo., U.S.	79	38°46'N	94°18'W
Pleasanton, Ca., U.S. (plĕz'ăn-tŭn)	74b	37°40'N	121°53'W
Pleasanton, Ks., U.S.	79	38°10'N	94°41'W
Pleasanton, Tx., U.S.	80	28°58'N	98°30'W
Pleasant Plain, Oh., U.S. (plĕz'ănt)	69f	39°17'N	84°06'W
Pleasant Ridge, Mi., U.S.	69b	42°28'N	83°09'W
Pleasant View, Ut., U.S. (plĕz'ănt vū)	75b	41°20'N	112°02'W
Pleasantville, N.Y., U.S. (plĕz'ănt-vĭl)	68a	41°08'N	73°47'W
Pleasure Ridge Park, Ky., U.S. (plĕzh'ēr rĭj)	69h	38°09'N	85°49'W
Plenty, Bay of, b., N.Z. (plĕn'tē)	175a	37°30'S	177°10'E
Plentywood, Mt., U.S. (plĕn'tē-wŏd)	73	48°47'N	104°38'W
Ples, Russia (plyĕs)	132	57°26'N	41°29'E
Pleshcheyevo, l., Russia (plĕsh-chā'yĕ-vô)	132	56°50'N	38°22'E
Plessisville, Can. (plĕ-sē'vēl')	59	46°12'N	71°47'W
Pleszew, Pol. (plĕ'zhĕf)	125	51°54'N	17°48'E
Plettenberg, Ger. (plĕ'tĕn-bĕrgh)	127c	51°13'N	7°53'E
Pleven, Blg. (plĕ'vĕn)	119	43°24'N	24°26'E
Pljevlja, Serb. (plĕv'lyä)	119	43°20'N	19°21'E
Płock, Pol. (pwôtsk)	117	52°32'N	19°44'E
Ploërmel, Fr. (plô-ĕr-mĕl')	126	47°56'N	2°25'W
Ploiești, Rom. (plô-yĕsht'')	110	44°56'N	26°01'E
Plomári, Grc.	131	38°51'N	26°24'E
Plomb du Cantal, mtn., Fr. (plôN'dükän-tăl')	117	45°30'N	2°49'E
Plonge, Lac la, l., Can. (plōnzh)	56	55°08'N	107°25'W
Plovdiv, Blg. (plôv'dĭf) (fĭl-ĭp-ŏp'ō-lĭs)	110	42°09'N	24°43'E
Pluma Hidalgo, Mex. (plōō'mä ē-däl'gō)	89	15°54'N	96°23'W
Plunge, Lith. (plŏn'gā)	123	55°56'N	21°45'E
Plymouth, Monts.	91b	16°43'N	62°12'W
Plymouth, Eng., U.K. (plĭm'ŭth)	117	50°25'N	4°14'W
Plymouth, In., U.S.	66	41°20'N	86°20'W
Plymouth, Ma., U.S.	67	42°00'N	70°45'W
Plymouth, Mi., U.S.	69b	42°23'N	83°27'W
Plymouth, N.C., U.S.	83	35°50'N	76°44'W
Plymouth, N.H., U.S.	67	43°50'N	71°40'W
Plymouth, Pa., U.S.	67	41°15'N	75°55'W
Plymouth, Wi., U.S.	71	43°45'N	87°59'W
Plyussa, r., Russia (plyōō'så)	132	58°33'N	28°30'E
Plzeň, Czech Rep.	110	49°45'N	13°23'E
Po, r., Italy	112	45°10'N	11°00'E
Pocahontas, Ar., U.S. (pō-kă-hŏn'tăs)	79	36°15'N	91°01'W
Pocahontas, Ia., U.S.	71	42°43'N	94°41'W
Pocatello, Id., U.S. (pō-kă-tĕl'ō)	64	42°54'N	112°30'W
Pochëp, Russia (pô-chĕp')	137	52°56'N	33°27'E
Pochinok, Russia (pô-chē'nôk)	132	54°14'N	32°27'E
Pochinski, Russia	136	54°40'N	44°50'E
Pochotitán, Mex. (pô-chô-tē-tá'n)	88	21°37'N	104°33'W
Pochutla, Mex.	89	15°46'N	96°28'W
Pocomoke City, Md., U.S. (pō-kō-mōk')	67	38°05'N	75°35'W
Pocono Mountains, mts., Pa., U.S. (pō-cō'nō)	67	41°10'N	75°30'W
Poços de Caldas, Braz. (pō-sôs-dĕ-käl'dás)	101	21°48'S	46°34'W
Poder, Sen. (pô-dôr')	184	16°35'N	15°04'W
Podgorica, Serb.	131	42°25'N	19°15'E
Podkamennaya Tunguska, r., Russia	135	61°43'N	93°45'E
Podol'sk, Russia (pô-dôl''sk)	136	55°26'N	37°33'E
Poggibonsi, Italy (pôd-jē-bôn'sē)	130	43°27'N	11°12'E
Pogodino, Bela. (pô-gô'dē-nô)	136	54°17'N	31°00'E
P'ohangdong, Kor., S.	166	35°57'N	129°23'E
Pointe-à-Pitre, Guad. (pwăNt' á pē-tr')	87	16°15'N	61°32'W
Pointe-aux-Trembles, Can. (pōō-ăNt' ō-träNbl)	62a	45°39'N	73°30'W
Pointe Claire, Can. (pōō-ăNt' klĕr)	62a	45°27'N	73°48'W
Pointe-des-Cascades, Can. (kăs-kădz')	62a	45°19'N	73°58'W

ng-sing; ŋ-baŋk; N-nasalized n; nŏd; cŏmmit; ōld; ôbey; ôrder; oi-boil; fōōd; ŏ-as oo in foot; ou-out; s-soft; sh-dish; th-thin; pūre; ûnite; ûrn; stŭd; circŭs; ü-as in French tu; '-indeterminate vowel.

290 Pointe Fortune – Poygan PRONOUNCING INDEX

PLACE (Pronunciation)	PAGE	LAT.	LONG.
Pointe Fortune, Can. (fôr´tūn)	62a	45°34'N	74°23'W
Pointe-Gatineau, Can. (po͞o-äɴt´gä-tē-nō´)	62c	45°28'N	75°42'W
Pointe Noire, Congo	186	4°48'S	11°51'E
Point Hope, Ak., U.S. (hōp)	63	68°18'N	166°38'W
Point Pleasant, W.V., U.S. (plĕz´ănt)	66	38°50'N	82°10'W
Point Roberts, Wa., U.S. (rŏb´ĕrts)	74d	48°59'N	123°04'W
Poissy, Fr. (pwȧ-sē´)	127b	48°55'N	2°02'E
Poitiers, Fr. (pwȧ-tyā´)	117	46°35'N	0°18'E
Pokaran, India (pō´kŭr-ŭn)	158	27°00'N	72°05'E
Pokrov, Russia (pô-krôf)	132	55°56'N	39°09'E
Pokrovskoye, Russia (pô-krôf´skô-yĕ)	133	47°27'N	38°54'E
Pola, r., Russia (pō´lä)	132	57°44'N	31°53'E
Pola de Laviana, Spain (dĕ-lä-vyä´nä)	128	43°15'N	5°29'W
Pola de Siero, Spain	128	43°24'N	5°39'W
Poland, nation, Eur. (pō´lănd)	110	52°37'N	17°01'E
Polangui, Phil. (pô-läŋ´gē)	169a	13°18'N	123°29'E
Polatsk, Bela.	136	55°30'N	28°48'E
Polazna, Russia (pô´läz-nȧ)	142a	58°18'N	56°25'E
Polessk, Russia (pô´lĕsk)	123	54°50'N	21°14'E
Polevskoy, Russia (pô-lĕ´vs-kô´ē)	142a	56°28'N	60°14'E
Polgár, Hung. (pôl´gär)	125	47°54'N	21°10'E
Policastro, Golfo di, b., Italy	130	40°00'N	13°23'E
Polichnítos, Grc.	131	39°05'N	26°11'E
Poligny, Fr. (pō-lē-nyē´)	127	46°48'N	5°42'E
Polillo, Phil. (pô-lēl´yō)	169a	14°42'N	121°56'E
Polillo Islands, is., Phil.	155	15°05'N	122°15'E
Polillo Strait, strt., Phil.	169a	15°02'N	121°40'E
Polist', r., Russia (pô´lĭst)	132	57°42'N	31°02'E
Polistena, Italy (pō-lĕs-tă´nä)	130	38°25'N	16°05'E
Polkan, Gora, mtn., Russia	135	60°18'N	92°08'E
Polochic, r., Guat. (pō-lō-chēk´)	90	15°19'N	89°45'W
Polonne, Ukr.	133	50°07'N	27°31'E
Polpaico, Chile (pôl-pá´y-kô)	99b	33°10'S	70°53'W
Polson, Mt., U.S. (pōl´sŭn)	73	47°40'N	114°10'W
Poltava, Ukr. (pôl-tä´vä)	134	49°35'N	34°33'E
Poltava, prov., Ukr.	133	49°53'N	32°58'E
Põltsamaa, Est. (pōlt´sȧ-mä)	123	58°39'N	26°00'E
Polunochnoye, Russia (pô-lo͞o-nô´ch-nô´yĕ)	142a	60°52'N	60°27'E
Poluy, r., Russia (pôl´wē)	140	65°45'N	68°15'E
Polyakovka, Russia (pŭl-yä´kôv-kȧ)	142a	54°38'N	59°42'E
Polyarnyy, Russia (pŭl-yär´nē)	134	69°10'N	33°30'E
Polygyros, Grc.	131	40°23'N	23°27'E
Polynesia, is., Oc.	194	4°00'S	156°00'W
Pomba, r., Braz. (pô´m-bȧ)	99a	21°28'S	42°28'W
Pomerania, hist. reg., Pol. (pŏm-ē-rā´nĭ-ȧ)	124	53°50'N	15°20'E
Pomeroy, S. Afr. (pŏm´ēr-roi)	187c	28°36'S	30°26'E
Pomeroy, Wa., U.S. (pŏm´ēr-oi)	72	46°28'N	117°35'W
Pomezia, Italy (pô-mĕ´t-zyä)	129d	41°41'N	12°31'E
Pomigliano d'Arco, Italy (pô-mē-lyá´nô-d-ä´r-kô)	129c	40°39'N	14°23'E
Pomme de Terre, Mn., U.S. (pôm dē tĕr´)	70	45°22'N	95°52'W
Pomona, Ca., U.S. (pô-mō´nȧ)	64	34°04'N	117°45'W
Pomorie, Blg.	119	42°24'N	27°41'E
Pompano Beach, Fl., U.S. (pŏm´pȧ-nō)	83a	26°12'N	80°07'W
Pompeii Ruins, hist., Italy	129c	40°31'N	14°29'E
Pompton Lakes, N.J., U.S. (pŏmp´tŏn)	68a	41°01'N	74°16'W
Pomuch, Mex. (pô-mo͞o´ch)	90a	20°12'N	90°10'W
Ponca, Ne., U.S. (pŏn´kȧ)	70	42°34'N	96°43'W
Ponca City, Ok., U.S.	79	36°42'N	97°07'W
Ponce, P.R. (pōn´sā)	87	18°01'N	66°43'W
Pondicherry, India	155	11°58'N	79°48'E
Pondicherry, state, India	155	11°50'N	74°50'E
Ponferrada, Spain (pôn-fĕr-rä´dhä)	118	42°33'N	6°38'W
Ponoka, Can. (pŏ-nō´kȧ)	50	52°42'N	113°35'W
Ponoy, Russia	136	66°58'N	41°00'E
Ponoy, r., Russia	136	67°00'N	39°00'E
Ponta Delgada, Port. (pôn´tȧ dĕl-gä´dȧ)	184a	37°40'N	25°45'W
Ponta Grossa, Braz. (grō´sȧ)	101	25°09'S	50°05'W
Pont-à-Mousson, Fr. (pôɴ´tȧ-mo͞o-sôɴ´)	127	48°55'N	6°02'E
Pontarlier, Fr. (pôɴ´tȧr-lyā´)	127	46°53'N	6°22'E
Pont-Audemer, Fr. (pôɴ´tōd´mâr´)	126	49°23'N	0°28'E
Pontchartrain Lake, l., La., U.S. (pôn-shȧr-trăn´)	81	30°10'N	90°10'W
Ponteareas, Spain	128	42°09'N	8°23'W
Pontedera, Italy (pōn-tā-dā´rä)	130	43°37'N	10°37'E
Ponte de Sor, Port.	128	39°14'N	8°03'W
Pontefract, Eng., U.K. (pŏn´tē-frăkt)	114a	53°41'N	1°18'W
Ponte Nova, Braz. (pô´n-tĕ-nô´vȧ)	101	20°26'S	42°52'W
Pontevedra, Spain (pôn-tĕ-vĕ-drä)	128	42°28'N	8°38'W
Ponthierville see Ubundi, D.R.C.	186	0°21'S	25°29'E
Pontiac, Il., U.S. (pŏn´tĭ-ăk)	66	40°55'N	88°35'W
Pontiac, Mi., U.S.	65	42°37'N	83°17'W
Pontianak, Indon. (pôn-tē-ä´nȧk)	168	0°04'S	109°20'E
Pontian Kechil, Malay.	153b	1°29'N	103°24'E
Pontic Mountains, mts., Tur.	137	41°20'N	34°30'E
Pontivy, Fr. (pôɴ-tē-vē´)	126	48°05'N	2°57'W
Pontoise, Fr. (pôɴ-twȧz´)	126	49°03'N	2°05'E
Pontonnyy, Russia (pôn´tôn-nyĭ)	142c	59°47'N	30°39'E
Pontotoc, Ms., U.S. (pŏn-tō-tŏk´)	82	34°11'N	88°59'W
Pontremoli, Italy (pôn-trĕm´ô-lē)	130	44°21'N	9°50'E
Ponziane, Isole, i., Italy (ĕ´sô-lĕ)	118	40°55'N	12°58'E
Poole, Eng., U.K. (po͞ol)	120	50°43'N	2°00'W
Poolesville, Md., U.S. (poolēs-vĭl)	68e	39°08'N	77°26'W
Pooley Island, i., Can. (po͞o´lē)	54	52°44'N	128°16'W
Poopó, Lago de, l., Bol.	100	18°45'S	67°07'W
Popayán, Col. (pō-pä-yän´)	100	2°21'N	76°43'W
Poplar, Mt., U.S. (pŏp´lēr)	73	48°08'N	105°10'W
Poplar, r., Mt., U.S.	73	48°34'N	105°20'W
Poplar, West Fork, r., Mt., U.S.	73	48°59'N	106°06'W
Poplar Bluff, Mo., U.S. (blŭf)	79	36°43'N	90°22'W
Poplar Plains, Ky., U.S. (plāns)	66	38°20'N	83°40'W
Poplar Point, Can.	62f	50°04'N	97°57'W
Poplarville, Ms., U.S. (pŏp´lēr-vĭl)	82	30°50'N	89°33'W
Popocatépetl Volcán, Mex. (pô-pô-kä-tā´pĕ´t'l)	86	19°01'N	98°38'W
Popokabaka, D.R.C. (pō´pō-kȧ-bä´kȧ)	186	5°42'S	16°35'E
Popovo, Blg. (pô´pô-vō)	131	43°23'N	26°17'E
Porbandar, India (pōr-bŭn´dŭr)	155	21°44'N	69°40'E
Porce, r., Col. (pôr-sĕ)	100a	7°11'N	74°55'W
Porcher Island, i., Can. (pôr´kēr)	54	53°57'N	130°30'W
Porcuna, Spain (pôr-ko͞o´nä)	128	37°54'N	4°10'W
Porcupine, r., N.A.	63	67°38'N	140°07'W
Porcupine Creek, r., Mt., U.S.	73	48°27'N	106°24'W
Porcupine Hills, hills, Can.	57	52°30'N	101°45'W
Pordenone, Italy (pōr-dā-nō´nā)	130	45°58'N	12°38'E
Pori, Fin. (pô´rē)	116	61°29'N	21°45'E
Poriúncula, Braz.	99a	20°58'S	42°02'W
Porkhov, Russia (pôr´κôf)	136	57°46'N	29°33'E
Porlamar, Ven. (pŏr-lä-mär´)	100	11°00'N	63°55'W
Pornic, Fr. (pōr-nēk´)	126	47°08'N	2°07'W
Poronaysk, Russia (pô´rô-nīsk)	135	49°21'N	143°23'E
Porrentruy, Switz. (pô-räɴ-trŭē´)	124	47°25'N	7°02'E
Porsgrunn, Nor. (pōrs´grōn´)	122	59°09'N	9°36'E
Portachuelo, Bol. (pôrt-ä-chwā´lō)	100	17°20'S	63°12'W
Portage, Pa., U.S. (pôr´tåj)	67	40°25'N	78°35'W
Portage, Wi., U.S.	71	43°33'N	89°29'W
Portage Des Sioux, Mo., U.S. (dē so͞o)	75e	38°56'N	90°21'W
Portage la Prairie, Can. (lä-prâ´rĭ)	50	49°57'N	98°25'W
Port Alberni, Can. (pōr äl-bēr-nē´)	50	49°14'N	124°48'W
Portalegre, Port. (pôr-tä-lā´grĕ)	118	39°18'N	7°26'W
Portales, N.M., U.S. (pôr-tä´lĕs)	78	34°10'N	103°11'W
Port Alfred, S. Afr.	186	33°36'S	26°55'E
Port Alice, Can. (ăl´ĭs)	50	50°23'N	127°27'W
Port Allegany, Pa., U.S. (ăl-ē-gā´nĭ)	67	41°50'N	78°10'W
Port Angeles, Wa., U.S. (ăn´jē-lĕs)	64	48°07'N	123°26'W
Port Antonio, Jam.	87	18°10'N	76°25'W
Portarlington, Austl.	173a	38°07'S	144°39'E
Port Arthur, Tx., U.S.	65	29°52'N	93°59'W
Port Augusta, Austl. (ô-gŭs´tȧ)	176	32°28'S	137°50'E
Port au Port Bay, b., Can. (pôr´tō pōr´)	61	48°41'N	58°45'W
Port-au-Prince, Haiti (prăɴs´)	87	18°35'N	72°20'W
Port Austin, Mi., U.S. (ôs´tĭn)	66	44°00'N	83°00'W
Port Blair, India (blâr)	168	12°07'N	92°45'E
Port Bolivar, Tx., U.S. (bŏl´ĭ-vȧr)	81a	29°22'N	94°46'W
Port Borden, Can. (bôr´dĕn)	60	46°15'N	63°42'W
Port-Bouët, C. Iv.	184	5°24'N	3°56'W
Port-Cartier, Can.	60	50°01'N	66°53'W
Port Chester, N.Y., U.S. (chĕs´tēr)	68a	40°59'N	73°40'W
Port Chicago, Ca., U.S. (shĭ-kô´gō)	74b	38°03'N	122°01'W
Port Clinton, Oh., U.S. (klĭn´tŭn)	66	41°30'N	83°00'W
Port Colborne, Can.	59	42°53'N	79°13'W
Port Coquitlam, Can. (kô-kwĭt´lȧm)	55	49°16'N	122°46'W
Port Credit, Can. (krĕd´ĭt)	62d	43°33'N	79°35'W
Port-de-Bouc, Fr. (pôr-dē-bo͞ok´)	126a	43°24'N	5°00'E
Port de Paix, Haiti (pĕ)	93	19°55'N	72°50'W
Port Dickson, Malay. (dĭk´sŭn)	153b	2°33'N	101°49'E
Port Discovery, b., Wa., U.S. (dĭs-kŭv´ēr-ĭ)	74a	48°05'N	122°55'W
Port Edward, S. Afr. (ĕd´wērd)	187c	31°04'S	30°14'E
Port Elgin, Can. (ĕl´jĭn)	60	46°03'N	64°05'W
Port Elizabeth, S. Afr. (ē-lĭz´ȧ-bĕth)	186	33°57'S	25°37'E
Porterdale, Ga., U.S. (pōr´tēr-dāl)	82	33°34'N	83°53'W
Porterville, Ca., U.S. (pōr´tēr-vĭl)	76	36°03'N	119°05'W
Port Francqui see Ilebo, D.R.C.	186	4°19'S	20°35'E
Port Gamble, Wa., U.S. (găm´bŭl)	74a	47°52'N	122°36'W
Port Gamble Indian Reservation, I.R., Wa., U.S.	74a	47°54'N	122°33'W
Port-Gentil, Gabon (zhäɴ-tē´)	186	0°43'S	8°47'E
Port Gibson, Ms., U.S.	82	31°56'N	90°57'W
Port Harcourt, Nig. (här´kŭrt)	184	4°43'N	7°05'E
Port Hardy, Can. (här´dĭ)	54	50°43'N	127°29'W
Port Hawkesbury, Can.	61	45°37'N	61°21'W
Port Hedland, Austl. (hĕd´lănd)	174	20°30'S	118°30'E
Porthill, Id., U.S.	72	49°00'N	116°30'W
Port Hood, Can. (ho͝od)	61	46°01'N	61°32'W
Port Hope, Can. (hōp)	59	43°55'N	78°10'W
Port Huron, Mi., U.S. (hū´rŏn)	65	43°00'N	82°30'W
Portici, Italy (pôr´tē-chē)	129c	40°34'N	14°20'E
Portillo, Chile (pôr-tē´l-yô)	99b	32°51'S	70°09'W
Portimão, Port. (pôr-tē-mouɴ)	128	37°09'N	8°34'W
Port Jervis, N.Y., U.S. (jûr´vĭs)	68a	41°22'N	74°41'W
Portland, Austl. (pōrt´lănd)	175	38°20'S	142°40'E
Portland, In., U.S.	66	40°25'N	85°00'W
Portland, Me., U.S.	65	43°40'N	70°16'W
Portland, Mi., U.S.	66	42°50'N	85°00'W
Portland, Or., U.S.	64	45°31'N	122°41'W
Portland, Tx., U.S.	81	27°53'N	97°20'W
Portland Bight, b., Jam.	92	17°45'N	77°05'W
Portland Canal, can., Ak., U.S.	54	55°10'N	130°08'W
Portland Inlet, b., Can.	54	54°50'N	130°15'W
Portland Point, c., Jam.	92	17°40'N	77°20'W
Port Lavaca, Tx., U.S. (lȧ-vä´kȧ)	81	28°36'N	96°38'W
Port Lincoln, Austl. (lĭŋ-kŭn)	174	34°39'S	135°50'E
Port Ludlow, Wa., U.S. (lŭd´lō)	74a	47°26'N	122°41'W
Port Macquarie, Austl. (mȧ-kwŏ´rĭ)	175	31°25'S	152°45'E
Port Madison Indian Reservation, I.R., Wa., U.S. (măd´ĭ-sŭn)	74a	47°46'N	122°38'W
Port Maria, Jam. (mȧ-rī´ȧ)	92	18°20'N	76°55'W
Port Moody, Can. (mo͞od´ĭ)	55	49°17'N	122°51'W
Port Moresby, Pap. N. Gui. (mōrz´bē)	169	9°34'S	147°20'E
Port Neches, Tx., U.S. (nĕch´ĕz)	81	29°59'N	93°57'W
Port Nelson, Can. (nĕl´sŭn)	57	57°03'N	92°36'W
Portneuf-Sur-Mer, Can. (pôr-nûf´sür mĕr)	60	48°36'N	69°06'W
Port Nolloth, S. Afr. (nŏl´ŏth)	186	29°10'S	17°00'E
Porto (Oporto), Port. (pōr´to)	110	41°10'N	8°38'W
Porto Acre, Braz. (ä´krĕ)	100	9°38'S	67°34'W
Porto Alegre, Braz. (ä-lā´grĕ)	102	29°58'S	51°11'W
Porto Amboim, Ang.	186	11°01'S	13°45'E
Portobelo, Pan. (pôr´tô-bā´lô)	87	9°32'N	79°40'W
Pôrto de Pedras, Braz. (pā´dräzh)	101	9°09'S	35°20'W
Pôrto Feliz, Braz. (fē-lē´s)	99a	23°12'S	47°30'W
Portoferraio, Italy (pōr´tô-fĕr-rä´yō)	130	42°47'N	10°20'E
Port of Spain, Trin. (spān)	100	10°44'N	61°24'W
Portogruaro, Italy (pōr´tô-grô-ä´rō)	130	45°48'N	12°49'E
Portola, Ca., U.S. (pôr´tô-lȧ)	76	39°47'N	120°29'W
Porto Mendes, Braz. (mĕ´n-dĕs)	101	24°41'S	54°13'W
Porto Murtinho, Braz. (mòr-tēn´yò)	101	21°43'S	57°43'W
Porto Nacional, Braz. (nä-syô-näl´)	101	10°43'S	48°14'W
Porto Novo, Benin (pōr´tô-nō´vō)	184	6°29'N	2°37'E
Port Orchard, Wa., U.S. (ôr´chērd)	74a	47°32'N	122°38'W
Port Orchard, b., Wa., U.S.	74a	47°40'N	122°39'W
Porto Santo, Ilha de, i., Port. (säɴ´tò)	184	32°41'N	16°15'W
Porto Seguro, Braz. (sā-go͞o´rò)	101	16°26'S	38°59'W
Porto Torres, Italy (tôr´rĕs)	130	40°49'N	8°25'E
Porto-Vecchio, Fr. (vĕk´ē-ô)	130	41°36'N	9°17'E
Pôrto Velho, Braz. (väl´yò)	100	8°45'S	63°43'W
Portoviejo, Ec. (pôr-tō-vyā´hō)	100	1°11'S	80°28'W
Port Phillip Bay, b., Austl. (fĭl´ĭp)	175	37°57'S	144°50'E
Port Pirie, Austl. (pĭ´rē)	174	33°10'S	138°00'E
Port Royal, b., Jam. (roi´ăl)	92	17°50'N	76°45'W
Port Said, Egypt	192d	31°15'N	32°19'E
Port Saint Johns, S. Afr. (sānt jŏnz)	186	31°37'S	29°32'E
Port Saint Lucie, Fl., U.S.	83a	27°20'N	80°20'W
Port Shepstone, S. Afr. (shĕps´tŭn)	186	30°45'S	30°23'E
Portsmouth, Dom.	91b	15°33'N	61°28'W
Portsmouth, Eng., U.K. (pôrts´mŭth)	110	50°45'N	1°03'W
Portsmouth, N.H., U.S.	65	43°05'N	70°50'W
Portsmouth, Oh., U.S.	65	38°45'N	83°00'W
Portsmouth, Va., U.S.	65	36°50'N	76°19'W
Port Sulphur, La., U.S. (sŭl´fēr)	82	29°28'N	89°41'W
Port Susan, b., Wa., U.S. (sū-zán´)	74a	48°11'N	122°25'W
Port Townsend, Wa., U.S. (tounz´ĕnd)	74a	48°07'N	122°46'W
Port Townsend, b., Wa., U.S.	74a	48°05'N	122°47'W
Portugal, nation, Eur. (pōr´tu-găl)	110	38°15'N	8°08'W
Portugalete, Spain (pōr-to͞o-gä-lā´tā)	128	43°18'N	3°05'W
Portuguese West Africa see Angola, nation, Ang.	186	14°15'S	16°00'E
Port Vendres, Fr.	126	42°32'N	3°07'E
Port Vila, Vanuatu	175	17°44'S	168°19'E
Port Wakefield, Austl. (wāk´fēld)	174	34°12'S	138°10'E
Port Washington, N.Y., U.S. (wôsh´ĭng-tŭn)	68a	40°49'N	73°42'W
Port Washington, Wi., U.S.	71	43°24'N	87°52'W
Posadas, Arg. (pō-sä´dhäs)	102	27°32'S	55°56'W
Posadas, Spain (pô-sä-däs)	128	37°48'N	5°09'W
Poshekhon'ye Volodarsk, Russia (pô-shyĕ´kōn-yĕ vôl´ô-dȧrsk)	132	58°31'N	39°07'E
Poso, Danau, l., Indon. (pō´sō)	168	2°00'S	119°40'E
Pospelokova, Russia (pŏs-pyĕl´kô-vȧ)	142a	59°25'N	60°50'E
Possession Sound, strt., Wa., U.S. (pŏ-zĕsh-ŭn)	74a	47°59'N	122°17'W
Possum Kingdom Reservoir, res., Tx., U.S. (pŏs´ŭm kĭng´dŭm)	80	32°58'N	98°12'W
Post, Tx., U.S. (pōst)	78	33°12'N	101°21'W
Postojna, Slvn. (pōs-tōynȧ)	130	45°45'N	14°13'E
Pos'yet, Russia (pos-yĕt´)	166	42°27'N	130°47'E
Potawatomi Indian Reservation, I.R., Ks., U.S. (pōt-ȧ-wā´tō mē)	79	39°30'N	96°11'W
Potchefstroom, S. Afr. (pŏch´ĕf-strōm)	186	26°42'S	27°06'E
Poteau, Ok., U.S. (pō-tō´)	79	35°03'N	94°37'W
Poteet, Tx., U.S. (pô-tēt)	80	29°05'N	98°35'W
Potenza, Italy (pô-tĕnt´sä)	119	40°39'N	15°49'E
Potenza, r., Italy	130	43°09'N	13°00'E
Potgietersrus, S. Afr. (pōt-kē´tĕrs-rûs)	186	24°09'S	29°04'E
Potholes Reservoir, res., Wa., U.S.	72	47°00'N	119°20'W
Poti, Geor. (pô´tē)	137	42°10'N	41°40'E
Potiskum, Nig.	184	11°43'N	11°05'E
Potomac, Md., U.S. (pō-tō´măk)	68e	39°01'N	77°13'W
Potomac, r., U.S. (pō-tō´măk)	65	38°15'N	76°55'W
Potosí, Bol.	100	19°35'S	65°45'W
Potosi, Mo., U.S. (pŏ-tō´sĭ)	79	37°56'N	90°46'W
Potosí, r., Mex. (pō-tō-sē´)	80	25°04'N	99°36'W
Potrerillos, Hond. (pō-trā-rēl´yôs)	90	15°13'N	87°58'W
Potsdam, Ger. (pôts´däm)	117	52°24'N	13°04'E
Potsdam, N.Y., U.S. (pŏts´dăm)	67	44°40'N	75°00'W
Pottenstein, Aus.	115e	47°58'N	16°06'E
Potters Bar, Eng., U.K. (pŏt´ĕz bär)	114b	51°41'N	0°12'W
Pottstown, Pa., U.S. (pŏts´toun)	67	40°15'N	75°40'W
Pottsville, Pa., U.S. (pôts´vĭl)	67	40°40'N	76°15'W
Poughkeepsie, N.Y., U.S. (pŏ-kĭp´sē)	65	41°45'N	73°55'W
Poulsbo, Wa., U.S. (pōlz´bō)	74a	47°44'N	122°38'W
Poulton-le-Fylde, Eng., U.K. (pōl´tŭn-lē-fīld´)	114a	53°52'N	2°59'W
Pouso Alegre, Braz. (pō´zò ä-lā´grĕ)	101	22°13'S	45°56'W
Póvoa de Varzim, Port. (pō-vō´ȧ dā vär´zēɴ)	118	41°23'N	8°44'W
Powder, r., Or., U.S.	72	44°55'N	117°35'W
Powder, r., U.S. (pou´dēr)	64	45°18'N	105°37'W
Powder, South Fork, r., Wy., U.S.	73	43°13'N	106°54'W
Powder River, Wy., U.S.	73	43°06'N	106°55'W
Powell, Wy., U.S. (pou´ĕl)	73	44°44'N	108°44'W
Powell, Lake, res., U.S.	64	37°26'N	110°25'W
Powell Lake, l., Can.	54	50°10'N	124°13'W
Powell Point, c., Bah.	92	24°50'N	76°20'W
Powell Reservoir, res., Ky., U.S.	82	36°30'N	83°35'W
Powell River, Can.	50	49°52'N	124°33'W
Poyang Hu, l., China	161	29°20'N	116°28'E
Poygan, r., Wi., U.S. (poi´găn)	71	44°10'N	89°05'W

ăt; finăl; rāte; senåte; ärm; åsk; sofȧ; fâre; ch-choose; dh-as th in other; bē; ēvent; bĕt; recĕnt; cratēr; g-gō; gh-guttural g; bĭt; ĭ-short neutral; rīde; κ-guttural k as ch in German ich;

PRONOUNCING INDEX Požarevac – Punta Gorda 291

PLACE (Pronunciation)	PAGE	LAT.	LONG.
Požarevac, Serb. (pô′zhȧ′rĕ-vȧts)	131	44°38′N	21°12′E
Poza Rica, Mex. (pô-zô-rē′kä)	89	20°32′N	97°25′W
Poznań, Pol.	110	52°25′N	16°55′E
Pozoblanco, Spain (pō-thō-bläŋ′kō)	128	38°23′N	4°50′W
Pozos, Mex. (pō′zōs)	88	22°05′N	100°50′W
Pozuelo de Alarcón, Spain (pō-thwā′lō dā ä-lär-kōn′)	129a	40°27′N	3°49′W
Pozzuoli, Italy (pôt-swô′lē)	130	40°34′N	14°08′E
Pra, r., Ghana (prà)	188	5°45′N	1°35′W
Pra, r., Russia	132	55°00′N	40°13′E
Prachin Buri, Thai. (prä′chēn)	168	13°59′N	101°15′E
Pradera, Col. (prä-dĕ′rä)	100a	3°24′N	76°13′W
Prades, Fr. (prȧd)	126	42°37′N	2°23′E
Prado, Col. (prä′dō)	100a	3°44′N	74°55′W
Prado Reservoir, res., Ca., U.S. (prä′dō)	75a	33°45′N	117°40′W
Prados, Braz. (prä′dōs)	99a	21°05′S	44°04′W
Prague, Czech Rep.	124	50°05′N	14°26′E
Praha *see* Prague, Czech Rep.	110	50°05′N	14°26′E
Praia, C.V. (prä′yȧ)	184b	15°00′N	23°30′W
Praia Funda, Ponta da, c., Braz. (pôn′tä-dȧ-prä′yȧ-fōō′n-dä)	102b	23°04′S	43°34′W
Prairie du Chien, Wi., U.S. (prā′rĭ dò shēn′)	71	43°02′N	91°10′W
Prairie Grove, Can. (prā′rĭ grōv)	62f	49°48′N	96°57′W
Prairie Island Indian Reservation, I.R., Mn., U.S.	71	44°42′N	92°32′W
Prairies, Rivière des, r., Can. (rē-vyâr′ dā prâ-rē′)	62a	45°40′N	73°34′W
Pratas Island, i., Asia	165	20°40′N	116°30′E
Prato, Italy (prä′tō)	130	43°53′N	11°03′E
Pratt, Ks., U.S. (prăt)	78	37°37′N	98°43′W
Prattville, Al., U.S. (prăt′vĭl)	82	32°28′N	86°27′W
Pravdinsk, Russia	123	54°26′N	21°00′E
Pravdinskiy, Russia (prȧv-dĕn′skĭ)	142b	56°03′N	37°52′E
Pravia, Spain (prä′vē-ä)	128	43°30′N	6°08′W
Pregolya, r., Russia (prĕ-gô′lä)	123	54°37′N	20°50′E
Premont, Tx., U.S. (prē-mŏnt′)	80	27°20′N	98°07′W
Prenzlau, Ger. (prĕnts′lou)	124	53°19′N	13°52′E
Přerov, Czech Rep. (przhĕ′rôf)	117	49°28′N	17°28′E
Prescot, Eng., U.K. (prĕs′kŭt)	114a	53°25′N	2°48′W
Prescott, Can. (prĕs′kŭt)	67	44°45′N	75°35′W
Prescott, Ar., U.S.	79	33°47′N	93°23′W
Prescott, Az., U.S. (prĕs′kŏt)	64	34°30′N	112°30′W
Prescott, Wi., U.S. (prĕs′kŏt)	75g	44°45′N	92°48′W
Presho, S.D., U.S. (prĕsh′ō)	70	43°56′N	100°04′W
Presidencia Rogue Sáenz Peña, Arg.	102	26°52′S	60°15′W
Presidente Epitácio, Braz. (prā-sē-dĕn′tĕ â-pē-tä′syō)	101	21°56′S	52°01′W
Presidio, Tx., U.S. (prē-sĭ′dĭ-ō)	80	29°33′N	104°23′W
Presidio, Río del, r., Mex. (rē′ō-dĕl-prē-sē′dyô)	88	23°54′N	105°44′W
Prešov, Slvk. (prĕ′shôf)	117	49°00′N	21°18′E
Prespa, Lake, l., Eur. (prĕs′pä)	131	40°49′N	20°50′E
Prespuntal, r., Ven.	101b	9°55′N	64°32′W
Presque Isle, Me., U.S. (prĕsk′ēl′)	60	46°41′N	68°03′W
Pressbaum, Aus.	115e	48°12′N	16°06′E
Prestea, Ghana	188	5°27′N	2°08′W
Preston, Austl.	173a	37°45′S	145°01′E
Preston, Eng., U.K. (prĕs′tŭn)	120	53°46′N	2°42′W
Preston, Id., U.S. (pres′tŭn)	73	42°05′N	111°54′W
Preston, Mn., U.S. (prĕs′tŭn)	71	43°42′N	92°06′W
Preston, Wa., U.S.	74a	47°31′N	121°56′W
Prestonburg, Ky., U.S. (prĕs′tŭn-bûrg)	66	37°35′N	82°50′W
Prestwich, Eng., U.K. (prĕst′wĭch)	114a	53°32′N	2°17′W
Pretoria, S. Afr. (prē-tō′rĭ-ȧ)	186	25°43′S	28°16′E
Pretoria North, S. Afr. (prĕ-tô′rĭ-ȧ nōōrd)	192c	25°41′S	28°11′E
Préveza, Grc. (prĕ′vȧ-zä)	131	38°58′N	20°44′E
Pribilof Islands, is., Ak., U.S. (prĭ′bĭ-lof)	63	57°00′N	169°20′W
Priboj, Serb. (prē′boi)	131	43°33′N	19°33′E
Price, Ut., U.S. (prīs)	77	39°35′N	110°50′W
Price, r., Ut., U.S.	77	39°21′N	110°35′W
Prichard, Al., U.S. (prĭt′chârd)	82	30°44′N	88°04′W
Priddis, Can. (prĭd′dĭs)	62e	50°53′N	114°20′W
Priddis Creek, r., Can.	62e	50°56′N	114°32′W
Priego, Spain (prē-ā′gō)	128	37°27′N	4°13′W
Prienai, Lith. (prē-ĕn′ī)	123	54°38′N	23°56′E
Prieska, S. Afr. (prē-ĕs′kȧ)	186	29°40′S	22°50′E
Priest Lake, l., Id., U.S. (prēst)	72	48°30′N	116°43′W
Priest Rapids Dam, Wa., U.S.	72	46°39′N	119°55′W
Priest Rapids Lake, res., Wa., U.S.	72	46°42′N	119°58′W
Priiskovaya, Russia (prĭ-ēs′kô-va-yȧ)	142a	60°50′N	58°55′E
Prijedor, Bos. (prē′yĕ-dôr)	130	44°58′N	16°43′E
Prijepolje, Serb. (prē′yĕ-pô′lyĕ)	131	43°22′N	19°41′E
Prilep, Mac. (prē′lĕp)	119	41°20′N	21°35′E
Primorsk, Russia (prē-môrsk′)	123	60°24′N	28°35′E
Primorsko-Akhtarskaya, Russia (prē-môr′skô äk-tär′skĭ-ê)	137	46°03′N	38°09′E
Primrose, S. Afr.	187b	26°11′S	28°11′E
Primrose Lake, l., Can.	56	54°55′N	109°45′W
Prince Albert, Can. (prĭns ăl′bērt)	50	53°12′N	105°46′W
Prince Albert National Park, rec., Can.	52	54°10′N	105°25′W
Prince Albert Sound, strt., Can.	52	70°23′N	116°57′W
Prince Charles Island, i., Can. (chärlz)	53	67°41′N	74°10′W
Prince Edward Island, prov., Can.	51	46°45′N	63°10′W
Prince Edward Islands, is., S. Afr.	178	46°36′S	37°57′E
Prince Edward National Park, rec., Can. (ĕd′wērd)	53	46°33′N	63°35′W
Prince Edward Peninsula, pen., Can.	67	44°00′N	77°15′W
Prince Frederick, Md., U.S. (prĭnce frĕdĕrĭk)	68e	38°33′N	76°35′W
Prince George, Can. (jôrj)	50	53°51′N	122°57′W
Prince of Wales, i., Austl.	175	10°47′S	142°15′E
Prince of Wales, i., Ak., U.S.	63	55°47′N	132°50′W

PLACE (Pronunciation)	PAGE	LAT.	LONG.
Prince of Wales, Cape, c., Ak., U.S. (wālz)	63	65°48′N	169°08′W
Prince Rupert, Can. (roo′pērt)	50	54°19′N	130°19′W
Princes Risborough, Eng., U.K. (prĭns′ĕz rĭz′brŭ)	114b	51°41′N	0°51′W
Princess Charlotte Bay, b., Austl. (shär′lŏt)	175	13°45′S	144°15′E
Princess Royal Channel, strt., Can. (roi′ăl)	54	53°10′N	128°37′W
Princess Royal Island, i., Can.	54	52°57′N	128°49′W
Princeton, Can. (prĭns′tŭn)	50	49°27′N	120°31′W
Princeton, Il., U.S.	66	41°20′N	89°25′W
Princeton, In., U.S.	66	38°20′N	87°35′W
Princeton, Ky., U.S.	82	37°07′N	87°52′W
Princeton, Mi., U.S.	71	46°16′N	87°33′W
Princeton, Mn., U.S.	71	45°34′N	93°36′W
Princeton, Mo., U.S.	79	40°23′N	93°34′W
Princeton, N.J., U.S.	67	40°21′N	74°40′W
Princeton, Wi., U.S.	71	43°50′N	89°09′W
Princeton, W.V., U.S.	83	37°21′N	81°05′W
Prince William Sound, strt., Ak., U.S. (wĭl′yăm)	63	60°40′N	147°10′W
Príncipe, i., S. Tom./P. (prēn′sē-pĕ)	184	1°37′N	7°25′E
Principe Channel, strt., Can. (prĭn′sĭ-pē)	54	53°28′N	129°45′W
Prineville, Or., U.S. (prīn′vĭl)	72	44°17′N	120°48′W
Prineville Reservoir, res., Or., U.S.	72	44°07′N	120°45′W
Prinzapolca, Nic. (prēn-zä-pōl′kä)	91	13°18′N	83°35′W
Prinzapolca, r., Nic.	91	13°23′N	84°23′W
Prior Lake, Mn., U.S. (prī′ēr)	75g	44°43′N	93°26′W
Priozërsk, Russia (prĭ-ô′zĕrsk)	123	61°03′N	30°08′E
Pripet, r., Eur.	137	51°50′N	29°45′E
Pripet Marshes, sw., Eur.	137	52°10′N	27°30′E
Priština, Serb. (prēsh′tĭ-nä)	119	42°39′N	21°12′E
Pritzwalk, Ger. (prĕts′välk)	124	53°09′N	12°12′E
Privas, Fr. (prē-väs′)	126	44°44′N	4°37′E
Prizren, Serb. (prē′zrĕn)	119	42°11′N	20°45′E
Procida, Italy (prô′chē-dä)	129c	40°31′N	14°02′E
Procida, Isola di, i., Italy	129c	40°32′N	13°57′E
Proctor, Mn., U.S. (prŏk′tēr)	75h	46°45′N	92°14′W
Proctor, Vt., U.S.	67	43°40′N	73°00′W
Proebstel, Wa., U.S. (prōb′stĕl)	74c	45°40′N	122°29′W
Proenca-a-Nova, Port. (prō-ān′sȧ-ä-nō′vȧ)	128	39°44′N	7°55′W
Progreso, Hond. (prō-grĕ′sō)	90	15°28′N	87°49′W
Progreso, Mex. (prō-grā′sō)	86	21°14′N	89°39′W
Progreso, Mex.	80	27°29′N	101°05′W
Prokhladnyy, Russia	138	43°46′N	44°00′E
Prokop′yevsk, Russia	140	53°53′N	86°45′E
Prokuplje, Serb. (prô′kōp′l-yĕ)	131	43°16′N	21°40′E
Prome, Mya.	168	18°46′N	95°15′E
Pronya, r., Bela. (prô′nyä)	132	54°08′N	30°58′E
Pronya, r., Russia	132	54°08′N	39°30′E
Prospect, Ky., U.S. (prŏs′pĕkt)	69h	38°21′N	85°36′W
Prospect Park, Pa., U.S. (prŏs′pĕkt pärk)	68f	39°53′N	75°18′W
Prosser, Wa., U.S. (prŏs′ēr)	72	46°10′N	119°46′W
Prostějov, Czech Rep. (prôs′tyĕ-yôf)	125	49°28′N	17°08′E
Protection, i., Wa., U.S. (prō-tĕk′shŭn)	74a	48°07′N	122°56′W
Protoka, r., Russia (prôt′ō-kȧ)	132	55°00′N	36°42′E
Provadiya, Blg. (prō-väd′ē-yȧ)	131	43°13′N	27°28′E
Providence, Ky., U.S. (prŏv′ĭ-dĕns)	66	37°25′N	87°45′W
Providence, R.I., U.S.	65	41°50′N	71°23′W
Providence, Ut., U.S.	73	41°42′N	111°50′W
Providencia, Isla de, i., Col.	91	13°21′N	80°55′W
Providenciales, i., T./C. Is.	93	21°50′N	72°15′W
Provideniya, Russia (prô-vĭ-dä′nĭ-yȧ)	63	64°30′N	172°54′W
Provincetown, Ma., U.S.	67	42°03′N	70°11′W
Provo, Ut., U.S. (prō′vō)	64	40°15′N	111°40′W
Prozor, Bos. (prō′zôr)	131	43°48′N	17°59′E
Prudence Island, i., R.I., U.S. (prōō′dĕns)	68b	41°38′N	71°20′W
Prudhoe Bay, b., Ak., U.S.	63	70°40′N	147°25′W
Prudnik, Pol. (prôd′nĭk)	125	50°19′N	17°34′E
Prussia, hist. reg., Eur. (prŭsh′ȧ)	124	50°43′N	8°35′E
Pruszków, Pol. (prôsh′kôf)	125	52°09′N	20°50′E
Prut, r., Eur. (prōōt)	112	48°05′N	27°07′E
Pryluky, Ukr.	137	50°36′N	32°21′E
Prymors′k, Ukr.	133	46°43′N	36°21′E
Pryor, Ok., U.S. (prī′ēr)	79	36°16′N	95°19′W
Pryvil′ne, Ukr.	133	47°30′N	32°21′E
Przedbórz, Pol.	125	51°05′N	19°53′E
Przemyśl, Pol. (pzhĕ′mĭsh′l)	110	49°47′N	22°45′E
Przheval′sk, Kyrg. (p′r-zhĭ-välsk′)	139	42°29′N	78°24′E
Psel, r., Eur.	137	49°45′N	33°42′E
Pskov, Russia (pskôf)	134	57°48′N	28°19′E
Pskov, prov., Russia	132	57°33′N	29°05′E
Pskovskoye Ozero, l., Eur. (p′skôv′skô′yĕ ôzĕ-rō)	136	58°05′N	28°15′E
Ptich′, r., Bela. (p′tĕch)	136	53°17′N	28°16′E
Ptuj, Slvn. (ptōō′ē)	130	46°24′N	15°54′E
Pucheng, China (pōō-chĕng′)	165	28°02′N	118°25′E
Pucheng, China (pōō-chŭŋ)	162	35°43′N	115°22′E
Puck, Pol. (pòtsk)	125	54°43′N	18°23′E
Pudozh, Russia (pōō′dôzh)	136	61°50′N	36°50′E
Puebla, Mex. (pwā′blä)	86	19°02′N	98°11′W
Puebla, state, Mex.	89	19°00′N	97°45′W
Puebla de Don Fadrique, Spain	128	37°55′N	2°55′W
Pueblo, Co., U.S. (pwā′blō)	64	38°15′N	104°36′W
Pueblo Nuevo, Mex. (nwā′vō)	88	23°23′N	105°21′W
Pueblo Viejo, Mex. (vyā′hō)	89	17°23′N	93°46′W
Puente Alto, Chile (pwĕ′n-tĕ äl′tō)	99b	33°36′S	70°34′W
Puentedeume, Spain (pwĕn-tā-dhā-ōō′mā)	128	43°28′N	8°09′W
Puente-Genil, Spain (pwĕn′tā-hā-nēl′)	128	37°25′N	4°18′W
Puerco, Rio, r., N.M., U.S. (pwĕr′kō)	77	35°15′N	107°05′W

PLACE (Pronunciation)	PAGE	LAT.	LONG.
Puerto Aisén, Chile (pwĕ′r-tō ä′y-sĕ′n)	102	45°28′S	72°44′W
Puerto Angel, Mex. (pwĕ′r-tō äŋ′hâl)	89	15°42′N	96°32′W
Puerto Armuelles, Pan. (pwĕ′r-tô är-mōō-ā′lyās)	91	8°18′N	82°52′W
Puerto Barrios, Guat. (pwĕ′r-tō bär′rē-ôs)	86	15°43′N	88°36′W
Puerto Bermúdez, Peru (pwĕ′r-tō bĕr-mōō′däz)	100	10°17′S	74°57′W
Puerto Berrío, Col. (pwĕ′r-tō bĕr-rē′ō)	100	6°29′N	74°27′W
Puerto Cabello, Ven. (pwĕ′r-tō kä-bĕl′yō)	100	10°28′N	68°01′W
Puerto Cabezas, Nic. (pwĕ′r-tō kä-bā′zäs)	91	14°01′N	83°26′W
Puerto Casado, Para. (pwĕ′r-tō kä-sä′dō)	102	22°16′S	57°57′W
Puerto Castilla, Hond. (pwĕ′r-tō käs-tēl′yō)	90	16°01′N	86°01′W
Puerto Chicama, Peru (pwĕ′r-tō chē-kä′mä)	100	7°46′S	79°18′W
Puerto Colombia, Col. (pwĕr′tô kô-lôm′bē-ä)	100	11°08′N	75°09′W
Puerto Cortés, C.R. (pwĕ′r-tō kôr-tās′)	91	9°00′N	83°37′W
Puerto Cortés, Hond. (pwĕ′r-tô kôr-tās′)	86	15°48′N	87°57′W
Puerto Cumarebo, Ven. (pwĕ′r-tō kōō-mä-rĕ′bô)	100	11°25′N	69°17′W
Puerto de Luna, N.M., U.S. (pwĕr′tô dâ lōō′nä)	78	34°49′N	104°36′W
Puerto de Nutrias, Ven. (pwĕ′r-tō dĕ nōō-trē-äs′)	100	8°02′N	69°19′W
Puerto Deseado, Arg. (pwĕ′r-tō dā-sā-ä′dhō)	102	47°38′S	66°00′W
Puerto de Somport, p., Eur.	129	42°51′N	0°25′W
Puerto Eten, Peru (pwĕ′r-tō ĕ-tĕ′n)	100	6°59′S	79°51′W
Puerto Jiménez, C.R. (pwĕ′r-tô κē-mĕ′nĕz)	91	8°35′N	83°23′W
Puerto La Cruz, Ven. (pwĕ′r-tō lä krōō′z)	100	10°14′N	64°38′W
Puertollano, Spain (pwĕ-tôl-yä′nō)	118	38°41′N	4°05′W
Puerto Madryn, Arg. (pwĕ′r-tō mä-drĕn′)	102	42°45′S	65°01′W
Puerto Maldonado, Peru (pwĕ′r-tō mäl-dō-nä′dô)	100	12°43′S	69°01′W
Puerto Miniso, Mex. (pwĕ′r-tō mē-nē′sô)	88	16°06′N	98°02′W
Puerto Montt, Chile (pwĕ′r-tō mô′nt)	102	41°29′S	73°00′W
Puerto Natales, Chile (pwĕ′r-tō nä-tä′lēs)	102	51°48′S	72°01′W
Puerto Niño, Col. (pwĕ′r-tō nē′n-yô)	100a	5°57′N	74°36′W
Puerto Padre, Cuba (pwĕ′r-tō pä′drä)	92	21°10′N	76°40′W
Puerto Peñasco, Mex. (pwĕ′r-tō pĕn-yä′s-kô)	86	31°39′N	113°15′W
Puerto Pinasco, Para. (pwĕ′r-tō pē-nä′s-kô)	102	22°31′S	57°50′W
Puerto Píritu, Ven. (pwĕ′r-tō pē′rē-tōō)	101b	10°05′N	65°04′W
Puerto Plata, Dom. Rep. (pwĕ′r-tō plä′tä)	87	19°50′N	70°40′W
Puerto Princesa, Phil. (pwĕr-tô prēn-sā′sä)	168	9°45′N	118°41′E
Puerto Rico, dep., N.A. (pwĕr′tô rē′kō)	87	18°16′N	66°50′W
Puerto Rico Trench, deep,	87	19°45′N	66°30′W
Puerto Salgar, Col. (pwĕ′r-tō säl-gär′)	100a	5°30′N	74°39′W
Puerto Santa Cruz, Arg. (pwĕ′r-tō sän′tä krōōz′)	102	50°04′S	68°32′W
Puerto Suárez, Bol. (pwĕ′r-tō swä′râz)	101	18°55′S	57°39′W
Puerto Tejada, Col. (pwĕ′r-tō tĕ-κä′dä)	100	3°13′N	76°23′W
Puerto Vallarta, Mex. (pwĕ′r-tō väl-yär′tä)	88	20°36′N	105°13′W
Puerto Varas, Chile (pwĕ′r-tō vä′räs)	102	41°16′S	73°03′W
Puerto Wilches, Col. (pwĕ′r-tō vēl′c-hĕs)	100	7°19′N	73°54′W
Pugachëv, Russia (pōō′gȧ-chyôf)	137	52°00′N	48°40′E
Puget, Wa., U.S. (pū′jĕt)	74c	46°10′N	123°23′W
Puget Sound, strt., Wa., U.S.	72	47°49′N	122°26′W
Puglia (Apulia), hist. reg., Italy (pōō′lyä) (ä-pōō′lyä)	130	41°13′N	16°10′E
Pukaskwa National Park, rec., Can.	53	48°22′N	85°55′W
Pukeashun Mountain, mtn., Can.	55	51°12′N	119°14′W
Pukin, r., Malay.	153b	2°53′N	102°54′E
Pula, Cro. (pōō′lä)	118	44°52′N	13°55′E
Pulacayo, Bol. (pōō-lä-kä′yō)	100	20°12′N	66°33′W
Pulaski, Tn., U.S. (pů-lăs′kĭ)	82	35°11′N	87°03′W
Pulaski, Va., U.S.	83	37°00′N	81°45′W
Puławy, Pol. (pò-wä′vĕ)	125	51°24′N	21°59′E
Pulicat, r., India	159	13°58′N	79°52′E
Pullman, Wa., U.S. (pòl′măn)	72	46°44′N	117°10′W
Pulog, Mount, mtn., Phil. (pōō′lôg)	169a	16°38′N	120°53′E
Puma Yumco, l., China (pōō-mä yōōm-tswo)	158	28°30′N	90°10′E
Pumpkin Creek, r., Mt., U.S. (pŭmp′kĭn)	73	45°47′N	105°35′W
Punakha, Bhu. (pōō-nŭk′ŭ)	155	27°45′N	89°59′E
Punata, Bol. (pōō-nä′tä)	100	17°43′S	65°43′W
Pune, India	155	18°38′N	73°53′E
Punjab, state, India (pŭn′jäb′)	155	31°00′N	75°30′E
Puno, Peru (pōō′nō)	100	15°58′S	70°02′W
Punta Arenas, Chile (pōō′n-tä-rĕ′näs)	102	53°09′S	70°48′W
Punta de Piedras, Ven. (pōō′n-tä dĕ pyĕ′dräs)	101b	10°54′N	64°06′W
Punta Gorda, Belize (pòn′tä gôr′dä)	90	16°07′N	88°50′W
Punta Gorda, Fl., U.S. (pŭn′tȧ gôr′dȧ)	83a	26°55′N	82°02′W
Punta Gorda, Río, r., Nic. (pōō′n-tä gô′r-dä)	91	11°34′N	84°13′W

ng-sing; ŋ-baŋk; ɴ-nasalized n; nŏd; cŏmmit; ōld; ôbey; ôrder; oi-boil; fōōd; ò-as oo in foot; ou-out; s-soft; sh-dish; th-thin; pūre; ûnite; ûrn; stŭd; circŭs; ü-as in French tu; ′-indeterminate vowel.

292 Punta Indio – Raipur PRONOUNCING INDEX

PLACE (Pronunciation)	PAGE	LAT.	LONG.
Punta Indio, Canal, strt., Arg. (po͞o′n-tä- ē′n-dyô)	99c	34°56′S	57°20′W
Puntarenas, C.R. (pōnt-ä-rā′näs)	87	9°59′N	84°49′W
Punto Fijo, Ven. (po͞o′n-tô fē′kô)	100	11°48′N	70°14′W
Punxsutawney, Pa., U.S. (pŭnk-sŭ-tô′nē)	67	40°55′N	79°00′W
Puquio, Peru (po͞o′kyō)	100	14°43′S	74°02′W
Pur, r., Russia	140	65°30′N	77°30′E
Purcell, Ok., U.S. (pûr-sĕl′)	79	35°01′N	97°22′W
Purcell Mountains, mts., N.A. (pûr-sĕl′)	55	50°00′N	116°30′W
Purdy, Wa., U.S. (pûr′dē)	74a	47°23′N	122°37′W
Purépero, Mex. (po͞o-rā′pâ-rō)	88	19°56′N	102°02′W
Purgatoire, r., Co., U.S. (pûr-gá-twär′)	78	37°25′N	103°53′W
Puri, India (po′rē)	155	19°52′N	85°51′E
Purial, Sierra de, mts., Cuba (sē-ĕ′r-rä-dĕ-po͞o-rē-äl′)	93	20°15′N	74°40′W
Purificación, Col. (po͞o-rē-fē-kä-syōn′)	100	3°52′N	74°54′W
Purificación, Mex. (po͞o-rē-fē-kä-syō′n)	88	19°44′N	104°38′W
Purificación, r., Mex.	88	19°30′N	104°54′W
Purkersdorf, Aus.	115e	48°13′N	16°11′E
Puruandiro, Mex. (pō-ro͞o-än′dē-rō)	88	20°04′N	101°33′W
Purús, r., S.A. (po͞o-ro͞o′s)	100	6°45′S	64°34′W
Pusan, Kor., S.	161	35°08′N	129°05′E
Pushkin, Russia (pŏsh′kĭn)	136	59°43′N	30°25′E
Pushkino, Russia (po͞osh′kē-nô)	132	56°01′N	37°51′E
Pustoshka, Russia (pûs-tôsh′kȧ)	132	56°20′N	29°33′E
Pustunich, Mex. (po͞os-to͞o′nēch)	89	19°10′N	90°29′W
Putaendo, Chile (po͞o-tä-ĕn-dô)	99b	32°37′S	70°42′W
Puteaux, Fr. (pü-tō′)	127b	48°52′N	2°12′E
Putfontein, S. Afr. (pōt′fōn-tān)	187b	26°08′S	28°24′E
Putian, China (po͞o-tĭĕn)	165	25°40′N	119°02′E
Putla de Guerrero, Mex. (po͞o′tlä-dē-gĕr-rĕ′rō)	89	17°03′N	97°55′W
Putnam, Ct., U.S. (pŭt′năm)	67	41°55′N	71°55′W
Putorana, Gory, mts., Russia	135	68°45′N	93°15′E
Puttalam, Sri L.	159	8°02′N	79°44′E
Putumayo, r., S.A. (po-to͞o-mä′yō)	100	1°02′S	73°50′W
Putung, Tanjung, c., Indon.	168	3°35′S	111°50′E
Putyvl′, Ukr.	133	51°21′N	33°52′E
Puulavesi, l., Fin.	123	61°49′N	27°10′E
Puyallup, Wa., U.S. (pū-ăl′ŭp)	74a	47°12′N	122°18′W
Puyang, China (po͞o-yäŋ)	164	35°42′N	114°58′E
Pweto, D.R.C. (pwā′tō)	186	8°29′S	28°58′E
Pyasina, r., Russia (pyä-sē′nȧ)	140	72°45′N	87°37′E
Pyatigorsk, Russia (pyȧ-tē-gôrsk′)	137	44°00′N	43°00′E
Pyetrykaw, Bela.	132	52°09′N	28°30′E
Pyhäjärvi, l., Fin.	123	60°57′N	21°50′E
Pyinmana, Mya. (pyēn-mä′nŭ)	155	19°47′N	96°15′E
Pymatuning Reservoir, res., Pa., U.S. (pī-mȧ-tŭn′ĭng)	66	41°40′N	80°30′W
Pyŏnggang, Kor., N. (pyŭng′gäng′)	166	38°21′N	127°18′E
P′yŏngyang, Kor., N.	161	39°03′N	125°48′E
Pyramid, l., Nv., U.S. (pĭ′ră-mĭd)	76	40°02′N	119°50′W
Pyramid Lake Indian Reservation, I.R., Nv., U.S.	76	40°17′N	119°52′W
Pyramids, hist., Egypt	192b	29°53′N	31°10′E
Pyrenees, mts., Eur. (pĭr-e-nēz′)	112	43°00′N	0°05′E
Pýrgos, Grc.	119	37°51′N	21°28′E
Pyriatyn, Ukr.	137	50°13′N	32°31′E
Pyrzyce, Pol. (pĕzhĭ′tsĕ)	124	53°09′N	14°53′E

Q

PLACE (Pronunciation)	PAGE	LAT.	LONG.
Qal′at Bishah, Sau. Ar.	154	20°01′N	42°30′E
Qamdo, China (chyäm-dwō)	160	31°06′N	96°30′E
Qandala, Som.	157	11°28′N	49°52′E
Qaraghandy (Karaganda), Kaz.	139	49°42′N	73°18′E
Qaraözen, r.	137	49°50′N	49°35′E
Qarqan see Qiemo, China	160	38°02′N	85°16′E
Qarqan, r., China	160	38°55′N	87°15′E
Qarqaraly, Kaz.	139	49°18′N	75°28′E
Qārūn, Birket, l., Egypt	185	29°34′N	30°34′E
Qaşr al Burayqah, Libya	185	30°25′N	19°20′E
Qasr al-Farāfirah, Egypt	185	27°04′N	28°13′E
Qaşr Banī Walīd, Libya	185	31°45′N	14°04′E
Qasr el Boukhari, Alg.	118	35°50′N	2°48′E
Qatar, nation, Asia (kä′tȧr)	154	25°00′N	52°45′E
Qaţţārah, Munkhafaḑ al, depr., Egypt	185	30°07′N	27°30′E
Qausuittuq (Resolute), Can.	49	74°41′N	95°00′W
Qāyen, Iran	154	33°45′N	59°08′E
Qazvīn, Iran	154	36°10′N	49°59′E
Qeshm, Iran	154	26°51′N	56°10′E
Qeshm, i., Iran	154	26°52′N	56°15′E
Qezel Owzan, r., Iran	154	36°30′N	49°00′E
Qezi′ot, Isr.	153a	30°53′N	34°28′E
Qianwei, China (chyĕn-wā)	162	40°11′N	120°05′E
Qi′anzhen, China (chyē-än-jŭn)	162	32°16′N	120°59′E
Qibao, China (chyē-bou)	163b	31°06′N	121°16′E
Qiblīyah, Jabal al Jalālat al, mts., Egypt	153a	28°49′N	32°21′E
Qijiang, China (chyē-jyäŋ)	165	29°05′N	106°40′E
Qikou, China (chyē-kō)	162	38°37′N	117°33′E
Qilian Shan, mts., China (chyē-lĭĕn shän)	160	38°43′N	98°00′E
Qiliping, China (chyē-lē-pĭŋ)	162	31°28′N	114°41′E
Qindao, China (chyĭn-dou)	161	36°05′N	120°10′E
Qing′an, China (chyĭŋ-än)	164	46°50′N	127°30′E
Qingcheng, China (chyĭŋ-chŭŋ)	162	37°12′N	117°43′E
Qingfeng, China (chyĭŋ-fŭŋ)	162	35°52′N	115°05′E
Qinghai, prov., China (chyĭŋ-hī)	160	36°14′N	95°30′E
Qinghai Hu see Koko Nor, l., China	160	37°26′N	98°30′E
Qinghe, China (chyĭŋ-hŭ)	164a	40°08′N	116°16′E
Qingjiang, China (chyĭŋ-jyäŋ)	165	28°00′N	115°30′E
Qingjiang, China	162	33°34′N	118°58′E
Qingliu, China (chyĭŋ-lĭô)	165	26°15′N	116°50′E
Qingningsi, China (chyĭŋ-nĭŋ-sz)	163b	31°16′N	121°33′E
Qingping, China (chyĭŋ-pĭŋ)	162	36°46′N	116°03′E
Qingpu, China (chyĭŋ-po͞o)	165	31°08′N	121°06′E
Qingxian, China (chyĭŋ shyĕn)	162	38°37′N	116°48′E
Qingyang, China (chyĭŋ-yäŋ)	160	36°02′N	107°42′E
Qingyuan, China (chyĭŋ-yôän)	165	23°43′N	113°10′E
Qingyuan, China	164	42°05′N	125°00′E
Qingyun, China (chyĭŋ-yôn)	162	37°52′N	117°26′E
Qingyundian, China (chĭŋ-yôn-dĭĕn)	164a	39°41′N	116°31′E
Qinhuangdao, China (chyĭn-huaŋ-dou)	161	39°57′N	119°34′E
Qin Ling, mts., China (chyĭn lĭŋ)	160	33°25′N	108°58′E
Qinyang, China (chyĭn-yäŋ)	164	35°00′N	112°55′E
Qinzhou, China (chyĭn-jō)	165	22°00′N	108°35′E
Qionghai, China (chyôŋ-hī)	165	19°10′N	110°28′E
Qiqian, China (chyē-chyĕn)	161	52°23′N	121°04′E
Qiqihar, China	161	47°18′N	124°00′E
Qiryat Gat, Isr.	153a	31°38′N	34°36′E
Qiryat Shemona, Isr.	153a	33°12′N	35°34′E
Qitai, China (chyē-tī)	160	44°07′N	89°04′E
Qiuxian, China (chyô shyĕn)	162	36°43′N	115°13′E
Qixian, China (chyē-shyĕn)	162	34°33′N	114°47′E
Qixian, China	164	35°36′N	114°13′E
Qiyang, China (chyē-yäŋ)	165	26°40′N	112°00′E
Qobda, r., Kaz. (kä-rȧ kôb′dȧ)	137	50°40′N	55°00′E
Qogir Feng see K2, mtn., Asia	155	36°06′N	76°38′E
Qom, Iran	154	34°28′N	50°53′E
Qongyrat, Kaz.	139	47°25′N	75°10′E
Qostanay, Kaz.	139	53°10′N	63°39′E
Quabbin Reservoir, res., Ma., U.S. (kwā′bĭn)	67	42°20′N	72°10′W
Quachita, Lake, l., Ar., U.S. (kwä shĭ′tô)	79	34°47′N	93°37′W
Quadra Island, i., Can.	54	50°08′N	125°16′W
Quakertown, Pa., U.S. (kwā′kēr-toun)	67	40°30′N	75°20′W
Quanah, Tx., U.S. (kwä′nȧ)	78	34°19′N	99°43′W
Quang Ngai, Viet. (kwäng n′gä′ē)	168	15°05′N	108°58′E
Quang Ngai, mtn., Viet.	165	15°10′N	108°20′E
Quanjiao, China (chyuän-jyou)	162	32°06′N	118°17′E
Quanzhou, China (chyuän-jō)	161	24°58′N	118°40′E
Quanzhou, China	165	25°58′N	111°02′E
Qu′Appelle, r., Can.	52	50°30′N	104°00′W
Qu′Appelle Dam, dam, Can.	56	51°00′N	106°25′W
Quartu Sant′Elena, Italy (kwär-to͞o′ sänt ä′lâ-nä)	130	39°16′N	9°12′E
Quartzsite, Az., U.S.	77	33°40′N	114°13′W
Quatsino Sound, strt., Can. (kwŏt-sē′nō)	54	50°25′N	128°10′W
Quba, Azer. (ko͞o′bä)	137	41°05′N	48°30′E
Qūchān, Iran	157	37°06′N	58°30′E
Qudi, China	162	37°06′N	117°15′E
Québec, Can. (kwē-bĕk′) (kȧ-bĕk′)	62b	46°49′N	71°13′W
Québec, prov., Can.	51	51°07′N	70°25′W
Quedlinburg, Ger. (kvĕd′lēn-bo͞orgh)	124	51°45′N	11°10′E
Queen Bess, Can.	54	51°16′N	124°34′W
Queen Charlotte Islands, is., Can. (kwēn shär′lŏt)	52	53°30′N	132°25′W
Queen Charlotte Ranges, mts., Can.	54	53°00′N	132°00′W
Queen Charlotte Sound, strt., Can.	54	51°30′N	129°30′W
Queen Charlotte Strait, strt., Can. (strāt)	52	50°40′N	127°25′W
Queen Elizabeth Islands, is., Can. (ê-lĭz′ȧ-bĕth)	49	78°20′N	110°00′W
Queen Maud Gulf, b., Can. (mäd)	52	68°27′N	102°55′W
Queen Maud Land, reg., Ant.	178	75°00′S	10°00′E
Queen Maud Mountains, mts., Ant.	178	85°00′S	179°00′W
Queens Channel, strt., Austl. (kwēnz)	174	14°25′S	129°10′E
Queenscliff, Austl.	173a	38°16′S	144°39′E
Queensland, state, Austl. (kwēnz′lănd)	175	22°45′S	141°01′E
Queenstown, Austl. (kwēnz′toun)	176	42°00′S	145°40′E
Queenstown, S. Afr.	187c	31°54′S	26°53′E
Queimados, Braz. (kā-mȧ′dôs)	102b	22°42′S	43°34′W
Quela, Ang.	190	9°16′S	17°02′E
Quelimane, Moz. (kā-lē-mä′nĕ)	187	17°48′S	37°05′E
Queluz, Port.	129b	38°45′N	9°15′W
Quemado de Güines, Cuba (kā-mä′dhä-dĕ-gwē′nĕs)	92	22°45′N	80°20′W
Quemoy, Tai.	165	24°30′N	118°20′E
Quemoy, i., Tai.	165	24°27′N	118°23′E
Quepos, C.R. (kā′pôs)	91	9°26′N	84°10′W
Quepos, Punta, c., C.R. (po͞o′n-tä)	91	9°23′N	84°20′W
Querétaro, Mex. (kå-rā′tä-rō)	86	20°37′N	100°25′W
Querétaro, state, Mex.	88	21°00′N	100°00′W
Quesada, Spain (kā-sä′dhä)	128	37°51′N	3°04′W
Quesnel, Can. (kā-nĕl′)	50	52°59′N	122°30′W
Quesnel, r., Can.	55	52°15′N	122°00′W
Quesnel Lake, l., Can.	52	52°32′N	121°05′W
Quetame, Col. (kĕ-tä′mĕ)	100a	4°20′N	73°50′W
Quetta, Pak. (kwĕt′ä)	155	30°19′N	67°01′E
Quezaltenango, Guat. (kå-zäl′tå-näŋ′gō)	86	14°50′N	91°30′W
Quezaltepeque, El Sal. (kĕ-zäl′tĕ′pĕ-kĕ)	90	13°50′N	89°17′W
Quezaltepeque, Guat. (kå-zäl′tå-pā′kå)	90	14°39′N	89°26′W
Quezon City, Phil. (kā-zōn)	168	14°40′N	121°02′E
Qufu, China (chyo͞o-fo͞o)	162	35°37′N	116°54′E
Quibdo, Col. (kēb′dō)	100	5°42′N	76°41′W
Quiberon, Fr. (kē-bē-rôn′)	126	47°29′N	3°08′W
Quiçama, Parque Nacional de, rec., Ang.	190	10°00′S	13°25′E
Quicksborn, Ger. (kvēks′bōrn)	115c	53°44′N	9°54′E
Quilcene, Wa., U.S. (kwĭl-sēn′)	74a	47°50′N	122°53′W
Quilimari, Chile (kē-lē-mä′rē)	99b	32°06′S	71°28′W
Quillan, Fr. (kē-yäɴ′)	126	42°53′N	2°13′E
Quillota, Chile (kēl-yō′tä)	102	32°52′S	71°14′W
Quilmes, Arg. (kēl′mās)	99c	34°43′S	58°16′W
Quilon, India (kwē-lōn′)	159	8°58′N	76°16′E
Quilpie, Austl. (kwĭl′pē)	175	26°34′S	149°20′E
Quimbaya, Col. (kēm-bä′yä)	100a	4°38′N	75°46′W
Quimbele, Ang.	190	6°28′S	16°13′E
Quimbonge, Ang.	190	8°36′S	18°30′E
Quimper, Fr. (kăɴ-pĕr′)	117	47°59′N	4°04′W
Quinalt, r., Wa., U.S.	72	47°23′N	124°10′W
Quinault Indian Reservation, I.R., Wa., U.S.	72	47°27′N	124°34′W
Quincy, Fl., U.S. (kwĭn′sē)	82	30°35′N	84°35′W
Quincy, Il., U.S.	65	39°55′N	91°23′W
Quincy, Ma., U.S.	61a	42°15′N	71°00′W
Quincy, Mi., U.S.	66	42°00′N	84°50′W
Quincy, Or., U.S.	74c	46°08′N	123°10′W
Qui Nhon, Viet. (kwĭnyôn)	168	13°51′N	109°03′E
Quinn, r., Nv., U.S. (kwĭn)	72	41°42′N	117°45′W
Quintanar de la Orden, Spain (kēn-tä-när′)	128	39°36′N	3°02′W
Quintana Roo, state, Mex. (rô′ô)	86	19°30′N	88°30′W
Quintero, Chile (kēn-tĕ′rô)	99b	32°48′S	71°30′W
Quionga, Moz.	191	10°37′S	40°30′E
Quiroga, Mex. (kē-rô′gä)	88	19°39′N	101°30′W
Quiroga, Spain (kē-rô′gä)	128	42°28′N	7°18′W
Quitman, Ga., U.S. (kwĭt′măn)	82	30°46′N	83°35′W
Quitman, Ms., U.S.	82	33°02′N	88°43′W
Quito, Ec. (kē′tō)	100	0°17′S	78°32′W
Qumbu, S. Afr. (kòm′bo͞o)	187c	31°10′S	28°48′E
Quorn, Austl. (kwôrn)	176	32°20′S	138°00′E
Qurayyah, Wādī, r., Egypt	153a	30°08′N	34°27′E
Qusmuryn köli, l., Kaz.	139	52°30′N	64°15′E
Qutang, China (chyo͞o-täŋ)	162	32°33′N	120°07′E
Quthing, Leso.	187c	30°35′S	27°42′E
Quxian, China (chyo͞o-shyĕn)	161	28°58′N	118°58′E
Quxian, China	165	30°40′N	106°48′E
Quzhou, China (chyoō-jō)	162	36°47′N	114°58′E
Qyzylorda, Kaz.	139	44°58′N	65°45′E

R

PLACE (Pronunciation)	PAGE	LAT.	LONG.
Raab (Raba), r., Eur. (räp)	124	46°55′N	15°55′E
Raahe, Fin. (rä′ĕ)	116	64°39′N	24°22′E
Rab, i., Serb. (räb)	130	44°45′N	14°40′E
Raba, Indon.	168	8°32′S	118°49′E
Raba (Raab), r., Eur.	125	47°28′N	17°12′E
Rabat, Mor. (rȧ-bät′)	184	33°59′N	6°47′W
Rabaul, Pap. N. Gui. (rä′boul)	169	4°15′S	152°19′E
Rābigh, Sau. Ar.	157	22°48′N	39°01′E
Raccoon, r., Ia., U.S. (ră-ko͞on′)	71	42°07′N	94°45′W
Raccoon Cay, i., Bah.	93	22°25′N	75°50′W
Race, Cape, c., Can. (rās)	61	46°40′N	53°10′W
Rachado, Cape, c., Malay.	153b	2°26′N	101°29′E
Racibórz, Pol. (rä-chē′bo͞ozh)	125	50°06′N	18°14′E
Racine, Wi., U.S. (rȧ-sēn′)	65	42°43′N	87°49′W
Raco, Mi., U.S. (rȧ cō)	75k	46°22′N	84°43′W
Rădăuţi, Rom.	119	47°53′N	25°55′E
Radcliffe, Eng., U.K. (răd′klĭf)	114a	53°34′N	2°20′W
Radevormwald, Ger. (rä′dĕ-fōrm-väld)	127c	51°12′N	7°22′E
Radford, Va., U.S. (răd′fērd)	83	37°06′N	81°33′W
Rādhanpur, India	158	23°57′N	71°38′E
Radium, S. Afr. (rā′dĭ-ŭm)	192c	25°06′S	28°18′E
Radom, Pol. (rä′dôm)	117	51°24′N	21°11′E
Radomir, Blg. (rä′dô-mēr)	131	42°33′N	22°58′E
Radomsko, Pol. (rä-dôm′skô)	117	51°04′N	19°27′E
Radomyshl, Ukr. (rä-dô-mēsh′′l)	137	50°30′N	29°13′E
Radul′, Ukr. (rȧ′do͞ol)	133	51°52′N	30°46′E
Radviliškis, Lith. (räd′vē-lēsh′kēs)	123	55°49′N	23°31′E
Radwah, Jabal, mtn., Sau. Ar.	154	24°44′N	38°14′E
Radzyń Podlaski, Pol. (räd′zēn-y′ pŭd-lä′skĭ)	125	51°49′N	22°40′E
Raeford, N.C., U.S. (rā′fērd)	83	34°57′N	79°15′W
Raesfeld, Ger. (răz′fĕld)	127c	51°46′N	6°50′E
Raeside, l., Austl. (rā′sīd)	174	29°20′S	122°30′E
Rae Strait, strt., Can. (rā)	52	68°40′N	95°03′W
Rafaela, Arg. (rä-fä-ā′lä)	102	31°15′S	61°21′W
Rafah, Pak. (rä′fä)	153a	31°14′N	34°12′E
Rafsanjān, Iran	154	30°45′N	56°30′E
Raft, r., Id., U.S. (răft)	73	42°20′N	113°17′W
Ragay, Phil. (rä-gī′)	169a	13°49′N	122°45′E
Ragay Gulf, b., Phil.	169a	13°44′N	122°38′E
Ragunda, Swe. (rä-gòn′dä)	122	63°07′N	16°24′E
Ragusa, Italy (rä-go͞o′sä)	118	36°58′N	14°41′E
Rahachow, Bela.	136	53°07′N	30°04′E
Rahway, N.J., U.S. (rô′wā)	68a	40°37′N	74°16′W
Rāichūr, India (rä′ē-cho͞or′)	155	16°23′N	77°18′E
Raigarh, India (rī′gŭr)	155	21°57′N	83°32′E
Rainbow Bridge National Monument, rec., Ut., U.S. (rān′bō)	77	37°05′N	111°00′W
Rainbow City, Pan.	86a	9°20′N	79°53′W
Rainier, Or., U.S.	74c	46°05′N	122°56′W
Rainier, Mount, mtn., Wa., U.S. (rā-nēr′)	64	46°52′N	121°46′W
Rainy, r., N.A.	65	48°50′N	94°41′W
Rainy Lake, l., N.A. (rān′ē)	53	48°43′N	94°29′W
Rainy River, Can.	51	48°43′N	94°29′W
Raipur, India (rä′jŭ-bo͞o-rĕ′)	158	21°25′N	81°37′E

ăt; finăl; rāte; senāte; ärm; åsk; sofȧ; fâre; ch-choose; dh-as th in other; bē; ēvent; bĕt; recĕnt; cratẽr; g-gō; gh-guttural g; bĭt; ĭ-short neutral; rīde; ĸ-guttural k as ch in German ich;

PRONOUNCING INDEX
Raisin – Resolution 293

PLACE (Pronunciation)	PAGE	LAT.	LONG.
Raisin, r., Mi., U.S. (rā'zĭn)	66	42°00'N	83°35'W
Raitan, N.J., U.S. (rā-tăn)	68a	40°34'N	74°40'W
Rājahmundry, India (räj-ŭ-mŭn'drē)	155	17°03'N	81°51'E
Rajang, r., Malay.	168	2°10'N	113°30'E
Rājapālaiyam, India	159	9°30'N	77°33'E
Rājasthān, state, India (rä'jŭs-tän)	155	26°00'N	72°00'E
Rājkot, India (räj'kôt)	155	22°20'N	70°48'E
Rājpur, India	158a	22°24'N	88°25'E
Rājshāhi, Bngl.	155	24°26'S	88°39'E
Rakhiv, Ukr.	125	48°02'N	24°13'E
Rakh'oya, Russia (räk'yȧ)	142c	60°06'N	30°50'E
Rakitnoye, Russia (rȧ-kēt'nô-yĕ)	137	50°51'N	35°53'E
Rakovník, Czech Rep.	124	50°07'N	13°45'E
Rakvere, Est. (rȧk'vĕ-rĕ)	136	59°22'N	26°14'E
Raleigh, N.C., U.S.	65	35°45'N	78°39'W
Ram, r., Can.	55	52°10'N	115°05'W
Rama, Nic. (rä'mä)	91	12°11'N	84°14'W
Ramallo, Arg. (rä-mä'l-yô)	99c	33°28'S	60°02'W
Ramanāthapuram, India	159	9°13'N	78°52'E
Rambouillet, Fr. (räɴ-bōō-yĕ')	126	48°39'N	1°49'E
Rame Head, c., S. Afr.	187c	31°48'S	29°22'E
Ramenskoye, Russia (rȧ'mĕn-skô-yĕ)	132	55°34'N	38°15'E
Ramlat as Sab'atayn, reg., Asia	154	16°08'N	45°15'E
Ramm, Jabal, mtn., Jord.	153a	29°37'N	35°32'E
Râmnicu Sărat, Rom.	119	45°24'N	27°06'E
Râmnicu Vâlcea, Rom.	131	45°07'N	24°22'E
Ramos, Mex. (rä'mōs)	88	22°46'N	101°52'W
Ramos, r., Nig.	189	5°10'N	5°40'E
Ramos Arizpe, Mex. (ä-rēz'pä)	80	25°33'N	100°57'W
Rampart, Ak., U.S. (răm'pärt)	63	65°28'N	150°18'W
Rampo Mountains, mts., N.J., U.S. (răm'pō)	68a	41°06'N	72°12'W
Rāmpur, India (räm'pōōr)	155	28°53'N	79°03'E
Ramree Island, i., Mya. (räm'rē')	168	19°01'N	93°23'E
Ramsayville, Can. (răm'zē vĭl)	62c	45°23'N	75°34'W
Ramsbottom, Eng., U.K. (rămz'bŏt-ŭm)	114a	53°39'N	2°20'W
Ramsey, I. of Man (răm'zē)	120	54°20'N	4°25'W
Ramsey, N.J., U.S.	68a	41°03'N	74°09'W
Ramsey Lake, l., Can.	58	47°15'N	82°16'W
Ramsgate, Eng., U.K. (rămz''gāt)	121	51°19'N	1°20'E
Ramu, r., Pap. N. Gui. (rä'mōō)	169	5°35'S	145°16'E
Rancagua, Chile (rän-kä'gwä)	102	34°10'S	70°43'W
Rance, r., Fr. (räɴs)	126	48°17'N	2°30'W
Rānchī, India	155	23°21'N	85°20'E
Rancho Boyeros, Cuba (rä'n-chô-bô-yĕ'rôs)	93a	23°00'N	82°23'W
Randallstown, Md., U.S. (răn'dȧlz-toun)	68e	39°22'N	76°48'W
Randers, Den. (rän'ĕrs)	116	56°28'N	10°03'E
Randfontein, S. Afr. (ränt'fŏn-tān)	187b	26°10'S	27°42'E
Randleman, N.C., U.S. (răn'd'l-măn)	83	35°49'N	79°50'W
Randolph, Ma., U.S. (răn'dôlf)	61a	42°10'N	71°03'W
Randolph, Ne., U.S.	70	42°22'N	97°22'W
Randolph, Vt., U.S.	67	43°55'N	72°40'W
Random Island, i., Can. (răn'dŭm)	61	48°12'N	53°25'W
Randsfjorden, Nor.	122	60°35'N	10°10'E
Randwick, Austl.	173b	33°55'S	151°15'E
Ranérou, Sen.	188	15°18'N	13°58'W
Rangeley, Me., U.S. (rānj'lē)	60	44°56'N	70°38'W
Rangeley, l., Me., U.S.	60	45°00'N	70°25'W
Ranger, Tx., U.S. (rān'jēr)	64	32°26'N	98°41'W
Rangia, India	158	26°32'N	91°39'E
Rangoon (Yangon), Mya. (răŋ-gōōn')	155	16°46'N	96°09'E
Rangpur, Bngl. (rŭng'pōōr)	155	25°48'N	89°19'E
Rangsang, i., Indon. (räng'säng')	153b	0°53'N	103°05'E
Rangsdorf, Ger. (rängs'dōrf)	115b	52°17'N	13°25'E
Rānīganj, India (rä-nē-gŭnj')	158	23°40'N	87°08'E
Rankin Inlet, b., Can. (răŋ'kĕn)	53	62°45'N	94°27'W
Ranova, r., Russia (rä'nô-vȧ)	132	53°55'N	40°03'E
Rantau, Malay.	153b	2°35'N	101°58'E
Rantekombola, Bulu, mtn., Indon.	168	3°22'S	119°50'E
Rantoul, Il., U.S. (răn-tōōl')	66	40°25'N	88°05'W
Raoyang, China (rou-yäŋ)	162	38°16'N	115°45'E
Rapallo, Italy (rä-päl'lō)	130	44°21'N	9°14'E
Rapel, r., Chile (rä-pål')	99b	34°05'S	71°30'W
Rapid, r., Mn., U.S. (răp'ĭd)	71	48°21'N	94°50'W
Rapid City, S.D., U.S.	64	44°06'N	103°14'W
Rapla, Est. (räp'lȧ)	123	59°02'N	24°46'E
Rappahannock, r., Va., U.S. (răp'ȧ-hăn'ŭk)	67	38°20'N	75°25'W
Raquette, l., N.Y., U.S. (răk'ĕt)	67	43°50'N	74°35'W
Raritan, r., N.J., U.S. (răr'ĭ-tăn)	68a	40°32'N	74°27'W
Rarotonga, Cook Is. (rä'rô-tŏŋ'gȧ)	2	20°40'S	163°00'W
Ra's an Naqb, Jord.	153a	30°00'N	35°29'E
Raşcov, Mol.	133	47°55'N	28°51'E
Ras Dashen Terara, mtn., Eth. (räs dä-shän')	185	12°49'N	38°14'E
Raseiniai, Lith. (rä-syā'nyī)	123	55°23'N	23°04'E
Rashayya, Leb.	153a	33°30'N	35°50'E
Rashīd, Egypt (rȧ-shēd') (rō-zĕt'ȧ)	156	31°22'N	30°25'E
Rashīd, Masabb, mth., Egypt	192b	31°30'N	29°58'E
Rashkina, Russia (rȧsh'kĭ-nȧ)	142a	59°57'N	61°30'E
Rasht, Iran	154	37°13'N	49°45'E
Raška, Serb. (räsh'kȧ)	131	43°16'N	20°40'E
Rasskazovo, Russia (räs-kä'sô-vô)	137	52°40'N	41°40'E
Rastatt, Ger. (rä-shtät)	124	48°51'N	8°12'E
Rastes, Russia (rȧs'tĕs)	142a	59°24'N	58°49'E
Rastunovo, Russia (rȧs-tōō'nô-vô)	142b	55°15'N	37°50'E
Ratangarh, India (rŭ-tŭn'gŭr)	158	28°10'N	74°30'E
Ratcliff, Tx., U.S. (răt'klĭf)	81	31°22'N	95°09'W
Rathenow, Ger. (rä'tĕ-nō)	124	52°36'N	12°20'E
Rathlin Island, i., N. Ire., U.K. (răth-lĭn)	120	55°18'N	6°13'W
Ratingen, Ger. (rä'tĕn-gĕn)	127c	51°18'N	6°51'E
Rat Islands, is., Ak., U.S. (răt)	63a	51°35'N	176°48'E
Ratlām, India	158	23°19'N	75°05'E
Ratnāgiri, India	159	17°04'N	73°24'E
Raton, N.M., U.S. (rȧ-tōn')	64	36°52'N	104°26'W
Rattlesnake Creek, r., Or., U.S. (răt''l snāk)	72	42°38'N	117°39'W
Rättvik, Swe. (rĕt'vēk)	122	60°54'N	15°07'E
Rauch, Arg. (rá'ōōch)	102	36°47'S	59°05'W
Raufoss, Nor. (rou'fôs)	122	60°44'N	10°30'E
Raúl Soares, Braz. (rä-ōō'l-sôä'rĕs)	99a	20°05'S	42°28'W
Rauma, Fin. (rä'ò-mȧ)	116	61°07'N	21°31'E
Rauna, Lat. (rȧū'nȧ)	123	57°21'N	25°31'E
Raurkela, India	155	22°15'N	84°53'E
Rautalampi, Fin. (rä'ōō-tĕ-läm'pò)	123	62°39'N	26°25'E
Rava-Rus'ka, Ukr.	125	50°14'N	23°40'E
Ravenna, Italy (rä-vĕn'nä)	118	44°27'N	12°13'E
Ravenna, Ne., U.S. (rȧ-vĕn'ȧ)	70	41°20'N	98°50'W
Ravenna, Oh., U.S.	66	41°10'N	81°20'W
Ravensburg, Ger. (rä'vĕns-bōōrgh)	124	47°48'N	9°35'E
Ravensdale, Wa., U.S. (rā'vĕnz-dāl)	74a	47°22'N	121°58'W
Ravensthorpe, Austl. (rā'vĕns-thôrp)	174	33°30'S	120°20'E
Ravenswood, W.V., U.S. (rā'vĕnz-wŏŏd)	66	38°55'N	81°50'W
Rāwalpindi, Pak. (rä-wŭl-pēn'dē)	155	33°40'N	73°10'E
Rawa Mazowiecka, Pol.	125	51°46'N	20°17'E
Rawandoz, Iraq	137	36°37'N	44°30'E
Rawicz, Pol. (rä'vēch)	124	51°36'N	16°51'E
Rawlina, Austl. (rôr-lēnȧ)	174	31°13'S	125°45'E
Rawlins, Wy., U.S. (rô'lĭnz)	64	41°46'N	107°15'W
Rawson, Arg. (rô'sŭn)	102	43°16'S	65°09'W
Rawson, Arg.	99c	34°36'S	60°03'W
Rawtenstall, Eng., U.K. (rô'tĕn-stôl)	114a	53°42'N	2°17'W
Ray, Cape, c., Can. (rā)	53a	47°40'N	59°18'W
Raya, Bukit, mtn., Indon.	168	0°45'S	112°11'E
Raychikinsk, Russia (rī'chĭ-kēnsk)	141	49°52'N	129°17'E
Rayleigh, Eng., U.K. (rā'lē)	114b	51°35'N	0°36'E
Raymond, Can. (rā'mŭnd)	55	49°27'N	112°39'W
Raymond, Wa., U.S.	72	46°41'N	123°42'W
Raymondville, Tx., U.S. (rā'mŭnd-vĭl)	79	26°30'N	97°46'W
Ray Mountains, mts., Ak., U.S.	63	65°40'N	151°45'W
Rayne, La., U.S. (rān)	81	30°12'N	92°15'W
Rayón, Mex. (rä-yōn')	88	21°49'N	99°39'W
Rayton, S. Afr. (rā'tŭn)	187b	25°45'S	28°33'E
Raytown, Mo., U.S. (rā'toun)	75f	39°01'N	94°48'W
Rayville, La., U.S. (rā-vĭl)	81	32°28'N	91°46'W
Raz, Pointe du, c., Fr. (pwăɴt dü rä)	117	48°02'N	4°43'W
Razdan, Arm.	138	40°30'N	44°46'E
Razdol'noye, Russia (räz-dôl'nô-yĕ)	166	43°38'N	131°58'E
Razgrad, Blg.	119	43°32'N	26°32'E
Razlog, Blg. (räz'lôk)	131	41°54'N	23°32'E
Razorback Mountain, mtn., Can. (rā'zēr-băk)	54	51°35'N	124°42'W
Rea, r., Eng., U.K. (rē)	114a	52°25'N	2°31'W
Reaburn, Can. (rā'bûrn)	62f	50°06'N	97°53'W
Reading, Eng., U.K. (rĕd'ĭng)	117	51°25'N	0°58'W
Reading, Ma., U.S.	61a	42°32'N	71°07'W
Reading, Mi., U.S.	66	41°45'N	84°45'W
Reading, Oh., U.S.	69f	39°14'N	84°26'W
Reading, Pa., U.S.	65	40°20'N	75°55'W
Reading, co., Eng., U.K.	114a	52°37'N	0°40'W
Realengo, Braz. (rĕ-ä-län-gô)	99a	23°50'S	43°25'W
Rebiana, Libya	185	24°10'N	22°03'E
Rebun, i., Japan (rē'bōōn)	166	45°25'N	140°54'E
Recanati, Italy (rā-kä-nä'tē)	130	43°25'N	13°35'E
Recherche, Archipelago of the, is., Austl. (rē-shârsh')	174	34°17'S	122°30'E
Rechytsa, Bela. (ryĕ'chĕt-sȧ)	137	52°22'N	30°24'E
Recife, Braz. (rȧ-sē'fĕ)	101	8°09'S	34°59'W
Recife, Kapp, c., S. Afr. (rȧ-sē'fĕ)	187c	34°03'S	25°43'E
Recklinghausen, Ger. (rĕk'lĭng-hou-zĕn)	127c	51°36'N	7°13'E
Reconquista, Arg. (rā-kôn-kēs'tä)	102	29°01'S	59°41'W
Rector, Ar., U.S. (rĕk'tēr)	79	36°16'N	90°21'W
Red, r., Asia	168	21°00'N	103°00'E
Red, r., N.A. (rĕd)	64	48°00'N	97°00'W
Red, r., Tn., U.S.	82	36°35'N	86°55'W
Red, r., U.S.	65	31°40'N	92°55'W
Red, North Fork, r., U.S.	78	35°20'N	100°08'W
Red, Prairie Dog Town Fork, r., U.S. (prā'rĭ)	78	34°54'N	101°31'W
Red, Salt Fork, r., U.S.	78	35°04'N	100°31'W
Redan, Ga., U.S. (rē-dăn') (rĕd'ăn)	68c	33°44'N	84°09'W
Red Bank, N.J., U.S. (băngk)	68a	40°21'N	74°06'W
Red Bluff Reservoir, res., Tx., U.S.	80	32°03'N	103°52'W
Redby, Mn., Mn., U.S. (rĕd'bē)	71	47°52'N	94°55'W
Red Cedar, r., Wi., U.S. (sē'dēr)	71	45°03'N	91°48'W
Redcliff, Can. (rĕd'clĭf)	50	50°05'N	110°47'W
Redcliffe, Austl. (rĕd'clĭf)	176	27°20'S	153°12'E
Red Cliff Indian Reservation, I.R., Wi., U.S.	71	46°48'N	91°22'W
Red Cloud, Ne., U.S. (kloud)	78	40°06'N	98°32'W
Red Deer, Can. (dēr)	50	52°16'N	113°48'W
Red Deer, r., Can.	52	51°00'N	111°00'W
Red Deer, r., Can.	57	52°55'N	102°10'W
Red Deer Lake, l., Can.	57	52°58'N	101°28'W
Reddick, Il., U.S. (rĕd'dĭk)	69a	41°06'N	88°16'W
Redding, Ca., U.S. (rĕd'ĭng)	72	40°36'N	122°25'W
Redenção da Serra, Braz. (rĕ-dĕn-souɴ-dä-sĕ'r-rä)	99a	23°17'S	45°31'W
Redfield, S.D., U.S. (rĕd'fēld)	70	44°53'N	98°30'W
Red Fish Bar, Tx., U.S.	81a	29°29'N	94°53'W
Red Indian Lake, l., Can. (ĭn'dĭ-ăn)	53a	48°40'N	56°50'W
Red Lake, Can. (lāk)	51	51°02'N	93°49'W
Red Lake, r., Mn., U.S.	70	48°02'N	96°04'W
Red Lake Falls, Mn., U.S. (lāk fôls)	70	47°52'N	96°17'W
Red Lake Indian Reservation, I.R., Mn., U.S.	70	48°09'N	95°55'W
Redlands, Ca., U.S. (rĕd'lăndz)	75a	34°04'N	117°11'W
Red Lion, Pa., U.S. (lī'ŭn)	67	39°55'N	76°30'W
Red Lodge, Mt., U.S.	73	45°13'N	107°16'W
Redmond, Wa., U.S. (rĕd'mŭnd)	74a	47°40'N	122°07'W
Rednitz, r., Ger. (rĕd'nētz)	124	49°10'N	11°00'E
Red Oak, Ia., U.S. (ōk)	70	41°00'N	95°12'W
Redon, Fr. (rē-dôɴ')	126	47°42'N	2°03'W
Redonda, Isla, i., Braz. (ē's-lä-rē-dô'n-dä)	102b	23°05'S	43°11'W
Redonda Island, i., Antig. (rĕ-dŏn'dȧ)	91b	16°55'N	62°28'W
Redondela, Spain (rā-dhôn-dā'lä)	128	42°16'N	8°34'W
Redondo, Port. (rå-dôn'dò)	128	38°40'N	7°32'W
Redondo, Wa., U.S. (rē-dŏn'dō)	74a	47°21'N	122°19'W
Redondo Beach, Ca., U.S.	75a	33°50'N	118°23'W
Red Pass, Can. (pås)	55	52°59'N	118°59'W
Red Rock, r., Mt., U.S.	73	44°54'N	112°44'W
Red Sea, sea,	154	23°15'N	37°00'E
Redstone, Can. (rĕd'stōn)	54	52°08'N	123°42'W
Red Sucker Lake, l., Can. (sŭk'ēr)	57	54°09'N	93°40'W
Redwater, r., Mt., U.S.	73	47°37'N	105°25'W
Red Willow Creek, r., Ne., U.S.	78	40°34'N	100°48'W
Red Wing, Mn., U.S.	71	44°34'N	92°35'W
Redwood City, Ca., U.S. (rĕd' wŏŏd)	74b	37°29'N	122°13'W
Redwood Falls, Mn., U.S.	70	44°32'N	95°06'W
Redwood National Park, rec., Ca., U.S.	72	41°20'N	124°00'W
Redwood Valley, Ca., U.S.	76	39°15'N	123°12'W
Ree, Lough, l., Ire. (lŏk'rē')	116	53°30'N	7°45'W
Reed City, Mi., U.S. (rēd)	66	43°50'N	85°35'W
Reed Lake, l., Can.	57	54°37'N	100°30'W
Reedley, Ca., U.S. (rēd'lē)	76	36°37'N	119°27'W
Reedsburg, Wi., U.S. (rēdz'bûrg)	71	43°32'N	90°01'W
Reedsport, Or., U.S. (rēdz'pōrt)	72	43°42'N	124°08'W
Reelfoot Lake, res., Tn., U.S. (rēl'fŏŏt)	82	36°18'N	89°20'W
Rees, Ger. (rĕz)	127c	51°46'N	6°25'E
Reeves, Mount, mtn., Austl. (rēv's)	176	33°50'S	149°56'E
Reform, Al., U.S. (rē-fôrm')	82	33°23'N	88°00'W
Refugio, Tx., U.S. (rȧ-fōō'hyō) (rē-fū'jō)	81	28°18'N	97°15'W
Rega, r., Pol. (rĕ-gä)	124	53°48'N	15°30'E
Regen, r., Ger. (rā'ghĕn)	124	49°09'N	12°21'E
Regensburg, Ger. (rā'ghĕns-bôrgh)	117	49°02'N	12°06'E
Reggio, La., U.S. (rĕg'jĭ-ō)	68d	29°50'N	89°46'W
Reggio di Calabria, Italy (rĕ'jô dē kä-lä'brē-ä)	119	38°07'N	15°42'E
Reggio nell' Emilia, Italy	118	44°43'N	10°34'E
Reghin, Rom. (rā-gēn')	125	46°47'N	24°44'E
Regina, Can. (rē-jī'nȧ)	56	50°25'N	104°39'W
Regla, Cuba (răg'lä)	92	23°08'N	82°20'W
Regnitz, r., Ger. (rĕg'nētz)	124	49°50'N	10°55'E
Reguengos de Monsaraz, Port.	128	38°26'N	7°30'W
Rehoboth, Nmb.	186	23°10'S	17°15'E
Reḥovot, Isr.	153a	31°53'N	34°49'E
Reichenbach, Ger. (rī'kĕn-bäk)	124	50°36'N	12°18'E
Reidsville, N.C., U.S. (rēdz'vĭl)	83	36°20'N	79°37'W
Reigate, Eng., U.K. (rī'gāt)	120	51°12'N	0°12'W
Reims, Fr. (răɴs)	110	49°16'N	4°00'E
Reina Adelaida, Archipiélago, is., Chile	102	52°00'S	74°15'W
Reinbeck, Ia., U.S. (rīn'bĕk)	71	42°22'N	92°34'W
Reindeer, l., Can. (rān'dēr)	52	57°36'N	101°23'W
Reindeer, r., Can.	56	55°45'N	103°30'W
Reindeer Island, i., Can.	57	52°25'N	98°00'W
Reinosa, Spain (rå-ē-nō'sä)	128	43°01'N	4°08'W
Reistertown, Md., U.S. (rēs'tēr-toun)	68e	39°28'N	76°50'W
Reitz, S. Afr.	192c	27°48'S	28°25'E
Rema, Jabal, mtn., Yemen	154	14°13'N	44°38'E
Rembau, Malay.	153b	2°36'N	102°06'E
Remedios, Col. (rĕ-mĕ'dyôs)	100a	7°03'N	74°42'W
Remedios, Cuba (rā-mā'dhē-ōs)	92	22°30'N	79°35'W
Remedios, Pan. (rĕ-mĕ'dyōs)	91	8°14'N	81°46'W
Remiremont, Fr. (rē-mēr-môɴ')	127	48°01'N	6°35'E
Rempang, i., Indon.	153b	0°51'N	104°04'E
Remscheid, Ger. (rĕm'shīt)	127c	51°10'N	7°11'E
Rena, Nor.	122	61°08'N	11°17'E
Rendova, i., Sol. Is. (rĕn'dô-vȧ)	175	8°38'S	156°26'E
Rendsburg, Ger. (rĕnts'bôrgh)	124	54°19'N	9°39'E
Renfrew, Can. (rĕn'frōō)	51	45°30'N	76°30'W
Rengam, Malay. (rĕn'gäm')	153b	1°53'N	103°24'E
Rengo, Chile (rĕn'gō)	99b	34°22'S	70°50'W
Reni, Ukr. (ran')	133	45°26'N	28°18'E
Renmark, Austl. (rĕn'märk)	175	34°10'S	140°50'E
Rennel, i., Sol. Is. (rĕn-nĕl')	175	11°50'S	160°38'E
Rennes, Fr. (rĕn)	110	48°07'N	1°02'W
Reno, Nv., U.S. (rē'nō)	64	39°32'N	119°49'W
Reno, r., Italy (rā'nō)	130	44°10'N	10°55'E
Renovo, Pa., U.S. (rē-nō'vō)	67	41°20'N	77°50'W
Renqiu, China (rŭn-chyō)	162	38°44'N	116°05'E
Rensselaer, In., U.S. (rĕn'sē-lâr)	66	41°00'N	87°10'W
Rensselaer, N.Y., U.S. (rĕn'sē-lâr)	67	42°40'N	73°45'W
Rentchler, Il., U.S. (rĕnt'chlēr)	75e	38°30'N	89°52'W
Renton, Wa., U.S. (rĕn'tŭn)	74a	47°29'N	122°13'W
Repentigny, Can.	62a	45°47'N	73°26'W
Republic, Al., U.S. (rē-pŭb'lĭk)	68h	33°37'N	86°54'W
Republic, Wa., U.S.	72	48°38'N	118°44'W
Republican, r., U.S.	64	40°15'N	100°00'W
Republican, South Fork, r., Co., U.S. (rē-pŭb'lĭ-kăn)	78	39°35'N	102°28'W
Repulse Bay, b., Austl. (rē-pŭls')	175	20°56'S	149°22'E
Requena, Spain (rå-kā'nä)	118	39°29'N	1°03'W
Resende, Braz. (rĕ-sĕ'n-dĕ)	99a	22°30'S	44°26'W
Resende Costa, Braz. (kôs-tä)	99a	20°55'S	44°12'W
Reshetylivka, Ukr.	133	49°34'N	34°04'E
Resistencia, Arg. (rä-sēs-tĕn'syä)	102	27°24'S	58°54'W
Reşiţa, Rom. (rå'shē-tä)	131	45°18'N	21°56'E
Resolute *see* Qausuittuq, Can.	49	74°41'N	95°00'W
Resolution, i., Can. (rĕz-ô-lū'shŭn)	53	61°30'N	63°58'W

ng-sing; ŋ-baŋk; ɴ-nasalized n; nŏd; cŏmmit; ōld; ôbey; ôrder; oi-boil; fōōd; ò-as oo in foot; ou-out; s-soft; sh-dish; th-thin; pūre; ůnite; ûrn; stŭd; circŭs; ü-as in French tu; ′-indeterminate vowel.

294 Resolution Island – Rockdale PRONOUNCING INDEX

PLACE (Pronunciation)	PAGE	LAT.	LONG.
Resolution Island, i., N.Z. (rĕz-ŏl-ûshŭn)	175a	45°43'S	166°20'E
Restigouche, r., Can.	60	47°35'N	67°35'W
Restrepo, Col. (rĕs-trĕ'pô)	100a	3°49'N	76°31'W
Restrepo, Col.	100a	4°16'N	73°32'W
Retalhuleu, Guat. (rā-täl-ōō-lān')	90	14°31'N	91°41'W
Rethel, Fr. (r-tl')	126	49°34'N	4°20'E
Réthimnon, Grc.	130a	35°21'N	24°30'E
Retie, Bel.	115a	51°16'N	5°08'E
Retsil, Wa., U.S. (rĕt'sĭl)	74a	47°33'N	122°37'W
Réunion, dep., Afr. (rā-ü-nyôn')	3	21°06'S	55°36'E
Reus, Spain (rā'ōōs)	118	41°08'N	1°05'E
Reutlingen, Ger. (roit'lĭng-ĕn)	124	48°29'N	9°14'E
Reutov, Russia (rĕ-ōō'ôf)	142b	55°45'N	37°52'E
Revda, Russia (ryâv'dá)	142a	56°48'N	59°57'E
Revelstoke, Can. (rĕv'ĕl-stōk)	50	51°00'N	118°12'W
Reventazón, Río, r., C.R. (rå-vĕn-tä-zōn')	91	10°10'N	83°30'W
Revere, Ma., U.S. (rē-vēr')	61a	42°24'N	71°01'W
Revillagigedo, Islas, is., Mex. (ē's-läs-rĕ-vêl-yä-hĕ'gĕ-dô)	86	18°45'N	111°00'W
Revillagigedo Chan, Ak., U.S. (rĕ-vĭl'á-gĭ-gē'dō)	54	55°10'N	131°13'W
Revillagigedo Island, i., Ak., U.S.	54	55°35'N	131°23'W
Revin, Fr. (rĕ-văN)	126	49°56'N	4°34'E
Rewa, India (rā'wä)	155	24°41'N	81°11'E
Rewāri, India	158	28°19'N	76°39'E
Rexburg, Id., U.S. (rĕks'bûrg)	73	43°50'N	111°48'W
Rey, Iran	157	35°35'N	51°25'E
Rey, I., Mex. (rā)	80	27°00'N	103°33'W
Rey, Isla del, i., Pan. (ē's-lä-dĕl-rā'ē)	91	8°20'N	78°40'W
Reyes, Bol. (rā'yĕs)	100	14°19'S	67°16'W
Reyes, Point, c., Ca., U.S.	76	38°00'N	123°00'W
Reykjanes, c., Ice. (rā'kyá-nĕs)	112	63°37'N	24°33'W
Reykjavik, Ice. (rā'kyá-vēk)	110	64°09'N	21°39'W
Reynosa, Mex. (rā-ē-nō'sä)	80	26°05'N	98°21'W
Rēzekne, Lat. (rå'zĕk-nĕ)	136	56°31'N	27°19'E
Rezh, Russia (rĕzh')	142a	57°22'N	61°23'E
Rezina, Mol. (ryĕzh'ĕ-nĭ)	133	47°44'N	28°56'E
Rhaetian Alps, mts., Eur.	124	46°30'N	10°00'E
Rhaetien Alps, mts., Eur.	130	46°22'N	10°33'E
Rheinberg, Ger. (rīn'bĕrgh)	127c	51°33'N	6°37'E
Rheine, Ger. (rī'nĕ)	124	52°16'N	7°26'E
Rheinkamp, Ger.	127c	51°30'N	6°37'E
Rheinland, hist. reg., Ger.	124	50°05'N	6°40'E
Rheydt, Ger. (rĕ'yt)	127c	51°10'N	6°28'E
Rhin, r., Ger. (rēn)	115b	52°52'N	12°49'E
Rhine, r., Eur.	112	50°34'N	7°21'E
Rhinelander, Wi., U.S. (rīn'lăn-dĕr)	71	45°39'N	89°25'W
Rhin Kanal, can., Ger. (rēn kä-näl')	115b	52°47'N	12°40'E
Rhiou, r., Alg.	129	35°45'N	1°18'E
Rhode Island, state, U.S. (rōd ī'lănd)	65	41°35'N	71°40'W
Rhode Island, i., R.I., U.S.	68b	41°31'N	71°14'W
Rhodes, S. Afr. (rôdz)	187c	30°48'S	27°56'E
Rhodes *see* Ródos, i., Grc.	112	36°00'N	28°29'E
Rhodesia *see* Zimbabwe, nation, Afr.	186	17°50'S	29°30'E
Rhodope Mountains, mts., Eur. (rô'dô-pĕ)	112	42°00'N	24°08'E
Rhondda, Wales, U.K. (rŏn'dhá)	120	51°40'N	3°40'W
Rhône, r., Fr. (rōn)	112	44°30'N	4°45'E
Rhoon, Neth.	115a	51°52'N	4°24'E
Rhum, i., Scot., U.K. (rŭm)	120	57°00'N	6°20'W
Riachão, Braz. (rē-ä-chouN')	101	7°15'S	46°30'W
Rialto, Ca., U.S. (rē-ăl'tō)	75a	34°06'N	117°23'W
Riau, prov., Indon.	153b	0°56'N	101°25'E
Riau, Kepulauan, i., Indon.	168	0°30'N	104°55'E
Riau, Selat, strt., Indon.	153b	0°40'N	104°27'E
Riaza, r., Spain (rē-ä'thä)	128	41°25'N	3°25'W
Ribadavia, Spain (rē-bä-dhä-vē-ä)	128	42°18'N	8°06'W
Ribadeo, Spain (rē-bä-dhā'ō)	128	43°32'N	7°05'W
Ribadesella, Spain (rē'bä-dā-säl'yä)	128	43°30'N	5°02'W
Ribe, Den. (rē'bĕ)	122	55°20'N	8°45'E
Ribeirão Prêto, Braz. (rē-bā-rouN-prĕ'tô)	101	21°11'S	47°47'W
Ribera, N.M., U.S. (rē-bĕ'rä)	78	35°23'N	105°27'W
Riberalta, Bol. (rē-bå-räl'tä)	100	11°06'S	66°02'W
Rib Lake, Wi., U.S. (rĭb lāk)	71	45°20'N	90°11'W
Rîbnita, Mol.	133	47°45'N	29°02'E
Rice, l., Can.	59	44°05'N	78°10'W
Rice Lake, Wi., U.S.	71	45°30'N	91°44'W
Rice Lake, l., Mn., U.S.	75g	45°10'N	93°09'W
Richards Island, i., Can. (rĭch'ĕrds)	63	69°45'N	135°30'W
Richards Landing, Can. (lănd'ĭng)	75k	46°18'N	84°02'W
Richardson, Tx., U.S. (rĭch'ĕrd-sŭn)	75c	32°56'N	96°44'W
Richardson, Wa., U.S.	74a	48°27'N	122°54'W
Richardson Mountains, mts., Can.	52	66°58'N	136°19'W
Richardson Mountains, mts., N.Z.	177	44°50'S	168°30'E
Richardson Park, De., U.S. (pärk)	67	39°45'N	75°35'W
Richelieu, r., Can. (rēsh'lyû')	59	45°05'N	73°25'W
Richfield, Mn., U.S.	75g	44°53'N	93°17'W
Richfield, Oh., U.S.	69d	41°14'N	81°38'W
Richfield, Ut., U.S.	77	38°45'N	112°05'W
Richford, Vt., U.S. (rĭch'fērd)	67	45°00'N	72°35'W
Rich Hill, Mo., U.S. (rĭch hĭl)	79	38°05'N	94°21'W
Richibucto, Can. (rĭ-chĭ-bŭk'tō)	51	46°41'N	64°52'W
Richland, Ga., U.S. (rĭch'lănd)	82	32°05'N	84°40'W
Richland, Wa., U.S.	72	46°17'N	119°19'W
Richland Center, Wi., U.S. (sĕn'tĕr)	71	43°20'N	90°25'W
Richmond, Austl. (rĭch'mŭnd)	175	20°47'S	143°14'E
Richmond, Austl.	173b	33°36'S	150°45'E
Richmond, Can.	62c	45°12'N	75°49'W
Richmond, Can.	59	45°40'N	72°07'W
Richmond, S. Afr.	187c	29°52'S	30°17'E
Richmond, Il., U.S.	69a	42°29'N	88°18'W
Richmond, In., U.S.	66	39°50'N	85°00'W
Richmond, Ky., U.S.	66	37°45'N	84°20'W
Richmond, Mo., U.S.	79	39°16'N	93°58'W
Richmond, Tx., U.S.	81	29°35'N	95°45'W
Richmond, Ut., U.S.	73	41°55'N	111°50'W
Richmond, Va., U.S.	65	37°35'N	77°30'W
Richmond Beach, Wa., U.S.	74a	47°47'N	122°23'W
Richmond Heights, Mo., U.S.	75e	38°38'N	90°20'W
Richmond Highlands, Wa., U.S.	74a	47°46'N	122°22'W
Richmond Hill, Can. (hĭl)	59	43°53'N	79°26'W
Richton, Ms., U.S. (rĭch'tŭn)	82	31°20'N	89°54'W
Richwood, W.V., U.S. (rĭch'wŏd)	66	38°10'N	80°30'W
Ridderkerk, Neth.	115a	51°52'N	4°35'E
Rideau, r., Can.	62c	45°17'N	75°41'W
Rideau Lake, l., Can. (rē-dō')	59	44°40'N	76°20'W
Ridgefield, Ct., U.S. (rĭj'fēld)	68a	41°16'N	73°30'W
Ridgefield, Wa., U.S.	74c	45°49'N	122°40'W
Ridgeway, Can. (rĭj'wā)	69c	42°53'N	79°02'W
Ridgewood, N.J., U.S. (rĭdj'wŏd)	68a	40°59'N	74°08'W
Ridgway, Pa., U.S.	67	41°25'N	78°40'W
Riding Mountain, mtn., Can. (rīd'ĭng)	57	50°37'N	99°37'W
Riding Mountain National Park, rec., Can. (rīd'ĭng)	52	50°59'N	99°19'W
Riding Rocks, is., Bah.	92	25°20'N	79°10'W
Riebeek-Oos, S. Afr.	187c	33°14'S	26°09'E
Ried, Aus. (rēd)	124	48°13'N	13°30'E
Riesa, Ger. (rē'zá)	124	51°17'N	13°17'E
Rieti, Italy (rē-ā'tē)	118	42°25'N	12°51'E
Rievleidam, res., S. Afr.	187b	25°52'S	28°18'E
Riffe Lake, res., Wa., U.S.	72	46°20'N	122°10'W
Rifle, Co., U.S. (rī'f'l)	77	39°35'N	107°50'W
Rīga, Lat. (rē'gá)	134	56°55'N	24°05'E
Riga, Gulf of, b., Eur.	136	57°56'N	23°05'E
Rīgān, Iran	154	28°45'N	58°55'E
Rigaud, Can. (rē-gō')	62a	45°29'N	74°18'W
Rigby, Id., U.S. (rĭg'bē)	73	43°40'N	111°55'W
Rigeley, W.V., U.S. (rĭj'lē)	67	39°40'N	78°45'W
Rīgestān, des., Afg.	154	30°53'N	64°42'E
Rigolet, Can. (rĭg-ō-lā')	51	54°10'N	58°40'W
Riihimäki, Fin.	123	60°44'N	24°44'E
Rijeka, Cro. (rĭ-yĕ'kä)	118	45°22'N	14°24'E
Rijkevorsel, Bel.	115a	51°21'N	4°46'E
Rijswijk, Neth.	115a	52°03'N	4°19'E
Rika, r., Ukr. (rē'ká)	125	48°21'N	23°37'E
Rima, r., Nig.	189	13°30'N	5°50'E
Rimavska Sobota, Slvk. (rē'máf-ská sô'bô-tá)	125	48°25'N	20°01'E
Rimbo, Swe. (rēm'bō)	122	59°45'N	18°22'E
Rimini, Italy (rē'mē-nē)	118	44°03'N	12°33'E
Rimouski, Can. (rē-mōōs'kē)	51	48°27'N	68°32'W
Rincón de Romos, Mex. (rēn-kōn dā rô-mōs')	88	22°13'N	102°21'W
Ringkøbing, Den. (rĭng'kûb-ĭng)	116	56°06'N	8°14'E
Ringkøbing Fjord, b., Den.	122	55°55'N	8°04'E
Ringsted, Den. (rĭng'stĕdh)	122	55°27'N	11°49'E
Ringvassøya, i., Nor. (rĭng'väs-ûê)	116	69°58'N	16°43'E
Ringwood, Austl.	173a	37°49'S	145°14'E
Rinjani, Gunung, mtn., Indon.	168	8°39'S	116°22'E
Río Abajo, Pan. (rē-ō-ä-bä'kō)	86a	9°01'N	78°30'W
Río Balsas, Mex. (rē'ō-bäl-säs)	88	17°59'N	99°45'W
Riobamba, Ec. (rē'ō-bäm-bä)	100	1°45'S	78°37'W
Rio Bonito, Braz. (rē'ō bō-nē'tō)	99a	22°44'S	42°38'W
Rio Branco, Braz. (rē'ō brän'kō)	100	9°57'S	67°50'W
Río Branco, Ur. (rīō brăncō)	102	32°33'S	53°29'W
Rio Casca, Braz. (rē'ō-ká's-kä)	99a	20°15'S	42°39'W
Río Chico, Ven. (rē'ō chē'kŏ)	101b	10°20'N	65°58'W
Rio Claro, Braz. (rē'ō klä'rō)	101	22°25'S	47°33'W
Río Cuarto, Arg. (rē'ō kwär'tō)	102	33°05'S	64°15'W
Rio das Flores, Braz. (rē'ō-däs-flō-rĕs)	99a	22°10'S	43°35'W
Rio de Janeiro, Braz. (rē'ō dā zhä-nā'ē-rō)	102b	22°50'S	43°20'W
Rio de Janeiro, state, Braz.	101	22°27'S	42°43'W
Río de Jesús, Pan.	91	7°54'N	80°59'W
Río Frío, Mex. (rē'ō-frē'ō)	89a	19°21'N	98°40'W
Río Gallegos, Arg. (rē'ō gä-lā'gōs)	102	51°43'S	69°15'W
Rio Grande, Braz. (rē'ō grän'dĕ)	102	31°04'S	52°14'W
Río Grande, Mex. (rē'ō grän'dā)	88	23°51'N	102°59'W
Riogrande, Tx., U.S. (rē'ō grån-dā)	80	26°23'N	98°48'W
Rio Grande do Norte, state, Braz.	101	5°26'S	37°20'W
Rio Grande do Sul, state, Braz. (rē'ō grän'dĕ-dô-sōō'l)	102	29°00'S	54°00'W
Ríohacha, Col. (rē'ō-ä'chä)	100	11°30'N	72°54'W
Río Hato, Pan. (rē-ō-ä'tō)	91	8°19'N	80°11'W
Riom, Fr. (rē-ôN')	126	45°54'N	3°08'E
Rio Muni, hist. reg., Eq. Gui. (rē'ō mōō'nē)	184	1°47'N	8°33'E
Ríonegro, Col. (rē'ō-nĕ'grō)	100a	6°09'N	75°22'W
Río Negro, prov., Arg. (rē'ō nā'grō)	102	40°15'S	68°15'W
Río Negro, dept., Ur. (rē'ō-nĕ'grō)	99c	32°48'S	57°45'W
Río Negro, Embalse del, res., Ur.	102	32°45'S	55°50'W
Rionero, Italy (rē-ō-nā'rō)	130	40°55'N	15°42'E
Rioni, r., Geor.	138	42°08'N	41°39'E
Rio Novo, Braz. (rē'ō-nō'vō)	99a	21°30'S	43°08'W
Rio Pardo de Minas, Braz. (rē'ō pär'dō-dĕ-mē'näs)	101	15°43'S	42°24'W
Rio Pombo, Braz. (rē'ō pôm'bä)	99a	21°17'S	43°09'W
Rio Sorocaba, Represa do, res., Braz.	99a	23°37'S	47°19'W
Ríosucio, Col. (rē'ō-sōō'syō)	100a	5°25'N	75°41'W
Río Tercero, Arg. (rē'ō dĕr-sĕ'rō)	102	32°12'S	63°59'W
Rio Verde, Braz. (vĕr'dĕ)	101	17°47'S	50°49'W
Rioverde, Mex. (rē'ō-vĕr'dā)	86	21°54'N	99°59'W
Ripley, Eng., U.K. (rĭp'lē)	114a	53°03'N	1°24'W
Ripley, Ms., U.S.	82	34°44'N	88°55'W
Ripley, Tn., U.S.	82	35°44'N	89°34'W
Ripoll, Spain (rē-pōl'')	129	42°10'N	2°10'E
Ripon, Wi., U.S. (rĭp'ŏn)	71	43°49'N	88°50'W
Ripon, i., Austl.	174	20°05'S	118°10'E
Ripon Falls, wtfl., Ug.	186	0°38'N	33°02'E
Risaralda, dept., Col.	100a	5°15'N	76°00'W
Risdon, Austl. (rĭz'dŭn)	175	42°37'S	147°32'E
Rishiri, i., Japan (rē-shē'rē)	166	45°10'N	141°08'E
Rishon le Ziyyon, Isr.	153a	31°57'N	34°48'E
Rishra, India	158a	22°42'N	88°22'E
Rising Sun, In., U.S. (rīz'ĭng sŭn)	66	38°55'N	84°55'W
Risør, Nor. (rēs'ûr)	116	58°44'N	9°10'E
Ritacuva, Alto, mtn., Col. (ä'l-tō-rē-tä-kōō'vä)	100	6°22'N	72°13'W
Rittman, Oh., U.S. (rĭt'năn)	69d	40°58'N	81°47'W
Ritzville, Wa., U.S. (rĭts'vĭl)	72	47°08'N	118°23'W
Riva, Dom. Rep. (rē'vä)	93	19°10'N	69°55'W
Riva, Italy (rē'vä)	130	45°54'N	10°49'E
Riva, Md., U.S. (rĭ'vă)	68e	38°57'N	76°36'W
Rivas, Nic. (rē'väs)	90	11°25'N	85°51'W
Rive-de-Gier, Fr. (rēv-dĕ-zhē-ā')	126	45°32'N	4°37'E
Rivera, Ur. (rē-vā'rä)	102	30°52'S	55°32'W
River Cess, Lib. (rĭv'ĕr sĕs)	184	5°46'N	9°52'W
Riverdale, Il., U.S. (rĭv'ĕr dāl)	69a	41°38'N	87°36'W
Riverdale, Ut., U.S.	75b	41°11'N	112°00'W
River Falls, Al., U.S.	82	31°20'N	86°25'W
River Falls, Wi., U.S.	71	44°48'N	92°38'W
Riverhead, N.Y., U.S. (rĭv'ĕr hĕd)	67	40°55'N	72°40'W
Riverina, reg., Austl. (rĭv-ĕr-ē'nä)	175	34°55'S	144°30'E
River Jordan, Can. (jôr'dăn)	74a	48°25'N	124°03'W
River Oaks, Tx., U.S. (ōkz)	75c	32°47'N	97°24'W
River Rouge, Mi., U.S. (rōōzh)	69b	42°16'N	83°09'W
Rivers, Can.	57	50°01'N	100°15'W
Riverside, Ca., U.S. (rĭv'ĕr-sīd)	64	33°59'N	117°21'W
Riverside, N.J., U.S.	68f	40°02'N	74°58'W
Rivers Inlet, Can.	54	51°45'N	127°15'W
Riverstone, Austl.	173b	33°41'S	150°52'E
Riverton, Va., U.S.	67	39°00'N	78°15'W
Riverton, Wy., U.S.	73	43°02'N	108°24'W
Rivesaltes, Fr. (rēv'zält')	126	42°48'N	2°48'E
Riviera Beach, Fl., U.S. (rĭv-ĭ-ĕr'á bēch)	83a	26°46'N	80°04'W
Riviera Beach, Md., U.S.	68e	39°10'N	76°32'W
Rivière-Beaudette, Can.	62a	45°14'N	74°20'W
Rivière-du-Loup, Can. (rē-vyâr' dü lōō')	51	47°50'N	69°32'W
Rivière Qui Barre, Can. (rē-vyĕr' kē-bär)	62g	53°47'N	113°51'W
Rivière-Trois-Pistoles, Can. (trwä'pēs-tôl')	60	48°07'N	69°10'W
Rivne, Ukr.	133	48°11'N	31°46'E
Rivne, Ukr.	137	50°37'N	26°17'E
Rivne, prov., Ukr.	133	50°55'N	27°00'E
Riyadh, Sau. Ar.	154	24°31'N	46°47'E
Rize, Tur. (rē'zĕ)	119	41°00'N	40°30'E
Rizhao, China (rē-jou)	164	35°27'N	119°28'E
Rizzuto, Cape, c., Italy (rēt-sōō'tô)	131	38°53'N	17°05'E
Rjukan, Nor. (ryōō'kän)	116	59°53'N	8°30'E
Roanne, Fr. (rō-än')	117	46°02'N	4°04'E
Roanoke, Al., U.S. (rō'á-nōk)	82	33°08'N	85°21'W
Roanoke, Va., U.S.	65	37°16'N	79°55'W
Roanoke, r., U.S.	65	36°17'N	77°22'W
Roanoke Rapids, N.C., U.S.	83	36°25'N	77°40'W
Roanoke Rapids Lake, res., N.C., U.S.	83	36°28'N	77°37'W
Roan Plateau, plat., Co., U.S. (rōn)	77	39°25'N	110°00'W
Roatan, Hond. (rō-ä-tän')	90	16°18'N	86°33'W
Roatán, i., Hond.	90	16°19'N	86°46'W
Robbeneiland, i., S. Afr.	186a	33°48'S	18°22'E
Robbins, Il., U.S. (rŏb'ĭnz)	69a	41°39'N	87°42'W
Robbinsdale, Mn., U.S. (rŏb'ĭnz-dāl)	75g	45°03'N	93°22'W
Robe, Wa., U.S. (rōb)	74a	48°06'N	121°50'W
Roberts, Mount, mtn., Austl. (rŏb'ĕrts)	175	28°05'S	152°30'E
Roberts, Point, c., Wa., U.S. (rŏb'ĕrts)	74d	48°58'N	123°05'W
Robertson, Lac, l., Can.	61	51°00'N	59°10'W
Robertsport, Lib. (rŏb'ĕrts-pōrt)	184	6°45'N	11°22'W
Roberval, Can. (rŏb'ĕr-văl) (rô-bĕr-vál')	51	48°32'N	72°15'W
Robinson, Can.	61	48°16'N	58°50'W
Robinson, Il., U.S. (rŏb'ĭn-sŭn)	66	39°00'N	87°45'W
Robinvale, Austl. (rŏb-ĭn'vāl)	176	34°45'S	142°45'E
Roblin, Can.	57	51°15'N	101°25'W
Robson, Mount, mtn., Can. (rŏb'sŭn)	55	53°07'N	119°09'W
Robstown, Tx., U.S. (rŏbz'toun)	81	27°46'N	97°41'W
Roca, Cabo da, c., Port. (ká'bō-dä-rō'kä)	128	38°47'N	9°30'W
Rocas, Atol das, atoll, Braz. (ä-tôl-däs-rō'kás)	101	3°50'S	33°46'W
Rocha, Ur. (rō'chás)	102	34°26'S	54°14'W
Rochdale, Eng., U.K. (rŏch'dāl)	120	53°37'N	2°09'W
Roche à Bateau, Haiti (rôsh á bá-tō')	93	18°10'N	74°00'W
Rochefort, Fr. (rôsh-fŏr')	117	45°55'N	0°57'W
Rochelle, Il., U.S. (rō-shĕl')	71	41°53'N	89°06'W
Rochester, Eng., U.K.	114a	51°24'N	0°30'E
Rochester, In., U.S. (rŏch'ĕs-tĕr)	66	41°05'N	86°20'W
Rochester, Mi., U.S.	69b	42°41'N	83°09'W
Rochester, Mn., U.S.	65	44°01'N	92°30'W
Rochester, N.H., U.S.	67	43°20'N	71°00'W
Rochester, N.Y., U.S.	65	43°15'N	77°35'W
Rochester, Pa., U.S.	69e	40°42'N	80°16'W
Rock, r., Ia., U.S.	70	43°17'N	96°13'W
Rock, r., Or., U.S.	74c	45°34'N	122°52'W
Rock, r., Or., U.S.	74c	45°52'N	123°14'W
Rock, r., U.S.	65	41°40'N	90°00'W
Rockaway, N.J., U.S. (rŏck'á-wā)	68a	40°54'N	74°30'W
Rockbank, Austl.	173a	37°44'S	144°40'E
Rockcliffe Park, Can. (rok'klĭf pärk)	62c	45°27'N	75°40'W
Rock Creek, r., Can. (rŏk)	73	49°01'N	107°00'W
Rock Creek, r., Il., U.S.	69a	41°16'N	87°54'W
Rock Creek, r., Mt., U.S.	73	46°25'N	113°40'W
Rock Creek, r., Or., U.S.	74c	45°30'N	120°06'W
Rock Creek, r., Wa., U.S.	72	47°09'N	117°50'W
Rockdale, Austl.	173b	33°57'S	151°08'E

ăt; finăl; rāte; senåte; ärm; åsk; sofá; fâre; ch-choose; dh-as th in other; bē; ēvent; bĕt; recĕnt; cratĕr; g-gō; gh-guttural g; bĭt; ĭ-short neutral; rīde; K-guttural k as ch in German ich;

PLACE (Pronunciation)	PAGE	LAT.	LONG.
Rockdale, Md., U.S.	68e	39°22'N	76°49'W
Rockdale, Tx., U.S. (rŏk'dāl)	81	30°39'N	97°00'W
Rock Falls, Il., U.S. (rŏk fôlz)	71	41°45'N	89°42'W
Rockford, Il., U.S. (rŏk'fẽrd)	65	42°16'N	89°07'W
Rockhampton, Austl. (rŏk-hămp'tŭn)	175	23°26'S	150°29'E
Rock Hill, S.C., U.S. (rŏk'hĭl)	65	34°55'N	81°01'W
Rockingham, N.C., U.S. (rŏk'ĭng-hăm)	83	34°54'N	79°45'W
Rockingham Forest, for., Eng., U.K. (rok'ĭng-hăm)	114a	52°29'N	0°43'W
Rock Island, Il., U.S.	65	41°31'N	90°37'W
Rock Island Dam, Wa., U.S. (ī lănd)	72	47°17'N	120°33'W
Rockland, Can. (rŏk'lănd)	62c	45°33'N	75°17'W
Rockland, Ma., U.S.	61a	42°07'N	70°55'W
Rockland, Me., U.S.	60	44°06'N	69°09'W
Rockland Reservoir, res., Austl.	176	36°55'S	142°20'E
Rockmart, Ga., U.S. (rŏk'märt)	82	33°58'N	85°00'W
Rockmont, Wi., U.S. (rŏk'mŏnt)	75h	46°34'N	91°54'W
Rockport, In., U.S. (rŏk'pōrt)	66	38°20'N	87°00'W
Rockport, Ma., U.S.	61a	42°39'N	70°37'W
Rockport, Mo., U.S.	79	40°25'N	95°30'W
Rockport, Tx., U.S.	81	28°03'N	97°03'W
Rock Rapids, Ia., U.S. (răp'ĭdz)	70	43°26'N	96°10'W
Rock Sound, strt., Bah.	92	24°50'N	76°05'W
Rocksprings, Tx., U.S. (rŏk sprĭngs)	80	30°02'N	100°12'W
Rock Springs, Wy., U.S.	64	41°35'N	109°13'W
Rockstone, Guy. (rŏk'stôn)	101	5°55'N	57°27'W
Rock Valley, Ia., U.S. (văl'ĭ)	70	43°13'N	96°17'W
Rockville, In., U.S. (rŏk'vĭl)	66	39°45'N	87°15'W
Rockville, Md., U.S.	68e	39°05'N	77°11'W
Rockville Centre, N.Y., U.S. (sĕn'tẽr)	68a	40°39'N	73°39'W
Rockwall, Tx., U.S. (rŏk'wôl)	79	32°55'N	96°23'W
Rockwell City, Ia., U.S. (rŏk'wĕl)	71	42°22'N	94°37'W
Rockwood, Can. (rŏk-wo͝od)	62d	43°37'N	80°08'W
Rockwood, Me., U.S.	60	45°39'N	69°45'W
Rockwood, Tn., U.S.	82	35°51'N	84°41'W
Rocky, East Branch, r., Oh., U.S.	69d	41°13'N	81°43'W
Rocky, West Branch, r., Oh., U.S.	69d	41°17'N	81°54'W
Rocky Boys Indian Reservation, I.R., Mt., U.S.	73	48°08'N	109°34'W
Rocky Ford, Co., U.S.	78	38°02'N	103°43'W
Rocky Hill, N.J., U.S. (hĭl)	68a	40°24'N	74°38'W
Rocky Island Lake, l., Can.	58	46°56'N	83°04'W
Rocky Mount, N.C., U.S.	83	35°55'N	77°47'W
Rocky Mountain House, Can.	55	52°22'N	114°55'W
Rocky Mountain National Park, rec., Co., U.S.	64	40°29'N	106°06'W
Rocky Mountains, mts., N.A.	49	50°00'N	114°00'W
Rocky River, Oh., U.S.	69d	41°29'N	81°51'W
Rodas, Cuba (rō'dhäs)	92	22°20'N	80°35'W
Roden, r., Eng., U.K. (rō'dĕn)	114a	52°49'N	2°38'W
Rodeo, Mex. (rō-dā'ō)	80	25°12'N	104°34'W
Rodeo, Ca., U.S. (rō'dēō)	74b	38°02'N	122°16'W
Roderick Island, i., Can. (rŏd'ẽ-rĭk)	54	52°40'N	128°22'W
Rodez, Fr. (rō-dĕz')	117	44°22'N	2°34'E
Rodnei, Munţii, mts., Rom.	125	47°41'N	24°05'E
Rodniki, Russia (rôd'nē-kē)	136	57°08'N	41°48'E
Rodonit, Kep I, c., Alb.	131	41°38'N	19°01'E
Ródos, Grc.	119	36°24'N	28°15'E
Ródos, i., Grc.	119	36°00'N	28°29'E
Roebling, N.J., U.S. (rōb'lĭng)	68f	40°07'N	74°48'W
Roebourne, Austl. (rō'bŭrn)	174	20°50'S	117°15'E
Roebuck Bay, b., Austl. (rō'bŭck)	174	18°15'S	121°10'E
Roedtan, S. Afr.	192c	24°37'S	29°08'E
Roeselare, Bel.	121	50°55'N	3°05'E
Roesiger, l., Wa., U.S. (rŏz'ĭ-gẽr)	74a	47°59'N	121°56'W
Roes Welcome Sound, strt., Can. (rōz)	53	64°10'N	87°23'W
Rogatica, Bos. (rō-gä'tē-tsä)	131	43°46'N	19°00'E
Rogers, Ar., U.S. (rŏj-ẽrz)	79	36°19'N	94°07'W
Rogers City, Mi., U.S.	66	45°30'N	83°50'W
Rogersville, Tn., U.S.	82	36°21'N	83°00'W
Rognac, Fr. (rōn-yäk')	126a	43°29'N	5°15'E
Rogoaguado, l., Bol. (rō'gō-ä-gwä-dō)	100	12°42'S	66°46'W
Rogovskaya, Russia (rô-gôf'ská-yá)	133	45°43'N	38°42'E
Rogózno, Pol. (rô'gôzh-nō)	124	52°44'N	16°53'E
Rogue, r., Or., U.S. (rōg)	72	42°32'N	124°13'W
Rohatyn, Ukr.	125	49°22'N	24°37'E
Rojas, Arg. (rō'häs)	99c	34°11'S	60°42'W
Rojo, Cabo, c., Mex. (rō'hō)	89	21°35'N	97°16'W
Rojo, Cabo, c., P.R. (rō'hō)	87b	17°55'N	67°14'W
Rokel, r., S.L.	188	9°00'N	11°55'W
Rokkō-Zan, mtn., Japan (rôk'kō zän)	167b	34°46'N	135°16'E
Rokycany, Czech Rep. (rō'kĭ'tsá-nĭ)	124	49°44'N	13°37'E
Roldanillo, Col. (rôl-dä-nē'l-yō)	100a	4°24'N	76°09'W
Rolla, Mo., U.S.	79	37°56'N	91°45'W
Rolla, N.D., U.S.	70	48°52'N	99°32'W
Rolleville, Bah.	92	23°40'N	76°00'W
Roma, Austl. (rō'mȧ)	175	26°30'S	148°48'E
Roma *see* Rome, Italy	110	41°52'N	12°37'E
Roma, Leso.	187c	29°28'S	27°43'E
Romaine, r., Can. (rō-mĕn')	53	51°22'N	63°23'W
Roman, Rom. (rō'män)	125	46°56'N	26°57'E
Romania, nation, Eur. (rō-mā'nē-ȧ)	110	46°18'N	22°53'E
Romano, Cape, c., Fl., U.S. (rō-mä'nō)	83a	25°48'N	82°00'W
Romano, Cayo, i., Cuba (kä'yō-rō-mä'nô)	92	22°15'N	78°00'W
Romanovo, Russia (rô-mä'nô-vô)	142a	59°09'N	61°24'E
Romans, Fr. (rô-mäN')	126	45°04'N	4°49'E
Romblon, Phil. (rōm-blōn')	169a	12°34'N	122°16'E
Romblon Island, i., Phil.	169a	12°33'N	122°17'E
Rome (Roma), Italy	110	41°52'N	12°37'E
Rome, Ga., U.S. (rōm)	65	34°14'N	85°10'W
Rome, N.Y., U.S.	67	43°15'N	75°25'W
Romeo, Mi., U.S. (rō'mē-ō)	66	42°50'N	83°00'W
Romford, Eng., U.K. (rŭm'fẽrd)	114b	51°35'N	0°11'E
Romilly-sur-Seine, Fr. (rô-mē-yē'sür-sān')	126	48°32'N	3°41'E
Romita, Mex. (rō-mē'tä)	88	20°53'N	101°32'W
Romny, Ukr. (rôm'nĭ)	137	50°46'N	33°31'E
Rømø, i., Den. (rûm'û)	122	55°08'N	8°17'E
Romoland, Ca., U.S. (rō'mō'lănd)	75a	33°44'N	117°11'W
Romorantin-Lanthenay, Fr. (rô-mô-räN-tăN')	126	47°24'N	1°46'E
Rompin, Malay.	153b	2°42'N	102°30'E
Rompin, r., Malay.	153b	2°54'N	103°10'E
Romsdalsfjorden, Nor.	122	62°40'N	7°05'W
Romulus, Mi., U.S. (rom'ū lŭs)	69b	42°14'N	83°24'W
Ron, Mui, c., Viet.	165	18°05'N	106°45'E
Ronan, Mt., U.S. (rō'năn)	73	47°28'N	114°03'W
Roncador, Serra do, mts., Braz. (sĕr'rȧ dō rôn-kä-dōr')	101	12°44'S	52°19'W
Ronceverte, W.V., U.S. (rŏn'sē-vûrt)	66	37°45'N	80°30'W
Ronda, Spain (rōn'dä)	137	36°45'N	5°10'W
Ronda, Sierra de, mts., Spain	128	36°35'N	5°03'W
Rondônia, state, Braz.	100	10°15'S	63°07'W
Ronge, Lac la, l., Can. (rōnzh)	52	55°10'N	105°00'W
Rongjiang, China (rôŋ-jyäŋ)	165	25°52'N	108°45'E
Rongxian, China	165	22°50'N	110°32'E
Rønne, Den. (rûn'ĕ)	116	55°08'N	14°46'E
Ronneby, Swe. (rōn'ĕ-bü)	122	56°13'N	15°17'E
Ronne Ice Shelf, ice, Ant.	178	77°30'S	38°00'W
Roodepoort, S. Afr. (rō'dĕ-pōrt)	187b	26°10'S	27°52'E
Roodhouse, Il., U.S. (ro͞od'hous)	79	39°29'N	90°21'W
Rooiberg, S. Afr.	192c	24°46'S	27°42'E
Roosendaal, Neth. (rō'zĕn-däl)	115a	51°32'N	4°27'E
Roosevelt, Ut., U.S. (rōz''vĕlt)	77	40°20'N	110°00'W
Roosevelt, r., Braz. (rō'sĕ-vĕlt)	101	9°22'S	60°28'W
Roosevelt Island, i., Ant.	178	79°30'S	168°00'W
Root, r., Wi., U.S.	69a	42°49'N	87°54'W
Roper, r., Austl. (rōp'ẽr)	174	14°50'S	134°00'E
Ropsha, Russia (rôp'shȧ)	142c	59°44'N	29°53'E
Roque Pérez, Arg. (rō'kĕ-pĕ'rĕz)	99c	35°23'S	59°22'W
Roques, Islas los, is., Ven.	100	12°25'N	67°40'W
Roraima, state, Braz. (rō'rīy-mä)	100	2°00'N	62°15'W
Roraima, Mount, mtn., S.A. (rô-rä-ē'mä)	101	5°12'N	60°52'W
Røros, Nor. (rûr'ôs)	116	62°36'N	11°25'E
Ros', r., Ukr. (rôs)	133	49°40'N	30°22'E
Rosa, Monte, mtn., Italy (mōn'tä rō'zä)	118	45°56'N	7°51'E
Rosales, Mex. (rō-zä'läs)	80	28°15'N	100°43'W
Rosales, Phil. (rō-sä'lĕs)	169a	15°54'N	120°38'E
Rosamorada, Mex. (rō'zä-mō-rä'dhä)	88	22°06'N	105°16'W
Rosaria, Laguna, l., Mex. (lä-gō'nä-rō-sä'ryä)	89	17°50'N	93°51'W
Rosario, Arg. (rō-zä'rē-ō)	102	32°58'S	60°42'W
Rosario, Braz. (rô-zä'rē-ô)	101	2°49'S	44°15'W
Rosario, Mex.	80	26°31'N	105°40'W
Rosario, Mex.	88	22°58'N	105°54'W
Rosario, Phil.	169a	13°49'N	121°13'W
Rosario, Ur.	99c	34°19'S	57°24'E
Rosario, Cayo, i., Cuba (kä'yō-rō-sä'ryō)	92	21°40'N	81°55'W
Rosário do Sul, Braz. (rō-zä'rē-ō-dō-so͞o'l)	102	30°17'S	54°52'W
Rosário Oeste, Braz. (ō'ĕst'ĕ)	101	14°47'S	56°20'W
Rosario Strait, strt., Wa., U.S.	74a	48°27'N	122°45'W
Rosbach, Ger. (rōz'bäĸ)	127c	50°47'N	7°38'E
Roscoe, Tx., U.S. (rôs'kō)	80	32°26'N	100°38'W
Roseau, Dom.	91b	15°17'N	61°23'W
Roseau, Mn., U.S. (rô-zō')	70	48°52'N	95°47'W
Roseau, r., Mn., U.S.	70	48°52'N	96°11'W
Roseberg, Or., U.S. (rōz'bûrg)	64	43°13'N	123°30'W
Rosebud, r., Can. (rōz'bŭd)	55	51°20'N	112°20'W
Rosebud Creek, r., Mt., U.S.	73	45°48'N	106°34'W
Rosebud Indian Reservation, I.R., S.D., U.S.	70	43°13'N	100°42'W
Rosedale, Ms., U.S.	82	33°49'N	90°56'W
Rosedale, Wa., U.S.	74a	47°20'N	122°39'W
Roseires Reservoir, res., Sudan	185	11°15'N	34°45'E
Roselle, Il., U.S. (rō-zĕl')	69a	41°59'N	88°05'W
Rosemère, Can. (rōz'mēr)	62a	45°38'N	73°48'W
Rosemount, Mn., U.S. (rōz'mount)	75g	44°44'N	93°08'W
Rosendal, S. Afr. (rô-sĕn'täl)	192c	28°32'S	27°56'E
Rosenheim, Ger. (rō'zĕn-hīm)	117	47°52'N	12°06'E
Roses, Golf de, b., Spain	129	42°10'N	3°20'E
Rosetown, Can. (rōz'toun)	50	51°33'N	108°00'W
Rosetta *see* Rashīd, Egypt	156	31°22'N	30°25'E
Rosettenville, neigh., S. Afr.	187b	26°15'S	28°04'E
Roseville, Ca., U.S. (rōz'vĭl)	76	38°44'N	121°19'W
Roseville, Mi., U.S.	69b	42°30'N	82°55'W
Roseville, Mn., U.S.	75g	45°01'N	93°10'W
Rosiclare, Il., U.S. (rōz'y-klâr)	66	37°30'N	88°15'W
Rosignol, Guy. (rŏs-ĭg-nōl)	101	6°16'N	57°37'W
Roşiori de Vede, Rom. (rô-shôr'ĕ dĕ vĕ-dĕ)	131	44°06'N	25°00'E
Roskilde, Den. (rôs'kēl-dĕ)	122	55°39'N	12°04'E
Roslavl', Russia (rôs'läv'l)	136	53°56'N	32°52'E
Roslyn, Wa., U.S. (rŏz'lĭn)	72	47°14'N	121°00'W
Rösrath, Ger. (rûz'rät)	127c	50°53'N	7°11'E
Ross, Oh., U.S. (rôs)	69f	39°19'N	84°39'W
Rossano, Italy (rôs-sä'nō)	119	39°34'N	16°38'E
Rossan Point, c., Ire.	120	54°45'N	8°30'W
Ross Creek, r., Can.	62g	53°40'N	113°08'W
Rosseau, l., Can. (rôs-sō')	59	45°15'N	79°30'W
Rossel, i., Pap. N. Gui. (rô-sĕl')	175	11°31'S	154°00'E
Rosser, Can. (rôs'sẽr)	62f	49°59'N	97°27'W
Ross Ice Shelf, ice, Ant.	178	81°30'S	175°00'W
Rossignol, Lake, l., Can.	60	44°10'N	65°10'W
Ross Island, i., Can.	57	54°14'N	97°45'W
Ross Lake, res., Wa., U.S.	72	48°40'N	121°07'W
Rossland, Can. (rôs'lănd)	50	49°05'N	118°48'W
Rossosh', Russia (rôs'sŭsh)	137	50°12'N	39°32'E
Rossouw, S. Afr.	187c	31°12'S	27°18'E
Ross Sea, sea, Ant.	178	76°00'S	178°00'W
Rossvatnet, l., Nor.	116	65°36'N	13°08'E
Rossville, Ga., U.S. (rôs'vĭl)	82	34°57'N	85°22'W
Rosthern, Can.	56	52°41'N	106°25'W
Rostock, Ger. (rôs'tŭk)	116	54°04'N	12°06'E
Rostov, Russia	136	57°13'N	39°23'E
Rostov, prov., Russia	133	47°38'N	39°15'E
Rostov-na-Donu, Russia (rôstôv-nȧ-dô-no͞o)	134	47°16'N	39°47'E
Roswell, Ga., U.S. (rôz'wĕl)	82	34°02'N	84°21'W
Roswell, N.M., U.S.	64	33°23'N	104°32'W
Rotan, Tx., U.S. (rô-tăn')	78	32°51'N	100°27'W
Rothenburg, Ger.	124	49°20'N	10°10'E
Rotherham, Eng., U.K. (rŏdh'ẽr-ăm)	114a	53°26'N	1°21'W
Rotherham, co., Eng., U.K.	114a	53°52'N	1°45'W
Rothesay, Can. (rôth'sȧ)	60	45°23'N	66°00'W
Rothesay, Scot., U.K.	120	55°50'N	3°14'W
Rothwell, Eng., U.K.	114a	53°44'N	1°30'W
Roti, Pulau, i., Indon. (rō'tē)	168	10°30'S	122°52'E
Roto, Austl. (rō'tō)	176	33°07'S	145°30'E
Rotorua, N.Z.	177	38°07'S	176°17'E
Rotterdam, Neth. (rôt'ẽr-däm')	110	51°55'N	4°27'E
Rottweil, Ger. (rōt'vīl)	124	48°10'N	8°36'E
Roubaix, Fr. (ro͞o-bĕ')	126	50°42'N	3°10'E
Rouen, Fr. (ro͞o-äN')	110	49°25'N	1°05'E
Rouge, r., Can. (ro͞ozh)	62d	43°53'N	79°21'W
Rouge, r., Can.	59	46°40'N	74°50'W
Rouge, r., Mi., U.S.	69b	42°30'N	83°15'W
Rough River Reservoir, res., Ky., U.S.	66	37°45'N	86°10'W
Round Lake, Il., U.S.	69a	42°21'N	88°05'W
Round Pond, l., Can.	61	48°15'N	55°57'W
Round Rock, Tx., U.S.	81	30°31'N	97°41'W
Round Top, mtn., Or., U.S. (tŏp)	74c	45°41'N	123°22'W
Roundup, Mt., U.S. (round'ŭp)	73	46°25'N	108°35'W
Rousay, i., Scot., U.K. (ro͞o'zȧ)	120a	59°10'N	3°04'W
Rouyn, Can. (ro͞on)	51	48°22'N	79°03'W
Rovaniemi, Fin. (rô'vȧ-nyĕ'mĭ)	116	66°29'N	25°45'E
Rovato, Italy (rô-vä'tō)	130	45°33'N	10°00'E
Roven'ki, Russia	133	49°54'N	38°54'E
Roven'ky, Ukr.	133	48°06'N	39°44'E
Rovereto, Italy (rō-vå-rā'tô)	130	45°53'N	11°05'E
Rovigo, Italy (rô-vē'gô)	130	45°05'N	11°48'E
Rovinj, Cro. (rō'ēn')	130	45°05'N	13°40'E
Rovira, Col. (rô-vē'rä)	100a	4°14'N	75°13'W
Rovuma (Ruvuma), r., Afr.	191	10°50'S	39°50'E
Rowley, Ma., U.S. (rou'lē)	61a	42°43'N	70°53'W
Roxana, Il., U.S. (rŏks'ăn-nȧ)	75e	38°51'N	90°05'W
Roxas, Phil. (rō-xäs)	168	11°30'N	122°47'E
Roxo, Cap, c., Sen.	188	12°20'N	16°43'W
Roy, N.M., U.S. (roi)	78	35°54'N	104°09'W
Roy, Ut., U.S.	75b	41°10'N	112°02'W
Royal, i., Bah.	92	25°30'N	76°50'W
Royal Canal, can., Ire. (roi-ăl)	120	53°28'N	6°45'W
Royal Natal National Park, rec., S. Afr.	187c	28°35'S	28°54'E
Royal Oak, Can. (roi'ăl ōk)	74a	48°30'N	123°24'W
Royal Oak, Mi., U.S.	69b	42°29'N	83°09'W
Royalton, Mi., U.S. (roi'ăl-tŭn)	66	42°00'N	86°25'W
Royan, Fr. (rwä-yäN')	126	45°40'N	1°02'W
Roye, Fr. (rwä)	126	49°43'N	2°40'E
Royersford, Pa., U.S. (rô' yẽrz-fẽrd)	68f	40°11'N	75°32'W
Royston, Ga., U.S. (roiz'tŭn)	82	34°15'N	83°06'W
Royton, Eng., U.K. (roi'tŭn)	114a	53°34'N	2°07'W
Rozay-en-Brie, Fr. (rō-zā-ĕN-brē')	127b	48°41'N	2°57'E
Rozdil'na, Ukr.	133	46°47'N	30°08'E
Rozhaya, r., Russia (rô'zhȧ-yȧ)	142b	55°20'N	37°37'E
Rozivka, Ukr.	133	47°14'N	36°35'E
Rožňava, Slvk. (rôzh'nyȧ-vȧ)	125	48°39'N	20°32'E
Rtishchevo, Russia ('r-tĭsh'chĕ-vô)	137	52°15'N	43°40'E
Ru, r., China (ro͞o)	162	33°07'N	114°18'E
Ruacana Falls, wtfl., Afr.	186	17°15'S	14°45'E
Ruaha National Park, rec., Tan.	191	7°15'S	34°50'E
Ruapehu, vol., N.Z. (rō-ȧ-pā'ho͞o)	175a	39°15'S	175°37'E
Rub' al Khali *see* Ar Rub' al Khālī, des., Asia	154	20°00'N	51°00'E
Rubeho Mountains, mts., Tan.	191	6°45'S	36°15'E
Rubidoux, Ca., U.S.	75a	33°59'N	117°24'W
Rubizhne, Ukr.	133	48°53'N	38°29'E
Rubondo Island, i., Tan.	191	2°10'S	31°55'E
Rubtsovsk, Russia	134	51°31'N	81°17'E
Ruby, Ak., U.S. (ro͞o'bē)	64a	64°38'N	155°22'W
Ruby, l., Nv., U.S.	76	40°11'N	115°20'W
Ruby, r., Mt., U.S.	73	45°06'N	112°10'W
Ruby Mountains, mts., Nv., U.S.	76	40°11'N	115°36'W
Rudkøbing, Den. (ro͞odh'kûb-ĭng)	122	54°56'N	10°44'E
Rüdnitz, Ger. (rüd'nētz)	115b	52°44'N	13°38'E
Rudolf, Lake, l., Afr. (ro͞o'dôlf)	185	3°30'N	36°05'E
Rufā'ah, Sudan (ro͞o-fä'ä)	185	14°52'N	33°30'E
Ruffec, Fr. (rü-fĕk')	126	46°03'N	0°11'E
Rufiji, r., Tan. (ro͞o-fē'jē)	187	8°00'S	38°00'E
Rufisque, Sen. (rü-fēsk')	184	14°43'N	17°17'W
Rufunsa, Zam.	191	15°05'S	29°40'E
Rufus Woods, Wa., U.S.	72	48°02'N	119°33'W
Rugao, China (ro͞o-gou)	164	32°24'N	120°33'E
Rugby, Eng., U.K. (rŭg'bē)	114a	52°22'N	1°15'W
Rugby, N.D., U.S.	70	48°22'N	100°00'W
Rugeley, Eng., U.K. (ro͞oj'lē)	114a	52°46'N	1°56'W
Rügen, i., Ger. (rü'ghĕn)	112	54°28'N	13°47'E
Ruhnu-Saar, i., Est. (ro͞onō-sä'âr)	123	57°46'N	23°15'E
Ruhr, r., Ger. (ro͝or)	124	51°18'N	8°17'E
Rui'an, China (rwā-än)	165	27°48'N	120°40'E
Ruiz, Mex. (ro͞oē'z)	88	21°55'N	105°09'W
Ruiz, Nevado del, vol., Col. (nĕ-vä'dô-dĕl-ro͞oē'z)	100a	4°52'N	75°20'W
Rūjiena, Lat. (ro͞o'yĭ-ă-nä)	123	57°54'N	25°19'E
Ruki, r., D.R.C.	190	0°05'S	18°55'E
Rukwa, Lake, l., Tan. (ro͞ok-wä')	186	8°00'S	32°25'E

ng-sing; ŋ-baŋk; N-nasalized n; nŏd; cŏmmit; ōld; ôbey; ôrder; oi-boil; fo͞od; ȯ-as oo in foot; ou-out; s-soft; sh-dish; th-thin; pūre; ûnite; ûrn; stŭd; circŭs; ü-as in French tu; '-indeterminate vowel.

296 Rum – Saint George

PRONOUNCING INDEX

PLACE (Pronunciation)	PAGE	LAT.	LONG.
Rum, r., Mn., U.S. (rŭm)	71	45°52'N	93°45'W
Ruma, Serb. (rōō'mà)	131	45°00'N	19°53'E
Rumbek, Sudan (rŭm'bĕk)	185	6°52'N	29°43'E
Rum Cay, i., Bah.	93	23°40'N	74°50'W
Rumford, Me., U.S. (rŭm'fẽrd)	60	44°32'N	70°35'W
Rummah, Wādī ar, val., Sau. Ar.	154	26°17'N	41°45'E
Rummānah, Egypt	153a	31°01'N	32°39'E
Runan, China (rōō-nän)	164	32°59'N	114°22'E
Runcorn, Eng., U.K. (rŭŋ'kôrn)	114a	53°20'N	2°44'W
Ruo, r., China (rwô)	160	41°15'N	100°46'E
Rupat, i., Indon. (rōō'pät)	153b	1°55'N	101°35'E
Rupat, Selat, strt., Indon.	153b	1°55'N	101°17'E
Rupert, Id., U.S. (rōō'pẽrt)	73	42°36'N	113°41'W
Rupert, Rivière de, r., Can.	53	51°35'N	76°30'W
Ruse, Blg. (rōō'sĕ) (rŏ'sĕ)	110	43°50'N	25°59'E
Rushan, China (rōō-shän)	162	36°54'N	121°31'E
Rush City, Mn., U.S.	71	45°40'N	92°59'W
Rushville, Il., U.S. (rŭsh'vĭl)	79	40°08'N	90°34'W
Rushville, In., U.S.	66	39°35'N	85°30'W
Rushville, Ne., U.S.	70	42°43'N	102°27'W
Rusizi, r., Afr.	191	3°00'S	29°05'E
Rusk, Tx., U.S. (rŭsk)	81	31°49'N	95°09'W
Ruskin, Can. (rŭs'kĭn)	74d	49°10'N	122°25'W
Russ, r., Aus.	115e	48°12'N	16°55'E
Russas, Braz. (rōō's-säs)	101	4°48'S	37°50'W
Russell, Can. (rŭs'ĕl)	50	50°47'N	101°15'W
Russell, Can.	62c	45°15'N	75°22'W
Russell, Ca., U.S.	74b	37°39'N	122°08'W
Russell, Ks., U.S.	78	38°51'N	98°51'W
Russell, Ky., U.S.	66	38°30'N	82°45'W
Russel Lake, l., Can.	57	56°15'N	101°30'W
Russell Islands, is., Sol. Is.	175	9°16'S	158°30'E
Russellville, Al., U.S. (rŭs'ĕl-vĭl)	82	34°29'N	87°44'W
Russellville, Ar., U.S.	79	35°16'N	93°08'W
Russellville, Ky., U.S.	82	36°48'N	86°51'W
Russia, nation, Eur., Asia	134	61°00'N	60°00'E
Russian, r., Ca., U.S. (rŭsh'ăn)	76	38°59'N	123°10'W
Rustavi, Geor.	138	41°33'N	45°02'E
Rustenburg, S. Afr. (rŭs'tĕn-bûrg)	192c	25°40'S	27°15'E
Ruston, La., U.S. (rŭs'tŭn)	81	32°32'N	92°39'W
Ruston, Wa., U.S.	74a	47°18'N	122°30'W
Rute, Spain (rōō'tā)	128	38°20'N	4°34'W
Ruth, Nv., U.S. (rōōth)	76	39°17'N	115°00'W
Ruthenia, hist. reg., Ukr.	125	48°25'N	23°00'E
Rutherfordton, N.C., U.S. (rŭdh'ẽr-fẽrd-tŭn)	83	35°23'N	81°58'W
Rutland, Vt., U.S.	67	43°35'N	72°55'W
Rutledge, Md., U.S. (rŭt'lĕdj)	68e	39°34'N	76°33'W
Rutog, China	160	33°29'N	79°26'E
Rutshuru, D.R.C. (rōōt-shōō'rōō)	186	1°11'S	29°27'E
Ruvo, Italy (rōō'vô)	130	41°07'N	16°32'E
Ruvuma, r., Afr.	186	11°30'S	37°00'E
Ruza, Russia (rōō'zà)	132	55°42'N	36°12'E
Ruzhany, Bela. (rò-zhän'ĭ)	125	52°49'N	24°54'E
Rwanda, nation, Afr.	186	2°10'S	29°37'E
Ryabovo, Russia (ryä'bô-vô)	142c	59°24'N	31°08'E
Ryazan', Russia (ryä-zän'')	134	54°37'N	39°43'E
Ryazan', prov., Russia	132	54°10'N	39°37'E
Ryazhsk, Russia (ryäzh'sk')	136	53°43'N	40°04'E
Rybachiy, Poluostrov, pen., Russia	136	69°50'N	33°20'E
Rybatskoye, Russia	142c	59°50'N	30°31'E
Rybinsk, Russia	134	58°02'N	38°52'E
Rybinskoye, res., Russia	134	58°23'N	38°15'E
Rybnik, Pol. (rĭb'nēk)	125	50°06'N	18°37'E
Ryde, Eng., U.K. (rīd)	120	50°43'N	1°16'W
Rye, N.Y., U.S. (rī)	68a	40°58'N	73°42'W
Ryl'sk, Russia (rēl''sk)	137	51°33'N	34°42'E
Ryōtsu, Japan (ryōt'sōō)	166	38°02'N	138°23'E
Rypin, Pol. (rĭ'pēn)	125	53°04'N	19°25'E
Rysy, mtn., Eur.	125	49°12'N	20°04'E
Ryukyu Islands *see* Nansei-shotō, is., Japan	161	27°30'N	127°00'E
Rzeszów, Pol. (zhå-shôf)	117	50°02'N	22°00'E
Rzhev, Russia ('r-zhĕf)	134	56°16'N	34°17'E
Rzhyshchiv, Ukr.	133	49°58'N	31°05'E

S

PLACE (Pronunciation)	PAGE	LAT.	LONG.
Saale, r., Ger. (sä-lĕ)	124	51°14'N	11°52'E
Saalfeld, Ger. (säl'fĕlt)	124	50°38'N	11°20'E
Saarbrücken, Ger. (zähr'brü-kĕn)	117	49°15'N	7°01'E
Saaremaa, i., Est.	136	58°25'N	22°30'E
Saavedra, Arg. (sä-ä-vā'drä)	102	37°45'S	62°23'W
Saba, i., Neth. Ant. (sä'bä)	91b	17°39'N	63°20'W
Šabac, Serb. (shä'bàts)	119	44°45'N	19°49'E
Sabadell, Spain (sä-bä-dhäl')	118	41°32'N	2°07'E
Sabah, hist. reg., Malay.	168	5°10'N	116°25'E
Sabana, Archipiélago de, is., Cuba	92	23°05'N	80°00'W
Sabana, Río, r., Pan. (sä-bä'nä)	91	8°40'N	78°02'W
Sabana de la Mar, Dom. Rep. (sä-bä'nä dā lä mär')	93	19°05'N	69°30'W
Sabana de Uchire, Ven. (sä-bà'nä dĕ ōō-chē'rĕ)	101b	10°02'N	65°32'W
Sabanagrande, Hond. (sä-bä'nä-grä'n-dĕ)	90	13°47'N	87°16'W
Sabanalarga, Col. (sä-bà'nä-lär'gä)	100	10°38'N	75°02'W
Sabanas Páramo, mtn., Col. (sä-bà'näs pà'rä-mô)	100a	6°28'N	76°08'W
Sabancuy, Mex. (sä-bäŋ-kwē')	89	18°58'N	91°09'W
Sabang, Indon. (sä'bäng)	168	5°52'N	95°26'E
Sabaudia, Italy (sà-bou'dē-ä)	130	41°19'N	13°00'E
Sabetha, Ks., U.S. (sȧ-bĕth'ȧ)	79	39°54'N	95°49'W
Sabi (Rio Save), r., Afr. (sä'bē)	186	20°18'S	32°07'E
Sabile, Lat. (sä'bē-lĕ)	123	57°03'N	22°34'E
Sabinal, Tx., U.S. (sȧ-bī'nȧl)	80	29°19'N	99°27'W
Sabinal, Cayo, i., Cuba (kä'yō sä-bē-näl')	92	21°40'N	77°20'W
Sabinas, Mex.	86	28°05'N	101°30'W
Sabinas, r., Mex. (sä-bē'näs)	80	26°37'N	99°52'W
Sabinas, Río, r., Mex. (rē'ō sä-bē'näs)	80	27°25'N	100°33'W
Sabinas Hidalgo, Mex. (ē-däl'gô)	80	26°30'N	100°10'W
Sabine, Tx., U.S. (sȧ-bēn')	81	29°44'N	93°54'W
Sabine, r., U.S.	65	32°00'N	94°30'W
Sabine, Mount, mtn., Ant.	178	72°05'S	169°10'E
Sabine Lake, l., La., U.S.	81	29°53'N	93°41'W
Sablayan, Phil. (säb-lä-yän')	169a	12°49'N	120°47'E
Sable, Cape, c., Can. (sā'b'l)	53	43°25'N	65°24'W
Sable, Cape, c., Fl., U.S.	65	25°12'N	81°10'W
Sables, Rivière aux, r., Can.	59	49°00'N	70°20'W
Sablé-sur-Sarthe, Fr. (säb-lā-sür-särt')	126	47°50'N	0°17'W
Sablya, Gora, mtn., Russia	136	64°50'N	59°00'E
Sàbor, r., Port. (sä-bōr')	128	41°18'N	6°54'W
Sabunchu, Azer.	138	40°26'N	49°56'E
Sabzevār, Iran	157	36°13'N	57°42'E
Sac, r., Mo., U.S. (sôk)	79	38°11'N	93°45'W
Sacandaga Reservoir, res., N.Y., U.S. (să-kăn-dȧ'gȧ)	67	43°10'N	74°15'W
Sacavém, Port. (sä-kä-vĕn')	129b	38°47'N	9°06'W
Sacavém, r., Port.	129b	38°52'N	9°06'W
Sac City, Ia., U.S. (sŏk)	70	42°25'N	95°00'W
Sachigo Lake, l., Can. (săch'ĭ-gō)	57	53°49'N	92°08'W
Sachsen, hist. reg., Ger. (zäk'sĕn)	124	50°45'N	12°17'E
Sacketts Harbor, N.Y., U.S. (săk'ĕts)	67	43°55'N	76°05'W
Sackville, Can. (săk'vĭl)	60	45°54'N	64°22'W
Saco, Me., U.S. (sô'kô)	60	43°30'N	70°28'W
Saco, r., Braz. (sä'kô)	102b	22°20'S	43°26'W
Saco, r., Me., U.S.	60	43°53'N	70°46'W
Sacramento, Mex.	80	25°45'N	103°22'W
Sacramento, Mex.	80	27°05'N	101°45'W
Sacramento, Ca., U.S. (săk-rȧ-mĕn'tō)	64	38°35'N	121°30'W
Sacramento, r., Ca., U.S.	76	40°20'N	122°07'W
Şa'dah, Yemen	154	16°50'N	43°45'E
Saddle Lake Indian Reserve, I.R., Can.	55	54°00'N	111°40'W
Saddle Mountain, mtn., Or., U.S. (săd''l)	74c	45°58'N	123°40'W
Sadiya, India (sŭ-dē'yä)	155	27°53'N	95°35'E
Sado, i., Japan (sä'dō)	161	38°05'N	138°26'E
Sado, r., Port. (sä'dō)	128	38°15'N	8°20'W
Saeby, Den. (sĕ'bü)	122	57°21'N	10°29'E
Saeki, Japan (sä'ā-kē)	166	32°56'N	131°51'E
Säffle, Swe.	122	59°10'N	12°55'E
Safford, Az., U.S. (săf'fẽrd)	77	32°50'N	109°45'W
Safi, Mor. (sä'fē) (ȧs'fē)	184	32°24'N	9°09'W
Safid Koh, Selseleh-ye, mts., Afg.	154	34°45'N	63°58'E
Saga, Japan (sä'gä)	167	33°15'N	130°18'E
Sagami-Nada, b., Japan (sä'gä'mē nä-dä)	167	35°06'N	139°24'E
Sagamore Hills, Oh., U.S. (săg'ȧ-môr hĭlz)	69d	41°19'N	81°34'W
Saganaga, l., N.A. (să-gȧ-nä'gȧ)	71	48°13'N	91°17'W
Sāgar, India	155	23°55'N	78°45'E
Saghyz, r., Kaz.	137	48°30'N	56°10'E
Saginaw, Mi., U.S. (săg'ĭ-nô)	65	43°25'N	84°00'W
Saginaw, Mn., U.S.	75h	46°51'N	92°26'W
Saginaw, Tx., U.S.	75c	32°52'N	97°22'W
Saginaw Bay, b., Mi., U.S.	65	43°50'N	83°40'W
Saguache, Co., U.S. (sȧ-wăch')	77	38°05'N	106°10'W
Saguache Creek, r., Co., U.S.	66	38°05'N	106°40'W
Sagua de Tánamo, Cuba (sä-gwä dĕ tä'nä-mō)	93	20°40'N	75°15'W
Sagua la Grande, Cuba (sä-gwä lä grä'n-dĕ)	92	22°45'N	80°05'W
Saguaro National Park, rec., Az., U.S. (säg-wä'rō)	77	32°12'N	110°40'W
Saguenay, r., Can. (săg-ē-nā')	53	48°20'N	70°15'W
Sagunt, Spain	129	38°58'N	1°29'E
Sagunto, Spain (sä-gōn'tō)	118	39°40'N	0°17'W
Sahara, des., Afr. (sȧ-hä'rȧ)	184	23°44'N	1°40'W
Saharan Atlas, mts., Afr.	118	32°51'N	1°02'W
Sahāranpur, India (sŭ-hä'rŭn-pōōr')	155	29°58'N	77°41'E
Sahara Village, Ut., U.S. (sȧ-hä'rȧ)	75b	41°06'N	111°58'W
Sahel *see* Sudan, reg., Afr.	184	15°00'N	7°00'E
Sāhīwāl, Pak.	158	30°43'N	73°04'E
Sahuayo de Dias, Mex.	88	20°03'N	102°43'W
Saigon *see* Ho Chi Minh City, Viet.	168	10°46'N	106°34'E
Saijō, Japan (sä'ē-jō)	167	33°55'N	133°13'E
Saimaa, l., Fin. (sä'ĭ-mä)	116	61°24'N	28°45'E
Sain Alto, Mex. (sä-ēn' äl'tō)	88	23°35'N	103°13'W
Saint Adolphe, Can. (sånt a'dŏlf) (săN' tȧ-dôlf')	62f	49°40'N	97°07'W
Saint Afrique, Fr. (săN' tȧ-frēk')	126	43°58'N	2°52'E
Saint Albans, Austl. (sånt ôl'bănz)	173a	37°44'S	144°47'E
Saint Albans, Eng., U.K.	120	51°44'N	0°20'W
Saint Albans, Vt., U.S.	67	44°50'N	73°05'W
Saint Albans, W.V., U.S.	66	38°20'N	81°50'W
Saint Albert, Can. (sånt ăl'bẽrt)	55	53°38'N	113°38'W
Saint Amand-Mont Rond, Fr. (săN't ȧ-mäN' môN-rôN')	126	46°44'N	2°28'E
Saint André-Est, Can.	62a	45°33'N	74°19'W
Saint Andrews, Can.	51	45°05'N	67°03'W
Saint Andrews, Scot., U.K.	120	56°20'N	2°40'W
Saint Andrew's Channel, strt., Can.	61	46°06'N	60°28'W
Saint Anicet, Can. (sĕNt ä-nē-sĕ')	62a	45°07'N	74°23'W
Saint Ann, Mo., U.S. (sånt ăn')	75e	38°44'N	90°23'W
Sainte Anne, Guad.	91b	16°15'N	61°23'W
Saint Anne, Il., U.S.	69a	41°01'N	87°44'W
Sainte Anne, r., Can. (sănt ȧn') (sånt ăn')	59	46°55'N	71°46'W
Sainte-Anne, r., Can.	62b	47°07'N	70°50'W
Sainte Anne-des-Plaines, Can. (dā plĕN)	62a	45°46'N	73°49'W
Saint Ann's Bay, Jam.	92	18°25'N	77°15'W
Saint Anns Bay, b., Can. (ănz)	61	46°20'N	60°30'W
Saint Anselme, Can. (săN' tăN-sĕlm')	62b	46°37'N	70°58'W
Saint Anthony, Can. (săn ăn'thô-nē)	51	51°24'N	55°35'W
Saint Anthony, Id., U.S. (sånt ăn'thô-nē)	73	43°59'N	111°42'W
Saint Antoine-de-Tilly, Can.	62b	46°40'N	71°31'W
Saint Apollinaire, Can. (săN' tȧ-pôl-ē-nâr')	62b	46°36'N	71°30'W
Saint Arnoult-en-Yvelines, Fr. (săN-tär-nōō'ĕN-nēv-lēn')	127b	48°33'N	1°55'E
Saint Augustin-de-Québec, Can. (sĕn tō-güs-tēn')	62b	46°45'N	71°27'W
Saint Augustin-Deux-Montagnes, Can.	62a	45°38'N	73°59'W
Saint Augustine, Fl., U.S. (sånt ô'gŭs-tēn)	65	29°53'N	81°21'W
Sainte Barbe, Can. (sånt bärb')	62a	45°14'N	74°12'W
Saint Barthélemy, i., Guad.	91b	17°55'N	62°32'W
Saint Bees Head, c., Eng., U.K. (sånt bēz' hĕd)	120	54°30'N	3°40'W
Saint Benoît, Can. (sĕN bē-nōō-ä')	62a	45°34'N	74°05'W
Saint Bernard, La., U.S. (bẽr-närd')	68d	29°52'N	89°52'W
Saint Bernard, Oh., U.S.	69f	39°10'N	84°30'W
Saint Bride, Mount, mtn., Can. (sånt brīd)	55	51°30'N	115°57'W
Saint Brieuc, Fr. (săN' brēs')	117	48°32'N	2°47'W
Saint Bruno, Can. (brū'nō)	62a	45°31'N	73°20'W
Saint Canut, Can. (săN' kȧ-nū')	62a	45°43'N	74°04'W
Saint Casimir, Can. (kȧ-zē-mēr')	59	46°45'N	72°34'W
Saint Catharines, Can. (kăth'ȧ-rĭnz)	51	43°10'N	79°14'W
Saint Catherine, Mount, mtn., Gren.	91b	12°10'N	61°42'W
Saint Chamas, Fr. (sàN-shä-mä')	126a	43°32'N	5°03'E
Saint Chamond, Fr. (săN' shȧ-môN')	117	45°30'N	4°17'E
Saint Charles, Can. (săN' shärlz')	62b	46°47'N	70°57'W
Saint Charles, Il., U.S. (sånt chärlz')	69a	41°55'N	88°19'W
Saint Charles, Mi., U.S.	66	43°20'N	84°10'W
Saint Charles, Mn., U.S.	71	43°56'N	92°05'W
Saint Charles, Mo., U.S.	75e	38°47'N	90°29'W
Saint Charles, Lac, l., Can.	62b	46°56'N	71°21'W
Saint Christopher-Nevis *see* Saint Kitts and Nevis, nation, N.A.	87	17°24'N	63°30'W
Saint Clair, Mi., U.S. (sånt klâr)	66	42°55'N	82°30'W
Saint Clair, l., Can.	65	42°25'N	82°30'W
Saint Clair, r., Can.	58	42°45'N	82°25'W
Sainte Claire, Can.	62b	46°36'N	70°52'W
Saint Clair Shores, Mi., U.S.	69b	42°30'N	82°54'W
Saint Claude, Fr. (săN' klōd')	127	46°24'N	5°53'E
Saint Clet, Can. (sănt' klā')	62a	45°22'N	74°21'W
Saint Cloud, Fl., U.S. (sånt kloud')	83a	28°13'N	81°17'W
Saint Cloud, Mn., U.S.	65	45°33'N	94°08'W
Saint Constant, Can. (kŏn'stănt)	62a	45°23'N	73°34'W
Saint Croix, i., V.I.U.S. (sånt kroi')	87	17°40'N	64°43'W
Saint Croix, r., N.A. (kroi')	60	45°28'N	67°32'W
Saint Croix, r., U.S. (sånt kroi')	65	45°45'N	93°00'W
Saint Croix Indian Reservation, I.R., Wi., U.S.	71	45°40'N	92°21'W
Saint Croix Island, i., S. Afr. (săn krwä)	187c	33°48'S	25°45'E
Saint Damien-de-Buckland, Can. (sånt dä'mē-ĕn)	62b	46°37'N	70°39'W
Saint David, Can. (dā'vĭd)	62b	46°47'N	71°11'W
Saint David's Head, c., Wales, U.K.	120	51°54'N	5°25'W
Saint-Denis, Fr. (săN'dē-nē')	126	48°26'N	2°22'E
Saint Dizier, Fr. (dē-zyā')	117	48°49'N	4°55'E
Saint Dominique, Can. (sĕN dō-mē-nēk')	62a	45°19'N	74°09'W
Saint Edouard-de-Napierville, Can. (sĕN-tē-dōō-är')	62a	45°14'N	73°31'W
Saint Elias, Mount, mtn., N.A. (sånt ē-lī'ăs)	52	60°25'N	141°00'W
Saint Étienne, Fr.	117	45°26'N	4°22'E
Saint Etienne-de-Lauzon, Can. (săN' tā-tyĕn')	62b	46°39'N	71°19'W
Sainte Euphémie, Can. (sĕNt û-fē-mē')	62b	46°47'N	70°27'W
Saint Eustache, Can. (săN' tû-stàsh')	62a	45°34'N	73°54'W
Saint Eustache, Can.	62f	49°58'N	97°47'W
Sainte Famille, Can. (săN't fȧ-mē'y')	62b	46°58'N	70°58'W
Saint Félicien, Can. (săN fā-lē-syăN')	51	48°39'N	72°28'W
Sainte Felicite, Can.	60	48°54'N	67°20'W
Sainte Féréol, Can. (fa-rā-ôl')	62b	47°07'N	70°52'W
Saint Florent-sur-Cher, Fr. (săN' flô-räN'sür-shâr')	126	46°58'N	2°15'E
Saint Flour, Fr. (săN flōōr')	126	45°02'N	3°09'E
Sainte Foy, Can. (sănt fwä)	59	46°47'N	71°18'W
Saint Francis, r., Ar., U.S.	79	35°56'N	90°27'W
Saint Francis Lake, l., Can. (săn frăn'sĭs)	59	45°00'N	74°20'W
Saint François, Can. (săN'frăN-swä')	62b	47°01'N	70°49'W
Saint François de Boundji, Congo	190	1°03'S	15°22'E
Saint Francois Xavier, Can.	62f	49°55'N	97°32'W
Saint Gaudens, Fr. (gō-dăNs')	126	43°07'N	0°43'E
Sainte Genevieve, Mo., U.S. (sånt jĕn'ĕ-vēv)	79	37°58'N	90°02'W
Saint George, Austl. (sånt jôrj')	175	28°02'S	148°40'E
Saint George, Can. (săN jôrj')	51	45°08'N	66°49'W
Saint George, Can. (săN'zhôrzh')	62d	43°14'N	80°15'W
Saint George, S.C., U.S. (sånt jôrj')	83	33°11'N	80°35'W
Saint George, Ut., U.S.	77	37°05'N	113°40'W
Saint George, i., Ak., U.S.	63	56°30'N	169°40'W
Saint George, Cape, c., Can.	53a	48°28'N	59°15'W

PRONOUNCING INDEX

Saint George – Salina 297

PLACE (Pronunciation)	PAGE	LAT.	LONG.
Saint George, Cape, c., Fl., U.S.	82	29°30'N	85°20'W
Saint George's, Can. (jôrj'ĕs)	51	48°26'N	58°29'W
Saint Georges, Fr. Gu.	101	3°48'N	51°47'W
Saint George's, Gren.	91b	12°02'N	61°57'W
Saint George's Bay, b., Can.	53a	48°20'N	59°00'W
Saint Georges Bay, b., Can.	61	45°49'N	61°45'W
Saint George's Channel, strt., Eur. (jôr-jĕz)	112	51°45'N	6°30'W
Saint Germain-en-Laye, Fr. (săN' zhĕr-măN-äN-lā')	126	48°53'N	2°05'E
Saint Gervais, Can. (zhĕr-vĕ')	62b	46°43'N	70°53'W
Saint Girons, Fr. (zhē-rôN')	126	42°58'N	1°08'E
Saint Gotthard Pass, p., Switz.	124	46°33'N	8°34'E
Saint Gregory, Mount, mtn., Can. (sȧnt grĕg'ẽr-ē)	61	49°19'N	58°13'W
Saint Helena, i., St. Hel.	183	16°01'S	5°16'W
Saint Helenabaai, b., S. Afr.	186	32°25'S	17°15'E
Saint Helens, Eng., U.K. (sȧnt hĕl'ĕnz)	114a	53°27'N	2°44'W
Saint Helens, Or., U.S. (hĕl'ĕnz)	74c	45°52'N	122°49'W
Saint Helens, Mount, vol., Wa., U.S.	72	46°13'N	122°10'W
Saint Helier, Jersey (hyĕl'yẽr)	126	49°12'N	2°06'W
Saint Henri, Can. (săN' hĕn'rē)	62b	46°41'N	71°04'W
Saint Hubert, Can.	62a	45°29'N	73°24'W
Saint Hyacinthe, Can.	51	45°35'N	72°55'W
Saint Ignace, Mi., U.S. (sȧnt ĭg'nȧs)	71	45°51'N	84°39'W
Saint Ignace, i., Can. (săN' ĭg'nȧs)	58	48°47'N	88°14'W
Saint Irenee, Can. (săN' tē-rā-nā')	59	47°34'N	70°15'W
Saint Isidore-de-Laprairie, Can.	62a	45°18'N	73°41'W
Saint Isidore-de-Prescott, Can. (săN' ĭz'ĭ-dôr-prĕs-kŏt)	62c	45°23'N	74°54'W
Saint Isidore-Dorchester, Can. (dŏr-chĕs'tẽr)	62b	46°35'N	71°05'W
Saint Jacob, Il., U.S. (jā-kŏb)	75e	38°43'N	89°46'W
Saint James, Mn., U.S. (sȧnt jāmz')	71	43°58'N	94°37'W
Saint James, Mo., U.S.	79	37°59'N	91°37'W
Saint James, Cape, c., Can.	54	51°58'N	131°00'W
Saint Janvier, Can. (săN' zhäN-vyā')	62a	45°43'N	73°56'W
Saint Jean, Can. (săN' zhäN')	51	45°20'N	73°15'W
Saint Jean, Can.	62b	46°55'N	70°54'W
Saint Jean, Lac, l., Can.	53	48°35'N	72°00'W
Saint Jean-Chrysostome, Can. (krĭ-zōs-tōm')	62b	46°43'N	71°12'W
Saint Jean-d'Angely, Fr. (däN-zhā-lē')	126	45°56'N	0°33'W
Saint Jean-de-Luz, Fr. (dē lüz')	126	43°23'N	1°40'W
Saint Jérôme, Can. (sȧnt jĕ-rōm') (săN zhā-rōm')	62a	45°47'N	74°00'W
Saint Joachim-de-Montmorency, Can. (sȧnt jō'ȧ-kĭm)	62b	47°04'N	70°51'W
Saint John, Can. (sȧnt jŏn)	51	45°16'N	66°03'W
Saint John, In., U.S.	69a	41°27'N	87°29'W
Saint John, Ks., U.S.	78	37°59'N	98°44'W
Saint John, N.D., U.S.	70	48°57'N	99°42'W
Saint John, i., V.I.U.S.	87b	18°16'N	64°48'W
Saint John, r., N.A.	53	47°00'N	68°00'W
Saint John, Cape, c., Can.	61	50°00'N	55°32'W
Saint Johns, Antig.	91b	17°07'N	61°50'W
Saint John's, Can. (jŏns)	53a	47°34'N	52°43'W
Saint Johns, Az., U.S. (jŏnz)	77	34°30'N	109°25'W
Saint Johns, Mi., U.S.	66	43°05'N	84°35'W
Saint Johns, r., Fl., U.S.	65	29°54'N	81°32'W
Saint Johnsbury, Vt., U.S. (jŏnz'bĕr-ē)	67	44°25'N	72°00'W
Saint Joseph, Dom.	91b	15°25'N	61°26'W
Saint Joseph, Mi., U.S.	66	42°05'N	86°30'W
Saint Joseph, Mo., U.S. (sȧnt jō-sĕf)	65	39°44'N	94°49'W
Saint Joseph, i., Can.	66	46°15'N	83°55'W
Saint Joseph, l., Can. (jō'zhûf)	53	51°31'N	90°40'W
Saint Joseph, r., Mi., U.S. (sȧnt jō'sĕf)	66	41°45'N	85°50'W
Saint Joseph Bay, b., Fl., U.S. (jō'zhûf)	82	29°48'N	85°26'W
Saint Joseph-de-Beauce, Can. (sĕN zhō-zĕf' dĕ bōs)	59	46°18'N	70°52'W
Saint Joseph-du-Lac, Can. (sĕN zhō-zĕf' dü läk)	62a	45°32'N	74°00'W
Saint Joseph Island, i., Tx., U.S. (sȧnt jō-sĕf)	81	27°58'N	96°50'W
Saint Junien, Fr. (săN'zhü-nyăN')	126	45°53'N	0°54'E
Sainte Justine-de-Newton, Can. (sȧnt jŭs-tēn')	62a	45°22'N	74°22'W
Saint Kilda, Austl.	173a	37°52'S	144°59'E
Saint Kilda, i., Scot., U.K. (kĭl'dȧ)	120	57°50'N	8°32'W
Saint Kitts, i., St. K./N. (sȧnt kĭtts)	87	17°24'N	63°30'W
Saint Kitts and Nevis, nation, N.A.	87	17°24'N	63°30'W
Saint Lambert, Can.	67	45°29'N	73°29'W
Saint Lambert-de-Lévis, Can.	62b	46°35'N	71°12'W
Saint Laurent, Can. (săN'lō-räN)	62a	45°31'N	73°41'W
Saint Laurent, Fr. Gu.	101	5°27'N	53°56'W
Saint Laurent-d'Orleans, Can.	62b	46°52'N	71°00'W
Saint Lawrence, Can. (sȧnt lô'rĕns)	61	46°55'N	55°23'W
Saint Lawrence, i., Ak., U.S. (sȧnt lô'rĕns)	64a	63°10'N	172°12'W
Saint Lawrence, r., N.A.	53	48°24'N	69°30'W
Saint Lawrence, Gulf of, b., Can.	53	48°00'N	62°00'W
Saint Lazare, Can. (săN'lá-zár')	62b	46°39'N	70°48'W
Saint Lazare-de-Vaudreuil, Can.	62a	45°24'N	74°08'W
Saint Léger-en-Yvelines, Fr. (săN-lā-zhĕ'ĕN-nēv-lēn')	127b	48°43'N	1°45'E
Saint Leonard, Can. (sȧnt lĕn'árd)	60	47°10'N	67°56'W
Saint Léonard, Can.	62a	45°36'N	73°35'W
Saint Leonard, Md., U.S.	68e	38°29'N	76°31'W
Saint Lô, Fr.	117	49°07'N	1°05'W
Saint-Louis, Sen.	184	16°02'N	16°30'W
Saint Louis, Mi., U.S. (sȧnt lo͞o'ĭs)	66	43°25'N	84°35'W
Saint Louis, Mo., U.S. (sȧnt lo͞o'ĭs) (lo͞o'ē)	65	38°39'N	90°15'W

PLACE (Pronunciation)	PAGE	LAT.	LONG.
Saint Louis, r., Mn., U.S. (sȧnt lo͞o'ĭs)	71	46°57'N	92°58'W
Saint Louis, Lac, l., Can. (săN' lo͞o-ē')	62a	45°24'N	73°51'W
Saint Louis-de-Gonzague, Can. (săN' lo͞o ē')	62a	45°13'N	74°00'W
Saint Louis Park, Mn., U.S.	75g	44°56'N	93°21'W
Saint Lucia, nation, N.A.	87	13°54'N	60°40'W
Saint Lucia Channel, strt., N.A. (lū'shĭ-ȧ)	91b	14°15'N	61°00'W
Saint Lucie Canal, can., Fl., U.S. (lū'sē)	83a	26°57'N	80°25'W
Saint Magnus Bay, b., Scot., U.K. (măg'nŭs)	120a	60°25'N	2°09'W
Saint Malo, Fr. (săN' má-lō')	117	48°40'N	2°02'W
Saint Malo, Golfe de, b., Fr. (gôlf-dĕ-săN-mä-lō')	117	48°50'N	2°49'W
Saint Marc, Haiti (săN' márk')	93	19°10'N	72°40'W
Saint-Marc, Canal de, strt., Haiti	93	19°05'N	73°15'W
Saint Marcellin, Fr. (mär-sĕ-lăN')	127	45°08'N	5°15'E
Saint Margarets, Md., U.S.	68e	39°02'N	76°30'W
Sainte Marie, Cap, c., Madag.	187	25°31'S	45°00'E
Sainte-Marie-aux-Mines, Fr. (săN'tĕ-mä-rē'ō-mēn')	127	48°14'N	7°08'E
Sainte Marie-Beauce, Can. (săNt'má-rē')	59	46°27'N	71°03'W
Saint Maries, Id., U.S. (sȧnt mā'rēs)	72	47°18'N	116°34'W
Saint Martin, i., N.A. (mär'tĭn)	91b	18°06'N	62°54'W
Sainte Martine, Can.	62a	45°14'N	73°37'W
Saint Martins, Can. (mär'tĭnz)	60	45°21'N	65°32'W
Saint Martinville, La., U.S. (mär'tĭn-vĭl)	81	30°08'N	91°50'W
Saint Mary, r., Can. (mā'rē)	55	49°25'N	113°00'W
Saint Mary, Cape, c., Gam.	188	13°28'N	16°40'W
Saint Mary Reservoir, res., Can.	55	49°30'N	113°00'W
Saint Marys, Austl. (mā'rēz)	176	41°40'S	148°10'E
Saint Marys, Can.	58	43°15'N	81°10'W
Saint Marys, Ga., U.S.	83	30°43'N	81°35'W
Saint Mary's, Ks., U.S.	79	39°12'N	96°03'W
Saint Mary's, Oh., U.S.	66	40°30'N	84°25'W
Saint Marys, Pa., U.S.	67	41°25'N	78°30'W
Saint Marys, W.V., U.S.	66	39°20'N	81°15'W
Saint Marys, r., N.A.	75k	46°27'N	84°33'W
Saint Marys, r., U.S.	83	30°37'N	82°05'W
Saint Mary's Bay, b., Can.	61	46°50'N	53°47'W
Saint Mary's Bay, b., Can.	60	44°20'N	66°10'W
Saint Mathew, S.C., U.S. (măth'ū)	83	33°40'N	80°46'W
Saint Matthew, i., Ak., U.S.	63	60°25'N	172°10'W
Saint Matthews, Ky., U.S. (măth'ūz)	69h	38°15'N	85°39'W
Saint Maur-des-Fossés, Fr.	127b	48°48'N	2°29'E
Saint Maurice, r., Can. (săN' mō-rēs') (sȧnt mô'rĭs)	53	47°20'N	72°55'W
Saint Michael, Ak., U.S. (sȧnt mī'kĕl)	63	63°22'N	162°20'W
Saint Michel, Can. (săN'mē-shĕl')	62b	46°52'N	70°54'W
Saint Michel, Bras, r., Can.	62b	46°47'N	70°51'W
Saint Michel-de-l'Atalaye, Haiti	93	19°25'N	72°20'W
Saint Michel-de-Napierville, Can.	62a	45°14'N	73°34'W
Saint Mihiel, Fr. (săN' mē-yĕl')	127	48°53'N	5°30'E
Saint Nazaire, Fr. (săN'ná-zâr')	110	47°18'N	2°13'W
Saint Nérée, Can. (nā-rā')	62b	46°43'N	70°43'W
Saint Nicolas, Can. (ne-kô-lä')	62b	46°42'N	71°22'W
Saint Nicolas, Cap, c., Haiti	93	19°45'N	73°35'W
Saint Omer, Fr. (săN'tô-mâr')	126	50°44'N	2°16'E
Saint Pascal, Can. (sĕN pă-skăl')	60	47°32'N	69°48'W
Saint Paul, Can. (sȧnt pôl')	50	53°59'N	111°17'W
Saint Paul, Mn., U.S.	65	44°57'N	93°05'W
Saint Paul, Ne., U.S.	70	41°13'N	98°28'W
Saint Paul, i., Can.	61	47°15'N	60°10'W
Saint Paul, i., Ak., U.S.	63	57°10'N	170°20'W
Saint Paul, r., Lib.	188	7°10'N	10°00'W
Saint Paul, Île, i., Afr.	3	38°43'S	77°31'E
Saint Paul Park, Mn., U.S. (pärk)	75g	44°51'N	93°00'W
Saint Pauls, N.C., U.S. (pôls)	83	34°47'N	78°57'W
Saint Peter, Mn., U.S. (pē tẽr)	71	44°20'N	93°56'W
Saint Peter Port, Guern.	126	49°27'N	2°35'W
Saint Petersburg (Sankt-Peterburg) (Leningrad), Russia	134	59°57'N	30°20'E
Saint Petersburg, Fl., U.S. (pē'tẽrz-bûrg)	65	27°47'N	82°38'W
Sainte Pétronille, Can. (sĕNt pĕt-rō-nēl')	62b	46°51'N	71°08'W
Saint Philémon, Can. (sĕN fĕl-mōN')	62b	46°41'N	70°28'W
Saint Philippe-d'Argenteuil, Can. (săN'fe-lēp')	62a	45°38'N	74°25'W
Saint Philippe-de-Laprairie, Can.	62a	45°20'N	73°28'W
Saint Pierre, Mart. (săN'pyâr')	91b	14°45'N	61°12'W
Saint Pierre, St. P./M.	61	46°47'N	56°11'W
Saint Pierre, i., St. P./M.	61	46°47'N	56°11'W
Saint Pierre, Lac, l., Can.	59	46°07'N	72°45'W
Saint Pierre and Miquelon, dep., N.A.	53a	46°53'N	56°40'W
Saint Pierre-d'Orléans, Can.	62b	46°53'N	71°04'W
Saint Pierre-Montmagny, Can.	62b	46°55'N	70°37'W
Saint Placide, Can. (plás'ĭd)	62a	45°32'N	74°11'W
Saint Pol-de-Léon, Fr. (săN-pô'dĕ-lā-ôN')	126	48°41'N	4°00'W
Saint Quentin, Fr. (săN'käN-tăN')	117	49°52'N	3°16'E
Saint Raphaël, Can. (rä-fä-ĕl')	62b	46°48'N	70°46'W
Saint Raymond, Can.	59	46°50'N	71°51'W
Saint Rédempteur, Can. (săN rā-dăNp-tûr')	62b	46°42'N	71°18'W
Saint Rémi, Can. (sĕN rĕ-mē')	62a	45°15'N	73°36'W
Saint Romuald-d'Etchemin, Can. (sĕN rō'mo͞o-äl)	59	46°45'N	71°14'W
Sainte Rose, Guad.	91b	16°19'N	61°45'W
Saintes, Fr.	126	45°44'N	0°41'W
Sainte Scholastique, Can. (skô-lȧs-tēk')	62a	45°39'N	74°05'W
Saint Siméon, Can.	59	47°51'N	69°55'W
Saint Stanislas-de-Kostka, Can.	62a	45°11'N	74°08'W

PLACE (Pronunciation)	PAGE	LAT.	LONG.
Saint Stephen, Can. (stē'vĕn)	51	45°12'N	66°17'W
Saint Sulpice, Can.	62a	45°50'N	73°21'W
Saint Thérèse-de-Blainville, Can. (tĕ-rĕz' dĕ blĕN-vēl')	59	45°38'N	73°51'W
Saint Thomas, Can. (tŏm'ȧs)	51	42°45'N	81°15'W
Saint Thomas, i., V.I.U.S.	87	18°22'N	64°57'W
Saint Thomas Harbor, b., V.I.U.S. (tŏm'ȧs)	87c	18°19'N	64°56'W
Saint Timothée, Can. (tē-mô-tā')	62a	45°17'N	74°03'W
Saint Tropez, Fr. (trô-pĕ')	127	43°15'N	6°42'E
Saint Valentin, Can. (văl-ĕn-tĭn)	62a	45°07'N	73°19'W
Saint Valéry-sur-Somme, Fr. (vá-lā-rē')	126	50°10'N	1°39'E
Saint Vallier, Can. (väl-yā')	62b	46°54'N	70°49'W
Saint Victor, Can. (vĭk'tẽr)	59	46°09'N	70°56'W
Saint Vincent, Gulf, b., Austl. (vĭn'sĕnt)	176	34°55'S	138°00'E
Saint Vincent and the Grenadines, nation, N.A.	87	13°20'N	60°50'W
Saint Vincent Passage, strt., N.A.	91b	13°35'N	61°10'W
Saint Walburg, Can.	50	53°39'N	109°12'W
Saint Yrieix-la-Perche, Fr. (ē-rē-ĕ)	126	45°30'N	1°08'E
Saitama, dept., Japan (sī'tä-mä)	167a	35°52'N	139°40'E
Saitbaba, Russia (sá-ĕt'bá-bá)	142a	54°06'N	56°42'E
Sajama, Nevada, mtn., Bol. (nĕ-vá'dä-sä-hä'mä)	100	18°13'S	68°53'W
Sakai, Japan (sä'kä-ē)	166	34°34'N	135°28'E
Sakaiminato, Japan	167	35°33'N	133°15'E
Sakākah, Sau. Ar.	154	29°58'N	40°03'E
Sakakawea, Lake, res., N.D., U.S.	64	47°49'N	101°58'W
Sakania, D.R.C. (sá-kä'nĭ-á)	186	12°45'S	28°34'E
Sakarya, r., Tur. (sá-kär'yá)	154	40°10'N	31°00'E
Sakata, Japan (sä'kä-tä)	161	38°56'N	139°57'E
Sakchu, Kor., N. (säk'chò)	166	40°29'N	125°09'E
Sakha (Yakutia), prov., Russia	141	65°21'N	117°13'E
Sakhalin, i., Russia (sá-ká-lēn')	135	52°00'N	143°00'E
Šakiai, Lith. (shä'kĭ-ī)	123	54°59'N	23°05'E
Sakishima-guntō, is., Japan (sä'kē-shē'ma gòn'tō')	161	24°25'N	125°00'E
Sakmara, r., Russia	137	52°00'N	56°10'E
Sakomet, r., R.I., U.S. (sä-kō'mĕt)	68b	41°32'N	71°11'W
Sakurai, Japan	167b	34°31'N	135°51'E
Sakwaso Lake, l., Can. (sȧ-kwä'sō)	57	53°01'N	91°55'W
Sal, i., C.V. (säal)	184b	16°45'N	22°39'W
Sal, r., Russia (sál)	137	47°30'N	43°00'E
Sal, Cay, i., Bah. (kē săl)	92	23°45'N	80°25'W
Sala, Swe. (sô'lä)	122	59°56'N	16°34'E
Sala Consilina, Italy (sä'lä kôn-sē-lē'nä)	130	40°24'N	15°38'E
Salada, Laguna, l., Mex. (lä-gò'nä-sä-lä'dä)	76	32°34'N	115°45'W
Saladillo, Arg. (sä-lä-dēl'yō)	102	35°38'S	59°48'W
Salado, Hond. (sä-lä'dhō)	90	15°44'N	87°03'W
Salado, r., Arg.	99c	35°53'S	58°12'W
Salado, r., Arg.	102	37°00'S	67°00'W
Salado, r., Arg. (sä-lä'dō)	102	26°05'S	63°35'W
Salado, r., Mex.	86	28°00'N	102°00'W
Salado, r., Mex. (sä-lä'dō)	89	18°30'N	97°29'W
Salado Creek, r., Tx., U.S.	75d	29°23'N	98°25'W
Salado de los Nadadores, Río, r., Mex. (dĕ-lôs-nä-dä-dō'rĕs)	80	27°26'N	101°35'W
Salal, Chad	189	14°51'N	17°13'E
Salamanca, Chile (sä-lä-mä'n-kä)	99b	31°48'S	70°57'W
Salamanca, Mex.	86	20°36'N	101°10'W
Salamanca, Spain (sä-lä-mä'n-ká)	110	40°54'N	5°42'W
Salamanca, N.Y., U.S. (săl-á-măŋ'kȧ)	67	42°10'N	78°45'W
Salamat, Bahr, r., Chad (bär sä-lä-mät')	185	10°06'N	19°16'E
Salamina, Col. (sä-lä-mē'-nä)	100a	5°25'N	75°29'W
Salamína, Grc.	131	37°58'N	23°30'E
Salat-la-Canada, Fr.	126	44°52'N	1°13'E
Salaverry, Peru (sä-lä-vā'rē)	100	8°16'S	78°54'W
Salawati, i., Indon. (sä-lä-wä'tē)	169	1°07'S	130°52'E
Salawe, Tan.	191	3°19'S	32°52'E
Sala y Gómez, Isla, i., Chile	195	26°50'S	105°50'W
Salcedo, Dom. Rep. (säl-sä'dō)	93	19°25'N	70°30'W
Saldaña, r., Col. (säl-dá'n-yä)	100a	3°42'N	75°16'W
Saldanha, S. Afr.	186	32°55'S	18°05'E
Saldus, Lat. (sál'dòs)	123	56°39'N	22°30'E
Sale, Austl. (sāl)	176	38°10'S	147°07'E
Sale, Eng., U.K.	114a	53°24'N	2°20'W
Sale, r., Can. (sál'rē-vyâr')	62f	49°44'N	97°11'W
Salekhard, Russia (sŭ-lyĭ-kärt)	136	66°35'N	66°50'E
Salem, India	155	11°39'N	78°11'E
Salem, S. Afr.	187c	33°29'S	26°30'E
Salem, Il., U.S. (sā'lĕm)	66	38°40'N	89°00'W
Salem, In., U.S.	66	38°35'N	86°00'W
Salem, Ma., U.S.	61a	42°31'N	70°54'W
Salem, Mo., U.S.	79	37°36'N	91°33'W
Salem, N.H., U.S.	61a	42°46'N	71°16'W
Salem, N.J., U.S.	67	39°35'N	75°30'W
Salem, Oh., U.S.	66	40°55'N	80°50'W
Salem, Or., U.S.	64	44°55'N	123°03'W
Salem, S.D., U.S.	70	43°43'N	97°23'W
Salem, Va., U.S.	83	37°16'N	80°05'W
Salem, W.V., U.S.	66	39°15'N	80°35'W
Salemi, Italy (sä-lā'mē)	130	37°49'N	12°48'E
Salerno, Italy (sä-lĕr'nō)	118	40°27'N	14°46'E
Salerno, Golfo di, b., Italy (gôl-fō-dē)	118	40°30'N	14°40'E
Salford, Eng., U.K. (săl'fẽrd)	120	53°26'N	2°19'W
Salgótarján, Hung. (shôl'gō-tôr-yän)	125	48°06'N	19°50'E
Salhyr, r., Ukr.	133	45°25'N	34°22'E
Salida, Co., U.S. (sá-lī'dá)	78	38°31'N	106°01'W
Salies-de-Béan, Fr.	126	43°27'N	0°58'W
Salima, Mwi.	191	13°47'S	34°26'E
Salina, Ks., U.S. (sȧ-lī'nȧ)	64	38°50'N	97°37'W
Salina, Ut., U.S.	77	39°00'N	111°55'W

298 Salina – San Francisco
PRONOUNCING INDEX

PLACE (Pronunciation)	PAGE	LAT.	LONG.
Salina, i., Italy (sä-lē'nä)	130	38°35'N	14°48'E
Salina Cruz, Mex. (sä-lē'nä krōōz')	86	16°10'N	95°12'W
Salina Point, c., Bah.	93	22°10'N	74°20'W
Salinas, Mex.	86	22°38'N	101°42'W
Salinas, P.R.	87b	17°58'N	66°16'W
Salinas, Ca., U.S. (så-lē'nås)	76	36°41'N	121°40'W
Salinas, r., Mex. (sä-lē'näs)	89	16°15'N	90°31'W
Salinas, r., Ca., U.S.	76	36°33'N	121°29'W
Salinas, Bahía de, b., N.A. (bä-ē'ä-dĕ-så-lē'nås)	90	11°05'N	85°55'W
Salinas National Monument, rec., N.M., U.S.	77	34°10'N	106°05'W
Salinas Victoria, Mex. (sä-lē'näs vēk-tō'rē-ä)	80	25°59'N	100°19'W
Saline, r., Ar., U.S. (så-lēn')	79	34°06'N	92°30'W
Saline, r., Ks., U.S.	78	39°05'N	99°43'W
Salins-les-Bains, Fr. (så-lăɴ'-lā-băɴ')	127	46°55'N	5°54'E
Salisbury, Can.	60	46°03'N	65°05'W
Salisbury, Eng., U.K. (sôlz'bē-rē)	117	50°35'N	1°51'W
Salisbury, Md., U.S.	67	38°20'N	75°40'W
Salisbury, Mo., U.S.	79	39°24'N	92°47'W
Salisbury, N.C., U.S.	83	35°40'N	80°29'W
Salisbury *see* Harare, Zimb.	186	17°50'S	31°03'E
Salisbury Island, i., Can.	53	63°36'N	76°20'W
Salisbury Plain, pl., Eng., U.K.	120	51°15'N	1°52'W
Salkehatchie, r., S.C., U.S. (sô-kē-hăch'ē)	83	33°09'N	81°10'W
Sallisaw, Ok., U.S. (săl'ĭ-sô)	79	35°27'N	94°48'W
Salmon, Id., U.S. (săm'ŭn)	73	45°11'N	113°54'W
Salmon, r., Can.	54	54°00'N	123°50'W
Salmon, r., Can.	60	46°19'N	65°36'W
Salmon, r., Id., U.S.	64	45°30'N	115°45'W
Salmon, r., N.Y., U.S.	67	44°35'N	74°15'W
Salmon, r., Wa., U.S.	74c	45°44'N	122°36'W
Salmon, Middle Fork, r., Id., U.S.	72	44°50'N	114°52'W
Salmon Arm, Can.	55	50°42'N	119°16'W
Salmon Falls Creek, r., Id., U.S.	73	42°22'N	114°53'W
Salmon Gums, Austl. (gŭmz)	174	33°00'S	122°00'E
Salmon River Mountains, mts., Id., U.S.	64	44°15'N	115°44'W
Salon-de-Provence, Fr. (så-lôɴ-dē-prô-väɴs')	127	43°48'N	5°09'E
Salonika *see* Thessaloníki, Grc.	110	40°38'N	22°59'E
Salonta, Rom. (sä-lôn'tä)	125	46°46'N	21°38'E
Saloum, r., Sen.	188	14°10'N	15°45'W
Salsette Island, i., India	159b	19°12'N	72°52'E
Sal'sk, Russia (sälsk)	137	46°30'N	41°20'E
Salt, r., Az., U.S. (sôlt)	64	33°28'N	111°35'W
Salt, r., Mo., U.S.	79	39°54'N	92°11'W
Salta, Arg. (säl'tä)	102	24°50'S	65°16'W
Salta, prov., Arg.	102	25°15'S	65°00'W
Saltair, Ut., U.S. (sôlt'âr)	75b	40°46'N	112°09'W
Salt Cay, i., T./C. Is.	93	21°20'N	71°15'W
Salt Creek, r., Il., U.S. (sôlt)	69a	42°01'N	88°01'W
Saltillo, Mex. (säl-tēl'yō)	86	25°24'N	100°59'W
Salt Lake City, Ut., U.S. (sôlt lāk sĭ'tĭ)	64	40°45'N	111°52'W
Salto, Arg. (säl'tō)	99c	34°17'S	60°15'W
Salto, Ur.	102	31°18'S	57°45'W
Salto, r., Mex.	88	22°16'N	99°18'W
Salto, Serra do, mtn., Braz. (sĕ'r-rä-dô)	99a	20°26'S	43°28'W
Salto Grande, Braz. (grän'dä)	101	22°57'S	49°58'W
Salton Sea, Ca., U.S. (sôlt'ŭn)	76	33°28'N	115°43'W
Salton Sea, l., Ca., U.S.	64	33°19'N	115°50'W
Saltpond, Ghana	184	5°16'N	1°07'W
Salt River Indian Reservation, I.R., Az., U.S. (sôlt rĭv'ẽr)	77	33°40'N	112°01'W
Saltsjöbaden, Swe. (sält'shû-bäd'ĕn)	122	59°15'N	18°20'E
Saltspring Island, i., Can. (sält'sprĭng)	54	48°47'N	123°30'W
Saltville, Va., U.S. (sôlt'vĭl)	83	36°50'N	81°45'W
Saltykovka, Russia (sål-tē'kôf-kå)	142b	55°45'N	37°56'E
Salud, Mount, mtn., Pan. (sä-lōō'th)	86a	9°14'N	79°42'W
Saluda, S.C., U.S. (så-lōō'då)	83	34°02'N	81°46'W
Saluda, r., S.C., U.S.	83	34°07'N	81°48'W
Saluzzo, Italy (sä-lōōt'sō)	130	44°39'N	7°31'E
Salvador, Braz. (säl-vä-dōr') (bä-ē'à)	101	12°59'S	38°27'W
Salvador Lake, l., La., U.S.	81	29°45'N	90°20'W
Salvador Point, c., Bah.	92	24°30'N	77°45'W
Salvatierra, Mex. (säl-vä-tyĕr'rä)	88	20°13'N	100°52'W
Salween, r., Asia	152	21°00'N	98°00'E
Salyan, Azer.	137	39°40'N	49°10'E
Salzburg, Aus. (sälts'bòrgh)	117	47°48'N	13°04'E
Salzwedel, Ger. (sälts-vä'dĕl)	124	52°51'N	11°10'E
Samālūt, Egypt (sä-mä-lōōt')	156	28°17'N	30°43'E
Samana, Cabo, c., Dom. Rep.	87	19°20'N	69°00'W
Samana or Atwood Cay, i., Bah.	93	23°05'N	73°45'W
Samar, i., Phil. (sä'mär)	169	11°30'N	126°07'E
Samara (Kuybyshev), Russia	136	53°10'N	50°05'E
Samara, r., Russia	137	52°50'N	50°35'E
Samara, r., Ukr. (så-mä'rå)	133	48°47'N	35°30'E
Samarai, Pap. N. Gui. (sä-mä-rä'ē)	169	10°45'S	150°49'E
Samarinda, Indon.	168	0°30'S	117°10'E
Samarkand, Uzb. (så-mår-känt')	139	39°42'N	67°00'E
Şamaxı, Azer.	137	40°35'N	48°40'E
Samba, D.R.C.	191	4°38'S	26°22'E
Sambalpur, India (sŭm'bŭl-pòr)	155	21°30'N	84°05'E
Sāmbhar, r., India	158	27°00'N	74°58'E
Sambir, Ukr.	125	49°31'N	23°12'E
Samborombón, r., Arg.	99c	35°20'S	57°52'W
Samborombón, Bahía, b., Arg. (bä-ē'ä-säm-bô-rôm-bô'n)	99c	35°57'S	57°05'W
Sambre, r., Eur. (säɴ'br')	121	50°20'N	4°15'E
Sambungo, Ang.	190	8°39'S	20°43'E
Sammamish, r., Wa., U.S.	74a	47°43'N	122°08'W
Sammamish, Lake, l., Wa., U.S. (så-măm'ĭsh)	74a	47°35'N	122°02'W
Samoa, nation, Oc.	2	14°30'S	172°00'W

PLACE (Pronunciation)	PAGE	LAT.	LONG.
Samoa Islands, is., Oc.	170a	14°00'S	171°00'W
Samokov, Blg. (sä'mô-kôf)	131	42°20'N	23°33'E
Samora Correia, Port. (sä-mō'rä-kôr-rĕ'yä)	129b	38°55'N	8°52'W
Samorovo, Russia (så-må-rô'vô)	140	60°47'N	69°13'E
Sámos, i., Grc. (să'mŏs)	119	37°53'N	26°35'E
Samothráki, i., Grc.	119	40°23'N	25°10'E
Sampaloc Point, c., Phil. (säm-pä'lôk)	169a	14°43'N	119°56'E
Sam Rayburn Reservoir, res., Tx., U.S.	81	31°10'N	94°15'W
Samson, Al., U.S. (săm'sŭn)	82	31°06'N	86°02'W
Samsu, Kor., N. (säm'sōō')	166	41°12'N	128°00'E
Samsun, Tur. (såm'sōōn')	154	41°20'N	36°05'E
Samtredia, Geor. (såm'trĕ-dē)	137	42°18'N	42°25'E
Samuel, i., Can. (săm'ū-ĕl)	74d	48°50'N	123°10'W
Samur, r., (sä-mōōr')	137	41°40'N	47°20'E
San, Mali (sän)	184	13°18'N	4°54'W
San, r., Eur.	117	50°33'N	22°12'E
Şan'ā', Yemen (sän'ä)	154	15°17'N	44°05'E
Sanaga, r., Cam. (sä-nä'gä)	184	4°30'N	12°00'E
San Ambrosio, Isla, i., Chile (ē's-lä-dĕ-sän äm-brō'zē-ō)	97	26°40'S	80°00'W
Sanana, Pulau, i., Indon.	169	2°15'S	126°38'E
Sanandaj, Iran	154	36°44'N	46°43'E
San Andreas, Ca., U.S. (săn ăn'drē-ăs)	76	38°10'N	120°42'W
San Andreas, l., Ca., U.S.	74b	37°36'N	122°26'W
San Andrés, Col. (sän-än-drĕ's)	100a	6°57'N	75°41'W
San Andrés, Mex. (sän än-dräs')	89a	19°15'N	99°10'W
San Andrés, i., Col.	91	12°32'N	81°34'W
San Andrés, Laguna de, l., Mex.	89	22°40'N	97°50'W
San Andres Mountains, mts., N.M., U.S. (săn ăn'drē-ăs)	64	33°00'N	106°40'W
San Andrés Tuxtla, Mex. (sän-än-drā's-tōōs'tlä)	86	18°27'N	95°12'W
San Angelo, Tx., U.S. (săn ăn-jĕ-lō)	64	31°28'N	100°22'W
San Antioco, Isola di, i., Italy (ē'sō-lä-dē-sän-än-tyŏ'kô)	130	39°00'N	8°25'E
San Antonio, Chile (sän-än-tō'nyō)	102	33°34'S	71°36'W
San Antonio, Col.	100a	2°57'N	75°06'W
San Antonio, Col.	100a	3°55'N	75°28'W
San Antonio, Phil.	169a	14°57'N	120°05'E
San Antonio, Tx., U.S. (săn ăn-tō'nē-ō)	64	29°25'N	98°30'W
San Antonio, r., Tx., U.S.	81	29°00'N	97°58'W
San Antonio, Cabo, c., Cuba (kå'bô-sän-än-tō'nyô)	87	21°55'N	84°55'W
San Antonio, Lake, res., Ca., U.S.	76	36°00'N	121°13'W
San Antonio Bay, b., Tx., U.S.	81	28°20'N	97°08'W
San Antonio de Areco, Arg. (dā ä-rä'kō)	99c	34°16'S	59°30'W
San Antonio de las Vegas, Cuba	93a	22°51'N	82°23'W
San Antonio de los Baños, Cuba (då lōs bän'yōs)	92	22°54'N	82°30'W
San Antonio de los Cobres, Arg. (då lōs kō'brås)	102	24°15'S	66°29'W
San Antônio de Pádua, Braz. (dĕ-på'dwä)	99a	21°32'S	42°09'W
San Antonio de Tamanaco, Ven.	101b	9°42'N	66°03'W
San Antonio Oeste, Arg. (sän-nä-tō'nyō ô-ĕs'tå)	102	40°49'S	64°56'W
San Antonio Peak, mtn., Ca., U.S. (săn ăn-tō'nĭ-ō)	75a	34°17'N	117°39'W
Sanarate, Guat. (sä-nä-rä'tĕ)	90	14°47'N	90°12'W
San Augustine, Tx., U.S. (săn ô'gŭs-tēn)	81	31°33'N	94°08'W
San Bartolo, Mex. (sän bär-tō'lō)	89a	19°36'N	99°43'W
San Bartolo, Mex.	80	24°43'N	103°12'W
San Bartolomeo, Italy (bär-tô-lô-mā'ō)	130	41°25'N	15°04'E
San Benedetto del Tronto, Italy (bā'nå-dĕt'tô dĕl trōn'tô)	130	42°58'N	13°54'E
San Benito, Tx., U.S. (săn bĕ-nē'tō)	81	26°07'N	97°37'W
San Benito, r., Ca., U.S.	76	36°40'N	121°20'W
San Bernardino, Ca., U.S. (bûr-när-dē'nō)	64	34°07'N	117°19'W
San Bernardino Mountains, mts., Ca., U.S.	76	34°05'N	116°23'W
San Bernardo, Chile (sän bĕr-när'dō)	99b	33°35'S	70°42'W
San Blas, Mex. (sän bläs')	86	21°33'N	105°19'W
San Blas, Cape, c., Fl., U.S.	65	29°38'N	85°38'W
San Blas, Cordillera de, mts., Pan.	91	9°17'N	78°20'W
San Blas, Golfo de, b., Pan.	91	9°33'N	78°42'W
San Blas, Punta, c., Pan.	91	9°35'N	78°55'W
San Bruno, Ca., U.S. (săn brū-nō)	74b	37°38'N	122°25'W
San Buenaventura, Mex. (bwā'nå-vĕn-tōō'rä)	80	27°07'N	101°30'W
San Carlos, Chile (sän-kå'r-lŏs)	102	36°23'S	71°58'W
San Carlos, Col.	100a	6°11'N	74°58'W
San Carlos, Eq. Gui.	190	3°27'N	8°33'E
San Carlos, Mex. (sän kär'lōs)	89	17°49'N	92°33'W
San Carlos, Mex.	80	24°36'N	98°52'W
San Carlos, Nic. (sän-kä'r-lôs)	91	11°08'N	84°48'W
San Carlos, Phil.	169a	15°56'N	120°20'E
San Carlos, Ca., U.S. (săn kär'lōs)	74b	37°30'N	122°15'W
San Carlos, Ven.	100	9°36'N	68°35'W
San Carlos, r., C.R.	91	10°36'N	84°18'W
San Carlos de Bariloche, Arg.	102	41°15'S	71°26'W
San Carlos Indian Reservation, I.R., Az., U.S. (săn kär'lōs)	77	33°27'N	110°15'W
San Carlos Lake, res., Az., U.S.	77	33°05'N	110°29'W
San Casimiro, Ven. (kä-sē-mē'rō)	101b	10°01'N	67°02'W
San Cataldo, Italy (kä-täl'dō)	130	37°30'N	13°59'E
Sánchez, Dom. Rep. (sän'chĕz)	87	19°15'N	69°40'W
Sanchez, Río de los, r., Mex. (rē'ō-dĕ-lôs)	88	20°31'N	102°29'W
Sánchez Román, Mex. (rô-må'n)	88	21°48'N	103°20'W

PLACE (Pronunciation)	PAGE	LAT.	LONG.
San Clemente, Spain (sän klå-mĕn'tå)	128	39°25'N	2°24'W
San Clemente Island, i., Ca., U.S.	64	32°54'N	118°29'W
San Cristóbal, Dom. Rep. (krēs-tō'bäl)	93	18°25'N	70°05'W
San Cristóbal, Guat.	90	15°22'N	90°26'W
San Cristóbal, Ven.	100	7°43'N	72°15'W
San Cristobal, i., Sol. Is.	175	10°47'S	162°17'E
San Cristóbal de las Casas, Mex.	86	16°44'N	92°39'W
Sancti Spíritus, Cuba (säŋk'tē spē'rē-tōōs)	87	21°55'N	79°25'W
Sancti Spíritus, prov., Cuba	92	22°05'N	79°20'W
Sancy, Puy de, mtn., Fr. (pwē-dĕ-säɴ-sē')	117	45°30'N	2°53'E
Sand, i., Or., U.S. (sănd)	74c	46°16'N	124°01'W
Sand, i., Wi., U.S.	71	46°03'N	91°09'W
Sand, r., S. Afr.	187c	28°30'S	29°30'E
Sand, r., S. Afr.	192c	28°09'S	26°46'E
Sanda, Japan (sän'dä)	167	34°53'N	135°14'E
Sandakan, Malay. (sån-dä'kån)	168	5°51'N	118°03'E
Sanday, i., Scot., U.K. (sănd'ā)	120a	59°17'N	2°25'W
Sandbach, Eng., U.K. (sănd'băch)	114a	53°08'N	2°22'W
Sandefjord, Nor. (sän'dĕ-fyôr')	122	59°09'N	10°14'E
San de Fuca, Wa., U.S. (de-fōō-cä)	74a	48°14'N	122°44'W
Sanders, Az., U.S.	77	35°13'N	109°20'W
Sanderson, Tx., U.S. (săn'dẽr-sŭn)	80	30°09'N	102°24'W
Sandersville, Ga., U.S. (săn'dẽrz-vĭl)	83	32°57'N	82°50'W
Sandhammaren, c., Swe. (sänt'häm-mår)	116	55°24'N	14°37'E
Sand Hills, reg., Ne., U.S. (sănd)	70	41°57'N	101°29'W
Sand Hook, N.J., U.S. (sănd hòk)	68a	40°29'N	74°05'W
Sandhurst, Eng., U.K. (sănd'hûrst)	114b	51°20'N	0°48'W
Sandia Indian Reservation, I.R., N.M., U.S.	77	35°15'N	106°30'W
San Diego, Ca., U.S. (săn dē-ā'gō)	64	32°43'N	117°10'W
San Diego, Tx., U.S.	78	27°47'N	98°13'W
San Diego, r., Ca., U.S.	76	32°53'N	116°57'W
San Diego de la Unión, Mex. (sän dĕ-å-gô då lä ōō-nyōn')	88	21°27'N	100°52'W
Sandies Creek, r., Tx., U.S. (sănd'ēz)	81	29°13'N	97°34'W
San Dimas, Mex. (dē-mäs')	88	24°08'N	105°57'W
San Dimas, Ca., U.S. (săn dē-mås)	75a	34°07'N	117°49'W
Sandnes, Nor. (sänd'nĕs)	122	58°52'N	5°44'E
Sandoa, D.R.C. (sän-dō'ä)	186	9°39'S	23°00'E
Sandomierz, Pol. (sån-dô'myĕzh)	125	50°39'N	21°45'E
San Doná di Piave, Italy (sän dô nä' dē pyä'vē)	130	45°38'N	12°34'E
Sandoway, Mya. (sän-dô-wī')	155	18°24'N	94°28'E
Sandpoint, Id., U.S. (sănd point)	72	48°17'N	116°34'W
Sandringham, Austl. (săn'drĭng-ăm)	173a	37°57'S	145°01'E
Sandrio, Italy (sä'n-dryô)	130	46°11'N	9°53'E
Sand Springs, Ok., U.S. (sănd sprĭnz)	79	36°08'N	96°06'W
Sandstone, Austl. (sănd'stōn)	174	28°00'S	119°25'E
Sandstone, Mn., U.S.	71	46°08'N	92°53'W
Sanduo, China (sän-dwô)	162	32°49'N	119°39'E
Sandusky, Al., U.S. (săn-dŭs'kē)	68h	33°32'N	86°50'W
Sandusky, Mi., U.S.	66	43°25'N	82°50'W
Sandusky, Oh., U.S.	65	41°25'N	82°45'W
Sandusky, r., Oh., U.S.	66	41°10'N	83°20'W
Sandwich, Il., U.S. (sănd'wĭch)	66	41°35'N	88°53'W
Sandy, Or., U.S. (sănd'ē)	74c	45°24'N	122°16'W
Sandy, Ut., U.S.	75b	40°36'N	111°53'W
Sandy, r., Or., U.S.	74c	45°28'N	122°17'W
Sandy Cape, c., Austl.	175	24°25'S	153°10'E
Sandy Hook, Ct., U.S. (hòk)	68a	41°25'N	73°17'W
Sandy Lake, l., Can.	62g	53°46'N	113°58'W
Sandy Lake, l., Can.	61	49°16'N	57°00'W
Sandy Lake, l., Can.	57	53°00'N	93°07'W
Sandy Point, Tx., U.S.	81a	29°22'N	95°27'W
Sandy Point, c., Wa., U.S.	74d	48°48'N	122°42'W
Sandy Springs, Ga., U.S. (springz)	68c	33°55'N	84°23'W
San Estanislao, Para. (ĕs-tä-nēs-lä'ô)	102	24°38'S	56°20'W
San Esteban, Hond. (ĕs-tĕ'bän)	90	15°13'N	85°53'W
San Fabian, Phil. (fä-byä'n)	169a	16°14'N	120°28'E
San Felipe, Chile (fĕ-lē'på)	102	32°45'S	70°43'W
San Felipe, Mex. (fĕ-lē'pĕ)	88	21°29'N	101°13'W
San Felipe, Mex.	88	22°21'N	105°26'W
San Felipe, Ven. (fĕ-lē'pĕ)	100	10°13'N	68°45'W
San Felipe, Cayos de, is., Cuba (kä'yōs-dĕ-sän-fĕ-lē'pĕ)	92	22°00'N	83°30'W
San Felipe Creek, r., Ca., U.S. (săn fĕ-lēp'ā)	76	33°10'N	116°03'W
San Felipe Indian Reservation, I.R., N.M., U.S.	77	35°26'N	106°26'W
San Félix, Isla, i., Chile (ē's-lä-dĕ-sän fā-lēks')	97	26°20'S	80°10'W
San Fernanda, Spain (fĕr-nä'n-dä)	128	36°28'N	6°13'W
San Fernando, Arg. (fĕr-nä'n-dô)	102a	34°26'S	58°34'W
San Fernando, Chile	99b	35°36'S	70°58'W
San Fernando, Mex. (fĕr-nän'dō)	80	24°52'N	98°10'W
San Fernando, Phil. (sän fĕr-nä'n-dô)	168	16°38'N	120°19'E
San Fernando, Ca., U.S. (fĕr-nän'dō)	75a	34°17'N	118°27'W
San Fernando, r., Mex. (sän fĕr-nän'dô)	80	25°07'N	98°25'W
San Fernando de Apure, Ven. (sän-fĕr-nä'n-dō-dĕ-ä-pōō'rä)	100	7°46'N	67°29'W
San Fernando de Atabapo, Ven. (dĕ-ä-tä-bä'pô)	100	3°58'N	67°41'W
San Fernando de Henares, Spain (dĕ-ā-nä'räs)	129a	40°23'N	3°31'W
Sånfjället, mtn., Swe.	116	62°19'N	13°30'E
Sanford, Can. (săn'fẽrd)	62f	49°41'N	97°27'W
Sanford, Fl., U.S. (săn'fôrd)	65	28°46'N	81°18'W
Sanford, Me., U.S. (săn'fẽrd)	60	43°26'N	70°47'W
Sanford, N.C., U.S.	83	35°26'N	79°10'W
San Francisco, Arg. (sän frăn'sĭs'kô)	102	31°23'S	62°09'W
San Francisco, El Sal.	90	13°48'N	88°11'W

ăt; finăl; rāte; senåte; ärm; åsk; sofà; fâre; ch-choose; dh-as th in other; bē; ēvent; bĕt; recĕnt; cratẽr; g-gō; gh-guttural g; bĭt; ĭ-short neutral; rīde; ᴋ-guttural k as ch in German ich;

PRONOUNCING INDEX San Francisco – Santa Clara 299

PLACE (Pronunciation)	PAGE	LAT.	LONG.
San Francisco, Ca., U.S.	64	37°45'N	122°26'W
San Francisco, r., N.M., U.S.	77	33°35'N	108°55'W
San Francisco Bay, b., Ca., U.S. (săn frăn'sĭs'kō)	76	37°45'N	122°21'W
San Francisco del Oro, Mex. (dĕl ō'rō)	86	27°00'N	106°37'W
San Francisco del Rincón, Mex. (dĕl rēn-kōn')	88	21°01'N	101°51'W
San Francisco de Macaira, Ven. (dĕ-mä-kī'rä)	101b	9°58'N	66°17'W
San Francisco de Macoris, Dom. Rep. (dä-mä-kō'rēs)	93	19°20'N	70°15'W
San Francisco de Paula, Cuba (dā pou'lä)	93a	23°04'N	82°18'W
San Gabriel, Ca., U.S. (săn gä-brē-ĕl') (gä'brē-ĕl)	75a	34°06'N	118°06'W
San Gabriel, r., Ca., U.S.	75a	33°47'N	118°06'W
San Gabriel Chilac, Mex. (sän-gä-brē-ĕl-chē-läk')	89	18°19'N	97°22'W
San Gabriel Mts, Ca., U.S.	75a	34°17'N	118°03'W
San Gabriel Reservoir, res., Ca., U.S.	75a	34°14'N	117°48'W
Sangamon, r., Il., U.S. (săn'gá-msion)	79	40°08'N	90°08'W
Sanger, Ca., U.S. (săng'ēr)	76	36°42'N	119°33'W
Sangerhausen, Ger. (säng'ĕr-hou-zĕn)	124	51°28'N	11°17'E
Sangha, r., Afr.	185	2°40'N	16°10'E
Sangihe, Pulau, i., Indon.	169	3°30'N	125°30'E
San Gil, Col. (sän-kē'l)	100	6°32'N	73°13'W
San Giovanni in Fiore, Italy (sän jō-vän'nē ēn fyō'rā)	130	39°15'N	16°40'E
San Giuseppe Vesuviano, Italy	129c	40°36'N	14°31'E
Sangju, Kor., S. (säng'jōō')	166	36°20'N	128°07'E
Sāngli, India	155	16°56'N	74°38'E
Sangmélima, Cam.	189	2°56'N	11°59'E
San Gorgonio Mountain, mtn., Ca., U.S. (săn gôr-gō'nĭ-ō)	75a	34°06'N	116°50'W
Sangre de Cristo Mountains, mts., U.S.	64	37°45'N	105°50'W
San Gregoria, Ca., U.S. (săn grĕ-gôr'ä)	74b	37°20'N	122°23'W
Sangro, r., Italy (säŋ'grō)	130	41°38'N	13°56'E
Sangüesa, Spain (säŋ-gwĕ'sä)	128	42°36'N	1°15'W
Sanhe, China (sän-hŭ)	162	39°59'N	117°06'E
Sanibel Island, i., Fl., U.S. (săn'ĭ-bĕl)	83a	26°26'N	82°15'W
San Ignacio, Belize	90a	17°11'N	89°04'W
San Ildefonso, Cape, c., Phil. (sän-ēl-dĕ-fōn-sō)	169a	16°03'N	122°10'E
San Ildefonso o la Granja, Spain (ō lä grän'khä)	128	40°54'N	4°02'W
San Isidro, Arg. (ē-sē'drō)	99c	34°28'S	58°31'W
San Isidro, C.R.	91	9°24'N	83°43'W
San Jacinto, Phil. (sän hä-sēn'tō)	169a	12°33'N	123°43'E
San Jacinto, Ca., U.S. (săn jȧ-sĭn'tô)	75a	33°47'N	116°57'W
San Jacinto, r., Ca., U.S. (săn jȧ-sĭn'tô)	75a	33°44'N	117°14'W
San Jacinto, r., Tx., U.S.	81	30°25'N	95°05'W
San Jacinto, West Fork, r., Tx., U.S.	81	30°35'N	95°37'W
San Javier, Chile (sän-hä-vē'ĕr)	99b	35°35'S	71°43'W
San Jerónimo, Mex.	89a	19°31'N	98°46'W
San Jerónimo de Juárez, Mex. (hā-rō'nē-mô dā hwä'râz)	88	17°08'N	100°30'W
San Joaquin, Ven.	101b	10°16'N	67°47'W
San Joaquin, r., Ca., U.S. (săn hwä-kēn')	76	37°10'N	120°51'W
San Joaquin Valley, Ca., U.S.	76	36°45'N	120°30'W
San Jorge, Golfo, b., Arg. (gôl-fô-sän-kô'r-kĕ)	102	46°15'S	66°45'W
San José, C.R. (sän hō-sā')	87	9°57'N	84°05'W
San Jose, Phil.	169a	12°22'N	121°04'E
San Jose, Phil.	169a	15°49'N	120°57'E
San Jose, Ca., U.S. (săn hō-zā')	64	37°20'N	121°54'W
San José, i., Mex. (kō-sĕ')	86	25°00'N	110°35'W
San José, Isla de, i., Pan. (ē's-lä-dĕ-sän hō-sā')	91	8°17'N	79°20'W
San Jose, Rio, r., N.M., U.S. (săn hō-zā')	77	35°15'N	108°10'W
San José de Feliciano, Arg. (dā lä ĕs-kē'nä)	102	30°26'S	58°44'W
San José de Gauribe, Ven. (sän-hō-sĕ'dĕ-gaōō-rē'bĕ)	101b	9°51'N	65°49'W
San José de las Lajas, Cuba (sän-kō-sĕ'dĕ-läs-lä'käs)	93a	22°58'N	82°10'W
San José Iturbide, Mex. (ē-tōōr-bē'dĕ)	88	21°00'N	100°24'W
San Juan, Arg. (hwän')	102	31°36'S	68°29'W
San Juan, Col. (hōä'n)	100a	3°23'N	73°48'W
San Juan, Dom. Rep. (sän hwän')	93	18°50'N	71°15'W
San Juan, Phil.	169a	16°41'N	120°20'E
San Juan, P.R. (sän hwän')	87	18°30'N	66°10'W
San Juan, prov., Arg.	102	31°00'S	69°30'W
San Juan, r., Mex. (sän-hōō-än')	89	18°10'N	95°23'W
San Juan, r., N.A.	87	10°58'N	84°18'W
San Juan, r., U.S.	64	36°30'N	109°00'W
San Juan, Cabezas de, c., P.R.	87b	18°29'N	65°30'W
San Juan, Cabo, c., Eq. Gui.	190	1°08'N	9°23'E
San Juan, Pico, mtn., Cuba (pē'kô-sän-kōä'n)	92	21°55'N	80°00'W
San Juan, Río, r., Mex. (rē'ō-sän-hwän)	80	25°35'N	99°15'W
San Juan Bautista, Para. (sän hwän' bou-tēs'tä)	102	26°48'S	57°09'W
San Juan Capistrano, Mex. (sän-hōō-än' kä-pēs-trä'nō)	88	22°41'N	104°07'W
San Juan Creek, r., Ca., U.S. (săn hwän')	76	35°24'N	120°12'W
San Juan de Guadalupe, Mex. (sän hwan dā gwä-dhä-lōō'pä)	80	24°37'N	102°43'W
San Juan del Norte, Nic.	91	10°55'N	83°44'W
San Juan del Norte, Bahía de, b., Nic.	91	11°12'N	83°40'W
San Juan de los Lagos, Mex. (sän-hōō-än'dā lōs lä'gōs)	88	21°15'N	102°18'W
San Juan de los Lagos, r., Mex. (dā lōs lä'gōs)	88	21°13'N	102°12'W
San Juan de los Morros, Ven. (dĕ-lōs-mô'r-rôs)	101b	9°54'N	67°22'W
San Juan del Río, Mex.	88	20°21'N	99°59'W
San Juan del Río, Mex. (sän hwän del rē'ō)	80	24°47'N	104°29'W
San Juan del Sur, Nic. (dĕl sōōr)	86	11°15'N	85°53'W
San Juan Evangelista, Mex. (sän-hōō-ä'n-å-väŋ-kå-lĕs'ta')	89	17°57'N	95°08'W
San Juan Island, i., Wa., U.S.	74a	48°28'N	123°08'W
San Juan Islands, is., Can. (săn hwän)	54	48°49'N	123°14'W
San Juan Islands, is., Wa., U.S.	142a	48°36'N	122°50'W
San Juan Ixtenco, Mex. (ēx-tĕ'n-kō)	89	19°14'N	97°52'W
San Juan Martínez, Cuba	92	22°15'N	83°50'W
San Juan Mountains, mts., Co., U.S. (san hwán')	64	37°50'N	107°30'W
San Julián, Arg. (sän hōō-lyá'n)	102	49°17'S	68°02'W
San Justo, Arg. (hōōs'tō)	102a	34°40'S	58°33'W
Sankanbiriwa, mtn., S.L.	188	8°56'N	10°48'W
Sankarani, r., Afr. (sän'kä-rä'nē)	184	11°10'N	8°35'W
Sankt Gallen, Switz.	117	47°25'N	9°22'E
Sankt Moritz, Switz. (sånt mō'rĭts) (zäŋkt mō'rēts)	124	46°31'N	9°50'E
Sankt Pölten, Aus. (zäŋkt-pûl'tĕn)	124	48°12'N	15°38'E
Sankt Veit, Aus. (zäŋkt vĭt')	124	46°46'N	14°20'E
Sankuru, r., D.R.C. (säŋ-kōō'rōō)	186	4°00'S	22°35'E
San Lázaro, Cabo, c., Mex. (sän-lá'zä-rō)	86	24°58'N	113°30'W
San Leandro, Ca., U.S. (săn lē-ăn'drō)	74b	37°43'N	122°10'W
Şanlıurfa, Tur.	154	37°20'N	38°45'E
San Lorenzo, Arg. (sän lō-rĕn'zō)	102	32°46'S	60°44'W
San Lorenzo, Hond. (sän lō-rĕn'zō)	90	13°24'N	87°24'W
San Lorenzo, Ca., U.S. (săn lō-rĕn'zō)	74b	37°41'N	122°08'W
San Lorenzo de El Escorial, Spain	128	40°36'N	4°09'W
Sanlúcar de Barrameda, Spain (sän-lōō'kär)	118	36°46'N	6°21'W
San Lucas, Bol. (lōō'kås)	100	20°12'S	65°06'W
San Lucas, Cabo, c., Mex.	86	22°45'N	109°45'W
San Luis, Arg. (lô-ēs')	102	33°16'S	66°15'W
San Luis, Col. (lôē's)	100a	6°03'N	74°57'W
San Luis, Cuba	93	20°15'N	75°50'W
San Luis, Guat.	90	14°38'N	89°42'W
San Luis, prov., Arg.	102	32°45'S	66°00'W
San Luis de la Paz, Mex. (dā lä päz')	88	21°17'N	100°32'W
San Luis del Cordero, Mex. (dĕl kōr-dā'rō)	80	25°25'N	104°20'W
San Luis Obispo, Ca., U.S. (ō-bĭs'pō)	64	35°18'N	120°40'W
San Luis Obispo Bay, b., Ca., U.S.	76	35°07'N	121°05'W
San Luis Potosí, Mex.	86	22°08'N	100°58'W
San Luis Potosí, state, Mex.	86	22°45'N	101°45'W
San Luis Rey, r., Ca., U.S. (rā'ē)	76	33°22'N	117°06'W
San Manuel, Az., U.S. (sän măn'ū-ĕl)	77	32°30'N	110°45'W
San Marcial, N.M., U.S. (săn mår-shäl')	77	33°40'N	107°00'W
San Marco, Italy (sän mär'kō)	130	41°53'N	15°50'E
San Marcos, Guat. (mär'kōs)	90	14°57'N	91°49'W
San Marcos, Mex.	88	16°46'N	99°23'W
San Marcos, Tx., U.S. (săn mär'kŏs)	81	29°53'N	97°56'W
San Marcos, r., Tx., U.S.	80	30°08'N	98°15'W
San Marcos de Colón, Hond. (sän-mä'r-kōs-dĕ-kō-lō'n)	90	13°17'N	86°50'W
San Maria di Léuca, Cape, c., Italy (dē-lĕ'ōō-kä)	119	39°47'N	18°20'E
San Marino, S. Mar. (sän mä-rē'nō)	130	44°55'N	12°26'E
San Marino, Ca., U.S. (săn mēr-ē'nō)	75a	34°07'N	118°06'W
San Marino, nation, Eur.	110	43°40'N	13°00'E
San Martín, Col. (sän mär-tē'n)	100a	3°42'N	73°44'W
San Martín, vol., Mex. (mär-tē'n)	89	18°36'N	95°11'W
San Martín, l., S.A.	102	48°15'S	72°30'W
San Martín Chalchicuautla, Mex.	88	21°22'N	98°39'W
San Martin de la Vega, Spain (sän mär ten' dā lä vā'gä)	129a	40°12'N	3°34'W
San Martín Hidalgo, Mex. (sän mär-tē'n-ē-däl'gō)	88	20°27'N	103°55'W
San Mateo, Mex.	89	16°59'N	97°04'W
San Mateo, Ca., U.S. (săn mä-tā'ō)	74b	37°34'N	122°20'W
San Mateo, Ven. (sän mä-tĕ'ō)	101b	9°45'N	64°34'W
San Matías, Golfo, b., Arg. (sän mä-tē'äs)	102	41°30'S	63°45'W
Sanmen Wan, b., China	165	29°00'N	122°15'E
San Miguel, El Sal. (sän mē-gāl')	86	13°28'N	88°11'W
San Miguel, Mex. (sän mē-gāl')	89	18°18'N	97°09'W
San Miguel, Pan.	91	8°26'N	78°55'W
San Miguel, Phil. (sän mē-gē'l)	169a	15°09'N	120°56'E
San Miguel, Ven. (sän mē-gē'l)	101b	9°56'N	64°58'W
San Miguel, vol., El Sal.	90	13°27'N	88°17'W
San Miguel, i., Ca., U.S.	76	34°03'N	120°23'W
San Miguel, r., Bol. (sän-mē-gĕl')	100	13°34'S	63°58'W
San Miguel, r., N.A. (sän mē-gāl')	89	15°27'N	92°00'W
San Miguel, r., Co., U.S. (săn mē-gĕl')	77	38°15'N	108°40'W
San Miguel, Bahía, b., Pan. (bä-ē'ä-sän mē-gāl')	91	8°17'N	78°26'W
San Miguel Bay, b., Phil.	169a	13°55'N	123°12'E
San Miguel de Allende, Mex. (dā ä-lyĕn'dā)	88	20°54'N	100°44'W
San Miguel el Alto, Mex. (ĕl äl'tō)	88	21°03'N	102°26'W
Sannär, Sudan	185	14°25'N	33°30'E
San Narcisco, Phil. (sän när-sē'sō)	169a	15°01'N	120°05'E
San Narcisco, Phil.	169a	13°34'N	122°33'E
San Nicolás, Arg. (sän nē-kô-lä's)	102	33°20'S	60°14'W
San Nicolas, Phil. (nē-kô-läs')	169a	16°05'N	120°45'E
San Nicolas, i., Ca., U.S. (săn nĭ'kô-lȧ)	76	33°14'N	119°10'W
San Nicolás, r., Mex.	88	19°40'N	105°08'W
Sanniquellie, Lib.	188	7°22'N	8°43'W
Sannūr, Wādī, Egypt	192b	28°48'N	31°12'E
Sanok, Pol. (sä'nōk)	125	49°31'N	22°13'E
San Pablo, Phil. (sän-pä-blō)	169a	14°05'N	121°20'E
San Pablo, Ca., U.S. (săn păb'lō)	74b	37°58'N	122°21'W
San Pablo, Ven. (sän-pá'blō)	101b	9°46'N	65°04'W
San Pablo, r., Pan. (sän päb'lō)	91	8°12'N	81°12'W
San Pablo Bay, b., Ca., U.S. (săn păb'lō)	74b	38°04'N	122°25'W
San Pablo Res, Ca., U.S.	74b	37°55'N	122°12'W
San Pascual, Phil. (päs-kwäl')	169a	13°08'N	122°59'E
San Pedro, Arg. (sän pā'drō)	102	24°15'S	64°15'W
San Pedro, Arg.	99c	33°41'S	59°42'W
San Pedro, Chile (sän pĕ'drō)	99b	33°54'S	71°27'W
San Pedro, El Sal. (sän pā'drō)	90	13°49'N	88°58'W
San Pedro, Mex. (sän pā'drō)	88	18°38'N	92°25'W
San Pedro, Para. (sän-pĕ'drō)	102	24°13'S	57°00'W
San Pedro, Ca., U.S. (săn pē'drō)	75a	33°44'N	118°17'W
San Pedro, r., Cuba (sän-pĕ'drō)	92	21°05'N	78°15'W
San Pedro, r., Mex. (sän pā'drō)	88	22°08'N	104°59'W
San Pedro, r., Mex.	80	27°56'N	105°50'W
San Pedro, r., Az., U.S.	77	32°48'N	110°37'W
San Pedro, Río de, r., Mex.	88	21°51'N	102°24'W
San Pedro, Río de, r., N.A.	89	18°23'N	92°13'W
San Pedro Bay, b., Ca., U.S. (săn pē'drō)	75a	33°42'N	118°12'W
San Pedro de las Colonias, Mex. (dĕ-läs-kō-lō'nyäs)	80	25°47'N	102°58'W
San Pedro de Macorís, Dom. Rep. (sän-pĕ'drō-då mä-kō-rēs')	93	18°30'N	69°30'W
San Pedro Lagunillas, Mex. (sän pā'drō lä-gōō-nēl'yäs)	88	21°12'N	104°47'W
San Pedro Sula, Hond. (sän pā'drō sōō'lä)	90	15°29'N	88°01'W
San Pietro, Isola di, i., Italy (ē'sō-lä-dē-sän pyā'trō)	130	39°09'N	8°15'E
San Quentin, Ca., U.S. (săn kwĕn-tēn')	74b	37°57'N	122°29'W
San Quintin, Phil. (sän kēn-tēn')	169a	15°59'N	120°47'E
San Rafael, Arg. (sän rä-fä-āl')	102	34°30'S	68°13'W
San Rafael, Col. (sän-rä-fä-ĕ'l)	100a	6°18'N	75°02'W
San Rafael, Ca., U.S. (săn rȧ-fĕl)	74b	37°58'N	122°31'W
San Rafael, r., Ut., U.S. (săn rȧ-fĕl')	77	39°05'N	110°50'W
San Rafael, Cabo, c., Dom. Rep. (ká'bō)	93	19°00'N	68°50'W
San Ramón, C.R.	91	10°07'N	84°30'W
San Ramon, Ca., U.S. (săn rȧ-mōn')	74b	37°47'N	122°59'W
San Remo, Italy (sän rā'mō)	130	43°48'N	7°46'E
San Roque, Col. (sän-rô'kĕ)	100a	6°29'N	75°00'W
San Roque, Spain	128	36°13'N	5°23'W
San Saba, Tx., U.S. (săn sä'bȧ)	80	31°12'N	98°43'W
San Saba, r., Tx., U.S.	80	30°58'N	99°12'W
San Salvador, El Sal. (sän säl-vä-dōr')	86	13°45'N	89°11'W
San Salvador (Watling), i., Bah. (săn săl'vȧ-dôr)	93	24°05'N	74°30'W
San Salvador, i., Ec.	100	0°14'S	90°50'W
San Salvador, r., Ur. (sän-säl-vä-dô'r)	99c	33°42'S	58°04'W
Sansanné-Mango, Togo (sän-så-nā' mäŋ'gō)	184	10°21'N	0°28'E
San Sebastian, Spain (sän så-bås-tyän')	184	28°09'N	17°11'W
San Sebastián *see* Donostia-San Sebastián, Spain	110	43°19'N	1°59'W
San Sebastián, Ven. (sän-sĕ-bäs-tyá'n)	101b	9°58'N	67°11'W
San Sebastiàn de los Reyes, Spain	129a	40°33'N	3°38'W
San Severo, Italy (sän sĕ-vâ'rō)	119	41°43'N	15°24'E
Sanshui, China (sän-shwā)	161	23°14'N	112°51'E
San Simon Creek, r., Az., U.S. (săn sī-mōn')	77	32°45'N	109°30'W
Santa Ana, El Sal.	86	14°02'N	89°35'W
Santa Ana, Mex. (sän'tä ä'nä)	88	19°18'N	98°10'W
Santa Ana, Ca., U.S. (săn'tȧ ăn'ȧ)	64	33°45'N	117°52'W
Santa Ana, r., Ca., U.S.	75a	33°41'N	117°57'W
Santa Ana Mountains, mts., Ca., U.S.	75a	33°44'N	117°36'W
Santa Anna, Tx., U.S.	80	31°44'N	99°18'W
Santa Antão, i., C.V. (sä-tä-ä'n-zhĕ-lō)	184b	17°20'N	26°05'W
Santa Bárbara, Braz. (sän-tä-bá'r-bä-rä)	101	19°57'S	43°25'W
Santa Bárbara, Hond.	90	14°52'N	88°20'W
Santa Bárbara, Mex.	80	26°48'N	105°50'W
Santa Barbara, Ca., U.S.	64	34°26'N	119°43'W
Santa Barbara, i., Ca., U.S.	76	33°30'N	118°44'W
Santa Barbara Channel, strt., Ca., U.S.	76	34°15'N	120°00'W
Santa Branca, Braz. (sän-tä-brä'n-kä)	99a	23°25'S	45°52'W
Santa Catalina, i., Ca., U.S.	64	33°29'N	118°37'W
Santa Catalina, Cerro de, mtn., Pan.	91	8°39'N	81°36'W
Santa Catalina, Gulf of, b., Ca., U.S. (săn'tȧ kă-tȧ-lē'nȧ)	76	33°00'N	117°58'W
Santa Catarina, Mex. (sän'tä kä-tä-rē'nä)	80	25°41'N	100°27'W
Santa Catarina, state, Braz. (sän-tä-kä-tä-rē'nä)	102	27°15'S	50°30'W
Santa Catarina, r., Mex.	88	16°31'N	98°39'W
Santa Clara, Cuba (sän'tȧ klä'rá)	87	22°25'N	80°00'W
Santa Clara, Mex.	80	24°29'N	103°22'W
Santa Clara, Ur.	102	32°46'S	54°51'W
Santa Clara, Ca., U.S. (săn'tȧ klârȧ)	72	37°21'N	121°56'W
Santa Clara, vol., Nic.	90	12°44'N	87°00'W
Santa Clara, r., Ca., U.S. (săn'tȧ klă'rȧ)	76	34°22'N	118°53'W
Santa Clara, Bahía de, b., Cuba (bä-ē'ä-dĕ-sän-tä-klä-rä)	92	23°05'N	80°50'W
Santa Clara, Sierra, mts., Mex. (sē-ĕ'r-rä-sän'tä klä'rä)	86	27°30'N	113°50'W

ng-sing; ŋ-baŋk; N-nasalized n; nŏd; cŏmmit; ōld; ôbey; ôrder; oi-boil; fo͞od; ŏ-as oo in foot; ou-out; s-soft; sh-dish; th-thin; pūre; ŭnite; ûrn; stŭd; circŭs; ü-as in French tu; '-indeterminate vowel.

PLACE (Pronunciation)	PAGE	LAT.	LONG.
Santa Clara Indian Reservation, I.R., N.M., U.S.	77	35°59'N	106°10'W
Santa Cruz, Bol. (sän'tä krōōz')	100	17°45'S	63°03'W
Santa Cruz, Braz. (sän-tä-krōō's)	102	29°43'S	52°15'W
Santa Cruz, Braz.	102b	22°55'S	43°41'W
Santa Cruz, Chile	99b	34°38'S	71°21'W
Santa Cruz, C.R.	90	10°16'N	85°37'W
Santa Cruz, Mex.	80	25°50'N	105°25'W
Santa Cruz, Phil.	169a	13°28'N	122°02'E
Santa Cruz, Phil.	169a	14°17'N	121°25'E
Santa Cruz, Phil.	169a	15°46'N	119°53'E
Santa Cruz, Ca., U.S.	64	36°59'N	122°02'W
Santa Cruz, prov., Arg.	102	48°00'S	70°00'W
Santa Cruz, i., Ec. (sän-tä-krōō'z)	100	0°38'S	90°20'W
Santa Cruz, r., Arg. (sän'tä krōōz')	102	50°05'S	71°00'W
Santa Cruz, r., Az., U.S. (săn'tȧ krōōz')	77	32°30'N	111°30'W
Santa Cruz Barillas, Guat. (sän-tä-krōō'z-bä-rē'l-yäs)	90	15°47'N	91°22'W
Santa Cruz del Sur, Cuba (sän-tä-krōō's-dĕl-sōō'r)	92	20°45'N	78°00'W
Santa Cruz de Tenerife, Spain (sän'tä krōōz dā tā-nå-rē'fā)	182	28°07'N	15°27'W
Santa Cruz Islands, is., Sol. Is.	175	10°58'S	166°47'E
Santa Cruz Mountains, mts., Ca., U.S. (săn'tȧ krōōz')	74b	37°30'N	122°19'W
Santa Domingo, Cay, i., Bah.	93	21°50'N	75°45'W
Santa Fe, Arg. (sän'tä fā')	102	31°33'S	60°45'W
Santa Fé, Cuba (sän-tä-fē')	92	21°45'N	82°40'W
Santa Fe, Spain (sän'tä-fā)	128	37°12'N	3°43'W
Santa Fe, N.M., U.S. (săn'tȧ fā')	64	35°40'N	106°00'W
Santa Fe, prov., Arg. (sän'tä fā')	102	32°00'S	61°15'W
Santa Fe de Bogotá *see* Bogotá, Col.	100	4°36'N	74°05'W
Santa Filomena, Braz. (sän-tä-fē-lō-mĕ'nä)	101	9°09'S	44°45'W
Santa Genoveva, mtn., Mex. (sän-tä-hĕ-nō-vĕ'vä)	86	23°30'N	110°00'W
Santai, China (san-tī)	160	31°02'N	105°02'E
Santa Inés, Ven. (sän'tä ē-nĕ's)	101b	9°54'N	64°21'W
Santa Inés, i., Chile (sän'tä ē-nās')	102	53°45'S	74°15'W
Santa Isabel, i., Sol. Is.	175	7°57'S	159°28'E
Santa Isabel, Pico de, mtn., Eq. Gui.	189	3°35'N	8°46'E
Santa Lucía, Cuba (sän'tä lōō-sē'ä)	92	21°15'N	77°30'W
Santa Lucía, Ur. (sän-tä-lōō-sē'ä)	102	34°27'S	56°23'W
Santa Lucía, Ven.	101b	10°18'N	66°40'W
Santa Lucía, r., Ur.	99c	34°19'S	56°13'W
Santa Lucía Bay, b., Cuba (sän'tä lōō-sē'ä)	92	22°55'N	84°20'W
Santa Margarita, i., Mex. (sän'tä mär-gä-rē'tä)	86	24°15'N	112°00'W
Santa Maria, Braz. (sän'tä mä-rē'ä)	102	29°40'S	54°00'W
Santa Maria, Italy (sän-tä mä-rē'ä)	130	41°05'N	14°15'E
Santa María, Phil. (sän-tä-mä-rē'ä)	169a	14°48'N	120°57'E
Santa Maria, Ca., U.S. (săn-tȧ mȧ-rē'ȧ)	76	34°57'N	120°28'W
Santa María, vol., Guat.	90	14°45'N	91°33'W
Santa María, r., Mex. (sän'tä mä-rē'ä)	88	21°33'N	100°17'W
Santa Maria, Cabo de, c., Port. (kä'bō-dĕ-sän-tä-mä-rē'ä)	128	36°58'N	7°54'W
Santa Maria, Cape, c., Bah.	93	23°45'N	75°30'W
Santa María, Cayo, i., Cuba	92	22°40'N	79°00'W
Santa María del Oro, Mex. (sän'tä-mä-rē'ä-dĕl-ô-rō)	88	21°21'N	104°35'W
Santa María de los Angeles, Mex. (dĕ-lôs-ä'n-hĕ-lĕs)	88	22°10'N	103°34'W
Santa María del Río, Mex.	88	21°46'N	100°43'W
Santa María de Ocotán, Mex.	88	22°56'N	104°30'W
Santa Maria Island, i., Port. (sän-tä-mä-rē'ä)	184a	37°09'N	26°02'W
Santa Maria Madalena, Braz.	99a	22°00'S	42°00'W
Santa Marta, Col. (sän'tä mär'tä)	100	11°15'N	74°13'W
Santa Marta, Cabo de, c., Ang.	190	13°52'S	12°25'E
Santa Monica, Ca., U.S. (săn'tȧ mŏn'ĭ-kȧ)	64	34°01'N	118°29'W
Santa Monica Mountains, mts., Ca., U.S.	75a	34°08'N	118°38'W
Santana, r., Braz. (sän-tä'nä)	102b	22°33'S	43°37'W
Santander, Col. (sän-tän-dĕr')	100a	3°00'N	76°25'W
Santander, Spain (sän-tän-dår')	110	43°27'N	3°50'W
Sant Antoni de Portmany, Spain	129	38°59'N	1°17'E
Santa Paula, Ca., U.S. (săn'tȧ pô'lȧ)	76	34°24'N	119°05'W
Santarém, Braz. (sän-tä-rĕN')	101	2°28'S	54°37'W
Santarém, Port.	128	39°18'N	8°48'W
Santaren Channel, strt., Bah. (săn-tȧ-rĕn')	92	24°15'N	79°30'W
Santa Rita do Sapucai, Braz. (sä-pō-kȧ'ē)	99a	22°15'S	45°41'W
Santa Rosa, Arg. (sän-tä-rō-sä)	102	36°45'S	64°10'W
Santa Rosa, Col. (sän-tä-rō-sä)	100a	6°38'N	75°26'W
Santa Rosa, Ec.	100	3°29'S	79°55'W
Santa Rosa, Guat. (sän'tä rō'sȧ)	90	14°21'N	90°16'W
Santa Rosa, Hond.	90	14°45'N	88°51'W
Santa Rosa, Ca., U.S. (săn'tȧ rō'zȧ)	64	38°27'N	122°42'W
Santa Rosa, N.M., U.S. (săn'tȧ rō'sȧ)	78	34°55'N	104°41'W
Santa Rosa, Ven. (sän-tä-rō-sä)	101b	9°37'N	64°10'W
Santa Rosa de Cabal, Col. (sän-tä-rō-sä-dĕ-kä-bä'l)	100a	4°53'N	75°38'W
Santa Rosa de Viterbo, Braz. (sän-tä-rō-sá-dĕ-vē-tĕr'-bō)	99a	21°30'S	47°21'W
Santa Rosa Indian Reservation, I.R., Ca., U.S. (săn'tȧ rō'zȧ')	76	33°28'N	116°50'W
Santa Rosalía, Mex. (sän'tä rō-zä'lē-ä)	86	27°13'N	112°15'W
Santa Rosa Range, mts., Nv., U.S. (săn'tȧ rō'zȧ)	72	41°33'N	117°50'W
Santa Susana, Ca., U.S. (săn'tȧ sōō-ză'nȧ)	75a	34°16'N	118°42'W
Santa Teresa, Arg. (sän-tä-tĕ-rĕ'sä)	99c	33°27'S	60°47'W
Santa Teresa, Ven.	101b	10°14'N	66°40'W
Santa Uxia, Spain	128	42°34'N	8°55'W
Santa Vitória do Palmar, Braz. (sän-tä-vē-tō'ryä-dō-päl-màr)	102	33°30'S	53°16'W
Santa Ynez, r., Ca., U.S. (săn'tȧ ē-nĕz')	76	34°40'N	120°20'W
Santa Ysabel Indian Reservation, I.R., Ca., U.S. (săn-tȧ ĭ-zȧ-bĕl')	76	33°05'N	116°46'W
Santee, Ca., U.S. (săn tē')	76a	32°50'N	116°58'W
Santee, r., S.C., U.S.	65	33°00'N	79°45'W
Sant' Eufemia, Golfo di, b., Italy (gôl-fō-dē-sän-tĕ'ō-fē'myä)	130	38°53'N	15°53'E
Sant Feliu de Guixols, Spain	129	41°45'N	3°01'E
Santiago, Braz. (sän-tyá'gô)	102	29°05'S	54°46'W
Santiago, Chile (sän-tē-ä'gō)	102	33°26'S	70°40'W
Santiago, Pan.	87	8°07'N	80°58'W
Santiago, Phil. (sän-tyä'gō)	169a	16°42'N	121°33'E
Santiago, prov., Chile (sän-tyá'gō)	99b	33°28'S	70°55'W
Santiago, i., Phil.	169a	16°29'N	120°03'E
Santiago de Compostela, Spain	118	42°52'N	8°32'W
Santiago de Cuba, Cuba (sän-tyá'gô-dā kōō'bä)	87	20°00'N	75°50'W
Santiago de Cuba, prov., Cuba	92	20°20'N	76°05'W
Santiago de las Vegas, Cuba (sän-tyá'gô-dĕ-läs-vĕ'gäs)	93a	22°58'N	82°23'W
Santiago del Estero, Arg.	102	27°50'S	64°14'W
Santiago del Estero, prov., Arg. (sän-tē-á'gō-dĕl ĕs-tā-rō)	102	27°15'S	63°30'W
Santiago de los Caballeros, Dom. Rep.	87	19°30'N	70°45'W
Santiago Mountains, mts., Tx., U.S. (săn-tē-ä'gō)	64	30°00'N	103°30'W
Santiago Reservoir, res., Ca., U.S.	75a	33°47'N	117°42'W
Santiago Rodríguez, Dom. Rep. (sän-tyá'gô-rô-drē'gĕz)	93	19°30'N	71°25'W
Santiago Tuxtla, Mex. (sän-tyá'gô-tōō'x-tlä)	89	18°28'N	95°18'W
Santiaguillo, Laguna de, l., Mex. (lä-ōō'nä-dĕ-sän-tê-a-gē'l-yō)	80	24°51'N	104°43'W
Santisteban del Puerto, Spain (sän'tē stä-bän'dĕl pwĕr'tō)	128	38°15'N	3°12'W
Sant Mateu, Spain	129	40°26'N	0°09'E
Santo Amaro, Braz. (sän'tō ä-mä'rō)	101	12°32'S	38°33'W
Santo Amaro de Campos, Braz.	99a	22°01'S	41°05'W
Santo André, Braz.	99a	23°40'S	46°31'W
Santo Angelo, Braz. (sän-tô-ä'n-zhĕ-lô)	102	28°16'S	53°59'W
Santo Antônio do Monte, Braz. (sän-tô-än-tô'nyô-dô-môn'tĕ)	99a	20°06'S	45°18'W
Santo Domingo, Cuba (sän'tô-dōmĭn'gō)	92	22°35'N	80°20'W
Santo Domingo, Dom. Rep. (sän'tô dô-mĭn'gô)	87	18°30'N	69°55'W
Santo Domingo, Nic. (sän-tô-dô-mē'n-gō)	91	12°15'N	84°56'W
Santo Domingo de la Caizada, Spain (dā lä käl-thä'dä)	128	42°27'N	2°55'W
Santoña, Spain (sän-tō'nyä)	128	43°25'N	3°27'W
Santos, Braz. (sän'tozh)	101	23°58'S	46°20'W
Santos Dumont, Braz. (sän'tôs-dô-mô'nt)	101	21°28'S	43°33'W
Sanuki, Japan (sä'nōō-kē)	167a	35°16'N	139°53'E
San Urbano, Arg. (sän-ōr-bä'nō)	99c	33°39'S	61°28'W
San Valentín, Monte, mtn., Chile (sän-vä-lĕn-tē'n)	102	46°41'S	73°30'W
San Vicente, Arg. (sän-vē-sĕn'tĕ)	99c	35°00'S	58°26'W
San Vicente, Chile	99b	34°25'S	71°06'W
San Vicente, El Sal. (sän vē-sĕn'tā)	90	13°41'N	88°43'W
San Vicente de Alcántara, Spain	128	39°24'N	7°08'W
San Vito al Tagliamento, Italy (san vē'tô)	130	45°53'N	12°52'E
San Xavier Indian Reservation, I.R., Az., U.S. (x-ä'vĭēr)	77	32°07'N	111°12'W
San Ysidro, Ca., U.S. (sän ysĭ-drō')	76a	32°33'N	117°02'W
Sanyuanli, China (sän-yüän-lē)	163a	23°11'N	113°16'E
São Bernardo do Campo, Braz. (souN-bĕr-nár'dō-dō-ká'm-pō)	99a	23°44'S	46°33'W
São Borja, Braz. (souN-bôr-zhä)	102	28°44'S	55°59'W
São Carlos, Braz. (souN kär'lōzh)	101	22°02'S	47°54'W
São Cristovão, Braz. (souN-krēs-tō-vouN)	101	11°04'S	37°11'W
São Fidélis, Braz. (souN-fē-dĕ'lĕs)	99a	21°41'S	41°45'W
São Francisco, Braz. (souN frän-sēsh'kō)	101	15°59'S	44°42'W
São Francisco, r., Braz. (sän-frän-sē's-kō)	101	8°56'S	40°20'W
São Francisco do Sul, Braz. (souN frän-sēsh'kō-dō-sōō'l)	102	26°15'S	48°42'W
São Gabriel, Braz. (souN'gä-brē-ĕl')	102	30°28'S	54°11'W
São Geraldo, Braz. (souN-zhĕ-rä'l-dō)	99a	21°01'S	42°49'W
São Gonçalo, Braz. (souN'gōŋ-sä'lō)	99a	22°55'S	43°04'W
Sao Hill, Tan.	191	8°20'S	35°12'E
São João, Gui.-B.	188	11°32'N	15°26'W
São João da Barra, Braz. (souN-zhōuN-dä-bä'rä)	99a	21°40'S	41°03'W
São João da Boa Vista, Braz. (souN-zhōuN-dä-bōä-vē's-tä)	99a	21°58'S	46°45'W
São João del Rei, Braz. (souN zhō-ouN'dĕl-rā)	102	21°08'S	44°14'W
São João de Meriti, Braz. (souN-zhōuN-dĕ-mĕ-rē-tē)	102b	22°47'S	43°22'W
São João do Araguaia, Braz. (souN zhō-ouN'dō-ä-rä-gwä'yä)	101	5°29'S	48°44'W
São João dos Lampas, Port. (souN' zhō-ouN' dōzh län-päzh')	129b	38°52'N	9°24'W
São João Nepomuceno, Braz. (souN-zhōuN-nĕ-pō-mōō-sĕ-nō)	99a	21°33'S	43°00'W
São Jorge Island, i., Port. (souN zhôr'zhĕ)	184a	38°28'N	27°34'W
São José do Rio Pardo, Braz. (souN-zhō-sĕ'dō-rē'ō-pá'r-dō)	99a	21°36'S	46°50'W
São José do Rio Prêto, Braz. (souN zhō-zĕ'dō-re'ō-prĕ-tō)	101	20°57'S	49°12'W
São José dos Campos, Braz. (souN zhō-zā'dōzh käN pōzh')	99a	23°12'S	45°53'W
São Leopoldo, Braz. (souN-lĕ-ô-pôl'dō)	102	29°46'S	51°09'W
São Luis, Braz.	101	2°31'S	43°14'W
São Luis do Paraitinga, Braz. (souN-lōōē's-dō-pä-rä-ē-tē'n-gä)	99a	23°15'S	45°18'W
São Manuel, r., Braz.	101	8°28'S	57°07'E
São Mateus, Braz. (souN mä-tā'ōzh)	101	18°44'S	39°45'W
São Mateus, Braz.	102b	22°49'S	43°23'W
São Miguel Arcanjo, Braz. (souN-mē-gĕ'l-år-kän-zhō)	99a	23°54'S	47°59'W
São Miguel Island, i., Port.	184a	37°59'N	26°38'W
Saona, i., Dom. Rep. (sä-ō'nä)	93	18°10'N	68°55'W
Saône, r., Fr. (sōn)	112	47°00'N	5°30'E
São Nicolau, i., C.V. (souN' nē-kō-louN')	184b	16°19'N	25°19'W
São Paulo, Braz. (souN' pou'lō)	101	23°34'S	46°38'W
São Paulo, state, Braz. (souN pou'lō)	101	21°45'S	50°47'W
São Paulo de Olivença, Braz. (souN'pou'lōdä ō-lē-vĕn'sȧ)	100	3°32'S	68°46'W
São Pedro, Braz. (souN-pĕ'drō)	99a	22°34'S	47°54'W
São Pedro de Aldeia, Braz. (souN-pĕ'drō-dĕ-äl-dĕ'yä)	99a	22°50'S	42°04'W
São Pedro e São Paulo, Rocedos, rocks, Braz.	97	1°50'N	30°00'W
São Raimundo Nonato, Braz. (souN' rī-mō'n-dō nō-nä'tō)	101	9°09'S	42°32'W
São Roque, Braz. (souN' rō'kĕ)	99a	23°32'S	47°08'W
São Roque, Cabo de, c., Braz. (kä'bo-dĕ-souN' rō'kĕ)	101	5°06'S	35°11'W
São Sebastião, Braz. (souN sȧ-bäs-tē-ouN')	99a	23°48'S	45°25'W
São Sebastião, Ilha de, i., Braz.	99a	23°52'S	45°22'W
São Sebastião do Paraíso, Braz.	99a	20°54'S	46°58'W
São Simão, Braz. (souN-sē-mouN)	99a	21°30'S	47°33'W
São Tiago, i., C.V. (souN tē-ä'gō)	184b	15°09'N	24°45'W
São Tomé, S. Tom./P.	184	0°20'N	6°44'E
Sao Tome and Principe, nation, Afr. (prēn'sē-pĕ)	184	1°00'N	6°00'E
Saoura, Oued, r., Alg.	184	29°39'N	1°42'W
São Vicente, Braz. (souN ve-se'n-tĕ)	101	23°57'S	46°25'W
São Vicente, i., C.V. (souN vē-sĕn'tā)	184b	16°51'N	24°35'W
São Vicente, Cabo de, c., Port. (ká'bō-dĕ-sän-vê-sĕ'n-tĕ)	112	37°03'N	9°31'W
Sapele, Nig. (sä-pā'lā)	184	5°54'N	5°41'E
Sapitwa, mtn., Mwi.	191	15°58'S	35°38'E
Sa Pobla, Spain	129	39°46'N	3°02'E
Sapozhok, Russia (sä-pō-zhôk')	132	53°58'N	40°44'E
Sapporo, Japan (säp-pō'rō)	161	43°02'N	141°29'E
Sapronovo, Russia (sáp-rō'nô-vô)	142b	55°13'N	38°25'E
Sapucaí, r., Braz. (sä-pōō-kȧ-ē')	99a	22°20'S	45°53'W
Sapucaia, Braz. (sä-pōō-kä'yȧ)	99a	22°01'S	42°54'W
Sapucaí Mirim, r., Braz. (sä-pōō-kȧ-ē'mē-rēN)	99a	21°06'S	47°03'W
Sapulpa, Ok., U.S. (sȧ-pŭl'pȧ)	79	36°01'N	96°05'W
Saqqez, Iran	157	36°14'N	46°16'E
Saquarema, Braz. (sä-kwä-rĕ-mä)	99a	22°56'S	42°32'W
Sara, Wa., U.S. (sä'rä)	74c	45°45'N	122°42'W
Sara, Bahr, r., Chad (bär)	185	8°19'N	17°44'E
Sarajevo, Bos. (sä-rá-yĕv'ō) (sá-rá'ya-vô)	110	43°50'N	18°26'E
Sarakhs, Iran	157	36°32'N	61°11'E
Sarana, Russia (sá-rä'ná)	142a	56°31'N	57°44'E
Saranac Lake, N.Y., U.S.	67	44°20'N	74°05'W
Saranac Lake, l., N.Y., U.S. (săr'ȧ-năk)	67	44°15'N	74°20'W
Sarandí, Arg. (sä-rän'dē)	102a	34°41'S	58°21'W
Sarandí Grande, Ur. (sä-rän'dē-grän'dĕ)	99c	33°42'S	56°21'W
Saranley, Som.	192a	2°28'N	42°15'E
Saransk, Russia (sá-ränsk')	134	54°10'N	45°10'E
Sarany, Russia (sá-rá'nĭ)	142a	58°33'N	58°48'E
Sara Peak, mtn., Nig.	189	9°37'N	9°25'E
Sarapul, Russia (sä-räpōl')	136	56°28'N	53°50'E
Sarasota, Fl., U.S. (săr-ȧ-sōtȧ)	83a	27°27'N	82°30'W
Saratoga, Tx., U.S. (săr-ȧ-tō'gȧ)	81	30°17'N	94°31'W
Saratoga, Wa., U.S.	74a	48°04'N	122°29'W
Saratoga Pass, Wa., U.S.	74a	48°09'N	122°33'W
Saratoga Springs, N.Y., U.S. (sprĭngz)	67	43°05'N	74°50'W
Saratov, Russia (sá rä'tôf)	134	51°30'N	45°30'E
Saravane, Laos	165	15°48'N	106°40'E
Sarawak, hist. reg., Malay. (sȧ-rä'wäk)	168	2°30'N	112°45'E
Sárbogárd, Hung. (shär'bō-gärd)	125	46°53'N	18°38'E
Sarcee Indian Reserve, I.R., Can. (sär'sē)	62e	50°58'N	114°23'W
Sarcelles, Fr.	127b	49°00'N	2°23'E
Sardalas, Libya	184	25°59'N	10°33'E
Sardinia, i., Italy (sär-dĭn'ĭȧ)	112	40°08'N	9°05'E
Sardis, Ms., U.S. (sär'dĭs)	82	34°26'N	89°55'W
Sardis Lake, res., Ms., U.S.	82	34°27'N	89°43'W
Sargent, Ne., U.S. (sär'jĕnt)	70	41°40'N	99°38'W
Sarh, Chad (är-chaN-bō')	185	9°09'N	18°23'E
Sarikamis, Tur.	137	40°30'N	42°40'E
Sariñena, Spain (sä-rēn-yĕ'nä)	129	41°46'N	0°11'W
Sark, i., Guern. (särk)	126	49°28'N	2°22'W
Şarköy, Tur. (shär'kû-ē)	131	40°39'N	27°07'E
Sarmiento, Monte, mtn., Chile (mô'n-tĕ-sär-myĕn'tō)	102	54°28'S	70°40'W
Sarnia, Can. (sär'nē-ȧ)	51	43°00'N	82°25'W

ăt; finăl; rāte; senåte; ärm; åsk; sofȧ; fâre; ch-choose; dh-as th in other; bē; ėvent; bĕt; recĕnt; cratĕr; g-gō; gh-guttural g; bĭt; ĭ-short neutral; rīde; K-guttural k as ch in German ich;

PRONOUNCING INDEX Sarno – Semichi Islands 301

PLACE (Pronunciation)	PAGE	LAT.	LONG.
Sarno, Italy (sä'r-nô)	129c	40°35'N	14°38'E
Sarny, Ukr. (sär'nê)	137	51°17'N	26°39'E
Saronikós Kólpos, b., Grc.	131	37°51'N	23°30'E
Saros Körfezi, b., Tur. (sä'rôs)	131	40°30'N	26°20'E
Sárospatak, Hung. (shä'rôsh-pô'tôk)	125	48°19'N	21°35'E
Šar Planina, mts., Serb. (shär plä'nê-na)	131	42°07'N	21°54'E
Sarpsborg, Nor. (särps'bôrg)	122	59°17'N	11°07'E
Sarrebourg, Fr. (sär-bōōr')	127	48°44'N	7°02'E
Sarreguemines, Fr. (sär-gē-mēn')	117	49°06'N	7°05'E
Sarria, Spain (sär'ē-ä)	118	42°14'N	7°17'W
Sarstun, r., N.A. (särs-tōō'n)	90	15°50'N	89°26'W
Sartène, Fr. (sär-těn')	130	41°36'N	8°59'E
Sarthe, r., Fr. (särt)	117	47°44'N	0°32'W
Şärur, Azer.	138	39°33'N	44°58'E
Sárvár, Hung. (shär'vär)	124	47°14'N	16°55'E
Sarych, Mys, c., Ukr. (mĭs sá-rēch')	137	44°25'N	33°00'E
Saryesik-Atyraū, des., Kaz.	139	45°30'N	76°00'E
Sary-Ishikotrau, Peski, des., Kyrg. (sä'rē ē' shěk-ō'trou)	139	46°12'N	75°30'E
Sarysū, r., Kaz. (sä'rē-sōō)	139	47°47'N	69°14'E
Sasarām, India (sŭs-ŭ-räm')	155	25°00'N	84°00'E
Sasayama, Japan (sä'sä-yä'mä)	167	35°05'N	135°14'E
Sasebo, Japan (sä'sâ-bô)	161	33°12'N	129°43'E
Saskatchewan, prov., Can.	50	54°46'N	107°40'W
Saskatchewan, r., Can. (săs-kăch'ê-wän)	52	53°45'N	103°20'W
Saskatoon, Can. (săs-*kà*-tōōn')	50	52°07'N	106°38'W
Sasolburg, S. Afr.	192c	26°52'S	27°47'E
Sasovo, Russia (sás'ô-vô)	136	54°20'N	42°00'E
Saspamco, Tx., U.S. (săs-păm'cō)	75d	29°13'N	98°18'W
Sassandra, C. Iv.	188	4°58'N	6°05'W
Sassandra, r., C. Iv. (sás-sän'drá)	184	5°35'N	6°25'W
Sassari, Italy (säs'sä-rê)	118	40°44'N	8°33'E
Sassnitz, Ger. (säs'nêts)	124	54°31'N	13°37'E
Satadougou, Mali (sä-tá-dōō-gōō')	188	12°21'N	12°07'W
Säter, Swe. (sĕ'tĕr)	122	60°21'N	15°50'E
Satilla, r., Ga., U.S. (sá-tĭl'*à*)	83	31°15'N	82°13'W
Satka, Russia (sät'kà)	136	55°03'N	59°02'E
Sátoraljaujhely, Hung. (shä'tô-rô-lyô-ōō'yěl')	125	48°24'N	21°40'E
Satu Mare, Rom. (sá'tōō-má'rě)	119	47°50'N	22°53'E
Saturna, Can. (să-tûr'ná)	74d	48°48'N	123°12'W
Saturna, i., Can.	74d	48°47'N	123°03'W
Sauda, Nor.	116	59°40'N	6°21'E
Saudárkrókur, Ice.	110	65°41'N	19°38'W
Saudi Arabia, nation, Asia (sá-ōō'dĭ á-rā'bĭ-á)	154	22°40'N	46°00'E
Sauerlach, Ger. (zou'ĕr-läk)	115d	47°58'N	11°39'E
Saugatuck, Mi., U.S. (sô'*gà*-tŭk)	66	42°40'N	86°10'W
Saugeen, r., Can.	58	44°20'N	81°20'W
Saugerties, N.Y., U.S. (sô'gĕr-tēz)	67	42°05'N	73°55'W
Saugus, Ma., U.S. (sô'gŭs)	61a	42°28'N	71°01'W
Sauk, r., Mn., U.S. (sôk)	71	45°30'N	94°45'W
Sauk Centre, Mn., U.S.	71	45°43'N	94°58'W
Sauk City, Wi., U.S.	71	43°16'N	89°45'W
Sauk Rapids, Mn., U.S. (răp'ĭd)	71	45°35'N	94°08'W
Sault Sainte Marie, Can.	51	46°31'N	84°20'W
Sault Sainte Marie, Mi., U.S. (sōō sānt mạ-rē')	65	46°29'N	84°21'W
Saumatre, Étang, l., Haiti	93	18°40'N	72°10'W
Saunders Lake, l., Can. (sän'dẽrs)	62g	53°18'N	113°25'W
Saurimo, Ang.	186	9°39'S	20°24'E
Sausalito, Ca., U.S. (sô-sá-lē'tô)	74b	37°51'N	122°29'W
Sausset-les-Pins, Fr. (sō-sĕ'lā-păɴ')	126a	43°20'N	5°08'E
Saútar, Ang.	190	11°06'S	18°27'E
Sauvie Island, i., Or., U.S. (sô'vē)	74c	45°43'N	123°49'W
Sava, r., Serb. (sä'vä)	112	44°50'N	18°30'E
Savage, Md., U.S. (să'věj)	68e	39°07'N	76°49'W
Savage, Mn., U.S.	75g	44°47'N	93°20'W
Savai'i, i., Samoa	170a	13°35'S	172°25'W
Savalen, l., Nor.	122	62°19'N	10°15'E
Savalou, Benin	184	7°56'N	1°58'E
Savanna, Il., U.S. (sá-văn'á)	71	42°05'N	90°09'W
Savannah, Ga., U.S. (sá-văn'á)	65	32°04'N	81°07'W
Savannah, Mo., U.S.	79	39°58'N	94°49'W
Savannah, Tn., U.S.	82	35°13'N	88°14'W
Savannah, r., U.S.	65	33°11'N	81°51'W
Savannakhét, Laos	168	16°33'N	104°45'E
Savanna la Mar, Jam. (sá-vän'á lá mär')	92	18°10'N	78°10'W
Save, r., Fr.	126	43°32'N	0°50'E
Save, Rio (Sabi), r., Afr. (rē'ō-sä'vĕ)	186	21°28'S	34°14'E
Sāveh, Iran	157	35°01'N	50°20'E
Saverne, Fr. (sá-vĕrn')	127	48°40'N	7°22'E
Savigliano, Italy (sä-vēl-yä'nô)	130	44°38'N	7°42'E
Savigny-sur-Orge, Fr.	127b	48°41'N	2°22'E
Savona, Italy (sä-nō'nä)	118	44°19'N	8°28'E
Savonlinna, Fin. (sä'vôn-lēn'nà)	123	61°53'N	28°49'E
Savran', Ukr. (säv-rän')	133	48°07'N	30°09'E
Sawahlunto, Indon.	168	0°37'S	100°50'E
Sawākin, Sudan	185	19°02'N	37°19'E
Sawda, Jabal as, mts., Libya	185	28°14'N	13°46'E
Sawhāj, Egypt	185	26°34'N	31°40'E
Sawknah, Libya	185	29°04'N	15°53'E
Sawu, Laut (Savu Sea), sea, Asia	168	9°15'S	122°15'E
Sawyer, l., Wa., U.S. (sô'yẽr)	74a	47°20'N	122°02'W
Saxony *see* Sachsen, hist. reg., Ger.	124	50°45'N	12°17'E
Say, Niger (sä'ē)	184	13°09'N	2°16'E
Sayan Khrebet, mts., Russia (sŭ-yän')	135	51°30'N	90°00'E
Sayhūt, Yemen	154	15°23'N	51°28'E
Sayre, Ok., U.S. (sâ'ẽr)	78	35°19'N	99°40'W
Sayre, Pa., U.S.	67	41°55'N	76°30'W
Sayreton, Al., U.S. (sâ'ẽr-tŭn)	68h	33°34'N	86°51'W
Sayreville, N.J., U.S. (sâr'vĭl)	68a	40°28'N	74°21'W
Sayr Usa, Mong.	160	44°15'N	107°00'E
Sayula, Mex. (sä-yōō'lä)	89	17°51'N	94°56'W
Sayula, Mex.	88	19°50'N	103°33'W
Sayula, Laguna de, l., Mex. (lä-gô'nä-dě)	88	20°00'N	103°33'W
Say'un, Yemen	154	16°00'N	48°59'E
Sayville, N.Y., U.S. (sā'vĭl)	67	40°45'N	73°10'W
Sazanit, i., Alb.	119	40°30'N	19°17'E
Sázava, r., Czech Rep.	124	49°36'N	15°24'E
Sazhino, Russia (sáz-hē'nô)	142a	56°20'N	58°15'E
Scandinavian Peninsula, pen., Eur.	152	62°00'N	14°00'E
Scanlon, Mn., U.S. (skăn'lôn)	75h	46°27'N	92°26'W
Scappoose, Or., U.S. (skă-pōōs')	74c	45°46'N	122°53'W
Scappoose, r., Or., U.S.	74c	45°47'N	122°57'W
Scarborough, Eng., U.K. (skär'bŭr-ô)	120	54°16'N	0°19'W
Scarsdale, N.Y., U.S. (skärz'dāl)	68a	41°01'N	73°47'W
Scatari I, Can. (skăt'*à*-rē)	61	46°00'N	59°44'W
Schaerbeek, Bel. (skär'bāk)	115a	50°50'N	4°23'E
Schaffhausen, Switz. (shäf'hou-zěn)	117	47°42'N	8°38'E
Schefferville, Can.	51	54°52'N	67°01'W
Schelde,, r., Eur.	121	51°04'N	3°55'E
Schenectady, N.Y., U.S. (skě-něk'tà-dê)	65	42°50'N	73°55'W
Scheveningen, Neth.	115a	52°06'N	4°15'E
Schiedam, Neth.	115a	51°55'N	4°23'E
Schiltigheim, Fr. (shĕl'tegh-hĭm)	127	48°48'N	7°47'E
Schio, Italy (skē'ô)	130	45°43'N	11°23'E
Schleswig, Ger. (shlĕs'vēgh)	116	54°32'N	9°32'E
Schleswig, hist. reg., Ger. (shlĕs'vēgh)	124	54°40'N	9°10'E
Schleswig-Holstein, state, Ger. (shlĕs'vēgh-hōl'shtīn)	115c	53°40'N	9°45'E
Schmalkalden, Ger. (shmäl'käl-děn)	124	50°41'N	10°25'E
Schneider, In., U.S. (schnīd'ẽr)	69a	41°12'N	87°26'W
Schofield, Wi., U.S. (skō'fēld)	71	44°52'N	89°37'W
Schönebeck, Ger. (shû'nē-bergh)	124	52°01'N	11°44'E
Schoonhoven, Neth.	115a	51°56'N	4°51'E
Schramberg, Ger. (shräm'bĕrgh)	124	48°14'N	8°24'E
Schreiber, Can.	58	48°50'N	87°10'W
Schroon, l., N.Y., U.S. (skrōōn)	67	43°50'N	73°50'W
Schultzendorf, Ger. (shōōl'tzĕn-dōrf)	115b	52°21'N	13°55'E
Schumacher, Can.	58	48°30'N	81°30'W
Schuyler, Ne., U.S. (slī'ler)	70	41°28'N	97°05'W
Schuylkill, r., Pa., U.S. (skōōl'kĭl)	68f	40°10'N	75°31'W
Schuylkill-Haven, Pa., U.S. (skōōl'kĭl hā-vĕn)	67	40°35'N	76°10'W
Schwabach, Ger. (shvä'bäk)	124	49°19'N	11°02'E
Schwäbische Alb, mts., Ger. (shvă'bē-shē älb)	124	48°11'N	9°09'E
Schwäbisch Gmünd, Ger. (shvă'bēsh gmünd)	124	48°47'N	9°49'E
Schwäbisch Hall, Ger. (häl)	124	49°08'N	9°44'E
Schwandorf, Ger. (shvän'dôrf)	124	49°19'N	12°08'E
Schwaner, Pegunungan, mts., Indon. (skvän'ẽr)	168	1°05'S	112°30'E
Schwarzwald, for., Ger. (shvärts'väld)	124	47°54'N	7°57'E
Schwaz, Aus.	124	47°20'N	11°45'E
Schwechat, Aus. (shvĕk'ät)	124	48°09'N	16°29'E
Schwedt, Ger. (shvĕt)	124	53°04'N	14°17'E
Schweinfurt, Ger. (shvīn'fōrt)	124	50°03'N	10°14'E
Schwelm, Ger. (shvĕlm)	127c	51°17'N	7°18'E
Schwerin, Ger. (shvĕ-rēn')	124	53°36'N	11°25'E
Schweriner See, l., Ger. (shvĕ'rē-nẽr zā)	124	53°40'N	11°06'E
Schwerte, Ger. (shvĕr'tĕ)	127c	51°26'N	7°34'E
Schwielowsee, l., Ger. (shvē'lōv zā)	115b	52°20'N	12°52'E
Schwyz, Switz. (shēts)	124	47°01'N	8°38'E
Sciacca, Italy (shē-äk'kä)	130	37°30'N	13°09'E
Scilly, Isles of, is., Eng., U.K. (sĭl'ē)	112	49°56'N	6°50'W
Scioto, r., Oh., U.S. (sī-ō'tō)	65	39°10'N	82°55'W
Scituate, Ma., U.S. (sĭt'ū-āt)	61a	42°12'N	70°45'W
Scobey, Mt., U.S. (skō'bē)	73	48°48'N	105°29'W
Scoggin, Or., U.S. (skō'gĭn)	74c	45°28'N	123°14'W
Scotch, r., Can. (skŏch)	62c	45°21'N	74°56'W
Scotia, Ca., U.S. (skō'shá)	72	40°29'N	124°06'W
Scotland, S.D., U.S.	70	43°08'N	97°43'W
Scotland, state, U.K. (skŏt'lánd)	110	57°05'N	5°10'W
Scotland Neck, N.C., U.S. (nĕk)	83	36°06'N	77°25'W
Scotstown, Can. (skŏts'toun)	67	45°35'N	71°15'W
Scott, r., Ca., U.S.	72	41°20'N	122°55'W
Scott, Cape, c., Can. (skŏt)	52	50°47'N	128°26'W
Scott, Mount, mtn., Or., U.S.	74c	45°27'N	122°33'W
Scott, Mount, mtn., Or., U.S.	72	42°55'N	122°00'W
Scott Air Force Base, Il., U.S.	75e	38°33'N	89°52'W
Scottburgh, S. Afr. (skŏt'bŭr-ō)	186	30°18'S	30°42'E
Scott City, Ks., U.S.	78	38°28'N	100°54'W
Scottdale, Ga., U.S. (skŏt'dāl)	68c	33°47'N	84°16'W
Scott Islands, is., Ant.	178	67°00'S	178°00'E
Scottsbluff, Ne., U.S. (skŏts'blŭf)	70	41°52'N	103°40'W
Scottsboro, Al., U.S. (skŏts'bŭro)	82	34°40'N	86°03'W
Scottsburg, In., U.S. (skŏts'bûrg)	66	38°40'N	85°50'W
Scottsdale, Austl. (skŏts'dāl)	176	41°12'S	147°37'E
Scottsville, Ky., U.S. (skŏts'vĭl)	82	36°45'N	86°10'W
Scottville, Mi., U.S.	66	44°00'N	86°20'W
Scranton, Pa., U.S. (skrăn'tŭn)	65	41°15'N	75°45'W
Scugog, l., Can. (skū'gôg)	59	44°05'N	78°55'W
Scunthorpe, Eng., U.K. (skŭn'thôrp)	114a	53°36'N	0°38'W
Scutari *see* Shkodër, Alb.	110	42°04'N	19°30'E
Scutari, Lake, l., Eur. (skōō'tä-rê)	119	42°14'N	19°33'E
Seabeck, Wa., U.S. (sē'bĕck)	74a	47°38'N	122°50'W
Sea Bright, N.J., U.S. (sē brīt)	68a	40°22'N	73°58'W
Seabrook, Tx., U.S. (sē'brŏŏk)	81	29°34'N	95°01'W
Seaford, De., U.S. (sē'fẽrd)	67	38°35'N	75°40'W
Seagraves, Tx., U.S. (sē'grāvs)	78	32°51'N	102°38'W
Sea Islands, is., Ga., U.S. (sē)	83	31°21'N	81°05'W
Seal, r., Can.	52	59°08'N	96°37'W
Seal Beach, Ca., U.S.	75a	33°44'N	118°06'W
Seal Cays, is., Bah.	93	22°40'N	75°55'W
Seal Cays, is., T./C. Is.	93	21°10'N	71°45'W
Seal Island, i., S. Afr. (sēl)	186a	34°07'S	18°36'E
Sealy, Tx., U.S. (sē'lê)	81	29°46'N	96°10'W
Searcy, Ar., U.S. (sûr'sê)	79	35°13'N	91°43'W
Searles, l., Ca., U.S. (sûrl's)	76	35°44'N	117°22'W
Searsport, Me., U.S. (sērz'pōrt)	60	44°28'N	68°55'W
Seaside, Or., U.S. (sē'sīd)	72	45°59'N	123°55'W
Seattle, Wa., U.S. (sê-ăt''l)	64	47°36'N	122°20'W
Sebaco, Nic. (sĕ-bä'kô)	90	12°50'N	86°03'W
Sebago, Me., U.S. (sê-bā'gō)	60	43°52'N	70°20'W
Sebastián Vizcaíno, Bahía, b., Mex.	86	28°45'N	115°15'W
Sebastopol, Ca., U.S. (sê-bás'tô-pôl)	76	38°27'N	122°50'W
Sebderat, Erit.	185	15°30'N	36°45'E
Sebewaing, Mi., U.S. (se'bê-wăng)	66	43°45'N	83°25'W
Sebezh, Russia (syĕ'bĕzh)	132	56°16'N	28°29'E
Sebinkarahisar, Tur.	119	40°15'N	38°10'E
Sebnitz, Ger. (zĕb'nêts)	124	51°01'N	14°16'E
Sebou, Oued, r., Mor.	184	34°23'N	5°18'W
Sebree, Ky., U.S. (sê-brē')	66	37°35'N	87°30'W
Sebring, Fl., U.S. (sē'brĭng)	83a	27°30'N	81°26'W
Sebring, Oh., U.S.	66	40°55'N	81°05'W
Secchia, r., Italy (sĕ'kyä)	130	44°25'N	10°25'E
Seco, r., Mex. (sĕ'kô)	89	18°11'N	93°18'W
Sedalia, Mo., U.S.	65	38°42'N	93°12'W
Sedan, Fr. (sẽ-däɴ)	117	49°49'N	4°55'E
Sedan, Ks., U.S. (sê-dăn')	79	37°07'N	96°08'W
Sedom, Isr.	153a	31°04'N	35°24'E
Sedro Woolley, Wa., U.S. (sē'drô-wŏŏl'ê)	74a	48°30'N	122°14'W
Šeduva, Lith. (shĕ'dŏŏ-vá)	123	55°46'N	23°45'E
Seestall, Ger. (zā'shtäl)	115d	47°58'N	10°52'E
Sefrou, Mor. (sĕ-frōō')	118	33°49'N	4°46'W
Seg, l., Russia (syĕgh)	136	63°20'N	33°30'E
Segamat, Malay. (sā'gá-mát)	153b	2°30'N	102°49'E
Segang, China (sŭ-gäŋ)	162	31°59'N	114°13'E
Segbana, Benin	189	10°56'N	3°42'E
Segorbe, Spain (sĕ-gôr-bĕ)	129	39°50'N	0°30'W
Ségou, Mali (sā-gōō')	184	13°27'N	6°16'W
Segovia, Col. (sĕ-gō'vēä)	100a	7°08'N	74°42'W
Segovia, Spain (sâ-gō'vê-ä)	118	40°58'N	4°05'W
Segre, r., Spain (sā'grä)	129	41°54'N	1°10'E
Seguam, i., Ak., U.S. (sē'gwäm)	63a	52°16'N	172°10'W
Seguam Passage, strt., Ak., U.S.	63a	52°20'N	173°00'W
Séguédine, Niger	189	20°12'N	12°59'E
Séguéla, C. Iv. (sā-gā-lä')	184	7°57'N	6°40'W
Seguin, Tx., U.S. (sẽ-gēn')	81	29°35'N	97°58'W
Segula, i., Ak., U.S. (sē-gū'lá)	63a	52°08'N	178°35'E
Segura, r., Spain	118	38°24'N	2°12'W
Segura, Sierra de, mts., Spain (sē-ĕ'r-rä-dĕ)	128	38°05'N	2°45'W
Sehwān, Pak.	158	26°33'N	67°51'E
Seibo, Dom. Rep. (sĕ'y-bō)	93	18°45'N	69°05'W
Seiling, Ok., U.S.	78	36°09'N	98°56'W
Seim, r., Eur.	137	51°23'N	33°22'E
Seinäjoki, Fin. (sâ'ē-nĕ-yô'kê)	123	62°47'N	22°50'E
Seine, r., Can. (sân)	62f	49°48'N	97°03'W
Seine, r., Can. (sân)	58	49°04'N	91°00'W
Seine, r., Fr.	112	48°00'N	4°30'E
Seine, Baie de la, b., Fr. (bī dĕ lä sân)	126	49°37'N	0°53'W
Seio do Venus, mtn., Braz. (sĕ-yô-dô-vĕ'nōōs)	102b	22°28'S	43°12'W
Seixal, Port. (sâ-ê-shäl')	129b	38°38'N	9°06'W
Sekenke, Tan.	191	4°16'S	34°10'E
Şeki, Azer.	138	41°12'N	47°12'E
Sekondi-Takoradi, Ghana (sĕ-kŏn'dê tä-kô-rä'dê)	184	4°59'N	1°43'W
Sekota, Eth.	185	12°47'N	38°59'E
Selangor, state, Malay. (sâ-län'gōr)	153b	2°53'N	101°29'E
Selanovtsi, Blg. (sĕl'á-nôv-tsĭ)	131	43°42'N	24°05'E
Selaru, Pulau, i., Indon.	169	8°30'S	130°30'E
Selatan, Tanjung, c., Indon. (sá-lä'tän)	168	4°09'S	114°40'E
Selawik, Ak., U.S. (sē-lá-wĭk)	63	66°30'N	160°09'W
Selayar, Pulau, i., Indon.	168	6°15'S	121°15'E
Selbusjøen, l., Nor. (sĕl'bōō)	122	63°18'N	11°55'E
Selby, Eng., U.K. (sĕl'bē)	114a	53°47'N	1°03'W
Seldovia, Ak., U.S. (sĕl-dō'vē-á)	63	59°26'N	151°42'W
Selemdzha, r., Russia (sâ-lĕmt-zhä')	141	52°28'N	131°50'E
Selenga (Selenge), r., Asia (sĕ lĕŋ gä')	135	49°00'N	102°00'E
Selenge, r., Asia	160	49°04'N	102°23'E
Selennyakh, r., Russia (sĕl-yĭn-yäk)	141	67°42'N	141°45'E
Sélestat, Fr. (sē-lē-stä')	127	48°16'N	7°27'E
Sélibaby, Maur. (sâ-lê-bà-bē')	184	15°21'N	12°11'W
Seliger, l., Russia (sĕl'lē-gĕr)	136	57°14'N	33°18'E
Selizharovo, Russia (sā'lê-zhä'rô-vô)	132	56°51'N	33°28'E
Selkirk, Can. (sĕl'kûrk)	50	50°09'N	96°52'W
Selkirk Mountains, mts., Can.	52	51°00'N	117°40'W
Selleck, Wa., U.S. (sĕl'ĕck)	74a	47°22'N	121°52'W
Sellersburg, In., U.S. (sĕl'ĕrs-bûrg)	69h	38°25'N	85°45'W
Sellya Khskaya, Guba, b., Russia (sĕl-yäk'ská-yá)	141	72°30'N	136°00'E
Selma, Al., U.S. (sĕl'má)	65	32°25'N	87°00'W
Selma, Ca., U.S.	76	36°34'N	119°37'W
Selma, N.C., U.S.	83	35°33'N	78°16'W
Selma, Tx., U.S.	75d	29°33'N	98°19'W
Selmer, Tn., U.S.	82	35°11'N	88°36'W
Selsingen, Ger. (zĕl'zēn-gĕn)	115c	53°22'N	9°13'E
Selway, r., Id., U.S. (sĕl'wâ)	72	46°07'N	115°12'W
Selwyn, l., Can. (sĕl'wĭn)	52	59°41'N	104°30'W
Seman, r., Alb.	131	40°48'N	19°53'E
Semarang, Indon. (sĕ-mä'räng)	168	7°03'S	110°27'E
Semenivka, Ukr.	137	52°10'N	32°34'E
Semeru, Gunung, mtn., Indon.	168	8°06'S	112°55'E
Semey (Semipalatinsk), Kaz.	139	50°28'N	80°29'E
Semiahmoo Indian Reserve, I.R., Can.	74d	49°01'N	122°43'W
Semiahmoo Spit, Wa., U.S. (sĕm'ĭ-á-mōō)	74d	48°59'N	122°52'W
Semichi Islands, is., Ak., U.S. (sē-mē'chī)	63a	52°40'N	174°50'E

302 Seminoe Reservoir – Shiawassee PRONOUNCING INDEX

PLACE (Pronunciation)	PAGE	LAT.	LONG.
Seminoe Reservoir, res., Wy., U.S. (sĕm'ĭ nô)	73	42°08'N	107°10'W
Seminole, Ok., U.S. (sĕm'ĭ-nōl)	79	35°13'N	96°41'W
Seminole, Tx., U.S.	80	32°43'N	102°39'W
Seminole, Lake, res., U.S.	82	30°57'N	84°46'W
Semipalatinsk *see* Semey, Kaz.	139	50°28'N	80°29'E
Semisopochnoi, i., Ak., U.S. (sĕ-mē-sá-pōsh' noi)	63a	51°45'N	179°25'E
Semliki, r., Afr. (sĕm'lê-kē)	185	0°45'N	29°36'E
Semmering Pass, p., Aus. (sĕm'ĕr-ĭng)	124	47°39'N	15°50'E
Senador Pompeu, Braz. (sĕ-nä-dōr-pôm-pĕ'ō)	101	5°34'S	39°18'W
Senaki, Geor.	138	42°17'N	42°04'E
Senatobia, Ms., U.S. (sĕ-ná-tō'bē-á)	82	34°36'N	89°56'W
Sendai, Japan (sĕn-dī')	161	38°18'N	141°02'E
Seneca, Ks., U.S. (sĕn'ĕ-ká)	79	39°49'N	96°03'W
Seneca, Md., U.S.	68e	39°04'N	77°20'W
Seneca, S.C., U.S.	83	34°40'N	82°58'W
Seneca, l., N.Y., U.S.	67	42°30'N	76°55'W
Seneca Falls, N.Y., U.S.	67	42°55'N	76°55'W
Sénégal, nation, Afr. (sĕn-ĕ-gôl')	184	14°53'N	14°58'W
Sénégal, r., Afr.	184	16°00'N	14°00'W
Senekal, S. Afr. (sĕn'ĕ-kál)	192c	28°20'S	27°37'E
Senftenberg, Ger. (zĕnf'tĕn-bĕrgh)	124	51°32'N	14°00'E
Sengunyane, r., Leso.	187c	29°35'S	28°08'E
Senhor do Bonfim, Braz. (sĕn-yôr dô bôn-fē'n)	101	10°21'S	40°09'W
Senigallia, Italy (så-nē-gäl'lyä)	130	43°42'N	13°16'E
Senj, Cro. (sĕn')	130	44°58'N	14°55'E
Senja, i., Nor. (sĕnyä)	116	69°28'N	16°10'E
Senlis, Fr. (sän-lēs')	127b	49°13'N	2°35'E
Sennar Dam, dam, Sudan	185	13°38'N	33°38'E
Senneterre, Can.	51	48°20'N	77°22'W
Sens, Fr. (säns)	126	48°05'N	3°18'E
Sensuntepeque, El Sal. (sĕn-sōōn-tå-pā'kå)	90	13°53'N	88°34'W
Senta, Serb. (sĕn'tä)	119	45°54'N	20°05'E
Senzaki, Japan (sĕn'zä-kē)	167	34°22'N	131°09'E
Seoul (Sŏul), Kor., S.	161	37°35'N	127°03'E
Sepang, Malay.	153b	2°43'N	101°45'E
Sepetiba, Baía de, b., Braz. (bäē'ä dĕ så-på-tē'bá)	102b	23°01'S	43°42'W
Sepik, r., (sĕp-ĕk')	169	4°07'S	142°40'E
Septentrional, Cordillera, mts., Dom. Rep.	93	19°50'N	71°15'W
Septeuil, Fr. (sĕ-tû')	127b	48°53'N	1°40'E
Sept-Îles, Can. (sĕ-tēl')	60	50°12'N	66°23'W
Sequatchie, r., Tn., U.S. (sē-kwăch'ē)	82	35°33'N	85°14'W
Sequim, Wa., U.S. (sē'kwĭm)	74a	48°05'N	123°07'W
Sequim Bay, b., Wa., U.S.	74a	48°04'N	122°58'W
Sequoia National Park, rec., Ca., U.S. (sē-kwoi'á)	64	36°34'N	118°37'W
Seraing, Bel. (sĕ-răn')	121	50°38'N	5°28'E
Serāmpore, India	158a	22°44'N	88°21'E
Serang, Indon. (så-räng')	168	6°13'S	106°10'E
Seranggung, Indon.	153b	0°49'N	104°11'E
Serbia and Montenegro (Yugoslavia), nation, Eur.	110	44°00'N	21°00'E
Serbia *see* Srbija, hist. reg., Serb.	131	44°05'N	20°35'E
Serdobsk, Russia (sĕr-dôpsk')	137	52°30'N	44°20'E
Sered', Slvk.	125	48°17'N	17°43'E
Seredyna-Buda, Ukr.	137	52°11'N	34°03'E
Seremban, Malay. (sĕr-ĕm-bän')	153b	2°44'N	101°57'E
Serengeti National Park, rec., Tan.	191	2°20'S	34°50'E
Serengeti Plain, pl., Tan.	191	2°40'S	34°55'E
Serenje, Zam. (sĕ-rĕn'yĕ)	186	13°12'S	30°49'E
Seret, r., Ukr. (sĕr'ĕt)	125	49°45'N	25°30'E
Sergeya Kirova, i., Russia (sĕr-gyē'yà kē'rô-vå)	140	77°30'N	86°10'E
Sergipe, state, Braz. (sĕr-zhē'pĕ)	101	10°27'S	37°04'W
Sergiyev Posad, Russia	142b	56°18'N	38°08'E
Sergiyevsk, Russia	136	53°58'N	51°00'E
Sérifos, Grc.	131	37°10'N	24°32'E
Sérifos, i., Grc.	131	37°42'N	24°17'E
Serodino, Arg. (sĕ-rô-dē'nō)	99c	32°36'S	60°56'W
Seropédica, Braz. (sĕ-rô-pĕ'dē-kä)	102b	22°44'S	43°43'W
Serov, Russia (syĕ-rôf')	140	59°36'N	60°30'E
Serowe, Bots. (sĕ-rō'wĕ)	186	22°18'S	26°39'E
Serpa, Port. (sĕr-pä)	128	37°56'N	7°38'W
Serpukhov, Russia (syĕr'pô-kôf)	134	54°53'N	37°27'E
Sérres, Grc. (sĕr'rĕ) (sĕr'ĕs)	119	41°06'N	23°36'E
Serrinha, Braz. (sĕr-rēn'yà)	101	11°43'S	38°49'W
Serta, Port. (sĕr'tá)	128	39°48'N	8°01'W
Sertânia, Braz. (sĕr-tá'nyä)	101	8°28'S	37°13'W
Sertãozinho, Braz. (sĕr-toun-zē'n-yô)	99a	21°10'S	47°58'W
Serting, r., Malay.	153b	3°01'N	102°32'E
Sese Islands, is., Ug.	191	0°30'S	32°30'E
Sesia, r., Italy (såz'yä)	130	45°33'N	8°25'E
Sesimbra, Port. (sĕ-sē'm-brä)	129b	38°27'N	9°06'W
Sesmyl, r., S. Afr.	187b	25°51'S	28°06'E
Ses Salines, Cap de, c., Spain	129	39°16'N	3°03'E
Sestri Levante, Italy (sĕs'trē lå-vän'tå)	130	44°15'N	9°24'E
Sestroretsk, Russia (sĕs-trô-rĕtsk)	136	60°06'N	29°58'E
Sestroretskiy Razliv, Ozero, l., Russia	142c	60°05'N	30°07'E
Seta, Japan (sĕ'tä)	167b	34°58'N	135°56'W
Sète, Fr. (sĕt)	117	43°24'N	3°42'E
Sete Lagoas, Braz. (sĕ-tĕ lä-gō'äs)	101	19°23'S	43°58'W
Sete Pontes, Braz.	102b	22°51'S	43°05'W
Seto, Japan (sĕ'tō)	167	35°11'N	137°07'E
Seto-Naikai, sea, Japan (sĕ'tô nī'kī)	167	33°50'N	132°25'E
Settat, Mor. (sĕt-ät') (sĕ-tá')	184	33°02'N	7°30'W
Sette-Cama, Gabon (sĕ-tĕ-kä-mä')	186	2°29'S	9°40'E
Settlement Point, c., Bah. (sĕt'l-mĕnt)	92	26°40'N	79°00'W
Settlers, S. Afr. (sĕt'lĕrs)	192c	24°57'S	28°33'E
Settsu, Japan	167b	34°46'N	135°33'E
Setúbal, Port. (så-tōō'bäl)	118	30°32'N	8°54'W

PLACE (Pronunciation)	PAGE	LAT.	LONG.
Setúbal, Baía de, b., Port.	128	38°27'N	9°08'W
Seul, Lac, l., Can. (lák sûl)	53	50°20'N	92°30'W
Sevan, l., Arm. (syĭ-vän')	137	40°10'N	45°20'E
Sevastopol', Ukr. (syĕ-vås-tô'pôl'')	134	44°34'N	33°34'E
Sevenoaks, Eng., U.K. (sĕ-vĕn-ōks')	114b	51°16'N	0°12'E
Severka, r., Russia (sâ'vĕr-kà)	142b	55°11'N	38°41'E
Severn, r., Can. (sĕv'ĕrn)	53	55°21'N	88°42'W
Severn, r., U.K.	120	51°50'N	2°25'W
Severna Park, Md., U.S. (sĕv'ĕrn-â)	68e	39°04'N	76°33'W
Severnaya Dvina, r., Russia	134	63°00'N	42°40'E
Severnaya Zemlya (Northern Land), is., Russia (sĕ-vyĭr-nīŭ zĭ-m'lyä')	135	79°33'N	101°15'E
Severoural'sk, Russia (sĕ-vyĭ-rŭ-ōō-rälsk')	140	60°08'N	59°53'E
Sevier, r., Ut., U.S.	64	39°25'N	112°20'W
Sevier, East Fork, r., Ut., U.S.	77	37°45'N	112°10'W
Sevier Lake, l., Ut., U.S. (sē-vēr')	77	38°55'N	113°10'W
Sevilla, Col. (sĕ-vē'l-yä)	100a	4°16'N	75°56'W
Sevilla, Spain (så-vēl'yä)	110	37°29'N	5°58'W
Seville, Oh., U.S. (sĕ'vĭl)	69d	41°01'N	81°45'W
Sevlievo, Blg. (sĕv'lyĕ-vô)	119	43°02'N	25°05'E
Sevsk, Russia (syĕfsk)	132	52°08'N	34°28'E
Seward, Ak., U.S. (sū'ârd)	64a	60°18'N	149°28'W
Seward, Ne., U.S.	79	40°55'N	97°06'W
Seward Peninsula, pen., Ak., U.S.	63	65°40'N	164°00'W
Sewell, Chile (sĕ'ō-ĕl)	102	34°01'S	70°18'W
Sewickley, Pa., U.S. (sē-wĭk'lē)	69e	40°33'N	80°11'W
Seybaplaya, Mex. (sā-ē-bä-plä'yä)	89	19°38'N	90°40'W
Seychelles, nation, Afr. (sā-shĕl')	3	5°20'S	55°10'E
Seydisfjördur, Ice. (sā'dĕs-fyûr-dŏr)	116	65°21'N	14°08'W
Seyhan, r., Tur.	119	37°28'N	35°40'E
Seylac, Som.	192a	11°19'N	43°20'E
Seymour, S. Afr. (sē'môr)	187c	32°33'S	26°48'E
Seymour, Ia., U.S.	71	40°41'N	93°03'W
Seymour, In., U.S. (sē'mōr)	66	38°55'N	85°55'W
Seymour, Tx., U.S.	78	33°35'N	99°16'W
Sezela, S. Afr.	187c	30°33'S	30°37'W
Sezze, Italy (sĕt'så)	130	41°32'N	13°00'E
Sfântu Gheorghe, Rom.	119	45°53'N	25°49'E
Sfax, Tun. (sfäks)	184	34°51'N	10°45'E
's-Gravenhage *see* The Hague, Neth. ('s krä'vĕn-hä'kĕ) (hāg)	110	52°05'N	4°16'E
Sha, r., China (shä)	161	33°33'N	114°30'E
Shaanxi, prov., China (shän-shyē)	160	35°30'N	109°10'E
Shabeelle (Shebele), r., Afr.	192a	1°38'N	43°50'E
Shache, China (shä-chŭ)	160	38°15'N	77°15'E
Shackleton Ice Shelf, ice, Ant. (shăk''l-tŭn)	178	65°00'S	100°00'E
Shades Creek, r., Al., U.S. (shādz)	68h	33°20'N	86°55'W
Shades Mountain, mtn., Al., U.S.	68h	33°22'N	86°51'W
Shagamu, Nig.	189	6°51'N	3°39'E
Shāhdād, Namakzār-e, l., Iran (nŭ-mŭk-zär')	154	31°00'N	58°30'E
Shāhjahānpur, India (shä-jŭ-hän'pōōr)	155	27°58'N	79°58'E
Shajing, China (shä-jyĭŋ)	163a	22°44'N	113°48'E
Shaker Heights, Oh., U.S. (shā'kĕr)	69d	41°28'N	81°34'W
Shakhty, Russia (shäk'tē)	134	47°41'N	40°11'E
Shaki, Nig.	189	8°39'N	3°25'E
Shakopee, Mn., U.S. (shăk'ō-pe)	75g	44°48'N	93°31'W
Shala Lake, l., Eth. (shä'là)	185	7°34'N	39°00'E
Shalqar, Kaz.	139	47°52'N	59°41'E
Shalqar köli, l., Kaz.	137	50°30'N	51°30'E
Shām, Jabal ash, mtn., Oman	154	23°01'N	57°45'E
Shambe, Sudan (shäm'bå)	185	7°08'N	30°46'E
Shammar, Jabal, mts., Sau. Ar. (jĕb'ĕl shŭm'år)	154	27°13'N	40°16'E
Shamokin, Pa., U.S. (shá-mō'kĭn)	67	40°45'N	76°30'W
Shamrock, Tx., U.S. (shăm'rŏk)	78	35°14'N	100°12'W
Shamva, Zimb. (shäm'vá)	186	17°18'S	31°35'E
Shandon, Oh., U.S. (shăn-dŭn)	69f	39°20'N	84°13'W
Shandong, prov., China (shän-dôŋ)	161	36°08'N	117°09'E
Shandong Bandao, pen., China (shän-dôŋ bän-dou)	161	37°00'N	120°10'E
Shangcai, China (shäŋ-tsī)	162	33°16'N	114°16'E
Shangcheng, China (shäŋ-chŭŋ)	162	31°47'N	115°22'E
Shangdu, China (shäŋ-dōō)	164	41°38'N	113°22'E
Shanghai, China (shäng'hī')	161	31°14'N	121°27'E
Shanghai Shi, prov., China (shäŋ-hī shr)	161	31°30'N	121°45'E
Shanghe, China (shäŋ-hŭ)	162	37°18'N	117°10'E
Shanglin, China (shäŋ-lĭn)	162	38°20'N	116°05'E
Shangqiu, China (shäŋ-chyô)	164	34°24'N	115°39'E
Shangrao, China (shäŋ-rou)	165	28°25'N	117°58'E
Shangzhi, China (shäŋ-jr)	164	45°18'N	127°52'E
Shanhaiguan, China	164	40°01'N	119°45'E
Shannon, Al., U.S. (shăn'ŭn)	68h	33°23'N	86°52'W
Shannon, r., Ire. (shăn'ŏn)	117	52°30'N	10°15'W
Shanshan, China (shän'shän')	160	42°51'N	89°53'E
Shantar, i., Russia (shän'tär)	141	55°13'N	138°42'E
Shantou, China (shän-tō)	161	23°20'N	116°40'E
Shanxi, prov., China (shän-shyē)	161	37°30'N	112°00'E
Shan Xian, China (shän shyĕn)	162	34°47'N	116°04'E
Shaobo, China (shou-bwo)	164	32°33'N	119°30'E
Shaobo Hu, l., China (shou-bwo hōō)	162	32°47'N	119°13'E
Shaoguan, China (shou-gŭän)	161	24°58'N	113°42'E
Shaoxing, China (shou-shyĭŋ)	161	30°00'N	120°40'E
Shaoyang, China	161	27°15'N	111°28'E
Shapki, Russia (shäp'kĭ)	142c	59°36'N	31°11'E
Shark Bay, b., Austl. (shärk)	174	25°30'S	113°00'E
Sharon, Ma., U.S. (shăr'ŏn)	61a	42°07'N	71°11'W
Sharon, Pa., U.S.	66	41°15'N	80°30'W
Sharon Springs, Ks., U.S.	78	38°51'N	101°45'W
Sharonville, Oh., U.S. (shăr'ŏn vĭl)	69f	39°16'N	84°24'W
Sharpsburg, Pa., U.S. (shärps'bûrg)	69e	40°30'N	79°54'W
Sharr, Jabal, mtn., Sau. Ar.	154	28°00'N	36°07'E
Shashi, China (shä-shē)	161	30°20'N	112°18'E
Shasta, Mount, mtn., Ca., U.S.	64	41°35'N	122°12'W

PLACE (Pronunciation)	PAGE	LAT.	LONG.
Shasta Lake, res., Ca., U.S. (shăs'tá)	64	40°51'N	122°32'W
Shatsk, Russia (shátsk)	136	54°00'N	41°40'E
Shattuck, Ok., U.S. (shăt'ŭk)	78	36°16'N	99°53'W
Shaunavon, Can.	50	49°40'N	108°25'W
Shaw, Ms., U.S. (shô)	82	33°36'N	90°44'W
Shawano, Wi., U.S. (shá-wô'nô)	71	44°41'N	88°13'W
Shawinigan, Can.	51	46°32'N	72°46'W
Shawnee, Ks., U.S. (shô-nē')	75f	39°01'N	94°43'W
Shawnee, Ok., U.S.	64	35°20'N	96°54'W
Shawneetown, Il., U.S. (shô'nē-toun)	66	37°40'N	88°05'W
Shayang, China	165	31°00'N	112°38'E
Shchara, r., Bela. (sh-chá'rá)	125	53°17'N	25°12'E
Shchëlkovo, Russia (shchĕl'kô-vô)	132	55°55'N	38°00'E
Shchigry, Russia (shchē'grē)	133	51°52'N	36°54'E
Shchors, Ukr. (shchôrs)	133	51°38'N	31°58'E
Shchuch'ye Ozero, Russia (shchōōch'yĕ ô'zĕ-rô)	142a	56°31'N	56°35'E
Sheakhala, India	158a	22°47'N	88°10'E
Shebele (Shabeelle), r., Afr. (shā'bå-lē)	192a	6°07'N	43°10'E
Sheboygan, Wi., U.S. (shē-boi'găn)	65	43°45'N	87°44'W
Sheboygan Falls, Wi., U.S.	71	43°43'N	87°51'W
Shechem, hist., W.B.	153a	32°15'N	35°22'E
Shedandoah, Pa., U.S.	67	40°50'N	76°15'W
Shediac, Can. (shĕ'dē-ăk)	60	46°13'N	64°32'W
Shedin Peak, mtn., Can. (shĕd'ĭn)	54	55°55'N	127°32'W
Sheerness, Eng., U.K. (shēr'nĕs)	114b	51°26'N	0°46'E
Sheffield, Can.	62d	43°20'N	80°13'W
Sheffield, Eng., U.K.	116	53°23'N	1°28'W
Sheffield, Al., U.S. (shĕf'fēld)	82	35°42'N	87°42'W
Sheffield, Oh., U.S.	69d	41°26'N	82°05'W
Sheffield, co., Eng., U.K.	114a	53°52'N	1°35'W
Sheffield Lake, Oh., U.S.	69d	41°30'N	82°03'W
Sheksna, r., Russia (shĕks'ná)	136	59°50'N	38°40'E
Shelagskiy, Mys, c., Russia (shĭ-läg'skē)	135	70°08'N	170°52'E
Shelbina, Ar., U.S. (shĕl-bī'ná)	79	39°41'N	92°03'W
Shelburn, In., U.S. (shĕl'bŭrn)	66	39°10'N	87°30'W
Shelburne, Can.	51	43°46'N	65°19'W
Shelburne, Can.	59	44°04'N	80°12'W
Shelby, In., U.S. (shĕl'bē)	69a	41°12'N	87°21'W
Shelby, Mi., U.S.	66	43°35'N	86°20'W
Shelby, Ms., U.S.	82	33°56'N	90°44'W
Shelby, Mt., U.S.	73	48°35'N	111°55'W
Shelby, N.C., U.S.	83	35°16'N	81°35'W
Shelby, Oh., U.S.	66	40°50'N	82°40'W
Shelbyville, Il., U.S. (shĕl'bē-vĭl)	66	39°20'N	88°45'W
Shelbyville, In., U.S.	66	39°30'N	85°45'W
Shelbyville, Ky., U.S.	66	38°10'N	85°15'W
Shelbyville, Tn., U.S.	82	35°30'N	86°28'W
Shelbyville Reservoir, res., Il., U.S.	66	39°30'N	88°45'W
Sheldon, Ia., U.S. (shĕl'dŭn)	70	43°10'N	95°50'W
Sheldon, Tx., U.S.	81a	29°52'N	95°07'W
Shelekhova, Zaliv, b., Russia	135	60°00'N	156°00'E
Shelikof Strait, strt., Ak., U.S. (shĕ'lē-kôf)	63	57°56'N	154°20'W
Shellbrook, Can.	56	53°15'N	106°22'W
Shelley, Id., U.S. (shĕl'lē)	73	43°24'N	112°06'W
Shellrock, r., Ia., U.S. (shĕl'rŏk)	71	43°25'N	93°19'W
Shelon', r., Russia (shå'lôn)	132	57°50'N	29°40'E
Shelton, Ct., U.S. (shĕl'tŭn)	67	41°15'N	73°05'W
Shelton, Ne., U.S.	78	40°46'N	98°41'W
Shelton, Wa., U.S.	72	47°14'N	123°05'W
Shemakha, Russia (shĕ-mà-kä')	142a	56°16'N	59°19'E
Shenandoah, Ia., U.S. (shĕn-ăn-dō'á)	79	40°46'N	95°23'W
Shenandoah, Va., U.S.	67	38°30'N	78°30'W
Shenandoah, r., Va., U.S.	67	38°55'N	78°05'W
Shenandoah National Park, rec., Va., U.S.	65	38°35'N	78°25'W
Shendam, Nig.	189	8°53'N	9°32'E
Shengfang, China (shengfāng)	162	39°05'N	116°40'E
Shenkursk, Russia (shĕn-kōōrsk')	134	62°10'N	43°08'E
Shenmu, China	164	38°55'N	110°35'E
Shenqiu, China	164	33°11'N	115°06'E
Shenxian, China (shŭn shyän)	162	38°02'N	115°33'E
Shenxian, China (shŭn shyĕn)	162	36°14'N	115°38'E
Shenyang, China (shŭn-yäŋ)	161	41°45'N	123°22'E
Shenze, China (shŭn-dzŭ)	162	38°12'N	115°12'E
Shenzhen, China	165	22°32'N	114°08'E
Sheopur, India	155	25°37'N	77°10'E
Shepard, Can. (shĕ'pärd)	62e	50°57'N	113°55'W
Shepetivka, Ukr.	137	50°10'N	27°01'E
Shepparton, Austl. (shĕp'ăr-tŭn)	176	36°15'S	145°25'E
Sherborn, Ma., U.S. (shûr'bŭrn)	61a	42°15'N	71°22'W
Sherbrooke, Can.	51	45°24'N	71°54'W
Sherburn, Eng., U.K. (shûr'bŭrn)	114a	53°47'N	1°15'W
Shereshevo, Bela. (shĕ-rĕ-shĕ-vô)	125	52°31'N	24°08'E
Sheridan, Ar., U.S. (shĕr'ĭ-dăn)	79	34°19'N	92°21'W
Sheridan, Or., U.S.	72	45°06'N	123°22'W
Sheridan, Wy., U.S.	64	44°48'N	106°56'W
Sherman, Tx., U.S. (shĕr'măn)	64	33°39'N	96°37'W
Sherna, r., Russia (shĕr'ná)	142b	56°08'N	38°45'E
Sherridon, Can.	57	55°10'N	101°10'W
's Hertogenbosch, Neth. (sĕr-tō'ghĕn-bôs)	121	51°41'N	5°19'E
Sherwood, Or., U.S.	74c	45°21'N	122°50'W
Sherwood Forest, for., Eng., U.K.	114a	53°11'N	1°07'W
Sherwood Park, Can.	55	53°31'N	113°19'W
Shetland Islands, is., Scot., U.K. (shĕt'lănd)	112	60°35'N	2°10'W
Shewa Gimira, Eth.	185	7°13'N	35°49'E
Shexian, China (shŭ shyĕn)	162	36°34'N	113°42'E
Sheyang, r., China (shē-yäŋ)	162	33°42'N	119°40'E
Sheyenne, r., N.D., U.S. (shī-ĕn')	70	46°42'N	97°52'W
Shi, r., China (shr)	162	31°58'N	115°50'E
Shi, r., China	162	32°09'N	114°11'E
Shiawassee, r., Mi., U.S. (shī-á-wôs'ē)	66	43°15'N	84°05'W

ăt; finăl; rāte; senåte; ärm; åsk; sofá; fâre; ch-choose; dh-as th in other; bē; ēvent; bĕt; recĕnt; cratẽr; g-gō; gh-guttural g; bĭt; ĭ-short neutral; rīde; ĸ-guttural k as ch in German ich;

PRONOUNCING INDEX
Shibām – Siracusa 303

PLACE (Pronunciation)	PAGE	LAT.	LONG.
Shibām, Yemen (shē'bäm)	154	16°02'N	48°40'E
Shibīn al Kawn, Egypt (shē-bēn'ĕl kōm')	192b	30°31'N	31°01'E
Shibīn al Qanāṭir, Egypt (kȧ-nä'tēr)	192b	30°18'N	31°21'E
Shicun, China (shr-tsòn)	162	33°47'N	117°18'E
Shields, r., Mt., U.S. (shēldz)	73	45°54'N	110°40'W
Shifnal, Eng., U.K. (shĭf'năl)	114a	52°40'N	2°22'W
Shijian, China (shr-jyĕn)	162	31°27'N	117°51'E
Shijiazhuang, China (shr-jyä-jůäŋ)	161	38°04'N	114°31'E
Shijiu Hu, l., China (shr-jyō hōō)	162	31°29'N	119°07'E
Shikārpur, Pak.	155	27°51'N	68°52'E
Shiki, Japan (shē'kē)	167a	35°50'N	139°35'E
Shikoku, i., Japan (shē'kō'kōō)	161	33°43'N	133°33'E
Shilka, r., Russia (shĭl'kȧ)	141	53°00'N	118°45'E
Shilla, mtn., India	158	32°18'N	78°17'E
Shillong, India (shēl-lông')	155	25°39'N	91°58'E
Shiloh, Il., U.S. (shī'lō)	75e	38°34'N	89°54'W
Shilong, China (shr-lôŋ)	165	23°05'N	113°58'E
Shilou, China	163a	22°58'N	113°29'E
Shimabara, Japan (shē'mä-bä'rä)	167	32°46'N	130°22'E
Shimada, Japan (shē'mä-dä)	167	34°49'N	138°13'E
Shimbiris, mtn., Som.	192a	10°40'N	47°23'E
Shimizu, Japan (shē'mē-zōō)	166	35°00'N	138°29'E
Shimminato, Japan (shēm'mē'nä-tō)	167	36°47'N	137°05'E
Shimoda, Japan (shē'mô-dȧ)	167	34°41'N	138°58'E
Shimoga, India	159	13°59'N	75°38'E
Shimoni, Kenya	191	4°39'S	39°23'E
Shimonoseki, Japan	161	33°58'N	130°55'E
Shimo-Saga, Japan (shē'mô sä'gä)	167b	35°01'N	135°41'E
Shin, Loch, l., Scot., U.K. (lŏk shĭn)	120	58°08'N	4°02'W
Shinagawa-Wan, b., Japan (shē'nä-gä'wä wän)	167a	35°37'N	139°49'E
Shinano-Gawa, r., Japan (shē-nä'nô gä'wä)	167	36°43'N	138°22'E
Shīndand, Afg.	157	33°18'N	62°08'E
Shinji, l., Japan (shĭn'jē)	167	35°23'N	133°05'E
Shinkolobwe, D.R.C.	191	11°02'S	26°35'E
Shinyanga, Tan. (shĭn-yäŋ'gä)	186	3°40'S	33°26'E
Shiono Misaki, c., Japan (shē-ō'nô mē'sä-kē)	166	33°20'N	136°10'E
Shipai, China (shr-pī)	163a	23°07'N	113°23'E
Ship Channel Cay, i., Bah. (shĭp chă-nĕl kē)	92	24°50'N	76°50'W
Shipley, Eng., U.K. (shĭp'lē)	114a	53°50'N	1°47'W
Shippegan, Can. (shĭ'pē-găn)	60	47°45'N	64°42'W
Shippegan Island, i., Can.	60	47°50'N	64°38'W
Shippenburg, Pa., U.S. (shĭp'ĕn bûrg)	67	40°00'N	77°30'W
Shipshaw, r., Can. (shĭp'shô)	59	48°50'N	71°03'W
Shiqma, r., Isr.	153a	31°31'N	34°40'E
Shirane-san, mtn., Japan (shē'rä'nȧ-sän')	167	35°44'N	138°14'E
Shirati, Tan. (shē-rä'tē)	186	1°15'S	34°02'E
Shīrāz, Iran (shē-räz')	154	29°32'N	52°27'E
Shire, r., Afr. (shē'rȧ)	186	15°00'S	35°00'E
Shiriya Saki, c., Japan (shē'rä sä'kē)	166	41°25'N	142°10'E
Shirley, Ma., U.S. (shûr'lē)	61a	42°33'N	71°39'W
Shishaldin Volcano, vol., Ak., U.S. (shī-shăl'dĭn)	63a	54°48'N	164°00'W
Shively, Ky., U.S. (shīv'lē)	69h	38°11'N	85°47'W
Shivpuri, India	155	25°31'N	77°46'E
Shivta, Horvot, hist., Isr.	153a	30°54'N	34°36'E
Shivwits Plateau, plat., Az., U.S.	77	36°13'N	113°42'W
Shiwan, China (shr-wän)	163a	23°01'N	113°04'E
Shiwan Dashan, mts., China (shr-wän dä-shän)	165	22°10'N	107°30'E
Shizuki, Japan (shĭ'zōō-kē)	167	34°29'N	134°51'E
Shizuoka, Japan (shē'zōō'ōkȧ)	166	34°58'N	138°24'E
Shklow, Bela.	132	54°11'N	30°23'E
Shkodër, Alb. (shkô'dûr) (skōō'tarē)	110	42°04'N	19°30'E
Shkotovo, Russia (shkô'tô-vô)	166	43°15'N	132°21'E
Shoal Creek, r., Il., U.S. (shōl)	73	38°37'N	89°25'W
Shoal Lake, l., Can.	57	49°32'N	95°00'W
Shoals, In., U.S. (shōlz)	66	38°40'N	86°45'W
Shōdo, i., Japan (shō'dō)	167	34°27'N	134°27'E
Sholāpur, India (shō'lä-pōōr)	155	17°42'N	75°51'E
Shorewood, Wi., U.S. (shōr'wŏŏd)	69a	43°05'N	87°54'W
Shoshone, Id., U.S. (shō-shōn'tē)	73	42°56'N	114°24'W
Shoshone, r., Wy., U.S.	73	44°35'N	108°50'W
Shoshone Lake, l., Wy., U.S.	73	44°17'N	110°50'W
Shoshoni, Wy., U.S.	73	43°14'N	108°05'W
Shostka, Ukr. (shôst'kȧ)	133	51°51'N	33°31'E
Shouguang, China (shō-gůäŋ)	162	36°53'N	118°45'E
Shouxian, China (shō shyĕn)	162	32°36'N	116°45'E
Shpola, Ukr. (shpô'lȧ)	137	49°01'N	31°36'E
Shreveport, La., U.S. (shrēv'pōrt)	65	32°30'N	93°46'W
Shrewsbury, Eng., U.K. (shrōōz'bēr-ĭ)	120	52°43'N	2°44'W
Shrewsbury, Ma., U.S.	61a	42°18'N	71°43'W
Shropshire, co., Eng., U.K.	114a	52°36'N	2°45'W
Shroud Cay, i., Bah.	92	24°20'N	76°40'W
Shuangcheng, China (shůäŋ-chŭŋ)	164	45°18'N	126°18'E
Shuanghe, China (shůäŋ-hŭ)	162	31°33'N	116°48'E
Shuangliao, China	161	43°37'N	123°30'E
Shuangyang, China	164	43°28'N	125°45'E
Shuhedun, China (shōō-hŭ-dòn)	162	31°33'N	117°01'E
Shuiye, China (shwā-yŭ)	162	36°08'N	114°07'E
Shule, r., China (shōō-lŭ)	160	40°53'N	94°55'E
Shullsburg, Wi., U.S. (shŭlz'bûrg)	71	42°35'N	90°16'W
Shumagin, is., Ak., U.S. (shōō'mȧ-gēn)	63	55°22'N	159°20'W
Shumen, Blg.	119	43°15'N	26°54'E
Shunde, China (shòn-dŭ)	163a	22°50'N	113°15'E
Shungnak, Ak., U.S. (shŭng'nȧk)	63	66°55'N	157°20'W
Shunut, Gora, mtn., Russia (gȧ-rä shōō'nòt)	142a	56°33'N	59°45'E
Shunyi, China (shòn-yē)	162	40°09'N	116°38'E
Shuqrah, Yemen	154	13°32'N	46°02'E
Shūrāb, r., Iran (shōō räb)	154	31°08'N	55°30'E
Shuri, Japan (shōō'rē)	166	26°10'N	127°48'E
Shurugwi, Zimb.	186	19°34'S	30°03'E
Shūshtar, Iran (shōōsh'tŭr)	154	31°50'N	48°46'E
Shuswap Lake, l., Can. (shōōs'wŏp)	55	50°57'N	119°15'W
Shuya, Russia (shōō'yȧ)	134	56°52'N	41°23'E
Shuyang, China (shōō yäng)	162	34°09'N	118°47'E
Shweba, Mya.	155	22°23'N	96°13'E
Shymkent, Kaz.	139	42°17'N	69°42'E
Shyroke, Ukr.	133	47°40'N	33°18'E
Siak Kecil, r., Indon.	153b	1°01'N	101°45'E
Siaksriinderapura, Indon. (sē-äks'rī ĕn'drȧ-pōō'rä)	153b	0°48'N	102°05'E
Siālkot, Pak. (sē-äl'kōt)	155	32°39'N	74°30'E
Siátista, Grc. (syä'tĭs-tȧ)	131	40°15'N	21°32'E
Siau, Pulau, i., Indon.	169	2°40'N	126°00'E
Šiauliai, Lith. (shē-ou'lē-ī)	136	55°57'N	23°19'E
Sibay, Russia (sē'bȧy)	142a	52°41'N	58°40'E
Šibenik, Cro. (shē-bä'nĕk)	119	43°44'N	15°55'E
Siberia, reg., Russia	152	57°00'N	97°00'E
Siberut, Pulau, i., Indon. (sē'bȧ-rōōt)	168	1°22'S	99°45'E
Sibiti, Congo (sē-bē-tē')	186	3°41'S	13°21'E
Sibiu, Rom. (sē-bē-ōō')	119	45°47'N	24°09'E
Sibley, Ia., U.S. (sĭb'lē)	70	43°24'N	95°33'W
Sibolga, Indon. (sē-bō'gȧ)	168	1°45'N	98°45'E
Sibsāgar, India (sĕb-sŭ'gŭr)	155	26°47'N	94°45'E
Sibutu Island, i., Phil.	168	4°40'N	119°30'E
Sibuyan, i., Phil. (sē-bōō-yän')	169a	12°19'N	122°25'E
Sibuyan Sea, sea, Phil.	168	12°43'N	122°38'E
Sichuan, prov., China (sz-chůän)	160	31°20'N	103°00'E
Sicily, i., Italy (sĭs'ĭ-lē)	112	37°38'N	13°30'E
Sico, r., Hond. (sē-kō)	90	15°32'N	85°42'W
Sidamo, hist. reg., Eth. (sē-dä'mō)	185	5°08'N	37°45'E
Siderno Marina, Italy (sē-dĕr'nô mä-rē'nä)	130	38°18'N	16°19'E
Sídheros, Ákra, c., Grc.	130a	35°19'N	26°20'E
Sidi Aïssa, Alg.	129	35°53'N	3°44'E
Sidi bel Abbès, Alg. (sē'dē-bĕl ȧ-bĕs')	184	35°15'N	0°43'W
Sidi Ifni, Mor. (ēf'nē)	184	29°22'N	10°15'W
Sidirókastro, Grc.	131	41°13'N	23°27'E
Sidley, Mount, mtn., Ant. (sĭd'lē)	178	77°25'S	129°00'W
Sidney, Can.	54	48°39'N	123°24'W
Sidney, Mt., U.S. (sĭd'nē)	73	47°43'N	104°07'W
Sidney, Ne., U.S.	70	41°10'N	103°00'W
Sidney, Oh., U.S.	66	40°20'N	84°10'W
Sidney Lanier, Lake, res., Ga., U.S. (lăn'yẽr)	65	34°27'N	83°56'W
Sido, Mali	188	11°40'N	7°36'W
Sidon *see* Saydā, Leb.	154	33°34'N	35°23'E
Sidr, Wādī, r., Egypt	153a	29°43'N	32°58'E
Sidra, Gulf of *see* Surt, Khalīj, b., Libya	185	31°30'N	18°28'E
Siedlce, Pol. (syĕd'l-tsĕ)	125	52°09'N	22°20'E
Siegburg, Ger. (zēg'bōōrgh)	124	50°48'N	7°13'E
Siegen, Ger. (zē'ghĕn)	124	50°52'N	8°01'E
Sieghartskirchen, Aus.	115e	48°16'N	16°00'E
Siemiatycze, Pol. (syĕm'yȧ'tē-chĕ)	125	52°26'N	22°52'E
Siemionówka, Pol. (sĕĕ-mēō'nôf-kä)	125	52°53'N	23°50'E
Siem Reap, Camb. (syĕm'rā'äp)	168	13°32'N	103°54'E
Siena, Italy (sē-ĕn'ä)	118	43°19'N	11°21'E
Sieradz, Pol. (syĕ'rädz)	125	51°35'N	18°45'E
Sierpc, Pol. (syĕrpts)	125	52°51'N	19°42'E
Sierra Blanca, Tx., U.S. (sē-ĕ'rȧ blaŋ-kä)	80	31°10'N	105°20'W
Sierra Blanca Peak, mtn., N.M., U.S. (blăn'kȧ)	64	33°25'N	105°50'W
Sierra Leone, nation, Afr. (sē-ĕr'rä lā-ō'nā)	184	8°48'N	12°30'W
Sierra Madre, Ca., U.S. (mä'drē)	75a	34°10'N	118°03'W
Sierra Mojada, Mex. (sē-ĕ'r-rä-mô-kä'dä)	80	27°22'N	103°42'W
Sífnos, i., Grc.	131	36°58'N	24°30'E
Sigean, Fr. (sē-zhŏn')	126	43°02'N	2°56'E
Sigeurney, Ia., U.S. (sē-gûr-nĭ)	71	41°16'N	92°10'W
Sighetu Marmaţiei, Rom.	125	47°57'N	23°55'E
Sighişoara, Rom. (sē-gē-shwä'rȧ)	125	46°11'N	24°48'E
Siglufjördur, Ice.	116	66°06'N	18°45'W
Signakhi, Geor.	137	41°45'N	45°50'E
Signal Hill, Ca., U.S. (sĭg'năl hĭl)	75a	33°48'N	118°11'W
Sigsig, Ec. (sēg-sēg')	100	3°04'S	78°44'W
Sigtuna, Swe. (sēgh-tōō'nȧ)	122	59°40'N	17°39'E
Siguanea, Ensenada de la, b., Cuba	92	21°45'N	83°15'W
Siguatepeque, Hond. (sē-gwä'tē-pē-kē)	90	14°33'N	87°51'W
Sigüenza, Spain (sē-gwĕ'n-zä)	118	41°03'N	2°38'W
Siguiri, Gui. (sē-gē-rē')	184	11°25'N	9°10'W
Sihong, China (sz-hôŋ)	162	33°25'N	118°13'E
Siirt, Tur. (sĭ-ērt')	137	38°00'N	42°00'E
Sikalongo, Zam.	191	16°46'S	27°07'E
Sikasso, Mali (sē-käs'sō)	184	11°19'N	5°40'W
Sikeston, Mo., U.S. (sīks'tŭn)	79	36°50'N	89°35'W
Sikhote Alin', Khrebet, mts., Russia (se-kô'ta a-lēn')	135	45°00'N	135°45'E
Sikinos, i., Grc. (sĭ'kĭ-nōs)	131	36°45'N	24°55'E
Sikkim, state, India	155	27°42'N	88°25'E
Siklós, Hung. (sĭ'klōsh)	125	45°51'N	18°18'E
Sil, r., Spain (sē'l)	128	42°20'N	7°13'W
Silang, Phil. (sē-läng')	169a	14°14'N	120°58'E
Silao, Mex. (sē-lä'ō)	88	20°56'N	101°25'W
Silchar, India (sĭl-chär')	155	24°52'N	92°50'E
Silent Valley, S. Afr. (sī'lĕnt vă'lē)	192c	24°32'S	26°40'E
Siler City, N.C., U.S. (sī'lẽr)	83	35°45'N	79°29'W
Silesia, hist. reg., Pol. (sĭ-lē'shȧ)	124	50°58'N	16°53'E
Silifke, Tur.	119	36°20'N	34°00'E
Siling Co, l., China	160	32°05'N	89°10'E
Silistra, Blg. (sē-lēs'trȧ)	119	44°01'N	27°13'E
Siljan, l., Swe. (sēl'yän)	116	60°48'N	14°28'E
Silkeborg, Den. (sĭl'kĕ-bôr')	122	56°10'N	9°33'E
Sillery, Can. (sēl'-re')	62b	46°46'N	71°15'W
Siloam Springs, Ar., U.S. (sī-lōm)	79	36°10'N	94°32'W
Siloana Plains, pl., Zam.	190	16°55'S	23°10'E
Silocayoápan, Mex. (sē-lô-kä-yô-ȧ'pän)	88	17°29'N	98°09'W
Silsbee, Tx., U.S. (sĭlz' bē)	81	30°19'N	94°09'W
Šilutė, Lith.	123	55°21'N	21°29'E
Silva Jardim, Braz. (sē'l-vä-zhär-dēN)	99a	22°40'N	42°24'W
Silvana, Wa., U.S. (sĭ-vän'ȧ)	74a	48°12'N	122°16'W
Silvânia, Braz. (sēl-vä'nyä)	101	16°43'S	48°33'W
Silvassa, India	158	20°10'N	73°00'E
Silver, l., Mo., U.S.	79	39°38'N	93°12'W
Silverado, Ca., U.S. (sĭl-vẽr-ä'dō)	75a	33°45'N	117°40'W
Silver Bank, bk.,	93	20°40'N	69°40'W
Silver Bank Passage, strt., N.A.	93	20°40'N	70°20'W
Silver Bay, Mn., U.S.	71	47°24'N	91°07'W
Silver City, Pan.	91	9°20'N	79°54'W
Silver City, N.M., U.S. (sĭl'vẽr sĭ'tĭ)	77	32°45'N	108°20'W
Silver Creek, N.Y., U.S. (crēk)	67	42°35'N	79°10'W
Silver Creek, r., Az., U.S.	77	34°30'N	110°05'W
Silver Creek, r., In., U.S.	69h	38°20'N	85°45'W
Silver Creek, Muddy Fork, r., In., U.S.	69h	38°26'N	85°52'W
Silverdale, Wa., U.S. (sĭl'vẽr-dāl)	74a	47°39'N	122°42'W
Silver Lake, Wi., U.S. (lāk)	69a	42°33'N	88°10'W
Silver Lake, l., Wi., U.S.	69a	42°35'N	88°08'W
Silver Spring, Md., U.S. (sprĭng)	68e	39°00'N	77°00'W
Silver Star Mountain, mtn., Wa., U.S.	74c	45°45'N	122°15'W
Silverthrone Mountain, mtn., Can. (sĭl'vẽr-thrōn)	54	51°31'N	126°06'W
Silverton, S. Afr.	192c	25°45'S	28°13'E
Silverton, Co., U.S. (sĭl'vẽr-tŭn)	77	37°50'N	107°40'W
Silverton, Oh., U.S.	69f	39°12'N	84°24'W
Silverton, Or., U.S.	72	45°02'N	122°46'W
Silves, Port. (sēl'vĕzh)	118	37°15'N	8°24'W
Silvies, r., Or., U.S. (sĭl'vēz)	72	43°44'N	119°15'W
Sim, Russia (sĭm)	142a	55°00'N	57°42'E
Sim, r., Russia	142a	54°50'N	56°50'E
Simao, China (sz-mou)	160	22°56'N	101°07'E
Simard, Lac, l., Can.	59	47°38'N	78°40'W
Simba, D.R.C.	190	0°36'N	22°55'E
Simcoe, Can. (sĭm'kō)	120	42°50'N	80°20'W
Simcoe, l., Can.	53	44°30'N	79°20'W
Simeulue, Pulau, i., Indon.	168	2°27'N	95°30'E
Simferopol', Ukr.	134	44°58'N	34°04'E
Similk Beach, Wa., U.S. (sē'mĭlk)	74a	48°27'N	122°35'W
Simla, India (sĭm'lȧ)	155	31°09'N	77°15'E
Şimleu Silvaniei, Rom.	119	47°14'N	22°46'E
Simms Point, c., Bah.	92	25°00'N	77°40'W
Simojovel, Mex. (sē-mô-hô-vĕl')	89	17°12'N	92°43'W
Simonésia, Braz. (sē-mô-nĕ'syä)	99a	20°04'S	41°53'W
Simonette, r., Can. (sī-mŏn-ĕt')	55	54°15'N	118°00'W
Simonstad, S. Afr.	186a	34°11'S	18°25'E
Simood Sound, Can.	54	50°45'N	126°25'W
Simplon Pass, p., Switz. (sĭm'plŏn) (săN-plôN')	124	46°13'N	7°53'E
Simpson, i., Can.	71	48°43'N	87°44'W
Simpson Desert, des., Austl. (sĭmp-sŭn)	174	24°40'S	136°40'E
Simrishamn, Swe. (sēm'rēs-häm'n)	122	55°35'N	14°19'E
Sims Bayou, Tx., U.S. (sĭmz bī-yōō')	81a	29°37'N	95°23'W
Simushir, i., Russia (se-mōō'shēr)	161	47°15'N	150°47'E
Sinaia, Rom. (sĭ-nä'yȧ)	131	45°20'N	25°30'E
Sinai Peninsula, pen., Egypt (sī'nī)	185	29°24'N	33°29'E
Sinaloa, state, Mex. (sē-nä-lô-ä)	86	25°15'N	107°45'W
Sinan, China (sz-nän)	160	27°50'N	108°30'E
Sinanju, Kor., N. (sĭ'nän-jŏ')	166	39°39'N	125°41'E
Sincelejo, Col. (sēn-sȧ-lā'hō)	100	9°12'N	75°30'W
Sinclair Inlet, Wa., U.S. (sĭn-klâr')	74a	47°31'N	122°41'W
Sinclair Mills, Can.	55	54°02'N	121°41'W
Sindi, Est. (sēn'dē)	123	58°20'N	24°40'E
Sines, Port. (sē'nȧzh)	128	37°57'N	8°50'W
Singapore, Sing. (sĭn'gȧ-pōr')	168	1°18'N	103°52'E
Singapore, nation, Asia	168	1°22'N	103°45'E
Singapore Strait, strt., Asia	153b	1°14'N	104°20'E
Singu, Mya. (sĭn'gŭ)	160	22°37'N	96°04'E
Siniye Lipyagi, Russia (sēn'ē lēp'yä-gē)	133	51°24'N	38°29'E
Sinj, Cro. (sēn')	130	43°42'N	16°39'E
Sinjah, Sudan	185	13°09'N	33°52'E
Sinkāt, Sudan	156	18°50'N	36°50'E
Sinkiang *see* Xinjiang, prov., China	160	40°15'N	82°15'E
Sin'kovo, Russia (sĭn-kô'vô)	142b	56°23'N	37°19'E
Sinnamary, Fr. Gu.	101	5°15'N	52°52'W
Sinni, r., Italy (sēn'nē)	130	40°05'N	16°15'E
Sinnūris, Egypt	192b	29°25'N	30°52'E
Sino, Pedra de, mtn., Braz. (pĕ'drä-dō-sē'nô)	102b	22°27'S	43°02'W
Sinop, Tur.	154	42°00'N	35°05'E
Sint Eustatius, i., Neth. Ant.	91b	17°32'N	62°45'W
Sint Niklaas, Bel.	115a	51°10'N	4°07'E
Sinton, Tx., U.S. (sĭn'tŭn)	81	28°03'N	97°30'W
Sintra, Port. (sēn'trȧ)	128	38°48'N	9°23'W
Sint Truiden, Bel.	115a	50°49'N	5°14'E
Sinŭiju, Kor., N. (sĭ'nòĭ-jōō)	161	40°04'N	124°33'E
Sinyavino, Russia (sĭn-yä'vĭ-nô)	142c	59°50'N	31°07'E
Sinyaya, r., Eur. (sēn'yȧ-yȧ)	132	56°40'N	28°20'E
Sion, Switz. (sē'ôN')	124	46°15'N	7°17'E
Sioux City, Ia., U.S. (sōō sĭ'tĭ)	65	42°30'N	96°25'W
Sioux Falls, S.D., U.S. (fôlz)	64	43°33'N	96°43'W
Sioux Lookout, Can.	51	50°06'N	91°55'W
Siping, China (sz-pĭŋ)	161	43°05'N	124°24'E
Sipiwesk, Can.	50	55°27'N	97°24'W
Sipsey, r., Al., U.S. (sĭp'sē)	82	33°26'N	87°42'W
Sipura, Pulau, i., Indon.	168	2°15'S	99°33'E
Siqueros, Mex. (sē-kā'rōs)	88	23°19'N	106°14'W
Siquía, Río, r., Nic. (sē-kē'ȧ)	91	12°23'N	84°36'W
Siracusa, Italy (sē-rä-kōō'sä)	119	37°02'N	15°19'E

304 Sirājganj - Songwe PRONOUNCING INDEX

PLACE (Pronunciation)	PAGE	LAT.	LONG.
Sirājganj, Bngl. (sĭ-räj′gŭnj)	155	24°23′N	89°43′E
Sirama, El Sal. (Sē-rä-mä)	90	13°23′N	87°55′W
Sir Douglas, Mount, mtn., Can. (sûr dŭg′lăs)	55	50°44′N	115°20′W
Sir Edward Pellew Group, is., Austl. (pĕl′ū)	174	15°15′S	137°15′E
Siret, Rom.	125	47°58′N	26°01′E
Siret, r., Eur.	119	47°00′N	27°00′E
Sirhān, Wādī, depr., Sau. Ar.	154	31°02′N	37°16′E
Sirsa, India	158	29°39′N	75°02′E
Sir Sandford, Mount, mtn., Can. (sûr sănd′fērd)	55	51°40′N	117°52′W
Sirvintos, Lith. (shēr′vĭn-tôs)	123	55°02′N	24°59′E
Sir Wilfrid Laurier, Mount, mtn., Can. (sûr wĭl′frĭd lôr′yĕr)	55	52°47′N	119°45′W
Sisak, Cro. (sē′sȧk)	119	45°29′N	16°20′E
Sisal, Mex. (sē-säl′)	86	21°09′N	90°03′W
Sishui, China (sz-shwä)	162	35°40′N	117°17′E
Sisquoc, r., Ca., U.S. (sĭs′kwŏk)	76	34°47′N	120°13′W
Sisseton, S.D., U.S. (sĭs′tŭn)	70	45°39′N	97°04′W
Sistān, Daryacheh-ye, l., Asia	154	31°45′N	61°15′E
Sisteron, Fr. (sēst′rôɴ′)	127	44°10′N	5°55′E
Sisterville, W.V., U.S. (sĭs′tēr-vĭl)	66	39°30′N	81°00′W
Sitía, Grc. (sē′tĭ-ȧ)	130a	35°09′N	26°10′E
Sitka, Ak., U.S. (sĭt′kȧ)	64a	57°08′N	135°18′W
Sittingbourne, Eng., U.K. (sĭt-ĭng-bôrn)	114b	51°20′N	0°44′E
Sittwe, Mya.	155	20°09′N	92°54′E
Sivas, Tur. (sē′väs)	154	39°50′N	36°50′E
Siverek, Tur. (sē′vĕ-rĕk)	154	37°50′N	39°20′E
Siverskaya, Russia (sē′vĕr-skȧ-yȧ)	123	59°17′N	30°03′E
Sivers'kyi Donets', r., Eur.	133	48°48′N	38°42′E
Sīwah, Egypt	156	29°12′N	25°31′E
Siwah, oasis, Egypt (sē′wä)	185	29°33′N	25°11′E
Sixaola, r., C.R.	91	9°31′N	83°07′W
Sixian, China (sz shyĕn)	162	33°37′N	117°51′E
Sixth Cataract, wtfl., Sudan	185	16°26′N	32°44′E
Siyang, China (sz-yäŋ)	162	33°43′N	118°42′E
Sjaelland, i., Den. (shĕl′lȧn′)	122	55°34′N	11°35′E
Sjenica, Serb. (syĕ′nē-tsȧ)	131	43°15′N	20°02′E
Skadovs'k, Ukr.	133	46°08′N	32°54′E
Skagen, Den. (skä′ghĕn)	122	57°43′N	10°32′E
Skagerrak, strt., Eur. (skä-ghĕ-räk′)	112	57°43′N	8°28′E
Skagit, r., Wa., U.S.	72	48°29′N	121°52′W
Skagit Bay, b., Wa., U.S. (skăg′ĭt)	74a	48°20′N	122°32′W
Skagway, Ak., U.S. (skăg-wā)	64a	59°30′N	135°28′W
Skälderviken, b., Swe.	122	56°20′N	12°25′E
Skalistyy, Golets, mtn., Russia	135	57°28′N	119°48′E
Skalistyy Khrebet, mts., Russia	138	43°15′N	43°00′E
Skamania, Wa., U.S. (skȧ-mā′nĭ-ȧ)	74c	45°37′N	112°03′W
Skamokawa, Wa., U.S.	74c	46°16′N	123°27′W
Skanderborg, Den. (skän-ĕr-bôr′)	122	56°04′N	9°55′E
Skaneateles, N.Y., U.S. (skăn-ĕ-ăt′lĕs)	67	42°55′N	76°25′W
Skaneateles, l., N.Y., U.S.	67	42°50′N	76°20′W
Skänninge, Swe. (shĕn′ĭng-ĕ)	122	58°24′N	15°02′E
Skanör-Falsterbo, Swe. (skän′ûr)	122	55°24′N	12°49′E
Skara, Swe. (skä′rȧ)	122	58°25′N	13°24′E
Skeena, r., Can. (skē′nä)	52	54°30′N	129°00′W
Skeena Mountains, mts., Can.	54	56°00′N	128°00′W
Skeerpoort, S. Afr.	187b	25°49′S	27°45′E
Skeerpoort, r., S. Afr.	187b	25°58′S	27°41′E
Skeldon, Guy. (skĕl′dŭn)	101	5°49′N	57°15′W
Skellefteå, Swe. (shĕl′ĕf-tĕ-ȧ′)	116	64°47′N	20°48′E
Skellefteälven, r., Swe.	116	65°15′N	19°30′E
Skhodnya, Russia (skôd′nyȧ)	142b	55°57′N	37°21′E
Skhodnya, r., Russia	142b	55°55′N	37°16′E
Skiathos, i., Grc. (skē′ȧ-thôs)	131	39°15′N	23°25′E
Skibbereen, Ire. (skĭb′ĕr-ēn)	120	51°32′N	9°25′W
Skidegate, b., Can. (skĭ′-dē-gāt′)	54	53°15′N	132°00′W
Skidmore, Tx., U.S. (skĭd′môr)	81	28°16′N	97°40′W
Skien, Nor. (skē′ĕn)	116	59°13′N	9°35′E
Skierniewice, Pol. (skyĕr-nyĕ-vēt′sĕ)	125	51°58′N	20°13′E
Skihist Mountain, mtn., Can.	55	50°11′N	121°54′W
Skikda, Alg.	184	36°58′N	6°51′E
Skilpadfontein, S. Afr.	192c	25°02′S	28°50′E
Skive, Den. (skē′vĕ)	122	56°34′N	8°56′E
Skjálfandafljót, r., Ice. (skyäl′fänd-ô)	116	65°24′N	16°40′W
Skjerstad, Nor. (skyĕr-städ)	116	67°12′N	15°37′E
Škofja Loka, Slvn. (shkôf′yȧ lō′kȧ)	130	46°10′N	14°20′E
Skokie, Il., U.S. (skō′kē)	69a	42°02′N	87°45′W
Skokomish Indian Reservation, I.R., Wa., U.S. (Skō-kō′mĭsh)	74a	47°22′N	123°07′W
Skole, Ukr. (skô′lĕ)	125	49°03′N	23°32′E
Skópelos, i., Grc. (skô′pä-lôs)	131	39°04′N	23°31′E
Skopin, Russia (skô′pĕn)	136	53°49′N	39°35′E
Skopje, Mac. (skôp′yĕ)	130	42°02′N	21°26′E
Skövde, Swe. (shûv′dĕ)	116	58°25′N	13°48′E
Skovorodino, Russia (skô′vô-rô′dĭ-nô)	135	53°53′N	123°56′E
Skowhegan, Me., U.S. (skou-hē′găn)	60	44°45′N	69°27′W
Skradin, Cro. (skrä′dĕn)	131	43°49′N	17°58′E
Skreia, Nor. (skrā′ȧ)	122	60°40′N	10°55′E
Skudeneshavn, Nor. (skōō′dĕ-nes-houn′)	122	59°10′N	5°19′E
Skull Valley Indian Reservation, I.R., Ut., U.S. (skŭl)	77	40°25′N	112°50′W
Skuna, r., Ms., U.S. (skū′nȧ)	82	33°57′N	89°36′W
Skunk, r., Ia., U.S. (skŭnk)	71	41°12′N	92°14′W
Skuodas, Lith. (skwô′dȧs)	123	56°16′N	21°32′E
Skurup, Swe. (skû′rŏp)	122	55°29′N	13°27′E
Skvyra, Ukr.	137	49°43′N	29°41′E
Skwierzyna, Pol. (skvē-ĕr′zhĭ-nȧ)	124	52°35′N	15°30′E
Skye, Island of, i., Scot., U.K. (skī)	116	57°25′N	6°17′W
Skykomish, r., Wa., U.S. (skī′kō-mĭsh)	74a	47°50′N	121°55′W
Skyring, Seno de, b., Chile (sĕ′nô-s-krē′ng)	102	52°35′S	72°30′W
Skýros, Grc.	131	38°53′N	24°32′E
Skýros, i., Grc.	119	38°50′N	24°43′E
Slagese, Den.	122	55°25′N	11°19′E
Slamet, Gunung, mtn., Indon. (slä′mĕt)	168	7°15′S	109°15′E
Slănic, Rom. (slŭ′nĕk)	131	45°13′N	25°56′E
Slater, Mo., U.S. (slāt′ēr)	79	39°13′N	93°03′W
Slatina, Rom. (slä′tē-nä)	131	44°26′N	24°21′E
Slaton, Tx., U.S. (slā′tŭn)	78	33°26′N	101°38′W
Slave, r., Can. (slāv)	52	59°40′N	111°21′W
Slavgorod, Russia (sláf′gô-rôt)	134	52°58′N	78°43′E
Slavonija, hist. reg., Serb. (slä-vô′nē-yä)	131	45°29′N	17°31′E
Slavonska Požega, Cro. (slä-vôn′skä pô′zhĕ-gä)	131	45°18′N	17°42′E
Slavonski Brod, Cro. (skä-vôn′skē brôd)	119	45°10′N	18°01′E
Slavuta, Ukr. (slȧ-vōō′tȧ)	133	50°18′N	27°01′E
Slavyanskaya, Russia (slȧv-yȧn′skȧ-yȧ)	133	45°14′N	38°09′E
Sławno, Pol. (swav′nō)	124	54°21′N	16°38′E
Slayton, Mn., U.S. (slā′tŭn)	70	44°00′N	95°44′W
Sleaford, Eng., U.K. (slē′fērd)	114a	53°00′N	0°25′W
Sleepy Eye, Mn., U.S. (slēp′ĭ ī)	71	44°17′N	94°44′W
Slidell, La., U.S. (slī-dĕl′)	81	30°17′N	89°47′W
Sliedrecht, Neth.	115a	51°49′N	4°46′E
Sligo, Ire. (slī′gō)	116	54°17′N	8°19′W
Slite, Swe. (slē′tĕ)	122	57°41′N	18°47′E
Sliven, Blg. (slē′vĕn)	119	42°41′N	26°20′E
Sloatsburg, N.Y., U.S. (slōts′bûrg)	68a	41°09′N	74°11′W
Slonim, Bela. (swō′nēm)	125	53°05′N	25°19′E
Slough, Eng., U.K. (slou)	114b	51°29′N	0°36′W
Slovakia, nation, Eur.	125	48°50′N	20°00′E
Slovenia, nation, Eur.	130	45°58′N	14°43′E
Slovians'k, Ukr.	137	48°52′N	37°34′E
Sluch, r., Ukr.	137	50°56′N	26°48′E
Slunj, Cro. (slòn′)	130	45°08′N	15°46′E
Słupsk, Pol. (swôpsk)	116	54°28′N	17°02′E
Slutsk, Bela. (slòtsk)	136	53°02′N	27°34′E
Slyne Head, c., Ire. (slīn)	116	53°25′N	10°05′W
Smackover, Ar., U.S. (smăk′ō-vēr)	79	33°22′N	92°42′W
Smederevo, Serb.	131	44°39′N	20°54′E
Smederevska Palanka, Serb. (smĕ-dĕ-rĕv′skä pä-län′kä)	131	44°21′N	21°00′E
Smedjebacken, Swe. (smĭ′tyĕ-bä-kĕn)	122	60°09′N	15°19′E
Smethport, Pa., U.S. (smĕth′pōrt)	67	41°50′N	78°25′W
Smethwick, Eng., U.K.	120	52°31′N	2°04′W
Smila, Ukr.	137	49°14′N	31°52′E
Smile, Ukr.	133	50°55′N	33°36′E
Smiltene, Lat. (smēl′tĕ-nĕ)	123	57°26′N	25°57′E
Smith, Can. (smĭth)	50	55°10′N	114°02′W
Smith, i., Wa., U.S.	74a	48°20′N	122°53′W
Smith, r., Mt., U.S.	73	47°00′N	111°20′W
Smith Center, Ks., U.S. (sĕn′tēr)	78	39°45′N	98°46′W
Smithers, Can. (smĭth′ērs)	50	54°47′N	127°10′W
Smithfield, N.C., U.S. (smĭth′fēld)	83	35°30′N	78°21′W
Smithfield, Ut., U.S.	73	41°50′N	111°49′W
Smithland, Ky., U.S. (smĭth′lănd)	66	37°10′N	88°25′W
Smith Mountain Lake, res., Va., U.S.	83	37°00′N	79°45′W
Smith Point, Tx., U.S.	81a	29°32′N	94°45′W
Smiths Falls, Can. (smĭths)	51	44°55′N	76°05′W
Smithton, Austl. (smĭth′tŭn)	176	40°55′S	145°12′E
Smithton, Il., U.S.	75e	38°24′N	89°59′W
Smithville, Tx., U.S. (smĭth′vĭl)	81	30°00′N	97°08′W
Smitswinkelvlakte, pl., S. Afr.	186a	34°16′S	18°25′E
Smoke Creek Desert, des., Nv., U.S. (smōk crēk)	76	40°28′N	119°40′W
Smoky, r., Can. (smōk′ĭ)	55	55°30′N	117°30′W
Smoky Hill, r., U.S. (smōk′ĭ hĭl)	64	38°40′N	100°00′W
Smøla, i., Nor. (smûlä)	116	63°16′N	7°40′E
Smolensk, Russia (smô-lyĕnsk′)	134	54°46′N	32°03′E
Smolensk, prov., Russia	132	55°00′N	32°18′E
Smyadovo, Blg.	131	43°04′N	27°00′E
Smyrna see İzmir, Tur.	154	38°25′N	27°05′E
Smyrna, De., U.S. (smûr′nȧ)	67	39°20′N	75°35′W
Smyrna, Ga., U.S.	68c	33°53′N	84°31′W
Snag, Can. (snăg)	63	62°18′N	140°30′W
Snake, r., Mn., U.S. (snāk)	71	45°58′N	93°20′W
Snake, r., U.S.	64	45°30′N	117°00′W
Snake Range, mts., Nv., U.S.	77	39°20′N	114°15′W
Snake River Plain, pl., Id., U.S.	73	43°08′N	114°46′W
Snap Point, c., Bah.	92	23°45′N	77°30′W
Sneffels, Mount, mtn., Co., U.S. (snĕf′ĕlz)	77	38°00′N	107°50′W
Snelgrove, Can. (snĕl′grōv)	62d	43°44′N	79°50′W
Sniardwy, Jezioro, l., Pol. (snyärt′vĭ)	125	53°46′N	21°59′E
Snöhetta, mtn., Nor. (snû-hĕttä)	116	62°18′N	9°12′E
Snohomish, Wa., U.S. (snô-hō′mĭsh)	74a	47°55′N	122°05′W
Snohomish, r., Wa., U.S.	74a	47°53′N	122°04′W
Snoqualmie, Wa., U.S. (snō qwäl′mē)	74a	47°32′N	121°50′W
Snoqualmie, r., Wa., U.S.	72	47°32′N	121°53′W
Snov, r., Eur. (snôf)	133	51°38′N	31°38′E
Snowdon, mtn., Wales, U.K.	120	53°05′N	4°04′W
Snow Hill, Md., U.S. (hĭl)	67	38°15′N	75°20′W
Snow Lake, Can.	57	54°50′N	100°10′W
Snowy Mountains, mts., Austl. (snō′ē)	175	36°17′S	148°30′E
Snyder, Ok., U.S. (snī′dēr)	78	34°40′N	98°57′W
Snyder, Tx., U.S.	80	32°48′N	100°53′W
Soar, r., Eng., U.K. (sōr)	114a	52°44′N	1°09′W
Sobat, r., Sudan (sō′bȧt)	185	9°04′N	32°02′E
Sobinka, Russia (sô-bĭŋ′kȧ)	132	55°59′N	40°02′E
Sobo Zan, mtn., Japan (sô′bô zän)	166	32°47′N	131°27′E
Sobral, Braz. (sô-brä′l)	101	3°39′S	40°16′W
Sochaczew, Pol. (sô-ĸä′chĕf)	125	52°14′N	20°18′E
Sochi, Russia (sôch′ĭ)	134	43°35′N	39°50′E
Society Islands, is., Fr. Poly. (sô-sī′ĕ-tē)	195	15°00′S	157°30′W
Socoltenango, Mex. (sô-kôl-tĕ-näŋ′gō)	89	16°17′N	92°20′W
Socorro, Braz. (sō-kō′r-rō)	99a	22°35′S	46°32′W
Socorro, Col. (sô-kôr′rō)	100	6°23′N	73°19′W
Socorro, N.M., U.S.	77	34°05′N	106°55′W
Socúellamos, Spain (sô-kōō-āl′yä-mōs)	128	39°18′N	2°48′W
Soda, l., Ca., U.S. (sō′dȧ)	76	35°12′N	116°25′W
Soda Peak, mtn., Wa., U.S.	74c	45°53′N	122°04′W
Soda Springs, Id., U.S. (sprĭngz)	73	42°39′N	111°37′W
Söderhamn, Swe. (sû-dĕr-häm′′n)	116	61°20′N	17°00′E
Söderköping, Swe.	122	58°30′N	16°14′E
Södertälje, Swe. (sû-dĕr-tĕl′yĕ)	116	59°12′N	17°35′E
Sodo, Eth.	185	7°03′N	37°46′E
Soest, Ger. (zōst)	124	51°35′N	8°05′E
Sofia (Sofiya), Blg. (sô′fē-yȧ) (sō′fē-ȧ)	110	42°43′N	23°20′E
Sofiïvka, Ukr.	133	48°03′N	33°53′E
Sofiya see Sofia, Blg.	110	42°43′N	23°20′E
Soga, Japan (sō′gä)	167a	35°35′N	140°08′E
Sogamoso, Col. (sô-gä-mô′sō)	100	5°42′N	72°51′W
Sognafjorden, b., Nor.	112	61°09′N	5°30′E
Sogozha, r., Russia (sô′gô-zhȧ)	132	58°35′N	39°08′E
Sohano, Pap. N. Gui.	170e	5°27′S	154°40′E
Soissons, Fr. (swä-sôɴ′)	126	49°23′N	3°17′E
Sōka, Japan (sō′kä)	167a	35°50′N	139°49′E
Sokal', Ukr. (sô′käl′)	125	50°28′N	24°20′E
Söke, Tur. (sû′kĕ)	119	37°40′N	27°10′E
Sokółka, Pol. (sô-kôl′kȧ)	125	53°23′N	23°30′E
Sokolo, Mali (sô-kô-lō′)	184	14°51′N	6°09′W
Sokołów Podlaski, Pol. (sô-kô-wôf′ pŭd-lä′skĭ)	125	52°24′N	22°15′E
Sokone, Sen.	188	13°53′N	16°22′W
Sokoto, Nig. (sō′kô-tō)	184	13°04′N	5°16′E
Sola de Vega, Mex.	89	16°31′N	96°58′W
Solander, Cape, c., Austl.	173b	34°03′S	151°16′E
Solano, Phil. (sō-lä′nō)	169a	16°31′N	121°11′E
Soledad, Col. (sô-lĕ-dä′d)	100	10°47′N	75°00′W
Soledad Díez Gutiérrez, Mex.	88	22°19′N	100°54′W
Soleduck, r., Wa., U.S. (sōl′dŭk)	72	47°59′N	124°28′W
Solentiname, Islas de, is., Nic. (ē′s-läs-dĕ-sô-lĕn-tē-nä′mȧ)	90	11°15′N	85°16′W
Solihull, Eng., U.K. (sō′lĭ-hŭl)	114a	52°25′N	1°46′W
Solihull, co., Eng., U.K.	114a	52°25′N	1°42′W
Solikamsk, Russia (sô-lē-kȧmsk′)	136	59°38′N	56°48′E
Sol'-Iletsk, Russia	134	51°10′N	55°05′E
Solimões see Amazon, r., Braz.	100	2°45′S	67°44′W
Solingen, Ger. (zō′lĭng-ĕn)	124	51°10′N	7°05′E
Sóller, Spain (sō′lyĕr)	129	39°45′N	2°40′E
Sologne, reg., Fr. (sō-lôn′yĕ)	126	47°36′N	1°53′E
Solola, Guat. (sô-lō′lä)	90	14°45′N	91°12′W
Solomon, r., Ks., U.S.	78	39°24′N	98°19′W
Solomon, North Fork, r., Ks., U.S.	78	39°34′N	99°52′W
Solomon, South Fork, r., Ks., U.S.	78	39°19′N	99°52′W
Solomon Islands, nation, Oc. (sŏ′lō-mŭn)	3	7°00′S	160°00′E
Solon, China (swo-lōōn)	161	46°32′N	121°18′E
Solon, Oh., U.S. (sō′lŭn)	69d	41°23′N	81°26′W
Solothurn, Switz. (zō′lō-thōōrn)	124	47°13′N	7°30′E
Solovetskiye Ostrova, is., Russia	136	65°10′N	35°40′E
Šolta, i., Serb. (shôl′tä)	130	43°20′N	16°15′E
Soltau, Ger. (sōl′tou)	124	53°00′N	9°50′E
Sol'tsy, Russia (sôl′tsē)	132	58°04′N	30°13′E
Solvay, N.Y., U.S. (sŏl′vā)	67	43°05′N	76°10′W
Sölvesborg, Swe. (sŏl′vĕs-bôrg)	122	56°04′N	14°35′E
Sol'vychegodsk, Russia (sôl′vē-chĕ-gôtsk′)	136	61°18′N	46°58′E
Solway Firth, b., U.K. (sŏl′wāfûrth′)	116	54°42′N	3°55′W
Solwezi, Zam.	191	12°11′S	26°25′E
Soly, Bela.	122	54°31′N	26°11′E
Somalia, nation, Afr. (sō-mä′lē-ȧ)	192a	3°28′N	44°47′E
Somanga, Tan.	191	8°24′S	39°17′E
Sombor, Serb. (sôm′bôr)	119	45°45′N	19°10′E
Sombrerete, Mex. (sōm-brā-rā′tȧ)	88	23°38′N	103°37′W
Sombrero, Cayo, i., Ven. (kä-yô-sôm-brē′rô)	101b	10°52′N	68°12′W
Somerset, Ky., U.S. (sŭm′ēr-sĕt)	82	37°05′N	84°35′W
Somerset, Ma., U.S.	68b	41°46′N	71°05′W
Somerset, Pa., U.S.	67	40°00′N	79°05′W
Somerset, Tx., U.S.	75d	29°13′N	98°39′W
Somerset East, S. Afr.	187c	32°44′S	25°36′E
Somersworth, N.H., U.S. (sŭm′ērz-wûrth)	60	43°16′N	70°53′W
Somerton, Az., U.S. (sŭm′ēr-tŭn)	77	32°36′N	114°43′W
Somerville, Ma., U.S. (sŭm′ēr-vĭl)	61a	42°23′N	71°06′W
Somerville, N.J., U.S.	68a	40°34′N	74°37′W
Somerville, Tn., U.S.	82	35°14′N	89°21′W
Somerville, Tx., U.S.	81	30°21′N	96°31′W
Someş, r., Eur.	125	47°43′N	23°09′E
Somma Vesuviana, Italy (sôm′mä vā-zōō-vē-ä′nä)	129c	40°38′N	14°27′E
Somme, r., Fr. (sôm)	126	50°02′N	2°04′E
Sommerfeld, Ger. (zō′mĕr-fĕld)	115b	52°48′N	13°02′E
Sommerville, Austl.	173a	38°14′S	145°10′E
Somoto, Nic. (sô-mō′tō)	90	13°28′N	86°37′W
Son, r., India (sōn)	155	24°40′N	82°35′E
Sŏnchŏn, Kor., N. (sŭn′shŭn)	166	39°49′N	124°56′E
Sondags, r., S. Afr.	187c	33°17′S	25°14′E
Sønderborg, Den. (sûn′′ĕr-bôrgh)	116	54°55′N	9°47′E
Sondershausen, Ger. (zōn′dĕrz-hou′zĕn)	124	51°17′N	10°45′E
Song Ca, r., Viet.	165	19°15′N	105°00′E
Songea, Tan. (sôn-gā′ȧ)	186	10°41′S	35°39′E
Songjiang, China	161	31°01′N	121°14′E
Sŏngjin, Kor., N. (sŭng′jĭn′)	166	40°38′N	129°10′E
Songkhla, Thai. (sông′ĸlä′)	168	7°09′N	100°34′E
Songwe, D.R.C.	191	12°25′S	29°40′E

ăt; finăl; rāte; senȧte; ärm; ȧsk; sofȧ; fâre; ch-choose; dh-as th in other; bē; ėvent; bĕt; recĕnt; cratẽr; g-gō; gh-guttural g; bĭt; ĭ-short neutral; rīde; ĸ-guttural k as ch in German ich;

PRONOUNCING INDEX
Sonneberg – Sri Jayewardenepura Kotte 305

PLACE (Pronunciation)	PAGE	LAT.	LONG.
Sonneberg, Ger. (sŏn'ĕ-bĕrgh)	124	50°20'N	11°14'E
Sonora, Ca., U.S. (sō-nō'rȧ)	76	37°58'N	120°22'W
Sonora, Tx., U.S.	80	30°33'N	100°38'W
Sonora, state, Mex.	86	29°45'N	111°15'W
Sonora, r., Mex.	86	28°45'N	111°35'W
Sonora Peak, mtn., Ca., U.S.	64	38°22'N	119°39'W
Sonseca, Spain (sôn-sā'kä)	128	39°41'N	3°56'W
Sonsón, Col. (sôn-sôn')	100	5°42'N	75°28'W
Sonsonate, El Sal. (sōn-sō-nä'tā)	90	13°46'N	89°43'W
Sonsorol Islands, is., Palau (sŏn-sō-rōl')	169	5°03'N	132°33'E
Sooke Basin, b., Can. (sŏŏk)	74a	48°21'N	123°47'W
Soo Locks, trans., Mi., U.S. (sōō lŏks)	75a	46°30'N	84°30'W
Sopetrán, Col. (sō-pĕ-trä'n)	100a	6°30'N	75°44'W
Sopot, Pol. (sô'pôt)	125	54°26'N	18°25'E
Sopron, Hung. (shŏp'rôn)	119	47°41'N	16°36'E
Sora, Italy (sō'rä)	130	41°43'N	13°37'E
Sorbas, Spain (sôr'bäs)	128	37°05'N	2°07'W
Sordo, r., Mex. (sô'r-dō)	89	16°39'N	97°33'W
Sorel, Can. (sō-rĕl')	51	46°01'N	73°07'W
Sorell, Cape, c., Austl.	176	42°10'S	144°50'E
Soresina, Italy (sō-rå-zē'nä)	130	45°17'N	9°51'E
Soria, Spain (sō'rē-ä)	118	41°46'N	2°28'W
Soriano, dept., Ur. (sō-rēä'nō)	99c	33°25'S	58°00'W
Soroca, Mol.	137	48°09'N	28°17'E
Sorocaba, Braz. (sō-rō-kä'bȧ)	101	23°29'S	47°27'W
Sorong, Indon. (sō-rông')	169	1°00'S	131°20'E
Sorot', r., Russia (sō-rō'tzh)	132	57°08'N	29°23'E
Soroti, Ug. (sō-rō'tē)	185	1°43'N	33°37'E
Sørøya, i., Nor.	116	70°37'N	20°58'E
Sorraia, r., Port. (sôr-rī'ȧ)	128	38°55'N	8°42'W
Sorrento, Italy (sôr-rĕn'tō)	130	40°23'N	14°23'E
Sorsogon, Phil. (sôr-sŏgōn')	169	12°51'N	124°02'E
Sortavala, Russia (sôr'tä-vä-lä)	133	61°43'N	30°40'E
Sosna, r., Russia (sôs'nȧ)	133	50°33'N	38°15'E
Sosnogorsk, Russia	134	63°13'N	54°09'E
Sosnowiec, Pol. (sôs-nô'vyĕts)	125	50°17'N	19°10'E
Sosnytsia, Ukr.	133	51°30'N	32°29'E
Sosunova, Mys, c., Russia (mĭs sô'sō-nôf'ȧ)	166	46°28'N	138°06'E
Sos'va, r., Russia (sôs'vȧ)	142a	59°55'N	60°40'E
Sos'va, r., Russia (sôs'vȧ)	136	63°10'N	63°30'E
Sota, r., Benin	189	11°10'N	3°20'E
Sota la Marina, Mex. (sô-tä-lä-mä-rē'nä)	88	23°45'N	98°11'W
Soteapan, Mex. (sō-tå-ä'pän)	89	18°14'N	94°51'W
Soto la Marina, Río, r., Mex. (rē'ō-so'tō lä mä-rē'nä)	88	23°55'N	98°30'W
Sotuta, Mex. (sō-tōō'tä)	90a	20°35'N	89°00'W
Soublette, Ven. (sō-ōō-blĕ'tĕ)	101b	9°55'N	66°06'W
Souflí, Grc.	131	41°12'N	26°17'E
Soufrière, St. Luc. (sōō-frē-âr')	91b	13°50'N	61°03'W
Soufrière, mtn., St. Vin.	91b	13°19'N	61°12'W
Soufrière, vol., Guad. (sōō-frē-är')	91b	16°06'N	61°42'W
Sŏul *see* Seoul, Kor., S.	161	37°35'N	127°03'E
Sounding Creek, r., Can. (soun'dĭng)	56	51°35'N	111°00'W
Souq Ahras, Alg.	117	36°23'N	8°00'E
Sources, Mount aux, mtn., Afr. (môn'tō sôrs')	186	28°47'S	29°04'E
Soure, Port. (sōr-ĕ)	128	40°04'N	8°37'W
Souris, Can. (sōō'rē')	61	46°20'N	62°17'W
Souris, Can.	50	49°38'N	100°15'W
Souris, r., N.A.	52	48°30'N	101°30'W
Sourlake, Tx., U.S. (sour'lāk)	81	30°09'N	94°24'W
Sousse, Tun. (sōōs)	184	36°00'N	10°39'E
South, r., Ga., U.S.	68c	33°40'N	84°15'W
South, r., N.C., U.S.	83	34°49'N	78°33'W
South Africa, nation, Afr.	186	28°00'S	24°50'E
South Amboy, N.J., U.S. (south'ăm'boi)	68a	40°28'N	74°17'W
South America, cont.	97	15°00'S	60°00'W
Southampton, Eng., U.K. (south-ămp'tŭn)	110	50°54'N	1°30'W
Southampton, N.Y., U.S.	67	40°53'N	72°24'W
Southampton Island, i., Can.	53	64°38'N	84°00'W
South Andaman Island, i., India (ăn-dȧ-măn')	168	11°57'N	93°24'E
South Australia, state, Austl. (ôs-trā'lĭ-ȧ)	174	29°45'S	132°00'E
South Bay, b., Bah.	93	20°55'N	73°35'W
South Bend, In., U.S. (bĕnd)	65	41°40'N	86°20'W
South Bend, Wa., U.S. (bĕnd)	72	46°39'N	123°48'W
South Bight, b., Bah.	92	24°20'N	77°35'W
South Bimini, i., Bah. (bē'mē-nē)	92	25°40'N	79°20'W
Southborough, Ma., U.S. (south'bŭr-ō)	61a	42°18'N	71°33'W
South Boston, Va., U.S. (bôs'tŭn)	83	36°41'N	78°55'W
Southbridge, Ma., U.S. (south'brĭj)	67	42°05'N	72°00'W
South Caicos, i., T./C. Is. (kī'kōs)	93	21°30'N	71°35'W
South Carolina, state, U.S. (kăr-ō-lī'nȧ)	65	34°15'N	81°10'W
South Cave, Eng., U.K. (cāv)	114a	53°45'N	0°35'W
South Charleston, W.V., U.S.	66	38°20'N	81°40'W
South China Sea, sea, Asia (chī'nȧ)	168	15°23'N	114°12'E
South Creek, r., Austl.	173b	33°43'S	150°50'E
South Dakota, state, U.S. (dȧ-kō'tȧ)	64	44°20'N	101°55'W
South Downs, Eng., U.K. (dounz)	120	50°55'N	1°13'W
South Dum-Dum, India	158a	22°36'N	88°25'E
South East Cape, c., Austl.	175	43°47'S	146°03'E
Southend-on-Sea, Eng., U.K. (south-ĕnd')	121	51°33'N	0°41'E
Southern Alps, mts., N.Z. (sŭ-thûrn ălps)	175a	43°35'S	170°00'E
Southern Cross, Austl.	174	31°13'S	119°30'E
Southern Indian, l., Can. (sŭth'ẽrn ĭn'dĭ-ăn)	52	56°46'N	98°57'W
Southern Pines, N.C., U.S. (sŭth'ẽrn pīnz)	83	35°10'N	79°23'W
Southern Ute Indian Reservation, I.R., Co., U.S. (ūt)	77	37°05'N	108°23'W
South Euclid, Oh., U.S. (ū'klĭd)	69d	41°30'N	81°34'W
South Fox, i., Mi., U.S. (fŏks)	66	45°25'N	85°55'W
South Gate, Ca., U.S. (gāt)	75a	33°57'N	118°13'W
South Georgia, i., S. Geor. (jôr'jȧ)	97	54°00'S	37°00'W
South Haven, Mi., U.S. (hāv''n)	66	42°25'N	86°15'W
South Hill, Va., U.S.	83	36°44'N	78°08'W
South Holston Lake, res., U.S.	83	36°35'N	82°00'W
South Indian Lake, Can.	57	56°50'N	99°00'W
Southington, Ct., U.S. (sŭdh'ĭng-tŭn)	67	41°35'N	72°55'W
South Island, i., N.Z.	175a	42°40'S	169°00'E
South Loup, r., Ne., U.S. (lōōp)	70	41°21'N	100°08'W
South Magnetic Pole, pt. of. i.,	178	65°18'S	139°30'E
South Merrimack, N.H., U.S. (mĕr'ĭ-măk)	61a	42°47'N	71°36'W
South Milwaukee, Wi., U.S. (mĭl-wô'kē)	69a	42°55'N	87°52'W
South Moose Lake, l., Can.	57	53°51'N	100°20'W
South Nation, r., Can.	59	45°00'N	75°25'W
South Negril Point, c., Jam. (nå-grĕl')	92	18°15'N	78°25'W
South Ogden, Ut., U.S. (ŏg'dĕn)	75b	41°12'N	111°58'W
South Orkney Islands, is., Ant.	97	57°00'S	45°00'W
South Ossetia, hist. reg., Geor.	138	42°20'N	44°00'E
South Paris, Me., U.S. (păr'ĭs)	60	44°13'N	70°32'W
South Park, Ky., U.S. (pärk)	69h	38°06'N	85°43'W
South Pasadena, Ca., U.S. (păs-ȧ-dē'nȧ)	75a	34°06'N	118°08'W
South Pease, r., Tx., U.S. (pēz)	78	33°54'N	100°45'W
South Pender, i., Can. (pĕn'dẽr)	74d	48°45'N	123°09'W
South Pittsburg, Tn., U.S. (pĭs'bûrg)	82	35°00'N	85°42'W
South Platte, r., U.S. (plăt)	64	40°40'N	102°40'W
South Point, c., Barb.	91b	13°00'N	59°43'W
South Point, c., Mi., U.S.	66	44°50'N	83°20'W
South Pole, pt. of. i., Ant.	178	90°00'S	0°00'
South Porcupine, Can.	58	48°28'N	81°13'W
Southport, Austl. (south'pōrt)	175	27°57'S	153°27'E
Southport, Eng., U.K. (south'pōrt)	120	53°38'N	3°00'W
Southport, In., U.S.	69g	39°40'N	86°07'W
Southport, N.C., U.S.	83	35°55'N	78°02'W
South Portland, Me., U.S. (pōrt-lănd)	60	43°37'N	70°15'W
South Prairie, Wa., U.S. (prā'rĭ)	74a	47°08'N	122°06'W
South Range, Wi., U.S. (rānj)	75h	46°37'N	91°59'W
South River, N.J., U.S. (rĭv'ẽr)	68a	40°27'N	74°23'W
South Ronaldsay, i., Scot., U.K. (rŏn'ăld-s'ā)	120a	58°48'N	2°55'W
South Saint Paul, Mn., U.S.	75g	44°54'N	93°02'W
South Salt Lake, Ut., U.S. (sôlt lāk)	75b	40°44'N	111°53'W
South Sandwich Islands, is., S. Geor. (sănd'wĭch)	97	58°00'S	27°00'W
South Sandwich Trench, deep,	97	55°00'S	27°00'W
South San Francisco, Ca., U.S. (săn frăn-sĭs'kō)	74b	37°39'N	122°24'W
South Saskatchewan, r., Can. (săs-kach'ĕ-wän)	52	50°30'N	110°30'W
South Shetland Islands, is., Ant.	97	62°00'S	70°00'W
South Shields, Eng., U.K. (shēldz)	116	55°00'N	1°22'W
South Sioux City, Ne., U.S. (sōō sĭt'ē)	70	42°48'N	96°26'W
South Taranaki Bight, b., N.Z. (tä-rä-nä'kē)	175a	39°35'S	173°50'E
South Thompson, r., Can. (tŏmp'sŭn)	55	50°41'N	120°21'W
Southton, Tx., U.S. (south'tŭn)	75d	29°18'N	98°26'W
South Uist, i., Scot., U.K. (ū'ĭst)	120	57°15'N	7°24'W
South Umpqua, r., Or., U.S. (ŭmp'kwȧ)	72	43°00'N	122°54'W
Southwell, Eng., U.K. (south'wĕl)	114a	53°04'N	0°56'W
South West Africa *see* Namibia, nation, Afr.	186	19°30'S	16°13'E
Southwest Miramichi, r., Can. (mĭr ȧ-mē'shē)	60	46°35'N	66°17'W
Southwest Point, c., Bah.	92	25°50'N	77°10'W
Southwest Point, c., Bah.	93	23°55'N	74°30'W
South Yorkshire, hist. reg., Eng., U.K.	114a	53°29'N	1°35'W
Sovetsk, Russia (sô-vyĕtsk')	136	55°04'N	21°54'E
Sovetskaya Gavan', Russia (sŭ-vyĕt'skī-u gä'vŭn')	135	48°59'N	140°14'E
Sow, r., Eng., U.K. (sou)	114a	52°45'N	2°12'W
Soya Kaikyō, strt., Asia	166	45°45'N	141°38'E
Sōya Misaki, c., Japan (sō'yȧ mē'sä-kē)	166	45°35'N	141°25'E
Soyo, Ang.	186	6°10'S	12°25'E
Sozh, r., Eur. (sôzh)	137	52°50'N	31°00'E
Sozopol, Blg. (sôz'ô-pôl')	131	42°18'N	27°50'E
Spa, Bel. (spä)	121	50°30'N	5°50'E
Spain, nation, Eur. (spān)	110	40°15'N	4°30'W
Spalding, Ne., U.S. (spôl'dĭng)	70	41°43'N	98°23'W
Spanaway, Wa., U.S. (spăn'ȧ-wā)	74a	47°06'N	122°26'W
Spangler, Pa., U.S. (spăng'lẽr)	67	40°40'N	78°50'W
Spanish Fork, Ut., U.S. (spăn'ĭsh fôrk)	77	40°10'N	111°40'W
Spanish Town, Jam.	87	18°00'N	76°55'W
Sparks, Nv., U.S. (spärks)	76	39°34'N	119°45'W
Sparrows Point, Md., U.S. (spăr'ōz)	68e	39°13'N	76°29'W
Sparta *see* Spárti, Grc.	131	37°07'N	22°28'E
Sparta, Ga., U.S. (spär'tȧ)	83	33°16'N	82°59'W
Sparta, Il., U.S.	79	38°07'N	89°42'W
Sparta, Mi., U.S.	66	43°10'N	85°45'W
Sparta, Tn., U.S.	82	35°54'N	85°26'W
Sparta, Wi., U.S.	71	43°56'N	90°50'W
Sparta Mountains, mts., N.J., U.S.	68a	41°00'N	74°38'W
Spartanburg, S.C., U.S. (spär'tăn-bûrg)	65	34°57'N	82°13'W
Spartel, Cap, c., Mor. (spär-tĕl')	128	35°48'N	5°50'W
Spárti (Sparta), Grc.	131	37°07'N	22°28'E
Spartivento, Cape, c., Italy (spär-tē-vĕn'tô)	130	37°55'N	16°09'E
Spartivento, Cape, c., Italy	112	38°54'N	8°52'E
Spas-Demensk, Russia (spȧs dyĕ-mĕnsk')	132	54°24'N	34°02'E
Spas-Klepiki, Russia (spȧs klĕp'ē-kē)	132	55°09'N	40°11'E
Spassk-Ryazanskiy, Russia (ryä-zän'skĭ)	132	54°24'N	40°21'E
Spassk-Dal'niy, Russia (spŭsk'däl'nyē)	135	44°30'N	133°00'E
Spátha, Ákra, c., Grc.	130a	35°42'N	23°45'E
Spaulding, Al., U.S. (spôl'dĭng)	68h	33°27'N	86°50'W
Spear, Cape, c., Can. (spēr)	61	47°32'N	52°32'W
Spearfish, S.D., U.S. (spēr'fĭsh)	70	44°28'N	103°52'W
Speed, In., U.S. (spēd)	69h	38°25'N	85°45'W
Speedway, In., U.S. (spēd'wā)	69g	39°47'N	86°14'W
Speichersee, l., Ger.	115d	48°12'N	11°47'E
Spencer, Ia., U.S.	70	43°09'N	95°08'W
Spencer, In., U.S. (spĕn'sẽr)	66	39°15'N	86°45'W
Spencer, N.C., U.S.	83	35°43'N	80°25'W
Spencer, W.V., U.S.	66	38°55'N	81°20'W
Spencer Gulf, b., Austl. (spĕn'sẽr)	174	34°20'S	136°55'E
Sperenberg, Ger. (shpĕ'rĕn-bĕrgh)	115b	52°09'N	13°22'E
Spey, l., Scot., U.K. (spā)	120	57°25'N	3°29'W
Speyer, Ger. (shpī'ẽr)	124	49°18'N	8°26'E
Sphinx, hist., Egypt (sfĭnks)	192b	29°57'N	31°08'E
Spijkenisse, Neth.	115a	51°51'N	4°18'E
Spinazzola, Italy (spē-nät'zō-lä)	130	40°58'N	16°05'E
Spirit Lake, Ia., U.S. (lāk)	70	43°25'N	95°08'W
Spirit Lake, Id., U.S. (spĭr'ĭt)	72	47°58'N	116°51'W
Spišská Nová Ves, Slvk. (spĕsh'skä nō'vä vĕs)	117	48°56'N	20°35'E
Spitsbergen *see* Svalbard, dep., Nor.	134	77°00'N	20°00'E
Split, Cro. (splēt)	110	43°30'N	16°28'E
Split Lake, l., Can.	57	56°08'N	96°15'W
Spokane, Wa., U.S. (spōkăn')	64	47°39'N	117°25'W
Spokane, r., Wa., U.S.	72	47°47'N	118°00'W
Spokane Indian Reservation, I.R., Wa., U.S.	72	47°55'N	118°00'W
Spoleto, Italy (spô-lā'tō)	130	42°44'N	12°44'E
Spoon, r., Il., U.S. (spōōn)	79	40°36'N	90°22'W
Spooner, Wi., U.S. (spōōn'ẽr)	71	45°50'N	91°53'W
Spotswood, N.J., U.S. (spŏtz'wŏŏd)	68a	40°23'N	74°22'W
Sprague, r., Or., U.S. (sprāg)	72	42°30'N	121°42'W
Spratly, i., Asia (sprăt'lē)	168	8°38'N	111°54'E
Spray, N.C., U.S. (sprā)	83	36°30'N	79°44'W
Spree, r., Ger. (sprā)	124	51°53'N	14°08'E
Spremberg, Ger. (shprĕm'bĕrgh)	124	51°35'N	14°23'E
Spring, r., Ar., U.S.	79	36°25'N	91°35'W
Springbok, S. Afr. (spring'bŏk)	186	29°35'S	17°55'E
Spring Creek, r., Nv., U.S. (sprĭng)	76	40°18'N	117°45'W
Spring Creek, r., Tx., U.S.	81	30°03'N	95°43'W
Spring Creek, r., Tx., U.S.	80	31°08'N	100°50'W
Springdale, Can.	61	49°30'N	56°05'W
Springdale, Ar., U.S. (sprĭng'dāl)	79	36°10'N	94°07'W
Springdale, Pa., U.S.	69e	40°33'N	79°46'W
Springer, N.M., U.S. (sprĭng'ẽr)	78	36°21'N	104°37'W
Springerville, Az., U.S.	77	34°08'N	109°17'W
Springfield, Co., U.S. (sprĭng'fēld)	78	37°24'N	102°04'W
Springfield, Il., U.S.	65	39°46'N	89°37'W
Springfield, Ky., U.S.	66	37°35'N	85°10'W
Springfield, Ma., U.S.	65	42°05'N	72°35'W
Springfield, Mn., U.S.	71	44°14'N	94°59'W
Springfield, Mo., U.S.	65	37°13'N	93°17'W
Springfield, Oh., U.S.	65	39°55'N	83°50'W
Springfield, Or., U.S.	72	44°01'N	123°02'W
Springfield, Tn., U.S.	82	36°30'N	86°53'W
Springfield, Vt., U.S.	67	43°20'N	72°35'W
Springfontein, S. Afr. (sprĭng'fŏn-tīn)	186	30°16'S	25°45'E
Springhill, Can. (sprĭng-hĭl')	51	45°39'N	64°03'W
Spring Mountains, mts., Nv., U.S.	76	36°18'N	115°49'W
Springs, S. Afr. (sprĭngs)	192c	26°16'S	28°27'E
Springstein, Can. (sprĭng'stīn)	62f	49°49'N	97°29'W
Springton Reservoir, res., Pa., U.S. (sprĭng-tŭn)	68f	39°57'N	75°26'W
Springvale, Austl.	173a	37°57'S	145°09'E
Spring Valley, Ca., U.S.	76a	32°46'N	117°01'W
Springvalley, Il., U.S. (sprĭng-văl'ĭ)	66	41°20'N	89°15'W
Spring Valley, Mn., U.S.	71	43°41'N	92°26'W
Spring Valley, N.Y., U.S.	68a	41°07'N	74°03'W
Springville, Ut., U.S. (sprĭng-vĭl)	77	40°10'N	111°40'W
Springwood, Austl.	173b	33°42'S	150°34'E
Spruce Grove, Can. (sprōōs grōv)	62g	53°32'N	113°55'W
Spur, Tx., U.S. (spûr)	78	33°29'N	100°51'W
Squam, l., N.H., U.S. (skwôm)	67	43°45'N	71°30'W
Squamish, Can. (skwō'mĭsh)	54	49°42'N	123°09'W
Squamish, r., Can.	54	50°10'N	123°30'W
Squillace, Golfo di, b., Italy (gōō'l-fō-dē skwēl-lä'chå)	130	38°44'N	16°47'E
Srbija (Serbia), hist. reg., Serb. (sr bē-yä) (sĕr'bē-ä)	131	44°05'N	20°35'E
Srbobran, Serb. (s'r'bô-brän')	131	45°32'N	19°50'E
Sredne-Kolymsk, Russia (s'rĕd'nyĕ kô-lêmsk')	135	67°49'N	154°55'E
Sredne Rogatka, Russia (s'red'nȧ-ya) (rô gär'tkä)	142c	59°49'N	30°20'E
Sredniy Ik, r., Russia (srĕd'nĭ ĭk)	142a	55°46'N	58°50'E
Sredniy Ural, mts., Russia (ō'rȧl)	142a	57°47'N	59°00'E
Šrem, Pol. (shrĕm)	125	52°06'N	17°01'E
Sremska Karlovci, Serb. (srĕm'skē kär'lov-tsē)	131	45°10'N	19°57'E
Sremska Mitrovica, Serb. (srĕm'skä mē'trō-vē-tsä')	131	44°59'N	19°39'E
Sretensk, Russia (s'rĕ'tĕnsk)	135	52°13'N	117°39'E
Sri Jayewardenepura Kotte, Sri L.	159	6°50'N	80°05'E

ng-sing; ŋ-baŋk; ɴ-nasalized n; nŏd; cŏmmit; ōld; ôbey; ôrder; oi-boil; fōōd; ŏŏ-as oo in foot; ou-out; s-soft; sh-dish; th-thin; pūre; ûnite; ûrn; stŭd; circŭs; ü-as in French tu; '-indeterminate vowel.

306 Sri Lanka – Suide PRONOUNCING INDEX

PLACE (Pronunciation)	PAGE	LAT.	LONG.
Sri Lanka, nation, Asia	159	8°45'N	82°30'E
Srīnagar, India (srē-nŭg'ŭr)	155	34°11'N	74°49'E
Środa, Pol. (shrŏ'dȧ)	125	52°14'N	17°17'E
Stabroek, Bel.	115a	51°20'N	4°21'E
Stade, Ger. (shtä'dĕ)	124	53°36'N	9°28'E
Städjan, mtn., Swe. (stĕd'yän)	122	61°53'N	12°50'E
Stafford, Eng., U.K. (stăf'fērd)	120	52°48'N	2°06'W
Stafford, Ks., U.S.	78	37°58'N	98°37'W
Staffordshire, co., Eng., U.K.	114a	52°45'N	2°00'W
Stahnsdorf, Ger. (shtäns'dôrf)	115b	52°22'N	13°10'E
Staines, Eng., U.K.	114b	51°26'N	0°13'W
Stakhanov, Ukr.	137	48°34'N	38°37'E
Stalingrad *see* Volgograd, Russia	134	48°40'N	42°20'E
Stalybridge, Eng., U.K.	114a	53°29'N	2°03'W
Stambaugh, Mi., U.S. (stăm'bô)	71	46°03'N	88°38'W
Stamford, Eng., U.K.	114a	52°39'N	0°28'W
Stamford, Ct., U.S. (stăm'fērd)	68a	41°03'N	73°32'W
Stamford, Tx., U.S.	78	32°57'N	99°48'W
Stammersdorf, Aus. (shtäm'ĕrs-dôrf)	115e	48°19'N	16°25'E
Stamps, Ar., U.S. (stămps)	79	33°22'N	93°31'W
Stanberry, Mo., U.S. (stan'bĕr-ē)	79	40°12'N	94°34'W
Standerton, S. Afr. (stăn'dēr-tŭn)	186	26°57'S	29°17'E
Standing Rock Indian Reservation, I.R., N.D., U.S. (stănd'ĭng rŏk)	70	47°07'N	101°05'W
Standish, Eng., U.K. (stăn'dĭsh)	114a	53°36'N	2°39'W
Stanford, Ky., U.S. (stăn'fērd)	82	37°29'N	84°40'W
Stanger, S. Afr. (stăŋ-ger)	187c	29°22'S	31°18'E
Staniard Creek, Bah.	92	24°50'N	77°55'W
Stanislaus, r., Ca., U.S. (stăn'ĭs-lô)	76	38°10'N	120°16'W
Stanley, Can. (stăn'lē)	60	46°17'N	66°44'W
Stanley, Falk. Is.	102	51°46'S	57°59'W
Stanley, N.D., U.S.	70	48°20'N	102°25'W
Stanley, Wi., U.S.	71	44°56'N	90°56'W
Stanley Pool, l., Afr.	186	4°07'S	15°40'E
Stanley Reservoir, res., India (stăn'lē)	159	12°07'N	77°27'E
Stanleyville *see* Kisangani, D.R.C.	185	0°30'S	25°12'E
Stann Creek, Belize (stăn krēk)	90a	17°01'N	88°14'W
Stanovoy Khrebet, mts., Russia (stŭn-ȧ-voi')	135	56°12'N	127°12'E
Stanton, Ca., U.S. (stăn'tŭn)	75a	33°48'N	118°00'W
Stanton, Ne., U.S.	70	41°57'N	97°15'W
Stanton, Tx., U.S.	80	32°08'N	101°46'W
Stanwood, Wa., U.S. (stăn'wŏŏd)	74a	48°14'N	122°23'W
Staples, Mn., U.S. (stā'p'lz)	71	46°21'N	94°48'W
Stapleton, Al., U.S.	82	30°45'N	87°48'W
Stara Planina, mts., Blg.	112	42°50'N	24°45'E
Staraya Kupavna, Russia (stä'rȧ-yȧ kû-päf'nȧ)	142b	55°48'N	38°10'E
Staraya Russa, Russia (stä'rȧ-yȧ rōōsȧ)	136	57°58'N	31°21'E
Stara Zagora, Blg. (zä'gô-rȧ)	119	42°26'N	25°37'E
Starbuck, Can. (stär'bŭk)	62f	49°46'N	97°36'W
Stargard Szczeciński, Pol. (shtär'gärt shchĕ-chyn'skē)	116	53°19'N	15°03'E
Staritsa, Russia (stä'rē-tsȧ)	132	56°29'N	34°58'E
Starke, Fl., U.S. (stärk)	83	29°55'N	82°07'W
Starkville, Co., U.S. (stärk'vĭl)	78	37°06'N	104°34'W
Starkville, Ms., U.S.	82	33°27'N	88°47'W
Starnberg, Ger. (shtärn-bĕrgh)	115d	47°59'N	11°20'E
Starnberger See, l., Ger.	124	47°58'N	11°30'E
Starobil's'k, Ukr.	137	49°19'N	38°57'E
Starodub, Russia (stä-rô-drŏop')	132	52°25'N	32°49'E
Starogard Gdański, Pol. (stä'rō-grad gdĕn'skē)	116	53°58'N	18°33'E
Starokostiantyniv, Ukr.	137	49°45'N	27°12'E
Staro-Minskaya, Russia (stä'rô mĭn'skȧ-yȧ)	137	46°19'N	38°51'E
Staro-Shcherbinovskaya, Russia	133	46°38'N	38°38'E
Staro-Subkhangulovo, Russia (stäro-sŏŏb-kan-gōō'lôvō)	142a	53°08'N	57°24'E
Staroutkinsk, Russia (stȧ-rô-ōōt'kĭnsk)	142a	57°14'N	59°21'E
Starovirivka, Ukr.	133	49°31'N	35°48'E
Start Point, c., Eng., U.K. (stärt)	117	50°14'N	3°34'W
Staryi Ostropil', Ukr.	133	49°48'N	27°32'E
Stary Sącz, Pol. (stä-rē sônch')	125	49°32'N	20°36'E
Staryy Oskol, Russia (stä'rē ôs-kôl')	137	51°18'N	37°51'E
Stassfurt, Ger. (shtäs'fōōrt)	124	51°52'N	11°35'E
Staszów, Pol. (stä'shŏf)	125	50°32'N	21°13'E
State College, Pa., U.S. (stāt kŏl'ĕj)	67	40°50'N	77°55'W
State Line, Mn., U.S. (līn)	75h	46°36'N	92°18'W
Staten Island, i., N.Y., U.S. (stăt'ĕn)	68a	40°35'N	74°10'W
Statesboro, Ga., U.S. (stāts'bŭr-ô)	83	32°26'N	81°47'W
Statesville, N.C., U.S. (stās'vĭl)	83	34°45'N	80°54'W
Staunton, Il., U.S. (stôn'tŭn)	75e	39°01'N	89°47'W
Staunton, Va., U.S.	67	38°10'N	79°05'W
Stavanger, Nor. (stä'väng'ĕr)	110	58°59'N	5°44'E
Stave, r., Can. (stāv)	74d	49°12'N	122°24'W
Staveley, Eng., U.K. (stāv'lē)	114a	53°17'N	1°21'W
Stavenisse, Neth.	115a	51°35'N	3°59'E
Stavropol', Russia	134	45°05'N	41°50'E
Steamboat Springs, Co., U.S. (stēm'bōt')	78	40°30'N	106°48'W
Stebliv, Ukr.	133	49°23'N	31°03'E
Steel, r., Can. (stēl)	58	49°08'N	86°55'W
Steelton, Pa., U.S. (stēl'tŭn)	67	40°15'N	76°45'W
Steenbergen, Neth.	115a	51°35'N	4°18'E
Steens Mountain, mts., Or., U.S. (stēnz)	72	42°15'N	118°52'W
Steep Point, c., Austl. (stēp)	174	26°15'N	112°05'E
Stefanie, Lake *see* Chew Bahir, l., Afr.	185	4°46'N	37°31'E
Steinbach, Can.	50	49°32'N	96°41'W
Steinkjer, Nor. (stĕĭn-kyĕr)	116	64°00'N	11°19'E
Stella, Wa., U.S. (stĕl'ȧ)	74c	46°11'N	123°12'W
Stellarton, Can. (stĕl'ȧr-tŭn)	51	45°34'N	62°40'W
Stendal, Ger. (shtĕn'däl)	124	52°37'N	11°51'E
Stepanakert *see* Xankändi, Azer.	136	39°50'N	46°40'E
Stephens, Port, b., Austl. (stē'fĕns)	176	32°43'N	152°55'E
Stephenville, Can. (stē'vĕn-vĭl)	53a	48°33'N	58°35'W
Stepnogorsk, Kaz.	139	52°20'N	72°05'E
Sterkrade, Ger. (shtĕr'krädĕ)	127c	51°31'N	6°51'E
Sterkstroom, S. Afr.	187c	31°33'S	26°36'E
Sterling, Co., U.S. (stûr'lĭng)	64	40°38'N	103°14'W
Sterling, Il., U.S.	66	41°48'N	89°42'W
Sterling, Ks., U.S.	78	38°11'N	98°11'W
Sterling, Ma., U.S.	61a	42°26'N	71°41'W
Sterling, Tx., U.S.	80	31°53'N	100°58'W
Sterlitamak, Russia (styĕr'lē-ta-mȧk')	134	53°38'N	55°56'E
Šternberk, Czech Rep. (shtĕrn'bĕrk)	125	49°44'N	17°18'E
Stettin *see* Szczecin, Pol.	110	53°25'N	14°35'E
Stettler, Can.	50	52°19'N	112°43'W
Steubenville, Oh., U.S. (stū'bĕn-vĭl)	66	40°20'N	80°40'W
Stevens, l., Wa., U.S. (stē'vĕnz)	74a	47°59'N	122°06'W
Stevens Point, Wi., U.S.	71	44°30'N	89°35'W
Stevensville, Mt., U.S. (stē'vĕnz-vĭl)	73	46°31'N	114°03'W
Stewart, r., Can. (stū'ērt)	52	63°27'N	138°48'W
Stewart Island, i., N.Z.	175a	46°56'S	167°40'E
Stewiacke, Can. (stū'wē-ăk)	60	45°08'N	63°21'W
Steynsrus, S. Afr. (stīns'rōōs)	192c	27°58'S	27°33'E
Steyr, Aus. (shtīr)	117	48°03'N	14°24'E
Stif, Alg.	184	36°18'N	5°21'E
Stikine, r., Can. (stĭ-kēn')	52	58°17'N	130°10'W
Stikine Ranges, Can.	50	59°05'N	130°00'W
Stillaguamish, r., Wa., U.S.	74a	48°11'N	122°18'W
Stillaguamish, South Fork, r., Wa., U.S. (stĭl-ȧ-gwä'mĭsh)	74a	48°05'N	121°59'W
Stillwater, Mn., U.S. (stĭl'wô-tēr)	75g	45°04'N	92°48'W
Stillwater, Mt., U.S.	73	45°23'N	109°45'W
Stillwater, Ok., U.S.	79	36°06'N	97°03'W
Stillwater, r., Mt., U.S.	73	48°47'N	114°40'W
Stillwater Range, mts., Nv., U.S.	76	39°43'N	118°11'W
Štip, Mac. (shtĭp)	131	41°43'N	22°07'E
Stirling, Scot., U.K. (stûr'lĭng)	120	56°05'N	3°59'W
Stittsville, Can. (stĭts'vĭl)	62c	45°15'N	75°54'W
Stizef, Alg. (mĕr-syä' lȧ-kôɴb)	129	35°18'N	0°11'W
Stjördalshalsen, Nor. (styûr-däls-hälsĕn)	122	63°26'N	11°00'E
Stockbridge Munsee Indian Reservation, I.R., Wi., U.S. (stŏk'brĭdj mŭn-sē)	71	44°49'N	89°00'W
Stockerau, Aus. (shtô'kĕ-rou)	124	48°24'N	16°13'E
Stockholm, Swe. (stôk'hôlm)	110	59°23'N	18°00'E
Stockholm, Me., U.S. (stŏk'hōlm)	60	47°05'N	68°08'W
Stockport, Eng., U.K. (stŏk'pôrt)	120	53°24'N	2°09'W
Stockton, Eng., U.K.	120	54°35'N	1°25'W
Stockton, Ca., U.S. (stŏk'tŭn)	64	37°56'N	121°16'W
Stockton, Ks., U.S.	78	39°26'N	99°16'W
Stockton, i., Wi., U.S.	71	46°56'N	90°25'W
Stockton Plateau, plat., Tx., U.S.	64	30°34'N	102°35'W
Stockton Reservoir, res., Mo., U.S.	79	37°40'N	93°45'W
Stöde, Swe. (stû'dĕ)	122	62°26'N	16°35'E
Stoeng Trêng, Camb. (stŏng'trĕng')	168	13°36'N	106°00'E
Stoke-on-Trent, Eng., U.K. (stōk-ŏn-trĕnt)	116	53°01'N	2°12'W
Stokhid, r., Ukr.	125	51°24'N	25°20'E
Stolac, Bos. (stô'lȧts)	131	43°03'N	17°59'E
Stolbovoy, is., Russia (stôl-bô-voi')	141	74°05'N	136°00'E
Stolin, Bela. (stô'lēn)	125	51°54'N	26°52'E
Stömstad, Swe.	122	58°58'N	11°09'E
Stone, Eng., U.K.	114a	52°54'N	2°09'W
Stoneham, Can. (stōn'ȧm)	62b	46°59'N	71°22'W
Stoneham, Ma., U.S.	61a	42°30'N	71°05'W
Stonehaven, Scot., U.K. (stōn'hā-v'n)	120	56°57'N	2°09'W
Stone Mountain, Ga., U.S. (stōn)	68c	33°49'N	84°10'W
Stonewall, Can. (stōn'wôl)	62f	50°09'N	97°21'W
Stonewall, Ms., U.S.	82	32°08'N	88°44'W
Stoney Creek, Can. (stō'nē)	62d	43°13'N	79°45'W
Stonington, Ct., U.S. (stōn'ĭng-tŭn)	67	41°20'N	71°55'W
Stony Indian Reserve, I.R., Can.	62e	51°10'N	114°45'W
Stony Mountain, Can.	62f	50°05'N	97°13'W
Stony Plain, Can. (stō'nē plān)	62g	53°32'N	114°00'W
Stony Plain Indian Reserve, I.R., Can.	62g	53°29'N	113°48'W
Stony Point, N.Y., U.S.	68a	41°13'N	73°58'W
Stora Sotra, i., Nor.	122	60°24'N	4°35'E
Stord, i., Nor. (stôrd)	122	59°54'N	5°15'E
Store Baelt, strt., Den.	122	55°25'N	10°50'E
Storfjorden, b., Nor.	122	62°17'N	6°19'E
Stormberg, mts., S. Afr. (stôrm'bûrg)	187c	31°28'S	26°35'E
Storm Lake, Ia., U.S.	70	42°39'N	95°12'W
Stormy Point, c., V.I.U.S. (stôr'mē)	87c	18°22'N	65°01'W
Stornoway, Scot., U.K. (stôr'nō-wā)	116	58°13'N	6°21'W
Storozhynets', Ukr.	125	48°10'N	25°44'E
Störsjo, Swe. (stôr'shû)	122	62°49'N	13°08'E
Störsjoen, l., Nor. (stôr-syûĕn)	122	61°32'N	11°30'E
Störsjon, l., Swe.	116	63°06'N	14°00'E
Storvik, Swe.	122	60°37'N	16°31'E
Stoughton, Wi., U.S.	71	42°54'N	89°15'W
Stour, r., Eng., U.K. (stour)	121	52°09'N	0°29'E
Stourbridge, Eng., U.K. (stour'brĭj)	114a	52°27'N	2°08'W
Stow, Ma., U.S. (stō)	61a	42°56'N	71°31'W
Stow, Oh., U.S.	69d	41°09'N	81°26'W
Straatsdrif, S. Afr.	192c	25°19'S	26°22'E
Strabane, N. Ire., U.K. (stră-băn')	120	54°59'N	7°27'W
Straelen, Ger. (shtrā'lĕn)	127c	51°26'N	6°16'E
Strahan, Austl. (strā'ăn)	175	42°08'S	145°28'E
Strakonice, Czech Rep. (strä'kô-nyē-tsĕ)	124	49°18'N	13°52'E
Straldzha, Blg. (sträl'dzhȧ)	131	42°37'N	26°44'E
Stralsund, Ger. (shräl'sŏŏnt)	116	54°18'N	13°04'E
Strangford Lough, l., N. Ire., U.K.	120	54°30'N	5°34'W
Strängnäs, Swe. (strĕng'nĕs)	122	59°23'N	16°59'E
Stranraer, Scot., U.K. (străn-rär')	120	54°55'N	5°05'W
Strasbourg, Fr. (strȧs-bōōr')	110	48°36'N	7°49'E
Stratford, Can. (străt'fērd)	58	43°20'N	81°05'W
Stratford, Ct., U.S.	67	41°10'N	73°05'W
Stratford, Wi., U.S.	71	44°16'N	90°02'W
Stratford-upon-Avon, Eng., U.K.	120	52°13'N	1°41'W
Straubing, Ger. (strou'bĭng)	124	48°52'N	12°36'E
Strausberg, Ger. (strous'bĕrgh)	124	52°35'N	13°50'E
Strawberry, r., Ut., U.S.	77	40°05'N	110°55'W
Strawn, Tx., U.S. (strôn)	80	32°38'N	98°28'W
Streator, Il., U.S. (strē'tēr)	66	41°05'N	88°50'W
Streeter, N.D., U.S.	70	46°40'N	99°22'W
Streetsville, Can. (strēts'vĭl)	62d	43°34'N	79°43'W
Strehaia, Rom. (strĕ-kä'yȧ)	131	44°37'N	23°13'E
Strel'na, Russia (strĕl'nȧ)	142c	59°52'N	30°01'E
Stretford, Eng., U.K. (strĕt'fērd)	114a	53°25'N	2°19'W
Strickland, r., Pap. N. Gui. (strĭk'lănd)	169	6°15'S	142°00'E
Strijen, Neth.	115a	51°44'N	4°32'E
Stromboli, Italy (strŏm'bô-lē)	119	38°46'N	15°16'E
Stromyn, Russia (strô'mĭn)	142b	56°02'N	38°29'E
Strong, r., Ms., U.S. (strông)	82	32°03'N	89°42'W
Strongsville, Oh., U.S. (strôngz'vĭl)	69d	41°19'N	81°50'W
Stronsay, i., Scot., U.K. (strŏn'sā)	120a	59°09'N	2°35'W
Stroudsburg, Pa., U.S. (stroudz'bûrg)	67	41°00'N	75°15'W
Struer, Den.	122	56°29'N	8°34'E
Strugi Krasnyye, Russia (strōō'gĭ krä's-ny'yĕ)	132	58°14'N	29°10'E
Struma, r., Eur. (strōō'mȧ)	131	41°55'N	23°05'E
Strumica, Mac. (strōō'mĭ-tsȧ)	131	41°26'N	22°38'E
Strunino, Russia	142b	56°23'N	38°34'E
Struthers, Oh., U.S. (strŭdh'ērz)	66	41°00'N	80°35'W
Struvenhütten, Ger. (shtrōō'vĕn-hü-tĕn)	115c	53°52'N	10°04'E
Strydpoortberge, mts., S. Afr.	192c	24°08'N	29°18'E
Stryi, Ukr.	125	49°16'N	23°51'E
Strzelce Opolskie, Pol. (stzhĕl'tsĕ o-pŏl'skyĕ)	125	50°31'N	18°20'E
Strzelin, Pol. (stzhĕ-lĭn)	125	50°48'N	17°06'E
Strzelno, Pol. (stzhål'nô)	125	52°37'N	18°10'E
Stuart, Fl., U.S. (stū'ērt)	83a	27°10'N	80°14'W
Stuart, Ia., U.S.	71	41°31'N	94°20'W
Stuart, i., Ak., U.S.	63	63°25'N	162°45'W
Stuart, i., Wa., U.S.	74d	48°42'N	123°10'W
Stuart Lake, l., Can.	54	54°32'N	124°35'W
Stuart Range, mts., Austl.	174	29°00'S	134°30'E
Sturgeon, r., Can.	62g	53°41'N	113°46'W
Sturgeon, r., Mi., U.S.	71	46°43'N	88°43'W
Sturgeon Bay, Wi., U.S.	71	44°50'N	87°22'W
Sturgeon Bay, b., Can.	57	52°00'N	98°00'W
Sturgeon Falls, Can.	51	46°19'N	79°49'W
Sturgis, Ky., U.S.	66	37°35'N	88°00'W
Sturgis, Mi., U.S.	66	41°45'N	85°25'W
Sturgis, S.D., U.S.	70	44°25'N	103°31'W
Sturt Creek, r., Austl.	174	19°40'S	127°40'E
Sturtevant, Wi., U.S. (stûr'tē-vănt)	69a	42°42'N	87°54'W
Stutterheim, S. Afr. (stŭrt'ĕr-hīm)	187c	32°34'S	27°27'E
Stuttgart, Ger. (shtŏŏt'gärt)	110	48°48'N	9°15'E
Stuttgart, Ar., U.S. (stŭt'gärt)	79	34°30'N	91°33'W
Stykkishólmur, Ice.	116	65°00'N	21°48'W
Styr', r., Eur. (stēr)	125	51°44'N	26°07'E
Suao, Tai. (sōō'ou)	165	24°35'N	121°45'E
Subarnarekha, r., India	158	22°38'N	86°26'E
Subata, Lat. (sŏŏ'bȧ-tȧ)	123	56°02'N	25°54'E
Subic, Phil. (sōō'bĭk)	169a	14°52'N	120°15'E
Subic Bay, b., Phil.	169a	14°41'N	120°11'E
Subotica, Serb. (sōō'bô'tē-tsȧ)	110	46°06'N	19°41'E
Subugo, mtn., Kenya	191	1°40'S	35°49'E
Succasunna, N.J., U.S. (sŭk'kȧ-sŭn'nȧ)	68a	40°52'N	74°37'W
Suceava, Rom. (sōō-chȧ-ä'vȧ)	125	47°39'N	26°17'E
Suceava, r., Rom.	125	47°45'N	26°10'E
Sucha, Pol. (sōō'kȧ)	125	49°44'N	19°40'E
Suchiapa, Mex. (sōō-chē-ä'pä)	89	16°38'N	93°08'W
Suchiapa, r., Mex.	89	16°27'N	93°26'W
Suchitoto, El Sal. (sōō-chē-tō'tō)	90	13°58'N	89°03'W
Sucio, r., Col. (sōō'syô)	100a	6°55'N	76°15'W
Suck, r., Ire. (sŭk)	120	53°34'N	8°16'W
Sucre, Bol. (sōō'krā)	100	19°06'S	65°16'W
Sucre, dept., Ven. (sōō'krē)	101b	10°18'N	64°12'W
Sud, Canal du, strt., Haiti	93	18°40'N	73°15'W
Sud, Rivière du, r., Can. (rē-vyâr'dü süd')	62b	46°56'N	70°35'W
Suda, Russia (sŏ'dȧ)	142a	56°58'N	56°45'E
Suda, r., Russia (sŏ'dȧ)	132	59°24'N	36°40'E
Sudair, Sau. Ar. (sū-dä'ĕr)	154	25°48'N	46°28'E
Sudalsvatnet, l., Nor.	122	59°35'N	6°59'E
Sudan, nation, Afr.	185	14°00'N	28°00'E
Sudan, reg., Afr. (sōō-dän')	184	15°00'N	7°00'E
Sudbury, Can. (sŭd'bĕr-ē)	51	46°28'N	81°00'W
Sudbury, Ma., U.S.	61a	42°23'N	71°25'W
Sudetes, mts., Eur.	112	50°41'N	15°37'E
Sudogda, Russia (sŏ'dôk-dȧ)	132	55°57'N	40°29'E
Sudost', r., Eur. (sŏ-dôst')	132	52°43'N	33°13'E
Sudzha, Russia (sŏd'zhȧ)	133	51°14'N	35°11'E
Sueca, Spain (swā'kä)	129	39°12'N	0°18'W
Suez, Egypt	185	29°58'N	32°34'E
Suez, Gulf of, b., Egypt (sōō-ĕz')	185	29°53'N	32°33'E
Suez Canal, can., Egypt	185	30°53'N	32°21'E
Suffern, N.Y., U.S. (sŭf'fērn)	68a	41°07'N	74°09'W
Suffolk, Va., U.S. (sŭf'ŭk)	68g	36°43'N	76°35'W
Sugar City, Co., U.S.	78	38°12'N	103°42'W
Sugar Creek, Mo., U.S.	75f	39°07'N	94°27'W
Sugar Creek, r., Il., U.S. (shŏŏg'ēr)	79	40°14'N	89°28'W
Sugar Creek, r., In., U.S.	66	39°55'N	87°10'W
Sugar Island, i., Mi., U.S.	75k	46°31'N	84°12'W
Sugarloaf Point, c., Austl. (sŏŏg'ēr-lôf)	176	32°19'S	153°04'E
Suggi Lake, l., Can.	57	54°22'N	102°47'W
Sühbaatar, Mong.	160	50°18'N	106°31'E
Suhl, Ger. (zōōl)	124	50°37'N	10°41'E
Suichuan, mtn., China	165	26°25'N	114°10'E
Suide, China (swā-dōō)	164	37°32'N	110°12'E

ăt; finăl; rāte; senåte; ärm; åsk; sofȧ; fâre; ch-choose; dh-as th in other; bē; ēvent; bĕt; recĕnt; cratēr; g-gō; gh-guttural g; bĭt; ĭ-short neutral; rīde; ᴋ-guttural k as ch in German ich;

T

ng-sing; ŋ-baŋk; N-nasalized n; nŏd; cŏmmit; ōld; ôbey; ôrder; oi-boil; fōōd; ò-as oo in foot; ou-out; s-soft; sh-dish; th-thin; pūre; ûnite; ûrn; stŭd; circŭs; ü-as in French tu; ′-indeterminate vowel.

308 Tablas Strait – Tarboro

PRONOUNCING INDEX

PLACE (Pronunciation)	PAGE	LAT.	LONG.
Tablas Strait, strt., Phil.	169a	12°17'N	121°41'E
Table Bay, b., S. Afr. (tā'b'l)	186a	33°41'S	18°27'E
Table Mountain, mtn., S. Afr.	186a	33°58'S	18°26'E
Table Rock Lake, Mo., U.S.	79	36°37'N	93°29'W
Tabligbo, Togo	188	6°35'N	1°30'E
Taboga, i., Pan. (tä-bō'gä)	86a	8°48'N	79°35'W
Taboguilla, i., Pan. (tä-bô-gē'l-yä)	86a	8°48'N	79°31'W
Tábor, Czech Rep. (tä'bôr)	124	49°25'N	14°40'E
Tabora, Tan. (tä-bō'rä)	186	5°01'S	32°48'E
Tabou, C. Iv. (tå-bōō')	184	4°25'N	7°21'W
Tabrīz, Iran (tå-brēz')	154	38°00'N	46°13'E
Tabuaeran, i., Kir.	2	3°52'N	159°20'W
Tabwémasana, Mont, mtn., Vanuatu	170f	15°20'S	166°44'E
Tacámbaro, r., Mex. (tä-käm'bä-rō)	88	18°55'N	101°25'W
Tacámbaro de Codallos, Mex.	88	19°12'N	101°28'W
Tacarigua, Laguna de la, l., Ven.	101b	10°18'N	65°43'W
Tacheng, China (tä-chŭŋ)	160	46°50'N	83°24'E
Tachie, r., Can.	54	54°30'N	125°00'W
Tacloban, Phil. (tä-klō'bän)	169	11°06'N	124°58'E
Tacna, Peru (täk'nä)	100	18°34'S	70°16'W
Tacoma, Wa., U.S. (tȧ-kō'mȧ)	64	47°14'N	122°27'W
Taconic Range, mts., N.Y., U.S. (tȧ-kŏn'ĭk)	67	41°55'N	73°40'W
Tacotalpa, Mex. (tä-kō-täl'pä)	89	17°37'N	92°51'W
Tacotalpa, r., Mex.	89	17°24'N	92°38'W
Tademaït, Plateau du, plat., Alg. (tä-dē-mä'ēt)	184	28°00'N	2°15'E
Tadio, Lagune, b., C. Iv.	188	5°20'N	5°25'W
Tadjoura, Dji. (tád-zhōō'rá)	192a	11°48'N	42°54'E
Tadley, Eng., U.K. (tăd'lē)	114b	51°19'N	1°08'W
Tadotsu, Japan (tä'dô-tsò)	167	34°14'N	133°43'E
Tadoussac, Can. (tá-dōō-sák')	59	48°09'N	69°43'W
Tadzhikistan *see* Tajikistan, nation, Asia	134	39°22'N	69°30'E
Taebaek Sanmaek, mts., Asia (tī-bĭk' sän-mīk')	166	37°20'N	128°50'E
Taedong, r., Kor., N. (tī-dŏng)	166	38°38'N	124°32'E
Taegu, Kor., S. (tī'gōō')	161	35°49'N	128°41'E
Taejŏn, Kor., S.	166	36°20'N	127°26'E
Tafalla, Spain (tä-fäl'yä)	128	42°30'N	1°42'W
Tafna, r., Alg. (täf'nä)	128	35°28'N	1°00'W
Taft, Ca., U.S. (tȧft)	76	35°09'N	119°27'W
Tagama, reg., Niger	189	15°50'N	6°30'E
Taganrog, Russia (tá-gán-rôk')	137	47°12'N	38°56'E
Taganrogskiy Zaliv, b., Eur. (tá-gán-rôk'skĭ zä'lĭf)	137	46°55'N	38°17'E
Tagula, i., Pap. N. Gui. (tä'gōō-lá)	175	11°45'S	153°46'E
Tagus (Tajo), r., Eur. (tā'gŭs)	112	39°40'N	5°07'W
Tahan, Gunong, mtn., Malay.	168	4°33'N	101°52'E
Tahat, mtn., Alg. (tä-hät')	184	23°22'N	5°21'E
Tahiti, i., Fr. Poly. (tä-hē'tē) (tä'ē-tē')	2	17°30'S	149°30'W
Tahkuna Nina, c., Est. (táh-kōō'ná nē'ná)	123	59°08'N	22°03'E
Tahlequah, Ok., U.S. (tä-lĕ-kwä')	79	35°54'N	94°58'W
Tahoe, l., U.S. (tä'hō)	64	39°09'N	120°18'W
Tahoua, Niger (tä'ōō-ä)	184	14°54'N	5°16'E
Tahtsa Lake, l., Can.	54	53°33'N	127°47'W
Tahuya, Wa., U.S. (tȧ-hū-yă')	74a	47°23'N	123°03'W
Tahuya, r., Wa., U.S.	74a	47°28'N	122°55'W
Tai'an, China (tī-än)	164	36°13'N	117°08'E
Taibai Shan, mtn., China (tī-bī shän)	164	33°42'N	107°25'E
Taibus Qi, China (tī-bōō-sz chyē)	164	41°52'N	115°25'E
Taicang, China (tī-tsäŋ)	162	31°26'N	121°06'E
T'aichung, Tai. (tī'chŏng)	161	24°10'N	120°42'E
Tai'erzhuang, China (tī-är-jůäŋ)	162	34°34'N	117°44'E
Taigu, China (tī-gōō)	164	37°25'N	112°35'E
Taihang Shan, mts., China (tī-häŋ shän)	164	35°45'N	112°00'E
Taihe, China (tī-hŭ)	162	33°10'N	115°38'E
Tai Hu, l., China (tī hōō)	161	31°13'N	120°00'E
Tailagoin, reg., Mong. (tī'lȧ-gän' kä'rä)	160	43°39'N	105°54'E
Tailai, China (tī-lī)	164	46°20'N	123°10'E
Tailem Bend, Austl. (tā-lĕm)	176	35°15'S	139°30'E
T'ainan, Tai. (tī'nan')	161	23°08'N	120°18'E
Taínaro, c., Grc.	118	37°45'N	22°00'E
Taining, China (tī'nĭng')	165	26°58'N	117°15'E
T'aipei, Tai. (tī'pā')	161	25°02'N	121°38'E
Taiping, pt. of. i., Malay.	168	4°56'N	100°39'E
Taiping Ling, mtn., China	164	47°03'N	120°30'E
Taisha, Japan (tī'shä)	167	35°23'N	132°40'E
Taishan, China (tī-shän)	165	22°15'N	112°50'E
Tai Shan, mts., China (tī shän)	164	36°16'N	117°05'E
Taitao, Península de, pen., Chile	102	46°20'S	77°15'W
T'aitung, Tai. (tī'tōōng')	165	22°45'N	121°02'E
Taiwan, nation, Asia (tī-wän) (fŏr-mō'sȧ)	161	23°30'N	122°20'E
Taiwan Strait, strt., Asia	161	24°30'N	120°00'E
Taixian, China (tī shyĕn)	162	32°31'N	119°54'E
Taixing, China (tī-shyĭŋ)	162	32°12'N	119°58'E
Taiyuan, China (tī-yůän)	161	37°32'N	112°38'E
Taizhou, China (tī-jō)	162	32°23'N	119°41'E
Ta'Izz, Yemen	157	13°38'N	44°04'E
Tajano de Morais, Braz. (tĕ-zhä'nô-dĕ-mô-rä'ēs)	99a	22°05'S	42°04'W
Tajikistan, nation, Asia	134	39°22'N	69°30'E
Tajumulco, vol., Guat. (tä-hōō-mōōl'kō)	90	15°03'N	91°53'W
Tajuña, r., Spain (tä-κōō'n-yä)	128	40°23'N	2°36'W
Tājūrā', Libya	118	32°56'N	13°24'W
Tak, Thai.	168	16°57'N	99°12'E
Taka, i., Japan (tä'kä)	167	30°47'N	130°23'E
Takada, Japan (tä'kà-dä)	166	37°08'N	138°30'E
Takahashi, Japan (tä'kä'hä-shĭ)	167	34°47'N	133°35'E
Takaishi, Japan	167b	34°32'N	135°27'E
Takamatsu, Japan (tä'kä'mä-tsōō')	161	34°20'N	134°02'E
Takamori, Japan (tä'kä'mō-rē')	167	32°50'N	131°08'E
Takaoka, Japan (ta'kä'ō-kä')	166	36°45'N	136°59'E
Takapuna, N.Z.	177	36°48'S	174°47'E
Takarazuka, Japan (tä'kä-rä-zōō'kä)	167b	34°48'N	135°22'E
Takasaki, Japan (tä'kät'sōō-kē')	166	36°20'N	139°00'E
Takatsu, Japan (tä-kät'sōō) (mĕ'zō-nô-kŏ'chē)	167a	35°36'N	139°37'E
Takatsuki, Japan (tä'kät'sōō-kē')	167b	34°51'N	135°38'E
Takayama, Japan (tä'kä'yä'mä)	167	36°11'N	137°16'E
Takefu, Japan (tä'kĕ-fōō)	166	35°57'N	136°09'E
Take-shima, is., Asia	166	37°15'N	131°51'E
Takla Lake, l., Can.	52	55°25'N	125°53'W
Takla Makan, des., China (mä-kán')	160	39°22'N	82°34'E
Takoma Park, Md., U.S. (tȧ'kōmȧ pärk)	68e	38°59'N	77°00'W
Takum, Nig.	189	7°17'N	9°59'E
Tala, Mex. (tä'lä)	88	20°39'N	103°42'W
Talagante, Chile (tä-lä-gá'n-tĕ)	99b	33°39'S	70°54'W
Talamanca, Cordillera de, mts., C.R.	91	9°37'N	83°55'W
Talanga, Hond. (tä-lä'n-gä)	90	14°21'N	87°09'W
Talara, Peru (tä-lä'rä)	100	4°32'S	81°17'W
Talasea, Pap. N. Gui. (tä-lä-sā'ä)	169	5°20'S	150°00'E
Talata Mafara, Nig.	189	12°35'N	6°04'E
Talaud, Kepulauan, is., Indon. (tä-lout')	169	4°17'N	127°30'E
Talavera de la Reina, Spain	118	39°58'N	4°51'W
Talca, Chile (täl'kä)	102	35°25'S	71°39'W
Talca, prov., Chile	99b	35°23'S	71°15'W
Talca, Punta, c., Chile (pōō'n-tä-täl'kä)	99b	33°25'S	71°42'W
Talcahuano, Chile (täl-kä-wä'nō)	102	36°41'S	73°05'W
Taldom, Russia (täl-dôm)	132	56°44'N	37°33'E
Taldyqorghan, Kaz.	139	45°03'N	77°18'E
Talea de Castro, Mex. (tä'lå-ä dä käs'trō)	89	17°22'N	96°14'W
Talibu, Pulau, i., Indon.	169	1°30'S	125°00'E
Talim, i., Phil. (tä-lēm')	169a	14°21'N	121°14'E
Talisay, Phil. (tä-lē'sī)	169a	14°08'N	122°56'E
Talkeetna, Ak., U.S. (tál-kēt'ná)	63	62°18'N	150°02'W
Talladega, Al., U.S. (täl-à-dē'gȧ)	82	33°25'N	86°06'W
Tallahassee, Fl., U.S. (tăl-ȧ-hăs'ē)	65	30°25'N	84°17'W
Tallahatchie, r., Ms., U.S. (tal-ȧ hăch'ē)	82	34°21'N	90°03'W
Tallapoosa, Ga., U.S. (tăl-ȧ-pōō'sȧ)	82	33°44'N	85°15'W
Tallapoosa, r., Al., U.S.	82	32°22'N	86°08'W
Tallassee, Al., U.S. (tăl'ȧ-sē)	82	32°30'N	85°54'W
Tallinn, Est. (tál'lĕn) (rĕ'väl)	134	59°26'N	24°44'E
Tallmadge, Oh., U.S. (tăl'mĭj)	69d	41°06'N	81°26'W
Tallulah, La., U.S. (tä-lōō'lȧ)	81	32°25'N	91°13'W
Tal'ne, Ukr.	133	48°52'N	30°43'E
Talo, mtn., Eth.	185	10°45'N	37°55'E
Taloje Budrukh, India	159b	19°05'N	73°05'E
Talpa de Allende, Mex. (täl'pä dā äl-yĕn'då)	88	20°25'N	104°48'W
Talquin, Lake, res., Fl., U.S.	82	30°26'N	84°33'W
Talsi, Lat. (tal'sĭ)	123	57°16'N	22°35'E
Taltal, Chile (täl-täl')	102	25°26'S	70°32'W
Taly, Russia (täl'ĭ)	133	49°51'N	40°07'E
Tama, Ia., U.S. (tā'mȧ)	71	41°57'N	92°36'W
Tama, r., Japan	167a	35°38'N	139°35'E
Tamale, Ghana (tä-mä'lå)	184	9°25'N	0°50'W
Taman', Russia (tå-män'')	133	45°13'N	36°46'E
Tamanaco, r., Ven. (tä-mä-nä'kō)	101b	9°32'N	66°00'W
Tamaqua, Pa., U.S. (tȧ-mô'kwȧ)	67	40°45'N	75°50'W
Tamar, r., Eng., U.K. (tā'mär)	120	50°35'N	4°15'W
Tamarite de Litera, Spain (tä-mä-rē'tā)	129	41°52'N	0°24'E
Tamaulipas, state, Mex. (tä-mä-ōō-lē'päs')	86	23°45'N	98°30'W
Tamazula de Gordiano, Mex.	88	19°44'N	103°09'W
Tamazulapan del Progreso, Mex.	89	17°41'N	97°34'W
Tamazunchale, Mex. (tä-mä-zòn-chä'lå)	88	21°16'N	98°46'W
Tambacounda, Sen. (täm-bä-kōōn'dä)	184	13°47'N	13°40'W
Tambador, Serra do, mts., Braz. (sĕ'r-rä-dô-täm'bä-dōr)	101	10°33'S	41°16'W
Tambelan, Kepulauan, is., Indon. (täm-bå-län')	168	0°38'N	107°38'E
Tambo, Austl. (tăm'bō)	175	24°50'S	146°15'E
Tambov, Russia (tám-bôf')	134	52°45'N	41°10'E
Tambov, prov., Russia	132	52°50'N	40°42'E
Tambre, r., Spain (täm'brā)	128	42°59'N	8°33'W
Tambura, Sudan (täm-bōō'rä)	185	5°34'N	27°30'E
Tame, r., Eng., U.K. (tām)	114a	52°41'N	1°42'W
Tâmega, r., Port. (tä-mā'gà)	128	41°30'N	7°45'W
Tamenghest, Alg.	184	22°34'N	5°34'E
Tamenghest, Oued, r., Alg.	184	22°15'N	2°51'E
Tamgak, Monts, mtn., Niger (tam-gäk')	184	18°40'N	8°40'E
Tamgué, Massif du, mtn., Gui.	184	12°15'N	12°35'W
Tamiahua, Mex. (tä-myä-wä)	89	21°17'N	97°26'W
Tamiahua, Laguna, l., Mex. (lä-gò'nä-tä-myä-wä)	89	21°38'N	97°33'W
Tamiami Canal, can., Fl., U.S. (tă-mī-ăm'ĭ)	83a	25°52'N	80°08'W
Tamil Nadu, state, India	155	11°30'N	78°00'E
Tampa, Fl., U.S. (tăm'pȧ)	65	27°57'N	82°25'W
Tampa Bay, b., Fl., U.S.	65	27°35'N	82°38'W
Tampere, Fin. (tám'pĕ-rĕ)	116	61°21'N	23°39'E
Tampico, Mex. (täm-pē'kō)	86	22°14'N	97°51'W
Tampico Alto, Mex. (täm-pē'kō äl'tō)	89	22°07'N	97°48'W
Tampin, Malay.	153b	2°28'N	102°15'E
Tam Quan, Viet.	165	14°20'N	109°10'E
Tamuín, Mex. (tä-mōō-ē'n)	88	22°04'N	98°47'W
Tamworth, Austl. (tăm'wûrth)	175	31°01'S	151°00'E
Tamworth, Eng., U.K.	114a	52°38'N	1°41'W
Tana, i., Vanuatu	175	19°32'S	169°27'E
Tana, r., Kenya (tä'nä)	187	0°30'S	39°30'E
Tanabe, Japan (tä-nä'bā)	166	33°45'N	135°21'E
Tanabe, Japan	167b	34°49'N	135°46'E
Tanacross, Ak., U.S. (tă'nȧ-crŏs)	63	63°20'N	143°30'W
Tanaga, i., Ak., U.S. (tä-nä'gä)	63a	51°28'N	178°10'W
Tanahbala, Pulau, i., Indon. (tá-nä-bä'lä)	168	0°30'S	98°22'E
Tanahmasa, Pulau, i., Indon. (tá-nä-mä'sä)	168	0°03'S	97°30'E
Tanakpur, India (tăn'ăk-pòr)	158	29°10'N	80°07'E
Tana Lake, l., Eth.	185	12°09'N	36°41'E
Tanami, Austl. (tá-nä'mē)	174	19°45'S	129°50'E
Tanana, Ak., U.S. (tă'ná-nô)	63	65°18'N	152°20'W
Tanana, r., Ak., U.S.	63	64°26'N	148°40'W
Tanaro, r., Italy (tä-nä'rô)	130	44°45'N	8°02'E
Tanashi, Japan	167a	35°44'N	139°34'E
Tanbu, China (tän-bōō)	163a	23°20'N	113°06'E
Tancheng, China (tän-chŭŋ)	164	34°37'N	118°22'E
Tanchŏn, Kor., N. (tän'chŭn)	166	40°29'N	128°50'E
Tancítaro, Mex. (tän-sē'tä-rō)	88	19°16'N	102°24'W
Tancítaro, Cerro de, mtn., Mex. (sĕ'r-rô-dĕ)	88	19°24'N	102°19'W
Tancoco, Mex. (tän-kō'kō)	89	21°16'N	97°45'W
Tandil, Arg. (tän-dēl')	102	36°16'S	59°01'W
Tandil, Sierra del, mts., Arg.	102	38°40'S	59°40'W
Tanega, i., Japan (tä'nȧ-gä')	161	30°36'N	131°11'E
Tanezrouft, reg., Alg. (tä'nĕz-rōft)	184	24°17'N	0°30'W
Tang, r., China (täŋ)	162	33°38'N	117°29'E
Tang, r., China	162	39°13'N	114°45'E
Tanga, Tan. (täŋ'gȧ)	187	5°04'S	39°06'E
Tangancícuaro, Mex. (täŋ-gän-sē'kwa-rō)	88	19°52'N	102°13'W
Tanganyika, Lake, l., Afr.	186	5°15'S	29°40'E
Tanger, Mor. (tăn-jĕr')	184	35°52'N	5°55'W
Tangermünde, Ger. (täŋ'ĕr-mün'de)	124	52°33'N	11°58'E
Tanggu, China (täŋ-gōō)	162	39°04'N	117°41'E
Tanggula Shan, mts., China (täŋ-gōō-lä shän)	160	33°15'N	89°07'E
Tanghe, China	164	32°40'N	112°50'E
Tangier *see* Tanger, Mor.	184	35°52'N	5°55'W
Tangipahoa, r., La., U.S. (tăn'jē-pȧ-hō'ȧ)	81	30°48'N	90°28'W
Tangra Yumco, l., China (täŋ-rä yōōm-tswo)	158	30°50'N	85°40'E
T'angshan, China	164	39°38'N	118°11'E
Tangxian, China (täŋ shyĕn)	162	38°49'N	115°00'E
Tangzha, China (täŋ-jä)	162	32°06'N	120°48'E
Tanimbar, Kepulauan, is., Indon.	169	8°00'S	132°00'E
Tanjong Piai, c., Malay.	153b	1°16'N	103°11'E
Tanjong Ramunia, c., Malay.	153b	1°27'N	104°44'E
Tanjungbalai, Indon. (tän'jông-bä'lå)	153b	1°00'N	103°26'E
Tanjungpandan, Indon.	168	2°47'S	107°51'E
Tanjungpinang, Indon. (tän'jông-pē'näng)	153b	0°55'N	104°29'E
Tannu-Ola, mts., Asia	135	51°00'N	94°00'E
Tannūrah, Ra's at, c., Sau. Ar.	154	26°45'N	49°59'E
Tano, r., Afr.	188	5°40'N	2°30'W
Tanquijo, Arrecife, i., Mex. (är-rĕ-sē'fĕ-tän-kē'kō)	89	21°07'N	97°16'W
Ṭanṭā, Egypt	185	30°47'N	31°00'E
Tantoyuca, Mex. (tän-tō-yōō'kä)	88	21°22'N	98°13'W
Tanyang, Kor., S.	166	36°53'N	128°20'E
Tanzania, nation, Afr.	186	6°48'S	33°58'E
Tao, r., China (tou)	164	35°30'N	103°40'E
Tao'an, China (tou-än)	161	45°15'N	122°45'E
Tao'er, r., China (tou-är)	161	45°40'N	122°00'E
Taormina, Italy (tä-ôr-mē'nä)	130	37°53'N	15°18'E
Taos, N.M., U.S. (tä'ōs)	77	36°25'N	105°35'W
Taoudenni, Mali (tä'ōō-dĕ-nē')	184	22°57'N	3°37'W
Taoussa, Mali	188	16°55'N	0°35'W
Taoyuan, China (tou-yůän)	165	29°00'N	111°15'E
Tapa, Est. (tá'pá)	123	59°16'N	25°56'E
Tapachula, Mex.	90	14°55'N	92°20'W
Tapajós, r., Braz. (tä-pä-zhô's)	101	3°27'S	55°33'W
Tapalque, Arg. (tä-päl-kĕ')	99c	36°22'S	60°05'W
Tapanatepec, Mex. (tä-pä-nä-tĕ-pĕk)	89	16°22'N	94°19'W
Tāpi, r., India	155	21°00'N	76°30'E
Tappi Saki, c., Japan (täp'pē sä'kē)	166	41°05'N	139°40'E
Tapps, l., Wa., U.S. (tăpz)	74a	47°20'N	122°12'W
Taquara, Serra de, mts., Braz. (sĕ'r-rä-dĕ-tä-kwä'rä)	101	15°28'S	54°33'W
Taquari, r., Braz. (tä-kwä'rĭ)	101	18°35'S	56°50'W
Tar, r., N.C., U.S. (tär)	83	35°58'N	78°06'W
Tara, Russia (tä'rá)	134	56°58'N	74°13'E
Tara, i., Phil. (tä'rä)	169a	12°18'N	120°28'E
Tara, r., Russia (tä'rá)	140	56°32'N	76°13'E
Ţarābulus, Leb. (tä-rä'bò-lōōs)	154	34°25'N	35°50'E
Ţarābulus (Tripolitania), hist. reg., Libya	184	31°00'N	12°26'E
Tarakan, Indon.	168	3°17'N	118°04'E
Taranaki, Mount, vol., N.Z.	177	39°18'S	174°04'E
Tarancón, Spain (tä-rän-kōn')	128	40°01'N	3°00'W
Taranto, Italy (tä'rän-tô)	119	40°30'N	17°15'E
Taranto, Golfo di, b., Italy (gôl-fō-dē tä'rän-tô)	112	40°03'N	17°10'E
Tarapoto, Peru (tä-rä-pô'tō)	100	6°29'S	76°26'W
Tarare, Fr. (tá-rär')	126	45°55'N	4°23'E
Tarascon, Fr. (tá-räs-kôN')	126	42°53'N	1°35'E
Tarascon, Fr. (tä-räs-kôN)	126	43°47'N	4°41'E
Tarashcha, Ukr. (tä'rȧsh-chá)	133	49°34'N	30°52'E
Tarata, Bol. (tä-rä'tä)	100	17°43'S	66°00'W
Taravo, r., Fr.	130	41°54'N	8°58'E
Tarazit, Massif de, mts., Niger	189	20°05'N	7°35'E
Tarazona, Spain (tä-rä-thō'nä)	128	41°54'N	1°45'W
Tarazona de la Mancha, Spain (tä-rä-zô'nä-dĕ-lä-mä'n-chä)	128	39°13'N	1°50'W
Tarbes, Fr. (tàrb)	117	43°04'N	0°05'E
Tarboro, N.C., U.S. (tär'bŭr-ō)	83	35°53'N	77°34'W

ăt; finăl; rāte; senāte; ärm; ȧsk; sofȧ; fâre; ch-choose; dh-as th in other; bē; ēvent; bĕt; recĕnt; cratĕr; g-gō; gh-guttural g; bĭt; ĭ-short neutral; rīde; κ-guttural k as ch in German ich;

PRONOUNCING INDEX

Taree – Testa del Gargano 309

PLACE (Pronunciation)	PAGE	LAT.	LONG.
Taree, Austl. (tă-rē')	176	31°52'S	152°21'E
Tarentum, Pa., U.S. (tȧ-rĕn'tŭm)	69e	40°36'N	79°44'W
Tarfa, Wādī at, val., Egypt	192b	28°14'N	31°00'E
Târgovişte, Rom.	119	44°54'N	25°29'E
Târgu Jiu, Rom.	119	45°02'N	23°17'E
Târgu Mureş, Rom.	119	46°33'N	24°33'E
Târgu Neamţ, Rom.	125	47°14'N	26°23'E
Târgu Ocna, Rom.	125	46°18'N	26°38'E
Târgu Secuiesc, Rom.	125	46°04'N	26°06'E
Tarhūnah, Libya	156	32°26'N	13°38'E
Tarija, Bol. (tär-rē'hä)	100	21°42'S	64°52'W
Tarīm, Yemen (tá-rĭm')	154	16°13'N	49°08'E
Tarim, r., China (tá-rĭm')	160	40°45'N	85°39'E
Tarim Basin, basin, China (tá-rĭm')	160	39°52'N	82°34'E
Tarka, r., S. Afr. (tä'kȧ)	187c	32°15'S	26°00'E
Tarkastad, S. Afr.	187c	32°01'S	26°18'E
Tarkhankut, Mys, c., Ukr. (mĭs tár-kän'kôt)	137	45°21'N	32°30'E
Tarkio, Mo., U.S. (tär'kĭ-ō)	79	40°27'N	95°22'W
Tarkwa, Ghana (tärk'wä)	184	5°19'N	1°59'W
Tarlac, Phil. (tär'läk)	168	15°29'N	120°36'E
Tarlton, S. Afr. (tärl'tŭn)	187b	26°05'S	27°38'E
Tarma, Peru (tär'mä)	100	11°26'S	75°40'W
Tarn, r., Fr. (tärn)	117	43°45'N	2°00'E
Târnăveni, Rom.	125	46°19'N	24°18'E
Tarnów, Pol. (tär'nôf)	117	50°02'N	21°00'E
Taro, r., Italy (tä'rō)	130	44°41'N	10°03'E
Taroudant, Mor. (tá-rōō-dänt')	184	30°39'N	8°52'W
Tarpon Springs, Fl., U.S. (tär'pŏn)	83a	28°07'N	82°44'W
Tarporley, Eng., U.K. (tär'pēr-lē)	114a	53°09'N	2°40'W
Tarpum Bay, b., Bah. (tär'pŭm)	92	25°05'N	76°20'W
Tarquinia, Italy (tär-kwē'nē-ä)	130	42°16'N	11°46'E
Tarragona, Spain (tär-rä-gō'nä)	110	41°05'N	1°15'E
Tarrant, Al., U.S. (tăr'ănt)	68h	33°35'N	86°46'W
Tárrega, Spain (tä rå-gä)	129	41°40'N	1°09'E
Tarrejón de Ardoz, Spain (tär-rĕ-kô'n-dĕ-är-dôz)	129a	40°28'N	3°29'W
Tarrytown, N.Y., U.S. (tăr'ĭ-toun)	68a	41°04'N	73°52'W
Tarsus, Tur. (tár'sòs) (tär'sŭs)	154	37°00'N	34°50'E
Tartagal, Arg. (tär-tä-gá'l)	102	23°31'S	63°47'W
Tartu, Est. (tär'tōō) (dôr'pät)	134	58°23'N	26°44'E
Ţarţūs, Syria	156	34°54'N	35°59'E
Tarumi, Japan (tä'rōō-mē)	167b	34°38'N	135°04'E
Tarusa, Russia (tä-rōōs'ȧ)	132	54°43'N	37°11'E
Tarzana, Ca., U.S. (tär-ză'ȧ)	75a	34°10'N	118°32'W
Tashkent, Uzb. (tásh'kĕnt)	139	41°23'N	69°04'E
Tasman Bay, b., N.Z. (tăz'măn)	175a	40°50'S	173°20'E
Tasmania, state, Austl.	175	41°28'S	142°30'E
Tasman Peninsula, pen., Austl.	176	43°00'S	148°30'E
Tasman Sea, sea, Oc.	195	29°30'S	155°00'E
Tasquillo, Mex. (täs-kē'lyō)	88	20°34'N	99°21'W
Tatarsk, Russia (tá-tärsk')	134	55°13'N	75°58'E
Tatarstan, prov., Russia	136	55°00'N	51°00'E
Tatar Strait, strt., Russia	135	51°00'N	141°45'E
Tater Hill, mtn., Or., U.S. (tāt'ēr hĭl)	74c	45°47'N	123°02'W
Tateyama, Japan (tä'tĕ-yä'mä)	167	35°04'N	139°52'E
Tatlow, Mount, mtn., Can.	54	51°23'N	123°52'W
Tau, Nor.	122	59°05'N	5°59'E
Tauern Tunnel, trans., Aus.	124	47°12'N	13°17'E
Taung, S. Afr. (tä'ông)	186	27°25'S	24°47'E
Taunton, Ma., U.S. (tän'tŭn)	67	41°54'N	71°03'W
Taunton, r., R.I., U.S.	68b	41°50'N	71°02'W
Taupo, Lake, l., N.Z. (tä'ōō-pō)	175a	38°42'S	175°55'E
Taurage, Lith. (tou'rá-gā)	123	55°15'N	22°18'E
Taurus Mountains *see* Toros Dağları, mts., Tur.	154	37°00'N	32°40'E
Tauste, Spain (tä-ōōs'tå)	128	41°55'N	1°15'W
Tavda, Russia (táv-dá')	134	58°00'N	64°44'E
Tavda, r., Russia	140	58°30'N	64°15'E
Taverny, Fr. (tá-vĕr-nē')	127b	49°02'N	2°13'E
Taviche, Mex. (tä-vē'chĕ)	89	16°43'N	96°35'W
Tavira, Port. (tä-vē'rȧ)	128	37°09'N	7°42'W
Tavşanlı, Tur. (táv'shän-lĭ)	137	39°30'N	29°30'E
Tawakoni, l., Tx., U.S.	81	32°51'N	95°59'W
Tawaramoto, Japan (tä'wä-rä-mô-tô)	167b	34°33'N	135°48'E
Tawas City, Mi., U.S.	66	44°15'N	83°30'W
Tawas Point, c., Mi., U.S. (tô'wȧs)	66	44°15'N	83°25'W
Tawitawi Group, is., Phil. (tä'wē-tä'wē)	168	4°52'N	120°35'E
Tawkar, Sudan	185	18°28'N	37°46'E
Taxco de Alarcón, Mex. (täs'kō dā ä-lär-kō'n)	88	18°34'N	99°37'W
Tay, r., Scot., U.K.	120	56°35'N	3°37'W
Tay, Loch, l., Scot., U.K.	120	56°25'N	4°07'W
Tayabas Bay, b., Phil. (tä-yä'bäs)	169a	13°44'N	121°40'E
Tayga, Russia (tī'gä)	140	56°12'N	85°47'E
Taygonos, Mys, c., Russia	135	60°37'N	160°17'E
Taylor, Tx., U.S.	81	30°35'N	97°25'W
Taylor, Mount, mtn., N.M., U.S.	64	35°20'N	107°40'W
Taylorville, Il., U.S. (tā'lēr-vĭl)	66	39°30'N	89°20'W
Taymyr, l., Russia (tī-mĭr')	135	74°13'N	100°45'E
Taymyr, Poluostrov, pen., Russia	135	75°15'N	95°00'E
Tayshet, Russia (tī-shĕt')	135	56°09'N	97°49'E
Tayug, Phil.	169a	16°01'N	120°45'E
Taz, r., Russia (táz)	140	67°15'N	80°45'E
Taza, Mor. (tä'zä)	184	34°08'N	4°00'W
Tazovskoye, Russia	134	66°58'N	78°28'E
Tbessa, Alg.	184	35°27'N	8°13'E
Tbilisi, Geor. ('tbĭl-yē'sē)	137	41°40'N	44°45'E
Tchentlo Lake, l., Can.	54	55°11'N	125°00'W
Tchibanga, Gabon (chē-bäŋ'gä)	186	2°51'S	11°02'E
Tchien, Lib.	188	6°04'N	8°08'W
Tchigai, Plateau du, plat., Afr.	189	21°20'N	14°50'E
Tczew, Pol. (t'chĕf')	116	54°06'N	18°48'E
Teabo, Mex. (tĕ-ä'bô)	90a	20°25'N	89°14'W
Teague, Tx., U.S.	81	31°39'N	96°16'W
Teapa, Mex. (tā-ä'pä)	89	17°35'N	92°56'W
Tebing Tinggi, i., Indon. (teb'ĭng-tĭng'gä)	153b	0°54'N	102°39'E
Tecalitlán, Mex. (tā-kä-lē-tlän')	88	19°28'N	103°17'W
Techiman, Ghana	188	7°35'N	1°56'W
Tecoanapa, Mex. (tāk-wä-nä-pä')	88	16°33'N	98°46'W
Tecoh, Mex. (tĕ-kô)	90a	20°46'N	89°27'W
Tecolotlán, Mex. (tā-kô-lô-tlän')	88	20°13'N	103°57'W
Tecolutla, Mex. (tā-kô-lōō'tlä)	89	20°33'N	97°00'W
Tecolutla, r., Mex.	89	20°16'N	97°14'W
Tecomán, Mex. (tā-kô-män')	88	18°53'N	103°53'W
Tecómitl, Mex. (tĕ-kŏ'mētl)	89a	19°13'N	98°59'W
Tecozautla, Mex. (tā'kô-zä-ōō'tlä)	88	20°33'N	99°38'W
Tecpan de Galeana, Mex. (tĕk-pän' dā gä-lā-ä'nä)	88	17°13'N	100°41'W
Tecpatán, Mex. (tĕk-pä-tá'n)	89	17°08'N	93°18'W
Tecuala, Mex. (tĕ-kwä-lä)	88	22°24'N	105°29'W
Tecuci, Rom. (tá-kōch')	119	45°51'N	27°30'E
Tecumseh, Can. (tē-kŭm'sĕ)	69b	42°19'N	82°53'W
Tecumseh, Mi., U.S.	66	42°00'N	84°00'W
Tecumseh, Ne., U.S.	79	40°21'N	96°09'W
Tecumseh, Ok., U.S.	79	35°18'N	96°55'W
Tees, r., Eng., U.K. (tēz)	120	54°40'N	2°10'W
Teganuna, l., Japan (tā'gä-nōō'nä)	167a	35°50'N	140°02'E
Tegucigalpa, Hond. (tå-gōō-sē-gäl'pä)	86	14°08'N	87°15'W
Tehachapi Mountains, mts., Ca., U.S. (tē-hă-shă'pĭ)	76	34°50'N	118°55'W
Tehrān, Iran (tē-hrän')	154	35°45'N	51°30'E
Tehuacán, Mex. (tā-wä-kän')	86	18°27'N	97°23'W
Tehuantepec, Mex.	86	16°20'N	95°14'W
Tehuantepec, r., Mex.	89	16°30'N	95°23'W
Tehuantepec, Golfo de, b., Mex. (gôl-fô dĕ)	86	15°45'N	95°00'W
Tehuantepec, Istmo de, isth., Mex. (ē'st-mô dĕ)	89	17°55'N	94°35'W
Tehuehuetla, Arroyo, r., Mex. (tĕ-wĕ-wĕ'tlä är-rô-yô)	88	17°54'N	100°26'W
Tehuitzingo, Mex. (tā-wē-tzĭŋ'gō)	88	18°21'N	98°16'W
Tejeda, Sierra de, mts., Spain (sē-ĕ'r-rä dĕ tĕ-kĕ'dä)	128	36°55'N	4°00'W
Tejúpan, Mex. (tĕ-kōō-pä'n) (sän-tyá'gô)	89	17°39'N	97°34'W
Tejúpan, Punta, c., Mex.	88	18°19'N	103°30'W
Tejupilco de Hidalgo, Mex. (tå-hōō-pēl'kô dā ē-dhäl'gō)	88	18°52'N	100°07'W
Tekamah, Ne., U.S. (tē-kā'mȧ)	70	41°46'N	96°13'W
Tekax de Alvaro Obregon, Mex.	90a	20°12'N	89°11'W
Tekeze, r., Afr.	185	13°38'N	38°00'E
Tekit, Mex. (tĕ-kē't)	90a	20°35'N	89°18'W
Tekoa, Wa., U.S. (tē-kō'ȧ)	72	47°15'N	117°03'W
Tela, Hond. (tā'lä)	86	15°45'N	87°25'W
Tela, Bahía de, b., Hond.	90	15°53'N	87°29'W
Telapa Burok, Gunong, mtn., Malay.	153b	2°51'N	102°04'E
Telavi, Geor.	137	42°00'N	45°20'E
Tel Aviv-Yafo, Isr. (tĕl-ä-vēv'já'já'fȧ)	154	32°03'N	34°46'E
Telegraph Creek, Can. (tĕl'ē-gráf)	50	57°59'N	131°22'W
Teleneşti, Mol.	133	47°31'N	28°22'E
Telescope Peak, mtn., Ca., U.S. (tĕl'ē skōp)	64	36°12'N	117°05'W
Telesung, Indon.	153b	1°07'N	102°53'E
Telica, vol., Nic. (tā-lē'kä)	90	12°38'N	86°52'W
Tell City, In., U.S. (tĕl)	66	38°00'N	86°45'W
Teller, Ak., U.S. (tĕl'ēr)	63	65°17'N	166°28'W
Tello, Col. (tĕ'l-yô)	100a	3°05'N	75°08'W
Telluride, Co., U.S. (tĕl'ū-rīd)	77	37°55'N	107°50'W
Telok Datok, Malay.	153b	2°51'N	101°33'E
Teloloapan, Mex. (tā'lō-lō-ä'pän)	88	18°19'N	99°54'W
Tel'pos-Iz, Gora, mtn., Russia (tyĕl'pôs-ēz')	134	63°50'N	59°20'E
Telšiai, Lith. (tĕl'sha'ē)	123	55°59'N	22°17'E
Teltow, Ger. (tĕl'tō)	115b	52°24'N	13°12'E
Teluklecak, Indon.	153b	1°53'N	101°45'E
Tema, Ghana	188	5°38'N	0°01'E
Temascalcingo, Mex. (tā'mäs-käl-sĭŋ'gō)	88	19°55'N	100°00'W
Temascaltepec, Mex. (tā'mäs-käl-tå pĕk)	88	19°00'N	100°03'W
Temax, Mex. (tĕ'mäx)	86	21°10'N	88°51'W
Temir, Kaz.	139	49°10'N	57°15'E
Temirtaū, Kaz.	139	50°08'N	73°13'E
Témiscouata, l., Can. (tĕ'mĭs-kō-ä'tä)	60	47°40'N	68°50'W
Témiskaming, Can. (tē-mĭs'kȧ-mĭng)	51	46°41'N	79°01'W
Temoaya, Mex. (tĕ-mô-a-um-yä)	89a	19°28'N	99°36'W
Tempe, Az., U.S.	77	33°24'N	111°54'W
Temperley, Arg. (tĕ'm-pĕr-lā)	102a	34°47'S	58°24'W
Tempio Pausania, Italy (tĕm'pē-ō pou-sä'nē-ä)	130	40°55'N	9°05'E
Temple, Tx., U.S. (tĕm'p'l)	81	31°06'N	97°20'W
Temple City, Ca., U.S.	75a	34°07'N	118°02'W
Templeton, Can. (tĕm'p'l-tŭn)	62c	45°29'N	75°37'W
Templin, Ger. (tĕm-plēn')	124	53°08'N	13°30'E
Tempoal, r., Mex. (tĕm-pô-ä'l)	88	21°38'N	98°23'W
Temryuk, Russia (tyĕm-ryŏŏk')	137	45°17'N	37°21'E
Temuco, Chile (tå-mōō'kō)	102	38°46'S	72°38'W
Temyasovo, Russia (tĕm-yä'sô-vô)	142a	53°00'N	58°06'E
Tenāli, India	159	16°10'N	80°32'E
Tenamaxtlán, Mex. (tā'nä-mäs-tlän')	88	20°13'N	104°06'W
Tenancingo, Mex. (tā-nän-sēŋ'gō)	88	18°54'N	99°36'W
Tenango, Mex. (tå-näŋ'gō)	89a	19°09'N	98°51'W
Tenasserim, Mya. (tĕn-ăs'ēr-ĭm)	168	12°09'N	99°01'E
Tendrivs'ka Kosa, ostriv, i., Ukr.	133	46°12'N	31°17'E
Tenerife Island, i., Spain (tå-nå-rē'fā) (tĕn-ēr-īf')	184	28°41'N	17°02'W
Ténès, Alg. (tā-nĕs')	117	36°28'N	1°22'E
Tengiz köli, l., Kaz.	139	50°45'N	68°39'E
Tengxian, China (tŭŋ shyĕn)	164	35°07'N	117°08'E
Tenjin, Japan (tĕn'jĕn)	167b	34°54'N	135°04'E
Tenke, D.R.C. (tĕŋ'kä)	186	11°26'S	26°45'E
Tenkiller Ferry Reservoir, res., Ok., U.S. (tĕn-kĭl'ēr)	79	35°42'N	94°47'W
Tenkodogo, Burkina (tĕŋ-kô-dō'gô)	184	11°47'N	0°22'W
Tenmile, r., Wa., U.S. (tĕn mīl)	74d	48°52'N	122°32'W
Tennant Creek, Austl. (tĕn'ănt)	174	19°45'S	134°00'E
Tennessee, state, U.S. (tĕn-ĕ-sē')	65	35°50'N	88°00'W
Tennessee, r., U.S.	65	35°35'N	88°20'W
Tennille, Ga., U.S. (tĕn'ĭl)	82	32°55'N	86°50'W
Teno, r., Chile (tĕ'nô)	99b	34°55'S	71°00'W
Tenora, Austl. (tĕn-ôrȧ)	176	34°23'S	147°33'E
Tenosique, Mex. (tā-nô-sē'kå)	89	17°27'N	91°25'W
Tenri, Japan	167b	34°36'N	135°50'E
Tenryū-Gawa, r., Japan (tĕn'ryōō'gä'wä)	167	35°16'N	137°54'E
Tensas, r., La., U.S. (tĕn'sô)	81	31°54'N	91°30'W
Tensaw, r., Al., U.S. (tĕn'sô)	82	30°45'N	87°52'W
Tenterfield, Austl. (tĕn'tēr-fēld)	175	29°00'S	152°06'E
Ten Thousand, Islands, is., Fl., U.S. (tĕn thou'zănd)	83a	25°45'N	81°35'W
Teocaltiche, Mex. (tā'ō-käl-tē'chå)	88	21°27'N	102°38'W
Teocelo, Mex. (tā-ō-sā'lō)	89	19°22'N	96°57'W
Teocuitatlán de Corona, Mex.	88	20°06'N	103°22'W
Teófilo Otoni, Braz. (tē-ô'fē-lō-tô'nē)	101	17°49'S	41°18'W
Teoloyucan, Mex. (tā'ō-lō-yōō'kän)	88	19°43'N	99°12'W
Teopisca, Mex. (tā-ō-pēs'kä)	89	16°30'N	92°33'W
Teotihuacán, Mex. (tĕ-ô-tē-wá-ká'n)	89a	19°40'N	98°52'W
Teotitlán del Camino, Mex. (tā-ô-tē-tlän' dĕl kä-mē'nô)	89	18°07'N	97°04'W
Tepalcatepec, Mex. (tā'päl-kä-tā'pĕk)	88	19°11'N	102°51'W
Tepalcatepec, r., Mex.	88	18°54'N	102°25'W
Tepalcingo, Mex. (tā-päl-sēŋ'gô)	88	18°34'N	98°49'W
Tepatitlán de Morelos, Mex. (tā-pä-tē-tlän' dā mô-rā'los)	86	20°55'N	102°47'W
Tepeaca, Mex. (tā-pā-ä'kä)	89	18°57'N	97°54'W
Tepecoacuilco de Trujano, Mex.	88	18°15'N	99°29'W
Tepeji del Río, Mex. (tā-pā-ĸe' dĕl rē'ō)	88	19°55'N	99°22'W
Tepelmeme, Mex. (tā'pĕl-mā'må)	89	17°51'N	97°23'W
Tepetlaoxtoc, Mex. (tā'pā-tlä'ōs-tōk')	88	19°34'N	98°49'W
Tepezala, Mex. (tā-pā-zä-lä')	88	22°12'N	102°12'W
Tepic, Mex. (tā-pēk')	86	21°32'N	104°53'W
Tëplaya Gora, Russia (tyôp'lá-yá gô-rá)	142a	58°32'N	59°08'W
Teplice, Czech Rep.	117	50°39'N	13°50'E
Teposcolula, Mex.	89	17°33'N	97°29'W
Tequendama, Salto de, wtfl., Col. (sá'l-tô dĕ tĕ-kĕn-dä'mä)	100	4°34'N	74°18'W
Tequila, Mex. (tå-kē'lä)	88	20°53'N	103°48'W
Tequisistlán, r., Mex. (tĕ-kē-sēs-tlá'n)	89	16°20'N	95°40'W
Tequisquiapan, Mex. (tå-kēs-kē-ä'pän)	88	20°33'N	99°57'W
Ter, r., Spain (tĕr)	129	42°04'N	2°52'E
Téra, Niger	188	14°01'N	0°45'E
Tera, r., Spain (tā'rä)	128	42°05'N	6°24'W
Teramo, Italy (tā'rä-mô)	130	42°40'N	13°41'E
Terborg, Neth. (tĕr-bōrg)	127c	51°55'N	6°23'E
Tercan, Tur. (tĕr'jän)	137	39°40'N	40°12'E
Terceira Island, i., Port. (tĕr-sā'rä)	184a	38°49'N	26°36'W
Terebovlia, Ukr.	125	49°18'N	25°43'E
Terek, r., Russia	137	43°30'N	45°10'E
Terenkul', Russia (tĕ-rĕn'kôl)	142a	55°38'N	62°18'E
Teresina, Braz. (tĕr-å-sē'nȧ)	101	5°04'S	42°42'W
Teresópolis, Braz. (tĕr-ā-sô'pō-lēzh)	99a	22°25'S	42°59'W
Teriberka, Russia (tyĕr-ē-byôr'kȧ)	136	69°00'N	35°15'E
Terme, Tur. (tĕr'mĕ)	137	41°05'N	37°00'E
Termez, Uzb. (tyĕr'mĕz)	139	37°19'N	67°20'E
Termini, Italy (tĕr'mē-nē)	130	37°58'N	13°39'E
Términos, Laguna de, l., Mex. (lä-gō'nä dĕ tĕ'r-mē-nôs)	86	18°37'N	91°32'W
Termoli, Italy (tĕr'mô-lē)	130	42°00'N	15°01'E
Tern, r., Eng., U.K. (tûrn)	114a	52°49'N	2°31'W
Ternate, Indon. (tĕr-nä'tä)	169	0°52'N	127°25'E
Terni, Italy (tĕr'nē)	118	42°38'N	12°41'E
Ternopil', Ukr.	137	49°32'N	25°36'E
Terpeniya, Mys, c., Russia	135	48°44'N	144°42'E
Terpeniya, Zaliv, b., Russia (zá'lĭf tĕr-pă'nĭ-yá)	166	49°10'N	143°05'E
Terrace, Can. (tĕr'ĭs)	50	54°31'N	128°35'W
Terracina, Italy (tĕr-rä-chē'nä)	118	41°18'N	13°14'E
Terra Nova National Park, rec., Can.	53a	48°37'N	54°15'W
Terrassa, Spain	129	41°34'N	2°01'E
Terrebonne, Can. (tĕr-bŏn')	67	45°42'N	73°38'W
Terrebonne Bay, b., La., U.S.	81	28°55'N	90°30'W
Terre Haute, In., U.S. (tĕr-ē hōt')	65	39°25'N	87°25'W
Terrell, Tx., U.S. (tĕr'ĕl)	81	32°44'N	96°15'W
Terrell, Wa., U.S.	74d	48°53'N	122°44'W
Terrell Hills, Tx., U.S. (tĕr'ĕl hĭlz)	75d	29°28'N	98°27'W
Terschelling, i., Neth. (tĕr-skĕl'ĭng)	121	53°25'N	5°12'E
Teruel, Spain (tå-rōō-ĕl')	118	40°20'N	1°05'W
Tešanj, Bos. (tĕ'shän')	131	44°36'N	17°59'E
Teschendorf, Ger. (tĕ'shĕn-dōrf)	115b	52°51'N	13°10'E
Tesecheacan, Mex. (tĕ-sĕ-chĕ-ä-ká'n)	89	18°10'N	95°41'W
Teshekpuk, l., Ak., U.S. (tē-shĕk'pŭk)	63	70°18'N	152°36'W
Teshio Dake, mtn., Japan (tĕsh'ē-ō-dä'kä)	166	44°00'N	142°50'E
Teshio Gawa, r., Japan (tĕsh'ē-ō gä'wä)	166	44°53'N	144°55'E
Tesiyn, r., Asia	160	49°45'N	96°00'E
Teslin, Can. (tĕs-lĭn)	63	60°10'N	132°30'W
Teslin, l., Can.	52	60°12'N	132°08'W
Teslin, r., Can.	52	61°18'N	134°14'W
Tessaoua, Niger (tĕs-sä'ō-ä)	184	13°53'N	7°53'E
Tessenderlo, Bel.	115a	51°04'N	5°08'E
Test, r., Eng., U.K. (tĕst)	120	51°10'N	1°30'W
Testa del Gargano, c., Italy (täs'tä dĕl gär-gä'nō)	130	41°48'N	16°13'E

ng-sing; ŋ-baŋk; ɴ-nasalized n; nŏd; cŏmmit; ōld; ôbey; ôrder; oi-boil; fōōd; ŏ-as oo in foot; ou-out; s-soft; sh-dish; th-thin; pūre; ûnite; ûrn; stŭd; circŭs; ü-as in French tu; '-indeterminate vowel.

ăt; finăl; rāte; senâte; ärm; àsk; sofá; fâre; ch-choose; dh-as th in other; bē; ĕvent; bĕt; recĕnt; cratĕr; g-gō; gh-guttural g; bĭt; ī-short neutral; rīde; ĸ-guttural k as ch in German ich;

PRONOUNCING INDEX Tlalixcoyán – Travnik 311

PLACE (Pronunciation)	PAGE	LAT.	LONG.
Tlalixcoyán, Mex. (tlä-lēs'kō-yän')	89	18°53'N	96°04'W
Tlalmanalco, Mex. (tläl-mä-nä'l-kō)	89a	19°12'N	98°48'W
Tlalnepantla, Mex.	89a	19°32'N	99°13'W
Tlalnepantla, Mex. (tläl-nå-pän'tlä)	89a	18°59'N	99°01'W
Tlalpan, Mex. (tläl-pä'n)	88	19°17'N	99°10'W
Tlalpujahua, Mex. (tläl-pōō-κä'wä)	88	19°50'N	100°10'W
Tlapa, Mex. (tlä'pä)	88	17°30'N	98°30'W
Tlapacoyán, Mex. (tlä-pä-kō-yä'n)	89	19°57'N	97°11'W
Tlapehuala, Mex. (tlä-på-wä'lä)	88	18°17'N	100°30'W
Tlaquepaque, Mex. (tlä-kē-pä'kē)	88	20°39'N	103°17'W
Tlatlaya, Mex. (tlä-tlä'yä)	88	18°36'N	100°14'W
Tlaxcala, Mex. (tläs-kä'lä)	86	19°16'N	98°14'W
Tlaxcala, state, Mex.	88	19°30'N	98°15'W
Tlaxco, Mex. (tläs'kō)	88	19°37'N	98°06'W
Tlaxiaco Santa María Asunción, Mex.	89	17°16'N	97°41'W
Tlayacapán, Mex. (tlä-yä-kä-pä'n)	89a	18°57'N	99°00'W
Tlevak Strait, strt., Ak., U.S.	54	53°03'N	132°58'W
Tlumach, Ukr. (t'lû-mäch')	125	48°47'N	25°00'E
Toa, r., Cuba (tō'ä)	93	20°25'N	74°35'W
Toamasina, Madag.	187	18°14'S	49°25'E
Toar, Cuchillas de, mts., Cuba (kōō-chē'l-lyäs-dĕ-tô-ä'r)	93	20°20'N	74°50'W
Tobago, i., Trin. (tô-bā'gō)	87	11°15'N	60°30'W
Toba Inlet, b., Can.	54	50°20'N	124°50'W
Tobarra, Spain (tô-bär'rä)	128	38°37'N	1°42'W
Tobol (Tobyl), r., Asia	140	56°00'N	66°30'E
Tobol'sk, Russia (tô-bôlsk')	140	58°09'N	68°28'E
Tobyl *see* Tobol, r., Asia	140	52°00'N	62°00'E
Tocaima, Col. (tô-kä'y-mä)	100a	4°28'N	74°38'W
Tocantinópolis, Braz. (tō-kän-tē-nō'pō-lēs)	101	6°27'S	47°18'W
Tocantins, state, Braz.	101	10°00'S	48°00'W
Tocantins, r., Braz. (tō-kän-tēNs')	101	3°28'S	49°22'W
Toccoa, Ga., U.S. (tŏk'ō-ȧ)	82	34°35'N	83°20'W
Toccoa, r., Ga., U.S.	82	34°53'N	84°24'W
Tochigi, Japan (tō'chē-gĭ)	167	36°25'N	139°45'E
Tocoa, Hond. (tō-kō'ä)	90	15°37'N	86°01'W
Tocopilla, Chile (tō-kō-pēl'yä)	102	22°03'S	70°08'W
Tocuyo de la Costa, Ven. (tô-kōō'yō-dĕ-lä-kôs'tä)	101b	11°03'N	68°24'W
Toda, Japan	167a	35°48'N	139°42'E
Todmorden, Eng., U.K. (tŏd'môr-dĕn)	114a	53°43'N	2°05'W
Tofino, Can. (tō-fē'nō)	54	49°09'N	125°54'W
Töfsingdalens National Park, rec., Swe.	122	62°09'N	13°05'E
Tōgane, Japan (tō'gä-nā)	167	35°29'N	140°16'E
Togian, Kepulauan, is., Indon.	168	0°20'S	122°00'E
Togo, nation, Afr. (tō'gō)	184	8°00'N	0°52'E
Toguzak, r., Russia (tō'gò-zák)	142a	53°40'N	61°42'E
Tohono O'odham Indian Reservation, I.R., Az., U.S.	77	32°33'N	112°12'W
Tohopekaliga, Lake, l., Fl., U.S. (tō'hō-pē'kȧ-lī'gȧ)	83a	28°16'N	81°09'W
Tohor, Tanjong, c., Malay.	153b	1°53'N	102°29'E
Toijala, Fin. (toi'yȧ-lȧ)	123	61°11'N	23°46'E
Toi-Misaki, c., Japan (toi mē'sä-kē)	166	31°20'N	131°20'E
Toiyabe, Nv., U.S. (toi'yă-bē)	76	38°59'N	117°22'W
Tokachi Gawa, r., Japan (tō-kä'chē gä'wä)	166	43°10'N	142°30'E
Tokaj, Hung. (tô'kô-ê)	125	48°06'N	21°24'E
Tokat, Tur. (tô-κät')	154	40°20'N	36°30'E
Tokelau, dep., Oc. (tō-kē-lä'ō)	2	8°00'S	176°00'W
Tokmak, Kyrg. (tôk'mák)	139	42°44'N	75°41'E
Tokmak, Ukr.	133	47°17'N	35°48'E
Tokorozawa, Japan (tō'kô-rô-zä'wä)	167a	35°47'N	139°29'E
Tok-to, atoll, Asia	166	37°15'N	131°51'E
Tokuno, i., Japan (tô-kōō'nō)	161	27°42'N	129°25'E
Tokushima, Japan (tō'kô'shē-mä)	161	34°06'N	134°31'E
Tokuyama, Japan (tō'kô'yä-mä)	167	34°04'N	131°49'E
Tōkyō, Japan	161	35°42'N	139°46'E
Tōkyō-Wan, b., Japan (tō'kyō wän)	167	35°56'N	139°56'E
Tolcayuca, Mex. (tôl-kä-yōō'kä)	88	19°55'N	98°54'W
Toledo, Spain (tô-lĕ'dô)	118	39°53'N	4°02'W
Toledo, Ia., U.S. (tō-lē'dō)	71	41°59'N	92°35'W
Toledo, Oh., U.S.	65	41°40'N	83°35'W
Toledo, Or., U.S.	72	44°37'N	123°58'W
Toledo, Montes de, mts., Spain (mô'n-tĕs-dĕ-tô-lĕ'dô)	128	39°33'N	4°40'W
Toledo Bend Reservoir, res., U.S.	65	31°30'N	93°30'W
Toliara, Madag.	187	23°16'S	43°44'E
Tolima, dept., Col. (tô-lē'mä)	100a	4°07'N	75°20'W
Tolima, Nevado del, mtn., Col. (nĕ-vä-dô-dĕl-tô-lē'mä)	100a	4°40'N	75°20'W
Tolimán, Mex. (tô-lê-män')	88	20°54'N	99°54'W
Tollesbury, Eng., U.K. (tōl'z-bĕrĭ)	114b	51°46'N	0°49'E
Tolmezzo, Italy (tôl-mĕt'zô)	130	46°25'N	13°03'E
Tolmin, Slvn. (tôl'mēn)	130	46°12'N	13°45'E
Tolna, Hung. (tôl'nô)	125	46°25'N	18°47'E
Tolo, Teluk, b., Indon. (tō'lō)	168	2°00'S	122°06'E
Tolosa, Spain (tô-lō'sä)	118	43°10'N	2°05'W
Tolt, r., Wa., U.S. (tōlt)	74a	47°13'N	121°49'W
Toluca, Mex. (tô-lōō'kä)	86	19°17'N	99°40'W
Toluca, Il., U.S. (tô-lōō'kȧ)	66	41°00'N	89°10'W
Toluca, Nevado de, mtn., Mex. (nĕ-vä-dô-dĕ-tô-lōō'kä)	86	19°09'N	99°42'W
Tolyatti, Russia	136	53°30'N	49°10'E
Tom', r., Russia	140	55°33'N	85°00'E
Tomah, Wi., U.S. (tō'mȧ)	71	43°58'N	90°31'W
Tomahawk, Wi., U.S. (tŏm'ȧ-hôk)	71	45°27'N	89°44'W
Tomakivka, Ukr.	133	47°49'N	34°43'E
Tomanivi, mtn., Fiji	170g	17°37'S	178°01'E
Tomar, Port. (tō-mär')	128	39°36'N	8°26'W
Tomashovka, Bela.	125	51°34'N	23°37'E
Tomaszów Lubelski, Pol. (tô-mä'shôf lōō-bĕl'skĭ)	125	50°20'N	23°27'E
Tomaszów Mazowiecki, Pol. (tô-mä'shôf mä-zô'vyĕt-skĭ)	125	51°33'N	20°00'E
Tomatlán, Mex. (tô-mä-tlä'n)	88	19°54'N	105°14'W
Tombadonkéa, Gui.	188	11°00'N	14°23'W
Tombador, Serra do, mts., Braz. (sĕr'rȧ dò tōm-bä-dôr')	101	11°31'S	57°33'W
Tombigbee, r., U.S. (tŏm-bĭg'bē)	65	33°00'N	88°30'W
Tombos, Braz. (tô'm-bōs)	99a	20°53'S	42°00'W
Tombouctou, Mali	184	16°46'N	3°01'W
Tombstone, Az., U.S. (tōōm'stōn)	77	31°40'N	110°00'W
Tombua, Ang. (ȧ-lē-zhän'drĕ)	186	15°49'S	11°53'E
Tomelilla, Swe. (tô'mĕ-lēl-lä)	122	55°34'N	13°55'E
Tomelloso, Spain (tō-mål-lyō'sō)	128	39°09'N	3°02'W
Tommot, Russia (tŏm-mŏt')	135	59°13'N	126°22'E
Tomsk, Russia (tŏmsk)	134	56°29'N	84°57'E
Tonala, Mex.	88	20°38'N	103°14'W
Tonalá, r., Mex.	89	18°05'N	94°08'W
Tonawanda, N.Y., U.S. (tôn-ȧ-wŏn'dȧ)	69c	43°01'N	78°53'W
Tonawanda Creek, r., N.Y., U.S.	69c	43°05'N	78°43'W
Tonbridge, Eng., U.K. (tŭn-brĭj)	114b	51°11'N	0°17'E
Tonda, Japan (tôn'dä)	167b	34°51'N	135°38'E
Tondabayashi, Japan (tôn-dä-bä'yä-shē)	167b	34°29'N	135°36'E
Tondano, Indon. (tôn-dä'nō)	169	1°15'N	124°50'E
Tønder, Den. (tûn'nĕr)	122	54°47'N	8°49'E
Tone-Gawa, r., Japan (tô'nē gä'wa)	167	36°12'N	139°19'E
Tonga, nation, Oc. (tŏŋ'gȧ)	194	18°50'S	175°20'W
Tong'an, China (tôŋ-än)	165	24°48'N	118°02'E
Tonga Trench, deep,	194	23°00'S	172°30'W
Tongbei, China (tôŋ-bā)	161	48°00'N	126°48'E
Tongguan, China (tôŋ-gūän)	161	34°48'N	110°25'E
Tonghe, China (tôŋ-hŭ)	164	45°58'N	128°40'E
Tonghua, China (tôŋ-hwä)	161	41°43'N	125°50'E
Tongjiang, China (tôŋ-jyäŋ)	161	47°38'N	132°54'E
Tongliao, China (tôŋ-lĭou)	164	43°30'N	122°15'E
Tongo, Cam.	189	5°11'N	14°00'E
Tongoy, Chile (tōn-goi')	102	30°16'S	71°29'W
Tongren, China (tôŋ-rŭn)	160	27°45'N	109°12'E
Tongshan, China (tôŋ-shän)	162	34°27'N	116°27'E
Tongtian, r., China (tôŋ-tĭĕn)	160	33°00'N	97°00'E
Tongue, r., Mt., U.S. (tŭng)	73	45°08'N	106°40'W
Tongxian, China (tôŋ shyĕn)	162	39°55'N	116°40'E
Tonj, r., Sudan (tônj)	185	6°18'N	28°33'E
Tonk, India (Tŏŋk)	155	26°13'N	75°45'E
Tonkawa, Ok., U.S. (tŏŋ kȧ-wô)	79	36°42'N	97°19'W
Tonkin, Gulf of, b., Asia (tôn-kăn')	168	20°30'N	108°10'E
Tonle Sap, l., Camb. (tôn'lā săp')	168	13°03'N	102°49'E
Tonneins, Fr. (tō-năN')	126	44°24'N	0°18'E
Tönning, Ger. (tû'nēng)	124	54°20'N	8°55'E
Tonopah, Nv., U.S. (tō-nô-pä')	64	38°04'N	117°15'W
Tönsberg, Nor. (tûns'bĕrgh)	116	59°19'N	10°25'E
Tonto, r., Mex. (tôn'tō)	89	18°15'N	96°13'W
Tonto Creek, r., Az., U.S.	77	34°05'N	111°15'W
Tonto National Monument, rec., Az., U.S. (tôn'tô)	77	33°33'N	111°08'W
Tooele, Ut., U.S. (tò-ĕl ê)	75b	40°33'N	112°17'W
Toowoomba, Austl. (tò wōōm'bȧ)	175	27°32'S	152°10'E
Topanga, Ca., U.S. (tō'păn-gȧ)	75a	34°05'N	118°36'W
Topeka, Ks., U.S. (tô-pē'kȧ)	65	39°02'N	95°41'W
Topilejo, Mex. (tô-pē-lĕ'hô)	89a	19°12'N	99°09'W
Topock, Az., U.S.	77	34°40'N	114°20'W
Topol'čany, Slvk. (tô-pôl'chä-nü)	125	48°38'N	18°10'E
Topolobampo, Mex. (tō-pō-lō-bä'm-pô)	86	25°45'N	109°00'W
Topolovgrad, Blg.	131	42°05'N	26°19'E
Toppenish, Wa., U.S. (tôp'ĕn-ĭsh)	72	46°22'N	120°00'W
Torbat-e Ḩeydarīyeh, Iran	157	35°16'N	59°13'E
Torbat-e Jām, Iran	157	35°14'N	60°36'E
Torbay, Can. (tôr-bā')	61	47°40'N	52°43'W
Torbay *see* Torquay, Eng., U.K.	120	50°30'N	3°26'W
Torbreck, Mount, mtn., Austl. (tôr-brĕk)	176	37°05'S	146°55'E
Torch, l., Mi., U.S. (tôrch)	66	45°00'N	85°30'W
Töreboda, Swe. (tŭ'rē-bō'dä)	122	58°44'N	14°04'E
Torhout, Bel.	121	51°01'N	3°04'E
Toribío, Col. (tô-rē-bē'ô)	100a	2°58'N	76°14'W
Toride, Japan (tō'rē-dā)	167a	35°54'N	104°04'E
Torino *see* Turin, Italy	110	45°05'N	7°44'E
Tormes, r., Spain (tōr'mās)	128	41°12'N	6°15'W
Torneälven, r., Eur.	112	67°00'N	22°30'E
Torneträsk, l., Swe. (tôr'nĕ trĕsk)	116	68°10'N	20°36'E
Torngat Mountains, mts., Can.	53	59°18'N	64°35'W
Tornio, Fin. (tôr'nĭ-ô)	110	65°55'N	24°09'E
Toro, Lac, l., Can.	59	46°53'N	73°46'W
Toronto, Can. (tô-rŏn'tō)	51	43°40'N	79°23'W
Toronto, Oh., U.S.	66	40°30'N	80°35'W
Toronto, res., Mex.	80	27°35'N	105°37'W
Toropets, Russia (tô'rô-pyĕts)	136	56°31'N	31°37'E
Toros Dağları, mts., Tur. (tô'rŭs)	154	37°00'N	32°40'E
Torote, r., Spain (tô-rō'tä)	129a	40°36'N	3°24'W
Torquay, Eng., U.K. (tôr-kē')	120	50°30'N	3°26'W
Torra, Cerro, mtn., Col. (sĕ'r-rô-tô'r-rä)	100a	4°41'N	76°22'W
Torrance, Ca., U.S. (tŏr'rănc)	75a	33°50'N	118°20'W
Torre Annunziata, Italy (tôr'rä ä-nōōn-tsē-ä'tä)	129c	40°31'N	14°27'E
Torreblanca, Spain	129	40°18'N	0°12'E
Torre del Greco, Italy (tôr'rā dĕl grā'kô)	130	40°32'N	14°23'E
Torrejoncillo, Spain (tôr'rā-hōn-thē'lyō)	128	39°54'N	6°26'W
Torrelavega, Spain (tôr-rā'lä-vä'gä)	128	43°22'N	4°02'W
Torre Maggiore, Italy (tôr'rā mäd-jō'rā)	130	41°41'N	15°18'E
Torrens, Lake, l., Austl. (tŏr-ĕns)	174	30°07'S	137°40'E
Torrent, Spain	129	39°25'N	0°28'W
Torreón, Mex. (tôr-rå-ōn')	86	25°32'N	103°26'W
Torres Islands, is., Vanuatu (tôr'rĕs) (tŏr'ĕz)	175	13°18'N	165°59'E
Torres Martinez Indian Reservation, I.R., Ca., U.S. (tôr'ĕz mär-tē'nĕz)	76	33°33'N	116°21'W
Torres Novas, Port. (tôr'rĕzh nō'väzh)	128	39°28'N	8°37'W
Torres Strait, strt., Austl. (tôr'rĕs)	175	10°30'S	141°30'E
Torres Vedras, Port. (tôr'rĕsh vā'dräzh)	128	39°08'N	9°18'W
Torrevieja, Spain (tôr-rā-vyā'hä)	129	37°58'N	0°40'W
Torrijos, Phil. (tôr-rē'hōs)	169a	13°19'N	122°06'E
Torrington, Ct., U.S. (tŏr'ĭng-tŭn)	67	41°50'N	73°10'W
Torrington, Wy., U.S.	70	42°04'N	104°11'W
Torro, Spain (tô'r-rō)	128	41°27'N	5°23'W
Torsby, Swe. (tôrs'bü)	122	60°07'N	12°56'E
Torshälla, Swe. (tôrs'hĕl-ä)	122	59°26'N	16°21'E
Tórshavn, Far. Is. (tôrs-houn')	110	62°00'N	6°55'W
Tortola, i., Br. Vir. Is. (tôr-tō'lä)	87b	18°34'N	64°40'W
Tortona, Italy (tôr-tō'nä)	130	44°52'N	8°52'W
Tortosa, Spain (tôr-tō'sä)	110	40°59'N	0°33'E
Tortosa, Cap de, c., Spain	129	40°42'N	0°55'E
Tortue, Canal de la, strt., Haiti (tôr-tü')	93	20°05'N	73°20'W
Tortue, Île de la, i., Haiti	93	20°10'N	73°00'W
Tortue, Rivière de la, r., Can. (lä tōr-tü')	62a	45°12'N	73°32'W
Toruń, Pol.	110	53°02'N	18°35'E
Tõrva, Est. (t'r'vá)	123	58°02'N	25°56'E
Torzhok, Russia (tôr'zhôk)	136	57°03'N	34°53'E
Toscana, hist. reg., Italy (tôs-kä'nä)	130	43°23'N	11°08'E
Tosna, r., Russia	142c	59°28'N	30°53'E
Tosno, Russia (tôs'nô)	132	59°32'N	30°52'E
Tostado, Arg. (tôs-tä'dô)	102	29°10'S	61°43'W
Tosya, Tur. (tōz'yȧ)	119	41°00'N	34°00'E
Totana, Spain (tô-tä-nä)	128	37°45'N	1°28'W
Tot'ma, Russia (tôt'mȧ)	136	60°00'N	42°20'E
Totness, Sur.	101	5°51'N	56°17'W
Totonicapán, Guat. (tôtō-nê-kä'pän)	86	14°55'N	91°20'W
Totoras, Arg. (tô-tô'räs)	99c	32°33'S	61°13'W
Totsuka, Japan (tōt'sōō-kä)	167a	35°24'N	139°32'E
Tottenham, Eng., U.K. (tŏt'ĕn-ȧm)	114b	51°35'N	0°06'W
Tottori, Japan (tô'tô-rê)	161	35°30'N	134°15'E
Touba, C. Iv.	188	8°17'N	7°41'W
Touba, Sen.	188	14°51'N	15°53'W
Toubkal, Jebel, mtn., Mor.	184	31°15'N	7°46'W
Tougan, Burkina	188	13°04'N	3°04'W
Touggourt, Alg. (tò-gōōrt') (tōō-gōōr')	184	33°09'N	6°07'E
Touil, Oued, r., Alg. (tōō-ēl')	118	34°42'N	2°16'E
Toul, Fr. (tōōl)	117	48°39'N	5°51'E
Toulon, Fr. (tōō-lôN')	110	43°09'N	5°54'E
Toulouse, Fr. (tōō-lōōz')	110	43°37'N	1°27'E
Toungoo, Mya. (tô-ôŋ-gōō')	168	19°00'N	96°29'E
Tourcoing, Fr. (tòr-kwaN')	117	50°44'N	3°06'E
Tournan-en-Brie, Fr. (tōōr-naN-ĕN-brē')	127b	48°45'N	2°47'E
Tours, Fr. (tōōr)	110	47°23'N	0°39'E
Touside, Pic, mtn., Chad (tōō-sē-dā')	185	21°10'N	16°30'E
Tovdalselva, r., Nor. (tôv-däls-ĕlvå)	122	58°23'N	8°16'E
Towanda, Pa., U.S. (tô-wän'dȧ)	67	41°45'N	76°30'W
Town Bluff Lake, l., Tx., U.S.	81	30°52'N	94°30'W
Towner, N.D., U.S. (tou'nĕr)	70	48°21'N	100°24'W
Townsend, Ma., U.S. (toun'zĕnd)	61a	42°41'N	71°42'W
Townsend, Mt., U.S.	73	46°19'N	111°35'W
Townsend, Mount, mtn., Wa., U.S.	74a	47°52'N	123°03'W
Townsville, Austl. (tounz'vĭl)	175	19°18'S	146°50'E
Towson, Md., U.S. (tou'sŭn)	68e	39°24'N	76°36'W
Towuti, Danau, l., Indon. (tô-wōō'tê)	168	3°00'S	121°45'E
Toxkan, r., China	160	40°34'N	77°15'E
Toyah, Tx., U.S. (tô'yȧ)	80	31°19'N	103°46'W
Toyama, Japan (tō'yä-mä)	161	36°42'N	137°14'E
Toyama-Wan, b., Japan	167	36°58'N	137°16'E
Toyohashi, Japan (tō'yō-hä'shē)	166	34°44'N	137°21'E
Toyonaka, Japan (tō'yō-nä'kȧ)	167b	34°47'N	135°28'E
Tozeur, Tun. (tô-zûr')	118	33°59'N	8°11'E
Trabzon, Tur. (tráb'zôn)	154	41°00'N	39°45'E
Tracy, Can.	59	46°00'N	73°13'W
Tracy, Ca., U.S. (trā'sē)	76	37°45'N	121°27'W
Tracy, Mn., U.S.	70	44°13'N	95°37'W
Tracy City, Tn., U.S.	82	35°15'N	85°44'W
Trafalgar, Cabo, c., Spain (kȧ'bô-trä-fäl-gä'r)	128	36°10'N	6°02'W
Trafonomby, mtn., Madag.	187	24°32'S	46°35'E
Trail, Can. (trāl)	50	49°06'N	117°42'W
Traisen, r., Aus.	115e	48°15'N	15°55'E
Traiskirchen, Aus.	115e	48°01'N	16°18'E
Trakai, Lith. (trá-kày)	123	54°38'N	24°59'E
Trakiszki, Pol. (trä-kē'-sh-kê)	125	54°16'N	23°07'E
Tralee, Ire. (trȧ-lē')	117	52°16'N	9°20'W
Tranås, Swe. (trän'ôs)	122	58°03'N	14°56'E
Trancoso, Port. (träŋ-kō'sō)	128	40°46'N	7°23'W
Trangan, Pulau, i., Indon. (träŋ'gän)	169	6°52'S	133°30'E
Trani, Italy (trä'nē)	130	41°15'N	16°25'E
Transylvania, hist. reg., Rom. (trăn-sĭl-vā'nĭ-ȧ)	125	46°30'N	22°35'E
Trapani, Italy	118	38°01'N	12°31'E
Trappes, Fr. (tràp)	127b	48°47'N	2°01'E
Traralgon, Austl. (trä'răl-gŏn)	176	38°15'S	146°33'E
Trarza, reg., Maur.	188	17°35'N	15°15'W
Trasimeno, Lago, l., Italy (lä'gō trä-sē-mä'nō)	130	43°00'N	12°12'E
Trás-os-Montes, hist. reg., Port. (träzh'ōzh mōn'tāzh)	118	41°33'N	7°13'W
Traun, r., Aus. (troun)	124	48°10'N	14°15'E
Traunstein, Ger. (troun'stīn)	124	47°52'N	12°38'E
Traverse, Lake, l., Mn., U.S. (trăv'ērs)	70	45°46'N	96°53'W
Traverse City, Mi., U.S.	66	44°45'N	85°40'W
Travnik, Bos. (träv'nēk)	131	44°13'N	17°43'E

312 Treasure Island – Turnov PRONOUNCING INDEX

PLACE (Pronunciation)	PAGE	LAT.	LONG.
Treasure Island, i., Ca., U.S. (trĕzh'ēr)	74b	37°49'N	122°22'W
Trebbin, Ger. (trĕ'bēn)	115b	52°13'N	13°13'E
Trebinje, Bos. (trå'bēn-yĕ)	131	42°43'N	18°21'E
Trebišov, Slvk. (trĕ'bē-shôf)	125	48°36'N	21°32'E
Tregrosse Islands, is., Austl. (trē-grōs')	175	18°08'S	150°53'E
Treinta y Tres, Ur. (trå-ēn'tä ē träs')	102	33°14'S	54°17'W
Trelew, Arg. (trē'lū)	102	43°15'S	65°25'W
Trelleborg, Swe.	122	55°24'N	13°07'E
Tremiti, Isole, is., Italy (ē'sō-lĕ trā-mē'tē)	130	42°07'N	16°33'E
Trenčín, Czech Rep. (trĕn'chēn)	117	48°52'N	18°02'E
Trenque Lauquén, Arg. (trĕn'kĕ-lâ'ŏ-kĕ'n)	102	35°50'S	62°44'W
Trent, r., Can. (trĕnt)	59	44°15'N	77°55'W
Trent, r., Eng., U.K.	114a	53°25'N	0°45'W
Trent and Mersey Canal, can., Eng., U.K. (trĕnt) (mûr zē)	114a	53°11'N	2°24'W
Trentino-Alto Adige, hist. reg., Italy	130	46°16'N	10°47'E
Trento, Italy (trĕn'tô)	118	46°04'N	11°07'E
Trenton, Can. (trĕn'tŭn)	51	44°05'N	77°35'W
Trenton, Can.	61	45°37'N	62°38'W
Trenton, Mi., U.S.	69b	42°08'N	83°12'W
Trenton, Mo., U.S.	79	40°05'N	93°36'W
Trenton, N.J., U.S.	65	40°13'N	74°46'W
Trenton, Tn., U.S.	82	35°57'N	88°55'W
Trepassey, Can. (trē-păs'ē)	61	46°44'N	53°22'W
Trepassey Bay, b., Can.	61	46°40'N	53°20'W
Tres Arroyos, Arg. (träs'är-rō'yōs)	102	38°18'S	60°16'W
Três Corações, Braz. (trĕ's kō-rä-zô'ĕs)	99a	21°41'S	45°14'W
Tres Cumbres, Mex. (trĕ's kōō'm-brĕs)	89a	19°03'N	99°14'W
Três Lagoas, Braz. (trĕ's lä-gô'äs)	101	20°48'S	51°42'W
Três Marias, Reprêsa, res., Braz.	101	18°15'S	45°30'W
Tres Morros, Alto de, mtn., Col. (ä'l-tō dĕ trĕ's mô'r-rôs)	100a	7°08'N	76°10'W
Três Pontas, Braz. (trĕ'pô'n-täs)	99a	21°22'S	45°30'W
Três Pontas, Cabo das, c., Ang.	190	10°23'S	13°32'E
Três Rios, Braz. (trĕ's rē'ōs)	99a	22°07'S	43°13'W
Très-Saint Rédempteur, Can. (săn rā-dänp-tûr')	62a	45°26'N	74°23'W
Treuenbrietzen, Ger. (troi'ĕn-brē-tzĕn)	115b	52°06'N	12°52'E
Treviglio, Italy (trā-vē'lyô)	130	45°30'N	9°34'E
Treviso, Italy (trĕ-vē'sō)	118	45°39'N	12°15'E
Trichardt, S. Afr. (trĭ-kärt')	192c	26°32'N	29°16'E
Trier, Ger.	117	49°45'N	6°38'E
Trieste, Italy (trē-ĕs'tä)	110	45°39'N	13°48'E
Triglav, mtn., Slvn.	130	46°23'N	13°50'E
Trigueros, Spain (trē-gā'rōs)	128	37°23'N	6°50'W
Tríkala, Grc.	119	39°33'N	21°49'E
Trikora, Puncak, mtn., Indon.	169	4°15'S	138°45'E
Trim Creek, r., Il., U.S. (trĭm)	69a	41°19'N	87°39'W
Trincomalee, Sri L. (trĭŋ-kō-má-lē')	159	8°39'N	81°12'E
Tring, Eng., U.K. (trĭng)	114b	51°46'N	0°40'W
Trinidad, Bol. (trē-nē-dhädh')	100	14°48'S	64°43'W
Trinidad, Cuba (trē-nē-dhädh')	87	21°50'N	80°00'W
Trinidad, Ur.	102	33°29'S	56°55'W
Trinidad, Co., U.S. (trĭn'ĭdăd)	64	37°11'N	104°31'W
Trinidad, i., Trin. (trĭn'ĭ-dăd)	101	10°00'N	61°00'W
Trinidad, r., Pan.	86a	8°55'N	80°01'W
Trinidad, Sierra de, mts., Cuba (sē-ĕ'r-rä dĕ trē-nē-dä'd)	92	21°50'N	79°55'W
Trinidad and Tobago, nation, N.A. (trĭn'ĭ-dăd) (tō-bā'gō)	87	11°00'N	61°00'W
Trinitaria, Mex. (trē-nē-tä'ryä)	89	16°09'N	92°04'W
Trinity, Can. (trĭn'ĭ-tē)	61	48°59'N	53°55'W
Trinity, Tx., U.S.	81	30°52'N	95°27'W
Trinity, is., Ak., U.S.	63	56°25'N	153°15'W
Trinity, r., Ca., U.S.	72	40°50'N	123°20'W
Trinity, r., Tx., U.S.	65	30°50'N	95°09'W
Trinity, East Fork, r., Tx., U.S.	79	33°24'N	96°42'W
Trinity, West Fork, r., Tx., U.S.	78	33°22'N	98°26'W
Trinity Bay, b., Can.	53	48°00'N	53°40'W
Trino, Italy (trē'nō)	130	45°11'N	8°16'E
Trion, Ga., U.S. (trī'ŏn)	82	34°32'N	85°18'W
Trípoli, Grc.	119	37°32'N	22°32'E
Tripoli (Tarābulus), Libya	185	32°50'N	13°13'E
Tripolitania *see* Tarābulus, hist. reg., Libya	184	31°00'N	12°26'E
Tripura, state, India	155	24°00'N	92°00'E
Tristan da Cunha Islands, is., St. Hel. (trēs-tän'dä kōōn'yà)	2	35°30'S	12°15'W
Triste, Golfo, b., Ven. (gôl-fô trē's-tĕ)	101b	10°40'N	68°05'W
Triticus Reservoir, res., N.Y., U.S. (trī tĭ-cŭs)	68a	41°20'N	73°36'W
Trnava, Slvk. (t'r'ná-vá)	125	48°22'N	17°34'E
Trobriand Islands, is., Pap. N. Gui. (trō-brē-ănd')	169	8°25'S	151°45'E
Trogir, Cro. (trō'gēr)	130	43°32'N	16°17'E
Trois Fourches, Cap des, c., Mor.	128	35°28'N	2°58'W
Trois-Rivières, Can. (trwä'rē-vyâ')	51	46°21'N	72°35'W
Troitsk, Russia (trô'ētsk)	140	54°06'N	61°35'E
Troits'ke, Ukr.	133	47°39'N	30°16'E
Troitsko-Pechorsk, Russia (trô'ĭtsk-ô-pyĕ-chôrsk')	134	62°18'N	56°07'E
Trollhättan, Swe. (trôl'hĕt-ĕn)	116	58°17'N	12°17'E
Trollheimen, mts., Nor. (trôll-hĕĭm)	122	62°48'N	9°05'E
Trona, Ca., U.S. (trō'nà)	76	35°49'N	117°20'W
Tronador, Cerro, mtn., S.A. (sĕ'r-rô trō-nä'dôr)	102	41°17'S	71°56'W
Troncoso, Mex. (trôn-kō'sō)	88	22°43'N	102°22'W
Trondheim, Nor. (trôn'hâm)	110	63°25'N	11°35'E
Trosa, Swe. (trô'sä)	122	58°54'N	17°25'E
Trout, l., Can.	53	51°16'N	92°46'W
Trout, l., Can.	52	61°10'N	121°30'W
Trout Creek, r., Or., U.S.	72	42°18'N	118°31'W
Troutdale, Or., U.S. (trout'dāl)	74c	45°32'N	122°23'W
Trout Lake, Mi., U.S.	71	46°20'N	85°02'W
Trouville, Fr. (trōō-vēl')	126	49°23'N	0°05'E
Troy, Al., U.S. (troi)	82	31°47'N	85°46'W
Troy, Il., U.S.	75e	38°44'N	89°53'W
Troy, Ks., U.S.	79	39°46'N	95°07'W
Troy, Mo., U.S.	78	38°56'N	99°57'W
Troy, Mt., U.S.	72	48°28'N	115°56'W
Troy, N.C., U.S.	83	35°21'N	79°58'W
Troy, N.Y., U.S.	65	42°45'N	73°45'W
Troy, Oh., U.S.	66	40°00'N	84°10'W
Troy, hist., Tur.	154	39°59'N	26°14'E
Troyes, Fr. (trwä)	117	48°18'N	4°03'E
Trstenik, Serb. (t'r'stĕ-nĕk)	119	43°36'N	21°00'E
Trubchëvsk, Russia (tròp'chĕfsk)	137	52°36'N	33°46'E
Trucial States *see* United Arab Emirates, nation, Asia	154	24°00'N	54°00'E
Truckee, Ca., U.S. (trŭk'ē)	76	39°20'N	120°12'W
Truckee, r., Ca., U.S.	76	39°25'N	120°07'W
Truganina, Austl.	173a	37°49'N	144°44'E
Trujillo, Col. (trò-κē'l-yō)	100a	4°10'N	76°20'W
Trujillo, Peru	100	8°08'S	79°00'W
Trujillo, Spain (trōō-κē'l-yô)	118	39°27'N	5°50'W
Trujillo, Ven.	100	9°15'N	70°28'W
Trujillo, r., Mex.	88	23°12'N	103°10'W
Trujín, Lago, l., Dom. Rep. (trōō-κēn')	93	17°45'N	71°25'W
Truk *see* Chuuk, is., Micron.	170c	7°25'N	151°47'E
Trumann, Ar., U.S. (trōō'măn)	79	35°41'N	90°31'W
Trŭn, Blg. (trŭn)	131	42°49'N	22°39'E
Truro, Can. (trōō'rō)	51	45°22'N	63°16'W
Truro, Eng., U.K.	120	50°17'N	5°05'W
Trussville, Al., U.S. (trŭs'vĭl)	68h	33°37'N	86°37'W
Truth or Consequences, N.M., U.S. (trōōth ŏr kŏn'sē-kwĕn-sĭs)	77	33°10'N	107°20'W
Trutnov, Czech Rep. (tròt'nôf)	124	50°36'N	15°36'E
Trzcianka, Pol. (tchyän'kà)	124	53°02'N	16°27'E
Trzebiatów, Pol. (tchĕ-byä'tò-v)	124	54°03'N	15°16'E
Tsaidam Basin, basin, China (tsī-däm)	160	37°19'N	94°08'E
Tsala Apopka Lake, r., Fl., U.S. (tsä'lá ă-pŏp'kà)	83	28°57'N	82°11'W
Tsast Bogd, mtn., Mong.	160	46°44'N	92°34'E
Tsavo National Park, rec., Kenya	191	2°35'S	38°45'E
Tsawwassen Indian Reserve, I.R., Can.	74d	49°03'N	123°11'W
Tsentral'nyy-Kospashskiy, Russia (tsĕn-träl'nyĭ-kôs-pash'skĭ)	142a	59°03'N	57°48'E
Tshela, D.R.C. (tshā'là)	186	4°59'S	12°56'E
Tshikapa, D.R.C. (tshē-kä'pä)	186	6°25'S	20°48'E
Tshofa, D.R.C.	191	5°14'S	25°15'E
Tshuapa, r., D.R.C.	186	0°30'S	22°00'E
Tsiafajovona, mtn., Madag.	187	19°17'S	47°27'E
Tsiribihina, r., Madag. (tsē'rē-bē-hē-nä')	187	19°45'S	43°30'E
Tsitsa, r., S. Afr. (tsē'tsá)	187c	31°28'S	28°53'E
Tskhinvali, Geor.	138	42°13'N	43°56'E
Tsolo, S. Afr. (tsŏ'lō)	187c	31°19'S	28°47'E
Tsomo, S. Afr.	187c	32°03'S	27°49'E
Tsomo, r., S. Afr.	187c	31°53'S	27°48'E
Tsu, Japan (tsōō)	166	34°42'N	136°31'E
Tsuchiura, Japan (tsōō'chē-ōō-rä)	167	36°04'N	140°09'E
Tsuda, Japan (tsōō'dä)	167b	34°48'N	135°43'E
Tsugaru Kaikyō, strt., Japan	161	41°25'N	140°20'E
Tsumeb, Nmb. (tsōō'mĕb)	186	19°10'S	17°45'E
Tsunashima, Japan (tsōō'nä-shē'mä)	167a	35°32'N	139°37'E
Tsuruga, Japan (tsōō'rò-gä)	166	35°39'N	136°04'E
Tsurugi San, mtn., Japan (tsōō'rò-gē sän)	166	33°52'N	134°07'E
Tsuruoka, Japan (tsōō'rò-ō'kä)	166	38°43'N	139°51'E
Tsurusaki, Japan (tsōō'rò-sä'kē)	167	33°15'N	131°42'E
Tsu Shima, is., Japan (tsōō shē'mä)	161	34°28'N	129°30'E
Tsushima Strait, strt., Asia	161	34°00'N	129°00'E
Tsuwano, Japan (tsōō'wà-nô')	167	34°28'N	131°47'E
Tsuyama, Japan (tsōō'yä-mä')	166	35°05'N	134°00'E
Tua, r., Port. (tōō'ä)	128	41°23'N	7°18'W
Tualatin, r., Or., U.S. (tōō'á-lă-tĭn)	74c	45°25'N	122°54'W
Tuamoto, Îles, Fr. Poly. (tōō-ä-mō'tōō)	195	19°00'S	141°20'W
Tuapse, Russia (tò'áp-sĕ)	137	44°00'N	39°10'E
Tuareg, hist. reg., Alg.	184	21°26'N	2°51'E
Tubarão, Braz. (tōō-bä-roun')	102	28°23'S	48°56'W
Tübingen, Ger. (tü'bĭng-ĕn)	124	48°33'N	9°05'E
Tubinskiy, Russia (tû bĭn'skĭ)	142a	52°53'N	58°15'E
Tubruq, Libya	185	32°03'N	24°04'E
Tucacas, Ven. (tōō-kä'käs)	100	10°48'N	68°20'W
Tucker, Ga., U.S. (tŭk'ēr)	68c	33°51'N	84°13'W
Tucson, Az., U.S. (tōō-sŏn')	64	32°15'N	111°00'W
Tucumán, Arg. (tōō-kōō-män')	102	26°52'S	65°08'W
Tucumán, prov., Arg.	102	26°30'S	65°30'W
Tucumcari, N.M., U.S. (tōō'kŭm-kâr-ē)	78	35°11'N	103°43'W
Tucupita, Ven. (tōō-kōō-pē'tä)	100	9°00'N	62°09'W
Tudela, Spain (tōō-dhā'lä)	118	42°03'N	1°37'W
Tugaloo, r., Ga., U.S. (tŭg'á-lōō)	82	34°35'N	83°05'W
Tugela, r., S. Afr. (tōō-gel'á)	187c	28°50'S	30°52'E
Tugela Ferry, S. Afr.	187c	28°44'S	30°27'E
Tug Fork, r., U.S. (tŭg)	66	37°50'N	82°30'W
Tuguegarao, Phil. (tōō-gā-gà-rä'ō)	168	17°37'N	121°44'E
Tuhai, r., China (tōō-hī)	162	37°05'N	116°56'E
Tui, Slvn.	128	42°03'N	8°38'W
Tuinplaas, S. Afr.	192c	24°54'S	28°46'E
Tujunga, Ca., U.S. (tōō-jŭŋ'gá)	75a	34°15'N	118°16'W
Tukan, Russia (tōō'kàn)	142a	53°52'N	57°25'E
Tukangbesi, Kepulauan, is., Indon.	169	6°00'S	124°15'E
Tūkrah, Libya	185	32°34'N	20°47'E
Tuktoyaktuk, Can.	50	69°32'N	132°37'W
Tuktut Nogait National Park, rec., Can.	52	69°00'N	122°00'W
Tukums, Lat. (tò'kòms)	136	56°57'N	23°09'E
Tukuyu, Tan. (tōō-kōō'yä)	186	9°13'S	33°43'E
Tukwila, Wa., U.S. (tŭk'wĭ-lá)	74a	47°28'N	122°16'W
Tula, Mex. (tōō'lä)	88	20°04'N	99°22'W
Tula, Russia (tōō'lá)	136	54°12'N	37°37'E
Tula, prov., Russia	132	53°45'N	37°19'E
Tula, r., Mex. (tōō'lä)	88	20°40'N	99°27'W
Tulagai, i., Sol. Is. (tōō-lä'gē)	175	9°15'S	160°17'E
Tulaghi, Sol. Is.	170e	9°06'S	160°09'E
Tulalip, Wa., U.S. (tū-lă'lĭp)	74a	48°04'N	122°18'W
Tulalip Indian Reservation, I.R., Wa., U.S.	74a	48°06'N	122°16'W
Tulancingo, Mex. (tōō-län-sĭŋ'gō)	86	20°04'N	98°24'W
Tulangbawang, r., Indon.	168	4°17'S	105°00'E
Tulare, Ca., U.S. (tōō-lä'rà) (tul-âr')	76	36°12'N	119°22'W
Tulare Lake Bed, l., Ca., U.S.	76	35°57'N	120°18'W
Tularosa, N.M., U.S. (tōō-lá-rō'zá)	77	33°05'N	106°05'W
Tulcán, Ec. (tōōl-kän')	100	0°44'N	77°52'W
Tulcea, Rom. (tól'chà)	119	45°10'N	28°47'E
Tul'chyn, Ukr.	137	48°42'N	28°53'E
Tulcingo, Mex. (tōōl-sĭŋ'gō)	88	18°03'N	98°27'W
Tule, r., Ca., U.S. (tōōl'lå)	76	36°08'N	118°50'W
Tule River Indian Reservation, I.R., Ca., U.S. (tōō'lå)	76	36°00'N	118°40'W
Tuli, Zimb. (tōō'lē)	186	20°58'S	29°12'E
Tulia, Tx., U.S. (tōō'lĭ-á)	78	34°32'N	101°46'W
Tulik Volcano, vol., Ak., U.S. (tò'lĭk)	63a	53°28'N	168°10'W
Tūlkarm, W.B. (tōōl kärm)	153a	32°19'N	35°02'E
Tullahoma, Tn., U.S. (tŭl-á-hō'má)	82	35°21'N	86°12'W
Tullamore, Ire. (tŭl-á-mōr')	120	53°15'N	7°29'W
Tulle, Fr. (tül)	126	45°15'N	1°45'E
Tulln, Aus. (tòln)	124	48°21'N	16°04'E
Tullner Feld, reg., Aus.	115e	48°20'N	15°59'E
Tulpetlac, Mex. (tōōl-på-tlàk')	89a	19°33'N	99°04'W
Tulsa, Ok., U.S. (tŭl'sá)	65	36°08'N	95°58'W
Tulum, Mex. (tōō-lò'm)	90a	20°17'N	87°26'W
Tulun, Russia (tò-lōōn')	135	54°29'N	100°43'E
Tuma, r., Nic. (tōō'mä)	90	13°07'N	85°32'W
Tumba, Lac, l., D.R.C. (tòm'bä)	186	0°50'S	17°45'E
Tumbes, Peru (tōō'm-bĕs)	100	3°39'S	80°27'W
Tumbiscatío, Mex. (tōōm-bē-ská-tē'ō)	88	18°32'N	102°23'W
Tumbo, i., Can.	74d	48°49'N	123°04'W
Tumacocori National Monument, rec., Az., U.S. (tōō-mä-kä'kà-rē)	77	31°36'N	110°20'W
Tumen, China (tōō-mŭn)	164	43°00'N	129°50'E
Tumen, r., Asia	166	42°08'N	128°40'E
Tumeremo, Ven. (tōō-må-rā'mō)	101	7°15'N	61°28'W
Tumkūr, India	159	13°22'N	77°05'E
Tumuc-Humac Mountains, mts., S.A. (tōō-mòk'ōō-mäk')	101	2°15'N	54°50'W
Tunas de Zaza, Cuba (tōō'näs dā zä'zä)	92	21°40'N	79°35'W
Tunbridge Wells, Eng., U.K. (tŭn'brĭj welz')	121	51°05'N	0°09'E
Tunduru, Tan.	191	11°07'S	37°21'E
Tungabhadra Reservoir, res., India	159	15°26'N	75°57'E
Tuni, India	159	17°29'N	82°38'E
Tunica, Ms., U.S. (tū'nĭ-kà)	82	34°41'N	90°23'W
Tunis, Tun. (tū'nĭs)	184	36°59'N	10°06'E
Tunis, Golfe de, b., Tun.	118	37°06'N	10°43'E
Tunisia, nation, Afr. (tu-nĭzh'ē-á)	184	35°00'N	10°11'E
Tunja, Col. (tōō'n-hä)	100	5°32'N	73°19'W
Tunkhannock, Pa., U.S. (tŭnk-hăn'ŭk)	67	41°35'N	75°55'W
Tunnel, r., Wa., U.S. (tŭn'ĕl)	74a	47°48'N	123°04'W
Tuoji Dao, i., China (twô-jyē dou)	162	38°11'N	120°45'E
Tuolumne, r., Ca., U.S. (twô-lŭm'nē)	76	37°35'N	120°37'W
Tuostakh, r., Russia	141	67°09'N	137°30'E
Tupelo, Ms., U.S. (tū'pē-lō)	82	34°14'N	88°43'W
Tupinambaranas, Ilha, i., Braz.	101	3°04'S	58°09'W
Tupiza, Bol. (tōō-pē'zä)	100	21°26'S	65°43'W
Tupper Lake, N.Y., U.S. (tŭp'ēr)	67	44°15'N	74°25'W
Tüpqaraghan tübegi, pen., Kaz.	137	44°30'N	50°40'E
Tupungato, Cerro, vol., S.A.	102	33°30'S	69°52'W
Tuquerres, Col. (tōō-kĕ'r-rĕs)	100	1°12'N	77°44'W
Tura, Russia (tòr'á)	135	64°08'N	99°58'E
Turbio, r., Mex. (tōōr-byô)	88	20°28'N	101°40'W
Turbo, Col. (tōō'bō)	100	8°02'N	76°43'W
Turda, Rom. (tòr'dá)	125	46°35'N	23°47'E
Turfan Depression, depr., China	160	42°16'N	90°00'E
Turffontein, neigh., S. Afr.	187b	26°15'S	28°02'E
Tŭrgovishte, Blg.	131	43°14'N	26°36'E
Turgutlu, Tur.	137	38°30'N	27°20'E
Türi, Est. (tü'rĭ)	123	58°49'N	25°29'E
Turia, r., Spain (tōō'ryä)	128	40°12'N	1°18'W
Turicato, Mex. (tōō-rē-kä'tō)	88	19°03'N	101°24'W
Turiguano, i., Cuba (tōō-rē-gwä'nō)	92	22°20'N	78°35'W
Turin, Italy	110	45°05'N	7°44'E
Turiya, r., Ukr.	125	51°18'N	24°55'E
Turka, Ukr. (tòr'kà)	125	49°10'N	23°02'E
Turkestan, hist. reg., Asia	134	43°27'N	62°14'E
Turkey, nation, Asia	111	38°45'N	32°00'E
Turkey, r., Ia., U.S. (tûrk'ē)	71	43°20'N	92°16'W
Türkistan, Kaz.	139	44°00'N	68°00'E
Turkmenbashy, Turkmen.	139	40°00'N	52°50'E
Turkmenistan, nation, Asia	134	40°46'N	56°01'E
Turks, is., T./C. Is. (tûrks)	87	21°40'N	71°45'W
Turks Island Passage, strt., T./C. Is.	93	21°15'N	71°25'W
Turku, Fin. (tòrgökò)	110	60°28'N	22°12'E
Turlock, Ca., U.S. (tûr'lŏk)	76	37°30'N	120°51'W
Turneffe, i., Belize	86	17°25'N	87°43'W
Turner, Ks., U.S. (tûr'nēr)	75f	39°05'N	94°42'W
Turner Sound, strt., Bah.	92	24°20'N	78°05'W
Turners Peninsula, pen., S.L.	188	7°20'N	12°40'W
Turnhout, Bel. (türn-hout')	121	51°19'N	4°58'E
Turnov, Czech Rep. (tòr'nôf)	124	50°36'N	15°12'E

ăt; finăl; rāte; senāte; ärm; åsk; sofá; fâre; ch-choose; dh-as th in other; bē; ēvent; bĕt; recĕnt; cratēr; g-gō; gh-guttural g; bĭt; ĭ-short neutral; rīde; κ-guttural k as ch in German ich;

PRONOUNCING INDEX

Turnu Măgurele – Urdaneta 313

PLACE (Pronunciation)	PAGE	LAT.	LONG.
Turnu Măgurele, Rom.	119	43°54′N	24°49′E
Turpan, China (to͞o-är-pän)	160	43°06′N	88°41′E
Turquino, Pico, mtn., Cuba (pē′kō dä to͞or-kē′nō)	92	20°00′N	76°50′W
Turrialba, C.R. (to͞or-ryä′l-bä)	91	9°54′N	83°41′W
Turtkul′, Uzb. (tòrt-kòl′)	139	41°28′N	61°02′E
Turtle, r., Can.	57	49°20′N	92°30′W
Turtle Bay, b., Tx., U.S.	81a	29°48′N	94°38′W
Turtle Creek, r., S.D., U.S.	70	44°40′N	98°53′W
Turtle Mountain Indian Reservation, I.R., N.D., U.S.	70	48°45′N	99°57′W
Turtle Mountains, mts., N.D., U.S.	70	48°57′N	100°11′W
Turukhansk, Russia (to͞o-ro͞o-känsk′)	134	66°03′N	88°39′E
Tuscaloosa, Al., U.S. (tŭs-kȧ-lo͞o′sȧ)	65	33°10′N	87°35′W
Tuscarora, Nv., U.S. (tŭs-kȧ-rō′rȧ)	72	41°18′N	116°15′W
Tuscarora Indian Reservation, I.R., N.Y., U.S.	69c	43°10′N	78°51′W
Tuscola, Il., U.S. (tŭs-kō-lȧ)	66	39°50′N	88°20′W
Tuscumbia, Al., U.S. (tŭs-kŭm′bĭ-ȧ)	82	34°41′N	87°42′W
Tushino, Russia (to͞o′shĭ-nô)	142b	55°51′N	37°24′E
Tuskegee, Al., U.S. (tŭs-kē′gē)	82	32°25′N	85°40′W
Tustin, Ca., U.S. (tŭs′tĭn)	75a	33°44′N	117°49′W
Tutayev, Russia (to͞o-tȧ-yĕf′)	136	57°53′N	39°34′E
Tutbury, Eng., U.K. (tŭt′bẽr-ė)	114a	52°52′N	1°51′W
Tuticorin, India (to͞o-tē-kô-rĭn′)	159	8°51′N	78°09′E
Tutitlán, Mex. (to͞o-tē-tlä′n)	89a	19°38′N	99°10′W
Tutóia, Braz. (to͞o-tō′yȧ)	101	2°42′S	42°21′W
Tutrakan, Blg.	119	44°02′N	26°36′E
Tuttle Creek Reservoir, res., Ks., U.S.	79	39°30′N	96°38′W
Tuttlingen, Ger. (tòt′lĭng-ĕn)	124	47°58′N	8°50′E
Tutuila, i., Am. Sam.	170a	14°18′S	170°42′W
Tutwiler, Ms., U.S. (tŭt′wĭ-lẽr)	82	34°01′N	90°25′W
Tuva, prov., Russia	140	51°15′N	90°45′E
Tuvalu, nation, Oc.	3	5°20′S	174°00′E
Tuwayq, Jabal, mts., Sau. Ar.	154	20°45′N	46°30′E
Tuxedo Park, N.Y., U.S. (tŭk-sē′dō pärk)	68a	41°11′N	74°11′W
Tuxford, Eng., U.K. (tŭks′fẽrd)	114a	53°14′N	0°54′W
Túxpan, Mex. (to͞os′pän)	88	19°34′N	103°22′W
Túxpan, Mex.	86	20°57′N	97°26′W
Túxpan, r., Mex. (to͞os′pän)	89	20°55′N	97°52′W
Túxpan, Arrecife, i., Mex. (är-rĕ-sē′fĕ-to͞o′x-pä′n)	89	21°01′N	97°12′W
Tuxtepec, Mex. (tòs-tȧ-pĕk′)	89	18°06′N	96°09′W
Tuxtla Gutiérrez, Mex. (tòs′tlä go͞o-tyâr′rĕs)	86	16°44′N	93°08′W
Tuy, r., Ven. (to͞o′ē)	101b	10°15′N	66°03′W
Tuyra, r., Pan. (to͞o-ē′rȧ)	91	7°55′N	77°37′W
Tuz Gölü, l., Tur.	136	38°45′N	33°25′E
Tuzigoot National Monument, rec., Az., U.S.	77	34°40′N	111°52′W
Tuzla, Bos. (tòz′lȧ)	119	44°33′N	18°46′E
Tvedestrand, Nor. (tvĭ′dhĕ-stränd)	122	58°39′N	8°54′E
Tveitsund, Nor. (tvåt′sònd)	122	59°03′N	8°29′E
Tver′, Russia	134	56°52′N	35°57′E
Tver′, prov., Russia	132	56°50′N	33°08′E
Tvertsa, r., Russia (tvẽr′tsȧ)	132	56°58′N	35°22′E
Tweed, r., U.K. (twēd)	120	55°32′N	2°35′W
Tweeling, S. Afr. (twē′lĭng)	192c	27°34′S	28°31′E
Twenty Mile Creek, r., Can. (twĕn′tĭ mīl)	62d	43°09′N	79°49′W
Twickenham, Eng., U.K. (twĭk′′n-ȧm)	114b	51°26′N	0°20′W
Twillingate, Can. (twĭl′ĭn-gāt)	53a	49°39′N	54°46′W
Twin Bridges, Mt., U.S. (twĭn brĭ-jĕz)	73	45°34′N	112°17′W
Twin Falls, Id., U.S. (fôls)	64	42°33′N	114°29′W
Twinsburg, Oh., U.S. (twĭnz′bŭrg)	69d	41°19′N	81°26′W
Twitchell Reservoir, res., Ca., U.S.	76	34°50′N	120°10′W
Two Butte Creek, r., Co., U.S. (to͞o būt)	78	37°39′N	102°45′W
Two Harbors, Mn., U.S.	71	47°00′N	91°42′W
Two Prairie Bay, Ar., U.S. (prâ′rĭ bī o͞o′)	79	34°48′N	92°07′W
Two Rivers, Wi., U.S. (rĭv′ẽrz)	71	44°09′N	87°36′W
Tyabb, Austl.	173a	38°16′S	145°11′E
Tylden, S. Afr. (tĭl-dĕn)	187c	32°08′S	27°06′E
Tyldesley, Eng., U.K. (tĭldz′lė)	114a	53°32′N	2°28′W
Tyler, Mn., U.S. (tī′lẽr)	70	44°18′N	96°08′W
Tyler, Tx., U.S.	65	32°21′N	95°19′W
Tylertown, Ms., U.S. (tī′lẽr-toun)	82	31°08′N	90°06′W
Tylihul, r., Ukr.	133	47°25′N	30°27′E
Tyndall, S.D., U.S. (tĭn′dȧl)	70	42°58′N	97°52′W
Tyndinskiy, Russia	135	55°22′N	124°45′E
Tyne, r., Eng., U.K. (tīn)	120	54°59′N	1°56′W
Tynemouth, Eng., U.K. (tīn′mŭth)	116	55°04′N	1°39′W
Tyngsboro, Ma., U.S. (tĭnj-bûr′ō)	61a	42°40′N	71°27′W
Tynset, Nor. (tün′sĕt)	116	62°17′N	10°45′E
Tyre *see* Şūr, Leb.	153a	33°16′N	35°13′E
Tyrifjorden, l., Nor.	122	60°03′N	10°25′E
Tyrnavos, Grc.	131	39°50′N	22°14′E
Tyrone, Pa., U.S.	67	40°40′N	78°15′W
Tyrrell, Lake, l., Austl. (tĭr′ĕll)	176	35°12′S	143°00′E
Tyrrhenian Sea, sea, Italy (tĭr-rē′nĭ-ȧn)	112	40°10′N	12°15′E
Tyukalinsk, Russia (työ-kȧ-lĭnsk′)	134	56°03′N	71°43′E
Tyukyan, r., Russia (työk′yȧn)	141	65°42′N	116°09′E
Tyuleniy, i., Russia	137	44°30′N	48°00′E
Tyumen′, Russia (tyo͞o-mĕn′)	134	57°02′N	65°28′E
Tzucacab, Mex. (tzo͞o-kä-kä′b)	90a	20°06′N	89°03′W
U			
Uaupés, Braz. (wä-o͞o′pās)	100	0°02′S	67°03′W
Ubangi, r., Afr. (o͞o-bäŋ′gē)	185	3°00′N	18°00′E
Ubatuba, Braz. (o͞o-bȧ-to͞o′bȧ)	99a	23°25′S	45°06′W
Ubeda, Spain (o͞o′bå-dä)	128	38°01′N	3°23′W
Uberaba, Braz. (o͞o-bå-rä′bȧ)	101	19°47′S	47°47′W
Uberlândia, Braz. (o͞o-bĕr-lä′n-dyä)	101	18°54′S	48°11′W
Ubombo, S. Afr. (o͞o-bôm′bō)	186	27°33′S	32°13′E
Ubon Ratchathani, Thai. (o͞o′bŭn rä′chätá-nē)	168	15°15′N	104°52′E
Ubort′, r., Eur. (o͞o-bôrt′)	133	51°18′N	27°43′E
Ubrique, Spain (o͞o-brē′kå)	128	36°43′N	5°36′W
Ubundu, D.R.C.	186	0°21′S	25°29′E
Ucayali, r., Peru (o͞o′kä-yä′lē)	100	8°58′S	74°13′W
Uccle, Bel. (ü′kl′)	115a	50°48′N	4°17′E
Uchaly, Russia (û-chä′lĭ)	142a	54°22′N	59°28′E
Uchiko, Japan (o͞o′chē-kō)	167	33°30′N	132°39′E
Uchinoura, Japan (o͞o′chē-nō-o͞o′rä)	167	31°16′N	131°03′E
Uchinskoye Vodokhranilishche, res., Russia	142b	56°08′N	37°44′E
Uchiura-Wan, b., Japan (o͞o′chē-o͞o′rä wän)	166	42°20′N	140°44′E
Uchur, r., Russia (ò-chòr′)	141	57°25′N	130°35′E
Uda, r., Russia	141	53°54′N	131°29′E
Uda, r., Russia (ò′dȧ)	141	52°28′N	110°51′E
Udai, r., Ukr.	133	50°45′N	32°13′E
Udaipur, India (ò-dŭ′ē-po͞or)	158	24°41′N	73°41′E
Uddevalla, Swe. (o͞od′dĕ-väl-ȧ)	116	58°21′N	11°55′E
Udine, Italy (o͞o′dē-nå)	118	46°05′N	13°14′E
Udmurtia, prov., Russia	136	57°00′N	53°00′E
Udon Thani, Thai.	168	17°31′N	102°51′E
Udskaya Guba, b., Russia	135	55°00′N	136°30′E
Ueckermünde, Ger.	124	53°43′N	14°01′E
Ueda, Japan (wä′dä)	166	36°26′N	138°16′E
Uele, r., D.R.C. (wā′lȧ)	185	3°55′N	23°30′E
Uelzen, Ger. (ült′sĕn)	124	52°58′N	10°34′E
Ufa, Russia (ò′fȧ)	134	54°45′N	55°57′E
Ufa, r., Russia	136	56°00′N	57°05′E
Ugab, r., Nmb. (o͞o′gäb)	186	21°10′S	14°00′E
Ugalla, r., Tan. (o͞o-gä′lä)	186	6°15′S	32°30′E
Uganda, nation, Afr. (o͞o-gän′dä) (û-găn′dȧ)	185	2°00′N	32°28′E
Ugashik Lake, l., Ak., U.S. (o͞o′gá-shĕk)	63	57°36′N	157°10′W
Ugie, S. Afr. (ò′jē)	187c	31°13′S	28°14′E
Uglegorsk, Russia (o͞o-glĕ-gôrsk)	135	49°00′N	142°31′E
Ugleural′sk, Russia (òg-lē-ò-rálsk′)	142a	58°58′N	57°35′E
Uglich, Russia (o͞og-lĕch′)	132	57°33′N	38°19′E
Uglitskiy, Russia (o͞og-lĭt′skĭ)	142a	53°50′N	60°18′E
Uglovka, Russia (o͞og-lôf′kȧ)	132	58°14′N	33°24′E
Ugra, r., Russia (o͞og′rȧ)	136	54°43′N	34°20′E
Ugŭrchin, Blg.	131	43°06′N	24°23′E
Uhrichsville, Oh., U.S. (ū′rĭks-vĭl)	66	40°25′N	81°20′W
Uíge, Ang.	186	7°37′S	15°03′E
Uiju, Kor., N. (ò′ējo͞o)	161	40°09′N	124°33′E
Uinkaret Plateau, plat., Az., U.S. (ū-ĭn′kâr-ĕt)	77	36°43′N	113°15′W
Uinskoye, Russia (ò-ĭn′skô-yĕ)	142a	56°53′N	56°25′E
Uinta, r., Ut., U.S. (ū-ĭn′tȧ)	77	40°25′N	109°55′W
Uintah and Ouray Indian Reservation, I.R., Ut., U.S.	77	40°20′N	110°20′W
Uinta Mountains, mts., Ut., U.S.	64	40°35′N	111°00′W
Uitenhage, S. Afr.	186	33°46′S	25°26′E
Uithoorn, Neth.	115a	52°13′N	4°49′E
Uji, Japan (o͞o′jē)	167b	34°53′N	135°49′E
Ujiji, Tan. (o͞o-jē′jē)	186	4°55′S	29°41′E
Ujjain, India (o͞o-jŭĕn)	155	23°18′N	75°37′E
Ujungpandang, Indon.	168	5°08′S	119°28′E
Ukerewe Island, i., Tan.	191	2°00′S	32°40′E
Ukhta, Russia (o͞ok′tȧ)	136	65°22′N	31°30′E
Ukhta, Russia	136	63°08′N	53°42′E
Ukiah, Ca., U.S. (ū-kī′ȧ)	76	39°09′N	123°12′W
Ukmerge, Lith. (òk′mĕr-ghå)	136	55°16′N	24°45′E
Ukraine, nation, Eur.	134	49°15′N	30°15′E
Uku, i., Japan (o͞ok′o͞o)	167	33°18′N	129°02′E
Ulaangom, Mong.	160	50°23′N	92°14′E
Ulan Bator (Ulaanbaatar), Mong.	160	47°56′N	107°00′E
Ulan-Ude, Russia (o͞o′län o͞o′dȧ)	135	51°59′N	107°41′E
Ulchin, Kor., S. (o͞ol′chĕn′)	166	36°57′N	129°26′E
Ulcinj, Serb. (o͞ol′tsĕn′)	119	41°56′N	19°15′E
Ulhās, r., India	159b	19°13′N	73°03′E
Ulhāsnagar, India	158	19°10′N	73°07′E
Uliastay, Mong.	160	47°49′N	97°00′E
Ulindi, r., D.R.C. (o͞o-lĭn′dė)	186	1°55′S	26°17′E
Ulla, Bela. (òl′ȧ)	132	55°14′N	29°15′E
Ulla, r., Bela.	132	54°58′N	29°03′E
Ulla, r., Spain (o͞o′lä)	128	42°45′N	8°33′W
Ullŭng, i., Kor., S. (o͞ol′lòng′)	166	37°29′N	130°50′E
Ulm, Ger. (òlm)	117	48°24′N	9°59′E
Ulmer, Mount, mtn., Ant. (ŭl′mûr′)	178	77°30′S	86°00′W
Ulricehamn, Swe. (òl-rē′sĕ-häm)	122	57°49′N	13°23′E
Ulsan, Kor., S. (o͞ol′sän′)	166	35°35′N	129°22′E
Ulster, hist. reg., Eur. (ŭl′stẽr)	120	54°41′N	7°10′W
Ulua, r., Hond. (o͞o-lo͞o′ȧ)	90	15°49′N	87°45′W
Ulubaria, India	158a	22°27′N	88°09′E
Ulukışla, Tur. (o͞o-lo͞o-kēsh′lȧ)	119	36°40′N	34°30′E
Ulunga, Russia (ò-lo͞on′gȧ)	166	46°16′N	136°29′E
Ulungur, r., China (o͞o-lo͞on-gŭr)	160	46°31′N	88°00′E
Uluru (Ayers Rock), mtn., Austl.	174	25°23′S	131°05′E
Ulu-Telyak, Russia (o͞o lò′tĕlyăk)	142a	54°54′N	57°01′E
Ulverstone, Austl. (ŭl′vẽr-stŭn)	175	41°20′S	146°22′E
Ul′yanovka, Russia	142c	59°38′N	30°47′E
Ul′yanovsk, Russia (o͞o-lyä′nôfsk)	134	54°20′N	48°24′E
Ulysses, Ks., U.S. (ū-lĭs′ēz)	78	37°34′N	101°25′W
Umán, Mex. (o͞o-män′)	90a	20°52′N	89°44′W
Uman′, Ukr. (ò-män′)	137	48°44′N	30°13′E
Umatilla Indian Reservation, I.R., Or., U.S. (ū-mȧ-tĭl′ȧ)	72	45°38′N	118°35′W
Umberpāda, India	159b	19°28′N	73°04′E
Umbria, hist. reg., Italy (ŭm′brĭ-ȧ)	130	42°53′N	12°22′E
Umeälven, r., Swe.	112	64°57′N	18°51′E
Umhlatuzi, r., S. Afr. (òm′hlä-to͞o′zĭ)	187c	28°47′S	31°17′E
Umiat, Ak., U.S. (o͞o′mĭ-ăt)	64a	69°20′N	152°28′W
Umkomaas, S. Afr. (òm-kō′mäs)	187c	30°12′S	30°48′E
Umnak, i., Ak., U.S. (o͞om′nák)	64b	53°10′N	169°08′W
Umnak Pass, Ak., U.S.	63a	53°10′N	168°04′W
Umniati, r., Zimb.	186	17°08′S	29°11′E
Umpqua, r., Or., U.S. (ŭmp′kwȧ)	72	43°42′N	123°50′W
Umtata, S. Afr. (òm-tä′tä)	186	31°36′S	28°47′E
Umtentweni, S. Afr.	187c	30°41′S	30°29′E
Umzimkulu, S. Afr. (òm-zēm-ko͞o′lo͞o)	187c	30°12′S	29°53′E
Umzinto, S. Afr. (òm-zĭn′tō)	187c	30°19′S	30°41′E
Una, r., Serb. (o͞o′nȧ)	130	44°38′N	16°10′E
Unalakleet, Ak., U.S. (ū-nȧ-lăk′lēt)	63	63°50′N	160°42′W
Unalaska, Ak., U.S. (ū-nȧ-lás′kȧ)	63a	53°30′N	166°20′W
Unare, r., Ven.	101b	9°45′N	65°12′W
Unare, Laguna de, l., Ven. (lä-gò′nä-de-o͞o-ná′rĕ)	101b	10°07′N	65°23′W
Unayzah, Sau. Ar.	154	25°50′N	44°02′E
Uncas, Can. (ŭŋ′kȧs)	62g	53°30′N	113°02′W
Uncia, Bol. (o͞on′sē-ä)	100	18°28′S	66°32′W
Uncompahgre, r., Co., U.S.	77	38°20′N	107°45′W
Uncompahgre Peak, mtn., Co., U.S. (ŭn-kŭm-pä′grē)	77	38°00′N	107°30′W
Uncompahgre Plateau, plat., Co., U.S.	77	38°40′N	108°40′W
Underberg, S. Afr. (ŭn′dẽr-bûrg)	187c	29°51′S	29°32′E
Unecha, Russia (ò-nĕ′chȧ)	132	52°51′N	32°44′E
Ungava, Péninsule d′, pen., Can.	53	59°55′N	74°00′W
Ungava Bay, b., Can. (ŭŋ-gā′vȧ)	53	59°46′N	67°18′W
União da Vitória, Braz. (o͞o-nē-ouN′ dä vē-tô′ryä)	102	26°17′S	51°13′W
Unije, i., Serb. (o͞o′nē-yĕ)	130	44°39′N	14°10′E
Unimak, i., Ak., U.S. (o͞o-nē-mák′)	63	54°30′N	163°35′W
Unimak Pass, Ak., U.S.	63a	54°22′N	165°22′W
Union, Mo., U.S.	79	38°28′N	90°59′W
Union, Ms., U.S. (ūn′yŭn)	82	32°35′N	89°07′W
Union, N.C., U.S.	83	34°42′N	81°40′W
Union, Or., U.S.	72	45°13′N	117°52′W
Union City, Ca., U.S.	74b	37°36′N	122°01′W
Union City, In., U.S.	66	40°10′N	85°00′W
Union City, Mi., U.S.	66	42°00′N	85°10′W
Union City, Pa., U.S.	67	41°50′N	79°50′W
Union City, Tn., U.S.	82	36°25′N	89°04′W
Unión de Reyes, Cuba	92	22°45′N	81°30′W
Unión de San Antonio, Mex.	88	21°07′N	101°56′W
Unión de Tula, Mex.	88	19°57′N	104°14′W
Union Grove, Wi., U.S. (ūn-yŭn grōv)	69a	42°41′N	88°03′W
Unión Hidalgo, Mex. (ē-dä′lgô)	89	16°29′N	94°51′W
Union Point, Ga., U.S.	82	33°37′N	83°08′W
Union Springs, Al., U.S. (sprĭngz)	82	32°08′N	85°43′W
Uniontown, Al., U.S. (ūn′yŭn-toun)	82	32°26′N	87°30′W
Uniontown, Oh., U.S.	69d	40°58′N	81°25′W
Uniontown, Pa., U.S.	67	39°55′N	79°45′W
Unionville, Mo., U.S. (ūn′yŭn-vĭl)	79	40°28′N	92°58′W
Unisan, Phil. (o͞o-nē′sän)	169a	13°50′N	121°59′E
United Arab Emirates, nation, Asia	154	24°00′N	54°00′E
United Kingdom, nation, Eur.	110	56°30′N	1°40′W
United States, nation, N.A.	64	38°00′N	110°00′W
Unity, Can.	56	52°27′N	109°10′W
Universal, In., U.S. (ū-nĭ-vûr′sȧl)	66	39°35′N	87°30′W
University City, Mo., U.S. (ū′nĭ-vûr′sĭ-tĭ)	75e	38°40′N	90°19′W
University Park, Tx., U.S.	75c	32°51′N	96°48′W
Unna, Ger. (o͞o′nä)	127c	51°32′N	7°41′E
Uno, Canal Numero, can., Arg.	99c	36°43′S	58°14′W
Unterhaching, Ger. (o͞on′tĕr-hä-kĕng)	115d	48°03′N	11°38′E
Ünye, Tur. (ün′yĕ)	119	41°00′N	37°10′E
Unzha, r., Russia (òn′zhȧ)	136	57°45′N	44°10′E
Upa, r., Russia (ò′pȧ)	132	53°54′N	36°48′E
Upata, Ven. (o͞o-pä′tä)	100	7°58′N	62°27′W
Upemba, Parc National de l′, rec., D.R.C.	191	9°10′S	26°15′E
Upington, S. Afr. (ŭp′ĭng-tŭn)	186	28°25′S	21°15′E
Upland, Ca., U.S. (ŭp′lȧnd)	75a	34°06′N	117°38′W
Upolu, i., Samoa	170a	13°55′S	171°45′W
Upolu Point, c., Hi., U.S. (o͞o-pô′lo͞o)	84a	20°15′N	155°48′W
Upper Arrow Lake, l., Can. (ăr′ō)	55	50°30′N	117°55′W
Upper Darby, Pa., U.S. (där′bĭ)	68f	39°58′N	75°16′W
Upper des Lacs, l., N.A. (dĕ läk)	70	48°58′N	101°55′W
Upper Kapuas Mountains, mts., Asia	168	1°45′N	112°06′E
Upper Klamath Lake, l., Or., U.S.	72	42°23′N	122°55′W
Upper Lake, l., Nv., U.S. (ŭp′ẽr)	72	41°42′N	119°59′W
Upper Marlboro, Md., U.S. (ŭpẽr mărl′bōrō)	68e	38°49′N	76°46′W
Upper Mill, Wa., U.S. (mĭl)	74a	47°11′N	121°55′W
Upper Red Lake, l., Mn., U.S. (rĕd)	71	48°14′N	94°53′W
Upper Sandusky, Oh., U.S. (săn-dŭs′kė)	66	40°50′N	83°20′W
Upper San Leandro Reservoir, res., Ca., U.S. (ŭp′ẽr săn lē-ăn′drō)	74b	37°47′N	122°04′W
Upper Volta *see* Burkina Faso, nation, Afr.	184	13°00′N	2°00′W
Uppingham, Eng., U.K. (ŭp′ĭng-ȧm)	114a	52°35′N	0°43′W
Uppsala, Swe. (o͞op′sȧ-lä)	110	59°53′N	17°39′E
Uptown, Ma., U.S. (ŭp′toun)	61a	42°10′N	71°36′W
Uraga, Japan (o͞o′rȧ-gȧ′)	167a	35°15′N	139°43′E
Ural, r., (ò-räl′′) (ū-rôl)	134	48°00′N	51°00′E
Urals, mts., Russia	134	56°28′N	58°13′E
Uran, India (o͞o-rän′)	159b	18°53′N	72°46′E
Uranium City, Can.	50	59°34′N	108°59′W
Urawa, Japan (o͞o′rȧ-wä′)	166	35°52′N	139°39′E
Urayasu, Japan (o͞o′rä-yä′so͞o)	167a	35°40′N	139°54′W
Urazovo, Russia (ò-rä′zô-vô)	133	50°08′N	38°03′E
Urbana, Il., U.S. (ûr-băn′ȧ)	66	40°10′N	88°15′W
Urbana, Oh., U.S.	66	40°05′N	83°50′W
Urbino, Italy (o͞or-bē′nô)	130	43°43′N	12°37′E
Urdaneta, Phil. (o͞or-dä-nä′tä)	169a	15°59′N	120°34′E

ng-sing; ŋ-baŋk; N-nasalized n; nŏd; cŏmmit; ōld; ôbey; ôrder; oi-boil; fo͞od; ò-as oo in foot; ou-out; s-soft; sh-dish; th-thin; pūre; ûnite; ûrn; stŭd; circŭs; ü-as in French tu; ′-indeterminate vowel.

314 Urdinarrain - Vega PRONOUNCING INDEX

PLACE (Pronunciation)	PAGE	LAT.	LONG.
Urdinarrain, Arg. (o͞or-dē-när-rāē'n)	99c	32°43'S	58°53'W
Uritsk, Russia (o͞o'rĭtsk)	142c	59°50'N	30°11'E
Urla, Tur. (ór'lȧ)	131	38°20'N	26°44'E
Urman, Russia (òr'mán)	142a	54°53'N	56°52'E
Urmi, r., Russia (òr'mē)	166	48°50'N	134°00'E
Uromi, Nig.	189	6°44'N	6°18'E
Urrao, Col. (o͞or-rä'ô)	100	6°19'N	76°11'W
Urshel'skiy, Russia (o͞or-shĕl'skē-ē)	132	55°50'N	40°11'E
Ursus, Pol.	125	52°12'N	20°53'E
Urubamba, r., Peru (o͞o-ro͞o-bäm'bä)	100	11°48'S	72°34'W
Uruguaiana, Braz.	102	29°45'S	57°00'W
Uruguay, nation, S.A. (o͞o-ro͞o-gwī') (ū'ro͞o-gwā)	102	32°45'S	56°00'W
Uruguay, r., S.A. (o͞o-ro͞o-gwī')	102	27°05'S	55°15'W
Ürümqi, China (ů-rům-chyē)	160	43°49'N	87°43'E
Urup, i., Russia (ò'ro͝op')	161	46°00'N	150°00'E
Uryupinsk, Russia (òr'yò-pēn-sk')	137	50°50'N	42°00'E
Ürzhar, Kaz.	139	47°28'N	82°00'E
Urziceni, Rom. (ò-zē-chěn'')	131	44°45'N	26°42'E
Usa, Japan	166	33°31'N	131°22'E
Usa, r., Russia (ò'sȧ)	136	66°00'N	58°20'E
Uşak, Tur. (o͞o'shȧk)	119	38°45'N	29°15'E
Usakos, Nmb. (o͞o-sä'kōs)	186	22°00'S	15°40'E
Usambara Mountains, mts., Tan.	191	4°40'S	38°25'E
Usangu Flats, sw., Tan.	191	8°10'S	34°00'E
Ushaki, Russia (o͞o'shá-kĭ)	142c	59°28'N	31°00'E
Ushakovskoye, Russia (ò-shá-kôv'skô-yĕ)	142a	56°18'N	62°23'E
Ushashi, Tan.	191	2°00'S	33°57'E
Ushiku, Japan (o͞o'shē-ko͞o)	167a	35°24'N	140°09'E
Ushimado, Japan (o͞o'shē-mä'dō)	167	34°37'N	134°09'E
Ushuaia, Arg. (o͞o-sho͞o-ī'ä)	102	54°46'S	68°24'W
Usman', Russia (o͞os-mán')	137	52°03'N	39°40'E
Usol'ye, Russia (ò-sô'lyĕ)	142a	59°24'N	56°40'E
Usol'ye-Sibirskoye, Russia (ò-sô'lyĕsĭ' bĕr'skô-yĕ)	140	52°44'N	103°46'E
Uspallata Pass, p., S.A. (o͞os-pä-lyä'tä)	102	32°47'S	70°08'W
Uspanapa, r., Mex. (o͞os-pä-nä'pä)	89	17°43'N	94°14'W
Ussel, Fr. (üs'ĕl)	126	45°33'N	2°17'E
Ussuri, r., Asia (o͞o-so͞o'rē)	141	47°30'N	134°00'E
Ussuriysk, Russia	135	43°48'N	132°09'E
Ust'-Bol'sheretsk, Russia	135	52°41'N	157°00'E
Ustica, Isola di, i., Italy	130	38°43'N	12°11'E
Ústí nad Labem, Czech Rep.	124	50°40'N	14°02'E
Ust'-Izhora, Russia (òst-ēz'hô-rȧ)	142c	59°49'N	30°35'E
Ustka, Pol. (o͞ost'kȧ)	124	54°34'N	16°52'E
Ust'-Kamchatsk, Russia	135	56°13'N	162°18'E
Ust'-Katav, Russia (òst kȧ'táf)	142a	54°55'N	58°12'E
Ust'-Kishert', Russia (òst kē'shĕrt)	142a	57°21'N	57°13'E
Ust'-Kulom, Russia (kò'lŭm)	134	61°38'N	54°00'E
Ust'-Maya, Russia (má'yȧ)	135	60°33'N	134°43'E
Ust' Olenëk, Russia	135	72°52'N	120°15'E
Ust-Ordynskiy, Russia (òst-ôr-dyēnsk'ĭ)	140	52°47'N	104°39'E
Ust' Penzhino, Russia	141	63°00'N	165°10'E
Ust' Port, Russia (òst'pôrt')	134	69°20'N	83°41'E
Ust'-Tsil'ma, Russia (tsĭl'mȧ)	134	65°25'N	52°10'E
Ust'-Tyrma, Russia (tur'mȧ)	135	50°27'N	131°17'E
Ust' Uls, Russia	142a	60°35'N	58°32'E
Ust-Urt, Plateau, plat., Asia	134	44°03'N	54°58'E
Ustynivka, Ukr.	133	47°59'N	32°31'E
Ustyuzhna, Russia (yo͞ozh'nȧ)	136	58°49'N	36°19'E
Usu, China (ů-so͞o)	160	44°28'N	84°07'E
Usuki, Japan (o͞o'so͞o-kē')	167	33°06'N	131°47'E
Usulutan, El Sal. (o͞o-so͞o-lä-tän')	90	13°22'N	88°25'W
Usumacinta, r., N.A. (o͞o'so͞o-mä-sēn'tō)	89	18°24'N	92°30'W
Us'va, Russia (o͞os'vȧ)	142a	58°41'N	57°38'E
Utah, state, U.S. (ū'tô)	64	39°25'N	112°40'W
Utah Lake, l., Ut., U.S.	77	40°10'N	111°55'W
Utan, India	159b	19°17'N	72°43'E
Ute Mountain Indian Reservation, I.R., N.M., U.S.	77	36°57'N	108°34'W
Utena, Lith. (o͞o'tā-nä)	123	55°32'N	25°40'E
Utete, Tan. (o͞o-tā'tȧ)	187	8°05'S	38°47'E
Utica, In., U.S. (ū'tĭ-kȧ)	69h	38°20'N	85°39'W
Utica, N.Y., U.S.	65	43°05'N	75°10'W
Utiel, Spain (o͞o-tyâl')	128	39°34'N	1°13'W
Utika, Mi., U.S. (ū'tĭ-kȧ)	69b	42°37'N	83°02'W
Utik Lake, l., Can.	57	55°16'N	96°00'W
Utikuma Lake, l., Can.	55	55°50'N	115°25'W
Utila, i., Hond. (o͞o-tē'lä)	90	16°07'N	87°05'W
Uto, Japan (o͞o'tō')	166	32°43'N	130°39'E
Utrecht, Neth. (ü'trĕkt) (ū'trĕkt)	117	52°05'N	5°06'E
Utrera, Spain (o͞o-trā'rä)	118	37°12'N	5°48'W
Utsunomiya, Japan (o͞ot'sò-nô-mē-yȧ')	161	36°35'N	139°52'E
Uttaradit, Thai.	168	17°47'N	100°10'E
Uttaranchal, state, India	155	29°30'N	78°30'E
Uttarpara-Kotrung, India	158a	22°40'N	88°21'E
Uttar Pradesh, state, India (òt-tär-prä-dēsh)	155	27°00'N	80°00'E
Uttoxeter, Eng., U.K. (ŭt-tôk'sē-tēr)	114a	52°54'N	1°52'W
Utuado, P.R. (o͞o-to͞o-ä'dhō)	87b	18°16'N	66°40'W
Uusikaupunki, Fin.	123	60°48'N	21°24'E
Uvalde, Tx., U.S. (ū-văl'dē)	80	29°14'N	99°47'W
Uvel'skiy, Russia (ò-vyĕl'skĭ)	142a	54°27'N	61°22'E
Uvinza, Tan.	191	5°06'S	30°22'E
Uvira, D.R.C. (o͞o-vē'rä)	186	3°28'S	29°03'E
Uvod', r., Russia (ò-vôd')	132	56°40'N	41°10'E
Uvongo Beach, S. Afr.	187c	30°49'S	30°23'E
Uvs Nuur, l., Asia	160	50°29'N	93°32'E
Uwajima, Japan (o͞o-wä'jē-mä)	166	33°12'N	132°35'E
Uxbridge, Ma., U.S. (ŭks'brĭj)	61a	42°05'N	71°38'W
Uxmal, hist., Mex. (o͞o'x-mä'l)	90a	20°22'N	89°44'W
Uy, r., Russia (o͞oy)	142a	54°05'N	62°11'E
Uyskoye, Russia (ûy'skô-yĕ)	142a	54°22'N	60°01'E
Uyuni, Bol. (o͞o-yo͞o'nē)	100	20°28'S	66°45'W
Uyuni, Salar de, pl., Bol. (sä-lär-dĕ)	100	20°58'S	67°09'W
Uzbekistan, nation, Asia	134	42°42'N	60°00'E
Uzh, r., Ukr. (òzh)	133	51°07'N	29°05'E
Uzhhorod, Ukr.	125	48°38'N	22°18'E
Užice, Serb. (o͞o'zhē-tsĕ)	131	43°51'N	19°53'E
Uzunköprü, Tur.	131	41°17'N	26°42'E

V

PLACE (Pronunciation)	PAGE	LAT.	LONG.
Vaal, r., S. Afr. (väl)	186	28°15'S	24°30'E
Vaaldam, res., S. Afr.	192c	26°58'S	28°37'E
Vaalplaas, S. Afr.	192c	25°39'S	28°56'E
Vaalwater, S. Afr.	192c	24°17'S	28°08'E
Vaasa, Fin. (vä'sȧ)	110	63°06'N	21°39'E
Vác, Hung. (väts)	125	47°46'N	19°10'E
Vache, Île à, i., Haiti	93	18°05'N	73°40'W
Vadstena, Swe. (väd'stĭ'nȧ)	122	58°27'N	14°53'E
Vaduz, Liech. (vä'dòts)	124	47°10'N	9°32'E
Vaga, r., Russia (vá'gȧ)	136	61°55'N	42°30'E
Vah, r., Slvk. (väk)	117	48°07'N	17°52'E
Vaigai, r., India	159	10°20'N	78°13'E
Vakh, r., Russia (våk)	140	61°30'N	81°33'E
Valachia, hist. reg., Rom.	131	44°45'N	24°17'E
Valcartier-Village, Can. (väl-kärt-yĕ'vē-läzh')	62b	46°56'N	71°28'W
Valdai Hills, hills, Russia (vál-dī' gô'rĭ)	136	57°50'N	32°35'E
Valday, Russia (vál-dī')	136	57°58'N	33°13'E
Valdecañas, Embalse de, res., Spain	128	39°45'N	5°30'W
Valdemārpils, Lat.	123	57°22'N	22°34'E
Valdemorillo, Spain (väl-dȧ-mô-rēl'yō)	129a	40°30'N	4°04'W
Valdepeñas, Spain (väl-dȧ-pān'yäs)	118	38°46'N	3°22'W
Valderaduey, r., Spain (väl-dĕ-rä-dwĕ'y)	128	41°39'N	5°35'W
Valdés, Península, pen., Arg. (väl-dĕ's)	102	42°15'S	63°15'W
Valdez, Ak., U.S. (văl'dēz)	63	61°10'N	146°18'W
Valdilecha, Spain (väl-dē-lā'chä)	129a	40°17'N	3°19'W
Valdivia, Chile (väl-dē'vä)	102	39°47'S	73°13'W
Valdivia, Col. (väl-dē'vēä)	100a	7°10'N	75°26'W
Val-d'Or, Can.	51	48°03'N	77°50'W
Valdosta, Ga., U.S. (văl-dŏs'tȧ)	65	30°50'N	83°18'W
Vale, Or., U.S. (vāl)	72	43°59'N	117°14'W
Valença, Braz. (vä-lĕn'sȧ)	101	13°43'S	38°58'W
Valença, Port.	128	42°03'N	8°36'W
Valence, Fr. (vä-lĕnns)	117	44°56'N	4°54'E
Valencia, Spain	110	39°26'N	0°23'W
Valencia, Ven. (vä-lĕn'syä)	100	10°11'N	68°00'W
Valencia, hist. reg., Spain	129	39°08'N	0°43'W
Valencia, Golf de, b., Spain	129	39°50'N	0°30'E
Valencia, Lago de, l., Ven.	101b	10°11'N	67°45'W
Valencia de Alcántara, Spain	128	39°34'N	7°13'W
Valenciennes, Fr. (vȧ-län-syĕn')	126	50°24'N	3°36'E
Valentine, Ne., U.S. (vă lăn-tē-nyē')	64	42°52'N	100°34'W
Valera, Ven. (vä-lē'rä)	100	9°12'N	70°45'W
Valerianovsk, Russia (vȧ-lĕ-rĭ-ä'nôvsk)	142a	58°47'N	59°34'E
Valga, Est. (väl'gȧ)	136	57°47'N	26°03'E
Valhalla, S. Afr. (väl-hăl-ȧ)	187b	25°49'S	28°09'E
Valier, Mt., U.S. (vă-lēr')	73	48°17'N	112°14'W
Valjevo, Serb. (väl'yȧ-vô)	131	44°17'N	19°57'E
Valky, Ukr.	133	49°49'N	35°40'E
Valladolid, Mex. (väl-yä-dhō-lēdh')	86	20°39'N	88°13'W
Valladolid, Spain (väl-yä-dhō-lēdh')	110	41°41'N	4°41'W
Valle, Arroyo del, Ca., U.S. (ä-rō'yō dĕl väl'yȧ)	76	37°36'N	121°43'W
Vallecas, Spain (väl-yā'käs)	129a	40°23'N	3°37'W
Valle de Allende, Mex. (väl'yä dā äl-yĕn'dȧ)	80	26°55'N	105°25'W
Valle de Bravo, Mex. (brä'vô)	88	19°12'N	100°07'W
Valle de Guanape, Ven. (vä'l-yĕ-dĕ-gwä-nä'pĕ)	101b	9°54'N	65°41'W
Valle de la Pascua, Ven. (lä-pä's-ko͞oä)	100	9°12'N	65°08'W
Valle del Cauca, dept., Col. (vä'l-yĕ del kou'kä)	100a	4°03'N	76°13'W
Valle de Santiago, Mex. (sän-tē-ä'gô)	88	20°23'N	101°11'W
Valledupar, Col. (do͞o-pär')	100	10°13'N	73°39'W
Valle Grande, Bol. (grän'dā)	100	18°27'S	64°03'W
Vallejo, Ca., U.S. (vä-yä'hō) (vă-lā'hō)	64	38°06'N	122°15'W
Vallejo, Sierra de, mts., Mex. (sē-ĕ'r-rä-dĕ-väl-yĕ'kô)	88	21°00'N	105°10'W
Vallenar, Chile (väl-yȧ-när')	102	28°39'S	70°52'W
Valles, Mex.	86	21°59'N	99°02'W
Valletta, Malta (väl-lĕt'ä)	118	35°50'N	14°29'E
Valle Vista, Ca., U.S. (väl'yȧ vĭs'tȧ)	75a	33°45'N	116°53'W
Valley City, N.D., U.S.	64	46°55'N	97°59'W
Valley City, Oh., U.S. (văl'ĭ)	69d	41°14'N	81°56'W
Valley Falls, Ks., U.S.	79	39°25'N	95°26'W
Valleyfield, Can. (văl'ē-fēld)	51	45°16'N	74°09'W
Valley Park, Mo., U.S. (văl'ē pärk)	75e	38°33'N	90°30'W
Valley Stream, N.Y., U.S. (văl'ĭ strēm)	68a	40°39'N	73°42'W
Valli di Comácchio, l., Italy (vä'lē-dē-kô-mä'chyô)	130	44°38'N	12°15'E
Vallière, Haiti (väl-yâr')	93	19°30'N	71°55'W
Vallimanca, r., Arg. (väl-yē-mä'n-kä)	99c	36°21'S	60°55'W
Valls, Spain (väls)	118	41°15'N	1°15'E
Valmiera, Lat. (väl'myĕ-rȧ)	136	57°34'N	25°54'E
Valognes, Fr. (vȧ-lôn'y')	126	49°32'N	1°30'W
Valona *see* Vlorë, Alb.	119	40°28'N	19°31'E
Valozhyn, Bela.	132	54°04'N	26°38'E
Valparaíso, Chile (väl'pä-rä-ē'sō)	102	33°02'S	71°32'W
Valparaíso, Mex.	88	22°49'N	103°33'W
Valparaiso, In., U.S. (văl-pȧ-rā'zō)	66	41°25'N	87°05'W
Valpariso, prov., Chile	99b	32°58'S	71°23'W
Valréas, Fr. (vȧl-rȧ-ä')	126	44°25'N	4°56'E
Vals, r., S. Afr.	192c	27°32'S	26°51'E
Vals, Tanjung, c., Indon.	169	8°30'S	137°15'E
Valsbaai, b., S. Afr.	186a	34°14'S	18°35'E
Valuyevo, Russia (vá-lo͞o'yĕ-vô)	142b	55°34'N	37°21'E
Valuyki, Russia (vá-lò-ē'kē)	137	50°14'N	38°04'E
Valverde del Camino, Spain (väl-vĕr-dĕ-dĕl-kä-mē'nō)	128	37°34'N	6°44'W
Vammala, Fin.	123	61°19'N	22°51'E
Van, Tur. (vȧn)	154	38°04'N	43°10'E
Van Buren, Ar., U.S. (văn bū'rĕn)	79	35°26'N	94°20'W
Van Buren, Me., U.S.	60	47°09'N	67°58'W
Vanceburg, Ky., U.S. (văns'bûrg)	66	38°35'N	83°20'W
Vancouver, Can. (văn-ko͞o'vēr)	50	49°16'N	123°06'W
Vancouver, Wa., U.S.	64	45°37'N	122°40'W
Vancouver Island, i., Can.	52	49°50'N	125°05'W
Vancouver Island Ranges, mts., Can.	54	49°25'N	125°25'W
Vandalia, Il., U.S. (văn-dā'lĭ-ȧ)	66	39°00'N	89°00'W
Vandalia, Mo., U.S.	79	39°19'N	91°30'W
Vanderbijlpark, S. Afr.	192c	26°43'S	27°50'E
Vanderhoof, Can.	50	54°01'N	124°01'W
Van Diemen, Cape, c., Austl. (văndē'mĕn)	174	11°05'S	130°15'E
Van Diemen Gulf, b., Austl.	174	11°50'S	131°30'E
Vanegas, Mex. (vä-nē'gäs)	86	23°54'N	100°54'W
Vänern, l., Swe.	112	58°52'N	13°17'E
Vänersborg, Swe. (vĕ'nĕrs-bôr')	116	58°24'N	12°15'E
Vanga, Kenya (väŋ'gä)	187	4°38'S	39°10'E
Vangani, India	159b	19°07'N	73°15'E
Van Gölü, l., Tur.	136	38°33'N	42°46'E
Van Horn, Tx., U.S.	80	31°03'N	104°50'W
Vanier, Can.	62c	45°27'N	75°39'W
Van Lear, Ky., U.S. (văn lēr')	66	37°45'N	82°50'W
Vannes, Fr. (vȧn)	117	47°42'N	2°46'W
Van Nuys, Ca., U.S. (văn nīz')	75a	34°11'N	118°27'W
Van Rees, Pegunungan, mts., Indon.	169	2°30'S	138°45'E
Vantaan, r., Fin.	123	60°25'N	24°43'E
Vanua Levu, i., Fiji	170g	16°33'S	179°15'E
Vanuatu, nation, Oc.	175	16°02'S	169°15'E
Van Wert, Oh., U.S. (văn wûrt')	66	40°50'N	84°35'W
Vara, Swe. (vä'rä)	122	58°17'N	12°55'E
Varakļāni, Lat.	123	56°38'N	26°46'E
Varallo, Italy (vä-räl'lô)	130	45°44'N	8°14'E
Vārānasi (Benares), India	155	25°25'N	83°00'E
Varangerfjorden, b., Nor.	113	70°05'N	30°20'E
Varano, Lago di, l., Italy (lä'gō-dē-vä-rä'nô)	130	41°52'N	15°55'E
Varaždin, Cro. (vä'räzh'dĕn)	119	46°17'N	16°20'E
Varazze, Italy (vä-rät'sā)	130	44°23'N	8°34'E
Varberg, Swe. (vär'bĕrg)	122	57°06'N	12°16'E
Vardar, r., Serb. (vär'där)	131	41°40'N	21°50'E
Varėna, Lith. (vä-rā'nȧ)	123	54°16'N	24°35'E
Varennes, Can. (vȧ-rĕn')	62a	45°41'N	73°27'W
Vareš, Bos. (vä'rĕsh)	131	44°10'N	18°20'E
Varese, Italy (vä-rā'sā)	130	45°45'N	8°49'E
Varginha, Braz. (vär-zhē'n-yä)	101	21°33'S	45°25'W
Varkaus, Fin. (vär'kous)	123	62°19'N	27°51'E
Varlamovo, Russia (vár-lá'mô-vô)	142a	54°37'N	60°41'E
Varna, Blg. (vär'nȧ)	110	43°14'N	27°58'E
Varna, Russia	142a	53°22'N	60°59'E
Värnamo, Swe. (vĕr'nȧ-mô)	122	57°11'N	13°45'E
Varnsdorf, Czech Rep. (värns'dôrf)	124	50°54'N	14°36'E
Varnville, S.C., U.S. (värn'vĭl)	83	32°49'N	81°05'W
Vasa, India	159b	19°20'N	72°47'E
Vascongadas *see* Basque Provinces, hist. reg., Spain	128	43°00'N	2°46'W
Vashka, r., Russia	136	64°00'N	48°00'E
Vashon, Wa., U.S. (văsh'ŭn)	74a	47°27'N	122°28'W
Vashon Heights, Wa., U.S. (hītz)	74a	47°30'N	122°28'W
Vashon Island, i., Wa., U.S.	74a	47°27'N	122°27'W
Vaslui, Rom. (vás-lo͞o'ē)	125	46°39'N	27°49'E
Vassar, Mi., U.S. (văs'ēr)	66	43°25'N	83°35'W
Vassouras, Braz. (väs-sō'räzh)	99a	22°25'S	43°40'W
Västerås, Swe. (vĕs'tĕr-ôs)	116	59°39'N	16°30'E
Västerdalälven, r., Swe.	116	61°06'N	13°10'E
Västervik, Swe. (vĕs'tĕr-vēk)	116	57°45'N	16°35'E
Vasto, Italy (väs'tô)	118	42°06'N	12°42'E
Vasyl'kiv, Ukr.	137	50°10'N	30°22'E
Vasyugan, r., Russia (vás-yo͞o-gán')	140	58°52'N	77°30'E
Vatican City, nation, Eur.	130	41°54'N	12°22'E
Vaticano, Cape, c., Italy (vä-tē-kä'nô)	130	38°38'N	15°52'E
Vatnajökull, ice, Ice. (vät'nȧ-yû-kôl)	116	64°34'N	16°41'W
Vatomandry, Madag.	187	18°53'S	48°13'E
Vatra Dornei, Rom. (vät'rȧ dôr'nȧ')	125	47°22'N	25°20'E
Vättern, l., Swe.	112	58°15'N	14°24'E
Vattholma, Swe.	122	60°01'N	17°40'E
Vaudreuil, Can. (vō-drû'y')	62a	45°24'N	74°02'W
Vaugh, Wa., U.S. (vôn)	74a	47°21'N	122°47'W
Vaughan, Can.	62d	43°47'N	79°36'W
Vaughn, N.M., U.S.	78	34°37'N	105°13'W
Vaupés, r., S.A. (vá'o͞o-pĕ's)	100	1°18'N	71°14'W
Vawkavysk, Bela. (vôl-kô-vēsk')	125	53°11'N	24°29'E
Vaxholm, Swe. (väks'hôlm)	122	59°26'N	18°19'E
Växjo, Swe. (vĕks'shû)	116	56°53'N	14°46'E
Vaygach, i., Russia (vī-gäch')	134	70°00'N	59°00'E
Veadeiros, Chapadas dos, hills, Braz. (shä-pä'däs-dôs-vĕ-ȧ-dā'rōs)	101	14°00'S	47°00'W
Vedea, r., Rom. (vȧ'dyȧ)	131	44°25'N	24°45'E
Vedia, Arg. (vĕ'dyä)	99c	34°29'S	61°30'W
Veedersburg, In., U.S. (vē'dĕrz-bûrg)	66	40°05'N	87°15'W
Vega, i., Nor.	116	65°38'N	10°51'E

ăt; finăl; rāte; senåte; ärm; ȧsk; sofȧ; fâre; ch-choose; dh-as th in other; bē; ēvent; bĕt; recĕnt; cratēr; g-gō; gh-guttural g; bĭt; ĭ-short neutral; rīde; κ-guttural k as ch in German ich;

PRONOUNCING INDEX Vega de Alatorre – Villa Hidalgo 315

PLACE (Pronunciation)	PAGE	LAT.	LONG.
Vega de Alatorre, Mex. (vä′gä dā ä-lä-tōr′rå)	89	20°02′N	96°39′W
Vega Real, reg., Dom. Rep. (vĕ′gä-rĕ-ä′l)	93	19°30′N	71°05′W
Vegreville, Can.	50	53°30′N	112°03′W
Vehär Lake, l., India	159b	19°11′N	72°52′E
Veinticinco de Mayo, Arg.	99c	35°26′S	60°09′W
Vejer de la Frontera, Spain	128	36°15′N	5°58′W
Vejle, Den. (vī′lĕ)	116	55°41′N	9°29′E
Velbert, Ger. (fĕl′bĕrt)	127c	51°20′N	7°03′E
Velebit, mts., Serb. (vā′lĕ-bēt)	119	44°25′N	15°23′E
Velen, Ger. (fĕ′lĕn)	127c	51°54′N	7°00′E
Vélez-Málaga, Spain (vā′lāth-mä′lä-gä)	128	36°48′N	4°05′W
Vélez-Rubio, Spain (rōō′bē-ō)	128	37°38′N	2°05′W
Velika Kapela, mts., Serb. (vĕ′lē-kä kä-pĕ′lä)	119	45°03′N	15°20′E
Velika Morava, r., Serb. (mô′rä-vä)	119	44°00′N	21°30′E
Velikaya, r., Russia (vå-lē′kå-yå)	132	57°25′N	28°07′E
Velikiye Luki, Russia (vyē-lē′-kyē lōō′ke)	134	56°19′N	30°32′E
Velikiy Ustyug, Russia (vå-lē′kĭ ōōs-tyōg′)	134	60°45′N	46°38′E
Veliko Tŭrnovo, Blg.	119	43°06′N	25°38′E
Velikoye, Russia (vå-lē′kô-yĕ)	132	57°21′N	39°45′E
Velikoye, l., Russia	132	57°00′N	36°53′E
Veli Lošinj, Cro. (lô′shēn′)	130	44°30′N	14°29′E
Velizh, Russia (vå′lēzh)	136	55°37′N	31°11′E
Vella Lavella, i., Sol. Is.	175	8°00′S	156°42′E
Velletri, Italy (vĕl-lā′trē)	130	41°42′N	12°48′E
Vellore, India (vĕl-lōr′)	155	12°57′N	79°09′E
Vels, Russia (vĕls)	142a	60°35′N	58°47′E
Vel'sk, Russia (vĕlsk)	134	61°00′N	42°18′E
Velten, Ger. (fel′tĕn)	115b	52°41′N	13°11′E
Velya, r., Russia (vĕl′yå)	142b	56°23′N	37°54′E
Velyka Lepetykha, Ukr.	133	47°11′N	33°58′E
Velykyi Bychkiv, Ukr.	125	47°59′N	24°01′E
Venadillo, Col. (vĕ-nä-dē′l-yō)	100a	4°43′N	74°55′W
Venado, Mex. (vå-mä′dō)	88	22°54′N	101°07′W
Venado Tuerto, Arg. (vĕ-nä′dô-tōōĕ′r-tô)	102	33°28′S	61°47′W
Vendôme, Fr. (väN-dōm′)	126	47°46′N	1°05′E
Veneto, hist. reg., Italy (vĕ-nĕ′tô)	130	45°58′N	11°24′E
Venëv, Russia (vĕn-ĕf′)	136	54°19′N	38°14′E
Venezia *see* Venice, Italy	110	45°25′N	12°18′E
Venezuela, nation, S.A. (vĕn-ē-zwē′lȧ)	100	8°00′N	65°00′W
Venezuela, Golfo de, b., S.A. (gôl-fô-dĕ)	100	11°34′N	71°02′W
Veniaminof, Mount, mtn., Ak., U.S.	63	56°12′N	159°20′W
Venice, Italy	110	45°25′N	12°18′E
Venice, Ca., U.S. (vĕn′ĭs)	75a	33°59′N	118°28′W
Venice, Il., U.S.	75e	38°40′N	90°10′W
Venice, Gulf of, b., Italy	118	45°23′N	13°00′E
Venlo, Neth.	127c	51°22′N	6°11′E
Venta, r., Eur. (vĕn′tå)	123	57°05′N	21°45′E
Ventana, Sierra de la, mts., Arg. (sē-ĕ-rä-dĕ-lä-vĕn-tä′nä)	102	38°00′S	63°00′W
Ventersburg, S. Afr. (vĕn-tĕrs′bûrg)	192c	28°06′S	27°10′E
Ventersdorp, S. Afr. (vĕn-tĕrs′dôrp)	192c	26°20′S	26°48′E
Ventimiglia, Italy (vĕn-tē-mēl′yä)	130	43°46′N	7°37′E
Ventnor, N.J., U.S. (vĕnt′nẽr)	67	39°20′N	74°25′W
Ventspils, Lat. (vĕnt′spēls)	136	57°24′N	21°41′E
Ventuari, r., Ven. (vĕn-tōōä′rē)	100	4°47′N	65°56′W
Ventura, Ca., U.S. (vĕn-tōō′rȧ)	76	34°18′N	119°18′W
Venukovsky, Russia (vĕ-nōō′kôv-skĭ)	142b	55°10′N	37°26′E
Venustiano Carranza, Mex. (vĕ-nōōs-tyä′nō-kär-rä′n-zä)	88	19°44′N	103°48′W
Venustiano Carranzo, Mex. (kär-rä′n-zô)	89	16°21′N	92°36′W
Vera, Arg. (vĕ-rä)	102	29°22′S	60°09′W
Vera, Spain (vā′rä)	128	37°18′N	1°53′W
Veracruz, Mex.	86	19°13′N	96°07′W
Veracruz, state, Mex. (vā-rä-krōōz′)	86	20°30′N	97°15′W
Verāval, India (vĕr′vŭ-väl)	155	20°59′N	70°49′E
Vercelli, Italy (vĕr-chĕl′lē)	130	45°18′N	8°27′E
Verchères, Can. (vĕr-shâr′)	62a	45°46′N	73°21′W
Verde, i., Phil. (vĕr′dā)	169a	13°34′N	121°11′E
Verde, r., Mex.	88	21°48′N	99°50′W
Verde, r., Mex.	88	20°50′N	103°00′W
Verde, r., Mex.	89	16°05′N	97°44′W
Verde, r., Az., U.S. (vûrd)	77	34°04′N	111°40′W
Verde, Cap, c., Bah.	93	22°50′N	75°00′W
Verde, Cay, i., Bah.	93	22°00′N	75°05′W
Verde Island Passage, strt., Phil. (vĕr′dē)	169a	13°36′N	120°39′E
Verdemont, Ca., U.S. (vûr′dē-mŏnt)	75a	34°12′N	117°22′W
Verden, Ger. (fĕr′dĕn)	124	52°55′N	9°15′E
Verdigris, r., Ok., U.S. (vûr′dē-grēs)	79	36°50′N	95°29′W
Verdun, Can. (vĕr′dŭN′)	59	45°27′N	73°34′W
Verdun, Fr. (vâr-dûN′)	117	49°09′N	5°21′E
Verdun, Fr.	127	43°48′N	1°10′E
Vereeniging, S. Afr. (vĕ-rä′nĭ-gĭng)	192c	26°40′S	27°56′E
Verena, S. Afr. (vĕr-ĕn ȧ)	192c	25°30′S	29°02′E
Vereya, Russia (vĕ-rå′yå)	132	55°21′N	36°08′E
Verín, Spain (vå-rēn′)	128	41°56′N	7°26′W
Verkhne-Kamchatsk, Russia (vyĕrk′nyĕ kåm-chatsk′)	135	54°42′N	158°41′E
Verkhne Neyvinskiy, Russia (nä-vĭn′skĭ)	142a	57°17′N	60°10′E
Verkhne Ural'sk, Russia (ö-ralsk′)	134	53°53′N	59°13′E
Verkhniy Avzyan, Russia (vyĕrk′nyĕ åv-zyän′)	142a	53°32′N	57°30′E
Verkhniye Kigi, Russia (vyĕrk′nĭ-yĕ kĭ′gĭ)	142a	55°23′N	58°37′E
Verkhniy Ufaley, Russia (ö-få′lā)	142a	56°04′N	60°15′E
Verkhnyaya Pyshma, Russia (vyĕrk′nyä-yä pōōsh′må)	142a	56°57′N	60°37′E
Verkhnyaya Salda, Russia (säl′då)	142a	58°03′N	60°33′E
Verkhnyaya Tunguska (Angara), r., Russia (tön-gōs′kå)	140	58°13′N	97°00′E
Verkhnyaya Tura, Russia (tö′rå)	142a	58°22′N	59°51′E
Verkhnyaya Yayva, Russia (yäy′vå)	142a	59°28′N	57°38′E
Verkhotur'ye, Russia (vyĕr-kô-tōōr′yĕ)	142a	58°52′N	60°47′E
Verkhoyansk, Russia (vyĕr-κô-yänsk′)	135	67°43′N	133°33′E
Verkhoyanskiy Khrebet, mts., Russia (vyĕr-κô-yänskĭ)	135	67°45′N	128°00′E
Vermilion, Can. (vẽr-mĭl′yŭn)	50	53°22′N	110°51′W
Vermilion, l., Mn., U.S.	71	47°49′N	92°35′W
Vermilion, r., Can.	59	47°30′N	73°15′W
Vermilion, r., Can.	56	53°30′N	111°00′W
Vermilion, r., Il., U.S.	66	41°05′N	89°00′W
Vermilion, r., Mn., U.S.	71	48°09′N	92°31′W
Vermilion Hills, hills, Can.	56	50°43′N	106°50′W
Vermilion Range, mts., Mn., U.S.	71	47°55′N	91°59′W
Vermillion, S.D., U.S.	70	42°46′N	96°56′W
Vermillion, r., S.D., U.S.	70	43°54′N	97°14′W
Vermillion Bay, b., La., U.S.	81	29°47′N	92°00′W
Vermont, state, U.S. (vẽr-mŏnt′)	65	43°50′N	72°50′W
Vernal, Ut., U.S. (vûr′nȧl)	73	40°29′N	109°40′W
Verneuk Pan, pl., S. Afr. (vẽr-nŭk′)	186	30°10′S	21°46′E
Vernon, Can. (vẽr-nôN′)	50	50°18′N	119°15′W
Vernon, Can.	62c	45°10′N	75°27′W
Vernon, Ca., U.S. (vûr′nŭn)	75a	34°01′N	118°12′W
Vernon, In., U.S. (vûr′nŭn)	66	39°00′N	85°40′W
Vernon, N.J., U.S.	68a	39°00′N	85°40′W
Vernon, Tx., U.S.	78	34°09′N	99°16′W
Vernonia, Or., U.S. (vûr-nō′nyȧ)	74c	45°52′N	123°12′W
Vero Beach, Fl., U.S. (vē′rō)	83a	27°36′N	80°25′W
Véroia, Grc.	131	40°30′N	22°13′E
Verona, Italy (vā-rō′nä)	118	45°28′N	11°02′E
Versailles, Fr. (vĕr-sī′y′)	117	48°48′N	2°07′E
Versailles, Ky., U.S. (vẽr-sālz′)	66	38°05′N	84°45′W
Versailles, Mo., U.S.	79	38°27′N	92°52′W
Vert, Cap, c., Sen.	184	14°43′N	17°30′W
Verulam, S. Afr. (vĕ-rōō-lăm)	187c	29°39′S	31°08′E
Verviers, Bel. (vĕr-vyā′)	121	50°35′N	5°57′E
Vesele, Ukr.	133	46°59′N	34°56′E
Vesijärvi, l., Fin.	123	61°09′N	25°10′E
Vesoul, Fr. (vē-sōōl′)	127	47°38′N	6°11′E
Vestavia Hills, Al., U.S.	68h	33°26′N	86°46′W
Vesterålen, is., Nor. (vĕs′tĕr ô′lĕn)	116	68°54′N	14°03′E
Vestfjord, b., Nor.	112	67°33′N	12°59′E
Vestmannaeyjar, Ice. (vĕst′män-ä-ā′yär)	116	63°12′N	20°17′W
Vesuvio, vol., Italy (vĕ-sōō′vyä)	112	40°35′N	14°26′E
Ves'yegonsk, Russia (vĕ-syĕ-gônsk′)	132	58°42′N	37°09′E
Veszprem, Hung. (vĕs′prām)	125	47°05′N	17°53′E
Vészto, Hung. (vĕs′tû)	125	46°55′N	21°18′E
Vet, r., S. Afr. (vĕt)	192c	28°25′S	26°37′E
Vetlanda, Swe. (vĕt-län′dä)	122	57°26′N	15°05′E
Vetluga, Russia (vyĕt-lōō′gå)	136	57°50′N	45°42′E
Vetluga, r., Russia	136	56°50′N	45°50′E
Vetovo, Blg. (vā′tô-vô)	131	43°42′N	26°18′E
Vetren, Blg. (vĕt′rĕn′)	131	42°16′N	24°04′E
Vevay, In., U.S. (vē′vā)	66	38°45′N	85°05′W
Veynes, Fr. (vān′′)	127	44°31′N	5°47′E
Vézère, r., Fr. (vā-zer′)	126	45°01′N	1°00′E
Viacha, Bol. (vēå′chå)	100	16°43′S	68°16′W
Viadana, Italy (vē-ä-dä′nä)	130	44°55′N	10°30′E
Vian, Ok., U.S. (vī′ăn)	79	35°30′N	95°00′W
Viana, Braz. (vē-ä′nä)	101	3°09′S	44°44′W
Viana do Alentejo, Port. (vē-ä′nȧ dö ä-lĕn-tā′hō)	128	38°20′N	8°02′W
Viana do Bolo, Spain	128	42°10′N	7°07′W
Viana do Castelo, Port. (dö käs-tā′lö)	118	41°41′N	8°45′W
Viangchan, Laos	168	18°07′N	102°33′E
Viar, r., Spain (vē-ä′rä)	128	38°15′N	6°08′W
Viareggio, Italy (vē-ä-rĕd′jô)	130	43°52′N	10°14′E
Viborg, Den. (vē′bôr)	122	56°27′N	9°22′E
Vibo Valentia, Italy (vē′bô-vä-lĕ′n-tyä)	130	38°47′N	16°06′E
Vic, Spain	129	41°55′N	2°14′E
Vicálvaro, Spain	129a	40°25′N	3°37′W
Vicente López, Arg. (vē-sĕ′n-tĕ-lō′pĕz)	102a	34°31′S	58°29′W
Vicenza, Italy (vē-chĕnt′sä)	118	45°33′N	11°33′E
Vichuga, Russia (vē-chōō′gå)	136	57°13′N	41°58′E
Vichy, Fr. (vē-shē′)	117	46°06′N	3°28′E
Vickersund, Nor.	122	60°00′N	9°59′E
Vicksburg, Mi., U.S. (vĭks′bûrg)	66	42°10′N	85°30′W
Vicksburg, Ms., U.S.	65	32°20′N	90°50′W
Viçosa, Braz. (vē-sô′sä)	99a	20°46′S	42°51′W
Victoria, Arg. (vēk-tō′rēä)	102	32°36′S	60°09′W
Victoria, Can. (vĭk-tō′rĭ-ȧ)	50	48°26′N	123°23′W
Victoria, Chile (vēk-tô-rēä)	102	38°15′S	72°16′W
Victoria, Col. (vēk-tô′rēä)	100a	5°19′N	74°54′W
Victoria, Phil. (vēk-tô-ryä)	169a	15°34′N	120°41′E
Victoria, Tx., U.S. (vĭk-tō′rĭ-ȧ)	81	28°48′N	97°00′W
Victoria, Va., U.S.	83	36°57′N	78°13′W
Victoria, state, Austl.	175	36°46′S	143°15′E
Victoria, l., Afr.	186	0°50′S	32°50′E
Victoria, r., Austl.	174	17°25′S	130°50′E
Victoria, Mount, mtn., Mya.	155	21°26′N	93°59′E
Victoria, Mount, mtn., Pap. N. Gui.	169	9°35′S	147°45′E
Victoria de las Tunas, Cuba (vēk-tō′rē-ä dā läs tōō′näs)	92	20°55′N	77°05′W
Victoria Falls, wtfl., Afr.	186	17°55′S	25°51′E
Victoria Island, i., Can.	49	70°13′N	107°45′W
Victoria Lake, l., Can.	61	48°20′N	57°40′W
Victoria Land, reg., Ant.	178	75°00′S	160°00′E
Victoria Nile, r., Ug.	191	2°20′N	31°35′E
Victoria Peak, mtn., Belize (vēk-tōrī′ȧ)	90a	16°47′N	88°40′W
Victoria Peak, mtn., Can.	54	50°03′N	126°06′W
Victoria River Downs, Austl. (vĭc-tŏr′ĭȧ)	174	16°30′S	131°10′E
Victoria Strait, strt., Can. (vĭk-tō′rĭ-ȧ)	52	69°10′N	100°58′W
Victoriaville, Can. (vĭk-tō′rĭ-ȧ-vĭl)	51	46°04′N	71°59′W
Victoria West, S. Afr. (wĕst)	186	31°25′S	23°10′E
Vidalia, Ga., U.S. (vĭ-dā′lĭ-ȧ)	83	32°10′N	82°26′W
Vidalia, La., U.S.	81	31°33′N	91°28′W
Vidin, Blg. (vĭ′dĕn)	119	44°00′N	22°53′E
Vidnoye, Russia	142b	55°33′N	37°41′E
Vidzy, Bela. (vê′dzĭ)	132	55°23′N	26°46′E
Viedma, Arg. (vyād′mä)	102	40°55′S	63°03′W
Viedma, l., Arg.	102	49°40′S	72°35′W
Viejo, r., Nic. (vyā′hō)	90	12°45′N	86°19′W
Vienna (Wien), Aus.	110	48°13′N	16°22′E
Vienna, Ga., U.S. (vē-ĕn′ȧ)	82	32°03′N	83°50′W
Vienna, Il., U.S.	79	37°24′N	88°50′W
Vienna, Va., U.S.	68e	38°54′N	77°16′W
Vienne, Fr. (vyĕn′)	117	45°31′N	4°54′E
Vienne, r., Fr.	126	47°06′N	0°20′E
Vientiane *see* Viangchan, Laos	168	18°07′N	102°33′E
Vieques, P.R. (vyā′kås)	87b	18°09′N	65°27′W
Vieques, i., P.R. (vyā′kås)	87b	18°05′N	65°28′W
Vierfontein, S. Afr. (vēr′fŏn-tān)	192c	27°06′S	26°45′E
Viersen, Ger. (fēr′zĕn)	127c	51°15′N	6°24′E
Vierwaldstätter See, l., Switz.	124	46°54′N	8°36′E
Vierzon, Fr. (vyâr-zôN′)	117	47°14′N	2°04′E
Viesca, Mex. (vē-ās′kä)	80	25°21′N	102°47′W
Viesca, Laguna de, l., Mex. (lä-ö′nä-dĕ)	80	25°30′N	102°40′W
Vieste, Italy (vyĕs′tå)	130	41°52′N	16°10′E
Vietnam, nation, Asia (vyĕt′näm′)	168	18°00′N	107°00′E
Vigan, Phil. (vēgän)	168	17°36′N	120°22′E
Vigevano, Italy (vē-jå-vä′nô)	130	45°18′N	8°52′E
Vigny, Fr. (vēn-y′ē′)	127b	49°05′N	1°54′E
Vigo, Spain (vē′gō)	110	42°18′N	8°42′W
Vihti, Fin. (vē′tĭ)	123	60°27′N	24°18′E
Vijayawāda, India	155	16°31′N	80°37′E
Viksøyri, Nor.	122	61°06′N	6°35′E
Vila Caldas Xavier, Moz.	191	15°59′S	34°12′E
Vila de Manica, Moz. (vē′lä dĕ mä-nē′kå)	186	18°48′S	32°49′E
Vila de Rei, Port. (vē′lȧ dā rā′ĭ)	128	39°42′N	8°03′W
Vila do Conde, Port. (vē′lä dö kôn′dĕ)	128	41°21′N	8°44′W
Vilafranca del Penedès, Spain	129	41°20′N	1°40′E
Vilafranca de Xira, Port. (frăn′kȧ dä shē′rȧ)	128	38°58′N	8°59′W
Vilaine, r., Fr. (vē-lân′)	126	47°34′N	2°15′W
Vilalba, Spain	128	43°18′N	7°43′W
Vilanculos, Moz. (vē-län-kōō′lōs)	186	22°03′S	35°13′E
Vilāni, Lat. (vē′lá-nĭ)	123	56°31′N	27°00′E
Vila Nova de Foz Côa, Port. (nō′vȧ dā fōz-kō′ȧ)	128	41°08′N	7°11′W
Vila Nova de Gaia, Port. (vē′lȧ nō′vȧ dā gä′yä)	128	41°08′N	8°40′W
Vila Nova de Milfontes, Port. (nō′vȧ dā mēl-fôn′tāzh)	128	37°44′N	8°48′W
Vila Real, Port. (rā-äl′)	118	41°18′N	7°48′W
Vila-real, Spain	129	39°55′N	0°07′W
Vila Real de Santo Antonio, Port.	128	37°14′N	7°25′W
Vila Viçosa, Port. (vē-sō′zȧ)	128	38°47′N	7°24′W
Vileyka, Bela. (vē-lā′ē-kå)	132	54°19′N	26°58′E
Vilhelmina, Swe.	116	64°37′N	16°30′E
Viljandi, Est. (vēl′yän-dē)	136	58°24′N	25°34′E
Viljoenskroon, S. Afr.	192c	27°13′S	26°58′E
Vilkaviškis, Lith. (vēl-kå-vēsh′kēs)	123	54°40′N	23°08′E
Vil'kitskogo, i., Russia (vyl-kēts-kōgō)	140	73°25′N	76°00′E
Villa Acuña, Mex. (vēl′yä-kōō′n-yä)	80	29°20′N	100°56′W
Villa Ahumada, Mex. (ä-ōō-mä′dä)	80	30°43′N	106°30′W
Villa Alta, Mex. (äl′tä)(sän ēl-då-fōn′sō)	89	17°20′N	96°08′W
Villa Angela, Arg. (vē′l-yä á′n-κē-lä)	102	27°31′S	60°42′W
Villa Ballester, Arg. (vē′l-yä-bál-yĕs-tĕr)	102a	34°33′S	58°33′W
Villa Bella, Bol. (bĕ′l-yä)	100	10°25′S	65°22′W
Villablino, Spain (vēl-yä-blē′nô)	128	42°58′N	6°18′W
Villacañas, Spain (vēl-yä-kän′yäs)	128	39°39′N	3°20′W
Villacarrillo, Spain (vēl-yä-kä-rēl′yō)	128	38°09′N	3°07′W
Villach, Aus. (fē′läκ)	117	46°38′N	13°50′E
Villacidro, Italy (vē-lä-chē′drō)	130	39°28′N	8°41′E
Villa Clara, prov., Cuba	92	22°40′N	80°10′W
Villa Constitución, Arg. (kōn-stē-tōō-syōn′)	99c	33°15′S	60°19′W
Villa Coronado, Mex. (kō-rō-nä′dhō)	80	26°45′N	105°10′W
Villa Cuauhtémoc, Mex. (vēl′yä-kōō-äö-tē′mōk)	89	22°11′N	97°50′W
Villa de Allende, Mex. (vēl′yä′dä äl-yĕn′dā)	80	25°18′N	100°01′W
Villa de Alvarez, Mex. (vēl′yä-dē-ä′l-vä-rēz)	88	19°17′N	103°44′W
Villa de Cura, Ven. (dē-kōō′rä)	101b	10°03′N	67°29′W
Villa de Guadalupe, Mex. (dē-gwä-dhä-lōō′pā)	88	23°22′N	100°44′W
Villa de Mayo, Arg.	102a	34°31′S	58°41′W
Villa Dolores, Arg. (vēl′yä dô-lō′räs)	102	31°50′S	65°05′W
Villa Escalante, Mex. (vēl′yä-ĕs-kä-lán′tē)	88	19°24′N	101°36′W
Villa Flores, Mex. (vēl′yä-flō′räs)	89	16°13′N	93°17′W
Villafranca, Italy (vēl-lä-frän′kä)	130	45°22′N	10°53′E
Villafranca del Bierzo, Spain	128	42°37′N	6°49′W
Villafranca de los Barros, Spain	128	38°34′N	6°22′W
Villafranche-de-Rouergue, Fr. (dĕ-rōō-ĕrg′)	126	44°21′N	2°02′E
Villa García, Mex. (gär-sē′ä)	88	22°07′N	101°55′W
Villagarcía, Spain	128	42°38′N	8°43′W
Villagrán, Mex.	80	24°28′N	99°30′W
Villa Grove, Il., U.S. (vĭl′ȧ grōv′)	66	39°55′N	88°15′W
Villaguay, Arg. (vē′l-yä-gwī)	102	31°47′S	58°53′W
Villa Hayes, Para. (vēl′yä āyås)(hāz)	102	25°07′S	57°31′W
Villahermosa, Mex. (vēl′yä-ĕr-mō′sä)	86	17°59′N	92°56′W
Villa Hidalgo, Mex. (vēl′yäē-däl′gō)	88	21°39′N	102°41′W

ng-sing; ŋ-baŋk; N-nasalized n; nŏd; cŏmmit; ōld; ôbey; ôrder; oi-boil; fōōd; ö-as oo in foot; ou-out; s-soft; sh-dish; th-thin; pūre; ûnite; ûrn; stŭd; circŭs; ü-as in French tu; ′-indeterminate vowel.

316 Villaldama – Waipahu PRONOUNCING INDEX

PLACE (Pronunciation)	PAGE	LAT.	LONG.
Villaldama, Mex. (vēl-yäl-dä'mä)	86	26°30'N	100°26'W
Villa Lopez, Mex. (vēl'yä lō'pĕz)	80	27°00'N	105°02'W
Villalpando, Spain (vēl-yäl-pän'dō)	128	41°54'N	5°24'W
Villa María, Arg. (vē'l-yä-mä-rē'ä)	102	32°17'S	63°08'W
Villamatín, Spain (vēl-yä-mä-tē'n)	128	36°50'N	5°38'W
Villa Mercedes, Arg. (mĕr-sā'dås)	102	33°38'S	65°16'W
Villa Montes, Bol. (vē'l-yä-mô'n-tĕs)	100	21°13'S	63°26'W
Villa Morelos, Mex. (mō-rē'lomcs)	88	20°01'N	101°24'W
Villanueva, Col. (vē'l-yä-nòĕ'vä)	100	10°44'N	73°08'W
Villanueva, Hond. (vēl'yä-nwā'vä)	90	15°19'N	88°02'W
Villanueva, Mex. (vēl'yä-nòĕ'vä)	88	22°25'N	102°53'W
Villanueva de Córdoba, Spain (vēl-yä-nwĕ'vä-dā kôr'dô-bä)	128	38°18'N	4°38'W
Villanueva de la Serena, Spain (lä sā-rā'nä)	128	38°59'N	5°56'W
Villa Obregón, Mex. (vē'l-yä-ō-brĕ-gô'n)	89a	19°21'N	99°11'W
Villa Ocampo, Mex. (ō-käm'pō)	80	26°26'N	105°30'W
Villa Pedro Montoya, Mex. (vēl'yä-pĕ'drô-môn-tô'yà)	88	21°38'N	99°51'W
Villard-Bonnot, Fr. (vēl-yär'bôn-nō')	127	45°15'N	5°53'E
Villarrica, Para. (vēl-yä-rē'kä)	102	25°55'S	56°23'W
Villarrobledo, Spain (vēl-yär-rô-blā'dhō)	118	39°15'N	2°37'W
Villa Unión, Mex. (vēl'yä-ōō-nyōn')	88	23°10'N	106°14'W
Villavicencio, Col. (vē'l-yä-vē-sĕ'n-syō)	100	4°09'N	73°38'W
Villaviciosa de Odón, Spain	129a	40°22'N	3°38'W
Villavieja, Col. (vē'l-yä-vē-ĕ'κä)	100a	3°13'N	75°13'W
Villazón, Bol. (vē'l-yä-zô'n)	100	22°02'S	65°42'W
Villefranche, Fr.	117	45°59'N	4°43'E
Villejuif, Fr. (vēl'zhüst')	127b	48°48'N	2°22'E
Ville-Marie, Can.	51	47°18'N	79°22'W
Villena, Spain (vē-lyā'nà)	118	38°37'N	0°52'W
Villeneuve, Can. (vēl'nûv')	62g	53°40'N	113°49'W
Villeneuve-Saint Georges, Fr. (säɴ-zhôrzh')	127b	48°43'N	2°27'E
Villeneuve-sur-Lot, Fr. (sür-lō')	126	44°25'N	0°41'E
Ville Platte, La., U.S. (vēl plát')	81	30°41'N	92°17'W
Villers Cotterêts, Fr. (vē-âr'kô-trā')	127b	49°15'N	3°05'E
Villerupt, Fr. (vēl'rüp')	127	49°28'N	6°16'E
Ville-Saint Georges, Can. (vĭl-sĕɴ-zhôrzh')	59	46°07'N	70°40'W
Villeta, Col. (vē'l-yĕ'tà)	100a	5°02'N	74°29'W
Villeurbanne, Fr. (vēl-ûr-bän')	117	45°43'N	4°55'E
Villiers, S. Afr. (vĭl'ĭ-ĕrs)	192c	27°03'S	28°38'E
Villingen-Schwenningen, Ger.	124	48°04'N	8°33'E
Villisca, Ia., U.S. (vĭ'lĭs'kà)	71	40°56'N	94°56'W
Villupuram, India	159	11°59'N	79°33'E
Vilnius, Lith. (vĭl'nĕ-ŏs)	134	54°40'N	25°26'E
Vilppula, Fin. (vĭl'pŭ-lå)	123	62°01'N	24°24'E
Vil'shanka, Ukr.	133	48°14'N	30°52'E
Vil'shany, Ukr.	133	50°02'N	35°54'E
Vilvoorde, Bel.	115a	50°56'N	4°25'E
Vilyuy, r., Russia (vēl'yĭ)	135	63°00'N	121°00'E
Vilyuysk, Russia (vē-lyōō'ĭsk')	135	63°41'N	121°47'E
Vimmerby, Swe. (vĭm'ĕr-bü)	122	57°41'N	15°51'E
Vimperk, Czech Rep. (vĭm-pĕrk')	124	49°04'N	13°41'E
Viña del Mar, Chile (vē'nyä dĕl mär')	102	33°00'S	71°33'W
Vinalhaven, Me., U.S. (vī-nal-hā'vĕn)	60	44°03'N	68°49'W
Vinaròs, Spain	129	40°29'N	0°27'E
Vincennes, Fr. (văɴ-sĕn')	127b	48°51'N	2°27'E
Vincennes, In., U.S. (vĭn-zĕnz')	65	38°40'N	87°30'W
Vincent, Al., U.S. (vĭn'sĕnt)	82	33°21'N	86°25'W
Vindelälven, r., Swe.	116	65°02'N	18°30'E
Vindeln, Swe. (vĭn'dĕln)	116	64°10'N	19°52'E
Vindhya Range, mts., India (vĭnd'yä)	155	22°30'N	75°50'E
Vineland, N.J., U.S. (vīn'lănd)	67	39°30'N	75°00'W
Vinh, Viet. (vĕn'y')	168	18°38'N	105°42'E
Vinhais, Port. (vēn-yä'ēzh)	128	41°51'N	7°00'W
Vinings, Ga., U.S. (vī'nĭngz)	68c	33°52'N	84°28'W
Vinita, Ok., U.S. (vĭ-nē'tà)	79	36°38'N	95°09'W
Vinkovci, Cro. (vēn'kôv-tsē)	131	45°17'N	18°47'E
Vinnytsia, Ukr.	134	49°13'N	28°31'E
Vinnytsya, prov., Ukr.	133	48°45'N	28°01'E
Vinogradovo, Russia (vĭ-nô-grä'do-vô)	142b	55°25'N	38°33'E
Vinson Massif, mtn., Ant.	178	77°40'S	87°00'W
Vinton, Ia., U.S. (vĭn'tŭn)	71	42°08'N	92°01'W
Vinton, La., U.S.	81	30°12'N	93°35'W
Violet, La., U.S. (vī'ō-lĕt)	68d	29°54'N	89°54'W
Virac, Phil. (vē-räk')	165	13°38'N	124°20'E
Virbalis, Lith. (vēr'bá-lēs)	123	54°38'N	22°55'E
Virden, Can. (vûr'dĕn)	50	49°51'N	101°55'W
Virden, Il., U.S.	79	39°28'N	89°46'W
Virgin, r., U.S.	77	36°51'N	113°50'W
Virginia, S. Afr.	192c	28°07'S	26°54'E
Virginia, Mn., U.S. (vĕr-jĭn'yà)	65	47°32'N	92°36'W
Virginia, state, U.S.	65	37°00'N	80°45'W
Virginia Beach, Va., U.S.	67	36°50'N	75°58'W
Virginia City, Nv., U.S.	76	39°18'N	119°40'W
Virgin Islands, is., N.A. (vûr'jĭn)	87	18°15'N	64°00'W
Viroqua, Wi., U.S. (vĭ-rō'kwà)	71	43°33'N	90°54'W
Virovitica, Cro. (vē-rô-vē'tē-tsä)	131	45°50'N	17°24'E
Virpazar, Serb. (vĕr'pä-zär')	131	42°16'N	19°06'E
Virrat, Fin. (vĭr'ät)	123	62°15'N	23°45'E
Virserum, Swe. (vĭr'sĕ-rŏm)	122	57°22'N	15°35'E
Vis, Cro. (vēs)	130	43°03'N	16°11'E
Vis, i., Serb.	119	43°00'N	16°10'E
Visalia, Ca., U.S. (vĭ-sā'lĭ-à)	76	36°20'N	119°18'W
Visby, Swe. (vĭs'bü)	110	57°39'N	18°19'E
Viscount Melville Sound, strt., Can.	49	74°00'N	110°00'W
Višegrad, Bos. (vē'shē-gräd)	131	43°48'N	19°17'E
Vishākhapatnam, India	155	17°48'N	83°21'E
Vishera, r., Russia (vĭ'shĕ-rà)	142a	60°40'N	58°46'E
Vishnyakovo, Russia	142b	55°44'N	38°10'E
Vishoek, S. Afr.	186a	34°13'S	18°26'E
Visim, Russia (vē'sĭm)	142a	57°38'N	59°32'E
Viskan, r., Swe.	122	57°20'N	12°25'E
Viški, Lat. (vēs'kĭ)	123	56°02'N	26°47'E
Visoko, Bos. (vē'sô-kô)	131	43°59'N	18°10'E
Vistula see Wisła, r., Pol.	112	52°30'N	20°00'E
Vitebsk, prov., Bela.	132	55°05'N	29°18'E
Viterbo, Italy (vē-tĕr'bō)	118	42°24'N	12°08'E
Viti Levu, i., Fiji	170g	18°00'S	178°00'E
Vitim, Russia (vē'tēm)	135	59°22'N	112°43'E
Vitim, r., Russia (vē'tēm)	135	54°00'N	115°00'E
Vitino, Russia (vē'tĭ-nô)	142c	59°40'N	29°51'E
Vitória, Braz. (vē-tō'rē-ä)	101	20°09'S	40°17'W
Vitoria, Spain (vē-tō-ryä)	118	42°43'N	2°43'W
Vitória de Conquista, Braz. (vē-tô'rē-ä-dä-kōn-kwē's-tä)	101	14°51'S	40°44'W
Vitry-le-François, Fr. (vē-trē'lē-frän-swá')	126	48°44'N	4°34'E
Vitsyebsk, Bela. (vē'tyĕpsk)	136	55°12'N	30°16'E
Vittorio, Italy (vē-tô'rē-ô)	130	45°59'N	12°17'E
Viveiro, Spain	128	43°39'N	7°37'W
Vivian, La., U.S. (vĭv'ĭ-ăn)	81	32°51'N	93°59'W
Vizianagaram, India	155	18°10'N	83°29'E
Vlaardingen, Neth. (vlär'dĭng-ĕn)	121	51°54'N	4°20'E
Vladikavkaz, Russia	137	43°05'N	44°35'E
Vladimir, Russia (vlá-dyē'mēr)	134	56°08'N	40°24'E
Vladimir, prov., Russia (vlä-dyē'mēr)	132	56°08'N	39°53'E
Vladimiro-Aleksandrovskoye, Russia	166	42°50'N	133°00'E
Vladivostok, Russia (vlá-dē-vôs-tŏk')	135	43°06'N	131°47'E
Vlasenica, Bos. (vlä'sĕ-nēt'sà)	131	44°11'N	18°58'E
Vlasotince, Serb. (vlä'sô-tēn-tsĕ)	131	42°58'N	22°08'E
Vlieland, i., Neth. (vlē'länt)	121	53°19'N	4°55'E
Vlissingen, Neth. (vlĭs'sĭng-ĕn)	121	51°30'N	3°34'E
Vlorë, Alb.	119	40°27'N	19°30'E
Vltava, r., Czech Rep.	124	49°24'N	14°18'E
Vodl, l., Russia (vôd''l)	136	62°20'N	37°20'E
Voerde, Ger.	127c	51°35'N	6°41'E
Voghera, Italy (vô-gā'rä)	130	44°58'N	9°02'E
Voight, r., Wa., U.S.	74a	47°03'N	122°08'W
Voinjama, Lib.	188	8°25'N	9°45'W
Voiron, Fr. (vwà-rôɴ')	127	45°23'N	5°48'E
Voisin, Lac, l., Can. (vô-zĭn)	56	54°13'N	107°15'W
Volchansk, Ukr. (vôl-chänsk')	137	50°18'N	36°56'E
Volga, r., Russia (vôl'gä)	134	47°30'N	46°20'E
Volga, Mouths of the, mth.	137	46°00'N	49°10'E
Volgograd, Russia (vŏl-gō-grā't)	134	48°40'N	42°20'E
Volgogradskoye, res., Russia (vôl-gô-grad'skô-yĕ)	134	51°10'N	45°10'E
Volkhov, Russia (vôl'kôf)	123	59°54'N	32°21'E
Volkhov, r., Russia	136	58°45'N	31°40'E
Volodarskiy, Russia (vô-lô-där'skĭ)	142c	59°49'N	30°06'E
Volodymyr-Volyns'kyi, Ukr.	125	50°50'N	24°20'E
Vologda, Russia (vô'lŏg-dà)	134	59°12'N	39°52'E
Vologda, prov., Russia	132	59°00'N	37°26'E
Volokolamsk, Russia (vô-lô-kô'lamsk)	132	56°02'N	35°58'E
Volokonovka, Russia (vô-lô-kô'nôf-kà)	133	50°28'N	37°52'E
Vol'sk, Russia (vôl'sk)	137	52°02'N	47°23'E
Volta, r., Ghana	188	6°05'N	0°30'E
Volta, Lake, res., Ghana (vōl'tà)	184	7°10'N	0°30'W
Volta Blanche (White Volta), r., Afr.	188	11°30'N	0°40'W
Volta Noire see Black Volta, r., Afr.	184	11°30'N	4°00'W
Volta Redonda, Braz. (vôl'tä-rā-dôn'dä)	101	22°32'S	44°05'W
Volterra, Italy (vôl-tĕr'rä)	130	43°22'N	10°51'E
Voltri, Italy (vōl'trē)	130	44°25'N	8°45'E
Volturno, r., Italy (vôl-tōōr'nô)	130	41°12'N	14°20'E
Vólvi, Límni, l., Grc.	131	40°41'N	23°23'E
Volzhskoye, l., Russia (vôl'sh-skô-yĕ)	132	56°43'N	36°18'E
Von Ormy, Tx., U.S. (vŏn ôr'mē)	75d	29°18'N	98°36'W
Vōōpsu, Est. (vōōp'sô)	123	58°06'N	27°30'E
Voorburg, Neth.	115a	52°04'N	4°21'E
Voortrekkerhoogte, S. Afr.	187b	25°48'S	28°10'E
Vop', r., Russia (vôp)	132	55°20'N	32°55'E
Vopnafjördur, Ice.	116	65°43'N	14°58'W
Vordingborg, Den. (vôr'dĭng-bôr)	122	55°10'N	11°55'E
Vóreioi Sporades, is., Grc.	131	38°55'N	24°05'E
Vóreios Evvoïkós Kólpos, b., Grc.	131	38°48'N	23°02'E
Vorkuta, Russia (vôr-kōō'tà)	134	67°28'N	63°40'E
Vormsi, i., Est. (vôrm'sĭ)	123	59°06'N	23°05'E
Vorona, r., Russia (vô-rô'na)	137	51°50'N	42°00'E
Voronava, Bela.	125	54°07'N	25°16'E
Voronezh, Russia (vô-rô'nyĕzh)	134	51°39'N	39°11'E
Voronezh, prov., Russia	133	51°10'N	39°13'E
Voronezh, r., Russia	137	52°17'N	39°32'E
Vorontsovka, Russia (vô-rônt'sôv-kà)	142a	59°40'N	60°14'E
Voron'ya, r., Russia (vô-rônyá)	136	68°20'N	35°20'E
Võrts-Järv, l., Est. (vôrts yärv)	123	58°15'N	26°12'E
Võru, Est. (vô'rû)	136	57°50'N	26°58'E
Vorya, r., Russia (vôr'yà)	142b	55°55'N	38°15'E
Vosges, mts., Fr. (vōzh)	117	48°09'N	6°57'E
Voskresensk, Russia (vôs-krĕ-sĕnsk')	142b	55°20'N	38°42'E
Voss, Nor. (vôs)	116	60°40'N	6°24'E
Vostryakovo, Russia	142b	55°23'N	37°49'E
Votkinsk, Russia (vôt-kēnsk')	136	57°00'N	54°00'E
Votkinskoye Vodokhranilishche, res., Russia	136	57°30'N	55°00'E
Vouga, r., Port. (vō'gá)	128	40°43'N	7°51'W
Vouziers, Fr. (vōō-zyā')	126	49°25'N	4°40'E
Voxnan, r., Swe.	122	61°30'N	15°24'E
Voyageurs National Park, rec., Mn., U.S.	71	48°30'N	92°40'W
Vozhe, l., Russia (vôzh'yĕ)	136	60°40'N	39°00'E
Voznesens'k, Ukr.	137	47°34'N	31°22'E
Vradiïvka, Ukr.	133	47°51'N	30°38'E
Vrangelya (Wrangel), i., Russia	135	71°25'N	178°30'W
Vranje, Serb. (vrän'yĕ)	131	42°33'N	21°55'E
Vratsa, Blg. (vrät'tsà)	119	43°12'N	23°31'E
Vrbas, Serb. (v'r'bäs)	131	45°34'N	19°43'E
Vrbas, r., Serb.	131	44°25'N	17°17'E
Vrchlabí, Czech Rep. (v'r'chlä-bē)	124	50°32'N	15°51'E
Vrede, S. Afr. (vrĭ'dĕ)(vrēd)	192c	27°25'S	29°11'E
Vredefort, S. Afr. (vrĭ'dĕ-fŏrt)(vrēd'fōrt)	192c	27°00'S	27°21'E
Vreeswijk, Neth.	115a	52°00'N	5°06'E
Vršac, Serb. (v'r'shäts)	119	45°08'N	21°18'E
Vrutky, Slvk. (vrōōt'kē)	125	49°09'N	18°55'E
Vryburg, S. Afr. (vrī'bûrg)	186	26°55'S	24°45'E
Vryheid, S. Afr. (vrī'hīt)	186	27°43'S	30°58'E
Vsetín, Czech Rep. (fsĕt'yēn)	125	49°21'N	18°01'E
Vsevolozhskiy, Russia (vsyĕ'vôlô'zh-skēē)	142c	60°01'N	30°41'E
Vuelta Abajo, reg., Cuba (vwĕl'tä ä-bä'hō)	92	22°20'N	83°45'W
Vught, Neth.	115a	51°38'N	5°18'E
Vukovar, Cro. (vò'kô-vär)	131	45°20'N	19°00'E
Vulcan, Mi., U.S. (vŭl'kăn)	66	45°45'N	87°50'W
Vulcano, i., Italy (vōōl-kä'nô)	130	38°23'N	15°00'E
Vûlchedrŭma, Blg.	131	43°43'N	23°29'E
Vuntut National Park, rec., Can.	52	68°27'N	139°58'W
Vyartsilya, Russia (vyár-tsē'lyá)	123	62°10'N	30°40'E
Vyatka, r., Russia (vyát'ká)	136	59°20'N	51°25'E
Vyazemskiy, Russia (vyá-zĕm'skĭ)	166	47°29'N	134°39'E
Vyaz'ma, Russia (vyáz'má)	136	55°12'N	34°17'E
Vyazniki, Russia (vyáz'nē-kē)	136	56°10'N	42°10'E
Vyborg, Russia (vwē'bôrk)	134	60°43'N	28°46'E
Vychegda, r., Russia (vē'chĕg-dà)	136	61°40'N	48°00'E
Vyerkhnyadzvinsk, Bela.	132	55°48'N	27°59'E
Vyetka, Bela. (vyĕt'ká)	132	52°36'N	31°05'E
Vylkove, Ukr.	137	45°24'N	29°36'E
Vym, r., Russia (vwēm)	136	63°15'N	51°20'E
Vyritsa, Russia (vē'rĭ-tsä)	142c	59°24'N	30°20'E
Vyshnevolotskoye, l., Russia (vŭy'sh-ņĕ'vôlôt's-kô'yĕ)	132	57°30'N	34°27'E
Vyshniy Volochëk, Russia (vĕsh'nyĭ vôl-ô-chĕk')	134	57°34'N	34°35'E
Vyškov, Czech Rep. (vêsh'kôf)	124	49°17'N	16°58'E
Vysoké Mýto, Czech Rep. (vû'sô-kä mû'tô)	124	49°58'N	16°07'E
Vysokovsk, Russia (vĭ-sô'kôfsk)	132	56°16'N	36°32'E
Vytegra, Russia (vû'tĕg-rà)	134	61°00'N	36°20'E
Vyzhnytsia, Ukr.	125	48°16'N	25°12'E

W

PLACE (Pronunciation)	PAGE	LAT.	LONG.
W, Parcs Nationaux du, rec., Niger	189	12°20'N	2°40'E
Waal, r., Neth. (väl)	121	51°46'N	5°00'E
Waalwijk, Neth.	115a	51°41'N	5°05'E
Wabamun, Grc.	119	39°23'N	22°56'E
Wabamuno, Can. (wŏ'bă-mŭn)	55	53°33'N	114°28'W
Wabasca, Can. (wō-bäs'kă)	55	56°00'N	113°53'W
Wabash, In., U.S. (wô'băsh)	66	40°45'N	85°50'W
Wabash, r., U.S.	65	38°00'N	88°00'W
Wabasha, Mn., U.S. (wä'bá-shô)	71	44°24'N	92°04'W
Wabe Gestro, r., Eth.	185	6°25'N	41°21'E
Wabowden, Can. (wä-bō'd'n)	57	54°55'N	98°38'W
Wąbrzeźno, Pol. (vôŋ-bzĕzh'nô)	125	53°17'N	18°59'E
Wabu Hu, l., China (wä-bōō hōō)	162	32°25'N	116°35'E
W. A. C. Bennett Dam, dam, Can.	55	56°01'N	122°10'W
Waccamaw, r., S.C., U.S. (wăk'à-mô)	83	33°47'N	78°55'W
Waccasassa Bay, b., Fl., U.S. (wă-kà-să'sà)	82	29°02'N	83°10'W
Wachow, Ger. (vä'κōv)	115b	53°32'N	12°46'E
Waco, Tx., U.S. (wā'kō)	64	31°35'N	97°06'W
Waconda Lake, res., Ks., U.S.	78	39°45'N	98°15'W
Wadayama, Japan (wä'dä'yä-mä)	167	35°19'N	134°49'E
Waddenzee, sea, Neth.	121	53°00'N	4°50'E
Waddington, Mount, mtn., Can. (wŏd'dĭng-tŭn)	52	51°23'N	125°15'W
Wadena, Can.	56	51°57'N	103°50'W
Wadena, Mn., U.S. (wŏ-dē'nà)	70	46°26'N	95°09'W
Wadesboro, N.C., U.S. (wādz'bŭr-ô)	83	34°57'N	80°05'W
Wadley, Ga., U.S. (wŭd'lē)	83	32°54'N	82°25'W
Wad Madani, Sudan (wäd mē-dä'nē)	185	14°27'N	33°31'E
Wadowice, Pol. (vá-dô'vēt-sĕ)	125	49°53'N	19°31'E
Wadsworth, Oh., U.S. (wŏdz'wûrth)	69d	41°01'N	81°44'W
Wager Bay, b., Can. (wā'jēr)	53	65°48'N	88°19'W
Wagga Wagga, Austl. (wôg'à wôg'à)	175	35°10'S	147°30'E
Wagoner, Ok., U.S. (wăg'ŭn-ĕr)	79	35°58'N	95°22'W
Wagon Mound, N.M., U.S. (wăg'ŭn mound)	78	35°59'N	104°45'W
Wągrowiec, Pol. (vôŋ-grô'vyĕts)	125	52°47'N	17°14'E
Waha, Libya	156	28°16'N	19°54'E
Wahiawā, Hi., U.S.	64d	21°30'N	158°03'W
Wahoo, Ne., U.S. (wä-hōō')	70	41°14'N	96°39'W
Wahpeton, N.D., U.S. (wô'pē-tŭn)	70	46°17'N	96°38'W
Waialua, Hi., U.S. (wä'ē-ä-lōō'ä)	84a	21°33'N	158°08'W
Wai'anae, Hi., U.S. (wä'ē-à-nä'ä)	84a	21°25'N	158°11'W
Waidhofen, Aus. (vīd'hôf-ĕn)	124	47°58'N	14°46'E
Waigeo, Pulau, i., Indon. (wä-ē-gä'ô)	169	0°07'N	131°00'E
Waikato, r., N.Z. (wä'ē-kä'to)	175a	38°10'S	175°35'E
Waikerie, Austl. (wā'kēr-ē)	176	34°15'S	140°00'E
Wailuku, Hi., U.S. (wä'ē-lōō'kōō)	64c	20°55'N	156°30'W
Waimānalo, Hi., U.S. (wä-ē-mä'nä-lo)	84a	21°19'N	157°43'W
Waimea, Hi., U.S. (wä-ē-mā'ä)	84a	21°56'N	159°38'W
Wainganga, r., India (wä-ēn-gŭn'gä)	155	20°30'N	80°15'E
Waingapu, Indon.	168	9°32'S	120°00'E
Wainwright, Can.	50	52°49'N	110°52'W
Wainwright, Ak., U.S. (wān-rīt)	63	74°40'N	159°00'W
Waipahu, Hi., U.S. (wä'ē-pä'hōō)	64d	21°20'N	158°02'W

ăt; finăl; rāte; senâte; ärm; åsk; sofà; fâre; ch-choose; dh-as th in other; bē; ēvent; bĕt; recĕnt; cratẽr; g-gō; gh-guttural g; bĭt; ĭ-short neutral; rīde; κ-guttural k as ch in German ich;

PRONOUNCING INDEX
Waiska – Webster Springs 317

PLACE (Pronunciation)	PAGE	LAT.	LONG.
Waiska, r., Mi., U.S. (wȧ-ĭz-kȧ)	75k	46°20'N	84°38'W
Waitsburg, Wa., U.S. (wāts'bûrg)	72	46°17'N	118°08'W
Wajima, Japan (wä'jē-mȧ)	167	37°23'N	136°56'E
Wajir, Kenya	191	1°45'N	40°04'E
Wakami, r., Can.	58	47°43'N	82°22'W
Wakasa-Wan, b., Japan (wä'kä-sä wän)	166	35°43'N	135°39'E
Wakatipu, l., N.Z. (wä-kä-tē'po͞o)	175a	45°04'S	168°30'E
Wakayama, Japan (wä-kä'yä-mä)	161	34°14'N	135°11'E
Wake, i., Oc. (wāk)	3	19°25'N	167°00'E
Wa Keeney, Ks., U.S. (wô-kē'nē)	78	39°01'N	99°53'W
Wakefield, Can. (wāk-fēld)	62c	45°39'N	75°55'W
Wakefield, Eng., U.K.	120	53°41'N	1°25'W
Wakefield, Ma., U.S.	61a	42°31'N	71°05'W
Wakefield, Mi., U.S.	71	46°28'N	89°55'W
Wakefield, Ne., U.S.	70	42°15'N	96°52'W
Wakefield, R.I., U.S.	68b	41°26'N	71°30'W
Wakefield, co., Eng., U.K.	114a	53°12'N	1°25'W
Wake Forest, N.C., U.S. (wāk fŏr'ĕst)	83	35°58'N	78°31'W
Waki, Japan (wä'kē)	167	34°05'N	134°10'E
Wakkanai, Japan (wä'kä-nä'ē)	161	45°19'N	141°43'E
Wakkerstroom, S. Afr. (vȧk'ēr-strōm)(wăk'ēr-stro͞om)	186	27°19'S	30°04'E
Wakonassin, r., Can.	58	46°35'N	82°10'W
Waku Kundo, Ang.	186	11°25'S	15°07'E
Wałbrzych, Pol. (väl'bzhûk)	124	50°46'N	16°16'E
Walcott, Lake, res., Id., U.S.	73	42°40'N	113°23'W
Wałcz, Pol. (välch)	124	53°11'N	16°30'E
Waldoboro, Me., U.S. (wôl'dô-bŭr-ô)	60	44°06'N	69°22'W
Waldo Lake, l., Or., U.S. (wôl'dō)	72	43°46'N	122°10'W
Waldorf, Md., U.S. (wăl'dôrf)	68e	38°37'N	76°57'W
Waldron, Mo., U.S.	75f	39°14'N	94°47'W
Waldron, i., Wa., U.S.	74d	48°42'N	123°02'W
Wales, Ak., U.S. (wālz)	63	65°35'N	168°14'W
Wales, state, U.K.	110	52°12'N	3°40'W
Walewale, Ghana	188	10°21'N	0°48'W
Walgett, Austl. (wôl'gĕt)	175	30°00'S	148°10'E
Walhalla, S.C., U.S. (wŭl-hăl'ȧ)	82	34°45'N	83°04'W
Walikale, D.R.C.	191	1°25'S	28°03'E
Walkden, Eng., U.K.	114a	53°32'N	2°24'W
Walker, Mn., U.S. (wôk'ēr)	71	47°06'N	94°37'W
Walker, r., Nv., U.S.	76	39°07'N	119°10'W
Walker, Mount, mtn., Wa., U.S.	74a	47°47'N	122°54'W
Walker Lake, l., Can.	57	54°42'N	96°57'W
Walker Lake, l., Nv., U.S.	76	38°46'N	118°30'W
Walker River Indian Reservation, I.R., Nv., U.S.	76	39°06'N	118°20'W
Walkerville, Mt., U.S. (wôk'ēr-vĭl)	73	46°20'N	112°32'W
Wallace, Id., U.S. (wŏl'ās)	72	47°27'N	115°55'W
Wallaceburg, Can.	58	42°39'N	82°25'W
Wallacia, Austl.	173b	33°52'S	150°40'E
Wallaroo, Austl. (wŏl-ȧ-ro͞o)	174	33°52'S	137°45'E
Wallasey, Eng., U.K. (wŏl'ȧ-sē)	114a	53°25'N	3°03'W
Walla Walla, Wa., U.S. (wŏl'ȧ wŏl'ȧ)	64	46°03'N	118°20'W
Walled Lake, Mi., U.S. (wôl'd lāk)	69b	42°32'N	83°29'W
Wallel, Tulu, mtn., Eth.	185	9°00'N	34°52'E
Wallingford, Eng., U.K. (wŏl'ĭng-fērd)	114b	51°34'N	1°08'W
Wallingford, Vt., U.S.	67	43°30'N	72°55'W
Wallis and Futuna Islands, dep., Oc.	194	13°00'S	176°10'E
Wallisville, Tx., U.S. (wŏl'ĭs-vĭl)	81a	29°50'N	94°44'W
Wallowa, Or., U.S. (wôl'ô-wȧ)	72	45°34'N	117°32'W
Wallowa, r., Or., U.S.	72	45°28'N	117°28'W
Wallowa Mountains, mts., Or., U.S.	72	45°10'N	117°22'W
Wallula, Wa., U.S.	72	46°08'N	118°55'W
Walnut, Ca., U.S. (wôl'nŭt)	75a	34°00'N	117°51'W
Walnut, r., Ks., U.S.	79	37°28'N	97°06'W
Walnut Canyon National Mon, rec., Az., U.S.	77	35°10'N	111°30'W
Walnut Creek, Ca., U.S.	74b	37°54'N	122°04'W
Walnut Creek, r., Tx., U.S.	75c	32°37'N	97°03'W
Walnut Ridge, Ar., U.S. (rĭj)	79	36°04'N	90°56'W
Walpole, Ma., U.S. (wôl'pōl)	61a	42°09'N	71°15'W
Walpole, N.H., U.S.	67	43°05'N	72°25'W
Walsall, Eng., U.K. (wôl-sôl)	120	52°35'N	1°58'W
Walsenburg, Co., U.S. (wôl'sĕn-bûrg)	78	37°38'N	104°46'W
Walsum, Ger.	127c	51°32'N	6°41'E
Walter F. George Reservoir, res., U.S.	82	32°00'N	85°00'W
Walters, Ok., U.S. (wôl'tērz)	78	34°21'N	98°19'W
Waltham, Ma., U.S. (wôl'thȧm)	61a	42°22'N	71°14'W
Walthamstow, Eng., U.K. (wôl'tȧm-stō)	114b	51°34'N	0°01'W
Walton, N.Y., U.S.	67	42°10'N	75°05'W
Walton-le-Dale, Eng., U.K. (lē-dāl')	114a	53°44'N	2°40'W
Walvis Bay, Nmb. (wôl'vĭs)	186	22°50'S	14°30'E
Walworth, Wi., U.S. (wôl'wŭrth)	71	42°33'N	88°39'W
Wama, Ang.	190	12°14'S	15°33'E
Wamba, r., D.R.C.	186	7°00'S	18°00'E
Wamego, Ks., U.S. (wŏ-mē'gō)	79	39°13'N	96°17'W
Wami, r., Tan. (wä'mē)	187	6°31'S	37°17'E
Wanapitei Lake, l., Can.	59	46°45'N	80°45'W
Wanaque, N.J., U.S. (wŏn'ȧ-kū)	68a	41°03'N	74°16'W
Wanaque Reservoir, res., N.J., U.S.	68a	41°06'N	74°20'W
Wanda Shan, mts., China (wän-dä shän)	161	45°54'N	131°45'E
Wandoan, Austl.	176	26°09'S	149°51'E
Wandsbek, Ger. (vänds'bĕk)	115c	53°34'N	10°07'E
Wandsworth, Eng., U.K. (wôndz'wûrth)	114b	51°26'N	0°12'W
Wanganui, N.Z. (wôŋ'gȧ-no͞o'ē)	175a	39°53'S	175°01'E
Wangaratta, Austl. (wŏŋ'gȧ-răt'ȧ)	176	36°23'S	146°18'E
Wangeroog, i., Ger. (vän'gĕ-rōg)	124	53°49'N	7°57'E
Wangqingtuo, China (wäŋ-chyĭŋ-twô)	162	39°14'N	116°56'E
Wangsi, China (wäŋ-sē)	162	37°59'N	116°57'E
Wantage, Eng., U.K. (wŏn'tȧj)	114b	51°33'N	1°26'W
Wantagh, N.Y., U.S.	68a	40°41'N	73°30'W
Wanxian, China (wän shyĕn)	162	38°51'N	115°10'E
Wanxian, China (wän-shyĕn)	160	30°48'N	108°22'E
Wanzai, China (wän-dzī)	165	28°05'N	114°25'E
Wanzhi, China (wän-jr)	162	31°11'N	118°31'E
Wapakoneta, Oh., U.S. (wä'pȧ-kô-nĕt'ȧ)	66	40°35'N	84°10'W
Wapawekka Hills, hills, Can. (wŏ'pă-wĕ'kă-hĭlz)	56	54°45'N	104°20'W
Wapawekka Lake, l., Can.	56	54°55'N	104°40'W
Wapello, Ia., U.S. (wŏ-pĕl'ō)	71	41°10'N	91°11'W
Wappapello Reservoir, res., Mo., U.S. (wă'pȧ-pĕl-lō)	65	37°07'N	90°10'W
Wappingers Falls, N.Y., U.S. (wŏp'ĭn-jērz)	67	41°35'N	73°55'W
Wapsipinicon, r., Ia., U.S. (wŏp'sĭ-pĭn'ĭ-kŏn)	71	42°16'N	91°35'W
Wapusk National Park, rec., Can.	52	58°00'N	94°15'W
Warabi, Japan (wä'rä-bē)	167a	35°50'N	139°41'E
Warangal, India (wŭ'raŋ-gál)	155	18°03'N	79°45'E
Warburton, The, r., Austl. (wŏr'bûr-tŭn)	174	27°30'S	138°45'E
Wardān, Wādī, r., Egypt	153a	29°22'N	33°00'E
Ward Cove, Ak., U.S.	54	55°24'N	131°43'W
Warden, S. Afr. (wôr'dĕn)	192c	27°52'S	28°59'E
Wardha, India (wŭr'dä)	155	20°46'N	78°42'E
War Eagle, W.V., U.S. (wôr ē'g'l)	66	37°30'N	81°50'W
Waren, Ger. (vä'rĕn)	124	53°32'N	12°43'E
Warendorf, Ger. (vä'rĕn-dōrf)	127c	51°57'N	7°59'E
Wargla, Alg.	184	32°00'N	5°18'E
Warialda, Austl.	176	29°32'S	150°34'E
Warmbad, Nmb. (värm'bäd) (wôrm'băd)	186	28°25'S	18°45'E
Warmbad, S. Afr.	192c	24°52'S	28°18'E
Warm Beach, Wa., U.S. (wôrm)	74a	48°10'N	122°22'W
Warm Springs Indian Reservation, I.R., Or., U.S. (wôrm sprĭnz)	72	44°55'N	121°30'W
Warm Springs Reservoir, res., Or., U.S.	72	43°42'N	118°40'W
Warner Mountains, mts., Ca., U.S.	64	41°30'N	120°17'W
Warner Robins, Ga., U.S.	82	32°37'N	83°36'W
Warnow, r., Ger. (vär'nō)	124	53°51'N	11°55'E
Warracknabeal, Austl.	176	36°20'S	142°28'E
Warragamba Reservoir, res., Austl.	176	33°40'S	150°00'E
Warrego, r., Austl. (wôr'ē-gô)	175	27°13'S	145°58'E
Warren, Can.	62f	50°08'N	97°32'W
Warren, Ar., U.S. (wŏr'ĕn)	79	33°37'N	92°03'W
Warren, In., U.S.	66	40°40'N	85°25'W
Warren, Mi., U.S.	69b	42°33'N	83°03'W
Warren, Mn., U.S.	70	48°11'N	96°44'W
Warren, Oh., U.S.	66	41°15'N	80°50'W
Warren, Or., U.S.	74c	45°49'N	122°51'W
Warren, Pa., U.S.	67	41°50'N	79°10'W
Warren, R.I., U.S.	68b	41°44'N	71°14'W
Warrendale, Pa., U.S. (wŏr'ĕn-dāl)	69e	40°39'N	80°04'W
Warrensburg, Mo., U.S. (wŏr'ĕnz-bûrg)	79	38°45'N	93°42'W
Warrenton, Ga., U.S. (wŏr'ĕn-tŭn)	83	33°26'N	82°37'W
Warrenton, Or., U.S.	74c	46°10'N	123°56'W
Warrenton, Va., U.S.	67	38°45'N	77°50'W
Warri, Nig. (wär'ē)	184	5°33'N	5°43'E
Warrington, Eng., U.K.	114a	53°22'N	2°30'W
Warrington, Fl., U.S. (wŏ'ĭng-tŭn)	82	30°21'N	87°15'W
Warrnambool, Austl. (wôr'nȧm-bo͞ol)	175	38°20'S	142°28'E
Warroad, Mn., U.S. (wôr'rōd)	70	48°55'N	95°20'W
Warrumbungle Range, mts., Austl. (wôr'ŭm-bŭŋ-g'l)	175	31°18'S	150°00'E
Warsaw, Pol.	110	52°15'N	21°05'E
Warsaw, Il., U.S. (wôr'sô)	79	40°21'N	91°26'W
Warsaw, In., U.S.	66	41°15'N	85°50'W
Warsaw, N.Y., U.S.	67	42°45'N	78°10'W
Warsaw, NC, N.C., U.S.	83	35°00'N	78°07'W
Warsop, Eng., U.K. (wôr'sŭp)	114a	53°13'N	1°05'W
Warszawa *see* Warsaw, Pol.	110	52°15'N	21°05'E
Warta, r., Pol. (vär'tá)	117	52°30'N	16°00'E
Wartburg, S. Afr.	187c	29°26'S	30°39'E
Warwick, Austl. (wŏr'ĭk)	175	28°05'S	152°10'E
Warwick, Can.	59	45°58'N	71°57'W
Warwick, Eng., U.K.	120	52°19'N	1°46'W
Warwick, N.Y., U.S.	68a	41°15'N	74°22'W
Warwick, R.I., U.S.	67	41°42'N	71°27'W
Warwickshire, co., Eng., U.K.	114a	52°30'N	1°35'W
Wasatch Mountains, mts., Ut., U.S. (wô'săch)	75b	40°45'N	111°46'W
Wasatch Plateau, plat., Ut., U.S.	77	38°55'N	111°40'W
Wasatch Range, mts., U.S.	64	39°10'N	111°30'W
Wasbank, S. Afr.	187c	28°27'S	30°09'E
Wasco, Or., U.S. (wăs'kō)	72	45°36'N	120°42'W
Waseca, Mn., U.S. (wô-sē'kȧ)	71	44°04'N	93°31'W
Wash, The, Eng., U.K. (wŏsh)	116	53°00'N	0°20'E
Washburn, Me., U.S. (wŏsh'bŭrn)	60	46°46'N	68°10'W
Washburn, Wi., U.S.	71	46°41'N	90°55'W
Washburn, Mount, mtn., Wy., U.S.	73	44°55'N	110°10'W
Washington, D.C., U.S. (wŏsh'ĭng-tŭn)	65	38°50'N	77°00'W
Washington, Ga., U.S.	83	33°43'N	82°46'W
Washington, Ia., U.S.	71	41°17'N	91°42'W
Washington, In., U.S.	66	38°40'N	87°10'W
Washington, Ks., U.S.	79	39°48'N	97°04'W
Washington, Mo., U.S.	79	38°33'N	91°00'W
Washington, N.C., U.S.	83	35°32'N	77°01'W
Washington, Pa., U.S.	66	40°10'N	80°14'W
Washington, state, U.S.	64	47°30'N	121°10'W
Washington, i., Wi., U.S.	71	45°18'N	86°42'W
Washington, Lake, l., Wa., U.S.	74a	47°34'N	122°12'W
Washington, Mount, mtn., N.H., U.S.	65	44°15'N	71°15'W
Washington Court House, Oh., U.S.	66	39°30'N	83°25'W
Washington Park, Il., U.S.	75e	38°38'N	90°06'W
Washita, r., Ok., U.S. (wŏsh'ĭ-tô)	78	35°33'N	99°16'W
Washougal, Wa., U.S. (wŏ-sho͞o'gál)	74c	45°35'N	122°21'W
Washougal, r., Wa., U.S.	74c	45°38'N	122°17'W
Wasilków, Pol. (vá-sēl'kōf)	125	53°12'N	23°13'E
Waskaiowaka Lake, l., Can. (wŏ'skă-yō'wō-kă)	57	56°30'N	96°20'W
Wassenberg, Ger. (vä'sĕn-bĕrgh)	127c	51°06'N	6°07'E
Wassuk Range, mts., Nv., U.S. (wăs'sŭk)	76	38°58'N	119°00'W
Waswanipi, Lac, l., Can.	59	49°35'N	76°15'W
Water, i., V.I.U.S. (wô'tēr)	87c	18°20'N	64°57'W
Waterberge, mts., S. Afr. (wôrtēr'bûrg)	192c	24°25'S	27°53'E
Waterboro, S.C., U.S. (wô'tēr-bûr-ō)	83	32°50'N	80°40'W
Waterbury, Ct., U.S. (wô'tēr-bĕr-ē)	67	41°30'N	73°00'W
Water Cay, i., Bah.	93	22°55'N	75°50'W
Waterdown, Can. (wô'tēr-doun)	62d	43°20'N	79°54'W
Wateree Lake, res., S.C., U.S. (wô'tēr-ē)	83	34°40'N	80°48'W
Waterford, Ire. (wô'tēr-fērd)	117	52°20'N	7°03'W
Waterford, Wi., U.S.	69a	42°46'N	88°13'W
Waterloo, Bel.	115a	50°44'N	4°24'E
Waterloo, Can. (wô-tēr-lo͞o')	59	43°30'N	80°40'W
Waterloo, Can.	59	45°25'N	72°30'W
Waterloo, Ia., U.S.	65	42°30'N	92°22'W
Waterloo, Il., U.S.	79	38°19'N	90°08'W
Waterloo, Md., U.S.	68e	39°11'N	76°50'W
Waterloo, N.Y., U.S.	67	42°55'N	76°50'W
Waterton-Glacier International Peace Park, rec., N.A. (wô'ter-tŭn-glā'shûr)	64	48°55'N	114°10'W
Waterton Lakes National Park, rec., Can.	55	49°05'N	113°50'W
Watertown, Ma., U.S. (wô'tēr-toun)	61a	42°22'N	71°11'W
Watertown, N.Y., U.S.	65	44°00'N	75°55'W
Watertown, S.D., U.S.	64	44°53'N	97°07'W
Watertown, Wi., U.S.	71	43°13'N	88°40'W
Water Valley, Ms., U.S. (văl'ē)	82	34°08'N	89°38'W
Waterville, Me., U.S.	60	44°34'N	69°37'W
Waterville, Mn., U.S.	71	44°10'N	93°35'W
Waterville, Wa., U.S.	72	47°38'N	120°04'W
Watervliet, N.Y., U.S. (wô'tēr-vlēt')	67	42°45'N	73°54'W
Watford, Eng., U.K. (wŏt'fōrd)	120	51°38'N	0°24'W
Wathaman Lake, l., Can.	56	56°55'N	103°43'W
Watlington, Eng., U.K.	114b	51°37'N	1°01'W
Watonga, Ok., U.S. (wŏ-tôŋ'gȧ)	79	35°50'N	98°26'E
Watsa, D.R.C. (wät'sä)	185	3°03'N	29°32'E
Watseka, Il., U.S. (wŏt-sē'kȧ)	66	40°45'N	87°45'W
Watson, In., U.S. (wŏt'sŭn)	69h	38°21'N	85°42'W
Watson Lake, Can.	50	60°18'N	128°50'W
Watsonville, Ca., U.S. (wŏt'sŭn-vĭl)	76	36°55'N	121°46'W
Wattenscheid, Ger. (vä'tĕn-shīd)	127c	51°30'N	7°07'E
Watts, Ca., U.S. (wŏts)	75a	33°56'N	118°15'W
Watts Bar Lake, res., Tn., U.S. (bär)	82	35°45'N	84°49'W
Waubay, S.D., U.S. (wô'bā)	70	45°19'N	97°18'W
Wauchula, Fl., U.S. (wô-cho͞o'lȧ)	83a	27°32'N	81°48'W
Wauconda, Il., U.S. (wô-kŏn'dȧ)	69a	42°15'N	88°08'W
Waukegan, Il., U.S. (wô-kē'gȧn)	65	42°22'N	87°51'W
Waukesha, Wi., U.S. (wô'kē-shô)	69a	43°01'N	88°13'W
Waukon, Ia., U.S. (wô kŏn)	71	43°15'N	91°30'W
Waupaca, Wi., U.S. (wô-păk'ȧ)	71	44°22'N	89°06'W
Waupun, Wi., U.S. (wô-pŭn')	71	43°37'N	88°45'W
Waurika, Ok., U.S. (wô-rē'kȧ)	79	34°09'N	97°59'W
Wausau, Wi., U.S. (wô'sô)	65	44°58'N	89°40'W
Wausaukee, Wi., U.S. (wô-sô'kē)	71	45°22'N	87°58'W
Wauseon, Oh., U.S. (wô'sē-ŏn)	66	41°30'N	84°10'W
Wautoma, Wi., U.S. (wô-tō'mȧ)	71	44°04'N	89°11'W
Wauwatosa, Wi., U.S. (wô-wȧ-t'ō'sȧ)	69a	43°03'N	88°00'W
Waveney, r., Eng., U.K. (wāv'nē)	121	52°27'N	1°17'E
Waverly, S. Afr.	187c	31°54'S	26°29'E
Waverly, Ia., U.S. (wā'vēr-lē)	71	42°43'N	92°29'W
Waverly, Tn., U.S.	82	36°04'N	87°46'W
Wāw, Sudan	185	7°41'N	28°00'E
Wawa, Can.	58	47°59'N	84°47'W
Wāw al-Kabir, Libya	185	25°23'N	16°52'E
Wawanesa, Can. (wŏ'wō-nē'sä)	57	49°36'N	99°41'W
Wawasee, l., In., U.S. (wô-wô-sē')	66	41°25'N	85°45'W
Waxahachie, Tx., U.S. (wăk-sȧ-hăch'ē)	81	32°23'N	96°50'W
Wayland, Ky., U.S. (wā'lănd)	83	37°25'N	82°47'W
Wayland, Ma., U.S.	61a	42°23'N	71°22'W
Wayne, Mi., U.S.	69b	42°17'N	83°23'W
Wayne, Ne., U.S.	70	42°13'N	97°03'W
Wayne, N.J., U.S.	68a	40°56'N	74°16'W
Wayne, Pa., U.S.	68f	40°03'N	75°22'W
Waynesboro, Ga., U.S. (wānz'bŭr-ô)	83	33°05'N	82°02'W
Waynesboro, Pa., U.S.	67	39°45'N	77°35'W
Waynesboro, Va., U.S.	67	38°05'N	78°50'W
Waynesburg, Pa., U.S. (wānz'bûrg)	66	39°55'N	80°10'W
Waynesville, N.C., U.S. (wānz'vĭl)	83	35°28'N	82°58'W
Waynoka, Ok., U.S. (wā-nō'kȧ)	78	36°34'N	98°52'W
Wayzata, Mn., U.S. (wā-zä-tȧ)	75g	44°58'N	93°31'W
Wazīrbad, Pak.	158	32°39'N	74°11'E
Weagamow Lake, l., Can. (wē'ăg-ȧ-mou)	57	52°53'N	91°22'W
Weald, The, reg., Eng., U.K. (wēld)	120	50°58'N	0°15'W
Weatherford, Ok., U.S. (wĕ-dhēr-fērd)	78	85°32'N	98°41'W
Weatherford, Tx., U.S.	81	32°45'N	97°46'W
Weaver, r., Eng., U.K. (wē'vēr)	114a	53°09'N	2°31'W
Weaverville, Ca., U.S. (wē'vēr-vĭl)	72	40°44'N	122°55'W
Webb City, Mo., U.S.	79	37°10'N	94°26'W
Weber, r., Ut., U.S.	75b	41°13'N	112°07'W
Webster, Ma., U.S.	61a	42°04'N	71°52'W
Webster, S.D., U.S.	70	45°19'N	97°30'W
Webster City, Ia., U.S.	71	42°28'N	93°49'W
Webster Groves, Mo., U.S. (grōvz)	75e	38°36'N	90°22'W
Webster Springs, W.V., U.S. (sprĭngz)	66	38°30'N	80°20'W

ng-sing; ŋ-baŋk; ɴ-nasalized n; nŏd; cŏmmit; ōld; ôbey; ôrder; oi-boil; fo͞od; ȯ-as oo in foot; ou-out; s-soft; sh-dish; th-thin; pūre; ůnite; ûrn; stŭd; circŭs; ü-as in French tu; '-indeterminate vowel.

318 Weddell Sea – White Rock PRONOUNCING INDEX

PLACE (Pronunciation)	PAGE	LAT.	LONG.
Weddell Sea, sea, Ant. (wĕd′ĕl)	178	73°00′S	45°00′W
Wedel, Ger. (vā′dĕl)	115c	53°35′N	9°42′E
Wedge Mountain, mtn., Can. (wĕj)	55	50°10′N	122°50′W
Wedgeport, Can. (wĕj′pōrt)	60	43°44′N	65°59′W
Wednesfield, Eng., U.K. (wĕd″nz-fēld)	114a	52°36′N	2°04′W
Weed, Ca., U.S. (wēd)	72	41°35′N	122°21′W
Weenen, S. Afr. (vā′nĕn)	187c	28°52′S	30°05′E
Weert, Neth.	121	51°16′N	5°39′E
Weesp, Neth.	115a	52°18′N	5°01′E
Węgorzewo, Pol. (vôn-gô′zhĕ-vô)	125	54°14′N	21°46′E
Węgrow, Pol. (vôn′grôf)	125	52°23′N	22°02′E
Wei, r., China (wā)	162	35°47′N	114°27′E
Wei, r., China (wā)	160	34°00′N	108°10′E
Weichang, China (wā-chäŋ)	161	41°50′N	118°00′E
Weiden, Ger.	124	49°41′N	12°09′E
Weifang, China	161	36°43′N	119°08′E
Weihai, China (wa′hăĭ′)	161	37°30′N	122°05′E
Weilheim, Ger. (vīl′hīm′)	124	47°50′N	11°06′E
Weimar, Ger. (vī′mär)	117	50°59′N	11°20′E
Weinan, China	164	34°32′N	109°40′E
Weipa, Austl.	175	12°25′S	141°54′E
Weir, r., Can. (wēr-rĭv-ēr)	57	56°49′N	94°04′W
Weirton, W.V., U.S.	66	40°25′N	80°35′W
Weiser, Id., U.S. (wē′zēr)	72	44°15′N	116°58′W
Weiser, r., Id., U.S.	72	44°26′N	116°40′W
Weishi, China (wā-shr)	164	34°23′N	114°12′E
Weissenburg, Ger.	124	49°04′N	11°20′E
Weissenfels, Ger. (vī′sĕn-fĕlz)	124	51°13′N	11°58′E
Weiss Lake, res., Al., U.S.	82	34°15′N	85°35′W
Weixi, China (wā-shyē)	160	27°27′N	99°30′E
Weixian, China (wā shyĕn)	162	36°59′N	115°17′E
Wejherowo, Pol. (vā-hĕ-rô′vô)	125	54°36′N	18°15′E
Welch, W.V., U.S. (wĕlch)	83	37°24′N	81°28′W
Weldon, N.C., U.S. (wĕl′dŭn)	83	36°24′N	77°36′W
Weldon, r., Mo., U.S.	79	40°22′N	93°39′W
Weleetka, Ok., U.S. (wĕ-lēt′kȧ)	79	35°19′N	96°08′W
Welford, Austl. (wĕl′fērd)	176	25°08′S	144°43′E
Welkom, S. Afr. (wĕl′kŏm)	186	27°57′S	26°45′E
Welland, Can. (wĕl′ănd)	59	42°59′N	79°13′W
Wellesley, Ma., U.S. (wĕlz′lē)	61a	42°18′N	71°17′W
Wellesley Islands, is., Austl.	174	16°15′S	139°25′E
Wellington, Austl. (wĕl′lĭng-tŭn)	176	32°40′S	148°50′E
Wellington, N.Z.	175a	41°15′S	174°45′E
Wellington, Eng., U.K.	114a	52°42′N	2°30′W
Wellington, Ks., U.S.	79	37°16′N	97°24′W
Wellington, Oh., U.S.	66	41°10′N	82°10′W
Wellington, Tx., U.S.	78	34°51′N	100°12′W
Wellington, i., Chile (ŏĕ′lēng-tŏn)	102	49°30′S	76°30′W
Wells, Can.	50	53°06′N	121°34′W
Wells, Mi., U.S.	66	45°50′N	87°00′W
Wells, Mn., U.S.	71	43°44′N	93°43′W
Wells, Nv., U.S.	72	41°07′N	115°04′W
Wells, l., Austl. (wĕlz)	174	26°35′S	123°40′E
Wellsboro, Pa., U.S. (wĕlz′bŭ-rô)	67	41°45′N	77°15′W
Wellsburg, W.V., U.S. (wĕlz′bûrg)	66	40°10′N	80°40′W
Wells Dam, dam, Wa., U.S.	72	48°00′N	119°39′W
Wellston, Oh., U.S. (wĕlz′tŭn)	66	39°05′N	82°30′W
Wellsville, Mo., U.S. (wĕlz′vĭl)	79	39°04′N	91°33′W
Wellsville, N.Y., U.S.	67	42°10′N	78°00′W
Wellsville, Oh., U.S.	66	40°35′N	80°40′W
Wellsville, Ut., U.S.	73	41°38′N	111°57′W
Wels, Aus. (vĕls)	117	48°10′N	14°01′E
Welshpool, Wales, U.K. (wĕlsh′pōōl)	120	52°44′N	3°10′W
Welverdiend, S. Afr. (vĕl-vĕr-dēnd′)	192c	26°23′S	27°16′E
Welwyn Garden City, Eng., U.K. (wĕlĭn)	114b	51°46′N	0°17′W
Wem, Eng., U.K. (wĕm)	114a	52°51′N	2°44′W
Wembere, r., Tan.	191	4°35′S	33°55′E
Wen, r., China (wŭn)	162	36°24′N	119°00′E
Wenan Wa, sw., China (wĕn′ān′ wä)	162	38°56′N	116°29′E
Wenatchee, Wa., U.S. (wĕ-năch′ē)	72	47°24′N	120°18′W
Wenatchee Mountains, mts., Wa., U.S.	72	47°28′N	121°10′W
Wenchang, China (wŭn-chäŋ)	165	19°32′N	110°42′E
Wenchi, Ghana	188	7°42′N	2°07′W
Wendeng, China (wŭn-dŭŋ)	162	37°14′N	122°03′E
Wendo, Eth.	185	6°37′N	38°29′E
Wendorer, Ut., U.S.	73	40°47′N	114°01′W
Wendover, Can. (wĕn-dōv′ēr)	62c	45°34′N	75°07′W
Wendover, Eng., U.K.	114b	51°44′N	0°45′W
Wenham, Ma., U.S. (wĕn′ăm)	61a	42°36′N	70°53′W
Wenquan, China (wŭn-chyůän)	161	47°10′N	120°00′E
Wenshan, China	160	23°20′N	104°15′E
Wenshang, China (wĕn′shäng)	162	35°43′N	116°31′E
Wensu, China (wĕn-sŏ)	160	41°45′N	80°30′E
Wentworth, Austl. (wĕnt′wûrth)	175	34°03′S	141°53′E
Wenzhou, China (wŭn-jō)	161	28°00′N	120°40′E
Wepener, S. Afr. (wē′pĕn-ēr) (vā′pĕn-ĕr)	186	29°43′S	27°04′E
Werder, Ger. (vĕr′dĕr)	115b	52°23′N	12°56′E
Were Ilu, Eth.	185	10°39′N	39°21′E
Werl, Ger. (vĕrl)	127c	51°33′N	7°55′E
Wermelskirchen, Ger.	127c	51°08′N	7°13′E
Werneuchen, Ger. (vĕr′hoi-kĕn)	115b	52°38′N	13°44′E
Werra, r., Ger. (vĕr′ä)	124	51°16′N	9°54′E
Werribee, Austl.	173a	37°54′S	144°40′E
Werribee, r., Austl.	173a	37°40′S	144°37′E
Wertach, r., Ger. (vĕr′täk)	124	48°12′N	10°40′E
Weseke, Ger. (vĕ′zĕ-kĕ)	127c	51°54′N	6°51′E
Wesel, Ger. (vā′zĕl)	127c	51°39′N	6°37′E
Weser, r., Ger. (vā′zĕr)	112	51°00′N	10°30′E
Weslaco, Tx., U.S. (wĕs-lā′kō)	81	26°10′N	97°59′W
Weslemkoon, l., Can.	59	45°02′N	77°25′W
Wesleyville, Can. (wĕs′lē-vĭl)	61	49°09′N	53°34′W
Wessel Islands, is., Austl. (wĕs′ĕl)	174	11°45′S	136°25′E
Wesselsbron, S. Afr. (wĕs′ĕl-brŏn)	192c	27°51′S	26°22′E
Wessington Springs, S.D., U.S. (wĕs′ĭng-tŭn)	70	44°06′N	98°35′W
West, Mount, mtn., Pan.	86a	9°10′N	79°52′W
West Allis, Wi., U.S. (wĕst-ăl′ĭs)	69a	43°01′N	88°01′W
West Alton, Mo., U.S. (ôl′tŭn)	75e	38°52′N	90°13′W
West Bay, b., Fl., U.S.	82	30°20′N	85°45′W
West Bay, b., Tx., U.S.	81a	29°11′N	95°03′W
West Bend, Wi., U.S. (wĕst bĕnd)	71	43°25′N	88°13′W
West Bengal, state, India (bĕn-gôl′)	155	23°30′N	87°30′E
West Blocton, Al., U.S. (blŏk′tŭn)	82	33°05′N	87°05′W
Westborough, Ma., U.S. (wĕst′bŭr-ô)	61a	42°17′N	71°37′W
West Boylston, Ma., U.S. (boil′stŭn)	61a	42°22′N	71°46′W
West Branch, Mi., U.S. (wĕst branch)	66	44°15′N	84°10′W
West Bridgford, Eng., U.K. (brĭj′fērd)	114a	52°55′N	1°08′W
West Bromwich, Eng., U.K. (wĕst brŭm′ĭj)	114a	52°32′N	1°59′W
Westbrook, Me., U.S. (wĕst′brŏŏk)	60	43°41′N	70°23′W
Westby, Wi., U.S. (wĕst′bē)	71	43°40′N	90°52′W
West Caicos, i., T./C. Is. (käē′kō) (kī′kōs)	93	21°40′N	72°30′W
West Cape Howe, c., Austl.	174	35°15′S	117°30′E
West Chester, Oh., U.S. (chĕs′tēr)	69f	39°20′N	84°24′W
West Chester, Pa., U.S.	68f	39°57′N	75°36′W
West Chicago, Il., U.S. (chĭ-kâ′gō)	69a	41°53′N	88°12′W
West Columbia, S.C., U.S. (cōl′ŭm-bē-ȧ)	83	33°58′N	81°05′W
West Columbia, Tx., U.S.	81	29°08′N	95°39′W
West Cote Blanche Bay, b., La., U.S.	81	29°30′N	92°17′W
West Covina, Ca., U.S. (wĕst kō-vē′nȧ)	75a	34°04′N	117°55′W
West Des Moines, Ia., U.S. (dē moin′)	71	41°35′N	93°42′W
West Des Moines, r., Ia., U.S.	71	42°52′N	94°32′W
West End, Bah.	92	26°40′N	78°55′W
Westerham, Eng., U.K. (wĕ′stēr′ŭm)	114b	51°15′N	0°05′E
Westerhörn, Ger. (vĕs′tĕr-hōrn)	115c	53°52′N	9°41′E
Westerlo, Bel.	115a	51°05′N	4°57′E
Westerly, R.I., U.S. (wĕs′tēr-lē)	67	41°25′N	71°50′W
Western Australia, state, Austl. (ôs-trā′lĭ-ȧ)	174	24°15′S	121°30′E
Western Dvina, r., Eur.	123	55°30′N	28°27′E
Western Ghāts, mts., India	155	17°35′N	74°00′E
Western Port, Md., U.S. (wĕs′tērn pōrt)	67	39°30′N	79°00′W
Western Sahara, dep., Afr. (sȧ-hä′rȧ)	184	23°05′N	15°33′W
Western Samoa *see* Samoa, nation, Oc.	2	14°30′S	172°00′W
Western Siberian Lowland, depr., Russia	134	63°37′N	72°45′E
Westerville, Oh., U.S. (wĕs′tēr-vĭl)	66	40°10′N	83°00′W
Westerwald, for., Ger. (vĕs′tĕr-väld)	124	50°35′N	7°45′E
Westfalen, hist. reg., Ger. (vĕst-fä-lĕn)	124	51°20′N	8°30′E
Westfield, Ma., U.S. (wĕst′fēld)	67	42°05′N	72°45′W
Westfield, N.J., U.S.	68a	40°39′N	74°21′W
Westfield, N.Y., U.S. (wĕst′fēld)	68a	42°20′N	79°40′W
Westford, Ma., U.S. (wĕst′fērd)	61a	42°35′N	71°26′W
West Frankfort, Il., U.S. (frăŋk′fŭrt)	66	37°55′N	88°55′W
West Ham, Eng., U.K.	114b	51°30′N	0°00′W
West Hartford, Ct., U.S. (härt′fērd)	67	41°45′N	72°45′W
West Helena, Ar., U.S. (hĕl′ĕn-ȧ)	79	34°32′N	90°39′W
West Indies, is., (ĭn′dēz)	87	19°00′N	78°30′W
West Jordon, Ut., U.S. (jôr′dăn)	75b	40°37′N	111°56′W
West Kirby, Eng., U.K. (kûr′bē)	114a	53°22′N	3°11′W
West Lafayette, In., U.S. (lä-fā-yĕt′)	66	40°25′N	86°55′W
Westlake, Oh., U.S.	69d	41°27′N	81°55′W
Westleigh, S. Afr. (wĕst-lē)	192c	27°39′S	27°18′E
West Liberty, Ia., U.S. (wĕst lĭb′ēr-tĭ)	71	41°34′N	91°15′W
West Linn, Or., U.S. (lĭn)	74c	45°22′N	122°37′W
Westlock, Can. (wĕst′lŏk)	55	54°09′N	113°52′W
West Memphis, Ar., U.S.	79	35°08′N	90°11′W
West Midlands, hist. reg., Eng., U.K.	114a	52°26′N	1°50′W
Westminster, Ca., U.S. (wĕst′mĭn-stēr)	75a	33°45′N	117°59′W
Westminster, Md., U.S.	67	39°40′N	76°55′W
Westminster, S.C., U.S.	82	34°38′N	83°10′W
Westmount, Can. (wĕst′mount)	62a	45°29′N	73°36′W
West Newbury, Ma., U.S. (nū′bēr-ē)	61a	42°47′N	70°57′W
West Newton, Pa., U.S. (nū′tŭn)	69e	40°12′N	79°45′W
West New York, N.J., U.S. (nů yôrk)	68a	40°47′N	74°01′W
West Nishnabotna, r., Ia., U.S. (nĭsh-nȧ-bŏt′nȧ)	70	40°56′N	95°37′W
Weston, Ma., U.S. (wĕs′tŭn)	61a	42°22′N	71°18′W
Weston, W.V., U.S.	66	39°00′N	80°30′W
Westonaria, S. Afr.	192c	26°19′S	27°38′E
Weston-super-Mare, Eng., U.K. (wĕs′tŭn sū′pēr-mā′rē)	120	51°23′N	3°00′W
West Orange, N.J., U.S. (wĕst ŏr′ĕnj)	68a	40°46′N	74°14′W
West Palm Beach, Fl., U.S. (pām bēch)	65	26°44′N	80°04′W
West Pensacola, Fl., U.S. (pĕn-sȧ-kō′lȧ)	82	30°24′N	87°18′W
West Pittsburg, Ca., U.S. (pĭts′bûrg)	74b	38°02′N	121°56′W
Westplains, Mo., U.S. (wĕst-plānz′)	79	36°42′N	91°51′W
West Point, Ga., U.S.	82	32°52′N	85°10′W
West Point, Ms., U.S.	82	33°36′N	88°39′W
Westpoint, Ne., U.S.	70	41°50′N	96°00′W
West Point, N.Y., U.S.	68a	41°23′N	73°58′W
West Point, Ut., U.S.	75b	41°07′N	112°05′W
West Point, Va., U.S.	67	37°25′N	76°50′W
West Point Lake, res., U.S.	82	33°00′N	85°10′W
Westport, Ire.	120	53°44′N	9°36′W
Westport, Ct., U.S. (wĕst′pōrt)	68a	41°07′N	73°22′W
Westport, Or., U.S. (wĕst′pōrt)	74c	46°08′N	123°22′W
Westray, i., Scot., U.K. (wĕs′trā)	120a	59°19′N	3°05′W
West Road, r., Can. (rōd)	54	53°00′N	124°00′W
West Saint Paul, Mn., U.S. (sānt pôl′)	75g	44°55′N	93°05′W
West Sand Spit, i., T./C. Is.	93	21°25′N	72°10′W
West Slope, Or., U.S.	74c	45°30′N	122°46′W
West Tavaputs Plateau, plat., Ut., U.S. (wĕst tăv′ȧ-pōts)	77	39°45′N	110°35′W
West Terre Haute, In., U.S. (tĕr-ē hōt′)	66	39°30′N	87°30′W
West Union, Ia., U.S. (ūn′yŭn)	71	42°58′N	91°48′W
West University Place, Tx., U.S.	81a	29°43′N	95°26′W
Westview, Oh., U.S. (wĕst′vū)	69d	41°21′N	81°54′W
West View, Pa., U.S.	69e	40°31′N	80°02′W
Westville, Can. (wĕst′vĭl)	61	45°35′N	62°43′W
Westville, Il., U.S.	66	40°00′N	87°40′W
West Virginia, state, U.S. (wĕst vẽr-jĭn′ĭ-ȧ)	65	39°00′N	80°50′W
West Walker, r., Ca., U.S. (wôk′ēr)	76	38°25′N	119°25′W
West Warwick, R.I., U.S. (wŏr′ĭk)	68b	41°42′N	71°31′W
Westwego, La., U.S. (wĕst-wē′gō)	68d	29°55′N	90°09′W
Westwood, Ca., U.S. (wĕst′wŏŏd)	76	40°18′N	121°00′W
Westwood, Ks., U.S.	75f	39°03′N	94°37′W
Westwood, Ma., U.S.	61a	42°13′N	71°14′W
Westwood, N.J., U.S.	68a	40°59′N	74°02′W
West Wyalong, Austl. (wīălông)	175	34°00′S	147°20′E
West Yorkshire, hist. reg., Eng., U.K.	114a	53°37′N	1°48′W
Wetar, Pulau, i., Indon. (wĕt′âr)	169	7°34′S	126°00′E
Wetaskiwin, Can. (wĕ-tăs′kē-wŏn)	50	52°58′N	113°22′W
Wetmore, Tx., U.S. (wĕt′mōr)	75d	29°34′N	98°25′W
Wetter, Ger.	127c	51°23′N	7°23′E
Wetumpka, Al., U.S. (wē-tŭmp′kȧ)	82	32°33′N	86°12′W
Wetzlar, Ger. (vets′lär)	124	50°35′N	8°30′E
Wewak, Pap. N. Gui. (wā-wăk′)	169	3°19′S	143°30′E
Wewoka, Ok., U.S. (wē-wō′kȧ)	79	35°09′N	96°30′W
Wexford, Ire. (wĕks′fērd)	117	52°20′N	6°30′W
Weybridge, Eng., U.K. (wā′brĭj)	114b	51°20′N	0°26′W
Weyburn, Can. (wā′bûrn)	50	49°41′N	103°52′W
Weymouth, Eng., U.K. (wā′mŭth)	120	50°37′N	2°34′W
Weymouth, Ma., U.S.	61a	42°44′N	70°57′W
Weymouth, Oh., U.S.	69d	41°11′N	81°48′W
Whale Cay, i., Bah.	92	25°20′N	77°45′W
Whale Cay Channels, strt., Bah.	92	26°45′N	77°10′W
Wharton, N.J., U.S. (hwôr′tŭn)	68a	40°54′N	74°35′W
Wharton, Tx., U.S.	81	29°19′N	96°06′W
What Cheer, Ia., U.S. (hwŏt chēr)	71	41°23′N	92°24′W
Whatcom, Lake, l., Wa., U.S. (hwät′kŭm)	74c	48°44′N	123°34′W
Whatshan Lake, l., Can. (wŏt′shăn)	55	50°00′N	118°03′W
Wheatland, Wy., U.S. (hwēt′lănd)	73	42°04′N	104°52′W
Wheatland Reservoir Number 02, res., Wy., U.S.	73	41°52′N	105°36′W
Wheaton, Il., U.S. (hwē′tŭn)	69a	41°52′N	88°06′W
Wheaton, Md., U.S.	68e	39°05′N	77°05′W
Wheaton, Mn., U.S.	70	45°48′N	96°29′W
Wheeler Peak, mtn., N.M., U.S.	78	36°34′N	105°25′W
Wheeler Peak, mtn., Nv., U.S.	64	38°58′N	114°15′W
Wheeling, Il., U.S. (hwēl′ĭng)	69a	42°08′N	87°54′W
Wheeling, W.V., U.S.	66	40°05′N	80°45′W
Wheelwright, Arg. (ōō′ĕl-rē′gt)	99c	33°46′S	61°14′W
Whidbey Island, i., Wa., U.S. (hwĭd′bē)	74a	48°13′N	122°50′W
Whippany, N.J., U.S. (hwĭp′á-nē)	68a	40°49′N	74°25′W
Whitby, Can. (hwĭt′bē)	51	43°50′N	79°00′W
Whitchurch, Eng., U.K. (hwĭt′chûrch)	114a	52°58′N	2°49′W
White, l., Can.	58	48°47′N	85°50′W
White, l., Can.	59	45°15′N	76°35′W
White, r., Can.	58	48°34′N	85°46′W
White, r., In., U.S.	66	39°15′N	86°45′W
White, r., S.D., U.S.	70	43°13′N	101°04′W
White, r., Tx., U.S.	78	36°25′N	102°20′W
White, r., Vt., U.S.	67	43°45′N	72°35′W
White, r., Wa., U.S.	72	47°07′N	121°48′W
White, r., U.S.	65	35°30′N	92°00′W
White, r., U.S.	70	43°41′N	99°48′W
White, r., U.S.	77	40°10′N	108°55′W
White, East Fork, r., In., U.S.	66	38°45′N	86°20′W
White Bay, b., Can.	53a	50°00′N	56°30′W
White Bear Indian Reserve, I.R., Can.	57	49°50′N	102°15′W
White Bear Lake, l., Mn., U.S.	75g	45°04′N	92°58′W
White Castle, La., U.S.	81	30°10′N	91°09′W
White Center, Wa., U.S.	74a	47°31′N	122°21′W
White Cloud, Mi., U.S.	66	43°35′N	85°45′W
Whitecourt, Can. (wīt′cōrt)	50	54°09′N	115°41′W
White Earth, r., N.D., U.S.	70	48°30′N	102°44′W
White Earth Indian Reservation, I.R., Mn., U.S.	70	47°18′N	95°42′W
Whiteface, r., Mn., U.S. (whīt′fās)	71	47°12′N	92°13′W
Whitefield, N.H., U.S. (hwīt′fēld)	67	44°20′N	71°35′W
Whitefish Bay, Wi., U.S.	69a	43°07′N	77°54′W
Whitefish Bay, b., Can.	57	49°26′N	94°14′W
Whitefish Bay, b., N.A.	71	46°36′N	84°50′W
White Hall, Il., U.S.	79	39°26′N	90°23′W
Whitehall, Mi., U.S. (hwīt′hôl)	66	43°20′N	86°20′W
Whitehall, N.Y., U.S.	67	43°30′N	73°25′W
Whitehaven, Eng., U.K. (hwīt′hā-vĕn)	120	54°35′N	3°30′W
Whitehorn, Point, c., Wa., U.S. (hwīt′hôrn)	74d	48°54′N	122°48′W
Whitehorse, Can. (whīt′hôrs)	50	60°39′N	135°01′W
White Lake, l., La., U.S.	81	29°40′N	92°35′W
White Mountain Peak, mtn., Ca., U.S.	76	37°38′N	118°13′W
White Mountains, mts., Me., U.S.	60	44°22′N	71°15′W
White Mountains, mts., N.H., U.S.	67	42°20′N	71°05′W
Whitemouth, l., Can.	57	49°14′N	95°40′W
White Nile (Al Bahr al Abyad), r., Sudan	185	12°30′N	32°30′E
White Otter, l., Can.	58	49°15′N	91°48′W
White Pass, p., N.A.	63	59°35′N	135°03′W
White Plains, N.Y., U.S.	68a	41°02′N	73°47′W
White River, Can.	58	48°38′N	85°23′W
White Rock, Can.	55	49°01′N	122°49′W

ăt; finăl; rāte; senāte; ärm; åsk; sofȧ; fâre; ch-choose; dh-as th in other; bē; ēvent; bĕt; recĕnt; cratẽr; g-gō; gh-guttural g; bĭt; ĭ-short neutral; rīde; ĸ-guttural k as ch in German ich;

PRONOUNCING INDEX

Whiterock Reservoir – Wolkersdorf 319

PLACE (Pronunciation)	PAGE	LAT.	LONG.
Whiterock Reservoir, res., Tx., U.S. (hwīt'rŏk)	75c	32°51'N	96°40'W
White Russia *see* Belarus, nation, Eur.	134	53°30'N	25°33'E
Whitesail Lake, l., Can. (whīt'sāl)	54	53°30'N	127°00'W
White Sands National Monument, rec., N.M., U.S.	77	32°50'N	106°20'W
White Sea, sea, Russia	134	66°00'N	40°00'E
White Settlement, Tx., U.S.	75c	32°45'N	97°28'W
White Sulphur Springs, Mt., U.S.	73	46°32'N	110°49'W
White Umfolzi, r., S. Afr. (ŭm-fô-lō'zē)	187c	28°12'S	30°55'E
Whiteville, N.C., U.S. (hwīt'vĭl)	83	34°18'N	78°45'W
White Volta (Volta Blanche), r., Afr.	188	9°40'N	1°10'W
Whitewater, Wi., U.S. (whīt-wŏt'ẽr)	71	42°49'N	88°40'W
Whitewater, l., Can.	57	49°14'N	100°39'W
Whitewater, r., In., U.S.	69f	39°19'N	84°55'W
Whitewater Bay, b., Fl., U.S.	83a	25°16'N	80°21'W
Whitewater Creek, r., Mt., U.S.	73	48°50'N	107°50'W
Whitewell, Tn., U.S. (hwīt'wĕl)	82	35°11'N	85°31'W
Whitewright, Tx., U.S. (hwīt'rīt)	79	33°33'N	96°25'W
Whitham, r., Eng., U.K. (wĭth'ŭm)	114a	53°08'N	0°15'W
Whiting, In., U.S. (hwīt'ĭng)	69a	41°41'N	87°30'W
Whitinsville, Ma., U.S. (hwīt'ĕns-vĭl)	61a	42°06'N	71°40'W
Whitman, Ma., U.S. (hwĭt'măn)	61a	42°05'N	70°57'W
Whitmire, S.C., U.S. (hwĭt'mīr)	83	34°30'N	81°40'W
Whitney, Mount, mtn., Ca., U.S.	64	36°34'N	118°18'W
Whitney Lake, l., Tx., U.S. (hwĭt'nē)	81	32°02'N	97°36'W
Whitstable, Eng., U.K. (wĭt'stăb'l)	114b	51°22'N	1°03'E
Whitsunday, i., Austl. (hwĭt's'n-dā)	175	20°16'S	149°00'E
Whittier, Ca., U.S. (hwĭt'ĭ-ẽr)	75a	33°58'N	118°02'W
Whittlesea, S. Afr. (wĭt'l'sē)	187c	32°11'S	26°51'E
Whitworth, Eng., U.K. (hwĭt'wŭrth)	114a	53°40'N	2°10'W
Whyalla, Austl. (hwī-ăl'ȧ)	174	33°00'S	137°32'E
Whymper, Mount, mtn., Can. (wĭm'pẽr)	54	48°57'N	124°10'W
Wiarton, Can. (wī'ăr-tŭn)	51	44°45'N	80°45'W
Wichita, Ks., U.S. (wĭch'ĭ-tô)	64	37°42'N	97°21'W
Wichita, r., Tx., U.S.	78	33°50'N	99°38'W
Wichita Falls, Tx., U.S. (fôls)	64	33°54'N	98°29'W
Wichita Mountains, mts., Ok., U.S.	64	34°48'N	98°43'W
Wick, Scot., U.K. (wĭk)	116	58°25'N	3°05'W
Wickatunk, N.J., U.S. (wĭk'á-tŭnk)	68a	40°21'N	74°15'W
Wickenburg, Az., U.S.	77	33°58'N	112°44'W
Wickiup Reservoir, res., Or., U.S.	72	43°40'N	121°43'W
Wickliffe, Oh., U.S. (wĭk'klĭf)	69d	41°37'N	81°29'W
Wicklow, Ire.	120	52°59'N	6°06'W
Wicklow Mountains, mts., Ire. (wĭk'lō)	120	52°49'N	6°20'W
Wickup Mountain, mtn., Or., U.S. (wĭk'ŭp)	74c	46°06'N	123°35'W
Wiconisco, Pa., U.S. (wĭ-kon'ĭs-kō)	67	43°35'N	76°45'W
Widen, W.V., U.S. (wī'dĕn)	66	38°25'N	80°55'W
Widnes, Eng., U.K. (wĭd'nĕs)	114a	53°21'N	2°44'W
Wieliczka, Pol. (vyĕ-lēch'kȧ)	125	49°58'N	20°06'E
Wien *see* Vienna, Aus.	110	48°13'N	16°22'E
Wien, state, Aus.	115e	48°11'N	16°23'E
Wiener Neustadt, Aus. (vē'nẽr noi'shtät)	117	47°48'N	16°15'E
Wiener Wald, for., Aus.	115e	48°09'N	16°05'E
Wieprz, r., Pol. (vyĕpzh)	125	51°25'N	22°45'E
Wiergate, Tx., U.S. (wẽr'gāt)	81	31°00'N	93°42'W
Wiesbaden, Ger. (vēs'bä-dĕn)	117	50°05'N	8°15'E
Wigan, Eng., U.K. (wĭg'ăn)	120	53°33'N	2°37'W
Wiggins, Ms., U.S. (wĭg'ĭnz)	82	30°51'N	89°05'W
Wight, Isle of, i., Eng., U.K. (wīt)	120	50°44'N	1°17'W
Wilber, Ne., U.S. (wĭl'bẽr)	79	40°29'N	96°57'W
Wilburton, Ok., U.S. (wĭl'bẽr-tŭn)	79	34°54'N	95°18'W
Wilcannia, Austl. (wĭl-cān-ĭȧ)	175	31°30'S	143°30'E
Wildau, Ger. (vēl'dou)	115b	52°20'N	13°39'E
Wildberg, Ger. (vēl'bĕrgh)	115b	52°52'N	12°39'E
Wildcat Hill, hill, Can. (wīld'kăt)	57	53°17'N	102°30'W
Wildhay, r., Can. (wīld'hā)	55	53°15'N	117°20'W
Wildomar, Ca., U.S. (wĭl'dō-mär)	75a	33°35'N	117°17'W
Wild Rice, r., Mn., U.S.	70	47°10'N	96°40'W
Wild Rice, r., N.D., U.S.	70	46°10'N	97°12'W
Wild Rice Lake, l., Mn., U.S.	75h	46°54'N	92°10'W
Wildspitze, mtn., Aus.	124	46°55'N	10°50'E
Wildwood, N.J., U.S.	67	39°00'N	74°50'W
Wiley, Co., U.S. (wī'lē)	78	38°08'N	102°41'W
Wilge, r., S. Afr. (wĭl'jĕ)	192c	25°38'S	29°09'E
Wilge, r., S. Afr.	192c	27°27'S	28°46'E
Wilhelm, Mount, mtn., Pap. N. Gui.	169	5°58'S	144°58'E
Wilhelmina Gebergte, mts., Sur.	101	4°30'N	57°00'W
Wilhelmina Kanaal, can., Neth.	115a	51°37'N	4°55'E
Wilhelmshaven, Ger. (vēl-hĕlms-hä'fĕn)	116	53°30'N	8°10'E
Wilkes-Barre, Pa., U.S. (wĭlks'băr-ē)	65	41°15'N	75°50'W
Wilkes Land, reg., Ant.	178	71°00'S	126°00'E
Wilkeson, Wa., U.S. (wĭl-kē'sŭn)	74a	47°06'N	122°03'W
Wilkie, Can. (wĭlk'ē)	50	52°25'N	108°43'W
Wilkinsburg, Pa., U.S. (wĭl'kĭnz-bûrg)	69e	40°26'N	79°53'W
Willamette, r., Or., U.S.	64	45°00'N	123°00'W
Willapa Bay, b., Wa., U.S.	72	46°37'N	124°00'W
Willard, Oh., U.S. (wĭl'ȧrd)	66	41°00'N	82°50'W
Willard, Ut., U.S.	75b	41°24'N	112°02'W
Willcox, Az., U.S. (wĭl'kŏks)	77	32°15'N	109°50'W
Willcox Playa, l., Az., U.S.	77	32°08'N	109°51'W
Willemstad, Neth. Ant.	100	12°12'N	68°58'W
Willesden, Eng., U.K. (wĭlz'dĕn)	114b	51°31'N	0°17'W
William "Bill" Dannelly Reservoir, res., Al., U.S.	82	32°10'N	87°15'W
William Creek, Austl. (wĭl'yăm)	174	28°45'S	136°20'E
Williams, Az., U.S. (wĭl'yămz)	77	35°15'N	112°15'W
Williams, i., Bah.	92	24°30'N	78°30'W
Williamsburg, Ky., U.S. (wĭl'yămz-bûrg)	82	36°42'N	84°09'W
Williamsburg, Oh., U.S.	69f	39°04'N	84°02'W
Williamsburg, Va., U.S.	83	37°15'N	76°41'W
Williams Lake, Can.	55	52°08'N	122°09'W
Williamson, W.V., U.S. (wĭl'yăm-sŭn)	66	37°40'N	82°15'W
Williamsport, Md., U.S.	67	39°35'N	77°45'W
Williamsport, Pa., U.S.	67	41°15'N	77°05'W
Williamston, N.C., U.S. (wĭl'yămz-tŭn)	83	35°50'N	77°04'W
Williamston, S.C., U.S.	83	34°36'N	82°30'W
Williamstown, Austl.	173a	37°52'S	144°54'E
Williamstown, W.V., U.S. (wĭl'yămz-toun)	66	39°20'N	81°30'W
Williamsville, N.Y., U.S. (wĭl'yăm-vĭl)	69c	42°58'N	78°46'W
Willimantic, Ct., U.S. (wĭl-ĭ-măn'tĭk)	67	41°40'N	72°10'W
Willis, Tx., U.S. (wĭl'ĭs)	81	30°24'N	95°29'W
Willis Islands, is., Austl.	175	16°15'S	150°30'E
Williston, N.D., U.S. (wĭl'ĭs-tŭn)	64	48°08'N	103°38'W
Williston, Lake, l., Can.	52	55°40'N	123°40'W
Willmar, Mn., U.S. (wĭl'mär)	70	45°07'N	95°05'W
Willoughby, Oh., U.S. (wĭl'ō-bē)	69d	41°39'N	81°25'W
Willow, Ak., U.S.	63	61°50'N	150°00'W
Willow Creek, r., Or., U.S.	72	44°21'N	117°34'W
Willow Grove, Pa., U.S.	68f	40°07'N	75°07'W
Willowick, Oh., U.S. (wĭl'ō-wĭk)	69d	41°39'N	81°28'W
Willowmore, S. Afr. (wĭl'ō-môr)	186	33°15'S	23°37'E
Willow Run, Mi., U.S. (wĭl'ō rŭn)	69b	42°16'N	83°34'W
Willows, Ca., U.S. (wĭl'ōz)	76	39°32'N	122°11'W
Willow Springs, Mo., U.S. (sprĭngz)	79	36°59'N	91°56'W
Willowvale, S. Afr. (wĭ-lō'vāl)	187c	32°17'S	28°32'E
Wills Point, Tx., U.S. (wĭlz point)	81	32°42'N	96°02'W
Wilmer, Tx., U.S. (wĭl'mẽr)	75c	32°35'N	96°40'W
Wilmette, Il., U.S. (wĭl-mĕt')	69a	42°04'N	87°42'W
Wilmington, Austl.	176	32°39'S	138°07'E
Wilmington, Ca., U.S. (wĭl'mĭng-tŭn)	75a	33°46'N	118°16'W
Wilmington, De., U.S.	65	39°45'N	75°33'W
Wilmington, Il., U.S.	69a	41°19'N	88°09'W
Wilmington, Ma., U.S.	61a	42°34'N	71°10'W
Wilmington, N.C., U.S.	65	34°12'N	77°56'W
Wilmington, Oh., U.S.	66	39°20'N	83°50'W
Wilmore, Ky., U.S. (wĭl'mŏr)	66	37°50'N	84°35'W
Wilmslow, Eng., U.K. (wĭlmz'lō)	114a	53°19'N	2°14'W
Wilno *see* Vilnius, Lith.	134	54°40'N	25°26'E
Wilpoort, S. Afr.	192c	26°57'S	26°17'E
Wilson, Ar., U.S. (wĭl'sŭn)	79	35°35'N	90°02'W
Wilson, N.C., U.S.	83	35°42'N	77°55'W
Wilson, Ok., U.S.	79	34°09'N	97°27'W
Wilson, r., Al., U.S.	82	34°53'N	87°28'W
Wilson, Mount, mtn., Ca., U.S.	75a	34°15'N	118°06'W
Wilson, Point, c., Austl.	173a	38°05'S	144°31'E
Wilson Lake, res., Al., U.S.	65	34°45'N	87°30'W
Wilson's Promontory, pen., Austl. (wĭl'sŭnz)	175	39°05'S	146°50'E
Wilsonville, Il., U.S. (wĭl'sŭn-vĭl)	75e	39°04'N	89°52'W
Wilstedt, Ger. (vēl'shtĕt)	115c	53°45'N	10°04'E
Wilster, Ger. (vēl'stẽr)	115c	53°55'N	9°23'E
Wilton, Ct., U.S. (wĭl'tŭn)	68a	41°11'N	73°25'W
Wilton, N.D., U.S.	70	47°09'N	100°47'W
Wiluna, Austl. (wĭ-lōō'nȧ)	174	26°35'S	120°25'E
Winamac, In., U.S. (wĭn'ȧ măk)	66	41°05'N	86°40'W
Winburg, S. Afr. (wĭm-bûrg)	192c	28°31'S	27°02'E
Winchester, Eng., U.K.	120	51°04'N	1°20'W
Winchester, Ca., U.S. (wĭn'chĕs-tẽr)	75a	33°41'N	117°06'W
Winchester, Id., U.S.	72	46°14'N	116°39'W
Winchester, In., U.S.	66	40°10'N	84°50'W
Winchester, Ky., U.S.	66	38°00'N	84°15'W
Winchester, Ma., U.S.	61a	42°28'N	71°09'W
Winchester, N.H., U.S.	67	42°45'N	72°25'W
Winchester, Tn., U.S.	82	35°11'N	86°06'W
Winchester, Va., U.S.	67	39°10'N	78°10'W
Wind, r., Wy., U.S.	73	43°17'N	109°02'W
Windber, Pa., U.S. (wĭnd'bẽr)	67	40°15'N	78°45'W
Wind Cave National Park, rec., S.D., U.S.	70	43°36'N	103°53'W
Winder, Ga., U.S. (wīn'dẽr)	82	33°58'N	83°43'W
Windermere, Eng., U.K. (wĭn'dẽr-mẽr)	120	54°25'N	2°59'W
Windham, Ct., U.S. (wĭnd'ăm)	67	41°45'N	72°05'W
Windham, N.H., U.S.	61a	42°49'N	71°21'W
Windhoek, Nmb. (vĭnt'hŏok)	186	22°05'S	17°10'E
Wind Lake, l., Wi., U.S.	69a	42°49'N	88°06'W
Wind Mountain, mtn., N.M., U.S.	80	32°02'N	105°30'W
Windom, Mn., U.S. (wĭn'dŭm)	70	43°50'N	95°04'W
Windorah, Austl. (wĭn-dō'rȧ)	175	25°15'S	142°50'E
Wind River Indian Reservation, I.R., Wy., U.S.	73	43°26'N	109°00'W
Wind River Range, mts., Wy., U.S.	64	43°19'N	109°47'W
Windsor, Austl. (wĭn'zẽr)	173b	33°37'S	150°49'E
Windsor, Can.	51	42°19'N	83°00'W
Windsor, Can.	53a	48°57'N	55°40'W
Windsor, Can.	51	44°59'N	64°08'W
Windsor, Eng., U.K.	120	51°27'N	0°37'W
Windsor, Co., U.S.	78	40°27'N	104°51'W
Windsor, Mo., U.S.	79	38°32'N	93°31'W
Windsor, N.C., U.S.	83	35°58'N	76°57'W
Windsor, Vt., U.S.	67	43°30'N	72°25'W
Windward Islands, is., N.A. (wĭnd'wẽrd)	87	12°45'N	61°40'W
Windward Passage, strt., N.A.	87	19°30'N	74°20'W
Winefred Lake, l., Can.	56	55°30'N	110°35'W
Winfield, Ks., U.S.	79	37°14'N	97°00'W
Winifred, Mt., U.S. (wĭn ĭ frĕd)	73	47°35'N	109°20'W
Winisk, r., Can.	53	54°30'N	86°30'W
Wink, Tx., U.S. (wĭŋk)	80	31°48'N	103°06'W
Winkler, Can. (wĭnk'lẽr)	57	49°11'N	97°56'W
Winneba, Ghana (wĭn'ē-bȧ)	188	5°25'N	0°36'W
Winnebago, Mn., U.S. (wĭn'ē-bā'gō)	71	43°45'N	94°08'W
Winnebago, Lake, l., Wi., U.S.	71	44°09'N	88°10'W
Winnebago Indian Reservation, I.R., Ne., U.S.	70	42°15'N	96°06'W
Winnemucca, Nv., U.S. (wĭn-ē-mŭk'ȧ)	64	40°59'N	117°43'W
Winnemucca, l., Nv., U.S.	76	40°06'N	119°07'W
Winner, S.D., U.S. (wĭn'ẽr)	70	43°22'N	99°50'W
Winnetka, Il., U.S. (wĭ-nĕt'kȧ)	69a	42°07'N	87°44'W
Winnett, Mt., U.S. (wĭn'ĕt)	73	47°01'N	108°20'W
Winnfield, La., U.S. (wĭn'fēld)	81	31°56'N	92°39'W
Winnibigoshish, l., Mn., U.S. (wĭn'ĭ-bĭ-gō'shĭsh)	71	47°30'N	93°45'W
Winnipeg, Can. (wĭn'ĭ-pĕg)	50	49°53'N	97°09'W
Winnipeg, r., Can.	52	50°30'N	95°00'W
Winnipeg, Lake, l., Can.	52	52°00'N	97°00'W
Winnipegosis, Can. (wĭn'ĭ-pē-gō'sĭs)	50	51°39'N	99°56'W
Winnipegosis, l., Can.	52	52°30'N	100°00'W
Winnipesaukee, l., N.H., U.S. (wĭn'ē-pē-sô'kē)	67	43°40'N	71°20'W
Winnsboro, La., U.S. (wĭnz'bŭr'ō)	81	32°09'N	91°42'W
Winnsboro, S.C., U.S.	83	34°29'N	81°05'W
Winnsboro, Tx., U.S.	79	32°56'N	95°15'W
Winona, Can. (wĭ-nō'nȧ)	62d	43°13'N	79°39'W
Winona, Mn., U.S.	65	44°03'N	91°40'W
Winona, Ms., U.S.	82	33°29'N	89°43'W
Winooski, Vt., U.S. (wĭ-nōōs-kē)	67	44°30'N	73°10'W
Winsen, Ger. (vēn'zĕn)	115c	53°22'N	10°13'E
Winsford, Eng., U.K. (wĭnz'fẽrd)	114a	53°11'N	2°30'W
Winslow, Az., U.S. (wĭnz'lō)	77	35°00'N	110°45'W
Winslow, Wa., U.S.	74a	47°38'N	122°31'W
Winsted, Ct., U.S. (wĭn'stĕd)	67	41°55'N	73°05'W
Winster, Eng., U.K. (wĭn'stẽr)	114a	53°08'N	1°38'W
Winston-Salem, N.C., U.S. (wĭn stŭn-sā'lĕm)	65	36°05'N	80°15'W
Winterberge, mts., Afr.	187c	32°18'S	26°25'E
Winter Garden, Fl., U.S. (wĭn'tẽr gär'd'n)	83a	28°32'N	81°35'W
Winter Haven, Fl., U.S. (hā'vĕn)	83a	28°01'N	81°38'W
Winter Park, Fl., U.S. (pärk)	83a	28°35'N	81°21'W
Winters, Tx., U.S. (wĭn'tẽrz)	80	31°59'N	99°58'W
Winterset, Ia., U.S. (wĭn'tẽr-sĕt)	71	41°19'N	94°03'W
Winterswijk, Neth.	127c	51°58'N	6°44'E
Winterthur, Switz. (vĭn'tẽr-tōōr)	124	47°30'N	8°32'E
Winterton, S. Afr.	187c	28°51'S	29°33'E
Winthrop, Ma., U.S.	61a	42°23'N	70°59'W
Winthrop, Me., U.S. (wĭn'thrŭp)	60	44°19'N	70°00'W
Winthrop, Mn., U.S.	71	44°31'N	94°20'W
Winton, Austl. (wĭn-tŭn)	175	22°17'S	143°08'E
Wipperfürth, Ger. (vē'pẽr-fürt)	127c	51°07'N	7°23'E
Wirksworth, Eng., U.K. (wûrks'wûrth)	114a	53°05'N	1°35'W
Wisconsin, state, U.S. (wĭs-kŏn'sĭn)	65	44°30'N	91°00'W
Wisconsin, r., Wi., U.S.	65	43°14'N	90°34'W
Wisconsin Dells, Wi., U.S.	71	43°38'N	89°46'W
Wisconsin Rapids, Wi., U.S.	71	44°24'N	89°50'W
Wishek, N.D., U.S. (wĭsh'ĕk)	70	46°15'N	99°34'W
Wisła, r., Pol. (vēs'wȧ)	112	52°30'N	20°00'E
Wisłoka, r., Pol. (vēs-wō'kȧ)	125	49°55'N	21°26'E
Wismar, Ger. (vĭs'mär)	116	53°53'N	11°28'E
Wismar, Guy. (wĭs'mȧr)	101	5°58'N	58°15'W
Wisner, Ne., U.S. (wĭz'nẽr)	70	42°00'N	96°55'W
Wissembourg, Fr. (vē-sän-bōōr')	127	49°03'N	7°58'E
Wister, Lake, l., Ok., U.S. (vĭs'tẽr)	79	35°02'N	94°52'W
Witbank, S. Afr. (wĭt-băŋk)	192c	25°53'S	29°14'E
Witberg, mtn., Afr.	187c	30°32'S	27°18'E
Witham, Eng., U.K. (wĭdh'ăm)	114b	51°48'N	0°37'E
Witham, r., Eng., U.K.	114a	53°11'N	0°20'W
Withamsville, Oh., U.S. (wĭdh'ămz-vĭl)	69f	39°04'N	84°16'W
Withlacoochee, r., Fl., U.S. (wĭth-lȧ-kōō'chē)	83a	28°58'N	82°30'W
Withlacoochee, r., Ga., U.S.	82	31°15'N	83°30'W
Withrow, Mn., U.S. (wĭdh'rō)	75g	45°08'N	92°54'W
Witney, Eng., U.K. (wĭt'nē)	114b	51°45'N	1°30'W
Witt, Il., U.S. (vĭt)	66	39°10'N	89°15'W
Witten, Ger. (vē'tĕn)	127c	51°26'N	7°19'E
Wittenberg, Ger. (vē'tĕn-bĕrgh)	124	51°53'N	12°40'E
Wittenberge, Ger. (vĭt-ĕn-bĕr'gĕ)	124	52°59'N	11°45'E
Wittlich, Ger. (vĭt'lĭk)	124	49°58'N	6°54'E
Witu, Kenya (wē'tōō)	187	2°18'S	40°28'E
Witu Islands, is., Pap. N. Gui.	169	4°45'S	149°50'E
Witwatersberg, mts., S. Afr. (wĭt-wôr-tẽrz-bûrg)	187b	25°58'S	27°53'E
Witwatersrand, mtn., S. Afr. (wĭt-wôr'tẽrs-rănd)	192c	25°55'S	26°27'E
Wkra, r., Pol. (f'krȧ)	125	52°40'N	20°35'E
Włocławek, Pol. (vwô-tswä'vĕk)	125	52°38'N	19°08'E
Włodawa, Pol. (vwô-dä'vä)	125	51°33'N	23°33'E
Włoszczowa, Pol. (vwôsh-chô'vä)	125	50°51'N	19°58'E
Woburn, Ma., U.S. (wō'bŭrn) (wō'bûrn)	61a	42°29'N	71°10'W
Woerden, Neth.	115a	52°05'N	4°52'E
Woking, Eng., U.K.	114b	51°18'N	0°33'W
Wokingham, Eng., U.K. (wô'kĭng-hăm)	114b	51°23'N	0°50'W
Wolcott, Ks., U.S. (wōl'kŏt)	75f	39°12'N	94°47'W
Wolf, i., Can. (wŏlf)	59	44°10'N	76°25'W
Wolf, r., Ms., U.S.	82	30°45'N	89°36'W
Wolf, r., Wi., U.S.	71	45°14'N	88°45'W
Wolfenbüttel, Ger. (vôl'fĕn-büt-ĕl)	124	52°10'N	10°32'E
Wolf Lake, l., Il., U.S.	69a	41°39'N	87°33'W
Wolf Point, Mt., U.S. (wŏlf point)	73	48°07'N	105°40'W
Wolfratshausen, Ger. (vōlf'räts-hou-zĕn)	115d	47°55'N	11°25'E
Wolfsburg, Ger. (vôlfs'bōōrgh)	124	52°30'N	10°37'E
Wolfville, Can. (wŏlf'vĭl)	60	45°05'N	64°22'W
Wolgast, Ger. (vôl'gäst)	124	54°04'N	13°46'E
Wolhuterskop, S. Afr.	187b	25°41'S	27°40'E
Wolkersdorf, Aus.	115e	48°24'N	16°31'E

ng-sing; ŋ-baŋk; N-nasalized n; nŏd; cŏmmit; ōld; ōbey; ôrder; oi-boil; fōōd; ŏ-as oo in foot; ou-out; s-soft; sh-dish; th-thin; pūre; ŭnite; ûrn; stŭd; circŭs; ü-as in French tu; '-indeterminate vowel.

320 Wollaston - Yancheng PRONOUNCING INDEX

PLACE (Pronunciation)	PAGE	LAT.	LONG.
Wollaston, l., Can. (wòl'ás-tŭn)	52	58°15'N	103°20'W
Wollaston Peninsula, pen., Can.	52	70°00'N	115°00'W
Wollongong, Austl. (wòl'ŭn-gông)	175	34°26'S	151°05'E
Wołomin, Pol. (vô-wō'mên)	125	52°19'N	21°17'E
Wolseley, Can.	56	50°25'N	103°15'W
Woltersdorf, Ger. (vōl'tĕs-dōrf)	115b	52°07'N	13°13'E
Wolverhampton, Eng., U.K. (wòl'vēr-hămp-tŭn)	117	52°35'N	2°07'W
Wolwehoek, S. Afr.	192c	26°55'S	27°50'E
Wŏnsan, Kor., N. (wŭn'sän')	161	39°08'N	127°24'E
Wonthaggi, Austl. (wônt-hăg'ê)	175	38°45'S	145°42'E
Wood, S.D., U.S. (wòd)	70	43°26'N	100°25'W
Woodbine, Ia., U.S. (wòd'bīn)	70	41°44'N	95°42'W
Woodbridge, N.J., U.S. (wòd'brĭj')	68a	40°33'N	74°18'W
Wood Buffalo National Park, rec., Can.	52	59°50'N	118°53'W
Woodburn, Il., U.S. (wòd'bûrn)	75e	39°03'N	90°01'W
Woodburn, Or., U.S.	72	45°10'N	122°51'W
Woodbury, N.J., U.S. (wòd'bĕr-ê)	68f	39°50'N	75°14'W
Woodcrest, Ca., U.S. (wòd'krĕst)	75a	33°53'N	117°18'W
Woodinville, Wa., U.S. (wòd'ĭn-vĭl)	74a	47°46'N	122°09'W
Woodland, Ca., U.S. (wòd'lănd)	76	38°41'N	121°47'W
Woodland, Wa., U.S.	74c	45°54'N	122°45'W
Woodland Hills, Ca., U.S.	75a	34°10'N	118°36'W
Woodlark Island, i., Pap. N. Gui. (wòd'lärk)	169	9°07'S	152°00'E
Woodlawn Beach, N.Y., U.S. (wòd'lôn bēch)	69c	42°48'N	78°51'W
Wood Mountain, mtn., Can.	56	49°14'N	106°20'W
Wood River, Il., U.S.	75e	38°52'N	90°06'W
Woodroffe, Mount, mtn., Austl. (wòd'rŭf)	174	26°05'S	132°00'E
Woodruff, S.C., U.S. (wòd'rŭf)	83	34°43'N	82°03'W
Woods, l., Austl. (wòdz)	174	18°00'S	133°18'E
Woods, Lake of the, l., N.A.	53	49°25'N	93°25'W
Woods Cross, Ut., U.S. (krôs)	75b	40°53'N	111°54'W
Woodsfield, Oh., U.S. (wòdz-fēld)	66	39°45'N	81°10'W
Woodson, Or., U.S. (wòdsŭn)	74c	46°07'N	123°20'W
Woodstock, Can. (wòd'stŏk)	59	43°10'N	80°50'W
Woodstock, Can.	51	46°09'N	67°34'W
Woodstock, Eng., U.K.	114b	51°48'N	1°22'W
Woodstock, Il., U.S.	71	42°20'N	88°29'W
Woodstock, Va., U.S.	67	38°55'N	78°25'W
Woodsville, N.H., U.S. (wòdz'vĭl)	67	44°10'N	72°00'W
Woodville, Ms., U.S. (wòd'vĭl)	82	31°06'N	91°11'W
Woodville, Tx., U.S.	81	30°48'N	94°25'W
Woodward, Ok., U.S. (wòd'wôrd)	78	36°25'N	99°24'W
Woolwich, Eng., U.K. (wòl'ĭj)	114b	51°28'N	0°05'E
Woomera, Austl. (wōōm'ĕrá)	174	31°15'S	136°43'E
Woonsocket, R.I., U.S. (wōōn-sŏk'ĕt)	68b	42°00'N	71°30'W
Woonsocket, S.D., U.S.	70	44°03'N	98°17'W
Wooster, Oh., U.S. (wòs'tēr)	66	40°50'N	81°55'W
Worcester, S. Afr. (wōōs'tēr)	186	33°35'S	19°31'E
Worcester, Eng., U.K. (wòs'tēr)	117	52°09'N	2°14'W
Worcester, Ma., U.S. (wòs'tēr)	65	42°16'N	71°49'W
Worcestershire, co., Eng., U.K.	114a	52°25'N	2°10'W
Worden, Il., U.S. (wôr'dĕn)	75e	38°56'N	89°50'W
Workington, Eng., U.K. (wûr'kĭng-tŭn)	120	54°40'N	3°30'W
Worksop, Eng., U.K. (wûrk'sŏp) (wûr'sŭp)	114a	53°18'N	1°07'W
Worland, Wy., U.S. (wûr'lănd)	73	44°02'N	107°56'W
Worms, Ger. (vōrms)	117	49°37'N	8°22'E
Worona Reservoir, res., Austl.	173b	34°12'S	150°55'E
Worth, Il., U.S. (wûrth)	69a	41°42'N	87°47'W
Wortham, Tx., U.S. (wûr'dhăm)	81	31°46'N	96°22'W
Worthing, Eng., U.K. (wûr'dhĭng)	120	50°48'N	0°29'W
Worthington, In., U.S. (wûr'dhĭng-tŭn)	66	39°05'N	87°00'W
Worthington, Mn., U.S.	70	43°38'N	95°36'W
Worth Lake, l., Tx., U.S.	75c	32°48'N	97°32'W
Wowoni, Pulau, i., Indon. (wō-wō'nê)	169	4°05'S	123°45'E
Wragby, Eng., U.K. (răg'bê)	114a	53°17'N	0°19'W
Wrangell, Ak., U.S. (răn'gĕl)	64a	56°28'N	132°25'W
Wrangell, Cape, c., Ak., U.S.	63a	52°55'N	172°30'E
Wrangell, Mount, mtn., Ak., U.S.	63	61°58'N	143°50'W
Wrangell Mountains, mts., Ak., U.S.	63	62°28'N	142°40'W
Wrangell-Saint Elias National Park, rec., Ak., U.S.	63	61°00'N	142°00'W
Wrath, Cape, c., Scot., U.K. (răth)	120	58°34'N	5°01'W
Wray, Co., U.S. (rā)	78	40°06'N	102°14'W
Wreak, r., Eng., U.K. (rĕk)	114a	52°45'N	0°59'W
Wreck Reefs, rf., Austl. (rĕk)	175	22°00'S	155°52'E
Wrekin, The, mtn., Eng., U.K. (rĕk'ĭn)	114a	52°40'N	2°33'W
Wrens, Ga., U.S. (rĕnz)	83	33°15'N	82°25'W
Wrentham, Ma., U.S.	61a	42°04'N	71°20'W
Wrexham, Wales, U.K. (rĕk'săm)	120	53°03'N	3°00'W
Wrexham, co., Wales, U.K.	114a	53°00'N	2°57'W
Wrights Corners, N.Y., U.S. (rīts kôr'nērz)	69c	43°14'N	78°42'W
Wrightsville, Ga., U.S. (rīts'vĭl)	83	32°44'N	82°44'W
Wrocław, Pol. (vrôtsläv) (brĕs'lou)	125	51°07'N	17°10'E
Wrotham, Eng., U.K. (rōōt'ŭm)	114b	51°18'N	0°19'E
Września, Pol. (vzhăsh'nyá)	125	52°19'N	17°33'E
Wu, r., China (wōō')	160	27°30'N	107°00'E
Wuchang, China	164	44°59'N	127°00'E
Wuchang, China (wōō-chäŋ)	161	30°32'N	114°25'E
Wucheng, China (wōō-chŭŋ)	162	37°14'N	116°03'E
Wuhan, China	161	30°30'N	114°15'E
Wuhu, China (wōō'hōō)	165	31°22'N	118°22'E
Wuji, China (wōō-jyĭ)	162	38°12'N	114°57'E
Wujiang, China (wōō-jyäŋ)	162	31°10'N	120°38'E
Wuleidao Wan, b., China (wōō-lā-dou wän)	162	36°55'N	122°00'E
Wulidian, China (wōō-lē-dĭĕn)	162	32°09'N	114°17'E
Wünsdorf, Ger. (vüns'dorf)	115b	52°10'N	13°29'E
Wupatki National Monument, rec., Az., U.S.	77	35°35'N	111°45'W
Wuping, China (wōō-pĭŋ)	165	25°05'N	116°01'E
Wuppertal, Ger. (vòp'ĕr-täl)	117	51°16'N	7°14'E
Wuqiao, China (wōō-chyou)	162	37°37'N	116°29'E
Würm, r., Ger. (vürm)	115d	48°07'N	11°20'E
Würselen, Ger. (vür'zĕ-lĕn)	127c	50°49'N	6°09'E
Würzburg, Ger. (vürts'bòrgh)	117	49°48'N	9°57'E
Wurzen, Ger. (vòrt'sĕn)	117	51°22'N	12°45'E
Wushi, China (wōō-shr)	160	41°13'N	79°08'E
Wusong, China (wōō-sôŋ)	162	31°23'N	121°29'E
Wustermark, Ger. (vōōs'tĕr-märk)	115b	52°33'N	12°57'E
Wustrau, Ger. (vōōst'rou)	115b	52°40'N	12°51'E
Wuustwezel, Bel.	115a	51°23'N	4°36'E
Wuwei, China (wōō'wā')	165	31°19'N	117°53'E
Wuxi, China (wōō-shyē)	161	31°36'N	120°17'E
Wuxing, China (wōō-shyĭŋ)	161	30°38'N	120°10'E
Wuyi Shan, mts., China (wōō-yē shän)	165	26°38'N	116°35'E
Wuyou, China (wōō-yō)	162	33°18'N	120°15'E
Wuzhi Shan, mtn., China (wōō-jr shän)	165	18°48'N	109°30'E
Wuzhou, China (wōō-jō)	161	23°32'N	111°25'E
Wyandotte, Mi., U.S. (wī'ăn-dŏt)	69b	42°12'N	83°10'W
Wye, Eng., U.K. (wī)	114b	51°12'N	0°57'E
Wye, r., Eng., U.K.	114a	53°14'N	1°46'W
Wylie, Lake, res., S.C., U.S.	83	35°02'N	81°21'W
Wymore, Ne., U.S. (wī'mōr)	79	40°09'N	96°41'W
Wynberg, S. Afr. (wīn'bĕrg)	186a	34°00'S	18°28'E
Wyndham, Austl. (wīnd'ăm)	174	15°30'S	128°15'E
Wynne, Ar., U.S. (wĭn)	79	35°12'N	90°46'W
Wynnewood, Ok., U.S. (wĭn'wòd)	79	34°39'N	97°10'W
Wynona, Ok., U.S. (wī-nō'ná)	79	36°33'N	96°19'W
Wynyard, Can. (wīn'yĕrd)	50	51°47'N	104°10'W
Wyoming, Oh., U.S. (wī-ō'mĭng)	69f	39°14'N	84°28'W
Wyoming, state, U.S.	64	42°50'N	108°30'W
Wyoming Range, mts., Wy., U.S.	64	42°43'N	110°35'W
Wyre Forest, for., Eng., U.K. (wīr)	114a	52°24'N	2°24'W
Wysokie Mazowieckie, Pol. (vê-sô'kyĕ mä-zô-vyĕts'kyĕ)	125	52°55'N	22°42'E
Wyszków, Pol. (vêsh'kôf)	125	52°35'N	21°29'E
Wytheville, Va., U.S. (wĭth'vĭl)	83	36°55'N	81°06'W

X

PLACE (Pronunciation)	PAGE	LAT.	LONG.
Xàbia, Spain	129	38°45'N	0°07'E
Xagua, Banco, bk., Cuba (bä'n-kō-sä'gwä)	92	21°35'N	80°50'W
Xai Xai, Moz.	186	25°00'S	33°45'E
Xalapa, Mex.	86	19°32'N	96°53'W
Xangongo, Ang.	186	16°50'S	15°05'E
Xankändi (Stepanakert), Azer. (styĕ'pän-á-kĕrt)	137	39°50'N	46°40'E
Xanten, Ger. (ksän'tĕn)	127c	51°40'N	6°28'E
Xánthi, Grc.	119	41°08'N	24°53'E
Xàtiva, Spain	118	38°58'N	0°31'W
Xau, Lake, l., Bots.	186	21°15'S	24°38'E
Xcalak, Mex. (sä-lä'k)	90a	18°15'N	87°50'W
Xelva, Spain	128	39°43'N	1°00'W
Xenia, Oh., U.S. (zē'nĭ-á)	66	39°40'N	83°55'W
Xi, r., China (shyē)	165	23°15'N	112°10'E
Xiajin, China (shyä-jyĭn)	164	36°58'N	115°59'E
Xiamen, China	161	24°30'N	118°10'E
Xiamen, i., Tai. (shyä-mŭn)	165	24°28'N	118°20'E
Xi'an, China (shyē-än)	160	34°20'N	109°00'E
Xiang, r., China (shyäŋ)	161	27°30'N	112°30'E
Xianghe, China (shyäŋ-hŭ)	162	39°46'N	116°59'E
Xiangtan, China (shyäŋ-tän)	161	27°55'N	112°45'E
Xianyang, China (shyĕn-yäŋ)	164	34°20'N	108°40'E
Xiaoxingkai Hu, l., China (shyou-shyĭŋ-kī hōō)	166	42°25'N	132°45'E
Xiapu, China (shyä-pōō)	161	27°00'N	120°00'E
Xiayi, China (shyä-yē)	162	34°15'N	116°07'E
Xicotencatl, Mex. (sē-kō-tĕn-kät'l)	88	23°00'N	98°58'W
Xifeng, China (shyē-fŭŋ)	164	42°40'N	124°40'E
Xiheying, China (shyē-hŭ-yĭŋ)	162	39°58'N	114°50'E
Xiliao, r., China (shyē-lĭou)	164	43°23'N	121°40'E
Xilitla, Mex. (sê-lē'tlä)	88	21°24'N	98°59'W
Xinchang, China (shyĭn-chäŋ)	163b	31°02'N	121°38'E
Xing'an, China (shyĭŋ-än)	165	25°44'N	110°32'E
Xingcheng, China (shyĭŋ-chŭŋ)	162	40°38'N	120°41'E
Xinghua, China (shyĭŋ-hwä)	162	32°58'N	119°48'E
Xingjiawan, China (shyĭŋ-jyä-wän)	162	37°16'N	114°54'E
Xingtai, China (shyĭŋ-tī)	164	37°04'N	114°33'E
Xingu, r., Braz. (zhĕŋ-gò')	101	6°20'S	52°34'W
Xinhai, China (shyĭn-hī)	162	36°59'N	117°33'E
Xinhua, China (shyĭn-hwä)	165	27°45'N	111°20'E
Xinhuai, r., China (shyĭn-hwī)	162	33°48'N	119°39'E
Xinhui, China (shyn-hwā)	165	22°40'N	113°08'E
Xining, China (shyē-nĭŋ)	160	36°52'N	101°36'E
Xinjiang (Sinkiang), prov., China (shyĭn-jyäŋ)	160	40°15'N	82°15'E
Xinjin, China (shyĭn-jyĭn)	164	39°23'N	121°57'E
Xinmin, China (shyĭn-mĭn)	164	42°00'N	122°42'E
Xintai, China (shyĭn-tī)	165	35°55'N	117°44'E
Xintang, China (shyĭn-täŋ)	163a	23°08'N	113°36'E
Xinxian, China (shyĭn shyĕn)	162	31°47'N	114°50'E
Xinxian, China	164	38°20'N	112°45'E
Xinxiang, China (shyĭn-shyäŋ)	164	35°17'N	113°49'E
Xinyang, China (shyĭn-yäŋ)	161	32°08'N	114°04'E
Xinye, China (shyĭn-yŭ)	164	32°40'N	112°20'E
Xinzao, China (shyĭn-dzou)	163a	23°01'N	113°25'E
Xinzheng, China (shyĭn-jŭŋ)	162	34°24'N	113°43'E
Xinzo de Limia, Spain	128	42°03'N	7°43'W
Xiongyuecheng, China (shyôŋ-yüĕ-chŭŋ)	162	40°10'N	122°08'E
Xiping, China (shyē-pĭŋ)	162	33°21'N	114°01'E
Xishui, China (shyē-shwā)	165	30°30'N	115°10'E
Xixian, China (shyē shyĕn)	162	32°20'N	114°42'E
Xixona, Spain	129	38°31'N	0°29'W
Xiyang, China (shyē-yäŋ)	162	37°37'N	113°42'E
Xiyou, China (shyē-yō)	162	37°21'N	119°59'E
Xizang (Tibet), prov., China (shyē-dzäŋ)	160	31°15'N	87°30'E
Xizhong Dao, i., China (shyē-jôŋ dou)	162	39°27'N	121°06'E
Xochihuehuetlán, Mex. (sô-chê-wĕ-wĕ-tlà'n)	89	17°53'N	98°29'E
Xochimilco, Mex. (sō-chê-mēl'kô)	89a	19°15'N	99°06'W
Xuancheng, China (shyůän-chŭŋ)	165	30°52'N	118°48'E
Xuanhua, China (shyůän-hwä)	164	40°35'N	115°05'E
Xuanhuadian, China (shyůän-hwä-dĭĕn)	162	31°42'N	114°29'E
Xuchang, China (shyōō-chäŋ)	164	34°02'N	113°49'E
Xudat, Azer.	138	41°38'N	48°42'E
Xuddur, Som.	192a	3°55'N	43°45'E
Xun, r., China (shyòn)	165	23°28'N	110°30'E
Xuzhou, China	161	34°17'N	117°10'E

Y

PLACE (Pronunciation)	PAGE	LAT.	LONG.
Ya'an, China (yä-än)	160	30°00'N	103°20'E
Yablonovyy Khrebet, mts., Russia (yá-blô-nô-vē')	135	51°15'N	111°30'E
Yablunivskyi, Pereval, p., Ukr.	125	48°20'N	24°25'E
Yacheng, China (yä-chŭŋ)	165	18°20'N	109°10'E
Yachiyo, Japan	167a	35°43'N	140°07'E
Yacolt, Wa., U.S. (yă'kôlt)	74c	45°52'N	122°24'W
Yacolt Mountain, mtn., Wa., U.S.	74c	45°52'N	122°27'W
Yacona, r., Ms., U.S. (yá'cō nä)	82	34°13'N	89°30'W
Yacuiba, Bol. (yá-kōō-ē'bá)	100	22°02'S	63°44'W
Yadkin, r., N.C., U.S. (yăd'kĭn)	83	36°12'N	80°40'W
Yafran, Libya	184	31°57'N	12°04'E
Yaguajay, Cuba (yä-guä-hä'ē)	92	22°20'N	79°20'W
Yahagi-Gawa, r., Japan (yä'hä-gê gä'wä)	167	35°16'N	137°22'E
Yahongqiao, China (yä-hôŋ-chyou)	162	39°45'N	117°52'E
Yahualica, Mex. (yä-wä-lē'kä)	88	21°08'N	102°53'W
Yajalón, Mex. (yä-hä-lōn')	89	17°16'N	92°20'W
Yakhroma, Russia (yäl'rô-ma)	142b	56°17'N	37°30'E
Yakhroma, r., Russia	142b	56°15'N	37°38'E
Yakima, Wa., U.S. (yăk'ĭmá)	64	46°35'N	120°30'W
Yakima, r., Wa., U.S. (yăk'ĭ-má)	72	46°48'N	120°22'W
Yakima Indian Reservation, I.R., Wa., U.S.	72	46°16'N	121°03'W
Yakoma, D.R.C.	190	4°05'N	22°27'E
Yaku, i., Japan (yä'kōō)	161	30°15'N	130°41'E
Yakutat, Ak., U.S. (yăk'ò-tăt)	63	59°32'N	139°35'W
Yakutsk, Russia (yá-kòtsk')	135	62°13'N	129°49'E
Yale, Mi., U.S.	66	43°05'N	82°45'W
Yale, Ok., U.S.	79	36°07'N	96°42'W
Yale Lake, res., Wa., U.S.	72	46°00'N	122°20'W
Yalinga, C.A.R. (yá-lĭŋ'gá)	185	6°56'N	23°22'E
Yalobusha, r., Ms., U.S. (yá-lô-bòsh'á)	82	33°48'N	90°02'W
Yalong, r., China (yä-lôŋ)	160	32°29'N	98°41'E
Yalta, Ukr. (yäl'tá)	137	44°29'N	34°12'E
Yalu, r., Asia	161	41°20'N	126°35'E
Yalutorovsk, Russia (yä-lōō-tô'rôfsk)	134	56°42'N	66°32'E
Yamada, Japan (yä'mà-dä)	167	33°37'N	133°39'E
Yamagata, Japan (yä-mà'gä-tá)	161	38°12'N	140°24'E
Yamaguchi, Japan (yä-mà'gōō-chê)	166	34°10'N	131°30'E
Yamal, Poluostrov, pen., Russia (yä-mäl')	134	71°15'N	70°00'E
Yamantau, Gora, mtn., Russia (gá-rä' yá'man-táw)	142a	54°16'N	58°08'E
Yamasaki, Japan (yä'mà'sä-kê)	167	35°01'N	134°33'E
Yamasaki, Japan	167b	34°53'N	135°41'E
Yamashina, Japan (yä'mä-shê'nä)	167b	34°59'N	135°50'E
Yamashita, Japan (yá'mä-shê'tá)	167b	34°53'N	135°25'E
Yamato, Japan	167a	35°28'N	139°28'E
Yamato-Kōriyama, Japan	167b	34°39'N	135°48'E
Yamato-takada, Japan (yä'mä-tô tä'kä-dä)	167b	34°31'N	135°45'E
Yambi, Mesa de, mtn., Col. (mĕ'sä-dĕ-yá'm-bē)	100	1°55'N	71°45'W
Yambol, Blg. (yäm'bôl)	119	42°28'N	26°31'E
Yamdena, i., Indon.	169	7°23'S	130°30'E
Yamethin, Mya. (yŭ-mē'thĕn)	155	20°14'N	96°27'E
Yamhill, Or., U.S. (yäm'hĭl)	74c	45°20'N	123°11'W
Yamkino, Russia (yäm'kĭ-nô)	142b	55°56'N	38°25'E
Yamma Yamma, Lake, l., Austl. (yăm'á yăm'á)	175	26°15'S	141°30'E
Yamoussoukro, C. Iv.	184	6°49'N	5°17'W
Yamsk, Russia (yämsk)	135	59°41'N	154°09'E
Yamuna, r., India	155	25°30'N	80°30'E
Yamzho Yumco, l., China (yäm-jwo yōōm-tswo)	160	29°11'N	91°26'E
Yana, r., Russia (yá'ná)	135	71°00'N	136°00'E
Yanac, Austl. (yăn'ăk)	175	36°10'S	141°30'E
Yanagawa, Japan (yä-nä'gä-wä)	167	33°11'N	130°24'E
Yanam, India (yŭnŭm')	155	16°48'N	82°15'E
Yan'an, China (yän-än)	160	36°46'N	109°15'E
Yanbu', Sau. Ar.	154	23°57'N	38°02'E
Yancheng, China (yän-chŭŋ)	164	33°23'N	120°11'E
Yancheng, China	164	33°38'N	113°59'E

ăt; finăl; rāte; senâte; ärm; àsk; sofá; fâre; ch-choose; dh-as th in other; bē; ėvent; bĕt; recĕnt; cratēr; g-gō; gh-guttural g; bĭt; ĭ-short neutral; rīde; κ-guttural k as ch in German ich;

PRONOUNCING INDEX

Yandongi – Zaltbommel 321

PLACE (Pronunciation)	PAGE	LAT.	LONG.
Yandongi, D.R.C.	190	2°51'N	22°16'E
Yangcheng Hu, l., China (yäŋ-chŭŋ hōō)	162	31°30'N	120°31'E
Yangchun, China (yäŋ-chòn)	165	22°08'N	111°48'E
Yang'erzhuang, China (yäŋ-är-jûäŋ)	162	38°18'N	117°31'E
Yanggezhuang, China (yäŋ-gŭ-jûäŋ)	164a	40°10'N	116°48'E
Yanggu, China (yäŋ-gōō)	162	36°06'N	115°46'E
Yanghe, China (yäŋ-hŭ)	162	33°48'N	118°23'E
Yangjiang, China (yäŋ-jyäŋ)	165	21°52'N	111°58'E
Yangjiaogou, China (yäŋ-jyou-gō)	162	37°17'N	118°53'E
Yangon *see* Rangoon, Mya.	155	16°46'N	96°09'E
Yangquan, China (yäŋ-chyůän)	162	37°52'N	113°36'E
Yangtze (Chang), r., China (yăng'tse) (chäŋ)	161	30°30'N	117°25'E
Yangxin, China (yäŋ-shyĭn)	162	37°39'N	117°34'E
Yangyang, Kor., S. (yäng'yäng')	166	38°02'N	128°38'E
Yangzhou, China (yäŋ-jō)	161	32°24'N	119°24'E
Yanji, China (yän-jyē)	161	42°55'N	129°35'E
Yanjiahe, China (yän-jyä-hŭ)	162	31°55'N	114°47'E
Yanjin, China (yän-jyĭn)	162	35°09'N	114°13'E
Yankton, S.D., U.S. (yănk'tŭn)	64	42°51'N	97°24'W
Yanling, China (yän-lĭŋ)	162	34°07'N	114°12'E
Yanshan, China (yän-shän)	164	38°05'N	117°15'E
Yanshou, China (yän-shō)	164	45°25'N	128°43'E
Yantai, China	161	37°32'N	121°22'E
Yanychi, Russia (yä'nĭ-chĭ)	142a	57°42'N	56°24'E
Yanzhou, China (yäŋ-jō)	161	35°35'N	116°50'E
Yanzhuang, China (yän-jûäŋ)	162	36°08'N	117°47'E
Yao, Chad (yä'ō)	185	13°00'N	17°38'E
Yao, Japan	167b	34°37'N	135°37'E
Yaoundé, Cam.	184	3°52'N	11°31'E
Yap, i., Micron. (yăp)	3	11°00'N	138°00'E
Yapen, Pulau, i., Indon.	169	1°30'S	136°15'E
Yaque del Norte, r., Dom. Rep. (yä'kå dĕl nôr'tå)	87	19°40'N	71°25'W
Yaque del Sur, r., Dom. Rep. (yä-kĕ-dĕl-sōō'r)	93	18°35'N	71°05'W
Yaqui, r., Mex. (yä'kē)	86	28°15'N	109°40'W
Yaracuy, dept., Ven. (yä-rä-kōō'ē)	101b	10°10'N	68°31'W
Yaraka, Austl. (yȧ-răk'ȧ)	175	24°50'S	144°08'E
Yaransk, Russia (yä-ränsk')	134	57°18'N	48°05'E
Yarda, oasis, Chad (yär'dȧ)	185	18°29'N	19°13'E
Yare, r., Eng., U.K.	121	52°40'N	1°32'E
Yarkand *see* Shache, China	160	38°15'N	77°15'E
Yarmouth, Can. (yär'mŭth)	60	43°50'N	66°07'W
Yaroslavka, Russia (yȧ-rô-släv'kȧ)	142a	55°52'N	57°59'E
Yaroslavl', Russia (yȧ-rô-släv''l)	134	57°37'N	39°54'E
Yaroslavl', prov., Russia	132	58°05'N	38°05'E
Yarra, r., Austl.	173a	37°51'S	144°54'E
Yarro-to, l., Russia (yȧ'rô-tô')	136	67°55'N	71°35'E
Yartsevo, Russia (yär'tsyĕ-vô)	136	55°04'N	32°38'E
Yartsevo, Russia	135	60°13'N	89°52'E
Yarumal, Col. (yä-rōō-mäl')	100	6°57'N	75°24'W
Yasawa Group, is., Fiji	170g	17°00'S	177°23'E
Yasel'da, r., Bela. (yä-syŭl'dȧ)	125	52°13'N	25°53'E
Yateras, Cuba (yä-tā'räs)	93	20°00'N	75°00'W
Yates Center, Ks., U.S. (yāts)	79	37°53'N	95°44'W
Yathkyed, l., Can. (yȧth-kī-ĕd')	52	62°41'N	98°00'W
Yatsuga-take, mtn., Japan (yät'sōō-gä dä'kä)	167	36°01'N	138°21'W
Yatsushiro, Japan (yät'sōō'shē-rô)	167	32°30'N	130°35'E
Yatta Plateau, plat., Kenya	191	1°55'S	38°10'E
Yautepec, Mex. (yä-ōō-tå-pĕk')	88	18°53'N	99°04'W
Yawata, Japan (yä'wä-tä)	167	34°52'N	135°43'E
Yawatahama, Japan (yä'wä'tä'hä-mä)	167	33°24'N	132°25'E
Yaxian, China (yä shyĕn)	165	18°10'N	109°32'E
Yayama, D.R.C.	190	1°16'S	23°07'E
Yayao, China (yä-you)	163a	23°10'N	113°40'E
Yazd, Iran	154	31°59'N	54°03'E
Yazoo, r., Ms., U.S. (yă'zōō)	65	32°32'N	90°40'W
Yazoo City, Ms., U.S.	82	32°50'N	90°18'W
Ydra, i., Grc.	131	37°20'N	23°30'E
Ye, Mya. (yā)	168	15°13'N	97°52'E
Yeadon, Pa., U.S. (yē'dŭn)	68f	39°56'N	75°16'W
Yecla, Spain (yā'klä)	128	38°35'N	1°09'W
Yefremov, Russia (yĕ-frā'môf)	132	53°08'N	38°04'E
Yegor'yevsk, Russia (yĕ-gôr'yĕfsk)	136	55°23'N	38°59'E
Yeji, China (yŭ-jyē)	162	31°52'N	115°57'E
Yekaterinburg, Russia	134	56°51'N	60°36'E
Yelabuga, Russia (yĕ-lä'bŏ-gȧ)	136	55°50'N	52°18'E
Yelan, Russia	137	50°50'N	44°00'E
Yelets, Russia (yĕ-lyĕts')	134	52°35'N	38°28'E
Yelizavetpol'skiy, Russia (yĕ'lĭ-za-vĕt-pôl-skĭ)	142a	52°51'N	60°38'E
Yelizavety, Mys, c., Russia (yĕ-lyē-sȧ-vyĕ'tĭ)	135	54°28'N	142°59'E
Yell, i., Scot., U.K. (yĕl)	120a	60°35'N	1°27'W
Yellow *see* Huang, r., China	161	35°06'N	113°39'E
Yellow, r., Fl., U.S. (yĕl'ô)	82	30°33'N	86°53'W
Yellowhead Pass, p., Can. (yĕl'ô-hĕd)	55	52°52'N	118°35'W
Yellowknife, Can. (yĕl'ô-nīf)	50	62°29'N	114°38'W
Yellow Sea, sea, Asia	161	35°20'N	122°15'E
Yellowstone, r., U.S.	64	46°00'N	108°00'W
Yellowstone, Clarks Fork, r., U.S.	73	44°55'N	109°05'W
Yellowstone Lake, l., Wy., U.S.	64	44°27'N	110°03'W
Yellowstone National Park, rec., U.S. (yĕl'ô-stōn)	64	44°45'N	110°35'W
Yel'nya, Russia (yĕl'nyȧ)	132	54°34'N	33°12'E
Yemanzhelinsk, Russia (yĕ-mán-zhâ'lĭnsk)	142a	54°47'N	61°24'E
Yemen, nation, Asia (yĕm'ĕn)	154	15°00'N	47°00'E
Yemetsk, Russia	136	63°28'N	41°28'E
Yenangyaung, Mya. (yā'nän-d oung)	155	20°27'N	94°59'E
Yencheng, China	160	37°30'N	79°26'E
Yendi, Ghana (yĕn'dē)	184	9°26'N	0°01'W
Yengisar, China (yŭn-gē-sär)	160	39°01'N	75°29'E
Yenice, r., Tur.	137	41°10'N	33°00'E
Yenisey, r., Russia (yĕ-nē-sĕ'ē)	134	71°00'N	82°00'E
Yeniseysk, Russia (yĕ-nĭĕsā'ĭsk)	135	58°27'N	90°28'E
Yeo, l., Austl. (yō)	174	28°15'S	124°00'E
Yerevan, Arm. (yĕ-rĕ-vän')	137	40°10'N	44°30'E
Yerington, Nv., U.S. (yē'rĭng-tŭn)	76	38°59'N	119°10'W
Yermak, i., Russia	136	66°45'N	71°30'E
Yeste, Spain (yĕs'tå)	128	38°23'N	2°19'W
Yeu, Île d', i., Fr. (ēl dyû)	117	46°43'N	2°45'W
Yevlax, Azer.	138	40°36'N	47°09'E
Yexian, China (yŭ-shyĕn)	162	37°09'N	119°57'E
Yeya, r., Russia (yä'yȧ)	133	46°25'N	39°17'E
Yeysk, Russia (yĕysk)	137	46°41'N	38°13'E
Yi, r., China	162	34°38'N	118°07'E
Yibin, China (yē-bĭn)	160	28°50'N	104°40'E
Yichang, China (yē-chäŋ)	161	30°38'N	111°22'E
Yidu, China (yē-dōō)	164	36°42'N	118°30'E
Yilan, China (yē-län)	161	46°10'N	129°40'E
Yinchuan, China (yĭn-chůän)	160	38°22'N	106°22'E
Yingkou, China (yĭŋ-kō)	161	40°35'N	122°10'E
Yining, China (yē-nĭŋ)	160	43°58'N	80°40'E
Yin Shan, mts., China (yĭng'shän')	164	40°50'N	110°30'E
Yishan, China (yē-shän)	160	24°32'N	108°42'E
Yishui, China (yē-shwā)	162	35°49'N	118°40'E
Yitong, China (yē-tôŋ)	161	43°15'N	125°10'E
Yixian, China (yē shyĕn)	164	41°30'N	121°15'E
Yixing, China	162	31°26'N	119°57'E
Yiyang, China (yē-yäŋ)	165	28°52'N	112°12'E
Yoakum, Tx., U.S. (yō'kŭm)	81	29°18'N	97°09'W
Yockanookany, r., Ms., U.S. (yŏk'á-nōō-kă-nĭ)	82	32°47'N	89°38'W
Yodo-Gawa, strt., Japan (yō'dō'gä-wä)	167b	34°46'N	135°35'E
Yog Point, c., Phil. (yōg)	165	14°00'N	124°30'E
Yogyakarta, Indon. (yŏg-yȧ-kär'tȧ)	168	7°50'S	110°20'E
Yoho National Park, rec., Can. (yō'hō)	50	51°26'N	116°30'W
Yojoa, Lago de, l., Hond. (lä'gô dĕ yô-hō'ä)	90	14°49'N	87°53'W
Yokkaichi, Japan (yô'kä'ē-chē)	166	34°58'N	136°35'E
Yokohama, Japan (yō'kō-hä'mȧ)	161	35°37'N	139°40'E
Yokosuka, Japan (yō-kō'sō-kä)	166	35°17'N	139°40'E
Yokota, Japan (yō-kō'tä)	167a	35°23'N	140°02'E
Yola, Nig. (yō'lä)	184	9°13'N	12°27'E
Yolaina, Cordillera de, mts., Nic.	91	11°34'N	84°34'W
Yomou, Gui.	188	7°34'N	9°16'W
Yonago, Japan (yō'nä-gō)	166	35°27'N	133°19'E
Yonezawa, Japan (yō'nĕ'zȧ-wä)	166	37°50'N	140°07'E
Yong'an, China (yôŋ-än)	165	26°00'N	117°22'E
Yongding, r., China (yôŋ-dĭŋ)	164	40°25'N	115°00'E
Yŏngdŏk, Kor., S. (yŭng'dŭk')	166	36°28'N	129°25'E
Yŏnghŭng, Kor., N. (yŭng'hòng')	166	39°31'N	127°11'E
Yonghŭng Man, b., Kor., N.	166	39°10'N	128°00'E
Yongnian, China (yôŋ-nĭĕn)	164	36°47'N	114°32'E
Yongqing, China (yôŋ-chyĭŋ)	164a	39°18'N	116°27'E
Yongshun, China (yôŋ-shòn)	160	29°05'N	109°58'E
Yonkers, N.Y., U.S. (yŏŋ'kĕrz)	68a	40°57'N	73°54'W
Yonne, r., Fr. (yôn)	126	48°18'N	3°15'E
Yono, Japan (yō'nō)	167a	35°53'N	139°36'E
Yorba Linda, Ca., U.S. (yôr'bȧ lĭn'dȧ)	75a	33°55'N	117°51'W
York, Austl.	174	32°00'S	117°00'E
York, Eng., U.K.	116	53°58'N	1°10'W
York, Al., U.S. (yôrk)	82	32°33'N	88°16'W
York, Ne., U.S.	79	40°52'N	97°36'W
York, Pa., U.S.	65	40°00'N	76°40'W
York, S.C., U.S.	83	34°59'N	81°14'W
York, Cape, c., Austl.	175	10°45'S	142°35'E
York, Kap, c., Grnld.	49	75°30'N	73°00'W
Yorke Peninsula, pen., Austl.	176	34°24'S	137°20'E
Yorketown, Austl.	176	35°00'S	137°28'E
York Factory, Can.	57	57°05'N	92°18'W
Yorkshire Wolds, Eng., U.K. (yôrk'shĭr)	120	54°00'N	0°35'W
Yorkton, Can. (yôrk'tŭn)	50	51°13'N	102°28'W
Yorktown, Tx., U.S. (yôrk'toun)	81	28°57'N	97°30'W
Yorktown, Va., U.S.	83	37°12'N	76°31'W
Yoro, Hond. (yō'rō)	90	15°09'N	87°05'W
Yoron, i., Japan	166	26°48'N	128°40'E
Yosemite National Park, rec., Ca., U.S. (yô-sĕm'ĭ-tē)	64	38°03'N	119°36'W
Yoshida, Japan (yō'shē-dȧ)	167	34°39'N	132°41'E
Yoshikawa, Japan (yō-shē'kä'wä')	167a	35°53'N	139°51'E
Yoshino, r., Japan (yō'shē-nō)	167	34°04'N	133°57'E
Yoshkar-Ola, Russia (yôsh-kär'ô-lä')	136	56°35'N	48°05'E
Yos Sudarsa, Pulau, i., Indon.	169	7°20'S	138°30'E
Yŏsu, Kor., S. (yŭ'sōō')	166	34°42'N	127°42'W
You, r., China (yō)	165	23°55'N	106°50'E
Youghal, Ire. (yōō'ôl) (yôl)	121	51°58'N	7°57'E
Youghal Bay, b., Ire.	120	51°52'N	7°46'W
Young, Austl. (yŭng)	176	34°15'S	148°18'E
Young, Ur. (yô-ōō'ng)	99c	32°42'S	57°38'W
Youngs, l., Wa., U.S. (yŭngz)	74a	47°25'N	122°08'W
Youngstown, N.Y., U.S.	69c	43°15'N	79°02'W
Youngstown, Oh., U.S.	66	41°05'N	80°40'W
Yozgat, Tur. (yôz'gȧd)	154	39°50'N	34°50'E
Ypsilanti, Mi., U.S. (ĭp-sĭ-lăn'tĭ)	69b	42°15'N	83°37'W
Yreka, Ca., U.S. (wī-rē'kȧ)	72	41°43'N	122°36'W
Yrghyz, Kaz.	139	48°30'N	61°17'E
Yrghyz, r., Kaz.	112	49°30'N	60°32'E
Ysleta, Tx., U.S. (ēz-lĕ'tȧ)	80	31°42'N	106°18'W
Yssingeaux, Fr. (ē-săN-zhō)	126	45°09'N	4°08'E
Ystad, Swe.	116	55°25'N	13°49'E
Ystādeh-ye Moqor, Āb-e, l., Afg.	158	32°35'N	68°00'E
Yu'alliq, Jabal, mts., Egypt	153a	30°12'N	33°42'E
Yuan, r., China (yůän)	161	28°50'N	110°50'E
Yuan'an, China (yůän-än)	165	31°08'N	111°28'E
Yuanling, China (yůän-lĭŋ)	165	28°30'N	110°18'E
Yuanshi, China (yůän-shr)	164	37°45'N	114°32'E
Yuasa, Japan	167	34°02'N	135°10'E
Yuba City, Ca., U.S. (yōō'bȧ)	76	39°08'N	121°38'W
Yucaipa, Ca., U.S. (yŭ-kȧ-ē'pȧ)	75a	34°02'N	117°02'W
Yucatán, state, Mex. (yōō-kä-tän')	86	20°45'N	89°00'W
Yucatán Channel, strt., N.A.	86	22°30'N	87°00'W
Yucatán Peninsula, pen., N.A.	90	19°30'N	89°00'W
Yucheng, China (yōō-chŭŋ)	162	34°31'N	115°54'E
Yucheng, China	164	36°55'N	116°39'E
Yuci, China (yōō-tsz)	164	37°32'N	112°40'E
Yudoma, r., Russia (yōō-dō'mȧ)	141	59°13'N	137°00'E
Yueqing, China (yůĕ-chyĭn)	165	28°02'N	120°40'E
Yueyang, China (yůĕ-yäŋ)	161	29°25'N	113°05'E
Yuezhuang, China (yůĕ-jûäŋ)	162	36°13'N	118°17'E
Yug, r., Russia (yòg)	136	59°50'N	45°55'E
Yugoslavia *see* Serbia and Montenegro, nation, Eur. (yōō-gô-slä-vĭ-ȧ)	110	44°00'N	21°00'E
Yukhnov, Russia (yòk'nof)	132	54°44'N	35°15'E
Yukon, ter., Can. (yōō'kŏn)	50	63°16'N	135°30'W
Yukon, r., N.A.	64a	64°00'N	159°30'W
Yukutat Bay, b., Ak., U.S. (yōō-kū tăt')	63	59°34'N	140°50'W
Yuldybayevo, Russia (yòld'bä'yĕ-vô)	142a	52°20'N	57°52'E
Yulin, China (yōō-lĭn)	165	22°38'N	110°10'E
Yulin, China	160	38°18'N	109°45'E
Yuma, Az., U.S. (yōō'mä)	64	32°40'N	114°40'W
Yuma, Co., U.S.	78	40°08'N	102°50'W
Yuma, r., Dom. Rep.	93	19°05'N	70°05'W
Yumbi, D.R.C.	191	1°14'S	26°14'E
Yumen, China (yōō-mŭn)	160	40°14'N	96°56'E
Yuncheng, China (yòn-chŭŋ)	164	35°00'N	110°40'E
Yunnan, prov., China (yun'nän')	160	24°23'N	101°03'E
Yunnan Plat, plat., China (yò-nän)	160	26°03'N	101°26'E
Yunxian, China (yòn shyĕn)	161	32°50'N	110°55'E
Yunxiao, China (yòn-shyou)	165	24°00'N	117°20'E
Yura, Japan (yōō'rä)	167	34°18'N	134°54'E
Yurécuaro, Mex. (yōō-rā'kwä-rō)	88	20°21'N	102°16'W
Yurimaguas, Peru (yōō-rē-mä'gwäs)	100	5°59'S	76°12'W
Yuriria, Mex. (yōō'rē-rē'ä)	88	20°11'N	101°08'W
Yurovo, Russia	142b	55°30'N	38°24'E
Yur'yevets, Russia	136	57°15'N	43°08'E
Yuscarán, Hond. (yōōs-kä-rän')	90	13°57'N	86°48'W
Yushan, China (yōō-shän)	165	28°42'N	118°20'E
Yü Shan, mtn., Tai.	161	23°38'N	121°05'E
Yushu, China (yōō-shōō)	164	44°58'N	126°32'E
Yutian, China (yōō-tĭĕn)	164	39°54'N	117°45'E
Yutian, China (yōō-tĭĕn) (kŭ-r-yä)	160	36°55'N	81°39'E
Yuty, Para. (yōō-tē')	102	26°45'S	56°13'W
Yuwangcheng, China (yü'wäng'chĕng)	162	31°32'N	114°26'E
Yuxian, China (yōō shyĕn)	164	39°40'N	114°38'E
Yuzha, Russia (yōō'zhȧ)	136	56°38'N	42°20'E
Yuzhno-Sakhalinsk, Russia (yōōzh'nô-sä-κä-lĭnsk')	135	47°11'N	143°04'E
Yuzhnoural'skiy, Russia (yōōzh-nô-ô-rál'skĭ)	142a	54°26'N	61°17'E
Yuzhnyy Ural, mts., Russia (yōō'zhnĭ ò-räl')	142a	52°51'N	57°48'E
Yverdon, Switz. (ē-vĕr-dôn)	124	46°46'N	6°35'E
Yvetot, Fr. (ēv-tō')	126	49°39'N	0°45'E

Z

PLACE (Pronunciation)	PAGE	LAT.	LONG.
Za, r., Mor.	118	34°19'N	2°23'W
Zaachila, Mex. (sä-ä-chē'lȧ)	89	16°56'N	96°45'W
Zaandam, Neth. (zän'dȧm)	121	52°25'N	4°49'E
Ząbkowice Śląskie, Pol.	124	50°35'N	16°48'E
Zabrze, Pol. (zäb'zhĕ)	117	50°18'N	18°48'E
Zacapa, Guat. (sä-kä'pä)	90	14°56'N	89°30'W
Zacapoaxtla, Mex. (sä-kä-pō-äs'tlä)	89	19°51'N	97°34'W
Zacatecas, Mex. (sä-kä-tā'käs)	86	22°44'N	102°32'W
Zacatecas, state, Mex.	86	24°00'N	102°45'W
Zacatecoluca, El Sal. (sä-kä-tå-kô-lōō'kä)	90	13°31'N	88°50'W
Zacatelco, Mex.	88	19°12'N	98°12'W
Zacatepec, Mex. (sä-kä-tå-pĕk') (sän-tē-ä'gô)	89	17°10'N	95°53'W
Zacatlán, Mex. (sä-kä-tlän')	89	19°55'N	97°57'W
Zacoalco de Torres, Mex. (sä-kô-äl'kô dä tōr'rĕs)	88	20°12'N	103°33'W
Zacualpan, Mex. (sä-kò-äl-pän')	88	18°43'N	99°46'W
Zacualtipan, Mex. (sá-kò-äl-tē-pän')	88	20°38'N	98°39'W
Zadar, Cro. (zä'där)	110	44°08'N	15°16'E
Zadonsk, Russia (zä-dônsk')	132	52°22'N	38°55'E
Žagare, Lat. (zhágárĕ)	123	56°21'N	23°14'E
Zagarolo, Italy (tzä-gä-rô'lô)	129d	41°51'N	12°53'E
Zaghouan, Tun. (zá-gwäN')	184	36°30'N	10°04'E
Zagreb, Cro. (zä'grĕb)	110	45°50'N	15°58'E
Zagros Mountains, mts., Iran	154	33°30'N	46°30'E
Zāhedān, Iran (zä'hȧ-dän)	154	29°37'N	60°31'E
Zahlah, Leb. (zä'lä')	153a	33°50'N	35°54'E
Zaire *see* Congo, Democratic Republic of the, nation, Afr.	186	1°00'S	22°15'E
Zaječar, Serb. (zä'yĕ-chär')	131	43°54'N	22°16'E
Zakhidnyi Buh (Bug), r., Eur.	124	52°29'N	21°20'E
Zakopane, Pol. (zá-kô-pä'nĕ)	125	49°18'N	19°57'E
Zakouma, Parc National de, rec., Chad	189	10°50'N	19°20'E
Zákynthos, Grc.	131	37°48'N	20°55'E
Zákynthos, i., Grc.	119	37°45'N	20°32'E
Zalaegerszeg, Hung. (zŏ'lô-ĕ'gĕr-sĕg)	124	46°50'N	16°50'E
Zalău, Rom. (zȧ-lŭ'ò)	125	47°11'N	23°06'E
Zaltan, Libya	185	28°20'N	19°40'E
Zaltbommel, Neth.	115a	51°48'N	5°15'E

ng-sing; ŋ-baŋk; N-nasalized n; nŏd; cŏmmit; ōld; ôbey; ôrder; oi-boil; fōōd; ò-as oo in foot; ou-out; s-soft; sh-dish; th-thin; pūre; ûnite; ûrn; stŭd; circŭs; ü-as in French tu; '-indeterminate vowel.

322 Zambezi – Zyryanovsk PRONOUNCING INDEX

PLACE (Pronunciation)	PAGE	LAT.	LONG.
Zambezi, r., Afr. (zȧm-bā′zẻ)	186	16°00′S	29°45′E
Zambia, nation, Afr. (zăm′bẻ-ȧ)	186	14°23′S	24°15′E
Zamboanga, Phil. (säm-bỏ-aŋ′gä)	168	6°58′N	122°02′E
Zambrów, Pol. (zäm′bro̊f)	125	52°29′N	22°17′E
Zamora, Mex. (sä-mō′rä)	86	19°59′N	102°16′W
Zamora, Spain (thä-mō′rä)	118	41°32′N	5°43′W
Zanatepec, Mex.	89	16°30′N	94°22′W
Zandvoort, Neth.	115a	52°22′N	4°30′E
Zanesville, Oh., U.S. (zānz′vĭl)	66	39°55′N	82°00′W
Zangasso, Mali	188	12°09′N	5°37′W
Zanjān, Iran	154	36°26′N	48°24′E
Zanzibar, Tan. (zăn′zĭ-bär)	187	6°10′S	39°11′E
Zanzibar, i., Tan.	187	6°20′S	39°37′E
Zanzibar Channel, strt., Tan.	191	6°05′S	39°00′E
Zaozhuang, China (dzou-jůäŋ)	162	34°51′N	117°34′E
Zapadnaya Dvina *see* Western Dvina, r., Eur.	123	55°30′N	28°27′E
Zapala, Arg. (zä-pä′lä)	102	38°53′S	70°02′W
Zapata, Tx., U.S. (sä-pä′tä)	80	26°52′N	99°18′W
Zapata, Ciénaga de, sw., Cuba (syĕ′nä-gä-dĕ-zä-pȧ′tä)	92	22°30′N	81°20′W
Zapata, Península de, pen., Cuba (pĕ-nẻ′n-so͞o-lä-dĕ-zä-pȧ′tä)	92	22°20′N	81°30′W
Zapatera, Isla, i., Nic. (ẻ′s-lä-sä-pä-tā′rō)	90	11°45′N	85°45′W
Zapopan, Mex. (sä-pō′pän)	88	20°42′N	103°23′W
Zaporizhzhia, Ukr.	134	47°50′N	35°10′E
Zaporizhzhia, prov., Ukr.	133	47°20′N	35°05′E
Zaporoshskoye, Russia (zȧ-pỏ-rôsh′skỏ-yĕ)	123	60°36′N	30°31′E
Zapotiltic, Mex. (sä-pỏ-tẻl-tēk′)	88	19°37′N	103°25′W
Zapotitlán, Mex. (sä-pỏ-tẻ-tlän′)	88	17°13′N	98°58′W
Zapotitlán, Punta, c., Mex.	89	18°34′N	94°48′W
Zapotlanejo, Mex. (sä-pỏ-tlä-nā′hỏ)	88	20°38′N	103°05′W
Zaragoza, Mex. (sä-rä-gō′sä)	88	23°59′N	99°45′W
Zaragoza, Mex.	88	22°02′N	100°45′W
Zaragoza, Spain (thä-rä-gō′thä)	110	41°39′N	0°53′W
Zarand, Munţii, mts., Rom.	125	46°07′N	22°21′E
Zaranda Hill, mtn., Nig.	189	10°15′N	9°35′E
Zaranj, Afg.	157	31°06′N	61°53′E
Zarasai, Lith. (zä-rä-sī′)	123	55°45′N	26°18′E
Zárate, Arg. (zä-rä′tā)	102	34°05′S	59°05′W
Zaraysk, Russia (zä-rä′ẻsk)	136	54°46′N	38°53′E
Zaria, Nig. (zä′rẻ-ä)	184	11°07′N	7°44′E
Zarqā′, r., Jord.	153a	32°13′N	35°43′E
Zarzal, Col. (zär-zȧ′l)	100a	4°23′N	76°04′W
Zashiversk, Russia (zȧ′shĭ-vĕrsk′)	135	67°08′N	144°02′E
Zastavna, Ukr. (zȧs-täf′nȧ)	125	48°32′N	25°50′E
Zastron, S. Afr. (zȧs′trŭn)	187c	30°19′S	27°07′E
Žatec, Czech Rep. (zhä′tĕts)	124	50°19′N	13°32′E
Zavitinsk, Russia	141	50°12′N	129°44′E
Zawiercie, Pol. (zȧ-vyĕr′tsyĕ)	125	50°28′N	19°25′E
Zāwiyat al-Baydā′, Libya	185	32°49′N	21°46′E
Zāyandeh, r., Iran	154	32°15′N	51°00′E
Zaysan, Kaz. (zī′sȧn)	139	47°43′N	84°44′E
Zaza, r., Cuba (zȧ′zȧ)	92	21°40′N	79°25′W
Zbarazh, Ukr. (zbä-räzh′)	125	49°39′N	25°48′E
Zbruch, r., Ukr. (zbro̊ch)	125	48°56′N	26°18′E
Zdolbuniv, Ukr.	125	50°31′N	26°17′E
Zduńska Wola, Pol. (zdo͞on′′skä vō′lä)	125	51°36′N	18°27′E
Zebediela, S. Afr.	192c	24°19′S	29°21′E
Zeeland, Mi., U.S. (zē′lănd)	66	42°50′N	86°00′W
Ẕefat, Isr.	153a	32°58′N	35°30′E
Zehdenick, Ger. (tsā′dē-nẻk)	124	52°59′N	13°20′E
Zehlendorf, Ger. (tsā′lĕn-dôrf)	115b	52°47′N	13°23′E
Zeist, Neth.	115a	52°05′N	5°14′E
Zelenogorsk, Russia (zĕ-lă′nỏ-gỏrsk)	123	60°13′N	29°39′E
Zella-Mehlis, Ger. (tsâl′ȧ-mā′lẻs)	124	50°40′N	10°38′E
Zémio, C.A.R. (za-myō′)	185	5°03′N	25°11′E
Zemlya Frantsa-Iosifa (Franz Josef Land), is., Russia	134	81°32′N	40°00′E
Zempoala, Punta, c., Mex. (po͞o′n-tä-sĕm-pỏ-ä′lä)	89	19°30′N	96°18′W
Zempoatlépetl, mtn., Mex. (sĕm-pỏ-ä-tlā′pĕt′l)	89	17°13′N	95°59′W
Zemun, Serb. (zĕ′mo͞on) (sĕm′lĭn)	119	44°50′N	20°25′E
Zengcheng, China (dzŭŋ-chŭŋ)	163a	23°18′N	113°49′E
Zenica, Bos. (zĕ′nẻt-sä)	131	44°10′N	17°54′E
Ẕeni-Su, is., Japan (zĕ′nē so͞o)	167	33°55′N	138°55′E
Žepče, Bos. (zhĕp′chĕ)	133	44°26′N	18°01′E
Zepernick, Ger. (tsĕ′pĕr-nēk)	115b	52°39′N	13°32′E
Zerbst, Ger. (tsĕrbst)	124	51°58′N	12°03′E
Zerpenschleuse, Ger. (tsĕr′pĕn-shloi-zĕ)	115b	52°51′N	13°30′E
Zeuthen, Ger. (tsoi′tĕn)	115b	52°21′N	13°38′E
Zevenaar, Neth.	127c	51°56′N	6°06′E
Zevenbergen, Neth.	115a	51°38′N	4°36′E
Zeya, Russia (zå′yȧ)	135	53°43′N	127°29′E
Zeya, r., Russia	141	52°31′N	128°30′E
Zeytun, Tur. (zā-to͞on′)	137	38°00′N	36°40′E
Zezere, r., Port. (zĕ′zå-rĕ)	128	39°54′N	8°12′W
Zgierz, Pol. (zgyĕzh)	125	51°51′N	19°26′E
Zhambyl, Kaz.	139	42°51′N	71°29′E
Zhangaqazaly, Kaz.	139	45°47′N	62°00′E
Zhangbei, China (jäŋ-bā)	161	41°12′N	114°50′E
Zhanggezhuang, China (jäŋ-gŭ-jůäŋ)	162	40°09′N	116°56′E
Zhangguangcai Ling, mts., China (jäŋ-gůäŋ-tsī lĭŋ)	164	43°50′N	127°55′E
Zhangjiakou, China	161	40°45′N	114°58′E
Zhangqiu, China (jäŋ-chyỏ)	162	36°50′N	117°29′E
Zhangye, China (jäŋ-yu)	160	38°46′N	101°00′E
Zhangzhou, China (jäŋ-jō)	161	24°35′N	117°45′E
Zhangzi Dao, i., China (jäŋ-dz dou)	162	39°02′N	122°44′E
Zhanhua, China (jän-hwä)	162	37°42′N	117°49′E
Zhanjiang, China (jän-jyäŋ)	161	21°20′N	110°28′E
Zhanyu, China (jän-yo͞o)	164	44°30′N	122°30′E
Zhao′an, China (jou-än)	165	23°48′N	117°10′E
Zhaodong, China (jou-dỏŋ)	164	45°58′N	126°00′E
Zhaotong, China (jou-tỏŋ)	160	27°18′N	103°50′E
Zhaoxian, China (jou shyĕn)	162	37°46′N	114°48′E
Zhaoyuan, China (jou-yuän)	162	37°22′N	120°23′E
Zharkent, Kaz.	139	44°12′N	79°58′E
Zhaysang köli, l., Kaz.	139	48°16′N	84°05′E
Zhecheng, China (jŭ-chŭŋ)	164	34°05′N	115°19′E
Zhegao, China (jŭ-gou)	162	31°47′N	117°44′E
Zhejiang, prov., China (jŭ-jyäŋ)	161	29°30′N	120°00′E
Zhelaniya, Mys, c., Russia (zhĕ′lä-nĭ-yȧ)	134	75°43′N	69°10′E
Zhem, r., Kaz.	137	46°50′N	54°10′E
Zhengding, China (jŭŋ-dĭŋ)	164	38°10′N	114°35′E
Zhengyang, China (jŭŋ-yäŋ)	162	32°34′N	114°22′E
Zhengzhou, China (jŭŋ-jō)	161	34°46′N	113°42′E
Zhenjiang, China (jŭn-jyäŋ)	161	32°13′N	119°24′E
Zhenyuan, China (jŭn-yůän)	165	27°08′N	108°30′E
Zhetiqara, Kaz.	139	52°12′N	61°18′E
Zhigalovo, Russia (zhẻ-gä′lỏ-vỏ)	135	54°52′N	105°05′E
Zhigansk, Russia (zhē-gȧnsk′)	135	66°45′N	123°20′E
Zhijiang, China (jr-jyäŋ)	165	27°25′N	109°45′E
Zhizdra, Russia (zhẻz′drȧ)	132	53°47′N	34°41′E
Zhizhitskoye, l., Russia (zhẻ-zhēt′skỏ-yĕ)	132	56°08′N	31°34′E
Zhmerynka, Ukr.	137	49°02′N	28°09′E
Zhongwei, China (jỏŋ-wā)	160	37°32′N	105°10′E
Zhongxian, China (jỏŋ shyĕn)	160	30°20′N	108°00′E
Zhongxin, China (jỏŋ-shyĭn)	163a	23°16′N	113°38′E
Zhoucun, China (jō-tso͞on)	164	36°49′N	117°52′E
Zhoukouzhen, China (jō-kō-jŭn)	162	33°39′N	114°40′E
Zhoupu, China (jō-po͞o)	162	31°07′N	121°33′E
Zhoushan Qundao, is., China (jō-shän-chyo̊n-dou)	161	30°00′N	123°00′E
Zhouxian, China (jō shyĕn)	164	39°30′N	115°59′E
Zhovkva, Ukr.	125	50°03′N	23°58′E
Zhu, r., China (jo͞o)	163a	22°48′N	113°36′E
Zhuanghe, China (jůäŋ-hŭ)	164	39°40′N	123°00′E
Zhuanqiao, China (jůäŋ-chyou)	163b	31°02′N	121°24′E
Zhucheng, China (jo͞o-chŭŋ)	164	36°01′N	119°24′E
Zhuji, China (jo͞o-jyē)	165	29°58′N	120°10′E
Zhujiang Kou, b., Asia (jo͞o-jyäŋ kō)	165	22°00′N	114°00′E
Zhukovskiy, Russia (zho̊-kôf′skĭ)	142b	55°33′N	38°09′E
Zhurivka, Ukr.	133	50°31′N	31°43′E
Zhytomyr, Ukr.	134	50°15′N	28°40′E
Zhytomyr, prov., Ukr.	133	50°40′N	28°07′E
Zi, r., China (dzē)	165	26°50′N	111°00′E
Zia Indian Reservation, I.R., N.M., U.S.	77	35°30′N	106°43′W
Zibo, China (dzē-bwo)	162	36°48′N	118°04′E
Ziel, Mount, mtn., Austl. (zēl)	174	23°15′S	132°45′E
Zielona Góra, Pol. (zhyĕ-lô′nä go͞o′rä)	124	51°56′N	15°30′E
Zigazinskiy, Russia (zĭ-gazinskēẻ)	142a	53°50′N	57°18′E
Ziguinchor, Sen.	184	12°35′N	16°16′W
Ẕile, Tur. (zẻ-lĕ′)	119	40°20′N	35°50′E
Žilina, Slvk. (zhẻ′lĭ-nä)	117	49°14′N	18°45′E
Zillah, Libya	185	28°26′N	17°52′E
Zima, Russia (zē′mȧ)	140	53°58′N	102°08′E
Zimapan, Mex. (sē-mä′pän)	88	20°43′N	99°23′W
Zimatlán de Alvarez, Mex.	89	16°52′N	96°47′W
Zimba, Zam.	191	17°19′S	26°13′E
Zimbabwe, nation, Afr. (rỏ-dē′zhĭ-ȧ)	186	17°50′S	29°30′E
Zimnicea, Rom. (zẻm-nē′chȧ)	131	43°39′N	25°22′E
Ẕin, r., Isr.	153a	30°50′N	35°12′E
Zinacatepec, Mex. (zē-nä-kä-tĕ′pĕk)	89	18°19′N	97°15′W
Zinapécuaro, Mex. (sē-nä-pā′kwȧ-rỏ)	88	19°50′N	100°49′W
Zinder, Niger (zĭn′dẽr)	184	13°48′N	8°59′E
Zin′kiv, Ukr.	133	50°13′N	34°23′E
Zion, Il., U.S. (zī′ŭn)	69a	42°27′N	87°50′W
Zion National Park, rec., Ut., U.S.	64	37°20′N	113°00′W
Zionsville, In., U.S. (zīŭnz-vĭl)	69g	39°57′N	86°15′W
Zirandaro, Mex. (sē-rän-dä′rō)	88	18°28′N	101°02′W
Zitacuaro, Mex. (sē-tȧ-kwä′rō)	88	19°25′N	100°22′W
Zitlala, Mex. (sẻ-tlä′lä)	88	17°38′N	99°09′W
Zittau, Ger. (tsẻ′tou)	124	50°55′N	14°48′E
Ziway, l., Eth.	185	8°08′N	39°11′E
Ziya, r., China (dzē-yä)	162	38°38′N	116°31′E
Zlatograd, Blg.	131	41°24′N	25°05′E
Zlatoust, Russia (zlȧ-tỏ-o̊st′)	134	55°13′N	59°39′E
Zlītan, Libya	185	32°27′N	14°33′E
Złoczew, Pol. (zwô′chĕf)	125	51°23′N	18°34′E
Zlynka, Russia (zlẻŋ′kȧ)	132	52°28′N	31°39′E
Znamensk, Russia (znä′mĕnsk)	123	54°37′N	21°13′E
Znamianka, Ukr.	133	48°43′N	32°35′E
Znojmo, Czech Rep. (znoi′mô)	117	48°52′N	16°03′E
Zoetermeer, Neth.	115a	52°08′N	4°29′E
Zoeterwoude, Neth.	115a	52°08′N	4°29′E
Zolochiv, Ukr.	125	49°48′N	24°55′E
Zolotonosha, Ukr. (zỏ′lỏ-tỏ-nô′shȧ)	137	49°41′N	32°03′E
Zolotoy, Mys, c., Russia (mĭs zỏ-lỏ-tôy′)	166	47°24′N	139°10′E
Zomba, Mwi. (zỏm′bȧ)	186	15°23′S	35°18′E
Zongo, D.R.C. (zŏŋ′gỏ)	185	4°19′N	18°36′E
Zonguldak, Tur. (zỏn′go͞ol′dȧk)	154	41°25′N	31°50′E
Zonhoven, Bel.	115a	50°59′N	5°24′E
Zoquitlán, Mex. (sỏ-kēt-län′)	89	18°09′N	97°02′W
Zorita, Spain (thō-rē′tä)	128	39°18′N	5°41′W
Zossen, Ger. (tsō′sĕn)	115b	52°13′N	13°27′E
Zouar, Chad	189	20°27′N	16°32′E
Zouxian, China (dzō shyĕn)	164	35°24′N	116°54′E
Zubtsov, Russia (zo̊p-tsôf′)	132	56°13′N	34°34′E
Zuera, Spain (thwā′rä)	129	41°40′N	0°48′W
Zugdidi, Geor.	138	42°30′N	41°53′E
Zuger See, l., Switz. (tso͞og)	124	47°10′N	8°40′E
Zugspitze, mtn., Eur.	124	47°25′N	11°00′E
Zuidelijk Flevoland, reg., Neth.	115a	52°22′N	5°20′E
Zújar, r., Spain (zo͞o′ĸär)	128	38°55′N	5°05′W
Zújar, Embalse del, res., Spain	128	38°50′N	5°20′W
Zulueta, Cuba (zo͞o-lỏ-ĕ′tä)	92	22°20′N	79°35′W
Zumbo, Moz. (zo͞om′bỏ)	186	15°36′S	30°25′E
Zumbro, r., Mn., U.S. (zŭm′brō)	71	44°18′N	92°14′W
Zumbrota, Mn., U.S. (zŭm-brō′tȧ)	71	44°16′N	92°39′W
Zumpango, Mex. (so̊m-päŋ-gō)	88	19°48′N	99°06′W
Zundert, Neth.	115a	51°28′N	4°39′E
Zungeru, Nig. (zo̊ŋ-gā′ro͞o)	184	9°48′N	6°09′E
Zunhua, China (dzo̊n-hwä)	164	40°12′N	117°55′E
Zuni, r., Az., U.S.	77	34°40′N	109°30′W
Zuni Indian Reservation, I.R., N.M., U.S. (zo͞o′nẻ)	77	35°10′N	108°40′W
Zuni Mountains, mts., N.M., U.S.	77	35°10′N	108°10′W
Zunyi, China	160	27°58′N	106°40′E
Zürich, Switz. (tsü′rĭk)	110	47°22′N	8°32′E
Zürichsee, l., Switz.	124	47°18′N	8°47′E
Zushi, Japan (zo͞o′shẻ)	167a	35°17′N	139°35′E
Zuwārah, Libya	184	32°58′N	12°07′E
Zuwayzā, Jord.	153a	31°42′N	35°55′E
Zvenigorod, Russia (zvā-nẻ′gô-rôt)	132	55°46′N	36°54′E
Zvenyhorodka, Ukr.	137	49°07′N	30°59′E
Zvishavane, Zimb.	186	20°15′S	30°28′E
Zvolen, Slvk. (zvô′lĕn)	125	48°35′N	19°10′E
Zvornik, Bos. (zvôr′nēk)	131	44°24′N	19°08′E
Zweibrücken, Ger. (tsvī-brük′ĕn)	124	49°16′N	7°20′E
Zwickau, Ger. (tsvĭkỏu)	117	50°43′N	12°30′E
Zwolle, Neth. (zvỏl′ĕ)	117	52°33′N	6°05′E
Żyradów, Pol. (zhẻ-rär′do̊f)	125	52°04′N	20°28′E
Zyryanka, Russia (zẻ-ryän′kȧ)	135	65°45′N	151°15′E
Zyryanovsk, Kaz.	139	49°43′N	84°20′E

ăt; finăl; rāte; senȧte; ärm; ȧsk; sofȧ; fâre; ch-choose; dh-as th in other; bē; ẻvent; bĕt; recĕnt; cratẽr; g-gō; gh-guttural g; bĭt; ĭ-short neutral; rīde; ĸ-guttural k as ch in German ich;

Listed below are major topics covered by the thematic maps, graphs and/or statistics.
Page citations are for world, continent and country maps and for world tables.

SOURCES

The following sources have been consulted during the process of creating and updating the thematic maps and statistics for the 21st Edition.

Air Carrier Traffic at Canadian Airports, Statistics Canada

Annual Coal Report, U.S. Dept. of Energy, Energy Information Administration

Armed Conflicts Report, Project Ploughshares

Atlas of Canada, Natural Resources Canada

Canadian Minerals Yearbook, Statistics Canada

Census of Canada, Statistics Canada

Census of Population, U.S. Census Bureau

Chromium Industry Directory, International Chromium Development Association

Coal Fields of the Conterminous United States, U.S. Geological Survey

Coal Quality and Resources of the Former Soviet Union, U.S. Geological Survey

Coal-Bearing Regions and Structural Sedimentary Basins of China and Adjacent Seas, U.S. Geological Survey

Commercial Service Airports in the United States with Percent Boardings Change, Federal Aviation Administration (FAA)

Completed Peacekeeping Operations, Center for Defense Information

Conventional Arms Transfers to Developing Nations, Library of Congress, Congressional Research Service

Current Status of the World's Major Episodes of Political Violence: Hot Wars and Hot Spots, Center for Systemic Peace

Dependencies and Areas of Special Sovereignty, U.S. Dept. of State, Bureau of Intelligence and Research

Earth's Seasons—Equinoxes, Solstices, Perihelion, and Aphelion, U.S. Naval Observatory

EarthTrends: The Environmental Information Portal, World Resources Institute and World Conservation Monitoring Centre 2003. Available at http://earthtrends.wri.org/ Washington, D.C.: World Resources Institute

Economic Census, U.S. Census Bureau

Employment, Hours, and Earnings from the Current Employment Statistics Survey, U.S. Dept. of Labor, Bureau of Labor Statistics

Energy Statistics Yearbook, United Nations Dept. of Economic and Social Affairs

Epidemiological Fact Sheets by Country, Joint United Nations Program on HIV/AIDS (UNAIDS), World Health Organization, United Nations Children's Fund (UNICEF)

Estimated Water Use in the United States, U.S. Geological Survey

Estimates of Health Personnel, World Health Organization

FAO Food Balance Sheet, Food and Agriculture Organization of the United Nations (FAO)

FAO Statistical Databases (FAOSTAT), Food and Agriculture Organization of the United Nations (FAO)

Fishstat Plus, Food and Agriculture Organization of the United Nations (FAO)

Geothermal Resources Council Bulletin, Geothermal Resources Bulletin

Geothermal Resources in China, Bob Lawrence and Associates, Inc.

Global Alcohol Database, World Health Organization

Global Forest Resources Assessment, Food and Agriculture Organization of the United Nations (FAO), Forest Resources Assessment Programme

Great Lakes Factsheet Number 1, U.S. Environmental Protection Agency

The Hop Atlas, Joh. Barth & Sohn GmbH & Co. KG

Human Development Report 2003, United Nations Development Programme, © 2003 by United Nations Development Programme. Used by permission of Oxford University Press, Inc.

Installed Generating Capacity, International Geothermal Association

International Database, U.S. Census Bureau

International Energy Annual, U.S. Dept. of Energy, Energy Information Administration

International Journal on Hydropower and Dams, International Commission on Large Dams

International Petroleum Encyclopedia, PennWell Publishing Co.

International Sugar and Sweetener Report, F.O. Licht, Licht Interactive Data

International Trade Statistics, World Trade Organization

International Water Power and Dam Construction Yearbook, Wilmington Publishing

Iron and Steel Statistics, U.S. Geological Survey, Thomas D. Kelly and Michael D. Fenton

Lakes at a Glance, LakeNet

Land Scan Global Population Database, U.S. Dept. of Energy, Oak Ridge National Laboratory (© 2003 UT-Battelle, LLC. All rights reserved. Notice: These data were produced by UT-Battelle, LLC under Contract No. DE-AC05-00OR22725 with the Department of Energy. The Government has certain rights in this data. Neither UT-Battelle, LLC nor the United States Department of Energy, nor any of their employees, makes any warranty, express or implied, or assumes any legal liability or responsibility for the accuracy, completeness, or usefulness of any data, apparatus, product, or process disclosed, or represents that its use would not infringe privately owned rights.)

Largest Rivers in the United States, U.S. Geological Survey

Lengths of the Major Rivers, U.S. Geological Survey

Likely Nuclear Arsenals Under the Strategic Offensive Reductions Treaty, Center for Defense Information

Major Episodes of Political Violence, Center for Systemic Peace

Maps of Nuclear Power Reactors, International Nuclear Safety Center

Mineral Commodity Summaries, U.S. Geological Survey, Bureau of Mines

Mineral Industry Surveys, U.S. Geological Survey, Bureau of Mines

Minerals Yearbook, U.S. Geological Survey, Bureau of Mines

National Priorities List, U.S. Environmental Protection Agency

National Tobacco Information Online System (NATIONS), U.S. Dept. of Health and Human Services, Centers for Disease Control and Prevention (CDC)

Natural Gas Annual, U.S. Dept. of Energy, Energy Information Administration

New and Recent Conflicts of the World, The History Guy

Nuclear Power Reactors in the World, International Atomic Energy Agency

Oil and Gas Journal DataBook, PennWell Publishing Co.

Oil and Gas Resources of the World, Oilfield Publications, Ltd.

Petroleum Supply Annual, U.S. Dept. of Energy, Energy Information Administration

Population of Capital Cities and Cities of 100,000 and More Inhabitants, United Nations Dept. of Economic and Social Affairs

Preliminary Estimate of the Mineral Production of Canada, Natural Resources Canada

Red List of Threatened Species, International Union for Conservation and Natural Resources

Significant Earthquakes of the World, U.S. Geological Survey

State of Food Insecurity in the World, Food and Agriculture Organization of the United Nations (FAO)

State of the World's Children, United Nations Children's Fund (UNICEF)

Statistical Abstract of the United States, U.S. Census Bureau

Statistics on Asylum-Seekers, Refugees and Others of Concern to UNHCR, United Nations High Commissioner for Refugees (UNHCR)

Survey of Energy Resources, World Energy Council

Tables of Nuclear Weapons Stockpiles, Natural Resources Defense Council

TeleGeography Research, PriMetrica, Inc. (www.primetrica.com)

Tobacco Atlas, World Health Organization

Tobacco Control Country Profiles, World Health Organization

Transportation in Canada, Minister of Public Works and Government Services, Transport Canada

UNESCO Statistical Tables, United Nations Educational, Scientific and Cultural Organization (UNESCO)

United Nations Commodity Trade Statistics (COMTRADE), United Nations Dept. of Economic and Social Affairs

United Nations Peacekeeping in the Service of Peace, United Nations Dept. of Peacekeeping Operations

United Nations Peacekeeping Operations, United Nations Dept. of Peacekeeping Operations

Uranium: Resources, Production and Demand, United Nations Organization for Economic Co-operation and Development (OECD)

Volcanoes of the World, Smithsonian National Museum of Natural History

Water Account for Australia, Australian Bureau of Statistics

Women in National Parliaments, Inter-Parliamentary Union

Women's Suffrage, Inter-Parliamentary Union

The World at War, Center for Defense Information, The Defense Monitor

The World at War, Federation of American Scientists, Military Analysis Network

World Conflict List, National Defense Council Foundation

World Contraceptive Use, United Nations Dept. of Economic and Social Affairs

The World Factbook, U.S. Dept. of State, Central Intelligence Agency (CIA)

World Facts and Maps, Rand McNally

World Lakes Database, International Lake Environment Committee

World Population Prospects, United Nations Dept. of Economic and Social Affairs

World Urbanization Prospects, United Nations Dept. of Economic and Social Affairs

World Water Resources and Their Use, State Hydrological Institute of Russia/UNESCO

The World's Nuclear Arsenal, Center for Defense Information

Special Acknowledgements

The American Geographical Society, for permission to use the Miller cylindrical projection.

The Association of American Geographers, for permission to use R. Murphy's landforms map.

The McGraw-Hill Book Company, for permission to use G. Trewartha's climatic regions map.

The University of Chicago Press, for permission to use Goode's Homolosine equal-area projection.